LYME DISEASE AND RELAPSING FEVER SPIROCHETES

Genomics, Molecular Biology, Host Interactions and Disease Pathogenesis

Editors:

Justin D. Radolf
Departments of Medicine, Pediatrics,
Molecular Biology and Biophysics,
Genetics and Genome Sciences,
and Immunology
UConn Health
263 Farmington Avenue
Farmington, CT 06030-3715
USA

D. Scott Samuels
Division of Biological Sciences
University of Montana
32 Campus Dr
Missoula MT 59812-4824
USA

Caister Academic Press

www.caister.com

ISBN: 978-1-913652-61-6 (paperback)
ISBN: 978-1-913652-62-3 (ebook)

DOI: https://doi.org/10.21775/9781913652616

Cover image

Spinning disk laser confocal micrograph showing endothelial activation in the murine cremaster microvasculature in *Cd1d*$^{-/-}$ mice 24 hours after intravenous injection of GFP-expressing *B. burgdorferi* (green). Adherent and crawling neutrophils, physically distanced from spirochetes, were stained were with phycoerythin-conjugated anti-Ly6G (red) and venules were stained with Alexa Fluor 647-conjugated anti-PECAM1 (blue). See Chapter 17. Image from Björn Petri, Xi Tan and George Chaconas. Imaging was supported by the Live Cell Imaging Laboratory, Snyder Institute for Chronic Diseases, The University of Calgary.

Dedication

This book is dedicated to Jorge Benach, Alan Barbour, and the late Willy Burgdorferi for their seminal contributions – beginning with the discovery that started it all – and leading to lifetimes of accomplishments in *Borrelia* research and training successive generations of spirochetologists. We also gratefully acknowledge our colleagues and trainees for their passion and commitment.

Contents

Preface

Remarkably, ten years have elapsed since the publication of ***Borrelia*: Molecular Biology, Host Interaction and Pathogenesis**. By 2010, Lyme disease (Lyme borreliosis) had become the prototype of an 'emerging infectious disease'. Once an obscure cause of arthritis in Southeastern Connecticut, in the span of three decades, Lyme disease was by then well recognized as the most prevalent vector-borne illness in the United States and Europe and a growing threat to global health. The first edition reflected two currents impacting the *Borrelia* field. One was the evolution of the study of pathogenic *Borrelia* from an arcane branch of microbiology to the mainstream of bacterial pathogenesis and clinical/translational research. The other was the advent of an unprecedented sociologic phenomenon – a coalition of self-proclaimed authorities accusing the biomedical community of conspiring to conceal the 'real truths' about the diagnosis and treatment of Lyme disease and the insidious nature of its etiologic agent. The book was the product of experts who pooled their energies and intellects to create the first state-of-the-art reference work covering the myriad, interlaced facets of the enzootic disorders caused by pathogenic *Borrelia*. The editors and authors had other intentions as well. We sought to showcase the extraordinary progress on a multitude of scientific and clinical fronts stemming from well-controlled, carefully executed, peer-reviewed research. And we conceived of the book as a vehicle for pushing back against the sophistry and shibboleths proffered to the public and legislators by our 'Lyme literate' detractors.

So, what, other than the title of the second edition, has changed during the last decade? Plenty. In reviewing the 24 chapters that comprise the volume, the editors have been impressed not just by the amount of new information generated by our colleagues but also its scope, technical virtuosity, and ingenuity. In fact, it should be evident to readers familiar with the first edition that the second edition is much more than an update; it is a reboot, and, in many instances, a re-evaluation of ideas and concepts prevalent at the time of the first edition. We now have much greater insight into the intricate ecologic factors that sustain *Borrelia* species in the wild and the environmental forces driving their alarming expansion throughout the Northern Hemisphere. With the availability of next-generation sequencing, genomic sequences for numerous *Borrelia* strains and species, sophisticated methods for genetic manipulation, transcriptomics, and structural data for many virulence-related proteins, our understanding of borrelial evolutionary biology, physiology, pathogenicity, and immune evasion has increased exponentially. Genomics and genetics have brought us to the cusp of unraveling the great mystery of how different species and strains of Lyme disease spirochetes elicit such varied clinical outcomes; indeed, it has become increasingly clear that the answer lies to a large extent in the evolutionary and immunologic mismatch between the spirochete and its incidental, reservoir-incompetent human host. In 2010, experimentation with ticks was limited to a few laboratories. Now, many investigators study the tick phases of the enzootic cycle and our coverage of vector biology and *Borrelia*-tick interactions has expanded to provide more guidance for this experimental approach. In addition, live imaging of spirochetes was just coming online a decade ago: the second edition contains an expansive treatment of the options available to investigators who need to track live spirochetes and evaluate gene expression *in situ* in ticks and mice. The importance of motility and chemotaxis for borrelial invasiveness has long been appreciated; however, cryoelectron microscopy, in combination with genetics and structural biology, has provided stunning visual insights into the workings of the flagellar motor that enables these serpentine organisms to migrate from the bite site to distal tissues where they inflict damage. At the time of the first edition, the signaling molecule c-di-GMP and the transcription factor BosR had only just been discovered in *B. burgdorferi* and there was essentially no mention of the stringent response. In the ensuing ten years, genetic and biochemical dissection of the molecular mechanisms modulating gene expression as the spirochete traverses its enzootic cycle has revealed hitherto unrecognized complexities and, considering the small size of the borrelial genome, a surprisingly sophisticated network of regulatory pathways. The composition of the field has changed as well. The author list contains the names of many gifted, enthusiastic early career investigators willing to take on the challenge of *Borrelia* for their life's work. Unfortunately, one thing has not changed – the Lyme Wars continue, led by individuals who espouse flawed studies purporting to show that Lyme disease spirochetes in humans adopt a persister state impervious to antimicrobial therapy that causes chronic symptomatology. In this edition, three authorities offer their analysis of these and other data as well as their perspectives on the chronic Lyme disease controversy that continues to ensnare our field.

This book was completed in the shadow of COVID-19. It seems ironic that one newly emergent infectious disease would nearly derail our efforts to bring to fruition ten years of progress studying another. Everyone's research and personal lives were upended by the pandemic; some of the authors served on the front lines. We, the editors, wish to express our deepest gratitude to our colleagues for their perseverance and commitment.

Justin D. Radolf and D. Scott Samuels

Chapter 1

The World-Wide Saga of Lyme Borreliosis

Jorge L. Benach[1]* and Juan Carlos García-Moncó[2]

[1]Department of Microbiology and Immunology, Stony Brook University, Stony Brook, New York 11794, USA
[2]Department of Neurology, Basurto University Hospital, Bilbao, Vizcaya, Spain
*Corresponding author: Jorge.benach@stonybrook.edu

DOI: https://doi.org/10.21775/9781913652616.01

Abstract
We have kept this chapter as a historical account of the early days of Lyme borreliosis in both Europe and the United States. Recent advances at all levels in this field are being updated in all the other chapters. Moreover, controversies that surround the diagnosis and treatment of this disease have been taken up in a new chapter 23 in this book. Thus, the saga of Lyme borreliosis in this chapter is told now with the same perspective as in the previous edition. The clinical-epidemiologic studies of a cluster of arthritis in Lyme, Connecticut intersected with ongoing studies on tick borne diseases in Long Island, New York to set the stage for the discovery of the spirochete agent of Lyme disease. Studies on the microbiology of spirochetes at the Rocky Mountain Laboratory in Montana were critical for the cultivation of the newly discovered organism. The multi-national clinical studies in Europe throughout the 20th century on the dermatologic and neurologic manifestations characterized and enlarged the syndrome that we now know as Lyme borreliosis. For reasons that are discussed in this chapter, controversies have arisen among advocate patient groups and the clinicians and scientists that are at the forefront of research in Lyme borreliosis. Polarized opinions on what constitutes this disease and how to treat it have resulted in unprecedented litigation involving medical societies and advocate patient groups with no obvious end in sight.

The epidemic of oligoarticular arthritis in Connecticut and the Northeast

The story of Lyme disease in Connecticut has been told on numerous occasions and has often been cited as an example of a very positive interplay among the lay community, academic medicine, and public health workers. In October 1975, two mothers, *Polly Murray* and *Judith Mensch*, from Lyme, Connecticut, contacted the Connecticut State Health Department to report that a number of children who lived in the same neighborhood in their town had developed what they had been told was juvenile rheumatoid arthritis. Soon thereafter, a joint investigation by the Connecticut State Health Department and the Yale University School of Medicine was begun to investigate the reports made by both mothers. This fortuitous occurrence developed into a dramatic epidemiologic study that resulted in the identification of a world-wide infection.

The possibility of a focal episode of juvenile rheumatoid arthritis in Lyme kindled the interest of Dr. Allen Steere, who had just joined the Division of Rheumatology at Yale, which in turn was headed by Dr. Stephen Malawista, a noted rheumatologist with a long track record studying inflammatory rheumatic diseases. Steere, joined by David Snydman of the Connecticut State Health Department, initiated an active surveillance system in the affected towns of Lyme, Old Lyme and East Haddam (Connecticut) to identify cases of the disorder reported by the two mothers.

The active community surveillance disclosed 39 children (Steere et al., 1977b) with recurrent attacks of swelling and pain in large joints, notably in the knees, and some for substantial periods of time. In addition, there were twelve adults, who were related to or lived near the arthritic children, with similar signs and symptoms. The surveillance showed an epidemic of oligoarticular arthritis in a focal geographic area. The focality of Lyme disease has remained a constant feature of this disease. In fact, even within these affected towns, clustering of cases was actually the norm in that many of the patients lived in several adjoining streets. Furthermore, there was also clustering within families, with six having more than one affected member, although the manifestations of the disease may have had onsets in different years. In the first study, the prevalence of the disease was 4.3 per 1,000 residents but was three times higher among children. The overrepresentation of children in Lyme disease has also remained a constant feature of this disease. This investigation essentially excluded juvenile rheumatoid arthritis as the cause of the epidemic given that the prevalence of arthritis in these Connecticut communities was 100 times greater than that expected prevalence for juvenile rheumatoid arthritis. Thus, with a high degree of certainty that the articular manifestations being reported in the three towns in Connecticut were not juvenile rheumatoid arthritis, the disease was named Lyme arthritis. With the enlarging clinical spectrum, this name was changed to Lyme disease.

The enlarging investigation revealed that approximately one quarter of the patients had developed an erythematous cutaneous lesion that, in some cases, appeared to expand into reddish, often annular, concentric rings. Interestingly, the skin lesion appeared weeks before the onset of the articular manifestations. This lesion had been previously described in Sweden in 1909 by Arvid Afzelius (see below) who named it "erythema migrans" because of its propensity to enlarge. Sometime later, the lesion became known as "erythema chronicum migrans" to emphasize its protracted duration. Afzelius associated the lesion with the bite of *Ixodes ricinus* ticks. During the summer of 1976, the Yale investigators followed prospectively the patients with the cutaneous lesions (Steere et al., 1977a). Many of these patients developed arthritis, while some also developed neurologic or cardiac abnormalities (Steere et al., 1980a). The neurologic manifestations among the prospective study group subsequently were defined by Reik and coworkers in 1979 (Reik et al., 1979). The addition of neurologic and cardiac involvement to the already known skin and articular manifestations underscored the complex, multi-organ-system disease that was being characterized in Connecticut.

The established association of erythema migrans with the bite of *Ixodes ricinus* ticks in Europe led to investigations showing that a local *Ixodes* species was associated with Lyme disease (Steere et al., 1978). Two studies provided indirect evidence for a role of *Ixodes scapularis* in the development of the erythema migrans. One study showed that the incidence of Lyme disease was 2.8 cases per 1,000 residents in the three communities of Lyme, Old Lyme, and East Haddam, on the east side of the Connecticut river. This was in marked contrast to the almost 30-fold lower incidence of 0.1 cases per 1,000 in several communities on the west side of the river (Steere and Malawista, 1979). The other study by Wallis et al . (Wallis et al., 1978) showed that *I. scapularis* were much more abundant on the east side (with a high incidence of erythema migrans and Lyme arthritis) of the Connecticut river than on the west where the incidence of the disease was significantly less. On the east side of the river, subadult *I. scapularis* were also more abundant on white-footed field mice, *Peromyscus leucopus*, while the adults were more abundant on white-tailed deer. Importantly, attempts to recover a possible etiologic agent from the ticks collected in the affected areas were unsuccessful. Nonetheless, the epidemiologic evidence strongly implicated these ixodid ticks in the natural history of Lyme disease.

One final important step was taken that led to successful therapy of Lyme disease with antibiotics. European erythema migrans had long been known to be responsive to antibiotic therapy, while studies demonstrating the efficacy of antibiotic therapy for this skin lesion had also appeared in the United States (Scrimenti, 1970). Therefore, Steere and collaborators gave antibiotics to patients with erythema migrans and compared the outcomes with those in untreated patients from their investigations. The results showed that penicillin, given early in the illness, shortened the duration of erythema migrans and appeared to prevent or lessen subsequent arthritis (Steere et al., 1980b).

Prior to the Connecticut epidemic, erythema migrans in North America was first described in 1970in a case report of a hunter from Wisconsin. The attending physician treated this patient successfully with penicillin (Scrimenti, 1970). A cluster of four cases of erythema migrans that occurred in the summer of 1975 in southeastern Connecticut was also reported before the Yale University studies (Mast and Burrows, 1976a, b). Although determining when erythema migrans first appeared in the United States is difficult, the lack of any reports prior to 1960 indicate that Lyme disease became epidemic in the 1970's coincident with the reports from Connecticut, Long Island, New York, and Massachusetts. In retrospect, the Wisconsin case was a harbinger of the subsequent emergence of Lyme disease in the upper Midwest. Moreover, it is likely that cases of Lyme disease in these endemic areas were not diagnosed as such before it gained widespread recognition. For example, in eastern Long Island, which is a major area endemic for Lyme disease, there are persistent anecdotes of a condition known as the "Montauk knee" that sounds suspiciously like the monoarticular arhtritis of Lyme. This entity evidently has been around for decades, and it is tempting to speculate that the Montauk knee may have been the late stage of Lyme disease.

The discovery of the agent of Lyme disease

A recent and detailed account of the discovery of the Lyme disease agent has been published, and it documents the recollection of events that led to the discovery. The narrative below mirrors closely the published account (Barbour and Benach, 2019).

The tick revolution of the 1970's

In the early 1970's, an increase in the number of cases of Rocky Mountain Spotted Fever (RMSF) occurred in the Southeast, South central and Northeastern states. This increase was amply documented and sparked a renewed interest in tick-borne diseases in this country (Benach et al., 1977; Burgdorfer, 1975). This was particularly true in Long Island and other coastal New England and the Massachusetts offshore islands that are at the northernmost limit of the geographic distribution of this disease. Parallel with the increase in the number of cases of RMSF were reports of large populations of *Dermacentor variabilis* ticks, the vector of the causative organism of RMSF, *Rickettsia rickettsii*. Although the reasons for this spike in the prevalence of RMSF have been debated, a general consensus has not developed. Nevertheless, it is generally believed that the increase in cases of RMSF in the early 1970's resulted from a number of environmental changes that may have been occurring during these years, and possibly even earlier. One such change was the banning in the late 1960s of DDT, a chlorinated hydrocarbon pesticide that had been widely used for agricultural, suburban pest, and public health insect control. The widespread use of this pesticide may have affected tick populations as non-target organisms and may have helped keep these populations in check; banning of this pesticide may have allowed for a gradual increase in the tick populations in large geographic areas of the country. Another environmental change reflects the increased suburbanization of rural land leading to a closer relationship of wildlife with new human residents. While this demographic trend is certainly well-documented and can explain the increase in the number of cases of tick-borne diseases, it does not explain the large increases in the tick populations. To explain the latter, increases in the populations of reservoirs have been implicated as the sources of large numbers of vectors. This is particularly true in the case of deer and the *Ixodes scapularis* ticks. As mentioned earlier, the increases in cases of RMSF and its vector, *D. variabilis*, were not an isolated event. In fact, another, possibly larger and of higher public health impact, increase in the numbers of human-biting *I. scapularis* led to the epidemic of Lyme disease in the northeast and upper midwest.

The entomological literature prior to the 1970's clearly documents the presence in the northeast of an *Ixodes* species whose subadult stages are ectoparasites of field rodents, notably *Peromyscus leucopus*, and whose females feed on white-tailed deer. Depending on the source, this tick species was identified as *I. muris* or as *I. scapularis*. Stepping into this taxonomic void, Spielman and coworkers described a new species using morphological characteristics of ticks collected in Massachusetts which they named *Ixodes dammini*, for Gustave Dammin, a pathologist at Harvard University (Spielman et al., 1979). This tick, once largely thought to be confined to feral hosts, assumed immediate importance as the vector to humans of *Babesia microti*, and, a few years later, as the vector of the etiologic agent of Lyme disease.

However, concerns that the newly named *Ixodes dammini* was not a new species prompted researchers to use genetic approaches to address this taxonomic question. Reciprocal crosses between *I. dammini* from Massachusetts and *I. scapularis* from Georgia produced offspring through the F3 generation whereas reciprocal crosses between *I. dammini* and *I. pacificus* from California (now known to be the vector of *B. burgdorferi* on the West coast (Burgdorfer et al., 1985) and *I. scapularis* and *I. pacificus* crosses produced sterile F1 progeny. Thorough assortive mating experiments, morphometric measurements, and chromosome and isoenzyme analyses led Oliver and coworkers (Oliver et al., 1993) to conclude that *I. dammini* was not a species separate from *I. scapularis*, and that the name *Ixodes scapularis*, which had been used for more than one hundred years to identify this species, had priority over the name *Ixodes dammini.* General acceptance of the name change resulted in the now near universal use of the *I. scapularis* nomenclature in the scientific literature.

Was there a "tick revolution" in the 1970's and possibly even earlier? On the face of the question, it appears improbable that virtually parallel increases in the populations of two separate tick species in such geographic proximity over such a short period of time are due to chance. Convincing arguments have been proposed, but single theories do not explain all of the increases in the tick populations and of the epidemics that developed thereafter. As with other environmental events, the causes are probably multifactorial, and their consequences certainly important enough to be debated for years to come. To be sure, significant multifactorial theories have been tested with an extraordinary degree of insight. One study linked concurrent defoliation of oak trees by gypsy moths (another environmental event in the northeast in the 1970's) with changing acorn production by the defoliated oaks, deer eating acorns, and rodent reservoirs eating the pupae of the gypsy moth (Jones et al., 1998; LoGiudice et al., 2003; Ostfeld et al., 2006). This theory was characterized as an environmental chain reaction using disparate animal and plant players as the basis for the changes in vectors and reservoirs that resulted in the tick-borne epidemics of the last 30 or so year.

Human babesiosis

The finding of human cases of babesiosis on Nantucket Island (Massachusetts) (Ruebush et al., 1977; Ruebush et al., 1981) and on Long Island (New York) (Benach et al., 1982; Benach et al., 1978; Benach et al., 1979) was a pivotal event in the discovery of the causative organism of Lyme disease. Human babesiosis in the northeastern United States is caused by the intraerythrocytic piroplasm *Babesia microti* (Homer et al., 2000; Rudzinska et al., 1976). Babesiosis is a malaria-like opportunistic infection with a broad spectrum of severity, ranging from subclinical in immunologically intact persons to fatal in elderly, immunocompromised, or cirrhotic patients. This protozoan pathogen is maintained enzootically in the white-footed mouse reservoir (*Peromyscus leucopus*) and is transmitted through the bite of *Ixodes scapularis* (Piesman and Spielman, 1980; Piesman et al., 1979; Spielman et al., 1981). The first cases of human babesiosis appeared in the 1970's in the offshore islands of Massachusetts and New York. The number of cases increased from a handful to more than one dozen in a short time, and more ominously, the severity of the illness in elderly patients, along with the detection of transfusion-transmitted and transplacental cases, required investigation by federal and state public health officials (Esernio-Jenssen et al., 1987; Grabowski et al., 1982). The finding of transfusion-acquired cases resulted in a ban on blood collection from the enzootic areas by the local blood banks. A very clear disease pattern emerged from the increasing number of cases (Dammin et al., 1981; Meldrum et al., 1992; Rosner et al., 1984; Ruebush et al., 1981). The typical patient was 50 years old or older and presented with non-periodic fever and chills, myalgia, fatigue, hemolytic anemia, and hematuria. A number of patients had undergone splenectomy, with a fatality rate of around 10% in this subset. In fact, studies with experimental animal models also have documented that advanced age is a risk factor for more severe disease (Habicht et al., 1983; Vannier et al., 2004). The parasitemia varied markedly and did not always correlate with clinical severity (Dammin et al., 1981; Filstein et al., 1980; Meldrum et al., 1992; Rosner et al., 1984; Ruebush et al., 1981). In some immunocompromised patients, reduction in parasitemia required red blood cell exchange (Cahill et al., 1981), and even in these cases, it persisted at low levels for months. Importantly, in the intervening years since the first cases of babesiosis were discovered, its geographic distribution has not changed markedly. While single cases of *B. microti* infection have been reported from areas other than coastal New England and New York, the large majority of the reported cases remains clustered geographically in the same endemic areas.

A serosurvey conducted in Shelter Island (New York) was carried out to detect asymptomatic individuals that could transmit *B. microti* through blood banking procedures and to determine the prevalence and incidence of serologically-reactive but asymptomatic babesiosis (Filstein et al., 1980). Healthy individuals at high risk of exposure to ticks were surveyed during a single transmission season. Paired sera obtained in June and October, 1978 were tested for antibodies to *B. microti* using the newly developed indirect immunofluorescence test (Chisholm et al., 1978). Point prevalence values of 4.4% in June and 6.9% in October were obtained. Six of 102 persons tested in both months showed at least a fourfold rise in titer of antibodies to *B. microti* for an incidence of 6% for the season. Control sera from inland areas did not show any significant reactivity to *B. microti*, but of the six persons who seroconverted, four had histories of tick bites during the transmission season. Of note, this serosurvey identified a number of individuals who reported a history of erythema migrans and also had antibodies to *B. microti*. These individuals did not have clinical babesiosis, but retrospective inquiries yielded histories compatible with early Lyme disease. Sera banked from these individuals were used several years later to strengthen the association between *B. burgdorferi* and Lyme disease (Burgdorfer et al., 1982). This association between subclinical exposure to *B. microti* and signs of Lyme disease proved to be a common occurrence that underscored the close natural history of both pathogens. This pattern also underscored the opportunistic nature of babesiosis and the high degree of susceptibility of healthy persons to infection with *B. burgdorferi*.

The sharing of reservoir and vector hosts by *B. microti* and *B. burgdorferi* leads to frequent coinfection in nature (Anderson et al., 1986; Piesman et al., 1987; Piesman et al., 1986) and in patients (Benach and Habicht, 1981; Krause et al., 1992b; Sweeney et al., 1998). Evidence for coinfection in humans was noted before the discovery of *B. burgdorferi*. Clinical histories consistent with Lyme disease were elicited from patients who had antibodies to *B. microti* in the serosurvey in 1978 (Filstein et al., 1980). Subsequently, in 1980 a number of patients with clinical babesiosis reported erythema migrans at the onset of the disease (Benach and Habicht, 1981), Although serologic confirmation of Lyme disease was not possible at the time, subsequent serological confirmation of coinfection or coexposure was obtained shortly after *B. burgdorferi* was discovered. Clinical coinfection was further documented in a series of patient studies (Benach et al., 1985; Grunwaldt et al., 1983). The existence of clearly established human cases of coinfection with *B. burgdorferi* and *B. microti* led to clinical and animal model studies to determine whether the course of illness is more severe following coinfection (Hilton et al., 1999; Krause et al., 1996; Wang et al., 2000). In the clinical arena, there were early conflicting results. One study reported greater severity of Lyme disease (Krause et al., 1996), while another did not report any synergistic interactions between the two diseases (Wang et al., 2000). Experimental studies of coinfection in mice did not show any synergy leading to greater severity of either infection (Coleman et al., 2005), but another study showed greater arthritis severity in double infections (Moro et al., 2002).

Borrelia research in Montana and Minnesota

The medium defined by Kelly for the *in vitro* cultivation of *Borrelia hermsii* greatly facilitated the studies of Herbert Stoenner on the biology of relapsing fever and made possible the subsequent discovery of the *Borrelia* agent of Lyme disease (Kelly, 1971). Burgdorfer also conducted relapsing fever research as early as 1954, and his work was instrumental in institutionalizing investigations on the tick–borne Borrelia at the Rocky Mountain Laboratory (Burgdorfer, 1954; Burgdorfer and Mavros, 1970; Davis and Burgdorfer, 1954). Following Stoenner, research on relapsing fever Borrelia was taken up by Alan G. Barbour, who focused on antigenic variation (Barbour et al., 1982; Stoenner, 1974; Stoenner et al., 1982). After several attempts, Barbour was able to grow Lyme disease spirochetes from *Ixodes* ticks in pure culture using what we know today as BSK medium (Barbour, Stoenner, Kelly). These organisms were then used as antigen determine if antibodies were present in the sera from the Long Island patients (see below).

The discovery of the spirochete

The discovery of the Lyme disease spirochete was the result of a convergence of events that were occurring in different geographic areas of the United States although not necessarily in isolation from each other: 1. The collaboration between Burgdorfer and Benach centered around the Long Island, New York endemic areas of several tick-borne diseases, and led to joint investigations that yielded the isolation of the organism from ticks collected in Shelter Island, NY. 2. The state of readiness at the Rocky Mountain Laboratory to culture spirochetes in an improved Kelly's medium was another factor in the discovery of the organism. This medium, improved by Stoenner and then Barbour, permitted the growth of the newly discovered *Borrelia*. 3. The description of an enlarging clinical spectrum for Lyme disease by Steere, Malawista, and coworkers permitted the clinical identification of cases outside of the Lyme, Connecticut area.

The serosurveys for babesiosis in 1978 and 1979 in Shelter Island, New York had provided serum specimens from individuals with clinically inapparent babesiosis, but who gave a history of erythema migrans and joint manifestations diagnosed as Lyme disease by Edgar Grunwaldt, the internist on Shelter Island. These sera, stored in our serum bank at Stony Brook University, were used for establishing the etiology of one of the organisms seen in *Ixodes scapularis* adults collected from Shelter Island. One such organism was a microfilaria seen in the hemolymph of the ticks. On darkfield microscopy, spirochetes were seen and later cultured in Kelly's medium. In addition to the banked sera, we provided the serum of a faculty member, a physician at Stony Brook University, who had suffered for some time from many of the manifestations of Lyme disease. These sera reacted with the newly cultured spirochetes, initially thought to be a species of *Treponema*, and these findings were reported in Science (Burgdorfer et al., 1982). In short order, spirochetes were isolated from patients with Lyme disease by two groups, thus fulfilling Koch's postulates; both studies were reported in the same issue of the New England Journal of Medicine in 1983 (Benach et al., 1983) (Steere et al., 1983). Spirochetes were subsequently isolated from the skin of patients with erythema migrans (Berger et al., 1985).

Continuing field investigations headed by Edward Bosler isolated the spirochetes from the blood of white-footed mice collected in Shelter Island and from the same areas where the serosurveys of 1978 and 1979 had been carried out (Bosler et al., 1983). The isolation of the *Borrelia* organisms from *Peromyscus leucopus* served to identify the main reservoir of this organism. This finding, also published in Science in 1983, was subsequently confirmed in several studies from other parts of the country (Burgdorfer et al., 1988). Interestingly, deer long thought to be involved in the zoonotic cycle of the spirochete, proved to be a relatively incompetent host (Telford et al., 1988). Canine Lyme disease was identified through our continuing studies on Long Island (Lissman et al., 1984). Dr. Barry Lissman, a veterinarian in private practice in our area, had been a prior collaborator in our studies on tick-borne disease. A few years earlier, Dr. Lissman and our group had characterized, for the first time, the canine cases of Rocky Mountain Spotted Fever (Lissman and Benach, 1980). Aware of the problem of high tick burdens in Long island dogs, Dr. Lissman had seen a series of animals with lameness and arthritis that were heavily infested by both *I. scapularis* and *D. variabilis*. The discovery of the spirochete permitted the testing of canine serum against this agent with strongly positive results. Moreover, we were able to grow the *Borrelia* from blood samples from dogs with this syndrome (Lissman et al., 1984). The transmission of the organism by the feeding of *I. scapularis* ticks was subsequently verified (Benach et al., 1987; Ribeiro et al., 1987). Robert Lane in California and Burgdorfer were able to grow the Borrelia organism from *Ixodes pacificus* ticks, thus identifying a third endemic area in the western United States (Burgdorfer et al., 1985).

The early *Borrelia* work at the Rocky Mountain Laboratory was paralleled by the seminal studies, also on relapsing fever, by Russell Johnson at the University of Minnesota. His manuscripts (Johnson, 1977) were required reading for all investigators who joined this field after the identification of the agent of Lyme disease. In 1983, at the first International Meeting on Lyme Disease at Yale University in New Haven, Connecticut, a group of investigators that had been instrumental in the discovery and characterization of the Lyme disease spirochete agreed that the new organism would be named *Borrelia burgdorferi*, after Willy Burgdorfer. Johnson wrote the taxonomic study proposing this name (Johnson et al., 1984). Years later, taxonomists in Europe determined that Lyme disease spirochetes in Europe were sufficiently diverse to diverse to warrant separate species designations (Belfaiza et al., 1993; Canica et al., 1993). This led to the naming of *B. garinii* and *B. afzelii* as well as the designation of the original isolate as *B. burgdorferi* sensu stricto, with the entire complex being *B. burgdorferi* sensu lato (Baranton et al., 1992).

More recently, *B. japonica*, and other *Borrelia* species of unknown infectivity were described (Kawabata et al., 1993). A taxonomic study of the genus Borrelia used three molecular markers (conserved signature inserts/deletions, conserved signature proteins, and average nucleotide identity analysis on Borrelia genomes. These analyses supported the existence of two separate Borrelia clades and led to a proposed a division of the genus Borrelia into two separate genera. In this proposal, there would be an emended genus Borrelia, containing the agents of relapsing fever and a new genus, Borreliella, that includes the causative agents of Lyme disease(Adeolu and Gupta, 2014). However, this generic separation was not supported unanimously. One study pointed out that there was inadequate evidence for splitting the genus, and cited insufficient use of other taxonomic tools(Margos et al., 2017). In turn, these criticisms were addressed, and refuted, in another publication (Barbour et al., 2017). Additional taxonomic arguments were presented against the separation of the genus Borrelia, and these resulted in a recommended re-unification of the genus supported by the results of an analysis of percentage of conserved proteins, a method that has been used for generic demarcation (Margos et al., 2018). Another multi-author publication cited public health and patient safety concerns if the new generic designation of "Borreliella" gained widespread use (Stevenson et al., 2019). This publication reflected the current state of the field of Lyme disease, with its polarized constituencies, and the consensus that renaming its etiological agent would lead to additional confusion.

A notable milestone in the therapy of Lyme disease was made by Dattwyler *et al.* (Dattwyler et al., 1987; Dattwyler et al., 1988) with the introduction of ceftriaxone as the choice antibiotic. Barthold developed a murine model of Lyme arthritis and carditis that has provided the basis for numerous studies on pathogenesis and host responses (Barthold et al., 1988). The genetic transformation barrier for *B. burgdorferi* was broken by the work of Samuels, (Samuels, 1995; Samuels et al., 1994a; Samuels et al., 1994b) and Rosa and coworkers (Rosa et al., 1996; Rosa et al., 2005). These studies have already permitted important research on gene product function in conjunction with the genome of *B. burgdorferi* (Fraser et al., 1997).

The European evidence of one hundred years

Although Lyme disease was named for the place where it was first described in the United States, the disorder had been present in Europe for a long time, mainly as a dermatological and neurological disorder, in contrast with the original rheumatologic presentation in the USA. Furthermore, many years before, several European scientists had linked this disorder to a presumed infectious agent transmitted by a tick bite. The historical hallmarks are presented in Table 1.

Erythema migrans and Azfelius

The first description of the most frequent and worldwide pathognomonic skin lesion of what we now know as Lyme borreliosis or Lyme disease, was given at the beginning of the 20th century by Arvid Azfelius (1857-1923), one of the founders of the Swedish Society of Dermatology. In 1909 he presented before the society an elderly woman with an expanding annular erythema with central fading that had developed at the site of a bite by the tick *Ixodes reduvius* (Azfelius, 1910). Because of its expansion pattern he named this lesion erythema migrans and suggested that it was due to a tick-transmitted infection. A few years later, the Austrian dermatologist, Lipschütz, described another patient with a similar condition that lasted for 7 months, and suggested naming it erythema chronicum migrans (ECM) on the basis of its duration (Lipschutz, 1918). He did not speculate on its etiology. In 1921 Azfelius reported six additional patients with a similar skin lesion, and two more were presented before the Swedish Society of Dermatology by other authors (Azfelius, 1921). All of the patients had a previous tick bite, a fact that was held by Azfelius as evidence that a tick bite (*Ixodes ricinus*) was implicated in this entity. In 1930, another Swedish dermatologist, Sven Hellerström, from the Karolinska Institute, reported a patient who developed lymphocytic meningitis three months after an ECM and suggested that both conditions were related (Hellerstrom, 1930). In the years to follow, other similar cases were discussed by Hellerström, who suggested a common etiology for the skin lesion and for the accompanying meningitis. Other European authors reported additional patients with meningitis following eythema migrans (Dalsgaard-Nielsen, 1948; Gelbjerg-Hansen, 1945; Zwelleger, 1946). In 1936, Askani from the Dermatology Clinic at the University of Heidelberg reviewed the significance of ticks as vectors of human and animal pathogens and concluded that ECM was caused either by a toxin or by a living pathogen from the salivary glands of ticks (Askani, 1936). Together with Hellerström's findings, this became the preliminary evidence for the salivary route of transmission during tick feeding.

The cause of ECM remained elusive, and Hellerström mentioned a possible role for spirochetes, at that time being investigated by Carl Lennhoff, also a member of the Dermatology Clinic at the Karolinska Institute. Lennhoffclaimed to have developed a staining technique (based on the interaction of mercuric chloride with spirochetes) that permitted him to identify spirochetal elements in many different skin lesions of unknown etiology, including psoriasis, zoster, varicella, lymphadenosis benigna cutis, and erythema migrans, among others (Lennhoff, 1948). Although his staining method appeared suitable for demonstrating spirochetes in culture, it was not suitable for identifying these microorganisms in smears or in tissue sections of ECM lesions. In fact, to Lennhoff's disappointment, the structures that he thought were spirochetes turned out to be artifacts.

In 1949, Schaltenbrand (Schaltenbrand, 1949) from the Neurological Clinic at the University of Würzburg showed that the nervous system disorders associated with ECM were different from those of the central European tick-borne encephalitis. He speculated on the relationship between the anatomical location of the tick bite and the subsequent neurologic manifestations and assumed that ECM meningitis was caused by an unidentified viral agent associated with *I. ricinus*. He also described prompt improvement after treatment with corticosteroids and tetracycline. In 1950, Hellerström, at the Annual Meeting of the Southem Medical Association in Cincinnati, Ohio, presented "ECM Afzelius with meningitis" (Hellesträm, 1950). He reviewed several cases in which both erythema migrans and subsequent meningocerebrospinal symptoms occurred after tick bites. He also discussed the still unsolved cause of ECM and wondered whether the sheep tick, *I. ricinus,* could be a vector of spirochetes. He pointed out that "spirochetes have been known to cause circular erythemas that subside in the center and advance peripherally and to affect the nervous system". Hellerström injected tick salivary gland extracts into the skin of patients and healthy people and noted an intense reaction in the former, whereas only a weak response was obtained in the later. In 1951, a colleague of Hellerström, Hollström, reported on the successful use of penicillin in treating ECM (Hollström, 1951). Despite the accumulating evidence, and perhaps because spirochetes had never been found in association with ticks of the genus *Ixodes*, the search for spirochetes in ticks was not pursued and the cause of ECM remained obscure. In 1955, Binder and coworkers (Binder et al., 1955) succeeded in transmitting ECM from human to human by transplanting pieces of skin taken from the peripheral zone of a lesion of a patient with ECM. In each of the three volunteers, typical lesions developed within 1 to 3 weeks, and a sub-passage was equally successful. In each case, treatment with penicillin caused the lesions to disappear within a few days. These experiments clearly suggested that the cause of ECM was an infectious agent susceptible to penicillin.

A rickettsial cause of ECM was suggested in 1962 by Degos *et al.* (Degos et al., 1962) in France. Based on results obtained by Giroud's slide agglutination test, they reported six of seven patients with significant titers against the epidemic typhus agent, *Rickettsia prowazekii,* the murine typhus rickettsia, *R. mooseri,* and the boutonneuse fever agent, *R. conorii.* Although these results could not be confirmed, the notion that ECM is caused by a rickettsia was kept alive by the electron microscopic demonstration of rickettsia-like microorganisms in macrophages of two ECM patients (Sandbank and Feuerman, 1979).

In 1974, Klaus Weber, from Munich, successfully treated a patient with ECM and meningitis with antibiotics (penicillin and chloramphenicol) and concluded that this nervous disorder was a bacterial illness (Weber, 1974). He ruled out spirochetes as the causative agent because "these microorganisms are transmitted by argasid and not ixodid ticks." He also referred to the negative results of S. Hard who attempted to infect larval, nymphal, and adult *I. ricinus* by allowing them to feed on the erythematous ring of an ECM lesion (Hard, 1966). Subsequent darkfield examination of tick tissues showed no spirochetes, and transmission experiments with nymphal ticks that had fed on the lesion as larvae proved negative. He pursued further the rickettsial hypothesis and examined the sera of 13 ECM patients for the detection of antibodies to 14 different rickettsial antigens (Sandbank and Feuerman, 1979; Weber, 1981). All tests were negative.

Garin-Bujadoux, Bannwarth, and the neurological spectrum

In 1922, two French neurologists, Charles Garin and Antoine Bujadoux, from Lyon, gave the first description of neurologic complications after a tick bite and of a skin lesion that, although not recognized as such, was compatible with erythema migrans (Garin and Bujadoux, 1922). They reported on a 58-year-old male who developed a redish, painful and expanding skin lesion, accompanied by inguinal lymphadenopathy, three weeks after having been bitten by a tick on his left buttock. This was followed by excruciating bilateral sciatic pain, intercostal pain, and finally a painful unilateral right brachial plexopathy with amyotrophy of the deltoid muscle. He required morphine, obtaining only partial relief. Four months after disease onset the pain had decreased significantly, and he was left with deltoid amyotrophy and weakness. Serologic test for syphilis (Wassermann test) was weakly positive, but the authors were aware that "in certain tick-transmitted diseases such as relapsing fever and Rocky Mountain Spotted Fever these test can be positive"[sic]. They ruled out syphilis as the cause and presented the case as an example of "paralysie par les tiques" (tick palsy), although they also argued

Table 1. Milestones in the history of Lyme borreliosis in Europe

Year	Author, country of origin	Hallmark
1883	Alfred Buchwald, Germany	Description of acrodermatitis chornica atrophicans (ACA)
1902	Herxheimer and Hartman	Further characterization of ACA
1910	Arvid Azfelius, Sweden	Description of the erythema migrans (EM)
1918	Lipschütz, Austria	Suggested the term erythema chronicum migrans
1922	Garin and Bujadoux, France	Description of meningopolyneuritis after tick bite and probable EM
1930	Hellerström, Sweden	Description of meningitis following EM. Suggested common etiology for both conditions
1943	Bo Bäfverstedt, Sweden	Description of lymphadenosis benigna cutis
1944	Bannwarth, Germany	Further descriptions of meningopolyneuritis after EM
1948	Carl Lennhoff, Germany	Reported seeing spirochetes in biopsy specimens (turned out to be artifacts)
1949	Hellerström, Sweden	Suggested that a spirochete might be involved
1949	Thyresson, Sweden	Successful treatment of ACA with penicillin
1951	Hollstrom, Sweden	Successful treatment of EM with penicillin
1954	Gotz, Germany	Transmission of ACA among humans
1955	Binder et al., Germany	Transmission of EM among humans
1974	Weber, Germany	Successful response of ECM to antibiotics
1985	Ackermann et al, Germany	Description of central nervous system involvement
1984-1985	Ackermann et al., Asbrink et al., Wislke et al., Germany and Sweden	Cultivation of *B. burgdorferi* in skin and fluids

that this disorder could be caused by a tick-borne virus. However, in contrast to their patient, they noted that those cases of tick palsy in the literature were not accompanied by a previous skin lesion. Furthermore, tick palsy is usually caused by Dermacentor tick bites, while their patient had been bitten by *Ixodes hexagonus*, the hedgehog tick.

Between 1941 and 1944, Bannwarth (Bannwarth, 1941, 1944) in Germany reported several patients who suffered meningitis with evidence of cranial or peripheral neuropathy. In his first report, he described 14 patients with lymphocytic meningitis and severe radicular pain in the trunk and extremities. Pain was migrating, neuralgic and more intense at night. Facial palsy was present in 6 patients, unilateral in two of them, and bilateral in four. Bannwarth, however, did not make the connection between prior tick bites or ECM, which one of the patients seemed to have had, and the neurological disturbances favored, instead, an allergic or rheumatic etiology.

In 1985 Ackermann et al. (Ackermann et al., 1985) reported on eight patients with chronic progressive encephalomyelitis and intrathecal synthesis of specific antibodies against *B. burgdorferi* who improved following parenteral antibiotic therapy. This is the first report of central nervous system involvement in Lyme borreliosis and was followed by larger series (Ackermann et al., 1987). The protean manifestations of the neurological disorder created some confusion with other neurological diseases of uncertain origin, including multiple sclerosis or Guillain-Barré syndrome, among others (Kristoferitsch, 1991; Schmutzhard et al., 1989). In the 1970s and 1980s, Horstrup and Ackerman reviewed the manifestations of this entity in Europe. They (Hörstrup and Ackermann, 1973), and others (Weber, 1974) insisted on an infectious etiology even before the spirochetal origin of Lyme disease was discovered.

Acrodermatitis chronica atrophicans (ACA) and Lymphadenosis benigna cutis

ACA was first described in the late 19th century by Buchwald (Buchwald, 1883) and then clearly characterized by Herxheimer and Hartmann in 1902 (Herxheimer and Hartman, 1902). Many patients with this intriguing disorder remembered having been bitten by the sheep tick, *I. ricinus,* and some reported subsequent development of ECM-like skin lesions. In 1942, R. H. Kahie, a German dermatologist noted that six of seven patients with ACA had a positive serologic response against *Treponema pallidum.* Ten years later, his mentor, Dr. Grüneberg, confirmed these results and speculated that the reactions must have been caused by a group-specific spirochete as the causative agent of ACA (Grüneberg, 1954). In the 1940s, the beneficial effect of penicillin on ACA was demonstrated (Thyresson, 1949), and Gotz (Gotz, 1954, 1955) showed in the 1950s that it was a transmissible condition. He successfully transplanted affected skin of a patient into four volunteers, including himself, and recorded a clinical picture similar to ACA (i.e., inflammation that lasted for as long as 312 days and promptly disappeared after treatment with penicillin). Gotz speculated the etiologic agent to be a large virus, but, lacking evidence, he concluded that the cause of ACÁ would continue to be an open question. In 1955, Lohel (Lohel, 1955) showed antibodies against spirochetes in mice inoculated with blood from ACA patients. His claim of ACA being a spirochetosis, however, was rejected. The geographical distribution in Europe and the observation that most patients had been from rural areas further strenghthened the suspicion that the tick *Ixodes ricinus* was the vector.

Lymphadenosis benigna cutis, or borrelial lymphocytoma, was first described by Bäfverstedt (Bäfverstedt, 1943), another Swedish dermatologist, in 1943 as a benign skin condition characterized by the appearance of lymphoreticular tissue in the epidermis and subcutaneous layers. He also mentioned a possible association with cases of ACA. In the ensuing years, it was successfully transmitted and treated with penicillin, again reinforcing the idea of an infectious etiology.

A missing articular manifestation in Europe?

The rheumatological aspects of Lyme disease became evident after the initial epidemic of arthritis in Lyme, Connecticut was first reported (see above). On the other hand, the European descriptions of this disease consisted primarily of a cutaneous and neurological disorder. This notion, that American Lyme disease primarily affects the joints, while European Lyme disease is primarily neurological, may represent an acquisition bias. That dermatologists and neurologists in Europe were well aware of this disease, while in the United States rheumatologists led by Stephen Malawista became involved in the arthritis epidemic may at least partially account for this notion. However, a German patient (Herzer et al. in 1976) developed an expanding skin lesion, neurological abnormalities and arthritis following a tick bite(Herzer et al., 1983). This patient insisted on her disease being related to the prior tick bite but was not diagnosed until she read an article on the lay press about the presentation of Lyme disease in the United States. She was then told that Lyme arthritis did not exist in Europe.

Although joint manifestations had been observed in European patients with erythema migrans, ACA, and lymphocytic meningitis (Bannwarth, 1941; Benjamowistch and Maschkilleisson, 1933; Erbslöh and Kohlmeyer, 1968; Gans, 1933, 1952; Hoövelborn, 1931; Hopf, 1966; Jessner, 1922; Jessner and Loewenstamm, 1924; Schaltenbrand, 1949), the initial reports on European Lyme disease after the discovery of its etiology basically neglected the rheumatological aspects to the point that they were regarded as curiosities (Ackermann et al., 1980; Hewitt et al., 1980). In fact, the Second International Symposium on Lyme Disease and Related Disorders, held in Vienna in 1985, did not include any studies by European rheumatologists. In the meantime, an increasing number of reports of Lyme arthritis from several European countries have accrued, suggesting that this joint manifestations may have been overlooked or at least underestimated (Herzer, 1983, 1991; Herzer et al., 1983).

Linking the etiology of the European syndromes to *Borrelia burgdorferi*

In the early 1980s, the clinical spectrum of a disorder of still unknown etiology had been clearly delineated, particularly in its cutaneous and neurological manifestations, and investigators had concluded that it was caused by an infectious agent, probably a spirochete, and that it was transmitted by Ixodid ticks. Soon after the discovery of the Lyme agent in the United States, *B. burgdorferi* was finally isolated from European patients with EM by Swedish and German investigators (Ackermann et al., 1984; Asbrink et al., 1984; Wilske et al., 1985). In 1983, Asbrink et al (Asbrink et al., 1984) from Sweden, succeeded for the first time in isolating *B. burgdorferi* from the skin of a patient with ACA , thus demonstrating that this cutaneous disease was also part of the spectrum of Lyme disease.

The availability of serologic testing following the discovery of *B. burgdorferi* enabled investigators to obtain evidence for the spirochetal etiology of the meningopolyneuritis of Garin-Bujadoux and Bannwarth and other less frequent neurological. Cultivation of the spirochete from the cerebrospinal fluid of some patients confirmed the conclusions drawn from serologic studies (Allal et al., 1986; Pfister et al., 1989; Preac Mursic et al., 1984).

Clinical research on Lyme borreliosis, and particularly, in neuroborreliosis since the early days has continued at a great pace. A recent review of the outcomes of neuroborreliosis cases in Europe and America (to 2019) disclosed that this disease can be treated effectively with antibiotics. However, this review also showed that there are some lingering issues that suggest either unusual severity or insufficient therapy in a minority of studies (Garcia-Monco and Benach, 2019). Since this review was published, additional, comprehensive studies have demonstrated the benefits of large-scale approaches to this disease. A meta-analysis of Lyme disease serology in Europe observed that heterogeneity of results and risk of bias complicated the extrapolation of the results to clinical practice(Leeflang et al., 2016). A French study of 9594 patients with Lyme disease from 2011-2016 disclosed a mean annual incidence of 53 cases per 100,000. The study also verified focality and a bimodal age distribution (children and eldery) (Septfons et al., 2019). A nationwide population-based cohort study in Denmark showed that neuroborreliosis patients did not have a long-term risk of dementia, Parkinson's, motor neuron diseases, epilepsy or Guillain-Barre, but it did show a short-term risk for the latter two diseases(Haahr et al., 2019).

The Rocky Road of Lyme disease

The vaccine

The scientific body of work that led to the development of a commercial vaccine for Lyme disease can be considered a major outcome of the interdisciplinary approach that uncovered the major biological features of *B. burgdorferi* and its natural history in both the vector and the vertebrate host. The cationic lipoproteins, OspA and OspB (Benach et al., 1988; Bergstrom et al., 1989; Brandt et al., 1990), which are prominent in the electrophoretic profile of cultured *B. burgdorferi*, were the first to be studied for their role in eliciting an immune response (Barbour et al., 1983; Johnson et al., 1986). The protective role of antibodies to OspA was revealed by passive immunization of previously infected *scid* mice with anti-OspA monoclonal antibodies (Barthold and Bockenstedt, 1993; Fikrig et al., 1992b; Fikrig et al., 1994). Using sera from patients with Lyme disease, passive protection of mice was accomplished but only if the sera contained antibodies to OspA (Zhong et al., 1997). Moreover, two additional studies provided evidence for a role for passive immunization in the attenuation and resolution of chronic disease (Fikrig et al., 1990; Schaible et al., 1990). The lipid moiety turned out to be critical for the immunogenicity of OspA, since only the full-length lipidated molecule protected mice from challenge (Bockenstedt et al., 1993; Brandt et al., 1990; Erdile et al., 1993; Weis et al., 1994). Clearly, the lipidation is important biologically but also for conferring immunogenicity to the molecule. In addition, antigens expressed by *B. burgdorferi* in culture could be different from those present in ticks or the mammalian host (Gern et al., 1993; Schaible et al., 1993a). This observation led to studies on the extent and nature of the protection that could be elicited following challenge with antigenically diverse strains of this organism. Furthermore, determining if vaccine preparations could protect from challenge with spirochetes transmitted by ticks became critical. The antibody response to tick or needle inoculated cultured spirochetes is dependent on the amount of antigen given. Inocula of 1000 cultured spirochetes or less, or by tick exposure, generated antibodies to a diverse group of antigens but not to OspA/B. In contrast, higher doses of cultured organisms developed antibodies to these two antigens (Gern et al., 1997; Schaible et al., 1993b).

There were also geographic differences in some areas of the American continent; protection could be achieved by OspA vaccination followed by challenge with infected ticks (Telford et al., 1993). Although OspA can induce a robust antibody response when administered as a purified protein, it had long been recognized that Lyme disease patients do not develop antibodies to OspA in the early stages of the disease. This ostensible paradox was subsequently explained on the basis that expression of OspA is downregulated as the tick begins to feed on a host (Schwan et al., 1995). In this manner, the OspA vaccine would induce an antibody to an antigen that is not present in the host as it is lost during transmission from the tick to the host. If this was the case, how can the OspA subunit vaccine be protective? Attempts to answer this question showed that the vaccine was protective of the host by eliminating the organisms that expressed OspA in the feeding tick (de Silva et al., 1996; Fikrig et al., 1992a). Thus, the mode of action of the OspA vaccine is to block transmission by inducing the formation of antibodies that eliminate spirochetes in the midgut of the feeding tick (de Silva et al., 1996). That antibodies had an effect in the vector underscores the uniqueness of this vaccine paradigm. Antibodies to OspA have a remote effect on *B. burgdorferi*, affecting the organisms outside of the host where the antibodies were developed. This was a novel approach to a vaccine (Connolly and Benach, 2005).

Based on the effectiveness of the antibody response to OspA, two commercial recombinant lipidated, adjuvant-containing, OspA vaccines were formulated, tested and marketed in the United States. The United States Food and Drug Administration (FDA) licensed a vaccine, LYMErix, on December 21, 1998, for persons 15 to 70 years of age. This licensed vaccine was a subunit formulation containing recombinant OspA (rOspA). Two vaccine studies demonstrated the effectiveness of a triple-dose vaccine regimen that protected 100% of the volunteers after the second year in one trial (Steere et al., 1998) and 92% in another (Sigal et al., 1998). Three doses of 0.5 mL (30 µg) of rOspA vaccine administered by intramuscular injection at 0, 1, and 12 months were required for optimal protection in subjects who were between 15 and 70 years of age. The duration of protective immunity for longer than 1 year after the third dose of immunization was not determined at that time. Unfortunately, safety and efficacy of the rOspA vaccine were not established for persons older than 70 years or younger than 15 years of age, two age groups that have been traditionally overrepresented in this disease (Steigbigel and Benach, 1998). Volunteers and patients that received this vaccine developed a positive serologic pattern that distinguished their response from that of patients.

As of February 2002, the LYMErix vaccine was withdrawn from the commercial market in the United States. The reasons for its removal from the market were poor market penetration, not issues related to the vaccine's safety, immunogenicity, or efficacy. Cross-reactivity of OspA with self-antigens may have caused some doubts about the possibility of an autoimmune response (Gross et al., 1998). IgG antibodies to OspA had been shown to correlate with severe and prolonged arthritis, furthering additional concerns about autoreactivity (Akin et al., 1999; Krause et al., 1992a).

On January 22, 2002, officers from the Lyme Disease Association (LDA) and International Lyme and Associated Disease Society (ILADS) met with both scientists and administrators from the Food and Drug Administration (FDA) to express concerns about the vaccine's track record on safety and effectiveness. This meeting was made possible at the request of a Congressman from New Jersey. The evidence presented included the occurrence of joint and muscle pain after receiving the vaccine. On February 25, a month or so after meeting, the LDA received written answers to their questions from FDA officials. Coincidentally, on or about the same time, Glaxo SmithKline withdrew LYMErix from the market. Although this chapter ends here, not so the history of Lyme Borreliosis, as told in the remaining 23 chapters of this volume.

References

Ackermann, R., Gollmer, E., and Rehse-Kupper, B. (1985). [Progressive Borrelia encephalomyelitis. Chronic manifestation of erythema chronicum migrans disease of the nervous system]. Dtsch Med Wochenschr *110*, 1039-1042.

Ackermann, R., Horstrup, P., and Schmidt, R. (1984). Tick-borne meningopolyneuritis (Garin-Bujadoux, Bannwarth). Yale J Biol Med *57*, 485-490.

Ackermann, R., Rehse-Küpper, B., Gollmer, E., and Schmidt, R. (1987). Chronic neurologic manifestations of erythema migrans borreliosis. Ann NY Acad Sci *539*, 16-23.

Ackermann, R., Runne, U., Klenk, W., and Dienst, C. (1980). [Erythema chronicum migrans with arthritis (author's transl)]. Dtsch Med Wochenschr *105*, 1779-1781.

Adeolu, M., and Gupta, R.S. (2014). A phylogenomic and molecular marker based proposal for the division of the genus Borrelia into two genera: the emended genus Borrelia containing only the members of the relapsing fever Borrelia, and the genus Borreliella gen. nov. containing the members of the Lyme disease Borrelia (Borrelia burgdorferi sensu lato complex). Antonie Van Leeuwenhoek *105*, 1049-1072. https://doi.org/10.1007/s10482-014-0164-x

Akin, E., McHugh, G.L., Flavell, R.A., Fikrig, E., and Steere, A.C. (1999). The immunoglobulin (IgG) antibody response to OspA and OspB correlates with severe and prolonged Lyme arthritis and the IgG response to P35 correlates with mild and brief arthritis. Infect Immun *67*, 173-181.

Allal, J., Thomas, P., and Mazzonelli, J. (1986). Borrelia isolated from cerebrospinal fluid in a French case of Lyme disease. Annals of the rheumatic diseases *45*, 789-790.

Anderson, J.F., Johnson, R.C., Magnarelli, L.A., Hyde, F.W., and Myers, J.E. (1986). Peromyscus leucopus and Microtus pennsylvanicus simultaneously infected with Borrelia burgdorferi and Babesia microti. J Clin Microbiol *23*, 135-137.

Asbrink, E., Hederstedt, B., and Hovmark, A. (1984). The spirochetal etiology of erythema chronicum migrans Afzelius. Acta Derm Venereol *64*, 291-295.

Askani, H. (1936). Zur Ätiologie des Erythema chronicum migrans. Dermatol Wochenschr *102*, 125.

Azfelius, A. (1910). Verhandlungen der dermatologischen gesellschaft zu Stockholm, 28 Oct 1909. Arch Derm Syph *101*, 404.

Azfelius, A. (1921). Erythema chronicum migrans. Acta Derm Venereol 2, 120-125.

Bäfverstedt, B. (1943). Über Lymphadenosis benigna cutis. Eine klinische und patologisch-anatomische Studie. (Stockholm).

Bannwarth, A. (1941). Chronische lymphocytare meningitis, entzundliche polyneuritis und "rheumatismus" eing beitrag zumn problem "allergic und nervensystem". Arch Psychiatr Nervenkr *113*, 284-376.

Bannwarth, A. (1944). Zur klinik und patohogenese der "chronischen lymphocytare meningitis". Arch Psychiatr Nervenkr *117*, 161-185.

Baranton, G., Postic, D., Saint Girons, I., Boerlin, P., Piffaretti, J.C., Assous, M., and Grimont, P.A. (1992). Delineation of Borrelia burgdorferi sensu stricto, Borrelia garinii sp. nov., and group VS461 associated with Lyme borreliosis. Int J Syst Bacteriol *42*, 378-383.

Barbour, A.G., Adeolu, M., and Gupta, R.S. (2017). Division of the genus Borrelia into two genera (corresponding to Lyme disease and relapsing fever groups) reflects their genetic and phenotypic distinctiveness and will lead to a better understanding of these two groups of microbes

(Margos et al. (2016) There is inadequate evidence to support the division of the genus Borrelia. Int. J. Syst. Evol. Microbiol. doi: 10.1099/ijsem.0.001717). Int J Syst Evol Microbiol *67*, 2058-2067. https://doi.org/10.1099/ijsem.0.001815

Barbour, A.G., and Benach, J.L. (2019). Discovery of the Lyme Disease Agent. MBio *10*. https://doi.org/10.1128/mBio.02166-19

Barbour, A.G., Tessier, S.L., and Stoenner, H.G. (1982). Variable major proteins of Borrellia hermsii. J Exp Med *156*, 1312-1324.

Barbour, A.G., Tessier, S.L., and Todd, W.J. (1983). Lyme disease spirochetes and ixodid tick spirochetes share a common surface antigenic determinant defined by a monoclonal antibody. Infect Immun *41*, 795-804.

Barthold, S.W., and Bockenstedt, L.K. (1993). Passive immunizing activity of sera from mice infected with Borrelia burgdorferi. Infect Immun *61*, 4696-4702.

Barthold, S.W., Moody, K.D., Terwilliger, G.A., Jacoby, R.O., and Steere, A.C. (1988). An animal model for Lyme arthritis. Ann N Y Acad Sci *539*, 264-273.

Belfaiza, J., Postic, D., Bellenger, E., Baranton, G., and Girons, I.S. (1993). Genomic fingerprinting of Borrelia burgdorferi sensu lato by pulsed-field gel electrophoresis. J Clin Microbiol *31*, 2873-2877.

Benach, J.L., Bosler, E.M., Hanrahan, J.P., Coleman, J.L., Habicht, G.S., Bast, T.F., Cameron, D.J., Ziegler, J.L., Barbour, A.G., Burgdorfer, W., *et al.* (1983). Spirochetes isolated from the blood of two patients with Lyme disease. N Engl J Med *308*, 740-742.

Benach, J.L., Coleman, J.L., and Golightly, M.G. (1988). A murine IgM monoclonal antibody binds an antigenic determinant in outer surface protein A, an immunodominant basic protein of the Lyme disease spirochete. J Immunol *140*, 265-272.

Benach, J.L., Coleman, J.L., Habicht, G.S., MacDonald, A., Grunwaldt, E., and Giron, J.A. (1985). Serological evidence for simultaneous occurrences of Lyme disease and babesiosis. J Infect Dis *152*, 473-477.

Benach, J.L., Coleman, J.L., Skinner, R.A., and Bosler, E.M. (1987). Adult Ixodes dammini on rabbits: a hypothesis for the development and transmission of Borrelia burgdorferi. J Infect Dis *155*, 1300-1306.

Benach, J.L., and Habicht, G.S. (1981). Clinical characteristics of human babesiosis. J Infect Dis *144*, 481.

Benach, J.L., Habicht, G.S., and Hamburger, M.I. (1982). Immunoresponsiveness in acute babesiosis in humans. J Infect Dis *146*, 369-380.

Benach, J.L., White, D.J., Burgdorfer, W., Keelan, T., Guirgis, S., and Altieri, R.H. (1977). Changing patterns in the incidence of Rocky Mountain Spotted Fever on Long Island (1971-1976). Am J Epidemiol *106*, 380-387.

Benach, J.L., White, D.J., and McGovern, J.P. (1978). Babesiosis in Long Island. Host-parasite relationships of rodent- and human-derived Babesia microti isolates in hamsters. Am J Trop Med Hyg *27*, 1073-1078.

Benach, J.L., White, D.J., McGovern, J.P., and Jacovina, M.M. (1979). Immunological relationships of Long Island isolates of Babesia microti. Am J Trop Med Hyg *28*, 643-648.

Benjamowistch, E., and Maschkilleisson, L.N. (1933). Witere Beiträge zur Präge über die Atrophie der Haut. III. Acrodermatitis cronica atrophicans und ihre Beziehungen zur Sklerodermia. Acta Derm Venereol (Stockh) *14*, 313-341.

Berger, B.W., Kaplan, M.H., Rothenberg, I.R., and Barbour, A.G. (1985). Isolation and characterization of the Lyme disease spirochete from the skin of patients with erythema chronicum migrans. J Am Acad Dermatol *13*, 444-449.

Bergstrom, S., Bundoc, V.G., and Barbour, A.G. (1989). Molecular analysis of linear plasmid-encoded major surface proteins, OspA and OspB, of the Lyme disease spirochaete Borrelia burgdorferi. Mol Microbiol *3*, 479-486.

Binder, E., Doepfmer, R., and Homstein, O. (1955). Experimentelle Übertragung des Erythema chronicum migrans von Mensch zu Mensch. Hautarzi *6*, 494.

Bockenstedt, L.K., Fikrig, E., Barthold, S.W., Kantor, F.S., and Flavell, R.A. (1993). Inability of truncated recombinant Osp A proteins to elicit protective immunity to Borrelia burgdorferi in mice. J Immunol *151*, 900-906.

Bosler, E.M., Coleman, J.L., Benach, J.L., Massey, D.A., Hanrahan, J.P., Burgdorfer, W., and Barbour, A.G. (1983). Natural Distribution of the Ixodes dammini spirochete. Science *220*, 321-322.

Brandt, M.E., Riley, B.S., Radolf, J.D., and Norgard, M.V. (1990). Immunogenic integral membrane proteins of Borrelia burgdorferi are lipoproteins. Infect Immun *58*, 983-991. https://doi.org/10.1128/IAI.58.4.983-991.1990

Buchwald, A. (1883). Ein fall von diffuser idiopathischcr Haunt-Atrofie. Arch Dermatol Syph *10*, 553-556.

Burgdorfer, W. (1954). On the occult infection in relapsing fevers. Bull Soc Pathol Exot Filiales *47*, 664-667.

Burgdorfer, W. (1975). A review of Rocky Mountain spotted fever (tick-borne typhus), its agent, and its tick vectors in the United States. J Med Entomol *12*, 269-278.

Burgdorfer, W., Barbour, A.G., Hayes, S.F., Benach, J.L., Grunwaldt, E., and Davis, J.P. (1982). Lyme disease-a tick-borne spirochetosis? Science *216*, 1317-1319.

Burgdorfer, W., Hayes, S.F., and Benach, J.L. (1988). Development of Borrelia burgdorferi in ixodid tick vectors. Ann N Y Acad Sci *539*, 172-179.

Burgdorfer, W., Lane, R.S., Barbour, A.G., Gresbrink, R.A., and Anderson, J.R. (1985). The western black-legged tick, Ixodes pacificus: a vector of Borrelia burgdorferi. Am J Trop Med Hyg *34*, 925-930.

Burgdorfer, W., and Mavros, A.J. (1970). Susceptibility of Various Species of Rodents to the Relapsing Fever Spirochete, Borrelia hermsii. Infect Immun *2*, 256-259.

Cahill, K.M., Benach, J.L., Reich, L.M., Bilmes, E., Zins, J.H., Siegel, F.P., and Hochweis, S. (1981). Red cell exchange: treatment of babesiosis in a splenectomized patient. Transfusion *21*, 193-198.

Canica, M.M., Nato, F., du Merle, L., Mazie, J.C., Baranton, G., and Postic, D. (1993). Monoclonal antibodies for identification of Borrelia afzelii sp. nov. associated with late cutaneous manifestations of Lyme borreliosis. Scand J Infect Dis *25*, 441-448.

Chisholm, E.S., Ruebush, T.K., 2nd, Sulzer, A.J., and Healy, G.R. (1978). Babesia microti infection in man: evaluation of an indirect immunofluorescent antibody test. Am J Trop Med Hyg *27*, 14-19.

Coleman, J.L., LeVine, D., Thill, C., Kuhlow, C., and Benach, J.L. (2005). Babesia microti and Borrelia burgdorferi follow independent courses of infection in mice. J Infect Dis *192*, 1634-1641.

Connolly, S.E., and Benach, J.L. (2005). The versatile roles of antibodies in Borrelia infections. Nat Rev Microbiol *3*, 411-420.

Dalsgaard-Nielsen, T. (1948). Allergic meningitis and chronic erythema migrans azfelii after bite by Ixodes reduvius. Acta Allerg (Kbh) *1*.

Dammin, G.J., Spielman, A., Benach, J.L., and Piesman, J. (1981). The rising incidence of clinical Babesia microti infection. Hum Pathol *12*, 398-400.

Dattwyler, R.J., Halperin, J.J., Pass, H., and Luft, B.J. (1987). Ceftriaxone as effective therapy in refractory Lyme disease. J Infect Dis *155*, 1322-1325.

Dattwyler, R.J., Halperin, J.J., Volkman, D.J., and Luft, B.J. (1988). Treatment of late Lyme borreliosis--randomised comparison of ceftriaxone and penicillin. Lancet *1*, 1191-1194.

Davis, G.E., and Burgdorfer, W. (1954). On the susceptibility of the guinea pig to the relapsing fever spirochete Borrelia duttonii. Bull Soc Pathol Exot Filiales *47*, 498-501.

de Silva, A.M., Telford, S.R., 3rd, Brunet, L.R., Barthold, S.W., and Fikrig, E. (1996). Borrelia burgdorferi OspA is an arthropod-specific transmission-blocking Lyme disease vaccine. J Exp Med *183*, 271-275.

Degos, R., Touraine, R., and Arouete, J. (1962). [Chronic erythema migrans. (Discussion of a rickettsial origin)]. Ann Dermatol Syphiligr (Paris) *89*, 247-260.

Erbslöh, F., and Kohlmeyer, K. (1968). Über polytope Erkrankungen des peripheren Nervensystems bei lymphozytärer Meningitis. Fortschr Neurol Psychiat *35*, 321-342.

Erdile, L.F., Brandt, M.A., Warakomski, D.J., Westrack, G.J., Sadziene, A., Barbour, A.G., and Mays, J.P. (1993). Role of attached lipid in immunogenicity of Borrelia burgdorferi OspA. Infect Immun *61*, 81-90.

Esernio-Jenssen, D., Scimeca, P.G., Benach, J.L., and Tenenbaum, M.J. (1987). Transplacental/perinatal babesiosis. J Pediatr *110*, 570-572.

Fikrig, E., Barthold, S.W., Kantor, F.S., and Flavell, R.A. (1990). Protection of mice against the Lyme disease agent by immunizing with recombinant OspA. Science *250*, 553-556.

Fikrig, E., Barthold, S.W., Kantor, F.S., and Flavell, R.A. (1992a). Long-term protection of mice from Lyme disease by vaccination with OspA. Infect Immun *60*, 773-777.

Fikrig, E., Barthold, S.W., Marcantonio, N., Deponte, K., Kantor, F.S., and Flavell, R.A. (1992b). Roles of OspA, OspB, and flagellin in protective immunity to Lyme borreliosis in laboratory mice. Infect Immun *60*, 657-661.

Fikrig, E., Bockenstedt, L.K., Barthold, S.W., Chen, M., Tao, H., Ali-Salaam, P., Telford, S.R., and Flavell, R.A. (1994). Sera from patients with chronic Lyme disease protect mice from Lyme borreliosis. J Infect Dis *169*, 568-574.

Filstein, M.R., Benach, J.L., White, D.J., Brody, B.A., Goldman, W.D., Bakal, C.W., and Schwartz, R.S. (1980). Serosurvey for human babesiosis in New York. J Infect Dis *141*, 518-521.

Fraser, C.M., Casjens, S., Huang, W.M., Sutton, G.G., Clayton, R., Lathigra, R., White, O., Ketchum, K.A., Dodson, R., Hickey, E.K., *et al.* (1997). Genomic sequence of a Lyme disease spirochaete, Borrelia burgdorferi. Nature *390*, 580-586. https://doi.org/10.1038/37551

Gans, O. (1933). Akrodermatitis atrophicans und Sklerodermie. Zbl Haut Geschl Kr *45*, 678-679.

Gans, O. (1952). Akrodermatitis atrophicans arthropathica. Hautarzt *3*, 151-155.

Garcia-Monco, J.C., and Benach, J.L. (2019). Lyme Neuroborreliosis: Clinical Outcomes, Controversy, Pathogenesis, and Polymicrobial Infections. Ann Neurol *85*, 21-31. https://doi.org/10.1002/ana.25389

Garin, C., and Bujadoux (1922). Paralysie par les tiques. J Med Lyon *71*, 765-767.

Gelbjerg-Hansen, G. (1945). Erythema chronicum migrans Azfelii and meningitis after a tick bite. Acta Derm Venereol (Stockh) *25*, 458-463.

Gern, L., Hu, C.M., Voet, P., Hauser, P., and Lobet, Y. (1997). Immunization with a polyvalent OspA vaccine protects mice against Ixodes ricinus tick bites infected by Borrelia burgdorferi ss, Borrelia garinii and Borrelia afzelii. Vaccine *15*, 1551-1557. https://doi.org/S0264-410X(97)00066-2 [pii]

Gern, L., Schaible, U.E., and Simon, M.M. (1993). Mode of inoculation of the Lyme disease agent Borrelia burgdorferi influences infection and immune responses in inbred strains of mice. J Infect Dis *167*, 971-975.

Gotz, H. (1954). Die Acrodermatitis chronica atrophicans Herx-heimer ais Infektionskrankheit. Hautarzt *5*, 491-504.

Gotz, H. (1955). Die Acrodermatitis chronica atrophicans Herx-heimer ais Infektionskrankheit—Erganzung zur I. Mittei-lung. Hautarzt *6*, 249-252.

Grabowski, E.F., Giardina, P.J., Goldberg, D., Masur, H., Read, S.E., Hirsch, R.L., and Benach, J.L. (1982). Babesiosis transmitted by a transfusion of frozen-thawed blood. Ann Intern Med *96*, 466-467.

Gross, D.M., Forsthuber, T., Tary-Lehmann, M., Etling, C., Ito, K., Nagy, Z.A., Field, J.A., Steere, A.C., and Huber, B.T. (1998). Identification of LFA-1 as a candidate autoantigen in treatment-resistant Lyme arthritis. Science *281*, 703-706.

Grüneberg, T. (1954). Auffällige serologische Refunde bei Akro-dermatitis chronica atrophicans Herxheimer Klin Wochenschr *32*, 935-936.

Grunwaldt, E., Barbour, A.G., and Benach, J.L. (1983). Simultaneous occurrence of babesiosis and Lyme disease. N Engl J Med *308*, 1166.

Haahr, R., Tetens, M.M., Dessau, R.B., Krogfelt, K.A., Bodilsen, J., Andersen, N.S., Moller, J.K., Roed, C., Christiansen, C.B., Ellermann-Eriksen, S., *et al.* (2019). Risk of neurological disorders in patients with European Lyme neuroborreliosis. A nationwide population-based cohort study. Clin Infect Dis. https://doi.org/10.1093/cid/ciz997

Habicht, G.S., Benach, J.L., Leichtling, K.D., Gocinski, B.L., and Coleman, J.L. (1983). The effect of age on the infection and immunoresponsiveness of mice to Babesia microti. Mech Ageing Dev *23*, 357-369.

Hard, S. (1966). Erythema chronicum migrans (Afzelii) associated with mosquito bite. Acta Derm Venereol *46*, 473-476.

Hellerstrom, S. (1930). Erythema chronicum migrans Azfelii. Acta Derm Venereol (Stockh) *11*, 315-321.

Hellström, S. (1950). Erythema chronicum migrans Azfelius with meningitis. South Med J *43*, 330-334.

Herxheimer , K., and Hartman, K. (1902). Über Acrodermatitis chronica atrophicans. Arch Derm Syph *61*, 255-300.

Herzer, P. (1983). [Lyme disease. A new or newly recognized clinical entity]. MMW Munch Med Wochenschr *125*, 737-738.

Herzer, P. (1991). Joint manifestations of Lyme borreliosis in Europe. Scand J Infect Dis Suppl *77*, 55-63.

Herzer, P., Schattenkirchner, M., and Zöllner, N. (1983). Lyme arthritis-eine zu selten bedachte Diagnose? . Verh Dtsch Ges Inn Med *89*, 299-302.

Hewitt, J., Illouz, G., Giroin, D., Dolivo, M., Leibowitch, M., and Gutman, L. (1980). Erythema chronicum migrans suivi d'une polyarthrite séronégative. Premier cas européen d'arthrite de Lyme? . In Journées de la Societé francaise de Dermatologie (Paris).

Hilton, E., DeVoti, J., Benach, J.L., Halluska, M.L., White, D.J., Paxton, H., and Dumler, J.S. (1999). Seroprevalence and seroconversion for tick-borne diseases in a high-risk population in the northeast United States. Am J Med *106*, 404-409. https://doi.org/S0002-9343(99)00046-7 [pii]

Hollström, S. (1951). Successful treatment of erythema migrans Afzelius. Acta Derm Venereol (Stockh) *31*, 235-234.

Homer, M.J., Aguilar-Delfin, I., Telford, S.R., 3rd, Krause, P.J., and Persing, D.H. (2000). Babesiosis. Clin Microbiol Rev *13*, 451-469.

Hoövelborn, C. (1931). Gelenkveränderungen bei Acrodermatitis cronica atrophicans. Arch Derm Syph (Berlin) *164*, 349-356.

Hopf, H.C. (1966). [Acrodermatitis chronica atrophicans (Herxheimer) and the nervous system. Analysis of clinical, physiologic, histologic and electromyographic findings]. Monographien aus dem Gesamtgebiete der Neurologie und Psychiatrie *114*, 1-130.

Hörstrup, P., and Ackermann, R. (1973). Durch zecken übertragene meningopolyneuritis (Garin-Bujadoux, Bannwarth). Fortschr Neurol Psychiat *41*.

Jessner, M. (1922). Weiterer Beitag zur Kenntnis der Acrodermatitis chronica Atrophican. Arch Derm Syph (Berlin) *164*, 349-356.

Jessner, M., and Loewenstamm, A. (1924). Bericht über 66 Fälle von Acrodermatitis cronica atrophican. Dermatol Wochenschr *79*, 1171-1177.

Johnson, R.C. (1977). The spirochetes. Annu Rev Microbiol *31*, 89-106. https://doi.org/10.1146/annurev.mi.31.100177.000513

Johnson, R.C., Hyde, F.W., and Brenner, D.J. (1984). Borrelia burgdorferi sp.nov.: etiological agent of Lyme disease. Int J Syst Bacteriol *34*, 496-497.

Johnson, R.C., Kodner, C., and Russell, M. (1986). Passive immunization of hamsters against experimental infection with the Lyme disease spirochete. Infect Immun *53*, 713-714.

Jones, C.G., Ostfeld, R.S., Richard, M.P., Schauber, E.M., and Wolff, J.O. (1998). Chain reactions linking acorns to gypsy moth outbreaks and Lyme disease risk. Science *279*, 1023-1026.

Kawabata, H., Masuzawa, T., and Yanagihara, Y. (1993). Genomic analysis of Borrelia japonica sp. nov. isolated from Ixodes ovatus in Japan. Microbiol Immunol *37*, 843-848.

Kelly, R. (1971). Cultivation of Borrelia hermsi. Science *173*, 443-444.

Krause, A., Burmester, G.R., Rensing, A., Schoerner, C., Schaible, U.E., Simon, M.M., Herzer, P., Kramer, M.D., and Wallich, R. (1992a). Cellular immune reactivity to recombinant OspA and flagellin from Borrelia burgdorferi in patients with Lyme borreliosis. Complexity of humoral and cellular immune responses. J Clin Invest *90*, 1077-1084. https://doi.org/10.1172/JCI115923

Krause, P.J., Telford, S.R., 3rd, Pollack, R.J., Ryan, R., Brassard, P., Zemel, L., and Spielman, A. (1992b). Babesiosis: an underdiagnosed disease of children. Pediatrics *89*, 1045-1048.

Krause, P.J., Telford, S.R., 3rd, Spielman, A., Sikand, V., Ryan, R., Christianson, D., Burke, G., Brassard, P., Pollack, R., Peck, J., *et al.* (1996). Concurrent Lyme disease and babesiosis. Evidence for increased severity and duration of illness. JAMA *275*, 1657-1660.

Kristoferitsch, W. (1991). Neurological manifestations of Lyme borreliosis: Clinical definition and differential diagnosis. Scand J Infect Dis *Suppl 77*, 64-73.

Leeflang, M.M., Ang, C.W., Berkhout, J., Bijlmer, H.A., Van Bortel, W., Brandenburg, A.H., Van Burgel, N.D., Van Dam, A.P., Dessau, R.B., Fingerle, V., *et al.* (2016). The diagnostic accuracy of serological tests for Lyme borreliosis in Europe: a systematic review and meta-analysis. BMC Infect Dis *16*, 140. https://doi.org/10.1186/s12879-016-1468-4

Lennhoff, C. (1948). Spirochetes in aetiologically obscure diseases. Acta Derm Venereol (Stockh) *28*, 295.

Lipschutz, B. (1918). Uber eine seltene erytheform (erythema chronicum migrans). Arch Dermatol Syph *118*, 349-356.

Lissman, B.A., and Benach, J.L. (1980). Rocky Mountain spotted fever in dogs. J Am Vet Med Assoc *176*, 994-995.

Lissman, B.A., Bosler, E.M., Camay, H., Ormiston, B.G., and Benach, J.L. (1984). Spirochete-associated arthritis (Lyme disease) in a dog. J Am Vet Med Assoc *185*, 219-220.

LoGiudice, K., Ostfeld, R.S., Schmidt, K.A., and Keesing, F. (2003). The ecology of infectious disease: effects of host diversity and community composition on Lyme disease risk. Proc Natl Acad Sci U S A *100*, 567-571.

Lohel, H. (1955). [Experimental studies on etiology of Herxheimer's acrodermatitis chronica atrophicans.]. Klin Wochenschr *33*, 185-186.

Margos, G., Fedorova, N., Kleinjan, J.E., Hartberger, C., Schwan, T.G., Sing, A., and Fingerle, V. (2017). Borrelia lanei sp. nov. extends the diversity of Borrelia species in California. Int J Syst Evol Microbiol *67*, 3872-3876. https://doi.org/10.1099/ijsem.0.002214

Margos, G., Gofton, A., Wibberg, D., Dangel, A., Marosevic, D., Loh, S.M., Oskam, C., and Fingerle, V. (2018). The genus Borrelia reloaded. PLoS One *13*, e0208432. https://doi.org/10.1371/journal.pone.0208432

Mast, W.E., and Burrows, W.M. (1976a). Erythema chronicum migrans and "lyme arthritis". JAMA *236*, 2392.

Mast, W.E., and Burrows, W.M., Jr. (1976b). Erythema chronicum migrans in the United States. JAMA *236*, 859-860.

Meldrum, S.C., Birkhead, G.S., White, D.J., Benach, J.L., and Morse, D.L. (1992). Human babesiosis in New York State: an epidemiological description of 136 cases. Clin Infect Dis *15*, 1019-1023.

Moro, M.H., Zegarra-Moro, O.L., Bjornsson, J., Hofmeister, E.K., Bruinsma, E., Germer, J.J., and Persing, D.H. (2002). Increased arthritis severity in mice coinfected with Borrelia burgdorferi and Babesia microti. J Infect Dis *186*, 428-431. https://doi.org/JID011441 [pii]

Ostfeld, R.S., Canham, C.D., Oggenfuss, K., Winchcombe, R.J., and Keesing, F. (2006). Climate, deer, rodents, and acorns as determinants of variation in lyme-disease risk. PLoS Biol *4*, e145. https://doi.org/05-PLBI-RA-1433R3 [pii] 10.1371/journal.pbio.0040145

Pfister, H.W., Preac-Mursic, V., Wilske, B., Einhaupl, K.M., and Weinberger, K. (1989). Latent Lyme neuroborreliosis: presence of Borrelia burgdorferi in the cerebrospinal fluid without concurrent inflammatory signs. Neurology *39*, 1118-1120.

Piesman, J., Hicks, T.C., Sinsky, R.J., and Obiri, G. (1987). Simultaneous transmission of Borrelia burgdorferi and Babesia microti by individual nymphal Ixodes dammini ticks. J Clin Microbiol *25*, 2012-2013.

Piesman, J., Mather, T.N., Telford, S.R., 3rd, and Spielman, A. (1986). Concurrent Borrelia burgdorferi and Babesia microti infection in nymphal Ixodes dammini. J Clin Microbiol *24*, 446-447.

Piesman, J., and Spielman, A. (1980). Human babesiosis on Nantucket Island: prevalence of Babesia microti in ticks. Am J Trop Med Hyg *29*, 742-746.

Piesman, J., Spielman, A., Etkind, P., Ruebush, T.K., 2nd, and Juranek, D.D. (1979). Role of deer in the epizootiology of Babesia microti in Massachusetts, USA. J Med Entomol *15*, 537-540.

Preac Mursic, V., Wilske, B., Schierz, G., Pfister, H.W., and Einhaupl, K. (1984). Repeated isolation of spirochetes from the cerebrospinal fluid of a patient with meningoradiculitis Bannwarth. European journal of clinical microbiology *3*, 564-565.

Reik, L., Steere, A.C., Bartenhagen, N.H., Shope, R.E., and Malawista, S.E. (1979). Neurologic abnormalities of Lyme disease. Medicine (Baltimore) *58*, 281-294.

Ribeiro, J.M., Mather, T.N., Piesman, J., and Spielman, A. (1987). Dissemination and salivary delivery of Lyme disease spirochetes in vector ticks (Acari: Ixodidae). J Med Entomol *24*, 201-205.

Rosa, P., Samuels, D.S., Hogan, D., Stevenson, B., Casjens, S., and Tilly, K. (1996). Directed insertion of a selectable marker into a circular plasmid of Borrelia burgdorferi. J Bacteriol *178*, 5946-5953.

Rosa, P.A., Tilly, K., and Stewart, P.E. (2005). The burgeoning molecular genetics of the Lyme disease spirochaete. Nat Rev Microbiol *3*, 129-143. https://doi.org/nrmicro1086 [pii] 10.1038/nrmicro1086

Rosner, F., Zarrabi, M.H., Benach, J.L., and Habicht, G.S. (1984). Babesiosis in splenectomized adults. Review of 22 reported cases. Am J Med *76*, 696-701. https://doi.org/0002-9343(84)90298-5 [pii]

Rudzinska, M.A., Trager, W., Lewengrub, S.J., and Gubert, E. (1976). An electron microscopic study of Babesia microti invading erythrocytes. Cell Tissue Res *169*, 323-334.

Ruebush, T.K., 2nd, Juranek, D.D., Chisholm, E.S., Snow, P.C., Healy, G.R., and Sulzer, A.J. (1977). Human babesiosis on Nantucket Island. Evidence for self-limited and subclinical infections. N Engl J Med *297*, 825-827.

Ruebush, T.K., 2nd, Juranek, D.D., Spielman, A., Piesman, J., and Healy, G.R. (1981). Epidemiology of human babesiosis on Nantucket Island. Am J Trop Med Hyg *30*, 937-941.

Samuels, D.S. (1995). Electrotransformation of the spirochete Borrelia burgdorferi. Methods Mol Biol *47*, 253-259. https://doi.org/10.1385/0-89603-310-4:253

Samuels, D.S., Mach, K.E., and Garon, C.F. (1994a). Genetic transformation of the Lyme disease agent Borrelia burgdorferi with coumarin-resistant gyrB. J Bacteriol *176*, 6045-6049.

Samuels, D.S., Marconi, R.T., Huang, W.M., and Garon, C.F. (1994b). gyrB mutations in coumermycin A1-resistant Borrelia burgdorferi. J Bacteriol *176*, 3072-3075.

Sandbank, M., and Feuerman, E.J. (1979). Ultrastructural observation of rickettsia-like bodies in erythema cronicum migrans. J Cutan Pathol *6*, 253-264.

Schaible, U.E., Gern, L., Wallich, R., Kramer, M.D., Prester, M., and Simon, M.M. (1993a). Distinct patterns of protective antibodies are generated against Borrelia burgdorferi in mice experimentally inoculated with high and low doses of antigen. Immunol Lett *36*, 219-226.

Schaible, U.E., Kramer, M.D., Eichmann, K., Modolell, M., Museteanu, C., and Simon, M.M. (1990). Monoclonal antibodies specific for the outer surface protein A (OspA) of Borrelia burgdorferi prevent Lyme borreliosis in severe combined immunodeficiency (scid) mice. Proc Natl Acad Sci U S A *87*, 3768-3772.

Schaible, U.E., Wallich, R., Kramer, M.D., Gern, L., Anderson, J.F., Museteanu, C., and Simon, M.M. (1993b). Immune sera to individual Borrelia burgdorferi isolates or recombinant OspA thereof protect SCID mice against infection with homologous strains but only partially or not at all against those of different OspA/OspB genotype. Vaccine *11*, 1049-1054.

Schaltenbrand, G. (1949). Chronische aseptische Meningitis. Nervenarzt *20*, 433-442.

Schmutzhard, E., Pohl, P., and Stanek, G. (1989). Failure to demonstrate B. burgdorferi as an etiologic agent in multiple sclerosis. Zbl Bakt *Suppl 18*, 219-226.

Schwan, T.G., Piesman, J., Golde, W.T., Dolan, M.C., and Rosa, P.A. (1995). Induction of an outer surface protein on Borrelia burgdorferi during tick feeding. Proc Natl Acad Sci U S A *92*, 2909-2913.

Scrimenti, R.J. (1970). Erythema chronicum migrans. Arch Dermatol *102*, 104-105.

Septfons, A., Goronflot, T., Jaulhac, B., Roussel, V., De Martino, S., Guerreiro, S., Launay, T., Fournier, L., De Valk, H., Figoni, J., *et al.* (2019). Epidemiology of Lyme borreliosis through two surveillance systems: the national Sentinelles GP network and the national hospital discharge database, France, 2005 to 2016. Euro Surveill *24*. https://doi.org/10.2807/1560-7917.ES.2019.24.11.1800134

Sigal, L.H., Zahradnik, J.M., Lavin, P., Patella, S.J., Bryant, G., Haselby, R., Hilton, E., Kunkel, M., Adler-Klein, D., Doherty, T., *et al.* (1998). A vaccine consisting of recombinant Borrelia burgdorferi outer-surface protein A to prevent Lyme disease. Recombinant Outer-Surface Protein A Lyme Disease Vaccine Study Consortium. N Engl J Med *339*, 216-222.

Spielman, A., Clifford, C.M., Piesman, J., and Corwin, M.D. (1979). Human babesiosis on Nantucket Island, USA: description of the vector, Ixodes (Ixodes) dammini, n. sp. (Acarina: Ixodidae). J Med Entomol *15*, 218-234.

Spielman, A., Etkind, P., Piesman, J., Ruebush, T.K., 2nd, Juranek, D.D., and Jacobs, M.S. (1981). Reservoir hosts of human babesiosis on Nantucket Island. Am J Trop Med Hyg *30*, 560-565.

Steere, A.C., Batsford, W.P., Weinberg, M., Alexander, J., Berger, H.J., Wolfson, S., and Malawista, S.E. (1980a). Lyme carditis: cardiac abnormalities of Lyme disease. Ann Intern Med *93*, 8-16.

Steere, A.C., Broderick, T.F., and Malawista, S.E. (1978). Erythema chronicum migrans and Lyme arthritis: epidemiologic evidence for a tick vector. Am J Epidemiol *108*, 312-321.

Steere, A.C., Grodzicki, R.L., Kornblatt, A.N., Craft, J.E., Barbour, A.G., Burgdorfer, W., Schmid, G.P., Johnson, E., and Malawista, S.E. (1983). The spirochetal etiology of Lyme disease. N Engl J Med *308*, 733-740.

Steere, A.C., and Malawista, S.E. (1979). Cases of Lyme disease in the United States: locations correlated with distribution of Ixodes dammini. Ann Intern Med *91*, 730-733.

Steere, A.C., Malawista, S.E., Hardin, J.A., Ruddy, S., Askenase, W., and Andiman, W.A. (1977a). Erythema chronicum migrans and Lyme arthritis. The enlarging clinical spectrum. Ann Intern Med *86*, 685-698.

Steere, A.C., Malawista, S.E., Newman, J.H., Spieler, P.N., and Bartenhagen, N.H. (1980b). Antibiotic therapy in Lyme disease. Ann Intern Med *93*, 1-8.

Steere, A.C., Malawista, S.E., Snydman, D.R., Shope, R.E., Andiman, W.A., Ross, M.R., and Steele, F.M. (1977b). Lyme arthritis: an epidemic of oligoarticular arthritis in children and adults in three connecticut communities. Arthritis Rheum *20*, 7-17.

Steere, A.C., Sikand, V.K., Meurice, F., Parenti, D.L., Fikrig, E., Schoen, R.T., Nowakowski, J., Schmid, C.H., Laukamp, S., Buscarino, C., *et al.* (1998). Vaccination against Lyme disease with recombinant Borrelia burgdorferi outer-surface lipoprotein A with adjuvant. Lyme Disease Vaccine Study Group. N Engl J Med *339*, 209-215.

Steigbigel, R.T., and Benach, J.L. (1998). Immunization against Lyme disease--an important first step. N Engl J Med *339*, 263-264.

Stevenson, B., Fingerle, V., Wormser, G.P., and Margos, G. (2019). Public health and patient safety concerns merit retention of Lyme borreliosis-associated spirochetes within the genus Borrelia, and rejection of the genus novum Borreliella. Ticks Tick Borne Dis *10*, 1-4. https://doi.org/10.1016/j.ttbdis.2018.08.010

Stoenner, H.G. (1974). Biology of Borrelia hermsii in Kelly medium. Appl Microbiol *28*, 540-543.

Stoenner, H.G., Dodd, T., and Larsen, C. (1982). Antigenic variation of Borrelia hermsii. J Exp Med *156*, 1297-1311.

Sweeney, C.J., Ghassemi, M., Agger, W.A., and Persing, D.H. (1998). Coinfection with Babesia microti and Borrelia burgdorferi in a western Wisconsin resident. Mayo Clin Proc *73*, 338-341.

Telford, S.R., 3rd, Fikrig, E., Barthold, S.W., Brunet, L.R., Spielman, A., and Flavell, R.A. (1993). Protection against antigenically variable Borrelia burgdorferi conferred by recombinant vaccines. J Exp Med *178*, 755-758.

Telford, S.R., 3rd, Mather, T.N., Moore, S.I., Wilson, M.L., and Spielman, A. (1988). Incompetence of deer as reservoirs of the Lyme disease spirochete. Am J Trop Med Hyg *39*, 105-109.

Thyresson, N. (1949). The penicillin treatment of acrodermatitis chronica atrophicans. Acta Derm Venereol (Stockh) *29*, 572-621.

Vannier, E., Borggraefe, I., Telford, S.R., 3rd, Menon, S., Brauns, T., Spielman, A., Gelfand, J.A., and Wortis, H.H. (2004). Age-associated decline in resistance to Babesia microti is genetically determined. J Infect Dis *189*, 1721-1728. https://doi.org/10.1086/382965 JID31784 [pii]

Wallis, R.C., Brown, S.E., Kloter, K.O., and Main, A.J., Jr. (1978). Erythema chronicum migrans and lyme arthritis: field study of ticks. Am J Epidemiol *108*, 322-327.

Wang, T.J., Liang, M.H., Sangha, O., Phillips, C.B., Lew, R.A., Wright, E.A., Berardi, V., Fossel, A.H., and Shadick, N.A. (2000). Coexposure to Borrelia burgdorferi and Babesia microti does not worsen the long-term outcome of lyme disease. Clin Infect Dis *31*, 1149-1154. https://doi.org/CID994396 [pii]

Weber, K. (1974). [Erythema chronicum migrans meningitis - a bacterial infectious disease? (author's transl))]. MMW Munch Med Wochenschr *116*, 1993-1998.

Weber, K. (1981). Serological study with rickettsial antigens in erythema chronicum migrans. Dermatologica *163*, 460-467.

Weis, J.J., Ma, Y., and Erdile, L.F. (1994). Biological activities of native and recombinant Borrelia burgdorferi outer surface protein A: dependence on lipid modification. Infect Immun *62*, 4632-4636.

Wilske, B., Preac-Mursic, V., and Schierz, G. (1985). Antigenic heterogeneity of European Borrelia burgdorferi strains isolated from patients and ticks. Lancet *1*, 1099. https://doi.org/S0140-6736(85)92396-7 [pii]

Zhong, W., Stehle, T., Museteanu, C., Siebers, A., Gern, L., Kramer, M., Wallich, R., and Simon, M.M. (1997). Therapeutic passive vaccination against chronic Lyme disease in mice. Proc Natl Acad Sci U S A *94*, 12533-12538.

Zwelleger, H. (1946). Uber die chronische allergische meningitis. Helv Paediatr Acta *1*, 417-426.

Chapter 2

Multipartite Genome of Lyme Disease *Borrelia*: Structure, Variation and Prophages

Ira Schwartz[1]*, Gabriele Margos[2], Sherwood R. Casjens[3], Wei-Gang Qiu[4] and Christian H. Eggers[5]

[1]Department of Microbiology and Immunology, New York Medical College, Valhalla, NY USA, [2]National Reference Centre for *Borrelia* and Bavarian Health and Food Safety Authority, Oberschleissheim, Germany,
[3]Division of Microbiology and Immunology, Pathology Department, University of Utah School of Medicine, Salt Lake City, UT USA,
[4]Department of Biological Sciences, Hunter College of the City University of New York, New York, NY USA
[5]Department of Biomedical Sciences, Quinnipiac University, Hamden, CT, USA
*Corresponding author: ira_schwartz@nymc.edu

DOI: https://doi.org/10.21775/9781913652616.02

Abstract

All members of the *Borrelia* genus that have been examined harbour a linear chromosome that is about 900 kbp in length, as well as a plethora of both linear and circular plasmids in the 5-220 kbp size range. Genome sequences for 27 Lyme disease *Borrelia* isolates have been determined since the elucidation of the *B. burgdorferi* B31 genome sequence in 1997. The chromosomes, which carry the vast majority of the housekeeping genes, appear to be very constant in gene content and organization across all Lyme disease *Borrelia* species. The content of the plasmids, which carry most of the genes that encode the differentially expressed surface proteins that interact with the spirochete's arthropod and vertebrate hosts, is much more variable. Lyme disease *Borrelia* isolates carry between 7-21 different plasmids, ranging in size from 5-84 kbp. All strains analyzed to date harbor three plasmids, cp26, lp54 and lp17. The plasmids are unusual, as compared to most bacterial plasmids, in that they contain many paralogous sequences, a large number of pseudogenes, and, in some cases, essential genes. In addition, a number of the plasmids have features indicating that they are prophages. Numerous methods have been developed for Lyme disease *Borrelia* strain typing. These have proven valuable for clinical and epidemiological studies, as well as phylogenomic and population genetic analyses. Increasingly, these approaches have been displaced by whole genome sequencing techniques. Some correlations between genome content and pathogenicity have been deduced, and comparative whole genome analyses promise future progress in this arena.

Introduction

The genus *Borrelia* forms a deeply separated lineage within the *Spirochaetes* branch of the bacterial tree of life (Paster et al., 1991). The organisms are genomically unique and not closely related to any other bacteria, including the other *Spirochaetes*. *Borrelia burgdorferi* and *Borrelia hermsii* are representative of the two major branches within the *Borrelia* genus, wherein *B. burgdorferi* typifies the Lyme disease *Borrelia* branch (abbreviated herein as LB) and *B. hermsii* typifies the relapsing fever agent branch (Paster et al., 1991; Ras et al., 1996; Schwan et al., 2007). The term '*B. burgdorferi*' denotes *B. burgdorferi sensu stricto* throughout this chapter. By analysing the genomes of both *B. burgdorferi* and *B. hermsii,* Barbour and coworkers originally showed that both have a topologically unusual genome with linear chromosomes and multiple linear plasmids (Plasterk et al., 1985; Barbour, 1988; Ferdows and Barbour, 1989; Hinnebusch et al., 1990; Kitten and Barbour, 1990; Hinnebusch and Barbour, 1991) (Figure 1).

Subsequent work has shown this to be a property of all members of the genus *Borrelia* that have been examined. This chapter will focus on the Lyme Disease *Borrelia* (abbreviated throughout as LB) branch of the genus; the genomics of the relapsing

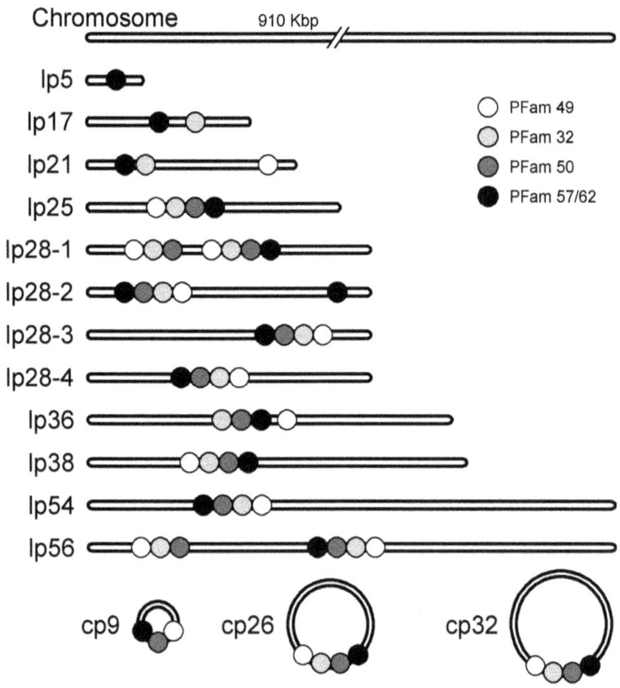

Figure 1. The genome of *B. burgdorferi* type strain B31. The linear and circular plasmids are shown with the genes thought to be involved in plasmid replication, partitioning and compatibility (see text) indicated by small circles with different shading. Different B31 cultures have been found to carry one or two related cp9 plasmids and up to eight different cp32 plasmids. The figure is modified from figure 8 of Casjens *et al.* (2000).

fever branch is covered in an accompanying chapter (Chapter 3).

The *B. burgdorferi* genome was among the first bacterial chromosomes to be sequenced (Fraser et al., 1997). Since then 27 complete LB genomes (including chromosome and plasmids) have been elucidated and numerous other less complete genome sequences have been deposited in the NCBI Genomes database (as of 5/12/20). Each LB species carries a linear chromosome about 900 kbp in length and multiple circular and linear low copy number plasmids that are usually, but not always, in the 5-84 kbp range. These features have been demonstrated directly for many of the defined LB species, including *Borrelia afzelii*, *Borrelia andersonii*, *Borrelia bavariensis, Borrelia bissettiae*, *B. burgdorferi, Borrelia filandensis*, *Borrelia garinii*, *Borrelia japonica, Borrelia lusitaniae*, *Borrelia maritima, Borrelia mayonii, Borrelia spielmanii*, *Borrelia turdi* and *Borrelia valaisiana* (Ferdows and Barbour, 1989; Davidson et al., 1992; Casjens and Huang, 1993; Ojaimi et al., 1994; Casjens et al., 1995; Vitorino et al., 2010; Mongodin et al., 2013; Kingry et al., 2016; Casjens et al., 2018; Margos et al., 2019a; Margos et al., 2019b; Margos et al., 2020) and all LB species likely will have genomes of this type. The chromosomes of these bacteria all have quite similar gene contents, whereas the plasmids are more variable (details below).

In light of their importance for disease pathogenesis and maintenance of the enzootic cycle (see below and Chapters 13 and 24), the number and diversity of the plasmids carried by LB species have been the focus of much research. This enterprise, however, has been hampered by their unusual features. First, individual LB cells harbor a more diverse plasmid complement than that of any other bacterium - the sequenced genome of *B. burgdorferi* type strain B31 has 21 plasmids, and several additional plasmids appear to have been lost between isolation of the strain in 1982 and the completion of the complete genome sequence (Casjens et al., 2000). Second, a number of the plasmids have very similar sizes, and thus cannot be separated by pulsed-field gel electrophoresis. Consequently, mapping a gene to a particular plasmid by Southern hybridization can be difficult. Third, different plasmids in the same cell often carry paralogous sequences that can be extremely similar (Casjens, 2000; Casjens et al., 2000). This can and does confuse both plasmid identification and assembly of plasmid sequences. Thus, the complete plasmid content of a particular LB strain is difficult to ascertain without complete genomic sequence analysis. Fortunately, advances in sequencing technologies has made such analysis simpler and is leading to a much better understanding of LB species plasmid content (Casjens et al., 2012; Mongodin et al., 2013; Margos et al., 2017b; Casjens et al., 2018). Finally, assessing the true plasmid content of an LB cell is further complicated by the fact that plasmids can be lost during primary isolation; knowing whether native plasmids have been lost from any strain isolated in the laboratory is essentially impossible at this time. Furthermore, many of the *Borrelia* plasmids are readily lost during passage or manipulation *in vitro*, and many clones of any given strain will have lost one or more (Schwan et al., 1988; Xu et al., 1996; Purser and Norris, 2000; Elias et al., 2002; Grimm et al., 2003).

Overall, the 900-kbp chromosomes (often called the "large" or "main" chromosome by workers in this field) carry the great majority of the genes that encode metabolic enzymes, and the 400 to 650 kbp (in different isolates) of plasmid DNAs carry the bulk of the surface lipoprotein encoding genes. With the notable exception of cp26 (see below), the plasmids are not required for growth in culture (Sadziene et al., 1993; Casjens et al., 1997b), but have often been found to be required for mouse infectivity or tick transmission in the laboratory (Schwan et al., 1988; Xu et al., 1996; Purser and Norris, 2000; Purser et al., 2003; Grimm et al., 2005; Revel et al., 2005; Stewart et al., 2005; Strother et al., 2005; Strother and de Silva, 2005; Lin et al., 2012; Radolf et al., 2012). The chromosome carries tightly packed genes, as is typical of bacteria, while many of the linear plasmids have substantially lower gene densities and many apparently decaying pseudogenes (Casjens, 2000; Casjens et al., 2000). Currently, the general impression is that of an evolutionarily quite stable chromosome that encodes the machinery required for existence as a bacterial cell along with a rather large menu of much more evolutionarily variable plasmids that encode most of the proteins that interact with the vertebrate and arthropod host environments encountered by the bacteria (there are, however, a small number of important metabolic enzymes encoded by the plasmids and surface-exposed proteins encoded by the chromosome; see below).

We note that references cited in this review were chosen to allow the reader access to the latest literature and not necessarily to credit discoverers, the originators of ideas, or the first to obtain particular kinds of data. Thus, cited articles are sometimes more recent articles instead of those in which the original discoveries were first reported.

Lyme Disease *Borrelia* Chromosomes

Borrelia burgdorferi B31 Chromosome

The complete 910,725-bp sequence of the isolate B31 (the type strain of *B. burgdorferi*) linear chromosome was published in 1997 (Fraser et al., 1997). Its nucleotide composition is 28.6% G+C, and it is predicted to contain 803 protein-encoding genes and 17 pseudogenes. The chromosomal genes have been shown to be under intense purifying selection (Tyler et al., 2018). The fraction of mutationally inactivated chromosomal genes is among the lowest of the analyzed bacterial genomes. Although smaller genomes of free-living bacteria are known, this chromosome size is near the small end of the spectrum (*e.g.* Casjens, 1998)). The protein-encoding genes occupy 93% of the chromosome, a typical value for a bacterial genome not undergoing current reduction in size (Lynch, 2006). About 67% of the genes are oriented such that they are transcribed away from the center of the chromosome. GC skew analysis and experimental evidence strongly suggest the presence of a replication origin near the center of the linear chromosome (Picardeau et al., 1999). The chromosome of B31 carries five rRNA genes (two 23S, two 5S and one 16S) and 32 tRNA genes. The former are clustered very near the center of the chromosome. The tRNAs, predicted to encode individuals specific for all 20 amino acids, are scattered across the chromosome in seven clusters and 13 single genes (Schwartz et al., 1992; Fraser et al., 1997). Variable numbers of tandem repeats were found in three genes: BB_0210, BB_0546 and BB_0801 (Mongodin et al., 2013). BB_0210 encodes a surface protein, Lmp1, which has been observed to hamper chromosome sequence assembly from short-read sequences such as Illumina likely due to difficulties in assembling the tandem repeats (Tyler et al., 2018).

Since the complete genome of *B. burgdorferi* strain B31 was sequenced (Fraser et al., 1997; Casjens et al., 2000), the *Borrelia* research community has usually used the 'locus tags' in its GenBank annotation as names for its genes and their encoded proteins (and their homologues discovered in other isolates). Thus, according to bacterial convention, chromosomal genes are named '*bb0xxx*' (lower case and italicized) in ascending order from *bb0001* through *bb0853* across the chromosome, and chromosomally encoded proteins are named 'BB_0xxx' (upper case and not italicized, *e.g.* BB_0364). The B31 plasmid locus tag names are similar but have the form 'BB$xx' in which '$' indicates a letter, A through U, denoting the plasmid which carries the gene (*e.g. bba74* encodes protein BBA74 and lies on lp54; *bbs09* lies on cp32-3, etc.; see LB plasmid section below).

Depending on the annotation pipeline, 60-85% of the chromosome's predicted genes have some similarity to a gene in another organism whose role or function is at least partly understood; about 10% are similar to known genes in other organisms whose roles are unknown; the remainder are unique to *Borrelia* and have unknown functions (Fraser et al., 1997; Margos et al., 2017c). The chromosome carries what appears to be a rather minimal set of genes required for cell maintenance and replication. These include genes for cell wall biosynthesis (but not synthesis of lipopolysaccharide); protein export and lipidation; DNA,

RNA and protein biosynthesis; DNA repair; nucleotide metabolism (but not *de novo* synthesis); membrane lipid and phospholipid biosynthesis; glycolysis and a few enzymes that provide substrates for the glycolytic pathway, the bacterium's sole mode for generating ATP. A large and complete set of the genes known to be required for motility and chemotaxis also is present, as are a number of genes for transport of small molecules across the cytoplasmic membrane. A major conclusion from this bioinformatic analysis is that the biosynthetic and intermediary metabolic capacity of *Borrelia* is very limited. Genes that encode enzymes that perform functions in respiration, amino acid synthesis, nucleotide synthesis, lipid synthesis and enzyme cofactor synthesis are almost completely lacking, consistent with the many and fastidious requirements for growth in culture, as well as its restricted enzootic, host-associated lifestyle. Although originally no requirement for iron or iron transporter genes were predicted (Posey and Gherardini, 2000), in a subsequent paper a ferritin-like Dps (DNA-binding protein from starved bacteria) protein with iron and copper binding properties was described (Wang et al., 2012). In most cases that have been examined manganese- or zinc-dependent enzymes are used instead of iron-dependent enzymes (Nguyen et al., 2007; Ouyang et al., 2009; Groshong and Blevins, 2014). Given this limited biosynthetic capacity, the spirochete must acquire nearly all carbohydrates, amino acids, nucleosides, lipids and cofactors from its environment. Numerous transporter and permease components are encoded on the chromosome (with several others encoded on plasmids); these include those for transport/uptake of carbohydrates (14 genes) and amino acids/peptides (15 genes) (Fraser et al., 1997; Corona and Schwartz, 2015; Groshong et al., 2017). (See also chapter 6 for additional discussion of metabolism).

The initial annotation of the B31 chromosome suggested a surprisingly small number of genes identified as encoding transcriptional regulators. This was not expected as numerous *B. burgdorferi* genes are known to be regulated by environmental signals (Samuels, 2011; Radolf et al., 2012). Only two alternative sigma factors (RpoS and RpoN) and two two-component systems (Hk1-Rrp1 and Hk2-Rrp2) initially were identified by sequence homology. Since then several additional transcriptional regulators have been identified and characterized, including BosR, BadR, CsrA, Rel_{Bbu} and DksA (Samuels, 2011; Radolf et al., 2012; Groshong and Blevins, 2014; Arnold et al., 2018) (see Chapter 5 for more details).

Chromosomes of Other Lyme Disease Borrelia species

The seminal publication of the genome of the *B. burgdorferi* type strain B31 (Fraser et al., 1997) revealed its complexity. It took several years before the full closure of all plasmids was achieved and the diversity of the plasmids, even within one species, was fully appreciated (Casjens et al., 2000; Casjens et al., 2012; Casjens et al., 2017; Casjens et al., 2018). The stability and co-linearity of chromosome and core plasmids (lp54, cp26) between *Borrelia* species was initially demonstrated by Glöckner et al. (2004; 2006). These authors sequenced the genomes of *B. bavariensis* PBi (at the time still designated *B. garinii*) and *B. afzelii* PKo and showed that the chromosomes of these two species were co-linear with that of *B. burgdorferi* B31. The chromosomes of PBi and PKo had only three insertions and six and nine deletions of >100 bp, respectively (Glockner et al., 2006). Two of these insertions were >1000 bp; one resided within the coding sequence of the basic membrane protein A and the other was near the origin or replication. A large indel of >600 bp in the genome of PBi was found in the coding region of LMP1 (BB_0210) which contains variable number of tandem repeats (Mongodin et al., 2013; Tyler et al., 2018).

Subsequently determined genome sequences for many of the 22 Lyme disease *Borrelia* species are now available (Table 1), either as completed genomes, draft genomes or in the form of raw reads in the NCBI sequence read archive (SRA) (https://www.ncbi.nlm.nih.gov/sra) (Casjens et al., 2011a; Casjens et al., 2011b; Schutzer et al., 2011; Schutzer et al., 2012; Mongodin et al., 2013; Ivanova et al., 2014; Jacquot et al., 2014; Gatzmann et al., 2015; Becker et al., 2016; Castillo-Ramirez et al., 2016; Kingry et al., 2016; Margos et al., 2017a; Margos et al., 2017b; Margos et al., 2017c; Margos et al., 2018; Margos et al., 2019a; Margos et al., 2020). The genome structure, i.e. its composition of a linear main chromosome and a number of circular and linear plasmids, is conserved in all species investigated (Schutzer et al., 2012; Mongodin et al., 2013; Ivanova et al., 2014; Becker et al., 2016; Kingry et al., 2016; Margos et al., 2019a; Margos et al., 2020). The chromosome of the different LB species varies in size, ranging from 900 kbp (900,694 in *B. chilensis*) to 922 kbp (922,901 in *B. burgdorferi* JD1) (Table 1). Right end extensions (plasmid fusion) of chromosomes have been described in *B. burgdorferi* , *B. mayonii* and *B. valaisiana* (Casjens et al., 1997a; Huang et al., 2004; Banik et al., 2011; Casjens et al., 2012; Casjens et al., 2018) and has most recently has been found in the *B. turdi* isolate TPT2017, in which an lp28-2-like plasmid was found fused to the right end of its chromosome (Margos et al., 2019a).

The rRNA gene profiles of LB spirochetes were described as consisting of one 16S rRNA locus that is located approximately 2 kb upstream of tandemly duplicated 5S-23S rRNA loci separated by an intergenic spacer of approximately 200-350 bp (Schwartz et al., 1992; Gazumyan et al., 1994); this arrangement appeared to be similar in other LB species (Ojaimi et al., 1994). However, differences in the rRNA gene profiles have been identified in *B. bavariensis*, *B. afzelii*, *B. spielmanii*, *B. turdi* and *B. yangtzensis* which contain two 16S rRNA genes. Other species, such as *B. americana* (3), *B. carolinensis* (4), *B. japonica* (4) and *B. kurtenbachii* (4) appear to have three or four 5S-23S rRNA loci (Table 1).

In two isolates of *B. andersonii,* 19857 and 21038, and the *B. japonica* isolate IKA2, a single 5S rRNA locus and insertions in the second copy of the 23S rRNA were reported (Marconi et al., 1995). Curiously, two genome assemblies of *B. bavariensis,* isolates NMJW1 and SZ, (erroneously named *B. garinii* in GenBank), seem to possess only one 5S and one 23S rRNA locus but two 16S rRNA loci. However, whether these deviations in the arrangement of the rRNA loci are correct or may in some cases be due to mis-assembly of sequence data needs to be confirmed. In *B. chilensis,* 31 tRNAs and 1 non-coding (nc) RNA are present (Ivanova et al., 2014), whereas the number of tRNAs and ncRNAs in the genomes of the majority of LB species with completed chromosome sequences in GenBank were 32 and 3, respectively (Table 1).

In general, the synteny of the main chromosome appears to be well conserved in most investigated LB species. The length of the constant chromosomal regions was reported to be approximately 903 kbp with only a few indels or differences between the chromosomes of LB species (although in some it may be shorter (e.g. in *B. bissettiae*, *B. chilensis* or *B. maritima* chromosomes are only 900 - 902 kbp; Table 1) (Mongodin et al., 2013; Ivanova et al., 2014; Margos et al., 2020). Interspecies differences in the three repeat-containing genes (*bb0210*, *bb0546* and *bb0801*) were found. In addition, indels larger than 25 bp differentiate the chromosomes of *B. burgdorferi* and *B. afzelii* (Glockner et al., 2006; Mongodin et al., 2013). Furthermore, compared to strain B31, *B. bavariensis* isolate PBi and *B. afzelii* PKo had duplications in the *bmp* gene region, five indels of <330 bp in intergenic regions and indels of <150 bp in genes *bb0309*, *bb0704* and *bb0749* (Glockner et al., 2006; Mongodin et al., 2013). Additional comparative genome analyses of *B. maritima* and its closest relative *B. bissettiae* isolate DN127 showed that *B. maritima* lacked five chromosomal protein coding sequences, including three hypothetical proteins, the orthologue of *bb0049* and the gene encoding competence protein ComEC/Rec2, but included an additional three genes in its repertoire --- a gene encoding an apolipoprotein N-acetyl-transferase, a hypothetical protein and the arginine/ ornithine antiporter ArcD (Margos et al., 2020).

Although different statistical methods were used for calculating genetic distances within and between species, it is apparent that the genetic divergence is much smaller within species than between species, confirming previous data using MLST/MLSA (Margos et al., 2009). Within *B. burgdorferi* only minor differences between chromosomes were noted, e.g. ZS7 and Bol26 differed by 0.084% and 94b and 29805 by 0.625% (Mongodin et al., 2013), whereas distances calculated between species ranged from 2.8% - 8 % (Mongodin et al., 2013; Jacquot et al., 2014). Consistent with this, the most recently described LB species, *B. mayonii* and *B. maritima* shared average nucleotide identities of 93.8% and 90.6%, respectively, with other LB species. Thus, phylogenetic analyses confirm the genetic distinctness of the different LB species, which is likely related to their ecology (see below).

Lyme Disease *Borrelia* plasmids

Members of the *Borrelia* genus carry more plasmids than any other bacteria. LB isolates that have been analyzed carry between 7 and 21 different plasmids (Figure 1). These plasmids are both linear and circular and range from 5 to 84 kbp in size (Casjens et al., 2000; Casjens et al., 2012; Kingry et al., 2016; Casjens et al., 2017; Margos et al., 2017b; Casjens et al., 2018; Jabbari et al., 2018; Margos et al., 2020). Most of the plasmids, except for cp26 (Byram et al., 2004), are not required for growth in culture (Sadziene et al., 1992; Casjens et al., 1997b). Table 2 lists the plasmid complement for the 32 LB isolates for which complete or nearly complete genome sequences have been reported to date.

Table 1. Chromosomal genetic features of the Lyme Disease *Borrelia* species complex

Species	Isolate name	No. of tRNA/ncRNA	No. of rRNAs rrs/rrf-rrl	Average G+C % *	Chromosome size°	No. of sequenced genomes (source)
B. afzelii	Pko		2/2	27.9	903690	11 (GB#)
	HLJ01	32/3	2/2	28.3	905471	12 (SRA##)
	K78	32/3	2/2	27.9	905949	
	Tom3107	32/3	2/2	28.1	905861	
	BO23	32/3	2/2	27.8	905349	
B. americana	SCW-41	nd	1/ 3	nd	nd	1 (SRA)
B. bavariensis	PBi	32/3	2/2	27.8	905817	1 (GB)
	BgVir	32/3	2/2	28.2	905534	25 (SRA)
	NMJW1	32/3	2/1	28.4	902789	
	SZ	32/3	2/1	28.4	902487	
B. bissettiae	DN-127	32/3	1/ 2	28.3	900755	2 (GB) 1 (SRA)
B. burgdorferi sensu stricto	B31	31/1?	1/ 2	28.1	910724	111 (GB, many draft)
	JD1	32/2	1/ 2	28.3	922801	16 (SRA)
	CA382	32/3	1/ 2	28.6	910736	
	PAbe	32/3	1/ 2	28.5	910728	
	PAli	32/3	1/ 2	28.3	909921	
	MM1	32/3	1/ 2	28.2	908512	
	ZS7	32/3	1/ 1	28.2	906701	
	B331	32/3	1/ 2	28.2	904573	
	N40	32/3	1/ 2	28.2	902191	
B. californiensis	CA446	32/?	1/ 2	28.1	nd	1 (SRA)
B. carolinensis	CA446	nd	1/ 4	nd	nd	1 (SRA)
B. chilensis	VA1	31/1	1/ 2	28.5	900694	1 (GB)
B. garinii	20047	32/3	1/ 2	28.1	906449	42 (GB, mostly draft)
	Far04	29/3	1/ 2 (1 partial 23S)	nd	nd	26 (SRA)
B. japonica	HO14	33/?	1/ 4	28.1	nd	1 (SRA)
B. kurtenbachii	25015	33/?	1/ 4	28.4	nd	1 (SRA)
B. lanei	CA28-91	Nd	Nd	Nd	Nd	
B. lusitaniae	PoTiB2		1/ 2			1 (SRA)
B. maritima	CA690	33/1	1/ 2	27	902176	1 (GB)
B. mayonii	M14-1420	32/3	1/ 2	27.8	904387	2 (GB)
	M14-1539	32/3	1/ 2	27.8	904387	
B. spielmanii	PMew	Nd	2/2	27.6	nd	1 (GB) 3 (SRA)
B. turdi	TPT2017	33/?	2/2	27.7	935973	3 (GB, draft)
	T1990A		2/2	27.7	906931	
	T2084	33?/	2/2	27.6	901717	
B. valaisiana	VS116	32/2	1/ 2	27.4	913294	2 (GB)
	Tom4006	32/3	1/ 2	28.0	912160	2 (SRA)
B. yangtzensis	Okinawa CW62	33/?	2/ 1	27.8	nd	1 (SRA)

*may depend on whether and which plasmids are included
°Telomere regions may not be included in all samples
#GB=GenBank
##SRA=Sequence Read Archive

Plasmid types and nomenclature

B. burgdorferi plasmids were originally named in the type strain B31 with "lp" for linear plasmid and "cp" for circular plasmid and numerical designations of their approximate kbp sizes as determined by agarose gel electrophoresis and sequencing that did not always reach the linear plasmid telomeres (Fraser et al., 1997; Casjens et al., 2000). Naming the plasmids in all strains according to their size, however, creates several difficulties: (i) most of the plasmids are in the 25 to 31 kbp size range, so the number of different names based only on size is limited, (ii) plasmids from different isolates with closely related segments can, nonetheless, harbor major differences in size and organization, and (iii) names based purely on size have no biological significance. Naming plasmids according to more biologically meaningful criteria would be more informative. It was noticed early on that all of the B31 plasmids larger than 15 kbp carry a set of four genes, usually in a contiguous cluster, that are predicted to be involved in plasmid replication and maintenance. These genes encode proteins that are members of the LB protein families (PFams) named PFam32, PFam49, PFam50 and PFam57/62 (originally defined as *paralogous* families within strain B31 in Casjens *et al.* (2000) and extended to other strains in Casjens *et al.* (2012) (Figure 1). Smaller plasmids lp5, lp21 and cp9 often lack one or more of these four gene types but always carry a PFam57/62 gene. The PFam32 proteins are homologues of the well-studied ParA proteins that play a central role in the partitioning of many bacterial plasmids (Shih and Rothfield, 2006); the other three families have no strong similarity to genes of known function. Experimental studies have strongly supported the notion that the PFam49, PFam50 and PFam57/62 protein types are required for plasmid maintenance (Stewart et al., 2001; Eggers et al., 2002). The PFam32 proteins of circular cp26 or linear lp17 are apparently not required for maintenance (Beaurepaire and Chaconas, 2005; Tilly et al., 2012); however, the mutational analysis of Lin et al. (2012) suggests that they are required for other plasmids. PFam32 proteins are quite diverse and can be robustly parsed into different sequence types; strikingly, no LB cell has been found that carries two plasmids that encode PFam32 proteins of the same sequence type. This PFam32-type mutual exclusivity currently holds true for the 669 sequenced plasmids in 50 completely or nearly completely sequenced isolates (423 plasmids in 29 isolates with completely sequenced and published genomes (Table 2) and 21 additional unpublished LB genomes that carry 246 plasmids; Casjens, unpublished results). This strongly suggests that two plasmids with the same PFam32 type gene cannot coexist in the same cell and so are incompatible in spite of the fact that this protein is not always essential for plasmid maintenance in the laboratory. Consequently, it has been proposed that, when possible, LB plasmids be named according to the PFam32 protein they encode (reviewed by Casjens *et al.*, 2006; 2010; 2018). Rather than inventing a completely new set of plasmid names (the B31 names had been in use for nearly a decade), each PFam32 type was given its already existing B31 name (*e.g.*, cp26, lp28-2, lp36, *etc.*). As new PFam types were discovered in other isolates, they were given new names of the same style (*e.g.,* lp28-5 discovered in *B. burgdorferi* strain N40; lp28-8 discovered in *B. burgdorferi* strain 94a). The various B31 lp28 plasmids as well as lp21, lp25, lp36, lp38 and lp56 plasmids are all very variable in size between strains, and all of these plasmids contain many overlapping paralogous gene types.

Thus, when a newly discovered plasmid type contains members of these same gene sets it is given the next number in the lp28 series which is currently up to lp28-13 (see below). Linear plasmids that carry a cp32 type PFam32 gene are given a name that corresponds to its cp32 type (*e.g,* a linear plasmid with a cp32-6 type PFam32 gene would be lp32-6) (Casjens et al., 2017; Casjens et al., 2018). Table 3 lists all the published plasmids and their size ranges.

This PFam32 type nomenclature scheme has the major advantage that plasmids of the same compatibility type have related names, but it also has disadvantages: (i) the PFam32 gene on a given plasmid must be sequenced in order to determine its final designation, and (ii) plasmids of the same PFam32 type often differ substantially in gene content (see below). PFam32 sequence types are not limited to single species as most types are found in multiple LB species (Casjens et al., 2018). Of course, any biological nomenclature system will have some complications, and LB isolates are no exception. First, several cases exist in which individual plasmids carry more than one apparently intact PFam32 gene (though adhering to the "only one copy of any PFam32 type gene per cell" rule). There are examples of circular plasmids integrated into linear plasmids (cp32-10 integrated into lp56 in *B. burgdorferi* strain B31 and cp32-11 into lp54 in *B. finlandensis* strain SV1), apparent end-to-end fusion of linear plasmids lp36 and lp28-4 in *B. burgdorferi* strain CA-11.2A, and fusion of two or more circular plasmids (*e.g.*, cp32-1 and cp32-5 in *B. burgdorferi* strain JD1; fusion of parts of four cp32s in cp32-quad in *B. bissettiae* strain DN127) (Casjens et al., 2012; Casjens et al., 2018). The names of fused plasmids have not always been consistent, but the recommendation is that they be named with both PFam32 types, for example "cp32-1+5". A second more serious complication is that it is not known how different the PFam32 proteins must be to define different plasmid compatibilities. Although the PFam32 proteins form well-defined clusters, there is

Table 2. Total plasmid numbers in 32 Lyme Disease *Borrelia* isolates

Species	Isolate (X[1])	cp9	cp26	cp32[2]	lp5	lp17	linears[3]	lp54	Total[4]	Reference
B. burgdorferi										
	64b (A)	-	1	9[8]	-	1	7	1	18	Casjens, 2017
	B31 (A)	1[5]	1	8[6]	1	1	10	1	21[2,5]	Casjens, 2000
	Bol26 (A)	-	1	6[7,8]	-	1	4	1	13	Casjens, 2017
	PAli (A[1])	-	1	7[6]	-	1	4	1	13	Margos, 2017
	PAbe (A[1])	-	1	7[6]	-	1	4[9]	1	11	Margos, 2017
	ZS7(A)	-	1	6[8]	-	1	6	1	14	Casjens, 2017
	29805 (B)	-	1	5[8]	-	1	7	1	15	Casjens, 2017
	N40 (B)	1	1	6[8]	-	1	6	1	16	Casjens, 2017
	WI91-23 (B)	2	1	7[8]	1	1	9	1	21	Casjens, 2017
	72a (C)	-	1	6[8]	-	1	4	1	13	Casjens, 2017
	94a (C)	1	1	5	-	1	5[7]	1	14	Casjens, 2017
	118a (C)	1	1	8[8]	-	1	8	1	19	Casjens, 2017
	CA11_2a (C)	-	1	5	-	1	4[10]	1	12	Casjens, 2017
	MM1 (C[1])	1	1	6	-	1	5	1	15	Jabbari, 2018
	156a (D)	-	1	8[8]	-	1	9	1	20	Casjens, 2017
	297 (D[11])	-	1	9[8]	-	1	7	1	19	Casjens, 2017
	JD1 (D)	-	1	9[8]	-	1	9	1	20	Casjens, 2017
B. "finlandensis"										
	SV1	-	1	4[8,12]	-	1	4	1	10	Casjens, 2018
B. bissettiae										
	DN127	1	1	9[8]	-	1	5	1	16	Casjens, 2018
B. mayonii										
	MN14-1420	1	1	5[8]	1	1	5	1	15	Kingry, 2016
	MN14-1539	1	1	5[8]	-	1	5	1	14	Kingry, 2016
B. bavariensis										
	PBi	-	1	4[13]	-	1	6[13]	1	11[13]	Glockner, 2006 Margos, 2018
B. garinii										
	20047	-	1	2[13]	-	1	5[13]	1	>6	Margos, 2018-[15]
	Far04	-	1	-	-	1	4	1	7	Casjens, 2018
	PBr	-	1	2[8]	-	1	6	1	11	Casjens, 2018
B. afzelii										
	ACA-1	-	1	4[8]	-	1	7	1	14	Casjens, 2018-[15]
	BO23	1	1	?[13]	-	1	5[13,14]	1	>9	
	K78	-	1	4	-	1	6	1	13	Schuler, 2015
	PKo	-	1	7[8]	-	1	7	1	17	Casjens, 2018
B. spielmanii										
	A14S	1[7]	1	4[7]	-	1	5[7]	1	13	Casjens, 2018
B. valaisiana										
	VS116	1	1	3[8]	-	1	4	1	11	Casjens, 2018
B. maritima										
	CA690	-	1	2	-	1	3	1	7	Margos, 2019 Casjens, unpub
B. turdi										
	T1990A	1	1	3[2]	-	1	3	1	10	Margos, 2019
	TPT2017	-	1	3[2]	-	1	4	1	10	Margos, 2019
	T2084	-	1	3[2]	-	1	3	1	9	Margos, 2019

[1]The sequences were analyzed in the following references (Kingry *et al.*, 2016; Casjens *et al.*, 2017; 2018; Margos *et al.*, 2017; 2019; Jabbari *et al.*, 2018). The Bbss chromosomal SNP type (Mongodin *et al.*, 2013) is indicated in parentheses; The SNP types were predicted for PAbe, PAli and MM1 chromosomes by our analysis of cp26, lp54 and chromosome right end (see text) (S. Casjens and W. Qiu, unpublished).

[2]Number of cp32 PFam32 compatibility types, not DNA molecules. Strain B31 may have had 9th cp32 (cp32-5) that was lost before genome sequencing (Zuckert and Meyer, 1996; Casjens *et al.*, 1997).

[3]This category includes all the linear plasmids except lp5, lp17 and lp54 (see text).

[4]The number of different plasmid DNA molecules; fused plasmids counted as one.

[5]B31 may have had a second cp9 that was lost before genome sequencing (Miller *et al.*, 2000).

[6]cp32-10 is integrated into the lp56 linear plasmid (Casjens *et al.*, 2000).

[7]Assembly of one or more of these plasmids sequences was incomplete.

[8]One or more cp32s either truncated or fused with other cp32s.

[9]Includes lp28-1 since Illumina sequencing methods found some lp28-1 sequences; it may have been lost in most cells in the culture.

[10]Lp36 and lp28-4 are fused end-to-end.

[11]The 297 chromosome has not been sequenced. Its chromosomal SNP type is inferred from the cp26 and lp54 plasmids (Mongodin *et al.*, 2013).

[12]Cp32-11 is integrated into lp54 (Casjens *et al.*, 2018).

[13]There may be additional plasmids in this group whose sequences were not completely assembled.

[14]Lp54 and lp38 are reported to be fused end-to-end in CA-11.2A.

[15]Sébastien Bontemps-Gallo, personal communication and unpublished; cp32s were not assembled; see relevant GenBank entries.

Table 3. Total Lyme Disease *Borrelia* plasmid numbers and sizes in 29 genomes from 9 species.

	burgdorferi	*finlandensis*	*bissettiae*	*mayonii*	*afzelii*	*garinii*	*spielmanii*	*valaisiana*	*maritima*	Size range
	17[1]	1	1	2	3	2	1	1	1	(kbp)[3]
Plasmid										
lp5	2	-	1	-	-	-	-	-	-	5
lp17	17	1	1	2	3	2	1	1	1	15-28
lp21	2	-	-	-	-	-	-	-	-	15-19
lp25	9[2]	-	1	2	-	2	-	1	-	22-33
lp28-1	8	-	-	-	-	-	-	-	-	24-30
lp28-2	5	1	-	-	3	-	-	-	1	21-36
lp28-3	14[2]	-	1	2	2	1	1	1	-	22-82
lp28-4	15	1	1	1	3	1	1	-	1	22-41[4]
lp28-5	6	-	-	-	-	-	-	-	-	24-28
lp28-6	6	-	-	-	-	-	-	-	-	27-31
lp28-7	1	-	1	-	2	1	-	-	-	27-34
lp28-8	1	-	-	2	2	1	1	1	-	18-23
lp28-9	1	-	-	-	2	2	-	-	-	27-39
lp32-3	2	-	-	-	-	-	-	-	-	17
lp32-6	-	1	-	-	-	-	-	-	-	48
lp32-10	-	-	-	-	2	1	-	-	-	30-42
lp32-12	-	1	-	-	-	-	-	-	-	53
lp36	17	-	-	2	-	2	2	2	1	17-46[4]
lp38	12	-	-	-	3	-	1	-	-	20-39
lp54	17	1	1	2	3	2	1	1	1	53-61[4]
lp56	6	-	1	-	-	-	-	-	-	21-30[4]
cp9	8[2]	-	1	2	-	-	1	1	-	8-11
cp26	17	1	1	2	3	2	1	1	1	26-27
cp32's	108[5]	3	7	10	15	2	4	3	1	27-32[6]

[1]Below the species name is the number of genomes analyzed in the following references (Kingry *et al.*, 2016; Casjens *et al.*, 2017; 2018; Margos *et al.*, 2017; 2019; Jabbari *et al.*, 2018); they are those in table 2 but not including strains PBi, 20047 and BO23 since their plasmids are likely incompletely described.

[2]Includes unsequenced 297 lp25 and N40 lp28-3 and second B31 cp9 type (see Casjens *et al.*, 2017).

[3]The upper size estimates can be slight underestimates since most sequences do not extend quite to the termini of the plasmids (see Casjens *et al.*, 2012).

[4]Sizes do not include neatly integrated ~30 kbp cp32s in B31 lp56 and SV1 lp54 or fused lp36 and lp28-4 in CA-11.2A

[5]Number of cp32 apparently intact PFam32 genes and not DNA molecules (see text).

[6]All apparently intact cp32s are between 27 and 32 kbp; however, a significant fraction of cp32s exhibit large deletions and these can be as small as 13.5 kbp.

insufficient experimental evidence to determine whether such clusters indeed always have different compatibilities. The most closely related PFam32 types that must have different compatibilities, since they reside in the in the same cell, are cp32-8, cp32-10 and cp32-12, all of which are present in *B. burgdorferi* strain JD1and are about 25% different in amino acid sequence (see figure S1 in Casjens *et al.*, 2018). lp28-1 and lp28-9 (17-18% different PFam32s) have been arbitrarily split into different PFam32 types (Casjens et al., 2017), whereas it has been argued that lp28-10 (Kingry et al., 2016) and lp28-8, at 12-15% different, may be the same compatibility group (Casjens et al., 2018). However, it should be emphasized that in the absence of finding a cell that contains both types, experimental tests of compatibility in these cases will be required to ascertain whether these groups have the same or different compatibilities. Finally, plasmids smaller than ~15 kbp, lp5 and cp9, often do not encode a PFam32 protein and so cannot be categorized in this fashion. For example, *B. burgdorferi* strain WI91-23 (Casjens et al., 2017) and perhaps the original *B. burgdorferi* B31 isolate (Miller et al., 2000) contain two different cp9 plasmids that lack PFam32 genes, indicating that there are multiple compatibility types for this plasmid. These may correlate with differences in their PFam57/62 proteins (which are encoded by *all* LB species plasmids), but this remains speculative.

Plasmid relationships and gene content

How many plasmid compatibility types exist in LB species? Thirty-two different PFam32 types have been reported (Casjens et al., 2018). These include two types that were previously noted as: (i) a second type of PFam32 gene on *B. burgdorferi* B31 lp28-1 (*bbf13*) and recently found as the only PFam32 gene on a *B. burgdorferi* strain B331 plasmid named lp28-11 (Schwartz et al., unpublished), and (ii) another type (unnamed because the plasmids they reside on have not been completely sequenced or described) present in contiguous sequences (contigs) in a *B. afzelii* PKo draft genome sequence (locus_tag BAPKO_2556) and a *B. japonica* HO14 draft genome sequence (locus_tag SAMN02983004_01117). In addition, two new types recently discovered in *B. turdi* have been named lp28-12 and lp28-13 (the former designated lp30 in (Margos et al., 2019a; S. Casjens and W. Qiu, unpublished). This brings the current total to 34 PFam32 types. Thus, with lp5 and *at least* two types of cp9, there appears to be a minimum of 37 LB plasmid compatibility types, suggesting that it would be theoretically possible that a single cell could harbor 37 different plasmids. The rate of discovery of new PFam32 types has greatly diminished as more genome sequences have been determined, so it is likely that nearly all extant types are now known.

For LB strains in which the plasmid copy numbers have been measured, the range is one to two copies per chromosome. The relative staining intensities/bp of all the plasmid and chromosome electrophoresis bands are typically quite similar, suggesting that plasmids are generally present in low copy number (Hinnebusch and Barbour, 1992; Casjens and Huang, 1993; Glockner et al., 2004; Tilly et al., 2012). The observation that sequence coverage was 10-fold higher for *B. burgdorferi* strain JD1 lp28-6 than for other plasmids suggests that there may be exceptions to this rule (Casjens et al., 2012). The LB plasmids also have a number of additional interesting and unusual features: (i) Only a small fraction of the plasmid genes encode proteins with recognizable similarity to proteins outside of the *Borrelia* genus; these include proteins related to previously known plasmid partitioning proteins (above), small molecule transporters, DNA restriction-modification systems, and nucleotide and DNA metabolism enzymes. (ii) The plasmids encode many N-terminally lipidated proteins most of which are targeted to the outer surface of the bacteria (*e.g.*, Casjens et al., 2000; Dowdell et al., 2017). These proteins are important mediators of interactions between spirochetes and their hosts and are potential vaccine and detection targets. (iii) A large number of paralogous gene families and paralogous intergenic sequences are present on the plasmids (Casjens et al., 2000). (iv) Like the main chromosome, the linear plasmids have covalently closed hairpin ends (Barbour and Garon, 1987; Hinnebusch et al., 1990; Hinnebusch and Barbour, 1991, 1992) (see Chapter 4). (v) Most of the LB plasmid types appear to have undergone structural rearrangements at a significant frequency in nature, judging from comparisons of cognate PFam32 type plasmids present in different isolates. Each resulting structural variant or "organizational subtype" is typically restricted to a single species. The different plasmid types and their rearrangements are discussed in more detail below.

Cp9

Six of the seventeen *B. burgdorferi* complete genome sequences and five of the fifteen genomes from nine other LB species include a cp9. Although the cp9s from *B. burgdorferi* do not carry a PFam32 gene, those from the other LB species encode a unique PFam32 type (Casjens et al., 2018). The different species harbor different cp9 organizational subtypes, and three are known in *B. burgdorferi*, including two in strain WI91-23. The *B. burgdorferi* cp9s encode PFam95 *eppA* genes (Champion et al., 1994; Miller and Stevenson, 2003), and all cp9s encode PFam96 and PFam165 proteins, also called BppC and BppA, respectively (El-Hage and Stevenson, 2002), which also are encoded by all intact cp32 plasmids; the roles of these proteins are not known. Cp9s may be readily lost during in vitro passage (Grimm et al., 2003), and there is no current evidence that they are involved in virulence.

Cp26

All LB isolates analyzed to date carry very similar, syntenic cp26 plasmids (Tilly et al., 1997; Terekhova et al., 2006; Casjens et al., 2012; 2017; 2018) and it has been shown to be essential for growth in culture in *B. burgdorferi* strain B31 (Byram et al., 2004; Jewett et al., 2007a). These plasmids encode proteins involved in GMP synthesis (Margolis et al.,

1994), chitobiose import (Tilly et al., 2004), host integrin binding (Behera et al., 2008), oligopeptide import (Bono et al., 1998), and the telomere resolvase that creates the hairpin ends of the linear replicons (Ravin et al., 2000; Kobryn and Chaconas, 2002) (see chapter 4). It also encodes the important surface antigen lipoprotein, OspC, that is expressed early in infection of a mammalian host and is essential for establishment of infection (Wilske et al., 1986; Grimm et al., 2004; Tilly et al., 2006; Fingerle et al., 2007). *ospC*, one of the most variable single-copy genes in the LB genome (Mongodin et al., 2013), appears to have undergone a number of short horizontal exchange events (Livey et al., 1995; Qiu et al., 1997; Wang et al., 1999a; Wang et al., 1999c; Barbour and Travinsky, 2010; Brisson et al., 2010; Haven et al., 2011) Single nucleotide polymorphism (SNP) tree analysis shows that, in the strains whose whole genome sequences have been determined, whole cp26 plasmids have not exchanged between chromosomal lineages, either within or between LB species (Mongodin et al., 2013).

Cp32s

The cp32 plasmids are largely syntenic bacteriophage prophages (see below) that can be quite similar throughout their lengths - in some cases 99% identical over several kbp (which makes unambiguous sequence assembly difficult without very long run sequence technology). Most LB isolates carry one or more cp32 plasmids (Casjens et al., 1997b; Iyer et al., 2003; Miller and Stevenson, 2003; Casjens et al., 2017; 2018). The two *B. garinii* isolates and one *B. maritima* isolate whose genomes have been sequenced have the fewest with two or fewer (*B. garinii* Far04 is the only isolate that carried no cp32s when its genome was sequenced) (Casjens et al., 2018; Margos et al., 2020). Their putative phage virion assembly proteins (see below) are presumably only expressed when the prophages are induced to lytic growth (Zhang and Marconi, 2005), but the cp32s also encode a number of surface exposed lipoproteins that are expressed from the prophage. The latter include the *revA* genes whose surface lipoprotein products bind fibronectin (Brissette et al., 2009a; Byram et al., 2015), the *mlp* encoded surface lipoproteins (Theisen, 1996; Yang et al., 1999; Porcella et al., 2000), the *bdr* (*Borrelia* direct repeat) genes whose functions are unknown (Zuckert et al., 1999; Zuckert and Barbour, 2000), and the complex family of *erp* (also called *ospE, ospF* or *elp*) genes whose various members have been shown to encode surface lipoproteins that bind to plasminogen (Brissette et al., 2009b), laminin (Brissette et al., 2009c), and factor H complement regulatory factor (Alitalo et al., 2002; Metts et al., 2003; Kraiczy et al., 2004), as well as having possible cell adhesin function (Antonara et al., 2007).

Twelve different types of PFam32 proteins are encoded by the cp32 group of circular plasmids, and these apparently can exist in the same cell (the name cp32-2 is not currently in use as it has the same PFam32 type as cp32-7 (Fraser et al., 1997; Casjens et al., 2000). Relatively frequent homologous recombination appears to shuffle sequences among the cp32s as well as allowing fusions between them (Brisson et al., 2013; Casjens et al., 2017; Margos et al., 2017b). In addition, about 15% of the cp32s that have been sequenced have undergone some apparently nonhomologous organizational rearrangements such as truncation, fusion and inversion.

Lp5

Lp5 is a relatively rare small plasmid that has been found only in two isolates of *B. burgdorferi* and one of *B. mayonii* (Casjens et al., 2000; Kingry et al., 2016; Casjens et al., 2017). They appear to encode only a PFam57/62 plasmid maintenance protein, a PFam137 protein of unknown function and several pseudogenes. There is no current indication that lp5 is involved in tick transmission or mouse virulence.

Lp17

Lp17, along with cp26 and lp54, is one of the three plasmids found in all natural LB isolates. It encodes unique PFam32 and PFam57/62 plasmid maintenance/compatibility proteins; in addition, all known lp17s encode homologues of *B. burgdorferi* B31 proteins BBD9, 10, 11, 15 and 18. Except for BBD18, a global regulator that affects the function of the alternative sigma factor RpoS and thereby controls the transition from mammalian host to tick vector (Dulebohn et al., 2014; Hayes et al., 2014), none of the other genes have a known function. Curiously, among the 32 lp17s in completely sequenced genomes, there are currently 16 different organizational variants (Figure 2) and nearly all are restricted to one species. Nearly all such variants are restricted to one species, but a single species can have up to seven different variants (as in the case of *B. burgdorferi* where 17 isolates have been sequenced (Casjens et al., 2017; Margos et al., 2017b; Casjens et al., 2018; Jabbari et al., 2018; Margos et al., 2020). Interestingly, all the LB lp17

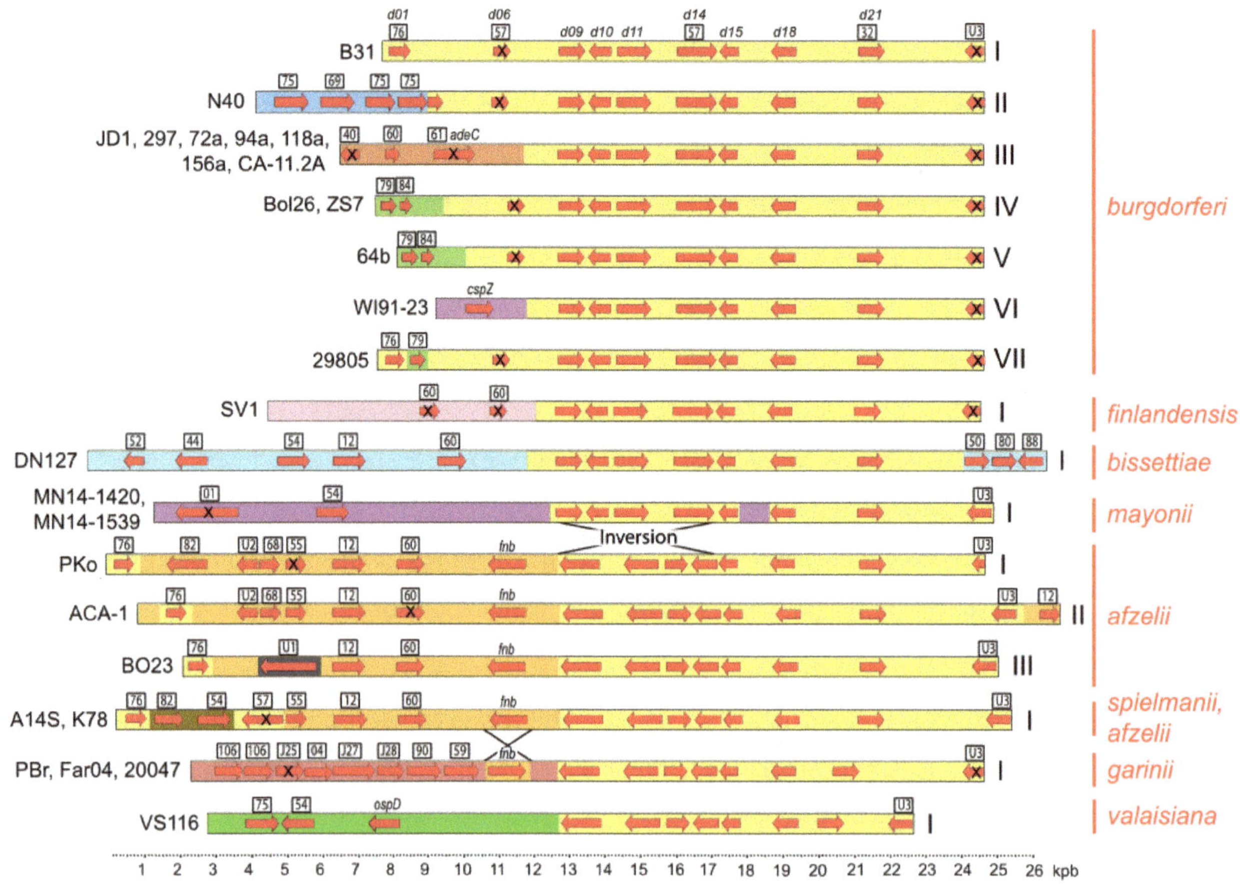

Figure 2. ORF maps of lp17 plasmids. The horizontal bars represent all the currently sequenced and annotated lp17 plasmids. identical background colors indicate regions of homologous DNA in the different plasmids. The bacterial species and organizational subtypes are indicated by Roman numerals on the right, and isolates that carry each subtype are indicated on the left. Selected genes are indicated by red arrows, and black "X"s mark pseudogenes. *B. burgdorferi* paralogous family (PFam) numbers (Casjens et al., 2012) are indicated in the boxes above; "U"s in boxes are proteins for which no intact gene is known in *B. burgdorferi*; "J's in boxes indicate strain B31 homologous proteins for which there is only one homologue in B31 (*i.e.*, no PFam number exists); *adeC, cspZ* and *fbn* refer to adenine deaminase, complement regulator-acquiring surface protein Z and fibronectin binding protein encoding genes, respectively. Selected strain B31 gene names are indicated above its map. This figure is reproduced with permission from Casjens *et al.* (2018).

variants include alternate sequences (3-21 kbp long) at their left ends, and all of these alternate left end sequences encode proteins that are homologous to proteins encoded by other linear plasmids.

Lp54

Lp54 plasmids are present in all LB isolates and are relatively uniform in gene content and organization. Two exceptions are apparent fusions with lp38 in strain BO23 and with cp32-11 in SV1. One variation among canonical, unfused lp54s is the presence of different 2-3 kbp left end extensions in (i) *B. chilensis,* (ii) *B. japonica* and (iii) *B. afzelii, B. garinii* and *B. spielmanii* that encode different proteins (Samuels et al., 1993; Casjens et al., 2018). The lack of repeated sequences on the lp54s allows easier sequence assembly, so more lp54 plasmid sequences have been reported than are present in the complete sets of plasmids that are listed in Tables 2 and 3.

Lp54 carries a number of genes that are important in mouse infection and tick transmission. The expression of one-third of lp54 genes is regulated by RpoS; these include genes required during both the tick and mammalian phases of the spirochete life cycle (Caimano et al., 2007; Iyer et al., 2015;

Caimano et al., 2019). OspA and OspB are major surface proteins that are expressed in ticks and are important for vector colonization (Pal et al., 2004; Yang et al., 2004; Tilly et al., 2012). Decorin binding proteins (DbpA and B) are important in mouse infection (Blevins et al., 2008; Shi et al., 2008; Weening et al., 2008; Salo et al., 2015). BBA34 (OppA5) is an oligopeptide binding protein that is required for mammalian infection (Bono et al., 1998; Groshong et al., 2017; Caimano et al., 2019) and BBA57 protein is critical for early infection success (Bernard et al., 2018). Bestor *et al.* (2010; 2012) and Promnares *et al.* (2009) used directed genetic analysis to show that the BBA03 and BBA62 (lp6.6) proteins, respectively, are important during tick transmission to mice. The PFam54 gene cluster on lp54, which encodes proteins BBA64-A73 in strain B31, includes genes that affect host complement function and/or bind host plasminogen (Wallich et al., 2005; Hallstrom et al., 2013; Koenigs et al., 2013; Hammerschmidt et al., 2016; Kraiczy, 2016). BBA64 is required for tick transmission (Gilmore et al., 2010). The central portion of this tandem gene array (*e.g.,* B31 genes *bba68-bba72*) varies in gene content and gene number both within and between species (Qiu et al., 1997; Wywial et al., 2009). Four conserved PFam54 genes make up the constant outside portions of this cluster, and there are 28 different full-length gene types in the variable regions of 28 LB genomes. The number of PFam54 genes in the central variable region ranges from two in *B. chilensis* to six or eight in *B. afzelii, B. bissettiae, B. garinii* and *B. valaisiana* isolates (Casjens et al., 2018). Essentially identical arrays are often present in different isolates of the same species, and any given array type is apparently restricted to a single species. This observation, along with single nucleotide polymorphism (SNP) tree analysis, shows that whole lp54 plasmids have not exchanged among the LB species analyzed (Mongodin et al., 2013).

Other linear plasmids

The remaining LB plasmids are linear and vary in size from 15 kbp to 82 kbp, although the vast majority are between 24 and 40 kbp. These include lp21, lp25, lp28-1, -2, -3, -4, -5, -6, -7, -8, -9, -11, lp32-3, -6, -10, -12, -13, lp36, lp38 and lp56 (lp28-10 is not used here because it was used by Kingery *et al.* (2016), and analyses of lp28-11, -12 and -13 mentioned above have not yet been published) (see Table 2). In addition, the left end of lp17 (above) and the right end of the chromosome (Banik et al., 2011; Casjens et al., 2012; 2018) have a number of different extensions that appear to be fragments derived from these "other" plasmids. These plasmids are discussed together since they have the following common properties: (i) They contain numerous similar sequence patches. (ii) Many of them have an exceptionally low (for bacteria) density of protein coding genes. (iii) They carry an unusually large (for bacteria) number of pseudogenes with reading frames that have been damaged by frame-disrupting mutations and/or large deletions and insertions relative to intact homologues. The ten *B. burgdorferi* B31 plasmids which contain most (87%) of this strain's pseudogenes have a total predicted protein coding fraction of 0.42, a much lower value than that (>0.90) typical of bacterial DNA (Casjens, 2000). (iv) Each PFam32 type includes plasmids from different isolates of quite different lengths (see Table 3) (Palmer et al., 2000; Casjens et al., 2012; 2017; 2018). (v) Finally, these plasmids have patchwork or mosaic relationships wherein patches of similar sequence appear to be more or less randomly scattered. When two "cognate" (*i.e.*, same PFam32 type) plasmids from this group are compared, they typically contain patches of high sequence similarity and yet may have immediately adjacent sequences that are completely different. The rearrangements that generated the mosaic relationship patterns appear to be largely non-homologous (Casjens et al., 2012; 2017; 2018). These observations suggest a tumultuous history of (often duplicative) random rearrangements followed by decay of broken and redundant genes. The reasons for this type of evolution remain mysterious.

Such apparently chaotic evolution has generated multiple organizational subtypes for all of this group of linear plasmid types. The number of such subtypes is not small (*e.g.,* see lp17 section above), but the fact that essentially identical cognate plasmids have been found in many strains suggests that nonetheless a limited number of subtypes exist for each plasmid type. The rearrangement process that generates these organizational subtypes is ongoing, since different paralogous segments have diverged to different degrees and individual subtypes are nearly always limited to one species or one lineage within a species (Casjens et al., 2017; 2018). On the other hand, the process is not so fast that every isolate or lineage has a completely unique set of organizational subtypes. The fact that no such rearrangement has been observed during laboratory cultivation indicates that such events are quite rare. Nonetheless, a small number of exceptionally similar plasmids have been found in more than one species. In particular, nearly identical lp5s in two *B. burgdorferi* lineages and *B. mayonii* and extremely similar lp17s in a *B. afzelii* strain and a *B. spielmanii* strain suggest that rare whole linear plasmid transfer may occur (Casjens et al., 2017; 2018).

Plasmid roles in Lyme disease

Numerous directed gene knockout and global scanning strategies have been used to identify plasmids or encoded genes in *B. burgdorferi* strain B31 that are required for infection of mice or survival in and transmission by ticks (*e.g.,* (Purser and Norris, 2000; Labandeira-Rey and Skare, 2001; Grimm et al., 2005; Strother et al., 2005; Lin et al., 2012; Ellis et al., 2013; Krupna-Gaylord et al., 2014; Troy et al., 2016; Ramsey et al., 2017; Phelan et al., 2019) (see Chapter 16). These studies found several plasmids, lp25, lp28-1, lp36, lp54 and cp26, to be essential for strain B31 pathogenesis in mice, while lp25, lp28-1 and lp28-4 are critical for tick transmission. Numerous individual genes on these plasmids have been identified as having essential or modulatory roles in these processes, including the following: lp28-1 carries the *vlsE* gene that encodes an outer surface protein and the cassettes with which *vlsE* exchanges sequences to produce antigenic variation (Norris, 2014; Verhey et al., 2019) and *arp* (Barthold et al., 2006). Lp25 contains the nicotinamidase-encoding *pncA* gene that is critical for mouse infection (Purser et al., 2003; Jewett et al., 2011) and *bptA* which is essential for tick transmission (Revel et al., 2005). Lp36 carries *bbk13*, which encodes a non-surface-exposed membrane protein important in early skin infection (Aranjuez et al., 2019); *bbk17 (adeC)*, which encodes adenine deaminase, important for mouse infectivity (Jewett et al., 2007b); *bbk32*, which encodes a surface exposed lipoprotein that binds fibronectin and certain glycosaminoglycans through its N-terminal domain and inhibits the classical complement pathway through its C-terminal domain (Probert and Johnson, 1998; Fischer et al., 2006; Seshu et al., 2006; Xie et al., 2019); and *bbk46*, which is critical for evasion of the humoral immune response in mice (Ellis et al., 2013). The roles of individual genes on lp54 and cp26 were discussed above. Numerous other candidates for plasmid genes that play roles during infection of mice or ticks have been identified by various global strategies, but have not yet been studied in detail (Lin et al., 2012; Dulebohn et al., 2013; Ellis et al., 2013; Phelan et al., 2019). Although they have not been shown to be directly important in pathogenesis, strain B31 lp25 and lp56 genes that encode restriction/modification proteins are technically important in that their inactivation is required for efficient transformation of *B. burgdorferi* strain B31 cells by exogenous DNA (Lawrenz et al., 2002; Kawabata et al., 2004; Rego et al., 2011). We note that, because of the past rearrangements among these plasmids, orthologs of these genes are not always carried on plasmids of the same PFam32 type. For example, the *vlsE* gene/*vls* cassette system is carried on lp28-1, -3, -8, -9, lp32-3, -6 or lp36 in different LB isolates, and *pncA* is found on lp25, lp28-2 or lp38 (Casjens et al., 2017; 2018).

Almost no genes with homology to previously known virulence genes in other bacterial families have been found in the LB. The one exception is a cluster of genes (so far, always on an lp28-8 plasmid) that encode, modify and probably export a short 25 amino acid peptide toxin (Molloy et al., 2015). This toxin, borreliolysin S, and its putative extensive modifications are very similar to streptolysin S, a cytolytic molecule produced by *Streptococcus pyogenes* strains, whose mechanism of action is not well understood. Members of this family of toxins are reported to target eukaryotic or prokaryotic cells (Quereda et al., 2017; Tsao et al., 2019). Interestingly, this set of toxin forming genes is present in some isolates of *B. afzelii, B. spielmanii, B. valaisiana, B. japonica* and *B. lusitaniae* (its presence in *B. garinii* is still uncertain), but they have not been found in *B. burgdorferi, B. finlandensis, B. mayonii*, or *B. bavariensis* isolates (Molloy et al., 2015; Casjens et al., 2018; S. Casjens, unpublished). Thus, the presence of the toxin genes does not correlate with the LB species that cause human Lyme disease and it is not a critical requirement for that disease, although it could be involved in acrodermatitis chronica atrophicans, a cell chronic inflammatory of Lyme disease that is most commonly correlated with *B. afzelii* infection (Maraspin et al., 2019).

Comparative Genomics

Typing methods

Typing of bacterial strains has great importance for clinical and epidemiological studies, for diagnostics, as well as for population genetic or evolutionary research of LB strains. While many PCR-based approaches, especially for diagnostic purposes, target single genes (based on time and financial considerations), a technique that was termed multilocus sequence typing (MLST) was invented in the late 1990s and has been used with great effect for bacterial epidemiology (Enright and Spratt, 1999; Spratt, 1999; Urwin and Maiden, 2003) In addition to methods described in this section, several other molecular typing methods, including DNA-DNA reassociation analysis, RNA gene restriction analysis (ribotyping), DNA sequencing of 16S rRNA or other conserved genes (e.g., *fla*, *ospA*, *hbb*), species-specific PCR, PCR-based restriction fragment length polymorphism (RFLP) analysis, pulsed-field gel electrophoresis (PFGE), randomly amplified polymorphic DNA (RAPD) fingerprinting, and variable-number tandem repeat (VNTR), have been developed and used for the identification and classification of LB species (Wang et al., 1999b; Farlow et al., 2002; Liveris, 2013). Except for OspC, these methods rely largely on chromosomal loci for typing determinations.

Discriminatory power varies among different molecular typing methods. In general, whole genome-based approaches such as PFGE, RAPD, MLST and whole genome sequencing (WGS) have the highest resolution. *ospC* typing was frequently used to differentiate strains within the same species, in particular in *B. burgdorferi*, and for disease association studies (Seinost et al., 1999; Wang et al., 1999c; Qiu et al., 2002; Brisson and Dykhuizen, 2004; Hanincova et al., 2008; Qiu et al., 2008; Wormser et al., 2008), but MLST is better suited than *ospC* for differentiation of disease associated sequence types (Hanincova et al., 2013) (see below). Depending on the study purpose (i.e., required discriminatory power) and available resources, appropriate selection of one or more molecular typing methods is critical to yield species, subtype and strain-specific information. Multilocus sequence typing/multilocus sequence analysis (MLST/MLSA) and WGS have been widely employed for new species delineation, epidemiology, population genetics and evolutionary studies of different LB species (Vitorino et al., 2008; Hoen et al., 2009; Margos et al., 2009; Margos et al., 2010; Ogden et al., 2010; Ogden et al., 2011; Vollmer et al., 2011; Margos et al., 2012; Hanincova et al., 2013; Vollmer et al., 2013; Jacquot et al., 2014; Carpi et al., 2015; Gatzmann et al., 2015; Mechai et al., 2015; Castillo-Ramirez et al., 2016; Walter et al., 2016; Walter et al., 2017; Tyler et al., 2018; Margos et al., 2020).

In this section, the focus will be on the following methods: *rrs-rrlA* (16S-23S) intergenic spacer (IGS) typing, *rrf-rrl* (5S-23S) IGS typing, *ospC* typing, MLST/MLSA and WGS analysis. These methods can be employed alone or in combination for molecular characterization of LB cultured isolates and directly or indirectly on uncultivated organisms in ticks, mammalian host reservoirs and human clinical specimens. Most PCR based methods can be used for molecular typing of *B. burgdorferi* directly in a variety of field-collected and clinical specimens, but mixed infections which have been reported in many sample types (Qiu et al., 2002; Vitorino et al., 2008; Hoen et al., 2009; Ogden et al., 2010; Vollmer et al., 2011; Mechai et al., 2015; Walter et al., 2016), may not be disentangled (Margos et al., 2011). Some molecular typing methods (i.e., ribotyping, PFGE, RAPD) do require pure cultured spirochetes. Selection of particular genotypes or strains of *B. burgdorferi* during *in vitro* cultivation has been demonstrated (Norris et al., 1997; Liveris et al., 1999) and this potential "culture bias" needs to be considered in data interpretation when cultured isolates are used for molecular typing studies.

rrs-rrlA (16S-23S) intergenic spacer (IGS)

As described earlier, LB species possess a unique rRNA gene organization that is distinct from that of other prokaryotes (Schwartz et al., 1992; Ojaimi et al., 1994). In *B. burgdorferi,* the rRNA gene locus consists of a single 16S rRNA gene (*rrs*) followed by a large intergenic spacer that varies in size among different LB species. This is followed by a tandem repeat of 23S rRNA-5S rRNA genes in most LB species that is separated by a short spacer of 200-350 bp (Schwartz et al., 1992; Gazumyan et al., 1994; Ojaimi et al., 1994; Fraser et al., 1997). Two different rRNA-based PCR-RFLP typing methods have been developed for *B. burgdorferi*, targeting either the *rrs-rrlA* spacer or the *rrfA-rrlB* spacer (Postic et al., 1994; Liveris et al., 1995; Rijpkema et al., 1995).

PCR amplification of the proximal 941 bp of the *rrs-rrlA* spacer followed by RFLP analysis using restriction endonuclease digestion distinguishes *B. burgdorferi* strains into three ribosomal spacer types referred to as RST1, RST2 and RST3. Studies have shown that RST types correlate with pathogenic potential in humans and mice (Liveris et al., 1999; Wormser et al., 1999; Wang et al., 2001; Wang et al., 2002; Wormser et al., 2008). Application of this typing scheme to uncultivated spirochetes from LD patients and field-collected ticks showed that both may be simultaneously infected with one or more distinct genotypes of *B. burgdorferi* (Liveris et al., 1999; Ranka et al., 2004). The *rrs-rrlA* spacer PCR product can be subjected to direct sequencing and sequence polymorphism within the first 250 nucleotides of this PCR product, resulting in delineation of 10 distinct Intragenic Sequence (IGS) types (Bunikis et al., 2004). Using this approach, 127 *B. burgdorferi* clinical isolates were classified into 16 IGS types and demonstrated linkage disequlibrium between the *rrs-rrlA* IGS and *ospC* loci, indicating a lack of random distribution of the plasmid that harbors *ospC* (Hanincova et al., 2008) (see section below on population genomics).

rrfA-rrlB (5S-23S) IGS

The 5S-23S IGS was used extensively for species determination via PCR-RFLP, sequencing or reverse line blot analysis (e.g., Postic et al., 1994; Rijpkema et al., 1995; Postic et al., 1998; Saint Girons et al., 1998; Kurtenbach et al., 2001). It is still used today for sample screening and LB species determination, as well as for resolving mixed LB species infection, (e.g., Jenkins et al., 2012; Coipan et al., 2016; Blazejak et al., 2018). While single locus approaches appear attractive and economical for sample screening or diagnostic purposes, the problems associated with single locus approaches for bacterial species determination or strain characterization are well known. Single loci provide a genealogy only for the targeted locus not for the organism, do not buffer against exchange of genetic material and PCR may show differential sensitivity (Urwin and Maiden, 2003; Lager et al., 2017). Phylogenetic analyses of IGS sequences may give deceptive results in species assignment, as exemplified by *B. afzelii, B. lanei* (previously named genomospecies 2), and *B. garinii/ B. bavariensis* (Postic et al., 2007; Margos et al., 2009; Coipan et al., 2016; Margos et al., 2017a; Sabitova et al., 2018).

ospC typing

As noted, OspC is essential for *B. burgdorferi* to establish a productive initial infection in mammals (Tilly et al., 2006). *ospC* was among the first single gene loci to be targeted for strain typing and has been used most widely for *B. burgdorferi* strain differentiation. *B. burgdorferi* samples are most commonly genotyped by amplifying a 617 bp region of *ospC* (Seinost et al., 1999; Wang et al., 1999c; Qiu et al., 2002). *ospC* typing has been employed to investigate *B. burgdorferi* genetic diversity in environmental samples (e.g. ticks) (Wang et al., 1999c; Qiu et al., 2002; Brisson and Dykhuizen, 2004; Barbour and Travinsky, 2010) and different allelic variants are associated with a differential capacity to cause disseminated infection in mammals (Seinost et al., 1999; Wormser et al., 1999; Wang et al., 2001; 2002; Wormser et al., 2008). *ospC* is one of the most highly polymorphic gene in LB species; by convention, *ospC* major groups are designated using the criteria that *ospC* alleles are <2% different within a group and >8% divergent between groups (Wang et al., 1999c). Approximately 25 *ospC* major groups have been identified in *B. burgdorferi* populations (Barbour and Travinsky, 2010; Tyler et al., 2018). Aspects of *ospC* in relation to population genomics are presented in later sections.

MLST and MLSA

MLST and MLSA (the application of MLST at the genus level) were developed from MultiLocus Enzyme Electrophosis (MLEE) and provide a sensitive method for bacterial strain characterization and species delimitation (Enright and Spratt, 1999; Spratt, 1999; Urwin and Maiden, 2003; Bishop et al., 2009). Typical MLST/MLSA schemes are based on evolutionary congruent (Loza Reyes, 2010) housekeeping loci and certain criteria should be considered for selecting such genes (Urwin and Maiden, 2003). These include that the chosen housekeeping genes: i) are scattered across the linear chromosome to avoid local bias, ii) should not be flanked by genes under strong positive selection pressure as this may influence the neighboring genes, and iii) all should have similar levels of genetic diversity so that each gene contributes to phylogenetic analyses and no single gene dominates a tree generated by use of the concatenated sequences.

Although several MLST/MLSA schemes have been introduced for LB, several of these depart from typical MLST in that they combine different categories of loci, such as genes encoding outer surface proteins, conserved housekeeping genes, or noncoding loci (Bunikis et al., 2004; Qiu et al., 2004; Richter et al., 2006; Postic et al., 2007). If loci that are not evolutionary congruent are combined in phylogenetic analyses, mixture models must be applied (Loza Reyes, 2010). The MLST scheme described here (Margos et al., 2008) adheres to the principles of MLST systems as originally intended (Spratt, 1999; Urwin and Maiden, 2003). It has been integrated into the MLST database maintained at the University of Oxford at https://pubmlst.org/borrelia (Jolley et al., 2004; Jolley, 2009; Margos et al., 2015a).

The MLST method developed for the *B. burgdorferi* sensu lato species complex (Margos et al., 2011) also has been utilized in modified form for relapsing-fever species (Margos et al., 2015a; Boden et al., 2016; Fingerle et al., 2016; Kingry et al., 2018; Stete et al., 2018) (see chapter 3). It permits characterization of isolates at the different levels required for evolutionary, epidemiological and population/landscape genetic studies. The method consists of three steps: i) data acquisition, ii) data assembly and iii) analysis (reviewed by (Margos et al., 2011)). Data acquisition involves sample collection, DNA extraction, targeted PCR of housekeeping loci and sequencing, either by Sanger sequencing or via Illumina MiSeq sequencing (Kingry et al., 2018). Comparison of sequence fragments with each other or to isolates in the MLST database allows species identification, determination of DNA variation in strains and species, as well as identification of recombination. It reflects the evolutionary history of bacterial isolates more realistically than single locus approaches.

As the method considers single point mutation differences, amplicons need to be sequenced in forward and reverse directions for comparison when using Sanger sequencing. Good quality sequences of the same fragment length as available through the MLST website (https://pubmlst.org/borrelia) permit obtaining allele numbers. The allele number for each locus results in a chain of eight integers which corresponds to the allelic profile of the isolate and defines the sequence type (ST). Novel alleles or novel sequence types must be submitted to the MLST database (via the curator, see https://pubmlst.org/borrelia/submission of data) to obtain sequential numbers. For analysis of MLST/MLSA data either concatenated sequences for all loci or allelic profiles can be used in downstream methods such as phylogenetic analysis, eBURST/goeBURST (Francisco et al., 2012), or population genetics methods. Currently, the *Borrelia* MLST database (https://pubmlst.org/borrelia) contains data for >3,100 isolates and >900 STs (as of 05.05.2020).

The species threshold level, adjusted to the MLST loci employed in this system, was determined to be 0.017 (Margos et al., 2009) and is comparable to the species threshold suggested by DNA-DNA hybridization (Postic et al., 2007). As for DNA-DNA hybridization, slight variations in genetic differences around the threshold may be observed using MLST. This poses the question whether such isolates should be regarded as new species. For example, when *B. yangtzensis* was described, some isolates were below the species threshold compared to the type strain but not compared to other isolates of the species. It was concluded that all isolates should belong to the same species (Margos et al., 2015b), indicating that slight variations around the species threshold should be considered when proposing new species. Also, when delineating new LB species by MLSA, ecological data should be taken into account whenever possible as this has been an important consideration when describing new bacterial species in general (Cohan, 2002).

Whole Genome Sequencing

Next generation sequencing (NGS) has advanced bacterial genome sequencing at a very fast pace (reviewed by Niedringhaus et al., 2011; Kulski, 2016; Levy and Myers, 2016). However, sequencing and assembly of LB genomes using NGS methods has shown that in addition to accurate short-read methods, technologies that provide long-reads covering several thousand bases (such as Oxford Nanopore Technology (ONT) or Pacific Bioscience single molecule real-time (SMRT) technology) are required for complete genome assembly, including plasmids (Kingry et al., 2016; Margos et al., 2017b; Gofton et al., 2018; Kuleshov et al., 2020). The value of completed bacterial genomes vs. high-quality draft status has been emphasized (Casjens et al., 2000; 2012) (see plasmid section above) and analyses that are contingent on multi-copy genes important for bacterial niche adaptation can only be achieved using completed genomes (Margos and Becker, unpublished).

Currently, the most commonly used technologies for bacterial genome sequencing are Illumina, Pacific Bioscience, and Oxford Nanopore technologies. The principles of these sequencing technologies are well described in several reviews (Niedringhaus et al., 2011; Kulski, 2016; Levy and Myers, 2016). All three technologies have been used for sequencing LB genomes. The accuracy of sequences is best with Illumina. However, due to the shortness of reads (600 bp max) difficult regions in the genome (in particular plasmid sequences) may not be completely assembled using only Illumina reads (Margos et al., 2017b; 2017c; Tyler et al., 2018; Margos et al., 2019a; 2020). For example, repetitive sequences on the chromosome (e.g. *lmp1*) prevent complete assembly of these genomic elements (Tyler et al., 2018; Margos et al., 2019a). Good results have been obtained with the PacBio RSII system sequencing often in combination with Illumina technology to correct small insertions/deletions. However, it has been noted that plasmids may appear incorrectly fused in the assembly or a single plasmid may consist of two or more contigs even using SMRT technology (Margos et al., 2017b; Gofton et al., 2018; Kuleshov et al., 2020). Another issue is that long sequence "overhangs" can be found at the ends of the resulting contigs due to the adapters added during SMRT library production and circular sequencing; these overhangs must be removed. On the other hand, SMRT sequencing provides valuable information on telomere regions of linear replicons which may be difficult or impossible to obtain with the other sequencing methods (Margos and Casjens, unpublished). Although ONT provides the longest reads, the error rate in DNA sequencing is high (Margos et al., 2017b). This may have an effect on genome assembly. Assembly software available for hybrid assembly such as SPAdes (Bankevich et al., 2012) or Canu (Koren et al., 2016) will not always assemble all plasmids accurately; there may be several copies of the same plasmid or plasmids may be lost during assembly (in particular cp32s) (Margos and Casjens, unpublished).

There is great utility of WGS for species and strain typing as well as for phylogenetic investigations. Without doubt, WGS provides the highest level of resolution for isolate discrimination. MLST data and sequences for other genes are easily extracted from WGS data and this provides an opportunity to compare isolates between studies (e.g.Gatzmann et al., 2015; Becker et al., 2016; Castillo-Ramirez et al., 2016; Walter et al., 2017; Tyler et al., 2018). Tools for MLST and genomic analyses are available via several websites, among them https://pubmlst.org/borrelia/ or the website of the Center for Genomic Epidemiology (http://www.genomicepidemiology.org/). Several studies have shown that phylogenies based on a larger number of conserved single copy genes (i.e., more than eight MLST genes) or genomic SNPs provide better node support even for internal nodes (Becker et al., 2016; Tyler et al., 2018). Some examples and details are given in the following section.

Genome phylogeny and biogeography

The application of next-generation sequencing methods to cultured LB species (Jacquot et al., 2014; Gatzmann et al., 2015; Becker et al., 2016; Tyler et al., 2018) and *B. burgdorferi*-infected ticks using an oligo-capture approach (Carpi et al., 2015; Walter et al., 2017) permitted assembly of the chromosome and contiguous sequences (contigs) for several, more or less completed, plasmids. These data allowed the re-construction of robust phylogenies based on chromosomal or genomic SNPs or large numbers of orthologous single copy genes with high support for external and even internal nodes (Mongodin et al., 2013; Becker et al., 2016; Tyler et al., 2018). The utility of genomics for phylogenetic/biogeographic analyses is briefly described here as the biogeography of species and their populations will be described in more detail in Chapter 9.

Investigations on Lyme disease risk in the United States revealed a pattern of distinct hotspots in the Northeast, the Midwest and California (Bacon et al., 2008; Schwartz et al., 2017). The pattern of Lyme disease occurrence is driven by the distribution of tick vectors (Diuk-Wasser et al., 2006) and wildlife hosts (Hanincova et al., 2006; Hamer et al., 2010; Ostfeld et al., 2018). Lyme disease risk also was affected by historical events impacting the distribution of *Ixodes* vectors and their reproductive hosts (Barbour and Fish, 1993; Spielman, 1994) followed by independent expansions of *Ixodes spp.* and *B. burgdorferi* in the Northeast and Midwest US (Hoen et al., 2009) (see also Chapters 10 and 21). The pattern of sequence type distribution demonstrated that the focal *B. burgdorferi* populations east (NE and MW USA) and west of the Rocky Mountains were genetically distinct but phylogenetically related (Hoen et al., 2009; Margos et al., 2012; Hanincova et al., 2013). These data also suggested that *B. burgdorferi* had been resident in North America for several thousand, if not millions of years (Hoen et al., 2009), findings that were confirmed by phylogenomic analyses of *B. burgdorferi* from North America (Walter et al., 2017). The predicted (Ogden et al., 2006) and recently observed population expansion of *I. scapularis* and *B. burgdorferi* into Canada (Ogden et al., 2010; 2011) indicated an influx of *B. burgdorferi* genotypes from the different focal populations of the USA likely driven by consequences of climate change (Ogden et al., 2006; Mechai et al., 2015; Tyler et al., 2018). Unique genotypes found in Canada, both by MLST and genome analyses (Mechai et al., 2015; Tyler et al., 2018), clustered next to rare genotypes from the USA (Walter et al., 2017) which suggested that these are also likely to originate from the USA (Tyler et al., 2018) (Figure 3). All phylogenetic/phylogenomic investigations conducted to date showed that *B. burgdorferi* populations do not cluster according to geography; that is, Northeast, Midwest and California isolates are intermingled in phylogenies and even European isolates form clades nested between North American clades (Figure 3) (Hoen et al., 2009; Hanincova et al., 2013; Castillo-Ramirez et al., 2016; Walter et al., 2017; Tyler et al., 2018; but see Qiu et al., 2008). While this is suggestive of migration events between Europe and North America and perhaps numerous migration events between populations within North America, the exact nature or mechanisms of these exchanges (and whether they are historical or contemporary) is currently unknown (Qiu et al., 2008; Castillo-Ramirez et al., 2016; Walter et al., 2017). Some *B. burgdorferi* genotypes isolated from patients in Europe were more closely related to North American genotypes than to tick-associated European genotypes; genome analyses suggested that these individuals may have acquired their infection in the USA (Castillo-Ramirez et al., 2016).

Approximately 90 genomes of various LB species were sequenced and used in combination with 18 previously published genomes for phylogenomic analyses (Becker et al., 2016). Phylogenetic

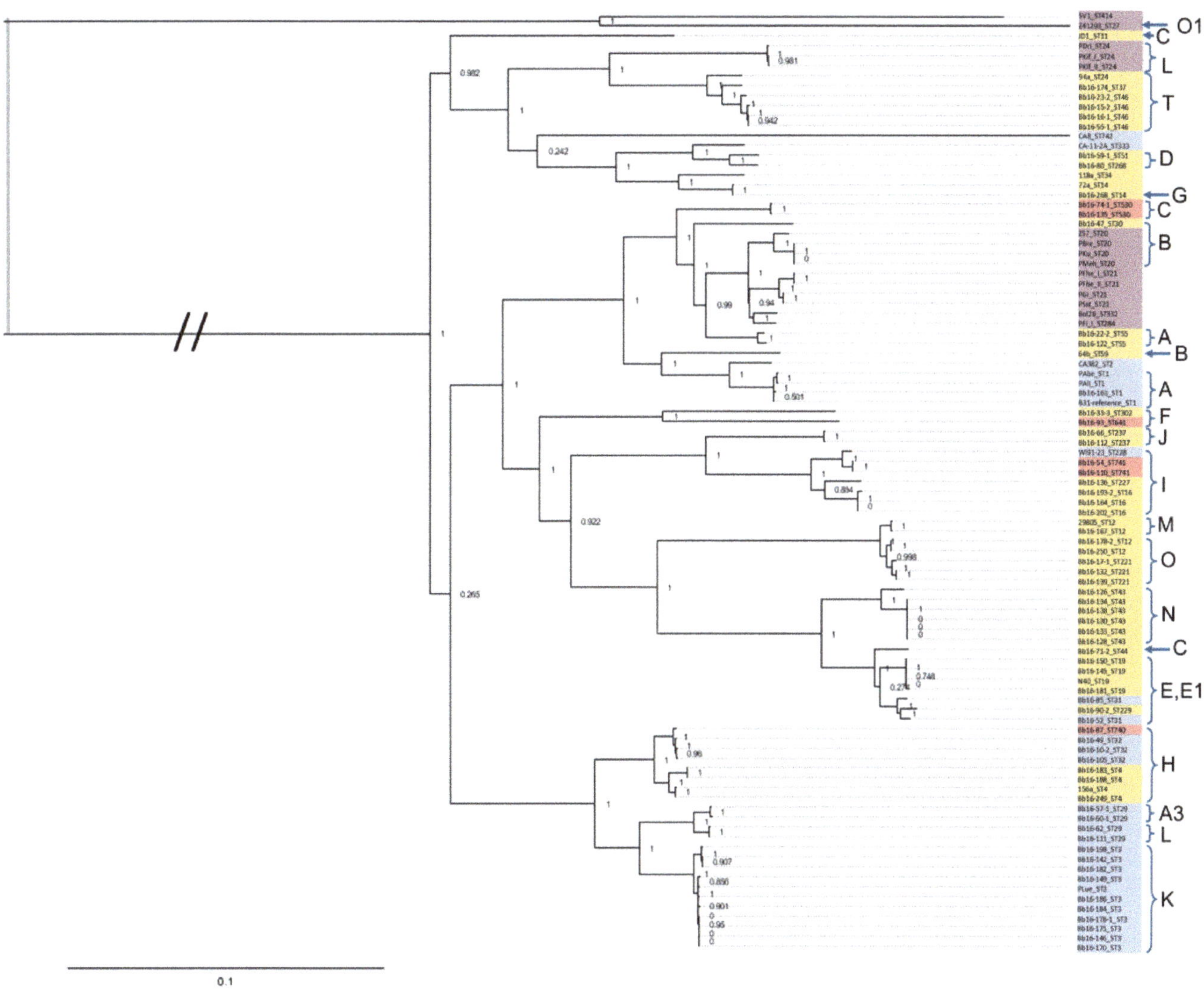

Figure 3. Population structure of *B. burgdorferi* inferred from genomic data. The figure shows a midpoint rooted phylogenetic tree of the *B. burgdorferi* chromosome generated via the SNVPhyl pipeline. All sequences are currently published full sequences from samples collected in North America and Europe. Color coding indicates the geographic occurrence of the samples: blue = USA only; red = Canada only, yellow = USA and Canada; violet = Europe. Letters to the right of the tree indicate the *ospC* major groups of the strains. Likelihood values for branches with a value less than 0.9 are shown. Reprinted with permission from (Tyler et al., 2018).

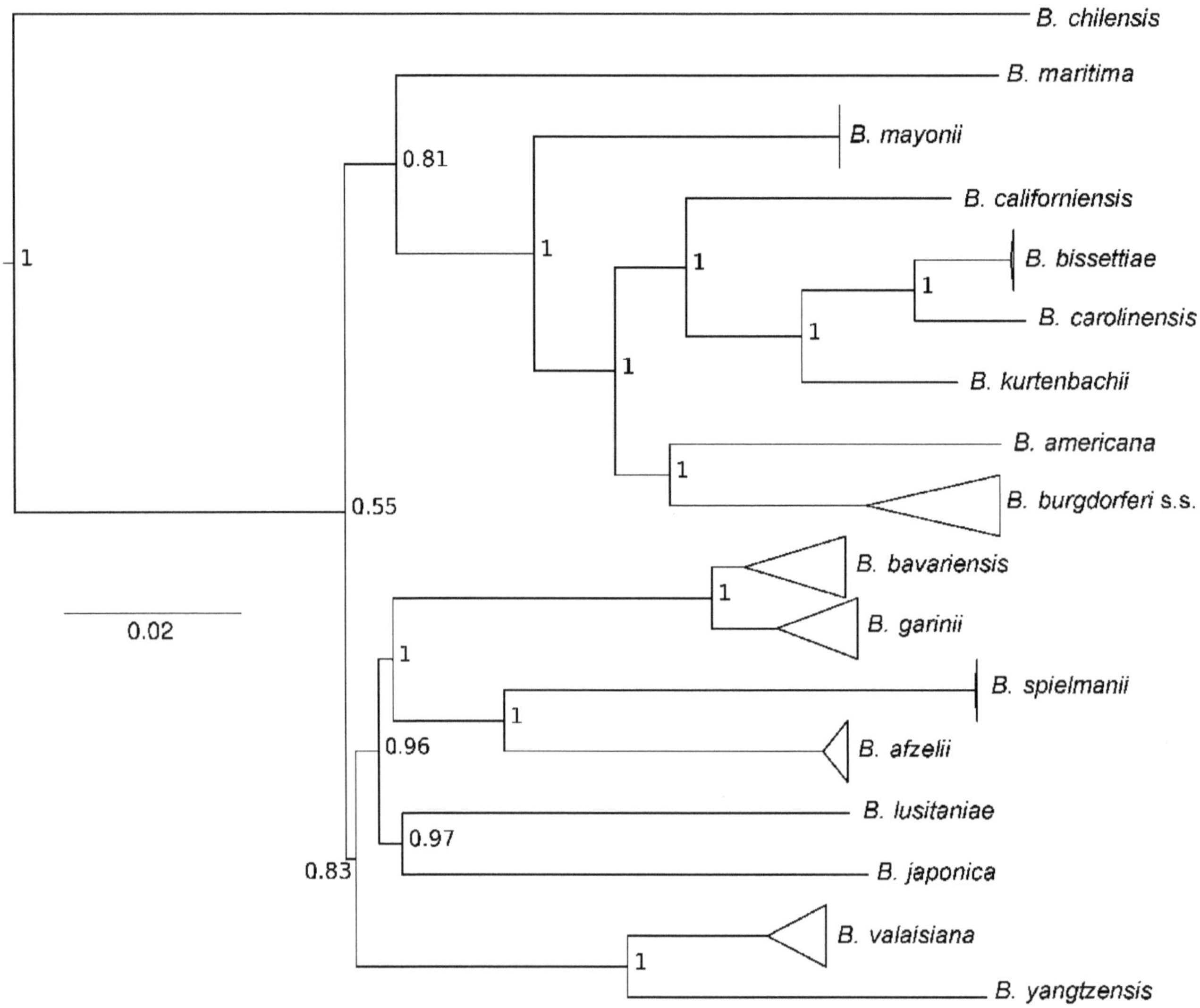

Figure 4. Phylogeny of 17 Lyme disease *Borrelia* species. Homologous sequences of chromosomal single copy genes were identified via BLASTn (v. 2.2.26) using *B. burgdorferi* B31 as reference (for details see Becker et al. 2016). DNA sequences of 113 genes were aligned in muscle (v. 3.8.31) and an unrooted maximum clade credibility tree was generated in BEAST (v.1.7.5). The tree contains 115 samples belonging to 17 of the currently known species. The nodes (branches) are very well supported; posterior probabilities are given next to the nodes. Scale bar corresponds to number of base substitutions per site. In the tree, *B. chinensis* takes a more basal position compared with other species. Two major lineages corresponding to a predominantly Eurasian clade and another predominantly North American clade are discernible. Figure modified from (Margos et al., 2020), with permission.

reconstruction based on >100 conserved chromosomal single copy genes revealed that the genospecies formed clearly distinguishable clusters corresponding to (predicted) host associations. Populations of only two species (as investigated by MLST), namely *B. lusitaniae* and *B. bavariensis*, formed clusters that corresponded to geography (Vitorino et al., 2008; Margos et al., 2013). These data implied that LB host associations are an important driver of diversification and speciation but that vector associations may also drive speciation (Gatzmann et al., 2015; Becker et al., 2016). There was strong node support throughout the tree and its topology, suggesting that host switching followed by speciation occurred several times in the evolutionary history of the LB species complex and is still ongoing. A phylogenetic tree for 17 LB species is presented in Figure 4.

A particularly interesting example relates to *B. garinii* and *B. bavariensis. B. garinii* is considered one of the most heterogeneous species within the LB species complex (Jacquot et al., 2014). Its wide distribution throughout Eurasia, in the Northern Hemisphere (including rarely the Atlantic coast of North America (Smith et al., 2006; Munro et al., 2019)) and Southern Hemisphere (Comstedt et al., 2011) is likely owing to its adaptation to terrestrial and sea birds as hosts and its ability to utilize several tick species as vectors (*Ixodes ricinus*, *Ixodes persulcatus*, *Ixodes. uriae* and *Ixodes pavlovsky*) (Gomez-Diaz et al., 2011; Vollmer et al., 2011; Mukhacheva and Kovalev, 2013; Vollmer et al., 2013; Norte et al., 2020). It has been speculated that the absence of *B. garinii* from terrestrial transmission cycles of North America may be due to vector incompetence of *I. scapularis* and *I. pacificus* (reviewed by Margos et al., 2019b). The population structure of *B. garinii* differs from that of *B. afzelii* (another widely distributed species throughout Eurasia) in that *B. garinii* MLST sequence types are more geographically mixed than those of *B. afzelii;* this has been attributed to migration patterns of their hosts (at least at a European scale) (Vollmer et al., 2011; Vollmer et al., 2013; Jacquot et al., 2014; Munro et al., 2019; Norte et al., 2020).

The closest related species to *B. garinii* is *B. bavariensis* with which it shares a most recent common ancestor (Margos et al., 2013; Gatzmann et al., 2015; Becker et al., 2016) (Figure 4). Speciation of *B. bavariensis* was likely due to a host switch from bird to rodent (Hu et al., 2001; Huegli et al., 2002). Phylogeographic analyses employing genetic data suggested that *B. bavariensis* has undergone a recent range expansion, invading Europe from Asia, its likely geographic origin. This range expansion was accompanied by a drastic genetic bottleneck, suggesting that the colonization of Europe was a single event, probably coinciding with a vector switch. Phylogenomics and biogeographic investigation confirmed the existence of two *B. bavariensis* populations, one in Eastern Europe/Asia that utilizes *I. persulcatus* as vector and a Western European population adapted to *I. ricinus*. Genome analyses of the chromosome and two plasmids (lp54 and cp26) confirmed a strong genetic bottleneck in, and supports the hypothesis for clonal expansion of, the European *B. bavariensis* population (reviewed in Margos et al., 2019b). Genes that were found under positive selection in genome analysis in the European population included nine chromosomal loci (BG_0142, BG_0275, BG_0298, BG_0327, BG_0359, BG_0368, BG_0706, BG_0748, BG_0827 although the latter showed a high value for recombination) and two loci on lp54 (BGA04, BGA37) (Gatzmann et al., 2015).

Genome analysis of *B. maritima*, a species described from California, revealed a small genome comprised of a chromosome and only seven plasmids, two circular and five linear (Margos et al., 2020). It remains to be investigated whether this is indicative of a narrow host and vector association. In the phylogenetic reconstruction from MLST data, *B. maritima* isolate CA690 was positioned close to the European species *B. afzelii,* whereas in the genomic-based phylogeny it formed a sister clade to *B. chilensis* and occupied a basal position compared to other LB species occurring in North America (Fedorova et al., 2014; Margos et al., 2020) (Figure 4). These data particularly highlight the utility of whole genome analyses for elucidation of phylogenetic relationships.

Lyme Disease *Borrelia* population genomics

Con-specific genomic groups in bacteria: selection or rare recombination?

Population genomics quantifies genome-scale genetic variations within natural populations of a single biological species and aims to understand underlying evolutionary processes including mutation, recombination, genetic drift and natural selection (Li et al., 2008; Ellegren, 2014). Population genomics can be studied through large-scale comparisons of DNA sequences of natural populations, and with the advent of high-throughput DNA sequencing technologies this term is replacing the more commonly used term "population genetics" (Luikart et al., 2003). For bacterial species, an absence of apparent genetic barriers to DNA exchange makes it challenging to analyze genetic diversification within populations as an evolutionary process distinct from genetic divergence between populations (Rocha, 2018). Like some other obligate and vector-borne parasites, however, LB bacteria show strong species-specific geographic distributions such that genomic variations within and between populations can be productively analyzed (Kurtenbach et al., 2006; Margos et al., 2012; Jacquot et al., 2014; Qiu and Martin, 2014; Seifert et al., 2015; Walter et al., 2016).

Unlike eukaryotes, bacteria can often exchange DNA across large phylogenetic distances. Recombination among con-specific bacterial strains is not negligible, but occurs at a rate that is not as high as in sexual eukaryotes (Smith et al., 1993; Rocha, 2018). The promiscuous nature and intermediate rates of bacterial recombination make the study of genomic variations challenging in practice. One apparent difficulty is the decreased statistical confidence in classification of species and strains based on phylogenetic reconstruction, which assumes an absence of recombination. Special phylogenetic methods are often necessary to either identify and exclude recombination hotspots prior to phylogenetic reconstruction or to estimate recombination rates in conjunction with strain phylogeny (Marttinen et al., 2012; Ansari and Didelot, 2014).

Furthermore, quantifying bacterial recombination rates is critical for identification of genes under adaptive evolution in bacterial pathogens. For example, loci encoding serotype-determining surface antigens often exhibit the highest recombination rates as well as the highest sequence variability in the genomes of pathogenic bacteria, including *E. coli* and LB bacteria (Haven et al., 2011; Bobay et al., 2015). Indeed, recombination is a universal genetic mechanism for acceleration of species adaptation, without which beneficial mutations may not be maintained due to interference with each other ("Hill-Robertson effect") and deleterious mutations cannot be easily purged ("Muller's Ratchet") (Muller, 1964; Hill and Robertson, 1966).

The limited but non-negligible levels of recombination in bacteria result in a mosaic genomic structure, whereby genetic variations are strongly linked across the entire chromosome yet are unlinked at recombination hotspots (Milkman and Bridges, 1990; Smith et al., 1993). As such, a limited number of distinct sequence types - far fewer than one would expect from random recombination but more than one would expect from strict clonality - exist within bacterial populations and are known as "clonal frames" (Milkman and Bridges, 1990). Here we refer to these clonal frames or genome-sequence clusters, representing recognizable evolutionary lineages within a con-specific LB population, as genome-wide linkage groups or "genomic groups" for short. Contrary to the original and long-enduring belief that the presence of genomic groups within a con-specific bacterial population is primarily a reflection of low recombination rates, recombination typically occurs at a higher frequency than *de novo* mutations in bacterial pathogens, supporting the critical role natural selection plays in the maintenance of genomic groups (Smith et al., 1993). Indeed, in the absence of natural selection such as adaptive diversification, a recombination rate on par with the rate of mutation is sufficient to render genomic groups within a bacterial population unrecognizable (Fraser et al., 2007). In the subsequent sections, a review of studies identifying the genomic groups within LB populations based on phylogenetic reconstruction of DNA sequences from across the main chromosome is presented. The role of natural selection targeting two highly variable surface antigen loci (*ospC* and *vls*) in maintaining con-specific genomic groups in LB populations is then examined.

LB populations contain well defined genomic groups

Diverse genomic groups co-exist within LB populations, often infecting a single tick and presumably a single vertebrate host (Rauter and Hartung, 2005; Andersson et al., 2013; Di et al., 2018). Multilocus linkage disequilibrium (nonrandom distribution of genetic alleles) among LB strains was first noted at three genetic loci (*p93*, *fla*, and *ospA*) based on a number of isolates which are now recognized as distinct LB species (Dykhuizen et al., 1993). Subsequently, linkage disequilibrium was observed within natural populations within the same LB species as well, between alleles at two distinct loci (Qiu et al., 2002), at multiple loci (Bunikis et al., 2004; Hanincova et al., 2008; Margos et al., 2008; Qiu et al., 2008; Hanincova et al., 2013) and, most importantly in terms of defining genomic groups, at the whole genome level (Mongodin et al., 2013; Walter et al., 2017; Tyler et al., 2018). It should be noted that genome-wide linkage disequilibrium is particularly strong within LB populations that have experienced recent expansion, such as those in the northeastern USA and southern Canada (Qiu et al., 2002; Margos et al., 2012; Tyler et al., 2018). When sampling is performed over a wider geographic area with a longer history of indigenous populations, such as those in Western and Midwestern US and Europe, decay of linkage disequilibrium is noted especially between plasmid-borne loci (e.g, *ospC*) and chromosomal loci (e.g., IGS) (Travinsky et al., 2010; Margos et al., 2012; Hanincova et al., 2013).

Con-specific genomic groups within LB populations displaying distinct evolutionary lineages are maintained to a greater extent by natural selection than by a low genetic recombination rate (Brisson and Dykhuizen, 2004; Haven et al., 2011). Indeed, rates of homologous recombination among coexisting

B. burgdorferi strains in the Northeast US is about three times the rate of point mutations based on a comparison of three co-existing genomes (Qiu et al., 2004). A subsequent simulation-based study showed that diversifying selection at a single surface-antigen locus was sufficient to overcome sequence-homogenizing effects of recombination and maintain the coexistence of within-population genomic groups (Haven et al., 2011). Recombination occurs more frequently within than between LB species in areas where they coexist, perhaps due to vector and host specialization providing more opportunities for strain mixing (Jacquot et al., 2014; Gatzmann et al., 2015). Nevertheless, cross-species horizontal transfers of whole plasmids in nature have been documented among coexisting LB species, including lp5 exchange between *B. mayonii* and *B. burgdorferi* in the Midwest US, lp56 exchange between *B. bissettiae* and *B. burgdorferi* in Western US, lp17 exchange between *B. spielmanii* and *B. afzelii* in Europe, and lp28-9 exchange between *B. burgdorferi* and *B. garinii* or *B. afzelii* in Europe (Casjens et al., 2018).

The genomic diversity of LB species and populations within them has profound public-health and clinical consequences. First, LB species differ in ecological prevalence, human pathogenicity, and clinical manifestations. In the USA, *B. burgdorferi* is the predominant pathogenic LB species, present in over 50% of adult *Ixodes scapularis* ticks in the Midwest and Northeast (Greay et al., 2018; Chauhan et al., 2019; Tokarz et al., 2019). While humans are incidental hosts, *B. burgdorferi* is pathogenic to humans and has caused over 275,000 cases of Lyme disease from 2008-2015 (Schwartz et al., 2017). On the other hand, *B. mayonii,* a more recently discovered LB species in the Midwest USA, has a prevalence of ~2% in ticks and causes acute spirochaetaemia (Pritt et al., 2016; Cross et al., 2018). No documented human disease has been associated with other recognized North American LB species (with the possible exception of *B. bissettiae* DNA in human serum; see Girard et al., 2011), including *B. americana, B. andersonii, B. californiensis, B. carolinensis, B. kurtenbachii, B. lanei* and the most recently named *B. maritima* (Barbour, 2019; Margos et al., 2020). In Eurasia, human infections are caused more frequently by LB species associated with mammalian hosts (*B. afzelii, B. burgdorferi*, *B. bavariensis*, and *B. spielmanii*) than by those associated with bird and lizard hosts (*B. garinii*, *B. lusitaniae*, and *B. valaisiana*) (Stanek and Reiter, 2011; Coipan et al., 2016). Remarkably, *B. bavariensis* is infrequently found in the tick vectors in Europe, but was frequently isolated from human patients (Margos et al., 2013; Springer et al., 2020). For *B. bissettiae*, a single human isolate has been obtained in Germany from a patient without travel history, whereas two human isolates exist for *B. lusitaniae* (Collares-Pereira et al., 2004; de Carvalho et al., 2008; Margos et al., 2016). In Europe, the prevalence of *B. bissettiae* in ticks is extremely low (Coipan et al., 2016) and *B. lusitaniae* is moderately prevalent in countries around the Mediterranean Sea (Baptista et al., 2004). It has been proposed that *B. valaisiana* is non-pathogenic to humans because no single human isolate has been obtained to date (Margos et al., 2017d), although the species is found as frequently in *Ixodes ricinus* ticks (the vector of *Borrelia* species in Europe) as other known human-pathogenic *Borrelia* species (Rauter and Hartung, 2005). It remains an open question whether some *Borrelia* species do not cause human disease because they are rare in the environment or utilize tick vectors that do not bite humans at any frequency.

Second, genomic groups within *B. burgdorferi* vary in disease propensities. At least 18 *B. burgdorferi* genomic groups, corresponding to major sequence variations at *ospC* (allelic types A-O, T, and U) co-segregate with chromosome-based phylogeny in northeast USA populations (Wang et al., 1999c; Barbour and Travinsky, 2010; Di et al., 2018). Initially, four groups (*ospC* types A, B, I, and K) were found to be strongly associated with disseminated (i.e. extracutaneous) Lyme disease (Seinost et al., 1999). This was confirmed and expanded to include ospC type H (Wormser et al., 2008). A subsequent MLST-based study demonstrated that there are *B. burgdorferi* sequence types (ST) with different capacities to cause disseminated infection in humans (Hanincova et al., 2013). Considering that all *ospC* groups vary in human virulence, invasiveness of different *ospC* type clonal groups has been ranked by scaling the frequency of their presence in blood and cerebrospinal fluid cultures by their natural prevalence in ticks (Figure 5) (Dykhuizen et al., 2008). Decreasing levels of *ospC* type diversity from tick to skin, blood, synovial fluid and cerebrospinal fluid as biological niches have been further confirmed by (Brisson et al., 2011).

Third, population genomics holds the promise of unbiased identification of molecular mechanisms underlying variations in clinical manifestations among LB strains (Seifert et al., 2015; Tufts et al., 2019). In the following sections, patterns of molecular polymorphisms at two highly variable surface antigen loci, *ospC* and *vlsE*, are summarized and the molecular and evolutionary mechanisms involved in sustaining LB prevalence in nature and pathogenicity in humans are discussed.

ospC variability: a serotype- and lineage-determinant of Lyme disease Borrelia

Two main selective forces have been proposed as evolutionary mechanisms that maintain the even and stable coexistence of diverse con-specific genomic groups within local LB populations despite the presence of recombination (Brisson et al., 2012; Andersson et al., 2013; Qiu and Martin, 2014; Seifert et al., 2015). First, the various LB genomic groups may prefer different vertebrate host species (Brisson and Dykhuizen, 2004; Mechai et al., 2016; Tufts et al., 2019). Second, diverse LB genomic groups may be maintained by negative frequency-dependent selection by which strains displaying rare antigenic variants have higher fitness than strains carrying more common antigen alleles (Haven et al., 2011; Durand et al., 2015; Durand et al., 2017a). The *ospC* locus encodes the major serotype determinant in LB (Wilske et al., 1993; Barbour and Travinsky, 2010). As noted earlier, it is also the most variable non-paralogous gene in the LB genome (Mongodin et al., 2013). The nearly one-to-one correspondence between the sequence allele types of OspC and *B. burgdorferi* genomic groups in the Northeast USA supports the mechanism of immune-mediated balancing selection for maintaining diverse coexisting genomic groups (Figures 3 and 5) (Attie et al., 2007; Hanincova et al., 2008); *ospC* alleles are highly divergent from one another (an average of ~15% pairwise differences in amino acid sequences of different OspC proteins) (Wang et al., 1999c). This pattern of genetic variation at immune-dominant loci is expected from, and strongly supports, the predominance of negative frequency-dependent selection in shaping microbial populations in nature, including those of LB (Haven et al., 2011; Andersson et al., 2013; Jacquot et al., 2014; Durand et al., 2015; Strandh and Raberg, 2015).

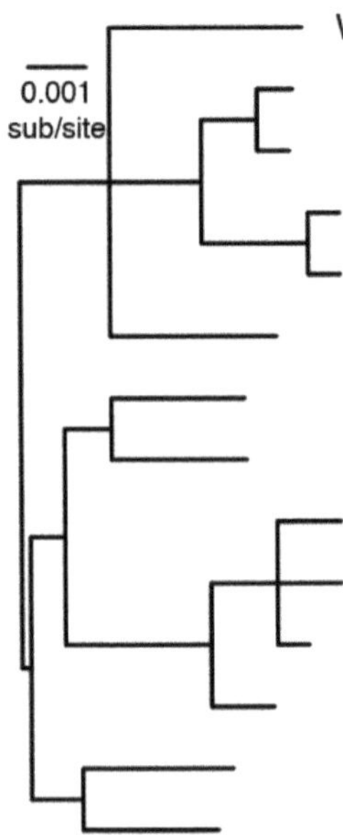

Strain	Group	OspC	Inv. Rank
WI91-23	B	I	1.8
72a	C	G	NA
118a	C	J	NA
94a	C	U	NA
MM1	C	U	NA
80a	B	N	4.0
JD1	E	C	5.3
156a	D	H	4.8
PAli	A	A	1.5
PAbe	A	A	1.5
B31	A	A	1.5
64b	A	B	3.9
N40	B	E	NA
29805	B	M	6.1

Figure 5. Genomic groups in *B. burgdorferi* from Northeast and Midwest United States. A phylogenetic tree of published complete *B. burgdorferi* genomes (n=14) (Mongodin et al., 2013) (left) was reconstructed based on nucleotide sequences at 20 loci, chosen randomly based on their approximate equal spacing on the main chromosome (*BB_0057, BB_0178, BB_0198, BB_0199, BB_0228, BB_0243, BB_0251, BB_0253, BB_0259, BB_0271, BB_0328, BB_0344, BB_0369, BB_0388, BB_0435, BB_0512, BB_0545, BB_0548, BB_0622, and BB_0809*). Sequences were translated, codon-aligned, and concatenated using MUSCLE (Edgar, 2004) and BpWrapper (Hernandez et al., 2018). An approximate maximum likelihood tree was inferred using FastTree (Price et al., 2010). All branches shown are supported by >0.9 bootstrap values. The tree was mid-point rooted and plotted using the APE package in R (Paradis et al., 2004). Phyogenetic groups (2nd column), major OspC allelic types (3rd column), and an invasiveness rank (4th column, "NA" for not available) were marked for each strain (Dykhuizen et al., 2008; Casjens et al., 2017). The nearly perfect linkage disequilibrium (and, in fact, the one-to-one correspondence) between major-group OspC alleles and distinct genomic groups suggests the critical role of OspC in maintaining within-species genomic diversity.

Critical ecological and evolutionary roles of *ospC* in shaping population structure of natural *B. burgdorferi* populations from the Northeast US are manifested, first, by a nearly complete linkage disequilibrium between sequence variations at *ospC* and genetic variations elsewhere in the genome (Figures 3 and 5) (Wang et al., 1999c; Hanincova et al., 2008; Wormser et al., 2008; Haven et al., 2011). The linkage disequilibrium is high, but less than complete at a larger geographic scale in the Western USA (Barbour and Travinsky, 2010; Travinsky et al., 2010; Tyler et al., 2018). On the other hand, population genomic studies on *B. afzelii, B. garinii* and *B. burgdorferi* in Europe showed lower level of linkage disequilibrium between *ospC* and chromosomal loci, likely caused by linkage decay in older populations (Jacquot et al., 2014). Second, as noted above, *ospC* is the most polymorphic non-paralogous locus in the genome, driven by a high rate of recombination that spills over to its genomic neighbors (Figure 6). Third, *B. burgdorferi* strains with major sequence variations at *ospC* are rather evenly represented in local populations among infected *Ixodes* ticks, consistent with expectations of negative frequency-dependent selection (Rannala et al., 2000; Qiu et al., 2002; Di et al., 2018). However, this was not found in a population of *B. afzelii* sampled at one habitat in Switzerland over a period of 11 years where one OspC major group dominated (Durand et al., 2017b).

A closer look at *ospC* sequence variability among 10 LB species reveals non-uniform selective pressures among gene regions associated with functional and structural domains (Figure 6). The first 18 codons (nucleotide positions 1-54) encode a signal peptide and are conserved not only for amino acid sequences (low evolution rates at 1st and 2nd codon positions), but also for synonymous nucleotides (low rates at the 3rd codon positions) (Wang et al., 1999c). Following the signal peptide, the N-terminal membrane-tethering motif (codons 19-30, or nucleotide positions 55-123) lacks sequence conservation (Kumru et al., 2011). The OspC molecules form dimeric α-helical bundles that are covalently anchored on the outer membrane (Kumaran et al., 2001; Eicken et al., 2002; Kumru et al., 2011). The α-helices (e.g., α1, α2, and α3) tend to be conserved at 1st and 2nd codon positions but not at the 3rd codon position, indicating influence of purifying selection. The surface-exposed regions (e.g, β1 and β2) tend to evolve at nearly the same rates among the three codon positions, indicating influence of positive selection driven by antagonistic co-evolution with host immune defense (Figure 6). LB strains differing in *ospC* alleles are immunologically distinct and wildlife hosts are frequently infected by multiple LB strains carrying distinct *ospC* alleles (Andersson et al., 2013; States et al., 2014; Bhatia et al., 2018). Serotype-determining antibody-binding epitopes have been mapped to hypervariable regions, making it challenging to design broadly effective OspC-based diagnostics or vaccines (Buckles et al., 2006; Ivanova et al., 2009; Baum et al., 2013; Izac et al., 2019). An OspC-based vaccine designed with a linear concatenation of eight serotype-specific epitopes has been approved for veterinary use against Lyme disease in dogs (Earnhart and Marconi, 2007; Izac and Marconi, 2019). The potential of conserved regions of *ospC* (Figure 6) as immune-protective vaccinogens against a broad array of coexisting LB strains within local endemic regions of Lyme disease remains to be explored (Ivanova et al., 2009; Baum et al., 2013).

vls variability: fast sequence and duplicative evolution

Lyme disease *Borrelia* express numerous lipoproteins in a highly regulated fashion during its enzootic life cycle (Radolf et al., 2012) (see Chapter 5). For example, OspA is highly expressed during tick infection and OspC is expressed during initial mammalian host invasion (Tilly et al., 2006; Mulay et al., 2009; Caimano et al., 2019). In contrast, the vls (*vmp* [variable major protein]-like sequences) family of lipoproteins is expressed during persistent infection of vertebrate hosts (Tilly et al., 2013; Norris, 2014; Chaconas et al., 2020), a key mechanism of antigenic variation in LB underlining its evasion host adaptive immunity. Comparative sequence analysis revealed distinct patterns of genomic variation, perhaps reflective their distinct biological functions. For example, OspA sequences vary greatly between LB species, but show minimal variation within species. In contrast, OspC sequences are substantially different both between and within LB species. The *vls* locus varies in both gene sequence, as well as in copy number, even within the same LB species (Mongodin et al., 2013; Tufts et al., 2019).

The *vls* system of antigenic variation in LB was discovered by sequence homology to the *vlp*/*vsp* (variable large/small protein) system in relapsing fever (RF) *Borrelia* (Zhang et al., 1997; Norris, 2014; Barbour, 2016) (See chapters 13 and 14). This system consists of a plasmid-borne expression locus, *vlsE*, for the variable surface antigen protein and an adjacent set of typically 10-20 tandem unexpressed "*vls*" cassettes containing variants of the *vlsE* sequence (Figure 7). Genetic variation at the *vlsE* expression locus is mediated by frequent non-homologous recombination between the silent *vls* cassettes and the expression locus (Coutte et al., 2009; Lin et al., 2009; Norris, 2014; Chaconas et al., 2020). Similar to the *vmp* system in RF *Borrelia*, capsule variability in *Neisseria* and the Variant Surface Glycoprotein (VSG) system in *Trypanosoma*, generation of random and abundant genetic variants mediated by intra-genomic recombination serve as a population-level adaptive mechanism (Vink et al., 2012; Graves et al., 2013; Chaconas et al., 2020).

The *vls* system in LB bacteria varies in both paralogous copy numbers and in cassette sequences (Graves et al., 2013) (Figure 8). Greater sequence similarity of *vls* cassettes within than between the

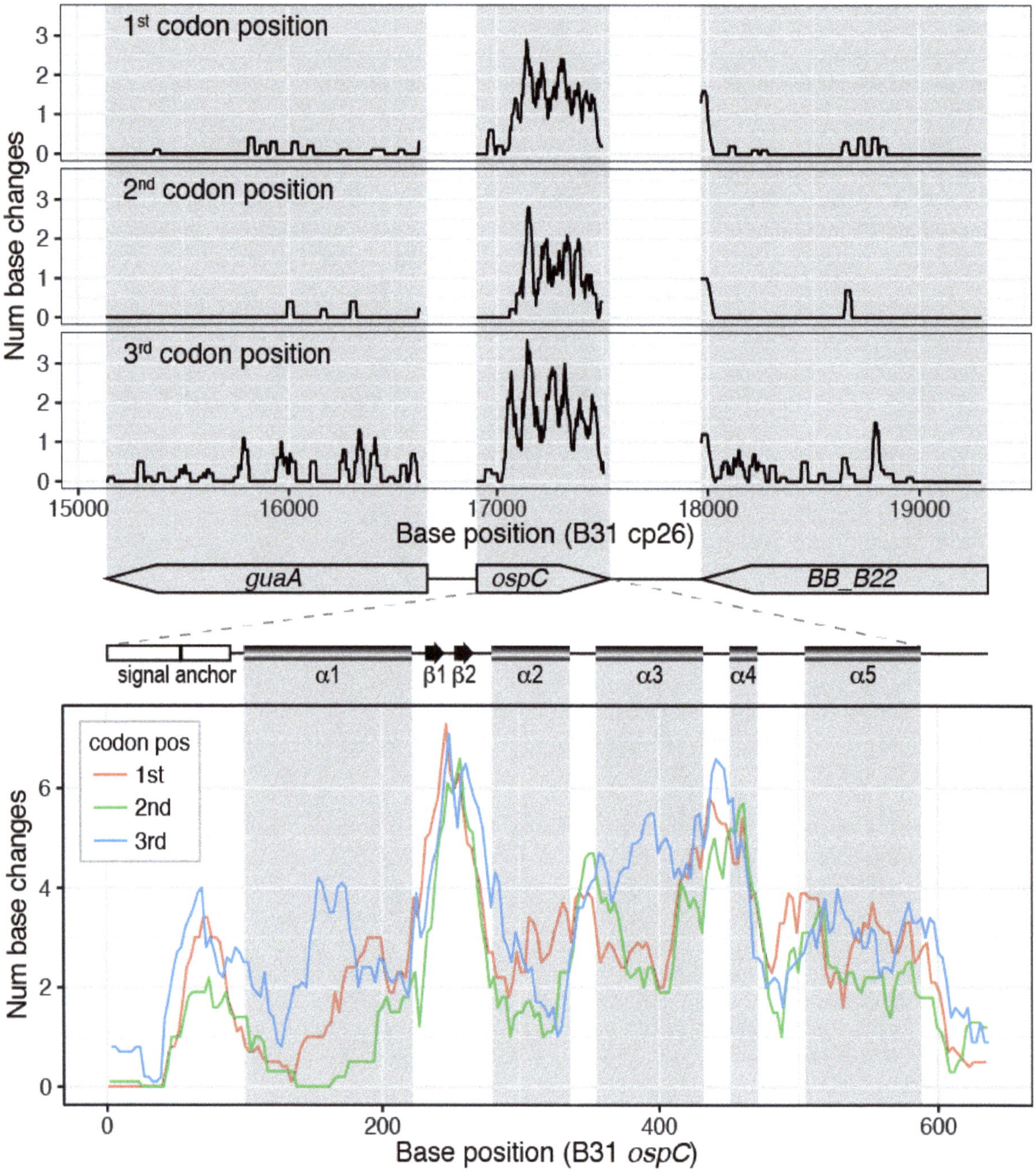

Figure 6. Association of *ospC* sequence variability with its molecular function and structure. (Top 3 Panels) Within-population evolutionary rates at *ospC* and its neighboring loci (*BB_B18* [*guaA*] and *BB_B22*). Genomic sequences at the three loci from 14 *B. burgdorferi* strains from Northeast USA (same strains as in Figure 5) were aligned by preserving codon positions using BIOALN of the BpWrapper sequence utilities (Hernandez et al., 2018). X-axis shows genomic coordinates of the B31 cp26 plasmid and a genome map of the region. Y-axis shows the number of nucleotide substitutions at the three nucleotide positions within each codon, calculated with the program DNACOMP of the PHYLIP package (Felsenstein, 1989). (Bottom panel) Evolutionary rates at the *ospC* locus among 32 *LB* strains of 10 LB species. The species include *B. burgdorferi* (16 strains; 118a, 156a, 29805, 64b, 72a, 94a, B31, BOL26, JD1, MM1, N40, PAbe, PAli, ZS7, CA-11-2A, and WI91-23), *B. spielmanii* (A14S), B. afzelii (ACA-1, K78, Tom3107, and PKo), *B.bavariensis* (BgVir and PBi), *B. bissettiae* (DN127), *B. garinii* (Far04 and PBr), *B. mayonii* (MN14-1420 and MN14-1539), *B. finlandensis* (SV1), *B. valaisiana* (Tom4006 and VS116), and *B. chinensis* (VA1). The positions of two functional domains (the signal peptide and the outer-membrane anchoring domain) and the secondary structural elements of OspC (Eicken et al., 2002; Kumru et al., 2011) are indicated above the panel.

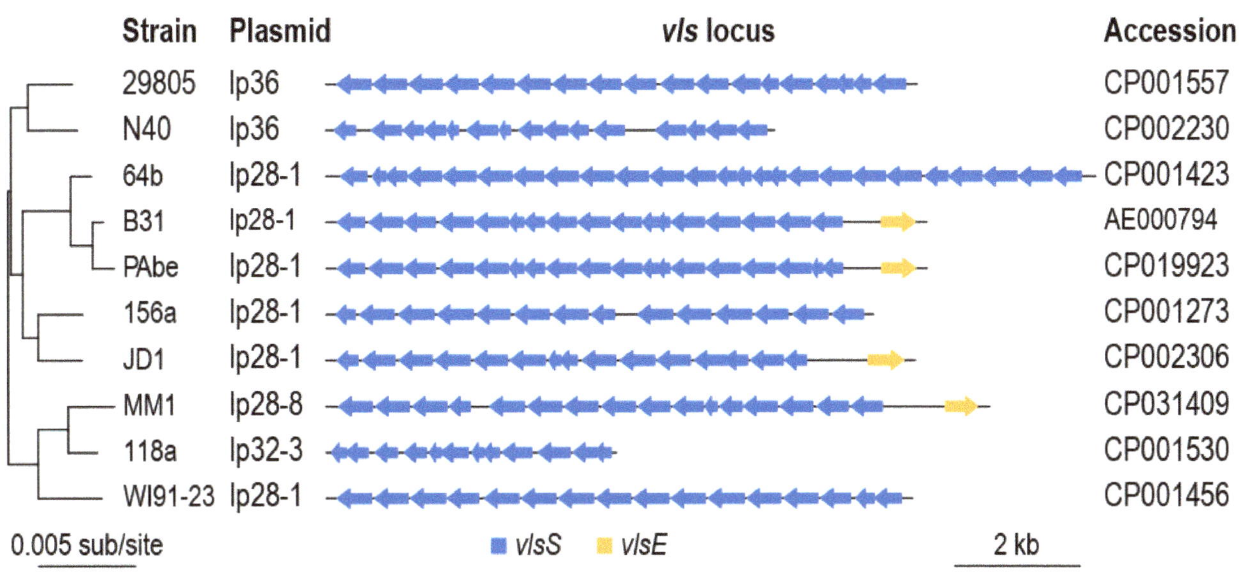

Figure 7. Copy-number and sequence variations at the *vls* locus. Genomic map of the *vls* locus in 10 *B. burgdorferi* strains. MLST phylogeny (left) was derived from Figure 5. Individual cassette sequences were originally identified manually (Casjens et al., 2017) and re-derived here by using HMMER (Potter et al., 2018) with B31 *vls* sequences (Zhang et al., 1997) as queries. *vlsE* expression locus is indicated in yellow. Note that the absence of *vlsE*, located in the telomeric regions of linear plasmids, in most strains is mostly likely due to incomplete genome assemblies.

closely related strains (Figure 8) could be due to either concerted evolution or rapid duplication and loss of gene copies (Nei and Rooney, 2005). Concerted evolution is the mechanism by which paralogous gene copies (e.g., ribosomal RNA genes) in a species evolve non-independently from each other and converge in DNA sequences by intra-genomic recombination (Elder and Turner, 1995). The gene birth-death hypothesis, on the other hand, proposes a mechanism by which sequences converge by independent gene duplications and losses while mutations accumulate, also independently, among paralogous gene copies (Nei and Rooney, 2005). The birth-death mechanism appears to be a more parsimonious model that does not require the assumption of the additional process of intra-genomic recombination. Gene tree analysis of *vls* cassettes from sequenced *B. burgdorferi* genomes supports the birth-death model of *vls* cassette evolution. First, the cassette copy number varies among the closely related *B. burgdorferi* strains, suggesting frequent gene duplications and losses (Figure 7).

Second, cassette sequences have diverged greatly within the same genome, suggesting that mutations accumulate independently among cassette copies, with a lack of (or infrequent) intra-genomic recombination (Figure 8). Third, *vls* cassettes among the most recently diverged sister-group *B. burgdorferi* strains do not sort into strain-specific sequence groups, as one would expect from concerted evolution. Examples of recently diverged sister groups include B31/64b/PAbe, 72a/118a, and 94a/MM1 (Figure 8). Fourth, although distantly related genomic groups tend to show strain-specific *vls* cassette sequences, exceptions exist. For example, SNP Groups D (156a) and E (JD1) are a pair of distantly related sister groups (Figure 5), but their vis sequences fail to evolve into distinct strain-specific lineages as one would expect from concerted evolution within each strain (Figure 8). It is possible that natural selection has kept the *vls* cassettes in these two strains from diverging from each other. Further comparative analysis of *vls* sequences among closely related genomes would enable quantification of precise rates of cassette duplication, gene loss, mutation, intra-genomic recombination,

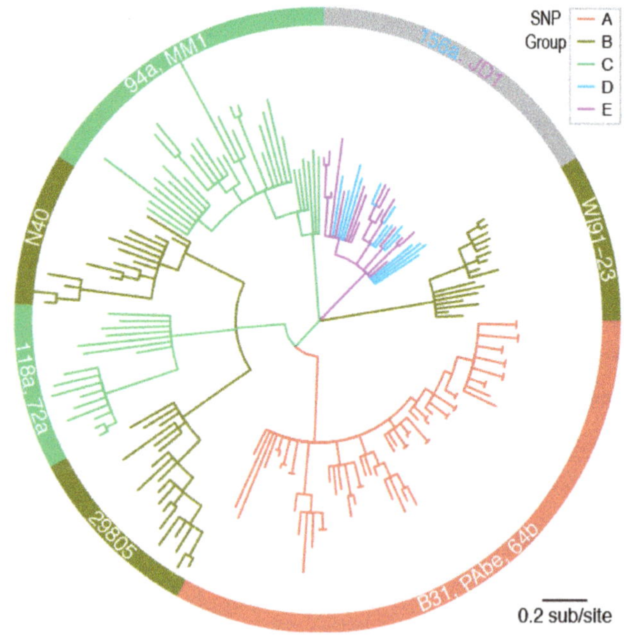

Figure 8. An unrooted maximum likelihood tree of *vls* cassette protein sequences. Genomes analyzed are those in Figure 3 plus 72a and 94a (Figure 5). Tree shows fast divergence of *vls* cassette sequences within and among con-specific genomic groups. Each tip represents a single cassette sequence. All branches are supported by a bootstrap value of 0.9 or greater, as determined by FastTree (Price et al., 2010). Clades (i.e., monophyletic groups) are colored by "SNP Group", referring to phylogenetic groups defined by chromosomal SNPs (Mongodin et al., 2013). This tree is an update of a previously published study using a similar set of genomes (Graves et al., 2013).

and natural selection in this key antigenic variation system responsible for LB persistence in hosts.

Deep homology between diverse mechanisms of antigenic variation in Borrelia

The antigenic variation systems in LB and the RF *Borrelia* apparently descended from a common ancestral molecular system (Norris, 2006). While the two systems share characteristics such as sequence homology, structural resemblance, plasmid-borne locations, and antigenic variation mediated by recombination between archival and expressed copies, they have diverged in genomic organization and molecular mechanisms of recombination. For example, the archival copies in RF *Borrelia* are dispersed among plasmids, whereas the *vls* sequences are arranged in tandem arrays of cassettes adjacent to *vlsE* on a single linear plasmid (Norris, 2006; Barbour, 2016; Casjens et al., 2017; Casjens et al., 2018) (Figure 7).

Furthermore, sequence and structural analyses revealed a deep molecular homology between *vls* and *ospC* systems of antigenic variation. VlsE, OspC, and Vsp share a similar, predominantly α-helical molecular structure (Kumaran et al., 2001; Eicken et al., 2002; Kumru et al., 2011). Genetic complementation experiments showed interchangeable roles for OspC and VlsE in maintaining infection in immune-deficient mice, but they have divergent molecular functions in early and persistent infections, respectively, in natural hosts (Tilly et al., 2013). Primers targeting the upstream and downstream conserved regions of the *ospC* locus could also be successfully used to amplify a full-length *vsp* gene in *B. miyamotoi* (an RF *Borrelia*), hinting at shared gene regulatory mechanisms (Di et al., 2018). In sum, the three main molecular systems of antigenic variations in *Borrelia*, including the *ospC* and *vls* systems in LB and the *vsp/vlp* system in RF *Borrelia,* likely share a common evolutionary origin. Comparative analyses among closely related LB genomes would further illuminate the origin, diversification, and functions of these highly adaptable molecular systems that are keys to LB survival as an obligate tick-borne parasite of vertebrates.

Population genomics is indispensable for reconstructing biogeographic histories of natural LB populations (Margos et al., 2012; Gatzmann et al., 2015; Walter et al., 2017; Tyler et al., 2018). In addition, population analyses of con-specific genomic groups are essential for uncovering mechanisms of LB evolution in nature including recombination, genome divergence driven by adaptation to host and vector, and genome diversification driven by host immunity (Kurtenbach et al., 2006; Qiu and Martin, 2014; Seifert et al., 2015; Becker et al., 2016). Furthermore, population genomics informs the investigation of key genes and genetic mechanisms (e.g., *ospC* and *vls* cassettes) associated with invasion of, and persistence in, hosts including humans (Mongodin et al., 2013; Jacquot et al., 2014; Seifert et al., 2015).

Bacteriophages of the Lyme disease *Borrelia*

The ϕBB-1 bacteriophage

A number of studies in which supernatants of *B. burgdorferi* cultures were examined by electron microscopy have reported the presence of tailed bacteriophage-like particles (Hayes et al., 1983; Barbour and Hayes, 1986; Neubert et al., 1993; Schaller and Neubert, 1994; Eggers and Samuels, 1999; Eggers et al., 2000). Sequence analysis also recognized that several *B. burgdorferi* plasmids could be prophages (see below). These observations have been borne out by more directed studies in strain CA-11.2A, where the supernatant yielded phage-like particles with symmetrical heads and apparently contractile tails (Eggers and Samuels, 1999; Eggers et al., 2000; 2001a). These particles were purified and shown to contain the cell's complement of cp32 DNA molecules in a linearized form (Eggers and Samuels, 1999) (Figure 9). This phage has been designated φBB-1.

The presence of cp32 DNA in the particles strongly suggests that these plasmids are prophage. With the exception of *B. garinii* strain Far04, cp32-like plasmids have been found in all other LB species that have been analyzed (see plasmid section above); however ϕBB-1 or similar phage particles have been detected in the supernatants of only a small number of *B. burgdorferi* strains and in one *B. bissettiae* strain following spontaneous release or after treatment with sub-lethal concentrations of 1-methyl-3-nitroso-nitroguanidine (MNNG), mitomycin C, or ethanol (Eggers and Samuels, 1999; Eggers et al., 2000; 2016). φBB-1 has not yet been propagated by lytic growth in the laboratory, and the reason(s) for this failure are not yet understood. Possibilities include: i) a high propensity to form a stable lysogen; ii) possible genetic defects in the machinery required for prophage induction; iii) a technical inability to generate a 'lawn' of *B. burgdorferi* cells in which to detect viral plaques; iv) a lack of sensitivity in the methods needed to detect phage from a small percentage of cells that are undergoing lysis; and/or (v) the absence *in vitro* of unique signals that induce fully lytic growth, such as might be found in either the tick vector or mammalian host.

Horizontal gene transfer (HGT) in *B. burgdorferi* appears to primarily involve the cp32s and other plasmid-borne loci consisting of fragments <2kb (Stevenson and Miller, 2003; Qiu et al., 2004; Schulte-Spechtel et al., 2006; Barbour and Travinsky, 2010; Haven et al., 2011; Brisson et al., 2012; 2013; Hanincova et al., 2013; Mongodin et al., 2013). The demonstration that ϕBB-1 can transduce a cp32 between two cells of the same *B. burgdorferi* strain and between different *B. burgdorferi* strains, suggests the possibility that φBB-1 is capable of enhancing the genetic diversity observed in the complement of cp32 plasmids found in different *B. burgdorferi* isolates (Casjens et al., 1997b; Caimano et al., 2000; Iyer et al., 2003; Stevenson and Miller, 2003; Casjens et al., 2018). Furthermore, the observation that ϕBB-1 can package small heterologous *E. coli/B. burgdorferi* shuttle vectors and transduce them between different *B. burgdorferi* strains (Eggers et al., 2016), suggests that this phage may act in generalized transduction, as well. With many bacteriophage, a small percentage (<1%) of the capsids are filled with bacterial host DNA rather than phage genomic DNA (Stanton, 2007; Thierauf et al., 2009; Muniesa et al., 2011; Penades et al., 2015). This mispackaging is thought to likely occur, at least in part, when the phage packaging mechanism recognizes packaging sequence homologs throughout the bacterial genome (Thierauf et al., 2009). Although the role of generalized transduction in HGT in naturally-occurring spirochetes is not yet well understood, "directed mispackaging" could be one factor as to why some non-phage genes, such as *ospC*, appear to be more likely than others to be subject to HGT and selective retainment in *B. burgdorferi* (Dykhuizen and Baranton, 2001; Brisson and Dykhuizen, 2004; Qiu et al., 2004; Attie et al., 2007).

cp32 as prophage

The cp32s share organizational properties with the genomes of other tailed bacteriophages. Comparative genomic studies of double-stranded DNA bacteriophage groups, such as the λ-like family of phages, have shown that their genomes are genetic mosaics in which each has an overall conserved gene order, with occasional non-homologous genes inserted into the genomes of different phages (Casjens et al., 1992; Hendrix, 2002; Casjens, 2005; Hendrix and Casjens, 2006; Casjens, 2008); this pattern is similar to that observed among the cp32s (Casjens et al., 1997b; Caimano et al., 2000; Casjens et al., 2000; Stevenson et al., 2000b; Stevenson and Miller, 2003). The 32-kb circular plasmids can have long stretches of nearly identical DNA, broken by three regions of significant variability; these regions correspond to the partitioning/compatibility regions that allow several cp32 isoforms to be stably maintained in a single isolate (see plasmid section above), and two regions that encode multiple families of differentially-expressed OspE/OspF/Elp and Mlp lipoproteins (Marconi et al., 1996; Porcella et al., 1996; Stevenson et al., 1996; Zuckert and Meyer, 1996; Casjens et al., 1997b; Akins et al., 1999; Yang et al., 1999; Caimano et al., 2000; Casjens et al., 2000; Porcella et al., 2000; Yang et al., 2003; Casjens et al., 2012).

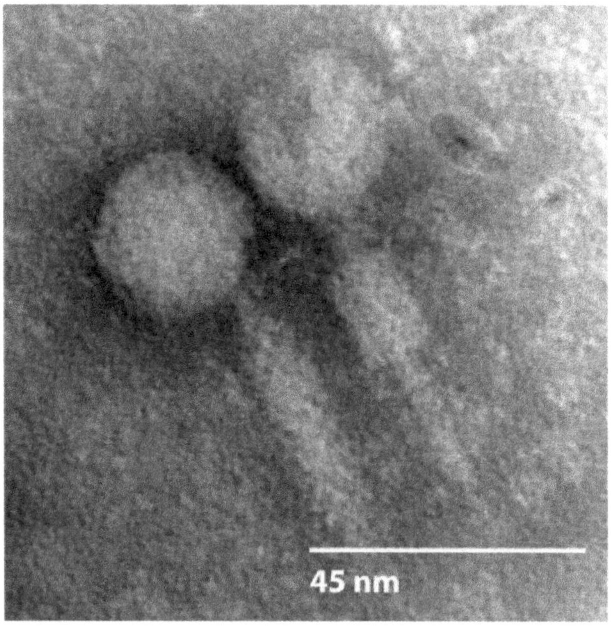

Figure 9. *B. burgdorferi* bacteriophage φBB-1 virions. Phage particles from the supernatant of an MNNG-treated culture of *B. burgdorferi* strain CA-11.2A were negatively stained for transmission electron microscopy with phosphotungstic acid; bar is 45 nm. Reproduced from Eggers et. al (Eggers et al., 2001b) with permission of the publisher.

cp32 phage-like genes

Two genes that encode proteins with homology to phage virion head assembly proteins are present on the cp32s. Genes L43 (and paralogues on other cp32s; L43 is on cp32-8 in *B. burgdorferi* strain B31) and L01 (and paralogues) encode a putative phage large subunit terminase and a portal protein, respectively (Casjens et al., 2000; Eggers et al., 2000; 2001a). Terminases form the ATP-dependent motors that translocate phage DNA into virions, and the portal proteins form the hole through which the DNA is packaged into the preformed phage heads (Catalano, 2005; Prevelige and Cortines, 2018). These two proteins are the most highly conserved tailed-phage proteins, and no homologues have been found that have different functions (Casjens, 2003). The presence of these genes strongly supports the notion that cp32 plasmids are in fact prophages. The positions of genes L43 and L01 (and cp32 paralogs) are consistent with the typical arrangement of virion head assembly genes in other dsDNA temperate bacteriophages, where the large terminase and portal are often the second and third genes, respectively, in the phage's late operon (see below) (Casjens, 2003; 2005; 2008). In contrast to L43 and L01, a majority of the predicted proteins encoded on the cp32s have no homologues in current databases outside of *B. burgdorferi*, while the structural proteins for φBB-1 have not yet been identified. This lack of homology may not be surprising, as virion structural proteins are known to be extremely variable, and many phage structural proteins have no recognizable homologues, even among proteins with identical functions in other phages (Casjens et al., 1992; Hendrix et al., 1999; 2000; Casjens, 2003; 2005). The known dsDNA tailed phages with genomes in the 20-60 kbp range typically have a single large operon that is expressed late in lytic infection and contains the 20-30 genes required for virion assembly and lysis (Casjens, 2005; 2008). The putative terminase and portal protein genes on the cp32s are followed by 25-30 contiguous genes, which are transcribed in the same direction, have no homologues outside the cp32s, and remain uncharacterized (Casjens et al., 2000). We previously have proposed that this region is likely the "late operon" of φBB-1 (Eggers et al., 2000; 2001a).

As part of their survival strategy, bacteriophages escape from their host cell by degrading the cell wall and lysing the cell after a cycle of lytic growth. To do this, all tailed bacteriophages encode an endolysin that degrades the cell wall and a holin whose permeabilization of the membrane allows the endolysin access to the cell wall thus controlling the timing of cell lysis (Young and Blasi, 1995). The genes that encode these proteins are sufficiently variable that they are often not recognizable in nucleotide sequence (Wang et al., 2000; Young, 2002). The function of these lysis genes necessitates that their expression be tightly controlled (holins in particular are lethal to bacteria (Young and Blasi, 1995), and they most often are clustered at the beginning or the end of the late operon (Casjens et al., 1992; Ford et al., 1998). On the cp32s, *blyA* (L23 and its paralogs) lies near the distal end of the putative phage late operon; the *blyA* protein product has membrane-disruptive activity and has been shown to mediate the release of both the phage lambda encoded endolysin (Damman et al., 2000) and an *E. coli* host-encoded cytolysin (Ludwig et al., 2007) in surrogate systems. These results suggest that *blyA* may encode a holin, although no cp32-encoded endolysin has yet been identified.

Late operons are not expressed from the uninduced prophage DNA and are only expressed during lytic growth of the phage; thus, the above model predicts that these genes will only be expressed during "inducing" conditions. A second, slightly weaker, prediction of the model is that these genes will in fact be co-transcribed as an operon. Zhang and Marconi (2005) used RT-PCR across the gene boundaries to demonstrate that the thirty genes from BBL42 to BBL28 (using cp32-8 as a typical cp32) are in fact co-transcribed after *B. burgdorferi* cells are exposed to the φBB-1-inducing agent MNNG, and that these genes are apparently not expressed in uninduced cells. Other DNA array transcription studies are largely in agreement that most of these genes are expressed minimally, if at all, in the absence of induction (Ojaimi et al., 2003; Anderton et al., 2004; Tokarz et al., 2004). The expression of these genes is influenced by the *B. burgdorferi* stringent response regulator, Rel_{Bbu}; a *B. burgdorferi* strain in which this protein was inactivated exhibited a marked increase in their expression (Drecktrah et al., 2015). These data are completely consistent with the proposal that this block of genes comprise a late phage operon. Curiously, the expression pattern of *blyA* is an exception to this rule and not what is expected for a lethal holin. An increase of the *blyA* transcript, in the absence of a concomitant increase in the expression of other genes within the late phage operon, has been observed with an increase in temperature (Ojaimi et al., 2003), after treatment with a bactericidal antibody (Anderton et al., 2004) and in the mammalian host (Zhang and Marconi, 2005). Thus, the *blyA* (and the adjacent *blyB*) gene appears to be under different regulatory control from the rest of the putative cp32 late operon genes. Consequently, a final determination as to whether BlyA is in fact a holin awaits further experimentation. Similarly, the cp32 *bdr* genes (R27 in cp32-4 and paralogs), the most promoter distal member of this gene cluster, appear to be expressed under non-inducing conditions and so may not be part of a late operon (Zuckert et al., 1999; Roberts et al., 2000).

There also are cp32 genes outside of the putative late operon that have homology to phage-related genes; these include those encoding the PFam161, PFam165, and PFam96 proteins. The *pfam96* gene is predicted to be a recombinase. The genes for PFam161 and PFam165 encode a single-stranded DNA binding protein and a DNA nuclease, respectively, and were renamed *ssbP* and *nucP* (Chenail et al., 2012). The products of these genes could play a role in the circularization of the linear φBB-1 genome into cp32 or in phage-mediated recombination. Tokarz et al. (2004) observed that members of all three of these PFams are upregulated in response to blood, a stimulus that *B. burgdorferi* encounters during the blood meal within the tick vector or during infection of the mammalian host. Chenail, et al. (2012) demonstrated that *ssbP* and *nucP* are up-regulated by the transcription factor, BpaB, which also represses gene expression from the cp32 *ospE/ospF/elp* loci (see below) and is itself encoded on each cp32 within the region responsible for plasmid replication and maintenance. Chromatin immunoprecipitation using an antibody specific for BpaB revealed other cp32 binding sites for this transcription factor, including several within the putative late phage operon near the genes encoding the putative terminase and portal protein genes and one near the *blyA* (holin) gene. The coordinated regulation of the "early" and "late" genes of many better studied plasmid prophages remains rather poorly understood (Ravin et al., 2000; Lobocka et al., 2004; Lehnherr, 2006), so an understanding of these findings in a larger context is difficult at this time.

Each cp32 also encodes a set of genes that are involved in their replication and stable maintenance, including the PFam57, PFam32, and PFam49 genes (see plasmid section above). PFam57 and the untranslated region directly upstream were found to be essential and sufficient for replication of the prophage plasmid (Eggers et al., 2002). The PFam32 paralogs have significant homology to the partitioning gene, *parA* (Zuckert and Meyer, 1996; Casjens et al., 1997b, 2000; Stevenson et al., 2000a). Homologues of *parA* are found on a number of well-studied low copy number plasmids, including two *E. coli* prophages, P1 and N15, which exist as autonomously replicating plasmids (Austin and Abeles, 1985; Grigoriev and Lobocka, 2001; Surtees and Funnell, 2003; Lobocka et al., 2004). Finally, the PFam49 genes found downstream of the PFam32 paralogs encode the transcription factor and predicted ParB homolog, BpaB (Stevenson et al., 1996; Stevenson et al., 2000b; Eggers et al., 2002; Burns et al., 2010; Jutras et al., 2012).

Lysogenic conversion

Most prophages alter their hosts, often by modifying the bacterial surface, through expression of their "host conversion" genes (Cheetham et al., 1995; Casjens and Hendrix, 2002; Canchaya et al., 2003; Casjens, 2003; Hendrix and Casjens, 2006; Dion et al., 2020), and the cp32s appear to be no exception. The cp32s express several families of proteins that have been well-studied. These include the Mlp, Bdr,

Rev, BapA and OspE/OspF/Elp proteins (Gilmore and Mbow, 1998; Skare et al., 1999; Yang et al., 1999; Zuckert et al., 1999; Porcella et al., 2000; Bauer et al., 2001; Miller and Stevenson, 2003; Zhang et al., 2005; Stevenson et al., 2006; Brissette et al., 2008; Brissette et al., 2009b; Kenedy and Akins, 2011; Lin et al., 2015). All these are membrane proteins, many demonstrate increased expression during the bloodmeal, and most have been shown to be on the cell surface. While the functional roles for many of these gene products are not known, an exception is OspE and related proteins that are able to bind complement factor H and so appear to play a role in the ability of *B. burgdorferi* to evade the host complement system (Hellwage et al., 2001; Alitalo et al., 2002; Stevenson et al., 2002; Metts et al., 2003; McDowell et al., 2004; Miller and Stevenson, 2004; Hovis et al., 2006; Kenedy and Akins, 2011). Additionally, OspF and related proteins may act as adhesins during *B. burgdorferi* infection and have been shown to bind to host cell heparin sulfate glycosaminoglycans on glial epithelial cells (Antonara et al., 2007; Antonara et al., 2011; Lin et al., 2015). Thus, these genes appear to be involved in the interactions between the pathogenic bacterium and its host, and so can be considered lysogenic conversion genes.

A multiplicity of prophages

In many bacterial systems, prophages have evolved ways to enhance the resistance of their bacterial hosts to superinfection by similar phages that might represent a threat (i.e., cause cell death) (Bondy-Denomy et al., 2016). φBB-1 represents an unusual group of bacteriophages, in that there can be as many as nine quite similar prophages resident in a single cell. Although the PFam32 proteins play a role in determining the maintenance and compatibility of prophage plasmids (see plasmid section above), it is not clear what role, if any, these proteins have in the mechanisms governing superinfection. Furthermore, Eggers, *et al.* (2016) reported differences in the efficiency with which ϕBB-1 transduces DNA packaged from one *B. burgdorferi* strain into either the same strain or into other *B. burgdorferi* strains. These data suggest that differences in surface receptors, restriction modification systems (Lawrenz et al., 2002; Kawabata et al., 2004; Rego et al., 2011), resident plasmid complements, or other as-yet-unidentified factors may influence the initial establishment of a new prophage. The entire complement of cp32s within a given isolate is represented in the extracellular supernatant after prophage induction and are all presumably packaged individually within φBB-1 capsids (Eggers et al., 2001b). Additionally, the putative structural and functional phage genes on different cp32s are highly conserved; thus, we anticipate that all cp32s might well be packaged within virtually identical capsids, whether or not all individual cp32s are still fully functional prophages. For example, two B31 plasmids, cp32-4 and cp32-9, have frameshift mutations in one and seven putative late phage operon genes, respectively (Casjens et al., 2000; Eggers et al., 2000); however we have observed, that cp32-4 and/or cp32-9 (the two plasmids were indistinguishable by the methods used) are packaged in abundance within φBB-1 capsids released from *B. burgdorferi* strain B31 (C. H. Eggers and D. S. Samuels, unpublished).

Several other *B. burgdorferi* plasmids (for example, cp9, cp18, lp54, lp56 and lp28-2) contain DNA that is homologous to virion structural protein genes and so may be phage-related. Cp9 plasmids in *B. burgdorferi* and other LB species appear to have evolved from cp32 plasmids by a series of inversion and deletion mutations (Champion et al., 1994; Dunn et al., 1994; Stevenson et al., 2000b). Similarly, cp18 plasmids present in strains *B. burgdorferi* 297 and N40 appear to be defective prophages that arose from deletions of the putative late phage operons on cp32s (Stevenson et al., 1997; Caimano et al., 2000). Both of these plasmid types may represent satellite phages that lack structural and functional phage genes but rely on the machinery encoded on cp32s to spread through a population. *B. burgdorferi* strain B31 lp56 contains an integrated cp32 (Casjens et al., 2000). This integration, which appears to be a rather recent event, occurred in the middle of a putative late operon gene, and two of its other late operon genes have frameshift mutations (Casjens et al., 2000), suggesting that the late operon of this cp32 is beginning to decay; such decay is characteristic of prophage genes that are no longer under selection (Canchaya et al., 2003; Casjens, 2003). lp54 contains twenty noncontiguous genes that are similar (but diverged) in sequence to the putative cp32 late operon genes (including the putative terminase and portal genes). However, in this region of lp54 (from approximately BB_A31 to BB_A55), there are two large insertions, one deletion and one substitution relative to the cp32 sequences (Casjens et al., 2000). Thus, whether lp54 is a productive prophage is unclear; neither this plasmid nor lp56 has been observed in preparations of phage from induced *B.*

burgdorferi cultures. One other linear plasmid, lp28-2 (and its relatives lp28-6, lp28-7 and lp28-9), is only very distantly related to the cp32s, but also carries a cluster of genes that may be phage-related (Eggers et al., 2001a; Casjens et al., 2018). These genes should encode a potential large subunit terminase (G21), a portal (G20), a tape-measure protein (G08) that is responsible for determining tail length in other bacteriophages (Hendrix, 1988), and several proteins (G10, G22, G23 and G27) with very distant homology to other phage proteins. Linear plasmids with these genes are found in other LB species as well, but an lp28-2-like genome has not yet been demonstrated to be packaged into phage virions. Given that a number of different bacteriophages have been observed in association with LB, however, one of these could potentially be derived from a more intact lp28-2 or related plasmid.

Future prospects

Determination of the complete genome sequence of the *B. burgdorferi* type strain B31, including the chromosome and all plasmids, in the year 2000 was a major step forward in the study of the Lyme disease bacteria as it opened the door to many new experimental avenues. Complete genome sequences of other *B. burgdorferi* isolates and a number of other LB species in the subsequent two decades has provided a much more detailed and refined picture of genetic variation and relationships within and among these LB lineages. This information also formed the basis for more accurate strain typing and lineage tracking, as well as disease diagnosis and detection of LB spirochetes. Since NextGen sequencing is much cheaper and faster, elucidation of many more complete LB genome sequences is expected in the near future. This information should make analyses of LB phylogeny, population biology, as well as host species and phylogeographic correlations, much more robust and informative. The genome sequences of LB species that do not cause human disease may point the way towards identification of genes important in human Lyme disease.

Although significant and ongoing progress has been made in understanding the roles of particular LB genes in mouse infection, much remains to be learned. Understanding of the molecular pathogenesis of Lyme disease, especially in humans, remains quite incomplete. Further, the possible loss of plasmids during the initial isolation of LB from their natural hosts still causes uncertainty in such studies. Genetic manipulation of LB and the use of mutant LB spirochetes in animal infection studies has been largely limited to the B31 strain of *B. burgdorferi* in laboratory mice. This is due to technical considerations and the current inability to modify DNA of most other isolates. Overcoming these difficulties in future studies of other *B. burgdorferi* lineages and LB species should lead to new insights in the molecular pathogenesis of Lyme disease.

Acknowledgements

The authors graciously thank current and former members of their research groups and their colleagues for their contributions to LB genomics. The authors research was supported by grants AR041511 and AI045801 (IS), AI049003 (SRC) and AI139782 (WQ) from the National Institutes of Health and a Quinnipiac University School of Health Sciences Faculty Research Award (CHE).

References

Akins, D.R., Caimano, M.J., Yang, X., Cerna, F., Norgard, M.V., and Radolf, J.D. (1999). Molecular and evolutionary analysis of *Borrelia burgdorferi* 297 circular plasmid-encoded lipoproteins with OspE- and OspF-like leader peptides. Infect Immun *67*, 1526-1532.

Alitalo, A., Meri, T., Lankinen, H., Seppala, I., Lahdenne, P., Hefty, P.S., Akins, D., and Meri, S. (2002). Complement inhibitor factor H binding to Lyme disease spirochetes is mediated by inducible expression of multiple plasmid-encoded outer surface protein E paralogs. J Immunol *169*, 3847-3853.

Andersson, M., Scherman, K., and Raberg, L. (2013). Multiple-strain infections of *Borrelia afzelii*: a role for within-host interactions in the maintenance of antigenic diversity? Am Nat *181*, 545-554. https://doi.org/10.1086/669905

Anderton, J.M., Tokarz, R., Thill, C.D., Kuhlow, C.J., Brooks, C.S., Akins, D.R., Katona, L.I., and Benach, J.L. (2004). Whole-genome DNA array analysis of the response of *Borrelia burgdorferi* to a bactericidal monoclonal antibody. Infect Immun *72*, 2035-2044.

Ansari, M.A., and Didelot, X. (2014). Inference of the properties of the recombination process from whole bacterial genomes. Genetics *196*, 253-265. https://doi.org/10.1534/genetics.113.157172

Antonara, S., Chafel, R.M., LaFrance, M., and Coburn, J. (2007). *Borrelia burgdorferi* adhesins identified using in vivo phage display. Mol Microbiol *66*, 262-276.

Antonara, S., Ristow, L., and Coburn, J. (2011). Adhesion mechanisms of *Borrelia burgdorferi*. Adv Exp Med Biol *715*, 35-49. https://doi.org/10.1007/978-94-007-0940-9_3

Aranjuez, G.F., Kuhn, H.W., Adams, P.P., and Jewett, M.W. (2019). *Borrelia burgdorferi bbk13* Is Critical for Spirochete Population Expansion in the Skin during Early Infection. Infect Immun *87*. https://doi.org/10.1128/iai.00887-18

Arnold, W.K., Savage, C.R., Lethbridge, K.G., Smith, T.C., 2nd, Brissette, C.A., Seshu, J., and Stevenson, B. (2018). Transcriptomic insights on the virulence-controlling CsrA, BadR, RpoN, and RpoS regulatory networks in the Lyme disease spirochete. PLoS One *13*, e0203286. https://doi.org/10.1371/journal.pone.0203286

Attie, O., Bruno, J.F., Xu, Y., Qiu, D., Luft, B.J., and Qiu, W.G. (2007). Co-evolution of the outer surface protein C gene (*ospC*) and intraspecific lineages of *Borrelia burgdorferi* sensu stricto in the northeastern United States. Infect Genet Evol *7*, 1-12. https://doi.org/10.1016/j.meegid.2006.02.008

Austin, S., and Abeles, A. (1985). The partition functions of P1, P7, and F miniplasmids. Basic Life Sci *30*, 215-226.

Bacon, R.M., Kugeler, K.J., and Mead, P.S. (2008). Surveillance for Lyme disease--United States, 1992-2006. MMWR Surveill Summ *57*, 1-9.

Banik, S., Terekhova, D., Iyer, R., Pappas, C.J., Caimano, M.J., Radolf, J.D., and Schwartz, I. (2011). BB0844, an RpoS-regulated protein, is dispensable for *Borrelia burgdorferi* infectivity and maintenance in the mouse-tick infectious cycle. Infect Immun *79*, 1208-1217. https://doi.org/10.1128/IAI.01156-10

Bankevich, A., Nurk, S., Antipov, D., Gurevich, A.A., Dvorkin, M., Kulikov, A.S., Lesin, V.M., Nikolenko, S.I., Pham, S., Prjibelski, A.D., *et al.* (2012). SPAdes: a new genome assembly algorithm and its applications to single-cell sequencing. J Comput Biol *19*, 455-477. https://doi.org/10.1089/cmb.2012.0021

Baptista, S., Quaresma, A., Aires, T., Kurtenbach, K., Santos-Reis, M., Nicholson, M., and Collares-Pereira, M. (2004). Lyme borreliosis spirochetes in questing ticks from mainland Portugal. Int J Med Microbiol *293 Suppl 37*, 109-116. https://doi.org/10.1016/s1433-1128(04)80016-0

Barbour, A.G. (1988). Plasmid analysis of *Borrelia burgdorferi*, the Lyme disease agent. J Clin Microbiol *26*, 475-478.

Barbour, A.G. (2016). Multiple and Diverse *vsp* and *vlp* Sequences in *Borrelia miyamotoi*, a Hard Tick-Borne Zoonotic Pathogen. PLoS One *11*, e0146283. https://doi.org/10.1371/journal.pone.0146283

Barbour, A.G., and Fish, D. (1993). The biological and social phenomenon of Lyme disease. Science *260*, 1610-1616.

Barbour, A.G., and Garon, C.F. (1987). Linear plasmids of the bacterium *Borrelia burgdorferi* have covalently closed ends. Science *237*, 409-411.

Barbour, A.G., and Hayes, S.F. (1986). Biology of *Borrelia* species. Microbiol Rev *50*, 381-400.

Barbour, A.G., and Travinsky, B. (2010). Evolution and distribution of the *ospC* Gene, a transferable serotype determinant of *Borrelia burgdorferi*. MBio *1*, e00153-00110. https://doi.org/10.1128/mBio.00153-10

Barbour, A.Q., W. (2019). Borreliella. In Bergey's Manual of Systematics of Archaea and Bacteria (John Wiley & Sons, Inc. in association with Bergey's Manual Trust), pp. 1-22. https://doi.org/https://doi.org/10.1002/9781118960608.gbm01525

Barthold, S.W., Hodzic, E., Tunev, S., and Feng, S. (2006). Antibody-mediated disease remission in the mouse model of lyme borreliosis. Infect Immun *74*, 4817-4825. https://doi.org/74/8/4817 [pii] 10.1128/IAI.00469-06

Bauer, Y., Hofmann, H., Jahraus, O., Mytilineos, J., Simon, M.M., and Wallich, R. (2001). Prominent T cell response to a selectively in vivo expressed *Borrelia burgdorferi* outer surface protein (pG) in patients with Lyme disease. Eur J Immunol *31*, 767-776.

Baum, E., Randall, A.Z., Zeller, M., and Barbour, A.G. (2013). Inferring epitopes of a polymorphic antigen amidst broadly cross-reactive antibodies using protein microarrays: a study of OspC proteins of *Borrelia burgdorferi*. PLoS One *8*, e67445. https://doi.org/10.1371/journal.pone.0067445

Beaurepaire, C., and Chaconas, G. (2005). Mapping of essential replication functions of the linear plasmid lp17 of *B. burgdorferi* by targeted deletion walking. Mol Microbiol *57*, 132-142. https://doi.org/MMI4688 [pii] 10.1111/j.1365-2958.2005.04688.x

Becker, N.S., Margos, G., Blum, H., Krebs, S., Graf, A., Lane, R.S., Castillo-Ramirez, S., Sing, A., and Fingerle, V. (2016). Recurrent evolution of host and vector association in bacteria of the *Borrelia burgdorferi* sensu lato species complex. BMC Genomics *17*, 734. https://doi.org/10.1186/s12864-016-3016-4

Behera, A.K., Durand, E., Cugini, C., Antonara, S., Bourassa, L., Hildebrand, E., Hu, L.T., and Coburn, J. (2008). *Borrelia burgdorferi* BBB07 interaction with integrin alpha3beta1 stimulates production of pro-inflammatory mediators in primary human chondrocytes. Cell Microbiol *10*, 320-331. https://doi.org/CMI1043 [pii] 10.1111/j.1462-5822.2007.01043.x

Bernard, Q., Smith, A.A., Yang, X., Koci, J., Foor, S.D., Cramer, S.D., Zhuang, X., Dwyer, J.E., Lin, Y.P., Mongodin, E.F., *et al.* (2018). Plasticity in early immune evasion strategies of a bacterial pathogen. Proc Natl Acad Sci U S A *115*, E3788-e3797. https://doi.org/10.1073/pnas.1718595115

Bestor, A., Rego, R.O., Tilly, K., and Rosa, P.A. (2012). Competitive advantage of *Borrelia burgdorferi* with outer surface protein BBA03 during tick-mediated infection of the mammalian host. Infect Immun *80*, 3501-3511. https://doi.org/10.1128/iai.00521-12

Bestor, A., Stewart, P.E., Jewett, M.W., Sarkar, A., Tilly, K., and Rosa, P.A. (2010). Use of the Cre-lox recombination system to investigate the lp54 gene requirement in the infectious cycle of *Borrelia burgdorferi*. Infect Immun *78*, 2397-2407. https://doi.org/IAI.01059-09 [pii] 10.1128/IAI.01059-09

Bhatia, B., Hillman, C., Carracoi, V., Cheff, B.N., Tilly, K., and Rosa, P.A. (2018). Infection history of the blood-meal host dictates pathogenic potential of the Lyme disease spirochete within the feeding tick vector. PLoS Pathogens *14*, e1006959. https://doi.org/10.1371/journal.ppat.1006959

Bishop, C.J., Aanensen, D.M., Jordan, G.E., Kilian, M., Hanage, W.P., and Spratt, B.G. (2009). Assigning strains to bacterial species via the internet. BMC Biol *7*, 3. https://doi.org/10.1186/1741-7007-7-3

Blazejak, K., Raulf, M.K., Janecek, E., Jordan, D., Fingerle, V., and Strube, C. (2018). Shifts in *Borrelia burgdorferi* (s.l.) geno-species infections in *Ixodes ricinus* over a 10-year surveillance period in the city of Hanover (Germany) and *Borrelia miyamotoi*-specific Reverse Line Blot detection. Parasit Vectors *11*, 304. https://doi.org/10.1186/s13071-018-2882-9

Blevins, J.S., Hagman, K.E., and Norgard, M.V. (2008). Assessment of decorin-binding protein A to the infectivity of *Borrelia burgdorferi* in the murine models of needle and tick infection. BMC Microbiol *8*, 82. https://doi.org/1471-2180-8-82 [pii] 10.1186/1471-2180-8-82

Bobay, L.M., Traverse, C.C., and Ochman, H. (2015). Impermanence of bacterial clones. Proc Natl Acad Sci U S A *112*, 8893-8900. https://doi.org/10.1073/pnas.1501724112

Boden, K., Lobenstein, S., Hermann, B., Margos, G., and Fingerle, V. (2016). *Borrelia miyamotoi*-Associated Neuroborreliosis in Immunocompromised Person. Emerg Infect Dis *22*, 1617-1620. https://doi.org/10.3201/eid2209.152034

Bondy-Denomy, J., Qian, J., Westra, E.R., Buckling, A., Guttman, D.S., Davidson, A.R., and Maxwell, K.L. (2016). Prophages mediate defense against phage infection through diverse mechanisms. ISME J *10*, 2854-2866. https://doi.org/10.1038/ismej.2016.79

Bono, J.L., Tilly, K., Stevenson, B., Hogan, D., and Rosa, P. (1998). Oligopeptide permease in *Borrelia burgdorferi*: putative peptide-binding components encoded by both chromosomal and plasmid loci. Microbiology *144*, 1033-1044.

Brissette, C.A., Bykowski, T., Cooley, A.E., Bowman, A., and Stevenson, B. (2009a). *Borrelia burgdorferi* RevA antigen binds host fibronectin. Infect Immun *77*, 2802-2812. https://doi.org/IAI.00227-09 [pii] 10.1128/IAI.00227-09

Brissette, C.A., Cooley, A.E., Burns, L.H., Riley, S.P., Verma, A., Woodman, M.E., Bykowski, T., and Stevenson, B. (2008). Lyme borreliosis spirochete Erp proteins, their known host ligands, and potential roles in mammalian infection. Int J Med Microbiol *298*, 257-267. https://doi.org/10.1016/j.ijmm.2007.09.004

Brissette, C.A., Haupt, K., Barthel, D., Cooley, A.E., Bowman, A., Skerka, C., Wallich, R., Zipfel, P.F., Kraiczy, P., and Stevenson, B. (2009b). *Borrelia burgdorferi* infection-associated surface proteins ErpP, ErpA, and ErpC bind human plasminogen. Infect Immun *77*, 300-306. https://doi.org/IAI.01133-08 [pii] 10.1128/IAI.01133-08

Brissette, C.A., Verma, A., Bowman, A., Cooley, A.E., and Stevenson, B. (2009c). The *Borrelia burgdorferi* outer-surface protein ErpX binds mammalian laminin. Microbiology *155*, 863-872. https://doi.org/155/3/863 [pii] 10.1099/mic.0.024604-0

Brisson, D., Baxamusa, N., Schwartz, I., and Wormser, G.P. (2011). Biodiversity of *Borrelia burgdorferi* strains in tissues of Lyme disease patients. PLoS One *6*, e22926. https://doi.org/10.1371/journal.pone.0022926

Brisson, D., Drecktrah, D., Eggers, C.H., and Samuels, D.S. (2012). Genetics of *Borrelia burgdorferi*. Annu Rev Genet *46*, 515-536. https://doi.org/10.1146/annurev-genet-011112-112140

Brisson, D., and Dykhuizen, D.E. (2004). *ospC* diversity in *Borrelia burgdorferi*: different hosts are different niches. Genetics *168*, 713-722.

Brisson, D., Vandermause, M.F., Meece, J.K., Reed, K.D., and Dykhuizen, D.E. (2010). Evolution of northeastern and midwestern *Borrelia burgdorferi*, United States. Emerg Infect Dis *16*, 911-917.

Brisson, D., Zhou, W., Jutras, B.L., Casjens, S., and Stevenson, B. (2013). Distribution of cp32 prophages among Lyme disease-causing spirochetes and natural diversity of their lipoprotein-encoding *erp* loci. Appl Environ Microbiol *79*, 4115-4128. https://doi.org/AEM.00817-13 [pii] 10.1128/AEM.00817-13

Buckles, E.L., Earnhart, C.G., and Marconi, R.T. (2006). Analysis of antibody response in humans to the type A OspC loop 5 domain and assessment of the potential utility of the loop 5 epitope in Lyme disease vaccine development. Clin Vaccine Immunol *13*, 1162-1165. https://doi.org/10.1128/CVI.00099-06

Bunikis, J., Garpmo, U., Tsao, J., Berglund, J., Fish, D., and Barbour, A.G. (2004). Sequence typing reveals extensive strain diversity of the Lyme borreliosis agents *Borrelia burgdorferi* in North America and *Borrelia afzelii* in Europe. Microbiology *150*, 1741-1755.

Burns, L.H., Adams, C.A., Riley, S.P., Jutras, B.L., Bowman, A., Chenail, A.M., Cooley, A.E., Haselhorst, L.A., Moore, A.M., Babb, K., *et al.* (2010). BpaB, a novel protein encoded by the Lyme disease spirochete's cp32 prophages, binds to erp Operator 2 DNA. Nucleic Acids Res *38*, 5443-5455. https://doi.org/10.1093/nar/gkq284

Byram, R., Gaultney, R.A., Floden, A.M., Hellekson, C., Stone, B.L., Bowman, A., Stevenson, B., Johnson, B.J., and Brissette, C.A. (2015). *Borrelia burgdorferi* RevA Significantly Affects Pathogenicity and Host Response in the Mouse Model of Lyme Disease. Infect Immun *83*, 3675-3683. https://doi.org/10.1128/iai.00530-15

Byram, R., Stewart, P.E., and Rosa, P. (2004). The essential nature of the ubiquitous 26-kilobase circular replicon of *Borrelia burgdorferi*. J Bacteriol *186*, 3561-3569. https://doi.org/10.1128/JB.186.11.3561-3569.2004 186/11/3561

Caimano, M.J., Groshong, A.M., Belperron, A., Mao, J., Hawley, K.L., Luthra, A., Graham, D.E., Earnhart, C.G., Marconi, R.T., Bockenstedt, L.K., *et al.* (2019). The RpoS Gatekeeper in *Borrelia burgdorferi*: An Invariant Regulatory Scheme That Promotes Spirochete Persistence in Reservoir Hosts and Niche Diversity. Front Microbiol *10*, 1923. https://doi.org/10.3389/fmicb.2019.01923

Caimano, M.J., Iyer, R., Eggers, C.H., Gonzalez, C., Morton, E.A., Gilbert, M.A., Schwartz, I., and Radolf, J.D. (2007). Analysis of the RpoS regulon in *Borrelia burgdorferi* in response to mammalian host signals provides insight into RpoS function during the enzootic cycle. Mol Microbiol *65*, 1193-1217. https://doi.org/MMI5860 [pii] 10.1111/j.1365-2958.2007.05860.x

Caimano, M.J., Yang, X., Popova, T.G., Clawson, M.L., Akins, D.R., Norgard, M.V., and Radolf, J.D. (2000). Molecular and evolutionary characterization of the cp32/18 family of supercoiled plasmids in *Borrelia burgdorferi* 297. Infect Immun *68*, 1574-1586.

Canchaya, C., Proux, C., Fournous, G., Bruttin, A., and Brussow, H. (2003). Prophage genomics. Microbiol Mol Biol Rev *67*, 238-276.

Carpi, G., Walter, K.S., Bent, S.J., Hoen, A.G., Diuk-Wasser, M., and Caccone, A. (2015). Whole genome capture of vector-borne pathogens from mixed DNA samples: a case study of *Borrelia burgdorferi*. BMC Genomics *16*, 434. https://doi.org/10.1186/s12864-015-1634-x

Casjens, S. (1998). The diverse and dynamic structure of bacterial genomes. Annu Rev Genet *32*, 339-377. https://doi.org/10.1146/annurev.genet.32.1.339

Casjens, S. (2000). Borrelia genomes in the year 2000. J Mol Microbiol Biotechnol *2*, 401-410.

Casjens, S. (2003). Prophages in bacterial genomics: What have we learned so far? Molec Microbiol *249*, 277-300.

Casjens, S., Delange, M., Ley, H.L., 3rd, Rosa, P., and Huang, W.M. (1995). Linear chromosomes of Lyme disease agent spirochetes: genetic diversity and conservation of gene order. J Bacteriol *177*, 2769-2780.

Casjens, S., Eggers, C.H., and Schwartz, I. (2010). *Borrelia* genomics: chromsosome, plasmids, bacteriophges and genetic variation. In *Borrelia* molecular biology, host interaction and pathogenicity, D.S. Samuels, and J. Radolf, eds. (Norfolk: Caister Academic Press), pp. 27-53.

Casjens, S., Hatfull, G., and Hendrix, R. (1992). Evolution of dsDNA tailed-bacteriophage genomes. Sem Virol *3*, 383-397.

Casjens, S., and Hendrix, R. (2002). Bacteriophage roles in bacterial chromosome evolution. In The Bacterial Chromosome, P. Higgins, ed. (Washington, D.C.: ASM Press), p. 39-52.

Casjens, S., and Huang, W.M. (1993). Linear chromosomal physical and genetic map of *Borrelia burgdorferi*, the Lyme disease agent. Mol Microbiol *8*, 967-980.

Casjens, S., Murphy, M., DeLange, M., Sampson, L., van Vugt, R., and Huang, W.M. (1997a). Telomeres of the linear chromosomes of Lyme disease spirochaetes: nucleotide sequence and possible exchange with linear plasmid telomeres. Mol Microbiol *26*, 581-596. https://doi.org/10.1046/j.1365-2958.1997.6051963.x

Casjens, S., Palmer, N., van, V.R., Huang, W.M., Stevenson, B., Rosa, P., Lathigra, R., Sutton, G., Peterson, J., Dodson, R.J., *et al.* (2000). A bacterial genome in flux: the twelve linear and nine circular extrachromosomal DNAs in an infectious isolate of the Lyme disease spirochete *Borrelia burgdorferi*. Mol Microbiol *35*, 490-516.

Casjens, S., van Vugt, R., Tilly, K., Rosa, P.A., and Stevenson, B. (1997b). Homology throughout the multiple 32-kilobase circular plasmids present in Lyme disease spirochetes. J Bacteriol *179*, 217-227.

Casjens, S.R. (2005). Comparative genomics and evolution of the tailed-bacteriophages. Curr Opin Microbiol *8*, 451-458.

Casjens, S.R. (2008). Diversity among the tailed-bacteriophages that infect the *Enterobacteriaceae*. Res Microbiol *159*, 340-348. https://doi.org/10.1016/j.resmic.2008.04.005

Casjens, S.R., Di, L., Akther, S., Mongodin, E.F., Luft, B.J., Schutzer, S.E., Fraser, C.M., and Qiu, W.G. (2018). Primordial origin and diversification of plasmids in Lyme disease agent bacteria. BMC Genomics *19*, 218. https://doi.org/10.1186/s12864-018-4597-x

Casjens, S.R., Fraser-Liggett, C.M., Mongodin, E.F., Qiu, W.G., Dunn, J.J., Luft, B.J., and Schutzer, S.E. (2011a). Whole genome sequence of an unusual *Borrelia burgdorferi* sensu lato isolate. J Bacteriol *193*, 1489-1490. https://doi.org/10.1128/JB.01521-10

Casjens, S.R., Gilcrease, E.B., Vujadinovic, M., Mongodin, E.F., Luft, B.J., Schutzer, S.E., Fraser, C.M., and Qiu, W.G. (2017). Plasmid diversity and phylogenetic consistency in the Lyme disease agent *Borrelia burgdorferi*. BMC Genomics *18*, 165. https://doi.org/10.1186/s12864-017-3553-5

Casjens, S.R., Huang, W.M., Gilcrease, E.B., Qiu, W.G., McCaig, W.D., Luft, B.J., and Schutzer, S.E. (2006). Comparative genomics of *Borrelia burgdorferi*. In Molecular Biology of Spirochetes, F.C. Cabello, D. Hulinska, and H.P. Godfrey, eds. (Amsterdam: IOS Press), pp. 79-95.

Casjens, S.R., Mongodin, E.F., Qiu, W.G., Dunn, J.J., Luft, B.J., Fraser-Liggett, C.M., and Schutzer, S.E. (2011b). Whole-genome sequences of two *Borrelia afzelii* and two *Borrelia garinii* Lyme disease agent isolates. J Bacteriol *193*, 6995-6996. https://doi.org/10.1128/JB.05951-11

Casjens, S.R., Mongodin, E.F., Qiu, W.G., Luft, B.J., Schutzer, S.E., Gilcrease, E.B., Huang, W.M., Vujadinovic, M., Aron, J.K., Vargas, L.C., *et al.* (2012). Genome stability of Lyme disease spirochetes: comparative genomics of *Borrelia burgdorferi* plasmids. PLoS One *7*, e33280.

Castillo-Ramirez, S., Fingerle, V., Jungnick, S., Straubinger, R.K., Krebs, S., Blum, H., Meinel, D.M., Hofmann, H., Guertler, P., Sing, A., *et al.* (2016). Trans-Atlantic exchanges have shaped the population structure of the Lyme disease agent *Borrelia burgdorferi* sensu stricto. Sci Rep *6*, 22794. https://doi.org/10.1038/srep22794

Catalano, C.E. (2005). Viral genome packaging machines: genetics, structure, and mechanism (Georgetown, TX: Landes Bioscience).

Chaconas, G., Castellanos, M., and Verhey, T.B. (2020). Changing of the guard: How the Lyme disease spirochete subverts the host immune response. J Biol Chem *295*, 301-313. https://doi.org/10.1074/jbc.REV119.008583

Champion, C.I., Blanco, D.R., Skare, J.T., Haake, D.A., Giladi, M., Foley, D., Miller, J.N., and Lovett, M.A. (1994). A 9.0-kilobase-pair circular plasmid of *Borrelia burgdorferi* encodes an exported protein: evidence for expression only during infection. Infect Immun *62*, 2653-2661.

Chauhan, G., McClure, J., Hekman, J., Marsh, P.W., Bailey, J.A., Daniels, R.F., Genereux, D.P., and Karlsson, E.K. (2019). Combining Citizen Science and Genomics to Investigate Tick, Pathogen, and Commensal Microbiome at Single-Tick Resolution. Front Genet *10*, 1322. https://doi.org/10.3389/fgene.2019.01322

Cheetham, B.F., Tattersall, D.B., Bloomfield, G.A., Rood, J.I., and Katz, M.E. (1995). Identification of a gene encoding a bacteriophage-related integrase in a vap region of the Dichelobacter nodosus genome. Gene *162*, 53-58.

Chenail, A.M., Jutras, B.L., Adams, C.A., Burns, L.H., Bowman, A., Verma, A., and Stevenson, B. (2012). *Borrelia burgdorferi* cp32 BpaB modulates expression of the prophage NucP nuclease and SsbP single-stranded DNA-binding protein. J Bacteriol *194*, 4570-4578. https://doi.org/10.1128/JB.00661-12

Cohan, F.M. (2002). What are bacterial species? Annu Rev Microbiol *56*, 457-487.

Coipan, E.C., Jahfari, S., Fonville, M., Oei, G.A., Spanjaard, L., Takumi, K., Hovius, J.W., and Sprong, H. (2016). Imbalanced presence of *Borrelia burgdorferi* s.l. multilocus sequence types in clinical manifestations of Lyme borreliosis. Infect Genet Evol *42*, 66-76. https://doi.org/10.1016/j.meegid.2016.04.019

Collares-Pereira, M., Couceiro, S., Franca, I., Kurtenbach, K., Schafer, S.M., Vitorino, L., Goncalves, L., Baptista, S., Vieira, M.L., and Cunha, C. (2004). First isolation of *Borrelia lusitaniae* from a human patient. J Clin Microbiol *42*, 1316-1318. https://doi.org/10.1128/jcm.42.3.1316-1318.2004

Comstedt, P., Jakobsson, T., and Bergstrom, S. (2011). Global ecology and epidemiology of *Borrelia garinii* spirochetes. Infect Ecol Epidemiol *1*. https://doi.org/10.3402/iee.v1i0.9545

Corona, A., and Schwartz, I. (2015). *Borrelia burgdorferi*: Carbon Metabolism and the Tick-Mammal Enzootic Cycle. Microbiol Spectr *3*. https://doi.org/10.1128/microbiolspec.MBP-0011-2014

Coutte, L., Botkin, D.J., Gao, L., and Norris, S.J. (2009). Detailed analysis of sequence changes occurring during *vlsE* antigenic variation in the mouse model of *Borrelia burgdorferi* infection. PLoS Pathogens *5*, e1000293. https://doi.org/10.1371/journal.ppat.1000293

Cross, S.T., Kapuscinski, M.L., Perino, J., Maertens, B.L., Weger-Lucarelli, J., Ebel, G.D., and Stenglein, M.D. (2018). Co-Infection Patterns in Individual *Ixodes scapularis* Ticks Reveal Associations between Viral, Eukaryotic and Bacterial Microorganisms. Viruses *10*. https://doi.org/10.3390/v10070388

Damman, C.J., Eggers, C.H., Samuels, D.S., and Oliver, D.B. (2000). Characterization of *Borrelia burgdorferi* BlyA and BlyB proteins: a prophage-encoded holin-like system. J Bacteriol *182*, 6791-6797.

Davidson, B.E., MacDougall, J., and Saint Girons, I. (1992). Physical map of the linear chromosome of the bacterium *Borrelia burgdorferi* 212, a causative agent of Lyme disease, and localization of rRNA genes. J Bacteriol *174*, 3766-3774.

de Carvalho, I.L., Fonseca, J.E., Marques, J.G., Ullmann, A., Hojgaard, A., Zeidner, N., and Nuncio, M.S. (2008). Vasculitis-like syndrome associated with Borrelia lusitaniae infection. Clin Rheumatol *27*, 1587-1591. https://doi.org/10.1007/s10067-008-1012-z

Di, L., Wan, Z., Akther, S., Ying, C., Larracuente, A., Li, L., Di, C., Nunez, R., Cucura, D.M., Goddard, N.L., *et al.* (2018). Genotyping and Quantifying Lyme Pathogen Strains by Deep Sequencing of the Outer Surface Protein C (*ospC*) Locus. J Clin Microbiol *56*. https://doi.org/10.1128/JCM.00940-18

Dion, M.B., Oechslin, F., and Moineau, S. (2020). Phage diversity, genomics and phylogeny. Nat Rev Microbiol *18*, 125-138. https://doi.org/10.1038/s41579-019-0311-5

Diuk-Wasser, M.A., Gatewood, A.G., Cortinas, M.R., Yaremych-Hamer, S., Tsao, J., Kitron, U., Hickling, G., Brownstein, J.S., Walker, E., Piesman, J., *et al.* (2006). Spatiotemporal patterns of host-seeking *Ixodes scapularis* nymphs (Acari: Ixodidae) in the United States. J Med Entomol *43*, 166-176.

Dowdell, A.S., Murphy, M.D., Azodi, C., Swanson, S.K., Florens, L., Chen, S., and Zuckert, W.R. (2017). Comprehensive Spatial Analysis of the *Borrelia burgdorferi* Lipoproteome Reveals a Compartmentalization Bias toward the Bacterial Surface. J Bacteriol *199*. https://doi.org/10.1128/JB.00658-16

Drecktrah, D., Lybecker, M., Popitsch, N., Rescheneder, P., Hall, L.S., and Samuels, D.S. (2015). The *Borrelia burgdorferi* RelA/SpoT homolog and stringent response regulate survival in the tick vector and global gene expression during starvation. PLoS Pathogens *11*, e1005160. https://doi.org/10.1371/journal.ppat.1005160

Dulebohn, D.P., Bestor, A., and Rosa, P.A. (2013). *Borrelia burgdorferi* linear plasmid 28-3 confers a selective advantage in an experimental mouse-tick infection model. Infect Immun *81*, 2986-2996. https://doi.org/10.1128/IAI.00219-13

Dulebohn, D.P., Hayes, B.M., and Rosa, P.A. (2014). Global repression of host-associated genes of the Lyme disease spirochete through post-transcriptional modulation of the alternative sigma factor RpoS. PLoS One *9*, e93141. https://doi.org/10.1371/journal.pone.0093141

Dunn, J.J., Buchstein, S.R., Butler, L.-L., Fisenne, S., Polin, D.S., Lade, B.N., and Luft, B.J. (1994). Complete nucleotide sequence of a circular plasmid from the Lyme disease spirochete, *Borrelia burgdorferi*. J Bacteriol *176*, 2706-2717.

Durand, J., Herrmann, C., Genne, D., Sarr, A., Gern, L., and Voordouw, M.J. (2017a). Multistrain Infections with Lyme Borreliosis Pathogens in the Tick Vector. Appl Environ Microbiol *83*. https://doi.org/10.1128/AEM.02552-16

Durand, J., Jacquet, M., Paillard, L., Rais, O., Gern, L., and Voordouw, M.J. (2015). Cross-Immunity and Community Structure of a Multiple-Strain Pathogen in the Tick Vector. Appl Environ Microbiol *81*, 7740-7752. https://doi.org/10.1128/AEM.02296-15

Durand, J., Jacquet, M., Rais, O., Gern, L., and Voordouw, M.J. (2017b). Fitness estimates from experimental infections predict the long-term strain structure of a vector-borne pathogen in the field. Sci Rep *7*, 1851. https://doi.org/10.1038/s41598-017-01821-1

Dykhuizen, D.E., and Baranton, G. (2001). The implications of a low rate of horizontal transfer in *Borrelia*. Trends Microbiol *9*, 344-350.

Dykhuizen, D.E., Brisson, D., Sandigursky, S., Wormser, G.P., Nowakowski, J., Nadelman, R.B., and Schwartz, I. (2008). The propensity of different *Borrelia burgdorferi* sensu stricto genotypes to cause disseminated infections in humans. Am J Trop Med Hyg *78*, 806-810.

Dykhuizen, D.E., Polin, D.S., Dunn, J.J., Wilske, B., Preac-Mursic, V., Dattwyler, R.J., and Luft, B.J. (1993). *Borrelia burgdorferi* is clonal: implications for taxonomy and vaccine development. Proc Natl Acad Sci U S A *90*, 10163-10167. https://doi.org/10.1073/pnas.90.21.10163

Earnhart, C.G., and Marconi, R.T. (2007). Construction and analysis of variants of a polyvalent Lyme disease vaccine: approaches for improving the immune response to chimeric vaccinogens. Vaccine *25*, 3419-3427. https://doi.org/10.1016/j.vaccine.2006.12.051

Edgar, R.C. (2004). MUSCLE: multiple sequence alignment with high accuracy and high throughput. Nucleic Acids Res *32*, 1792-1797. https://doi.org/10.1093/nar/gkh340

Eggers, C.H., Caimano, M.J., Clawson, M.L., Miller, W.G., Samuels, D.S., and Radolf, J.D. (2002). Identification of loci critical for replication and compatibility of a *Borrelia burgdorferi* cp32 plasmid and use of a cp32-based shuttle vector for the expression of fluorescent reporters in the Lyme disease spirochaete. Mol Microbiol *43*, 281-295.

Eggers, C.H., Casjens, S., Hayes, S.F., Garon, C.F., Damman, C.J., Oliver, D.B., and Samuels, D.S. (2000). Bacteriophages of spirochetes. J Mol Microbiol Biotechnol *2*, 365-373.

Eggers, C.H., Casjens, S., and Samuels, D.S. (2001a). Bacteriophges of *Borrelia burgdorferi* and other spirochetes. In The spirochetes Molecilar and celullar biology, M. Saier, and J. Garcia-Lara, eds. (Wiltshire: Hozison Scientific Press), pp. 35-44.

Eggers, C.H., Gray, C.M., Preisig, A.M., Glenn, D.M., Pereira, J., Ayers, R.W., Alshahrani, M., Acabbo, C., Becker, M.R., Bruenn, K.N., *et al.* (2016). Phage-mediated horizontal gene transfer of both prophage and heterologous DNA by fBB-1, a bacteriophage of *Borrelia burgdorferi*. Pathog Dis *74*. https://doi.org/10.1093/femspd/ftw107

Eggers, C.H., Kimmel, B.J., Bono, J.L., Elias, A.F., Rosa, P., and Samuels, D.S. (2001b). Transduction by fBB-1, a bacteriophage of *Borrelia burgdorferi*. J Bacteriol *183*, 4771-4778.

Eggers, C.H., and Samuels, D.S. (1999). Molecular evidence for a new bacteriophage of *Borrelia burgdorferi*. J Bacteriol *181*, 7308-7313.

Eicken, C., Sharma, V., Klabunde, T., Lawrenz, M.B., Hardham, J.M., Norris, S.J., and Sacchettini, J.C. (2002). Crystal structure of Lyme disease variable surface antigen VlsE of *Borrelia burgdorferi*. J Biol Chem *277*, 21691-21696. https://doi.org/10.1074/jbc.M201547200

El-Hage, N., and Stevenson, B. (2002). Simultaneous coexpression of *Borrelia burgdorferi* Erp proteins occurs through a specific, *erp* locus-directed regulatory mechanism. J Bacteriol *184*, 4536-4543. https://doi.org/10.1128/jb.184.16.4536-4543.2002

Elder, J.F., Jr., and Turner, B.J. (1995). Concerted evolution of repetitive DNA sequences in eukaryotes. Q Rev Biol *70*, 297-320. https://doi.org/10.1086/419073

Elias, A.F., Stewart, P.E., Grimm, D., Caimano, M.J., Eggers, C.H., Tilly, K., Bono, J.L., Akins, D.R., Radolf, J.D., Schwan, T.G., *et al.* (2002). Clonal polymorphism of *Borrelia burgdorferi* strain B31 MI: implications for mutagenesis in an infectious strain background. Infect Immun *70*, 2139-2150.

Ellegren, H. (2014). Genome sequencing and population genomics in non-model organisms. Trends Ecol Evol *29*, 51-63. https://doi.org/10.1016/j.tree.2013.09.008

Ellis, T.C., Jain, S., Linowski, A.K., Rike, K., Bestor, A., Rosa, P.A., Halpern, M., Kurhanewicz, S., and Jewett, M.W. (2013). In vivo expression technology identifies a novel virulence factor critical for *Borrelia burgdorferi* persistence in mice. PLoS Pathogens *9*, e1003567. https://doi.org/10.1371/journal.ppat.1003567

Enright, M.C., and Spratt, B.G. (1999). Multilocus sequence typing. Trends Microbiol *7*, 482-487.

Farlow, J., Postic, D., Smith, K.L., Jay, Z., Baranton, G., and Keim, P. (2002). Strain typing of *Borrelia burgdorferi*, *Borrelia afzelii*, and *Borrelia garinii* by using multiple-locus variable-number tandem repeat analysis. J Clin Microbiol *40*, 4612-4618.

Fedorova, N., Kleinjan, J.E., James, D., Hui, L.T., Peeters, H., and Lane, R.S. (2014). Remarkable diversity of tick or mammalian-associated Borreliae in the metropolitan San Francisco Bay Area, California. Ticks Tick Borne Dis *5*, 951-961. https://doi.org/10.1016/j.ttbdis.2014.07.015

Felsenstein, J. (1989). PHYLIP - Phylogeny Inference Package. Cladistics *5*, 164-166.

Ferdows, M.S., and Barbour, A.G. (1989). Megabase-sized linear DNA in the bacterium *Borrelia burgdorferi*, the Lyme disease agent. Proc Natl Acad Sci USA *86*, 5969-5973.

Fingerle, V., Goettner, G., Gern, L., Wilske, B., and Schulte-Spechtel, U. (2007). Complementation of a *Borrelia afzelii* OspC mutant highlights the crucial role of OspC for dissemination of *Borrelia afzelii* in *Ixodes ricinus*. Int J Med Microbiol *297*, 97-107. https://doi.org/10.1016/j.ijmm.2006.11.003

Fingerle, V., Pritsch, M., Wachtler, M., Margos, G., Ruske, S., Jung, J., Loscher, T., Wendtner, C., and Wieser, A. (2016). "*Candidatus Borrelia kalaharica*" Detected from a Febrile Traveller Returning to Germany from Vacation in Southern Africa. PLoS Neg Trop Dis *10*, e0004559. https://doi.org/10.1371/journal.pntd.0004559

Fischer, J.R., LeBlanc, K.T., and Leong, J.M. (2006). Fibronectin binding protein BBK32 of the Lyme disease spirochete promotes bacterial attachment to glycosaminoglycans. Infect Immun *74*, 435-441. https://doi.org/74/1/435 [pii] 10.1128/IAI.74.1.435-441.2006

Ford, M.E., Sarkis, G.J., Belanger, A.E., Hendrix, R.W., and Hatfull, G.F. (1998). Genome structure of mycobacteriophage D29: implications for phage evolution. J Mol Biol *279*, 143-164.

Francisco, A.P., Vaz, C., Monteiro, P.T., Melo-Cristino, J., Ramirez, M., and Carrico, J.A. (2012). PHYLOViZ: phylogenetic inference and data visualization for sequence based typing methods. BMC Bioinformatics *13*, 87. https://doi.org/10.1186/1471-2105-13-87

Fraser, C., Hanage, W.P., and Spratt, B.G. (2007). Recombination and the nature of bacterial speciation. Science *315*, 476-480. https://doi.org/10.1126/science.1127573

Fraser, C.M., Casjens, S., Huang, W.M., Sutton, G.G., Clayton, R., Lathigra, R., White, O., Ketchum, K.A., Dodson, R., Hickey, E.K., *et al.* (1997). Genomic sequence of a Lyme disease spirochaete, *Borrelia burgdorferi*. Nature *390*, 580-586.

Gatzmann, F., Metzler, D., Krebs, S., Blum, H., Sing, A., Takano, A., Kawabata, H., Fingerle, V., Margos, G., and Becker, N.S. (2015). NGS population genetics analyses reveal divergent evolution of a Lyme Borreliosis agent in Europe and Asia. Ticks Tick Borne Dis *6*, 344-351. https://doi.org/10.1016/j.ttbdis.2015.02.008

Gazumyan, A., Schwartz, J.J., Liveris, D., and Schwartz, I. (1994). Sequence analysis of the ribosomal RNA operon of the Lyme disease spirochete, *Borrelia burgdorferi*. Gene *146*, 57-65.

Gilmore, R.D., Jr., Howison, R.R., Dietrich, G., Patton, T.G., Clifton, D.R., and Carroll, J.A. (2010). The *bba64* gene of *Borrelia burgdorferi*, the Lyme disease agent, is critical for mammalian infection via tick bite transmission. Proc Natl Acad Sci U S A *107*, 7515-7520. https://doi.org/10.1073/pnas.1000268107

Gilmore, R.D., Jr., and Mbow, M.L. (1998). A monoclonal antibody generated by antigen inoculation via tick bite is reactive to the *Borrelia burgdorferi* Rev protein, a member of the 2.9 gene family locus. Infect Immun *66*, 980-986.

Girard, Y.A., Fedorova, N., and Lane, R.S. (2011). Genetic diversity of *Borrelia burgdorferi* and detection of B. *bissettii*-like DNA in serum of north-coastal California residents. J Clin Microbiol *49*, 945-954. https://doi.org/10.1128/JCM.01689-10

Glockner, G., Lehmann, R., Romualdi, A., Pradella, S., Schulte-Spechtel, U., Schilhabel, M., Wilske, B., Suhnel, J., and Platzer, M. (2004). Comparative analysis of the *Borrelia garinii* genome. Nucleic

Acids Res *32*, 6038-6046. https://doi.org/32/20/6038 [pii] 10.1093/nar/gkh953

Glockner, G., Schulte-Spechtel, U., Schilhabel, M., Felder, M., Suhnel, J., Wilske, B., and Platzer, M. (2006). Comparative genome analysis: selection pressure on the *Borrelia vls* cassettes is essential for infectivity. BMC Genomics *7*, 211. https://doi.org/10.1186/1471-2164-7-211

Gofton, A.W., Margos, G., Fingerle, V., Hepner, S., Loh, S.M., Ryan, U., Irwin, P., and Oskam, C.L. (2018). Genome-wide analysis of *Borrelia turcica* and '*Candidatus Borrelia tachyglossi*' shows relapsing fever-like genomes with unique genomic links to Lyme disease *Borrelia*. Infect Genet Evol *66*, 72-81. https://doi.org/10.1016/j.meegid.2018.09.013

Gomez-Diaz, E., Boulinier, T., Sertour, N., Cornet, M., Ferquel, E., and McCoy, K.D. (2011). Genetic structure of marine *Borrelia garinii* and population admixture with the terrestrial cycle of Lyme borreliosis. Environ Microbiol *13*, 2453-2467. https://doi.org/10.1111/j.1462-2920.2011.02515.x

Graves, C.J., Ros, V.I., Stevenson, B., Sniegowski, P.D., and Brisson, D. (2013). Natural selection promotes antigenic evolvability. PLoS Pathogens *9*, e1003766. https://doi.org/10.1371/journal.ppat.1003766

Greay, T.L., Gofton, A.W., Paparini, A., Ryan, U.M., Oskam, C.L., and Irwin, P.J. (2018). Recent insights into the tick microbiome gained through next-generation sequencing. Parasit Vectors *11*, 12. https://doi.org/10.1186/s13071-017-2550-5

Grigoriev, P.S., and Lobocka, M.B. (2001). Determinants of segregational stability of the linear plasmid-prophage N15 of *Escherichia coli*. Molec Microbiol *42*, 355-368.

Grimm, D., Elias, A.F., Tilly, K., and Rosa, P.A. (2003). Plasmid stability during in vitro propagation of *Borrelia burgdorferi* assessed at a clonal level. Infect Immun *71*, 3138-3145.

Grimm, D., Tilly, K., Bueschel, D.M., Fisher, M.A., Policastro, P.F., Gherardini, F.C., Schwan, T.G., and Rosa, P.A. (2005). Defining plasmids required by *Borrelia burgdorferi* for colonization of tick vector *Ixodes scapularis* (Acari: Ixodidae). J Med Entomol *42*, 676-684.

Grimm, D., Tilly, K., Byram, R., Stewart, P.E., Krum, J.G., Bueschel, D.M., Schwan, T.G., Policastro, P.F., Elias, A.F., and Rosa, P.A. (2004). Outer-surface protein C of the Lyme disease spirochete: a protein induced in ticks for infection of mammals. Proc Natl Acad Sci U S A *101*, 3142-3147.

Groshong, A.M., and Blevins, J.S. (2014). Insights into the biology of *Borrelia burgdorferi* gained through the application of molecular genetics. Advances in applied microbiology *86*, 41-143. https://doi.org/10.1016/B978-0-12-800262-9.00002-0

Groshong, A.M., Dey, A., Bezsonova, I., Caimano, M.J., and Radolf, J.D. (2017). Peptide Uptake Is Essential for *Borrelia burgdorferi* Viability and Involves Structural and Regulatory Complexity of its Oligopeptide Transporter. mBio *8*. https://doi.org/10.1128/mBio.02047-17

Hallstrom, T., Siegel, C., Morgelin, M., Kraiczy, P., Skerka, C., and Zipfel, P.F. (2013). CspA from *Borrelia burgdorferi* inhibits the terminal complement pathway. MBio *4*. https://doi.org/10.1128/mBio.00481-13

Hamer, S.A., Tsao, J.I., Walker, E.D., and Hickling, G.J. (2010). Invasion of the Lyme Disease Vector *Ixodes scapularis*: Implications for *Borrelia burgdorferi* Endemicity. Ecohealth *DOI: 10.1007/s10393-010-0287-0*.

Hammerschmidt, C., Klevenhaus, Y., Koenigs, A., Hallstrom, T., Fingerle, V., Skerka, C., Pos, K.M., Zipfel, P.F., Wallich, R., and Kraiczy, P. (2016). BGA66 and BGA71 facilitate complement resistance of *Borrelia bavariensis* by inhibiting assembly of the membrane attack complex. Mol Microbiol *99*, 407-424. https://doi.org/10.1111/mmi.13239

Hanincova, K., Kurtenbach, K., Diuk-Wasser, M., Brei, B., and Fish, D. (2006). Epidemic spread of Lyme borreliosis, northeastern United States. Emerg Infect Dis *12*, 604-611.

Hanincova, K., Liveris, D., Sandigursky, S., Wormser, G.P., and Schwartz, I. (2008). *Borrelia burgdorferi* sensu stricto is clonal in patients with early Lyme borreliosis. Appl Environ Microbiol *74*, 5008-5014.

Hanincova, K., Mukherjee, P., Ogden, N.H., Margos, G., Wormser, G.P., Reed, K.D., Meece, J.K., Vandermause, M.F., and Schwartz, I. (2013). Multilocus sequence typing of *Borrelia burgdorferi* suggests existence of lineages with differential pathogenic properties in humans. PLoS One *8*, e73066.

Haven, J., Vargas, L.C., Mongodin, E.F., Xue, V., Hernandez, Y., Pagan, P., Fraser-Liggett, C.M., Schutzer, S.E., Luft, B.J., Casjens, S.R., *et al.* (2011). Pervasive Recombination and Sympatric Genome Diversification Driven by Frequency-Dependent Selection in *Borrelia burgdorferi*, the Lyme Disease Bacterium. Genetics *189*, 951-966. https://doi.org/genetics.111.130773 [pii] 10.1534/genetics.111.130773

Hayes, B.M., Dulebohn, D.P., Sarkar, A., Tilly, K., Bestor, A., Ambroggio, X., and Rosa, P.A. (2014). Regulatory protein BBD18 of the lyme disease spirochete: essential role during tick acquisition? MBio *5*, e01017-01014. https://doi.org/10.1128/mBio.01017-14

Hayes, S.F., Burgdorfer, W., and Barbour, A.G. (1983). Bacteriophage in the *Ixodes dammini* spirochete, etiological agent of Lyme disease. J Bacteriol *154*, 1436-1439.

Hellwage, J., Meri, T., Heikkila, T., Alitalo, A., Panelius, J., Lahdenne, P., Seppala, I.J., and Meri, S. (2001). The complement regulator factor H binds to the surface protein OspE of *Borrelia burgdorferi*. J Biol Chem *276*, 8427-8435.

Hendrix, R., and Casjens, S. (2006). Bacteriophage l and its genetic neighborhood. In The Bacteriophages, 2nd Edition, R. Calendar, ed. (New York City, N.Y.: Oxford Press), pp. 409-447.

Hendrix, R.W. (1988). Tail length determination in double-stranded DNA bacteriophages. Curr Top Microbiol Immunol *136*, 21-29.

Hendrix, R.W. (2002). Bacteriophages: evolution of the majority. Theor Popul Biol *61*, 471-480.

Hendrix, R.W., Lawrence, J.G., Hatfull, G.F., and Casjens, S. (2000). The origins and ongoing evolution of viruses. Trends Microbiol *8*, 504-508.

Hendrix, R.W., Smith, M.C., Burns, R.N., Ford, M.E., and Hatfull, G.F. (1999). Evolutionary relationships among diverse bacteriophages and prophages: all the world's a phage. Proc Natl Acad Sci USA *96*, 2192-2197.

Hernandez, Y., Bernstein, R., Pagan, P., Vargas, L., McCaig, W., Ramrattan, G., Akther, S., Larracuente, A., Di, L., Vieira, F.G., *et al.* (2018). BpWrapper: BioPerl-based sequence and tree utilities for rapid prototyping of bioinformatics pipelines. BMC bioinformatics *19*, 76. https://doi.org/10.1186/s12859-018-2074-9

Hill, W.G., and Robertson, A. (1966). The effect of linkage on limits to artificial selection. Genet Res *8*, 269-294.

Hinnebusch, J., and Barbour, A.G. (1991). Linear plasmids of *Borrelia burgdorferi* have a telomeric structure and sequence similar to those of a eukaryotic virus. J Bacteriol *173*, 7233-7239.

Hinnebusch, J., and Barbour, A.G. (1992). Linear- and circular-plasmid copy numbers in *Borrelia burgdorferi*. J Bacteriol *174*, 5251-5257.

Hinnebusch, J., Bergstrom, S., and Barbour, A.G. (1990). Cloning and sequence analysis of linear plasmid telomeres of the bacterium *Borrelia burgdorferi*. Mol Microbiol *4*, 811-820.

Hoen, A.G., Margos, G., Bent, S.J., Kurtenbach, K., and Fish, D. (2009). Phylogeography of *Borrelia burgdorferi* in the eastern United States reflects multiple independent Lyme disease emergence events. Proc Natl Acad Sci U S A *106*, 15013-15018.

Hovis, K.M., Tran, E., Sundy, C.M., Buckles, E., McDowell, J.V., and Marconi, R.T. (2006). Selective binding of *Borrelia burgdorferi* OspE paralogs to factor H and serum proteins from diverse animals: possible expansion of the role of OspE in Lyme disease pathogenesis. Infect Immun *74*, 1967-1972.

Hu, C.M., Wilske, B., Fingerle, V., Lobet, Y., and Gern, L. (2001). Transmission of *Borrelia garinii* OspA serotype 4 to BALB/c mice by *Ixodes ricinus* ticks collected in the field. J Clin Microbiol *39*, 1169-1171.

Huang, W.M., Robertson, M., Aron, J., and Casjens, S. (2004). Telomere exchange between linear replicons of *Borrelia burgdorferi*. J Bacteriol *186*, 4134-4141. https://doi.org/10.1128/JB.186.13.4134-4141.2004

Huegli, D., Hu, C.M., Humair, P.F., Wilske, B., and Gern, L. (2002). *Apodemus* species mice are reservoir hosts of *Borrelia garinii* OspA serotype 4 in Switzerland. J Clin Microbiol *40*, 4735-4737.

Ivanova, L., Christova, I., Neves, V., Aroso, M., Meirelles, L., Brisson, D., and Gomes-Solecki, M. (2009). Comprehensive seroprofiling of sixteen *B. burgdorferi* OspC: implications for Lyme disease diagnostics design. Clin Immunol *132*, 393-400. https://doi.org/10.1016/j.clim.2009.05.017

Ivanova, L.B., Tomova, A., Gonzalez-Acuna, D., Murua, R., Moreno, C.X., Hernandez, C., Cabello, J., Cabello, C., Daniels, T.J., Godfrey, H.P., *et al.* (2014). *Borrelia chilensis*, a new member of the *Borrelia burgdorferi* sensu lato complex that extends the range of this genospecies in the Southern Hemisphere. Environ Microbiol *16*, 1069-1080. https://doi.org/10.1111/1462-2920.12310

Iyer, R., Caimano, M.J., Luthra, A., Axline, D., Jr., Corona, A., Iacobas, D.A., Radolf, J.D., and Schwartz, I. (2015). Stage-specific global alterations in the transcriptomes of Lyme disease spirochetes during tick feeding and following mammalian host adaptation. Mol Microbiol *95*, 509-538. https://doi.org/10.1111/mmi.12882

Iyer, R., Kalu, O., Purser, J., Norris, S., Stevenson, B., and Schwartz, I. (2003). Linear and circular plasmid content in *Borrelia burgdorferi* clinical isolates. Infect Immun *71*, 3699-3706.

Izac, J.R., Camire, A.C., Earnhart, C.G., Embers, M.E., Funk, R.A., Breitschwerdt, E.B., and Marconi, R.T. (2019). Analysis of the antigenic determinants of the OspC protein of the Lyme disease spirochetes: Evidence that the C10 motif is not immunodominant or required to elicit bactericidal antibody responses. Vaccine *37*, 2401-2407. https://doi.org/10.1016/j.vaccine.2019.02.007

Izac, J.R., and Marconi, R.T. (2019). Diversity of the Lyme Disease Spirochetes and its Influence on Immune Responses to Infection and Vaccination. Vet Clin North Am Small Anim Pract *49*, 671-686. https://doi.org/10.1016/j.cvsm.2019.02.007

Jabbari, N., Glusman, G., Joesch-Cohen, L.M., Reddy, P.J., Moritz, R.L., Hood, L., and Lausted, C.G. (2018). Whole genome sequence and comparative analysis of *Borrelia burgdorferi* MM1. PLoS One *13*, e0198135. https://doi.org/10.1371/journal.pone.0198135

Jacquot, M., Gonnet, M., Ferquel, E., Abrial, D., Claude, A., Gasqui, P., Choumet, V., Charras-Garrido, M., Garnier, M., Faure, B., *et al.* (2014). Comparative population genomics of the *Borrelia burgdorferi* species complex reveals high degree of genetic isolation among species and underscores benefits and constraints to studying intra-specific epidemiological processes. PLoS One *9*, e94384. https://doi.org/10.1371/journal.pone.0094384

Jenkins, A., Hvidsten, D., Matussek, A., Lindgren, P.E., Stuen, S., and Kristiansen, B.E. (2012). *Borrelia burgdorferi* sensu lato in *Ixodes ricinus* ticks from Norway: evaluation of a PCR test targeting the chromosomal *flaB* gene. Exp Appl Acarol *58*, 431-439. https://doi.org/10.1007/s10493-012-9585-2

Jewett, M.W., Byram, R., Bestor, A., Tilly, K., Lawrence, K., Burtnick, M.N., Gherardini, F., and Rosa, P.A. (2007a). Genetic basis for retention of a critical virulence plasmid of *Borrelia burgdorferi*. Mol Microbiol *66*, 975-990. https://doi.org/MMI5969 [pii] 10.1111/j.1365-2958.2007.05969.x

Jewett, M.W., Jain, S., Linowski, A.K., Sarkar, A., and Rosa, P.A. (2011). Molecular characterization of the *Borrelia burgdorferi* in vivo-essential protein PncA. Microbiology *157*, 2831-2840. https://doi.org/mic.0.051706-0 [pii] 10.1099/mic.0.051706-0

Jewett, M.W., Lawrence, K., Bestor, A.C., Tilly, K., Grimm, D., Shaw, P., VanRaden, M., Gherardini, F., and Rosa, P.A. (2007b). The critical role of the linear plasmid lp36 in the infectious cycle of *Borrelia burgdorferi*. Mol Microbiol *64*, 1358-1374.

Jolley, K.A. (2009). Internet-based sequence-typing databases for bacterial molecular epidemiology. Methods Mol Biol *551*, 305-312. https://doi.org/10.1007/978-1-60327-999-4_21

Jolley, K.A., Chan, M.S., and Maiden, M.C. (2004). mlstdbNet - distributed multi-locus sequence typing (MLST) databases. BMC bioinformatics *5*, 86. https://doi.org/10.1186/1471-2105-5-86

Jutras, B.L., Verma, A., Adams, C.A., Brissette, C.A., Burns, L.H., Whetstine, C.R., Bowman, A., Chenail, A.M., Zuckert, W.R., and Stevenson, B. (2012). BpaB and EbfC DNA-binding proteins regulate production of the Lyme disease spirochete's infection-associated Erp surface proteins. J Bacteriol *194*, 778-786. https://doi.org/10.1128/JB.06394-11

Kawabata, H., Norris, S.J., and Watanabe, H. (2004). BBE02 disruption mutants of *Borrelia burgdorferi* B31 have a highly transformable, infectious phenotype. Infect Immun *72*, 7147-7154. https://doi.org/72/12/7147 [pii] 10.1128/IAI.72.12.7147-7154.2004

Kenedy, M.R., and Akins, D.R. (2011). The OspE-related proteins inhibit complement deposition and enhance serum resistance of *Borrelia burgdorferi*, the Lyme disease spirochete. Infect Immun *79*, 1451-1457. https://doi.org/10.1128/IAI.01274-10

Kingry, L.C., Anacker, M., Pritt, B., Bjork, J., Respicio-Kingry, L., Liu, G., Sheldon, S., Boxrud, D., Strain, A., Oatman, S., *et al.* (2018). Surveillance for and Discovery of *Borrelia* Species in US Patients Suspected of Tickborne Illness. Clin Infect Dis *66*, 1864-1871. https://doi.org/10.1093/cid/cix1107

Kingry, L.C., Batra, D., Replogle, A., Rowe, L.A., Pritt, B.S., and Petersen, J.M. (2016). Whole Genome Sequence and Comparative Genomics of the Novel Lyme Borreliosis Causing Pathogen, *Borrelia mayonii*. PLoS One *11*, e0168994. https://doi.org/10.1371/journal.pone.0168994

Kitten, T., and Barbour, A.G. (1990). Juxtaposition of expressed variable antigen genes with a conserved telomere in the bacterium *Borrelia hermsii*. Proc Natl Acad Sci U S A *87*, 6077-6081.

Kobryn, K., and Chaconas, G. (2002). ResT, a telomere resolvase encoded by the Lyme disease spirochete. Mol Cell *9*, 195-201. https://doi.org/S1097276501004336

Koenigs, A., Hammerschmidt, C., Jutras, B.L., Pogoryelov, D., Barthel, D., Skerka, C., Kugelstadt, D., Wallich, R., Stevenson, B., Zipfel, P.F., *et al.* (2013). BBA70 of *Borrelia burgdorferi* is a novel plasminogen-binding protein. J Biol Chem *288*, 25229-25243. https://doi.org/10.1074/jbc.M112.413872

Koren, S., Walenz, B.P., Berlin, K., Miller, J.R., and Phillippy, A.M. (2016). Canu: scalable and accurate long-read assembly via adaptive k-mer weighting and repeat separation. bioRxiv.

Kraiczy, P. (2016). Hide and Seek: How Lyme Disease Spirochetes Overcome Complement Attack. Frontiers in immunology *7*, 385. https://doi.org/10.3389/fimmu.2016.00385

Kraiczy, P., Hartmann, K., Hellwage, J., Skerka, C., Kirschfink, M., Brade, V., Zipfel, P.F., Wallich, R., and Stevenson, B. (2004). Immunological characterization of the complement regulator factor H-binding CRASP and Erp proteins of *Borrelia burgdorferi*. Int J Med Microbiol *293 Suppl 37*, 152-157.

Krupna-Gaylord, M.A., Liveris, D., Love, A.C., Wormser, G.P., Schwartz, I., and Petzke, M.M. (2014). Induction of type I and type III interferons by *Borrelia burgdorferi* correlates with pathogenesis and requires linear plasmid 36. PLoS One *9*, e100174. https://doi.org/10.1371/journal.pone.0100174

Kuleshov, K.V., Margos, G., Fingerle, V., Koetsveld, J., Goptar, I.A., Markelov, M.L., Kolyasnikova, N.M., Sarksyan, D.S., Kirdyashkina, N.P., Shipulin, G.A., *et al.* (2020). Whole genome sequencing of *Borrelia miyamotoi* isolate Izh-4: reference for a complex bacterial genome. BMC Genomics *21*, 16. https://doi.org/10.1186/s12864-019-6388-4

Kulski, J.K. (2016). Next-Generation Sequencing — An Overview of the History, Tools, and "Omic" Applications. In Next Generation Sequencing - Advances, Applications and Challenges, J.K. Kulski, ed. (London UK: IntechOpen). https://doi.org/DOI: 10.5772/61964

Kumaran, D., Eswaramoorthy, S., Luft, B.J., Koide, S., Dunn, J.J., Lawson, C.L., and Swaminathan, S. (2001). Crystal structure of outer surface protein C (OspC) from the Lyme disease spirochete, *Borrelia burgdorferi*. EMBO J *20*, 971-978. https://doi.org/10.1093/emboj/20.5.971

Kumru, O.S., Schulze, R.J., Rodnin, M.V., Ladokhin, A.S., and Zuckert, W.R. (2011). Surface localization determinants of Borrelia OspC/Vsp family lipoproteins. J Bacteriol *193*, 2814-2825. https://doi.org/10.1128/JB.00015-11

Kurtenbach, K., De Michelis, S., Sewell, H.S., Etti, S., Schafer, S.M., Hails, R., Collares-Pereira, M., Santos-Reis, M., Hanincova, K., Labuda, M., *et al.* (2001). Distinct combinations of *Borrelia burgdorferi* sensu lato genospecies found in individual questing ticks from Europe. Appl Environ Microbiol *67*, 4926-4929.

Kurtenbach, K., Hanincova, K., Tsao, J.I., Margos, G., Fish, D., and Ogden, N.H. (2006). Fundamental processes in the evolutionary ecology of Lyme borreliosis. Nat Rev Microbiol *4*, 660-669. https://doi.org/10.1038/nrmicro1475

Labandeira-Rey, M., and Skare, J.T. (2001). Decreased infectivity in *Borrelia burgdorferi* strain B31 is associated with loss of linear plasmid 25 or 28-1. Infect Immun *69*, 446-455. https://doi.org/10.1128/IAI.69.1.446-455.2001

Lager, M., Faller, M., Wilhelmsson, P., Kjelland, V., Andreassen, A., Dargis, R., Quarsten, H., Dessau, R., Fingerle, V., Margos, G., *et al.* (2017). Molecular detection of *Borrelia burgdorferi* sensu lato - An analytical comparison of real-time PCR protocols from five different Scandinavian laboratories. PloS ONE *12*, e0185434. https://doi.org/10.1371/journal.pone.0185434

Lawrenz, M.B., Kawabata, H., Purser, J.E., and Norris, S.J. (2002). Decreased electroporation efficiency in *Borrelia burgdorferi* containing linear plasmids lp25 and lp56: impact on transformation of infectious *B. burgdorferi*. Infect Immun *70*, 4798-4804.

Lehnherr, H. (2006). Bacteriophage P1. In The Bacteriophages 2nd edition, S.T. Calendar, ed. (New York: Oxford University Press), pp. 350-364.

Levy, S.E., and Myers, R.M. (2016). Advancements in Next-Generation Sequencing. Annu Rev Genom Hum Gen *17*, 95-115. https://doi.org/10.1146/annurev-genom-083115-022413

Li, Y.F., Costello, J.C., Holloway, A.K., and Hahn, M.W. (2008). "Reverse ecology" and the power of population genomics. Evolution *62*, 2984-2994. https://doi.org/10.1111/j.1558-5646.2008.00486.x

Lin, T., Gao, L., Edmondson, D.G., Jacobs, M.B., Philipp, M.T., and Norris, S.J. (2009). Central role of the Holliday junction helicase RuvAB in vlsE recombination and infectivity of *Borrelia burgdorferi*. PLoS Pathogens *5*, e1000679. https://doi.org/10.1371/journal.ppat.1000679

Lin, T., Gao, L., Zhang, C., Odeh, E., Jacobs, M.B., Coutte, L., Chaconas, G., Philipp, M.T., and Norris, S.J. (2012). Analysis of an ordered, comprehensive STM mutant library in infectious *Borrelia burgdorferi*: insights into the genes required for mouse infectivity. PLoS One 7, e47532. https://doi.org/10.1371/journal.pone.0047532

Lin, Y.P., Bhowmick, R., Coburn, J., and Leong, J.M. (2015). Host cell heparan sulfate glycosaminoglycans are ligands for OspF-related proteins of the Lyme disease spirochete. Cell Microbiol *17*, 1464-1476. https://doi.org/10.1111/cmi.12448

Liveris, D., Gazumyan, A., and Schwartz, I. (1995). Molecular typing of *Borrelia burgdorferi* sensu lato by PCR-restriction fragment length polymorphism analysis. J Clin Microbiol *33*, 589-595.

Liveris, D., Varde, S., Iyer, R., Koenig, S., Bittker, S., Cooper, D., McKenna, D., Nowakowski, J., Nadelman, R.B., Wormser, G.P., *et al.* (1999). Genetic diversity of *Borrelia burgdorferi* in Lyme disease patients as determined by culture versus direct PCR with clinical specimens. J Clin Microbiol *37*, 565-569.

Liveris, D., Hanincova, K. and Schwartz, I. (2013). Borreliae. In Molecular Typing in Bacterial Infections, I.M. de Fillipis, M.L. McKee, ed. (New York: Humana Press), pp. 353-369. https://doi.org/10.1007/978-1-62703-185-1

Livey, I., Gibbs, C.P., Schuster, R., and Dorner, F. (1995). Evidence for lateral transfer and recombination in OspC variation in Lyme disease *Borrelia*. Mol Microbiol *18*, 257-269.

Lobocka, M.B., Rose, D.J., Plunkett, G., 3rd, Rusin, M., Samojedny, A., Lehnherr, H., Yarmolinsky, M.B., and Blattner, F.R. (2004). Genome of bacteriophage P1. J Bacteriol *186*, 7032-7068. https://doi.org/10.1128/JB.186.21.7032-7068.2004

Loza Reyes, E. (2010). Classification of phylogenetic data via Bayesian mixture modelling. PhD thesis University of Bath, UK.

Ludwig, A., von Rhein, C., Mischke, A., and Brade, V. (2007). Release of latent ClyA cytolysin from Escherichia coli mediated by a bacteriophage-associated putative holin (BlyA) from *Borrelia burgdorferi*. Int J Med Microbiol 298, 473-481. DOI: 10.1016/j.ijmm.2007.07.014

Luikart, G., England, P.R., Tallmon, D., Jordan, S., and Taberlet, P. (2003). The power and promise of population genomics: from genotyping to genome typing. Nat Rev Genet *4*, 981-994. https://doi.org/10.1038/nrg1226

Lynch, M. (2006). Streamlining and simplification of microbial genome architecture. Annu Rev Microbiol *60*, 327-349. https://doi.org/10.1146/annurev.micro.60.080805.142300

Maraspin, V., Mrvic, T., Ruzic-Sabljic, E., Jurcic, V., and Strle, F. (2019). Acrodermatitis chronica atrophicans in children: Report on two cases and review of the literature. Ticks Tick Borne Dis *10*, 180-185. https://doi.org/10.1016/j.ttbdis.2018.10.009

Marconi, R.T., Liveris, D., and Schwartz, I. (1995). Identification of novel insertion elements, restriction fragment length polymorphism patterns, and discontinuous 23S rRNA in Lyme disease spirochetes: phylogenetic analyses of rRNA genes and their intergenic spacers in *Borrelia japonica* sp. nov. and genomic group 21038 (*Borrelia andersonii* sp. nov.) isolates. J Clin Microbiol *33*, 2427-2434.

Marconi, R.T., Sung, S.Y., Hughes, C.A., and Carlyon, J.A. (1996). Molecular and evolutionary analyses of a variable series of genes in *Borrelia burgdorferi* that are related to *ospE* and *ospF*, constitute a gene family, and share a common upstream homology box. J Bacteriol *178*, 5615-5626.

Margolis, N., Hogan, D., Tilly, K., and Rosa, P.A. (1994). Plasmid location of *Borrelia* purine biosynthesis gene homologs. J Bacteriol *176*, 6427-6432.

Margos, G., Becker, N.S., Fingerle, V., Sing, A., Ramos, J.A., Carvalho, I.L., and Norte, A.C. (2019a). Core genome phylogenetic analysis of the avian associated Borrelia turdi indicates a close relationship to *Borrelia garinii*. Mol Phylogenet Evol *131*, 93-98. https://doi.org/10.1016/j.ympev.2018.10.044

Margos, G., Binder, K., Dzaferovic, E., Hizo-Teufel, C., Sing, A., Wildner, M., Fingerle, V., and Jolley, K.A. (2015a). PubMLST.org - The new home for the Borrelia MLSA database. Ticks Tick Borne Dis *6*, 869-871. https://doi.org/10.1016/j.ttbdis.2015.06.007

Margos, G., Chu, C.Y., Takano, A., Jiang, B.G., Liu, W., Kurtenbach, K., Masuzawa, T., Fingerle, V., Cao, W.C., and Kawabata, H. (2015b). *Borrelia yangtzensis* sp. nov. a rodent associated species in Asia is related to *B. valaisiana*. Int J Syst Evol Microbiol. https://doi.org/10.1099/ijsem.0.000491

Margos, G., Fedorova, N., Becker, N.S., Kleinjan, J.E., Marosevic, D., Krebs, S., Hui, L., Fingerle, V., and Lane, R.S. (2020). *Borrelia maritima* sp. nov., a novel species of the *Borrelia burgdorferi* sensu lato complex, occupying a basal position to North American species. Int J Syst Evol Microbiol *70*, 849-856. https://doi.org/10.1099/ijsem.0.003833

Margos, G., Fedorova, N., Kleinjan, J.E., Hartberger, C., Schwan, T.G., Sing, A., and Fingerle, V. (2017a). *Borrelia lanei* sp. nov. extends the diversity of *Borrelia* species in California. Int J Syst Evol Microbiol *67*, 3872-3876. https://doi.org/10.1099/ijsem.0.002214

Margos, G., Fingerle, V., and Reynolds, S. (2019b). *Borrelia bavariensis*: Vector Switch, Niche Invasion, and Geographical Spread of a Tick-Borne Bacterial Parasite. Frontiers in Ecology and Evolution 7. https://doi.org/10.3389/fevo.2019.00401

Margos, G., Gatewood, A.G., Aanensen, D.M., Hanincova, K., Terekhova, D., Vollmer, S.A., Cornet, M., Piesman, J., Donaghy, M., Bormane, A., *et al.* (2008). MLST of housekeeping genes captures geographic population structure and suggests a European origin of *Borrelia burgdorferi*. Proc Natl Acad Sci U S A *105*, 8730-8735. https://doi.org/10.1073/pnas.0800323105

Margos, G., Gofton, A., Wibberg, D., Dangel, A., Marosevic, D., Loh, S.M., Oskam, C., and Fingerle, V. (2018). The genus *Borrelia* reloaded. PLoS One *13*, e0208432. https://doi.org/10.1371/journal.pone.0208432

Margos, G., Hepner, S., Mang, C., Marosevic, D., Reynolds, S.E., Krebs, S., Sing, A., Derdakova, M., Reiter, M.A., and Fingerle, V. (2017b). Lost in plasmids: next generation sequencing and the complex genome of the tick-borne pathogen *Borrelia burgdorferi*. BMC Genomics *18*, 422. https://doi.org/10.1186/s12864-017-3804-5

Margos, G., Hepner, S., Mang, C., Sing, A., Liebl, B., and Fingerle, V. (2017c). Completed Genome Sequences of *Borrelia burgdorferi* Sensu Stricto B31(NRZ) and Closely Related Patient Isolates from Europe. Genome Announc *5*. https://doi.org/10.1128/genomeA.00637-17

Margos, G., Hojgaard, A., Lane, R.S., Cornet, M., Fingerle, V., Rudenko, N., Ogden, N., Aanensen, D.M., Fish, D., and Piesman, J. (2010). Multilocus sequence analysis of *Borrelia bissettii* strains from North America reveals a new *Borrelia* species, *Borrelia kurtenbachii*. Ticks Tick Borne Dis *1*, 151-158.

Margos, G., Lane, R.S., Fedorova, N., Koloczek, J., Piesman, J., Hojgaard, A., Sing, A., and Fingerle, V. (2016). *Borrelia bissettiae* sp. nov. and *Borrelia californiensis* sp. nov. prevail in diverse enzootic transmission cycles. Int J Syst Evol Microbiol *66*, 1447-1452. https://doi.org/10.1099/ijsem.0.000897

Margos, G., Sing, A., and Fingerle, V. (2017d). Published data do not support the notion that *Borrelia valaisiana* is human pathogenic. Infection *45*, 567-569. https://doi.org/10.1007/s15010-017-1032-1

Margos, G., Tsao, J.I., Castillo-Ramirez, S., Girard, Y.A., Hamer, S.A., Hoen, A.G., Lane, R.S., Raper, S.L., and Ogden, N.H. (2012). Two boundaries separate *Borrelia burgdorferi* populations in North America. Appl Environ Microbiol *78*, 6059-6067. https://doi.org/10.1128/AEM.00231-12

Margos, G., Vollmer, S.A., Cornet, M., Garnier, M., Fingerle, V., Wilske, B., Bormane, A., Vitorino, L., Collares-Pereira, M., Drancourt, M., *et al.* (2009). A new *Borrelia* species defined by Multilocus Sequence Analysis of Housekeeping Genes. Appl Environ Microbiol *75*, 5410-5416.

Margos, G., Vollmer, S.A., Ogden, N.H., and Fish, D. (2011). Population genetics, taxonomy, phylogeny and evolution of *Borrelia burgdorferi* sensu lato. Infect Genet Evol *11*, 1545-1563.

Margos, G., Wilske, B., Sing, A., Hizo-Teufel, C., Cao, W.C., Chu, C., Scholz, H., Straubinger, R.K., and Fingerle, V. (2013). *Borrelia bavariensis* sp. nov. is widely distributed in Europe and Asia. Int J Syst Evol Microbiol *63*, 4284-4288.

Marttinen, P., Hanage, W.P., Croucher, N.J., Connor, T.R., Harris, S.R., Bentley, S.D., and Corander, J. (2012). Detection of recombination events in bacterial genomes from large population samples. Nucleic Acids Res *40*, e6. https://doi.org/10.1093/nar/gkr928

McDowell, J.V., Wolfgang, J., Senty, L., Sundy, C.M., Noto, M.J., and Marconi, R.T. (2004). Demonstration of the involvement of outer surface protein E coiled coil structural domains and higher order structural elements in the binding of infection-induced antibody and the complement-regulatory protein, factor H. J Immunol *173*, 7471-7480.

Mechai, S., Margos, G., Feil, E.J., Barairo, N., Lindsay, L.R., Michel, P., and Ogden, N.H. (2016). Evidence for Host-Genotype Associations of *Borrelia burgdorferi* Sensu Stricto. PLoS One *11*, e0149345. https://doi.org/10.1371/journal.pone.0149345

Mechai, S., Margos, G., Feil, E.J., Lindsay, L.R., and Ogden, N.H. (2015). Complex population structure of *Borrelia burgdorferi* in southeastern and south central Canada as revealed by phylogeographic analysis. Appl Environ Microbiol *81*, 1309-1318. https://doi.org/10.1128/AEM.03730-14

Metts, M.S., McDowell, J.V., Theisen, M., Hansen, P.R., and Marconi, R.T. (2003). Analysis of the OspE determinants involved in binding of factor H and OspE-targeting antibodies elicited during *Borrelia burgdorferi* infection in mice. Infect Immun *71*, 3587-3596.

Milkman, R., and Bridges, M.M. (1990). Molecular evolution of the *Escherichia coli* chromosome. III. Clonal frames. Genetics *126*, 505-517.

Miller, J., Bono, J., Babb, K., El-Hage, N., Casjens, S., and Stevenson, B. (2000). A second allele of *eppA* in *Borrelia burgdorferi* strain B31 is located on the previously undetected circular plasmid cp9-2. J Bacteriol *182*, 6254-6258.

Miller, J.C., and Stevenson, B. (2003). Immunological and genetic characterization of *Borrelia burgdorferi* BapA and EppA proteins. Microbiology *149*, 1113-1125.

Miller, J.C., and Stevenson, B. (2004). Increased expression of *Borrelia burgdorferi* factor H-binding surface proteins during transmission from ticks to mice. Int J Med Microbiol *293 Suppl 37*, 120-125.

Molloy, E.M., Casjens, S.R., Cox, C.L., Maxson, T., Ethridge, N.A., Margos, G., Fingerle, V., and Mitchell, D.A. (2015). Identification of the minimal cytolytic unit for streptolysin S and an expansion of the toxin family. BMC Microbiol *15*, 141. https://doi.org/10.1186/s12866-015-0464-y

Mongodin, E.F., Casjens, S.R., Bruno, J.F., Xu, Y., Drabek, E.F., Riley, D.R., Cantarel, B.L., Pagan, P.E., Hernandez, Y.A., Vargas, L.C., *et al.* (2013). Inter- and intra-specific pan-genomes of *Borrelia burgdorferi* sensu lato: genome stability and adaptive radiation. BMC Genomics *14*, 693. https://doi.org/10.1186/1471-2164-14-693

Mukhacheva, T.A., and Kovalev, S.Y. (2013). Multilocus sequence analysis of *Borrelia burgdorferi* s.l. in Russia. Ticks Tick Borne Dis *4*, 275-279.

Mulay, V.B., Caimano, M.J., Iyer, R., Dunham-Ems, S., Liveris, D., Petzke, M.M., Schwartz, I., and Radolf, J.D. (2009). *Borrelia burgdorferi bba74* is expressed exclusively during tick feeding and is regulated by both arthropod- and mammalian host-specific signals. J Bacteriol *191*, 2783-2794. https://doi.org/10.1128/JB.01802-08

Muller, H.J. (1964). The Relation of Recombination to Mutational Advance. Mutat Res *106*, 2-9. https://doi.org/10.1016/0027-5107(64)90047-8

Muniesa, M., Imamovic, L., and Jofre, J. (2011). Bacteriophages and genetic mobilization in sewage and faecally polluted environments. Microb Biotechnol *4*, 725-734. https://doi.org/10.1111/j.1751-7915.2011.00264.x

Munro, H.J., Ogden, N.H., Mechai, S., Lindsay, L.R., Robertson, G.J., Whitney, H., and Lang, A.S. (2019). Genetic diversity of *Borrelia garinii* from *Ixodes uriae* collected in seabird colonies of the northwestern Atlantic Ocean. Ticks Tick Borne Dis *10*, 101255. https://doi.org/10.1016/j.ttbdis.2019.06.014

Nei, M., and Rooney, A.P. (2005). Concerted and birth-and-death evolution of multigene families. Annu Rev Genet *39*, 121-152. https://doi.org/10.1146/annurev.genet.39.073003.112240

Neubert, U., Schaller, M., Januschke, E., Stolz, W., and Schmieger, H. (1993). Bacteriophages induced by ciprofloxacin in a *Borrelia burgdorferi* skin isolate. Zentralbl Bakteriol *279*, 307-315.

Nguyen, K.T., Wu, J.C., Boylan, J.A., Gherardini, F.C., and Pei, D. (2007). Zinc is the metal cofactor of *Borrelia burgdorferi* peptide deformylase. Arch Biochem Biophys *468*, 217-225. https://doi.org/S0003-9861(07)00483-3 [pii] 10.1016/j.abb.2007.09.023

Niedringhaus, T.P., Milanova, D., Kerby, M.B., Snyder, M.P., and Barron, A.E. (2011). Landscape of next-generation sequencing technologies. Anal Chem *83*, 4327-4341. https://doi.org/10.1021/ac2010857

Norris, D.E., Johnson, B.J., Piesman, J., Maupin, G.O., Clark, J.L., and Black, W.C.t. (1997). Culturing selects for specific genotypes of *Borrelia burgdorferi* in an enzootic cycle in Colorado. J Clin Microbiol *35*, 2359-2364.

Norris, S.J. (2006). Antigenic variation with a twist--the *Borrelia* story. Mol Microbiol *60*, 1319-1322. https://doi.org/10.1111/j.1365-2958.2006.05204.x

Norris, S.J. (2014). *vls* Antigenic Variation Systems of Lyme Disease *Borrelia*: Eluding Host Immunity through both Random, Segmental Gene Conversion and Framework Heterogeneity. Microbiol Spectr *2*, 1. https://doi.org/10.1128/microbiolspec.MDNA3-0038-2014

Norte, A.C., Margos, G., Becker, N.S., Albino Ramos, J., Nuncio, M.S., Fingerle, V., Araujo, P.M., Adamik, P., Alivizatos, H., Barba, E., *et al.* (2020). Host dispersal shapes the population structure of a tick-borne bacterial pathogen. Mol Ecol *29*, 485-501. https://doi.org/10.1111/mec.15336

Ogden, N.H., Bouchard, C., Kurtenbach, K., Margos, G., Lindsay, L.R., Trudel, L., Nguon, S., and Milord, F. (2010). Active and passive surveillance and phylogenetic analysis of *Borrelia burgdorferi* elucidate the process of Lyme disease risk emergence in Canada. Environ Health Perspect *118*, 909-914.

Ogden, N.H., Maarouf, A., Barker, I.K., Bigras-Poulin, M., Lindsay, L.R., Morshed, M.G., O'Callaghan C, J., Ramay, F., Waltner-Toews, D., and Charron, D.F. (2006). Climate change and the potential for range expansion of the Lyme disease vector *Ixodes scapularis* in Canada. Int J Parasitol *36*, 63-70.

Ogden, N.H., Margos, G., Aanensen, D.M., Drebot, M.A., Feil, E.J., Hanincova, K., Schwartz, I., Tyler, S., and Lindsay, L.R. (2011). Investigation of genotypes of *Borrelia burgdorferi* in *Ixodes scapularis* ticks collected during surveillance in Canada. Appl Environ Microbiol *77*, 3244-3254. https://doi.org/10.1128/AEM.02636-10

Ojaimi, C., Brooks, C., Casjens, S., Rosa, P., Elias, A., Barbour, A., Jasinskas, A., Benach, J., Katona, L., Radolf, J., *et al.* (2003). Profiling of temperature-induced changes in *Borrelia burgdorferi* gene expression by using whole genome arrays. Infect Immun *71*, 1689-1705.

Ojaimi, C., Davidson, B.E., Saint Girons, I., and Old, I.G. (1994). Conservation of gene arrangement and an unusual organization of rRNA genes in the linear chromosomes of the Lyme disease spirochaetes *Borrelia burgdorferi*, B. garinii and B. afzelii. Microbiology *140*, 2931-2940. https://doi.org/10.1099/13500872-140-11-2931

Ostfeld, R.S., Levi, T., Keesing, F., Oggenfuss, K., and Canham, C.D. (2018). Tick-borne disease risk in a forest food web. Ecology *99*, 1562-1573. https://doi.org/10.1002/ecy.2386

Ouyang, Z., He, M., Oman, T., Yang, X.F., and Norgard, M.V. (2009). A manganese transporter, BB0219 (BmtA), is required for virulence by the Lyme disease spirochete, *Borrelia burgdorferi*. Proc Natl Acad Sci U S A *106*, 3449-3454. https://doi.org/10.1073/pnas.0812999106

Pal, U., Li, X., Wang, T., Montgomery, R.R., Ramamoorthi, N., Desilva, A.M., Bao, F., Yang, X., Pypaert, M., Pradhan, D., *et al.* (2004). TROSPA, an *Ixodes scapularis* receptor for *Borrelia burgdorferi*. Cell *119*, 457-468. https://doi.org/10.1016/j.cell.2004.10.027

Palmer, N., Fraser, C., and Casjens, S. (2000). Distribution of twelve linear extrachromosomal DNAs in natural isolates of Lyme disease spirochetes. J Bacteriol *182*, 2476-2480. https://doi.org/10.1128/jb.182.9.2476-2480.2000

Paradis, E., Claude, J., and Strimmer, K. (2004). APE: Analyses of Phylogenetics and Evolution in R language. Bioinformatics *20*, 289-290. https://doi.org/10.1093/bioinformatics/btg412

Paster, B.J., Dewhirst, F.E., Weisburg, W.G., Tordoff, L.A., Fraser, G.J., Hespell, R.B., Stanton, T.B., Zablen, L., Mandelco, L., and Woese, C.R. (1991). Phylogenetic analysis of the spirochetes. J Bacteriol *173*, 6101-6109.

Penades, J.R., Chen, J., Quiles-Puchalt, N., Carpena, N., and Novick, R.P. (2015). Bacteriophage-mediated spread of bacterial virulence genes. Curr Opin Microbiol *23*, 171-178. https://doi.org/10.1016/j.mib.2014.11.019

Phelan, J.P., Kern, A., Ramsey, M.E., Lundt, M.E., Sharma, B., Lin, T., Gao, L., Norris, S.J., Hyde, J.A., Skare, J.T., *et al.* (2019). Genome-wide screen identifies novel genes required for *Borrelia burgdorferi* survival in its *Ixodes* tick vector. PLoS Pathogens *15*, e1007644. https://doi.org/10.1371/journal.ppat.1007644

Picardeau, M., Lobry, J.R., and Hinnebusch, B.J. (1999). Physical mapping of an origin of bidirectional replication at the centre of the *Borrelia burgdorferi* linear chromosome. Mol Microbiol *32*, 437-445. https://doi.org/10.1046/j.1365-2958.1999.01368.x

Plasterk, R.H., Simon, M.I., and Barbour, A.G. (1985). Transposition of structural genes to an expression sequence on a linear plasmid causes antigenic variation in the bacterium *Borrelia hermsii*. Nature *318*, 257-263.

Porcella, S.F., Fitzpatrick, C.A., and Bono, J.L. (2000). Expression and immunological analysis of the plasmid-borne *mlp* genes of *Borrelia burgdorferi* strain B31. Infect Immun *68*, 4992-5001.

Porcella, S.F., Popova, T.G., Akins, D.R., Li, M., Radolf, J.D., and Norgard, M.V. (1996). *Borrelia burgdorferi* supercoiled plasmids encode multicopy tandem open reading frames and a lipoprotein gene family. J Bacteriol *178*, 3293-3307.

Posey, J.E., and Gherardini, F.C. (2000). Lack of a role for iron in the Lyme disease pathogen. Science *288*, 1651-1653.

Postic, D., Assous, M.V., Grimont, P.A., and Baranton, G. (1994). Diversity of *Borrelia burgdorferi* sensu lato evidenced by restriction fragment length polymorphism of *rrf* (5S)-*rrl* (23S) intergenic spacer amplicons. Int J Syst Bacteriol *44*, 743-752.

Postic, D., Garnier, M., and Baranton, G. (2007). Multilocus sequence analysis of atypical *Borrelia burgdorferi* sensu lato isolates - description of *Borrelia californiensis* sp. nov., and genomospecies 1 and 2. Int J Med Microbiol *297*, 263-271.

Postic, D., Ras, N.M., Lane, R.S., Hendson, M., and Baranton, G. (1998). Expanded diversity among Californian *Borrelia* isolates and description of *Borrelia bissettii* sp. nov. (formerly *Borrelia* group DN127). J Clin Microbiol *36*, 3497-3504.

Potter, S.C., Luciani, A., Eddy, S.R., Park, Y., Lopez, R., and Finn, R.D. (2018). HMMER web server: 2018 update. Nucleic Acids Res *46*, W200-W204. https://doi.org/10.1093/nar/gky448

Prevelige, P.E., Jr., and Cortines, J.R. (2018). Phage assembly and the special role of the portal protein. Curr Opin Virol *31*, 66-73. https://doi.org/10.1016/j.coviro.2018.09.004

Price, M.N., Dehal, P.S., and Arkin, A.P. (2010). FastTree 2--approximately maximum-likelihood trees for large alignments. PLoS One *5*, e9490. https://doi.org/10.1371/journal.pone.0009490

Pritt, B.S., Mead, P.S., Johnson, D.K.H., Neitzel, D.F., Respicio-Kingry, L.B., Davis, J.P., Schiffman, E., Sloan, L.M., Schriefer, M.E., Replogle, A.J., *et al.* (2016). Identification of a novel pathogenic *Borrelia* species causing Lyme borreliosis with unusually high spirochaetaemia: a descriptive study. Lancet Infect Dis *16*, 556-564. https://doi.org/10.1016/S1473-3099(15)00464-8

Probert, W., and Johnson, B. (1998). Identification of a 47 kd fibronectin-binding protein expressed by *Borrelia burgdorferi* isolate B31. Molecular Microbiology *30*, 1003-1015.

Promnares, K., Kumar, M., Shroder, D.Y., Zhang, X., Anderson, J.F., and Pal, U. (2009). *Borrelia burgdorferi* small lipoprotein Lp6.6 is a member of multiple protein complexes in the outer membrane and facilitates pathogen transmission from ticks to mice. Mol Microbiol *74*, 112-125. https://doi.org/10.1111/j.1365-2958.2009.06853.x

Purser, J.E., Lawrenz, M.B., Caimano, M.J., Howell, J.K., Radolf, J.D., and Norris, S.J. (2003). A plasmid-encoded nicotinamidase (PncA) is essential for infectivity of *Borrelia burgdorferi* in a mammalian host. Mol Microbiol *48*, 753-764. https://doi.org/3452

Purser, J.E., and Norris, S.J. (2000). Correlation between plasmid content and infectivity in *Borrelia burgdorferi*. Proc Natl Acad Sci USA *97*, 13865-13870. https://doi.org/Doi 10.1073/Pnas.97.25.13865

Qiu, W.G., Bosler, E.M., Campbell, J.R., Ugine, G.D., Wang, I.N., Luft, B.J., and Dykhuizen, D.E. (1997). A population genetic study of *Borrelia burgdorferi* sensu stricto from eastern Long Island, New York, suggested frequency-dependent selection, gene flow and host adaptation. Hereditas *127*, 203-216.

Qiu, W.G., Bruno, J.F., McCaig, W.D., Xu, Y., Livey, I., Schriefer, M.E., and Luft, B.J. (2008). Wide distribution of a high-virulence *Borrelia burgdorferi* clone in Europe and North America. Emerg Infect Dis *14*, 1097-1104.

Qiu, W.G., Dykhuizen, D.E., Acosta, M.S., and Luft, B.J. (2002). Geographic uniformity of the Lyme disease spirochete (*Borrelia burgdorferi*) and its shared history with tick vector (*Ixodes scapularis*) in the Northeastern United States. Genetics *160*, 833-849.

Qiu, W.G., and Martin, C.L. (2014). Evolutionary genomics of *Borrelia burgdorferi* sensu lato: findings, hypotheses, and the rise of hybrids. Infect Genet Evol *27*, 576-593. https://doi.org/10.1016/j.meegid.2014.03.025

Qiu, W.G., Schutzer, S.E., Bruno, J.F., Attie, O., Xu, Y., Dunn, J.J., Fraser, C.M., Casjens, S.R., and Luft, B.J. (2004). Genetic exchange and plasmid transfers in *Borrelia burgdorferi* sensu stricto revealed by three-way genome comparisons and multilocus sequence typing. Proc Natl Acad Sci U S A *101*, 14150-14155.

Quereda, J.J., Nahori, M.A., Meza-Torres, J., Sachse, M., Titos-Jimenez, P., Gomez-Laguna, J., Dussurget, O., Cossart, P., and Pizarro-Cerda, J. (2017). Listeriolysin S Is a Streptolysin S-Like Virulence Factor That Targets Exclusively Prokaryotic Cells In Vivo. MBio *8*. https://doi.org/10.1128/mBio.00259-17

Radolf, J.D., Caimano, M.J., Stevenson, B., and Hu, L.T. (2012). Of ticks, mice and men: understanding the dual-host lifestyle of Lyme disease spirochaetes. NatRevMicrobiol *10*, 87-99.

Ramsey, M.E., Hyde, J.A., Medina-Perez, D.N., Lin, T., Gao, L., Lundt, M.E., Li, X., Norris, S.J., Skare, J.T., and Hu, L.T. (2017). A high-throughput genetic screen identifies previously uncharacterized *Borrelia burgdorferi* genes important for resistance against reactive oxygen and nitrogen species. PLoS Pathogens *13*, e1006225. https://doi.org/10.1371/journal.ppat.1006225

Ranka, R., Bormane, A., Salmina, K., and Baumanis, V. (2004). Identification of three clinically relevant *Borrelia burgdorferi* sensu lato genospecies by PCR-restriction fragment length polymorphism analysis of 16S-23S ribosomal DNA spacer amplicons. J Clin Microbiol *42*, 1444-1449. https://doi.org/10.1128/jcm.42.4.1444-1449.2004

Rannala, B., Qiu, W.G., and Dykhuizen, D.E. (2000). Methods for estimating gene frequencies and detecting selection in bacterial populations. Genetics *155*, 499-508.

Ras, N.M., Lascola, B., Postic, D., Cutler, S.J., Rodhain, F., Baranton, G., and Raoult, D. (1996). Phylogenesis of relapsing fever *Borrelia* spp. Int J Syst Bacteriol *46*, 859-865. https://doi.org/10.1099/00207713-46-4-859

Rauter, C., and Hartung, T. (2005). Prevalence of *Borrelia burgdorferi* sensu lato genospecies in *Ixodes* ricinus ticks in Europe: a metaanalysis. Appl Environ Microbiol *71*, 7203-7216. https://doi.org/10.1128/AEM.71.11.7203-7216.2005

Ravin, V., Ravin, N., Casjens, S., Ford, M.E., Hatfull, G.F., and Hendrix, R.W. (2000). Genomic sequence and analysis of the atypical temperate bacteriophage N15. J Mol Biol *299*, 53-73.

Rego, R.O., Bestor, A., and Rosa, P.A. (2011). Defining the plasmid-borne restriction-modification systems of the Lyme disease spirochete *Borrelia burgdorferi*. J Bacteriol *193*, 1161-1171. https://doi.org/JB.01176-10 [pii] 10.1128/JB.01176-10

Revel, A.T., Blevins, J.S., Almazan, C., Neil, L., Kocan, K.M., de la Fuente, J., Hagman, K.E., and Norgard, M.V. (2005). *bptA* (bbe16) is essential for the persistence of the Lyme disease spirochete, *Borrelia burgdorferi*, in its natural tick vector. Proc Natl Acad Sci U S A *102*, 6972-6977. https://doi.org/0502565102 [pii] 10.1073/pnas.0502565102

Richter, D., Postic, D., Sertour, N., Livey, I., Matuschka, F.R., and Baranton, G. (2006). Delineation of *Borrelia burgdorferi* sensu lato species by multilocus sequence analysis and confirmation of the delineation of *Borrelia spielmanii* sp. nov. Int J Syst Evol Microbiol *56*, 873-881.

Rijpkema, S.G., Molkenboer, M.J., Schouls, L.M., Jongejan, F., and Schellekens, J.F. (1995). Simultaneous detection and genotyping of three genomic groups of *Borrelia burgdorferi* sensu lato in Dutch *Ixodes ricinus* ticks by characterization of the amplified intergenic spacer region between 5S and 23S rRNA genes. J Clin Microbiol *33*, 3091-3095.

Roberts, D.M., Theisen, M., and Marconi, R.T. (2000). Analysis of the cellular localization of Bdr paralogs in *Borrelia burgdorferi*, a causative agent of lyme disease: evidence for functional diversity. J Bacteriol *182*, 4222-4226.

Rocha, E.P.C. (2018). Neutral Theory, Microbial Practice: Challenges in Bacterial Population Genetics. Mol Biol Evol *35*, 1338-1347. https://doi.org/10.1093/molbev/msy078

Sabitova, Y., Fomenko, N., Tikunov, A., Stronin, O., Khasnatinov, M., Abmed, D., Danchinova, G., Golovljova, I., and Tikunova, N. (2018). Multilocus sequence analysis of *Borrelia burgdorferi* sensu lato isolates from Western Siberia, Russia and Northern Mongolia. Infect Genet Evol *62*, 160-169. https://doi.org/10.1016/j.meegid.2018.04.015

Sadziene, A., Rosa, P., Thompson, P., Hogan, D., and Barbour, A. (1992). Antibody-resistant mutants of *Borrelia burgdorferi*: in vitro selection and characterization. J Exp Med *176*, 799-809.

Sadziene, A., Wilske, B., Ferdows, M.S., and Barbour, A.G. (1993). The cryptic ospC gene of *Borrelia burgdorferi* B31 is located on a circular plasmid. Infect Immun *61*, 2192-2195.

Saint Girons, I., Gern, L., Gray, J.S., Guy, E.C., Korenberg, E., Nuttall, P.A., Rijpkema, S.G., Schonberg, A., Stanek, G., and Postic, D. (1998). Identification of *Borrelia burgdorferi* sensu lato species in Europe. Zentralbl Bakteriol *287*, 190-195.

Salo, J., Jaatinen, A., Soderstrom, M., Viljanen, M.K., and Hytonen, J. (2015). Decorin binding proteins of *Borrelia burgdorferi* promote arthritis development and joint specific post-treatment DNA persistence in mice. PLoS One *10*, e0121512. https://doi.org/10.1371/journal.pone.0121512

Samuels, D., Marconi, R., and Garon, C. (1993). Variation in the size of the *ospA*-containing linear plasmid, but not the linear chromosome, among the three *Borrelia* species associated with Lyme disease. J Gen Microbiol *139 (Pt 10)*, 2445-2449.

Samuels, D.S. (2011). Gene regulation in *Borrelia burgdorferi*. Annu Rev Microbiol *65*, 479-499. https://doi.org/10.1146/annurev.micro.112408.134040

Schaller, M., and Neubert, U. (1994). Bacteriophages and ultrastructural alterations of *Borrelia burgdorferi* induced by ciprofloxacin. J Spirochetal Tick-borne Dis *1*, 37-40.

Schulte-Spechtel, U., Fingerle, V., Goettner, G., Rogge, S., and Wilske, B. (2006). Molecular analysis of decorin-binding protein A (DbpA) reveals five major groups among European *Borrelia burgdorferi sensu lato* strains with impact for the development of serological assays and indicates lateral gene transfer of the *dbpA* gene. Int J Med Microbiol *296 Suppl 40*, 250-266.

Schutzer, S.E., Fraser-Liggett, C.M., Casjens, S.R., Qiu, W.G., Dunn, J.J., Mongodin, E.F., and Luft, B.J. (2011). Whole-genome sequences of thirteen isolates of *Borrelia burgdorferi*. J Bacteriol *193*, 1018-1020. https://doi.org/JB.01158-10 [pii] 10.1128/JB.01158-10

Schutzer, S.E., Fraser-Liggett, C.M., Qiu, W.G., Kraiczy, P., Mongodin, E.F., Dunn, J.J., Luft, B.J., and Casjens, S.R. (2012). Whole-genome sequences of *Borrelia bissettii*, *Borrelia valaisiana*, and *Borrelia spielmanii*. J Bacteriol *194*, 545-546. https://doi.org/10.1128/JB.06263-11

Schwan, T.G., Burgdorfer, W., and Garon, C.F. (1988). Changes in infectivity and plasmid profile of the Lyme disease spirochete, *Borrelia burgdorferi*, as a result of in vitro cultivation. Infect Immun *56*, 1831-1836.

Schwan, T.G., Raffel, S.J., Schrumpf, M.E., and Porcella, S.F. (2007). Diversity and distribution of *Borrelia hermsii*. Emerg Infect Dis *13*, 436-442.

Schwartz, A.M., Hinckley, A.F., Mead, P.S., Hook, S.A., and Kugeler, K.J. (2017). Surveillance for Lyme Disease - United States, 2008-2015. MMWR Surveill Summ *66*, 1-12. https://doi.org/10.15585/mmwr.ss6622a1

Schwartz, J.J., Gazumyan, A., and Schwartz, I. (1992). rRNA gene organization in the Lyme disease spirochete, *Borrelia burgdorferi*. J Bacteriol *174*, 3757-3765.

Seifert, S.N., Khatchikian, C.E., Zhou, W., and Brisson, D. (2015). Evolution and population genomics of the Lyme borreliosis pathogen, *Borrelia burgdorferi*. Trends Genet *31*, 201-207. https://doi.org/10.1016/j.tig.2015.02.006

Seinost, G., Dykhuizen, D.E., Dattwyler, R.J., Golde, W.T., Dunn, J.J., Wang, I.N., Wormser, G.P., Schriefer, M.E., and Luft, B.J. (1999). Four clones of *Borrelia burgdorferi* sensu stricto cause invasive infection in humans. Infect Immun *67*, 3518-3524.

Seshu, J., Esteve-Gassent, M.D., Labandeira-Rey, M., Kim, J.H., Trzeciakowski, J.P., Hook, M., and Skare, J.T. (2006). Inactivation of the fibronectin-binding adhesin gene *bbk32* significantly attenuates the infectivity potential of *Borrelia burgdorferi*. Mol Microbiol *59*, 1591-1601. https://doi.org/MMI5042 [pii] 10.1111/j.1365-2958.2005.05042.x

Shi, Y., Xu, Q., McShan, K., and Liang, F.T. (2008). Both decorin-binding proteins A and B are critical for the overall virulence of *Borrelia burgdorferi*. Infect Immun *76*, 1239-1246. https://doi.org/IAI.00897-07 [pii] 10.1128/IAI.00897-07

Shih, Y.L., and Rothfield, L. (2006). The bacterial cytoskeleton. Microbiol Mol Biol Rev *70*, 729-754.

Skare, J.T., Foley, D.M., Hernandez, S.R., Moore, D.C., Blanco, D.R., Miller, J.N., and Lovett, M.A. (1999). Cloning and molecular characterization of plasmid-encoded antigens of *Borrelia burgdorferi*. Infect Immun *67*, 4407-4417.

Smith, J.M., Smith, N.H., O'Rourke, M., and Spratt, B.G. (1993). How clonal are bacteria? Proc Natl Acad Sci U S A *90*, 4384-4388. https://doi.org/10.1073/pnas.90.10.4384

Smith, R.P., Jr., Muzaffar, S.B., Lavers, J., Lacombe, E.H., Cahill, B.K., Lubelczyk, C.B., Kinsler, A., Mathers, A.J., and Rand, P.W. (2006). *Borrelia garinii* in seabird ticks (*Ixodes uriae*), Atlantic Coast, North America. Emerg Infect Dis *12*, 1909-1912.

Spielman, A. (1994). The emergence of Lyme disease and human babesiosis in a changing environment. Ann N Y Acad Sci *740*, 146-156.

Spratt, B.G. (1999). Multilocus sequence typing: molecular typing of bacterial pathogens in an era of rapid DNA sequencing and the Internet. Curr Opin Microbiol *2*, 312-316. https://doi.org/http:/Ibiomednet.com/elecref/1369527400200312

Springer, A., Raulf, M.K., Fingerle, V., and Strube, C. (2020). Borrelia prevalence and species distribution in ticks removed from humans in Germany, 2013-2017. Ticks Tick Borne Dis *11*, 101363. https://doi.org/10.1016/j.ttbdis.2019.101363

Stanek, G., and Reiter, M. (2011). The expanding Lyme Borrelia complex--clinical significance of genomic species? Clin Microbiol Infect *17*, 487-493. https://doi.org/10.1111/j.1469-0691.2011.03492.x

Stanton, T.B. (2007). Prophage-like gene transfer agents-novel mechanisms of gene exchange for *Methanococcus*, *Desulfovibrio*, *Brachyspira*, and *Rhodobacter* species. Anaerobe *13*, 43-49. https://doi.org/10.1016/j.anaerobe.2007.03.004

States, S.L., Brinkerhoff, R.J., Carpi, G., Steeves, T.K., Folsom-O'Keefe, C., DeVeaux, M., and Diuk-Wasser, M.A. (2014). Lyme disease risk not amplified in a species-poor vertebrate community: similar *Borrelia burgdorferi* tick infection prevalence and OspC genotype frequencies. Infect Genet Evol *27*, 566-575. https://doi.org/10.1016/j.meegid.2014.04.014

Stete, K., Rieg, S., Margos, G., Hacker, G., Wagner, D., Kern, W.V., and Fingerle, V. (2018). Case Report and Genetic Sequence Analysis of *Candidatus Borrelia kalaharica*, Southern Africa. Emerg Infect Dis *24*, 1659-1664. https://doi.org/10.3201/eid2409.171381

Stevenson, B., Bykowski, T., Cooley, A.E., Babb, K., Miller, J.C., Woodman, M.E., von Lackum, K., and Riley, S.P. (2006). The Lyme disease spirochete Erp protein family: structure, function and regulation of expression. In Moleculr biology of spirochetes, F.C. Cabello, D. Hulinska, and H.P. Godfrey, eds. (Amsterdam: IOS Press), pp. 354-372.

Stevenson, B., Casjens, S., van Vugt, R., Porcella, S.F., Tilly, K., Bono, J.L., and Rosa, P. (1997). Characterization of cp18, a naturally truncated member of the cp32 family of *Borrelia burgdorferi* plasmids. J Bacteriol *179*, 4285-4291.

Stevenson, B., El-Hage, N., Hines, M.A., Miller, J.C., and Babb, K. (2002). Differential binding of host complement inhibitor factor H by *Borrelia burgdorferi* Erp surface proteins: a possible mechanism underlying the expansive host range of Lyme disease spirochetes. Infect Immun *70*, 491-497.

Stevenson, B., and Miller, J.C. (2003). Intra- and interbacterial genetic exchange of Lyme disease spirochete *erp* genes generates sequence identity amidst diversity. J Mol Evol *57*, 309-324.

Stevenson, B., Porcella, S.F., Oie, K.L., Fitzpatrick, C.A., Raffel, S.J., Lubke, L., Schrumpf, M.E., and Schwan, T.G. (2000a). The relapsing fever spirochete *Borrelia hermsii* contains multiple, antigen-encoding circular plasmids that are homologous to the cp32 plasmids of Lyme disease spirochetes. Infect Immun *68*, 3900-3908.

Stevenson, B., Tilly, K., and Rosa, P. (1996). A family of genes located on four separate 32-kilobase circular plasmids in *Borrelia burgdorferi* B31. J Bacteriol *178*, 3508-3516.

Stevenson, B., Zückert, W., and Akins, D. (2000b). Repetition, conservation, and variation: the multiple cp32 plasmids of *Borrelia* species. J Molec Microbiol Biotech, in press.

Stewart, P.E., Byram, R., Grimm, D., Tilly, K., and Rosa, P.A. (2005). The plasmids of *Borrelia burgdorferi*: essential genetic elements of a pathogen. Plasmid *53*, 1-13. https://doi.org/S0147-619X(04)00122-2 [pii] 10.1016/j.plasmid.2004.10.006

Stewart, P.E., Thalken, R., Bono, J.L., and Rosa, P. (2001). Isolation of a circular plasmid region sufficient for autonomous replication and transformation of infectious *Borrelia burgdorferi*. Mol Microbiol *39*, 714-721. https://doi.org/10.1046/j.1365-2958.2001.02256.x

Strandh, M., and Raberg, L. (2015). Within-host competition between *Borrelia afzelii ospC* strains in wild hosts as revealed by massively parallel amplicon sequencing. Philos Trans R Soc Lond B Biol Sci *370*. https://doi.org/10.1098/rstb.2014.0293

Strother, K.O., Broadwater, A., and De Silva, A. (2005). Plasmid requirements for infection of ticks by *Borrelia burgdorferi*. Vector Borne Zoonotic Dis *5*, 237-245. https://doi.org/10.1089/vbz.2005.5.237

Strother, K.O., and de Silva, A. (2005). Role of *Borrelia burgdorferi* linear plasmid 25 in infection of *Ixodes scapularis* ticks. J Bacteriol *187*, 5776-5781. https://doi.org/10.1128/JB.187.16.5776-5781.2005

Surtees, J.A., and Funnell, B.E. (2003). Plasmid and chromosome traffic control: how ParA and ParB drive partition. Curr Top Dev Biol *56*, 145-180.

Terekhova, D., Iyer, R., Wormser, G.P., and Schwartz, I. (2006). Comparative genome hybridization reveals substantial variation among clinical isolates of *Borrelia burgdorferi* sensu stricto with different pathogenic properties. J Bacteriol *188*, 6124-6134. https://doi.org/188/17/6124 [pii] 10.1128/JB.00459-06

Theisen, M. (1996). Molecular cloning and characterization of *nlpH*, encoding a novel, surface-exposed, polymorphic, plasmid-encoded 33-kilodalton lipoprotein of *Borrelia afzelii*. J Bacteriol *178*, 6435-6442.

Thierauf, A., Perez, G., and Maloy, A.S. (2009). Generalized transduction. Methods Mol Biol *501*, 267-286. https://doi.org/10.1007/978-1-60327-164-6_23

Tilly, K., Bestor, A., and Rosa, P.A. (2013). Lipoprotein succession in *Borrelia burgdorferi*: similar but distinct roles for OspC and VlsE at different stages of mammalian infection. Mol Microbiol *89*, 216-227. https://doi.org/10.1111/mmi.12271

Tilly, K., Casjens, S., Stevenson, B., Bono, J.L., Samuels, D.S., Hogan, D., and Rosa, P. (1997). The *Borrelia burgdorferi* circular plasmid cp26: conservation of plasmid structure and targeted inactivation of the *ospC* gene. Mol Microbiol *25*, 361-373.

Tilly, K., Checroun, C., and Rosa, P.A. (2012). Requirements for *Borrelia burgdorferi* plasmid maintenance. Plasmid. https://doi.org/S0147-619X(12)00013-3 [pii] 10.1016/j.plasmid.2012.01.009

Tilly, K., Grimm, D., Bueschel, D.M., Krum, J.G., and Rosa, P. (2004). Infectious cycle analysis of a *Borrelia burgdorferi* mutant defective in transport of chitobiose, a tick cuticle component. Vector Borne Zoonotic Dis *4*, 159-168.

Tilly, K., Krum, J.G., Bestor, A., Jewett, M.W., Grimm, D., Bueschel, D., Byram, R., Dorward, D., Vanraden, M.J., Stewart, P., *et al.* (2006). *Borrelia burgdorferi* OspC protein required exclusively in a crucial early stage of mammalian infection. Infect Immun *74*, 3554-3564.

Tokarz, R., Anderton, J.M., Katona, L.I., and Benach, J.L. (2004). Combined effects of blood and temperature shift on *Borrelia burgdorferi* gene expression as determined by whole genome DNA array. Infect Immun *72*, 5419-5432.

Tokarz, R., Tagliafierro, T., Sameroff, S., Cucura, D.M., Oleynik, A., Che, X., Jain, K., and Lipkin, W.I. (2019). Microbiome analysis of *Ixodes scapularis* ticks from New York and Connecticut. Ticks Tick Borne Dis *10*, 894-900. https://doi.org/10.1016/j.ttbdis.2019.04.011

Travinsky, B., Bunikis, J., and Barbour, A.G. (2010). Geographic differences in genetic locus linkages for *Borrelia burgdorferi*. Emerg Infect Dis *16*, 1147-1150. https://doi.org/10.3201/eid1607.091452

Troy, E.B., Lin, T., Gao, L., Lazinski, D.W., Lundt, M., Camilli, A., Norris, S.J., and Hu, L.T. (2016). Global Tn-seq analysis of carbohydrate utilization and vertebrate infectivity of *Borrelia burgdorferi*. Mol Microbiol *101*, 1003-1023. https://doi.org/10.1111/mmi.13437

Tsao, N., Kuo, C.F., Cheng, M.H., Lin, W.C., Lin, C.F., and Lin, Y.S. (2019). Streptolysin S induces mitochondrial damage and macrophage death through inhibiting degradation of glycogen synthase kinase-3beta in *Streptococcus pyogenes* infection. Sci Rep *9*, 5371. https://doi.org/10.1038/s41598-019-41853-3

Tufts, D.M., Hart, T.M., Chen, G.F., Kolokotronis, S.O., Diuk-Wasser, M.A., and Lin, Y.P. (2019). Outer surface protein polymorphisms linked to host-spirochete association in Lyme *Borreliae*. Mol Microbiol *111*, 868-882. https://doi.org/10.1111/mmi.14209

Tyler, S., Tyson, S., Dibernardo, A., Drebot, M., Feil, E.J., Graham, M., Knox, N.C., Lindsay, L.R., Margos, G., Mechai, S., *et al.* (2018). Whole genome sequencing and phylogenetic analysis of strains of the agent of Lyme disease *Borrelia burgdorferi* from Canadian emergence zones. Sci Rep *8*, 10552. https://doi.org/10.1038/s41598-018-28908-7

Urwin, R., and Maiden, M.C. (2003). Multi-locus sequence typing: a tool for global epidemiology. Trends Microbiol *11*, 479-487.

Verhey, T.B., Castellanos, M., and Chaconas, G. (2019). Antigenic variation in the Lyme spirochete: detailed functional assessment of recombinational switching at *vlsE* in the JD1 strain of *Borrelia burgdorferi*. Mol Microbiol *111*, 750-763. https://doi.org/10.1111/mmi.14189

Vink, C., Rudenko, G., and Seifert, H.S. (2012). Microbial antigenic variation mediated by homologous DNA recombination. FEMS Microbiol Rev *36*, 917-948. https://doi.org/10.1111/j.1574-6976.2011.00321.x

Vitorino, L., Margos, G., Ze-Ze, L., Kurtenbach, K., and Collares-Pereira, M. (2010). Plasmid profile analysis of Portuguese *Borrelia lusitaniae* strains. Ticks Tick Borne Dis *1*, 125-128. https://doi.org/10.1016/j.ttbdis.2010.07.001

Vitorino, L.R., Margos, G., Feil, E.J., Collares-Pereira, M., Ze-Ze, L., and Kurtenbach, K. (2008). Fine-scale Phylogeographic Structure of *Borrelia lusitaniae* Revealed by Multilocus Sequence Typing. PloS ONE *3*, e4002.

Vollmer, S.A., Bormane, A., Dinnis, R.E., Seelig, F., Dobson, A.D., Aanensen, D.M., James, M.C., Donaghy, M., Randolph, S.E., Feil, E.J., *et al.* (2011). Host migration impacts on the phylogeography of Lyme Borreliosis spirochaete species in Europe. Environ Microbiol *13*, 184-192. https://doi.org/10.1111/j.1462-2920.2010.02319.x

Vollmer, S.A., Feil, E.J., Chu, C.Y., Raper, S.L., Cao, W.C., Kurtenbach, K., and Margos, G. (2013). Spatial spread and demographic expansion of Lyme borreliosis spirochaetes in Eurasia. Infect Genet Evol *14C*, 147-155. https://doi.org/10.1016/j.meegid.2012.11.014

Wallich, R., Pattathu, J., Kitiratschky, V., Brenner, C., Zipfel, P.F., Brade, V., Simon, M.M., and Kraiczy, P. (2005). Identification and functional characterization of complement regulator-acquiring surface protein 1 of the Lyme disease spirochetes *Borrelia afzelii* and *Borrelia garinii*. Infect Immun *73*, 2351-2359. https://doi.org/10.1128/IAI.73.4.2351-2359.2005

Walter, K.S., Carpi, G., Caccone, A., and Diuk-Wasser, M.A. (2017). Genomic insights into the ancient spread of Lyme disease across North America. Nat Ecol Evol *1*, 1569-1576. https://doi.org/10.1038/s41559-017-0282-8

Walter, K.S., Carpi, G., Evans, B.R., Caccone, A., and Diuk-Wasser, M.A. (2016). Vectors as Epidemiological Sentinels: Patterns of Within-Tick *Borrelia burgdorferi* Diversity. PLoS Pathogens *12*, e1005759. https://doi.org/10.1371/journal.ppat.1005759

Wang, G., Ojaimi, C., Iyer, R., Saksenberg, V., McClain, S.A., Wormser, G.P., and Schwartz, I. (2001). Impact of genotypic variation of *Borrelia burgdorferi* sensu stricto on kinetics of dissemination and severity of disease in C3H/HeJ mice. Infect Immun *69*, 4303-4312. https://doi.org/10.1128/IAI.69.7.4303-4312.2001

Wang, G., Ojaimi, C., Wu, H., Saksenberg, V., Iyer, R., Liveris, D., McClain, S.A., Wormser, G.P., and Schwartz, I. (2002). Disease severity in a murine model of lyme borreliosis is associated with the genotype of the infecting *Borrelia burgdorferi* sensu stricto strain. J Infect Dis *186*, 782-791. https://doi.org/10.1086/343043

Wang, G., van Dam, A.P., and Dankert, J. (1999a). Evidence for frequent OspC gene transfer between *Borrelia valaisiana* sp. nov. and other Lyme disease spirochetes. FEMS Microbiol Lett *177*, 289-296. https://doi.org/10.1111/j.1574-6968.1999.tb13745.x

Wang, G., van Dam, A.P., Schwartz, I., and Dankert, J. (1999b). Molecular typing of *Borrelia burgdorferi* sensu lato: taxonomic, epidemiological, and clinical implications. Clin Microbiol Rev *12*, 633-653.

Wang, I.N., Dykhuizen, D.E., Qiu, W., Dunn, J.J., Bosler, E.M., and Luft, B.J. (1999c). Genetic diversity of *ospC* in a local population of *Borrelia burgdorferi* sensu stricto. Genetics *151*, 15-30.

Wang, I.N., Smith, D.L., and Young, R. (2000). Holins: the protein clocks of bacteriophage infections. Annu Rev Microbiol *54*, 799-825.

Wang, P., Lutton, A., Olesik, J., Vali, H., and Li, X. (2012). A novel iron- and copper-binding protein in the Lyme disease spirochaete. Mol Microbiol *86*, 1441-1451. https://doi.org/10.1111/mmi.12068

Weening, E.H., Parveen, N., Trzeciakowski, J.P., Leong, J.M., Hook, M., and Skare, J.T. (2008). *Borrelia burgdorferi* lacking DbpBA exhibits an early survival defect during experimental infection. Infect Immun *76*, 5694-5705. https://doi.org/IAI.00690-08 [pii] 10.1128/IAI.00690-08

Wilske, B., Preac-Mursic, V., Jauris, S., Hofmann, A., Pradel, I., Soutschek, E., Schwab, E., Will, G., and Wanner, G. (1993). Immunological and molecular polymorphisms of OspC, an immunodominant major outer surface protein of *Borrelia burgdorferi*. Infect Immun *61*, 2182-2191.

Wilske, B., Preac-Mursic, V., Schierz, G., and Busch, K.V. (1986). Immunochemical and immunological analysis of European *Borrelia burgdorferi* strains. Zentralbl Bakteriol Mikrobiol Hyg [A] *263*, 92-102.

Wormser, G.P., Brisson, D., Liveris, D., Hanincova, K., Sandigursky, S., Nowakowski, J., Nadelman, R.B., Ludin, S., and Schwartz, I. (2008). *Borrelia burgdorferi* genotype predicts the capacity for hematogenous dissemination during early Lyme disease. J Infect Dis *198*, 1358-1364.

Wormser, G.P., Liveris, D., Nowakowski, J., Nadelman, R.B., Cavaliere, L.F., McKenna, D., Holmgren, D., and Schwartz, I. (1999). Association of specific subtypes of *Borrelia burgdorferi* with hematogenous dissemination in early Lyme disease. J Infect Dis *180*, 720-725. https://doi.org/JID981202 [pii]10.1086/314922

Wywial, E., Haven, J., Casjens, S.R., Hernandez, Y.A., Singh, S., Mongodin, E.F., Fraser-Liggett, C.M., Luft, B.J., Schutzer, S.E., and Qiu, W.G. (2009). Fast, adaptive evolution at a bacterial host-resistance locus: The PFam54 gene array in *Borrelia burgdorferi*. Gene *445*, 26-37. https://doi.org/S0378-1119(09)00325-4 [pii] 10.1016/j.gene.2009.05.017

Xie, J., Zhi, H., Garrigues, R.J., Keightley, A., Garcia, B.L., and Skare, J.T. (2019). Structural determination of the complement inhibitory domain of *Borrelia burgdorferi* BBK32 provides insight into classical pathway complement evasion by Lyme disease spirochetes. PLoS Pathogens *15*, e1007659. https://doi.org/10.1371/journal.ppat.1007659

Xu, Y., Kodner, C., Coleman, L., and Johnson, R.C. (1996). Correlation of plasmids with infectivity of *Borrelia burgdorferi* sensu stricto type strain B31. Infect Immun *64*, 3870-3876.

Yang, X., Popova, T.G., Hagman, K.E., Wikel, S.K., Schoeler, G.B., Caimano, M.J., Radolf, J.D., and Norgard, M.V. (1999). Identification, characterization, and expression of three new members of the *Borrelia burgdorferi* Mlp (2.9) lipoprotein gene family. Infect Immun *67*, 6008-6018.

Yang, X.F., Hubner, A., Popova, T.G., Hagman, K.E., and Norgard, M.V. (2003). Regulation of expression of the paralogous Mlp family in *Borrelia burgdorferi*. Infect Immun *71*, 5012-5020.

Yang, X.F., Pal, U., Alani, S.M., Fikrig, E., and Norgard, M.V. (2004). Essential role for OspA/B in the life cycle of the Lyme disease spirochete. J Exp Med *199*, 641-648.

Young, R. (2002). Bacteriophage holins: deadly diversity. J Molec Microbiol Biotechnol *4*, 21-36.

Young, R., and Blasi, U. (1995). Holins: form and function in bacteriophage lysis. FEMS Microbiol Rev *17*, 191-205.

Zhang, H., and Marconi, R.T. (2005). Demonstration of cotranscription and 1-methyl-3-nitroso-nitroguanidine induction of a 30-gene operon of *Borrelia burgdorferi*: evidence that the 32-kilobase circular plasmids are prophages. J Bacteriol *187*, 7985-7995.

Zhang, H., Raji, A., Theisen, M., Hansen, P.R., and Marconi, R.T. (2005). bdrF2 of Lyme disease spirochetes is coexpressed with a series of cytoplasmic proteins and is produced specifically during early infection. J Bacteriol *187*, 175-184.

Zhang, J.R., Hardham, J.M., Barbour, A.G., and Norris, S.J. (1997). Antigenic variation in Lyme disease borreliae by promiscuous recombination of VMP-like sequence cassettes. Cell *89*, 275-285. https://doi.org/10.1016/s0092-8674(00)80206-8

Zuckert, W.R., and Barbour, A.G. (2000). Stability of *Borrelia burgdorferi bdr* loci *in vitro* and *in vivo*. Infect Immun *68*, 1727-1730.

Zuckert, W.R., and Meyer, J. (1996). Circular and linear plasmids of Lyme disease spirochetes have extensive homology: characterization of a repeated DNA element. J Bacteriol *178*, 2287-2298.

Zuckert, W.R., Meyer, J., and Barbour, A.G. (1999). Comparative analysis and immunological characterization of the *Borrelia* Bdr protein family. Infect Immun *67*, 3257-3266.

Chapter 3

Genomics of Relapsing Fever Spirochetes

Luke C. Kingry

Division of Vector-Borne Diseases, Centers for Disease Control and Prevention, USA
Corresponding author: vtx8@cdc.gov

DOI: https://doi.org/10.21775/9781913652616.03

Abstract

The relapsing fever *Borrelia* are composed of a genetically diverse array of bacterial species found in many areas of the world. All species studied to date share a complex genomic structure including a long linear chromosome, an array of linear plasmids, and in some cases circular plasmids similar to the Lyme disease *Borrelia*. While this complex genomic structure is shared with the Lyme disease *Borrelia*, the relapsing fever species possess distinct genetic elements that have been shown to be important for their unique pathogenesis as compared to the Lyme disease spirochete species. Genomic characterization of relapsing fever *Borrelia* species has greatly accelerated over the last decade. This chapter aims to discuss those advances and the role genomics will play in the understanding of this expanding group of bacterial species.

Introduction

Relapsing fever (RF) spirochetes comprise those bacterial species in the *Borrelia* genus that fall into a phylogenetically related clade associated with mammalian, reptilian, and avian borreliosis. Amongst this group are numerous pathogenic species transmitted by both soft and hard ticks as well as body lice that cause human infections throughout the world and some of which have been implicated in significant human epidemics since the medieval times (Guellil et al., 2018) (see Chapter 14). The genomics of RF spirochetes is a field behind that of the Lyme disease spirochete genomics field as the first published genome was made available in 2008, eleven years after publication of the first Lyme disease group spirochete genome (Fraser et al., 1997; Lescot et al., 2008). Table 1 provides an overview of geographic distribution, arthropod vector, and the number of genomic sequences available for the different RF species discussed in this chapter. Genomic characterization of RF species is currently occurring at a rapid pace and the goal of this chapter is to describe the state of the field according to current literature. Indeed, during the writing of this chapter whole genomes sequences continue to be added to publicly available databases that will further the genetic understanding of this group of bacteria that has long been associated with human illness.

Recently, the *Borrelia* genus was separated with the RF group retaining the *Borrelia* genus designation and those falling into the Lyme Borreliosis group being designated as the *Borreliella* genus (Adeolu and Gupta, 2014). Comparison of 38 genome sequences identified 44 unique conserved signature insertions or deletions in conserved proteins that distinguish the two genera and 25 conserved signature proteins that differentiate the two genera. This was countered by a study that used proteomic sequence comparisons derived from 30 *Borrelia* type strain genomes and 54 additional proteomes from *Brachyspira, Leptospira, Spirochaeta, and Treponema* species (Margos et al., 2018). By comparing the percentage of conserved proteins (POCP), the authors argued that there was strong genetic support for a single genus designation. This was further supported by the inclusion of two recently characterized species, *B. turcica* and *B. tachyglossi*, which displayed genomic composition that shared genetic similarities within the conserved signature proteins of both Lyme disease and RF spirochetes. Both studies highlight the importance of comparative genomics for true understanding of phylogenetic relationships that are constantly being updated due to the availability of affordable next generation sequencing. While there have been genomic sequence based arguments for and against this division of the genus, the *Borrelia* and *Borreliella* genus designations recently were formally rejected; consequently, this chapter will refer to Lyme disease (LD) spirochetes and relapsing fever (RF) spiro-

Table 1. Relapsing fever species and their geographic distribution, vector, and genomics sequences available (as of 2020).

RF *Borrelia* species	Geographic distribution	Arthropod vector	Genomes available
B. crocidurae	Northern Africa	*Ornithodoros sonrai*	3
B. duttonii	Central Africa	*Ornithodoros moubata/ Ornithodoros porcinus*	2
B. hermsii	Western USA	*Ornithodoros hermsi*	7
B. hispanica	Northern Africa, Southwestern Europe	*Ornithodoros eraticus*	1
B. miyamotoi	Northern Hemisphere	*Ixodes* spp.	21
B. parkeri	West Coast USA	*Ornithodoros parkeri*	2
B. persica	Eurasia, Middle East	*Ornithodoros tholozani*	1
B. recurrentis	Eastern Africa	*Pediculus humanus*	1
B. turicatae	Central and Southern USA	*Ornithodoros turicata*	2

chetes when making comparisons between the two (Margos et al., 2020). The phylogenetic makeup of the *Borreliaceae* continues to expand with the discovery of new species and as such, comparative genomics will be key to defining the true genetic relationships within this family. This chapter will focus on the genomics of the classic RF clade of the *Borrelia* genus and not the recently characterized genomes of the reptile and echidna associated clade within the RF group of the *Borrelia* genus (Margos et al., 2018).

Early comparisons of nucleotide sequence between the known species of RF spirochetes found a genetic distinction between "Old World" and "New World" RF spirochetes with those species causing disease in the Afrotropical and Palearctic ecological regions in the former category and those causing disease in the Nearctic in the latter category. Recent discovery of *Borrelia* species that group with the "New World" RFs have since been discovered in Zambia and Tanzania, Africa, and some "Old World" *Borrelia* have global distribution (Mitani et al., 2004; Barbour, 2005; Qiu et al., 2019).

The RF spirochetes, like the Lyme disease group spirochetes, possess one of the most segmented and complex genomes known in the bacterial domain. As the genomic era has transformed our understanding of many infectious agents, so too has it begun to transform our understanding of the *Borrelia* genus. In the last decade the amount of available sequence data and genome assemblies of RF species have expanded drastically. The number of genome assemblies publicly available has increased from four strains and four species in 2010 to 41 strains and 11 species in 2020. There is still much to be learned about the genomics of RF spirochetes, as the most recent Bergey's Manual description of RF spirochetes lists 33 potentially unique species, 22 of which do not currently have genome assemblies publicly available (Barbour et al., 2018).

All RF spirochete genomes studied to date possess a linear chromosome of approximately 900 kbp, a large linear plasmid of approximately 72 to 200 kbp, an array of linear plasmids ranging from 5 to 30 kbp, with some species possessing circular plasmids ranging from 5 to 40 kbp. Each extrachromosomal replicon encodes distinct partitioning genes responsible for plasmid partitioning during growth and cell-division; these RF spirochete replicons have considerable stability unlike the Lyme disease spirochetes that readily lose several of extra-chromosomal replicons during laboratory passage (Lopez et al., 2008). Encoded throughout the linear plasmids are archived genes encoding variable large and small proteins (Vlp and Vsp). These variable gene sequences are periodically recombined to an expression site where they are responsible for the antigenic phase variations that coincide with the tell-tale febrile relapses during human infection (Dai et al., 2006) (see Chapter 14).

Description of the RF Spirochete chromosome

RF spirochete chromosomes sequenced to date range from 906 kbp to 935 kbp and have an average GC content of 27-30% (Barbour, 2014). GC skew analysis suggests an origin of replication at the center of the chromosome with most genes transcribed away from the center of the chromosome (Lescot et al., 2008). The 5′ and 3′ ends of the chromosome and linear plasmids are covalently closed with hairpin telomeres (Kitten and Barbour, 1990). The chromosome encodes a single copy of 16S, 23S and 5S ribosomal RNA genes in contrast to the Lyme disease spirochete chromosomes that have two copies of 23S and 5S rRNAs (Fraser et al., 1997; Lescot et al., 2008). RF spirochete species with complete chromosome sequences reliably encode 32 transfer RNAs, although *B. miyamotoi* encodes 31 tRNAs, with predicted specificity for all 20 amino acids (Hue et al., 2013). *B. hermsii* has been shown to carry multiple copies of its chromosome in a single cell in both diffuse and discrete nucleoid forms distributed throughout the length of the cell (Kitten and Barbour, 1992). RF *Borrelia* species that have whole chromosome assemblies share highly colinear chromosomes with very few whole gene differences (Elbir et al., 2014a) and average nucleotide identity ranging from 82 to 99% (Table 2). This overall chromosomal synteny is largely shared with Lyme disease spirochete chromosomes with notable differences. Unique chromosomal genes in RF spirochetes include three genes involved in purine salvage, *hpt, purA*, and *purB* (mentioned below) and two genes involved in glycerol biosynthesis, *glpQ* and *glpT* (Fraser et al., 1997; Schwan et al., 2003; Guyard et al., 2006; Pettersson et al., 2007; Lescot et al., 2008). Coding density in RF spirochete chromosomes ranges from 91.5-94.7%.

As has been described for Lyme disease spirochetes, the RF spirochetes chromosome encodes limited metabolic capacity with few exceptions that are attributed to the differing pathogenicity and tissue tropism of the RF spirochetes. The limited metabolic capacity is likely a product of its evolution through complex interactions with vertebrate hosts and arthropod vectors that have resulted in both parasitic and mutualistic outcomes depending on the host or vector (Samaddar et al., 2020). Around 30% of the RF spirochete chromosome encodes genes of unknown function. Translation, cell wall and membrane biogenesis, and cell motility are the most abundant functional categories encoded in the chromosome as predicted by the clusters of orthologous groups (COG) database (Elbir et al., 2014a).

Metabolic differences between Lyme disease and RF spirochetes

One of the primary phenotypic differences between the RF and Lyme disease group species is the ability to reach high spirochete numbers in the blood, which has been attributed in part to genomic differences, including intact metabolic pathways for purine salvage and glycerol biosynthesis. Notable, exceptions include the LD spirochete *B. mayonii* and the hard tick-transmitted RF species *B. miyamotoi* (Molloy et al., 2015; Pritt et al., 2016). Pettersson et al. found three chromosomal (*hpt, purA, purB*) and three plasmid-borne (*nrdIEF*) genes involved in purine salvage that are not present in Lyme disease spirochetes (Pettersson et al., 2007). They tested the ability of RF versus Lyme disease spirochetes to salvage purines and found RF spirochetes to be more efficient at purine salvage than Lyme disease spirochetes for several biologically relevant purines. They also observed a difference in purine salvage efficiency between the two genetic groups of *B. hermsii* GGI and GGII (discussed below). The GGII strain salvaged purines more efficiently than GGI, which was genetically attributed to an intact adenine deaminase in the GGII strain that was disrupted in the GGI strain (Pettersson et al., 2007). All RF strains tested appeared to most efficiently salvage and incorporate hypoxanthine, which is the most abundant purine in mammalian blood. The close association of RF spirochetes with red blood cells in infected mice and humans has been theorized to be a mechanism to salvage hypoxanthine released from red blood cells (Pettersson et al., 2007). Another genetic feature thought to be responsible for RF spirochetes to reach high bacterial numbers in mammalian blood is a completely intact glycerol metabolism pathway that

Table 2. Average nucleotide identity between chromosome sequences of relapsing fever spirochetes.

	B. hispanica	B. persica	B. anserina	B. coriaceae	B. crocidurae	B. duttonii	B. hermsii_DAH	B. miyamotoi_CA17-2241	B. miyamotoi_Fr64b	B. miyamotoi_Izh5	B. miyamotoi_LB-2001	B. parkeri	B. recurrentis
B. hispanica													
B. persica	87.54												
B. anserina	82.74	82.61											
B. coriaceae	83.01	82.92	84.91										
B. crocidurae	95.97	87.24	82.54	83.1									
B. duttonii	95.97	87.32	82.67	83.13	98.84								
B. hermsii_DAH	83.85	83.85	86.8	87.1	83.76	83.85							
B. miyamotoi_CA17-2241	82.26	82.1	84.15	84.01	82.28	82.26	85.64						
B. miyamotoi_Fr64b	82.33	82.13	84.21	84.12	82.31	82.21	85.84	97.42					
B. miyamotoi_Izh5	82.1	82.1	84.16	84.1	82.26	82.32	85.68	97.44	99.98				
B. miyamotoi_LB-2001	82.3	82.13	84.07	84.01	82.25	82.21	85.57	98.56	97.51	97.53			
B. parkeri	83.96	83.85	86.63	87.8	83.8	83.81	89.64	85.5	85.57	85.6	85.52		
B. recurrentis	95.88	87.38	82.61	83	98.66	99.35	83.76	82.24	82.34	82.34	82.22	83.77	
B. turicatae	83.97	83.81	86.67	87.92	83.77	83.79	89.72	85.51	85.69	85.61	85.59	97.36	83.86

includes the previously mentioned glycerophosphodiester phosphodiesterase (*glpQ*) and glycerol 3-phosphate transporter (*glpT*) absent in Lyme disease group spirochetes. Both gene products have been shown to allow RF spirochetes to efficiently hydrolyze deacylated phospholipids and transport G3P across the cytoplasmic membrane (Schwan et al., 2003). Finally, the thymidylate synthase (*thyX*) gene encoded adjacent to the *nrdIEF* gene cluster on the large linear plasmid has been shown to allow RF spirochetes to be significantly more efficient for *de novo* pyrimidine biosynthesis than the *thyX* encoded in the Lyme disease spirochete genomes, which carries a substitution of a conserved amino acid (Zhong et al., 2006). Rapid growth requires readily available nucleic acids and recent plasmid assemblies have also identified multiple *thyX* loci encoded throughout the plasmidome of *B. miyamotoi* and *B. turicatae*, perhaps also responsible for more efficient *de novo* pyrimidine biosynthesis (Kingry et al., 2016; Kingry et al., 2017b).

Plasmids

In contrast to the chromosome, the RF *Borreliae* plasmid sequences encode very few genes that have orthologs outside of the *Borrelia* genus (~10%). The plasmids encode metabolic, structural, regulatory, and virulence related proteins. The archived Vsp/Vlp proteins for antigenic switching encoded throughout multiple linear plasmids are unique amongst all bacteria that are known to use antigen variation for host immune evasion. All sequenced strains of RF *Borreliae* have at least one plasmid sequence available and most if not all assemblies from individual strains could be considered draft assemblies as it is very difficult to know that all extrachromosomal DNA is represented and properly assembled due to redundant sequence and similar plasmid sizes that make confirmation of the exact number of molecules nearly impossible by traditional methods (i.e., Southern blot). Nomenclature for naming plasmids is also not standardized in part owing to the vast number of different sized molecules and non-conserved nature of their coding within and across species. Many assemblies simply name plasmids lp (linear plasmid) and cp (circular plasmid) with size in kilobases (i.e., lp165 and cp28), although this nomenclature is not universally used. Like Lyme disease group spirochetes, each plasmid has unique plasmid partitioning genes (Pfam 32, 49, 50, and 57) commonly grouped together on a given plasmid, likely for proper plasmid partitioning during replication. Examples of plasmid fusion events have been observed and recombination amongst linear plasmids is thought to be relatively common.

Large linear plasmid

The large linear plasmid of RF spirochetes varies widely in size from ~72 kbp in *B. miyamotoi* to ~180 kbp in *B. hermsii*. This plasmid is characterized by the *thyX* and *nrdIEF* open reading frames. In many species the large linear plasmid has numerous colinear repeat regions with the exception of *B. miyamotoi* (Miller et al., 2013; Kingry et al., 2017a; Kingry et al., 2017b). The absence of repetitive sequences in *B. miyamotoi* may contribute to the smaller size of the large linear plasmid. Genetic comparisons of the large linear plasmids from RF spirochetes and plasmid sequence from Lyme disease group spirochetes suggests there are orthologous sequence blocks corresponding to regions of lp54 and cp26 of *B. burgdorferi*, both of which appear to encode essential virulence loci and have been identified in all Lyme disease group species (Miller et al., 2013). Only 5-10% of the large linear plasmid open reading frames encode for genes that share homology with bacterial species outside of *Borrelia* species (Miller et al., 2013). Two open reading frames on the large linear plasmid encode proteins that have sequence homology and predicted structure similar to the CRASP family proteins in Lyme disease group spirochetes that have been shown to be important for complement resistance (Miller et al., 2013). Genes unique to the RF spirochetes on the large linear plasmid include the *Borrelia* immunogenic protein A (*bipA*), a factor H binding protein (*fhbA*) associated with complement resistance, and an arthropod associated lipoprotein (*alp*) (Hovis et al., 2004; Lopez et al., 2010; Marcsisin et al., 2012). Species-specific differences in these proteins may be interesting candidates to understand differences in vector/host specificity as well as virulence differences in humans amongst the RF species.

Other linear plasmids

Each RF species that has plasmid assemblies available also possesses a linear plasmid with orthologous sequence to the cp26 plasmid of the Lyme disease group spirochetes that encodes the variable tick protein (*vtp*), an *ospC* ortholog, and the telomere resolvase gene, *resT*. The *vtp* gene product has been shown to be important for persistence in the tick vector and for tick transmission (Raffel et al., 2014). The *resT* gene product is essential for resolving the covalently closed telomeric ends of the chromosome and linear plasmids during cell division; this has only been shown in Lyme disease group spirochetes but it is thought to occur in RF spirochete species considering the conservation of the gene and existence of linear molecules with hairpin telomeres in both genera (Chaconas, 2005). Another common feature amongst the linear RF spirochete plasmids includes multiple copies of genetically distinct forms of the variable major protein (VMP) family consisting of variable large and small proteins (Vlp and Vsp) encoded on numerous different linear replicons. Despite the multiple copies of the gene throughout the linear plasmids, only a single copy of *vsp/vlp* is expressed at a time while the other copies exist in "archived" form. *B hermsii* has been shown to possess a discrete expression site at the 3′ terminus of a 27-kbp linear plasmid where the active *vlp/vsp* copy is expressed. The surface exposed antigenic proteins are altered by recombination between the expression site and one of the archived copies of the gene throughout the multiple linear plasmids (Dai et al., 2006). The change to the antigenic surface of the bacterial cell results in the distinctive relapsing febrile episodes during infection as the bacteria evades the immune response. The *vlp/vsp* expression site *B. miyamotoi* is thought to be in the center of a 41-kbp "virulence-associated" plasmid (Kuleshov et al., 2020). More studies are required to understand the species-specific mechanisms of surface antigen variation that is characteristic of RF spirochetes as this mechanism has primarily been characterized in *B. hermsii*.

Circular plasmids

In contrast to the limited gene similarities between the linear plasmids of RF and Lyme disease group spirochetes, circular plasmids identified in RF are highly syntenic with the cp32 circular plasmids identified in Lyme disease spirochetes (Casjens et al., 2018). Analysis of the gene content present on the circular plasmids suggests these replicons represent prophage genomes that encode surface exposed proteins that decorate the spirochete outer membrane (Eggers and Samuels, 1999). Sequence comparison of the *B. burgdorferi* cp32 plasmids and the circular plasmids of *B. hermsii* showed homologs of the antigenic major lipoprotein L and K (*mlpL* and *mlpK*) genes, the *Borrelia* direct repeat A and B (*bdrA* and *bdrB*) genes, and the phage holin system encoding *blyA* and *blyB* genes (Damman et al., 2000; Stevenson et al., 2000). Based on the analysis of *B. hermsii* circular plasmid sequence, the notable difference in RF and Lyme disease spirochete circular plasmids appears to be the absence of *ospE, ospF,* and *elp* genes found on the cp32 circular plasmids of Lyme disease spirochetes (see chapter 2). Genes identified in *B. hermsii* circular plasmids not previously observed in Lyme disease spirochetes included a single strand DNA binding protein (*ssb*), a nucleotide methylase, and a protein of unknown function. Both *ssb* and DNA methylase genes have been associated with phage packaging and segregation. Indeed, early culture work with *B. hermsii* suggested the presence of phage-like particles in the culture supernatant (Barbour and Hayes, 1986). Evidence for circular plasmids has not been observed *B. anserina* and *B. parkeri,* although low-stringency hybridization probes suggested similar gene content to the circular plasmids may exist in linear form in *B. parkeri* (Stevenson et al., 2000).

Comparative genomics

Intraspecies genetic diversity

Old World RF *Borrelia* (*B. duttonii, B. recurrentis, B. hispanica,* and *B. crocidurae*) possess a high degree of relatedness based on whole chromosome nucleotide identity with higher ANI between the individual species than in usually considered for species designation (<95%). This is also true for New World RF *Borrelia* (*B. hermsii, B. turicatae*, and *B. parkeri*). However, differences in host reservoir, tick vector, and geographic distribution have been generally acceptable for species differentiation in the RF *Borrelia* field. Whole genome comparisons of the Old World species *B. recurrentis* and *B. duttonii* determined that *B. recurrentis* is a sub-strain of *B. duttonii* in an advanced state of genomic decay. *B. recurrentis* was found to have lost crucial DNA repair genes (*recA and mutS*) allowing for gene loss and niche restriction as a louse-borne rather than tick-borne pathogen with its only known reservoir being humans, unlike other tick-borne RF spirochetes that exist in enzootic cycles with humans being terminal hosts (Elbir et al., 2014a).

B. persica has the lowest average chromosomal nucleotide identity to other species of RF spirochetes of the species that have whole chromosome sequences available (Elbir et al., 2014b) (Table 2). Molecular characterization of strains using *flaB*, *glpQ*, and IGS amplicon sequences from ticks, cats, dogs, and human blood specimens collected in Israel and the Palestinian Authority have demonstrated limited genetic diversity between strains collected from distinct geographical areas and over a period of years (Safdie et al., 2010; Baneth et al., 2016).

Concatenated sequence comparisons showed a clear distinction between *B. persica* and the African RF species. Sequence of the 16S rRNA gene found that strains of *B. persica* can possess an A to G polymorphism at position 1443 that was previously a difference only observed between Old and New World RF spirochete species (Safdie et al., 2010).

A genomic comparison of five RF species found that the New World species *B. hermsii* and *B. turicatae* possess colinear chromosomal structure while Old World species *B. duttonii, B. recurrentis,* and *B. crocidurae* have an 1,134-bp insertional sequence element in all three species or a 414-bp transposase only in *B. crocidurae* that disrupts their colinearity with North American species (Elbir et al., 2014a). Further disruptions in colinearity for Old World *Borrelia* include a 5-kbp duplication in the *B. duttonii* chromosome and a duplication of the genes *rpsU, ftsK*, and *bacA* in *B. recurrentis*. Two genes, encoding RNA polymerase σ^{54} and antigen P35-like protein, were found to be absent from New World RF species and one gene, *radA*, was absent in the Old World *Borrelia* species examined. Overall, the 5′ portion of the Old World *Borrelia* species was found to be more variable than the same region in New World *Borrelia* species (Elbir et al., 2014a). To date, no functional differences have been determined due to these differences between Old and New World RF species.

Interspecies genetic diversity

RF spirochetes species are generally thought to be more genetically stable than the Lyme disease group spirochetes with very low interspecies diversity. Whole genome comparisons of multiple strains from the same RF spirochete species are limited due, in part, to culture not routinely being utilized for RF diagnostics. Much of what is currently known about interspecies diversity of the RF spirochetes has relied on culture-free approaches for genetic characterization. Multi-locus sequence analysis (MLSA) can be a powerful approach in the absence of culture isolates and whole genome sequence data to infer phylogenetic relationships and has proven to be an efficacious tool for both the RF and Lyme disease group spirochetes. Using concatenated sequence from multiple PCR amplicon targets and performing multi-sequence alignments can provide phylogenetic comparisons between and within bacterial species. Two main MLSA schemes have been utilized to characterize RF species designation and genetic diversity. Gene sequence for 16S rRNA, *flaB, gyrB,* IGS (portions of sequence between the 23S and 16S rRNAs), and *glpQ* have traditionally been used to place a given species in the RF group, as there is the most publicly available sequence data for these genes. An eight-gene MLST approach targeting the housekeeping genes *clpA, clpX, pepX, pyrG, nifS, recG, rplB*, and *uvrA* has been shown to be a robust approach for differentiating RF and Lyme disease species but may be less than ideal for determining interspecies diversity (Margos et al., 2008). Studies utilizing the eight housekeeping MLST approach have had trouble inferring interspecies population structures for both *B. recurrentis* and *B. miyamotoi*, with these species showing a higher degree of stability within these genes than the Lyme disease group spirochetes (Marosevic et al., 2017; Kingry et al., 2018).

B. recurrentis genomic analyses confirmed earlier work based on MLSA comparisons showing low genetic variability amongst *B. recurrentis* strains across their whole genome sequences (Elbir et al., 2012). Two archived clinical isolates from 1985 and 2004 and four clinical isolates from 2015 were sequenced and the reads mapped to the original reference sequence *B. recurrentis* A1 (Marosevic et al., 2017). The four isolates from 2015 possessed the same 17 chromosomal SNPs compared to the A1 reference sequence. The archived isolates from 1985 and 2004 possessed 12 and 16 chromosomal SNPs, respectively, compared to the A1 reference suggesting a high degree of genomic conservation over time for *B. recurrentis*. Six chromosomal SNPs were shared among all six isolates. Mapping sequence reads to the seven known plasmid sequences from the A1 reference showed that all six isolates possessed at least the same number of plasmids as the reference strain with only 17-21 total SNPs identified, the four 2015 isolates did not share the same SNP pattern in the plasmid sequence as they did for the chromosome sequence. The *B. recurrentis* A1 reference sequence for the large linear plasmid was 124 kbp in length; however, *de novo* assembly of the large linear plasmid for all six clinical isolates showed a large linear plasmid of 165 kbp similar to both *B. duttoni* and *B. crocidurae*. In addition, a 6-kbp linear plasmid identified in the A1 reference was missing 1 kbp at the 3′ end for all six isolates. Whether these differences are due to errors in the original reference sequence, the result of laboratory culture adaptation, or true structural differences in *B. recurrentis* is unknown. The 7-kbp plasmid was also identified in *B. crocidurae* but is not represented in the *B. duttonii* reference sequence. The low genetic variability identified in this study also highlighted that the two traditional MLSA approaches would not have differentiated any of the six isolates from each other or the reference. In the event of *B. recurrentis* outbreak/epidemic investigations, whole genome sequencing would likely be required to implicate a common source of infection. This highlights the utility of whole genome sequencing for genetically stable RF species to understand genetic variability and identify potential strain-specific markers that can be targeted for diagnostic/strain typing purposes.

Early comparisons of *B. hermsii* strains using MLSA analysis of 16S rRNA, *glpQ, gyrB, flaB*, and IGS sequences determined that there are two main genomic groups, GGI and GGII (Porcella et al., 2005; Schwan et al., 2007). Further MLSA analysis of additional *B. hermsii* isolates identified two subgroups within GGI, GGIa and GGIb (Fischer et al., 2009; Johnson et al., 2016). Whole chromosome sequence comparisons of 21 GGI isolates and 28 GGII isolates found that both *B. hermsii* genomic groups displayed low genetic diversity across the full chromosome sequence detecting 3,425 and 1,833 SNPs for GGI and GGII, respectively. Due to the higher degree genetic diversity found in *B. hermsii* GGI, it is thought that GGI is the ancestral lineage of *B. hermsii* in North America (Johnson, 2012). This is the most comprehensive genomic comparison amongst the RF species.

Using the eight housekeeping gene scheme to type *B. miyamotoi* strains worldwide appears to differentiate *B. miyamotoi* strains by the *Ixodes* tick vector that they have been identified in (Cook et al., 2016). Whole chromosome sequencing confirmed MLSA analysis and showed *B. miyamotoi* strains transmitted by the same tick share 99.9% nucleotide identity, whereas strains transmitted by different tick species share 97-98% nucleotide identity (Hue et al., 2013; Kingry et al., 2017a; Kingry et al., 2017b; Kuleshov et al., 2018; Kuleshov et al., 2019). Understanding *B. miyamotoi* sequence diversity will likely require whole genome sequence comparisons to identify the best approach for differentiating individual strains. The vector specificity of *B. miyamotoi* genotypes presents a unique opportunity for genome-wide comparisons to identify the genetic basis for vector-specific adaptations within a single species. Although only recognized as a human pathogen since 2011, there are currently 21 whole genome assemblies (many with plasmid assemblies) publicly available for *B. miyamotoi*, representing the most whole genome sequence data for any given RF species.

Genetic variation of several RF spirochete species has only been explored using MLSA approaches rather than whole genome comparisons to date. Many of these species have only one or two sequenced genomes limiting the utility of whole genome comparisons currently. Current understanding of species diversity with limited whole genome sequence are summarized in Table 3.

Future directions in the genomics of RF spirochetes

Advances in culture methods and continued discovery of *Borrelia* species will undoubtedly change our understanding of the genetic diversity and comparative genomics of the RF spirochetes. Available genome sequence data are largely focused on those species that cause human disease but this does not diminish the importance of generating a more complete understanding of RF spirochete genomics from those species that either do not cause human illness or have yet to be associated with human illness. In addition, culture-free methods for genetic characterization of bacterial species will continue to expand our understanding of the diversity of the *Borrelia* genus. Whole genome amplification and bait and capture sequencing approaches may one day prove to be a straightforward way to obtain whole genome data without the culture step that requires expertise and highly specific media formulations for *Borrelia*. A thorough genetic understanding of a bacterial species is required for the development of accurate molecular assays for diagnosis and a comprehensive description of the molecular epidemiology of a species. Less work has been done to understand the RF *Borrelia* plasmid repertoire as compared to those of Lyme disease group spirochetes. Due to the repetitive nature of *Borrelia* plasmid sequences, accurate assembly remains difficult. Advances in long-read sequencing in the last decade has increased our understanding of *Borrelia* plasmids greatly and will likely continue to expand our ability to achieve a truly complete *Borrelia* genome.

Questions remaining to be answered through comparative genomics

What is the genetic basis for vector preference: individual soft tick species, hard tick versus soft

Table 3. Summary of nucleotide comparisons of relapsing fever spirochete species with limited genomic sequence available.

Species	Genetic Content Analyzed	Observation	References
B. hispanica	16S, *flaB*, *glpQ*, *groEL*, *p66*, *recC*	99.4 % nucleotide identity between 3 strains from Spain and 1 strain from Morocco	(Sarih et al., 2009; Toledo et al., 2010)
B. crocidurae	5 intragenic loci	98.4-99.1 pairwise identity between 5 loci amplified from 35 *B. crocidurae* infected ticks and humans from Mali, Mauritania, Morocco, and Senegal	(Elbir et al., 2015)
B. crocidurae	16S, IGS, *flaB*, *glpQ*	99.5% nucleotide identity between 6 strains from Mali	(Schwan et al., 2012)
B. duttoni	5 intragenic loci	98-100% pairwise nucleotide identity between 5 loci amplified 18 *B. duttoni* infected humans from Tanzania	(Elbir et al., 2012)
B. parkeri	16S, *flaB*, *glpQ*, *gyrB*	99.8% nucleotide identity between 6 tick isolates from CA, USA	(Schwan et al., 2005)
B. turicatae	16S, *flaB*, *glpQ*, *gyrB*	99.9% nucleotide identity between 8 tick isolates from TX, KS, FL, USA	(Schwan et al., 2005)
B. persica	16S, IGS, *flaB*, *glpQ*, *purA*, *purB*	>99% identity amongst 3 distinct genotypes from several studies from Eurasia and the Middle East	(Safdie et al., 2010; Baneth et al., 2016; Kleinerman et al., 2018)

tick, and tick versus body louse? Genomic sequence data of other RF spirochetes that are transmitted by hard ticks, i.e., *B. lonestari* and *B. theileri*, will help address this question.

What is the genetic basis for alternative tissue tropism in the tick vector between the RF and Lyme disease spirochetes: tick salivary glands versus midgut?

What is the genetic basis for transovarial transmission of the RF spirochetes, which does not occur in Lyme disease spirochetes?

Understanding these gaps in our knowledge of RF spirochetes would greatly benefit from population level comparative genomic studies using a large number of sequenced strains for genome-wide association studies that are more capable of differentiating between population diversity and diversity that can actually be linked to phenotypic differences, including host/vector preference and disease manifestation in humans (i.e. *Borrelia miyamotoi* disease versus classic relapsing fever).

Disclaimer

The views expressed are those of the author alone and do not represent the Centers for Disease Control and Prevention.

References

Adeolu, M., and Gupta, R.S. (2014). A phylogenomic and molecular marker based proposal for the division of the genus *Borrelia* into two genera: the emended genus *Borrelia* containing only the members of the relapsing fever *Borrelia*, and the genus *Borreliella* gen. nov. containing the members of the Lyme disease *Borrelia* (*Borrelia* burgdorferi sensu lato complex). Antonie Van Leeuwenhoek *105*, 1049-1072. https://doi.org/10.1007/s10482-014-0164-x

Baneth, G., Nachum-Biala, Y., Halperin, T., Hershko, Y., Kleinerman, G., Anug, Y., Abdeen, Z., Lavy, E., Aroch, I., and Straubinger, R.K. (2016). *Borrelia persica* infection in dogs and cats: clinical manifestations, clinicopathological findings and genetic characterization. Parasit Vectors *9*, 244. https://doi.org/10.1186/s13071-016-1530-5

Barbour, A.G. (2005). Relapsing Fever. In Tick-Borne Diseases of Humans, J.L. Goodman, ed. (Washington, D.C.: ASM Press).

Barbour, A.G. (2014). Phylogeny of a relapsing fever *Borrelia* species transmitted by the hard tick *Ixodes scapularis*. Infect Genet Evol *27*, 551-558. https://doi.org/10.1016/j.meegid.2014.04.022

Barbour, A.G., and Hayes, S.F. (1986). Biology of *Borrelia* species. Microbiol Rev *50*, 381-400.

Barbour, A.G., Schwan, T.G.J.B.s.M.o.S.o.A., and Bacteria (2018). *Borrelia*. 1-22.

Casjens, S.R., Di, L., Akther, S., Mongodin, E.F., Luft, B.J., Schutzer, S.E., Fraser, C.M., and Qiu, W.G. (2018). Primordial origin and diversification of plasmids in Lyme disease agent bacteria. BMC Genomics *19*, 218. https://doi.org/10.1186/s12864-018-4597-x

Chaconas, G. (2005). Hairpin telomeres and genome plasticity in *Borrelia*: all mixed up in the end. Mol Microbiol *58*, 625-635. https://doi.org/10.1111/j.1365-2958.2005.04872.x

Cook, V.J., Fedorova, N., Macdonald, W.P., Lane, R.S., and Barbour, A.G. (2016). Unique Strain of *Borrelia* miyamotoi in *Ixodes pacificus* Ticks, California, USA. Emerg Infect Dis *22*, 2205-2207. https://doi.org/10.3201/eid2212.152046

Dai, Q., Restrepo, B.I., Porcella, S.F., Raffel, S.J., Schwan, T.G., and Barbour, A.G. (2006). Antigenic variation by *Borrelia hermsii* occurs through recombination between extragenic repetitive elements on linear plasmids. Mol Microbiol *60*, 1329-1343. https://doi.org/10.1111/j.1365-2958.2006.05177.x

Damman, C.J., Eggers, C.H., Samuels, D.S., and Oliver, D.B. (2000). Characterization of *Borrelia burgdorferi* BlyA and BlyB proteins: a prophage-encoded holin-like system. J Bacteriol *182*, 6791-6797. https://doi.org/10.1128/jb.182.23.6791-6797.2000

Eggers, C.H., and Samuels, D.S. (1999). Molecular evidence for a new bacteriophage of *Borrelia burgdorferi*. J Bacteriol *181*, 7308-7313. https://doi.org/10.1128/JB.181.23.7308-7313.1999

Elbir, H., Abi-Rached, L., Pontarotti, P., Yoosuf, N., and Drancourt, M. (2014a). African relapsing Fever *Borreliae* genomospecies revealed by comparative genomics. Front Public Health *2*, 43. https://doi.org/10.3389/fpubh.2014.00043

Elbir, H., FotsoFotso, A., Diatta, G., Trape, J.F., Arnathau, C., Renaud, F., and Durand, P. (2015). Ubiquitous bacteria *Borrelia crocidurae* in Western African ticks *Ornithodoros sonrai*. Parasit Vectors *8*, 477. https://doi.org/10.1186/s13071-015-1089-6

Elbir, H., Gimenez, G., Sokhna, C., Bilcha, K.D., Ali, J., Barker, S.C., Cutler, S.J., Raoult, D., and Drancourt, M. (2012). Multispacer sequence typing relapsing fever *Borreliae* in Africa. PLoS Negl Trop Dis *6*, e1652. https://doi.org/10.1371/journal.pntd.0001652

Elbir, H., Larsson, P., Normark, J., Upreti, M., Korenberg, E., Larsson, C., and Bergstrom, S. (2014b). Genome Sequence of the Asiatic Species *Borrelia persica*. Genome Announc 2. https://doi.org/10.1128/genomeA.01127-13

Fischer, R.J., Johnson, T.L., Raffel, S.J., and Schwan, T.G. (2009). Identical strains of *Borrelia hermsii* in mammal and bird. Emerg Infect Dis *15*, 2064-2066. https://doi.org/10.3201/eid1512.090792

Fraser, C.M., Casjens, S., Huang, W.M., Sutton, G.G., Clayton, R., Lathigra, R., White, O., Ketchum, K.A., Dodson, R., Hickey, E.K., *et al.* (1997). Genomic sequence of a Lyme disease spirochaete, *Borrelia burgdorferi*. Nature *390*, 580-586. https://doi.org/10.1038/37551

Guellil, M., Kersten, O., Namouchi, A., Bauer, E.L., Derrick, M., Jensen, A.O., Stenseth, N.C., and Bramanti, B. (2018). Genomic blueprint of a relapsing fever pathogen in 15th century Scandinavia. Proc Natl Acad Sci U S A *115*, 10422-10427. https://doi.org/10.1073/pnas.1807266115

Guyard, C., Battisti, J.M., Raffel, S.J., Schrumpf, M.E., Whitney, A.R., Krum, J.G., Porcella, S.F., Rosa, P.A., DeLeo, F.R., and Schwan, T.G. (2006). Relapsing fever spirochaetes produce a serine protease that provides resistance to oxidative stress and killing by neutrophils. Mol Microbiol *60*, 710-722. https://doi.org/10.1111/j.1365-2958.2006.05122.x

Hovis, K.M., McDowell, J.V., Griffin, L., and Marconi, R.T. (2004). Identification and characterization of a linear-plasmid-encoded factor H-binding protein (FhbA) of the relapsing fever spirochete *Borrelia hermsii*. J Bacteriol *186*, 2612-2618. https://doi.org/10.1128/jb.186.9.2612-2618.2004

Hue, F., Ghalyanchi Langeroudi, A., and Barbour, A.G. (2013). Chromosome Sequence of *Borrelia miyamotoi*, an Uncultivable Tick-Borne Agent of Human Infection. Genome Announc *1*. https://doi.org/10.1128/genomeA.00713-13

Johnson, T.L. (2012). The ecology of tick-borne relapsing fever in western North America.

Johnson, T.L., Fischer, R.J., Raffel, S.J., and Schwan, T.G. (2016). Host associations and genomic diversity of *Borrelia hermsii* in an endemic focus of tick-borne relapsing fever in western North America. Parasit Vectors *9*, 575. https://doi.org/10.1186/s13071-016-1863-0

Kingry, L.C., Anacker, M., Pritt, B., Bjork, J., Respicio-Kingry, L., Liu, G., Sheldon, S., Boxrud, D., Strain, A., Oatman, S., *et al.* (2018). Surveillance for and Discovery of *Borrelia* Species in US Patients Suspected of Tickborne Illness. Clin Infect Dis *66*, 1864-1871. https://doi.org/10.1093/cid/cix1107

Kingry, L.C., Batra, D., Replogle, A., Sexton, C., Rowe, L., Stermole, B.M., Christensen, A.M., and Schriefer, M.E. (2016). Chromosome and Linear Plasmid Sequences of a 2015 Human Isolate of the Tick-Borne Relapsing Fever Spirochete, *Borrelia turicatae*. Genome Announc *4*. https://doi.org/10.1128/genomeA.00655-16

Kingry, L.C., Replogle, A., Batra, D., Rowe, L.A., Sexton, C., Dolan, M., Connally, N., Petersen, J.M., and Schriefer, M.E. (2017a). Toward a Complete North American *Borrelia miyamotoi* Genome. Genome Announc *5*. https://doi.org/10.1128/genomeA.01557-16

Kingry, L.C., Replogle, A., Dolan, M., Sexton, C., Padgett, K.A., and Schriefer, M.E. (2017b). Chromosome and Large Linear Plasmid Sequences of a *Borrelia miyamotoi* Strain Isolated from *Ixodes pacificus* Ticks from California. Genome Announc *5*. https://doi.org/10.1128/genomeA.00960-17

Kitten, T., and Barbour, A.G. (1990). Juxtaposition of expressed variable antigen genes with a conserved telomere in the bacterium *Borrelia hermsii*. Proc Natl Acad Sci U S A *87*, 6077-6081. https://doi.org/10.1073/pnas.87.16.6077

Kitten, T., and Barbour, A.G. (1992). The relapsing fever agent *Borrelia hermsii* has multiple copies of its chromosome and linear plasmids. Genetics *132*, 311-324.

Kleinerman, G., King, R., Nachum-Biala, Y., and Baneth, G. (2018). *Borrelia persica* infection in rock hyraxes. Ticks Tick Borne Dis *9*, 382-388. https://doi.org/10.1016/j.ttbdis.2017.12.003

Kuleshov, K.V., Hoornstra, D., Sprong, H., Platonov, A.E., and Hovius, J.W. (2019). Draft Whole-Genome Sequences of Two Western European *Borrelia miyamotoi* Isolates. Microbiol Resour Announc *8*. https://doi.org/10.1128/MRA.01314-19

Kuleshov, K.V., Koetsveld, J., Goptar, I.A., Markelov, M.L., Kolyasnikova, N.M., Sarksyan, D.S., Toporkova, M.G., Kirdyashkina, N.P., Shipulin, G.A., Hovius, J.W., *et al.* (2018). Whole-Genome Sequencing of Six *Borrelia miyamotoi* Clinical Strains Isolated in Russia. Genome Announc *6*. https://doi.org/10.1128/genomeA.01424-17

Kuleshov, K.V., Margos, G., Fingerle, V., Koetsveld, J., Goptar, I.A., Markelov, M.L., Kolyasnikova, N.M., Sarksyan, D.S., Kirdyashkina, N.P., Shipulin, G.A., *et al.* (2020). Whole genome sequencing of *Borrelia miyamotoi* isolate Izh-4: reference for a complex bacterial genome. BMC Genomics *21*, 16. https://doi.org/10.1186/s12864-019-6388-4

Lescot, M., Audic, S., Robert, C., Nguyen, T.T., Blanc, G., Cutler, S.J., Wincker, P., Couloux, A., Claverie, J.M., Raoult, D., *et al.* (2008). The genome of *Borrelia* recurrentis, the agent of deadly louse-borne relapsing fever, is a degraded subset of tick-borne *Borrelia duttonii*. PLoS Genet *4*, e1000185. https://doi.org/10.1371/journal.pgen.1000185

Lopez, J.E., Schrumpf, M.E., Nagarajan, V., Raffel, S.J., McCoy, B.N., and Schwan, T.G. (2010). A novel surface antigen of relapsing fever spirochetes can discriminate between relapsing fever and Lyme borreliosis. Clin Vaccine Immunol *17*, 564-571. https://doi.org/10.1128/CVI.00518-09

Lopez, J.E., Schrumpf, M.E., Raffel, S.J., Policastro, P.F., Porcella, S.F., and Schwan, T.G. (2008). Relapsing fever spirochetes retain infectivity after prolonged in vitro cultivation. Vector Borne Zoonotic Dis *8*, 813-820. https://doi.org/10.1089/vbz.2008.0033

Marcsisin, R.A., Campeau, S.A., Lopez, J.E., and Barbour, A.G. (2012). Alp, an arthropod-associated outer membrane protein of *Borrelia* species that cause relapsing fever. Infect Immun *80*, 1881-1890. https://doi.org/10.1128/IAI.06419-11

Margos, G., Castillo-Ramirez, S., Cutler, S., Dessau, R.B., Eikeland, R., Estrada-Pena, A., Gofton, A., Grana-Miraglia, L., Hunfeld, K.P., Krause, A., *et al.* (2020). Rejection of the name *Borreliella* and all proposed species comb. nov. placed therein. Int J Syst Evol Microbiol *70*, 3577-3581. https://doi.org/10.1099/ijsem.0.004149

Margos, G., Gatewood, A.G., Aanensen, D.M., Hanincova, K., Terekhova, D., Vollmer, S.A., Cornet, M., Piesman, J., Donaghy, M., Bormane, A., *et al.* (2008). MLST of housekeeping genes captures geographic population structure and suggests a European origin of *Borrelia burgdorferi*. Proc Natl Acad Sci U S A *105*, 8730-8735. https://doi.org/10.1073/pnas.0800323105

Margos, G., Gofton, A., Wibberg, D., Dangel, A., Marosevic, D., Loh, S.M., Oskam, C., and Fingerle, V. (2018). The genus *Borrelia* reloaded. PLoS One *13*, e0208432. https://doi.org/10.1371/journal.pone.0208432

Marosevic, D., Margos, G., Wallich, R., Wieser, A., Sing, A., and Fingerle, V. (2017). First insights in the variability of *Borrelia recurrentis* genomes. PLoS Negl Trop Dis *11*, e0005865. https://doi.org/10.1371/journal.pntd.0005865

Miller, S.C., Porcella, S.F., Raffel, S.J., Schwan, T.G., and Barbour, A.G. (2013). Large linear plasmids of *Borrelia* species that cause relapsing fever. J Bacteriol *195*, 3629-3639. https://doi.org/10.1128/JB.00347-13

Mitani, H., Talbert, A., and Fukunaga, M. (2004). New World relapsing fever *Borrelia* found in *Ornithodoros porcinus* ticks in central Tanzania. Microbiol Immunol *48*, 501-505. https://doi.org/10.1111/j.1348-0421.2004.tb03545.x

Molloy, P.J., Telford, S.R., 3rd, Chowdri, H.R., Lepore, T.J., Gugliotta, J.L., Weeks, K.E., Hewins, M.E., Goethert, H.K., and Berardi, V.P. (2015). *Borrelia miyamotoi* Disease in the Northeastern United States: A Case Series. Ann Intern Med *163*, 91-98. https://doi.org/10.7326/M15-0333

Pettersson, J., Schrumpf, M.E., Raffel, S.J., Porcella, S.F., Guyard, C., Lawrence, K., Gherardini, F.C., and Schwan, T.G. (2007). Purine salvage pathways among *Borrelia* species. Infect Immun *75*, 3877-3884. https://doi.org/10.1128/IAI.00199-07

Porcella, S.F., Raffel, S.J., Anderson, D.E., Jr., Gilk, S.D., Bono, J.L., Schrumpf, M.E., and Schwan, T.G. (2005). Variable tick protein in two genomic groups of the relapsing fever spirochete *Borrelia hermsii* in western North America. Infect Immun *73*, 6647-6658. https://doi.org/10.1128/IAI.73.10.6647-6658.2005

Pritt, B.S., Mead, P.S., Johnson, D.K.H., Neitzel, D.F., Respicio-Kingry, L.B., Davis, J.P., Schiffman, E., Sloan, L.M., Schriefer, M.E., Replogle, A.J., *et al.* (2016). Identification of a novel pathogenic *Borrelia* species causing Lyme borreliosis with unusually high spirochaetaemia: a descriptive study. Lancet Infect Dis *16*, 556-564. https://doi.org/10.1016/S1473-3099(15)00464-8

Qiu, Y., Nakao, R., Hang'ombe, B.M., Sato, K., Kajihara, M., Kanchela, S., Changula, K., Eto, Y., Ndebe, J., Sasaki, M., *et al.* (2019). Human Borreliosis Caused by a New World Relapsing Fever *Borrelia*-like Organism in the Old World. Clin Infect Dis *69*, 107-112. https://doi.org/10.1093/cid/ciy850

Raffel, S.J., Battisti, J.M., Fischer, R.J., and Schwan, T.G. (2014). Inactivation of genes for antigenic variation in the relapsing fever spirochete *Borrelia hermsii* reduces infectivity in mice and transmission by ticks. PLoS Pathog *10*, e1004056. https://doi.org/10.1371/journal.ppat.1004056

Safdie, G., Farrah, I.Y., Yahia, R., Marva, E., Wilamowski, A., Sawalha, S.S., Wald, N., Schmiedel, J., Moter, A., Gobel, U.B., *et al.* (2010). Molecular characterization of *Borrelia persica*, the agent of tick borne relapsing fever in Israel and the Palestinian Authority. PLoS One *5*, e14105. https://doi.org/10.1371/journal.pone.0014105

Samaddar, S., Marnin, L., Butler, L.R., and Pedra, J.H.F. (2020). Immunometabolism in Arthropod Vectors: Redefining Interspecies Relationships. Trends Parasitol. https://doi.org/10.1016/j.pt.2020.07.010

Sarih, M., Garnier, M., Boudebouch, N., Bouattour, A., Rihani, A., Hassar, M., Gern, L., Postic, D., and Cornet, M. (2009). *Borrelia* hispanica relapsing fever, Morocco. Emerg Infect Dis *15*, 1626-1629. https://doi.org/10.3201/eid1510.090403

Schwan, T.G., Anderson, J.M., Lopez, J.E., Fischer, R.J., Raffel, S.J., McCoy, B.N., Safronetz, D., Sogoba, N., Maiga, O., and Traore, S.F. (2012). Endemic foci of the tick-borne relapsing fever spirochete *Borrelia crocidurae* in Mali, West Africa, and the potential for human infection. PLoS Negl Trop Dis *6*, e1924. https://doi.org/10.1371/journal.pntd.0001924

Schwan, T.G., Battisti, J.M., Porcella, S.F., Raffel, S.J., Schrumpf, M.E., Fischer, E.R., Carroll, J.A., Stewart, P.E., Rosa, P., and Somerville, G.A. (2003). Glycerol-3-phosphate acquisition in spirochetes: distribution and biological activity of glycerophosphodiester phosphodiesterase (GlpQ) among *Borrelia* species. J Bacteriol *185*, 1346-1356. https://doi.org/10.1128/jb.185.4.1346-1356.2003

Schwan, T.G., Raffel, S.J., Schrumpf, M.E., Policastro, P.F., Rawlings, J.A., Lane, R.S., Breitschwerdt, E.B., and Porcella, S.F. (2005). Phylogenetic analysis of the spirochetes *Borrelia parkeri* and *Borrelia turicatae* and the potential for tick-borne relapsing fever in Florida. J Clin Microbiol *43*, 3851-3859. https://doi.org/10.1128/JCM.43.8.3851-3859.2005

Schwan, T.G., Raffel, S.J., Schrumpf, M.E., and Porcella, S.F. (2007). Diversity and distribution of *Borrelia hermsii*. Emerg Infect Dis *13*, 436-442. https://doi.org/10.3201/eid1303.060958

Stevenson, B., Porcella, S.F., Oie, K.L., Fitzpatrick, C.A., Raffel, S.J., Lubke, L., Schrumpf, M.E., and Schwan, T.G. (2000). The relapsing fever spirochete *Borrelia hermsii* contains multiple, antigen-encoding circular plasmids that are homologous to the cp32 plasmids of Lyme disease spirochetes. Infect Immun *68*, 3900-3908. https://doi.org/10.1128/iai.68.7.3900-3908.2000

Toledo, A., Anda, P., Escudero, R., Larsson, C., Bergstrom, S., and Benach, J.L. (2010). Phylogenetic analysis of a virulent *Borrelia* species isolated from patients with relapsing fever. J Clin Microbiol *48*, 2484-2489. https://doi.org/10.1128/JCM.00541-10

Zhong, J., Skouloubris, S., Dai, Q., Myllykallio, H., and Barbour, A.G. (2006). Function and evolution of plasmid-borne genes for pyrimidine biosynthesis in *Borrelia* spp. J Bacteriol *188*, 909-918. https://doi.org/10.1128/JB.188.3.909-918.2006

Chapter 4

Replication of the *Borrelia burgdorferi* Genome

Kerri Kobryn*

Department of Microbiology and Immunology, College of Medicine, University of Saskatchewan, Saskatoon, SK, Canada
*Corresponding author: kerri.kobryn@usask.ca

DOI: https://doi.org/10.21775/9781913652616.04

Abstract

Borrelia species possess unusual genomes that harbor multiple circular and linear replicons. The termini of the linear replicons are covalently closed DNA hairpins, structures referred to as hairpin (hp) telomeres. Hairpin telomeres are formed from replicated intermediates by a specialized DNA cleavage and rejoining reaction called telomere resolution. Current knowledge of the mechanisms of the DNA replication, hairpin telomere metabolism and replicon maintenance of these unusual, segmented genomes will be reviewed. We will also discuss proposals on how hairpin telomere metabolism may contribute to rearrangements within the linear component of the genome.

Introduction

All organisms with linear genomes face the twin challenges of avoiding the loss of genetic material from replicon ends due to problems with priming lagging strand synthesis at the termini and degradation of the terminal sequences by cellular nucleases. Additionally, there is the need to prevent inappropriate genetic fusions with other replicons. These challenges are referred to as the 'end-replication' and 'end-protection' problems (Olovnikov, 1973; Ohki et al., 2001; Soudet et al., 2014). It was recognized early that specialized structures or functions seem to exist at the end of eukaryotic chromosomes (Muller, 1938; McClintock, 1941). Chromosome termini were named 'telomeres' by Herman Muller (from the Greek *telos* and *meros* meaning end-part). Perhaps the conceptually simplest telomere structure discovered since the coining of the term is the hp telomere, consisting of a covalently closed DNA hairpin that eliminates the free DNA end that causes the end-replication and -protection problems. The hp telomeres of *Borrelia burgdorferi* have provided a model system for study of this simplest type of telomere.

Replication of the linear chromosome

The replication machinery

Bacterial chromosomal replication has been studied in detail only for a small number of model systems. The overarching fundamentals of DNA replication appear well conserved, but there is greater diversity in the details of regulation and segregation than at first appreciated when *Escherichia coli* was the primary system studied (Reyes-Lamothe et al., 2012). *B. burgdorferi* appears to accomplish chromosome replication with a restricted set of replication proteins relative to *E. coli* (Fraser et al., 1997). Table 1 compares the replication machinery of *E. coli* to that of *B. burgdorferi* inferred from annotation of the genome. *B. burgdorferi* appears to have a simpler DNA polymerase III holoenzyme lacking the ε and θ subunits of the replicative polymerase associated with proofreading and the ψ and χ subunits of the clamp loader. *B. burgdorferi*'s replicative polymerase has an N-terminal Polymerase and Histidinol Phosphatase (PHP) domain that likely provides the proofreading exonuclease, normally provided by ε and stimulated by θ (Studwell-Vaughan and O'Donnell, 1993). The δ and δ′ subunits of the clamp loader also appeared to be absent in the initial annotation of the *B. burgdorferi* genome but the Spirochete Genome Browser (http://sgb.leibniz-fli.de) now annotates a δ-like subunit in several *Borrelia* species (the BB0455 locus is the *B. burgdorferi* B31 strain gene equivalent). Interestingly, the χ and ψ subunits of the DNA polymerase III holoenzyme that interact with the SSB and the τ/γ subunits, respectively, seem to be absent from the *B. burgdorferi* genome (see Table 1; (Fraser et al., 1997)). The χ / ψ subcomplex stimulates clamp loading, contributes to replisome stability and aids coordination of leading and lagging strand synthesis (Glover and McHenry, 1998; Simonetta et al., 2009; Marceau et al., 2011). Viable null *holC* (χ) and a

Table 1. Replication machinery of *E. coli* and *B. burgdorferi.*

Protein function	*E. coli*	*B. burgdorferi*
Origin activator	DnaA	DnaA
Helicase loader	DnaC	absent
Helicase	DnaB	DnaB
Primase	DnaG	DnaG
Replicative polymerase	Pol III $\alpha\varepsilon\theta$	Pol III α
Clamp loader	$(\tau/\gamma)_3\ \delta\delta'\psi\ \chi$	τ/γ δ-like?
Sliding clamp	β	β
ssDNA binding protein	SSB	SSB

mutant allele of *holD* (ψ) have been recovered in *E. coli* but they cause aberrant replication and genome instability (Saveson and Lovett, 1997; Flores et al., 2002). *B. burgdorferi*'s clamp loader is either simpler than *E. coli*'s or has undiscovered, non-orthologous versions of the 'missing' components. Precedent for a simpler clamp loader is provided by the well-characterized T4 phage replisome where the clamp loader is made up of a heteropentamer of two proteins (gp44/62) (Jarvis et al., 1989).

Replication is regulated primarily at the initiation stage; the key step in building a replisome is the loading of the replicative helicase, DnaB, onto the ssDNA generated at *oriC* that will be the templates for leading and lagging strand synthesis (Reyes-Lamothe et al., 2012). In *E. coli* this involves DnaA-ATP melting origin DNA and the DnaC-DnaB complex loading onto the exposed SSB-coated ssDNA via interactions of the helicase loader, DnaC, with DnaA and DnaB (Mott et al., 2008). However, this mechanism of replicative helicase loading is not well conserved (Reyes-Lamothe et al., 2012). In some cases, it even seems the replicative helicases can self-load (Soni et al., 2005). The *B. burgdorferi* genome does not encode obvious helicase loading homologues but it does encode the PriA helicase (Fraser et al., 1997). PriA is involved in loading DnaB at abandoned replication fork structures (Heller and Marians, 2005). This suggests that at least control of DnaB loading during replication restart is conserved.

Topoisomerases required for replisome elongation and chromosome segregation

Elongating replisomes generate positive supercoiling ahead of the replication fork. For continued elongation this excess positive supercoiling must be removed ahead of the fork by the action of the type II topoisomerase, DNA gyrase (Postow et al., 2001). The *B. burgdorferi* genome encodes loci for the two subunits of DNA gyrase (*gyrA* and *gyrB*). Positive supercoiling can also diffuse behind the replication fork, producing 'pre-catenane' intertwining of the replicated daughters. This form of positive supercoiling is removed by the action of a second type II topoisomerase enzyme referred to as topoisomerase IV (Deibler et al., 2001; Postow et al., 2001). The *B. burgdorferi* genome also encodes loci for the two subunits of topoisomerase IV (*parC* and *parE*); *parC*, as expected, is the target of fluoroquinolones (Fraser et al., 1997(Fraser et al., 1997; Galbraith et al., 2005). Catenation of daughter chromosomes remains a barrier for successful chromosome segregation, even in bacteria with a linear chromosome. This is demonstrated by *Borrelia*'s possession of topoisomerase IV and by the nonviability of topoisomerase IV mutants in *E. coli* strains harboring a chromosome linearized by introduction of the reaction site for the N15 telomere resolvase, TelN, into the terminus region and subsequent lysogenization of the strain with N15 (Cui et al., 2007).

Clearing the chromosome from the division septum

The DNA translocase, FtsK, was discovered as a division septum-located, accessory factor for the Xer-*dif* site-specific recombination system that resolves chromosome dimers into monomers (Recchia et al., 1999). FtsK acts as a directional DNA pump that brings *dif* sites together for recombination by XerCD, removes catenation interlinks, positions *dif* sites at the septum and is thought to pump DNA away from the division septum (Grainge et al., 2007; Lesterlin et al., 2008; Reyes-Lamothe et al., 2012). FtsK incidence in a genome usually correlates with the presence of the Xer-*dif* resolution system (Reyes-Lamothe et al., 2012). *Streptomyces coelicolor* and *Borrelia* spp. with linear chromosomes are notable exceptions. The high rate of DNA loss from S. *coelicolor* chromosome ends in an *ftsK* mutant and the filamentous phenotype of a transposon insertion mutant into *B. burgdorferi ftsK* are consistent with a requirement for FtsK to clear the chromosome away from the developing division septum to allow cell division (Stewart et al., 2004; Flardh and Buttner, 2009).

Bidirectional replication from an internal origin of replication

The initial report of the genomic sequence of *B. burgdorferi* B31 revealed an orientation pattern of transcriptional units and CG skew on the chromosome suggestive of an origin of replication (*oriC*) located almost exactly in the middle of the linear chromosome (Fraser et al., 1997). Subsequent analysis of nascent strand production mapped *oriC* to a 2-kb region in the middle of the chromosome (Picardeau et al., 1999). Importantly, this study failed to detect replication initiated near the hp telomeres. The 240-bp intergenic region between *dnaA* and *dnaN* was the position of the switch in CG skew polarity and was predicted to be the most likely position of *oriC*. Subsequent discovery of an Hbb-binding site in the *dnaA* and *dnaN* intergenic region provided further evidence that this may be the location of *oriC* (Kobryn et al., 2000). Hbb is an architectural protein with structural and functional similarity to the IHF and HU family of DNA-bending proteins (Tilly et al., 1996; Mouw and Rice, 2007). Such architectural proteins are often required to aid DnaA in origin melting (Dixon and Kornberg, 1984; Bonnefoy and Rouviere-Yaniv, 1992). While only three candidate DnaA-binding sites were found in the putative origin region, DnaA from *B. burgdorferi* has yet to be purified and its binding site defined. This would help define the sequence of 'DnaA boxes' and to identify all the DnaA-dependent origins in the genome.

Replication termination: circular vs. linear chromosomes

Bacteria with circular chromosomes send out a pair of replisomes from *oriC* that eventually converge in the terminus region (*ter*) of the chromosome. In order to properly duplicate the chromosome, it is important that the converging forks do not pass each other to continue replication, leading to over-replication of *ter* associated sequences. Recent genetic analyses of termination in *E. coli* suggest that converging forks can lead to the production of 3′-flap structures that trigger PriA-dependent re-initiation events in the *ter* region that lead to over-replication of *ter* sequences. The RecG DNA translocase and several 3′-exonucleases have been implicated in suppressing such over-replication (Rudolph et al., 2009; Rudolph et al., 2013). The *E. coli* chromosome was converted into a linear form with hp telomeres to assess the effect of eliminating fork convergence in the terminus (Rudolph et al., 2013). Unexpectedly, *ter* over-replication in *recG* mutants was not completely abolished by chromosome linearization (Rudolph et al., 2013). This hints at unanticipated complexities of replication at hp telomeres.

Replication model

The discovery that chromosome replication featured bidirectional replication from an internal origin strongly constrained possible replication models (Figure 1) (Picardeau et al., 1999; Chaconas et al., 2001). Models wherein the replication machinery rounds the hp telomeres to produce inverted repeat circular dimers with replicated telomere junctions (*rTels*) came to the fore (Figure 1). Here, the chromosome has been copied but is present in a dimeric form that cannot be segregated to daughter cells. A specialized DNA breakage and reunion event to process the replicated telomeres into a pair of hp telomeres, releasing linear copies from the circular dimer was hypothesized and shown to occur *in vivo* (Chaconas et al., 2001). Subsequent work in which a telomere resolvase from *B. burgdorferi* was purified and characterized confirmed the general outline of the replication pathway presented in Figure 1 (Kobryn and Chaconas, 2002). Other general replication issues such as that of copy number of the chromosome and plasmids in *B. burgdorferi* was previously reviewed in the previous edition of this book (Chaconas, 2010).

Replication of the plasmids

Sequence analysis of the plasmid component of the *B. burgdorferi* genome revealed that most plasmids carry a common cluster of two to four genes that fall into five paralogous gene families (PF; see Table 2). It was hypothesized that these may constitute plasmid replication and partition functions (Casjens et al., 2000). Paralogous gene family 32 (PF32) members have some sequence similarity to the plasmid partition ATPase, ParA, by virtue of possessing a Walker box type ATPase and Mg^{2+} binding regions. However, the other paralogous gene family members have no homology to genes in other genera.

Several studies that have constructed shuttle vectors have confirmed that DNA segments containing the putative plasmid maintenance functions act to support autonomous replication (Stewart et al., 2001; Eggers et al., 2002; Stewart et al., 2003). Circular replicons can be linearized by introduction of synthetic telomeres (Chaconas et al., 2001) and spontaneous inter-conversion between linear and circular plasmid forms has been inferred (Ferdows et al., 1996). This suggests that the mechanism of plasmid maintenance is conserved, irrespective of the topological form of the replicon. Deletion analyses of some shuttle vectors have also defined possession, in *cis*, of a PF62 or PF57 family member as sufficient for plasmid replication, suggesting that these encode related versions of a plasmid origin activator (Eggers et al., 2002; Beaurepaire and Chaconas, 2005). Similar deletion analyses of other shuttle vectors have suggested a minimum, in *cis*, requirement of PF57/62 and PF50 for autonomous replication (Stewart et al., 2003). However, a small change in a presumed *cis* acting site was found to change the minimum complement of required proteins (Stewart et al., 2003).

The paralogous gene family members that define plasmid maintenance functions seem to come in replicon-specific flavors suggesting that they define plasmid incompatibility groups (Casjens et al., 2000; Casjens et al., 2012). Experimental support for this hypothesis has been provided by studies where shuttle vectors constructed from cp9 or cp32-derived plasmid maintenance functions were used to displace the endogenous version of the plasmids (Stewart et al., 2001; Eggers et al., 2002). Though its role in plasmid compatibility is unknown, the most consistent scheme for defining plasmid incompatibility groups is to define it by the PF32 gene present on a replicon (Casjens et al., 2010; Casjens et al., 2017). Complicating this notion is the persistent suspicion that some plasmid maintenance functions can be donated in *trans* from one plasmid to another.

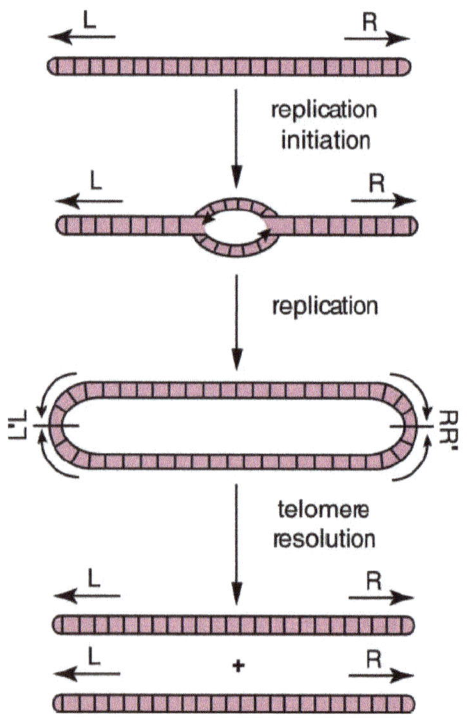

Figure 1. Replication pathway for linear DNA. L and R, with accompanying arrows, denote the left and right hp telomeres, respectively. Lines in the replicated telomere junctions of the circular head-to-head dimer intermediate mark the symmetry axis of the replicated telomeres. For simplicity, only the scenario where both replicated telomeres are formed before telomere resolution starts is shown. The simplest case of one initiation event per telomere resolution cycle is shown. The dynamics of replication and telomere resolution timing and are not currently known. This figure has been adapted, with permission from (Kobryn and Chaconas, 2002).

In an effort to eliminate concerns about *trans* complementation effects, a study was undertaken in *E. coli* to determine the ability of cp26 plasmid maintenance genes to complement plasmid stability defects of a mini-F plasmid lacking its partition system (Tilly et al., 2012). In this experimental set-up, the *B. burgdorferi* proteins are not required for plasmid replication, only stability. The cp26 PF62, 50 and 49 functions, but not the PF32 function were required for plasmid stability. The dispensability of PF32 was confirmed by successful disruption of this gene in *B. burgdorferi* (Tilly et al., 2012). These results suggest that despite its apparent homology to the partition protein, ParA, PF32 proteins may not be involved in partition but could act, instead, as an accessory factor for replication. PF62 has been assumed based on minimal replicon data to be the origin activator but instead was required for stability. It seems likely that a role for plasmid replication functions in partitioning, separate from their roles in replication, may explain these apparent contradictions. Replication/partition of the well-studied pSC101 plasmid provides an example where plasmid and host replication proteins play genetically separable roles in these processes (Conley and Cohen, 1995; Miller et al., 1995; Miller and Cohen, 1999).

Further evidence defining plasmid maintenance requirements came from a study mapping the insertion sites of a high-density transposon insertion library (Lin et al., 2012; Chaconas and Norris, 2013). As expected for a putative origin activator, PF57/62 genes did not tolerate transposon insertion unless the affected gene was a second 'orphan' initiator gene present on the replicon. The minimal complement of maintenance genes seemed to be context-dependent, plasmids with an incomplete set of maintenance genes (or *cis* acting sites) tolerated further loss better than plasmids with a full

Table 2. Paralogous gene families implicated in plasmid maintenance.

Paralogous family	Proposed function	Evidence
62/57	Origin activator/stability	Ubiquity, minimal replicon requirement, required for mini-F stability, no transposon insertions recovered
32	Unknown/replication	P-loop NTPase homology, interaction with putative origin activator, dispensable for mini-F stability
49	Regulation/stability	Expression/*in vitro* studies, mini-F stability
50	Unknown/stability	Mini-F stability

complement. Finally, loss of PF32 function on linear plasmids was well tolerated but was never observed amongst the cp32 plasmids (Chaconas and Norris, 2013).

Exemplars of the PF 62, 32 and 49 gene families have been purified and characterized in an effort to gain greater insight into their function (Deneke and Chaconas, 2008; Chenail et al., 2012). PF62/32 proteins from lp17 (B31 strain) were purified. The PF32 protein was expected to be an ATPase based on its ParA homology and this was confirmed. It was also found to possess non-specific DNA-binding activity. The putative initiator (PF62) was purified but found not to bind specifically to the putative origin of replication. Unexpectedly, PF62 was found to interact with PF32 (Deneke and Chaconas, 2008). PF32 proteins have proven amenable to crystallization and there are apo- and ATP-bound structures deposited in the Protein Structural Database that are not currently published with any biological context (accession numbers 3K9H, 3K9G, 5TVK) (Abendroth et al., 2011). PF49 protein from a cp32 plasmid was unexpectedly recovered as a protein that binds the operator region of *erp* genes and was found to regulate *erp* transcription (Burns et al., 2010). Subsequent characterization uncovered binding sites that regulate transcription of genes that are putative single strand annealing and exonuclease functions (Chenail et al., 2012). PF49, therefore, clearly plays roles in transcription and plasmid maintenance. Such multifunctionality has precedents in that plasmid maintenance proteins often autoregulate and DnaA has been found to play a role in activating transcription needed for origin opening during λ replication (Szalewska-Palasz et al., 1998; Flatten et al., 2009).

Telomere resolution

Bidirectional replication from an internal *oriC* would eventually produce fully replicated DNA present in copies linked through replicated telomere (*rTel*) junctions (Figure1). The hp telomeres have overcome the end-replication problem but have created the alternate problem of a head-to-head inverted repeat dimeric replication intermediate that cannot be segregated to daughter cells. This necessitates a specialized DNA breakage and reunion reaction, referred to as telomere resolution, to generate linear monomers with hp telomeres. Figure 1 shows the simplest case of a single replication initiation event per pair of telomere resolution reactions and the case where both *rTel* junctions are resolved together from a circular dimer form. Uncoordinated telomere resolution events and multiple replication initiation events would produce complex, branched forms with higher copy number near *oriC* than at hp telomeres (Ravin et al., 2003). Replication/resolution dynamics are not presently understood but deviations from this simple model could explain the higher copy number of origin proximal sequences relative to telomere proximal sequences and the highly complex form of *B. burgdorferi* nucleoids recovered from the wells of pulsed field gels (Hinnebusch and Bendich, 1997; Beaurepaire and Chaconas, 2007).

Telomere resolution was shown to occur *in vivo* by introduction of *rTel* junctions into lp17. Hairpin telomeres were introduced at the site of the artificially introduced *rTel* (Chaconas et al., 2001). Targeting *rTel* junctions into linear replicons allows programmed deletions to be produced, a technique that has been used to define minimal replicon requirements of lp17 and to produce lp28-1 variants lacking the *vls* locus (Beaurepaire and Chaconas, 2005; Bankhead and Chaconas, 2007). Targeting *rTel* junctions into circular replicons allows conversion of the circular replicon into a linear form, though there is evidence that plasmid initiator functions may have adapted for optimal expression in the native topological form (Chaconas et al., 2001; Beaurepaire and Chaconas, 2007).

The enzyme that promotes telomere resolution has been purified and characterized. The telomere resolvase was named ResT for Resolvase of Telomeres. Telomere resolvases have also been purified and characterized from a few phage systems and from *Agrobacterium tumefaciens* (Deneke et al., 2000; Hertwig et al., 2003; Huang et al., 2004a; Huang et al., 2012; Kobryn and Chaconas, 2014). In these systems, the enzyme that promotes telomere resolution is often called a protelomerase (a discussion of nomenclature can be found in Chaconas, 2005 ref).

The hairpin telomeres

Initially, the sequence of only a few hp telomeres was available. The hp telomeres of lp17 were sequenced around the hairpin turn around by the labor intensive method of Maxam-Gilbert sequencing (Hinnebusch and Barbour, 1991). Subsequent studies employed mung bean nuclease opening of the hairpins followed by cloning and Sanger sequencing (Hinnebusch et al., 1990; Hinnebusch and Barbour, 1991; Casjens et al., 1997; Fraser et al., 1997; Zhang et al., 1997; Huang et al., 2004b). The mung bean nuclease approach does raise the possible technical issue of missing nucleotides at the hp turnaround. Nuclease opening of the hp telomeres followed by direct genomic Sanger sequencing captured the sequence of a strain's full complement of hp telomeres (excepting lp5) (Tourand et al., 2009). The issue of possibly missing terminal nucleotides was addressed by constructing *rTel* junctions from each sequence, including -1 and +1 variants, and testing their reactivity *in vitro* with ResT. The results of this study are summarized in Figure 2. There is a surprising diversity of hp telomere sequences that can be grouped into three telomere types by the sequence characteristics (Chaconas, 2005; Tourand et al., 2009).

Each telomere possesses a conserved sequence called box 3 located at nucleotide positions 14-21 that serves as the main signal for site-specific binding mediated by ResT (164-449) (Tourand et al., 2003; Tourand et al., 2007). Also conserved in most telomeres is an AT-rich box 1 sequence that comes in TATAAT and TATTAT flavors with alternate spacing relative to box 3 and the hp turnaround that defines the type 1 and type 2 telomeres. The box 1 sequence does not seem to contribute to site-specific binding of ResT but instead seems to be an important determinant of the telomere's *in vitro* reactivity, since its absence defines the type 3 telomeres that have generally poor *in vitro* reactivity (Tourand et al., 2009). Box 1's contribution to telomere reactivity is, however, not straightforward since experiments with ResT from the relapsing fever spirochete, *B. hermsii*, show that this telomere resolvase can only process type 1 telomeres *in vitro* and that possession of a histidine instead of a proline at position 326 in BhResT is the main determinant of this phenotype (Tourand et al., 2006; Moriarty and Chaconas, 2009). BhResT *trans*-complementation allows knockout of endogenous ResT in *B. burgdorferi*. The telomeres present within a single strain possess a wide range of reactivity *in vitro*, nonetheless, all the telomeres are resolved *in vivo* (Bankhead et al., 2006; Tourand et al., 2006; Tourand et al., 2009). This implies that there may be accessory factors *in vivo* that promote telomere resolution by ResT. The fact that positive DNA supercoiling strongly stimulates telomere resolution raises the possibility that a cellular process like replication, transcription or DNA transport that transiently generates positive supercoiling may promote telomere resolution rather than a particular accessory factor (Bankhead et al., 2006).

The telomere resolvase, ResT

ResT domain organization

ResT is thought to have three domains that divide the protein into N-, catalytic and C-terminal domains (see Figure 3) (Tourand et al., 2007; Chaconas, 2010).

ResT (1-163) and ResT (164-449) domains were defined by partial chymotrypsin cleavage and independent expression of ResT (1-163) and ResT (164-449). ResT (1-163) possesses non-specific DNA binding activity and the ability to anneal free ssDNA (Tourand et al., 2007; Mir et al., 2013). ResT (164-449) possesses the telomere resolvase catalytic domain that shares homology with other telomere resolvases and with the catalytic domain of tyrosine recombinases. ResT (164-449) also carries the determinants of site-specific telomere binding. ResT (341-449) is predicted, *in silico*, to fold into an independent domain (Tourand et al., 2007; Kobryn and Chaconas, 2014). The function of the C-terminal domain is currently unknown.

General characteristics of the telomere resolution reaction

ResT, and other telomere resolvases, catalyze a unique reaction in which a replicated telomere

Type	Telomere	Sequence (box 1, box 3)	Initial Rate (fmol/min)
2	lp28-4R	AATTTATTATCTTTTAGTATAATGA	48.8 ± 2.5
2	lp36R	TATTTATTATCTTTTAGTATAATGA	43.0 ± 7.8
1	ChromR	ATATAATTTTTAATTAGTATAGAAT	41.7 ± 7.3
2	lp28-1R	TATTTATTATCTTTTAGTATATATA	41.4 ± 0.9
1	lp28-1L	ATATAATTTTTAATTAGTATAGATA	40.4 ± 7.4
2	lp28-2L, 36L	AATTTATTATTAATTAGTATAAATA	39.3 ± 13.8
1	lp17L, 28-3L, 28-4L, 21L	ATATAATTTTTTATTAGTATAGAGT	33.7 ± 6.1
2	lp28-3R	ATATTATTATTACTTAGTATAAATT	29.2 ± 6.8
1	lp56R, lp17R	ATATAATATTTATTTAGTACAAAGT	28.4 ± 2.8
2	lp38R	AATTTATTATCTTTTAGTATAATAG	27.4 ± 0.1
2	lp56L	AATTTATTATCTTTTAGTATAATGC	26.5 ± 5.4
3	lp21R	AAATTAGTTTTTTTTAGTATAAAGC	26.5 ± 6.1
2	ChromL	TAAATATAATTTAATAGTATAAAAA	25.6 ± 0.4
3	lp25L	TATAAATTTTTAAATAGTATAGTTA	18.3± 3.1
3	lp25R	AAATATTTTTTTATTAGTATAGAGA	13.6± 2.1
3	lp28-2R	ATACAATTATTAATTAGTATAGAAA	4.7± 2.2
3	lp38L	AATTTACAATTTTTTAGTATAAAAA	<0.3
3	lp54L	TGAAGATAATCTATTAGTATACTAA	<0.3
3	lp54R	TAATAAGAGTTTATTAGTATACTAA	<0.3

Figure 2. Hairpin telomere sequence alignment. Telomere sequences are arranged in descending order of activity, as assessed by in vitro reactions with replicated telomere (rTel) junctions constructed from the hp telomere sequences shown. The sequences are aligned with the hairpin termini, or symmetry axis of the replicated telomere, positioned to the left. Boxes 1 and 3 (shaded blue) are conserved sequence elements identified in (Tourand et al., 2003). Three types of telomere are designated based on the relative spacing of boxes 1 and 3 (types 1 & 2) or by the absence of a box 1 sequence element (type 3). The sequences of lp17L, lp17R, lp21R, lp28-1R, lp56R, ChromL and ChromR have been reported previously (Hinnebusch et al., 1990; Hinnebusch and Barbour, 1991; Casjens et al., 1997; Fraser et al., 1997; Zhang et al., 1997; Huang et al., 2004b), while the complete set of hp telomere sequences present in a B. burgdorferi strain was reported in (Tourand et al., 2009). This figure has been adapted, with permission, from (Tourand et al., 2009).

substrate is cleaved 6 bp apart on opposing strands followed by refolding of the cleaved strands and rejoining the DNA into a pair of hp telomeres (Deneke et al., 2000; Kobryn and Chaconas, 2002; Huang et al., 2004a; Kobryn, 2007; Huang et al., 2012). The reaction chemistry and catalytic domain used to promote DNA cleavage and rejoining is similar to that employed by the type IB topoisomerases and tyrosine recombinase class of enzymes and proceeds by a pair of transesterifications (Deneke et al., 2004; Aihara et al., 2007; Shi et al., 2013). DNA cleavage involves the formation of a transient 3′-phosphotyrosyl cleavage intermediate in which ResT becomes covalently linked to the cleaved DNA and DNA rejoining occurs when the 5′-OH of the refolding strands attack the 3′-phosphotyrosyl bonds of the cleaved intermediate (Kobryn et al., 2005). The similarity to the tyrosine recombinase family of enzymes even extends to ResT's ability to promote Holliday junction formation between a pair of *rTel*s under conditions that inhibit telomere resolution (Kobryn et al., 2009). Telomere resolvases have been hypothesized to have originally been a phage integrase or chromosomal dimer resolvase enzyme that subsequently linearized the *Borrelia* genome when it acquired the ability to form hairpin structures on symmetrized substrates (Kobryn et al., 2009; Chaconas and Kobryn, 2010).

The unique aspect of telomere resolution is the refolding of the cleaved strands into hairpin telomere structures (Figure 4). ResT effects this refolding in the context of an intact dimer that stays together until both hp telomeres are formed. This prevents formation of broken telomeres capped by covalently linked ResT (Briffotaux and Kobryn, 2010). Because

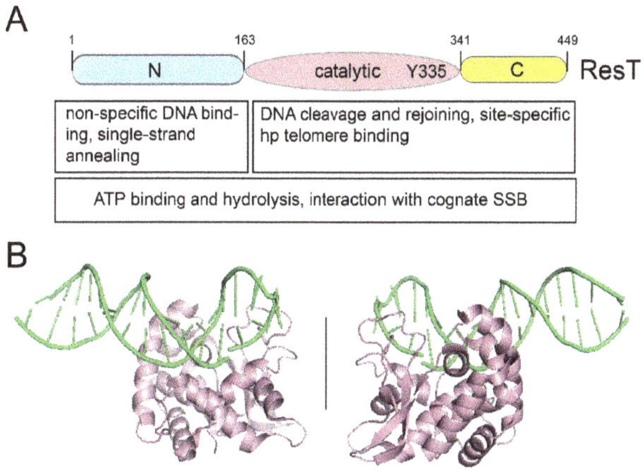

Figure 3. ResT domain organization. A) Graphic presenting the three domains of ResT. Noted beneath the graphic are boxes with listed functions ascribed to the individual ResT (1-163) and ResT (164-449) domains. The box that spans the entire protein indicates activities requiring full-length protein. Where the full-length protein is required for an activity this is noted in the box that spans the entire protein. This graphic is adapted, with permission, from (Kobryn and Chaconas, 2014). B) A structural view of the telomere resolvase catalytic domain bound to an hp telomere is shown. The structure to the right has been rotated horizontally by 180° around the indicated axis. The structure is derived from the crystallization of the Agrobacterium tumefaciens telomere resolvase, TelA, bound to its hairpin telomere (Shi et al., 2013). Only the protein residues that align with the ResT catalytic domain, TelA (218-410), are shown. The graphic was produced with PyMOL (Schrodinger, LLC; http://www.pymol.org/) using PDB ID 4e0g.

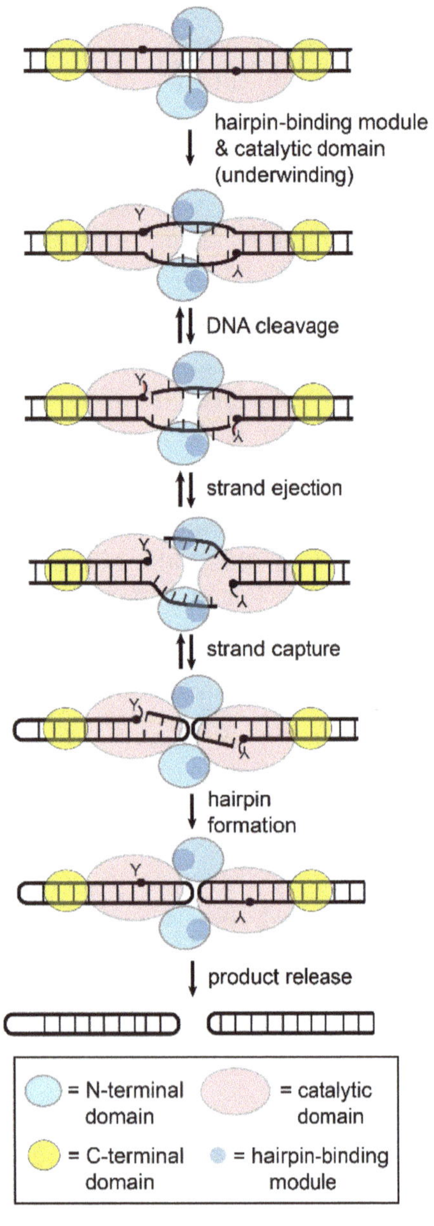

Figure 4. Telomere resolution promoted by ResT (Lucyshyn et al., 2015).

the strand rejoining reaction of topoisomerase-type enzymes is faster than the DNA cleavage step, telomere resolvases need a mechanism to ensure that the 5′-OH containing cleaved strands are ejected away from the phosphotyrosyl bonds to prevent strand resealing in the parental configuration, regenerating substrate instead of forming products. For ResT, the forward trajectory of the reaction is promoted by a pre-cleavage underwinding of the substrate DNA between the scissile phosphates (Bankhead and Chaconas, 2004; Lucyshyn et al., 2015). Numerous residues in the N-terminal and catalytic domains are implicated in forming and stabilizing this pre-cleavage intermediate. The integrity of the developing hp telomeres is tested at the strand-rejoining step, allowing some reaction reversibility to prevent hairpin formation at sequences lacking the ability to basepair the first two bases in the hairpin product (Lucyshyn et al., 2015). There seems to be some diversity of hairpin formation mechanisms for other telomere resolvases (reviewed in (Kobryn and Chaconas, 2014).

Phenotype of ResT depletion

As expected for the enzyme that forms the hp telomeres, ResT has been shown to be an essential function (Byram et al., 2004; Tourand et al., 2006). ResT's essential nature had made study of its *in vivo* roles difficult until it became possible to make strains with conditional expression of essential genes in *B. burgdorferi* (Gilbert et al., 2007; Liang et al., 2010; Groshong et al., 2012). This allowed the development of *B. burgdorferi* strains whose only functional copy of ResT is provided on a shuttle vector that allows inducible expression of ResT depending on whether IPTG is present in the growth media or not (Bandy et al., 2014). The conditional *resT* strain recapitulates the unexpectedly high abundance of ResT *in vivo* (15,000 monomers/cell), and IPTG removal from the media causes growth arrest after three ResT half-lives have elapsed (48 h). However, the phenotype of ResT depletion is unexpectedly complex. Upon ResT depletion, unit-sized linear replicon forms disappear, becoming progressively more complex forms that have an equal mix of hp telomeres and unresolved *rTel* junctions. Importantly, the accumulated *rTel* junctions can be resolved *in vitro* with ResT (Bandy et al., 2014). There was the strong expectation that ResT depletion would cause the cells to filament because the failure to decatenate replicated chromosomes or to resolve chromosome dimers leads to inhibition of cell division and subsequent filamentation in *E. coli* (Reyes-Lamothe et al., 2012). However, ResT-depleted cells do not filament. Furthermore, DNA replication ceases in ResT-depleted cells indicating that initiation or elongation steps of replication may require ResT in some direct or indirect capacity (Bandy et al., 2014). Another unexpected phenotype of ResT depletion is the survival, in a quiescent state, of 15-24% of the cells for as long as 14 days post-IPTG removal. Re-expression of ResT in these survivors 'reanimates' them (Bandy et al., 2014). The reason for the survival of a fraction of the cells is unknown. Conditional depletion of ResT confirmed the general outline of

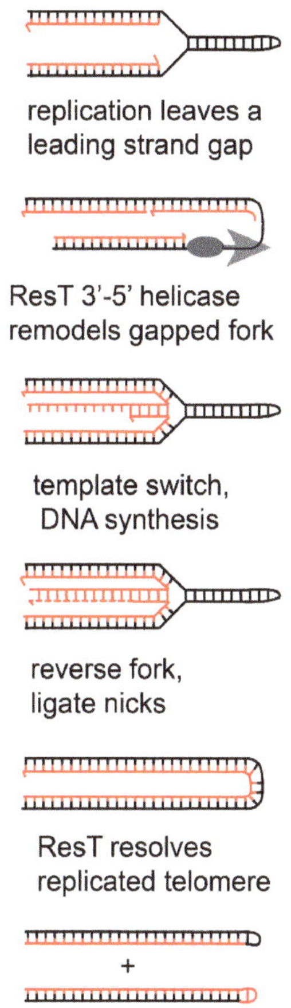

Figure 5. Proposed role of ResT in replication fork remodeling at hairpin telomeres. A hypothesized conflict between the replicative helicase and the leading strand polymerase as the replisome approaches a hp telomere could cause a leading strand gap to be left at hp telomeres. The ATP-dependent fork regression activity of ResT remodels the resulting stalled or abandoned fork to allow a template switch and synthesis across the gap using the nascent lagging strand as template. Reversal of the fork, followed by ligation produces the replicated telomere junction that is then resolved by ResT into hp telomeres (Huang et al., 2017). Black lines represent the parental DNA strands, solid red lines indicate nascent strands produced prior to fork stalling and dashed red lines denote DNA synthesized after the hypothesized template switch.

ResT's essential role in telomere resolution presented in Figure1 but also hinted at unappreciated complexities of the replication pathway and suggested that ResT may be multifunctional.

Evidence that ResT is multifunctional

Biochemical evidence has accumulated that ResT is multifunctional. ResT was demonstrated to promote single-strand DNA annealing reactions and limited strand exchange promoted by its annealing activity (Figure 3) (Mir et al., 2013). These activities are similar to those of the β recombinase of phage λ and of RecO(R) acting on free DNA strands (Luisi-DeLuca and Kolodner, 1994; Li et al., 1998). Since single-strand annealing proteins can promote recombination reactions directly or by acting as recombinational mediators for RecA-like recombinases, it was important to determine if ResT has mediator properties. A key property of recombinational mediators is the ability to promote annealing of complementary single-stranded DNA complexed with ssDNA binding protein (SSB) (Kantake et al., 2002). ResT shares this key mediator property with RecO(R) (Figure 3) (Huang and Kobryn, 2016). ResT could serve as a functional analogue of RecO(R) since the *B. burgdorferi* genome lacks these and other components of the RecF pathway (Fraser et al., 1997). ResT-facilitated annealing of complementary single strands requires a specific physical interaction between ResT and the amphipathic C-terminal tail of SSB (Huang and Kobryn, 2016). Bacterial SSBs employ this conserved C-terminal tail to recruit the replisome, and many recombination and repair functions, to their sites of action on ssDNA (Shereda et al., 2008).

Further evidence of ResT's multifunctionality came with the discovery that ResT promotes ATP-dependent DNA unwinding and branch migration reactions (Huang et al., 2017). Despite lacking characteristic sequence signature motifs of a helicase, ResT is able to bind ATP, to hydrolyze ATP in response to ssDNA binding and to use the energy of ATP hydrolysis to power DNA unwinding reactions that proceed with a 3′-5′ polarity. The preferred substrates for ResT unwinding are branched substrates with 3-arms, including D-loops and replication fork mimics. The combination of annealing and unwinding activities allows ResT to promote regression of replication fork mimics into Holliday junction structures, a process that allows template switching and repair of blocked replication forks (Huang et al., 2017).

Discovery of ResT's unanticipated DNA annealing/unwinding activities, combined with the complexity of the phenotype of ResT depletion, prompted a reevaluation of the model of the replication cycle for the linear DNAs. The revised model hypothesizes a need for ResT regression of replication forks stalled at hp telomeres, allowing the completion of replication at the hp telomeres. The final fully replicated telomere junctions are then resolved by

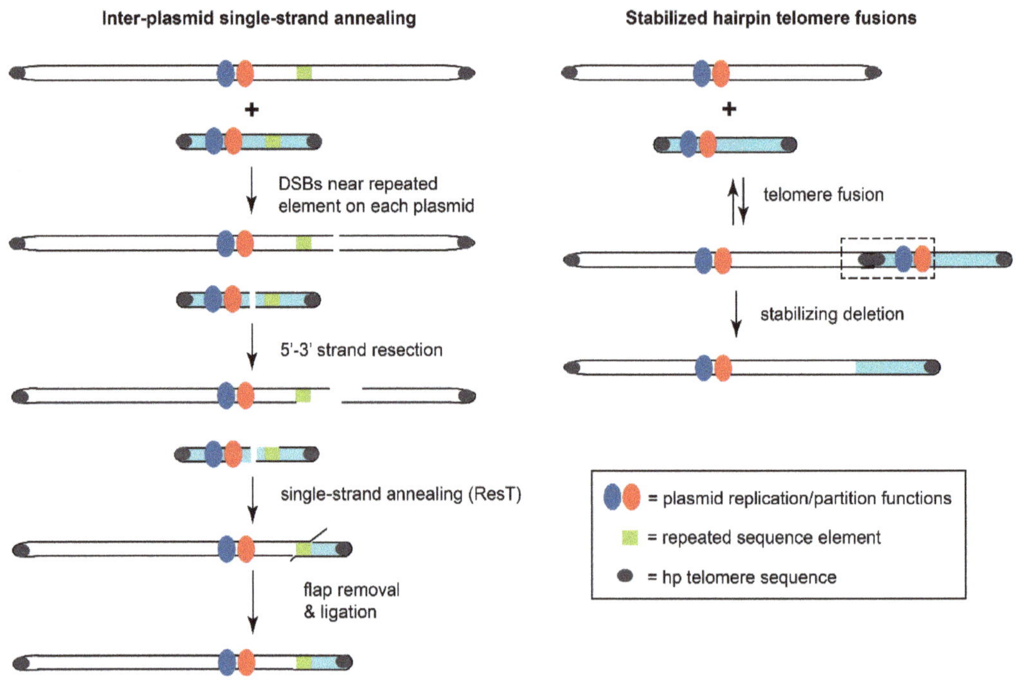

Figure 6. Proposed role of ResT in linear replicon rearrangements. ResT has been shown to promote single-strand annealing and telomere fusion reactions (Kobryn and Chaconas, 2005; Mir et al., 2013; Huang and Kobryn, 2016). Both activities could promote DNA rearrangements of the type documented for the linear component of the *B. burgdorferi* genome (Casjens et al., 2000; Casjens et al., 2012; Casjens et al., 2017). Single-strand annealing can repair DNA double-strand breaks using limited regions of sequence homology but uses a non-conservative method that deletes the homologous DNA sequence in one of the recombining partner DNAs. How this can produce inter-plasmidic replicon rearrangements by an apparently non-homologous mechanism is detailed in the left panel. cp32 plasmids may encode single-strand annealing and exonuclease proteins (Aravind et al., 2000; Iyer et al., 2002; Chenail et al., 2012). Telomere fusions promoted by ResT may produce plasmid fusion events that become permanent when accompanied by a stabilizing deletion event (Kobryn and Chaconas, 2005; Chaconas, 2010; Chaconas and Kobryn, 2010).

ResT back into hp telomeres (Figure 5). Investigation of the generality of these findings to other telomere resolvases as well as isolation and characterization ResT mutants that separate the telomere resolution and annealing/unwinding activities will be needed to test this revised replication model and to determine the biological relevance of these activities.

Proposed role of ResT in linear replicon rearrangements

The linear component of the *B. burgdorferi* genome appears to be undergoing a high level of DNA rearrangement relative to the circular component. The plasticity of the linear genome compartment is revealed when multiple, independently isolated *B. burgdorferi* strains are sequenced and compared (Casjens et al., 2000; Casjens et al., 2012; Casjens et al., 2017). Notable rearrangements include transfer of linear plasmid sequences to the right end of the chromosome and production of mosaic variants of the linear plasmids. In most cases, no obvious sequence homology exists at the breakpoint, suggesting that most events occur by a non-homologous mechanism (Casjens et al., 2017). Additionally, homologous targeting for genetic manipulation of *B. burgdorferi* is stimulated more than tenfold by the presence of a replicated telomere in the gene targeting vector (Chaconas et al., 2001). These observations have led to the suggestion that properties intrinsic to hp telomeres and ResT may promote DNA rearrangements in the linear component of the genome (Chaconas, 2010; Chaconas and Kobryn, 2010). While this contention is difficult to test *in vivo*, biochemical study of ResT has revealed two activities that could plausibly contribute to DNA rearrangements: hp telomere fusion and single-strand annealing (Kobryn and Chaconas, 2005; Mir et al., 2013; Huang and Kobryn, 2016). Both activities would promote DNA rearrangements by a non-conservative, mutagenic pathway that would mimic a non-homologous event (Figure 6). In the case of hp telomere fusion between two distinct linear replicons, any stabilizing deletion could freeze the fusion event by destroying the *rTel* junction produced by reaction reversal (Figure 6; Kobryn and Chaconas, 2005). Plasmid rearrangements produced by single-strand annealing depend upon short (20-35 bp) stretches of homology that can be annealed by ResT; processing with a hypothetical 5′-exonuclease would produce a recombinant with only one of the original homologous repeats, mimicking a non-homologous event (Figure 6) (Mir et al., 2013). This is essentially the reaction promoted by phage recombination systems typified by the λ Red and RecE/T systems. In line with the discovery that the cp32 plasmids are prophages it should be noted that the cp32 plasmids encode potential annealing/exonuclease genes that could promote the reaction depicted in the left panel of Figure 3.6 (Eggers and Samuels, 1999; Eggers et al., 2001). To date, it has been confirmed that a purified member of this paralogous gene family binds ssDNA (Chenail et al., 2012).

References

Abendroth, J., Gardberg, A.S., Robinson, J.I., Christensen, J.S., Staker, B.L., Myler, P.J., Stewart, L.J., and Edwards, T.E. (2011). SAD phasing using iodide ions in a high-throughput structural genomics environment. J Struct Funct Genomics *12*, 83-95. https://doi.org/10.1007/s10969-011-9101-7

Aihara, H., Huang, W.M., and Ellenberger, T. (2007). An interlocked dimer of the protelomerase TelK distorts DNA structure for the formation of hairpin telomeres. Mol Cell *27*, 901-913.

Aravind, L., Makarova, K.S., and Koonin, E.V. (2000). Survey and summary: holliday junction resolvases and related nucleases: identification of new families, phyletic distribution and evolutionary trajectories. Nucleic Acids Res *28*, 3417-3432.

Bandy, N.J., Salman-Dilgimen, A., and Chaconas, G. (2014). Construction and characterization of a *Borrelia burgdorferi* strain with conditional expression of the essential telomere resolvase, ResT. J Bacteriol *196*, 2396-2404. https://doi.org/10.1128/JB.01435-13

Bankhead, T., and Chaconas, G. (2004). Mixing active site components: A recipe for the unique enzymatic activity of a telomere resolvase. Proc Natl Acad Sci USA *101*, 13768-13773.

Bankhead, T., Kobryn, K., and Chaconas, G. (2006). Unexpected twist: harnessing the energy in positive supercoils to control telomere resolution. Mol Microbiol *62*, 895-905.

Bankhead, T., and Chaconas, G. (2007). The role of VlsE antigenic variation in the Lyme disease spirochete: persistence through a mechanism that differs from other pathogens. Mol Microbiol *65*, 1547-1558. https://doi.org/10.1111/j.1365-2958.2007.05895.x

Beaurepaire, C., and Chaconas, G. (2005). Mapping of essential replication functions of the linear plasmid lp17 of B. burgdorferi by targeted deletion walking. Mol Microbiol *57*, 132-142. https://doi.org/10.1111/j.1365-2958.2005.04688.x

Beaurepaire, C., and Chaconas, G. (2007). Topology-dependent transcription in linear and circular plasmids of the segmented genome of *Borrelia burgdorferi*. Mol Microbiol *63*, 443-453.

Bonnefoy, E., and Rouviere-Yaniv, J. (1992). HU, the major histone-like protein of *E. coli*, modulates the binding of IHF to *oriC*. Embo J *11*, 4489-4496.

Briffotaux, J., and Kobryn, K. (2010). Preventing broken borrelia telomeres: ResT couples dual hairpin telomere formation to product release. J Biol Chem *285*, 4100-41018. https://doi.org/doi: 10.1074/jbc.M110.150060.

Burns, L.H., Adams, C.A., Riley, S.P., Jutras, B.L., Bowman, A., Chenail, A.M., Cooley, A.E., Haselhorst, L.A., Moore, A.M., Babb, K., *et al.* (2010). BpaB, a novel protein encoded by the Lyme disease spirochete's cp32 prophages, binds to *erp* Operator 2 DNA. Nucleic acids research *38*, 5443-5455.

Byram, R., Stewart, P.E., and Rosa, P. (2004). The essential nature of the ubiquitous 26-kilobase circular replicon of *Borrelia burgdorferi*. J Bacteriol *186*, 3561-3569.

Casjens, S., Murphy, M., DeLange, M., Sampson, L., van Vugt, R., and Huang, W.M. (1997). Telomeres of the linear chromosomes of Lyme disease spirochaetes: nucleotide sequence and possible exchange with linear plasmid telomeres. Mol Microbiol *26*, 581-596.

Casjens, S., Palmer, N., van Vugt, R., Huang, W.H., Stevenson, B., Rosa, P., Lathigra, R., Sutton, G., Peterson, J., Dodson, R.J., *et al.* (2000). A bacterial genome in flux: the twelve linear and nine circular extrachromosomal DNAs in an infectious isolate of the Lyme disease spirochete *Borrelia burgdorferi*. Mol Microbiol *35*, 490-516.

Casjens, S.R., Eggers, C.H., and Schwartz, I. (2010). *Borrelia* Genomics: Chromosome, Plasmids, Bacteriophages and Genetic Variation. In *Borrelia*: Molecular Biology, Host Interaction and Pathogenesis, D.S. Samuels, and J.D. Radolf, eds. (Norwich, UK: Horizon Scientific Press), pp. 21-47.

Casjens, S.R., Mongodin, E.F., Qiu, W.G., Luft, B.J., Schutzer, S.E., Gilcrease, E.B., Huang, W.M., Vujadinovic, M., Aron, J.K., Vargas, L.C., *et al.* (2012). Genome stability of Lyme disease spirochetes: comparative genomics of *Borrelia burgdorferi* plasmids. PLoS One *7*, e33280. https://doi.org/10.1371/journal.pone.0033280

Casjens, S.R., Gilcrease, E.B., Vujadinovic, M., Mongodin, E.F., Luft, B.J., Schutzer, S.E., Fraser, C.M., and Qiu, W.G. (2017). Plasmid diversity and phylogenetic consistency in the Lyme disease agent *Borrelia burgdorferi*. BMC Genomics *18*, 165. https://doi.org/10.1186/s12864-017-3553-5

Chaconas, G., Stewart, P.E., Tilly, K., Bono, J.L., and Rosa, P. (2001). Telomere resolution in the Lyme disease spirochete. EMBO J *20*, 3229-3237. https://doi.org/10.1093/emboj/20.12.3229

Chaconas, G. (2005). Hairpin telomeres and genome plasticity in *Borrelia*: all mixed up in the end. Molecular Microbiology *58*, 625-635.

Chaconas, G. (2010). Replication of the *B. burgdorferi* genome and scrambling of the linear replicons through reverse telomere resolution. In *Borrelia*: Molecular Biology, Host Interaction and Pathogenesis

D.S. Samuels, and J.D. Radolf, eds. (Norwich, UK: Horizon Scientific Press), pp. 49-60.

Chaconas, G., and Kobryn, K. (2010). Structure, function, and evolution of linear replicons in *Borrelia*. Annual review of microbiology *64*, 185-202.

Chaconas, G., and Norris, S.J. (2013). Peaceful coexistence amongst *Borrelia* plasmids: getting by with a little help from their friends? Plasmid *70*, 161-167. https://doi.org/10.1016/j.plasmid.2013.05.002

Chenail, A.M., Jutras, B.L., Adams, C.A., Burns, L.H., Bowman, A., Verma, A., and Stevenson, B. (2012). *Borrelia burgdorferi* cp32 BpaB modulates expression of the prophage NucP nuclease and SsbP single-stranded DNA-binding protein. J Bacteriol *194*, 4570-4578. https://doi.org/10.1128/JB.00661-12

Conley, D.L., and Cohen, S.N. (1995). Isolation and characterization of plasmid mutations that enable partitioning of pSC101 replicons lacking the partition (*par*) locus. J Bacteriol *177*, 1086-1089.

Cui, T., Moro-oka, N., Ohsumi, K., Kodama, K., Ohshima, T., Ogasawara, N., Mori, H., Wanner, B., Niki, H., and Horiuchi, T. (2007). *Escherichia coli* with a linear genome. EMBO reports *8*, 181-187.

Deibler, R.W., Rahmati, S., and Zechiedrich, E.L. (2001). Topoisomerase IV, alone, unknots DNA in E. coli. Genes Dev *15*, 748-761. https://doi.org/10.1101/gad.872301

Deneke, J., Ziegelin, G., Lurz, R., and Lanka, E. (2000). The protelomerase of temperate *Escherichia coli* phage N15 has cleaving-joining activity. Proc Natl Acad Sci U S A *97*, 7721-7726.

Deneke, J., Burgin, A.B., Wilson, S.L., and Chaconas, G. (2004). Catalytic residues of the telomere resolvase ResT: a pattern similar to, but distinct from tyrosine recombinases and type IB topoisomerases. J Biol Chem *279*, 53699-53706.

Deneke, J., and Chaconas, G. (2008). Purification and properties of the plasmid maintenance proteins from the Borrelia burgdorferi linear plasmid lp17. J Bacteriol *190*, 3992-4000. https://doi.org/10.1128/JB.00057-08

Dixon, N.E., and Kornberg, A. (1984). Protein HU in the enzymatic replication of the chromosomal origin of *Escherichia coli*. Proc Natl Acad Sci U S A *81*, 424-428.

Eggers, C.H., and Samuels, D.S. (1999). Molecular evidence for a new bacteriophage of *Borrelia burgdorferi*. J Bacteriol *181*, 7308-7313.

Eggers, C.H., Kimmel, B.J., Bono, J.L., Elias, A.F., Rosa, P., and Samuels, D.S. (2001). Transduction by fBB-1, a bacteriophage of *Borrelia burgdorferi*. J Bacteriol *183*, 4771-4778.

Eggers, C.H., Caimano, M.J., Clawson, M.L., Miller, W.G., Samuels, D.S., and Radolf, J.D. (2002). Identification of loci critical for replication and compatibility of a *Borrelia burgdorferi* cp32-based shuttle vector for the expression of fluorescent reporters in the Lyme disease spirochete. Mol Microbiol *43*, 281-295.

Ferdows, M.S., Serwer, P., Griess, G.A., Norris, S.J., and Barbour, A.G. (1996). Conversion of a linear to a circular plasmid in the relapsing fever agent *Borrelia hermsii*. J Bacteriol *178*, 793-800.

Flardh, K., and Buttner, M.J. (2009). Streptomyces morphogenetics: dissecting differentiation in a filamentous bacterium. Nat Rev Microbiol *7*, 36-49. https://doi.org/nrmicro1968 [pii] 10.1038/nrmicro1968

Flatten, I., Morigen, and Skarstad, K. (2009). DnaA protein interacts with RNA polymerase and partially protects it from the effect of rifampicin. Mol Microbiol *71*, 1018-1030. https://doi.org/10.1111/j.1365-2958.2008.06585.x

Flores, M.J., Ehrlich, S.D., and Michel, B. (2002). Primosome assembly requirement for replication restart in the *Escherichia coli holD*^{G10} replication mutant. Mol Microbiol *44*, 783-792.

Fraser, C.M., Casjens, S., Huang, W.M., Sutton, G.G., Clayton, R., Lathigra, R., White, O., Ketchum, K.A., Dodson, R., Hickey, E.K., *et al.* (1997). Genomic sequence of a Lyme disease spirochaete, *Borrelia burgdorferi*. Nature *390*, 580-586.

Galbraith, K.M., Ng, A.C., Eggers, B.J., Kuchel, C.R., Eggers, C.H., and Samuels, D.S. (2005). parC mutations in fluoroquinolone-resistant Borrelia burgdorferi. Antimicrob Agents Chemother *49*, 4354-4357. https://doi.org/10.1128/AAC.49.10.4354-4357.2005

Gilbert, M.A., Morton, E.A., Bundle, S.F., and Samuels, D.S. (2007). Artificial regulation of *ospC* expression in *Borrelia burgdorferi*. Mol Microbiol *63*, 1259-1273.

Glover, B.P., and McHenry, C.S. (1998). The χ ψ subunits of DNA polymerase III holoenzyme bind to single-stranded DNA-binding protein (SSB) and facilitate replication of an SSB-coated template. J Biol Chem *273*, 23476-23484.

Grainge, I., Bregu, M., Vazquez, M., Sivanathan, V., Ip, S.C., and Sherratt, D.J. (2007). Unlinking chromosome catenanes *in vivo* by site-specific recombination. EMBO J *26*, 4228-4238. https://doi.org/7601849 [pii] 10.1038/sj.emboj.7601849

Groshong, A.M., Gibbons, N.E., Yang, X.F., and Blevins, J.S. (2012). Rrp2, a prokaryotic enhancer-like binding protein, is essential for viability of *Borrelia burgdorferi*. J Bacteriol *194*, 3336-3342.

Heller, R.C., and Marians, K.J. (2005). The disposition of nascent strands at stalled replication forks dictates the pathway of replisome loading during restart. Mol Cell *17*, 733-743. https://doi.org/S1097-2765(05)01084-1 [pii] 10.1016/j.molcel.2005.01.019

Hertwig, S., Klein, I., Lurz, R., Lanka, E., and Appel, B. (2003). PY54, a linear plasmid prophage of *Yersinia enterocolitica* with covalently closed ends. Mol Microbiol *48*, 989-1003.

Hinnebusch, B.J., and Bendich, A.J. (1997). The bacterial nucleoid visualized by fluorescence microscopy of cells lysed within agarose: comparison of *Escherichia coli* and spirochetes of the genus *Borrelia*. J Bacteriol *179*, 2228-2237.

Hinnebusch, J., Bergstrom, S., and Barbour, A.G. (1990). Cloning and sequence analysis of linear plasmid telomeres of the bacterium *Borrelia burgdorferi*. Mol Microbiol *4*, 811-820.

Hinnebusch, J., and Barbour, A.G. (1991). Linear plasmids of *Borrelia burgdorferi* have a telomeric structure and sequence similar to those of a eukaryotic virus. J Bacteriol *173*, 7233-7239.

Huang, S.H., and Kobryn, K. (2016). The *Borrelia burgdorferi* telomere resolvase, ResT, anneals ssDNA complexed with its cognate ssDNA-binding protein. Nucleic acids research *44*, 5288-5298. https://doi.org/10.1093/nar/gkw344

Huang, S.H., Cozart, M.R., Hart, M.A., and Kobryn, K. (2017). The *Borrelia burgdorferi* telomere resolvase, ResT, possesses ATP-dependent DNA unwinding activity. Nucleic acids research *45*, 1319-1329. https://doi.org/10.1093/nar/gkw1243

Huang, W.M., Joss, L., Hsieh, T., and Casjens, S. (2004a). Protelomerase uses a topoisomerase IB/Y-recombinase type mechanism to generate DNA hairpin ends. J Mol Biol *337*, 77-92. https://doi.org/10.1016/j.jmb.2004.01.012

Huang, W.M., Robertson, M., Aron, J., and Casjens, S. (2004b). Telomere exchange between linear replicons of *Borrelia burgdorferi*. J Bacteriol *186*, 4134-4141.

Huang, W.M., DaGloria, J., Fox, H., Ruan, Q., Tillou, J., Shi, K., Aihara, H., Aron, J., and Casjens, S. (2012). Linear chromosome-generating system of *Agrobacterium tumefaciens* C58: protelomerase generates and protects hairpin ends. J Biol Chem *287*, 25551-25563.

Iyer, L.M., Koonin, E.V., and Aravind, L. (2002). Classification and evolutionary history of the single-strand annealing proteins, RecT, Redbeta, ERF and RAD52. BMC Genomics *3*, 8.

Jarvis, T.C., Paul, L.S., and von Hippel, P.H. (1989). Structural and enzymatic studies of the T4 DNA replication system. I. Physical characterization of the polymerase accessory protein complex. J Biol Chem *264*, 12709-12716.

Kantake, N., Madiraju, M.V., Sugiyama, T., and Kowalczykowski, S.C. (2002). *Escherichia coli* RecO protein anneals ssDNA complexed with its cognate ssDNA-binding protein: A common step in genetic recombination. Proc Natl Acad Sci U S A *99*, 15327-15332. https://doi.org/10.1073/pnas.252633399 252633399 [pii]

Kobryn, K., Naigamwalla, D.Z., and Chaconas, G. (2000). Site-specific DNA binding and bending by the *Borrelia burgdorferi* Hbb protein. Mol Microbiol *37*, 145-155.

Kobryn, K., and Chaconas, G. (2002). ResT, a telomere resolvase encoded by the Lyme disease spirochete. Mol Cell *9*, 195-201.

Kobryn, K., Burgin, A.B., and Chaconas, G. (2005). Uncoupling the chemical steps of telomere resolution by ResT. J Biol Chem *280*, 26788-26795.

Kobryn, K., and Chaconas, G. (2005). Fusion of hairpin telomeres by the *B. burgdorferi* telomere resolvase ResT: Implications for shaping a genome in flux. Mol Cell *17*, 783-791.

Kobryn, K. (2007). The linear hairpin replicons of *Borrelia burgdorferi*. In Microbial Linear Plasmids, F. Meinhardt, and R. Klassen, eds. (Berlin Heidelberg: Springer), pp. 117-140.

Kobryn, K., Briffotaux, J., and Karpov, V. (2009). Holliday junction formation by the *Borrelia burgdorferi* telomere resolvase, ResT: implications for the origin of genome linearity. Mol Microbiol *71*, 1117-1130. https://doi.org/MMI6584 [pii] 10.1111/j.1365-2958.2008.06584.x

Kobryn, K., and Chaconas, G. (2014). Hairpin telomere resolvases. In Microbiol Spectrum, pp. MDNA-0023-2014. https://doi.org/doi:10.1128/microbiolspec.MDNA3-0023-2014

Lesterlin, C., Pages, C., Dubarry, N., Dasgupta, S., and Cornet, F. (2008). Asymmetry of chromosome replichores renders the DNA translocase activity of FtsK essential for cell division and cell shape maintenance in *Escherichia coli*. PLoS Genet *4*, e1000288. https://doi.org/10.1371/journal.pgen.1000288

Li, Z., Karakousis, G., Chiu, S.K., Reddy, G., and Radding, C.M. (1998). The beta protein of phage λ promotes strand exchange. J Mol Biol *276*, 733-744. https://doi.org/S0022-2836(97)91572-2 [pii] 10.1006/jmbi.1997.1572

Liang, F.T., Xu, Q., Sikdar, R., Xiao, Y., Cox, J.S., and Doerrler, W.T. (2010). BB0250 of *Borrelia burgdorferi* is a conserved and essential inner membrane protein required for cell division. J Bacteriol *192*, 6105-6115.

Lin, T., Gao, L., Zhang, C., Odeh, E., Jacobs, M.B., Coutte, L., Chaconas, G., Philipp, M.T., and Norris, S.J. (2012). Analysis of an ordered, comprehensive STM mutant library in infectious *Borrelia burgdorferi*: insights into the genes required for mouse infectivity. PLoS One *7*, e47532. https://doi.org/10.1371/journal.pone.0047532

Lucyshyn, D., Huang, S.H., and Kobryn, K. (2015). Spring loading a pre-cleavage intermediate for hairpin telomere formation. Nucleic acids research *43*, 6062-6074. https://doi.org/10.1093/nar/gkv497

Luisi-DeLuca, C., and Kolodner, R. (1994). Purification and characterization of the *Escherichia coli* RecO protein. Renaturation of complementary single-stranded DNA molecules catalyzed by the RecO protein. J Mol Biol *236*, 124-138. https://doi.org/10.1006/jmbi.1994.1123

Marceau, A.H., Bahng, S., Massoni, S.C., George, N.P., Sandler, S.J., Marians, K.J., and Keck, J.L. (2011). Structure of the SSB-DNA polymerase III interface and its role in DNA replication. EMBO J *30*, 4236-4247. https://doi.org/10.1038/emboj.2011.305

McClintock, B. (1941). The stability of broken ends of chromosomes in *Zea mays*. Genetics *26*, 234-282.

Miller, C., and Cohen, S.N. (1999). Separate roles of *Escherichia coli* replication proteins in synthesis and partitioning of pSC101 plasmid DNA. J Bacteriol *181*, 7552-7557.

Miller, C.A., Ingmer, H., and Cohen, S.N. (1995). Boundaries of the pSC101 minimal replicon are conditional. J Bacteriol *177*, 4865-4871.

Mir, T., Huang, S.H., and Kobryn, K. (2013). The telomere resolvase of the Lyme disease spirochete, *Borrelia burgdorferi*, promotes DNA single-strand annealing and strand exchange. Nucleic acids research *41*, 10438-10448.

Moriarty, T.J., and Chaconas, G. (2009). Identification of the determinant conferring permissive substrate usage in the telomere resolvase, ResT. J Biol Chem *284*, 23293-23301.

Mott, M.L., Erzberger, J.P., Coons, M.M., and Berger, J.M. (2008). Structural synergy and molecular crosstalk between bacterial helicase loaders and replication initiators. Cell *135*, 623-634. https://doi.org/S0092-8674(08)01254-3 [pii] 10.1016/j.cell.2008.09.058

Mouw, K.W., and Rice, P.A. (2007). Shaping the *Borrelia burgdorferi* genome: crystal structure and binding properties of the DNA-bending protein Hbb. Mol Microbiol *63*, 1319-1330.

Muller, H. (1938). The remaking of chromosomes. Collecting Net (Woods Hole) *13*, 181-198.

Ohki, R., Tsurimoto, T., and Ishikawa, F. (2001). In vitro reconstitution of the end replication problem. Mol Cell Biol *21*, 5753-5766. https://doi.org/10.1128/mcb.21.17.5753-5766.2001

Olovnikov, A.M. (1973). A theory of marginotomy. The incomplete copying of template margin in enzymic synthesis of polynucleotides and biological significance of the phenomenon. J Theor Biol *41*, 181-190. https://doi.org/10.1016/0022-5193(73)90198-7

Picardeau, M., Lobry, J.R., and Hinnebusch, B.J. (1999). Physical mapping of an origin of bidirectional replication at the centre of the *Borrelia burgdorferi* linear chromosome. Mol Microbiol *32*, 437-445.

Postow, L., Crisona, N.J., Peter, B.J., Hardy, C.D., and Cozzarelli, N.R. (2001). Topological challenges to DNA replication: conformations at the fork. Proc Natl Acad Sci U S A *98*, 8219-8226. https://doi.org/10.1073/pnas.11100699898/15/8219 [pii]

Ravin, N.V., Kuprianov, V.V., Gilcrease, E.B., and Casjens, S.R. (2003). Bidirectional replication from an internal ori site of the linear N15 plasmid prophage. Nucleic acids research *31*, 6552-6560.

Recchia, G.D., Aroyo, M., Wolf, D., Blakely, G., and Sherratt, D.J. (1999). FtsK-dependent and -independent pathways of Xer site-specific recombination. Embo J *18*, 5724-5734.

Reyes-Lamothe, R., Nicolas, E., and Sherratt, D.J. (2012). Chromosome replication and segregation in bacteria. Annual review of genetics *46*, 121-143.

Rudolph, C.J., Upton, A.L., and Lloyd, R.G. (2009). Replication fork collisions cause pathological chromosomal amplification in cells lacking RecG DNA translocase. Mol Microbiol *74*, 940-955. https://doi.org/MMI6909 [pii] 10.1111/j.1365-2958.2009.06909.x

Rudolph, C.J., Upton, A.L., Stockum, A., Nieduszynski, C.A., and Lloyd, R.G. (2013). Avoiding chromosome pathology when replication forks collide. Nature *500*, 608-611. https://doi.org/nature12312 [pii] 10.1038/nature12312

Saveson, C.J., and Lovett, S.T. (1997). Enhanced deletion formation by aberrant DNA replication in Escherichia coli. Genetics *146*, 457-470.

Shereda, R.D., Kozlov, A.G., Lohman, T.M., Cox, M.M., and Keck, J.L. (2008). SSB as an organizer/mobilizer of genome maintenance complexes. Crit Rev Biochem Mol Biol *43*, 289-318. https://doi.org/904496150 [pii] 10.1080/10409230802341296

Shi, K., Huang, W.M., and Aihara, H. (2013). An enzyme-catalyzed multistep DNA refolding mechanism in hairpin telomere formation. PLoS biology *11*, e1001472.

Simonetta, K.R., Kazmirski, S.L., Goedken, E.R., Cantor, A.J., Kelch, B.A., McNally, R., Seyedin, S.N., Makino, D.L., O'Donnell, M., and Kuriyan, J. (2009). The mechanism of ATP-dependent primer-template recognition by a clamp loader complex. Cell *137*, 659-671. https://doi.org/10.1016/j.cell.2009.03.044

Soni, R.K., Mehra, P., Mukhopadhyay, G., and Dhar, S.K. (2005). Helicobacter pylori DnaB helicase can bypass Escherichia coli DnaC function in vivo. Biochem J *389*, 541-548. https://doi.org/BJ20050062 [pii] 10.1042/BJ20050062

Soudet, J., Jolivet, P., and Teixeira, M.T. (2014). Elucidation of the DNA end-replication problem in Saccharomyces cerevisiae. Mol Cell *53*, 954-964. https://doi.org/10.1016/j.molcel.2014.02.030

Stewart, P.E., Thalken, R., Bono, J.L., and Rosa, P. (2001). Isolation of a circular plasmid region sufficient for autonomous replication and transformation of infectious Borrelia burgdorferi. Mol Microbiol *39*, 714-721.

Stewart, P.E., Chaconas, G., and Rosa, P. (2003). Conservation of plasmid maintenance functions between linear and circular plasmids in *Borrelia burgdorferi*. J Bacteriol *185*, 3202-3209.

Stewart, P.E., Hoff, J., Fischer, E., Krum, J.G., and Rosa, P.A. (2004). Genome-wide transposon mutagenesis of *Borrelia burgdorferi* for identification of phenotypic mutants. Appl Environ Microbiol *70*, 5973-5979.

Studwell-Vaughan, P.S., and O'Donnell, M. (1993). DNA polymerase III accessory proteins. V. θ encoded by *holE*. J Biol Chem *268*, 11785-11791.

Szalewska-Palasz, A., Wegrzyn, A., Blaszczak, A., Taylor, K., and Wegrzyn, G. (1998). DnaA-stimulated transcriptional activation of orilambda: *Escherichia coli* RNA polymerase βbo subunit as a transcriptional activator contact site. Proc Natl Acad Sci U S A *95*, 4241-4246.

Tilly, K., Fuhrman, J., Campbell, J., and Samuels, D.S. (1996). Isolation of *Borrelia burgdorferi* genes encoding homologues of DNA- binding protein HU and ribosomal protein S20. Microbiology *142*, 2471-2479.

Tilly, K., Checroun, C., and Rosa, P.A. (2012). Requirements for *Borrelia burgdorferi* plasmid maintenance. Plasmid *68*, 1-12. https://doi.org/10.1016/j.plasmid.2012.01.009

Tourand, Y., Kobryn, K., and Chaconas, G. (2003). Sequence-specific recognition but position-dependent cleavage of two distinct telomeres by the *Borrelia burgdorferi* telomere resolvase, ResT. Mol Microbiol *48*, 901-911.

Tourand, Y., Bankhead, T., Wilson, S.L., Putteet-Driver, A.D., Barbour, A.G., Byram, R., Rosa, P.A., and Chaconas, G. (2006). Differential telomere processing by *Borrelia* telomere resolvases *in vitro* but not *in vivo*. J Bacteriol *188*, 7378-7386.

Tourand, Y., Lee, L., and Chaconas, G. (2007). Telomere resolution by *Borrelia burgdorferi* ResT through the collaborative efforts of tethered DNA binding domains. Mol Microbiol *64*, 580-590.

Tourand, Y., Deneke, J., Moriarty, T.J., and Chaconas, G. (2009). Characterization and *in vitro* reaction properties of 19 unique hairpin telomeres from the linear plasmids of the Lyme disease spirochete. J Biol Chem *284*, 7264-7272.

Zhang, J.R., Hardham, J.M., Barbour, A.G., and Norris, S.J. (1997). Antigenic variation in Lyme disease borreliae by promiscuous recombination of VMP-like sequence cassettes. Cell *89*, 275-285.

Chapter 5

Gene Regulation and Transcriptomics

D. Scott Samuels[1]*, Meghan C. Lybecker[2], X. Frank Yang[3], Zhiming Ouyang[4], Travis J. Bourret[5], William K. Boyle[5], Brian Stevenson[6], Dan Drecktrah[1] and Melissa J. Caimano[7]

[1]Division of Biological Sciences, University of Montana, Missoula, MT 59812, USA
[2]Department of Biology, University of Colorado, Colorado Springs, CO 80918, USA
[3]Department of Microbiology and Immunology, Indiana University School of Medicine, Indianapolis, IN4 6202, USA
[4]Department of Molecular Medicine, University of South Florida, Tampa, FL 33612, USA
[5]Department of Medical Microbiology and Immunology, Creighton University, Omaha, NE, 68105 USA
[6]Department of Microbiology, Immunology and Molecular Genetics, University of Kentucky School of Medicine, Lexington, KY 40536, USA
[7]Departments of Medicine, Pediatrics, and Molecular Biology and Biophysics, UConn Health, Farmington, CT, USA
*Corresponding author: samuels@mso.umt.edu

DOI: https://doi.org/10.21775/9781913652616.05

Abstract

Borrelia (*Borreliella*) *burgdorferi*, along with closely related species, is the etiologic agent of Lyme disease. The spirochete subsists in an enzootic cycle that encompasses acquisition from a vertebrate host to a tick vector and transmission from a tick vector to a vertebrate host. To adapt to its environment and persist in each phase of its enzootic cycle, *B. burgdorferi* wields three systems to regulate the expression of genes: the RpoN-RpoS alternative sigma (σ) factor cascade, the Hk1/Rrp1 two-component system and its product c-di-GMP, and the stringent response mediated by Rel_{Bbu} and DksA. These regulatory systems respond to enzootic phase-specific signals and are controlled or fine-tuned by transcription factors, including BosR and BadR, as well as small RNAs, including $DsrA_{Bb}$ and Bb6S RNA. In addition, several other DNA-binding and RNA-binding proteins have been identified, although their functions have not all been defined. Global changes in gene expression revealed by high-throughput transcriptomic studies have elucidated various regulons, albeit technical obstacles have mostly limited this experimental approach to cultivated spirochetes. Regardless, we know that the spirochete, which carries a relatively small genome, regulates the expression of a considerable number of genes required for the transitions between the tick vector and the vertebrate host as well as the adaptation to each.

Introduction

The Lyme disease spirochete *Borrelia* (*Borreliella*) *burgdorferi*, and closely related species *B. afzelii, B. garinii*, *B. bissettii*, and *B. mayonii*, are maintained in an enzootic cycle involving a tick vector and a vertebrate host (Figure 1) (Lane et al., 1991; Radolf et al., 2012; Caimano et al., 2016). The spirochetes regulate the expression of their genes in a phase-specific fashion as they traverse this cycle in nature, moving back and forth between the tick and a vertebrate. *Ixodes* larvae feed on an infected host and acquire the spirochetes, which take up residence in the midgut. Following the molt, nymphs feed on a vertebrate and the spirochetes exit the midgut, migrate through the hemocoel, reach the salivary glands, and transmit to the naïve host (Ribeiro et al., 1987; Dunham-Ems et al., 2009). The spirochete survives in these two disparate habitats, and moves between them at the proper time, by sensing biological cues that alter the expression of the requisite collection of genes. Different 'modulons' are expressed as *B. burgdorferi* cycles from tick to vertebrate and vertebrate to tick, and adapts to these different environments (Iyer et al., 2015). The physiological signposts of the enzootic cycle, molecular cues prompting the spirochete to regulate its gene expression, have not been fully delineated nor have the mechanisms of signal integration and transduction to the regulatory machinery been elucidated. Schwan et al. (1995) uncovered temperature as an environmental signal; these seminal findings suggested that the increased temperature of the warm blood meal regulates the levels of outer membrane lipoproteins during tick feeding. *B. burgdorferi* is exposed to a variety of other factors that vary during the enzootic cycle and have been shown to affect gene expression, including pH (Carroll et al., 1999; Yang et al., 2000), dissolved oxygen (Seshu et al., 2004a) and carbon dioxide (Hyde et al., 2007), transition metals (Troxell et al., 2013), nutrients (Bugrysheva et al., 2015; Drecktrah et al., 2015), osmolarity (Bontemps-Gallo et al., 2016), weak organic acids (Dulebohn et al., 2017), and short-chain fatty acids (Lin et al., 2018). Nevertheless, uncertainty remains regarding the relative roles of each and their interactions.

This chapter covers our current understanding of gene regulation during the enzootic cycle; since the original chapter in the first edition of this book (Skare et al., 2010), a number of recommended reviews have been published (Samuels, 2011; Radolf et al., 2012; Kung et al., 2013; Troxell and Yang, 2013; Groshong and Blevins, 2014; Caimano et al., 2016; Iyer and Schwartz, 2016; Samuels and Samuels, 2016; Ye et al., 2016; Lybecker and Samuels, 2017; Stevenson and Seshu, 2018). We begin the chapter

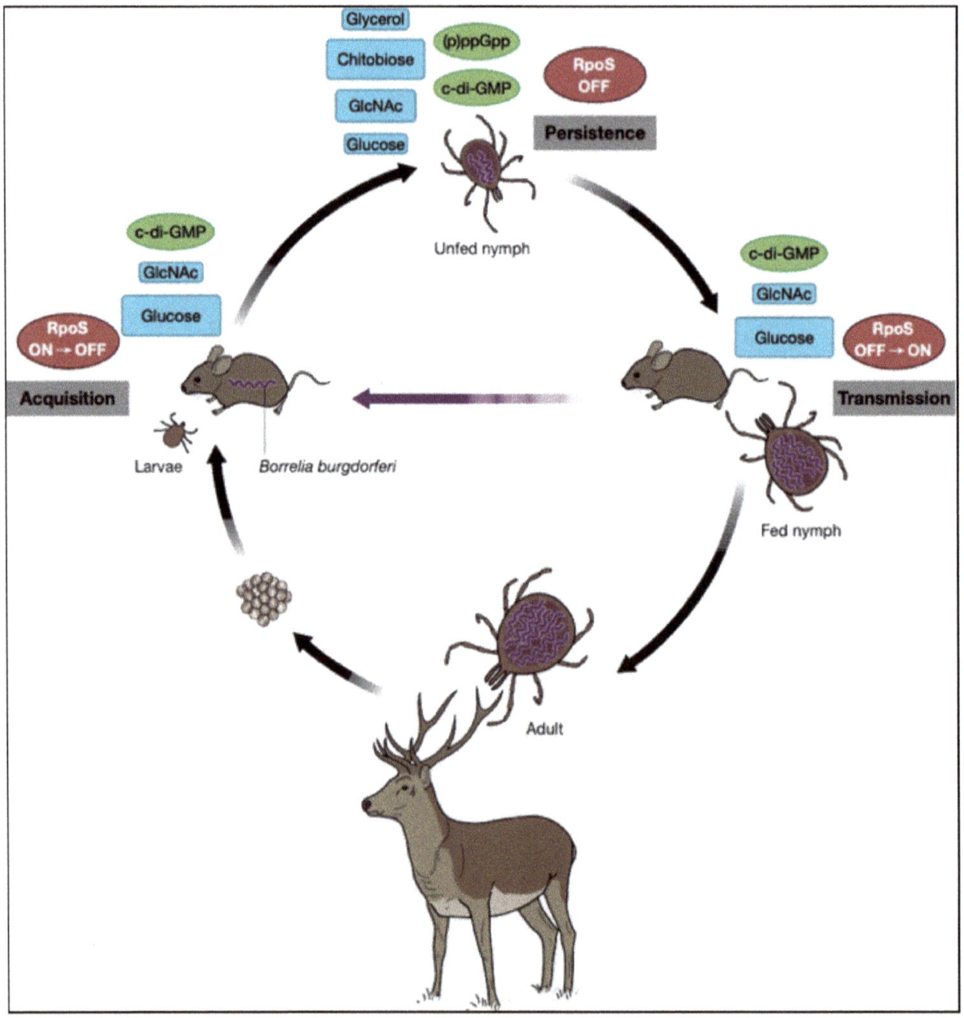

Figure 1. The enzootic cycle of *B. burgdorferi* showing milestones indicating the action of gene regulatory systems. Acquisition: larval ticks acquire *B. burgdorferi* by feeding on an infected vertebrate. Persistence: intracellular second messengers (p)ppGpp and c-di-GMP regulate persistence in the tick by various mechanisms, including control of genes encoding carbohydrate utilization, while the molecular gatekeeper RpoS is absent in unfed ticks. Transmission: following the molt into nymphs, infected ticks can transmit *B. burgdorferi* to uninfected vertebrates, completing the cycle. Reprinted with permission from Caimano et al. (2016). GlcNAc, *N*-acetyl glucosamine.

by introducing the enzyme catalyzing transcription, and invoke the maxim *spirochetes do it differently*, as the *B. burgdorferi* RNA polymerase is unusually, albeit not uniquely, dependent on manganese (Boyle et al., 2020). We know of three main gene regulatory systems that are in play throughout the enzootic cycle: the RpoN-RpoS alternative sigma (σ) factor cascade; the Hk1/Rrp1 two-component system and its second messenger c-di-GMP; and stringent response regulators, Rel_{Bbu} and DksA, and the alarmone (p)ppGpp. Orthologs of several transcriptional regulators that have been thoroughly characterized in model microorganisms are involved in modulating these gene regulatory networks, although in ways that, more often than not, are utterly idiosyncratic for this deeply branching group of bacteria (Samuels and Radolf, 2009). We discuss nucleic-acid binding proteins and mention a relatively recently described signaling molecule. We then take up post-transcriptional mechanisms, focusing on small regulatory RNAs, before turning our attention to what has been learned via high-throughput transcriptomic approaches.

RNA polymerase

The transcription of genes in *B. burgdorferi* is carried out by the RNA polymerase holoenzyme consisting of subunits encoded by *rpoA* (α subunit), *rpoB* (β subunit), *rpoC* (β′ subunit), and *rpoZ* (ω subunit), which constitute the core RNA polymerase, along with one of three σ factors encoded by *rpoD* (housekeeping sigma factor σ^{70} or σ^{D}), *rpoN* (σ^{N} or σ^{54}), or *rpoS* (σ^{S} or σ^{38}). It has the same $\alpha_2\beta\beta'\omega$ plus σ architecture as other bacterial RNA polymerases. The σ subunit is required for recognition of gene promoters and separation of the two strands of the double helix (Marchetti et al., 2017). The σ subunit associates with the $\alpha_2\beta\beta'\omega$ core to initiate transcription and then elongation proceeds, often after the σ subunit dissociates from the holoenzyme. The largest subunit, β′, houses the active site that catalyzes RNA synthesis. The ω subunit promotes RNA polymerase activity in other bacteria (Mathew and Chatterji, 2006; Kurkela et al., 2020); however, its role in the *B. burgdorferi* RNA polymerase has yet to be characterized.

Unlike prototypical bacterial RNA polymerases that require magnesium for activity (Sosunov et al., 2003), the enzymatic activity of the *B. burgdorferi* RNA polymerase is manganese-dependent (Boyle et al., 2020). Manganese-dependent RNA polymerases have been found in other bacteria, including *Clostridium acetobutylicum*, *Bacillus subtilis*, and *Lactobacillus curvatus* (Rutberg and Armentrout, 1972; Stetter and Zillig, 1974; Pich and Bahl, 1991). The molecular basis of the manganese requirement for the *B. burgdorferi* RNA polymerase is unclear, as the conserved catalytic site amino acids within the β′ subunit are identical to those found in both magnesium- and manganese-utilizing bacteria. The spirochete regulates the cellular concentrations of manganese through the BmtA transporter (Ouyang et al., 2009a; Wagh et al., 2015), and alterations in extracellular levels of manganese affect the expression of a variety of virulence genes (Troxell and Yang, 2013; Troxell et al., 2013). Changes in *B. burgdorferi* gene expression in response to lower concentrations of manganese may be the consequence of reducing the bioavailability of manganese required for RNA polymerase activity, resulting in transcriptional stalling as described in other bacteria (Rutberg and Armentrout, 1972; Stetter and Zillig, 1974; Borbely and Schneider, 1988; Pich and Bahl, 1991; Sosunov et al., 2003; Poranen et al., 2008; Agapov et al., 2017).

The enzymatic activity of the *B. burgdorferi* RNA polymerase also responds to temperature and pH (Boyle et al., 2020), which is notable because the environmental temperature and pH vary as *B. burgdorferi* traverses its enzootic cycle (Carroll et al., 1999; Schwan and Piesman, 2000; Yang et al., 2000). *In vitro* transcription assays initiated from RpoD (σ^{70})-dependent promoters using the *B. burgdorferi* RNA polymerase core and recombinant RpoD σ factor showed RNA polymerase activity is affected by temperature with maximal activity at 37°C. Similarly, RNA polymerase activity is pH-dependent within the physiological range experienced by *B. burgdorferi* with the lowest activity observed at pH 6.8 and the highest activity at pH 7.5 (Boyle et al., 2020). These *in vitro* experiments suggest regulation of *B. burgdorferi* gene expression in response to vector and host environments might be a consequence, at least partially, of biochemical changes in RNA polymerase activity.

During growth of *B. burgdorferi in vitro*, RpoD (σ^{70}) is the primary σ factor responsible for the initiation of transcription, as the alternative σ factors, RpoN (σ^{N}) and RpoS (σ^{S}), are either inactive or present at low levels in the cell during logarithmic growth of *B. burgdorferi* (Hübner et al., 2001; Burtnick et al., 2007; Bontemps-Gallo et al., 2016). Sites of RpoD-dependent transcriptional initiation were mapped by *in vitro* transcription reactions initiated by recombinant RpoD (Adams et al., 2017; Boyle et al., 2020). The ability of *B. burgdorferi* to complete its enzootic cycle is dependent on RNA polymerase utilization of RpoN and RpoS (see the next section on the RpoN-RpoS alternative σ factor cascade). The expression of *rpoN* and *rpoS*, along with corresponding changes in RpoN- and RpoS-dependent gene expression, occur in response to environmental signals. While much of the work on gene regulation has focused on the impact of transcription factors, alternative σ factors and small RNAs in *B. burgdorferi*, the intracellular environment, including pH, temperature, and manganese availability, may directly alter RNA polymerase function along with the interactions of gene regulatory systems with RNA polymerase.

RpoN-RpoS alternative σ factor cascade

The RpoN-RpoS alternative σ factor cascade has undoubtedly been the most studied gene regulatory pathway in *B. burgdorferi* (Samuels, 2011; Radolf et al., 2012; Samuels and Samuels, 2016; Ye et al., 2016; Stevenson and Seshu, 2018) (Figure 2). It plays a vital role in the enzootic cycle of *B. burgdorferi* by activating transcription of genes that are essential for transmission and vertebrate host infection, while repressing genes that are required for spirochete survival in the tick vector (Hübner et al., 2001; Caimano et al., 2004; Caimano et al., 2005; Fisher et al., 2005; Caimano et al., 2007; Ouyang et al., 2008; Dunham-Ems et al., 2012; Grove et al., 2017; Arnold et al., 2018). The housekeeping σ factor (e.g., σ^{70} in *E. coli*) is responsible for transcription of most genes (see the section on RNA polymerase above). On the other hand, alternative σ factors recognize different promoters and allow for transcription of a subpopulation of genes in response to environmental or developmental signals (Kazmierczak et al., 2005; Österberg et al., 2011). RpoN (σ^N or σ^{54}; *bb0450*) and RpoS (σ^S or σ^{38}; *bb0771*) are the only two alternative σ factors encoded in the *B. burgdorferi* genome (Fraser et al., 1997). RpoN directly controls *rpoS* expression, which constitutes the RpoN-RpoS σ factor cascade (or RpoN-RpoS pathway).

RpoN is a unique σ factor that does not belong to the σ^{70} (RpoD) family (Reitzer and Schneider, 2001; Wigneshweraraj et al., 2008). RpoN-dependent gene expression possesses several distinctive features. First, RpoN-holoenzyme ($E\sigma^{54}$) exclusively requires a transcriptional activator, called bacterial enhancer binding protein (bEBP) (Bush and Dixon, 2012). *B. burgdorferi* has a single bEBP, Rrp2 (Yang et al., 2003). Second, RpoN recognizes a highly conserved –24/–12 promoter sequence that differs dramatically from the –35/–10 promoter sequence recognized by σ^{70} family σ factors (Studholme and Buck, 2000). Third, RpoN-dependent gene activation requires ATP binding (although not hydrolysis). As such, RpoN controls the expression of only a few genes in most bacteria (Reitzer and Schneider, 2001; Kazmierczak et al., 2005). In *B. burgdorferi*, the only gene with an RpoN-type promoter identified is *rpoS*, which has a perfect –24/–12 promoter sequence (Hübner et al., 2001; Smith et al., 2007).

RpoS from *B. burgdorferi* is evolutionarily distinct from the prototypical RpoS of γ-proteobacteria (Chiang and Schellhorn, 2010). However, it retains key features of its namesake, including the presence of a region σ1.2 and promoter recognition via an extended –10 (Eggers et al., 2006; Caimano et al., 2019). In most bacteria, RpoS serves to globally regulate gene expression to adapt to environmental stresses, such as entering stationary growth phase (reviewed in Hengge-Aronis, 2002a). RpoS accumulates following low or high temperature, low pH, high osmolarity, oxidative stress, UV exposure, and carbon starvation, resulting in regulation of genes to address the stressful conditions and prevent cellular damage. Yet, in *B. burgdorferi*, RpoS controls expression of enzootic phase-specific genes rather than the general stress response, and is required for infection of the vertebrate host (Caimano et al., 2004; Caimano et al., 2019) as well as transmission during tick feeding (Fisher et al., 2005; Dunham-Ems et al., 2012). RpoS in *B. burgdorferi* induces vertebrate host-specific genes and represses tick vector-specific genes (Hübner et al., 2001; Caimano et al., 2004; Caimano et al., 2005; Fisher et al., 2005; Caimano et al., 2007; Ouyang et al., 2008; Grove et al., 2017; Arnold et al., 2018) and, thus, RpoS has been designated the "gatekeeper" of the enzootic cycle (Figure 1) (Caimano et al., 2007; Mulay et al., 2009; Caimano et al., 2019).

RpoN and RpoS constitute a σ factor cascade

The discovery of the RpoN-RpoS σ factor cascade by Hübner et al. (2001) was made possible by inactivation of the *rpoS* and *rpoN* genes in an infectious *B. burgdorferi* strain. Both *rpoS* and *rpoN* mutants showed a striking phenotype: loss of production of OspC and DbpB/A, two of the abundant surface antigens. Abrogation of OspC and DbpB/A in

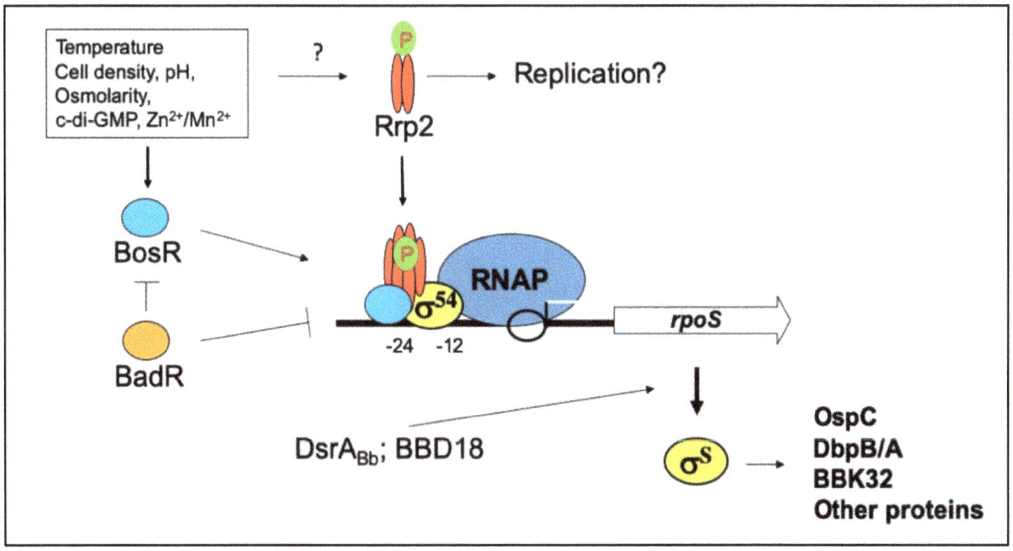

Figure 2. A model of the RpoN-RpoS σ factor cascade. During tick feeding and mammalian infection, environmental and host signals activate the RpoN-RpoS σ factor cascade. Activation of RpoN requires phosphorylation of Rrp2 and accumulation of BosR. Rrp2 is the sole prokaryotic enhancer-binding protein present in *B. burgdorferi* that is required for RpoN (σ^{54}) activation. Phosphorylation of Rrp2 not only is required for RpoN-RpoS activation, but also is indispensable for cell survival, presumably replication. Levels of BosR respond to environmental signals and accumulation of BosR activates *rpoS* at its RpoN-dependent promoter via an unknown mechanism. BadR represses *rpoS* transcription by directly binding near the RpoN-dependent promoter region. In addition to the major *rpoS* mRNA species transcribed from the RpoN-dependent promoter, a longer *rpoS* transcript is produced at low cell density from an RpoN-independent promoter located within the upstream *flgJ* gene. The sRNA $DsrA_{Bb}$ regulates the efficiency of long *rpoS* mRNA species translation in response to temperature. DDB18 can regulate RpoS (σ^S) levels at the post-transcriptional level. Accumulation of *rpoS* transcript leads to the production of OspC, DbpA, DbpB, BBK32, and other mammalian infection-associated proteins.

the *rpoN* mutant was due to abolished production of RpoS, and constitutive expression of *rpoS* in the *rpoN* mutant rescued OspC and DbpB/A production. RpoN controlling *rpoS* activation appeared to be direct, as the *rpoS* gene has a perfect consensus –24/–12 RpoN-dependent promoter sequence (Hübner et al., 2001; Burtnick et al., 2007; Lybecker and Samuels, 2007; Smith et al., 2007). Further biochemical and genetic analyses indicated that the –24/–12 RpoN-dependent promoter is the major promoter responsible for cell density-dependent *rpoS* expression (Burtnick et al., 2007; Lybecker and Samuels, 2007; Smith et al., 2007).

To date, *rpoS* is the only gene identified whose transcription has been demonstrated to be directly controlled by RpoN in *B. burgdorferi*. Virtually all differentially expressed genes in the *rpoS* and *rpoN* mutants are overlapping (Ouyang et al., 2008), supporting the hypothesis that RpoN modulates *B. burgdorferi* genes mainly through RpoS; however, other data suggest that RpoN and RpoS can regulate different groups of genes under certain conditions (Fisher et al., 2005; Arnold et al., 2018). In addition, a second, RpoD (σ^{70})-dependent, promoter (–35/–10) has been reported upstream that transcribes a longer *rpoS* mRNA (Lybecker and Samuels, 2007). Translation of this long *rpoS* mRNA is regulated by the small RNA (sRNA) $DsrA_{Bb}$ (Lybecker and Samuels, 2007) and may explain the controversy regarding the RpoN-dependent and RpoS-dependent transcriptomes (see section below on Small regulatory RNAs and RNA chaperones). There is evidence that RpoS is post-transcriptionally regulated by BBD18 encoded on the linear plasmid lp17, although the mechanistic details are unknown (Dulebohn et al., 2014) (Figure 2).

Genes controlled by RpoN-RpoS σ factor cascade

The first genes identified as being under the control of RpoN-RpoS σ factor cascade were *ospC* and *dbpBA* (Hübner et al., 2001); microarray and RNA-seq analyses have revealed that the RpoN-RpoS σ factor cascade influences the expression of more than 100 genes (Fisher et al., 2005; Caimano et al., 2007; Boardman et al., 2008; Ouyang et al., 2008; Arnold et al., 2018; Caimano et al., 2019) (see the section on Transcriptomics below). Many of the vertebrate phase-specific are under the control of the RpoN-RpoS σ factor cascade and include, in addition to *ospC* and *dbpBA, bbk32, oppA5, bba64, bba66, ospF* and *mlp* (Fischer et al., 2003; Eggers et al., 2004; Grimm et al., 2004; Clifton et al., 2006; Fischer et al., 2006; Seshu et al., 2006; Gilmore et al., 2007; Medrano et al., 2007; Gautam et al., 2008; Maruskova et al., 2008; Shi et al., 2008). Many of these genes are differentially regulated under a variety of environmental conditions including temperature, pH, blood, and host adaptation (Revel et al., 2002; Brooks et al., 2003; Ojaimi et al., 2003; Tokarz et al., 2004; Caimano et al., 2007). These genes often encode surface lipoproteins, but many are hypothetical genes with no known function. Several chromosomal genes, including genes involved in chemotaxis and metabolism are also controlled by the RpoN-RpoS σ factor cascade (Fisher et al., 2005; Caimano et al., 2007; Ouyang et al., 2008). Recently, Caimano et al. (2019) compared the RpoS regulons of two *B. burgdorferi* strains and postulated that the RpoN-RpoS σ factor cascade controls two cohorts of genes: a core group that is required for maintenance of *B. burgdorferi* in the enzootic cycle and a group of plasmid-encoded genes encoding variable surface lipoproteins important for adaptation to diverse vertebrate reservoir hosts.

RpoS controls *B. burgdorferi* gene expression both directly and indirectly. Genetic evidence indicates that the –35/–10 promoters of *ospC*, *ospF* and several other RpoS-regulated genes are recognized exclusively by RpoS; in other words, RpoS directly governs *ospC* expression by binding to the *ospC* promoter (Eggers et al., 2004; Yang et al., 2005). RpoD (σ^{70}) and RpoS (σ^{S}) are closely related and recognize similar core promoter elements (Gaal et al., 2001; Hengge-Aronis, 2002b). Promoter selectivity by the two σ factors in *E. coli* is complicated, involving minor sequence differences, architectural DNA-binding proteins, or DNA supercoiling (Kusano et al., 1996; Colland et al., 2000; Becker and Hengge-Aronis, 2001; Gaal et al., 2001; Hengge-Aronis, 2002b; Typas et al., 2007). Alternative σ factor utilization by RNA polymerase is also influenced by 6S RNA (Wassarman and Storz, 2000; Cavanagh and Wassarman, 2014; Steuten et al., 2014a; Steuten et al., 2014b; Wassarman, 2018) (see section below on Small regulatory RNAs and RNA chaperones). In *B. burgdorferi*, the sequence within the extended –10 region of the promoter contributes to RpoS recognition and a point mutation of C to G at position –13 abolished RpoS-dependent expression of *ospC* (Eggers et al., 2004; Yang et al., 2005; Caimano et al., 2019). Replacing the –10 region of the RpoD-dependent promoter of the *ospE* gene with the –10 region of the RpoS-dependent promoter of *ospF* converted *ospE* to an RpoS-dependent gene (Eggers et al., 2006). A putative RpoS-dependent consensus extended –10 region sequence of TG(G/A)(G/A)ATA(T/A)ATT has been proposed based on analyzing the promoters of ten RpoS-dependent genes of *B. burgdorferi* (Caimano et al., 2007). Although *ospC* transcription is chiefly controlled by RpoS, low levels of *ospC* mRNA have been detected in the absence of RpoS: He et al. (2008) reported that *ospC* expression was greatly reduced, but not completely eliminated in a *rpoS ospAB* double mutant. A recent transcriptome study detected *ospC* expression in the *rpoN* and *rpoS* mutants (Arnold et al., 2018). Given that RpoD and RpoS recognize similar promoter sequences, this *ospC* mRNA is likely due to leaky transcription mediated by RpoD.

RpoS appears also to regulate *B. burgdorferi* gene expression in an indirect manner. The *bba66* genes has a 29-bp inverted repeat (IR) element upstream of the promoter that is required for transcription, suggesting that RpoS may influence *bba66* indirectly via an unidentified factor that binds to the IR motif (Clifton et al., 2006). In addition, some genes are only partially controlled by RpoS, such as *gltP* (encoding a glutamate transporter) and *cdr* (encoding NAD-dependent coenzyme A disulphide reductase). These genes likely have a promoter that is recognized by both RpoS and RpoD (Caimano et al., 2007; Eggers et al., 2011).

In addition to activating transcription, RpoS represses expression of some *B. burgdorferi* genes (Caimano et al., 2005; Caimano et al., 2007; Grove et al., 2017; Arnold et al., 2018). Several of these RpoS-repressed genes are important for tick colonization and are downregulated following transmission, including *ospAB*, *lp6.6*, *bba74,* and *bb0365* (de Silva et al., 1996; Schwan and Piesman, 2000; Pal et al., 2008; Mulay et al., 2009). Some *bdr* paralogs, which encode members of the *B*orrelia direct repeat protein family with no known function (Zückert et al., 1999), are also repressed by RpoS (Roberts et al., 2002). While the mechanism of RpoS-dependent repression of this regulon remains unclear, the T-rich track upstream of the promoter was proposed to be involved (Sohaskey et al., 1999; Brooks et al., 2003;

Caimano et al., 2007). Recently, using a GFP reporter and the dialysis membrane chambers (DMC) model, Grove et al. (2017) uncovered two distinct mechanisms of RpoN-dependent gene repression: one is direct repression (*ospA*, *glp*) in which RpoS competes with RpoD at the –35/–10 sequence for promoter binding and the other is indirect repression (*bba74*) via an unknown *trans*-acting repressor whose expression is controlled by RpoS. However, other data suggest that BosR, a transcriptional activator of *rpoS*, not RpoS, directly represses *ospAB* and other lipoprotein genes (Wang et al., 2013; Shi et al., 2014) (see next section on Transcription factors BosR and BadR).

Factors regulating the RpoN-RpoS σ factor cascade
Rrp2, encoded by *bb0763*, is the sole bEBP present in *B. burgdorferi* (Yang et al., 2003). Rrp2 belongs to the well-studied NtrC protein family (Studholme and Buck, 2000). Members of this family contain three putative functional domains: an N-terminal response regulator domain, a central RpoN-activation domain, and a C-terminal helix-turn-helix (HTH) DNA-binding domain. The central domain becomes activated upon phosphorylation at a conserved Asp residue within the N-terminal domain. The activated central domain then contacts the RpoN-holoenzyme through DNA looping, hydrolyzes ATP, and promotes open promoter complex formation for transcriptional initiation. Yang et al. (2003) first demonstrated that Rrp2 is the upstream activator for the RpoN-RpoS alternative σ factor cascade (Figure 2): a single G239C point mutation within the central activation domain of Rrp2 abolished expression of *rpoS*, *ospC*, and other RpoN-RpoS pathway-dependent genes. The G239C mutation did not abolish the ATPase activity of Rrp2, suggesting that the mutation affects interaction of Rrp2 with RpoN necessary for initiation of open complex formation (Ouyang and Zhou, 2017). Artificial induction of an inducible *rrp2* construct also resulted in *rpoS* transcription (Ouyang et al., 2014c). Interestingly, Rrp2 activates *rpoS* transcription without binding to a specific enhancer binding site (Burtnick et al., 2007; Blevins et al., 2009), despite having a predicted HTH-containing C-terminal domain and being capable of non-specific binding to DNA (Burtnick et al., 2007; Blevins et al., 2009). Surprisingly, *rpoN* is not required for *B. burgdorferi* growth and is readily inactivated, but *rrp2* is essential (Yang et al., 2003; Burtnick et al., 2007; Blevins et al., 2009; Groshong et al., 2012; Yin et al., 2016). This suggests that Rrp2 has an additional role besides serving as the activator for the RpoN-RpoS σ factor cascade. Phosphorylation, ATPase activity, and the C-terminal-mediated DNA-binding are all required for the essential function of Rrp2, but the mechanistic details remain unclear (Yin et al., 2016; Ouyang and Zhou, 2017). Further work, such as chromatin immunoprecipitation (ChIP), will be required to identify Rrp2-dependent but RpoN-independent genes to gain insight the essential nature of Rrp2.

Given that Rrp2 is a two-component response regulator, its cognate histidine kinase Hk2 (BB0763) predicted in the genome was hypothesized to receive upstream signals to phosphorylate and activate Rrp2, which, in turn, activates the RpoN-RpoS σ factor cascade (Yang et al., 2003; Burtnick et al., 2007; Liu et al., 2020). However, unlike *rrp2*, *hk2* is not essential for cell survival and, unexpectedly, disruption of *hk2* does not affect the expression of *rpoS* and *ospC*, suggesting that Hk2 is not required for Rrp2 phosphorylation. One the other hand, Hk2 can function as a phosphatase to dephosphorylate Rrp2 and a strain that overexpresses Hk2 has reduced level of phosphorylated Rrp2 and impaired growth (Liu et al., 2020). Further study indicated that acetyl phosphate, a high-energy metabolic intermediate that can donate phosphoryl or acetyl groups, activates Rrp2 via phosphorylation (Xu et al., 2010); the hypothesis was compelling as *B. burgdorferi* senses the blood meal, begins to replicate and activates the RpoN-RpoS σ factor cascade upon tick feeding, so acetyl phosphate would serve to connect metabolic status and enzootic phase-specific gene expression. However, inactivation of the AckA-Pta pathway that produces acetyl phosphate proved to be lethal to spirochete survival but did not affect the expression of *rpoS* and *ospC* (Richards et al., 2015), arguing against a role for acetyl phosphate in Rrp2 phosphorylation. Therefore, the source of phosphate and the mechanism of Rrp2 phosphorylation are enigmatic, although acetyl phosphate, Hk2 and other small phosphate donors might have overlapping and complementary functions.

Another two-component system present in the *B. burgdorferi* genome, Hk1-Rrp1, appears to engage in cross-talk with the Rrp2-controlled RpoN-RpoS σ factor cascade (Rogers et al., 2009; He et al., 2011; Sze et al., 2013; He et al., 2014; Caimano et al., 2015) (see section below on Hk1/Rrp1 two-component system and c-di-GMP). The interplay between the Hk1-Rrp1 and Hk2-Rrp2 signaling systems has only begun to be revealed. Sze et al. (2013) reported that Rrp1 affects *rpoS* via regulation of *bosR*. c-di-GMP, synthesized by Rrp1, controls *rpoS* expression through the c-di-GMP binding protein PlzA and PlzA links the two sets of two-component systems (He et al., 2014; Caimano et al., 2015; Zhang et al., 2018).

Transcription factors BosR and BadR

BosR

BosR (BB0647; *Borrelia* oxidative stress regulator) is a transcriptional regulator that belongs to the ferric uptake regulator (Fur) family (Fraser et al., 1997; Boylan et al., 2003; Katona et al., 2004; Samuels and Radolf, 2009). It shares 22.5% and 21.8% identity, and 43% and 49% similarity, respectively, with the Fur protein in *Vibrio cholerae* and the stress-responsive Fur family protein PerR in *Bacillus subtilis*. *B. burgdorferi* does not rely on iron (Posey and Gherardini, 2000; Wang et al., 2012b), so BosR seemed unlikely to regulate iron homeostasis in the spirochete. Instead, BosR is analogous to PerR and was first reported to regulate the expression of *dps/napA/bicA* as well as other genes involved in oxidative stress response in *B. burgdorferi* (Boylan et al., 2003), but BosR was eventually shown to activate *rpoS* transcription and, thus, the RpoS-dependent regulon (Hyde et al., 2009; Ouyang et al., 2009b; Hyde et al., 2010; Ouyang et al., 2011; Ouyang et al., 2014a; Katona, 2015).

Expression and regulation of *bosR*

Transcription of *bosR* is driven by two promoters: a strong RpoD-dependent promoter located directly upstream of *bosR* (Katona et al., 2004; Hyde et al., 2010; Katona, 2015; Ouyang et al., 2016) and a second RpoD-dependent promoter upstream of an operon in which *bosR* is co-transcribed with *bb0648* (encoding a putative serine/threonine kinase) and *bb0646* (encoding a putative lipase) (Ouyang et al., 2009b; Ouyang et al., 2016). While *bosR* is expressed throughout the entire enzootic cycle (Medrano et al., 2007; Ouyang et al., 2016), environmental factors, such as temperature and pH, modulate its transcription (Katona et al., 2004; Ouyang et al., 2016). The intracellular second messengers c-di-GMP (Sze et al., 2013; He et al., 2014; Novak et al., 2014; Ye et al., 2014) and the alarmones (p)ppGpp (Bugrysheva et al., 2015; Drecktrah et al., 2015; Drecktrah et al., 2018) induce *bosR* expression at both transcriptional and post-transcriptional levels. Dissolved carbon dioxide (Hyde et al., 2007) and the transition metals manganese, zinc and copper (Troxell and Yang, 2013; Troxell et al., 2013; Wang et al., 2017) influence *bosR* expression at the post-transcriptional level. BadR, a ROK repressor, negatively affects *bosR* expression at both transcriptional and post-transcriptional levels (Miller et al., 2013; Ouyang and Zhou, 2015). Transcription of *bosR* also is activated by the BosR protein, indicating that BosR is an auto-activator (Ouyang et al., 2016), which is different from classical Fur family proteins as auto-repressors. The expression and regulatory patterns of *bosR* highlight the crucial role of BosR in the enzootic cycle of *B. burgdorferi*.

BosR is an atypical Fur family member

Fur family proteins consist of two structural domains including a N-terminal winged helix-turn-helix DNA-binding domain and a C-terminal dimerization domain, connected by a flexible hinge region. Structural analyses of numerous Fur proteins have revealed two sites involved in metal binding (Helmann, 2014; Pinochet-Barros and Helmann, 2018; Sarvan et al., 2018; Nader et al., 2019). The first site (S1), located in the dimerization domain, is usually occupied by a zinc ion and is crucial for protein dimerization, so it is called the structural site. The second metal binding site (S2) is located in the hinge region between the DNA-binding domain and the dimerization domain. The metal ion (e.g., iron, manganese, zinc, or nickel) occupying the S2 site determines the specific regulatory function of the Fur protein; thus, it is called the regulatory site. Occasionally, Fur proteins such as those in *Helicobacter pylori* and *Campylobacter jejuni* contain a third metal binding site (S3) in the dimerization domain, which is also important for the regulatory function of the protein (Dian et al., 2011; Butcher et al., 2012). The regulatory functions of Fur family members are responsive to the availability of transition metal ions such as iron, manganese, zinc, and nickel (Hantke, 1981, 2001; Carpenter et al., 2009; Fillat, 2014). In the absence of metal cofactors, the DNA-binding domain of Fur typically has an "open" conformation that does not bind DNA. Occupancy of S1 with a zinc ion induces Fur dimerization, which brings the DNA-binding domains of two monomers into closer proximity. Binding of metal ions to S2 triggers additional conformational changes, which in turn stimulates the formation of a "closed" DNA-binding domain. As a result, the "closed" Fur binds to promoter DNA with high affinity and represses gene expression. In addition, metal-dependent Fur family proteins also sense non-metal signals. For example, PerR contains the zinc-binding structural S1 site and a manganese- or iron-coordinating regulatory S2 site (Traoré et al., 2006; Jacquamet et al., 2009; Traoré et al., 2009). Manganese or iron occupying S2 triggers the formation of a "closed" PerR (i.e., PerR-Zn-Mn and PerR-Zn-Fe) that binds to the Per box and blocks gene transcription. When iron in the S2 site detects peroxide, such as H_2O_2, histidine residues of the PerR-Zn-Fe are oxidized. This leads to the release of iron and changes PerR to the "open" conformation. The "open" PerR then dissociates from the Per box, which results in the derepression of genes involved in the peroxide stress response (Traoré et al., 2006; Jacquamet et al., 2009; Traoré et al., 2009).

Recombinant BosR contains zinc, but no other metal ions (including iron, copper, manganese, chromium, cadmium, nickel, magnesium, or lead) (Ouyang et al., 2011; Wang et al., 2012b; Mason et al., 2019). Two CXXC motifs were identified in the dimerization domain of BosR that constitute the zinc-coordination structural S1 site of BosR (Mason et al., 2019). This site is similar to the S1 site in many Fur family proteins including PerR (Figure 3). The ability of BosR to bind Zn was dramatically reduced when the CXXC motifs were mutated (Mason et al., 2019). The CXXC motifs also contribute to the ability of BosR to form dimers and are crucial for the regulatory function of BosR. However, BosR lacks a recognizable regulatory S2 site. Of the residues constituting the iron, manganese, zinc, or nickel-coordinating regulatory S2 site of Fur proteins, BosR retains only a single conserved residue (His111) in the hinge region (Figure 3). The vicinity surrounding His111 lacks metal coordinating residues other than the known CXXC motifs. Therefore, the DNA-binding conformation of BosR may not be induced by metal binding, other than at the S1 site. Instead, the His111 may be modified through an as yet unknown mechanism to facilitate DNA binding, which is consistent with the presence of only zinc in recombinant BosR (Ouyang et al., 2011; Wang et al., 2012b; Mason et al., 2019). Alternatively, BosR may use a noncanonical S2 site for signal sensing. Recombinant BosR is able to bind copper, zinc and iron when incubated with unusually high amounts of these metal ions, so perhaps BosR contains two

```
                                                            S2
BbBosR MNDNIIDVHSALEKVGITNDPVLLKNLTSELGMKASHSRNRIILYIASNPKEYFTAKEVY   60
BsPerR ----------------MAAHELKEALETLKETGVRITPQRHAILEYLVNSM-AHPTADDIY   44
EcFUR  --------------------MTDNNTALKKAGLKVTLPRLKILEVLQEPDNHHVSAEDLY   40
VcFur  --------------------MSDNNQALKDAGLKVTLPRLKILEVLQQPECQHISAEELY   40
CjFur  ----------MLIENVEYDVLLERFKKILRQGGLKYTKQREVLLKTLYHSD-THYTPESLY   50
HpFur  -----------MKRLETLESILERLRMSIKKNGLKNSKQREEVVSVLYRSG-THLSPEEIT   49
EcZur  ------------MEKTTTQELLAQAEKICAQRNVRLTPQRLEVLRLMSLQD-GAISAYDLL   48
              DNA binding       S2              S2      S3    S2 S1
BbBosR NKLIKE--IPSLSKATVYNTLNILKERNILKDIKTTDQKETKFYLSLASTI--AHFKCNK  116
BsPerR KALEGK--FPNMSVATVYNNLRVFRESGLVKELTYGDASS-RFDFVTSD-H--YHAICEN   98
EcFUR  KRLIDM--GEEIGLATVYRVLNQFDDAGIVTRHNFEGGKS-VFELTQQHHH--DHLICLD   95
VcFur  KKLIDL--GEEIGLATVYRVLNQFDDAGIVTRHHFEGGKS-VFELSTQHHH--DHLVCLD   95
CjFur  MEIKQAEPDLNVGIATVYRTLNLLEEAEMVTSISFGSAGK-KYELANKPHH--DHMICKN  107
HpFur  HSIRQK--DKNTSISSVYRILNFLEKENFICVLETSKSGR-RYEIAAKEHH--DHIICLH  104
EcZur  DLLREA--EPQAKPPTVYRALDFLLEQGFVHKVESTNSYV-LCHLFDQPTHTSAMFICDR  105
       S1  S2    S3                S3       S1    S1
BbBosR CNQVHPIQLDDIKDIL---KDKLGENWETKSIEIIYSGHCNN---CYKKDTHNNNNVPDE  170
BsPerR CGKIVDFHYPGLDEVEQLAAHVTGFKVSH--HRLEIYGVCQE---CSKKENH--------  145
EcFUR  CGKVIEFSDDSIEARQREIAAKHGIRLTN--HSLYLYGHCAE-GDCREDEHAHEGK----  148
VcFur  CGEVIEFSDDVIEQRQKEIAAKYNVQLTN--HSLYLYGKCGSDGSCKDNPNAHKPKK---  150
CjFur  CGKIIEFENPIIERQQALIAKEHGFKLTG--HLMQLYGVCGD---CNNQKAKVKI-----  157
HpFur  CGKIIEFADPEIENRQNEVVKKYQAKLIS--HDMKMFVWCKE---CQESEC---------  150
EcZur  CGAVKEECAEGVEDIMHTLAAKMGFALRH--NVIEAHGLCAA---CVEVEACRHPEQCQH  160
```

Figure 3. Alignment of BosR and Fur proteins using Clustal Omega. Bs, *B. subtilis*; Ec, *E. coli*; Vc, *V. cholera*; Cj, *C. jejuni*; Hp, *H. pylori*. The residues constituting the S1, S2 and S3 metal binding sites are indicated in red, green, and blue, respectively. Residues potentially important for DNA binding are indicated in purple.

regulatory sites coordinating zinc, iron, or copper in addition to the structural S1 site (Wang et al., 2017). Surprisingly, adding zinc or copper, but not iron, inhibited the binding of BosR to the *ospA* promoter region (Wang et al., 2017). However, we do not know which metal is bound to BosR *in vivo*.

There remain considerable knowledge gaps regarding BosR. The role of this regulatory protein in the oxidative stress response and transition metal homeostasis is, frankly, uncertain. In addition, the contributions of the BosR regulon to the enzootic cycle need to be addressed. The binding of Fur protein from other bacteria to DNA results in gene repression; however, the binding of BosR to the promoter region of *rpoS* and *bosR* is known to activate transcription, so the mechanistic basis for this unusual activity should be explored. Furthermore, whether BosR has a second regulatory metal-binding site like other Fur proteins is unknown. Dissecting the unusual DNA binding activity and regulatory function of BosR is fundamental toward determining the pathway by which BosR senses the host- and vector-specific signals to control gene expression during the enzootic cycle.

BosR function

Seshu et al. (2004b) constructed the first *bosR* mutant in a high-passage, non-infectious strain of *B. burgdorferi*. However, the *bosR* allele (*bosRR39K*) in the parental high-passage strain possessed a point mutation (R39K) relative to the wild-type *bosR* in infectious strains (Seshu et al., 2004a; Hyde et al., 2006); the R39 residue of BosR is essential for protein function (Seshu et al., 2004a; Katona, 2015). *bosR* deletion mutants were then generated in low-passage, infectious *B. burgdorferi* strains. Hyde et al. (2009, 2010) reported that *bosR* contributes to the *in vitro* growth of *B. burgdorferi,* and the levels of Dps and SodA were decreased in the *bosR* mutant. However, neither the growth defect nor the change in Dps synthesis were observed in the *bosR* mutant generated by Ouyang et al (2009b; 2011). Despite these and other discrepancies, the strategic *bosR* mutants yielded breakthroughs in our understanding of BosR function. First, BosR is essential for tick to vertebrate transmission and vertebrate infection by *B. burgdorferi*. Second, BosR is required for the transcriptional activation of *rpoS*. The expression of *rpoS* and RpoS-dependent OspC and DbpA was abolished in *bosR* mutants. BosR directly activates transcription of *rpoS* by binding to the *rpoS* promoter region (Ouyang et al., 2011; Ouyang et al., 2014a; Katona, 2015). Specifically, BosR binds to three sites in the *rpoS* promoter region as a homodimer, including a site overlapping the -24/-12 RpoN-dependent promoter (Ouyang et al., 2011). This suggests that BosR interacts with RpoN (σ^{54}) to facilitate the recruitment of RpoN-RNA polymerase holoenzyme to the *rpoS* promoter to activate gene expression. Further analyses revealed that BosR recognizes the *rpoS* promoter region via a core sequence called the BosR box, a novel DNA element consisting of a 13-bp palindromic sequence (ATTTAANTTAAAT) with dyad symmetry (Ouyang et al., 2014a). Different from the 19-bp Fur binding consensus (i.e., Fur box, GATAATGATAATCATTATC) and the 15-bp PerR binding consensus (i.e., Per box, TTATAATTATTATAA), the BosR box probably comprises a 6–1–6 inverted repeat composed of two hexamers (ATTTAA) in a head-to-tail orientation. The requirement of BosR in the activation of RpoN-dependent *rpoS* transcription was unexpected, as activation of bacterial RpoN (σ^{54}) systems generally only requires the bacterial enhancer binding protein (bEBP) (Bush and Dixon, 2012), and suggests that BosR and Rrp2 together mimic the bEBP.

The BosR regulon

Recombinant BosR binds to many DNA sequences including the Per box (Katona et al., 2004; Ouyang et al., 2014a) and the promoter regions of *dps* (Boylan et al., 2003; Katona et al., 2004; Ouyang et al., 2011), *sodA* (Seshu et al., 2004a), *cdr* (Boylan et al., 2006), *oppA4* (Medrano et al., 2007), *ospA* (Wang et al., 2013; Shi et al., 2014; Wang et al., 2017), *ospD* (Wang et al., 2013), and *bosR* itself (Katona et al., 2004; Ouyang et al., 2011). Such versatile DNA recognition capacity is not typical of canonical Fur family proteins. Contradictory results were also obtained regarding the DNA binding activity of BosR. Boylan et al. (2003) reported that the binding of BosR to the *dps* promoter was inhibited by manganese and required zinc as well as the reduction with DTT. Katona et al. (2004), however, found that binding of BosR to the Per box and the *dps* promoter was not affected by manganese, but was inhibited by zinc. Wang et al. (2017) showed that the binding of BosR to the *ospA* promoter was not affected by manganese or iron but was inhibited by copper and zinc. These discrepancies may be due to proteins with different conformations used in the *in vitro* binding studies. Analytical ultracentrifugation analyses revealed a monomer-dimer equilibrium of recombinant BosR in free solution (Ouyang et al., 2014a). Metal ions and oxidizing reagents may induce conformational changes, and thus affect the DNA binding activity of recombinant protein. In an attempt to discern the global regulatory effects of BosR, Katona et al. (2004) searched the *B. burgdorferi* genome for the Per box by using the BlastN program and found 26 genes possibly regulated by BosR. Ouyang et al. (2014a) searched the genome (Fraser et al., 1997) for putative BosR boxes using the Regulatory Sequence Analysis Tools, and found 47 genes carrying one or more BosR box sequences in their promoter regions. However, identification of a BosR binding site in a promoter does not mean that BosR regulates the gene. Microarray-based transcriptomic analysis of the *bosR* mutant identified 137 and 62 genes that were upregulated or downregulated, respectively, by BosR, although the function of most them is unknown (Ouyang et al., 2009b). A comparison of the *bosR* and *rpoS* microarray data (Ouyang et al., 2008) found that 50 and 57 genes were upregulated or downregulated, respectively, only by BosR, which suggests that BosR is a dual-functional global regulator in *B. burgdorferi*.

BadR

BadR (BB0693; *Borrelia* host adaption regulator) is a DNA-binding transcriptional regulator that represses *rpoS* transcription in *B. burgdorferi* (Fraser et al., 1997; Miller et al., 2013; Ouyang and Zhou, 2015). BadR synthesis is induced at low temperature, mimicking the tick environment (Miller et al., 2013). *badR* is expressed throughout the entire enzootic cycle of *B. burgdorferi*, reaching a maximal level in intermolt larvae (Ouyang and Zhou, 2015). BadR is required by *B. burgdorferi* to infect vertebrates; *badR* mutants are incapable of infecting mice through needle injection (Miller et al., 2013; Ouyang and Zhou, 2015). In addition, BadR is crucial for optimal growth of *B. burgdorferi in vitro*; *badR* mutants display a marked growth defect in culture compared to the wild-type parental strain. Expression of *rpoS* is enhanced in *badR* mutants, indicating that BadR is a repressor of *rpoS* transcription. Recombinant BadR binds to multiple sites in the *rpoS* promoter region; several residues in the putative sugar-binding domain of BadR were crucial for binding. A merodiploid with *badR* expression controlled by an inducible promoter was constructed to overcome the inability to complement the *badR* null mutant: induction of *badR* inhibited transcription of *rpoS* (Ouyang and Zhou, 2015). Some differences exist between the two genetic studies of *badR* (Miller et al., 2013; Ouyang and Zhou, 2015) including whether expression of *rpoS* in the *badR* mutants is growth-phase dependent. In addition, Miller et al. (2013) reported that binding of recombinant BadR to the *rpoS* promoter region was inhibited by the addition of phosphorylated sugars, while this inhibition was not observed by Ouyang et al (2015), which may be due to the presence of the maltose-binding protein tag on the recombinant BadR in the first study.

BadR inhibits expression of *bosR* at both transcriptional and post-transcriptional levels. Binding sites for BadR have been identified in the sequences upstream of *bosR* (Ouyang and Zhou, 2015). Both BadR and BosR bind to the promoter regions of *rpoS* and *bosR*, although they recognize different core sequences. BadR recognizes DNA at TAAAATAT motifs. Expression of *rpoS* is abolished in a *bosR-badR* double mutant (Ouyang and Zhou, 2015), which suggests that BosR functions to activate *rpoS* transcription rather than to derepress BadR. BadR represses or induces the expression of 127 or 80 genes, respectively (Miller et al., 2013). Subsequent RNA-seq analyses showed that expression of 134 and 100 genes were decreased and increased, respectively, in the *badR* mutant (Arnold et al., 2018). Many of these genes are associated with sugar transport and utilization.

How BadR modulates gene expression in *B. burgdorferi* remains elusive. BadR was originally annotated as a xylose operon regulatory protein XylR-1 involved in the utilization and degradation of D-xylose (Fraser et al., 1997). BadR, however, likely does not function as a xylose-responsive transcriptional regulator like XylR, as *B. burgdorferi* does not use xylose as a carbon source (von Lackum and Stevenson, 2005) and its genome does not encode proteins for xylose transport and catabolism (Fraser et al., 1997). Rather, BadR has ~21% amino acid sequence identity with the *E. coli* transcriptional regulator NagC. Three-dimensional protein modeling indicated that BadR had a structure similar to the Mlc protein in *V. cholerae* and *E. coli* (Ouyang and Zhou, 2015). Mlc is a homolog of NagC and both proteins belong to the ROK (Repressor, ORF, Kinase) protein family. ROK repressors contain an N-terminal helix-turn-helix (HTH) DNA-binding motif and a C-terminal sugar-binding domain. The DNA-binding activity and regulatory function of ROK repressors are responsive to the availability of specific sugars. Despite the homology to NagC and Mlc, BadR likely functions differently from other ROK repressors. Typically, ROK repressors bind a specific sugar or effector through specificity-determining residues (SDRs) (Titgemeyer et al., 1994; Kazanov et al., 2013). However, BadR lacks the SDRs found in NagC, Mlc, and other ROK repressors, consistent with the observation that the DNA-binding of recombinant BadR was not affected by phosphorylated sugars (Ouyang and Zhou, 2015), and the expression of genes involved in GlcNAc transport and utilization were not affected in the *badR* mutant (Miller et al., 2013), supporting that BadR does not function analogously to NagC or Mlc and is yet another example of an atypical transcriptional regulator influencing the expression of enzootic cycle-specific genes (Samuels and Radolf, 2009).

YebC

YebC (BB0025) is a transcription factor that belongs to a putative YebC/PmpR family of DNA-binding transcriptional regulators (TACO1 family, pfam PF01709) (Shin et al., 2002; Zhang et al., 2012) and controls *vlsE* expression (Zhang et al., 2020). VlsE undergoes antigenic variation in the vertebrate host and *vlsE* expression increases concomitantly with the decrease in *ospC* expression in response to the host adaptive immune response, but there are few mechanistic details on the regulation of *vlsE* expression (Norris, 2014; Chaconas et al., 2020). Zhang et al. (2020) very recently demonstrated that deletion of *yebC* dramatically reduced *vlsE* mRNA levels *in vitro*. Expression of *yebC* and *vlsE* is co-regulated in response to culture temperature, and is inversely regulated with expression of *ospC* during vertebrate infection. The *yebC* mutant has a similar phenotype as the *vlsE* mutant: both are infectious in immunocompromised mice but unable to infect immunocompetent mice. YebC binds to the *vlsE* promoter, suggesting that YebC directly regulates *vlsE* transcription; however, how YebC is activated during the vertebrate phase remains unknown.

Hk1/Rrp1 two-component system and c-di-GMP

Since its discovery as a novel allosteric activator of cellulose synthase in *Komagataeibacter xylinus* (formerly *Acetobacter xylinus*) more than 30 years ago (Ross et al., 1987), bis-(3′-5′)-cyclic dimeric guanosine monophosphate (c-di-GMP) has emerged as a ubiquitous global secondary messenger associated with a wide range of lifestyle control networks, including the transition from planktonic to sessile states, biofilm formation, cell cycle progression, and virulence (Cotter and Stibitz, 2007; Wolfe and Visick, 2008; Hengge, 2009; Römling et al., 2013; Valentini and Filloux, 2019). Diguanylate cyclases (DGCs), the enzymes responsible for synthesis of c-di-GMP from two GTP molecules, are readily identified by the presence of an ~170 amino acid GGDEF (Gly-Gly-Asp-Glu-Phe) conserved domain (Pfam PF00990) (Pei and Grishin, 2001; Schirmer and Jenal, 2009). GGDEF domain-containing proteins are omnipresent in bacteria (Galperin, 2004; Ryjenkov et al., 2005), with most bacterial genomes encoding multiple DGCs. The potential relevance of c-di-GMP for environmental signaling in Lyme disease spirochetes first came to light in a bioinformatics study by Galperin et al. (2001). Using purified recombinant protein, Ryjenkov et al. (2005) provided direct biochemical evidence that Rrp1 (BB0419), the only identifiable GGDEF domain-containing protein encoded by *Borrelia* spp. (Galperin et al., 2001), functions as a diguanylate cyclase. Intracellular concentrations of c-di-GMP in *B. burgdorferi* are controlled by the activity of two evolutionarily distinct phosphodiesterases (PDEs), PdeA and PdeB, belonging to the EAL and HD-GYP families of non-specific PDEs, respectively (Sultan et al., 2010; Sultan et al., 2011; Novak et al., 2014). While both borrelial PDEs hydrolyze c-di-GMP with high affinity (Sultan et al., 2010; Sultan et al., 2011; Novak et al., 2014), PdeA- and PdeB-deficient spirochetes display different motility- and virulence-related phenotypes (Sultan et al., 2010; Sultan et al., 2011), suggesting that these PDEs are part of a complex regulatory system that involves multiple effectors and control mechanisms.

Rrp1 along with Hk1 form a prototypical bacterial sensory two-component signal transduction system (TCS). Hk1, the sensory component, is composed of four modular domains: a periplasmic sensor domain of three ligand-binding subdomains, each with homology to bacterial periplasmic solute-binding proteins (PBPs), a histidine kinase domain, a phosphoacceptor receiver (REC) domain, and a histidine phosphotransferase (Hpt) domain (Bauer et al., 2015). Ligand binding by Hk1 initiates a signal transduction cascade that results in phosphorylation of the Rrp1 REC domain and activation of c-di-GMP synthesis (Caimano et al., 2011; Bauer et al., 2015). Although the small molecule(s) responsible for activating Hk1 in nature have yet to be identified, bioinformatics and structural analyses of the Hk1 sensory PBP domains point to exogenously-derived amino acids and/or their derivatives as potential activating ligands (Caimano et al., 2011; Caimano et al., 2015). Consistent with this prediction, the Hk1/Rrp1 TCS is active when spirochetes are grown in

BSK II (Rogers et al., 2009; Caimano et al., 2011), a semi-defined medium that contains abundant free amino acids (Barbour, 1984). The genes encoding Hk1 and Rrp1 (*bb0420* and *bb0419*, respectively) are co-transcribed as a bicistronic message (Caimano et al., 2011). *hk1/rrp1* is expressed more or less constitutively *in vitro* and throughout the enzootic cycle (Rogers et al., 2009; Caimano et al., 2011).

Phenotypic characterization of *hk1* and *rrp1* mutants demonstrated that this TCS functions exclusively during the tick phase of the enzootic cycle (Caimano et al., 2011; He et al., 2011; Kostick et al., 2011). Hk1- and Rrp1-deficient organisms display wild-type burdens in murine tissues following needle inoculation but are rapidly destroyed in the midguts of feeding larvae and nymphs as early as 36 hours into the acquisition blood meal (Caimano et al., 2011; He et al., 2011; Kostick et al., 2011). The phenotypic similarities between *hk1* and *rrp1* mutants in larvae and nymphs suggest that the ligand(s) responsible for activating c-di-GMP signaling are not tick life-stage-specific (Caimano et al., 2011; He et al., 2011; Kostick et al., 2011; Caimano et al., 2015). Little to no killing was observed in the absence of tick feeding, indicating that the destruction of both mutants is mediated by the feeding process. An important clue regarding this unusual tick phenotype was provided by Bontemps-Gallo et al. (2016), who found that *hk1* and *rrp1* mutant *B. burgdorferi* grown at 34°C in BSK II were substantially more osmo-sensitive than their wild-type isogenic parent under conditions that mimic the high osmolality (~550-650 mOsm) within fed midguts. Additional insight into the contribution of the Hk1/Rrp1 TCS to tick adaptation comes from genome-wide transcriptomics analyses by microarray (Rogers et al., 2009; He et al., 2011) and RNA-seq (Caimano et al., 2015). Results from these analyses revealed two broad categories of c-di-GMP-regulated genes that could contribute to survival within tick midguts during the blood meal. The first includes genes involved in utilization of alternate carbon sources for energy generation (glycolysis) and biosynthesis of phospholipids and/or peptidoglycan. In addition to the glycerol utilization (Glp) system, c-di-GMP upregulates expression of PTS transporters (*malX2* and *chbC*) for *N*-acetylglucosamine (GlcNAc) and chitobiose (a GlcNAc dimer derived from chitin, the major component of tick cuticle), respectively (Tilly et al., 2001; Tilly et al., 2004; Rhodes et al., 2009; Gherardini et al., 2010; Pappas et al., 2011; Sze et al., 2013; Caimano et al., 2015; Corona and Schwartz, 2015; Caimano et al., 2016; Zhang et al., 2018). The phenotypes for *hk1* and *rrp1* mutants in feeding ticks, however, are more pronounced than those of either *glp* or *chbC* mutants. Presumably, utilization of multiple alternative carbon sources enables spirochetes to expand exponentially and maintain cell envelope homeostasis as blood meal-derived glucose becomes increasingly limited due to competition from midgut epithelial cells. Additional c-di-GMP-regulated genes contribute to maintenance of the cell envelope: the Rrp1 regulon includes *ackA*, encoding the acetate kinase required for generation of acetyl phosphate, which in turn gives rise to acetyl-CoA, the starting point for synthesis of peptidoglycan via the mevalonate pathway (Xu et al., 2010; Van Laar et al., 2012). The second category of c-di-GMP regulated genes includes differentially-expressed lipoproteins, including the cp32-encoded OspE/BbCRASP lipoproteins, which have been shown to inhibit complement-mediated lysis by binding complement factor H (CFH) and CFH-related proteins (Radolf et al., 2012; Bhattacharjee et al., 2013; Kraiczy and Stevenson, 2013). Other lipoproteins regulated by c-di-GMP (e.g., Mlps) also may protect *B. burgdorferi* from hazards encountered in the fed midgut (e.g., antimicrobial peptides and reactive oxygen species).

c-di-GMP modulates cellular processes in bacteria by a variety of mechanisms, including transcriptional, translational, allosteric inhibition of enzymatic activity (I-sites), protein-protein interactions, protein secretion, and protein stability (Cotter and Stibitz, 2007; Wolfe and Visick, 2008; Römling et al., 2013; Valentini and Filloux, 2019). A wide range of effector molecules involved in c-di-GMP signaling have been identified, including two types of riboswitches (Sudarsan et al., 2008; Lee et al., 2010; Smith et al., 2011) and numerous c-di-GMP receptor proteins (Schirmer and Jenal, 2009; Dahlstrom and O'Toole, 2017; Valentini and Filloux, 2019). One of the most well-studied group of c-di-GMP effector proteins are those containing an ~110 amino acid PilZ domain (PF07238) (Amikam and Galperin, 2006; Schirmer and Jenal, 2009), first identified at the C-terminus of the cellulose synthase catalytic subunit of BcsA from *K. xylinum* (Ross et al., 1987). To date, two c-di-GMP effector PilZ domain proteins, PlzA (BB0733) and PlzB, have been described in *B. burgdorferi* (Freedman et al., 2010; Pitzer et al., 2011; He et al., 2014; Mallory et al., 2016). A third PilZ-domain protein, PlzC, is found only in relapsing fever spirochetes (Mallory et al., 2016). PlzA, encoded on the chromosome, and PlzB, encoded on a linear plasmid, share only 63% amino acid identity and appear to carry out distinct functions (Mallory et al., 2016). The majority of Lyme disease spirochetes encode only PlzA. c-di-GMP binding by borrelial Plz proteins is mediated by a signature PilZ domain motif: RIHER and DzSYGG (z=hydrophobic residue) (Amikam and Galperin, 2006; Ryjenkov et al., 2006; Freedman et al., 2010; Pitzer et al., 2011; Mallory et al., 2016). In independent studies, Motaleb, Marconi and colleagues demonstrated that recombinant PlzA binds c-di-GMP at 1:1 stoichiometry with relatively high affinity (~1 – 6 μM) (Freedman et al., 2010; Pitzer et al., 2011; Mallory et al., 2016). Binding of c-di-GMP by PlzA is thought to induce a conformational change enabling a 'switch mechanism' (Mallory et al., 2016). Interestingly, the PilZ motifs in borrelial Plz proteins are separated by a variable 32 amino acid spacer that is significantly longer than the 19 to 23 residues typically found in prototypical PilZ domains, perhaps allowing spirochetal Plz proteins to engage a more extensive collection of downstream effector molecules (Mallory et al., 2016). Consistent with its c-di-GMP effector function, expression of *plzA* is increased during tick feeding (Freedman et al., 2010; Pitzer et al., 2011). However, there are conflicting reports regarding the extent to which PlzA is required for survival during tick feeding. Pitzer et al. (2011) reported that *plzA* mutants survive poorly in naturally-infected ticks during the acquisition and transmission blood meals, while Kostick-Dunn et al. (2018) found that PlzA was not strictly required to establish infection in ticks. Nonetheless, both groups demonstrated that *plzA*-infected ticks are unable to transmit spirochetes to naïve mice. Moreover, overexpression of PlzA is able to partially compensate for the loss of Rrp1 in feeding ticks (Zhang et al., 2018). This phenotype stems, in part, from the requirement for both c-di-GMP and PlzA for transcription of the *glp* operon (*bb0240-bb0243*) (Caimano et al., 2015; Zhang et al., 2018), which encodes a system for the uptake and utilization of glycerol (Pappas et al., 2011). However, unlike *rrp1* mutants, which are infectious in mice, spirochetes lacking PlzA are substantially less virulent by needle-inoculation (Pitzer et al., 2011; He et al., 2014; Kostick-Dunn et al., 2018; Zhang et al., 2018). Decreased virulence of *plzA* mutants has been attributed, at least in part, to aberrant motility (i.e., decreased swarming, reduced translational motility and flex frequency) (Pitzer et al., 2011; Kostick-Dunn et al., 2018) and/or dysregulation of RpoS-dependent gene regulation (Sze et al., 2013; He et al., 2014). He et al. (2014) reported that PlzA influences the RpoS pathway via its effects on BosR, a Fur family protein required for expression of *rpoS* (see the section above on BosR and BadR transcription factors) (Samuels and Radolf, 2009); *plzA* mutants express two- to threefold lower levels of *bosR* mRNA compared to wild type and little to no BosR protein *in vitro* (He et al., 2014). Given that the Hk1/Rrp1 TCS is not required and presumably 'OFF' in vertebrates, any virulence-related aspects of the *plzA* mutant phenotype, including dysregulation of BosR and/or RpoS, are likely mediated by apo-PlzA (Pitzer et al., 2011; Zhang et al., 2018).

Genome-wide comparison of the RpoS and Hk1/Rrp1 regulons identified only a handful of genes that are regulated by both RpoS and c-di-GMP (Caimano et al., 2015). The expression profiles of these co-regulated genes also were strikingly different from those of the prototypical RpoS-regulated genes *ospC* and *dbpA* in fed nymphs (i.e., when both pathways are active) (Dunham-Ems et al., 2012; Caimano et al., 2019). Thus, while c-di-GMP may enhance expression of some RpoS-upregulated genes, these two pathways likely function in parallel rather than cooperatively. As part of its gatekeeper function, RpoS also represses the expression of RpoD (σ^{70})-dependent tick-phase genes, including *ospA*, *bba62*, and the *glp* operon, within the vertebrate (Caimano et al., 2005; Caimano et al., 2007; Dunham-Ems et al., 2012; Grove et al., 2017; Caimano et al., 2019) (see the section above on the RpoN-RpoS σ factor cascade). Although transcription of RpoS-upregulated genes begins during the transmission blood meal (Schwan et al., 1995; Dunham-Ems et al., 2012; Ouyang et al., 2012; Caimano et al., 2019), RpoS-mediated repression does not occur to any appreciable extent until spirochetes have entered the vertebrate host (Dunham-Ems et al., 2012; Caimano et al., 2019). Overlap between the Hk1/Rrp1 and RpoS regulons includes a number of tick phase-specific genes that are repressed by RpoS during infection. These data led Caimano et al. (2015) to propose that c-di-GMP antagonizes RpoS-mediated repression either directly or indirectly in feeding ticks. Once *B. burgdorferi* is in a vertebrate, the absence of c-di-GMP would allow unfettered repression of these tick phase-specific genes.

Stringent response and (p)ppGpp

Rel_{Bbu} and DksA mediate adaptation to nutrient stress

B. burgdorferi adapts to environmental nutritional changes encountered during its enzootic cycle by globally altering gene expression through the stringent response, which is particularly pivotal during persistence in the tick vector between blood meals. The stringent response steels the spirochete by regulating the expression of numerous pathways, including replication, motility, morphology, stress responses, and virulence factors. These changes are driven by modulating the levels of two intracellular alarmones, guanosine pentaphosphate and guanosine tetraphosphate, collectively known as (p)ppGpp, which directly bind RNA polymerase and indirectly affect numerous regulatory systems to alter transcription (Potrykus and Cashel, 2008; Dalebroux and Swanson, 2012; Gaca et al., 2015; Hauryliuk et al., 2015; Liu et al., 2015; Steinchen and Bange, 2016; Gourse et al., 2018). In *B. burgdorferi*, (p)ppGpp levels are regulated by a single gene product, Rel$_{Bbu}$, containing both a canonical synthetase (RelA) domain induced by amino acid starvation and a bifunctional synthetase/hydrolase (SpoT) domain (Concepcion and Nelson, 2003; Bugrysheva et al., 2005; Drecktrah et al., 2015). In addition to the catalytic domains, Rel$_{Bbu}$ also contains the regulatory TGS (threonyl-tRNA synthetases, GTPases and SpoT) and DC (aspartate-cysteine) oligomerization domains, but lacks the ACT regulatory domain in the C-terminal region (Grant, 2006; Hauryliuk et al., 2015; Ronneau and Hallez, 2019). (p)ppGpp levels rise during nutrient deprivation and fall upon return to nutrient-rich conditions in a Rel$_{Bbu}$-dependent manner (Concepcion and Nelson, 2003; Bugrysheva et al., 2005; Drecktrah et al., 2015). The specific triggers of the stringent response in *B. burgdorferi* remain unknown, but amino acid starvation may not serve as the classic signal since the *B. burgdorferi* genome is predicted to contain few amino acid transporters and may rely more on peptide transport and proteolysis (Fraser et al., 1997; Groshong et al., 2017); instead, the spirochete may sense changes in levels of lipids or phosphate.

The Rel$_{Bbu}$-mediated transcriptional changes of the stringent response play an important role in survival during nutrient stress both in culture and within the nymphal tick midgut during the transmission blood meal (Drecktrah et al., 2015). Survival in the midgut and transmission to mice via tick bite are not abolished in a *rel*$_{Bbu}$ mutant incapable of synthesizing (p)ppGpp, but are significantly curtailed (Drecktrah et al., 2015). Growth in nutrient-rich media at 23°C, the temperature used to represent the tick environment, is also Rel$_{Bbu}$-dependent (Bugrysheva et al., 2015). The stringent response also regulates *B. burgdorferi* morphology as the *rel*$_{Bbu}$ mutant is more likely to form round bodies, a spherical condensed form within an outer membrane observed *in vitro* and within ticks between blood meals, in response to nutrient stress (Brorson and Brorson, 1997, 1998; Alban et al., 2000; Dunham-Ems et al., 2012; Drecktrah et al., 2015). These studies illustrate an important role for Rel$_{Bbu}$ and the stringent response in persistence in the tick and subsequent transmission to the host.

The transcription factor DksA (DnaK suppressor), which potentiates the transcriptional effects of (p)ppGpp on RNA polymerase activity (Gourse et al., 2018), also globally reprograms the *B. burgdorferi* transcriptome in response to nutrient stress (Boyle et al., 2019; Mason et al., 2020). There is considerable overlap between the DksA and Rel$_{Bbu}$ regulons, but differences exist (see below), illustrating that DksA is a cooperating partner in the stringent response but also exerts an independent influence. Recent studies have found that, similar to Rel$_{Bbu}$, DksA regulates genes that affect *B. burgdorferi* survival under nutrient and osmotic stresses (Boyle et al., 2019). Surprisingly, DksA is essential for *B. burgdorferi* infectivity in the murine model of Lyme disease, at least by needle inoculation (Mason et al., 2020), while Rel$_{Bbu}$ is dispensable for murine infectivity (Drecktrah et al., 2015). Tick transmission studies await further investigation as do experiments examining the role of DksA in persistence in the tick.

The Rel$_{Bbu}$ regulon

The Rel$_{Bbu}$-dependent regulon is characterized by broad patterns of gene regulation typical of bacteria adapting to nutrient stress by reducing growth, including repression of many genes involved in metabolism (glycolysis and mevalonate biosynthesis), translation (ribosomal proteins) and transcription (RNA polymerase subunits), as revealed by RNA-seq and microarray analyses (Bugrysheva et al., 2015; Drecktrah et al., 2015). Rel$_{Bbu}$-dependent regulation of gene products whose function is not fully understood may provide clues to their role in the enzootic cycle. For example, the expression of the genes encoding the oligopeptide binding proteins (*oppA1*, *oppA2*, *oppA3* and *oppA5*) are Rel$_{Bbu}$-upregulated suggesting that peptide uptake functions during persistence of the spirochete in the tick, as this is the phase where survival of the *rel*$_{Bbu}$ mutant is compromised (Drecktrah et al., 2015). Rel$_{Bbu}$ also regulates the glycerol metabolic (*glp*) operon (Bugrysheva et al., 2015; Drecktrah et al., 2015). The *glp* operon contributes to persistence of *B.*

burgdorferi in the tick where the spirochete is thought to utilize glycerol as a carbon source and environmental signal, as has been demonstrated *in vitro* (He et al., 2011; Pappas et al., 2011; Bugrysheva et al., 2015). The defect in persistence in the tick of the *rel*$_{Bbu}$ mutant is likely due, at least in part, to compromised glycerol metabolism. The metabolic fate of glycerol may be directed by the stringent response as well: Rel$_{Bbu}$ upregulates expression of genes encoding the glycerol uptake facilitator (*glpF*) and glycerol kinase (*glpK*), but studies disagree on the effect on the last gene in the operon, glycerol 3-phosphate dehydrogenase (*glpD*) (Bugrysheva et al., 2015; Drecktrah et al., 2015). If GlpD is upregulated then glycerol 3-phosphate would be trafficked to the glycolytic pathway, while downregulation of GlpD would instead shuttle phosphorylated glycerol to membrane and lipoprotein biosynthesis. This suggests that the stringent response coordinates transcriptional changes to affect metabolism to adapt the spirochete to nutritional flux in the enzootic cycle.

These transcriptomic studies also illuminate potential functions for Rel$_{Bbu}$ in processes not traditionally thought to be regulated by the stringent response. Expression of *vlsE,* a Vmp-like surface lipoprotein used to evade the adaptive host response (Norris, 2014; Chaconas et al., 2020) (see Chapter 13), is Rel$_{Bbu}$-upregulated (Drecktrah et al., 2015). Although *vlsE* recombination only occurs in the vertebrate host, the stringent response is the first signaling system identified that regulates *vlsE* expression. In addition, Rel$_{Bbu}$ regulates expression of the *bosR* gene (Drecktrah et al., 2015) (see section above on Transcription factors BosR and BadR). The reduction in *bosR* expression is not enough to render the *rel*$_{Bbu}$ mutant non-infectious by needle inoculation, but could play a role in the tick transmission defect. Rel$_{Bbu}$ also upregulates a number of genes encoding extracellular matrix-binding proteins important in the vertebrate host, such as *dbpBA* and *bbk32* (Drecktrah et al., 2015). These findings suggest that *B. burgdorferi* may encounter nutrient-limited niches in the vertebrate, such as the synovial regions in joints, that trigger the stringent response or there may be subtler, as of yet unidentified, signals inducing (p)ppGpp production in the spirochete to modulate gene expression in the vertebrate host.

Rel$_{Bbu}$ also seems to coordinate the life cycle of *B. burgdorferi* bacteriophages by controlling numerous genes located on the cp32 prophages (Eggers and Samuels, 1999; Zhang and Marconi, 2005). Rel$_{Bbu}$ represses expression of many more cp32 genes than it activates, with most located in the putative late operon (Drecktrah et al., 2015). Furthermore, expression of the essential gene *ftsH* (Chu et al., 2016) was also Rel$_{Bbu}$-downregulated (Bugrysheva et al., 2015; Drecktrah et al., 2015); the *ftsH* homolog in *E. coli* encodes a protease that regulates the λ phage life cycle (Kihara et al., 1997; Shotland et al., 1997). Thus, changes in nutrient availability in the tick midgut or vertebrate may signal, through the stringent response, conditions for prophage induction.

Rel$_{Bbu}$ regulates sRNAs

The stringent response regulates expression of a third of all identified small RNAs (sRNAs) in *B. burgdorferi*, including intragenic sRNAs, antisense sRNAs and 5′ UTR sRNAs (Popitsch et al., 2017; Drecktrah et al., 2018) (see the section below on Small regulatory RNAs and RNA chaperones). Threefold more sRNAs were Rel$_{Bbu}$-upregulated than downregulated and were mapped throughout the genome (Drecktrah et al., 2015). Antisense sRNAs and 5′ UTR sRNAs associated with genes of known or predicted function mostly targeted genes encoding translation, transporter and cell envelope functions. One intriguing example is that Rel$_{Bbu}$ downregulates an antisense sRNA within the *glpF* ORF (SR0186) while concurrently upregulating the *glpF* mRNA (Drecktrah et al., 2018). The sense:antisense pairing may favor degradation of the *glpF* transcript in response to vertebrate-specific signals (Drecktrah et al., 2015). Thus, one mechanism *B. burgdorferi* may employ to survive in the tick involves sensing nutrient stress between blood meals to induce the stringent response that represses the *glpF* antisense sRNA, which, in turn, stabilizes the *glpF* mRNA and increases GlpF production to enhance transport and utilization of glycerol.

Rel$_{Bbu}$ may also indirectly modulate RNA polymerase activity to alter the global transcriptome through the Bb6S RNA (see the section below on Small regulatory RNAs and RNA chaperones). Bb6S RNA is a highly expressed sRNA that binds directly to the RpoD (σ^{70})-RNA polymerase (Drecktrah et al., 2020). Rel$_{Bbu}$ upregulates Bb6S RNA, potentially providing another mechanism to shape the transcriptional landscape in addition to the direct interaction of (p)ppGpp with RNA polymerase (Drecktrah et al., 2018). The vast contingent of sRNAs controlled by the stringent response represent multiple pathways for regulating gene expression in response to nutrient stress.

The DksA regulon

The transcription factor DksA collaborates with (p)ppGpp to globally regulate gene expression (Gourse et al., 2018). DksA targets many of the same genes as Rel_{Bbu}, but also affects a distinct set (Boyle et al., 2019), a phenomenon that has been reported in other bacteria (Potrykus and Cashel, 2008; Dalebroux et al., 2010). Downregulation of genes encoding ribosomal proteins, DNA replication proteins, and transcriptional and translational machinery was DksA-dependent in addition to Rel_{Bbu}-dependent (Bugrysheva et al., 2015; Drecktrah et al., 2015; Boyle et al., 2019). DksA-dependent genes were similarly found throughout the genome and spanned all functional categories, reinforcing the extensive genomic reach of this stringent response-associated transcription factor. As with Rel_{Bbu}, DksA upregulated expression of factors important for persistence in the tick, such as the glycerol metabolic genes *glpF* and *glpK*, the bacterioferritin ortholog gene *dps/napA/bicA*, as well as the tick phase-specific lipoprotein-encoding genes *ospA* and *lp6.6* (Boyle et al., 2019). Further transcriptional overlap with the Rel_{Bbu} regulon includes the repression of many cp32 genes in the late phage operon and upregulation of a number of genes encoding lipoproteins associated with the vertebrate phase of the enzootic cycle, including *dbpA* and *ospC* (Drecktrah et al., 2015; Boyle et al., 2019).

Interplay between (p)ppGpp and DksA is evident as levels of *dksA* transcript increase in the rel_{Bbu} null mutant (Bugrysheva et al., 2015; Drecktrah et al., 2015) while (p)ppGpp levels are increased in a *dksA* null mutant (Boyle et al., 2019). Thus, the spirochete may be compensating for the lack of one component of the stringent response by increasing the levels of its partner, illustrating the cooperative and complex relationship of (p)ppGpp and DksA in remodeling the transcriptome to adapt to nutrient flux in the dynamic environments of the enzootic cycle.

That both Rel_{Bbu} and DksA upregulate expression of gene products controlled by the RpoN-RpoS σ factor cascade mediating vertebrate phase-specific gene expression and host infectivity was unexpected, and suggests a more intricate role for these factors than merely mediating the response to nutrient limitations. DksA, as well as Rel_{Bbu}, does not alter the level of *rpoS* transcript during nutrient stress, but rather appears to affect RpoS levels post-transcriptionally (Drecktrah et al., 2015; Boyle et al., 2019; Mason et al., 2020), a function that has been previously reported in *E. coli* (Brown et al., 2002). The mechanism for post-transcriptional RpoS regulation in these mutants has not been determined but may be driven, at least in part, by Rel_{Bbu}-upregulation of $DsrA_{Bb}$, an sRNA known to post-transcriptionally increase RpoS levels (Lybecker and Samuels, 2007; Drecktrah et al., 2018). Thus, reduction of *ospC* and *dbpA* transcripts is likely due to the lack of RpoS protein in the DksA mutant and may account for the non-infectious phenotype observed when mice are challenged by needle inoculation (Mason et al., 2020). The finding that the Rel_{Bbu} mutant is infectious indicates a marked divergence of the two regulators that have similar regulons and are thought to work hand-in-hand. Future studies are required to discern the subtle functional differences between Rel_{Bbu} and DksA, and elucidate the induction mechanisms of genes encoding lipoproteins involved in *B. burgdorferi* infectivity as well as the complexities of the stringent response.

Small regulatory RNAs and RNA chaperones

Small non-coding RNAs have been described as the "dark matter of the cell" (Riddihough, 2005) because,

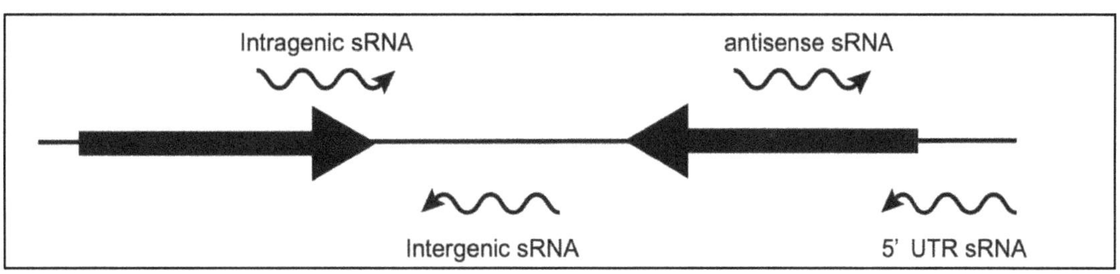

Figure 4. Classification of sRNAs based on their genomic location. sRNAs (black wavy arrows) were categorized based on their relation to annotated open reading frames (black solid arrows).

despite their plethora, they remained mostly undiscovered in the pre-genomic world. Of course, many of these small non-coding RNAs were known since the early days of molecular biology, such as 6S RNA, the ubiquitous and, in certain situations, predominant global regulator of bacterial transcription whose function was outright mysterious for over 30 years (Wassarman and Storz, 2000). Aside from a few functional small RNAs (sRNAs), such as tmRNA (SsrA), annotated in the genome sequence (Fraser et al., 1997), the first small RNA identified in *B. burgdorferi* was $DsrA_{Bb}$, a post-transcriptional regulator of the RpoN-RpoS alternative σ factor cascade (Lybecker and Samuels, 2007). The constellation of non-coding sRNAs in *B. burgdorferi* exponentially expanded a decade later via high-throughput transcriptome studies (Arnold et al., 2016; Adams et al., 2017; Popitsch et al., 2017; Drecktrah et al., 2018). The bevy of sRNAs was classified based on their genomic location: intergenic sRNAs, intragenic sRNAs, antisense sRNAs, and 5′ UTR-associated sRNAs (Figure 4) (Popitsch et al., 2017).

Several mechanisms of riboregulation by sRNAs have been defined, including those employed by pathogenic bacteria to modulate virulence strategies (Fröhlich and Vogel, 2009; Papenfort and Vanderpool, 2015; Svensson and Sharma, 2016; Hör et al., 2018; Lejars et al., 2019; Adams and Storz, 2020; González Plaza, 2020). Many sRNAs regulate gene expression via base-pairing with target mRNAs, affecting their stability, translation, transcription, or processing: *trans*-acting sRNAs base-pair with their target mRNAs and *cis*-acting antisense sRNAs are completely complementarity to their cognate mRNA (Waters and Storz, 2009; Storz et al., 2011; Caldelari et al., 2013). Other sRNAs interact with proteins, titrating their levels or otherwise affecting their activities. As of 2020, only three sRNAs have been functionally characterized in *B. burgdorferi*. (i) $DsrA_{Bb}$ is a *trans*-acting sRNA that base-pairs with *rpoS* mRNA regulating its translation (Lybecker and Samuels, 2007). (ii) Bb6S RNA is a protein-interacting sRNA that directly influences σ factor selectivity by RNA polymerase and globally regulates gene expression (Drecktrah et al., 2020). (iii) *ittA* is an sRNA that affects expression of several genes and is required for dissemination in the vertebrate host via an unknown molecular mechanism (Medina-Pérez et al., 2020). To date, there is no experimental evidence in *B. burgdorferi* for riboswitches, *cis*-acting regulatory RNAs usually found in the 5′ UTR of the mRNA they regulate that function by altering their secondary structure in response to binding intracellular metabolites (Roth and Breaker, 2009; Serganov and Nudler, 2013; Sherwood and Henkin, 2016; McCown et al., 2017); however, the class of 5′ UTR-associated sRNAs (Popitsch et al., 2017) may represent heretofore unrecognized riboswitches.

sRNAs

$DsrA_{Bb}$

In *Escherichia coli*, DsrA is one of three regulatory sRNAs that affect levels of the σ factor RpoS (Hengge-Aronis, 2002a; Repoila et al., 2003; Majdalani et al., 2005; Narberhaus et al., 2006; Waters and Storz, 2009). The sRNA disrupts a hairpin stem-loop that blocks access of the ribosome to the Shine-Dalgarno sequence on the *rpoS* mRNA and enhances its translation at low temperature (Hengge-Aronis, 2002a; Repoila et al., 2003). DsrA is considered to be "a thermometer for the RpoS regulon" in *E. coli* (Repoila et al., 2003) because both its transcription and stability increase at low temperature, causing an accumulation of RpoS protein during the cold stress response (Repoila and Gottesman, 2001).

An sRNA was discovered in *B. burgdorferi* that has extensive complementarity to the upstream region of the *rpoS* mRNA, which appears to assume a hairpin stem-loop structure akin to that of *E. coli*, implying a similar mode of regulation (Lybecker and Samuels, 2007). Several experiments suggested that the sRNA functioned in post-transcriptional temperature-mediated regulation of RpoS, so it was called $DsrA_{Bb}$; however, there are substantial differences between the *E. coli* and *B. burgdorferi* DsrA analogs (Lybecker and Samuels, 2007). Notably, $DsrA_{Bb}$ senses an increase in temperature, presumably when warm vertebrate blood enters the ambient feeding tick, while DsrA in *E. coli* responds to a decrease in temperature. In addition, $DsrA_{Bb}$, which is about four times longer than DsrA from *E. coli*, has more extensive base-pairing potential (31 of 34 nucleotides, or up to 51 of 68 nucleotides, compared to 26 of 38 nucleotides) with its target, the *rpoS* mRNA in *B. burgdorferi* (Lybecker and Samuels, 2007). Furthermore, via an as yet unknown mechanism, $DsrA_{Bb}$ requires translation of an upstream and overlapping open reading frame encoded by *bb0577* (Lybecker and Samuels, 2007). Lastly, the steady-state levels of DsrA in *E. coli* are influenced by temperature, while $DsrA_{Bb}$ is unaffected by temperature (Lybecker and Samuels, 2007) and instead is upregulated by the stringent response

(Drecktrah et al., 2018) (see the section above on the Stringent response and (p)ppGpp). Other than the potential crosstalk between the stringent response (which regulates gene expression in response to starvation in the unfed tick midgut) and the RpoN-RpoS alternative σ factor cascade (which activates gene expression required in the vertebrate host and represses gene expression required in the tick vector), almost nothing is known about the temperature-sensing modus operandi of $DsrA_{Bb}$.

$DsrA_{Bb}$ is hypothesized to serve as a molecular thermometer regulating gene expression during the enzootic cycle of *B. burgdorferi* by modulating the RpoN-RpoS σ factor cascade. Yet, the increase in *rpoS* mRNA levels (Caimano et al., 2004) does not require $DsrA_{Bb}$ (Lybecker and Samuels, 2007), which suggests the presence of two independent mechanisms that regulate RpoS. Two *rpoS* mRNAs have been identified (Lybecker and Samuels, 2007; Smith et al., 2007): a short *rpoS* mRNA that is transcribed from an RpoN-dependent promoter (Smith et al., 2007) at high cell density (Lybecker and Samuels, 2007) and a long *rpoS* mRNA that is transcribed from an RpoD-dependent promoter at low cell density (Lybecker and Samuels, 2007). Of note, $DsrA_{Bb}$ stimulates the translation of *rpoS* mRNA exclusively at low cell density in culture (Lybecker and Samuels, 2007). Only the long *rpoS* mRNA contains the sequence complementary to $DsrA_{Bb}$ (Lybecker and Samuels, 2007) and, therefore, $DsrA_{Bb}$ regulates translation of the RpoD-dependent long *rpoS* mRNA, but not translation of the RpoN-dependent short *rpoS* mRNA. The seminal studies (Hübner et al., 2001; Smith et al., 2007) on *rpoS* regulation missed the long *rpoS* mRNA presumably because the spirochetes were grown to a high cell density, culture conditions in which the short *rpoS* mRNA is transcribed (Lybecker and Samuels, 2007). The long *rpoS* transcript has been observed in both an *rpoN* null mutant and in a strain with the RpoN-dependent *rpoS* promoter mutated (Hall et al., 2021). What, then, is the regulatory role of the two *rpoS* mRNA species, and the $DsrA_{Bb}$ sRNA, in the enzootic cycle? The proposed model is that the pathway controlling RpoS and, subsequently, enzootic cycle phase-specific gene expression, bifurcates: the $DsrA_{Bb}$-RpoS branch activates RpoS during transmission from the tick to the vertebrate and the RpoN-RpoS branch maintains RpoS levels, and RpoS-dependent gene expression, in the vertebrate (Hall et al., 2021). This accounts for the apparently discrepant observations that *rpoN* mutants migrate from the midgut to the salivary glands (Fisher et al., 2005), but *rpoS* mutants cannot exit the midgut (Dunham-Ems et al., 2012).

Bb6S RNA

6S RNA, encoded by *ssrS*, directly interacts with RpoD-holoenzyme (E), the RNA polymerase carrying the RpoD ($E\sigma^{70}$) σ factor (Wassarman and Storz, 2000; Cavanagh and Wassarman, 2014; Steuten et al., 2014a; Steuten et al., 2014b; Burenina et al., 2015; Wassarman, 2018), and serves as a global regulator of transcription (Waters and Storz, 2009; Storz et al., 2011). The primary sequences of 6S RNAs from different bacterial species are not conserved, but rather share a conserved secondary structure that imitates the open promoter complex (Wassarman and Storz, 2000; Barrick et al., 2005; Trotochaud and Wassarman, 2005; Chen et al., 2017), and 6S RNA prevents binding of $E\sigma^{70}$ to RpoD-dependent promoters, resulting in decreased transcription of RpoD-dependent genes and increased transcription of many RpoS-dependent genes (Wassarman and Storz, 2000; Trotochaud and Wassarman, 2004, 2006; Cavanagh et al., 2008; Cavanagh et al., 2010; Neusser et al., 2010). In general, the function of 6S RNA is to shift expression from the RpoD-dependent regulon to the RpoS-dependent regulon (Cavanagh et al., 2008). In *E. coli* and other bacteria, 6S RNA has been associated with adaptation to environmental stresses, but 6S RNA also affects bacterial virulence and infectivity (Faucher et al., 2010; Yan et al., 2013; Warrier et al., 2014; Ren et al., 2017).

Bb6S RNA, the canonical 6S RNA in *B. burgdorferi*, was first predicted by computational genome mining (Barrick et al., 2005) and was then found in the noncoding sRNA transcriptomes of *B. burgdorferi* (Arnold et al., 2016; Popitsch et al., 2017; Drecktrah et al., 2018). Recently, Bb6S RNA was shown to bind RNA polymerase *in vivo*, be processed by RNase Y, regulate expression of lipoprotein genes, and play a role in infectivity in the murine model (Drecktrah et al., 2020). Mapping and modeling of Bb6S RNA suggest that the secondary structure is conserved with a single-stranded central bubble surrounded by double-stranded stem regions (Barrick et al., 2005; Trotochaud and Wassarman, 2005; Chen et al., 2017; Drecktrah et al., 2020).

Unlike 6S RNA from *E. coli* and some other bacteria (Wassarman and Storz, 2000; Cavanagh and Wassarman, 2014; Steuten et al., 2014b), steady-state levels of Bb6S RNA were not affected in cultured *B. burgdorferi* by nutrient stress or growth phase (Drecktrah et al., 2020), which is consistent with RNA-seq studies showing temperature and growth phase had no effect on Bb6S RNA levels (Arnold et al., 2016; Popitsch et al., 2017). However, the amount of Bb6S RNA increased as fed larvae molted into nymphs and decreased after nymphs fed, indicating a function for Bb6S RNA during spirochete persistence in the tick vector (Drecktrah et al., 2020). Bb6S RNA also is upregulated by the stringent response (Drecktrah et al., 2018), which facilitates survival of *B. burgdorferi* between the larval and nymph blood meals (Drecktrah et al., 2015) (see section above on the Stringent response and (p)ppGpp). However, Bb6S RNA was not required for spirochetes to persist through the larval and nymph stages (Drecktrah et al., 2020). Note, though, that in other bacteria, the phenotypes of 6S RNA mutants are typically subtle (Wassarman and Storz, 2000; Trotochaud and Wassarman, 2004, 2006; Cavanagh et al., 2012; Cavanagh and Wassarman, 2013) and Bb6S RNA was required for full infectivity of mice by needle inoculation, although, the infectivity defect of the Bb6S RNA mutant was overcome by either an increased inoculum or tick transmission (Drecktrah et al., 2020). Bb6S RNA regulates expression of the RpoS-dependent genes *ospC* and *dbpBA* without significantly changing RpoS protein levels, results consistent with Bb6S RNA sequestering $E\sigma^{70}$ and promoting transcription by the RpoS-holoenzyme ($E\sigma^{S}$) as found in model microorganisms (Drecktrah et al., 2020).

ittA

The intergenic sRNA SR0736, encoded between *bbd18* and *bbd21* on lp17, was recently identified as attenuating vertebrate infectivity and dissemination of *B. burgdorferi* by filtering data from a global transposon insertion (Lin et al., 2012) sequencing screen with the output of an sRNA transcriptome (Popitsch et al., 2017). This sRNA was renamed *ittA*, for infectivity-associated and tissue-tropic sRNA locus A (Medina-Pérez et al., 2020). An *ittA* mutant strain differentially expresses 19 genes compared to the wild-type parental strain, including increased levels of *bba66* and *vra*, as well as 11 other mRNAs, and decreased levels of *ospD*, *ospA* and *bba74/oms28*, as well as three other mRNAs (Medina-Pérez et al., 2020). The molecular mechanism(s) *ittA* employs to regulate gene expression are currently unknown. Curiously, despite restoring the *in vivo* infectivity phenotype, the *ittA* complement did not rescue the gene expression defect of many of the validated differentially regulated genes.

RNA chaperones

Hfq

Hfq is an RNA chaperone that facilitates the interaction between sRNAs and their targets (Gottesman, 2004; Valentin-Hansen et al., 2004; Majdalani et al., 2005; Brennan and Link, 2007; Waters and Storz, 2009). The Hfq of *B. burgdorferi* was difficult to identify because it has limited homology with the Hfq from *E. coli* and other bacteria, but it is able to heterologously complement an *hfq* mutant of *E. coli* (Lybecker et al., 2010). Conversely, the *E. coli hfq* gene heterologously complements several phenotypes of the pleiotropic *B. burgdorferi hfq* mutant *in vitro*, albeit not the vertebrate infectivity defect (Lybecker et al., 2010). *B. burgdorferi* Hfq is required for infectivity in mice, as well as induction of RpoS and its regulon (Lybecker et al., 2010); therefore, the phenotype is not due solely to a defect in facilitating the $DsrA_{Bb}$ interaction with *rpoS* mRNA, as RpoS can be induced at high cell density in the *dsrA*$_{Bb}$ mutant (Lybecker and Samuels, 2007), suggesting an Hfq-dependent sRNA network in the spirochete that regulates infectivity (Lybecker et al., 2010).

CsrA

CsrA is an RNA-binding protein that inhibits translation of several genes by binding the ribosome-binding sites of their mRNAs and blocking the ribosome; CsrB and CsrC are sRNAs that bind to and sequester CsrA, inhibiting its activity on mRNAs (Romeo, 1998; Majdalani et al., 2005; Hör et al., 2018; Romeo and Babitzke, 2018). CsrA regulates a variety of physiological processes in *E. coli* and other bacteria, including carbon metabolism and motility (Romeo, 1998; Majdalani et al., 2005; Hör et al., 2018; Romeo and Babitzke, 2018), as well as bacterial infectivity (Lucchetti-Miganeh et al., 2008; Vakulskas et al., 2015). Levels of CsrA in *B. burgdorferi* increase in response to elevated temperature and decreased pH (Sanjuan et al., 2009). Initial reports suggested that CsrA affects the expression of *ospC* as well as other RpoS-dependent genes and is required for murine infectivity in *B. burgdorferi* (Sanjuan et al., 2009; Karna et al., 2011; Sze and Li, 2011; Sze et al., 2011; Karna et al., 2013; Arnold et al., 2018; Lin et al., 2018); however, a subsequent study reported that CsrA does not activate RpoS or its regulon and that *csrA* mutants are infectious in the murine model (Ouyang et al.,

2014b). As of 2020, the role of CsrA in the physiology of *B. burgdorferi* remains enigmatic and its RNA targets remain undefined.

RNA turnover and ribonucleases

The steady-state levels of RNAs are controlled not just by the rate of their transcription, but also by the rate of their degradation, which is catalyzed by a variety of ribonucleases (Mohanty and Kushner, 2016; Bechhofer and Deutscher, 2019; Trinquier et al., 2020). Compared to other bacteria, *B. burgdorferi* encodes a sparse quiver of ribonucleases, which includes the endoribonucleases RNase III (BB0705), RNase P (BB0441), RNase Z (BB0755), YbeY (BB0060), RNase Y (BB0504), RNase M5 (BB0626), and RNase HII (BB0046) and the exoribonuclease PNPase (BB0805) (Fraser et al., 1997; Archambault et al., 2013; Anacker et al., 2018; Drecktrah et al., 2020). Lyme disease spirochetes lack detectable orthologs of RNase E/G and RNase J1/J2; RNase Y functions in RNA processing and decay in *Bacillus subtilis*, *Staphylococcus aureus*, and *Streptococcus pyogenes* in an analogous fashion as RNase E or G in *E. coli*, so *B. burgdorferi* appears to be more similar in its suite of ribonucleases to the gram-positive bacteria (Shahbabian et al., 2009; Yao and Bechhofer, 2010; Bechhofer, 2011; Lehnik-Habrink et al., 2011; Marincola et al., 2012; Chen et al., 2013; Bechhofer and Deutscher, 2019). Two ribonucleases have been genetically characterized in *B. burgdorferi*: RNase III, encoded by the *rnc* gene (Anacker et al., 2018) and RNase Y, encoded by the essential *rny* gene (Drecktrah et al., 2020). RNase III, which recognizes and cleaves double-stranded RNA, is required for proper rRNA processing (Anacker et al., 2018) and for mRNA decay, at least for several transcripts (Snow et al., 2020), in the spirochete. RNase Y is involved in the biogenesis of 6S RNA in *B. burgdorferi* (Drecktrah et al., 2020), which has not been shown in any other bacterium, but the ribonuclease surely has additional roles in the cell, likely in the turnover of mRNA based on precedence in other bacteria (Lehnik-Habrink et al., 2011; Durand et al., 2012; Chen et al., 2013), as Bb6S RNA null mutants are viable but *rny* is essential for growth of the spirochete (Drecktrah et al., 2020).

Schlax and colleagues have assayed mRNA turnover kinetics *in vitro* and demonstrated that half-lives vary from a minute to almost an hour, which is considerably longer than the half-lives measured for the majority of mRNAs in most bacteria (Archambault et al., 2013). Note that these studies are more complicated in *B. burgdorferi* because mRNA turnover in bacteria is generally quantified following inhibition of transcription with rifampicin, but the unusual RNA polymerase of *B. burgdorferi* is resistant to this antibiotic, so, instead, actinomycin D is used to arrest transcription (Archambault et al., 2013). The half-lives and steady-state levels of several mRNAs, as well as the steady-state levels of the stable rRNAs, were significantly increased in an *rnc* mutant, indicating a role for RNase III in RNA degradation (Snow et al., 2020). However, the steady-state levels of a couple mRNAs decreased in the *rnc* mutant, suggesting a more nuanced regulatory scheme involving RNase III. At this point, the mechanism by which RNA degradation and RNase activity is regulated to effect post-transcriptional levels of gene expression in *B. burgdorferi* is unknown; however, RNase III is known to regulate the interaction between antisense sRNAs and their target mRNAs in several bacteria, including *E. coli*, *B. subtilis*, *S. aureus*, *S. pyogenes* and *Streptomyces antibioticus* (Lasa et al., 2011; Lasa et al., 2012; Lioliou et al., 2012; Lee et al., 2013; Lybecker et al., 2014a; Lybecker et al., 2014b; Le Rhun et al., 2016), and a similar mechanism may be employed by the spirochete.

Nucleic acid-binding proteins

The genome of *B. burgdorferi* encodes several proteins that resemble known nucleic acid-binding proteins of other bacteria, such as BosR and BadR (see the section above). However, most of those predicted nucleic acid-binding proteins have yet to be characterized. In addition, novel proteins have been identified that do not possess structures previously associated with nucleic acid interactions. Two of the most notable of these borrelial proteins, EbfC and SpoVG, are also encoded by a wide spectrum of other bacterial species. Thus, studies of *B. burgdorferi* are providing insights on the physiology of bacteria ranging from *E. coli* to *S. aureus*.

At least two novel *B. burgdorferi* proteins, SpoVG and BpuR, initially discovered due to their DNA-binding activities, have since been found to have greater affinities for RNA than for DNA. Combined with data from other species, this finding raises the possibility that many other "DNA-binding" proteins might also preferentially bind to RNAs. Additionally, the known activities of BpuR and SpoVG support a paradigm that RNA-binding proteins can have functions beyond serving as RNA chaperones, such as operating solo to directly affect mRNA translation.

EbfC

This 99-amino acid protein forms a homodimer in solution; the alpha helical N- and C-termini of both subunits form a "pincer" that fits over the DNA double helix (Lim et al., 2003; Riley et al., 2009). Consistent with binding by a homodimer, EbfC preferentially interacts with a palindromic sequence, 5′-GTnAC-3′. This sequence occurs approximately every 1.1 kb throughout the *B. burgdorferi* genome. EbfC also forms higher-order oligomers, including tetramers and octamers, an activity that could serve to bridge separate DNA strands (Riley et al., 2009). Microscopic analyses of GFP-tagged EbfC revealed that it associates with the borrelial nucleoid (Jutras et al., 2012a). These are all characteristics of nucleoid-associated proteins, leading to a model that a major function of EbfC is to organize the structure of borrelial DNA. Moreover, the *ebfC* gene is co-expressed with *dnaX*, which encodes a subunit of DNA polymerase, suggesting that DNA-binding by EbfC plays an indirect role in replication (Jutras et al., 2012a). Almost all known species of bacteria also encode an EbfC homolog (variously named YbaB, ORF12, and other terms.), implying a conserved function (Cooley et al., 2009; Wang et al., 2012a).

EbfC initially was identified due to its high affinity binding to the upstream region of *B. burgdorferi erp* operons (Babb et al., 2006). The majority of *erp* operators contain two to three complete EbfC-binding sites, a density that has not been identified elsewhere in the *B. burgdorferi* genome. EbfC acts as an anti-repressor of *erp* transcription by competing with the BpaB repressor for binding to the operator (Jutras et al., 2012b). Erp protein expression is repressed during colonization of an unfed tick, but is induced when the tick begins to feed, and those outer surface lipoproteins are expressed during transmission and throughout the vertebrate phase of the enzootic cycle. Studies of *B. burgdorferi* cultivated *in vitro* found that *erp* transcript levels increase as bacterial replication rates increase, a phenomenon that also occurs as a tick begins to feed and the spirochetes rapidly multiply due to nutrients in the blood (Jutras et al., 2013c). The current model for *erp* regulation posits that the rapid bacterial replication rate in feeding ticks necessitates DnaX and EbfC for DNA replication, and that *erp* operons have evolved to recognize elevated EbfC levels as a signal that the tick is feeding and the time is appropriate to produce the Erp proteins (Jutras et al., 2012a; Jutras et al., 2013c).

BpaB

All of the plasmids in the segmented genome of *B. burgdorferi* encode a unique BpaB protein. BpaB and two other ORFs are essential for plasmid maintenance and partitioning, with BpaB apparently performing a function comparable to the ParB proteins of other bacterial plasmids. The unique sequences of BpaB from each plasmid appear to be involved with compatibility (Zückert and Meyer, 1996; Casjens et al., 1997; Fraser et al., 1997; Casjens et al., 2000; Stewart et al., 2001; Eggers et al., 2002; Stewart et al., 2003; Brisson et al., 2013). The BpaB proteins of the cp32 family of prophages are the best studied to date. While BpaB presumably binds DNA near the origin of replication, as do ParB proteins from other species, three other binding sites have been identified on cp32s: one each in the *erp* operator, adjacent to the *nucP* promoter, and adjacent to the *ssbP* promoter (Burns et al., 2010; Chenail et al., 2012). In *erp* operators, BpaB binds to a specific sequence 5′ of the promoter, then oligomerizes along the DNA, stabilized by protein-protein interactions (Burns et al., 2010). Occlusion of the promoter by BpaB prevents RNA polymerase binding, thereby inhibiting transcription (Jutras et al., 2012b). BpaB appears to enhance transcription of *nucP* and *ssbP*, through unknown mechanisms (Chenail et al., 2012).

BpuR

BpuR is a homodimer of two 122-amino acid subunits, which fold together into a "PUR" domain (Graebsch et al., 2010; Jutras et al., 2013a). This domain is named because it preferentially binds to <u>pur</u>ine-rich stretches of nucleic acids. PUR domains are found in a few other bacterial genera, such as *Treponema* and *Bacteroides*, and in all multicellular eukaryotes. This curious distribution suggests that there was an exchange of a PUR motif between Domains over the course of evolution.

PUR domain proteins bind to double-stranded and single-stranded DNAs and to RNA, with BpuR exhibiting its greatest affinity for RNA. A specific nucleotide sequence has yet to be identified for BpuR binding, a feature shared with eukaryotic PUR domain proteins. This suggests that PUR domain proteins interact with higher-order structural determinants.

BpuR was first identified due to its ability to bind DNA, as it has a high affinity for sequences in *erp* operator sites. It acts as a co-repressor of *erp* transcription, enhancing the repressor activity of BpaB through an undefined mechanism. BpuR has also been found to bind several borrelial mRNAs, including *bpuR* itself, where binding to the 5′ end inhibits translation (Jutras et al., 2013d). Comparisons of transcriptomic and proteomic data indicate that BpuR is also likely to affect translation of numerous mRNAs (Jutras et al., 2019).

SpoVG

Homologs of SpoVG are produced by many species of spirochetes and firmicutes. Its name comes from early studies of *Bacillus subtilis*, where *spoVG* mutants are defective in stage V of sporulation (Rosenbluh et al., 1981). SpoVG proteins do not contain previously known nucleic acid-binding motifs, and their functions were a mystery until investigations of a DNA-binding protein of *B. burgdorferi* resulted in fishing out SpoVG (Jutras et al., 2013b). That initial study further demonstrated that the SpoVG proteins of *Listeria monocytogenes* and *Staphylococcus aureus* also bind to DNA, and mapped nucleic acid-binding specificity to the C-terminal alpha helix. Subsequent studies found that the *L. monocytogenes* and *B. burgdorferi* SpoVG proteins exhibit higher affinities for RNA than for DNA (Burke and Portnoy, 2016; Savage et al., 2018). A defined nucleotide sequence for SpoVG-binding has yet to be identified, suggesting that it interacts with conformational aspects of the nucleic acid sugar-phosphate backbone. *B. burgdorferi spoVG* mutants exhibit pleotropic defects, evidently due to both transcriptional and post-transcriptional regulatory activities (Savage et al., 2018).

Hbb

Integration host factor (IHF) was discovered on the basis of its role in the integration of bacteriophage λ into the *E. coli* chromosome (Drlica and Rouviere-Yaniv, 1987; Friedman, 1988; Nash, 1996). IHF binds to DNA at specific sites and induces bending. The origin of replication on the *E. coli* chromosome contains an IHF-binding site, which is conserved in several bacteria, and IHF is essential for replication (Nash, 1996). Another small DNA-binding protein in *E. coli* is HU, which is homologous to IHF (Drlica and Rouviere-Yaniv, 1987; Schmid, 1990; Nash, 1996). HU also bends DNA, but it lacks sequence specificity. Many bacteria have homologs of both IHF and HU, although several species have only a single homolog. Tilly et al. (1996) isolated a gene encoding an IHF/HU homolog, termed Hbb, from *B. burgdorferi* in a genetic screen based on complementation of the λ DNA packaging defect of *E. coli* IHF/HU mutants. Hbb is most similar to homologs from bacteria that encode a single IHF/HU homolog and is 27 to 30% identical to IHF and HU from *E. coli* (Tilly et al., 1996). Hbb is unable to complement an *E. coli* HU mutant for the defect in bacteriophage Mu growth (Tilly et al., 1996) and it binds DNA in a sequence-specific fashion (Kobryn et al., 2000; Mouw and Rice, 2007), suggesting that Hbb is more like IHF than HU. Kobryn et al. (2000) demonstrated that Hbb binds to a specific sequence that is related to the IHF consensus binding site. In addition, IHF binds to this Hbb-binding site with high affinity and Hbb will bind to an IHF consensus site (Mouw and Rice, 2007). However, Hbb affects DNA topology, which is a property of HU, so Hbb has a substantial non-specific DNA-binding activity and is not a straightforward IHF homolog (Kobryn et al., 2000).

Mouw and Rice (2007) solved the structure of an Hbb-DNA complex. The structure and mechanism of DNA bending is similar to that of IHF and HU proteins. Hbb binds DNA as a homodimer with two β-ribbon "arms" that wrap in the minor groove, with a pair of conserved prolines intercalating between adjacent base pairs (Mouw and Rice, 2007), which results in a DNA bend of more than 180° (Kobryn et al., 2000; Mouw and Rice, 2007). The binding site identified by Kobryn et al. (2000) is near the origin of replication on the chromosome (between *dnaA* and *dnaN*). Thus, Hbb may be involved in replication initiation (Chaconas, 2005), although results in an *E. coli* surrogate system show that recombinant Hbb increases transcription of *oppA2* and *oppA4* promoter fusions (Medrano et al., 2007).

Gac

Knight and Samuels (1999) identified Gac in a biochemical screen for proteins that bound to a sub-telomeric sequence from linear plasmid lp17. Gac has HU-like non-specific DNA-binding activity (Knight and Samuels, 1999; Corbett et al., 2004), but it is a novel protein that completely lacks sequence similarity to HU or other members of the architectural DNA-binding protein family: it has only been identified to date in *B. burgdorferi*. Gac is the C-terminal domain of GyrA, the A subunit of DNA gyrase, and it is produced as an independent protein by a mechanism unprecedented in bacteria. *B. burgdorferi* transcribes two mRNAs carrying *gac* sequences: the larger one encodes the entire *gyrBA* operon that presumably generates DNA gyrase, the essential type II DNA topoisomerase in bacteria that introduces negative DNA supercoiling (Reece and Maxwell, 1991; Corbett and Berger, 2004), and the smaller, more abundant one, encodes Gac. The *gac* mRNA is predicted to be transcribed from its own promoter embedded in the *gyrA* gene (Knight and Samuels, 1999). This unusual architectural DNA-binding protein was termed Gac (for GyrA C-terminal domain), and its gene *gac*, because, although it is translated from a subset of the GyrA ORF, it is a different gene product of a unique genetic locus (Knight et al., 2000). In other words, mutations can disrupt the synthesis of Gac, but have no phenotypic effect on GyrA. In fact, genetic disruption of Gac was accomplished by mutating the start codon and Met 5 codon (to Leu and Ile, respectively) (Knight et al., 2000). Disappointingly, the *gac* mutant had no detectable phenotype. The phenotype was hypothesized to be suppressed by the coumermycin A_1 resistance marker in the upstream *gyrB* gene used to introduce the site-specific mutations (Samuels, 2006). Although reverse genetics has so far failed to reveal the function of Gac in *B. burgdorferi*, other experiments suggest that Gac is an architectural DNA-binding protein with HU activity. HU is essential for replicative transposition of bacteriophage Mu in *E. coli*: generating the Type I complex, an early intermediate, requires HU (Craigie et al., 1985). Gac was able to substitute for HU, albeit not as efficiently, in promoting formation of the Type I complex in an *in vitro* donor-cleavage reaction (Knight and Samuels, 1999). Notably, Gac was not able to substitute in the assay for IHF. In addition, Gac, expressed from its own promoter, was able to complement the defect in Mu growth, as assayed by efficiency of plating, in an *E. coli* HU mutant (Knight and Samuels, 1999). Like HU, IHF, and Hbb, Gac bends DNA, which was demonstrated by FRET measurements (Corbett et al., 2004). However, the mechanism by which Gac bends DNA is completely different than the other architectural DNA-binding proteins that involve a β-sheet in the minor groove. Instead, Gac has a novel structure, determined by Corbett et al. (2004) and termed a β-pinwheel, which is similar to a β-propeller; the DNA wraps around the perimeter of four of the six blades of the pinwheel, resulting in a profound bend. This structure also provided insight into the enzymatic mechanism of DNA gyrase (Corbett et al., 2004), as Gac, and the C-terminal domains of DNA gyrases from other bacteria, impart positive supercoiling on the wrapped DNA, which is converted to negative supercoiling following the strand-passage reaction (Ruthenburg et al., 2005; Huang et al., 2006; Kramlinger and Hiasa, 2006).

DNA supercoiling and *cis*-acting sequences

DNA supercoiling, which is known to regulate gene expression in other bacteria (López-García and Forterre, 2000; Cheung et al., 2003; Rui and Tse-Dinh, 2003; Peter et al., 2004), has been hypothesized to serve as a molecular thermometer for *ospC* expression (Alverson et al., 2003). There are several mechanisms by which DNA supercoiling affects gene expression, including the regulation of promoter recognition by RpoS. *E. coli* RpoS preferentially recognizes promoters on a relaxed template (Kusano et al., 1996) and DNA relaxation in *E. coli* is required for transcription by RNA polymerase holoenzyme containing RpoS during the cellular response to osmotic stress (Bordes et al., 2003). In fact, DNA supercoiling has been proposed to be a post-translational regulator of RpoS activity (Hengge-Aronis, 2002b; Bordes et al., 2003).

In *B. burgdorferi*, *ospC* gene is transcribed from an RpoS-dependent promoter (see the section above on the RpoN-RpoS alternative σ factor cascade), but transcription is repressed early during the infection because OspC elicits a potent immune response (Liang et al., 2002a; Liang et al., 2002b; Crother et al., 2004; Liang et al., 2004; Tilly et al., 2006). Two overlapping inverted repeats (IRs) are present upstream of the promoter (Margolis et al., 1994) that serve as an operator to mediate repression (Xu et al., 2007). The inverted repeats probably either bind a *trans*-acting repressor or alter secondary structure in response to changes in DNA supercoiling, or both. Surprisingly, mutating the inverted repeats has little effect on *ospC* transcription in *trans* from a shuttle vector (Eggers et al., 2004; Yang et al., 2005). Drecktrah et al. (2013) obtained different results using site-directed mutagenesis of the inverted repeats in *cis* on cp26: induction of *ospC* by DNA supercoiling, temperature, and pH required the distal inverted repeat, but not the proximal inverted repeat. In addition, the sequence of the IRs was less important for the regulation of *ospC* expression than the base-pairing potential of the two halves of the IR element (Drecktrah et al., 2013).

Cyclic-di-AMP

Cyclic-di-AMP is a recently identified second messenger in bacteria (Römling, 2008; Corrigan and Gründling, 2013; Commichau et al., 2019). It is synthesized from two molecules of ATP by diadenylate cyclase (DAC), and is hydrolyzed by phosphodiesterases (PDEs). Virtually all organisms identified to date that are endowed with c-di-AMP are gram-positive bacteria. *B. burgdorferi* is one of a few exceptions (Barker et al., 2013; Ye et al., 2014; Savage et al., 2015; Rubin et al., 2018). *B. burgdorferi* encodes a diadenylate cyclase, CdaA (BB0008) (Savage et al., 2015), and a phosphodiesterase, DhhP (BB0619) (Ye et al., 2014). In gram-positive bacteria, genes encoding c-di-AMP phosphodiesterase are often non-essential: mutants lacking phosphodiesterase were readily obtained and those mutants have been key to our understanding of c-di-AMP signaling (Römling, 2008; Corrigan and Gründling, 2013; Commichau et al., 2019). On the other hand, genes encoding c-di-AMP cyclase appear to be essential in most bacteria. In *B. burgdorferi*, the phenotypes of c-di-AMP mutants are unique. Inactivation of *dhhP* in *B. burgdorferi* could only be achieved in the presence of a plasmid encoded inducible *dhhP* gene (Ye et al., 2014). The conditional *dhhP* mutant failed to grow in the absence of the inducer. This mutant had a decreased, not increased, resistance to β-lactamase antibiotics, which also differs from gram-positive bacteria.

Wild-type *B. burgdorferi* cultured *in vitro* has very low levels of intracellular c-di-AMP (~9 nM), and overproduction of CdaA does not increase c-di-AMP levels (Ye et al., 2014; Savage et al., 2015), suggesting that CdaA may not be active under standard *in vitro* culture conditions. However, this may not be true as inactivation of *dhhP* resulted in over fiftyfold increase of intracellular c-di-AMP levels, suggesting that CdaA is active to produce c-di-AMP in cultured spirochetes. c-di-AMP appears to be involved in several cellular processes in other bacteria, including potassium transport, cell wall and membrane homeostasis, DNA damage repair, sporulation, and osmoregulation (Römling, 2008; Corrigan and Gründling, 2013; Commichau et al., 2019). In *B. burgdorferi*, the essential nature of the *dhhP* mutant strongly indicates that c-di-AMP plays an important role that is yet to be elucidated.

Transcriptomics

Transcriptomics, or genome-wide expression profiling, is a powerful tool that comparatively quantifies RNAs produced from the genome. Methods of transcriptome analyses are continuously evolving, but two of the most commonly used approaches to date are microarrays and RNA-sequencing (RNA-seq) technologies. Microarray-based studies quantify transcript levels of predetermined RNAs, while RNA-seq can identify all expressed RNAs. Microarray-based transcriptomic data from multiple *B. burgdorferi* studies was recently reviewed by Iyer et al. (2016). RNA-seq has more recently been utilized for differential gene expression analysis, and to globally identify transcription start sites (TSSs) and small RNAs (sRNAs).

The vast majority of transcriptomic studies have been performed on spirochetes cultivated *in vitro* under conditions attempting, perhaps not successfully, to mimic either the vertebrate host or tick vector (Table 1). Performing transcriptomics on spirochetes within infected ticks or vertebrate tissues is challenging due to the paucity of organisms present within these host environments. Using an enrichment-based microarray approach, Iyer et al. (2015) identified an *in vivo* transcriptome during acquisition (fed larvae) and transmission (fed nymphs), while Narasimhan et al. (2003) examined *in vivo* gene expression in the central nervous system (CNS) and heart tissues of non-human primates (NHP). Notably, the genome coverage is low in both the tick and NHP transcriptomes, perhaps because the transcriptome is indeed reduced *in vivo* compared to *in vitro*, or because of limitations of the microarray technology to detect low-level transcripts. In addition, differentially-expressed genes may have been missed due to the limited dynamic range of microarray technology. Several groups have taken advantage of a model developed by Akins et al. (1998) in which spirochetes are cultivated within dialysis membrane chambers (DMCs) implanted into the peritoneal cavity of a rat or rabbit to generated large numbers of spirochetes in a mammalian 'host-adapted' state. These 'host-adapted' organisms undergo many, but not all, of the transcriptomic and proteomic changes associated with vertebrate infection, including repression of tick-phase genes, such as *ospA* and the *glp* operon.

Comparative transcriptome analyses using either microarray or RNA-seq have identified cohorts of genes regulated by specific environmental conditions as well as important regulatory networks in *B. burgdorferi*. Strikingly, there is often limited overlap between published data sets for seemingly identical strains (i.e., wild type) grown under ostensibly the same environmental condition(s). The lack of concordance between different data sets is likely due to a combination of biological and technical factors, including variations in growth media components,

Table 1. Transcriptome studies in *B. burgdorferi.*

Experimental Condition	Strain	Platform	Reference
Increased temperature	B31	Membrane microarray	Ojaimi et al., Infect. Immun., 2003
Comparative analysis of clinical isolates	B31	Membrane microarray	Ojaimi et al., Infect. Immun., 2005
Temperature-shift *in vitro* vs. host adapted DMC	B31	Glass slide microarray	Revel et al., PNAS, 2002
In vitro vs. host adapted DMC	B31	Membrane microarray	Brooks et al. Infect. Immun., 2003
Host mimic: Influence of whole blood *in vitro*	B31	Membrane microarray	Tokarz et al., Infect. Immun., 2004
Monoclonal OspB antibody co-cultivation	B31	Membrane microarray	Anderton et al., Infect. Immun., 2004
Neurological cell co-cultivation	B31	Affymetrix slide microarray	Livengood et al., Infect. Immun., 2008
RpoS regulon (*in vitro* vs. DMC)	297	70 m oligo glass slide microarray	Caimano et al., Mol. Microbiol., 2007
RpoS regulon (*in vitro* vs. DMC)	B31, 297	Illumina RNA sequencing	Caimano et al., Front Microbiol., 2019
RpoN/RpoS regulon (*in vitro*)	B31	70 m oligo glass slide microarray	Fisher et al., PNAS, 2005
BosR regulon (*in vitro*)	B31	Membrane microarray	Hyde et al., Microbiology, 2006
BosR regulon (*in vitro*)	B31	70 m oligo glass slide microarray	Ouyang et al., Mol. Microbiol., 2009
Rrp2 regulon (*in vitro*)	B31	70 m oligo glass slide microarray	Boardman et al., Infect. Immun.,2008
Rrp2/RpoN/RpoS regulon (*in vitro*)	B31	70 m oligo glass slide microarray	Ouyang et al., Microbiology, 2008
CsrA, BadR, RpoN and RpoS	B31	Illumina RNA sequencing	Arnold et al., PLOS One, 2018
Rrp1 regulon (*in vitro*)	B31	70 m oligo glass slide microarray	Rogers et al. Mol. Microbiol., 2009
Rrp1 regulon (*in vitro*)	B31	70 m oligo glass slide microarray	He et al., PLOS Pathog., 2011
Rrp1 regulon (*in vitro*)	B31	Illumina RNA sequencing	Caimano et al., Infect. Immun., 2015
Rel_{Bbu} regulon (*in vitro*)	297	70 m oligo glass slide microarray	Bugrysheva et al., PLOS One, 2015
BadR regulon (*in vitro*)	B31	Nimblegen	Miller et al., Mol. Microbiol., 2013
HrpA regulon (*in vitro*)	B31	Nimblegen	Salman-Dilgimen et al., PLOS Pathog., 2013
Non-human primate tissues	N40, JD1	Glass slide microarray	Narasimhan et al., PNAS, 2003
Fed nymphs	N40	Glass slide microarray	Narasimhan et al., J. Bacteriol., 2002
Infected murine tissues	B31	70 m oligo glass slide microarray	Pal et al., J. Infect. Dis., 2008
Fed larvae and nymphs vs DMCs and *in vitro*	B31	70 m oligo glass slide microarray	Iyer et al., Mol. Microbiol., 2015
GlcNAc starvation	B31	Illumina RNA sequencing	Schneider et al., FEMS Microbiol Lett., 2018
Global 5' end identification	B31	Illumina RNA sequencing	Adams et al., Nucleic Acid Res, 2016
In vitro growth phases	B31	Illumina RNA sequencing	Arnold et al., PLOS One, 2016
Temperature response small RNA regulon	B31	Illumina RNA sequencing	Popitsch et al., BMC genomics, 2016
DNA methylation	B31	Illumina RNA sequencing	Casselli et al., J. Bacteriol., 2018
Rel_{Bbu} regulon and starvation	B31	Illumina RNA sequencing	Drecktrah et al., PLOS Pathogens, 2015
Rel_{Bbu} small RNA regulon	B31	Illumina RNA sequencing	Drecktrah et al., Front. Cell. Infect. Microbiol., 2018
Doxycycline treatment	B31	Ion Torrent RNA sequencing	Caskey et al., Front Microbiol., 2019
EbfC regulon	B31	Illumina RNA sequencing	Jutras et al., J. Bacteriol., 2012
Outer membrane vesicle transcriptome	B31	Illumina RNA sequencing	Malge et al., FEMS Microbiol. Lett., 2018
ittA regulon	B31	Illumina RNA sequencing	Medina et al., PLOS Pathogens, 2020
DksA and starvation regulon	B31	Affymetrix slide microarray	Boyle et al., J. Bacteriol., 2019

phase of growth at the time of harvest, methods used for RNA extraction, cDNA synthesis and library preparation, approach used to identify differentially expressed genes (i.e., microarray or RNA-seq), and data analyses. Population and cell-to-cell heterogeneity likely account for some of the differential gene expression observed in transcriptome studies, which is further confounded by the complex plasmid profiles of *B. burgdorferi.* Plasmid content and the repertoires of plasmid-encoded genes vary between strains. In addition, plasmids frequently are lost during *in vitro* cultivation. Several studies have compared gene expression profiles of different strains of *B. burgdorferi*, revealing strain-specific gene regulation (Ojaimi et al., 2005; Caimano et al., 2019).

Despite the lack of concordance among transcriptome studies in *B. burgdorferi*, several trends have emerged. Genes encoding outer surface proteins induced during transmission and vertebrate infection (OspC-like) often are induced by a temperature upshift, cultivation in DMCs and the RpoN/RpoS alternative σ factor cascade (Revel et al., 2002; Brooks et al., 2003; Ojaimi et al., 2003; Tokarz et al., 2004; Fisher et al., 2005; Hyde et al., 2006; Caimano et al., 2007; Boardman et al., 2008; Livengood et al., 2008; Ouyang et al., 2008; Ouyang et al., 2011; Caimano et al., 2019). Starvation, and the signaling molecules (p)ppGpp and c-di-GMP regulate genes encoding outer surface proteins, chemotactic proteins and transporters and enzymes required to utilize different carbon sources. Strikingly, differential expression of the *glp* operon in many of the transcriptome studies substantiate its role in the tick. This operon encodes three genes, *glpF*, *glpK*, and *glpD*, required for uptake and utilization of glycerol as an alternative carbon source in ticks; *glp*-deficient spirochetes survive at much lower levels during the tick blood meal and intermolt (He et al., 2011; Pappas et al., 2011). The *glp* operon or specific genes in the operon are upregulated by tick-associated regulators Rrp1, $RelA_{Bbu}$, and DksA as well as starvation, while it is repressed by a vertebrate phase-specific environmental cues (temperature upshift, recovery from starvation, co-cultivation with neuroglial cells, and cultivation within DMCs) along with vertebrate phase regulators BosR, BadR and CsrA (Revel et al., 2002; Fisher et al., 2005; Hyde et al., 2006; Livengood et al., 2008; Ouyang et al., 2009b; Rogers et al., 2009; He et al., 2011; Miller et al., 2013; Ouyang et al., 2014b; Bugrysheva et al., 2015; Caimano et al., 2015; Drecktrah et al., 2015; Iyer et al., 2015; Ouyang and Zhou, 2015; Iyer and Schwartz, 2016; Schneider and Rhodes, 2018; Boyle et al., 2019).

Temperature response

Growth at different temperatures is often employed to mimic the environmental change spirochetes undergo as *B. burgdorferi* transitions between the tick vector and vertebrate host (23°C and 34-37°C, respectively). Revel et al. (2002) altered both the pH and temperature of *in vitro B. burgdorferi* cultures to simulate the unfed and fed tick environments (23°C/ pH 7.5 and 37°C/pH 6.8, respectively). Ojaimi et al. (2003) and Iyer et al. (2015) also compared differential gene expression of cells grown at 23°C and 35°C, but did not alter the pH. To simulate nymphal transmission, Tokarz et al. (2004) compared the transcriptomes of spirochetes grown at 35°C in BSK II medium with and without the addition of 6% human blood.

Starvation response

In order to persist in the tick midgut in between blood meals, *B. burgdorferi* must adapt to and survive nutrient deprivation. Three transcriptomic analyses of *B. burgdorferi* cultured in nutrient-limited media provide insight into gene expression during spirochete persistence (Drecktrah et al., 2015; Schneider and Rhodes, 2018; Boyle et al., 2019). Drecktrah et al. (2015) shifted *B. burgdorferi* cultures growing in BSK II with rabbit serum (BSK II + RS) at stationary phase to RPMI without rabbit serum for 6 hours to represent nutrient limitation and starvation conditions in the tick. RPMI lacks many of the nutrients present in BSK II including neopeptone, yeastolate, *N*-acetylglucosamine and bovine serum albumin. To mimic the influx of nutrients that accompanies an incoming blood meal during transmission, cultures were subsequently shifted back to BSK II + RS for 2 hours. Boyle et al. (2019) also examined differential gene expression in response to starvation in RPMI media using microarray analyses instead of RNA-seq.

N-acetylglucosamine (GlcNAc) starvation also has been used to imitate tick overwintering *in vitro* (Schneider and Rhodes, 2018). GlcNAc is required for cell wall synthesis and, once phosphorylated, the sugar can be shuttled into the glycolytic pathway. *B. burgdorferi* is not able to synthesize GlcNAc from glucose but acquires it from the breakdown of chitobiose, a major component of tick cuticle, via the phosphoenolpyruvate-phosphotransferase systems (PEP-PTS) (Tilly et al., 2001; Corona and Schwartz, 2015; Caimano et al., 2016; Troy et al., 2016). Schneider and Rhodes (2018) used RNA-seq to compare the gene expression profiles of wild-type *B. burgdorferi* cultivated in the presence or absence of GlcNAc.

In vivo *tick transcriptomics*

As previously discussed, *in vivo* transcriptomic analyses are challenging due to the low numbers of spirochetes that can be recovered from the tick vector or infected host tissues. To date, the only genome-wide transcriptomic study of *B. burgdorferi* within ticks compared fed larvae (acquisition) and fed nymphs (transmission). Using an RNA enrichment pre-amplification step prior to microarray analysis, Iyer et al. (2015) detected ~500 transcripts in the fed nymphs and larvae. The transcriptomes of *B. burgdorferi* in both tick life stages were also compared to *B. burgdorferi* grown *in vitro* following temperature-shift or cultivated within DMCs.

Hierarchal clustering revealed that the expression profile of temperature-shifted *B. burgdorferi* was more similar to that of fed larvae than fed nymphs, although the *in vivo* transcriptomes were very different than the *in vitro* transcriptome.

Host-adapted transcriptomics

Several studies also sought to demonstrate the transcriptomic expression profile of mammalian host-adapted organisms. Narasimhan et al. (2003) examined *in vivo* gene expression in the central nervous system (CNS) and heart tissues of non-human primates (NHP), which included a comparison between steroid treated and immunocompromised NHPs. An enrichment technique to selectively amplify specific bacterial transcriptomes was used to enhance detection of spirochetal mRNA in NHP tissues. While the detectable transcriptome was limited (~500 transcripts), the authors identified over 150 differentially expressed genes in the heart compared to CNS tissue. To date, this is the only *in vivo* transcriptome analysis performed using infected vertebrate tissues. However, multiple groups have attempted to recapitulate the tissue environment by cultivating spirochetes within DMCs or by co-cultivating spirochetes with human neuroglial cells (Revel et al., 2002; Brooks et al., 2003; Tokarz et al., 2004; Caimano et al., 2007; Livengood et al., 2008; Iyer et al., 2015; Caimano et al., 2019). Three microarray analyses have directly compared the transcriptomes of host-adapted *B. burgdorferi* within DMCs compared to *in vitro* cultured spirochetes at 37°C. These studies have demonstrated that gene expression varies substantially between the two growth environments (Revel et al., 2002; Brooks et al., 2003; Iyer et al., 2015). Livengood et al. (2008) identified genes related to spirochete-host interactions by comparing the transcriptomes of *B. burgdorferi* co-cultivated with human neuroglial cells to those cultured in cell-free medium.

Regulators of gene expression

The RpoN/RpoS alternative σ factor cascade (see section above) is activated within ticks during the nymphal blood meal resulting in a sea change of gene expression to allow for transmission and vertebrate infection. RpoS, an alternate σ factor, induces expression of genes required for transmission and maintenance of *B. burgdorferi* in the vertebrate host and represses many tick phase-specific genes. Transcription of *rpoS* is induced during the nymphal blood meal by RpoN, as part of a complex that also includes Rrp2 and the Fur family regulator BosR (see sections above on the RpoN/RpoS alternative σ factor cascade and the Transcription factors BosR and BadR).

The BosR regulon has been examined by microarray by two groups (Hyde et al., 2006; Ouyang et al., 2009b). Hyde et al. (2006) compared the transcriptomes of a *bosR* point mutant sensitive to oxidative stress and corresponding complemented strain. Ouyang et al. (2009b) analyzed the BosR-dependent transcriptome by comparing the transcriptomes of a wild-type strain and *bosR* null mutant.

Fisher et al. (2005) was first to define the RpoN- and RpoS-dependent transcriptome. Their comparative transcriptomics revealed a set of genes regulated through the RpoN/RpoS alternative σ factor cascade as well as subsets of genes that are uniquely regulated by either RpoN or RpoS alone. Ouyang et al. (2008) examined the Rrp2-, RpoN- and RpoS-dependent transcriptomes utilizing microarray analyses. They also identified a large number of genes coordinately regulated by all three factors suggesting a hierarchical coordination of Rrp2, RpoN and RpoS. In contrast to the Fisher et al. (2005) study, however, Ouyang et al. (2008) found only a limited number of genes regulated by RpoN or RpoS alone. In the latter study, the Rrp2 regulon contained >100 genes that were differentially expressed outside of the RpoN-RpoS pathway. Another Rrp2-dependent regulon was identified by Boardman et al. (2008).

Caimano et al. (2007) initially described the RpoS regulon from *B. burgdorferi* strain 297 grown *in vitro* at 37°C and in DMCs, by comparing the transcriptomes of wild type and a *rpoS* mutant via microarray analysis. The *in vitro* and DMC RpoS regulons overlapped significantly with many genes being induced by RpoS under both growth conditions. However, they also identified a large cohort of genes that were repressed by RpoS only in DMC, demonstrating that mammalian host-specific signals are necessary for RpoS repression. More recently, Caimano et al. (2019) compared the RpoS regulons of two phylogenetically distinct *B. burgdorferi* strains, 297 and B31, with different origins (human and tick, respectively): both *in vitro* temperature-shifted, and DMC-cultivated spirochetes were used for transcriptome analyses. These studies identified a core set of genes that were differentially regulated in both strains *in vitro* and/or in DMCs as well as RpoS-regulated genes that were unique to each strain. Notably, the core set of genes was primarily encoded on the chromosome, while the strain-specific genes were predominately plasmid-borne. Importantly, RpoS-mediated repression of tick phase-specific genes in both strains occurred only in response to mammalian host-specific signals.

Arnold et al. (2018) employed RNA-seq to examine the RpoN and RpoS regulons and, unexpectedly, detected only one differentially expressed transcript in the *rpoS* mutant, and two in the *rpoN* mutant compared to the wild type. The spirochetes were cultured under conditions that do not induce synthesis of RpoS and therefore levels of the alternate σ factor were assumed to be low in the wild-type strain. This study underscores the impact of growth phase and culture conditions on comparative transcriptome analyses.

The two-component system Hk1/Rrp1 appears to function primarily in ticks during transmission and acquisition (see section above on the Hk1/Rrp1 two-component system and c-di-GMP). In fact, *B. burgdorferi* lacking either Hk1 or Rrp1 do not survive within tick midguts during the blood meal (Caimano et al., 2011; He et al., 2011; Kostick et al., 2011; Caimano et al., 2015). The response regulator Rrp1 is a diguanylate cyclase that synthesizes c-di-GMP, a small signaling molecule that regulates many cellular processes in bacteria (Cotter and Stibitz, 2007; Wolfe and Visick, 2008; Hengge, 2009; Römling et al., 2013; Valentini and Filloux, 2019). Three groups have examined the Rrp1 regulon using both microarray and RNA-seq (Rogers et al., 2009; He et al., 2011; Caimano et al., 2015). Rogers et al. (2009) first identified the Rrp1 regulon using microarray technology and a deletion mutant of *rrp1* in a non-infectious *B. burgdorferi* strain (B31 5A13). He et al. (2011) identified Rrp1-dependent differentially expressed transcripts also using microarray technology but in an infectious wild-type strain. The Rrp1 regulon was also examined by RNA-seq (Caimano et al., 2015).

B. burgdorferi also utilizes the stringent response mechanism to sense and adapt to nutrient limitation by adjusting gene expression that modifies their growth and physiology (Potrykus and Cashel, 2008; Dalebroux and Swanson, 2012; Gaca et al., 2015; Hauryliuk et al., 2015; Liu et al., 2015; Steinchen and Bange, 2016; Gourse et al., 2018) (see section above on the Stringent response and (p)ppGpp). The stringent response is mediated by levels of the alarmones, guanosine pentaphosphate and guanosine tetraphosphate, abbreviated (p)ppGpp. In *B. burgdorferi,* (p)ppGpp levels are controlled by Rel_{Bbu}, a bifunctional synthetase/hydrolase containing domains homologous to RelA and SpoT (Concepcion and Nelson, 2003; Bugrysheva et al., 2005; Drecktrah et al., 2015). RelA typically synthesizes (p)ppGpp in response to amino acid starvation, while SpoT induces accumulation of (p)ppGpp in response to limiting fatty acids, phosphate, carbon, or iron. The Rel_{Bbu} regulon has been characterized by RNA sequencing and microarray analyses (Bugrysheva et al., 2015; Drecktrah et al., 2015). Bugrysheva et al. (2015) used microarray analyses to characterize the Rel_{Bbu} transcriptome from *in vitro* cultivated *B. burgdorferi* at exponential and stationary growth phase. Drecktrah et al. (2015) used RNA sequencing analyses to describe the Rel_{Bbu} regulon at stationary growth phase, starvation conditions and recovery from starvation.

DksA works in conjunction with (p)ppGpp to affect RNA polymerase activity in *E. coli* (Gourse et al., 2018) (see section above on the Stringent response and (p)ppGpp). Boyle et al. (2019) identified the DksA regulon at mid-logarithmic growth phase and under starvation conditions using microarray analyses in *B. burgdorferi*. The *dksA* mutant has a significant increase in (p)ppGpp at both mid-logarithmic and starvation cultivation conditions, which thwarted an attempt to delineate the role of *dksA* changes in gene expression, independent of (p)ppGpp concentration.

Borrelia host adapted regulator (BadR) is a DNA-binding protein that is necessary for *B. burgdorferi* to colonize mice (Miller et al., 2013; Ouyang and Zhou, 2015) (see section above on the Transcription factors BosR and BadR). BadR binds the DNA upstream of both the *rpoS* and *bosR* genes (Miller et al., 2013; Ouyang and Zhou, 2015). Two transcriptomic studies have examined the BadR regulon. Miller et al. (2013) utilized microarray analyses, while Arnold et al. (2018) performed RNA-seq to compare wild-type and mutant *badR* strains.

The function of the carbon storage regulator A (*csrA*) homolog in *B. burgdorferi* remains unknown and there are contradictory reports as to its influence on the RpoS-RpoN σ factor cascade (Sanjuan et al., 2009; Karna et al., 2011; Sze and Li, 2011; Sze et al., 2011; Karna et al., 2013; Ouyang et al., 2014b; Lin et al., 2018) (see section above on Small regulatory RNAs and RNA chaperones). In other organisms, CsrA is an RNA-binding protein that regulates carbon storage, virulence factors, and other cellular processes (Romeo, 1998; Majdalani et al., 2005; Lucchetti-Miganeh et al., 2008; Vakulskas et al., 2015; Hör et al., 2018; Romeo and Babitzke, 2018). Arnold et al. (2018) utilized RNA-seq to assay the CsrA regulon and compared it to the BadR regulon revealing a cohort of genes regulated by both factors.

Differential gene expression on plasmids

Several plasmids appear to be hotbeds of differential gene expression. Genes on lp54 have been implicated in transmission and borrelial survival in the vertebrate host. Strikingly, many genes on lp54 are reported as differentially expressed in nearly all transcriptome studies. Many genes on lp28-2 are upregulated by the RpoN/RpoS alternative σ factor cascade and downregulated by $RelA_{Bbu}$ in response to starvation (Caimano et al., 2007; Boardman et al., 2008; Ouyang et al., 2008; Ouyang et al., 2009b; Drecktrah et al., 2015; Caimano et al., 2019). These data indicate that induction of genes on lp28-2 may be important for transmission and vertebrate survival or repression of these genes may be needed for adaptation to the tick vector. However, lp28-2 is not necessary for the murine infection via needle inoculation (Purser and Norris, 2000). The cp32 family of circular plasmids also harbor many genes that are differentially expressed under several environmental conditions and by several regulators, including the RpoN/RpoS alternative σ factor cascade, $RelA_{Bbu}$ and Rrp1 (Caimano et al., 2005; Caimano et al., 2007; Boardman et al., 2008; Ouyang et al., 2008; Ouyang et al., 2009b; Rogers et al., 2009; He et al., 2011; Drecktrah et al., 2015; Caimano et al., 2019). Genes on lp36 are downregulated by RpoS in the DMC and are primarily upregulated by $RelA_{Bbu}$, suggesting phase-specific expression may be important as spirochetes traverse their enzootic cycle (Caimano et al., 2007; Drecktrah et al., 2015; Caimano et al., 2019). Jewett et al. (2007) reported that the lp36 plasmid is not required for survival in the tick but does play a role in mammalian infectivity. As discussed above, the complex plasmid profiles of *B. burgdorferi* pose a significant challenge to identifying bona fide differentially expressed genes harbored on plasmids. In comparative transcriptomic analyses, if all genes on a plasmid are either upregulated or downregulated, then the cause may be plasmid loss in one of the strains.

Identification of sRNAs

The advent of RNA-seq allowed researchers to globally identify sRNAs in *B. burgdorferi* (Arnold et al., 2016; Popitsch et al., 2017) (see section above on Small regulatory RNAs and RNA chaperones). Popitsch et al. (2017) size-selected for sRNAs 50-500 nucleotides in length, and sequenced the sRNA transcriptome from *B. burgdorferi* cultivated *in vitro* at 23°C and after a temperature shift to 37°C, identifying 1005 sRNAs. Arnold et al. (2016) sequenced total RNA from three phases of growth *in vitro* at 37°C and found putative sRNAs from short predicted transcriptional units (<425 nucleotide in length) in non-coding regions of the genome, or antisense to annotated protein coding genes. The Arnold et al. (2016) data set identified 351 sRNAs, but the type of RNA-seq and subsequent analyses do not allow for the identification of sRNAs that overlap an annotated open reading frame. As such, the data set excludes intragenic sRNAs and 5′ UTR-associated sRNAs. In addition, the size selection was performed *in silico* and included predicted transcriptional units. If the intragenic sRNAs and 5′ UTR-associated sRNAs are excluded, then the Popitsch et al. (2017) data set identifies 550 intergenic sRNAs and antisense sRNAs, similar to the findings of Arnold et al. (2016). Further comparison and overlap of the two data sets is difficult due to differences in the sRNA annotations and reference annotations for the genes. In addition, the spirochetes were cultivated in different media and at different growth phases, as well as the two studies used different cDNA library types, construction, and analyses.

The Rel_{Bbu}-dependent sRNA transcriptome revealed that the stringent response regulates the expression of a large number of sRNAs (Drecktrah et al., 2018). Approximately one third of the annotated sRNAs, sans the intragenic sRNA class, were differentially expressed during starvation of *in vitro* cultivated *B. burgdorferi.* 187 sRNAs are upregulated by Rel_{Bbu}, while 54 are downregulated. The non-coding RNA (ncRNA) regulons of CsrA and BadR were also identified using the predicted ncRNAs annotations from Arnold et al. (2016). Antisense RNAs associated with *glpF* and *rpoN* were differentially regulated by both regulators.

Identification of transcriptional start sites (TSSs) and 5′ ends

5′ end-specific RNA-sequencing (5′ RNA-seq) can globally identify the 5′ ends and transcriptional start sites (TSSs) of transcripts. This technique exploits some of the biochemical properties of transcription (Adams et al., 2017), namely that the first nucleotide transcribed in an RNA contains a triphosphate at the 5′ end, while processed or degraded 5′ ends of RNAs contain a monophosphate or a hydroxyl group. Adams et al. (2017) used 5′ RNA-seq on *in vitro* cultivated *B. burgdorferi* and identified 6042 TSSs. 63% of the TSSs were termed 'internal', denoting that the promoter and 5′ end is within an annotated gene in the same orientation, while 13% were antisense TSSs with the 5′ ends of RNAs encoded opposite to annotated genes. 22% of the TSSs were within 300 nucleotides of the start codon of an annotated open reading frame. Arnold et al. (2016) predicted 5′ ends of transcripts by analyzing total RNA sequencing data from *B. burgdorferi* cultivated *in vitro* at three different growth phases. A custom script was used to determine putative transcriptional units and their transcript boundaries. 600 to 800 putative 5′ ends were reported in each of the growth phases examined. The predicted 5′ ends identified could be associated with TSSs or processed and/or degraded RNAs.

Future perspectives

Transcriptomic studies have significantly contributed toward understanding global gene regulation in *B. burgdorferi*. The accumulated data from studies performed over the last twenty years are laudable, but many challenges still remain to more accurately understand global gene expression during the enzootic cycle of *B. burgdorferi*. More robust *in vivo* transcriptome analyses with higher genomic coverage in both the tick victor and the vertebrate host would paint a more accurate big picture of the changes in gene expression. Specifically, tick tissue-specific (salivary gland and midguts) and mouse tissue-specific (joint, heart, bladder, and others) transcriptomes will shed light on gene expression patterns necessary for *B. burgdorferi* to transmit and persist in the tick and vertebrate. RNA-sequencing technologies continue to evolve with increased sensitivity and coverage yields from significantly less input RNA. Single-cell RNA-seq is an advancing technology that will uncover cell-to-cell heterogeneity in populations of spirochetes that may determine infection and persistence outcomes. Dual RNA-seq analyzes host or vector transcriptomes simultaneously with the bacterial transcriptome to reveal correlations between bacterial and host-specific responses in gene expression. RNA-seq can be coupled to other biochemical approaches to identify RNA targets of RNA-binding proteins (via CLIP-seq, RIP-seq, RIL-seq, and CLASH and Grad-seq), novel ribo-regulators (via Term-seq) and functional antisense RNAs (via dsRNA-seq) (Lybecker et al., 2014b; Lybecker and Samuels, 2017; Saliba et al., 2017). Using and adapting these '-seq' methodologies to *in vitro* and *in vivo* models of infection should rapidly advance our understanding of global gene expression patterns in the Lyme disease spirochete.

Concluding remarks

Gene expression is regulated in *B. burgdorferi* and related species as the spirochetes traverse their natural enzootic cycles, which requires acquisition by a naïve tick from an infected vertebrate, adaptation to and persistence in the tick, transmission from an infected tick to a naïve host, and adaptation to and survival in the vertebrate. Many of the molecular mechanisms controlling gene expression as the spirochete transitions between and adapts to these phases of the enzootic cycle have been experimentally uncovered and genetically dissected, albeit the majority of investigative approaches are restricted to cultured organisms or an emulation of the genuine environments *B. burgdorferi* experiences. The spirochete employs three regulatory systems to control gene expression, which include a sequential cascade of a pair of alternative σ factors, a couple of two-component systems, and, at the most recent count, three purine second messengers. A trio of transcription factors and three sRNAs are known to activate or fine-tune gene expression at the transcriptional and post-transcriptional levels. High-throughput transcriptomic approaches have identified the genes and defined the phase-specific regulons that are activated or repressed by environmental signals, although their precise roles have yet to be fully elucidated. Transcriptomic studies have revealed a surprising swath of previously unrecognized small RNAs, the vast majority of which have no known function, but are likely responsible for fine-tuning the expression of specific genes throughout the enzootic cycle of the spirochete. Many of the gene products controlled by the RpoN-RpoS σ factor cascade and the Hk1-Rrp1 two-component system function to move the spirochete between the vector and host or to facilitate survival by altering the metabolism of the spirochete or its surface that interfaces directly with the vector or host; however, many of these regulated genes have no homology to genes outside of the deeply branching spirochete phylum (or even genus), so their physiological functions remain a mystery. *B. burgdorferi* is renowned for doing things its own way, but, curiously, the spirochete wields the canonical bacterial response to nutrient deprivation, the stringent response, to survive in the tick between blood meals, which entails shifting the expression of genes encoding various metabolic functions, carbohydrate utilization and surface lipoproteins. Our hope is that this chapter has shed light into the current state-of-the-art and built a conceptual framework for continuing studies on gene regulation and transcriptomics in the Lyme disease spirochetes.

Acknowledgements
This chapter is a reboot of the Gene Regulation, Transcriptomics, and Proteomics chapter written by Jon Skare, Jay Carroll, Frank Yang, Scott Samuels, and Darrin Akins for the first edition of the book in 2010. DSS thanks JDR for his passion, mentorship and forbearance…and apologizes for the agita.

Research in our laboratories on gene regulation and transcriptomics is supported by Public Health Service grants R01 AI051486 to DSS, R01 AI130247 to DSS, DD, and ML, R21 AI133334 to DD and DSS, R03 AI146909 to ZO, R01 AI083640 to XFY, R21 AI128379 to MJC, R01 AI029735 to MJC, R21 AI126146 to MJC, R21 AI139940 to MJC, R01 AI144126 to BS, R21 AI147139 to BS, and R21 AI139956 to BS as well as Connecticut Children's Medical Center to MJC.

References

Adams, P.P., Flores Avile, C., Popitsch, N., Bilusic, I., Schroeder, R., Lybecker, M., and Jewett, M.W. (2017). *In vivo* expression technology and 5' end mapping of the *Borrelia burgdorferi* transcriptome identify novel RNAs expressed during mammalian infection. Nucleic Acids Res *45*, 775-792. https://doi.org/10.1093/nar/gkw1180

Adams, P.P., and Storz, G. (2020). Prevalence of small base-pairing RNAs derived from diverse genomic loci. Biochim Biophys Acta Gene Regul Mech *1863*, 194524. https://doi.org/10.1016/j.bbagrm.2020.194524

Agapov, A., Olina, A., Esyunina, D., and Kulbachinskiy, A. (2017). Gfh factors and NusA cooperate to stimulate transcriptional pausing and termination. FEBS Lett *591*, 946-953. https://doi.org/10.1002/1873-3468.12609

Akins, D.R., Bourell, K.W., Caimano, M.J., Norgard, M.V., and Radolf, J.D. (1998). A new animal model for studying Lyme disease spirochetes in a mammalian host-adapted state. J Clin Invest *101*, 2240-2250.

Alban, P.S., Johnson, P.W., and Nelson, D.R. (2000). Serum-starvation-induced changes in protein synthesis and morphology of *Borrelia burgdorferi*. Microbiology *146*, 119-127.

Alverson, J., Bundle, S.F., Sohaskey, C.D., Lybecker, M.C., and Samuels, D.S. (2003). Transcriptional regulation of the *ospAB* and *ospC* promoters from *Borrelia burgdorferi*. Mol Microbiol *48*, 1665-1677. https://doi.org/10.1046/j.1365-2958.2003.03537.x

Amikam, D., and Galperin, M.Y. (2006). PilZ domain is part of the bacterial c-di-GMP binding protein. Bioinformatics *22*, 3-6. https://doi.org/10.1093/bioinformatics/bti739

Anacker, M.L., Drecktrah, D., LeCoultre, R.D., Lybecker, M., and Samuels, D.S. (2018). RNase III processing of rRNA in the Lyme disease spirochete *Borrelia burgdorferi*. J Bacteriol *200*, e00035-18. https://doi.org/10.1128/JB.00035-18

Archambault, L., Borchert, J.S., Bergeron, J., Snow, S., and Schlax, P.J. (2013). Measurements of mRNA degradation in *Borrelia burgdorferi*. J Bacteriol *195*, 4879-4887. https://doi.org/10.1128/JB.00659-13

Arnold, W.K., Savage, C.R., Brissette, C.A., Seshu, J., Livny, J., and Stevenson, B. (2016). RNA-seq of *Borrelia burgdorferi* in multiple phases of growth reveals insights into the dynamics of gene expression, transcriptome architecture, and noncoding RNAs. PLoS One *11*, e0164165. https://doi.org/10.1371/journal.pone.0164165

Arnold, W.K., Savage, C.R., Lethbridge, K.G., Smith, T.C., 2nd, Brissette, C.A., Seshu, J., and Stevenson, B. (2018). Transcriptomic insights on the virulence-controlling CsrA, BadR, RpoN, and RpoS regulatory networks in the Lyme disease spirochete. PLoS One *13*, e0203286. https://doi.org/10.1371/journal.pone.0203286

Babb, K., Bykowski, T., Riley, S.P., Miller, M.C., Demoll, E., and Stevenson, B. (2006). *Borrelia burgdorferi* EbfC, a novel, chromosomally encoded protein, binds specific DNA sequences adjacent to *erp* loci on the spirochete's resident cp32 prophages. J Bacteriol *188*, 4331-4339. https://doi.org/10.1128/JB.00005-06

Barbour, A.G. (1984). Isolation and cultivation of Lyme disease spirochetes. Yale J Biol Med *57*, 521-525.

Barker, J.R., Koestler, B.J., Carpenter, V.K., Burdette, D.L., Waters, C.M., Vance, R.E., and Valdivia, R.H. (2013). STING-dependent recognition of cyclic di-AMP mediates type I interferon responses during *Chlamydia trachomatis* infection. mBio *4*, e00018-13. https://doi.org/10.1128/mBio.00018-13

Barrick, J.E., Sudarsan, N., Weinberg, Z., Ruzzo, W.L., and Breaker, R.R. (2005). 6S RNA is a widespread regulator of eubacterial RNA polymerase that resembles an open promoter. RNA *11*, 774-784. https://doi.org/10.1261/rna.7286705

Bauer, W.J., Luthra, A., Zhu, G., Radolf, J.D., Malkowski, M.G., and Caimano, M.J. (2015). Structural characterization and modeling of the *Borrelia burgdorferi* hybrid histidine kinase Hk1 periplasmic sensor: a system for sensing small molecules associated with tick feeding. J Struct Biol *192*, 48-58. https://doi.org/10.1016/j.jsb.2015.08.013

Bechhofer, D.H. (2011). *Bacillus subtilis* mRNA decay: new parts in the toolkit. Wiley Interdiscip Rev RNA 2, 387-394. https://doi.org/10.1002/wrna.66

Bechhofer, D.H., and Deutscher, M.P. (2019). Bacterial ribonucleases and their roles in RNA metabolism. Crit Rev Biochem Mol Biol *54*, 242-300. https://doi.org/10.1080/10409238.2019.1651816

Becker, G., and Hengge-Aronis, R. (2001). What makes an *Escherichia coli* promoter σ^S dependent? Role of the –13/–14 nucleotide promoter positions and region 2.5 of σ^S. Mol Microbiol *39*, 1153-1165.

Bhattacharjee, A., Oeemig, J.S., Kolodziejczyk, R., Meri, T., Kajander, T., Lehtinen, M.J., Iwai, H., Jokiranta, T.S., and Goldman, A. (2013). Structural basis for complement evasion by Lyme disease pathogen *Borrelia burgdorferi*. J Biol Chem *288*, 18685-18695. https://doi.org/10.1074/jbc.M113.459040

Blevins, J.S., Xu, H., He, M., Norgard, M.V., Reitzer, L., and Yang, X.F. (2009). Rrp2, a σ^{54}-dependent transcriptional activator of *Borrelia burgdorferi*, activates *rpoS* in an enhancer-independent manner. J Bacteriol *191*, 2902-2905. https://doi.org/10.1128/JB.01721-08

Boardman, B.K., He, M., Ouyang, Z., Xu, H., Pang, X., and Yang, X.F. (2008). Essential role of the response regulator Rrp2 in the infectious cycle of *Borrelia burgdorferi*. Infect Immun *76*, 3844-3853. https://doi.org/10.1128/IAI.00467-08

Bontemps-Gallo, S., Lawrence, K., and Gherardini, F.C. (2016). Two different virulence-related regulatory pathways in *Borrelia burgdorferi* are directly affected by osmotic fluxes in the blood meal of feeding *Ixodes* ticks. PLoS Pathog *12*, e1005791. https://doi.org/10.1371/journal.ppat.1005791

Borbely, G., and Schneider, G.J. (1988). Cyanobacterial DNA-dependent RNA polymerase. Methods Enzymol *167*, 592-599. https://doi.org/10.1016/0076-6879(88)67067-4

Bordes, P., Conter, A., Morales, V., Bouvier, J., Kolb, A., and Gutierrez, C. (2003). DNA supercoiling contributes to disconnect σ^S accumulation from σ^S-dependent transcription in *Escherichia coli*. Mol Microbiol *48*, 561-571.

Boylan, J.A., Hummel, C.S., Benoit, S., Garcia-Lara, J., Treglown-Downey, J., Crane, E.J., III, and Gherardini, F.C. (2006). *Borrelia burgdorferi bb0728* encodes a coenzyme A disulphide reductase whose function suggests a role in intracellular redox and the oxidative stress response. Mol Microbiol *59*, 475-486. https://doi.org/10.1111/j.1365-2958.2005.04963.x

Boylan, J.A., Posey, J.E., and Gherardini, F.C. (2003). *Borrelia* oxidative stress response regulator, BosR: a distinctive Zn-dependent transcriptional activator. Proc Natl Acad Sci USA *100*, 11684-11689. https://doi.org/10.1073/pnas.2032956100

Boyle, W.K., Groshong, A.M., Drecktrah, D., Boylan, J.A., Gherardini, F.C., Blevins, J.S., Samuels, D.S., and Bourret, T.J. (2019). DksA controls the response of the Lyme disease spirochete *Borrelia burgdorferi* to starvation. J Bacteriol *201*, e00582-18. https://doi.org/10.1128/JB.00582-18

Boyle, W.K., Hall, L.S., Armstrong, A.A., Dulebohn, D.P., Samuels, D.S., Gherardini, F.C., and Bourret, T.J. (2020). Establishment of an *in vitro* RNA polymerase transcription system: a new tool to study transcriptional activation in *Borrelia burgdorferi*. Sci Rep *10*, 8246. https://doi.org/10.1038/s41598-020-65104-y

Brennan, R.G., and Link, T.M. (2007). Hfq structure, function and ligand binding. Curr Opin Microbiol *10*, 125-133. https://doi.org/10.1016/j.mib.2007.03.015

Brisson, D., Zhou, W., Jutras, B.L., Casjens, S., and Stevenson, B. (2013). Distribution of cp32 prophages among Lyme disease-causing spirochetes and natural diversity of their lipoprotein-encoding *erp* loci. Appl Environ Microbiol *79*, 4115-4128. https://doi.org/10.1128/AEM.00817-13

Brooks, C.S., Hefty, P.S., Jolliff, S.E., and Akins, D.R. (2003). Global analysis of *Borrelia burgdorferi* genes regulated by mammalian host-specific signals. Infect Immun *71*, 3371-3383.

Brorson, Ø., and Brorson, S.-H. (1997). Transformation of cystic forms of *Borrelia burgdorferi* to normal, mobile spirochetes. Infection *25*, 240-246.

Brorson, Ø., and Brorson, S.-H. (1998). A rapid method for generating cystic forms of *Borrelia burgdorferi*, and their reversal to mobile spirochetes. APMIS *106*, 1131-41.

Brown, L., Gentry, D., Elliott, T., and Cashel, M. (2002). DksA affects ppGpp induction of RpoS at a translational level. J Bacteriol *184*, 4455-4465. https://doi.org/10.1128/jb.184.16.4455-4465.2002

Bugrysheva, J.V., Bryksin, A.V., Godfrey, H.P., and Cabello, F.C. (2005). *Borrelia burgdorferi rel* is responsible for generation of guanosine-3'-diphosphate-5'-triphosphate and growth control. Infect Immun *73*, 4972-4981.

Bugrysheva, J.V., Pappas, C.J., Terekhova, D.A., Iyer, R., Godfrey, H.P., Schwartz, I., and Cabello, F.C. (2015). Characterization of the Rel_{Bbu} regulon in *Borrelia burgdorferi* reveals modulation of glycerol metabolism by (p)ppGpp. PLoS One *10*, e0118063. https://doi.org/10.1371/journal.pone.0118063

Burenina, O.Y., Elkina, D.A., Hartmann, R.K., Oretskaya, T.S., and Kubareva, E.A. (2015). Small noncoding 6S RNAs of bacteria. Biochemistry (Mosc) *80*, 1429-1446. https://doi.org/10.1134/S0006297915110048

Burke, T.P., and Portnoy, D.A. (2016). SpoVG Is a conserved RNA-binding protein that regulates *Listeria monocytogenes* lysozyme resistance, virulence, and swarming motility. mBio *7*, e00240. https://doi.org/10.1128/mBio.00240-16

Burns, L.H., Adams, C.A., Riley, S.P., Jutras, B.L., Bowman, A., Chenail, A.M., Cooley, A.E., Haselhorst, L.A., Moore, A.M., Babb, K., *et al.* (2010). BpaB, a novel protein encoded by the Lyme disease spirochete's cp32 prophages, binds to *erp* operator 2 DNA. Nucleic Acids Res *38*, 5443-5455. https://doi.org/10.1093/nar/gkq284

Burtnick, M.N., Downey, J.S., Brett, P.J., Boylan, J.A., Frye, J.G., Hoover, T.R., and Gherardini, F.C. (2007). Insights into the complex regulation of *rpoS* in *Borrelia burgdorferi*. Mol Microbiol *65*, 277-293. https://doi.org/10.1111/j.1365-2958.2007.05813.x

Bush, M., and Dixon, R. (2012). The role of bacterial enhancer binding proteins as specialized activators of σ^{54}-dependent transcription. Microbiol Mol Biol Rev *76*, 497-529. https://doi.org/10.1128/MMBR.00006-12

Butcher, J., Sarvan, S., Brunzelle, J.S., Couture, J.-F., and Stintzi, A. (2012). Structure and regulon of *Campylobacter jejuni* ferric uptake regulator Fur define apo-Fur regulation. Proc Natl Acad Sci USA *109*, 10047-10052. https://doi.org/10.1073/pnas.1118321109

Caimano, M.J., Drecktrah, D., Kung, F., and Samuels, D.S. (2016). Interaction of the Lyme disease spirochete with its tick vector. Cell Microbiol *18*, 919-927. https://doi.org/10.1111/cmi.12609

Caimano, M.J., Dunham-Ems, S., Allard, A.M., Cassera, M.B., Kenedy, M., and Radolf, J.D. (2015). Cyclic di-GMP modulates gene expression in Lyme disease spirochetes at the tick-mammal interface to promote spirochete survival during the blood meal and tick-to-mammal transmission. Infect Immun *83*, 3043-3060. https://doi.org/10.1128/IAI.00315-15

Caimano, M.J., Eggers, C.H., Gonzalez, C.A., and Radolf, J.D. (2005). Alternate sigma factor RpoS is required for the in vivo-specific repression of *Borrelia burgdorferi* plasmid lp54-borne *ospA* and *lp6.6* genes. J Bacteriol *187*, 7845-7852.

Caimano, M.J., Eggers, C.H., Hazlett, K.R.O., and Radolf, J.D. (2004). RpoS is not central to the general stress response in *Borrelia burgdorferi* but does control expression of one or more essential virulence determinants. Infect Immun *72*, 6433-6445.

Caimano, M.J., Groshong, A.M., Belperron, A., Mao, J., Hawley, K.L., Luthra, A., Graham, D.E., Earnhart, C.G., Marconi, R.T., Bockenstedt, L.K., *et al.* (2019). The RpoS gatekeeper in *Borrelia burgdorferi*: an invariant regulatory scheme that promotes spirochete persistence in reservoir hosts and niche diversity. Front Microbiol *10*, 1923. https://doi.org/10.3389/fmicb.2019.01923

Caimano, M.J., Iyer, R., Eggers, C.H., Gonzalez, C., Morton, E.A., Gilbert, M.A., Schwartz, I., and Radolf, J.D. (2007). Analysis of the RpoS regulon in *Borrelia burgdorferi* in response to mammalian host signals provides insight into RpoS function during the enzootic cycle. Mol Microbiol *65*, 1193-1217. https://doi.org/10.1111/j.1365-2958.2007.05860.x

Caimano, M.J., Kenedy, M.R., Kairu, T., Desrosiers, D.C., Harman, M., Dunham-Ems, S., Akins, D.R., Pal, U., and Radolf, J.D. (2011). The hybrid histidine kinase Hk1 is part of a two-component system that is essential for survival of *Borrelia burgdorferi* in feeding *Ixodes scapularis* ticks. Infect Immun *79*, 3117-3130. https://doi.org/10.1128/IAI.05136-11

Caldelari, I., Chao, Y., Romby, P., and Vogel, J. (2013). RNA-mediated regulation in pathogenic bacteria. Cold Spring Harb Perspect Med *3*, a010298. https://doi.org/10.1101/cshperspect.a010298

Carpenter, B.M., Whitmire, J.M., and Merrell, D.S. (2009). This is not your mother's repressor: the complex role of Fur in pathogenesis. Infect Immun *77*, 2590-2601. https://doi.org/10.1128/IAI.00116-09

Carroll, J.A., Garon, C.F., and Schwan, T.G. (1999). Effects of environmental pH on membrane proteins in *Borrelia burgdorferi*. Infect Immun *67*, 3181-3187.

Casjens, S., Palmer, N., van Vugt, R., Huang, W.M., Stevenson, B., Rosa, P., Lathigra, R., Sutton, G., Peterson, J., Dodson, R.J., *et al.* (2000). A bacterial genome in flux: the twelve linear and nine circular extrachromosomal DNAs in an infectious isolate of the Lyme disease spirochete *Borrelia burgdorferi*. Mol Microbiol *35*, 490-516.

Casjens, S., van Vugt, R., Tilly, K., Rosa, P.A., and Stevenson, B. (1997). Homology throughout the multiple 32-kilobase circular plasmids present in Lyme disease spirochetes. J Bacteriol *179*, 217-227.

Cavanagh, A.T., Chandrangsu, P., and Wassarman, K.M. (2010). 6S RNA regulation of *relA* alters ppGpp levels in early stationary phase. Microbiology *156*, 3791-3800. https://doi.org/10.1099/mic.0.043992-0

Cavanagh, A.T., Klocko, A.D., Liu, X., and Wassarman, K.M. (2008). Promoter specificity for 6S RNA regulation of transcription is determined by core promoter sequences and competition for region 4.2 of σ^{70}. Mol Microbiol *67*, 1242-1256. https://doi.org/10.1111/j.1365-2958.2008.06117.x

Cavanagh, A.T., Sperger, J.M., and Wassarman, K.M. (2012). Regulation of 6S RNA by pRNA synthesis is required for efficient recovery from stationary phase in *E. coli* and *B. subtilis*. Nucleic Acids Res *40*, 2234-2246. https://doi.org/10.1093/nar/gkr1003

Cavanagh, A.T., and Wassarman, K.M. (2013). 6S-1 RNA function leads to a delay in sporulation in *Bacillus subtilis*. J Bacteriol *195*, 2079-2086. https://doi.org/10.1128/JB.00050-13

Cavanagh, A.T., and Wassarman, K.M. (2014). 6S RNA, a global regulator of transcription in *Escherichia coli*, *Bacillus subtilis*, and beyond. Annu Rev Microbiol *68*, 45-60. https://doi.org/10.1146/annurev-micro-092611-150135

Chaconas, G. (2005). Hairpin telomeres and genome plasticity in *Borrelia*: all mixed up in the end. Mol Microbiol *58*, 625-635. https://doi.org/10.1111/j.1365-2958.2005.04872.x

Chaconas, G., Castellanos, M., and Verhey, T.B. (2020). Changing of the guard: How the Lyme disease spirochete subverts the host immune response. J Biol Chem *295*, 301-313. https://doi.org/10.1074/jbc.REV119.008583

Chen, J., Wassarman, K.M., Feng, S., Leon, K., Feklistov, A., Winkelman, J.T., Li, Z., Walz, T., Campbell, E.A., and Darst, S.A. (2017). 6S RNA mimics B-form DNA to regulate *Escherichia coli* RNA polymerase. Mol Cell *68*, 388-397. https://doi.org/10.1016/j.molcel.2017.09.006

Chen, Z., Itzek, A., Malke, H., Ferretti, J.J., and Kreth, J. (2013). Multiple roles of RNase Y in *Streptococcus pyogenes* mRNA processing and degradation. J Bacteriol *195*, 2585-2594. https://doi.org/10.1128/JB.00097-13

Chenail, A.M., Jutras, B.L., Adams, C.A., Burns, L.H., Bowman, A., Verma, A., and Stevenson, B. (2012). *Borrelia burgdorferi* cp32 BpaB modulates expression of the prophage NucP nuclease and SsbP single-stranded DNA-binding protein. J Bacteriol *194*, 4570-4578. https://doi.org/10.1128/JB.00661-12

Cheung, K.J., Badarinarayana, V., Selinger, D.W., Janse, D., and Church, G.M. (2003). A microarray-based antibiotic screen identifies a regulatory role for supercoiling in the osmotic stress response of *Escherichia coli*. Genome Res *13*, 206-215.

Chiang, S.M., and Schellhorn, H.E. (2010). Evolution of the RpoS regulon: origin of RpoS and the conservation of RpoS-dependent regulation in bacteria. J Mol Evol *70*, 557-571. https://doi.org/10.1007/s00239-010-9352-0

Chu, C.-Y., Stewart, P.E., Bestor, A., Hansen, B., Lin, T., Gao, L., Norris, S.J., and Rosa, P.A. (2016). Function of the *Borrelia burgdorferi* FtsH homolog is essential for viability both *in vitro* and *in vivo* and independent of HflK/C. mBio *7*, e00404-16. https://doi.org/10.1128/mBio.00404-16

Clifton, D.R., Nolder, C.L., Hughes, J.L., Nowalk, A.J., and Carroll, J.A. (2006). Regulation and expression of *bba66* encoding an immunogenic infection-associated lipoprotein in *Borrelia burgdorferi*. Mol Microbiol *61*, 243-258. https://doi.org/10.1111/j.1365-2958.2006.05224.x

Colland, F., Barth, M., Hengge-Aronis, R., and Kolb, A. (2000). σ factor selectivity of *Escherichia coli* RNA polymerase: role for CRP, IHF and Lrp transcription factors. EMBO J *19*, 3028-3037.

Commichau, F.M., Heidemann, J.L., Ficner, R., and Stülke, J. (2019). Making and breaking of an essential poison: the cyclases and phosphodiesterases that produce and degrade the essential second messenger cyclic di-AMP in bacteria. J Bacteriol *201*, e00462-18. https://doi.org/10.1128/JB.00462-18

Concepcion, M.B., and Nelson, D.R. (2003). Expression of *spoT* in *Borrelia burgdorferi* during serum starvation. J Bacteriol *185*, 444-452.

Cooley, A.E., Riley, S.P., Kral, K., Miller, M.C., DeMoll, E., Fried, M.G., and Stevenson, B. (2009). DNA-binding by *Haemophilus influenzae* and *Escherichia coli* YbaB, members of a widely-distributed bacterial protein family. BMC Microbiol *9*, 137. https://doi.org/10.1186/1471-2180-9-137

Corbett, K.D., and Berger, J.M. (2004). Structure, molecular mechanisms, and evolutionary relationships in DNA topoisomerases. Annu Rev Biophys Biomol Struct *33*, 95-118.

Corbett, K.D., Shultzaberger, R.K., and Berger, J.M. (2004). The C-terminal domain of DNA gyrase A adopts a DNA-bending β-pinwheel fold. Proc Natl Acad Sci USA *101*, 7293-7298.

Corona, A., and Schwartz, I. (2015). *Borrelia burgdorferi*: carbon metabolism and the tick-mammal enzootic cycle. Microbiol Spectr *3*, MBP-0011-2014. https://doi.org/10.1128/microbiolspec.MBP-0011-2014

Corrigan, R.M., and Gründling, A. (2013). Cyclic di-AMP: another second messenger enters the fray. Nat Rev Microbiol *11*, 513-524. https://doi.org/10.1038/nrmicro3069

Cotter, P.A., and Stibitz, S. (2007). c-di-GMP-mediated regulation of virulence and biofilm formation. Curr Opin Microbiol *10*, 17-23. https://doi.org/10.1016/j.mib.2006.12.006

Craigie, R., Arndt-Jovin, D.J., and Mizuuchi, K. (1985). A defined system for the DNA strand-transfer reaction at the initiation of bacteriophage Mu transposition: protein and DNA substrate requirements. Proc Natl Acad Sci USA *82*, 7570-7574.

Crother, T.R., Champion, C.I., Whitelegge, J.P., Aguilera, R., Wu, X.-Y., Blanco, D.R., Miller, J.N., and Lovett, M.A. (2004). Temporal analysis of the antigenic composition of *Borrelia burgdorferi* during infection in rabbit skin. Infect Immun *72*, 5063-5072.

Dahlstrom, K.M., and O'Toole, G.A. (2017). A symphony of cyclases: specificity in diguanylate cyclase signaling. Annu Rev Microbiol *71*, 179-195. https://doi.org/10.1146/annurev-micro-090816-093325

Dalebroux, Z.D., and Swanson, M.S. (2012). ppGpp: magic beyond RNA polymerase. Nat Rev Microbiol *10*, 203-212. https://doi.org/10.1038/nrmicro2720

Dalebroux, Z.D., Yagi, B.F., Sahr, T., Buchrieser, C., and Swanson, M.S. (2010). Distinct roles of ppGpp and DksA in *Legionella pneumophila* differentiation. Mol Microbiol *76*, 200-219. https://doi.org/10.1111/j.1365-2958.2010.07094.x

de Silva, A.M., Telford, S.R., 3rd, Brunet, L.R., Barthold, S.W., and Fikrig, E. (1996). *Borrelia burgdorferi* OspA is an arthropod-specific transmission-blocking Lyme disease vaccine. J Exp Med *183*, 271-275.

Dian, C., Vitale, S., Leonard, G.A., Bahlawane, C., Fauquant, C., Leduc, D., Muller, C., de Reuse, H., Michaud-Soret, I., and Terradot, L.

(2011). The structure of the *Helicobacter pylori* ferric uptake regulator Fur reveals three functional metal binding sites. Mol Microbiol *79*, 1260-1275. https://doi.org/10.1111/j.1365-2958.2010.07517.x

Drecktrah, D., Hall, L.S., Brinkworth, A.J., Comstock, J.R., Wassarman, K.M., and Samuels, D.S. (2020). Characterization of 6S RNA in the Lyme disease spirochete. Mol Microbiol *113*, 399-417. https://doi.org/10.1111/mmi.14427

Drecktrah, D., Hall, L.S., Hoon-Hanks, L.L., and Samuels, D.S. (2013). An inverted repeat in the *ospC* operator is required for induction in *Borrelia burgdorferi*. PLoS One *8*, e68799. https://doi.org/10.1371/journal.pone.0068799

Drecktrah, D., Hall, L.S., Rescheneder, P., Lybecker, M., and Samuels, D.S. (2018). The stringent response-regulated sRNA transcriptome of *Borrelia burgdorferi*. Front Cell Infect Microbiol *8*, 231. https://doi.org/10.3389/fcimb.2018.00231

Drecktrah, D., Lybecker, M., Popitsch, N., Rescheneder, P., Hall, L.S., and Samuels, D.S. (2015). The *Borrelia burgdorferi* RelA/SpoT homolog and stringent response regulate survival in the tick vector and global gene expression during starvation. PLoS Pathog *11*, e1005160. https://doi.org/10.1371/journal.ppat.1005160

Drlica, K., and Rouviere-Yaniv, J. (1987). Histonelike proteins of bacteria. Microbiol Rev *51*, 301-319.

Dulebohn, D.P., Hayes, B.M., and Rosa, P.A. (2014). Global repression of host-associated genes of the Lyme disease spirochete through post-transcriptional modulation of the alternative sigma factor RpoS. PLoS One *9*, e93141. https://doi.org/10.1371/journal.pone.0093141

Dulebohn, D.P., Richards, C.L., Su, H., Lawrence, K.A., and Gherardini, F.C. (2017). Weak organic acids decrease *Borrelia burgdorferi* cytoplasmic pH, eliciting an acid stress response and impacting RpoN- and RpoS-dependent gene expression. Front Microbiol *8*, 1734. https://doi.org/10.3389/fmicb.2017.01734

Dunham-Ems, S.M., Caimano, M.J., Eggers, C.H., and Radolf, J.D. (2012). *Borrelia burgdorferi* requires the alternative sigma factor RpoS for dissemination within the vector during tick-to-mammal transmission. PLoS Pathog *8*, e1002532. https://doi.org/10.1371/journal.ppat.1002532

Dunham-Ems, S.M., Caimano, M.J., Pal, U., Wolgemuth, C.W., Eggers, C.H., Balic, A., and Radolf, J.D. (2009). Live imaging reveals a biphasic mode of dissemination of *Borrelia burgdorferi* within ticks. J Clin Invest *119*, 3652-3665. https://doi.org/10.1172/JCI39401

Durand, S., Gilet, L., and Condon, C. (2012). The essential function of *B. subtilis* RNase III is to silence foreign toxin genes. PLoS Genet *8*, e1003181. https://doi.org/10.1371/journal.pgen.1003181

Eggers, C.H., Caimano, M.J., Clawson, M.L., Miller, W.G., Samuels, D.S., and Radolf, J.D. (2002). Identification of loci critical for replication and compatibility of a *Borrelia burgdorferi* cp32 plasmid and use of a cp32-based shuttle vector for expression of fluorescent reporters in the Lyme disease spirochaete. Mol Microbiol *43*, 281-296.

Eggers, C.H., Caimano, M.J., Malizia, R.A., Kariu, T., Cusack, B., Desrosiers, D.C., Hazlett, K.R.O., Claiborne, A., Pal, U., and Radolf, J.D. (2011). The coenzyme A disulphide reductase of *Borrelia burgdorferi* is important for rapid growth throughout the enzootic cycle and essential for infection of the mammalian host. Mol Microbiol *82*, 679-697. https://doi.org/10.1111/j.1365-2958.2011.07845.x

Eggers, C.H., Caimano, M.J., and Radolf, J.D. (2004). Analysis of promoter elements involved in the transcriptional initiation of RpoS-dependent *Borrelia burgdorferi* genes. J Bacteriol *186*, 7390-402.

Eggers, C.H., Caimano, M.J., and Radolf, J.D. (2006). Sigma factor selectivity in *Borrelia burgdorferi*: RpoS recognition of the *ospE*/*ospF*/*elp* promoters is dependent on the sequence of the –10 region. Mol Microbiol *59*, 1859-1875.

Eggers, C.H., and Samuels, D.S. (1999). Molecular evidence for a new bacteriophage of *Borrelia burgdorferi*. J Bacteriol *181*, 7308-7313. https://doi.org/10.1128/JB.181.23.7308-7313.1999

Faucher, S.P., Friedlander, G., Livny, J., Margalit, H., and Shuman, H.A. (2010). *Legionella pneumophila* 6S RNA optimizes intracellular multiplication. Proc Natl Acad Sci USA *107*, 7533-7538. https://doi.org/10.1073/pnas.0911764107

Fillat, M.F. (2014). The FUR (ferric uptake regulator) superfamily: diversity and versatility of key transcriptional regulators. Arch Biochem Biophys *546*, 41-52. https://doi.org/10.1016/j.abb.2014.01.029

Fischer, J.R., LeBlanc, K.T., and Leong, J.M. (2006). Fibronectin binding protein BBK32 of the Lyme disease spirochete promotes bacterial attachment to glycosaminoglycans. Infect Immun *74*, 435-441. https://doi.org/10.1128/IAI.74.1.435-441.2006

Fischer, J.R., Parveen, N., Magoun, L., and Leong, J.M. (2003). Decorin-binding proteins A and B confer distinct mammalian cell type-specific attachment by *Borrelia burgdorferi*, the Lyme disease spirochete. Proc Natl Acad Sci USA *100*, 7307-7312. https://doi.org/10.1073/pnas.1231043100

Fisher, M.A., Grimm, D., Henion, A.K., Elias, A.F., Stewart, P.E., Rosa, P.A., and Gherardini, F.C. (2005). *Borrelia burgdorferi* σ^{54} is required for mammalian infection and vector transmission but not for tick colonization. Proc Natl Acad Sci USA *102*, 5162-5167.

Fraser, C.M., Casjens, S., Huang, W.M., Sutton, G.G., Clayton, R., Lathigra, R., White, O., Ketchum, K.A., Dodson, R., Hickey, E.K., *et al.* (1997). Genomic sequence of a Lyme disease spirochete, *Borrelia burgdorferi*. Nature *390*, 580-586.

Freedman, J.C., Rogers, E.A., Kostick, J.L., Zhang, H., Iyer, R., Schwartz, I., and Marconi, R.T. (2010). Identification and molecular characterization of a cyclic-di-GMP effector protein, PlzA (BB0733): additional evidence for the existence of a functional cyclic-di-GMP regulatory network in the Lyme disease spirochete, *Borrelia burgdorferi*. FEMS Immunol Med Microbiol *58*, 285-294. https://doi.org/10.1111/j.1574-695X.2009.00635.x

Friedman, D.I. (1988). Integration host factor: a protein for all reasons. Cell *55*, 545-554.

Fröhlich, K.S., and Vogel, J. (2009). Activation of gene expression by small RNA. Curr Opin Microbiol *12*, 674-682. https://doi.org/10.1016/j.mib.2009.09.009

Gaal, T., Ross, W., Estrem, S.T., Nguyen, L.H., Burgess, R.R., and Gourse, R.L. (2001). Promoter recognition and discrimination by Eσ^S RNA polymerase. Mol Microbiol *42*, 939-954.

Gaca, A.O., Colomer-Winter, C., and Lemos, J.A. (2015). Many means to a common end: the intricacies of (p)ppGpp metabolism and its control of bacterial homeostasis. J Bacteriol, 1146-1156. https://doi.org/10.1128/JB.02577-14

Galperin, M.Y. (2004). Bacterial signal transduction network in a genomic perspective. Environ Microbiol *6*, 552-567. https://doi.org/10.1111/j.1462-2920.2004.00633.x

Galperin, M.Y., Nikolskaya, A.N., and Koonin, E.V. (2001). Novel domains of the prokaryotic two-component signal transduction systems. FEMS Microbiol Lett *203*, 11-21. https://doi.org/10.1111/j.1574-6968.2001.tb10814.x

Gautam, A., Hathaway, M., McClain, N., Ramesh, G., and Ramamoorthy, R. (2008). Analysis of the determinants of *bba64* (*P35*) gene expression in *Borrelia burgdorferi* using a *gfp* reporter. Microbiology *154*, 275-285. https://doi.org/10.1099/mic.0.2007/011676-0

Gherardini, F., Boylan, J., Lawrence, K., and Skare, J. (2010). Metabolism and physiology of *Borrelia*. In *Borrelia*: Molecular Biology, Host Interaction and Pathogenesis, D.S. Samuels, and J.D. Radolf, eds. (Norfolk, UK: Caister Academic Press), pp. 103-138.

Gilmore, R.D., Jr., Howison, R.R., Schmit, V.L., Nowalk, A.J., Clifton, D.R., Nolder, C., Hughes, J.L., and Carroll, J.A. (2007). Temporal expression analysis of the *Borrelia burgdorferi* paralogous gene family 54 genes BBA64, BBA65, and BBA66 during persistent infection in mice. Infect Immun *75*, 2753-2764. https://doi.org/10.1128/IAI.00037-07

González Plaza, J.J. (2020). Small RNAs as fundamental players in the transference of information during bacterial infectious diseases. Front Mol Biosci 7, 101. https://doi.org/10.3389/fmolb.2020.00101

Gottesman, S. (2004). The small RNA regulators of *Escherichia coli*: roles and mechanisms. Annu Rev Microbiol *58*, 303-328.

Gourse, R.L., Chen, A.Y., Gopalkrishnan, S., Sanchez-Vazquez, P., Myers, A., and Ross, W. (2018). Transcriptional responses to ppGpp and DksA. Annu Rev Microbiol *72*, 163-184. https://doi.org/10.1146/annurev-micro-090817-062444

Graebsch, A., Roche, S., Kostrewa, D., Söding, J., and Niessing, D. (2010). Of bits and bugs–on the use of bioinformatics and a bacterial crystal structure to solve a eukaryotic repeat-protein structure. PLoS One *5*, e13402. https://doi.org/10.1371/journal.pone.0013402

Grant, G.A. (2006). The ACT domain: a small molecule binding domain and its role as a common regulatory element. J Biol Chem *281*, 33825-33829. https://doi.org/10.1074/jbc.R600024200

Grimm, D., Tilly, K., Byram, R., Stewart, P.E., Krum, J.G., Bueschel, D.M., Schwan, T.G., Policastro, P.F., Elias, A.F., and Rosa, P.A. (2004). Outer-surface protein C of the Lyme disease spirochete: a protein induced in ticks for infection of mammals. Proc Natl Acad Sci USA *101*, 3142-3147.

Groshong, A.M., and Blevins, J.S. (2014). Insights into the biology of *Borrelia burgdorferi* gained through the application of molecular genetics. Adv Appl Microbiol *86*, 41-143. https://doi.org/10.1016/B978-0-12-800262-9.00002-0

Groshong, A.M., Dey, A., Bezsonova, I., Caimano, M.J., and Radolf, J.D. (2017). Peptide uptake is essential for *Borrelia burgdorferi* viability and involves structural and regulatory complexity of its oligopeptide transporter. mBio *8*, e02047-17. https://doi.org/10.1128/mBio.02047-17

Groshong, A.M., Gibbons, N.E., Yang, X.F., and Blevins, J.S. (2012). Rrp2, a prokaryotic enhancer-like binding protein, is essential for viability of *Borrelia burgdorferi*. J Bacteriol *194*, 3336-3342. https://doi.org/10.1128/JB.00253-12

Grove, A.P., Liveris, D., Iyer, R., Petzke, M., Rudman, J., Caimano, M.J., Radolf, J.D., and Schwartz, I. (2017). Two distinct mechanisms govern RpoS-mediated repression of tick-phase genes during mammalian host adaptation by *Borrelia burgdorferi*, the Lyme disease spirochete. mBio *8*, e01204-17. https://doi.org/10.1128/mBio.01204-17

Hall, L.S., Simonich, C.A., Brinkworth, A.J., Lybecker, M.C., Drecktrah, D., and Samuels, D.S. (2021). A novel RpoN-independent *rpoS* transcript in *Borrelia burgdorferi*. Manuscript in preparation.

Hantke, K. (1981). Regulation of ferric iron transport in *Escherichia coli* K12: isolation of a constitutive mutant. Mol Gen Genet *182*, 288-292. https://doi.org/10.1007/BF00269672

Hantke, K. (2001). Iron and metal regulation in bacteria. Curr Opin Microbiol *4*, 172-177. https://doi.org/10.1016/s1369-5274(00)00184-3

Hauryliuk, V., Atkinson, G.C., Murakami, K.S., Tenson, T., and Gerdes, K. (2015). Recent functional insights into the role of (p)ppGpp in bacterial physiology. Nat Rev Microbiol *13*, 298-309. https://doi.org/10.1038/nrmicro3448

He, M., Oman, T., Xu, H., Blevins, J., Norgard, M.V., and Yang, X.F. (2008). Abrogation of *ospAB* constitutively activates the Rrp2-RpoN-RpoS pathway (sigmaN-sigmaS cascade) in *Borrelia burgdorferi*. Mol Microbiol *70*, 1453-1464. https://doi.org/10.1111/j.1365-2958.2008.06491.x

He, M., Ouyang, Z., Troxell, B., Xu, H., Moh, A., Piesman, J., Norgard, M.V., Gomelsky, M., and Yang, X.F. (2011). Cyclic di-GMP is essential for the survival of the Lyme disease spirochete in ticks. PLoS Pathog *7*, e1002133.

He, M., Zhang, J.-J., Ye, M., Lou, Y., and Yang, X.F. (2014). Cyclic di-GMP receptor PlzA controls virulence gene expression through RpoS in *Borrelia burgdorferi*. Infect Immun *82*, 445-452. https://doi.org/10.1128/IAI.01238-13

Helmann, J.D. (2014). Specificity of metal sensing: iron and manganese homeostasis in *Bacillus subtilis*. J Biol Chem *289*, 28112-28120. https://doi.org/10.1074/jbc.R114.587071

Hengge, R. (2009). Principles of c-di-GMP signalling in bacteria. Nat Rev Microbiol *7*, 263-273. https://doi.org/10.1038/nrmicro2109

Hengge-Aronis, R. (2002a). Signal transduction and regulatory mechanisms involved in control of the σ^S (RpoS) subunit of RNA polymerase. Microbiol Mol Biol Rev *66*, 373-395.

Hengge-Aronis, R. (2002b). Stationary phase gene regulation: what makes an *Escherichia coli* promoter σ^S-selective? Curr Opin Microbiol *5*, 591-595.

Hör, J., Gorski, S.A., and Vogel, J. (2018). Bacterial RNA biology on a genome scale. Mol Cell *70*, 785-799. https://doi.org/10.1016/j.molcel.2017.12.023

Huang, Y.-Y., Deng, J.-Y., Gu, J., Zhang, Z.-P., Maxwell, A., Bi, L.-J., Chen, Y.-Y., Zhou, Y.-F., Yu, Z.-N., and Zhang, X.-E. (2006). The key DNA-binding residues in the C-terminal domain of *Mycobacterium tuberculosis* DNA gyrase A subunit (GyrA). Nucleic Acids Res *34*, 5650-5659. https://doi.org/10.1093/nar/gkl695

Hübner, A., Yang, X., Nolen, D.M., Popova, T.G., Cabello, F.C., and Norgard, M.V. (2001). Expression of *Borrelia burgdorferi* OspC and DbpA is controlled by a RpoN-RpoS regulatory pathway. Proc Natl Acad Sci USA *98*, 12724-12729.

Hyde, J.A., Seshu, J., and Skare, J.T. (2006). Transcriptional profiling of *Borrelia burgdorferi* containing a unique *bosR* allele identifies a putative oxidative stress regulon. Microbiology *152*, 2599-2609. https://doi.org/10.1099/mic.0.28996-0

Hyde, J.A., Shaw, D.K., Smith, R., III, Trzeciakowski, J.P., and Skare, J.T. (2009). The BosR regulatory protein of *Borrelia burgdorferi* interfaces with the RpoS regulatory pathway and modulates both homeostatic and pathogenic properties of the Lyme disease spirochete. Mol Microbiol *74*, 1344-1355. https://doi.org/10.1111/j.1365-2958.2009.06951.x

Hyde, J.A., Shaw, D.K., Smith, R., III, Trzeciakowski, J.P., and Skare, J.T. (2010). Characterization of a conditional *bosR* mutant in *Borrelia burgdorferi*. Infect Immun *78*, 265-274. https://doi.org/10.1128/IAI.01018-09

Hyde, J.A., Trzeciakowski, J.P., and Skare, J.T. (2007). *Borrelia burgdorferi* alters its gene expression and antigenic profile in response to CO_2 levels. J Bacteriol *189*, 437-445.

Iyer, R., Caimano, M.J., Luthra, A., Axline, D., Jr., Corona, A., Iacobas, D.A., Radolf, J.D., and Schwartz, I. (2015). Stage-specific global alterations in the transcriptomes of Lyme disease spirochetes during tick feeding and following mammalian host adaptation. Mol Microbiol *95*, 509-538. https://doi.org/10.1111/mmi.12882

Iyer, R., and Schwartz, I. (2016). Microarray-based comparative genomic and transcriptome analysis of *Borrelia burgdorferi*. Microarrays *5*, 9. https://doi.org/10.3390/microarrays5020009

Jacquamet, L., Traoré, D.A.K., Ferrer, J.-L., Proux, O., Testemale, D., Hazemann, J.-L., Nazarenko, E., El Ghazouani, A., Caux-Thang, C., Duarte, V., *et al.* (2009). Structural characterization of the active form of PerR: insights into the metal-induced activation of PerR and Fur proteins for DNA binding. Mol Microbiol *73*, 20-31. https://doi.org/10.1111/j.1365-2958.2009.06753.x

Jewett, M.W., Lawrence, K., Bestor, A.C., Tilly, K., Grimm, D., Shaw, P., VanRaden, M., Gherardini, F., and Rosa, P.A. (2007). The critical role of the linear plasmid lp36 in the infectious cycle of *Borrelia burgdorferi*. Mol Microbiol *64*, 1358-1374. https://doi.org/10.1111/j.1365-2958.2007.05746.x

Jutras, B.L., Bowman, A., Brissette, C.A., Adams, C.A., Verma, A., Chenail, A.M., and Stevenson, B. (2012a). EbfC (YbaB) is a new type of bacterial nucleoid-associated protein and a global regulator of gene expression in the Lyme disease spirochete. J Bacteriol *194*, 3395-3406. https://doi.org/10.1128/JB.00252-12

Jutras, B.L., Chenail, A.M., Carroll, D.W., Miller, M.C., Zhu, H., Bowman, A., and Stevenson, B. (2013a). Bpur, the Lyme disease spirochete's PUR domain protein: identification as a transcriptional modulator and characterization of nucleic acid interactions. J Biol Chem *288*, 26220-26234. https://doi.org/10.1074/jbc.M113.491357

Jutras, B.L., Chenail, A.M., Rowland, C.L., Carroll, D., Miller, M.C., Bykowski, T., and Stevenson, B. (2013b). Eubacterial SpoVG homologs constitute a new family of site-specific DNA-binding proteins. PLoS One *8*, e66683. https://doi.org/10.1371/journal.pone.0066683

Jutras, B.L., Chenail, A.M., and Stevenson, B. (2013c). Changes in bacterial growth rate govern expression of the *Borrelia burgdorferi* OspC and Erp infection-associated surface proteins. J Bacteriol *195*, 757-764. https://doi.org/10.1128/JB.01956-12

Jutras, B.L., Jones, G.S., Verma, A., Brown, N.A., Antonicello, A.D., Chenail, A.M., and Stevenson, B. (2013d). Posttranscriptional self-regulation by the Lyme disease bacterium's BpuR DNA/RNA-binding protein. J Bacteriol *195*, 4915-4923. https://doi.org/10.1128/JB.00819-13

Jutras, B.L., Savage, C.R., Arnold, W.K., Lethbridge, K.G., Carroll, D.W., Tilly, K., Bestor, A., Zhu, H., Seshu, J., Zückert, W.R., *et al.* (2019). The Lyme disease spirochete's BpuR DNA/RNA-binding protein is differentially expressed during the mammal-tick infectious cycle, which affects translation of the SodA superoxide dismutase. Mol Microbiol *112*, 973-991. https://doi.org/10.1111/mmi.14336

Jutras, B.L., Verma, A., Adams, C.A., Brissette, C.A., Burns, L.H., Whetstine, C.R., Bowman, A., Chenail, A.M., Zückert, W.R., and Stevenson, B. (2012b). BpaB and EbfC DNA-binding proteins regulate production of the Lyme disease spirochete's infection-associated Erp surface proteins. J Bacteriol *194*, 778-786. https://doi.org/10.1128/JB.06394-11

Karna, S.L.R., Prabhu, R.G., Lin, Y.-H., Miller, C.L., and Seshu, J. (2013). Contributions of environmental signals and conserved residues to the functions of carbon storage regulator A of *Borrelia burgdorferi*. Infect Immun *81*, 2972-2985. https://doi.org/10.1128/IAI.00494-13

Karna, S.L.R., Sanjuan, E., Esteve-Gassent, M.D., Miller, C.L., Maruskova, M., and Seshu, J. (2011). CsrA modulates levels of lipoproteins and key regulators of gene expression critical for pathogenic mechanisms of *Borrelia burgdorferi*. Infect Immun *79*, 732-744. https://doi.org/10.1128/IAI.00882-10

Katona, L.I. (2015). The Fur homologue BosR requires Arg39 to activate *rpoS* transcription in *Borrelia burgdorferi* and thereby direct spirochaete infection in mice. Microbiology *161*, 2243-2255. https://doi.org/10.1099/mic.0.000166

Katona, L.I., Tokarz, R., Kuhlow, C.J., Benach, J., and Benach, J.L. (2004). The Fur homologue in *Borrelia burgdorferi*. J Bacteriol *186*, 6443-6456. https://doi.org/10.1128/JB.186.19.6443-6456.2004

Kazanov, M.D., Li, X., Gelfand, M.S., Osterman, A.L., and Rodionov, D.A. (2013). Functional diversification of ROK-family transcriptional regulators of sugar catabolism in the Thermotogae phylum. Nucleic Acids Res *41*, 790-803. https://doi.org/10.1093/nar/gks1184

Kazmierczak, M.J., Wiedmann, M., and Boor, K.J. (2005). Alternative sigma factors and their roles in bacterial virulence. Microbiol Mol Biol Rev *69*, 527-543.

Kihara, A., Akiyama, Y., and Ito, K. (1997). Host regulation of lysogenic decision in bacteriophage λ: transmembrane modulation of FtsH (HflB), the cII degrading protease, by HflKC (HflA). Proc Natl Acad Sci USA *94*, 5544-5549.

Knight, S.W., Kimmel, B.J., Eggers, C.H., and Samuels, D.S. (2000). Disruption of the *Borrelia burgdorferi gac* gene, encoding the naturally synthesized GyrA C-terminal domain. J Bacteriol *182*, 2048-2051. https://doi.org/10.1128/jb.182.7.2048-2051.2000

Knight, S.W., and Samuels, D.S. (1999). Natural synthesis of a DNA-binding protein from the C-terminal domain of DNA gyrase A in *Borrelia burgdorferi*. EMBO J *18*, 4875-4881. https://doi.org/10.1093/emboj/18.17.4875

Kobryn, K., Naigamwalla, D.Z., and Chaconas, G. (2000). Site-specific DNA binding and bending by the *Borrelia burgdorferi* Hbb protein. Mol Microbiol *37*, 145-155.

Kostick, J.L., Szkotnicki, L.T., Rogers, E.A., Bocci, P., Raffaelli, N., and Marconi, R.T. (2011). The diguanylate cyclase, Rrp1, regulates critical steps in the enzootic cycle of the Lyme disease spirochetes. Mol Microbiol *81*, 219-231. https://doi.org/10.1111/j.1365-2958.2011.07687.x

Kostick-Dunn, J.L., Izac, J.R., Freedman, J.C., Szkotnicki, L.T., Oliver, L.D., Jr., and Marconi, R.T. (2018). The *Borrelia burgdorferi* c-di-GMP binding receptors, PlzA and PlzB, are functionally distinct. Front Cell Infect Microbiol *8*, 213. https://doi.org/10.3389/fcimb.2018.00213

Kraiczy, P., and Stevenson, B. (2013). Complement regulator-acquiring surface proteins of *Borrelia burgdorferi*: Structure, function and regulation of gene expression. Ticks Tick Borne Dis *4*, 26-34. https://doi.org/10.1016/j.ttbdis.2012.10.039

Kramlinger, V.M., and Hiasa, H. (2006). The "GyrA-box" is required for the ability of DNA gyrase to wrap DNA and catalyze the supercoiling reaction. J Biol Chem *281*, 3738-3742. https://doi.org/10.1074/jbc.M511160200

Kung, F., Anguita, J., and Pal, U. (2013). *Borrelia burgdorferi* and tick proteins supporting pathogen persistence in the vector. Future Microbiol *8*, 41-56. https://doi.org/10.2217/fmb.12.121

Kurkela, J., Fredman, J., Salminen, T.A., and Tyystjärvi, T. (2020). Revealing secrets of the enigmatic omega subunit of bacterial RNA polymerase. Mol Microbiol. https://doi.org/10.1111/mmi.14603

Kusano, S., Ding, Q., Fujita, N., and Ishihama, A. (1996). Promoter selectivity of *Escherichia coli* RNA polymerase $E\sigma^{70}$ and $E\sigma^{38}$ holoenzymes: effect of DNA supercoiling. J Biol Chem *271*, 1998-2004.

Lane, R.S., Piesman, J., and Burgdorfer, W. (1991). Lyme borreliosis: relation of its causative agent to its vectors and hosts in North America and Europe. Annu Rev Entomol *36*, 587-609.

Lasa, I., Toledo-Arana, A., Dobin, A., Villanueva, M., Ruiz de los Mozos, I., Vergara-Irigaray, M., Segura, V., Fagegaltier, D., Penadés, J.R., Valle, J., *et al.* (2011). Genome-wide antisense transcription drives mRNA processing in bacteria. Proc Natl Acad Sci USA *108*, 20172-20177. https://doi.org/10.1073/pnas.1113521108

Lasa, I., Toledo-Arana, A., and Gingeras, T.R. (2012). An effort to make sense of antisense transcription in bacteria. RNA Biol *9*, 1039-1044. https://doi.org/10.4161/rna.21167

Le Rhun, A., Beer, Y.Y., Reimegård, J., Chylinski, K., and Charpentier, E. (2016). RNA sequencing uncovers antisense RNAs and novel small RNAs in *Streptococcus pyogenes*. RNA Biol *13*, 177-195. https://doi.org/10.1080/15476286.2015.1110674

Lee, E.R., Baker, J.L., Weinberg, Z., Sudarsan, N., and Breaker, R.R. (2010). An allosteric self-splicing ribozyme triggered by a bacterial second messenger. Science *329*, 845-848. https://doi.org/10.1126/science.1190713

Lee, J.-H., Gatewood, M.L., and Jones, G.H. (2013). RNase III is required for actinomycin production in *Streptomyces antibioticus*. Appl Environ Microbiol *79*, 6447-6451. https://doi.org/10.1128/AEM.02272-13

Lehnik-Habrink, M., Schaffer, M., Mäder, U., Diethmaier, C., Herzberg, C., and Stülke, J. (2011). RNA processing in *Bacillus subtilis*: identification of targets of the essential RNase Y. Mol Microbiol *81*, 1459-1473. https://doi.org/10.1111/j.1365-2958.2011.07777.x

Lejars, M., Kobayashi, A., and Hajnsdorf, E. (2019). Physiological roles of antisense RNAs in prokaryotes. Biochimie *164*, 3-16. https://doi.org/10.1016/j.biochi.2019.04.015

Liang, F.T., Jacobs, M.B., Bowers, L.C., and Philipp, M.T. (2002a). An immune evasion mechanism for spirochetal persistence in Lyme borreliosis. J Exp Med *195*, 415-422.

Liang, F.T., Nelson, F.K., and Fikrig, E. (2002b). Molecular adaptation of *Borrelia burgdorferi* in the murine host. J Exp Med *196*, 275-280.

Liang, F.T., Yan, J., Mbow, M.L., Sviat, S.L., Gilmore, R.D., Mamula, M., and Fikrig, E. (2004). *Borrelia burgdorferi* changes its surface antigenic expression in response to host immune responses. Infect Immun *72*, 5759-5767.

Lim, K., Tempczyk, A., Parsons, J.F., Bonander, N., Toedt, J., Kelman, Z., Howard, A., Eisenstein, E., and Herzberg, O. (2003). Crystal structure of YbaB from *Haemophilus influenzae* (HI0442), a protein of unknown function coexpressed with the recombinational DNA repair protein RecR. Proteins *50*, 375-379. https://doi.org/10.1002/prot.10297

Lin, T., Gao, L., Zhang, C., Odeh, E., Jacobs, M.B., Coutte, L., Chaconas, G., Philipp, M.T., and Norris, S.J. (2012). Analysis of an ordered, comprehensive STM mutant library in infectious *Borrelia burgdorferi*: insights into the genes required for mouse infectivity. PLoS One *7*, e47532. https://doi.org/10.1371/journal.pone.0047532

Lin, Y.H., Chen, Y., Smith, T.C., II, Karna, S.L.R., and Seshu, J. (2018). Short-chain fatty acids alter metabolic and virulence attributes of *Borrelia burgdorferi*. Infect Immun *86*, e00217-18. https://doi.org/10.1128/IAI.00217-18

Lioliou, E., Sharma, C.M., Caldelari, I., Helfer, A.-C., Fechter, P., Vandenesch, F., Vogel, J., and Romby, P. (2012). Global regulatory functions of the *Staphylococcus aureus* endoribonuclease III in gene expression. PLoS Genet *8*, e1002782. https://doi.org/10.1371/journal.pgen.1002782

Liu, K., Bittner, A.N., and Wang, J.D. (2015). Diversity in (p)ppGpp metabolism and effectors. Curr Opin Microbiol *24*, 72-79. https://doi.org/10.1016/j.mib.2015.01.012

Liu, Q., Xu, H., Zhang, Y., Yang, J., Du, J., Zhou, Y., Yang, X.F., and Lou, Y. (2020). Role of Hk2 in the enzootic cycle of *Borrelia burgdorferi*. Front Med *7*, 652. https://doi.org/10.3389/fmed.2020.573648

Livengood, J.A., Schmit, V.L., and Gilmore, R.D., Jr. (2008). Global transcriptome analysis of *Borrelia burgdorferi* during association with human neuroglial cells. Infect Immun *76*, 298-307. https://doi.org/10.1128/IAI.00866-07

López-García, P., and Forterre, P. (2000). DNA topology and the thermal stress response, a tale from mesophiles and hyperthermophiles. Bioessays *22*, 738-746.

Lucchetti-Miganeh, C., Burrowes, E., Baysse, C., and Ermel, G. (2008). The post-transcriptional regulator CsrA plays a central role in the adaptation of bacterial pathogens to different stages of infection in animal hosts. Microbiology *154*, 16-29. https://doi.org/10.1099/mic.0.2007/012286-0

Lybecker, M., Bilusic, I., and Raghavan, R. (2014a). Pervasive transcription: detecting functional RNAs in bacteria. Transcription *5*, e944039. https://doi.org/10.4161/21541272.2014.944039

Lybecker, M., Zimmermann, B., Bilusic, I., Tukhtubaeva, N., and Schroeder, R. (2014b). The double-stranded transcriptome of *Escherichia coli*. Proc Natl Acad Sci USA *111*, 3134-3139. https://doi.org/10.1073/pnas.1315974111

Lybecker, M.C., Abel, C.A., Feig, A.L., and Samuels, D.S. (2010). Identification and function of the RNA chaperone Hfq in the Lyme disease spirochete *Borrelia burgdorferi*. Mol Microbiol *78*, 622-635. https://doi.org/10.1111/j.1365-2958.2010.07374.x

Lybecker, M.C., and Samuels, D.S. (2007). Temperature-induced regulation of RpoS by a small RNA in *Borrelia burgdorferi*. Mol Microbiol *64*, 1075-1089. https://doi.org/10.1111/j.1365-2958.2007.05716.x

Lybecker, M.C., and Samuels, D.S. (2017). Small RNAs of *Borrelia burgdorferi*: characterizing functional regulators in a sea of sRNAs. Yale J Biol Med *90*, 317-323.

Majdalani, N., Vanderpool, C.K., and Gottesman, S. (2005). Bacterial small RNA regulators. Crit Rev Biochem Mol Biol *40*, 93-113.

Mallory, K.L., Miller, D.P., Oliver, L.D., Jr., Freedman, J.C., Kostick-Dunn, J.L., Carlyon, J.A., Marion, J.D., Bell, J.K., and Marconi, R.T. (2016). Cyclic-di-GMP binding induces structural rearrangements in the PlzA and PlzC proteins of the Lyme disease and relapsing fever spirochetes: a possible switch mechanism for c-di-GMP-mediated effector functions. Pathog Dis *74*, ftw105. https://doi.org/10.1093/femspd/ftw105

Marchetti, M., Malinowska, A., Heller, I., and Wuite, G.J.L. (2017). How to switch the motor on: RNA polymerase initiation steps at the single-molecule level. Protein Sci *26*, 1303-1313. https://doi.org/10.1002/pro.3183

Margolis, N., Hogan, D., Tilly, K., and Rosa, P.A. (1994). Plasmid location of *Borrelia* purine biosynthesis gene homologs. J Bacteriol *176*, 6427-6432.

Marincola, G., Schäfer, T., Behler, J., Bernhardt, J., Ohlsen, K., Goerke, C., and Wolz, C. (2012). RNase Y of *Staphylococcus aureus* and its role in the activation of virulence genes. Mol Microbiol *85*, 817-832. https://doi.org/10.1111/j.1365-2958.2012.08144.x

Maruskova, M., Esteve-Gassent, M.D., Sexton, V.L., and Seshu, J. (2008). Role of the *bbA64* locus of *Borrelia burgdorferi* in early stages of infectivity in a murine model of Lyme disease. Infect Immun *76*, 391-402. https://doi.org/10.1128/IAI.01118-07

Mason, C., Liu, X., Prabhudeva, S., and Ouyang, Z. (2019). The CXXC motifs are essential for the function of BosR in *Borrelia burgdorferi*. Front Cell Infect Microbiol *9*, 109. https://doi.org/10.3389/fcimb.2019.00109

Mason, C., Thompson, C., and Ouyang, Z. (2020). DksA plays an essential role in regulating the virulence of *Borrelia burgdorferi*. Mol Microbiol *114*, 172-183. https://doi.org/10.1111/mmi.14504

Mathew, R., and Chatterji, D. (2006). The evolving story of the omega subunit of bacterial RNA polymerase. Trends Microbiol *14*, 450-455. https://doi.org/10.1016/j.tim.2006.08.002

McCown, P.J., Corbino, K.A., Stav, S., Sherlock, M.E., and Breaker, R.R. (2017). Riboswitch diversity and distribution. RNA *23*, 995-1011. https://doi.org/10.1261/rna.061234.117

Medina-Pérez, D.N., Wager, B., Troy, E., Gao, L., Norris, S.J., Lin, T., Hu, L., Hyde, J.A., Lybecker, M., and Skare, J.T. (2020). The intergenic small non-coding RNA *ittA* is required for optimal infectivity and tissue tropism in *Borrelia burgdorferi*. PLoS Pathog *16*, e1008423. https://doi.org/10.1371/journal.ppat.1008423

Medrano, M., Ding, Y., Wang, X.-G., Lu, P., Coburn, J., and Hu, L.T. (2007). Regulators of expression of the oligopeptide permease A proteins of *Borrelia burgdorferi*. J Bacteriol *189*, 2653-2659.

Miller, C.L., Karna, S.L.R., and Seshu, J. (2013). *Borrelia* host adaptation Regulator (BadR) regulates *rpoS* to modulate host adaptation and virulence factors in *Borrelia burgdorferi*. Mol Microbiol *88*, 105-124. https://doi.org/10.1111/mmi.12171

Mohanty, B.K., and Kushner, S.R. (2016). Regulation of mRNA decay in bacteria. Annu Rev Microbiol *70*, 25-44. https://doi.org/10.1146/annurev-micro-091014-104515

Mouw, K.W., and Rice, P.A. (2007). Shaping the *Borrelia burgdorferi* genome: crystal structure and binding properties of the DNA-bending protein Hbb. Mol Microbiol *63*, 1319-1330. https://doi.org/10.1111/j.1365-2958.2007.05586.x

Mulay, V.B., Caimano, M.J., Iyer, R., Dunham-Ems, S., Liveris, D., Petzke, M.M., Schwartz, I., and Radolf, J.D. (2009). *Borrelia burgdorferi bba74* is expressed exclusively during tick feeding and is regulated by both arthropod- and mammalian host-specific signals. J Bacteriol *191*, 2783-2794. https://doi.org/10.1128/JB.01802-08

Nader, S., Pérard, J., Carpentier, P., Arnaud, L., Crouzy, S., and Michaud-Soret, I. (2019). New insights into the tetrameric family of the Fur metalloregulators. Biometals *32*, 501-519. https://doi.org/10.1007/s10534-019-00201-8

Narasimhan, S., Caimano, M.J., Liang, F.T., Santiago, F., Laskowski, M., Philipp, M.T., Pachner, A.R., Radolf, J.D., and Fikrig, E. (2003). *Borrelia burgdorferi* transcriptome in the central nervous system of non-human primates. Proc Natl Acad Sci USA *100*, 15953-15958. https://doi.org/10.1073/pnas.2432412100

Narberhaus, F., Waldminghaus, T., and Chowdhury, S. (2006). RNA thermometers. FEMS Microbiol Rev *30*, 3-16.

Nash, H.A. (1996). The HU and IHF proteins: accessory factors for complex protein-DNA assemblies. In Regulation of Gene Expression in *Escherichia coli*, E.C.C. Lin, and A.S. Lynch, eds. (Austin, Texas: R. G. Landes Co), pp. 149-179.

Neusser, T., Polen, T., Geissen, R., and Wagner, R. (2010). Depletion of the non-coding regulatory 6S RNA in *E. coli* causes a surprising reduction in the expression of the translation machinery. BMC Genomics *11*, 165. https://doi.org/10.1186/1471-2164-11-165

Norris, S.J. (2014). *vls* antigenic variation systems of Lyme disease *Borrelia*: eluding host immunity through both random, segmental gene conversion and framework heterogeneity. Microbiol Spectr 2, MDNA3-0038-2014. https://doi.org/10.1128/microbiolspec.MDNA3-0038-2014

Novak, E.A., Sultan, S.Z., and Motaleb, M.A. (2014). The cyclic-di-GMP signaling pathway in the Lyme disease spirochete, *Borrelia burgdorferi*. Front Cell Infect Microbiol *4*, 56. https://doi.org/10.3389/fcimb.2014.00056

Ojaimi, C., Brooks, C., Casjens, S., Rosa, P., Elias, A., Barbour, A., Jasinskas, A., Benach, J., Katona, L., Radolf, J., *et al.* (2003). Profiling of temperature-induced changes in *Borrelia burgdorferi* gene expression by using whole genome arrays. Infect Immun *71*, 1689-1705.

Ojaimi, C., Mulay, V., Liveris, D., Iyer, R., and Schwartz, I. (2005). Comparative transcriptional profiling of *Borrelia burgdorferi* clinical isolates differing in capacities for hematogenous dissemination. Infect Immun *73*, 6791-6802. https://doi.org/10.1128/IAI.73.10.6791-6802.2005

Österberg, S., del Peso-Santos, T., and Shingler, V. (2011). Regulation of alternative sigma factor use. Annu Rev Microbiol *65*, 37-55. https://doi.org/10.1146/annurev.micro.112408.134219

Ouyang, Z., Blevins, J.S., and Norgard, M.V. (2008). Transcriptional interplay among the regulators Rrp2, RpoN and RpoS in *Borrelia burgdorferi*. Microbiology *154*, 2641-2658. https://doi.org/10.1099/mic.0.2008/019992-0

Ouyang, Z., Deka, R.K., and Norgard, M.V. (2011). BosR (BB0647) controls the RpoN-RpoS regulatory pathway and virulence expression in *Borrelia burgdorferi* by a novel DNA-binding mechanism. PLoS Pathog *7*, e1001272. https://doi.org/10.1371/journal.ppat.1001272

Ouyang, Z., He, M., Oman, T., Yang, X.F., and Norgard, M.V. (2009a). A manganese transporter, BB0219 (BmtA), is required for virulence by the Lyme disease spirochete, *Borrelia burgdorferi*. Proc Natl Acad Sci USA *106*, 3449-3454. https://doi.org/10.1073/pnas.0812999106

Ouyang, Z., Kumar, M., Kariu, T., Haq, S., Goldberg, M., Pal, U., and Norgard, M.V. (2009b). BosR (BB0647) governs virulence expression in *Borrelia burgdorferi*. Mol Microbiol *74*, 1331-1343. https://doi.org/10.1111/j.1365-2958.2009.06945.x

Ouyang, Z., Narasimhan, S., Neelakanta, G., Kumar, M., Pal, U., Fikrig, E., and Norgard, M.V. (2012). Activation of the RpoN-RpoS regulatory pathway during the enzootic life cycle of *Borrelia burgdorferi*. BMC Microbiol *12*, 44. https://doi.org/10.1186/1471-2180-12-44

Ouyang, Z., and Zhou, J. (2015). BadR (BB0693) controls growth phase-dependent induction of *rpoS* and *bosR* in *Borrelia burgdorferi* via recognizing TAAAATAT motifs. Mol Microbiol *98*, 1147-1167. https://doi.org/10.1111/mmi.13206

Ouyang, Z., and Zhou, J. (2017). The putative Walker A and Walker B motifs of Rrp2 are required for the growth of *Borrelia burgdorferi*. Mol Microbiol *103*, 86-98. https://doi.org/10.1111/mmi.13545

Ouyang, Z., Zhou, J., Brautigam, C.A., Deka, R., and Norgard, M.V. (2014a). Identification of a core sequence for the binding of BosR to the *rpoS* promoter region in *Borrelia burgdorferi*. Microbiology *160*, 851-862. https://doi.org/10.1099/mic.0.075655-0

Ouyang, Z., Zhou, J., and Norgard, M.V. (2014b). CsrA (BB0184) is not involved in activation of the RpoN-RpoS regulatory pathway in *Borrelia burgdorferi*. Infect Immun *82*, 1511-1522. https://doi.org/10.1128/IAI.01555-13

Ouyang, Z., Zhou, J., and Norgard, M.V. (2014c). Synthesis of RpoS is dependent on a putative enhancer binding protein Rrp2 in *Borrelia burgdorferi*. PLoS One *9*, e96917. https://doi.org/10.1371/journal.pone.0096917

Ouyang, Z., Zhou, J., and Norgard, M.V. (2016). Evidence that BosR (BB0647) Is a positive autoregulator in *Borrelia burgdorferi*. Infect Immun *84*, 2566-2574. https://doi.org/10.1128/IAI.00297-16

Pal, U., Dai, J., Li, X., Neelakanta, G., Luo, P., Kumar, M., Wang, P., Yang, X., Anderson, J.F., and Fikrig, E. (2008). A differential role for BB0365 in the persistence of *Borrelia burgdorferi* in mice and ticks. J Infect Dis *197*, 148-155. https://doi.org/10.1086/523764

Papenfort, K., and Vanderpool, C.K. (2015). Target activation by regulatory RNAs in bacteria. FEMS Microbiol Rev *39*, 362-378. https://doi.org/10.1093/femsre/fuv016

Pappas, C.J., Iyer, R., Petzke, M.M., Caimano, M.J., Radolf, J.D., and Schwartz, I. (2011). *Borrelia burgdorferi* requires glycerol for maximum fitness during the tick phase of the enzootic cycle. PLoS Pathog *7*, e1002102. https://doi.org/10.1371/journal.ppat.1002102

Pei, J., and Grishin, N.V. (2001). GGDEF domain is homologous to adenylyl cyclase. Proteins *42*, 210-216. https://doi.org/10.1002/1097-0134(20010201)42:2<210::aid-prot80>3.0.co;2-8

Peter, B.J., Arsuaga, J., Breier, A.M., Khodursky, A.B., Brown, P.O., and Cozzarelli, N.R. (2004). Genomic transcriptional response to loss of chromosomal supercoiling in *Escherichia coli*. Genome Biol *5*, R87.

Pich, A., and Bahl, H. (1991). Purification and characterization of the DNA-dependent RNA polymerase from *Clostridium acetobutylicum*. J Bacteriol *173*, 2120-2124. https://doi.org/10.1128/jb.173.6.2120-2124.1991

Pinochet-Barros, A., and Helmann, J.D. (2018). Redox Sensing by Fe^{2+} in Bacterial Fur Family Metalloregulators. Antioxid Redox Signal *29*, 1858-1871. https://doi.org/10.1089/ars.2017.7359

Pitzer, J.E., Sultan, S.Z., Hayakawa, Y., Hobbs, G., Miller, M.R., and Motaleb, M.A. (2011). Analysis of the *Borrelia burgdorferi* cyclic-di-GMP-binding protein PlzA reveals a role in motility and virulence. Infect Immun *79*, 1815-1825. https://doi.org/10.1128/IAI.00075-11

Popitsch, N., Bilusic, I., Rescheneder, P., Schroeder, R., and Lybecker, M. (2017). Temperature-dependent sRNA transcriptome of the Lyme disease spirochete. BMC Genomics *18*, 28. https://doi.org/10.1186/s12864-016-3398-3

Poranen, M.M., Salgado, P.S., Koivunen, M.R., Wright, S., Bamford, D.H., Stuart, D.I., and Grimes, J.M. (2008). Structural explanation for the role of Mn^{2+} in the activity of ϕ6 RNA-dependent RNA polymerase. Nucleic Acids Res *36*, 6633-6644. https://doi.org/10.1093/nar/gkn632

Posey, J.E., and Gherardini, F.C. (2000). Lack of a role for iron in the Lyme disease pathogen. Science *288*, 1651-1653. https://doi.org/10.1126/science.288.5471.1651

Potrykus, K., and Cashel, M. (2008). (p)ppGpp: still magical? Annu Rev Microbiol *62*, 35-51. https://doi.org/10.1146/annurev.micro.62.081307.162903

Purser, J.E., and Norris, S.J. (2000). Correlation between plasmid content and infectivity in *Borrelia burgdorferi*. Proc Natl Acad Sci USA *97*, 13865-13870.

Radolf, J.D., Caimano, M.J., Stevenson, B., and Hu, L.T. (2012). Of ticks, mice and men: understanding the dual-host lifestyle of Lyme disease spirochaetes. Nat Rev Microbiol *10*, 87-99. https://doi.org/10.1038/nrmicro2714

Reece, R.J., and Maxwell, A. (1991). DNA gyrase: structure and function. Crit Rev Biochem Mol Biol *26*, 335-375.

Reitzer, L., and Schneider, B.L. (2001). Metabolic context and possible physiological themes of σ^{54}-dependent genes in *Escherichia coli*. Microbiol Mol Biol Rev *65*, 422-444. https://doi.org/10.1128/MMBR.65.3.422-444.2001

Ren, J., Sang, Y., Qin, R., Cui, Z., and Yao, Y.-F. (2017). 6S RNA is involved in acid resistance and invasion of epithelial cells in *Salmonella enterica* serovar Typhimurium. Future Microbiol *12*, 1045-1057. https://doi.org/10.2217/fmb-2017-0055

Repoila, F., and Gottesman, S. (2001). Signal transduction cascade for regulation of RpoS: temperature regulation of DsrA. J Bacteriol *183*, 4012-4023.

Repoila, F., Majdalani, N., and Gottesman, S. (2003). Small non-coding RNAs, co-ordinators of adaptation processes in *Escherichia coli*: the RpoS paradigm. Mol Microbiol *48*, 855-861.

Revel, A.T., Talaat, A.M., and Norgard, M.V. (2002). DNA microarray analysis of differential gene expression in *Borrelia burgdorferi*, the Lyme disease spirochete. Proc Natl Acad Sci USA *99*, 1562-1567.

Rhodes, R.G., Coy, W., and Nelson, D.R. (2009). Chitobiose utilization in *Borrelia burgdorferi* is dually regulated by RpoD and RpoS. BMC Microbiol *9*, 108. https://doi.org/10.1186/1471-2180-9-108

Ribeiro, J.M.C., Mather, T.N., Piesman, J., and Spielman, A. (1987). Dissemination and salivary delivery of Lyme disease spirochetes in vector ticks (Acari: Ixodidae). J Med Entomol *24*, 201-205. https://doi.org/10.1093/jmedent/24.2.201

Richards, C.L., Lawrence, K.A., Su, H., Yang, Y., Yang, X.F., Dulebohn, D.P., and Gherardini, F.C. (2015). Acetyl-phosphate is not a global regulatory bridge between virulence and central metabolism in *Borrelia burgdorferi*. PLoS One *10*, e0144472. https://doi.org/10.1371/journal.pone.0144472

Riddihough, G. (2005). In the forests of RNA dark matter. Science *309*, 1507-1507. https://doi.org/10.1126/science.309.5740.1507

Riley, S.P., Bykowski, T., Cooley, A.E., Burns, L.H., Babb, K., Brissette, C.A., Bowman, A., Rotondi, M., Miller, M.C., DeMoll, E., *et al.* (2009). *Borrelia burgdorferi* EbfC defines a newly-identified, widespread family of bacterial DNA-binding proteins. Nucleic Acids Res *37*, 1973-1983. https://doi.org/10.1093/nar/gkp027

Roberts, D.M., Caimano, M., McDowell, J., Theisen, M., Holm, A., Orff, E., Nelson, D., Wikel, S., Radolf, J., and Marconi, R.T. (2002). Environmental regulation and differential production of members of the Bdr protein family of *Borrelia burgdorferi*. Infect Immun *70*, 7033-7041.

Rogers, E.A., Terekhova, D., Zhang, H.-M., Hovis, K.M., Schwartz, I., and Marconi, R.T. (2009). Rrp1, a cyclic-di-GMP-producing response regulator, is an important regulator of *Borrelia burgdorferi* core cellular functions. Mol Microbiol *71*, 1551-1573. https://doi.org/10.1111/j.1365-2958.2009.06621.x

Romeo, T. (1998). Global regulation by the small RNA-binding protein CsrA and the non-coding RNA molecule CsrB. Mol Microbiol *29*, 1321-1330.

Romeo, T., and Babitzke, P. (2018). Global regulation by CsrA and its RNA antagonists. Microbiol Spectr *6*, RWR-0009-2017. https://doi.org/10.1128/microbiolspec.RWR-0009-2017

Römling, U. (2008). Great times for small molecules: c-di-AMP, a second messenger candidate in Bacteria and Archaea. Sci Signal *1*, pe39. https://doi.org/10.1126/scisignal.133pe39

Römling, U., Galperin, M.Y., and Gomelsky, M. (2013). Cyclic di-GMP: the first 25 years of a universal bacterial second messenger. Microbiol Mol Biol Rev *77*, 1-52. https://doi.org/10.1128/MMBR.00043-12

Ronneau, S., and Hallez, R. (2019). Make and break the alarmone: regulation of (p)ppGpp synthetase/hydrolase enzymes in bacteria. FEMS Microbiol Rev *43*, 389-400. https://doi.org/10.1093/femsre/fuz009

Rosenbluh, A., Banner, C.D., Losick, R., and Fitz-James, P.C. (1981). Identification of a new developmental locus in *Bacillus subtilis* by construction of a deletion mutation in a cloned gene under sporulation control. J Bacteriol *148*, 341-351. https://doi.org/10.1128/JB.148.1.341-351.1981

Ross, P., Weinhouse, H., Aloni, Y., Michaeli, D., Weinberger-Ohana, P., Mayer, R., Braun, S., de Vroom, E., van der Marel, G.A., van Boom, J.H., *et al.* (1987). Regulation of cellulose synthesis in *Acetobacter xylinum* by cyclic diguanylic acid. Nature *325*, 279-281. https://doi.org/10.1038/325279a0

Roth, A., and Breaker, R.R. (2009). The structural and functional diversity of metabolite-binding riboswitches. Annu Rev Biochem *78*, 305-334. https://doi.org/10.1146/annurev.biochem.78.070507.135656

Rubin, B.E., Huynh, T.N., Welkie, D.G., Diamond, S., Simkovsky, R., Pierce, E.C., Taton, A., Lowe, L.C., Lee, J.J., Rifkin, S.A., *et al.* (2018). High-throughput interaction screens illuminate the role of c-di-AMP in cyanobacterial nighttime survival. PLoS Genet *14*, e1007301. https://doi.org/10.1371/journal.pgen.1007301

Rui, S., and Tse-Dinh, Y.-C. (2003). Topoisomerase function during bacterial responses to environmental challenge. Front Biosci *8*, d256-263.

Rutberg, L., and Armentrout, R.W. (1972). Deoxyribonucleic acid polymerase activity in a deoxyribonucleic acid polymerase I-deficient mutant of *Bacillus subtilis* infected with temperature bacteriophage SPO2. J Virol *10*, 658-660. https://doi.org/10.1128/JVI.10.4.658-660.1972

Ruthenburg, A.J., Graybosch, D.M., Huetsch, J.C., and Verdine, G.L. (2005). A superhelical spiral in the *Escherichia coli* DNA gyrase A C-terminal domain imparts unidirectional supercoiling bias. J Biol Chem *280*, 26177-26184. https://doi.org/10.1074/jbc.M502838200

Ryjenkov, D.A., Simm, R., Römling, U., and Gomelsky, M. (2006). The PilZ domain is a receptor for the second messenger c-di-GMP: the PilZ domain protein YcgR controls motility in enterobacteria. J Biol Chem *281*, 30310-30314. https://doi.org/10.1074/jbc.C600179200

Ryjenkov, D.A., Tarutina, M., Moskvin, O.V., and Gomelsky, M. (2005). Cyclic diguanylate is a ubiquitous signaling molecule in bacteria: insights into biochemistry of the GGDEF protein domain. J Bacteriol *187*, 1792-1798. https://doi.org/10.1128/JB.187.5.1792-1798.2005

Saliba, A.-E., Santos, S.C., and Vogel, J. (2017). New RNA-seq approaches for the study of bacterial pathogens. Curr Opin Microbiol *35*, 78-87. https://doi.org/10.1016/j.mib.2017.01.001

Samuels, D.S. (2006). Antibiotic resistance in *Borrelia burgdorferi*: applications for genetic manipulation and implications for evolution. In Molecular Biology of Spirochetes, F.C. Cabello, D. Hulinska, and H.P. Godfrey, eds. (Amsterdam, Netherlands: IOS Press), pp. 56-70.

Samuels, D.S. (2011). Gene regulation in *Borrelia burgdorferi*. Annu Rev Microbiol *65*, 479-499. https://doi.org/10.1146/annurev.micro.112408.134040

Samuels, D.S., and Radolf, J.D. (2009). Who is the BosR around here anyway? Mol Microbiol *74*, 1295-1299. https://doi.org/10.1111/j.1365-2958.2009.06971.x

Samuels, D.S., and Samuels, L.R.N. (2016). Gene regulation during the enzootic cycle of the Lyme disease spirochete. For Immunopathol Dis Therap *7*, 205-212. https://doi.org/10.1615/ForumImmunDisTher.2017019469

Sanjuan, E., Esteve-Gassent, M.D., Maruskova, M., and Seshu, J. (2009). Overexpression of CsrA (BB0184) alters the morphology and antigen profiles of *Borrelia burgdorferi*. Infect Immun *77*, 5149-5162. https://doi.org/10.1128/IAI.00673-09

Sarvan, S., Butcher, J., Stintzi, A., and Couture, J.-F. (2018). Variation on a theme: investigating the structural repertoires used by ferric uptake regulators to control gene expression. Biometals *31*, 681-704. https://doi.org/10.1007/s10534-018-0120-8

Savage, C.R., Arnold, W.K., Gjevre-Nail, A., Koestler, B.J., Bruger, E.L., Barker, J.R., Waters, C.M., and Stevenson, B. (2015). Intracellular concentrations of *Borrelia burgdorferi* cyclic di-AMP are not changed by altered expression of the CdaA synthase. PLoS One *10*, e0125440. https://doi.org/10.1371/journal.pone.0125440

Savage, C.R., Jutras, B.L., Bestor, A., Tilly, K., Rosa, P.A., Tourand, Y., Stewart, P.E., Brissette, C.A., and Stevenson, B. (2018). *Borrelia burgdorferi* SpoVG DNA- and RNA-binding protein modulates the physiology of the Lyme disease spirochete. J Bacteriol *200*, e00033-18. https://doi.org/10.1128/JB.00033-18

Schirmer, T., and Jenal, U. (2009). Structural and mechanistic determinants of c-di-GMP signalling. Nat Rev Microbiol *7*, 724-735. https://doi.org/10.1038/nrmicro2203

Schmid, M.B. (1990). More than just "histone-like" proteins. Cell *63*, 451-453.

Schneider, E.M., and Rhodes, R.G. (2018). N-acetylmannosamine (ManNAc) supports the growth of *Borrelia burgdorferi* in the absence of N-acetylglucosamine (GlcNAc). FEMS Microbiol Lett *365*, fny243. https://doi.org/10.1093/femsle/fny243

Schwan, T.G., and Piesman, J. (2000). Temporal changes in outer surface proteins A and C of the Lyme disease-associated spirochete, *Borrelia burgdorferi*, during the chain of infection in ticks and mice. J Clin Microbiol *38*, 382-388.

Schwan, T.G., Piesman, J., Golde, W.T., Dolan, M.C., and Rosa, P.A. (1995). Induction of an outer surface protein on *Borrelia burgdorferi* during tick feeding. Proc Natl Acad Sci USA *92*, 2909-2913.

Serganov, A., and Nudler, E. (2013). A decade of riboswitches. Cell *152*, 17-24. https://doi.org/10.1016/j.cell.2012.12.024

Seshu, J., Boylan, J.A., Gherardini, F.C., and Skare, J.T. (2004a). Dissolved oxygen levels alter gene expression and antigen profiles in *Borrelia burgdorferi*. Infect Immun *72*, 1580-1586.

Seshu, J., Boylan, J.A., Hyde, J.A., Swingle, K.L., Gherardini, F.C., and Skare, J.T. (2004b). A conservative amino acid change alters the function of BosR, the redox regulator of *Borrelia burgdorferi*. Mol Microbiol *54*, 1352-1363.

Seshu, J., Esteve-Gassent, M.D., Labandeira-Rey, M., Kim, J.H., Trzeciakowski, J.P., Höök, M., and Skare, J.T. (2006). Inactivation of the fibronectin-binding adhesin gene *bbk32* significantly attenuates the infectivity potential of *Borrelia burgdorferi*. Mol Microbiol *59*, 1591-1601.

Shahbabian, K., Jamalli, A., Zig, L., and Putzer, H. (2009). RNase Y, a novel endoribonuclease, initiates riboswitch turnover in *Bacillus subtilis*. EMBO J *28*, 3523-3533. https://doi.org/10.1038/emboj.2009.283

Sherwood, A.V., and Henkin, T.M. (2016). Riboswitch-mediated gene regulation: novel RNA architectures dictate gene expression responses. Annu Rev Microbiol *70*, 361-374. https://doi.org/10.1146/annurev-micro-091014-104306

Shi, Y., Dadhwal, P., Li, X., and Liang, F.T. (2014). BosR functions as a repressor of the *ospAB* operon in *Borrelia burgdorferi*. PLoS One *9*, e109307. https://doi.org/10.1371/journal.pone.0109307

Shi, Y., Xu, Q., McShan, K., and Liang, F.T. (2008). Both decorin-binding proteins A and B are critical for the overall virulence of *Borrelia burgdorferi*. Infect Immun *76*, 1239-1246. https://doi.org/10.1128/IAI.00897-07

Shin, D.H., Yokota, H., Kim, R., and Kim, S.-H. (2002). Crystal structure of conserved hypothetical protein Aq1575 from *Aquifex aeolicus*. Proc Natl Acad Sci USA *99*, 7980-7985. https://doi.org/10.1073/pnas.132241399

Shotland, Y., Koby, S., Teff, D., Mansur, N., Oren, D.A., Tatematsu, K., Tomoyasu, T., Kessel, M., Bukau, B., Ogura, T., *et al.* (1997). Proteolysis of the phage λ CII regulatory protein by FtsH (HflB) of *Escherichia coli*. Mol Microbiol *24*, 1303-1310.

Skare, J.T., Carroll, J.A., Yang, X.F., Samuels, D.S., and Akins, D.R. (2010). Gene regulation, transcriptomics and proteomics. In *Borrelia*: Molecular Biology, Host Interaction and Pathogenesis, D.S. Samuels, and J.D. Radolf, eds. (Norfolk, UK: Caister Academic Press), pp. 67-101.

Smith, A.H., Blevins, J.S., Bachlani, G.N., Yang, X.F., and Norgard, M.V. (2007). Evidence that RpoS (σ^S) in *Borrelia burgdorferi* is controlled directly by RpoN (σ^{54}/σ^N). J Bacteriol *189*, 2139–2144.

Smith, K.D., Shanahan, C.A., Moore, E.L., Simon, A.C., and Strobel, S.A. (2011). Structural basis of differential ligand recognition by two classes of bis-(3'-5')-cyclic dimeric guanosine monophosphate-binding riboswitches. Proc Natl Acad Sci USA *108*, 7757-7762. https://doi.org/10.1073/pnas.1018857108

Snow, S., Bacon, E., Bergeron, J., Katzman, D., Wilhelm, A., Lewis, O., Syangtan, D., Calkins, A., Archambault, L., Anacker, M.L., *et al.* (2020). Transcript decay mediated by RNase III in *Borrelia burgdorferi*. Biochem Biophys Res Commun *529*, 386-391. https://doi.org/10.1016/j.bbrc.2020.05.201

Sohaskey, C.D., Zückert, W.R., and Barbour, A.G. (1999). The extended promoters for two outer membrane lipoprotein genes of *Borrelia* spp. uniquely include a T-rich region. Mol Microbiol *33*, 41-51.

Sosunov, V., Sosunova, E., Mustaev, A., Bass, I., Nikiforov, V., and Goldfarb, A. (2003). Unified two-metal mechanism of RNA synthesis and degradation by RNA polymerase. EMBO J *22*, 2234-2244. https://doi.org/10.1093/emboj/cdg193

Steinchen, W., and Bange, G. (2016). The magic dance of the alarmones (p)ppGpp. Mol Microbiol *101*, 531-544. https://doi.org/10.1111/mmi.13412

Stetter, K.O., and Zillig, W. (1974). Transcription in Lactobacillaceae: DNA-dependent RNA polymerase from *Lactobacillus curvatus*. Eur J Biochem *48*, 527-540. https://doi.org/10.1111/j.1432-1033.1974.tb03794.x

Steuten, B., Hoch, P.G., Damm, K., Schneider, S., Köhler, K., Wagner, R., and Hartmann, R.K. (2014a). Regulation of transcription by 6S RNAs: insights from the *Escherichia coli* and *Bacillus subtilis* model systems. RNA Biol *11*, 508-521. https://doi.org/10.4161/rna.28827

Steuten, B., Schneider, S., and Wagner, R. (2014b). 6S RNA: recent answers–future questions. Mol Microbiol *91*, 641-648. https://doi.org/10.1111/mmi.12484

Stevenson, B., and Seshu, J. (2018). Regulation of gene and protein expression in the Lyme disease spirochete. In Spirochete Biology: The Post Genomic Era, B. Adler, ed. (Cham, Switzerland: Springer), pp. 83-112. https://doi.org/10.1007/82_2017_49

Stewart, P., Thalken, R., Bono, J., and Rosa, P. (2001). Isolation of a circular plasmid region sufficient for autonomous replication and transformation of infectious *Borrelia burgdorferi*. Mol Microbiol *39*, 714-721.

Stewart, P.E., Chaconas, G., and Rosa, P. (2003). Conservation of plasmid maintenance functions between linear and circular plasmids in *Borrelia burgdorferi*. J Bacteriol *185*, 3202-3209.

Storz, G., Vogel, J., and Wassarman, K.M. (2011). Regulation by small RNAs in bacteria: expanding frontiers. Mol Cell *43*, 880-891. https://doi.org/10.1016/j.molcel.2011.08.022

Studholme, D.J., and Buck, M. (2000). The biology of enhancer-dependent transcriptional regulation in bacteria: insights from genome sequences. FEMS Microbiol Lett *186*, 1-9.

Sudarsan, N., Lee, E.R., Weinberg, Z., Moy, R.H., Kim, J.N., Link, K.H., and Breaker, R.R. (2008). Riboswitches in eubacteria sense the second messenger cyclic di-GMP. Science *321*, 411-3. https://doi.org/10.1126/science.1159519

Sultan, S.Z., Pitzer, J.E., Boquoi, T., Hobbs, G., Miller, M.R., and Motaleb, M.A. (2011). Analysis of the HD-GYP domain cyclic-di-GMP phosphodiesterase reveals a role in motility and enzootic life cycle of *Borrelia burgdorferi*. Infect Immun *79*, 3273-3283. https://doi.org/10.1128/IAI.05153-11

Sultan, S.Z., Pitzer, J.E., Miller, M.R., and Motaleb, M.A. (2010). Analysis of a *Borrelia burgdorferi* phosphodiesterase demonstrates a role for cyclic-di-guanosine monophosphate in motility and virulence. Mol Microbiol *77*, 128-142. https://doi.org/10.1111/j.1365-2958.2010.07191.x

Svensson, S.L., and Sharma, C.M. (2016). Small RNAs in bacterial virulence and communication. Microbiol Spectr *4*. https://doi.org/10.1128/microbiolspec.VMBF-0028-2015

Sze, C.W., and Li, C. (2011). Inactivation of *bb0184,* which encodes carbon storage regulator A, represses the infectivity of *Borrelia burgdorferi*. Infect Immun *79*, 1270-1279. https://doi.org/10.1128/IAI.00871-10

Sze, C.W., Morado, D.R., Liu, J., Charon, N.W., Xu, H., and Li, C. (2011). Carbon storage regulator A ($CsrA_{Bb}$) is a repressor of *Borrelia burgdorferi* flagellin protein FlaB. Mol Microbiol *82*, 851-864. https://doi.org/10.1111/j.1365-2958.2011.07853.x

Sze, C.W., Smith, A., Choi, Y.H., Yang, X., Pal, U., Yu, A., and Li, C. (2013). Study of the response regulator Rrp1 reveals its regulatory role in chitobiose utilization and virulence of *Borrelia burgdorferi*. Infect Immun *81*, 1775-1787. https://doi.org/10.1128/IAI.00050-13

Tilly, K., Elias, A.F., Errett, J., Fischer, E., Iyer, R., Schwartz, I., Bono, J.L., and Rosa, P. (2001). Genetics and regulation of chitobiose utilization in *Borrelia burgdorferi*. J Bacteriol *183*, 5544-5553.

Tilly, K., Fuhrman, J., Campbell, J., and Samuels, D.S. (1996). Isolation of *Borrelia burgdorferi* genes encoding homologues of DNA-binding protein HU and ribosomal protein S20. Microbiology *142*, 2471-2479. https://doi.org/10.1099/00221287-142-9-2471

Tilly, K., Grimm, D., Bueschel, D.M., Krum, J.G., and Rosa, P. (2004). Infectious cycle analysis of a *Borrelia burgdorferi* mutant defective in transport of chitobiose, a tick cuticle component. Vector Borne Zoonotic Dis *4*, 159-168. https://doi.org/10.1089/1530366041210738

Tilly, K., Krum, J.G., Bestor, A., Jewett, M.W., Grimm, D., Bueschel, D., Byram, R., Dorward, D., Vanraden, M.J., Stewart, P., *et al.* (2006). *Borrelia burgdorferi* OspC protein required exclusively in a crucial early stage of mammalian infection. Infect Immun *74*, 3554-3564.

Titgemeyer, F., Reizer, J., Reizer, A., and Saier, M.H., Jr. (1994). Evolutionary relationships between sugar kinases and transcriptional repressors in bacteria. Microbiology *140 (Pt 9)*, 2349-2354. https://doi.org/10.1099/13500872-140-9-2349

Tokarz, R., Anderton, J.M., Katona, L.I., and Benach, J.L. (2004). Combined effects of blood and temperature shift on *Borrelia burgdorferi* gene expression as determined by whole genome DNA array. Infect Immun *72*, 5419-32.

Traoré, D.A.K., El Ghazouani, A., Ilango, S., Dupuy, J., Jacquamet, L., Ferrer, J.-L., Caux-Thang, C., Duarte, V., and Latour, J.-M. (2006). Crystal structure of the apo-PerR-Zn protein from *Bacillus subtilis*. Mol Microbiol *61*, 1211-1219. https://doi.org/10.1111/j.1365-2958.2006.05313.x

Traoré, D.A.K., El Ghazouani, A., Jacquamet, L., Borel, F., Ferrer, J.-L., Lascoux, D., Ravanat, J.-L., Jaquinod, M., Blondin, G., Caux-Thang, C., *et al.* (2009). Structural and functional characterization of 2-oxo-histidine in oxidized PerR protein. Nat Chem Biol *5*, 53-59. https://doi.org/10.1038/nchembio.133

Trinquier, A., Durand, S., Braun, F., and Condon, C. (2020). Regulation of RNA processing and degradation in bacteria. Biochim Biophys Acta Gene Regul Mech *1863*, 194505. https://doi.org/10.1016/j.bbagrm.2020.194505

Trotochaud, A.E., and Wassarman, K.M. (2004). 6S RNA function enhances long-term cell survival. J Bacteriol *186*, 4978-4985. https://doi.org/10.1128/JB.186.15.4978-4985.2004

Trotochaud, A.E., and Wassarman, K.M. (2005). A highly conserved 6S RNA structure is required for regulation of transcription. Nat Struct Mol Biol *12*, 313-319. https://doi.org/10.1038/nsmb917

Trotochaud, A.E., and Wassarman, K.M. (2006). 6S RNA regulation of *pspF* transcription leads to altered cell survival at high pH. J Bacteriol *188*, 3936-3943. https://doi.org/10.1128/JB.00079-06

Troxell, B., and Yang, X.F. (2013). Metal-dependent gene regulation in the causative agent of Lyme disease. Front Cell Infect Microbiol *3*, 79. https://doi.org/10.3389/fcimb.2013.00079

Troxell, B., Ye, M., Yang, Y., Carrasco, S.E., Lou, Y., and Yang, X.F. (2013). Manganese and zinc regulate virulence determinants in *Borrelia burgdorferi*. Infect Immun *81*, 2743-2752. https://doi.org/10.1128/IAI.00507-13

Troy, E.B., Lin, T., Gao, L., Lazinski, D.W., Lundt, M., Camilli, A., Norris, S.J., and Hu, L.T. (2016). Global Tn-seq analysis of carbohydrate utilization and vertebrate infectivity of *Borrelia burgdorferi*. Mol Microbiol *101*, 1003-1023. https://doi.org/10.1111/mmi.13437

Typas, A., Becker, G., and Hengge, R. (2007). The molecular basis of selective promoter activation by the σ^S subunit of RNA polymerase. Mol Microbiol *63*, 1296-1306.

Vakulskas, C.A., Potts, A.H., Babitzke, P., Ahmer, B.M., and Romeo, T. (2015). Regulation of bacterial virulence by Csr (Rsm) systems. Microbiol Mol Biol Rev *79*, 193-224. https://doi.org/10.1128/MMBR.00052-14

Valentin-Hansen, P., Eriksen, M., and Udesen, C. (2004). The bacterial Sm-like protein Hfq: a key player in RNA transactions. Mol Microbiol *51*, 1525-1533.

Valentini, M., and Filloux, A. (2019). Multiple roles of c-di-GMP sgnaling in bacterial pathogenesis. Annu Rev Microbiol *73*, 387-406. https://doi.org/10.1146/annurev-micro-020518-115555

Van Laar, T.A., Lin, Y.-H., Miller, C.L., Karna, S.L.R., Chambers, J.P., and Seshu, J. (2012). Effect of levels of acetate on the mevalonate pathway of *Borrelia burgdorferi*. PLoS One *7*, e38171. https://doi.org/10.1371/journal.pone.0038171

von Lackum, K., and Stevenson, B. (2005). Carbohydrate utilization by the Lyme borreliosis spirochete, *Borrelia burgdorferi*. FEMS Microbiol Lett *243*, 173-179.

Wagh, D., Pothineni, V.R., Inayathullah, M., Liu, S., Kim, K.-M., and Rajadas, J. (2015). Borreliacidal activity of *Borrelia* metal transporter A (BmtA) binding small molecules by manganese transport inhibition. Drug Des Devel Ther *9*, 805-816. https://doi.org/10.2147/DDDT.S77063

Wang, H., Wang, F., Hua, X., Ma, T., Chen, J., Xu, X., Wang, L., Tian, B., and Hua, Y. (2012a). Genetic and biochemical characteristics of the histone-like protein DR0199 in *Deinococcus radiodurans*. Microbiology *158*, 936-943. https://doi.org/10.1099/mic.0.053702-0

Wang, P., Dadhwal, P., Cheng, Z., Zianni, M.R., Rikihisa, Y., Liang, F.T., and Li, X. (2013). *Borrelia burgdorferi* oxidative stress regulator BosR directly represses lipoproteins primarily expressed in the tick during mammalian infection. Mol Microbiol *89*, 1140-1153. https://doi.org/10.1111/mmi.12337

Wang, P., Lutton, A., Olesik, J., Vali, H., and Li, X. (2012b). A novel iron- and copper-binding protein in the Lyme disease spirochaete. Mol Microbiol *86*, 1441-1451. https://doi.org/10.1111/mmi.12068

Wang, P., Yu, Z., Santangelo, T.J., Olesik, J., Wang, Y., Heldwein, E., and Li, X. (2017). BosR is a novel Fur family member responsive to copper and regulating copper homeostasis in *Borrelia burgdorferi*. J Bacteriol *199*, e00276-17. https://doi.org/10.1128/JB.00276-17

Warrier, I., Hicks, L.D., Battisti, J.M., Raghavan, R., and Minnick, M.F. (2014). Identification of novel small RNAs and characterization of the 6S RNA of *Coxiella burnetii*. PLoS One *9*, e100147. https://doi.org/10.1371/journal.pone.0100147

Wassarman, K.M. (2018). 6S RNA, a global regulator of transcription. Microbiol Spectr *6*, RWR-0019-2018. https://doi.org/10.1128/microbiolspec.RWR-0019-2018

Wassarman, K.M., and Storz, G. (2000). 6S RNA regulates *E. coli* RNA polymerase activity. Cell *101*, 613-623.

Waters, L.S., and Storz, G. (2009). Regulatory RNAs in bacteria. Cell *136*, 615-628. https://doi.org/10.1016/j.cell.2009.01.043

Wigneshweraraj, S., Bose, D., Burrows, P.C., Joly, N., Schumacher, J., Rappas, M., Pape, T., Zhang, X., Stockley, P., Severinov, K., *et al.* (2008). *Modus operandi* of the bacterial RNA polymerase containing the σ^{54} promoter-specificity factor. Mol Microbiol *68*, 538-546. https://doi.org/10.1111/j.1365-2958.2008.06181.x

Wolfe, A.J., and Visick, K.L. (2008). Get the message out: cyclic-di-GMP regulates multiple levels of flagellum-based motility. J Bacteriol *190*, 463-475. https://doi.org/10.1128/JB.01418-07

Xu, H., Caimano, M.J., Lin, T., He, M., Radolf, J.D., Norris, S.J., Gheradini, F., Wolfe, A.J., and Yang, X.F. (2010). Role of acetyl-phosphate in activation of the Rrp2-RpoN-RpoS pathway in *Borrelia burgdorferi*. PLoS Pathog *6*, e1001104. https://doi.org/10.1371/journal.ppat.1001104

Xu, Q., McShan, K., and Liang, F.T. (2007). Identification of an *ospC* operator critical for immune evasion of *Borrelia burgdorferi*. Mol Microbiol *64*, 220-231. https://doi.org/10.1111/j.1365-2958.2007.05636.x

Yan, Y., Su, S., Meng, X., Ji, X., Qu, Y., Liu, Z., Wang, X., Cui, Y., Deng, Z., Zhou, D., *et al.* (2013). Determination of sRNA expressions by RNA-seq in *Yersinia pestis* grown *in vitro* and during infection. PLoS One *8*, e74495. https://doi.org/10.1371/journal.pone.0074495

Yang, X., Goldberg, M.S., Popova, T.G., Schoeler, G.B., Wikel, S.K., Hagman, K.E., and Norgard, M.V. (2000). Interdependence of environmental factors influencing reciprocal patterns of gene expression in virulent *Borrelia burgdorferi*. Mol Microbiol *37*, 1470-1479.

Yang, X.F., Alani, S.M., and Norgard, M.V. (2003). The response regulator Rrp2 is essential for the expression of major membrane lipoproteins in *Borrelia burgdorferi*. Proc Natl Acad Sci USA *100*, 11001-11006.

Yang, X.F., Lybecker, M.C., Pal, U., Alani, S.M., Blevins, J., Revel, A.T., Samuels, D.S., and Norgard, M.V. (2005). Analysis of the *ospC* regulatory element controlled by the RpoN-RpoS regulatory pathway in *Borrelia burgdorferi*. J Bacteriol *187*, 4822-4829. https://doi.org/10.1128/JB.187.14.4822-4829.2005

Yao, S., and Bechhofer, D.H. (2010). Initiation of decay of *Bacillus subtilis rpsO* mRNA by endoribonuclease RNase Y. J Bacteriol *192*, 3279-3286. https://doi.org/10.1128/JB.00230-10

Ye, M., Zhang, J.-J., Fang, X., Lawlis, G.B., Troxell, B., Zhou, Y., Gomelsky, M., Lou, Y., and Yang, X.F. (2014). DhhP, a cyclic di-AMP phosphodiesterase of *Borrelia burgdorferi*, is essential for cell growth and virulence. Infect Immun *82*, 1840-1849. https://doi.org/10.1128/IAI.00030-14

Ye, M., Zhou, Y., Lou, Y., and Yang, X.F. (2016). Genome reduction of *Borrelia burgdorferi*: two TCS signaling pathways for two distinct host habitats. Sci China Life Sci *59*, 19-21. https://doi.org/10.1007/s11427-015-4996-z

Yin, Y., Yang, Y., Xiang, X., Wang, Q., Yang, Z.-N., Blevins, J., Lou, Y., and Yang, X.F. (2016). Insight into the dual functions of bacterial enhancer-binding protein Rrp2 of *Borrelia burgdorferi*. J Bacteriol *198*, 1543-1552. https://doi.org/10.1128/JB.01010-15

Zhang, H., and Marconi, R.T. (2005). Demonstration of cotranscription and 1-methyl-3-nitroso-nitroguanidine induction of a 30-gene operon of *Borrelia burgdorferi*: evidence that the 32-kilobase circular plasmids are prophages. J Bacteriol *187*, 7985-7995. https://doi.org/10.1128/JB.187.23.7985-7995.2005

Zhang, J.-J., Chen, T., Yang, Y., Du, J., Li, H., Troxell, B., He, M., Carrasco, S.E., Gomelsky, M., and Yang, X.F. (2018). Positive and negative regulation of glycerol utilization by the c-di-GMP binding protein PlzA in *Borrelia burgdorferi*. J Bacteriol *200*, e00243-18. https://doi.org/10.1128/JB.00243-18

Zhang, Y., Chen, T., Raghunandanan, S., Xiang, X., Yang, J., Liu, Q., Edmondson, D.G., Norris, S., Yang, X.F., and Lou, Y. (2020). YebC regulates variable surface antigen VlsE expression and is required for host immune evasion in *Borrelia burgdorferi*. PLoS Pathog *16, e1008953. https://doi.org/10.1371/journal.ppat.1008953*

Zhang, Y., Lin, J., and Gao, Y. (2012). *In silico* identification of a multi-functional regulatory protein involved in Holliday junction resolution in bacteria. BMC Syst Biol *6 Suppl 1*, S20. https://doi.org/10.1186/1752-0509-6-S1-S20

Zückert, W.R., and Meyer, J. (1996). Circular and linear plasmids of Lyme disease spirochetes have extensive homology: characterization of a repeated DNA element. J Bacteriol *178*, 2287-2298.

Zückert, W.R., Meyer, J., and Barbour, A.G. (1999). Comparative analysis and immunological characterization of the *Borrelia* Bdr protein family. Infect Immun *67*, 3257-3266.

Chapter 6

Metabolism and Physiology of *Borrelia*

Frank C. Gherardini[1]*, Daniel P. Dulebohn[1], Travis J. Bourret[2] and Crystal L. Richards[1]

[1]Laboratory of Bacteriology, Gene Regulation Section, Division of Intramural Research, Rocky Mountain Laboratories, National Institute of Allergy and Infectious Diseases, National Institutes of Health, Hamilton, Montana, USA
[2]Department of Medical Microbiology and Immunology, Creighton University, Omaha, Nebraska, USA
*Corresponding author: fgherardini@nih.gov

DOI: https://doi.org/10.21775/9781913652616.06

Abstract
In this chapter, we define key biochemical pathways and metabolic systems that are the underpinning of the physiology of *Borrelia*. Energy extracted from the fermentation of a few simple sugars fuels these biochemical reactions and also energizes a V-type ATPase (V-ATPase), establishing a membrane potential that drives motility and the transport of most solutes. After transport, metabolites and biochemical intermediates, such as simple sugars, fatty acids, purines and pyrimidines, peptides and metal ions, are chemically utilized and/or modified to provide an intracellular pool of compounds necessary for protein, nucleic acid, membrane and cell wall biosynthesis. The extremely limited *de novo* biosynthetic capacity of *Borrelia* restricts members of this genus to a host-dependent lifestyle but conserves energy and reflects a reduction of the genome that is an interesting example of adaptive biology. It is fascinating how well the metabolism and physiology of *Borrelia* spp. dovetail with the physiology of their vertebrate and arthropod hosts.

Introduction

Like other bacterial pathogens with limited genomes, *Borrelia* sp. have significantly reduced their metabolic capabilities and adapted to their particular hosts and vectors. It is not surprising, therefore, that they have taken advantage of metabolites that are provided in excess in the host to eliminate 'expensive' energy requiring biochemical processes. For example, by utilizing amino acids, nucleotides and fatty acids from the host, energy required for the synthesis of these essential cellular building blocks is conserved for the synthesis of macromolecular structures (e.g., membranes, proteins, nucleic acids, flagella) and to fuel transport, motility and cell division. Naturally, this survival "strategy" imposes limitations on the bacterium. To become completely host-dependent, some bacterial pathogens, of which *B. burgdorferi* is a prime example, must totally abandon the ability to exist in a free-living form. While evolutionarily limiting, this strategy has been quite successful for a number of symbiotic, commensal, parasitic and pathogenic bacteria, although irreversibly linking them to their specific host(s).

Analysis of the genome and preliminary biochemical characterization of enzymes and pathways in *Borrelia burgdorferi* provides some insight into its biosynthetic/metabolic capacity. First, its genome lacks the genes encoding enzymes for biosynthesis of amino acids, nucleotides, fatty acids, enzymes of the TCA cycle (e.g., succinate dehydrogenase, fumarase, aconitase, etc.), a respiratory chain (e.g., cytochromes, etc.) or lipopolysaccharide (LPS) (Fraser et al., 1997). While harboring the genes encoding enzymes/proteins for DNA, RNA, and protein synthesis, as well as those for membrane assembly, cell wall biosynthesis, protein synthesis, cell wall assembly, cell division, chemotaxis and motility, import of critical building blocks becomes essential for the survival of *Borrelia* sp. (Fraser et al., 1997). Second, it indicates that *Borrelia* is a homofermenter that degrades a limited number of sugars (e.g., glucose, glucosamine) via the Embden-Meyerhof pathway to lactate as the sole electron acceptor. Also, some 3 carbon compounds (e.g., glycerol and glycerol-3-phosphate, G3P) can be used to supplement central metabolism, as well as lipid and lipoprotein biosynthetic pathways (Posey and Gherardini, 2000; Schwan et al., 2003). Third, experimental data from two laboratories indicate that *B. burgdorferi* does not utilize Fe. The Fe content of the organism was less than 10 Fe atoms per cell (Posey and Gherardini, 2000), while a more recent study measured the levels at <0.01nmoles/1e^10 cells (Aguirre et al., 2013). In another study, Wang et al. detected levels of Fe ($2x10^5$ atoms/cell), Cu ($4x10^5$ atoms/cell), Ni ($5x10^3$ atoms/cell), and Co ($2x10^2$ atoms/cell), indicting much higher levels of these metals than had been detected in the previous studies (Wang et al., 2012). Since the bacterium lacks known Fe, Ni, Cu, or Co containing metalloproteins, that are at the core of many biochemical synthetic pathways, it is hard to determine the biological relevance of these metals in *B. burgdorferi*. This has had a dramatic effect on the intracellular targets of reactive oxygen species since they don't sustain DNA damage as a result of Fenton chemistry (Boylan and Gherardini, 2008).

In the following sections, we will attempt to outline the genes/proteins which impact biochemical pathways that play a role in energy utilization, transport of precursor molecules/compounds (fatty acids, amino acids, nucleotides etc.) and the modification of those precursors to yield essential components for cell growth, membrane/lipoprotein assembly, nucleic acid synthesis and protein synthesis. Molecular architecture and the assembly of cellular macromolecular fraction in *Borrelia* (proteins, membranes assembly, DNA and RNAs) are very similar to those described in other bacteria and will not be addressed in these sections.

Transport

Very little is known about the mechanisms by which solutes are transported across the cell envelopes of spirochetes. Like other bacteria, however, membrane transport systems presumably provide the necessary exchange of sufficient amounts of essential metabolites from the environment to the appropriate intracellular location for biochemical processes. This is particularly important in *B. burgdorferi* since it has such limited biosynthetic capacity. In fact, ~100 of the ~850 genes located on the chromosome encode proteins potentially involved in transport of required biosynthetic intermediates and ions (Fraser et al., 1997). To date, a limited number of these proteins have been specifically characterized biochemically. However, by comparing genomic, genetic, and uptake/incorporation data combined with biochemical analyses of specific pathways, one can speculate as to the potential role for some of the putative transporters or transport systems. As expected, the uptake of certain ions or solutes indicates that transport systems are active in *B. burgdorferi* and some genes/proteins have been yet identified to explain their accumulation within the cell. For example, Mn has been shown to be actively transported in *B. burgdorferi* in an energy dependent fashion and a Mn transport system has been identified (Posey and Gherardini, 2000; Ouyang et al., 2009a).

The first step in uptake of essential molecules in *B. burgdorferi* is transport of these moieties across the outer membrane. In Gram-negative bacteria, three different mechanisms are used to pass various substrates into the periplasm: (i) nonspecific porins where small molecular weight substrates (<650 Da) diffuse through general porins such as OmpF or OmpC (Nakae, 1976; Heller and Wilson, 1981), (ii) specific or specialized porins where larger substrates, which exceed the size limits of the general porins, are transported into the cell. Examples of these types of transporters are LamB in *E. coli* which is essential for the transport of maltodextrins made up of more than three glucose subunits (Wandersman et al., 1979) and ScrY, the sucrose porin from *Klebsiella pneumoniae* (Hardesty et al., 1991). These porins do not bind substrates tightly but rather facilitate transport, and (iii) high affinity receptors where substrates, which are excluded by size, charge, or availability, are specifically transported across the outer membrane. These receptors bind specific substrates with high affinity and the subsequent release/translocation into the periplasm requires the expenditure of energy. In Gram-negative bacteria, electrochemical energy (Proton Motive Force, PMF) transduction from the cytoplasmic membrane to the receptor is achieved by three inner membrane proteins: ExbB, ExbD, and TonB (Eick-Helmerich et al., 1987; Eick-Helmerich and Braun, 1989; Fischer et al., 1989). In *E. coli*, the transport of vitamin B_{12} and Fe-siderophore complexes across the outer membrane is mediated by this energy-dependent mechanism. In *B. burgdorferi*, the absence of homologs for TonB, ExbB or ExbD, suggests that this mechanism does not play a role in transport of substrates across its outer membrane (Fraser et al., 1997).

Bacteria typically utilize a variety of classes of transport systems: i) ion channels which function as highly selective pores that open and close in response to specific stimuli allowing ions and solutes to rapidly flow down a gradient (Ferguson and Deisenhofer, 2004) and ii) electrochemical transport systems, specifically uniporters, symporters and antiporters, that couple electrochemical proton gradients to move across the cytoplasmic membrane (Jap and Walian, 1996). Electrical gradient potential cannot be maintained across outer membranes; therefore, active transport depends upon the ability of cells to generate electrochemical gradients across cytoplasmic membranes (Bradbeer, 1993; Ferguson and Deisenhofer, 2004). In *B. burgdorferi*, larger solutes, such as fatty acids and nucleotides, appear to follow the similar path. (iii) Finally, active transporters, including phosphoenolpyruvate sugar phosphotransferase systems (PEP:PTS), the ATP-binding cassette (ABC) transporters and the V-type ATPases, pump solutes across the inner membrane against concentration gradients. These general mechanisms define the transport of metabolites in *B. burgdorferi*.

Porins

To date, three putative outer membrane porins have been identified in the outer leaflet of *B. burgdorferi*. Initially, the approach taken was to isolate/purify the outer membranes, assay the purity of the membrane fraction using inner membrane markers and measure porin activity using a planar lipid bi-layer assay to demonstrate porin activity (Skare et al., 1997). Using this methodology, Skare et al. were able to show that the purified outer membranes of *B. burgdorferi* contained two putative porin activities (16). Another putative porin, Oms28 (encoded by *bba74*), is a 28-kDa protein that was originally characterized as an outer membrane porin with a single channel conductance (SSC) of 0.6 nano-Siemen (nS) suggesting a relatively small channel (0.64 nm) which could accommodate small molecular weight molecules and ions (Skare et al., 1997). Recently, however, BBA74 has been demonstrated as a soluble periplasmic protein (Mulay et al., 2007). These conflicting data leave the function of BBA74 to be determined. It is interesting to note that the gene encoding Oms28/BBA74 resides on plasmid lp54 which has been shown to be lost in certain high-passage, avirulent strains (Purser and Norris, 2000). Since these isolates have no apparent growth defects, it appears that Oms66 or other unidentified porins are able to compensate for the loss of Oms28 during in vitro growth. A second porin, designated Oms66 (P66, encoded by *bb0603*), is a 66-kDa outer membrane protein also with single channel conductance measuring 12.6 nS (Skare et al., 1997). This would represent a large channel which could accommodate much larger solutes than Oms28.

P66 is an interesting case. As mentioned in other sections of this chapter, larger molecular weight solutes such as peptides, long chain fatty acids, and deoxynucleotides have no obvious outer membrane transport systems (e.g., fatty acid facilitator, FadL) and are proposed to diffuse across the outer membrane of *B. burgdorferi*. Because of the relative sizes of channels formed by Oms28 and P66, it seems likely that P66 may play an important role in the passive transport of some of these solutes. P66 assembles into a homo-oligomeric complex with an apparent molecular weight of ~460 kDa (Kenedy et al., 2014). Additionally, using a phage display library of *B. burgdorferi,* Coburn and Leong were able to identify P66 as an outer membrane adhesion that bound integrin $\alpha_3\beta_3$ (Coburn et al., 1999). Subsequently, Coburn and Cugini disrupted *bb0603* and were able to show that the loss of P66 abolished integrin binding activity (Coburn and Cugini, 2003). Importantly, further studies showed that P66 integrin binding activity is essential for endothelial transmigration in vitro and in vivo and ΔP66 mutants lost infectivity in mice and ticks (LaFrance et al., 2011; Ristow et al., 2015; Curtis et al., 2018). Therefore, the P66 porin activity is not essential for *in vitro* growth while integrin binding activities (plus other unidentified functions) are essential for infectivity. This same dual functionality was demonstrated for the 64-kDa and 53-kDa (Msp) proteins from *T. denticola.* Both proteins have porin activity while the 64-kDa protein is also involved in binding of *T. denticola* cells to gingival fibroblasts or in the case of the Msp, in binding to various extracellular matrix proteins (Fenno et al., 1996; Fenno et al., 1997). It is clear that a great deal remains to be determined about transport across the outer membrane of *B. burgdorferi.*

A similar approach was used to identify P13, encoded by *bb0034*, as a potential outer membrane porin (Noppa et al., 2001; Pinne et al., 2004). Protease treatment of mutant strain B313 (ΔOspA-D, ΔDbpA-B), confirmed the surface localization of P13. Interestingly P13 is both N- and C-terminally processed and forms a complex of ~300 kDa in the outer membrane. SSC analysis measured a voltage independent, cannel forming activity of 0.6 nS that is selective for cations and anions (Bárcena-Uribarri et al., 2014). Because of this ion selectivity, P13 was thought to be involved in resistance to osmotic stress, however, analyses of a ΔP13 mutant did not show an increased sensitivity osmotic stress induced by 7% glycerol, while this treatment reduced the growth rate of this mutant. In fact, *bb0034* and *bb0603* are down-regulated when wild-type strain B31-A cells are exposed to this stress condition, suggesting that these porins are not involved in the osmotic stress response in *B. burgdorferi*. Conversely, the expression of *bba01*, which encodes a paralog of P13, is induced under these conditions. An in-depth study of the osmotic stress response in strain B31-A3 showed that ion uptake (e.g., KtrA-B), ion antiporters (e.g., BB0447, NhaC-1 and NhaC-2), and the transport of osmoprotectant molecules, such as glutamate, choline, proline, and glycine betaine across the inner membrane, are more relevant to resistance to osmotic stress than outer membrane porins, P13 and P66.

Inner membrane-associated transport systems

Transport across the inner membrane of bacteria relies on electrochemical transport systems and ATP-driven import or export systems. *B. burgdorferi* apparently uses similar mechanisms to import/export substrates across its inner membrane and a list of these putative transporters is shown in Table 1. The first major category of inner membrane transporters, the electrochemical transporters, can be divided into subgroups that include uniporters, symporters, antiporters and resistance-nodulation-division (RND) transporters that use electrochemical proton gradients to move substrates across the cytoplasmic membrane. Uniporters transport one solute and are driven directly by the substrate gradient (e.g., BB0724 and BB0725, putative K^+ transporters; BB0380, putative Mg^{2++} transporter; and BB0219, Mn transport (Fraser et al., 1997; Ouyang et al., 2009a). Symporters that pump two or more types of solutes in the same direction, are driven by the electrochemical gradient of one solute (e.g., BB0729, BB0401 and BB0814, putative Na^+/dicarboxylate symporters (Fraser et al., 1997) while antiporters, driven by the single solute gradient, pump substrates in the opposite direction (e.g., BB0447, BB0637 and BB0638, putative Na^+/H^+ antiporters; BB0164, putative Na^+/Ca^+ transporter, BB0616, putative cation efflux pump; BB0843, putative arginine/ornithine antiporter) (Fraser et al., 1997). Jain et al. characterized two key purine uniport transporters, BBB22 and BBB23, encoded on an essential plasmid (cp26) that are required for the transport of hypoxanthine, adenine and guanine (Jain et al., 2012). These nucleotides are channeled into a novel purine salvage pathway (Lawrence et al., 2009). Additionally, mutant strains lacking *bbb22* and *bbb23,* are noninfectious in mice (Jewett et al., 2009). These transporters should derive the energy required for transport from the inner membrane proton motive force (PMF) established by the hydrolysis of ATP by the V_1/V_0 ATPase. These types of transport systems are essential for the physiology, metabolism and virulence in *B. burgdorferi* and have been shown to be critical for resistance to acid and osmotic stresses (Bontemps-Gallo et al., 2016; Dulebohn et al., 2017).

RND transporters are specialized uniporters that, in Gram-negative bacteria, export toxic substances across both membranes driven by an electrochemical gradient, not by ATP hydrolysis (Li and Nikaido, 2004). The best characterized of these is the AcrAB-TolC drug efflux system in *E. coli* (Nikaido and Zgurskaya, 2001). AcrA and AcrB form an inner membrane complex that is made up of 3 monomers of each protein. These bind specific substrates and couple export with an inner membrane proton gradient. The TolC component, a membrane fusion protein, is linked to the AcrA-B efflux pump, forming a complex that spans the periplasmic space and the outer membrane. Because of this structural bridge between the inner and outer membranes, substrates can be translocated from the cytoplasm directly to the external medium. Bunikis et al. identified an RND transport system in *B. burgdorferi* designated BesA (BB0141), BesB (BB0140) and BesC (BB0142; TolC homolog) (Bunikis et al., 2008). They showed that the Bes protein complex was involved in resistance of *B. burgdorferi* to various antibiotics and was potentially driven by an electrochemical gradient. In addition, disruption of the Bes operon yielded mutants that were avirulent (Bunikis et al., 2008). It seems very clear that electrochemical transport systems described above are very important to physiology, internal biochemistry and survival of *B. burgdorferi*.

Bacteria use group translocation of carbohydrates mediated by the phosphoenolpyruvate phospho-transferase system (PEP:PTS) to transport carbohydrates across the cytoplasmic membrane. These systems consist of two cytoplasmic proteins, Enzyme I (EI) and HPr, and a variety of membrane-associated transport complexes, designated as Enzyme II, that are specific for certain sugars (EII^S) (Table 1). These proteins function in concert to transport and phosphorylate incoming sugars for central metabolism. Specifically, EI transfers a phosphoryl group from PEP to HPr, which subsequently transfers a phosphoryl moiety to the various EII transport complexes. EII complexes are made up of three functional units, EIIA, EIIB and

Table 1. Putative transporter systems in *Borrelia burgdorferi*.

Enzyme Name	Gene Designation	Open Reading Frame
ABC transporter	—	*bb0753, bb0754, bb0742, bb0466*
ABC transporter	—	*bb0573, bb0574, bbj26, bbj27*
Oligopeptide transport system	*oppB, oppC, oppD, oppF*	*bb0332, bb0333, bb0334, bb0335*
Oligopeptide binding protein	*oppA-1, -2, -3, -4, -5*	*bb0328, bb0329, bb0330, bbb16, bba34*
Ribose/galactose ABC transporter system	*mglA, mglC1,mglC2*	*bb0677, bb0678, bb0679*
Ribose/galactose ABC transporter system	*bmpB, bmpA, bmpC, bmpD*	*bb0382, bb0383, bb0384, bb0385*
Nucleoside transporter system	—	*bb0316, bb0317, bb0318*
Spermidine/putrescine transport system	*potC, potB, potA, potD*	*bb0639, bb0640, bb0641, bb0642*
Mn uptake transporter	*bmtA*	*bb0219*
Phosphate transport system	—, *pstC, pstA, pstB*	*bb0215, bb0216, bb0217, bb0218*
Glycine betaine transport system	—	*bb0144, bb0145, bb0146*
Lipoprotein releasing factor	—	*bb0078, bb0079, bb0080, bb0081*
Phosphate transporter	*phoU*	*bb0042*
Mg transporter	*mgtE*	*bb0380*
Chromate efflux	—	*bb0451, bb0452*
Guanine-hypoxanthine-adenine permease	*pbuG*	*bbb22, bbb23*
L-lactate permease	—	*bb0604*
Potassium transport system	*ktrB, ktrA*	*bb0724, bb0725*
Glycerol uptake facilitator	*glpF*	*bb0240*
Cysteine transport	*tcyP*	*bb0729*
Sodium:dicarboxylate symporter	—	*bb0401*
Sodium:pantothenate symporter	—	*bb0814*
PTS system, phosphoenolpyruvate, Enzyme I	—	*bb0558*
PTS system, phosphocarrier protein, HPr	—	*bb0557*
PTS system, phosphocarrier protein, HPr	—	*bb0448*
PTS system, fructose-specific Enzyme IIABC	—	*bb0629*
PTS system, fructose-specific Enzyme IIABC	—	*bb0408*
PTS system, glucose-specific Enzyme IIABC	*malX2*	*bbb29*
PTS system, glucose-specific Enzyme IIABC	*malX1*	*bb0116*
PTS system, glucose-specific Enzyme IIAB	*ptsG*	*bb0645*
PTS system, glucose-specific Enzyme IIA	*crr*	*bb0559*
PTS system, chitobiose-specific Enzyme IIC, AB	*chbC, chbA, chbB*	*bbb04, bbb05, bbb06*
Multidrug resistance protein B	—	*bbq69, bbi26*
Na+ driven multidrug efflux pump	—	*bb0583, bb0584*
Na+ driven multidrug efflux pump	—	*bb0473*
Virulence factor MviN	—	*bb0810*
Met, Try or Lys transport	*nhaC*	*bb0638, bb0637*
Arginine/ornithine antiporter	*argD*	*bb0843*
Mechanosensitive ion channel	—	*bb0453*
Sodium-calcium exchanger	—	*bb0164*
Cation efflux	—	*bb0616*
Na+/H+ antiporter	—	*bb0447*
Methicillin resistance factor	—	*bb0586*
Bacitracin resistance factor	—	*bb0258*
60 kDa inner membrane protein, YidC homolog	—	*bb0442*
Borrelia efflux system	*besB, besA, besC*	*bb0140, bb0141, bb0142*
Antibiotic efflux	—	*bb0252*
Protein export (SecDF)	—	*bb0652, bb0653*
V-type ATPase	subunits K, I, D, B, A, E, X, Y	*bb0090, bb0091, bb0092, bb0093, bb0094, bb0096*
Outer membrane porin (oms66)	—	*bb0603*

EIIC, which occur as three separate protein subunits or as domains within the same polypeptide. EIIA and EIIB proteins/domains receive the phosphoryl group from HPr and transfer it to the incoming sugar, while EIIC, an integral inner membrane protein, functions as the sugar binding protein/domain. The genome of *B. burgdorferi* contains only a single EI homolog, encoded by *bb0558* and two HPr homologs, encoded by *bb0557* and *bb0448* (Fraser et al., 1997). Also, there are homologs for six EIIABC complexes that transport and phosphorylate specific sugars for utilization in central metabolism and RNA synthesis. These include four EIIABC transporters that exist as single polypeptides: BB0116, a putative glucose transporter, BB0408 and BB0629, putative mannose/galactose transporters, and BBB29, a putative glucose transporter. A fifth EIIABC sugar transport system is made up of two proteins encoded by *bb0559* (EIIA) and *bb0645* (EIIBC) where EIIA exists as a single protein and, EIIB and EIIC exist as domains within the same polypeptide (Fraser et al., 1997). This PTS system is thought to be involved in glucose transport. The sixth, encoded by *bbb04* (EIIC), *bbb05* (EIIA) and *bbb06* (EIIB) is composed of three distinct proteins and has been implicated in chitobiose uptake and utilization (Tilly et al., 2004). Clearly these PEP:PTS transport systems are important for energy metabolism in *B. burgdorferi* (Khajanchi et al., 2015).

ABC transporters are specific systems that transport substrates against a concentration gradient driven by the hydrolysis of ATP. These can function as importers (ABC_{IM}) or exporters (ABC_{EX}), and have been implicated in the virulence of some bacterial pathogens. ABC_{IM} mediate the transport of nutrients including oligosaccharides, amino acids, peptides, Fe-siderophores and polyamines (Davidson et al., 2008). ABC_{EX} function as secretion systems for a variety of molecules including lipids, polysaccharides and protein toxins, such as hemolysins (Davidson et al., 2008). Despite the difference in substrate specificity and the direction of transport, ABC_{IM} and ABC_{EX} are structurally organized in a similar fashion. Generally, most ABC_{IM} are made up of four proteins that form a complex composed of hydrophobic, integral membrane protein(s) (homodimers or heterodimers) and two hydrophilic monomers that are associated with the cytoplasmic side of the membrane complex. The final components of ABC_{IM} are periplasmic binding proteins (PBP) that deliver substrates to the periplasmic side of the cytoplasmic inner membrane complex. The completed complex then mediates ATP hydrolysis-dependent import of the substrate to the cytoplasm. Conversely, in most ABC_{EX} systems, the protein subunits of the complex are fused into a single polypeptide, and export of substrates across the outer membrane is coupled to TolC (Zgurskaya and Nikaido, 2000; Poole, 2001; Murakami and Yamaguchi, 2003).

Several ABC_{IM} systems have been identified in the genome of *B. burgdorferi* (see Table 1). The best characterized of these is the oligopeptide transport system, designated as the oligopeptide permease (Opp) operon, which is similar to the Opp and dipeptide permease (Dpp) systems described in *E. coli* and *Salmonella enterica* (Abouhamad et al., 1991). The homologous Opp system in *E. coli* is composed of a periplasmic binding protein OppA, two inner membrane transporters, OppB and OppC, and two ATP binding proteins, OppD and OppF, which hydrolyze ATP to drive peptide transport. The homologous component proteins of the Opp system in *B. burgdorferi* are encoded by *bb0332 (oppB), bb0333 (oppC), bb0334 (oppD), bb0335 (oppF)* and five periplasmic binding proteins encoded by *bb0328 (oppA-1), bb0329 (oppA-2), bb0330 (oppA-3), bbb16 (oppA-4*, encoded on plasmid cp26*)* and *bba34 (oppA-5*, encoded on plasmid lp54) (Fraser et al., 1997). It should be noted that two additional Opp homologs have been identified in the chromosome of *B. burgdorferi,* encoded by *bb0746 (oppC)* and *bb0747(oppB)* but their role in peptide transport remains to be determined (Fraser et al., 1997). Lin et al. demonstrated that *B. burgdorferi* OppA-1, -2, -4 or -5 with *E. coli* OppB, OppC, OppD and OppF encoded on recombinant plasmids were able to functionally complement an *E. coli oppABCDF* deletion mutant (Lin et al., 2001). Further, Wang et al., using a combinatorial phage display heptapeptide library and competitive binding assays, were able to show that OppA-1 and -2 have a preference for tripeptides over dipeptides or longer peptides (Wang et al., 2004). In these assays, OppA-1 had broad substrate specificity, including peptides with bulky hydrophobic residues, while OppA-2 and -3 preferred peptides composed of smaller, nonpolar amino acids. Studies of the regulation of the *B. burgdorferi* OppA proteins suggest that these differences in substrate specificity may be important at different stages of the infective cycle. For example, OppA-4 and OppA-5 are up regulated in a mouse while OppA-1 is constitutively expressed in vivo or in vitro (Wang et al., 2002). Medrano et al. showed that *oppA-4* was potentially regulated by BosR while they and others have shown that *oppA-5* is regulated by RpoN and RpoS, and is a candidate transcription factor, EbfC (Caimano et al., 2007; Medrano et al., 2007). Importantly, Groshong et al. showed that the oligopeptide binding proteins are differentially regulated in fed ticks and DMCs while the permeases are constitutively expressed under similar conditions (Groshong et al., 2017). Since it cannot synthesize or transport most amino acids (except possibly arginine, glutamate and cysteine), Opp-mediated peptide transport is critical for amino acid metabolism, in vitro survival, and infectivity of *B. burgdorferi* (Groshong et al., 2017).

A second ABC transport system has been characterized in *B. burgdorferi*. Bontemps-Gallo et al. characterized an essential, spermidine specific ABC transport system, encoded by *bb0639-bb0642* (Bontemps-Gallo et al., 2018). In other bacteria, polyamines (*e.g.* putrescine, spermidine, spermine) are involved in the efficiency polymerase function as well as the defenses and adaptations to ROS and osmotic stresses (Munro et al., 1972; Cohen, 1998; Chattopadhyay et al., 2003; Shah and Swiatlo, 2008; Altendorf et al., 2009). Polyamine content in bacterial cells is regulated by both biosynthesis and transport (Igarashi and Kashiwagi, 1999). In *E. coli*, the most commonly used bacterial model for polyamine research, the intracellular concentration of putrescine (22.1 nmol/10^9 cells) is higher than spermidine (3.8 nmol/10^9 cells) (Morris and Jorstad, 1970; Jorstad and Morris, 1974; Fillingame et al., 1975). Uptake experiments in *E. coli* demonstrate that spermidine/putrescine transport (via PotABCD) is a spermidine-preferential uptake system (Kashiwagi et al., 1993; Kashiwagi et al., 1995; Igarashi and Kashiwagi, 1999; Terui et al., 2013) that also transports putrescine at low affinity. Using an inducible LacI expression system, Bontemps-Gallo et al. were able to show that PotABCD in *B. burgdorferi* is required for cell viability, regulated by RpoS at low osmolarity (< 300 mOsM) and is spermidine specific (Bontemps-Gallo et al., 2018). In addition, Boyle et al. showed that spermidine is required for optimal activity of DNA-dependent RNA polymerase using an in vitro transcription system (Boyle et al., 2020). Clearly, the PotABCD and oligopeptide ABC transport systems are critical for *B. burgdorferi* physiology and viability.

V-type ATPase

Vacuole-type ATPases (V-ATPases) are found in both eukaryotic and prokaryotic cells. In eukaryotes, they are complex structures with at least 13 subunits that function as proton pumps driven by energy derived from the hydrolysis of ATP (Lolkema et al., 2003; Yokoyama and Imamura, 2005; Qi et al., 2007; Murata et al., 2008; Minagawa et al., 2013). To date, genome comparisons have identified some 37 V-ATPases in 16 bacteria and 21 archaea and more are likely to be found as more genomes are completed. In archaea, these ATPases are common and function in ATP synthesis. In bacteria, V-type ATPases are rare while the vast majority harbor F-type ATPases which function as ATP synthetases (Grüber et al., 2014). V-Type ATPases are found almost exclusively in low-GC Gram-positive bacteria but some have been identified in members of the phyla Fusobacteria, Chlamydiae and Spirochaetae, but not in Protobacteria. For example, the structure of a V-type ATPase from *Enterococcus hirae* has been determined experimentally and, based upon the subunit homology, it seems very similar to those in chlamydial and spirochetal species in function and structure. Detailed analyses of these bacteria, based upon the organization of the genes in the V-ATPases operons, protein homology and the protein subunits contain therein, show that spirochetal V-ATPases are most closely related to those found in Chlamydial species. [See review: (Lolkema et al., 2003)]. It is interesting to note that *T. pallidum* and *Clostridium tetani* are the only bacteria that contain two functionally distinct V-ATPases with either proton pumping or putative synthetase activity in their genomes (Deka et al., 2015).

V-type ATPases from prokaryotes are less complex than their eukaryotic counterparts and contain 8 to 9 subunits arranged in two heterologous domains designated as V_1 and V_0. V_1 is a hydrophilic, cytosolic domain that is responsible for ATP hydrolysis, while the V_0 is a membrane embedded ion-translocating domain. In *E. hirae*, the V-ATPase which contains 9 subunits, has been characterized biochemically and the proposed structure is shown in Figure 1, panel B. The V_1 is a 500-kDa protein complex made up of subunits A, B, C, D, E, F, and G which function to hydrolyze ATP and transfer that energy to V_0. Subunits A and B bind ATP, with subunit A functioning as the enzymatic site for hydrolysis. Three monomers each of A and B are arranged around subunits D, G and C which form the central stem. The hydrolysis of ATP by subunit A causes a rotational shift in the stem complex that "activates" the V_0 complex to pump ions/protons. The E and F subunits are thought to form an auxiliary stem that might function to stabilize the central stem, but their exact function has not been demonstrated experimentally. The final subunit in the central stem is subunit C which is considered essential while its exact function is not known. The V_0 is composed of subunits K and I with a stoichiometry of 10 to 1, respectively. Each of the K subunits bind a Na^+ ion and as the central stem rotates, it drives the rotation of the K ring bringing those bound ions into contact with the ion channels in protein I. The bound Na^+ is "released" into that channel for translocation into the periplasm (Kakinuma et al., 1999; Lolkema et al., 2003).

B. burgdorferi harbors the genes encoding a V-type ATPase like those observed in low-GC Gram-positive

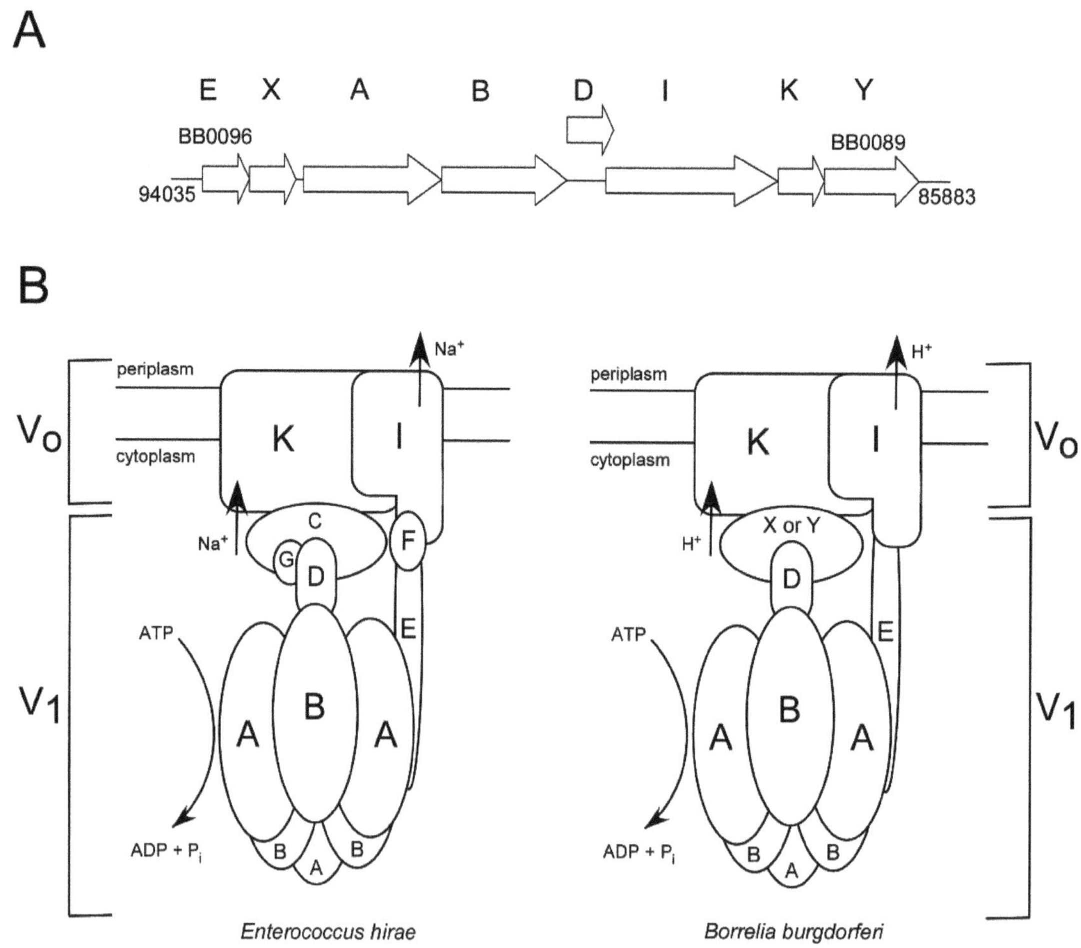

Figure 1. Panel A. Gene organization of the V-ATPase operon from *B. burgdorferi*. Panel B. Model for the proposed structure of the V-ATPase transporter in *Enterococcus hirae* compared to the model predicted for *B. burgdorferi*. Figure modified with permission from Gherardini et al., 2010.

bacteria, Chlamydial species and other Spirochetes (Fraser et al., 1997). The genes are organized in an operon with the gene order E, X, A, B, D, I, K and Y (Figure 1, panel A). Based upon the well characterized V-type ATPase from *E. hirae,* we propose a model for the structure of the V-type ATPase from *B. burgdorferi* (Figure1, panel B). In the V_1 domain, the A and B subunits are proposed to also function in ATP binding and hydrolysis, while D and E function as major and minor components of the central and auxiliary stems respectively. Conspicuously absent is a gene coding a C subunit which is thought to perform an essential function in the interaction between the central stem and V_0 domain. Also absent are homologs for the G and F subunits (Fraser et al., 1997). Lolkema et al. observed that these subunits where missing from some genomes (Spirochetes and Chlamydia species) and proposed that other, unidentified genes in their V-ATPase operons might perform the enzymatic functions of C, F, and/or G (see Figure 1, panels A and B, designated X and Y) (Lolkema et al., 2003). Homologs of the I and K proteins are contained in the V-ATPase operon of *B. burgdorferi* and presumably perform functions like those proposed for the V_0 domain *E. hirae*. Murata et al. recently showed that binding of Na^+ to the purified K subunit of the V-type ATPase from *E. hirae* was competitively inhibited by Li^+ or H^+, suggesting that those ions could also be transported by this complex (Murata et al., 2008). Based upon these observations and experimental data (Dulebohn et al., 2017), we propose that the V-type ATPase in *B. burgdorferi* functions primarily as a H^+ pump that establishes a proton motive force across the inner membrane thereby driving some proton coupled transport systems (e.g., Na^+, BB0637,

BB0638 and BB0447) and, most importantly, *B. burgdorferi* motility. It is important to note that some V-type-ATPases function as proton transport systems that drive ATP synthesis. Because of the similarities to the V-ATPase from *E. hirae,* this seems unlikely in *B. burgdorferi.*

Recently, Dulebohn et al. demonstrated that the V-ATPase from *B. burgdorferi* is essential during acid stress to pump protons from the cytosol when cells are exposed to high levels of weak organic acids (Dulebohn et al., 2017). Membrane-permeable weak monocarboxylic organic acids, like acetate and benzoate, have been shown to trigger acid/proton stress in some bacteria and *B. burgdorferi* (Kirkpatrick et al., 2001). This stress occurs when excess, free protons accumulate in the cytoplasm (Castanie-Cornet et al., 1999; Castanié-Cornet et al., 2010) (Castanie-Cornet et al., 1999; Foster, 1999). Protonated weak acids can diffuse across the cell membrane, dissociate in the cytoplasm and increase the concentration of intracellular protons. This generates differences in the pH between the external environment (pHe) and the cytoplasm (pH_i), acidifying the cytosol, decreasing the activity of pH-sensitive enzymes, and disrupting the proton motive force (PMF) (Bearson et al., 1997; Foster, 1999; Richard and Foster, 2004; Kanjee and Houry, 2013). To restore pH homeostasis, many bacteria use dissimilar processes to reestablish a favorable pH_i including: (i) coupling proton exchange with cations or anions (antiporters); (ii) generating or transporting neutralizing buffering molecules including arginine, lysine and/or glutamine, polyamines, poly-phosphates, and inorganic phosphates; (iii) utilizing amino acid decarboxylases and/or deaminases; (iv) synthesizing acid resistant fatty acids and lipids; and (v) reversing proton import via an F1/Fo ATPase by hydrolyzing ATP to pump protons out of the cytosol to restore intracellular pH_i. (Bearson et al., 1997; Foster, 1999; Slonczewski et al., 2009). Most of these mechanisms/systems are not present in *B. burgdorferi* (Fraser et al., 1997). Dulebohn et al. showed that Lyme disease spirochetes re-establish a neutral pH_i using BB0843 (arginine/ornithine antiporter), BB0447, BB0637 and BB0638 (Na^+/H^+ antiporters) and BB0090-BB0096 (Vo/V1 vacuolar-type ATPase) (Dulebohn et al., 2017). Clearly, the V-ATPase from *B. burgdorferi* is critical for establishing and maintaining a stable PMF for cell viability.

Central metabolism

B. burgdorferi is an obligate bacterial pathogen with no known ability to survive outside of either an arthropod vector or mammalian host. This apparent inability to survive outside of a host implies that *B. burgdorferi* is restricted in terms of its metabolic capabilities and is consistent with its fastidious nature. On the other hand, the ability to survive in such disparate hosts as ticks and warm-blooded mammals implies that *B. burgdorferi* is adept at incorporating metabolites found in the vectors and hosts that they infect. The genome sequence of *B. burgdorferi* indicates that the primary source for energy production is by homolactic fermentation via the Embden-Meyerhof pathway (Fraser et al., 1997; Das et al., 2000; von Lackum and Stevenson, 2005). No homologs of TCA cycle enzymes (e.g., succinate dehydrogenase, fumarase, aconitase etc.), respiratory chain enzymes (e.g., cytochromes etc.), or enzymes involved in oxidation phosphorylation are encoded in the *B. burgdorferi* genome (Fraser et al., 1997; Das et al., 2000; von Lackum and Stevenson, 2005). Despite a paucity of ATP-generating and biosynthetic pathways, *B. burgdorferi* thrives and persists in the arthropod vectors and reservoir hosts, suggesting a very successful evolutionary, adaptive metabolism. This section will address the fundamental mechanisms and pathways *B. burgdorferi* utilizes to generate sufficient energy to survive in the disparate environments of the arthropod vector and reservoir host.

Carbon utilization

One mechanism used by *B. burgdorferi* that aids in ATP generation in glycolysis is its apparent ability to transport several different six-carbon sugars via phosphotransferase systems (PTS). The PEP:PTS system in *B. burgdorferi* transports substrates ultimately directed into glycolysis for the production of ATP (see PEP:PTS systems above in "Transport" section). The PEP:PTS systems of *B. burgdorferi* are hypothesized to transport glucose, N-acetyl-glucosamine (GlcNAc), chitobiose, mannose, and possibly other sugars and disaccharides that can be shuttled into glycolysis directly or following conversion into fructose-6-P or fructose 1,6-bisphosphate (Figure 1) (Fraser et al., 1997; Saier and Paulsen, 2000; von Lackum and Stevenson, 2005). Analysis of the genome sequence of *B. burgdorferi* suggests that additional carbon sources including fructose, galactose, and ribose may be transported to support growth. However, growth was restricted on these sugars while glucose, maltose, N-acetylglucosamine, N-acetylmannosamine, chitobiose, mannose, trehalose and glycerol support growth *in vitro* (von Lackum and Stevenson, 2005; Hoon-Hanks et al., 2012). Glucose is thought to be the preferred carbon source for *B. burgdorferi* since it can be fed directly into glycolysis (Figure 2). The phosphorylation of glucose takes place at the inner membrane during transport by the PEP:PTS system; glucose-6-P then is isomerized to fructose-6-P by glucose-6-P isomerase (BB0730). *B. burgdorferi* encodes two phosphofructokinases (BB0020/BB0727), and the phosphofructokinase (BB0020) is pyrophosphate-dependent (Deng et al., 1999), generating fructose-1,6-bisP while conserving ATP. The importance of glucose transport is clear since strains lacking the EIIBC component PtsG (encoded by BB0645), hypothesized to be involved in glucose transport, aberrantly regulated key virulence factors and were unable to survive in mice by needle inoculation or by tick transmission (Khajanchi et al., 2015). However, spirochetes lacking PtsG were able to survive long-term in ticks, suggesting an important role for this transport system during transmission and mouse infection. Additionally, a TnSeq study has identified several factors that contribute to the bacterial fitness of *B. burgdorferi* when grown in the presence of mannose, maltose, trehalose, and GlcNAc (Troy et al., 2016). Although the enzymatic function of these factors has not been confirmed, carbohydrate utilization and energy metabolism are critical for *B. burgdorferi* to survive in their complex infectious cycle when limited to glycolysis as the major source of ATP generation. Clearly, more studies are warranted to define the specific roles of these carbon sources, their availability during the infectious cycle, and their role in energy metabolism of *B. burgdorferi*.

Analyses of other biochemical pathways may explain why galactose and ribose are transported in *B. burgdorferi*. As described in section "Phosphatidylcholine (PC), phosphatidylglycerol (PG) and α-monogalactosyl diacylglycerol (α-MgalDAG) synthesis", galactose is required for the synthesis of the major membrane glycolipid, α-MgalDAG. It has been proposed that this glycolipid is required for membrane stability and function (Mannock et al., 2001). In the case of ribose, homologs of all of the enzymes required for the oxidative phase of the pentose phosphate pathway are present within *B. burgdorferi* (Fraser et al., 1997), while several enzymes are missing from the non-oxidative phase of the pentose phosphate pathway (notably ribulose-phosphate 3-epimerase, transketolase, and transaldolase). Given that the oxidative phase of the pentose phosphate pathway pushes ribose toward RNA biosynthesis while the non-oxidative phase yields glyceraldehyde 3-phosphate and fructose 6-phosphate that fuel glycolysis and lipid/lipoprotein biosynthesis (Figure 2), it seems likely that ribose is required for RNA synthesis and not for generating energy. This would explain the presence of a putative ribose transport system in *B. burgdorferi* (Fraser et al., 1997) and the observation of von Lackum and Stevenson that *B. burgdorferi* cannot use ribose as a carbon source during in vitro growth (von Lackum and Stevenson, 2005).

The transport of chitobiose is of particular interest as it is a derivative of the chitin exoskeleton of the arthropod vector, specifically a β-1-4 linked disaccharide of N-acetyl glucosamine (NAG) not found in mammals. The ability of *B. burgdorferi* to transport chitobiose was originally shown by Tilly et al. (Tilly et al., 2001). A subsequent study demonstrated that genetic inactivation of *chbC* (the IIC component of the chitobiose PTS transporter) did not affect growth in either the tick vector or mammalian hosts, indicating that other sugars present within the tick can compensate for the loss of this transport system (Tilly et al., 2004). It is curious that while NAG is sufficient as a carbon source to promote growth, the non-acetylated form, glucosamine, is not (von Lackum and Stevenson, 2005). This observation suggests that glucosamine cannot be funneled into the glycolytic pathway because *B. burgdorferi* does not encode a kinase that converts glucosamine to glucosamine-6-phosphate, which following deamination, forms fructose 6-phosphate, a critical intermediate substrate in the glycolytic pathway (Figure 2). In contrast, after NAG is phosphorylated, it can then either be incorporated into cell wall biosynthesis or deacetylated to generate glucosamine-6-phosphate that is then converted to fructose 6-phosphate for use in glycolysis (Figure 2) (von Lackum and Stevenson, 2005). Again, the observation that NAG can support *in vitro* growth of *B. burgdorferi* (von Lackum and Stevenson, 2005) suggests that it can serve an alternate role as a carbon source in addition to its more widely accepted function as a building block in cell wall biosynthesis.

Finally, the utilization of glycerol is critical for the metabolism of *B. burgdorferi*. An operon encoded by *bb0240-243* is predicted to include a glycerol

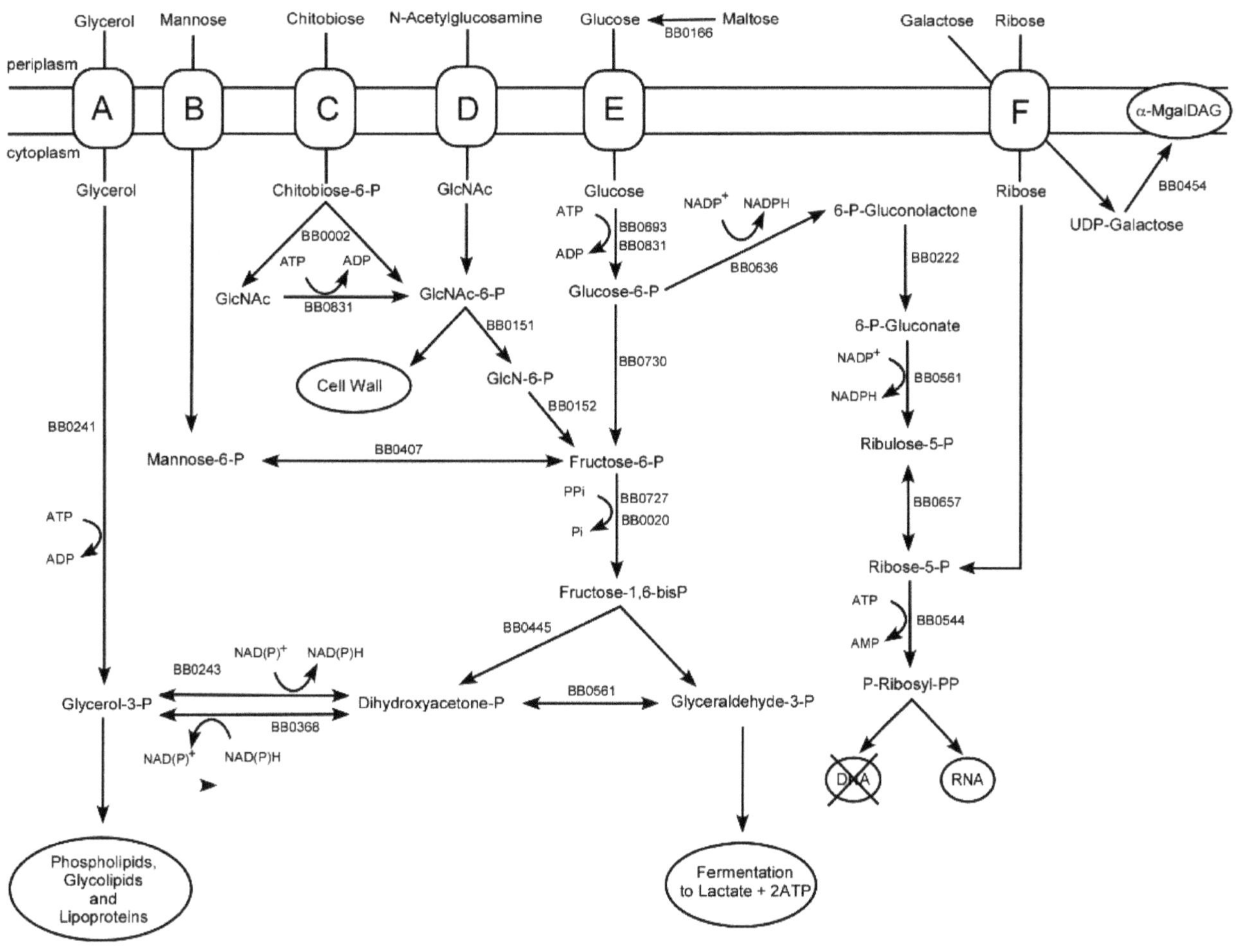

Enzymes Involved in Transport

Protein	Function	Gene Designation
	TRANSPORTER A	
BB0240	Glycerol uptake facilitator	*glpF*
	TRANSPORTER B	
BB0408	PTS, mannose-specific Enzyme IIABC	—
BB0629	PTS, mannose-specific Enzyme IIABC	—
	TRANSPORTER C	
BBB04	PTS, chitobiose-specific Enzyme IIC	*chbC*
BBB05	PTS, chitobiose-specific Enzyme IIA	*chbA*
BBB06	PTS, chitobiose-specific Enzyme IIB	*chbB*
	TRANSPORTER D/E	
BB0559	PTS, glucose-specific Enzyme IIA	*crr*
BB0116	PTS, glucose-specific Enzyme IIABC	*malX1*
BBB29	PTS, glucose-specific Enzyme IIABC	*malX2*
BB0645	PTS, glucose-specific Enzyme IIAB	*ptsG*
	TRANSPORTER F	
BB0382	Ribose/galactose ABC transporter, basic membrane protein B	*bmpB*
BB0383	Ribose/galactose ABC transporter, basic membrane protein A	*bmpA*
BB0384	Ribose/galactose ABC transporter, basic membrane protein C	*bmpC*
BB0385	Ribose/galactose ABC transporter, basic membrane protein D	*bmpD*
BB0677	Ribose/galactose ABC transporter, ATP-binding protein	*mglA*
BB0678	Ribose/galactose ABC transporter, permease protein	*mglC1*
BB0679	Ribose/galactose ABC transporter, permease protein	*mglC2*

Enzymes Involved in Carbon Metabolism

Protein	Function	Gene Designation
	TRANSPORTER F (cont.)	
BB0002	beta-N-acetylhexosaminidase	*cht*
BB0020	pyrophosphate-fructose-6-P 1-phosphotransferase	*pfkA*
BB0151	N-acetylglucosamine-6-P deacetylase	*nagA*
BB0166	4-alpha-glucanotransferase	*malQ*
BB0222	6-phosphoglucanolactonase	*pgl*
BB0241	glycerol kinase	*glpK*
BB0243	glycerol-3-P dehydrogenase	*glpD*
BB0368	glycerol-3-P dehydrogenase	*gpsA*
BB0407	mannose-6-P isomerase	*manA*
BB0445	fructose-bi-P adolase	*fbaA*
BB0544	ribose phosphate pyrophosphokinase	*prsA*
BB0454	monogalactosyl-1,2-diacylglycerol synthase	-
BB0561	6-phosphogluconate dehydrogenase	*gnd*
BB0636	glucose-6-P 1-dehydrogenase	*zwf*
BB0657	ribose-5-P isomerase	*rpi*
BB0727	pyrophosphate-fructose-6-P 1-phosphotransferase	*pfpB*
BB0730	glucose-6-P isomerase	*pgi*
BB0831	glucokinase	*nagC*

Figure 2. Carbohydrate transport and metabolism *in B. burgdorferi.* Enzymes are identified by ORF number in the diagram and correspond to the function and gene designation given in the inserted tables. Figure modified with permission from von Lackum et al., 2005.

transporter (GlpF; *bb0240*), a glycerol kinase (GlpK; *bb0241*), and a reversible glycerol-3-phosphate dehydrogenase (GlpD; *bb0243*) which sequentially bring glycerol into the cell and convert it to dihydroxyacetone phosphate (Fraser et al., 1997; Saier and Paulsen, 2000; von Lackum and Stevenson, 2005; Corona and Schwartz, 2015). Once dihydroxyacetone phosphate is formed, it can be isomerized into glyceraldehyde-3-phosphate by triphosphate isomerase, encoded by *bb0055,* and enter the glycolytic pathway to generate ATP (Figure 2). The exact role of the protein encoded by *bb0242* is not known. A second reversible glycerol-3-phosphate dehydrogenase, GpsA (encoded by *bb0368*), catalyzes the reversible redox conversion of dihydroxyacetone phosphate to glycerol-3-phosphate which serves as a crucial link between carbohydrate and lipid metabolism, as well as lipoprotein biosynthesis. This link is central to maintain a maximum yield of ATP generated by carbohydrate catabolism, redox balance, and the synthesis of essential lipids, glycolipids, and lipoproteins for cell growth and viability. Because of the importance of this pathway, GlpD and GpsA appear to be coordinately regulated by RpoS through DksA, (p)ppGpp, and c-di-GMP to modulate the flow of carbon between glycerol catabolism and central metabolism (Bugrysheva et al., 2002; Bugrysheva et al., 2003; Bugrysheva et al., 2005; Pappas et al., 2011; Bugrysheva et al., 2015; Drecktrah et al., 2015; Caimano et al., 2016; Boyle et al., 2019; Mason et al., 2020). While glycerol metabolism is important in a mammalian infection, it seems to be critically important for metabolism, long term survival and gene regulation in the tick-phase of the infective cycle.

Arginine as a potential carbon source for relapsing fever spirochetes

In some bacteria, amino acid catabolism can supplement primary metabolic pathways, providing an alternate source for ATP synthesis when other carbon sources are limited. In *Borrelia* species, amino acid catabolism is extremely limited, and the arginine deiminase (ADI) pathway is potentially the only secondary pathway capable of generating ATP (Figure 3). The ADI pathway has been studied in a variety of organisms including both Gram-positive and Gram-negative bacterial species (Elias et al., 2002; Elias et al., 2003; Vrancken et al., 2009; Rechnitzer et al., 2013; Lindgren et al., 2014; Noens and Lolkema, 2016; Majsnerowska et al., 2018). This amino acid catabolism system is generally comprised of three enzymes and a membrane bound antiporter. Arginine deiminase (ArcA), which converts arginine to citrulline, was originally identified in *Streptococcus pyogenes* cellular extracts due to its antitumor activity (Elias et al., 2002). Since then it has been the subject of many studies attempting to define the role of the enzyme in energy acquisition, survival of acid stress and infectivity (Champomier Verges et al., 1999; Cusumano et al., 2013; Rechnitzer et al., 2013; Thurlow et al., 2013; Cusumano and Caparon, 2015; Geiger et al., 2016). Ornithine carbamoyltransferase (ArcB) converts citrulline to ornithine and carbamoyl-phosphate and has been shown to contribute to bacterial growth, infectivity and survival of acid stress (Frank et al., 2003; Cusumano et al., 2013; Noens and Lolkema, 2016; Jutras et al., 2019). Carbamate kinase (ArcC) uses substrate level phosphorylation to generate an ATP from carbamoyl-phosphate + ADP and has been shown to contribute to growth in some bacteria (Jutras et al., 2019). The arginine-ornithine antiporter, ArcD, is a membrane bound protein which transports one arginine into the bacterial cytosol while exporting ornithine to the extracellular milieu (Rimaux et al., 2012; Rechnitzer et al., 2013; Majsnerowska et al., 2018). The genes encoding a putative arginine/ornithine antiporter (*bb0843*) as well as the catabolic enzymes arginine deiminase (*bb0841*) and ornithine carbamoyltransferase (*bb0842*) were identified in both Lyme disease and relapsing fever spirochetes (ERGO). However, the gene encoding the final enzyme, carbamate kinase, responsible for generating ATP from carbamoyl phosphate, is missing from the Lyme disease spirochetes but is found in all relapsing fever spirochetes (*bh0844*) (Figure 3, panel B). This one difference in the operons, while seemly trivial, might suggest that the Lyme disease and relapsing fever spirochetes are using the ADI pathway for entirely different metabolic roles. Preliminary experimental data indicates that might be the case.

In *B. burgdorferi,* the ADI pathway has been shown to be involved in the acid stress response (See section: Acid stress) (Dulebohn et al., 2017). When cells are exposed to the weak organic acids, acetate, benzoate or propionate, intracellular pH dropped, and growth stopped. The addition of arginine to the growth media restored the intracellular pH (pH_i) and cell viability. BB0842, the arginine/ornithine antiporter and BB0841, arginine deiminase, were central to this effect by transporting arginine and converting it to citrulline + NH_3. Arginine and NH_3 serve as intracellular buffers by scavenging excess H^+ to

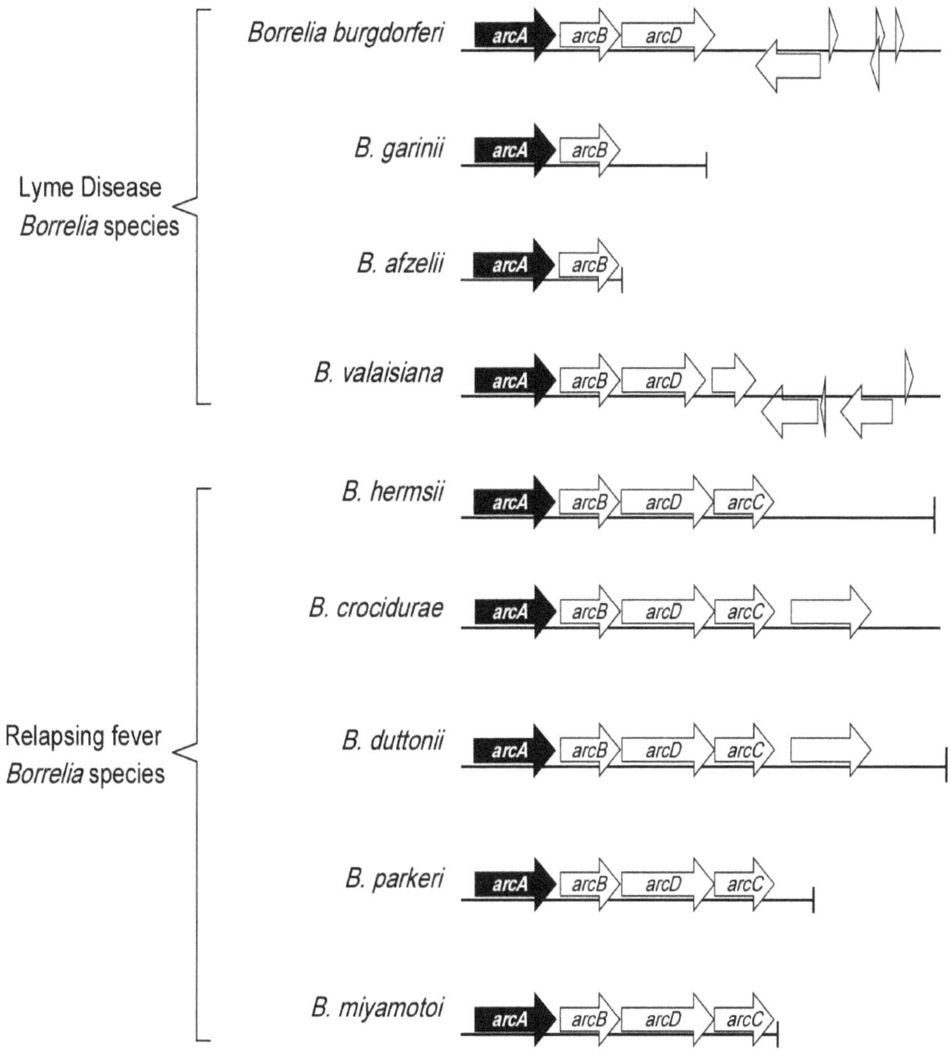

Figure 3. Genetic organization of the Arc operon from Lyme and Relapsing Fever spirochetes.

restore intracellular pH_i during acid stress. These same effects were observed when *B. hermsii* cells were exposed to weak organic acids. However, excess arginine is toxic to *B. burgdorferi* in the absence of exogenous organic acids, possibly due to the accumulation of intracellular NH_3 and the exclusion of ornithine, an essential amino acid for cell wall biosynthesis, from the cytoplasm. Conversely, excess arginine is not toxic to *B. hermsii* cells and, in fact, stimulates in vitro growth. These data suggest that *B. hermsii* and other relapsing fever spirochetes, which harbor a gene encoding a carbamate kinase (*bh0844*), are generating ATP from their ADI pathway. It is important to note that BB0843 and BH0843 have significant differences in amino acid sequence; experimental data suggest that BB0843 has a preference for ornithine for cell wall biosynthesis, while BH0843 preferentially transports arginine to generate ATP. The significance of the ADI pathways in Lyme disease and relapsing fever spirochetes, and their potential effects on disease outcome and the transmission cycle, need to be determined experimentally.

ATP synthesis

As indicated above, *B. burgdorferi* lacks genes homologous to those encoding proteins involved in oxidative phosphorylation or the TCA cycle (Fraser et al., 1997; Das et al., 2000; von Lackum and Stevenson, 2005). They also do not appear to make energy via the pentose phosphate pathway but, instead, use it to generate building blocks for DNA and RNA (von Lackum and Stevenson, 2005; Bedlack et al., 2012). However, they do have all the genes required for the glycolytic pathway (Fraser et al., 1997; Das et al., 2000; von Lackum and Stevenson, 2005). It appears the major metabolic pathway for ATP generation in *B. burgdorferi* is substrate level phosphorylation mediated by the phosphoglycerate kinase and pyruvate kinase enzymatic steps during fermentation (Fraser et al., 1997; von Lackum and Stevenson, 2005; Bedlack et al., 2012). Because oxidative phosphorylation is not operative in *B. burgdorferi*, the ability to harness the energy stored in these reduced substrates is not possible. Thus, reduced $NADH^+$ is oxidized independent of oxidative phosphorylation and, instead, is dependent on reduction of pyruvate to lactate through lactate dehydrogenase and other redox enzymes that oxidize $NADH^+$, including coenzyme disulfide reductase (i.e., CoADR, encoded by *bb0728*) (Boylan et al., 2006).

Most bacteria generate a proton gradient across their cytoplasmic membrane using electron transport systems that result in a greater concentration of protons outside the cytoplasm. The protons move down this concentration gradient, through an F-type ATPase, in a process that generates ATP within the cytoplasm of the bacterium. However, *B. burgdorferi* does not encode an F-type ATPase. Therefore, this concentration gradient must be established independently. The maintenance of the proton gradient is likely mediated by the function of the V-type ATPase in a process that hydrolyzes ATP (see section: V-Type ATPase) (Fraser et al., 1997; Fraser et al., 1998). Effectively, V-type ATPases function in reverse of F-type ATPases and, thus, utilize energy instead of synthesizing ATP. Previously, V-type ATPases were associated with vacuolar membranes but were more recently found in plasma membranes of most eukaryotes (Beyenbach and Wieczorek, 2006) and select bacteria, including *B. burgdorferi* and *T. pallidum* (Fraser et al., 1997; Fraser et al., 1998). Although, in theory, the pumping of protons in reverse through its V-type ATPase could be used to generate ATP, no evidence exists that this occurs in spirochetes or any other prokaryote.

Purine salvage

Nucleotide metabolism serves an essential function in all living organisms as a means of providing critical precursors for nucleic acid biosynthesis. The complexity of these biochemical pathways varies among organisms and spans the range from rudimentary salvage pathways, which rely upon

Table 2. Enzymes in *Borrelia burgdorferi* nucleotide salvage pathway.

Enzyme Name	Gene Designation	Open Reading Frame
Adenine deaminase	*adeC*	*bbk17*
Adenine phosphoribosyltransferase	*apt*	*bb0777*
Adenylate kinase	*adk*	*bb0417*
Deoxynucleoside kinase	*dgk*	*bb0239*
Deoxyribosyl transferase	*ptd*	*bb0426*
GMP synthase	*guaA*	*bbb18*
IMP dehydrogenase	*guaB*	*bbb17*
Guanine-hypoxanthine permease	*pbuG*	*bbb22, bbb23*
Nucleoside diphosphate kinase	*ndk*	*bb0463*
Xanthine-guanine phosphoribosyltransferase	*gpt*	*bb0103*
Cytidine deaminase	*cda*	*bb0618*
Thymidine kinase	*tdk*	*bb0791*
CTP synthase	*pyrG*	*bb0575*
Uridylate kinase	*pyrH*	*bb0571*
Uridine kinase	*udk*	*bb0015*
Cytidylate kinase	*cmk*	*bb0819, bb0128*
Thymidylate synthase	*thyX*	*bba76*

exogenous sources of nucleotides, to more complex *de novo* biosynthetic pathways. The degree to which an organism is dependent upon exogenous sources of nucleotide precursors is the direct result of the complexity of the organism's nucleotide biosynthetic pathway and is an important determinant in the niches where this organism can survive. Differences in nucleotide salvage pathways differ among *Borrelia* species (Pettersson et al., 2007). The genomes of Lyme disease spirochetes, (*e.g.*, *B. burgdorferi, B. garinii, B. afzelii*) are missing genes encoding enzymes such as ribonucleotide reductase, adenylosuccinate lyase, adenylosuccinate synthase and hypoxanthine phosphoribosyltransferase (Table 2). Conversely, in the genomes of relapsing fever species (*B. hermsii, B. turicatae*), these genes are present. *B. hermsii* encodes a classical nucleotide salvage pathway, like those described for other organisms (Barbour et al., 2005; Pettersson et al., 2007). This type of pathway allows for the uptake of purine and pyrimidine bases, which can undergo enzymatic transformation to ribonucleotides for incorporation into RNA or be reduced to deoxyribonucleotides by ribonucleotide reductase (RNR) for incorporation into DNA.

Ribonucleotide reductase

Because *B. hermsii* has a functional RNR, this classical salvage pathway does not require an exogenous source of deoxynucleotides for DNA synthesis and, as a result, the availability of deoxynucleotides in host tissues does not limit where this spirochete can survive. Hypoxanthine phosphoribosyltransferase in *B. hermsii* is responsible for the production of inosine monophosphate through the transfer of a ribose-5'-monophosphate from PRPP to hypoxanthine (Pettersson et al., 2007). The presence of this enzyme provides for the direct utilization of hypoxanthine to produce GMP and dGMP through inosine monophosphate dehydrogenase (GuaB) and GMP synthase (GuaA). Furthermore, adenylosuccinate lyase (PurB) and adenylosuccinate synthase (PurA) could provide an additional advantage to *B. hermsii* by performing the conversion of IMP to AMP. This pathway may allow for the utilization of hypoxanthine for the synthesis of AMP and dAMP. Without PurAB, hypoxanthine can serve only to produce GMP and dGMP, thereby imposing a requirement on the bacterium for host derived sources of adenine nucleosides.

In silico analysis of *B. burgdorferi*'s genome failed to identify any sequence with homology to known RNR enzymes (Fraser et al., 1997). RNRs catalyze the reduction of ribonucleotides to their deoxy-analogs as essential precursors for DNA synthesis. RNRs are widely believed to be an absolute requirement for all living organisms due to their role in DNA synthesis (Jordan and Reichard, 1998), although the genome of *Ureaplasma urealyticum* also has been reported to be missing an RNR homolog (Carnrot et al., 2003). Additional evidence for the absence of an RNR in *B. burgdorferi* was illustrated by the lack of cross hybridization in a Southern blot when *B. burgdorferi* DNA was probed with *nrdF* from *B. hermsii* (Pettersson et al., 2007). Furthermore, the absence of RNR enzyme activity was demonstrated in a radio-labeled nucleotide incorporation assay when *B. burgdorferi* was unable to convert ribonucleotides to deoxyribonucleotides (Boylan et al., 2008). The consequences to an organism deficient in this enzyme are: i) the organism must depend on exogenous sources of deoxynucleosides from host tissue, and/or ii) the organism must have evolved an alternative means of producing and regulating deoxynucleotide pools. *B. burgdorferi* takes advantage of both approaches and has evolved a novel biochemical approach to generate deoxynucleotides for DNA synthesis which is outlined in Figure 4. Thus, the availability of exogenous deoxynucleosides, which are required to sustain these pathways, is a determining factor in the survival of *B. burgdorferi* in different host tissues/sites (i.e. tick midgut or mammalian tissue).

Since deoxynucleotides and various nucleic acid precursors can be found in mammalian tissue, *B. burgdorferi* can take advantage of this by efficiently transporting and incorporating free deoxynucleosides into DNA. Purine bases, such as hypoxanthine and adenine, are also available in blood and other tissues for incorporation into RNA or for conversion by the deoxyribosyltransferase (Ptd/*bb0426*) to deoxynucleosides (Boylan et al., 2008). Although the specific transporters that are responsible for this uptake have not been characterized, the genome of *B. burgdorferi* contains two ORFs that are believed to encode guanine-hypoxanthine permeases (PbuG) (Byram et al., 2004) and an operon encoding a putative nucleoside permease system, BB0677-679 (Overbeek et al.2003). These transporters may also be responsible for the uptake of several additional nucleotide precursors. Given these few building blocks and a unique purine salvage pathway, *B. burgdorferi* can produce DNA in sufficient quantity

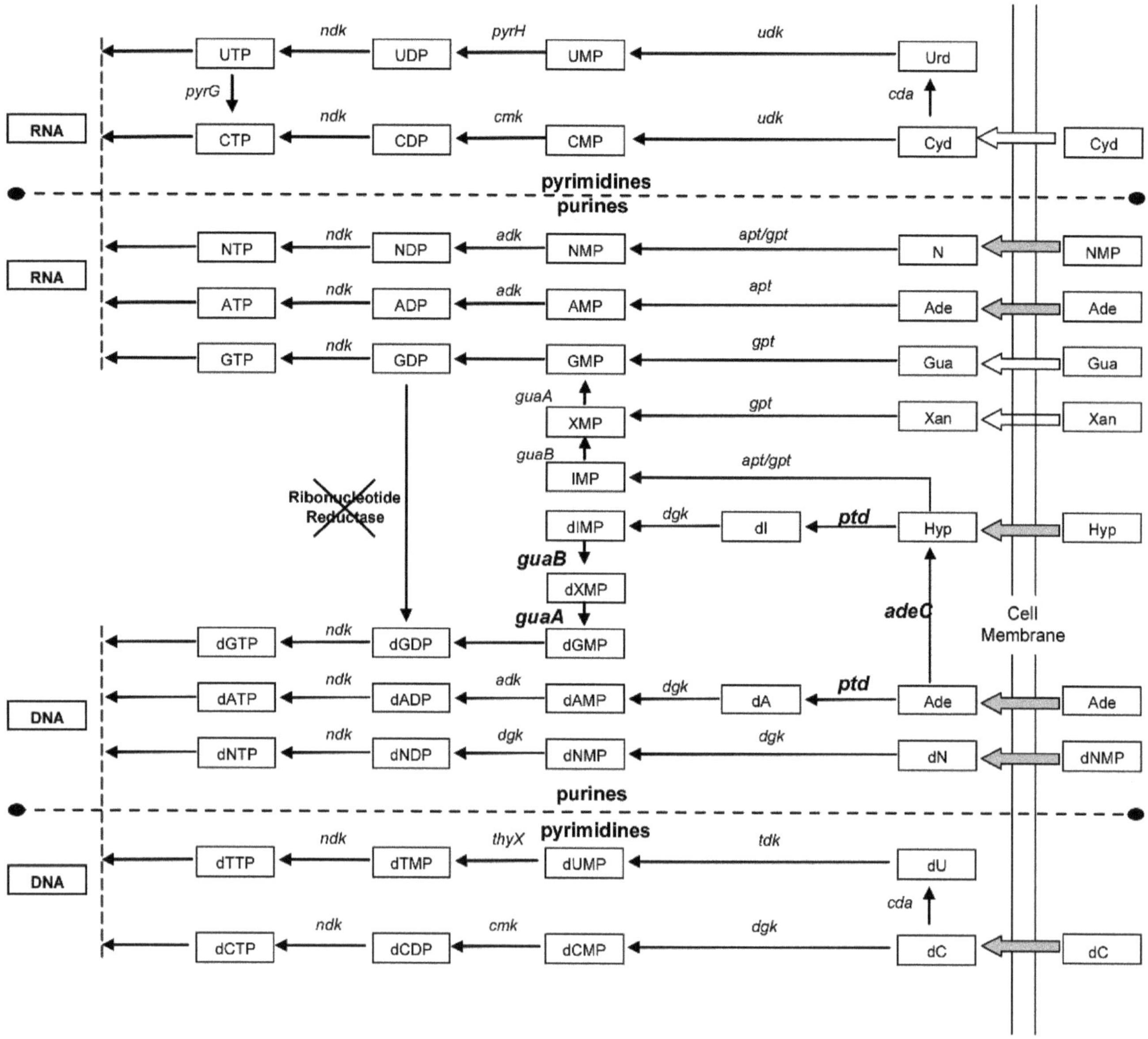

Figure 4. The purine and pyrimidine salvage pathway in *Borrelia burgdorferi*. Figure modified with permission from Gherardini et al., 2010.

and with suitable efficiency to support cell division without the requirement for an RNR.

2'-Deoxyribosyltransferase

In *B. burgdorferi*, a key player that is crucial to its purine salvage pathway is the 2'-deoxyribosyltransferase (Ptd), a 21-kDa protein that is encoded by *bb0426* (Boylan et al., 2008). This enzyme functions by catalyzing the transfer of a 2'-deoxyribose moiety from a donor purine deoxynucleoside to an acceptor purine base. Two types of deoxyribosyltransferases have been described for several *Lactobacillius* species and characterization of both proteins has included elucidation of catalytic mechanisms and crystal structures (Kaminski, 2002; Anand et al., 2004). Kinetic studies using a recombinant BB0426 have demonstrated that *B. burgdorferi* harbors a purine specific 2'-deoxyribosyltransferase designated Ptd (Boylan et al., 2008). Enzymatic activity was not detected when pyrimidine pairs or pyrimidine-purine pairs were tested as substrates. Due to the absence of the reductive capability of an RNR, host tissues serve as the source of the deoxynucleosides substrates *in vivo*. Once a deoxynucleoside has been transported into the cell, Ptd can transfer its deoxyribose to another purine base and, after phosphorylation, incorporation into DNA (Figure 4). In particular, this function benefits *B. burgdorferi* in the production of dGMP for DNA synthesis by binding hypoxanthine by Ptd, as a deoxyribose donor, thereby forming deoxyinosine (dI). With subsequent phosphorylation to dIMP, and the catalytic action of GuaA and GuaB (discussed below), dGMP is produced. Utilizing Ptd as a substitute for an RNR provides *B. burgdorferi* with the ability to regulate the composition of deoxynucleotide pools when the availability of one or more deoxynucleotides is limited from exogenous sources.

Inosine monophosphate dehydrogenase and GMP synthase

Another critical component to *B. burgdorferi's* purine salvage pathway is inosine-5'-monophosphate dehydrogenase (GuaB). This enzyme is a 44-kDa gene product of *bbb17* (*guaB*) located on circular plasmid cp-26 in *B. burgdorferi's* genome. GuaB has been shown to catalyze the oxidation of IMP to XMP through the corresponding reduction of NAD^+ (Zhou et al., 1997). This activity has been the primary focus of enzymatic studies due to the importance of GuaB in the pathway for GMP production. In *B. burgdorferi*, recombinant GuaB also has been shown to catalyze the conversion of dIMP to dXMP (Jewett et al., 2009; Lawrence et al., 2009). This activity, along with GMP synthase (GuaA), satisfies the essential role of dGMP production for DNA synthesis in the absence of a functional RNR. The advantage of this mechanism is the utilization of host derived hypoxanthine to produce deoxyinosine (dI) through the action of Ptd. Subsequent phosphorylation provides dIMP which serves as a substrate for GuaB and GuaA, arriving at dGMP for DNA synthesis. The absence of *guaAB* through site directed mutagenesis was reported to result in loss of virulence in the standard mouse model (Jewett et al., 2009).

Adenine deaminase

Adenine deaminase (AdeC) is a 61-kDa protein encoded by *bbk17* in *B. burgdorferi,* whose function has been shown to catalyze the deamination of adenine to produce hypoxanthine (Jewett et al., 2007). This mechanism provides a distinct advantage to *B. burgdorferi's* purine salvage pathway by allowing for the catabolism of adenine nucleotides when exogenous sources of hypoxanthine are limited in host cell tissue. This secondary source of hypoxanthine can then be utilized in the production of GMP and dGMP as described above. Mutation of *adeC* in *B. burgdorferi* results in an attenuated virulence phenotype in vivo but like the *guaAB* mutant, did not have a growth defect when cultured in vitro (Jewett et al., 2007).

Cofactor biosynthesis

Enzymes are biological molecules which catalyze chemical reactions; many require the presence of an additional, non-protein cofactor to promote catalysis. There are 2 types of cofactors: essential ions (mostly metals) and coenzymes (organic compounds) which act as group-transfer reagents. The latter can be divided into two groups based on their metabolic origin: metabolites and vitamin-derived coenzymes. Metabolite coenzymes include a number of nucleoside triphosphates, of which ATP is by far the most abundant. The vitamin-derived coenzymes are derivatives of vitamins, which are compounds that cannot be synthesized by animals and must be obtained as nutrients. This section will focus on the three major vitamin-derived coenzymes of *B. burgdorferi*: Coenzyme A (CoA), nicotinamide adenine dinucleotide (NAD^+), nicotinamide adenine dinucleotide phosphate ($NADP^+$) and flavin adenine dinucleotide (FAD).

CoA

Coenzyme A (CoA) and acyl carrier protein (ACP) are the two predominant acyl group carriers in cells. These compounds are essential cofactors in all cells and participate in numerous reactions in intermediary metabolism. In addition, reduced coenzyme A likely plays other roles in *B. burgdorferi*. The high levels of reduced CoA inside the cell suggest a role in supporting the reduced intracellular environment and the thiol/disulfide distribution of other thiol-containing molecules. Although putative genes have been identified in *B. burgdorferi* that may be involved in the synthesis and regulation of CoA (Eggers et al., 2011), most of what is known about the biochemistry and genetics of the enzymes involved in the formation of pantothenate and CoA has been derived from research with *E. coli* and *S. typhimurium*.

Pantothenate transport

The first step in the synthesis of coenzyme A is the production of pantothenate. Most bacteria are capable of synthesizing pantothenate using intermediates from the amino acid pathways. However, *B. burgdorferi* is missing key enzymes in this biosynthetic pathway and therefore, must acquire pantothenate from its environment via a transport system. Pantothenic acid is widely distributed in mammalian tissues (liver, kidney, heart, skin and muscle) and available for transport into the bacterium. The best-characterized transport system is the high-affinity pantothenate permease of *E. coli* that catalyzes the concentrative uptake of pantothenate by a sodium ion cotransport mechanism (Nakamura and Tamura, 1973; Vallari and Rock, 1985). Pantothenate permease belongs to the group of inner membrane permeases that catalyze active cation-dependent symport. It has been shown that in bacteria that lack a pantothenate biosynthetic pathway (e.g., Lactobacillus spp.), this transport activity is required for growth. *B. burgdorferi* harbors the gene (BB0814) encoding a putative pantothenate permease.

CoA biosynthesis

Pantothenate phosphorylation by pantothenate kinase (BB0527) to form 4'-phosphopantothenate is the first step in the biosynthesis of CoA. In *E. coli* this is the most highly regulated step in the process and is regulated through feedback inhibition by CoA and its thioesters (Davis, 1950; Maas and Davis, 1950; Jackowski and Rock, 1981). Non-esterified CoA was significantly more potent than CoA thioesters in inhibiting the enzyme in extracts from *E. coli*. CoA is a competitive inhibitor with respect to ATP, thus providing a mechanism to coordinate CoA production with the energy state of the cell. In *E. coli*, mutants have been isolated in which the pantothenate kinase activity in crude extracts is refractory to feedback inhibition by CoA. These strains have CoA levels that are significantly higher than in strains containing the wild-type kinase. Also, the effects of pantothenate kinase overexpression on the size of the CoA pool have been studied. Strains expressing significantly more wild-type kinase exhibited only a small increase in the steady-state CoA level, corroborating the conclusion that modulation of pantothenate kinase activity by feedback regulation is the critical factor controlling the intracellular CoA concentration. At this time, there is no information on the regulation of the putative pantothenate kinase (BB0527) in *B. burgdorferi*.

Following pantothenate phosphorylation, a cysteine is then added to 4'-phosphopantothenate to generate 4'-phosphopantothenoylcysteine and then decarboxylated by 4'-phosphopantetheine decarboxylases to generate 4'-phosphopantetheine. This 2-step enzymatic process is predicted to be carried out by the phosphopantothenate-cysteine ligase and phosphopantothenoylcysteine decarboxylase activities of the protein encoded by BB0812 (Fraser et al., 1997). Basically, nothing is known about the biochemistry or genetics of this enzyme. The next step requires two enzymes that convert 4-phosphopantetheine to CoA. First, phosphopantetheine-phosphate adenylyltransferase (BB0702) transfers the AMP moiety to 4'-phosphopantetheine, with the release of PP_i to form dephospho-CoA. Next, a phosphate group is added to the 3'-hydroxyl of dephospho-CoA by dephospho-CoA kinase (BB0547) to yield CoA. Again, little is known about the biochemistry or genetics of these two enzymes. Interestingly, in mammals, these enzymes are co-purified and exist as a bifunctional protein that is designated CoA synthase (Worrall and Tubbs, 1983). However, these two activities have been separated by chromatographic fractionation of *Brevibacterium ammoniagenes* extracts, suggesting that individual enzymes catalyze the two steps in bacteria (Martin and Drueckhammer, 1993). This step has been identified as a secondary regulatory point in CoA biosynthesis (Jackowski and Rock, 1981, 1984). Double-label experiments indicate that extracellular 4'- phosphopantetheine is derived from the degradation of ACP, and reutilization of this intermediate is regulated at the adenylyltransferase step. Excretion of 4'-phosphopantetheine is an irreversible mechanism for reduction of the intracellular CoA and ACP pools since *E. coli* is unable to assimilate exogenous 4'-phosphopantetheine into CoA. However, it is possible that *B. burgdorferi* may use a similar mechanism to modulate intracellular levels of CoA and ACP. Understanding the biochemical mechanisms that regulate the activity of 4'-phosphopantetheine adenylyltransferase will be necessary to establish the role of this enzyme in controlling the intracellular CoA concentration.

The composition of the intracellular CoA pool has been examined in *B. burgdorferi* by using the thiol-specific labelling reagent monobromobimane to label small thiols, which are then separated and identified by their retention time on high-performance liquid chromatography (HPLC) (Boylan et al., 2006). Aerobic cells grown in the presence of pantothenic acid showed CoA levels of 3.02 ± 0.19 µmol per gram of cells (dry weight), while those grown without pantothenic acid were 2.77 ± 0.23. *Borrelia* cells grown in the presence of reduced CoA contained levels of CoA equivalent to cells grown without supplements (2.59 ± 0.48 µmol per gram of cells), indicating that exogenous CoA does not alter the CoA level within the cell. The CoA present in the cell was almost entirely in the thiol form, as the disulfide content was below the detection limit of this assay, determined by calculating: (thiol + disulfide) − thiol. This thiol form is maintained in the cell by the specific disulfide reductase, CoADR (encoded by *bb0728*) (Boylan et al., 2006).

CoA is used 4'-phosphopantetheine as a donor for the synthesis of ACP from apo-ACP catalyzed by ACP synthase (BB0010). In *E. coli*, metabolic labeling, following the starvation of pantothenate auxotrophs, shows that the level of active ACP is maintained at the expense of CoA (Alberts and Vagelos, 1966; Jackowski and Rock, 1981, 1984). The ACP prosthetic group is removed by the action of ACP phosphodiesterase (EC 3.1.4.14) (Vagelos and Larrabes, 1967), an enzyme lacking in *B. burgdorferi*. ACP prosthetic group turnover is rapid compared with the stability of the protein moiety (Powell et al., 1969; Powell et al., 1973; Jackowski and Rock, 1984), thus generating a cycle of ACP synthesis and degradation. The physiological significance of this cycle remains a mystery, although it seems unlikely to be related to the regulation of fatty acid biosynthesis (Jackowski and Rock, 1983). The 4'-phosphopantetheine released from ACP can be either reincorporated into CoA or excreted from the cell (Jackowski and Rock, 1984). When the CoA level is low, however, turnover of the ACP prosthetic group is faster and excretion of 4'-phosphopantetheine is minimal, suggesting that the cycle participates in the metabolic maintenance of a basal CoA concentration.

NAD^+

NAD^+ is an essential cofactor that is ubiquitous in nature. It contains a reactive nicotinamide ring that accepts H^+ and electrons during the oxidation of molecules (Westheimer, 1987). In addition to its role in oxidation-reduction reactions in glycolysis, NAD^+ can also be used as a substrate for the modification of proteins by ADP-ribosyltransferases and for the synthesis and repair of DNA by prokaryotic DNA ligase. Therefore, NAD^+ has an impact, directly or indirectly, on most pathways within the cell. In most prokaryotes, NAD can be synthesized *de novo* from aspartate through a series of five reactions (White, 1982). However, of the 5 enzymes in this pathway, *B. burgdorferi* harbors only 1, nicotinate-nucleotide adenylytransferase, *nadD* (BB0782). This enzyme catalyzes the conversion of nicotinate mononucleotide (NAMN) to Deamido-NAD^+ and the conversion of nicotinamide D-ribonucleotide (NMN) to NAD both at the expense of ATP. Therefore, for *de novo* synthesis to occur, *B. burgdorferi* must be transporting NMN or nicotinate mononucleotide through an unidentified transporter. *E. coli* and *S. typhimurium* can scavenge from their environments a variety of exogenous pyridines, including NMN, through a membrane-associated NAD pyrophosphatase, PnuE, or other NMN uptake systems, PnuA, PnuB, PnuC, or PnuD. However, the *B. burgdorferi* genome does not contain any orthologs for these transport systems. Given the numerous transporters that have been identified in the *B. burgdorferi* genome with no assigned function, it is possible that one functions as a specific NMN transporter or a generalized pyridine nucleotide uptake system.

In addition to *de novo* synthesis, prokaryotes such as *E. coli* can scavenge NAD^+ from environmental sources through the pyridine nucleotide cycles (PNCs). NAD-dependent oxidation-reduction reactions do not consume NAD. However, NAD is degraded and the half-life during aerobic growth of *Salmonella typhimurium* determined to be approximately 90 minutes (Park et al., 1989). Therefore, systems functioning to recycle the pyridine moiety are advantageous to the cell. These cycles have been well-characterized in both *S. typhimurium and E. coli*. An important feature of the salvage pathways of these organisms is the energetically favorable conservation of the pyridine ring. When recycling is possible, it takes precedence over *de novo* synthesis (McLaren et al., 1973). The existence of various NAD-recycling pathways consisting of four-, five-, and six-membered cycles, referred to as PNCs IV, V, and VI, respectively. PNCs IV and V function predominantly for intracellular recycling, while PNC VI plays an important role in scavenging preformed pyridines from the environment.

The majority of intracellular NAD recycling begins with hydrolysis of the PPi bond, releasing NMN and AMP (Hillyard et al., 1981). One enzyme that can degrade NAD to NMN is prokaryotic DNA ligase, which utilizes the energy released from the PPi bond to reseal nicks in double-stranded DNA. This reaction is important in the repair, recombination, and replication of DNA. This degradation by DNA ligase was thought to be the primary cause of NAD recycling, but this was shown to be false. Park et al. demonstrated that *S. typhimurium* cells containing a

T4 phage ATP-dependent DNA ligase do not exhibit any significant decrease in NAD turnover (Park et al., 1989). Interestingly, the NAD turnover rate is much higher (three- to four-fold) under aerobic growth conditions than under anaerobic conditions. This may possibly be due to some unknown NAD- dependent enzyme that protects membranes or DNA from oxygen damage. The enzyme responsible for initiating most intracellular NAD degradation has yet to be identified. Once NMN has been generated it can be converted to NAD through the actions of *nadD*.

In addition to NAD recycling, *B. burgdorferi* may be capable of scavenging nicotinamide from the environment to synthesize NAD^+ through the PNC VI pathway. Nicotinamide (Nam) is found in various tissues including the skin and liver and therefore, would be available for transport into the cell. Both *S. typhimurium and E. coli* have been shown to uptake Nam, however, uptake systems have not been identified in *B. burgdorferi* (Fraser et al., 1997). Nicotinamide is then deaminated to nicotinic acid (NA) by the nicotinamidase, PncA (encoded by *bbe22*), encoded on the linear plasmid lp25. Interestingly, numerous studies have demonstrated that lp25 is an essential plasmid for mammalian infection. *B. burgdorferi* cells lacking lp25 lost infectivity in mice which was restored by introducing a shuttle vector containing BBE22 (Purser et al., 2003). This indicated that the nicotinamidase encoded by BBE22 is essential for *B. burgdorferi* infectivity and suggests it is important in the production of NAD in a mammalian host. The next member step is the formation of NAMN through action of a NA phosphoribosyltransferase, PncB (encoded by *bb0635*) (Foster et al., 1979), which links the recycling/scavenging systems to the *de novo* pathway. The final series of reactions in the synthesis of NAD, catalyzed by NadD and NadE, are essential for cell viability. The NAMN produced by recycling or scavenging is adenylated by NadD (Namn adenylytransferase, BB0782), producing NA adenine dinucleotide, which is then amidated by NadE (NAD synthetase, BB0522) to form NAD^+.

$NADP^+$

B. burgdorferi also encodes a putative homolog of the NAD^+ kinase, NadF (BB0311). NadF transfers a phosphate from ATP, ADP or polyphosphate to the 2'moiety of the adenosine to generate $NADP^+$ with the most common phosphate donor being ATP (Cheng and Roth, 1994; Shi et al., 2009). Some NAD^+ kinases phosphorylate NAD^+ or NADH to generate $NADP^+$ and NADPH respectively (Shi et al., 2009) and it is unclear which substrates or phosphate sources are preferred by BB0311. It is likely that the production of $NADP^+$ is required for the oxidative stage of the pentose phosphate pathway and ultimately the generation of ribose-5-phosphate (Fraenkel, 1986; Sprenger, 1995; Fitzpatrick et al., 2007). In the pentose phosphate pathway, the conversion of glucose-6-phosphate to 6-phosphogluconolactone by glucose 6-phosphate dehydrogenase (BB0636) and the conversion of 6-phosphogluconate to ribulose-5-phosphate by 6-phosphogluconate dehydrogenase (BB0561) are likely $NADP^+$-dependent reactions (Figure 2). Another potential metabolic role for NADPH is by a putative FAD-dependent thymidylate synthase (BBA76) which uses 5,10-methylenetetrahydrofolate, dUMP and $FADH_2$ to generate deoxythymidine monophosphate. Sequence analysis of these enzymes support the hypothesis that both enzymes require $NADP^+$ for activity and generate reduced NADPH in the process. However, direct experimental analyses of these enzymes would be necessary to confirm their cofactor requirements.

The type II HMG-CoA reductase (HMG-CoAR, BB0685) in *B. burgdorferi* converts HMG-CoA to mevalonate (Richards et al., 2015) and has been shown to oxidize NADPH to $NADP^+$ (Km 132uM/ Vmax, 1.92uM NADPH $minute^{-1}$ mg^{-1}) in *B. burgdorferi* (Figure 6)(Van Laar et al., 2016). The sequence similarities between BB0685 and other type II HMG-CoA reductases (EC 1.1.1.88) suggests that BB0685 may be able to utilize either NADH or NADPH like the HMG-CoAR from *Pseudomonas mevalonii.* This enzyme has a kinetic preference for NAD^+ in the oxidative acetylation of mevalonate in catabolism. In contrast, the enzyme prefers NADPH for the reduction of HMG-CoA to mevalonate for the synthesis of isoprenoids (Friesen et al., 1996). Since the HMG-CoAR from *B. burgdorferi* synthesizes Lipid II, an intermediate in the synthesis of peptidoglycan (See *Undecaprenyl-pyrophosphate* (C55-P) *synthesis* below), from acetate via mevalonate and does not have a degradative pathway for isoprenoids, it seems very likely that HMG-CoAR uses NADPH. The type II HMG-CoAR in *Archaeoglobus fulgidus* also has a complex allosteric regulatory mechanism controlling its activity (Kim et al., 2000), and further analysis of the HMG-CoA reductase in *B. burgdorferi* may reveal similar mechanisms that play a role in regulating the synthesis of mevalonate and, ultimately, the cell wall of *B. burgdorferi*.

In addition to these metabolic roles, the reductive potential of NADPH may also be utilized to maintain intracellular redox and for oxidative defense mechanisms in *B. burgdorferi*. While CoADR (BB0728) exclusively uses NADH for enzyme activity (Boylan et al., 2006), the activities of the thioredoxin reductase (BB0515) are predicted to require NADPH (Fraser et al., 1997) to reduce protein disulfides. Although experimental evidence is required to support this role for NADPH in oxidative stress defense, it is interesting to note that the transcript of *nadF (bb0311)* increases in fed nymphs compared to fed larvae which also corresponds with high level expression of thioredoxin (Iyer et al., 2015). The concentration and relative ratios of NAD^+/NADH and $NADP^+$/NADPH have not been characterized but are likely to fluctuate in *B. burgdorferi* under certain environmental conditions such as oxidative stress or under certain nutrient rich or nutrient limiting conditions. How these levels fluctuate is unknown. More work is required to understand the function these key cofactors play in metabolism, oxidative defense and their critical roles in maintenance of *B. burgdorferi* during its complex life cycle.

Riboflavin

Riboflavin is the central component of the cofactors FAD and FMN, and is, therefore, required by all flavoproteins. The biosynthesis of one riboflavin molecule requires one molecule of GTP and two molecules of ribulose 5-phosphate as substrates (Bacher et al., 2000). This pathway has been well characterized in *E. coli*. The imidazole ring of GTP is hydrolytically opened, yielding a 4,5-diamino-pyrimidine which is converted to 5-amino-6-ribitylamino-2,4(1H,3H)-pyrimidinedione by a sequence of deamination, side chain reduction and dephosphorylation. Condensation of 5-amino-6-ribitylamino-2,4(1H,3H)-pyrimidinedione with 3,4-dihydroxy-2-butanone-4-phosphate obtained from ribulose 5-phosphate affords 6,7-dimethyl-8-ribityllumazine. Dismutation of the lumazine derivative yields riboflavin and 5-amino-6-ribitylamino-2,4(1H,3H)-pyrimidinedione, which is recycled in the biosynthetic pathway.

To perform its metabolic role, riboflavin must be converted to the cofactors FMN and FAD. This is accomplished through the phosphorylation of the 5'hydroxyl group of riboflavin generating FMN and is catalyzed by riboflavin kinase. FMN is then converted to FAD by FAD synthetase. However, genome analysis indicates that *B. burgdorferi* does not contain any of the enzymes responsible for *de novo* synthesis of riboflavin, FMN, or FAD (Fraser et al., 1997). Yet, the genome does contain predicted flavoproteins including BB0812 (Phosphopanto-thenoylcysteine decarboxylase/phosphopanto-thenate-cysteine ligase), BBA76 (FAD-dependent thymidylate synthase), BB0515, (thioredoxin reductase) and BB0728 (CoA reductase). The CoA reductase (CoADR), purified from *B. burgdorferi* cells, contained FAD and the cofactor was required for activity (Boylan et al., 2006). Additionally, Eggers et al., using *in silico* analyses, confirmed an FAD binding motif in CoADR. Also, HPLC analyses of *B. burgdorferi* cell extracts revealed that intracellular concentrations of FAD approached 1 micromolar, indicating that the bacterium possibly transports FAD (unpublished data).

Recently, an ABC-type riboflavin transporter system was identified which would provide a mechanism for riboflavin acquisition by *B. burgdorferi* (Showman et al., 2016a; Ramsey et al., 2017). The riboflavin uptake transport (Rfu) operon (*bb0316-bb0319*) encodes two putative permeases, RfuD (BB0316) and RfuC (BB0317), an ATP-binding component, RfuB (BB0318) and a solute binding protein, RfuA (BB00319) whose homolog in *T. pallidum* binds riboflavin (Showman et al., 2016a). *B. burgdorferi* strains lacking either RfuB (BB0318) or one of the permeases, RfuC (BB0317) were more sensitive to oxidative stress when exposed to inorganic (H_2O_2) and organic peroxides (*t*-butyl hydroperoxide) (Showman et al., 2016a; Ramsey et al., 2017; Phelan et al., 2019). Additionally, spirochetes lacking RfuB (BB0318) were more sensitive to killing by murine macrophages and were highly attenuated in their ability to infect mice (Showman et al., 2016). Since FAD is a cofactor for CoADR and thioredoxin reductase, it seems likely that a reduction in the activities of these enzymes would make these strains more sensitive to reactive oxygen species (ROS). Although RufA binds riboflavin (Deka et al., 2013) and is found in an ABC-type transporter operon, direct transport of riboflavin by the Ruf system or a decrease in riboflavin, FMN or FAD concentrations in strains lacking Ruf components has not been demonstrated. Further research on this system will help to identify what compounds are transported by Ruf (riboflavin, FMN and FAD). Also, it will be important to identify enzymes that could convert riboflavin to FMN and FAD. It is clear that riboflavin, FMN and FAD are key cofactors that promote the survival of *B. burgdorferi* during its infective cycle.

Lipid, undecaprenyl-pyrophosphate (C55-P) and lipoprotein biosynthesis

Membrane lipids, glycolipids and lipoproteins are particularly important, not only for cell growth and nutrient acquisition, but for formation of a structural scaffold and physiochemical environment for efficient membrane and membrane protein function. This is important for cells as they adapt to various environments and environmental conditions and, in the case of bacterial pathogens, to various hosts and/ or different conditions within a particular host.

Phosphatidylcholine (PC), phosphatidylglycerol (PG) and α-monogalactosyl diacylglycerol (α-MgalDAG) synthesis

Lipid and lipoprotein biochemistry in bacteria involve the synthesis of fatty acids and the attachment of these hydrophobic moieties to glycerol-3-phosphate (G3P) for the final synthesis of membrane phospholipids, glycolipids, or modification of lipoproteins. In prokaryotes and eukaryotes, there are a host of divergent enzymes to catalyze reactions that follow a basic biochemical pathway in which either acetyl-ACP or acetyl-CoA is condensed to malonyl-ACP for conversion to acetoacetyl-ACP or condensation with the growing acyl chain. Subsequent reactions catalyze the reduction, dehydration, and final reduction to acyl-ACP for more rounds of elongation. In addition, β-hydroxydecanoyl-dehydrase can catalytically trigger the synthesis of unsaturated fatty acids (Pugh and Wakil, 1965). Except for photosynthetic prokaryotes (Tasaka et al., 1996), most bacteria do not synthesize polyunsaturated fatty acids. In *B. burgdorferi,* there are no genes encoding any enzymes involved in fatty acid biosynthesis, so the bacterium must acquire fatty acids from its environment for lipoprotein and lipid biosynthesis. Therefore, Belisle et al. and Boylan et al. characterized the fatty acid composition of these molecules in *B. burgdorferi* (Belisle et al., 1994; Boylan et al., 2011). In fact, both studies detected polyunsaturated fatty acids (18:1, 18:2) in their membranes and lipoproteins, strongly suggesting that there were able to acquire these fatty acids from the growth media.

Several groups have shown that the membranes of *B. burgdorferi* are composed primarily of phosphatidylcholine (PC; ~11%), phosphatidylglycerol (PG; ~10%) and monogalactosyl diacylglycerol (α-MGalDAG; ~36) as well as an immunogenic sterol galactoside (Belisle et al., 1994; Cox and Radolf, 2001; Hossain et al., 2001). The acyl subcomponents of these moieties have been shown to be predominately palmitic acid (~55%) while 35% of the remaining fatty acids were unsaturated (oleic acid; ~22%) and polyunsaturated (linoleic or linolenic acids; 12% and 1% respectively) fatty acids. It has been shown that *B. burgdorferi* selectively incorporates these fatty acids from the growth media and that the polyunsaturated moieties are the primary targets of reactive oxygen species (ROS) in these cells (see Oxidative Stress Response, Targets of ROS) (Boylan and Gherardini, 2008). It is not known how the composition is affected during growth in the arthropod vector or the various mammalian hosts. Clearly, these varying environments might require alterations in the composition of the membranes due to physiochemical effects (temperature) or any potential changes may simply be the result ofavailability.

The synthesis of lipids, glycolipids and lipoproteins in *B. burgdorferi* require: (i) the transport of fatty acids, (ii) the activation of apo-ACP by reduced CoA (See Section Biosynthesis of co-factors), (iii) the covalent linkage of fatty acids to CoA or holo-ACP, (iv) the synthesis of CDP-1,2 diacylglycerol (CDP-DAG) from G-3-P, acyl-ACP or acyl-CoA and CTP, (v) the addition of choline, G-3-P, or galactose to CDP-DAG (or DAG) yielding PC, PG or α-MgalDAG, (vi), and finally, the addition of PG to proteins to generate lipoproteins. The first step in the process involves the transport of fatty acids which has been extensively studied in *E. coli.* Initially, it was proposed that fatty acids passively diffuse across the outer membrane and did not require a protein carrier. This seems to be the case for short-chain fatty acids (SCFAs) since specific transport proteins (C_4-C_6) for these acyl moieties have not been identified. In the absence of specific transport systems, it has been proposed that SCFAs diffuse through porins in the outer membrane. However, it was first shown in *E. coli* that transport of long chain fatty acids (LCFAs: C_{12}-C_{18}) across the outer membrane is facilitated by the protein encoded by *fadL* which is up-regulated by changes in osmolarity (Nunn and Simons, 1978; Black, 1991; Higashitani et al., 1993). FadL has been shown to bind LCFAs with high affinity and that binding opens a LCFA-specific channel that allows the movement of the fatty acid into the periplasmic space (Hearn et al., 2009; Lepore et al., 2011). When *fadL* was genetically inactivated in *E. coli*, long-chain fatty acids could still enter the cell but at a rate that could not support growth. No FadL homolog has been identified in *B. burgdorferi* and experimental data suggests that no specific LCFA transport system maybe be required in these and some other spirochetes. Cox and Radolf showed that FA-aminofluorescein conjugates could readily insert into the outer membrane of *B. burgdorferi* and *T. pallidum* (Cox and Radolf, 2001). They hypothesized that the lack of LPS in the spirochete membranes made them more permeable to lipophilic moieties promoting the uptake of FAs that could not be synthesized by these bacteria. It seems very likely that this mechanism explains SCFA and LCFA transport across the outer membrane of *B. burgdorferi.*

LCFAs in *E. coli* are transported across the cytoplasmic membrane by long chain acyl-CoA synthetase (ACAS) which is encoded by *fadD* (Black et al., 1992). ACASs are multifunctional proteins that bind LCFAs, covalently link the acyl moiety to reduced CoA and release the activated form into the cytosol. In addition to CoA, the enzyme requires ATP and it has been proposed that a soluble form of the protein binds ATP and this ATP- ACAS bound form inserts into the inner membrane. Once in the membrane, ATP-ACAS retrieves a LCFA, binds CoA, and enzymatically transfers the acyl moiety to CoA with the release of AMP and acyl-CoA into the cell. The *B. burgdorferi* genome harbors two genes which encode putative FadD homologs, *bb0137* and *bb0593* (Fraser et al., 1997). These proteins have ~36 percent identity and presumable function to transport fatty acids across the inner membrane increasing the intracellular pools of acyl-CoA for lipid, glycolipid, and lipoprotein biosynthesis (see below). In *E. coli*, acyl-CoA is the substrate for β-oxidation when fatty acids are utilized as a carbon source, however, *B. burgdorferi* does not contain any of the enzymes required to catalyze these reactions. This suggests that acyl-CoA is a substrate/precursor for only lipid, glycolipid, and lipoprotein biosynthetic pathways in *B. burgdorferi.* It should be noted that acyl-ACP plays a pivotal role in lipid biosynthesis in bacteria, and it is interesting to speculate that the transfer of the acyl moiety from acyl-CoA to ACP could target fatty acids to biosynthesis of membrane lipids. Yet, it is not clear if this transfer is taking place or required since acyl-CoA could potentially channel fatty acids into the same biosynthetic pathway(s). The exact role of acyl-CoA or acyl-ACP in lipid and lipoprotein synthesis *B. burgdorferi* remains to be determined experimentally. The process for the synthesis of membrane lipids from G-3-P and acyl-ACP is outlined in Figure 5. BB0037 (glycerol-3-phosphate acyltransferase) begins the process by catalyzing the addition of 2 acyl moieties to G-3-P generating phosphatidic acid (PA) which is subsequently converted to CDP-DAG by BB0119 (CDP-diacylglycerol synthase). It should be noted that acyl-ACP or acyl-CoA may serve as the acyl donor to G-3-P since either can potentially be generated by the transfer of an acyl moiety between CoA and ACP in a reversible reaction catalyzed by CoA acyltransferase (BB0137). Currently, it is not known which carrier serves as the major acyl donor for G-3-P. At this stage, there are three potential fates for CDP-diacylglycerol (CDP-DAG).

In one pathway, phosphatidylcholine synthase (BB0249) converts CDP-DAG to phosphatidylcholine (PC) using choline which has to be transported into the cell (Sohlenkamp et al., 2000; López-Lara and Geiger, 2001; Wang et al., 2004) (Figure 5, left branch of the pathway). In the second pathway, CDP-DAG is condensed with G-3-P to phosphatidyl-glycerol phosphate (PGP) by phosphatidylglycerol phosphate synthase (BB0721) (Wang et al., 2004). The next step requires the dephosphorylation of PGP. In *E. coli*, this reaction is catalyzed by PGP phosphatase (PgpA) which specifically hydrolyses PGP to PG. A second enzyme, encoded by *pgpB*, has similar activity against PGP but also has activity against PA suggesting that PgpA was the primary enzyme involved in PG synthesis. However, when both *pgpA* and *pgpB* were disrupted, phosphatase activity was reduced but PG synthesis remained at wild-type levels, suggesting that one or more additional phosphatases were able to keep the PG biosynthetic pathway functioning at normal levels in *E. coli* (Funk et al., 1992). The enzyme(s) responsible for this activity has not been identified. While no homologs for PgpA or PgpB have been identified in *B. burgdorferi,* the presence of high levels of PG in its membranes suggests that other, unidentified phosphatases are catalyzing this final step in the biosynthesis of PG (Figure 5, right branch of the pathway).

The major component of membrane lipids in *B. burgdorferi* is the glycolipid α-MgalDAG (Radolf et al., 1995; Ostberg et al., 2007). This important component of the outer and inner membranes of *B. burgdorferi* has been shown to be immunogenic (Kinjo et al., 2006). Glycolipids, like α-MgalDAG, are most often found in

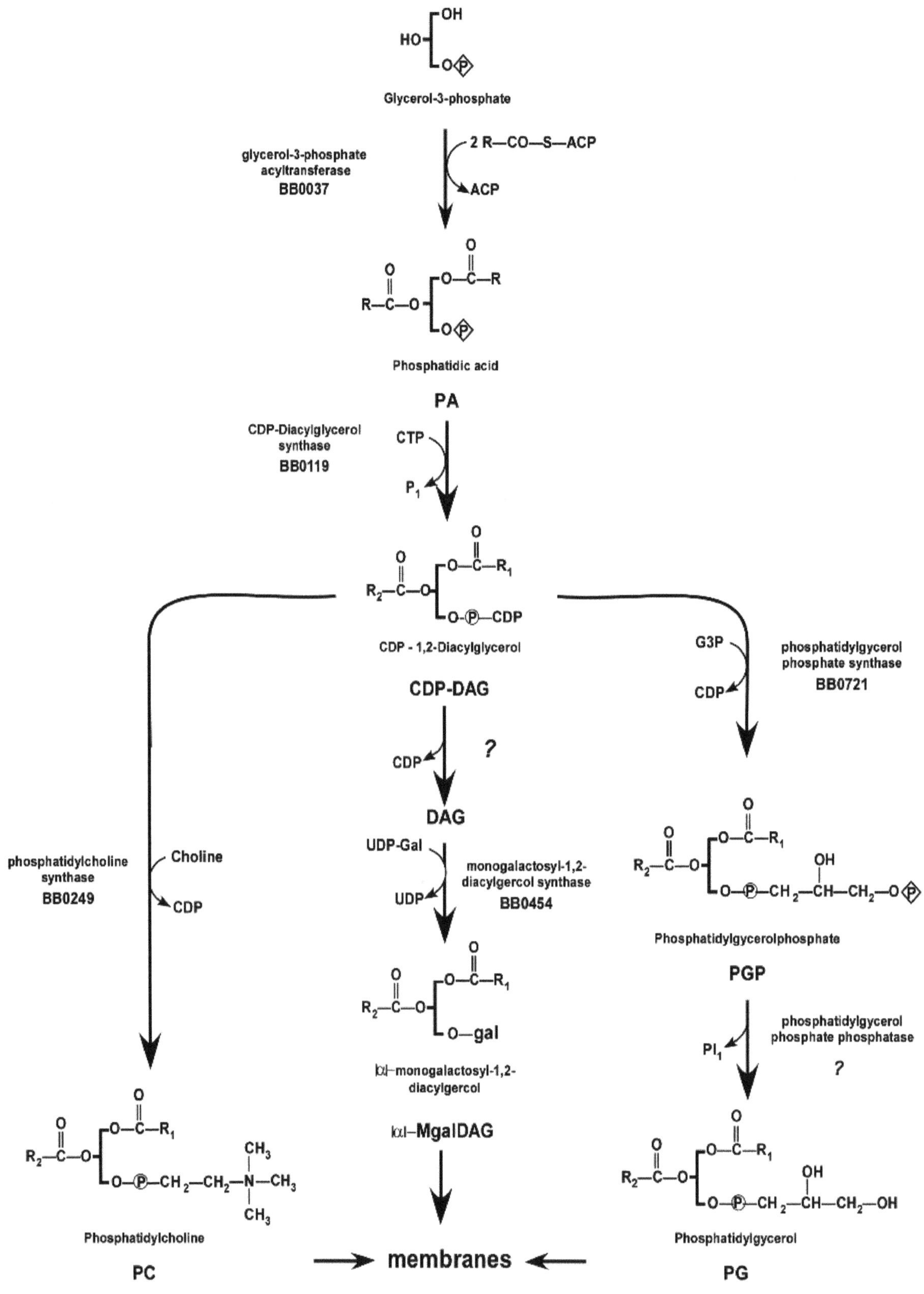

Figure 5. Putative lipid and glycolipid biosynthetic pathways in *Borrelia burgdorferi*. Figure modified with permission from Gherardini et al., 2010.

the membranes of Gram-positive bacteria and rarely in those of Gram-negative bacteria (exceptions: photosynthetic bacteria and some *Pseudomonas* species) whose most abundant glycolipid is lipopolysaccharide (Shaw, 1970). The synthesis of α-MgalDAG in *B. burgdorferi* from DAG and UDP-galactose (UDP-gal) is catalyzed by α-MgalDAG synthase, encoded by *bb0454*. This enzyme has been purified and characterized and was shown to transfer the galactose moiety from UDP-gal to DAG (Ostberg et al., 2007). UDP-Gal is essential for this reaction and is believed to be generated from UDP-Glu, an epimerase encoded by *bb0444,* which is adjacent to the gene encoding α-MgalDAG synthase. The second substrate in this reaction is DAG which was shown to be the only diacyl donor for the synthesis of α-MgalDAG. How DAG is generated in *B. burgdorferi* remains to be determined. Typically, DAG is synthesized from PA by PA phosphatase (PAS) (e.g., in *B. subtilis*), however, no homolog for PAS has been identified in *B. burgdorferi.* It is possible that an uncharacterized lipid phosphatase is catalyzing this reaction or that DAG is generated from CDP-DAG by an unknown mechanism (see Figure 5,

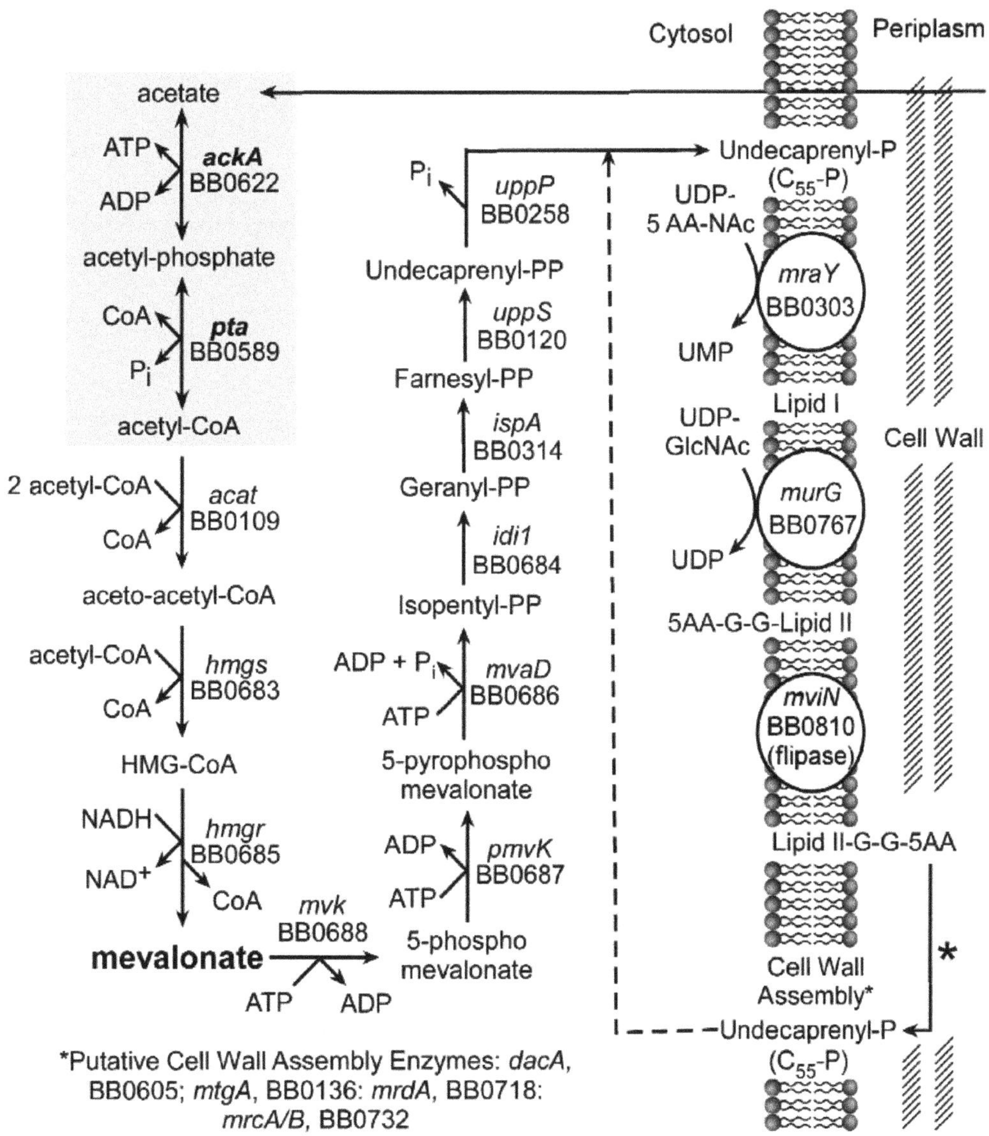

Figure 6. Schematic representation of the mevalonate pathway. Figure modified with permission from Richards et al., 2015.

central pathway). This remains to be determined experimentally.

Undecaprenyl-pyrophosphate (C55-P) synthesis
Critical for bacterial peptidoglycan biosynthesis is the translocation of sugar and amino acid components which are assembled in the cytoplasm and transferred across the plasma membrane for glycan polymerization. Half of the cycle occurs at the cytoplasmic face of the inner membrane leaflet where sugar and sugar-peptides are linked together on a carrier lipid molecule, undecaprenyl-pyrophosphate (C55-P). The complete form of the carrier lipid, called Lipid II, is transferred across the membrane by an integral membrane enzyme (flippase) where it is anchored in the outer leaflet. The disaccharide-peptide moiety is incorporated into the cell wall and crosslinked via the pentapeptide chain with the release of C55-PP. C55-PP in the outer leaflet is converted to C55-P by a pyrophosphatase and C55-P is returned to the cytoplasmic side of the membrane where the cycle is repeated. These steps are essential for cell wall biosynthesis. The linear-chain polyprenyl-phosphate lipids are pivotal for this process and are very highly conserved in bacteria (Bouhss et al., 2008; Teo and Roper, 2015; Workman and Strynadka, 2020) Bacteria require significant amounts of lipid II for cell wall biosynthesis and growth, synthesizing up to 2 x 10^5 copies of C_{55}-P per cell (Storm and Strominger, 1974; Kramer et al., 2004).

C55-P biosynthesis requires the production of isoprenoids that are synthesized by condensations of isopentyl-di-phosphate (IPP) and its isomer dimethylallyl diphosphate (DMAPP). Two pathways lead to the biosynthesis of these intermediates: (a) the methylerythritol phosphate (MEP) or non-mevalonate pathway synthesizes DMAPP and (b) the mevalonate (MVA) pathway which produces IPP. DMAPP is synthesized from pyruvate and D-glyceraldehyde-3-phosphate derived from central metabolism, while IPP is derived from the condensation of 3 molecules of acetyl-CoA (AcCoA) derived via multiple degradative and biosynthetic pathways (Kawakami and Fujisaki, 2018). The MEP pathway is found in plant chloroplasts and most bacteria while the MVA pathway is much more commonly used by Archaea, eukaryotes and some prokaryotes. Borrelia spp. use the MVA pathway to generate IPP but produce the precursor molecule, AcCoA, from a single pathway that converts acetate → acetyl-P → AcCoA (Richards et al., 2015) (Figure 6).

Acetate, AcP, and AcCoA are key metabolites in central metabolism, cell signaling and gene regulation in many eukaryotes, prokaryotes and some species of *Archaea* (Wolfe et al., 2003; Wolfe, 2010; Muller et al., 2012; Ferry, 2015). Most commonly, these molecules are produced during fermentation (acetogenesis). Typically, pyruvate dehydrogenase converts pyruvate to AcCoA and NAD^+, with AcCoA being converted to AcP by phosphate acetyltransferase (Pta) and to acetate by acetate kinase (AckA). While *Borrelia* species have *ackA* and *pta* homologs, they do not appear to use Pta and AckA for acetogenesis. There are several lines of evidence to support this conclusion. First, *B. burgdorferi* are homo-fermenters catabolizing glucose to lactate, not acetate (Johnson et al., 1984). Second, they do not harbor a gene encoding pyruvate dehydrogenase. Third, they do not harbor genes encoding pyruvate oxidase or pyruvate formate lyase, which are enzymes capable of generating AcCoA from pyruvate. Finally, they harbor no genes encoding enzymes for the β-oxidation of fatty acids or for a TCA cycle to generate and consume AcCoA for energy (Fraser et al., 1997). Taken together, these data suggest a different role for *ackA* and *pta* in *B. burgdorferi*.

The utilization or activation of acetate in *Archaea* or *Bacteria* involves four possible pathways. In acetoclastic methanogens (e.g., genera *Methnosarcina*), AckA, Pta, CO dehydrogenase, and AcCoA synthetase convert acetate to CO_2 and CH_4 (Kurth et al., 2020). Acetate uptake/utilization in bacteria can be accomplished using: (i) irreversible AcCoA synthetase (AMP-forming) that generates AcCoA in a two-step reaction, (ii) irreversible AcCoA synthetase (ADP-forming) in a single step, or (iii) via the Ack-Pta pathway (Kurth et al., 2020). The *B. burgdorferi* genome does not contain genes annotated as AcCoA synthetase, indicating that acetate activation requires AckA and Pta. Due to the limited metabolic capabilities of *B. burgdorferi*, AcCoA, generated by these enzymes, appears to be used exclusively for the synthesis of mevalonate and activated isoprenoids. Ultimately, these isoprene units are condensed to C_{55}-P for cell wall biogenesis (see Figure 6). The genome of *B. burgdorferi* harbors genes encoding enzymes for the complete synthesis of C_{55}-P, lipid I, lipid II and peptidoglycan (Figure 6). While the enzymes required for the completion of the final steps of this pathway have not been characterized experimentally, evidence suggests that the downstream portion of the pathway is intact and functions as annotated. Lipid II molecules are susceptible to certain antibiotics, such as vancomycin, which binds and disrupts lipid II function, leading to cell death in bacteria containing this structure (Martin and Breukink, 2007). *B. burgdorferi* is sensitive to vancomycin and exposure to the drug disrupts cell integrity (Dever et al., 1993; Kazragis et al., 1996). Using the pathway described above, *B. burgdorferi* uses an estimated 66 ATPs to synthesize one lipid II molecule representing a huge investment in energy to produce the required amount of this essential lipid (Storm and Strominger, 1974; Kramer et al., 2004; Martin and Breukink, 2007).

Lipoprotein synthesis

The synthesis of lipoproteins is essential in *Borrelia spp.* Genome analysis of *Borrelia* suggests that as many as >150 genes encode lipoproteins with many of these being directed toward the bacterial surface (Dowdell et al., 2017). While the function of most of these are not known, several are thought to function; (i) as transporters for sugars, peptides, fatty acids, nucleotides, metals, etc., (ii) as adhesins, (iii) for complement evasion, and (iv) as cell surface receptors. The potential pathway for synthesis of lipoproteins in *B. burgdorferi* is outlined in Figure 7. This putative pathway is believed to be similar to that outlined for *E. coli* and other Gram-negative bacteria. Diacylglycerol transferase (BB0362) mediates the transfer of DAG from PG to apo-lipoproteins (Chen and Zückert, 2011). It should be noted that the acyl side chains in DAG are heterogeneous and contain unsaturated as well as saturated fatty acid moieties. In fact, Belisle et al. have demonstrated that the lipoproteins of *B. burgdorferi* contain palmitic, oleic, linoleic and linolenic acid (Belisle et al., 1994) similar to those found in the membrane lipids. By deriving the lipid moieties on lipoproteins from PG, the holo-lipoproteins directly reflect the composition of bacterial membranes providing a conducive environment for proper lipoprotein function.

After the DAG is added to the apo-lipoprotein, the signal sequence is removed by signal peptidase II (BB0469) (Figure 7) (Regue et al., 1984). The last step in the biosynthesis is the addition of the last acyl moiety by lipoprotein acyltransferase to the NH_2 terminus of the apo-lipoprotein. The genome of *B. burgdorferi* harbors no apparent homolog to lipoprotein acyltransferase so it is not known if and how the last acyl moiety is added to the apo-lipoprotein (Fraser et al., 1997). Assembly and processing of lipoproteins in bacteria take place on the cytosolic membrane and, in Gram-negative bacteria, targeting these proteins to the correct intracellular location is critical for proper function.

Metal utilization

Approximately a third of all structurally characterized proteins are metalloproteins. Several of these metalloproteins function as essential enzymes; therefore, the metal co-factors (usually divalent transition metals) associated with them are essential. When present in excess, transition metals can be extremely toxic (Rosen, 2002; Finney and O'Halloran, 2003). Therefore, both eukaryotic and prokaryotic cells must carefully regulate the uptake and efflux of metals across biological membranes to maintain an equilibrium at sub-toxic levels (Hall, 2002; Rosen, 2002). Traditionally, essential metals have included the alkali metals potassium (K) and sodium (Na), the alkaline earth metals calcium (Ca) and magnesium (Mg), and the transition metals iron (Fe), manganese (Mn), zinc (Zn) and copper (Cu). Other transition metals which have less well-defined or more obscure biological functions include cobalt (Co), molybdenum (Mo), and nickel (Ni). Metals such as Fe are particularly important to biological systems because of their need as an essential co-factor for critical enzymes (e.g., ribonucleotide reductase and aconitase) as well as its involvement in the damaging side effects inherent to its redox biology under physiological conditions. Metalloproteins serve as biocatalysts or electron carriers and thus play an important role in the homeostasis of most living systems.

Metals are important for many biological processes including respiration, nitrogen fixation, methanogenesis, nucleic acid synthesis, oxygen transport, oxygen tolerance, proteolysis and hydrogen production/consumption since the various reaction steps in these processes are catalyzed by metalloenzymes (Butler, 1998; Andrews et al., 2003; Klebba, 2003). There are two main classes of metal-requiring enzymes that can be distinguished from one another based upon the strengths of their metal binding interactions. Metalloenzymes contain tightly bound metals, most commonly the transition metals Fe, Mn, Zn, Co, or Cu, while metal-activated enzymes loosely bind metal ions such as the alkali and alkaline earth metals Ca, K, Na, and Mg. Metals are known to participate in the catalytic process in

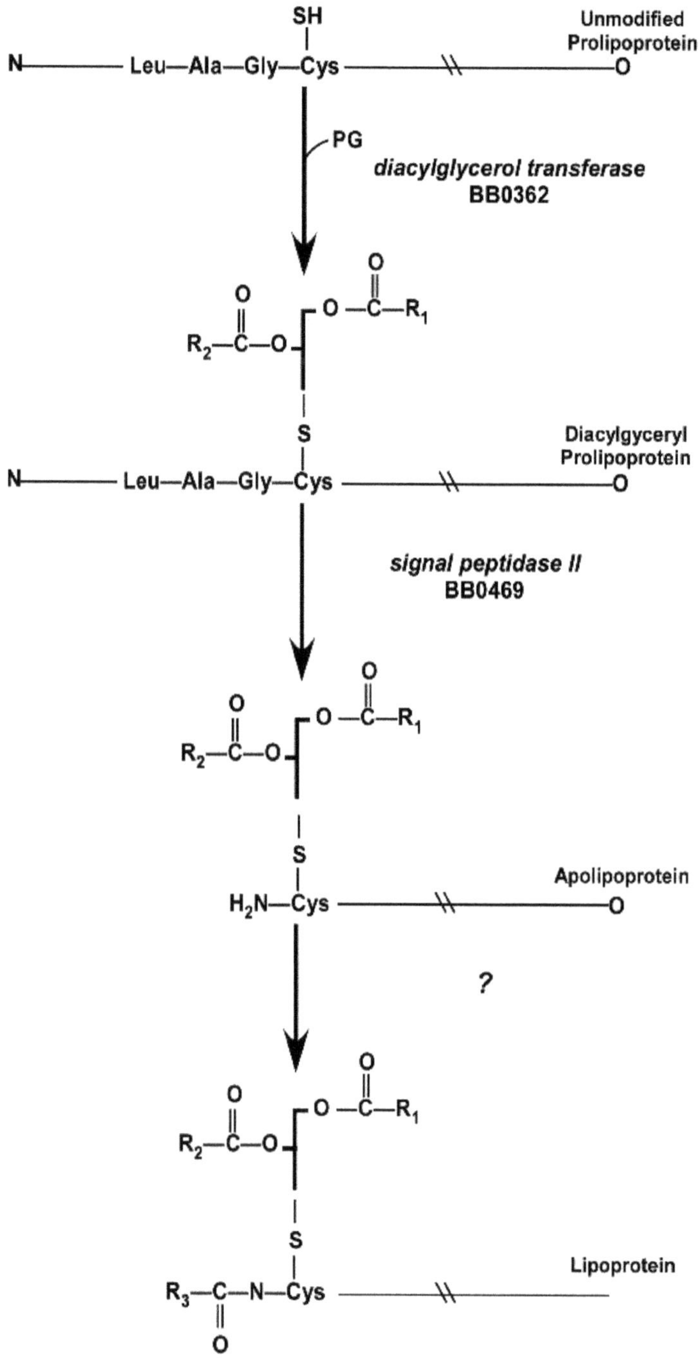

Figure 7. Putative lipoprotein biosynthetic pathway in *Borrelia burgdorferi*. Modified with permission from Gherardini et al., 2010.

three ways: i) by binding to substrates so as to orient them properly for reaction, ii) by mediating redox reactions through reversible changes in the metal's oxidation state, and iii) by electrostatically stabilizing or shielding negative charges.

Although the prior dogma dictated that all living organisms required Fe for survival, a number of studies demonstrated that there were exceptions. Specifically, the Lactobacilli have been shown to grow in Fe-depleted media (Archibald, 1983; Bruyneel et al., 1990). Under these conditions, *Lactobacillus plantarum* contains only one to two Fe atoms per cell. This extremely low Fe content is consistent with the ability of lactobacilli to grow in highly Fe-limited environments such as milk (Weinberg, 1997). It is believed that under these conditions, *L. plantarum* substitutes Mn for Fe in key metalloproteins. The genome sequence of *B. burgdorferi* indicated that this pathogen encoded few orthologues of known Fe-requiring proteins (Fraser et al., 1997). In addition, Fe limitation had no effect on growth in defined media and an independent evaluation of the total metal composition indicated that, while Mg, Mn, Zn, and Ca were readily detected, the organism carried <Fe atoms per cell (Posey and Gherardini, 2000). Thus, host-imposed Fe restriction has no apparent effect on *B. burgdorferi* growth.

The ability of bacterial pathogens, such as *B. burgdorferi*, to acquire essential elements from their host during an infection is critical for their survival (Outten and O'Halloran, 2001; Blencowe and Morby, 2003; Cornelissen, 2003; Mulrooney and Hausinger, 2003; Rensing and Grass, 2003). Although host tissues, fluids and secretions are seemingly rich in nutrients, transition metals are not as readily available. Bacterial pathogens have evolved both passive and active mechanisms for acquiring metals from their host environments (Ferguson and Deisenhofer, 2004). Passive mechanisms are associated with high rates of transport and low metal binding affinities, while active mechanisms demonstrate slower transport rates with higher binding affinities (Nikaido and Saier, 1992). With the exception of Fe, Gram-negative organisms appear to acquire cations via the use of outer membrane porins (Blencowe and Morby, 2003; Rensing and Grass, 2003). These proteins form barrel-shaped, fluid-filled channels in the outer membrane of cells allowing for the transport of small, hydrophilic solutes including divalent metals into their periplasmic space (Jap and Walian, 1996). Since porin-based mechanisms are a passive means by which to acquire metals, cations transported by this method must be present at concentrations far higher outside of the cell than inside. Potential host sources of these cations include ceruloplasmin, albumin, α2-macroglobulin and β1-globulin as well as other host serum proteins (de la Calle Guntiñas et al., 2002).

Metal transport

Analyses of the *B. burgdorferi* genome demonstrates the presence of very few predicted metal transporters (Fraser et al., 1997). Interestingly, none of those that were originally annotated by TIGR were predicted to participate in the transport of transition metals. After the publication of the *B. burgdorferi* genome, several investigators noted that the *B. burgdorferi* genome included a gene (*bb0219*) encoding a member of the ZIP family of metal transporters which are widely distributed among prokaryotes, eukaryotes, and archaea (Guerinot, 2000; Begum et al., 2002). Some members of the ZIP family are promiscuous, effectively transporting many different metals including Fe^{2+}, Zn^{2+}, Cd^{2+}, and Mn^{2+}, while others show a preference for specific metals (Grass et al., 2002; Grass et al., 2005). The prokaryotic homologues include the ZupT proteins which are found in several bacteria including *E. coli*, *Campylobacter jejuni,* and *B. burgdorferi* (Begum et al., 2002). Based on studies in *E. coli*, BB0219 appeared to be a divalent metal transporter that could transport Zn^{2+}, Mn^{2+}, Fe^{2+}, Co^{2+}, Cu^{2+}, or Cd^{2+}, although the affinities for different metals remained to be determined (Grass et al., 2002; Grass et al., 2005). Given that *B. burgdorferi* can transport detectable amounts of Zn^{2+} and Mn^{2+}, but does not accumulate Fe (Posey and Gherardini, 2000), it was proposed (Gherardini et al., 2010) that BB0219 was involved in the transport of Zn^{2+} and/or Mn^{2+}. Ouyang et al. confirmed that BB0219 is a Mn transporter (Ouyang et al., 2009a). Using a BB0219 mutant strain, they were able to show that intercellular concentrations of Mn were significantly reduced during in vitro growth of that mutant in BSKII. In addition, the strain was more sensitive to *t*-butyl hydroperoxide suggesting a role for Mn in the oxidative stress response in *B. burgdorferi* (Ouyang, 2009). This was predicable since it has been shown that the SOD enzyme from *B. burgdorferi* requires Mn for activity (Brown et al., 2019). This was the first metal transporter to be experimentally characterized in *B. burgdorferi* and was designated as Borrelia Mn transporter A (BmtA) (Ouyang et al., 2009a).

The requirement of Zn^{2+} in *B. burgdorferi* is supported by several studies indicating that this metal serves as a co-factor for important metalloproteins which are Fe^{2+}-dependent proteins in other bacteria. Specifically, BosR, the redox regulator of *B. burgdorferi* (a Ferric uptake regulator homolog) (Boylan et al., 2003; Seshu et al., 2004a; Boylan et al., 2006) and peptide deformylase (Pdf)(Nguyen et al., 2007) both require Zn^{2+} for DNA binding and enzymatic activity, respectively. In the case of BosR, the binding of Zn^{2+} is believed to be coordinated by two conserved cysteine residues and is required for efficient DNA binding (Boylan et al., 2003; Seshu et al., 2004b; Boylan et al., 2006). The exact configuration of this coordination is not known and the ability to bind additional metals, as is seen for the PerR protein from *Bacillus subtilis*, which can bind either Fe^{2+} and Zn^{2+} or Mn^{2+} and Zn^{2+} (Herbig and Helmann, 2001), remains to be determined. The Pdf from *B. burgdorferi* is a very interesting metalloprotein. All other Pdfs characterized from eukaryotes and prokaryotes have been shown to require Fe^{2+} for maximum activity in vivo. The enzyme purified from *B. burgdorferi* contains Zn^{2+} as a cofactor and, in this form, has enzyme activity levels which approach those measured for the Fe^{2+} Pdf purified from *E. coli* (Nguyen et al., 2007). Taken together, these data indicate that *B. burgdorferi* has eliminated the need for Fe^{2+} by substituting other metals for Fe^{2+} in key metalloproteins (e.g., BosR, Pdf, see section: Oxidative stress, SodA) and by eliminating biosynthetic pathways that normally require Fe^{2+} containing enzymes (see section: Ribonucleotide reductase).

With regard to other divalent metal transporters, *B. burgdorferi* also encodes both Mg^{2+} (*bb0380*) and Ca^{2+} (*bb0164*) transporters which are homologues of MgtE and CabD transporters, respectively (Saier and Paulsen, 2000). These same transporters are also found in *Treponema pallidum* (Fraser et al., 1998). Unlike *B. burgdorferi*, however, *T. pallidum* encodes other ABC transporters that are involved in the uptake of divalent metals, including Mn^{2+}, Zn^{2+}, Fe, and Co^{2+} (Fraser et al., 1998). The most well characterized *T. pallidum* ABC transporter is encoded by the *tro* operon (*troRABCD*) that transports several divalent metals including Mn^{2+}, Zn^{2+}, and perhaps Fe^{2+} (Posey et al., 1999; Hazlett et al., 2003; Desrosiers et al., 2007). In contrast to *T. pallidum*, key ABC transporters in *B. burgdorferi* are associated with peptide and ribose/galactose uptake (Fraser et al., 1997); none of the *Borrelia* ABC transporters are predicted to transport metals.

Although there is only one apparent transition metal transporter in *B. burgdorferi*, there are several secondary transporters that appear to transport monovalent cations (i.e., Na^+ and K^+). Briefly, *B. burgdorferi* encodes three Na^+/H^+ antiporters (*b0447*, *bb0637*, and *bb0638*), two dicarboxylic acid/Na^+ (or H^+) symporters (*bb0401* and *bb0729*), a pantothenate/Na^+ symporter (*bb0814*), and two K^+ transporters, *ktrA* and *ktrB* (*bb0724, bb0725*). When compared against *T. pallidum*, *B. burgdorferi* encodes three times more monovalent cation transporters (Fraser et al., 1997; Fraser et al., 1998; Saier and Paulsen, 2000); the significance of this observation is not known. *B. burgdorferi* also contains a cation diffusion facilitator (i.e., CDF family member; *bb0616*) that presumably is used to pump out excess metal to promote the proper cytosolic homeostasis to prevent intracellular levels from reaching toxic levels. Finally, genes encoding proteins putatively involved in chromate transporter/efflux (*bb0451* and *bb0452*) have been identified in the *Borrelia* genome (Table 1). Given that chromate usually pairs with Na^+ or K^+, this may serve as another modality for *B. burgdorferi* to obtain monovalent alkali metals. It is also interesting to note that the chromate transporters are genetically linked to a MscS homologue (*bb0453*), a mechanosensitive ion channel protein that helps to maintain the osmotic balance of cells as the osmolarity of their environment changes. This may be the function of this system in specific infection/colonization sites during the infective cycle (e.g., tick midgut).

Oxidative stress

Oxidative stress occurs when reactive oxygen species (ROS) exceeds the detoxification capacity of the cell. Active oxygen species are a potentially damaging consequence of life in an oxygen-rich environment and are generated as a by-product of normal aerobic metabolism (e.g., respiration or fatty acid oxidation) and/or when redox enzymes are oxidized by O_2. However, because *B. burgdorferi* generates ATP via the Embden-Meyerhof pathway, the generation of ROS does not come from the metabolism of oxygen. Pathogenic bacteria are exposed to exogenous ROS and/or reactive nitrogen species (RNS) which are used as cytotoxic weapons of the immune system and inflammatory response against invading microorganisms. The key reactive species are superoxide radicals (O_2^-), hydrogen peroxide (H_2O_2), hydroxyl radicals ($^{\bullet}OH$), and singlet oxygen, which result from the sequential univalent reductions of molecular oxygen and nitric oxide (NO). The ability to neutralize toxic ROS and RNS constitutes a formidable challenge that the bacterium must overcome to successfully establish and maintain an infection. Presumably, the conditions in which *B. burgdorferi* would need to respond to ROS and RNS include, but are not limited to: (i) in the tick following a bloodmeal and (ii) in the host due to the inflammatory response that occurs early in the infectious process (iii) Interaction of O_2 with electron shuttling molecules (e.g., FAD containing proteins)

B. burgdorferi is metabolically limited since they encode no enzymes involved in the TCA cycle or respiration via the electron transport system (Fraser et al., 1997) and oxidative phosphorylation is absent, therefore the source of the majority of endogenous O_2^- is absent. However, due to its infective cycle, *B. burgdorferi* is exposed to anaerobic and microaerobic environments and may encounter O_2 at levels that could generate biological relevant levels of cytosolic O_2^-. Hydrogen peroxide readily diffuses across biological membranes such that the intracellular concentration is easily influenced by both exogenous and endogenous sources. Intracellular production of peroxides can occur via a non-enzymatic process or by superoxide dismutase (SOD)-catalyzed dismutation. Also, *B. burgdorferi* encounters external H_2O_2 due to the inflammatory response mediated by the infection, e.g., the erythema migrans rash as well as the inflammation associated with carditis and arthritis. A proposed model for the oxidative stress response is shown in Figure 8.

Targets for ROS

The biological targets for ROS are DNA, RNA, proteins, and lipids. Superoxide can inactivate enzymes and proteins containing Fe-S clusters (e.g., aconitase). The most deleterious effects of O_2^- are related to its propensity to be converted to more potent oxidants like H_2O_2 or $^{\bullet}OH$. Hydrogen peroxide is poorly reactive with most biomolecules such as DNA, lipids, and proteins. However, H_2O_2 is capable

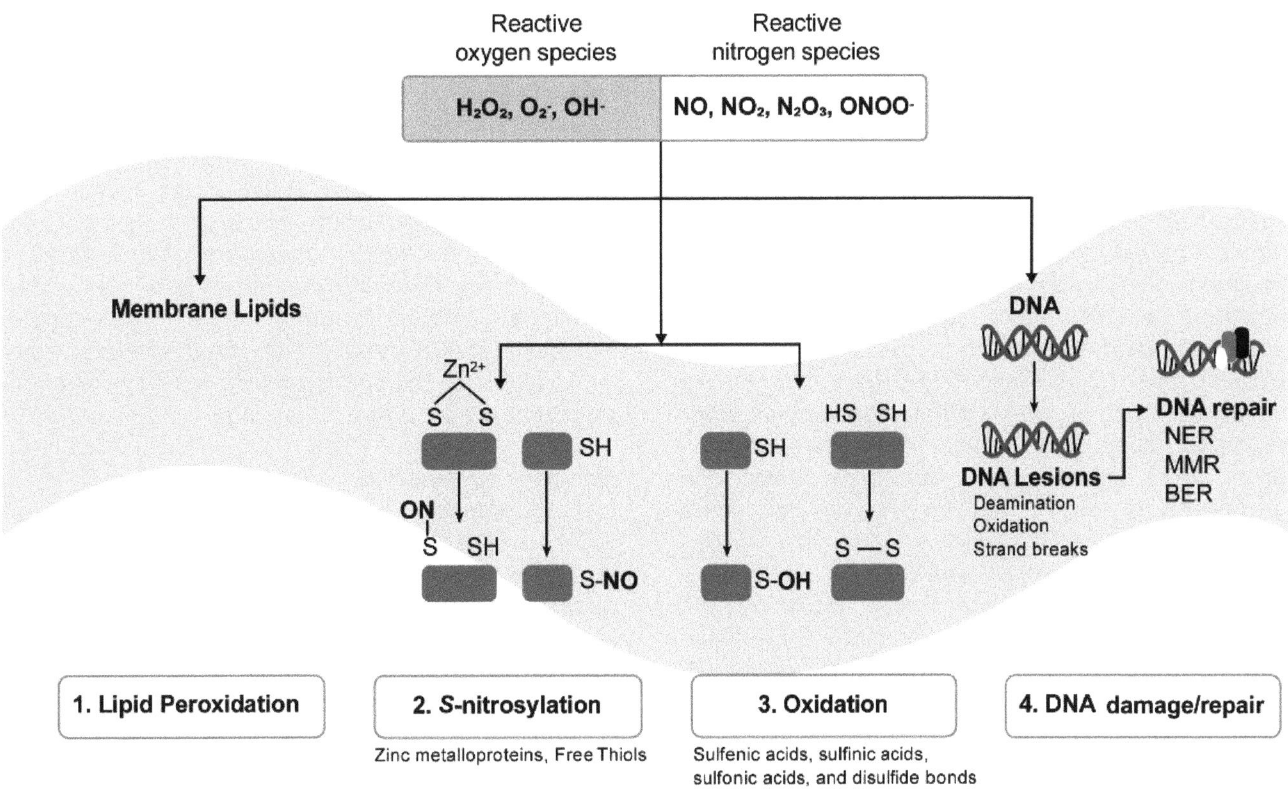

Figure 8. Reactive oxygen (ROS) and nitrogen species (RNS) encountered by *Borrelia burgdorferi*. Cellular lipids and proteins are the likely targets of ROS/RNS. Figure modified with permission from Bourret et al., 2011.

of directly oxidizing sulfhydryl groups of cysteines, particularly with ionized (thiolate, S^-) forms. H_2O_2 is capable of directly inactivating enzymes such as glyceraldhyde-3-phosphate dehydrogenases and peroxiredoxins that rely upon active-site cysteine residues for catalytic activity.

Although inactivation of proteins through the oxidation of cysteine residues may occur during oxidative stress in bacteria, they do not appear to be a major target of ROS (Imlay, 2002). The most damaging effects of H_2O_2 result from its interactions with reduced Fe (and to a lesser extent, Cu) centers in proteins, generating $^{\bullet}OH$ via the Fenton reaction (Imlay, 2003). $^{\bullet}OH$ reacts with most biomolecules and, because it is so reactive, will oxidize structures within cells depending largely upon proximity. Because Fe can localize along the phosphodiester backbone of nucleic acid, DNA is a major target of $^{\bullet}OH$. Several recent studies suggest that DNA serves as the major target for host toxicity for ROS and RNS assuming that Fe accumulates within the cell and intracellular concentrations of H_2O_2 approach 0.2-1 μM (Keyer et al., 1995; Keyer and Imlay, 1996, 1997; Seaver and Imlay, 2001). However, *B. burgdorferi* has been shown to not require Fe for growth and has intracellular Fe concentrations estimated to be <10 atoms per cell (Posey and Gherardini, 2000). At these low levels, it appeared unlikely that *B. burgdorferi* DNA is a target for ROS. This was supported by the fact that *B. burgdorferi* cells exposed to high concentrations of ROS (e.g., H_2O_2) showed no increase in the spontaneous mutation rate or the number of DNA lesions (AP sites or 8-oxoguanine) (Boylan et al., 2008). However, the apparent lack of detectable DNA damage in *B. burgdorferi* treated with ROS is most likely due to the efficient repair of ROS-induced DNA lesions by the nucleotide excision repair (NER) and methyl-directed mismatch repair (MMR) pathways (Salman-Dilgimen et al., 2013; Bourret et al., 2016), as strains with mutations in either pathway show increased susceptibility to killing by H_2O_2. These observations suggest H_2O_2 undergoes Fenton or Fenton-like chemistry within *B. burgdorferi* to produce DNA damaging $^{\bullet}OH$. Whether $^{\bullet}OH$ production is the result of H_2O_2 reacting with trace Fe or Cu within *B. burgdorferi* remains to be investigated.

Bacteria that incorporate polyunsaturated fatty acids into membranes risk having these structures targeted for peroxidation (Tasaka et al., 1996; Beermann et al., 2000; Ben-Menachem et al., 2003; Imlay, 2003). In this situation, free radicals can attack the polyunsaturated fatty acids and initiate lipid peroxidation. A primary effect of lipid peroxidation is the loss of membrane fluidity that can alter the physical properties of the membrane and the function of membrane-associated proteins. The operative question is whether bacterial cells contain polyunsaturated fatty acids and, more pointedly, whether *Borrelia* spp. contain these molecules. Since *B. burgdorferi* cannot synthesize its own lipids, it must scavenge them from its host so that their membrane composition reflects the host's lipid profile or their growth medium (Barbour, 1984; Fraser et al., 1997). Several publications demonstrated that *B. burgdorferi* could incorporate polyunsaturated fatty acids, specifically linoleic and linolenic acids, into their membranes (Beermann et al., 2000; Ben-Menachem et al., 2003; Boylan et al., 2008), suggesting that their membranes could undergo peroxidation if exposed to ROS (Boylan and Gherardini, 2008; Boylan et al., 2008). Based on the previously mentioned lack of Fe accumulation and the lack of molecules to generate endogenous superoxide, it is possible that polyunsaturated lipids serve as the most likely target for oxidative damage in *B. burgdorferi*. In fact, when *Borrelia* cells were treated with oxidants, including H_2O_2, *t*-butyl hydroperoxide, and lipoxidase, the polyunsaturated lipid composition, specifically linoleic acid, decreased in amount suggesting that they were substrates for oxidation while other fatty acids remained unaffected (Boylan et al., 2008). In addition, as linoleic acid concentrations decreased, aldehydes accumulated. Furthermore, the overall morphology of *B. burgdorferi* was adversely affected by exposure to oxygen and to lipoxidase. Taken together, these observations indicate that the presence of polyunsaturated fatty acids puts *B. burgdorferi* lipids containing these targets at risk for oxidative damage.

Defenses against ROS

Cellular defenses against the damaging effects of ROS involve both enzymatic and nonenzymatic components. The enzymatic components may directly scavenge ROS or may act by producing the nonenzymatic antioxidants. There are five categories of enzymes that provide the bulk of protection against deleterious reactions involving ROS in bacteria: superoxide dismutases (SODs)/superoxide reductases (SORs), hydroperoxidases, catalases, AhpR homologues, and Dps/Dpr proteins. Superoxide dismutases, ubiquitous among aerobic and most aerotolerant organisms, provide a defense against oxidative stress by catalyzing the dismutation of superoxide into oxygen and hydrogen peroxide through alternate oxidation and reduction of their catalytic metal ions (McCord and Fridovich, 1969; McCord et al., 1971). To date, four types of superoxide dismutases have been identified: Mn-dependent (MnSODs), Fe-dependent (FeSODs), Cu/Zn-dependent (Cu/ZnSODs), and Ni-dependent (NiSOD) (Youn et al., 1996; Palenik et al., 2003). The MnSOD was the first identified form and is encoded by *sodA* (Touati, 1983). Austin *et al* (Austin et al., 1981; Whitehouse et al., 1997) cloned and characterized a gene encoding a SOD from *B. burgdorferi*. This gene had 68% similarity to the *sodA* from *Haemophilus influenzae* and the partially purified enzyme had superoxide dismutase activity. The subsequent genome sequence confirmed that *B. burgdorferi* encoded a SOD designated as *sodA* (*bb0153*) (Fraser et al., 1997). Recent studies have confirmed SodA is a Mn-dependent superoxide dismutase, whose activity is supported by the manganese-rich environment within the cytosol of *B. burgdorferi* (Seshu et al., 2004a; Seshu et al., 2004b). SodA is required for the virulence of *B. burgdorferi* in a murine model of infection, suggesting *B. burgdorferi* encounters potentially lethal levels of ROS during infection (Sanjuan et al., 2009). However, ROS-hypersensitive *B. burgdorferi* strains harboring mutations in the NER or MMR pathways are capable of infecting mice (Salman-Dilgimen et al., 2013; Bourret et al., 2016), suggesting the inability of *sodA* mutants to establish infection may be independent of the effects of host-derived oxidants. This is supported by the observation that *sodA*-deficient *B. burgdorferi* strains show reduced expression of proteins required for mammalian infection such as OspC (Esteve-Gassent et al., 2015). This may be the consequence of the endogenous generation of ROS within *B. burgdorferi* that are capable of oxidizing and inactivating a variety of proteins required for infection in the absence of SodA.

The product of SOD is H_2O_2, which itself is a reactive species. Cellular catalases, which destroy H_2O_2 with remarkable rapidity, and peroxidases, which require NADH or NADPH as an electron source, are both capable of ridding the cells of H_2O_2. Since *B. burgdorferi* does not encode for either a catalase or peroxidase, it was unclear how *B. burgdorferi* contended with the H_2O_2 that was generated by the SodA. A potential explanation comes from the study of CoADR (encoded by *bb0728*) (Boylan et al., 2006). It was demonstrated that reduced CoA could react to reduce H_2O_2, thereby acting as a weak peroxidase.

Borrelia cells contain a large amount of the low molecular weight thiol coenzyme A (CoA) that could function as a glutathione equivalent in *B. burgdorferi* and thus serve as a molecular sink for toxic oxidizing compounds within the cytoplasm (Boylan et al., 2006). Specifically, CoA and CoADR function as a target for redox recycling and an enzyme to regenerate the reduced form of CoA (CoASH) from oxidized CoA, respectively (Boylan et al., 2006). This, coupled with the absence of any identifiable gene encoding a glutathione synthase or glutathione reductase within the borrelial genome, strongly suggested that CoA and CoADR are responsible for establishing and maintaining the redox balance in *B. burgdorferi* (Fraser et al., 1997; Boylan et al., 2006). The genes encoding thioredoxin (*trxA*; *bb0061*) and thioredoxin reductase (*trxB*; *bb0515*) may represent an additional redox active system in *B. burgdorferi*; however, these proteins have not yet been characterized. Recent studies have shown a riboflavin ABC transporter encoded by *bb0316 –319* is involved in the resistance of *B. burgdorferi* to ROS (Showman et al., 2016b; Ramsey et al., 2017), which may be due to the fact that other antioxidant defense systems, including CoA/CoADR and TrxA/B, require cofactors FMN and FAD that are synthesized from riboflavin.

In addition to the few detoxification genes found in the *B. burgdorferi* genome, a homologue of *dps*/*dpr* is encoded by open reading frame *bb0690, napA*. Dps/Dpr proteins have been associated with several functions related to oxidative stress in other bacterial systems. One activity is to non-specifically bind to and protect DNA (Almirón et al., 1992; Martinez and Kolter, 1997; Wolf et al., 1999; Frenkiel-Krispin et al., 2001; Morikawa et al., 2006) or to protect the DNA indirectly without physical interaction (Ceci et al., 2003). The other activity attributed to Dps/Dpr proteins is that of a ferritin-like molecule that can sequester Fe (Grant et al., 1998; Ceci et al., 2003; Ren et al., 2003; Roy et al., 2004) such that the cell is protected from Fenton based chemistry and subsequent damage to its DNA. Given that *B. burgdorferi* does not accumulate and has no apparent requirement for Fe (Posey and Gherardini, 2000), it is unclear as to how the ferritin-like activity would be relevant to this spirochete unless it could also sequester other toxic transition metals that *B.*

burgdorferi obtained. Yet, the recombinant form of the *B. burgdorferi* NapA protein does attain a dodecyl stoichiometric configuration that binds iron (Li et al., 2007). However, the NapA protein did not bind DNA relative to the *E. coli* homologue (Li et al., 2007). Given the lack of Fe present in *B. burgdorferi* and the inability of NapA to non-specifically protect DNA, it has been unclear how NapA contributes in a detoxification scheme to protect *B. burgdorferi* from oxidative stress. It is important to note that other Dps homologues protect DNA in the absence of direct binding (Ceci et al., 2003), therefore it is possible that *B. burgdorferi* utilizes a similar strategy. A recent study shed new insights into the functional role of NapA, showing that the protein may serve as a metallothionein through the iron- and copper-binding activity of its C-terminal poly-Cys tail (Wang et al., 2012). These observations have led to the renaming of NapA to BicA (*Borrelia* iron- and copper-binding protein A). *B. burgdorferi* strains lacking *napA/bicA* had lower cellular concentrations of both Fe and Cu compared to wild-type controls. The presence of Fe in *B. burgdorferi* strains used in this study contrast with previous studies that have found significantly less iron in *B. burgdorferi* (Posey and Gherardini, 2000; Troxell et al., 2013). Moreover, *napA/bicA*-deficient *B. burgdorferi* have shown increased resistance to oxidative stress produced by H_2O_2 compared to wild-type controls, suggesting that the bound metal in NapA/BicA may contribute an antioxidant defense of *Borrelia* (Wang et al., 2012).

In addition to the putative antioxidant defenses mentioned above, several additional genes have been implicated in protecting *B. burgdorferi* against ROS. In a recent study, a transposon library of *B. burgdorferi* mutants was screened for resistance to ROS leading to the identification of 37 genes implicated in resistance to H_2O_2 and 16 genes in resistance to tertiary-butyl hydroperoxide (Ramsey et al., 2017). Three of the candidate antioxidant defense genes encode predicted membrane proteins (*bb0017*, *bb0164*, and *bb0202*) of which, *bb0164*, is required for regulating cellular levels of manganese. The observation that *bb0164* was required for resistance to ROS and maintaining cellular manganese concentrations is consistent with the phenotypes observed in a *B. burgdorferi* strain harboring a mutation in the manganese transporter, BmtA (Ouyang et al., 2009a). Together, these studies revealed an essential role for manganese homeostasis in the antioxidant defenses of *B. burgdorferi*.

Regulation of the oxidative stress response

The most extensively studied organisms for understanding the genetic regulation in response to reactive oxygen/nitrogen species have been the enteric bacteria *E. coli* and *Salmonella typhimurium* as well as the Gram-positive bacterium *Bacillus subtilis*. These bacteria can activate the expression of some 30-40 proteins in response to ROS (Pomposiello et al., 2001; Zheng et al., 2001). The key regulators of these adaptive responses are the SoxRS and OxyR transcription factors of *E. coli* and the peroxide-sensitive repressor PerR from *Bacillus subtilis*. OxyR is a transcription factor, which when oxidized, forms an intramolecular disulfide bond. It is this oxidized form that is capable of activating the transcription of genes involved in the oxidative stress response (Zheng et al., 1998). SoxR contains a redox active Fe-S cluster that regulates DNA binding (Hidalgo et al., 1997). The PerR protein is a metalloprotein which binds one or more copies of a conserved 14-bp inverted repeat to repress expression of target genes. The oxidation of the Zn^{2+}-Fe^{2+} form of PerR (the most redox sensitive form of this regulator) by H_2O_2 abolishes binding, causing derepression of target genes (Herbig and Helmann, 2001). Of all of these regulators, *B. burgdorferi* does not contain a SoxRS or OxyR homologue, but does contain a PerR ortholog (annotated as Fur by TIGR and is a member of the Fur family of regulatory proteins) that responds to ROS, albeit in a manner more consistent with functional OxyR, and has been designated as BosR (Boylan et al., 2003).

Several studies have shown that BosR functions to regulate genes involved in the oxidative stress response, nitrosative stress response and dramatically influences the Rrp2-RpoN-RpoS regulatory cascade (for a detailed discussion of the effects of BosR on this regulatory cascade see Samuels et al. Chapter). Most of these analyses are based on; (i) protein/DNA binding studies, (ii) analyses of BosR mutants, (iii) RNA sequencing, and (iv) exposure to various stressors.

BosR is capable of binding DNA under both oxidizing and reducing conditions, requires Zn^{2+} for efficient binding, and has an unusually large (45 bp to 50 bp) target sequence binding domain based on DNase I footprint analysis (Boylan et al., 2003). BosR binds target sequences upstream of *sodA* (*bb0153*) (Seshu et al., 2004a), *napA/bicA* (*bb0690*) (Boylan et al., 2003; Seshu et al., 2004a), *cdr (Boylan et al., 2006)*, *oppA4 (Zhou et al., 2018) rpoS* (*bb0771*) (Ouyang et al., 2011; Samuels, 2011; Ouyang et al., 2014), and *bosR* (*bb0647*) (Katona et al., 2004). BosR acts as a transcriptional activator of oxidative stress genes, including *cdr*, *napA,* and *oppA4* (Boylan et al., 2003; Medrano et al., 2007), although other investigators have suggested that this regulator functions as a repressor (Katona et al., 2004). Several members of the Fur family of regulators can function as both an activator and a repressor (Delany et al., 2004; Grifantini et al., 2004; Gao et al., 2008), therefore it is likely that BosR also functions in this manner. Several recent studies have shed light on the functional role of BosR on *B. burgdorferi* gene expression and virulence (Boylan et al., 2003; Ouyang et al., 2009b; Hyde et al., 2010). Consistent with its hypothesized role in regulating the antioxidant defenses of *B. burgdorferi,* strains harboring mutations of *bosR* are more susceptible to killing by ROS compared to wild-type controls (Hyde et al., 2010; Phelan et al., 2019). BosR encodes four cysteine residues that are arranged in two distinct CXXC motifs (Cys-114/117 and Cys-153/156) that were previously speculated to serve as a Zn-binding motif and a redox sensor, respectively (Boylan et al., 2003). Results from a recent study support a role for both CXXC motifs in binding Zn, BosR dimerization, and BosR binding to cognate promoters (Ouyang et al., 2011; Ouyang et al., 2014). The ability of BosR to regulate the expression of antioxidant defense genes likely depends on the redox status of the cysteines encoded within its two CXXC motifs. Previous studies have shown disparate effects of ROS on the ability of BosR to bind to its cognate promoters with one study purportedly showing increased DNA binding in tertiary-butyl peroxide, while the other showed ROS-dependent inhibition of BosR binding (Boylan et al., 2003; Katona et al., 2004). Further investigations will be required to determine the role of BosR's putative redox-active cysteines in regulating the expression of its antioxidant defense genes.

Nitrosative stress

Nitric oxide (NO) is a diatomic free radical typically produced in living organisms through the oxidation of L-arginine to L-citrulline by nitric oxide synthases (NOS). Despite being a free radical, most of the biochemical effects of NO are due to its oxidative congeners, collectively termed reactive nitrogen species (RNS) (e.g., $NO_2^{\bullet}$, N_2O_3 and $ONOO^-$). Some RNS are membrane permeable (NO, $NO_2^{\bullet}$ and N_2O_3) while others, due to their charge, are not (e.g., $ONOO^-$). RNS act both as signaling molecules and as effector molecules in antimicrobial defenses of plants, arthropods, and mammals (Fang, 2004; Fang and Vázquez-Torres, 2019). RNS catalyze oxidative, nitrative, and nitrosative modifications to an array of biomolecules including DNA, protein thiols, secondary amines, tyrosine residues, cytochromes, thiol-bound metals, and polyunsaturated fatty acids of lipid membranes (Fang, 2004; Fang and Vázquez-Torres, 2019). Much attention has been given to characterizing the effects of RNS on microbial pathogens such as *Salmonella enterica* serovar Typhimurium, *Escherichia coli* and *Mycobacterium tuberculosis* (Fang, 2004). The redox biochemistry of *B. burgdorferi* is unique compared to these well-studied microbial pathogens as it lacks both intracellular iron and iron-containing proteins (i.e. cytochromes, etc.) (Posey and Gherardini, 2000), which are well-appreciated targets of RNS.

Studies on the effects of RNS during *in vitro* growth of *B. burgdorferi* have primarily relied on the use of NO donating molecules such as diethylamine NONOate (DEA/NO) (Bourret et al., 2011; Salman-Dilgimen et al., 2013; Bourret et al., 2016; Ramsey et al., 2017), which release NO at distinct rates depending on environmental growth conditions such as pH and temperature. Under microaerobic growth conditions (5% CO_2, 3% O_2), the release of NO by DEA/NO promotes a primarily nitrosative environment where the production of N_2O_3 is associated with widespread *S*-nitrosylation of free or zinc-bound cysteine thiols of *B. burgdorferi* proteins (Bourret et al., 2011). In contrast to studies on the effects of ROS on *B. burgdorferi*, RNS do not appear to trigger peroxidation of polyunsaturated fatty acids within *B. burgdorferi* cell membranes, which is in accord with the ability of NO and N_2O_3 to terminate lipid peroxidation reactions. The impact of RNS on *B. burgdorferi* extends beyond the *S*-nitrosylation of cysteine thiols. Several additional studies have revealed RNS are capable of damaging *B. burgdorferi* DNA, however, these DNA lesions are efficiently repaired by the nucleotide excision repair (NER) pathway (Bourret et al., 2011; Salman-Dilgimen et al., 2013; Bourret et al., 2016; Ramsey et al., 2017).

In contrast to *B. burgdorferi*, relatively little is known about the impact of RNS on the relapsing fever group *Borrelia*. One recent study revealed *Borrelia turicatae* is much more sensitive to killing by RNS compared to *B. burgdorferi* when challenged with DEA/NO during in vitro growth (Bourret et al., 2019). It is unclear if other relapsing fever group spirochetes exhibit similar patterns of sensitivity to RNS as was observed in *B. turicatae*. The increased sensitivity of *B. turicatae* to RNS suggests it may not encounter a significant degree of nitrosative stress through the course of its infectious cycle as has been shown for *B. burgdorferi*.

RNS in the infective cycle

Mammals express three NOS isoforms: endothelial NOS (eNOS), neuronal NOS (nNOS), and inducible NOS (iNOS). While RNS generated by eNOS and nNOS serve a primary function in cell signaling, the antimicrobial activity of RNS in mammalian systems is due to the production of NO by iNOS (Tsutsui et al., 2015). While the susceptibility of *B. burgdorferi* to RNS generated chemically during *in vitro* culture has been established by several studies (Ma et al., 1994; Modolell et al., 1994; Bourret et al., 2011; Salman-Dilgimen et al., 2013; Bourret et al., 2016; Ramsey et al., 2017), there does not appear to be a role for iNOS-generated RNS in the innate immune defenses against *B. burgdorferi* (Seiler et al., 1995; Brown and Reiner, 1999). This is evidenced by no differences in the burden of *B. burgdorferi*, or in disease phenotypes (such as arthritis) in wild-type and iNOS-deficient mice, or in mice treated with the NOS competitive inhibitor NG-L-monomethyl-L-arginine. Furthermore, RNS-hypersensitive *B. burgdorferi* strains harboring mutations in the NER pathway genes are able to successfully infect mice following intradermal inoculation or following transmission by *I. scapularis* (Bourret et al., 2016).

Hematophagous arthropods, including ticks, use RNS to induce dilation of their salivary ducts promoting the delivery of saliva enriched with a diverse cocktail of anti-inflammatory proteins to the bite site (Anderson et al., 2017; Narasimhan et al., 2019). Salivary RNS are also hypothesized to induce vasodilation at the host bite site promoting acquisition of the blood meal. Both Ixodid and Argasid ticks harbor a single NOS isoform that shares greater homology to eNOS and nNOS compared to iNOS, which contributes to the innate host defenses of mammals. It is, therefore, likely that tick-borne RNS contribute to cellular signaling rather than to the innate host defenses of ticks against *Borrelia* or other tick-borne pathogens. Nevertheless, the levels of RNS produced in the salivary glands and midguts of *I. scapularis* are substantial as shown by the decreased survival of RNS-hypersensitive *B. burgdorferi* strains with mutations in the NER pathway genes in *I. scapularis* nymphs following a blood meal (Bourret et al., 2016).

Little is known about whether relapsing fever group spirochetes encounter a significant level of nitrosative stress during their infectious cycle. A recent study highlighted the expression of NOS and the production of RNS within the midguts of the soft-bodied tick *Ornithodoros turicatae* (Bourret et al., 2019), suggesting *B. turicatae* may have to withstand RNS while residing in the midgut of unfed ticks. Additionally, it is unclear from this study whether the production of RNS within tick tissues increases during blood meal acquisition as was previously shown for *I. scapularis* (Bourret et al., 2016). Further investigations will be required to determine whether RNS pose a significant challenge to *B. turicatae* and other relapsing fever group spirochetes during infection.

Defenses against RNS

Studies investigating the effects of RNS on *B. burgdorferi* have focused primarily on their effects on viability and what genes are involved in defense against RNS. This is illustrated by the fact that *S*-nitrosylation of free and zinc-bound cysteine thiols in *B. burgdorferi* is associated with RNS-dependent cytotoxicity (Bourret et al., 2011), and studies showing the NER pathway plays a primary role in alleviating the genotoxicity inflicted by RNS and promotes survival of *B. burgdorferi* in the nitrosative environment of the *I. scapularis* midgut (Salman-Dilgimen et al., 2013; Bourret et al., 2016; Ramsey et al., 2017). Several genes encode candidates for thiol-based detoxification of RNS including the low molecular weight thiols homocysteine (Hcy) and coenzyme A (CoA), as well as thioredoxin and thioredoxin reductase (TrxA/TrxB), and NapA/BicA. *B. burgdorferi* encodes an incomplete methionine biosynthetic pathway that should result in the accumulation of high levels of Hcy (Babb et al., 2005; Riley et al., 2007), which plays a crucial role in the antioxidant defenses of other pathogens like *Salmonella* (De Groote et al., 1995). Presently, the role of the Hcy-synthesizing LuxS enzyme in *B. burgdorferi* infectivity is unclear. Although not strictly required for infectivity (Hübner et al., 2003; Blevins et al., 2004; Arnold et al., 2015), *luxS*-deficient *B. burgdorferi* have been shown to have a competitive disadvantage in vivo compared to wild-type controls (Hübner et al., 2003; Blevins et al., 2004; Arnold et al., 2015).

Coenzyme A is present in millimolar concentrations in *B. burgdorferi* and may directly detoxify RNS or maintain redox homeostasis through the activity of CoADR (Boylan et al., 2006). Moreover, CoADR is necessary for infection of mice and promotes survival in feeding nymphs where RNS are highly produced (Eggers et al., 2011). Thioredoxin and thioredoxin reductase contribute to the anti-nitrosative defenses of both *Salmonella* and *Neisseria gonorrhoeae* (Bjur et al., 2006; Potter et al., 2009), however, their role in *B. burgdorferi* physiology or virulence has yet to be determined. The function of NapA/BicA has been elusive. However, a recent study indicated that NapA/BicA may serve as a metallothionein through the iron- and copper-binding activity of its C-terminal, poly-Cys tail (Wang et al., 2012). NapA/BicA has also been suggested to have a role in the protection of the inner membrane of *B. burgdorferi* from lipid peroxidation. Similarly, NapA/BicA could act as a sink for RNS, protecting the *B. burgdorferi* cell envelope from nitro-oxidative damage through oxidation or S-nitrosylation of its C-terminal poly-Cys tail. Recently, a screen of a transposon library of *B. burgdorferi* mutants identified additional genes associated with the anti-nitrosative defenses of *B. burgdorferi* including *bb0839* encoding a putative ribonuclease H1 and *bb0431* encoding a putative nucleotide binding protein that shares homology to CobQ (Ramsey et al., 2017). The mechanisms by which these putative anti-nitrosative defenses protect *B. burgdorferi* against RNS during in vitro growth or during the course of its infectious cycle remains to be determined.

Osmotic stress

The tick midgut and salivary glands are sites where *B. burgdorferi* encounters different environmental and physiological conditions. These conditions change before, during and after feeding, but most would not be considered extreme compared to conditions encountered by free-living bacteria. For example, temperature (23–34°C), oxygen (mostly anaerobic to ~2–3% O_2 during feeding), pH (6.8 in flattened or feeding ticks) do not vary significantly while nutrients (nutrient rich to starvation), ROS or RNS may be considered more significant challenges to cell viability. Another physiochemical parameter, osmolarity, undergoes moderate fluctuations during tick feeding and *B. burgdorferi* cells regulate two regulatory networks (Hk1-Rrp1, Rrp2-RpoN-RpoS) that are required for survival and virulence. What is noteworthy is that *B. burgdorferi* is very well adapted to these conditions and senses minor changes in the tick "environment" to adapt, grow, and regulate the expression of protective/ mechanisms, and, importantly, key virulence factors (e.g., OspC, DbpA, etc.) required for successful transmission.

The osmolarity (mOsm) in the bloodmeal (~300-600 mOsm), saliva (~323.5±6.4 mOsm) and hemolymph (~311.5±33 mOsm) has been measured from feeding ticks (Bontemps-Gallo et al., 2016). Unpredictably, *B. burgdorferi* was only able to grow normally between 250 mOsm and 650 mOsm, which closely matches the range of osmolarity observed in the bloodmeal during tick feeding (~300–600 mOsm). The bacterial growth, morphology, and motility were dramatically affected by osmolarity outside this narrow range. Interestingly, at 250 to 300 mOsm, *B. burgdorferi* cells upregulated the levels of virulence factors (OspC, DbpA) which are required for infection. Following osmoadaptation, genes encoding proteins potentially involved in osmoregulation, *proV* (*bb0146*), *proX* (*bb144*), and *gltP* (*bb0729*) were induced at low osmolarity. The proteins encoded by these genes are involved in osmolyte transport. *gltP* encodes a putative cysteine transporter while the others are part of the ProU locus which includes *proV* (ATP binding subunit, *bb0146*), proW (integral membrane protein, *bb0145* and *proX* (periplasmic glycine betaine binding protein, *bb0144*). Mutants that did not express the putative L-glutamate transporter (gltP) or the ProU system were more sensitive to changes in osmolarity than wild-type cells, suggesting that L-glutamate and proline are osmoprotectant molecules in *B. burgdorferi*. Ion transport has also been shown to be involved in osmotolerance and osmoadaptation. Gene expression analyses of *B. burgdorferi* cells grown at different osmolarity identified specific osmo-responsive transporters: (a) three Na^+/H^+ antiporters, (*bb0447*), *nhaC-1* (*bb0637*) and *nhaC-2* (*bb638*), (b) a potassium uptake system, *ktrAB* (*bb0724*-5), (c) a $K^+/Na^+/Ca^{2+}$ antiporter (*bb0164*), and (d) a Mg^{2+} transporter, *mgtE*, (*bb0380*). The expression of both *nhaC-1* and *nhaC-2* increased 10-fold at 250 mOsm, suggesting an import of H^+ and export of Na $^+$ is involved in osmoadaptation. The expression of *bb0164* ($K^+/Na^+/Ca^{2+}$ antiporter), *ktrA,* and *ktrB* increased 3-fold at the lower osmolarity, suggesting an adaptive flux of K^+, Na^+ and/or Ca^{2+}. These systems are involved in osmotic adaption despite the narrow range of osmolarity to which *B. burgdorferi* is exposed.

Other interesting trends occurred in *B. burgdorferi* cells at physiologically relevant osmolarities. Immunoblots of cell lysates isolated from cells grown at low osmolarity (250 mOsm), indicated that the cells increased the expression of virulence related proteins, such as OspC, DbpA, and BBA66, that are required for the successful transmission and survival of *B. burgdorferi* in mammalian hosts (Schwan et al., 1995; Schwan and Piesman, 2000; Fischer et al., 2003; Grimm et al., 2004; Yang et al., 2004; Gilmore et al., 2007; Blevins et al., 2009; Samuels, 2011). Analyses by qRT-PCR of RNA isolated from these cells showed increased transcription of *ospC, dbpA,* and *bba66* correlating with the increase in expression of these proteins in immunoblots. An increase in the expression of *rpoN* and *rpoS* also was observed at low osmolarity. Additionally, immunoblot analyses of mutants B31-A3Δ*rpoN* and B31-A3Δ*rpoS* indicated that OspC, DbpA, and BBA66 were not induced in these mutants at low osmolarity. Since it has been shown that OspC, DbpA and other virulence factors are controlled by the Rrp2-RpoN-RpoS regulatory cascade (Hübner et al., 2001), it seems likely that low osmolarity is directly affecting this regulatory network.

High osmolarity not only affected motility and cell viability but also had another interesting effect on *B. burgdorferi*. Analyses of a B31-5A4Δ*rrp1* mutant indicated that this strain was sensitive to osmolarities >500 mOsm compared to the parent strain, B31-5A4, and the cells rapidly lysed after <4h of exposure to increased osmolarity. Rrp1, the response regulator in the Hk1-Rrp1 two component system, functions as a di-guanylate cyclase (Caimano et al., 2011; He et al., 2011; Kostick et al., 2011; Iyer et al., 2015). C-di-GMP acts as a secondary messenger for signal transduction in bacteria and the levels of c-di-GMP increased dramatically at high osmolarity (Bontemps-Gallo et al., 2016). Rrp1 has also been shown to be required for tick colonization, motility, and the regulation of genes involved in glycerol metabolism (Caimano et al., 2011; Kostick et al., 2011; Caimano et al., 2015; Iyer et al., 2015). Caimano et al. showed that B31-5A4Δ*rrp1* was virulent in mice. However, this mutant lysed after acquisition by ticks that were fed on mice infected with this strain or when the mutant was introduced into ticks by artificial feeding (Caimano et al., 2015). Collectively, these data suggest that at high osmolarity, Rrp1: i) was required for survival in ticks, ii) had increased diguanylate cyclase activity, iii) was required for tick colonization, and iv) could regulate *B. burgdorferi* motility. This indicated that, over the narrow range of osmolarities encountered in feeding ticks, *B. burgdorferi* adapts and regulates the genes involved in acquisition and long-term survival through Hk1-Rrp1. More importantly, the drop in osmolarity measured during tick feeding (600-300 mOsm), along with multiple other cues (temperature, nutrient excess, etc.), triggers the Rrp2-RpoN-RpoS regulatory cascade and the expression of multiple virulence factors.

Acid stress

Bacterial pathogens experience acid stress when exposed to a drastic decrease in the extracellular pH or when weak organic acids diffuse across cellular membranes, and rapidly deprotonate in the cytoplasm, leading to an increase in the cytoplasmic H^+ ion concentration and a decrease in the intracellular pH(pH_i)(Foster, 1999). The decrease in pH_i can affect proteins, enzymatic activity, nucleic acids, and destroy the proton motive force (PMF) (Bearson et al., 1997; Castanie-Cornet et al., 1999; Foster, 1999; Kanjee and Houry, 2013). The acid stress response works to stabilize pH_i and retain the PMF (Bearson et al., 1997). The acid stress response mechanisms have been best characterized in enteric bacteria which encounter drastic changes in extracellular pH as well as encounter organic carboxylic acids as they transit from the stomach into the lower digestive tract. (Bearson et al., 1997; Castanie-Cornet et al., 1999; Foster, 1999; Richard and Foster, 2003; Richard and Foster, 2004; Wilks and Slonczewski, 2007; Slonczewski et al., 2009).

In *B. burgdorferi*, the addition of weak organic acids to the growth medium promotes the expression of RpoS and RpoS-dependent genes (e.g., OspC, BBA66, DbpA, etc.) (Xu et al., 2010; Dulebohn et al., 2017; Lin et al., 2018). Weak acid-triggered induction was originally presumed to be a result of increased levels of intracellular acetyl-phosphate (Ac-P), which in turn, initiated the Rrp2-RpoN-RpoS cascade (Xu et al., 2010). However, in the absence of Ac-P, weak acid-triggered induction still occurs (Richards et al., 2015). This observation led to the discovery of an acid stress response in *B. burgdorferi* that activates the Rrp2-RpoN-RpoS cascade as well as proteins (e.g. NapA/BicA, CoADR) in the oxidative stress response (Dulebohn et al., 2017).

The weak organic acid, benzoate has been used to study the acid stress response in *E. coli* due to its ability to decrease pH_i and induce a robust acid stress response (Wilks and Slonczewski, 2007; Slonczewski et al., 2009). When benzoate is added to the growth medium of *B. burgdorferi*, a significant decrease in pH_i occurs with the concomitant increase in the levels of the regulators RpoS, BosR, and the downstream effectors OspC, BBA66, and NapA/BicA (Dulebohn et al., 2017). The induction of the acid stress response in *B. burgdorferi*, following a decrease in pH_i, was not limited to the weak acid benzoate and a correlation between the proportion of protonated acids available and the level of expression of RpoS and RpoS-dependent factors was observed (Dulebohn et al., 2017). The higher the proportion of protonated weak acid available to diffuse across the membrane, the greater the induction of the acid stress response. Additionally, when the acidification of the cytoplasm was neutralized with exogenous arginine, it prevented the activation of the Rrp2-RpoN-RpoS cascade and decreased OspC, BBA66, and NapA/BicA production. The acid stress response also increased the production of proteins involved in the response to oxidative stress, including BosR. Interestingly, the acid and oxidative stress responses are linked in other bacteria, most likely though reactive thiols in the sensing residues of key regulatory proteins (Mols et al., 2010; Chi et al., 2019). This also appears to be the case for *B. burgdorferi*.

The membrane-permeable weak acids acetate and lactate have both been identified in millimolar quantities in the blood meal of fed ticks (Dulebohn et al., 2017). The weak acids were detected at significant levels late in the feeding cycle and likely elicit an acid stress response in *B. burgdorferi* to protect spirochetes from multiple stressors found in the tick midgut (e.g., reactive nitrogen and reactive oxygen species) and promote long-term survival. To protect against the deleterious effects of acid stress, bacteria must maintain both the pH_i and their PMF to survive (Bearson et al., 1997; Foster, 1999; Richard and Foster, 2004; Slonczewski et al., 2009). *B. burgdorferi* lacks many of the systems utilized by other bacteria to neutralize excess protons in the cytoplasm, including amino acid synthesis, amino acid decarboxylases, fatty acid biosynthesis, most neutralizing metabolites, an F1/FO ATPase, and some proton pumping transporters (Bearson et al., 1997; Foster, 1999; Slonczewski et al., 2009); (Fraser et al., 1997). However, homologs of some *B. burgdorferi* enzymes and transporters have been previously shown in other bacteria to restore a favorable pH_i. These include: (i) putative coupled proton exchange with cations (BB0637-BB0638, BB0447, BB0724-BB0725, BB0380) and Na^+/H^+ antiporter (BB0637-38), (ii) transporting neutralizing buffering molecules (polyamines: BB0639-BB0640), (iii) amino acid deaminases (BB0841-BB0843), and (iv) proton export via a V_0V_1 vacuolar-type ATPase (BB090-BB096). Consistent with this role, the transcription of *bb0637-bb0638* (Na+/H+), *bb0724-bb0725* (K^+ transport), and *bb0380* (Mg^{++} transport) are all induced under acid stress conditions. The gene encoding the arginine transporter, ArcD (*bb0843*), a putative arginine/ornithine symport system, is also upregulated when *B. burgdorferi* cells were exposed to acid stress and exogenous arginine neutralized the effects of the monocarboxylic acid benzoate. After transport, arginine can be hydrolyzed to citrulline and NH_3 by ArcA (BB0841) and ultimately to ornithine and carbamoyl-P by ArcB (BB0842). NH_3, arginine, and ornithine would be excellent cytosolic buffering molecules. Additionally, the genes encoding a proton export V_0V_1 vacuolar-type ATPase (*bb090-bb096*) were induced under acid stress. Without respiratory enzyme complexes, *B. burgdorferi* utilizes this ATPase to hydrolyze ATP and pump protons across the cytoplasmic membrane to maintain the PMF and it is likely that the vacuolar ATPase is a major defense against acid stress (See section: V-type ATPase).

Acknowledgements

This work was supported by the Division of Intramural Research, National Institute of Allergy and Infectious Diseases, National Institutes of Health.

References

Abouhamad, W.N., Manson, M., Gibson, M.M., and Higgins, C.F. (1991). Peptide transport and chemotaxis in *Escherichia coli* and *Salmonella typhimurium*: characterization of the dipeptide permease (Dpp) and the dipeptide-binding protein. Mol microbio *5*, 1035-1047. https://doi.org/10.1111/j.1365-2958.1991.tb01876.x

Aguirre, J.D., Clark, H.M., McIlvin, M., Vazquez, C., Palmere, S.L., Grab, D.J., Seshu, J., Hart, P.J., Saito, M., and Culotta, V.C. (2013). A manganese-rich environment supports superoxide dismutase activity in a Lyme disease pathogen, *Borrelia burgdorferi*. J Biol Chem *288*, 8468-8478. https://doi.org/10.1074/jbc.M112.433540

Alberts, A.W., and Vagelos, P.R. (1966). Acyl carrier protein. J Biol Chem, 5201-5204.

Almirón, M., Link, A.J., Furlong, D., and Kolter, R. (1992). A novel DNA-binding protein with regulatory and protective roles in starved *Escherichia coli*. Genes Dev *6*, 2646-2654. https://doi.org/10.1101/gad.6.12b.2646

Altendorf, K., Booth, I.R., Gralla, J., Greie, J.C., Rosenthal, A.Z., and Wood, J.M. (2009). Osmotic stress. EcoSal Plus *3*. https://doi.org/10.1128/ecosalplus.5.4.5

Anand, R., Kaminski, P.A., and Ealick, S.E. (2004). Structures of purine 2'-deoxyribosyltransferase, substrate complexes, and the ribosylated enzyme intermediate at 2.0 A resolution. Biochem *43*, 2384-2393. https://doi.org/10.1021/bi035723k

Anderson, J.M., Moore, I.N., Nagata, B.M., Ribeiro, J.M.C., Valenzuela, J.G., and Sonenshine, D.E. (2017). Ticks, *Ixodes scapularis*, feed repeatedly on white-footed mice despite strong inflammatory response: An expanding paradigm for understanding tick-host interactions. Front Immunol *8*, 1784. https://doi.org/10.3389/fimmu.2017.01784

Andrews, S.C., Robinson, A.K., and Rodríguez-Quiñones, F. (2003). Bacterial iron homeostasis. FEMS Microbiol Rev *27*, 215-237. https://doi.org/10.1016/S0168-6445(03)00055-X

Archibald, F. (1983). *Lactobacillus plantarum*, an organism not requiring iron. FEMS Microbiol Lett *19*, 29-32. https://doi.org/10.1111/j.1574-6968.1983.tb00504.x

Arnold, W.K., Savage, C.R., Antonicello, A.D., and Stevenson, B. (2015). Apparent role for *Borrelia burgdorferi* LuxS during mammalian infection. Infect Immun *83*, 1347-1353. https://doi.org/10.1128/iai.00032-15

Austin, F.E., Barbieri, J.T., Corin, R.E., Grigas, K.E., and Cox, C.D. (1981). Distribution of Superoxide Dismutase, Catalase, and Peroxidase activities among *Treponema pallidum* and other spirochetes. Infect immun *33*, 372-379.

Babb, K., von Lackum, K., Wattier, R.L., Riley, S.P., and Stevenson, B. (2005). Synthesis of autoinducer 2 by the lyme disease spirochete, *Borrelia burgdorferi*. J Bacteriol *187*, 3079-3087. https://doi.org/10.1128/jb.187.9.3079-3087.2005

Bacher, A., Eberhardt, S., Fischer, M., Kis, K., and Richter, G. (2000). Biosynthesis of vitamin B2 (Riboflavin). Annu rev of nutr *20*, 153-167. https://doi.org/10.1146/annurev.nutr.20.1.153

Barbour, A.G. (1984). Isolation and cultivation of Lyme disease spirochetes. Yale J Biol Med *57*, 521-525.

Barbour, A.G., Putteet-Driver, A.D., and Bunikis, J. (2005). Horizontally acquired genes for purine salvage in *Borrelia* Spp. causing Relapsing Fever. Infect Immun *73*, 6165-6168. https://doi.org/10.1128/IAI.73.9.6165-6168.2005

Bárcena-Uribarri, I., Thein, M., Barbot, M., Sans-Serramitjana, E., Bonde, M., Mentele, R., Lottspeich, F., Bergström, S., and Benz, R. (2014). Study of the protein complex, pore diameter, and pore-forming activity of the *Borrelia burgdorferi* P13 porin. J Biol Chem *289*, 18614-18624. https://doi.org/10.1074/jbc.M113.539528

Bearson, S., Bearson, B., and Foster, J.W. (1997). Acid stress responses in enterobacteria. FEMS Microbiol Lett *147*, 173-180. https://doi.org/10.1111/j.1574-6968.1997.tb10238.x

Bedlack, R., Hardiman, O., Heiman-Patterson, T., Gutmann, L., Bromberg, M., Ostrow, L., Carter, G., Kabashi, E., Bertorini, T., Mozaffar, T., *et al.* (2012). ALS Untangled No. 17: "When ALS Is Lyme". Amyotroph Lateral Scler *13*, 487-491. https://doi.org/10.3109/17482968.2012.717796

Beermann, C., Lochnit, G., Geyer, R., Groscurth, P., and Filgueira, L. (2000). The lipid component of lipoproteins from *Borrelia burgdorferi*: structural analysis, antigenicity, and presentation via human dendritic cells. Biochem Biophys Res Commun *267*, 897-905. https://doi.org/10.1006/bbrc.1999.2057

Begum, N.A., Kobayashi, M., Moriwaki, Y., Matsumoto, M., Toyoshima, K., and Seya, T. (2002). *Mycobacterium Bovis* BCG cell wall and lipopolysaccharide induce a novel gene, BIGM103, encoding a 7-TM protein: identification of a new protein family having Zn-transporter and Zn-metalloprotease signatures. Genom *80*, 630-645. https://doi.org/10.1006/geno.2002.7000

Belisle, J.T., Brandt, M.E., Radolf, J., and Norgard, M.V. (1994). Fatty acids of *Treponema pallidum* and *Borrelia burgdorferi* lipoproteins. J Bacteriol *176*, 2151-2157.

Ben-Menachem, G., Kubler-Kielb, J., Coxon, B., Yergey, A., and Schneerson, R. (2003). A newly discovered cholesteryl galactoside from *Borrelia burgdorferi*. Proc Natl Acad Sci U S A *100*, 7913-7918.

Beyenbach, K.W., and Wieczorek, H. (2006). The V-type H+ ATPase: molecular structure and function, physiological roles and regulation. J Exp Biol *209*, 577-589. https://doi.org/10.1242/jeb.02014

Bjur, E., Eriksson-Ygberg, S., Aslund, F., and Rhen, M. (2006). Thioredoxin 1 promotes intracellular replication and virulence of *Salmonella enterica* serovar *typhimurium*. Infect Immun *74*, 5140-5151. https://doi.org/10.1128/iai.00449-06

Black, P.N. (1991). Primary sequence of the *Escherichia coli fadL* gene encoding an outer membrane protein required for long-chain fatty acid transport. J Bacteriol *173*, 435-442. https://doi.org/10.1128/jb.173.2.435-442.1991

Black, P.N., DiRusso, C.C., Metzger, A.K., and Heimert, T.L. (1992). Cloning, sequencing, and expression of the *fadD* gene of *Escherichia coli* encoding Acyl Coenzyme A Synthetase. J Biol Chem *267*, 25513-25520.

Blencowe, D.K., and Morby, A.P. (2003). Zn(II) metabolism in prokaryotes. FEMS Microbiol Rev *27*, 291-311. https://doi.org/10.1016/S0168-6445(03)00041-X

Blevins, J.S., Revel, A.T., Caimano, M.J., Yang, X.F., Richardson, J.A., Hagman, K.E., and Norgard, M.V. (2004). The *luxS* gene is not required for *Borrelia burgdorferi* tick colonization, transmission to a mammalian host, or induction of disease. Infect Immun *72*, 4864-4867. https://doi.org/10.1128/iai.72.8.4864-4867.2004

Blevins, J.S., Xu, H., He, M., Norgard, M.V., Reitzer, L., and Yang, X.F. (2009). Rrp2, a sigma54-dependent transcriptional activator of *Borrelia burgdorferi*, activates rpoS in an enhancer-independent manner. J Bacteriol *191*, 2902-2905. https://doi.org/10.1128/jb.01721-08

Bontemps-Gallo, S., Gaviard, C., Richards, C.L., Kentache, T., Raffel, S.J., Lawrence, K.A., Schindler, J.C., Lovelace, J., Dulebohn, D.P., Cluss, R.G., *et al.* (2018). Global profiling of lysine acetylation in *Borrelia burgdorferi* B31 reveals its role in central metabolism. Front Microbiol *9*, 2036. https://doi.org/10.3389/fmicb.2018.02036

Bontemps-Gallo, S., Lawrence, K., and Gherardini, F.C. (2016). Two different virulence-related regulatory pathways in *Borrelia burgdorferi* are directly affected by osmotic fluxes in the blood meal of feeding *Ixodes* ticks. PLoS Pathog *12*, e1005791. https://doi.org/10.1371/journal.ppat.1005791 eCollection 2016 Aug.

Bouhss, A., Trunkfield, A.E., Bugg, T.D., and Mengin-Lecreulx, D. (2008). The biosynthesis of peptidoglycan lipid-linked intermediates. FEMS Microbiol Rev *32*, 208-233. https://doi.org/10.1111/j.1574-6976.2007.00089.x

Bourret, T.J., Boylan, J.A., Lawrence, K.A., and Gherardini, F.C. (2011). Nitrosative damage to free and zinc-bound cysteine thiols underlies nitric oxide toxicity in wild-type *Borrelia burgdorferi*. Mol Microbiol *81*, 259-273. https://doi.org/10.1111/j.1365-2958.2011.07691.x

Bourret, T.J., Boyle, W.K., Zalud, A.K., Valenzuela, J.G., Oliveira, F., and Lopez, J.E. (2019). The relapsing fever spirochete *Borrelia turicatae* persists in the highly oxidative environment of its soft-bodied tick vector. Cell Microbiol *21*, e12987. https://doi.org/10.1111/cmi.12987

Bourret, T.J., Lawrence, K.A., Shaw, J.A., Lin, T., Norris, S.J., and Gherardini, F.C. (2016). The nucleotide excision repair pathway protects *Borrelia burgdorferi* from nitrosative stress in *Ixodes scapularis* ticks. Front microbiol *7*, 1397. https://doi.org/10.3389/fmicb.2016.01397 eCollection 2016.

Boylan, J.A., and Gherardini, F.C. (2008). Determining the cellular targets of reactive oxygen species in *Borrelia burgdorferi*. Methods Mol Biol (Clifton, NJ) *431*, 213-221. https://doi.org/10.1007/978-1-60327-032-8_17

Boylan, J.A., Hummel, C.S., Benoit, S., Garcia-Lara, J., Treglown-Downey, J., Crane, E.J., and Gherardini, F.C. (2006). *Borrelia Burgdorferi bb0728* encodes a Coenzyme A Disulphide Reductase whose function suggests a role in intracellular redox and the oxidative stress response. Mol Microbiol *59*, 475-486. https://doi.org/10.1111/j.1365-2958.2005.04963.x

Boylan, J.A., Lawrence, K.A., Downey, J.S., and Gherardini, F.C. (2008). *Borrelia burgdorferi* membranes are the primary targets of reactive oxygen species. Mol Microbiol *68*, 786-799. https://doi.org/10.1111/j.1365-2958.2008.06204.x

Boylan, J.A., Posey, J.E., and Gherardini, F.C. (2003). *Borrelia* oxidative stress response regulator, BosR: A distinctive Zn-dependent transcriptional activator. Proc Natl Acad Sci U S A *100*, 11684-11689. https://doi.org/10.1073/pnas.2032956100

Boyle, W.K., Groshong, A.M., Drecktrah, D., Boylan, J.A., Gherardini, F.C., Blevins, J.S., Samuels, D.S., and Bourret, T.J. (2019). DksA controls the response of the Lyme disease spirochete *Borrelia burgdorferi* to starvation. J Bacteriol *201*. https://doi.org/10.1128/jb.00582-18

Boyle, W.K., Hall, L.S., Armstrong, A.A., Dulebohn, D.P., Samuels, D.S., Gherardini, F.C., and Bourret, T.J. (2020). Establishment of an in vitro RNA polymerase transcription system: a new tool to study transcriptional activation in *Borrelia burgdorferi*. Sci Rep *10*, 8246. https://doi.org/10.1038/s41598-020-65104-y

Bradbeer, C. (1993). The proton motive force drives the outer membrane transport of cobalamin in Escherichia coli. J Bacteriol *175*, 3146-3150. https://doi.org/10.1128/jb.175.10.3146-3150.1993

Brown, C.R., and Reiner, S.L. (1999). Development of lyme arthritis in mice deficient in inducible nitric oxide synthase. J Infect Dis *179*, 1573-1576. https://doi.org/10.1086/314774

Brown, G., Broxham, A.H., Cherrington, S.E., Thomas, D.C., Dyer, A., Stejskal, L., and Bingham, R.J. (2019). Expression, purification and metal utilization of recombinant SodA from *Borrelia burgdorferi*. Protein Expr Purif *163*, 105447. https://doi.org/10.1016/j.pep.2019.105447

Bruyneel, B., Vande Woestyne, M., and Verstraete, W. (1990). Iron complexation as a tool to direct mixed Clostridia-Lactobacilli fermentations. J Applied Bacteriol *69*, 80-85. https://doi.org/10.1111/j.1365-2672.1990.tb02914.x

Bugrysheva, J., Dobrikova, E.Y., Godfrey, H.P., Sartakova, M.L., and Cabello, F.C. (2002). Modulation of *Borrelia burgdorferi* stringent response and gene expression during extracellular growth with tick cells. Infect Immun *70*, 3061-3067. https://doi.org/10.1128/iai.70.6.3061-3067.2002

Bugrysheva, J., Dobrikova, E.Y., Sartakova, M.L., Caimano, M.J., Daniels, T.J., Radolf, J.D., Godfrey, H.P., and Cabello, F.C. (2003). Characterization of the stringent response and rel(Bbu) expression in *Borrelia burgdorferi*. J Bacteriol *185*, 957-965. https://doi.org/10.1128/jb.185.3.957-965.2003

Bugrysheva, J.V., Bryksin, A.V., Godfrey, H.P., and Cabello, F.C. (2005). *Borrelia burgdorferi rel* is responsible for generation of guanosine-3'-diphosphate-5'-triphosphate and growth control. Infect Immun *73*, 4972-4981. https://doi.org/10.1128/iai.73.8.4972-4981.2005

Bugrysheva, J.V., Pappas, C.J., Terekhova, D.A., Iyer, R., Godfrey, H.P., Schwartz, I., and Cabello, F.C. (2015). Characterization of the Rel_{Bbu} regulon in *Borrelia burgdorferi* reveals modulation of glycerol metabolism by (p)ppGpp. PLoS One *10*, e0118063. https://doi.org/10.1371/journal.pone.0118063 10.1371/journal.pone.0118063. eCollection 2015.

Bunikis, I., Denker, K., Ostberg, Y., Andersen, C., Benz, R., and Bergström, S. (2008). An RND-type efflux system in *Borrelia burgdorferi* is involved in virulence and resistance to antimicrobial compounds. PLoS Pathog *4*, e1000009. https://doi.org/10.1371/journal.ppat.1000009

Butler, A. (1998). Acquisition and utilization of transition metal ions by marine organisms. Sci *281*, 207-210. https://doi.org/10.1126/science.281.5374.207

Byram, R., Stewart, P.E., and Rosa, P.A. (2004). The essential nature of the ubiquitous 26-kilobase circular replicon of *Borrelia burgdorferi*. J Bacteriol *186*, 3561-3569. https://doi.org/10.1128/JB.186.11.3561-3569.2004

Caimano, M.J., Drecktrah, D., Kung, F., and Samuels, D.S. (2016). Interaction of the Lyme disease spirochete with its tick vector. Cell Microbiol *18*, 919-927. https://doi.org/10.1111/cmi.12609

Caimano, M.J., Dunham-Ems, S., Allard, A.M., Cassera, M.B., Kenedy, M., and Radolf, J.D. (2015). Cyclic di-GMP modulates gene expression in Lyme disease spirochetes at the tick-mammal interface to promote spirochete survival during the blood meal and tick-to-mammal transmission. Infect Immun *83*, 3043-3060. https://doi.org/10.1128/iai.00315-15

Caimano, M.J., Iyer, R., Eggers, C.H., Gonzalez, C., Morton, E.A., Gilbert, M.A., Schwartz, I., and Radolf, J.D. (2007). Analysis of the RpoS regulon in *Borrelia burgdorferi* in response to mammalian host signals provides insight into RpoS function during the enzootic cycle. Mol Microbiol *65*, 1193-1217. https://doi.org/10.1111/j.1365-2958.2007.05860.x

Caimano, M.J., Kenedy, M.R., Kairu, T., Desrosiers, D.C., Harman, M., Dunham-Ems, S., Akins, D.R., Pal, U., and Radolf, J.D. (2011). The hybrid histidine kinase Hk1 is part of a two-component system that is essential for survival of *Borrelia burgdorferi* in feeding *Ixodes scapularis* ticks. Infect Immun *79*, 3117-3130. https://doi.org/10.1128/iai.05136-11

Carnrot, C., Wehelie, R., Eriksson, S., Bölske, G., and Wang, L. (2003). Molecular characterization of thymidine kinase from *Ureaplasma urealyticum*: nucleoside analogues as potent inhibitors of *Mycoplasma* growth. Mol microbiol *50*, 771-780. https://doi.org/10.1046/j.1365-2958.2003.03717.x

Castanié-Cornet, M.P., Cam, K., Bastiat, B., Cros, A., Bordes, P., and Gutierrez, C. (2010). Acid stress response in *Escherichia coli*: mechanism of regulation of *gadA* transcription by RcsB and GadE. Nucleic Acids Res *38*, 3546-3554. https://doi.org/10.1093/nar/gkq097

Castanie-Cornet, M.P., Penfound, T.A., Smith, D., Elliott, J.F., and Foster, J.W. (1999). Control of acid resistance in *Escherichia coli*. J Bacteriol *181*, 3525-3535.

Ceci, P., Ilari, A., Falvo, E., and Chiancone, E. (2003). The Dps protein of *Agrobacterium tumefaciens* does not bind to DNA but protects it toward oxidative cleavage: X-Ray crystal structure, iron binding, and hydroxyl-radical scavenging properties. J Biol Chem *278*, 20319-20326. https://doi.org/10.1074/jbc.M302114200

Champomier Verges, M.C., Zuniga, M., Morel-Deville, F., Perez-Martinez, G., Zagorec, M., and Ehrlich, S.D. (1999). Relationships between arginine degradation, pH and survival in *Lactobacillus sakei*. FEMS Microbiol Lett *180*, 297-304.

Chattopadhyay, M.K., Tabor, C.W., and Tabor, H. (2003). Polyamines protect *Escherichia coli* cells from the toxic effect of oxygen. Proc Natl Acad Sci U S A *100*, 2261-2265. https://doi.org/10.1073/pnas.2627990100

Chen, S., and Zückert, W.R. (2011). Probing the *Borrelia burgdorferi* surface lipoprotein secretion pathway using a conditionally folding protein domain. J Bacteriol *193*, 6724-6732. https://doi.org/10.1128/jb.06042-11

Cheng, W., and Roth, J.R. (1994). Evidence for two NAD kinases in *Salmonella typhimurium*. J Bacteriol *176*, 4260-4268. https://doi.org/10.1128/jb.176.14.4260-4268.1994

Chi, B.K., Huyen, N.T.T., Loi, V.V., Gruhlke, M.C.H., Schaffer, M., Mäder, U., Maaß, S., Becher, D., Bernhardt, J., Arbach, M., *et al.* (2019). The disulfide stress response and protein S-thioallylation caused by allicin and diallyl polysulfanes in *Bacillus subtilis* as revealed by transcriptomics and proteomics. Antioxidants (Basel) *8*. https://doi.org/10.3390/antiox8120605

Coburn, J., Chege, W., Magoun, L., Bodary, S.C., and Leong, J.M. (1999). Characterization of a candidate *Borrelia burgdorferi* beta3-chain integrin ligand identified using a phage display library. Mol Microbiol *34*, 926-940. https://doi.org/10.1046/j.1365-2958.1999.01654.x 10.1046/j.1365-2958.1999.01654.x.

Coburn, J., and Cugini, C. (2003). Targeted mutation of the outer membrane protein P66 disrupts attachment of the Lyme disease agent, *Borrelia burgdorferi*, to integrin alphavbeta3. Proc Natl Acad Sci U S A *100*, 7301-7306. https://doi.org/10.1073/pnas.1131117100

Cohen, S.S. (1998). Bacterial metabolism and polyamines: . In a guide to the polyamines, S.S. Cohen, ed. (New York: Oxford University Press), pp. 94-121.

Cornelissen, C.N. (2003). Transferrin-iron uptake by Gram-negative bacteria. Front Biosci: J Virt Lib *8*, 836-847. https://doi.org/10.2741/1076

Corona, A., and Schwartz, I. (2015). *Borrelia burgdorferi*: Carbon metabolism and the tick-mammal enzootic cycle. Microbiol Spectr *3*. https://doi.org/10.1128/microbiolspec.MBP-0011-2014

Cox, D.L., and Radolf, J.D. (2001). Insertion of fluorescent fatty acid probes into the outer membranes of the pathogenic spirochaetes *Treponema pallidum* and *Borrelia burgdorferi*. Microbiol *147*, 1161-1169. https://doi.org/10.1099/00221287-147-5-1161

Curtis, M.W., Hahn, B.L., Zhang, K., Li, C., Robinson, R.T., and Coburn, J. (2018). Characterization of stress and innate immunity resistance of wild-type and Δp66 *Borrelia burgdorferi*. Infect Immun *86*. https://doi.org/10.1128/iai.00186-17

Cusumano, Z.T., and Caparon, M.G. (2015). Citrulline protects *Streptococcus pyogenes* from acid stress using the arginine deiminase pathway and the F1Fo-ATPase. J Bacteriol *197*, 1288-1296. https://doi.org/10.1128/jb.02517-14

Cusumano, Z.T., Watson, M.E., Jr., and Caparon, M.G. (2013). *Streptococcus pyogenes* arginine and citrulline catabolism promotes infection and modulates innate immunity. Infect Immun *82*, 233-242. https://doi.org/10.1128/iai.00916-13

Das, R., Hegyi, H., and Gerstein, M. (2000). Genome analyses of spirochetes: A study of the protein structures, functions and metabolic pathways in *Treponema pallidum* and *Borrelia burgdorferi*. J Mol Microbiol Biotechnol 2, 387-392.

Davidson, A.L., Dassa, E., Orelle, C., and Chen, S. (2008). Structure, function, and evolution of bacterial ATP-binding cassette systems. Microbiol Mol Biol Rev *72*, 317-364. https://doi.org/10.1128/MMBR.00031-07

Davis, B.D. (1950). Studies on nutritionally deficient bacterial mutants isolated by means of penicillin. Experientia *6*, 41-50. https://doi.org/10.1007/BF02174810

De Groote, M.A., Granger, D., Xu, Y., Campbell, G., Prince, R., and Fang, F.C. (1995). Genetic and redox determinants of nitric oxide cytotoxicity in a *Salmonella typhimurium* model. Proc Natl Acad Sci U S A *92*, 6399-6403. https://doi.org/10.1073/pnas.92.14.6399

de la Calle Guntiñas, M.B., Bordin, G., and Rodriguez, A.R. (2002). Identification, characterization and determination of metal-binding proteins by liquid chromatography. A review. Analytical and bioanalytical chemistry *374*, 369-378. https://doi.org/10.1007/s00216-002-1508-3

Deka, R.K., Brautigam, C.A., Liu, W.Z., Tomchick, D.R., and Norgard, M.V. (2015). Evidence for posttranslational protein flavinylation in the syphilis spirochete *Treponema pallidum*: structural and biochemical insights from the catalytic core of a periplasmic flavin-trafficking protein. mBio *6*, e00519-00515. https://doi.org/10.1128/mBio.00519-15

Delany, I., Rappuoli, R., and Scarlato, V. (2004). Fur functions as an activator and as a repressor of putative virulence genes in *Neisseria meningitidis*. Mol Microbiol *52*, 1081-1090. https://doi.org/10.1111/j.1365-2958.2004.04030.x

Deng, Z., Roberts, D., Wang, X., and Kemp, R.G. (1999). Expression, characterization, and crystallization of the pyrophosphate-dependent phosphofructo-1-kinase of *Borrelia burgdorferi*. Arch Biochem Biophys *371*, 326-331. https://doi.org/10.1006/abbi.1999.1446

Desrosiers, D.C., Sun, Y.C., Zaidi, A.A., Eggers, C.H., Cox, D.L., and Radolf, J.D. (2007). The general transition metal (Tro) and Zn2+ (Znu) transporters in *Treponema pallidum*: analysis of metal specificities and expression profiles. Mol Microbiol *65*, 137-152. https://doi.org/10.1111/j.1365-2958.2007.05771.x

Dever, L.L., Jorgensen, J.H., and Barbour, A.G. (1993). In vitro activity of vancomycin against the spirochete *Borrelia burgdorferi*. Antimicrob Agents Chemother *37*, 1115-1121.

Dowdell, A.S., Murphy, M.D., Azodi, C., Swanson, S.K., Florens, L., Chen, S., and Zückert, W.R. (2017). Comprehensive spatial analysis of the *Borrelia burgdorferi* lipoproteome reveals a compartmentalization bias toward the bacterial surface. J Bacteriol *199*. https://doi.org/10.1128/jb.00658-16

Drecktrah, D., Lybecker, M., Popitsch, N., Rescheneder, P., Hall, L.S., and Samuels, D.S. (2015). The *Borrelia burgdorferi* RelA/SpoT homolog and stringent response regulate survival in the tick vector and global gene expression during starvation. PLoS Pathog *11*, e1005160. https://doi.org/10.1371/journal.ppat.1005160 eCollection 2015 Sep.

Dulebohn, D.P., Richards, C.L., Su, H., Lawrence, K.A., and Gherardini, F.C. (2017). Weak organic acids decrease *Borrelia burgdorferi* cytoplasmic pH, eliciting an acid stress response and impacting RpoN- and RpoS-dependent gene expression. Front Microbiol *8*, 1734. https://doi.org/10.3389/fmicb.2017.01734 eCollection 2017.

Eggers, C.H., Caimano, M.J., Malizia, R.A., Kariu, T., Cusack, B., Desrosiers, D.C., Hazlett, K.R., Claiborne, A., Pal, U., and Radolf, J.D. (2011). The coenzyme A disulphide reductase of *Borrelia burgdorferi* is important for rapid growth throughout the enzootic cycle and essential for infection of the mammalian host. Mol Microbiol *82*, 679-697. https://doi.org/10.1111/j.1365-2958.2011.07845.x

Eick-Helmerich, K., and Braun, V. (1989). Import of biopolymers into *Escherichia coli*: nucleotide sequences of the *exbB* and *exbD* genes are homologous to those of the *tolQ* and *tolR* genes, respectively. J Bacteriol *171*, 5117-51126. https://doi.org/10.1128/jb.171.9.5117-5126.1989

Eick-Helmerich, K., Hantke, K., and Braun, V. (1987). Cloning and expression of the *exbB* gene of Escherichia coli K-12. Mol Gen Genet *206*, 246-251. https://doi.org/10.1007/bf00333580

Elias, A.F., Bono, J.L., Kupko, J.J., 3rd, Stewart, P.E., Krum, J.G., and Rosa, P.A. (2003). New antibiotic resistance cassettes suitable for genetic studies in *Borrelia burgdorferi*. J Mol Microbiol Biotechnol *6*, 29-40. https://doi.org/10.1159/000073406

Elias, A.F., Stewart, P.E., Grimm, D., Caimano, M.J., Eggers, C.H., Tilly, K., Bono, J.L., Akins, D.R., Radolf, J.D., Schwan, T.G., *et al.* (2002). Clonal polymorphism of *Borrelia burgdorferi* strain B31 MI: implications for mutagenesis in an infectious strain background. Infect Immun *70*, 2139-2150.

Esteve-Gassent, M.D., Smith, T.C., 2nd, Small, C.M., Thomas, D.P., and Seshu, J. (2015). Absence of *sodA* increases the levels of oxidation of key metabolic determinants of *Borrelia burgdorferi*. PLoS One *10*, e0136707. https://doi.org/10.1371/journal.pone.0136707 eCollection 2015.

Fang, F.C. (2004). Antimicrobial reactive oxygen and nitrogen species: concepts and controversies. Nat Rev Microbiol 2, 820-832. https://doi.org/10.1038/nrmicro1004

Fang, F.C., and Vázquez-Torres, A. (2019). Reactive nitrogen species in host-bacterial interactions. Curr Opin Immunol *60*, 96-102. https://doi.org/10.1016/j.coi.2019.05.008

Fenno, J.C., Müller, K.H., and McBride, B.C. (1996). Sequence analysis, expression, and binding activity of recombinant major outer sheath protein (Msp) of *Treponema denticola*. J Bacteriol *178*, 2489-2497. https://doi.org/10.1128/jb.178.9.2489-2497.1996

Fenno, J.C., Wong, G.W., Hannam, P.M., Müller, K.H., Leung, W.K., and McBride, B.C. (1997). Conservation of Msp, the gene encoding the major outer membrane protein of oral *Treponema* Spp. J Bacteriol *179*, 2489-2497. https://doi.org/10.1128/jb.179.4.1082-1089.

Ferguson, A.D., and Deisenhofer, J. (2004). Metal import through microbial membranes. Cell *116*, 15-24. https://doi.org/10.1016/s0092-8674(03)01030-4

Ferry, J.G. (2015). Acetate metabolism in anaerobes from the Domain *Archaea*. Life (Basel, Switzerland) *5*, 1454-1471. https://doi.org/10.3390/life502145410.3390/life5021454.

Fillingame, R.H., Jorstad, C.M., and Morris, D.R. (1975). Increased cellular levels of spermidine or spermine are required for optimal DNA synthesis in lymphocytes activated by concanavalin A. Proc Natl Acad Sci U S A *72*, 4042-4045. https://doi.org/10.1073/pnas.72.10.4042

Finney, L.A., and O'Halloran, T.V. (2003). Transition metal speciation in the cell: insights from the chemistry of metal ion receptors. Science *300*, 931-936. https://doi.org/10.1126/science.1085049

Fischer, E., Gunter, K., and Braun, V. (1989). Involvement of ExbB and TonB in transport across the outer membrane of *Escherichia coli*: phenotypic complementation of exb mutants by overexpressed tonB and physical stabilization of TonB by ExbB. J Bacteriol *171*, 5127-5134. https://doi.org/10.1128/jb.171.9.5127-5134.1989

Fischer, J.R., Parveen, N., Magoun, L., and Leong, J.M. (2003). Decorin-binding proteins A and B confer distinct mammalian cell type-specific attachment by *Borrelia burgdorferi*, the Lyme disease spirochete Proc Natl Acad Sci U S A *100*, 7307-7312. https://doi.org/10.1073/pnas.1231043100

Fitzpatrick, T.B., Amrhein, N., Kappes, B., Macheroux, P., Tews, I., and Raschle, T. (2007). Two independent routes of de novo vitamin B6 biosynthesis: not that different after all. Biochem J *407*, 1-13. https://doi.org/10.1042/bj20070765

Foster, J.W. (1999). When protons attack: microbial strategies of acid adaptation. Curr Opin Microbiol *2*, 170-174. https://doi.org/10.1016/s1369-5274(99)80030-7

Foster, J.W., Kinney, D.M., and Moat, A.G. (1979). Pyridine nucleotide cycle of *Salmonella typhimurium*: regulation of nicotinic acid phosphoribosyltransferase and nicotinamide deamidase. J Bacteriol *138*, 957-961.

Fraenkel, D.G. (1986). Mutants in glucose metabolism. Annu Rev Biochem *55*, 317-337. https://doi.org/10.1146/annurev.bi.55.070186.001533

Frank, K.L., Bundle, S.F., Kresge, M.E., Eggers, C.H., and Samuels, D.S. (2003). *aadA* confers streptomycin resistance in *Borrelia burgdorferi*. J Bacteriol *185*, 6723-6727.

Fraser, C.M., Casjens, S., Huang, W.M., Sutton, G.G., Clayton, R., Lathigra, R., White, O., Ketchum, K.A., Dodson, R., Hickey, E.K., *et al.* (1997). Genomic sequence of a Lyme disease spirochaete, *Borrelia Burgdorferi*. Nature *390*, 375-388. https://doi.org/10.1038/37551

Fraser, C.M., Richardson, D., Norris, S.J., Weinstock, G.M., White, O., Sutton, G.G., Dodson, R., Gwinn, M., Hickey, E.K., Clayton, R., *et al.* (1998). Complete genome sequence of *Treponema Pallidum*, the syphilis spirochete. Science *281*, 375-388. https://doi.org/10.1126/science.281.5375.375

Frenkiel-Krispin, D., Levin-Zaidman, S., Shimoni, E., Wolf, S.G., Wachtel, E.J., Arad, T., Finkel, S.E., Kolter, R., and Minsky, A. (2001). Regulated phase transitions of bacterial chromatin: a non-enzymatic pathway for generic DNA protection. EMBO J *20*, 1184-1191. https://doi.org/10.1093/emboj/20.5.1184

Friesen, J.A., Lawrence, C.M., Stauffacher, C.V., and Rodwell, V.W. (1996). Structural determinants of nucleotide coenzyme specificity in the distinctive dinucleotide binding fold of HMG-CoA reductase from *Pseudomonas mevalonii*. Biochemistry *35*, 11945-11950. https://doi.org/10.1021/bi9609937

Funk, C.R., Zimniak, L., and Dowhan, W. (1992). The *pgpA* and *pgpB* Genes of *Escherichia coli* are not essential: evidence for a third phosphatidylglycerophosphate phosphatase. J Bacteriol *174*, 205-213. https://doi.org/10.1128/jb.174.1.205-213

Gao, H., Zhou, D., Li, Y., Guo, Z., Han, Y., Song, Y., Zhai, J., Du, Z., Wang, X., Lu, J., *et al.* (2008). The iron-responsive Fur regulon in *Yersinia pestis*. J Bacteriol *190*, 1259-1273. https://doi.org/10.1128/JB.01910-07

Geiger, R., Rieckmann, J.C., Wolf, T., Basso, C., Feng, Y., Fuhrer, T., Kogadeeva, M., Picotti, P., Meissner, F., Mann, M., *et al.* (2016). L-Arginine modulates T cell metabolism and enhances survival and anti-tumor activity. Cell *167*, 829-842.e813. https://doi.org/10.1016/j.cell.2016.09.031

Gilmore, R.D., Jr., Howison, R.R., Schmit, V.L., Nowalk, A.J., Clifton, D.R., Nolder, C., Hughes, J.L., and Carroll, J.A. (2007). Temporal expression analysis of the *Borrelia burgdorferi* paralogous gene family 54 genes BBA64, BBA65, and BBA66 during persistent infection in mice. Infect Immun *75*, 2753-2764. https://doi.org/10.1128/iai.00037-07

Grant, R.A., Filman, D.J., Finkel, S.E., Kolter, R., and Hogle, J.M. (1998). The crystal structure of Dps, a ferritin homolog that binds and protects DNA. Nat Struct Biol *5*, 294-303. https://doi.org/10.1038/nsb0498-294

Grass, G., Franke, S., Taudte, N., Nies, D.H., Kucharski, L.M., Maguire, M.E., and Rensing, C. (2005). The metal permease ZupT from *Escherichia coli* is a transporter with a broad substrate spectrum. J Bacteriol *187*, 1604-1611. https://doi.org/10.1128/JB.187.5.1604-1611.2005

Grass, G., Wong, M.D., Rosen, B.P., Smith, R.L., and Rensing, C. (2002). ZupT Is a Zn(II) uptake system in *Escherichia coli*. J Bacteriol *184*, 864-866. https://doi.org/10.1128/jb.184.3.864-866.2002

Grifantini, R., Frigimelica, E., Delany, I., Bartolini, E., Giovinazzi, S., Balloni, S., Agarwal, S., Galli, G., Genco, C., and Grandi, G. (2004). Characterization of a novel *Neisseria meningitidis* Fur and Iron-Regulated operon required for protection from oxidative stress: utility of DNA microarray in the assignment of the biological role of hypothetical genes. Mol Microbiol *54*, 962-979. https://doi.org/10.1111/j.1365-2958.2004.04315.x

Grimm, D., Tilly, K., Byram, R., Stewart, P.E., Krum, J.G., Bueschel, D.M., Schwan, T.G., Policastro, P.F., Elias, A.F., and Rosa, P.A. (2004). Outer-surface protein C of the Lyme disease spirochete: a protein induced in ticks for infection of mammals. Proc Natl Acad Sci U S A *101*, 3142-3147. https://doi.org/10.1073/pnas.0306845101

Groshong, A.M., Dey, A., Bezsonova, I., Caimano, M.J., and Radolf, J.D. (2017). Peptide uptake is essential for *Borrelia burgdorferi* viability and involves structural and regulatory complexity of its oligopeptide transporter. mBio *8*. https://doi.org/10.1128/mBio.02047-17

Grüber, G., Manimekalai, M.S., Mayer, F., and Müller, V. (2014). ATP synthases from archaea: the beauty of a molecular motor. Biochim Biophys Acta *1837*, 940-952. https://doi.org/10.1016/j.bbabio.2014.03.004

Guerinot, M.L. (2000). The ZIP family of metal transporters. Biochimica et biophysica acta *1465*, 190-198. https://doi.org/10.1016/s0005-2736(00)00138-3

Hall, J.L. (2002). Cellular mechanisms for heavy metal detoxification and tolerance. J Exp Bot *53*, 1-11.

Hardesty, C., Ferran, C., and DiRienzo, J.M. (1991). Plasmid-mediated sucrose metabolism in *Escherichia coli*: characterization of *scrY*, the structural gene for a phosphoenolpyruvate-dependent sucrose phosphotransferase system outer membrane porin. J Bacteriol *173*, 449-456. https://doi.org/10.1128/jb.173.2.449-456.1991

Hazlett, K.R., Rusnak, F., Kehres, D.G., Bearden, S.W., La Vake, C.J., La Vake, M.E., Maguire, M.E., Perry, R.D., and Radolf, J.D. (2003). The *Treponema pallidum* Tro operon encodes a multiple metal transporter, a Zinc-dependent transcriptional repressor, and a semi-autonomously expressed phosphoglycerate mutase. J Biological Chem *278*, 20687-20694. https://doi.org/10.1074/jbc.M300781200

He, M., Ouyang, Z., Troxell, B., Xu, H., Moh, A., Piesman, J., Norgard, M.V., Gomelsky, M., and Yang, X.F. (2011). Cyclic di-GMP is essential for the survival of the lyme disease spirochete in ticks. PLoS Pathog *7*, e1002133. https://doi.org/10.1371/journal.ppat.1002133

Hearn, E.M., Patel, D.R., Lepore, B.W., Indic, M., and van den Berg, B. (2009). Transmembrane passage of hydrophobic compounds through a protein channel wall. Nature *458*, 367-370. https://doi.org/10.1038/nature07678

Heller, K.B., and Wilson, T.H. (1981). Selectivity of the *Escherichia coli* outer membrane porins *ompC* and *ompF*. FEBS Lett *129*, 253-255. https://doi.org/10.1016/0014-5793(81)80177-9

Herbig, A.F., and Helmann, J.D. (2001). Roles of metal ions and hydrogen peroxide in modulating the interaction of the *Bacillus subtilis* PerR peroxide regulon repressor with operator DNA. Mol Microbiol *41*, 849-859. https://doi.org/10.1046/j.1365-2958.2001.02543.x

Hidalgo, E., Ding, H., and Demple, B. (1997). Redox signal transduction via iron-sulfur clusters in the SoxR transcription activator. Trends Biochem Sci *22*, 207-210. https://doi.org/10.1016/s0968-0004(97)01068-2

Higashitani, A., Nishimura, Y., Hara, H., Aiba, H., Mizuno, T., and Horiuchi, K. (1993). Osmoregulation of the fatty acid receptor gene *fadL* in *Escherichia coli*. Mol Gen Genet *240*, 339-347. https://doi.org/10.1007/BF00280384

Hillyard, D., Rechsteiner, M., Manlapaz-Ramos, P., Imperial, J.S., Cruz, L.J., and Olivera, B.M. (1981). The pyridine nucleotide cycle. Studies in *Escherichia coli* and the human cell line D98/AH2. J Biological Chem *256*, 8491-8497.

Hoon-Hanks, L.L., Morton, E.A., Lybecker, M.C., Battisti, J.M., Samuels, D.S., and Drecktrah, D. (2012). *Borrelia burgdorferi malQ* mutants utilize disaccharides and traverse the enzootic cycle. FEMS Immunol Med Microbiol *66*, 157-165. https://doi.org/10.1111/j.1574-695X.2012.00996.x

Hossain, H., Wellensiek, H.J., Geyer, R., and Lochnit, G. (2001). Structural analysis of glycolipids from *Borrelia burgdorferi*. Biochimie *83*, 683-692. https://doi.org/10.1016/s0300-9084(01)01296-2

Hübner, A., Revel, A.T., Nolen, D.M., Hagman, K.E., and Norgard, M.V. (2003). Expression of a luxS gene is not required for *Borrelia burgdorferi* infection of mice via needle inoculation. Infect Immun *71*, 2892-2896. https://doi.org/10.1128/iai.71.5.2892-2896.2003

Hübner, A., Yang, X., Nolen, D.M., Popova, T.G., Cabello, F.C., and Norgard, M.V. (2001). Expression of *Borrelia burgdorferi* OspC and DbpA is controlled by a RpoN-RpoS regulatory pathway. Proc Natl Acad Sci U S A *98*, 12724-12729. https://doi.org/10.1073/pnas.231442498

Hyde, J.A., Shaw, D.K., Smith, R., Trzeciakowski, J.P., and Skare, J.T. (2010). Characterization of a conditional *bosR* Mutant in *Borrelia burgdorferi*. Infect Immun *78*, 265-274. https://doi.org/10.1128/iai.01018-09

Igarashi, K., and Kashiwagi, K. (1999). Polyamine transport in bacteria and yeast. Biochem J *344 Pt 3*, 633-642.

Imlay, J.A. (2002). How oxygen damages microbes: oxygen tolerance and obligate anaerobiosis. Adv Micro Phys *46*, 111-153. https://doi.org/10.1016/s0065-2911(02)46003-1

Imlay, J.A. (2003). Pathways of oxidative damage. Annu Rev Microbiol *57*, 395-418. https://doi.org/10.1146/annurev.micro.57.030502.090938

Iyer, R., Caimano, M.J., Luthra, A., Axline, D., Jr., Corona, A., Iacobas, D.A., Radolf, J.D., and Schwartz, I. (2015). Stage-specific global alterations in the transcriptomes of Lyme disease spirochetes during tick feeding and following mammalian host adaptation. Mol Microbiol *95*, 509-538. https://doi.org/10.1111/mmi.12882

Jackowski, S., and Rock, C.O. (1981). Regulation of coenzyme A biosynthesis. J Bacteriol *148*, 926-932.

Jackowski, S., and Rock, C.O. (1983). Ratio of active to inactive forms of acyl carrier protein in *Escherichia coli*. J Biol Chem *258*, 15186-15191.

Jackowski, S., and Rock, C.O. (1984). Metabolism of 4'-phosphopantetheine in *Escherichia coli*. J Bacteriol *158*, 115-120.

Jain, S., Sutchu, S., Rosa, P.A., Byram, R., and Jewett, M.W. (2012). *Borrelia burgdorferi* harbors a transport system essential for purine salvage and mammalian infection. Infect Immun *80*, 3086-3093. https://doi.org/10.1128/iai.00514-12

Jap, B.K., and Walian, P.J. (1996). Structure and functional mechanism of porins. Phys Rev *76*, 1073-1088. https://doi.org/10.1152/physrev.1996.76.4.1073

Jewett, M.W., Byram, R., Bestor, A., Tilly, K., Lawrence, K., Burtnick, M.N., Gherardini, F., and Rosa, P.A. (2007). Genetic basis for retention of a critical virulence plasmid of Borrelia burgdorferi. Mol Microbiol *66*, 975-990. https://doi.org/10.1111/j.1365-2958.2007.05969.x

Jewett, M.W., Lawrence, K.A., Bestor, A., Byram, R., Gherardini, F., and Rosa, P.A. (2009). GuaA and GuaB are essential for *Borrelia Burgdorferi* survival in the tick-mouse infection cycle. J Bacteriol *191*, 6231-6241. https://doi.org/10.1128/JB.00450-09

Johnson, R.C., Hyde, F.W., and Rumpel, C.M. (1984). Taxonomy of the Lyme disease spirochetes. Yale J Biol Med *57*, 529-537.

Jordan, A., and Reichard, P. (1998). Ribonucleotide reductases. Annu Rev Biochem *67*, 71-98. https://doi.org/10.1146/annurev.biochem.67.1.71

Jorstad, C.M., and Morris, D.R. (1974). Polyamine limitation of growth slows the rate of polypeptide chain elongation in Escherichia coli. J Bacteriol *119*, 857-860. https://doi.org/10.1128/jb.119.3.857-860.1974

Jutras, B.L., Lochhead, R.B., Kloos, Z.A., Biboy, J., Strle, K., Booth, C.J., Govers, S.K., Gray, J., Schumann, P., Vollmer, W., *et al.* (2019). *Borrelia burgdorferi* peptidoglycan is a persistent antigen in patients with Lyme arthritis. Proc Natl Acad Sci U S A *116*, 13498-13507. https://doi.org/10.1073/pnas.1904170116

Kakinuma, Y., Yamato, I., and Murata, T. (1999). Structure and function of vacuolar Na+-translocating ATPase in Enterococcus hirae. J Bioenerg Biomembr *31*, 7-14. https://doi.org/10.1023/a:1005499126939

Kaminski, P.A. (2002). Functional cloning, heterologous expression, and purification of two different N-deoxyribosyltransferases from *Lactobacillus helveticus*. J Biol Chem *277*, 14400-14407. https://doi.org/10.1074/jbc.M111995200

Kanjee, U., and Houry, W.A. (2013). Mechanisms of acid resistance in *Escherichia coli*. Annu Rev Microbiol *67*, 65-81. https://doi.org/10.1146/annurev-micro-092412-155708

Kashiwagi, K., Endo, H., Kobayashi, H., Takio, K., and Igarashi, K. (1995). Spermidine-preferential uptake system in *Escherichia coli*. ATP hydrolysis by PotA protein and its association with membrane. J Biol Chem *270*, 25377-25382.

Kashiwagi, K., Miyamoto, S., Nukui, E., Kobayashi, H., and Igarashi, K. (1993). Functions of *potA* and *potD* proteins in spermidine-preferential uptake system in *Escherichia coli*. J Biol Chem *268*, 19358-19363.

Katona, L.I., Tokarz, R., Kuhlow, C.J., Benach, J., and Benach, J.L. (2004). The Fur homologue in *Borrelia burgdorferi*. J Bacteriol *186*, 6443-6456. https://doi.org/10.1128/JB.186.19.6443-6456.2004

Kawakami, N., and Fujisaki, S. (2018). Undecaprenyl phosphate metabolism in Gram-negative and Gram-positive bacteria. Biosci Biotechnol Biochem *82*, 940-946. https://doi.org/10.1080/09168451.2017.1401915

Kazragis, R.J., Dever, L.L., Jorgensen, J.H., and Barbour, A.G. (1996). In vivo activities of ceftriaxone and vancomycin against *Borrelia* spp. in the mouse brain and other sites. Antimicrob Agents Chemother *40*, 2632-2636. https://doi.org/10.1128/aac.40.11.2632

Kenedy, M.R., Luthra, A., Anand, A., Dunn, J.P., Radolf, J.D., and Akins, D.R. (2014). Structural modeling and physicochemical characterization provide evidence that P66 forms a β-barrel in the *Borrelia burgdorferi* outer membrane. J Bacteriol *196*, 859-872. https://doi.org/10.1128/jb.01236-13

Keyer, K., Gort, A.S., and Imlay, J.A. (1995). Superoxide and the production of oxidative DNA damage. J Bacteriol *177*, 6782-6790. https://doi.org/10.1128/jb.177.23.6782-6790.1995

Keyer, K., and Imlay, J.A. (1996). Superoxide accelerates DNA damage by elevating free-iron levels. Proc Natl Acad Sci U S A *93*, 13635-13640. https://doi.org/10.1073/pnas.93.24.13635

Keyer, K., and Imlay, J.A. (1997). Inactivation of dehydratase [4Fe-4S] clusters and disruption of iron homeostasis upon cell exposure to peroxynitrite. J Biol Chem *272*, 27652-27659. https://doi.org/10.1074/jbc.272.44.27652

Khajanchi, B.K., Odeh, E., Gao, L., Jacobs, M.B., Philipp, M.T., Lin, T., and Norris, S.J. (2015). Phosphoenolpyruvate phosphotransferase system components modulate gene Ttanscription and virulence of *Borrelia burgdorferi*. Infect Immun *84*, 754-764. https://doi.org/10.1128/iai.00917-15

Kim, D.Y., Stauffacher, C.V., and Rodwell, V.W. (2000). Dual coenzyme specificity of Archaeoglobus fulgidus HMG-CoA reductase. Protein Sci *9*, 1226-1234. https://doi.org/10.1110/ps.9.6.1226

Kinjo, Y., Tupin, E., Wu, D., Fujio, M., Garcia-Navarro, R., Benhnia, M.R., Zajonc, D.M., Ben-Menachem, G., Ainge, G.D., Painter, G.F., *et al.* (2006). Natural killer T cells recognize diacylglycerol antigens from pathogenic bacteria. Nature Immunol *7*, 978-986. https://doi.org/10.1038/ni1380

Kirkpatrick, C., Maurer, L.M., Oyelakin, N.E., Yoncheva, Y.N., Maurer, R., and Slonczewski, J.L. (2001). Acetate and formate stress: opposite responses in the proteome of *Escherichia coli*. J Bacteriol *183*, 6466-6477. https://doi.org/10.1128/jb.183.21.6466-6477.2001

Klebba, P.E. (2003). Three paradoxes of ferric enterobactin uptake. Front Biosci *8*, s1422-1436. https://doi.org/10.2741/1233

Kostick, J.L., Szkotnicki, L.T., Rogers, E.A., Bocci, P., Raffaelli, N., and Marconi, R.T. (2011). The diguanylate cyclase, Rrp1, regulates critical steps in the enzootic cycle of the Lyme disease spirochetes. Mol Microbiol *81*, 219-231. https://doi.org/10.1111/j.1365-2958.2011.07687.x

Kramer, N.E., Smid, E.J., Kok, J., de Kruijff, B., Kuipers, O.P., and Breukink, E. (2004). Resistance of Gram-positive bacteria to nisin is not determined by lipid II levels. FEMS Microbiol Lett *239*, 157-161. https://doi.org/10.1016/j.femsle.2004.08.033

Kurth, J.M., Op den Camp, H.J.M., and Welte, C.U. (2020). Several ways one goal-methanogenesis from unconventional substrates. Appl Microbiol Biotechnol *104*, 6839-6854. https://doi.org/10.1007/s00253-020-10724-7

LaFrance, M.E., Pierce, J.V., Antonara, S., and Coburn, J. (2011). The *Borrelia burgdorferi* integrin ligand P66 affects gene expression by human cells in culture. Infect Immun *79*, 3249-3261. https://doi.org/10.1128/iai.05122-11

Lawrence, K.A., Jewett, M.W., Rosa, P.A., and Gherardini, F.C. (2009). *Borrelia Burgdorferi bb0426* encodes a 2'-deoxyribosyltransferase that plays a central role in purine salvage. Mol Microbiol *72*, 1517-1529. https://doi.org/10.1111/j.1365-2958.2009.06740.x

Lepore, B.W., Indic, M., Pham, H., Hearn, E.M., Patel, D.R., and van den Berg, B. (2011). Ligand-gated diffusion across the bacterial outer membrane. Proc Natl Acad Sci U S A *108*, 10121-10126. https://doi.org/10.1073/pnas.1018532108

Li, X., Pal, U., Ramamoorthi, N., Liu, X., Desrosiers, D.C., Eggers, C.H., Anderson, J.F., Radolf, J.D., and Fikrig, E. (2007). The Lyme disease agent *Borrelia Burgdorferi* requires BB0690, a Dps homologue, to persist within ticks. Mol Microbiol *63*, 694-710. https://doi.org/10.1111/j.1365-2958.2006.05550.x

Li, X.Z., and Nikaido, H. (2004). Efflux-mediated drug resistance in Bacteria. Drugs *64*, 159-204. https://doi.org/10.2165/00003495-200464020-00004

Lin, B., Short, S.A., Eskildsen, M., Klempner, M.S., and Hu, L.T. (2001). Functional testing of putative Oligopeptide Permease (Opp) Proteins of *Borrelia burgdorferi*: a complementation model in Opp(-) *Escherichia coli*. Biochimica et biophysica acta *1499*, 222-231. https://doi.org/10.1016/s0167-4889(00)00121-x

Lin, Y.H., Chen, Y., Smith, T.C., 2nd, Karna, S.L.R., and Seshu, J. (2018). Short-Chain Fatty Acids Alter Metabolic and Virulence Attributes of *Borrelia burgdorferi*. Infect Immun *86*. https://doi.org/10.1128/iai.00217-18

Lindgren, J.K., Thomas, V.C., Olson, M.E., Chaudhari, S.S., Nuxoll, A.S., Schaeffer, C.R., Lindgren, K.E., Jones, J., Zimmerman, M.C., Dunman, P.M., *et al.* (2014). Arginine deiminase in *Staphylococcus epidermidis* functions to augment biofilm maturation through pH homeostasis. J Bacteriol *196*, 2277-2289. https://doi.org/10.1128/jb.00051-14

Lolkema, J.S., Chaban, Y., and Boekema, E.J. (2003). Subunit composition, structure, and distribution of bacterial V-type ATPases. J Bioenerg Biomembr *35*, 323-335. https://doi.org/10.1023/a:1025776831494

López-Lara, I.M., and Geiger, O. (2001). Novel pathway for phosphatidylcholine biosynthesis in bacteria associated with eukaryotes. J Biotechnol *91*, 211-221. https://doi.org/10.1016/s0168-1656(01)00331-5

Ma, Y., Seiler, K.P., Tai, K.F., Yang, L., Woods, M., and Weis, J.J. (1994). Outer surface lipoproteins of *Borrelia burgdorferi* stimulate nitric oxide production by the cytokine-inducible pathway. Infect Immun *62*, 3663-3671.

Maas, W.K., and Davis, B.D. (1950). Pantothenate studies. I. Interference by D-serine and L-aspartic acid with pantothenate synthesis in *Escherichia coli*. J Bacteriol *60*, 733-745.

Majsnerowska, M., Noens, E.E.E., and Lolkema, J.S. (2018). Arginine and citrulline catabolic pathways encoded by the *arc* gene cluster of *Lactobacillus brevis* ATCC367. J Bacteriol 200(14), e00182-18. https://doi.org/10.1128/jb.00182-18

Mannock, D.A., Lewis, R.N., McElhaney, R.N., Harper, P.E., Turner, D.C., and Gruner, S.M. (2001). An analysis of the relationship between fatty acid composition and the lamellar gel to liquid-crystalline and the lamellar to inverted nonlamellar phase transition temperatures of phosphatidylethanolamines and diacyl-alpha-D-glucosyl glycerols. Eur Biophys J *30*, 537-554. https://doi.org/10.1007/s00249-001-0185-z

Martin, D.P., and Drueckhammer, D.G. (1993). Separate enzymes catalyze the final two steps of coenzyme A biosynthesis in *Brevibacterium ammoniagenes*: purification of pantetheine phosphate adenylyltransferase. Biochem Biophys Res Commun *192*, 1155-1161. https://doi.org/10.1006/bbrc.1993.1537

Martin, N.I., and Breukink, E. (2007). Expanding role of lipid II as a target for lantibiotics. Future Microbiol *2*, 513-525. https://doi.org/10.2217/17460913.2.5.513

Martinez, A., and Kolter, R. (1997). Protection of DNA during oxidative stress by the nonspecific DNA-binding protein Dps. J Bacteriol *179*, 5188-5194. https://doi.org/10.1128/jb.179.16.5188-5194.1997

Mason, C., Thompson, C., and Ouyang, Z. (2020). DksA plays an essential role in regulating the virulence of *Borrelia burgdorferi*. Mol Microbiol *114*, 172-183. https://doi.org/10.1111/mmi.14504

McCord, J.M., and Fridovich, I. (1969). The utility of superoxide dismutase in studying free radical reactions. I. Radicals generated by the interaction of sulfite, dimethyl sulfoxide, and oxygen. J Biol Chem *244*, 6056-6063.

McCord, J.M., Keele, B.B., and Fridovich, I. (1971). An enzyme-based theory of obligate anaerobiosis: The physiological function of Superoxide Dismutase. Proc Natl Acad Sci U S A *68*, 1024-1027. https://doi.org/10.1073/pnas.68.5.1024

McLaren, J., Ngo, D.T., and Olivera, B.M. (1973). Pyridine nucleotide metabolism in *Escherichia coli*. 3. Biosynthesis from alternative precursors in vivo. J Biol Chem *248*, 5144-5149.

Medrano, M.S., Ding, Y., Wang, X.G., Lu, P., Coburn, J., and Hu, L.T. (2007). Regulators of expression of the oligopeptide permease A proteins of Borrelia burgdorferi. J Bacteriol *189*, 2653-2659. https://doi.org/10.1128/jb.01760-06

Minagawa, Y., Ueno, H., Hara, M., Ishizuka-Katsura, Y., Ohsawa, N., Terada, T., Shirouzu, M., Yokoyama, S., Yamato, I., Muneyuki, E., *et al.* (2013). Basic properties of rotary dynamics of the molecular motor *Enterococcus hirae* V1-ATPase. J Biol Chem *288*, 32700-32707. https://doi.org/10.1074/jbc.M113.506329

Modolell, M., Schaible, U.E., Rittig, M., and Simon, M.M. (1994). Killing of *Borrelia burgdorferi* by macrophages is dependent on oxygen radicals and nitric oxide and can be enhanced by antibodies to outer surface proteins of the spirochete. Immunol Lett *40*, 139-146.

Mols, M., van Kranenburg, R., van Melis, C.C., Moezelaar, R., and Abee, T. (2010). Analysis of acid-stressed Bacillus cereus reveals a major oxidative response and inactivation-associated radical formation. Environ Microbiol *12*, 873-885. https://doi.org/10.1111/j.1462-2920.2009.02132.x

Morikawa, K., Ohniwa, R.L., Kim, J., Maruyama, A., Ohta, T., and Takeyasu, K. (2006). Bacterial nucleoid dynamics: oxidative stress response in *Staphylococcus aureus*. Genes Cells *11*, 409-423. https://doi.org/10.1111/j.1365-2443.2006.00949.x

Morris, D.R., and Jorstad, C.M. (1970). Isolation of conditionally putrescine-deficient mutants of Escherichia coli. J Bacteriol *101*, 731-737. https://doi.org/10.1128/jb.101.3.731-737.1970

Mulay, V., Caimano, M.J., Liveris, D., Desrosiers, D.C., Radolf, J.D., and Schwartz, I. (2007). *Borrelia burgdorferi* BBA74, a periplasmic protein associated with the outer membrane, lacks porin-like properties. J Bacteriol *189*, 2063-2068. https://doi.org/10.1128/jb.01239-06

Muller, M., Mentel, M., van Hellemond, J.J., Henze, K., Woehle, C., Gould, S.B., Yu, R.Y., van der Giezen, M., Tielens, A.G., and Martin, W.F. (2012). Biochemistry and evolution of anaerobic energy metabolism in eukaryotes. Microbiol Mol Biol Rev *76*, 444-495. https://doi.org/10.1128/mmbr.05024-11

Mulrooney, S.B., and Hausinger, R.P. (2003). Nickel uptake and utilization by microorganisms. FEMS Microbiol Rev *27*, 239-261. https://doi.org/10.1016/S0168-6445(03)00042-1

Munro, G.F., Hercules, K., Morgan, J., and Sauerbier, W. (1972). Dependence of the putrescine content of *Escherichia coli* on the osmotic strength of the medium. J Biol Chem *247*, 1272-1280.

Murakami, S., and Yamaguchi, A. (2003). Multidrug-exporting secondary transporters. Curr Opin Struct Biol *13*, 443-452. https://doi.org/10.1016/s0959-440x(03)00109-x

Murata, T., Yamato, I., Kakinuma, Y., Shirouzu, M., Walker, J.E., Yokoyama, S., and Iwata, S. (2008). Ion binding and selectivity of the rotor ring of the Na+-transporting V-ATPase. Proc Natl Acad Sci U S A *105*, 8607-8612. https://doi.org/10.1073/pnas.0800992105

Nakae, T. (1976). Outer membrane of *Salmonella*. Isolation of protein complex that produces transmembrane channels. J Biol Chem *251*, 2176-2178.

Nakamura, H., and Tamura, Z. (1973). Pantothenate uptake in *Escherichia coli* K-12. J Nutr Sci Vitaminol *19*, 389-400. https://doi.org/10.3177/jnsv.19.389

Narasimhan, S., Booth, C.J., DePonte, K., Wu, M.J., Liang, X., Mohanty, S., Kantor, F., and Fikrig, E. (2019). Host-specific expression of Ixodes scapularis salivary genes. Ticks Tick Borne Dis *10*, 386-397. https://doi.org/10.1016/j.ttbdis.2018.12.001

Nguyen, K.T., Wu, J.C., Boylan, J.A., Gherardini, F.C., and Pei, D. (2007). Zinc is the metal cofactor of *Borrelia burgdorferi* peptide deformylase. Arch Biochem Biophys *468*, 217-225. https://doi.org/10.1016/j.abb.2007.09.023

Nikaido, H., and Saier, M.H. (1992). Transport proteins in bacteria: common themes in their design. Science *258*, 936-942. https://doi.org/10.1126/science.1279804

Nikaido, H., and Zgurskaya, H.I. (2001). AcrAB and related multidrug efflux pumps of *Escherichia coli*. J Mol Microbiol Biotechnol *3*, 215-218.

Noens, E.E., and Lolkema, J.S. (2016). Convergent evolution of the arginine deiminase pathway: the ArcD and ArcE arginine/ornithine exchangers. Microbiologyopen 6(1):e00412. https://doi.org/10.1002/mbo3.412

Noppa, L., Ostberg, Y., Lavrinovicha, M., and Bergström, S. (2001). P13, an integral membrane protein of *Borrelia burgdorferi*, is C-terminally processed and contains surface-exposed domains. Infect Immun *69*, 3323-3334. https://doi.org/10.1128/iai.69.5.3323-3334.2001

Nunn, W.D., and Simons, R.W. (1978). Transport of long-chain fatty acids by *Escherichia coli*: Mapping and characterization of mutants in the *fadL* gene. Proc Natl Acad Sci U S A *75*, 3377-3381. https://doi.org/10.1073/pnas.75.7.3377

Ostberg, Y., Berg, S., Comstedt, P., Wieslander, A., and Bergström, S. (2007). Functional analysis of a lipid galactosyltransferase synthesizing the major envelope lipid in the Lyme disease spirochete *Borrelia burgdorferi*. FEMS Microbiol Lett *272*, 22-29. https://doi.org/10.1111/j.1574-6968.2007.00728.x

Outten, C.E., and O'Halloran, T.V. (2001). Femtomolar sensitivity of metalloregulatory proteins controlling zinc homeostasis. Science *292*. https://doi.org/10.1126/science.1060331

Ouyang, Z., Deka, R.K., and Norgard, M.V. (2011). BosR (BB0647) controls the RpoN-RpoS regulatory pathway and virulence expression in *Borrelia burgdorferi* by a novel DNA-binding mechanism. PLoS Pathog *7*, e1001272. https://doi.org/10.1371/journal.ppat.1001272

Ouyang, Z., He, M., Oman, T., Yang, X.F., and Norgard, M.V. (2009a). A manganese transporter, BB0219 (BmtA), is required for virulence by the Lyme disease spirochete, Borrelia burgdorferi. Proc Natl Acad Sci U S A *106*, 3449-3454. https://doi.org/10.1073/pnas.0812999106

Ouyang, Z., Zhou, J., Brautigam, C.A., Deka, R., and Norgard, M.V. (2014). Identification of a core sequence for the binding of BosR to the *rpoS* promoter region in *Borrelia burgdorferi*. Microbiol *160*, 851-862. https://doi.org/10.1099/mic.0.075655-0

Ouyang, Z.M., Kumar, M., Kariu, T., Haq, S., Goldberg, M., Pal, U., and Norgard, M.V. (2009b). BosR (BB0647) governs virulence expression in *Borrelia burgdorferi*. Mol Microbiol *74*, 1331-1343. https://doi.org/10.1111/j.1365-2958.2009.06945.x

Overbeek, R., Larsen, N., Walunas, T., D'Souza, M., Pusch, G., Selkov, E., Liolios, K., Joukov, V., Kaznadzey, D., Anderson, I., *et al.* (2003). The ERGO genome analysis and discovery system. Nuc acids res *31*, 164-171. https://doi.org/10.1093/nar/gkg148

Palenik, B., Brahamsha, B., Larimer, F.W., Land, M., Hauser, L., Chain, P., Lamerdin, J., Regala, W., Allen, E.E., McCarren, J., *et al.* (2003). The genome of a motile marine *Synechococcus*. Nature *424*, 1037-1042. https://doi.org/10.1038/nature01943

Pappas, C.J., Iyer, R., Petzke, M.M., Caimano, M.J., Radolf, J.D., and Schwartz, I. (2011). *Borrelia burgdorferi* requires glycerol for maximum fitness during the tick phase of the enzootic cycle. PLoS Pathog *7*, e1002102. https://doi.org/10.1371/journal.ppat.1002102

Park, U.E., Olivera, B.M., Hughes, K.T., Roth, J.R., and Hillyard, D.R. (1989). DNA ligase and the pyridine nucleotide cycle in *Salmonella typhimurium*. J Bacteriol *171*, 2173-2180. https://doi.org/10.1128/jb.171.4.2173-2180.1989

Pettersson, J., Schrumpf, M.E., Raffel, S.J., Porcella, S.F., Guyard, C., Lawrence, K., Gherardini, F.C., and Schwan, T.G. (2007). Purine salvage pathways among *Borrelia* species. Infect immun *75*, 3877-3884. https://doi.org/10.1128/IAI.00199-07

Phelan, J.P., Kern, A., Ramsey, M.E., Lundt, M.E., Sharma, B., Lin, T., Gao, L., Norris, S.J., Hyde, J.A., Skare, J.T., *et al.* (2019). Genome-wide screen identifies novel genes required for *Borrelia burgdorferi* survival in its *Ixodes* tick vector. PLoS Pathog *15*, e1007644. https://doi.org/10.1371/journal.ppat.1007644

Pinne, M., Östberg, Y., Comstedt, P., and Bergström, S. (2004). Molecular analysis of the channel-forming protein P13 and its paralogue family 48 from different Lyme disease *Borrelia* species. Microbiol *150*, 549-559. https://doi.org/10.1099/mic.0.26728-0

Pomposiello, P.J., Bennik, M.H., and Demple, B. (2001). Genome-wide transcriptional profiling of the *Escherichia coli* responses to superoxide stress and sodium salicylate. J Bacteriol *183*, 3890-3902. https://doi.org/10.1128/JB.183.13.3890-3902.2001

Poole, K. (2001). Multidrug resistance in Gram-negative bacteria. Curr Opin Microbiol *4*, 500-508. https://doi.org/10.1016/s1369-5274(00)00242-3

Posey, J.E., and Gherardini, F.C. (2000). Lack of a role for iron in the Lyme disease pathogen. Science *288*, 1651-1653. https://doi.org/10.1126/science.288.5471.1651
10.1126/science.288.5471.1651.

Posey, J.E., Hardham, J.M., Norris, S.J., and Gherardini, F.C. (1999). Characterization of a manganese-dependent regulatory protein, TroR, From *Treponema pallidum*. Proc Natl Acad Sci U S A *96*, 10887-10892. https://doi.org/10.1073/pnas.96.19.10887

Potter, A.J., Kidd, S.P., Edwards, J.L., Falsetta, M.L., Apicella, M.A., Jennings, M.P., and McEwan, A.G. (2009). Thioredoxin reductase is essential for protection of Neisseria gonorrhoeae against killing by nitric oxide and for bacterial growth during interaction with cervical epithelial cells. J Infect Dis *199*, 227-235. https://doi.org/10.1086/595737

Powell, G.L., Bauza, M., and Larrabee, A.R. (1973). The stability of acyl carrier protein in *Escherichia coli*. J Biol Chem *248*, 4461-4466.

Powell, G.L., Elovson, J., and Vagelos, P.R. (1969). Acyl carrier protein. XII. Synthesis and turnover of the prosthetic group of acyl carrier protein in vivo. J Biol Chem *244*, 5616-5624.

Pugh, E.L., and Wakil, S.J. (1965). Studies on the mechanism of fatty acid synthesis. XIV. The prosthetic group of acyl carrier protein and the mode of its attachment to the protein. J Biol Chem *240*, 4727-4733.

Purser, J.E., Lawrenz, M.B., Caimano, M.J., Howell, J.K., Radolf, J.D., and Norris, S.J. (2003). A plasmid-encoded nicotinamidase (PncA) is essential for infectivity of *Borrelia burgdorferi* in a mammalian host. Mol Microbiol *48*. https://doi.org/10.1046/j.1365-2958.2003.03452.x

Purser, J.E., and Norris, S.J. (2000). Correlation between plasmid content and infectivity in Borrelia burgdorferi. Proc Natl Acad Sci U S A *97*, 13865-13870. https://doi.org/10.1073/pnas.97.25.13865

Qi, J., Wang, Y., and Forgac, M. (2007). The vacuolar (H+)-ATPase: subunit arrangement and in vivo regulation. J Bioenerg Biomembr *39*, 423-426. https://doi.org/10.1007/s10863-007-9116-8

Radolf, J.D., Robinson, E.J., Bourell, K.W., Akins, D.R., Porcella, S.F., Weigel, L.M., Jones, J.D., and Norgard, M.V. (1995). Characterization of outer membranes isolated from *Treponema pallidum*, the syphilis spirochete. Infect Immun *63*, 4244-4252. https://doi.org/10.1128/iai.63.11.4244-4252.1995

Ramsey, M.E., Hyde, J.A., Medina-Perez, D.N., Lin, T., Gao, L., Lundt, M.E., Li, X., Norris, S.J., Skare, J.T., and Hu, L.T. (2017). A high-throughput genetic screen identifies previously uncharacterized *Borrelia burgdorferi* genes important for resistance against reactive oxygen and nitrogen species. PLoS Pathog *13*, e1006225. https://doi.org/10.1371/journal.ppat.1006225

Rechnitzer, H., Rottem, S., and Herrmann, R. (2013). Reconstitution of an active arginine deiminase pathway in *Mycoplasma pneumoniae* M129. Infect Immun *81*, 3742-3749. https://doi.org/10.1128/iai.00441-13

Regue, M., Remenick, J., Tokunaga, M., Mackie, G.A., and Wu, H.C. (1984). Mapping of the lipoprotein signal peptidase gene (*Lsp*). J Bacteriol *158*, 632-635.

Ren, B., Tibbelin, G., Kajino, T., Asami, O., and Ladenstein, R. (2003). The multi-layered structure of Dps with a novel di-nuclear ferroxidase center. J Mol Biol *329*, 467-477. https://doi.org/10.1016/s0022-2836(03)00466-2

Rensing, C., and Grass, G. (2003). *Escherichia coli* mechanisms of copper homeostasis in a changing environment. FEMS Microbiology Reviews 27, 197-213. https://doi.org/10.1016/S0168-6445(03)00049-4

Richard, H., and Foster, J.W. (2004). *Escherichia coli* glutamate- and arginine-dependent acid resistance systems increase internal pH and reverse transmembrane potential. J Bacteriol *186*, 6032-6041. https://doi.org/10.1128/jb.186.18.6032-6041.2004

Richard, H.T., and Foster, J.W. (2003). Acid resistance in *Escherichia coli*. Adv Appl Microbiol *52*, 167-186.

Richards, C.L., Lawrence, K.A., Su, H., Yang, Y., Yang, X.F., Dulebohn, D.P., and Gherardini, F.C. (2015). Acetyl-phosphate is not a global regulatory bridge between virulence and central metabolism in *Borrelia burgdorferi*. PLoS One *10*, e0144472. https://doi.org/10.1371/journal.pone.0144472

Riley, S.P., Bykowski, T., Babb, K., von Lackum, K., and Stevenson, B. (2007). Genetic and physiological characterization of the *Borrelia burgdorferi* ORF BB0374-*pfs-metK-luxS* operon. Microbiol *153*, 2304-2311. https://doi.org/10.1099/mic.0.2006/004424-0

Rimaux, T., Riviere, A., Illeghems, K., Weckx, S., De Vuyst, L., and Leroy, F. (2012). Expression of the arginine deiminase pathway genes in *Lactobacillus sakei* is strain dependent and is affected by the environmental pH. Appl Environ Microbiol *78*, 4874-4883. https://doi.org/10.1128/aem.07724-11

Ristow, L.C., Bonde, M., Lin, Y.P., Sato, H., Curtis, M., Wesley, E., Hahn, B.L., Fang, J., Wilcox, D.A., Leong, J.M., *et al.* (2015). Integrin binding by *Borrelia burgdorferi* P66 facilitates dissemination but is not required for infectivity. Cell Microbiol *17*, 1021-1036. https://doi.org/10.1111/cmi.12418

Rosen, B.P. (2002). Transport and detoxification systems for transition metals, heavy metals and metalloids in eukaryotic and prokaryotic microbes. Comp Biochem Phys Part A, Mol Int Phys *133*, 689-693. https://doi.org/10.1016/s1095-6433(02)00201-5

Roy, S., Gupta, S., Das, S., Sekar, K., Chatterji, D., and Vijayan, M. (2004). X-ray analysis of *Mycobacterium smegmatis* Dps and a comparative study involving other Dps and Dps-like molecules. J Mol Biol *339*, 1103-1113. https://doi.org/10.1016/j.jmb.2004.04.042

Saier, M.H., and Paulsen, I.T. (2000). Whole genome analyses of transporters in spirochetes: *Borrelia burgdorferi* and *Treponema pallidum*. J Mol Microbiol Biotechnol 2, 393-399.

Salman-Dilgimen, A., Hardy, P.O., Radolf, J.D., Caimano, M.J., and Chaconas, G. (2013). HrpA, an RNA helicase involved in RNA processing, is required for mouse infectivity and tick transmission of the Lyme disease spirochete. PLoS Pathog *9*, e1003841. https://doi.org/10.1371/journal.ppat.1003841

Samuels, D.S. (2011). Gene regulation in *Borrelia burgdorferi*. Annu Rev Microbiol *65*, 479-499. https://doi.org/10.1146/annurev.micro.112408.134040

Sanjuan, E., Esteve-Gassent, M.D., Maruskova, M., and Seshu, J. (2009). Overexpression of CsrA (BB0184) alters the morphology and antigen profiles of *Borrelia burgdorferi*. Infect Immun *77*, 5149-5162. https://doi.org/10.1128/iai.00673-09

Schwan, T.G., Battisti, J.M., Porcella, S.F., Raffel, S.J., Schrumpf, M.E., Fischer, E.R., Carroll, J.A., Stewart, P.E., Rosa, P., and Somerville, G.A. (2003). Glycerol-3-phosphate acquisition in spirochetes: distribution and biological activity of glycerophosphodiester phosphodiesterase (GlpQ) among *Borrelia* species. J Bacteriol *185*, 1346-1356.

Schwan, T.G., and Piesman, J. (2000). Temporal changes in outer surface proteins A and C of the Lyme disease-associated spirochete, *Borrelia burgdorferi*, during the chain of infection in ticks and mice. J Clin Microbiol *38*, 382-388.

Schwan, T.G., Piesman, J., Golde, W.T., Dolan, M.C., and Rosa, P.A. (1995). Induction of an outer surface protein on *Borrelia burgdorferi* during tick feeding. Proc Natl Acad Sci U S A *92*, 2909-2913.

Seaver, L.C., and Imlay, J.A. (2001). Hydrogen peroxide fluxes and compartmentalization inside growing *Escherichia coli*. J Bacteriol *183*, 7182-7189. https://doi.org/10.1128/JB.183.24.7182-7189.2001

Seiler, K.P., Vavrin, Z., Eichwald, E., Hibbs, J.B., Jr., and Weis, J.J. (1995). Nitric oxide production during murine Lyme disease: lack of involvement in host resistance or pathology. Infect Immun *63*, 3886-3895.

Seshu, J., Boylan, J.A., Gherardini, F.C., and Skare, J.T. (2004a). Dissolved oxygen levels alter gene expression and antigen profiles in *Borrelia burgdorferi*. Infect Immun *72*, 1580-1586. https://doi.org/10.1128/iai.72.3.1580-1586.2004

Seshu, J., Boylan, J.A., Hyde, J.A., Swingle, K.L., Gherardini, F.C., and Skare, J.T. (2004b). A conservative amino acid change alters the function of BosR, the redox regulator of *Borrelia Burgdorferi*. Mol Microbiol *54*, 1352-1363. https://doi.org/10.1111/j.1365-2958.2004.04352.x

Shah, P., and Swiatlo, E. (2008). A multifaceted role for polyamines in bacterial pathogens. Mol Microbiol *68*, 4-16. https://doi.org/10.1111/j.1365-2958.2008.06126.x

Shaw, N. (1970). Bacterial glycolipids. Bacteriol Rev *34*, 365-377.

Shi, F., Li, Y., Li, Y., and Wang, X. (2009). Molecular properties, functions, and potential applications of NAD kinases. Acta Biochim Biophys Sin *41*, 352-361. https://doi.org/10.1093/abbs/gmp029

Showman, A.C., Aranjuez, G., Adams, P.P., and Jewett, M.W. (2016a). Gene bb0318 Is Critical for the Oxidative Stress Response and Infectivity of Borrelia burgdorferi. Infect Immun *84*, 3141-3151. https://doi.org/10.1128/iai.00430-16

Showman, A.C., Aranjuez, G., Adams, P.P., and Jewett, M.W. (2016b). Gene *bb0318* is critical for the oxidative stress response and infectivity of *Borrelia burgdorferi*. Infect Immun. https://doi.org/10.1128/iai.00430-16

Skare, J.T., Mirzabekov, T.A., Shang, E.S., Blanco, D.R., Erdjument-Bromage, H., Bunikis, J., Bergstrom, S., Tempst, P., Kagan, B.L., Miller, J.N., *et al.* (1997). The Oms66 (p66) protein is a *Borrelia burgdorferi* porin. Infect Immun *65*, 3654-3661.

Slonczewski, J.L., Fujisawa, M., Dopson, M., and Krulwich, T.A. (2009). Cytoplasmic pH measurement and homeostasis in *Bacteria* and *Archaea*. Adv Micro Phys *55*, 1-79, 317. https://doi.org/10.1016/s0065-2911(09)05501-5

Sohlenkamp, C., de Rudder, K.E., Rohrs, V., Lopez-Lara, I.M., and Geiger, O. (2000). Cloning and characterization of the gene for phosphatidylcholine synthase. J Biol Chem *275*, 18919-18925. https://doi.org/10.1074/jbc.M000844200

Sprenger, G.A. (1995). Genetics of pentose-phosphate pathway enzymes of *Escherichia coli* K-12. Arch Microbiol *164*, 324-330. https://doi.org/10.1007/bf02529978

Storm, D.R., and Strominger, J.L. (1974). Binding of bacitracin to cells and protoplasts of *Micrococcus lysodeikticus*. J Biol Chem *249*, 1823-1827.

Tasaka, Y., Gombos, Z., Nishiyama, Y., Mohanty, P., Ohba, T., Ohki, K., and Murata, N. (1996). Targeted mutagenesis of acyl-lipid desaturases in *Synechocystis*: Evidence for the important roles of polyunsaturated membrane lipids in growth, respiration and photosynthesis. The EMBO J *15*, 6416-6425.

Teo, A.C., and Roper, D.I. (2015). Core Steps of Membrane-Bound Peptidoglycan Biosynthesis: Recent Advances, Insight and Opportunities. Antibiotics *4*, 495-520. https://doi.org/10.3390/antibiotics4040495

Terui, Y., Saroj, S.D., Sakamoto, A., Yoshida, T., Higashi, K., Kurihara, S., Suzuki, H., Toida, T., Kashiwagi, K., and Igarashi, K. (2013). Properties of putrescine uptake by PotFGHI and PuuP and their physiological significance in *Escherichia coli*. Amino Acids *46*, 661-670. https://doi.org/10.1007/s00726-013-1517-x

Thurlow, L.R., Joshi, G.S., Clark, J.R., Spontak, J.S., Neely, C.J., Maile, R., and Richardson, A.R. (2013). Functional modularity of the arginine catabolic mobile element contributes to the success of USA300 methicillin-resistant *Staphylococcus aureus*. Cell Host Microbe *13*, 100-107. https://doi.org/10.1016/j.chom.2012.11.012

Tilly, K., Elias, A.F., Errett, J., Fischer, E., Iyer, R., Schwartz, I., Bono, J.L., and Rosa, P. (2001). Genetics and regulation of chitobiose utilization in *Borrelia burgdorferi*. J Bacteriol *183*, 5544-5553. https://doi.org/10.1128/jb.183.19.5544-5553.2001

Tilly, K., Grimm, D., Bueschel, D.M., Krum, J.G., and Rosa, P.A. (2004). Infectious cycle analysis of a *Borrelia burgdorferi* mutant defective in transport of chitobiose, a tick cuticle component. Vector Borne Zoonotic Dis *4*, 159-168. https://doi.org/10.1089/1530366041210738

Touati, D. (1983). Cloning and mapping of the manganese Superoxide Dismutase gene (*sodA*) of *Escherichia coli* K-12. J Bacteriol *155*, 1078-1087.

Troxell, B., Ye, M., Yang, Y., Carrasco, S.E., Lou, Y., and Yang, X.F. (2013). Manganese and zinc regulate virulence determinants in Borrelia burgdorferi. Infect Immun *81*, 2743-2752. https://doi.org/10.1128/iai.00507-13

Troy, E.B., Lin, T., Gao, L., Lazinski, D.W., Lundt, M., Camilli, A., Norris, S.J., and Hu, L.T. (2016). Global Tn-seq analysis of carbohydrate utilization and vertebrate infectivity of *Borrelia burgdorferi*. Mol Microbiol *101*, 1003-1023. https://doi.org/10.1111/mmi.13437

Tsutsui, M., Tanimoto, A., Tamura, M., Mukae, H., Yanagihara, N., Shimokawa, H., and Otsuji, Y. (2015). Significance of nitric oxide synthases: Lessons from triple nitric oxide synthases null mice. J Pharmacol Sci *127*, 42-52. https://doi.org/10.1016/j.jphs.2014.10.002

Vagelos, P.R., and Larrabes, A.R. (1967). Acyl carrier protein. IX. Acyl carrier protein hydrolase. J Biol Chem *242*, 1776-1781.

Vallari, D.S., and Rock, C.O. (1985). Pantothenate transport in *Escherichia coli*. J Bacteriol *162*, 1156-1161.

Van Laar, T.A., Hole, C., Rajasekhar Karna, S.L., Miller, C.L., Reddick, R., Wormley, F.L., and Seshu, J. (2016). Statins reduce spirochetal burden and modulate immune responses in the C3H/HeN mouse model of Lyme disease. Microbes Infect 18(6):430-435. https://doi.org/10.1016/j.micinf.2016.03.004

von Lackum, K., and Stevenson, B. (2005). Carbohydrate utilization by the Lyme borreliosis spirochete, *Borrelia burgdorferi*. FEMS Microbiol Lett *243*, 173-179. https://doi.org/10.1016/j.femsle.2004.12.002

Vrancken, G., Rimaux, T., Weckx, S., De Vuyst, L., and Leroy, F. (2009). Environmental pH determines citrulline and ornithine release through the arginine deiminase pathway in *Lactobacillus fermentum* IMDO 130101. Int J Food Microbiol *135*, 216-222. https://doi.org/10.1016/j.ijfoodmicro.2009.07.035

Wandersman, C., Schwartz, M., and Ferenci, T. (1979). *Escherichia coli* mutants impaired in maltodextrin transport. J Bacteriol *140*, 1-13.

Wang, P., Lutton, A., Olesik, J., Vali, H., and Li, X. (2012). A novel iron- and copper-binding protein in the Lyme disease spirochaete. Mol Microbiol *86*, 1441-1451. https://doi.org/10.1111/mmi.12068

Wang, X.G., Kidder, J.M., Scagliotti, J.P., Klempner, M.S., Noring, R., and Hu, L.T. (2004). Analysis of differences in the functional properties of the substrate binding proteins of the *Borrelia burgdorferi* oligopeptide permease (Opp) operon. J Bacteriol *186*, 51-60. https://doi.org/10.1128/jb.186.1.51-60.2004

Wang, X.G., Lin, B., Kidder, J.M., Telford, S., and Hu, L.T. (2002). Effects of environmental changes on expression of the oligopeptide permease (opp) genes of *Borrelia burgdorferi*. J Bacteriol *184*, 6198-6206. https://doi.org/10.1128/jb.184.22.6198-6206.2002

Weinberg, E.D. (1997). The *Lactobacillus* Anomaly: Total Iron Abstinence. Perspect Biol Med *40*, 578-583. https://doi.org/10.1353/pbm.1997.0072

Westheimer, F.H. (1987). Mechanism of action of the pyridine nucleotides. In Coenzymes and Cofactors, D. Dolphin, R. Poulson, and O. Avramovic, eds. (New York: John Wiley & Sons, Inc), pp. 253-322.

White, B.H. (1982). Biosynthetic and salvage pathways of pyridine nucleotide coenzymes. In The pyridine nucleotide coenzymes, J. Everse, B. Anderson, and K.-S. You, eds. (New York: Academic Press), pp. 225-248.

Whitehouse, C.A., Williams, L.R., and Austin, F.E. (1997). Identification of superoxide dismutase activity in *Borrelia burgdorferi*. Infect Immun *65*, 4865-4868.

Wilks, J.C., and Slonczewski, J.L. (2007). pH of the cytoplasm and periplasm of *Escherichia coli*: rapid measurement by green fluorescent protein fluorimetry. J Bacteriol *189*, 5601-5607. https://doi.org/10.1128/jb.00615-07

Wolf, S.G., Frenkiel, D., Arad, T., Finkel, S.E., Kolter, R., and Minsky, A. (1999). DNA protection by stress-induced biocrystallization. Nature *400*, 83-85. https://doi.org/10.1038/21918

Wolfe, A.J. (2010). Physiologically relevant small phosphodonors link metabolism to signal transduction. Curr Opin Microbiol *13*, 204-209. https://doi.org/10.1016/j.mib.2010.01.002

Wolfe, A.J., Chang, D.E., Walker, J.D., Seitz-Partridge, J.E., Vidaurri, M.D., Lange, C.F., Pruss, B.M., Henk, M.C., Larkin, J.C., and Conway, T. (2003). Evidence that acetyl phosphate functions as a global signal during biofilm development. Mol Microbiol *48*, 977-988. https://doi.org/10.1046/j.1365-2958.2003.03457.x

Workman, S.D., and Strynadka, N.C.J. (2020). A Slippery Scaffold: Synthesis and Recycling of the Bacterial Cell Wall Carrier Lipid. J Mol Biol. https://doi.org/10.1016/j.jmb.2020.03.025

Worrall, D.M., and Tubbs, P.K. (1983). A bifunctional enzyme complex in coenzyme A biosynthesis: purification of pantetheine phosphate adenylyltransferase and dephospho-CoA kinase. Biochem J *215*, 153-157. https://doi.org/10.1042/bj2150153

Xu, H., Caimano, M., Lin, T., He, M., Radolf, J.D., Norris, S.J., Gherardini, F., Wolfe, A.J., and Yang, X.F. (2010). Role of acetyl-phosphate in activation of the Rrp2-RpoN-RpoS pathway in *Borrelia burgdorferi*. PLoS Pathogens *6*, e1001104. https://doi.org/10.1371

Xu, H., Caimano, M.J., Lin, T., He, M., Radolf, J.D., Norris, S.J., Gherardini, F., Wolfe, A.J., and Yang, X.F., Pal, U., Alani, S.M., Fikrig, E., and Norgard, M.V. (2004). Essential role for OspA/B in the life cycle of the Lyme disease spirochete. J Exp Med *199*, 641-648. https://doi.org/10.1084/jem.20031960

Yokoyama, K., and Imamura, H. (2005). Rotation, structure, and classification of prokaryotic V-ATPase. J Bioenerg Biomemb *37*, 405-410. https://doi.org/10.1007/s10863-005-9480-1

Youn, H.D., Kim, E.J., Roe, J.H., Hah, Y.C., and Kang, S.O. (1996). A novel nickel-containing superoxide dismutase from *Streptomyces* spp. Biochem J *318 (Pt 3)*, 889-896. https://doi.org/10.1042/bj3180889

Zgurskaya, H.I., and Nikaido, H. (2000). Multidrug resistance mechanisms: drug efflux across two membranes. Mol Microbiol *37*, 219-225. https://doi.org/10.1046/j.1365-2958.2000.01926.x

Zheng, M., Aslund, F., and Storz, G. (1998). Activation of the OxyR transcription factor by reversible disulfide bond formation. Science *279*, 1718-1721. https://doi.org/10.1126/science.279.5357.1718

Zheng, M., Wang, X., Templeton, L.J., Smulski, D.R., LaRossa, R.A., and Storz, G. (2001). DNA microarray-mediated transcriptional profiling of the *Escherichia coli* response to hydrogen peroxide. J Bacteriol *183*, 4562-4570. https://doi.org/10.1128/JB.183.15.4562-4570.2001

Zhou, B., Yang, Y., Chen, T., Lou, Y., and Yang, X.F. (2018). The oligopeptide ABC transporter OppA4 negatively regulates the virulence factor OspC production of the Lyme disease pathogen. Ticks Tick Borne Dis *9*, 1343-1349. https://doi.org/10.1016/j.ttbdis.2018.06.006

Zhou, X., Cahoon, M., Rosa, P., and Hedstrom, L. (1997). Expression, purification, and characterization of inosine 5'-monophosphate dehydrogenase from *Borrelia burgdorferi*. J Biol Chem *272*, 21977-21981. https://doi.org/10.1074/jbc.272.35.21977

Chapter 7

Structure, Function, Biogenesis and Maintenance of the *Borrelia* Cell Envelope

Wolfram R. Zückert[1]*, Brandon L. Jutras[2], Alvaro M. Toledo[3] and Sven Bergström[4]

[1]Department of Microbiology, Molecular Genetics and Immunology, University of Kansas School of Medicine, Kansas City, KS, USA
[2]Department of Biochemistry, Virginia Polytechnic Institute, Blacksburg, VA, USA
[3]Department of Entomology, Rutgers University, New Brunswick, NJ, USA
[4]Department of Molecular Biology, Umeå University, Umeå, Sweden
*Corresponding author: wzueckert@kumc.edu

DOI: https://doi.org/10.21775/9781913652616.07

Abstract

Although *Borrelia* spirochaetes are often, but mistakenly, described as Gram-negative bacteria due to their diderm, i.e. double-membrane envelopes, a closer examination reveals significant differences in composition and architecture. Particularly striking is the lack of classical endotoxin/lipopolysaccharide. Instead, glycolipids and surface lipoproteins dominate the the host-pathogen interface, where they play important roles during transmission, persistence and ensuing pathogenic processes. A modified peptidoglycan cell wall is also emerging as a potent pathogenicity determinant, in addition to contributing to cell shape in concert with periplasmic flagella. While surface lipoproteins such as the Osps interact with a variety of ligands in different organ tissues, they are also targets of the immune response and several have emerged as vaccine candidates. Some of the identified periplasmic lipoproteins, i.e., the OppAs, are components of substrate transport complexes. Investigations into integral membrane proteins led to the identification of several *Borrelia* porins: P13, whose structure and function is unknown, DipA, which is specific for dicarboxylates, and P66, which has a dual role as a pore-forming outer membrane protein and adhesin. Tol homologs BesA, -B, and -C appear to form a Type I 'channel' to export exogenous toxic agents such as antibiotics and maintain infectivity by an unknown mechanism. Initial studies on envelope biogenesis pathways and mechanisms based on diderm proteobacterial model organisms have revealed significant deviations from the Gram-negative norm, further bolstering the unique status of *Borrelia* among microbial pathogens.

Introduction

The complex life style of *Borrelia* spirochaetes, i.e. their ability to shuttle between hematophagous arthropods and various vertebrates, exposes them to a variety of niches differing in nutritional content and immunological pressure. Survival, transmission and ultimately pathogenesis, therefore, requires these organisms to possess a large degree of adaptive biological capacity. First, the surface of the pathogen needs to be optimized repeatedly for interaction with environments and tissues as different as the tick midgut, the mammalian skin or joint cartilage. Diverse outer surface-exposed proteins play a critical role in this adaptive process by providing ligands for receptor-mediated adhesion, mechanisms of host immune response avoidance, as well as pathways for the acquisition of nutrients. As they are antigens evoking immune reactions in the mammalian host, they represent pre-eminent vaccine candidates. The spatiotemporal regulation of borrelial outer surface proteins is being understood in ever increasing depth, and structure-function relationships have begun to emerge. Genomics, proteomics, novel genetic tools and sophisticated visualization techniques now aim to reveal a more complete picture of the structure, function, biogenesis and maintenance of the *Borrelia* cell envelope. Of particular interest are (i) transport and assembly of its abundant lipoproteins, (ii) the uptake of essential nutrients and the efflux of toxic and undesired compounds, (ii) the role of glycolipids and other lipid bilayer components in the formation and biological function of lipid rafts, and (iv) the emerging importance of a unique peptidoglycan cell wall in pathogenesis, cell shape and cell division. In this chapter, we will take an inward journey through the *Borrelia* envelope, using *Borrelia burgdorferi* as a well-studied guide, highlighting common and unique features, reviewing structure-function relationships of instructive examples, and exploring open questions and hypotheses.

Structural features and components of the *Borrelia* envelope

Borrelia has a diderm (i.e. two lipid bilayer) ultrastructure that is distinct from that of gram-negative bacteria, such as *Escherichia coli*. An overview of the *Borrelia* cell envelope is depicted in Figure 1. As with other spirochaetes, a fragile outer membrane surrounds the protoplasmic cylinder, which consists of a peptidoglycan layer, a cytoplasmic (inner) membrane and the enclosed cytoplasmic contents (Johnson et al., 1984; Barbour and Hayes, 1986; Kudryashev et al., 2009). A ribbon of 7-11 flagella is located in the periplasmic space where it is attached to the poles and wraps around the cell cylinder, giving the bacterium its characteristic flat wave shape (see Chapter 8 by Sze *et al.*) (Barbour and Hayes, 1986; Goldstein et al., 1994; Motaleb et al., 2000; Charon et al., 2009).

The outer membrane of all bacteria forms a selective barrier between the cell and the environment that excludes certain molecules due to its physico-chemical properties and simultaneously allows entry of nutrient substances by diffusion through the outer membrane, either via porins (Benz, 2001) or

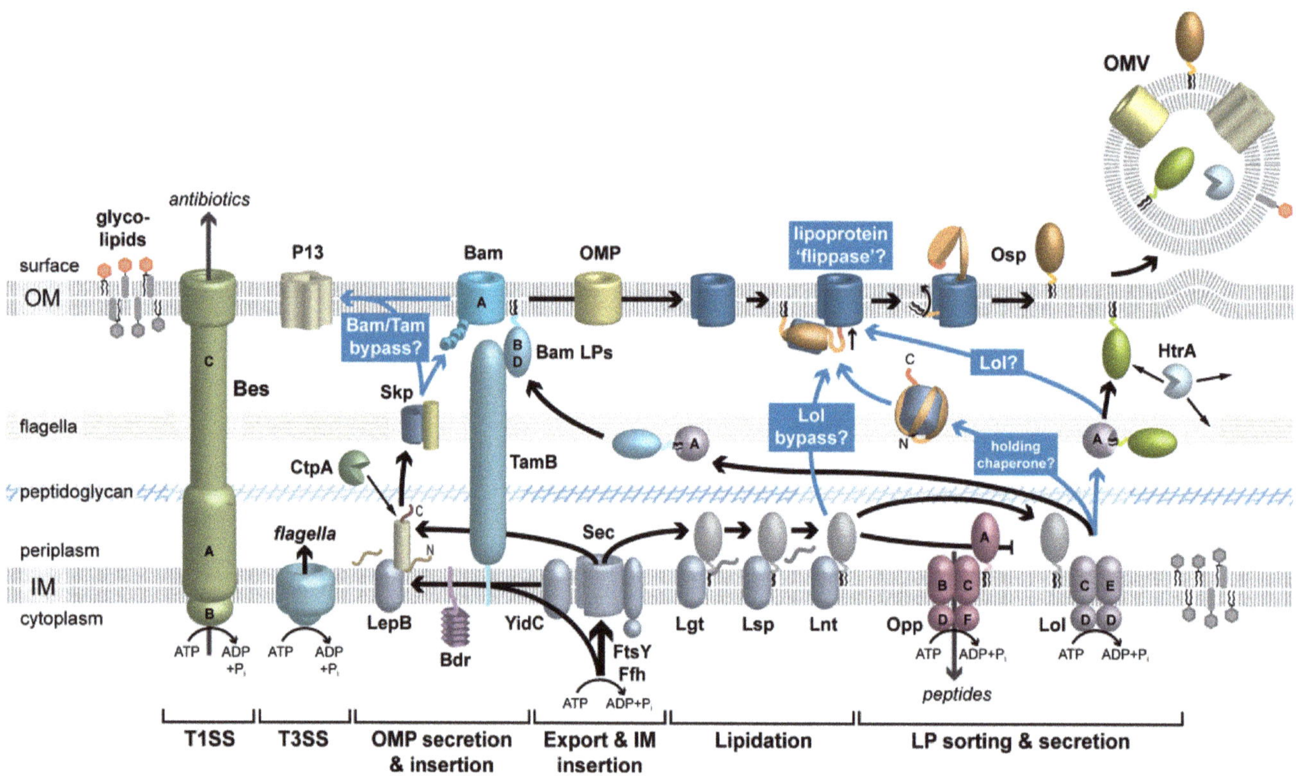

Figure 1. Major components of the diderm *B. burgdorferi* envelope and a model of spirochetal protein secretion. Displayed on the LPS-deficient bacterial surface are abundant and diverse surface lipoproteins, such as the Osps, a variety of glycolipids, and relatively few OM porins (OMPs). A few lipoproteins are anchored in the periplasmic leaflet of the OM. Outer membrane vesicles (OMVs) containing OM and periplasmic contents can be released from bacterial cells. Flagella are confined to the periplasm, atop a thin peptidoglycan layer. Envelope-spanning complexes include the Type I secretion system (T1SS) Bes machinery and the modular Bam/Tam complex involved in OMP assembly. Soluble periplasmic proteins include protein transport chaperones such as Skp and LolA, and proteases like CtpA or HtrA. Shown IM proteins are involved in protein export and processing, e.g., the flagellar T3SS, the general secretory (Sec) pathway components, lipoprotein modification (Lgt, Lsp, Lnt) and localization (Lol) pathways, or peptide import (OppA and OppBCDF) complexes. The IM also contains glycolipids. A depiction of lipid rafts containing glycolipids and membrane proteins has been omitted for clarity. See also text. (Modified from (Zückert, 2019) and used with permission).

receptor-mediated uptake (Nikaido, 2003). Additionally, outer membrane proteins (OMPs) play a critical role in the virulence of bacterial pathogens by responding to different environments, by mediating interactions with the host, or by competing with specific host defense mechanisms, thereby facilitating invasion and colonization of various tissues.

The borrelial outer cell membrane is fluid and consists of 45-62% protein, 23-50% lipid and 3-4% carbohydrate (Barbour and Hayes, 1986). Its composition differs significantly from those of gram-negative bacteria. First, a particularly distinguishing feature is the absence of phosphatidylethanolamine (Belisle et al., 1994) and lipopolysaccharide (LPS) (Takayama et al., 1987) and the presence of non-LPS glycolipid antigens (Eiffert et al., 1991; Wheeler et al., 1993; Belisle et al., 1994). These glycolipids represent about 50% of the total lipids and contain only galactose as the monosaccharide constituent (Berg, 2001; Hossain et al., 2001). It is unclear whether the glycolipids are asymmetrically distributed in the borrelial outer membrane, as with LPS in gram-negative bacteria (Kamio and Nikaido, 1976). Secondly, the *B. burgdorferi* outer membrane exhibits a relatively low density of transmembrane-spanning proteins as initially determined by freeze fracture EM studies (Walker et al., 1991; Radolf et al., 1994; Jones et al., 1995). This may explain why *Borrelia* is more susceptible than Gram negative bacteria to detergents or to disruption by routine physical manipulations such as centrifugation and resuspension. Thirdly, the outer membrane contains an unusually large number of lipoproteins (Brandt et al., 1990) many of which are on the bacterial surface, the host-pathogen interface (Dowdell et al., 2017), where they act as (i) adhesins, (ii) targets for bactericidal antibodies, (iii) or receptors for various molecules.

Outer membrane lipoproteins

An initial analysis of the genome sequence of *B. burgdorferi* sensu stricto predicted 132 open reading frames that contain the signal sequence characteristic for lipoproteins and 32 additional open reading frames that have other similarities to lipoproteins (Hayashi and Wu, 1990; Fraser et al., 1997; Casjens et al., 2000). A re-evaluation using a modified prediction algorithm indicated that the genome encodes up to 127 lipoproteins, corresponding to 7.8% of open reading frames (Setubal et al., 2006), a proportion significantly higher than in other bacterial genomes (Fraser et al., 1997; Fraser et al., 1998; Casjens et al., 2000). A comprehensive *in vitro* spatial assessment of the *B. burgdorferi* lipoproteome indicated that about two thirds of the encoded lipoproteins localize to the surface (Dowdell et al., 2017). This abundance of surface-localized lipoproteins, together with the general importance of outer surface proteins in bacterial adaptation and persistence, has led many researchers to focus their investigations on this class of surface molecules. Furthermore, the unusual membrane architecture of spirochaetes and the ancient phylogeny of these bacteria suggest that the export, structure and function of spirochetal lipoproteins are unique to these organisms.

Borrelial lipoproteins play a major role in the activation of the host inflammatory response and elicitation of inflammation in tissues at sites of infection during Lyme borreliosis. The immuno-potentiating activities of bacterial lipoproteins are related to their N-termini based on two lines of evidence: (i) acylation is essential for their inflammatory activities and (ii) synthetic N-terminal lipopeptides reproduce the proinflammatory activities of the native molecules (Bessler et al., 1985; Lex et al., 1986; Hoffmann et al., 1988; Deres et al., 1989; Erdile et al., 1993; Radolf et al., 1995). While activation of the innate immune response by lipoproteins is crucial for defense against *B. burgdorferi* infection, surface exposed lipoproteins also play essential roles in adaptive responses and pathogenicity of *Borrelia* spirochaetes. Lipoproteins dominate pathogenic mechanisms exploited by *B. burgdorferi,* including those of antigenic variation (Zhang et al., 1997; Zhang and Norris, 1998a), evasion of complement killing (Kraiczy et al., 2001; Alitalo et al., 2002; Stevenson et al., 2002) and adherence mechanisms (Coburn et al., 2005). The role of surface (lipo)proteins in pathogenesis is comprehensively discussed in a recent review by Caine and Coburn (Caine and Coburn, 2016) and chapter 13 by Coburn et al. We will, therefore, discuss only selected lipoprotein examples.

Successful succession: OspA/B and OspC

The reciprocal expression of OspA/B and OspC during *Borrelia* transition from its tick vector to the mammalian host represents the prototypical paradigm of how stage-specific protein expression can contribute to pathogenesis during the natural cycle of spirochetal transmission. In ticks, *Borrelia* expresses large amounts of OspA and OspB and almost no OspC. The production of OspA and OspB proteins decreases once withiin the mammalian host and, instead, the spirochaete produces OspC (Schwan et al., 1995; Ohnishi et al., 2001). Expression of OspC is induced 36-48 hours into the blood meal, while OspA remains expressed throughout the tick phase and is repressed once the bacteria reach the mammalian host (Barthold et al., 1995; Schwan et al., 1995; de Silva et al., 1996; Fingerle et al., 1998; Schwan and Piesman, 2000; Ohnishi et al., 2001; Iyer et al., 2015; Aranjuez et al., 2019; Caimano et al., 2019).

OspA, whose coding sequence was the first one cloned and sequenced from *B. burgdorferi*, remains one of the most studied proteins in the bacterium (Howe et al., 1986; Bergström et al., 1989). It is encoded by the same operon as OspB on the 54-kb linear plasmid lp54 (Barbour and Garon, 1988; Bergström et al., 1989). OspA consists of four β-sheets, formed by 21 anti-parallel β-strands and a C-terminal α-helix (Li and Lawson, 1995). The crystal structure of the OspB C-terminus (residues 152-296) was identical to that of OspA (Becker et al., 2005). OspA and OspB have been shown to play important roles in the pathogenesis of Lyme borreliosis in a large number of *in vivo* and *in vitro* experiments. OspA is preferentially expressed by the spirochaetes in the unfed tick midgut (Pal et al., 2000) where it mediates adhesion to the midgut epithelium by interacting with a tick receptor, TROSPA (Pal et al., 2004a). The OspA structure revealed a hydrophobic cavity buried in a positively charged cleft in the carboxyterminal domain that might be a binding site the receptor (Li et al., 1997). Providing an example into the emerging multifunctionality of *Borrelia* surface lipoproteins, the tick-specific expression of OspA (and its close homolog OspB) were later shown to shield the bacteria from bactericidal antibodies in the blood meal that are directed against otherwise accessible host-specific *Borrelia* antigens (Battisti et al., 2008; Tilly et al., 2016).

OspC is encoded on the 26-kb circular plasmid cp26 (Fraser et al., 1997) and forms a genus-wide, structurally conserved protein family with the variable small proteins (Vsps) in RF *Borrelia*. The common OspC/Vsp structural fold consists of α-helices, organized as dimer of five parallel α-helices and two short β-strands (Carter et al., 1994; Eicken et al., 2001; Kumaran et al., 2001; Zückert et al., 2001; Lawson et al., 2006). OspC initially was thought to facilitate the migration of *Borrelia* to tick salivary glands, which is initiated by a blood meal promoting the transmission of spirochaetes to the mammalian host (Gilmore and Piesman, 2000; Pal et al., 2004b). Yet, in subsequent studies, OspC was dispensable for localization or migration within the tick but was strictly required to efficiently infect mice (Grimm et al., 2004; Tilly et al., 2006; Dunham-Ems et al., 2012; Tilly et al., 2013). Further studies consolidated OspC's impressive role as a multifunctional essential early host colonization and dissemination factor, showing binding of both an immunosuppressive tick salivary protein (Ramamoorthi et al., 2005) and host complement component C4b (Caine et al., 2017), as well as protection from phagocytosis by macrophages (Carrasco et al., 2015).

Initial evaluation of the *ospC* genes from different *Borrelia* genospecies revealed a large degree of sequence polymorphism (Jauris-Heipke et al., 1995; Livey et al., 1995), resulting in the classification of strains into 19 major OspC groups (Wang et al., 1999). Molecular analysis suggested that recombination between different *ospC* alleles occurs frequently and that this genetic exchange is mediated by lateral transfer of *ospC* sequences within the same genospecies or between different *Borrelia* species (Livey et al., 1995; Seinost et al., 1999; Barbour and Travinsky, 2010). Of particular interest for the further functional definition of OspC was the finding that certain invasive *Borrelia* isolates belonged to defined OspC groups, implying that the OspC protein was also an important driver of invasiveness and dissemination of Lyme borreliosis (Seinost et al., 1999; Lagal et al., 2003). Indeed, OspC variants were later shown to differ in their affinity for plasminogen and fibrinogen (Lagal et al., 2006; Bierwagen et al., 2019), and a recent well-controlled study demonstrated that differential OspC binding to fibronectin and/or dermatan sulfate directly affects tissue tropism in an OspC variant-dependent manner (Lin et al., 2020b).

Adhesin and anti-complement factors: multifunctionality and redundancy

Attachment to tissues and cellular matrix is an important prerequisite for colonization of the host by pathogenic bacteria. Therefore, it is not surprising that in addition to OspA/B and OspC, additional borrelial lipoproteins have been shown to function as adhesins (see also Chapter 13). Decorin-binding proteins A and B (DbpA/B) are early-identified examples of this functional group of lipoproteins, facilitating the colonization of collagenous tissues and tissue tropism to the joints by adhering to the collagen-associated proteoglycan decorin (Guo et al., 1995; Guo et al., 1998; Hanson et al., 1998; Brown et al., 2001; Fischer et al., 2003). Antibodies against DbpA can prevent infection via needle infection only (Hagman et al., 1998; Hanson et al., 1998), nevertheless illustrating the importance of adhesion in the pathogenesis of Lyme disease.

Bgp (*Borrelia* glycosaminoglycan-binding protein) is a *B. burgdorferi* hemagglutinin with the capacity to bind heparan sulphate, dermatan sulphate, and aggrecan (Parveen and Leong, 2000; Russell and Johnson, 2013). Bgp is surface exposed, but its classification as a lipoprotein remains unresolved (Parveen and Leong, 2000). Bgp production was upregulated during conditions that mimicked tick feeding (Ramamoorthy and Philipp, 1998), while secretion of Bgp into extracellular environment also has been observed, indicating a possible role as a immune response decoy (Cluss et al., 2004). Since the adaptation of spirochaetes to the mammalian host environment leads to enhanced binding of glycosaminoglycan (Parveen et al., 2003) and aggrecan is a major proteoglycan found in joints, Bgp is likely contributing to efficient dissemination and colonization.

A diverse family of surface lipoproteins contributes to the bacterium's ability to evade clearance by the complement cascade, a system of plasma proteins that can be activated by antibody or lectins bound to the bacterial surface or by the pathogen surface itself (Medzhitov and Janeway, 2000). Several of the lipoproteins recruit complement regulatory factors to the bacterial surface and are therefore called CRASPs, for complement regulatory factor acquiring surface proteins. CspA (CRASP-1), expressed in the unfed tick environment, binds Factor H (FH), Factor H-like 1 (FHL-1) as well as complement components C7 and C9; CspZ (CRASP-2), expressed in the mammalian environment, binds FH and FHL-1; and some of the the OspE/F-related proteins, ErpA (CRASP-5), ErpC (CRASP-4) and ErpP (CRASP-3), bind FH, several FHL protein variants, as well as plasminogen (reviewed in (Lin et al., 2020a). CspA is a homodimer that forms a novel α-helical fold (Cordes et al., 2005) and binds FH, C7 and C9 at distinct sites (Kenedy et al., 2009; Hallstrom et al., 2013; Hammerschmidt et al., 2014), indicating that serum resistance remains important in the tick environment.

Another example of a multifunctional surface lipoprotein is BBK32. BBK32 (P35) initially was shown to bind fibronectin (Probert and Johnson, 1998) and later also to promote attachment of *Borrelia* to GAGs. Fibronectin binding proved to be important for initiating microvascular interactions, while binding of GAG via a discrete domain was shown to drive joint colonization (Fischer et al., 2006; Norman et al., 2008; Lin et al., 2015). A recent study demonstrated BBK32-mediated binding to complement factor C1r, thereby blocking the classical complement pathway. Thus, BBK32 further contributes to the bacterium's serum resistance and hematogenous spread (Garcia et al., 2016).

Multiphasic antigenic variation and persistence

The VlsE proteins are of particular interest because they were shown to be essential for persistence of *B. burgdorferi* in the natural enzootic cycle (Rogovskyy et al., 2015). VlsE belongs to the VMP (variable major proteins) family also found in RF *Borrelia* (Barbour, 1990; Burman et al., 1990; Barbour, 1993) (see Chapter 14 by Lopez et al.). VlsE is encoded on lp28-1 of *B. burgdorferi* and undergoes antigenic vajriation through an elaborate unidirectional gene conversion mechanism (Zhang et al., 1997; Zhang and Norris, 1998a; Coutte et al., 2009) via a mechanism similar to that responsible for recombination between the expressed pilin gene *pilE* and silent copies of *pilS* in *Neisseria gonorrhoeae* (Hagblom et al., 1985; Haas and Meyer, 1986; Zhang et al., 1992; Koomey, 1997; Zhang et al., 1997). Fifteen silent, non-expressed *vls* cassettes can be recombined into a functional, expressed *vlsE* gene, resulting in multiphasic antigenic variation of the expressed lipoprotein (Zhang and Norris, 1998a, b). The Holliday junction RuvAB helicase was shown to play a central role in the required recombinatory processes (Dresser et al., 2009; Lin et al., 2009).

VlsE contains invariable N- and C-terminal domains and a central variable domain that includes six variable and six invariable regions. Little is known about the signals that trigger recombination events at the *vls* locus. Recombination occurs *in vivo* as soon as four days after experimental infection of mice but not *in vitro*, suggesting that the mammalian host provides the signal for Vls recombination (Zhang and Norris, 1998b). The role of the mammalian host is important since the VlsE recombination rate of *B. burgdorferi* s.l. was found to be lower in IFN-γ R-deficient mice than in control animals. These results suggest that the murine immune response can promote the *in vivo* adaptation of *B. burgdorferi* s.l. (Anguita et al., 2001). The solved 3-D structure of VlsE is unlike OspA-C, consisting of eleven α-helices and four short β-strands (Eicken et al., 2002). This is in spite of the fact that VlsE and OspC/Vlps share overall structural features, such as the close proximity of their N and C termini and a hypervariable sequence corresponding to a surface exposed region (Cullen et al., 2004; Lawson et al., 2006; Radolf and Caimano, 2008; Zückert, 2013). A recent report indicates that VlsE masks epitopes of another surface protein, Arp (Lone and Bankhead, 2020), a phenomenon that may be similar to OspA blocking access to the outer membrane porin P66 and other surface proteins (Bunikis and Barbour, 1999; Tilly et al., 2016).

Another protein important for host adaptation is Lp6.6, a small but abundant 6.6-kDa lipoprotein localizing to the periplasmic face of the outer membrane in large protein complexes (Katona et al., 1992; Lahdenne et al., 1997; Promnares et al., 2009). Lp6.6 was shown to be important for transmission from the tick to the vertebrate host (Promnares et al., 2009) and is significantly downregulated during mammalian infection (Lahdenne et al., 1997). This downregulation is initiated during tick feeding through an RpoS-dependent mechanism (Yang et al., 2000; Caimano et al., 2005).

Outer membrane porins

The molecular sieving properties of bacterial cell envelopes are the result of channel forming proteins in their outer membranes, called porins. Porins are integral membrane proteins that form large water filled pores through the outer membrane (Benz, 1994; Achouak et al., 2001). Their function is mainly for uptake of substances from the environment and can be subdivided into two classes. The porins of the first class can be described as general diffusion pores. These porins sort mainly according the molecular mass of the solutes and show a linear relation between translocation rate and solute concentration gradient. The second class contains pores with a binding site inside the channel. These specific porins are responsible for the rapid uptake of classes of solutes such as carbohydrates, phoshate or nucleosides (Ferenci et al., 1980; Benz et al., 1986; Hancock and Benz, 1986; Benz et al., 1988).

Borrelia are limited in their metabolic and biosynthetic capacities and, therefore, highly dependent on nutrients provided by their hosts (Fraser et al., 1997; Casjens et al., 2000; Thein et al., 2008) (see Chapter 6 by Gherardini et al.). Consequently, these bacteria need to have an efficient regulation of the nutrient uptake across the cell envelope. The first indication of porins in *B. burgdorferi* came through investigation of channel-forming activities in the planar lipid bilayer assay of outer membrane vesicles (OMV). Two porin activities of 0.6 nS and 12.6 nS were found (Skare et al., 1995). Subsequent work characterized three possible porins in *Borrelia burgdorferi*: P13 (Östberg et al., 2002; Pinne et al., 2004; Barcena-Uribarri et al., 2014), P66 (Skare et al., 1997; Pinne et al., 2007; Barcena-Uribarri et al., 2010; Barcena-Uribarri et al., 2013), and Oms38/Dip A (Thein et al., 2008).

There are also indications of uncharacterized channel-forming activities present in *B. burgdorferi* as determined by black lipid bilayer analysis of outer membrane fractions (Östberg et al., 2002). Using computer-based algorithms, Kenedy and coworkers identified two novel borrelial OMPs with porin activities, BB0405 and BB0406 (Kenedy et al., 2016). BB0405 and BB0406 are cotranscribed and share a high degree of amino acid homology. Yet, BB0405 is accessible to proteases whereas BB0406 is not (Kenedy et al., 2016). Further analysis of these putative porins revealed that BB0405 appears to be essential for infection and that BB0406 is interacting with laminin and important for the dissemination into distant organs (Bista et al., 2020). In contrast to LD *Borrelia* species, knowledge of porin activities in RF *Borrelia* is rather limited. There are, however, indications of several pore-forming activities in outer membrane preparations of RF spirochaetes (Shang et al., 1998; Thein et al., 2008), and genes with a high degree of homology to *B. burgdorferi p13, p66,* and *dipA* can be found in the published genomes of the RF agents *B. duttonii, B. recurrentis* and *B. hermsii.*

The P13 porin family

The outer membrane protein P13 is encoded by ORF BB0034 which belongs to the paralogous gene family (PF) 48. PF48 has eight additional plasmid-encoded genes or pseudo-genes (Fraser et al., 1997; Noppa et al., 2001; Pinne et al., 2006). P13 was discovered in a high passage derivative of *B. burgdorferi* strain B31, B313, which lacked major Osps (OspA-D and DbpA/B) (Sadziene et al., 1995). Mice infected with B313 were used to generate MAbs against P13, which, in turn, inhibited the growth of the Osp-deficient strain but not the wild-type strain (Sadziene et al., 1995). This implied that surface exposed lipoproteins must hide the P13 epitope, in a similar manner to OspA protecting P66 (Bunikis and Barbour, 1999) and VlsE shielding Arp (Lone and Bankhead, 2020). Surface exposure of P13 was later confirmed by a variety of techniques, while the presence of an N-terminal signal sequence, a signal peptidase I cleavage site and three transmembrane spanning α-helices were predicted by computer analysis (Noppa et al., 2001). Furthermore, peptide sequencing revealed that P13 is processed at the C-terminus, but the N-terminal sequence was blocked by possible modification at the N-terminus (Noppa et al., 2001). Combined mass-spectrometry revealed this modification to be pyroglutamylation, i.e., pyroglutamate formation at an N-terminal glutamine (Nilsson et al., 2002). Pyroglutamate modification is either catalysed by the enzyme glutamine cyclotransferase or can occur spontaneously (Busby et al., 1987; Khandke et al., 1989; Awade et al., 1994). The latter might occur in P13 since no glutamine cyclotransferase homologue was found in the *B. burgdorferi* genome. Although formation of pyroglutamate at N-terminal glutamine residues has been suggested to protect some proteins from proteolytic degradation (Strobl et al., 1997), the role of pyroglutamylation of P13 remains to be elucidated.

The surface exposure and predicted transmembrane spanning domains of P13 indicated that it might possess channel-forming properties. The pore-forming activity of P13 was first described in 2002 using protein purified by FPLC from an outer membrane protein preparation. In a black lipid bilayer (BLB) analysis, it displayed an average single channel conductance (SCC) of 3.5 nS (Östberg et al., 2002). However, in a subsequent study using a different purification method, the SCC was revised and confirmed to be 0,6 nS (Barcena-Uribarri et al., 2014). Most porins exist in either open or closed states, depending on the transmembrane potential, a phenomenon known as voltage gating. Whether this is physiologically significant is questionable, since porin voltage dependence measurements in planar lipid bilayer assays suggest that the critical voltage (*Vc*), above which general porins close far exceeds the naturally occurring Donnan potential across the outer membrane (Koebnik et al., 2000). However, this strategy is still widely used to characterize porins. The channel formed by P13 was not influenced by voltage increases, reflecting a pore stably incorporated in the lipid bilayer. Since substrate-specific channels tend to be voltage-independent (Koebnik et al., 2000), perhaps P13 is a specific porin. Single channel and selectivity measurements showed that P13 had no preference for either cations or anions and showed no voltage-gating up to +/- 100 mV (Barcena-Uribarri et al., 2014).

Blue-native PAGE using P13 in its native state revealed that the P13 complex, consisting only of P13 monomers, has a molecular mass of roughly 300 kDa (Barcena-Uribarri et al., 2014). Using non-electrolytes to estimate channel diameter, the P13 homo-oligomer was estimated to be approximately 1,4 nM with a 400-Da molecular mass cut-off (Barcena-Uribarri et al., 2014). While numerous substrates have been tested, no substrate specificity for the P13 channel has yet been demonstrated.

The channel-forming activity of P13 also was confirmed by genetic experiments. The *p13* gene was inactivated by allelic-exchange mutagenesis, this being the first reported example of disruption of a gene encoding an integral outer membrane protein of *B. burgdorferi.* The porin activities of OM protein preparations derived from knock-out and wild-type strains were compared using the planar lipid bilayer assay. The SCC activity corresponding to P13 was eliminated in the *p13* knock-out strain. Evidence of additional porin activities present in OM of *B. burgdorferi*, distinct from the activities of P66, P13 and DipA, have been suggested (Östberg et al., 2002; Pinne et al., 2007; Thein et al., 2008).
Several interesting characteristics of P13, such as its small size, unusual amounts of subunits in an oligomer, hydrophobicity and predicted membrane-spanning α-helices are all in disagreement with common features of conventional porins (Östberg et al., 2002; Pinne et al., 2006; Barcena-Uribarri et al., 2014).

To further investigate the unusual processing of P13, an attempt to identify the C-terminal protease was performed. Using the amino acid sequence of the known carboxyl-terminal protease A (CtpA) from *Bartonella bacilliformis* (Mitchell and Minnick, 1997) to BLAST search the *B. burgdorferi* B31 MI genome (Fraser et al., 1997; Casjens et al., 2000), a homologous gene (*bb0359*) was found (Östberg et al., 2004). To elucidate if the activities of P13 and/or other *Borrelia* proteins are affected by CtpA, the *ctpA* gene was inactivated by allelic exchange mutagenesis (Östberg et al., 2004). Immunoblot analysis revealed that P13 was larger and had a more acidic pI in the *ctpA* knock-out, consistent with the theoretical size and pI of P13 if it were not processed at the C-terminus. 2-D gel electrophoresis of *B. burgdorferi* total proteins revealed several proteins present only in the *ctpA* mutant. Mass spectroscopy analysis of these potential CtpA substrates revealed that P13 and BB0323 are processed by CtpA, while BBA74 is up-regulated and produced in multiple isoforms in the absence of CtpA. These effects were due to loss of CtpA because complementation with a wild type copy of *ctpA* restored wild type-like protein expression profiles (Östberg et al., 2004). However, BBA74 is not a substrate for CtpA, so that its up-regulation is probably a secondary effect of *ctpA* inactivation (Östberg et al., 2004). Interestingly, the expression level of CtpA in the complemented strain is higher than in wild type, reducing the levels of BBA74 below its detection limit (Östberg et al., 2004). Thus, CtpA has pleiotropic effects, processing P13 and influencing the appearance of several other proteins. The importance of this unusual C-terminal processing of P13 is not fully known, but it may play a role in translocation of the protein *in vivo*. In a study by Kumru and coworkers, the P13 C-terminal peptide was altered to investigate its function (Kumru et al., 2011a). These studies suggests that the C-terminal peptide likely acts as a safeguard against misfolding or mislocalization in the outer membrane prior to its removal by CtpA (Kumru et al., 2011a). The connection of this C-terminal processing to the channel forming activity needs to be elucidated.

Being extremely hydrophobic (hydropathicity index, 0.47) and spanning the membrane with α-helices, one would have assumed that P13 would insert into the inner membrane by the conventional SRP-dependent protein secretion machinery. Nevertheless, the unusual feature of C-terminal cleavage CtpA led to the hypothesis that this event initiates translocation to the outer membrane (Östberg et al., 2004). Yet, P13 is present in OM protein preparations from the *ctpA* knock-out mutant. Therefore, the 28-amino acid C-terminal extension of P13 appears not to be required for transportation to the outer membrane, but cleavage may be necessary for correct P13 assembly. This mechanism is supported by the function of CtpA in photosynthetic organisms. A substrate for CtpA in photosystem II, the hydrophobic peptide D1 that spans the thylakoid membranes of chloroplasts (Hankamer et al., 2001), must be cleaved at its C-terminus for correct assembly and translocation and to ensure a functional photosystem II (Nixon et al., 1992; Ivleva et al., 2000). The C-terminus of P13 is processed directly after the alanine residue removing the extreme C-terminal 28 aa (Noppa et al., 2001). Interestingly, the C-terminal extension removed in D1 from different organisms varies from between 8 to 16 aa but is always cleaved after an alanine (Ivleva et al., 2000). Clearly, similarities between P13 and D1 exist and imply that CtpA might play a critical role in the translocation and assembly of functional P13. Moreover, CtpA of *B. burgdorferi* contains a predicted N-terminal signal sequence reminiscent of the need to transport CtpA across the thylakoid membrane in chloroplasts (Karnauchov et al., 1997) . By analogy, we interpret that CtpA is transported through the *Borrelia* inner membrane and that C-terminal processing of P13 occurs in the periplasmic space. In summary, P13 is able to form channels in the borrelial outer membrane despite its small molecular mass and α-helical secondary structure (Östberg et al., 2002; Barcena-Uribarri et al., 2014). This homooligomeric porin with a 400 kDa cut-off possibly acts as a general diffusion channel for small molecules into *Borrelia*.

P66 (Oms66)

P66 is a well-studied dual-function, integral outer membrane protein that is encoded by a chromosomal gene in both Lyme disease and relapsing fever spirochaetes. Its function as a pore-forming outer membrane protein has been clearly demonstrated by studies in planar lipid membranes and liposomes (Skare et al., 1997; Pinne et al., 2007; Thein et al., 2008; Barcena-Uribarri et al., 2010; Kenedy et al., 2014). Skare and colleagues initially showed that P66 forms pores in artificial membranes with an atypical huge single channel conductance of 9.6 nS in 1 M KCl (Skare et al., 1997). In later experiments by Pinne et al. (Pinne et al., 2007) using FPLC-purified protein, P66 exhibited voltage-dependent closure of the channels although at higher voltages. The zero-current membrane potential was –5 mV for an five-fold KCl-gradient, which means that the permeability ratio *PK/PCl* was about 0.77 as calculated from the Goldman–Hodgkin–Katz equation (Benz et al., 1979). This suggests that P66 is permeable to anions and cations. Incubation of purified P66 with corresponding polyclonal rabbit antiserum resulted in total loss of the 11 nS channel-forming activity (Pinne et al., 2007).

A channel sizing analysis, examining single-channel conductance in black lipid bilayers in the presence of different non-electrolytes with known hydrodynamic radii, showed that non-electrolytes with hydrodynamic radii smaller than 0.34 nm entered the pore, whereas non-charged molecules with a greater radius were non-permeable. The effective diameter of the P66 channel lumen was determined to be ~1.9 nm, with a central constriction of about 0.8 nm (Barcena-Uribarri et al., 2013). As derived from single- and multichannel experiments, the P66-induced membrane conductance could be blocked by 80–90% after addition of the non-electrolytes PEG 400, PEG 600 and maltohexaose in the low millimolar range (Barcena-Uribarri et al., 2013). The blockage of one P66 single-channel conductance unit of 11 nS occurred by seven or eight subconducting states, indicating a heptameric or octameric constitution of P66. This could be confirmed by Blue native PAGE and immunoblot analysis which revealed that P66 forms a complex with a corresponding mass of approximately 440 kDa (Barcena-Uribarri et al., 2013). Of note, other *Borrelia* species, including the relapsing fever spirochaetes *B. duttonii*, *B. recurrentis* and *B. hermsii*, also carry P66 homologs (Lescot et al., 2008) with inter-species amino acid identities of 41% but similar physicochemical properties; all form large pores with single-channel conductances between 9 and 11 nS (Barcena-Uribarri et al., 2010).

The second known function of P66, as a ligand for β3-chain integrins, was initially revealed using a phage display library of total genomic *B. burgdorferi* DNA (Coburn et al., 1999; Antonara et al., 2007). In follow-up studies, a synthetic peptide corresponding to P66 amino acids 203–209, i.e, within the region panned from the phage display library, inhibited *B. burgdorferi* attachment to integrin αIIbβ3 (Defoe and Coburn, 2001), while targeted mutagenesis of the same domain led to P66 mutants that were deficient in binding to αvβ3 integrin (Coburn and Cugini, 2003). About a decade later, P66-mediated integrin binding was shown to play a role in bacterial escape from the site of entry and further dissemination into the tissues in the mammalian host (Ristow et al., 2012; Ristow et al., 2015), likely by interactions with integrins found on different immune cells, blood platelets and endothelial cells. Most recently, intravital imaging of *B. burgdorferi* in a mouse model of infection demonstrated the importance of P66 for vascular transmigration (Kumar et al., 2015). Together, these studies firmly establish the role of P66 in vascular spread of the bacterium. Of note, site-directed mutagenesis of the P66 integrin binding domain revealed that binding activity is distinct from porin activity (Ristow et al., 2015).

The precise structure of P66 remains undetermined, although available data point to the formation of a large beta-barrel. Early experiments already predicted and then confirmed surface-exposed epitopes (Bunikis et al., 1995; Probert et al., 1995), with a major antigenic surface loop being protected by Osp lipoproteins, in particular OspA, from access of both antibodies and trypsin (Bunikis and Barbour, 1999) - whether the role of this interaction extends beyond this shielding function remains to be determined. Structural modeling based on the surface accessibility of this loop and circular dichroism spectrometry suggests that P66 forms a β-barrel with either 22 or 24 transmembrane domains (Kenedy et al., 2014).

DipA

Borrelia spirochaetes are – due to their small genomes – metabolically and biosynthetically deficient, thereby making them highly dependent on nutrients provided by their various hosts (see Chapter 6 by Gherardini et al.). With neither P13 or P66 displaying any substrate specificity, the search for nutrient-specific porins continued. The OM porin Oms38 was discovered in the OM fractions of RF *Borrelia* and is highly conserved in *B. duttonii*, *B. hermsii* and *B. recurrentis* (Thein et al., 2008). In black lipid bilayers, purified Oms38 formed water-filled pores of small conductance (80 pS in 1 M KCl) that were selective for anions. Further analyses of analogous proteins in LD borreliae revealed a 36-kDa protein (BB0418) with sequence and functional similarity to Oms36 in RF *Borrelia* (Thein et al., 2012). The protein, designated DipA (for dicarboxylate-specific porin A), has an average single-channel conductance of 50 pS in 1 M KCl. DipA is selective for anions with a ratio of permeability for cations over anions of 0.57 in KCl and does not show voltage-dependent closure. The permeation of KCl through the channel could be partly blocked by titrating the DipA-mediated membrane conductance with increasing concentrations of different organic di- and tricarboxylic anions. Particular high stability constants up to 28,320 l/mol (in 0.1 M KCl) were obtained among the 11 tested anions such as oxaloacetate, 2-oxoglutarate and citrate. The obtained results imply that DipA does not form a general diffusion pore, but a porin with a binding site specific for dicarboxylates which play important key roles in the deficient metabolic and biosynthetic pathways of *Borrelia* species (Thein et al., 2008; Thein et al., 2012).

Outer membrane glycolipids

Lipids constitute roughly 25-30 % of *B. burgdorferi* dry weight, of which >36% correspond to glycolipids (Hossain et al., 2001; Ben-Menachem et al., 2003). Early studies on glycolipids focused on their antigenic role in Lyme disease (Eiffert et al., 1991; Wheeler et al., 1993) while their structural analysis (Hossain et al., 2001; Ben-Menachem et al., 2003; Schroder et al., 2003; Stubs et al., 2009; Stubs et al., 2011) and biophysical properties (Huang et al., 2016) were later studied.

While lipoproteins are significant contributors to the immune response in Lyme disease patients and therefore, useful markers for serological diagnostic tests, several studies also noted the contribution of *Borrelia* lipids to the immune response. For example, *Borrelia* fatty acids and carbohydrates strongly reacted with sera containing anti-*B. burgdorferi* antibodies (Eiffert et al., 1991), which supported that *B. burgdorferi* lipids could mediate in antibody responses. Further studies showed that two lipid fractions from *B. burgdorferi* were recognized by sera collected from patients with Lyme disease (Wheeler et al., 1993) and some of the components of these antigenic fractions were identified as glycolipids (Wheeler et al., 1993). Interestingly, a subset of Lyme disease patients with neuroborreliosis had IgM antibodies that reacted to gangliosides with a Gal (β1-3) GalNac terminal sequence (Garcia Monco et al., 1993). Subsequent studies in rats proved that purified lipid fractions of *B. burgdorferi* elicit high levels of IgM antibodies that cross-reacted with asialo-GM1 and GM1 (Garcia-Monco et al., 1995). Conversely, sera of rats immunized with asialo-GM1 produced antibodies that cross-reacted with *B. burgdorferi* antigens (Garcia-Monco et al., 1995). These studies showed the antigenic potential of *Borrelia* lipids and opened the possibility, not yet resolved, that the pathogenesis of Lyme disease could be driven, at least in part, by cross-reaction between *Borrelia* glycolipids and host lipids. More recently, it was shown that glycolipids also produce a strong IgG antibody response in patients with Lyme arthritis (Jones et al., 2009), which supports that host antibody responses to glycolipids are not limited to Lyme disease patients with central nervous involvement. Several studies have purified, characterized and confirmed their immunogenicity (Hossain et al., 2001; Ben-Menachem et al., 2003; Schroder et al., 2003; Stubs et al., 2009; Stubs et al., 2011).

It was not until the beginning of the 21st century that several studies shed light on the structure of lipids in *B. burgdorferi*. A study using high-performance thin liquid chromatography (TLC) revealed the presence of 11 components, including two phospholipids, phosphatidylcholine (PC) and phosphatidylglycerol (PG), as well as a glycolipid, mono α-galactosyl-diacylglycerol (MGalD), which contains galactose and two fatty acids (Hossain et al., 2001). Shortly thereafter, two additional glycolipids containing cholesterol were identified, cholesteryl-β-D-galactopyranoside (CGal) and cholesteryl 6-O-palmitoyl-β-D-galactopyranoside (ACGal) (Ben-Menachem et al., 2003; Schroder et al., 2003). Both cholesterol glycolipids are very similar in their structure, with ACGal having an extra acyl chain bound to the galactose group (Ben-Menachem et al., 2003; Schroder et al., 2003). Glycolipids MGalD, CGal, and ACGal are ubiquitous within the *B. burgdorferi* sensu lato complex and are present in both pathogenic and non-pathogenic genospecies. (Stubs et al., 2009; Stubs et al., 2011). The cholesterol glycolipid ACGal is immunogenic and recognized by sera from patients with Lyme disease but not syphilis (Stubs et al., 2011). Other glycolipids that can be found in *B. burgdorferi* in trace amounts are CGlu and ACGlu (Stubs et al., 2009), which are the only cholesterol glycolipids present in the relapsing fever species *Borrelia hermsii* (Livermore et al., 1978). Although both glycolipids are very similar and only differ in the sugar group, ACGal is recognized by sera from Lyme disease patients while ACGlu is not (Stubs et al., 2011). The specificity and immunogenicity of *Borrelia* glycolipids MGalD and ACGal led to their synthesis and evaluation as potential vaccine antigens as well as diagnostic applications (Pozsgay et al., 2011; Twibanire et al., 2012; Jager et al., 2019). Glycolipids are present in both the outer and inner membranes of *Borrelia* (Toledo et al., 2018a). Nonetheless, the relative amount of the non-cholesterol glycolipid MGalD is higher in the inner membrane, whereas ACGal and CGal are relatively more abundant in the outer membrane (Toledo et al., 2018b). A proton nuclear magnetic resonance (^{1}H-NMR) spectrum showed that the three glycolipids had an equivalent double bond content, which supports that the structure and biophysical properties of *Borrelia* glycolipids in the outer and inner membranes are similar (Toledo et al., 2018b)

Outer membrane lipid rafts

Lipid rafts are membrane microdomains that are rich in cholesterol and sphingolipids and have a subset of specific proteins associated with them (London, 2002). These microdomains are highly dynamic and float freely within the liquid-disorder bilayer of cellular membranes but also can cluster together to form larger ordered domains (Simons and Ehehalt, 2002). Until recently, lipid rafts were associated with eukaryotic cells since sphingolipids and cholesterol are rare among prokaryotes (London, 2002; Bieberich, 2018). Pioneer studies on *Bacillus subtilis* and *B. burgdorferi* showed that prokaryotes also have membrane domains that resemble eukaryotic lipid rafts (LaRocca et al., 2010; Lopez and Kolter, 2010; LaRocca et al., 2013). The presence of lipid rafts in *B. burgdorferi* was first identified while studying the mechanism of action of CB2, a complement-independent bactericidal antibody (LaRocca et al., 2009; LaRocca et al., 2010). The CB2 antibody has a very high affinity for its antigen, OspB protein, and disrupts the membrane of the spirochaete leading to an osmotic shock and subsequent death of the bacterium (LaRocca et al., 2009). The bactericidal action of CB2 is specific to *Borrelia* and depends on the presence of cholesterol (LaRocca et al., 2010). Partial depletion of cholesterol from the membrane using methyl-β-cyclodextrin protects the spirochaete from the bactericidal effect of the antibody. The bactericidal CB2 antibody also recognizes, but does not lyse, an *E.coli* strain that expresses a surface-exposed OspB.

Transmission electron microscopy showed the presence of cholesterol-rich microdomains from *in vitro*- and *in vivo*-derived spirochaetes after tagging cholesterol glycolipids with 6 nm colloidal gold particles (LaRocca et al., 2010; LaRocca et al., 2013; Toledo et al., 2014). One of the characteristics of lipid rafts is that they form fluid microdomains enriched in cholesterol and sphingolipids that are more ordered and tightly packed than the surrounding bilayer (London, 2002; Simons and Ehehalt, 2002). The properties of *Borrelia* cholesterol-rich domains were studied using biophysical approaches, including anisotropy and fluorescence resonance energy transfer (FRET). Anisotropy experiments conducted on vesicles made out of total lipid extracts, as well as on live spirochaetes, confirmed the presence of temperature-dependent ordered domains. (LaRocca et al., 2010; LaRocca et al., 2013; Toledo et al., 2014). Similarly, weak FRET, measured as the ratio of fluorescence intensity with acceptor to that without acceptor (F/F_0), supported the co-existence of temperature-dependent ordered (rafts) and disordered domains in the membrane of the spirochaete (LaRocca et al., 2010; LaRocca et al., 2013; Toledo et al., 2014)

In contrast to eukaryotic cells, *Borrelia* does not have sphingolipids. Sphingolipids are critical components of eukaryotic lipid rafts (Brown and London, 2000; Bieberich, 2018), which raised questions about the biophysical properties of *Borrelia* microdomains and whether the same principles of lipid raft formation apply to prokaryotic and eukaryotic organisms. Cholesterol substitution experiments performed in live spirochaetes using sterols with different biophysical properties, ranging from raft-supporting to raft-inhibiting when mixed with eukaryotic sphingolipids, proved that sterols are necessary and sufficient for the formation of lipid rafts in the spirochaete (Crowley et al., 2013). Raft-supporting sterols also formed detergent-resistant membranes, and saturated acyl chains were required for biotinylated probes to partition within lipid rafts, supporting that cholesterol-rich domains in *Borrelia* have the hallmarks of eukaryotic lipid rafts (LaRocca et al., 2013).

The role of different lipids in the formation of rafts is crucial in understanding the organization of the outer membrane of *B. burgdorferi*. The most abundant lipids in *Borrelia*, including ACGal, MGalD, and PC, were purified and used to make vesicles in order to study the principles that govern raft-formation. Experiments using vesicles made of ACGal, MGalD and PC and their mixtures, analyzed by FRET and anisotropy, proved that ACGal was the primary lipid responsible for order (raft) domain formation and that its role was similar to that observed for cholesterol in eukaryotic rafts (Huang et al., 2016). In contrast to unsaturated PC or *Borrelia* MGalD, the level of saturation detected in *Borrelia* PC by NMR was relatively high, which facilitates domain formation by taking a similar role as to sphingolipids in eukaryotic rafts (Huang et al., 2016).

The first evidence that lipid rafts in *Borrelia* were molecular platforms that contain a subset of specific proteins came from co-localization experiments using transmission electron microscopy. Cholesterol glycolipids formed discrete domains on the surface of the spirochaete associated with lipoproteins OspA and OspB (LaRocca et al., 2010). Depletion of cholesterol from the spirochetal outer membrane using methyl-β-cyclodextrin disrupted lipid rafts as well as the proteins associated with them (LaRocca et al., 2010; Toledo et al., 2015). The protein content of lipid rafts was determined by mass spectrometry (Toledo et al., 2015; Toledo et al., 2018a).

The proteome of outer membrane lipid rafts provided a consistent set of selected lipoproteins, proteases and sensing molecules, some of which are prokaryotic homologs of eukaryotic lipid raft markers (Toledo et al., 2015; Toledo et al., 2018b). It is not surprising that the proteome was enriched in acylated proteins since fatty acylation is one mechanism for targeting proteins to lipid rafts (Moffett et al., 2000). Two of the most abundant proteins found in rafts were OspA and OspB. Both lipoproteins are predominantly expressed during the tick phase (Fingerle et al., 1995; Schwan et al., 1995) and mediate in the adhesion of the spirochaete to the tick midgut (Yang et al., 2004; Battisti et al., 2008). Nonetheless, a lipid tail is not sufficient to drive proteins to rafts since some lipoproteins (*e.g.*, OspC) are not associated with these microdomains. The raft proteome also included protease HtrA. The Deg/HtrA protein family is involved in adaptation to stress responses acting either as a chaperone or protease to maintain tigh protein quality control (Clausen et al., 2011). The periplasmic serine protease HtrA is present in different subcellular locations, including in the IM and OM (Coleman et al., 2013; Russell et al., 2013; Toledo et al., 2018b) where it is associated with lipid rafts (Toledo et al., 2018b). In the periplasm, the enzyme degrades unfolded proteins such as flagellin (Zhang et al., 2019). Nonetheless, *in vitro* experiments showed that a recombinant BbHtrA also degrades membrane proteins including P66 and BB0323 that co-localize with the protease in outer and inner membrane rafts, respectively (Kariu et al., 2015; Toledo et al., 2015; Coleman et al., 2016). In fact, a *B. burgdorferi* strain that overexpresses HtrA showed a two-fold decrease in the levels of P66 (Coleman et al., 2016), suggesting that co-localization of proteases and their substrates in rafts could facilitate protein degradation. Other proteins with important biological functions that were enriched in outer membrane rafts included plasminogen binding protein ErpA (Brissette et al., 2009) and Factor H binding protein CspA (Kenedy et al., 2009).

The selective association of lipoproteins and their contributions to raft formation was assessed using OspA and OspB mutants. Single OspA and OspB mutants showed no defect in lipid raft formation compared to the WT strain. However, *B. burgdorferi* B313, a strain that lacks OspA and OspB as well as most of the linear and circular plasmid, showed lower fluorescence values by fluorescence resonance energy transfer (FRET), which is compatible with less or smaller lipid rafts. The WT phenotype was recovered when B313 was transformed with a plasmid containing OspA but not when it was transformed by a plasmid expressing OspC, which is not found on lipid rafts. Collectively, these findings support a selective association of lipoproteins and lipid rafts that favors the formation of these microdomains (Toledo et al., 2014).

The peptidoglycan cell wall

Sandwiched between outer and inner membranes of the *B. burgdorferi* cell envelope is the peptidoglycan (PG) cell wall. PG is a mesh-like, stress-bearing, biological sac (i.e. sacculus). The primary role of PG is to act as an osmoprotectant, a dynamic bio-fabric that prevents the cell from bursting due to the high solute concentration in the cytoplasm (Holtje, 1998). Apart from combating turgor, PG has additional functions. PG is, in essence, the exoskeleton of most bacteria as well as the major cell envelope component that imparts cell shape (Young, 2006). Cell envelope organization is largely dictated by PG in that the sacculus often acts as a scaffold and provides continuity between layers (Cava and de Pedro, 2014). Because the PG layer is porous, it acts as a nondiscriminatory sieve for molecules up to 24kDa (Demchick and Koch, 1996). Apart from playing a structural role, PG organization is also central to the sub-cellular organization of many cell cycle events (Turner et al., 2014). Fragments of PG, or muropeptides, are even capable of acting as a chemical means of communication between bacteria and even kingdoms (Dworkin, 2014). Despite these exciting and versatile functions, until recently, the PG of *B. burgdorferi* has been largely ignored. Here we consider the unique chemical and structural features of *B. burgdorferi* PG and how these peculiarities may be involved in the biology and pathogenesis of Lyme disease.

The PG mesh is composed of two salient features – roughly parallel (Verwer et al., 1978) glycan strands that are connected by short peptides. Each glycan strand contains the repeating disaccharide *N*-Acetyl Glucosamine (Glc*N*Ac) and *N*-Acetyl Muramic acid (Mur*N*Ac). The latter is a sugar that is only found in bacteria (Vollmer et al., 2008). Disaccharides are joined by a ß-1,4 glycosydic bond. Virtually all bacteria follow this glycan strand organization; however, modifications that make the linkage resistant to lysozyme degradation, such as *O*-acetylation, are known to occur in several pathogens (Vollmer, 2008). While glycan organization and structure are well-conserved among all bacteria, some variability exists in the bridging peptides. The most common amino acids in the first and second position are L-Alanine and D-Glutamine, respectively. Variations occur at the third position, but chemical constraints require that this position is occupied by a dibasic amino acid to complete the cross-bridge with the neighboring peptide chain. Most gram-negative bacteria incorporate diaminopimelic acid (DAP) as the diamine, in the third position, while gram-positive bacteria carry Lysine (Lys) at the analogous position (Desmarais et al., 2013). Deviations from this dichotomy are rare (Vollmer, 2008). Di-D-Alanine terminate the pentapeptide and are the substrate(s) for Penicillin-binding protein (PBP) transpeptidase reactions, reactions that are required for remodeling the overall PG structure during sacculus elongation.

B. burgdorferi PG has several unusual features that deviate from convention. For instance, L-Ornithine (L-Orn) occupies the third position of the stem peptide, linked to a single Glycine (Gly) residue (Beck et al., 1990; Jutras et al., 2019). Several earlier reports show that some spirochaetes, including the relapsing fever agent *Borrelia hermsii* and treponemes incorporate L-Orn into PG (Azuma et al., 1975; Yanagihara et al., 1984; Caimano et al., 1999). The amidation of DAP – a common PG modification (Vollmer, 2008) – has the exact same mass as L-Orn-Gly. Thus, a more indepth analysis may be required to validate earlier findings ((Azuma et al., 1975; Yanagihara et al., 1984). Nevertheless, recent radiolabeling studies, coupled with tandem liquid chromatography/liquid scintillation (Jutras et al., 2019), confirmed the presence of L-Orn-Gly in *B. burgdorferi* PG. The same study uncovered a surprising alteration in glycan strand composition, providing evidence for a yet to be identified *N*-acetylated Hexose (Hex*N*Ac). Our high-resolution analysis of *B. burgdorferi* PG has both confirmed the presence of a third Hex*N*Ac, while identifying amidated Glutamine as another modification (DeHart and Jutras, unpublished data). The biological relevance of these uncommon PG alterations remains to be determined.

The structural architecture of *B. burgdorferi* PG is beginning to be elucidated. Composition and abundance of muropeptide profiles indicate that, on average, *B. burgdorferi* glycan strands are 30 disaccharides long (Jutras et al., 2019), which equates to ~30nm. In comparison, this is roughly similar to the glycan strand length of *Escherichia coli* grown under normal conditions. Glycan strands lie perpendicular to the long axis (Gan et al., 2008). When we consider the diameter of the spirochetal cell— 5 times smaller than *E. coli*— the number of strands required to cover the *B. burgdorferi* cell circumference (~20) is miniscule relative to other PGs previously analyzed (Vollmer and Seligman, 2010). Of the peptides present in *B. burgdorferi* PG, approximately 40% are participating in cross-linking glycan strands (Jutras et al., 2019). In line with this observation, free stem peptides are devoid of terminal D-Alanine(s); their detection required a high-resolution, targeted MS approach (Jutras et al., 2016). The degree of cross-linking may be involved in the seemingly non-uniform distribution of *B. burgdorferi* PG. Cryo-electron tomography images of intact *B. burgdorferi* cells show that the thickness of the PG layer varies from ~6 nm at mid-cell to ~40 nm at the poles (Figure 3A). Preliminary analysis of purified sacculi suggests that poles may be reinforced with multiple layers of PG (Brock, Liu, and Jutras, unpublished data) – yet another unusual feature and paradigm shift from the ~1-layer model for diderms (Gan et al., 2008).

Why do *Borrelia*, and seemingly other spirochaetes (DeHart and Jutras, unpublished data), produce PG with unusual chemical and structural features? Unlike most bacteria, *Borreliae* and *Treponema* possess periplasmic flagella form a ribbon that wrap around the cell cylinder (Charon et al., 2009) and are inserted, sub-terminally, at each pole. The natural flat-wave morphology is created by the inherent helicity of the collective flagellar filaments (Motaleb et al., 2000). Together, many studies support an unusual paradigm as it relates to PG and cell morphology – PG doesn't impart cell shape in *Borrelia* and treponemes (Charon et al., 2012). Instead, the PG – a straight rod when purified (Jutras et al., 2016) – pushes against the helical pitch created by the filaments, indicating that *Borrelia* flagella are the actual cytoskeletal structure. Flagellar motors are anchored into the PG layer and their rotation (and thus movement) produces immense torque on the cell cylinder. To protect cytoplasmic contents, PG produced by most spirochaete must be exceptionally flexible in order to exert counter force against the torque of the flagellar filaments (Charon et al., 2009). PG chemistry and structure are likely to be important in the elasticity of the biopolymer. Several lines of evidence support this notion. First, in order to create flat-wave morphology, mathematical modeling suggests that the PG of *B. burgdorferi* be more flexible than other, helical-shaped spirochaetes that do not possess L-Orn-type PG (Wolgemuth et al., 2006; Dombrowski et al., 2009; Slamti et al., 2011). Treating *B. burgdorferi* with vancomycin, an inhibitor of PG transpeptidation and cell wall architecture, alters both PG stiffness and bacterial motility (Harman et al., 2017). Finally, recent data indicate that mutations in PG synthesis enzymes that result in altered PG peptide composition cause a dramatic change in cell morphology (MecClune and Jutras, unpublished data), which raises the notion that that *B. burgdorferi* PG may have evolved to strike the perfect biophysical balance with the spirochaete motility machinery (DeHart and Jutras, unpublished data).

There may be unintended consequences of having atypical PG. *B. burgdorferi* is considered by many to be a stealth pathogen. Many of the ill effects during the course of LD infection are propagated by the human host response. As *B. burgdorferi* elongates its PG sacculus, it sheds muropeptides – 45% of all PG inside the cell is released per generation (Jutras et al., 2019). Released muropeptides are known to act as host immune modulatory molecules in other pathogens. For instance, *Neisseria gonorrhoeae* release many different muropeptides (Sinha and Rosenthal, 1980; Garcia and Dillard, 2008), most of which are cytotoxic and play a critical role in pathogenesis (Melly et al., 1984; Cloud-Hansen et al., 2008; Chan and Dillard, 2016; Schaub et al., 2016; Chan and Dillard, 2017; Lenz et al., 2017). Another gram negative, *Helicobacter pylori* both releases and injects PG breakdown products during mammalian infection to modulate host response (Viala et al., 2004; Gorrell et al., 2013; Irving et al., 2014). Tracheal Cytotoxin (TCT), one of the first bacterial molecules to demonstrate toxin activity (Goldman et al., 1982) is indeed a released muropeptide produced by *Bordetella pertussis* involved in whooping cough pathology (Goldman et al., 1982; Heiss et al., 1993; Heiss et al., 1994; Flak and Goldman, 1999).

Transport proteins and complexes

Transport proteins catalyze the translocation of solutes by various systems that differ with respect to energy coupling, substrate specificity and transmembrane topology. So far, four distinct types of functionally characterized transport systems have been described. These are (i) bacterial channel proteins, (ii) secondary transporters that utilize chemo-osmotic energy, i.e. proton gradient, (iii) primary transporters utilizing chemical energy, typically in the form of ATP, and (iv) group translocators that phosphorylate their substrates during transport (PTS). From a comparative genome analysis of transporters from 18 different prokaryotes by Paulsen and co-workers (Paulsen et al., 2000), it appeared that, relative to genome size, the spirochaetes *B. burgdorferi* and *T. pallidum* have few transporters, particularly secondary transporters (Saier and Paulsen, 2000). Interestingly, both spirochaetes contained the lowest numbers of transmembrane segments and transporters as compared to other bacteria (Paulsen et al., 2000). In another comparative study of 215 protein-translocation outer membrane porins of Gram-negative bacteria, only the TolC homologue (BB0142) that belongs to the outer membrane factor (OMF) family was identified in *B. burgdorferi* (Yen et al., 2002). Among the bacteria compared, *B. burgdorferi* has the largest percentage of phosphotransferase system (PTS) permeases. Notably *B. burgdorferi* also encodes a large number of peptide transporters (Paulsen et al., 2000).

Within the OMF family are the so called channel-tunnels which are engaged in exporting substances across the membranes of Gram- negative cell envelopes. This family incorporates the efflux pumps, which confer resistance to various harmful substances like antibiotics, heavy metals, dyes, bile salts and detergents. Efflux pumps consist of an inner membrane transporter, a periplasmic accessory protein and an outer membrane channel-tunnel. The well-studied *E. coli* channel-tunnel, TolC, is involved in protein secretion and can also be part of diverse efflux pumps (Andersen et al., 2000). The RND (resistance-nodulation division) transporters (Li and Nikaido, 2004) exist in all kingdoms of living organisms but seem to be involved in drug resistance, especially in Gram-negative bacteria, in which they export toxic substances across the cell envelope in a single energy-coupled step (Koronakis et al., 2000). The TolC crystal structure has been solved and consists of channel, equatorial, and tunnel domains (Koronakis et al., 2000). The channel domain is embedded in the outer membrane and consists of non-polar residues directed towards lipids and polar residues that locate to the interior resulting in typical amphipathic β-strands. The β-barrel of channel-tunnel is different from those of typical porins formed by single peptide chain. To form a TolC channel domain, three protomers are necessary in order to generate a single 12-stranded β-barrel. The tunnel domain of TolC consists entirely of α-helices assembled in an α-barrel, representing a unique structure. In the *B. burgdorferi* outer membrane, a channel-forming activity corresponding to a TolC homologue, BesC, has been identified, which is part of a putative export system comprising the components BesA, BesB and BesC (Bunikis et al., 2008). BesC forms channels in planar lipid bilayers. A structural model of the efflux apparatus was described showing the putative spatial orientation of BesC with respect to the AcrAB homologues BesAB. This structural modelling showed both similarities and differences between TolC of *E. coli* and BesC of *B. burgdorferi* (Bunikis et al., 2008). For TolC, it is well known that residues lining the periplasmic entrance are critical for the electrophysiological behaviour. A ring consists of six aspartate residues at the TolC tunnel entrance, viewed from the periplasm, is believed to be responsible for its high cation selectivity (Andersen et al., 2002). In BesC, the aspartate ring is replaced by positively charged lysine residues (K366) and negatively charged aspartate residues (D363), resulting in no net charge at the tunnel entrance and, thus, a non-selective BesC channel (Bunikis et al., 2008). Another difference between BesC and TolC is that the BesC channel has a nearly four-fold higher conductance in 1 M KCl than TolC (80 pS in 1 M KCl) (Benz et al., 1993). Bunikis and coworkers further demonstrated that the BesC protein was necessary for *B. burgdorferi* to establish infection in mice and also is involved in antibiotic resistance. This efflux system might be part of a Type I secretion machinery for maintenance of cellular homeostasis or export of exogenous toxic agents, perhaps also necessary for survival in vastly different host environments (Bunikis et al., 2008). An additional finding from the structural modelling of BesA is the lack of the α-hairpin domain creating a different adaptor structure as compared to *E. coli* (Bunikis et al., 2008). This modelling was subsequently verified by a high resolution structure analysis of BesA that confirmed the missing α-hairpin domain of the periplasmic adaptor protein (Greene et al., 2013). Furthermore, a primary sequence comparison revealed that this paradigm shift of adaptor structure is not unique to *Borrelia* as it also was found in other spirochaetes, including *Treponema* and *Spirochaeta* (Greene et al., 2013).

The genome of *B. burgdorferi* encodes a large amount of homologous carbohydrate transporters that belong to the PTS group (Fraser et al., 1997). A number of studies have attempted to characterize some of these transporters. Chitobiose is a component of the tick cuticle and peritrophic matrix. The products of *chbA*, *chbB*, and *chbC* allow *Borrelia* to transport and utilize chitobiose in the tick environment (Tilly et al., 2001; Tilly et al., 2004). Different *chb* genes may respond differently to various temperatures – *chbB* transcription is up-regulated at 23°C, whereas *chbC* transcription increases at 35°C (Ojaimi et al., 2003). Glucose is the major energy source for *B. burgdorferi* involving several PTS glucose transporters (von Lackum and Stevenson, 2005). BBB29 is one of the components of a glucose transporter system of *B. burgdorferi* (Byram et al., 2004). BBB29 expression is elevated in a *B. burgdorferi* non-disseminating clinical isolate compared to a disseminating isolate (Ojaimi et al., 2005) and at 23°C versus 35°C (Ojaimi et al., 2003). Mannose-modified host proteins are also a likely carbohydrate source for *B. burgdorferi,* with the BB0408 and BB0629 proteins suggested to be mannose transporters (von Lackum and Stevenson, 2005).

The members of major intrinsic protein (MIP) group consists of essential channel forming proteins that maintain an osmotic cell equilibrium and include aquaporins and glycerol facilitators (GlpF) (Paulsen et al., 2000; Froger et al., 2001). The GlpF homologue in *B. burgdorferi* (BB0240) is involved in glycerol uptake, which can either be used as an energy source or for phospholipid and lipoprotein biosynthesis (Schwan et al., 2003). The process of glycerol uptake (*glpF*) and utilization (*glpK/bb0241* and *glpA/bb0243*) might be required when *Borrelia* resides in its arthropod host since these *glp* genes are significantly up-regulated in *B. burgdorferi* grown at 23°C versus 35°C (Ojaimi et al., 2003).

Inner membrane transmembrane proteins

The oligopeptide permeases (Opps) belong to the family of ATP-binding cassette (ABC) transporters used for transportation of peptides into bacteria to be subsequently used as carbon and nitrogen sources (Lazazzera, 2001). Although the major role of bacterial Opp is in nutrient uptake, they may also be involved in sensing environmental signals in the form of peptides leading to diverse responses, including expression of virulence determinants, quorum sensing, chemotaxis, sporulation, conjugation, and competence (Perego et al., 1991; Rudner et al., 1991; Leonard et al., 1996). It is not known if the oligopeptide permease has a sensing capability in *B. burgdorferi s.l.* similar to other bacteria although an interesting feature to note is that Opp-like proteins are involved in expression of virulence determinants in other bacteria (Payne and Gilvarg, 1968; Goodell and Higgins, 1987; Finlay and Falkow, 1989; Gominet et al., 2001). The Opp complex in *B. burgdorferi* consists of an oligopeptide-binding lipoprotein OppA, a heterodimeric cytoplasmic membrane permease (OppB and OppC), and the nucleotide binding domain (NBD) heterodimer OppD and OppF that drives the transport of oligopeptides using ATPases (Das et al., 1996; Bono et al., 1998; Kornacki and Oliver, 1998; Monnet, 2003; Groshong et al., 2017). Interestingly, *B. burgdorferi* has five *oppA* genes each controlled by promoters which respond differently to various environmental signals (Wang et al., 2002; Ojaimi et al., 2003). However, *oppA* genes can also be coordinately up-regulated in *B. burgdorferi* clinical isolates with different capacities for haematogenous dissemination (Ojaimi et al., 2005). Perhaps each OppA possesses a different substrate binding affinity that is active during different stages of *B. burgdorferi* life cycle (Wang et al., 2004a), which would attribute to the targeting and penetration of specific tissues in invertebrate and vertebrate hosts (Ojaimi et al., 2005). The Opp system is particularly interesting in *B. burgdorferi s.l.* since it was evident from the genome sequence that only rudimentary machinery for synthesis or transport of amino acids exists. In fact, only a single intact peptide transporter operon could be found. From the genome analysis of *B. burgdorferi s.l.* it is also apparent that the *oppA* operon differs from that of *E. coli* as it has three separate chromosomally encoded substrate binding proteins (Lin et al., 2001; Wang et al., 2002), OppA-1, 2 and 3 and two additional plasmid encoded homologues OppA-4 and 5 (Wang et al., 2002). Interestingly, all of the *B. burgdorferi s.l.* OppA proteins were found to complement *E. coli* OppBCDF to form a functional peptide transport system for nutritional purposes (Lin et al., 2001). All five *oppA* genes have their own active promoters, and the apparent difference in regulation of gene expression implies that the *oppA* genes respond by different environmental signals for the adaptation of the *Borrelia* spirochaetes to various hosts (Wang et al., 2002). There was also a considerable difference in the substrate specificities of the five different OppA proteins, which suggest that they may have separate functions in the spirochaete (Wang et al., 2004a). These findings led to an in depth investigation of the factors that regulate the different *oppA* promoters. Thus, the alternative sigma factors, RpoS and RpoN, regulate the expression of *oppA5* (Caimano et al., 2007), the BosR/Fur homologue of *Borrelia* interacts with the *oppA4* promoter, and another candidate transcription factor, EbfC, interacts with the *oppA5* promoter (Medrano et al., 2007) (see chapter five by Samuels et al.). In a very elegantly executed study by Groshong and coworkers, structural analyses and modeling explain how the five oligopeptide binding proteins (OBPs) interact with variable peptide fragments. Transcriptional analyses of the different OBPs further indicated that they are both independently regulated and sensing different signals whereas the permeases and NBDs are constitutively expressed (Groshong et al., 2017). This study further showed the importance of peptide uptake for virulence and physiology of *Borrelia* spirochaetes.

Inner membrane lipid rafts

The presence of inner membrane proteins, including HflC, HflC, FtsH, among others, in the proteome of lipid rafts, was the first indirect evidence that these microdomains exist in both membranes (Toledo et al., 2015). The lipid profiles of the inner and outer membranes are very similar and contain the same lipid species, including cholesterol, cholesterol esters, ACGal, CGal, MGalD, PC and PG (Toledo et al., 2018b). The main difference between the lipid composition of the two membranes is the relative amount of glycolipids. Cholesterol glycolipids, ACGal and CGal, are relatively more abundant in the outer membrane than in the inner membrane, whereas the non-cholesterol glycolipid, MGalD, is more abundant in the inner membrane. Multilamellar vesicles (MLVs) prepared from isolated lipids from both membranes showed ordered (rafts) and disordered domains, as measured by FRET and anisotropy. However, they exhibited different levels of membrane order and domain segregation (Toledo et al., 2018b). These differences could be attributed, at least in part, to the presence of higher quantities of ACGal in the outer membrane, since ACGal is the primary lipid implicated in raft formation (Huang et al., 2016). Another factor that might contribute to the differences observed between both membranes is the level of saturation of PC, which is significantly higher in the outer membrane compared to the inner membrane (Toledo et al., 2018b).

The proteome of inner membrane lipid rafts is enriched in proteins involved in trafficking, signaling and contains proteases, including HtrA and FtsH, as well as lipid rafts markers. The proteome included HflC/K proteins, which have Stomatin Prohibitin Flotillin HflC/K (SPFH) domains (Rivera-Milla et al., 2006). Proteins with SPFH domains are very diverse and have different functions, but they partition in lipid rafts and are commonly used as lipid raft markers (Browman et al., 2007). In contrast to HtrA, FtsH is only associated with the inner membrane, although it is unclear whether FtsH partition in lipid rafts or is recruited by HflC/K, which are transmembrane regulators of FtsH (Kihara et al., 1997). FtsH is essential for the viability of the spirochaete *in vitro* and *in vivo* and its independent of regulators HflC/K (Chu et al., 2016). The substrates for FtsH in *Borrelia* are unknown, but in *E. coli* FtsH degrades membrane-embedded proteins such as SecY and SecD as well as soluble substrates (Kihara et al., 1995; Chiba et al., 2002). Interestingly, Sec proteins, including SecA and SecD, are enriched in lipid rafts in the inner membrane (Toledo et al., 2018b) and are potential substrates of FtsH. Similarly, inner membrane lipid rafts also were enriched in ABC transporters (BB0330, BB0334, BB0335, BB0466, BB0754 and BBB16, among others) and methyl chemotaxis proteins (BB0596, BB0597), core sensing elements in prokaryotes (Salah Ud-Din and Roujeinikova, 2017). In summary, inner membrane lipids rafts, based on their protein content, are associated with protein trafficking and signaling processes, whereas outer membrane lipid rafts discussed earlier are rich in proteins associated with adaptation and survival to tick and vertebrate milieus.

The cell cycle and cell division

The bacterial cell cycle – the life of a single cell from birth to division – is riddled with challenging events that need to be coordinated in both time and space. If one were to think of the cell like a water balloon - the envelope being the elastic balloon, and the cytoplasmic contents the water - we can start to understand the challenges. To grow, the balloon needs to physically expand and divide, not just stretch, while controlling the distribution and displacement of the water. To divide, the balloon needs to be cut in half, and each new end needs to be fabricated, all without spilling the water. This seems like an insurmountable feat, and yet, all bacteria perform these tasks rapidly and virtually mistake free. The key is controlling the timing and spatial arrangement of enzymes involved in cell envelope elongation with the machinery necessary for division. Here, we discuss these processes, highlight some of the recent findings in understanding these events, and discuss the implications in the context of Lyme disease treatment.

The PG sacculus is akin to the aforementioned balloon – a big molecular bag that surrounds the cytoplasm. Cell growth requires PG elongation. Cell division means that the *sac* is cut and PG poles are formed. Thus, by tracking PG synthesis we can understand all facets of the *B. burgdorferi* cell cycle. Recent advances in chemical synthesis have made tracking PG biogenesis possible. The covalent attachment of fluorophores, with different physical and spectral properties (Hsu et al., 2017), to the ubiquitous PG component D-Alanine (D-Ala), has furthered our understanding of *how* bacteria grow (Kuru et al., 2012). While there are always incredible exceptions in nature (Caccamo and Brun, 2018), most bacteria grow using one of the three following strategies. (i) Many rod-shaped bacteria elongate their cell wall by lateral synthesis, which involves near homogenous growth throughout the cell cylinder prior to septal synthesis at division (Figure 2A). (ii) Some rod-shaped bacteria, like *Agrobacterium tumefaciens* and *Mycobacterium tuberculosis*, have an alternative mode of growth – elongating from either one or both poles, respectively (Figure 2B) (Kuru et al., 2012; Garcia-Heredia et al., 2018). (iii) Most coccoids replicate by simply producing two halves of each daughter cell, also known as septal synthesis (Figure 2C). Once again, *B. burgdorferi* has deviated from convention. Feeding *B. burgdorferi* fluorescent D-Ala resulted in an unmistakably unique pattern of PG synthesis that seems to combine several forms of growth, simultaneously. With the exception of 2-4 um at the poles, *B. burgdorferi* elongates by both lateral synthesis and discrete zones of intense growth (Jutras et al., 2016). Zones are organized in both time and space – a central zone of growth is followed by secondary and tertiary sites. Despite distances that can exceed 30 um in length, each zone precisely forms at the ¼, ½, and ¾ positions along the cell length (Figure 2D). The primary zone at the ½ position becomes the site of cell division in the F_0 generation, while the ¼ and ¾ positioned zones are the site of cell division in the F_1 generation. In essence, how the mother cell grows determines where the daughter cells will divide. Multi-zonal growth appears to be a unique feature to the *Borrelia* genus. All *Borrelia spp.* tested, including those responsible for relapsing fever, display a similar PG growth pattern. Other spirochaetes, such as *Treponema denticola* and *Leptospira interrogans,* grow through lateral PG synthesis. The mechanism(s) responsible for zone placement, across a distance 30-50 times longer than the average rod-shaped cell, requires exquisite spatial and temporal balance between the elongation and division machinery.

We know very little about multi-zone, PG synthesis in *B. burgdorferi*. Co-incubation with fluorescent D-Ala and D-cycloserine (Kuru et al., 2019), a putative inhibitor of cytoplasmic D-Ala-D-Ala synthesis (Strominger et al., 1960), leads to the disappearance of growth zones in *B. burgdorferi,* unlike in other bacteria. Together, these findings suggest that zones of growth are in essence PG synthesis machinery hubs that are pre-established. These focal points extend into the periplasm, and orchestrate the incorporation of muropeptides into nascent PG, since there is a near perfect correlation between zones of PG synthesis and the localization of Penicillin-Binding Proteins (PBPs) in *B. burgdorferi* (Jutras et al., 2016). In depth analysis of D-Ala signal, relative to PBP signal, suggests that active sites of incorporation are more finite and lie inside the expanded area of each zone (Brock and Jutras, unpublished data, and Figure 3B). These observations provide clues as to a possible mechanism for zone placement (see below). Wu and colleagues made a case for an alternative mode of *B. burgdorferi* growth, which only occurs in stationary phase cultures, whereby cells grow by polar synthesis (Wu et al., 2018). In exponential, or stationary-phase growth, vancomycin – a glycopeptide inhibitor of periplasmic PG incorporation – eliminated both zonal and *polar* synthesis (Wu et al., 2018). These findings weaken the argument for polar growth in stationary phase and provide for an alternative, more parsimonious explanation for their observation. Late-stage cultures grow slowly, due to a depletion of nutrients in the culture media. Recently divided bacteria have a single zone of growth and polar signal from the previous division event (Figure 2E). As stationary phase and vancomycin treatment continue, slow lateral synthesis continues without further zonal growth. Upon reaching the total cessation of growth, the previously established zone becomes the site of division, which is forced by washing and centrifugation prior to image acquisition. The resulting cells have either a single pole labeled, or both poles, but one with more recent/intense signal (Figure 2E). In short, zonal growth and experimental conditions, can explain the recent observation without inferring a secondary strategy for PG synthesis.

The mechanism(s) that drive zonal growth and its placement are not known. Absolute cell length is not the impetus since different strains, which achieve average lengths that vary by >20um still form zones at the same relative ¼, ½, and ¾ positions (Jutras et al., 2016). Researchers in the field have long recognized the association between motility and cell division, as if the forceful torque applied by flagella on the cell cylinder is required for final cell separation (Motaleb et al. 2000). Be that as it may, a mutant that is unable to produce the major flagellar filament protein FlaB still faithfully positions zones that are correctly placed along the long axis of the cell. *B. burgdorferi* PG zones are truly finite, but extended regions of elongation. Photobleaching experiments demonstrated that the *B. burgdorferi* cytoplasm is contiguous in cells with three zones, indicating that these wide regions (2-4um) are not septa, produced by three discrete cells that are chained together (Jutras et al., 2016). That said, zones clearly become sites of cell division, implicating the cell division machinery in zone establishment and placement.

FtsZ is the most likely candidate for determining both the sites of zonal growth, and septal PG synthesis, in *B. burgdorferi.* FtsZ is a tubulin homologue that is essential for cell division in virtually all bacteria (Du and Lutkenhaus, 2019). FtsZ forms a ring-like structure and coordinates the placement and activity of nearly 20 proteins in most bacteria; many of which are essential for viability under typical growth conditions (Egan et al., 2020). The dynamic treadmilling behavior of FtsZ (Yang et al., 2017) coordinates the location and activity of PG septal synthesis machinery (Bisson-Filho et al., 2017), and ultimately, cell division. Many of the factors that influence FtsZ activity, placement, and cell division related processes are present in the *B. burgdorferi* genome, including putative homologues of *fts* genes A, B, E, H, I, K, Q, W, Y, Z. However, the negative regulators, whose role is to prevent FtsZ ring formation, too early or in erroneous places (e.g. Min and SlmA/Noc systems, EzrA, MipZ, etc) (Levin et al., 1999; Errington et al., 2003; Wu and Errington, 2004; Thanbichler and Shapiro; Cho et al., 2011) are conspicuously absent. Homologues of so-called positive regulators of Z-ring assembly, such as MapZ, PomZ, and SsgA/B (Willemse et al., 2011; Treuner-Lange et al., 2013; Fleurie et al., 2014), are also not found in the *B. burgdorferi* genome. Given the complex genome, in combination with the absolute cell length of *B. burgdorferi*, there are likely analogous systems yet to be identified.

Efforts to produce bacteria with fluorescent protein fusions to *B. burgdorferi* FtsZ, including in *cis* and in *trans*, have been futile (Pichoff, Zückert and Lutkenhaus; Brock and Jutras, unpublished data). This includes a strategy whereby a fluorescent protein is sandwiched between disordered domains in FtsZ (Moore et al., 2017), a hybrid fusion that is fully functional in *E. coli*. These observations suggest that *B. burgdorferi* FtsZ may have unique properties, which could explain both the unusual mode of *Borrelia* growth and the unsuccessful attempts to localize the protein in live cells. Indeed, unpublished data indicate that *B. burgdorferi* produces two isoforms of FtsZ (Brock and Jutras, unpublished data). Analysis of the *B. burgdorferi ftsZ* locus provides clues. A primary Ribosome Binding Site (RBS), upstream of the annotated Methionine (Met) start codon, is followed by an internal RBS and an additional Met in the same reading frame. Thus, *in vitro* cultured bacteria produce two variants of FtsZ.

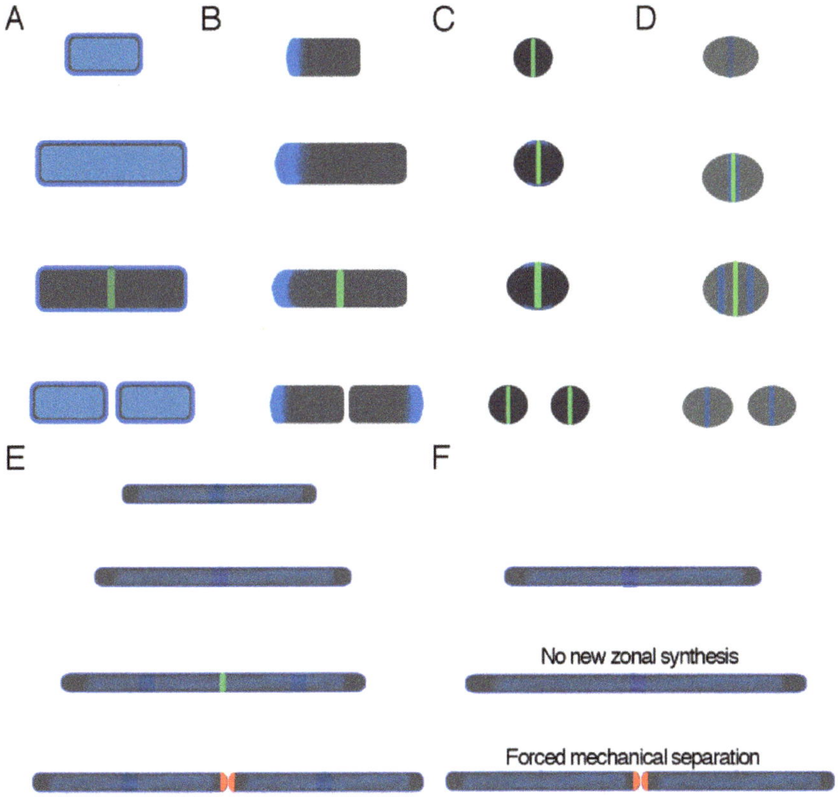

Figure 2. Different modes of bacterial growth. New fluorescent probes that track PG synthesis have illuminated different bacterial strategies for self-replication. (A) Lateral growth involves near ubiquitous PG synthesis (blue) throughout the cell cylinder, prior to septal PG synthesis (green) at division. (B) Polar growth occurs at one or both ends, only, prior to division. (C) Many spherical bacteria grow by producing two halves of a circle, almost exclusively synthesizing a septum. Note some pre-septal (blue) growth, adjacent to the septum, does occur in some species. (D) Ovoccoid bacteria (e.g. *Streptococcus pneumonia*) simultaneously elongate by producing pre-septal zones of PG synthesis, while producing a septum. Prior to septation, pre-septal zones are apparent in a mother cell. (E) In all *Borreliae* tested thus far, PG synthesis occurs by both lateral synthesis, and discrete zones of elongation. A single zone, at mid-cell, becomes the site of septal PG synthesis. Prior to membrane invagination, secondary and tertiary zones are asynchronously established, which will become the sites of division in the following generation. Note that lateral synthesis in *B. burgdorferi* does not occur (gray) at the poles, but post-division, residual signal becomes trapped (red). (F) An alternative explanation for apparent polar growth in *B. burgdorferi*. Lateral signal, at any stage of growth, has never been observed at poles. Upon vancomycin treatment, in stationary phase, primary zonal growth is reduced, no other zones are great, and lateral synthesis slowly continues. Both the zonal and lateral PG synthesis are, eventually, halted. After washing and mechanical manipulation of the cells, cell separation occurs, producing two daughter cells with trapped signal (red).

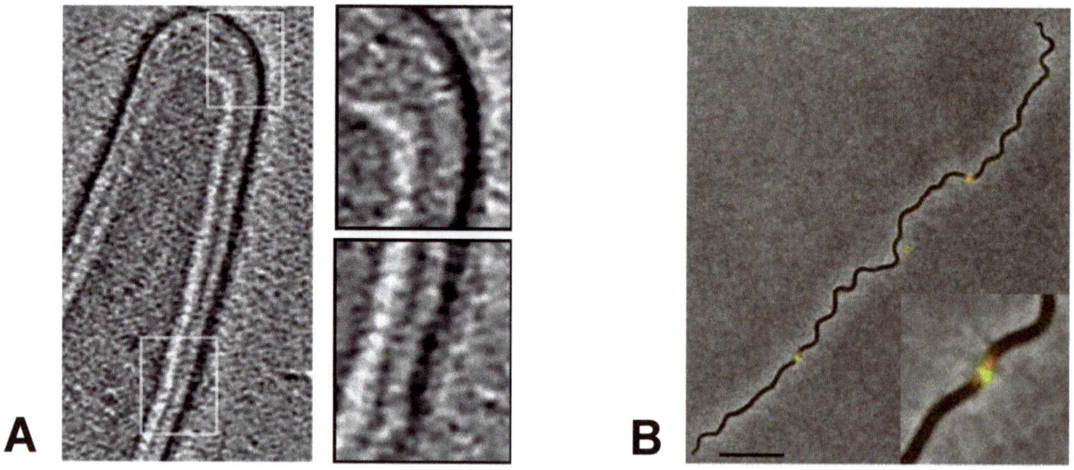

Figure 3. A. Cryo-electron tomography images of intact *B. burgdorferi* cells. PG thickness from mid cell to pole changes from ~6nm to ~40nm. **B. Penicillin-Binding Protein probes identify foci within PG synthesis zones.** Merged phase contrast and epifluorescence micrograph of *B. burgdorferi* co-labeled with fluorescent D-Alanine HADA (red) and PBP probe bocillin (green) suggest that the PG synthesis machinery is positioned in finite positions, inside zones of growth. Scale bar 10 µm, inset 4X.

The consequences of producing two isoforms of FtsZ are not known.

The pattern of *B. burgdorferi* PG synthesis, and the priming of division sites, bares resemblance to the cell cycle of the ovacoccoid bacterium *Streptococcus pneumoniae.* Three discrete zones of PG synthesis are apparent in *S. pneumoniae* prior to cell division (Peters et al., 2014). Each zone is equidistant and delineate future division sites (Peters et al., 2014; Tsui et al., 2014). These similarities are contrasted by some notable differences that truly make *B. burgdorferi* growth unique. For instance, *S. pneumoniae* PG zones are peripheral to the site of septal synthesis and both modes of growth occur simultaneously but do so in the absence of lateral synthesis. The latter can be attributed to *S. pneunmonia* not producing MreB, the central player in lateral PG elongation (Vollmer et al., 2019). Peripheral and septal PG zones are driven by Pbp2a and Pbp2x, respectively (Pinho et al., 2013). In contrast, *B. burgdorferi* zonal and lateral PG biogenesis occur simultaneously, but are discrete from septal synthesis. *B. burgdorferi* does not appear to encode a Pbp2a homologue but does produce an unusual Pbp analogous to FtsI with an additional PASTA (penicillin-binding protein and serine/threonine kinase associated) domain –a domain found within Pbp2x (Beilharz et al., 2012) and important regulator cell division (Fleurie 2014}.

Lipid acquisition/biosynthesis and modification

Phospholipids are the main components of cellular membranes; they form permeable bilayers that act as a barrier and support critical biological processes, including sensing and signaling. Most bacteria possess anionic phospholipids such as PG and cardiolipin (CL) and the zwitterionic phosphatidylethanolamine (PE). Like most bacteria, *B. burgdorferi* has PG, but it also possesses PC (Belisle et al., 1994), which is the most common phospholipid in eukaryotic membranes but is far less common among prokaryotes (Sohlenkamp and Geiger, 2016). The levels of PC in the membrane of prokaryotes range from small amounts in *Pseudomonas aeruginosa* to up to 73% of total membrane lipids in *Acetobacter aceti* (Sohlenkamp et al., 2003). In *Borrelia*, PC is the most abundant phospholipid (68%-74%), while PG constitutes 26-32% of the membrane phospholipids. The structures of PE and PC are very similar, and the only difference is the methylation of the nitrogen in the choline of PC. Nonetheless, PC is a bilayer forming phospholipid while PE prefers to form lipid structures such as inverse hexagonal phase (Popova and Hincha, 2011). There are two biosynthesis pathways in bacteria that produce PC, the methylation pathway and the phosphatidylcholine synthase (PCs) pathway. The methylation pathway is a stepwise process that converts phosphatidylethanolamine (PE) to PC and involves one or more phospholipid Nmethyltransferases (Geiger et al., 2013). The second pathway present in bacteria is the PCs pathway, in which choline condenses directly with CDP-diacylglycerol to form PC and CMP (Geiger et al., 2013; Sohlenkamp and Geiger, 2016). Some bacteria species possess both pathways, including *Legionella pneumophila* and *Agrobacterium tumefaciens* (Wessel et al., 2006; Geiger et al., 2013), but the majority rely on one of these pathways. The genome of *B. burgdorferi* is one of the smallest genomes among bacteria and lacks homologs of N-methyltransferase genes (Fraser et al., 1997; Geiger et al., 2013). In addition, PE is required to produce PC via the methylation pathway, but it is not found in the membrane of *Borrelia* (Belisle et al., 1994). The lack of PE in *Borrelia*, which is required in many prokaryotes, suggests that other lipids, possibly glycolipids, could compensate for it (Wang et al., 2004b). The spirochaete possesses a phosphatidyl-choline synthase, encoded in the gene *bb0249*, that is involved in the production of PC via the PCs pathway (Wang et al., 2004b). *bb0249* was cloned in *Escherichia coli*, which does not contain PC, resulting in the production of PC in the presence of exogenous choline (Wang et al., 2004b). Choline is readily available in the host and is likely transported into the cell through the ATP dependent transporter system ProU (Bontemps-Gallo et al., 2016)

The second phospholipid present in the membrane of *B. burgdorferi* is PG. The spirochaete encodes a phosphatidylglycerol synthase (PGs) gene (*bb0721*) in the chromosome that is critical for the production of PG from CDP-DAG. *E. coli* HD38/pHD102 was transformed with a plasmid containing *bb0721* (pTAC56) to assess the PGs activity of this gene. *E. coli* HD38 lacks a PGs, which is essential for growth, and therefore requires a functional copy expressed by the temperature-sensitive plasmid pHD102. As a result, at temperatures above 30°C, the growth of *E. coli* HD38/pHD102 stops. In contrast, transformed *E. coli* HD38/pTAC56 grew at 43°C, supporting the role of BB0721 as a PGs. This was further confirmed by analyzing the contribution of PG to the total phospholipid composition in both strains. The PG content in HD38/pTAC56 did not change, whereas in HD38/pHD102 dropped from 22% at 30°C to less than 5% at 43°C. (Wang et al., 2004b)

Overall, the presence of functional PCs and PGs indicates that PC and PG are synthesized in a pathway(s) involving CDP-DAG. The spirochaete has a *cdsA* homolog (*bb0119*) that mediates in the transfer of CMP from CTP to phosphatidic acid (PA) to form CDP-DAG, which is used by PCs that condense choline directly with CDP-DAG to form PC and releases CMP (Wang et al., 2004b). An unusual trait of phospholipids in *Borrelia* is the low double bond/acyl chain ratio (~0.25) that results in saturated acyl chain levels close to 75% (Hossain et al., 2001; Huang et al., 2016). Thus, at least 50% of PC molecules have two saturated acyl chains since some of the remaining PC molecules may have two unsaturated or one polyunsaturated acyl chains (Huang et al., 2016). Nonetheless, *B. burgdorferi* has outer and inner membranes. The phospholipid composition in both membranes is similar, but the saturated acyl chain levels of PC from the inner membrane is significantly lower (Toledo et al., 2018b). It is important to note that differences in the saturation of acyl chains confer different biophysical properties to lipids and is vital for the formation of tightly packed ordered domains (rafts) in eukaryotic cells (London, 2002) and *B. burgdorferi* (Huang et al., 2016). Therefore, this difference suggests that IM PC has a lower ability to support rafts relative to PC in the OM. (Toledo et al., 2018b).

As mentioned above, *B. burgdorferi* has three glycolipids. ACGal, CGal, and MGalD. The presence of glycolipids, similar to MGalD in Gram-negative organisms, is unusual. In contrast, Gram-negative bacteria possess LPS, a glycolipid that is absent in *Borrelia*. On the other hand, the presence of glycolipids similar to MGalD is frequent in Gram-positive bacteria, where major glycolipids are usually di-hexose DAG. The monogalacto-syldiacylglycerol synthase from *B. burgdorferi* (bbMGS) was identified by homology using the amino acid sequence of the monoglucosyl diacyglycerol synthase from the *Acholeplasma laidlawii* (alMGS) (Berg et al., 2001; Östberg et al., 2007). The activity of bbMGS was addressed by expressing the recombinant protein in *E. coli* and supplementing the cell extract with potential acceptor and donor substrates. The addition of radiolabeled Gal generated MGalD, whereas the addition of radiolabeled glucose did not render any product, which supports that the glycosylation is galactose-specific (Östberg et al., 2007).

Interestingly, the activity of bbMGS was influenced by the phospholipid composition in the mixture. The activity of bbMGS increased 10-fold in the presence of PG compared to PC (Östberg et al., 2007). In Gram-positives, glycosyltransferases generally yield di-hexose and sometimes tri-hexose DAG products.

Nonetheless, bbMGS only produce a mono-hexose DAG, MGalD, which is the only hexose DAG present in the membrane of Lyme disease and relapsing fever *Borrelia* (Livermore et al., 1978; Hossain et al., 2001). Two other glycolipids in *B. burgdorferi*, CGal and ACGal, require a glycosyltransferase. However, bbMGS is not involved in the glycosylation of these lipids since no product was detected when dyacylglycerol was substituted with cholesterol. Nevertheless, the addition of UDP-Gal/UDP-Glu to *Borrelia* cell extracts confirmed the synthesis of a cholesteryl galactoside (Östberg et al., 2007). There are three other glycosyltransferases in *B. burgdorferi* (Fraser et al., 1997), and the most suitable candidate involved in the glycosylation of cholesterol is *bb0572*, a putative inverted (α→β) galactosyltransferase since both cholesterol glycolipids have β-glycosidic linkages between the galactose residue and cholesterol (Ben-Menachem et al., 2003; Schroder et al., 2003).

Cholesterol is present in the membranes of *B. burgdorferi* as cholesterol, cholesterol esters, and also, it is an essential component of cholesterol glycolipids (Ben-Menachem et al., 2003; Schroder et al., 2003). The spirochaete is auxotrophic for cholesterol and relies on the host to acquire the sterol. *In vitro* experiments demonstrated that *B. burgdorferi* takes up cholesterol directly from HeLa cells to make cholesterol glycolipids (Crowley et al., 2013). During the transfer, the spirochaete exchanges lipids and antigens with HeLa cells, including cholesterol glycolipids. In addition, *B. burgdorferi* can also produce outer membrane vesicles that mediate in the antigen transfer to HeLa cells (Crowley et al., 2013). The implications of this mechanism in the pathogenesis of Lyme disease has not been elucidated yet, but the transfer of antigens from the spirochaete to vertebrate cells could contribute to an exacerbated and long-lasting immune response.

Protein export and secretion

The biogenesis of the *Borrelia* cell envelope is still very much a mystery. Inference based on known mechanisms and pathways in other diderm bacteria provides some intriguing clues, but few of these clues have been experimentally tested. The working model of *Borrelia burgdorferi* envelope biogenesis that we develop below (see also Figure 2) is, therefore, largely speculative in nature. It should serve as a working model to delineate how these two important components of the borrelial envelope are synthesized and transported to the sites where they are biologically active.

Lipoprotein transport

As already stated above and also evidenced by other chapters in this book, lipoproteins represent a class of membrane-anchored proteins that are crucial for the natural maintenance, transmission and virulence of the spirochaetes. In contrast to integral membrane proteins, lipoproteins are only peripherally tethered to the lipid bilayer leaflets of the inner (IM) or outer membranes (OM) via their acylated amino terminal cysteine residues. This not only mirrors their variable functions as flexible components of the periplasmic space or the host-pathogen interface, but also has implications for their thermodynamically different translocation through both membranes and the aqueous periplasmic space in between. Detailed studies on lipoprotein transport in Gram-negative bacteria have revealed a general three-step process: (i) Sec-dependent transport of prolipoproteins through the IM, (ii) lipid modification on the periplasmic face of the IM and cleavage of the signal peptide by signal peptidase II, and (iii) retention or release of acylated proteins and transport to the inner or outer leaflet of the OM. The latter sorting step is currently known to be mediated by either the Lol pathway (Tokuda and Matsuyama, 2004), a type II (T2SS) (Pugsley, 1993) or a type V/autotransporter (T5SS/AT) secretion system (Coutte et al., 2003; van Ulsen et al., 2003).

The enzymes responsible for lipoprotein maturation were first discovered in *E. coli* (Sankaran et al., 1995). The modification of lipoproteins occurs in three steps. First, the signal sequence LXYC specifies the addition of diacyl glycerol in a thioether linkage to the sulfur of the cysteine side chain. Then signal peptidase II cleaves at the amino side of the cysteine residue and a third fatty acid is added in an amide linkage to the new amino terminus (Hayashi and Wu, 1990). Homologues for these three enzymes have been identified in the *B. burgdorferi* genome (Fraser et al., 1997), suggesting that spirochetal lipoproteins are processed in an identical way.

Lipoproteins entering the *Escherichia coli* Lol pathway (Tokuda and Matsuyama, 2004) first interact with an IM ABC transporter-like complex, LolCDE, which releases OM-targeted lipoproteins from the IM in an ATP-dependent manner (Yakushi et al., 2000). Released lipoproteins then form a water-soluble complex with the periplasmic chaperone LolA (Yokota et al., 1999). After crossing the periplasm, this complex interacts with the OM receptor LolB, which mediates anchoring of lipoproteins to the inner leaflet of the OM (Yokota et al., 1999). The initial '+2 lipoprotein sorting rule' established in *E. coli* indicated that Asp at the penultimate (+2) N-terminal position of the mature lipoprotein leads to IM retention (Yamaguchi et al., 1988), while Ser or other amino acids at the +2 position allow the lipoproteins to be released to the OM. One potential mechanistic explanation is that the interaction of the negatively charged aspartate with the zwitterionic phosphatidylethanolamine blocks an interaction with the LolCDE sortase (Masuda et al., 2002; Harada et al., 2003). Yet, its general applicability remains to be determined as it has become increasingly clear that the '+2 rule' is far from universal. In *E. coli*, amino acids at position +3 can also affect localization (Seydel et al., 1999; Terada et al., 2001; Masuda et al., 2002; Lewenza et al., 2006). Furthermore, lipoprotein sorting in other Gram-negative pathogens such as *Pseudomonas aeruginosa* (Lewenza et al., 2008) and *Yersinia pestis* (Silva-Herzog et al., 2008) appears to rely on N-terminal signals at positions +3/+4 and beyond.

While the Lol machinery operates within the periplasm, established pathways of lipoprotein translocation through the OM involve either a T2SS or T5SS/AT mechanism. The prototypical T2SS, the *Klebsiella oxytoca* Pul secretion (Pugsley, 1993), secretes the surface-anchored enzyme PulA. Its secretion signal is located within at least three non-adjacent regions of the structural protein (Sauvonnet and Pugsley, 1996; Francetic and Pugsley, 2005). In the absence of a type II secretion system, PulA's penultimate Asp residue functions as IM retention/Lol avoidance signal (Pugsley et al., 1990). Lipoproteins exported via a T5SS/AT are translocated through the OM via a C-terminal autotransporter domain and include the *Bordetella pertussis* subtilisin SphB1 (Coutte et al., 2003) and the *Neisseria meningitidis* serine protease NalP (van Ulsen et al., 2003).

A database homology search for analogous spirochetal lipoprotein export pathways showed that orthologs of all essential components of the Sec translocase complex and the three enzymes required for lipoprotein modification are present in *Borrelia* (Figure 1), *Leptospira* and *Treponema* (Fraser et al., 1997; Fraser et al., 1998; Haake, 2000; Nascimento et al., 2004; Seshadri et al., 2004). Yet, further Sec-dependent or -independent bacterial protein secretion pathways are missing in spirochaetes, with the exception of the described Type I Bes secretion system in *Borrelia burgdorferi* (Bunikis et al., 2008) or potential twin-arginine translocation (Tat) and type II secretion orthologs in *Leptospira* (Nascimento et al., 2004). Lol pathway orthologs were identified as well.

In diderm bacteria, the Lol pathway is responsible for ensuring proper sorting of lipoproteins within the periplasm. LolCDE forms an IM ABC transporter-like complex that uses the energy from ATP hydrolysis by LolD to release fully modified lipoproteins from the IM via LolCE to the periplasmic carrier protein LolA (Matsuyama et al., 1995; Matsuyama et al., 1997; Tajima et al., 1998; Yakushi et al., 1998; Yakushi et al., 2000; Miyamoto et al., 2001; Tanaka et al., 2001; Fukuda et al., 2002; Masuda et al., 2002; Hara et al., 2003; Takeda et al., 2003; Taniguchi et al., 2005; Ito et al., 2007; Oguchi et al., 2008; Okuda et al., 2008; Watanabe et al., 2008; Okuda and Tokuda, 2009; Tsukahara et al., 2009; Yasuda et al., 2009; Remans et al., 2010; Tao et al., 2010; Narita and Tokuda, 2011; Okuda and Tokuda, 2011; Mizutani et al., 2013; Hayashi et al., 2014; Zückert, 2014; Grabowicz and Silhavy, 2017). Orthologs of LolCDE (Yakushi et al., 2000), LolA (Yokota et al., 1999), but not the OM lipoprotein receptor LolB (Yokota et al., 1999), were detected in the *B. burgdorferi* genome. The absence of LolB is not surprising, as LolB, an OM periplasmic lipoprotein and structural homolog of LolA (Takeda et al., 2003) is found only in β- and γ-proteobacteria (Wilson et al., 2015).

Conditional *B. burgdorferi* type strain B31 mutants in the LolD and the periplasmic lipoprotein carrier LolA were shown to have a significant growth defect, indicating that the pathway is essential (Bridges, Dowdell, Chen, Kueker, Liu, Blevins & Zückert, unpublished results). A working model of the *Borrelia* lipoprotein transport pathway therefore includes LolCDE (BB0078-81) and LolA (BB0346) orthologs, but diverges from that of other diderm bacteria by utilizing a so far unidentified additional OM module to facilitate translocation, or 'flipping', of surface lipoproteins through the OM (Figure 1). This mechanism might be analogous to the Slam machinery in *Neisseria meningitidis* (Hooda et al., 2016), although those proteins have restricted substrate specificities, and Slam homologs appear absent from *Borrelia* genomes. Genetically engineered *B. burgdorferi* express and present on their surface biologically active VMPs of RF *Borrelia*

(Zückert et al., 2004). This indicates that lipoprotein export machineries in different species of the genus *Borrelia* are functionally conserved.

Several independent lines of evidence point toward a periplasmic mechanism that maintains the large and diverse set of *B. burgdorferi* surface lipoproteins (Dowdell et al., 2017) in a translocation-competent conformation by preventing them from folding prematurely in the periplasm. (Schulze et al., 2010; Chen et al., 2011; Chen and Zückert, 2011; Kumru et al., 2011b). Site-directed mutagenesis studies of model surface lipoproteins OspA, OspC and Vsp1, including fusions to a faithful localization reporter variant of monomeric red fluorescent protein mRFP1 or conditionally folding calmodulin, also have repeatedly implied that the N-terminal tether peptides of the lipoproteins are intimately involved in this process - tether deletion and substitution mutants were specifically deficient in OM translocation (Schulze and Zückert, 2006; Kumru et al., 2010; Schulze et al., 2010), and surface localization of lipoprotein-calmodulin fusions required either the presence of a wild-type tether or calcium chelation-induced unfolding of calmodulin (Chen and Zückert, 2011). One potential mechanism would be that surface-targeted pro-lipoprotein peptides interact with a periplasmic "holding" chaperone (Figure 1) as they emerge tether-first from the IM Sec complex on the periplasmic side of the IM. This would be analogous to the "high affinity, low specificity" interaction of the proteobacterial chaperone SecB with the diverse peptides released from the ribosome (Park et al., 1988; Liu et al., 1989; Randall and Hardy, 2002; Lilly et al., 2009). The holding chaperone would then deliver surface lipoproteins in an at least partially unfolded conformation to an OM lipoprotein "flippase", which facilitates translocation through the OM and leads to the demonstrated ultimate anchoring of surface lipoproteins in the surface leaflet of the OM (Schulze et al., 2010; Chen et al., 2011; Chen and Zückert, 2011) (Figure 1).

The *B. burgdorferi* version of the Lol pathway may function either like a purely periplasmic lipoprotein sorting pathway as in other diderm bacteria, albeit with a LolA that can operate in absence of a cognate LolB. This would require an additional branch within a dichotomous lipoprotein secretion machinery where surface lipoproteins bypass the Lol machinery altogether to reach the surface. Alternatively, *B. burgorferi* LolA could have evolved to become fully integrated into the surface lipoprotein secretion pathway, potentially taking on the added role of a "holding" chaperone (Figure 1)

Integral membrane protein transport

While other bacteria commonly ensure correct localization of their membrane proteins by including transmembrane-spanning domains, *Borrelia* seemingly prefer N-terminal lipidation as a membrane anchoring method. Proteins containing transmembrane-spanning domains appear to be rare compared to Gram-negative bacteria. Biosynthetic labeling of *Borrelia* cultures with [^{14}C] amino acids identified only a few amphiphilic polypeptides that did not co-migrate with lipoproteins (Brandt et al., 1990) and freeze-fracture electron microscopy visualized only few outer membrane pore structures (Walker et al., 1991; Radolf et al., 1994).

Integral transmembrane proteins (TMPs) are important components of both the cytoplasmic inner (IM) and outer membrane (OM) lipid bilayers and required for their selective permeability. Studies on the mechanisms of processing, folding and membrane insertion in model systems have identified several IM, periplasmic and OM components involved in the assembly of TMPs in both the IM and OM. Figure 1 shows potential functional *B. burgdorferi* orthologs identified in sequence homology searches. The long-standing textbook dogma on membrane protein structures stated that α-helical transmembrane (TM) domains span the IM and β-barrels span the OM. Yet, recent data suggest that this might be an oversimplification. The Wza translocon for *E. coli* polysaccharides forms an α-helical barrel in the OM (Dong et al., 2006). P13, a multimeric *B. burgdorferi* OM porin, is predicted to have largely α-helical secondary structure and might form similar tertiary structure in the OM (Noppa et al., 2001; Nilsson et al., 2002).

TMPs, like LPs, are also predominantly targeted to the Sec translocon, but their signal peptides are cleaved by a separate periplasmic enzyme, signal peptidase I. Intriguingly, the *B. burgdorferi* genome encodes for three potential signal I pepdidases that – like sets of multiple LepBs in some gram-positive bacteria (Bonnemain et al., 2004) – may have different substrate specificities. *B. burgdorferi* LepB2 (BB0031), is closest to a canonical signal I peptidase, with an N-terminal two-TM anchor domain and an appropriately localized periplasmic active site. LepB3 (BB0263) similarly is predicted to have all enzymatically relevant residues located in the periplasm, anchored by a single N-terminal TM domain. LepB1 (BB0030) immediately upstream of LepB2, however, appears to be a signal peptidase I-like protein with a four-TM domain topology that lacks the canonical active site residues. A conditional knockout of *lepB1* led to a marked growth defect with signs of significant envelope instability (You, Whetstine and Zückert, unpublished results), but the protein's precise function remains unknown.

Subsequent to being threaded through, the TM domains of IM-localizing TMPs exit the Sec channel through a lateral gate into the hydrophobic environment of the lipid bilayer. This process is catalyzed by YidC, an IM insertase (Xie and Dalbey, 2008). Interestingly, YidC can also mediate a Sec-independent insertion of IM TMPs, particularly those with only one to two TM domains (Samuelson et al., 2000). The Bdr proteins, a family of bitopic type II membrane proteins with cytoplasmic N-termini and single C-terminal TM domains (Zückert et al., 1999), might, therefore, insert into the IM in a YidC/BB0442-dependent, but Sec-independent manner.

TMPs destined to the OM somehow avoid IM insertion and enter the periplasmic space where they interact with two chaperones, Skp and SurA. A double *skp surA* mutation is synthetically lethal in *E. coli* (Rizzitello et al., 2001). Yet, it remains unresolved whether Skp and SurA act in two parallel, redundant pathways, or rather sequentially. In the parallel scenario, SurA is the primary chaperone involved in periplasmic transit of OM TMPs, Skp attempts to rescue proteins that fall off the SurA pathway, and DegP degrades proteins abandoned by Skp (Sklar et al., 2007). In the sequential scenario, Skp prevents aggregation of nonnative TMPs, and SurA then promotes structural folding (Bos et al., 2007). A Skp homolog (BB0796) is expressed downstream from BamA, but its function remains to be determined (Iqbal et al., 2016). The *B. burgdorferi* HtrA/DegP chaperone/protease homolog (BB0104) appears to be a multifunctional protein with activity in the *Borrelia* periplasm, at the OM and as a secreted protein (Gherardini, 2013). Together with CtpA, HtrA is involved in the proteolytic maturation of BB0323, a periplasmic OM lipoprotein that is required for envelope stability and contains a C-terminal peptidoglycan-binding LysM domain (Stewart et al., 2004; Kariu et al., 2013). Additional HtrA substrates are *Borrelia* chemotaxis and motility factors (Coleman et al., 2013). Secreted HtrA has been shown to degrade components of host extracellular matrix components (Russell et al., 2013).

The insertion of OM TMPs is catalyzed by the Bam complex (Hagan et al., 2011; Konovalova et al., 2017). The OM β-barrel TMP BamA (formerly YaeT or Omp85) (Voulhoux et al., 2003) displays several periplasmic polypeptide-transport-associated (POTRA) domains (Sanchez-Pulido et al., 2003). These N-terminal domains have been shown to recognize C-terminal motifs of OM TMPs and might initiate the structural assembly process of TMPs (Robert et al., 2006). They are also thought to provide a scaffold for the interaction with four associated lipoproteins, BamB (YfgL), BamC (NlpB), BamD (YfiO) and BamE (SmpA) (Kim et al., 2007). The exact roles of these periplasmic OM lipoproteins in OM biogenesis and their interaction with BamA and substrate proteins are under intense investigation. So far, only BamD appears to be essential (Onufryk et al., 2005; Malinverni et al., 2006), indicating that BamAD forms the core Bam machinery in *E. coli*.

The *B. burgdorferi* BamA homolog (BB0795) is associated with only two structural Bam lipoprotein homologs, BamB (BB0324) and BamD (BB0028). Depletion of BamA led to a noticeable decrease of OMPs and OM lipoproteins (Lenhart and Akins, 2010). Of note, that study did not assess surface exposure, but the data are consistent with a likely indirect Bam complex dependency of the surface lipoprotein flipping process (Figure 1). BamB and BamD are the only two of the 8 lipoproteins anchored in the periplasmic leaflet of the OM (Dowdell et al.) that could not be disrupted by transposon mutagenesis (Lenhart et al., 2012; Lin et al., 2012; Lin et al., 2014); as mentioned above, a deletion mutant of BB0323 remains viable, albeit with a severe envelope structure defect (Stewart et al., 2004). *B. burgdorferi* mutants lacking BamB and BamD exhibited altered membrane permeability and enhanced sensitivity to various antimicrobials. BamB mutants also exhibited significantly impaired *in vitro* growth (Dunn et al., 2015). Thus, in deviation from the gram-negative/diderm dogma, BamB might replace BamD in the core Bam complex of *B. burgdorferi*.

Recently, the *B. burgdorferi* Bam machinery was shown to interact with a TamB (BB0794) homolog (Iqbal et al., 2016) that in other bacteria is part of a parallel OMP Translocation and Assembly Module

(TAM) pathway. This indicates that *B. burgdorferi* has adapted modules of separate secretion machineries to function in a single hybrid pathway that allows OMPs to efficiently traverse the periplasm (Stubenrauch et al., 2016). It is tempting to speculate that the *Borrelia* Bam/Tam machinery has evolved to also catalyze the insertion of outer membrane porins with predicted α-helical TM domains such as the outer membrane porin P13 (Sadziene et al., 1995; Noppa et al., 2001). If this were indeed the case, P13 would have to somehow avoid recognition and sorting to the IM by the Sec/YidC complex, possibly through the 28 amino acid C-terminal domain that is ultimately removed by CtpA (BB0359) (Östberg et al. 2004).

Conclusions and future directions

Since the discovery of the first outer membrane components, the borrelial cell envelope has given up many secrets, but many more await unraveling. The completed genome sequences of several species have also aided in the understanding of the common biology of *Borrelia* spirochaetes. In future studies, structure-function relationships of lipoprotein virulence factors need to be determined beyond their antibody epitopes. Recent advances in protein structure determination need to be leveraged to reveal the likely unusual conformations of *Borrelia* OM transmembrane proteins and to further understand the adaptive capacity of the bacterium in the various ecological niches it encounters. Envelope biogenesis pathways have to be further tested experimentally. A better understanding of the *Borrelia* cell cycle and envelope homeostasis are paramount to a better understanding and improved treatments for Lyme disease and relapsing fever. For example, quantitative, high-resolution reporters for PG synthesis can address issues related to so called persister organisms, and the overall status of an active infection. Identification and characterization of essential structural proteins and enzymes that can be targeted by of small molecule inhibitors can lead to novel therapeutics. We continue to discover that this phylogenetic ancient spirochetal microorganism has developed solutions of its own and often does not conform to the dogmatic structure and function of the cell envelope of Gram-negative bacteria - be it with a bacterial surface dominated by lipoproteins, the formation of lipid rafts in both membranes, an intriguing interplay of its unique peptidoglycan with the nearby motility apparatus, and the modular shuffling and adaptive salvaging of proven molecular machineries to maintain an envelope under conditions where change is constant. We continue to anticipate a bright future with many challenges and unsolved mysteries for several generations of *Borrelia* researchers.

Acknowledgements

This work was supported by grants from the National Institutes of Health (NIH) to WRZ (R21 AI144624) and AMT (R21 AI125806), by the Steven & Alexandra Cohen Foundation to BLJ, and by the Swedish Research Council to SB (VR-MH 01885). WRZ wishes to dedicate his contributions to the chapter to his parents.

References

Achouak, W., Heulin, T., and Pages, J.M. (2001). Multiple facets of bacterial porins. FEMS Microbiol Lett *199*, 1-7. https://doi.org/10.1111/j.1574-6968.2001.tb10642.x

Alitalo, A., Meri, T., Lankinen, H., Seppala, I., Lahdenne, P., Hefty, P.S., Akins, D., and Meri, S. (2002). Complement inhibitor factor H binding to Lyme disease spirochetes is mediated by inducible expression of multiple plasmid-encoded outer surface protein E paralogs. J Immunol *169*, 3847-3853. https://doi.org/10.4049/jimmunol.169.7.3847

Andersen, C., Hughes, C., and Koronakis, V. (2000). Chunnel vision. Export and efflux through bacterial channel-tunnels. EMBO Rep *1*, 313-318. https://doi.org/10.1093/embo-reports/kvd075

Andersen, C., Koronakis, E., Hughes, C., and Koronakis, V. (2002). An aspartate ring at the TolC tunnel entrance determines ion selectivity and presents a target for blocking by large cations. Mol Microbiol *44*, 1131-1139. https://doi.org/10.1046/j.1365-2958.2002.02898.x

Anguita, J., Thomas, V., Samanta, S., Persinski, R., Hernanz, C., Barthold, S.W., and Fikrig, E. (2001). *Borrelia burgdorferi*-induced inflammation facilitates spirochete adaptation and variable major protein-like sequence locus recombination. J Immunol *167*, 3383-3390. https://doi.org/10.4049/jimmunol.167.6.3383

Antonara, S., Chafel, R.M., LaFrance, M., and Coburn, J. (2007). *Borrelia burgdorferi* adhesins identified using in vivo phage display. Mol Microbiol *66*, 262-276. https://doi.org/10.1111/j.1365-2958.2007.05924.x

Aranjuez, G.F., Kuhn, H.W., Adams, P.P., and Jewett, M.W. (2019). *Borrelia burgdorferi* bbk13 Is Critical for Spirochete Population Expansion in the Skin during Early Infection. Infect Immun *87*. https://doi.org/10.1128/IAI.00887-18

Awade, A.C., Cleuziat, P., Gonzales, T., and Robert-Baudouy, J. (1994). Pyrrolidone carboxyl peptidase (Pcp): an enzyme that removes pyroglutamic acid (pGlu) from pGlu-peptides and pGlu-proteins. Proteins *20*, 34-51. https://doi.org/10.1002/prot.340200106

Azuma, I., Taniyama, T., Yamamura, Y., Yanagihara, Y., and Hattori, Y. (1975). Chemical studies on the cell walls of *Leptospira biflexa* strain Urawa and Treponema pallidum strain Reiter. Jpn J Microbiol *19*, 45-51. https://doi.org/10.1111/j.1348-0421.1975.tb00846.x

Barbour, A.G. (1990). Antigenic variation of a relapsing fever *Borrelia* species. Annual Reviews in Microbiology *44*, 155-171. https://doi.org/10.1146/annurev.mi.44.100190.001103

Barbour, A.G. (1993). Linear DNA of *Borrelia* species and antigenic variation. Trends Microbiol *1*, 236-239. https://doi.org/10.1016/0966-842x(93)90139-i

Barbour, A.G., and Garon, C.F. (1988). The genes encoding major surface proteins of *Borrelia burgdorferi* are located on a plasmid. Ann N Y Acad Sci *539*, 144-153. https://doi.org/10.1111/j.1749-6632.1988.tb31847.x

Barbour, A.G., and Hayes, S.F. (1986). Biology of *Borrelia* species. Microbiol Rev *50*, 381-400.

Barbour, A.G., and Travinsky, B. (2010). Evolution and distribution of the *ospC* gene, a transferable serotype determinant of *Borrelia burgdorferi*. mBio *1*. https://doi.org/10.1128/mBio.00153-10

Barcena-Uribarri, I., Thein, M., Barbot, M., Sans-Serramitjana, E., Bonde, M., Mentele, R., Lottspeich, F., Bergström, S., and Benz, R. (2014). Study of the protein complex, pore diameter, and pore-forming activity of the *Borrelia burgdorferi* P13 porin. J Biol Chem *289*, 18614-18624. https://doi.org/10.1074/jbc.M113.539528

Barcena-Uribarri, I., Thein, M., Maier, E., Bonde, M., Bergström, S., and Benz, R. (2013). Use of nonelectrolytes reveals the channel size and oligomeric constitution of the *Borrelia burgdorferi* P66 porin. PLoS One *8*, e78272. https://doi.org/10.1371/journal.pone.0078272

Barcena-Uribarri, I., Thein, M., Sacher, A., Bunikis, I., Bonde, M., Bergström, S., and Benz, R. (2010). P66 porins are present in both Lyme disease and relapsing fever spirochetes: a comparison of the biophysical properties of P66 porins from six *Borrelia* species. Biochim Biophys Acta *1798*, 1197-1203. https://doi.org/10.1016/j.bbamem.2010.02.011

Barthold, S.W., Fikrig, E., Bockenstedt, L.K., and Persing, D.H. (1995). Circumvention of outer surface protein A immunity by host-adapted *Borrelia burgdorferi*. Infect Immun *63*, 2255-2261. https://doi.org/10.1128/IAI.63.6.2255-2261.1995

Battisti, J.M., Bono, J.L., Rosa, P.A., Schrumpf, M.E., Schwan, T.G., and Policastro, P.F. (2008). Outer surface protein A protects Lyme disease spirochetes from acquired host immunity in the tick vector. Infect Immun *76*, 5228-5237. https://doi.org/10.1128/IAI.00410-08

Beck, G., Benach, J.L., and Habicht, G.S. (1990). Isolation, preliminary chemical characterization, and biological activity of *Borrelia burgdorferi* peptidoglycan. Biochem Biophys Res Commun *167*, 89-95. https://doi.org/10.1016/0006-291x(90)91734-a

Becker, M., Bunikis, J., Lade, B.D., Dunn, J.J., Barbour, A.G., and Lawson, C.L. (2005). Structural investigation of *Borrelia burgdorferi* OspB, a bactericidal Fab target. J Biol Chem *280*, 17363-17370. https://doi.org/10.1074/jbc.M412842200

Beilharz, K., Novakova, L., Fadda, D., Branny, P., Massidda, O., and Veening, J.W. (2012). Control of cell division in *Streptococcus pneumoniae* by the conserved Ser/Thr protein kinase StkP. Proc Natl Acad Sci U S A *109*, E905-913. https://doi.org/10.1073/pnas.1119172109

Belisle, J.T., Brandt, M.E., Radolf, J.D., and Norgard, M.V. (1994). Fatty acids of *Treponema pallidum* and *Borrelia burgdorferi* lipoproteins. J Bacteriol *176*, 2151-2157. https://doi.org/10.1128/jb.176.8.2151-2157.1994

Ben-Menachem, G., Kubler-Kielb, J., Coxon, B., Yergey, A., and Schneerson, R. (2003). A newly discovered cholesteryl galactoside from *Borrelia burgdorferi*. Proc Natl Acad Sci U S A *100*, 7913-7918. https://doi.org/10.1073/pnas.1232451100

Benz, R. (1994). Uptake of solutes through bacterial outer membranes. In Bacterial Cell Wall, M. Ghuysen, and R. Hackenbek, eds. (Amsterdam: Elsevier), pp. 397 - 423. https://doi.org/https://doi.org/10.1016/S0167-7306(08)60422-6

Benz, R. (2001). Porins - structure and function. In: Microbial Transport Systems. (Winkelmann, G, ed), WILEY-VCH Verlag GmbH, Weinheim/Germany, pp. 227-246.

Benz, R., Janko, K., and Lauger, P. (1979). Ionic selectivity of pores formed by the matrix protein (porin) of *Escherichia coli*. Biochim Biophys Acta *551*, 238-247. https://doi.org/10.1016/0005-2736(89)90002-3

Benz, R., Maier, E., and Gentschev, I. (1993). TolC of *Escherichia coli* functions as an outer membrane channel. Zentralbl Bakteriol *278*, 187-196. https://doi.org/10.1016/s0934-8840(11)80836-4

Benz, R., Schmid, A., Maier, C., and Bremer, E. (1988). Characterization of the nucleoside-binding site inside the Tsx channel of *Escherichia coli* outer membrane. Reconstitution experiments with lipid bilayer membranes. Eur J Biochem *176*, 699-705. https://doi.org/10.1111/j.1432-1033.1988.tb14333.x

Benz, R., Schmid, A., Nakae, T., and Vos-Scheperkeuter, G.H. (1986). Pore formation by LamB of *Escherichia coli* in lipid bilayer membranes. J Bacteriol *165*, 978-986. https://doi.org/10.1128/jb.165.3.978-986.1986

Berg, S. (2001). The glucolipid pathway in *Acholeplasma laidawii*: Functional and structural properties of three consequetively-acting enzymes and similarities to orthologs in prokaryotes. Doctoral thesis, Umeå University, Umeå, Sweden ISBN 91-7305-025-3.

Berg, S., Edman, M., Li, L., Wikstrom, M., and Wieslander, A. (2001). Sequence properties of the 1,2-diacylglycerol 3-glucosyltransferase from *Acholeplasma laidlawii* membranes. Recognition of a large group of lipid glycosyltransferases in eubacteria and archaea. J Biol Chem *276*, 22056-22063. https://doi.org/10.1074/jbc.M102576200

Bergström, S., Bundoc, V.G., and Barbour, A.G. (1989). Molecular analysis of linear plasmid-encoded major surface proteins, OspA and OspB, of the Lyme disease spirochaete *Borrelia burgdorferi*. Mol Microbiol *3*, 479-486. https://doi.org/10.1111/j.1365-2958.1989.tb00194.x

Bessler, W.G., Cox, M., Lex, A., Suhr, B., Wiesmuller, K.H., and Jung, G. (1985). Synthetic lipopeptide analogs of bacterial lipoprotein are potent polyclonal activators for murine B lymphocytes. J Immunol *135*, 1900-1905.

Bieberich, E. (2018). Sphingolipids and lipid rafts: Novel concepts and methods of analysis. Chem Phys Lipids *216*, 114-131. https://doi.org/10.1016/j.chemphyslip.2018.08.003

Bierwagen, P., Szpotkowski, K., Jaskolski, M., and Urbanowicz, A. (2019). *Borrelia* outer surface protein C is capable of human fibrinogen binding. FEBS J *286*, 2415-2428. https://doi.org/10.1111/febs.14810

Bisson-Filho, A.W., Hsu, Y.P., Squyres, G.R., Kuru, E., Wu, F., Jukes, C., Sun, Y., Dekker, C., Holden, S., VanNieuwenhze, M.S., *et al.* (2017). Treadmilling by FtsZ filaments drives peptidoglycan synthesis and bacterial cell division. Science *355*, 739-743. https://doi.org/10.1126/science.aak9973

Bista, S., Singh, P., Bernard, Q., Yang, X., Hart, T., Lin, Y.P., Kitsou, C., Singh Rana, V., Zhang, F., Linhardt, R.J., *et al.* (2020). A novel laminin-binding protein mediates microbial-endothelial cell interactions and facilitates dissemination of Lyme disease pathogens. Journal of Infectious Diseases *221*, 1438-1447. https://doi.org/10.1093/infdis/jiz626

Bonnemain, C., Raynaud, C., Reglier-Poupet, H., Dubail, I., Frehel, C., Lety, M.A., Berche, P., and Charbit, A. (2004). Differential roles of multiple signal peptidases in the virulence of *Listeria monocytogenes*. Mol Microbiol *51*, 1251-1266. https://doi.org/10.1111/j.1365-2958.2004.03916.x

Bono, J., Tilly, K., Stevenson, B., Hogan, D., and Rosa, P. (1998). Oligopeptide permease in *Borrelia burgdorferi*: putative peptide-binding components encoded by both chromosomal and plasmid loci. Microbiology *144* 1033-1044.

Bontemps-Gallo, S., Lawrence, K., and Gherardini, F.C. (2016). Two different virulence-related regulatory pathways in *Borrelia burgdorferi* are directly affected by osmotic fluxes in the blood meal of feeding *Ixodes* ticks. PLoS Pathog *12*, e1005791. https://doi.org/10.1371/journal.ppat.1005791

Bos, M.P., Robert, V., and Tommassen, J. (2007). Biogenesis of the gram-negative bacterial outer membrane. Annual Reviews in Microbiology *61*, 191-214. https://doi.org/10.1146/annurev.micro.61.080706.093245

Brandt, M.E., Riley, B.S., Radolf, J.D., and Norgard, M.V. (1990). Immunogenic integral membrane proteins of *Borrelia burgdorferi* are lipoproteins. Infect Immun *58*, 983-991. https://doi.org/10.1128/IAI.58.4.983-991.1990

Brissette, C.A., Haupt, K., Barthel, D., Cooley, A.E., Bowman, A., Skerka, C., Wallich, R., Zipfel, P.F., Kraiczy, P., and Stevenson, B.

(2009). *Borrelia burgdorferi* infection-associated surface proteins ErpP, ErpA, and ErpC bind human plasminogen. Infect Immun *77*, 300-306. https://doi.org/10.1128/IAI.01133-08

Browman, D.T., Hoegg, M.B., and Robbins, S.M. (2007). The SPFH domain-containing proteins: more than lipid raft markers. Trends Cell Biol *17*, 394-402. https://doi.org/10.1016/j.tcb.2007.06.005

Brown, D.A., and London, E. (2000). Structure and function of sphingolipid- and cholesterol-rich membrane rafts. J Biol Chem *275*, 17221-17224. https://doi.org/10.1074/jbc.R000005200

Brown, E.L., Wooten, R.M., Johnson, B.J., Iozzo, R.V., Smith, A., Dolan, M.C., Guo, B.P., Weis, J.J., and Hook, M. (2001). Resistance to Lyme disease in decorin-deficient mice. J Clin Invest *107*, 845-852. https://doi.org/10.1172/JCI11692

Bunikis, I., Denker, K., Östberg, Y., Andersen, C., Benz, R., and Bergström, S. (2008). An RND-type efflux system in *Borrelia burgdorferi* is involved in virulence and resistance to antimicrobial compounds. PLoS Pathog *4*, e1000009. https://doi.org/10.1371/journal.ppat.1000009

Bunikis, J., and Barbour, A. (1999). Access of antibody or trypsin to an integral outer membrane protein (P66) of *Borrelia burgdorferi* is hindered by Osp lipoproteins. Infect Immun *67*, 2874-2883.

Bunikis, J., Noppa, L., and Bergström, S. (1995). Molecular analysis of a 66-kDa protein associated with the outer membrane of Lyme disease *Borrelia*. FEMS Microbiol Lett *131*, 139-145. https://doi.org/10.1111/j.1574-6968.1995.tb07768.x

Burman, N., Bergström, S., Restrepo, B.I., and Barbour, A.G. (1990). The variable antigens Vmp7 and Vmp21 of the relapsing fever bacterium *Borrelia hermsii* are structurally analogous to the VSG proteins of the African trypanosome. Mol Microbiol *4*, 1715-1726. https://doi.org/10.1111/j.1365-2958.1990.tb00549.x

Busby, W.H., Jr., Quackenbush, G.E., Humm, J., Youngblood, W.W., and Kizer, J.S. (1987). An enzyme(s) that converts glutaminyl-peptides into pyroglutamyl-peptides. Presence in pituitary, brain, adrenal medulla, and lymphocytes. J Biol Chem *262*, 8532-8536.

Byram, R., Stewart, P.E., and Rosa, P. (2004). The essential nature of the ubiquitous 26-kilobase circular replicon of *Borrelia burgdorferi*. J Bacteriol *186*, 3561-3569. https://doi.org/10.1128/JB.186.11.3561-3569.2004

Caccamo, P.D., and Brun, Y.V. (2018). The Molecular Basis of Noncanonical Bacterial Morphology. Trends Microbiol *26*, 191-208. https://doi.org/10.1016/j.tim.2017.09.012

Caimano, M.J., Bourell, K.W., Bannister, T.D., Cox, D.L., and Radolf, J.D. (1999). The *Treponema denticola* major sheath protein is predominantly periplasmic and has only limited surface exposure. Infect Immun *67*, 4072-4083. https://doi.org/10.1128/IAI.67.8.4072-4083.1999

Caimano, M.J., Eggers, C.H., Gonzalez, C.A., and Radolf, J.D. (2005). Alternate sigma factor RpoS is required for the in vivo-specific repression of *Borrelia burgdorferi* plasmid lp54-borne ospA and lp6.6 genes. J Bacteriol *187*, 7845-7852. https://doi.org/10.1128/JB.187.22.7845-7852.2005

Caimano, M.J., Groshong, A.M., Belperron, A., Mao, J., Hawley, K.L., Luthra, A., Graham, D.E., Earnhart, C.G., Marconi, R.T., Bockenstedt, L.K., *et al.* (2019). The RpoS Gatekeeper in *Borrelia burgdorferi*: An Invariant Regulatory Scheme That Promotes Spirochete Persistence in Reservoir Hosts and Niche Diversity. Front Microbiol *10*, 1923. https://doi.org/10.3389/fmicb.2019.01923

Caimano, M.J., Iyer, R., Eggers, C.H., Gonzalez, C., Morton, E.A., Gilbert, M.A., Schwartz, I., and Radolf, J.D. (2007). Analysis of the RpoS regulon in *Borrelia burgdorferi* in response to mammalian host signals provides insight into RpoS function during the enzootic cycle. Mol Microbiol *65*, 1193-1217. https://doi.org/10.1111/j.1365-2958.2007.05860.x

Caine, J.A., and Coburn, J. (2016). Multifunctional and Redundant Roles of *Borrelia burgdorferi* Outer Surface Proteins in Tissue Adhesion, Colonization, and Complement Evasion. Front Immunol *7*, 442. https://doi.org/10.3389/fimmu.2016.00442

Caine, J.A., Lin, Y.P., Kessler, J.R., Sato, H., Leong, J.M., and Coburn, J. (2017). *Borrelia burgdorferi* outer surface protein C (OspC) binds complement component C4b and confers bloodstream survival. Cell Microbiol *19*. https://doi.org/10.1111/cmi.12786

Carrasco, S.E., Troxell, B., Yang, Y., Brandt, S.L., Li, H., Sandusky, G.E., Condon, K.W., Serezani, C.H., and Yang, X.F. (2015). Outer surface protein OspC is an antiphagocytic factor that protects *Borrelia burgdorferi* from phagocytosis by macrophages. Infect Immun *83*, 4848-4860. https://doi.org/10.1128/IAI.01215-15

Carter, C.J., Bergström, S., Norris, S.J., and Barbour, A.G. (1994). A family of surface-exposed proteins of 20 kilodaltons in the genus *Borrelia*. Infect Immun *62*, 2792-2799. https://doi.org/10.1128/IAI.62.7.2792-2799.1994

Casjens, S., Palmer, N., van Vugt, R., Huang, W.M., Stevenson, B., Rosa, P., Lathigra, R., Sutton, G., Peterson, J., Dodson, R.J., *et al.* (2000). A bacterial genome in flux: the twelve linear and nine circular extrachromosomal DNAs in an infectious isolate of the Lyme disease spirochete *Borrelia burgdorferi*. Mol Microbiol *35*, 490-516. https://doi.org/10.1046/j.1365-2958.2000.01698.x

Cava, F., and de Pedro, M.A. (2014). Peptidoglycan plasticity in bacteria: emerging variability of the murein sacculus and their associated biological functions. Curr Opin Microbiol *18*, 46-53. https://doi.org/10.1016/j.mib.2014.01.004

Chan, J.M., and Dillard, J.P. (2016). *Neisseria gonorrhoeae* Crippled Its Peptidoglycan Fragment Permease To Facilitate Toxic Peptidoglycan Monomer Release. J Bacteriol *198*, 3029-3040. https://doi.org/10.1128/JB.00437-16

Chan, J.M., and Dillard, J.P. (2017). Attention Seeker: Production, Modification, and Release of Inflammatory Peptidoglycan Fragments in *Neisseria* Species. J Bacteriol *199*. https://doi.org/10.1128/JB.00354-17

Charon, N.W., Cockburn, A., Li, C., Liu, J., Miller, K.A., Miller, M.R., Motaleb, M.A., and Wolgemuth, C.W. (2012). The unique paradigm of spirochete motility and chemotaxis. Annual Reviews in Microbiology *66*, 349-370. https://doi.org/10.1146/annurev-micro-092611-150145

Charon, N.W., Goldstein, S.F., Marko, M., Hsieh, C., Gebhardt, L.L., Motaleb, M.A., Wolgemuth, C.W., Limberger, R.J., and Rowe, N. (2009). The flat-ribbon configuration of the periplasmic flagella of *Borrelia burgdorferi* and its relationship to motility and morphology. J Bacteriol *191*, 600-607. https://doi.org/10.1128/JB.01288-08

Chen, S., Kumru, O.S., and Zückert, W.R. (2011). Determination of *Borrelia* surface lipoprotein anchor topology by surface proteolysis. J Bacteriol *193*, 6379-6383. https://doi.org/10.1128/JB.05849-11

Chen, S., and Zückert, W.R. (2011). Probing the *Borrelia burgdorferi* surface lipoprotein secretion pathway using a conditionally folding protein domain. J Bacteriol *193*, 6724-6732. https://doi.org/10.1128/JB.06042-11

Chiba, S., Akiyama, Y., and Ito, K. (2002). Membrane protein degradation by FtsH can be initiated from either end. J Bacteriol *184*, 4775-4782. https://doi.org/10.1128/jb.184.17.4775-4782.2002

Cho, H., McManus, H.R., Dove, S.L., and Bernhardt, T.G. (2011). Nucleoid occlusion factor SlmA is a DNA-activated FtsZ polymerization antagonist. Proc Natl Acad Sci U S A *108*, 3773-3778. https://doi.org/10.1073/pnas.1018674108

Chu, C.Y., Stewart, P.E., Bestor, A., Hansen, B., Lin, T., Gao, L., Norris, S.J., and Rosa, P.A. (2016). Function of the *Borrelia burgdorferi* FtsH Homolog Is Essential for Viability both In Vitro and In Vivo and Independent of HflK/C. mBio *7*, e00404-00416. https://doi.org/10.1128/mBio.00404-16

Clausen, T., Kaiser, M., Huber, R., and Ehrmann, M. (2011). HTRA proteases: regulated proteolysis in protein quality control. Nat Rev Mol Cell Biol *12*, 152-162. https://doi.org/10.1038/nrm3065

Cloud-Hansen, K.A., Hackett, K.T., Garcia, D.L., and Dillard, J.P. (2008). *Neisseria gonorrhoeae* uses two lytic transglycosylases to produce cytotoxic peptidoglycan monomers. J Bacteriol *190*, 5989-5994. https://doi.org/10.1128/JB.00506-08

Cluss, R.G., Silverman, D.A., and Stafford, T.R. (2004). Extracellular secretion of the *Borrelia burgdorferi* Oms28 porin and Bgp, a glycosaminoglycan binding protein. Infect Immun *72*, 6279-6286. https://doi.org/10.1128/IAI.72.11.6279-6286.2004

Coburn, J., Chege, W., Magoun, L., Bodary, S.C., and Leong, J.M. (1999). Characterization of a candidate *Borrelia burgdorferi* beta3-chain integrin ligand identified using a phage display library. Mol Microbiol *34*, 926-940. https://doi.org/10.1046/j.1365-2958.1999.01654.x

Coburn, J., and Cugini, C. (2003). Targeted mutation of the outer membrane protein P66 disrupts attachment of the Lyme disease agent, *Borrelia burgdorferi*, to integrin alphavbeta3. Proc Natl Acad Sci U S A *100*, 7301-7306. https://doi.org/10.1073/pnas.1131117100

Coburn, J., Fischer, J.R., and Leong, J.M. (2005). Solving a sticky problem: new genetic approaches to host cell adhesion by the Lyme disease spirochete. Mol Microbiol *57*, 1182-1195. https://doi.org/10.1111/j.1365-2958.2005.04759.x

Coleman, J.L., Crowley, J.T., Toledo, A.M., and Benach, J.L. (2013). The HtrA protease of *Borrelia burgdorferi* degrades outer membrane protein BmpD and chemotaxis phosphatase CheX. Mol Microbiol *88*, 619-633. https://doi.org/10.1111/mmi.12213

Coleman, J.L., Toledo, A., and Benach, J.L. (2016). *Borrelia burgdorferi* HtrA: evidence for twofold proteolysis of outer membrane protein p66. Mol Microbiol *99*, 135-150. https://doi.org/10.1111/mmi.13221

Cordes, F.S., Roversi, P., Kraiczy, P., Simon, M.M., Brade, V., Jahraus, O., Wallis, R., Skerka, C., Zipfel, P.F., Wallich, R., *et al.* (2005). A novel fold for the factor H-binding protein BbCRASP-1 of *Borrelia burgdorferi*. Nat Struct Mol Biol *12*, 276-277. https://doi.org/10.1038/nsmb902

Coutte, L., Botkin, D.J., Gao, L., and Norris, S.J. (2009). Detailed analysis of sequence changes occurring during vlsE antigenic variation in the mouse model of *Borrelia burgdorferi* infection. PLoS Pathog *5*, e1000293. https://doi.org/10.1371/journal.ppat.1000293

Coutte, L., Willery, E., Antoine, R., Drobecq, H., Locht, C., and Jacob-Dubuisson, F. (2003). Surface anchoring of bacterial subtilisin important for maturation function. Mol Microbiol *49*, 529-539. https://doi.org/10.1046/j.1365-2958.2003.03573.x

Crowley, J.T., Toledo, A.M., LaRocca, T.J., Coleman, J.L., London, E., and Benach, J.L. (2013). Lipid exchange between *Borrelia burgdorferi* and host cells. PLoS Pathog *9*, e1003109. https://doi.org/10.1371/journal.ppat.1003109

Cullen, P.A., Haake, D.A., and Adler, B. (2004). Outer membrane proteins of pathogenic spirochetes. FEMS Microbiol Rev *28*, 291-318. https://doi.org/10.1016/j.femsre.2003.10.004

Das, S., Shraga, D., Gannon, C., Lam, T.T., Feng, S., Brunet, L.R., Telford, S.R., Barthold, S.W., Flavell, R.A., and Fikrig, E. (1996). Characterization of a 30-kDa *Borrelia burgdorferi* substrate-binding protein homologue. Res Microbiol *147*, 739-751. https://doi.org/10.1016/s0923-2508(97)85121-2

de Silva, A.M., Telford, S.R., 3rd, Brunet, L.R., Barthold, S.W., and Fikrig, E. (1996). *Borrelia burgdorferi* OspA is an arthropod-specific transmission-blocking Lyme disease vaccine. J Exp Med *183*, 271-275. https://doi.org/10.1084/jem.183.1.271

Defoe, G., and Coburn, J. (2001). Delineation of *Borrelia burgdorferi* p66 sequences required for integrin alpha(IIb)beta(3) recognition. Infect Immun *69*, 3455-3459. https://doi.org/10.1128/IAI.69.5.3455-3459.2001

Demchick, P., and Koch, A.L. (1996). The permeability of the wall fabric of *Escherichia coli* and *Bacillus subtilis*. J Bacteriol *178*, 768-773. https://doi.org/10.1128/jb.178.3.768-773.1996

Deres, K., Schild, H., Wiesmuller, K.H., Jung, G., and Rammensee, H.G. (1989). In vivo priming of virus-specific cytotoxic T lymphocytes with synthetic lipopeptide vaccine. Nature *342*, 561-564. https://doi.org/10.1038/342561a0

Desmarais, S.M., De Pedro, M.A., Cava, F., and Huang, K.C. (2013). Peptidoglycan at its peaks: how chromatographic analyses can reveal bacterial cell wall structure and assembly. Mol Microbiol *89*, 1-13. https://doi.org/10.1111/mmi.12266

Dombrowski, C., Kan, W., Motaleb, M.A., Charon, N.W., Goldstein, R.E., and Wolgemuth, C.W. (2009). The elastic basis for the shape of *Borrelia burgdorferi*. Biophys J *96*, 4409-4417. https://doi.org/10.1016/j.bpj.2009.02.066

Dong, C., Beis, K., Nesper, J., Brunkan-Lamontagne, A.L., Clarke, B.R., Whitfield, C., and Naismith, J.H. (2006). Wza, the translocon for *E. coli* capsular polysaccharides, defines a new class of membrane protein. Nature *444*, 226-229. https://doi.org/10.1038/nature05267

Dowdell, A.S., Murphy, M.D., Azodi, C., Swanson, S.K., Florens, L., Chen, S., and Zückert, W.R. (2017). Comprehensive Spatial Analysis of the *Borrelia burgdorferi* Lipoproteome Reveals a Compartmentalization Bias toward the Bacterial Surface. J Bacteriol *199*. https://doi.org/10.1128/JB.00658-16

Dresser, A.R., Hardy, P.O., and Chaconas, G. (2009). Investigation of the genes involved in antigenic switching at the vlsE locus in *Borrelia burgdorferi*: an essential role for the RuvAB branch migrase. PLoS Pathog *5*, e1000680. https://doi.org/10.1371/journal.ppat.1000680

Du, S., and Lutkenhaus, J. (2019). At the Heart of Bacterial Cytokinesis: The Z Ring. Trends Microbiol *27*, 781-791. https://doi.org/10.1016/j.tim.2019.04.011

Dunham-Ems, S.M., Caimano, M.J., Eggers, C.H., and Radolf, J.D. (2012). *Borrelia burgdorferi* requires the alternative sigma factor RpoS for dissemination within the vector during tick-to-mammal transmission. PLoS Pathog *8*, e1002532. https://doi.org/10.1371/journal.ppat.1002532

Dunn, J.P., Kenedy, M.R., Iqbal, H., and Akins, D.R. (2015). Characterization of the beta-barrel assembly machine accessory lipoproteins from *Borrelia burgdorferi*. BMC Microbiol *15*, 70. https://doi.org/10.1186/s12866-015-0411-y

Dworkin, J. (2014). The medium is the message: interspecies and interkingdom signaling by peptidoglycan and related bacterial glycans. Annual Reviews in Microbiology *68*, 137-154. https://doi.org/10.1146/annurev-micro-091213-112844

Egan, A.J.F., Errington, J., and Vollmer, W. (2020). Regulation of peptidoglycan synthesis and remodelling. Nat Rev Microbiol *18*, 446-460. https://doi.org/10.1038/s41579-020-0366-3

Eicken, C., Sharma, V., Klabunde, T., Lawrenz, M.B., Hardham, J.M., Norris, S.J., and Sacchettini, J.C. (2002). Crystal structure of Lyme disease variable surface antigen VlsE of *Borrelia burgdorferi*. J Biol Chem *277*, 21691-21696. https://doi.org/10.1074/jbc.M201547200

Eicken, C., Sharma, V., Klabunde, T., Owens, R.T., Pikas, D.S., Hook, M., and Sacchettini, J.C. (2001). Crystal structure of Lyme disease antigen outer surface protein C from *Borrelia burgdorferi*. J Biol Chem *276*, 10010-10015. https://doi.org/10.1074/jbc.M010062200

Eiffert, H., Lotter, H., Jarecki-Khan, K., and Thomssen, R. (1991). Identification of an immunoreactive non-proteinaleous component in *Borrelia burgdorferi*. Med Microbiol Immunol *180*, 229-237. https://doi.org/10.1007/BF00202557

Erdile, L.F., Brandt, M.A., Warakomski, D.J., Westrack, G.J., Sadziene, A., Barbour, A.G., and Mays, J.P. (1993). Role of attached lipid in immunogenicity of *Borrelia burgdorferi* OspA. Infect Immun *61*, 81-90. https://doi.org/10.1128/IAI.61.1.81-90.1993

Errington, J., Daniel, R.A., and Scheffers, D.J. (2003). Cytokinesis in bacteria. Microbiol Mol Biol Rev *67*, 52-65, table of contents. https://doi.org/10.1128/mmbr.67.1.52-65.2003

Ferenci, T., Schwentorat, M., Ullrich, S., and Vilmart, J. (1980). Lambda receptor in the outer membrane of *Escherichia coli* as a binding protein for maltodextrins and starch polysaccharides. J Bacteriol *142*, 521-526. https://doi.org/10.1128/JB.142.2.521-526.1980

Fingerle, V., Hauser, U., Liegl, G., Petko, B., Preac-Mursic, V., and Wilske, B. (1995). Expression of outer surface proteins A and C of *Borrelia burgdorferi* in Ixodes ricinus. J Clin Microbiol *33*, 1867-1869. https://doi.org/10.1128/JCM.33.7.1867-1869.1995

Fingerle, V., Liegl, G., Munderloh, U., and Wilske, B. (1998). Expression of outer surface proteins A and C of *Borrelia burgdorferi* in Ixodes ricinus ticks removed from humans. Med Microbiol Immunol *187*, 121-126. https://doi.org/10.1007/s004300050083

Finlay, B.B., and Falkow, S. (1989). Common themes in microbial pathogenicity. Microbiol Rev *53*, 210-230.

Fischer, J.R., LeBlanc, K.T., and Leong, J.M. (2006). Fibronectin binding protein BBK32 of the Lyme disease spirochete promotes bacterial attachment to glycosaminoglycans. Infect Immun *74*, 435-441. https://doi.org/10.1128/IAI.74.1.435-441.2006

Fischer, J.R., Parveen, N., Magoun, L., and Leong, J.M. (2003). Decorin-binding proteins A and B confer distinct mammalian cell type-specific attachment by *Borrelia burgdorferi*, the Lyme disease spirochete. Proc Natl Acad Sci U S A *100*, 7307-7312. https://doi.org/10.1073/pnas.1231043100

Flak, T.A., and Goldman, W.E. (1999). Signalling and cellular specificity of airway nitric oxide production in pertussis. Cell Microbiol *1*, 51-60. https://doi.org/10.1046/j.1462-5822.1999.00004.x

Fleurie, A., Lesterlin, C., Manuse, S., Zhao, C., Cluzel, C., Lavergne, J.P., Franz-Wachtel, M., Macek, B., Combet, C., Kuru, E., *et al.* (2014). MapZ marks the division sites and positions FtsZ rings in *Streptococcus pneumoniae*. Nature *516*, 259-262. https://doi.org/10.1038/nature13966

Francetic, O., and Pugsley, A.P. (2005). Towards the identification of type II secretion signals in a nonacylated variant of pullulanase from *Klebsiella oxytoca*. J Bacteriol *187*, 7045-7055. https://doi.org/10.1128/JB.187.20.7045-7055.2005

Fraser, C.M., Casjens, S., Huang, W.M., Sutton, G.G., Clayton, R., Lathigra, R., White, O., Ketchum, K.A., Dodson, R., Hickey, E.K., *et al.* (1997). Genomic sequence of a Lyme disease spirochaete, *Borrelia burgdorferi*. Nature *390*, 580-586. https://doi.org/10.1038/37551

Fraser, C.M., Norris, S.J., Weinstock, G.M., White, O., Sutton, G.G., Dodson, R., Gwinn, M., Hickey, E.K., Clayton, R., Ketchum, K.A., *et al.* (1998). Complete genome sequence of *Treponema pallidum*, the syphilis spirochete. Science *281*, 375-388. https://doi.org/10.1126/science.281.5375.375

Froger, A., Rolland, J.P., Bron, P., Lagree, V., Caherec, F.L., Deschamps, S., Hubert, J.F., Pellerin, I., Thomas, D., and Delamarche, C. (2001). Functional characterization of a microbial aquaglyceroporin. Microbiology (Reading) *147*, 1129-1135. https://doi.org/10.1099/00221287-147-5-1129

Fukuda, A., Matsuyama, S., Hara, T., Nakayama, J., Nagasawa, H., and Tokuda, H. (2002). Aminoacylation of the N-terminal cysteine is essential for Lol-dependent release of lipoproteins from membranes but does not depend on lipoprotein sorting signals. J Biol Chem *277*, 43512-43518.

Gan, L., Chen, S., and Jensen, G.J. (2008). Molecular organization of Gram-negative peptidoglycan. Proc Natl Acad Sci U S A *105*, 18953-18957. https://doi.org/10.1073/pnas.0808035105

Garcia, B.L., Zhi, H., Wager, B., Hook, M., and Skare, J.T. (2016). *Borrelia burgdorferi* BBK32 Inhibits the Classical Pathway by Blocking Activation of the C1 Complement Complex. PLoS Pathog *12*, e1005404. https://doi.org/10.1371/journal.ppat.1005404

Garcia, D.L., and Dillard, J.P. (2008). Mutations in ampG or ampD affect peptidoglycan fragment release from *Neisseria gonorrhoeae*. J Bacteriol *190*, 3799-3807. https://doi.org/10.1128/JB.01194-07

Garcia Monco, J.C., Wheeler, C.M., Benach, J.L., Furie, R.A., Lukehart, S.A., Stanek, G., and Steere, A.C. (1993). Reactivity of neuroborreliosis patients (Lyme disease) to cardiolipin and gangliosides. J Neurol Sci *117*, 206-214. https://doi.org/10.1016/0022-510x(93)90175-x

Garcia-Heredia, A., Pohane, A.A., Melzer, E.S., Carr, C.R., Fiolek, T.J., Rundell, S.R., Lim, H.C., Wagner, J.C., Morita, Y.S., Swarts, B.M., *et al.* (2018). Peptidoglycan precursor synthesis along the sidewall of pole-growing mycobacteria. Elife *7*. https://doi.org/10.7554/eLife.37243

Garcia-Monco, J.C., Seidman, R.J., and Benach, J.L. (1995). Experimental immunization with *Borrelia burgdorferi* induces development of antibodies to gangliosides. Infect Immun *63*, 4130-4137. https://doi.org/10.1128/IAI.63.10.4130-4137.1995

Geiger, O., Lopez-Lara, I.M., and Sohlenkamp, C. (2013). Phosphatidylcholine biosynthesis and function in bacteria. Biochim Biophys Acta *1831*, 503-513. https://doi.org/10.1016/j.bbalip.2012.08.009

Gherardini, F.C. (2013). *Borrelia burgdorferi* HtrA may promote dissemination and irritation. Mol Microbiol *90*, 209-213. https://doi.org/10.1111/mmi.12390

Gilmore, R.D., Jr., and Piesman, J. (2000). Inhibition of *Borrelia burgdorferi* migration from the midgut to the salivary glands following feeding by ticks on OspC-immunized mice. Infect Immun *68*, 411-414. https://doi.org/10.1128/iai.68.1.411-414.2000

Goldman, W.E., Klapper, D.G., and Baseman, J.B. (1982). Detection, isolation, and analysis of a released *Bordetella pertussis* product toxic to cultured tracheal cells. Infect Immun *36*, 782-794. https://doi.org/10.1128/IAI.36.2.782-794.1982

Goldstein, S.F., Charon, N.W., and Kreiling, J.A. (1994). *Borrelia burgdorferi* swims with a planar waveform similar to that of eukaryotic flagella. Proceedings of the National Academy of Sciences of the United States of America *91*, 3433-3437.

Gominet, M., Slamti, L., Gilois, N., Rose, M., and Lereclus, D. (2001). Oligopeptide permease is required for expression of the *Bacillus thuringiensis* plcR regulon and for virulence. Mol Microbiol *40*, 963-975. https://doi.org/10.1046/j.1365-2958.2001.02440.x

Goodell, E.W., and Higgins, C.F. (1987). Uptake of cell wall peptides by *Salmonella typhimurium* and *Escherichia coli*. J Bacteriol *169*, 3861-3865. https://doi.org/10.1128/jb.169.8.3861-3865.1987

Gorrell, R.J., Guan, J., Xin, Y., Tafreshi, M.A., Hutton, M.L., McGuckin, M.A., Ferrero, R.L., and Kwok, T. (2013). A novel NOD1- and CagA-independent pathway of interleukin-8 induction mediated by the *Helicobacter pylori* type IV secretion system. Cell Microbiol *15*, 554-570. https://doi.org/10.1111/cmi.12055

Grabowicz, M., and Silhavy, T.J. (2017). Redefining the essential trafficking pathway for outer membrane lipoproteins. Proc Natl Acad Sci U S A *114*, 4769-4774. https://doi.org/10.1073/pnas.1702248114

Greene, N.P., Hinchliffe, P., Crow, A., Ababou, A., Hughes, C., and Koronakis, V. (2013). Structure of an atypical periplasmic adaptor from a multidrug efflux pump of the spirochete *Borrelia burgdorferi*. FEBS Lett *587*, 2984-2988. https://doi.org/10.1016/j.febslet.2013.06.056

Grimm, D., Tilly, K., Byram, R., Stewart, P.E., Krum, J.G., Bueschel, D.M., Schwan, T.G., Policastro, P.F., Elias, A.F., and Rosa, P.A. (2004). Outer-surface protein C of the Lyme disease spirochete: a protein induced in ticks for infection of mammals. Proc Natl Acad Sci U S A *101*, 3142-3147. https://doi.org/10.1073/pnas.0306845101

Groshong, A.M., Dey, A., Bezsonova, I., Caimano, M.J., and Radolf, J.D. (2017). Peptide uptake is essential for *Borrelia burgdorferi* viability and involves structural and regulatory complexity of its oligopeptide transporter. mBio *8*. https://doi.org/10.1128/mBio.02047-17

Guo, B.P., Brown, E.L., Dorward, D.W., Rosenberg, L.C., and Hook, M. (1998). Decorin-binding adhesins from *Borrelia burgdorferi*. Mol Microbiol *30*, 711-723. https://doi.org/10.1046/j.1365-2958.1998.01103.x

Guo, B.P., Norris, S.J., Rosenberg, L.C., and Hook, M. (1995). Adherence of *Borrelia burgdorferi* to the proteoglycan decorin. Infect Immun *63*, 3467-3472. https://doi.org/10.1128/IAI.63.9.3467-3472.1995

Haake, D.A. (2000). Spirochaetal lipoproteins and pathogenesis. Microbiology (Reading) *146 (Pt 7)*, 1491-1504. https://doi.org/10.1099/00221287-146-7-1491

Haas, R., and Meyer, T.F. (1986). The repertoire of silent pilus genes in *Neisseria gonorrhoeae*: evidence for gene conversion. Cell *44*, 107-115. https://doi.org/10.1016/0092-8674(86)90489-7

Hagan, C.L., Silhavy, T.J., and Kahne, D. (2011). β-Barrel membrane protein assembly by the Bam complex. Annu Rev Biochem *80*, 189-210. https://doi.org/10.1146/annurev-biochem-061408-144611

Hagblom, P., Segal, E., Billyard, E., and So, M. (1985). Intragenic recombination leads to pilus antigenic variation in *Neisseria gonorrhoeae*. Nature *315*, 156-158. https://doi.org/10.1038/315156a0

Hagman, K.E., Lahdenne, P., Popova, T.G., Porcella, S.F., Akins, D.R., Radolf, J.D., and Norgard, M.V. (1998). Decorin-binding protein of *Borrelia burgdorferi* is encoded within a two-gene operon and is protective in the murine model of Lyme borreliosis. Infect Immun *66*, 2674-2683. https://doi.org/10.1128/IAI.66.6.2674-2683.1998

Hallstrom, T., Siegel, C., Morgelin, M., Kraiczy, P., Skerka, C., and Zipfel, P.F. (2013). CspA from *Borrelia burgdorferi* inhibits the terminal complement pathway. mBio *4*. https://doi.org/10.1128/mBio.00481-13

Hammerschmidt, C., Koenigs, A., Siegel, C., Hallstrom, T., Skerka, C., Wallich, R., Zipfel, P.F., and Kraiczy, P. (2014). Versatile roles of CspA orthologs in complement inactivation of serum-resistant Lyme disease spirochetes. Infect Immun *82*, 380-392. https://doi.org/10.1128/IAI.01094-13

Hancock, R.E., and Benz, R. (1986). Demonstration and chemical modification of a specific phosphate binding site in the phosphate-starvation-inducible outer membrane porin protein P of *Pseudomonas aeruginosa*. Biochim Biophys Acta *860*, 699-707. https://doi.org/10.1016/0005-2736(86)90569-9

Hankamer, B., Morris, E., Nield, J., Gerle, C., and Barber, J. (2001). Three-dimensional structure of the photosystem II core dimer of higher plants determined by electron microscopy. J Struct Biol *135*, 262-269. https://doi.org/10.1006/jsbi.2001.4405

Hanson, M.S., Cassatt, D.R., Guo, B.P., Patel, N.K., McCarthy, M.P., Dorward, D.W., and Hook, M. (1998). Active and passive immunity against *Borrelia burgdorferi* decorin binding protein A (DbpA) protects against infection. Infect Immun *66*, 2143-2153. https://doi.org/10.1128/IAI.66.5.2143-2153.1998

Hara, T., Matsuyama, S., and Tokuda, H. (2003). Mechanism underlying the inner membrane retention of *Escherichia coli* lipoproteins caused by Lol avoidance signals. J Biol Chem *278*, 40408-40414. https://doi.org/10.1074/jbc.M307836200

Harada, K., Ando, M., Fujimura-Ito, T., Kasahara-Imamura, M., Sako, Y., Uchida, A., Ishida, Y., Kadota, H., Ohnishi, T., and Moseley, B.E. (2003). Mechanism of thermotolerance induction by split-dose hyperthermia in *Deinococcus radiodurans* DNA repair deficient mutants. Int J Mol Med *12*, 741-747.

Harman, M.W., Hamby, A.E., Boltyanskiy, R., Belperron, A.A., Bockenstedt, L.K., Kress, H., Dufresne, E.R., and Wolgemuth, C.W. (2017). Vancomycin reduces cell wall stiffness and slows swim speed of the Lyme disease bacterium. Biophys J *112*, 746-754. https://doi.org/10.1016/j.bpj.2016.12.039

Hayashi, S., and Wu, H.C. (1990). Lipoproteins in bacteria. J Bioenerg Biomembr *22*, 451-471. https://doi.org/10.1007/BF00763177

Hayashi, Y., Tsurumizu, R., Tsukahara, J., Takeda, K., Narita, S., Mori, M., Miki, K., and Tokuda, H. (2014). Roles of the protruding loop of factor B essential for the localization of lipoproteins (LolB) in the anchoring of bacterial triacylated proteins to the outer membrane. J Biol Chem *289*, 10530-10539. https://doi.org/10.1074/jbc.M113.539270

Heiss, L.N., Lancaster, J.R., Jr., Corbett, J.A., and Goldman, W.E. (1994). Epithelial autotoxicity of nitric oxide: role in the respiratory cytopathology of pertussis. Proc Natl Acad Sci U S A *91*, 267-270. https://doi.org/10.1073/pnas.91.1.267

Heiss, L.N., Moser, S.A., Unanue, E.R., and Goldman, W.E. (1993). Interleukin-1 is linked to the respiratory epithelial cytopathology of pertussis. Infect Immun *61*, 3123-3128. https://doi.org/10.1128/IAI.61.8.3123-3128.1993

Hoffmann, P., Heinle, S., Schade, U.F., Loppnow, H., Ulmer, A.J., Flad, H.D., Jung, G., and Bessler, W.G. (1988). Stimulation of human and murine adherent cells by bacterial lipoprotein and synthetic lipopeptide analogues. Immunobiology *177*, 158-170. https://doi.org/10.1016/S0171-2985(88)80036-6

Holtje, J.V. (1998). Growth of the stress-bearing and shape-maintaining murein sacculus of *Escherichia coli*. Microbiol Mol Biol Rev *62*, 181-203.

Hooda, Y., Lai, C.C., Judd, A., Buckwalter, C.M., Shin, H.E., Gray-Owen, S.D., and Moraes, T.F. (2016). Slam is an outer membrane protein that is required for the surface display of lipidated virulence factors in *Neisseria*. Nat Microbiol *1*, 16009. https://doi.org/10.1038/nmicrobiol.2016.9

Hossain, H., Wellensiek, H.J., Geyer, R., and Lochnit, G. (2001). Structural analysis of glycolipids from *Borrelia burgdorferi*. Biochimie *83*, 683-692. https://doi.org/10.1016/s0300-9084(01)01296-2

Howe, T.R., LaQuier, F.W., and Barbour, A.G. (1986). Organization of genes encoding two outer membrane proteins of the Lyme disease agent *Borrelia burgdorferi* within a single transcriptional unit. Infect Immun *54*, 207-212.

Hsu, Y.P., Rittichier, J., Kuru, E., Yablonowski, J., Pasciak, E., Tekkam, S., Hall, E., Murphy, B., Lee, T.K., Garner, E.C., *et al.* (2017). Full color palette of fluorescent d-amino acids for in situ labeling of bacterial cell walls. Chem Sci *8*, 6313-6321. https://doi.org/10.1039/c7sc01800b

Huang, Z., Toledo, A.M., Benach, J.L., and London, E. (2016). Ordered Membrane Domain-Forming Properties of the Lipids of *Borrelia burgdorferi*. Biophys J *111*, 2666-2675. https://doi.org/10.1016/j.bpj.2016.11.012

Iqbal, H., Kenedy, M.R., Lybecker, M., and Akins, D.R. (2016). The TamB ortholog of *Borrelia burgdorferi* interacts with the beta-barrel assembly machine (BAM) complex protein BamA. Mol Microbiol *102*, 757-774. https://doi.org/10.1111/mmi.13492

Irving, A.T., Mimuro, H., Kufer, T.A., Lo, C., Wheeler, R., Turner, L.J., Thomas, B.J., Malosse, C., Gantier, M.P., Casillas, L.N., *et al.* (2014). The immune receptor NOD1 and kinase RIP2 interact with bacterial peptidoglycan on early endosomes to promote autophagy and inflammatory signaling. Cell Host Microbe *15*, 623-635. https://doi.org/10.1016/j.chom.2014.04.001

Ito, H., Ura, A., Oyamada, Y., Yoshida, H., Yamagishi, J., Narita, S., Matsuyama, S., and Tokuda, H. (2007). A new screening method to identify inhibitors of the Lol (localization of lipoproteins) system, a novel antibacterial target. Microbiol Immunol *51*, 263-270. https://doi.org/10.1111/j.1348-0421.2007.tb03906.x

Ivleva, N.B., Shestakov, S.V., and Pakrasi, H.B. (2000). The carboxyl-terminal extension of the precursor D1 protein of photosystem II is required for optimal photosynthetic performance of the cyanobacterium *Synechocystis sp.* PCC 6803. Plant Physiol *124*, 1403-1412. https://doi.org/10.1104/pp.124.3.1403

Iyer, R., Caimano, M.J., Luthra, A., Axline, D., Jr., Corona, A., Iacobas, D.A., Radolf, J.D., and Schwartz, I. (2015). Stage-specific global alterations in the transcriptomes of Lyme disease spirochetes during tick feeding and following mammalian host adaptation. Mol Microbiol *95*, 509-538. https://doi.org/10.1111/mmi.12882

Jager, S.N., Porta, E.O.J., and Labadie, G.R. (2019). Convenient synthesis of the immunogenic glycolipid BbGL1. Steroids *141*, 41-45. https://doi.org/10.1016/j.steroids.2018.11.012

Jauris-Heipke, S., Liegl, G., Preac-Mursic, V., Rossler, D., Schwab, E., Soutschek, E., Will, G., and Wilske, B. (1995). Molecular analysis of genes encoding outer surface protein C (OspC) of *Borrelia burgdorferi* sensu lato: relationship to ospA genotype and evidence of lateral gene exchange of ospC. J Clin Microbiol *33*, 1860-1866. https://doi.org/10.1128/JCM.33.7.1860-1866.1995

Johnson, R.C., Schmid, G.P., Hyde, F.W., Steigerwalt, A.G., and Brenner, D.J. (1984). *Borrelia burgdorferi* sp. nov.: Ethiologic agent og Lyme disease. International Journal of SysteaticBacteriology, 496-497.

Jones, J.D., Bourell, K.W., Norgard, M.V., and Radolf, J.D. (1995). Membrane topology of *Borrelia burgdorferi* and *Treponema pallidum* lipoproteins. Infect Immun *63*, 2424-2434. https://doi.org/10.1128/IAI.63.7.2424-2434.1995

Jones, K.L., Seward, R.J., Ben-Menachem, G., Glickstein, L.J., Costello, C.E., and Steere, A.C. (2009). Strong IgG antibody responses to *Borrelia burgdorferi* glycolipids in patients with Lyme arthritis, a late manifestation of the infection. Clin Immunol *132*, 93-102. https://doi.org/10.1016/j.clim.2009.03.510

Jutras, B.L., Lochhead, R.B., Kloos, Z.A., Biboy, J., Strle, K., Booth, C.J., Govers, S.K., Gray, J., Schumann, P., Vollmer, W., *et al.* (2019). *Borrelia burgdorferi* peptidoglycan is a persistent antigen in patients with Lyme arthritis. Proc Natl Acad Sci U S A *116*, 13498-13507. https://doi.org/10.1073/pnas.1904170116

Jutras, B.L., Scott, M., Parry, B., Biboy, J., Gray, J., Vollmer, W., and Jacobs-Wagner, C. (2016). Lyme disease and relapsing fever *Borrelia* elongate through zones of peptidoglycan synthesis that mark division sites of daughter cells. Proc Natl Acad Sci U S A *113*, 9162-9170. https://doi.org/10.1073/pnas.1610805113

Kamio, Y., and Nikaido, H. (1976). Outer membrane of *Salmonella typhimurium*: accessibility of phospholipid head groups to phospholipase c and cyanogen bromide activated dextran in the external medium. Biochemistry *15*, 2561-2570. https://doi.org/10.1021/bi00657a012

Kariu, T., Sharma, K., Singh, P., Smith, A.A., Backstedt, B., Buyuktanir, O., and Pal, U. (2015). BB0323 and novel virulence determinant BB0238: *Borrelia burgdorferi* proteins that interact with and stabilize each other and are critical for infectivity. Journal of Infectious Diseases *211*, 462-471. https://doi.org/10.1093/infdis/jiu460

Kariu, T., Yang, X., Marks, C.B., Zhang, X., and Pal, U. (2013). Proteolysis of BB0323 results in two polypeptides that impact physiologic and infectious phenotypes in *Borrelia burgdorferi*. Mol Microbiol *88*, 510-522. https://doi.org/10.1111/mmi.12202

Karnauchov, I., Herrmann, R.G., Pakrasi, H.B., and Klosgen, R.B. (1997). Transport of CtpA protein from the cyanobacterium *Synechocystis* 6803 across the thylakoid membrane in chloroplasts. Eur J Biochem *249*, 497-504. https://doi.org/10.1111/j.1432-1033.1997.t01-1-00497.x

Katona, L.I., Beck, G., and Habicht, G.S. (1992). Purification and immunological characterization of a major low-molecular-weight lipoprotein from *Borrelia burgdorferi*. Infect Immun *60*, 4995-5003.

Kenedy, M.R., Luthra, A., Anand, A., Dunn, J.P., Radolf, J.D., and Akins, D.R. (2014). Structural modeling and physicochemical characterization provide evidence that P66 forms a beta-barrel in the *Borrelia burgdorferi* outer membrane. J Bacteriol *196*, 859-872. https://doi.org/10.1128/JB.01236-13

Kenedy, M.R., Scott, E.J., 2nd, Shrestha, B., Anand, A., Iqbal, H., Radolf, J.D., Dyer, D.W., and Akins, D.R. (2016). Consensus computational network analysis for identifying candidate outer membrane proteins from *Borrelia* spirochetes. BMC Microbiol *16*, 141. https://doi.org/10.1186/s12866-016-0762-z

Kenedy, M.R., Vuppala, S.R., Siegel, C., Kraiczy, P., and Akins, D.R. (2009). CspA-mediated binding of human factor H inhibits complement deposition and confers serum resistance in *Borrelia burgdorferi*. Infect Immun *77*, 2773-2782. https://doi.org/10.1128/IAI.00318-09

Khandke, K.M., Fairwell, T., Chait, B.T., and Manjula, B.N. (1989). Influence of ions on cyclization of the amino terminal glutamine residues of tryptic peptides of streptococcal PepM49 protein. Resolution of cyclized peptides by HPLC and characterization by mass spectrometry. Int J Pept Protein Res *34*, 118-123. https://doi.org/10.1111/j.1399-3011.1989.tb01499.x

Kihara, A., Akiyama, Y., and Ito, K. (1995). FtsH is required for proteolytic elimination of uncomplexed forms of SecY, an essential protein translocase subunit. Proc Natl Acad Sci U S A *92*, 4532-4536. https://doi.org/10.1073/pnas.92.10.4532

Kihara, A., Akiyama, Y., and Ito, K. (1997). Host regulation of lysogenic decision in bacteriophage lambda: transmembrane modulation of FtsH (HflB), the cII degrading protease, by HflKC (HflA). Proc Natl Acad Sci U S A *94*, 5544-5549. https://doi.org/10.1073/pnas.94.11.5544

Kim, S., Malinverni, J.C., Sliz, P., Silhavy, T.J., Harrison, S.C., and Kahne, D. (2007). Structure and function of an essential component of the outer membrane protein assembly machine. Science *317*, 961-964. https://doi.org/10.1126/science.1143993

Koebnik, R., Locher, K.P., and Van Gelder, P. (2000). Structure and function of bacterial outer membrane proteins: barrels in a nutshell. Mol Microbiol *37*, 239-253. https://doi.org/10.1046/j.1365-2958.2000.01983.x

Konovalova, A., Kahne, D.E., and Silhavy, T.J. (2017). Outer membrane biogenesis. Annual Reviews in Microbiology *71*, 539-556. https://doi.org/10.1146/annurev-micro-090816-093754

Koomey, M. (1997). Bacterial pathogenesis: a variation on variation in Lyme disease. Curr Biol *7*, R538-540. https://doi.org/10.1016/s0960-9822(06)00275-2

Kornacki, J.A., and Oliver, D.B. (1998). Lyme disease-causing *Borrelia* species encode multiple lipoproteins homologous to peptide-binding proteins of ABC-type transporters. Infect Immun *66*, 4115-4122.

Koronakis, V., Sharff, A., Koronakis, E., Luisi, B., and Hughes, C. (2000). Crystal structure of the bacterial membrane protein TolC central to multidrug efflux and protein export. Nature *405*, 914-919. https://doi.org/10.1038/35016007

Kraiczy, P., Skerka, C., Kirschfink, M., Zipfel, P.F., and Brade, V. (2001). Mechanism of complement resistance of pathogenic *Borrelia burgdorferi* isolates. Int Immunopharmacol *1*, 393-401. https://doi.org/10.1016/s1567-5769(00)00041-2

Kudryashev, M., Cyrklaff, M., Baumeister, W., Simon, M.M., Wallich, R., and Frischknecht, F. (2009). Comparative cryo-electron tomography of pathogenic Lyme disease spirochetes. Mol Microbiol *71*, 1415-1434. https://doi.org/10.1111/j.1365-2958.2009.06613.x

Kumar, D., Ristow, L.C., Shi, M., Mukherjee, P., Caine, J.A., Lee, W.Y., Kubes, P., Coburn, J., and Chaconas, G. (2015). Intravital Imaging of Vascular Transmigration by the Lyme Spirochete: Requirement for the integrin binding residues of the *B. burgdorferi* P66 protein. PLoS Pathog *11*, e1005333. https://doi.org/10.1371/journal.ppat.1005333

Kumaran, D., Eswaramoorthy, S., Luft, B.J., Koide, S., Dunn, J.J., Lawson, C.L., and Swaminathan, S. (2001). Crystal structure of outer surface protein C (OspC) from the Lyme disease spirochete, *Borrelia burgdorferi*. EMBO J *20*, 971-978. https://doi.org/10.1093/emboj/20.5.971

Kumru, O.S., Bunikis, I., Sorokina, I., Bergström, S., and Zückert, W.R. (2011a). Specificity and role of the *Borrelia burgdorferi* CtpA protease in outer membrane protein processing. J Bacteriol *193*, 5759-5765. https://doi.org/10.1128/JB.05622-11

Kumru, O.S., Schulze, R.J., Rodnin, M.V., Ladokhin, A.S., and Zückert, W.R. (2011b). Surface localization determinants of *Borrelia* OspC/Vsp family lipoproteins. J Bacteriol *193*, 2814-2825. https://doi.org/10.1128/JB.00015-11

Kumru, O.S., Schulze, R.J., Slusser, J.G., and Zückert, W.R. (2010). Development and validation of a FACS-based lipoprotein localization screen in the Lyme disease spirochete *Borrelia burgdorferi*. BMC Microbiol *10*, 277. https://doi.org/10.1186/1471-2180-10-277

Kuru, E., Hughes, H.V., Brown, P.J., Hall, E., Tekkam, S., Cava, F., de Pedro, M.A., Brun, Y.V., and VanNieuwenhze, M.S. (2012). *In situ* probing of newly synthesized peptidoglycan in live bacteria with fluorescent D-amino acids. Angew Chem Int Ed Engl *51*, 12519-12523. https://doi.org/10.1002/anie.201206749

Kuru, E., Radkov, A., Meng, X., Egan, A., Alvarez, L., Dowson, A., Booher, G., Breukink, E., Roper, D.I., Cava, F., *et al.* (2019). Mechanisms of incorporation for D-Amino acid probes that target peptidoglycan biosynthesis. ACS Chem Biol *14*, 2745-2756. https://doi.org/10.1021/acschembio.9b00664

Lagal, V., Portnoi, D., Faure, G., Postic, D., and Baranton, G. (2006). *Borrelia burgdorferi* sensu stricto invasiveness is correlated with OspC-plasminogen affinity. Microbes Infect *8*, 645-652. https://doi.org/10.1016/j.micinf.2005.08.017

Lagal, V., Postic, D., Ruzic-Sabljic, E., and Baranton, G. (2003). Genetic diversity among *Borrelia* strains determined by single-strand conformation polymorphism analysis of the *ospC* gene and its association with invasiveness. J Clin Microbiol *41*, 5059-5065. https://doi.org/10.1128/jcm.41.11.5059-5065.2003

Lahdenne, P., Porcella, S., Hagman, K., Akins, D., Popova, T., Cox, D., Katona, L., Radolf, J., and Norgard, M. (1997). Molecular characterization of a 6.6-kilodalton *Borrelia burgdorferi* outer membrane-associated lipoprotein (lp6.6) which appears to be downregulated during mammalian infection. Infect Immun *65*, 412-421.

LaRocca, T.J., Crowley, J.T., Cusack, B.J., Pathak, P., Benach, J., London, E., Garcia-Monco, J.C., and Benach, J.L. (2010). Cholesterol lipids of *Borrelia burgdorferi* form lipid rafts and are required for the bactericidal activity of a complement-independent antibody. Cell Host Microbe *8*, 331-342. https://doi.org/10.1016/j.chom.2010.09.001

LaRocca, T.J., Holthausen, D.J., Hsieh, C., Renken, C., Mannella, C.A., and Benach, J.L. (2009). The bactericidal effect of a complement-independent antibody is osmolytic and specific to *Borrelia*. Proc Natl Acad Sci U S A *106*, 10752-10757. https://doi.org/10.1073/pnas.0901858106

LaRocca, T.J., Pathak, P., Chiantia, S., Toledo, A., Silvius, J.R., Benach, J.L., and London, E. (2013). Proving lipid rafts exist: membrane domains in the prokaryote *Borrelia burgdorferi* have the same properties as eukaryotic lipid rafts. PLoS Pathog *9*, e1003353. https://doi.org/10.1371/journal.ppat.1003353

Lawson, C.L., Yung, B.H., Barbour, A.G., and Zückert, W.R. (2006). Crystal structure of neurotropism-associated variable surface protein 1 (Vsp1) of *Borrelia turicatae*. J Bacteriol *188*, 4522-4530. https://doi.org/10.1128/JB.00028-06

Lazazzera, B.A. (2001). The intracellular function of extracellular signaling peptides. Peptides *22*, 1519-1527. https://doi.org/10.1016/s0196-9781(01)00488-0

Lenhart, T.R., and Akins, D.R. (2010). *Borrelia burgdorferi* locus BB0795 encodes a BamA orthologue required for growth and efficient localization of outer membrane proteins. Mol Microbiol *75*, 692-709. https://doi.org/10.1111/j.1365-2958.2009.07015.x

Lenhart, T.R., Kenedy, M.R., Yang, X., Pal, U., and Akins, D.R. (2012). BB0324 and BB0028 are constituents of the *Borrelia burgdorferi* beta-barrel assembly machine (BAM) complex. BMC Microbiol *12*, 60. https://doi.org/10.1186/1471-2180-12-60

Lenz, J.D., Hackett, K.T., and Dillard, J.P. (2017). A Single Dual-Function Enzyme Controls the Production of Inflammatory NOD Agonist Peptidoglycan Fragments by *Neisseria gonorrhoeae*. mBio *8*. https://doi.org/10.1128/mBio.01464-17

Leonard, B.A., Podbielski, A., Hedberg, P.J., and Dunny, G.M. (1996). *Enterococcus faecalis* pheromone binding protein, PrgZ, recruits a chromosomal oligopeptide permease system to import sex pheromone cCF10 for induction of conjugation. Proc Natl Acad Sci U S A *93*, 260-264. https://doi.org/10.1073/pnas.93.1.260

Lescot, M., Audic, S., Robert, C., Nguyen, T.T., Blanc, G., Cutler, S.J., Wincker, P., Couloux, A., Claverie, J.M., Raoult, D., *et al.* (2008). The genome of *Borrelia recurrentis*, the agent of deadly louse-borne relapsing fever, is a degraded subset of tick-borne *Borrelia duttonii*. PLoS Genet *4*, e1000185. https://doi.org/10.1371/journal.pgen.1000185

Levin, P.A., Kurtser, I.G., and Grossman, A.D. (1999). Identification and characterization of a negative regulator of FtsZ ring formation in *Bacillus subtilis*. Proc Natl Acad Sci U S A *96*, 9642-9647. https://doi.org/10.1073/pnas.96.17.9642

Lewenza, S., Mhlanga, M.M., and Pugsley, A.P. (2008). Novel inner membrane retention signals in *Pseudomonas aeruginosa* lipoproteins. J Bacteriol *190*, 6119-6125. https://doi.org/10.1128/JB.00603-08

Lewenza, S., Vidal-Ingigliardi, D., and Pugsley, A.P. (2006). Direct visualization of red fluorescent lipoproteins indicates conservation of the membrane sorting rules in the family *Enterobacteriaceae*. J Bacteriol *188*, 3516-3524. https://doi.org/10.1128/JB.188.10.3516-3524.2006

Lex, A., Wiesmuller, K.H., Jung, G., and Bessler, W.G. (1986). A synthetic analogue of *Escherichia coli* lipoprotein, tripalmitoyl pentapeptide, constitutes a potent immune adjuvant. J Immunol *137*, 2676-2681.

Li, H., Dunn, J.J., Luft, B.J., and Lawson, C.L. (1997). Crystal structure of Lyme disease antigen outer surface protein A complexed with an Fab. Proc Natl Acad Sci U S A *94*, 3584-3589. https://doi.org/10.1073/pnas.94.8.3584

Li, H., and Lawson, C.L. (1995). Crystallization and preliminary X-ray analysis of *Borrelia burgdorferi* outer surface protein A (OspA) complexed with a murine monoclonal antibody Fab fragment. J Struct Biol *115*, 335-337. https://doi.org/10.1006/jsbi.1995.1058

Li, X.Z., and Nikaido, H. (2004). Efflux-mediated drug resistance in bacteria. Drugs *64*, 159-204. https://doi.org/10.2165/00003495-200464020-00004

Lilly, A.A., Crane, J.M., and Randall, L.L. (2009). Export chaperone SecB uses one surface of interaction for diverse unfolded polypeptide ligands. Protein Sci *18*, 1860-1868. https://doi.org/10.1002/pro.197

Lin, B., Short, S.A., Eskildsen, M., Klempner, M.S., and Hu, L.T. (2001). Functional testing of putative oligopeptide permease (Opp) proteins of *Borrelia burgdorferi*: a complementation model in *opp(-) Escherichia coli*. Biochim Biophys Acta *1499*, 222-231. https://doi.org/10.1016/s0167-4889(00)00121-x

Lin, T., Gao, L., Edmondson, D.G., Jacobs, M.B., Philipp, M.T., and Norris, S.J. (2009). Central role of the Holliday junction helicase RuvAB in vlsE recombination and infectivity of *Borrelia burgdorferi*. PLoS Pathog *5*, e1000679. https://doi.org/10.1371/journal.ppat.1000679

Lin, T., Gao, L., Zhang, C., Odeh, E., Jacobs, M.B., Coutte, L., Chaconas, G., Philipp, M.T., and Norris, S.J. (2012). Analysis of an ordered, comprehensive STM mutant library in infectious *Borrelia burgdorferi*: insights into the genes required for mouse infectivity. PLoS One *7*, e47532. https://doi.org/10.1371/journal.pone.0047532

Lin, T., Troy, E.B., Hu, L.T., Gao, L., and Norris, S.J. (2014). Transposon mutagenesis as an approach to improved understanding of *Borrelia* pathogenesis and biology. Front Cell Infect Microbiol *4*, 63. https://doi.org/10.3389/fcimb.2014.00063

Lin, Y.P., Chen, Q., Ritchie, J.A., Dufour, N.P., Fischer, J.R., Coburn, J., and Leong, J.M. (2015). Glycosaminoglycan binding by *Borrelia burgdorferi* adhesin BBK32 specifically and uniquely promotes joint colonization. Cell Microbiol *17*, 860-875. https://doi.org/10.1111/cmi.12407

Lin, Y.P., Frye, A.M., Nowak, T.A., and Kraiczy, P. (2020a). New Insights Into CRASP-Mediated Complement Evasion in the Lyme Disease Enzootic Cycle. Front Cell Infect Microbiol *10*, 1. https://doi.org/10.3389/fcimb.2020.00001

Lin, Y.P., Tan, X., Caine, J.A., Castellanos, M., Chaconas, G., Coburn, J., and Leong, J.M. (2020b). Strain-specific joint invasion and colonization by Lyme disease spirochetes is promoted by outer surface protein C. PLoS Pathog *16*, e1008516. https://doi.org/10.1371/journal.ppat.1008516

Liu, G., Topping, T.B., and Randall, L.L. (1989). Physiological role during export for the retardation of folding by the leader peptide of maltose-binding protein. Proc Natl Acad Sci U S A *86*, 9213-9217. https://doi.org/10.1073/pnas.86.23.9213

Livermore, B.P., Bey, R.F., and Johnson, R.C. (1978). Lipid metabolism of *Borrelia hermsii*. Infect Immun *20*, 215-220. https://doi.org/10.1128/IAI.20.1.215-220.1978

Livey, I., Gibbs, C.P., Schuster, R., and Dorner, F. (1995). Evidence for lateral transfer and recombination in OspC variation in Lyme disease *Borrelia*. Mol Microbiol *18*, 257-269. https://doi.org/10.1111/j.1365-2958.1995.mmi_18020257.x

London, E. (2002). Insights into lipid raft structure and formation from experiments in model membranes. Curr Opin Struct Biol *12*, 480-486. https://doi.org/10.1016/s0959-440x(02)00351-2

Lone, A.G., and Bankhead, T. (2020). The *Borrelia burgdorferi* VlsE Lipoprotein Prevents Antibody Binding to an Arthritis-Related Surface Antigen. Cell Rep *30*, 3663-3670 e3665. https://doi.org/10.1016/j.celrep.2020.02.081

Lopez, D., and Kolter, R. (2010). Functional microdomains in bacterial membranes. Genes Dev *24*, 1893-1902. https://doi.org/10.1101/gad.1945010

Malinverni, J.C., Werner, J., Kim, S., Sklar, J.G., Kahne, D., Misra, R., and Silhavy, T.J. (2006). YfiO stabilizes the YaeT complex and is essential for outer membrane protein assembly in *Escherichia coli*. Mol Microbiol *61*, 151-164. https://doi.org/10.1111/j.1365-2958.2006.05211.x

Masuda, K., Matsuyama, S., and Tokuda, H. (2002). Elucidation of the function of lipoprotein-sorting signals that determine membrane localization. Proc Natl Acad Sci U S A *99*, 7390-7395.

Matsuyama, S., Tajima, T., and Tokuda, H. (1995). A novel periplasmic carrier protein involved in the sorting and transport of *Escherichia coli* lipoproteins destined for the outer membrane. EMBO J *14*, 3365-3372.

Matsuyama, S., Yokota, N., and Tokuda, H. (1997). A novel outer membrane lipoprotein, LolB (HemM), involved in the LolA (p20)-dependent localization of lipoproteins to the outer membrane of *Escherichia coli*. EMBO J *16*, 6947-6955. https://doi.org/10.1093/emboj/16.23.6947

Medrano, M.S., Ding, Y., Wang, X.G., Lu, P., Coburn, J., and Hu, L.T. (2007). Regulators of expression of the oligopeptide permease A proteins of *Borrelia burgdorferi*. J Bacteriol *189*, 2653-2659. https://doi.org/10.1128/JB.01760-06

Medzhitov, R., and Janeway, C., Jr. (2000). Innate immune recognition: mechanisms and pathways. Immunol Rev *173*, 89-97. https://doi.org/10.1034/j.1600-065x.2000.917309.x

Melly, M.A., McGee, Z.A., and Rosenthal, R.S. (1984). Ability of monomeric peptidoglycan fragments from *Neisseria gonorrhoeae* to damage human fallopian-tube mucosa. Journal of Infectious Diseases *149*, 378-386. https://doi.org/10.1093/infdis/149.3.378

Mitchell, S.J., and Minnick, M.F. (1997). A carboxy-terminal processing protease gene is located immediately upstream of the invasion-associated locus from *Bartonella bacilliformis*. Microbiology (Reading) *143 (Pt 4)*, 1221-1233. https://doi.org/10.1099/00221287-143-4-1221

Miyamoto, A., Matsuyama, S., and Tokuda, H. (2001). Mutant of LolA, a lipoprotein-specific molecular chaperone of *Escherichia coli*, defective in the transfer of lipoproteins to LolB. Biochem Biophys Res Commun *287*, 1125-1128. https://doi.org/10.1006/bbrc.2001.5705

Mizutani, M., Mukaiyama, K., Xiao, J., Mori, M., Satou, R., Narita, S., Okuda, S., and Tokuda, H. (2013). Functional differentiation of structurally similar membrane subunits of the ABC transporter LolCDE complex. FEBS Lett *587*, 23-29. https://doi.org/10.1016/j.febslet.2012.11.009

Moffett, S., Brown, D.A., and Linder, M.E. (2000). Lipid-dependent targeting of G proteins into rafts. J Biol Chem *275*, 2191-2198. https://doi.org/10.1074/jbc.275.3.2191

Monnet, V. (2003). Bacterial oligopeptide-binding proteins. Cell Mol Life Sci *60*, 2100-2114. https://doi.org/10.1007/s00018-003-3054-3

Moore, D.A., Whatley, Z.N., Joshi, C.P., Osawa, M., and Erickson, H.P. (2017). Probing for Binding Regions of the FtsZ Protein Surface through Site-Directed Insertions: Discovery of Fully Functional FtsZ-Fluorescent Proteins. J Bacteriol *199*. https://doi.org/10.1128/JB.00553-16

Motaleb, M.A., Corum, L., Bono, J.L., Elias, A.F., Rosa, P., Samuels, D.S., and Charon, N.W. (2000). *Borrelia burgdorferi* periplasmic flagella have both skeletal and motility functions. Proc Natl Acad Sci U S A *97*, 10899-10904. https://doi.org/10.1073/pnas.200221797

Narita, S., and Tokuda, H. (2011). Overexpression of LolCDE allows deletion of the *Escherichia coli* gene encoding apolipoprotein N-acyltransferase. J Bacteriol *193*, 4832-4840. https://doi.org/10.1128/JB.05013-11

Nascimento, A.L., Ko, A.I., Martins, E.A., Monteiro-Vitorello, C.B., Ho, P.L., Haake, D.A., Verjovski-Almeida, S., Hartskeerl, R.A., Marques, M.V., Oliveira, M.C., *et al.* (2004). Comparative genomics of two *Leptospira interrogans* serovars reveals novel insights into physiology and pathogenesis. J Bacteriol *186*, 2164-2172. https://doi.org/10.1128/jb.186.7.2164-2172.2004

Nikaido, H. (2003). Molecular basis of bacterial outer membrane permeability revisited. Microbiol Mol Biol Rev *67*, 593-656. https://doi.org/10.1128/mmbr.67.4.593-656.2003

Nilsson, C.L., Cooper, H.J., Hakansson, K., Marshall, A.G., Östberg, Y., Lavrinovicha, M., and Bergström, S. (2002). Characterization of the P13 membrane protein of *Borrelia burgdorferi* by mass spectrometry. J Am Soc Mass Spectrom *13*, 295-299. https://doi.org/10.1016/S1044-0305(01)00365-8

Nixon, P.J., Trost, J.T., and Diner, B.A. (1992). Role of the carboxy terminus of polypeptide D1 in the assembly of a functional water-oxidizing manganese cluster in photosystem II of the cyanobacterium *Synechocystis* sp. PCC 6803: assembly requires a free carboxyl group at C-terminal position 344. Biochemistry *31*, 10859-10871. https://doi.org/10.1021/bi00159a029

Noppa, L., Östberg, Y., Lavrinovicha, M., and Bergström, S. (2001). P13, an integral membrane protein of *Borrelia burgdorferi*, is C-terminally processed and contains surface-exposed domains. Infect Immun *69*, 3323-3334. https://doi.org/10.1128/IAI.69.5.3323-3334.2001

Norman, M.U., Moriarty, T.J., Dresser, A.R., Millen, B., Kubes, P., and Chaconas, G. (2008). Molecular mechanisms involved in vascular interactions of the Lyme disease pathogen in a living host. PLoS Pathog *4*, e1000169. https://doi.org/10.1371/journal.ppat.1000169

Oguchi, Y., Takeda, K., Watanabe, S., Yokota, N., Miki, K., and Tokuda, H. (2008). Opening and closing of the hydrophobic cavity of LolA coupled to lipoprotein binding and release. J Biol Chem *283*, 25414-25420. https://doi.org/10.1074/jbc.M804736200

Ohnishi, J., Piesman, J., and de Silva, A.M. (2001). Antigenic and genetic heterogeneity of *Borrelia burgdorferi* populations transmitted by ticks. Proc Natl Acad Sci U S A *98*, 670-675. https://doi.org/10.1073/pnas.98.2.670

Ojaimi, C., Brooks, C., Casjens, S., Rosa, P., Elias, A., Barbour, A., Jasinskas, A., Benach, J., Katona, L., Radolf, J., *et al.* (2003). Profiling of temperature-induced changes in *Borrelia burgdorferi* gene expression by using whole genome arrays. Infect Immun *71*, 1689-1705. https://doi.org/10.1128/iai.71.4.1689-1705.2003

Ojaimi, C., Mulay, V., Liveris, D., Iyer, R., and Schwartz, I. (2005). Comparative transcriptional profiling of *Borrelia burgdorferi* clinical isolates differing in capacities for hematogenous dissemination. Infect Immun *73*, 6791-6802. https://doi.org/10.1128/IAI.73.10.6791-6802.2005

Okuda, S., and Tokuda, H. (2009). Model of mouth-to-mouth transfer of bacterial lipoproteins through inner membrane LolC, periplasmic LolA, and outer membrane LolB. Proc Natl Acad Sci U S A *106*, 5877-5882. https://doi.org/10.1073/pnas.0900896106

Okuda, S., and Tokuda, H. (2011). Lipoprotein sorting in bacteria. Annual Reviews in Microbiology *65*, 239-259. https://doi.org/10.1146/annurev-micro-090110-102859

Okuda, S., Watanabe, S., and Tokuda, H. (2008). A short helix in the C-terminal region of LolA is important for the specific membrane localization of lipoproteins. FEBS Lett *582*, 2247-2251. https://doi.org/10.1016/j.febslet.2008.05.022

Onufryk, C., Crouch, M.L., Fang, F.C., and Gross, C.A. (2005). Characterization of six lipoproteins in the sigmaE regulon. J Bacteriol *187*, 4552-4561. https://doi.org/10.1128/JB.187.13.4552-4561.2005

Östberg, Y., Berg, S., Comstedt, P., Wieslander, A., and Bergström, S. (2007). Functional analysis of a lipid galactosyltransferase synthesizing the major envelope lipid in the Lyme disease spirochete *Borrelia burgdorferi*. FEMS Microbiol Lett *272*, 22-29. https://doi.org/10.1111/j.1574-6968.2007.00728.x

Östberg, Y., Carroll, J.A., Pinne, M., Krum, J.G., Rosa, P., and Bergström, S. (2004). Pleiotropic effects of inactivating a carboxyl-terminal protease, CtpA, in *Borrelia burgdorferi*. J Bacteriol *186*, 2074-2084. https://doi.org/10.1128/jb.186.7.2074-2084.2004

Östberg, Y., Pinne, M., Benz, R., Rosa, P., and Bergström, S. (2002). Elimination of channel-forming activity by insertional inactivation of the p13 gene in *Borrelia burgdorferi*. J Bacteriol *184*, 6811-6819. https://doi.org/10.1128/jb.184.24.6811-6819.2002

Pal, U., de Silva, A., Montgomery, R., Fish, D., Anguita, J., Anderson, J., Lobet, Y., and Fikrig, E. (2000). Attachment of *Borrelia burgdorferi* within Ixodes scapularis mediated by outer surface protein A. J Clin Invest *106*, 561-569.

Pal, U., Li, X., Wang, T., Montgomery, R.R., Ramamoorthi, N., Desilva, A.M., Bao, F., Yang, X., Pypaert, M., Pradhan, D., *et al.* (2004a). TROSPA, an *Ixodes scapularis* receptor for *Borrelia burgdorferi*. Cell *119*, 457-468. https://doi.org/10.1016/j.cell.2004.10.027

Pal, U., Yang, X., Chen, M., Bockenstedt, L.K., Anderson, J.F., Flavell, R.A., Norgard, M.V., and Fikrig, E. (2004b). OspC facilitates *Borrelia burgdorferi* invasion of *Ixodes scapularis* salivary glands. J Clin Invest *113*, 220-230. https://doi.org/10.1172/JCI19894

Park, S., Liu, G., Topping, T.B., Cover, W.H., and Randall, L.L. (1988). Modulation of folding pathways of exported proteins by the leader sequence. Science *239*, 1033-1035. https://doi.org/10.1126/science.3278378

Parveen, N., Caimano, M., Radolf, J.D., and Leong, J.M. (2003). Adaptation of the Lyme disease spirochaete to the mammalian host environment results in enhanced glycosaminoglycan and host cell binding. Mol Microbiol *47*, 1433-1444. https://doi.org/10.1046/j.1365-2958.2003.03388.x

Parveen, N., and Leong, J.M. (2000). Identification of a candidate glycosaminoglycan-binding adhesin of the Lyme disease spirochete *Borrelia burgdorferi*. Mol Microbiol *35*, 1220-1234. https://doi.org/10.1046/j.1365-2958.2000.01792.x

Paulsen, I.T., Nguyen, L., Sliwinski, M.K., Rabus, R., and Saier, M.H., Jr. (2000). Microbial genome analyses: comparative transport capabilities in eighteen prokaryotes. J Mol Biol *301*, 75-100. https://doi.org/10.1006/jmbi.2000.3961

Payne, J.W., and Gilvarg, C. (1968). Size restriction on peptide utilization in *Escherichia coli*. J Biol Chem *243*, 6291-6299.

Perego, M., Higgins, C.F., Pearce, S.R., Gallagher, M.P., and Hoch, J.A. (1991). The oligopeptide transport system of *Bacillus subtilis* plays a role in the initiation of sporulation. Mol Microbiol *5*, 173-185. https://doi.org/10.1111/j.1365-2958.1991.tb01838.x

Peters, K., Schweizer, I., Beilharz, K., Stahlmann, C., Veening, J.W., Hakenbeck, R., and Denapaite, D. (2014). *Streptococcus pneumoniae* PBP2x mid-cell localization requires the C-terminal PASTA domains and is essential for cell shape maintenance. Mol Microbiol *92*, 733-755. https://doi.org/10.1111/mmi.12588

Pinho, M.G., Kjos, M., and Veening, J.W. (2013). How to get (a)round: mechanisms controlling growth and division of coccoid bacteria. Nat Rev Microbiol *11*, 601-614. https://doi.org/10.1038/nrmicro3088

Pinne, M., Denker, K., Nilsson, E., Benz, R., and Bergström, S. (2006). The BBA01 protein, a member of paralog family 48 from *Borrelia burgdorferi*, is potentially interchangeable with the channel-forming protein P13. J Bacteriol *188*, 4207-4217. https://doi.org/10.1128/JB.00302-06

Pinne, M., Östberg, Y., Comstedt, P., and Bergström, S. (2004). Molecular analysis of the channel-forming protein P13 and its paralogue family 48 from different Lyme disease *Borrelia* species. Microbiology (Reading) *150*, 549-559. https://doi.org/10.1099/mic.0.26728-0

Pinne, M., Thein, M., Denker, K., Benz, R., Coburn, J., and Bergström, S. (2007). Elimination of channel-forming activity by insertional inactivation of the p66 gene in *Borrelia burgdorferi*. FEMS Microbiol Lett *266*, 241-249. https://doi.org/10.1111/j.1574-6968.2006.00529.x

Popova, A.V., and Hincha, D.K. (2011). Thermotropic phase behavior and headgroup interactions of the nonbilayer lipids phosphatidylethanolamine and monogalactosyldiacylglycerol in the dry state. BMC Biophys *4*, 11. https://doi.org/10.1186/2046-1682-4-11

Pozsgay, V., Kubler-Kielb, J., Coxon, B., Marques, A., Robbins, J.B., and Schneerson, R. (2011). Synthesis and antigenicity of BBGL-2 glycolipids of *Borrelia burgdorferi*, the causative agent of Lyme disease. Carbohydr Res *346*, 1551-1563. https://doi.org/10.1016/j.carres.2011.04.045

Probert, W.S., Allsup, K.M., and LeFebvre, R.B. (1995). Identification and characterization of a surface-exposed, 66-kilodalton protein from *Borrelia burgdorferi*. Infect Immun *63*, 1933-1939. https://doi.org/10.1128/IAI.63.5.1933-1939.1995

Probert, W.S., and Johnson, B.J. (1998). Identification of a 47 kDa fibronectin-binding protein expressed by *Borrelia burgdorferi* isolate B31. Mol Microbiol *30*, 1003-1015. https://doi.org/10.1046/j.1365-2958.1998.01127.x

Promnares, K., Kumar, M., Shroder, D.Y., Zhang, X., Anderson, J.F., and Pal, U. (2009). *Borrelia burgdorferi* small lipoprotein Lp6.6 is a member of multiple protein complexes in the outer membrane and facilitates pathogen transmission from ticks to mice. Mol Microbiol *74*, 112-125. https://doi.org/10.1111/j.1365-2958.2009.06853.x

Pugsley, A.P. (1993). The complete general secretory pathway in gram-negative bacteria. Microbiol Rev *57*, 50-108.

Pugsley, A.P., Kornacker, M.G., and Ryter, A. (1990). Analysis of the subcellular location of pullulanase produced by *Escherichia coli* carrying the pulA gene from Klebsiella pneumoniae strain UNF5023. Mol Microbiol *4*, 59-72. https://doi.org/10.1111/j.1365-2958.1990.tb02015.x

Radolf, J.D., Arndt, L.L., Akins, D.R., Curetty, L.L., Levi, M.E., Shen, Y., Davis, L.S., and Norgard, M.V. (1995). *Treponema pallidum* and *Borrelia burgdorferi* lipoproteins and synthetic lipopeptides activate monocytes/macrophages. J Immunol *154*, 2866-2877.

Radolf, J.D., Bourell, K.W., Akins, D.R., Brusca, J.S., and Norgard, M.V. (1994). Analysis of *Borrelia burgdorferi* membrane architecture by freeze-fracture electron microscopy. J Bacteriol *176*, 21-31. https://doi.org/10.1128/jb.176.1.21-31.1994

Radolf, J.D., and Caimano, M.J. (2008). The long strange trip of *Borrelia burgdorferi* outer-surface protein C. Mol Microbiol *69*, 1-4. https://doi.org/10.1111/j.1365-2958.2008.06226.x

Ramamoorthi, N., Narasimhan, S., Pal, U., Bao, F., Yang, X.F., Fish, D., Anguita, J., Norgard, M.V., Kantor, F.S., Anderson, J.F., *et al.* (2005). The Lyme disease agent exploits a tick protein to infect the mammalian host. Nature *436*, 573-577. https://doi.org/10.1038/nature03812

Ramamoorthy, R., and Philipp, M.T. (1998). Differential expression of *Borrelia burgdorferi* proteins during growth in vitro. Infect Immun *66*, 5119-5124. https://doi.org/10.1128/IAI.66.11.5119-5124.1998

Randall, L.L., and Hardy, S.J. (2002). SecB, one small chaperone in the complex milieu of the cell. Cell Mol Life Sci *59*, 1617-1623. https://doi.org/10.1007/pl00012488

Remans, K., Pauwels, K., van Ulsen, P., Buts, L., Cornelis, P., Tommassen, J., Savvides, S.N., Decanniere, K., and Van Gelder, P. (2010). Hydrophobic surface patches on LolA of *Pseudomonas aeruginosa* are essential for lipoprotein binding. J Mol Biol *401*, 921-930. https://doi.org/10.1016/j.jmb.2010.06.067

Ristow, L.C., Bonde, M., Lin, Y.P., Sato, H., Curtis, M., Wesley, E., Hahn, B.L., Fang, J., Wilcox, D.A., Leong, J.M., *et al.* (2015). Integrin binding by *Borrelia burgdorferi* P66 facilitates dissemination but is not required for infectivity. Cell Microbiol *17*, 1021-1036. https://doi.org/10.1111/cmi.12418

Ristow, L.C., Miller, H.E., Padmore, L.J., Chettri, R., Salzman, N., Caimano, M.J., Rosa, P.A., and Coburn, J. (2012). The beta(3)-integrin ligand of *Borrelia burgdorferi* is critical for infection of mice but not ticks. Mol Microbiol *85*, 1105-1118. https://doi.org/10.1111/j.1365-2958.2012.08160.x

Rivera-Milla, E., Stuermer, C.A., and Malaga-Trillo, E. (2006). Ancient origin of reggie (flotillin), reggie-like, and other lipid-raft proteins: convergent evolution of the SPFH domain. Cell Mol Life Sci *63*, 343-357. https://doi.org/10.1007/s00018-005-5434-3

Rizzitello, A.E., Harper, J.R., and Silhavy, T.J. (2001). Genetic evidence for parallel pathways of chaperone activity in the periplasm of *Escherichia coli*. J Bacteriol *183*, 6794-6800. https://doi.org/10.1128/JB.183.23.6794-6800.2001

Robert, V., Volokhina, E.B., Senf, F., Bos, M.P., Van Gelder, P., and Tommassen, J. (2006). Assembly factor Omp85 recognizes its outer membrane protein substrates by a species-specific C-terminal motif. PLoS Biol *4*, e377. https://doi.org/10.1371/journal.pbio.0040377

Rogovskyy, A.S., Casselli, T., Tourand, Y., Jones, C.R., Owen, J.P., Mason, K.L., Scoles, G.A., and Bankhead, T. (2015). Evaluation of the Importance of VlsE Antigenic Variation for the Enzootic Cycle of *Borrelia burgdorferi*. PLoS One *10*, e0124268. https://doi.org/10.1371/journal.pone.0124268

Rudner, D.Z., LeDeaux, J.R., Ireton, K., and Grossman, A.D. (1991). The spo0K locus of *Bacillus subtilis* is homologous to the oligopeptide permease locus and is required for sporulation and competence. J Bacteriol *173*, 1388-1398. https://doi.org/10.1128/jb.173.4.1388-1398.1991

Russell, T.M., Delorey, M.J., and Johnson, B.J. (2013). *Borrelia burgdorferi* BbHtrA degrades host ECM proteins and stimulates release of inflammatory cytokines in vitro. Mol Microbiol *90*, 241-251. https://doi.org/10.1111/mmi.12377

Russell, T.M., and Johnson, B.J. (2013). Lyme disease spirochaetes possess an aggrecan-binding protease with aggrecanase activity. Mol Microbiol *90*, 228-240. https://doi.org/10.1111/mmi.12276

Sadziene, A., Thomas, D.D., and Barbour, A.G. (1995). *Borrelia burgdorferi* mutant lacking Osp: biological and immunological characterization. Infect Immun *63*, 1573-1580. https://doi.org/10.1128/IAI.63.4.1573-1580.1995

Saier, M.H., Jr., and Paulsen, I.T. (2000). Whole genome analyses of transporters in spirochetes: *Borrelia burgdorferi* and *Treponema pallidum*. J Mol Microbiol Biotechnol 2, 393-399.

Salah Ud-Din, A.I.M., and Roujeinikova, A. (2017). Methyl-accepting chemotaxis proteins: a core sensing element in prokaryotes and archaea. Cell Mol Life Sci *74*, 3293-3303. https://doi.org/10.1007/s00018-017-2514-0

Samuelson, J.C., Chen, M., Jiang, F., Moller, I., Wiedmann, M., Kuhn, A., Phillips, G.J., and Dalbey, R.E. (2000). YidC mediates membrane protein insertion in bacteria. Nature *406*, 637-641. https://doi.org/10.1038/35020586

Sanchez-Pulido, L., Devos, D., Genevrois, S., Vicente, M., and Valencia, A. (2003). POTRA: a conserved domain in the FtsQ family and a class of beta-barrel outer membrane proteins. Trends Biochem Sci *28*, 523-526. https://doi.org/10.1016/j.tibs.2003.08.003

Sankaran, K., Gupta, S.D., and Wu, H.C. (1995). Modification of bacterial lipoproteins. Methods Enzymol *250*, 683-697. https://doi.org/10.1016/0076-6879(95)50105-3

Sauvonnet, N., and Pugsley, A.P. (1996). Identification of two regions of *Klebsiella oxytoca* pullulanase that together are capable of promoting beta-lactamase secretion by the general secretory pathway. Mol Microbiol *22*, 1-7. https://doi.org/10.1111/j.1365-2958.1996.tb02650.x

Schaub, R.E., Chan, Y.A., Lee, M., Hesek, D., Mobashery, S., and Dillard, J.P. (2016). Lytic transglycosylases LtgA and LtgD perform distinct roles in remodeling, recycling and releasing peptidoglycan in *Neisseria gonorrhoeae*. Mol Microbiol *102*, 865-881. https://doi.org/10.1111/mmi.13496

Schroder, N.W., Schombel, U., Heine, H., Gobel, U.B., Zahringer, U., and Schumann, R.R. (2003). Acylated cholesteryl galactoside as a novel immunogenic motif in *Borrelia burgdorferi* sensu stricto. J Biol Chem *278*, 33645-33653. https://doi.org/10.1074/jbc.M305799200

Schulze, R.J., Chen, S., Kumru, O.S., and Zückert, W.R. (2010). Translocation of *Borrelia burgdorferi* surface lipoprotein OspA through the outer membrane requires an unfolded conformation and can initiate at the C-terminus. Mol Microbiol *76*, 1266-1278. https://doi.org/10.1111/j.1365-2958.2010.07172.x

Schulze, R.J., and Zückert, W.R. (2006). *Borrelia burgdorferi* lipoproteins are secreted to the outer surface by default. Mol Microbiol *59*, 1473-1484. https://doi.org/10.1111/j.1365-2958.2006.05039.x

Schwan, T., Battisti, J., Porcella, S., Raffel, S., Schrumpf, M., Fischer, E., Carroll, J., Stewart, P., Rosa, P., and Somerville, G. (2003). Glycerol-3-phosphate acquisition in spirochetes: distribution and biological activity of glycerophosphodiester phosphodiesterase (GlpQ) among *Borrelia* species. J Bacteriol *185*, 1346-1356.

Schwan, T.G., and Piesman, J. (2000). Temporal changes in outer surface proteins A and C of the lyme disease-associated spirochete, *Borrelia burgdorferi*, during the chain of infection in ticks and mice. J Clin Microbiol *38*, 382-388.

Schwan, T.G., Piesman, J., Golde, W.T., Dolan, M.C., and Rosa, P.A. (1995). Induction of an outer surface protein on *Borrelia burgdorferi* during tick feeding. Proc Natl Acad Sci U S A *92*, 2909-2913. https://doi.org/10.1073/pnas.92.7.2909

Seinost, G., Dykhuizen, D.E., Dattwyler, R.J., Golde, W.T., Dunn, J.J., Wang, I.N., Wormser, G.P., Schriefer, M.E., and Luft, B.J. (1999). Four clones of *Borrelia burgdorferi* sensu stricto cause invasive infection in humans. Infect Immun *67*, 3518-3524. https://doi.org/10.1128/IAI.67.7.3518-3524.1999

Seshadri, R., Myers, G.S., Tettelin, H., Eisen, J.A., Heidelberg, J.F., Dodson, R.J., Davidsen, T.M., DeBoy, R.T., Fouts, D.E., Haft, D.H., *et al.* (2004). Comparison of the genome of the oral pathogen *Treponema denticola* with other spirochete genomes. Proc Natl Acad Sci U S A *101*, 5646-5651. https://doi.org/10.1073/pnas.0307639101

Setubal, J.C., Reis, M., Matsunaga, J., and Haake, D.A. (2006). Lipoprotein computational prediction in spirochaetal genomes. Microbiology (Reading) *152*, 113-121. https://doi.org/10.1099/mic.0.28317-0

Seydel, A., Gounon, P., and Pugsley, A. (1999). Testing the '+2 rule' for lipoprotein sorting in the *Escherichia coli* cell envelope with a new genetic selection. Mol Microbiol *34*, 810-821.

Shang, E.S., Skare, J.T., Exner, M.M., Blanco, D.R., Kagan, B.L., Miller, J.N., and Lovett, M.A. (1998). Isolation and characterization of the outer membrane of *Borrelia hermsii*. Infect Immun *66*, 1082-1091. https://doi.org/10.1128/IAI.66.3.1082-1091.1998

Silva-Herzog, E., Ferracci, F., Jackson, M.W., Joseph, S.S., and Plano, G.V. (2008). Membrane localization and topology of the *Yersinia pestis* YscJ lipoprotein. Microbiology (Reading) *154*, 593-607. https://doi.org/10.1099/mic.0.2007/013045-0

Simons, K., and Ehehalt, R. (2002). Cholesterol, lipid rafts, and disease. J Clin Invest *110*, 597-603. https://doi.org/10.1172/JCI16390

Sinha, R.K., and Rosenthal, R.S. (1980). Release of soluble peptidoglycan from growing conococci: demonstration of anhydro-muramyl-containing fragments. Infect Immun *29*, 914-925.

Skare, J.T., Mirzabekov, T.A., Shang, E.S., Blanco, D.R., Erdjument-Bromage, H., Bunikis, J., Bergström, S., Tempst, P., Kagan, B.L., Miller, J.N., *et al.* (1997). The Oms66 (p66) protein is a *Borrelia burgdorferi* porin. Infect Immun *65*, 3654-3661. https://doi.org/10.1128/IAI.65.9.3654-3661.1997

Skare, J.T., Shang, E.S., Foley, D.M., Blanco, D.R., Champion, C.I., Mirzabekov, T., Sokolov, Y., Kagan, B.L., Miller, J.N., and Lovett, M.A. (1995). Virulent strain associated outer membrane proteins of *Borrelia burgdorferi*. J Clin Invest *96*, 2380-2392. https://doi.org/10.1172/JCI118295

Sklar, J.G., Wu, T., Kahne, D., and Silhavy, T.J. (2007). Defining the roles of the periplasmic chaperones SurA, Skp, and DegP in *Escherichia coli*. Genes Dev *21*, 2473-2484. https://doi.org/10.1101/gad.1581007

Slamti, L., de Pedro, M.A., Guichet, E., and Picardeau, M. (2011). Deciphering morphological determinants of the helix-shaped *Leptospira*. J Bacteriol *193*, 6266-6275. https://doi.org/10.1128/JB.05695-11

Sohlenkamp, C., and Geiger, O. (2016). Bacterial membrane lipids: diversity in structures and pathways. FEMS Microbiol Rev *40*, 133-159. https://doi.org/10.1093/femsre/fuv008

Sohlenkamp, C., Lopez-Lara, I.M., and Geiger, O. (2003). Biosynthesis of phosphatidylcholine in bacteria. Prog Lipid Res *42*, 115-162. https://doi.org/10.1016/s0163-7827(02)00050-4

Stevenson, B., El-Hage, N., Hines, M.A., Miller, J.C., and Babb, K. (2002). Differential binding of host complement inhibitor factor H by *Borrelia burgdorferi* Erp surface proteins: a possible mechanism underlying the expansive host range of Lyme disease spirochetes. Infect Immun *70*, 491-497. https://doi.org/10.1128/iai.70.2.491-497.2002

Stewart, P.E., Hoff, J., Fischer, E., Krum, J.G., and Rosa, P.A. (2004). Genome-wide transposon mutagenesis of *Borrelia burgdorferi* for identification of phenotypic mutants. Appl Environ Microbiol *70*, 5973-5979. https://doi.org/10.1128/AEM.70.10.5973-5979.2004

Strobl, S., Gomis-Ruth, F.X., Maskos, K., Frank, G., Huber, R., and Glockshuber, R. (1997). The alpha-amylase from the yellow meal worm: complete primary structure, crystallization and preliminary X-ray analysis. FEBS Lett *409*, 109-114. https://doi.org/10.1016/s0014-5793(97)00451-1

Strominger, J.L., Ito, E., and Threnn, R.H. (1960). Competitive inhibition of enzymatic reactions by oxamycin J Am Chem Soc *82*, 998-999.

Stubenrauch, C., Grinter, R., and Lithgow, T. (2016). The modular nature of the beta-barrel assembly machinery, illustrated in *Borrelia burgdorferi*. Mol Microbiol *102*, 753-756. https://doi.org/10.1111/mmi.13527

Stubs, G., Fingerle, V., Wilske, B., Gobel, U.B., Zahringer, U., Schumann, R.R., and Schroder, N.W. (2009). Acylated cholesteryl galactosides are specific antigens of *Borrelia* causing lyme disease and frequently induce antibodies in late stages of disease. J Biol Chem *284*, 13326-13334. https://doi.org/10.1074/jbc.M809575200

Stubs, G., Fingerle, V., Zahringer, U., Schumann, R.R., Rademann, J., and Schroder, N.W. (2011). Acylated cholesteryl galactosides are ubiquitous glycolipid antigens among *Borrelia burgdorferi* sensu lato. FEMS Immunol Med Microbiol *63*, 140-143. https://doi.org/10.1111/j.1574-695X.2011.00827.x

Tajima, T., Yokota, N., Matsuyama, S., and Tokuda, H. (1998). Genetic analyses of the in vivo function of LolA, a periplasmic chaperone involved in the outer membrane localization of *Escherichia coli* lipoproteins. FEBS Lett *439*, 51-54. https://doi.org/10.1016/s0014-5793(98)01334-9

Takayama, K., Rothenberg, R.J., and Barbour, A.G. (1987). Absence of lipopolysaccharide in the Lyme disease spirochete, *Borrelia burgdorferi*. Infect Immun *55*, 2311-2313. https://doi.org/10.1128/IAI.55.9.2311-2313.1987

Takeda, K., Miyatake, H., Yokota, N., Matsuyama, S., Tokuda, H., and Miki, K. (2003). Crystal structures of bacterial lipoprotein localization factors, LolA and LolB. EMBO J *22*, 3199-3209. https://doi.org/10.1093/emboj/cdg324

Tanaka, K., Matsuyama, S., and Tokuda, H. (2001). Deletion of *lolB*, encoding an outer membrane lipoprotein, is lethal for *Escherichia coli* and causes accumulation of lipoprotein localization intermediates in the periplasm. J Bacteriol *183*, 6538-6542.

Taniguchi, N., Matsuyama, S., and Tokuda, H. (2005). Mechanisms underlying energy-independent transfer of lipoproteins from LolA to LolB, which have similar unclosed beta-barrel structures. J Biol Chem *280*, 34481-34488.

Tao, K., Watanabe, S., Narita, S., and Tokuda, H. (2010). A periplasmic LolA derivative with a lethal disulfide bond activates the Cpx stress response system. J Bacteriol *192*, 5657-5662. https://doi.org/10.1128/JB.00821-10

Terada, M., Kuroda, T., Matsuyama, S., and Tokuda, H. (2001). Lipoprotein sorting signals evaluated as the LolA-dependent release of lipoproteins from the cytoplasmic membrane of *Escherichia coli*. J Biol Chem *276*, 47690-47694.

Thanbichler, M., and Shapiro, L. (2006). MipZ, a spatial regulator coordinating chromosome segregation with cell division in *Caulobacter*. Cell *126*, 147-162. https://doi.org/10.1016/j.cell.2006.05.038

Thein, M., Bonde, M., Bunikis, I., Denker, K., Sickmann, A., Bergström, S., and Benz, R. (2012). DipA, a pore-forming protein in the outer membrane of Lyme disease spirochetes exhibits specificity for the permeation of dicarboxylates. PLoS One *7*, e36523. https://doi.org/10.1371/journal.pone.0036523

Thein, M., Bunikis, I., Denker, K., Larsson, C., Cutler, S., Drancourt, M., Schwan, T.G., Mentele, R., Lottspeich, F., Bergström, S., *et al.* (2008). Oms38 is the first identified pore-forming protein in the outer membrane of relapsing fever spirochetes. J Bacteriol *190*, 7035-7042. https://doi.org/10.1128/JB.00818-08

Tilly, K., Bestor, A., and Rosa, P.A. (2013). Lipoprotein succession in *Borrelia burgdorferi*: similar but distinct roles for OspC and VlsE at different stages of mammalian infection. Mol Microbiol *89*, 216-227. https://doi.org/10.1111/mmi.12271

Tilly, K., Bestor, A., and Rosa, P.A. (2016). Functional Equivalence of OspA and OspB, but Not OspC, in Tick Colonization by *Borrelia burgdorferi*. Infect Immun *84*, 1565-1573. https://doi.org/10.1128/IAI.00063-16

Tilly, K., Elias, A.F., Errett, J., Fischer, E., Iyer, R., Schwartz, I., Bono, J.L., and Rosa, P. (2001). Genetics and regulation of chitobiose utilization in *Borrelia burgdorferi*. J Bacteriol *183*, 5544-5553. https://doi.org/10.1128/JB.183.19.5544-5553.2001

Tilly, K., Grimm, D., Bueschel, D.M., Krum, J.G., and Rosa, P. (2004). Infectious cycle analysis of a *Borrelia burgdorferi* mutant defective in transport of chitobiose, a tick cuticle component. Vector Borne Zoonotic Dis *4*, 159-168. https://doi.org/10.1089/1530366041210738

Tilly, K., Krum, J.G., Bestor, A., Jewett, M.W., Grimm, D., Bueschel, D., Byram, R., Dorward, D., Vanraden, M.J., Stewart, P., *et al.* (2006). *Borrelia burgdorferi* OspC protein required exclusively in a crucial early stage of mammalian infection. Infect Immun *74*, 3554-3564. https://doi.org/10.1128/IAI.01950-05

Tokuda, H., and Matsuyama, S. (2004). Sorting of lipoproteins to the outer membrane in E. coli. Biochim Biophys Acta *1694*, IN1-9.

Toledo, A., Crowley, J.T., Coleman, J.L., LaRocca, T.J., Chiantia, S., London, E., and Benach, J.L. (2014). Selective association of outer surface lipoproteins with the lipid rafts of *Borrelia burgdorferi*. mBio *5*, e00899-00814. https://doi.org/10.1128/mBio.00899-14

Toledo, A., Huang, Z., Benach, J.L., and London, E. (2018a). Analysis of Lipids and Lipid Rafts in *Borrelia*. Methods Mol Biol *1690*, 69-82. https://doi.org/10.1007/978-1-4939-7383-5_6

Toledo, A., Huang, Z., Coleman, J.L., London, E., and Benach, J.L. (2018b). Lipid rafts can form in the inner and outer membranes of *Borrelia burgdorferi* and have different properties and associated proteins. Mol Microbiol *108*, 63-76. https://doi.org/10.1111/mmi.13914

Toledo, A., Perez, A., Coleman, J.L., and Benach, J.L. (2015). The lipid raft proteome of *Borrelia burgdorferi*. Proteomics *15*, 3662-3675. https://doi.org/10.1002/pmic.201500093

Treuner-Lange, A., Aguiluz, K., van der Does, C., Gomez-Santos, N., Harms, A., Schumacher, D., Lenz, P., Hoppert, M., Kahnt, J., Munoz-Dorado, J., *et al.* (2013). PomZ, a ParA-like protein, regulates Z-ring formation and cell division in *Myxococcus xanthus*. Mol Microbiol *87*, 235-253. https://doi.org/10.1111/mmi.12094

Tsui, H.T., Boersma, M.J., Vella, S.A., Kocaoglu, O., Kuru, E., Peceny, J.K., Carlson, E.E., VanNieuwenhze, M.S., Brun, Y.V., Shaw, S.L., *et al.* (2014). Pbp2x localizes separately from Pbp2b and other peptidoglycan synthesis proteins during later stages of cell division of *Streptococcus pneumoniae* D39. Mol Microbiol *94*, 21-40. https://doi.org/10.1111/mmi.12745

Tsukahara, J., Mukaiyama, K., Okuda, S., Narita, S., and Tokuda, H. (2009). Dissection of LolB function--lipoprotein binding, membrane targeting and incorporation of lipoproteins into lipid bilayers. FEBS J *276*, 4496-4504. https://doi.org/10.1111/j.1742-4658.2009.07156.x

Turner, R.D., Vollmer, W., and Foster, S.J. (2014). Different walls for rods and balls: the diversity of peptidoglycan. Mol Microbiol *91*, 862-874. https://doi.org/10.1111/mmi.12513

Twibanire, J.D., Omran, R.P., and Grindley, T.B. (2012). Facile synthesis of a library of Lyme disease glycolipid antigens. Org Lett *14*, 3909-3911. https://doi.org/10.1021/ol301697c

van Ulsen, P., van Alphen, L., ten Hove, J., Fransen, F., van der Ley, P., and Tommassen, J. (2003). A Neisserial autotransporter NalP modulating the processing of other autotransporters. Mol Microbiol *50*, 1017-1030. https://doi.org/10.1046/j.1365-2958.2003.03773.x

Verwer, R.W., Nanninga, N., Keck, W., and Schwarz, U. (1978). Arrangement of glycan chains in the sacculus of *Escherichia coli*. J Bacteriol *136*, 723-729. https://doi.org/10.1128/JB.136.2.723-729.1978

Viala, J., Chaput, C., Boneca, I.G., Cardona, A., Girardin, S.E., Moran, A.P., Athman, R., Memet, S., Huerre, M.R., Coyle, A.J., *et al.* (2004). Nod1 responds to peptidoglycan delivered by the *Helicobacter pylori cag* pathogenicity island. Nat Immunol *5*, 1166-1174. https://doi.org/10.1038/ni1131

Vollmer, W. (2008). Structural variation in the glycan strands of bacterial peptidoglycan. FEMS Microbiol Rev *32*, 287-306. https://doi.org/10.1111/j.1574-6976.2007.00088.x

Vollmer, W., Blanot, D., and de Pedro, M.A. (2008). Peptidoglycan structure and architecture. FEMS Microbiol Rev *32*, 149-167. https://doi.org/10.1111/j.1574-6976.2007.00094.x

Vollmer, W., Massidda, O., and Tomasz, A. (2019). The Cell Wall of *Streptococcus pneumoniae*. Microbiol Spectr *7*. https://doi.org/10.1128/microbiolspec.GPP3-0018-2018

Vollmer, W., and Seligman, S.J. (2010). Architecture of peptidoglycan: more data and more models. Trends Microbiol *18*, 59-66. https://doi.org/10.1016/j.tim.2009.12.004

von Lackum, K., and Stevenson, B. (2005). Carbohydrate utilization by the Lyme borreliosis spirochete, *Borrelia burgdorferi*. FEMS Microbiol Lett *243*, 173-179. https://doi.org/10.1016/j.femsle.2004.12.002

Voulhoux, R., Bos, M.P., Geurtsen, J., Mols, M., and Tommassen, J. (2003). Role of a highly conserved bacterial protein in outer membrane protein assembly. Science *299*, 262-265. https://doi.org/10.1126/science.1078973

Walker, E.M., Borenstein, L.A., Blanco, D.R., Miller, J.N., and Lovett, M.A. (1991). Analysis of outer membrane ultrastructure of pathogenic *Treponema* and *Borrelia* species by freeze-fracture electron microscopy. J Bacteriol *173*, 5585-5588. https://doi.org/10.1128/jb.173.17.5585-5588.1991

Wang, G., van Dam, A.P., and Dankert, J. (1999). Evidence for frequent OspC gene transfer between *Borrelia valaisiana sp. nov.* and other Lyme disease spirochetes. FEMS Microbiol Lett *177*, 289-296. https://doi.org/10.1111/j.1574-6968.1999.tb13745.x

Wang, X.G., Kidder, J.M., Scagliotti, J.P., Klempner, M.S., Noring, R., and Hu, L.T. (2004a). Analysis of differences in the functional properties of the substrate binding proteins of the *Borrelia burgdorferi* oligopeptide permease (Opp) operon. J Bacteriol *186*, 51-60. https://doi.org/10.1128/jb.186.1.51-60.2004

Wang, X.G., Lin, B., Kidder, J.M., Telford, S., and Hu, L.T. (2002). Effects of environmental changes on expression of the oligopeptide permease (*opp*) genes of *Borrelia burgdorferi*. J Bacteriol *184*, 6198-6206. https://doi.org/10.1128/jb.184.22.6198-6206.2002

Wang, X.G., Scagliotti, J.P., and Hu, L.T. (2004b). Phospholipid synthesis in *Borrelia burgdorferi*: BB0249 and BB0721 encode functional phosphatidylcholine synthase and phosphatidylglycerolphosphate synthase proteins. Microbiology (Reading) *150*, 391-397. https://doi.org/10.1099/mic.0.26752-0

Watanabe, S., Oguchi, Y., Takeda, K., Miki, K., and Tokuda, H. (2008). Introduction of a lethal redox switch that controls the opening and closing of the hydrophobic cavity in LolA. J Biol Chem *283*, 25421-25427. https://doi.org/10.1074/jbc.M804737200

Wessel, M., Klusener, S., Godeke, J., Fritz, C., Hacker, S., and Narberhaus, F. (2006). Virulence of *Agrobacterium tumefaciens* requires phosphatidylcholine in the bacterial membrane. Mol Microbiol *62*, 906-915. https://doi.org/10.1111/j.1365-2958.2006.05425.x

Wheeler, C.M., Garcia Monco, J.C., Benach, J.L., Golightly, M.G., Habicht, G.S., and Steere, A.C. (1993). Nonprotein antigens of *Borrelia burgdorferi*. Journal of Infectious Diseases *167*, 665-674. https://doi.org/10.1093/infdis/167.3.665

Willemse, J., Borst, J.W., de Waal, E., Bisseling, T., and van Wezel, G.P. (2011). Positive control of cell division: FtsZ is recruited by SsgB during sporulation of *Streptomyces*. Genes Dev *25*, 89-99. https://doi.org/10.1101/gad.600211

Wilson, M.M., Anderson, D.E., and Bernstein, H.D. (2015). Analysis of the outer membrane proteome and secretome of *Bacteroides fragilis* reveals a multiplicity of secretion mechanisms. PLoS One *10*, e0117732. https://doi.org/10.1371/journal.pone.0117732

Wolgemuth, C.W., Charon, N.W., Goldstein, S.F., and Goldstein, R.E. (2006). The flagellar cytoskeleton of the spirochetes. J Mol Microbiol Biotechnol *11*, 221-227. https://doi.org/10.1159/000094056

Wu, L.J., and Errington, J. (2004). Coordination of cell division and chromosome segregation by a nucleoid occlusion protein in *Bacillus subtilis*. Cell *117*, 915-925. https://doi.org/10.1016/j.cell.2004.06.002

Wu, X., Sharma, B., Niles, S., O'Connor, K., Schilling, R., Matluck, N., D'Onofrio, A., Hu, L.T., and Lewis, K. (2018). Identifying Vancomycin as an Effective Antibiotic for Killing *Borrelia burgdorferi*. Antimicrob Agents Chemother *62*. https://doi.org/10.1128/AAC.01201-18

Xie, K., and Dalbey, R.E. (2008). Inserting proteins into the bacterial cytoplasmic membrane using the Sec and YidC translocases. Nat Rev Microbiol *6*, 234-244. https://doi.org/10.1038/nrmicro3595

Yakushi, T., Masuda, K., Narita, S., Matsuyama, S., and Tokuda, H. (2000). A new ABC transporter mediating the detachment of lipid-modified proteins from membranes. Nat Cell Biol *2*, 212-218. https://doi.org/10.1038/35008635

Yakushi, T., Yokota, N., Matsuyama, S., and Tokuda, H. (1998). LolA-dependent release of a lipid-modified protein from the inner membrane of *Escherichia coli* requires nucleoside triphosphate. J Biol Chem *273*, 32576-32581. https://doi.org/10.1074/jbc.273.49.32576

Yamaguchi, K., Yu, F., and Inouye, M. (1988). A single amino acid determinant of the membrane localization of lipoproteins in *E. coli*. Cell *53*, 423-432. https://doi.org/10.1016/0092-8674(88)90162-6

Yanagihara, Y., Kamisango, K., Yasuda, S., Kobayashi, S., Mifuchi, I., Azuma, I., Yamamura, Y., and Johnson, R.C. (1984). Chemical compositions of cell walls and polysaccharide fractions of spirochetes. Microbiol Immunol *28*, 535-544. https://doi.org/10.1111/j.1348-0421.1984.tb00706.x

Yang, X., Goldberg, M.S., Popova, T.G., Schoeler, G.B., Wikel, S.K., Hagman, K.E., and Norgard, M.V. (2000). Interdependence of environmental factors influencing reciprocal patterns of gene expression in virulent *Borrelia burgdorferi*. Mol Microbiol *37*, 1470-1479. https://doi.org/10.1046/j.1365-2958.2000.02104.x

Yang, X., Lyu, Z., Miguel, A., McQuillen, R., Huang, K.C., and Xiao, J. (2017). GTPase activity-coupled treadmilling of the bacterial tubulin FtsZ organizes septal cell wall synthesis. Science *355*, 744-747. https://doi.org/10.1126/science.aak9995

Yang, X.F., Pal, U., Alani, S.M., Fikrig, E., and Norgard, M.V. (2004). Essential role for OspA/B in the life cycle of the Lyme disease spirochete. J Exp Med *199*, 641-648. https://doi.org/10.1084/jem.20031960

Yasuda, M., Iguchi-Yokoyama, A., Matsuyama, S., Tokuda, H., and Narita, S. (2009). Membrane topology and functional importance of the periplasmic region of ABC transporter LolCDE. Biosci Biotechnol Biochem *73*, 2310-2316. https://doi.org/10.1271/bbb.90451

Yen, M.R., Peabody, C.R., Partovi, S.M., Zhai, Y., Tseng, Y.H., and Saier, M.H. (2002). Protein-translocating outer membrane porins of Gram-negative bacteria. Biochim Biophys Acta *1562*, 6-31. https://doi.org/10.1016/s0005-2736(02)00359-0

Yokota, N., Kuroda, T., Matsuyama, S., and Tokuda, H. (1999). Characterization of the LolA-LolB system as the general lipoprotein localization mechanism of *Escherichia coli*. J Biol Chem *274*, 30995-30999.

Young, K.D. (2006). The selective value of bacterial shape. Microbiol Mol Biol Rev *70*, 660-703. https://doi.org/10.1128/MMBR.00001-06

Zhang, J.R., Hardham, J.M., Barbour, A.G., and Norris, S.J. (1997). Antigenic variation in Lyme disease borreliae by promiscuous recombination of VMP-like sequence cassettes. Cell *89*, 275-285. https://doi.org/10.1016/s0092-8674(00)80206-8

Zhang, J.R., and Norris, S.J. (1998a). Genetic variation of the *Borrelia burgdorferi* gene *vlsE* involves cassette-specific, segmental gene conversion. Infect Immun *66*, 3698-3704. https://doi.org/10.1128/IAI.66.8.3698-3704.1998

Zhang, J.R., and Norris, S.J. (1998b). Kinetics and in vivo induction of genetic variation of vlsE in *Borrelia burgdorferi*. Infect Immun *66*, 3689-3697. https://doi.org/10.1128/IAI.66.8.3689-3697.1998

Zhang, K., Qin, Z., Chang, Y., Liu, J., Malkowski, M.G., Shipa, S., Li, L., Qiu, W., Zhang, J.R., and Li, C. (2019). Analysis of a flagellar filament cap mutant reveals that HtrA serine protease degrades unfolded flagellin protein in the periplasm of *Borrelia burgdorferi*. Mol Microbiol *111*, 1652-1670. https://doi.org/10.1111/mmi.14243

Zhang, Q.Y., DeRyckere, D., Lauer, P., and Koomey, M. (1992). Gene conversion in *Neisseria gonorrhoeae*: evidence for its role in pilus antigenic variation. Proc Natl Acad Sci U S A *89*, 5366-5370. https://doi.org/10.1073/pnas.89.12.5366

Zückert, W.R. (2013). A call to order at the spirochaetal host-pathogen interface. Mol Microbiol *89*, 207-211. https://doi.org/10.1111/mmi.12286

Zückert, W.R. (2014). Secretion of bacterial lipoproteins: through the cytoplasmic membrane, the periplasm and beyond. Biochim Biophys Acta *1843*, 1509-1516. https://doi.org/10.1016/j.bbamcr.2014.04.022

Zückert, W.R. (2019). Protein Secretion in Spirochetes. Microbiol Spectr *7*. https://doi.org/10.1128/microbiolspec.PSIB-0026-2019

Zückert, W.R., Kerentseva, T.A., Lawson, C.L., and Barbour, A.G. (2001). Structural conservation of neurotropism-associated VspA within the variable *Borrelia* Vsp-OspC lipoprotein family. J Biol Chem *276*, 457-463. https://doi.org/10.1074/jbc.M008449200

Zückert, W.R., Lloyd, J.E., Stewart, P.E., Rosa, P.A., and Barbour, A.G. (2004). Cross-species surface display of functional spirochetal lipoproteins by recombinant *Borrelia burgdorferi*. Infect Immun *72*, 1463-1469. https://doi.org/10.1128/iai.72.3.1463-1469.2004

Zückert, W.R., Meyer, J., and Barbour, A.G. (1999). Comparative analysis and immunological characterization of the *Borrelia* Bdr protein family. Infect Immun *67*, 3257-3266. https://doi.org/10.1128/IAI.67.7.3257-3266.1999

Chapter 8

Dancing with the Star: *Borrelia burgdorferi*, a Solo Dancer with All the Right Moves

Ching Wooen Sze[1], Hui Xu[2], Md A. Motaleb[2], Charles W. Wolgemuth[3], Jun Liu[4], Nyles W. Charon[5], and Chunhao Li[1]*

[1]Philips Institute for Oral Health Research, Virginia Commonwealth University, Richmond, VA 23298, USA
[2]Department of Microbiology and Immunology, Brody School of Medicine, East Carolina University, Greenville, NC 27834, USA
[3]Department of Physics and Molecular Cell Biology, University of Arizona, Tucson, AZ 85719, USA
[4]Department of Microbial Pathogenesis, Yale University School of Medicine, New Haven, CT 06516, USA
[5]Department of Microbiology, Immunology, and Cell Biology, West Virginia University, Morgantown, WV 26506, USA
*Corresponding author: cli5@vcu.edu

DOI: https://doi.org/10.21775/9781913652616.08

Abstract
Spirochetes are unique in many features, from the physical morphology down to the molecular mechanism used for gene regulation. As the model organism discussed in this chapter, the Lyme disease bacterium *Borrelia burgdorferi* shows us how spirochetes do it differently when it comes to the placement of flagella, coordination of two PF bundles during motility, a spirochete-specific flagellar collar protein in the basal body/motor complex, and the absolute reliance on motility and chemotaxis for disease transmission and colonization. Motility and chemotaxis are one of the most common traits adapted by different bacterial species, yet the 'classical' paradigms of hierarchical transcriptional regulation of flagellar genes and coordination of motility and chemotaxis derived from study of model bacteria do not fit in the world of spirochetes. As the genetic and molecular tool advances in the field of *B. burgdorferi*, the significance of more motility and chemotaxis genes begin to unravel making it possible to establish a working model best suited for spirochetes.

Introduction
When Anton van Leeuenhoek looked through a single-lens microscope in 1676 and obtained man's first recorded glimpse of bacteria, it was their motion that most delighted him: "I must say, for my part, that no more pleasant sight has ever yet come before my eye than these many thousands of living creatures, seen all alive in a little drop of water, moving among one another, each several creature having its own proper motion" (Berg, 1976).

The motility of bacteria, including the spirochetes, has fascinated biologists for centuries. Those who work on spirochetes are enticed to study these bacteria not only because of the diseases they cause and their roles in nature, but also because of their remarkable repertoire of moving and dancing, and their distinctive morphology. With the recent advances in targeted mutagenesis in spirochetes and advanced imaging technology, such as cryo-electron tomography (cryo-ET), there has been a burgeoning of information concerning their enigmatic mechanisms of motility. There have been a number of recent reviews on this topic, and the reader is referred to the following references: Charon and Goldstein, 2002; Goldstein, 2010; Charon et al., 2012; Zhao et al., 2014; Motaleb et al., 2015; Wolgemuth, 2015; Nakamura, 2020.

General structure of spirochetes
Spirochetes are a diverse group of diderm bacteria from the phylum of Spirochetes, the vast majority with long and helically coiled or flat wave shaped cells (Canale-Parola, 1978; Holt, 1978; Harwood and Canale-Parola, 1984; Goldstein et al., 1994; Charon and Goldstein, 2002; BJ, 2011; Charon et al., 2012). Spirochetes can be found throughout the environment as saprophytic organism in mud sediments, lakes, or deep sea hydrothermal vents to name a few, or in association with an array of

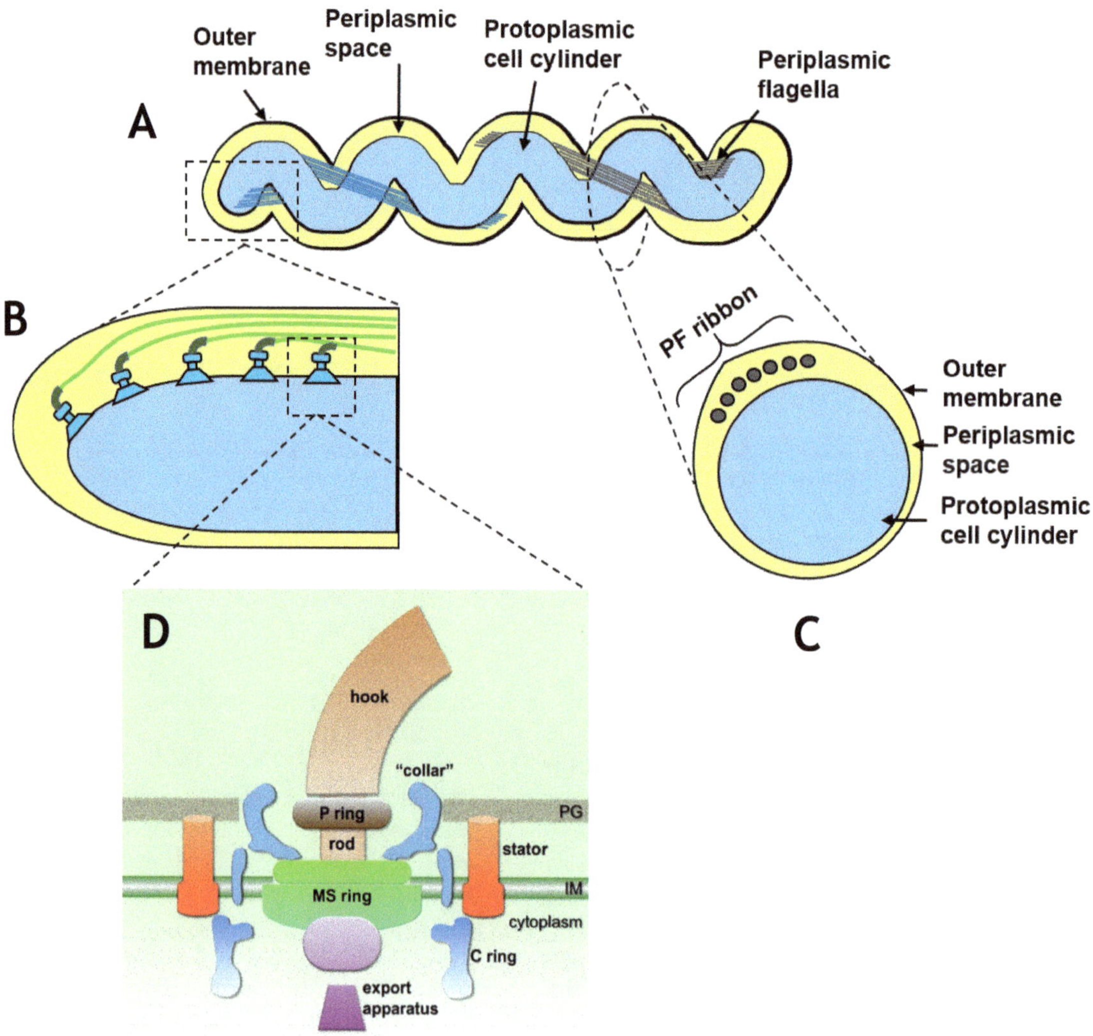

Figure 1. (A) Longitudinal diagram of a typical *B. burgdorferi* cell. Seven periplasmic flagella (PF) are depicted here at each cell pole inside the periplasmic space between the outer membrane and protoplasmic cell cylinder where each PF bundle is represented by different colors. Note that the PF overlap in the cell center. **(B)** Diagram depicting the arrangement of the flagellar basal bodies along the cell end of *B. burgdorferi*. **(C)** Cross-section diagram of *B. burgdorferi* illustrating the component parts. Seven PF form a tightly packed ribbon that causes the outer membrane to bulge at the vicinity of the PF ribbon. **(D)** Schematic model of the PF of *B. burgdorferi*. The core component of a hook-basal body complex is depicted here, including the export apparatus, the MS ring, the C ring, the stator, the rod, and the hook. The spirochete-specific "collar" is depicted. Reprinted and modified with permission from reference (Chang et al., 2020).

eukaryotic hosts such as termites, ticks, and mammals (Henry and Johnson, 1978; Canale-Parola, 1981; Harwood et al., 1982; Harwood and Canale-Parola, 1984; BJ, 2011). Given their affiliation with a diverse range of hosts, some species form a pathogenic relationship and cause severe infectious diseases in both humans and animals. The most medically significant and well known human diseases and their causative spirochetes include Lyme disease (*Borrelia burgdorferi*, also referred to as *Borreliella burgdorferi*), syphilis (*Treponema pallidum* subspecies *pallidum*), Yaws (*Treponema pallium* subspecies *pertenue*), leptospirosis (*Leptospira interrogans*), and periodontal disease (numerous *Treponema* species, including *Treponema denticola*) (Taber and Feigin, 1979; Benach et al., 1983; Cheng et al., 1985; Riviere et al., 1991; Ashimoto et al., 1996; BJ, 2011; Charon et al., 2012; Mitja et al., 2015; Wolgemuth, 2015). Most spirochetes have the distinct spiral or flat-wave morphology. Some rare ones are spherical (Troshina et al., 2015). Unlike other flagellated bacteria, the flagella of spirochetes are located within the periplasmic space (PS) between its outer membrane (OM) and the cell cylinder, hence they are designated as periplasmic flagella (PFs) (Figure 1). The number of PFs varies between species and can range from one to several hundred, but, in general, each PF is sub-terminally attached near the end of only one cell pole and wraps around the protoplasmic cell cylinder (PC), extending toward the center of the cell body (Figure 1) (Canale-Parola, 1978; Holt, 1978; Goldstein and Charon, 1988; Charon et al., 1992b; Li et al., 2000b; Lux et al., 2000; Charon and Goldstein, 2002; Wolgemuth et al., 2006; Radolf and Samuels, 2010; Charon et al., 2012; Nakamura, 2020). Rotation of the PF causes the cell cylinder to rotate and/or to gyrate its ends (to gyrate is to move in a circular motion without necessarily rotating), enabling the bacterium to swim in a given direction (Goldstein and Charon, 1988; Charon et al., 1992a; Charon and Goldstein, 2002; Goldstein, 2010; Charon et al., 2012; Wolgemuth, 2015).

General structure of *B. burgdorferi*

B. burgdorferi is the etiological agent of Lyme disease (LD) or Lyme borreliosis. In North America, *B. burgdorferi* is mainly transmitted by blacklegged ticks *Ixodes scapularis* and *Ixodes pacificus* to humans from its natural reservoirs, such as white-footed mice and other small animals (Piesman et al., 1999; Rosa et al., 2005; Goldstein, 2010; Stanek et al., 2012; MacDonald et al., 2017). In humans, 70-80% of infected patients present with the pathognomonic skin lesion erythema migrans (EM) during early infection, which can later *develop* into multi-systemic inflammatory disorders such as cardiac, neurological, and arthritic complications when left untreated (Murray and Shapiro, 2010; Schwartz et al., 2017). In contrast to most other spirochetes, *Borrelia* species, and *T. pallidum* as well, have a flat-wave morphology rather than being helical (Goldstein et al., 1994; Goldstein et al., 1996; Izard et al., 2009). An analysis of swimming *B. burgdorferi* cells revealed that planar waves are propagated from the anterior to the posterior end. In addition, high voltage electron microscopy confirmed the flat wave configuration of *B. burgdorferi* cells (Goldstein et al., 1994; Goldstein et al., 1996). Early electron microscopy work showed the presence of 7 to 11 PF in *B. burgdorferi* sub-terminally attached at each cell end (Figure 1) (Hovind-Hougen, 1984). In addition to being essential for the locomotion of *B. burgdorferi*, analysis of various flagella-related mutants also revealed that PF have a skeletal function responsible for the characteristic flat-wave morphology (see reviews (Charon et al., 2012; Motaleb et al., 2015; Nakamura, 2020)). For example, mutants with defects in forming intact PFs are rod shaped (Motaleb et al., 2000; Sartakova et al., 2001; Sal et al., 2008). In addition, cells immobilized at low temperature or with the protonophore carbonyl cyanide-mchlorophenyl-hydrazone (CCCP), which is known to inhibit flagellar rotation, still had a flat wave morphology (Motaleb et al., 2000). Thus, flagellar rotation is not necessary to form the flat wave in cells containing intact PFs. However, an analysis of a *motB* mutant, in which the PFs fail to rotate, revealed that the mutant cells have an irregular morphology (Sultan et al., 2015). These results suggest that flagellar rotation is necessary for the PFs to wrap around the cell cylinder to yield the flat wave morphology. In addition, the length of the filament also can influence the cell morphology (Sze et al., 2011), as will be discussed in another section.

Periplasmic flagella – a sophisticated nanomachine

PF structure and function

Similar to the external flagella from other bacterial species, such as *Escherichia coli* and *Salmonella enterica* serovar Typhimurium, the PFs of *B. burgdorferi* are composed of a basal body/motor complex, flagellar hook, and filament (Hovind-Hougen, 1984; Barbour and Hayes, 1986; Sal et al., 2008). The subterminal insertion of the PFs at the cell poles of *B. burgdorferi* was first visualized by Hovind-Hougen in 1984 (Hovind-Hougen, 1984), but the detailed configuration of PF was not solved until 12 years later. Using high voltage electron microscope, significant findings regarding the cellular architecture of *B. burgdorferi* were brought to light (Goldstein et al., 1996). First, the diameter of the spirochete was 0.33 μm, with the cell length between 10 to 20 μm. Second, the PFs formed as a continuous tight bundle with a diameter of 67 nm that extended from one cell end to the other. Third, the bundle of PFs wrapped around the body axis of the PC in a right-handed sense (the body axis is defined as a line through the center of the cell cylinder). Lastly, the PFs wrap around the cell axis in a left-handed sense (the cell axis is equivalent to the x axis of a sine wave, given the shape of the cell being a flat wave). The defined peak-to-peak amplitude of the cell is 0.78 μm, with a wavelength of 2.83 μm, with the latter identical to the helix pitch of the PFs in situ. The interaction between PF and PC is complex. For example, purified PFs, which are not constrained by the PC, have a smaller helical pitch and diameter than the intact cell (Charon et al., 1992a; Goldstein et al., 1996). In addition, Dombrowski et al. found two polymorphic shapes of purified PFs, the most common form had a helix pitch of 1.48 μm with a helix diameter of 0.4 μm (Dombrowski et al., 2009). The less common form had a helix pitch of 2.0 μm and helix diameter of 0.8 μm diameter. Mathematically modeling of the interaction of the PFs and the cell cylinder led to the finding that the most common configuration of the PFs fits best with the resulting flat-wave shape (Charon et al., 1992a; Goldstein et al., 1996; Dombrowski et al., 2009). As the helix pitch and diameter of the unconstrained PFs do not equal to those of the intact cell, the PFs must bend and stretch to fit within the PS. Since the PC is rod shaped, whereas the PFs are helical, the PFs exert force and cause elastic bending of the PC. Concomitantly, the PC in turn exerts an equal but opposite force back onto the flagella, causing the PFs to deform as well. The mathematical model developed by Wolgemuth and coworkers provides evidence that the balance between these two forces results in a cell with flat-wave architecture (Dombrowski et al., 2009).

Using cryo-ET we now know that PF form an elegant tight ribbon in the PS with roughly 3 nm spacing between each flagellum as they wrap around the body axis in a right handed sense (Charon et al., 2009). The formation of this tight ribbon structure is dependent on the rotation of the PFs (Sultan et al., 2015). The basal bodies are arranged linearly at each end along the long axis of the cell with spacing of approximately 90 to 120 nm apart from each other (Figure 1B) (Charon et al., 2009; Xu et al., 2011). The OM bulges in the vicinity of the PFs with the flagellar ribbon increasing the thickness of the PS (Figure 1C) (Charon et al., 2009; Kudryashev et al., 2009; Liu et al., 2009). Specifically, in the region without PFs, the PS thickness is roughly 16 nm whereas in the domains containing PFs, the PS thickness is between 42~49 nm (Charon et al., 2009; Liu et al., 2009).

Components and function of the flagellar motor

The flagellar motor is the rotary machinery responsible for flagellar assembly, rotation and directional switching. The major components include the export apparatus, the MS ring, the C-ring, the stator complex, and the rod (Figure 2). The stator complex is the torque generator composed of two membrane proteins MotA and MotB (Chang et al., 2019). The C-ring is the switch complex comprised of three proteins (FliG, FliM, and FliN) on the cytoplasmic side of the motor. The rotation of the motor is driven by the torque generated by the stator complexes. The rod serves as a drive shaft and consists of multiple different rod proteins (FlgB, FlgC, FlgF and FlgG) (Zhao et al., 2013). A unique spirochete-specific "collar" contributes significantly to the motor structures (Figure 1D and Figure 2) and is present in *B. burgdorferi* (Liu et al., 2009; Moon et al., 2016b) and all other spirochetes characterized to date (Murphy et al., 2006; Liu et al., 2010; Chen et al., 2011; Raddi et al., 2012). In addition, the flagellar motors of spirochetes are considerably larger than those in *E. coli* and *S. enterica* (~80 nm vs ~45 nm in diameter) (Murphy et al., 2006; Liu et al., 2009; Beeby et al., 2016).

Flagellar export apparatus

The export apparatus, which is responsible for the assembly of the rod, hook, and filament, is a flagellar-specific type III secretion system (T3SS) (Aldridge and Hughes, 2001). It is evolutionarily related to the injectisome, which is widely used by many bacterial pathogens to deliver virulence effectors into eukaryotic cells (Hu et al., 2015; Hu et al., 2017; Galan and Waksman, 2018). The export apparatus is composed of a membrane-bound export gate and a cytosolic ATPase complex. Cryo-ET imaging of the *B. burgdorferi* motor revealed a novel structure of its export apparatus (Qin et al., 2018). Specifically, the cytosolic ATPase complex is attached to the flagellar C-ring through multiple spokes to form a 'spoke and hub' structure. This structure not only strengthens the round-shaped C-ring but also appears to rotate with the C-ring (Qin et al., 2018). Further characterization of the intact export apparatus in PF will yield the mechanistic insights into flagellar assembly.

Stator-rotor interaction

Powered by the electrochemical gradient across the cytoplasmic membrane, flagellar rotation is mediated by the interaction between the stator and the C-ring. Comparison of the motors from wild-type cells and stator-deletion mutants in *B. burgdorferi* revealed that the stator–rotor interaction induces a conformational change in the flagellar C-ring (Chang et al., 2019). Analysis of the motors of a less motile *motB-D24E* mutant and a non-motile *motB-D24N* mutant, in which the proton translocation is reduced and blocked, respectively, provides direct evidence that the conformational change of the C-ring results from the proton-driven torque generated by the stator

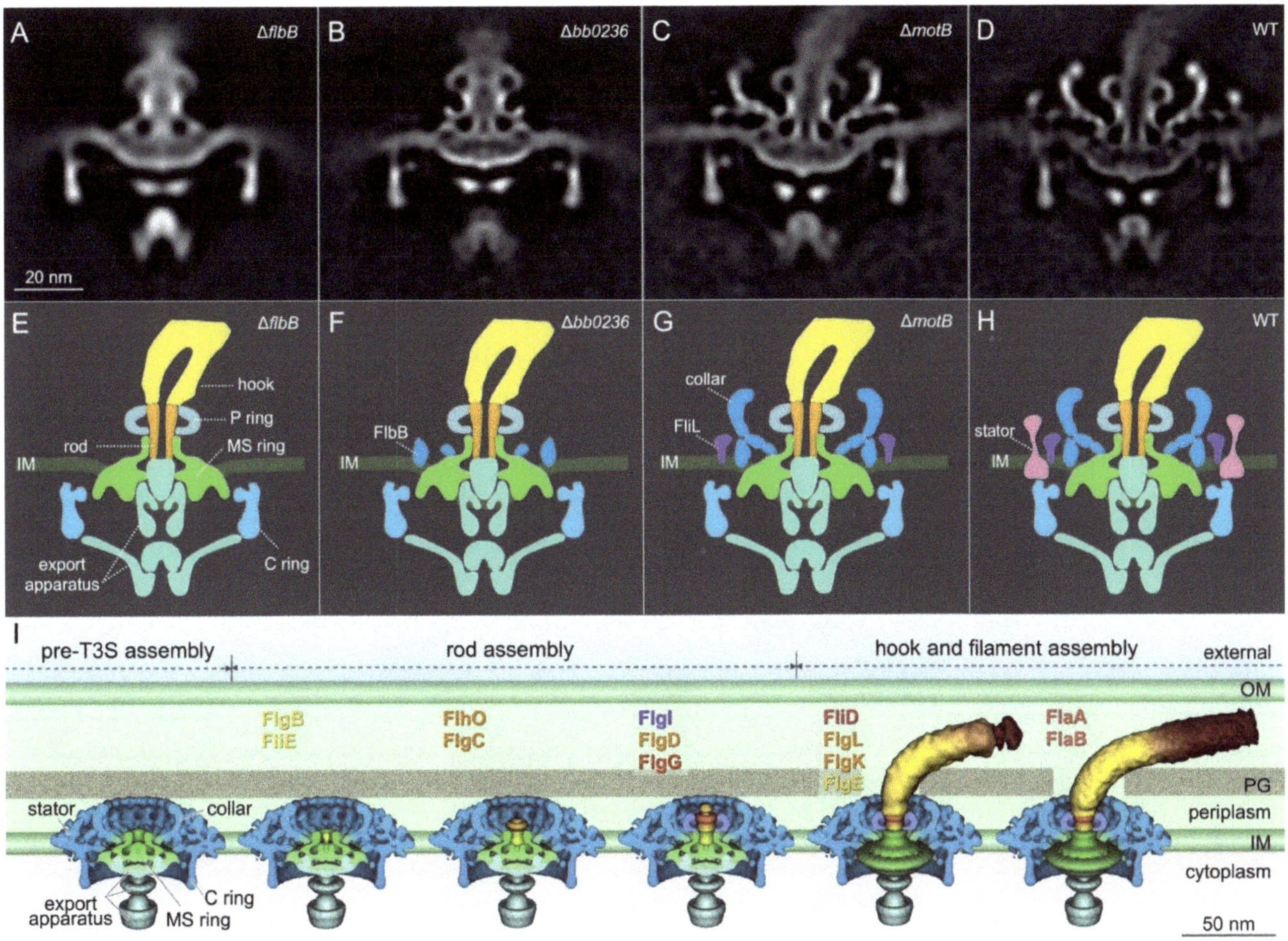

Figure 2. Visualizing the assembly of periplasmic flagella. (A) Motor structure of *flbB* mutant (*ΔflbB*) lacking the collar. **(B)** Motor structure of *bb0236* mutant (*Δbb0236*). The collar structure is largely absent in this mutant. **(C)** Motor structure of *motB* mutant (*ΔmotB*) lacking the stator complexes. **(D)** Motor structure of the wild type (WT). The intact collar, stator, and FliL are present. The corresponding models in each strain are shown in panels **(E-H)**, respectively. **(I)** T3SS mediated assembly of the rod, the hook, and the filament. Reprinted and modified with permission from reference (Chang et al., 2020).

complexes. These studies provide unprecedented structural information about the stator-rotor interactions and reveal the proton-driven conformational changes essential for flagellar rotation. Given that flagellum is the main organelle for motility in bacteria, the new insights into the molecular mechanisms underlying the motility and flagellar rotation in *B. burgdorferi* should be applicable not just to other spirochetes but motile bacteria in general.

Flagellar collar

It is becoming increasingly evident that the PFs have evolved as a unique nanomachine that drives the distinctive morphology and motility of spirochetes. The spirochete-specific collar clearly differentiates the PFs from other bacterial flagellar motors, as it constitutes a large, turbine-like complex ~80 nm in diameter and ~24 nm in height (Liu et al., 2009). FlbB (BB0286), the first collar protein identified, is located at the base of the collar (Moon et al., 2016b). Subsequently, the protein-protein interaction map developed in T. pallidum was used to identify BB0236 as a second collar protein (Rajagopala et al., 2007; Moon et al., 2018). Mutants deficient in either FlbB or BB0236 are non-motile and lack collars (Figure 2A-H). More recently, a novel protein FlcA (BB0326) was identified on the periphery of the collar, interacting directly with other flagellar proteins, such as the stator protein MotB (Xu et al., 2020). Although

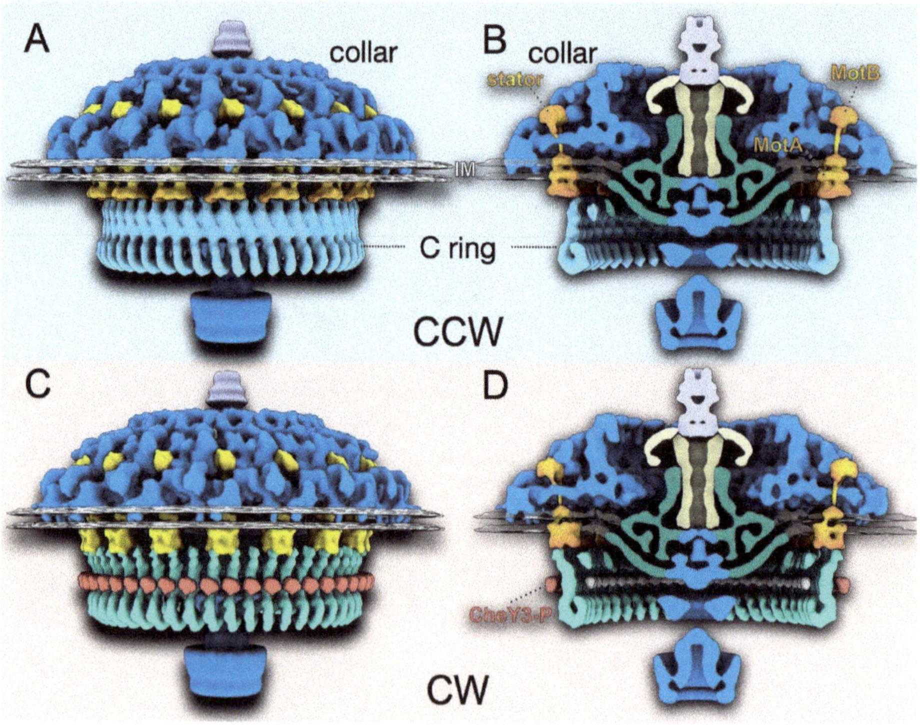

Figure 3. In situ structures of the flagellar motors in CCW and CW rotational states. (A) A side view of the motor in CCW rotation shows one conformation of the C-ring (light blue). **(B)** A section of the motor shows the stator (MotA/MotB) complex. **(C)** A side view of the motor in CW rotation shows another conformation of the C-ring (cyan). **(D)** A section of the motor shows the signaling protein CheY3-P. Note that the C-ring has undergone a profound conformational change from CCW to CW due to the binding of the CheY3-P, significantly altering its interactions with the stator complexes. Reprinted and modified with permission from reference (Chang et al., 2020).

additional proteins are likely involved in the collar assembly, it is evident that the periplasmic portion of the collar provides a framework to promote the recruitment of sixteen stator complexes, which are essential to generate the high torque required for the rotation of PFs and drive motility.

Motor rational switching mechanism

A hallmark of the flagellar motor is its ability to rotate in both counterclockwise (CCW) and clockwise (CW) directions and switch rapidly between the two rotational states. Regulation of rotational orientation is key for bacterial chemotaxis, a behavior that enables cells to move toward attractants or away from repellents. A sophisticated chemotaxis signaling system allows the cell to sense chemical stimuli and transmit this information via a signaling protein, CheY-P, to control the direction of flagellar rotation. Recent in situ high resolution of motor structures in *B. burgdorferi cheX* phosphatase and cheY3 response regulator mutants, in which motors are genetically locked in either CCW or CW rotation, provide mechanistic insights into the control mechanism for rotational switching. Specifically, CheY3 interacts directly with the FliM protein of the switch complex in a phosphorylation-dependent manner (Chang, 2020). The binding of CheY3-P to FliM induces a profound remodeling of the switch protein FliG2. As a result, the diameter of the FliG2 ring, which is composed of 46 units, expands from 55 nm to 62 nm upon binding of CheY3-P (Figure 3). Importantly, the large rearrangement of FliG2 during directional switching allows it to engage distinct interactions with the stator complexes, analogous to throwing an automobile transmission into reverse. Therefore, a novel model was proposed to explain the generation of torque and the switching of motor rotational direction: 1) Proton flux through the stator causes the cytoplasmic region to rotate CW (viewed from the hook to the C-ring). 2) Interaction with the outer part of the stator cytoplasmic region causes the C-ring to rotate CW, while interaction with the inner part of the stator cytoplasmic region causes the C-ring to rotate CCW. 3) Control of the direction of flagellar rotation consists of aligning the interaction sites of the stator and the switch complex through conformational changes in FliG2 to achieve the desired direction of flagellar rotation. Given that the flagellar core components, including the switch complex and stator complex, are highly conserved, this molecular mechanism for flagellar rotational switching is likely utilized, with some modifications, across a wide spectrum of bacterial species.

Unique structure of the flagellar hook

The hook is a flexible structure between the flagellar filament and the basal body/motor complex; it serves as a universal joint coupling the rotation of the inner membrane-bound motor to the filament (Figure 1D). In most bacteria, the hook consists of a polymer of roughly 120-130 copies of the hook protein FlgE (DePamphilis and Adler, 1971a, b; Silverman and Simon, 1977; Samatey et al., 2004). Compared to the flagellar filament, the hook is considerably more stable. Specifically, under conditions of low pH, high concentrations of urea, organic solvents, or detergent that completely dissociate the filaments, the hooks remain intact (Silverman and Simon, 1977). In contrast to other bacterial species, spirochetes have a unique feature in their hook architecture, as spirochete hook FlgE proteins are covalently cross-linked to one another. Consequently, the hooks form a high molecular weight complex (HMWC) on polyacrylamide gel (Limberger et al., 1994; Chi et al., 2002; Sal et al., 2008; Miller et al., 2016; Lynch et al., 2019). The flagellar hook in *B. burgdorferi* is approximately 61 nm long with 133 FlgE subunits (Sal et al., 2008; Miller et al., 2016; Zhang et al., 2019). Biochemical analysis of *B. burgdorferi* cell lysates and purified hook-basal bodies complex revealed that this HMWC is stable to boiling and even to treatments with β-mercaptoethanol, dithiothreitol, 8 M guanidine-HCl, or 8 M urea(Sal et al., 2008; Miller et al., 2016). Immunoblot analysis using anti-serum against FlgE showed a laddering effect with the strongest reacting protein migrated at molecular weight greater than 160 kDa (the monomeric form of FlgE is 46 kDa). Interestingly, in a *B. burgdorferi fliF* mutant where the MS ring of the basal body is absent, the formation of this HMWC is inhibited leaving trace amount of monomeric FlgE (Sal et al., 2008). As the flagellar subunits are assembled in a sequential manner, the flagellar components in the periplasmic region are secreted through the T3SS (except for FlaA, as described in the following section). In the absence of the MS ring in a *fliF* mutant, which is required for the secretion of the other basal body proteins, the hook protein fails to be exported into the PS for assembling, suggesting, as one might anticipate, that cross-linking occurs in the periplasm after FlgE monomers are secreted (Sal et al., 2008).

In analyzing the chemistry of FlgE cross-linking, the major breakthrough occurred when it was found that FlgE cross-linking occurred in vitro under specific conditions (Miller et al., 2016). *T. denticola* FlgE serves as the model system, as its in vitro cross-linking is the most robust. The interpeptide cross-link is between a conserved lysine on one FlgE subunit and a cysteine on another. In the reaction, the cysteine thiol group is eliminated, leading to the formation of the intermediate dehydroalanine, which then reacts with lysine forming the rare lysinoalanine (Lal) adduct. Lal is responsible for the cross-link and polymerization (Miller et al., 2016; Lynch et al., 2019). Importantly, *B. burgdorferi* hooks have the identical Lal cross-link as *T. denticola* FlgE (Miller et al., 2016) and can cross-link its FlgE protein in vitro with concomitant Lal formation (N. Charon, M. Miller, M. James, B. M. Lynch, B, Crane, unpublished). The unique Lal cross-link is completely self-catalytic and requires the conserved residues K165 and C178 (In *Leptospira*, S178 substitutes for C178, which can also lead to Lal synthesis). Phylogenetic analysis revealed that these residues are invariant in spirochetes but absent in the hook proteins of most other bacterial species (Miller et al., 2016; Lynch et al., 2019). Why do spirochetes need to cross-link their hook proteins? One possible explanation is that cross-linking provides extra stability and flexibility to the hook similar to the chain-mail type cross-linking of the pili of *Streptococcus pyogenes* (Kang et al., 2007) or the bacteriophage head proteins of HK97 (Hendrix, 2005). As compared to external flagella, the PFs wrap around the PC as they rotate causing the PC to deform. Under such intense stress, cross-linked FlgE should result in hooks that are flexible yet sufficiently strong to carry out this unique function.

Flagellar filament

The filament of *B. burgdorferi* consists of one major flagellin protein, FlaB, with a minor sheath protein, FlaA (Wallich et al., 1990; Ge and Charon, 1997a; Ge et al., 1998). The filament cap, FliD, is located at the distal end of the polymerized filament (Figure 2I) (Zhao et al., 2013; Zhang et al., 2019). The amino acid sequence of *B. burgdorferi* FlaB is highly homologous to the sequences of other bacterial flagellin proteins with conserved N and C terminal domains, indicating that they form similar structures (Wilson and Beveridge, 1993). FlaB monomers non-covalently self-polymerize in situ and assemble into a helical left-handed filament, even in the absence of FlaA sheath protein (Li et al., 2000a; Motaleb et al., 2004). FlaA shares no sequence homology to FlaB (Ge et al., 1998; Li et al., 2000a; Charon and Goldstein, 2002) and contains a typical peptidase I cleavage site at its N-terminal region, which is absent in the final FlaA protein, suggesting that FlaA is likely secreted via the type II SecA-dependent system instead of the flagellar specific T3SS (Ge and Charon, 1997a). Analyses of the flagella of other spirochetal species indicate that FlaA forms a sheath around the FlaB filament core, but the precise role of FlaA is unknown (Li et al., 2000a; Charon and Goldstein, 2002). Previous work on the swine dysentery spirochete, *Brachyspira hyodysenteriae*, showed that FlaA is present in excess relative to each of its three FlaB subunits. In the absence of FlaA, an intact PF can still form but with significantly smaller helix pitch and diameter, supporting FlaA as the filament sheath protein and indicating that its presence affects overall flagellar structure, perhaps by stabilizing FlaB core into an optimal helical configuration for rotation (Li et al., 2000a). However, there is considerably less FlaA protein in *B. burgdorferi* relative to FlaB (Ge et al., 1998; Motaleb et al., 2004). Preliminary evidence indicates that the small amount of FlaA in *B. burgdorferi* resides on the surface of the PF proximal to the basal body rather than encompassing the entire filament core, which may help to stabilize the structure of the filament to an optimal helical configuration near the basal body region (S. Satoshi, M. Motaleb, S. Aizawa, N.W. Charon, unpublished). Site directed mutagenesis of *flaA* led to the formation of a motile mutant capable of forming PFs but with compromised motility (M.A. Motaleb and N.W. Charon, unpublished). These results further reinforce the idea that FlaA plays a scaffolding role for the helical shape of PFs in *B. burgdorferi*.

As noted above, at the tip of the polymerized filament core sits the filament capping protein FliD (also known as the hook associated protein 2, HAP2) (Ikeda et al., 1985; Ikeda et al., 1987). As previously mentioned, flagellin monomers are secreted via the T3SS and self-polymerized, and in the case of spirochetes, within the PS. The capping FliD oligomer associates with the growing end of the filament and regulates the flagellin molecules which are transported through the hollow filament core to the tip of the flagella, thereby preventing leakage of flagellin subunits. The result of such interaction is a controlled incorporation of flagellin subunits and the elongation of a long helical filament (Homma et al., 1984; Ikeda et al., 1985; Ikeda et al., 1987; Yonekura et al., 2000). The capping protein FliD recently was identified in *B. burgdorferi* as a 665 amino acid (aa) protein roughly 20 kDa larger than the cap proteins of other bacteria (Zhang et al., 2019). Sequence alignment showed that the FliD of *B. burgdorferi* carries an extra variable region of 185 aa (VR) in its D3 domain. Phylogenetic analysis further revealed that such VR regions are exclusively found in spirochetes. This VR region forms a disordered loop along the peripheral region of the cap plate, which may serve to protect FliD from enzymatic degradation by proteases residing in the PS (Coleman et al., 2013; Zhang et al., 2019). Construction and analysis of a FliD variant with no VR region should clarify the function of the loop in FliD. Using cryo-ET, the structure of the FliD capping complex forms a pentamer-shaped oligomer with a 10 nm diameter. Surprisingly, depletion of FliD negatively impacts the expression levels of the two filament components, FlaA and FlaB, but no other PF-related structural proteins. Like other flagellated bacteria, the PFs of *B. burgdorferi* are assembled in a sequential manner, starting with the basal body/motor complex, next the hook, and finally the flagellar filament (Zhao et al., 2013). Deletion of *fliD* leads to the formation of rod-shaped and non-motile cells, as well as prematurely truncated, curved, and disoriented flagellar hooks, which also are observed in *flaB* mutants (Zhang et al., 2019). In the absence of the cap protein, newly synthesized unpolymerized flagellin subunits leak into the PS via T3SS and are degraded by proteases such as the serine protease HtrA (Coleman et al., 2013). Hence, we propose that FliD acts as a "plug" to stop the leakage of FlaB molecules and governs the assembly of the filament in the PS (Zhang et al., 2019). As mentioned, it would be interesting to determine if the VR region is necessary for this function.

3D Visualization of flagellar assembly

The bacterial flagellum is built from the inside out, from proximal to distal structures, in a temporally and spatially regulated fashion (Suzuki et al., 2004). By arresting assembly with a series of genetic mutations in *B. burgdorferi*, cryo-ET imaging of the motor provided a unique structural blueprint for the assembly process of this intricate molecular machine in intact cells (Figure 2I) (Zhao et al., 2013).

Secretion of rod substrates triggers remodeling of the central channel in the flagellar secretion apparatus from a closed to an open conformation. This open channel then serves as both a gateway and a template for flagellar rod assembly. The individual rod proteins (FlgB, FlgC, FlhO, FlgG) assemble sequentially to form a modular rod. The hook cap initiates hook assembly upon completion of the rod, and the filament cap facilitates filament assembly after the formation of the mature hook (Figure 2I). Given the intricate step-wise construction of flagella, disruption at any steps of this assembly process often results in mutants with defective flagellar structure, motility, and altered cell morphology (Charon et al., 2012).

Perspective for future study

Although the structure and function of bacterial flagellum have been studied for several decades, many important questions remain to be addressed. For example, how does the stator couple the proton gradient to generation of torque? How does the proton gradient power the export apparatus and facilitate protein transport across the inner membrane? Why is the spirochete hook covalently cross-linked? PFs raise additional questions: How do PFs coordinate their rotation from the two cell poles? How does the spirochete specific collar assemble?

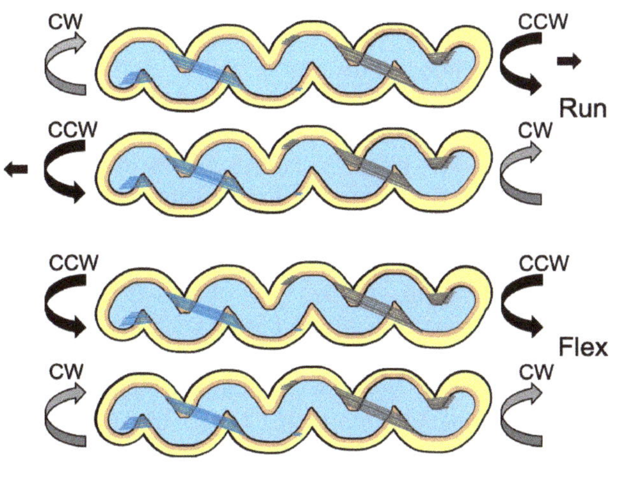

Figure 4. The four swimming configurations of *B. burgdorferi* as a function of the rotational direction of the PFs. Straight arrows at the cell ends depict the swimming direction of the spirochetes. Curved arrows depict the direction of the PF rotations. Top two panels show the two translational forms of *B. burgdorferi* or "run", while bottom two panels depict the non-translational forms or "flex". Reprinted and modified with permission from reference (Charon et al., 2012). Abbreviations: CW, clockwise; CCW, counterclockwise.

Given that flagella play critical roles in many bacterial pathogens, it will be important to understand not only the conserved structure and function among bacterial flagella, but also the specific features that set PFs apart from others.

Dynamics of Flagellar Motility

Asymmetrical rotation of PF

Bacterial motility is defined as the ability of a bacterium to move in an environment. For flagellated bacteria, this movement is primarily powered through the flagellar organelles. In most externally flagellated bacteria, there are two different swimming modalities controlled by the rotational mode of the flagellar motor: runs and tumbles (Berg, 1975; Macnab, 1996; Armitage, 1999; Berg, 2000). The motor can rotate either CW or CCW, which determines the mode. For example, in E. coli the flagella are externally attached around the entire cell body (peritrichous flagella). A run occurs when all the motors rotate CCW. During a run, the flagellar filaments form a bundle behind the cell, acting as a propeller. The bundle pushes the cell body forward (the flagellum and its direction of rotation is determined by viewing from the filament tip towards its insertion into the cell). When one or more of the rotors reverse direction to CW, the flagella bundle unravels and the cell body reorients and moves in an erratic fashion with little net displacement, and the cell "tumbles" (Macnab, 1996; Armitage, 1999; Berg, 2000; Turner et al., 2000). In a uniform environment with no chemical gradient, the cell runs and tumbles, performing a three-dimensional random walk (Berg, 2018). In other species, such as *Rhodobacter sphaeroides* and *Sinorhizobium meliloti*, instead of changing the direction of the flagella rotation, the organelles stop or reduce the speed of rotation, respectively, allowing the cells to reorient their motility mode (Armitage and Schmitt, 1997). Because of the complex geometry resulting from having periplasmic flagella and bundles of PFs at each cell pole, the swimming behavior of spirochetes is considerably more complicated than that of most bacteria (Charon et al., 2012). *B. burgdorferi* has four swimming configurations, depending on the direction of the flagella rotation: two translational/run modes (with each cell end as the leading end) and two non-translational modes (Figure 4) (Charon et al., 1984; Fosnaugh and Greenberg, 1988; Goldstein et al., 1994; Li et al., 2002). During run, the ribbons of PFs rotate asymmetrically, i.e. the anterior bundle rotates in one direction (e.g., CCW), and the posterior bundle rotates in the opposite direction CW (as a frame of reference, the PFs are viewed from the distal tip to their sites of insertion in the PC). Reversal occurs when both bundles switch their orientation of rotation. When both ribbons rotate in the same direction (i.e. both CCW or both CW), the cells are in non-translational mode. During this non-translational interval, many spirochetes become distorted and may bend in mid-cell, forming what is termed as a "flex", which is analogous to the "tumble" in exo-flagellated bacteria (Figure 4) (Fosnaugh and Greenberg, 1988; Goldstein et al., 1994). For most spirochete species, the two non-translating modes are morphologically identical except for *Leptospiraceae*, in which CCW and CW rotation result in spiral-shaped and hook-shaped cell ends, respectively; hence the two non-translating forms are hook-hook and spiral-spiral ends (Noguchi, 1918; Berg, 1978; Charon et al., 1984; Goldstein and Charon, 1988, 1990). Given that spirochetes have bipolar PFs, one could argue that only one PF bundle is actively rotating during translational motility. On the other hand, the other possibility is that both bundles are capable of simultaneously rotating both bundles of PFs during translational motility. To distinguish between these two possibilities, wild-type *B. burgdorferi* and a chemotaxis mutant cheA1cheA2 cells were tethered to a glass surface, and each cell end was monitored by tracking its beat frequency (i.e., number of gyrations per second) and direction of gyration (Li et al., 2002). The cheA1cheA2 mutant was chosen as it only runs in one direction and does not reverse as does the wild-type. The results indicate that both ends of the wild-type cell actively gyrate and independently change their gyrational directions (either CW or CCW). In contrast, the *cheA1cheA2* mutant, which gyrated both ends, could only gyrate its ends in one direction. These results indicate that the running phenotype of this mutant results in both ends simultaneously gyrating, with the cell swimming in one direction. Furthermore, in both the wild-type and mutant cells with actively gyrating ends, the beat frequency of one end was occasionally different from the other end, further suggesting that each cell end is capable of independent gyration. Analyses of other spirochete species yielded similar results (Charon et al., 1984; Charon et al., 1991; Nakamura, 2020).

Swimming behavior of B. burgdorferi in vitro

The swimming behavior of *B. burgdorferi* has been studied in detail. During translational motility, the rotation of PC is in the opposite direction of the rotating PFs. For example, if the PFs rotate in a CCW fashion as viewed from the back of the cell, the torque of the rotating bundle produces a CW counter rotation of the cell body. In addition, the rotating PFs cause the cell to gyrate with a concomitant flat wave moving backward along the length of the cell. The beat frequency of the wave is 5-10 Hz in liquid medium at room temperature (Goldstein et al., 1994; Li et al., 2002). The propagating waves in *B. burgdorferi* have the same wavelength as the helix pitch of the PF bundle and are propagated at a speed of 34 µm/sec relative to the cell and the mean swim speed of the cell is 4.25 µm/s in liquid medium (Goldstein et al., 1994). The ratio of the swim speed to wave propagation is 0.12, which means that the cell advances 12% of its length through the medium in the time it takes a wave to travel the entire length of the cell. Thus, backward moving waves are evident as cells swim in a given direction.

Recent mathematical modeling indicates that during translation the rotating PF bundles are not in direct contact with the PC, rather with a thin layer of lubricating fluid separating the two (Yang et al., 2011). Thus, the forces between PF and PC are, therefore, mediated by viscous forces in the fluid inside the periplasm. Based on this mathematical modeling, the thin layer of lubricating fluid is essential for the smooth backward propagating waves observed in translating cells. Alternately, if the bundles were in direct contact with the cell cylinder, the PF would become tangled and the propagating waves would not be as smooth and even as observed. In its natural habitat, *B. burgdorferi* constantly encounters environments that do not behave like liquid medium under laboratory conditions. When deposited into the dermis of mammal via tick bites, *B. burgdorferi* is in a gel-like surrounding composed largely of cross-linked collagen and glycosaminoglycans (GAG) network. The concomitant counter rotation of the cell cylinder as the PFs rotate enables spirochetes to penetrate highly viscoelastic surroundings, such as the connective tissue (Harman et al., 2012). Within the dermis layer, *B. burgdorferi* actively interacts with various extracellular matrix (ECM) components, such as fibronectin, decorin, and collagen, via its OM proteins [see(Coburn et al., 2013; Brissette and Gaultney, 2014; Petzke and Schwartz, 2015; Hyde, 2017; Lin et al., 2017) for recent reviews]. To better simulate the in vivo environment of dermis, gelatin matrices were utilized as an in vitro equivalent of ECM to study the relationships between borrelial motility and adhesion and dissemination (Zambrano et al., 2004; Dunham-Ems et al., 2009). Four motility states have been noted using gelatin, as determined by transient adhesion of the cell to the matrix, similar to what has been observed in mouse dermis. The four stages of motility are wriggling (where the cells undulate with propagating wave but not translating), lunging (where one cell end is fixed with the remainder of the cell body bent and relaxed as if trying to break free from its tether), translocating, and non-motile (stationary phase). Under such conditions, *B. burgdorferi* is penetrating and migrating through matrices with pore sizes that are much smaller than the diameter of the spirochete. However, the spirochete is swimming slower in gelatin matrix than in liquid media (Harman et al., 2012). Therefore, the unique motility mechanism of *B. burgdorferi* empowers it to penetrate dense tissues in mammals, but cell velocity is not enhanced during penetration.

Swimming behavior of B. burgdorferi *in vivo*

Recently, the movement of *B. burgdorferi* within the mammalian host has been analyzed. The first direct insight into the swimming behavior of *B. burgdorferi* in vivo came in 2008, when Moriarty et al. successfully imaged the dissemination of this spirochete in the vasculature of mice (Moriarty et al., 2008). Using green fluorescent protein (GFP)-tagged *B. burgdorferi* and spinning disk confocal intravital microscopy, real-time images were obtained to document the dissemination of fluorescent spirochetes from the microvasculature of living mice. Soon after injection, fluorescent spirochetes were seen interacting with capillaries and veins but not arterioles, most likely due to the elevated shear forces in arteries as compared to capillaries and veins. Such interactions only occurred with a virulent infectious *B. burgdorferi* strain, but not with an avirulent strain. These results suggest the involvement of surface-exposed virulence factors that are missing in the avirulent strain background. Three types of movements were documented before the spirochetes extravasate. The cells are at first transiently tethered to the endothelium, followed by short term dragging along the vasculature endothelium before coming to a stationary adhesion, where the spirochetes were partially or entirely embedded in the endothelium with one end gyrating before escaping the microvasculature. Roughly 90% of the interactions with endothelium were less than 1 sec and about 10% were 1-20 sec. In those stationary attached cells to the endothelium, approximately 70% occurred at the endothelial cell junctures, which is the preferred location for *B. burgdorferi* extravasation (Moriarty et al., 2008). BBK32, a fibronectin and GAG binding protein, augments the early interaction with vasculature endothelium (Norman et al., 2008; Moriarty et al., 2012), which subsequently facilitates P66 mediated integrin adhesion leading to transmigration of the spirochete into the connective tissues (Kumar et al., 2015). When *B. burgdorferi* is in the bloodstream, it encounters various innate immune cells and molecules whose sole purpose is to eliminate pathogens. In the liver microvasculature, Kupffer cells are the first line of defense against *B. burgdorferi*, which trap, immobilize, and engulf spirochetes before presenting processed antigens to resident invariant natural killer T cells. Thus, Kupffer cells play an important role in the clearance of blood-borne *B. burgdorferi* from the vasculature (Lee et al., 2010). The role of motility and chemotaxis in both the adherence and penetration through blood vessels, and in the interaction with immune cells in the liver, is likely important for the pathogenesis of *B. burgdorferi*. Future experiments using specific motility and chemotaxis mutants would help delineate the definitive role of each factor in the virulence and immune evasion of *B. burgdorferi*. As a tick-borne pathogen, *B. burgdorferi* spends half of its enzootic life cycle in the tick vector. Using isolated tick midguts and salivary glands before and after feeding on mice infected with GFP-tagged infectious strain of *B. burgdorferi*, analysis of the routes and dissemination mechanisms within ticks are rendered possible. Detailed swimming behavior of *B. burgdorferi* within tick vector will be discussed in detail later in this chapter on the role of motility and chemotaxis in LD.

Chemotaxis

Unique chemotaxis in B. burgdorferi

Bacterial chemotaxis is a complex sensory transduction pathway that enables cells to respond to environmental stimuli. During chemotaxis, bacteria undergo a biased random walk, which is achieved via modulation of the speed and direction of flagella rotation (Wadhams and Armitage, 2004; Sourjik and Armitage, 2010; Porter et al., 2011). The most studied organisms are *E. coli* and *S. enterica*, which serve as the prototypes for understanding the molecular mechanism of chemotaxis (Armitage, 1999; Bren and Eisenbach, 2000; Hazelbauer et al., 2008). A two-component phospho-relay system mediates the biased walk allowing cells to move toward a favorable condition or away from toxic one. A family of chemoreceptor proteins, referred to as the methyl-accepting chemotaxis proteins (MCPs), is coupled to CheA, a histidine kinase, by CheW.

During chemotactic responses, MCPs sense attractants or repellents and accordingly regulate the activity of CheA via autophosphorylation. Activated CheA relays the phosphoryl group to CheY (a response regulator) producing CheY-P, which, in turn, interacts with the motor switch complex (the C-ring), causing CW rotation. As noted above, CCW rotation results in smoothly swimming cells (runs), while CW rotation leads to tumbling. When MCPs bind attractants, runs are prolonged, yielding a biased random walk toward attractants (Wadhams and Armitage, 2004). Null mutants in either *cheY* or *cheA* are non-chemotactic and the flagella continuously rotate CCW; cells, therefore, run in one direction. The phosphatase CheZ dephosphorylates CheY-P; thus, mutants of *cheZ* constantly tumble (Boesch et al., 2000; Zhao et al., 2002). Bacteria quickly reset to the pre-stimulus state once they reach an environment with the same concentration of attractants. This process, known as adaptation, is controlled by CheR, a methyltransferase that specifically methylates MCPs. Conversely, CheB, a methylesterase, promotes demethylation of the MCPs so that the receptors can be excited again to sense chemical gradients (Porter et al., 2011).

The chemotaxis of *B. burgdorferi* is enigmatic (Goldstein et al., 1994; Charon and Goldstein, 2002; Li et al., 2002; Wolgemuth et al., 2006; Charon et al., 2012; Motaleb et al., 2015; Wolgemuth, 2015). There are some obvious differences between *B. burgdorferi* and the *E. coli-S. enterica* paradigm. One major difference is the copy number of chemotaxis genes. Whereas *E. coli* and *S. enterica* have a single copy of genes in the general chemotaxis pathway, *B. burgdorferi* has two *cheAs*, three *cheWs*, three *cheYs*, two *cheBs*, and two *cheRs* (Fraser et al., 1997). Second, *B. burgdorferi* lacks CheZ but encodes a CheX, which, as discussed below, is a phosphatase active on CheY-P (Fraser et al., 1997; Charon et al., 2012). Finally, *B. burgdorferi* has three modes of motility in vitro, namely, run, reverse, and flex, instead of the two presented in *E. coli* and *S. enterica*. *B. burgdorferi* must be able to control the direction of the PF at each cell end to modulate its chemotactic response. For instance, to run, the two PF bundles in any given cell must rotate in the opposite orientation, as viewed from the center of the cell (Figure 4). In addition, it only takes less than 0.3 s for a cell to reverse directions; in other words, during a reversal, both groups of PF bundles change direction almost simultaneously (Wolgemuth, 2015). The molecular mechanism that controls the asymmetric rotation of PF still remains unknown, though several models have been proposed (Fosnaugh and Greenberg, 1988; Charon et al., 2012).

Identification of chemoattractants and localization of chemoreceptors

A rapid and quantitative chemotaxis assay was developed to identify chemo-attractants to which *B. burgdorferi* responds. Initially, rabbit serum and tick saliva were found to be attractants by using chemotaxis capillary assays in context with manually cell counting (Shi et al., 1998; Shih et al., 2002). This method was modified by enumerating cells migrating into capillary tubes using flow cytometry (Bakker et al., 2007; Motaleb et al., 2007a). Using this method, several defined chemicals and host components were identified as chemo-attractants including glucosamine, N-acetylglucosamine (NAG), glucosamine dimers (chitosan), glutamate, and tick salivary gland protein Salp12 (Bakker et al., 2007; Motaleb et al., 2007a; Murfin et al., 2019). A microscopic agar drop assay (MAA) was recently developed to study the chemotaxis of Leptospira, which could potentially be adopted to measure *B. burgdorferi* chemotaxis and motility (Islam et al., 2014; Affroze et al., 2016). GFP fusions, immunofluorescence microscopy, and cryo-ET revealed that the chemoreceptors of *B. burgdorferi* form long, thin arrays near both cell poles in close proximity to the flagellar motors (Figure 5) (Xu et al., 2011). Genetic studies further demonstrate that an internal MCP (BB0569) and CheW3 are essential for the array formation (Zhang et al., 2012a; Zhang et al., 2016).

Functional studies of chemotaxis genes

B. burgdorferi encodes two putative chemotaxis histidine kinases CheA1 and CheA2 (Li et al., 2002). Amino acid sequence alignments indicate that *B. burgdorferi* CheA2 is more closely related than CheA1 to the CheA proteins of other spirochetes, such as *T. pallidum* (40% identity) and *T. denticola* (42% identity). In contrast, *B. burgdorferi* CheA1 is more closely related to CheA proteins from non-spirochetes, such as *R. sphaeroides* (35% identity) and *Vibrio cholerae* (35% identity). These results suggest that CheA1 is likely to be a recent acquisition from the Proteobacteria, whereas CheA2 is the ancestral protein (Li et al., 2002). Targeted mutagenesis revealed that deletion of cheA2 led to mutant cells constantly running in one direction and being unable to sense attractants such as rabbit

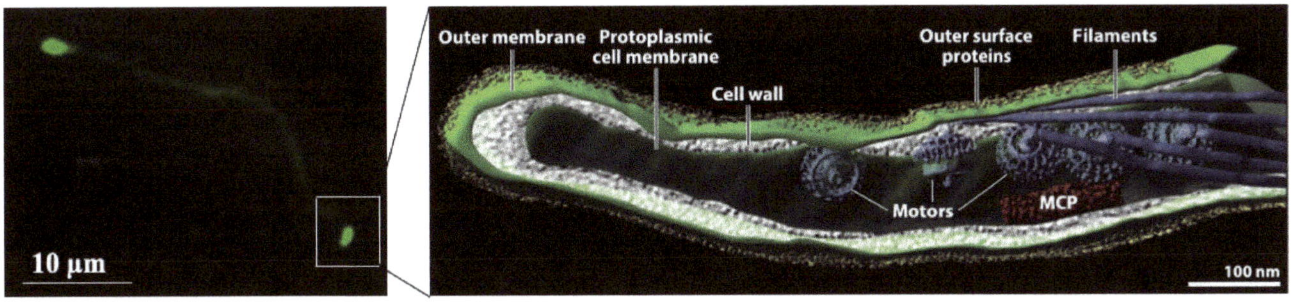

Figure 5. Cellular localization and architecture of chemoreceptors (MCPs) in *B. burgdorferi*. Left: Localization of MCPs by using GFP fusion protein. Right: Cellular architecture of one end of *B. burgdorferi* revealed by cryo-ET. A three- dimensional model was generated by manually segmenting the outer membrane, protoplasmic cell membrane, flagellar motors and filaments, methyl-accepting chemotaxis protein (MCP) array, and outer surface proteins. Reprinted and modified with permission from reference (Xu et al., 2011).

serum and NAG (Li et al., 2002; Sze et al., 2012). In contrast, deletion of cheA1 had no impact on the cell swimming behavior and chemotactic responses, indicating that cheA2, but not cheA1, is essential for the chemotaxis of *B. burgdorferi* in vitro (Li et al., 2002). Interestingly, in vitro phosphotransfer assays demonstrated that both CheA1 and CheA2 could be activated via autophosphorylation and transfer the phosphoryl group to all three CheYs (Motaleb et al., 2005; Motaleb et al., 2007b; Motaleb et al., 2011). The role of CheA1 still remains elusive. One possibility is that CheA1 may play a role in vivo, e.g., in a specific stage during the tick-mouse infection cycle which cannot be recapitulated in vitro.

The three chemotaxis response regulators share significant amino acid sequence identity with the CheYs from *E. coli* and *Bacillus subtilis*. While *B. burgdorferi* CheY2 is more similar to CheY of *E. coli*, CheY1 and CheY3 share highest amino acid sequence homology to their counterpart in *B. subtilis* (Motaleb et al., 2011). CheY has been well characterized in *E. coli*, and the key residues required for its function has been mapped. Sequence alignments reveal that all three *B. burgdorferi* CheYs contain those key residues but with some variations (Motaleb et al., 2011). Specifically, N59 and E89 of *E. coli* CheY, important for CheY auto-dephosphorylation (Silversmith et al., 2001; Silversmith et al., 2003), are present in CheY2 but not in CheY1 or CheY3; T81 of CheY3, which is involved in the CheX-mediated dephosphorylation of CheY3-P (Pazy et al., 2010), is found in CheY2 but not in CheY1 (Motaleb et al., 2011). In addition, phosphorylation assays suggested that the half-life of CheY2-P was remarkably shorter than those of CheY1-P and CheY3-P (H. Xu and M. Motaleb, unpublished). Moreover, structural analyses using crystallography showed that the helix (N76-L80) between helix α3 and strand β4 of CheY1 is absent in CheY2 and CheY3 (M. Lynch, B. Crane, H. Xu and M. Motaleb, unpublished). Targeted mutagenesis in a non-infectious *B. burgdorferi* strain revealed that only *cheY3* is essential for the chemotaxis of *B. burgdorferi* in vitro (Motaleb et al., 2011).

CheR and CheB play important roles in the chemotaxis of other bacteria including the spirochete Leptospira (Dong et al., 2010; Zhang et al., 2013); however, very little is known about these signaling regulators in *B. burgdorferi*. A transposon mutagenesis study indicated that *cheR2* and *mcp5* mutants had reduced motility (Lin et al., 2012). An atypical MCP protein, BB0569, was recently identified and reported to be essential for the motility and chemotaxis in *B. burgdorferi*. A *bb0569* mutant constantly ran in one direction and was also deficient in response to NAG, which is similar to the phenotype of *cheA2* and *cheY3* mutants. BB0569 was shown to directly interact with other MCPs affecting the polar localization of those receptors, and, therefore, it was concluded that BB0569 contributes to the stabilization of other MCPs (Zhang et al., 2016).

CheW couples MCPs to CheA forming a ternary complex (MCP-CheW-CheA) to facilitate signal transduction during chemotaxis (Liu and Parkinson, 1989; Gegner et al., 1992; Boukhvalova et al., 2002; Wadhams and Armitage, 2004; Hazelbauer et al., 2008). The three CheWs of *B. burgdorferi* share 28-30% amino acid sequence identity with *E. coli* CheW. Interestingly, CheW3 contains two domains: a N-terminus CheW domain (aa 26-165) and a C-terminus (aa 196-466) CheR–like domain (Zhang et al., 2012a). Targeted mutagenesis studies indicate that CheW1 and CheW3, but not CheW2, are required for the chemotaxis and chemoreceptor formation at the cell pole of *B. burgdorferi* (Xu et al., 2011; Zhang et al., 2012a). Biochemical, genetic and in vivo analyses reveal that those multiple chemotaxis proteins form two different chemotaxis signaling pathways: CheA2-CheW1/ CheW3-CheY3, which is essential for chemotaxis in vitro and CheA1-CheW2-CheY1/CheY2, which is not required for chemotaxis in vitro but affects infectivity of *B. burgdorferi* (Li et al., 2002; Motaleb et al., 2011; Sze et al., 2012; Zhang et al., 2012a; Novak et al., 2016; Xu et al., 2017).

As mentioned earlier, *cheZ* phosphatase mutants of *E. coli* and *S. enterica* constantly tumble. The explanation for this phenomenon is that in a *cheZ* mutant, the pool of CheY-P is relatively high, which, in turn, increases its probability of CW flagellar rotation. Similar to CheZ, the CheX of *B. burgdorferi* functions as a phosphatase that dephosphorylates CheY-P (Motaleb et al., 2005; Motaleb et al., 2007b). Targeted mutagenesis revealed that deletion of *cheX* led to mutant cells that are constantly flexing and unable to respond to attractants (Motaleb et al., 2005). *B. burgdorferi* possesses a *cheD* homolog (Fraser et al., 1997). In *B. subtilis*, CheD is directly involved in motility and chemotaxis through deamination of MCPs and by enhancing the activity of phosphatase CheC (Kristich and Ordal, 2002; Chao et al., 2006; Glekas et al., 2012; Yuan et al., 2012). In *B. burgdorferi*, CheD enhances the activity of CheX and also directly interacts with MCPs even though its function as a deaminase activity has not been established due to technical difficulties (Moon et al., 2016a). As *B. burgdorferi cheX* mutant constantly flexes, one would imagine that a cheD mutant should also be flexing due to reduced phosphatase activity of CheX. Interestingly, the swimming pattern of a *cheD* mutant is indistinguishable to that of the wild type, though its swarming ability is reduced (Moon et al., 2016a). The underlying mechanism remains unknown.

Structural analysis of chemotaxis proteins

Recently, there have been some notable advances in elucidating the structure and enzymatic mechanism of CheX. CheX is a homolog of CheC and FliY, two chemotaxis phosphatases present in other bacteria; however, it bears no sequence similarity to CheZ, a chemotaxis phosphatase of *E. coli*. Although the overall structure of CheX is distinct from that of CheZ (Zhao et al., 2002), the two proteins share some striking structural similarities. *E. coli* CheZ is a homodimer with two identical CheY binding sites (Silversmith et al., 2008). Each CheY binding site on CheZ has two conserved residues (an aspartate and a glutamine) exactly one alpha helical turn apart that directly insert into the CheY active site (Zhao et al., 2002). Co-crystal structural analysis of CheX·CheY3·BeF_3·Mg^{2+} revealed that CheX also contains two conserved CheY binding sites (Glu96 and Asn99) that are inserted into the active site of CheY3 (Pazy et al., 2010). Structural analyses of CheZ·CheY·BeF_3· Mg^{2+} and *B. burgdorferi* CheX·CheY3·BeF_3·Mg^{2+} further revealed that CheZ and *B. burgdorferi* CheX share virtually identical catalytic mechanisms despite having distinct binding modes with CheYs (Pazy et al., 2010). *B. burgdorferi* CheX is not present as a homodimer in the co-crystal complex, even though it is dimeric in solution (Motaleb et al., 2005; Pazy et al., 2010). Evidently, CheY3 induces dissociation of CheX, but the mechanism to cause this dissociation remains elusive. In *E. coli*, CheZ accelerates the intrinsic auto-dephosphorylation rate of CheY-P by about 100-fold, which consequently decreases the half-life of CheY-P to correlate with the chemotactic response time (~200 milliseconds) (Segall et al., 1982; Wadhams and Armitage, 2004; Silversmith et al., 2008; Pazy et al., 2010). The catalytic activity of CheZ is suppressed at low CheY-P concentrations and enhanced at high CheY-P levels (Silversmith et al., 2008). This type of allosteric regulation may be mediated by the dimeric formation of CheZ; it is not known if it occurs with *B. burgdorferi* CheX.

Regulation of motility chemotaxis genes

Motility and chemotaxis gene organization

B. burgdorferi devotes at least 5% of its genes to motility and chemotaxis, which emphasizes the importance to the lifecycle of the spirochete (Fraser et al., 1997; Charon and Goldstein, 2002). At least 57 putative motility and chemotaxis genes have been identified in *B. burgdorferi* via targeted mutagenesis, sequence homology to known motility and chemotaxis genes in other bacteria, or by virtue of their clustering with other known motility genes (Figure 6). In analyzing the motility genes of *B. burgdorferi*, the deduced amino acid sequences and the order of genes are most similar to those of other spirochete species, followed by *B. subtilis.* Since spirochete motility genes have high homology to other known motility genes, the basic architecture of

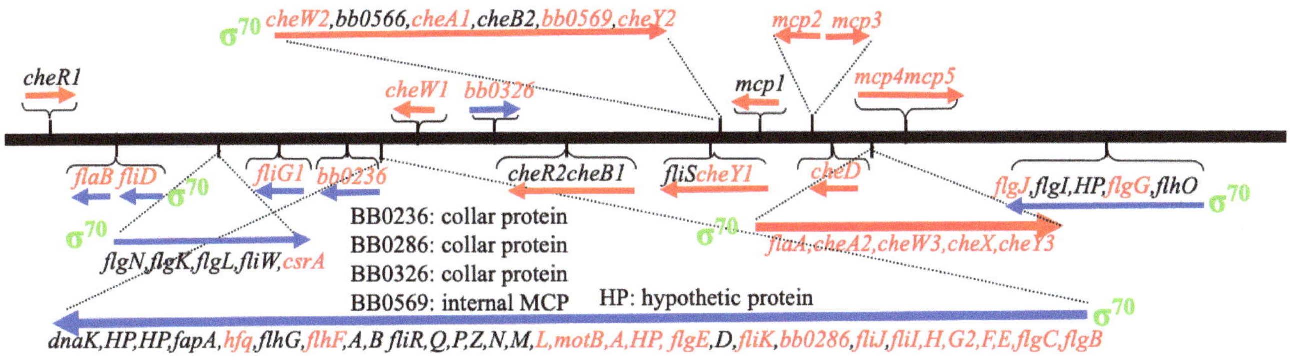

Figure 6. A genetic map of *B. burgdorferi* motility and chemotaxis genes. The functions of genes in red color are established and published (Motaleb et al., 2000; Li et al., 2002; Motaleb et al., 2005; Li et al., 2010; Sze and Li, 2011; Sze et al., 2011; Lin et al., 2012; Sze et al., 2012; Zhang et al., 2012a; Zhang et al., 2012b; Zhao et al., 2013; Lin et al., 2015; Moon et al., 2016b; Novak et al., 2016; Zhang et al., 2016; Xu et al., 2017; Moon et al., 2018; Qin et al., 2018; Chang et al., 2019; Zhang et al., 2019; Xu et al., 2020; Zhang et al., 2020). σ^{70}: sigma^{70}-like promoter sequences are mapped by either 5'RACE or primer extension (Ge and Charon, 1997b; Ge et al., 1997a; Ge et al., 1997b; Li et al., 2002; Yang and Li, 2009; Sze et al., 2011; Zhang et al., 2012b).

the flagella and the mechanism of flagellar rotation are well conserved. It is estimated that *B. burgdorferi* has roughly 16 motility operons with some genes co-transcribing with non-motility or chemotaxis-related genes (Figure 6). Most motility-related genes, aside from the chemotaxis genes, are encoded as single copy except for *fliG*, the flagellar motor switch protein as previously iterated (Fraser et al., 1997). One large motility operon comprising of 31 open reading frames (*orf*) resembles a gene cluster found in *B. subtilis* (Ge et al., 1997b; Zhang et al., 2020). Using primer extension and reverse transcription PCR (RT-PCR) analysis, a single σ^{70}-like promoter was mapped upstream of the first gene in this cluster, *flgB* (*bb0294*), with a single transcript of 26.3 kb in size, extending through the last gene in this operon, *dnaK* (*bb0264*). Many genes in this cluster were initially designated as *flb* genes despite their having no obvious homologs in non-spirochetes (Fraser et al., 1997; Ge et al., 1997b). Because *B. burgdorferi* has differences in comparison to most flagellated bacteria with respect to motility gene regulation (see below), the requirement for coordination of the two bundles of PFs, flagellar assembly and motor structure, some of these *flb* genes may eventually be identified as serving in one of these functions. For example, some of the *flb* genes were recently found to encode the motor collar proteins (Moon et al., 2016b; Xu et al., 2020).

With respect to the chemotaxis genes, it is presently unclear why multiple *cheA*, *cheY*, *cheW*, *cheB*, and *cheR* homologs are needed. Most of the chemotaxis genes are located within two gene clusters, *flaA-cheA2-cheW3-cheX-cheY3* and *cheW2-bb0566-cheA1-cheB2-bb0569-cheY2*. The remainder of the chemotaxis genes are scattered around the genome (Figure 6) (Fraser et al., 1997; Ge and Charon, 1997b). As previously mentioned, *B. burgdorferi* could conceivably utilize different chemosensory machineries under different environment during its infectious cycle, i.e., CheA2 pathway modulates chemotaxis during transmission/persistence in mammalian hosts, whereas the CheA1 pathway constitutes a second chemotaxis pathway to guide chemotactic responses in the tick vector, although there is no evidence for the latter. Along these lines, a body of microarray and transcriptomic analyses indicates that some of these chemotaxis genes are indeed differentially expressed under various culture conditions and during transmission or in response to different stimuli (Revel et al., 2002; Brooks et al., 2003; Ojaimi et al., 2003; Tokarz et al., 2004; Ouyang et al., 2008; Iyer et al., 2015).

Unique regulatory mechanism

The regulation of motility gene expression has been extensively studied in model organisms such as *E. coli* and *S. enterica*. Motility gene expression and flagella synthesis and assembly require tight coordination and are subject to a cascade of transcriptional events involving sequential expression of class1, class 2, and class 3 genes(Chilcott and Hughes, 2000; Aldridge and Hughes, 2002).

Expression of the class 1 *flhDC* master operon directs the transcription of the class 2 genes encoding structural proteins for the basal body as well as specific regulatory proteins which control the expression of the class 3 genes, encoding flagellin and chemotaxis proteins. Among the class 2 regulatory proteins are FliA, the flagellar alternative sigma factor σ^{28} and FlgM, the anti- σ^{28} protein. During basal body and hook synthesis and assembly, FliA and FlgM remain bound as a complex in the cytoplasm. Upon completion of the hook structure, the hook-basal body complex allows the export of FlgM, relieving inhibition on FliA and enabling initiation of transcription from the σ^{28}–dependent class 3 genes. This elaborate hierarchy control of gene expression enables bacteria to set up checkpoints to ensure proper assembly of hook-basal body structure prior to the synthesis of late flagellar proteins (Hughes et al., 1993). Similar schemes of transcriptional cascade in motility gene expression is found in various species of bacteria, with some utilizing an additional sigma factor, σ^{54} RpoN, in the hierarchical transcriptional control of motility gene expression (Anderson et al., 1995; McCarter, 2001; Aizawa S, 2002; Dasgupta et al., 2003; Niehus et al., 2004). In *B. burgdorferi*, the transcriptional cascade regulatory mechanism in PF synthesis is absent as no homologs of class 2 regulatory genes *fliA* and *flgM* have been identified. In addition, no σ^{28} promoter consensus sequences are evident in its genome. All of the chemotaxis and motility genes identified thus far are under the regulation of the house-keeping σ^{70}, emphasizing that *B. burgdorferi* lacks a transcriptional cascade control in its gene expression (Fraser et al., 1997; Ge et al., 1997a; Li et al., 2000b). Instead, *B. burgdorferi* employs a post-transcriptional regulatory mechanism for its flagellar synthesis as revealed by studies using various flagellar mutants. The *flaB* mutant of *B. burgdorferi* is aflagelated, non-motile and has a rod-shaped morphology (Motaleb et al., 2000). In this mutant, only the minor filament protein FlaA is synthesized but expressed at only 13% of its wild type level despite wild-type levels of the *flaA* transcript. These results were the first key evidence that the decrease of FlaA in *flaB* mutant occurs post-transcriptionally (Motaleb et al., 2004).

In the hierarchical control of gene expression, the hook provides a regulatory checkpoint during flagella synthesis. Thus, mutants that fail to synthesize the hook protein lack flagellar filament or class 3 chemotaxis gene transcripts. When the hook gene *flgE* is inactivated by targeted mutagenesis in *B. burgdorferi,* the level of *flaA* and *flaB* mRNA remain unchanged even though the accumulation of FlaA and FlaB proteins is markedly inhibited. In addition, inactivation of *flgE* does not impact the synthesis of CheA2, CheX, and CheY3 motility proteins (Sal et al., 2008). These results further reinforce the proposition that post-transcriptional control is involved in flagellar gene expression and that *B. burgdorferi* lacks the canonical cascade control of gene transcription found in other bacteria. Lastly, a protease residing in the PS was identified that is associated with FlaB turnover in mutants unable to form flagella, as in the *flgE* mutant described above. In the functional analysis of the filament cap, a *fliD* mutant expressed wild-type levels of *flaB* transcript despite a significant reduction in FlaB protein. The PS serine HtrA protease was responsible for FlaB degradation, as a deletion of *htrA* in *fliD* mutant background successfully restored the level of FlaB protein (Zhang et al., 2019). In sum, these mutational studies bolster the concept of post-transcriptional or translational regulation of chemotaxis and motility gene expression in *B. burgdorferi*.

The Csr system was first identified in *E. coli* as the global RNA-binding regulatory element for various biological processes, including motility and virulence. In *E. coli*, CsrA functions by binding to the consensus sequence RUACARGGAUGU, which is present within the leader region of its targeted transcripts and subsequently regulates gene expression post-transcriptionally [see reviews (Romeo, 1998; Romeo et al., 2013; Vakulskas et al., 2015; Romeo and Babitzke, 2018)]. Also, CsrA is a positive regulator of the *flhDC* master operon which encodes a DNA binding activator of flagella biosynthesis and chemotaxis genes(Wei et al., 2001; Yakhnin et al., 2013). The involvement of the carbon storage regulator A (CsrA) in the translational regulation of the flagellin protein FlaB marked the first known governance mechanism of flagellar synthesis in *B. burgdorferi* (Sze et al., 2011). In *B. burgdorferi*, *csrA* is clustered within several flagellar related genes in the *flgK* operon (Ge et al., 1997a). In contrast to *E. coli*, CsrA expression in *B. burgdorferi* negatively influences the synthesis of the major flagellin protein. Specifically, over-expression of CsrA (CsrA+) leads to reduced accumulation of FlaB protein. Such cells lack PFs and are rod-shaped. On the other hand, the inhibition of CsrA synthesis in a csrA mutant results in a 4-fold increment in FlaB accumulation with longer PFs and more inter-digitation (overlap) of the PFs in the center of cells. Such cells have shorter wavelengths (2.16 µm) than the wild-type (3.07 µm), These results indicate that PF length influences the cellular length of the flat waves, and reinforce the concept that the PFs have a skeletal function. Remarkably, both the CsrA+ and *csrA* mutants produce similar levels of *flaB* transcripts as the wild type, indicating that the control of FlaB occurs at the translational level. Two CsrA binding sites are present in the leader region of the *flaB* transcript, and one overlaps with the Shine-Dalgarno (SD) sequence. Modeling of the transcript suggests that CsrA interacts with these two regions to inhibit the translation by blocking the access of the ribosome to the SD sequence in *flaB* mRNA (Sze et al., 2011).

B. burgdorferi expresses many enzootic stage-specific genes controlled by the two-component regulatory system (TCS), HK1-Rrp1, and Rrp2-RpoN-RpoS pathway [see review (Samuels, 2011; Radolf et al., 2012; Stevenson and Seshu, 2018)]. Acetyl-phosphate, an intermediate product from the acetate metabolism, was initially thought to autophosphorylate the response regulator Rrp2 and upregulate the RpoN-RpoS TCS (Xu et al., 2010). However, subsequent results indicated that acetyl-phosphate was not involved in Rrp2 phosphorylation (Richards et al., 2015). Nevertheless, a linkage between CsrA and acetate metabolism was identified whereby CsrA serves as a repressor of phosphor-acetyltransferase (Pta), a key protein in the acetate metabolism by binding to the leader sequence of this *pta* transcript (Sze and Li, 2011). As mentioned previously, many chemotaxis- and motility-related genes are among the differentially expressed genes controlled by TCS (Revel et al., 2002; Brooks et al., 2003; Ojaimi et al., 2003; Tokarz et al., 2004; Ouyang et al., 2008; Angel et al., 2010; Iyer et al., 2015). For instance, eight chemotaxis genes are regulated by the Rrp2-RpoN-RpoS network (Ouyang et al., 2008) whose activity is highly induced in the mammalian host (Yang et al., 2000). The crosstalk between CsrA and the activation of Rrp2 coupled with the regulation of chemotaxis proteins expression via the Rrp2-RpoN-RpoS TCS may allow the spirochete to modulate its chemotactic activity during transmission to aid in colonization as well as dissemination.

In addition to motility and chemotaxis-related proteins, the second messenger, 3',5'-cyclic-diguanosine monophosphate (c-di-GMP), has emerged as a major factor influencing motility and possibly chemotaxis [see reviews(Novak et al., 2014; Orr and Lee, 2016; Baker and O'Toole, 2017; Jenal et al., 2017)]. C-di-GMP is synthesized by diguanylate cyclases (DGC) and degraded by phosphor-diesterases (PDE). C-di-GMP exerts its role via binding to its effectors such as PilZ domain containing proteins. In bacteria, such as *E. coli* and *S. enterica*, c-di-GMP negatively influences the speed and the direction of motor rotation using the c-di-GMP effector protein YcgR (Boehm et al., 2010; Fang and Gomelsky, 2010; Paul et al., 2010). C-di-GMP bound YcgR interacts with the flagellar switch-complex proteins FliG and FliM to reduce torque generation efficiency and induce CCW bias rotation. This mechanism is described as a "backstop brake" for regulation of motility and chemotaxis in bacteria (Paul et al., 2010). In addition, YcgR interacts with the stator protein, MotA, to reduce motor output and finely tune swimming velocity in response to environmental cues (Boehm et al., 2010). In *Pseudomonas aeruginosa*, c-di-GMP and its various effector proteins regulate motility along three tiers: transcription of flagellar genes via the flagellar transcriptional regulator FleQ (Dasgupta et al., 2002; Hickman and Harwood, 2008), adjustment of flagellar motor output via interaction of FlgZ with the stator protein MotC (Baker et al., 2016), and inhibition of chemotaxis signaling via interaction of MapZ with CheR (Xu et al., 2016; Yan et al., 2018). In *B. burgdorferi*, c-di-GMP is synthesized by Rrp1, a DGC (Rogers et al., 2009), and is degraded by two PDEs, PdeA and PdeB (Sultan et al., 2010; Sultan et al., 2011). A single PilZ domain containing effector protein was identified in the genome of *B. burgdorferi*, designated as PlzA (Freedman et al., 2010). Transcriptomic studies by microarray indicated that c-di-GMP regulates numerous chemotaxis and motility genes, including *mcp1*, *cheA1*, *cheY2*, *fliG1*, and *flaB* (Rogers et al., 2009). Depletion of c-di-GMP via deletion of *rrp1* led to mutants that are in constant straight run with attenuated chemotactic response (Kostick et al., 2011). Conversely, when PlzA the sole c-di-GMP effector protein was deleted, the *plzA* mutant exhibited reduced swarming ability (Pitzer et al., 2011). Perhaps PlzA in *B. burgdorferi* can differentially alter the activity of the motor complex depending on its ligand binding status. In congruence with this, it was recently revealed that PlzA can act as both the positive and negative regulator for the expression of glycerol utilization genes, *glp* operon (Zhang et al., 2018). The ability of PlzA to exert different modes of action in the presence and absence of c-di-GMP provides the cells with additional flexibility and capacity to regulate multiple cellular processes using a single effector protein. In sum, these studies demonstrate that c-di-GMP and PlzA are essential for the efficient functioning of the *B. burgdorferi* flagellar motor.

C-di-GMP can be hydrolyzed into 5'-phosphoguanylyl-(3',5')-guanosine (pGpG) and GMP [see review (Jenal et al., 2017)] by PDE enzymes. As mentioned, two PDE enzymes are encoded in the genome of *B. burgdorferi*, *pdeA* and *pdeB*. Mutants of PDE exhibit different phenotype from one another. Inactivation of *pdeA* resulted in cells that are unable to reverse direction (Sultan et al., 2010), similar to *cheA2* and *cheY3* chemotaxis mutants. On the other hand, mutation of *pdeB* led to cells with the wild-type swimming pattern but with increased flexing (Sultan et al., 2011). The hydrolyzed products of the two PDEs (pGpG vs GMP) could be responsible for the phenotypic differences that were observed in the *pdeA* and *pdeB* mutants. In addition, a double mutant of *pdeBplzA* constantly flexes, which is similar to *cheX* mutant, indicating that PlzA may be able to modulate the function of CheX or CheY3 to regulate the chemotaxis of *B. burgdorferi*. At this point, it is still difficult to determine how c-di-GMP influences the swimming behavior and chemotaxis of *B. burgdorferi* at the molecular level. Understanding how PlzA interacts with the motor complex or other motility/chemotaxis related proteins would be key to solving the puzzle behind the contribution of c-di-GMP and its effector protein in the regulation of motility/chemotaxis of *B. burgdorferi*. In summary, translational control of motility gene expression seems to be a crucial mechanism adopted by *B. burgdorferi*, with CsrA and c-di-GMP emerging as the preferred regulators of chemotaxis and motility.

Role of motility and chemotaxis in Lyme disease

The contribution of motility and chemotaxis during the infection and disease processes are well established in several pathogenic bacteria (Young et al., 2000; Josenhans and Suerbaum, 2002; Butler and Camilli, 2005; McGee et al., 2005; Duan et al., 2013; Matilla and Krell, 2018). Among the pathogenic spirochetes, defects in motility and/or chemotaxis have been linked to attenuated pathogenicity. For instance, motility-deficient mutants of *B. hyodysenteriae* are severely attenuated in their ability to colonize and infect a mouse model of swine dysentery (Rosey et al., 1996). In the oral spirochete *T. denticola*, motility and chemotaxis-deficient mutants are less invasive than the wild-type strain, with reduced ability to penetrate the oral epithelial cell layers (Lux et al., 2000; Lux et al., 2001). *B. burgdorferi* dedicates a significant proportion of its genome and cellular energy for the synthesis and operation of its flagella suggesting that motility and chemotaxis are likely important for the organism's infectious life cycle. Additionally, *B. burgdorferi* resides in different hosts such as the Ixodes ticks and many different vertebrates (Radolf et al., 2012). Experimental data under conditions mimicking these disparate environments demonstrate that the expression of many motility and chemotaxis genes is modulated, which likely enhances the overall fitness and survival of the spirochetes (Revel et al., 2002; Brooks et al., 2003; Ojaimi et al., 2003; Tokarz et al., 2004; Iyer et al., 2015). *B. burgdorferi* is highly invasive (Szczepanski et al., 1990) as it can traverse the extracellular matrix, penetrate the endothelial cell lining, and colonize joints, ears, heart and other tissues in the mammalian host (Niddam et al., 2017). In addition, motility mutants of *B. burgdorferi* are less able to penetrate tissue cell monolayers and bind to type I collagen lattices (Sadziene et al., 1991; Zambrano et al., 2004). Thus, motility and the ability to swim in highly viscous gel-like environments, such as connective tissue, is thought to play a critical role in the attachment and invasion processes of *B. burgdorferi*.

Recent advances of genetic tools in *B. burgdorferi* such as targeted gene inactivation, transposon mutagenesis, in trans and in cis complementation, and experimental tick-mouse infection models have facilitated the virulence studies of *B. burgdorferi* (Bono et al., 2000; Elias et al., 2002; Elias et al., 2003; Policastro and Schwan, 2003; Stewart et al., 2004; Li et al., 2007; Kariu et al., 2011; Oliver et al., 2016; Koci et al., 2018). A *fliG1* mutant constructed using the Himar1 transposon was the first motility mutant analyzed for its role in infection using a mouse model. Needle inoculation of the mutant into immunocompetent C3H/HeN mice was reported to be noninfectious (Botkin et al., 2006). Furthermore, a targeted mutant of *fliG1* failed to establish infections in both an immune-competent and an immune-deficient mouse model, whereas the complemented strain restored the infectivity to the wild-type level (Li et al., 2010). Subsequently, signature-tagged mutagenesis (STM) studies found that genes involved in chemotaxis including *mcp1*, *mcp3*, *mcp4*, *mcp5*, *cheA1*, *cheA2*, *cheB1*, *cheB2*, *cheW2*, *cheW3*, *cheR2*, *cheY2*, and *cheX* and genes important for motility *fliZ*, *flbA*, *fliH*, *fliI*, *flaA*, and *flgI* are required for the virulence in mouse, and in some instances, in tick colonization (Lin et al., 2012). However, results obtained from any STM study should be considered preliminary as the potential occurrence of cross-contamination within even well-isolated colonies may result in false positive results, and mutant polar effects need to be considered. Consequently, in vivo studies with individual clones obtained by targeted mutagenesis and complementation of mutants should be included to verify the importance of each gene in infection (Lin et al., 2012). To this end, several motility and chemotaxis mutants (e.g., *flaB*, *fliH*, *fliI*, *motB*, *cheA2*, *cheY1*, *cheY2*, *cheY3* and *cheD*) were reconstructed by targeted mutagenesis and resultant mutants were further complemented and examined in vitro and in vivo using mouse and/or tick infectious models (Sze et al., 2012; Sultan et al., 2013; Lin et al., 2015; Sultan et al., 2015; Moon et al., 2016a; Novak et al., 2016; Xu et al., 2017). Below are some examples.

Investigation of a non-chemotactic *cheA2* mutant revealed that chemotaxis is required for the spirochete to establish infection in mice, as the mutant failed to infect both wild-type and immunocompromised mice (Sze et al., 2012). In addition, it was eliminated from the inoculation site within 72 hours. Furthermore, although the mutant was able to survive in ticks, transmission from ticks to mice was unsuccessful, possibly due to its failure to sense and respond to chemo-attractants (e.g., serum) released during tick blood meal (Sze et al., 2012). Analyses of three response regulators CheYs reveal their different roles in the pathogenicity of *B. burgdorferi*. Like the cheA2 mutant, the cheY3 mutant failed to establish persistent infection in the murine host when the cells were inoculated by needle or by artificially infected ticks. Intravital imaging revealed that the mutant cells could not perform the backward-forward wild-type motility and, consequently, were cleared from the host within 96 hours after inoculation (Novak et al., 2016). In addition, this mutant showed significantly compromised fitness in fed ticks, although its survival in unfed ticks was unaffected, suggesting that a blood or tick-borne factor induced by feeding is responsible for the viability of the spirochetes in fed ticks. As mentioned above, CheY1 and CheY2 are not required for chemotaxis in vitro; however, in vivo studies reveal that they are required for the pathogenicity of *B. burgdorferi*. It was found that the *cheY2* mutant was severely attenuated in murine infection and compromised in dissemination from the skin inoculation site to the distant tissues. In addition, the mice fed upon by artificially *cheY2*-infected ticks did not develop a persistent infection. The ability of *cheY2* mutant to survive in the Ixodes ticks suggests that CheY2 is dispensable for persistence in ticks but important for transmission from ticks to mice and/or dissemination within the murine tissues, which is consistent with the elevated expression level of CheY2 in transmitting nymphs (Iyer et al., 2015; Xu et al., 2017). CheY1 was recently found to be important for the acquisition of the spirochete from mouse to ticks (H. Xu and M. Motaleb, unpublished). Taken together, these results indicate that three CheYs perform distinct functions in different stages of the enzootic life cycle of *B. burgdorferi*.

As previously mentioned, c-di-GMP is an emerging new factor with a regulatory role in the motility and chemotaxis of *B. burgdorferi*. Mutational analysis of genes involved in the c-di-GMP pathways resulted in mutants with altered motility and/or chemotaxis behaviors (Sultan et al., 2010; Kostick et al., 2011; Sultan et al., 2011). While it is unknown how c-di-GMP regulates motility and/or chemotaxis, experimental evidence indicates that PlzA is crucial for infectivity in mice. In addition, both rrp1 and *plzA* mutants showed significant reduction of viability in fed ticks (He et al., 2011; Kostick et al., 2011; Pitzer et al., 2011; Sze et al., 2013; Kostick-Dunn et al., 2018). A *pdeA* mutant with an altered swimming pattern (run-pause-no reverse) was unable to infect mouse either via needle inoculations or artificially infected ticks despite its normal colonization and viability in ticks (Sultan et al., 2010). On the other hand, a *pdeB* mutant with an increased flexing phenotype could infect mice normally via needle injection, but not via tick bite, which is likely due to the mutant's significantly reduced survivability in ticks (Sultan et al., 2011).

As discussed above, motility and chemotaxis are crucial for *B. burgdorferi* to complete its enzootic life cycle. One could speculate at which stages of the cycle motility and chemotaxis are crucial. First, within the tick, motility and chemotaxis could be essential for *B. burgdorferi* to migrate from the intestinal gut to the salivary glands and initiate a new infection. Second, once deposited in the skin of the mammalian host, chemotaxis and motility could be important for the spirochetes to migrate from the site of deposition into the surrounding skin tissue and then the circulatory system of the host. Several studies support the role of motility and chemotaxis at this stage (Li et al., 2010; Sze et al., 2012; Sultan et al., 2013; Sultan et al., 2015). Third, dissemination and subsequent colonization of the spirochetes into certain tissues, such as the joints, could also be dependent on chemotaxis and motility. Finally, when the tick takes a blood meal from a mammalian host, the spirochetes are likely attracted to the feeding pit by recognizing components of the extracellular matrix or tick saliva, which is a known attractant, and are taken up by tick to complete the enzootic cycle (Shih et al., 2002; Murfin et al., 2019). Additionally, the 2-year life cycle of the Ixodes tick involves two molts from larvae to nymph and nymph to adult, of which each only occurs after one blood meal. Therefore, along with increased body size following each molt and sustaining normal metabolism for a long period in each developmental stage, ticks encounter nutrient-deprived condition during the molting and unfed periods (Kurtenbach et al., 2006; Radolf et al., 2012; Troxell and Yang, 2013; Schotthoefer and Frost, 2015). *B. burgdorferi* is non-motile during these starvation stages (Dunham-Ems et al., 2009). The turn-off mode of chemotaxis and motility in *B. burgdorferi* during the molts and unfed periods should be conducive to its maintenance in tick midgut and transstadial transmission, especially when chemotaxis and motility are energy-consuming and unnecessary.

Although the detailed mechanism of how chemotaxis and motility are turned on or off is poorly understood, recent advances in imaging technologies have made it possible to visualize the behavior of the spirochete in conditions mimicking host environments in vitro or in the murine host and tick vector. To some degree, these advances provide mechanistic insights into the spatiotemporal regulation of chemotaxis and motility (Pal et al., 2004b; Moriarty et al., 2008; Dunham-Ems et al., 2009; Harman et al., 2012). Transmission of *B. burgdorferi* from its arthropod tick vector into a mammalian host requires the spirochete cells to penetrate through the epithelium to exit the tick midgut, then reach the salivary glands, and subsequently to be deposited into the dermis of a mammalian host (Dunham-Ems et al., 2009; Radolf et al., 2012). Intravital imaging data in the feeding tick demonstrated a novel biphasic mode of dissemination within ticks. In the first phase, replicating non-motile *B. burgdorferi* cells are passively transported towards the basement membrane through adhering to the differentiating epithelial cells, a process called "adherence-mediated migration." In the second phase, the non-motile *B. burgdorferi* cells transition into motile spirochetes that penetrate the basement membrane, enter the hemocoel, and then migrate into tick salivary glands(Dunham-Ems et al., 2009). This biphasic mode of dissemination provides direct evidence that the tick-to-mammal transmission involves the inhibition of the motility and/or chemotaxis of *B. burgdorferi* in the first phase and activation in the second phase. This transition is likely regulated by factors from tick and/or blood components from the host, and that motility and chemotaxis are essential for this biphasic dissemination (Dunham-Ems et al., 2009; Dunham-Ems et al., 2012; Sze et al., 2012; Sultan et al., 2013; Sultan et al., 2015). Once B. burgdorferi cells are inoculated into the mammalian skin, they hematogenously disseminate to distant tissues such as joints, ears, and heart (Coburn et al., 2013; Hyde, 2017). Intravital imaging showed that the back-and-forth motility of *B. burgdorferi* is important for vascular transmigration, dissemination and persistent infection, even though other regulatory proteins such as BBK32 and small regulatory RNAs (e.g. *ittA*) also play roles in these processes (Moriarty et al., 2008; Norman et al., 2008; Casselli et al., 2012; Moriarty et al., 2012; Coburn et al., 2013; Ebady et al., 2016; Novak et al., 2016; Lin et al., 2020; Medina-Perez et al., 2020). Consistent with this viewpoint, non-motile mutants (e.g., *ΔflaB* and *ΔmotB*) were cleared from the skin inoculation sites, and those spirochetes failed to disseminate (Sultan et al., 2013; Sultan et al., 2015). Intravital microscopic studies of nonchemotactic *cheY3* mutant revealed that this this mutant could not perform the back-and-forth movement and consequently was rapidly cleared from the site of inoculation (Novak et al., 2016). Surprisingly, long-term intravital microscopic studies showed that wild-type *B. burgdorferi* cells could remain motile for more than 720 days post-inoculation in immunocompetent mice (R. M. Wooten and M. Motaleb, unpublished), suggesting that motility and chemotaxis genes are being constitutively expressed during the mammalian infection, which is consistent with previous observations that these genes are controlled by the housekeeping σ^{70} (Ge et al., 1997a; Li et al., 2002; Motaleb et al., 2011). These results also suggest that there are sufficient nutrients present in the mammalian hosts to power the constant flagellar rotation and motility of the spirochete. When naïve questing ticks take a blood meal from an infected mouse, *B. burgdorferi* directly migrate toward the tick or tick-feeding site (Bockenstedt et al., 2014); this targeted movement is likely driven by chemotaxis.

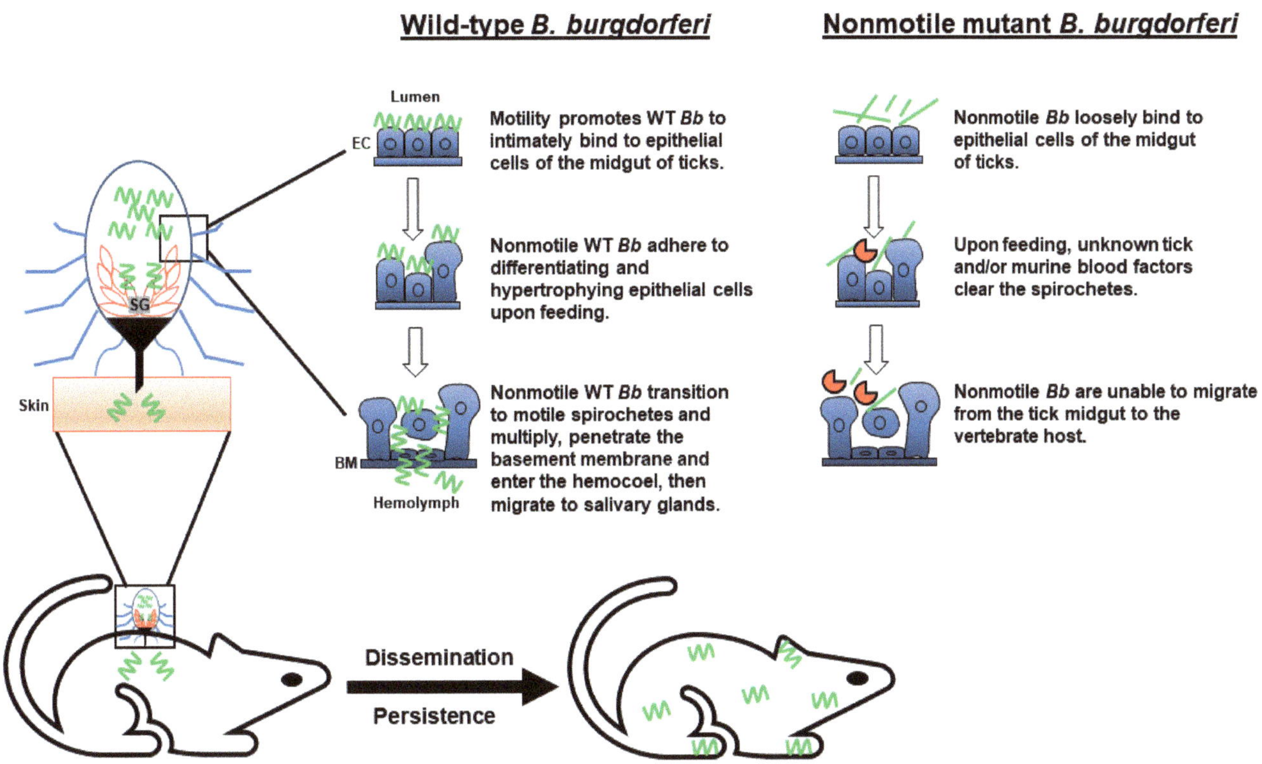

Figure 7. Proposed roles of *B. burgdorferi* motility in bacterial transmission, dissemination and persistence within and between the hosts. As ticks take a blood meal, wild-type *B. burgdorferi* cells replicate and use their run-and-reverse type of motility to cross the junction of the midgut basement membrane, and subsequently traverse through the salivary glands into the skin of the mammalian host (left diagram). Dissemination from the host skin tissues to the distant colonization sites also requires bacterial motility (Moriarty et al., 2008). During tick feeding, wild-type motility enables spirochetes to bind to the midgut using protein-protein interactions (OspA-TROSPA), which is believed to be important for bacterial viability in the tick (Pal et al., 2000; Pal et al., 2004a; Figlerowicz et al., 2013). Wild-type spirochetes then undergo a biphasic mode of dissemination during tick feeding (Dunham-Ems et al., 2009). However, it has not been investigated if the paralyzed cells also undergo a similar biphasic mode. During tick feeding, non-motile spirochetes in the midgut are unable to migrate, promoting clearance by tick and/or murine host factors and failed transmission (right diagram). Green curves and rod-shapes represent wild-type *B. burgdorferi* and non-motile mutant *B. burgdorferi*, respectively. Strings of oval shapes (◌) represent salivary glands (SG). Red three-fourths circles (◌) represent a tick and/or murine host blood-borne detrimental factor. BM: basement membrane; EC, epithelial cells. Adapted from reference (Motaleb et al., 2015) (diagram is not in scale).

Ticks, during feeding, excrete chemoattractants such as Salp12 to the feeding site, as this protein as well as tick salivary gland extracts were found to be attractants for *B. burgdorferi* (Shih et al., 2002; Murfin et al., 2019).

In summary, accumulating experimental evidence suggests that if a *B. burgdorferi* clone shows any defect in motility or chemotaxis, it also exhibits deficient in infectivity in vertebrate host, dissemination and/or migration within and between hosts (vertebrate and vector) or compromised viability in ticks. A working model is proposed to describe the role of *B. burgdorferi* motility/chemotaxis in the spirochete transmission, dissemination and persistence in host and vector (Figure 7). As the in vivo tracking systems of *B. burgdorferi* become available to monitor the spirochetes' migrations in the mammalian host and tick vector, couple with new fluorescence labeling for live-cell imaging of spirochetes (Moriarty et al., 2008; Dunham-Ems et al., 2009; Novak et al., 2016; Hillman et al., 2019), it will be interesting to determine how *B. burgdorferi* migrates within and between the tick and vertebrate hosts. Future experiments to identify and characterize the motility inhibitory factors present in the tick midgut would be enlightening to fully decipher the intricate communication that exists between *B. burgdorferi* and its host during transmission.

References

Affroze, S., Islam, M.S., Takabe, K., Kudo, S., and Nakamura, S. (2016). Characterization of Leptospiral Chemoreceptors Using a Microscopic Agar Drop Assay. Curr Microbiol *73*, 202-205.

Aizawa S, Z.I., Márquez-Magaña L, Ordal G. (2002). Chemotaxis and Motility. In *Bacillus subtilis* and Its Closest Relatives, L.R. Sonenshein A, Hoch J, ed. (Washington, DC: ASM Press), pp. 437-452.

Aldridge, P., and Hughes, K.T. (2001). How and when are substrates selected for type III secretion? Trends Microbiol *9*, 209-214. https://doi.org/10.1016/s0966-842x(01)02014-5

Aldridge, P., and Hughes, K.T. (2002). Regulation of flagellar assembly. Curr Opin Microbiol *5*, 160-165. https://doi.org/10.1016/s1369-5274(02)00302-8

Anderson, D.K., Ohta, N., Wu, J., and Newton, A. (1995). Regulation of the Caulobacter crescentus rpoN gene and function of the purified sigma 54 in flagellar gene transcription. Mol Gen Genet *246*, 697-706. https://doi.org/10.1007/bf00290715

Angel, T.E., Luft, B.J., Yang, X., Nicora, C.D., Camp, D.G., 2nd, Jacobs, J.M., and Smith, R.D. (2010). Proteome analysis of Borrelia burgdorferi response to environmental change. PLoS One *5*, e13800. https://doi.org/10.1371/journal.pone.0013800

Armitage, J.P., and Schmitt, R. (1997). Bacterial chemotaxis: Rhodobacter sphaeroides and Sinorhizobium meliloti--variations on a theme? Microbiology *143 (Pt 12)*, 3671-3682. https://doi.org/10.1099/00221287-143-12-3671

Armitage, J.P. (1999). Bacterial tactic responses. Adv Microb Physiol *41*, 229-289. https://doi.org/10.1016/s0065-2911(08)60168-x

Ashimoto, A., Chen, C., Bakker, I., and Slots, J. (1996). Polymerase chain reaction detection of 8 putative periodontal pathogens in subgingival plaque of gingivitis and advanced periodontitis lesions. Oral Microbiol Immunol *11*, 266-273. https://doi.org/10.1111/j.1399-302x.1996.tb00180.x

Baker, A.E., Diepold, A., Kuchma, S.L., Scott, J.E., Ha, D.G., Orazi, G., Armitage, J.P., and O'Toole, G.A. (2016). PilZ Domain Protein FlgZ Mediates Cyclic Di-GMP-Dependent Swarming Motility Control in Pseudomonas aeruginosa. J Bacteriol *198*, 1837-1846. https://doi.org/10.1128/JB.00196-16

Baker, A.E., and O'Toole, G.A. (2017). Bacteria, Rev Your Engines: Stator Dynamics Regulate Flagellar Motility. J Bacteriol *199*. https://doi.org/10.1128/JB.00088-17

Bakker, R.G., Li, C., Miller, M.R., Cunningham, C., and Charon, N.W. (2007). Identification of specific chemoattractants and genetic complementation of a Borrelia burgdorferi chemotaxis mutant: flow cytometry-based capillary tube chemotaxis assay. Appl Environ Microbiol *73*, 1180-1188. https://doi.org/10.1128/AEM.01913-06

Barbour, A.G., and Hayes, S.F. (1986). Biology of *Borrelia* species. Microbiol Rev *50*, 381-400.

Beeby, M., Ribardo, D.A., Brennan, C.A., Ruby, E.G., Jensen, G.J., and Hendrixson, D.R. (2016). Diverse high-torque bacterial flagellar motors assemble wider stator rings using a conserved protein scaffold. Proc Natl Acad Sci U S A *113*, E1917-1926. https://doi.org/10.1073/pnas.1518952113

Benach, J.L., Bosler, E.M., Hanrahan, J.P., Coleman, J.L., Habicht, G.S., Bast, T.F., Cameron, D.J., Ziegler, J.L., Barbour, A.G., Burgdorfer, W., *et al.* (1983). Spirochetes isolated from the blood of two patients with Lyme disease. N Engl J Med *308*, 740-742. https://doi.org/10.1056/NEJM198303313081302

Berg, H.C. (1975). Chemotaxis in bacteria. Annu Rev Biophys Bioeng *4*, 119-136. https://doi.org/10.1146/annurev.bb.04.060175.001003

Berg, H.C. (1976). How spirochetes may swim. J Theor Biol *56*, 269-273. https://doi.org/10.1016/s0022-5193(76)80074-4

Berg, H.C. (2000). Motile Behavior of Bacteria. Physics Today *53*, 24-29. https://doi.org/10.1063/1.882934

Berg, H.C. (2018). Random Walks in Biology: New and Expanded Edition (Princeton University Press).

Berg, H.C., Bromley,D.B., and Charon,N.W. (1978). Leptospiral motility. Symp Soc Gen Microbiol *28*, 285-294.

BJ, P. (2011). Phylum XV. Spirochaetes Vol 4 (New York: Springer).

Bockenstedt, L.K., Gonzalez, D., Mao, J., Li, M., Belperron, A.A., and Haberman, A. (2014). What ticks do under your skin: two-photon intravital imaging of Ixodes scapularis feeding in the presence of the lyme disease spirochete. Yale J Biol Med *87*, 3-13.

Boehm, A., Kaiser, M., Li, H., Spangler, C., Kasper, C.A., Ackermann, M., Kaever, V., Sourjik, V., Roth, V., and Jenal, U. (2010). Second messenger-mediated adjustment of bacterial swimming velocity. Cell *141*, 107-116. https://doi.org/10.1016/j.cell.2010.01.018

Boesch, K.C., Silversmith, R.E., and Bourret, R.B. (2000). Isolation and characterization of nonchemotactic CheZ mutants of Escherichia coli. J Bacteriol *182*, 3544-3552.

Bono, J.L., Elias, A.F., Kupko, J.J., 3rd, Stevenson, B., Tilly, K., and Rosa, P. (2000). Efficient targeted mutagenesis in Borrelia burgdorferi. J Bacteriol *182*, 2445-2452.

Botkin, D.J., Abbott, A.N., Stewart, P.E., Rosa, P.A., Kawabata, H., Watanabe, H., and Norris, S.J. (2006). Identification of potential virulence determinants by Himar1 transposition of infectious Borrelia burgdorferi B31. Infect Immun *74*, 6690-6699. https://doi.org/10.1128/IAI.00993-06

Boukhvalova, M.S., Dahlquist, F.W., and Stewart, R.C. (2002). CheW binding interactions with CheA and Tar. Importance for chemotaxis signaling in Escherichia coli. J Biol Chem *277*, 22251-22259. https://doi.org/10.1074/jbc.M110908200

Bren, A., and Eisenbach, M. (2000). How signals are heard during bacterial chemotaxis: protein-protein interactions in sensory signal propagation. J Bacteriol *182*, 6865-6873. https://doi.org/10.1128/jb.182.24.6865-6873.2000

Brissette, C.A., and Gaultney, R.A. (2014). That's my story, and I'm sticking to it--an update on B. burgdorferi adhesins. Front Cell Infect Microbiol *4*, 41. https://doi.org/10.3389/fcimb.2014.00041

Brooks, C.S., Hefty, P.S., Jolliff, S.E., and Akins, D.R. (2003). Global analysis of *Borrelia burgdorferi* genes regulated by mammalian host-specific signals. Infect Immun *71*, 3371-3383. https://doi.org/10.1128/iai.71.6.3371-3383.2003

Butler, S.M., and Camilli, A. (2005). Going against the grain: chemotaxis and infection in Vibrio cholerae. Nat Rev Microbiol *3*, 611-620. https://doi.org/10.1038/nrmicro1207

Canale-Parola, E. (1978). Motility and chemotaxis of spirochetes. Annu Rev Microbiol *32*, 69-99. https://doi.org/10.1146/annurev.mi.32.100178.000441

Canale-Parola, E. (1981). Free-Living Anaerobic and Facultatively Anaerobic Spirochetes: The Genus Spirochaeta. In The Prokaryotes: A Handbook on Habitats, Isolation, and Identification of Bacteria, M.P. Starr, H. Stolp, H.G. Trüper, A. Balows, and H.G. Schlegel, eds. (Berlin, Heidelberg: Springer Berlin Heidelberg), pp. 538-547. https://doi.org/10.1007/978-3-662-13187-9_46

Casselli, T., Tourand, Y., and Bankhead, T. (2012). Altered murine tissue colonization by Borrelia burgdorferi following targeted deletion of linear plasmid 17-carried genes. Infect Immun *80*, 1773-1782. https://doi.org/10.1128/IAI.05984-11

Chang, Y., Moon, K.H., Zhao, X., Norris, S.J., Motaleb, M.A., and Liu, J. (2019). Structural insights into flagellar stator-rotor interactions. Elife *8*. https://doi.org/10.7554/eLife.48979

Chang, Y.Z., K; Carroll, B.L; Zhao, X; Charon, N.W; Norris, S.J; Motaleb, M.A; Li, C; and Liu, J (2020). Molecular Mechanism for Rotational Switching of the Bacterial Flagellar Motor

Nature Structural and Molecular Biology *In press*.

Chao, X., Muff, T.J., Park, S.Y., Zhang, S., Pollard, A.M., Ordal, G.W., Bilwes, A.M., and Crane, B.R. (2006). A receptor-modifying deamidase in complex with a signaling phosphatase reveals reciprocal regulation. Cell *124*, 561-571. https://doi.org/10.1016/j.cell.2005.11.046

Charon, N.W., Daughtry, G.R., McCuskey, R.S., and Franz, G.N. (1984). Microcinematographic analysis of tethered Leptospira illini. J Bacteriol *160*, 1067-1073.

Charon, N.W., Goldstein, S.F., Curci, K., and Limberger, R.J. (1991). The bent-end morphology of Treponema phagedenis is associated with short, left-handed, periplasmic flagella. J Bacteriol *173*, 4820-4826. https://doi.org/10.1128/jb.173.15.4820-4826.1991

Charon, N.W., Goldstein, S.F., Block, S.M., Curci, K., Ruby, J.D., Kreiling, J.A., and Limberger, R.J. (1992a). Morphology and dynamics of protruding spirochete periplasmic flagella. J Bacteriol *174*, 832-840. https://doi.org/10.1128/jb.174.3.832-840.1992

Charon, N.W., Greenberg, E.P., Koopman, M.B., and Limberger, R.J. (1992b). Spirochete chemotaxis, motility, and the structure of the spirochetal periplasmic flagella. Res Microbiol *143*, 597-603. https://doi.org/10.1016/0923-2508(92)90117-7

Charon, N.W., and Goldstein, S.F. (2002). Genetics of motility and chemotaxis of a fascinating group of bacteria: the spirochetes. Annu Rev Genet *36*, 47-73. https://doi.org/10.1146/annurev.genet.36.041602.134359

Charon, N.W., Goldstein, S.F., Marko, M., Hsieh, C., Gebhardt, L.L., Motaleb, M.A., Wolgemuth, C.W., Limberger, R.J., and Rowe, N. (2009). The flat-ribbon configuration of the periplasmic flagella of *Borrelia burgdorferi* and its relationship to motility and morphology. J Bacteriol *191*, 600-607. https://doi.org/10.1128/JB.01288-08

Charon, N.W., Cockburn, A., Li, C., Liu, J., Miller, K.A., Miller, M.R., Motaleb, M.A., and Wolgemuth, C.W. (2012). The unique paradigm of spirochete motility and chemotaxis. Annu Rev Microbiol *66*, 349-370. https://doi.org/10.1146/annurev-micro-092611-150145

Chen, S., Beeby, M., Murphy, G.E., Leadbetter, J.R., Hendrixson, D.R., Briegel, A., Li, Z., Shi, J., Tocheva, E.I., Muller, A., *et al.* (2011). Structural diversity of bacterial flagellar motors. The EMBO journal *30*, 2972-2981. https://doi.org/10.1038/emboj.2011.186

Cheng, S.L., Siboo, R., Quee, T.C., Johnson, J.L., Mayberry, W.R., and Chan, E.C. (1985). Comparative study of six random oral spirochete isolates. Serological heterogeneity of Treponema denticola. J Periodontal Res *20*, 602-612. https://doi.org/10.1111/j.1600-0765.1985.tb00844.x

Chi, B., Limberger, R.J., and Kuramitsu, H.K. (2002). Complementation of a Treponema denticola flgE mutant with a novel coumermycin A1-resistant T. denticola shuttle vector system. Infect Immun *70*, 2233-2237. https://doi.org/10.1128/iai.70.4.2233-2237.2002

Chilcott, G.S., and Hughes, K.T. (2000). Coupling of flagellar gene expression to flagellar assembly in Salmonella enterica serovar typhimurium and Escherichia coli. Microbiol Mol Biol Rev *64*, 694-708. https://doi.org/10.1128/mmbr.64.4.694-708.2000

Coburn, J., Leong, J., and Chaconas, G. (2013). Illuminating the roles of the *Borrelia burgdorferi* adhesins. Trends Microbiol *21*, 372-379. https://doi.org/10.1016/j.tim.2013.06.005

Coleman, J.L., Crowley, J.T., Toledo, A.M., and Benach, J.L. (2013). The HtrA protease of Borrelia burgdorferi degrades outer membrane protein BmpD and chemotaxis phosphatase CheX. Mol Microbiol *88*, 619-633. https://doi.org/10.1111/mmi.12213

Dasgupta, N., Ferrell, E.P., Kanack, K.J., West, S.E., and Ramphal, R. (2002). fleQ, the gene encoding the major flagellar regulator of Pseudomonas aeruginosa, is sigma70 dependent and is downregulated by Vfr, a homolog of Escherichia coli cyclic AMP receptor protein. J Bacteriol *184*, 5240-5250. https://doi.org/10.1128/jb.184.19.5240-5250.2002

Dasgupta, N., Wolfgang, M.C., Goodman, A.L., Arora, S.K., Jyot, J., Lory, S., and Ramphal, R. (2003). A four-tiered transcriptional regulatory circuit controls flagellar biogenesis in Pseudomonas aeruginosa. Mol Microbiol *50*, 809-824. https://doi.org/10.1046/j.1365-2958.2003.03740.x

DePamphilis, M.L., and Adler, J. (1971a). Fine structure and isolation of the hook-basal body complex of flagella from Escherichia coli and Bacillus subtilis. J Bacteriol *105*, 384-395.

DePamphilis, M.L., and Adler, J. (1971b). Purification of intact flagella from Escherichia coli and Bacillus subtilis. J Bacteriol *105*, 376-383.

Dombrowski, C., Kan, W., Motaleb, M.A., Charon, N.W., Goldstein, R.E., and Wolgemuth, C.W. (2009). The elastic basis for the shape of Borrelia burgdorferi. Biophys J *96*, 4409-4417. https://doi.org/10.1016/j.bpj.2009.02.066

Dong, K., Li, Q., Liu, C., Zhang, Y., Zhao, G., and Guo, X. (2010). Cloning and characterization of three cheB genes in Leptospira interrogans. Acta Biochim Biophys Sin (Shanghai) *42*, 216-223. https://doi.org/10.1093/abbs/gmq007

Duan, Q., Zhou, M., Zhu, L., and Zhu, G. (2013). Flagella and bacterial pathogenicity. J Basic Microbiol *53*, 1-8. https://doi.org/10.1002/jobm.201100335

Dunham-Ems, S.M., Caimano, M.J., Pal, U., Wolgemuth, C.W., Eggers, C.H., Balic, A., and Radolf, J.D. (2009). Live imaging reveals a biphasic mode of dissemination of *Borrelia burgdorferi* within ticks. J Clin Invest *119*, 3652-3665. https://doi.org/10.1172/JCI39401

Dunham-Ems, S.M., Caimano, M.J., Eggers, C.H., and Radolf, J.D. (2012). *Borrelia burgdorferi* requires the alternative sigma factor RpoS for dissemination within the vector during tick-to-mammal transmission. PLoS Pathog *8*, e1002532. https://doi.org/10.1371/journal.ppat.1002532

Ebady, R., Niddam, A.F., Boczula, A.E., Kim, Y.R., Gupta, N., Tang, T.T., Odisho, T., Zhi, H., Simmons, C.A., Skare, J.T., *et al.* (2016). Biomechanics of *Borrelia burgdorferi* vascular interactions. Cell Rep *16*, 2593-2604. https://doi.org/10.1016/j.celrep.2016.08.013

Elias, A.F., Stewart, P.E., Grimm, D., Caimano, M.J., Eggers, C.H., Tilly, K., Bono, J.L., Akins, D.R., Radolf, J.D., Schwan, T.G., *et al.* (2002). Clonal polymorphism of Borrelia burgdorferi strain B31 MI: implications for mutagenesis in an infectious strain background. Infect Immun *70*, 2139-2150.

Elias, A.F., Bono, J.L., Kupko, J.J., 3rd, Stewart, P.E., Krum, J.G., and Rosa, P.A. (2003). New antibiotic resistance cassettes suitable for genetic studies in Borrelia burgdorferi. J Mol Microbiol Biotechnol *6*, 29-40. https://doi.org/73406

Fang, X., and Gomelsky, M. (2010). A post-translational, c-di-GMP-dependent mechanism regulating flagellar motility. Mol Microbiol *76*, 1295-1305. https://doi.org/10.1111/j.1365-2958.2010.07179.x

Figlerowicz, M., Urbanowicz, A., Lewandowski, D., Jodynis-Liebert, J., and Sadowski, C. (2013). Functional insights into recombinant TROSPA protein from Ixodes ricinus. PLoS One *8*, e76848. https://doi.org/10.1371/journal.pone.0076848

Fosnaugh, K., and Greenberg, E.P. (1988). Motility and chemotaxis of Spirochaeta aurantia: computer-assisted motion analysis. J Bacteriol *170*, 1768-1774. https://doi.org/10.1128/jb.170.4.1768-1774.1988

Fraser, C.M., Casjens, S., Huang, W.M., Sutton, G.G., Clayton, R., Lathigra, R., White, O., Ketchum, K.A., Dodson, R., Hickey, E.K., *et al.* (1997). Genomic sequence of a Lyme disease spirochaete, *Borrelia burgdorferi*. Nature *390*, 580-586. https://doi.org/10.1038/37551

Freedman, J.C., Rogers, E.A., Kostick, J.L., Zhang, H., Iyer, R., Schwartz, I., and Marconi, R.T. (2010). Identification and molecular characterization of a cyclic-di-GMP effector protein, PlzA (BB0733): additional evidence for the existence of a functional cyclic-di-GMP regulatory network in the Lyme disease spirochete, Borrelia burgdorferi. FEMS Immunol Med Microbiol *58*, 285-294. https://doi.org/10.1111/j.1574-695X.2009.00635.x

Galan, J.E., and Waksman, G. (2018). Protein-Injection Machines in Bacteria. Cell *172*, 1306-1318. https://doi.org/10.1016/j.cell.2018.01.034

Ge, Y., and Charon, N.W. (1997a). An unexpected flaA homolog is present and expressed in Borrelia burgdorferi. J Bacteriol *179*, 552-556. https://doi.org/10.1128/jb.179.2.552-556.1997

Ge, Y., and Charon, N.W. (1997b). Molecular characterization of a flagellar/chemotaxis operon in the spirochete Borrelia burgdorferi. FEMS Microbiol Lett *153*, 425-431. https://doi.org/10.1111/j.1574-6968.1997.tb12606.x

Ge, Y., Old, I.G., Girons, I.S., and Charon, N.W. (1997a). The flgK motility operon of Borrelia burgdorferi is initiated by a sigma 70-like promoter. Microbiology *143 (Pt 5)*, 1681-1690. https://doi.org/10.1099/00221287-143-5-1681

Ge, Y., Old, I.G., Saint Girons, I., and Charon, N.W. (1997b). Molecular characterization of a large Borrelia burgdorferi motility operon which is initiated by a consensus sigma70 promoter. J Bacteriol *179*, 2289-2299. https://doi.org/10.1128/jb.179.7.2289-2299.1997

Ge, Y., Li, C., Corum, L., Slaughter, C.A., and Charon, N.W. (1998). Structure and expression of the FlaA periplasmic flagellar protein of Borrelia burgdorferi. J Bacteriol *180*, 2418-2425.

Gegner, J.A., Graham, D.R., Roth, A.F., and Dahlquist, F.W. (1992). Assembly of an MCP receptor, CheW, and kinase CheA complex in the bacterial chemotaxis signal transduction pathway. Cell *70*, 975-982. https://doi.org/10.1016/0092-8674(92)90247-a

Glekas, G.D., Plutz, M.J., Walukiewicz, H.E., Allen, G.M., Rao, C.V., and Ordal, G.W. (2012). Elucidation of the multiple roles of CheD in Bacillus subtilis chemotaxis. Mol Microbiol *86*, 743-756. https://doi.org/10.1111/mmi.12015

Goldstein, S.F., and Charon, N.W. (1988). Motility of the spirochete Leptospira. Cell Motil Cytoskeleton *9*, 101-110. https://doi.org/10.1002/cm.970090202

Goldstein, S.F., and Charon, N.W. (1990). Multiple-exposure photographic analysis of a motile spirochete. Proc Natl Acad Sci U S A *87*, 4895-4899. https://doi.org/10.1073/pnas.87.13.4895

Goldstein, S.F., Charon, N.W., and Kreiling, J.A. (1994). Borrelia burgdorferi swims with a planar waveform similar to that of eukaryotic flagella. Proc Natl Acad Sci U S A *91*, 3433-3437.

Goldstein, S.F., Buttle, K.F., and Charon, N.W. (1996). Structural analysis of the Leptospiraceae and Borrelia burgdorferi by high-voltage electron microscopy. J Bacteriol *178*, 6539-6545. https://doi.org/10.1128/jb.178.22.6539-6545.1996

Goldstein, S.F., C. Li, J. Liu, M.R. Miller, M. A. Motaleb, S. J. Norris, R. E. Silversmith, C. W. Wolgemuth, C. and N. W. Charon (2010). The chic motility and chemotaxis of Borrelia burgdorferi. (Norfolk, UK: Caister Academic Press).

Harman, M.W., Dunham-Ems, S.M., Caimano, M.J., Belperron, A.A., Bockenstedt, L.K., Fu, H.C., Radolf, J.D., and Wolgemuth, C.W. (2012). The heterogeneous motility of the Lyme disease spirochete in gelatin mimics dissemination through tissue. Proc Natl Acad Sci U S A *109*, 3059-3064. https://doi.org/10.1073/pnas.1114362109

Harwood, C.S., Jannasch, H.W., and Canale-Parola, E. (1982). Anaerobic spirochete from a deep-sea hydrothermal vent. Appl Environ Microbiol *44*, 234-237.

Harwood, C.S., and Canale-Parola, E. (1984). Ecology of spirochetes. Annu Rev Microbiol *38*, 161-192. https://doi.org/10.1146/annurev.mi.38.100184.001113

Hazelbauer, G.L., Falke, J.J., and Parkinson, J.S. (2008). Bacterial chemoreceptors: high-performance signaling in networked arrays. Trends Biochem Sci *33*, 9-19. https://doi.org/10.1016/j.tibs.2007.09.014

He, M., Ouyang, Z., Troxell, B., Xu, H., Moh, A., Piesman, J., Norgard, M.V., Gomelsky, M., and Yang, X.F. (2011). Cyclic di-GMP is essential for the survival of the Lyme disease spirochete in ticks. PLoS Pathog *7*, e1002133. https://doi.org/10.1371/journal.ppat.1002133

Hendrix, R.W. (2005). Bacteriophage HK97: assembly of the capsid and evolutionary connections. Adv Virus Res *64*, 1-14. https://doi.org/10.1016/S0065-3527(05)64001-8

Henry, R.A., and Johnson, R.C. (1978). Distribution of the genus Leptospira in soil and water. Appl Environ Microbiol *35*, 492-499.

Hickman, J.W., and Harwood, C.S. (2008). Identification of FleQ from Pseudomonas aeruginosa as a c-di-GMP-responsive transcription factor. Mol Microbiol *69*, 376-389. https://doi.org/10.1111/j.1365-2958.2008.06281.x

Hillman, C., Stewart, P.E., Strnad, M., Stone, H., Starr, T., Carmody, A., Evans, T.J., Carracoi, V., Wachter, J., and Rosa, P.A. (2019). Visualization of Spirochetes by Labeling Membrane Proteins With Fluorescent Biarsenical Dyes. Front Cell Infect Microbiol *9*, 287. https://doi.org/10.3389/fcimb.2019.00287

Holt, S.C. (1978). Anatomy and chemistry of spirochetes. Microbiol Rev *42*, 114-160.

Homma, M., Fujita, H., Yamaguchi, S., and Iino, T. (1984). Excretion of unassembled flagellin by Salmonella typhimurium mutants deficient in hook-associated proteins. J Bacteriol *159*, 1056-1059.

Hovind-Hougen, K. (1984). Ultrastructure of spirochetes isolated from Ixodes ricinus and Ixodes dammini. Yale J Biol Med *57*, 543-548.

Hu, B., Morado, D.R., Margolin, W., Rohde, J.R., Arizmendi, O., Picking, W.L., Picking, W.D., and Liu, J. (2015). Visualization of the type III secretion sorting platform of *Shigella flexneri*. Proc Natl Acad Sci U S A *112*, 1047-1052. https://doi.org/10.1073/pnas.1411610112

Hu, B., Lara-Tejero, M., Kong, Q., Galan, J.E., and Liu, J. (2017). In Situ Molecular Architecture of the Salmonella Type III Secretion Machine. Cell *168*, 1065-1074 e1010. https://doi.org/10.1016/j.cell.2017.02.022

Hughes, K.T., Gillen, K.L., Semon, M.J., and Karlinsey, J.E. (1993). Sensing structural intermediates in bacterial flagellar assembly by export of a negative regulator. Science *262*, 1277-1280. https://doi.org/10.1126/science.8235660

Hyde, J.A. (2017). *Borrelia burgdorferi* keeps moving and carries on: a review of borrelial dissemination and invasion. Front Immunol *8*, 114. https://doi.org/10.3389/fimmu.2017.00114

Ikeda, T., Asakura, S., and Kamiya, R. (1985). "Cap" on the tip of Salmonella flagella. J Mol Biol *184*, 735-737. https://doi.org/10.1016/0022-2836(85)90317-1

Ikeda, T., Homma, M., Iino, T., Asakura, S., and Kamiya, R. (1987). Localization and stoichiometry of hook-associated proteins within Salmonella typhimurium flagella. J Bacteriol *169*, 1168-1173. https://doi.org/10.1128/jb.169.3.1168-1173.1987

Islam, M.S., Takabe, K., Kudo, S., and Nakamura, S. (2014). Analysis of the chemotactic behaviour of Leptospira using microscopic agar-drop assay. FEMS Microbiol Lett *356*, 39-44. https://doi.org/10.1111/1574-6968.12495

Iyer, R., Caimano, M.J., Luthra, A., Axline, D., Jr., Corona, A., Iacobas, D.A., Radolf, J.D., and Schwartz, I. (2015). Stage-specific global alterations in the transcriptomes of Lyme disease spirochetes during tick feeding and following mammalian host adaptation. Mol Microbiol *95*, 509-538. https://doi.org/10.1111/mmi.12882

Izard, J., Renken, C., Hsieh, C.E., Desrosiers, D.C., Dunham-Ems, S., La Vake, C., Gebhardt, L.L., Limberger, R.J., Cox, D.L., Marko, M., *et al.* (2009). Cryo-electron tomography elucidates the molecular architecture of Treponema pallidum, the syphilis spirochete. J Bacteriol *191*, 7566-7580. https://doi.org/10.1128/JB.01031-09

Jenal, U., Reinders, A., and Lori, C. (2017). Cyclic di-GMP: second messenger extraordinaire. Nat Rev Microbiol *15*, 271-284. https://doi.org/10.1038/nrmicro.2016.190

Josenhans, C., and Suerbaum, S. (2002). The role of motility as a virulence factor in bacteria. Int J Med Microbiol *291*, 605-614. https://doi.org/10.1078/1438-4221-00173

Kang, H.J., Coulibaly, F., Clow, F., Proft, T., and Baker, E.N. (2007). Stabilizing isopeptide bonds revealed in gram-positive bacterial pilus structure. Science *318*, 1625-1628. https://doi.org/10.1126/science.1145806

Kariu, T., Coleman, A.S., Anderson, J.F., and Pal, U. (2011). Methods for rapid transfer and localization of lyme disease pathogens within the tick gut. J Vis Exp. https://doi.org/10.3791/2544

Koci, J., Bernard, Q., Yang, X., and Pal, U. (2018). Borrelia burgdorferi surface protein Lmp1 facilitates pathogen dissemination through ticks as studied by an artificial membrane feeding system. Sci Rep *8*, 1910. https://doi.org/10.1038/s41598-018-20208-4

Kostick-Dunn, J.L., Izac, J.R., Freedman, J.C., Szkotnicki, L.T., Oliver, L.D., Jr., and Marconi, R.T. (2018). The *Borrelia burgdorferi* c-di-GMP binding receptors, PlzA and PlzB, are functionally distinct. Front Cell Infect Microbiol *8*, 213. https://doi.org/10.3389/fcimb.2018.00213

Kostick, J.L., Szkotnicki, L.T., Rogers, E.A., Bocci, P., Raffaelli, N., and Marconi, R.T. (2011). The diguanylate cyclase, Rrp1, regulates critical steps in the enzootic cycle of the Lyme disease spirochetes. Mol Microbiol *81*, 219-231. https://doi.org/10.1111/j.1365-2958.2011.07687.x

Kristich, C.J., and Ordal, G.W. (2002). Bacillus subtilis CheD is a chemoreceptor modification enzyme required for chemotaxis. J Biol Chem *277*, 25356-25362. https://doi.org/10.1074/jbc.M201334200

Kudryashev, M., Cyrklaff, M., Baumeister, W., Simon, M.M., Wallich, R., and Frischknecht, F. (2009). Comparative cryo-electron tomography of pathogenic Lyme disease spirochetes. Mol Microbiol *71*, 1415-1434. https://doi.org/10.1111/j.1365-2958.2009.06613.x

Kumar, D., Ristow, L.C., Shi, M., Mukherjee, P., Caine, J.A., Lee, W.Y., Kubes, P., Coburn, J., and Chaconas, G. (2015). Intravital Imaging of Vascular Transmigration by the Lyme Spirochete: Requirement for the Integrin Binding Residues of the B. burgdorferi P66 Protein. PLoS Pathog *11*, e1005333. https://doi.org/10.1371/journal.ppat.1005333

Kurtenbach, K., Hanincova, K., Tsao, J.I., Margos, G., Fish, D., and Ogden, N.H. (2006). Fundamental processes in the evolutionary ecology of Lyme borreliosis. Nat Rev Microbiol *4*, 660-669. https://doi.org/10.1038/nrmicro1475

Lee, W.Y., Moriarty, T.J., Wong, C.H., Zhou, H., Strieter, R.M., van Rooijen, N., Chaconas, G., and Kubes, P. (2010). An intravascular immune response to *Borrelia burgdorferi* involves Kupffer cells and iNKT cells. Nat Immunol *11*, 295-302. https://doi.org/10.1038/ni.1855

Li, C., Corum, L., Morgan, D., Rosey, E.L., Stanton, T.B., and Charon, N.W. (2000a). The spirochete FlaA periplasmic flagellar sheath protein impacts flagellar helicity. J Bacteriol *182*, 6698-6706. https://doi.org/10.1128/jb.182.23.6698-6706.2000

Li, C., Motaleb, A., Sal, M., Goldstein, S.F., and Charon, N.W. (2000b). Spirochete periplasmic flagella and motility. J Mol Microbiol Biotechnol *2*, 345-354.

Li, C., Bakker, R.G., Motaleb, M.A., Sartakova, M.L., Cabello, F.C., and Charon, N.W. (2002). Asymmetrical flagellar rotation in Borrelia burgdorferi nonchemotactic mutants. Proc Natl Acad Sci U S A *99*, 6169-6174. https://doi.org/10.1073/pnas.092010499

Li, C., Xu, H., Zhang, K., and Liang, F.T. (2010). Inactivation of a putative flagellar motor switch protein FliG1 prevents Borrelia burgdorferi from swimming in highly viscous media and blocks its infectivity. Mol Microbiol *75*, 1563-1576. https://doi.org/10.1111/j.1365-2958.2010.07078.x

Li, X., Pal, U., Ramamoorthi, N., Liu, X., Desrosiers, D.C., Eggers, C.H., Anderson, J.F., Radolf, J.D., and Fikrig, E. (2007). The Lyme disease agent Borrelia burgdorferi requires BB0690, a Dps homologue, to persist within ticks. Mol Microbiol *63*, 694-710. https://doi.org/10.1111/j.1365-2958.2006.05550.x

Limberger, R.J., Slivienski, L.L., and Samsonoff, W.A. (1994). Genetic and biochemical analysis of the flagellar hook of Treponema phagedenis. J Bacteriol *176*, 3631-3637. https://doi.org/10.1128/jb.176.12.3631-3637.1994

Lin, T., Gao, L., Zhang, C., Odeh, E., Jacobs, M.B., Coutte, L., Chaconas, G., Philipp, M.T., and Norris, S.J. (2012). Analysis of an ordered, comprehensive STM mutant library in infectious *Borrelia burgdorferi*: insights into the genes required for mouse infectivity. PLoS One *7*, e47532. https://doi.org/10.1371/journal.pone.0047532

Lin, T., Gao, L., Zhao, X., Liu, J., and Norris, S.J. (2015). Mutations in the Borrelia burgdorferi Flagellar Type III Secretion System Genes fliH and fliI Profoundly Affect Spirochete Flagellar Assembly, Morphology, Motility, Structure, and Cell Division. mBio *6*, e00579-00515. https://doi.org/10.1128/mBio.00579-15

Lin, Y.P., Li, L., Zhang, F., and Linhardt, R.J. (2017). Borrelia burgdorferi glycosaminoglycan-binding proteins: a potential target for new therapeutics against Lyme disease. Microbiology *163*, 1759-1766. https://doi.org/10.1099/mic.0.000571

Lin, Y.P., Tan, X., Caine, J.A., Castellanos, M., Chaconas, G., Coburn, J., and Leong, J.M. (2020). Strain-specific joint invasion and colonization by Lyme disease spirochetes is promoted by outer surface protein C. PLoS Pathog *16*, e1008516. https://doi.org/10.1371/journal.ppat.1008516

Liu, J., Lin, T., Botkin, D.J., McCrum, E., Winkler, H., and Norris, S.J. (2009). Intact flagellar motor of Borrelia burgdorferi revealed by cryo-electron tomography: evidence for stator ring curvature and rotor/C-ring assembly flexion. J Bacteriol *191*, 5026-5036. https://doi.org/10.1128/JB.00340-09

Liu, J., Howell, J.K., Bradley, S.D., Zheng, Y., Zhou, Z.H., and Norris, S.J. (2010). Cellular architecture of *Treponema pallidum*: novel flagellum, periplasmic cone, and cell envelope as revealed by cryo electron tomography. J Mol Biol *403*, 546-561. https://doi.org/S0022-2836(10)00996-4 [pii] 10.1016/j.jmb.2010.09.020

Liu, J.D., and Parkinson, J.S. (1989). Role of CheW protein in coupling membrane receptors to the intracellular signaling system of bacterial chemotaxis. Proc Natl Acad Sci U S A *86*, 8703-8707. https://doi.org/10.1073/pnas.86.22.8703

Lux, R., Moter, A., and Shi, W. (2000). Chemotaxis in pathogenic spirochetes: directed movement toward targeting tissues? J Mol Microbiol Biotechnol 2, 355-364.

Lux, R., Miller, J.N., Park, N.H., and Shi, W. (2001). Motility and chemotaxis in tissue penetration of oral epithelial cell layers by Treponema denticola. Infect Immun *69*, 6276-6283. https://doi.org/10.1128/IAI.69.10.6276-6283.2001

Lynch, M.J., Miller, M., James, M., Zhang, S., Zhang, K., Li, C., Charon, N.W., and Crane, B.R. (2019). Structure and chemistry of lysinoalanine crosslinking in the spirochaete flagella hook. Nat Chem Biol *15*, 959-965. https://doi.org/10.1038/s41589-019-0341-3

MacDonald, A.J., Hyon, D.W., Brewington, J.B., 3rd, O'Connor, K.E., Swei, A., and Briggs, C.J. (2017). Lyme disease risk in southern California: abiotic and environmental drivers of *Ixodes pacificus* (Acari: Ixodidae) density and infection prevalence with *Borrelia burgdorferi*. Parasit Vectors *10*, 7. https://doi.org/10.1186/s13071-016-1938-y

Macnab, R.M. (1996). Flagella and motility. In Escherichia coli and Salmonella: Cellular and Molecular Biology, F.C. Neidhardt, ed. (Washington: American Society for Microbiology), pp. 123-145.

Matilla, M.A., and Krell, T. (2018). The effect of bacterial chemotaxis on host infection and pathogenicity. FEMS Microbiol Rev *42*. https://doi.org/10.1093/femsre/fux052

McCarter, L.L. (2001). Polar flagellar motility of the Vibrionaceae. Microbiol Mol Biol Rev *65*, 445-462, table of contents. https://doi.org/10.1128/MMBR.65.3.445-462.2001

McGee, D.J., Langford, M.L., Watson, E.L., Carter, J.E., Chen, Y.T., and Ottemann, K.M. (2005). Colonization and inflammation deficiencies in Mongolian gerbils infected by Helicobacter pylori chemotaxis mutants. Infect Immun *73*, 1820-1827. https://doi.org/10.1128/IAI.73.3.1820-1827.2005

Medina-Perez, D.N., Wager, B., Troy, E., Gao, L., Norris, S.J., Lin, T., Hu, L., Hyde, J.A., Lybecker, M., and Skare, J.T. (2020). The intergenic small non-coding RNA ittA is required for optimal infectivity and tissue tropism in Borrelia burgdorferi. PLoS Pathog *16*, e1008423. https://doi.org/10.1371/journal.ppat.1008423

Miller, M.R., Miller, K.A., Bian, J., James, M.E., Zhang, S., Lynch, M.J., Callery, P.S., Hettick, J.M., Cockburn, A., Liu, J., *et al.* (2016). Spirochaete flagella hook proteins self-catalyse a lysinoalanine covalent crosslink for motility. Nat Microbiol *1*, 16134. https://doi.org/10.1038/nmicrobiol.2016.134

Mitja, O., Marks, M., Konan, D.J., Ayelo, G., Gonzalez-Beiras, C., Boua, B., Houinei, W., Kobara, Y., Tabah, E.N., Nsiire, A., *et al.* (2015). Global epidemiology of yaws: a systematic review. Lancet Glob Health *3*, e324-331. https://doi.org/10.1016/S2214-109X(15)00011-X

Moon, K.H., Hobbs, G., and Motaleb, M.A. (2016a). Borrelia burgdorferi CheD Promotes Various Functions in Chemotaxis and the Pathogenic Life Cycle of the Spirochete. Infect Immun *84*, 1743-1752. https://doi.org/10.1128/IAI.01347-15

Moon, K.H., Zhao, X., Manne, A., Wang, J., Yu, Z., Liu, J., and Motaleb, M.A. (2016b). Spirochetes flagellar collar protein FlbB has astounding effects in orientation of periplasmic flagella, bacterial shape, motility, and assembly of motors in Borrelia burgdorferi. Mol Microbiol *102*, 336-348. https://doi.org/10.1111/mmi.13463

Moon, K.H., Zhao, X., Xu, H., Liu, J., and Motaleb, M.A. (2018). A tetratricopeptide repeat domain protein has profound effects on assembly of periplasmic flagella, morphology and motility of the lyme disease spirochete Borrelia burgdorferi. Mol Microbiol *110*, 634-647. https://doi.org/10.1111/mmi.14121

Moriarty, T.J., Norman, M.U., Colarusso, P., Bankhead, T., Kubes, P., and Chaconas, G. (2008). Real-time high resolution 3D imaging of the Lyme disease spirochete adhering to and escaping from the vasculature of a living host. PLoS Pathog *4*, e1000090. https://doi.org/10.1371/journal.ppat.1000090

Moriarty, T.J., Shi, M., Lin, Y.P., Ebady, R., Zhou, H., Odisho, T., Hardy, P.O., Salman-Dilgimen, A., Wu, J., Weening, E.H., *et al.* (2012). Vascular binding of a pathogen under shear force through mechanistically distinct sequential interactions with host macromolecules. Mol Microbiol *86*, 1116-1131. https://doi.org/10.1111/mmi.12045

Motaleb, M.A., Corum, L., Bono, J.L., Elias, A.F., Rosa, P., Samuels, D.S., and Charon, N.W. (2000). *Borrelia burgdorferi* periplasmic flagella have both skeletal and motility functions. Proc Natl Acad Sci U S A *97*, 10899-10904. https://doi.org/10.1073/pnas.200221797

Motaleb, M.A., Sal, M.S., and Charon, N.W. (2004). The decrease in FlaA observed in a flaB mutant of Borrelia burgdorferi occurs posttranscriptionally. J Bacteriol *186*, 3703-3711. https://doi.org/10.1128/JB.186.12.3703-3711.2004

Motaleb, M.A., Miller, M.R., Li, C., Bakker, R.G., Goldstein, S.F., Silversmith, R.E., Bourret, R.B., and Charon, N.W. (2005). CheX is a phosphorylated CheY phosphatase essential for Borrelia burgdorferi chemotaxis. J Bacteriol *187*, 7963-7969. https://doi.org/10.1128/JB.187.23.7963-7969.2005

Motaleb, M.A., Miller, M.R., Bakker, R.G., Li, C., and Charon, N.W. (2007a). Isolation and characterization of chemotaxis mutants of the Lyme disease Spirochete Borrelia burgdorferi using allelic exchange mutagenesis, flow cytometry, and cell tracking. Methods Enzymol *422*, 421-437. https://doi.org/10.1016/S0076-6879(06)22021-4

Motaleb, M.A., Miller, M.R., Li, C., and Charon, N.W. (2007b). Phosphorylation assays of chemotaxis two-component system proteins in Borrelia burgdorferi. Methods Enzymol *422*, 438-447. https://doi.org/10.1016/S0076-6879(06)22022-6

Motaleb, M.A., Sultan, S.Z., Miller, M.R., Li, C., and Charon, N.W. (2011). CheY3 of Borrelia burgdorferi is the key response regulator essential for chemotaxis and forms a long-lived phosphorylated intermediate. J Bacteriol *193*, 3332-3341. https://doi.org/10.1128/JB.00362-11

Motaleb, M.A., Liu, J., and Wooten, R.M. (2015). Spirochetal motility and chemotaxis in the natural enzootic cycle and development of Lyme disease. Curr Opin Microbiol *28*, 106-113. https://doi.org/10.1016/j.mib.2015.09.006

Murfin, K.E., Kleinbard, R., Aydin, M., Salazar, S.A., and Fikrig, E. (2019). *Borrelia burgdorferi* chemotaxis toward tick protein Salp12 contributes to acquisition. Ticks Tick Borne Dis *10*, 1124-1134. https://doi.org/10.1016/j.ttbdis.2019.06.002

Murphy, G.E., Leadbetter, J.R., and Jensen, G.J. (2006). In situ structure of the complete *Treponema primitia* flagellar motor. Nature *442*, 1062-1064. https://doi.org/10.1038/nature05015

Murray, T.S., and Shapiro, E.D. (2010). Lyme disease. Clin Lab Med *30*, 311-328. https://doi.org/10.1016/j.cll.2010.01.003

Nakamura, S. (2020). Spirochete Flagella and Motility. Biomolecules *10*. https://doi.org/10.3390/biom10040550

Niddam, A.F., Ebady, R., Bansal, A., Koehler, A., Hinz, B., and Moriarty, T.J. (2017). Plasma fibronectin stabilizes Borrelia burgdorferi-endothelial interactions under vascular shear stress by a catch-bond mechanism. Proc Natl Acad Sci U S A *114*, E3490-E3498. https://doi.org/10.1073/pnas.1615007114

Niehus, E., Gressmann, H., Ye, F., Schlapbach, R., Dehio, M., Dehio, C., Stack, A., Meyer, T.F., Suerbaum, S., and Josenhans, C. (2004). Genome-wide analysis of transcriptional hierarchy and feedback regulation in the flagellar system of Helicobacter pylori. Mol Microbiol *52*, 947-961. https://doi.org/10.1111/j.1365-2958.2004.04006.x

Noguchi, H. (1918). Morphological Characteristics and Nomenclature of Leptospira (Spirochaeta) Icterohaemorrhagiae (Inada and Ido). J Exp Med *27*, 575-592. https://doi.org/10.1084/jem.27.5.575

Norman, M.U., Moriarty, T.J., Dresser, A.R., Millen, B., Kubes, P., and Chaconas, G. (2008). Molecular mechanisms involved in vascular interactions of the Lyme disease pathogen in a living host. PLoS Pathog *4*, e1000169. https://doi.org/10.1371/journal.ppat.1000169

Novak, E.A., Sultan, S.Z., and Motaleb, M.A. (2014). The cyclic-di-GMP signaling pathway in the Lyme disease spirochete, *Borrelia burgdorferi*. Front Cell Infect Microbiol *4*, 56. https://doi.org/10.3389/fcimb.2014.00056

Novak, E.A., Sekar, P., Xu, H., Moon, K.H., Manne, A., Wooten, R.M., and Motaleb, M.A. (2016). The Borrelia burgdorferi CheY3 response regulator is essential for chemotaxis and completion of its natural infection cycle. Cell Microbiol. https://doi.org/10.1111/cmi.12617

Ojaimi, C., Brooks, C., Casjens, S., Rosa, P., Elias, A., Barbour, A., Jasinskas, A., Benach, J., Katona, L., Radolf, J., *et al.* (2003). Profiling of temperature-induced changes in *Borrelia burgdorferi* gene expression by using whole genome arrays. Infect Immun *71*, 1689-1705.

Oliver, J.D., Lynn, G.E., Burkhardt, N.Y., Price, L.D., Nelson, C.M., Kurtti, T.J., and Munderloh, U.G. (2016). Infection of Immature Ixodes scapularis (Acari: Ixodidae) by Membrane Feeding. J Med Entomol *53*, 409-415. https://doi.org/10.1093/jme/tjv241

Orr, M.W., and Lee, V.T. (2016). A PilZ domain protein for chemotaxis adds another layer to c-di-GMP-mediated regulation of flagellar motility. Sci Signal *9*, fs16. https://doi.org/10.1126/scisignal.aai8859

Ouyang, Z., Blevins, J.S., and Norgard, M.V. (2008). Transcriptional interplay among the regulators Rrp2, RpoN and RpoS in Borrelia burgdorferi. Microbiology *154*, 2641-2658. https://doi.org/10.1099/mic.0.2008/019992-0

Pal, U., de Silva, A.M., Montgomery, R.R., Fish, D., Anguita, J., Anderson, J.F., Lobet, Y., and Fikrig, E. (2000). Attachment of Borrelia burgdorferi within Ixodes scapularis mediated by outer surface protein A. J Clin Invest *106*, 561-569. https://doi.org/10.1172/JCI9427

Pal, U., Li, X., Wang, T., Montgomery, R.R., Ramamoorthi, N., Desilva, A.M., Bao, F., Yang, X., Pypaert, M., Pradhan, D., *et al.* (2004a). TROSPA, an Ixodes scapularis receptor for Borrelia burgdorferi. Cell *119*, 457-468. https://doi.org/10.1016/j.cell.2004.10.027

Pal, U., Yang, X., Chen, M., Bockenstedt, L.K., Anderson, J.F., Flavell, R.A., Norgard, M.V., and Fikrig, E. (2004b). OspC facilitates *Borrelia burgdorferi* invasion of Ixodes scapularis salivary glands. J Clin Invest *113*, 220-230. https://doi.org/10.1172/JCI19894

Paul, K., Nieto, V., Carlquist, W.C., Blair, D.F., and Harshey, R.M. (2010). The c-di-GMP binding protein YcgR controls flagellar motor direction and speed to affect chemotaxis by a "backstop brake" mechanism. Mol Cell *38*, 128-139. https://doi.org/10.1016/j.molcel.2010.03.001

Pazy, Y., Motaleb, M.A., Guarnieri, M.T., Charon, N.W., Zhao, R., and Silversmith, R.E. (2010). Identical phosphatase mechanisms achieved through distinct modes of binding phosphoprotein substrate. Proc Natl Acad Sci U S A *107*, 1924-1929. https://doi.org/10.1073/pnas.0911185107

Petzke, M., and Schwartz, I. (2015). *Borrelia burgdorferi* pathogenesis and the immune response. Clin Lab Med *35*, 745-764. https://doi.org/10.1016/j.cll.2015.07.004

Piesman, J., Clark, K.L., Dolan, M.C., Happ, C.M., and Burkot, T.R. (1999). Geographic survey of vector ticks (Ixodes scapularis and Ixodes pacificus) for infection with the Lyme disease spirochete, Borrelia burgdorferi. J Vector Ecol *24*, 91-98.

Pitzer, J.E., Sultan, S.Z., Hayakawa, Y., Hobbs, G., Miller, M.R., and Motaleb, M.A. (2011). Analysis of the *Borrelia burgdorferi* cyclic-di-GMP-binding protein PlzA reveals a role in motility and virulence. Infect Immun *79*, 1815-1825. https://doi.org/10.1128/IAI.00075-11

Policastro, P.F., and Schwan, T.G. (2003). Experimental infection of Ixodes scapularis larvae (Acari: Ixodidae) by immersion in low passage cultures of Borrelia burgdorferi. J Med Entomol *40*, 364-370.

Porter, S.L., Wadhams, G.H., and Armitage, J.P. (2011). Signal processing in complex chemotaxis pathways. Nat Rev Microbiol *9*, 153-165. https://doi.org/10.1038/nrmicro2505

Qin, Z., Tu, J., Lin, T., Norris, S.J., Li, C., Motaleb, M.A., and Liu, J. (2018). Cryo-electron tomography of periplasmic flagella in Borrelia burgdorferi reveals a distinct cytoplasmic ATPase complex. PLoS Biol *16*, e3000050. https://doi.org/10.1371/journal.pbio.3000050

Raddi, G., Morado, D.R., Yan, J., Haake, D.A., Yang, X.F., and Liu, J. (2012). Three-dimensional structures of pathogenic and saprophytic *Leptospira* species revealed by cryo-electron tomography. J Bacteriol *194*, 1299-1306. https://doi.org/JB.06474-11 [pii] 10.1128/JB.06474-11

Radolf, J.D., and Samuels, D.S. (2010). Borrelia: Molecular Biology, Host Interaction and Pathogenesis (Caister Academic Press).

Radolf, J.D., Caimano, M.J., Stevenson, B., and Hu, L.T. (2012). Of ticks, mice and men: understanding the dual-host lifestyle of Lyme disease spirochaetes. Nat Rev Microbiol *10*, 87-99. https://doi.org/10.1038/nrmicro2714

Rajagopala, S.V., Titz, B., Goll, J., Parrish, J.R., Wohlbold, K., McKevitt, M.T., Palzkill, T., Mori, H., Finley, R.L., Jr., and Uetz, P. (2007). The protein network of bacterial motility. Mol Syst Biol *3*, 128. https://doi.org/10.1038/msb4100166

Revel, A.T., Talaat, A.M., and Norgard, M.V. (2002). DNA microarray analysis of differential gene expression in *Borrelia burgdorferi*, the Lyme disease spirochete. Proc Natl Acad Sci U S A *99*, 1562-1567. https://doi.org/10.1073/pnas.032667699

Richards, C.L., Lawrence, K.A., Su, H., Yang, Y., Yang, X.F., Dulebohn, D.P., and Gherardini, F.C. (2015). Acetyl-Phosphate Is Not a Global Regulatory Bridge between Virulence and Central Metabolism in Borrelia burgdorferi. PLoS One *10*, e0144472. https://doi.org/10.1371/journal.pone.0144472

Riviere, G.R., Wagoner, M.A., Baker-Zander, S.A., Weisz, K.S., Adams, D.F., Simonson, L., and Lukehart, S.A. (1991). Identification of spirochetes related to Treponema pallidum in necrotizing ulcerative gingivitis and chronic periodontitis. N Engl J Med *325*, 539-543. https://doi.org/10.1056/NEJM199108223250803

Rogers, E.A., Terekhova, D., Zhang, H.M., Hovis, K.M., Schwartz, I., and Marconi, R.T. (2009). Rrp1, a cyclic-di-GMP-producing response regulator, is an important regulator of *Borrelia burgdorferi* core cellular functions. Mol Microbiol *71*, 1551-1573. https://doi.org/10.1111/j.1365-2958.2009.06621.x

Romeo, T. (1998). Global regulation by the small RNA-binding protein CsrA and the non-coding RNA molecule CsrB. Mol Microbiol *29*, 1321-1330. https://doi.org/10.1046/j.1365-2958.1998.01021.x

Romeo, T., Vakulskas, C.A., and Babitzke, P. (2013). Post-transcriptional regulation on a global scale: form and function of Csr/Rsm systems. Environ Microbiol *15*, 313-324. https://doi.org/10.1111/j.1462-2920.2012.02794.x

Romeo, T., and Babitzke, P. (2018). Global Regulation by CsrA and Its RNA Antagonists. Microbiol Spectr *6*. https://doi.org/10.1128/microbiolspec.RWR-0009-2017

Rosa, P.A., Tilly, K., and Stewart, P.E. (2005). The burgeoning molecular genetics of the Lyme disease spirochaete. Nat Rev Microbiol *3*, 129-143. https://doi.org/10.1038/nrmicro1086

Rosey, E.L., Kennedy, M.J., and Yancey, R.J., Jr. (1996). Dual flaA1 flaB1 mutant of Serpulina hyodysenteriae expressing periplasmic flagella is severely attenuated in a murine model of swine dysentery. Infect Immun *64*, 4154-4162.

Sadziene, A., Thomas, D.D., Bundoc, V.G., Holt, S.C., and Barbour, A.G. (1991). A flagella-less mutant of Borrelia burgdorferi. Structural, molecular, and in vitro functional characterization. J Clin Invest *88*, 82-92. https://doi.org/10.1172/JCI115308

Sal, M.S., Li, C., Motalab, M.A., Shibata, S., Aizawa, S., and Charon, N.W. (2008). Borrelia burgdorferi uniquely regulates its motility genes and has an intricate flagellar hook-basal body structure. J Bacteriol *190*, 1912-1921. https://doi.org/10.1128/JB.01421-07

Samatey, F.A., Matsunami, H., Imada, K., Nagashima, S., Shaikh, T.R., Thomas, D.R., Chen, J.Z., Derosier, D.J., Kitao, A., and Namba, K. (2004). Structure of the bacterial flagellar hook and implication for the molecular universal joint mechanism. Nature *431*, 1062-1068. https://doi.org/10.1038/nature02997

Samuels, D.S. (2011). Gene regulation in *Borrelia burgdorferi*. Annu Rev Microbiol *65*, 479-499. https://doi.org/10.1146/annurev.micro.112408.134040

Sartakova, M.L., Dobrikova, E.Y., Motaleb, M.A., Godfrey, H.P., Charon, N.W., and Cabello, F.C. (2001). Complementation of a nonmotile flaB mutant of Borrelia burgdorferi by chromosomal integration of a plasmid containing a wild-type flaB allele. J Bacteriol *183*, 6558-6564. https://doi.org/10.1128/JB.183.22.6558-6564.2001

Schotthoefer, A.M., and Frost, H.M. (2015). Ecology and epidemiology of Lyme borreliosis. Clin Lab Med *35*, 723-743. https://doi.org/10.1016/j.cll.2015.08.003

Schwartz, A.M., Hinckley, A.F., Mead, P.S., Hook, S.A., and Kugeler, K.J. (2017). Surveillance for Lyme Disease - United States, 2008-2015. MMWR Surveill Summ *66*, 1-12. https://doi.org/10.15585/mmwr.ss6622a1

Segall, J.E., Manson, M.D., and Berg, H.C. (1982). Signal processing times in bacterial chemotaxis. Nature *296*, 855-857. https://doi.org/10.1038/296855a0

Shi, W., Yang, Z., Geng, Y., Wolinsky, L.E., and Lovett, M.A. (1998). Chemotaxis in Borrelia burgdorferi. J Bacteriol *180*, 231-235.

Shih, C.M., Chao, L.L., and Yu, C.P. (2002). Chemotactic migration of the Lyme disease spirochete (Borrelia burgdorferi) to salivary gland extracts of vector ticks. Am J Trop Med Hyg *66*, 616-621. https://doi.org/10.4269/ajtmh.2002.66.616

Silverman, M., and Simon, M.I. (1977). Bacterial flagella. Annu Rev Microbiol *31*, 397-419. https://doi.org/10.1146/annurev.mi.31.100177.002145

Silversmith, R.E., Smith, J.G., Guanga, G.P., Les, J.T., and Bourret, R.B. (2001). Alteration of a nonconserved active site residue in the chemotaxis response regulator CheY affects phosphorylation and interaction with CheZ. J Biol Chem *276*, 18478-18484. https://doi.org/10.1074/jbc.M011418200

Silversmith, R.E., Guanga, G.P., Betts, L., Chu, C., Zhao, R., and Bourret, R.B. (2003). CheZ-mediated dephosphorylation of the

Escherichia coli chemotaxis response regulator CheY: role for CheY glutamate 89. J Bacteriol *185*, 1495-1502.

Silversmith, R.E., Levin, M.D., Schilling, E., and Bourret, R.B. (2008). Kinetic characterization of catalysis by the chemotaxis phosphatase CheZ. Modulation of activity by the phosphorylated CheY substrate. J Biol Chem *283*, 756-765. https://doi.org/10.1074/jbc.M704400200

Sourjik, V., and Armitage, J.P. (2010). Spatial organization in bacterial chemotaxis. The EMBO journal *29*, 2724-2733. https://doi.org/10.1038/emboj.2010.178

Stanek, G., Wormser, G.P., Gray, J., and Strle, F. (2012). Lyme borreliosis. Lancet *379*, 461-473. https://doi.org/10.1016/S0140-6736(11)60103-7

Stevenson, B., and Seshu, J. (2018). Regulation of gene and protein expression in the Lyme disease spirochete. Curr Top Microbiol Immunol *415*, 83-112. https://doi.org/10.1007/82_2017_49

Stewart, P.E., Hoff, J., Fischer, E., Krum, J.G., and Rosa, P.A. (2004). Genome-wide transposon mutagenesis of Borrelia burgdorferi for identification of phenotypic mutants. Appl Environ Microbiol *70*, 5973-5979. https://doi.org/10.1128/AEM.70.10.5973-5979.2004

Sultan, S.Z., Pitzer, J.E., Miller, M.R., and Motaleb, M.A. (2010). Analysis of a Borrelia burgdorferi phosphodiesterase demonstrates a role for cyclic-di-guanosine monophosphate in motility and virulence. Mol Microbiol *77*, 128-142. https://doi.org/10.1111/j.1365-2958.2010.07191.x

Sultan, S.Z., Pitzer, J.E., Boquoi, T., Hobbs, G., Miller, M.R., and Motaleb, M.A. (2011). Analysis of the HD-GYP domain cyclic dimeric GMP phosphodiesterase reveals a role in motility and the enzootic life cycle of Borrelia burgdorferi. Infect Immun *79*, 3273-3283. https://doi.org/10.1128/IAI.05153-11

Sultan, S.Z., Manne, A., Stewart, P.E., Bestor, A., Rosa, P.A., Charon, N.W., and Motaleb, M.A. (2013). Motility is crucial for the infectious life cycle of Borrelia burgdorferi. Infect Immun *81*, 2012-2021. https://doi.org/10.1128/IAI.01228-12

Sultan, S.Z., Sekar, P., Zhao, X., Manne, A., Liu, J., Wooten, R.M., and Motaleb, M.A. (2015). Motor rotation is essential for the formation of the periplasmic flagellar ribbon, cellular morphology, and Borrelia burgdorferi persistence within Ixodes scapularis tick and murine hosts. Infect Immun *83*, 1765-1777. https://doi.org/10.1128/IAI.03097-14

Suzuki, H., Yonekura, K., and Namba, K. (2004). Structure of the Rotor of the Bacterial Flagellar Motor Revealed by Electron Cryomicroscopy and Single-particle Image Analysis. Journal of Molecular Biology *337*, 105-113. https://doi.org/https://doi.org/10.1016/j.jmb.2004.01.034

Szczepanski, A., Furie, M.B., Benach, J.L., Lane, B.P., and Fleit, H.B. (1990). Interaction between *Borrelia burgdorferi* and endothelium *in vitro*. J Clin Invest *85*, 1637-1647. https://doi.org/10.1172/JCI114615

Sze, C.W., and Li, C. (2011). Inactivation of bb0184, which encodes carbon storage regulator A, represses the infectivity of Borrelia burgdorferi. Infect Immun *79*, 1270-1279. https://doi.org/10.1128/IAI.00871-10

Sze, C.W., Morado, D.R., Liu, J., Charon, N.W., Xu, H., and Li, C. (2011). Carbon storage regulator A (CsrA(Bb)) is a repressor of Borrelia burgdorferi flagellin protein FlaB. Mol Microbiol *82*, 851-864. https://doi.org/10.1111/j.1365-2958.2011.07853.x

Sze, C.W., Zhang, K., Kariu, T., Pal, U., and Li, C. (2012). Borrelia burgdorferi needs chemotaxis to establish infection in mammals and to accomplish its enzootic cycle. Infect Immun *80*, 2485-2492. https://doi.org/10.1128/IAI.00145-12

Sze, C.W., Smith, A., Choi, Y.H., Yang, X., Pal, U., Yu, A., and Li, C. (2013). Study of the response regulator Rrp1 reveals its regulatory role in chitobiose utilization and virulence of *Borrelia burgdorferi*. Infect Immun *81*, 1775-1787. https://doi.org/10.1128/IAI.00050-13

Taber, L.H., and Feigin, R.D. (1979). Spirochetal infections. Pediatr Clin North Am *26*, 377-413. https://doi.org/10.1016/s0031-3955(16)33713-0

Tokarz, R., Anderton, J.M., Katona, L.I., and Benach, J.L. (2004). Combined effects of blood and temperature shift on *Borrelia burgdorferi* gene expression as determined by whole genome DNA array. Infect Immun *72*, 5419-5432. https://doi.org/10.1128/IAI.72.9.5419-5432.2004

Troshina, O., Oshurkova, V., Suzina, N., Machulin, A., Ariskina, E., Vinokurova, N., Kopitsyn, D., Novikov, A., and Shcherbakova, V. (2015). Sphaerochaeta associata sp. nov., a spherical spirochaete isolated from cultures of Methanosarcina mazei JL01. Int J Syst Evol Microbiol *65*, 4315-4322. https://doi.org/10.1099/ijsem.0.000575

Troxell, B., and Yang, X.F. (2013). Metal-dependent gene regulation in the causative agent of Lyme disease. Front Cell Infect Microbiol *3*, 79. https://doi.org/10.3389/fcimb.2013.00079

Turner, L., Ryu, W.S., and Berg, H.C. (2000). Real-time imaging of fluorescent flagellar filaments. J Bacteriol *182*, 2793-2801. https://doi.org/10.1128/jb.182.10.2793-2801.2000

Vakulskas, C.A., Potts, A.H., Babitzke, P., Ahmer, B.M., and Romeo, T. (2015). Regulation of bacterial virulence by Csr (Rsm) systems. Microbiol Mol Biol Rev *79*, 193-224. https://doi.org/10.1128/MMBR.00052-14

Wadhams, G.H., and Armitage, J.P. (2004). Making sense of it all: bacterial chemotaxis. Nat Rev Mol Cell Biol *5*, 1024-1037. https://doi.org/10.1038/nrm1524

Wallich, R., Moter, S.E., Simon, M.M., Ebnet, K., Heiberger, A., and Kramer, M.D. (1990). The Borrelia burgdorferi flagellum-associated 41-kilodalton antigen (flagellin): molecular cloning, expression, and amplification of the gene. Infect Immun *58*, 1711-1719.

Wei, B.L., Brun-Zinkernagel, A.M., Simecka, J.W., Pruss, B.M., Babitzke, P., and Romeo, T. (2001). Positive regulation of motility and flhDC expression by the RNA-binding protein CsrA of Escherichia coli. Mol Microbiol *40*, 245-256. https://doi.org/10.1046/j.1365-2958.2001.02380.x

Wilson, D.R., and Beveridge, T.J. (1993). Bacterial flagellar filaments and their component flagellins. Can J Microbiol *39*, 451-472. https://doi.org/10.1139/m93-066

Wolgemuth, C.W., Charon, N.W., Goldstein, S.F., and Goldstein, R.E. (2006). The flagellar cytoskeleton of the spirochetes. J Mol Microbiol Biotechnol *11*, 221-227. https://doi.org/10.1159/000094056

Wolgemuth, C.W. (2015). Flagellar motility of the pathogenic spirochetes. Semin Cell Dev Biol *46*, 104-112. https://doi.org/10.1016/j.semcdb.2015.10.015

Xu, H., Caimano, M.J., Lin, T., He, M., Radolf, J.D., Norris, S.J., Gherardini, F., Wolfe, A.J., and Yang, X.F. (2010). Role of acetyl-phosphate in activation of the Rrp2-RpoN-RpoS pathway in Borrelia burgdorferi. PLoS Pathog *6*, e1001104. https://doi.org/10.1371/journal.ppat.1001104

Xu, H., Raddi, G., Liu, J., Charon, N.W., and Li, C. (2011). Chemoreceptors and flagellar motors are subterminally located in close proximity at the two cell poles in spirochetes. J Bacteriol *193*, 2652-2656. https://doi.org/10.1128/JB.01530-10

Xu, H., Sultan, S., Yerke, A., Moon, K.H., Wooten, R.M., and Motaleb, M.A. (2017). Borrelia burgdorferi CheY2 Is Dispensable for Chemotaxis or Motility but Crucial for the Infectious Life Cycle of the Spirochete. Infect Immun *85*, e00264. https://doi.org/10.1128/IAI.00264-16

Xu, H., He, J., Liu, J., and Motaleb, M.A. (2020). BB0326 is responsible for the formation of periplasmic flagellar collar and assembly of the stator complex in Borrelia burgdorferi. Mol Microbiol *113*, 418-429. https://doi.org/10.1111/mmi.14428

Xu, L., Xin, L., Zeng, Y., Yam, J.K., Ding, Y., Venkataramani, P., Cheang, Q.W., Yang, X., Tang, X., Zhang, L.H., *et al.* (2016). A cyclic di-GMP-binding adaptor protein interacts with a chemotaxis methyltransferase to control flagellar motor switching. Sci Signal *9*, ra102. https://doi.org/10.1126/scisignal.aaf7584

Yakhnin, A.V., Baker, C.S., Vakulskas, C.A., Yakhnin, H., Berezin, I., Romeo, T., and Babitzke, P. (2013). CsrA activates flhDC expression by protecting flhDC mRNA from RNase E-mediated cleavage. Mol Microbiol *87*, 851-866. https://doi.org/10.1111/mmi.12136

Yan, X.F., Xin, L., Yen, J.T., Zeng, Y., Jin, S., Cheang, Q.W., Fong, R., Chiam, K.H., Liang, Z.X., and Gao, Y.G. (2018). Structural analyses unravel the molecular mechanism of cyclic di-GMP regulation of bacterial chemotaxis via a PilZ adaptor protein. J Biol Chem *293*, 100-111. https://doi.org/10.1074/jbc.M117.815704

Yang, J., Huber, G., and Wolgemuth, C.W. (2011). Forces and torques on rotating spirochete flagella. Phys Rev Lett *107*, 268101. https://doi.org/10.1103/PhysRevLett.107.268101

Yang, X., Goldberg, M.S., Popova, T.G., Schoeler, G.B., Wikel, S.K., Hagman, K.E., and Norgard, M.V. (2000). Interdependence of environmental factors influencing reciprocal patterns of gene expression in virulent Borrelia burgdorferi. Mol Microbiol *37*, 1470-1479. https://doi.org/10.1046/j.1365-2958.2000.02104.x

Yang, Y., and Li, C. (2009). Transcription and genetic analyses of a putative N-acetylmuramyl-L-alanine amidase in Borrelia burgdorferi. FEMS Microbiol Lett *290*, 164-173. https://doi.org/10.1111/j.1574-6968.2008.01416.x

Yonekura, K., Maki, S., Morgan, D.G., DeRosier, D.J., Vonderviszt, F., Imada, K., and Namba, K. (2000). The bacterial flagellar cap as the rotary promoter of flagellin self-assembly. Science *290*, 2148-2152. https://doi.org/10.1126/science.290.5499.2148

Young, G.M., Badger, J.L., and Miller, V.L. (2000). Motility is required to initiate host cell invasion by Yersinia enterocolitica. Infect Immun *68*, 4323-4326. https://doi.org/10.1128/iai.68.7.4323-4326.2000

Yuan, W., Glekas, G.D., Allen, G.M., Walukiewicz, H.E., Rao, C.V., and Ordal, G.W. (2012). The importance of the interaction of CheD with CheC and the chemoreceptors compared to its enzymatic activity during chemotaxis in Bacillus subtilis. PLoS One *7*, e50689. https://doi.org/10.1371/journal.pone.0050689

Zambrano, M.C., Beklemisheva, A.A., Bryksin, A.V., Newman, S.A., and Cabello, F.C. (2004). *Borrelia burgdorferi* binds to, invades, and colonizes native type I collagen lattices. Infect Immun *72*, 3138-3146. https://doi.org/10.1128/IAI.72.6.3138-3146.2004

Zhang, J.J., Chen, T., Yang, Y., Du, J., Li, H., Troxell, B., He, M., Carrasco, S.E., Gomelsky, M., and Yang, X.F. (2018). Positive and negative regulation of glycerol utilization by the c-di-GMP binding protein PlzA in *Borrelia burgdorferi*. J Bacteriol *200*. https://doi.org/10.1128/JB.00243-18

Zhang, K., Liu, J., Tu, Y., Xu, H., Charon, N.W., and Li, C. (2012a). Two CheW coupling proteins are essential in a chemosensory pathway of Borrelia burgdorferi. Mol Microbiol *85*, 782-794. https://doi.org/10.1111/j.1365-2958.2012.08139.x

Zhang, K., Tong, B.A., Liu, J., and Li, C. (2012b). A single-domain FlgJ contributes to flagellar hook and filament formation in the Lyme disease spirochete Borrelia burgdorferi. J Bacteriol *194*, 866-874. https://doi.org/10.1128/JB.06341-11

Zhang, K., Liu, J., Charon, N.W., and Li, C. (2016). Hypothetical Protein BB0569 Is Essential for Chemotaxis of the Lyme Disease Spirochete Borrelia burgdorferi. J Bacteriol *198*, 664-672. https://doi.org/10.1128/JB.00877-15

Zhang, K., Qin, Z., Chang, Y., Liu, J., Malkowski, M.G., Shipa, S., Li, L., Qiu, W., Zhang, J.R., and Li, C. (2019). Analysis of a flagellar filament cap mutant reveals that HtrA serine protease degrades unfolded flagellin protein in the periplasm of Borrelia burgdorferi. Mol Microbiol *111*, 1652-1670. https://doi.org/10.1111/mmi.14243

Zhang, K., He, J., Cantalano, C., Guo, Y., Liu, J., and Li, C. (2020). FlhF regulates the number and configuration of periplasmic flagella in Borrelia burgdorferi. Mol Microbiol. https://doi.org/10.1111/mmi.14482

Zhang, Y., Dong, K., Zeng, L., Li, Q., Liu, C., Wang, J., Guo, X., and Zhao, G.P. (2013). Genetic and molecular biological characterization of two homologous cheR genes from Leptospira interrogans. Acta Biochim Biophys Sin (Shanghai) *45*, 806-816. https://doi.org/10.1093/abbs/gmt081

Zhao, R., Collins, E.J., Bourret, R.B., and Silversmith, R.E. (2002). Structure and catalytic mechanism of the E. coli chemotaxis phosphatase CheZ. Nat Struct Biol *9*, 570-575. https://doi.org/10.1038/nsb816

Zhao, X., Zhang, K., Boquoi, T., Hu, B., Motaleb, M.A., Miller, K.A., James, M.E., Charon, N.W., Manson, M.D., Norris, S.J., *et al.* (2013). Cryoelectron tomography reveals the sequential assembly of bacterial flagella in Borrelia burgdorferi. Proc Natl Acad Sci U S A *110*, 14390-14395. https://doi.org/10.1073/pnas.1308306110

Zhao, X., Norris, S.J., and Liu, J. (2014). Molecular architecture of the bacterial flagellar motor in cells. Biochemistry *53*, 4323-4333. https://doi.org/10.1021/bi500059y

Chapter 9

Evolutionary Genetics of *Borrelia*

Zachary J. Oppler[1]*, Kayleigh R. O'Keeffe[1], Karen D. McCoy[2] and Dustin Brisson[1]

[1]Department of Biology, University of Pennsylvania, 433 South University Ave, Philadelphia, PA 19104, USA
[2]Centre for Research on the Ecology and Evolution of Diseases (CREES), MiVEGEC, University of Montpellier – CNRS - IRD, Montpellier, France
*Corresponding author: zoppler@sas.upenn.edu

DOI: https://doi.org/10.21775/9781913652616.09

Abstract

The genus *Borrelia* consists of evolutionarily and genetically diverse bacterial species that cause a variety of diseases in humans and domestic animals. These vector-borne spirochetes can be classified into two major evolutionary groups, the Lyme borreliosis clade and the relapsing fever clade, both of which have complex transmission cycles during which they interact with multiple host species and arthropod vectors. Molecular, ecological, and evolutionary studies have each provided significant contributions towards our understanding of the natural history, biology and evolutionary genetics of *Borrelia* species; however, integration of these studies is required to identify the evolutionary causes and consequences of the genetic variation within and among *Borrelia* species. For example, molecular and genetic studies have identified the adaptations that maximize fitness components throughout the *Borrelia* lifecycle and enhance transmission efficacy but provide limited insights into the evolutionary pressures that have produced them. Ecological studies can identify interactions between *Borrelia* species and the vertebrate hosts and arthropod vectors they encounter and the resulting impact on the geographic distribution and abundance of spirochetes but not the genetic or molecular basis underlying these interactions. In this chapter we discuss recent findings on the evolutionary genetics from both of the evolutionarily distinct clades of *Borrelia* species. We focus on connecting molecular interactions to the ecological processes that have driven the evolution and diversification of *Borrelia* species in order to understand the current distribution of genetic and molecular variation within and between *Borrelia* species.

Introduction

Zoonotic pathogens, like those that comprise the *Borrelia* genus, are a major cause of emerging and re-emerging infectious diseases worldwide and constitute a significant threat to public health (Jones et al., 1994; Taylor et al., 2001). Like many of the bacterial and viral species that cause zoonotic diseases, most *Borrelia* species have complex transmission cycles in which they interact with multiple vertebrate hosts and vectors over the course of their life cycles (Taylor et al., 2001; Lloyd-Smith et al., 2009; Huang et al., 2019). The bacterial spirochetes belonging to the genus *Borrelia* are vectored by arthropods and infect a wide range of vertebrate hosts; many cause diseases in humans and domestic animals (Burgdorfer et al., 1982; Parker and White, 1992; Pritt et al., 2016) (please see chapters 11 and 12). Species within the genus *Borrelia* are classified into two major evolutionary groups, the Lyme borreliosis clade (LB) and the relapsing fever clade (RF). Members of the Lyme borreliosis clade are vectored exclusively by hard-bodied ticks in the *Ixodes* genus and have been investigated to a much greater depth than members of the relapsing fever clade, most of which are vectored by soft-bodied *Argasid* ticks. Although these divergent evolutionary clades recently have been proposed as separate genera (Adeolu and Gupta, 2014), we will continue to refer to them as the Lyme borreliosis clade and the relapsing fever clade until the new classifications have additional consensus (Adeolu and Gupta, 2014; Barbour et al., 2017; Margos et al., 2017, 2020). In this chapter, we review the ecological interactions underlying the evolutionary forces that have produced the current distribution of genetic and genomic variation in both the LB and RF *Borrelia* clades (Table 1). In addition, we focus on the evolutionary and ecological

Table 1

		All *Borrelia*	Lyme *Borrelia*	RF *Borrelia*
Vector	Vector specialization	• Geographic range shaped by vector associations (Margos et al., 2011, 2019a) • Vector switch can result in genetic divergence among subpopulations, expansion into novel geographic range (Lescot et al., 2008; Gatzmann et al., 2015; Becker et al., 2016)	• Characterized molecular factors contributing to tick specialization and persistence (e.g. ospA) (Pal et al., 2004; Battisti et al., 2008; Konnai et al., 2012) • Long feeding time of *Ixodid* ticks provides opportunities for migration and gene flow (Sonenshine, 1979; Margos et al., 2011; Vollmer et al., 2011b; Norte et al., 2020b) • Rate and direction of gene flow only weakly correlated with that of ticks (Walter et al., 2017)	• Transovarial transmission in some species, although molecular mechanisms poorly characterized (Barbour and Hayes, 1986; Cutler, 2010; Rollend et al., 2013) • Short feeding time of *Argasid* ticks limits opportunities for migration and gene flow (Vial et al., 2006; Vial, 2009)
	Life inside vector environment	• Exposure to environmental conditions experienced by vectors (Kung et al., 2013)	• Many genes regulated in temperature-dependent manner likely due to exposure to fluctuating environmental conditions in hard *Ixodid* ticks (Ojaimi et al., 2003; Tokarz et al., 2004; Popitsch et al., 2017; Phelan et al., 2019)	• Sheltered microhabitats of *Argasid* ticks allow spirochetes to survive and remain infectious for several years (Francıs, 1938)
	Microbial interactions within the vector	• Limited number of genes involved in interbacterial interaction pathways owing to limited density and diversity of resident microbial communities in ticks (Ross et al., 2018; Couper et al., 2019)	• Frequent coinfection with other human pathogens (Hersh et al., 2014; Diuk-Wasser et al., 2016; Moutailler et al., 2016; Ross et al., 2018) • Competition with coinfecting strains reduces spirochete load in ticks (Rego et al., 2014; Moutailler et al., 2016; Walter et al., 2016; Durand et al., 2017; Genne et al., 2018; Genné et al., 2019)	• *Limited data have been collected on coinfections of RF* Borrelia *with conspecific strains or other human pathogens*
Host	Host specialization	• *Limited data have been collected on host associations within RF species, so it is unknown if the molecular mechanisms or evolutionary genetic impacts found among LB species apply to RF* Borrelia *species*	• Well characterized genetic and molecular mechanisms contributing to host specialization (e.g. CspA, OspC, OspE, etc.) (Marconi et al., 1993; Livey et al., 1995; Lagal et al., 2006; Önder et al., 2012; Hammerschmidt et al., 2014b; Hart et al., 2018; Tufts et al., 2019; Lin et al., 2020) • Limited HGT among species with different host associations (Jacquot et al., 2014) • Host associations shape population genetic structure (e.g. bird-associated species have much less genetic structure than rodent-associated species) (Victorino et al., 2008; Vollmer et al., 2011b, 2013; Šimo et al., 2017; Norte et al., 2020b) • Host switch results in genetic divergence among subpopulations, expansion into novel geographic range (Margos et al., 2009; Becker et al., 2016) • Wide geographic range of vectors permits hosts to drive changes in geographic range (Stanek et al., 2012; Šimo et al., 2017) • Diverse host communities maintain genetic diversity of *Borrelia* populations (Brisson and Dykhuizen, 2004; Mechai et al., 2016; Vuong et al., 2017)	• Limited geographic range of vectors limits extent to which hosts can drive changes in geographic range (Vial, 2009; Trape et al., 2013)
	Interactions with host immune system	• Host immune memory drives negative frequency dependent selection on *Borrelia* populations (Marcsisin et al., 2016; Gomez-Chamorro et al., 2019) • Well characterized genetic and molecular mechanisms that enable rapid immune evasion within the host (Zhang and Norris, 1998; Barbour et al., 2006; Coutte et al., 2009; Graves et al., 2013; Norris, 2015)	• vlsE system generates antigenic variability to evade host immunity (Zhang and Norris, 1998; Coutte et al., 2009; Graves et al., 2013; Norris, 2015; Tufts et al., 2019)	• Vmp system generates antigenic variability to evade host immunity (Barbour et al., 2006)

consequences arising from these past evolutionary changes.

Our current understanding of the evolutionary genetics of *Borrelia* species can be informed by investigations into the ecological interactions encountered by *Borrelia* species during their life cycle. These interactions are the product of past evolutionary processes that have shaped the observable genotypic and phenotypic diversity. Many investigations have focused separately on either the evolutionary history or the ecology of several *Borrelia* species, both of which have been invaluable to our understanding of the biology and history of *Borrelia*. Studies focusing on the evolutionary history of *Borrelia* have detailed genetic and phenotypic variation across space and among host species, genetic and genomic diversity and divergence as well as the processes occurring over millennia that have resulted in the variation within and between species observed today (De Michelis et al., 2000; Dykhuizen et al., 2008; Graves et al., 2013; Coipan et al., 2018). These evolutionary patterns and processes have resulted in molecular factors associated with host and vector specialization or mechanisms of escaping immune responses, differences in the rates and processes leading to diversification, and the identification of virulence factors that enhance survival or transmission (Pal et al., 2001; Fingerle et al., 2007; Neelakanta et al., 2007; de Silva et al., 2009; Kenedy et al., 2012; Ogden et al., 2015; Tufts et al., 2019); many of these factors are directly linked to host and vector specialization. Ecological studies have, on the other hand, characterized processes and interactions affecting the distribution and abundance of *Borrelia* species (Paul et al., 2016; Kilpatrick et al., 2017; Vuong et al., 2017). Connecting these ecological processes and interactions with the resulting evolutionary adaptations can reveal how these evolutionary changes have shaped and will continue to shape current and future ecological interactions and species distributions.

The common ancestor of all *Borrelia* species appears to be an obligate parasite that interacted exclusively with or within vector and host environments. Thus, vectors and hosts constitute the selective environment to which *Borrelia* must adapt to maximize survival and transmission efficacy (Haven et al., 2011). The disparate selective pressures experienced by species of *Borrelia* or populations within a species that interact with different vector or host species have resulted in divergent molecular adaptations that can lead to increased specialization on a subset of vectors or hosts. In many cases, the increased specialization feeds back, increasing selective pressures on other molecules to increase the efficacy of their interactions with molecules in some hosts or vectors at the cost of decreasing interaction efficacy in other host or vector species (Wang et al., 2001; Tufts et al., 2019). Associations with a limited subset of host or vector species further drive micro-evolutionary changes in *Borrelia* populations and species by shaping patterns of gene flow and geographic distributions (McCoy et al., 2013; Khatchikian et al., 2015; Seifert et al., 2015) as well as altering the effects of genetic drift and mutation (Brisson et al., 2012).

Investigating the complex web of ecological interactions among *Borrelia*, their vectors, and their hosts, facilitates detection of both the selective pressures that have given rise to the present-day genetic variation and the downstream evolutionary changes that are a product of that variation. Although the evolutionary and ecological history of Lyme borreliae have been the subject of extensive investigation (Kurtenbach et al., 2006; Margos et al., 2011; Brisson et al., 2012; Ogden et al., 2015; Kilpatrick et al., 2017), significantly less research has focused on the relapsing fever group (Cutler, 2010; Talagrand-Reboul et al., 2018). Here, we synthesize and compare recent findings on the evolutionary genetics from both groups of *Borrelia*, with a focus on identifying ecological processes that have driven adaptations and the current distribution of genetic variants within and between *Borrelia* species and the evolutionary consequences of that variation. Understanding the feedbacks between the evolution of these species and their ecological interactions is critical for predicting and controlling future disease epidemics.

Interactions between *Borrelia* and their vectors

Evolutionary impact of vector specialization

All *Borrelia* species depend on at least one vector species for the completion of their life cycle. Interactions between *Borrelia* and its vectors select for molecules that enable successful uptake from infected hosts, persistence within the vector, migration through vector tissues, and transmission and persistence in subsequent vertebrate hosts (Pal et al., 2001; Fingerle et al., 2007; Neelakanta et al., 2007; de Silva et al., 2009; Kenedy et al., 2012). Additionally, many RF *Borrelia* species have molecular machinery that enables transovarial transmission (Barbour and Hayes, 1986; Cutler, 2010; Rollend et al., 2013), although these molecules are not well characterized. For example, all studied LB species express the Outer Surface Protein A (OspA) which binds to the *Ixodid* tick midgut receptor, TROSPA, and enables colonization and persistence within the tick (Pal et al., 2004; Battisti et al., 2008). *Borrelia* specializing on different tick species experience different selective pressures at the molecular level owing to variation in tick proteins (like TROSPA) that *Borrelia* must exploit to complete its enzootic cycle (Konnai et al., 2012). That is, different tick environments impose different selective regimes on *Borrelia* traits, resulting in refinement by natural selection to maximize efficacy of each *Borrelia* species in a single vector species (Couper et al., 2020). The molecular variants resulting from this natural selection create variation in vector competence and vector specialization which further influences downstream interactions between *Borrelia* and vertebrate hosts. Molecules mediating persistence and specialization of RF *Borrelia* species within vectors have not been investigated to a similar depth. Nevertheless, most LB and RF *Borrelia* species are transmitted by a single species or related group of ticks such that the genes that mediate these interactions are expected to vary little within *Borrelia* species (Wilske et al., 1993; Eisen, 2020).

Vector associations are important drivers of the geographic range and migration rates of *Borrelia* species (Margos et al., 2011, 2019a). Variation in home ranges and rates of gene flow among *Borrelia* species can be attributed both to the feeding patterns of their vectors and to the range and mobility of host species parasitized by their vectors (Vial et al., 2006; Vial, 2009; Margos et al., 2011; Vollmer et al., 2011a; Stanek et al., 2012; Trape et al., 2013; Norte et al., 2020a). As an adaptation to their endophilic lifestyle, *Argasid* ticks like *Ornithodoros sonrai* typically take rapid bloodmeals to avoid predators and exposure to harsh environmental conditions outside of their sheltered microhabitats (Vial, 2009). Shorter feeding times in combination with feeding preferentially on territorial, burrowing vertebrates significantly constrains opportunities for migration and gene flow of both *O. sonrai* and *B. crocidurae*, the RF species that it vectors (Vial et al., 2006). Geographically isolated subpopulations of many RF *Borrelia* species vectored by ticks that parasitize less-mobile host species are genetically divergent due to both neutral evolution and to natural selection favoring specialization to local environmental conditions (Vial et al., 2006; Hoen et al., 2009; Humphrey et al., 2010). In contrast, *Ixodid* ticks remain attached to their hosts for several days and some, like *I. dentatus* or *I. uriae*, preferentially feed on mobile host species like birds (Sonenshine, 1979; Eveleigh and Threlfall, 1974.). LB species associated with vectors that feed on highly-mobile host species exhibit limited population genetic structure as a result of frequent migration among subpopulations (Margos et al., 2011; Vollmer et al., 2011b; Gómez-Díaz et al, 2011). Although these *Borrelia* species may experience a wider range of ecological interactions with more diverse host communities, they rarely show divergent features across their geographic range (Norte et al., 2020b). Thus, despite experiencing more diverse ecological interactions, *Borrelia* species with broad geographic ranges tend to harbor less genetic variation than *Borrelia* species exposed to fewer ecological interactions due to the homogenizing effect of gene flow among subpopulations.

Borrelia that colonize a novel vector species are exposed to different selective pressures derived from distinct ecological interactions in dissimilar geographic ranges. This can result in genetic divergence between populations of the same *Borrelia* species vectored by different arthropod species. For example, as populations of the RF species *B. duttonii* adapted to the body louse, *Pediculus humanus,* they diverged from the ancestral populations vectored by *O. moubata* (Lescot et al., 2008). Following this vector switch, the louse-borne populations experienced a massive geographic expansion due to their close association with human populations. Further, genetic data revealed sufficient evolutionary divergence and genome degradation to classify these populations as a distinct species, *B. recurrentis* (Lescot et al., 2008). Similarly, a vector switch by *B. bavariensis* from *I. persulcatus* to *I. ricinus* provided an opportunity for the *Borrelia* species to expand into Europe (Gatzmann et al., 2015; Becker et al., 2016). Vector switching not only allows *Borrelia* species to invade novel ecological niches, the resulting genomic diversification can result in speciation.

Despite the generally tight associations between *Borrelia* and their vectors, the direction and rate of gene flow of *Borrelia* species typically correlates only weakly with those of their vectors at fine temporal and geographic scales (Walter et al., 2017). This is at least partially due to the fact that vertebrate hosts frequently transport *Borrelia* spirochetes in the

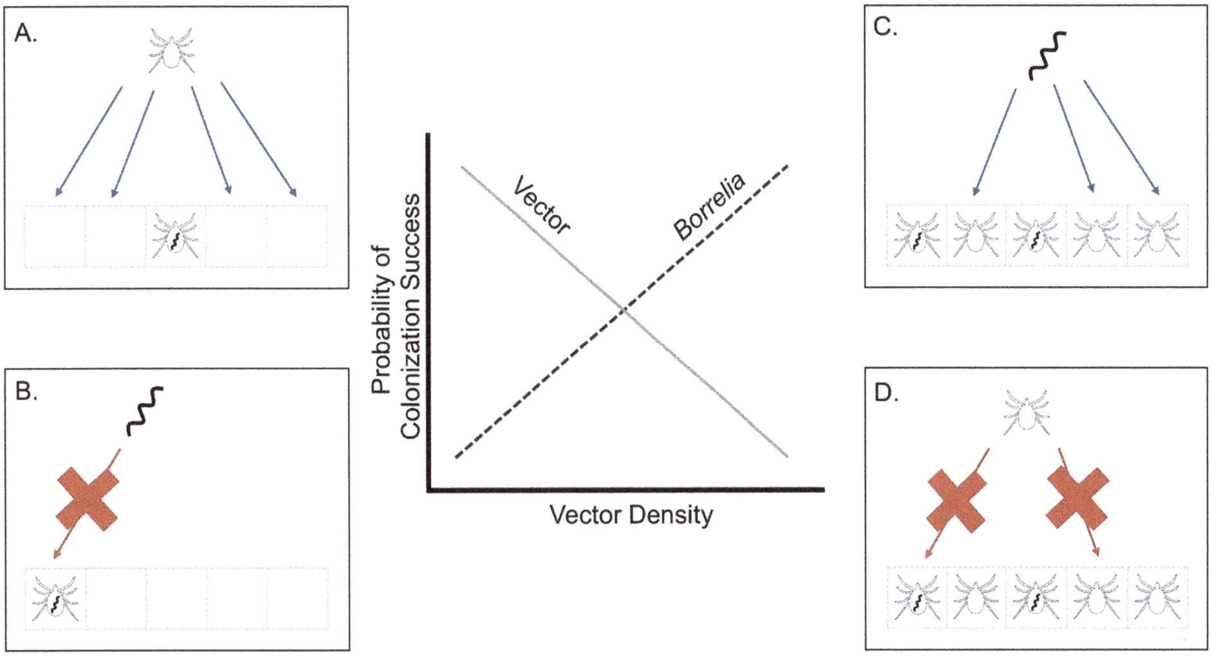

Figure 1. Differences in the rates of gene flow between *Borrelia* and their vectors may be impacted by differences in colonization efficiency. The rate at which *Borrelia* species and their vectors colonize new sites could be dependent, in part, on the size of the resident vector population. **A.** Vectors dispersing into areas with small resident vector populations may colonize with high efficiency due to limited competition with resident vectors if intraspecies competition imparts colonization efficiency, **B.** while migrant *Borrelia* spirochetes are likely to colonize with low efficiency in these areas due to the limited opportunities for transmission. **C.** By contrast, there are many transmission opportunities for *Borrelia* at locations with large vector populations that could result in higher colonization efficiency, **D.** whereas large resident vector populations could limit the probability that dispersing ticks will colonize due to competition with resident vectors. These differences in colonization efficiencies between *Borrelia* and their vectors, mediated by the size of resident vector populations, may have contributed to the observed differences in the rate and direction of gene flow between *Borrelia* species and their primary vectors. Additional research on the impact of intraspecies competition on population dynamics and colonization efficiency is necessary to determine the impacts of competition on the evolutionary history of both *Borrelia* and vector species.

absence of their vectors and vice versa. Namely, infected hosts can transport *Borrelia* even when they are not being parasitized by vectors, and uninfected hosts can transport uninfected vectors in the absence of *Borrelia*. For example, while soft ticks like *O. turicata* typically take short blood meals and have limited inter-burrow movement, *B. turicatae*-infected hosts may easily migrate between burrows (Adeyeye and Butler, 1989). and leave their bacteria in a new location. Likewise, population genetic structure among circulating *B. garinii* strains has been shown to correlate strongly with the local genetic structure of their tick vector *I. uriae* that diverge into distinct host races exploiting different sympatric seabird species at the scale of a colony (McCoy et al. 2013). However, no structure in *B. garinii* strains is evident at the scale of the ocean basin due to the large-scale movements and local colonization dynamics of the bacterium with the seabird host (Gómez-Díaz et al. 2011). The rates and direction of gene flow may differ between *Borrelia* species and vectors even if they disperse together because the bacterium and vector colonize locations at different efficiencies (Figure 1). That is, *Borrelia* may colonize new sites with high efficiency when the local density of vectors is high, while competition with resident vectors may limit the efficiency by which immigrating vectors can establish at these new sites. Additional work is needed to reveal how migration within vertebrates and differential colonization efficiencies contribute to the observed differences in the rate and direction of gene flow between *Borrelia* and their vectors (Walter et al., 2017).

Evolutionary genetic consequences arising from interactions with the vector environment and co-infecting pathogens

The evolutionary genetics of *Borrelia* populations are shaped by many of the environmental features experienced by their vectors. Survival of *Borrelia* within non-homeothermic ticks poses a significant challenge owing to exposure to daily and seasonal temperature fluctuations in addition to the physiological changes caused by environmental fluctuations (Kung et al., 2013). Experimental evidence from other bacterial species has demonstrated that fluctuating conditions select for microbes that are tolerant of a wider range of environments due to the strong selective pressure for tolerance resulting from extreme environmental variation (Chevin and Hoffmann, 2017; Saarinen et al., 2018). The extraordinary environmental fluctuations with which *Borrelia* interacts have even led some to hypothesize that *Borrelia* may have adapted to infect vertebrate hosts as a means of escaping environmentally hostile conditions that negatively impact vector populations (Cutler, 2010). Interestingly, RF *Borrelia* species vectored by nidicolous ticks have been shown to survive and remain infectious within unfed ticks for several years, possibly owing to the fact that their vectors predominantly reside within burrows, caves, and human and livestock habitats, buffered against environmental fluctuations (Francıs, 1938). By contrast, all LB species within *Ixodid* ticks remain exposed to environmental extremes of both the summers and winters in temperate climates and experimental studies have shown that LB survival in ticks may strongly decline at high temperatures (Shih et al, 1995). Understanding the mechanisms by which *Borrelia* withstand wide-ranging environmental fluctuations may provide insights into how recent and future changes in climate will ultimately impact their geographic distribution and evolutionary trajectories.

Borrelia and other microbes within their vectors may facilitate or compromise one another's ability to colonize, grow, and be transmitted to hosts (Diuk-Wasser et al., 2016; Moutailler et al., 2016; Ross et al., 2018). For example, several *Borrelia* species have been found to co-occur with a variety of other human pathogens, including the causal agent of human babesiosis, *Babesia microti* (Diuk-Wasser et al., 2016). *B. burgdorferi* sensu stricto (Bbss) is thought to facilitate *B. microti* within tick vectors as the number of ticks coinfected with these two pathogens is higher than would be expected by chance (Hersh et al., 2014; Diuk-Wasser et al., 2016). Coincidentally, the range and prevalence of *B. microti* has increased significantly in the northeastern United States where Bbss was already highly prevalent, suggesting important epidemiological consequences of this interaction (Dunn et al., 2014; Diuk-Wasser et al., 2016). However, for now, it remains unknown as to whether interactions between these pathogens within the tick vector are neutral or positive nor whether *B. microti* may reciprocally facilitate Bbss transmission. Further work on the nature of interactions between co-infecting infectious agents in the tick vector is now needed to better understand the role of the vector in determining the epidemiological dynamics of co-infecting pathogens.

Borrelia may also encounter other *Borrelia* species or conspecific strains within their vectors (Moutailler et al., 2016). Ticks infected by multiple strains of LB species have the same overall spirochete load as single-strain infections, suggesting there is competition among coinfecting strains which results in lower infection intensity on each strain within a tick (Walter et al., 2016; Durand et al., 2017; Genne et al., 2018; Genné et al., 2019). Since *Borrelia* density within ticks positively correlates with the probability of successful transmission to vertebrate hosts (Rego et al., 2014), competition among coinfecting strains is expected to reduce the evolutionary fitness of each strain. While inter-strain competition is expected to select for traits like increased growth rates or the production of spiteful molecules to suppress the growth of other strains (Cattadori et al., 2008; Telfer et al., 2010; Alizon et al., 2013; Susi et al., 2015), there is no evidence from genomic surveys nor from controlled experiments that such traits have evolved. More work is needed to identify the evolutionary consequences of interactions between co-infecting strains, both within vectors and within vertebrate hosts.

The limited density and diversity of microbial communities that naturally reside within *Ixodid* ticks seems likely to exert only minimal selective pressure on *Borrelia* species (Ross et al., 2018). Contrary to initial reports of a highly diverse *Ixodid* microbiome (van Treuren et al., 2015; Zolnik et al., 2016; Abraham et al., 2017), recent studies controlling for bacterial biomass demonstrated that the diversity of tick microbiomes is actually quite low (Ross et al., 2018; Couper et al., 2019). Recent investigations of both LB and RF *Borrelia* genomes revealed far fewer genes known to be involved in interbacterial

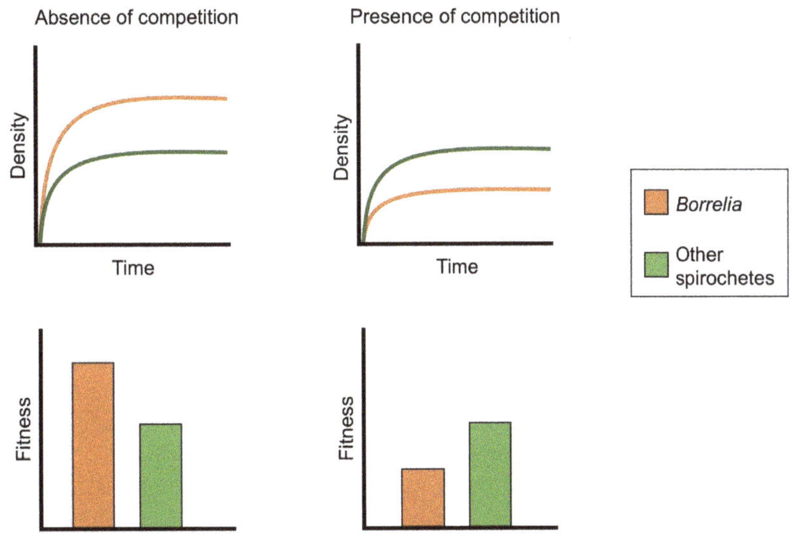

Figure 2. The presence and absence of genes involved in interbacterial interactions may have current and future fitness consequences. Many spirochetes have maintained genetic pathways for the production of molecules involved in interbacterial interaction. Most of these pathways have been lost in *Borrelia* species, likely because *Borrelia* rarely encounter rich microbiota within hosts or vectors that can reduce the fitness of *Borrelia* through competitive interactions. In the absence of competitive interbacterial interactions, the metabolic costs associated with the production of interbacterial-interaction molecules may result in negative fitness consequences for producers. The evolved loss of these interbacterial interaction pathways in *Borrelia* resulted in an increase in fitness in the absence of competition with other microbes in low diversity environments (**A**). In the presence of competition with other microbes (**B**), producers should be relatively more fit than non-producers as interbacterial interaction products provide a competitive advantage that offsets the metabolic costs of producing spiteful molecules; non-producers like *Borrelia* suffer fitness consequences as a result of their limited competitive abilities in higher diversity environments. The negative fitness consequences experienced by non-producers, like *Borrelia,* in higher microbial diversity environments may limit their ability to survive in novel environments where they are likely to encounter novel microbial species.

interaction pathways than have been found in many other surveyed bacterial species (Ross et al., 2018). Genes mediating interactions with other microbes likely have been lost in *Borrelia* owing to their limited encounters with other microbes and consequently weak natural selection for their maintenance (Zhang et al., 2012; Ross et al., 2018). This limited number of genes utilized for interbacterial interactions is likely to have negative fitness consequences for *Borrelia* species that engage in novel ecological interactions with microbes within new vectors or host species as geographic ranges change due to changes in climate or human land use (Figure 2).

Interactions between *Borrelia* and vertebrate hosts

Genetic basis and evolutionary consequences of host specialization

Interactions between *Borrelia* and their vertebrate hosts, similar to *Borrelia*-vector interactions, underlie selective regimes that favor a set of finely-tuned molecular mechanisms. These include molecules that enable vector-to-host transmission, colonization and spread to distal vertebrate tissues, evasion of the host immune system, and transmission back to the feeding vector (Kenedy et al., 2012; Ogden et al., 2015; Tufts et al., 2019). *Borrelia* species that interact with different sets of host species experience different selective pressures at the molecular level owing to variation in vertebrate immune components and other host molecules with which *Borrelia* must interact to complete its enzootic cycle (Roberts et al., 1998; Hammerschmidt et al., 2014a; Hart et al., 2018; Tufts et al., 2019). The sequence variation and molecular determinants of host associations in the LB group have been studied at much greater depth than those in the RF group, although it is likely that similar processes have occurred in both groups. Among LB species, even minor genetic differences between orthologous proteins can substantially impact resulting interactions with vertebrate host molecules. For example, molecular divergence at the complement regulator acquiring surface protein A (CspA) results in differential ability to bind immune components and establish infection within various vertebrate species (Hart et al., 2018; Lin et al., 2020).

The genetic sequences and related molecular processes in each *Borrelia* species ultimately feedback to shape the strength and frequency of subsequent interactions between *Borrelia* and the vertebrate host community which then impacts their geographic distribution, population genetic structure, and ensuing opportunities for interactions.

Host associations occur when vertebrate host species are infected more often than expected given their frequency in the host community. These associations are common among both LB and RF *Borrelia* species and are even emerging among *Borrelia* species previously considered to be host generalists (Hanincová et al., 2006; Kurtenbach et al., 2006; Mechai et al., 2016). For example, while Bbss infects a wide range of vertebrate species, individual strains clearly have host associations, as they do not all colonize and survive within the full range of vertebrate hosts (Anderson et al., 1990; Barthold et al., 1991; Norris et al., 1995; Wang et al., 2002; Barbour et al., 2009; Baum et al., 2012; Chan et al., 2012), and they vary in their capacity to infect and cause disease in humans (Dykhuizen et al., 2008; Wormser et al., 2008; Hanincova et al., 2013). The sequence variation among vertebrate species at molecules such as plasminogen and complement regulator proteins, molecules with which Bbss must interact to successfully colonize a host, limits the range of host species each Bbss strain can infect (Ripoche et al., 1988). As a result, sequence variation among Bbss strains at the molecules mediating these interactions, like OspC and OspE, produce species-specific differences in binding efficacy to plasminogen and complement regulator molecules (Lagal et al., 2006; Önder et al., 2012; Tufts et al., 2019). Since each Bbss strain maintains just a single OspC sequence, it can effectively bind plasminogen from only a narrow range of host species (Marconi et al., 1993; Livey et al., 1995; Casjens et al., 2018). The strong selective pressure to maximize efficacy of immune evasion imposes a major barrier to the long-term maintenance of host generalism in *Borrelia*.

The evolution of host associations is shaped by the strength and frequency of interactions between each *Borrelia* species, or individual strains within species, and each vertebrate species during its evolutionary history. As noted above, variation in host immune components creates disparate selective pressures for invading *Borrelia* species (Brisson and Dykhuizen, 2004; Bäumler and Fang, 2013). Hosts which are frequently infected and from which *Borrelia* achieve high host-to-tick transmission rates exert significantly stronger selective pressure than hosts that *Borrelia* rarely encounter or from which *Borrelia* are rarely transmitted to feeding ticks (Ripoche et al., 1988; Tufts et al., 2019). Among both LB and RF *Borrelia* species, the frequency of interactions with each vertebrate host is heavily influenced by the range and feeding preferences of their primary vectors (Margos et al., 2011; Talagrand-Reboul et al., 2018). As a *Borrelia* species or strain becomes increasingly well-adapted to survive and transmit from a particular host species, its ability to survive and transmit from other vertebrate species tends to diminish, constraining downstream ecological interactions (Anderson et al., 1990; Barthold et al., 1991; Norris et al., 1995; Wang et al., 2001, 2002; Barbour et al., 2009; Baum et al., 2012; Chan et al., 2012; Tufts et al., 2019).

While host associations limit the geographic range of *Borrelia* species and constrain the range of ecological interactions, adaptation to novel host species – often termed a host range expansion or a host switch – can expand the diversity and range of ecological interactions *Borrelia* experience. This has resulted in genetic divergence and even speciation among LB species. Novel host species can impose distinct selective pressures while also providing opportunities for expansion of geographic range and alternate routes and rates of gene flow. For example, evolutionary divergence between the sister species *B. garinii* and *B. bavariensis* likely resulted from the evolution of different host associations, with *B. garinii* specializing on birds and *B. bavariensis* specializing on small mammals (Margos et al., 2009; Becker et al., 2016). The divergent selective regimes imposed by the different sets of vertebrates infected by each species, along with neutral evolutionary divergence in populations isolated in different sets of host species, led to in the genetic differentiation and speciation among these specialized LB species.

Genetic differentiation among *Borrelia* adapted to different host species is a product of both a reduction in gene flow and alternate selective pressures. That is, opportunities for genetic exchange among *Borrelia* species with different host associations will be limited by a lack of proximity as members of different *Borrelia* species are rarely found infecting the same host (Dykhuizen and Baranton, 2001; Bunikis et al., 2004). In fact, rates of horizontal gene transfer among LB species are 50 times lower than the rates of horizontal gene transfer among strains within LB species (Jacquot et al., 2014), possibly as a result of limited proximity of different *Borrelia* species associated with different host species which can result in different *Borrelia* species infecting different individual ticks (Hildebrandt et al., 2003; Brisson and Dykhuizen, 2006; Herrmann et al., 2013). *Borrelia* species also experience very limited selective pressure from vertebrate species that they encounter infrequently or those to which they are not well-adapted and thus are not competent reservoirs. The lack of selective pressure to maintain the genetic and molecular variants necessary for a *Borrelia* species to infect these non-reservoir vertebrate host species eliminates evolutionary constraints, allowing specialization on the limited set of species that serve as regular hosts (Kurtenbach et al., 2006; Becker et al., 2016; Margos et al., 2019b). This positive feedback loop, along with the limited homogenizing effect of horizontal gene transfer among *Borrelia* species, has likely contributed to the high frequency of *Borrelia* species that utilize only a limited set of vertebrate species to complete their enzootic cycles.

Host associations shape the distribution and population genetic structure of Borrelia *species*

Vertebrate migration patterns shape the population genetic structure and geographic distribution of LB species. LB associated with small rodents and other less mobile host species tend to have well-defined population genetic structure as a result of limited opportunities for gene flow between isolated subpopulations (Victorino et al., 2008; Vollmer et al., 2011b; Norte et al., 2020b). In contrast, LB associated with highly-mobile host species, like birds, have limited population genetic structure and few distinct subpopulations as host movement and migration facilitate genetic exchange between geographically distant regions (Norte et al., 2020b). For example, population genetic structure can be observed even at very fine spatial scales for the rodent-associated *B. afzelii*, whereas genetic structure only began to emerge at inter-continental scales for the bird-associated *B. valaisiana* and *B. garinii* (Vollmer et al., 2011b, 2013; Gómez-Díaz et al., 2011). Thus, *Borrelia* species associated with mobile host species maintain genetic cohesion across their geographic range, while *Borrelia* associated with less mobile host species become genetically fragmented as a result of natural selection and adaptation to local conditions or host communities as well as genetic drift within the small effective subpopulations (Vollmer et al., 2011b; Norte et al., 2020b).

The extent to which vertebrate hosts can drive changes in the geographic range of *Borrelia* species is, of course, dependent upon the geographic distribution of competent vector species. As infected *Ixodid* ticks feed for up to several days (Šimo et al., 2017), they have many opportunities to migrate in tandem with the vertebrate hosts to which they are attached. On the other hand, most *Argasid* ticks detach from their hosts within minutes, limiting the opportunities to migrate to new sites on their hosts (Vial, 2009). As expected, *Argasid* ticks seem to have more constrained geographic ranges than their *Ixodid* relatives (Stanek et al., 2012; Trape et al., 2013), but ecological data on this tick family is still fragmentary. Since all *Borrelia* species depend on both vectors and vertebrate hosts for the completion of their enzootic cycle, the limited range and faster feeding time of *Argasid* vectors likely limits the opportunities for vertebrates to drive geographic range expansion of RF *Borrelia* species.

Interactions with diverse host species can promote the maintenance of genetic diversity within *Borrelia* populations. For example, many strains of Bbss coexist at geographic locations where the vertebrate community is highly diverse (Brisson and Dykhuizen, 2004). This is because antigenically distinct Bbss strains differ in their ability to successfully infect and be transmitted from each vertebrate species and no strain has the highest evolutionary fitness in all of the vertebrate species (Brisson and Dykhuizen, 2004; Cobey and Lipsitch, 2012). Variation among Bbss strains in host associations drives niche separation by allowing different subtypes to maximize transmission success in different hosts, thereby reducing within-host competition between strains (Brisson and Dykhuizen, 2004; Hanincová et al., 2006; Mechai et al., 2016; Vuong et al., 2017; Brisson, 2018). The presence of a diverse host community can thus promote and maintain an array of genetically divergent subtypes over the long-term.

Evolutionary genetic consequences arising from interactions with the vertebrate immune system

Interactions between *Borrelia* and the vertebrate adaptive immune system drive natural selection for antigenic variability. The selective pressure for *Borrelia* to persist within its host for the duration necessary to be transmitted back to another feeding vector has resulted in the evolution of genetic and molecular mechanisms that enable immune evasion with sufficient variation to remain effective throughout a months- or years-long infection (Tufts et al., 2019).

As infection proceeds, genetic loci encoding the immunodominant surface proteins Vmp in RF *Borrelia* species and Vmp-like sequences (VlsE) in LB species undergo frequent recombination with silent cassettes, generating novel antigens that are unrecognized by the host immune system (Zhang and Norris, 1998; Barbour et al., 2006; Coutte et al., 2009; Graves et al., 2013; Norris, 2015). Further, the genetic sequences at the silent cassettes evolve much more rapidly than the genes at other loci. This is due in part to the conservation of highly-mutable tandem repeat structures in the otherwise highly-diverged cassettes, suggesting that the ability to continuously generate novel sequence variation at this locus provides a significant fitness advantage (Lawrenz et al., 2004; Bankhead and Chaconas, 2007; Graves et al., 2013).

The adaptive immune systems of infected vertebrates also may drive balancing selection contributing to the maintenance of genetic diversity within *Borrelia* populations. For example, immune memory can create negative frequency dependent selection on *Borrelia* populations if rare genotypes have a selective advantage over common genotypes due to prior infections in the host community. That is, if immune memory to specific genotypes protects hosts from future infections of the same genotype but does not prevent infection with alternate genotypes, rare genotypes will have a selective advantage over more common genotypes. The underlying immune mechanisms necessary for negative frequency dependent selection to operate in natural populations have considerable empirical support in both RF and LB *Borrelia* species (Preac-Mursic et al., 1992; Gilmore et al., 1996; Probert et al., 1997; Barthold, 1999; Marcsisin et al., 2016). For example, mice previously infected with *B. hermsii* or *B. burgdorferi* were shown to be immune to antigenically homologous strains, but remained susceptible to infection by antigenically distinct strains (Probert et al., 1997; Marcsisin et al., 2016). Further, a recent study demonstrated that *B. afzelii* strain-specific antibodies can also be passed from mother to offspring, suggesting that the effects of immune memory may last beyond a single generation (Gomez-Chamorro et al., 2019). This type of negative frequency dependent process may enable antigenically distinct strains to be maintained within *Borrelia* populations, although no direct empirical evidence of this process operating in natural populations has been reported despite laboratory evidence that the underlying immune mechanisms necessary to select for rare genotypes are present in several natural host species (Preac-Mursic et al., 1992; Gilmore et al., 1996; Probert et al., 1997; Barthold, 1999; Marcsisin et al., 2016).

Concluding remarks

In this chapter, we summarized recent findings on the evolutionary genetics of both LB and RF *Borrelia* species. While we show that much work has been done to elucidate the biology and natural history of *Borrelia*, it is clear that there are still open questions about the causes and consequences of evolutionary change in these species, many of which are poised to be answered by leveraging existing knowledge on molecular mechanisms and ecological interactions. For example, molecular studies have played a critical role in identifying the mechanisms utilized by several *Borrelia* species to establish infection, evade immune responses, and transmit between hosts and vectors (Pal et al., 2001; Fingerle et al., 2007; Neelakanta et al., 2007; de Silva et al., 2009; Kenedy et al., 2012; Ogden et al., 2015; Tufts et al., 2019). Investigations incorporating the ecological interactions between species and their hosts and vectors have revealed the selective pressures that have given rise to the observable variation in the genetic and molecular components underlying these mechanisms (De Michelis et al., 2000; Dykhuizen et al., 2008; Graves et al., 2013; Coipan et al., 2018). Further, ecological and genetic studies have revealed how variation in these molecular mechanisms has continued to shape *Borrelia* evolution, driving changes in gene flow and geographic distribution, and altering the effects of genetic drift and mutation (Brisson et al., 2012; Khatchikian et al., 2015; Seifert et al., 2015). By integrating these studies, we can move beyond cataloging variation in molecular mechanisms and ecological relationships and understand how these factors are a consequence of historical evolutionary processes and how they will continue to drive evolutionary change.

Uncovering the evolutionary genetics of a system depends critically on a firm understanding of the molecular mechanisms and ecological interactions of the species being studied. While these aspects of *Borrelia* have been the subject of extensive investigation in many LB species, they remain to be investigated in even a single RF *Borrelia* species despite their greater diversity and medical importance worldwide (Table 1). For example, little is known about the molecular and genetic underpinnings of vector or host associations in any

RF species nor the pattern or causes of population genetic structure, both of which have been central to studies of many LB species. Investigations into the molecular mechanisms and the genetic underpinnings of those mechanisms from a range of RF *Borrelia* species will permit many questions about the evolutionary genetics of RF species to be addressed. Comparing evolutionary processes among species and among the LB and RF clades will provide a better understanding of the interactions and constraints that have led to the current assemblage of species and diversity of characteristics that are observable today. Without these data, identifying the major drivers of evolution in these species will remain a significant challenge.

References

Abraham, N.M., Liu, L., Jutras, B.L., Yadav, A.K., Narasimhan, S., Gopalakrishnan, V., Ansari, J.M., Jefferson, K.K., Cava, F., Jacobs-Wagner, C., and Fikrig, E. (2017). Pathogen-mediated manipulation of arthropod microbiota to promote infection. Proc. Natl. Acad. Sci. U.S.A. *114,* E781-E790 DOI: 10.1073/pnas.1613422114

Adeolu, M., and Gupta, R.S. (2014). A phylogenomic and molecular marker based proposal for the division of the genus *Borrelia* into two genera: the emended genus *Borrelia* containing only the members of the relapsing fever *Borrelia*, and the genus *Borreliella* gen. nov. containing the members of the Lyme disease *Borrelia* (*Borrelia burgdorferi* sensu lato complex). Antonie Van Leeuwenhoek *105,* 1049-1072. DOI: 10.1007/s10482-014-0164-x

Adeyeye, O.A., and Butler, J.F. (1989). Population structure and seasonal intra-burrow movement of *Ornithodoros turicata* (Acari: Argasidae) in gopher tortoise burrows. J. Med. Entomol. *26,* 279-283. DOI: 10.1093/jmedent/26.4.279

Alizon, S., de Roode, J.C., and Michalakis, Y. (2013). Multiple infections and the evolution of virulence. Ecol. Lett. *16,* 556-567. DOI: 10.1111/ele.12076

Anderson, J.F., Barthold, S.W., and Magnarelli, L.A. (1990). Infectious but nonpathogenic isolate of *Borrelia burgdorferi*. J. Clin. Microbiol. *28,* 2693-2699 DOI: 10.1128/JCM.28.12.2693-2699.1990

Bankhead, T., and Chaconas, G. (2007). The role of VlsE antigenic variation in the Lyme disease spirochete: persistence through a mechanism that differs from other pathogens. Mol. Microbiol. *65,* 1547-1558. DOI: 10.1111/j.1365-2958.2007.05895.x

Barbour, A.G., Adeolu, M., and Gupta, R.S. (2017). Division of the genus *Borrelia* into two genera (corresponding to Lyme disease and relapsing fever groups) reflects their genetic and phenotypic distinctiveness and will lead to a better understanding of these two groups of microbes (Margos et al. (2016) There is inadequate evidence to support the division of the genus *Borrelia. Int. J. Syst. Evol. Microbiol.*) Int. J. Syst. Evol. Microbiol. *67,* 2058-2067. DOI: 10.1099/ijsem.0.001815

Barbour, A.G., Bunikis, J., Travinsky, B., Hoen, A.G., Diuk-Wasser, M.A., Fish, D., and Tsao, J.I. (2009). Niche partitioning of *Borrelia burgdorferi* and *Borrelia miyamotoi* in the same tick vector and mammalian reservoir species. Am. J. Trop. Med. Hyg. *81,* 1120-1131. DOI: 10.4269/ajtmh.2009.09-0208

Barbour, A.G., Dai, Q., Restrepo, B.I., Stoenner, H.G., and Frank, S.A. (2006). Pathogen escape from host immunity by a genome program for antigenic variation. Proc. Natl. Acad. Sci. U.S.A. *103,* 18290-18295 DOI: 10.1073/pnas.0605302103

Barbour, A.G., and Hayes, S.F. (1986). Biology of *Borrelia* species. Microbiol. Rev. *50,* 381-400.

Barthold, S.W. (1999). Specificity of infection-induced immunity among *Borrelia burgdorferi* sensu lato species. Infect. Immun. *67,* 36-42 DOI: 10.1128/IAI.67.1.36-42.1999

Barthold, S.W., Persing, D.H., Armstrong, A.L., and Peeples, R.A. (1991). Kinetics of *Borrelia burgdorferi* dissemination and evolution of disease after intradermal inoculation of mice. Am. J. Pathol. *139,* 263-273.

Battisti, J.M., Bono, J.L., Rosa, P.A., Schrumpf, M.E., Schwan, T.G., and Policastro, P.F. (2008). Outer surface protein A protects Lyme disease spirochetes from acquired host immunity in the tick vector. Infect. Immun. *76,* 5228-5237 DOI: 10.1128/IAI.00410-08

Baum, E., Hue, F., and Barbour, A.G. (2012). Experimental infections of the reservoir species *Peromyscus leucopus* with diverse strains of *Borrelia burgdorferi*, a Lyme disease agent. Mbio *3,* e00434-12 DOI: 10.1128/mBio.00434-12

Baumler, A., and Fang, F.C. (2013). Host specificity of bacterial pathogens. Cold Spring Harb Perspect. Med. *3,* a010041 DOI: 10.1101/cshperspect.a010041

Becker, N.S., Margos, G., Blum, H., Krebs, S., Graf, A., Lane, R.S., Castillo-Ramírez, S., Sing, A., and Fingerle, V. (2016). Recurrent evolution of host and vector association in bacteria of the *Borrelia burgdorferi* sensu lato species complex. BMC Genomics *17,* 734. DOI: 10.1186/s12864-016-3016-4

Brisson, D. (2018). Negative frequency-dependent selection is frequently confounding. Frontiers in Ecology and Evolution *6,* 10. DOI: 10.3389/fevo.2018.00010

Brisson, D., Drecktrah, D., Eggers, C.H., and Samuels, D.S. (2012). Genetics of *Borrelia burgdorferi*. Annu. Rev. Genet. *46,* 515-536. DOI: 10.1146/annurev-genet-011112-112140

Brisson, D., and Dykhuizen, D.E. (2006). A modest model explains the distribution and abundance of *Borrelia burgdorferi* strains. Am. J. Trop. Med. Hyg. *74,* 615-622. DOI: 10.4269/ajtmh.2006.74.615

Brisson, D., and Dykhuizen, D.E. (2004). *ospC* diversity in *Borrelia burgdorferi*: different hosts are different niches. Genetics *168,* 713-722 DOI: 10.1534/genetics.104.028738

Bunikis, J., Garpmo, U., Tsao, J., Berglund, J., Fish, D., and Barbour, A.G. (2004). Sequence typing reveals extensive strain diversity of the Lyme borreliosis agents *Borrelia burgdorferi* in North America and *Borrelia afzelii* in Europe. Microbiology *150,* 1741-1755. DOI: 10.1099/mic.0.26944-0

Burgdorfer, W., Barbour, A.G., Hayes, S.F., Benach, J.L., Grunwaldt, E., and Davis, J.P. (1982). Lyme disease-a tick-borne spirochetosis? Science *216,* 1317-1319 DOI: 10.1126/science.7043737

Casjens, S.R., Di, L., Akther, S., Mongodin, E.F., Luft, B.J., Schutzer, S.E., Fraser, C.M., and Qiu, W. (2018). Primordial origin and diversification of plasmids in Lyme disease agent bacteria. BMC Genomics *19,* 1-24. DOI: 10.1186/s12864-018-4597-x

Cattadori, I., Boag, B., and Hudson, P.J. (2008). Parasite co-infection and interaction as drivers of host heterogeneity. Int. J. Parasitol. *38,* 371-380. DOI: 10.1016/j.ijpara.2007.08.004

Chan, K., Awan, M., Barthold, S.W., and Parveen, N. (2012). Comparative molecular analyses of *Borrelia burgdorferi* sensu stricto strains B31 and N40D10/E9 and determination of their pathogenicity. BMC Microbiology *12,* 157. DOI: 10.1186/1471-2180-12-157

Chevin, L., and Hoffmann, A.A. (2017). Evolution of phenotypic plasticity in extreme environments. Philosophical Transactions of the Royal Society B: Biological Sciences *372,* 20160138. DOI: 10.1098/rstb.2016.0138

Cobey, S., and Lipsitch, M. (2012). Niche and neutral effects of acquired immunity permit coexistence of pneumococcal serotypes. Science *335,* 1376-1380 DOI: 10.1126/science.1215947

Coipan, C.E., van Duijvendijk, G.L., Hofmeester, T.R., Takumi, K., and Sprong, H. (2018). The genetic diversity of *Borrelia afzelii* is not maintained by the diversity of the rodent hosts. Parasites & Vectors *11*, 454. DOI: 10.1186/s13071-018-3006-2

Couper, L.I., Kwan, J.Y., Ma, J., and Swei, A. (2019). Drivers and patterns of microbial community assembly in a Lyme disease vector. Ecology and Evolution *9*, 7768-7779. DOI: 10.1002/ece3.5361

Couper, L.I., Yang, Y., Yang, X.F., and Swei, A. (2020). Comparative vector competence of North American Lyme disease vectors. Parasites & Vectors *13*, 29. DOI: 10.1186/s13071-020-3893-x

Coutte, L., Botkin, D.J., Gao, L., and Norris, S.J. (2009). Detailed analysis of sequence changes occurring during *vlsE* antigenic variation in the mouse model of *Borrelia burgdorferi* infection. PLoS Pathog *5*, e1000293. DOI: 10.1371/journal.ppat.1000293

Cutler, S. (2010). Relapsing fever–a forgotten disease revealed. J. Appl. Microbiol. *108*, 1115-1122. DOI: 10.1111/j.1365-2672.2009.04598.x

De Michelis, S., Sewell, H.S., Collares-Pereira, M., Santos-Reis, M., Schouls, L.M., Benes, V., Holmes, E.C., and Kurtenbach, K. (2000). Genetic diversity of *Borrelia burgdorferi* sensu lato in ticks from mainland Portugal. J. Clin. Microbiol. *38*, 2128-2133. DOI: 10.1128/JCM.38.6.2128-2133.2000

de Silva, A.M., Tyson, K.R., and Pal, U. (2009). Molecular characterization of the tick-*Borrelia* interface. Front. Biosci. (Landmark Ed) *14*, 3051-3063 DOI: 10.2741/3434

Diuk-Wasser, M.A., Vannier, E., and Krause, P.J. (2016). Coinfection by *Ixodes* tick-borne pathogens: ecological, epidemiological, and clinical consequences. Trends Parasitol. *32*, 30-42. DOI: 10.1016/j.pt.2015.09.008

Dunn, J.M., Krause, P.J., Davis, S., Vannier, E.G., Fitzpatrick, M.C., Rollend, L., Belperron, A.A., Stacey, A., Bockenstedt, L.K., and Fish, D. (2014). *Borrelia burgdorferi* promotes the establishment of *Babesia microti* in the northeastern United States. PLoS One *9*, e115494. DOI: 10.1371/journal.pone.0115494

Durand, J., Herrmann, C., Genne, D., Sarr, A., Gern, L., and Voordouw, M.J. (2017). Multistrain infections with Lyme borreliosis pathogens in the tick vector. Appl. Environ. Microbiol. *83*, e02552-16 DOI: 10.1128/AEM.02552-16

Dykhuizen, D.E., and Baranton, G. (2001). The implications of a low rate of horizontal transfer in *Borrelia*. Trends Microbiol. *9*, 344-350. DOI: 10.1016/S0966-842X(01)02066-2

Dykhuizen, D.E., Brisson, D., Sandigursky, S., Wormser, G.P., Nowakowski, J., Nadelman, R.B., and Schwartz, I. (2008). The propensity of different *Borrelia burgdorferi* sensu stricto genotypes to cause disseminated infections in humans. Am. J. Trop. Med. Hyg. *78*, 806-810. DOI: 10.4269/ajtmh.2008.78.806

Eisen, L. (2020). Vector competence studies with hard ticks and *Borrelia burgdorferi* sensu lato spirochetes: A review. Ticks and Tick-borne Diseases 11: 101359 DOI: 10.1016/j.ttbdis.2019.101359

Eveleigh, E.S., and Threlfall, W. (1974). The biology of *Ixodes* (Ceratixodes) *uriae* White, 1852 in Newfoundland. Acarologia 16: 62135.

Fingerle, V., Goettner, G., Gern, L., Wilske, B., and Schulte-Spechtel, U. (2007). Complementation of a *Borrelia afzelii* OspC mutant highlights the crucial role of OspC for dissemination of *Borrelia afzelii* in *Ixodes ricinus*. International Journal of Medical Microbiology *297*, 97-107. DOI: 10.1016/j.ijmm.2006.11.003

Francis, E. (1938). Longevity of the tick *Ornithodoros turicata* and of *Spirochaeta recurrentis* within this tick. Public Health Reports (1896-1970) 2220-2241. DOI: 10.2307/4582740

Gatzmann, F., Metzler, D., Krebs, S., Blum, H., Sing, A., Takano, A., Kawabata, H., Fingerle, V., Margos, G., and Becker, N.S. (2015). NGS population genetics analyses reveal divergent evolution of a Lyme Borreliosis agent in Europe and Asia. Ticks and Tick-Borne Diseases *6*, 344-351. DOI: 10.1016/j.ttbdis.2015.02.008

Genné, D., Sarr, A., Gomez-Chamorro, A., Durand, J., Cayol, C., Rais, O., and Voordouw, M.J. (2018). Competition between strains of *Borrelia afzelii* inside the rodent host and the tick vector. Proceedings of the Royal Society B *285*, 20181804. DOI: 10.1098/rspb.2018.1804

Genné, D., Sarr, A., Rais, O., and Voordouw, M.J. (2019). Competition between strains of *Borrelia afzelii* in immature *Ixodes ricinus* ticks is not affected by season. Frontiers in Cellular and Infection Microbiology *9*, 431. DOI: 10.3389/fcimb.2019.00431

Gilmore, R.D., Jr, Kappel, K.J., Dolan, M.C., Burkot, T.R., and Johnson, B.J. (1996). Outer surface protein C (OspC), but not P39, is a protective immunogen against a tick-transmitted *Borrelia burgdorferi* challenge: evidence for a conformational protective epitope in OspC. Infect. Immun. *64*, 2234-2239 DOI: 10.1128/iai.64.6.2234-2239.1996

Gomez-Chamorro, A., Battilotti, F., Cayol, C., Mappes, T., Koskela, E., Boulanger, N., Genné, D., Sarr, A., and Voordouw, M.J. (2019). Susceptibility to infection with *Borrelia afzelii* and TLR2 polymorphism in a wild reservoir host. Scientific Reports *9*, 1-12. DOI: 10.1038/s41598-019-43160-3

Gómez-Díaz, E., Boulinier, T., Sertour, N., Cornet, M., Ferquel, E., and McCoy, K.D. (2011) Genetic structure of marine *Borrelia garinii* and population admixture with the terrestrial cycle of Lyme borreliosis. Environmental Microbiology 13: 245367 DOI: 10.1111/j.1462-2920.2011.02515

Gómez-Díaz, E., Jordà, M., Peinado, M.A., and Rivero, A. (2012). Epigenetics of host–pathogen interactions: the road ahead and the road behind. PLoS Pathog *8*, e1003007. DOI: 10.1371/journal.ppat.1003007

Graves, C.J., Ros, V.I., Stevenson, B., Sniegowski, P.D., and Brisson, D. (2013). Natural selection promotes antigenic evolvability. PLoS Pathog *9*, e1003766. DOI: 10.1371/journal.ppat.1003766

Hammerschmidt, C., Koenigs, A., Siegel, C., Hallstrom, T., Skerka, C., Wallich, R., Zipfel, P.F., and Kraiczy, P. (2014). Versatile roles of CspA orthologs in complement inactivation of serum-resistant Lyme disease spirochetes. Infect. Immun. *82*, 380-392 DOI: 10.1128/IAI.01094-13

Hanincova, K., Mukherjee, P., Ogden, N.H., Margos, G., Wormser, G.P., Reed, K.D., Meece, J.K., Vandermause, M.F., and Schwartz, I. (2013). Multilocus sequence typing of *Borrelia burgdorferi* suggests existence of lineages with differential pathogenic properties in humans. PLoS One *8*, e73066. DOI: 10.1371/journal.pone.0073066

Hanincova, K., Kurtenbach, K., Diuk-Wasser, M., Brei, B., and Fish, D. (2006). Epidemic spread of Lyme borreliosis, northeastern United States. Emerg. Infect. Dis. *12*, 604-611 DOI: 10.3201/eid1204.051016

Hart, T., Yang, X., Pal, U., and Lin, Y. (2018). Identification of Lyme borreliae proteins promoting vertebrate host blood-specific spirochete survival in *Ixodes scapularis* nymphs using artificial feeding chambers. Ticks and Tick-Borne Diseases *9*, 1057-1063. DOI: 10.1016/j.ttbdis.2018.03.033

Haven, J., Vargas, L.C., Mongodin, E.F., Xue, V., Hernandez, Y., Pagan, P., Fraser-Liggett, C.M., Schutzer, S.E., Luft, B.J., Casjens, S.R., and Qiu, W.G. (2011). Pervasive recombination and sympatric genome diversification driven by frequency-dependent selection in *Borrelia burgdorferi*, the Lyme disease bacterium. Genetics *189*, 951-966 DOI: 10.1534/genetics.111.130773

Herrmann, C., Gern, L., and Voordouw, M.J. (2013). Species co-occurrence patterns among Lyme borreliosis pathogens in the tick vector *Ixodes ricinus*. Appl. Environ. Microbiol. *79*, 7273-7280 DOI: 10.1128/AEM.02158-13

Hersh, M.H., Ostfeld, R.S., McHenry, D.J., Tibbetts, M., Brunner, J.L., Killilea, M.E., LoGiudice, K., Schmidt, K.A., and Keesing, F. (2014). Co-infection of blacklegged ticks with *Babesia microti* and *Borrelia burgdorferi* is higher than expected and acquired from small mammal hosts. PloS One *9*, e99348. DOI: 10.1371/journal.pone.0099348

Hildebrandt, A., Schmidt, K., Wilske, B., Dorn, W., Straube, E., and Fingerle, V. (2003). Prevalence of four species of *Borrelia burgdorferi* sensu lato and coinfection with *Anaplasma phagocytophila* in *Ixodes ricinus* ticks in central Germany. European Journal of Clinical

Microbiology and Infectious Diseases *22*, 364-367. DOI: 10.1007/s10096-003-0926-2

Hoen, A.G., Margos, G., Bent, S.J., Diuk-Wasser, M.A., Barbour, A., Kurtenbach, K., and Fish, D. (2009). Phylogeography of *Borrelia burgdorferi* in the eastern United States reflects multiple independent Lyme disease emergence events. Proc. Natl. Acad. Sci. U. S. A. *106*, 15013-15018 DOI: 10.1073/pnas.0903810106

Huang, Y.S., Higgs, S., and Vanlandingham, D.L. (2019). Arbovirus-mosquito vector-host interactions and the impact on transmission and disease pathogenesis of arboviruses. Frontiers in Microbiology *10*, 22. DOI: 10.3389/fmicb.2019.00022

Humphrey, P.T., Caporale, D.A., and Brisson, D. (2010). Uncoordinated phylogeography of *Borrelia burgdorferi* and its tick vector, *Ixodes scapularis*. Evolution: International Journal of Organic Evolution *64*, 2653-2663. DOI: 10.1111/j.1558-5646.2010.01001.x

Jacquot, M., Gonnet, M., Ferquel, E., Abrial, D., Claude, A., Gasqui, P., Choumet, V., Charras-Garrido, M., Garnier, M., and Faure, B. (2014). Comparative population genomics of the *Borrelia burgdorferi* species complex reveals high degree of genetic isolation among species and underscores benefits and constraints to studying intra-specific epidemiological processes. PloS One *9*, e94384. DOI: 10.1371/journal.pone.0094384

Jones, C.G., Lawton, J.H., and Shachak, M. (1994). Organisms as ecosystem engineers. In Ecosystem management, Springer) pp. 130-147. DOI: 10.1007/978-1-4612-4018-1_14

Kenedy, M.R., Lenhart, T.R., and Akins, D.R. (2012). The role of *Borrelia burgdorferi* outer surface proteins. FEMS Immunology & Medical Microbiology *66*, 1-19. DOI: 10.1111/j.1574-695x.2012.00980.x

Khatchikian, C.E., Prusinski, M.A., Stone, M., Backenson, P.B., Wang, I., Foley, E., Seifert, S.N., Levy, M.Z., and Brisson, D. (2015). Recent and rapid population growth and range expansion of the Lyme disease tick vector, *Ixodes scapularis*, in North America. Evolution *69*, 1678-1689. DOI: 10.1111/evo.12690

Kilpatrick, A.M., Dobson, A.D., Levi, T., Salkeld, D.J., Swei, A., Ginsberg, H.S., Kjemtrup, A., Padgett, K.A., Jensen, P.M., and Fish, D. (2017). Lyme disease ecology in a changing world: consensus, uncertainty and critical gaps for improving control. Philosophical Transactions of the Royal Society B: Biological Sciences *372*, 20160117. DOI: 10.1098/rstb.2016.0117

Konnai, S., Yamada, S., Imamura, S., Nishikado, H., Githaka, N., Ito, T., Takano, A., Kawabata, H., Murata, S., and Ohashi, K. (2012). Identification of TROSPA homologue in *Ixodes persulcatus* Schulze, the specific vector for human Lyme borreliosis in Japan. Ticks and Tick-Borne Diseases *3*, 75-77. DOI: 10.1016/j.ttbdis.2012.02.001

Kung, F., Anguita, J., and Pal, U. (2013). *Borrelia burgdorferi* and tick proteins supporting pathogen persistence in the vector. Future Microbiology *8*, 41-56. DOI: 10.2217/fmb.12.121

Kurtenbach, K., Hanincová, K., Tsao, J.I., Margos, G., Fish, D., and Ogden, N.H. (2006). Fundamental processes in the evolutionary ecology of Lyme borreliosis. Nature Reviews Microbiology *4*, 660-669. DOI: 10.1038/nrmicro1475

Lagal, V., Portnoï, D., Faure, G., Postic, D., and Baranton, G. (2006). *Borrelia burgdorferi* sensu stricto invasiveness is correlated with OspC–plasminogen affinity. Microb. Infect. *8*, 645-652. DOI: 10.1016/j.micinf.2005.08.017

Lawrenz, M.B., Wooten, R.M., and Norris, S.J. (2004). Effects of *vlsE* complementation on the infectivity of *Borrelia burgdorferi* lacking the linear plasmid lp28-1. Infect. Immun. *72*, 6577-6585 DOI: 10.1128/IAI.72.11.6577-6585.2004

Lescot, M., Audic, S., Robert, C., Nguyen, T.T., Blanc, G., Cutler, S.J., Wincker, P., Couloux, A., Claverie, J., and Raoult, D. (2008). The genome of *Borrelia recurrentis*, the agent of deadly louse-borne relapsing fever, is a degraded subset of tick-borne *Borrelia duttonii*. PLoS Genet *4*, e1000185. DOI: 10.1371/journal.pgen.1000185

Lin, Y., Frye, A.M., Nowak, T.A., and Kraiczy, P. (2020). New insights into CRASP-Mediated complement evasion in the Lyme disease enzootic cycle. Frontiers in Cellular and Infection Microbiology *10*, 1. DOI: 10.3389/fcimb.2020.00001

Livey, I., Gibbs, C., Schuster, R., and Dorner, F. (1995). Evidence for lateral transfer and recombination in OspC variation in Lyme disease *Borrelia*. Mol. Microbiol. *18*, 257-269. DOI: 10.1111/j.1365-2958.1995.mmi_18020257.x

Lloyd-Smith, J.O., George, D., Pepin, K.M., Pitzer, V.E., Pulliam, J.R., Dobson, A.P., Hudson, P.J., and Grenfell, B.T. (2009). Epidemic dynamics at the human-animal interface. Science *326*, 1362-1367 DOI: 10.1126/science.1177345

Marconi, R.T., Samuels, D.S., and Garon, C.F. (1993). Transcriptional analyses and mapping of the *ospC* gene in Lyme disease spirochetes. J. Bacteriol. *175*, 926-932 DOI: 10.1128/jb.175.4.926-932.1993

Marcsisin, R.A., Lewis, E.R., and Barbour, A.G. (2016). Expression of the tick-associated Vtp protein of *Borrelia hermsii* in a murine model of relapsing fever. PloS One *11*, e0149889. DOI: 10.1371/journal.pone.0149889

Margos, G., Marosevic, D., Cutler, S., Derdakova, M., Diuk-Wasser, M., Emler, S., Fish, D., Gray, J., Hunfeldt, K., and Jaulhac, B. (2017). There is inadequate evidence to support the division of the genus *Borrelia*. Int. J. Syst. Evol. Microbiol. *67*, 1081-1084. DOI: 10.1099/ijsem.0.001717

Margos, G., Becker, N.S., Fingerle, V., Sing, A., Ramos, J.A., de Carvalho, I.L., and Norte, A.C. (2019). Core genome phylogenetic analysis of the avian associated *Borrelia turdi* indicates a close relationship to *Borrelia garinii*. Mol. Phylogenet. Evol. *131*, 93-98. DOI: 10.1016/j.ympev.2018.10.044

Margos, G., Fingerle, V., Cutler, S., Gofton, A., Stevenson, B., and Estrada-Peña, A. (2020). Controversies in bacterial taxonomy: The example of the genus *Borrelia*. Ticks and Tick-Borne Diseases *11*, 101335. DOI: 10.1016/j.ttbdis.2019.101335

Margos, G., Fingerle, V., and Reynolds, S.E. (2019). *Borrelia bavariensis*: vector switch, niche invasion and geographical spread of a tick-borne bacterial parasite. Frontiers in Ecology and Evolution *7*, 401. DOI: 10.3389/fevo.2019.00401

Margos, G., Vollmer, S.A., Ogden, N.H., and Fish, D. (2011). Population genetics, taxonomy, phylogeny and evolution of *Borrelia burgdorferi* sensu lato. Infection, Genetics and Evolution *11*, 1545-1563. DOI: 10.1016/j.meegid.2011.07.022

Margos, G., Vollmer, S.A., Cornet, M., Garnier, M., Fingerle, V., Wilske, B., Bormane, A., Vitorino, L., Collares-Pereira, M., Drancourt, M., and Kurtenbach, K. (2009). A new *Borrelia* species defined by multilocus sequence analysis of housekeeping genes. Appl. Environ. Microbiol. *75*, 5410-5416 DOI: 10.1128/AEM.00116-09

McCoy, K. D., Leger, E., and Dietrich, M. (2013). Host specialization in ticks and transmission of tick-borne diseases: A review. Frontiers in Cellular and Infection Microbiology 3: 12 DOI: 10.3389/fcimb.2013.00057

Mechai, S., Margos, G., Feil, E.J., Barairo, N., Lindsay, L.R., Michel, P., and Ogden, N.H. (2016). Evidence for host-genotype associations of *Borrelia burgdorferi* sensu stricto. PLoS One *11*, e0149345. DOI: 10.1371/journal.pone.0149345

Moutailler, S., Valiente Moro, C., Vaumourin, E., Michelet, L., Tran, F.H., Devillers, E., Cosson, J., Gasqui, P., Van, V.T., and Mavingui, P. (2016). Co-infection of ticks: the rule rather than the exception. PLoS Neglected Tropical Diseases *10*, e0004539. DOI: 10.1371/journal.pntd.0004539

Neelakanta, G., Li, X., Pal, U., Liu, X., Beck, D.S., DePonte, K., Fish, D., Kantor, F.S., and Fikrig, E. (2007). Outer surface protein B is critical for *Borrelia burgdorferi* adherence and survival within *Ixodes* ticks. PLoS Pathog *3*, e33. DOI: 10.1371/journal.ppat.0030033

Norris, S.J. (2015). *vls* antigenic variation systems of Lyme disease *Borrelia*: eluding host immunity through both random, segmental gene conversion and framework heterogeneity. Mobile DNA III 471-489. DOI: 10.1128/9781555819217.ch22

Norris, S.J., Howell, J.K., Garza, S.A., Ferdows, M.S., and Barbour, A.G. (1995). High- and low-infectivity phenotypes of clonal populations of in vitro-cultured *Borrelia burgdorferi*. Infect. Immun. *63*, 2206-2212 DOI: 10.1128/IAI.63.6.2206-2212.1995

Norte, A.C., Margos, G., Becker, N.S., Albino Ramos, J., Núncio, M.S., Fingerle, V., Araújo, P.M., Adamík, P., Alivizatos, H., and Barba, E. (2020). Host dispersal shapes the population structure of a tick-borne bacterial pathogen. Mol. Ecol. *29*, 485-501. DOI: 10.1111/mec.15336

Ogden, N.H., Feil, E.J., Leighton, P.A., Lindsay, L.R., Margos, G., Mechai, S., Michel, P., and Moriarty, T.J. (2015). Evolutionary aspects of emerging Lyme disease in Canada. Appl. Environ. Microbiol. *81*, 7350-7359 DOI: 10.1128/AEM.01671-15

Ojaimi, C., Brooks, C., Casjens, S., Rosa, P., Elias, A., Barbour, A., Jasinskas, A., Benach, J., Katona, L., Radolf, J., *et al.* (2003). Profiling of temperature-induced changes in *Borrelia burgdorferi* gene expression by using whole genome arrays. Infect. Immun. *71*, 1689-1705 DOI: 10.1128/IAI.71.4.1689-1705.2003

Önder, Ö., Humphrey, P.T., McOmber, B., Korobova, F., Francella, N., Greenbaum, D.C., and Brisson, D. (2012). OspC is potent plasminogen receptor on surface of *Borrelia burgdorferi*. J. Biol. Chem. *287*, 16860-16868 DOI: 10.1074/jbc.M111.290775

Pal, U., Li, X., Wang, T., Montgomery, R.R., Ramamoorthi, N., DeSilva, A.M., Bao, F., Yang, X., Pypaert, M., and Pradhan, D. (2004). TROSPA, an *Ixodes scapularis* receptor for *Borrelia burgdorferi*. Cell *119*, 457-468. DOI: 10.1016/j.cell.2004.10.027

Pal, U., Montgomery, R.R., Lusitani, D., Voet, P., Weynants, V., Malawista, S.E., Lobet, Y., and Fikrig, E. (2001). Inhibition of *Borrelia burgdorferi*-tick interactions in vivo by outer surface protein A antibody. J. Immunol. *166*, 7398-7403 DOI: 10.4049/jimmunol.166.12.7398

Parker, J.L., and White, K.K. (1992). Lyme borreliosis in cattle and horses: a review of the literature. Cornell Vet. *82*, 253-274.

Paul, R.E., Cote, M., Le Naour, E., and Bonnet, S.I. (2016). Environmental factors influencing tick densities over seven years in a French suburban forest. Parasites & Vectors *9*, 309. DOI: 10.1186/s13071-016-1591-5

Phelan, J.P., Kern, A., Ramsey, M.E., Lundt, M.E., Sharma, B., Lin, T., Gao, L., Norris, S.J., Hyde, J.A., and Skare, J.T. (2019). Genome-wide screen identifies novel genes required for *Borrelia burgdorferi* survival in its *Ixodes* tick vector. PLoS Pathogens *15*, e1007644. DOI: 10.1371/journal.ppat.1007644

Popitsch, N., Bilusic, I., Rescheneder, P., Schroeder, R., and Lybecker, M. (2017). Temperature-dependent sRNA transcriptome of the Lyme disease spirochete. BMC Genomics *18*, 28. DOI: 10.1186/s12864-016-3398-3

Preac-Mursic, V., Wilske, B., Jauris, S., Will, G., Reinhardt, S., Lehnert, G., Patsouris, E., Mehraein, P., Soutschek, E., and Klockmann, U. (1992). Active immunization with pC protein of *Borrelia burgdorferi* protects gerbils against *B. burgdorferi* infection. Infection *20*, 342-349. DOI: 10.1007/BF01710681

Pritt, B.S., Mead, P.S., Johnson, D.K.H., Neitzel, D.F., Respicio-Kingry, L.B., Davis, J.P., Schiffman, E., Sloan, L.M., Schriefer, M.E., and Replogle, A.J. (2016). Identification of a novel pathogenic *Borrelia* species causing Lyme borreliosis with unusually high spirochaetaemia: a descriptive study. The Lancet Infectious Diseases *16*, 556-564. DOI: 10.1016/S1473-3099(15)00464-8

Probert, W.S., Crawford, M., Cadiz, R.B., and LeFebvre, R.B. (1997). Immunization with outer surface protein (Osp) A, but not OspC, provides cross-protection of mice challenged with North American isolates of *Borrelia burgdorferi*. J. Infect. Dis. *175*, 400-405. DOI: 10.1093/infdis/175.2.400

Rego, R.O., Bestor, A., Štefka, J., and Rosa, P.A. (2014). Population bottlenecks during the infectious cycle of the Lyme disease spirochete *Borrelia burgdorferi*. PLoS One *9*, e101009. DOI: 10.1371/journal.pone.0101009

Ripoche, J., Day, A., Harris, T.J., and Sim, R. (1988). The complete amino acid sequence of human complement factor H. Biochem. J. *249*, 593-602. DOI: 10.1042/bj2490593

Roberts, E.D., Bohm Jr, R.P., Lowrie Jr, R.C., Habicht, G., Katona, L., Piesman, J., and Philipp, M.T. (1998). Pathogenesis of Lyme neuroborreliosis in the rhesus monkey: the early disseminated and chronic phases of disease in the peripheral nervous system. J. Infect. Dis. *178*, 722-732. DOI: 10.1086/515357

Rollend, L., Fish, D., and Childs, J.E. (2013). Transovarial transmission of *Borrelia* spirochetes by *Ixodes scapularis*: a summary of the literature and recent observations. Ticks and Tick-Borne Diseases *4*, 46-51. DOI: 10.1016/j.ttbdis.2012.06.008

Ross, B.D., Hayes, B., Radey, M.C., Lee, X., Josek, T., Bjork, J., Neitzel, D., Paskewitz, S., Chou, S., and Mougous, J.D. (2018). *Ixodes scapularis* does not harbor a stable midgut microbiome. The ISME Journal *12*, 2596-2607. DOI: 10.1038/s41396-018-0161-6

Saarinen, K., Laakso, J., Lindström, L., and Ketola, T. (2018). Adaptation to fluctuations in temperature by nine species of bacteria. Ecology and Evolution *8*, 2901-2910. DOI: 10.1002/ece3.3823

Seifert, S.N., Khatchikian, C.E., Zhou, W., and Brisson, D. (2015). Evolution and population genomics of the Lyme borreliosis pathogen, *Borrelia burgdorferi*. Trends in Genetics *31*, 201-207. DOI: 10.1016/j.tig.2015.02.006

Shih, C. M., Telford, S. R., and Spielman, A. (1995). Effect of ambient temperature on competence of deer ticks as hosts for Lyme disease spirochetes. Journal of clinical microbiology, 33: 958-961. DOI: 10.1128/JCM.33.4.958-961.1995

Šimo, L., Kazimirova, M., Richardson, J., and Bonnet, S.I. (2017). The essential role of tick salivary glands and saliva in tick feeding and pathogen transmission. Frontiers in Cellular and Infection Microbiology *7*, 281. DOI: 10.3389/fcimb.2017.00281

Sonenshine, D. (1979). Ticks of Virginia. Virginia Polytechnic Institute and State University, College of Agriculture and Life Sciences, Blacksburg, VA 42.

Stanek, G., Wormser, G.P., Gray, J., and Strle, F. (2012). Lyme borreliosis. The Lancet *379*, 461-473. DOI: 10.1016/S0140-6736(11)60103-7

Susi, H., Barrès, B., Vale, P.F., and Laine, A. (2015). Co-infection alters population dynamics of infectious disease. Nature Communications *6*, 1-8. DOI: 10.1038/ncomms6975

Talagrand-Reboul, E., Boyer, P.H., Bergström, S., Vial, L., and Boulanger, N. (2018). Relapsing fevers: neglected tick-borne diseases. Frontiers in Cellular and Infection Microbiology *8*, 98. DOI: 10.3389/fcimb.2018.00098

Taylor, L.H., Latham, S.M., and Woolhouse, M.E. (2001). Risk factors for human disease emergence. Philosophical Transactions of the Royal Society of London.Series B: Biological Sciences *356*, 983-989. DOI: 10.1098/rstb.2001.0888

Telfer, S., Lambin, X., Birtles, R., Beldomenico, P., Burthe, S., Paterson, S., and Begon, M. (2010). Species interactions in a parasite community drive infection risk in a wildlife population. Science *330*, 243-246 DOI: 10.1126/science.1190333

Tokarz, R., Anderton, J.M., Katona, L.I., and Benach, J.L. (2004). Combined effects of blood and temperature shift on *Borrelia burgdorferi* gene expression as determined by whole genome DNA array. Infect. Immun. *72*, 5419-5432 DOI: 10.1128/IAI.72.9.5419-5432.2004

Trape, J., Diatta, G., Arnathau, C., Bitam, I., Sarih, M., Belghyti, D., Bouattour, A., Elguero, E., Vial, L., and Mane, Y. (2013). The epidemiology and geographic distribution of relapsing fever borreliosis in West and North Africa, with a review of the *Ornithodoros erraticus* complex (Acari: Ixodida). PLoS One *8,* e78473. DOI: 10.1371/journal.pone.0078473

Tufts, D.M., Hart, T.M., Chen, G.F., Kolokotronis, S., Diuk–Wasser, M.A., and Lin, Y. (2019). Outer surface protein polymorphisms linked to host–spirochete association in Lyme borreliae. Mol. Microbiol. *111,* 868-882. DOI: 10.1111/mmi.14209

Van Treuren, W., Ponnusamy, L., Brinkerhoff, R.J., Gonzalez, A., Parobek, C.M., Juliano, J.J., Andreadis, T.G., Falco, R.C., Ziegler, L.B., Hathaway, N., *et al.* (2015). Variation in the microbiota of *Ixodes* ticks with regard to geography, species, and sex. Appl. Environ. Microbiol. *81,* 6200-6209 DOI: 10.1128/AEM.01562-15

Vial, L. (2009). Biological and ecological characteristics of soft ticks (Ixodida: Argasidae) and their impact for predicting tick and associated disease distribution. Parasite *16,* 191-202. DOI: 10.1051/parasite/2009163191

Vial, L., Durand, P., Arnathau, C., Halos, L., Diatta, G., Trape, J., and Renaud, F. (2006). Molecular divergences of the *Ornithodoros sonrai* soft tick species, a vector of human relapsing fever in West Africa. Microb. Infect. *8,* 2605-2611. DOI: 10.1016/j.micinf.2006.07.012

Vitorino, L.R., Margos, G., Feil, E.J., Collares-Pereira, M., Ze-Ze, L., and Kurtenbach, K. (2008). Fine-scale phylogeographic structure of *Borrelia lusitaniae* revealed by multilocus sequence typing. PLoS One *3,* e4002. DOI: 10.1371/journal.pone.0004002

Vollmer, S.A., Bormane, A., Dinnis, R.E., Seelig, F., Dobson, A.D., Aanensen, D.M., James, M.C., Donaghy, M., Randolph, S.E., and Feil, E.J. (2011). Host migration impacts on the phylogeography of Lyme Borreliosis spirochaete species in Europe. Environ. Microbiol. *13,* 184-192. DOI: 10.1111/j.1462-2920.2010.02319.x

Vollmer, S.A., Feil, E.J., Chu, C., Raper, S.L., Cao, W., Kurtenbach, K., and Margos, G. (2013). Spatial spread and demographic expansion of Lyme borreliosis spirochaetes in Eurasia. Infection, Genetics and Evolution *14,* 147-155. DOI: 10.1016/j.meegid.2012.11.014

Vuong, H.B., Chiu, G.S., Smouse, P.E., Fonseca, D.M., Brisson, D., Morin, P.J., and Ostfeld, R.S. (2017). Influences of host community characteristics on *Borrelia burgdorferi* infection prevalence in blacklegged ticks. PloS One *12,* e0167810. DOI: 10.1371/journal.pone.0167810

Walter, K.S., Carpi, G., Caccone, A., and Diuk-Wasser, M.A. (2017). Genomic insights into the ancient spread of Lyme disease across North America. Nature Ecology & Evolution *1,* 1569-1576. DOI: 10.1038/s41559-017-0282-8

Walter, K.S., Carpi, G., Evans, B.R., Caccone, A., and Diuk-Wasser, M.A. (2016). Vectors as epidemiological sentinels: patterns of within-tick *Borrelia burgdorferi* diversity. PLoS Pathogens *12,* e1005759. DOI: 10.1371/journal.ppat.1005759

Wang, G., Ojaimi, C., Wu, H., Saksenberg, V., Iyer, R., Liveris, D., McClain, S.A., Wormser, G.P., and Schwartz, I. (2002). Disease severity in a murine model of Lyme borreliosis is associated with the genotype of the infecting *Borrelia burgdorferi* sensu stricto strain. J. Infect. Dis. *186,* 782-791. DOI: 10.1086/343043

Wang, G., Ojaimi, C., Iyer, R., Saksenberg, V., McClain, S.A., Wormser, G.P., and Schwartz, I. (2001). Impact of genotypic variation of *Borrelia burgdorferi* sensu stricto on kinetics of dissemination and severity of disease in C3H/HeJ mice. Infect. Immun. *69,* 4303-4312 DOI: 10.1128/IAI.69.7.4303-4312.2001

Wilske, B., Fingerle, V., Herzer, P., Hofmann, A., Lehnert, G., Peters, H., Pfister, H., Preac-Mursic, V., Soutschek, E., and Weber, K. (1993). Recombinant immunoblot in the serodiagnosis of Lyme borreliosis. Med. Microbiol. Immunol. *182,* 255-270. DOI: 10.1007/BF00579624

Wormser, G.P., Brisson, D., Liveris, D., Hanincová, K., Sandigursky, S., Nowakowski, J., Nadelman, R.B., Ludin, S., and Schwartz, I. (2008). *Borrelia burgdorferi* genotype predicts the capacity for hematogenous dissemination during early Lyme disease. J. Infect. Dis. *198,* 1358-1364. DOI: 10.1086/592279

Zhang, D., de Souza, R.F., Anantharaman, V., Iyer, L.M., and Aravind, L. (2012). Polymorphic toxin systems: comprehensive characterization of trafficking modes, processing, mechanisms of action, immunity and ecology using comparative genomics. Biology Direct *7,* 18. DOI: 10.1186/1745-6150-7-18

Zhang, J.R., and Norris, S.J. (1998). Kinetics and in vivo induction of genetic variation of *vlsE* in *Borrelia burgdorferi*. Infect. Immun. *66,* 3689-3697 DOI: 10.1128/IAI.66.8.3689-3697.1998

Zolnik, C.P., Prill, R.J., Falco, R.C., Daniels, T.J., and Kolokotronis, S. (2016). Microbiome changes through ontogeny of a tick pathogen vector. Mol. Ecol. *25,* 4963-4977. DOI: 10.1111/mec.1383

Chapter 10

Perpetuation of Borreliae

Sam R. Telford III* and Heidi K. Goethert

Dept of Infectious Disease and Global Health, Tufts University, Cummings School of Veterinary Medicine, 200 Westboro Road, North Grafton, MA 01536, USA
*Corresponding author: sam.telford@tufts.edu

DOI: https://doi.org/10.21775/9781913652616.10

Abstract
With one exception (epidemic relapsing fever), borreliae are obligately maintained in nature by ticks. Although some *Borrelia* spp. may be vertically transmitted to subsequent generations of ticks, most require amplification by a vertebrate host because inheritance is not stable. Enzootic cycles of borreliae have been found globally; those receiving the most attention from researchers are those whose vectors have some degree of anthropophily and, thus, cause zoonoses such as Lyme disease or relapsing fever. To some extent, our views on the synecology of the borreliae has been dominated by an applied focus, viz., analyses that seek to understand the elements of human risk for borreliosis. But, the elements of borrelial perpetuation do not necessarily bear upon risk, nor do our concepts of risk provide the best structure for analyzing perpetuation. We identify the major global themes for the perpetuation of borreliae, and summarize local variations on those themes, focusing on key literature to outline the factors that serve as the basis for the distribution and abundance of borreliae.

Introduction
Whereas the term "borreliosis" comprises relapsing fever, Lyme borreliosis, and the veterinary borrelioses, great global diversity of the causative agents, as well as related spirochetes not associated with disease, has been recognized during the last three decades. In part, borrelial diversity has been recognized more easily due to profound advances in genetic characterization but also because of increased attention by researchers. Is there a similar diversity in life cycles of these borreliae? In this review, we try to describe the essential character of the borrelioses at the ecological scale, with the objective of determining whether there are a limited number of orchestral pieces or a large oeuvre of distinct pieces. Are there many different themes (Beethoven's oeuvre) or simply variations on a few themes (kazoo vs. orchestral interpretations of the 5th Symphony)? Are the local vector-pathogen-host systems unique, or are they simply local variations? In what follows, we attempt to identify the main contributors to the perpetuation of the borreliae, with the aim of analyzing the factors that serve as the basis for their distribution and abundance. This is not a comprehensive review, nor do we recapitulate other reviews on the subject, but rather emphasize specific primary literature that make points that we consider to be critical to advancing the field. The perpetuation of borreliae, of course, depends on that of their obligate vectors and vertebrate hosts, but this review is not focused on tick or animal ecology, but to the extent possible, on the spirochetes in relation to their hosts.

Perpetuation and the zoonotic condition
The population biology of infectious agents is centered on understanding how the basic reproduction number (BRN), in its simplest form the number of secondary infections that derive from one infection, exceeds 1 from generation to generation(May, 1984). BRN<1 implies local extinction. BRN=1 is maintenance of the agent. BRN> 1 allows for perpetuation in local space and time. The enzootic cycle refers to the dynamic biotic and abiotic associations allowing for perpetuation. (For vector borne infections and those of ticks in particular, such associations necessitate conceptualizing BRN in terms of production of new infections in the next generation, averaging between infected vectors and vertebrate hosts (Hartemink et al., 2008). We use the simple heuristic concept of BRN in this chapter). Some spirochetes of the genus *Borrelia* (herein termed borreliae) can cause infection and disease in humans (e.g., relapsing fevers and

Lyme disease) and are zoonotic, that is, they are infections maintained by animals other than humans that can infect and cause disease in humans. The term "anthroponotic" designates the maintenance of an infection that solely has humans as the critical and only vertebrate host. A simplified view of the zoonotic condition is that BRN greatly exceeds that required for perpetuation, allowing spillover or overflow from the enzootic cycle and increased probability that humans encounter infection. Of course, even in the presence of large numbers of infected ticks, there might be no human infection unless the tick vector is anthropophilic and human behavior allowed for exposure to the ticks. Not all borreliae affect human health; some are veterinary agents. Some borreliae may even be elements of a tick or vertebrate's microbiome with no effects on the fitness of their hosts. There are also borreliae in enzootic cycles that are very unlikely to become zoonotic, or borreliae very closely related to zoonotic ones to which people are certainly exposed but evidence to date fails to confirm infectivity to humans. Other than the anthroponotic borreliae (African tick borne relapsing fever and louse borne relapsing fever), the capacity to infect humans has no bearing on perpetuation (i.e.,humans are dead end hosts for the borreliae).

Ecology and epidemiology

Analyses of the ecology of the borrelioses, particularly those caused by *Borrelia burgdorferi* sensu lato, have mainly had the objective of understanding human risk for acquiring infection and sustaining disease. (In practical terms, funding for ecological research is more easily justified by the risk implications.) Risk, however, has a strong human behavioral component that may skew our way of studying the natural history of the causative agents. For example, an epidemiologist would ask "what are the faunal attributes of a yard that are associated with risk to those exposed to that yard?" but an ecologist would ask "are the factors that contribute to the basic reproduction number of the agent different in a yard as opposed to the woodland?" with little interest in whether answering the question might have implications for risk. The field measurements would differ in emphasis: in the former, the entomological inoculation rate (also known as "density of infected nymphs") is determined by drag sampling, the fauna and its tick burden at the time of sampling is described (usually by limited cross-sectional or convenience trapping), and either or both are compared with measures of human disease incidence. Added layers of complexity might include description of pathogen genetic diversity or microbial interactions ("coinfection"). The kind of vertebrate host that is most abundant, with the greatest tick burden and greatest infectivity to ticks is identified within a site, again by limited cross-sectional or convenience trapping at specific time points and is called the "reservoir". All too often, such a "snapshot" is considered representative of the "ecology". Such snapshots, however, undoubtedly miss the complexity of borrelial perpetuation. With ecologists, the dynamic nature of an ecosystem is assumed: vertebrate hosts and their symbionts, as well as the animals and plants with which they interact, have varying demography within and between transmission seasons and, hence, longitudinal observations are needed. Sites, even those separated by a few hundred metres or less, may greatly differ in vegetation structure as a function of historic landscape use, local weather, geology or connectivity with other habitat patches and thus the fauna and flora may differ; such microhabitat heterogeneities comprise additional vessels in which natural selection may operate to promote perpetuation. Genetic diversity of the host, pathogen, and vector is axiomatic; natural selection operates continually.

Borreliae are obligately maintained by ticks, with one exception (louse borne relapsing fever). In some cases, they may be perpetuated by inheritance by the vector alone (in which case the vector is the reservoir) but in most cases a vertebrate amplifying host (typically called a reservoir) is required due to unstable inheritance in which infection is gradually lost with successive generations in the absence of new infection during a bloodmeal. The ecology of borreliae, therefore, comprises that of 3 distinct organisms (borrelia, arthropod, and vertebrate) and all the possible interactions between them, as well as those with other animals and plants in the environment and the physical environment itself. Even the bacterium within its tick vector might be indirectly affected by environmental stressors (e.g., prolonged exposure of an infected tick to dessication leading to its increased activity to rehydrate, thereby depleting tick energy reserves, which might lead to downregulating innate immunity or perturbation of microbiota). Borreliae would certainly be affected by the environment via the success of ticks finding and feeding on hosts, as well as anything else affecting the survival of their tick hosts, including weather or predation. Any single year's effects may exert changes for years to come: a decrease in hosts during the months when larvae are most active (in the example of the typical phenology of some hard ticks) within one transmission season might imply fewer resulting infected nymphs in the next year. This would lead to less spirochetal transmission (fewer infected vertebrate hosts); if there was no compensation for fewer infected vertebrates (e.g., by clustering of larvae and massive feeding on the few existing infected hosts...hence the term "amplifying hosts"), then one poor transmission year could imply poor transmission for many following years due to a chain of dependent events. In the absence of human-directed intervention, though, there has been no report of stable reduced transmission over time in any site, suggesting generally effective buffering of the transmission cycle.

There are two main phylogenetic groups of borreliae, comprising those related to the main agents of Lyme borreliosis (*Borrelia burgdorferi* sensu lato) and those causing relapsing fevers; these have classically been termed "hard tick" (ixodid) and "soft tick" (argasid) transmitted borreliae. The two groups have distinct genetic and biological characteristics(Barbour et al., 2017),and some have argued that they each merit their own genus (Gupta, 2019). Since the first studies of borreliae, relapsing fever-like spirochetes have been known to infect and be transmitted by hard ticks (the agent of bovine borreliosis), but few "hard tick" borreliae have been found in soft ticks (Lane et al., 2010) even as we have recognized additional "soft tick" spirochetes within "hard ticks". There is now evidence fora third phylogenetic group that does not cluster within the two classical clades, that of borreliae infecting reptiles and specific reptile feeding ticks or monotremes and their ticks(Loh et al., 2016; Takano et al., 2010), which are generally hard ticks.

The general paradigm of borrelial life cycles is that Lyme disease spirochetes are maintained by rodents, eulipotyphlans (formerly "insectivores" including shrews and hedgehogs), birds and *Ixodes* ticks and that relapsing fever spirochetes are maintained by nest or burrow inhabiting mammals (rodents or eulipotyphlans) and argasid (mainly *Ornithodoros*) ticks. There are major differences in these life cycles: with the former, vertebrates are generally infectious for extended durations; *Ixodes* ticks take several days to take a large bloodmeal, drop off, develop over a long time, often several months, to the next bloodfeeding stage with periods of dormancy. Many different kinds of vertebrates might be fed upon. Adult stages of *Ixodes* ticks may feed on larger hosts that are less susceptible to spirochetal infection, if at all.

With relapsing fever spirochetes, infectivity is thought to be of limited duration due to eventual immune control even with antigenic variation; soft ticks, with a few exceptions, take a few hours at most to feed and may become ready to feed again within a few weeks, with there being as many as 5 nymphal stages (and even multiple small bloodmeals within a nymphal stage) and multiple smaller bloodmeals by the female ticks resulting in many small egg batches. Soft ticks may feed on the same kind of animal, even the same individual, at any stage of their life, and their host range is restricted by whatever inhabits or visits infested burrows or sleeps near the ticks' hiding places. However, unlike hard ticks, soft ticks may survive for years in the absence of a bloodmeal, as does their capacity to transmit borreliae. Both the multihost (typically known as "three host") tick life strategy and the nidicolous (nest-inhabiting) "one host"strategy have obviously allowed for the extensive adaptive radiation and perpetuation of the borreliae.

Borrelia burgdorferi sensu lato: are there as many modes of perpetuation as there are genospecies or just variations on a theme?

The *Borrelia* entry in the NCBI taxonomy server https://tinyurl.com/y3bb7q7f lists 23 species of *B. burgdorferi* s.l. ("Borreliella") with additional *Candidatus* and as yet undescribed cognate sequences. We do not examine each of these but analyze those that provide examples of concepts that are the basis for understanding the perpetuation of these spirochetes.

Is there a general enzootic paradigm?

The universal association is that of *B. burgdorferi* s.l. with a tick in the genus *Ixodes*. Although *B. burgdorferi* s.l. has been reported to infect other genera of ticks, subsequent molecular analyses of the agent that was detected or experimental transmission experiments do not support the role of other tick vectors. Metastriate ticks such as *Dermacentor variabilis* do not support *B. burgdorferi* due to upregulation of hemolymph antimicrobial peptides and increased phagocytic activity (Johns et al., 2001). Indeed, the obligate binding of the major *B. burgdorferi* outer surface protein (OspA) with the *Ixodes* gut cell receptor TROSPA, promoting spirochetal infection of the vector(Pal et al., 2004), suggests that there is a very specific association of

B. burgdorferi with *Ixodes* ticks. It is noted, however, that TROSPA homologs occur in other tick genera (Urbanowicz et al., 2016) and thus their role as possible vectors cannot be completely dismissed.

Transovarial transmission (TOT)

Some vector-borne agents may be inherited by progeny from an infected female, a process known as transovarial or vertical transmission. Early studies of host seeking *I. scapularis* and *I.ricinus* larvae detected spirochetes by immunofluorescence and, in the absence of evidence that larvae might briefly attach to a host and become infected, then detach and seek hosts again, suggested that *B. burgdorferi* s.l. could be inherited (Piesman et al., 1986; Rijpkema et al., 1994). The comprehensive report demonstrating that *B. miyamotoi* was sympatric in most sites with *B. burgdorferi* s.l. and that it was maintained by TOT (Scoles et al., 2001) led to a reexamination of transovarial transmission. Over 100 pools of questing *I.ricinus* larvae yielded none with evidence of *B. burgdorferi*s.l. from Czech and German study sites (Richter et al., 2012)finding only *B. miyamotoi.* Rollend et al. (Rollend et al., 2013) analyzed 1214 pools of larval ticks (50 larvae per pool, each pool representing a single fed female ticks collected from field sites) by PCR and did not find any evidence of *B. burgdorferi* s.s. infection, but 1.4% contained evidence of *B. miyamotoi.* However, Van Duijvendijk et al. (van Duijvendijk et al., 2016) demonstrated that host seeking *I. ricinus* were infected by *B. afzelii* (0.62% of 1456 larvae) and that the larvae were infectious to mice, with *B. afzelii* recovered from skin of the mice. In their study sites, based on estimated larval and nymphal burdens on mice, it appears that larvae could account for as much as a quarter of all rodent infections due to the sheer number that feed on these hosts, assuming the same 0.62% prevalence of infection. TOT, thus, could contribute to BRN and might facilitate the establishment of new enzootic sites when infected ticks are introduced.

Is there a general vector?

What is clear is that for zoonotic transmission, the most important vectors are ticks in the *Ixodes persulcatus/ricinus* species complex, which comprises the species *I. persulcatus*, *I. ricinus*, *I. scapularis*, and *I. pacificus*, among others (Charrier et al., 2019). These ticks tend to share a general life cycle with adults feeding on larger animals, mainly deer, and the subadults on a wide range of small mammals, birds, and even reptiles. The full life cycle may be completed in 2 years, although those of *I. persulcatus* may extend for 4-6 years, depending on climatic zone (Shashina, 1985). There is marked seasonality for the activity of each life stage. Adult ticks are most common in colder months, from fall to spring. Larvae seek hosts in the summer, nymphs in the spring and early summer. There are species specific variations to this general temporal sequence reflecting the wide range of sites and habitats for this group of ticks. Temperature and photoperiod drive the life cycle via effects on development and host seeking behavior. These extended developmental or behavioral delays (diapauses, Greek "diapausis, = pause) between stages are a mechanism that allows adaptation to difficult environmental conditions. In the laboratory, all of these ticks may be induced to develop from egg to adult within 7 months at temperatures much greater than they would experience in nature (Ogden et al., 2004). At 21°C, a typical "room temperature" at which many tick biologists keep their colonies, *I.scapularis* eggs are deposited by the replete female and eclose (hatch) within 3 months; they may feed within the month after eclosion; the engorged larvae molt to nymphs in a month; the nymph may feed within the month after molting; and the engorged nymph can develop into adults within 2 months.

The critical attribute of a zoonotic vector is anthropophily. Because *B. burgdorferi* undergoes reactivation (Schwan and Piesman, 2000) and, hence, the probability of transmission increases with duration of attachment (Piesman et al., 1987), the prompt removal of an attached tick modifies apparent risk, implying that larger size ticks (adults) are less effective vectors than are smaller ones (nymphs). Interestingly, there is variation among the species complex as to the stage during which anthropophily is greatest, which influences zoonotic risk. In Russia, the main vectors are adult female *I. persulcatus* because the nymphal stage rarely attaches to humans (Korenberg, 1994).

Although the vector in the eastern U.S.is universally considered to be *I. scapularis*, there are at least 2 genetically distinct lineages or mitochondrial haplogroups (Qiu et al., 2002; Rich et al., 1995; Sakamoto et al., 2014; Xu et al., 2020), and likely as many as 5 (Gulia-Nuss et al., 2016). We remain the only research group continuing to refer to the "northern" lineage as *I. dammini* (considered the junior subjective synonym of *I. scapularis*) to remind workers that the more southerly populations of *I.*

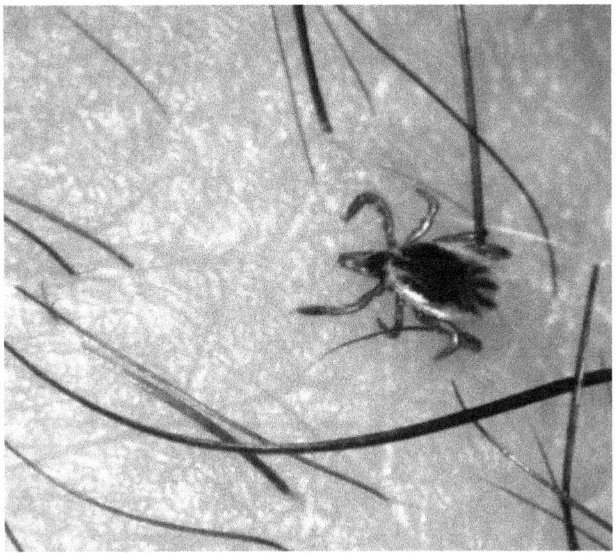

Figure 1. Nymphal *Ixodes dammini* seeking an attachment site, human skin.

scapularis rarely bite humans in the nymphal stage (Goddard and Piesman, 2006; Merten and Durden, 2000). Lyme disease risk in central and southern states is notably small even with known enzootic transmission (Gulia-Nuss et al., 2016). By contrast, infection by nymphal *I. dammini* (Figure 1) is the hallmark of the epidemiology of Lyme disease in New England and Upper Midwest sites of intense zoonotic transmission. Whether this behavioral difference (anthropophily), which is the basis for zoonotic transmission of *B. burgdorferi*, is genetically based remains to be explored (Telford III, 1998), although some SNP-defined clades of ticks are not represented in Lyme disease endemic states (Gulia-Nuss et al., 2016). During the last 30 years, "*I.scapularis*" has greatly expanded its range in part due to bird transport (Eisen et al., 2016), and both major lineages are now present in many central and southern U.S. sites (Sakamoto et al., 2014). A critical test of whether there is merit in distinguishing between the lineages is to determine whether human infestation increases where *I.scapularis* was previously known to only rarely bite humans; we predict that those ticks removed from people, if genetically characterized, will comprise the apparently more anthropophilic invasive northern lineages.

Of course, these considerations are not relevant to the perpetuation of *B. burgdorferi* unless anthropophily covaries with host choice or vector competence. The two lineages do not differ in vector competence (Dolan et al., 1997; Jacobs et al., 2003). Although *I.scapularis* appears to mainly feed on lizards in more southerly sites, it is not clear whether this is a specific choice or opportunism inasmuch as the diversity and abundance of lizards is significantly greater in more southerly sites than in middle Atlantic or northeastern American sites. *I. dammini* from Massachusetts feeds well on different kinds of lizards in the laboratory (unpublished). Interestingly, in experimental outdoor microcosms, nymphal ticks from northern populations were 8 times more likely to seek hosts above the leaf litter than those from southern populations, suggesting an adaptation of the latter to infesting *Scincella lateralis* lizards (Rogers, 1953)that dwell within leaf litter in those southern sites (Arsnoe et al., 2019). Warmer temperatures in more southerly sites, as well as less dense small mammal populations, may have influenced host seeking behavior of *I.scapularis* (Ginsberg et al., 2017), thereby helping to explain the generally lower force of *B. burgdorferi* transmission there (Bowman et al., 2009; Little et al., 2010; Xu et al., 2020). American lizard species are thought to be poorly competent for *B. burgdorferi* infection (Lane and Loye, 1989; Lane and Quistad, 1998), and this suggestion has been used to help explain the relatively low enzootic and zoonotic transmission in southerly sites where lizards are common. However, there are diverse reports that provide exceptions to the "lizards are incompetent reservoirs" dogma (Clark et al., 2005; Levin et al., 1996; Swei et al., 2011), and it may be that lizards serving as major hosts for *I.scapularis* subadults has only a minor or indirect role in transmission of *B. burgdorferi* in more southerly sites. The Eurasian *B. lusitaniae* is maintained by lacertid lizards and *I. ricinus* (Richter and Matuschka, 2006), demonstrating that lizards are not generally incompetent for *B. burgdorferi* s.l.

Early paradigms (the Northeastern U.S. model) for B. burgdorferi *.s.s.*

By the time an association of Lyme arthritis with a specific tick had been determined (Wallis et al., 1978), studies of the epidemiology and ecology of human babesiosis due to *Babesia microti* (Nantucket Fever) were well underway on Nantucket Island, Massachusetts by Andrew Spielman and Joseph Piesman. They incriminated *I. dammini* as the vector, and the white footed mouse, *Peromyscus leucopus* (Figure 2), as the main reservoir of the protozoal agent. Piesman and Spielman (Piesman and Spielman, 1979) identified the main host associations and basic phenology (larvae in late summer, nymphs

Figure 2. White footed mouse, *Peromyscus leucopus*, infested by larval deer ticks (ear).

in late spring, adults from fall to spring) and by the time *I.dammini* was also incriminated as the vector of Lyme arthritis, the major features of the tick's life cycle and the ecology of *B. microti* had been determined. It was not long thereafter that the suggestion was made that the main elements of the ecology of both infections were essentially the same (Spielman et al., 1984). Subsequent reports out of the Harvard lab on the reservoir capacity of this mouse (Donahue et al., 1987; Levine et al., 1985) for the "*Ixodes dammini* spirochete", soon to be named *B. burgdorferi*, established the paradigm of a mouse-driven zoonosis. The prevalence of *P. leucopus* on Nantucket Island, where there were fewer kinds of small mammals compared to mainland sites, and the strength of the evidence of its role in the ecology of *B. microti* there, biased early investigators as to how they thought of the perpetuation of *B. burgdorferi* in northeastern North America. White footed mice were considered to be the main reservoir.

Spirochetal isolates were obtained from most kinds of small mammals endemic to mainland southern New England, as well as from rabbits, mustelids, carnivores, and passerine birds (Anderson, 1988). The Harvard group tended to dismiss such findings as merely evidence of infection but not of reservoir capacity, insisting on the gold standard of xenodiagnosis (the feeding of uninfected laboratory-reared larval deer ticks on reservoir candidates) to confirm a reservoir. Subsequent work demonstrated the reservoir competence of shrews and rabbits (Telford III and Spielman, 1989; Telford et al., 1990), certain passerine birds, but not others (Mather et al., 1989a), and even voles and chipmunks (Mather et al., 1989b). Nevertheless, the Harvard group continued to insist that *P. leucopus* was the main reservoir and that others contributed considerably less to the overall population of infected ticks. Other hosts fed fewer subadult deer ticks (average larval burdens of >10 for mice vs. <2 for chipmunks or shrews) and were less infectious (typically >70% of xenodiagnostic ticks feeding on mice became infected but <50% of those feeding on shrews or chipmunks). Mice were also as much as 10 times more abundant than any of these other species in sites where deer ticks were common. In addition, mice were immunotolerant to tick infestation whereas other hosts such as voles or chipmunks tended to reject subadult *I. dammini* after an initial infestation (Davidar et al., 1989; Mather et al., 1989b), hence would fail to feed sufficient numbers of ticks to contribute to BRN. It should be noted, though, that the snapshot nature of such analyses fail to examine the full range of age of potential hosts; voles or chipmunks may be more tolerant of infestation or more infectious as recruits. Then too, immunotolerance to tick infestation has been measured with large numbers of ticks being fed on a host, as opposed to the likely "trickle" infestations (a few ticks every few days) that the majority of wild animals sustain. Rodents in general have rapid demographic turnover, so the extended duration of reservoir capacity for mice may not necessarily be as influential as expected.

On the other hand, deer had been determined by experimental infection attempts to be insusceptible to infection by *B.microti* (Piesman et al., 1979) and their reservoir incompetence for *B. burgdorferi* was demonstrated by xenodiagnosis (collecting replete larvae that naturally infested deer in intensely enzootic sites and allowing them to molt to nymphs, which were assayed for infection (Telford III et al., 1988). Such quantitative arguments were generated, in part, by trying to understand the conditions required for the excess BRN that leads to zoonotic overflow.

The central role for *P. leucopus* in the enzootic cycle of *B. burgdorferi* in the northeastern U.S. has its basis in a theoretical argument. Because ixodid ticks feed but once in each developmental stage, in the absence of inherited infection, a pathogen's transmission would depend on a vector focusing its feeding on a single kind of competent host (larvae that becomes infected must transmit as a nymph to a competent reservoir, otherwise the infection does not contribute to BRN). Narrowness of mosquito host range is the most influential of the variables in the malarial vectorial capacity model (Spielman and Rossignol, 1984). This seminal model greatly influenced Spielman, who used considerations of narrowness of host range to argue for a single most important reservoir for *B. microti* and *B. burgdorferi* (Spielman et al., 1984; Spielman, 1988). Such considerations directly led to the concept of a zooprophylactic (Russell, 1934) or dilution effect, that community host diversity will dilute the effect of white footed mice in producing infected ticks (Spielman et al., 1984; Spielman, 1988; Ostfeld and Keesing, 2000) and thereby human risk. Although data from insular sites (particularly Nantucket Island) tended to support this concept (Spielman et al., 1984), narrowness of vector host range does not seem to characterize the perpetuation of *B. burgdorferi* s.l. whose global distribution and abundance may be related more to adaptive plasticity: the spirochete and the tick are opportunistic. Indeed, the role of other species such as rabbits, shrews or chipmunks in buffering the perpetuation of eastern North American *B. burgdorferi* during times when or where mice might be scarce has been amply documented (Telford III and Spielman, 1989; Telford et al., 1990; LoGiudice et al., 2003; Brisson et al.; 2008).

Then, too, an adaptationist perspective is common among biologists, with natural selection caricatured as producing the most optimal solution (Gould and Lewontin, 1979). Enzootic and zoonotic transmission of *B. burgdorferi* should be optimal only with a tick that focuses its subadult feeding on a highly reservoir competent host. Such a host would allow the production of a great number of infected ticks, which made it more likely that BRN>>1 and the more likely that humans might be exposed to infection. But, is perpetuation really compromised by suboptimality? One might argue that if infection rates were small for host-seeking nymphs, the probability of encountering a susceptible individual mouse would be correspondingly diminished relative to when more infected nymphs were available. The 20-80 rule of parasitology, an empirical and theoretically based principle that posits 20% of the hosts are infested or infected by 80% of the individual parasites (Woolhouse et al., 1997) applies to ticks (Randolph et al., 1999) and undoubtedly helps maintenance of *B. burgdorferi*. Small numbers of mice in our Nantucket field sites feed most of the *I.dammini* larvae and nymphs (Figure 3). But, the basis for why an individual animal becomes heavily infested and the majority do not has not been explored.

Complexity is increasingly recognized: California

Evidence of complexity and of local differences in perpetuation of *B. burgdorferi* was provided in other regions nearly concurrently with the early New England studies. The western blacklegged tick, *Ixodes pacificus*, is the main vector for *B. burgdorferi* in the western USA and British Columbia (Lane et al., 1991). Lyme disease is more focal in its distribution in these areas, with most cases reported from the northern coastal counties of California. Seminal studies by Robert Lane and colleagues identified *I. pacificus* and *I. spinipalpis* (at the time, *I. neotomae* was implicated (Brown and Lane, 1992), but the infesting ticks were subsequently identified as *I. spinipalpis*) and relatively long lived rodents such woodrats and kangaroo rats, but only rarely *Peromyscus* spp. The study site was mainly the University of California Hopland Field Station in Mendocino County, which comprises chaparral habitat. *I. spinipalpis* only rarely bites humans, but *I. pacificus* was thought to serve as the zoonotic bridge by feeding on infected woodrats and kangaroo rats as well as humans. At the time, this paradigm served as a stark contrast to the *P. leucopus* driven transmission by an aggressive *I. dammini* vector and served as evidence that *B. burgdorferi* enzootic cycles vary geographically. Epidemiological studies determined that California Lyme disease cases were exposed mainly in dense woodlands (Clover and Lane, 1995) where woodrats and *Peromyscus* were infrequently infected by *B. burgdorferi* (Wright et al., 2000).

California is the most biodiverse state in the U.S., with 7 different biomes comprising 21 distinct ecological communities (Schoenherr, 1992). An analysis of mtDNA identified great haplotype diversity in California populations of *I. pacificus*, with many mtDNA haplotypes unique to sampling sites (Kain et al., 1999). In this context of small local lineages of vector ticks within diverse ecological communities, the finding that *B. burgdorferi* comprises a species complex in California should not be surprising. *B. burgdorferi* s.s., *B. bissettiae*, *B. americana*, *B. lanei*, *B. californiensis*, and other as yet unnamed genospecies have been isolated from *I. pacificus* and *I. spinipalpis* (Fedorova et al., 2014; Postic et al., 2007), but patterns of association between borreliael genospecies, vertebrate hosts, and tick lineages remain to be described. Most Lyme disease in California is caused by *B. burgdorferi* s.s., although there are case reports that suggest *B. bissettiae* may be an etiologic agent. Woodrats in chaparral and

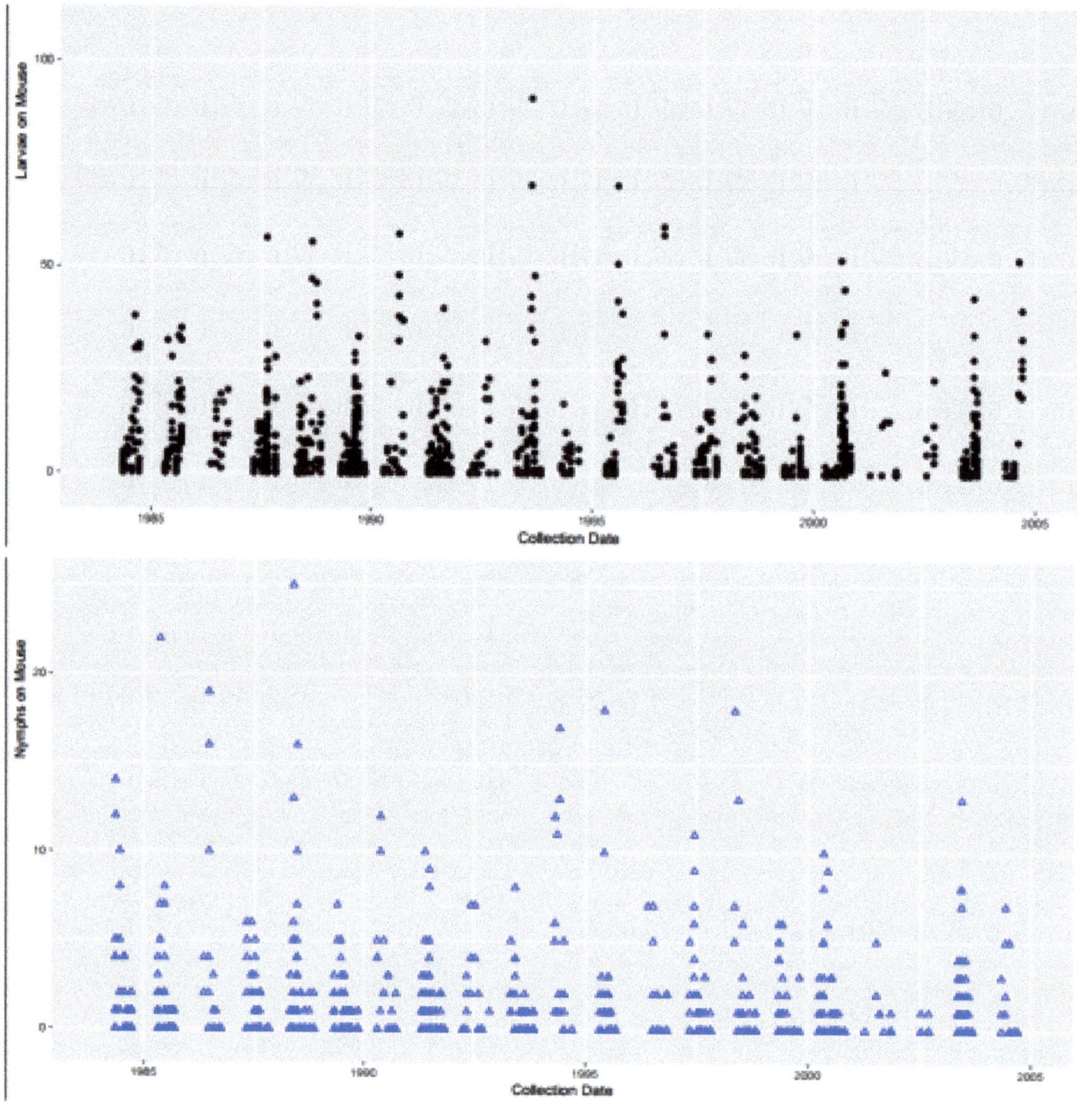

Figure 3. 20:80 rule. Infestation indices on *P. leucopus,* Nantucket Island, 1985-2004. Mice were trapped monthly (AMJJAS) on one acre grids. X axis, time. Y axis, number of ticks. Each dot represents the tick burden of a single mouse. Top graph is *I.dammini* larvae, bottom nymphs.

grassland sites maintain *B. bissettiae* or other *B. burgdorferi* s.l. (Postic et al., 1998, 2007; Vredevoe et al., 2004). Kangaroo rats maintain *B. californiensis* likely with *I. jellisonii* as the vector (Lane and Brown, 1991; Postic et al., 2007) although *I. spinipalpis* and *I. pacificus* also feed on this heteromyid rodent in the chaparral habitats of California. Of particular interest is the fact that western gray squirrels (*Sciurus griseus)* maintain *B. burgdorferi* s.s. in oak woodlands (Lane et al., 2005), suggesting that human risk might be promoted by the known tendency of this species to inhabit periurban and suburban habitats.

I. pacificus is found mainly in the summer Mediterranean climate biomes in California, but ranges into Washington, British Columbia and western Canada, with isolated populations in Arizona,

Utah, and Nevada. In California, the life cycle is extended, with adults seeking hosts in the fall and winter, fed females ovipositing in late winter through spring, and eggs eclosing in mid-late summer. The larvae do not immediately seek hosts, but overwinter and become active in the early spring, with a peak in May and April. Engorged larvae molt immediately, but the nymphs are inactive until late winter and feed through the spring. Fed nymphs will molt to adults in the late summer. Hence, the life cycle is a minimum of 3 years (Padgett and Lane, 2001).

Eurasian Lyme disease spirochetes

The main vector of Lyme disease spirochetes in Europe is the castor bean (sheep) tick, *I. ricinus*. This widely distributed tick, ranging from Ireland in the west to Scandinavia and the Russian Federation to St. Petersburg in the east, to the Mediterranean including parts of western North Africa, can be found in diverse habitats, even pasture, although it prefers heavily forested sites that provide the humid microhabitat required for development and survival. *I. ricinus* is recorded as infesting 44 different species of mammals, birds, and reptiles, but the majority of these ticks in continental Europe feed on roe deer, small rodents, and blackbirds (thrush family) (Hofmeester et al., 2016). In the southern part of its range, *I. ricinus*, like *I. scapularis*, may heavily infest lizards (Amore et al., 2007).

The life cycle of *I. ricinus* varies across its range. In the British Isles, where classic studies were undertaken of its natural history (Gray, 1984; Lees and Milne, 1951; Milne, 1943),there is a bimodal activity period for each of the stages, corresponding to spring and fall feeding cohorts. Adults seek hosts once temperatures reach 7ºC in the spring, with peak activity in May and June. A smaller peak is observed in August and September. Eggs from the spring cohort eclose in the summer, but this may be delayed until the following spring as it is for those from females feeding in the fall. Larvae feed in May and June, with a smaller peak in August and September. Nymphs generally follow the same activity pattern. Engorged larvae and nymphs will molt in the summer, but diapause until the following spring. Throughout its range, there is large overlap in seasonal activity between the stages, but such overlap may vary with latitude.

The identification of *B. burgdorferi* as the etiologic agent of Lyme disease included *I. ricinus* collected from Switzerland (Burgdorfer et al., 1982), and comprehensive studies there found infection in 5%-34% of host seeking ticks. In addition, *Apodemus* spp. mice were found to be infectious to laboratory reared ticks (xenodiagnosis), incriminating them as reservoirs (Aeschlimann et al., 1987). Strain heterogeneity was quickly recognized by European workers, distinguishing patient isolates from Europe from those of the U.S. by monoclonal antibody typing and by ultrastructure (Hovind-Hougen, 1984; Wilske et al., 1988). We now know that in Eurasia, *B. burgdorferi* s.l. includes over 20 genospecies, but only 3 cause the vast majority of human disease (*B. burgdorferi* s.s., *B. afzelii*, and *B. garinii*).

Population genetics analyses demonstrated little population structure for *I.ricinus* across northern and central Europe (Meeûs et al., 2002). Some isolated populations may be distinguishable (e.g., Great Britain from Latvia and Western Norway (Dinnis et al., 2014; Røed et al., 2016). Although the English Channel is a physical barrier to tick movement, birds may easily transport ticks great distances and, thus, the low degree of apparent divergence was not expected. This finding has been attributed to the very small likelihood that introduced subadults from birds would generally fail to reproduce; one estimate for developmental success between stages is <10% (*I.*e., 90% of all ticks at each stage die (Randolph et al., 2002). One exception to the rather homogenous population structure across continental Europe is that *I. ricinus* from southwest Mediterranean and North African sites appeared to be partially genetically isolated. These analyses led to re-examination of the Mediterranean "*I. ricinus*" and the recognition of a new cryptic species, *I. inopinatus* (Estrada-Peña et al., 2014). *I. inopinatus* has been found in northern and southern Germany and, in fact, is as frequently infected as *I. ricinus*, long known to be the only vector there (Hauck et al. 2019).

The diverse genospecies in Europe may be grouped into 3 categories with respect to their reservoir host: generalist, rodent specialized, and bird specialized. These categories derive from data on isolation (or DNA evidence) from or xenodiagostic studies of wild-caught animals, associations of animal species with particular sites, prevalence of infection in host-seeking ticks, and experimental infection. Of note is that only *B. burgdorferi* s.s. is considered a generalist, infecting rodents as well as birds. *B. burgdorferi* s.s. is the dominant genospecies in North America, but is less frequently encountered in Europe (relative to *B. afzelii* and *B. garinii*; (Strnad et al., 2017) for as yet undescribed reasons. *B. burgdorferi* s.s. is more frequently detected in adult *I. ricinus* than in nymphs (Mysterud et al., 2019), suggesting that this spirochete is likely associated with a host such as squirrels that feeds more nymphs than larvae (Humair and Gern, 1998).

In much of Eurasia, the vector for *B. burgdorferi* s.l. is *I. persulcatus*. The taiga tick ranges from the Baltic Sea to Japan, with the center of infestation within the Russian Federation, although half the provinces of China have known infestations (Chen et al., 2010). Infestations in Japan are most common in Hokkaido, although it exists in montane sites of Honshu and Kyushu (Takada et al., 2001). As the name suggests, the tick is most common in the taiga (boreal coniferous forests, typically with pine, larch and spruce) biome, but in the western part of the range deciduous forests are infested. The duration of the life cycle depends on temperature as well as photoperiod, adaptations that are not unexpected given its large latitudinal range (from roughly 35 to 65 N latitude, although the more southerly sites are in mountains). All stages seek hosts from spring through summer, with the peak during late May and mid-June. Subadults that engorge during May and June will molt during the summer but not seek hosts, overwintering to the next year. Subadults engorging during June and July will overwinter before molting in the spring. For all stages, a decreasing photoperiod will trigger diapause, either developmental or behavioral (Babenko, 1985; Belozerov, V.N., 1985).

The host range, as for all the *I. persulcatus* complex ticks, includes a wide range of mammals (>100 species), birds (175 species), and even some lizards (Labzin, 1985; Naumov, 1985). Larvae tend to feed on mice, voles and shrews; nymphs on chipmunks, hedgehogs, and birds; and adults on larger mammals such as hare or deer, although cattle also help maintain *I. persulcatus* infestations. During times when typical larval hosts are scarce, they may densely infest hares (Uspensky and Rubina, 1992). Hence, *I. persulcatus* is an opportunistic feeder.

Analyses of *I. persulcatus* population structure reveal low diversity, 1.14% within 12S rDNA, across its range from the west to Far East, including central Russia (Kovalev and Mukhacheva, 2012). As with *I. ricinus*, this may be explained by severe bottlenecking of genetic diversity during glaciation with rapid expansion of the populations during the 10,000 years after glacial retreat. The rapid expansion was certainly facilitated by intensive agriculture and the movement of large domestic animals. *I. persulcatus* appears to be increasing its range, first appearing in Finland in the early 2000s and then in the Norbotten archipelago (northern Sweden) 10 years later (Jaenson et al., 2016). This is the most northern infestation known for *I.persulcatus* and is interpreted by some as evidence for the effects of increasingly warmer temperatures in northern latitudes.

I. pavlovskyi, a tick that is parapatric and sympatric with *I. persulcatus* in many sites across Russia, was found to be infected by *B. afzelii* at a prevalence greater than that of the latter in Tomsk (Korenberg et al., 2010). Subsequent analysis of the ticks around Tomsk demonstrated that 10% of 783 ticks represented natural hybrids of *I. pavlovskyi* and *I. persulcatus*. These ticks apparently form fertile F1 and may be backcrossed (Kovalev et al., 2015), suggesting that it is simplistic to consider formation of fertile F1 hybrids between *Ixodes* ticks evidence for conspecificity. Additional research is required to determine whether hybrid *Ixodes* spp. have any implications for enzootic or zoonotic transmission of *B. burgdorferi* s.l. Subadult *I. persulcatus* and *I. pavlovskyi* feed on small mammals, often the same individual host, but the adults feed on different animals, viz., ungulates for *I. persulcatus* and birds for *I. pavlovskyi* (Korenberg et al., 2010), which suggests a means of coexistence via niche partitioning.

A Borrelia *for every kind of* Ixodes?

There are 235 species of ticks comprising at least 15 subgenera within the genus *Ixodes* (Camicas et al., 1998). Eight of these subgenera have been implicated as enzootic vectors for at least one *B. burgdorferi* s.l. (*Ceratixodes*, *Eschatocephalus*, *Pholeioixodes*, *Amerixodes*, *Exopalpiger*, *Ixodes*, *Trichotoixodes*, and *Partipalpiger*). To our knowledge, borreliae have not been reported from the particularly diverse *Scaphixodes* and *Afrixodes* (31 and 60 species, respectively), nor from the less speciose subgenera. The natural history of the vast majority of *Ixodes* spp. has not been completely described, although many are very host-specific (or at least circumstantially host specific) or nidicolous, thereby cryptically maintaining enzootic cycles. Although the subgenus *Ixodes* (which includes the *I.ricinus* species complex) has been the focus of much attention, enzootic *B. burgdorferi* s.l. cycles have been recognized for relatively host specific ticks such as the pelagic bird infesting *I. uriae* or the hedgehog tick *I. hexagonus* (Gern et al., 1997; Olsen et al., 1995). The nidicolous *I. trianguliceps* was found to be infected by *B. burgdorferi* s.l. 5-10 times less frequently than co-infesting *I. persulcatus* (the study was performed by analyzing ticks removed from trapped rodents), suggesting that it is less competent in pathogen uptake and/or spirochetal gut colonization (Korenberg et al., 2015). In Europe, *I. canisuga, I. frontalis, I. arboricola,* and *I. ventalloi* have all been found to be infected by *B. burgdorferi* s.l. but their role in maintenance is not clear. *I. ventalloi* is widely distributed in the southwestern Mediterranean, and feeds on diverse animals even though it is considered a "rabbit tick" (Santos and Santos-Silva, 2018; Tomassone et al., 2013). This tick is frequently found on cats, which implies that it could be common in peri-domestic sites and, thus, serve as a zoonotic bridge. However, even though *I. frontalis* and *I. arboricola*, which are bird-infesting species, are competent to acquire infection from birds and *B. burgdorferi* is transstadially maintained, they failed to actually transmit (Heylen et al., 2014). Thus, finding infection in a tick, even demonstration of positive xenodiagnosis, does not necessarily suggest it has a role in maintaining *B. burgdorferi* s.l. Nonetheless, it is almost a certainty that additional new *B. burgdorferi* s.l. enzootic cycles will be identified as the natural histories of the 200 or more valid understudied *Ixodes* species are examined.

Factors influencing host relationships

One might argue whether ticks are the reservoir (infection is maintained for long durations, given the extended tick life cycle) or the vertebrate is the reservoir because it can amplify infection (a single infected animal could give rise to dozens or hundreds of infected ticks) and thereby influence the BRN. Early in the studies of Lyme disease ecology, vertebrate hosts were classified as incompetent or competent reservoirs (Donahue et al., 1987; Telford III et al., 1988), usually based on whether xenodiagnostic ticks became infected either on experimentally infected or naturally exposed animals. Reservoir capacity comprised those with high reservoir competence (if a host is infected, most ticks feeding on that host become infected; and, the host does not become immune). They were densely infested by the relevant tick stage and did not become sensitized to the bites (immune rejection of the ticks). Theoretical considerations suggested that there must be "main" and ancillary reservoir hosts. The increasing recognition of borrelial genetic diversity has also been so influenced, with adaptation of genospecies to specific hosts by means of evading species-specific complement lysis. These paradigms need to be reexamined.

Cofeeding transmission

The BRN of *B. burgdorferi* s.l. depends on horizontal or transstadial transmission (Hartemink et al., 2008), i.e., a larva becomes infected by feeding on an infectious host that is systemically infected; spirochetes are maintained through the molt to the nymphal stage and are transmitted to a new host by nymphal feeding. However, nonsystemic or cofeeding transmission may play a secondary role in perpetuation. In this mode of transmission, uninfected ticks feeding next to infected ones may become infected without the agent needing to establish a persistent infection resulting in a bloodstream or skin infection (Jones et al., 1987). Ticks are highly clustered on hosts (Randolph et al., 1999). Cofeeding is the main mechanism for the perpetuation of tick borne encephalitis virus(Labuda et al., 1993) which only produces a transient viremia in the vertebrate host. The great overlap in seasonal activity of *I. ricinus* drives the force of TBEV transmission (Randolph et al., 1999), with infected nymphs serving as a source for larvae feeding on the same host. Cofeeding is thought to be of lesser importance for *B. burgdorferi* s.l. which generally produces a chronic infection in the vertebrate host. Sheep, which do not become systemically infected by *B. burgdorferi*, nonetheless may infect ticks via cofeeding (Ogden et al., 1997). Our recent host bloodmeal analyses suggest that some ticks may indeed become infected from deer (Goethert et al. submitted) which have long been thought to be incompetent hosts. The significance of cofeeding also includes the maintenance of *B. burgdorferi* diversity within a site where synchronous feeding of *I.dammini* larvae and nymphs is common(States et al., 2017)by promoting the survival of strains that are more rapidly cleared from the host.

Complement sensitivity or evasion as the main determinant for B. burgdorferi *s.l. host associations.*

A pervasive paradigm in the ecology of *B. burgdorferi* s.l. is that the adaptive radiation of these spirochetes into the great diversity of genospecies that we recognize today was driven and is maintained by adaptation or lack thereof to attack by host complement. In a seminal paper, Kurtenbach and colleagues (Kurtenbach et al., 1998) demonstrated specific borreliacidal activity when serum of diverse animals was incubated with *B. burgdorferi* s.l. strains. This borreliacidal activity was eliminated by heating serum, or supplementation of the serum with EDTA, suggesting the role of the alternative complement pathway. Mouse serum lysed *B. garinii*, bird serum lysed *B. afzelii*, and deer serum lysed all genospecies that were tested. The pattern of genospecies specific lysis is consistent with the prevailing views of reservoir competence, e.g., that deer were incompetent, that mice were never infected by *B.garinii*, and that birds were not infected by *B. afzelii*. Cattle, whose serum lyses all genospecies, are zooprophylactic (Gassner et al., 2008; Richter and Matuschka, 2010).

Spirochetes may evade complement lysis by binding the complement regulatory protein factor H (fH) (Kurtenbach et al., 1998). Factor H is a cofactor for factor I-mediated cleavage of C3b and bound Factor H will promote the degradation of C3b and thus inhibits the remainder of the complement cascade and formation of the membrane attack complex (MAC). *B. burgdorferi* s.l. are now known to have an array of outer surface proteins that target different modulators of the 3 different complement pathways (Kraiczy et al., 2004). The *B. burgdorferi* s.l. proteins involved in modulating the alternative complement pathway are called FH/FHL-1-binding complement acquiring surface proteins (CRASPs), and comprise CspA, CspZ, and OspE-related proteins (members of the Erp family) (Kraiczy et al., 2004). BBK32 and OspC also bind elements of the classical and lectin complement pathways and inhibit the formation of MAC (Garcia et al. 2016; Caine et al. 2017). CRASPS, OspC, and BBK32, among other *B. burgdorferi* s.l. proteins, may provide functional redundancy that would allow for some degree of plasticity in host associations and may be the basis for the leakiness of the complement lysis theory for host adaptation.

There are natural exceptions to the expected host associations: grey squirrels, for example, may be infected by *B. garinii* (Millins et al., 2015), and some inconsistencies from predicted host ranges for genospecies were detected when host seeking nymphal *I. ricinus* were analyzed by host bloodmeal analysis (Cadenas et al., 2007). The theory of differential sensitivity to host complement as a driver for the known genospecies/host associations should be viewed in the context of the fact that analyses of complement lysis (e.g., (Bhide et al., 2005) do not demonstrate 100% lysis but rather report the data as >90% lysed, or with varying degrees of lysis with an overall average of majority of the spirochetes lysed. In addition, the studies were done, by necessity, with established strains of *B. burgdorferi* s.l. One does not expect all or none phenomena in biology (we should not expect 100% lysis), but even the survival of 1% of spirochetes in a strain that has likely lost much of its variability via isolation and maintenance in BSK medium implies that the original strain in nature might have sufficient diversity for a variant to be selected. Experiments that serially passage surviving spirochetes in deer serum, or the use of representative sets of primary *B. burgdorferi* s.l. isolates, for example, might demonstrate such a possibility. The point is that this theory, based on *in vitro* cultivated isolates and general consistency with field observations, needs to be reexamined in the light of a dynamic environment in which natural selection is constantly operating.

Host bloodmeal remnant analysis: the key to analyzing host associations

To date, our understanding of the enzootic cycle has relied on indirect measures such as estimated host abundance, degree of tick infestation, and infectivity to ticks. Mosquito biologists have long used the forage ratio, the proportion of bloodmeals from a host relative to the proportion that host comprises relative to all possible hosts in the site where the sampling was undertaken (Hess et al., 1968) to determine whether a mosquito species prefers a host or is opportunistic. Given their wide host range, measurement of the forage ratio for the ticks that serve as vectors for *B. burgdorferi* s.l. is not straightforward. Trappability differs for individual animals within a species and also between species. Small mammal censuses using standard Sherman or Longworth traps tend to bias the results towards small rodents, such as mice, while squirrels, chipmunks, and shrews are more efficiently trapped with other kinds of traps (Huggins and Gee, 1995; Maddock, 1992; Stephens and Anderson, 2014). Birds are rarely concurrently sampled. There have been few comprehensive attempts to evaluate the reservoir capacity of all the possible hosts within a *B. burgdorferi* enzootic site; the most detailed of such studies (Brunner et al., 2008; LoGiudice et al., 2003) have required extrapolation from the literature or modeling to compensate for not being able to sample all kinds of animals. The logistical issues are large. Although one might own and easily set hundreds of mouse traps in one afternoon, few have and can set more than a few dozen larger box traps to sample

Table 1. Reports of bloodmeal analyses on host-seeking *Ixodes ricinus*.

Study	Method	Sensitivity	No. ticks identified	Proportion feeding on Rodents (mice/voles)	Most common host identified
Kirstein and Gray, 1999	RLB-cytb	63%	50	6%	deer
Pichon et al., 2003	RLB- 18S	53%	26	8%	bird, ruminant
Estrada-Peña et al., 2005	RLB-18S	41%	25	8%	bird
Pichon et al., 2005	RLB- 18S	49%	159	5%	songbirds, ruminants
Cadenas et al., 2007	RLB-12S	44%	712	8%	artiodactyls, red squirrel, boar
Humair et al., 2007	RLB-12S	50%	53	4%	red squirrel, deer
Wodecka et al., 2014	nested PCR-12S	63%	327	9%	roe and red deer boar
Collini et al., 2015	real-time PCR with HRM d-loop	65%	34	9%	domestic dog, deer
Collini et al., 2016	Real time PCR with HRM	55%	239	29%	rodents,dogs
Wodecka and Skotarczak, 2016	Nested PCR with RE digest-12S	65%	210	8%	boar, fox, red and red deer
Honig et al., 2017	RLB-12S	61%	219	28%	artiodactyls, rodents

raccoons. Tending mist nets requires continuous attendance to remove captures and depends on careful selection of the net sites. It would be almost impossible to conduct a longitudinal study of tick and spirochete dynamics within the entire fauna of a site, let alone over several sites. However, the use of game cameras allowed analysis of the possible contribution of medium sized and large mammals in 19 Dutch study sites (Takumi et al., 2019)and, combined with estimates of infection from host seeking nymphs, provided a good picture of the likely main contributors to *B. burgdorferi* BRN there.

Given the issues with estimating the complete diversity and relative abundance of possible hosts within the site and to sample often enough to describe temporal variation in host demography (particularly for larger mammals) within any site, host indices are more commonly used. In malarial entomology, the human blood index (HBI), the proportion of bloodmeals taken from humans by a mosquito population (Garrett-Jones, 1964), was used to demonstrate the effect of spraying DDT or dieldrin on the endophilic cohort of *Anopheles gambiae*, among other uses. Blood meal analysis provides a means to directly identify the host upon which a vector fed. This technique has been successfully utilized for mosquitoes since the 1950s (Weitz, 1956). Sufficient protein remains in the mosquito gut after digestion that hosts may be identified by immunological means. Bloodmeal analysis has proven to be much more difficult for ticks, due to their longer life cycle and complete intracellular digestion of their blood meal. Deer ticks feeding as larvae in August will not emerge as host-seeking nymphs until June the next year; what remains of the blood meal then is almost a full year old and little has survived the physiologic processes of molting.

The reported success of host bloodmeal remnant analysis in field-collected ticks has been variable, averaging about 50% success for tick samples, thus making the data difficult to interpret (Table 1): if the remaining samples had yielded identifications, would the distribution of host prevalences differ? Or is the failure to amplify from a large proportion of hosts because a host was not anticipated in a site and the correct assay was not used? Most of the assays have been based on amplifying mammalian mitochondrial gene sequences using conserved primers, and then identifying the host by reverse line blot using species-specific probes or by melting temperatures of the amplicon. To date, studies utilizing host bloodmeal remnant analysis on known Lyme disease vectors have been conducted only on *I. ricinus* in European sites (Table 1).

These valuable studies demonstrate that, for most of the sites examined, rodents play a very minor role as hosts of *I. ricinus*. Instead, ticks fed on roe and red deer, wild boar, and birds. Only Cadenas et al attempted to identify the blood meal host in a large number (>1300) of ticks (Cadenas et al., 2007). Their data clearly indicate that the majority of ticks in their sites fed on wild boar, deer and red squirrels. However, their efforts to incriminate reservoir hosts was unsuccessful due to low sample size; only 19% of all ticks tested positive for borreliae; half had successful bloodmeal identification, divided over 2 field sites, while there were fewer than 15 identified ticks for each *B. burgdorferi* s..l. genospecies. Only *B. afzelii* had a significant number of host-identified ticks, and these had fed on virtually every type of host for which assays were available; this finding is inconsistent with the current paradigm that this genospecies is maintained by voles. Of course, for those studies analyzing adult ticks, infection could have been acquired as a larva, but evidence of the identity of the infecting host would have not been retained into the adult stage.

Our laboratory is currently refining a new DNA-based method for blood meal remnant analysis that has the potential to overcome the sensitivity issues observed to date. Retrotransposons are genetic elements that integrate into the genome and can replicate themselves, leaving a copy behind in the genome after every jump. The most common forms are LINEs (long interspersed repetitive elements) and SINEs (short interspersed repetitive elements). As they have an ancient origin, over time they have accumulated extremely high copy numbers and have coevolved with their host genomes creating family-specific lineages. Previously called junk DNA, they are thought to make-up 40%of the human genome and similarly large portions of many other mammalian genomes. This extremely high copy number maximizes the probability that a fragment that can be targeted for amplification by PCR will remain after digestion in the tick gut. To date, infection status (using a multiplex real time PCR for *B. burgdorferi*, *B. microti*, and *Anaplasma phagocytophilum*) and host identity can be assayed from single host seeking nymphal *I.dammini.* In our Nantucket field site, during 2018, all the host seeking nymphs that were assayed had fed on *P. leucopus*, but in 2019 only 20% had done so. Interestingly, no change in prevalence of *B. burgdorferi* infection was observed between the years, confirming that even in insular sites, diverse animals may maintain the enzootic cycle. Even more surprising was that in two mainland sites in the same years, *P. leucopus* served as infection source in one while an as yet unidentified host was evident in the other (Goethert et al., submitted). As host bloodmeal remnant analysis is more widely used, we expect to see significant spatiotemporal differences in host contributions to infected ticks, perhaps even to the point that we can no longer say that there is a "main" reservoir host at a scale larger than the specific study site.

Ultimately, we seek to measure the *Peromyscus* blood index (PBI) or that of other species derived from host bloodmeal remnant analysis as a means of comparing the infected vs. noninfected tick samples from sites. An approximate forage ratio may thus be calculated to determine the contribution of mice relative to that of all other hosts identified as sources of infection from the same collection (approximated forage ratio = PBI divided by proportion of bloodmeals from all other hosts; if >1, mice contribute more infected ticks, if 1 then their contribution is no greater than all other hosts; and <1, other hosts contribute more to the force of transmission). Direct measurement of host contributions from host seeking ticks would eliminate the need for time intensive correlative trapping studies combined with xenodiagnosis.

Factors that influence the abundance of *B. burgdorferi* s.l.

The classical biotic and abiotic factors that affect abundance of animals and plants are not easily translated to the borreliae-tick-vertebrate system (Figure 4). Such factors would largely affect the hosts that are infected (vertebrate and tick), but it is possible that there are effects that directly bear on the spirochete. With respect to climate, a major abiotic factor, there is no question that temperature impacts the perpetuation of *B. burgdorferi* s.l., mainly with effects on the development and behavior of the tick, but also likely on its microbiota. For example, *B. burgdorferi* survives longer in ticks held in the laboratory at 21°C vs. those held at 26°C (Shih et al., 1995), suggesting infection stability is temperature dependent. The quantum of spirochetal infection within a tick may be associated with the latitude where the tick is found, perhaps by direct effect of temperature on spirochetal replication within the tick after molting (Sirotkin and Korenberg, 2018). It is also possible that borreliae become senescent faster if temperatures are warmer while the tick is seeking hosts. It is possible that temperature could act on tick innate immunity or its microbiome, which in turn would act on the spirochete, instead of directly on it.

Temperature, particularly in association with relative humidity (ticks are affected by "saturation deficit", a combined temperature and relative humidity index below which their activity and survival are curtailed), may be the most influential of the factors affecting

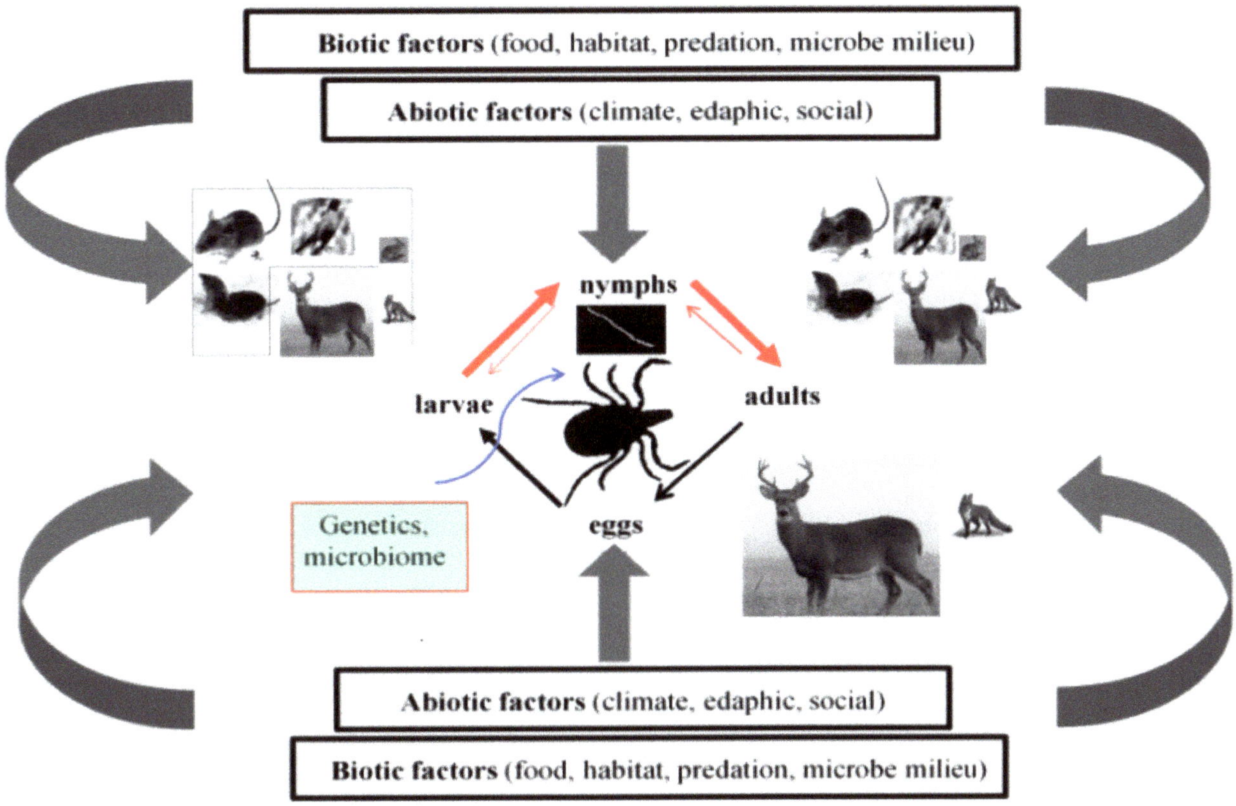

Figure 4. General factors operating on enzootic transmission of *B. burgdorferi.* Red arrow = flow of infection.

developmental rates as well as host seeking behavior and longevity in tick populations. Near ground temperatures, day length, and relative humidity are good predictors for nymphal *I. ricinus* activity (Daniel et al., 2015). Host seeking tick abundance, as measured by drag sampling, is reduced by the number of adverse moisture events (Berger et al., 2014; Perret et al., 2000) that is, periods of >8 hours with relative humidity <82% within the leaf litter within a transmission season. In our sites in coastal Massachusetts, the seasonal activity of nymphal *I. dammini* terminates in mid-July, coincident with mid-summer drought conditions, but when conditions remain cool and wet, nymphs may seek hosts well into August. Extreme variability of weather (*e.g.*, episodes of dryness) was associated with a decline in host seeking nymphal *I. ricinus* density, as opposed to overall temperature trends over a 15 year observation period (Hauser et al., 2018). Much has been speculated with respect to climate change and risk of tick transmitted infections, but any effects are likely to be locally dependent. The prevalence of infection of *B. burgdorferi* s.l. in host seeking *I. ricinus* seems remarkably stable across a continental scale, on the order of 11-14% in metanalysis studies with no strong evidence for increase or decrease over 30 years (Rauter and Hartung, 2005; Strnad et al., 2017), arguing against a role for climate change on a continental scale (global temperatures have increased during that time). Of course, there may be site-specific longitudinal changes, particularly where tick density changes: prevalence depends on density of ticks (Coipan et al., 2013; Randolph, 2001), with lesser prevalence where tick numbers are unusually great (>100 per square meter).

Lab studies poorly replicate natural conditions in which ticks are found. For example, it should be noted that the average annual temperature for coastal Massachusetts is 9ºC, taking into account daily maxima and minima and seasonal periods of cold and warm. At a constant 9ºC temperature, tick development does not proceed in the laboratory; in nature, development and activity is associated with accumulation of heating degree days (Ogden et al., 2004; Rand et al., 2004). At 24°C, engorged larvae may molt to nymphs within 3 weeks, as do engorged nymphs (unpublished); constant temperatures >28-30°C greatly reduce daily survival even at constant appropriate humidity (unpublished; (Ogden et al., 2004). As with many other facets of the ecology of Lyme disease, such studies examine a snapshot in time: natural selection may modify traits to allow for specific adaptation to a changing environment. For example, in the early 1990s, stable *I. dammini* infestations started to appear in west-central Massachusetts when previously they had only been found within 30km of the coast. The hypothesis at that time was that it was "too cold" to allow for the full life cycle to proceed because those from coastal sites would poorly molt (taking 50% longer with less success) from larvae to nymphs at 15°C. Ticks from the Quabbin reservoir (west central Massachusetts) would do so without delay. When the Quabbin ticks were mated with ticks from Ipswich (northeastern coast), an "intermediate" molting temperature (one month at 18°C) was noted for the progeny (unpublished). This single experiment, not pursued further, suggests a genetically based trait upon which selection could act to allow for invasion of colder sites. It follows that upper Midwestern U.S. populations of *I. dammini* might be cold adapted and thus be able to develop at lesser temperatures. Range expansion of *Ixodes* spp. could be due to local selection for lineages that can adapt to differing temperatures, as opposed to being presumed as evidence for climate change's effects.

Edaphic factors comprise those related to soil, including plant communities that are regulated by and in turn influence soil composition. Early GIS studies suggested that *I. dammini* abundance was associated with well drained sandy soils with a low water table, and negatively associated with wetlands or saturated soils (Glass et al., 1994). Moder or duff mull humus, in which there is accumulation of organic residues due to slow action of litter dwelling animals and fungi, was associated with nymphal *I. ricinus* density (Goldstein et al., 2018). However, in a more comprehensive analysis, macro- and microhabitat drivers determined nymphal *I. ricinus* abundance, with distance between trees and abundance of trees and shrubs serving as influential explanatory variables (Ehrmann et al., 2017). Functional properties of the habitat (e.g. presence of berries) was more influential as a predictor than soil quality. Interestingly, the most influential single driver for nymphal abundance was larval abundance, which suggests that the reproductive hosts for these ticks (larger mammals, mainly deer) are a critical determinant for transmission (nymphs being the pivotal stage of the tick for maintaining *B. burgdorferi* s.l.). An excellent review of the role of edaphic factors in tick ecology (Burtis et al., 2019) makes the astute point that, although most research on ticks focuses on host seeking stages, the majority of the tick life cycle occurs while the tick is sessile (engorged, molting, diapaused, resting) and takes place within the soil or leaf litter.

A "social" component of abiotic factors, comprising landscape disturbance (natural disasters) and anthropogenic factors (sometimes classified in a category of its own, e.g.,(Morris et al., 2020) creates opportunity for range expansion for *B. burgdorferi*-infected ticks. The emergence of Lyme disease in eastern North America was due to reforestation, increased abundance of deer, and suburbanization (Spielman et al., 1985). 100 years ago, many of the northeastern U.S. sites were characterized by a treeless landscape. Abandonment of farms has led to reforestation, with shrubby early to mid-successional habitats giving way to forests on their way to maturity (Foster et al., 1998). Successional forests may differ with respect to fauna and flora, depending on the duration of succession; mature forests may even be less biodiverse than those in early to mid-succession (Brooks et al., 2012). Risk due to *I. persulcatus*-transmitted infections may relate to cycles of timber harvesting in taiga forests (E. Korenberg, personal communication; (Tupikova and Korenberg, 1965)). The taiga is clear cut and development of successional habitat promotes greater densities of ticks, due to increased activity of ruminants (tick reproductive hosts) that browse shrubs as well as activity by people that exposes them to ticks (Korenberg, 1994). Once the taiga is regenerated, the *Borrelia*-tick biocenosis is stable, but people are less exposed because they do not visit within the deep forest as frequently and such sites have fewer ungulates. Such a scenario may characterize the suburban sites that comprise human risk in the eastern U.S., where new suburban developments create early successional habitat that can promote dense tick infestations coincident with those of the generalist rodent invaders such as white footed mice, chipmunks, and squirrels. Deer benefit from the nitrogen rich plantings in suburbia, as well as protection from hunting (which is usually discouraged). Suburban sites are thus prone to develop dense deer tick infestations due to being plagued by abundant deer, which deposit engorged female ticks. Some of the range expansion of *I.dammini* and risk of Lyme disease in the northeastern and Middle Atlantic states in the past 30 years may relate to increased development; in Massachusetts, 40 acres of land (90% in forested sites) was lost each day between 1985 and 1999 (Breunig, 2003). The majority of the development was for large lot, new residential construction.

Biotic factors include food, habitat, predation, and microbial communities. Of these, habitat and predation would influence ticks directly; microbial communities would influence the spirochete, tick, and vertebrate hosts; and all 4 would influence the vertebrate hosts. The biotic factors influencing vertebrate hosts is beyond the scope of this review. Habitat interacts with temperature and relative humidity as well as host abundance to help determine tick abundance, *vide supra*. Little is known about predation on ticks; other than parasitic chalcid wasps (Collatz et al., 2011; Hu et al., 1993), there are no predators adapted to ticks. Ants, beetles, spiders and other generalist arthropods may opportunistically prey on ticks as may ground foraging birds. None of these are thought to regulate tick abundance. Indeed, on Naushon Island, Massachusetts (which was the site of a chalcid wasp release in 1928, (Larrousse et al., 1928), as many as 40% of the host seeking nymphal *I. dammini* was infected each transmission season by *Ixodiphagus hookeri*, the chalcid wasp, suggesting that the wasp could kill a large fraction of the nymphs without negative effects on the tick population (T.N. Mather, personal communication).

It is likely that stable enzootic cycles may not require many ticks. *B. burgdorferi* was enzootic at least a generation before it became zoonotic in the U.S. given evidence of infected mice from 1896 and ticks from 1942 (Marshall et al., 1994; Persing et al., 1990). At the turn of the 20th century, the northeastern U.S. landscape was largely farmland (Cronon, 1983). Other than small interspersed herds, deer were largely extirpated Kilpatrick and LaBonte, 2007). The emergence of Lyme disease as a public health problem during the latter half of the 20th century coincided with the abandonment of farmland and reforestation; an expansion of deer herds in density and distribution; and suburbanization (Spielman et al., 1985). Deer, in particular, were central to the development of the zoonotic condition because they serve as the main reproductive hosts for *I. dammini* (Wilson et al., 1988), and risk is associated with tick density. However, for perpetuation of *B. burgdorferi*, it seems likely that dense infestations of ticks may not be required. The Great Island deer reduction experiment, which demonstrated causality between deer abundance and that of *I. dammini*, also provides evidence that an 80% reduction in tick density did not alter the seroprevalence of *B. burgdorferi* in white-footed mice, although seroconversion of mice was delayed by a month (Brunet et al., 1995a). The less efficiently transmitted *B.microti*, however, became less prevalent on Great Island among mice and its genetic diversity declined (unpublished). Prior to its emergence as a zoonotic infection, *B. burgdorferi* must have been maintained in the northeastern U.S. in a generally deer-deficient landscape that could be perpetuated in the absence of hosts that allow the production of dense tick infestations.

Mixed borrelial infections

The actual dynamics of spirochetal infection pose questions for analyses of BRN. The quantum of infection (density of spirochetes) within ticks is not homogeneous. The estimated quantum within host seeking ticks ranges from a few hundred to tens of thousands of spirochetes per tick, with an average of 3,000 or so for adult ticks (Brunet et al., 1995b; Kovalevskii and Korenberg, 1995). However, individual ticks may contain as many as 20,000 (Burkot et al., 1994). The significance of such heavily infected outliers for BRN is not clear: humans might suffer greater pathology (assuming a dose dependence for disease), but reservoir hosts might be rendered more infectious to ticks or perhaps a greater inoculum might induce more rapid immune selection. The routine use of ultra-sensitive PCR assays also confounds parameterizing and validating models for BRN. The infectious dose for a mouse is thought to be on the order of a few dozen to a hundred spirochetes (Wang et al., 2003; Kern et al., 2011). Would a tick with a spirochete burden at the limit of detection have any relevance to BRN?

B. burgdorferi s.l. may exist only rarely nature as a single clone within a reservoir or tick (Nakao and Miyamoto, 1995; Postic et al., 1998; Qiu et al., 2002; Pérez et al., 2011). When two different strains of *B. burgdorferi* s.s. infected white footed mice, only one of the two was efficiently transmitted to xenodiagnostic ticks (Derdáková et al., 2004). Competition between *B.afzelii* strains manifested as reduced efficiency of infection of xenodiagnostic ticks as well as quantum of infection for either compared to single infections (Genné et al., 2018, 2019).The average quantum of infection for individual strains diminished as strain richness increased (Strandh and Råberg, 2015), suggesting that competition may act to limit the overall spirochetal burden to the tick (Durand et al., 2017). However, those strains that achieve greater infection intensity in either the reservoir or vector are more likely to be transmitted (Rego et al., 2014). The contribution of those "lesser" strains to BRN requires further analyses: although apparently less fit, nonetheless they exist in natural populations. Balancing selection may maintain such strains; they serve as variation upon which selection may act in a changing environment (new hosts or abiotic influences), ensuring BRN >1 over time.

Does infection influence the fitness of its hosts?

White footed mice infected experimentally with *B. burgdorferi* s.s. generally failed to demonstrate pathology (Wright and Nielson, 1990; Schwanz et al., 2011), although young mice sustained carditis and multifocal arthritis (Moody et al., 1994). Infected *P. leucopus* had 20% more larvae infesting them than non-infected mice (Ostfeld et al., 2018), suggesting that infection altered grooming efforts. Infection did not influence mouse survival in a capture mark release study (Voordouw et al., 2015). With *B. afzelii*, infection did not influence survival of *Myodes glareolus* (bank vole), a main host for this genospecies, and although infection did not affect reproductive success of female voles, males were less likely to pass on their genes (Cayol et al., 2018). Kittiwakes infected by *B. burgdorferi* s.l., presumably *B. garinii*, did not differ in survival rates from those treated with an antibiotic (Chambert et al., 2012). *B. burgdorferi* s.l. does not appreciably alter the fitness of its vertebrate host, which may have promoted its ecological success across the Holarctic. Additional analyses might focus on whether very young animals might be burdened given the carditis and arthritis that have been reported; recruitment influences host population density and demography. There is some data that *B. burgdorferi* s.l. may actually enhance the fitness of its vector: infected ticks were more capable of surviving saturation deficit than those that were not (Herrmann and Gern, 2010).

Concurrent infections

Like all animals, ticks have a microbiome. A diverse guild of microbial agents may be found in the *Ixodes* spp. that serve as vectors for *B. burgdorferi* s.l. in any site (Carpi et al., 2011; Vayssier-Taussat et al., 2013; Couper et al., 2019; Li et al., 2020). These microbiomes may differ among ticks sampled from diverse sites (Van Treuren et al., 2015) and may differ between developmental stages, with much of the diversity originating from the environment as opposed to the bloodmeal (Zolnik et al., 2016), reflecting the close contact that a tick has with soil and leaf litter. Some reassessment of the reported microbiomes may be required given the possibility of unconsidered procedural artefacts during whole genome analyses (Lejal et al., 2020): about half of the operational taxonomic units (OTUs) identified in host seeking *I. ricinus* could be attributed to contamination of laboratory reagents. Nonetheless, *B. burgdorferi* s.s. less efficiently colonized the guts of subadult *I. dammini* held within sterile containers compared to the same colony held in nonsterile containers (Narasimhan et al., 2014), and the same effect was seen when larvae were allowed to feed on mice treated with gentamicin and infected with a gentamicin resistant strain of *B. burgdorferi* s.s. Dysbiosed ticks engorged better (weighed more) than those that were held in nonsterile conditions or not treated with gentamicin but were less likely to be infected, suggesting that perturbation of the microbiota modified vector competence. Accordingly, differences in prevalence of infection as well as individual quanta among host seeking ticks may need interpreting in the context of the microbial environment. There are few reports of whether members of typical small mammal parasite guilds may influence the reservoir competence for *B. burgdorferi* s.l., but there may only be rare effects given that all animals in nature are concurrently infected by diverse viruses, protozoa, helminths, and ectoparasites ("polymicrobial"). For example, the effect of *Heligmosomoides* sp. (a common nematode infection of wild rodents) was analyzed by experimental infection. Nematode infection along with *I.ricinus* subadult infestation led to greater Th2 responses in a mouse model, but this did not affect susceptibility to *B.afzelii* (Maaz et al., 2016).

The "coinfecting" agents that have received the most attention are ones that may infect and cause disease in humans, viz. *B. miyamotoi*, *B. microti*, *B. divergens* species complex, *Anaplasma phagocytophilum*, *Ehrlichia muris*, and the tickborne encephalitis complex of viruses, among others. Although "coinfection" with *B. microti* may modify the course of Lyme disease (Krause et al., 1996), the ecological effects of concurrent or sequential infection with the zoonotic ones remain debatable, particularly when one considers that wild animals and ticks contain diverse microbiota that may be more influential than those that we focus on because of their zoonotic impact. Concurrent infection by *B. microti* and *B. burgdorferi* did not appreciably change the transmission efficiency of spirochetes in an experimental *P. leucopus* model (Mather et al., 1990). However, *B. burgdorferi* s.s. appears to alter the susceptibility and infectivity of *P. leucopus* for *B. microti* and *A. phagocytophilum* (Dunn et al., 2014; Levin and Fish, 2001). Trying to determine whether there are fitness effects of concurrent infection within naturally infected ticks (analyzing host-seeking ticks) would be confounded: if there were fewer borreliae within a tick that also contained *B. microti*, is this because the host upon which the larva fed had few spirochetes or is it because there was some inhibitory influence of *B. microti* on the spirochete within the tick? Analyzing the naturally infected vertebrate hosts is equally difficult to interpret because of the complex mix of variables (immunogenetics, age, maternal effect, nutritional status).

Factors that influence the distribution of B. burgdorferi *s.l.*

What limits the distribution of the borreliae-tick-vertebrate system? *B. burgdorferi* s.l. are found on all continents except Antarctica, Australia, and Africa (enzootic cycles are present in North Africa). Relatively recent transport by humans of infected hosts, e.g., by livestock or rats (Matuschka et al., 1996; Smith et al., 1993), is possible; *R. norvegicus* is reservoir competent for *B. burgdorferi* and can maintain the enzootic cycle. Mammals and their ectoparasites could have moved North to South (or perhaps the other way) during the Great American Biotic Interchange (Panamanian isthmus) in the late Miocene through Pleistocene. *I. scapularis* infests deer in Panama (Fairchild, G.B. et al., 1966). It is also possible that mammals introduced infection during the several Beringian migrations during the last glaciations (Goethert et al., 2006). The main explanation, though, may be the transport of ticks or infection by birds. The pelagic bird tick *I. uriae* maintains *B. garinii* across the world, and even in colder sites in the northern (Iceland, Bering Straits) and southern (Crozet Islands, Falkland Islands) hemispheres (Olsen et al., 1995), demonstrating the importance of bird transport. *B. burgdorferi* has greatly extended its range across the eastern U.S., with strong evidence for the role of migratory birds in transporting ticks, as well as infection (within ticks and with birds themselves being reservoir competent) (Battaly et al., 1987; Ogden et al., 2008; Smith et al., 1996). In fact, *B. afzelii* reactivates in birds that are stressed from migration (Gylfe et al., 2000), suggesting a mechanism for long distance transport of infection. In a recent comprehensive European study, 37% of over 2000 ticks removed from more than 800 birds of 28 species were infected by *B. burgdorferi* s.l., demonstrating not only their critical role in dispersal of the bacterium but also, interestingly, their role in spatial mixing of spirochetal genotypes. The rodent-associated genospecies, in contrast seem less diverse and more likely to be geographically restricted (Norte et al., 2020).

The evidence for enzootic *B. burgdorferi* s.l. in tropical sites includes that from Taiwan and southern mainland China, as well as the southern cone countries of South America. These mainly comprise associations of newly recognized genospecies and *Ixodes* spp. that are generally rodent-infesting but also will attach to humans (e.g., *B. valaisiana* and *I. granulatus* in Taiwan, (Chao et al., 2010); *B.yangtze* and *I. granulatus* in southwestern China, (Chu et al., 2008); *B. chilensis* and *I. stilesi* (Ivanova et al., 2014); *Borrelia* sp. and *I. pararicinus* (SarachoBottero et al., 2017). The influence of temperature on the stability of vector infection by the diverse *B. burgdorferi* s.l. may be an interesting focus of future inquiry, given that most attention has been placed on the *B. burgdorferi* s.l. of northern latitudes, which is less likely to persist in the vector at greater temperatures (Shih et al., 1995).

The ecology of B. burgdorferi *s.l.: a synopsis*

Although there is a main theme in zoonotic sites (*I.ricinus/persulcatus* group ticks feeding on small mammals and birds maintain the enzootic cycle), the complexity that is evident across its range cannot simply be characterized as a range of performances of a symphonic piece by a range of instruments from kazoo (Massachusetts islands) to full orchestra (central Europe). The adaptive plasticity of the spirochete and tick has generated unusual enzootic cycles such as *B. lusitaniae*, *I. inopinatus*, and sand lizards; *B. garinii*, *I. uriae*, and pelagic seabirds; and *B. californiae*, *I. jellisoni* and kangaroo rats, among others. These have developed into distinct symphonies, and it should not be unexpected when other interesting *Borrelia*-*Ixodes*-vertebrate host combinations are reported.

Soft tick transmitted relapsing fever spirochetes

Although as diverse as *B. burgdorferi* s.l., the relapsing fever borreliae have one general enzootic cycle, comprising that of nest dwelling (nidicolous), cryptic soft ticks and the small mammals that they feed on.There are, however, some variations on the general theme. In addition, there are two exceptions to the variations on a theme characterization: louse borne relapsing fever, and hard tick transmitted relapsing fever borreliae.

Diversity of tick-relapsing fever group borreliae associations.

The systematics of the soft ticks have recently been revised, with two major groupings being recognized, the Argasinae and the Ornithodorinae (Mans et al., 2019). Relapsing fever spirochetes are mainly maintained by ticks of the Ornithodorinae (bold type indicates genera/subgenera including a known vector), although this may reflect the research effort related to human or veterinary health. The Argasinae comprise the genera *Alveonasus*, ***Argas*** (includingthe subgenus *Persicargas*), *Navis*, *Ogadenus*, *Proknekalia* and *Secretargas*. The Ornithodorinae include *Carios* (including the subgenera *Alectorobius*, *Antricola*, ***Carios***, *Nothoaspis*, *Parantricola*, *Reticulinasus*, *Subparmatus*), *Chiropterargas*, *Ornithodoros* (subgenera *Microargas*, *Ornamentum*, ***Ornithodoros***, ***Pavlovskyella***, *Theriodoros*) and *Otobius*. Many of these genera are monotypic or comprise a small number of species that have very specialized host associations. On the one hand, such specialization does not lend itself to host switches, a main mode of diversification for microbial parasites, but, the focused feeding on unique hosts promotes vectorial capacity and perpetuation if a microbial agent becomes transmitted during bloodfeeding. It seems likely that novel borreliae-soft tick associations will be recognized when more of these ticks (particularly with adequate sampling) are analyzed.

Felsenfeld (1965, 1971) listed 11 Old World (includes *B. anserina*) and 6 New World species of soft-tick transmitted *Borrelia* spp., along with 6 others that he felt were either too incompletely described to warrant recognition or those that he considered as belonging to another valid species. Today, 16 Old World and 8 New World soft tick transmitted *Borrelia* spp. are recognized (Talagrand-Reboul et al., 2018). Whole genome analyses support the classical Old World/ New World dichotomy but suggest some diversity simply represents ecotypes of a single genospecies (Elbir et al., 2014). As with other parasitic organisms (e.g., *Babesia* spp. or trypanosomes), a "one vector-one species" concept influenced the early taxonomy of the soft tick-transmitted borreliae (Davis, 1942) and thus a spirochete detected in a newly examined species of tick was assumed to be novel and worthy of a name. Field collected ticks from diverse Western U.S. sites transmitted spirochetes by feeding on mice. Only colony derived ticks of the species from which an infection originated (*O. turicata*, *O. hersmii*, and *O. parkeri* were fed at the same time) became infected by those spirochetes. The specificity of these associations remain to be fully analyzed, although a simple experiment suggests that there may be a salivary gland barrier: *O. turicata* and *O. parkeri* became infected (*B. hermsii* colonized the gut and disseminated to the hemolymph of both these species) but failed to be transmitted by bite whereas with the homologous association *B. hermsii* was efficiently transmitted by *O. hermsii* (Schwan, 1996). The barrier, thus, may not be just the capacity for spirochetes to attach to and invade the salivary glands, but could involve the salivary pharmacopeia and its influence at the site of inoculation (Krishnavajhala et al., 2018). Some borreliae, such as *B. duttonii,* may be transmitted not only by bite (as subadults) but also by contact with fluid from the coxal glands (an excretory organ that facilitates diuresis after a bloodmeal).

Burrows, logs, bird or bat roosts, mammal nests, caves, huts, and cabins are the venues of transmission. In general, the soft tick-transmitted borreliae are perpetuated in small natural foci (Pavlovsky, 1966), which can comprise just a few burrows in a larger area that contains many burrows or one hut in a village. The limited mobility of the soft tick vector (typically nidicolous, or nest dwelling), and its short feeding durations means that transport to new burrows or huts is a rare event. The limited capacity of these ticks to move or be transported, however, is compensated for by their long lives, often years. If transported, there is a great probability that the introduction will persist due to broad feeding preferences and TOT.. Soft ticks have a great capacity to adapt to diverse vertebrate hosts (Mans et al., 2008): critical salivary anti-hemostatic anti-thrombin and anti-platelet aggregation activities are highly conserved between the phylogenetically basal *A. monolakensis* and *Ornithodoros* spp. However, the argasid salivary pharmacopeia appears to be less elegant than that of ixodid ticks; a prominent

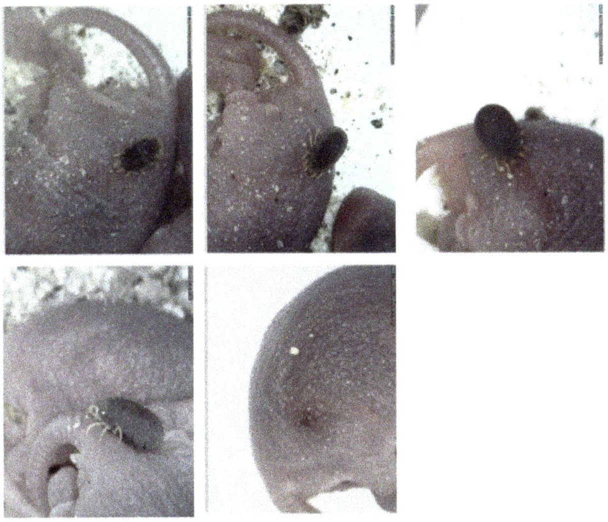

Figure 5. *O. hermsii* feeding on suckling mouse over 20 minutes, culminating in hemorrhagic feeding lesion.

hemorrhage or feeding lesion is left behind after many species of soft ticks feed (Figure 5). Then too, inheritance (TOT) tends to be the rule for the soft tick-transmitted borreliae, which suggests the possibility that a single infected fertilized female tick could initiate a natural focus. Thus, although transport to expand range is likely a rare event, when it happens introduction and development of a new soft tick-relapsing fever borreliae focus could be efficient.

As with studies of TOT for other vector-pathogen relationships, published reports of the efficiency of the process range from documenting that it does happen, to a more comprehensive analysis. The transovarial transmission rate (TOTR) is the percentage of infected females that transmit an agent to their progeny, without reference to how many of the progeny are actually infected. The filial infection rate (FIR) refers to the proportion of an infected female's progeny that are actually infected. The product of the FIR and TOTR is the vertical transmission rate (VTR), which is a measure of population prevalence and thus a more useful estimator for models of BRN (TOTR could simply be based on a few ticks that laid a few infected eggs, which might not survive to contribute to a new generation of infected ticks). There are very few reports that present data to estimate the VTR for the relapsing fever group borreliae.

Factors that influence the distribution of relapsing fever borreliae.

The ecology of soft ticks is not easily studied given their cryptic niche. They occupy crevices and burrows and are found under logs, rocks and bark; they are nocturnal or at least are active only under reduced light conditions. Few species have extended feeding durations and thus most are not to be collected from hosts. The standard methods for collecting hard ticks (dragging or flagging; trapping and examining vertebrate hosts) are not relevant. Some species, such as *O. coriaceus*, may be attracted by carbon dioxide and will accumulate nearby, to be collected by hand. Sentinel animals may be placed near potential tick hiding places and inspected frequently to find feeding ticks (Adeyeye and Butler, 1989). Most sampling methods comprise digging out or vacuuming a burrow and sifting the soil to look for ticks (soft ticks fluoresce under ultraviolet light, so sand collections may be easily screened by the use of a Woods lamp in a darkened room). Crevices or other hiding places within a house may be visually inspected and ticks collected with forceps. In short, the effort required for studying soft ticks is much more than for most hard ticks, leading to a paucity of published studies of their ecology. Surveys that serve as the basis for distribution maps tend to be biased by the inclusion of sites that come to our attention because of investigation of reports of human infection, which reflects zoonotic spillover due to local tick abundance; intense human exposureto sites where soft ticks are common; or investigator interest (e.g., the many contributions of the Rocky Mountain Laboratories to relapsing fever in the western U.S.).

A recent analysis using species distribution modeling of environmental variables suggests that the range of *O. turicata* (and potentially *B. turicatae*) in the U.S. is likely underestimated (Donaldson et al., 2016). Such an approach may allow for more targeted sampling efforts to fill in distributional gaps. Indeed, the distributional modeling results indicated possible overlap in the distributions *O. turicata* and *O. talaje*, and, in fact, the same woodrat and burrowing owl nests in central and southern Texas have yielded both species (Lopez et al., 2016). How the two species can occupy the same niche remains to be described.

One of the few long-term studies of the ecology of borreliae in relation to human risk for relapsing fever provides some general observations about their

Figure 6. *O. tartakovskyi* feeding on suckling mouse. All stages may feed simultaneously but may differ in feeding duration on a host.

distribution. Cases spatially clustered in southwestern Senegal near homes where *O. sonrai* had been collected and were most common in March(Vial et al., 2006a). Most cases were diagnosed about 4 months before the rainy season, not because the ticks were affected by the external conditions within their cryptic resting sites, but because small mammal populations were at their lowest abundance during the end of the dry season (Talagrand-Reboul et al., 2018). When small mammals are available, the ticks are more likely to feed on them, but when their density is low, humans become the source of bloodmeals. This pattern of small mammal scarcity (due to either seasonality or other factors leading to local declines) serving to focus bites on humans is seen with the paradigm of relapsing fever in the western U.S., where cabins left empty for the winter are suddenly occupied by human visitors (e.g., the Grand Canyon outbreaks (Paul et al., 2002). In such vacation cabins, rodents that nested there during the winter move to outdoor sites during the spring, leaving the long-lived ticks behind to attack new occupants (Wynns, 1942). The relevance of these epidemiological observations is that there is strong spatial clustering of transmission and the enzootic cycle is typically cryptic unless the small mammal hosts become scarce. This is as expected from the theory of natural focality, in which infectious agents such as the soft tick transmitted borreliae, generally exist in "nidi" or nests such as the burrows of rodents, which are "biocenoses" with optimal microhabitat and temperatures (Pavlovsky, 1966). Such nidi are relatively isolated from each other if the vector tick has limited mobility (due to rapid feeding, not being transported from a burrow) and humans never see the tick or infection unless they trespass into the nidus.

At the level of the deme, soft ticks generally have limited capacity to move between burrows or other nidi, although this is influenced by the duration of bloodfeeding attachment that characterizes a species. All stages (larvae, several nymphal stages, adults) are found in the same burrow (Figure 6). A capture-mark-release study of *O. turicata americana* in Florida demonstrated that marked ticks were never found to move from the gopher tortoise burrows into which they were released (Adeyeye and Butler, 1989); burrows within visual distance in a site could remain completely uninfested and in fact in one of the two study sites only 4 of 10 burrows contained ticks. *O. maritimus* infesting yellow-legged gulls, however, appeared to be able to move and recolonize empty nests, a conclusion based on microbial prevalence, which did not demonstrate spatial structure within the breeding site (Dupraz et al., 2017). Only 0.9% of marked *O. coriaceus* were found 15m away from where they were released (Garcia, 1963). At a regional scale, populations of *O. coriaceus* across California, Oregon and Nevada were highly structured with little evidence of gene flow as determined by mtDNA analysis (Teglas et al., 2005), suggesting that this tick has limited dispersal capacity despite the fact that its larvae may attach for many days to its highly vagile deer host. *O. sonrai* in Mauritania and Senegal demonstrated large diversity in 16S rDNA sequences, with 4 major subclades being detected; although there was some geographical clustering of subclades, in general genetic divergence did not correlate with increasing geographical distance and hence was interpreted as being inconsistent with the hypothesis that isolation is a function of low dispersal capacity (Vial et al., 2006b). Nonetheless, the main pattern of distribution of soft ticks is exemplified by long term micronidi that are determined by animal host resting behavior and microhabitat.

Fitness effects of infection

There is some evidence that the vertebrate hosts infested by soft ticks may be affected by tick feeding, but the influence of spirochetal infection on fitness remains poorly studied. Balearic storm petrel nestling survival was less in deeper aspects of the breeding cave where *O. maritimus* were dense, and individual nestling body weight was reduced for those with greater tick burdens (Sanz-Aguilar et al., 2020). Dense infestations of chickens by *A. persicus* reduce their egg laying capacity (Khan et al., 2001) and acute infections of fowl spirochetosis (*B. anserina*) can cause mortality. Sensitization of hosts by repeated feeding reduced *O. turicata* engorgement weights (Need and Butler, 1991), and fecundity of *O. moubata* was reduced (Chinzei and Minoura, 1988), suggesting possibilities for density dependent modulation of tick populations. Although *B. turicata* infected ticks remained infectious after 5 years of starvation (Francis, 1942), demonstrating remarkable longevity of both tick and spirochetes and general lack of effect on fitness, the study did not determine whether uninfected ticks could have lived even longer.

A poorly studied phenomenon, interference, could in some instances enhance the fitness of a vertebrate host that is infected by relapsing fever group borreliae. Trautman (Trautman, 1907) demonstrated that co-inoculation of a spirochete, presumably *B. duttonii*, influenced the parasitemia and disease due to *Trypanosoma vivax* and *T. gambiense* in experimentally infected mice. The effects were not observed when killed spirochetes were administered. Subsequent studies confirmed these results using other trypanosomes, but no effects were seen with plasmodia, rickettsiae, or leptospira. Trautmann did not specifically analyze the effects of trypanosome infection on borreliosis but statedhe had no doubt there was "egalement une action sur les spirilles". The generality of interference has been demonstrated by sporadic reports, for example, interference of a mouse bartonella by a hemoplasma (Eperythrozoon) was subsequently demonstrated by Tyzzer (Tyzzer, 1941), with the former disappearing from circulation when the latter was detected. This poorly studied phenomenon may have implications for the perpetuation of relapsing fever borreliae given the ubiquity of trypanosomes as vertebrate parasites.

East Africa and endemic tick borne relapsing fever

Obermeier found "a most delicate and peculiarly motile spirillum in the blood of relapsing fever patients", seen in samples taken and examined microscopically at bedside in Virchow's Berlin clinic in 1867. Berlin had been experiencing an outbreak of what is now known as epidemic or louse borne relapsing fever (Birkhaug, 1942). The discovery of the cause of endemic relapsing fever (African tick borne relapsing fever, ATBRF) is a story from the times of colonial exploration, with David Livingstone noting in 1857 that people bitten by ticks in Angola and Mozambique sustained an irregular fever (Geigy, 1968a).The vector of *B. duttoni*, the eyeless tampan *O. moubata* was identified in 1904 by Joseph E. Dutton who with John L. Todd were able to infect monkeys with relapsing fever by the bites of soft ticks; Dutton and Todd both became infected during the course of autopsies of relapsing fever fatalities and Dutton died of the disease. Robert Koch, at the same time, independently infected monkeys by the bites of soft ticks. It is not known whether any of these investigators were aware of the report (Marchaux and Salimbeni, 1903) demonstrating that *Argas miniatus* transmitted fowl spirochetosis.

It is possible, though, that the local inhabitants of East Africa recognized cause and effect long before scientific inquiry, as suggested by Geigy's (Geigy, 1968a) anecdote about soft ticks being carried by native travelers: "The author of this chapter can confirm this observation from personal experience. In 1957 in Tanzania, he met some native laborers who had come from Rwanda to work on a sisal estate near Morogoro. They carried with them *O. moubata* ticks from their own huts that they kept in matchboxes and allowed them to feed regularly on their skin...they explained that they did so in order not to contract relapsing fever on returning to their homes in Rwanda after a fairly long absence. They obviously knew from experience that a certain degree of protection against their local strain of borreliae could be achieved in this way. Unfortunately, the author could not find out whether their ticks were really infected or not because they refused to hand over these precious creatures." Thus, the distribution of *O. moubata* could have been heavily influenced by human transport. On the other hand, Geigy discounted human transport as a factor in the distribution of *O. moubata* and endemic tick borne relapsing fever, and speculated that this tick adapted from inhabiting the burrows of warthog and aardvark to human dwellings because "infested huts are practically always found where warthogs also occur and are hunted" and "conditions in a mud hut are strikingly similar to those in an aardvark burrow".

O. moubata is considered a superspecies or species complex with "races" adapted to wild or human domestic life cycles: to tortoises (*O. compactus*), to porcupines (*O. apertus*), and two races (*O. porcinus* and *O. domesticus*) found with warthogs and domestic pigs, respectively (Walton, 1962). The aardvark may be the original host for the *O. moubata* complex given that warthog abundance is apparently associated with that of aardvark burrows (Whittington-Jones et al., 2011). The great interchange of animals between burrows may have led to specific adaptations by the tick (*e.g.*, extended feeding on poikilothermic tortoises). The most parsimonious hypothesis to explain the apparent anthropophily of *O. moubata* sensu stricto is that the proximity of humans and their pigs to warthogs has selected tick lineages that adapted to living within huts, and were subsequently transported by humans, either incidentally or intentionally(Heisch and Grainger, 1950).

Although *O. moubata* has a wide host range, feeding on tortoises, birds (including domestic chickens), aardvark, porcupine, and warthogs or domestic pigs, ATBRF is not a zoonosis, but an anthroponosis, with humans as an amplification host. The classification of ATBRF as an anthroponosis rests on a general failure over the first half of the 20th century to detect *B.duttoni* in any animal other than humans. Experimental studies failed to infect warthogs with laboratory strains of *B. duttonii*; ticks collected from warthog burrows were never found to be infected, and subinoculation of warthog tissue samples into mice failed to demonstrate infection (Geigy, 1968b). This longstanding dogma has been challenged based on PCR detection of *B. duttoni* sequences in chickens and pigs sampled from ATBRF endemic villages (McCall et al., 2007). One might argue that PCR is a significantly more sensitive method to determine infection status of animals, certainly more so than microscopy of blood smears, but the gold standard for older workers was inoculation of diagnostic blood or tissues (often 0.5-1mL volumes) into laboratory rodents and assessing them for evidence of infection. At least with infection, the question of anthroponosis vs. zoonosis bears on interpreting the existing studies of the ecology of ATBRF as well as on planning future studies: a zoonosis is more complicated than is an anthroponosis, particularly to probe for critical BRN factors.

The life cycle of *O. moubata* has been studied by a number of researchers, and was nicely summarized by Hoogstraal (Hoogstraal, 1956). A female may lay as many as 7 egg batches (7 bloodmeals), totaling 500-1200 eggs during its lifetime. The larva that emerges from an egg is nonparasitic and nonmotile and molts to the first nymphal stage, without feeding, within days. Apparently, there is a reddish, hematin-like residue within larvae that may serve as a nutrient, suggesting that elements of the female's bloodmeal have been incorporated during oogenesis. There are 4-8 nymphal instars (a bloodmeal is required for each) before they become adults. The duration of feeding for the nymph ranges from 13-87 minutes, with later instars generally taking longer to feed. The entire life cycle may take as few as 62-73 days in the laboratory. The longevity of an adult tick is usually on the order of a year and a half, but maximal durations of 4-13 years have been reported.

Parthenogenetic colonies of *O. moubata* have been reported, with one well documented report (Davis, 1951) that assured a lack of fertilization by males. Progeny were all female, and all life cycle events (eclosion, molting, oviposition) greatly delayed relative to those that were sexually reproducing. The parthenogenetic females, when mated, produced progeny of both sexes. Whether all populations of *O. moubata* are capable of parthenogenesis remains undescribed. Given efficient inheritance of *B. duttonii*, which may last through 5 generations, albeit with reductions in transovarial transmission rate (Geigy and Aeschlimann, 1964), there is much potential for introduction to new sites as well as maintenance within sites with episodic presence of amplifying hosts.

O. moubata is apparently absent from areas with high humidity, although human living conditions may allow for colonization of sites where they would otherwise be excluded. In the laboratory, a saturated atmosphere as well as greater temperature reduces fitness (Cunliffe and Nuttall, 1921). In Kenya, some tribes construct houses with dry raised mud beds, and fires near the beds would dry the ground enough to promote tick survival (Walton, 1950). Similarly, *O. moubata* may be found in the Transvaal Highveld where there are very cold winters, simply because people have fires within their huts to provide heat (Hoogstraal, 1956). Accordingly, these ticks have interacted with humans to adapt locally and, thus, ATBRF is not only an anthroponosis but also an anthropogenic infection.

Fowl spirochetosis

B. anserina seems to be restricted to birds, with successful experimental infections of diverse birds but not to rodents or primates (Lisboa et al., 2009; McNeil et al., 1949). Human serum lyses *B. anserina*, helping to explain why zoonotic infection has not been reported even though fowl spirochetosis and *A. persicus* are widely distributed and can readily attain dense infestations within chicken coops, thereby potentially exposing the human caretakers of the birds. Although other *Argas* spp. are known to bite humans, *A. persicus* rarely bite people (Hoogstraal, 1956). The ecology of natural infestations of *A. persicus* and *B.anserina* (as opposed to agricultural settings) is dependent on breeding sites for their hosts, for example, birds that construct large permanent nests near where chickens are actively produced may sustain dense infestations. Heron rookeries near Cairo were found to have tens of thousands of *A. persicus* hiding in crevices and under bark of the fig trees where the herons roosted (Hoogstraal, 1956). Sporadic specimens have been found on diverse birds (e.g., quail, guinea fowl, vultures, cranes) but reports from mammals are rare. Why *B. anserina* has not been reported from wild birds is not known, particularly given the fact that many other *Argas* spp. such as *A. miniatus* are competent vectors (Lisboa et al., 2009).

A. persicus larvae have an extended feeding duration of 5-10 days, during which they could be transported; nymphs and adults take fewer than 2 hours to engorge. All feeding is done under reduced light conditions (Hooker et al., 1912). Unfed adults generally live a year, but larvae may survive over 200 days at 21-24°C and >85% relative humidity, although longevity was reduced to 50 days at 37°C at the same humidity. One estimate for the vertical transmission rate (VTR) for *B. anserine* was 67% (Zaher et al., 1977). Allowing infested chicken coops to remain uninhabited as a means of eliminating an infestation, therefore, would require an extended duration dependent on temperature. This would imply that peri-domestic infestations (*e.g.*, in a heron rookery) would likelypersist if nests were abandoned for the migratory season. The global distribution of this tick seems limited only by a minimum 20°C temperature for development. As with other pest infestations of domesticated animals, the increasing availability and use of antiparasitics as well as improved biosecurity and a trend for large-scale poultry production have been reducing the agricultural impact of fowl spirochetosis. *B. anserina* continues to affect poultry production in subsistence settings (Lisboa et al., 2009).

Borrelia coriaceae and O. coriaceus

Spirochetes were discovered in *O. coriaceae* during early investigations of Lyme disease in northern California (Lane et al., 1985), differing morphologically and behaviorally from the Lyme borreliosis agent and was described as *B. coriaceae* Johnson et al., 1987). At that time, *B. coriaceae* was considered a likely agent of epizootic bovine abortion, a condition that had long been epidemiologically linked to bites of this tick; the agent is now suggested to be a deltaproteobacterium transmitted by *O. coriaceae*. *O. coriaceae*had previously been demonstrated to be incompetent vectors for relapsing fever borreliae (Herms and Wheeler, 1935) and was known only for its painful bites. The 'pajaroello', as it was commonly known (Herms, 1916), was the subject of a comprehensive ecology-focused doctoral dissertation by Garcia (Garcia, 1963), and the following life history details are taken from that source. This is an unusual soft tick that feeds on large mammals such as the black tailed deer as well as cattle along the California coast down into Mexico. Although most feeding is thought to occur when deer or other large mammals bed down under shrubs such as manzanita, these ticks will move up to 25 feet along carbon dioxide gradients to the source. Body heat causes them to accumulate. Unlike most soft ticks, *O. coriaceus* is only weakly negatively phototropic and will move in sunlight. In one of the very few direct observations of predation on ticks, Garcia observed fence lizards (*Sceloporus occidentalis*) eating nymphs and confirmed this by examining gut contents of the lizards. The ticks have a characteristic response to being touched, which was to become immobile with the legs tucked underneath; he speculated that reduces predation by animals that targeted moving prey. The lower temperature limit to eggs eclosing (hatching) is 15°C and an upper limit 45°C, with high mortality at higher temperatures and low relative humidity. The larval stage has optimal survival between 20-25°C and >70% relative humidity. All stages may be active during all times of the year; rain and low temperatures inhibit activity. Deer are the main bloodmeal sources, and the ticks rapidly feed (an average of 20 or so minutes). Larvae, however, may be associated with the host for longer durations (an average of 9 days) and, thus, could be dispersed to new sites by deer. Coxal fluid is expressed off the host and thus transmission of *B. coriaceae* is via the saliva. Transovarial transmission appears to be inefficient, with a VTR of only 1.5%, indicating the requirement for a vertebrate amplifying host (Lane and Manweiler, 1988); more than half of deer that were sampled during the course of Lyme borreliosis studies in northern California demonstrated spirochetemias, likely due to *B. coriaceae* (Lane and Burgdorfer, 1988). Infection did not affect longevity or fecundity of the ticks (Lane and Manweiler, 1988). Like many soft tick species, adult *O. coriaceus* could survive as long as 5 years (Smith, 1944).

Hard tick transmitted relapsing fever spirochetes

Sir Arnold Theiler was the first to demonstrate that a spirochete was transmitted by hard ticks (Theiler, 1902); he demonstrated its presence in cattle during studies of theileriosis. He apparently thought little of the finding, particularly given that there was no pathology attributable to this spirochete (bovine borreliosis comprises fever, anemia, and transient malaise). Laveran named the organism *Spirillum theileri* from slides that Theiler had sent him. Theiler suspected that ticks served as vectors in his original report, noting that the borreliae accompanied piroplasm infections, and subsequently (Theiler, 1905) transmitted this infection to cattle using *Rhipicephalus (Boophilus) decoloratus*. The spirochete must be inherited (TOT) because all 3 motile life stages (larva, nymph, adult) of *R. decoloratus* develop on the same individual animal. Fed female ticks dropped from the host and eggs were laid in the environment; larvae emerged, attached to a new cow, fed to repletion, and molted in situ within a matter of days to the nymphal stage. The nymph, without moving (or moving very little) attached, fed, and developed in situ to the adult stage. Male ticks may move on the host to find a female. Theiler was working on theileriosis and babesiosis and knew about Theobald Smith and Fred Kilborne's pioneering incrimination of *R. annulatus*, a very similar one host cattle tick, as the vector of Texas cattle fever (due to *Babesia bigemina*). Inheritance was critical to perpetuating that parasite and Theiler also found that the borreliae were passed from the infected female through the egg to the larva. TOTR ranged from 70%-80% over 3 generations (Smith et al., 1978), confirming that inheritance is likely to be the main mode of perpetuation. Interestingly, although Theiler stated that larvae were infectious, subsequent work (Callow, 1967; Smith and Rogers, 1998) failed to confirm transmission by larvae. Smith and Rogers argued that even though larvae were infected, as demonstrated by detecting spirochetes within larval tissues, perhaps only nymphs were able to deliver sufficientinocula.

B.theileri has since been reported from many parts of Africa, Australia, and South America and undoubtedly has been transported globally with cattle as have their vectors, mainly *R. microplus* and *R. decoloratus*. Sheep, but not goats, horses, dogs or rodents were experimentally infected by blood subinoculation (Theiler, 1902). Deer can be infected by *B. theileri* (Callow, 1967), although spirochetemias are not detectable by microscopy. Evidence of infection was obtained by subinoculating blood into cattle, which became infected.The possibility that deer may become infected is of relevance given that they are naturally infested by *R..annulatus* in Latin America, and thus might serve as wildlife reservoirs for the agent of bovine borreliosis. Thus, although perpetuation of *B. theileri* mainly depends on lack of tick control in domestic animal production, an ancillary enzootic cycle between ticks and deer would perpetuate *B. theileri.* Indeed, in American sites (Texas, Florida) where *R. microplus* and *R. annulatus* were eradicated from cattle by the U.S. Department of Agriculture, infestations continued to appear due to maintenance of these ticks in the wild by deer (Pound et al., 2009). Whether deer maintain *B. theileri* remains to be explored.

Lyme disease-like illnesses of undocumented etiology have been confused with Lyme disease in sites where Lone Star ticks (*Amblyomma americanum*) are found (Campbell et al., 1995; Kirkland et al., 1997). This syndrome, termed Masters' disease (Armstrong et al., 1996) or southern tick-associated rash-illness, STARI, is characterized by an erythema migrans rash from which no evidence of *B. burgdorferi* may be detected, accompanied by generally mild constitutional symptoms with no chronic sequelae. Reactivity to *B. burgdorferi* antigens seems rare in sera from these patients (Campbell et al., 1995) but see Master and Donnell (Masters and Donnell, 1995). Lone Star ticks are not competent vectors for *B. burgdorferi*

(Piesman et al., 1979; Piesman and Sinsky, 1988) but are infected by *B. lonestari* (Armstrong et al., 1996; Barbour et al., 1996), which are closely related to *B.theileri* (Rich et al., 2001). The hypothesis that remains to be adequately tested is that *B. lonestari* is the etiologic agent of Masters' disease; definitive evidence for or against the hypothesis has not been reported. The prevalence of infection in Lone Star ticks ranges from 0-5% (Armstrong et al., 1996; Barbour et al., 1996; Kocan et al., 1992; Schulze et al., 1986), but given the proportionally greater densities of these ticks in endemic sites, compared with deer ticks, transmission risk indices for *B. burgdorferi* and *B. lonestari* differ only by 50% (Armstrong et al., 1996).

Deer serve as the main hosts for all stages of Lone Star ticks (Hair and Bowman, 1986), although raccoons, opossums, fox, coyote, rabbit, and turkeys are infested by subadult stages. Experimental inoculation of deer with in vitro cultivated *B. lonestari* produced spirochetemia that was detectable by blood smears and PCR (Moyer et al., 2006). About 15% of deer contained *B. lonestari* antibodies as detected by indirect immunofluorescence antibody testing, with a greater proportion seropositive from southern sites than in northern sites, corresponding to relative infestations of Lone Star ticks (Murdock et al., 2009). In a survey of deer in Arkansas, 11% of blood samples from hunter killed deer contained *B.lonestari* DNA (Fryxell et al., 2012). Although it is possible that deer are simply sentinels (exposed to bites of Lone Star ticks), the available evidence is suggestive of their role as an amplification host. *B.lonestari* was detected by PCR in turkeys, suggesting another possible amplification host (Jordan et al., 2009). *B. lonestari* infects host-seeking larval *A. americanum* (Stromdahl et al., 2003), demonstrating that TOT contributes to perpetuation. The stability of *B. lonestari* inheritance, however, has not been measured, and, thus, it is possible that amplifying hosts such as deer or turkeys, both important hosts for maintaining Lone Star tick infestations, are required for perpetuation.

In the early 1990s, BSK cultivated isolates from *I. persulcatus* collected from vegetation in Hokkaido, Japan were determined to differ from *B. burgdorferi* s.l. by DNA-DNA hybridization and by sequencing the 16S rDNA. The newly recognized spirochete, which grouped with the relapsing fever borreliae, was named *B. miyamotoi* in honor of Kenji Miyamoto, who had conducted the tick surveys in Lyme disease endemic sites in Hokkaido, and who had isolated the spirochetes (Fukunaga et al., 1995). This spirochete subsequently was found to infect *I.dammini* ticks used for experimental purposes (Scoles et al., 2001), and the comprehensive studies that followed determined that *P. leucopus* were competent reservoirs. 1.9%-2.5% of host seeking *I.dammini* from Lyme disease endemic sites in New York, Connecticut, Rhode Island and New Jersey were infected and infected female ticks provided filial infection rates of 6%-73%. A study of engorged female ticks removed from deer in 4 different states demonstrated a TOTR of 57%-100%, with FIR of 68%-100%, resulting in a VTR of 44%-93% (Han et al., 2019). Thus, inheritance is likely the main mode of perpetuation if VTR remains stable through additional generations.

Human infection and disease due to *B. miyamotoi* have been reported from Russia, Japan, the northeastern U.S., upper Midwestern U.S., and E. U. countries. *B. miyamotoi* has been detected in host seeking *Ixodes* spp. wherever surveys have been done in Lyme disease endemic sites e.g. (Crowder et al., 2014; Dibernardo et al., 2014; Fomenko et al., 2010; Hamšíková Svitálková et al., 2017; Richter et al., 2003; Wagemakers et al., 2017). Prevalence of infection in host seeking ticks has been determined to be on the order of 1%-2%, although there are reports of greater prevalences, which may represent sampling bias ("luck" of collecting many individuals that may have fed on a single spirochetemic animal or were the progeny of a single infected female).

At least 3 genotypes have been identified, suggesting that, like *B. burgdorferi* s.l., *B. miyamotoi* may comprise a species complex: Asian/Siberian, European, and American, mainly from ticks of the *I.ricinus* complex. *I.* (*Partipalpiger*) *ovatus*, a bird feeding eastern Asian tick, has been reported to be infected by the Asian genotype and what appears to be a fourth genotype of *B. miyamotoi* (Iwabu-Itoh et al., 2017). The Asian genotype of *B. miyamotoi* has also been reported from a single host-seeking *Haemaphysalis concinna* (Jiang, 2018). The detection of *B. miyamotoi* from *H. longicornis* or *H. inermis* removed from hosts (Pere David deer, woodmouse) (Heglasová et al., 2020; Yang, 2018) does not necessarily imply their contribution to perpetuation because these ticks may have acquired infection from the host from which they were removed.

Like *B. burgdorferi* s.l., the reported host range for *B. miyamotoi* is diverse. Deer likely serve as amplification hosts (Han et al., 2016), and the spirochete has been detected from wild boar (Wodecka and Skotarczak, 2016). Surveys of animals demonstrated that rodents (woodmice, deer mice, voles, chipmunk and squirrel (Hamer et al., 2012; Burri et al., 2014; Hamšíková Svitálková et al., 2017; Wagemakers et al., 2017), hedgehog (Jahfari et al., 2017) and raccoon (Barbour et al., 2009) were infected, as were diverse passerine birds (Hamer et al., 2012; Heylen et al., 2017; Wagemakers et al., 2017). A large proportion (60%) of turkeys from Tennessee (Scott et al., 2010) were considered to be infected. The wide range of possible amplification hosts reflects the wide host range of the *I. ricinus* species complex that serve as the main vectors for *B. burgdorferi* s.l. and *B. miyamotoi*. Concurrent infection of the two kinds of spirochetes appears to be rare, on the order of 0.1%-1% and no more than expected by chance (Hamer et al., 2012; Lynn et al., 2018). Indeed, the case has been made that the two spirochetes coexist using one vector by means of niche partitioning (Barbour et al., 2009). In the enzootic cycle with *I. dammini* and *P. leucopus* in New England, *B. burgdorferi* establishes chronic infection of the skin and the pivotal event in the enzootic cycle is transmission by nymphal ticks. *B. miyamotoi* has transient spirochetemia (blood) and transmission is by larvae (Barbour et al., 2009). Even a single *B.miyamotoi*-infected larva is sufficient to infect a host (Breuner et al., 2018).

Louse borne relapsing fever

Only two *Borrelia* spp. are anthroponotic, one causing louse borne relapsing fever (also known as epidemic relapsing fever) and one endemic tick-borne relapsing fever in Africa. The former is transmitted by the human louse (*Pediculus humanus*) and derives from endemic (African) tick borne relapsing fever spirochetes (*B. duttoni*) maintained by *O. moubata*-like ticks as an anthroponosis. The human louse has classically been distinguished into head louse (*Pediculus humanus capitis*) and body louse (*P. humanus corporis* or *P. humanus humanus*) based on morphology and feeding behavior. Body lice are not found in the hair but head lice could sometimes be found on the body. The two types were capable of interbreeding and both could be infected by *B. recurrentis* (Buxton, 1946).

As for the other louse-transmitted bacterial infections (trench fever and epidemic typhus), the agent of epidemic relapsing fever (*B.recurrentis*) contains a reduced genome relative to that of its tick maintained parent: 20.4% of the genome has been lost compared with *B.duttoni*, including the genes for the antigenic lipoproteins vlp and vsp as well as DNA repair genes *recA*, *mutS*, and *smf* (Elbir et al., 2014). The two borreliae apparently diverged 6-14 million years ago as determined by 16S rDNA comparison, consistent with the divergence of *P. humanus* from the chimpanzee infesting louse *P. schaeffi* (Reed et al., 2007). This estimate conflicts with the current evidence that body lice have originated multiple times, even within recent history, from head lice (Li et al., 2010), and that new populations may be constantly evolving from head lice. Body lice and head lice are the same species but are different ecotypes (subpopulations that specialize in different niches). It may be that as more *B.recurrentis* isolates are analyzed, the theory of an older divergence will become modified by the findings of more recent generation of local lineages from *B. duttoni.* Indeed, body lice were demonstrated to transmit *B. duttoni* (Heisch and Garnham, 1948), and the spirochete could be passaged indefinitely between lice by enema or intracoelomic inoculation (Mooser and Weyer 1954, as cited in (Geigy, 1968a) with no loss of infectivity to mice.

The perpetuation of *B.recurrentis* depends solely on its human reservoir. The spirochete's distribution, abundance, and relationships with the environment are regulated by those of its louse vector, which, in turn, is regulated by human ecology. Lice become infected only by feeding on people who have active spirochetemia. Apparently, a minimum of 4 spirochetes per microliter (1-2 per oil immersion field; (Heisch et al., 1960) is required to infect a louse. Only about a fifth of lice feeding on infectious people become infected. Within a week of an infectious bloodmeal, spirochetes may be observed in the louse hemolymph. They do not invade the ovarial tissues, hence there is no inherited infection (no transovarial transmission (Nicolle et al., 1912). There is a fitness cost to infection by *B. recurrentis* (and *Rickettsia prowazekii*) in that infected lice tend to die quickly (Houhamdi and Raoult, 2005). Individual lice live 3-7 days (at 30°C and 10°C respectively (Nuttall, 1917) in the absence of a bloodmeal and do not enter any dormancy that might prolong spirochetal survival in the absence of a live host. Thus, the louse is not the reservoir, and maintenance depends on humans.

The main mode of transmission is by crushing the infected louse, followed by the penetration of excoriated skin by the liberated spirochetes; *B. recurrentis* does not appear to invade the salivary glands and, thus, is not transmitted by bite. Although classic reviews on the biology of relapsing fever (Felsenfeld, 1971) state that louse feces are not infectious, rabbit adapted human lice infected by feeding on a rabbit that had been intravenously injected with 6 logs of *B. recurrentis* as they fed excreted viable spirochetes (Houhamdi and Raoult, 2005). The duration of infectivity for excreta is not clearly defined, and so, in the absence of extended survival of the spirochetes in dried louse excreta, the general role of fomites in maintaining *B. recurrentis* is likely minor; healthcare workers attending to victims of epidemic typhus often got infected by exposure to louse feces, whereas those attending relapsing fever patients did not (Zinsser, 1935).

Lice can be transferred between people by the social touching of heads (Rózsa and Apari, 2012), although there is some confusion as to whether this pertains to head lice or body lice), sharing clothing or bedding, by migration of lice from "verminous people" (Nuttall, 1917), or perhaps by deliberate infestation. Malabar holy men once apparently voluntarily let themselves be infested; young women in northern Siberia threw lice at visitors (Zinsser, 1935).Other than interesting cultural practices, prolonged close contact is required to maintain lice, such as under conditions of war, migration or during winter. Louse-borne relapsing fever is less common and more geographically restricted due to improved hygiene. Once upon a time, louse borne relapsing fever was likely cosmopolitan in distribution, traveling with people on ships and on war marches as did epidemic typhus (Zinsser, 1935). The Horn of Africa continues to remain endemic for this infection(de Jong et al., 1995; Nordmann et al., 2018; Yimer et al., 2014), with sporadic outbreaks. Large outbreaks may continue to occur where there are large accumulations of displaced people and a breakdown in hygiene and healthcare. In Sudan refugee camps, a 10% mortality was estimated to have occurred in 1999-2000 (Raoult and Roux, 1999). No country is immune from potential outbreaks: the EU has reported numerous louse borne relapsing fever cases in immigrants from the Horn of Africa, with spillover infections in people from other countries sharing the same migrant camps (ECDC(2015).

Comments on how to advance our knowledge of the "ecology of borreliae".

An ecologist will use an intervention to probe hypotheses about perpetuation; an epidemiologist will seek to discover the main elements of perpetuation as it relates to risk to either explain the public health burden (distribution and abundance of human infections), or formulate interventions, with the expectation that the results will reduce risk. Much of the past and current "Lyme disease ecology" literature is based on observational studies driven by epidemiology. We anticipate that the next version of this chapter will have a greater number of experimental ecology reports from which to draw. The new tools that are available will facilitate perturbation of nature. For example, Mice Against Ticks (MAT), a project that seeks to replace reservoir competent *P. leucopus* in zoonotic sites with those that are genetically modified to be less infectious to ticks (Buchthal et al., 2019), will provide the means to analyze whether there are essential species for enzootic *B. burgdorferi* transmission, as well as reduce BRN overflow. Other possibilities include the introduction into sites of ticks containing borreliae with genetic markers (e.g., GFP), although (as with MAT) there would be substantive ethical and regulatory issues that would require resolution.

One of the most vexing aspects of quantifying enzootic transmission (and entomological risk for zoonotic transmission) of *B. burgdorferi* s.l., as well as other hard tick transmitted borreliae, is the reliance on a single index: the entomological inoculation rate (EIR), or density of infected nymphs. This is the product of the prevalence of infection and the density of nymphs; much energy and resources are expended in the measurement of these indices without much thought into what they mean. EIR has its origins in malarial epidemiology and with caveats has some uses in measuring zoonotic risk of tick borne infections, mainly at larger scales and not at that of yards or small study sites (Eisen and Eisen, 2016). However, the utility of EIR for analyzing enzootic transmission is not axiomatic. Measuring the density of the tick vectors for *B. burgdorferi* is not straightforward. Although dragging (flagging, sweeping) to collect host seeking ticks provides an index of abundance, such an index is prone to error because the availability of ticks to be sampled depends on the abundance of hosts (Ginsberg and Ewing, 1989), and is affected by temperature and humidity at the time of sampling, accessibility of suitable habitat, and even inter-investigator prowess.

Indeed, sites are often selected because they yield many ticks as opposed to others (sample size is critical for measuring prevalence of infection), and are easier to traverse. Many sites would not be representative of what a mouse, shrew or bird would encounter; such animals go where we cannot. Standardizing transect selection, conditions and methods can reduce the variability that otherwise plagues measuring tick density, but drag sampling is a human risk-biased snapshot of the venue of enzootic transmission. Then too, all ticks from the transect (typically 500m or so, with a minimum of 20-30 minutes, usually 60 minutes of sampling) are placed into a single collection vial, ignoring the clustering of ticks and even infected ticks (Foppa et al., 2002; Telford et al., 1992); most 20m drags yield 0 or 1 ticks, and a small number >3, and an even smaller number may have dozens per drag. The clusters themselves provide a clue to the venue of transmission: subadult ticks engorge and drop from their hosts at the start of the host's activity (Mather and Spielman, 1986) and may accumulate in a mouse burrow, for example. By ignoring the clusters and averaging across a transect, we may be missing even more of the complexity of enzootic transmission, as we found in a study of focality for *Francisella tularensis* infection of *Dermacentor variabilis*(Goethert and Telford, 2009); bacterial diversity appeared to originate from a 260m diameter focus along a 1500m transect. We need to find better and complementary ways of estimating and interpreting infected and noninfected host seeking tick abundance, particularly where animals are most commonly encountering ticks.

Infestation of animals provides less biased indices of abundance for ixodid ticks, although the indices depend on number of animals examined as well as investigator experience in counting ticks on animals. Capture-mark-release is the standard method for censusing animal populations but to date has not been effectively used for *Ixodes* ticks (methods of marking that minimize fitness effects but are durable are expensive, e.g., the use of rare metals such as rubidium ingested in a bloodmeal, and would only provide estimates for host-seeking ticks but not those that are in the process of molting (which are difficult to collect). Microcosm studies (e.g., (Arsnoe et al., 2019) can provide good data on survival and questing behavior, but it may be logistically difficult to use microcosms to study the abundance of ticks or transmission of borreliae. The force of enzootic spirochetal transmission may be estimated by seroprevalence (or better, seroconversion), or direct detection of the agent by culture or PCR of animal skin samples. Xenodiagnosis is the best of all, but extremely labor intensive. Of all the current methods for analyzing BRN, measuring the infection status of enzootic hosts is the least biased if performed longitudinally. More objective means of sampling host seeking ticks would provide the complementary aspects of modeling BRN.

Acknowledgements

We are humbled by the opportunity to replace the previous version of this chapter, written by Joe Piesman and Tom Schwan, who have contributed so much basic knowledge on the transmission of borreliae. Joe is to blame for mentoring the senior author as a beginning graduate student and introducing him to borreliology as well as babesiology. Tom is the authority on the interaction of relapsing fever borreliae with soft ticks, has served as a role model for the ideal mix of lab and field investigator, and in particular, has emphasized the view that ecological considerations must inform cutting edge laboratory studies. We have deliberately eschewed reading and referring to many fine reviews on "Lyme disease ecology"; the interpretations of the literature presented in this chapter undoubtedly overlap but our intent was to present a personal view. Not all facets of borrelial perpetuation could be covered; many superb studies may not have been referenced. These choices do not reflect on merit and we apologize to our many colleagues and friends if their work has been omitted. Tucker Taylor generated the graphs for Figure 3. Our work on ticks has been supported over 3 decades by the National Institutes of Health Institute of Allergy and Infectious Diseases (R37 AI 19693 to the late Andrew Spielman; R29 AI 37993; R01 AI 39002; R21 AI 53411; R01 AI 064218; we are currently supported by R01 AI 137424 and R01 AI 130105). NIAID has strongly supported Lyme disease ecology research from the beginnings of the field and continues to do so; much of our collective knowledge on North American enzootic and zoonotic transmission derives from their support. This is a contribution of the University of Massachusetts Nantucket Field Station.

References

Adeyeye OA, Butler JF (1989) Population Structure and Seasonal Intra-Burrow Movement of Ornithodoros turicata (Acari: Argasidae) in Gopher Tortoise Burrows. *J Med Entomol* 26: 279–283, doi:10.1093/jmedent/26.4.279.

Aeschlimann A, Chamot E, Gigon F, Jeanneret JP, Kesseler D, Walther C (1987) B. burgdorferi in Switzerland. *Zentralbl Bakteriol Mikrobiol Hyg [A]* 263: 450–458, doi:10.1016/s0176-6724(87)80107-4.

Amore G, Tomassone L, Grego E, Ragagli C, Bertolotti L, Nebbia P, Rosati S, Mannelli A (2007) Borrelia lusitaniae in Immature Ixodes ricinus (Acari: Ixodidae) Feeding on Common Wall Lizards in Tuscany, Central Italy. *J Med Entomol* 44: 303–307, doi:10.1093/jmedent/44.2.303.

Anderson JF (1988) Mammalian and avian reservoirs for Borrelia burgdorferi. *Ann N Y Acad Sci* 539: 180–191.

Armstrong P, Rich SM, Smith R, Hartl DL, Spielman A, Telford 3rd SR (1996) A new Borrelia infecting Lone Star ticks. *The Lancet* 347: 67.

Arsnoe I, Tsao JI, Hickling GJ (2019) Nymphal Ixodes scapularis questing behavior explains geographic variation in Lyme borreliosis risk in the eastern United States. *Ticks Tick-Borne Dis* 10: 553–563, doi:10.1016/j.ttbdis.2019.01.001.

Babenko LV (1985) Periods of life cycle. In Taiga Tick Ixodes Persulcatus: Morphology, Systematics, Ecology, Medical Importance, (Leningrad: Nauka), pp. 213–214.

Barbour AG, Adeolu M, Gupta RS (2017) Division of the genus Borrelia into two genera (corresponding to Lyme disease and relapsing fever groups) reflects their genetic and phenotypic distinctiveness and will lead to a better understanding of these two groups of microbes (Margos et al. (2016) There is inadequate evidence to support the division of the genus Borrelia. Int. J. Syst. Evol. Microbiol. doi: 10.1099/ijsem.0.001717). *Int J Syst Evol Microbiol* 67: 2058–2067, doi:10.1099/ijsem.0.001815.

Barbour AG, Bunikis J, Travinsky B, Hoen AG, Diuk-Wasser MA, Fish D, Tsao JI (2009) Niche Partitioning of Borrelia burgdorferi and Borrelia miyamotoi in the Same Tick Vector and Mammalian Reservoir Species. *Am J Trop Med Hyg* 81: 1120–1131, doi:10.4269/ajtmh.2009.09-0208.

Barbour AG, Maupin GO, Teltow GJ, Carter CJ, Piesman J (1996) Identification of an Uncultivable Borrelia Species in the Hard Tick Amblyomma americanum: Possible Agent of a Lyme Disease-like Illness. *J Infect Dis* 173: 403–409, doi:10.1093/infdis/173.2.403.

Battaly GR, Fish D, Dowler RC (1987) The seasonal occurrence of Ixodes dammini and Ixodes dentatus (Acari: Ixodidae) on birds in a Lyme disease endemic area of southeastern New York State. *J N Y Entomol Soc* 461–468.

Belozerov, V.N. (1985) Diapause, its role in the life cycle, mechanism. In Taiga Tick Ixodes Persulcatus: Morphology, Systematics, Ecology, Medical Importance, (Leningrad), pp. 214–219.

Berger KA, Ginsberg HS, Dugas KD, Hamel LH, Mather TN (2014) Adverse moisture events predict seasonal abundance of Lyme disease vector ticks (Ixodes scapularis). *Parasit Vector* 7: 181.

Bhide MR, Travnicek M, Levkutova M, Curlik J, Revajova V, Levkut M (2005) Sensitivity of Borrelia genospecies to serum complement from different animals and human: a host—pathogen relationship. *FEMS Immunol Med Microbiol* 43: 165–172, doi:10.1016/j.femsim.2004.07.012.

Birkhaug K (1942) Otto H. F. Obermeier. In A Symposium on Relapsing Fever in the Americas, (Washington: American Association for the Advancement of Science), pp. 5–9.

Bowman D, Little SE, Lorentzen L, Shields J, Sullivan MP, Carlin EP (2009) Prevalence and geographic distribution of Dirofilaria immitis, Borrelia burgdorferi, Ehrlichia canis, and Anaplasma phagocytophilum in dogs in the United States: Results of a national clinic-based serologic survey. *Vet Parasitol* 160: 138–148, doi:10.1016/j.vetpar.2008.10.093.

Breuner NE, Hojgaard A, Replogle AJ, Boegler KA, Eisen L (2018) Transmission of the relapsing fever spirochete, Borrelia miyamotoi, by single transovarially-infected larval Ixodes scapularis ticks. *Ticks Tick-Borne Dis* 9: 1464–1467, doi:10.1016/j.ttbdis.2018.07.006.

Breunig K (2003) Changes in Land Use and Their Impact on Habitat, Biodiversity, and Ecosystem Services in Massachusetts.

Brisson D, Dykhuizen DE, Ostfeld RS (2008) Conspicuous impacts of inconspicuous hosts on the Lyme disease epidemic. *Proc R Soc B Biol Sci* 275: 227–235, doi:10.1098/rspb.2007.1208.

Brooks RT, Nislow KH, Lowe WH, Wilson MK, King DI (2012) Forest succession and terrestrial-aquatic biodiversity in small forested watersheds: a review of principles, relationships and implications for management. *Forestry* 85: 315–328, doi:10.1093/forestry/cps031.

Brown RN, Lane RS (1992) Lyme disease in California: a novel enzootic transmission cycle of Borrelia burgdorferi. *Science* 256: 1439–1442, doi:10.1126/science.1604318.

Brunet LR, Sellitto C, Spielman A, Telford SR (1995a) Antibody response of the mouse reservoir of Borrelia burgdorferi in nature. *Infect Immun* 63: 3030–3036.

Brunet LR, Spielman A, Iii SRT (1995b) Short Report: Density of Lyme Disease Spirochetes within Deer Ticks Collected from Zoonotic Sites. *Am J Trop Med Hyg* 53: 300–302, doi:10.4269/ajtmh.1995.53.300.

Brunner JL, LoGiudice K, Ostfeld RS (2008) Estimating Reservoir Competence of Borrelia burgdorferi; Hosts: Prevalence and Infectivity, Sensitivity, and Specificity. *J Med Entomol* 45: 139–147, doi: 10.1603/0022-2585(2008)45[139:ERCOBB]2.0.CO;2.

Buchthal J, Evans SW, Lunshof J, Telford SR, Esvelt KM (2019) Mice Against Ticks: an experimental community-guided effort to prevent tick-borne disease by altering the shared environment. *Philos Trans R Soc B-Biol Sci* 374: 20180105, doi:10.1098/rstb.2018.0105.

Burgdorfer W, Barbour A, Hayes SF, Benach JL, Grunwaldt E, Davis JP (1982) Lyme disease-a tick-borne spirochetosis? | Science http://science.sciencemag.org/content/216/4552/1317.full.pdf+html (accessed: 07/07/2016).

Burkot TR, Patrican L, Piesman J (1994) Field trial of an outer surface protein A (OspA) antigen-capture enzyme-linked immunosorbent assay (ELISA) to detect Borrelia burgdorferi in Ixodes scapularis. *Am J Trop Med Hyg* 50: 354–358, doi:10.4269/ajtmh.1994.50.354.

Burri C, Schumann O, Schumann C, Gern L (2014) Are Apodemus spp. mice and Myodes glareolus reservoirs for Borrelia miyamotoi, Candidatus Neoehrlichia mikurensis, Rickettsia helvetica, R. monacensis and Anaplasma phagocytophilum? *Ticks Tick-Borne Dis* 5: 245–251, doi:10.1016/j.ttbdis.2013.11.007.

Burtis JC, Yavitt JB, Fahey TJ, Ostfeld RS (2019) Ticks as Soil-Dwelling Arthropods: An Intersection Between Disease and Soil Ecology. *J Med Entomol* 56: 1555–1564, doi:10.1093/jme/tjz116.

Buxton P (1946) The Louse. An Account of the Lice Which Infest Man, Their Medical Importance and Control (Baltimore: Williams & Wilkins).

Cadenas FM, Rais O, Humair P-F, Douet V, Moret J, Gern L (2007) Identification of host bloodmeal source and Borrelia burgdorferi sensu lato in field-collected Ixodes ricinus ticks in Chaumont (Switzerland). *J Med Entomol* 44: 1109–1117.

Callow LL (1967) Observations on tick-transmitted spirochaetes of cattle in Australia and South Africa. *Br Vet J* 123: 492–497, doi:10.1016/s0007-1935(17)39704-x.

Camicas J-L, Hervy J-P, Adam F, Morel PC (1998) The ticks of the world (Acarida, Ixodida): nomenclature, described stages, hosts, distribution (including new species described before 1/01/96).

Campbell GL, Paul WS, Schriefer ME, Craven RB, Robbins KE, Dennis DT (1995) Epidemiologic and Diagnostic Studies of Patients with Suspected Early Lyme Disease, Missouri, 1990–1993. *J Infect Dis* 172: 470–480, doi:10.1093/infdis/172.2.470.

Cayol C, Giermek A, Gomez-Chamorro A, Hytönen J, Kallio ER, Mappes T, Salo J, Voordouw MJ, Koskela E (2018) Borrelia afzelii alters reproductive success in a rodent host. *Proc R Soc B Biol Sci* 285: 20181056, doi:10.1098/rspb.2018.1056.

Chambert T, Staszewski V, Lobato E, Choquet R, Carrie C, McCoy KD, Tveraa T, Boulinier T (2012) Exposure of black-legged kittiwakes to Lyme disease spirochetes: dynamics of the immune status of adult hosts and effects on their survival: Exposure and effects of Borrelia on kittiwakes. *J Anim Ecol* 81: 986–995, doi:10.1111/j.1365-2656.2012.01979.x.

Chao L-L, Wu W-J, Shih C-M (2010) Molecular detection of Borrelia valaisiana-related spirochetes from Ixodes granulatus ticks in Taiwan. *Exp Appl Acarol* 52: 393–407, doi:10.1007/s10493-010-9372-x.

Charrier NP, Hermouet A, Hervet C, Agoulon A, Barker SC, Heylen D, Toty C, McCoy KD, Plantard O, Rispe C (2019) A transcriptome-based phylogenetic study of hard ticks (Ixodidae). *Sci Rep* 9: 12923, doi:10.1038/s41598-019-49641-9.

Chen Z, Yang X, Bu F, Yang X, Yang X, Liu J (2010) Ticks (acari: ixodoidea: argasidae, ixodidae) of China. *Exp Appl Acarol* 51: 393–404, doi:10.1007/s10493-010-9335-2.

Chinzei Y, Minoura H (1988) Reduced Oviposition in Ornithodoros moubata (Acari: Argasidae) Fed on Tick-Sensitized and Vitellin-Immunized Rabbits. *J Med Entomol* 25: 26–31, doi:10.1093/jmedent/25.1.26.

Chu C-Y, Liu W, Jiang B-G, Wang D-M, Jiang W-J, Zhao Q-M, Zhang P-H, Wang Z-X, Tang G-P, Yang H, Cao W-C (2008) Novel Genospecies of Borrelia burgdorferi Sensu Lato from Rodents and Ticks in Southwestern China. *J Clin Microbiol* 46: 3130–3133, doi: 10.1128/JCM.01195-08.

Clark K, Hendricks A, Burge D (2005) Molecular Identification and Analysis of Borrelia burgdorferi Sensu Lato in Lizards in the Southeastern United States. *Appl Environ Microbiol* 71: 2616–2625, doi:10.1128/AEM.71.5.2616-2625.2005.

Clover JR, Lane RS (1995) Evidence implicating nymphal Ixodes pacificus (Acari: ixodidae) in the epidemiology of Lyme disease in California. *Am J Trop Med Hyg* 53: 237–240, doi:10.4269/ajtmh.1995.53.237.

Coipan EC, Jahfari S, Fonville M, Maassen CB, van der Giessen J, Takken W, Takumi K, Sprong H (2013) Spatiotemporal dynamics of emerging pathogens in questing Ixodes ricinus. *Front Cell Infect Microbiol* 3: doi:10.3389/fcimb.2013.00036.

Collatz J, Selzer P, Fuhrmann A, Oehme RM, Mackenstedt U, Kahl O, Steidle JLM (2011) A hidden beneficial: biology of the tick-wasp Ixodiphagus hookeri in Germany. *J Appl Entomol* 135: 351–358, doi: 10.1111/j.1439-0418.2010.01560.x.

Collini M, Albonico F, Hauffe HC, Mortarino M (2015) Identifying the last bloodmeal of questing sheep tick nymphs (Ixodes ricinus L.) using high resolution melting analysis. *Vet Parasitol* 210: 194–205, doi: 10.1016/j.vetpar.2015.04.007.

Collini M, Albonico F, Rosà R, Tagliapietra V, Arnoldi D, Conterno L, Rossi C, Mortarino M, Rizzoli A, Hauffe HC (2016) Identification of Ixodes ricinus blood meals using an automated protocol with high resolution melting analysis (HRMA) reveals the importance of domestic dogs as larval tick hosts in Italian alpine forests. *Parasit Vectors* 9: doi:10.1186/s13071-016-1901-y.

Cronon W (1983) Changes in the Land (New York: Hill and Wang).

Crowder CD, Carolan HE, Rounds MA, Honig V, Mothes B, Haag H, Nolte O, Luft BJ, Grubhoffer L, Ecker DJ, Schutzer SE, Eshoo MW (2014) Prevalence of *Borrelia miyamotoi* in *Ixodes* Ticks in Europe and the United States. *Emerg Infect Dis* 20: doi:10.3201/eid2010.131583.

Cunliffe N, Nuttall GHF (1921) Some Observations on the Biology and Structure of Ornithodorus moubata, Murray. Together with a Note on the External Characters which serve to differentiate the Sexes. *Parasitology* 13: 327–347, doi:10.1017/S0031182000012555.

Daniel M, Malý M, Danielová V, Kříž B, Nuttall P (2015) Abiotic predictors and annual seasonal dynamics of Ixodes ricinus, the major disease vector of Central Europe. *Parasit Vectors* 8: doi:10.1186/s13071-015-1092-y.

Davidar P, Wilson M, Ribeiro JM (1989) Differential distribution of immature Ixodes dammini (Acari: Ixodidae) on rodent hosts. *J Parasitol* 75: 898–904.

Davis G (1942) Species unity or plurality of the relapsing fever Spirochetes. *Publ Am Assoc Adv Sci* 18: 41–47.

Davis GE (1951) Parthenogenesis in the argasid tick Ornithodoros moubata (Murray, 1877). *J Parasitol* 37: 99–101.

Derdáková M, Dudiòák V, Brei B, Brownstein JS, Schwartz I, Fish D (2004) Interaction and transmission of two Borrelia burgdorferi sensu stricto strains in a tick-rodent maintenance system. *Appl Environ Microbiol* 70: 6783–6788, doi:10.1128/AEM.70.11.6783-6788.2004.

Dibernardo A, Cote T, Ogden NH, Lindsay LR (2014) The prevalence of Borrelia miyamotoi infection, and co-infections with other Borrelia spp. in Ixodes scapularis ticks collected in Canada. *Parasit Vectors* 7: 183, doi:10.1186/1756-3305-7-183.

Dinnis RE, Seelig F, Bormane A, Donaghy M, Vollmer SA, Feil EJ, Kurtenbach K, Margos G (2014) Multilocus sequence typing using mitochondrial genes (mtMLST) reveals geographic population structure of Ixodes ricinus ticks. *Ticks Tick-Borne Dis* 5: 152–160, doi: 10.1016/j.ttbdis.2013.10.001.

Dolan MC, Maupin GO, Panella NA, Golde WT, Piesman J (1997) Vector competence of Ixodes scapularis, I. spinipalpis, and Dermacentor andersoni (Acari:Ixodidae) in transmitting Borrelia burgdorferi, the etiologic agent of Lyme disease. *J Med Entomol* 34: 128–135, doi:10.1093/jmedent/34.2.128.

Donahue J, Piesman J, Spielman A (1987) Reservoir competence of white-footed mice for Lyme disease spirochetes. *Am J Trop Med Hyg* 36: 92–96.

Donaldson TG, León AAP de, Li AI, Castro-Arellano I, Wozniak E, Boyle WK, Hargrove R, Wilder HK, Kim HJ, Teel PD, Lopez JE (2016) Assessment of the Geographic Distribution of Ornithodoros turicata (Argasidae): Climate Variation and Host Diversity. *PLoS Negl Trop Dis* 10: e0004383, doi:10.1371/journal.pntd.0004383.

van Duijvendijk G, Coipan C, Wagemakers A, Fonville M, Ersöz J, Oei A, Földvári G, Hovius J, Takken W, Sprong H (2016) Larvae of Ixodes ricinus transmit Borrelia afzelii and B. miyamotoi to vertebrate hosts. *Parasit Vectors* 9: doi:10.1186/s13071-016-1389-5.

Dunn JM, Krause PJ, Davis S, Vannier EG, Fitzpatrick MC, Rollend L, Belperron AA, States SL, Stacey A, Bockenstedt LK, Fish D, Diuk-Wasser MA (2014) Borrelia burgdorferi Promotes the Establishment of Babesia microti in the Northeastern United States. *PLoS ONE* 9: e115494, doi:10.1371/journal.pone.0115494.

Dupraz M, Toty C, Devillers E, Blanchon T, Elguero E, Vittecoq M, Moutailler S, McCoy KD (2017) Population structure of the soft tick Ornithodoros maritimus and its associated infectious agents within a colony of its seabird host Larus michahellis. *Int J Parasitol Parasites Wildl* 6: 122–130, doi:10.1016/j.ijppaw.2017.05.001.

Durand J, Herrmann C, Genné D, Sarr A, Gern L, Voordouw MJ (2017) Multistrain Infections with Lyme Borreliosis Pathogens in the Tick Vector. *Appl Environ Microbiol* 83: doi:10.1128/AEM.02552-16.

Ehrmann S, Liira J, Gärtner S, Hansen K, Brunet J, Cousins SAO, Deconchat M, Decocq G, De Frenne P, De Smedt P, Diekmann M, Gallet-Moron E, Kolb A, Lenoir J, Lindgren J, Naaf T, Paal T, Valdés A, Verheyen K, Wulf M, Scherer-Lorenzen M (2017) Environmental drivers of Ixodes ricinus abundance in forest fragments of rural European landscapes. *BMC Ecol* 17: 31, doi:10.1186/s12898-017-0141-0.

Eisen L, Eisen RJ (2016) Critical Evaluation of the Linkage Between Tick-Based Risk Measures and the Occurrence of Lyme Disease Cases. *J Med Entomol* 53: 1050–1062, doi:10.1093/jme/tjw092.

Eisen RJ, Eisen L, Beard CB (2016) County-Scale Distribution of *Ixodes scapularis* and *Ixodes pacificus* (Acari: Ixodidae) in the Continental United States. *J Med Entomol* 53: 349–386, doi:10.1093/jme/tjv237.

Elbir H, Abi-Rached L, Pontarotti P, Yoosuf N, Drancourt M (2014) African Relapsing Fever Borreliae Genomospecies Revealed by Comparative Genomics. *Front Public Health* 2: doi:10.3389/fpubh.2014.00043.

Estrada-Peña A, Nava S, Petney T (2014) Description of all the stages of Ixodes inopinatus n. sp. (Acari: Ixodidae). *Ticks Tick-Borne Dis* 5: 734–743, doi:10.1016/j.ttbdis.2014.05.003.

Estrada-Peña A, Osácar JJ, Pichon B, Gray JS (2005) Hosts and Pathogen Detection for Immature Stages of Ixodes ricinus (Acari: Ixodidae) in North-Central Spain. *Exp Appl Acarol* 37: 257–268, doi: 10.1007/s10493-005-3271-6.

Fairchild, G.B., Kohls GM, Tipton VJ (1966) The ticks of Panama. In Ectoparasites of Panama, (Chicago: Field Museum of Natural History), pp. 167–219.

Fedorova N, Kleinjan JE, James D, Hui LT, Peeters H, Lane RS (2014) Remarkable diversity of tick or mammalian-associated Borreliae in the metropolitan San Francisco Bay Area, California. *Ticks Tick-Borne Dis* 5: 951–961, doi:10.1016/j.ttbdis.2014.07.015.

Felsenfeld O (1965) Borreliae, Human Relapsing Fever, and Parasite-Vector-Host Relationships. *Bacteriol Rev* 29: 46–74.

Felsenfeld O (1971) Borrelia. Strains, vectors, human and animal borreliosis. (St. Louis).

Fomenko NV, Livanova NN, Borgoyakov VYu, Kozlova IV, Shulaykina IV, Pukhovskaya NM, Tokarevich NK, Livanov SG, Doroschenko EK, Ivanov LI (2010) Detection of Borrelia miyamotoi in Ixodes persulcatus ticks in Russia. *Entomol Rev* 90: 1088–1094, doi: 10.1134/S0013873810080129.

Foppa IM, Krause PJ, Spielman A, Goethert H, Gern L, Brand B, Telford SR (2002) Entomologic and Serologic Evidence of Zoonotic Transmission of Babesia microti, Eastern Switzerland. *Emerg Infect Dis* 8: 722–726, doi:10.3201/eid0807.010459.

Foster DR, Motzkin G, Slater B (1998) Land-Use History as Long-Term Broad-Scale Disturbance: Regional Forest Dynamics in Central New England. *Ecosystems* 1: 96–119, doi:10.1007/s100219900008.

Francis E (1942) The longevity of fasting and non-fasting Ornithodoros turicata and the survival of Spirochaeta obermeieri within them. In N A Symposium on Relapsing Fever in the Americas, (Washington, DC: American Association for the Advancement of Science), pp. 85–88.

Fryxell RTT, Steelman CD, Szalanski AL, Kvamme KL, Billingsley PM, Williamson PC (2012) Survey of Borreliae in ticks, canines, and white-tailed deer from Arkansas, U.S.A. *Parasit Vectors* 5: 139, doi: 10.1186/1756-3305-5-139.

Fukunaga M, Takahashi Y, Tsuruta Y, Matsushita O, Ralph D, McClelland M, Nakao M (1995) Genetic and phenotypic analysis of Borrelia miyamotoi sp. nov., isolated from the ixodid tick Ixodes persulcatus, the vector for Lyme disease in Japan. *Int J Syst Bacteriol* 45: 804–810, doi:10.1099/00207713-45-4-804.

Garcia R (1963) Studies on the ecology of Ornithodoros coriaceus Koch (Acarina: Argasidae). Univercity of California.

Garrett-Jones C (1964) The human blood index of malaria vectors in relation to epidemiological assessment. *Bull World Health Organ* 30: 241–261.

Gassner F, Verbaarschot P, Smallegange RC, Spitzen J, Wieren SEV, Takken W (2008) Variations in Ixodes ricinus Density and Borrelia Infections Associated with Cattle Introduced into a Woodland in The Netherlands. *Appl Environ Microbiol* 74: 7138–7144, doi:10.1128/AEM.00310-08.

Geigy R (1968a) Relapsing Fevers. In Infectious Blood Diseases of Man and Animals, (New York: Academic Press), pp. 175–216.

Geigy R (1968b) Relapsing fevers. Infectious blood of man and animals (New York: Academic Press).

Geigy R, Aeschlimann A (1964) [Long-term observations on the transovarian transmission of Borrelia duttoni by Ornithodorus moubata]. *Acta Trop* 21: 87–91.

Genné D, Sarr A, Gomez-Chamorro A, Durand J, Cayol C, Rais O, Voordouw MJ (2018) Competition between strains of Borrelia afzelii inside the rodent host and the tick vector. *Proc R Soc B Biol Sci* 285: doi:10.1098/rspb.2018.1804.

Genné D, Sarr A, Rais O, Voordouw MJ (2019) Competition Between Strains of Borrelia afzelii in Immature Ixodes ricinus Ticks Is Not Affected by Season. *Front Cell Infect Microbiol* 9: doi:10.3389/fcimb.2019.00431.

Gern L, Rouvinez E, Toutoungi LN, Godfroid E (1997) Transmission cycles of Borrelia burgdorferi sensu lato involving Ixodes ricinus and/or I. hexagonus ticks and the European hedgehog. Erinaceus europaeus, in suburban and urban areas in Switzerland. *Folia Parasitol (Praha)* 44: 309–314.

Ginsberg HS, Albert M, Acevedo L, Dyer MC, Arsnoe IM, Tsao JI, Mather TN, LeBrun RA (2017) Environmental Factors Affecting Survival of Immature Ixodes scapularis and Implications for Geographical Distribution of Lyme Disease: The Climate/Behavior Hypothesis. *PLOS ONE* 12: e0168723, doi:10.1371/journal.pone.0168723.

Ginsberg HS, Ewing CP (1989) Comparison of flagging, walking, trapping, and collecting from hosts as sampling methods for northern deer ticks,Ixodes dammini, and lone-star ticks,Amblyomma americanum (Acari: Ixodidae). *Exp Appl Acarol* 7: 313–322, doi: 10.1007/BF01197925.

Glass GE, Amerasinghe FP, Morgan JM, Scott TW (1994) Predicting Ixodes scapularis abundance on white-tailed deer using geographic information systems. *Am J Trop Med Hyg* 51: 538–544, doi:10.4269/ajtmh.1994.51.538.

Goddard J, Piesman J (2006) New records of immature Ixodes scapularis from Mississippi. *J Vector Ecol J Soc Vector Ecol* 31: 421–422, doi:10.3376/1081-1710(2006)31[421:nroiis]2.0.co;2.

Goethert HK, Cook JA, Lance EW, Telford SR (2006) Fay and rausch 1969 revisited: babesia microti in alaskan small mammals. *J Parasitol* 92: 826–831, doi:10.1645/GE-584R.1.

Goethert HK, Mather TN, Buchthal J, Telford III S (submitted) Bloodmeal analysis of nymphal deer ticks demonstrates spatiotemporal diversity of Borrelia and Babesia reservoirs. *Appl Environ Microbiol.*

Goethert HK, Telford SR (2009) Nonrandom Distribution of Vector Ticks (Dermacentor variabilis) Infected by Francisella tularensis. *PLoS Pathog* 5: e1000319, doi:10.1371/journal.ppat.1000319.

Goldstein V, Boulanger N, Schwartz D, George J-C, Ertlen D, Zilliox L, Schaeffer M, Jaulhac B (2018) Factors responsible for Ixodes ricinus nymph abundance: Are soil features indicators of tick abundance in a French region where Lyme borreliosis is endemic? *Ticks Tick-Borne Dis* 9: 938–944, doi:10.1016/j.ttbdis.2018.03.013.

Gould SJ, Lewontin RC (1979) The Spandrels of San Marco and the Panglossian Paradigm: A Critique of the Adaptationist Programme. *Proc R Soc Lond B Biol Sci* 205: 581–598, doi:10.1098/rspb.1979.0086.

Gray JS (1984) Studies on the dynamics of active populations of the sheep tick, Ixodes ricinus L. in Co. Wicklow, Ireland. *Acarologia* 25: 167–178.

Gulia-Nuss M, Nuss AB, Meyer JM, Sonenshine DE, Roe RM, Waterhouse RM, Sattelle DB, de la Fuente J, Ribeiro JM, Megy K, Thimmapuram J, Miller JR, Walenz BP, Koren S, Hostetler JB, Thiagarajan M, Joardar VS, Hannick LI, Bidwell S, Hammond MP, Young S, Zeng Q, Abrudan JL, Almeida FC, Ayllón N, Bhide K, Bissinger BW, Bonzon-Kulichenko E, Buckingham SD, Caffrey DR, Caimano MJ, Croset V, Driscoll T, Gilbert D, Gillespie JJ, Giraldo-Calderón GI, Grabowski JM, Jiang D, Khalil SMS, Kim D, Kocan KM, Koči J, Kuhn RJ, Kurtti TJ, Lees K, Lang EG, Kennedy RC, Kwon H, Perera R, Qi Y, Radolf JD, Sakamoto JM, Sánchez-Gracia A, Severo MS, Silverman N, Šimo L, Tojo M, Tornador C, Van Zee JP, Vázquez J, Vieira FG, Villar M, Wespiser AR, Yang Y, Zhu J, Arensburger P, Pietrantonio PV, Barker SC, Shao R, Zdobnov EM, Hauser F, Grimmelikhuijzen CJP, Park Y, Rozas J, Benton R, Pedra JHF, Nelson DR, Unger MF, Tubio JMC, Tu Z, Robertson HM, Shumway M, Sutton G, Wortman JR, Lawson D, Wikel SK, Nene VM, Fraser CM, Collins FH, Birren B, Nelson KE, Caler E, Hill CA (2016) Genomic insights into the Ixodes scapularis tick vector of Lyme disease. *Nat Commun* 7: 10507, doi:10.1038/ncomms10507.

Gupta RS (2019) Distinction between Borrelia and Borreliella is more robustly supported by molecular and phenotypic characteristics than all other neighbouring prokaryotic genera: Response to Margos' et al. 'The genus Borrelia reloaded' (PLoS ONE 13(12): e0208432). *PLOS ONE* 14: e0221397, doi:10.1371/journal.pone.0221397.

Gylfe Å, Bergström S, Lundstróm J, Olsen B (2000) Reactivation of Borrelia infection in birds. *Nature* 403: 724–725, doi: 10.1038/35001663.

Hair JA, Bowman JL (1986) Behavioral ecology of Amblyomma americanum. In Morphology, Physiology, and Behavioral Biology of Ticks, (Chichester, West Sussex, England: E. Horwood; New York: Halsted Press), pp. 406–427.

Hamer SA, Hickling GJ, Keith R, Sidge JL, Walker ED, Tsao JI (2012) Associations of passerine birds, rabbits, and ticks with Borrelia miyamotoi and Borrelia andersonii in Michigan, USA. *Parasit Vectors* 5: 231.

Hamšíková Svitálková Z, Coipan E, Mahríková L, Minichová Berthová L, Sprong H, Kazimirova M (2017) Borrelia miyamotoi and Co-Infection with Borrelia afzelii in Ixodes ricinus Ticks and Rodents from Slovakia. *Microb Ecol* 73: 1000–1008, doi:10.1007/s00248-016-0918-2.

Han S, Hickling GJ, Tsao JI (2016) High Prevalence of *Borrelia miyamotoi* among Adult Blacklegged Ticks from White-Tailed Deer. *Emerg Infect Dis* 22: 316–318, doi:10.3201/eid2202.151218.

Han S, Lubelczyk C, Hicklingg GJ, Belperron AA, Bockenstedt LK, Tsao J (2019) Vertical transmission rates of Borrelia miyamotoi in Ixodes scapularis collected from white-tailed deer. *Ticks Tick-Borne Dis* 10: 682–689, doi:10.1016/j.ttbdis.2019.02.014.

Hartemink NA, Randolph SE, Davis SA, Heesterbeek JAP (2008) The Basic Reproduction Number for Complex Disease Systems: Defining R0 for Tick–Borne Infections. *Am Nat* 171: 743–754, doi: 10.1086/587530.

Hauck D, Springer A, Pachnicke S, Schunack B, Fingerle V, Strube C (2019) Ixodes inopinatus in northern Germany: occurrence and potential vector role for Borrelia spp., Rickettsia spp., and Anaplasma phagocytophilum in comparison with Ixodes ricinus. *Parasitol Res* 118: 3205–3216, doi:10.1007/s00436-019-06506-4.

Hauser G, Rais O, Morán Cadenas F, Gonseth Y, Bouzelboudjen M, Gern L (2018) Influence of climatic factors on Ixodes ricinus nymph abundance and phenology over a long-term monthly observation in Switzerland (2000–2014). *Parasites Amp Vectors* 11: 289, doi: 10.1186/s13071-018-2876-7.

Heglasová I, Rudenko N, Golovchenko M, Zubriková D, Miklisová D, Stanko M (2020) Ticks, fleas and rodent-hosts analyzed for the presence of Borrelia miyamotoi in Slovakia: the first record of Borrelia miyamotoi in a Haemaphysalis inermis tick. *Ticks Tick-Borne Dis* 11: 101456,doi:10.1016/j.ttbdis.2020.101456.

Heisch RB, Garnham PCC (1948) The transmission of Spirochaeta duttoni Novy & Knapp by Pediculus humanus corporis de Geer. *Parasitology* 38: 247–252, doi:10.1017/s0031182000023192.

Heisch RB, Grainger WE (1950) On the Occurrence of Ornithodoros Moubata Murray in Burrows. *Ann Trop Med Parasitol* 44: 153–155, doi:10.1080/00034983.1950.11685437.

Heisch RB, Sparrow H, Harvey A (1960) Behaviour of Sp. recurrentis Lebert in lice. *Bull Soc Pathol Exot* 53: 140–143.

Herms WB (1916) The Pajaroello Tick (Ornithodorous coriaceus Kock) with special reference to life history and biting habits. *J Parasitol* 2: 137–142.

Herms WB, Wheeler CM (1935) Tick Transmission of California Relapsing Fever. *J Econ Entomol* 28: 846–855, doi:10.1093/jee/28.6.846.

Herrmann C, Gern L (2010) Survival of Ixodes ricinus (Acari: Ixodidae) Under Challenging Conditions of Temperature and Humidity Is Influenced by Borrelia burgdorferi sensu lato Infection. *J Med Entomol* 47: 1196–1204, doi:10.1603/ME10111.

Hess AD, Haves RO, Tempelis CH (1968) The use of the forage ratio technique in mosquito host preference studies. *Mosq News* 28:.

Heylen D, Fonville M, van Leeuwen AD, Stroo A, Duisterwinkel M, van Wieren S, Diuk-Wasser M, de Bruin A, Sprong H (2017) Pathogen communities of songbird-derived ticks in Europe's low countries. *Parasit Vectors* 10: 497, doi:10.1186/s13071-017-2423-y.

Heylen D, Sprong H, Oers K van, Fonville M, Leirs H, Matthysen E (2014) Are the specialized bird ticks, Ixodes arboricola and I. frontalis, competent vectors for Borrelia burgdorferi sensu lato? *Environ Microbiol* 16: 1081–1089, doi:10.1111/1462-2920.12332.

Hofmeester TR, Coipan EC, Wieren SE van, Prins HHT, Takken W, Sprong H (2016) Few vertebrate species dominate theBorrelia burgdorferis.l. life cycle. *Environ Res Lett* 11: 043001, doi: 10.1088/1748-9326/11/4/043001.

Honig V, Carolan HE, Vavruskova Z, Massire C, Mosel MR, Crowder CD, Rounds MA, Ecker DJ, Ruzek D, Grubhoffer L, Luft BJ, Eshoo MW (2017) Broad-range survey of vector-borne pathogens and tick host identification of Ixodes ricinus from Southern Czech Republic. *FEMS Microbiol Ecol* 93: doi:10.1093/femsec/fix129.

Hoogstraal H (1956) Ticks of the Sudan. In African Ixodoidea Vol.1, (Washington, DC: Depatiment of Navy, Bureau of Medical Surgery), p.

Hooker W, Bishop F, Wood H (1912) The life history of some North American ticks. In Bull. Bur. Entomol, (USDA), pp. 1–239.

Houhamdi L, Raoult D (2005) Excretion of Living Borrelia recurrentis in Feces of Infected Human Body Lice. *J Infect Dis* 191: 1898–1906, doi:10.1086/429920.

Hovind-Hougen K (1984) Ultrastructure of spirochetes isolated from Ixodes ricinus and Ixodes dammini. *Yale J Biol Med* 57: 543–548.

Hu R, Hyland KE, Mather TN (1993) Occurrence and Distribution in Rhode Island of Hunterellus hookeri (Hymenoptera: Encyrtidae), a Wasp Parasitoid of Ixodes dammini. *J Med Entomol* 30: 277–280, doi: 10.1093/jmedent/30.1.277.

Huggins JG, Gee KL (1995) Efficiency and Selectivity of Cage Trap Sets for Gray and Fox Squirrels. *Wildl Soc Bull 1973-2006* 23: 204–207.

Humair P-F, Douet V, Cadenas FM, Schouls LM, Van De Pol I, Gern L (2007) Molecular identification of bloodmeal source in Ixodes ricinus ticks using 12S rDNA as a genetic marker. *J Med Entomol* 44: 869–880.

Humair P-F, Gern L (1998) Relationship between Borrelia burgdorferi sensu lato species, red squirrels (Sciurus vulgaris) and Ixodes ricinus in enzootic areas in Switzerland. *Acta Trop* 69: 213–227, doi:10.1016/S0001-706X(97)00126-5.

Ivanova LB, Tomova A, González–Acuña D, Murúa R, Moreno CX, Hernández C, Cabello J, Cabello C, Daniels TJ, Godfrey HP, Cabello FC (2014) Borrelia chilensis, a new member of the Borrelia burgdorferi sensu lato complex that extends the range of this genospecies in the Southern Hemisphere. *Environ Microbiol* 16: 1069–1080, doi:10.1111/1462-2920.12310.

Iwabu-Itoh Y, Bazartseren B, Naranbaatar O, Yondonjamts E, Furuno K, Lee K, Sato K, Kawabata H, Takada N, Andoh M, Kajita H, Oikawa Y, Nakao M, Ohnishi M, Watarai M, Shimoda H, Maeda K, Takano A (2017) Tick surveillance for Borrelia miyamotoi and phylogenetic analysis of isolates in Mongolia and Japan. *Ticks Tick-Borne Dis* 8: 850–857, doi:10.1016/j.ttbdis.2017.06.011.

Jacobs MB, Purcell JE, Philipp MT (2003) Ixodes scapularis ticks (Acari: Ixodidae) from Louisiana are competent to transmit Borrelia burgdorferi, the agent of Lyme borreliosis. *J Med Entomol* 40: 964–967, doi:10.1603/0022-2585-40.6.964.

Jaenson TGT, Värv K, Fröjdman I, Jääskeläinen A, Rundgren K, Versteirt V, Estrada-Peña A, Medlock JM, Golovljova I (2016) First evidence of established populations of the taiga tick Ixodes persulcatus (Acari: Ixodidae) in Sweden. *Parasit Vectors* 9: 377, doi: 10.1186/s13071-016-1658-3.

Jahfari S, Ruyts SC, Frazer-Mendelewska E, Jaarsma R, Verheyen K, Sprong H (2017) Melting pot of tick-borne zoonoses: the European hedgehog contributes to the maintenance of various tick-borne diseases in natural cycles urban and suburban areas. *Parasit Vectors* 10: doi:10.1186/s13071-017-2065-0.

Jiang B (2018) Borrelia miyamotoi Infections in Humans and Ticks, Northeastern China - Volume 24, Number 2—February 2018 - Emerging Infectious Disease journal - CDC. *Emerg Infect Dis* 24: doi: 10.3201/eid2402.160378.

Johns R, Ohnishi J, Broadwater A, Sonenshine DE, De Silva AM, Hynes WL (2001) Contrasts in Tick Innate Immune Responses to Borrelia burgdorferi Challenge: Immunotolerance in Ixodes scapularis Versus Immunocompetence in Dermacentor variabilis (Acari: Ixodidae). *J Med Entomol* 38: 99–107, doi:10.1603/0022-2585-38.1.99.

Johnson RC, Burgdorfer W, Lane RS, Barbour AG, Hayes SF, Hyde FW (1987) Borrelia coriaceae sp. nov.: Putative Agent of Epizootic Bovine Abortion. *Int J Syst Bacteriol* 37: 72–74.

Jones LD, Davies CR, Steele GM, Nuttall PA (1987) A novel mode of arbovirus transmission involving a nonviremic host. *Science* 237: 775–777, doi:10.1126/science.3616608.

de Jong J, Wilkinson RJ, Schaeffers P, Sondorp HE, Davidson RN (1995) Louse-borne relapsing fever in southern Sudan. *Trans R Soc Trop Med Hyg* 89: 621–621, doi:10.1016/0035-9203(95)90414-X.

Jordan BE, Onks KR, Hamilton SW, Hayslette SE, Wright SM (2009) Detection of Borrelia burgdorferi and Borrelia lonestari in birds in Tennessee. *J Med Entomol* 46: 131–138, doi:10.1603/033.046.0117.

Kain DE, Sperling FAH, Daly HV, Lane RS (1999) Mitochondrial DNA sequence variation in Ixodes pacificus (Acari: Ixodidae). *Heredity* 83: 378–386, doi:10.1038/sj.hdy.6886110.

Kern A, Collin E, Barthel C, Michel C, Jaulhac B, Boulanger N (2011) Tick Saliva Represses Innate Immunity and Cutaneous Inflammation in a Murine Model of Lyme Disease. *Vector-Borne Zoonotic Dis* 11: 1343–1350, doi:10.1089/vbz.2010.0197.

Khan MN, Khan LA, Mahmood S, Qudoos A (2001) Argas persicus infestation: prevalence and economic significance in poultry. *Pak 1 Agri Sei* 38: 3–4.

Kilpatrick HJ, LaBonte AM (2007) Managing urban deer in Connecticut: A guide for residents and communities. *Hartford CT Dep Environ Prot Bur Nat Resour Wildl Div.*

Kirkland KB, Klimko TB, Meriwether RA, Schriefer M, Levin M, Levine J, Kenzie WRM, Dennis DT (1997) Erythema Migrans—like Rash Illness at a Camp in North Carolina: A New Tick-Borne Disease? *Arch Intern Med* 157: 2635–2641, doi:10.1001/archinte.1997.00440430117014.

Kirstein F, Gray J (1999) Blood Meal Identification in Ticks: a Promising Tool in Ecological Research on Tick-borne Diseases. *Zentbl Bakteriol* 289: 760–764.

Kocan AA, Mukolwe SW, Murphy GL, Barker RW, Kocan KM (1992) Isolation of Borrelia burgdorferi (Spirochaetales: Spirochaetaceae) from Ixodes scapularis and Dermacentor albipictus Ticks (Acari: Ixodidae) in Oklahoma. *J Med Entomol* 29: 630–633, doi:10.1093/jmedent/29.4.630.

Korenberg EI (1994) Comparative ecology and epidemiology of lyme disease and tick-borne encephalitis in the former Soviet Union. *Parasitol Today Pers Ed* 10: 157–160, doi:10.1016/0169-4758(94)90269-0.

Korenberg EI, Kovalevskii YV, Gorelova NB, Nefedova VV (2015) Comparative analysis of the roles of Ixodes persulcatus and I. trianguliceps ticks in natural foci of ixodid tick-borne borrelioses in the Middle Urals, Russia. *Ticks Tick-Borne Dis* 6: 316–321, doi:10.1016/j.ttbdis.2015.02.004.

Korenberg EI, Nefedova VV, Romanenko VN, Gorelova NB (2010) The tick Ixodes pavlovskyi as a host of spirochetes pathogenic for humans and its possible role in the epizootiology and epidemiology of borrelioses. *Vector Borne Zoonotic Dis Larchmt N* 10: 453–458, doi: 10.1089/vbz.2009.0033.

Kovalev SY, Mikhaylishcheva MS, Mukhacheva TA (2015) Natural hybridization of the ticks Ixodes persulcatus and Ixodes pavlovskyi in their sympatric populations in Western Siberia. *Infect Genet Evol* 32: 388–395, doi:10.1016/j.meegid.2015.04.003.

Kovalev SY, Mukhacheva TA (2012) Phylogeographical structure of the tick Ixodes persulcatus: a novel view. *Ticks Tick-Borne Dis* 3: 212–218, doi:10.1016/j.ttbdis.2012.03.005.

Kovalevskii YV, Korenberg EI (1995) Differences in Borrelia infections in adult Ixodes persulcatus and Ixodes ricinus ticks (Acari: Ixodidae) in populations of north-western Russia. *Exp Appl Acarol* 19: 19–29, doi: 10.1007/BF00051934.

Kraiczy P, Hartmann K, Hellwage J, Skerka C, Kirschfink M, Brade V, Zipfel PF, Wallich R, Stevenson B (2004) Immunological characterization of the complement regulator factor H-binding CRASP and Erp proteins of Borrelia burgdorferi. *Int J Med Microbiol Suppl* 293: 152–157, doi:10.1016/S1433-1128(04)80029-9.

Krause PJ, Telford SR, Spielman A, Sikand V, Ryan R, Christianson D, Burke G, Brassard P, Pollack R, Peck J, Persing DH (1996) Concurrent Lyme disease and babesiosis. Evidence for increased severity and duration of illness. *JAMA* 275: 1657–1660.

Krishnavajhala A, Armstrong BA, Lopez JE (2018) Vector Competence of Geographical Populations of *Ornithodoros turicata* for the Tick-Borne Relapsing Fever Spirochete *Borrelia turicatae*. *Appl Environ Microbiol* 84: e01505-18, /aem/84/21/e01505-18.atom, doi:10.1128/AEM.01505-18.

Kurtenbach K, Sewell H-S, Ogden NH, Randolph SE, Nuttall PA (1998) Serum Complement Sensitivity as a Key Factor in Lyme Disease Ecology. *Infect Immun* 66: 1248–1251.

Labuda M, Jones LD, Williams T, Danielova V, Nuttall PA (1993) Efficient transmission of tick-borne encephalitis virus between cofeeding ticks. *J Med Entomol* 30: 295–299, doi:10.1093/jmedent/30.1.295.

Labzin VV (1985) Parasitizing on mammals. In Taiga Tick Ixodes Persulcatus: Morphology, Systematics, Ecology, Medical Importance, (Leningrad: Nauka), pp. 291–306.

Lane RS, Brown RN (1991) Wood Rats and Kangaroo Rats: Potential Reservoirs of the Lyme Disease Spirochete in California. *J Med Entomol* 28: 299–302, doi:10.1093/jmedent/28.3.299.

Lane RS, Burgdorfer W (1988) Spirochetes in mammals and ticks (Acari: Ixodidae) from a focus of Lyme borreliosis in California. *J Wildl Dis* 24: 1–9, doi:10.7589/0090-3558-24.1.1.

Lane RS, Burgdorfer W, Hayes SF, Barbour AG (1985) Isolation of a spirochete from the soft tick, Ornithodoros coriaceus: a possible agent of epizootic bovine abortion. *Science* 230: 85–87, doi:10.1126/science.3898367.

Lane RS, Loye JE (1989) Lyme disease in California: interrelationship of Ixodes pacificus (Acari: Ixodidae), the western fence lizard (Sceloporus occidentalis), and Borrelia burgdorferi. *J Med Entomol* 26: 272–278, doi:10.1093/jmedent/26.4.272.

Lane RS, Manweiler SA (1988) Borrelia coriaceae in its Tick Vector, Ornithodoros coriaceus (Acari: Argasidae), with Emphasis on Transstadial and Transovarial Infection. *J Med Entomol* 25: 172–177, doi:10.1093/jmedent/25.3.172.

Lane RS, Mun J, Eisen RJ, Eisen L (2005) Western gray squirrel (Rodentia: Sciuridae): a primary reservoir host of Borrelia burgdorferi in Californian oak woodlands? *J Med Entomol* 42: 388–396, doi: 10.1093/jmedent/42.3.388.

Lane RS, Mun J, Peribáñez MA, Fedorova N (2010) Differences in prevalence of Borrelia burgdorferi and Anaplasma spp. infection among host-seeking Dermacentor occidentalis, Ixodes pacificus, and Ornithodoros coriaceus ticks in northwestern California. *Ticks Tick-Borne Dis* 1: 159–167, doi:10.1016/j.ttbdis.2010.09.004.

Lane RS, Piesman J, Burgdorfer W (1991) Lyme borreliosis: relation of its causative agent to its vectors and hosts in North America and Europe. *Annu Rev Entomol* 36: 587–609.

Lane RS, Quistad GB (1998) Borreliacidal Factor in the Blood of the Western Fence Lizard (Sceloporus occidentalis). *J Parasitol* 84: 29–34, doi:10.2307/3284524.

Larrousse F, King AG, Wolbach SB (1928) The overwintering in Massachusetts of Ixodiphagus caucurtei. *Science* 67: 351–353.

Lees AD, Milne A (1951) The seasonal and diurnal activities of individual sheep ticks (Ixodes ricinus L). *Parasitology* 41: 189–208, doi:10.1017/s0031182000084031.

Lejal E, Estrada-Peña A, Marsot M, Cosson J-F, Rué O, Mariadassou M, Midoux C, Vayssier-Taussat M, Pollet T (2020) Taxon Appearance From Extraction and Amplification Steps Demonstrates the Value of Multiple Controls in Tick Microbiota Analysis. *Front Microbiol* 11: doi:10.3389/fmicb.2020.01093.

Levin M, Levine JF, Yang S, Howard P, Apperson CS (1996) Reservoir competence of the southeastern five-lined skink (Eumeces inexpectatus) and the green anole (Anolis carolinensis) for Borrelia burgdorferi. *Am J Trop Med Hyg* 54: 92–97, doi:10.4269/ajtmh.1996.54.92.

Levin ML, Fish D (2001) Interference Between the Agents of Lyme Disease and Human Granulocytic Ehrlichiosis in a Natural Reservoir Host. *Vector-Borne Zoonotic Dis* 1: 139–148, doi:10.1089/153036601316977741.

Levine JF, Wilson ML, Spielman A (1985) Mice as Reservoirs of the Lyme Disease Spirochete. *Am J Trop Med Hyg* 34: 355–360, doi:10.4269/ajtmh.1985.34.355.

Li L-F, Wei R, Liu H-B, Jiang B-G, Cui X-M, Wei W, Yuan T-T, Wang Q, Zhao L, Xia L-Y, Li J, Jiang J-F, Jia N, Hu Y-L (2020) Characterization of Microbial Communities in Ixodes persulcatus (Ixodida: Ixodidae), a Veterinary and Medical Important Tick Species in Northeastern China. *J Med Entomol* 57: 1270–1276, doi:10.1093/jme/tjaa013.

Li W, Ortiz G, Fournier P-E, Gimenez G, Reed DL, Pittendrigh B, Raoult D (2010) Genotyping of Human Lice Suggests Multiple Emergences of Body Lice from Local Head Louse Populations. *PLoS Negl Trop Dis* 4: e641, doi:10.1371/journal.pntd.0000641.

Lisboa RS, Teixeira RC, Rangel CP, Santos HA, Massard CL, Fonseca AH (2009) Avian Spirochetosis in Chickens Following Experimental Transmission of Borrelia anserina by Argas (Persicargas) miniatus. *Avian Dis* 53: 4.

Little SE, Heise SR, Blagburn BL, Callister SM, Mead PS (2010) Lyme borreliosis in dogs and humans in the USA. *Trends Parasitol* 26: 213–218, doi:10.1016/j.pt.2010.01.006.

LoGiudice K, Ostfeld RS, Schmidt KA, Keesing F (2003) The ecology of infectious disease: effects of host diversity and community composition on Lyme disease risk. *Proc Natl Acad Sci* 100: 567–571.

Loh S-M, Gofton AW, Lo N, Gillett A, Ryan UM, Irwin PJ, Oskam CL (2016) Novel Borrelia species detected in echidna ticks, Bothriocroton concolor, in Australia. *Parasit Vectors* 9: 339, doi:10.1186/s13071-016-1627-x.

Lopez J, Krishnavajhala A, Garcia M, Bermúdez S (2016) Tick-Borne Relapsing Fever Spirochetes in the Americas. *Vet Sci* 3: 16, doi:10.3390/vetsci3030016.

Lynn GE, Graham CB, Horiuchi K, Eisen L, Johnson TL, Lane RS, Eisen RJ (2018) Prevalence and Geographic Distribution of Borrelia miyamotoi in Host-Seeking Ixodes pacificus (Acari: Ixodidae) Nymphs in Mendocino County, California. *J Med Entomol* 55: 711–716, doi:10.1093/jme/tjx258.

Maaz D, Rausch S, Richter D, Krücken J, Kühl AA, Demeler J, Blümke J, Matuschka F-R, von Samson-Himmelstjerna G, Hartmann S (2016) Susceptibility to Ticks and Lyme Disease Spirochetes Is Not Affected in Mice Coinfected with Nematodes. *Infect Immun* 84: 1274–1286, doi:10.1128/IAI.01309-15.

MacArthur RH (1958) Population Ecology of Some Warblers of Northeastern Coniferous Forests. *Ecology* 39: 599–619, doi:10.2307/1931600.

Maddock AH (1992) Comparison of two methods for trapping rodents and shrews. *Isr J Ecol Evol* 38: 333–340, doi:10.1080/00212210.1992.10688680.

Mans BJ, Andersen JF, Schwan TG, Ribeiro JMC (2008) Characterization of anti-hemostatic factors in the argasid, Argas monolakensis: Implications for the evolution of blood-feeding in the soft tick family. *Insect Biochem Mol Biol* 38: 22–41, doi:10.1016/j.ibmb.2007.09.002.

Mans BJ, Featherston J, Kvas M, Pillay K-A, de Klerk DG, Pienaar R, de Castro MH, Schwan TG, Lopez JE, Teel P, Pérez de León AA, Sonenshine DE, Egekwu NI, Bakkes DK, Heyne H, Kanduma EG, Nyangiwe N, Bouattour A, Latif AA (2019) Argasid and ixodid systematics: Implications for soft tick evolution and systematics, with a new argasid species list. *Ticks Tick-Borne Dis* 10: 219–240, doi:10.1016/j.ttbdis.2018.09.010.

Marchaux E, Salimbeni A (1903) La Spirillose des Poules. *Ann Inst Pasteur* 17: 569–580.

Marshall WF, Telford SR, Rys PN, Rutledge BJ, Mathiesen D, Malawista SE, Spielman A, Persing DH (1994) Detection of Borrelia burgdorferi DNA in museum specimens of Peromyscus leucopus. *J Infect Dis* 170: 1027–1032, doi:10.1093/infdis/170.4.1027.

Masters EJ, Donnell HD (1995) Lyme and/or Lyme-like disease in Missouri. *Mo Med* 92: 346–353.

Mather TN, Iii SRT, MacLachlan AB, Spielman A (1989a) Incompetence of Catbirds as Reservoirs for the Lyme Disease Spirochete (Borrelia burgdorferi). *J Parasitol* 75: 66, doi:10.2307/3282938.

Mather TN, Spielman A (1986) Diurnal Detachment of Immature Deer Ticks (Ixodes dammini) from Nocturnal Hosts. *Am J Trop Med Hyg* 35: 182–186, doi:10.4269/ajtmh.1986.35.182.

Mather TN, Telford III SR, Moore SI, Spielman A (1990) Borrelia burgdorferi and Babesia microti: Efficiency of transmission from reservoirs to vector ticks (Ixodes dammini). *Exp Parasitol* 70: 55–61, doi:10.1016/0014-4894(90)90085-Q.

Mather TN, Wilson ML, Moore SI, Ribeiro JMC, Spielman A (1989b) Comparing the relative potential of rodents as reservoirs of the Lyme disease spirochete (Borrelia burgdorferi). *Am J Epidemiol* 130: 143–150;doi:10.1093/oxfordjournals.aje.a115306.

Matuschka F-R, Endepols S, Richter D, Ohlenbusch A, Eiffert H, Spielman A (1996) Risk of Urban Lyme Disease Enhanced by the Presence of Rats. *J Infect Dis* 174: 1108–1111, doi:10.1093/infdis/174.5.1108.

May RM (1984) Ecology and population biology. In Tropical and Geographic Medicine, (McGraw-Hill), pp. 152–166.

McCall PJ, Hume JCC, Motshegwa K, Pignatelli P, Talbert A, Kisinza W (2007) Does Tick-Borne Relapsing Fever Have an Animal Reservoir in East Africa? *Vector-Borne Zoonotic Dis* 7: 659–666, doi:10.1089/vbz.2007.0151.

McNeil E, Hinshaw WR, Kissling RE (1949) A study of Borrelia anserina infection (spirochetosis) in turkeys. *J Bacteriol* 57: 191–206, doi:10.1128/JB.57.2.191-206.1949.

Meeûs T de, Béati L, Delaye C, Aeschlimann A, Renaud F (2002) Sex-biased genetic structure in the vector of Lyme disease, Ixodes ricinus. *Evolution* 56: 1802–1807, doi:10.1554/0014-3820(2002)056[1802:SBGSIT]2.0.CO;2.

Merten HA, Durden LA (2000) A state-by-state survey of ticks recorded from humans in the United States. *J Vector Ecol J Soc Vector Ecol* 25: 102–113.

Millins C, Magierecka A, Gilbert L, Edoff A, Brereton A, Kilbride E, Denwood M, Birtles R, Biek R (2015) An Invasive Mammal (the Gray Squirrel, Sciurus carolinensis) Commonly Hosts Diverse and Atypical Genotypes of the Zoonotic Pathogen Borrelia burgdorferi Sensu Lato. *Appl Environ Microbiol* 81: 4236–4245, doi:10.1128/AEM.00109-15.

Milne A (1943) The comparison of sheep-tick populations (Ixodes ricinus L.). *Ann Appl Biol* 30: 240–250, doi:10.1111/j.1744-7348.1943.tb06195.x.

Moody KD, Terwilliger GA, Hansen GM, Barthold SW (1994) Experimental Borrelia burgdorferi infection in Peromyscus leucopus. *J Wildl Dis* 30: 155–161, doi:10.7589/0090-3558-30.2.155.

Morris WF, Ehrlén J, Dahlgren JP, Loomis AK, Louthan AM (2020) Biotic and anthropogenic forces rival climatic/abiotic factors in determining global plant population growth and fitness. *Proc Natl Acad Sci* 117: 1107–1112, doi:10.1073/pnas.1918363117.

Moyer PL, Varela AS, Luttrell MP, Moore VA, Stallknecht DE, Little SE (2006) White-tailed deer (Odocoileus virginianus) develop spirochetemia following experimental infection with Borrelia lonestari. *Vet Microbiol* 115: 229–236, doi:10.1016/j.vetmic.2005.12.020.

Murdock JH, Yabsley MJ, Little SE, Chandrashekar R, O'Connor TP, Caudell JN, Huffman JE, Langenberg JA, Hollamby S (2009) Distribution of Antibodies Reactive to Borrelia lonestari and Borrelia burgdorferi in White-Tailed Deer (Odocoileus virginianus) Populations in the Eastern United States. *Vector Borne Zoonotic Dis* 9: 729–736, doi:10.1089/vbz.2008.0144.

Mysterud A, Stigum VM, Jaarsma R, Sprong H (2019) Genospecies of Borrelia burgdorferi sensu lato detected in 16 mammal species and questing ticks from northern Europe. *Sci Rep* 9: 5088, doi:10.1038/s41598-019-41686-0.

Nakao M, Miyamoto K (1995) Mixed infection of different Borrelia species among Apodemus speciosus mice in Hokkaido, Japan. *J Clin Microbiol* 33: 490–492.

Narasimhan S, Rajeevan N, Liu L, Zhao YO, Heisig J, Pan J, Eppler-Epstein R, DePonte K, Fish D, Fikrig E (2014) Gut Microbiota of the Tick Vector Ixodes scapularis Modulate Colonization of the Lyme Disease Spirochete. *Cell Host Microbe* 15: 58–71, doi:10.1016/j.chom.2013.12.001.

Naumov RL (1985) Range of hosts. In Taiga Tick Ixodes Persulcatus: Morphology, Systematics, Ecology, Medical Importance, (Leningrad: Nauka), p. 277.

Need JT, Butler JF (1991) Sequential Feedings by Two Species of Argasid Tick on Laboratory Mice: Effects on Tick Survival, Weight Gain, and Attachment Time. *J Med Entomol* 28: 37–40, doi:10.1093/jmedent/28.1.37.

Nicolle C, Blaizot L, Conseil E (1912) Étiologie de la fièvre récurrente. Son mode de transmission par le pou. *Comptes Rendus Hebd Séances L'Académie Sci* 154: 1636–1638.

Nordmann T, Feldt T, Bosselmann M, Tufa TB, Lemma G, Holtfreter M, Häussinger D (2018) Outbreak of Louse-Borne Relapsing Fever among Urban Dwellers in Arsi Zone, Central Ethiopia, from July to November 2016. *Am J Trop Med Hyg* 98: 1599–1602, doi:10.4269/ajtmh.17-0470.

Norte AC, Margos G, Becker NS, Ramos JA, Núncio MS, Fingerle V, Araújo PM, Adamík P, Alivizatos H, Barba E, Barrientos R, Cauchard L, Csörgő T, Diakou A, Dingemanse NJ, Doligez B, Dubiec A, Eeva T, Flaisz B, Grim T, Hau M, Heylen D, Hornok S, Kazantzidis S, Kováts D, Krause F, Literak I, Mänd R, Mentesana L, Morinay J, Mutanen M, Neto JM, Nováková M, Sanz JJ, Silva LP da, Sprong H, Tirri I-S, Török J, Trilar T, Tyller Z, Visser ME, Carvalho IL de (2020) Host dispersal shapes the population structure of a tick-borne bacterial pathogen. *Mol Ecol* 29: 485–501, doi:10.1111/mec.15336.

Nuttall GHF (1917) The Biology of Pediculus humanus. *Parasitology* 10: 80–185, doi:10.1017/S0031182000003747.

Ogden NH, Lindsay LR, Beauchamp G, Charron D, Maarouf A, O'Callaghan CJ, Waltner-Toews D, Barker IK (2004) Investigation of Relationships Between Temperature and Developmental Rates of Tick Ixodes scapularis (Acari: Ixodidae) in the Laboratory and Field. *J Med Entomol* 41: 622–633, doi:10.1603/0022-2585-41.4.622.

Ogden NH, Lindsay LR, Hanincová K, Barker IK, Bigras-Poulin M, Charron DF, Heagy A, Francis CM, O'Callaghan CJ, Schwartz I, Thompson RA (2008) Role of Migratory Birds in Introduction and Range Expansion of Ixodes scapularis Ticks and of Borrelia burgdorferi and Anaplasma phagocytophilum in Canada. *Appl Environ Microbiol* 74: 1780–1790, doi:10.1128/AEM.01982-07.

Ogden NH, Nuttall PA, Randolph SE (1997) Natural Lyme disease cycles maintained via sheep by co-feeding ticks. *Parasitology* 115: 591–599, doi:10.1017/S0031182097001868.

Olsen B, Duffy DC, Jaenson TG, Gylfe A, Bonnedahl J, Bergström S (1995) Transhemispheric exchange of Lyme disease spirochetes by seabirds. *J Clin Microbiol* 33: 3270–3274.

Ostfeld RS, Brisson D, Oggenfuss K, Devine J, Levy MZ, Keesing F (2018) Effects of a zoonotic pathogen, Borrelia burgdorferi, on the behavior of a key reservoir host. *Ecol Evol* 8: 4074–4083, doi:10.1002/ece3.3961.

Ostfeld RS, Keesing F (2000) Biodiversity and Disease Risk: the Case of Lyme Disease. *Conserv Biol* 14: 722–728, doi:10.1046/j.1523-1739.2000.99014.x.

Padgett KA, Lane RS (2001) Life Cycle of Ixodes pacificus (Acari: Ixodidae): Timing of Developmental Processes Under Field and Laboratory Conditions. *J Med Entomol* 38: 684–693, doi:10.1603/0022-2585-38.5.684.

Pal U, Li X, Wang T, Montgomery RR, Ramamoorthi N, deSilva AM, Bao F, Yang X, Pypaert M, Pradhan D, Kantor FS, Telford S, Anderson JF, Fikrig E (2004) TROSPA, an Ixodes scapularis Receptor for Borrelia burgdorferi. *Cell* 119: 457–468, doi:10.1016/j.cell.2004.10.027.

Paul WS, Maupin G, Scott-Wright AO, Craven RB, Dennis DT (2002) Outbreak of tick-borne relapsing fever at the north rim of the Grand Canyon: evidence for effectiveness of preventive measures. *Am J Trop Med Hyg* 66: 71–75, doi:10.4269/ajtmh.2002.66.71.

Pavlovsky E (1966) Natural nidality of transmissible diseases (Urbana: Univeristy of Ilinois Press).

Pérez D, Kneubühler Y, Rais O, Jouda F, Gern L (2011) Borrelia afzelii ospC genotype diversity in Ixodes ricinus questing ticks and ticks from rodents in two Lyme borreliosis endemic areas: Contribution of co-feeding ticks. *Ticks Tick-Borne Dis* 2: 137–142, doi:10.1016/j.ttbdis.2011.06.003.

Perret JL, Guigoz E, Rais O, Gern L (2000) Influence of saturation deficit and temperature on Ixodes ricinus tick questing activity in a Lyme borreliosis-endemic area (Switzerland). *Parasitol Res* 86: 554–557, doi:10.1007/s004360000209.

Persing D, Telford S, Rys P, Dodge D, White T, Malawista S, Spielman A (1990) Detection of Borrelia burgdorferi DNA in museum specimens of Ixodes dammini ticks. *Science* 249: 1420–1423, doi:10.1126/science.2402635.

Pichon B, Egan D, Rogers M, Gray J (2003) Detection and identification of pathogens and host DNA in unfed host-seeking Ixodes ricinus L. (Acari: Ixodidae). *J Med Entomol* 40: 723–731.

Pichon B, Rogers M, Egan D, Gray J (2005) Blood-meal analysis for the identification of reservoir hosts of tick-borne pathogens in Ireland. *Vector-Borne Zoonotic Dis* 5: 172–180.

Piesman J, Donahue JG, Mather TN, Spielman A (1986) Transovarially acquired Lyme disease spirochetes (Borrelia burgdorferi) in field-collected larval Ixodes dammini (Acari: Ixodidae). *J Med Entomol* 23: 219.

Piesman J, Mather TN, Sinsky RJ, Spielman A (1987) Duration of tick attachment and Borrelia burgdorferi transmission. *J Clin Microbiol* 25: 557–558.

Piesman J, Sinsky RJ (1988) Ability of Ixodes scapularis, Dermacentor variabilis, and Amblyomma americanum (Acari: Ixodidae) to Acquire, Maintain, and Transmit Lyme Disease Spirochetes (Borrelia burgdorferi). *J Med Entomol* 25: 336–339, doi:10.1093/jmedent/25.5.336.

Piesman J, Spielman A (1979) Host-Associations and Seasonal Abundance of Immature Ixodes dammini in Southeastern Massachusetts. *Ann Entomol Soc Am* 72: 829–832, doi:10.1093/aesa/72.6.829.

Piesman J, Spielman A, Etkind P, Ruebush TK 2nd, Juranek DD (1979) Role of deer in the epizootiology of Babesia microti in Massachusetts, USA. *J Med Entomol* 15: 537–540.

Postic D, Garnier M, Baranton G (2007) Multilocus sequence analysis of atypical Borrelia burgdorferi sensu lato isolates--description of Borrelia californiensis sp. nov., and genomospecies 1 and 2. *Int J Med Microbiol IJMM* 297: 263–271, doi:10.1016/j.ijmm.2007.01.006.

Postic D, Ras NM, Lane RS, Hendson M, Baranton G (1998) Expanded Diversity among CalifornianBorrelia Isolates and Description of Borrelia bissettii sp. nov. (Formerly Borrelia Group DN127). *J Clin Microbiol* 36: 3497–3504.

Pound JM, Miller JA, George JE, Fish D, Carroll JF, Schulze TL, Daniels TJ, Falco RC, Stafford III KC, Mather TN (2009) The United States Department of Agriculture's Northeast area-wide tick control project: summary and conclusions. *Vector-Borne Zoonotic Dis* 9: 439–448.

Qiu W-G, Dykhuizen DE, Acosta MS, Luft BJ (2002) Geographic Uniformity of the Lyme Disease Spirochete (Borrelia burgdorferi) and Its Shared History With Tick Vector (Ixodes scapularis) in the Northeastern United States. *Genetics* 160: 833–849.

Rand PW, Lubelczyk C, Holman MS, Lacombe EH, Smith RP (2004) Abundance of Ixodes scapularis (Acari: Ixodidae) After the Complete Removal of Deer from an Isolated Offshore Island, Endemic for Lyme Disease. *J Med Entomol* 41: 779–784, doi:10.1603/0022-2585-41.4.779.

Randolph SE (2001) The shifting landscape of tick-borne zoonoses: tick-borne encephalitis and Lyme borreliosis in Europe. *Philos Trans R Soc Lond B Biol Sci* 356: 1045–1056, doi:10.1098/rstb.2001.0893.

Randolph SE, Green RM, Hoodless AN, Peacey MF (2002) An empirical quantitative framework for the seasonal population dynamics of the tick Ixodes ricinus. *Int J Parasitol* 32: 979–989.

Randolph SE, Miklisová D, Lysy J, Rogers DJ, Labuda M (1999) Incidence from coincidence: patterns of tick infestations on rodents facilitate transmission of tick-borne encephalitis virus. *Parasitology* 118: 177–186.

Raoult D, Roux V (1999) The Body Louse as a Vector of Reemerging Human Diseases. *Clin Infect Dis* 29: 888–911, doi:10.1086/520454.

Rapid Risk Assessment: Louse-borne relapsing fever in the EU, 19 November 2015 (2015) https://www.ecdc.europa.eu/en/publications-data/rapid-risk-assessment-louse-borne-relapsing-fever-eu-19-november-2015 (accessed: 03/08/2020).

Rauter C, Hartung T (2005) Prevalence of Borrelia burgdorferi Sensu Lato Genospecies in Ixodes ricinus Ticks in Europe: a Metaanalysis. *Appl Environ Microbiol* 71: 7203–7216, doi:10.1128/AEM.71.11.7203-7216.2005.

Reed DL, Light JE, Allen JM, Kirchman JJ (2007) Pair of lice lost or parasites regained: the evolutionary history of anthropoid primate lice. *BMC Biol* 5: 7, doi:10.1186/1741-7007-5-7.

Rego ROM, Bestor A, Štefka J, Rosa PA (2014) Population Bottlenecks during the Infectious Cycle of the Lyme Disease Spirochete Borrelia burgdorferi. *PLoS ONE* 9: e101009, doi:10.1371/journal.pone.0101009.

Rich SM, Armstrong PM, Smith RD, Telford SR (2001) Lone Star Tick-Infecting Borreliae Are Most Closely Related to the Agent of Bovine Borreliosis. *J Clin Microbiol* 39: 494–497, doi:10.1128/JCM.39.2.494-497.2001.

Rich SM, Caporale DA, Telford SR, Kocher TD, Hartl DL, Spielman A (1995) Distribution of the Ixodes ricinus-like ticks of eastern North America. *Proc Natl Acad Sci* 92: 6284–6288.

Richter D, Debski a, Hubalek Z, Matuschka F-R (2012) Absence of Lyme Disease Spirochetes in Larval Ixodes ricinus Ticks. *Vector Borne Zoonotic Dis* 12: 21–27.

Richter D, Matuschka F-R (2006) Perpetuation of the Lyme Disease Spirochete Borrelia lusitaniae by Lizards. *Appl Environ Microbiol* 72: 4627–4632, doi:10.1128/AEM.00285-06.

Richter D, Matuschka F-R (2010) Elimination of Lyme Disease Spirochetes from Ticks Feeding on Domestic Ruminants. *Appl Environ Microbiol* 76: 7650–7652, doi:10.1128/AEM.01649-10.

Richter D, Schlee DB, Matuschka F-R (2003) Relapsing Fever–Like Spirochetes Infecting European Vector Tick of Lyme Disease Agent. *Emerg Infect Dis* 9: 697–701, doi:10.3201/eid0906.020459.

Rijpkema S, Nieuwenhuijs J, Franssen FFJ, Jongejan F (1994) Infection rates of Borrelia burgdorferi in different instars of Ixodes ricinus ticks from the Dutch North Sea Island of Ameland. *Exp Appl Acarol* 18: 531–542, doi:10.1007/BF00058936.

Røed KH, Kvie KS, Hasle G, Gilbert L, Leinaas HP (2016) Phylogenetic Lineages and Postglacial Dispersal Dynamics Characterize the Genetic Structure of the Tick, Ixodes ricinus, in Northwest Europe. *PLOS ONE* 11: e0167450, doi:10.1371/journal.pone.0167450.

Rogers AJ (1953) A Study of the Ixodid Ticks of Northern Florida, Including the Biology and Life History of Ixodes scapularis Say (Ixodidae: Acarina). Thesis.

Rollend L, Fish D, Childs JE (2013) Transovarial transmission of Borrelia spirochetes by Ixodes scapularis: A summary of the literature and recent observations. *Ticks Tick-Borne Dis* 4: 46–51, doi:10.1016/j.ttbdis.2012.06.008.

Rózsa L, Apari P (2012) Why infest the loved ones--inherent human behaviour indicates former mutualism with head lice. *Parasitology* 139: 696–700, doi:10.1017/S0031182012000017.

Russell PF (1934) Zooprophylaxis failure: an experiment in the Philippines. *Riv Malariol* 13: 610–616.

Sakamoto JM, Goddard J, Rasgon JL (2014) Population and Demographic Structure of Ixodes scapularis Say in the Eastern United States. *PLoS ONE* 9: e101389, doi:10.1371/journal.pone.0101389.

Santos AS, Santos-Silva MM (2018) Ixodes ventalloi Gil Collado, 1936: A Vector Role to be Explored. *Vectors Vector-Borne Zoonotic Dis* doi:10.5772/intechopen.81615.

Sanz-Aguilar A, Payo-Payo A, Rotger A, Yousfi L, Moutailler S, Beck C, Dumarest M, Igual JM, Miranda MÁ, Viñas Torres M, Picorelli V, Gamble A, Boulinier T (2020) Infestation of small seabirds by Ornithodoros maritimus ticks: Effects on chick body condition, reproduction and associated infectious agents. *Ticks Tick-Borne Dis* 11: 101281, doi:10.1016/j.ttbdis.2019.101281.

Saracho Bottero MN, Sebastian PS, Carvalho LA, Claps LG, Mastropaolo M, Mangold AJ, Venzal JM, Nava S (2017) Presence of Borrelia in different populations of Ixodes pararicinus from northwestern Argentina. *Ticks Tick-Borne Dis* 8: 488–493, doi:10.1016/j.ttbdis.2017.02.008.

Schoenherr AA (1992) A Natural History of California (University of California Press).

Schulze TL, Lakat MF, Parkin WE, Shisler JK, Charette DJ, Bosler EM (1986) Comparison of rates of infection by the Lyme disease spirochete in selected populations of Ixodes dammini and Amblyomma americanum (Acari: Ixodidae). *Zentralblatt Für Bakteriol Mikrobiol Hyg Ser Med Microbiol Infect Dis Virol Parasitol* 263: 72–78, doi:10.1016/S0176-6724(86)80105-5.

Schwan TG (1996) Ticks and Borrelia: model systems for investigating pathogen-arthropod interactions. *Infect Agents Dis* 5: 167–181.

Schwan TG, Piesman J (2000) Temporal Changes in Outer Surface Proteins A and C of the Lyme Disease-Associated Spirochete, Borrelia burgdorferi, during the Chain of Infection in Ticks and Mice. *J Clin Microbiol* 38: 382–388.

Schwanz LE, Voordouw MJ, Brisson D, Ostfeld RS (2011) Borrelia burgdorferi has minimal impact on the Lyme disease reservoir host Peromyscus leucopus. *Vector-Borne Zoonotic Dis* 11: 117–124.

Scoles GA, Papero M, Beati L, Fish D (2001) A relapsing fever group spirochete transmitted by Ixodes scapularis ticks. *Vector Borne Zoonotic Dis Larchmt N* 1: 21–34, doi:10.1089/153036601750137624.

Scott MC, Rosen ME, Hamer SA, Baker E, Edwards H, Crowder C, Tsao JI, Hickling GJ (2010) High-Prevalence Borrelia miyamotoi Infection Among Wild Turkeys (Meleagris gallopavo) in Tennessee. *J Med Entomol* 47: 1238–1242, doi:10.1603/ME10075.

Shashina N (1985) Total duration of life cycle (Russian). In Taiga Tick Ixodes Persulcatus Schulze (Acarina, Ixodidae): Morphology, Systematics, Ecology, and Medical Significance, (Leningrad: Nauka), pp. 275–277.

Shih C, Liu L, Spielman A (1995) Differential Spirochetal Infectivities to Vector Ticks of Mice Chronically Infected by the Agent of Lyme-Disease. *J Clin Microbiol* 33: 3164–3168, doi:10.1128/JCM.33.12.3164-3168.1995.

Sirotkin M, Korenberg E (2018) Influence of Abiotic Factors on Different Developmental Stages of the Taiga Tick Ixodes persulcatus and the Sheep Tick Ixodes ricinus. *Entomol Rev* 98: 496–513, doi:10.1134/S0013873818040115.

Smith CN (1944) The Life History of the Tick Ornithodoros Coriaceus Koch (Argasidae). *Ann Entomol Soc Am* 37: 325–335, doi:10.1093/aesa/37.3.325.

Smith RD, Brener J, Osorno M, Ristic M (1978) Pathobiology of Borrelia theileri in the tropical cattle tick, Boophilus microplus. *J Invertebr Pathol* 32: 182–190, doi:10.1016/0022-2011(78)90028-9.

Smith RD, Rogers AB (1998) Borrelia theileri: a review. *J Spirochetal Tick Borne Dis* 5: 63–68.

Smith RP, Rand PW, Lacombe EH, Morris SR, Holmes DW, Caporale DA (1996) Role of bird migration in the long-distance dispersal of Ixodes dammini, the vector of Lyme disease. *J Infect Dis* 174: 221–224.

Smith RP, Rand PW, Lacombe EH, Telford SR, Rich SM, Piesman J, Spielman A (1993) Norway Rats as Reservoir Hosts for Lyme Disease Spirochetes on Monhegan Island, Maine. *J Infect Dis* 168: 687–691, doi:10.1093/infdis/168.3.687.

Spielman A (1988) Lyme disease and human babesiosis: Evidence incriminating vector and reservoir hosts. In Biology of Parasites, pp. 147–165.

Spielman A, Clifford CM, Piesman J, Corwin MD (1979) Human babesiosis on Nantucket Island, USA: description of the vector, Ixodes (Ixodes) dammini, n. sp. (Acarina: Ixodidae). *J Med Entomol* 15: 218–234.

Spielman A, Levine JF, Wilson ML (1984) Vectorial capacity of North American Ixodes ticks. *Yale J Biol Med* 57: 507.

Spielman A, Rossignol P (1984) Insect Vectors. In Tropical and Geographic Medicine, (New York: McGraw-Hill), pp. 167–183.

Spielman A, Wilson ML, Levine JF, Piesman J (1985) Ecology of Ixodes dammini-borne human babesiosis and Lyme disease. *Annu Rev Entomol* 30: 439–460.

States SL, Huang CI, Davis S, Tufts DM, Diuk-Wasser MA (2017) Co-feeding transmission facilitates strain coexistence in Borrelia burgdorferi, the Lyme disease agent. *Epidemics* 19: 33–42, doi:10.1016/j.epidem.2016.12.002.

Stephens RB, Anderson EM (2014) Effects of trap type on small mammal richness, diversity, and mortality. *Wildl Soc Bull* 38: 619–627, doi:10.1002/wsb.418.

Strandh M, Råberg L (2015) Within-host competition between Borrelia afzelii ospC strains in wild hosts as revealed by massively parallel amplicon sequencing. *Philos Trans R Soc B Biol Sci* 370: doi:10.1098/rstb.2014.0293.

Strnad M, Hönig V, Růžek D, Grubhoffer L, Rego ROM (2017) Europe-Wide Meta-Analysis of Borrelia burgdorferi Sensu Lato Prevalence in Questing Ixodes ricinus Ticks. *Appl Environ Microbiol* 83: doi:10.1128/AEM.00609-17.

Stromdahl EY, Williamson PC, Kollars TM, Evans SR, Barry RK, Vince MA, Dobbs NA (2003) Evidence of Borrelia lonestari DNA in Amblyomma americanum (Acari: Ixodidae) Removed from Humans. *J Clin Microbiol* 41: 5557–5562, doi:10.1128/JCM.41.12.5557-5562.2003.

Swei Andrea, Ostfeld Richard S., Lane Robert S., Briggs Cheryl J. (2011) Impact of the experimental removal of lizards on Lyme disease risk. *Proc R Soc B Biol Sci* 278: 2970–2978, doi:10.1098/rspb.2010.2402.

Takada N, Nakao M, Ishiguro F, Fujita H, Yano Y, Masuzawa T (2001) Prevalence of Lyme disease Borrelia in ticks and rodents in northern Kyushu, Japan. *Med Entomol Zool* 52: 117–123, doi:10.7601/mez.52.117_1.

Takano A, Goka K, Une Y, Shimada Y, Fujita H, Shiino T, Watanabe H, Kawabata H (2010) Isolation and characterization of a novel Borrelia group of tick-borne borreliae from imported reptiles and their associated ticks. *Environ Microbiol* 12: 134–146, doi:10.1111/j.1462-2920.2009.02054.x.

Takumi K, Sprong H, Hofmeester TR (2019) Impact of vertebrate communities on Ixodes ricinus-borne disease risk in forest areas. *Parasit Vectors* 12: 434, doi:10.1186/s13071-019-3700-8.

Talagrand-Reboul E, Boyer PH, Bergström S, Vial L, Boulanger N (2018) Relapsing Fevers: Neglected Tick-Borne Diseases. *Front Cell Infect Microbiol* 8: doi:10.3389/fcimb.2018.00098.

Teglas MB, May B, Crosbie PR, Stephens MR, Boyce WM (2005) Genetic Structure of the Tick Ornithodoros coriaceus (Acari: Argasidae) in California, Nevada, and Oregon. *J Med Entomol* 42: 7.

Telford III S (1998) The Name Ixodes dammini Epidemiologically Justified. *Emerg Infect Dis* 4: 132–134, doi:10.3201/eid0401.980126.

Telford III S, Spielman A (1989) The enzootic transmission of the agent of Lyme disease in rabbits. *Am J Trop Med Hyg* 41: 482–490.

Telford III SR, Mather TN, Moore SI, Wilson ML, Spielman A (1988) Incompetence of deer as reservoirs of the Lyme disease spirochete. *Am J Trop Med Hyg* 39: 105–109.

Telford SR 3rd, Mather TN, Adler GH, Spielman A (1990) Short-tailed shrews as reservoirs of the agents of Lyme disease and human babesiosis. *J Parasitol* 76: 681–683.

Telford SR 3rd, Urioste SS, Spielman A (1992) Clustering of Host-Seeking Nymphal Deer Ticks (Ixodes Dammini) Infected by Lyme Disease Spirochetes (Borrelia Burgdorferi). *Am J Trop Med Hyg* 47: 55–60, doi:10.4269/ajtmh.1992.47.55.

Theiler A (1905) Transmission and inoculability of Spirillum Theileri (Laveran). *Proc R Soc B-Biol Sci* 76: 505–506.

Theiler J (1902) Equine malaria and its sequelae. *J Comp Pathol Ther* 15: 40–54.

Tomassone L, Grego E, Auricchio D, Iori A, Giannini F, Rambozzi L (2013) Lyme borreliosis spirochetes and spotted fever group rickettsiae in ixodid ticks from Pianosa island, Tuscany Archipelago, Italy. *Vector Borne Zoonotic Dis Larchmt N* 13: 84–91, doi:10.1089/vbz.2012.1046.

Trautman R (1907) Etude experimentelle de l'association du spirille de la tick-fever et de divers trypanosomes. *Ann Inst Pasteur* 21: 808–824.

Tupikova NV, Korenberg EI (1965) The effect of concentrated deforestation on certain components of a natural focus of tick borne encephalitis in East European parts of the southern taiga forests. In Theoretical Questions of Natural Foci of Diseases,Proceedings of a Symposium, (Czechoslovak Academy of Sciences), pp. 319–324.

Tyzzer EE (1941) 'Interference' in mixed infections of bartonella and eperythrozoon in mice. *Am J Pathol* 17: 141–153.

Urbanowicz A, Lewandowski D, Szpotkowski K, Figlerowicz M (2016) Tick receptor for outer surface protein A from Ixodes ricinus — the first intrinsically disordered protein involved in vector-microbe recognition. *Sci Rep* 6: 25205, doi:10.1038/srep25205.

Uspensky I, Rubina M (1992) Host substitution by Ixodes persulcatus (Acari: Ixodidae) larvae in the years of deep depression in the abundance of small mammals. *Folia Parasitol (Praha)* 39: 171–176.

Van Treuren W, Ponnusamy L, Brinkerhoff RJ, Gonzalez A, Parobek CM, Juliano JJ, Andreadis TG, Falco RC, Ziegler LB, Hathaway N, Keeler C, Emch M, Bailey JA, Roe RM, Apperson CS, Knight R,

Meshnick SR (2015) Variation in the Microbiota of Ixodes Ticks with Regard to Geography, Species, and Sex. *Appl Environ Microbiol* 81: 6200–6209, doi:10.1128/AEM.01562-15.

Vial L, Diatta G, Tall A, Hadj Ba E, Bouganali H, Durand P, Sokhna C, Rogier C, Renaud F, Trape J-F (2006a) Incidence of tick-borne relapsing fever in west Africa: longitudinal study. *The Lancet* 368: 37–43, doi:10.1016/S0140-6736(06)68968-X.

Vial L, Durand P, Arnathau C, Halos L, Diatta G, Trape JF, Renaud F (2006b) Molecular divergences of the Ornithodoros sonrai soft tick species, a vector of human relapsing fever in West Africa. *Microbes Infect* 8: 2605–2611, doi:10.1016/j.micinf.2006.07.012.

Voordouw MJ, Lachish S, Dolan MC (2015) The Lyme Disease Pathogen Has No Effect on the Survival of Its Rodent Reservoir Host. *PLOS ONE* 10: e0118265, doi:10.1371/journal.pone.0118265.

Vredevoe LK, Stevens JR, Schneider BS (2004) Detection and characterization of Borrelia bissettii in rodents from the central California coast. *J Med Entomol* 41: 736–745, doi: 10.1603/0022-2585-41.4.736.

Wagemakers A, Jahfari S, de Wever B, Spanjaard L, Starink MV, de Vries HJC, Sprong H, Hovius JW (2017) Borrelia miyamotoi in vectors and hosts in The Netherlands. *Ticks Tick-Borne Dis* 8: 370–374, doi: 10.1016/j.ttbdis.2016.12.012.

Wallis RC, Brown SE, Kloter KO, Main AJ (1978) Erythema chronicum migrans and lyme arthritis: field study of ticks. *Am J Epidemiol* 108: 322–327, doi:10.1093/oxfordjournals.aje.a112626.

Walton G (1950) Relapsing fever in the Meru District of Kenya. *East Afr Med J* 27: 94–98.

Walton G (1962) The Ornithodoros moubata subspecies problem in relation to human relapsing fever epidemiology. *Symp Zool Soc Lond* 6: 83–156.

Wang G, Liveris D, Brei B, Wu H, Falco RC, Fish D, Schwartz I (2003) Real-Time PCR for Simultaneous Detection and Quantification of Borrelia burgdorferi in Field-Collected Ixodes scapularis Ticks from the Northeastern United States. *Appl Environ Microbiol* 69: 4561–4565, doi:10.1128/AEM.69.8.4561-4565.2003.

Weitz B (1956) Identification of blood meals of blood-sucking arthropods. *Bull World Health Organ* 15: 473–490.

Whittington-Jones GM, Bernard RTF, Parker DM (2011) Aardvark burrows: a potential resource for animals in arid and semi-arid environments. *Afr Zool* 46: 362–370, doi:10.1080/15627020.2011.11407509.

Wilske B, Preac–Mursic V, Schierz G, Kühbeck R, Barbour AG, Kramer M (1988) Antigenic Variability of Borrelia burgdorferi. *Ann N Y Acad Sci* 539: 126–143, doi:10.1111/j.1749-6632.1988.tb31846.x.

Wilson ML, Telford SR, Piesman J, Spielman A (1988) Reduced Abundance of Immature Ixodes dammini (Acari: Ixodidae) Following Elimination of Deer. *J Med Entomol* 25: 224–228, doi:10.1093/jmedent/25.4.224.

Wodecka B, Rymaszewska A, Skotarczak B (2014) Host and pathogen DNA identification in blood meals of nymphal Ixodes ricinus ticks from forest parks and rural forests of Poland. *Exp Appl Acarol* 62: 543–555, doi:10.1007/s10493-013-9763-x.

Wodecka B, Skotarczak B (2016) Identification of host blood-meal sources and Borrelia in field-collected Ixodes ricinus ticks in north-western Poland. *Ann Agric Environ Med* 23: 59–63, doi: 10.5604/12321966.1196853.

Woolhouse MEJ, Dye C, Etard J-F, Smith T, Charlwood JD, Garnett GP, Hagan P, Hii JLK, Ndhlovu PD, Quinnell RJ, Watts CH, Chandiwana SK, Anderson RM (1997) Heterogeneities in the transmission of infectious agents: Implications for the design of control programs. *Proc Natl Acad Sci* 94: 338–342, doi:10.1073/pnas.94.1.338.

Wright SA, Thompson MA, Miller MJ, Knerl KM, Elms SL, Karpowicz JC, Young JF, Kramer VL (2000) Ecology of Borrelia burgdorferi in ticks (Acari: Ixodidae), rodents, and birds in the Sierra Nevada foothills, Placer County, California. *J Med Entomol* 37: 909–918, doi: 10.1603/0022-2585-37.6.909.

Wright SD, Nielson SW (1990) Experimental infection of the white-footed mouse with Borrelia burgdorferi. *Am J Vet Res* 51: 1980–1987.

Wynns H (1942) The epidemiology of relapsing fever. In A Symposium on Relapsing Fever in the Americas, (Washington, DC: American Association for the Advancement of Science), pp. 100–105.

Xu G, Wielstra B, Rich SM (2020) Northern and southern blacklegged (deer) ticks are genetically distinct with different histories and Lyme spirochete infection rates. *Sci Rep* 10: 10289, doi:10.1038/s41598-020-67259-0.

Yang Y (2018) Borrelia miyamotoi sensu lato in Père David Deer and Haemaphysalis longicornis Ticks - Volume 24, Number 5—May 2018 - Emerging Infectious Disease journal - CDC. *Emerg Infect Dis* 24: doi:10.3201/eid2405.171355.

Yimer M, Mulu W, Ayalew W, Abera B (2014) Louse-borne relapsing fever profile at Felegehiwot referral hospital, Bahir Dar city, Ethiopia: a retrospective study. *BMC Res Notes* 7: 250, doi: 10.1186/1756-0500-7-250.

Zaher MA, Soliman ZR, Diab FM (1977) An experimental study of Borrelia anserina in four species of Argas ticks. *Z Für Parasitenkd* 53: 213–223, doi:10.1007/BF00380466.

Zinsser H (1935) Rats, lice and history (Boston: Little, Brown & Co).

Zolnik CP, Prill RJ, Falco RC, Daniels TJ, Kolokotronis S-O (2016) Microbiome changes through ontogeny of a tick pathogen vector. *Mol Ecol* 25: 4963–4977, doi:10.1111/mec.13832.

Chapter 11

Interactions Between Ticks and Lyme Disease Spirochetes

Utpal Pal[1,2]*, Chrysoula Kitsou[1], Dan Drecktrah[3], Özlem Büyüktanir Yaş[4] and Erol Fikrig[5]

[1]Department of Veterinary Medicine, University of Maryland, 8075 Greenmead Drive, College Park, MD 20742, USA
[2]Virginia-Maryland College of Veterinary Medicine, 8075 Greenmead Drive, College Park, MD 20742, USA
[3]Division of Biological Sciences, University of Montana, Missoula, MT, 59812, USA
[4]Department of Microbiology and Clinical Microbiology, Faculty of Medicine, Istinye University, Zeytinburnu, İstanbul, 34010, Turkey
[5]Section of Infectious Diseases, Department of Internal Medicine, Yale University School of Medicine, New Haven, CT 06520, USA
*Corresponding author: upal@umd.edu

DOI: https://doi.org/10.21775/9781913652616.11

Abstract
Borrelia burgdorferi sensu lato causes Lyme borreliosis in a variety of animals and humans. These atypical bacterial pathogens are maintained in a complex enzootic life cycle that primarily involves a vertebrate host and *Ixodes* spp. ticks. In the Northeastern United States, *I. scapularis* is the main vector, while wild rodents serve as the mammalian reservoir host. As *B. burgdorferi* is transmitted only by *I. scapularis* and closely related ticks, the spirochete-tick interactions are thought to be highly specific. Various borrelial and arthropod proteins that directly or indirectly contribute to the natural cycle of *B. burgdorferi* infection have been identified. Discrete molecular interactions between spirochetes and tick components also have been discovered, which often play critical roles in pathogen persistence and transmission by the arthropod vector. This chapter will focus on the past discoveries and future challenges that are relevant to our understanding of the molecular interactions between *B. burgdorferi* and *Ixodes* ticks. This information will not only impact scientific advancements in the research of tick-transmitted infections but will also contribute to the development of novel preventive measures that interfere with the *B. burgdorferi* life cycle.

Introduction
Ixodes ticks transmit a wide range of infections, including Lyme disease, to animals and humans (Nadelman and Wormser, 1998; Radolf et al., 2012; Nelder et al., 2016; Steere et al., 2016; Stanek and Strle, 2018). Lyme disease is a prominent tick-borne illness (Benach et al., 1983) that features widespread distribution (Mead, 2015) and is caused by a group of spirochetes belonging to the *Borrelia burgdorferi* sensu lato complex (Burgdorfer et al., 1988). Recently, more virulent (albeit less prevalent) strains of *Borrelia* pathogens have been identified, including new clinical isolates of *B. burgdorferi* sensu lato, *B. bissettii*, *B. mayonii*, and *B. miyamotoi*, which are also transmitted by *Ixodes* ticks and are associated with severe human diseases (see chapter 24) (Krause et al., 2015; Pritt et al., 2016; Rudenko et al., 2016). Globally, the *Borrelia* species that are commonly associated with Lyme disease include *B. burgdorferi* sensu stricto, which is prevalent throughout the United States and Europe, and *B. afzelii* and *B. garinii*, which are distributed throughout Eurasia (Nadelman and Wormser, 1998; Piesman and Gern, 2004; Mead, 2015).

In North America and Europe, Lyme disease spirochetes typically are maintained in nature through a complex tick-rodent infection cycle. Larval ticks acquire the pathogen while feeding on infected wild hosts, such as white-footed mice (Anderson et al., 1987; Anderson, 1989), and then transstadially maintain *B. burgdorferi* before transmitting the pathogen back to a naïve host during the subsequent blood meal (Steere, 2001; Steere et al., 2004). Humans and domesticated animals that develop Lyme disease are incidental hosts and do not play a role in the natural transmission cycle (Radolf et al., 2012).

B. burgdorferi does not readily infect most tick species, suggesting that its interactions with *I.*

scapularis are highly specific. Limited studies have shed light on these complex pathogen-tick interactions (Munderloh and Kurtti, 1995; Fikrig and Narasimhan, 2006); a recent review summarizes our most current state of knowledge about the interactions between *B. burgdorferi* and *I. scapularis* ticks (Kurokawa et al., 2020). Vector competence, which depends on genetic determinants influencing the vector's ability to transmit a pathogen, also shapes interactions involving the tick, pathogen, and host (de la Fuente et al., 2017). With the advent of genetic tools for creating *Borrelia* mutants (Samuels et al., 2018) (also see chapter 16) and the current availability of the *I. scapularis* genome (Gulia-Nuss et al., 2016; Miller et al., 2018), investigators will have increasingly robust tools for studying *Borrelia*-tick interactions, both during the arthropod phase of the spirochete life cycle and at the interface with the mammalian host. A few borrelial proteins already have been demonstrated as prerequisites for spirochete-vector interactions, particularly with receptors in select tick organs, such as the gut or salivary gland (Ramamoorthi et al., 2005; Rosa, 2005; Dunham-Ems et al., 2009; Zhang et al., 2011; Radolf et al., 2012). The adaptation of *Borrelia* to the vector or host also requires the tightly-controlled regulation of bacterial gene expression (Hubner et al., 2001; Caimano et al., 2005; Fisher et al., 2005; Caimano et al., 2007; Samuels, 2011; Caimano et al., 2016; Stevenson and Seshu, 2018). Exciting new studies have elucidated the complexities of gene regulation by *B. burgdorferi* in ticks (see Chapter 5). The RpoN/RpoS pathway is active in feeding nymphs (Caimano et al., 2007) and induces a set of proteins that are mandatory for *B. burgdorferi* transmission from ticks, as well as for host infection (Fisher et al., 2005). Additional proteins and nucleotide factors, like Rel_{Bbu} or 5'-triphosphate-guanosine-3'-diphosphate and 5'-diphosphate-guanosine-3'-diphosphate (collectively known as (p)ppGpp), are required for borrelial persistence, while the Hk1/Rrp1 two-component system and c-di-GMP are required for host transitions and the evasion of tick defenses (Bugrysheva et al., 2015; Caimano et al., 2015; Caimano et al., 2016). More recently, a large number of small noncoding RNAs (sRNAs) were discovered to be encoded in the borrelial genome, some of which are regulators of gene expression (Drecktrah et al., 2018). Future studies of *Borrelia*-tick interactions will allow us to understand the unique adaptive strategies of the bacterium for its survival in an extremely diverse range of host tissues. This information will contribute to the development of novel preventive measures against borrelial infection. The following sections will review how the molecular interactions of select *B. burgdorferi* and *I. scapularis* gene products support the pathogen's life cycle in nature. We also will comment on recent advances in *Ixodes* genomics, as well as other "omics" technologies (Hasin et al., 2017), and discuss how tick interaction studies may contribute to the development of a transmission-blocking Lyme disease vaccine.

The *B. burgdorferi* life cycle in ticks

B. burgdorferi survives in nature in a tick-mammal infection cycle (Anguita et al., 2003; Pal and Fikrig, 2003; Caimano et al., 2016). Given the lack of transovarial transmission, the pathogen needs to be acquired during one of the life stages of its vector at the time of engorgement on infected mammals, primarily wild rodents (Anderson, 1989; Munderloh and Kurtti, 1995; Pal and Fikrig, 2003). As the arthropod prepares for its blood meal ingestion within the first 12 hours of tick attachment to the host, the transfer of spirochetes to the vector already has begun. The details remain uncertain as to how *B. burgdorferi* transmits between the arthropod and vertebrate host, either by the active chemotactic migration of spirochetes or via passive transfer along with host fluid. During acquisition, spirochetes enter the tick gut from the dermis of an infected host and presumably continue to migrate until the tick is fully engorged, which usually takes 72-96 hours (Piesman et al., 1990; de Silva and Fikrig, 1995). *B. burgdorferi* intimately associates with the midgut tissues within 48 hours of feeding (Yang et al., 2004), and can persist in the gut throughout the life span of the arthropod (Pal et al., 2001). When the tick ingests a subsequent blood meal, the spirochetes multiply in the gut; at this time, an unknown fraction of the *B. burgdorferi* population exits the gut, enters the hemocoel, disseminates towards the salivary glands, and is transmitted through the salivary stream to the new mammalian host (de Silva and Fikrig, 1995; Coleman et al., 1997; Piesman et al., 2001; Schwan and Piesman, 2002; Dunham-Ems et al., 2009). The following section will highlight the varied environments and formidable challenges that *B. burgdorferi* encounters during its entry, long-term persistence, and subsequent transmission through ticks.

B. burgdorferi encounters diverse tick environments

The feeding gut during acquisition

When acquired from mammals, *B. burgdorferi* adapts to an entirely new environment in the vector. Despite recent breakthroughs (Smith and Pal, 2014; Smith et al., 2016; Oliva Chavez et al., 2017; Shaw et al., 2017; Shaw et al., 2018), relatively little is known about the immune system of the tick. Nevertheless, the feeding tick gut can be presumed to present a hostile environment, which *B. burgdorferi* encounters immediately following acquisition (Smith et al., 2016) (also see chapter 12). As spirochetes migrate into the luminal spaces of the gut and colonize it, they must avoid being digested with the blood meal and also must bypass the tick's innate immune defense mechanisms (Hynes et al., 2005). Phagocytosis allows the tick gut cells to capture constituents of the blood meal, such as cellular remnants and hemoglobin crystals (Sonenshine, 1993), although the ingestion of *B. burgdorferi* is rarely observed.

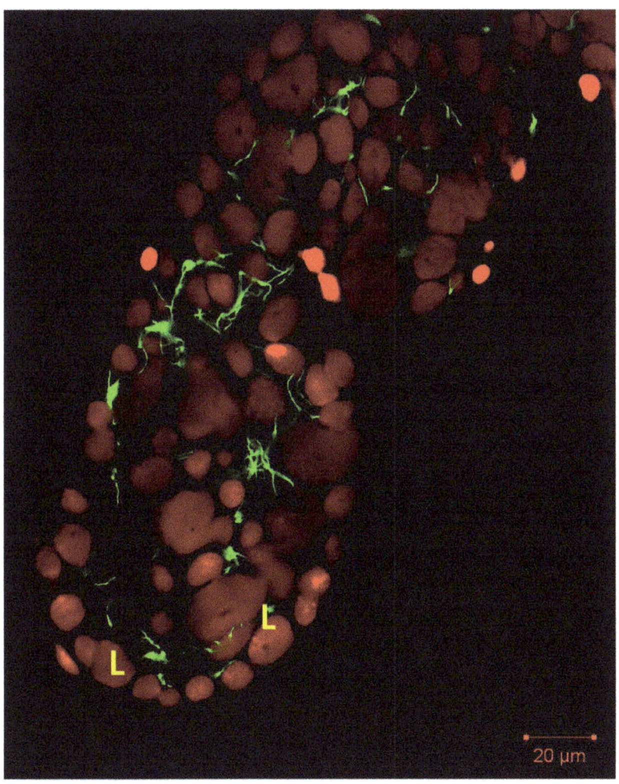

Figure 1. Colonized *B. burgdorferi* within the gut diverticulum of a nymphal *I. scapularis*. The gut of an unfed nymphal *I. scapularis* tick was dissected and processed for confocal immunofluorescence microscopy, as detailed (Pal et al., 2004a). The image shows a portion of the gut diverticulum, where tick nuclei (red) were labeled with propidium iodide, and spirochetes towards the luminal side (green) were detected using FITC-labeled anti-*B. burgdorferi* antibodies. L, Lumen of the gut, Bar = 20 µm.

While gastric digestion in ticks is primarily intracellular (Sonenshine, 1993), with intraluminal proteases virtually absent, the degradation of blood meal components could lead to peptides that have antimicrobial activities (Fogaca et al., 1999; Nakajima et al., 2003). In addition, ticks also may produce conventional antimicrobial peptides in the gut (Broadwater et al., 2002; Hynes et al., 2005). An *I. ricinus* defensin-like gene is upregulated in the gut after infection with *B. burgdorferi* (Grubhoffer et al., 2005); however, the mechanism by which the spirochetes evade this innate immune defense is unknown. *Ixodes* ticks also form a transient acellular peritrophic membrane (PM) that surrounds the blood meal (elaborated upon below). It remains unclear as to how this barrier influences *B. burgdorferi* acquisition and colonization of the gut epithelium in ticks. Invading spirochetes could rapidly colonize the gut epithelial cells prior to the formation of a peritrophic barrier or, alternatively, spirochetes could cross the peritrophic membrane after its formation and colonize the underlying gut epithelium. In either case, the peritrophic barrier could protect the bacteria that are bound to the midgut epithelium against the toxic contents of the lumen.

The unfed gut

In its natural cycle, *B. burgdorferi* persists in its arthropod vectors for months, confined to the guts of unfed ticks. *Ixodes* have a relatively long life cycle spanning at least 18 months (Ullmann et al., 2005), a significant part of which is dominated by phases of nutrient deprivation and limited metabolic activity (Sonenshine, 1993). Little is known about the mechanisms by which *B. burgdorferi* persists in unfed ticks (Figure 1), although more recent studies have suggested that the alteration of metabolic pathways may play a critical role in spirochete persistence. For example, studies have shown that the Hk1/Rrp1/c-di-GMP and Rel$_{Bbu}$/(p)ppGpp pathways have essential functions in spirochete adaptation and survival, include a necessary shift in carbon utilization from glucose to glycerol (He et al., 2011; Pappas et al., 2011; Drecktrah et al., 2015; Caimano et al., 2016). Residing within the gut of the poikilotherm, spirochetes also must survive the extremes of seasonal changes, as well as daily temperature fluctuations in temperate climates. The lumen of the unfed gut (Figure 1) is poor in residual nutrients, primarily because gut epithelia store nutrient macromolecules within cellular endosomes, leaving the lumen devoid of nutrients (Sonenshine, 1993). Genes expressed by *B. burgdorferi* in the unfed gut, or in "tick-like" growth conditions *in vitro*, could be associated with the maintenance of the most elemental functions for persistence in the nutrient-depleted and anoxic environment of the arthropod (Ojaimi et al., 2002; Revel et al., 2002). Spirochetes are known to upregulate specific gene products that protect DNA against starvation or oxidative stress-induced damage during the long intermolt period between blood meals (described below). Alternate forms of *B. burgdorferi*, called round bodies (RBs), also have been reported, especially under unfavorable conditions *in vitro* (Brorson and Brorson, 1998; Alban et al., 2000; Brorson and Brorson, 2004; Drecktrah et al., 2015), in ticks *in vivo* (Dunham-Ems et al., 2012), and in incidental hosts *in vivo* (Brorson and Brorson, 1998; Duray et al., 2005).

Feeding gut during transmission

Tick attachment to a host initiates a remarkable series of metabolic changes in the gut (Sonenshine, 1993). During the first 36 hours of host attachment, blood enters the tick slowly, and there is little digestive activity (Sauer, 1986; Sonenshine, 1993). Spirochetes in the gut must sense the appropriate physiochemical stimuli before they begin multiplying. Following 24 hours of host attachment, *B. burgdorferi* is principally confined within the lumen of the tick gut; by 72 hours, it is dispersed throughout the entire gut tissue (Pal and Fikrig, 2003). Although the density of *B. burgdorferi* in the gut increases exponentially as feeding progresses (Piesman et al., 1990; de Silva and Fikrig, 1995), thereby generating a phenotypically diverse population of spirochetes (Ohnishi et al., 2001; Srivastava and de Silva, 2008), the majority of the bacteria do not exit the gut. Only a fraction of spirochetes cross the gut epithelial barrier and disseminate to the hemocoel for transmission to the host via the salivary gland (Coleman et al., 1997). A seminal study provided a detailed mechanistic view of how *B. burgdorferi* travels through the feeding gut during transmission from infected ticks, using an unprecedented biphasic mode of dissemination (Dunham-Ems et al., 2009). In the initial phase of dissemination, the replicating spirochetes in the midgut form networks of nonmotile organisms, which later advance toward the basolateral gut surface via adherence to the resident gut epithelial cells. In the subsequent phase, the spirochetes become motile, invade the gut basement membrane, and migrate to the salivary glands through the hemocoel (Dunham-Ems et al., 2009). Future studies are warranted to decipher the identities of the tick and microbial components, or additional mechanisms that influence *B. burgdorferi* dissemination from the gut. Host-derived plasminogen has been implicated in the efficient dissemination of *Borrelia* within the tick, and in the enhancement of spirochetemia in vertebrate hosts (Coleman et al., 1995; Coleman et al., 1997). As previously mentioned, an ephemeral and porous peritrophic membrane (PM) is formed in the infected tick gut during feeding, which engulfs the blood meal and *B. burgdorferi* in the lumen (Rudzinska et al., 1982; Sonenshine, 1993), and is clearly visible within 36 hours of the onset of feeding. This membrane is a mechanical barrier, and particulate matters from the ingested blood meal can be retained in the lumen proper (Rudzinska et al., 1982). *B. burgdorferi* can be detected in the lumen of the feeding gut (Pal and Fikrig, 2003; Dunham-Ems et al., 2009); however, it is unknown as to how the physical division of the gut luminal spaces and epithelial tissues, as imposed by the peritrophic membrane, influences spirochete transmission. The integrity of the peritrophic membrane, which is essential for *B. burgdorferi* to efficiently colonize the gut epithelium, is influenced by extrinsic factors, such as the gut microbiota, or intrinsic proteins, like the transcription factor STAT (Kariu et al., 2013; Narasimhan et al., 2014b). Although studies have yet to define precisely how *B. burgdorferi* penetrates through the PM, another tick-borne pathogen, *Babesia microti*, is reported to develop a highly specialized organelle, the arrowhead, which helps the microbe to cross the peritrophic barrier of *Ixodes* ticks (Rudzinska et al., 1982). Nevertheless, the peritrophic membrane and its constituents have been shown to influence pathogen persistence or dissemination in many other arthropod-borne pathogens, including *B. burgdorferi* (Pimenta et al., 1997; Beerntsen et al., 2000; Kariu et al., 2013).

The hemocoel

During transmission, *B. burgdorferi* needs to traverse the hemocoel to access the salivary glands. In arthropods, the cellular constituents of the hemocoel are dominated by actively phagocytic blood cells or hemocytes that are the primary mediators of the arthropod innate immune responses (Sauer, 1986; Sonenshine, 1993). Little is known about how *B. burgdorferi* travels through the tick hemolymph, avoids phagocytosis and the innate immune defense of hemocytes, and selectively recognizes and enters the salivary gland. Studies indicate that a significant percentage (13 ± 5%) of isolated hemocytes in feeding *I. scapularis* could be associated with disseminated spirochetes (Coleman et al., 1997), although the biological significance of the *Ixodes* hemocyte-spirochete association is not clear. In another study, when *B. burgdorferi* was experimentally injected into the hemocoels of *I. scapularis* and *Dermacentor variabilis*, a non-competent tick vector of *Borrelia*, the spirochetes readily survived in the *Ixodes* hemolymph, whereas they were rapidly killed by hemocytes in the *D. variabilis* hemolymph, suggesting that *B. burgdorferi* is resistant to killing by *Ixodes* hemocytes (Johns et al., 2001). The recent availability of the *I. ricinus* hemocytome, as analyzed by next-generation sequencing technologies (Kotsyfakis et al., 2015), will contribute to our understanding of *Ixodes* hemocyte biology, including its role in spirochete interactions (Kurokawa et al., 2020). While the number of *B. burgdorferi* in the infected tick gut increases exponentially (Piesman et al., 1990; Piesman et al., 2001), the exact proportion of spirochetes that cross the gut epithelial barrier and systemically disseminate to the salivary gland through the hemocoel is unknown. Studies of infected *I. scapularis* during feeding indicate that approximately 5% and 0.7% of the gut spirochetes can be detected in the hemolymph and salivary glands, respectively (Coleman et al., 1997). Future studies on the kinetics and mechanisms of *B. burgdorferi* dissemination through the *Ixodes* hemocoel should provide important insights into the development of novel inhibitory measures against spirochete transmission from ticks to mammals.

The salivary gland

The salivary gland represents the other major tick organ (Sauer, 1986; Sonenshine, 1993) that spirochetes must traverse during transmission to a mammalian host. The saliva from an engorging tick, which is deposited in the tick bite site of the host dermis, plays a pivotal role in pathogen transmission (Simo et al., 2017; Nuttall, 2019). Once *B. burgdorferi* is acquired by the tick vector from an infected host, the pathogen resides in the tick gut and does not infect the salivary glands (Piesman, 1993; Piesman et al., 2001; Pal and Fikrig, 2003; Fikrig and Narasimhan, 2006). The spirochetes migrate from the infected tick gut to the salivary glands during a subsequent blood meal, which lasts for 3-7 days (Piesman, 1993; Piesman et al., 2001; Pal and Fikrig, 2003; Fikrig and Narasimhan, 2006). In comparison, *B. hermsii*, the relapsing fever spirochetes, readily colonize the salivary glands of unfed soft ticks, *Ornithodoros hermsi* (Sonenshine, 1993; Schwan and Piesman, 2002); soft ticks are fast feeders, and these particular spirochetes are transmitted within seconds of vector attachment to a mammalian host (Schwan and Piesman, 2002; Boyle et al., 2014). Regardless of the kinetics of transmission, tick salivary glands potentially act as the second distinct barrier that most arthropod-borne pathogens must cross for efficient transmission. Other tick-borne bacterial diseases, such as *Anaplasma marginale*, require active replication within the salivary gland for effective transmission (Ueti et al., 2007), yet there is no evidence of a similar event for *B. burgdorferi* within the tick. Borrelial spirochetes invade the salivary gland by an undefined mechanism, after which they may be passively carried to the host dermis, along with tick saliva. As several salivary gland genes are upregulated in infected *I. scapularis* nymphs, in comparison to uninfected nymphs (Ribeiro et al., 2006), salivary gland proteins may play an important role in *B. burgdorferi* transmission and infection.

Regulation of gene expression in ticks

B. burgdorferi alters its gene expression profile as it cycles between arthropods and mammals (de Silva and Fikrig, 1997; Anguita et al., 2003; Schwan, 2003; Rosa et al., 2005; Stevenson and Seshu, 2018), and specific genes maintain the spirochete infection cycle in nature (Coburn and Cugini, 2003; Purser et al., 2003; Grimm et al., 2004b; Pal et al., 2004b; Yang et al., 2004; Revel et al., 2005; Rosa et al., 2005; Li et al., 2007b; Pal et al., 2008b). Gene regulation in *B. burgdorferi* is the subject of several reviews, such as (Samuels, 2011; Caimano et al., 2016; Stevenson and Seshu, 2018) (also see chapter 6). The diverse microenvironments within vector tissues likely induce the changes in pathogen gene expression that are required for survival. In contrast to many pathogenic bacteria, *B. burgdorferi* devotes large portions of its genome (more than 8% of its coding genes) toward the production of approximately 150 lipoproteins (Fraser et al., 1997; Casjens et al., 2000; Schutzer et al., 2011), many of which are predicted to be exposed on the bacterial surface and be differentially expressed. A number of studies have attempted to identify *B. burgdorferi* genes that are selectively upregulated in the environments of either unfed or fed ticks *in vivo* (Narasimhan et al., 2002; Kumar et al., 2011; Dunham-Ems et al., 2012; Iyer et al., 2015; Groshong et al., 2017). The comparative microarray study revealed wide-spread global changes in expression of genes encoding lipoproteins, carbon utilization, nutrient uptake, intermediate metabolism and chemotaxis when comparing the transcriptomes of fed larvae and fed nymphs, as well as to those of *B. burgdorferi* cultured *in vitro* and in dialysis membrane chambers in the host (Iyer et al., 2015). This all-encompassing view of the transcriptomic modulation in the spirochetes grown in culture or in the vector provides an appreciable picture of the actual changes in the *in vivo* model of Lyme disease, while dissecting individual mechanisms regulating expression have mostly relied on specific *in vitro* conditions. To that end, a number of microarray and RNA-seq studies have provided indirect but important clues about the regulation and function of *B. burgdorferi* genes in *in vitro* culture models, in some cases partly mimicking tick-specific environments (Narasimhan et al., 2002; Ojaimi et al., 2002; Revel et al., 2002; Brooks et al., 2003; Ojaimi et al., 2003; Tokarz et al., 2004; Fisher et al., 2005; Bugrysheva et al., 2015; Caimano et al., 2015; Drecktrah et al., 2015; Arnold et al., 2016; Arnold et al., 2018; Boyle et al., 2019; Caimano et al., 2019). Temperature changes used to mimic the tick environment (23°C) and incoming blood meal during tick feeding and host environment (37°C) have identified this variable as an important regulator of global gene expression, albeit by unknown mechanisms (Ojaimi et al., 2002). Such studies have important implications, as the temperature in the tick gut changes rapidly during feeding, rising from the ambient arthropod temperature to mammalian body temperature, resulting in the modulation of the expression of genes necessary to establish infection, notably those encoding lipoproteins, especially outer surface lipoprotein C (*ospC*) (Schwan et al., 1995). Another simulation of the conditions of feeding ticks was accomplished by the addition of blood to spirochete cultures, which also induced major remodeling of the transcriptome with some similarities to temperature shifting, but also significant distinctions (Tokarz et al., 2004).

Induction of the RpoN-RpoS alternative sigma factor pathway, where RpoN controls the expression of RpoS (Hübner et al., 2001), is regarded as a key genetic regulatory network in *B. burgdorferi*, particularly during transmission from ticks to the host. The RpoN-RpoS pathway activates many spirochete genes in feeding ticks, implying a role in *B. burgdorferi* transmission through ticks. Both *rpoN* null and *rpoS* null mutants successfully colonized tick guts, but were unable to invade salivary glands or infect the vertebrate host (Fisher et al., 2005; Dunham-Ems et al., 2012). These studies indicate that genes regulated by RpoN and RpoS are involved in transmission from the vector to the host. The expression of the alternative sigma factor, *rpoS*, is selectively induced during the nymphal blood meal, suggesting a potential role of RpoS as an activator of genes that are relevant to spirochete transmission and the establishment of early infection in a mammalian host (Caimano et al., 2004; Caimano et al., 2007; Dunham-Ems et al., 2012; Caimano et al., 2019). Individual spirochete genes also have been identified that show dramatic expression in ticks, such as *ospA*, *ospB*, *ospC*, *ospD*, *ospE*, *ospF*, *elp*, *bptA*, *dps*, *cspA*, and *la7*, and that may play important roles in tick tissue colonization, transmission, and infection of a new host (Schwan et al., 1995; Schwan and Piesman, 2000; Ojaimi et al., 2002; Brooks et al., 2003; Grimm et al., 2004b; Tokarz et al., 2004; Yang et al., 2004; Revel et al., 2005; Bykowski et al., 2007a; Bykowski et al., 2007b; Caimano et al., 2007; Li et al., 2007b; Neelakanta et al., 2007; Pal et al., 2008a). Although many of these genes belong to *B. burgdorferi* paralogous gene families, such as Erps (OspE/F-related proteins), individual members have presented distinct expression patterns in the infected tick during feeding and transmission to the mammalian host. For example, whereas *ospE*, *ospF*, and *bbk2.11* display progressive induction in the gut while the arthropod is being engorged on the blood meal, certain members of the *elp* family, such as *bbk2.10* and *ospE/F* homolog *p21*, are expressed only at some later points of feeding and dissemination (Hefty et al., 2001).

The two intracellular second messengers guanosine tetraphosphate and pentaphosphate, collectively known as (p)ppGpp, and cyclic-di-GMP (c-di-GMP) are important signaling molecules that globally alter gene expression to adapt *B. burgdorferi* for persistence in the tick (He et al., 2011; Bugrysheva et al., 2015; Caimano et al., 2015; Drecktrah et al., 2015). The levels of (p)ppGpp are controlled by the bifunctional enzyme Rel_{Bbu}, which acts as both a synthetase and hydrolase to regulate the stringent response (Fraser et al., 1997; Concepcion and Nelson, 2003; Bugrysheva et al., 2005; Drecktrah et al., 2015). When *B. burgdorferi* is nutrient stressed *in vitro*, conditions used to simulate the tick environment between bloodmeals, levels of (p)ppGpp increase, leading to global changes in gene expression and sRNA levels (Bugrysheva et al., 2015; Drecktrah et al., 2015; Drecktrah et al., 2018). Genes encoding specific proteins involved in several pathways are targeted for regulation, including carbohydrate metabolism and persistence in the tick, such as the glycerol metabolism (*glp*) operon (Bugrysheva et al., 2015; Drecktrah et al., 2015). The Rel_{Bbu}-mediated transcriptional remodeling is important for persistence in the tick as a null mutant is compromised for survival between bloodmeals (Drecktrah et al., 2015). Additionally, the transcriptional regulator DksA also alters gene expression during nutrient stress both cooperatively and independently of (p)ppGpp, targeting similar functional pathways, such as the *glp* operon, and distinct ones as well (Boyle et al., 2019). Together, Rel_{Bbu} and DksA regulate the stringent response to control gene expression to adapt *B. burgdorferi* for persistence in the tick.

A series of studies described the identification and molecular characterization of a c-di-GMP regulatory system, which revealed a critical role in spirochete gene expression, motility, and survival in the tick (Freedman et al., 2010; Pitzer et al., 2011; He et al., 2014; Novak et al., 2014; Mallory et al., 2016; Kostick-Dunn et al., 2018; Zhang et al., 2018). Rrp1, a diguanulate cyclase of the two-component regulatory system Hk1-Rrp1, synthesizes c-di-GMP in response to as of yet unidentified signals to globally modulate gene expression, including those encoding genes involved in carbon utilization, notably, like the stringent response, the *glp* operon, and surface lipoproteins (Rogers et al., 2009; Sze et al., 2013; Caimano et al., 2015). The *rrp1* mutants unable to synthesize c-di-GMP have a dramatic survival defect in feeding larvae and nymphs, indicating that the c-di-GMP-mediated transcriptional changes likely adapt the spirochete for persistence (He et al., 2011; Kostick et al., 2011; Caimano et al., 2015). At least some of the effects of c-di-GMP on transcription may be mediated by the c-di-GMP binding protein PlzA, and in some strains PlzB (Freedman et al., 2010; Pitzer et al., 2011; He et al., 2014; Kostick-Dunn et al., 2018; Zhang et al., 2018). Overexpression of PlzA partially compensates for the survival defect of *rrp1* mutants (He et al., 2014).

The remarkable convergence of transcriptional regulatory mechanisms on the *glp* operon suggests a finely tuned, central role for these gene products in the tick. This operon is regulated during conditions that mimic the tick environment by both Rel_{Bbu} and DksA, Rrp1 and PlzA, as well as RpoS, the DNA-binding protein BadR, temperature and glycerol (Ojaimi et al., 2003; Caimano et al., 2007; He et al., 2011; Bugrysheva et al., 2015; Caimano et al., 2015; Drecktrah et al., 2015; Arnold et al., 2018; Zhang et al., 2018; Boyle et al., 2019). In fact, mutant strains with compromised glycerol metabolism, likely provided by the bloodmeal or tick itself, show a significant survival defect in ticks (Pappas et al., 2011). The utilization of glycerol within the tick phase of the *B. burgdorferi* life cycle is a prime example of the multiple mechanisms of gene regulation invoked by the spirochete in response to the arthropod environment in order to transit through the enzootic cycle (Caimano et al., 2016).

Overall, a substantial fraction of our current knowledge of gene regulation is supported by the initial microarray and RNA-seq studies, as these significantly contributed to our understanding of differentially expressed *B. burgdorferi* genes that might function in the adaptation of spirochetes to diverse environments *in vivo*. A collection of molecular tools has been developed (Gilbert et al., 2007; Brisson et al., 2012; Samuels et al., 2018), which should foster new studies delineating the regulatory mechanisms of *B. burgdorferi* genes important for interaction and persistence in the tick environment (Alverson et al., 2003; Caimano et al., 2005; Caimano et al., 2007; Lybecker and Samuels, 2007; Xu et al., 2007)(also see chapter 16). Further work is certainly needed to understand the intrinsic signal transduction pathways in spirochetes, as well as the details of the environmental factors that influence borrelial gene expression in ticks.

B. burgdorferi genes known to function in ticks

The *B. burgdorferi* genome encodes approximately 1,780 genes that are distributed among a single chromosome and 21 plasmids (Fraser et al., 1997; Casjens et al., 2000; Casjens et al., 2011; Schutzer et al., 2011; Casjens et al., 2012)(also see chapter 2). The vast majority of these genes have been annotated as "hypothetical proteins", as they have little resemblance to known proteins in existing databases, and could therefore possibly carry out specialized functions in the spirochete's enzootic life cycle (Fraser et al., 1997; Casjens et al., 2000). Previous efforts have focused on identifying the borrelial genes that are selectively upregulated at specific phases of the spirochete life cycle. As discussed above, microarray studies of the transcriptome profiling of cultured *B. burgdorferi* could identify genes that exhibit differential expression in response to environmental changes and are relevant to its life cycle and transmission through ticks. Indeed, mutagenesis studies indicate that preferentially regulated *B. burgdorferi* gene products (Grimm et al., 2004b; Pal et al., 2004b; Yang et al., 2004; Li et al., 2007b; Neelakanta et al., 2007; Radolf et al., 2012) could serve important functions in supporting the natural cycle of the spirochete. The following sections describe *B. burgdorferi* genes that are known to be highly expressed in the arthropod vector, many of which are thought to be essential in ticks (Grimm et al., 2004b; Pal et al., 2004b; Yang et al., 2004; Munderloh and Kurtti, 2005; Revel et al., 2005; Rosa et al., 2005; Li et al., 2007a; Li et al., 2007b; Pal et al., 2008a; Radolf et al., 2012; Kung et al., 2013). A recent study utilizing a genome-wide screen also identified a novel set of genes required for *B. burgdorferi* survival in the *Ixodes* vector (Phelan et al., 2019); many of these genes are encoded on plasmids, including *ospAB*, *ospC*, *ospD*, and *bbe16*, or on the chromosome, such as *dps* and *la7*.

B. burgdorferi *plasmids*

The *B. burgdorferi* genome contains nine circular (cp) and 12 linear (lp) plasmids (Fraser et al., 1997; Casjens et al., 2000)(see chapter 2). Several studies have indicated that specific plasmids are indispensable for *B. burgdorferi* survival in ticks (Purser and Norris, 2000; Grimm et al., 2005; Rosa et al., 2005; Stewart et al., 2005). As noted earlier, microarray studies have demonstrated that, compared to chromosomal genes, the expression of plasmid-encoded genes is more responsive to environmental changes, suggesting that they have pivotal roles in spirochete survival in diverse environments (Ojaimi et al., 2002; Brooks et al., 2003; Tokarz et al., 2004; Caimano et al., 2007; Iyer et al., 2015). *B. burgdorferi* isolates that lack certain plasmids reinforce the roles of plasmid-encoded genes in the spirochete life cycle in nature. While two linear plasmids, lp25 and lp28-1, were required for borrelial persistence in the mammalian host, only lp25 was found to be essential for tick infection (Purser et al., 2003; Grimm et al., 2004a; Strother et al., 2005; Strother and de Silva, 2005). *bbe22* on lp25 encodes for a nicotinamidase that is most likely essential to the biosynthesis of nicotinamide adenine dinucleotide. The complementation of *bbe22* alone was sufficient to restore the ability of spirochetes lacking lp25 to infect mice (Purser et al., 2003), whereas *bbe22* complementation only partially restored tick infectivity (Strother and de Silva, 2005). *B. burgdorferi* that was deficient in lp28-4 also displayed reduced abilities to persist in ticks and transmit to mice (Strother et al., 2005). The same study also determined that the plasmids lp5, lp28-1, and cp9 were not required for spirochete infectivity in the tick gut. Additionally, *B. burgdorferi* lacking the plasmid lp36 showed no survival defects in ticks, while murine infectivity was significantly compromised (Jewett et al., 2007). Individual genes carried on the same plasmid may have selective host- or vector-specific roles. For example, genes on the plasmid lp54 include *ospA*, which is essential in ticks (Yang et al., 2004), and *dbpAB*, which is important in mice (Liang et al., 2004; Shi et al., 2006; Shi et al., 2008). Microarray studies also indicated the differential expression of genes from many of the *B. burgdorferi* plasmids, most notably lp54, when spirochetes were grown in the presence or absence of mammal-specific factors, such as blood, or when cultured within a DMC (Brooks et al., 2003; Tokarz et al., 2004; Caimano et al., 2007; Iyer et al., 2015).

ospA and *ospB*

Borrelial genes *ospA* and *ospB* constitute a single operon and encode some of the most abundant outer membrane lipoproteins of *B. burgdorferi.* OspA in particular is shown to interact with multiple host receptors, such as tick receptor TROSPA (Pal et al., 2004a) and plasminogen (Fuchs et al., 1994). Both *ospA* and *ospB* show vector-specific expression in the *Borrelia* life cycle, with both exhibiting upregulation when spirochetes exit an infected mammalian host and migrate to feeding *Ixodes* ticks (de Silva et al., 1996; de Silva and Fikrig, 1997; Schwan and Piesman, 2000; Schwan, 2003; Fikrig et al., 2004; Neelakanta et al., 2007). Recombinant OspA and OspB specifically adhere to the tick gut, suggesting that these outer surface lipoproteins are important for spirochete adaptation within the gut (Pal et al., 2000; Fikrig et al., 2004). The experimental transfer of non-borreliacidal OspA or OspB antibodies into the murine host prevented the tissue adherence of spirochetes in feeding ticks *in vivo,* further implicating these proteins in the colonization of *Borrelia* in the tick gut (Pal et al., 2001). An *ospAB* null mutant has also been generated and used in tick-mouse infection studies (Yang et al., 2004), where it retained full murine infectivity and pathogenicity, but was unable to infect ticks. OspA/B-deficient *B. burgdorferi* migrated from infected mice into ticks, but failed to survive in the feeding tick gut, reinforcing the critical role of the *ospAB* operon in the maintenance of *Borrelia* in ticks. An OspB-deficient *B. burgdorferi* isolate was generated, and infection studies further demonstrated that OspB is also involved in the colonization and persistence of spirochetes in the arthropod vector (Neelakanta et al., 2007). These experiments demonstrate the importance of both OspA and OspB in *B. burgdorferi* colonization in the tick gut.

ospC

OspC is a prominent outer membrane lipoprotein of *B. burgdorferi* that is induced in ticks during the transmission of spirochetes from the feeding vector to the host (Schwan et al., 1995). *ospC* is the archetype of RpoS-dependent genes that are associated with tick transmission and host infection (Hubner et al., 2001; Yang et al., 2003; Fisher et al., 2005). Studies with *ospC* deletion mutants have demonstrated that the gene product is essential for infecting the murine host (Grimm et al., 2004b; Pal et al., 2004b; Dunham-Ems et al., 2012). However, the exact role of OspC in the borrelial transmission cycle through ticks warrants further investigation, as there are conflicting results about the phenotype of *ospC* mutants in ticks. In one study, the *ospC* null mutant failed to invade *I. scapularis* salivary glands in significant numbers, and was unable to establish infection in the mammalian host (Pal et al., 2004b). Subsequent studies showed that OspC binds a tick salivary gland protein, Salp15, and that the OspC-tick protein interaction facilitated the establishment of spirochete infection in mice (further discussed below). A study involving *B. afzelii,* which is prevalent in Europe and Asia and is transmitted by *Ixodes* ticks, also supports the notion that OspC is critical for spirochete dissemination from the tick gut to the salivary gland (Fingerle et al., 2007). The complementation of a naturally-occurring *ospC* mutant with a wild-type copy of the *ospC* gene resulted in more efficient migration to the salivary glands (Fingerle et al., 2007). Another group independently created a series of *ospC* mutants and assayed them in tick-mouse infection studies (Grimm et al., 2004b; Stewart et al., 2006; Tilly et al., 2006; Tilly et al., 2007). Their results showed that OspC is required for establishing infection in the murine host; however, *ospC* deletion had no influence on the migration of *B. burgdorferi* from the tick gut to the salivary gland. A more recent study also demonstrated that *Borrelia* requires one or more RpoS-dependent genes to migrate through the midgut, although OspC function is not required for the process, as it exclusively functions within the mammal (Dunham-Ems et al., 2012). Therefore, results indicating disparate phenotypes of OspC deletion are likely due to the variations in the experimental approaches to tick infection (microinjection versus immersion infection), spirochete detection modalities, and the use of different isolates of *B. burgdorferi* (strain 297 versus B31), all of which may have contributed to the overall divergent results. The kinetics of pathogen dissemination from the tick gut to the salivary gland could also vary among isolates. Finally, host-derived plasminogen, which is required for efficient spirochete dissemination through ticks (Coleman et al., 1995; Coleman et al., 1997), also is known to bind OspC (Lagal et al., 2006; Onder et al., 2012). Nonetheless, further studies are needed to clarify how the specificity of interactions between the same spirochete antigen, such as OspC, with multiple host or vector molecules, such as plasminogen or Salp15, contributes to borrelial infection. OspC has recently been implicated in various additional functions within mammals that support spirochete survival, such as an antiphagocytic factor (Carrasco et al., 2015), binding to the complement component C4b (Caine et al., 2017), or joint invasion and colonization in a strain-specific manner (Lin et al., 2020).

dps/napA/bicA

One of the *B. burgdorferi* chromosomal genes that displays regulated expression in the borrelial enzootic cycle is *bb0690,* which encodes a protein similar to the DNA-binding protein of the starved bacteria (Dps) family of proteins (Li et al., 2007b), also known as NapA and BicA (Wang et al., 2012). The protein represents the ortholog of bacterioferritin (Rivera, 2017) fused to a copper-binding metallothionein-like domain, as described in pathogenic Mycobacteria (Gold et al., 2008). *dps* expression is low throughout *B. burgdorferi* infection in mice but is high during tick intermolt periods. Notably, its expression is repressed in spirochetes during recovery from starvation *in vitro* (Drecktrah et al., 2015). Dps-deficient spirochetes retained full murine infectivity, readily migrated to feeding arthropods, and persisted in replete ticks. The mutant, however, failed to survive throughout the extended intermolt periods of resting *I. scapularis*, indicating that Dps is critical to the persistence of spirochetes in unfed ticks (Li et al., 2007b). The protein was later suggested to sequester iron and copper, thus promoting spirochete survival *in vivo* via the detoxification of metal ions, which the pathogen may encounter during its natural life cycle (Wang et al., 2012).

ospD

OspD is an outer membrane lipoprotein of *B. burgdorferi* that shows tightly regulated expression in the spirochete life cycle. *ospD*, in general, exhibits elevated expression in ticks as compared to mice (Li et al., 2007a). Two studies examined *ospD* expression in select stages of the tick-mouse infection cycle of *B. burgdorferi* (Li et al., 2007a; Stewart et al., 2008). Although the results from these studies differ in their kinetics of *ospD* induction, both show that *ospD* transcripts are significantly expressed in a narrow time window that corresponds to the phase of tick detachment from the host after feeding. An OspD-deficient isolate of *B. burgdorferi* was created, and subsequent examination of the mutant phenotype in mice indicated that OspD has no obvious influences on *B. burgdorferi* survival in the murine host, or on spirochete acquisition by ticks (Li et al., 2007a). Recombinant OspD binds to tick gut lysates *in vitro,* suggesting the protein's role in spirochete adherence to tick tissue (Li et al., 2007a). However, while the *ospD* mutant has a reduced capacity for attachment to the tick gut *in vivo*, this does not influence subsequent spirochete transmission during a second blood meal. Taken together, these data suggest that OspD may have a secondary or supportive role in *B. burgdorferi* persistence within ticks (Stewart et al., 2008).

bptA

BptA, annotated as *bbe16,* is a lipoprotein that is possibly surface-exposed. *bptA* is located on the plasmid lp25, which has been closely associated with *Borrelia* infectivity (Purser and Norris, 2000; Purser et al., 2003; Strother et al., 2005; Strother and de Silva, 2005). A microarray analysis first identified that *bbe16* is a differentially expressed gene, with upregulation when *B. burgdorferi* was cultivated in DMC, as compared to spirochetes grown in culture similar to the environment of fed ticks (Revel et al., 2002). Later studies showed that *bptA* is selectively expressed in culture conditions mimicking the tick, suggesting its functional role in spirochete persistence in the vector (Revel et al., 2005). *In vivo* analyses of infectious isolates of *bptA* mutants and complemented isolates demonstrated a requirement for BptA in borrelial persistence within ticks (Revel et al., 2005). The exact function of the protein in the *B. burgdorferi* natural cycle, however, remains to be elucidated.

la7

B. burgdorferi gene *la7*, annotated as *bb0365,* encodes P22, which is also known as lipoprotein LA7 (Lam et al., 1994; Grewe and Nuske, 1996). *la7* is produced in the mammalian host (von Lackum et al., 2007) and is highly upregulated in spirochetes upon their entry into feeding ticks (Pal et al., 2008a). Mutant spirochetes lacking *la7* do not differ from the parental isolates in the establishment of murine infection (Pal et al., 2008a). When naïve ticks engorge on spirochete-infected mice, the *la7* mutant enters ticks, but displays markedly decreased survival within feeding ticks, as compared to wild-type *B. burgdorferi*. LA7 may, therefore, be important for spirochete survival in feeding arthropods. A recent analysis of the high-resolution structure of LA7 will likely contribute to our understanding of the biological function of the protein (Brangulis et al., 2020a).

Additional genes

Specific sets of genes, located either on plasmids or the chromosome, show induced expression in ticks; a few of these have already proven to be of functional significance. Many of these proteins may be involved, either directly or indirectly, in interactions with tick molecules, thereby supporting spirochete survival in the vector, as discussed below. Additional borrelial genes have also been identified as being expressed *in vivo*, although they lack essential supportive roles. *B. burgdorferi* mutants were generated with a lack of *luxS* (Hubner et al., 2003; Tilly et al., 2004) or *chbC* (Hubner et al., 2003; Tilly et al., 2004), whose gene products are implicated in a quorum sensing pathway and other potential biochemical roles (Hubner et al., 2003; Tilly et al., 2004) or the transport of the tick cuticle component chitobiose, respectively. The functions of *luxS* or *chbC* are, however, not essential for spirochete persistence in ticks, nor during the natural infection cycle (Hubner et al., 2003; Tilly et al., 2004). More recently, a CdnL-type CarD regulator in *B. burgdorferi*, termed as LtpA, has been shown to support pathogen survival in ticks (Chen et al., 2018). A *Borrelia* protein annotated as BBE31, the crystal structure of which recently was solved (Brangulis et al., 2020b), is known to aid spirochete movement from the tick gut to the hemolymph during spirochete transmission (Zhang et al., 2011). A surface-exposed borrelial protein, enolase, has been shown to support the interaction of *B. burgdorferi* with plasminogen (Floden et al., 2011; Toledo et al., 2011; Nogueira et al., 2012), contributing to pathogen survival within feeding ticks (Nogueira et al., 2012). Another regulatory protein, annotated as BBD18, is thought to promote spirochete acquisition in ticks, via the regulation of RpoS (Hayes et al., 2014). A polyamine uptake system, termed as PotABCD and encoded by *bb0639-0642*, likely supports the vector phase of borrelial replication and survival (Bontemps-Gallo et al., 2018). The BpuR DNA/RNA-binding protein in *B. burgdorferi* (Jutras et al., 2013), which influences the translation of superoxide dismutase, has been shown to undergo preferential expression in ticks and to enhance spirochete survival in the vector (Jutras et al., 2019). One of the borrelial virulence determinants, annotated as Lmp1, is a surface-exposed and potentially multi-domain protein that is involved in various functions in spirochete infectivity (Yang et al., 2009; Yang et al., 2010). Recent studies indicate that Lmp1 contributes to host-pathogen interactions, including spirochete transmission from ticks to mammals (Zhuang et al., 2018), and possibly spirochete dissemination through tick tissues (Koci et al., 2018), which requires essential contributions from both the N-terminal and middle regions of Lmp1. The gene encoding BBA57 is constitutively expressed during the *Borrelia* infectious cycle involving both mice and ticks (Yang et al., 2013). Although it is unclear whether or how the protein supports *B. burgdorferi* infection in ticks, it is known to facilitate early spirochete infection in mice through the impairment of the host complement and neutrophil-derived antimicrobial defense mechanisms (Bernard et al., 2018).

The recent development of genetic tools has allowed researchers to create specific mutants of *B. burgdorferi* and to identify borrelial proteins that function at different stages in the vector phase of the spirochete life cycle (Radolf et al., 2012; Samuels et al., 2018). As *B. burgdorferi* spends a significant part of its life inside the arthropod, additional genes that are important in ticks are likely to be identified. High-throughput expression analyses, such as quantitative RT-PCR and RNA amplification-based microarray assays, have the potential to reveal genes that are highly expressed in ticks and may perform specialized functions in the complex enzootic life cycle of *B. burgdorferi.* In fact, with an ingenious use of genetic tools and a high-throughput screen, a recent seminal study detailed the identification of a set of novel genes that are required for borrelial survival in *Ixodes* ticks (Phelan et al., 2019). It was shown that many of these essential genes encode proteins of unknown function, membrane-associated proteins, or lipoproteins, which could support host-pathogen interactions or other important aspects of spirochete biology. *B. burgdorferi* harbors a homolog of the hibernation promotion factor, *bb0449*, although the gene function has been characterized as redundant for spirochete survival *in vivo*, including within the tick (Fazzino et al., 2015). The Rel_{Bbu}, the RelA/SpoT homolog in *B. burgdorferi,* regulates the stringent response and global gene expression in spirochetes and is required for pathogen survival in the tick (Drecktrah et al., 2015). Finally, the *ssrS* gene encoding the Bb6S RNA of *B. burgdorferi*, which is considered as one of the first regulatory RNAs, has been shown to control the expression of multiple lipoproteins involved in host infectivity (Drecktrah et al., 2020).

Molecular interfaces: interactions of microbial and vector gene products

Regardless of the nature of the ecological relationship between a microbe and its host, the development of specific molecular interactions is an essential prerequisite for the survival of the microbe *in vivo* (Finlay and Cossart, 1997; Dale and Moran, 2006). Mainly based on the vector-specific expression of limited *B. burgdorferi* gene products, scientists have uncovered a handful of discrete interactions between the spirochete and tick vector that are summarized below. These studies have established that certain borrelial proteins contribute to the maintenance of the spirochete infection cycle within ticks, either directly or via molecular interactions with specific vector receptors.

Interaction of B. burgdorferi *with tick gut proteins*

TROSPA

The tick receptor for OspA (TROSPA), which is predominantly expressed in the gut of *I. scapularis*, specifically interacts with *B. burgdorferi* OspA (Pal et al., 2004a). This receptor-ligand interaction is further explained by the findings that TROSPA co-localizes with OspA in the tick gut *in vivo*, and that OspA specifically bound recombinant TROSPA in a series of *in vitro* assays (Pal et al., 2004a). It was later shown that the N-terminal portion of TROSPA, which contains a putative transmembrane domain, is not involved in the OspA interaction, and that charges on the TROSPA protein influence its interaction with OspA (Figlerowicz et al., 2013). Interestingly, *TROSPA* expression within the tick was upregulated during spirochete infection and decreased during engorgement, suggesting its regulation by extrinsic cues. When uninfected ticks parasitized infected mice that had been passively administered TROSPA antisera, the acquisition of *B. burgdorferi* by the ticks was significantly reduced, as compared to those that fed on infected mice treated with pre-immune sera. TROSPA-immunized ticks also demonstrated a reduced capacity for pathogen transmission. Similarly, the RNA interference of *TROSPA* in nymphal ticks significantly diminished their ability to acquire *B. burgdorferi* from infected mice (Pal et al., 2004). The spirochetes appear to colonize the tick gut by adhering to TROSPA through OspA, which highlights the importance of this interaction in supporting the borrelial infection cycle in ticks. Nevertheless, the native function of TROSPA in the arthropod is currently unknown. *TROSPA* is not consistently expressed in all stages of *I. scapularis*, and its expression is highly influenced by environmental signals, such as blood meal engorgement or *B. burgdorferi* infection. These observations indicate a developmental function, rather than a housekeeping role, of TROSPA in tick biology.

Dystroglycan-like protein (ISDLP)

I. scapularis dystroglycan-like protein (ISDLP) is a *B. burgdorferi*-interacting tick gut protein with similarities to the dystroglycan family of proteins (Coumou et al., 2016). ISDLP is expressed on the gut epithelial surface and is induced during tick feeding and by borrelial infection. Additionally, this tick protein can bind to *B. burgdorferi*, and its knockdown by RNA interference has been shown to impair pathogen transmission, suggesting an important role in spirochete dissemination from ticks to mice.

TRE31 and Ixofin3D

Two tick gut proteins were described that support the dissemination of *B. burgdorferi* from the tick gut, particularly during pathogen transmission to the host. TRE31 was characterized as a tick gut protein that specifically interacts with BBE31, one of the borrelial outer surface lipoproteins (Zhang et al., 2011). The gene is preferentially expressed in the tissues of fed ticks, including the gut, hemolymph, and salivary glands. BBE31 antiserum impaired spirochete migration from the tick gut to the salivary glands, consequently attenuating murine infection by tick-transmitted *B. burgdorferi*. The silencing of *tre31* also lowered the pathogen burden in the tick hemolymph. These data indicate that TRE31 plays a role in enabling spirochete movement by interacting with BBE31. The crystal structure of BBE31 was recently solved, which will be important for a better understanding of the specific ligand-receptor interactions that support *B. burgdorferi* movement in the tick (Brangulis et al., 2020b). Additionally, Ixofin3D represents another tick gut protein that is relevant to pathogen dissemination, with similarities to putative fibronectin type III domain-containing proteins (Narasimhan et al., 2014a). As with TRE31, the antibody or RNA interference-mediated impairment of Ixofin3D resulted in decreased spirochete burdens in the tick salivary glands and in the murine host.

Dual oxidase

A dual oxidase in the tick gut (Duox), which is a member of the NADPH oxidase family, is involved in the maintenance of gut microbial homeostasis and the genesis of an acellular gut barrier called the dityrosine network (DTN) (Yang et al., 2014). The DTN develops from the tyrosine cross-linking of the extracellular matrix in *I. scapularis* ticks during blood meal engorgement. This Duox-mediated barrier influences the survival of *B. burgdorferi* in the gut, as an impaired DTN in *duox*-knockdown ticks can result in reduced levels of pathogen persistence within ticks. The absence of complete DTN formation in the tick gut activates a specific tick innate immune pathway, via the induction of nitric oxide synthase, which ultimately reduces the spirochete burden in the arthropods (Yang et al., 2014). Deeper understandings of the tick innate immune responses and the related vector-pathogen interactions, such as the one orchestrated by Duox, along with their contributions to microbial survival, are critical to the development of new interventions against tick-borne infections, including Lyme borreliosis.

PIXR

A Reeler domain-containing tick gut protein was described that influences the colonization of *B. burgdorferi* in the tick gut (Narasimhan et al., 2017). The gene product, which was termed "Protein of *I. scapularis* with a Reeler domain" (PIXR), is induced by spirochete infection. Both the knockdown of *pixr* via RNA interference and the antibody-mediated inhibition of PIXR reduced the ability of *B. burgdorferi* to infect the tick gut. Notably, PIXR was shown to be involved in biofilm formation, both *in vitro* and *in vivo*. The impairment of the PIXR function in ticks also had diverse effects on intrinsic and extrinsic elements within the arthropods, including their gut microbiome, metabolome, and immune responses. Additionally, PIXR has a proposed role in tick developmental biology, as the abrogation of its function impaired larval molting (Narasimhan et al., 2017). Subsequent analyses also revealed that the *Ixodes* microbiota is sensitive to anti-tick immunity, with biofilm formation potentially serving as the microbiome's defensive response to this immunity (Estrada-Pena et al., 2020).

IsCDA

One of the first detailed characterizations of the *Ixodes* peritrophic membrane (PM) involved a mass spectrometric analysis of the structure in ticks (Kariu et al., 2013). This revealed information about the *I. scapularis* CDA-like protein, or IsCDA, which is an immunogenic, antigenic, and predominant (60-kDa) component of the PM, with homology to arthropod chitin deacetylase. IsCDA is mainly expressed in the gut and is induced during early tick feeding. Interestingly, the knockdown of *IsCDA* via RNA interference failed to influence PM formation or spirochete persistence, suggesting nonessential roles of the protein in tick biology and host-pathogen interaction. However, IsCDA immunization selectively impaired *B. burgdorferi* survival in the gut, but not the rest of the microbial community, implying that cross-reactive IsCDA antibodies may influence PM functions relevant to spirochete persistence in ticks. It is therefore speculated that the tick PM plays a preferential role in limiting the persistence of *B. burgdorferi* within the vector.

Interaction of B. burgdorferi with tick salivary gland proteins

Since the discovery of immunodominant antigens that are deposited in the host dermis by feeding ticks, *I. scapularis* salivary gland proteins (Salps) have been an intense focus of *Borrelia* research (Das et al., 2000; Das et al., 2001; Francischetti et al., 2005; Ribeiro et al., 2006). Indeed, such so-called saliva-assisted transmission (SAT) (Nuttall, 2019) events have been studied in detail for many other vector-borne diseases, and novel salivary gland vaccine targets have been identified, such as for the leishmaniasis infection (Morris et al., 2001; Valenzuela et al., 2001). Many Salps are soluble, feeding-induced, and co-transmitted to the host along with *B. burgdorferi*; thus, these proteins have an opportunity to influence spirochete migration between the vector and host, either directly or indirectly (Ramamoorthi et al., 2005; Rosa, 2005; Radolf et al., 2012). The generation of host immunity against *I. scapularis* salivary gland proteins, particularly those expressed as early as 24 hours after feeding, is sufficient for the impairment of *B. burgdorferi* transmission between ticks and hosts (Narasimhan et al., 2007). The following are examples of tick Salps that influence spirochete infection.

Salp15

As discussed earlier, OspC is a prominent borrelial outer membrane lipoprotein that is swiftly upregulated during tick transmission (Schwan et al., 1995) and binds the tick salivary gland protein Salp15 (Anguita et al., 2002; Ramamoorthi et al., 2005). The OspC-Salp15 interaction has important functional consequences that impact spirochete transmission and pathogenesis in a multipartite manner, involving the pathogen, vector, and host (Ramamoorthi et al., 2005). *B. burgdorferi* carries Salp15 (via OspC binding) during its transmission from the tick salivary gland to the murine host, thereby facilitating *Borrelia* survival within the infected mammal (Anguita et al., 2002). Salp15 specifically binds *B. burgdorferi* OspC and prevents the killing of spirochetes by borreliacidal antibodies *in vitro*. Thus, the coating of the spirochetes via binding to Salp15 in the tick salivary gland could have important implications for pathogen survival in the host. The binding of Salp15 to *Borrelia* could potentially mask the exposure of OspC to components of the host immune system, as OspC is shown to be highly immunogenic and is a target of host borreliacidal antibodies (Wilske et al., 1993; Fung et al., 1994; Wilske et al., 1996; Mbow et al., 1999; Liang et al., 2002; Wilske, 2005; Buckles et al., 2006; Earnhart et al., 2007; Earnhart and Marconi, 2007b, a). Additionally, by itself, Salp15 carries host immunosuppressive properties by inhibiting $CD4^+$ T cell activation (Anguita et al., 2002). Thus, OspC-bound Salp15 provides further benefits to spirochetes, potentially affording protection against immune cells that infiltrate the tick bite site. Interestingly, the OspC-Salp15 interaction could be beneficial to both the vector and pathogen (Ramamoorthi et al., 2005). *B. burgdorferi* induces *salp15* expression in the salivary glands of feeding arthropods; therefore, borrelial infection may provide survival advantages to the arthropod that produces this protein, as it could promote effective feeding via the binding of Salp15 and plasminogen at the bite site (Coleman et al., 1995; Coleman et al., 1997; Ramamoorthi et al., 2005; Lagal et al., 2006).

Salp25D

One of the abundant tick Salps, Salp25D (Das et al., 2001), plays an important role in the mammalian host in regards to the acquisition of *B. burgdorferi* by the vector (Narasimhan et al., 2007). The silencing of *salp25D* in *Ixodes* salivary glands has been revealed to block spirochete acquisition by ticks from infected mice. Salp25D functions as an antioxidant; thus, the Salp25D-mediated detoxification of reactive oxygen radicals at the tick-pathogen-host interface potentially provides a survival advantage to *B. burgdorferi* at the bite site in mice.

Salp20

In addition to Salp15, which binds to *B. burgdorferi* and contributes to the protection of spirochetes against host immunity, Salp20 has also been shown to protect serum-sensitive strains of *Borrelia* from lysis in *in vitro* assays (Tyson et al., 2007). Accordingly, Salp20 may also protect spirochetes from the host complement system within the vector blood meal and during transmission. Salp20 is related to the *I. scapularis* anti-complement *(Isac)* family of proteins that may block the host alternative complement pathway (Soares et al., 2005; Ribeiro et al., 2006). RNA interference studies have established that *Isac* and related family members of tick proteins promote efficient tick feeding and spirochete growth in ticks (Soares et al., 2005).

Tick histamine release factor

Histamine release factor (tHRF) was identified as an *Ixodes* salivary gland protein induced by *B. burgdorferi* infection (Dai et al., 2010). Recombinant tHRF binds host basophils and stimulates histamine release. Tick saliva, from which tHRF is secreted, is known to induce specific host chemokines, including histamine-releasing cytokine, that favor feeding by ticks (Langhansova et al., 2015). As the expression of tHRF varies in correlation to the rapid feeding phase of the tick, it was speculated that the protein contributes to blood meal engorgement and likely influences spirochete transmission from the arthropod. The RNA interference-mediated knockdown of tHRF impaired tick feeding and also decreased *Borrelia* levels in the murine hosts (Dai et al., 2010). Vaccination with tHRF reduced both the tick feeding efficiency and *B. burgdorferi* burden in mice. Additionally, it was reasoned that tHRF in ticks serves to facilitate vascular permeability and to increase blood flow to the tick bite site, ultimately favoring their blood meal engorgement (Dai et al., 2010). These data highlight the vaccine potential of tHRF and its ability to interfere with tick feeding and pathogen transmission.

Tick salivary lectin pathway inhibitor

A tick protein designated as tick salivary lectin pathway inhibitor (TSLPI) was found to impair the host complement-mediated killing of *B. burgdorferi* (Schuijt et al., 2011). TSLPI was shown to interfere with the human lectin complement cascade (Wagemakers et al., 2016), which reduced neutrophil phagocytosis and chemotaxis, thus decreasing the lysis of *B. burgdorferi* cells. TSLPI-knockdown ticks, or ticks that fed on TSLPI-vaccinated mice, were unable to transmit the pathogen efficiently. These studies underscore the importance of the lectin complement cascade and associated immune responses (Coumou et al., 2019) in borrelial transmission through *Ixodes* ticks and the establishment of early infection in mammals.

Additional salivary gland proteins

In a recent study, researchers identified another *Ixodes* Salp, termed Salp12, which is involved in the vector chemoattractant for the spirochete (Murfin et al., 2019). The gene encoding Salp12 is expressed in the salivary glands and gut. Interference with the function of Salp12 impacted spirochete acquisition by ticks, but not pathogen transmission. In addition to *I. scapularis* Salp15, as detailed in earlier paragraphs, novel Salp15-like immunosuppressant genes were also shown to be produced from the salivary glands of other *Ixodes* ticks, such as *I. persulcatus* (Mori et al., 2010). Additionally, salivary lipocalins that are differentially expressed during infection with *B. afzelii* are thought to support pathogen transmission to the host (Valdes et al., 2016). A panel of immunogenic proteins in the saliva of 24h-fed female *I. scapularis* was also identified, some of which may be involved in interactions with *B. burgdorferi*, consequently influencing spirochete transmission through ticks (Lewis et al., 2015).

Interactions with additional tick proteins

Dae2

A novel antibacterial enzyme was discovered to be acquired by *Ixodes* ticks through lateral gene transfer from a bacterium during the early evolution of ticks and mites. The enzyme, termed Dae2 (domesticated amidase effector 2), shares similarities to peptidoglycan-degrading toxins and specifically limits the proliferation of *B. burgdorferi* in ticks (Chou et al., 2015). Dae2 was also identified as one of the *Ixodes* antimicrobial effector molecules that are downstream of a newly discovered cross-species interferon signaling pathway. This pathway, which is triggered by the mammalian IFNγ that is acquired in the tick blood meal, induces a tick Rho-like GTPase in a STAT-dependent manner, ultimately reducing the levels of *B. burgdorferi* via the induction of select antimicrobial proteins, including Dae2 (Smith et al., 2016).

Tick antimicrobial peptides (defensin and others)

Defensins are a well-known class of antimicrobial peptides of innate immunity that are effective against a wide range of pathogens (Ganz, 2003); not surprisingly, *Ixodes* ticks encode multiple defensin-like genes (Hynes et al., 2005; Smith and Pal, 2014; Gulia-Nuss et al., 2016; Smith et al., 2016), some of which are induced in the gut after infection with *B. burgdorferi* (Sonenshine et al., 2002; Rudenko et al., 2005; Smith et al., 2016)(also see chapter 12). Two isoforms (Chrudimska et al., 2011) and six novel putative defensins from *I. ricinus* were identified that may have antimicrobial activity (Tonk et al., 2014b). It is unclear whether defensins play a role in controlling *B. burgdorferi* infection in ticks (Smith et al., 2016); however, two chemically-synthesized defensin peptides based on *I. scapularis* tick defensins (Tonk et al., 2014b), termed as Scapularisin-3 and Scapularisin-6, carry potent anti-*Listeria* and antifungal activities (Tonk et al., 2014a). Codon-substitution models support the neutral evolution of *Ixodes* defensins, resulting in its vast gene expansion (Wang and Zhu, 2011). Other than these defensins, the *Ixodes* tick genome also contains genes that encode a diverse array of antimicrobial proteins or peptides (Smith et al., 2016), some with potential anti-*B. burgdorferi* activities. Tick saliva is reported to possess anti-alarmin effects during the transmission of *Borrelia*, thereby aiding in the establishment of infection in the host (Marchal et al., 2011). Furthermore, events like blood meal digestion in ticks generate novel peptides with antimicrobial activities (Sonenshine, 1993; Kopacek et al., 2010), and the recent deep sequencing analysis of the *I. ricinus* hemocytome suggests additional genes that may encode still more antimicrobial peptides (Kotsyfakis et al., 2015).

Other *Ixodes* tick proteins, even ones that are intracellular, are shown to be induced upon *B. burgdorferi* infection in the vector. One such example is a selenoprotein that is involved in the regulation of oxidative and endoplasmic reticulum stress. However, the mechanisms by which intracellular *Ixodes* proteins support pathogen survival in ticks remain enigmatic (Kumar et al., 2019). Nevertheless, these studies collectively indicate the existence of highly discrete molecular interfaces where specific microbial and vector gene products interact, either directly or indirectly, thus contributing to *B. burgdorferi* survival in a natural enzootic cycle.

Interface between tick immune system and *B. burgdorferi*

Despite substantial existing knowledge of arthropod innate immunity (Hoffmann et al., 1999; Irving et al., 2001; Hoffmann and Reichhart, 2002; Hoffmann, 2003; Ferrandon et al., 2007; Lemaitre and Hoffmann, 2007; Clayton et al., 2014), the molecular mechanisms by which the *Ixodes* immune system interfaces with and recognizes invading microbes remain poorly understood. Recent studies suggest that *B. burgdorferi* levels in ticks are affected by the independent knockdown of components from specific immune pathways, providing the first evidence of an active interface between the tick immune system and the agents of Lyme disease (Smith et al., 2016; Shaw et al., 2017; Kitsou and Pal, 2018). Research has also revealed that the gut microbiota associates with the *Ixodes* immune system, playing critical and influential roles in arthropod biology, including the ability to maintain and transmit *B. burgdorferi* (Narasimhan et al., 2014b; Narasimhan et al., 2017). These studies highlight novel concepts of tick immune responses that probably involve direct and indirect methods of pathogen recognition, including that of *B. burgdorferi* (Kitsou and Pal, 2018), and therefore incorporate multifaceted interactions between invading microbes and defense responses, as summarized below.

The arthropod immune system likely recognizes microorganisms directly by germ line-encoded, non-rearranging cellular receptors, and controls infection through the rapid generation of microbicidal effector responses (Hoffmann and Reichhart, 2002; Hoffmann, 2003; Ferrandon et al., 2007; Valanne et al., 2011; Myllymaki et al., 2014; Smith and Pal, 2014; Shaw et al., 2018). Recent studies show that the *Ixodes* immune deficiency (IMD) pathway senses microbial molecular patterns, like that of infection-derived lipids, such as 1-palmitoyl-2-oleoyl-*sn*-glycero-3-phosphoglycerol and 1-palmitoyl-2-oleoyl diacylglycerol, or diaminopimelic-type peptido-glycans, which are present in diverse bacterial pathogens including *B. burgdorferi*, ultimately leading to the generation of potent microbicidal responses in ticks (Shaw et al., 2017). The details of such molecular interactions between the tick immune system and spirochete cells or antigens, as well as the identity of any tick receptors that recognize microbial molecules, remain unknown. The tick E3 ubiquitin ligase X-linked inhibitor of apoptosis (XIAP) has recently been demonstrated to act as a molecular rheostat for the IMD pathway (McClure Carroll et al., 2019).

Studies of *Drosophila* and mammals also suggest the existence of alternative immune mechanisms, such as indirect or more host-centric immune recognition pathways, for interactions with virulent bacteria. These indirect immune cascades, including effector-triggered immunity (Cui et al., 2015), can sense pathogens or pathogenic-derived effector molecules that modify a host target molecule (Boyer et al., 2011). While examining the molecular mechanism of *B. burgdorferi* acquisition in *Ixodes* ticks, Smith et al. identified the existence of a novel cross-kingdom signaling cascade, incorporating both the mammalian host and tick vector (Smith et al., 2016). This cascade enables *I. scapularis* to recognize a mammalian cytokine, identified as IFNγ, which arrives in ticks via the ingested blood meal and functions as a potent infection cue by interacting with currently unknown tick receptor(s), thus activating microbicidal responses that limit the proliferation of invading spirochetes (de Silva, 2016; Smith et al., 2016). This novel IFNγ signaling involves the tick STAT protein, which induces an *Ixodes* GTPase (termed as IGTPase), as well as multiple borreliacidal effector molecules, including Dae2 (Chou et al., 2015), thereby reducing *B. burgdorferi* persistence in the arthropod (Smith et al., 2016). More recently, tick extracellular vesicles, resulting from exosome and microvesicle biogenesis, have been proposed to manipulate cross-kingdom signaling networks, potentially impacting pathogen spread and colonization in the host (Chavez et al., 2019). While these indirect or host-centric recognition pathways of infection and interaction between the tick immune system and invading microbe are emerging concepts, additional studies are required to enrich our fundamental knowledge of tick immunobiology and host-pathogen interactions.

Microbiome interactions

In *Ixodes* ticks, *B. burgdorferi* resides in the gut, which is also inhabited by diverse microbiota, as indicated by a series of recent studies using either laboratory or wild-collected ticks (Narasimhan and Fikrig, 2015; Bonnet et al., 2017; Swei and Kwan, 2017; Couper and Swei, 2018; Greay et al., 2018; Hernandez-Jarguin et al., 2018; Aivelo et al., 2019; Landesman et al., 2019; Tokarz et al., 2019). The role of gut microbiota, its interactions with invading pathogens, and the maintenance of immune homeostasis are now well-established concepts in many organisms (Ley et al., 2008; Molina-Cruz et al., 2008; Round and Mazmanian, 2009; Hooper et al., 2012; Buchon et al., 2013; Kamada et al., 2013). The *Ixodes* gut microbiota was recently described as having an important role in the persistence of tick-borne pathogens like *B. burgdorferi* (Narasimhan et al., 2014b). The presence of pathogens in the tick gut, along with their possible interactions with resident gut microbiota, could impair tick structural barriers, such as the peritrophic matrix, via the modulation of tick proteins and/or specific immune signaling pathways. These complex interactions, involving diverse pathogens, gut microbiota, and tick proteins, essentially shape the survival of spirochetes in the vector (Narasimhan et al., 2014b; Abraham et al., 2017). Further studies of tick-pathogen and immunome-gut microbiome interactions, particularly how they impact spirochete persistence, will be necessary to enhance our knowledge of tick-borne infections. The use of new experimental approaches, such as dysbiosis (Narasimhan et al., 2014b), aposymbiotic ticks, and the artificial feeder system, including ones recently developed (Oliver et al., 2016; Koci et al., 2018), which can eliminate unwanted host factors and allow an unprecedented degree of empirical manipulation, will be especially valuable for these investigations.

Ixodes genomics

Two separate studies provided initial information about the complete genome sequence of *I. scapularis*, both in ticks (Gulia-Nuss et al., 2016) and an *Ixodes* cell line (Miller et al., 2018). The exploration of the *Ixodes* genome offers new avenues for scientific investigation and should reveal novel features of the microbe-vector interaction. However, further research on this subject largely awaits major developments in gene manipulation tools and the pursuit of additional genome sequencing efforts, which are summarized below.

Gene manipulation

Stable methods of gene manipulation are well-developed in other arthropod vectors, such as in mosquitoes (Catteruccia et al., 2000; Catteruccia, 2007), and are influenced by the development of advanced genetic tools in the *Drosophila* (Venken and Bellen, 2005). Gene manipulation in ticks remains in its infancy because few of the available genetic tools are readily applicable. RNA interference (RNAi) is now viewed as the only possible gene-silencing technique in ticks (de la Fuente et al., 2007a), although the knockdown effects are transient. Unfortunately, unlike many other arthropods (Bucher et al., 2002; He et al., 2006; Khila and Grbic, 2007), the use of parental RNAi to introduce stable RNAi effects in offspring has brought limited success in ticks. Additional formidable challenges include the lack of information about tick embryology and the tick genome; further progress in these areas would stimulate the development of more advanced gene manipulation methodologies. However, the long life span and slow post-embryonic development of *Ixodes* ticks will pose serious problems for the phenotypic analysis of mutants after the genome has been manipulated. These efforts will be further complicated by the sensitivity of *I. scapularis* embryonic development, including premolt and preoviposition development, to environmental conditions, such as temperature (Heath, 1979; Ogden et al., 2004). Nevertheless, *I. scapularis* cell lines (Munderloh et al., 1994) are available to the research community, which should allow us to optimize the manipulation of the tick genome in cell culture.

Whole genome sequence

The *I. scapularis* Genome Project (IGP), a collaborative effort between the international community of tick researchers and genome sequencing centers, was initiated more than a decade ago with the support of the National Institutes of Health (Hill and Wikel, 2005; Pagel Van Zee et al., 2007), with the intention of performing whole genome sequencing of *I. scapularis*. More than 18 million trace reads, representing approximately a five-fold coverage of the genome, have been deposited thus far at the National Center for Biotechnology Information (NCBI) Trace Archive (Pagel Van Zee et al., 2007). In addition, 20 *Ixodes* bacterial artificial chromosome (BAC) clones, 370,000 BAC-end reads, and more than 80,000 expressed sequence tags (ESTs) have been sequenced. The sequencing, assembly, and annotation of a six-fold coverage of the *Ixodes* genome were expected by 2008, but these efforts took longer due to the large genome size, genomic complexity, and presence of extensive repetitive DNA in *I. scapularis* (Ullmann et al., 2005; Geraci et al., 2007). In 2016, the annotated whole genome sequencing of the *I. scapularis* tick was published (Gulia-Nuss et al., 2016), which was subsequently followed by a draft genome sequence for the *I. scapularis* ISE6 cell line (Miller et al., 2018). The study that addressed the whole genome analysis of ticks annotated approximately 57% of the genome, indicating the presence of at least 20,486 protein-coding genes, with a notable occurrence of gene families associated with tick-host interactions. Despite these seminal efforts, our knowledge of the tick genome remains limited, although more sophisticated sequencing tools are increasingly available and can be utilized to construct a complete tick genome atlas. More recently, high-throughput long-read DNA/RNA sequencing (Amarasinghe et al., 2020), single-cell and single-nucleus RNA-sequencing methods (Ding et al., 2020), and chromosome conformation capture methods (Lieberman-Aiden et al., 2009) have provided unprecedented views of the genome, including that of arthropod vectors, particularly the three-dimensional organization of chromosomes, genomes, and genome-wide chromatin contacts (Dudchenko et al., 2017; Wang et al., 2018), which are recognized as crucial aspects of gene regulation (Bonev et al., 2017). The wealth of information from these genome studies is expected to have a high impact on future gene manipulation technologies and will contribute to our understanding of vector biology, including pathogen survival and transmission through *Ixodes* ticks (de la Fuente et al., 2016).

Ixodes transcriptomics

The availability of next-generation sequencing technologies, such as genome-wide RNA-sequencing (RNA-Seq) (Wang et al., 2009; Stark et al., 2019) and single cell transcriptomics (La Manno, 2019; Skinnider et al., 2019), has provided a unique opportunity to study gene expression in *Ixodes* ticks, including its relevance to vector-pathogen interactions, not only on a high-throughput basis, but also at the levels of the single cell, tissue, or whole organism. Conventional and targeted approaches, like subtractive hybridization (Narasimhan et al., 2017) or quantitative RT-PCR (Smith et al., 2016), have previously been utilized for the identification of tick transcripts that are modulated during borrelial infection. The recent progress of next-generation sequencing, now with capabilities for short or long reads and *de novo* genome transcriptome analysis (Schwarz et al., 2013), has allowed for a robust and global analysis of the *Ixodes* genes that are modulated during the entry, persistence, and exit of *B. burgdorferi* through ticks. Over the past decades, a number of studies have addressed *Ixodes* transcriptomics in the presence or absence of tick-borne pathogens, including in the gut (Perner et al., 2016), salivary glands (Francischetti et al., 2005; McNally et al., 2012; Schwarz et al., 2013), hemocytes (Kotsyfakis et al., 2015), and whole ticks (Charrier et al., 2018) or cell lines (Weisheit et al., 2015; Alberdi et al., 2016; Mansfield et al., 2017), which is important for understanding the unusual biology of ticks, such as their blood meal digestion, development, and host-pathogen interactions. A recent meta-analysis of published RNA-Seq studies compared the genetic variations between wild strains and laboratory strains of *I. ricinus* (Charrier et al., 2018). Together, these data will enrich our knowledge of the biology of an ancient arthropod vector and will inform the development of novel control strategies against tick bites or tick-transmitted infections.

The integration of transcriptomics data with other available information, like tick biology and proteomics (Ribeiro and Francischetti, 2003), as well as experimental vaccination trials, will aid in the identification of novel tick protective antigens and possibly future vaccines (Contreras et al., 2016; Contreras et al., 2019). This is particularly important, as *Ixodes* ticks can parasitize almost any vertebrate class across the globe and are second only to mosquitoes as vectors of serious human infectious diseases. The tick salivary gland transcriptome (Ribeiro and Francischetti, 2003; Francischetti et al., 2005; Ribeiro et al., 2006) will foster research on the roles of additional tick salivary gland proteins that support *B. burgdorferi* infection and transmission to the host. The salivary gland transcriptome has identified transcripts encoding proteins with signature functional domains, including basic-tail and Kunitz, as well as proline-rich peptides, metalloproteases, and novel proteins with similarities to disintegrins, histamine-binding proteins, anti-complement proteins, and a neuropeptide-like protein (nlp-31) that may have antimicrobial activity (Francischetti et al., 2005). The further characterization of these proteins and their functions will greatly contribute to our understanding of host immunity and the development of vaccines against both ticks and the pathogens that they transmit.

Ixodes proteomics and metabolomics

Similar to transcriptomics, other high-throughput "omics" approaches, such as proteomics and metabolomics, provide useful information about the overall protein diversity and steady-state levels of small molecules in cells or tissues, which is insightful in regards to tick biology and the discovery of vaccine or therapeutic targets (Contreras et al., 2019). High resolution LC-MS/MS (Seger and Salzmann, 2020), iTRAQ (isobaric Tags for Relative and Absolute Quantitation) (Luo and Zhao, 2012), and shotgun proteomics (Lisacek et al., 2006) have allowed for more sensitive and accurate analyses of proteomics and metabolomics in tick-borne infections. These techniques have been applied to the study of pathogens like *B. burgdorferi*, including the characterization of the proteome in the borrelial outer membrane and the classification of protein-protein interactions (Yang et al., 2011; Yang et al., 2018), in addition to tracking the sources of blood meals (Onder et al., 2014) and analyzing the proteome in the tick vector (Di Venere et al., 2015; Iovinella et al., 2016) or in infected tick cells (Weisheit et al., 2015; Grabowski et al., 2016), as well as the immuno-proteomic identification of tick antigens (Vu Hai et al., 2013). These experimental approaches have been applied to a system-level analysis of proteome complexity in the *Ixodes* gut and salivary glands (Schwarz et al., 2014), which are the organs relevant to blood meal engorgement, vector interactions, and the colonization of tick-borne pathogens. Another study used iTRAQ to identify at least 12 salivary gland proteins in *I. ricinus* that were modulated in the presence of various *Borrelia* strains (Cotte et al., 2014). As the post-translational modification of proteins is critical to protein function, including host-pathogen interactions, bioanalytical methods for glycobiology were developed, such as those based on mass spectrometry (Vechtova et al., 2018), although they have not yet been used in studies addressing tick interactions with the spirochete. Proteomic comparative studies of various tick cell lines also were reported, which proved to be indispensable for tick research (Munderloh et al., 1994; Munderloh and Kurtti, 1995), as the tools allow for cell lines to be used as controlled experimental systems in the study of tick-pathogen interactions (Loginov et al., 2019). Using gas chromatography coupled to a mass spectrometry-based approach, a recent study explored the differences between the metabolic profiles of *B. burgdorferi*-infected and naïve *I. scapularis* nymphal ticks during blood meal engorgement (Hoxmeier et al., 2017). Additional studies on tick-pathogen metabolic interactions (see review (Cabezas-Cruz et al., 2019)) are certainly warranted, as they will contribute both to our understanding of the complex interactions between *B. burgdorferi* and *Ixodes* ticks, and to the development of novel interventions against tick-borne infections.

Interaction-blocking Lyme disease vaccines

During infection in mammals, including humans (see chapter 25), *B. burgdorferi* enters the skin at the site of the tick bite, typically resulting in a dermal lesion called erythema migrans (Nadelman and Wormser, 1998; Wormser et al., 2006), although almost 30% of human subjects fail to display the lesion. The diagnosis of Lyme disease (Marques, 2015) (see chapter 23), especially during early infection, still remains a challenging task, due to the clinical manifestations it has in common with other febrile diseases, in addition to variations in the immune responses of infected individuals. The available antibiotic treatment usually resolves clinical symptoms, if administered in the early stages of infection. However, persistent or relapsing symptoms (e.g., fatigue, musculoskeletal pain, cognitive difficulties, etc.) later develop in a subset of patients; these symptoms are collectively referred to as chronic Lyme disease or antibiotic treatment-resistant Lyme arthritis, a condition otherwise termed as post-treatment Lyme disease syndrome (PTLDS) (Marques, 2008; Aucott et al., 2013). The pathogenesis and treatment of PTLDS remain obscure (see chapter 24). These factors, combined with a remarkably high incidence of Lyme disease (Mead, 2015) (see chapter 22), further underscore the need for the development of effective preventive measures, such as vaccines to combat infection (Gomes-Solecki et al., 2020), which is the research focus in many laboratories. Such studies aim to investigate the protective efficacy of potential vaccine candidates in animal models of Lyme borreliosis (chapter 16), targeting specific spirochete and/or tick antigens that are essential for pathogen transmission and/or persistence, as exemplified in the following sections (also see chapter 20).

Antigens based on spirochetes

Understanding the molecular details of the *B. burgdorferi* life cycle, particularly in the arthropod vector, is likely to support the development of novel Lyme disease vaccines that are based on spirochete antigens. This notion is supported by the success and mechanism of protection afforded by the original human vaccine against Lyme disease (Fikrig et al., 1990; Sigal et al., 1998; Steere et al., 1998), as well as many other potential vaccine candidates (for a detailed review, see Chapter 19).

Antigens based on tick targets

In addition to bacterial antigens, proteins from the vector also represent promising vaccine candidates (de la Fuente et al., 2007b; Rego et al., 2019). Vaccines targeting tick gut or salivary gland proteins that interact with *Borrelia*, such as Salp15 (Anguita et al., 2002; Ramamoorthi et al., 2005; Dai et al., 2009; Sultana et al., 2016) or TROSPA (Pal et al., 2004a), could inhibit pathogen persistence in ticks and block pathogen transmission to vertebrates. Several *I. scapularis* proteins that impact borrelial persistence and transmission through ticks (Yang et al., 2014; Narasimhan et al., 2017) could be used as novel vaccination strategies. For example, a gut Dual oxidase (Duox) that supports spirochete persistence within ticks was identified as a *B. burgdorferi*-inducible *I. scapularis* gene product (Yang et al., 2014). As previously demonstrated by mosquitoes that were infected with malarial parasites (Kumar et al., 2010), the supportive role of Duox in spirochete survival in ticks was attributed to its involvement in the formation of the dityrosine network, a molecular barrier surrounding the gut lumen. This network has been shown to dampen the epithelial immune response by maintaining homeostasis within the gut microbial community, while also protecting invading pathogens from the vector immune system (Kumar et al., 2010). Therefore, vector proteins like TROSPA or Duox, especially the regions that are unique to ticks, could serve as components of novel transmission-blocking Lyme disease vaccines. Additional *Ixodes* proteins that have been studied as potential anti-tick vaccines have also demonstrated their impacts on the acquisition or transmission of *B. burgdorferi* in ticks, including subolesin (Almazan et al., 2003; Bensaci et al., 2012), TSLPI, a tick mannose-binding lectin inhibitor (Schuijt et al., 2011), and tHRF, a tick histamine release factor (Dai et al., 2010).

Vaccination studies using a combination of tick and spirochete antigens, such as Salp15 and OspA or OspC, were also described as markedly enhancing the protective efficacy of antibodies against *B. burgdorferi* antigens (Dai et al., 2009). A recent review highlighted the current state of knowledge about anti-tick vaccine candidates that can prevent the transmission of tick-borne pathogens (Rego et al., 2019; Bhowmick and Han, 2020).

Concluding remarks

With more than 800 tick species in existence globally, these ancient ectoparasites have evolved a remarkable tolerance to extreme climates, as well as survivability for prolonged periods, and the abilities to maintain and transmit varied sets of pathogens. *Ixodes* ticks, which arguably transmit the most diverse range of human and animal pathogens amongst all ticks, including a striking array of Lyme borreliosis agents, are remarkably adapted to engorge on a wide collection of vertebrate hosts. The interactions between a distinct set of ligand receptors contributes to the successful persistence of human pathogens, such as *B. burgdorferi*, through a complex life cycle in nature involving *Ixodes* ticks. Studies have established the clear evolution of an intimate relationship between the pathogen and its arthropod vector. *B. burgdorferi* can induce the expression of specific tick genes, such as *TROSPA* (Pal et al., 2004a), *salp15* (Ramamoorthi et al., 2005; Dai et al., 2009), *duox* (Yang et al., 2014), and *pixr* (Narasimhan et al., 2017). Experimental evidence suggests that at least one of these genes, *salp15*, can potentially modulate host immunity in favor of the tick feeding process; therefore, *B. burgdorferi* infection may provide survival advantages to ticks. Further work is warranted to determine the exact ecological relationship between ticks and *B. burgdorferi,* particularly to ascertain whether it is parasitic or mutualistic. Studies are also needed to address the interactions between *Ixodes* ticks and organisms that have been identified as potential symbionts (Stewart and Bloom, 2020), as well as how polymicrobial interactions shape *B. burgdorferi* infection in ticks. The identification of additional microbial and vector gene products that are actively engaged in spirochete acquisition, maintenance, and transmission through the arthropod should also be explored in future research.

The long life span and slow post-embryonic development of *Ixodes* (Sauer, 1986; Sonenshine, 1993), along with the relatively limited availability of high-resolution genomic information, complicate the application of gene manipulation tools to tick research and hinder the analyses of the tick transcriptome and proteome. Future *Ixodes* studies are required on embryology (Sonenshine, 1993), the three-dimensional configuration of the genome (Kempfer and Pombo, 2020), gene manipulation and transgenesis (Sun et al., 2017), and paratransgenesis (Wilke and Marrelli, 2015), as these would certainly expand upon experimental opportunities and foster a deeper understanding of tick biology and development. Such efforts will also contribute to our knowledge of the molecular interactions between *Borrelia* and tick gene products. Specific questions to be addressed include the molecular details of how *B. burgdorferi* evades tick immunity, the gut digestive activities that occur upon spirochete arrival in feeding ticks, and the mechanism by which the pathogen continues to persist in ticks throughout the intermolt periods that are dominated by temperature extremities and starvation. Studies should also distinguish how *B. burgdorferi* crosses physical barriers, such as the chitinous gut peritrophic membrane and newly-recognized acellular barriers like the dityrosine network, and then selectively migrates through the tick gut, hemocoel, and salivary gland epithelia during transmission to mammals. *Ixodes* and *Borrelia* have probably co-evolved numerous but distinct molecular interfaces that support their coexistence.

Globally, the incidences of tick-borne diseases, including Lyme borreliosis, are increasing. At least 13 new tick-borne pathogens were recognized in the Western Hemisphere during the last 20 years (Paddock et al., 2016). It is likely that ticks evolved robust genetic and epigenetic mechanisms that promote their survival in environmental extremities and their unparalleled tolerance to a wide range of microbial pathogens. Future investigations of the genetic and epigenetic bases of tick vectorial capacity will enrich our knowledge of the interactions between *Ixodes* and diverse pathogens, contributing to a better understanding of tick-borne infections. In particular, more thorough knowledge about the biology of the tick-*B. burgdorferi* interaction (Kurokawa et al., 2020) not only will contribute to the development of preventive measures against Lyme disease but also serve as a paradigm for novel control measures against additional tick-borne infections.

Acknowledgement
The authors are thankful to Kathryn Nassar for assistance with the preparation of this chapter. We are grateful to Aravinda M. de Silva, Deborah Shroder, and Ireen Dryburgh-Barry for assistance with the preparation of the original manuscript. Lyme disease research in our laboratories is supported by several past and ongoing funding supports from the National Institutes of Health (including currently active grants, P01AI138949 and R01AI154542 to U.P), the Department of Defense, Global Lyme Alliance, Merck & Co. Inc., the Arthritis Foundation, the American Heart Association, and the Steven and Alexandra Cohen Foundation. Erol Fikrig is an Investigator of the Howard Hughes Medical Institute. Chrysoula Kitsou is the recipient of the Deborah and Mark Blackman Postdoctoral Fellowship from Global Lyme Alliance.

References

Abraham, N.M., Liu, L., Jutras, B.L., Yadav, A.K., Narasimhan, S., Gopalakrishnan, V., Ansari, J.M., Jefferson, K.K., Cava, F., Jacobs-Wagner, C., *et al.* (2017). Pathogen-mediated manipulation of arthropod microbiota to promote infection. Proc Natl Acad Sci U S A *114*, E781-E790. https://doi.org/10.1073/pnas.1613422114

Aivelo, T., Norberg, A., and Tschirren, B. (2019). Bacterial microbiota composition of *Ixodes ricinus* ticks: the role of environmental variation, tick characteristics and microbial interactions. PeerJ *7*, e8217. https://doi.org/10.7717/peerj.8217

Alban, P.S., Johnson, P.W., and Nelson, D.R. (2000). Serum-starvation-induced changes in protein synthesis and morphology of *Borrelia burgdorferi*. Microbiology *146 (Pt 1)*, 119-127. doi: 10.1099/00221287-146-1-119.

Alberdi, P., Mansfield, K.L., Manzano-Roman, R., Cook, C., Ayllon, N., Villar, M., Johnson, N., Fooks, A.R., and de la Fuente, J. (2016). Tissue-Specific Signatures in the Transcriptional Response to *Anaplasma phagocytophilum* Infection of *Ixodes scapularis* and *Ixodes ricinus* Tick Cell Lines. Frontiers in cellular and infection microbiology *6*, 20. https://doi.org/10.3389/fcimb.2016.00020

Almazan, C., Kocan, K.M., Bergman, D.K., Garcia-Garcia, J.C., Blouin, E.F., and de la Fuente, J. (2003). Identification of protective antigens for the control of *Ixodes scapularis* infestations using cDNA expression library immunization. Vaccine *21*, 1492-1501. https://doi.org/S0264410X02006837

Alverson, J., Bundle, S.F., Sohaskey, C.D., Lybecker, M.C., and Samuels, D.S. (2003). Transcriptional regulation of the ospAB and ospC promoters from *Borrelia burgdorferi*. Mol Microbiol *48*, 1665-1677. doi: 10.1046/j.1365-2958.2003.03537.x.

Amarasinghe, S.L., Su, S., Dong, X., Zappia, L., Ritchie, M.E., and Gouil, Q. (2020). Opportunities and challenges in long-read sequencing data analysis. Genome Biol *21*, 30. https://doi.org/10.1186/s13059-020-1935-5

Anderson, J.F. (1989). Epizootiology of *Borrelia* in *Ixodes* tick vectors and reservoir hosts. Rev Infect Dis *11 Suppl 6*, S1451-1459. doi: 10.1093/clinids/11.supplement_6.s1451.

Anderson, J.F., Duray, P.H., and Magnarelli, L.A. (1987). Prevalence of *Borrelia burgdorferi* in white-footed mice and Ixodes dammini at Fort McCoy, Wis. J Clin Microbiol *25*, 1495-1497. doi: 10.1128/JCM.25.8.1495-1497.1987.

Anguita, J., Hedrick, M.N., and Fikrig, E. (2003). Adaptation of *Borrelia burgdorferi* in the tick and the mammalian host. FEMS Microbiol Rev *27*, 493-504. doi: 10.1016/S0168-6445(03)00036-6.

Anguita, J., Ramamoorthi, N., Das, S., Conze, D., Askenase, P., Rincón, M., Kantor, F.S., and Fikrig, E. (2002). Salp15, an *Ixodes scapularis* saliva protein, inhibits CD4+ T cell activation. Immunity *16*, 849-859. doi: 10.1016/s1074-7613(02)00325-4.

Arnold, W.K., Savage, C.R., Brissette, C.A., Seshu, J., Livny, J., and Stevenson, B. (2016). RNA-Seq of *Borrelia burgdorferi* in Multiple Phases of Growth Reveals Insights into the Dynamics of Gene Expression, Transcriptome Architecture, and Noncoding RNAs. PLoS One *11*, e0164165. https://doi.org/10.1371/journal.pone.0164165

Arnold, W.K., Savage, C.R., Lethbridge, K.G., Smith, T.C., 2nd, Brissette, C.A., Seshu, J., and Stevenson, B. (2018). Transcriptomic insights on the virulence-controlling CsrA, BadR, RpoN, and RpoS regulatory networks in the Lyme disease spirochete. PLoS One *13*, e0203286. https://doi.org/10.1371/journal.pone.0203286

Aucott, J.N., Crowder, L.A., and Kortte, K.B. (2013). Development of a foundation for a case definition of post-treatment Lyme disease syndrome. International journal of infectious diseases : IJID : *17*, e443-449. https://doi.org/10.1016/j.ijid.2013.01.008

Beerntsen, B.T., James, A.A., and Christensen, B.M. (2000). Genetics of mosquito vector competence. Microbiol Mol Biol Rev *64*, 115-137. doi: 10.1128/mmbr.64.1.115-137.2000.

Benach, J.L., Bosler, E.M., and Hanrahan, J.P. (1983). Spirochetes isolated from the blood of two patients with Lyme disease. N Engl J Med *308*, 740-742.

Bensaci, M., Bhattacharya, D., Clark, R., and Hu, H.T. (2012). Oral Vaccination With Vaccinia Virus Expressing the Tick Antigen Subolesin Inhibits Tick Feeding and Transmission of *Borrelia burgdorferi* Vaccination. Vaccine *30(42)*:6040-6. doi: 10.1016/j.vaccine.2012.07.053.

Bernard, Q., Smith, A.A., Yang, X., Koci, J., Foor, S.D., Cramer, S.D., Zhuang, X., Dwyer, J.E., Lin, Y.P., Mongodin, E.F., *et al.* (2018). Plasticity in early immune evasion strategies of a bacterial pathogen. Proc Natl Acad Sci U S A. https://doi.org/10.1073/pnas.1718595115

Bhowmick, B., and Han, Q. (2020). Understanding Tick Biology and Its Implications in Anti-tick and Transmission Blocking Vaccines Against Tick-Borne Pathogens. Front Vet Sci *7*, 319. https://doi.org/10.3389/fvets.2020.00319

Bonev, B., Mendelson Cohen, N., Szabo, Q., Fritsch, L., Papadopoulos, G.L., Lubling, Y., Xu, X., Lv, X., Hugnot, J.P., Tanay, A., *et al.* (2017). Multiscale 3D Genome Rewiring during Mouse Neural Development. Cell *171*, 557-572 e524. https://doi.org/10.1016/j.cell.2017.09.043

Bonnet, S.I., Binetruy, F., Hernandez-Jarguin, A.M., and Duron, O. (2017). The Tick Microbiome: Why Non-pathogenic Microorganisms Matter in Tick Biology and Pathogen Transmission. Frontiers in cellular and infection microbiology *7*, 236. https://doi.org/10.3389/fcimb.2017.00236

Bontemps-Gallo, S., Lawrence, K.A., Richards, C.L., and Gherardini, F.C. (2018). *Borrelia burgdorferi* genes, bb0639-0642, encode a putative putrescine/spermidine transport system, PotABCD, that is spermidine specific and essential for cell survival. Mol Microbiol *108*, 350-360. https://doi.org/10.1111/mmi.13940

Boyer, L., Magoc, L., Dejardin, S., Cappillino, M., Paquette, N., Hinault, C., Charriere, G.M., Ip, W.K., Fracchia, S., Hennessy, E., *et al.* (2011). Pathogen-derived effectors trigger protective immunity via activation of the Rac2 enzyme and the IMD or Rip kinase signaling pathway. Immunity *35*, 536-549. https://doi.org/10.1016/j.immuni.2011.08.015

Boyle, W.K., Groshong, A.M., Drecktrah, D., Boylan, J.A., Gherardini, F.C., Blevins, J.S., Samuels, D.S., and Bourret, T.J. (2019). DksA Controls the Response of the Lyme Disease Spirochete *Borrelia burgdorferi* to Starvation. J Bacteriol *201*. https://doi.org/10.1128/JB.00582-18

Boyle, W.K., Wilder, H.K., Lawrence, A.M., and Lopez, J.E. (2014). Transmission dynamics of *Borrelia turicatae* from the arthropod vector. PLoS Negl Trop Dis *8*, e2767. https://doi.org/10.1371/journal.pntd.0002767

Brangulis, K., Akopjana, I., Petrovskis, I., Kazaks, A., Jekabsons, A., Jaudzems, K., Viksna, A.,

Bertins, M., and Tars, K. (2020a). Structural analysis of *Borrelia burgdorferi* periplasmic lipoprotein BB0365 involved in Lyme disease infection. FEBS Lett *594*, 317-326. https://doi.org/10.1002/1873-3468.13594

Brangulis, K., Akopjana, I., Petrovskis, I., Kazaks, A., Zelencova, D., Jekabsons, A., Jaudzems, K., and Tars, K. (2020b). BBE31 from the Lyme disease agent *Borrelia burgdorferi,* known to play an important role in successful colonization of the mammalian host, shows the ability to bind glutathione. Biochim Biophys Acta Gen Subj *1864*, 129499. https://doi.org/10.1016/j.bbagen.2019.129499

Brisson, D., Drecktrah, D., Eggers, C.H., and Samuels, D.S. (2012). Genetics of *Borrelia burgdorferi*. Annual review of genetics *46*, 515-536. https://doi.org/10.1146/annurev-genet-011112-112140

Broadwater, A.H., Sonenshine, D.E., Hynes, W.L., Ceraul, S., and De, S.A. (2002). Glass capillary tube feeding: a method for infecting nymphal *Ixodes scapularis* (Acari: Ixodidae) with the Lyme disease spirochete B*orrelia burgdorferi*. J Med Entomol *39*, 285-292. doi: 10.1603/0022-2585-39.2.285.

Brooks, C.S., Hefty, P.S., Jolliff, S.E., and Akins, D.R. (2003). Global analysis of *Borrelia burgdorferi* genes regulated by mammalian host-specific signals. Infect Immun *71*, 3371-3383. doi: 10.1128/iai. 71.6.3371-3383.2003.

Brorson, O., and Brorson, S.H. (1998). In vitro conversion of *Borrelia burgdorferi* to cystic forms in spinal fluid, and transformation to mobile spirochetes by incubation in BSK-H medium. Infection *26*, 144-150. doi: 10.1007/BF02771839.

Brorson, O., and Brorson, S.H. (2004). An in vitro study of the susceptibility of mobile and cystic forms of Borrelia burgdorferi to tinidazole. Int Microbiol 7, 139-142.

Bucher, G., Scholten, J., and Klingler, M. (2002). Parental RNAi in Tribolium (Coleoptera). Curr Biol *12*, R85-86. doi: 10.1016/ s0960-9822(02)00666-8.

Buchon, N., Broderick, N.A., and Lemaitre, B. (2013). Gut homeostasis in a microbial world: insights from Drosophila melanogaster. Nat Rev Microbiol *11*, 615-626. https://doi.org/10.1038/nrmicro3074

Buckles, E.L., Earnhart, C.G., and Marconi, R.T. (2006). Analysis of antibody response in humans to the type A OspC loop 5 domain and assessment of the potential utility of the loop 5 epitope in Lyme disease vaccine development. Clin Vaccine Immunol *13*, 1162-1165. doi: 10.1128/CVI.00099-06.

Bugrysheva, J.V., Bryksin, A.V., Godfrey, H.P., and Cabello, F.C. (2005). *Borrelia burgdorferi* rel is responsible for generation of guanosine-3'-diphosphate-5'-triphosphate and growth control. Infect Immun *73*, 4972-4981. doi: 10.1128/IAI.73.8.4972-4981.2005

Bugrysheva, J.V., Pappas, C.J., Terekhova, D.A., Iyer, R., Godfrey, H.P., Schwartz, I., and Cabello, F.C. (2015). Characterization of the Rel_{Bbu} Regulon in *Borrelia burgdorferi* Reveals Modulation of Glycerol Metabolism by (p)ppGpp. PLoS One *10*, e0118063. https://doi.org/ 10.1371/journal.pone.0118063

Burgdorfer, W., Hayes, S., and Benach, J. (1988). Development of *Borrelia burgdorferi* in ixodid tick vectors. Ann NY Acad Sci *539*, 172-179. doi: 10.1111/j.1749-6632.1988.tb31851.x.

Bykowski, T., Woodman, M.E., Cooley, A.E., Brissette, C.A., Brade, V., Wallich, R., Kraiczy, P., and Stevenson, B. (2007a). Coordinated expression of *Borrelia burgdorferi* complement regulator-acquiring surface proteins during the Lyme disease spirochete's mammal-tick infection cycle. Infect Immun *75*, 4227-4236. doi: 10.1128/IAI.00604-07

Bykowski, T., Woodman, M.E., Cooley, A.E., Brissette, C.A., Wallich, R., Brade, V., Kraiczy, P., and Stevenson, B. (2007b). *Borrelia burgdorferi* complement regulator-acquiring surface proteins (BbCRASPs): Expression patterns during the mammal-tick infection cycle. Int J Med Microbiol. 298 Suppl 1(Suppl 1):249-56. doi: 10.1016/j.ijmm. 2007.10.002

Cabezas-Cruz, A., Espinosa, P., Alberdi, P., and de la Fuente, J. (2019). Tick-Pathogen Interactions: The Metabolic Perspective. Trends Parasitol *35*, 316-328. https://doi.org/10.1016/j.pt.2019.01.006

Caimano, M.J., Drecktrah, D., Kung, F., and Samuels, D.S. (2016). Interaction of the Lyme disease spirochete with its tick vector. Cell Microbiol *18*, 919-927. https://doi.org/10.1111/cmi.12609

Caimano, M.J., Dunham-Ems, S., Allard, A.M., Cassera, M.B., Kenedy, M., and Radolf, J.D. (2015). Cyclic di-GMP modulates gene expression in Lyme disease spirochetes at the tick-mammal interface to promote spirochete survival during the blood meal and tick-to-mammal transmission. Infect Immun *83*, 3043-3060. https://doi.org/10.1128/IAI. 00315-15

Caimano, M.J., Eggers, C.H., Gonzalez, C.A., and Radolf, J.D. (2005). Alternate sigma factor RpoS is required for the in vivo-specific repression of *Borrelia burgdorferi* plasmid lp54-borne ospA and lp6.6 genes. J Bacteriol *187*, 7845-7852. doi: 10.1128/JB. 187.22.7845-7852.2005.

Caimano, M.J., Eggers, C.H., Hazlett, K.R., and Radolf, J.D. (2004). RpoS is not central to the general stress response in *Borrelia burgdorferi* but does control expression of one or more essential virulence determinants. Infect Immun *72*, 6433-6445. doi: 10.1128/IAI. 72.11.6433-6445.2004.

Caimano, M.J., Groshong, A.M., Belperron, A., Mao, J., Hawley, K.L., Luthra, A., Graham, D.E., Earnhart, C.G., Marconi, R.T., Bockenstedt, L.K., *et al.* (2019). The RpoS Gatekeeper in *Borrelia burgdorferi:* An Invariant Regulatory Scheme That Promotes Spirochete Persistence in Reservoir Hosts and Niche Diversity. Front Microbiol *10*, 1923. https:// doi.org/10.3389/fmicb.2019.01923

Caimano, M.J., Iyer, R., Eggers, C.H., Gonzalez, C., Morton, E.A., Gilbert, M.A., Schwartz, I., and Radolf, J.D. (2007). Analysis of the RpoS regulon in *Borrelia burgdorferi* in response to mammalian host signals provides insight into RpoS function during the enzootic cycle. Mol Microbiol *65*, 1193-1217. doi: 10.1111/j.1365-2958.2007.05860.x.

Caine, J.A., Lin, Y.P., Kessler, J.R., Sato, H., Leong, J.M., and Coburn, J. (2017). *Borrelia burgdorferi* outer surface protein C (OspC) binds complement component C4b and confers bloodstream survival. Cell Microbiol *19*. https://doi.org/10.1111/cmi.12786

Carrasco, S.E., Troxell, B., Yang, Y., Brandt, S.L., Li, H., Sandusky, G.E., Condon, K.W., Serezani, C.H., and Yang, X.F. (2015). Outer surface protein OspC is an antiphagocytic factor that protects *Borrelia burgdorferi* from phagocytosis by macrophages. Infect Immun *83*, 4848-4860. https://doi.org/10.1128/IAI.01215-15

Casjens, S., Palmer, N., van Vugt, R., Huang, W.M., Stevenson, B., Rosa, P., Lathigra, R., Sutton, G., Peterson, J., Dodson, R.J., *et al.* (2000). A bacterial genome in flux: the twelve linear and nine circular extrachromosomal DNAs in an infectious isolate of the Lyme disease spirochete *Borrelia burgdorferi.* Mol Microbiol *35*, 490-516. doi: 10.1046/j.1365-2958.2000.01698.x.

Casjens, S.R., Mongodin, E.F., Qiu, W.G., Dunn, J.J., Luft, B.J., Fraser-Liggett, C.M., and Schutzer, S.E. (2011). Whole-genome sequences of two *Borrelia afzelii* and two *Borrelia garinii* Lyme disease agent isolates. J Bacteriol *193*, 6995-6996. https://doi.org/193/24/6995

Casjens, S.R., Mongodin, E.F., Qiu, W.G., Luft, B.J., Schutzer, S.E., Gilcrease, E.B., Huang, W.M., Vujadinovic, M., Aron, J.K., Vargas, L.C., *et al.* (2012). Genome stability of Lyme disease spirochetes: comparative genomics of *Borrelia burgdorferi* plasmids. PLoS ONE *7*, e33280. https://doi.org/10.1371/journal.pone.0033280

Catteruccia, F. (2007). Malaria vector control in the third millennium: progress and perspectives of molecular approaches. Pest Manag Sci *63*, 634-640. doi: 10.1002/ps.1

Catteruccia, F., Nolan, T., Loukeris, T.G., Blass, C., Savakis, C., Kafatos, F.C., and Crisanti, A. (2000). Stable germline transformation of the malaria mosquito *Anopheles stephensi.* Nature *405*, 959-962. doi: 10.1038/35016096.

Charrier, N.P., Couton, M., Voordouw, M.J., Rais, O., Durand-Hermouet, A., Hervet, C., Plantard, O., and Rispe, C. (2018). Whole body transcriptomes and new insights into the biology of the tick Ixodes ricinus. Parasites & vectors *11*, 364. https://doi.org/10.1186/ s13071-018-2932-3

Chavez, A.S.O., O'Neal, A.J., Santambrogio, L., Kotsyfakis, M., and Pedra, J.H.F. (2019). Message in a vesicle - trans-kingdom intercommunication at the vector-host interface. Journal of cell science *132*. https://doi.org/10.1242/jcs.224212

Chen, T., Xiang, X., Xu, H., Zhang, X., Zhou, B., Yang, Y., Lou, Y., and Yang, X.F. (2018). LtpA, a CdnL-type CarD regulator, is important for the enzootic cycle of the Lyme disease pathogen. Emerg Microbes Infect *7*, 126. https://doi.org/10.1038/s41426-018-0122-1

Chou, S., Daugherty, M.D., Peterson, S.B., Biboy, J., Yang, Y., Jutras, B.L., Fritz-Laylin, L.K., Ferrin, M.A., Harding, B.N., Jacobs-Wagner, C., *et al.* (2015). Transferred interbacterial antagonism genes augment eukaryotic innate immune function. Nature *518*, 98-101. https://doi.org/ 10.1038/nature13965

Chrudimska, T., Slaninova, J., Rudenko, N., Ruzek, D., and Grubhoffer, L. (2011). Functional characterization of two defensin isoforms of the hard tick Ixodes ricinus. Parasites & vectors *4*, 63. https://doi.org/10.1186/1756-3305-4-63

Clayton, A.M., Dong, Y., and Dimopoulos, G. (2014). The Anopheles innate immune system in the defense against malaria infection. J Innate Immun *6*, 169-181. https://doi.org/10.1159/000353602

Coburn, J., and Cugini, C. (2003). Targeted mutation of the outer membrane protein P66 disrupts attachment of the Lyme disease agent, *Borrelia burgdorferi*, to integrin alpha-v-beta-3. Proc Natl Acad Sci U S A *100*, 7301-7306. doi: 10.1073/pnas.1131117100

Coleman, J.L., Gebbia, J.A., Piesman, J., Degen, J.L., Bugge, T.H., and Benach, J.L. (1997). Plasminogen is required for efficient dissemination of *Borrelia burgdorferi* in ticks and for enhancement of spirochetemia in mice. Cell *89*, 1111-1119. doi: 10.1016/s0092-8674(00)80298-6.

Coleman, J.L., Sellati, T.J., Testa, J.E., Kew, R.R., Furie, M.B., and Benach, J.L. (1995). Borrelia burgdorferi binds plasminogen, resulting in enhanced penetration of endothelial monolayers. Infect Immun *63*, 2478-2484. doi: 10.1128/IAI.63.7.2478-2484.1995.

Concepcion, M.B., and Nelson, D.R. (2003). Expression of spoT in *Borrelia burgdorferi* during serum starvation. J Bacteriol *185*, 444-452. https://doi.org/10.1128/jb.185.2.444-452.2003

Contreras, M., Villar, M., Alberdi, P., and de la Fuente, J. (2016). Vaccinomics Approach to Tick Vaccine Development. Methods Mol Biol *1404*, 275-286. https://doi.org/10.1007/978-1-4939-3389-1_19

Contreras, M., Villar, M., and de la Fuente, J. (2019). A Vaccinomics Approach for the Identification of Tick Protective Antigens for the Control of Ixodes ricinus and *Dermacentor reticulatus* Infestations in Companion Animals. Front Physiol *10*, 977. https://doi.org/10.3389/fphys.2019.00977

Cotte, V., Sabatier, L., Schnell, G., Carmi-Leroy, A., Rousselle, J.C., Arsene-Ploetze, F., Malandrin, L., Sertour, N., Namane, A., Ferquel, E., *et al.* (2014). Differential expression of *Ixodes ricinus* salivary gland proteins in the presence of the Borrelia burgdorferi sensu lato complex. J Proteomics *96*, 29-43. https://doi.org/10.1016/j.jprot.2013.10.033

Coumou, J., Narasimhan, S., Trentelman, J.J., Wagemakers, A., Koetsveld, J., Ersoz, J.I., Oei, A., Fikrig, E., and Hovius, J.W. (2016). Ixodes scapularis dystroglycan-like protein promotes Borrelia burgdorferi migration from the gut. J Mol Med (Berl) *94*, 361-370. https://doi.org/10.1007/s00109-015-1365-0

Coumou, J., Wagemakers, A., Narasimhan, S., Schuijt, T.J., Ersoz, J.I., Oei, A., de Boer, O.J., Roelofs, J., Fikrig, E., and Hovius, J.W. (2019). The role of Mannose Binding Lectin in the immune response against *Borrelia burgdorferi* sensu lato. Scientific reports *9*, 1431. https://doi.org/10.1038/s41598-018-37922-8

Couper, L., and Swei, A. (2018). Tick Microbiome Characterization by Next-Generation 16S rRNA Amplicon Sequencing. J Vis Exp. https://doi.org/10.3791/58239

Cui, H., Tsuda, K., and Parker, J.E. (2015). Effector-triggered immunity: from pathogen perception to robust defense. Annu Rev Plant Biol *66*, 487-511. https://doi.org/10.1146/annurev-arplant-050213-040012

Dai, J., Narasimhan, S., Zhang, L., Liu, L., Wang, P., and Fikrig, E. (2010). Tick histamine release factor is critical for *Ixodes scapularis* engorgement and transmission of the lyme disease agent. PLoS Pathog *6*, e1001205. https://doi.org/10.1371/journal.ppat.1001205

Dai, J., Wang, P., Adusumilli, S., Booth, C.J., Narasimhan, S., Anguita, J., and Fikrig, E. (2009). Antibodies against a tick protein, Salp15, protect mice from the Lyme disease agent. Cell Host Microbe *6*, 482-492. https://doi.org/S1931-3128(09)00348-5 [pii]10.1016/j.chom.2009.10.006

Dale, C., and Moran, N.A. (2006). Molecular interactions between bacterial symbionts and their hosts. Cell *126*, 453-465. doi: 10.1016/j.cell.2006.07.014.

Das, S., Banerjee, G., DePonte, K., Marcantonio, N., Kantor, F.S., and Fikrig, E. (2001). Salp25D, an Ixodes scapularis antioxidant, is 1 of 14 immunodominant antigens in engorged tick salivary glands. J Infect Dis *184*, 1056-1064. doi: 10.1086/323351.

Das, S., Marcantonio, N., Deponte, K., Telford, S.R., 3rd, Anderson, J.F., Kantor, F.S., and Fikrig, E. (2000). SALP16, a gene induced in Ixodes scapularis salivary glands during tick feeding. Am J Trop Med Hyg *62*, 99-105. doi: 10.4269/ajtmh.2000.62.99.

de la Fuente, J., Antunes, S., Bonnet, S., Cabezas-Cruz, A., Domingos, A.G., Estrada-Pena, A., Johnson, N., Kocan, K.M., Mansfield, K.L., Nijhof, A.M., *et al.* (2017). Tick-Pathogen Interactions and Vector Competence: Identification of Molecular Drivers for Tick-Borne Diseases. Frontiers in cellular and infection microbiology 7, 114. https://doi.org/10.3389/fcimb.2017.00114

de la Fuente, J., Kocan, K.M., Almazan, C., and Blouin, E.F. (2007a). RNA interference for the study and genetic manipulation of ticks. Trends Parasitol *23*, 427-433. doi: 10.1016/j.pt.2007.07.002.

de la Fuente, J., Kocan, K.M., and Blouin, E.F. (2007b). Tick vaccines and the transmission of tick-borne pathogens. Vet Res Commun *31 Suppl 1*, 85-90.

de la Fuente, J., Waterhouse, R.M., Sonenshine, D.E., Roe, R.M., Ribeiro, J.M., Sattelle, D.B., and Hill, C.A. (2016). Tick Genome Assembled: New Opportunities for Research on Tick-Host-Pathogen Interactions. Frontiers in cellular and infection microbiology *6*, 103. https://doi.org/10.3389/fcimb.2016.00103

de Silva, A.M. (2016). Ticks Take Cues from Mammalian Interferon. Cell Host Microbe *20*, 3-4. https://doi.org/10.1016/j.chom.2016.06.016

de Silva, A.M., and Fikrig, E. (1995). Growth and migration of *Borrelia burgdorferi* in Ixodes ticks during blood feeding. Am J Trop Med Hyg *53*, 397-404. doi: 10.4269/ajtmh.1995.53.397.

de Silva, A.M., and Fikrig, E. (1997). Arthropod- and host-specific gene expression by *Borrelia burgdorferi*. J Clin Invest *99*, 377-379. doi: 10.1172/JCI119169.

de Silva, A.M., Telford, S.R., Brunet, L.R., Barthold, S.W., and Fikrig, E. (1996). *Borrelia burgdorferi* OspA is an arthropod-specific transmission-blocking Lyme disease vaccine. J Exp Med *183*, 271-275. doi: 10.1084/jem.183.1.271.

Di Venere, M., Fumagalli, M., Cafiso, A., De Marco, L., Epis, S., Plantard, O., Bardoni, A., Salvini, R., Viglio, S., Bazzocchi, C., *et al.* (2015). Ixodes ricinus and Its Endosymbiont Midichloria mitochondrii: A Comparative Proteomic Analysis of Salivary Glands and Ovaries. PLoS One *10*, e0138842. https://doi.org/10.1371/journal.pone.0138842

Ding, J., Adiconis, X., Simmons, S.K., Kowalczyk, M.S., Hession, C.C., Marjanovic, N.D., Hughes, T.K., Wadsworth, M.H., Burks, T., Nguyen, L.T., *et al.* (2020). Systematic comparison of single-cell and single-nucleus RNA-sequencing methods. Nat Biotechnol *38*, 737-746. https://doi.org/10.1038/s41587-020-0465-8

Drecktrah, D., Hall, L.S., Brinkworth, A.J., Comstock, J.R., Wassarman, K.M., and Samuels, D.S. (2020). Characterization of 6S RNA in the Lyme disease spirochete. Mol Microbiol *113*, 399-417. https://doi.org/10.1111/mmi.14427

Drecktrah, D., Hall, L.S., Rescheneder, P., Lybecker, M., and Samuels, D.S. (2018). The Stringent Response-Regulated sRNA Transcriptome of *Borrelia burgdorferi*. Frontiers in cellular and infection microbiology *8*, 231. https://doi.org/10.3389/fcimb.2018.00231

Drecktrah, D., Lybecker, M., Popitsch, N., Rescheneder, P., Hall, L.S., and Samuels, D.S. (2015). The Borrelia burgdorferi RelA/SpoT Homolog and Stringent Response Regulate Survival in the Tick Vector and Global Gene Expression during Starvation. PLoS Pathog *11*, e1005160. https://doi.org/10.1371/journal.ppat.1005160

Dudchenko, O., Batra, S.S., Omer, A.D., Nyquist, S.K., Hoeger, M., Durand, N.C., Shamim, M.S., Machol, I., Lander, E.S., Aiden, A.P., *et al.* (2017). De novo assembly of the *Aedes aegypti* genome using Hi-C yields chromosome-length scaffolds. Science *356*, 92-95. https://doi.org/10.1126/science.aal3327

Dunham-Ems, S.M., Caimano, M.J., Eggers, C.H., and Radolf, J.D. (2012). *Borrelia burgdorferi* requires the alternative sigma factor RpoS for dissemination within the vector during tick-to-mammal transmission.

PLoS Pathog *8*, e1002532. https://doi.org/10.1371/journal.ppat.1002532

Dunham-Ems, S.M., Caimano, M.J., Pal, U., Wolgemuth, C.W., Eggers, C.H., Balic, A., and Radolf, J.D. (2009). Live imaging reveals a biphasic mode of dissemination of *Borrelia burgdorferi* within ticks. J Clin Invest *119*, 3652-3665. https://doi.org/39401

Duray, P.H., Yin, S.R., Ito, Y., Bezrukov, L., Cox, C., Cho, M.S., Fitzgerald, W., Dorward, D., Zimmerberg, J., and Margolis, L. (2005). Invasion of human tissue ex vivo by *Borrelia burgdorferi*. J Infect Dis *191*, 1747-1754.

Earnhart, C.G., Buckles, E.L., and Marconi, R.T. (2007). Development of an OspC-based tetravalent, recombinant, chimeric vaccinogen that elicits bactericidal antibody against diverse Lyme disease spirochete strains. Vaccine *25*, 466-480. doi: 10.1016/j.vaccine.2006.07.052.

Earnhart, C.G., and Marconi, R.T. (2007a). An octavalent lyme disease vaccine induces antibodies that recognize all incorporated OspC type-specific sequences. Hum Vaccin *3*, 281-289. doi: 10.4161/hv.4661.

Earnhart, C.G., and Marconi, R.T. (2007b). OspC phylogenetic analyses support the feasibility of a broadly protective polyvalent chimeric Lyme disease vaccine. Clin Vaccine Immunol *14*, 628-634. doi: 10.1128/CVI.00409-06.

Estrada-Pena, A., Cabezas-Cruz, A., and Obregon, D. (2020). Resistance of Tick Gut Microbiome to Anti-Tick Vaccines, Pathogen Infection and Antimicrobial Peptides. Pathogens *9*. https://doi.org/10.3390/pathogens9040309

Fazzino, L., Tilly, K., Dulebohn, D.P., and Rosa, P.A. (2015). Long-term survival of Borrelia burgdorferi lacking the hibernation promotion factor homolog in the unfed tick vector. Infect Immun *83*, 4800-4810. https://doi.org/10.1128/IAI.00925-15

Ferrandon, D., Imler, J.L., Hetru, C., and Hoffmann, J.A. (2007). The *Drosophila* systemic immune response: sensing and signalling during bacterial and fungal infections. Nature reviews Immunology *7*, 862-874. https://doi.org/10.1038/nri2194

Figlerowicz, M., Urbanowicz, A., Lewandowski, D., Jodynis-Liebert, J., and Sadowski, C. (2013). Functional insights into recombinant TROSPA protein from *Ixodes ricinus*. PLoS One *8*, e76848. https://doi.org/10.1371/journal.pone.0076848

Fikrig, E., Barthold, S.W., Kantor, F.S., and Flavell, R.A. (1990). Protection of mice against the Lyme disease agent by immunizing with recombinant OspA. Science *250*, 553-556. doi: 10.1126/science.2237407

Fikrig, E., and Narasimhan, S. (2006). Borrelia burgdorferi--traveling incognito? Microbes Infect *8*, 1390-1399. https://doi.org/S1286-4579(06)00046-3 [pii] 10.1016/j.micinf.2005.12.022

Fikrig, E., Pal, U., Chen, M., Anderson, J.F., and Flavell, R.A. (2004). OspB antibody prevents *Borrelia burgdorferi* colonization of Ixodes scapularis. Infect Immun *72*, 1755-1759.

Fingerle, V., Goettner, G., Gern, L., Wilske, B., and Schulte-Spechtel, U. (2007). Complementation of a Borrelia afzelii OspC mutant highlights the crucial role of OspC for dissemination of *Borrelia afzelii* in *Ixodes ricinus*. Int J Med Microbiol *297*, 97-107. doi: 10.1016/j.ijmm.2006.11.003

Finlay, B.B., and Cossart, P. (1997). Exploitation of mammalian host cell functions by bacterial pathogens. Science *276*, 718-725. doi: 10.1126/science.276.5313.718.

Fisher, M.A., Grimm, D., Henion, A.K., Elias, A.F., Stewart, P.E., Rosa, P.A., and Gherardini, F.C. (2005). *Borrelia burgdorferi* sigma54 is required for mammalian infection and vector transmission but not for tick colonization. Proc Natl Acad Sci U S A *102*, 5162-5167. https://doi.org/0408536102 [pii] 10.1073/pnas.0408536102

Floden, A.M., Watt, J.A., and Brissette, C.A. (2011). *Borrelia burgdorferi* enolase is a surface-exposed plasminogen binding protein. PLoS ONE *6*, e27502. https://doi.org/10.1371/journal.pone.0027502 PONE-D-11-11262 [pii]

Fogaca, A.C., da Silva, P.I., Jr., Miranda, M.T., Bianchi, A.G., Miranda, A., Ribolla, P.E., and Daffre, S. (1999). Antimicrobial activity of a bovine hemoglobin fragment in the tick *Boophilus microplus*. J Biol Chem *274*, 25330-25334. doi: 10.1074/jbc.274.36.25330.

Francischetti, I.M., My Pham, V., Mans, B.J., Andersen, J.F., Mather, T.N., Lane, R.S., and Ribeiro, J.M. (2005). The transcriptome of the salivary glands of the female western black-legged tick *Ixodes pacificus* (Acari: Ixodidae). Insect Biochem Mol Biol *35*, 1142-1161. doi: 10.1016/j.ibmb.2005.05.007.

Fraser, C.M., Casjens, S., Huang, W.M., Sutton, G.G., Clayton, R., Lathigra, R., White, O., Ketchum, K.A., Dodson, R., Hickey, E.K., *et al.* (1997). Genomic sequence of a Lyme disease spirochaete, *Borrelia burgdorferi*. Nature *390*, 580-586. doi: 10.1038/37551.

Freedman, J.C., Rogers, E.A., Kostick, J.L., Zhang, H., Iyer, R., Schwartz, I., and Marconi, R.T. (2010). Identification and molecular characterization of a cyclic-di-GMP effector protein, PlzA (BB0733): additional evidence for the existence of a functional cyclic-di-GMP regulatory network in the Lyme disease spirochete, *Borrelia burgdorferi*. FEMS Immunol Med Microbiol *58*, 285-294. https://doi.org/FIM635

Fuchs, H., Wallich, R., Simon, M.M., and Kramer, M.D. (1994). The outer surface protein A of the spirochete *Borrelia burgdorferi* is a plasmin(ogen) receptor. Proc Natl Acad Sci (U S A) *91*, 12594-12598. doi: 10.1073/pnas.91.26.12594.

Fung, B.P., McHugh, G.L., Leong, J.M., and Steere, A.C. (1994). Humoral immune response to outer surface protein C of *Borrelia burgdorferi* in Lyme disease: role of the immunoglobulin M response in the serodiagnosis of early infection. Infect Immun *62*, 3213-3221. doi: 10.1128/IAI.62.8.3213-3221.

Ganz, T. (2003). Defensins: antimicrobial peptides of innate immunity. Nature reviews Immunology *3*, 710-720. https://doi.org/10.1038/nri1180

Geraci, N.S., Spencer Johnston, J., Paul Robinson, J., Wikel, S.K., and Hill, C.A. (2007). Variation in genome size of argasid and ixodid ticks. Insect Biochem Mol Biol *37*, 399-408. doi: 10.1016/j.ibmb.2006.12.007.

Gilbert, M.A., Morton, E.A., Bundle, S.F., and Samuels, D.S. (2007). Artificial regulation of ospC expression in *Borrelia burgdorferi*. Mol Microbiol *63*, 1259-1273. doi: 10.1111/j.1365-2958.2007.05593.x.

Gold, B., Deng, H., Bryk, R., Vargas, D., Eliezer, D., Roberts, J., Jiang, X., and Nathan, C. (2008). Identification of a copper-binding metallothionein in pathogenic mycobacteria. Nat Chem Biol *4*, 609-616. https://doi.org/10.1038/nchembio.109

Gomes-Solecki, M., Arnaboldi, P.M., Backenson, P.B., Benach, J.L., Cooper, C.L., Dattwyler, R.J., Diuk-Wasser, M., Fikrig, E., Hovius, J.W., Laegreid, W., *et al.* (2020). Protective Immunity and New Vaccines for Lyme Disease. Clin Infect Dis *70*, 1768-1773. https://doi.org/10.1093/cid/ciz872

Grabowski, J.M., Perera, R., Roumani, A.M., Hedrick, V.E., Inerowicz, H.D., Hill, C.A., and Kuhn, R.J. (2016). Changes in the Proteome of Langat-Infected *Ixodes scapularis* ISE6 Cells: Metabolic Pathways Associated with Flavivirus Infection. PLoS Negl Trop Dis *10*, e0004180. https://doi.org/10.1371/journal.pntd.0004180

Greay, T.L., Gofton, A.W., Paparini, A., Ryan, U.M., Oskam, C.L., and Irwin, P.J. (2018). Recent insights into the tick microbiome gained through next-generation sequencing. Parasites & vectors *11*, 12. https://doi.org/10.1186/s13071-017-2550-5

Grewe, C., and Nuske, J.H. (1996). Immunolocalization of a 22 kDa protein (IPLA7, P22) of *Borrelia burgdorferi*. FEMS Microbiol Lett *138*, 215-219. doi: 10.1111/j.1574-6968.1996.tb08160.x.

Grimm, D., Eggers, C.H., Caimano, M.J., Tilly, K., Stewart, P.E., Elias, A.F., Radolf, J.D., and Rosa, P.A. (2004a). Experimental assessment of the roles of linear plasmids lp25 and lp28-1 of *Borrelia burgdorferi* throughout the infectious cycle. Infect Immun *72*, 5938-5946.

Grimm, D., Tilly, K., Bueschel, D.M., Fisher, M.A., Policastro, P.F., Gherardini, F.C., Schwan, T.G., and Rosa, P.A. (2005). Defining plasmids required by *Borrelia burgdorferi* for colonization of tick vector

Ixodes scapularis (Acari: Ixodidae). J Med Entomol *42*, 676-684. doi: 10.1128/IAI.72.10.5938-5946.2004.

Grimm, D., Tilly, K., Byram, R., Stewart, P.E., Krum, J.G., Bueschel, D.M., Schwan, T.G., Policastro, P.F., Elias, A.F., and Rosa, P.A. (2004b). Outer-surface protein C of the Lyme disease spirochete: a protein induced in ticks for infection of mammals. Proc Natl Acad Sci U S A *101*, 3142-3147. doi: 10.1073/pnas.0306845101.

Groshong, A.M., Dey, A., Bezsonova, I., Caimano, M.J., and Radolf, J.D. (2017). Peptide Uptake Is Essential for *Borrelia burgdorferi* Viability and Involves Structural and Regulatory Complexity of its Oligopeptide Transporter. mBio *8*. https://doi.org/10.1128/mBio.02047-17

Grubhoffer, L., Golovchenko, M., Vancova, M., Zacharovova-Slavickova, K., Rudenko, N., and Oliver, J.H., Jr. (2005). Lyme borreliosis: insights into tick-/host-borrelia relations. Folia parasitologica *52*, 279-294. doi: 10.14411/fp.2005.039.

Gulia-Nuss, M., Nuss, A.B., Meyer, J.M., Sonenshine, D.E., Roe, R.M., Waterhouse, R.M., Sattelle, D.B., de la Fuente, J., Ribeiro, J.M., Megy, K., *et al.* (2016). Genomic insights into the *Ixodes scapularis* tick vector of Lyme disease. Nat Commun *7*, 10507. https://doi.org/10.1038/ncomms10507

Hasin, Y., Seldin, M., and Lusis, A. (2017). Multi-omics approaches to disease. Genome Biol *18*, 83. https://doi.org/10.1186/s13059-017-1215-1

Hayes, B.M., Dulebohn, D.P., Sarkar, A., Tilly, K., Bestor, A., Ambroggio, X., and Rosa, P.A. (2014). Regulatory protein BBD18 of the lyme disease spirochete: essential role during tick acquisition? mBio *5*, e01017-01014. https://doi.org/10.1128/mBio.01017-14

He, M., Ouyang, Z., Troxell, B., Xu, H., Moh, A., Piesman, J., Norgard, M.V., Gomelsky, M., and Yang, X.F. (2011). Cyclic di-GMP is essential for the survival of the lyme disease spirochete in ticks. PLoS Pathog *7*, e1002133. https://doi.org/10.1371/journal.ppat.1002133

He, M., Zhang, J.J., Ye, M., Lou, Y., and Yang, X.F. (2014). Cyclic Di-GMP receptor PlzA controls virulence gene expression through RpoS in *Borrelia burgdorferi*. Infect Immun *82*, 445-452. https://doi.org/10.1128/IAI.01238-13

He, Z.B., Cao, Y.Q., Yin, Y.P., Wang, Z.K., Chen, B., Peng, G.X., and Xia, Y.X. (2006). Role of hunchback in segment patterning of Locusta migratoria manilensis revealed by parental RNAi. Dev Growth Differ *48*, 439-445. doi: 10.1111/j.1440-169X.2006.00881.x.

Heath, A.C. (1979). The temperature and humidity preferences of *Haemaphysalis longicornis*, *Ixodes Holocyclus* and *Rhipicephalus sanguineus* (Ixodidae): studies on eggs. Int J Parasitol *9*, 33-39. doi: 10.1016/0020-7519(79)90063-8.

Hefty, P.S., Jolliff, S.E., Caimano, M.J., Wikel, S.K., Radolf, J.D., and Akins, D.R. (2001). Regulation of OspE-related, OspF-related, and Elp lipoproteins of *Borrelia burgdorferi* strain 297 by mammalian host-specific signals. Infect Immun *69*, 3618-3627. doi: 10.1128/IAI.69.6.3618-3627.2001.

Hernandez-Jarguin, A., Diaz-Sanchez, S., Villar, M., and de la Fuente, J. (2018). Integrated metatranscriptomics and metaproteomics for the characterization of bacterial microbiota in unfed Ixodes ricinus. Ticks Tick Borne Dis *9*, 1241-1251. https://doi.org/10.1016/j.ttbdis.2018.04.020

Hill, C.A., and Wikel, S.K. (2005). The Ixodes scapularis Genome Project: an opportunity for advancing tick research. Trends Parasitol *21*, 151-153. https://doi.org/10.1016/j.pt.2005.02.004

Hoffmann, J.A. (2003). The immune response of Drosophila. Nature *426*, 33-38. https://doi.org/10.1038/nature02021

Hoffmann, J.A., Kafatos, F.C., Janeway, C.A., and Ezekowitz, R.A. (1999). Phylogenetic perspectives in innate immunity. Science *284*, 1313-1318. doi: 10.1126/science.284.5418.1313.

Hoffmann, J.A., and Reichhart, J.M. (2002). Drosophila innate immunity: an evolutionary perspective. Nat Immunol *3*, 121-126. https://doi.org/10.1038/ni0202-121

Hooper, L.V., Littman, D.R., and Macpherson, A.J. (2012). Interactions between the microbiota and the immune system. Science *336*, 1268-1273. https://doi.org/10.1126/science.1223490

Hoxmeier, J.C., Fleshman, A.C., Broeckling, C.D., Prenni, J.E., Dolan, M.C., Gage, K.L., and Eisen, L. (2017). Metabolomics of the tick-Borrelia interaction during the nymphal tick blood meal. Scientific reports *7*, 44394. https://doi.org/10.1038/srep44394

Hubner, A., Revel, A.T., Nolen, D.M., Hagman, K.E., and Norgard, M.V. (2003). Expression of a luxS gene is not required for *Borrelia burgdorferi* infection of mice via needle inoculation. Infect Immun *71*, 2892-2896.

Hubner, A., Yang, X., Nolen, D.M., Popova, T.G., Cabello, F.C., and Norgard, M.V. (2001). Expression of *Borrelia burgdorferi* OspC and DbpA is controlled by a RpoN-RpoS regulatory pathway. Proc Natl Acad Sci U S A *98*, 12724-12729.

Hynes, W.L., Ceraul, S.M., Todd, S.M., Seguin, K.C., and Sonenshine, D.E. (2005). A defensin-like gene expressed in the black-legged tick, Ixodes scapularis. Med Vet Entomol *19*, 339-344. doi: 10.1111/j.1365-2915.2005.00579.x.

Iovinella, I., Ban, L., Song, L., Pelosi, P., and Dani, F.R. (2016). Proteomic analysis of castor bean tick Ixodes ricinus: a focus on chemosensory organs. Insect Biochem Mol Biol *78*, 58-68. https://doi.org/10.1016/j.ibmb.2016.09.004

Irving, P., Troxler, L., Heuer, T.S., Belvin, M., Kopczynski, C., Reichhart, J.M., Hoffmann, J.A., and Hetru, C. (2001). A genome-wide analysis of immune responses in Drosophila. Proc Natl Acad Sci U S A *98*, 15119-15124. https://doi.org/10.1073/pnas.261573998

Iyer, R., Caimano, M.J., Luthra, A., Axline, D., Jr., Corona, A., Iacobas, D.A., Radolf, J.D., and Schwartz, I. (2015). Stage-specific global alterations in the transcriptomes of Lyme disease spirochetes during tick feeding and following mammalian host adaptation. Mol Microbiol *95*, 509-538. https://doi.org/10.1111/mmi.12882

Jewett, M.W., Lawrence, K., Bestor, A.C., Tilly, K., Grimm, D., Shaw, P., VanRaden, M., Gherardini, F., and Rosa, P.A. (2007). The critical role of the linear plasmid lp36 in the infectious cycle of *Borrelia burgdorferi*. Mol Microbiol *64*, 1358-1374. doi: 10.1111/j.1365-2958.2007.05746.x.

Johns, R., Ohnishi, J., Broadwater, A., Sonenshine, D.E., De Silva, A.M., and Hynes, W.L. (2001). Contrasts in tick innate immune responses to *Borrelia burgdorferi* challenge: immunotolerance in *Ixodes scapularis* versus immunocompetence in *Dermacentor variabilis* (Acari: Ixodidae). J Med Entomol *38*, 99-107. doi: 10.1603/0022-2585-38.1.99.

Jutras, B.L., Jones, G.S., Verma, A., Brown, N.A., Antonicello, A.D., Chenail, A.M., and Stevenson, B. (2013). Posttranscriptional self-regulation by the Lyme disease bacterium's BpuR DNA/RNA-binding protein. J Bacteriol *195*, 4915-4923. https://doi.org/10.1128/JB.00819-13

Jutras, B.L., Savage, C.R., Arnold, W.K., Lethbridge, K.G., Carroll, D.W., Tilly, K., Bestor, A., Zhu, H., Seshu, J., Zuckert, W.R., *et al.* (2019). The Lyme disease spirochete's BpuR DNA/RNA-binding protein is differentially expressed during the mammal-tick infectious cycle, which affects translation of the SodA superoxide dismutase. Mol Microbiol *112*, 973-991. https://doi.org/10.1111/mmi.14336

Kamada, N., Seo, S.U., Chen, G.Y., and Nunez, G. (2013). Role of the gut microbiota in immunity and inflammatory disease. Nature reviews Immunology *13*, 321-335. https://doi.org/10.1038/nri3430

Kariu, T., Smith, A., Yang, X., and Pal, U. (2013). A Chitin Deacetylase-Like Protein Is a Predominant Constituent of Tick Peritrophic Membrane That Influences the Persistence of Lyme Disease Pathogens within the Vector. PLoS ONE *8*, e78376. https://doi.org/10.1371/journal.pone.0078376

Kempfer, R., and Pombo, A. (2020). Methods for mapping 3D chromosome architecture. Nat Rev Genet *21*, 207-226. https://doi.org/10.1038/s41576-019-0195-2

Khila, A., and Grbic, M. (2007). Gene silencing in the spider mite Tetranychus urticae: dsRNA and siRNA parental silencing of the Distal-less gene. Dev Genes Evol *217*, 241-251. doi: 10.1007/s00427-007-0132-9.

Kitsou, C., and Pal, U. (2018). Ixodes Immune Responses Against Lyme Disease Pathogens. Frontiers in cellular and infection microbiology *8*, 176. https://doi.org/10.3389/fcimb.2018.00176

Koci, J., Bernard, Q., Yang, X., and Pal, U. (2018). *Borrelia burgdorferi* surface protein Lmp1 facilitates pathogen dissemination through ticks as studied by an artificial membrane feeding system. Scientific reports *8*, 1910. https://doi.org/10.1038/s41598-018-20208-4

Kopacek, P., Hajdusek, O., Buresova, V., and Daffre, S. (2010). Tick innate immunity. Adv Exp Med Biol *708*, 137-162.

Kostick, J.L., Szkotnicki, L.T., Rogers, E.A., Bocci, P., Raffaelli, N., and Marconi, R.T. (2011). The diguanylate cyclase, Rrp1, regulates critical steps in the enzootic cycle of the Lyme disease spirochetes. Mol Microbiol *81*, 219-231. https://doi.org/10.1111/j.1365-2958.2011.07687.x

Kostick-Dunn, J.L., Izac, J.R., Freedman, J.C., Szkotnicki, L.T., Oliver, L.D., Jr., and Marconi, R.T. (2018). The *Borrelia burgdorferi* c-di-GMP Binding Receptors, PlzA and PlzB, Are Functionally Distinct. Frontiers in cellular and infection microbiology *8*, 213. https://doi.org/10.3389/fcimb.2018.00213

Kotsyfakis, M., Kopacek, P., Franta, Z., Pedra, J.H., and Ribeiro, J.M. (2015). Deep Sequencing Analysis of the *Ixodes ricinus* Haemocytome. PLoS Negl Trop Dis *9*, e0003754. https://doi.org/10.1371/journal.pntd.0003754

Krause, P.J., Fish, D., Narasimhan, S., and Barbour, A.G. (2015). *Borrelia miyamotoi* infection in nature and in humans. Clinical microbiology and infection : the official publication of the European Society of Clinical Microbiology and Infectious Diseases *21*, 631-639. https://doi.org/10.1016/j.cmi.2015.02.006

Kumar, D., Embers, M., Mather, T.N., and Karim, S. (2019). Is selenoprotein K required for *Borrelia burgdorferi* infection within the tick vector Ixodes scapularis? Parasites & vectors *12*, 289. https://doi.org/10.1186/s13071-019-3548-y

Kumar, M., Kaur, S., Kariu, T., Yang, X., Bossis, I., Anderson, J.F., and Pal, U. (2011). *Borrelia burgdorferi* BBA52 is a potential target for transmission blocking Lyme disease vaccine. Vaccine *29(48)*:9012-9, doi:10.1016/j.vaccine.2011.1009.1035.

Kumar, S., Molina-Cruz, A., Gupta, L., Rodrigues, J., and Barillas-Mury, C. (2010). A peroxidase/dual oxidase system modulates midgut epithelial immunity in Anopheles gambiae. Science *327*, 1644-1648. https://doi.org/10.1126/science.1184008

Kung, F., Anguita, J., and Pal, U. (2013). *Borrelia burgdorferi* and tick proteins supporting pathogen persistence in the vector. Future Microbiol *8*, 41-56. https://doi.org/10.2217/fmb.12.121

Kurokawa, C., Lynn, G.E., Pedra, J.H.F., Pal, U., Narasimhan, S., and Fikrig, E. (2020). Interactions between Borrelia burgdorferi and ticks. Nat Rev Microbiol. https://doi.org/10.1038/s41579-020-0400-5

La Manno, G. (2019). From single-cell RNA-seq to transcriptional regulation. Nat Biotechnol *37*, 1421-1422. https://doi.org/10.1038/s41587-019-0327-4

Lagal, V., Portnoi, D., Faure, G., Postic, D., and Baranton, G. (2006). *Borrelia burgdorferi* sensu stricto invasiveness is correlated with OspC-plasminogen affinity. Microbes Infect *8*, 645-652. doi: 10.1016/j.micinf.2005.08.017.

Lam, T.T., Nguyen, T.P., Fikrig, E., and Flavell, R.A. (1994). A chromosomal *Borrelia burgdorferi* gene encodes a 22-kilodalton lipoprotein, P22, that is serologically recognized in Lyme disease. J Clin Microbiol *32*, 876-883. doi: 10.1128/JCM.32.4.876-883.1994.

Landesman, W.J., Mulder, K., Fredericks, L.P., and Allan, B.F. (2019). Cross-kingdom analysis of nymphal-stage *Ixodes scapularis* microbial communities in relation to Borrelia burgdorferi infection and load. FEMS Microbiol Ecol *95*. https://doi.org/10.1093/femsec/fiz167

Langhansova, H., Bopp, T., Schmitt, E., and Kopecky, J. (2015). Tick saliva increases production of three chemokines including monocyte chemoattractant protein-1, a histamine-releasing cytokine. Parasite Immunol *37*, 92-96. https://doi.org/10.1111/pim.12168

Lemaitre, B., and Hoffmann, J. (2007). The host defense of *Drosophila melanogaster*. Annual review of immunology *25*, 697-743. https://doi.org/10.1146/annurev.immunol.25.022106.141615

Lewis, L.A., Radulovic, Z.M., Kim, T.K., Porter, L.M., and Mulenga, A. (2015). Identification of 24h Ixodes scapularis immunogenic tick saliva proteins. Ticks Tick Borne Dis *6*, 424-434. https://doi.org/10.1016/j.ttbdis.2015.03.012

Ley, R.E., Lozupone, C.A., Hamady, M., Knight, R., and Gordon, J.I. (2008). Worlds within worlds: evolution of the vertebrate gut microbiota. Nat Rev Microbiol *6*, 776-788. https://doi.org/10.1038/nrmicro1978

Li, X., Neelakanta, G., Liu, X., Beck, D.S., Kantor, F.S., Fish, D., Anderson, J.F., and Fikrig, E. (2007a). Role of outer surface protein D in the *Borrelia burgdorferi* life cycle. Infect Immun *75*, 4237-4244. doi: 10.1128/IAI.00632-07.

Li, X., Pal, U., Ramamoorthi, N., Liu, X., Desrosiers, D.C., Eggers, C.H., Anderson, J.F., Radolf, J.D., and Fikrig, E. (2007b). The Lyme disease agent *Borrelia burgdorferi* requires BB0690, a Dps homologue, to persist within ticks. Mol Microbiol *63(3)*, 694-710. doi: 10.1111/j.1365-2958.2006.05550.x.

Liang, F.T., Brown, E.L., Wang, T., Iozzo, R.V., and Fikrig, E. (2004). Protective niche for *Borrelia burgdorferi* to evade humoral immunity. The American journal of pathology *165*, 977-985. doi: 10.1016/S0002-9440(10)63359-7.

Liang, F.T., Jacobs, M.B., Bowers, L.C., and Philipp, M.T. (2002). An immune evasion mechanism for spirochetal persistence in Lyme borreliosis. J Exp Med *195*, 415-422. doi: 10.1084/jem.20011870.

Lieberman-Aiden, E., van Berkum, N.L., Williams, L., Imakaev, M., Ragoczy, T., Telling, A., Amit, I., Lajoie, B.R., Sabo, P.J., Dorschner, M.O., *et al.* (2009). Comprehensive mapping of long-range interactions reveals folding principles of the human genome. Science *326*, 289-293. https://doi.org/10.1126/science.1181369

Lin, Y.P., Tan, X., Caine, J.A., Castellanos, M., Chaconas, G., Coburn, J., and Leong, J.M. (2020). Strain-specific joint invasion and colonization by Lyme disease spirochetes is promoted by outer surface protein C. PLoS Pathog *16*, e1008516. https://doi.org/10.1371/journal.ppat.1008516

Lisacek, F., Cohen-Boulakia, S., and Appel, R.D. (2006). Proteome informatics II: bioinformatics for comparative proteomics. Proteomics *6*, 5445-5466. https://doi.org/10.1002/pmic.200600275

Loginov, D.S., Loginova, Y.F., Dycka, F., Bottinger, K., Vechtova, P., and Sterba, J. (2019). Tissue-specific signatures in tick cell line MS profiles. Parasites & vectors *12*, 212. https://doi.org/10.1186/s13071-019-3460-5

Luo, R., and Zhao, H. (2012). Protein quantitation using iTRAQ: Review on the sources of variations and analysis of nonrandom missingness. Stat Interface *5*, 99-107. https://doi.org/10.4310/sii.2012.v5.n1.a9

Lybecker, M.C., and Samuels, D.S. (2007). Temperature-induced regulation of RpoS by a small RNA in Borrelia burgdorferi. Mol Microbiol *64*, 1075-1089. doi: 10.1111/j.1365-2958.2007.05716.x.

Mallory, K.L., Miller, D.P., Oliver, L.D., Jr., Freedman, J.C., Kostick-Dunn, J.L., Carlyon, J.A., Marion, J.D., Bell, J.K., and Marconi, R.T. (2016). Cyclic-di-GMP binding induces structural rearrangements in the PlzA and PlzC proteins of the Lyme disease and relapsing fever spirochetes: a possible switch mechanism for c-di-GMP-mediated effector functions. Pathogens and disease *74*. https://doi.org/10.1093/femspd/ftw105

Mansfield, K.L., Cook, C., Ellis, R.J., Bell-Sakyi, L., Johnson, N., Alberdi, P., de la Fuente, J., and Fooks, A.R. (2017). Tick-borne pathogens induce differential expression of genes promoting cell survival and host resistance in Ixodes ricinus cells. Parasites & vectors *10*, 81. https://doi.org/10.1186/s13071-017-2011-1

Marchal, C., Schramm, F., Kern, A., Luft, B.J., Yang, X., Schuijt, T.J., Hovius, J.W., Jaulhac, B., and Boulanger, N. (2011). Antialarmin effect of tick saliva during the transmission of Lyme disease. Infect Immun *79*, 774-785. https://doi.org/IAI.00482-10 [pii] 10.1128/IAI.00482-10

Marques, A. (2008). Chronic Lyme disease: a review. Infect Dis Clin North Am *22*, 341-360, vii-viii. https://doi.org/10.1016/j.idc.2007.12.011

Marques, A.R. (2015). Laboratory diagnosis of Lyme disease: advances and challenges. Infect Dis Clin North Am *29*, 295-307. https://doi.org/10.1016/j.idc.2015.02.005

Mbow, M.L., Gilmore, R.D., Jr., and Titus, R.G. (1999). An OspC-specific monoclonal antibody passively protects mice from tick-transmitted infection by *Borrelia burgdorferi* B31. Infect Immun *67*, 5470-5472. doi: 10.1128/IAI.67.10.5470-5472.1999.

McClure Carroll, E.E., Wang, X., Shaw, D.K., O'Neal, A.J., Oliva Chavez, A.S., Brown, L.J., Boradia, V.M., Hammond, H.L., and Pedra, J.H.F. (2019). p47 licenses activation of the immune deficiency pathway in the tick *Ixodes scapularis*. Proc Natl Acad Sci U S A *116*, 205-210. https://doi.org/10.1073/pnas.1808905116

McNally, K.L., Mitzel, D.N., Anderson, J.M., Ribeiro, J.M., Valenzuela, J.G., Myers, T.G., Godinez, A., Wolfinbarger, J.B., Best, S.M., and Bloom, M.E. (2012). Differential salivary gland transcript expression profile in *Ixodes scapularis* nymphs upon feeding or flavivirus infection. Ticks Tick Borne Dis *3*, 18-26. https://doi.org/10.1016/j.ttbdis.2011.09.003

Mead, P.S. (2015). Epidemiology of Lyme disease. Infect Dis Clin North Am *29*, 187-210. https://doi.org/10.1016/j.idc.2015.02.010

Miller, J.R., Koren, S., Dilley, K.A., Harkins, D.M., Stockwell, T.B., Shabman, R.S., and Sutton, G.G. (2018). A draft genome sequence for the Ixodes scapularis cell line, ISE6. F1000Res *7*, 297. https://doi.org/10.12688/f1000research.13635.1

Molina-Cruz, A., DeJong, R.J., Charles, B., Gupta, L., Kumar, S., Jaramillo-Gutierrez, G., and Barillas-Mury, C. (2008). Reactive oxygen species modulate *Anopheles gambiae* immunity against bacteria and Plasmodium. J Biol Chem *283*, 3217-3223. https://doi.org/10.1074/jbc.M705873200

Mori, A., Konnai, S., Yamada, S., Hidano, A., Murase, Y., Ito, T., Takano, A., Kawabata, H., Onuma, M., and Ohashi, K. (2010). Two novel Salp15-like immunosuppressant genes from salivary glands of *Ixodes persulcatus* Schulze tick. Insect Mol Biol *19*, 359-365. https://doi.org/10.1111/j.1365-2583.2010.00994.x

Morris, R.V., Shoemaker, C.B., David, J.R., Lanzaro, G.C., and Titus, R.G. (2001). Sandfly maxadilan exacerbates infection with *Leishmania major* and vaccinating against it protects against L. major infection. J Immunol *167*, 5226-5230. doi: 10.4049/jimmunol.167.9.5226.

Munderloh, U.G., and Kurtti, T.J. (1995). Cellular and molecular interrelationships between ticks and prokaryotic tick-borne pathogens. Annu Rev Entomol *40*, 221-243. doi: 10.1146/annurev.en.40.010195.001253.

Munderloh, U.G., and Kurtti, T.J. (2005). The ABCs of Lyme disease spirochaetes in ticks. Lancet *366*, 962-964. doi: 10.1016/S0140-6736(05)67350-3.

Munderloh, U.G., Liu, Y., Wang, M., Chen, C., and Kurtti, T.J. (1994). Establishment, maintenance and description of cell lines from the tick Ixodes scapularis. J Parasitol *80*, 533-543.

Murfin, K.E., Kleinbard, R., Aydin, M., Salazar, S.A., and Fikrig, E. (2019). Borrelia burgdorferi chemotaxis toward tick protein Salp12 contributes to acquisition. Ticks Tick Borne Dis *10*, 1124-1134. https://doi.org/10.1016/j.ttbdis.2019.06.002

Myllymaki, H., Valanne, S., and Ramet, M. (2014). The Drosophila imd signaling pathway. J Immunol *192*, 3455-3462. https://doi.org/10.4049/jimmunol.1303309

Nadelman, R.B., and Wormser, G.P. (1998). Lyme borreliosis. Lancet *352*, 557-565.

Nakajima, Y., Ogihara, K., Taylor, D., and Yamakawa, M. (2003). Antibacterial hemoglobin fragments from the midgut of the soft tick, Ornithodoros moubata (Acari: Argasidae). J Med Entomol *40*, 78-81. doi: 10.1603/0022-2585-40.1.78.

Narasimhan, S., Coumou, J., Schuijt, T.J., Boder, E., Hovius, J.W., and Fikrig, E. (2014a). A tick gut protein with fibronectin III domains aids *Borrelia burgdorferi* congregation to the gut during transmission. PLoS Pathog *10*, e1004278. https://doi.org/10.1371/journal.ppat.1004278

Narasimhan, S., Deponte, K., Marcantonio, N., Liang, X., Royce, T.E., Nelson, K.F., Booth, C.J., Koski, B., Anderson, J.F., Kantor, F., *et al.* (2007). Immunity against *Ixodes scapularis* salivary proteins expressed within 24 hours of attachment thwarts tick feeding and impairs Borrelia transmission. doi: 10.1371/journal.pone.0000451. PLoS ONE 2, e451.

Narasimhan, S., and Fikrig, E. (2015). Tick microbiome: the force within. Trends Parasitol *31*, 315-323. https://doi.org/10.1016/j.pt.2015.03.010

Narasimhan, S., Rajeevan, N., Liu, L., Zhao, Y.O., Heisig, J., Pan, J., Eppler-Epstein, R., Deponte, K., Fish, D., and Fikrig, E. (2014b). Gut Microbiota of the Tick Vector *Ixodes scapularis* Modulate Colonization of the Lyme Disease Spirochete. Cell Host Microbe *15*, 58-71. https://doi.org/10.1016/j.chom.2013.12.001

Narasimhan, S., Santiago, F., Koski, R.A., Brei, B., Anderson, J.F., Fish, D., and Fikrig, E. (2002). Examination of the *Borrelia burgdorferi* transcriptome in Ixodes scapularis during feeding. J Bacteriol *184*, 3122-3125.

Narasimhan, S., Schuijt, T.J., Abraham, N.M., Rajeevan, N., Coumou, J., Graham, M., Robson, A., Wu, M.J., Daffre, S., Hovius, J.W., *et al.* (2017). Modulation of the tick gut milieu by a secreted tick protein favors *Borrelia burgdorferi* colonization. Nat Commun *8*, 184. https://doi.org/10.1038/s41467-017-00208-0

Neelakanta, G., Li, X., Pal, U., Liu, X., Beck, D.S., Deponte, K., Fish, D., Kantor, F.S., and Fikrig, E. (2007). Outer Surface Protein B Is Critical for *Borrelia burgdorferi* Adherence and Survival within Ixodes Ticks. PLoS Pathog *3*, e33. doi: 10.1371/journal.ppat.0030033.

Nelder, M.P., Russell, C.B., Sheehan, N.J., Sander, B., Moore, S., Li, Y., Johnson, S., Patel, S.N., and Sider, D. (2016). Human pathogens associated with the blacklegged tick *Ixodes scapularis*: a systematic review. Parasites & vectors *9*, 265. https://doi.org/10.1186/s13071-016-1529-y

Nogueira, S.V., Smith, A.A., Qin, J.H., and Pal, U. (2012). A surface enolase participates in *Borrelia burgdorferi*-plasminogen interaction and contributes to pathogen survival within feeding ticks. Infect Immun *80*, 82-90. https://doi.org/IAI.05671-11 [pii] 10.1128/IAI.05671-11

Novak, E.A., Sultan, S.Z., and Motaleb, M.A. (2014). The cyclic-di-GMP signaling pathway in the Lyme disease spirochete, *Borrelia burgdorferi*. Frontiers in cellular and infection microbiology *4*, 56. https://doi.org/10.3389/fcimb.2014.00056

Nuttall, P.A. (2019). Tick saliva and its role in pathogen transmission. Wien Klin Wochenschr. https://doi.org/10.1007/s00508-019-1500-y

Ogden, N.H., Lindsay, L.R., Beauchamp, G., Charron, D., Maarouf, A., O'Callaghan, C.J., Waltner-Toews, D., and Barker, I.K. (2004). Investigation of relationships between temperature and developmental rates of tick *Ixodes scapularis* (Acari: Ixodidae) in the laboratory and field. J Med Entomol *41*, 622-633. doi: 10.1603/0022-2585-41.4.622.

Ohnishi, J., Piesman, J., and de Silva, A.M. (2001). Antigenic and genetic heterogeneity of *Borrelia burgdorferi* populations transmitted by ticks. Proc Natl Acad Sci U S A *98*, 670-675. doi: 10.1073/pnas.98.2.670.

Ojaimi, C., Brooks, C., Akins, D., Casjens, S., Rosa, P., Elias, A., Barbour, A., Jasinskas, A., Benach, J., Katonah, L., *et al.* (2002). *Borrelia burgdorferi* gene expression profiling with membrane-based arrays. Methods Enzymol *358*, 165-177. doi: 10.1016/s0076-6879(02)58088-5.

Ojaimi, C., Brooks, C., Casjens, S., Rosa, P., Elias, A., Barbour, A., Jasinskas, A., Benach, J., Katona, L., Radolf, J., *et al.* (2003). Profiling of temperature-induced changes in *Borrelia burgdorferi* gene expression by using whole genome arrays. Infect Immun *71*, 1689-1705. doi: 10.1128/iai.71.4.1689-1705.2003.

Oliva Chavez, A.S., Shaw, D.K., Munderloh, U.G., and Pedra, J.H. (2017). Tick humoral responses: Marching to the beat of a different drummer. Front Microbiol *8*, 223. https://doi.org/10.3389/fmicb.2017.00223

Oliver, J.D., Lynn, G.E., Burkhardt, N.Y., Price, L.D., Nelson, C.M., Kurtti, T.J., and Munderloh, U.G. (2016). Infection of Immature *Ixodes scapularis* (Acari: Ixodidae) by Membrane Feeding. J Med Entomol *53*, 409-415. https://doi.org/10.1093/jme/tjv241

Onder, O., Humphrey, P.T., McOmber, B., Korobova, F., Francella, N., Greenbaum, D.C., and Brisson, D. (2012). OspC is potent plasminogen receptor on surface of *Borrelia burgdorferi*. J Biol Chem *287*, 16860-16868. https://doi.org/M111.290775 [pii] 10.1074/jbc.M111.290775

Onder, O., Shao, W., Lam, H., and Brisson, D. (2014). Tracking the sources of blood meals of parasitic arthropods using shotgun proteomics and unidentified tandem mass spectral libraries. Nat Protoc *9*, 842-850. https://doi.org/10.1038/nprot.2014.048

Paddock, C.D., Lane, R.S., Staples, J.E., and Labruna, M.B. (2016). CHANGING PARADIGMS FOR TICK-BORNE DISEASES IN THE AMERICAS. National Academies of Sciences, Engineering, and Medicine Washington (DC): National Academies Press (US) *Forum on Microbial Threats; Board on Global Health*.

Pagel Van Zee, J., Geraci, N.S., Guerrero, F.D., Wikel, S.K., Stuart, J.J., Nene, V.M., and Hill, C.A. (2007). Tick genomics: the *Ixodes* genome project and beyond. Int J Parasitol *37*, 1297-1305. https://doi.org/10.1016/j.ijpara.2007.05.011

Pal, U., Dai, J., Li, X., Neelakanta, G., Luo, P., Kumar, M., Wang, P., Yang, X., Anderson, J.F., and Fikrig, E. (2008a). A Differential Role for BB0365 in the Persistence of *Borrelia burgdorferi* in Mice and Ticks. J Infect Dis *197*, 148-155. doi: 10.1086/523764.

Pal, U., de Silva, A.M., Montgomery, R.R., Fish, D., Anguita, J., Anderson, J.F., Lobet, Y., and Fikrig, E. (2000). Attachment of *Borrelia burgdorferi* within *Ixodes scapularis* mediated by outer surface protein A. J Clin Invest *106*, 561-569.

Pal, U., and Fikrig, E. (2003). Adaptation of *Borrelia burgdorferi* in the vector and vertebrate host. Microbes Infect *5*, 659-666. doi: 10.1016/s1286-4579(03)00097-2.

Pal, U., Li, X., Wang, T., Montgomery, R.R., Ramamoorthi, N., Desilva, A.M., Bao, F., Yang, X.,

Pypaert, M., Pradhan, D., *et al.* (2004a). TROSPA, an *Ixodes scapularis* receptor for *Borrelia burgdorferi*. Cell *119*, 457-468. https://doi.org/10.1016/j.cell.2004.10.027

Pal, U., Montgomery, R.R., Lusitani, D., Voet, P., Weynants, V., Malawista, S.E., Lobet, Y., and Fikrig, E. (2001). Inhibition of *Borrelia burgdorferi*-tick Interactions *in vivo* by outer surface protein A antibody. J Immunol *166*, 7398-7403. doi: 10.4049/jimmunol.166.12.7398.

Pal, U., Wang, P., Bao, F., Yang, X., Samanta, S., Schoen, R., Wormser, G.P., Schwartz, I., and Fikrig, E. (2008b). *Borrelia burgdorferi* basic membrane proteins A and B participate in the genesis of Lyme arthritis. J Exp Med *205*, 133-141. doi: 10.1084/jem.20070962.

Pal, U., Yang, X., Chen, M., Bockenstedt, L.K., Anderson, J.F., Flavell, R.A., Norgard, M.V., and Fikrig, E. (2004b). OspC facilitates *Borrelia burgdorferi* invasion of *Ixodes scapularis* salivary glands. J Clin Invest *113*, 220-230. doi: 10.1172/JCI19894.

Pappas, C.J., Iyer, R., Petzke, M.M., Caimano, M.J., Radolf, J.D., and Schwartz, I. (2011). Borrelia burgdorferi requires glycerol for maximum fitness during the tick phase of the enzootic cycle. PLoS Pathog *7*, e1002102. https://doi.org/10.1371/journal.ppat.1002102PPATHOGENS-D-11-00446 [pii]

Perner, J., Provaznik, J., Schrenkova, J., Urbanova, V., Ribeiro, J.M., and Kopacek, P. (2016). RNA-seq analyses of the midgut from blood- and serum-fed Ixodes ricinus ticks. Scientific reports *6*, 36695. https://doi.org/10.1038/srep36695

Phelan, J.P., Kern, A., Ramsey, M.E., Lundt, M.E., Sharma, B., Lin, T., Gao, L., Norris, S.J., Hyde, J.A., Skare, J.T., *et al.* (2019). Genome-wide screen identifies novel genes required for *Borrelia burgdorferi* survival in its Ixodes tick vector. PLoS Pathog *15*, e1007644. https://doi.org/10.1371/journal.ppat.1007644

Piesman, J. (1993). Dynamics of *Borrelia burgdorferi* transmission by nymphal *Ixodes dammini* ticks. J Infect Dis *167*, 1082-1085. doi: 10.1093/infdis/167.5.1082.

Piesman, J., and Gern, L. (2004). Lyme borreliosis in Europe and North America. Parasitology *129 Suppl*, S191-220. doi: 10.1017/s0031182003004694.

Piesman, J., Oliver, J., and Sinsky, R. (1990). Growth kinetics of the Lyme disease spirochete (*Borrelia burgdorferi*) in vector ticks (Ixodes dammini). Am J Trop med Hyg *42*, 352-357. doi: 10.4269/ajtmh.1990.42.352.

Piesman, J., Schneider, B.S., and Zeidner, N.S. (2001). Use of quantitative PCR to measure density of *Borrelia burgdorferi* in the midgut and salivary glands of feeding tick vectors. J Clin Microbiol *39*, 4145-4148. doi: 10.1128/JCM.39.11.4145-4148.2001.

Pimenta, P.F., Modi, G.B., Pereira, S.T., Shahabuddin, M., and Sacks, D.L. (1997). A novel role for the peritrophic matrix in protecting Leishmania from the hydrolytic activities of the sand fly midgut. Parasitology *115 (Pt 4)*, 359-369.

Pitzer, J.E., Sultan, S.Z., Hayakawa, Y., Hobbs, G., Miller, M.R., and Motaleb, M.A. (2011). Analysis of the *Borrelia burgdorferi* cyclic-di-GMP-binding protein PlzA reveals a role in motility and virulence. Infect Immun *79*, 1815-1825. https://doi.org/IAI.00075-11

Pritt, B.S., Mead, P.S., Johnson, D.K., Neitzel, D.F., Respicio-Kingry, L.B., Davis, J.P., Schiffman, E., Sloan, L.M., Schriefer, M.E., Replogle, A.J., *et al.* (2016). Identification of a novel pathogenic *Borrelia* species causing Lyme borreliosis with unusually high spirochaetaemia: a descriptive study. Lancet Infect Dis. https://doi.org/10.1016/S1473-3099(15)00464-8

Purser, J.E., Lawrenz, M.B., Caimano, M.J., Howell, J.K., Radolf, J.D., and Norris, S.J. (2003). A plasmid-encoded nicotinamidase (PncA) is essential for infectivity of *Borrelia burgdorferi* in a mammalian host. Mol Microbiol *48*, 753-764. doi: 10.1046/j.1365-2958.2003.03452.x.

Purser, J.E., and Norris, S.J. (2000). Correlation between plasmid content and infectivity in *Borrelia burgdorferi*. Proc Natl Acad Sci U S A *97*, 13865-13870. doi: 10.1073/pnas.97.25.13865.

Radolf, J.D., Caimano, M.J., Stevenson, B., and Hu, L.T. (2012). Of ticks, mice and men: understanding the dual-host lifestyle of Lyme disease spirochaetes. Nat Rev Microbiol *10*, 87-99. https://doi.org/nrmicro2714 [pii] 10.1038/nrmicro2714

Ramamoorthi, N., Narasimhan, S., Pal, U., Bao, F., Yang, X.F., Fish, D., Anguita, J., Norgard, M.V., Kantor, F.S., Anderson, J.F., *et al.* (2005). The Lyme disease agent exploits a tick protein to infect the mammalian host. Nature *436*, 573-577. https://doi.org/10.1038/nature03812

Rego, R.O.M., Trentelman, J.J.A., Anguita, J., Nijhof, A.M., Sprong, H., Klempa, B., Hajdusek, O., Tomas-Cortazar, J., Azagi, T., Strnad, M., *et al.* (2019). Counterattacking the tick bite: towards a rational design of anti-tick vaccines targeting pathogen transmission. Parasites & vectors *12*, 229. https://doi.org/10.1186/s13071-019-3468-x

Revel, A.T., Blevins, J.S., Almazan, C., Neil, L., Kocan, K.M., de la Fuente, J., Hagman, K.E., and Norgard, M.V. (2005). *bptA* (bbe16) is essential for the persistence of the Lyme disease spirochete, *Borrelia burgdorferi*, in its natural tick vector. Proc Natl Acad Sci U S A *102*, 6972-6977. doi: 10.1073/pnas.0502565102.

Revel, A.T., Talaat, A.M., and Norgard, M.V. (2002). DNA microarray analysis of differential gene expression in *Borrelia burgdorferi*, the Lyme disease spirochete. Proc Natl Acad Sci U S A *99*, 1562-1567. doi: 10.1073/pnas.032667699.

Ribeiro, J.M., Alarcon-Chaidez, F., Francischetti, I.M., Mans, B.J., Mather, T.N., Valenzuela, J.G., and Wikel, S.K. (2006). An annotated catalog of salivary gland transcripts from *Ixodes scapularis* ticks. Insect Biochem Mol Biol *36*, 111-129.

Ribeiro, J.M., and Francischetti, I.M. (2003). Role of arthropod saliva in blood feeding: sialome and post-sialome perspectives. Annu Rev Entomol *48*, 73-88. doi: 10.1146/annurev.ento.48.060402.102812.

Rivera, M. (2017). Bacterioferritin: Structure, Dynamics, and Protein-Protein Interactions at Play in Iron Storage and Mobilization. Acc Chem Res *50*, 331-340. https://doi.org/10.1021/acs.accounts.6b00514.

Rogers, E.A., Terekhova, D., Zhang, H.M., Hovis, K.M., Schwartz, I., and Marconi, R.T. (2009). Rrp1, a cyclic-di-GMP-producing response regulator, is an important regulator of *Borrelia burgdorferi* core cellular functions. Mol Microbiol *71*, 1551-1573. https://doi.org/MMI6621 [pii]10.1111/j.1365-2958.2009.06621.x

Rosa, P. (2005). Lyme disease agent borrows a practical coat. Nat Med *11*, 831-832. doi: 10.1038/nm0805-831.

Rosa, P.A., Tilly, K., and Stewart, P.E. (2005). The burgeoning molecular genetics of the Lyme disease spirochaete. Nat Rev Microbiol *3*, 129-143. doi: 10.1038/nrmicro1086.

Round, J.L., and Mazmanian, S.K. (2009). The gut microbiota shapes intestinal immune responses during health and disease. Nature reviews Immunology *9*, 313-323. https://doi.org/10.1038/nri2515

Rudenko, N., Golovchenko, M., Edwards, M.J., and Grubhoffer, L. (2005). Differential expression of *Ixodes ricinus* tick genes induced by blood feeding or Borrelia burgdorferi infection. J Med Entomol *42*, 36-41. doi: 10.1093/jmedent/42.1.36.

Rudenko, N., Golovchenko, M., Vancova, M., Clark, K., Grubhoffer, L., and Oliver, J.H., Jr. (2016). Isolation of live *Borrelia burgdorferi* sensu lato spirochaetes from patients with undefined disorders and symptoms not typical for Lyme borreliosis. Clinical microbiology and infection *22*, 267 e269-267 e215. https://doi.org/10.1016/j.cmi.2015.11.009

Rudzinska, M.A., Spielman, A., Lewengrub, S., Piesman, J., and Karakashian, S. (1982). Penetration of the peritrophic membrane of the tick by *Babesia microti*. Cell Tissue Res *221*, 471-481. doi: 10.1007/BF00215696.

Samuels, D.S. (2011). Gene regulation in *Borrelia burgdorferi*. Annu Rev Microbiol *65*, 479-499. https://doi.org/10.1146/annurev.micro.112408.134040

Samuels, D.S., Drecktrah, D., and Hall, L.S. (2018). Genetic Transformation and Complementation. Methods Mol Biol *1690*, 183-200. https://doi.org/10.1007/978-1-4939-7383-5_15

Sauer, J.R.a.H., A.J (1986). Morphology, Physiology and Biology of Ticks, Vol Vol-1 (New York: Ellis Horwood Limited and John Wiley and Sons).

Schuijt, T.J., Coumou, J., Narasimhan, S., Dai, J., Deponte, K., Wouters, D., Brouwer, M., Oei, A., Roelofs, J.J., van Dam, A.P., *et al.* (2011). A tick mannose-binding lectin inhibitor interferes with the vertebrate complement cascade to enhance transmission of the lyme disease agent. Cell Host Microbe *10*, 136-146. https://doi.org/10.1016/j.chom.2011.06.010

Schutzer, S.E., Fraser-Liggett, C.M., Casjens, S.R., Qiu, W.G., Dunn, J.J., Mongodin, E.F., and Luft, B.J. (2011). Whole-genome sequences of thirteen isolates of *Borrelia burgdorferi*. J Bacteriol *193*, 1018-1020. https://doi.org/JB.01158-10

Schwan, T.G. (2003). Temporal regulation of outer surface proteins of the Lyme-disease spirochaete *Borrelia burgdorferi*. Biochemical Society transactions *31*, 108-112. DOI: 10.1042/bst0310108

Schwan, T.G., and Piesman, J. (2000). Temporal changes in outer surface proteins A and C of the Lyme disease- associated spirochete, *Borrelia burgdorferi*, during the chain of infection in ticks and mice. J Clin Microbiol *38*, 382-388

Schwan, T.G., and Piesman, J. (2002). Vector interactions and molecular adaptations of Lyme disease and relapsing fever spirochetes associated with transmission by ticks. Emerg Infect Dis *8*, 115-121. DOI: 10.3201/eid0802.010198

Schwan, T.G., Piesman, J., Golde, W.T., Dolan, M.C., and Rosa, P.A. (1995). Induction of an outer surface protein on *Borrelia burgdorferi* during tick feeding. Proc Natl Acad Sci U S A *92*, 2909-2913. doi: 10.1073/pnas.92.7.2909.

Schwarz, A., Tenzer, S., Hackenberg, M., Erhart, J., Gerhold-Ay, A., Mazur, J., Kuharev, J., Ribeiro, J.M., and Kotsyfakis, M. (2014). A systems level analysis reveals transcriptomic and proteomic complexity in Ixodes ricinus midgut and salivary glands during early attachment and feeding. Mol Cell Proteomics *13*, 2725-2735. https://doi.org/10.1074/mcp.M114.039289

Schwarz, A., von Reumont, B.M., Erhart, J., Chagas, A.C., Ribeiro, J.M., and Kotsyfakis, M. (2013). De novo Ixodes ricinus salivary gland transcriptome analysis using two next-generation sequencing methodologies. FASEB J *27*, 4745-4756. https://doi.org/10.1096/fj.13-232140

Seger, C., and Salzmann, L. (2020). After another decade: LC-MS/MS became routine in clinical diagnostics. Clinical biochemistry. https://doi.org/10.1016/j.clinbiochem.2020.03.004

Shaw, D.K., Tate, A.T., Schneider, D.S., Levashina, E.A., Kagan, J.C., Pal, U., Fikrig, E., and Pedra, J.H.F. (2018). Vector Immunity and Evolutionary Ecology: The Harmonious Dissonance. Trends Immunol *39*, 862-873. https://doi.org/10.1016/j.it.2018.09.003

Shaw, D.K., Wang, X., Brown, L.J., Chavez, A.S., Reif, K.E., Smith, A.A., Scott, A.J., McClure, E.E., Boradia, V.M., Hammond, H.L., *et al.* (2017). Infection-derived lipids elicit an immune deficiency circuit in arthropods. Nat Commun *8*, 14401. https://doi.org/10.1038/ncomms14401

Shi, Y., Xu, Q., McShan, K., and Liang, F.T. (2008). Both decorin-binding proteins A and B are critical for the overall virulence of *Borrelia burgdorferi*. Infect Immun *76*, 1239-1246. doi: 10.1128/IAI.00897-07. Epub 2008 Jan 14.

Shi, Y., Xu, Q., Seemanapalli, S.V., McShan, K., and Liang, F.T. (2006). The dbpBA locus of Borrelia burgdorferi is not essential for infection of mice. Infect Immun *74*, 6509-6512. doi: 10.1128/IAI.00740-06

Sigal, L.H., Zahradnik, J.M., Lavin, P., Patella, S.J., Bryant, G., Haselby, R., Hilton, E., Kunkel, M., Adler-Klein, D., Doherty, T., *et al.* (1998). A vaccine consisting of recombinant *Borrelia burgdorferi* outer-surface protein A to prevent Lyme disease. N Engl J Med *339*, 216-222. doi: 10.1056/NEJM199807233390402.

Simo, L., Kazimirova, M., Richardson, J., and Bonnet, S.I. (2017). The Essential Role of Tick Salivary Glands and Saliva in Tick Feeding and Pathogen Transmission. Frontiers in cellular and infection microbiology *7*, 281. https://doi.org/10.3389/fcimb.2017.00281

Skinnider, M.A., Squair, J.W., and Foster, L.J. (2019). Evaluating measures of association for single-cell transcriptomics. Nat Methods *16*, 381-386. https://doi.org/10.1038/s41592-019-0372-4

Smith, A.A., Navasa, N., Yang, X., Wilder, C.N., Buyuktanir, O., Marques, A., Anguita, J., and Pal, U. (2016). Cross-Species Interferon Signaling Boosts Microbicidal Activity within the Tick Vector. Cell Host Microbe *20*, 91-98. https://doi.org/10.1016/j.chom.2016.06.001

Smith, A.A., and Pal, U. (2014). Immunity-related genes in Ixodes scapularis--perspectives from genome information. Frontiers in cellular and infection microbiology *4*, 116. https://doi.org/10.3389/fcimb.2014.00116

Soares, C.A., Lima, C.M., Dolan, M.C., Piesman, J., Beard, C.B., and Zeidner, N.S. (2005). Capillary feeding of specific dsRNA induces silencing of the isac gene in nymphal *Ixodes scapularis* ticks. Insect Mol Biol *14*, 443-452. doi: 10.1111/j.1365-2583.2005.00575.x.

Sonenshine, D.E. (1993). Biology of Ticks, Vol Vol-1 (New York: Oxford University Press).

Sonenshine, D.E., Ceraul, S.M., Hynes, W.E., Macaluso, K.R., and Azad, A.F. (2002). Expression of defensin-like peptides in tick hemolymph and midgut in response to challenge with *Borrelia burgdorferi*, *Escherichia coli* and *Bacillus subtilis*. Exp Appl Acarol *28*, 127-134. doi: 10.1023/a:1025354326877.

Srivastava, S.Y., and de Silva, A.M. (2008). Reciprocal expression of ospA and ospC in single cells of *Borrelia burgdorferi*. J Bacteriol *190*, 3429-3433. doi: 10.1128/JB.00085-08

Stanek, G., and Strle, F. (2018). Lyme borreliosis-from tick bite to diagnosis and treatment. FEMS Microbiol Rev *42*, 233-258. https://doi.org/10.1093/femsre/fux047

Stark, R., Grzelak, M., and Hadfield, J. (2019). RNA sequencing: the teenage years. Nat Rev Genet *20*, 631-656. https://doi.org/10.1038/s41576-019-0150-2

Steere, A.C. (2001). Lyme disease. N Engl J Med *345*, 115-125. doi: 10.1056/NEJM200107123450207.

Steere, A.C., Coburn, J., and Glickstein, L. (2004). The emergence of Lyme disease. J Clin Invest *113*, 1093-1101. doi: 10.1172/JCI21681.

Steere, A.C., Sikand, V.K., Meurice, F., Parenti, D.L., Fikrig, E., Schoen, R.T., Nowakowski, J., Schmid, C.H., Laukamp, S., Buscarino, C., *et al.* (1998). Vaccination against Lyme disease with recombinant *Borrelia burgdorferi* outer-surface lipoprotein A with adjuvant. N Engl J Med *339*, 209-215. doi: 10.1056/NEJM199807233390401.

Steere, A.C., Strle, F., Wormser, G.P., Hu, L.T., Branda, J.A., Hovius, J.W., Li, X., and Mead, P.S. (2016). Lyme borreliosis. Nat Rev Dis Primers *2*, 16090. https://doi.org/10.1038/nrdp.2016.90

Stevenson, B., and Seshu, J. (2018). Regulation of gene and protein expression in the Lyme disease spirochete. Curr Top Microbiol Immunol *415*, 83-112. https://doi.org/10.1007/82_2017_49

Stewart, P.E., Bestor, A., Cullen, J.N., and Rosa, P.A. (2008). A tightly regulated surface protein of Borrelia burgdorferi is not essential to the mouse-tick infectious cycle. Infect Immun *76*, 1970-1978. doi: 10.1128/IAI.00714-07.

Stewart, P.E., and Bloom, M.E. (2020). Sharing the Ride: Ixodes scapularis Symbionts and Their Interactions. Frontiers in cellular and infection microbiology *10*, 142. https://doi.org/10.3389/fcimb.2020.00142

Stewart, P.E., Byram, R., Grimm, D., Tilly, K., and Rosa, P.A. (2005). The plasmids of *Borrelia burgdorferi*: essential genetic elements of a pathogen. Plasmid *53*, 1-13. doi: 10.1016/j.plasmid.2004.10.006.

Stewart, P.E., Wang, X., Bueschel, D.M., Clifton, D.R., Grimm, D., Tilly, K., Carroll, J.A., Weis, J.J., and Rosa, P.A. (2006). Delineating the requirement for the *Borrelia burgdorferi* virulence factor OspC in the mammalian host. Infect Immun *74*, 3547-3553. doi: 10.1128/IAI.00158-06.

Strother, K.O., Broadwater, A., and De Silva, A. (2005). Plasmid requirements for infection of ticks by *Borrelia burgdorferi*. Vector Borne Zoonotic Dis *5*, 237-245. doi: 10.1089/vbz.2005.5.237.

Strother, K.O., and de Silva, A. (2005). Role of Borrelia burgdorferi linear plasmid 25 in infection of *Ixodes scapularis* ticks. J Bacteriol *187*, 5776-5781. doi: 10.1128/JB.187.16.5776-5781.2005.

Sultana, H., Patel, U., Toliver, M., Maggi, R.G., and Neelakanta, G. (2016). Molecular identification and bioinformatics analysis of a potential anti-vector vaccine candidate, 15-kDa salivary gland protein (Salp15), from *Ixodes affinis* ticks. Ticks Tick Borne Dis *7*, 46-53. https://doi.org/10.1016/j.ttbdis.2015.08.003

Sun, D., Guo, Z., Liu, Y., and Zhang, Y. (2017). Progress and Prospects of CRISPR/Cas Systems in Insects and Other Arthropods. Front Physiol *8*, 608. https://doi.org/10.3389/fphys.2017.00608

Swei, A., and Kwan, J.Y. (2017). Tick microbiome and pathogen acquisition altered by host blood meal. The ISME journal *11*, 813-816. https://doi.org/10.1038/ismej.2016.152

Sze, C.W., Smith, A., Choi, Y.H., Yang, X., Pal, U., Yu, A., and Li, C. (2013). Study of the response regulator Rrp1 reveals its regulatory role in chitobiose utilization and virulence of *Borrelia burgdorferi*. Infect Immun *81*, 1775-1787. https://doi.org/10.1128/IAI.00050-13

Tilly, K., Bestor, A., Jewett, M.W., and Rosa, P. (2007). Rapid clearance of Lyme disease spirochetes lacking OspC from skin. Infect Immun *75*, 1517-1519. doi: 10.1128/IAI.01725-06

Tilly, K., Grimm, D., Bueschel, D.M., Krum, J.G., and Rosa, P. (2004). Infectious cycle analysis of a *Borrelia burgdorferi* mutant defective in transport of chitobiose, a tick cuticle component. Vector Borne Zoonotic Dis *4*, 159-168. doi: 10.1089/1530366041210738

Tilly, K., Krum, J.G., Bestor, A., Jewett, M.W., Grimm, D., Bueschel, D., Byram, R., Dorward, D., Vanraden, M.J., Stewart, P., *et al.* (2006). *Borrelia burgdorferi* OspC protein required exclusively in a crucial early stage of mammalian infection. Infect Immun *74*, 3554-3564. doi: 10.1128/IAI.01950-05.

Tokarz, R., Anderton, J.M., Katona, L.I., and Benach, J.L. (2004). Combined effects of blood and temperature shift on Borrelia burgdorferi gene expression as determined by whole genome DNA array. Infect Immun *72*, 5419-5432. doi: 10.1128/IAI.72.9.5419-5432

Tokarz, R., Tagliafierro, T., Sameroff, S., Cucura, D.M., Oleynik, A., Che, X., Jain, K., and Lipkin, W.I. (2019). Microbiome analysis of *Ixodes scapularis* ticks from New York and Connecticut. Ticks Tick Borne Dis *10*, 894-900. https://doi.org/10.1016/j.ttbdis.2019.04.011

Toledo, A., Coleman, J.L., Kuhlow, C.J., Crowley, J.T., and Benach, J.L. (2011). The enolase of *Borrelia burgdorferi* is a plasminogen receptor released in outer membrane vesicles. Infect Immun *80*, 359-368. https://doi.org/IAI.05836-11 [pii] 10.1128/IAI.05836-11

Tonk, M., Cabezas-Cruz, A., Valdes, J.J., Rego, R.O., Chrudimska, T., Strnad, M., Sima, R., Bell-Sakyi, L., Franta, Z., Vilcinskas, A., *et al.* (2014a). Defensins from the tick *Ixodes scapularis* are effective against phytopathogenic fungi and the human bacterial pathogen Listeria grayi. Parasites & vectors *7*, 554. https://doi.org/10.1186/s13071-014-0554-y

Tonk, M., Cabezas-Cruz, A., Valdes, J.J., Rego, R.O., Rudenko, N., Golovchenko, M., Bell-Sakyi, L., de la Fuente, J., and Grubhoffer, L. (2014b). Identification and partial characterisation of new members of the *Ixodes ricinus* defensin family. Gene *540*, 146-152. https://doi.org/10.1016/j.gene.2014.03.002

Tyson, K., Elkins, C., Patterson, H., Fikrig, E., and de Silva, A. (2007). Biochemical and functional characterization of Salp20, an *Ixodes scapularis* tick salivary protein that inhibits the complement pathway. Insect Mol Biol *16*, 469-479. doi: 10.1111/j.1365-2583.2007.00742.x.

Ueti, M.W., Reagan, J.O., Jr., Knowles, D.P., Jr., Scoles, G.A., Shkap, V., and Palmer, G.H. (2007). Identification of midgut and salivary glands as specific and distinct barriers to efficient tick-borne transmission of *Anaplasma marginale*. Infect Immun *75*, 2959-2964.

Ullmann, A.J., Lima, C.M., Guerrero, F.D., Piesman, J., and Black, W.C.t. (2005). Genome size and organization in the blacklegged tick, *Ixodes scapularis* and the Southern cattle tick, *Boophilus microplus*. Insect Mol Biol *14*, 217-222. doi: 10.1111/j.1365-2583.2005.00551.x.

Valanne, S., Wang, J.H., and Ramet, M. (2011). The Drosophila Toll signaling pathway. J Immunol *186*, 649-656. https://doi.org/10.4049/jimmunol.1002302

Valdes, J.J., Cabezas-Cruz, A., Sima, R., Butterill, P.T., Ruzek, D., and Nuttall, P.A. (2016). Substrate prediction of Ixodes ricinus salivary lipocalins differentially expressed during Borrelia afzelii infection. Scientific reports *6*, 32372. https://doi.org/10.1038/srep32372

Valenzuela, J.G., Belkaid, Y., Garfield, M.K., Mendez, S., Kamhawi, S., Rowton, E.D., Sacks, D.L., and Ribeiro, J.M. (2001). Toward a defined anti-Leishmania vaccine targeting vector antigens: characterization of a protective salivary protein. J Exp Med *194*, 331-342. doi: 10.1084/jem.194.3.331.

Vechtova, P., Sterbova, J., Sterba, J., Vancova, M., Rego, R.O.M., Selinger, M., Strnad, M., Golovchenko, M., Rudenko, N., and Grubhoffer, L. (2018). A bite so sweet: the glycobiology interface of tick-host-pathogen interactions. Parasites & vectors *11*, 594. https://doi.org/10.1186/s13071-018-3062-7

Venken, K.J., and Bellen, H.J. (2005). Emerging technologies for gene manipulation in *Drosophila melanogaster*. Nat Rev Genet *6*, 167-178. doi: 10.1038/nrg1553.

von Lackum, K., Ollison, K.M., Bykowski, T., Nowalk, A.J., Hughes, J.L., Carroll, J.A., Zuckert, W.R., and Stevenson, B. (2007). Regulated synthesis of the *Borrelia burgdorferi* inner-membrane lipoprotein IpLA7 (P22, P22-A) during the Lyme disease spirochaete's mammal-tick infectious cycle. Microbiology *153*, 1361-1371. https://doi.org/153/5/1361

Vu Hai, V., Pages, F., Boulanger, N., Audebert, S., Parola, P., and Almeras, L. (2013). Immunoproteomic identification of antigenic salivary biomarkers detected by *Ixodes ricinus*-exposed rabbit sera. Ticks Tick Borne Dis *4*, 459-468. https://doi.org/10.1016/j.ttbdis.2013.06.001

Wagemakers, A., Coumou, J., Schuijt, T.J., Oei, A., Nijhof, A.M., van 't Veer, C., van der Poll, T., Bins, A.D., and Hovius, J.W. (2016). An Ixodes ricinus Tick Salivary Lectin Pathway Inhibitor Protects *Borrelia burgdorferi* sensu lato from Human Complement. Vector Borne Zoonotic Dis *16*, 223-228. https://doi.org/10.1089/vbz.2015.1901

Wang, P., Lutton, A., Olesik, J., Vali, H., and Li, X. (2012). A novel iron- and copper-binding protein in the Lyme disease spirochaete. Mol Microbiol *86*, 1441-1451. https://doi.org/10.1111/mmi.12068

Wang, Q., Sun, Q., Czajkowsky, D.M., and Shao, Z. (2018). Sub-kb Hi-C in D. melanogaster reveals conserved characteristics of TADs between insect and mammalian cells. Nat Commun *9*, 188. https://doi.org/10.1038/s41467-017-02526-9

Wang, Y., and Zhu, S. (2011). The defensin gene family expansion in the tick *Ixodes scapularis*. Dev Comp Immunol *35*, 1128-1134. https://doi.org/10.1016/j.dci.2011.03.030

Wang, Z., Gerstein, M., and Snyder, M. (2009). RNA-Seq: a revolutionary tool for transcriptomics. Nat Rev Genet *10*, 57-63. https://doi.org/10.1038/nrg2484

Weisheit, S., Villar, M., Tykalova, H., Popara, M., Loecherbach, J., Watson, M., Ruzek, D., Grubhoffer, L., de la Fuente, J., Fazakerley, J.K., *et al.* (2015). *Ixodes scapularis* and *Ixodes ricinus* tick cell lines respond to infection with tick-borne encephalitis virus: transcriptomic and proteomic analysis. Parasites & vectors *8*, 599. https://doi.org/10.1186/s13071-015-1210-x

Wilke, A.B., and Marrelli, M.T. (2015). Paratransgenesis: a promising new strategy for mosquito vector control. Parasites & vectors *8*, 342. https://doi.org/10.1186/s13071-015-0959-2

Wilske, B. (2005). Epidemiology and diagnosis of Lyme borreliosis. Ann Med *37*, 568-579.

Wilske, B., Busch, U., Fingerle, V., Jauris-Heipke, S., Preac Mursic, V., Rossler, D., and Will, G. (1996). Immunological and molecular variability of OspA and OspC. Implications for *Borrelia* vaccine development. Infection *24*, 208-212. doi: 10.1007/BF01713341.

Wilske, B., Fingerle, V., Herzer, P., Hofmann, A., Lehnert, G., Peters, H., Pfister, H.W., Preac-Mursic, V., Soutschek, E., and Weber, K. (1993). Recombinant immunoblot in the serodiagnosis of Lyme borreliosis. Comparison with indirect immunofluorescence and enzyme-linked immunosorbent assay. Med Microbiol Immunol *182*, 255-270. DOI: 10.1007/BF00579624

Wormser, G.P., Dattwyler, R.J., Shapiro, E.D., Halperin, J.J., Steere, A.C., Klempner, M.S., Krause, P.J., Bakken, J.S., Strle, F., Stanek, G., *et al.* (2006). The clinical assessment, treatment, and prevention of lyme disease, human granulocytic anaplasmosis, and babesiosis: clinical practice guidelines by the Infectious Diseases Society of America. Clin Infect Dis *43*, 1089-1134. https://doi.org/10.1086/508667

Xu, Q., McShan, K., and Liang, F.T. (2007). Identification of an ospC operator critical for immune evasion of *Borrelia burgdorferi*. Mol Microbiol *64*, 220-231. doi: 10.1111/j.1365-2958.2007.05636.x.

Yang, X., Coleman, A.S., Anguita, J., and Pal, U. (2009). A chromosomally encoded virulence factor protects the Lyme disease pathogen against host-adaptive immunity. PLoS Pathog *5*, e1000326. doi: 10.1371/journal.ppat.1000326.

Yang, X., Lenhart, T.R., Kariu, T., An*guita, J., Akins, D.R., and Pal, U. (2010). Characterization of unique regions of* Borrelia burgdorferi surface-located membrane protein 1. Infect Immun *78*, 4477-4487. https://doi.org/IAI.00501-10

Yang, X., Promnares, K., Qin, J., He, M., Shroder, D.Y., Kariu, T., Wang, Y., and Pal, U. (2011). Characterization of Multiprotein Complexes of the *Borrelia burgdorferi* Outer Membrane Vesicles. J Proteome Res *10*, 4556-4566. https://doi.org/10.1021/pr200395b

Yang, X., Qin, J., Promnares, K., Kariu, T., Anderson, J.F., and Pal, U. (2013). Novel microbial virulence factor triggers murine lyme arthritis. J Infect Dis *207*, 907-918. https://doi.org/10.1093/infdis/jis930

Yang, X., Smith, A.A., Williams, M.S., and Pal, U. (2014). A dityrosine network mediated by dual oxidase and peroxidase influences the persistence of Lyme disease pathogens within the vector. J Biol Chem *289*, 12813-12822. https://doi.org/10.1074/jbc.M113.538272

Yang, X., Thakur, M., Koci, J., Smith, A.A., Singh, P., Zhuang, X., Promnares, K., Wang, Y., Buyuktanir, O., and Pal, U. (2018). Analysis of *Borrelia burgdorferi* proteome and protein-protein interactions. Methods Mol Biol *1690*, 259-277. https://doi.org/10.1007/978-1-4939-7383-5_19

Yang, X.F., Alani, S.M., and Norgard, M.V. (2003). The response regulator Rrp2 is essential for the expression of major membrane lipoproteins in *Borrelia burgdorferi*. Proc Natl Acad Sci U S A *100*, 11001-11006. doi: 10.1073/pnas.1834315100.

Yang, X.F., Pal, U., Alani, S.M., Fikrig, E., and Norgard, M.V. (2004). Essential role for OspA/B in the life cycle of the Lyme disease spirochete. J Exp Med *199*, 641-648. doi: 10.1084/jem.20031960

Zhang, J.J., Chen, T., Yang, Y., Du, J., Li, H., Troxell, B., He, M., Carrasco, S.E., Gomelsky, M., and Yang, X.F. (2018). Positive and negative regulation of glycerol utilization by the c-di-GMP binding protein PlzA in *Borrelia burgdorferi*. J Bacteriol *200*. https://doi.org/10.1128/JB.00243-18

Zhang, L., Zhang, Y., Adusumilli, S., Liu, L., Narasimhan, S., Dai, J., Zhao, Y.O., and Fikrig, E. (2011). Molecular interactions that enable movement of the Lyme disease agent from the tick gut into the hemolymph. PLoS Pathog *7*, e1002079. https://doi.org/10.1371/journal.ppat.1002079

Zhuang, X., Yang, X., Altieri, A.S., Nelson, D.C., and Pal, U. (2018). *Borrelia burgdorferi* surface-located Lmp1 protein processed into region-specific polypeptides that are critical for microbial persistence. Cell Microbiol, e12855. https://doi.org/10.1111/cmi.12855

Chapter 12

Biology and Molecular Biology of *Ixodes scapularis*

Daniel E. Sonenshine[1,2]* and Ladislav Šimo[3]

[1]Guest Researcher, Vector Biology Section, Laboratory of Malaria and Vector Research, National Institutes of Allergy and Infectious Diseases, NIH, Rockville, MD 20852, USA
[2]Department of Biological Sciences, Old Dominion University, Norfolk, VA 23529, USA
[3]UMR BIPAR, INRAE, Ecole Nationale Vétérinaire d'Alfort, ANSES, Université Paris-Est, Maisons-Alfort, 94700 France
*Corresponding author: dsonensh@odu.edu

DOI: https://doi.org/10.21775/9781913652616.12

Abstract
This chapter describes the biology of the tick *Ixodes scapularis* in relation to its role as the vector of the Lyme disease agent, *Borrelia burgdorferi*. Following a review of the internal anatomy of the tick, we review basic molecular processes that contribute to an understanding of the dynamics of the tick's specialized parasitic processes, including attachment behavior, salivation; silencing of host anti-inflammatory responses to enable blood ingestion at the dermal feeding site; hemoglobin digestion and reproduction. The chapter is divided into three parts: 1) systematic and anatomical characteristics of ticks; 2) host finding, attachment, salivary disruption of host defenses, blood feeding and digestion; and 3) molecular regulation of tick bodily functions and reproduction. In the first part, we review the systematics of ticks and the taxonomic position of the vector of Lyme disease, *I. scapularis*, compared to other tick species. Next, we review the general organization of the tick body, including (a) the mouthparts essential for sucking blood, (b) the powerful sucking pharynx, (c) the midgut and its role in blood and hemoglobin digestion, (d) the salivary glands and their complex cellular organization, (e) the synganglion (a fused central nervous system) responsible for controlling all body functions, (f) the reproductive organs, and (g) the tracheal system that facilitates air intake and removal of CO_2. In the second part, we highlight the role of the tick's salivary glands in secreting a remarkably complex array of anti-hemostatic molecules that modulate the bite site in the host skin and how these salivary molecules facilitate the lengthy blood-sucking process. We also describe how ticks capture hemoglobin and internalize it in midgut epithelial cells for intracellular digestion, followed by the sequestration of heme into specialized hemosomes for disposal as hematin. We also will review the neural control of regulation of tick salivary glands, blood uptake, hemoglobin digestion, blood meal concentration, water/salt elimination, vitellogenesis and receptor mediated vitellogenin uptake in the developing oocytes and their oviposition.

Introduction
Ticks represent one of the most successful ectoparasites, and certainly one of the most feared, because of the numerous disease-causing pathogens that they transmit to humans and animals. Ticks transmit a greater variety of pathogens than any other arthropod vector. In addition, they are feared as pests, causing irritation, inflammation and, in some cases, severe toxic reactions, paralysis and even death. The world-wide public health impact due to tick-borne diseases, including the costs of vaccination and treatment for these illnesses, is estimated to be in the billions of dollars. Annual cost estimates for Lyme disease in the U.S. are ~786 million USD per year and tens of millions USD in several European countries (Mac et al., 2019).

Among the factors that explain the success of ticks as parasites and vectors of infectious diseases are their unique feeding style, remarkably long survival periods, high reproductive potential and well-developed innate immune system. In contrast to all other ectoparasitic arthropods that bite quickly and soon flee with small amounts of stolen blood, hard ticks remain attached for long periods and consume large amounts of blood, up to 10- to 100-fold greater than their original body mass. This is especially the case for the hard ticks (family *Ixodidae*) that cement themselves into the skin of their vertebrate hosts and remain attached for many days, often consuming up 200 – 300 times their pre-feeding body weight (including excess blood meal water that they secrete back into the host) (Sonenshine and Roe, 2014). While feeding, ticks secrete a remarkable cocktail of anti-hemostatic salivary proteins, prostaglandins, and other small molecules that virtually silence the host's awareness of its presence by disabling pain and itch responses, preventing coagulation, increasing blood supply and other actions that disable the host's normal responses to injury (Nuttall, 2019). Nevertheless, silencing host defenses against tick bites does not always succeed, especially after repeated tick attacks on the same hosts (e.g., laboratory mice) or when ticks attempt to feed on non-permissive, e.g., guinea pigs, or other non-native host species.

Over the course of thousands or even millions of years, evolutionary pressures have led to increasing degrees of host specificity among ticks for birds, bats, other small mammals, and so on. *I. scapularis* is an outlier--feeding on diverse vertebrate hosts that include a wide range of reptiles, birds, and mammals. Incidentally, the ways in which tick saliva alters the host skin surrounding the feeding site enhances the survival and invasion of tick-borne pathogens, a form of saliva-assisted pathogen transmission (Bonnet et al., 2018; Šimo et al., 2017). Following successful blood feeding, ticks drop off their hosts and survive in leaf litter or other ground cover in woodland habitats, rotting vegetation and soil in meadows (non-nidicolous species) or in caves or nests of their hosts (nidicolous species) (Gray et al., 2014; Randolph, 2014). Ticks are remarkable in their ability to survive for long periods without fresh nutrients. Some species, especially among the soft ticks (family Argasidae), can survive for many months or even years between blood meals (Sonenshine and Roe, 2014). Finally, ticks hold the record among blood-feeding arthropods for the numbers of eggs produced. An *I. scapularis* adult female typically produces 1,000 – 1500 eggs after engorgement. Other ixodid ticks produce many more (e.g., ~ 5,400 in the case of the American dog tick *Dermacentor variabilis*) (Sonenshine, 1970). Although much smaller egg yields are produced by soft ticks following each blood meal, these species may feed and lay eggs multiple times, rather than all at once, as with ixodid ticks. Finally, the tick's natural immune system, controlled by the Toll and Imd pathways, ensures that most microbes harbored by ticks do not harm their tick hosts, although some tick-borne pathogens, e.g., *Anaplasma* sp., may affect tick fitness, either negatively or positively (Dumler, 2010; Harris et al., 2017). Following thousands and perhaps even millions of years of tick-pathogen association, a balance has evolved such that these pathogens are unlikely to disrupt the tick's physiological and biochemical processes even though these same microbes are pathogenic once transmitted to their reservoir competent vertebrate hosts.

This chapter will review the many unique physical, biological, and genetic attributes of ticks, primarily the black-legged tick, *I. scapularis*, to provide a framework for understanding their role in the transmission of disease-causing microorganisms to humans and animals.

Part I: Systematic and anatomical characteristics of ticks

Types of ticks (systematics)

Ticks are chelicerate arthropods (phylum Chelicerata), which includes animals that have chelicerae, rather than mandibles, as their cutting and biting appendages. Ticks are arachnids (class Arachnida), the same subphylum that includes spiders, scorpions, and other similar taxa, and are grouped with mites in the subclass Acari. They are classified in their own distinctive order, the order Ixodida, within the superorder Parasitiformes (Keirans, 2009). Their evolutionary origin remains controversial, with recent studies based on molecular systematics only partially supporting earlier hypotheses based on morphology. The chelicerates are believed to have diverged from the Pancrustacea, the group that eventually evolved into the insects, crustaceans and myriapods as early as the Cambrian period (Dunlop, 2010). Ticks are now believed to have evolved in a region of the ancient supercontinent Gondwana land (in the area which eventually became Antarctica and Australia) sometime during the Paleozoic era (Beati and Klompen, 2019; Beati et al., 2008; Jeyaprakash and Hoy, 2009), likely during the Carboniferous period (Beati and Klompen, 2019; Mans et al., 2016), although others predict a more recent origin, sometime during the early Cretaceous period (Klompen et al., 2007). The genus *Ixodes* is believed to have evolved early during this period, placing it close to the base of the ixodid phylogenetic tree (Klompen et al., 2000). Hypotheses concerning tick evolution, including *I. scapularis,* the focus of our book, are important since they may help to explain how *I. scapularis* remains a non-host-specific species, feeding on an exceptionally diverse array of vertebrate hosts including amphibians, reptiles, birds and mammals. They may also contribute to a better appreciation of why *I. scapularis* is so remarkably permissive for diverse microbial symbionts, including prokaryotic and eukaryotic symbionts pathogenic for humans and other animals.

Characteristics of ticks: external body features and similarities with other acarines

Knowledge of the external appearance of the black-legged tick, *I. scapularis,* is important, since only this tick species transmits the microbial pathogens that cause Lyme disease in humans. It is also a vector of human granulocytic anaplasmosis, human babesiosis and several microbes that cause other lesser known maladies. People, including many scientists, often mistake other human biting ticks, e.g., lone star ticks or American dog ticks as vectors of *B. burgdorferi*, leading to unnecessary expenditure of time and resources.

The tick body plan is similar to that found in many other parasitiform acarines. The body is organized into three major regions: the anterior capitulum, (gnathosoma) which bears the chelicerae, palps and hypostome; the idiosoma, or body region, and the walking legs. The idiosoma is divided further into the anterior podosoma, which bears the four pairs of walking legs (three pairs in larvae), and the posterior opisthosoma, which bears the paired respiratory spiracles and anal aperture. Some authors also combine the capitulum and podosoma into the "prosoma" to distinguish this region from the opisthosoma. Like all other acarines, the ancestral segmentation is obscured by the extreme fusion that occurred during tick evolution. As exemplified by *I. scapularis*, , on the ventral side, adult ixodid ticks have a genital pore located between the first and second pairs of legs, crescent-shaped in females, subtriangular in males, but absent in juveniles; the genital pore is open females but covered with a movable plate in males. A large subtriangular plate, the scutum, is located on the dorsal side of the body in females, nymphs and larvae; the scutum covers the entire body in males (and, therefore, limits expansion of the body during blood feeding). In ticks of the genus *Ixodes,* including *I. scapularis*, sclerotized plates cover the ventral part of the body in males but is absent in females. The ventral plates of the male include the large sternal plate, posterior to the genital pore, the lateral epimeral plates and the adanal plates adjacent to the anus. A large spiracular plate, containing the spiracular aperture, occurs on either side of the body just posterior to the coxae of the fourth pair of legs. The anal aperture is located near the posterior end of the opisthosoma. These anatomical features are shown in Figure 1.

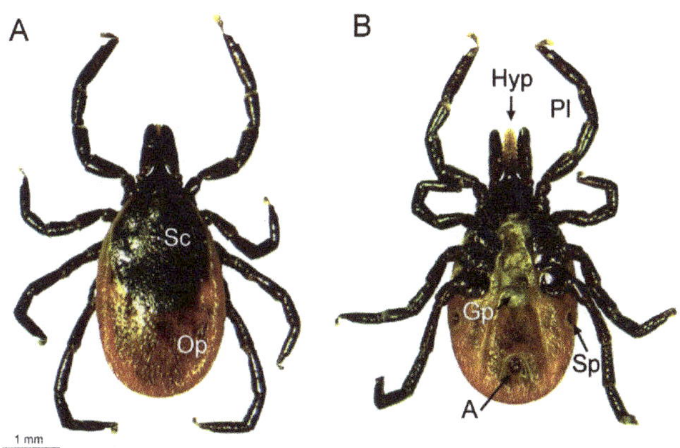

Figure 1. Photomicrographs of an unfed female *Ixodes ricinus* (castor bean ticks), a species closely related to *Ixodes scapularis* and the vector of Lyme disease in Europe. (A) Dorsal view, showing the black-colored scutum (Sc) and the reddish-brown opisthosomal region (Op). (B) Ventral view showing the mouthparts, including the hypostome (Hyp) with its rows of recurved denticles, palps (Pl), female genital pore (Gp) between the hindlegs and the spiracular plates (Sp) on the lateral margins of the body and the anus (A). (Photo authors: D. E. Sonenshine and L. Šimo).

The tick mouthparts (Figure 2), located on the capitulum, comprise the four-segmented palps, chelicerae with sharp cutting teeth, and the elongated, central toothed hypostome which bears the food canal for sucking blood. It is important to recognize that the capitulum is not a "head", since the synganglion, the tick's fused brain and central nerve chord, are located deep inside the body. The capitulum comprises the basis capitulum, the large, flexible structure which contains the pharynx. A pair of ring-like depressions containing numerous tiny pores occurs on the dorsal surface of females. The chelicerae on the dorsal side of the capitulum can be everted or retracted as needed. At their anterior ends, each chelicera has a set of cheliceral digits, bearing sharp spines for cutting into host skin. Instead of cutting like a pair of scissors, the digits move from side to side, ripping and tearing the skin surface. Extending anteriorly from the sides of the basis capituli are the 4-segemented palps which

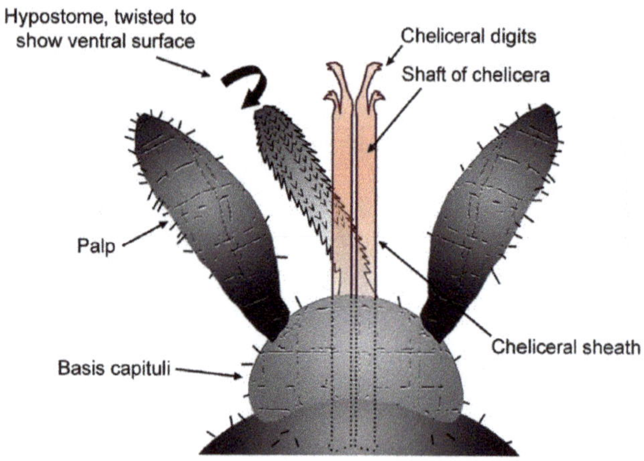

Figure 2. Diagrammatic representation of the mouthparts of an ixodid tick illustrating the locations of the major structures that form the capitulum (gnathosoma). The hypostome is shown teased out of its normal location between palps and the chelicerae. (Graphic authors: D. E. Sonenshine and L. Šimo).

provide gustatory (touch) sensory information that contributes to decisions the tick can make about whether to attach and where.

Finally, the elongated hypostome extending from the center of the basis capitula serves as the anchoring structure, penetrating deep into the host skin. Numerous sharp recurved denticles located all across the ventral side of the hypostome (Figure 2) help hold the tick in place during the initial period of attachment until sufficient cement is secreted (Sonenshine and Anderson, 2014). A deep groove (not shown in the figure) on the dorsal surface of the hypostome serves as both the food channel and the conduit for delivering saliva. During feeding, blood is sucked into this hypostomal food channel by the pumping action of the pharynx, alternating with periods of salivary secretion (3D animations are available in Vancová et al., 2020).

Internal body organization
The internal anatomy of ticks is similar to that of all other parasitiform acarines (Alberti and Coons, 1999) but very different from those of insects and other mandibulate arthropods. In hard ticks, such as *I. scapularis,* it comprises 1) the alimentary canal, beginning with the food canal in the hypostome, pharynx, esophagus, midgut and the hindgut; 2) paired salivary glands; 3) the synganglion and peripheral nerves; 4) heart; 5) Malpighian tubules; and, finally, 6) the male or female reproductive organs. Knowledge of the tick's internal anatomy is essential to understanding how ticks feed, reproduce and survive in their various natural habitats. In this review of the tick's internal anatomy, the major emphasis is directed to the salivary glands and the midgut. Images of the tick's body organs and tissues can be found online by going to the website www.vectorbase.org, selecting tools, controlled vocabulary, and then selecting TADS from the drop down list. This anatomical study is based on *D. variabilis*, but most features are similar to those found in *I. scapularis*.

Pharynx
Ticks are pool feeders. They suck up extravasated blood that oozes from the damaged dermal blood vessels into the tissues around their mouthparts. Since such fluids have low hydrostatic pressure compared to intact blood vessels, imbibing these fluids requires a powerful sucking pharynx. In *ixodid* ticks, the pharynx is an elongated organ located inside the base of the capitulum on its ventral side. The interior of the pharynx is lined with a delicate, thin cuticle. Externally, rings of constrictor muscles are wrapped around the pharynx alternate with dilator muscles that extend from the pharynx to the capitulum wall (Figure 3). Contractions of the dilator muscles pull the pharynx open, allowing blood sucked into the food channel to flow into the pharynx. Contractions of the constrictor muscles, coordinated with relaxation of the dilator muscles (3D animation available in Vancová et al., 2020), constricts the pharynx and forces imbibed blood posteriorly into the esophagus and midgut. Opening and closing of tiny valves at either end of the pharynx synchronizes with the contractions and relaxations of the pharyngeal muscles; the anterior valve closes when the constrictor muscles contract, while the posterior valve opens, insuring that blood flows posteriorly and does not regurgitate (for a more detailed review see (Sonenshine and Anderson, 2014; Vancová et al., 2020).

Esophagus and midgut
Blood imbibed from the host is pumped out of the pharynx into the esophagus, an elongated narrow tube that passes through the synganglion to the midgut. At the esophageal-midgut junction, fluid entry is regulated by the proventricular valve, which allows blood to pass into the midgut (Sonenshine and Anderson, 2014).The midgut is the first body organ in which pathogenic microbes imbibed with the host blood can survive and, in most cases, invade the

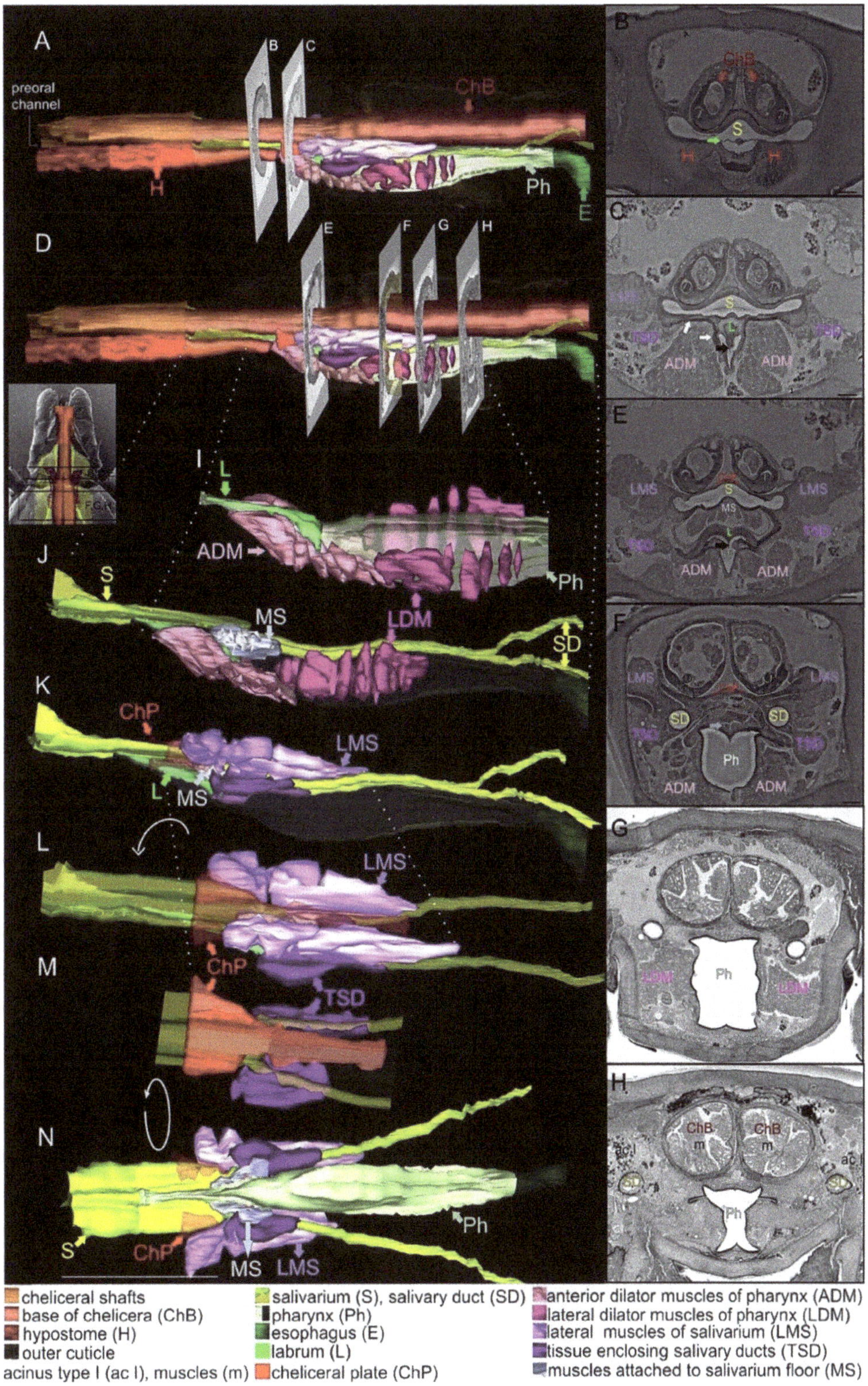

Figure 3. The feeding apparatus of the unfed nymphal stage of the European "castor bean tick", *Ixodes ricinus*. 3D models (A,D,I–N) were reconstructed from serial transverse sections (B,C,E–H) through capitulum part imaged by back-scattered electrons (inset). Bar: 25 µm. The 3D animations are available in original article. From Vancová et al., (2020) Three-dimensional reconstruction of the feeding apparatus of the tick *Ixodes ricinus* (Acari: *Ixodidae*): a new insight into the mechanism of blood-feeding. Sci. Rep. 10:165.

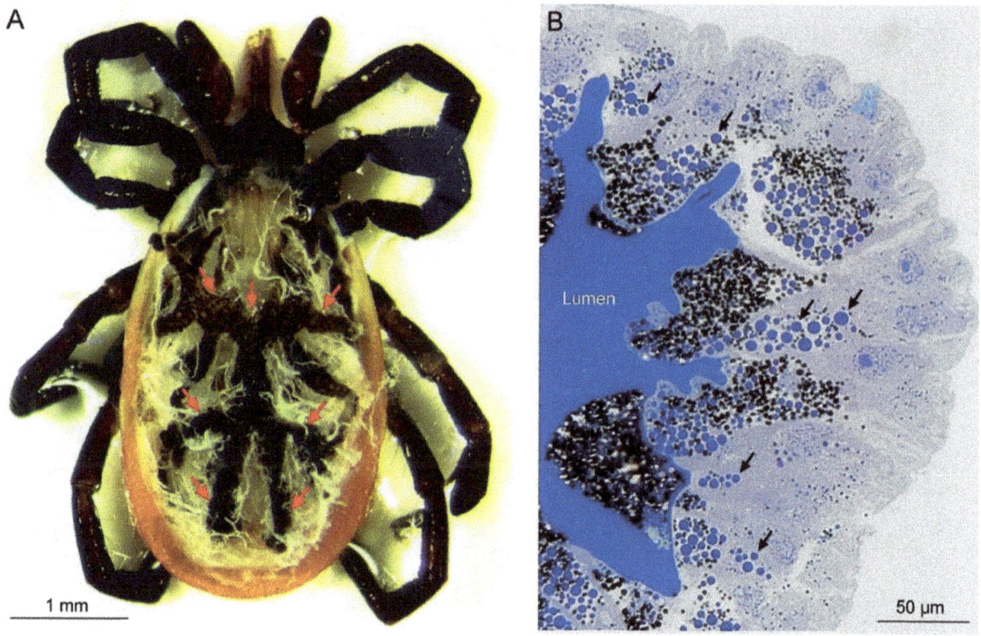

Figure 4. (A) An unfed *Ixodes ricinus* adult female with dorsum removed, highlighting the midgut (red arrows identifying the central ventriculus and lateral diverticulae). (B) Photomicrograph of a histological section of the midgut of a female *Ixodes scapularis* taken on day 5 of blood feeding on its mammalian host (rabbit, *Oryctolagus cuniculus*). Note the masses of hematin-filled vesicles (hemosomes) near the luminal side of many of the midgut cells. The blue-stained vesicles (black arrows in B) are believed to be phagolysosomes containing hemoglobin in various stages of digestion. (Photo authors: D. E. Sonenshine and L. Šimo).

tick's internal tissues. Consequently, knowledge of the morphology and function of the midgut is essential for readers to understand and appreciate its unusual method of intracellular hemoglobin digestion, heme sequestration and detoxification, as well as its role as a long-term food storage organ. These features have profound implications for the survival and development of tick-borne pathogens (Sonenshine and Macaluso, 2017).

The midgut is the largest organ in the tick's body. In *I. scapularis*, it consists of multiple pouch-like diverticula that extend outward from the central portion (stomach) and fill all the available spaces (Figure 4A). During blood feeding by the attached tick, the midgut fills with blood, and the diverticulae enlarge enormously, displacing other internal organs or compressing them (e.g., trachea which must stretch in order to continue to support gaseous exchanges). Digestion of all blood cells takes place in the midgut lumen (Figure 4B). However, except for hemoglobin and serum albumin, which are digested inside the midgut cells (discussed below), no further digestion takes place and serum proteins remain intact and are eventually expelled in the feces (see below) (Lara et al., 2005). Even some hemoglobin, acquired during the larval blood meal, can be found in *I. scapularis* nymphs more than 300 days later (Laskay et al., 2013). These findings show that blood digestion is a gradual process, often requiring weeks or even months, depending on environmental temperatures.

The midgut consists of a simple epithelium surrounded by a single layer of smooth muscle cells. Although previous studies have described different midgut epithelial cell types specialized for ingestion, digestion, secretion and elimination of waste, it is now generally accepted that these specialized cells from a single population of progenitor stem cells stimulated to divide and differentiate by the influx of fresh blood (Starck et al., 2018). The lining epithelial cells vary greatly in size, with some reaching up to 100 µm from the basal lamina to the midgut lumen. A thin semipermeable matrix, the peritrophic matrix, forms adjacent to the midgut cells within the lumen during blood feeding. This membrane protects the delicate epithelial cells from harsh materials (e.g., particulates) and also excludes Lyme disease spirochetes on the luminal side. On their luminal sides, midgut epithelial cells have numerous microvilli which greatly expand their surface area. Small subcellular pockets form at the cell surfaces, likely sites of hemoglobin uptake where these molecules bind to specific receptors in minute clathrin-coated pits and then are internalized by receptor mediated endocytosis. The pinocytotic vesicles are internalized (endosomes) and fuse with lysosomes to form digestive vesicles. Hemoglobin digestion is intracellular; many cells are filled with digestion vesicles, while other vesicles, termed hemosomes, are filled with hematin resulting from the detoxification of hemoglobin (Gulia-Nuss et al., 2016). Thus, the reader should recognize that one of the unique features of tick blood digestion is that, in contrast to all other blood-feeding arthropods, ticks are mostly hemoglobin feeders, although host serum albumin also is digested by the same or similar multienzyme pathway (Sojka et al., 2016). Also, in contrast to all other blood-feeders, digestion in ticks proceeds slowly, so that the midgut serves as a long-term storage organ, enabling these ectoparasites to survive for many months or even years between blood meals.

During engorgement, the midgut swells enormously. Feeding female ixodid ticks may increase their body size over 100 times or more as the blood meal progresses. During the rapid phase, the cuticle is stretched and thins considerably. However, this phenomenon results in tremendous hydrostatic pressure. One wonders how the internal body organs, especially the air-filled tracheal trunks and tracheae, can avoid displacement and compression. According to Kaufman *et al.* (Kaufman et al., 2016), ticks have evolved a process of pulsatile blood ingestion, "using pulses of high pressure rather than a continuous "squeeze", to maintain respiratory function. "The pulses of pressure are not highly regular, but average about 50 Hz…". These authors believe that this is a higher frequency than the "snap back" feature of the cuticle, so that stretch can be achieved by pulses without recovery (contractions) between pulses. These pulses are believed to reflect the periodic contractions and relaxations of the dorso-ventral muscles, since these large and numerous muscles are the ones most likely to produce pressures in the range of 55-60 kPa and frequencies in the range of 30–80 Hz. The frequent pulsations also have the advantage of enabling periodic air flow through the spiracles and tracheal system.

Salivary glands

The salivary glands are the primary organs responsible for transmission of most tick-borne pathogens (Šimo et al., 2017), although exceptions occur in argasid ticks in which pathogens disseminate via excretion of excess body water from their coxal glands (Sonenshine, 2014). Consequently, knowledge of the anatomy and physiology of these organs is essential for readers to understand how they function in support of the tick's feeding and long-term survival. Tick saliva is undoubtedly one of the most important reasons for their evolutionary success as ectoparasites, enabling them to consume enormous quantities of blood while remaining inconspicuous until they complete their bloodmeals.

In *I. scapularis,* the salivary glands are complex multi-functional organs. They secrete cement (Gulia-Nuss et al., 2016; Kim et al., 2016b; Suppan et al., 2018), essential for attaching the tick to its vertebrate host, as well as a complex assortment of anti-hemostatic proteins, prostaglandins and other agents that enable the tick to suck blood while the host remains unaware of its presence (Nuttall, 2019; Šimo et al., 2017). They also excrete excess water consumed from the blood, estimated to be as high as 70% of the blood meal volume (Bowman et al., 2008; Kaufman and Phillips, 1973; Kaufman and Sauer, 1982), concentrating the blood meal while the tick remains attached, thereby greatly increasing the volume of blood engorged. Exactly how they accomplish these remarkable functions, unique among all blood-feeding arthropods, is the subject of this and later sections (see *Regulation of salivary glands* and *Osmoregulation during the blood meal* in Part III) of this chapter.

The *I. scapularis* salivary gland is a multi-lobed organ (Figure 5), resembling a cluster of grapes, located in the anterior portion of the body (Alarcon-Chaidez, 2014; Bowman and Sauer, 2004; Fawcett et al., 1981). There are three types of alveoli, known as acini, namely, I, II and III (Figures 5 and 6A) in *ixodid* tick female salivary gland (Binnington, 1978).

The agranular type I acini are located at the anterior end of the gland and are situated adjacent to the main salivary duct (Figure 5). The cells of these acini are large with a central vacuole and numerous lamellae along their margins. The central lamellate cell abuts the acinus lumen and is believed to secrete directly into the main salivary duct. Other smaller cell types, such as peripheral lamellate cells, peritubular

Figure 5. (A) Microphotograph of dissected, intact salivary gland (upper panel) of partially fed *I. scapularis* female, and separated acini types I, II and III (lower panels). (B) Schematic illustration of unfed *Ixodes* female salivary gland highlighting the position of acini types. MD main salivary duct. (Photo and graphic authors: D. E. Sonenshine and L. Šimo).

cells and constrictor cells, are located adjacent to a short lobular duct that directly opens into the main salivary duct. The primary role of the type I acinus is to rehydrate the ticks by uptake of water during periods of fasting (Needham et al., 1990; Rudolph and Knülle, 1974; Rudolph and Knulle, 1978). They accomplish this by absorbing water from hygroscopic saliva secreted by type II and III acini (Figure 7B; Kim et al., 2017, 2019) onto the hypostome, which is subsequently diluted with water from the humid atmosphere of the tick's microenvironment. Typical of cells involved in active transport of water and water regulation, there is a tremendous amount of basal membrane infoldings all along the margin of the acinus (Needham et al., 1990). These salivary acini also appear to be active during tick feeding and in some cases also been implicated in cement secretion (Barker et al., 1984).

The structurally similar type II and III acini comprise the major portions of the salivary glands and lie towards the middle and distal (posterior) regions, respectively (Figure 5). Besides their prominent granular cells (discussed below), types II and III acini contain non-secretory support cells (e.g., basal, adlumenal and ablumenal epithelial (having interstitial projections) and neck cells) (Figures 6B and 12). Contractions of the single adlumenal (= myoepithelial) cell (Figures 6B, C and 12) drives the secretory contents of the acini into the connecting ducts for passage into the main salivary gland ducts (Alarcon-Chaidez, 2014; Vancová et al., 2019). The basal extensions of the myoepithelial cell, along with the apical extension of the basal epithelial cells overlay the acinar valve (Figures 6B, D and 12), serve to control expulsion of the acinar contents to the adjacent duct (Vancová et al., 2019). In the type II and III acini, various granule-rich cells, divided into different cell types within each acinus, namely, types *a-f* (differing in number depending on the tick species) have been recognized (Binnington, 1978). The type II acini have been reported to contain up to six different secretory cell types containing numerous intracellular electron-dense granules (Denardi et al., 2011). Among the most important are 1) the type "*a*" cells, believed to be rich in compounds that contribute to the cement, since they deplete their granules soon after attachment; 2) types "*b*" and "*c*" cells, which contain granules believed to contribute glycoproteins that likely serve in disrupting host

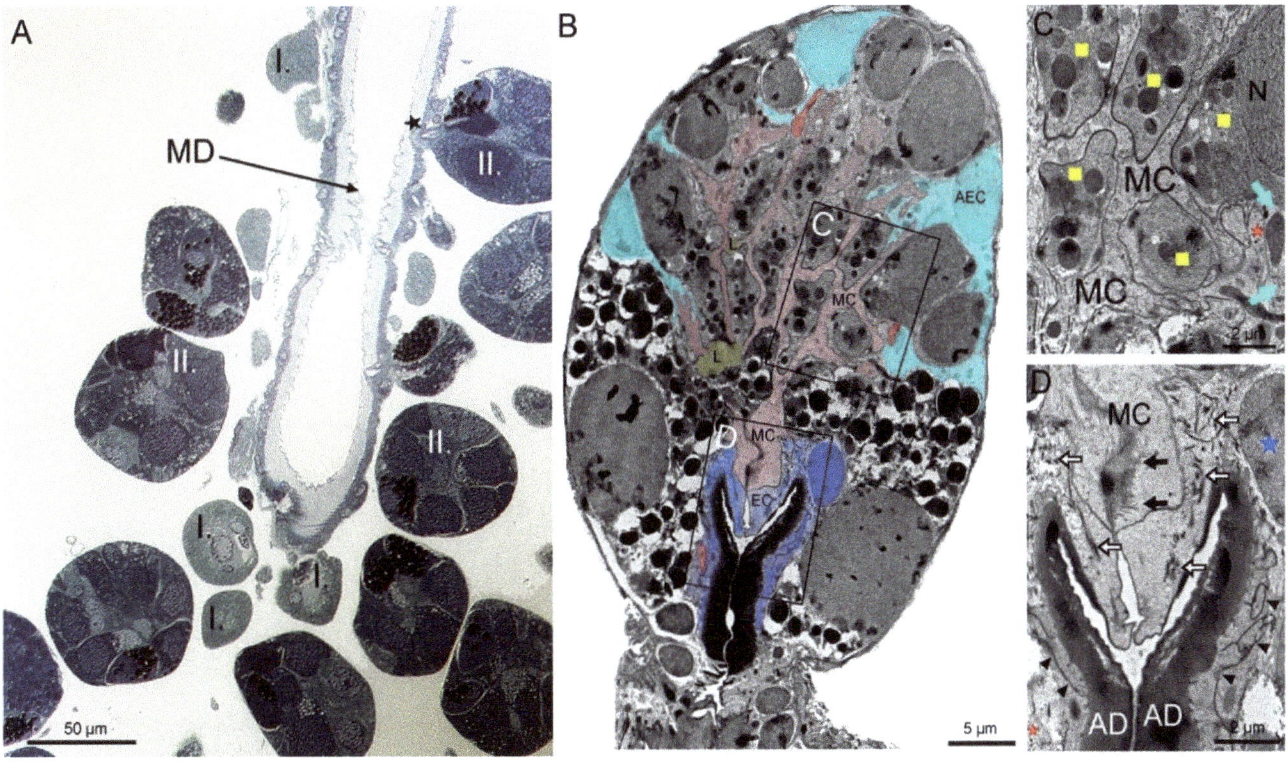

Figure 6. Detailed structure of the salivary gland and the different acinar types. (A) Photomicrographs of histological sections through the type I and type II acini of the salivary gland of a feeding female *Ixodes scapularis*. Image showing the gland acini adjacent to the main salivary duct (MD) in differential interference contrast (Roman number indicate acinus type). Asterisk shows small acinar duct directly attaching an acinus to the main duct. (B) Transmission electron microscopy images of the type III acinus from an unfed *I. ricinus* female salivary gland highlighting its cellular composition and associated axons. Cross-section through the middle of the acinus reveals common axon positions (red), basal epithelial cells (ECs, dark blue), adlumenal myoepithelial cell (MC, pink), ablumenal epithelial cells (AECs, aqua-blue) and lumen (L, yellow). Different granular cells at both basal and apical regions of the acinus are left in black and white. Insets in B are magnified in C and D. (C) The axons (red asterisks) are in direct contact with the MC's finger-like projections which radiate between granular cells (yellow squares) and fine projections of AECs (aqua-blue arrows). (D) The acinar valve region. Apparent microtubules (black arrows) in the MC and (white empty arrows) in ECs are shown. EC labyrinthine junctions (black arrowheads) are closely associated with the acinar duct (AD). The blue asterisk shows the nucleus of the EC. The basal axon in D is indicated by the red asterisk. Panel A photo author: D. E. Sonenshine; panels B – D are from Vancová et al., (2019) Ultrastructural mapping of salivary gland innervation in the tick *Ixodes ricinus*. Sci. Rep. 9:6860.

hemostatic and immune challenges to tick attack. In *Ixodes ricinus*, the secretory granules of the *a*, *b* and *c* cells are glycosylated; most contain mannose, galactose, n-acetyl-D-glucosamine and sialic acid (Vancová et al., 2006). In *D. variabilis* tick, the acinus III-specific "*d*" and "*e*" cells are reported to secrete cement compounds that bind the feeding tick to the host skin (Jaworski et al., 1992). However, it is not known whether the same cells types in *I. scapularis* produce the same compounds. In contrast to the type I acini, these large, complex acini type II and III have their own acinar valves (Figures 6B, D and 12), which open to secondary ducts that collect and transport the protein-rich saliva to the main salivary duct (Binnington, 1978; Vancová et al., 2019). The rich array of salivary proteins secreted by the salivary glands and their roles in tick feeding will be discussed below in *the tick sialome* in Part II of this chapter.

The specific activities of individual acini of tick salivary glands mediate diverse functions that ensure the tick's biological success during both on- and off-host periods (Figure 7)(Bowman et al., 2008; Kim et al., 2019). Studies examining the physiology of tick salivary glands have focused primarily on the on-host tick feeding period, during which the secretory activities of the types I and II acini are most apparent. When feeding, these structures produce hyperosmolar primary saliva that passes via ducts to

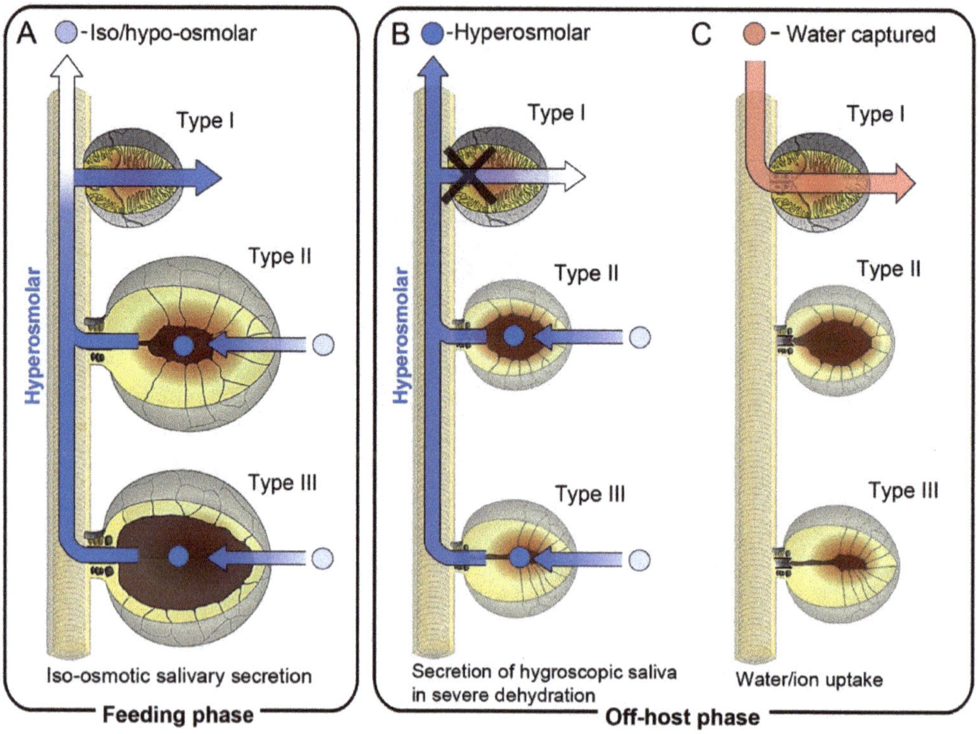

Figure 7. A model explaining roles of specific acini types in *Ixodes scapularis*. (A) The process for iso/hypo osmolar salivary excretion in feeding ticks. (B) Production of hyperosmolar saliva for capturing water molecules. (C) Uptake of water captured in the hygroscopic salivary secretion. From Kim et al., 2018. Neural and endocrine regulation of osmoregulatory organs in tick: Recent discoveries and implications. Gen. Comp. Endocrinol. 278: 42-49, with permission from Elsevier.

the anterior part of the gland where type I acini absorb ions resulting in iso-osmolar saliva (Figure 7A). During the fasting period, when the tick undergoes severe stress due to desiccation, the absorptive function of type I acini is shut down and hygroscopic hyperosmolar saliva is secreted (Figure 7B) onto the hypostome surface to from a crystalized matrix. In a more favorable environment, this deposit is deliquesced by atmospheric water, and the type I acini absorb water from this diluted saliva (Figure 7C), thereby keeping the ticks rehydrated (Kim et al., 2016a, 2017, 2019). The regulation mechanism of tick salivary gland activities seems to be unique among the invertebrate taxa. We discuss its characteristics in the *Regulation of salivary glands* in Part III of this chapter.

The tick bite site has often been viewed as a closed environment with a defined volume. However, as Mans (2019) noted, "the feeding site itself may be dynamic with constant changes in concentration of both host and tick proteins....since ticks alternately salivate and ingest blood meal". These repetitive salivating and sucking activities, as well as protein dispersal into the general systemic system, make it difficult to estimate the concentrations of the anti-hemostatic molecules active during the feeding process. The bite site is subject to host inflammatory responses, including edema, while salivary secretion of excess blood meal water back into the wound site all contribute to a changing volume. Estimates of the tick bite site (also described, inappropriately, as a feeding pool), range from 10 – 50 µl (Mans, 2019). However, more research is needed on this topic, especially how it differs between different animal hosts.

Reproductive system

Reproduction in females among the hard ticks is of the univoltine type (i.e., mating, insemination, engorgement, and ovary development occur with a single feeding). This results in deposition of thousands of eggs in a single oviposition cycle. The female dies following oviposition. This contrasts with the soft ticks, in which feeding, mating, ovary development and oviposition occur multiple times (multivoltine reproduction). As a result, hard ticks are among the most prolific egg layers of all blood

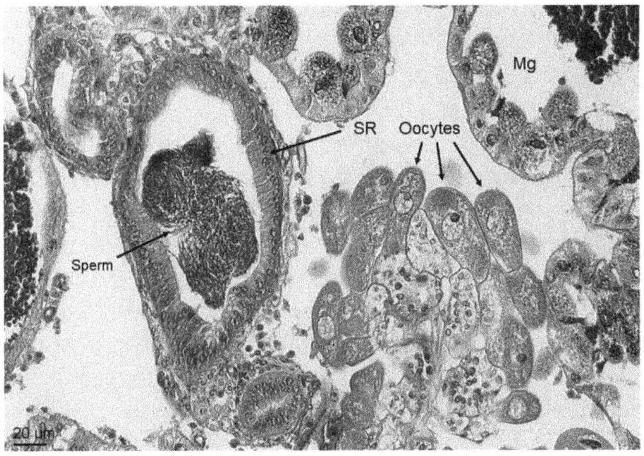

Figure 8. Photomicrograph of the feeding-mated female, *Ixodes scapularis*, fed 5 days, showing a mass of spermatozoa in the seminal receptacle, adjacent to oocytes in the ovary. Mg = midgut; Sperm = spermatozoa; SR = seminal receptacle. (Photo author: D. E. Sonenshine).

feeding arthropods, with some species laying up to 20,000 eggs in a single ovipositing cycle (Arthur, 1962). The biological significance of this phenomenon for ticks and parasites and vectors of infectious diseases will be considered later in the *Reproduction* section in Part III of this chapter.

In *I. scapularis*, as in all other species of the ixodid family, the female reproductive system comprises the U-shaped ovary located across the opisthosomal region of the body; a pair of narrow, coiled oviducts, each extending from either side of the ovary; a single uterus; a muscular vagina, and the cuticle-lined vestibular vagina which opens into the genital pore. As described by Ogihara and Taylor (Ogihara and Taylor, 2014), the ovary is panoistic, producing ova without nutritive cells (i.e., without nurse or follicle cells). In the *I. scapularis* nymphal ovary, the ovary wall consists of a simple epithelium and occasional elongated smooth muscle cells studded with oogonia and interstitial cells. Following the molt from nymph to adult female, many primary oocytes have developed, and ovary maturation progresses, especially after mating and insemination during feeding and engorgement. The oocytes occur on the hemocoel side of the ovary wall, connected to the epithelium by pedicels of funicular cells and bathed by hemolymph as they develop. In partially fed females, the oocytes are supported by funicular cells. One section of the ovary contains a longitudinal groove, replete with undifferentiated oocytes, while the developing oocytes occur in other parts of the ovary, attached to the ovary wall. The luminal side of the ovary is lined with interstitial cells (Ogihara and Taylor, 2014). After mating, the female stores the spermatozoa in the seminal receptacle where they remain available to fertilize the eggs following the transition to vitellogenesis and rapid development of the oocytes for oviposition (Figure 8).

The ecdysteroid hormone (20-hydroxyecdysone) stimulates the production of the yolk precursor; vitellogenin (Vg) is essential for ovarian maturation (the sites and molecular biology of vitellogenesis will be discussed in the *Reproduction* section of Part III). Following mating and the onset of vitellogenesis, waves of oocyte growth occur in preparation for and during oogenesis. The vitellogenic ova (i.e., ova that have incorporated Vg) appear brown in color due to the heme component of the Vg molecule. As they reach maturity, including fertilization and shell formation, the mature eggs are ovulated into the ovary lumen (Ogihara et al., 2019). Except for the ovary and parts of the oviducts, the other components of the reproductive system (i.e., uterus, muscular vagina, and vestibular vagina) are located in the podosomal region of the tick body. In addition, a pair of blind-ending tubules, the tubular accessory glands, extends from the vestibular vagina near its junction with the muscular region of the vagina. During blood-feeding, the cells of the vestibular vagina proliferate externally and form a large gland, the lobular accessory gland. A sac-like structure, the seminal receptacle, which stores incoming spermatozoa, is located above the vagina. Lastly, an unusual feature of ticks is the presence of a separate egg waxing organ, the Gené's organ, which extends its arms out in the space (the camerostome) created by downward flexing of the capitulum, enabling these structures to coat the eggs during oviposition. The reproductive organs are not static structures. Rather, they undergo extensive changes during blood-feeding as the oocytes expand and extend into the surrounding hemocoel (Ogihara and Taylor, 2014).

Synganglion

The nervous system consists of a synganglion and the peripheral nerves. In ticks, as well as in all other acarines, the synganglion is the entire central nervous system, the result of fusion of the ancestral "brain" and the ventral nerve cord (Šimo et al., 2014a). It is located in the ventro-anterior region of the body (Figure 9), immediately anterior to the muscular vagina. The synganglion is surrounded by a delicate sheath, the periganglionic membrane, enclosing the periganglionic sinus containing filtered hemolymph from the dorsal aorta and heart. There are two major regions, supraesophageal and subesophageal, so-called due to their positions in relation to the esophagus, which extends from the pharynx through this organ to the midgut. The tissue structure of the synganglion is divided into two separate zones, the outer cortical zone with numerous neuronal cell bodies, and the interior neuropile, filled with innumerable neural fibers, axons and dendrites that crisscross the synganglion in all directions. The synganglion contains numerous ganglia, named for the body organs with which they are associated (e.g., protocerebrum, cheliceral ganglia, pedal ganglia, neurosecretory centers, etc.) recognized only by special histochemical staining or electron microscopy (Obenchain and Oliver, 1975; Prullage et al., 1992).

The peripheral nervous system is comprised of the numerous nerves that innervate the legs, palps, chelicerae, and internal organs. Four large pedal trunk nerves extend from the lateral edges of the synganglion to the four walking legs. Several smaller nerves extend from the anterior end of the synganglion to innervate the salivary glands, chelicerae, pharynx, eyes, and other body structures in this region. Other nerves, especially the opisthosomal nerves, extend from the posterior end of the synganglion to innervate the internal organs in the posterior region of the body. Peripheral nerves also innervate several sensory organs, including the chemosensory sensilla (Ferreira et al., 2015; Sonenshine et al., 1984) and infrared-sensing sensilla in the Haller's organ on the leg I tarsi; gustatory sensilla on the terminal segments of the palps; gustatory sensilla on the foreleg tarsi (Phillips and Sonenshine, 1993); the chelicerae and cheliceral digits and the ocelli (eyes) on the dorsal body surface. For sensory perception of their external environment, ticks use a unique organ on their forelegs, the Haller's organ. Located on the foreleg tarsi, this organ comprises a cluster of olfactory sensilla in the anterior pit region of the organ (Carr et al., 2017) and a posterior capsule with pleomorphic sensilla-like elements. Transmembrane receptors associated with the Haller's organ sensilla detect animal odors that facilitate host recognition and attachment for feeding. The role of the posterior capsule is not clearly known, although a recent study in the American dog tick, *D. variabilis* demonstrated it functions as an infrared receptor (Mitchell et al., 2017). Whether it performs a similar function in *I. scapularis* is unknown.

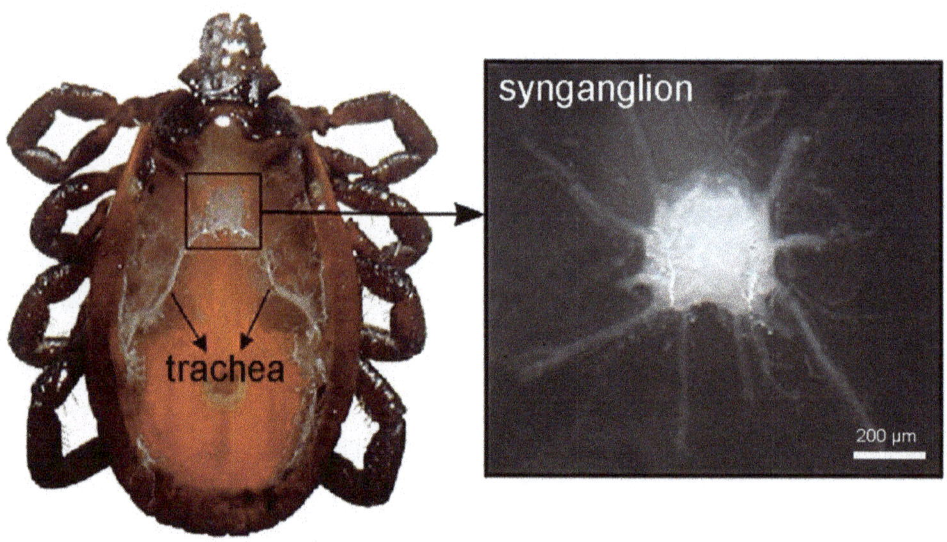

Figure 9. Tick central nervous system, showing the synganglion in the unfed female of the tick *Rhipicephalus appendiculatus*. Left - whole body image, with dorsal cuticle removed, showing the position of the synganglion, and tracheae leading to the spiracles on the lateral margins of the body. Right – enlargement of the ventral body region showing the synganglion and associated nerves. From Šimo et al., (2009) Identification of a complex peptidergic neuroendocrine network in the hard tick, *Rhipicephalus appendiculatus*, Cell Tissue Res. 335(3):639-55 with permission from Springer Nature.

Tracheal system
Respiration in adult and nymphal ticks, including *I. scapularis,* is accomplished via a system of tracheal tubes emanating from a pair of spiracles located on the lateral sides of the opisthosoma just behind the legs. Tracheal tubes are absent in tick larvae, which exchange oxygen and waste gases via the body cuticle. Each spiracle is surrounded by a large, multi-porous spiracular plate containing numerous aeropyles. The ostium of each spiracle abuts a ridge, the macula. Although one would assume that the spiracles can open or close, in ixodid ticks, the macula is immovable and does not seal the ostium. Instead, air exchange occurs via the aeropyles, not the spiracles. A large tracheal atrium is located internally below each spiracular plate and joins several large tracheal trunks. Increases in hemolymph hydrostatic pressure (due to contractions and relaxations of the dorso-ventral body wall muscles) can invert the wall of the atrium, forming a plug that blocks the opening between the ostium and the adjacent tracheal trunks, sealing the space between the ostium and the adjacent tracheal trunks, effectively closing the spiracle. Contraction of muscles attached to the outer surface of the atrium opens the chamber and draws air into the respiratory system. This mechanism acts like a bellows, driving air into the connecting tracheae or expelling waste gases to the exterior (Fielden and Duncan, 2014).

Numerous tracheae extend from the several large tracheal trunks and anastomose throughout the body, reaching into every organ and tissue, eventually differentiating into tiny tracheoles that carry air to individual cells. Each trachea is an elongated tube comprised of a simple epithelium surrounded by a thin layer of folded cuticle, forming the distinctive taenidia, structures that enable one to easily distinguish the tracheae (Fielden and Duncan, 2014).

Part II. Host finding, attachment, salivary disruption of host defenses, blood feeding and digestion.

Host finding behavior
Ticks, as obligate blood feeding ectoparasites, must find vertebrate hosts for their blood meal (hematophagy). Ticks use a unique organ on the tarsal segments of their forelegs, the Haller's organ, to detect common host-derived gases (e.g., NH_3, CO_2) and odors more characteristic of their preferred hosts (Carr et al., 2017). The Haller's organ (posterior capsule) also serves as a heat detector, capable of detecting heat emitting hosts at distances of at least 1 m (Carr and Salgado, 2019). *I. scapularis* is an "ambush" tick that waits in the vegetation for passing hosts, then sensing odors and minor heat emissions to recognize animals onto which it attaches and feeds. Following physical contact with their hosts, ticks also use sensilla on their palps to detect non-volatile compounds on the host skin to assess suitability for attachment and skin penetration (Foelix and Chu-Wang, 1972).

The tick sialome
Mosquitoes, fleas, bugs, and many other blood-sucking insects feed rapidly, often within seconds or at most a few minutes. These biting insects present the familiar "insect bite", and they secrete saliva to facilitate blood flow (Boulanger, 2018; Ribeiro et al., 2010). Ticks, however, have adopted a long-term feeding strategy, attaching to their hosts and imbibing blood over a lengthy time period. Indeed, hard ticks are unique among the vast array of blood-feeding arthropods by their ability to remain attached to host skin and continue blood feeding for many days (Wikel, 2013). Tick saliva is a veritable pharmacopeia of diverse bioactive molecules, mostly proteins and peptides, but also adenosine, prostaglandins, neurotransmitters, endocannabionoids and even microRNAs (Nuttall, 2019; Šimo et al., 2017). More than 500 proteins and peptides have been reported to occur in the salivary glands. The majority of these molecules are secreted proteins and peptides that are released into the saliva (Ribeiro et al., 2006, 2017).

Production of saliva is orchestrated by dopamine, neuropeptides, and their receptors, via both granular type II and III salivary gland acini (for a better understanding of their roles and control see *Salivary glands* and *Regulation of salivary glands* in Parts II and III, respectively) (Kim et al., 2019; Šimo et al., 2009a, 2011, 2014b, 2014a). Saliva at the stage of tick attachment is rich in cement compounds (including *I. scapularis*, although absent in some other species of this genus) that anchor the mouthparts to the host skin. Much has been learned about the composition of tick cement. Ticks begin secreting cement compounds within a few minutes after they attach to the host skin. The first layer, known as the cortex, forms a "cement cone" around the hypostome. Feeding ticks continue to secrete additional cement compounds to form an additional layer, known as the cortical cement layer, that spreads out and further strengthens the cement cone. In addition to binding the mouthparts to the host's skin, tick cement functions as a sealing agent, filling in any gaps between the epidermal and dermal layers and the inserted hypostome and chelicerae (Suppan et al., 2018). Several authors who have studied the composition of tick cement report glycine-rich proteins (GRPs) as important elements of its structure. In *A. americanum*, the cement is a complex mixture including GRPs, lipocalins, protease inhibitors, detoxification proteins, mucins, storage proteins and others; almost 50% of the saliva consists of proteins of unknown function (Hollmann et al., 2018). Earlier studies by Jaworski et al. (1992) showed evidence of a highly immunogenic 90 kDa protein secreted by specialized cells of the salivary gland, specifically the type d and "*e*" cells found in the type III acini in *D. variabilis* as well as the "*a*" cells of the type II acini in several other species (Coons and Alberti, 1999). Other workers (Antunes et al., 2018) reported a GRP as a major component of the cement composition and a candidate for a vaccine. A large (> 1700 amino acids) GRP also was found in the *I. scapularis* genome (GenBank: EEC15669) and is likely one of the major components of the salivary gland cement mixture. Another important component, 64TRP, was found to be highly immunogenic and provided cross-protection among diverse ixodid tick species (Trimnell et al., 2005). However, we could not find any evidence of this protein in *I. scapularis,* although it is present in the closely related European tick *I. ricinus* (Trimnel et al 2005).

Once the host's awareness of the biting tick has been compromised, the attached tick, soon to be cemented in the skin, can begin its blood meal. Hematophagy is a remarkably well-orchestrated process, controlled by downregulation of genes coding for molecules no longer needed and upregulation of new genes coding for hemolysis of incoming blood cells, the cascade of hemoglobin digesting enzymes, new cuticle growth to help maximize engorgement and water elimination to help concentrate the blood meal. Beginning with the ground-breaking discoveries by Ribeiro and colleagues (Ribeiro et al., 1985, 2006), investigators have identified approximately 500 peptides and proteins in the tick's salivary gland repertoire, most of which are secreted to facilitate feeding (Nuttall, 2019; Šimo et al., 2017). Among these are molecules that silence the host's itch and pain receptors, prevent blood coagulation, dilate blood vessels, disrupt immune rejection, and prevent tissue repair. A brief synopsis of these salivary molecules is described below.

Insertion of the tick's hypostome and chelicerae damages cells in both the epidermis and dermis. Damaged skin cells release adenosine diphosphate (ADP), activate the factor XII coagulation pathway, and elicit kallikreins and other inflammatory mediators causing release of bradykinin. Bradykinin, in turn, binds to nociceptor neuron that can respond within seconds or minutes to initiate the pain response and alert the host to respond (Pinho-Ribeiro et al., 2017). Other expected responses to the initial tick bite injury include itching, vasoconstriction, blood coagulation and inflammation (Šimo et al., 2017), all of which would alert the host to the presence of a foreign invader. However, tick saliva contains a variety of anti-clotting, anti-platelet, vasodilatory, anti-inflammatory and immuno-modulatory proteins that effectively disarm the host defense, at least in tick-naïve animals, allowing *I. scapularis* and other hard ticks to feed before an effective immune defense can be mounted to reject them.

To prevent pain, ticks secrete a salivary metalloprotease, specifically, an angiotensin converting enzyme (ACE) to block the pain-inducing hormone, bradykinin (Francischetti et al., 2009; Jelinski, 2016; Nuttall, 2019). To silence itching and irritation due to histamine released from basophils and mast cells, tick saliva also contains a histamine-binding protein (lipocalin) secreted soon after the initial attachment phase, which blocks host histamine at the wound site (Dai et al., 2010). Lipocalins have been reported to be among the most abundant salivary proteins, in some cases (e.g., *Ornithodoros* sp.) comprising as much as 20% of the total soluble salivary protein. However, the estimate for hard ticks, including *I. scapularis,* is much less, <1% of the total soluble protein mass. Other possible analgesics secreted in the saliva of some tick species include endocannabinoids and even adenosine (Nuttall, 2019). Collectively, these salivary agents effectively anesthetize the wound site, enabling the tick to continue feeding without alerting its host.

Tick bites damage blood vessels, triggering vasoconstriction and blood coagulation. To continue feeding, the tick's saliva contains compounds that counteract the host response to such injury (i.e., formation of a platelet plug, blood coagulation and vasoconstriction of vessels surrounding the bite site).

Anticoagulants in saliva (e.g., ixolaris and penthalaris) block clotting factor VIIa, disrupting clotting by the extrinsic pathway, while basic tail peptides inhibit clotting factor FXa of the intrinsic pathway. These same salivary proteins also activate plasminogen, a protein acquired from the host (Francischetti et al., 2009). Metalloproteases in saliva breakdown fibrinogen and fibrin, inhibiting clot formation by preventing platelet aggregation (Francischetti et al., 2005). Apyrases are enzymes that hydrolyze ATP, liberating ADP and AMP, which inhibit platelet aggregation and subsequent blood coagulation (I.e., acting as anti-coagulants) (Ribeiro et al., 1985). Serpins (serine protease inhibitors) found in the saliva of *Ixodes ricinus* (Chmelař et al., 2017; Šimo et al., 2017), as well as *I. scapularis*, also are important in inhibiting thrombin (i.e., preventing coagulation). They also are reported to function as vasodilators. Diverse serpins, differentiated by their Kunitz domains, occur in the different tick species and target different components of the coagulation cascade (Blisnick et al., 2017). In *I. scapularis* saliva, the basic tail protein Salp 14 is a specific thrombin inhibitor (Narasimhan et al., 2002), believed to work in conjunction with a different Kunitz domain protein, ixolaris, to prevent blood coagulation. In summary, since blood clotting is a complex process, beginning with release of thrombin from blood platelets, it is not surprising that *I. scapularis* ticks target several different elements of the clotting cascade to ensure continued blood flow.

Ensuring continued blood flow during tick feeding is dependent upon open (i.e., dilated) blood vessels. Prominent among the salivary compounds that function as vasodilators are the lipids prostacyclin and prostaglandin (PGE2) (Ribeiro et al., 1992). Another compound identified for this purpose is the apyrase, ATP-dihydrophosphohydrolase, which blocks vasoconstriction by blocking the agonist ADP (Nuttall, 2019). Other participants in this process are the cystatins, an important class of salivary proteins. Cystatins were shown to be secreted early in the attachment process. They are likely to function as among the earliest anti-hemostatic mediators (Ibelli et al., 2013). Cystatins also are known to affect cytokine production, especially IL-1β, a co-stimulator of the cytokine IL-9 (Nuttall, 2019), suggesting a role in blocking cytokine signaling that leads to early influx of T-lymphocytes into the bite site. Other unknown vasodilators present in tick saliva may be needed in the later phases of the feeding process. Blocking host complement is another important function of tick saliva. *I. scapularis* ticks secrete an anti-complement protein, Isac (*I. scapularis* anti-complement) (Valenzuela et al., 2000) as well as a suite of similar anti-complement proteins (Nuttall, 2019).

Tick salivary secretion is a dynamic process; the composition of salvia changes during the different phases of feeding from attachment to repletion (Nuttall, 2019)(Perner et al., 2018). During the first 24 hours, a large number of proteins are secreted into the host skin, including many of the same proteins described above, but also including a putative Kunitz BPTI protein, ferritin, a hemelipoglycoprotein, a glutathione-S-transferase, heat shock proteins, antimicrobial peptides (including defensin) and a variety of ribosomal proteins. Collectively, they contribute to preparing the feeding site, preventing pain and itch responses, enhancing blood flow, preventing blood coagulation, minimizing oxidative stress and combating reactive oxygen species (ROS) and also preventing bacterial contamination (Lewis et al., 2015). Later in the feeding process, numerous genes upregulated during the early stages (e.g., genes coding for lipocalins, cement compounds, etc.) are downregulated while others (e.g., the gene coding for ticks secrete histamine release factor (tHRF), Isac and others) are upregulated (Nuttall, 2019; Šimo et al., 2017).

When ready to begin the final rapid engorgement phase of blood-feeding, *I. scapularis* tHRF binds to basophils, inducing them to release histamine and induce vasodilation (Dai et al., 2010). This is the opposite effect of the early feeding process when blockade of histamine by lipocalins and other secreted proteins minimizes itching and/or pain that would have alerted the host to the tick's presence.

In addition to the complex array of secreted salivary proteins, tick salivary glands also secrete prostaglandins (PG), as noted previously, including PGE_2, $PGF_{2\alpha}$, and PGD_2. Tick salivary PGs have been reported to inhibit the cytokines interleukin-1 and tumor necrosis factor α (TNFα) by macrophages, interleukin-2 and interferon-γ. In addition, PGs, especially $PGF_{2\alpha}$ were reported to prevent platelet aggregation, increase vasodilation and also inhibit mast cell degranulation and release of mediators of inflammation (Sauer et al., 1995).

Despite the tick's well-defined armamentarium for countering host defenses, some reservoir species e.g., the white-footed mouse *Peromyscus leucopus*, are well equipped to respond to the tick's parasitic attack. When *I. scapularis* nymphs feed on white-footed mice, a permissive host, they provoke a strong inflammatory response. The skin of the bite site becomes replete with T-cells, macrophages, and neutrophils (but few if any eosinophils and basophils). This response is weakest in tick-naïve animals, typically newborns that have just begun foraging, but increases in severity in mice subjected to frequent tick attacks. Nevertheless, the ticks could continue feeding and engorge fully despite the inflammation, largely because the dermal architecture of the feeding lesion remains intact, appearing only as a broad, flattened (sessile) lesion. This phenomenon allows the ticks to remain anchored while they complete their blood meal. Other, unknown factors may also enable tick feeding despite the host's immune response. In contrast, when guinea pigs (*Cavia porcellus*), a non-permissive host native to South America, are subjected to repeated tick infestations, the inflammation is similar in severity to that seen in *P. leucopus,* although the immune cell population is especially rich in histamine-secreting basophils (Karasuyama et al., 2018). However, the dermis at the bite site degenerates as a large but narrow cavitary lesion mostly filled with leucocytes instead of erythrocytes, compromising the quality of the tick's blood meal. Observations of tick's attempts to blood feed from this unstable, cavity centered lesion and the associated hyperkeratotic epidermis showed that it was more difficult for the ticks to remain attached. Ticks on those animals fed poorly and were rejected before they could engorge (Anderson et al., 2017).

Digestion of the blood meal

Ticks are obligate blood feeding parasites, and hemoglobin is their primary nutritive food source. Ticks are unique among all other blood feeding arthropods in their sequestration and absorption of hemoglobin. When blood enters the midgut, midgut cells begin to proliferate, some differentiating into secretory cells, others into digestive cells (Coons et al., 1886). Subsequently, digestion of the blood meal begins with the lysis of blood cells, especially erythrocytes (hemolysis) by hemolysins (Ribeiro, 1988). In *Haemaphysalis longicornis*, trypsin-like serine proteases secreted by the midgut disrupts the membranes of ingested erythrocytes (Miyoshi et al., 2007), leading to lysis and liberating hemoglobin. Transcripts for serine peptidases and cysteine peptidases also were found in the midgut transcriptomes of *I. scapularis and D. variabilis*, implicating these enzymes in hemolysis of red cells. Instead of a single peptidase, multiple families of peptidases have been found in the midgut transcriptomes of these ticks, including serine peptidases, cysteine peptidases, aspartyl peptidases, and metallopeptidases, although it is likely that not all are involved in blood cell hemolysis (Anderson et al., 2008; Sonenshine and Anderson, 2014). Varying amounts of hemoglobin liberated from the hemolyzed erythrocytes bind to specific receptors in clathrin-coated pits on the luminal side of the midgut digestive cells. Subsequently, hemoglobin is internalized by receptor mediated endocytosis. The resulting endosomes fuse with lysosomes to form heterolysosomes (also known as "phagolysosomes") containing abundant hemoglobinases (i.e., enzymes that digest hemoglobin). Albumin also is absorbed, presumably by other, uncoated membrane pits, but almost all other proteins and potential nutrients in blood are excluded, remaining within the midgut lumen and eventually excreted (Sonenshine and Anderson, 2014).

Over the course of the next several hours, the globin moieties are digested while the highly toxic heme, also liberated during digestion of hemoglobin, is detoxified by heme-binding proteins to hematin-like aggregates, similar to hemozoin of *Plasmodium* sp. (Graça-Souza et al., 2006) or the hematin-like waste products in the blood flukes Schistosoma sp. (Brindley et al., 1997; Koehler et al., 2007), and shuttled to hemosomes that accumulate in the luminal side of the hypertrophied digestive cell for later disposal; this process is known as heterophagy. Small amounts of free heme bind to ferritin and are transcytosed into the hemolymph for eventual incorporation into storage proteins or vitellogenin (this is the reason why tick hemolymph and eggs are brown). Hypertrophied midgut epithelial cells containing hemosomes eventually rupture and dislodge their contents (exocytosis) into the midgut lumen along with other excretory wastes. This intracellular mechanism for detoxification of heme is believed to be unique among blood-feeding parasites (Lara et al., 2005). Hemoglobin digestion is a continuous process that progresses slowly over many days or even weeks, depending upon the life stage. Excess blood meal water accumulated during feeding (note that the tick's osmotic pressure is ~ 1.3 to 1.4 times higher than in typical vertebrate hosts) is collected by specialized cells in the salivary glands and secreted back into the host. The blood meal is digested slowly, often after many months (except in

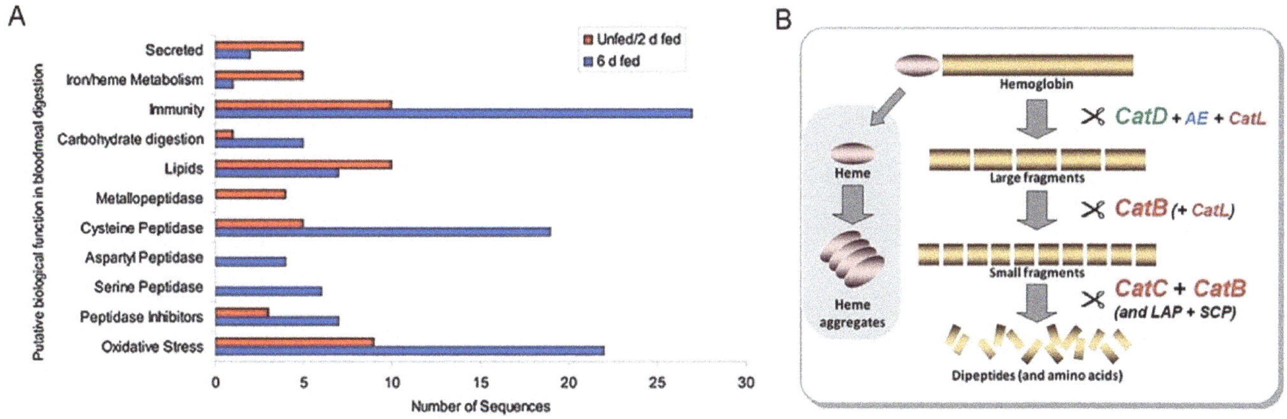

Figure 10. Diagrammatic illustrations of the protein families and/or specific proteins involved in blood meal digestion in feeding female ixodid ticks. (A) Changes in protein families involved in blood meal digestion comparing the early unfed – day 2 attachment period versus the day 6 post-host feeding period in *Dermacentor variabilis.* (B) The hypothetical proteolytic pathway of hemoglobin digestion in heterolysosomes in midgut epithelial cells in *Ixodes ricinus*. The initial breakdown of hemoglobin is done by Cathepsins D and L plus legumain, liberating large fragments, followed in turn by cathepsin B (and secondarily by cathepsin L), liberating even smaller fragments. Finally, cathepsins C and B, with secondary digestion by leucine amino peptidase (LAP) and serine carboxypeptidase (SCP), digest the small fragments into dipeptides and free amino acids. Panel A is from Anderson et al., (2008) Exploring the mialome of ticks: an annotated catalogue of midgut transcripts from the hard tick, *Dermacentor variabilis* (Acari: Ixodidae), BMC Genomics, 9:552. Panel B is from Horn et al. (2009) Hemoglobin digestion in blood-feeding ticks: mapping a multipeptidase pathway by functional proteomics, Chem. Biol. 16:1053–1063, with permission from Elsevier.

mated, engorged females that begin egg laying). This slow digestion allows the midgut to serve as a food reservoir. However, when the juvenile stage ticks feed again, the remnants of the old blood meal is passed out in the feces while fresh blood is ingested. This unique form of blood digestion, with few extracellular enzymes also provides a relatively permissive environment that shelters diverse microbes, including tick-borne pathogens.

Stages of blood meal and hemoglobin digestion

During the lengthy feeding period, the midgut undergoes extensive changes (Figure 10). Some of the genes coding for proteins associated with the early stages of blood meal digestion and hemoglobin sequestration are down-regulated later in the process, while others are more strongly upregulated or remain unchanged. Figure 10A illustrates midgut transcriptomic changes during the 6-day feeding period in a feeding *I. scapularis* female. During the brief attachment period, when the female tick is cementing itself and establishing the feeding site, genes coding for iron metabolism, immunity, lipids, metalloproteases, cysteine, aspartyl and serine proteases, peptidase inhibitors and oxidative stress are expressed, many rather strongly. However, by day 6, genes coding for immunity, carbohydrate digestion, cysteine peptidase, peptidase inhibitors and oxidative stress are even more strongly expressed while genes coding for iron metabolism and lipid metabolism were down regulated and genes coding for metallopeptidase, aspartyl peptidase and serine peptidase remained the same (Sonenshine and Anderson, 2014).

Once sequestered in the heterolysosomes of the midgut cells, hemoglobin molecules are disrupted so as to liberate the globin for breakdown into smaller peptides and amino acids. The process is carried out as a sequential cascade of different proteolytic events, each by a different proteolytic enzyme targeting different links in the amino acid chain of the globin moieties. These processes are done entirely within the small intracellular vesicles (hemosomes), not in the cellular cytoplasm. In the European sheep tick, *I. ricinus,* two types of proteolytic enzymes, cysteine proteases known as cathepsins D and L and legumain cleave the globins, releasing several large peptide fragments. Next, other cysteine proteases, cathepsin B and cathepsin L, digest the large fragments into smaller ones. Subsequently, cathepsins C and B, leucine amino peptidase (LAP) and serine carboxypeptidase (SCP) complete the process, cleaving the fragments into dipeptides and free amino acids (Figure 10 B). The end products liberating by these enzymatic steps escape into the cytoplasm and are transcytosed into the surrounding hemolymph (Horn et al., 2009; Sojka et al., 2011). It

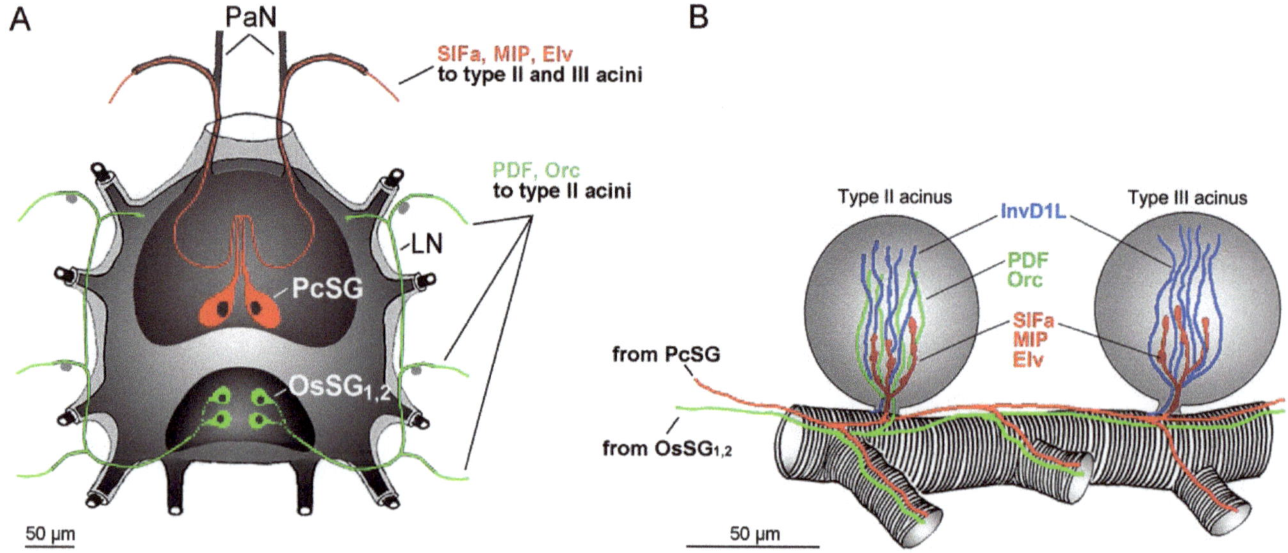

Figure 11. Neuronal control of the *Ixodes scapularis* salivary gland. (A) Tick synganglion with peptidergic neurons identified as sources of salivary gland innervation. Giant protocerebral neurons PcSG (red) expressing three different neuropeptides SIFa (SIFamide), MIP (myoinhibitory peptide) and Elv (elevenin) innervate basal parts of types II and III acini shown in (B). Another set of opisthosomal neurons $OsSG_{1,2}$ (green), reacting with antibodies against PDF (pigment dispersing factor) and Orc (orcokinin) innervate exclusively type II acini shown in (B). Abundant axonal projections expressing InvD1L (invertebrate-specific D1-like dopamine receptor) and innervating (blue in B) both type II and III acini have an unknown origin. Note that image B shows only one single type II and type III acinus attached to the salivary duct. Axons running along the main duct and different axon terminals entering the acini are shown in different colors. PaN – palpal nerve, LN – lateral nerve. Panel A graphic authors: D. E. Sonenshine and L. Šimo. Panel B is modified from Vancová et al., (2019) Ultrastructural mapping of salivary gland innervation in the tick *Ixodes ricinus*. Sci. Rep. 9:6860.

is likely that the same proteolytic cascade occurs in the midgut cells of *I. scapularis*, since homologues of the *I. ricinus* genes have been identified in this tick (Gulia-Nuss et al., 2016, Supplementary Table 21).

Part III. Molecular regulation of tick bodily functions and reproduction

Neuropeptides, neurotransmitters, and their receptors regulating physiologic processes

The *I. scapularis* genome contains at least 39 neuropeptide genes (encoding more than 100 mature neuropeptides), most of which are expressed during different phases of blood feeding and subsequent reproductive events (Gulia-Nuss et al., 2016). In invertebrates, neuropeptides and protein hormones (collectively, neuropeptides), along with their G-protein coupled receptors (GPCRs), are among the most important regulators of biological processes, such as development, behavior, host seeking, reproduction and feeding (Meeusen et al., 2003; Nässel, 2009). Although, within invertebrates, the number of neuropeptide genes is more or less the same, their specific functions may differ among taxa (Schoofs et al., 2017). Alerting the tick's physiological state for successful feeding and subsequent development involves neuropeptides produced primarily by their central nervous system, the synganglion (Christie, 2008; Šimo et al., 2009b). In ticks, functional inferences of neuropeptides has been based, thus far, almost exclusively on the identification of their release sites and/or localization of their receptors in specific target cells/organs (Šimo and Park, 2014; Šimo et al., 2009a; Vancová et al., 2006). In addition speculation about their putative functions has been based on the regulation of neuropeptide and receptor transcripts during feeding (Egekwu et al., 2016). In addition, the GPCRs recognizing a suite of neurotransmitters (e.g., dopamine, γ-aminobutyric acid (GABA), glutamate, etc.) also are expressed during these physiological events (Kim et al., 2014; Šimo et al., 2011, 2012, 2014b). Valuable knowledge of these receptors has been acquired from studies of adult females, although similar gene activities are assumed to apply to larvae, the stage responsible for acquisition of *Borrelia* spirochetes, and nymphs, the life cycle most often responsible for transmission of spirochetes to

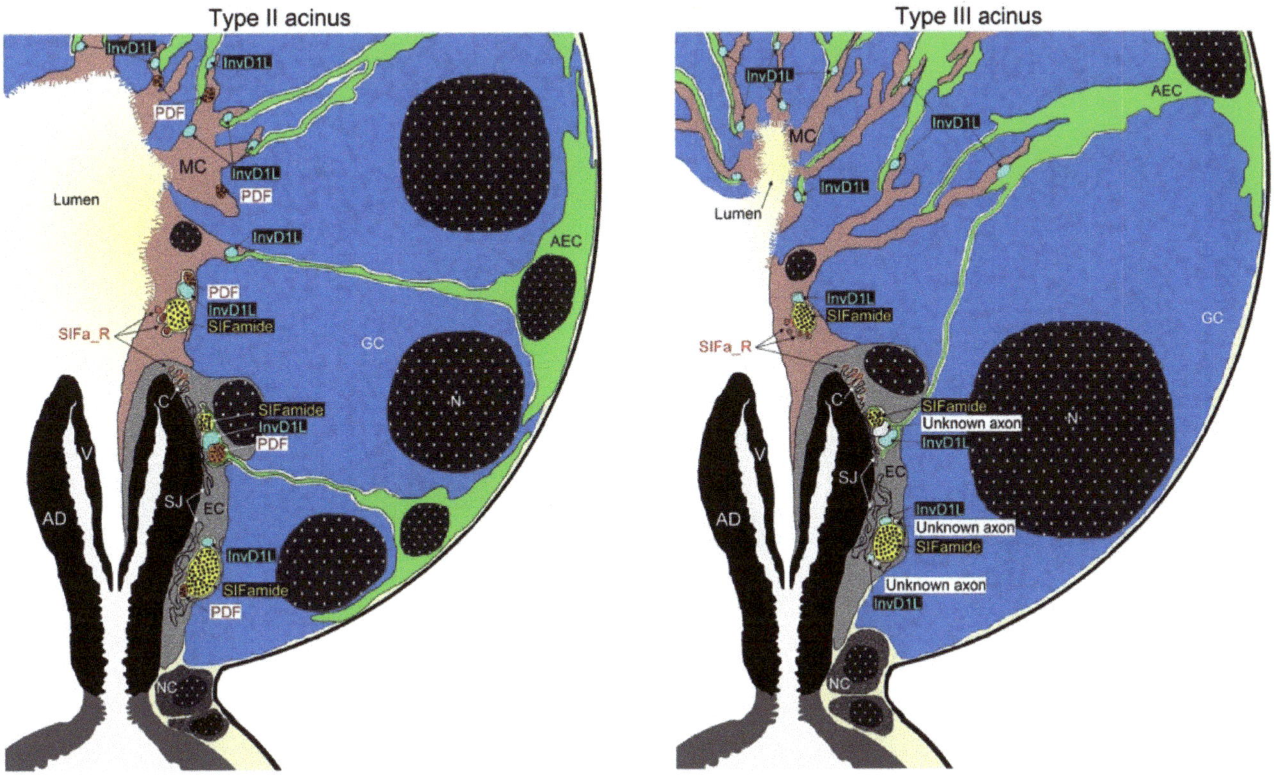

Figure 12. Schematic showing cross sections of type II and III acini from unfed *I. ricinus* female salivary glands highlighting the organization of the acinar cells and their putative associations with different axonal projections. Similar organization is believed to occur in the salivary glands of *Ixodes scapularis*. SIFamide-positive axons (yellow), pigment dispersing factor-positive axons (PDF, brown/orange), invertebrate-specific D1-like dopamine receptor-positive axons (InvD1L, aqua-blue). SIFamide receptor (SIFa_R, red): on plasma membrane and vesicle-like structure (upper arrow) of the myoepithelial cell (MC, pink), axons accompanying the SIFa-axon surrounded by the MC (middle arrow) and canaliculi (lower arrow) of basal epithelial cell (EC, grey). Note that SIFamide-axons also co-express myoinhibitory peptide (MIP) and elevenin (Elv) likely acting in the same acinar region. The acinar valve (V) is covered by EC extensions and overlaid by the MC. AEC, ablumenal epithelial cell: green; GC, granular cell: blue; NC, neck cells: dark grey; N, nucleus: dotted black; AD: acinar duct; C: canaliculi of ECs; SJ: septate junctions. For a complete description of the different granular cell types in hard tick salivary glands see Binnington et al, 1978 and Fawcet et al, 1986. From Vancová et al., (2019) Ultrastructural mapping of salivary gland innervation in the tick Ixodes ricinus. Sci. Rep. 9:6860.

humans and reservoir hosts (Egekwu et al., 2016; Gulia-Nuss et al., 2016). The most recent discoveries in the field of tick neurobiology are discussed below highlighting the unique functions of neuropeptides and neurotransmitters in the tick lineage.

Regulation of salivary glands

Among all of the internal tick organs, the control of salivary glands represents the best-known subject in terms of signaling molecules and their receptors. The activities of this organ are regulated primarily by direct upstream neural signaling pathways arising from the tick synganglion (Figure 11). The post-genomic era provided solid evidence that saliva production is orchestrated by dopamine, as well as neuropeptides and their receptors, via the primary saliva producers, the granular type II and III acini (Kim et al., 2014, 2019; Šimo et al., 2009a, 2011, 2014b). The neural connection of the tick salivary gland with specific cells in the synganglion has been reported in various tick species (Šimo et al., 2012, 2014a), with the most detailed information from the black-legged tick *I. scapularis*.

Specifically, in *Ixodes* females, three different axonal projections, arising from several distinct neurons, extend into the type II and III acini and arborize within them (Figure 11). These projections include a) neuropeptidergic axon terminals arising from giant protocerebral neurons (PcSG) co-expressing myoinhibitory peptide (MIP), SIFamide and elevenin (Elv) terminate at the basal regions in both type II

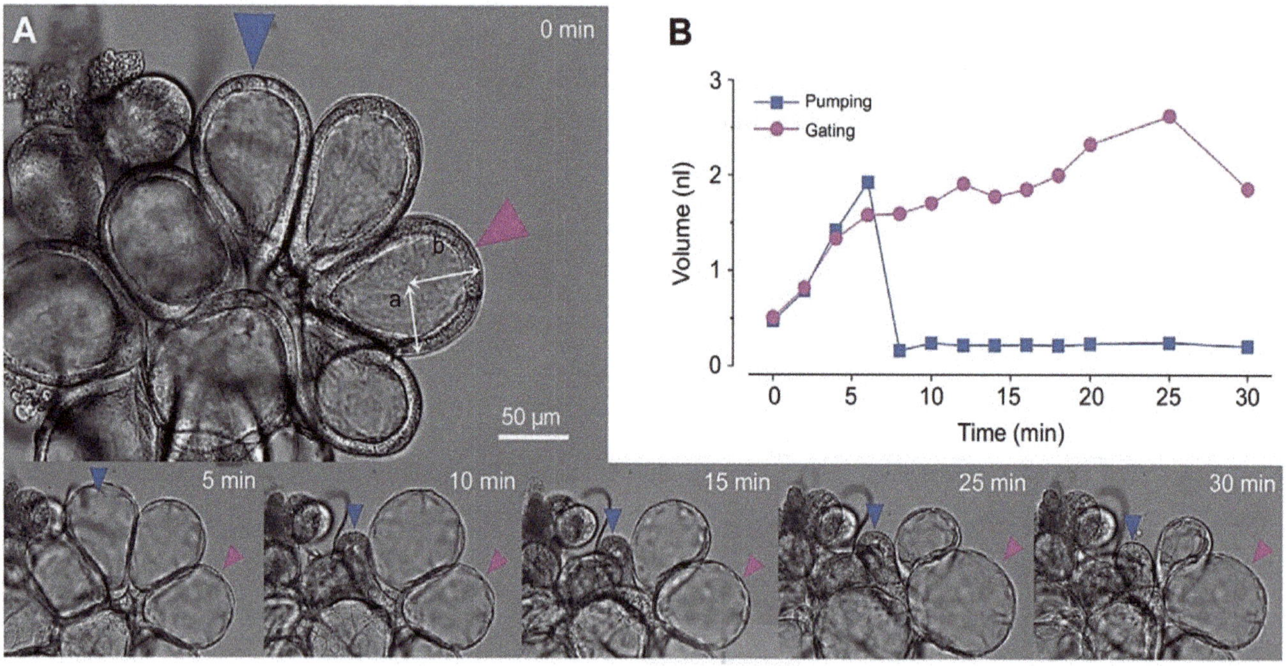

Figure 13. Changes in type III acini size following dopamine treatment, showing pumping and gating actions of individual acini in *I. scapularis* female. (A) Type III acini under the microscope, demonstrating changes in the acini size over time following dopamine treatment. (B) Changes in the acini volume after dopamine treatment. Video available in original article. From Kim et al., (2014) Orchestration of salivary secretion mediated by two different dopamine receptors in the blacklegged tick *Ixodes scapularis*, J. Exp. Biol. 217: 3656-3663.

and III acini (Kim et al., 2018; Šimo et al., 2009a); and b) neuropeptidergic axon terminals arising from two pairs of opisthosomal neurons ($OsSG_{1,2}$) positively immunoreacting to antibodies against both pigment dispersing factor (PDF) and orcokinin (Orc) reach the apical parts of exclusively type II acini (Roller et al., 2015; Šimo et al., 2009b); and c) axonal projections with unknown origin expressing invertebrate - specific D1-like dopamine receptor (InvD1L) reaching the apical regions of both type II and III acini (Šimo et al., 2014b).

The specific role of these molecules in type II and III acini has been inferred from immunogold labeling in *I. ricinus* (Vancová et al., 2019). Specifically, the basal axon terminals co-releasing MIP/SIFamide/Elv are thought to control the activities of the acinar valves via the regulation of myoepithelial and basal epithelial cells in *I. ricinus*, both of which overlay the valvular arms (Figure 12). The exclusive presence of PDF/Orc-axon terminals in type II acini suggests that ticks are capable of selectively controlling particular types of acini via different neuropeptidergic networks (Figures 11B and 12). The direct association of InvD1L - axons with the numerous zones of the myoepithelial cell (MC, Figure 6B, C and 12) supports a proposed model in which activation of InvD1L triggers Ca^{++}-mediated MC contraction and expulsion of acinar contents (Kim et al., 2014; Šimo et al., 2014b). In addition, another dopamine receptor, the epithelial D1, located on the luminal surface of both type II and III acini triggers the cAMP pathway for transport of fluid into the acini (Šimo et al., 2011). Thus, the autocrine/paracrine dopamine (Koči et al., 2014) in type II and III acini orchestrates the influx of fluids and pumping/gating actions (Figure 13) through the D1 and InvD1L dopamine receptors, respectively (Kim et al., 2014). Besides the direct action of dopamine, several other chemical agents, such as octopamine, norepinephrine, GABA, ergot alkaloids and pilocarpine, have been directly or indirectly implicated in salivary gland secretion; their precise mode of action, however, remains enigmatic (Lindsay and Kaufman, 1986; Needham and Pannabecker, 1983; Pannabecker and Needham, 1985).

Osmoregulation during the blood meal

Ixodid ticks, including *I. scapularis,* are unique among blood-feeding arthropods with regard to the

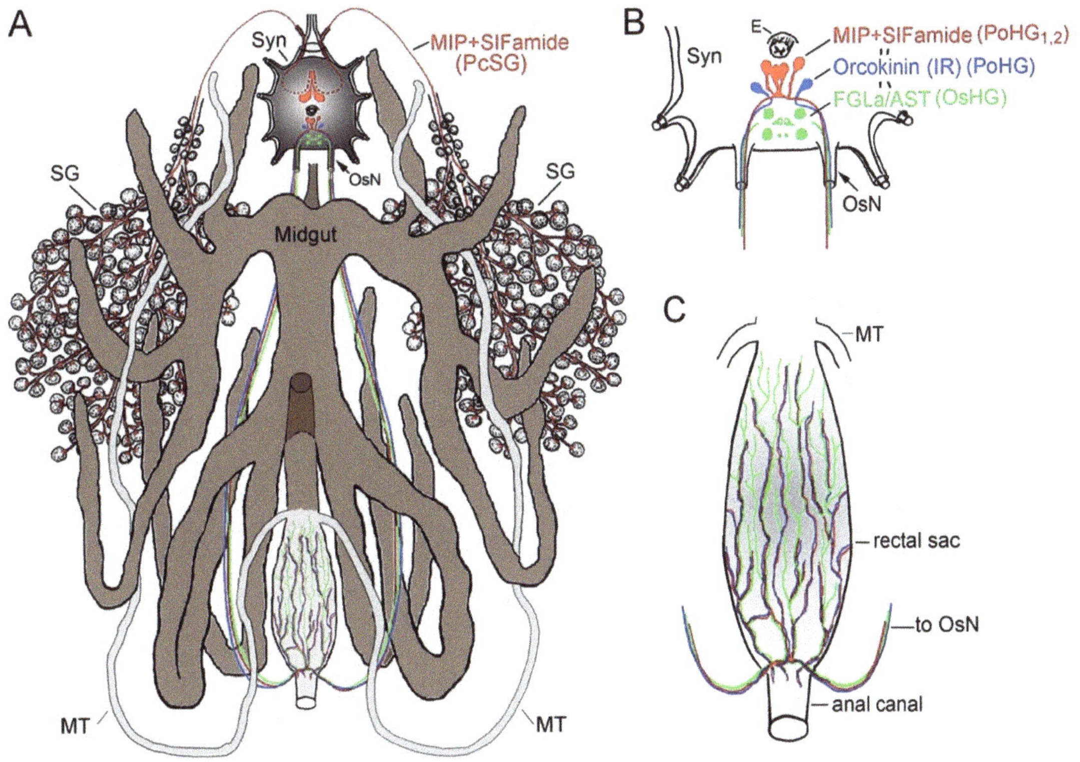

Figure 14. Simplified diagrams showing peptidergic innervation of *Ixodes scapularis* osmoregulatory organs--the salivary glands (SG) and hindgut. (A) Complex view of synganglion (Syn) neurons innervating the hindgut SG. (B) Magnified view of dorsal-post-esophageal region of the synganglion with a focus on putative peptidergic neurons innervating the hindgut. Note that immunoreactive axonal projections from all detected neurons exit the synganglion via the opisthosomal nerves (OsN) and reach the surface of the rectal sac and the anal canal as shown in (C). (C) Immunoreactive axonal projections reaching the surface of the rectal sac and anal canal. Protocerebral salivary gland (PcSG) neurons producing myoinhibitory peptide and SIFamide in innervation of the SG (see Figures 11, 12). E, oesophagus; MT, Malpighian tubule; OsHG, opisthosomal hindgut neurons; PoHG, postoesophageal hindgut neurons. From Šimo et al., (2014) Neuropeptidergic control of the hindgut in the black-legged tick *Ixodes scapularis*, 819-826, Int. J. Parasitol., with the permission from Elsevier.

enormous volumes of blood they consume, as much as 100 times their pre-fed body weight (Tatchell, 1967). Following the attachment phase, a strong pulsing current, achieved by contraction/relaxation of muscles associated with pharynx, drives the ingested blood into the pharynx, then into the esophagus and subsequently fills the central midgut (stomach) and its numerous diverticula (Vancová et al., 2020). To accomplish this remarkable feat, the tick remains attached for many days, while a suite of signaling molecules (neuropeptides and neurotransmitters) along with their receptors are hypothesized to regulate the rate of blood volume intake, periodically accelerating (e.g., FMRFamide neuropeptide and/or metabotropic glutamate receptor, (Audsley and Weaver, 2009; Dillon et al., 2015) or slowing (e.g., sulfakinin neuropeptide and/or GABA) the process (Moran, 2000). Upregulation of certain synganglion neuropeptides, their receptors and neurotransmitter receptors, that are expressed in varying degrees during feeding, is speculated to regulate this process since their activities are consistent with their reported roles in regulating the rates of blood imbibition in other arthropod vectors (Galun, 1987). A major reason for this exquisitely well-regulated feeding process is the tick's need to synthesize new integument as it feeds, effectively enlarging its body to accommodate the enormous fluid volume it will eventually consume. Here too, this unusual process of integument and cuticle growth also is suggested to be regulated by increased expression of genes, e.g., corazonin, eclosion hormone and bursicon, etc. (discussed below and illustrated in Figure 16). The large volume of imbibed blood induces over-hydration and causes an enormous osmoregulatory stress on the ticks due to the significantly lower

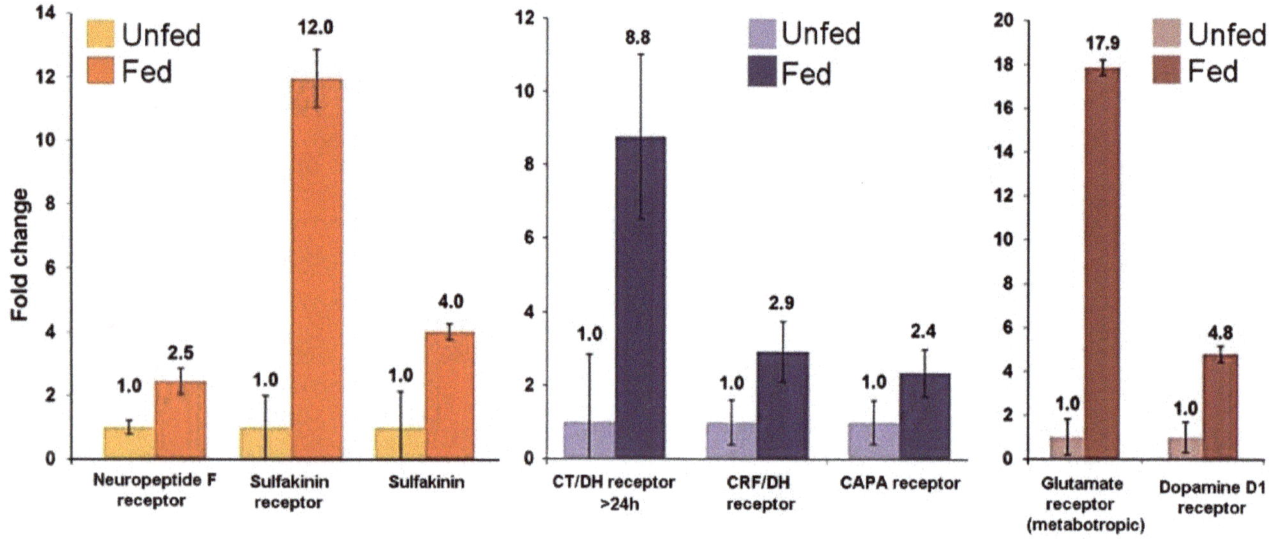

Figure 15. Quantitative PCR (qPCR) assays comparing expression of synganglion neuropeptides, neuropeptides, and neurotransmitters receptors between unfed and partially fed *I. scapularis* females. CT/DH receptor: calcitonin-like diuretic hormone receptor, CRF/DH receptor: corticotropin-releasing-factor related diuretic hormone. From Egekwu et al., (2016) Comparison of synganglion neuropeptides, neuropeptide receptors and neurotransmitter receptors and their gene expression in response to feeding in *Ixodes scapularis* (*Ixodidae*) vs. *Ornithodoros turicata* (*Argasidae*), 25:72-95, Insect Mol. Biol., with permission from Wiley Online Library.

osmotic pressure of vertebrate blood, ~ 300 milliosmoles, versus the much higher concentration of tick hemolymph, ~ 340 milliosmoles (summarized by Egekwu et al. 2016).

Ticks are known to utilize a unique osmoregulatory mechanism for substantial elimination of excess water, ions and metabolic waste essential to maintaining homeostasis during blood feeding (Kaufman and Phillips, 1973; Kim et al., 2019). In contrast to insects, in which the Malpighian tubules carry out these processes (Park, 2012), in ixodid ticks, the salivary glands are the primary osmoregulatory organs (Bowman and Sauer, 2004; Tatchell, 1967). An earlier study in *Dermacentor andersoni* found that 70% of water and 96% of sodium ions from the digested blood meal are transferred via the gut wall into the hemolymph where they are taken up by salivary glands and secreted back to the host through the saliva (Kaufman and Phillips, 1973). The epithelium of the type III acini (and likely type II as well) has been suggested as the specific transport site for water and electrolytes from hemolymph into the salivary glands (Fawcett et al., 1981; Meredith and Kaufman, 1973). Furthermore, it has been proposed that dopaminergic signaling mediates the formation of sodium-rich primary saliva by the potassium pump (Na/K-ATPase) located in the epithelial cells in the types II and III acini (Kim et al., 2016a). These same authors also suggested that, besides the ions transferred via an electrochemical gradient, Na/K-ATPase may activate aquaporins channels to regulate sequestration and elimination of water through the gut and salivary glands and its secretion back into the host (Campbell et al., 2010). By eliminating such a large proportion of the ingested blood meal water during the final phases of blood feeding, ticks are able to concentrate their blood meal by as much 3 times the amount they originally ingested (Koch and Sauer, 1984). In addition, the influx/efflux (Figure 13) of solute by type II and III acini is regulated by autocrine/paracrine dopamine signaling via D1 and InvD1L dopamine receptors respectively (Kim et al., 2014; Šimo et al., 2011, 2014b). Axons co-expressing the neuropeptides SIFamide, MIP and elevenin are thought to be involved in controlling myoepithelial cells in these structures to regulate salivary secretion (for more details see section *Regulation of salivary glands* in Part III of this chapter; also see Figures 11, 13; (Kim et al., 2019; Šimo et al., 2012; Vancová et al., 2019).

Another important excretory/osmoregulatory organ playing a significant physiological role in maintaining homoeostasis and waste elimination is the tick's

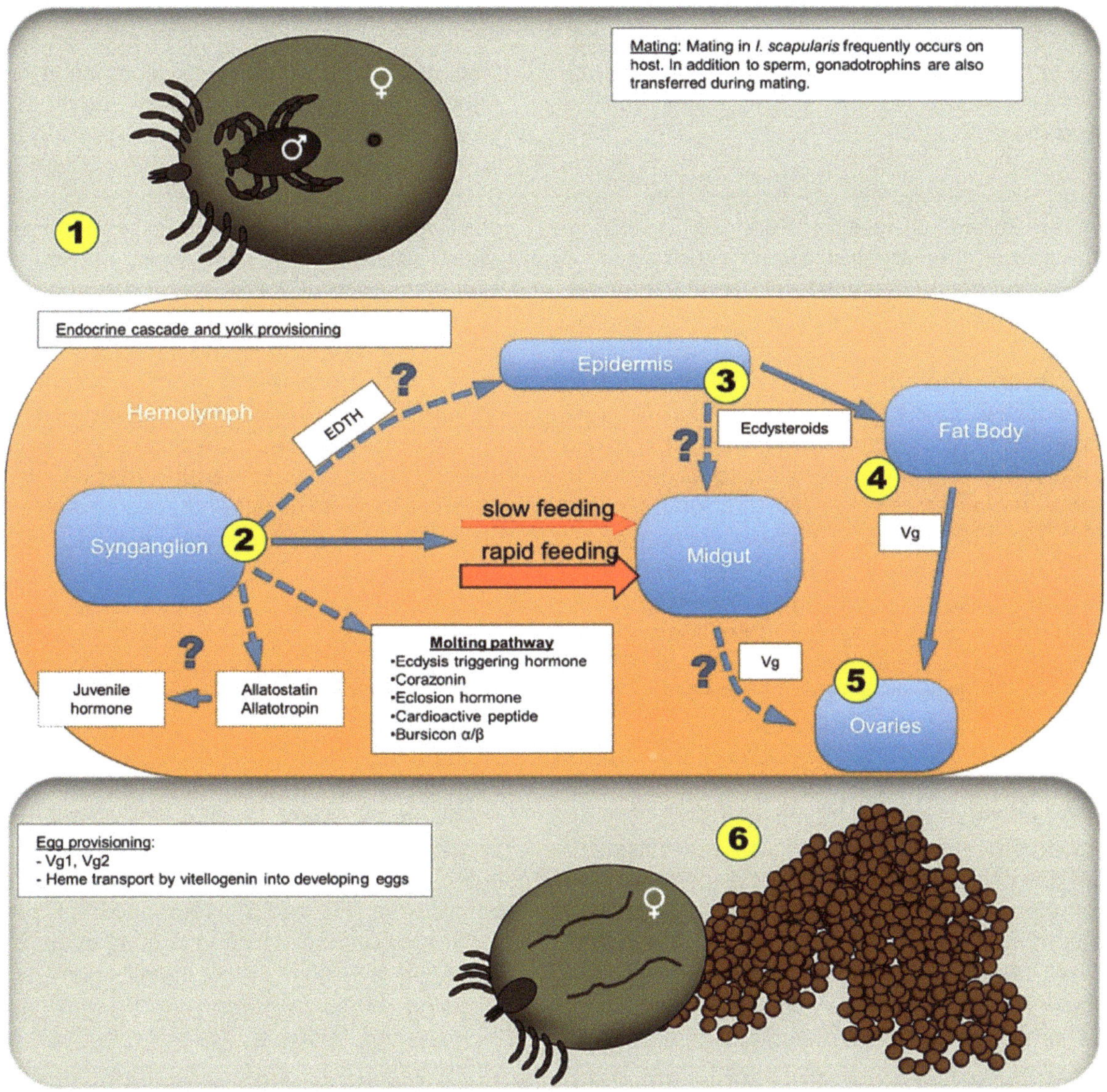

Figure 16. A putative model of neuroendocrine processes controlling mating and egg production in *Ixodes scapularis*. (1) Mating takes place off or on the host, required for rapid blood feeding during which the male transfers sperm and gonadotropins (unidentified at present), among other seminal components, including the spermatophore, (2) Gonadotropins stimulate the synganglion to release EDTH, which in turn stimulates rapid engorgement. The unknown gonadotropins also initiate synthesis and release of allatostatins and allatotropins (which may stimulate or inhibit the mevalonate-farnesal pathway), (3) EDTH initiates production of ecdysteroids by the epidermis, (4) rapidly rising ecdysteroid titers activate transcription factors for vitellogenin synthesis (Vg) in the fat body and specialized cells of the midgut activate transcription factors, as well as vitellogenin receptor (VgR) in the ovaries (5) Vg is taken up via VgR-receptor mediated endocytosis by developing oocytes and incorporated into the yolk as vitellin, and (6) The female produces a single batch of up to 3,000 mature eggs. Dashed lines indicate proposed pathways and factors. 20-E, 20-hydroxyecdysone; CAP, cardioactive peptide; EDTH, hypothesized epidermal trophic hormone; Vg, vitellogenin; VgR, vitellogenin receptor. Modified rom Gulia-Nuss et al., (2016) Genomic insights into the *Ixodes scapularis* tick vector of Lyme disease, Nat. Commun., 7:10507.

hindgut, comprised of Malpighian tubules, the rectal sac and the short anal canal (Sonenshine, 2014). During feeding, up to 74% of potassium and about 25% of water is excreted via tick's hindgut (Kaufman and Phillips, 1973). The feces, or wastes, from the midgut, consists of hematin, other nutrients and undigested hemoglobin. These fecal wastes are passed to the rectal sac. In addition, nitrogenous wastes in the form of guanine crystals accumulate in the Malpighian tubules and also are passed into the rectal sac (Sonenshine, 2014). During or immediately after feeding (or after molting of juvenile stages), masses of dark fecal material and white powdery-like wastes are passed to the external environment (Sonenshine et al., 2003).

Although the regulation of diuresis has been extensively studied in insects (Park, 2012), little is known about the mechanisms in ticks. In *I. scapularis*, a rich neuropeptidergic axonal network arising from specific neurons in the tick synganglion arborize on the surface of the tick rectal sac and anal canal (Šimo and Park, 2014). Specifically, immunopositive axons reacting with four different neuropeptides were described: FGLamide related allatostatin, MIP and SIFamide (both co-expressed in the same axonal projections) and orcokinin (Figure 14). Receptors for MIP and SIFamide were observed on the visceral muscles of the *Ixodes* rectal sac. These data suggest that these neuropeptides, through their cognate receptors, may regulate ion and water transport via the rectal sac (Šimo and Park, 2014, Campbell et al., 2014). An *in vitro* experimental setting for testing activities of these neuropeptides showed strong myostimulatory effects of SIFamide on hindgut motility, whereas MIP antagonized these actions (Šimo and Park, 2014). The bioactivity of SIFamide/MIP on the dissected hindgut demonstrated the first direct evidence for the role of these tick neuropeptides. Along with their proposed functions in the control of salivary glands (Figures 11 and 12), these two peptides are hypothesized to function as important regulatory factors of both primary osmoregulatory organs in ticks (Kim et al., 2019). A pictorial view of the target organs and the neuropeptides that regulate their function is shown below (Figure 14).

Although, the functions of most of the tick neuropeptides await experimental evidence, sequencing of the *I. scapularis* genome uncovered interesting features in tick neuropeptidergic signaling system and its potential involvement in processes relevant to their feeding biology (Gulia-Nuss et al., 2016; Šimo et al., 2014a). Most of the tick neuropeptides have clear orthologues to those of other invertebrates, including corazonin, eclosion hormone, cardioactive peptide and bursicon α and β (Figure 16), all known to regulate ecdysis, cuticle synthesis, hardening and tanning in insects, respectively (Nässel, 2009). The identification of specific functions of these neuropeptides in adult ticks, which do not molt, may help to uncover novel pathways for regulation of cuticle synthesis during the ingestion of large quantities of blood meal during feeding as well as other developmental processes. Another interesting feature reported in the *Ixodes* genome was the remarkable increase (up to 10-fold) in the number of certain neuropeptide receptors (Gulia-Nuss et al., 2016), among which were insect orthologs known to regulate satiety (sulfakinin) and water balance (kinin, tachykinin, and calcitonin-like diuretic hormone CT/DH, and corticotropin-releasing-factor related diuretic hormone CRF-DH) (Nässel, 2009). Whether all of these receptors are functional genes, rather that pseudogenes, awaits experimental confirmation; however, this finding supports the hypothesis that crucial functions regulating massive body increase size and osmoregulation are accomplished via greatly increased expression of this repertoire of specific neuropeptide receptors (Gulia-Nuss et al., 2016).

An interesting study by Egekwu *et al.* (2016) compared the levels of transcripts encoding neuropeptide/neuropeptide receptors and neuro-transmitter receptors in *I. scapularis* synganglion during blood feeding. Especially noteworthy was that transcripts for neuropeptide F and sulfakinin/sulfakinin receptor in the *I. scapularis* female synganglion were significantly upregulated in partially fed *I. scapularis* females when compared to unfed female ticks (Figure 15). Neuropeptide F plays a role in *Drosophila* food signaling (Chung et al., 2017), and sulfakinin is known as the satiety gene, slowing or stopping blood ingestion in various arthropods (Nässel and Williams, 2014), a function similar to cholecystokinin, which reduces food intake in vertebrates (Moran, 2000). Whether these neuropeptides play a similar role in ticks remains to be confirmed by further investigations. In the same study, another set of neuropeptides receptors for CT/DH, CRF/DH and CAPA was significantly upregulated in the synganglion of partially fed *I. scapularis* females. These are well-known for their contributions to water balance and diuresis in insects (Nässel, 2009; Park, 2012), and current data suggests they may control similar physiological processes in ticks. Very strong upregulation of the metabotropic glutamate receptor during feeding was hypothesized to correlate with the activities of the *I. scapularis* pharyngeal pump (Egekwu et al., 2016), but experimental proof is currently lacking (Figure 15). In addition, upregulation of the D1 dopamine receptor supports the involvement of this neurotransmitter in salivary gland regulation as described in previous studies (Šimo et al., 2011, 2014b).

Extensive additional investigations are required to fully understand the role of these signaling molecules and their receptors in tick biology. Current achievements in genomics, peptidomics, transcriptomics, immunolocalization or direct bioassays represent an excellent starting point to elucidate the different signal transductions at specific body sites. The tick neuropeptidergic system is an ideal target for the development of novel control measures against tick-borne diseases.

Reproduction

In *Ixodes*, mating occurs both on and off the host, but the other hard tick genera mate exclusively on the host. In these other genera, the males and females begin feeding separately from one another and both undergo development of their reproductive organs as they suck blood. Mating normally occurs within several days after slow feeding on their vertebrate hosts. However, mating and insemination are required for the females to commence rapid engorgement, vitellogenesis and oviposition. The transition of the mated adult female to the reproductive phase that occurs soon after blood feeding is a complex process (Figure 16) controlled by the ecdysteroid hormone, 20-hydroxyecdysone (20-E), in contrast to insects, where it is regulated by the juvenile hormone, JHIII (Chang, 1993). Exhaustive studies by (Neese et al., 2000) failed to reveal evidence for the occurrence of any of the known juvenile hormones in any life stages of a tick (*D. variabilis)* or soft tick (*Ornithodoros parkeri).* However, although all of the enzymes for the mevalonate pathway (homologous to the JH pathway) were found in different tick species, the final enzymatic steps that convert farnesoic acid to JH were absent. The authors suggest that this highly conserved pathway may function in regulating reproduction (Zhu et al., 2016). Activation and regulation of the pathway may be controlled by neuropeptides allatotropin and allatostatin (see description below).

In mated *I. scapularis* and most other hard tick females, unknown factors in the male's seminal fluid (gonadotropins noted above) trigger a rapid increase in blood feeding (Donohue et al., 2009; Weiss and Kaufman, 2004). This signaling event stimulates unknown elements in the synganglion that initiate production and secretion of ecdysone in the epidermis (Brown et al., 2009; Zhu et al., 1991), subsequently converted to 20-E in the fat body and tissues. Synthesis of 20-E is regulated by the ecdysiotropic hormone (Roe et al., 2008). Increasing concentrations of 20-E stimulate certain organs, primarily the fat body and midgut, to commence biosynthesis and secretion of vitellogenin (Vg), the precursor of the yolk protein, as well as upregulation of the Vg receptor (VgR) in the ovary. Discovery of the gene for the Vg receptor in the *I. scapularis* genome suggests that a similar 20-E regulated process also occurs in the tick species. However, it is not known precisely when vitellogenesis occurs in *I. scapularis* females (i.e, when insemination occurs) since the males remain in copula with their female partners through the blood feeding period, departing only after female repletion and drop off has occurred. All of these events have been thoroughly documented by numerous studies over many years (reviews by (Cabrera et al., 2009; Ogihara and Taylor, 2014).

It is not known whether *I. scapularis* and other ixodid ticks regulate reproductive activity by means of the same neuropeptides, allatotropin and allatostatin, that regulate it in insects (Klowden, 2013). Both of these neuropeptides, along with their receptors, were identified in the *I. scapularis* genome as well as in the synganglion transcriptome (Egekwu et al., 2016; Gulia-Nuss et al., 2016). In addition, gonadotropin-releasing hormone (GnRH), which is known to stimulate ovary (Zandawala et al., 2018) and prothoracicotropic hormone responsible for secretion of ecdysone (molting hormone) in insects (Schwedes and Carney, 2012), also has been reported in the *I. scapularis* genome (Gulia-Nuss et al., 2016), suggesting a similar pathway. Although the direct action of these neuropeptides in ticks awaits experimental confirmation, Egekwu et al., (2016) reported significant up-regulation of transcripts encoding allatostatin and its receptor in the synganglion of partially fed *I. scapularis* female. They speculated that inhibition of oocyte development by allatostatin signaling prevents oviposition before the full supply of nutrients needed is available. No enlarged oocytes were noted by one of the authors (Sonenshine, unpublished) during the dissections of partially fed *I. scapularis* females, supporting this hypothesis.

Much remains to be discovered regarding these complex biosynthetic pathways and their regulation by synganglion signaling (neuropeptides, neurotransmitters, and their receptors) that enable blood feeding ticks to produce thousands of eggs to ensure survival of their populations.

Acknowledgements
The authors are grateful for the comments concerning physical stresses on tick body tissues during blood feeding from Dr. Peter Flynn and Dr. Reuben Kaufman, both from the University of Alberta, Canada.

References

Alarcon-Chaidez, F. (2014). Salivary glands: Structure, Physiology, and Molecular Biology. In Biology of Ticks, D.E. Sonenshine, and R.M. Roe, eds. (New York: Oxford University Press), pp. 163–205.

Anderson, J.M., Sonenshine, D.E., and Valenzuela, J.G. (2008). Exploring the mialome of ticks: an annotated catalogue of midgut transcripts from the hard tick, Dermacentor variabilis (Acari: Ixodidae). BMC Genomics *9*, 552. https://doi.org/10.1186/1471-2164-9-552

Anderson, J.M., Moore, I.N., Nagata, B.M., Ribeiro, J.M.C., Valenzuela, J.G., and Sonenshine, D.E. (2017). Ticks, Ixodes scapularis, Feed Repeatedly on White-Footed Mice despite Strong Inflammatory Response: An Expanding Paradigm for Understanding Tick–Host Interactions. Front Immunol *8*. https://doi.org/10.3389/fimmu.2017.01784

Antunes, S., Couto, J., Ferrolho, J., Rodrigues, F., Nobre, J., Santos, A.S., Santos-Silva, M.M., de la Fuente, J., and Domingos, A. (2018). Rhipicephalus bursa Sialotranscriptomic Response to Blood Feeding and Babesia ovis Infection: Identification of Candidate Protective Antigens. Front Cell Infect Microbiol *8*, 116. https://doi.org/10.3389/fcimb.2018.00116

Arthur, D.R. (1962). Ticks and Disease. Oxford, U.K. Pergamon Press, 445 p.

Audsley, N., and Weaver, R.J. (2009). Neuropeptides associated with the regulation of feeding in insects. General and Comparative Endocrinology *162*, 93–104. https://doi.org/10.1016/j.ygcen.2008.08.003

Barker, D.M., Ownby, C.L., Krolak, J.M., Claypool, P.L., and Sauer, J.R. (1984). The Effects of Attachment, Feeding, and Mating on the Morphology of the Type I Alveolus of Salivary Glands of the Lone Star Tick, Amblyomma americanum (L.). The Journal of Parasitology *70*, 99–113.

Beati, L., and Klompen, H. (2019). Phylogeography of Ticks (Acari: Ixodida). Annu. Rev. Entomol. *64*, 379–397. https://doi.org/10.1146/annurev-ento-020117-043027

Beati, L., Keirans, J.E., Durden, L.A., and Opiang, M.D. (2008). Bothriocroton oudemansi (Neumann, 1910) n. comb. (Acari: Ixodida: Ixodidae), an ectoparasite of the western long-beaked echidna in Papua New Guinea: redescription of the male and first description of the female and nymph. Syst. Parasitol. *69*, 185–200. https://doi.org/10.1007/s11230-007-9115-5

Binnington, K.C. (1978). Sequential changes in salivary gland structure during attachment and feeding of the cattle tick, Boophilus microplus. Int. J. Parasitol. *8*, 97–115. https://doi.org/10.1016/0020-7519(78)90004-8

Blisnick, A.A., Foulon, T., and Bonnet, S.I. (2017). Serine Protease Inhibitors in Ticks: An Overview of Their Role in Tick Biology and Tick-Borne Pathogen Transmission. Front Cell Infect Microbiol 7. https://doi.org/10.3389/fcimb.2017.00199

Bonnet, S., Kazimírová, M., Richardson, J., and Šimo, L. (2018). Chapter 5 - Tick Saliva and Its Role in Pathogen Transmission. In Skin and Arthropod Vectors, N. Boulanger, ed. (Academic Press), pp. 121–191. https://doi.org/10.1016/B978-0-12-811436-0.00005-8

Boulanger, N. (2018). Skin and Arthropod Vectors - 1st Edition (London, U.K. Academic Press).

Bowman, A.S., and Sauer, J.R. (2004). Tick salivary glands: function, physiology and future. Parasitology *129 Suppl*, S67-81.

Bowman, A.S., Ball, A., and Sauer, J.R. (2008). Tick salivary glands: the physiology of tick water balance and their role in pathogen trafficking and transmission. In Ticks: Biology, Disease and Control, A.S. Bowman, and P. Nuttal, eds. (Cambridge, UK: Cambridge University Press), pp. 73–91. https://doi.org/10.1017/CBO9780511551802.004

Brindley, P.J., Kalinna, B.H., Dalton, J.P., Day, S.R., Wong, J.Y.M., Smythe, M.L., and McManus, D.P. (1997). Proteolytic degradation of host hemoglobin by schistosomes. Molecular and Biochemical Parasitology *89*, 1–9. https://doi.org/10.1016/S0166-6851(97)00098-4

Brown, M.R., Sieglaff, D.H., and Rees, H.H. (2009). Gonadal ecdysteroidogenesis in arthropoda: occurrence and regulation. Annu. Rev. Entomol. *54*, 105–125. https://doi.org/10.1146/annurev.ento.53.103106.093334

Cabrera, A.R., Donohue, K.V., and Roe, R.M. (2009). Regulation of female reproduction in mites: A unifying model for the Acari. Journal of Insect Physiology *55*, 1079–1090. https://doi.org/10.1016/j.jinsphys.2009.08.007

Campbell, E.M., Burdin, M., Hoppler, S., and Bowman, A.S. (2010). Role of an aquaporin in the sheep tick Ixodes ricinus: assessment as a potential control target. Int. J. Parasitol. *40*, 15–23. https://doi.org/10.1016/j.ijpara.2009.06.010

Carr, A.L., and Salgado, V.L. (2019). Ticks home in on body heat: A new understanding of Haller's organ and repellent action. PLoS ONE *14*, e0221659. https://doi.org/10.1371/journal.pone.0221659

Carr, A.L., Mitchell, R.D., Dhammi, A., Bissinger, B.W., Sonenshine, D.E., and Roe, R.M. (2017). Tick Haller's Organ, a New Paradigm for Arthropod Olfaction: How Ticks Differ from Insects. Int J Mol Sci *18*. https://doi.org/10.3390/ijms18071563

Chang, E.S. (1993). Comparative endocrinology of molting and reproduction: insects and crustaceans. Annu. Rev. Entomol. *38*, 161–180. https://doi.org/10.1146/annurev.en.38.010193.001113

Chmelař, J., Kotál, J., Langhansová, H., and Kotsyfakis, M. (2017). Protease Inhibitors in Tick Saliva: The Role of Serpins and Cystatins in Tick-host-Pathogen Interaction. Front Cell Infect Microbiol *7*. https://doi.org/10.3389/fcimb.2017.00216

Christie, A.E. (2008). Neuropeptide discovery in Ixodoidea: an in silico investigation using publicly accessible expressed sequence tags. Gen. Comp. Endocrinol. *157*, 174–185. https://doi.org/10.1016/j.ygcen.2008.03.027

Chung, B.Y., Ro, J., Hutter, S.A., Miller, K.M., Guduguntla, L.S., Kondo, S., and Pletcher, S.D. (2017). Drosophila Neuropeptide F Signaling Independently Regulates Feeding and Sleep-Wake Behavior. Cell Rep *19*, 2441–2450. https://doi.org/10.1016/j.celrep.2017.05.085

Coons, L.B., and Alberti, G. (1999). Acari: Ticks. In Microscopic Anatomy of Invertebrates, F.W. Harrison, ed. (New York: Wiley-Liss), pp. 267–514.

Coons, L.B., Rossell-Davis, R., and Tarnowski, B.I. (1886). Blood meal digestion in ticks. In Morphology, Physiology, and Behavioral Biology of Ticks, J.R. Sauer, and J.A. Hair, eds. (Chichester, UK: Ellis Horwood Limited), pp. 248–279.

Dai, J., Narasimhan, S., Zhang, L., Liu, L., Wang, P., and Fikrig, E. (2010). Tick histamine release factor is critical for Ixodes scapularis engorgement and transmission of the lyme disease agent. PLoS Pathog. *6*, e1001205. https://doi.org/10.1371/journal.ppat.1001205

Denardi, S.E., Furquim, K.C.S., Bechara, G.H., and Camargo-Mathias, M.I. (2011). Salivary glands of Amblyomma cajennense (Acari: Ixodidae): a histological and an ultrastructural overview. Exp. Appl. Acarol. *54*, 177–189. https://doi.org/10.1007/s10493-010-9421-5

Dillon, J., Franks, C.J., Murray, C., Edwards, R.J., Calahorro, F., Ishihara, T., Katsura, I., Holden-Dye, L., and O'Connor, V. (2015). Metabotropic Glutamate Receptors: Modulators Of Context-

Dependent Feeding Behaviour In C. Elegans. J. Biol. Chem. *290*, 15052–15065. https://dx.doi.org/10.1074%2Fjbc.M114.606608

Donohue, K.V., Khalil, S.M.S., Ross, E., Mitchell, R.D., Roe, R.M., and Sonenshine, D.E. (2009). Male engorgement factor: Role in stimulating engorgement to repletion in the ixodid tick, Dermacentor variabilis. J. Insect Physiol. *55*, 909–918. https://doi.org/10.1016/j.jinsphys.2009.05.019

Dumler, J.S. (2010). Fitness and freezing: vector biology and human health. J. Clin. Invest. *120*, 3087–3090. https://doi.org/10.1172/JCI44402

Dunlop, J.A. (2010). Geological history and phylogeny of Chelicerata. Arthropod Struct Dev *39*, 124–142. https://doi.org/10.1016/j.asd.2010.01.003

Egekwu, N., Sonenshine, D.E., Garman, H., Barshis, D.J., Cox, N., Bissinger, B.W., Zhu, J., and M Roe, R. (2016). Comparison of synganglion neuropeptides, neuropeptide receptors and neurotransmitter receptors and their gene expression in response to feeding in Ixodes scapularis (Ixodidae) vs. Ornithodoros turicata (Argasidae). Insect Mol. Biol. *25*, 72–92. https://doi.org/10.1111/imb.12202

Fawcett, D.W., Doxsey, S., and Büscher, G. (1981). Salivary gland of the tick vector (R. appendiculatus) of East Coast fever. II. Cellular basis for fluid secretion in the type III acinus. Tissue Cell *13*, 231–253. https://doi.org/10.1016/0040-8166(81)90003-3

Ferreira, L.L., Soares, S.F., de Oliveira Filho, J.G., Oliveira, T.T., Pérez de León, A.A., and Borges, L.M.F. (2015). Role of Rhipicephalus microplus cheliceral receptors in gustation and host differentiation. Ticks Tick Borne Dis *6*, 228–233. https://doi.org/10.1016/j.ttbdis.2014.12.008

Fielden, L.J., and Duncan, F.D. (2014). Respiratory system. In Biology of Ticks, D.E. Sonenshine, and R.M. Roe, eds. (New York: Oxford University Press), pp. 240–257. ISBN 9780199744053.

Foelix, R.F., and Chu-Wang, I.-W. (1972). Fine structural analysis of palpal receptors in the tick Amblyomma americanum (L.). Z.Zellforsch *129*, 548–560. https://doi.org/10.1007/BF00316749

Francischetti, I.M.B., My Pham, V., Mans, B.J., Andersen, J.F., Mather, T.N., Lane, R.S., and Ribeiro, J.M.C. (2005). The transcriptome of the salivary glands of the female western black-legged tick Ixodes pacificus (Acari: Ixodidae). Insect Biochem. Mol. Biol. *35*, 1142–1161. https://doi.org/10.1016/j.ibmb.2005.05.007

Francischetti, I.M.B., Sa-Nunes, A., Mans, B.J., Santos, I.M., and Ribeiro, J.M.C. (2009). The role of saliva in tick feeding. Front Biosci (Landmark Ed) *14*, 2051–2088.

Galun, R. (1987). Regulation of blood gorging. Int. J. Trop. Insect Sci. 8: 623-625 . doi.org/10.1017/S1742758400022700

Graça-Souza, A.V., Maya-Monteiro, C., Paiva-Silva, G.O., Braz, G.R.C., Paes, M.C., Sorgine, M.H.F., Oliveira, M.F., and Oliveira, P.L. (2006). Adaptations against heme toxicity in blood-feeding arthropods. Insect Biochem. Mol. Biol. *36*, 322–335. https://doi.org/10.1016/j.ibmb.2006.01.009

Gray, S.J., Estrada-Peña, A., and Vial, L. (2014). Ecology of nidicolous tick. In Biology of Ticks, D.E. Sonenshine, and R.M. Roe, eds. (New York: Oxford University Press), pp. 39–60. ISBN 9780195084313

Gulia-Nuss, M., Nuss, A.B., Meyer, J.M., Sonenshine, D.E., Roe, R.M., Waterhouse, R.M., Sattelle, D.B., de la Fuente, J., Ribeiro, J.M., Megy, K., et al. (2016). Genomic insights into the Ixodes scapularis tick vector of Lyme disease. Nat Commun *7*, 10507. https://doi.org/10.1038/ncomms10507

Harris, E.K., Verhoeve, V.I., Banajee, K.H., Macaluso, J.A., Azad, A.F., and Macaluso, K.R. (2017). Comparative vertical transmission of Rickettsia by Dermacentor variabilis and Amblyomma maculatum. Ticks Tick Borne Dis *8*, 598–604. https://dx.doi.org/10.1016%2Fj.ttbdis.2017.04.003

Hollmann, T., Kim, T.K., Tirloni, L., Radulović, Ž.M., Pinto, A.F.M., Diedrich, J.K., Yates, J.R., da Silva Vaz, I., and Mulenga, A. (2018). Identification and characterization of proteins in the Amblyomma americanum tick cement cone. Int. J. Parasitol. *48*, 211–224. https://doi.org/10.1016/j.ijpara.2017.08.018

Horn, M., Nussbaumerová, M., Sanda, M., Kovárová, Z., Srba, J., Franta, Z., Sojka, D., Bogyo, M., Caffrey, C.R., Kopácek, P., et al. (2009). Hemoglobin digestion in blood-feeding ticks: mapping a multipeptidase pathway by functional proteomics. Chem. Biol. *16*, 1053–1063. https://dx.doi.org/10.1016%2Fj.chembiol.2009.09.009

Ibelli, A.M.G., Hermance, M.M., Kim, T.K., Gonzalez, C.L., and Mulenga, A. (2013). Bioinformatics and expression analyses of the Ixodes scapularis tick cystatin family. Exp Appl Acarol *60*, 41–53. https://dx.doi.org/10.1007%2Fs10493-012-9613-2

Jaworski, D.C., Rosell, R., Coons, L.B., and Needham, G.R. (1992). Tick (Acari: Ixodidae) attachment cement and salivary gland cells contain similar immunoreactive polypeptides. J. Med. Entomol. *29*, 305–309. https://doi.org/10.1093/jmedent/29.2.305

Jelinski, J. (2016). Painless Hematophagy: The Functional Role of Novel Tick Metalloproteases in Pain Suppression. Honors Theses. https://aquila.usm.edu/honors_theses

Jeyaprakash, A., and Hoy, M.A. (2009). First divergence time estimate of spiders, scorpions, mites and ticks (subphylum: Chelicerata) inferred from mitochondrial phylogeny. Exp. Appl. Acarol. *47*, 1–18. https://doi.org/10.1007/s10493-008-9203-5

Karasuyama, H., Tabakawa, Y., Ohta, T., Wada, T., and Yoshikawa, S. (2018). Crucial Role for Basophils in Acquired Protective Immunity to Tick Infestation. Front Physiol *9*. https://dx.doi.org/10.3389%2Ffphys.2018.01769

Kaufman, W.R., and Phillips, J.E. (1973). Ion and water balance in the ixodid tick Dermacentor andersoni. I. Routes of ion and water excretion. Journal of Experimental Biology.

Kaufman, W.R., and Sauer, J.R. (1982). CHAPTER 7 - Ion and Water Balance in Feeding Ticks: Mechanisms of Tick Excretion. In Physiology of Ticks, F.D. Obenchain, and R. Galun, eds. (Oxford, U.K. Pergamon Press), pp. 213–244.

Kaufman, W.R., Kaufman, S., and Flynn, P.C. (2016). Cuticle expansion during feeding in the tick Amblyomma hebraeum (Acari: Ixodidae): The role of hydrostatic pressure. J. Insect Physiol. *88*, 10–14.

Keirans, J.E. (2009). Order Ixodida. In A Manual of Acarology, G.W. Krantz, and D.E. Walter, eds. (Lubbock: Texas Tech University Press), pp. 11–123.

Kim, D., Šimo, L., and Park, Y. (2014). Orchestration of salivary secretion mediated by two different dopamine receptors in the blacklegged tick Ixodes scapularis. J. Exp. Biol. *217*, 3656–3663. https://dx.doi.org/10.1242%2Fjeb.109462

Kim, D., Urban, J., Boyle, D.L., and Park, Y. (2016a). Multiple functions of Na/K-ATPase in dopamine-induced salivation of the Blacklegged tick, Ixodes scapularis. Sci Rep *6*, 1–13. https://doi.org/10.1038/srep21047

Kim, D., Maldonado-Ruiz, P., Zurek, L., and Park, Y. (2017). Water absorption through salivary gland type I acini in the blacklegged tick, Ixodes scapularis. PeerJ *5*, e3984. https://doi.org/10.7717/peerj.3984

Kim, D., Šimo, L., and Park, Y. (2018). Molecular characterization of neuropeptide elevenin and two elevenin receptors, IsElevR1 and IsElevR2, from the blacklegged tick, Ixodes scapularis. Insect Biochem. Mol. Biol. *101*, 66–75. https://doi.org/10.1016/j.ibmb.2018.07.005

Kim, D., Šimo, L., Vancová, M., Urban, J., and Park, Y. (2019). Neural and endocrine regulation of osmoregulatory organs in tick: Recent discoveries and implications. Gen. Comp. Endocrinol. *278*, 42–49. https://doi.org/10.1016/j.ygcen.2018.08.004

Kim, T.K., Tirloni, L., Pinto, A.F.M., Moresco, J., Iii, J.R.Y., Jr, I. da S.V., and Mulenga, A. (2016b). Ixodes scapularis Tick Saliva Proteins Sequentially Secreted Every 24 h during Blood Feeding. PLOS Neglected Tropical Diseases *10*, e0004323. https://doi.org/10.1371/journal.pntd.0004323

Klompen, H., Lekveishvili, M., and Black, W.C. (2007). Phylogeny of parasitiform mites (Acari) based on rRNA. Mol. Phylogenet. Evol. *43*, 936–951. https://doi.org/10.1016/j.ympev.2006.10.024

Klompen, J.S.H., Black, W.C., Keirans, J.E., and Norris, D.E. (2000). Systematics and Biogeography of Hard Ticks, a Total Evidence Approach. Cladistics *16*, 70–102. https://doi.org/10.1006/clad.1999.0126

Klowden, M.J. (2013). Physiological Systems in Insects - 3rd Edition (San Diego, CA: Academic Press).

Koch, H.G., and Sauer, J.R. (1984). Quantity of Blood Ingested by Four Species of Hard Ticks (Acari: Ixodidae) Fed on Domestic Dogs. Ann Entomol Soc Am *77*, 142–146. https://doi.org/10.1093/aesa/77.2.142

Koči, J., Šimo, L., and Park, Y. (2014). Autocrine/paracrine dopamine in the salivary glands of the blacklegged tick Ixodes scapularis. J. Insect Physiol. *62*, 39–45. https://dx.doi.org/10.1016%2Fj.jinsphys.2014.01.007

Koehler, J.W., Morales, M.E., Shelby, B.D., and Brindley, P.J. (2007). Aspartic protease activities of schistosomes cleave mammalian hemoglobins in a host-specific manner. Mem. Inst. Oswaldo Cruz *102*, 83–85. https://doi.org/10.1590/S0074-02762007000100014

Lara, F.A., Lins, U., Bechara, G.H., and Oliveira, P.L. (2005). Tracing heme in a living cell: hemoglobin degradation and heme traffic in digest cells of the cattle tick Boophilus microplus. J. Exp. Biol. *208*, 3093–3101. https://doi.org/10.1242/jeb.01749

Laskay, U.A., Breci, L., Vilcins, I.-M.E., Dietrich, G., Barbour, A.G., Piesman, J., and Wysocki, V.H. (2013). Survival of host blood proteins in Ixodes scapularis (Acari: Ixodidae) ticks: a time course study. J. Med. Entomol. *50*, 1282–1290. https://doi.org/10.1603/ME12125

Lewis, L.A., Radulović, Ž.M., Kim, T.K., Porter, L.M., and Mulenga, A. (2015). Identification of 24h Ixodes scapularis immunogenic tick saliva proteins. Ticks Tick Borne Dis *6*, 424–434. https://doi.org/10.1016/j.ttbdis.2015.03.012

Lindsay, P.J., and Kaufman, W.R. (1986). Potentiation of salivary fluid secretion in ixodid ticks: a new receptor system for gamma-aminobutyric acid. Can. J. Physiol. Pharmacol. *64*, 1119–1126. https://doi.org/10.1139/y86-191

Mac, S., da Silva, S.R., and Sander, B. (2019). The economic burden of Lyme disease and the cost-effectiveness of Lyme disease interventions: A scoping review. PLoS ONE *14*, e0210280. https://doi.org/10.1371/journal.pone.0210280

Mans, B.J. (2019). Chemical Equilibrium at the Tick–Host Feeding Interface:A Critical Examination of Biological Relevance in Hematophagous Behavior. Front Physiol *10*. https://doi.org/10.3389/fphys.2019.00530

Mans, B.J., de Castro, M.H., Pienaar, R., de Klerk, D., Gaven, P., Genu, S., and Latif, A.A. (2016). Ancestral reconstruction of tick lineages. Ticks Tick Borne Dis *7*, 509–535. https://doi.org/10.1016/j.ttbdis.2016.02.002

Meeusen, T., Mertens, I., Loof, A.D., and Schoofs, L. (2003). G Protein-Coupled Receptors in Invertebrates: A State of the Art. In International Review of Cytology, K. J. Leon, ed. (San Diego, U.S., Academic Press), pp. 189–261. https://doi.org/10.1016/s0074-7696(03)30004-x

Meredith, J., and Kaufman, W.R. (1973). A proposed site of fluid secretion in the salivary gland of the ixodid tick Dermacentor andersoni. Parasitology *67*, 205–217. https://doi.org/10.1017/S0031182000046424

Mitchell, R.D., Zhu, J., Carr, A.L., Dhammi, A., Cave, G., Sonenshine, D.E., and Roe, R.M. (2017). Infrared light detection by the haller's organ of adult american dog ticks, Dermacentor variabilis (Ixodida: Ixodidae). Ticks Tick Borne Dis *8*, 764–771. https://doi.org/10.1016/j.ttbdis.2017.06.001

Miyoshi, T., Tsuji, N., Islam, M.K., Huang, X., Motobu, M., Alim, M.A., and Fujisaki, K. (2007). Molecular and reverse genetic characterization of serine proteinase-induced hemolysis in the midgut of the ixodid tick Haemaphysalis longicornis. J. Insect Physiol. *53*, 195–203. https://doi.org/10.1016/j.jinsphys.2006.12.001

Moran, T.H. (2000). Cholecystokinin and satiety: current perspectives. Nutrition *16*, 858–865. https://doi.org/10.1016/s0899-9007(00)00419-6

Narasimhan, S., Koski, R.A., Beaulieu, B., Anderson, J.F., Ramamoorthi, N., Kantor, F., Cappello, M., and Fikrig, E. (2002). A novel family of anticoagulants from the saliva of Ixodes scapularis. Insect Mol. Biol. *11*, 641–650. https://doi.org/10.1046/j.1365-2583.2002.00375.x

Nässel, D.R. (2009). Neuropeptide Signaling in Invertebrates. In Encyclopedia of Neuroscience, L.R. Squire, ed. (Oxford, U.K.: Academic Press), pp. 821–828. https://doi.org/10.1016/B978-008045046-9.01446-7

Nässel, D.R., and Williams, M.J. (2014). Cholecystokinin-Like Peptide (DSK) in Drosophila, Not Only for Satiety Signaling. Front Endocrinol (Lausanne) *5*:219https://dx.doi.org/10.3389%2Ffendo.2014.00219

Needham, G.R., and Pannabecker, T.L. (1983). Effects of octopamine, chlordimeform, and demethylchlordimeform on amine-controlled tick salivary glands isolated from feeding Amblyomma americanum (L.). Pesticide Biochemistry and Physiology *19*, 133–140. https://doi.org/10.1016/0048-3575(83)90132-3

Needham, G.R., Rosell, R., and Greenwald, L. (1990). Ultrastructure of type-I salivary-gland acini in four species of ticks and the influence of hydration states on the type-I acini ofAmblyomma americanum. Exp Appl Acarol *10*, 83–104. https://doi.org/10.1007/BF01194085

Neese, P.A., E Sonenshine, D., Kallapur, V.L., Apperson, C.S., and Roe, R.M. (2000). Absence of insect juvenile hormones in the American dog tick, Dermacentor variabilis (Say) (Acari:Ixodidae), and in Ornithodoros parkeri Cooley (Acari:Argasidae). J. Insect Physiol. *46*, 477–490. https://doi.org/10.1016/S0022-1910(99)00134-1

Nuttall, P.A. (2019). Wonders of tick saliva. Ticks Tick Borne Dis *10*, 470–481. https://doi.org/10.1016/j.ttbdis.2018.11.005

Obenchain, F.D., and Oliver, J.H. (1975). Neurosecretory system of the American Dog Tick, Dermacentor variabilis (Acari: Ixodidae). ii. Distribution of secretory cell types, axonal pathways and putative neurohemal-neuroendocrine associations; comparative histological and anatomical implications. J. Morphol. *145*, 269–293. https://doi.org/10.1002/jmor.1051450303

Ogihara, M.H., and Taylor, D. (2014). Female reproductive system: Structure, Physiology, and Molecular Biology. In Biology of Ticks, D.E. Sonenshine, and R.M. Roe, eds. (New York: Oxford University Press), pp. 449 – 483.

Ogihara, M.H., Taylor, D., and Kataoka, H. (2019). Steroid hormones in Acari, their functions and synthesis. Appl Entomol Zool *54*, 323–338. https://doi.org/10.1007/s13355-019-00631-3

Pannabecker, T., and Needham, G.R. (1985). Effects of octopamine on fluid secretion by isolated salivary glands of a feeding ixodid tick. Arch. Insect Biochem. Physiol. *2*, 217–226. https://doi.org/10.1002/arch.940020209

Park, Y. (2012). Endocrine regulation of insect diuresis in the early postgenomic era1. Can. J. Zool. *90*, 507–520. https://doi.org/10.1139/z2012-013

Perner, J., Kropáčková, S., Kopáček, P., and Ribeiro, J.M.C. (2018). Sialome diversity of ticks revealed by RNAseq of single tick salivary glands. PLoS Negl Trop Dis *12*, e0006410. https://doi.org/10.1371/journal.pntd.0006410

Phillips, J.S., and Sonenshine, D.E. (1993). Role of the male claw sensilla in perception of female mounting sex pheromone in Dermacentor variabilis, Dermacentor andersoni and Amblyomma americanum. Exp Appl Acarol *17*, 631–653. https://doi.org/10.1007/BF00058504

Pinho-Ribeiro, F.A., Verri, W.A., and Chiu, I.M. (2017). Nociceptor Sensory Neuron-Immune Interactions in Pain and Inflammation. Trends Immunol. *38*, 5–19. https://doi.org/10.1016/j.it.2016.10.001

Prullage, J.B., Pound, J.M., and Meola, S.M. (1992). Synganglial morphology and neurosecretory centers of adult Amblyomma americanum (L.) (Acari: Ixodidae). J. Med. Entomol. *29*, 1023–1034. https://doi.org/10.1093/jmedent/29.6.1023

Randolph, S.E. (2014). Ecology of nidicolous tick. In Biology of Ticks, D.E. Sonenshine, and R.M. Roe, eds. (New York,:U.S. Oxford University Press), pp. 3–38.

Ribeiro, J.M.C. (1988). The Midgut Hemolysin of Ixodes dammini (Acari: Ixodidae). The Journal of Parasitology *74*, 532–537. https://www.jstor.org/stable/3282168

Ribeiro, J.M., Makoul, G.T., Levine, J., Robinson, D.R., and Spielman, A. (1985). Antihemostatic, antiinflammatory, and immunosuppressive properties of the saliva of a tick, Ixodes dammini. J. Exp. Med. *161*, 332–344. https://doi.org/10.1084/jem.161.2.332

Ribeiro, J.M., Evans, P.M., MacSwain, J.L., and Sauer, J. (1992). Amblyomma americanum: characterization of salivary prostaglandins E2 and F2 alpha by RP-HPLC/bioassay and gas chromatography-mass spectrometry. Exp. Parasitol. *74*, 112–116. https://doi.org/10.1016/0014-4894(92)90145-z

Ribeiro, J.M.C., Alarcon-Chaidez, F., B. Francischetti, I.M., Mans, B.J., Mather, T.N., Valenzuela, J.G., and Wikel, S.K. (2006). An annotated catalog of salivary gland transcripts from Ixodes scapularis ticks. Insect Biochemistry and Molecular Biology *36*, 111–129. https://doi.org/10.1016/j.ibmb.2005.11.005

Ribeiro, J.M.C., Mans, B.J., and Arcà, B. (2010). An insight into the sialome of blood-feeding Nematocera. Insect Biochem. Mol. Biol. *40*, 767–784. https://doi.org/10.1016/j.ibmb.2010.08.002

Ribeiro, J.M.C., Slovák, M., and Francischetti, I.M.B. (2017). An insight into the sialome of Hyalomma excavatum. Ticks Tick Borne Dis *8*, 201–207. https://doi.org/10.1016/j.ttbdis.2016.08.011

Roe, R.M., Donohue, K.V., Khalil, S.M.S., and Sonenshine, D.E. (2008). Hormonal regulation of metamorphosis and reproduction in ticks. Front. Biosci. *13*, 7250–7268. https://doi.org/10.2741/3226

Roller, L., Šimo, L., Mizoguchi, L., Slovák, M., Park, Y., and Žitňan, D. (2015). Orcokinin-like immunoreactivity in central neurons innervating the salivary glands and hindgut of ixodid ticks. Cell Tissue Res. *360*, 209–222. https://doi.org/10.1007/s00441-015-2121-z

Rudolph, D., and Knülle, W. (1974). Site and mechanism of water vapour uptake from the atmosphere in ixodid ticks. Nature *249*, 84–85. https://doi.org/10.1038/249084a0

Rudolph, D., and Knulle, W. (1978). Uptake of water vapour from the air: process, site and mechanism in ticks. In Comparative Physiology: Water Ions and Fluid Mechanics, K. Schmidt-Nielsen, L. Bolis, and S.H.P. Maddrell, eds. (Cambridge: Cambridge University Press), pp. 97–113.

Sauer, J.R., McSwain, J.L., Bowman, A.S., and Essenberg, R.C. (1995). Tick salivary gland physiology. Annu. Rev. Entomol. *40*, 245–267. https://doi.org/10.1146/annurev.en.40.010195.001333

Schoofs, L., De Loof, A., and Van Hiel, M.B. (2017). Neuropeptides as Regulators of Behavior in Insects. Annu. Rev. Entomol. *62*, 35–52.

Schwedes, C.C., and Carney, G.E. (2012). Ecdysone signaling in adult Drosophila melanogaster. J. Insect Physiol. *58*, 293–302. https://doi.org/10.1146/annurev-ento-031616-035500

Šimo, L., and Park, Y. (2014). Neuropeptidergic control of the hindgut in the black-legged tick Ixodes scapularis. Int J Parasitol *44*, 819–826. https://doi.org/10.1016/j.ijpara.2014.06.007

Šimo, L., Žitňan, D., and Park, Y. (2009a). Two novel neuropeptides in innervation of the salivary glands of the black-legged tick, Ixodes scapularis: Myoinhibitory peptide and SIFamide. J Comp Neurol *517*, 551–563. https://doi.org/10.1002/cne.22182

Šimo, L., Slovák, M., Park, Y., and Zitnan, D. (2009b). Identification of a complex peptidergic neuroendocrine network in the hard tick, Rhipicephalus appendiculatus. Cell Tissue Res. *335*, 639–655. https://doi.org/10.1007/s00441-008-0731-4

Šimo, L., Koči, J., Žitňan, D., and Park, Y. (2011). Evidence for D1 Dopamine Receptor Activation by a Paracrine Signal of Dopamine in Tick Salivary Glands. PLoS One *6*. https://dx.doi.org/10.1371%2Fjournal.pone.0016158

Šimo, L., Zitňan, D., and Park, Y. (2012). Neural control of salivary glands in ixodid ticks. J. Insect Physiol. *58*, 459–466. https://doi.org/10.1016/j.jinsphys.2011.11.006

Šimo, L., Daniel, S., E., Park, Y., and Žitňan, D. (2014a). The Nervous and Sensory Systems: Structure, Function, Proteomics and Genomics. In Biology of Ticks, D.E. Sonenshine, and R.M. Roe, eds. (New York: Oxford University Press), pp. 309–367.

Šimo, L., Koči, J., Kim, D., and Park, Y. (2014b). Invertebrate Specific D1-like Dopamine Receptor in Control of Salivary Glands in the Black-Legged Tick Ixodes scapularis. J Comp Neurol *522*, 2038–2052. https://doi.org/10.1002/cne.23515

Šimo, L., Kazimirova, M., Richardson, J., and Bonnet, S.I. (2017). The Essential Role of Tick Salivary Glands and Saliva in Tick Feeding and Pathogen Transmission. Front Cell Infect Microbiol *7*. https://dx.doi.org/10.3389%2Ffcimb.2017.00281

Sojka, D., Francischetti, I.M.B., Calvo, E., and Kotsyfakis, M. (2011). Cysteine proteases from bloodfeeding arthropod ectoparasites. Adv. Exp. Med. Biol. *712*, 177–191. https://doi.org/10.1007/978-1-4419-8414-2_11

Sojka, D., Pytelková, J., Perner, J., Horn, M., Konvičková, J., Schrenková, J., Mareš, M., and Kopáček, P. (2016). Multienzyme degradation of host serum albumin in ticks. Ticks Tick Borne Dis *7*, 604–613. https://doi.org/10.1016/j.ttbdis.2015.12.014

Sonenshine, D.E. (1970). Fertility and fecundity of Dermacentor variabilis reared from radioisotope--taggd larvae. J. Econ. Entomol. *63*, 1675–1676. https://doi.org/10.1093/jee/63.5.1675

Sonenshine, D.E. (2014). Excretion and water balance Hindgut, Malpighian Tubules, and Coxal Glands. In Biology of Ticks, D.E. Sonenshine, and R.M. Roe, eds. (New York, U.S. Oxford University Press), pp. 206–2019.

Sonenshine, D.E., and Anderson, J.M. (2014). Mouthparts and digestive system: Anatomy and Molecular Biology of Feeding and Digestion. In Biology of Ticks, D.E. Sonenshine, and R.M. Roe, eds. (New York, U.S.: Oxford University Press), pp. 122–162.

Sonenshine, D.E., and Macaluso, K.R. (2017). Microbial Invasion vs. Tick Immune Regulation. Front Cell Infect Microbiol *7*, 390. https://doi.org/10.3389/fcimb.2017.00390

Sonenshine, D.E., and Roe, R.M. (2014). Overview: Ticks, People, and Animals. In Biology of Ticks, D.E. Sonenshine, ed. (New York: Oxford University Press), pp. 3–16.

Sonenshine, D.E., Homsher, P.J., Carson, K.A., and Wang, V.D. (1984). Evidence of the role of the cheliceral digits in the perception of genital sex pheromones during mating in the American dog tick, Dermacentor variabilis (Acari: Ixodidae). J. Med. Entomol. *21*, 296–306. https://doi.org/10.1093/jmedent/21.3.296

Sonenshine, D.E., Adams, T., Allan, S.A., McLaughlin, J., and Webster, F.X. (2003). Chemical composition of some components of the arrestment pheromone of the black-legged tick, Ixodes scapularis (Acari: Ixodidae) and their use in tick control. J. Med. Entomol. *40*, 849–859. https://doi.org/10.1603/0022-2585-40.6.849

Starck, J.M., Mehnert, L., Biging, A., Bjarsch, J., Franz-Guess, S., Kleeberger, D., and Hörnig, M. (2018). Morphological responses to feeding in ticks (Ixodes ricinus). Zoological Letters *4*, 20. https://doi.org/10.1186/s40851-018-0104-0

Suppan, J., Engel, B., Marchetti-Deschmann, M., and Nürnberger, S. (2018). Tick attachment cement - reviewing the mysteries of a biological skin plug system. Biol Rev Camb Philos Soc *93*, 1056–1076. https://doi.org/10.1111/brv.12384

Tatchell, R.J. (1967). Salivary Secretion in the Cattle Tick as a Means of Water Elimination. Nature *213*, 940–941. https://doi.org/10.1038/213940a0

Trimnell, A.R., Davies, G.M., Lissina, O., Hails, R.S., and Nuttall, P.A. (2005). A cross-reactive tick cement antigen is a candidate broad-spectrum tick vaccine. Vaccine *23*, 4329–4341. https://doi.org/10.1016/j.vaccine.2005.03.041

Valenzuela, J.G., Charlab, R., Mather, T.N., and Ribeiro, J.M. (2000). Purification, cloning, and expression of a novel salivary anticomplement protein from the tick, Ixodes scapularis. J. Biol. Chem. *275*, 18717–18723. https://doi.org/10.1074/jbc.M001486200

Vancová, M., Zacharovová, K., Grubhoffer, L., and Nebesárová, J. (2006). Ultrastructure and lectin characterization of granular salivary cells from Ixodes ricinus females. J. Parasitol. *92*, 431–440. https://doi.org/10.1645/GE-648R.1

Vancová, M., Bílý, T., Nebesářová, J., Grubhoffer, L., Bonnet, S., Park, Y., and Šimo, L. (2019). Ultrastructural mapping of salivary gland innervation in the tick Ixodes ricinus. Sci Rep *9*, 1–13. https://doi.org/10.1038/s41598-019-43284-6

Vancová, M., Bílý, T., Šimo, L., Touš, J., Horodyský, P., Růžek, D., Novobilský, A., Salát, J., Strnad, M., Sonenshine, D.E., et al. (2020). Three-dimensional reconstruction of the feeding apparatus of the tick Ixodes ricinus (Acari: Ixodidae): a new insight into the mechanism of blood-feeding. Sci Rep *10*, 1–7. https://doi.org/10.1038/s41598-019-56811-2

Weiss, B.L., and Kaufman, W.R. (2004). Two feeding-induced proteins from the male gonad trigger engorgement of the female tick Amblyomma hebraeum. Proc. Natl. Acad. Sci. U.S.A. *101*, 5874–5879. https://doi.org/10.1073/pnas.0307529101

Wikel, S. (2013). Ticks and tick-borne pathogens at the cutaneous interface: host defenses, tick countermeasures, and a suitable environment for pathogen establishment. Front Microbiol *4*. https://dx.doi.org/10.3389%2Ffmicb.2013.00337

Zandawala, M., Tian, S., and Elphick, M.R. (2018). The evolution and nomenclature of GnRH-type and corazonin-type neuropeptide signaling systems. Gen. Comp. Endocrinol. *264*, 64–77. https://doi.org/10.1016/j.ygcen.2017.06.007

Zhu, J., Khalil, S.M., Mitchell, R.D., Bissinger, B.W., Egekwu, N., Sonenshine, D.E., and Roe, R.M. (2016). Mevalonate-Farnesal Biosynthesis in Ticks: Comparative Synganglion Transcriptomics and a New Perspective. PLoS ONE *11*, e0141084. https://doi.org/10.1371/journal.pone.0141084

Zhu, X.X., Oliver, J.H., and Dotson, E.M. (1991). Epidermis as the source of ecdysone in an argasid tick. Proc Natl Acad Sci U S A *88*, 3744–3747. https://dx.doi.org/10.1073%2Fpnas.88.9.3744

Chapter 13

Lyme Disease Pathogenesis

Jenifer Coburn[1]*, Brandon Garcia[2], Linden T. Hu[3], Mollie W. Jewett[4], Peter Kraiczy[5], Steven J. Norris[6] and Jon Skare[7]

[1]Center For Infectious Disease Research, Medical College of Wisconsin, 8701 Watertown Plank Rd., TBRC C3980, Milwaukee, WI 53226, USA
[2]Department of Microbiology and Immunology, East Carolina University, Brody School of Medicine, Greenville, NC 27858, USA
[3]Department of Molecular Biology and Microbiology, Vice Dean of Research, Tufts University School of Medicine, 136 Harrison Ave., Boston, MA 02111, USA
[4]Immunity and Pathogenesis Division Head, Burnett School of Biomedical Sciences, University of Central Florida College of Medicine, 6900 Lake Nona Blvd. Orlando, FL 32827, USA
[5]Institute of Medical Microbiology and Infection Control, University Hospital Frankfurt, Goethe University Frankfurt, Paul-Ehrlich-Str. 40, 60596 Frankfurt, Germany
[6]Department of Pathology and Laboratory Medicine, University of Texas Medical School at Houston, P.O. Box 20708, Houston, TX 77225, USA
[7]Professor and Associate Head, Texas A&M University, 8447 Riverside Pkwy, Bryan, TX 77807, USA
*Corresponding author: jcoburn@mcw.edu

DOI: https://doi.org/10.21775/9781913652616.13

Abstract
Lyme disease *Borrelia* are obligately parasitic, tick-transmitted, invasive, persistent bacterial pathogens that cause disease in humans and non-reservoir vertebrates primarily through the induction of inflammation. During transmission from the infected tick, the bacteria undergo significant changes in gene expression, resulting in adaptation to the mammalian environment. The organisms multiply and spread locally and induce inflammatory responses that, in humans, result in clinical signs and symptoms. *Borrelia* virulence involves a multiplicity of mechanisms for dissemination and colonization of multiple tissues and evasion of host immune responses. Most of the tissue damage, which is seen in non-reservoir hosts, appears to result from host inflammatory reactions, despite the low numbers of bacteria in affected sites. This host response to the Lyme disease *Borrelia* can cause neurologic, cardiovascular, arthritic, and dermatologic manifestations during the disseminated and persistent stages of infection. The mechanisms by which a paucity of organisms (in comparison to many other infectious diseases) can cause varied and in some cases profound inflammation and symptoms remains mysterious but are the subjects of diverse ongoing investigations. In this chapter, we provide an overview of virulence mechanisms and determinants for which roles have been demonstrated *in vivo*, primarily in mouse models of infection.

Introduction
Lyme disease was first recognized in 1976 when a cluster of cases of juvenile arthritis was recognized in Old Lyme, CT. Many of these patients also reported cutaneous skin lesions that were similar to those reported in Europe that were previously associated with tick bites. There was a strong suspicion that an infectious agent was the underlying cause of both cases in Old Lyme, CT and in Europe, but it was not until 1982 that a spirochete found in *Ixodes* ticks was suggested to be the cause (Burgdorfer et al., 1982). The role of this bacterium, named *Borrelia (B.) burgdorferi*, as the causative agent of Lyme disease was quickly established as the bacterium was recovered from patients as well as from reservoir hosts, such as the white-footed mouse. Although first recognized because of its association with arthritis, it became clear that arthritis was a late stage manifestation and that acute infection manifested with a characteristic erythema migrans rash and, in some cases, with cardiac or neurologic involvement.

Prokaryotes evolved approximately 3.5 billion years ago (Schopf, 2000) and, by definition, were initially free-living. The evolution of microbial communities and symbiotic relationships about 1 billion years ago eventually resulted in the development of multicellular eukaryotic organisms, creating new ecologic niches. Bacteria colonized these

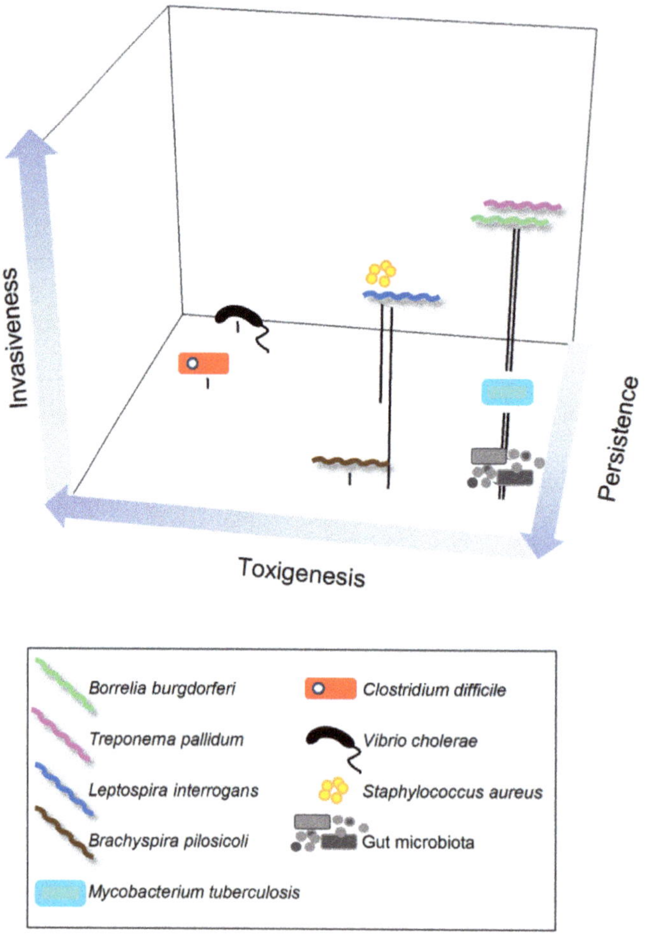

Figure 1. Characteristics of select bacterial pathogens. Schematic depiction of general properties of bacterial pathogens illustrated. Note that for many of the bacteria shown, the outcome of the interaction may depend on the host and even particular sites within the host. *B. burgdorferi* and *T. pallidum* are highly invasive and persistent, but non-toxigenic, while *M. tuberculosis* is usually less invasive but persistent and non-toxigenic; gut microbiota are persistent but neither invasive nor toxigenic. *S. aureus* is often a commensal that as opportunity arises can be invasive and toxigenic. *L. interrogans* is invasive and persistent, but unlike *B. burgdorferi* and *T. pallidum*, synthesizes lipopolysaccharide (LPS), a highly inflammatory glycolipid. *B. pilosicoli* causes persistent but not invasive infection, synthesizes LPS and hemolysins. *V. cholerae* and *C. difficile* are both toxigenic organisms, which cause disease largely attributable to toxin action; *C. difficile* is more persistent, at least in part, due to spore formation. The authors thank Valery Lozada-Fernandez, PhD candidate at the Medical College of Wisconsin, for generation of the 3D plot in this figure. Concept originally depicted in (Botkin et al., 2006).

ecologic niches as commensal, symbiotic and pathogenic organisms. Development of host-associated pathogens resulted from genetic adaptation to protect themselves, parasitize, and become harmful to other organisms. This adaptation involved modification, acquisition, and loss of a variety of genes, thereby acquiring, creating, and preserving functions that assure survival and losing those that are nonessential.

Lyme disease *Borrelia* developed an unusual lifestyle in which they must alternate between vertebrate and arthropod hosts; they are never free-living (*i.e.*, they do not "hit the ground"). In addition, they belong to a group of highly invasive organisms, the spirochetes, capable of invading multiple tissues in susceptible vertebrates and causing infection and manifestations for months to years (Figure 1). In this manner, infection with Lyme disease *Borrelia* in some respects parallels that of another spirochete, *Treponema pallidum* subsp. *pallidum*, the causative agent of syphilis. Both diseases exhibit local, disseminated and persistent manifestations, and tissue pathology appears to be due primarily to host inflammatory reactions. However, there are important differences in the mode of transmission (*i.e.* vectored by an infected tick vs. direct human to human contact). In addition, while Lyme disease and syphilis are both persistent infections with pathology predominantly caused by inflammatory reactions, each has distinct manifestations. Overlaid on these differences is the fact that the clinical manifestations of Lyme disease vary in different geographic regions. This is believed to be, at least in part, due to differences in the predominant Lyme disease agents in particular regions. For example, *Borrelia garinii*, which is found in Eurasia, has been linked to neurologic disease, while *B. burgdorferi*, the major species found in the U.S. is associated with arthritis. Even among strains of *B. burgdorferi*, there have been differences reported in manifestations and in the severity of inflammation associated with particular strains. The specific differences between species and strains responsible for the varying clinical presentations are only beginning to be identified. Despite a high degree of genetic similarity among pathogenic treponemes, greater clinical distinctions are observed among the diseases they cause (syphilis, bejel, yaws, pinta). The dissimilarities observed in these spirochetal infections reflect important differences in infection patterns, tissue tropisms, and host-pathogen interactions at the molecular level.

In this chapter, we examine the factors involved in the pathogenesis of Lyme disease in experimentally tractable animal models. In some cases, results obtained in animal models can be directly related to human disease (also see chapter 24). It is important to remember that the majority of animal models may not fully reflect *B. burgdorferi* infections of either humans (accidental) or of reservoir hosts in the natural tick-reservoir-tick cycle. Nevertheless, studying the pathogenesis of invasive, nontoxigenic organisms requires animal models for advancing understanding of mechanisms of invasion, dissemination, and persistence. Such investigations are relatively complex, requiring analysis of gene products (e.g. adhesins, or adhesive proteins, and other surface molecules) that are best studied in the context of intact host and pathogen organisms. Only in the setting of an intact animal can the roles of critical interactions be fully evaluated. It is also important to note that low passage, infectious strains of *B. burgdorferi* have low transformation rates, associated with the presence of putative restriction-modification systems encoded on genomic plasmids frequently lost during *in vitro* culture (Lawrenz et al., 2002; Kawabata et al., 2004). At least one of these plasmids is required for infection of mice and ticks, making genetic manipulations more challenging than with many other organisms. Thus, our knowledge of the genes important in *B. burgdorferi* pathogenesis is still expanding rapidly and is certainly not complete.

Additional complications in understanding *B. burgdorferi* pathogenesis are the differences in the immune responses between different hosts. In its natural reservoir host, *Peromyscus leucopus*, the white footed mouse, no inflammatory response is seen with infection. In contrast, different strains of inbred *Mus* mice manifest varying responses to the organism. With some strains of inbred mice, the manifestations may mimic those seen in human Lyme disease (e.g. carditis, arthritis). However, it is unclear whether the underlying mechanisms for generating these responses are similar. Because of the complexity of host-pathogen interactions, this chapter can include only a portion of relevant studies; the reader is asked to refer to the other chapters in this book and the primary literature for additional detail regarding gene regulation, motility, interactions with the tick vectors, animal models, host responses, and other factors that contribute to pathogenesis and disease.

The infection process

Virulence determinants of Borrelia burgdorferi

Current knowledge regarding the virulence determinants of Lyme disease *Borrelia* is summarized in Table 1, which appears at the end of this chapter. Herein, we define virulence determinants to include all genes that affect the infection process. This designation, therefore, includes genes involved in basic metabolic processes required for survival in a host animal, but that do not necessarily directly or actively engage the host. In Table 1, we focus on genes that encode functions required for *B. burgdorferi* to (i) survive in, and (ii) invade the mammalian host (disseminate), and (iii) cause disease, as described in greater detail below. The focus on *B. burgdorferi* reflects the volume of work done with this species, although most of the genes and proteins included in Table 1 have orthologs in the other Lyme disease spirochetes. One important consideration is that any virulence determinant should fulfill molecular Koch's postulates (Falkow, 1988), which essentially require that (i) a phenotype or gene be specific to pathogenic members of a genus, (ii) inactivation of a gene results in mutant with reduced infectivity (the ability to cause infection) or pathogenicity (the ability to cause disease), and (iii) complementation of the mutant restores the properties of the wild-type. Note that that all *Borrelia* isolated from ticks or wildlife could be

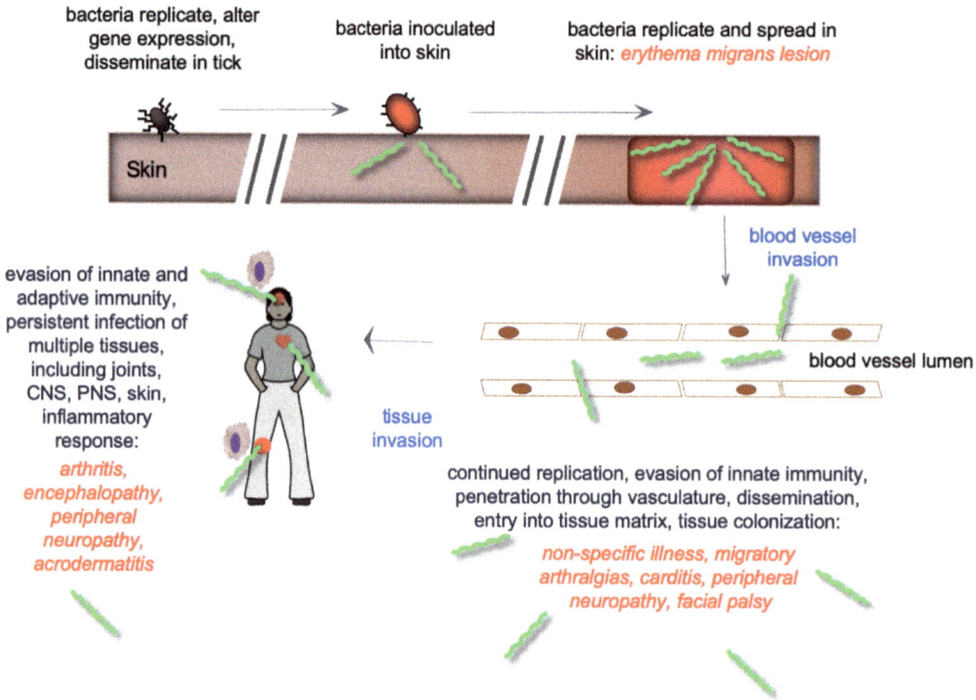

Figure 2. Overview of *Borrelia* infection and Lyme disease in humans. Schematic representation of the infection process (in black text) and common associated disease manifestations (in red text) in Lyme borreliosis. For detailed descriptions of Lyme disease clinical manifestations, stages, and associations with different Lyme spirochete species, we refer the reader to chapter 24 by Radolf *et al.* in this volume and to Steere, A.C. (2001) "Lyme Disease" N Engl J Med; 345:115-125. Adapted from Coburn, J., Leong, J. and Chaconas, G. (2013). "Illuminating the roles of the *Borrelia burgdorferi* adhesins" Trends Microbiol 21(8): 372-379, with permission from Elsevier.

considered infectious due to their obligate parasitic their lifecycles but that pathogenicity also depends on the host animal. For example, it is now clear that not all members of the Lyme disease taxonomic clade of *Borrelia* can cause human infection, so cannot be considered as human pathogens. In *Borrelia* species, attenuation of infectivity during laboratory culture is frequently associated with loss of genomic plasmids maintained in the natural life cycle.

Overview of the infection process

An overview of the basic steps of the infection process and host responses in Lyme borreliosis is presented in Figure 2. Table 1 summarizes specific *B. burgdorferi* genes and gene products that have been implicated as virulence determinants or factors *in vivo*. We recognize that additional *Borrelia* species cause Lyme borreliosis, but genetic tools that allow testing of hypotheses experimentally have been developed most extensively for use and deployed in, *B. burgdorferi* (chapter 16 by Rosa and Jewett).

During tick transmission, the bacterium undergoes changes in gene expression that permit it to adapt to the mammalian environment (see chapter 5 by Samuels *et al.*). Survival and replication in the dermis involve adaptation to the increased temperature, altered pH, and differences in nutrient availability, as well as fending off host innate immune mechanisms. Binding of complement regulators by a number of different *B. burgdorferi* proteins likely plays a key role in survival throughout infection and in diverse vertebrate hosts. Motility and chemotaxis (chapter 8 by Sze *et al.*) also are important for local and systemic dissemination. Binding to a variety of host tissue components provides activities, such as adherence to the vascular wall and extravasation, while adherence to extravascular tissue components facilitates persistent colonization of diverse tissue sites. The acquisition of host plasmin may aid in the tissue invasion process. Immune responses, both innate and adaptive (chapter 18 by Bockenstedt *et al.*), contribute to control of the numbers of the organisms but also contribute to tissue pathology and disease manifestations. Lyme disease *Borrelia* have established mechanisms, including antigenic variation, that permit survival for months to years in the mammalian host; indeed, experimentally infected mice appear to be infected for life. This persistence is likely required to maintain the bacteria in the natural tick-vertebrate infection cycle. Some of the gene products and mechanisms involved in invasion, dissemination, tissue colonization, and evasion of the host immune system are addressed in the remainder of this chapter, but other critical aspects (e.g. interactions of *B. burgdorferi* with the tick vector, regulation of gene expression, motility and chemotaxis, and immune responses) are covered in more detail elsewhere in this book. Emerging themes regarding the pathogenesis of Lyme disease include (i) *B. burgdorferi* encodes several multi-function proteins, (ii) *B. burgdorferi* encodes several distinct proteins that may inhibit host complement activity at different stages of the cascade, (iii) immune responses to *B. burgdorferi* are not only responsible for controlling bacterial numbers, but also for disease manifestations, and (iv) *B. burgdorferi* encodes multiple adhesins that play distinct roles in the infection process.

Overview of methods used to identify and evaluate candidate virulence determinants in *Borrelia*

Evaluation of the roles of candidate virulence determinants in model hosts

While the significance of candidate virulence determinants in many bacterial pathogens can be determined using cell culture and animal models, in the case of *B. burgdorferi*, ticks also must be considered. Each model type can be informative in different ways. Cell culture models can be informative regarding specific and detailed mechanisms of how the bacteria interact with different host cell types, and how individual host cell types respond to the spirochete. The strengths of cell culture models include lower cost and potential for higher throughput, but only in animals can virulence determinants be evaluated in the context of tissue architecture and host responses. Laboratory mice (usually *Mus musculus*) are the favored model host because of the relatively low cost and the abundance of mutant mice and mouse-specific reagents to facilitate evaluation of damage and pathology. However, a significant caveat to the use of mice is that they are not humans, nor even primates, and so are fundamentally different in terms of the host cells and molecules that are relevant in *Borrelia* infections. A second significant caveat is the possibility of off-target effects in mutant mouse lines that are not carefully and completely characterized. The primate (Rhesus macaque, *Macaca mulatta*) model is more reflective of human disease, but more challenging due to cost, availability of appropriate facilities and expertise, and the lack of genetic tractability. We refer the reader to the chapters in this volume covering animal models (chapter 15), host response (chapter 18), and veterinary Lyme disease (chapter 20) for in-depth information. The development of the dialysis membrane chamber implant model allowed, for the first time, harvest of sufficient numbers of *B. burgdorferi* cells for further studies on gene expression and protein production while the bacteria are in a host animal (Akins et al., 1998). Tick work requires available hosts for feeding and therefore progression of the ticks through their life stages, and appropriate containment, but is now performed by many laboratories. For further detail on the ticks and the tick phase of the *B. burgdorferi* life cycle, we refer the reader to (chapters 11 by Pal *et al.* and 12 by Sonenshine and Simo).

Genetic approaches used to identify virulence determinants of B. burgdorferi

Genome-wide analysis of the *B. burgdorferi* genes that are expressed during mammalian infection was historically challenging because the spirochete loads in blood and tissues are too low to allow use of older technologies. This obstacle has been partially overcome through technological advances, and the transcriptomic and proteomic analysis of *B. burgdorferi* grown under *in vitro* conditions that mimic certain aspects of the tick and mouse environments (Carroll et al., 1999; Revel et al., 2002; Brooks et al., 2003; Ojaimi et al., 2003; Seshu et al., 2004; Tokarz et al., 2004; Hyde et al., 2007; Angel et al., 2010) and mammalian host adapted spirochetes grown in dialysis membrane chambers implanted in the rat peritoneal cavity (Akins et al., 1998; Revel et al., 2002; Brooks et al., 2003; Caimano, 2005; Iyer et al., 2015). More recently, advances in the development of genetic tools for *B. burgdorferi* (see chapter 16 by Rosa and Jewett) and in new technologies have contributed significantly to understanding the importance of particular *B. burgdorferi* genes and likely, their products, during infection.

In Vivo Expression Technology (IVET)

In Vivo Expression Technology is a gene discovery method that allows for identification of transcriptionally active regions of a microbial genome when the microorganism is in an environment of interest and does not require isolation of microbial RNA (Mahan et al., 1993a; Mahan et al., 1993b). IVET has been applied to *B. burgdorferi* to identify DNA sequences that function as active promoters during murine infection (Ellis et al., 2013; Casselli and Bankhead, 2015). The details of the molecular genetics of *B. burgdorferi* IVET approaches are discussed in chapter 16. Application of IVET to *B. burgdorferi* has led to the discovery of novel genes important for murine infection, as well as previously unannotated sequences that are expressed during an active infection (Ellis et al., 2013; Jain et al., 2015; Showman et al., 2016; Adams et al., 2017).

Signature tagged mutagenesis (STM)
Signature tagged mutagenesis is a transposon-based approach to the generation of mutations on a genomic scale. The transposon (Tn) incorporates short unique DNA sequences, or "signature tags", to help discriminate between genetically identical "siblings" arising from the transformation vs. truly independent transposon insertion mutants. The library of Tn mutants can then be challenged in a number of ways to identify mutants that do not survive the challenge, i.e. mutants that were present in the starting pool but missing after the selection. In a *tour de force* effort, a large Tn library was generated in an infectious *B. burgdorferi* strain (Botkin et al., 2006; Lin et al., 2012), and used to infect mice. While some mutants were recovered at frequencies reflective of the starting pool, many were recovered at much lower frequencies or not at all. This latter group was considered attenuated. Sequencing of the Tn insertion sites in the clones in the latter group identified candidate virulence determinant-encoding genes at a scale that was unprecedented in the field. Some of these genes had already been shown to encode virulence determinants, validating the approach, but many other genes identified using these strategies opened new avenues of investigation for multiple research groups.

Transposon sequencing (Tn-seq)
The signature tagged mutagenesis library also has been used in transposon sequencing protocols (Tn-seq) to identify the fitness of insertional mutants in the STM library under specific conditions. Here, the STM library is placed under a specific selection pressure and the frequency of mutants before and after the selection pressure is compared using quantitative massively parallel sequencing to identify the location of the transposon insertion. This technique has been used to identify genes involved in carbon utilization, evasion of reactive oxygen species, and survival in the tick host (Troy et al., 2013; Troy et al., 2016; Ramsey et al., 2017; Phelan et al., 2019). The technique is highly reproducible when used under *in vitro* conditions due to the large numbers of mutants that can be tested in a single experiment. When using Tn-seq under *in vivo* conditions in a mouse or tick host, "bottle-neck" issues must be considered. Bottlenecks occur when there is stochastic loss of mutants for reasons that are unrelated to the specific transposon insertion. During *in vivo* infections, a large percentage of an inoculum does not survive the initial transition into a new host. Losses in the bottleneck can lead to false discovery of insertions that are not truly related to fitness and survival in the host. Bottleneck issues can be overcome either by studying smaller numbers of mutants or by increasing the sample size (Chao et al., 2016). One of the major advantages of Tn-seq is that it allows the potential for unbiased identification of possible function encoded by genes with unknown function.

Biochemical approaches
Characterization of the function of a protein or other molecule often requires thorough analyses at the biochemical level. For example, adhesins would have to be localized on the bacterial surface to interact with host substrates, and proof of their localization in or on the *Borrelia* cell has been approached in multiple ways. Since *Borrelia* species are diderm (*i.e.*, have outer and inner membranes with a periplasmic space in between), multiple fractionation approaches have been tested (Magnarelli et al., 1989; Bledsoe et al., 1994; Radolf et al., 1995; Jacobs et al., 2005; Nowalk et al., 2006b), although the results have at times been controversial as spirochete components may fractionate incorrectly using protocols developed for Gram-negative enteric organisms. An initial approach is evaluation of whether a secretion signal peptide (Emr et al., 1980a; Emr et al., 1980b; Hall et al., 1980; Beckwith and Silhavy, 1983) is located at the amino terminal end of a predicted protein sequence, although some properties of spirochete lipoprotein secretion signals differ from those found in *E. coli* (Setubal et al., 2006). Even if there is a signal peptide, the protein is not necessarily localized to the surface in diderm bacteria, so additional biochemical approaches are warranted. Interestingly, it appears that lipoproteins encoded by *B. burgdorferi* are, by default, localized to the outer membrane (Schulze and Zuckert, 2006; Dowdell et al., 2017), but experimental proof of surface localization always is required for any putative surface-exposed protein. One approach is determination of whether the protein is accessible using a protein-specific antibody, with detection of the antibody using a secondary antibody conjugated to a fluorochrome and visualization by fluorescence microscopy. In the case of *Borrelia*, a control

antibody directed against a periplasmic component, e.g. flagellin, is required to demonstrate that the outer membrane is intact (e.g. (Kraiczy et al., 2001b; Hefty et al., 2002; Kenedy et al., 2014)). A variation on this technique, designed to help protect the integrity of the outer membrane, is probing with antibodies after suspension of individual bacteria in agarose gel microdroplets (Cox et al., 1996). A second method employs protease (e.g. proteinase K, trypsin) digestion of intact *Borrelia* cells, followed by gel electrophoresis and immunoblotting to determine whether intact protein or fragments of the protein are present after protease treatment (e.g. (Bunikis et al., 2001; Zuckert et al., 2004; Revel et al., 2005; Jewett et al., 2007a)). Again, controls for the integrity of the outer membrane are essential. For both immunofluorescence and surface proteolysis approaches, a control used frequently in the *Borrelia* field is flagellin, as the flagella in *Borrelia* are in the periplasmic space, and not accessible to either antibody or protease if the outer membrane is intact (reviewed in Rosa, 1997; Charon et al., 2012; Zuckert, 2019).

Determination of binding affinities (K_Ds) is another approach used to assess the potential physiologic relevance of interactions of *Borrelia* proteins with DNA or RNA (e.g. regulators or gene expression) or with mammalian substrates (e.g. adhesins). What is considered high vs. low affinity binding is, in part, determined on the basis of physiologically relevant concentrations of the interacting molecules. K_D values for *B. burgdorferi* adhesin-ligand interactions vary considerably (nM to µM ranges) (Brissette et al., 2009a; Fortune et al., 2014; Lin et al., 2015; Ristow et al., 2015), but even lower affinity interactions may still be physiologically relevant, as bacterial cell interactions with host cells or extracellular matrix components tend to be multivalent. While the *in vitro* determination of the cellular locations and the ligand specificities and affinities of adhesins is of value, the ultimate measure of relevance is the assessment of adhesin expression and function in the infection cycle, using mutational analysis and other approaches.

Imaging of Borrelia burgdorferi in murine hosts

In this section, we provide a very brief overview of live imaging approaches for investigating *B. burgdorferi*-host interactions. More in-depth information is provided elsewhere in this volume (chapter 17 by Chaconas *et al.*).

In vivo whole animal imaging of *B. burgdorferi* in living mice has illuminated gene expression patterns and roles of several *Borrelia* proteins in infection (Hyde et al., 2011; Skare et al., 2016). In this biophotonic imaging approach, *B. burgdorferi* strains that do or do not produce a protein of interest are transformed with a *Borrelia*-codon optimized luciferase gene (Blevins et al., 2007), then injected into mice. A major advantage of this approach is that the *Borrelia* can be followed over time using the same mice for repeated measurements. Detection requires injection of luciferin, which, upon oxidation by luciferase, emits light. Quantification of the light emitted can be correlated with other methods of *Borrelia* quantification (*e.g.*, quantitative PCR). Beyond the whole animal approach, *Borrelia* burdens in tissues/organs also can be quantified after euthanasia of the mice.

Two photon microscopy has been used in the *Borrelia* field to image the bacteria the feeding tick, and host skin tissue components in living mice (Harman et al., 2012; Bockenstedt et al., 2014). This method takes advantage of near-infrared excitation of fluorescent molecules, which results in reduced scattering of light in mammalian tissues, providing a stronger signal to noise ratio and greater depth of imaging in tissues than is possible with confocal imaging. In addition, second harmonic generation resulting from absorption of photons by certain tissue components allows imaging of *Borrelia* movement in the context of mammalian tissue in real time.

Spinning disc confocal microscopy provides high-resolution live imaging of living, moving, targets of interest. In the world of bacterial pathogenesis, this approach was first used to image *Borrelia* interactions with the vasculature in living mice (Moriarty et al., 2008). The high resolution images are collected with background fluorescence removed due to filtration of the excitation and emission light through tiny pinholes in a rapidly spinning disc. This increases resolution, particularly for moving targets such as *Borrelia* and living mouse tissue, because of the high image acquisition rate and the fact that light outside the plane of focus is deflected by the disc unless it passes precisely through the pinholes. Several different types of interactions were discerned, and roles for particular *B. burgdorferi* proteins in each of these types of interactions have been defined (Moriarty et al., 2008; Norman et al., 2008; Lee et al., 2014; Kumar et al., 2015).

Genes and Proteins involved in infection and pathogenesis

Adaptation to the mammalian host

When an *Ixodes* vector tick begins feeding on a mammalian host, there is a dramatic alteration in the tick midgut environment (see chapter 11 by Pal *et al.*). The blood meal and proximity to the mammal results in changes in temperature, pH, availability of nutrients, and exposure to antibodies, complement and other host protective factors. In response to changing environmental conditions, *B. burgdorferi* replicates rapidly in the midgut during and following tick feeding, with the number of *Borrelia* increasing from <500 prior to feeding to about 1.7 x 10^5 per tick after 72 hours of feeding, a >300-fold increase (De Silva and Fikrig, 1995). Along with this population expansion, *B. burgdorferi* undergoes dramatic changes in gene expression, primarily but not exclusively *via* the Rrp2-RpoN-RpoS regulatory cascade (Table 1 and chapter 5 by Samuels *et al.*).

Some of these changes are reflected at the mRNA level *in vitro* with changes in culture temperature and pH, as well as incubation in dialysis membrane chambers or in the presence of blood. These *in vitro* environmental changes have been used as models to study gene regulation during the tick to mammal transition, and in part reflect the tick to mammal transition, with upregulation of mammalian infection-associated surface proteins such as OspC, DbpA, DbpB *via* the Rrp2-RpoN-RpoS sigma factor pathway. However, changes in the *in vitro* culture conditions do not affect expression of other genes, including *p66*, or result in downregulation of tick-related gene products OspA and OspB.

Survival in the mammalian host: Nutrient acquisition and metabolism

Once the bacteria have been introduced into the dermis of a mammalian host, they must replicate and disseminate in order to facilitate the tick-vertebrate-tick cycle that maintains the genus in nature. The biosynthetic capabilities of Lyme disease *Borrelia* are limited, so the bacteria require complex media for growth in the laboratory, and are dependent on acquisition of multiple nutrient classes from the host. Uptake of nutrients by cytoplasmic membrane transporters requires that the nutrients permeate through the outer membrane, likely through porins. Although the *in vivo* functional roles of *B. burgdorferi* porins have not been addressed experimentally, indirect data suggest that the porin, or some other function, of the adhesin/porin P66 may be required *in vivo* (Ristow et al., 2015). BB0406, which like P66 potentially may serve as a porin and an adhesin, facilitates *B. burgdorferi* infection (Shrestha et al., 2017; Bista et al., 2020). Several nutrient salvage proteins required for survival in mammalian tissue have been identified (Table 1). PncA, a nicotinamidase that converts nicotinamide (niacin) to nicotinic acid, is needed to provide a precursor for biosynthesis of NAD (Purser et al., 2003). AdeC, an adenine deaminase, which converts adenine to hypoxanthine and, thus, facilitates purine interconversion, and GuaA and GuaB, GMP synthase and IMP dehydrogenase, respectively, are enzymes involved in nucleotide metabolism and are required for infection of mice (Jewett et al., 2007b; Jewett et al., 2009). BBB22 and BBB23 are purine permeases critical for uptake of hypoxanthine, adenine and guanine and also are essential for *B. burgdorferi* infection of mice (Jain et al., 2012; Jain et al., 2015). OppAV, an oligopeptide permease component of the oligopeptide uptake system, is critical for *B. burgdorferi* persistence in the mammalian host (Caimano et al., 2019). BB0318, the ATPase component of a putative riboflavin ABC transport system (Deka et al., 2013), contributes to resistance to oxidative stress and macrophage killing as well as infection in mice (Showman et al., 2016), suggesting an important link between *B. burgdorferi* nutrient acquisition and immune evasion mechanisms. The genes encoding the majority of these proteins are located on genomic plasmids, reinforcing the essential roles of these replicons in the enzootic cycle. The chromosomally-encoded putative glucose specific EIIBC component of a phosphoenolpyruvate phosphotransferase system PtsG is critical for mouse infection (Lin et al., 2012; Khajanchi et al., 2015). The manganese transporter BmtA is also required for infection of both mice and ticks (Ouyang et al., 2009a). It also is possible that the lipolytic/hemolytic enzyme BB0646 (Shaw et al., 2012) serves a nutrient scavenging function as opposed to an intoxication function, as the pathologies seen in Lyme disease are not consistent with those seen in toxin-mediated diseases. Given that many genes in *B. burgdorferi* have not yet been assigned a role or function, it is likely that additional factors and their activities remain to be identified *in vivo*. If, however, their functions also are required for *in vitro* cultivation, they would not be covered in Tn mutant libraries, as the mutants would be lost from the population.

Survival in the mammalian host: Evasion of innate immune responses

B. burgdorferi components are recognized by multiple members of the toll like receptor (TLR) family of receptors and by Nod-like receptors (NLRs). More extensive descriptions of the functions of these receptors can be found in chapter 18 by Bockenstedt *et al.* Upon encountering their cognate ligands, TLRs and NLRs activate signaling pathways that lead to the release of both pro- and anti-inflammatory cytokines and chemokines. Depending upon the host, the initial response typically leads to recruitment of cells of the innate immune system (neutrophils, T cells) that have specific capabilities for killing the organism. *B. burgdorferi* has evolved mechanisms for evading these host defenses as detailed below and elsewhere in this volume.

Most pathogenic microorganisms have developed mechanisms for evading the innate and adaptive immune responses, permitting their survival and proliferation in a hostile host environment (reviewed in e.g. (van der Woude and Baumler, 2004; Norris, 2006; Vink et al., 2012; Norris, 2014; Foley, 2015; Knirel et al., 2015; Obergfell and Seifert, 2015; Phillips et al., 2019; Chaconas et al., 2020)). Immune evasion mechanisms are particularly important in pathogens that cause persistent disseminated infection. These infectious agents include Lyme disease and relapsing fever *Borrelia*, *T. pallidum* (syphilis and related illnesses), *Mycobacterium tuberculosis* and *M. leprae* (tuberculosis and leprosy), protozoa such as *Plasmodium* and *Babesia spp.* (malaria and related diseases), *Leishmania spp.* (leishmaniasis), and *Trypanosoma* (sleeping sickness and Chagas disease), the Herpesvirus family (Herpes Virus I and II, varicella zoster virus, Epstein-Barr virus, and others), and Human Immunodeficiency Virus. Mechanisms of immune evasion are myriad, and several are key contributors to the success of the Lyme disease *Borrelia* as pathogens in humans and in the natural enzootic cycle.

Borrelia Resistance to Reactive Oxygen Species (ROS) and Reactive Nitrogen Species (RNS)

As a way to neutralize infectious agents, hosts utilize a battery of innate and adaptive immune systems to protect themselves. The great majority of infectious agents that cause infection in immunocompetent hosts are equipped, at least to some extent, to circumvent these host responses through a variety of mechanisms. For an extracellular parasite and pathogen like *B. burgdorferi,* the ability to neutralize innate killing mechanisms is critical. One such protective system that an infected host deploys to protect against infection, be it an arthropod or a mammal, is the induction of reactive oxygen species (ROS) or reactive nitrogen species (RNS). Unlike many bacteria, *B. burgdorferi* does not encode a large collection of obvious detoxification enzymes to combat ROS or RNS, but independent groups have demonstrated roles for DNA repair enzymes, a ribonuclease, and transport proteins in resistance to ROS and RNS (Bourret et al., 2011; Bourret et al., 2016; Showman et al., 2016; Ramsey et al., 2017). Consistent with these findings, *B. burgdorferi* infection of mice in which either nitric oxide (NO) production was inhibited or inducible nitric oxide synthase (iNOS) was missing had no effect on the disease process, suggesting that the spirochetes have a robust way of combating this potent form of host defense (Seiler et al., 1995; Brown and Reiner, 1999). Other important factors include the role of tick saliva-specific factors that provide antioxidants and other immune modulatory compounds that promote the establishment of *B. burgdorferi* colonization (Ribeiro et al., 1990; Kuthejlova et al., 2001; Narasimhan et al., 2007). Please note that these topics are covered in more depth in chapters 18 by Bockenstedt *et al.* and 11 by Pal *et al.* in this volume.

B. burgdorferi responds differently to oxidative stress relative to most living systems due, in part, to an apparent lack of any requirement for iron (Fe) (Posey and Gherardini, 2000), at least in laboratory culture and as assessed using methodology available at the time. To date, no enzymes have been identified that require Fe as a cofactor. *B. burgdorferi* uses manganese as a cofactor instead of the iron used by other bacteria (Ouyang et al., 2009a; Troxell et al., 2012; Aguirre et al., 2013; Troxell and Yang, 2013; Troxell et al., 2013; Wagh et al., 2015) and chapter 6 by Gherardini *et al.*). The lack of Fe places *B. burgdorferi* at an advantage relative to other Fe-dependent pathogens (i.e. most) for two reasons. First, DNA damage mediated *via* Fenton-based chemistry is not likely to occur, particularly following exposure to ROS. Prior studies support this contention given that DNA damage was not detected upon the addition of ROS (Boylan and Gherardini, 2008; Boylan et al., 2008), although as stated above, DNA repair enzymes are important for survival of ROS and RNS challenge. Second, a defense strategy termed nutritional immunity, in which particular nutrients (*e.g.* Fe) are tightly sequestered by the host, is largely irrelevant to *B. burgdorferi* with regard to Fe.

B. burgdorferi do not synthesize their own fatty acids and thus are dependent on the host they infect to provide these essential lipid components (Fraser et al., 1997; Casjens et al., 2000). The incorporation of polyunsaturated fatty acids and lipids from hosts they infect into *B. burgdorferi* lipids (Crowley et al., 2013) places these structures at risk of oxidative damage and the toxicity that occurs as a result. Previous studies demonstrated that the oxidation of lipids occurs when *B. burgdorferi* is exposed to ROS under conditions in which DNA remains recalcitrant to oxidative damage (Boylan and Gherardini, 2008; Boylan et al., 2008). During the blood meal, where tick respiration may generate ROS, or under conditions where the host immune response is elevated, represent times that polyunsaturated lipids would place *B. burgdorferi* at risk of oxidative damage and/or death. As such, it behooves *B. burgdorferi* to neutralize this attack. Exactly how *B. burgdorferi* quells lipid oxidation and the deleterious effects that ensue is not known.

The collection of detoxification enzymes that *B. burgdorferi* encodes would appear to be incomplete based on those found in other pathogenic bacteria. For example, no catalases or peroxidases are encoded in the *B. burgdorferi* genome. A single manganese (Mn)-dependent superoxide dismutase (*sodA*), a Mn transporter (*bmtA*), and a riboflavin transporter (*bb0318*), all contribute to resistance to added ROS *in vitro* and *B. burgdorferi* infectivity in mice (Esteve-Gassent et al., 2009; Ouyang et al., 2009a; Showman et al., 2016). A *napA* homologue (also designated *bicA* or *dps*) plays a role in metal homeostasis with distinct roles presented in different studies (Li et al., 2007; Wang et al., 2012). A more recent report suggested that mutants in *napA*/*bicA* were more resistant to added oxidizing compounds when metals were added, particularly Fe and Cu, due to the ability of NapA/BicA/Dps to trap these metals and serve as a sink for them within the spirochete (Wang et al., 2012). The latter study, which used a more refined approach to measuring metal contents of *B. burgdorferi* than that used earlier (Posey and Gherardini, 2000), posited that the sequestered metals contribute to damage when ROS is encountered. Consistent with this, *bicA* mutants, which take up less Fe and Cu, were more resistant to H_2O_2 than their wild-type isogenic parent (Wang et al., 2012).

Other genes whose products are predicted to be associated with resistance to oxidative stress include thioredoxin, thioredoxin reductase, and sulfoxide reductase, although none of these have been functionally characterized. The genome predicts no glutathione or glutathione reductase equivalents to provide redox balance or resistance to oxidizing compounds. However, coenzyme A, in concert with coenzyme A disulphide reductase (denoted as Cdr and encoded by *bb0728*), reduces hydrogen peroxide and thus protects *B. burgdorferi* from damaging ROS (Boylan et al., 2006). These studies also determined that reduced coenzyme A was present in high concentrations within *B. burgdorferi* cells and that Cdr was specific for the thiol-containing coenzyme A, with coenzyme A serving as a functional analogue to glutathione in *B. burgdorferi* to maintain homeostatic redox balance within the cytoplasm of the spirochete.

In contrast to the effect of oxidative exposure, the consequence of nitrosative stress, due to NO exposure, results in significant damage to DNA that then requires nucleoside excision repair genes *uvrB* and *uvrC* to repair the lesions that accrue (Bourret et al., 2011; Bourret et al., 2016). The production of NO is as likely in the tick vector as it is in mammals (*via* iNOS) given the presence of *Ixodes*-specific nitric oxide synthase and nitrophorins. Further, the production of NO also contributes to vasodilation that seemingly benefits the blood meal process (Yang et al., 2014).

With the goal of identifying novel *B. burgdorferi* proteins involved in resistance to ROS and RNS, an unbiased screen was employed using the existing transposon library (Lin et al., 2012) coupled with Tn-seq analysis (Ramsey et al., 2017). From this analysis, 66 genes were identified that contributed to RNS and ROS resistance. Notably, several genes already known to be important for resistance to ROS and RNS were observed in this study. For example, it is known that nucleotide excision repair, mediated by *B. burgdorferi uvrB* and *uvrC*, is needed in response to RNS (Bourret et al., 2011; Bourret et al., 2016), and Tn mutants in *uvrC* and *uvrD* were disproportionally lower in abundance in the Tn-seq data following *B. burgdorferi* exposure to RNS (Ramsey et al., 2017). In addition, several genes of unknown function were identified in the RNS as well as the ROS screens (Ramsey et al., 2017). Two distinct ROS compounds were used, hydrogen peroxide and tertiary-butyl hydroperoxide, with several targets overlapping between these samples (Ramsey et al., 2017). In addition, unique targets were found for both reagents, suggesting that *B. burgdorferi* produces proteins with unique mechanisms that specifically combat either hydrogen peroxide or alkyl peroxides.

Several of the novel targets identified encode transmembrane protein transporters, most of which have not been characterized in any context (Ramsey et al., 2017). One of these proteins, designated BB0164, appears to transport Mn^{2+} similar to BmtA (Ouyang et al., 2009a; Ramsey et al., 2017). The need to accumulate divalent cations, particularly Mn^{2+}, is also seen in *Lactobacillus* species, which, similar to *B. burgdorferi,* also lack many detoxification enzymes (Archibald and Fridovich, 1981). Prior work suggests that Mn, in complex with other molecules, forms anti-oxidant compounds that scavenge ROS that protect the cell from damage (Archibald and Fridovich, 1981). It is also noteworthy that the majority of the genes identified in the initial *in vitro* RNS and ROS screen were also required for experimental infection, suggesting that the proteins they encode are important in the pathogenic capabilities of *B. burgdorferi* (Ramsey et al., 2017). Along these lines, another independent screen designed to track genes essential for survival in the tick identified several genes that were also seen in the RNS/ROS screen, implying that *B. burgdorferi* must negate RNS and ROS exposure in the tick to maintain its ability to survive in order to ultimately transmit to mammalian hosts (Ramsey et al., 2017; Phelan et al., 2019).

Complement evasion by Lyme disease Borrelia

The complement system is a primary arm of the innate immune system comprised of dozens of soluble and membrane-attached host proteins (Figure 3). Complement activation is initiated by surface-associated molecular recognition events that trigger a proteolytic cascade mediated by the action of specialized serine proteases. Complement activation occurs through one of three canonical pathways known as classical, lectin, or alternative (Figure 3). Regardless of the initiating pathway, complement activation (i) leads to the recruitment of professional phagocytes via release of powerful chemotactic peptides (*i.e.*, C3a and C5a), (ii) opsonizes surfaces near the site of activation *via* various complement component fragments (i.e., C4b, C3b, iC3b, C3dg, and C3d), and (iii) results in the formation of a pore-like structure, the membrane attack complex (MAC or C5b-9), that integrates into target membranes, leading to lysis of the invading pathogen. Vector-transmitted microorganisms like *Borrelia* that traffic in fluids where complement proteins are present at high concentrations (i.e. blood, interstitial fluids, etc.) must overcome destruction by complement.

Lyme disease spirochetes have evolved an arsenal of outer surface proteins with demonstrated anti-complement function *in vivo* (Figure 4 and Table 1). We also refer the reader to three recent reviews of Lyme disease *Borrelia* interactions with the host complement system (Lin et al., 2020a; Lin et al., 2020b; Skare and Garcia, 2020). Each of these *Borrelia* outer surface proteins exerts its inhibitory activity at proximal (i.e. initiation pathways), central (i.e. C3b amplification loop), or distal (i.e. terminal sequence) positions of the cascade. Complement evasion mechanisms associated with *Borrelia* outer surface proteins can be broadly characterized by one of three activities: (i) proteins that recruit endogenous complement regulators (CspA, CspZ, ErpP, ErpC, ErpA), (ii) proteins that recruit endogenous proteases and target their proteolytic activities towards complement protein substrates (CspA, CspZ, ErpA, ErpC, ErpP, BBA70), or (iii) proteins that inhibit the activity of a complement protein by directly binding to it (BBK32, BAD16, OspC, BGA66, BGA71). The collective activity of these proteins affords *Borrelia* one of the most diverse and functionally overlapping complement evasion repertoires known.

Interestingly, Lyme disease spirochetes vary in their susceptibility to killing by nonimmune human and various mammalian sera *in vitro*, indicating a differential ability of these spirochetes to prevent MAC formation (reviewed by Kurtenbach et al., 2002; Lin et al., 2020a; Lin et al., 2020b). All human-pathogenic species including *B. burgdorferi*, *B. afzelii*, *B. spielmanii*, *B. bavariensis* (formerly classified as *B. garinii* OspA serotype 4), and *B. mayonii,* as well as *B. japonica* and *B. bissettiae,* resist complement-mediated killing *in vitro* (reviewed by (Kraiczy, 2016b, a; Walter et al., 2019). Although pathogenic to humans, *B. garinii* predominantly displays a susceptible phenotype *in vitro* like the non-pathogenic genospecies *B. valaisiana* and *B. lusitaniae*. How *B. garinii* overcomes the cidal activity of human complement remains unresolved, but one possibility is that protein(s) critical for complement evasion are produced *in vivo* but not *in vitro.* In addition, recruitment of host-derived (plasminogen) or tick-derived proteins during transmission (Tick Salivary Lectin Pathway Inhibitor, TSLPI; Salp15, Salp20) (Schuijt et al., 2008; Schuijt et al., 2011) or specific pathogen-associated factors may protect this particular genospecies from complement attack during natural infection. More recently, complement resistance has

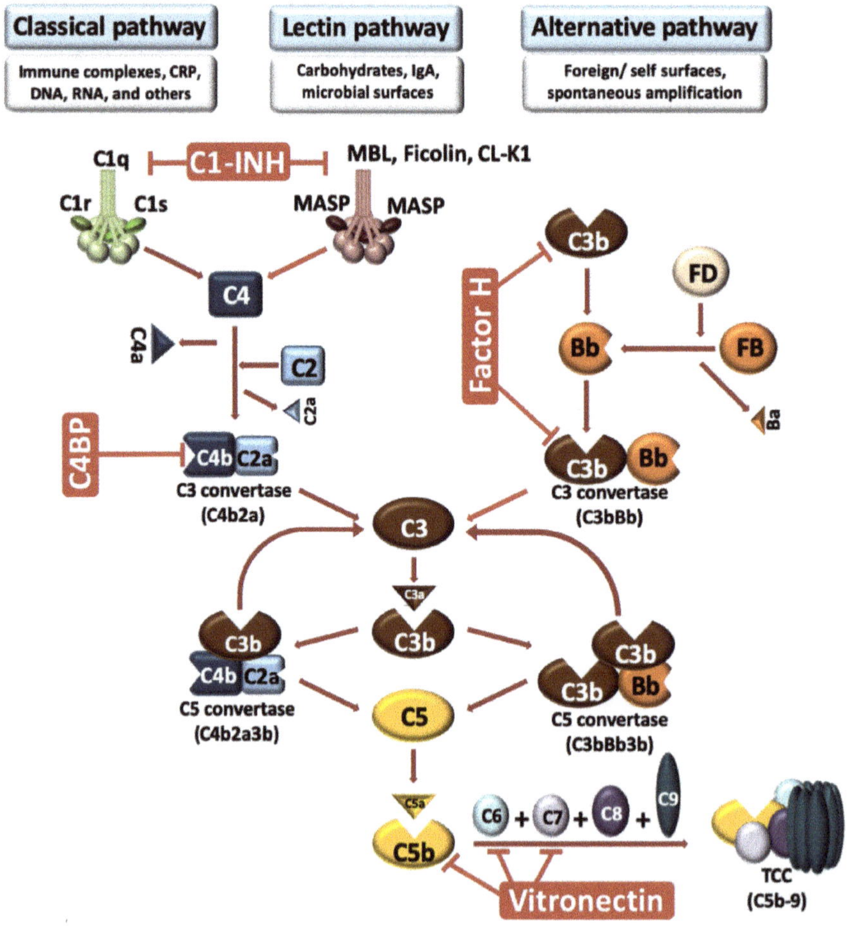

Figure 3. Schematic representation of the human complement cascades. Complement is activated by three pathways: the classical, lectin, and alternative pathway. In each pathway, a series of initial binding steps followed by proteolytic steps ultimately results in activation of the terminal pathway and lysis of the target cells. CRP, C-reactive protein; C1-INH, C1 esterase inhibitor; MBL, Mannose-binding lectin; MASP, MBL-associated serine protease; C4BP, C4b-binding protein; TCC, terminal complement complex. Parts of this figure are adapted from Kraiczy, P. (2016), "Travelling between two worlds: complement as a gatekeeper for an expanded host range of Lyme disease spirochetes" Vet. Sci., 3:12; doi:10.3390/vetsci3020012. Permission to reproduce under CC 4.0 (https://creativecommons.org/licenses/by/4.0).

been linked to alleles from distinct *Borrelia* species. The best example is CspA. CspA from *B. burgdorferi* and *B. afzelii* confers resistance to mammalian complement, whereas CspA from *B. garinii* provides resistance to avian serum (Hart et al., 2018).

Borrelia factors affecting activation of the classical and lectin pathways

The classical and lectin pathways of complement are activated by pattern recognition molecules capable of distinguishing specific molecular structures on the surfaces of microbes, including *Borrelia*. The classical pathway is controlled by the action of a multi-protein complex known as the first component of complement, C1. C1 consists of the pattern recognition subcomponent, C1q, in complex with a heterotetramer of two serine proteases ($C1r_2C1s_2$). Following binding of C1q to activating ligands such as IgG and IgM immune complexes, C1r is converted from a zymogen form into an active protease where it then cleaves and activates the C1s zymogen. The classical pathway is, therefore, linked to the adaptive immune response, as specific antibodies are developed in response to a pathogen. Likewise, the lectin pathway is triggered by the pattern recognition molecules mannose-binding lectin (MBL), ficolins, or collectins, which recognize foreign carbohydrates resulting in activation of the MBL-associated proteases (MASPs). Both C1s and MASP-2 cleave C4 and C2 and, thus, the classical and lectin pathways of complement intersect at the formation of the C3 convertases (i.e. C4bC2b).

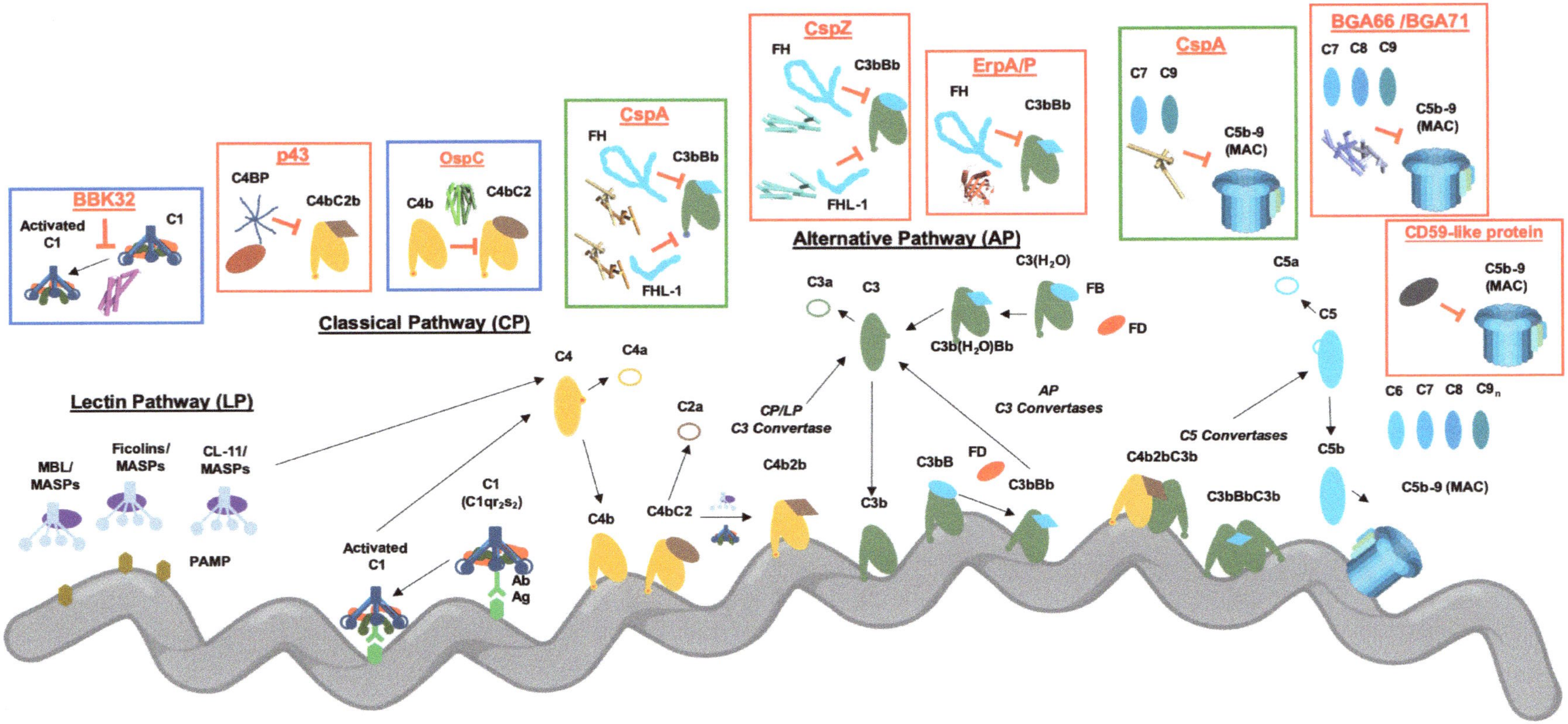

Figure 4. ***Borrelia burgdorferi*** **subversion of the complement system**. To evade complement, Lyme disease spirochetes produce outer surface lipoproteins that interfere with host complement by inhibiting complement activity directly or binding to host-derived regulators of complement activity (RCA). Several *Borrelia* inhibitors block the upstream initiation steps, including the CP-specific inhibitor BBK32, which binds to C1r and traps C1 in a zymogen state. OspC binds C4b and interferes with the activation of both the CP and LP. Lyme disease spirochetes also produce p43, which downregulates the CP and LP by recruiting the RCA called C4b-binding protein (C4BP). CspA and CspZ bind the negative regulators of the AP known as factor H (FH) and factor H-like protein 1 (FHL-1). ErpA and ErpP also bind to FH, but not to FHL-1. *Borrelia* produce at least four proteins that block the formation of the MAC complex. CspA binds C7 and C9 in a FH-independent manner and blocks C9 polymerization. Two homologous proteins from *B. bavariensis*, BGA66 and BGA71 also block C9 polymerization by binding to C7, C8, and C9. Finally, a surface protein that exhibits activity comparable to CD59 (i.e., CD59-like) was investigated, but the gene encoding this protein has not been identified. Proteins known to affect experimental infection by *B. burgdorferi* are outlined in blue, those that are important in transmission from *Ixodes* ticks are framed in green, and those whose functions *in vivo* are not defined are shown in red. PAMP: pathogen-associated molecular pattern; Ab: antibody; Ag: antigen; CL-11: collectin-11. This figure is reproduced from Skare J.T. and Garcia B.L. (2020) “Complement Evasion by Lyme Disease Spirochetes” Trends Microbiol. S0966-842X(20)30129-3. doi: 10.1016/j.tim.2020.05.004 with permission conveyed through Elsevier and Copyright Clearance Center.

Relatively few examples of pathogen-derived molecules with specific activity against the upstream activation steps of the classical and lectin pathways have been described. Lyme disease *Borrelia* spirochetes are outliers in this regard, however, as they target both pathways through two unrelated outer surface proteins, OspC and Bbk32. OspC was shown to bind C4b and prevent formation of the classical/lectin pathway C3 proconvertase by blocking binding of C2 (Caine et al., 2017). In this manner, OspC inhibits both pathways simultaneously. Consistent with a role in complement evasion, an isogenic *ospC* mutant of *B. burgdorferi* strain B31-A3 (i.e. B31-A3Δ*ospC*) exhibited reduced survival in the blood of intravenously (i.v.) inoculated mice by 30 minutes post-inoculation relative to wild-type. B31-A3 derived strains that constitutively produced OspC from B31-A3 or N40 D10/E9 showed prolonged survival. Yet to be determined is whether this complement-inhibiting activity of OspC is mechanistically tied to its antiphagocytic activity (Carrasco et al., 2015). OspC from *B. garinii* PBr also was protective in this system albeit to a lesser extent than OspCs of B31-A3 and N40 (Caine et al., 2017).

BBK32 is a multi-functional *B. burgdorferi* outer surface protein with roles in adhesion to fibronectin (Fn) and glycosaminoglycans (GAGs) conferred by distinct domains of the protein (Lin et al., 2015; Xie et al., 2019) (*see* adhesin section below). More recently, BBK32 was shown to form high-affinity interactions with the human complement C1 complex by binding specifically to the C1r protease subcomponent, thereby inhibiting the classical complement pathway. The C1r-binding activity of BBK32 is mediated by a third domain of the protein distinct from those that mediate binding to GAGs and fibronectin (Garcia et al., 2016; Xie et al., 2019). Evidence that BBK32 promotes bloodstream survival of *B. burgdorferi* in mice was reported independently (Caine and Coburn, 2015), supporting the *in vivo* relevance of its complement-inhibiting activity. Using surface plasmon resonance binding assays and human serum-based assays of complement function, it was shown that the C-terminal region of BBK32 (i.e. BBK32-C) is responsible for all BBK32-mediated C1/C1r interaction and anti-complement activities (Garcia et al., 2016; Xie et al., 2019). Thus, BBK32 possesses a modular architecture whereby its interactions with glycosaminoglycans, fibronectin, and C1r are mediated by non-overlapping binding sites. Strengthening the case that BBK32 binds C1, a serum-sensitive, high passage strain of *B. burgdorferi* missing linear plasmid-encoded outer surface proteins, including BBK32 (*i.e.*, strain B314), was protected from classical pathway-mediated lysis when transformed with a shuttle vector containing *bbk32* (Garcia et al., 2016). However, an isogenic *bbk32* mutant in an otherwise infectious strain of *B. burgdorferi* was serum-resistant, pointing to additional factors (*e.g.*, OspC) responsible for evasion of complement by Lyme disease spirochetes.

The complement inhibitory activities of BBK32 homologs in the Lyme spirochetes *B. afzelii* (*i.e.*, BAD16) and *B. garinii* (*i.e.*, BGD19) also were evaluated (Xie et al., 2019). While BAD16 exhibited similar properties to BBK32 *in vitro*, recombinant *B. garinii* BGD19 failed to inhibit the classical pathway. BGD19 also did not inhibit the classical pathway when expressed on the surface of *B. burgdorferi* B314. The three-dimensional structure of the complement inhibitory region of BBK32, BBK32-C, was solved by x-ray crystallography, and the binding site on C1r was mapped to the serine protease domain (Xie et al., 2019). A recombinant BBK32 chimeric protein with site-directed changes designed to mimic BDG19 at three surface-exposed residues exhibited BGD19-like activities in all assays, indicating that one or more of these residues in native BBK32 contributes to complement inhibition.

Borrelia factors affecting activation of the alternative pathway

Unlike the lectin and classical pathways, which rely on pattern recognition molecules to initiate the cascade, alternative pathway complement activation occurs by a mechanism known as tick-over. Tick-over is a process whereby spontaneously hydrolyzed C3 forms fluid phase C3 convertases of the form C3(H_2O)Bb, leading to surface-bound alternative pathway C3 convertases (C3bBb). The main negative regulators of C3 convertase formed by the alternative pathway are Factor H (FH) and FH-like protein (FHL-1) (Zipfel and Skerka, 2009). FH and FHL-1 bind directly to C3b and stimulate the release of the protease fragment (i.e. complement component Bb) through a process known as decay accelerating activity. Additionally, FH and FHL-1 each act as cofactors for factor-I mediated degradation of C3b. These inhibitory activities protect self surfaces from excessive activation and harmful attack by activated effector molecules (Zipfel and Skerka, 2009).

Borrelia produce outer surface proteins collectively called complement regulator-acquiring surface proteins or CRASPs that recruit FH and FHL-1 to the

Borrelia surface, thereby hijacking their anti-complement activities. CRASPs are members of at least two surface protein families and bind different proteins of the FH protein family (see above) (Kraiczy et al., 2001b; Kraiczy et al., 2001c). In this model of complement evasion, *Borrelia* cells usurp the host regulators of the innate immune system by binding FH and FHL-1 *via* CRASPs, thereby terminating the accumulation of activated C3b molecules on the bacterial cell surface. Resistant spirochetes with less C3b deposited on their surfaces would have decreased opsonization (reducing phagocytosis) and a significantly reduced formation of the complement membrane attack complex (decreasing complement-dependent lysis).

Most data accumulated to date support the importance of CspA (also CRASP-1 and BBA68) and its orthologous proteins as inhibitors of the complement cascades in diverse *Borrelia* species. Disruption of *bba68* (encoding CspA) in the complement resistant *B. burgdorferi* strain B31cF resulted in sensitivity to killing by human complement, while complementation with the shuttle vector pKFSS-1::CRASP-1 restored resistance *in vitro* (Brooks et al., 2005). Similarly, transformation of the serum-sensitive *B. garinii* strain Bg50 with pKFSS-1::CRASP-1 rendered the strain resistant to killing by human complement (Brooks et al., 2005). Further gain-of-function studies revealed that the FH and FHL-1 binding CspA proteins from *B. afzelii*, *B. spielmanii,* and *B. mayonii* also promote complement resistance in a serum sensitive *B. garinii* background (Hammerschmidt et al., 2014; Walter et al., 2019). RT-PCR and western blot analysis indicated that CspA is not expressed during experimental infection of mice with different North American *B. burgdorferi* strains, and that human Lyme disease patients do not produce antibodies against CspA (McDowell et al., 2006), bringing into question whether CspA is expressed during mammalian infection. However, a different group found that antibodies from Lyme disease patients react specifically with non-denatured, recombinant CspA in a line blot assay but not with the denatured protein in western blots (Rossmann et al., 2006). Further studies also provided evidence that CspA is expressed by spirochetes 72 hours post infection at the tick bite site but not at later time points (von Lackum et al., 2005; Hart et al., 2018). The collective results support the notion that CspA-mediated binding of FH during the blood meal permits survival of *Borrelia* in the feeding tick and during transit to the host (Hart et al., 2018). Binding of mouse FH to CspA of *B. burgdorferi* B31 and *B. afzelii* PKo allows the bacterium to survive in ticks fed on mice, while the parental *ΔcspA* or a derivative complemented with *cspA* from *B. garinii* ZQ1 are not detectable in infected nymphs. These findings suggested that the *Borrelia* binding to host-derived FH determines survival in fed ticks, a key prerequisite for transmission to the mammalian host.

Another member of the CRASP family, CspZ (CRASP-2, BBH06), binds FH and FHL-1 and also facilitates complement resistance in gain-of-function experiments with non-infectious strains (*i.e., B. burgdorferi* B313 and *B. garinii* G1) (Hartmann et al., 2006; Siegel et al., 2008). CspZ has been surface-localized on intact cells by immunofluorescence microscopy and FACS analyses (Hartmann et al., 2006; Marcinkiewicz et al., 2019), and by trypsin digestion (Hartmann et al., 2006; Coleman et al., 2008; Dowdell et al., 2017). In studies conducted with different infectious clones of *B. burgdorferi* B31 (Bykowski et al., 2007; Coleman et al., 2008), the *cspZ* gene is expressed by *B. burgdorferi* in all mouse tissues examined but not in unfed or feeding ticks. However, deletion of the *cspZ* gene did not affect mouse infectivity or disease, and did not reduce resistance to complement-mediated killing *in vitro* (Coleman et al., 2008). In contrast, it was recently shown that addition of human blood to the culture medium increased hematogenous dissemination and colonization of wild-type *B. burgdorferi* strain B31-A3 but not B31-A3 *ΔcspZ* in mice and *C. coturnix* quails (Marcinkiewicz et al., 2019). The addition of blood to the medium increased CspZ levels, suggesting that the pre-induction of CspZ might allow *B. burgdorferi* survival immediately after injection, whereas those grown conventionally cannot escape clearance by the host.

Three additional surface lipoproteins encoded on cp32 plasmids, BbCRASP-3, BbCRASP-4, and BbCRASP-5 have been shown to bind FH and certain factor H related (FHR) proteins (Kraiczy et al., 2001a; Alitalo et al., 2002; Stevenson et al., 2002; Kraiczy et al., 2003; Metts et al., 2003). *In vitro*, recombinant versions of these proteins bound FH in ELISA and surface plasmon resonance analyses, but the same proteins exposed on the *Borrelia* surface did not bind FH and gain-of-function strains remained complement sensitive (Siegel et al., 2010). No specific role for any of these proteins *in vivo* has been demonstrated, but analyses of these genes are complicated by the high degree of redundancy in the *B. burgdorferi* genome.

Borrelia factors affecting activation of the terminal pathway

Different *Borrelia* factors involved in inactivation of the terminal pathway have been described, including CspA of *B. burgdorferi*, the CspA orthologs of *B. afzelii*, *B. spielmanii*, and *B. mayonii*, and BGA66 and BGA71 of *B. bavariensis* (Hallstrom et al., 2013; Hammerschmidt et al., 2014; Hammerschmidt et al., 2016; Walter et al., 2019). Common to all of these proteins are: (i) binding to the late complement components C7, C8 or C9, and the terminal complement complex (TCC), (ii) termination of auto-polymerization of C9, (iii) inhibition of the formation of the TCC and (iv) promotion of complement resistance in gain-of-function strains (Hallstrom et al., 2013; Hammerschmidt et al., 2014; Hammerschmidt et al., 2016; Walter et al., 2019). Despite being members of the PFam54 protein family (the largest paralogous gene family in the Lyme disease spirochete genome), which includes CspA, BGA66 and BGA71 did not bind complement regulators FH or FHL-1. To date, their possible roles in a mouse infection model have not been investigated.

Role of host complement in control of Borrelia infectivity

The involvement of the classical and lectin pathways in Lyme borreliosis recently has been studied in animal models of infection using complement-deficient mice. The lectin pathway was evaluated in *B. burgdorferi* infection of mannose-binding lectin (MBL) deficient mice (Coumou et al., 2019). An early protective role was shown for MBL as MBL-deficient mice exhibited significantly higher bacterial loads in skin tissues of both needle- and *I. scapularis* infection models (Coumou et al., 2019). However, later infection time points showed similar bacterial loads in heart, joints, and bladder in WT vs. MBL-deficient mice. The role of classical pathway initiation in response to *B. burgdorferi* infection was investigated using C1q deficient mice (Zhi et al., 2018). As measured using bioluminescent *in vivo* imaging, C1q deficient mice harbored higher bacterial loads at 10 days relative to WT mice. Significantly increased bacterial loads also were measured by quantitative PCR in C1q deficient vs. WT mice at day 21 in the skin, joint, and heart but not lymph nodes. At day 28, loads were higher in joints and hearts, but not skin and lymph nodes in C1q-deficient mice. Collectively, these studies suggest that the upstream pattern recognition molecules of the complement system play roles in controlling *B. burgdorferi* infections *in vivo*, although individually, the roles are relatively minor.

The role of the alternative pathway in host response to Lyme borreliosis has been studied using mouse models of disease. No significant differences in *B. burgdorferi* loads could be detected at any time post-infection in mice genetically deficient in complement factor B relative to WT mice (Woodman et al., 2007). In the same study, the use of FH deficient mice demonstrated that FH is not required for *B. burgdorferi* to infect mice. However, a confounder in these studies is that FH-deficient mice have extremely low levels of C3 due to the high rate of spontaneous C3 consumption (Pickering et al., 2002). Infection of C3 deficient mice resulted in significantly increased bacterial loads in the ear tissue during the first two weeks of infection, but not in heart or joint tissue. As C3 activation is central to the complement cascade, C3 deficiency affects all three complement initiation pathways. Thus, while the factor B and FH deficient mouse studies suggested a muted role for the alternative pathway in control of *B. burgdorferi* burden in the murine host, the C3-deficient mouse infection experiments may indicate a role for the classical or lectin pathways, even prior to a strong humoral response (Woodman et al., 2007). Indeed, others have noted a role for C3 in experimental Lyme borreliosis. For example, C3 deficient mice showed generally increased bacterial loads compared to wild-type mice when needle infected with *B. burgdorferi* (Lawrenz et al., 2003). The same study also showed that a complement-sensitive strain of *B. garinii* exhibited a lower median infectious dose in C3 deficient mice compared to wild-type. These results indicate that complement is a potential barrier for *B. garinii* infection in mouse models and that C3 may play an important role in early dissemination of *B. burgdorferi*. In contrast, infections of C3-deficient mice deficient with *B. burgdorferi*, *B. afzelii*, and *B. bavariensis* did not show significant differences compared to WT mice, with the one exception of increased *B. burgdorferi* loads in joints (van Burgel et al., 2011). In the same study, complement-susceptible strains of *B. burgdorferi* remained noninfectious in C3 deficient mice. Finally, the involvement of the host terminal pathway in borreliosis was evaluated *in vivo* using mice deficient in C5. These studies showed that activation of C5 is not necessary to control murine borreliosis (Bockenstedt et al., 1993). Collectively, these studies suggest that complement regulation does play a modest role in determination of outcome in experimental Lyme borreliosis, but that additional *Borrelia* virulence determinants contribute to the host-pathogen standoff of persistent infection.

Regulation of expression of complement inhibiting factors
Several of the genes that encode complement-inhibitory proteins are modulated by both well-characterized and novel regulatory systems, as reviewed in (Kraiczy and Stevenson, 2013). CRASPs have different patterns or expression, but all are produced by *B. burgdorferi* in skin at the tick feeding site, and none in the unfed tick (Kraiczy and Stevenson, 2013). Expression of the multifunctional OspC and BBk32, along with that of many other gene products, is transcriptionally regulated by the RpoS/RpoN/Rrp2 and BosR systems (Hubner et al., 2001; Yang et al., 2003; Fisher et al., 2005; Caimano et al., 2007; Ouyang et al., 2008; Hyde et al., 2009; Ouyang et al., 2009b; Samuels, 2011; Caimano et al., 2019). Both *ospC* and *bbk32* are known to be expressed during the blood meal and early in mammalian infection. This regulatory scheme is well studied and is described in greater detail in chapter 5 by Samuels *et al.*

Evasion of Adaptive Immunity: Antigenic Variation of VlsE

The vertebrate host mounts a continually expanding (Craft et al., 1986), yet non-protective, response to *B. burgdorferi* during infection, indicating that immune evasion is a critical virulence determinant in this organism. Antigenic variation as a means to escape clearance by the adaptive immune system has been noted in multiple bacterial, protozoan, and fungal pathogens (reviewed in (Deitsch et al., 2009)). Antigenic variation was known to occur in the relapsing fever *Borrelia* species (e.g. (Plasterk et al., 1985; Burman et al., 1990; Schwan and Hinnebusch, 1998; Dai et al., 2006)), and accounts for the relapses for which the disease was named. The relapsing fever *Borrelia* antigenic variation entails a gene conversion type of recombination, copying a complete silent gene encoding a variable membrane protein (Vmp), into a single expression site near the telomere of one of the genomic linear plasmids (Barbour et al., 2000). Multiple *vmp* genes, which encode two classes of lipoproteins designated large (Vlp) or small (Vsp), are maintained as silent copies elsewhere in the genome. In most recombination events, a *vlp* or *vsp* gene recombines randomly *via* upstream and downstream homology sites into the single expression site (Barbour, 1991; Barbour et al., 2000; Dai et al., 2006; Norris, 2006). In some instances, chimeric genes are formed. The number of possible variants is, thus, largely dependent on the number of *vmp* genes in the genome. One particular *vmp* gene, designated *vtp* or *vsp33*, has its own expression site and does not undergo recombination. Vsp33 is required for transmission of spirochetes to a new vertebrate host (Raffel et al., 2014; Schwan et al., 2020).

In 1997, the ground-breaking discovery and description of the *vls* (*vmp*-like sequence) locus (Zhang et al., 1997) brought the Lyme disease *Borrelia* into the group of pathogens able to vary a surface protein. This locus as characterized in *B. burgdorferi* strain B31 consists of a single expression site (*vlsE*) near the telomere of a linear genomic plasmid, lp28-1, plus an adjacent set of 15 silent cassettes in the opposite orientation (*vls2-16*). The silent cassettes have a high degree of sequence similarity (76.9 to 91.4% at the amino acid level) to the central cassette region of the expression site (Figure 5). Although the *vls* system has been detected in all strains and species of Lyme disease *Borrelia* examined, the replicon in which the *vls* locus is contained and the number of silent cassettes vary between isolates (Kawabata et al., 1998; Wang et al., 2001; Iyer et al., 2003; Wang et al., 2003; Norris, 2014; Casjens et al., 2017). Notably, the locus is always near a telomere of a linear plasmid, as is true for the *vmp* system in the relapsing fever *Borrelia*, but in other ways it is quite distinct. *Vls* has been calculated to theoretically generate on the order of 10^{40} possible variants in *B. burgdorferi* strain B31 (Verhey et al., 2018a), many orders of magnitude greater than *vmp*. In contrast to the *vmp* system, *vlsE* undergoes random, segmental gene conversion (also termed "switching", or "templated changes") in which portions of any of the silent cassettes are copied into the variable domain of the *vlsE* expression site, resulting in cellular-level unique mosaics of VlsE due to continuous independent gene conversion events throughout infection (Figure 5) (Zhang and Norris, 1998a, b; Verhey et al., 2018b, a, 2019). Unfortunately for researchers investigating the molecular mechanisms involved in antigenic variation of VlsE, no switching has been detected during *in vitro* culture, and the *in vivo* signals that activate this process are currently unknown. Two excellent reviews provide a comparison of the antigenic variation systems of the relapsing fever and Lyme disease *Borrelia* (Norris, 2014) and a compilation of the unique features of the *vls* locus (Chaconas et al., 2020).

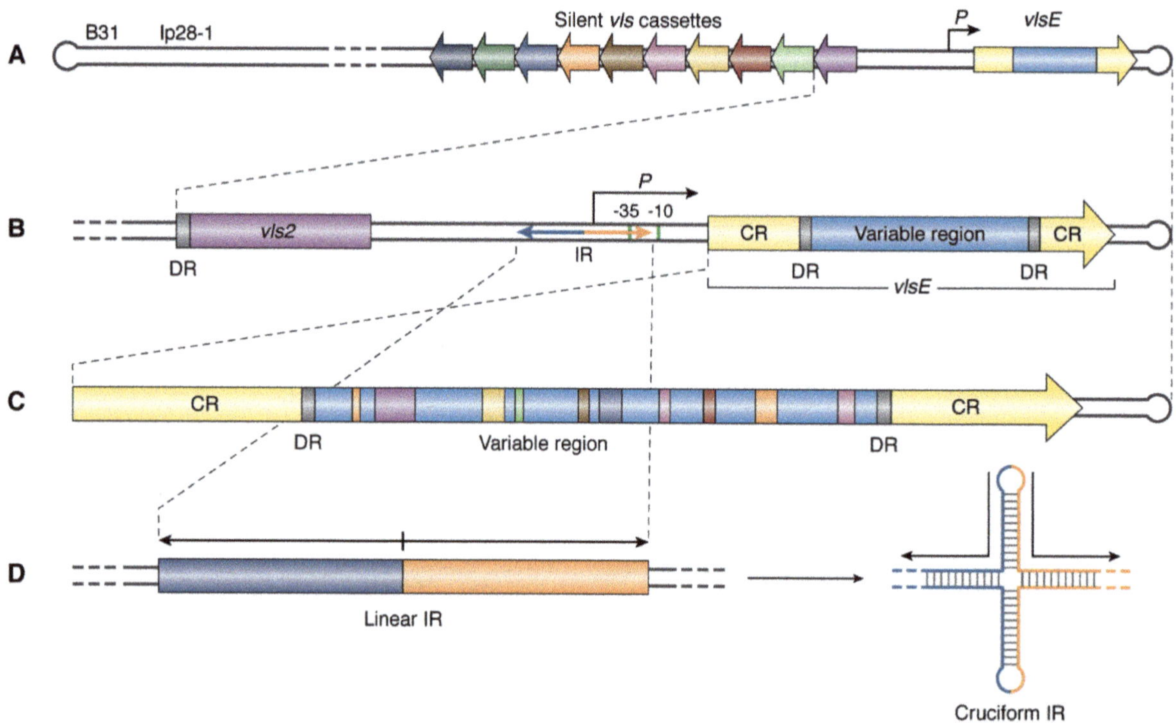

Figure 5. Schematic of the *vls* antigenic variation locus of *B. burgdorferi* strain B31. A, the *vls* expression locus (*vlsE*) with its promoter (P) is located 82 bp from the right covalently closed hairpin end of the linear plasmid lp28-1. To the left of the promoter and intergenic region (gray) are 15 silent cassettes carrying information corresponding to the variable region of *vlsE* and situated in the opposite orientation. B, the *vlsE* region is shown in greater detail, with the constant regions (CR) shown in yellow and the variable region, which corresponds to the information carried in the *vls* cassettes, shown in blue. The variable region is flanked by 17-bp direct repeats (DR)s. To the left of *vlsE* is its promoter, with the -10 and -35 sequences shown as green bars. Also shown by the bidirectional arrow is a 100-bp perfect inverted repeat (IR) that partially overlaps the promoter. C, an enlargement of the vlsE gene shows the product of multiple recombinational switching events that result in the copying of genetic information from the silent cassettes into the expression locus, producing a mosaic *vlsE* carrying information from a number of the silent cassettes. D, an IR found in the promoter region of the *vlsE* gene is shown in its normal linear configuration and as an extruded cruciform promoted by negative supercoiling or DNA unwinding from replication or transcription. This figure is republished from Chaconas, G., Castellanos, M. and Verhey, T. B. (2020). "Changing of the guard: How the Lyme disease spirochete subverts the host immune response" J Biol Chem 295(2): 301-313, with permission of The Journal of Biological Chemistry, © the American Society for Biochemistry and Molecular Biology, permission conveyed through Copyright Clearance Center, Inc.

B. burgdorferi clones lacking lp28-1 are eliminated from infected immunocompetent mice within 2-3 weeks following inoculation, but survive in the absence of adaptive immune system pressure in SCID mice (Labandeira-Rey et al., 2003; Purser et al., 2003); this result is consistent with a role in protection against adaptive immunity. As expected, transformation of a strain missing lp28-1 with lp28-1 restores full infectivity in immunocompetent mice (Grimm et al., 2004a). Refined analysis of the specific role of the *vls* system was achieved using engineered versions of lp28-1 that lacked either the *vls* sequences or the remaining portion of the plasmid (Bankhead and Chaconas, 2007). The Δ*vls* clones exhibited decreased infectivity in WT mice similar to that seen with a lp28-1 negative clone, with no positive cultures obtained at 21 or 28 days post infection. In contrast, a clone lacking the remainder of lp28-1 but carrying an intact *vls* locus was fully infectious in WT mice by needle inoculation. The Δ*vls* clones were recovered from every tissue site tested 28 days after infection of SCID mice; this result paralleled the infectivity of lp28-1 minus clones in SCID mice observed previously (Labandeira-Rey et al., 2003; Purser et al., 2003; Lawrenz et al., 2004). Since SCID mice cannot elicit an adaptive immune response, but have intact innate immune responses, these results corroborated other evidence that the role of *vls* is primarily evasion of adaptive immunity. VlsE also protects (masks) other surface-exposed proteins that do not vary (possibly due to important functional constraints) from host antibodies (Rogovskyy and Bankhead, 2013; Lone and Bankhead, 2020), and this variable protein appears to serve as an immune-evasive barrier critical to maintenance of *B. burgdorferi* in the natural tick-vertebrate infection cycle (Rogovskyy and Bankhead, 2013; Rogovskyy et al., 2015). Of note, the invariable major lipoprotein produced by *B. burgdorferi* in the tick appears to have a role in protecting *B. burgdorferi* from acquired immunity in the tick (which does not have an adaptive immune system), which also is relevant to maintenance in the natural cycle (Battisti et al., 2008).

The timing of the requirement for *vlsE* also is consistent with a role in evasion of adaptive immunity in that antibody responses are first detectable at approximately 2 weeks of *B. burgdorferi* infection. This is also the time frame at which OspC production begins to be detrimental in immunocompetent mice (Tilly et al., 2006; Xu et al., 2006; Tilly et al., 2013), as the *ospC* sequence of an individual *Borrelia* cell and its progeny does not vary and so is targeted by the adaptive immune response. VlsE previously was shown to succeed OspC in immunocompetent mice, but in SCID mice infected with a *vlsE*-deficient strain, *ospC* expression was maintained (Tilly et al., 2013). Since OspC has been shown to inhibit the innate complement host defense system, while VlsE renders the adaptive immune response essentially irrelevant to clearance of infection, the OspC-Vls succession is consistent with *B. burgdorferi*'s requirements for immune evasion to establish and maintain persistence. Intriguingly, both OspC and VlsE were selected for *in vivo* vascular adhesion using a phage display library of *B. burgdorferi* DNA (Antonara et al., 2007). This has been verified for OspC (Lin et al., 2020c), and hints that VlsE may also have an adhesin function in mice.

Within the *vlsE* expression site, there are invariant or "constant" regions at the 5' (amino terminal) and 3' (carboxy terminal) ends flanking a central variable region. Additional features of the best-characterized *vls* locus, that of the B31 isolate, include direct repeats flanking the variable region of *vlsE* and the silent copies of *vls* genes, but this feature is not conserved among all strains (Chaconas et al., 2020). A ~100 base pair inverted repeat upstream of the coding sequence and overlapping with the -35 promoter element is present in all *vls* loci sequenced to date (Hudson et al., 2001; Castellanos et al., 2018). The Vls protein is the target of antibody development during infection, with antibody responses targeting both variable and invariable regions (McDowell et al., 2002). Small regions scattered within the variable region are not variable, and one of these segments encodes the "C6" peptide that is a target of serologic testing used for diagnosis of Lyme disease (see, for example, (Branda et al., 2017; Marques, 2018; Wormser et al., 2018; Lipsett et al., 2019)). The variable region largely maps to the membrane-distal protein surface in the crystal structure of VlsE, while the invariant amino and carboxy termini include the major alpha-helices that comprise the framework of the surface-localized protein that is important for dimerization (Eicken et al., 2002; Verhey et al., 2018a, 2019).

Although point mutations may also occur (Sung et al., 2001; Coutte et al., 2009; Verhey et al., 2018b), most sequence changes can be attributed directly to substitution of silent cassette sequences of varying lengths of ~1-140 base pairs (Verhey et al., 2018b), and so are templated changes. These events can be detected as early as 4 days post infection in mice, and by 28 days each clone re-isolated from mice exhibited a different sequence resulting from separate recombination events (Zhang and Norris, 1998b). Recent advances in DNA sequencing technologies and, more importantly, comprehensive analysis of the enormous volumes of data generated by sequencing reads, have enabled the generation of vast datasets that illuminate *vlsE* changes to the single nucleotide level in *B. burgdorferi* isolated from multiple tissue sites in independently infected mice at multiple time points (Verhey et al., 2018a, b, 2019).

The use of long-read sequencing combined with multiple reads for each template strand enabled reliable assembly and analyses of the sequence data that had previously been impossible with older technologies, a particularly daunting task in light of the multicopy *vls* genes arranged in linear fashion on a single genomic replicon. These data revealed that essentially all *vlsE* sequences isolated from infected mice were unique, i.e. that no single sequence change predominated, or was strongly selected for, in murine infection (Verhey et al., 2018a, b, 2019), similar to earlier findings (Zhang and Norris, 1998b). The deep sequencing used in the more recent studies also suggested that, once disseminated, individual bacterial cells do not frequently re-enter the circulation to seed additional sites (Verhey et al., 2018a).

The data also revealed that gene conversion events (templated changes) were not randomly distributed in the variable region of *vlsE*, but instead occurred more frequently proximal to sites of previous events (Verhey et al., 2018b). Homology does appear to play an important role in segmental gene conversion, as particular silent cassettes flanked by direct repeats with the highest similarity to those in *vlsE* also have the highest frequency of being used as templates for segmental gene conversion (Verhey et al., 2018b). As noted by the authors, however, it is possible that a physical constraint in generating the recombination structure favors usage of the same silent cassettes. It is possible that maintenance of the similarities in the direct repeats is mutually selected for based on the physical restraints placed on the DNA by the recombination structure and/or machinery. Interestingly, however, studies using strain B31, which has direct repeats, and strain JD1, which does not, revealed that the direct repeats are not required for segmental gene conversion to occur (Verhey et al., 2018a, 2019). Nevertheless, even without the direct repeats, the *vls* cassettes, by definition, have significant homology, and stretches of highest homology at the base pair level were seen to recombine most frequently (Verhey et al., 2018b). A window of $\geq$ 20 homologous base pairs flanking a switch site was found to be optimal.

The frequency of "switch" events was estimated to be on the order of 3.3 x 10^{-2} per cell per generation based on the vast data sets generated in these studies (Verhey et al., 2018b). This rate is one to two logs higher than that estimated for *vmp* switching in the relapsing fever *Borrelia*, and it should be noted that this is based on data at the nucleotide, not the amino acid level. Templated changes would be expected to be either silent or missense, while non-templated changes, which occur most frequently in regions in which templated changes are most frequent (Verhey et al., 2018b), could be silent, missense, or nonsense. The cells in which silent or nonsense mutations occur in *vlsE* would more likely be cleared by the host immune system and not available for analysis after infection of wild-type mice. The clustering of non-templated changes with regions of templated switches, which were particularly biased to the 5' end of the templated regions, led the authors to speculate that error-prone repairs may participate in the changes at the *vlsE* locus (Verhey et al., 2018b). It is important to point out that switching of *vlsE* also occurs in SCID mice, but the elimination of clones expressing the parental *vlsE* takes longer, presumably because of the lack of immune clearance of invariant clones (Zhang and Norris, 1998b; Coutte et al., 2009; Verhey et al., 2018a). This observation was exploited in the use of SCID mice to characterize and quantify all types of change events in the *vls* locus, which permits study of *vls* changes independent of selection by the host adaptive immune response.

The potential roles of both the inverted repeat upstream of *vlsE* and the telomere-proximal location of the gene were explored using a mini-*vls* system that had to be constructed due to the genetic intractability of the full-length *vls* locus (Castellanos et al., 2018). In this system, the *vlsE* locus was cloned with or without an adjacent silent cassette, with or without the inverted repeat, and in plasmids that would be maintained either as circular replicons in *B. burgdorferi* or would resolve into linear replicons due to the inclusion of telomeric sequences. Multiple steps, some performed by direct transformation of ligation products into *B. burgdorferi*, were required to generate the constructs. The switching frequency assessed after infection of mice was very low (considered background) in strains carrying plasmids with no silent cassette and in those in which the replicon was circular and lacked the inverted repeat. In contrast, with the silent cassette AND the inverted repeat present, the switching frequencies on both circular and linear replicons were all similar and higher, though approximately three logs less than observed with the intact *vls* locus. At this point, it is not clear how the inverted repeats and the replicon topology affect *vls* variation, but this system will facilitate further investigations of these questions.

Together, all of the results thus far reinforce the concept that antigenic variation is a critical function of VlsE, and that VlsE is a critical virulence determinant of *B. burgdorferi*. Variation in *vlsE* sequences has not been detected during *in vitro* culture (Zhang and Norris, 1998b) or in infected ticks (Indest et al., 2001), indicating that as yet unidentified environmental signals in vertebrate hosts (in all of these experiments, mice) regulate the gene conversion process.

The genes critical to the segmental gene conversion at the *vls* locus were identified independently by two groups (Dresser et al., 2009; Lin et al., 2009). Interestingly, and perhaps surprisingly, RecA expression was shown to be NOT required for *vlsE* recombination during mouse infection (Liveris et al., 2008; Dresser et al., 2009), indicating that the process is not dependent upon 'standard' homologous recombination. This is somewhat surprising given the overall homology of the locus and the presence of direct repeats flanking the variable region in many isolates. Two different genetic approaches were employed to the gene(s) critical to *vls* switching; both found that the process depends on the activity of the RuvA-RuvB Holliday junction helicase/branch migrase. In one study, transposon (Tn) mutants (Lin et al., 2009) in genes involved in DNA repair and recombination were screened in wild-type mice by infection over several time points up to 28 days. Tn mutants in *ruvA* and *ruvB* both were attenuated in wild-type mice and had reduced *vls* recombination when isolates were analyzed after recovery from infected mice. A targeted *ruvA* mutant showed reduced infection in wild-type, but not SCID, mice (Lin et al., 2009). Sequence analyses of *B. burgdorferi* clones isolated after murine infection again showed that minimal *vlsE* change had occurred in the *ruvA* and *ruvB* mutants during infection of wild-type mice, and that the parental sequence was detected in only a small proportion of clones by 28 days post-inoculation. In contrast, the parental sequence was found in multiple isolates after infection of SCID mice with the *ruvA* mutant. The recombination events that did occur in the *ruvA* mutant during infection of wild-type mice were not only less frequent but also used fewer silent loci as templates, indicating that the fine-scale segmental gene conversion that generates multiple changes of few amino acids in the parental bacteria is not active in the mutant.

In the same journal issue, an independent group, using a different approach, also identified RuvA and RuvB as critical to *vlsE* segmental gene conversion by generating 17 *B. burgdorferi* strains with targeted mutations that disrupted genes predicted to encode proteins involved in DNA repair, recombination, and replication (Dresser et al., 2009). A strength of this targeted approach was that it facilitated analysis of independent clones. Two individual mutant strains for each targeted locus were tested for their ability to cause infection in mice, again over a time course, and for *vls* recombination during infection. Again, *ruvA* and *ruvB* mutants were the only strains reproducibly attenuated by both measures.

VlsE and the flanking silent cassettes therefore comprise an elegant system required for Lyme disease *Borrelia* to maintain persistent infection in immunocompetent reservoir animals, an essential element of the natural lifecycle. Although variation in the surface-exposed portions of the protein keeps antibody-mediated clearance in check, invariant domains that are not surface-exposed are targeted by antibody responses that are not effective at clearing the organism, but are useful in diagnosis of infection in accidental hosts including humans. Finally, there is some evidence that additional proteins help *B. burgdorferi* survive in the face of the host adaptive immune response, e.g. BB0210, also known as Lmp1 (Yang et al., 2009). As yet, the specific roles that other such proteins play remain to be determined.

Dissemination and tissue colonization: A plethora of adhesins

Colonization of the tick is mediated by several *B. burgdorferi* proteins, including OspA, an adhesin that binds to the tick midgut epithelium (also see chapter 11 by Pal *et al.*). In addition, OspA blocks antibody access to the bacterial cell when *B. burgdorferi* resides in the tick midgut, serving as a shield as well (Battisti et al., 2008). In this chapter, however, we focus on interactions that contribute to infection of the mammalian host. *B. burgdorferi* has been known for decades to bind to a variety of mammalian cell types in culture and to extracellular matrix components, as reviewed in (Benach et al., 1991; Coburn et al., 2005; Antonara et al., 2011; Coburn et al., 2013; Brissette and Gaultney, 2014), but the relevance of these interactions *in vivo* has not always been clear. Recent advances in genetic manipulation of the bacterium and in imaging have enabled studies of the roles of individual *B. burgdorferi*-host protein interactions *in vivo*. Here we focus on *B. burgdorferi* proteins that facilitate survival, dissemination, and tissue colonization in the mouse model of infection and that bind to specific mammalian substrates *in vitro*. Several adhesins also serve to combat host immune defenses, and, at least in some cases, these activities are mediated by distinct domains. Of note, adhesive activities found only *in vitro* warrant some caution regarding importance *in vivo*, as many of the substrate proteins are inherently "sticky". Affinity and specificity are important considerations, but the gold standard regarding any candidate adhesin is the fulfillment of molecular Koch's postulates (Falkow, 1988). We begin with the adhesins that bind to components of the extracellular matrix (ECM), then move on to adhesins that bind to mammalian integral membrane proteins that have transmembrane signaling potential.

Adhesins that bind to extracellular matrix components

Laminins constitute a family of widely distributed heterotrimeric extracellular matrix glycoproteins that serve a variety of functions; many bacteria produce laminin-binding adhesins. *B. burgdorferi* encodes at least three laminin-binding adhesins: ErpX, BmpA and BB0406. ErpX was shown to bind laminin (stated to be primarily $\alpha_1\beta_1\gamma_1$ by the vendor) produced by a murine sarcoma cell line

(Brissette et al., 2009c), but no assessments of a potential *in vivo* role have been reported.

Similar *in vitro* approaches were used to identify the Bmp proteins as candidate laminin-binding adhesins, although only BmpA was investigated in depth (Verma et al., 2009). *In vivo*, there is evidence that BmpA and B contribute to joint colonization and arthritis development (Pal et al., 2008), and BmpD was selected *in vivo* for vascular adhesion (Antonara et al., 2007). More recent structure based predictions suggest a role for Bmp proteins in nutrient uptake (Astrand et al., 2019; Cuellar et al., 2020), although this activity has not been experimentally verified. Furthermore, BmpB, C, and D all are associated with the cytoplasmic membrane, *i.e.* not available to serve as adhesins (Dowdell et al., 2017). Therefore, potential roles of Bmp proteins as adhesins are in serious doubt, as independent lines of evidence (in silico and experimental) suggest that they are not located on the *B. burgdorferi* surface.

BB0406 is a recent addition to the group of laminin-binding adhesins as well as a putative porin (Shrestha et al., 2017; Bista et al., 2020). Mutants in *bb0406* are infectious but attenuated in mice and show reduced bacterial burdens in tissues compared to the parental strain. The data presented support a role for this protein in dissemination and bacterial colonization in mouse tissues (Bista et al., 2020). The precise roles of the laminin-binding proteins in murine infection remain to be determined.

OspC is one case in which a protein known to be required for *B. burgdorferi* to establish infection in mice was subsequently found to have adhesin functions and a role in evasion of innate immunity. It is not yet clear whether the same or distinct domains of the protein are involved in each activity. *ospC* mutants are unable to initiate infection in either immunocompetent or MyD88-/- and SCID mice (Table 1), which are deficient in innate and adaptive immunity, respectively. These results are consistent with the more recent identification of OspC as an antiphagocytic factor (Carrasco et al., 2015), and as having a role in disabling the complement cascade (Caine et al., 2017) (Table 1). Expression is required only during initial infection, as *ΔospC* organisms complemented with *ospC* encoded on a plasmid had consistently lost the plasmid 6 weeks post infection (Tilly et al., 2006). Constitutive expression of OspC results in elimination of *B. burgdorferi* in immunocompetent but not SCID mice (Xu et al., 2006), suggesting that appropriate downregulation of this major surface protein is critical to evasion of the adaptive immune response. The documented succession of OspC to VlsE production by *B. burgdorferi* during mammalian infection (Tilly et al., 2013) is interesting in this regard, as OspC contributes to complement resistance in innate and adaptive immunity, while VlsE undergoes continual antigenic variation throughout mammalian infection, protecting the bacteria from the adaptive immune response (Table 1).

OspC also was identified as a candidate adhesin using *in vivo* phage display (Antonara et al., 2007) and was shown to bind to mammalian cells in culture. Subsequent work has demonstrated that OspC binds to mammalian ECM components fibronectin and/or dermatan sulfate, with binding to each substrate varying with the OspC sequence type (Lin et al., 2020c). OspC also contributes to extravasation of bacteria (Lin et al., 2020c). VlsE also was selected for vascular adhesive activities using *in vivo* phage display (Antonara et al., 2007) (Table 1), providing hints of possible adhesion-based functional overlap of these proteins in addition to their roles in subverting or evading host immune defenses.

An operon encoding two related adhesins, decorin-binding proteins A and B (DbpA and DbpB; Table 1), also is activated when the bacteria are transiting to the mammalian host. Decorin is a proteoglycan with a core protein modified by addition of a dermatan sulfate (chondroitin sulfate B) glycosaminoglycan, which "decorates" type 1 collagen fibrils. Decorin is known to have several structural and regulatory functions and is widely available as a substrate for bacterial adhesion. DbpA has been shown to bind the proteoglycan decorin, and to dermatan sulfate glycosaminoglycans in the absence of the core protein. In the first genetic study of the role of *B. burgdorferi* attachment to host molecules, decorin-deficient mice showed colonization defects by *B. burgdorferi* in heart, bladder, and joint at low doses; at higher inocula, decorin-deficient mice harbored fewer *B. burgdorferi* specifically in joint tissue and exhibited reduced arthritis severity and frequency (Brown et al., 2001). Mutant *B. burgdorferi* that do not produce DbpA and DbpB have significant attenuations in virulence as measured by bacterial loads in disseminated sites (Hyde et al., 2011; Fortune et al., 2014; Lin et al., 2014), suggesting that a key role for these proteins is in facilitating dissemination and/or colonization of multiple tissue sites. Various DbpA sequences produced in an isogenic strain background showed different levels of binding to decorin and dermatan sulfate *in vitro* and different levels of colonization in a variety of tissues *in vivo* (Lin et al., 2014), For example, DbpA sequences from *B. garinii* and *B. afzelii* strains colonized joint tissues in mice at significantly lower levels than did DbpA from a *B. burgdorferi* strain, reflecting the different propensities of the species to cause arthritis in humans (Lin et al., 2014).

The search for *B. burgdorferi* collagen-binding adhesins began long before the identification of BBA33 as a collagen-binging adhesin (Zhi et al., 2015). The interest in collagen was initially fueled by pathological observations of the bacteria residing primarily in perivascular connective tissues and frequently associated with collagen (Barthold et al., 1988; Barthold et al., 1991; Barthold et al., 1993). Although *B. burgdorferi* was found to bind to and invade biologically intact type 1 collagen (Zambrano et al., 2004), which is the most abundant collagen and is widely distributed, BBA33 was found to bind to types IV and VI collagen (Zhi et al., 2015). Type IV collagen is widely distributed, primarily found in the basement membrane (basal lamina) of endothelial and epithelial cell layers. Type VI collagen is abundant in skeletal muscle and nervous system tissue types and appears to serve as a connection between basal lamina and other ECM components. Since collagens are increasingly appreciated as signaling as well as structural proteins, it is intriguing to contemplate potential effects of *B. burgdorferi* binding to collagen not only in terms of tissue colonization but also in effects on host responses to the bacteria. Considerations of the complexities of *B. burgdorferi* interactions with ECM components are complicated by the complexities of interactions between the ECM components themselves (*e.g.*, binding of decorin to type I collagen). Interestingly, BBA33 does not bind to collagen type I (Zhi et al., 2015), suggesting that additional collagen-binding adhesins await discovery.

B. burgdorferi encodes several fibronectin-binding adhesins RevB, RevA (2 genomic copies), BB0347, and BBK32, in order of increasing affinity for fibronectin (decreasing K_D). All were identified originally using *in vitro* approaches, but further work has identified roles for some in the mouse model of infection. The fibronectin-binding adhesin with the lowest affinity for fibronectin, RevB (encoded by *bbc10*), does not yet have a known *in vivo* function. The only test reported thus far is in gain of function studies in which no increases in binding to skin vasculature were seen as compared to the parental strain (Moriarty et al., 2012).

Studies on RevA are inherently complicated by the fact that there are two copies of the gene on different replicons in the genome. Nevertheless, a double knockout strain was generated, and while there was no difference in the ID_{50} from the wild-type parental strain, a defect in colonization of heart tissue was identified (Byram et al., 2015). Complementation of the mutant restored heart colonization. A transposon insertion near the 5' end of one of the RevA genes, *bbm27*, revealed an infectivity defect, while a transposon insertion nearer the 3' end of the other RevA-encoding gene, *bbp27*, did not have an apparent virulence defect in mice (Lin et al., 2012). This provides a basis for future studies of the precise roles of the RevA proteins during *B. burgdorferi* infection. For example, no role for RevA was found in gain of function *B. burgdorferi* strain interactions with the vasculature in the skin in living mice (Moriarty et al., 2012), and the same gain of function strain showed reduced binding to heart and ear tissues in a different infection model (Caine and Coburn, 2015). BB0347 is a third fibronectin-binding adhesin of *B. burgdorferi* that was not found to have any significant effect on *B. burgdorferi* interactions with the skin vasculature (Moriarty et al., 2012), but in gain of function studies was found to increase *B. burgdorferi* interactions with the vasculature in tibiotarsal joint tissue (Caine and Coburn, 2015).

Finally, although not essential to the ability of *B. burgdorferi* to cause infection in mice, BBK32 is the highest-affinity fibronectin-binding protein produced by *B. burgdorferi* and has separable dermatan sulfate glycosaminoglycan (GAG)-binding activity as well. Both of the extracellular matrix-binding activities contribute to vascular interactions of *B. burgdorferi* in living mice (Moriarty et al., 2012). Initial, more transient, *B. burgdorferi*-endothelium interactions are mediated by fibronectin binding, while the more stable interactions are mediated by GAG binding. The GAG-binding activity of BBK32 also contributes to *B. burgdorferi* tropism for joint tissue during murine infection (Lin et al., 2015). As described earlier, BBK32 also binds C1r, blocking its activation and thereby the classical complement pathway (Garcia et al., 2016). Live imaging of *B. burgdorferi* infection in mice using bioluminescence also supports a role for BBK32 in the full virulence of the bacteria (Hyde et al., 2011).

Several questions remain to be resolved regarding the potential roles of the fibronectin-binding adhesins of *B. burgdorferi*. In addition to those described above, there are reports that CspA, a complement regulatory protein, also binds other host molecules, including fibronectin (Hallstrom et al., 2010). One interesting consideration is whether the different fibronectin-binding proteins might have different affinities for soluble (plasma/serum) fibronectin vs. insoluble matrix fibronectin. Several pathogens are known to coat themselves with host molecules as a mechanism to evade host recognition, and it is possible that *B. burgdorferi* does this as well. However, this possibility remains to be experimentally addressed, a task made more difficult by the number of adhesins with apparently overlapping or redundant functions produced by *B. burgdorferi* and other spirochetes.

Bgp (*Borrelia* GAG-binding protein, BB0588) was the first *B. burgdorferi* GAG-binding adhesin identified, based on biochemical purification from the spirochetes (Leong et al., 1995; Parveen and Leong, 2000). Like intact *B. burgdorferi*, Bgp bound heparin and dermatan sulfate (Parveen and Leong, 2000). Bgp is a secreted protein that is found in the culture medium as well as on the surface of the spirochete (Parveen and Leong, 2000; Cluss et al., 2004). In addition to its GAG-binding properties, Bgp is a 5'-methylthioadenosine/*S*-adenosyl homocysteine nucleosidase (Pfs), which in other bacteria is important in elimination of toxic metabolic products and in nucleotide salvage (Winzer and Williams, 2001). Recombinant Bgp partially inhibits *B. burgdorferi* attachment to purified GAGs and to cultured mammalian cells, supporting it as a candidate adhesin, but the nonadherent *B. burgdorferi* strain B314 produces Bgp, indicating that Bgp is not sufficient to promote bacterial binding. *B. burgdorferi bgp* mutants retain the ability to infect mice (Parveen et al., 2006; Schlachter et al., 2018). A different protein, BB0210, also termed Lmp1, was selected for adhesion to the vasculature in living mice using phage display (Antonara et al., 2007), and subsequently shown to bind to the GAG chondroitin-6-sulfate (Yang et al., 2016). The middle region of this protein contains several repeats, at least some of which were included in all selected phage clones. This middle region is responsible for the adhesin activity, which can in itself promote *B. burgdorferi* infectivity (Yang et al., 2016).

Among the *B. burgdorferi* mutants that do not produce known GAG-binding adhesins, those that do not produce *dbpAB* or *ospC* are most attenuated. Mutants in *bbK32* and *lmp1* are less attenuated, and *bgp* appears to play a very minor role in murine infection. Many of these proteins are multifunctional, and the roles of GAG binding, specifically, have been investigated only for DbpA, OspC, and BBK32 (Moriarty et al., 2012; Lin et al., 2014; Lin et al., 2015; Lin et al., 2020c). It remains possible that the other GAG-binding adhesins play more significant roles in other vertebrate hosts.

Integrin binding proteins

Integrins are heterodimeric, integral membrane proteins produced by all mammalian cell types except erythrocytes. They perform a number of critical activities for the cell, including regulation of growth and morphology, adhesion to extracellular matrix, and bidirectional signaling in response to a variety of stimuli (Dustin and Springer; Larson and Springer, 1990; Hynes, 1992; Ruoslahti, 1996, 1997; Hynes, 2002; Bachmann et al., 2019; Humphries et al., 2019; Mohammed et al., 2019). Each integrin is composed of one alpha (α) subunit plus one beta (β) subunit, with the primary subdivisions into families based on the β subunit. There are two known β_3 chain integrins, $\alpha_{IIb}\beta_3$ and $\alpha_v\beta_3$, and *B. burgdorferi* has been shown to bind to both (Coburn et al., 1993; Coburn et al., 1994; Coburn et al., 1998; Coburn et al., 1999). Both of these integrins are rather promiscuous, meaning that they bind to several different mammalian proteins (ligands). *B. burgdorferi* is also known to bind to at least two β_1 chain integrins, $\alpha_3\beta_1$ and $\alpha_5\beta_1$, but interactions with other β_1 chain integrins and other integrin β chain families have not been carefully investigated.

The integral outer membrane protein P66, which is also a porin, was the first integrin binding protein identified in *B. burgdorferi*. This was done using a functional phage display selection as opposed to homology or motif searching (Coburn et al., 1999). P66 binds primarily to the β_3 chain integrins, but also binds to integrins $\alpha_3\beta_1$ and $\alpha_5\beta_1$, showing that, like its integrin partners, P66 is promiscuous (Coburn et al., 1999; Coburn and Cugini, 2003). Subsequent studies demonstrated that P66 is essential for mammalian but not tick infection (Ristow et al., 2012), consistent with the observation that P66 is not detectable in the unfed tick (Cugini et al., 2003). The integrin binding activity has a specific role in facilitating dissemination and extravasation of the spirochete out of the vasculature and into perivascular connective tissue but is not essential for murine infection (Kumar et al., 2015; Ristow et al., 2015).

Leukocyte extravasation in the inflammatory response is a step-wise process in which leukocyte carbohydrates bind to endothelial cell selectins to capture the circulating cells, and leukocyte integrins then bind to endothelial cell ligands to form stationary adhesions (Lawrence and Springer, 1991; Lawrence et al., 1995). Further interactions and cellular changes at the chemical and mechanical levels promote leukocyte migration across the endothelium and basement membrane, which is critical to homing to sites of infection and inflammation. It is intriguing to note some conceptual similarities between leukocyte and *B. burgdorferi* extravasation, with the requirement for BBK32-host GAG interactions slowing the movement of the spirochetes in circulation and P66-host integrin and OspC-GAG interactions promoting extravasation.

Several candidate *B. burgdorferi* integrin ligands were identified by searching the genome for predicted proteins containing the RGD tripeptide, which is commonly found in mammalian β_1 and β_3 integrin ligands, and for possible secretion signals. Purified recombinant BBB07 promoted pro-inflammatory responses in synoviocytes through integrin $\alpha_3\beta_1$ (Behera et al., 2008), and transposon insertions showed mild defects in infectivity in mice (Lin et al., 2012). However, these results have not been connected experimentally, and at this point it is not clear that BBB07 is exposed on the surface of the *B. burgdorferi* cell and available to interact with host molecules. Further work regarding a different *B. burgdorferi* protein, BB0172, which in recombinant form also binds integrin $\alpha_3\beta_1$ (Wood et al., 2013) is also needed to determine relevance to *B. burgdorferi* infection and pathogenesis.

It is clear that *B. burgdorferi* encodes many adhesive proteins, and the roles in infection of mice for some are diagrammed in Figure 6. One might ask several questions: do all of the putative adhesins actually function as adhesins, and if so, why are so many needed? Several known adhesins have been shown to be essential to the ability of *B. burgdorferi* to cause infection in mice (e.g. OspC, P66) (Grimm et al., 2004b; Tilly et al., 2006; Ristow et al., 2012), while others contribute significantly to infection (BBK32, DbpA), but are not absolutely essential (Li et al., 2006; Seshu et al., 2006; Blevins et al., 2008; Hyde et al., 2011; Imai et al., 2013; Fortune et al., 2014). Others (e.g. Bgp) appear to make relatively minor contributions to infection and pathogenesis in laboratory mice (Parveen et al., 2006; Schlachter et al., 2018). It is certainly possible that, given the host range of *B. burgdorferi* in the natural infection cycle, some adhesins contribute more significantly to *B. burgdorferi* infection in natural reservoir hosts. It is also important to note that several of the *B. burgdorferi* adhesins (*e.g.*, OspC, BBK32, P66, BB0406) are multifunctional, and it is important to recognize that different functions may contribute to infectivity. In fact, this can and has been experimentally addressed (Seshu et al., 2006; Hyde et al., 2011; Ristow et al., 2012; Kumar et al., 2015; Lin et al., 2015; Ristow et al., 2015; Garcia et al., 2016; Caine et al., 2017; Lin et al., 2020c), but much remains to be done to fully understand the roles of each adhesin in *B. burgdorferi* infection.

As is the case for other pathogenic bacteria, it has been speculated that *B. burgdorferi* adhesins may play different roles in infection (Figure 6). In fact, evidence to date suggests that GAG binding by BBK32 is important for transient interactions with the vascular endothelium *in vivo* (Norman et al., 2008; Moriarty et al., 2012) and for colonization of joint tissue (Lin et al., 2015), while OspC GAG binding activity is required for extravasation and colonization of joint tissue, with associated joint swelling (Lin et al., 2020c). DbpA is also involved in colonization of joint tissue and promotes arthritis development in an allele-specific manner (Lin et al., 2014). P66 integrin-binding activity and OspC GAG-binding activity are required for extravasation *in vivo* (Kumar et al., 2015; Lin et al., 2020c). Since many of the *B. burgdorferi* adhesins bind glycosaminoglycans, one would think that they might be functionally redundant. However, since phenotypes seen in deletion mutant strains are decipherable, they are non-redundant, but in many cases the specific interactions that reveal the essential nature of each remain to be elucidated. The recognition of ECM components, such as Fn, collagen, or GAGs, which consist of repeating subunits of either protein or carbohydrate origin, may lead one to suspect that the corresponding *B. burgdorferi* adhesins might be recognizing diverse repeat structures with limited specificity, although specific roles for each have been demonstrated. Even interactions of relatively low affinity and specificity may promote bacterial binding to host components, which may then facilitate formation of more specific interactions mediated by other bacterial and host molecules. It is also possible that the

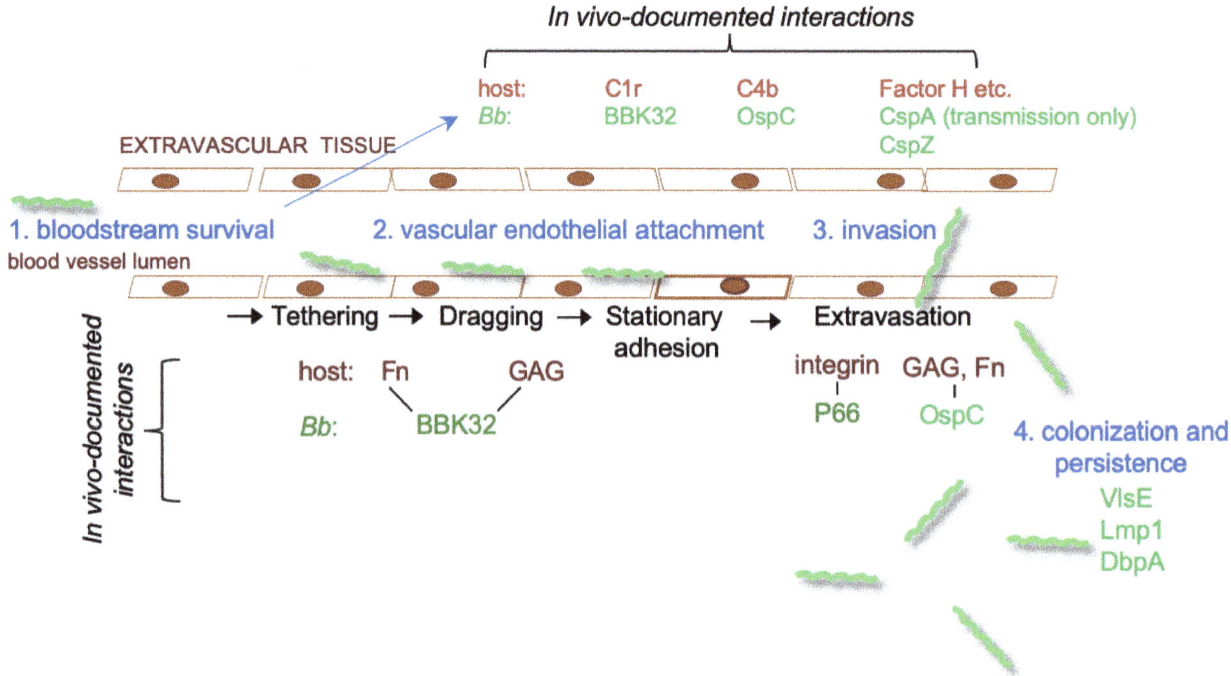

Figure 6. *Borrelia burgdorferi*-host interactions that contribute to infection. Shown are interactions between host molecules and *B. burgdorferi* virulence determinants that have been documented to influence infection *in vivo* in animal model(s). Many more await further characterization or have not yet been shown to make significant contributions in the mouse model of *B. burgdorferi* infection. Additional factors such as those involved in nutrient scavenging are not included here. Refer to Table 1 for a more complete summary of factors known to participate in *B. burgdorferi* survival in mice. Complement factors: C1r, C4b, Factor H (and related proteins). Adhesion substrates: GAG = glycosaminoglycan, Fn = fibronectin.

relatively low affinities for human molecules may be higher for comparable molecules of other species. It is also important to remember that some of the extracellular matrix substrates for adhesion by *B. burgdorferi* also modulate immune responses by the host (e.g. (Iozzo and Schaefer, 2010; Moreth et al., 2012; Frevert et al., 2018)), so the effects of *B. burgdorferi*-extracellular matrix interactions on the nature of the immune response remain to be explored. The same is true of *B. burgdorferi*-integrin interactions, as integrins are also involved in shaping the immune response (Striz and Costabel, 1992; Gorfu et al., 2009; Nishimura, 2009; Verma and Kelleher, 2017; Thome et al., 2018). A difficult challenge, particularly in an experimental system with limited genetic tools, will be to systematically characterize each of the adherence pathways of *B. burgdorferi*, with respect to their biochemical functions, their effects on host cell biology and immune responses, and their roles during infection of mammals and ticks.

Roles of bacterial and host proteinases in dissemination and inflammation

In addition to binding to extracellular matrix proteins, *B. burgdorferi* needs a mechanism for breaking down the molecules in the extracellular matrix that impede its movement and that promote the inflammatory response that might clear the infection. *B. burgdorferi* appears to have adopted several strategies for accomplishing this, as reviewed in (Hu et al., 2001).

B. burgdorferi encodes a homolog of the high temperature requirement A (HtrA) family of proteins. HtrA in *B. burgdorferi*, an ortholog of DegP, a periplasmic chaperone and quality control protein, is bound to outer membranes and released into the milieu (Coleman et al., 2013; Russell et al., 2013; Backert et al., 2018). It has been shown to play a role in degradation of *B. burgdorferi* proteins such as the basic membrane protein, BmpD, the chemotaxis phosphatase CheX, and outer membrane porin, P66, the monomeric but not polymerized form of flagellin, and the cellular fission protein BB0323 (Coleman et al., 2013; Kariu et al., 2013; Coleman et al., 2016; Ye et al., 2016; Coleman et al., 2018; Zhang et al., 2019). *In vitro*, HtrA also has aggrecanase activity and is able to degrade the extracellular matrix components fibronectin and aggrecan (Russell et al., 2013). An HtrA mutant is non-infectious in mice, but also exhibits significant growth defects *in vitro* (Ye et al., 2016), has decreased motility, and produces less pyruvate (Coleman et al., 2018). It is unclear whether the activity of HtrA on *B. burgdorferi* proteins and/or degradation of extracellular matrix proteins are important to the loss of infectivity of the HtrA mutant (Ye et al., 2016).

The Lon proteases are a group of ATP-dependent serine proteases ubiquitously distributed in the genomes sequenced to date. They fall into two major classes: LonA, encoded primarily in bacterial and eukaryotic genomes, and LonB, present in archaea. LonA proteases have generally been found to be localized in the cytoplasm and LonBs attached to the cytoplasmic membrane, but both groups appear to assure protein quality control. Most bacteria encode a single LonA, but *Borrelia* species encode two, Lon-1 (encoded by *bb0253* in *B. burgdorferi*) and Lon-2 (encoded by *bb0613* in *B. burgdorferi*). Both genes are expressed during mammalian infection (Mason et al., 2020; Thompson et al., 2020). The two *B. burgdorferi* proteins are functionally distinct, with different N-terminal substrate-binding domains and different abilities to complement an *E. coli lon* mutant (Coleman et al., 2009), suggesting distinct activities. Individual knockout mutants have been generated and assessed for both *in vitro* and *in vivo* phenotypes. Inactivation of *lon-1* results in reduced growth rate, reduced resilience to osmotic stress and oxidative stress, increased levels of OspC and RpoS *in vitro*, and markedly decreased infectivity (Thompson et al., 2020). The *lon-2* mutant did not display an *in vitro* growth defect but did show reduced resistance to osmotic and oxidative stress *in vitro* and, as with lon-1 mutants, significantly reduced burdens in tissue samples from mice infected with a single dose (Mason et al., 2020). In this mutant, however, some reductions in *in vitro* levels of OspC and RpoS protein levels were noted.

Another strategy utilized by *B. burgdorferi* for breaking down extracellular matrix proteins is to exploit host proteases. *B. burgdorferi* has been shown to bind and stabilize plasmin, an important protease in the host fibrinolytic cascade (Coleman et al., 1995; Hu et al., 1995; Klempner et al., 1995; Klempner et al., 1996). A number of *B. burgdorferi* proteins have been found to bind plasminogen *in vitro* (Fuchs et al., 1994; Hu et al., 1997; Brissette et al., 2009b; Hallstrom et al., 2010; Floden et al., 2011; Onder et al., 2012; Toledo et al., 2012; Koenigs et al., 2013). The role that each plays *in vivo* has generally not been experimentally established, with one exception (Earnhart et al., 2014). The C-terminal 10 amino acids of OspC were shown to bind plasminogen but were found not to be required for *B. burgdorferi* infection. Plasmin is a broad-spectrum serine protease that, in addition to degrading fibrin, is capable of degrading extracellular matrix components such as elastin and laminin. Plasmin is also an important activator of other extracellular matrix degrading proteases such as matrix metalloproteinases (MMPs). Plasmin is produced by activation of its circulating zymogen form, plasminogen, through the action of activating proteases such as urokinase-type plasminogen activator (u-PA) and tissue-type plasminogen activator (t-PA). The presence of *B. burgdorferi* induces production of u-PA and its receptor (CD87) by circulating monocytes (Fuchs et al., 1996; Coleman et al., 2001; Haile et al., 2006). CD87 can stabilize u-PA against degradation and inactivation by circulating inhibitors (e.g. plasminogen activator inhibitor-2, PAI-2), allowing it to efficiently convert plasminogen to plasmin. Urokinase also may bind directly to *B. burgdorferi,* which also protects it against inactivation (Klempner et al., 1996). The binding of plasmin to *B. burgdorferi* has been shown to increase the ability of the organism to pass through cell culture monolayers and to avoid inhibition by circulating inhibitors such as α_2-antiplasmin (Coleman et al., 1995; Perides et al., 1996). In studies of plasminogen deficient mice, it was shown that plasmin was required for efficient dissemination of *B. burgdorferi* within the tick (i.e. from the midgut to the salivary glands) (Coleman et al., 1997). The presence of plasmin also enhanced spirochetemia following feeding of infected ticks on uninfected mice. However, it did not affect the spread of bacteria to distant sites in mice.

Other proteases appear to be involved in extracellular matrix degradation and dissemination by *B. burgdorferi.* MMPs are the major host enzymes in degradation of extracellular matrix proteins. They are involved in normal host processes such as tissue remodeling and wound healing and, as a family, can degrade every component of the extracellular matrix. Infection with *B. burgdorferi* has been shown to induce production of numerous MMPs (including MMP-1,

MMP-3, MMP-9, MMP-13 and MMP-19) from different host tissues (keratinocytes, fibroblasts, astrocytes, PBMCs and chondrocytes) in cell culture (Perides et al., 1999; Gebbia et al., 2001; Lin et al., 2001a; Zhao et al., 2003; Behera et al., 2004; Gebbia et al., 2004; Behera et al., 2005; Grab et al., 2007). MMP-9 is up-regulated in erythema migrans skin lesions from infected patients (Zhao et al., 2003), and promotes development of arthritis but not *B. burgdorferi* dissemination, and does not increase bacterial loads in the joints (Heilpern et al., 2009). *In vitro* studies showed that MMP-9 can be activated by plasmin bound to *B. burgdorferi* and that activation of MMP-9 results in increased penetration of spirochetes through artificial barriers, including artificial blood-brain barrier models (Gebbia et al., 2001; Grab et al., 2005; Grab et al., 2007). MMP-9 also was found to be increased in the cerebrospinal fluid of patients with neuroborreliosis (Kirchner et al., 2000). Animal studies have shown that MMP-9 plays an important role in penetration of *Streptococcus pneumonia* through the blood brain barrier (Paul et al., 1998). However, studies of MMP-9 knockout mice in Lyme disease models have not shown any difference in dissemination of *B. burgdorferi* to distant sites, although it was important to the development of arthritis (Heilpern et al., 2009). However, because the MMPs show considerable overlap in their substrate specificity and because *B. burgdorferi* induces multiple MMPs from host tissues, it is possible that deficiency of any single MMP is compensated for by other induced enzymes.

While *B. burgdorferi* may co-opt host proteases to assist with dissemination, the induction and activation of MMPs resulting from the presence of *B. burgdorferi* contributes to some of the pathology associated with Lyme disease. In particular, MMPs have been closely associated with degradation of bone and cartilage in other forms of arthritis (e.g. rheumatoid arthritis). The collagenases MMP-1 and MMP-13, as well as stromelysins (MMP-3 and MMP-19), were found to be elevated in the synovial fluids of patients with Lyme arthritis and/or induced from cultured chondrocytes by the presence of *B. burgdorferi* (Hu et al., 2001; Lin et al., 2001a). Studies with monkey cartilage explants have shown that MMPs and a related metalloproteinase from the ADAMTS family (A Disintegrin-like And Metalloprotease domain with ThromboSpondin type I motifs), aggrecanase-1/ADAMTS-4, play a key role in the degradation of articular cartilage after *B. burgdorferi* infection (Hu et al., 2001; Behera et al., 2006). The specific MMPs induced by *B. burgdorferi* differ significantly between animals, as well as from tissue to tissue, and may be responsible for some of the differences in tissue-specific pathology. For example, the MMP-1, MMP-8 and MMP-13 collagenases are increased in human Lyme arthritis but not in joints of mouse strains infected with *B. burgdorferi* (Behera et al., 2005; Crandall et al., 2006). Irreparable cartilage damage occurs with the degradation of interstitial collagens by collagenases but not with degradation of other cartilage components (e.g. aggrecans, hyaluronans) mediated through MMPs in the non-collagenase families. It is possible that these differences explain the lack of permanent joint erosions in mouse models of Lyme arthritis compared with humans.

Summary: *Borrelia burgdorferi* Biology Requires Multiple Virulence Determinants for Infection, Dissemination, Persistence, and Pathogenesis

Dissemination of *Borrelia* species from the inoculation site to additional sites in an animal host and persistence in the skin are necessary for propagation in the natural tick-vertebrate infection cycle. While the focus in this chapter has been on consideration of these properties as virulence determinants, they are fundamental to the biology of *B. burgdorferi*. It is also important to remember that the focus of this chapter has been on features and questions that are experimentally tractable in mouse models of infection. *B. burgdorferi* has been shown by multiple groups to migrate in the skin following inoculation in humans (Berger et al., 1985; Berger et al., 1992; Mitchell et al., 1993), but not all *B. burgdorferi* strains cause disseminated infection in humans (Wang et al., 1999; Earnhart et al., 2005; Jones et al., 2006). In contrast, *B. burgdorferi* does generally spread to multiple diverse tissue sites as the infection proceeds in the laboratory mouse. The local spread, which is associated with the leading edge of the erythema migrans lesion in humans (Berger et al., 1985; Berger et al., 1992; Mitchell et al., 1993), is detectable in mice using microscopic imaging techniques including two photon and intravital microscopy (see the imaging section of this chapter and chapter 17 by Chaconas *et al.*), although mice do not develop erythema migrans. Dissemination through an entire animal (*e.g.* mouse) has been reliably followed over time and quantified using *in vivo* whole animal imaging, culture and qPCR, but these approaches are not possible, unethical, or more difficult in human patients. Once infection of most vertebrate hosts is established, it is maintained over long periods of time in the absence of antibiotic therapy.

Many researchers have contributed significantly to our understanding of *Borrelia*-host interactions in the mouse model. Spread through tissue matrices is thought to occur even between different tissue types, but the most efficient means of spread early after infection is likely through transient dissemination through the blood and/or lymphatic vascular systems. Imaging of lymphatic vessel involvement and blood vessel invasion has not yet been achieved, but i.v. inoculation of the bacteria has permitted intravital microscopic imaging of interactions with and extravasation from blood vessels (the vasculature) (Moriarty et al., 2008; Norman et al., 2008; Lee et al., 2010; Moriarty et al., 2012; Lee et al., 2014; Kumar et al., 2015; Lin et al., 2020c), and elucidation of specific roles in these processes for some adhesins. In order to even image such events, the bacteria must survive the innate host defenses in the circulation and in tissues, e.g. complement, ROS, and RNS, as discussed earlier in this chapter. In addition, the biosynthetically feeble *Borrelia* require nutrients supplied by the vertebrate host or through salvage pathways to survive and replicate, also as described earlier.

Despite the incredible amount of information clinical and basic science researchers have gained about *B. burgdorferi* and roles of some of its virulence determinants in infection, there are many interesting questions to be addressed in future research on the fundamental biology and virulence determinants of *B, burgdorferi*. For example, what is the mammalian host signal that activates switching at the *vls* locus, and what signaling pathways in the *Borrelia* cell are involved? What are the specific mechanisms by which different adhesins contribute to dissemination, tissue colonization, intravascular interactions, and persistence in host animals? Are these the same or distinct in different vertebrate hosts? How is it that multiple adhesins that have similar binding activities *in vitro* have discernable roles *in vivo*? The same questions could be applied to complement regulatory proteins of Lyme disease *Borrelia*. How do the adhesins, complement regulators, and proteases collaborate during infection without inactivating each other? Why is *B. burgdorferi* infection associated with arthritis in humans and certain mouse strains, while *B. garinii* is associated with neuroborreliosis in humans? Are differences in bacterial components and host responses to these components each important in determining differential disease manifestations? What are the critical nutritional factors required by *B. burgdorferi* to replicate in the vertebrate host? Fortunately, there is an ever-expanding and increasingly elegant genetic toolbox that facilitates research on the biology of *B. burgdorferi*, but the segmented genome and strain-to-strain and species-to-species variations remain vexing. Importantly, mice are not tiny humans, and differences in how *Borrelia* species replicate, disseminate, are potentially cleared, or are pathogenic in different host species remain underexplored. Human infection is likely best recapitulated in non-human primate models of infection, but these require knowledge and resources not available to most researchers.

The human pathogenic spirochetes, *Borrelia*, *Treponema*, *Leptospira*, and *Brachyspira*, are all capable of causing long-term infections, although the sites of persistence and the means of acquisition vary dramatically. Perhaps further analysis of the expanding genomic information, and the expansion of genetic tools for all four genera will shed further light on commonalities and differences among the pathogens, and comparisons with the spirochete genera that are not pathogenic will illuminate convergent vs. divergent evolution and the features critical to maintenance of the unique lifestyles of each.

Acknowledgements

Work in the authors' laboratories was supported by the National Institutes of Health (US) awards K22 AI081730 (MWJ), R01 AI099094 (MWJ), R01 AI37277 (SJN), R01 AI59048 (SJN), R21 AI103905 (LTH), R21 AI111317 (LTH), R01 AI118799 (JC), R21 AI119821 (to JS), R01 AI121217 (JC), R01 AI121401 (JC and PK), R01 AI131656 (LTH and JS), R21 AI133367 (BG and JS), R21 AI140510 (JC), R56 AI146930 (BG and JS), R01 AI146930 (BG and JS), R21 NS102766 (BG), R21 NS104767 (BG), The National Research for Tick-Borne Diseases (MWJ), and the LOEWE Center DRUID Novel Drug Targets against Poverty-Related and Neglected Tropical Infectious Diseases, project C3 (PK).

The authors thank Dr. George Chaconas of the University of Calgary for insightful suggestions on the *vlsE* section, Valery Lozada-Fernandez, PhD candidate at the Medical College of Wisconsin, for generation of the 3D plot in Figure 1, Dr. Theresa C. Maatman at the Medical College of Wisconsin for assistance with the artwork in Figure 2, and Alexandra D. Powell, a PhD candidate at Texas A&M University for assistance with Figure 4.

Table 1. Information regarding known and potential virulence determinants of *Borrelia burgdorferi* (partial listing). Genes are sorted first according to their apparent major functions during the infectious cycle, then by genetic locus designation. In each section, those encoded on the chromosome are listed before those encoded on genomic plasmids. Note that some appear in more than one major function section. Chr= chromosome

Protein(s) and *gene(s)*[a]	Gene location(s)	Findings / evidence *in vitro*	Findings/evidence *in vivo*
Regulation			
CsrA (*bb0184*)	chr	Carbon storage regulator A controls expression of a number of infection-relevant genes (Karna et al., 2011; Arnold et al., 2018).	A mutant lacking *csrA* is non-infectious in mice; however, complementation of the mutant in *trans* does not restore the infectious phenotype (Karna et al., 2011).
LtpA (*bb0355*)	chr	CarD-like transcriptional regulator important for *in vitro* growth and cold stress (Yang et al., 2008; Chen et al., 2018).	An *ltpA* mutant demonstrates delayed infection in mice and reduced survival in ticks (Chen et al., 2018).
BB0363 (*bb0363*)	chr	Cyclic di-GMP phosphodiesterase important for motility (Sultan et al., 2010).	Essential for infection of mice but dispensable for survival within ticks (Sultan et al., 2010).
LuxS (*bb0377*)	chr	Potential involvement in quorum sensing (Stevenson and Babb, 2002; Hubner et al., 2003; von Lackum et al., 2006; Riley et al., 2007).	Provides competitive advantage during mammalian infection (Arnold et al., 2015).
Rrp1 (*bb0419*)	chr	Two component response regulator, diguanylate cyclase (Rogers et al., 2009)	Required for survival in ticks (Caimano et al., 2011; He et al., 2011; Kostick et al., 2011), inactivation of gene leads to attenuated virulence but overexpression leads to avirulence in mice (Kostick et al., 2011)
HK1 (*bb0420*)	chr	Two component response regulator, histidine kinase (Caimano et al., 2011)	Required for survival in ticks (Caimano et al., 2011)
RpoN (*bb0450*)	chr	Sigma factor that regulates temperature- and pH-dependent gene expression (Ouyang et al., 2008); RpoN activates *rpoS* (Smith et al., 2007; Blevins et al., 2009), RpoS activates *ospC* promoter (Yang et al., 2003)	Inactivation results in loss of tick to mouse transmission, mouse infectivity and mammal-arthropod transmission, but does not affect survival in ticks (Caimano et al., 2004; Ouyang et al., 2008).
BosR (*bb0647*)	chr	Regulatory protein that affects RpoS production and thus promotes production of several known virulence determinants during transmission from tick to mammal and early in mammalian infection, including OspC, DbpAB (Hyde et al., 2009; Ouyang et al., 2011); involved in the oxidative stress response (Boylan et al., 2003; Hyde et al., 2009); binds to regulatory domains upstream of rpoS (Ouyang et al., 2011); exhibits post-transcriptional regulation (Hyde et al., 2007)	Required for infection in mice (Hyde et al., 2009; Ouyang et al., 2009b), including transmission by tick or needle inoculation (Ouyang et al., 2009b)
PlzA (*bb0733*)	chr	PilZ-domain containing protein that binds c-di-GMP, controls virulence gene expression through RpoS and is important for motility (Pitzer et al., 2011; He et al., 2014; Kostick-Dunn et al., 2018).	PlzA contributes to survival in ticks and mouse infection by needle inoculation (Pitzer et al., 2011; Kostick-Dunn et al., 2018). *plzA* mutant *B. burgdorferi* are non-infectious by tick bite transmission (Pitzer et al., 2011).
Rrp2 (*bb0763*)	chr	Two component response regulator required for activation of *rpoN* expression (Ouyang et al., 2008), is activated by acetyl phosphate (Xu et al., 2010a)	Inactivation by point mutation abrogates mouse infection (Boardman et al., 2008; Ouyang et al., 2008).
RpoS (*bb0771*)	chr	Sigma factor that promotes production of several known virulence determinants during transmission from tick to mammal and early in mammalian infection, including OspC, DbpAB (Hubner et al., 2001; Eggers et al., 2004; Yang et al., 2005), is regulated by small non-coding RNA DsrABb (Lybecker and Samuels, 2007) and by BBD18 (Hayes et al., 2014)	Required for infection and persistence in mice (Caimano et al., 2004; Xu et al., 2012; Caimano et al., 2019), and for dissemination within feeding tick (Dunham-Ems et al., 2012)
HrpA (*bb0827*)	chr	ATP-dependent DEAH-box RNA helicase involved in RNA processing and global gene regulation (Salman-Dilgimen et al., 2011; Salman-Dilgimen et al., 2013).	Required for mouse infection and tick transmission (Salman-Dilgimen et al., 2011; Salman-Dilgimen et al., 2013).
Basic cellular functions			
RuvB (*bb0022*)	chr	With RuvA, Holliday junction helicase/branch migrase (Dresser et al., 2009; Lin et al., 2009)	Required for *vlsE* recombination, therefore antigenic variation (Dresser et al., 2009; Lin et al., 2009)
RuvA (*bb0023*)	chr	With RuvB, Holliday junction helicase/branch migrase (Dresser et al., 2009; Lin et al., 2009)	Required for *vlsE* recombination, therefore antigenic variation (Dresser et al., 2009; Lin et al., 2009)
HtrA (*bb0104*)	chr	A serine protease with a number of infection relevant target proteins (Coleman et al., 2013; Gherardini, 2013; Kariu et al., 2013; Russell et al., 2013; Russell and Johnson, 2013; Coleman et al., 2016, 2018). Induced during stationary phase growth and at elevated temperatures (Ye et al., 2016). Sensitive to inactivation by salt and zinc (Russell et al., 2016). Important for spirochete growth and cellular organization at 37°C (Ye et al., 2016).	Essential for establishment of mammalian infection (Ye et al., 2016).
BmtA (*bb0219*)	chr	Manganese transporter (Ouyang et al., 2009a)	Required for mammalian and tick infection (Ouyang et al., 2009a)
Lon−1 (*bb0253*)	chr	ATP− and Mg-dependent protease, does not complement *E. coli lon* mutant (Coleman et al., 2009), slower growth in vitro and reduced resistance to osmotic and oxidative stress, increased levels of OspC, RpoS, BosR (Thompson et al., 2020)	Significantly attenuated in mice at single dose and time point evaluated (Thompson et al., 2020)
BB0318 (*bb0318*)	chr	ATPase component of a putative riboflavin ABC transport system (Deka et al., 2013; Showman et al., 2016).	Identified in IVET screen and is critical for resistance to oxidative stress, macrophage killing and mouse infectivity (Ellis et al., 2013; Showman et al., 2016).

Lon-2 (*bb0613*)	chr	ATP- and Mg-dependent protease complements *E. coli lon* mutant (Coleman et al., 2009), cotranscribed with *tig*, *clpP*, *clpX*, not required for growth in culture, decreased resistance to osmotic and oxidative stress, decreased OspC and RpoS levels (Mason et al., 2020).	Attenuated in mice at single dose and time point tested (Mason et al., 2020).
PtsG (*bb0645*)	chr	Putative glucose specific EIIBC component of a phosphoenolpyruvate phosphotransferase system and affects *B. burgdorferi* gene expression (PEP-PTS) (Khajanchi et al., 2015).	Critical for mouse infection by needle inoculation and tick bite transmission but dispensible for survival in ticks (Khajanchi et al., 2015).
CoADR (*bb0728*)	chr	Coenzyme A disulphide reductase likely important for intracellular redox and the oxidative stress response(Boylan et al., 2006; Eggers et al., 2011)	Required for infection of mice and important for survival in feeding nymphs (Eggers et al., 2011)
OppAV (*bba34*)	lp56	Oligopeptide substrate binding protein component of the Opp ABC transport system, regulated by RpoS (Kornacki and Oliver, 1998; Lin et al., 2001b; Wang et al., 2002; Wang et al., 2004; Caimano et al., 2007; Medrano et al., 2007; Dunham-Ems et al., 2012; Groshong et al., 2017; Caimano et al., 2019).	Important for persistence in mice (Caimano et al., 2019).
GuaA and B (*bbb17,18*)	cp26	GMP synthase and IMP dehydrogenase, respectively, involved in purine salvage (Margolis et al., 1994; Jewett et al., 2009)	Transposon mutagenesis suggested *guaB* is required for mouse infection (Botkin et al., 2006), promote replication in tick, required for mouse infection *via* needle inoculation and tick bite (Jewett et al., 2009)
BBB23, BBB23 (*bbb22,23*)	cp26	Purine permeases essential for hypoxanthine transport and critical for adenine and guanine transport (Jain et al., 2012)	Spirochetes lacking both *bbb22* and *bbb23* are non-infectious in mice. *bbb22* alone is sufficient to restore infectivity, but *bbb23* alone is not (Jain et al., 2012).
PncA (*bbe22*)	lp25	Nicotinamidase involved in NAD metabolism (Purser et al., 2003)	Inactivation results in loss of infectivity in mice (Purser et al., 2003)
AdeC (*bbk17*)	lp36	Adenine deaminase; converts adenine to hypoxanthine. Involved in purine interconversion (Jewett et al., 2007b).	Required for full infectivity in mice, but not in ticks (Jewett et al., 2007b)
Hypothetical proteins of unknown function			
BB0238 (*bb0238*)	chr	A tetratricopeptde repeat (TPR) domain containing protein that localizes to the inner membrane (Groshong et al., 2014)). Interacts with the N-terminus of BB0323 to promote posttranslational protein stability (Kariu et al., 2015).	Critical for mammalian infection and persistence by promoting spirochete burden in mouse tissues as well as heart and joint inflammation (Groshong et al., 2014; Kariu et al., 2015).
BB0323 (*bb0323*)	chr	A membrane-associated immunogenic protein that contains a C-terminal LysM domain and is important for spirochete morphology and cell division (Stewart et al., 2004; Zhang et al., 2009). The protein undergoes multiple proteolysis events mediated by HtrA (BB0104) and interacts with BB0238 (Kariu et al., 2015).	Essential for infectivity and persistence in the mouse and transmission throughout the enzootic cycle (Zhang et al., 2009).
BB0646 (*bb0646*)	chr	Lipolytic and hemolytic activities (Shaw et al., 2012)	Inactivation leads to reduced bacterial loads in mice (Shaw et al., 2012)
BB0744 (*bb0744*)	chr	Identified as possible adhesin but is subsurface (Wager et al., 2015)	Mutant is attenuated in mice (Wager et al., 2015)
BBA03 (*bba03*)	lp54	Induced during tick feeding (Bestor et al., 2012).	Provides a competitive advantage during tick transmission (Bestor et al., 2012).
BBA07 (*bba07*)	lp54	A surface exposed lipoprotein that is induced during tick feeding (Xu et al., 2010b).	Contribution to transmission by tick (Xu et al., 2010b).
BBA64 (*bba64*)	lp54	A surface exposed, immunogenic lipoprotein (Brooks et al., 2006; Nowalk et al., 2006a; Barbour et al., 2008).	Contributes to transmission by tick (Gilmore et al., 2010; Patton et al., 2011).
BBA66 (*bba66*)	lp54	Surface exposed, immunogenic lipoprotein (Brooks et al., 2006; Nowalk et al., 2006a; Barbour et al., 2008)..	Contributes to transmission by tick (Patton et al., 2013).
bbd07	lp17	317 bp intergenic transcript (Casselli et al., 2019)	Contributes to colonization of mouse tissues as well as induction of carditis and arthritis (Casselli et al., 2019).
BBK13 (*bbk13*)	lp36	Non-surface exposed immunogenic protein (Aranjuez et al., 2019).	Promotes *B. burgdorferi* proliferation and dissemination, especially in skin (Aranjuez et al., 2019)
Adhesion			
Lmp1 (*bb0210*)	chr	A surface localized membrane protein (Yang et al., 2009) that binds chondroitin-6-sulfate (Yang et al., 2016) and undergoes proteolytic processing mediated, in part, by HtrA (BB0104) (Zhuang X et al., 2018 doi: 10.1111/cmi.12855).	Selected using *in vivo* phage display for binding to the vascular endothelium (Antonara et al., 2007), protects against host acquired immunity (Yang et al., 2009), facilitates *B. burgdorferi* dissemination through ticks (Koci J. et al., 2018 doi: 10.1038/s41598-018-20208-4)
BB0172 (*bb0172*)	chr	Recombinant protein binds integrin α3β1 (Wood et al., 2013)	Immunization with BB0172 peptides protects mice against *B. burgdorferi* infection (Small et al., 2014; Hassan et al., 2019)
BB0347 (*bb0347*)	chr	Binds fibronectin (Gaultney et al., 2013)	Does not have a role in vascular interactions in skin (Moriarty et al., 2012), but increases binding in mouse joint tissue 1 hr post-inoculation (Caine and Coburn, 2015)
BmpA (*bb0383*), BmpD (*bb0385*)	chr	BmpA binds laminin (Verma et al., 2009)	BmpA/B contribute to joint persistence and arthritis (Pal et al., 2008), BmpD selected using *in vivo* phage display for binding to the vascular endothelium (Antonara et al., 2007)
BB0405, BB0406 (*bb0405, bb0406*)	chr	Closely related, BB0406 binds laminin (Bista et al., 2020)	Each is required for optimal infection in mice (Shrestha et al., 2017; Bista et al., 2020)

Bgp (*bb0588*)	chr	Binds heparin, dermatan sulfate GAGs (Parveen et al., 2006), aggrecan (Russell and Johnson, 2013), has nucleosidase activity (Parveen et al., 2006)	Not essential for infection of mice, but promotes colonization (Parveen et al., 2006; Schlachter et al., 2018)
P66 (*bb0603*)	chr	Porin (Skare et al., 1997; Pinne et al., 2007; Barcena-Uribarri et al., 2010; Barcena-Uribarri et al., 2013), binds certain integrins (Coburn et al., 1999; Coburn and Cugini, 2003). Inactivation in noninfectious *B. burgdorferi* reduces binding to integrins and eliminates specific channel forming activity (Coburn and Cugini, 2003; Pinne et al., 2007).	Selected using *in vivo* phage display for binding to the vascular endothelium (Antonara et al., 2007), inactivation in infectious *B. burgdorferi* results in loss of infectivity in mice (Ristow et al., 2012); disruption of integrin binding activity decreases dissemination in mice (Ristow et al., 2015) and vascular transmigration in mice (Kumar et al., 2015).
DbpA and B (*bba24, bba25*)	lp54	Binding to extracellular matrix (ECM) component decorin, GAGs (Guo et al., 1998; Fischer et al., 2003), 3 critical lysines required for binding (Brown et al., 1999), allelic variation influences binding to ECM and cells (Benoit et al., 2011)	Lack of expression of either *dbpA* or *dbpB* results in reduced virulence, (Blevins et al., 2008; Hyde et al., 2011), less so when delivered by tick bite than by needle inoculation of cultured bacteria (Blevins et al., 2008), effects seen primarily early in infection (Weening et al., 2008; Imai et al., 2013) 3 critical lysines required for full virulence (Fortune et al., 2014), promotes joint colonization and arthritis in mice (Lin et al., 2014; Salo et al., 2015), allelic variation influences tissue tropism (Shi et al., 2008; Lin et al., 2014)
BBA33 (*bba33*)	lp54	Binds to collagen types IV and VI (Zhi et al., 2015)	Transposon insertion attenuated in mice (Lin et al., 2012), targeted mutant cleared from mice by day 4 (Zhi et al., 2015)
OspC *(bbb19)*	cp26	Increased expression under mammalian tissue conditions (37°C) (Schwan and Piesman, 2000), binds a tick salivary protein that inhibits complement (Ramamoorthi et al., 2005; Schuijt et al., 2008), binds plasminogen (Lagal et al., 2006), binds complement factor C4b and inhibits complement cascade (Caine et al., 2017); binds fibrinogen (Bierwagen et al., 2019), binds fibronectin and/or dermatan sulfate GAGs depending on type (Lin et al., 2020c)	Induced in midgut of feeding tick (Schwan et al., 1995; Schwan and Piesman, 2000); required for early stages of infection (Grimm et al., 2004b; Stewart et al., 2006; Tilly et al., 2006), antiphagocytic (Carrasco et al., 2015), promotes bloodstream survival (Caine and Coburn, 2015; Caine et al., 2017), selected using *in vivo* phage display for binding to the vascular endothelium (Antonara et al., 2007); binding to dermatan sulfate GAG required for joint colonization; binding to Fn and GAG required for infectivity (Lin et al., 2020c)
BBK32 (*bbk32*)	lp36	Binds to fibronectin and GAGs (Probert and Johnson, 1998; Probert et al., 2001; Fischer et al., 2006); binds complement protein C1r and blocks activation of complement C1r and thus C1s (Garcia et al., 2016)	Quantifiable attenuation of *bbk32* mutant on overall infectivity in mice (Hyde et al., 2011) (Li et al., 2006; Seshu et al., 2006); mediates interactions with the vasculature in mice (Moriarty et al., 2008; Norman et al., 2008; Moriarty et al., 2012), affects tissue tropism in mice (Lin et al., 2015), GAG binding domain confers bloodstream survival in mice (Caine and Coburn, 2015) and promotes joint colonization (Lin et al., 2015)
RevA (*bbm27, bbp27*)	cp32s	Binds fibronectin, laminin (Brissette et al., 2009a; Moriarty et al., 2012)	Promotes heart colonization (Lin et al., 2012; Byram et al., 2015); but gain of function decreases bacterial binding in heart (Caine and Coburn, 2015), does not have a role in vascular interactions in skin (Moriarty et al., 2012);
Immune evasion			
BBA57 (*bba57*)	lp54	Lipoprotein originally of unknown function (Yang et al., 2013), modulates expression of multiple infection relevant *B. burgdorferi* proteins (Bernard et al., 2018)	Supports survival during early infection by protecting against neutrophils, suppressing host complement-mediated killing and regulating type I interferon and triggers neutrophil chemotaxis and inflammation during late disseminated infection (Yang et al., 2013; Bernard et al., 2018)
CspA (aka CRASP-1) (*bba68*)	lp54	Complement regulator-acquiring surface protein-1, binds to complement inhibitory factor H and FHL-1 (Kraiczy et al., 2001b; Kraiczy et al., 2001c); binds terminal complement factor C7 and C9, blocks C9 polymerization (Hallstrom et al., 2013), confers serum resistance (Brooks et al., 2005; Kenedy et al., 2009; Hammerschmidt et al., 2014); inactivation of gene decreases resistance to killing by normal human serum (Brooks et al., 2005; Kenedy et al., 2009), binds plasminogen (Hallstrom et al., 2010)	Produced during transmission/adaptation to mammal but not in tick (von Lackum et al., 2005; Bykowski et al., 2007) required to survive in feeding nymphal ticks and for transmission to mice (Hart et al., 2018)
OspC (*bbb19*)	cp26	Increased expression under mammalian tissue conditions (37°C) (Schwan and Piesman, 2000), binds a tick salivary protein that inhibits complement (Ramamoorthi et al., 2005; Schuijt et al., 2008), binds plasminogen (Lagal et al., 2006), binds complement factor C4b and inhibits complement cascade (Caine et al., 2017); binds fibrinogen (Bierwagen et al., 2019), binds fibronectin and/or dermatan sulfate GAGs depending on type (Lin et al., 2020c)	Induced in midgut of feeding tick (Schwan et al., 1995; Schwan and Piesman, 2000); required for early stages of infection (Grimm et al., 2004b; Stewart et al., 2006; Tilly et al., 2006), antiphagocytic (Carrasco et al., 2015), promotes bloodstream survival (Caine and Coburn, 2015; Caine et al., 2017), selected using *in vivo* phage display for binding to the vascular endothelium (Antonara et al., 2007)
VlsE (*bbf32*)	lp28-1	Antigenic variation system (Zhang et al., 1997; Zhang and Norris, 1998a, b), unusual recombination requirements for RuvA-RuvB Holliday junction helicase/branch migrase (Dresser et al., 2009; Lin et al., 2009; Castellanos et al., 2018; Verhey et al., 2018b, a, 2019) (reviewed in (Norris, 2006; Chaconas et al., 2020))	Antigenic variation system active during infection (Zhang and Norris, 1998b), complete *vls* locus required for full capacity for recombination and therefore evasion of adaptive immunity and persistence in immunocompetent mice (Zhang et al., 1997; Zhang and Norris, 1998a, b; Bankhead and Chaconas, 2007), changes recently tracked and quantified (Castellanos et al., 2018; Verhey et al., 2018b, a, 2019) and locus calculated to encode on the order of 10^{40} variants, selected using *in vivo* phage display for binding to the vascular endothelium (Antonara et al., 2007)
Crasp2 (aka CspZ) (*bbh06*)	lp28-3	Complement regulator-acquiring surface protein-2, binds to complement inhibitory factors H and FHL-1 (Hartmann et al., 2006; Haupt et al., 2007), confers serum resistance (Hartmann et al., 2006). CspZ production is induced by human blood leading to complement resistance (Marcinkiewicz et al., 2019).	Produced during mammalian infection but not in ticks (Bykowski et al., 2007; Marcinkiewicz et al., 2019), inactivation reduces dissemination and colonization in vertebrate and avian hosts (Marcinkiewicz et al., 2019)

References

Adams, P.P., Flores Avile, C., Popitsch, N., Bilusic, I., Schroeder, R., Lybecker, M., and Jewett, M.W. (2017). In vivo expression technology and 5' end mapping of the Borrelia burgdorferi transcriptome identify novel RNAs expressed during mammalian infection. Nucleic Acids Research 45, 775-792. https://doi.org/10.1093/nar/gkw1180

Aguirre, J.D., Clark, H.M., McIlvin, M., Vazquez, C., Palmere, S.L., Grab, D.J., Seshu, J., Hart, P.J., Saito, M., and Culotta, V.C. (2013). A manganese-rich environment supports superoxide dismutase activity in a Lyme disease pathogen, Borrelia burgdorferi. The Journal of Biological Chemistry 288, 8468-8478. https://doi.org/10.1074/jbc.M112.433540

Akins, D.R., Bourell, K.W., Caimano, M.J., Norgard, M.V., and Radolf, J.D. (1998). A new animal model for studying Lyme disease spirochetes in a mammalian host-adapted state. J Clin Invest 101, 2240-2250. https://doi.org/10.1172/JCI2325

Alitalo, A., Meri, T., Lankinen, H., Seppala, I., Lahdenne, P., Hefty, P.S., Akins, D., and Meri, S. (2002). Complement inhibitor factor H binding to Lyme disease spirochetes is mediated by inducible expression of multiple plasmid-encoded outer surface protein E paralogs. J Immunol 169, 3847-3853.

Angel, T.E., Luft, B.J., Yang, X., Nicora, C.D., Camp, D.G., 2nd, Jacobs, J.M., and Smith, R.D. (2010). Proteome analysis of Borrelia burgdorferi response to environmental change. PLoS ONE 5, e13800. https://doi.org/10.1371/journal.pone.0013800

Antonara, S., Chafel, R.M., Lafrance, M., and Coburn, J. (2007). Borrelia burgdorferi adhesins identified using in vivo phage display. Molecular microbiology 66, 262-276.

Antonara, S., Ristow, L., and Coburn, J. (2011). Adhesion mechanisms of Borrelia burgdorferi. Adv Exp Med Biol 715, 35-49. https://doi.org/10.1007/978-94-007-0940-9_3

Aranjuez, G.F., Kuhn, H.W., Adams, P.P., and Jewett, M.W. (2019). Borrelia burgdorferi bbk13 Is Critical for Spirochete Population Expansion in the Skin during Early Infection. Infect Immun 87. https://doi.org/10.1128/IAI.00887-18

Archibald, F.S., and Fridovich, I. (1981). Manganese, superoxide dismutase, and oxygen tolerance in some lactic acid bacteria. Journal of bacteriology 146, 928-936.

Arnold, W.K., Savage, C.R., Antonicello, A.D., and Stevenson, B. (2015). Apparent role for Borrelia burgdorferi LuxS during mammalian infection. Infection and immunity 83, 1347-1353. https://doi.org/10.1128/IAI.00032-15

Arnold, W.K., Savage, C.R., Lethbridge, K.G., Smith, T.C., 2nd, Brissette, C.A., Seshu, J., and Stevenson, B. (2018). Transcriptomic insights on the virulence-controlling CsrA, BadR, RpoN, and RpoS regulatory networks in the Lyme disease spirochete. PLoS ONE 13, e0203286. https://doi.org/10.1371/journal.pone.0203286

Astrand, M., Cuellar, J., Hytonen, J., and Salminen, T.A. (2019). Predicting the ligand-binding properties of Borrelia burgdorferi s.s. Bmp proteins in light of the conserved features of related Borrelia proteins. J Theor Biol 462, 97-108. https://doi.org/10.1016/j.jtbi.2018.11.004

Bachmann, M., Kukkurainen, S., Hytonen, V.P., and Wehrle-Haller, B. (2019). Cell Adhesion by Integrins. Physiological reviews 99, 1655-1699. https://doi.org/10.1152/physrev.00036.2018

Backert, S., Bernegger, S., Skorko-Glonek, J., and Wessler, S. (2018). Extracellular HtrA serine proteases: An emerging new strategy in bacterial pathogenesis. Cell Microbiol 20, e12845. https://doi.org/10.1111/cmi.12845

Bankhead, T., and Chaconas, G. (2007). The role of VlsE antigenic variation in the Lyme disease spirochete: persistence through a mechanism that differs from other pathogens. Mol Microbiol 65, 1547-1558. https://doi.org/10.1111/j.1365-2958.2007.05895.x

Barbour, A.G. (1991). Molecular biology of antigenic variation in Lyme borreliosis and relapsing fever: a comparative analysis. Scand J Infect Dis - Supplementum 77, 88-93.

Barbour, A.G., Carter, C.J., and Sohaskey, C.D. (2000). Surface protein variation by expression site switching in the relapsing fever agent Borrelia hermsii. Infection and immunity 68, 7114-7121.

Barbour, A.G., Jasinskas, A., Kayala, M.A., Davies, D.H., Steere, A.C., Baldi, P., and Felgner, P.L. (2008). A genome-wide proteome array reveals a limited set of immunogens in natural infections of humans and white-footed mice with Borrelia burgdorferi. Infect Immun 76, 3374-3389. https://doi.org/10.1128/IAI.00048-08

Barcena-Uribarri, I., Thein, M., Sacher, A., Bunikis, I., Bonde, M., Bergstrom, S., and Benz, R. (2010). P66 porins are present in both Lyme disease and relapsing fever spirochetes: a comparison of the biophysical properties of P66 porins from six Borrelia species. Biochim Biophys Acta 1798, 1197-1203. https://doi.org/10.1016/j.bbamem.2010.02.011

Barcena-Uribarri, I., Thein, M., Maier, E., Bonde, M., Bergstrom, S., and Benz, R. (2013). Use of nonelectrolytes reveals the channel size and oligomeric constitution of the Borrelia burgdorferi P66 porin. PLoS One 8, e78272. https://doi.org/10.1371/journal.pone.0078272

Barthold, S.W., Moody, K.D., Terwilliger, G.A., Duray, P.H., Jacoby, R.O., and Steere, A.C. (1988). Experimental Lyme arthritis in rats infected with Borrelia burgdorferi. J Infect Dis 157, 842-846.

Barthold, S.W., Persing, D.H., Armstrong, A.L., and Peeples, R.A. (1991). Kinetics of Borrelia burgdorferi dissemination and evolution of disease after intradermal inoculation of mice. Am J Pathol 139, 263-273.

Barthold, S.W., de Souza, M.S., Janotka, J.L., Smith, A.L., and Persing, D.H. (1993). Chronic Lyme borreliosis in the laboratory mouse. Am J Pathol 143, 959-971.

Battisti, J.M., Bono, J.L., Rosa, P.A., Schrumpf, M.E., Schwan, T.G., and Policastro, P.F. (2008). Outer surface protein A protects Lyme disease spirochetes from acquired host immunity in the tick vector. Infection and immunity 76, 5228-5237.

Beckwith, J., and Silhavy, T.J. (1983). Genetic analysis of protein export in Escherichia coli. Methods in Enzymology 97, 3-11. https://doi.org/10.1016/0076-6879(83)97114-8

Behera, A.K., Thorpe, C.M., Kidder, J.M., Smith, W., Hildebrand, E., and Hu, L.T. (2004). Borrelia burgdorferi-induced expression of matrix metalloproteinases from human chondrocytes requires mitogen-activated protein kinase and Janus kinase/signal transducer and activator of transcription signaling pathways. Infect Immun 72, 2864-2871. https://doi.org/10.1128/iai.72.5.2864-2871.2004

Behera, A.K., Hildebrand, E., Scagliotti, J., Steere, A.C., and Hu, L.T. (2005). Induction of host matrix metalloproteinases by Borrelia burgdorferi differs in human and murine lyme arthritis. Infect Immun 73, 126-134. https://doi.org/10.1128/IAI.73.1.126-134.2005

Behera, A.K., Hildebrand, E., Szafranski, J., Hung, H.H., Grodzinsky, A.J., Lafyatis, R., Koch, A.E., Kalish, R., Perides, G., Steere, A.C., et al. (2006). Role of aggrecanase 1 in Lyme arthritis. Arthritis Rheum 54, 3319-3329. https://doi.org/10.1002/art.22128

Behera, A.K., Durand, E., Cugini, C., Antonara, S., Bourassa, L., Hildebrand, E., Hu, L.T., and Coburn, J. (2008). Borrelia burgdorferi BBB07 interaction with integrin α3β1 stimulates production of pro-inflammatory mediators in primary human chondrocytes. Cell Microbiol 10, 320-331.

Benach, J.L., Fernandez Villar, B., Szczepanski, A., and Garcia Monco, J. (1991). Lyme borreliosis: non-specific interactions of the organism with the host. Scandinavian journal of infectious diseases Supplementum 77, 130-135.

Benoit, V.M., Fischer, J.R., Lin, Y.P., Parveen, N., and Leong, J.M. (2011). Allelic variation of the Lyme disease spirochete adhesin DbpA influences spirochetal binding to decorin, dermatan sulfate, and mammalian cells. Infect Immun 79, 3501-3509. https://doi.org/10.1128/IAI.00163-11

Berger, B.W., Kaplan, M.H., Rothenberg, I.R., and Barbour, A.G. (1985). Isolation and characterization of the Lyme disease spirochete from the skin of patients with erythema chronicum migrans. Journal of the American Academy of Dermatology 13, 444-449.

Berger, B.W., Johnson, R.C., Kodner, C., and Coleman, L. (1992). Cultivation of Borrelia burgdorferi from erythema migrans lesions and perilesional skin. Journal of clinical microbiology 30, 359-361.

Bernard, Q., Smith, A.A., Yang, X., Koci, J., Foor, S.D., Cramer, S.D., Zhuang, X., Dwyer, J.E., Lin, Y.P., Mongodin, E.F., et al. (2018). Plasticity in early immune evasion strategies of a bacterial pathogen. Proc Natl Acad Sci U S A 115, E3788-E3797. https://doi.org/10.1073/pnas.1718595115

Bestor, A., Rego, R.O., Tilly, K., and Rosa, P.A. (2012). Competitive advantage of Borrelia burgdorferi with outer surface protein BBA03 during tick-mediated infection of the mammalian host. Infection and immunity 80, 3501-3511. https://doi.org/10.1128/IAI.00521-12

Bierwagen, P., Szpotkowski, K., Jaskolski, M., and Urbanowicz, A. (2019). Borrelia outer surface protein C is capable of human fibrinogen binding. FEBS J 286, 2415-2428. https://doi.org/10.1111/febs.14810

Bista, S., Singh, P., Bernard, Q., Yang, X., Hart, T., Lin, Y.P., Kitsou, C., Singh Rana, V., Zhang, F., Linhardt, R.J., et al. (2020). A Novel Laminin-Binding Protein Mediates Microbial-Endothelial Cell Interactions and Facilitates Dissemination of Lyme Disease Pathogens. J Infect Dis 221, 1438-1447. https://doi.org/10.1093/infdis/jiz626

Bledsoe, H.A., Carroll, J.A., Whelchel, T.R., Farmer, M.A., Dorward, D.W., and Gherardini, F.C. (1994). Isolation and partial characterization of Borrelia burgdorferi inner and outer membranes by using isopycnic centrifugation. Journal of bacteriology 176, 7447-7455.

Blevins, J.S., Revel, A.T., Smith, A.H., Bachlani, G.N., and Norgard, M.V. (2007). Adaptation of a luciferase gene reporter and lac expression system to Borrelia burgdorferi. Appl Environ Microbiol 73, 1501-1513. https://doi.org/10.1128/AEM.02454-06

Blevins, J.S., Hagman, K.E., and Norgard, M.V. (2008). Assessment of decorin-binding protein A to the infectivity of Borrelia burgdorferi in the murine models of needle and tick infection. BMC Microbiol 8, 82. https://doi.org/10.1186/1471-2180-8-82

Blevins, J.S., Xu, H., He, M., Norgard, M.V., Reitzer, L., and Yang, X.F. (2009). Rrp2, a sigma54-dependent transcriptional activator of Borrelia burgdorferi, activates rpoS in an enhancer-independent manner. Journal of bacteriology 191, 2902-2905.

Boardman, B.K., He, M., Ouyang, Z., Xu, H., Pang, X., and Yang, X.F. (2008). Essential role of the response regulator Rrp2 in the infectious cycle of Borrelia burgdorferi. Infection and immunity 76, 3844-3853. https://doi.org/10.1128/IAI.00467-08

Bockenstedt, L., Barthold, S., Deponte, K., Marcantonio, N., and Kantor, F. (1993). Borrelia burgdorferi infection and immunity in mice deficient in the fifth component of complement. Infection and immunity 61, 2104-2107.

Bockenstedt, L.K., Gonzalez, D., Mao, J., Li, M., Belperron, A.A., and Haberman, A. (2014). What ticks do under your skin: two-photon intravital imaging of Ixodes scapularis feeding in the presence of the lyme disease spirochete. Yale J Biol Med 87, 3-13.

Botkin, D.J., Abbott, A.N., Stewart, P.E., Rosa, P.A., Kawabata, H., Watanabe, H., and Norris, S.J. (2006). Identification of potential virulence determinants by Himar1 transposition of infectious Borrelia burgdorferi B31. Infection and immunity 74, 6690-6699. https://doi.org/10.1128/IAI.00993-06

Bourret, T.J., Boylan, J.A., Lawrence, K.A., and Gherardini, F.C. (2011). Nitrosative damage to free and zinc-bound cysteine thiols underlies nitric oxide toxicity in wild-type Borrelia burgdorferi. Mol Microbiol 81, 259-273. https://doi.org/10.1111/j.1365-2958.2011.07691.x

Bourret, T.J., Lawrence, K.A., Shaw, J.A., Lin, T., Norris, S.J., and Gherardini, F.C. (2016). The nucleotide excision repair pathway protects Borrelia burgdorferi from nitrosative stress in Ixodes scapularis ticks. Front Microbiol 7, 1397. https://doi.org/10.3389/fmicb.2016.01397

Boylan, J.A., Posey, J.E., and Gherardini, F.C. (2003). Borrelia oxidative stress response regulator, BosR: a distinctive Zn-dependent transcriptional activator. Proc Natl Acad Sci U S A 100, 11684-11689. https://doi.org/10.1073/pnas.2032956100

Boylan, J.A., Hummel, C.S., Benoit, S., Garcia-Lara, J., Treglown-Downey, J., Crane, E.J., 3rd, and Gherardini, F.C. (2006). Borrelia burgdorferi bb0728 encodes a coenzyme A disulphide reductase whose function suggests a role in intracellular redox and the oxidative stress response. Molecular microbiology 59, 475-486. https://doi.org/10.1111/j.1365-2958.2005.04963.x

Boylan, J.A., and Gherardini, F.C. (2008). Determining the cellular targets of reactive oxygen species in Borrelia burgdorferi. Methods in molecular biology (Clifton, NJ 431, 213-221. https://doi.org/10.1007/978-1-60327-032-8_17

Boylan, J.A., Lawrence, K.A., Downey, J.S., and Gherardini, F.C. (2008). Borrelia burgdorferi membranes are the primary targets of reactive oxygen species. Mol Microbiol 68, 786-799. https://doi.org/10.1111/j.1365-2958.2008.06204.x

Branda, J.A., Strle, K., Nigrovic, L.E., Lantos, P.M., Lepore, T.J., Damle, N.S., Ferraro, M.J., and Steere, A.C. (2017). Evaluation of Modified 2-Tiered Serodiagnostic Testing Algorithms for Early Lyme Disease. Clin Infect Dis 64, 1074-1080. https://doi.org/10.1093/cid/cix043

Brissette, C.A., Bykowski, T., Cooley, A.E., Bowman, A., and Stevenson, B. (2009a). Borrelia burgdorferi RevA antigen binds host fibronectin. Infection and immunity 77, 2802-2812.

Brissette, C.A., Haupt, K., Barthel, D., Cooley, A.E., Bowman, A., Skerka, C., Wallich, R., Zipfel, P.F., Kraiczy, P., and Stevenson, B. (2009b). Borrelia burgdorferi infection-associated surface proteins ErpP, ErpA, and ErpC bind human plasminogen. Infect Immun 77, 300-306. https://doi.org/10.1128/IAI.01133-08

Brissette, C.A., Verma, A., Bowman, A., Cooley, A.E., and Stevenson, B. (2009c). The Borrelia burgdorferi outer-surface protein ErpX binds mammalian laminin. Microbiology (Reading, England) 155, 863-872.

Brissette, C.A., and Gaultney, R.A. (2014). That's my story, and I'm sticking to it--an update on B. burgdorferi adhesins. Front Cell Infect Microbiol 4, 41. https://doi.org/10.3389/fcimb.2014.00041

Brooks, C.S., Hefty, P.S., Jolliff, S.E., and Akins, D.R. (2003). Global analysis of Borrelia burgdorferi genes regulated by mammalian host-specific signals. Infect Immun 71, 3371-3383. https://doi.org/10.1128/iai.71.6.3371-3383.2003

Brooks, C.S., Vuppala, S.R., Jett, A.M., Alitalo, A., Meri, S., and Akins, D.R. (2005). Complement regulator-acquiring surface protein 1 imparts resistance to human serum in Borrelia burgdorferi. J Immunol 175, 3299-3308.

Brooks, C.S., Vuppala, S.R., Jett, A.M., and Akins, D.R. (2006). Identification of Borrelia burgdorferi outer surface proteins. Infect Immun 74, 296-304. https://doi.org/10.1128/IAI.74.1.296-304.2006

Brown, C.R., and Reiner, S.L. (1999). Development of lyme arthritis in mice deficient in inducible nitric oxide synthase. J Infect Dis 179, 1573-1576. https://doi.org/10.1086/314774

Brown, E.L., Guo, B.P., O'Neal, P., and Hook, M. (1999). Adherence of Borrelia burgdorferi. Identification of critical lysine residues in DbpA required for decorin binding. The Journal of biological chemistry 274, 26272-26278.

Brown, E.L., Wooten, R.M., Johnson, B.J., Iozzo, R.V., Smith, A., Dolan, M.C., Guo, B.P., Weis, J.J., and Hook, M. (2001). Resistance to Lyme disease in decorin-deficient mice. J Clin Invest 107, 845-852. https://doi.org/10.1172/JCI11692

Bunikis, J., Mirian, H., Bunikiene, E., and Barbour, A.G. (2001). Non-heritable change of a spirochaete's phenotype by decoration of the cell surface with exogenous lipoproteins. Molecular microbiology 40, 387-396.

Burgdorfer, W., Barbour, A.G., Hayes, S.F., Benach, J.L., Grunwaldt, E., and Davis, J.P. (1982). Lyme disease-a tick-borne spirochetosis? Science (New York, NY 216, 1317-1319.

Burman, N., Bergstrom, S., Restrepo, B.I., and Barbour, A.G. (1990). The variable antigens Vmp7 and Vmp21 of the relapsing fever bacterium Borrelia hermsii are structurally analogous to the VSG proteins of the African trypanosome. Molecular microbiology 4, 1715-1726.

Bykowski, T., Woodman, M.E., Cooley, A.E., Brissette, C.A., Brade, V., Wallich, R., Kraiczy, P., and Stevenson, B. (2007). Coordinated expression of Borrelia burgdorferi complement regulator-acquiring surface proteins during the Lyme disease spirochete's mammal-tick infection cycle. Infect Immun 75, 4227-4236. https://doi.org/10.1128/IAI.00604-07

Byram, R., Gaultney, R.A., Floden, A.M., Hellekson, C., Stone, B.L., Bowman, A., Stevenson, B., Johnson, B.J., and Brissette, C.A. (2015). Borrelia burgdorferi RevA Significantly Affects Pathogenicity and Host Response in the Mouse Model of Lyme Disease. Infection and immunity 83, 3675-3683. https://doi.org/10.1128/IAI.00530-15

Caimano, M.J., Eggers, C.H., Hazlett, K.R., and Radolf, J.D. (2004). RpoS is not central to the general stress response in Borrelia burgdorferi but does control expression of one or more essential virulence determinants. Infect Immun 72, 6433-6445. https://doi.org/10.1128/IAI.72.11.6433-6445.2004

Caimano, M.J. (2005). Cultivation of Borrelia burgdorferi in dialysis membrane chambers in rat peritonea. Current protocols in microbiology Chapter 12, Unit 12C 13. https://doi.org/10.1002/9780471729259.mc12c03s00

Caimano, M.J., Iyer, R., Eggers, C.H., Gonzalez, C., Morton, E.A., Gilbert, M.A., Schwartz, I., and Radolf, J.D. (2007). Analysis of the RpoS regulon in Borrelia burgdorferi in response to mammalian host signals provides insight into RpoS function during the enzootic cycle. Mol Microbiol 65, 1193-1217. https://doi.org/10.1111/j.1365-2958.2007.05860.x

Caimano, M.J., Kenedy, M.R., Kairu, T., Desrosiers, D.C., Harman, M., Dunham-Ems, S., Akins, D.R., Pal, U., and Radolf, J.D. (2011). The hybrid histidine kinase Hk1 is part of a two-component system that is essential for survival of Borrelia burgdorferi in feeding Ixodes scapularis ticks. Infect Immun 79, 3117-3130. https://doi.org/10.1128/IAI.05136-11

Caimano, M.J., Groshong, A.M., Belperron, A., Mao, J., Hawley, K.L., Luthra, A., Graham, D.E., Earnhart, C.G., Marconi, R.T., Bockenstedt, L.K., et al. (2019). The RpoS Gatekeeper in Borrelia burgdorferi: An Invariant Regulatory Scheme That Promotes Spirochete Persistence in Reservoir Hosts and Niche Diversity. Front Microbiol 10, 1923. https://doi.org/10.3389/fmicb.2019.01923

Caine, J.A., and Coburn, J. (2015). A short-term Borrelia burgdorferi infection model identifies tissue tropisms and bloodstream survival conferred by adhesion proteins. Infect Immun 83, 3184-3194. https://doi.org/10.1128/IAI.00349-15

Caine, J.A., Lin, Y.P., Kessler, J.R., Sato, H., Leong, J.M., and Coburn, J. (2017). Borrelia burgdorferi outer surface protein C (OspC) binds complement component C4b and confers bloodstream survival. Cell Microbiol 19. https://doi.org/10.1111/cmi.12786

Carrasco, S.E., Troxell, B., Yang, Y., Brandt, S.L., Li, H., Sandusky, G.E., Condon, K.W., Serezani, C.H., and Yang, X.F. (2015). Outer surface protein OspC is an antiphagocytic factor that protects Borrelia burgdorferi from phagocytosis by macrophages. Infect Immun 83, 4848-4860. https://doi.org/10.1128/IAI.01215-15

Carroll, J.A., Garon, C.F., and Schwan, T.G. (1999). Effects of environmental pH on membrane proteins in Borrelia burgdorferi. Infection & Immunity 67, 3181-3187.

Casjens, S., Palmer, N., van Vugt, R., Huang, W.M., Stevenson, B., Rosa, P., Lathigra, R., Sutton, G., Peterson, J., Dodson, R.J., et al. (2000). A bacterial genome in flux: the twelve linear and nine circular extrachromosomal DNAs in an infectious isolate of the Lyme disease spirochete Borrelia burgdorferi. Molecular microbiology 35, 490-516.

Casjens, S.R., Gilcrease, E.B., Vujadinovic, M., Mongodin, E.F., Luft, B.J., Schutzer, S.E., Fraser, C.M., and Qiu, W.G. (2017). Plasmid diversity and phylogenetic consistency in the Lyme disease agent Borrelia burgdorferi. BMC Genomics 18, 165. https://doi.org/10.1186/s12864-017-3553-5

Casselli, T., and Bankhead, T. (2015). Use of in vivo Expression Technology for the Identification of Putative Host Adaptation Factors of the Lyme Disease Spirochete. Journal of molecular microbiology and biotechnology 25, 349-361. https://doi.org/10.1159/000439305

Casselli, T., Crowley, M.A., Highland, M.A., Tourand, Y., and Bankhead, T. (2019). A small intergenic region of lp17 is required for evasion of adaptive immunity and induction of pathology by the Lyme disease spirochete. Cell Microbiol 21, e13029. https://doi.org/10.1111/cmi.13029

Castellanos, M., Verhey, T.B., and Chaconas, G. (2018). A Borrelia burgdorferi mini-vls system that undergoes antigenic switching in mice: investigation of the role of plasmid topology and the long inverted repeat. Molecular microbiology 109, 710-721. https://doi.org/10.1111/mmi.14071

Chaconas, G., Castellanos, M., and Verhey, T.B. (2020). Changing of the guard: How the Lyme disease spirochete subverts the host immune response. The Journal of biological chemistry 295, 301-313. https://doi.org/10.1074/jbc.REV119.008583

Chao, M.C., Abel, S., Davis, B.M., and Waldor, M.K. (2016). The design and analysis of transposon insertion sequencing experiments. Nature reviews 14, 119-128. https://doi.org/10.1038/nrmicro.2015.7

Charon, N.W., Cockburn, A., Li, C., Liu, J., Miller, K.A., Miller, M.R., Motaleb, M.A., and Wolgemuth, C.W. (2012). The unique paradigm of spirochete motility and chemotaxis. Annu Rev Microbiol 66, 349-370. https://doi.org/10.1146/annurev-micro-092611-150145

Chen, T., Xiang, X., Xu, H., Zhang, X., Zhou, B., Yang, Y., Lou, Y., and Yang, X.F. (2018). LtpA, a CdnL-type CarD regulator, is important for the enzootic cycle of the Lyme disease pathogen. Emerg Microbes Infect 7, 126. https://doi.org/10.1038/s41426-018-0122-1

Cluss, R.G., Silverman, D.A., and Stafford, T.R. (2004). Extracellular secretion of the Borrelia burgdorferi Oms28 porin and Bgp, a glycosaminoglycan binding protein. Infection and immunity 72, 6279-6286.

Coburn, J., Leong, J., and Erban, J. (1993). Integrin αIIbβ3 mediates binding of the Lyme disease agent, Borrelia burgdorferi, to human platelets. Proc Natl Acad Sci USA 90, 7058-7063.

Coburn, J., Barthold, S.W., and Leong, J.M. (1994). Diverse Lyme disease spirochetes bind integrin αIIbβ3 on human platelets. Infect Immunity 62, 5559-5567.

Coburn, J., Magoun, L., Bodary, S.C., and Leong, J.M. (1998). Integrins αvβ3 and α5β1 mediate attachment of lyme disease spirochetes to human cells. Infection and immunity 66, 1946-1952.

Coburn, J., Chege, W., Magoun, L., Bodary, S.C., and Leong, J.M. (1999). Characterization of a candidate Borrelia burgdorferi β3-chain integrin ligand identified using a phage display library. Molecular microbiology 34, 926-940.

Coburn, J., and Cugini, C. (2003). Targeted mutation of the outer membrane protein P66 disrupts attachment of the Lyme disease spirochete, Borrelia burgdorferi, to integrin αvβ3. Proc Natl Acad Sci USA 100, 7301-7306.

Coburn, J., Fischer, J.R., and Leong, J.M. (2005). Solving a sticky problem: new genetic approaches to host cell adhesion by the Lyme disease spirochete. Mol Microbiol 57, 1182-1195. https://doi.org/10.1111/j.1365-2958.2005.04759.x

Coburn, J., Leong, J., and Chaconas, G. (2013). Illuminating the roles of the Borrelia burgdorferi adhesins. Trends Microbiol 21, 372-379. https://doi.org/10.1016/j.tim.2013.06.005

Coleman, A.S., Yang, X., Kumar, M., Zhang, X., Promnares, K., Shroder, D., Kenedy, M.R., Anderson, J.F., Akins, D.R., and Pal, U. (2008). Borrelia burgdorferi complement regulator-acquiring surface protein 2 does not contribute to complement resistance or host infectivity. PLoS One 3, 3010e. https://doi.org/10.1371/journal.pone.0003010

Coleman, J., Gebbia, J., Piesman, J., Degan, J., Bugge, T., and Benach, J. (1997). Plasminogen is required for efficient dissemination of Borrelia burgdorferi in ticks and for enhancement of spirochetemia in the vertebrate host. Cell 90, 1111-1119.

Coleman, J.L., Sellati, T.J., Testa, J.E., Kew, R.R., Furie, M.B., and Benach, J.L. (1995). Borrelia burgdorferi binds plasminogen, resulting in enhanced penetration of endothelial monolayers. Infection & Immunity 63, 2478-2484.

Coleman, J.L., Gebbia, J.A., and Benach, J.L. (2001). Borrelia burgdorferi and other bacterial products induce expression and release of the urokinase receptor (CD87). J Immunol 166, 473-480. https://doi.org/10.4049/jimmunol.166.1.473

Coleman, J.L., Katona, L.I., Kuhlow, C., Toledo, A., Okan, N.A., Tokarz, R., and Benach, J.L. (2009). Evidence that two ATP-dependent (Lon) proteases in Borrelia burgdorferi serve different functions. PLoS Pathog 5, e1000676. https://doi.org/10.1371/journal.ppat.1000676

Coleman, J.L., Crowley, J.T., Toledo, A.M., and Benach, J.L. (2013). The HtrA protease of Borrelia burgdorferi degrades outer membrane protein BmpD and chemotaxis phosphatase CheX. Molecular microbiology 88, 619-633. https://doi.org/10.1111/mmi.12213

Coleman, J.L., Toledo, A., and Benach, J.L. (2016). Borrelia burgdorferi HtrA: evidence for twofold proteolysis of outer membrane protein p66. Mol Microbiol 99, 135-150. https://doi.org/10.1111/mmi.13221

Coleman, J.L., Toledo, A., and Benach, J.L. (2018). HtrA of Borrelia burgdorferi Leads to Decreased Swarm Motility and Decreased Production of Pyruvate. mBio 9. https://doi.org/10.1128/mBio.01136-18

Coumou, J., Wagemakers, A., Narasimhan, S., Schuijt, T.J., Ersoz, J.I., Oei, A., de Boer, O.J., Roelofs, J., Fikrig, E., and Hovius, J.W. (2019). The role of mannose binding lectin in the immune response against Borrelia burgdorferi sensu lato. Sci Rep 9, 1431. https://doi.org/10.1038/s41598-018-37922-8

Coutte, L., Botkin, D.J., Gao, L., and Norris, S.J. (2009). Detailed analysis of sequence changes occurring during vlsE antigenic variation in the mouse model of Borrelia burgdorferi infection. PLoS Pathog 5, e1000293.

Cox, D.L., Akins, D.R., Bourell, K.W., Lahdenne, P., Norgard, M.V., and Radolf, J.D. (1996). Limited surface exposure of Borrelia burgdorferi outer surface lipoproteins. Proc Natl Acad Sci U S A 93, 7973-7978.

Craft, J.E., Fischer, D.K., Shimamoto, G.T., and Steere, A.C. (1986). Antigens of Borrelia burgdorferi recognized during Lyme disease. Appearance of a new immunoglobulin M response and expansion of the immunoglobulin G response late in the illness. Journal of Clinical Investigation 78, 934-939.

Crandall, H., Dunn, D.M., Ma, Y., Wooten, R.M., Zachary, J.F., Weis, J.H., Weiss, R.B., and Weis, J.J. (2006). Gene expression profiling reveals unique pathways associated with differential severity of lyme arthritis. J Immunol 177, 7930-7942. https://doi.org/10.4049/jimmunol.177.11.7930

Crowley, J.T., Toledo, A.M., LaRocca, T.J., Coleman, J.L., London, E., and Benach, J.L. (2013). Lipid exchange between Borrelia burgdorferi and host cells. PLoS Pathog 9, e1003109. https://doi.org/10.1371/journal.ppat.1003109

Cuellar, J., Astrand, M., Elovaara, H., Pietikainen, A., Siren, S., Liljeblad, A., Guedez, G., Salminen, T.A., and Hytonen, J. (2020). Structural and Biomolecular Analyses of Borrelia burgdorferi BmpD Reveal a Substrate-Binding Protein of an ABC-Type Nucleoside Transporter Family. Infection and immunity 88. https://doi.org/10.1128/IAI.00962-19

Cugini, C., Medrano, M., Schwan, T.G., and Coburn, J. (2003). Regulation of expression of the Borrelia burgdorferi β3-chain integrin ligand, P66, in ticks and in culture. Infection and immunity 71, 1001-1007.

Dai, Q., Restrepo, B.I., Porcella, S.F., Raffel, S.J., Schwan, T.G., and Barbour, A.G. (2006). Antigenic variation by Borrelia hermsii occurs through recombination between extragenic repetitive elements on linear plasmids. Molecular microbiology 60, 1329-1343. https://doi.org/10.1111/j.1365-2958.2006.05177.x

De Silva, A.M., and Fikrig, E. (1995). Growth and migration of Borrelia burgdorferi in Ixodes ticks during blood feeding. American Journal of Tropical Medicine & Hygiene 53, 397-404.

Deitsch, K.W., Lukehart, S.A., and Stringer, J.R. (2009). Common strategies for antigenic variation by bacterial, fungal and protozoan pathogens. Nature reviews 7, 493-503. https://doi.org/10.1038/nrmicro2145

Deka, R.K., Brautigam, C.A., Biddy, B.A., Liu, W.Z., and Norgard, M.V. (2013). Evidence for an ABC-type riboflavin transporter system in pathogenic spirochetes. mBio 4, e00615-00612. https://doi.org/10.1128/mBio.00615-12

Dowdell, A.S., Murphy, M.D., Azodi, C., Swanson, S.K., Florens, L., Chen, S., and Zuckert, W.R. (2017). Comprehensive spatial analysis of the Borrelia burgdorferi lipoproteome reveals a compartmentalization bias toward the bacterial surface. J Bacteriol 199. https://doi.org/10.1128/JB.00658-16

Dresser, A.R., Hardy, P.O., and Chaconas, G. (2009). Investigation of the genes involved in antigenic switching at the vlsE locus in Borrelia burgdorferi: an essential role for the RuvAB branch migrase. PLoS Pathog 5, e1000680.

Dunham-Ems, S.M., Caimano, M.J., Eggers, C.H., and Radolf, J.D. (2012). Borrelia burgdorferi requires the alternative sigma factor RpoS for dissemination within the vector during tick-to-mammal transmission. PLoS Pathog 8, e1002532. https://doi.org/10.1371/journal.ppat.1002532

Dustin, M.L., and Springer, T.A. Role of lymphocyte adhesion receptors in transient interactions and cell locomotion. [Review]. Annual Review of Immunology 9:27-66, 1991.

Earnhart, C.G., Buckles, E.L., Dumler, J.S., and Marconi, R.T. (2005). Demonstration of OspC type diversity in invasive human Lyme disease isolates and identification of previously uncharacterized epitopes that define the specificity of the OspC murine antibody response. Infection and immunity 73, 7869-7877.

Earnhart, C.G., Rhodes, D.V., Smith, A.A., Yang, X., Tegels, B., Carlyon, J.A., Pal, U., and Marconi, R.T. (2014). Assessment of the potential contribution of the highly conserved C-terminal motif (C10) of Borrelia burgdorferi outer surface protein C in transmission and infectivity. Pathogens and disease 70, 176-184. https://doi.org/10.1111/2049-632X.12119

Eggers, C.H., Caimano, M.J., and Radolf, J.D. (2004). Analysis of promoter elements involved in the transcriptional initiation of RpoS-dependent Borrelia burgdorferi genes. J Bacteriol 186, 7390-7402. https://doi.org/10.1128/JB.186.21.7390-7402.2004

Eggers, C.H., Caimano, M.J., Malizia, R.A., Kariu, T., Cusack, B., Desrosiers, D.C., Hazlett, K.R., Claiborne, A., Pal, U., and Radolf, J.D. (2011). The coenzyme A disulphide reductase of Borrelia burgdorferi is important for rapid growth throughout the enzootic cycle and essential for infection of the mammalian host. Mol Microbiol 82, 679-697. https://doi.org/10.1111/j.1365-2958.2011.07845.x

Eicken, C., Sharma, V., Klabunde, T., Lawrenz, M.B., Hardham, J.M., Norris, S.J., and Sacchettini, J.C. (2002). Crystal structure of Lyme disease variable surface antigen VlsE of Borrelia burgdorferi. J Biol Chem 277, 21691-21696. https://doi.org/10.1074/jbc.M201547200

Ellis, T.C., Jain, S., Linowski, A.K., Rike, K., Bestor, A., Rosa, P.A., Halpern, M., Kurhanewicz, S., and Jewett, M.W. (2013). In vivo expression technology identifies a novel virulence factor critical for Borrelia burgdorferi persistence in mice. PLoS Pathog 9, e1003567. https://doi.org/10.1371/journal.ppat.1003567

Emr, S.D., Hall, M.N., and Silhavy, T.J. (1980a). A mechanism of protein localization: the signal hypothesis and bacteria. The Journal of cell biology 86, 701-711. https://doi.org/10.1083/jcb.86.3.701

Emr, S.D., Hedgpeth, J., Clement, J.M., Silhavy, T.J., and Hofnung, M. (1980b). Sequence analysis of mutations that prevent export of lambda receptor, an Escherichia coli outer membrane protein. Nature 285, 82-85. https://doi.org/10.1038/285082a0

Esteve-Gassent, M.D., Elliott, N.L., and Seshu, J. (2009). sodA is essential for virulence of Borrelia burgdorferi in the murine model of Lyme disease. Molecular microbiology 71, 594-612.

Falkow, S. (1988). Molecular Koch's postulates applied to microbial pathogenicity. Rev Infect Dis 10 Suppl 2, S274-276.

Fischer, J.R., Parveen, N., Magoun, L., and Leong, J.M. (2003). Decorin-binding proteins A and B confer distinct mammalian cell type-specific attachment by Borrelia burgdorferi, the Lyme disease spirochete. Proc Natl Acad Sci U S A 100, 7307-7312. https://doi.org/10.1073/pnas.1231043100

Fischer, J.R., LeBlanc, K.T., and Leong, J.M. (2006). Fibronectin binding protein BBK32 of the Lyme disease spirochete promotes bacterial attachment to glycosaminoglycans. Infection and immunity 74, 435-441.

Fisher, M.A., Grimm, D., Henion, A.K., Elias, A.F., Stewart, P.E., Rosa, P.A., and Gherardini, F.C. (2005). Borrelia burgdorferi sigma54 is required for mammalian infection and vector transmission but not for tick colonization. Proceedings of the National Academy of Sciences of the United States of America 102, 5162-5167. https://doi.org/10.1073/pnas.0408536102

Floden, A.M., Watt, J.A., and Brissette, C.A. (2011). Borrelia burgdorferi enolase is a surface-exposed plasminogen binding protein. PLoS One 6, e27502. https://doi.org/10.1371/journal.pone.0027502

Foley, J. (2015). Mini-review: Strategies for Variation and Evolution of Bacterial Antigens. Comput Struct Biotechnol J 13, 407-416. https://doi.org/10.1016/j.csbj.2015.07.002

Fortune, D.E., Lin, Y.P., Deka, R.K., Groshong, A.M., Moore, B.P., Hagman, K.E., Leong, J.M., Tomchick, D.R., and Blevins, J.S. (2014). Identification of lysine residues in the Borrelia burgdorferi DbpA adhesin required for murine infection. Infect Immun 82, 3186-3198. https://doi.org/10.1128/IAI.02036-14

Fraser, C.M., Casjens, S., Huang, W.M., Sutton, G.G., Clayton, R., Lathigra, R., White, O., Ketchum, K.A., Dodson, R., Hickey, E.K., et al. (1997). Genomic sequence of a Lyme disease spirochaete, Borrelia burgdorferi Nature 390, 580-586.

Frevert, C.W., Felgenhauer, J., Wygrecka, M., Nastase, M.V., and Schaefer, L. (2018). Danger-Associated Molecular Patterns Derived From the Extracellular Matrix Provide Temporal Control of Innate Immunity. J Histochem Cytochem 66, 213-227. https://doi.org/10.1369/0022155417740880

Fuchs, H., Wallich, R., Simon, M.M., and Kramer, M.D. (1994). The outer surface protein A of the spirochete Borrelia burgdorferi is a plasmin(ogen) receptor. Proceedings of the National Academy of Sciences of the United States of America 91, 12594-12598.

Fuchs, H., Simon, M.M., Wallich, R., Bechtel, M., and Kramer, M.D. (1996). Borrelia burgdorferi induces secretion of pro-urokinase-type plasminogen activator by human monocytes. Infection and immunity 64, 4307-4312.

Garcia, B.L., Zhi, H., Wager, B., Hook, M., and Skare, J.T. (2016). Borrelia burgdorferi BBK32 inhibits the classical pathway by blocking activation of the C1 complement complex. PLoS Pathog 12, e1005404. https://doi.org/10.1371/journal.ppat.1005404

Gaultney, R.A., Gonzalez, T., Floden, A.M., and Brissette, C.A. (2013). BB0347, from the lyme disease spirochete Borrelia burgdorferi, is surface exposed and interacts with the CS1 heparin-binding domain of human fibronectin. PLoS One 8, e75643. https://doi.org/10.1371/journal.pone.0075643

Gebbia, J.A., Coleman, J.L., and Benach, J.L. (2001). Borrelia spirochetes upregulate release and activation of matrix metalloproteinase gelatinase B (MMP-9) and collagenase 1 (MMP-1) in human cells. Infect Immun 69, 456-462. https://doi.org/10.1128/IAI.69.1.456-462.2001

Gebbia, J.A., Coleman, J.L., and Benach, J.L. (2004). Selective induction of matrix metalloproteinases by Borrelia burgdorferi via toll-like receptor 2 in monocytes. J Infect Dis 189, 113-119. https://doi.org/10.1086/380414

Gherardini, F.C. (2013). Borrelia burgdorferi HtrA may promote dissemination and irritation. Molecular microbiology 90, 209-213. https://doi.org/10.1111/mmi.12390

Gilmore, R.D., Jr., Howison, R.R., Dietrich, G., Patton, T.G., Clifton, D.R., and Carroll, J.A. (2010). The bba64 gene of Borrelia burgdorferi, the Lyme disease agent, is critical for mammalian infection via tick bite transmission. Proc Natl Acad Sci U S A 107, 7515-7520. https://doi.org/10.1073/pnas.1000268107

Gorfu, G., Rivera-Nieves, J., and Ley, K. (2009). Role of β7 integrins in intestinal lymphocyte homing and retention. Curr Mol Med 9, 836-850. https://doi.org/10.2174/156652409789105525

Grab, D.J., Perides, G., Dumler, J.S., Kim, K.J., Park, J., Kim, Y.V., Nikolskaia, O., Choi, K.S., Stins, M.F., and Kim, K.S. (2005). Borrelia burgdorferi, host-derived proteases, and the blood-brain barrier. Infection and immunity 73, 1014-1022.

Grab, D.J., Nyarko, E., Barat, N.C., Nikolskaia, O.V., and Dumler, J.S. (2007). Chemokine, cytokine and matrix metalloprotease expression by human brain microvascular endothelial cells is enhanced with Anaplasma phagocytophilum-Borrelia burgdorferi coinfection. Clin Vaccine Immunol.

Grimm, D., Eggers, C.H., Caimano, M.J., Tilly, K., Stewart, P.E., Elias, A.F., Radolf, J.D., and Rosa, P.A. (2004a). Experimental assessment of the roles of linear plasmids lp25 and lp28-1 of Borrelia burgdorferi throughout the infectious cycle. Infect Immun 72, 5938-5946. https://doi.org/10.1128/IAI.72.10.5938-5946.2004

Grimm, D., Tilly, K., Byram, R., Stewart, P.E., Krum, J.G., Bueschel, D.M., Schwan, T.G., Policastro, P.F., Elias, A.F., and Rosa, P.A. (2004b). Outer-surface protein C of the Lyme disease spirochete: a protein induced in ticks for infection of mammals. Proc Natl Acad Sci U S A 101, 3142-3147. https://doi.org/10.1073/pnas.0306845101

Groshong, A.M., Fortune, D.E., Moore, B.P., Spencer, H.J., Skinner, R.A., Bellamy, W.T., and Blevins, J.S. (2014). BB0238, a presumed tetratricopeptide repeat-containing protein, is required during Borrelia burgdorferi mammalian infection. Infection and immunity 82, 4292-4306. https://doi.org/10.1128/IAI.01977-14

Groshong, A.M., Dey, A., Bezsonova, I., Caimano, M.J., and Radolf, J.D. (2017). Peptide Uptake Is Essential for Borrelia burgdorferi Viability and Involves Structural and Regulatory Complexity of its Oligopeptide Transporter. mBio 8. https://doi.org/10.1128/mBio.02047-17

Guo, B.P., Brown, E.L., Dorward, D.W., Rosenberg, L.C., and Hook, M. (1998). Decorin-binding adhesins from Borrelia burgdorferi. Molecular microbiology 30, 711-723.

Haile, W.B., Coleman, J.L., and Benach, J.L. (2006). Reciprocal upregulation of urokinase plasminogen activator and its inhibitor, PAI-2, by Borrelia burgdorferi affects bacterial penetration and host-inflammatory response. Cell Microbiol 8, 1349-1360. https://doi.org/10.1111/j.1462-5822.2006.00717.x

Hall, M.N., Emr, S.D., and Silhavy, T.J. (1980). Genetic studies on mechanisms of protein localization in Escherichia coli K-12. J Supramol Struct 13, 147-163. https://doi.org/10.1002/jss.400130203

Hallstrom, T., Haupt, K., Kraiczy, P., Hortschansky, P., Wallich, R., Skerka, C., and Zipfel, P.F. (2010). Complement regulator-acquiring surface protein 1 of Borrelia burgdorferi binds to human bone morphogenic protein 2, several extracellular matrix proteins, and plasminogen. J Infect Dis 202, 490-498. https://doi.org/10.1086/653825

Hallstrom, T., Siegel, C., Morgelin, M., Kraiczy, P., Skerka, C., and Zipfel, P.F. (2013). CspA from Borrelia burgdorferi inhibits the terminal complement pathway. MBio 4. https://doi.org/10.1128/mBio.00481-13

Hammerschmidt, C., Koenigs, A., Siegel, C., Hallstrom, T., Skerka, C., Wallich, R., Zipfel, P.F., and Kraiczy, P. (2014). Versatile roles of CspA orthologs in complement inactivation of serum-resistant Lyme disease spirochetes. Infect Immun 82, 380-392. https://doi.org/10.1128/IAI.01094-13

Hammerschmidt, C., Klevenhaus, Y., Koenigs, A., Hallstrom, T., Fingerle, V., Skerka, C., Pos, K.M., Zipfel, P.F., Wallich, R., and Kraiczy, P. (2016). BGA66 and BGA71 facilitate complement resistance of Borrelia bavariensis by inhibiting assembly of the membrane attack complex. Mol Microbiol 99, 407-424. https://doi.org/10.1111/mmi.13239

Harman, M.W., Dunham-Ems, S.M., Caimano, M.J., Belperron, A.A., Bockenstedt, L.K., Fu, H.C., Radolf, J.D., and Wolgemuth, C.W. (2012). The heterogeneous motility of the Lyme disease spirochete in gelatin mimics dissemination through tissue. Proc Natl Acad Sci U S A 109, 3059-3064. https://doi.org/10.1073/pnas.1114362109

Hart, T., Nguyen, N.T.T., Nowak, N.A., Zhang, F., Linhardt, R.J., Diuk-Wasser, M., Ram, S., Kraiczy, P., and Lin, Y.P. (2018). Polymorphic factor H-binding activity of CspA protects Lyme borreliae from the host complement in feeding ticks to facilitate tick-to-host transmission. PLoS Pathog 14, e1007106. https://doi.org/10.1371/journal.ppat.1007106

Hartmann, K., Corvey, C., Skerka, C., Kirschfink, M., Karas, M., Brade, V., Miller, J.C., Stevenson, B., Wallich, R., Zipfel, P.F., et al. (2006). Functional characterization of BbCRASP-2, a distinct outer membrane protein of Borrelia burgdorferi that binds host complement regulators factor H and FHL-1. Mol Microbiol 61, 1220-1236. https://doi.org/10.1111/j.1365-2958.2006.05318.x

Hassan, W.S., Giaretta, P.R., Rech, R., Ollivault-Shiflett, M., and Esteve-Gasent, M.D. (2019). Enhanced protective efficacy of Borrelia burgdorferi BB0172 derived-peptide based vaccine to control Lyme disease. Vaccine 37, 5596-5606. https://doi.org/10.1016/j.vaccine.2019.07.092

Haupt, K., Kraiczy, P., Wallich, R., Brade, V., Skerka, C., and Zipfel, P.F. (2007). Binding of human factor H-related protein 1 to serum-resistant Borrelia burgdorferi is mediated by borrelial complement regulator-acquiring surface proteins. J Infect Dis 196, 124-133.

Hayes, B.M., Dulebohn, D.P., Sarkar, A., Tilly, K., Bestor, A., Ambroggio, X., and Rosa, P.A. (2014). Regulatory protein BBD18 of the lyme disease spirochete: essential role during tick acquisition? mBio 5, e01017-01014. https://doi.org/10.1128/mBio.01017-14

He, M., Ouyang, Z., Troxell, B., Xu, H., Moh, A., Piesman, J., Norgard, M.V., Gomelsky, M., and Yang, X.F. (2011). Cyclic di-GMP is essential for the survival of the Lyme disease spirochete in ticks. PLoS Pathog 7, e1002133. https://doi.org/10.1371/journal.ppat.1002133

He, M., Zhang, J.J., Ye, M., Lou, Y., and Yang, X.F. (2014). Cyclic Di-GMP receptor PlzA controls virulence gene expression through RpoS in Borrelia burgdorferi. Infect Immun 82, 445-452. https://doi.org/10.1128/IAI.01238-13

Hefty, P.S., Jolliff, S.E., Caimano, M.J., Wikel, S.K., and Akins, D.R. (2002). Changes in temporal and spatial patterns of outer surface lipoprotein expression generate population heterogeneity and antigenic diversity in the Lyme disease spirochete, Borrelia burgdorferi. Infect Immun 70, 3468-3478.

Heilpern, A.J., Wertheim, W., He, J., Perides, G., Bronson, R.T., and Hu, L.T. (2009). Matrix metalloproteinase 9 plays a key role in Lyme arthritis but not in dissemination of Borrelia burgdorferi. Infect Immun 77, 2643-2649. https://doi.org/10.1128/IAI.00214-09

Hu, L.T., Perides, G., Noring, R., and Klempner, M.S. (1995). Binding of human plasminogen to Borrelia burgdorferi. Infection & Immunity 63, 3491-3496.

Hu, L.T., Pratt, S.D., Perides, G., Katz, L., Rogers, R.A., and Klempner, M.S. (1997). Isolation, cloning, and expression of a 70-kilodalton plasminogen binding protein of Borrelia burgdorferi. Infection and immunity 65, 4989-4995.

Hu, L.T., Eskildsen, M.A., Masgala, C., Steere, A.C., Arner, E.C., Pratta, M.A., Grodzinsky, A.J., Loening, A., and Perides, G. (2001). Host metalloproteinases in Lyme arthritis. Arthritis Rheum 44, 1401-1410. https://doi.org/10.1002/1529-0131(200106)44:6<1401::AID-ART234>3.0.CO;2-S

Hubner, A., Yang, X., Nolen, D.M., Popova, T.G., Cabello, F.C., and Norgard, M.V. (2001). Expression of Borrelia burgdorferi OspC and DbpA is controlled by a RpoN-RpoS regulatory pathway. Proc Natl Acad Sci U S A 98, 12724-12729. https://doi.org/10.1073/pnas.231442498

Hubner, A., Revel, A.T., Nolen, D.M., Hagman, K.E., and Norgard, M.V. (2003). Expression of a luxS gene is not required for Borrelia burgdorferi infection of mice via needle inoculation. Infect Immun 71, 2892-2896. https://doi.org/10.1128/iai.71.5.2892-2896.2003

Hudson, C.R., Frye, J.G., Quinn, F.D., and Gherardini, F.C. (2001). Increased expression of Borrelia burgdorferi vlsE in response to human endothelial cell membranes. Molecular microbiology 41, 229-239.

Humphries, J.D., Chastney, M.R., Askari, J.A., and Humphries, M.J. (2019). Signal transduction via integrin adhesion complexes. Current opinion in cell biology 56, 14-21. https://doi.org/10.1016/j.ceb.2018.08.004

Hyde, J.A., Trzeciakowski, J.P., and Skare, J.T. (2007). Borrelia burgdorferi alters its gene expression and antigenic profile in response to CO2 levels. Journal of bacteriology 189, 437-445. https://doi.org/10.1128/JB.01109-06

Hyde, J.A., Shaw, D.K., Smith Iii, R., Trzeciakowski, J.P., and Skare, J.T. (2009). The BosR regulatory protein of Borrelia burgdorferi interfaces with the RpoS regulatory pathway and modulates both the oxidative stress response and pathogenic properties of the Lyme disease spirochete. Mol Microbiol 74, 1344-1355. https://doi.org/10.1111/j.1365-2958.2009.06951.x

Hyde, J.A., Weening, E.H., Chang, M., Trzeciakowski, J.P., Hook, M., Cirillo, J.D., and Skare, J.T. (2011). Bioluminescent imaging of Borrelia burgdorferi in vivo demonstrates that the fibronectin-binding protein BBK32 is required for optimal infectivity. Mol Microbiol 82, 99-113. https://doi.org/10.1111/j.1365-2958.2011.07801.x

Hynes, R.O. (1992). Integrins: versatility, modulation, and signaling in cell adhesion. Cell 69, 11-25.

Hynes, R.O. (2002). Integrins: bidirectional, allosteric signaling machines. Cell 110, 673-687.

Imai, D.M., Samuels, D.S., Feng, S., Hodzic, E., Olsen, K., and Barthold, S.W. (2013). The early dissemination defect attributed to disruption of decorin-binding proteins is abolished in chronic murine Lyme borreliosis. Infection and immunity 81, 1663-1673. https://doi.org/10.1128/IAI.01359-12

Indest, K.J., Howell, J.K., Jacobs, M.B., Scholl-Meeker, D., Norris, S.J., and Philipp, M.T. (2001). Analysis of Borrelia burgdorferi vlsE gene expression and recombination in the tick vector. Infection and immunity 69, 7083-7090. https://doi.org/10.1128/IAI.69.11.7083-7090.2001

Iozzo, R.V., and Schaefer, L. (2010). Proteoglycans in health and disease: novel regulatory signaling mechanisms evoked by the small leucine-rich proteoglycans. FEBS J 277, 3864-3875. https://doi.org/10.1111/j.1742-4658.2010.07797.x

Iyer, R., Kalu, O., Purser, J., Norris, S., Stevenson, B., and Schwartz, I. (2003). Linear and circular plasmid content in Borrelia burgdorferi clinical isolates. Infect Immun 71, 3699-3706. https://doi.org/10.1128/iai.71.7.3699-3706.2003

Iyer, R., Caimano, M.J., Luthra, A., Axline, D., Jr., Corona, A., Iacobas, D.A., Radolf, J.D., and Schwartz, I. (2015). Stage-specific global alterations in the transcriptomes of Lyme disease spirochetes during tick feeding and following mammalian host adaptation. Mol Microbiol 95, 509-538. https://doi.org/10.1111/mmi.12882

Jacobs, J.M., Yang, X., Luft, B.J., Dunn, J.J., Camp, D.G., 2nd, and Smith, R.D. (2005). Proteomic analysis of Lyme disease: global protein comparison of three strains of Borrelia burgdorferi. Proteomics 5, 1446-1453.

Jain, S., Sutchu, S., Rosa, P.A., Byram, R., and Jewett, M.W. (2012). Borrelia burgdorferi harbors a transport system essential for purine salvage and mammalian infection. Infection and immunity 80, 3086-3093. https://doi.org/10.1128/IAI.00514-12

Jain, S., Showman, A.C., and Jewett, M.W. (2015). Molecular dissection of a Borrelia burgdorferi in vivo essential purine transport system. Infection and immunity 83, 2224-2233. https://doi.org/10.1128/IAI.02859-14

Jewett, M.W., Byram, R., Bestor, A., Tilly, K., Lawrence, K., Burtnick, M.N., Gherardini, F., and Rosa, P.A. (2007a). Genetic basis for retention of a critical virulence plasmid of Borrelia burgdorferi. Molecular microbiology 66, 975-990.

Jewett, M.W., Lawrence, K., Bestor, A.C., Tilly, K., Grimm, D., Shaw, P., VanRaden, M., Gherardini, F., and Rosa, P.A. (2007b). The critical role of the linear plasmid lp36 in the infectious cycle of Borrelia burgdorferi. Molecular microbiology 64, 1358-1374.

Jewett, M.W., Lawrence, K.A., Bestor, A., Byram, R., Gherardini, F., and Rosa, P.A. (2009). GuaA and GuaB are essential for Borrelia burgdorferi survival in the tick-mouse infection cycle. Journal of bacteriology 191, 6231-6241.

Jones, K.L., Glickstein, L.J., Damle, N., Sikand, V.K., McHugh, G., and Steere, A.C. (2006). Borrelia burgdorferi genetic markers and disseminated disease in patients with early Lyme disease. J Clin Microbiol 44, 4407-4413. https://doi.org/10.1128/JCM.01077-06

Kariu, T., Yang, X., Marks, C.B., Zhang, X., and Pal, U. (2013). Proteolysis of BB0323 results in two polypeptides that impact physiologic and infectious phenotypes in Borrelia burgdorferi. Molecular microbiology 88, 510-522. https://doi.org/10.1111/mmi.12202

Kariu, T., Sharma, K., Singh, P., Smith, A.A., Backstedt, B., Buyuktanir, O., and Pal, U. (2015). BB0323 and novel virulence determinant BB0238: Borrelia burgdorferi proteins that interact with and stabilize each other and are critical for infectivity. J Infect Dis 211, 462-471. https://doi.org/10.1093/infdis/jiu460

Karna, S.L., Sanjuan, E., Esteve-Gassent, M.D., Miller, C.L., Maruskova, M., and Seshu, J. (2011). CsrA modulates levels of lipoproteins and key regulators of gene expression critical for pathogenic mechanisms of Borrelia burgdorferi. Infection and immunity 79, 732-744. https://doi.org/10.1128/IAI.00882-10

Kawabata, H., Myouga, F., Inagaki, Y., Murai, N., and Watanabe, H. (1998). Genetic and immunological analyses of Vls (VMP-like sequences) of Borrelia burgdorferi. Microbial pathogenesis 24, 155-166. https://doi.org/10.1006/mpat.1997.0183

Kawabata, H., Norris, S.J., and Watanabe, H. (2004). BBE02 disruption mutants of Borrelia burgdorferi B31 have a highly transformable, infectious phenotype. Infection and immunity 72, 7147-7154.

Kenedy, M.R., Vuppala, S.R., Siegel, C., Kraiczy, P., and Akins, D.R. (2009). CspA-mediated binding of human factor H inhibits complement deposition and confers serum resistance in Borrelia burgdorferi. Infect Immun 77, 2773-2782. https://doi.org/10.1128/IAI.00318-09

Kenedy, M.R., Luthra, A., Anand, A., Dunn, J.P., Radolf, J.D., and Akins, D.R. (2014). Structural modeling and physicochemical characterization provide evidence that P66 forms a beta-barrel in the Borrelia burgdorferi outer membrane. J Bacteriol 196, 859-872. https://doi.org/10.1128/JB.01236-13

Khajanchi, B.K., Odeh, E., Gao, L., Jacobs, M.B., Philipp, M.T., Lin, T., and Norris, S.J. (2015). Phosphoenolpyruvate Phosphotransferase System Components Modulate Gene Transcription and Virulence of Borrelia burgdorferi. Infect Immun 84, 754-764. https://doi.org/10.1128/IAI.00917-15

Kirchner, A., Koedel, U., Fingerle, V., Paul, R., Wilske, B., and Pfister, H.W. (2000). Upregulation of matrix metalloproteinase-9 in the cerebrospinal fluid of patients with acute Lyme neuroborreliosis. J Neurol Neurosurg Psychiatry 68, 368-371. https://doi.org/10.1136/jnnp.68.3.368

Klempner, M.S., Noring, R., Epstein, M.P., McCloud, B., Hu, R., Limentani, S.A., and Rogers, R.A. (1995). Binding of human plasminogen and urokinase-type plasminogen activator to the Lyme disease spirochete, Borrelia burgdorferi. J Infect Dis 171, 1258-1265.

Klempner, M.S., Noring, R., Epstein, M.P., McCloud, B., and Rogers, R.A. (1996). Binding of human urokinase type plasminogen activator and plasminogen to Borrelia species. Journal of Infectious Diseases 174, 97-104.

Knirel, Y.A., Sun, Q., Senchenkova, S.N., Perepelov, A.V., Shashkov, A.S., and Xu, J. (2015). O-antigen modifications providing antigenic diversity of Shigella flexneri and underlying genetic mechanisms. Biochemistry Biokhimiia 80, 901-914. https://doi.org/10.1134/S0006297915070093

Koenigs, A., Hammerschmidt, C., Jutras, B.L., Pogoryelov, D., Barthel, D., Skerka, C., Kugelstadt, D., Wallich, R., Stevenson, B., Zipfel, P.F., et al. (2013). BBA70 of Borrelia burgdorferi is a novel plasminogen-binding protein. J Biol Chem 288, 25229-25243. https://doi.org/10.1074/jbc.M112.413872

Kornacki, J.A., and Oliver, D.B. (1998). Lyme disease-causing Borrelia species encode multiple lipoproteins homologous to peptide-binding proteins of ABC-type transporters. Infection and immunity 66, 4115-4122.

Kostick-Dunn, J.L., Izac, J.R., Freedman, J.C., Szkotnicki, L.T., Oliver, L.D., Jr., and Marconi, R.T. (2018). The Borrelia burgdorferi c-di-GMP binding receptors, PlzA and PlzB, are functionally distinct. Front Cell Infect Microbiol 8, 213. https://doi.org/10.3389/fcimb.2018.00213

Kostick, J.L., Szkotnicki, L.T., Rogers, E.A., Bocci, P., Raffaelli, N., and Marconi, R.T. (2011). The diguanylate cyclase, Rrp1, regulates critical steps in the enzootic cycle of the Lyme disease spirochetes. Mol Microbiol 81, 219-231. https://doi.org/10.1111/j.1365-2958.2011.07687.x

Kraiczy, P., Skerka, C., Brade, V., and Zipfel, P.F. (2001a). Further characterization of complement regulator-acquiring surface proteins of Borrelia burgdorferi. Infection and immunity 69, 7800-7809.

Kraiczy, P., Skerka, C., Kirschfink, M., Brade, V., and Zipfel, P.F. (2001b). Immune evasion of Borrelia burgdorferi by acquisition of human complement regulators FHL-1/reconectin and Factor H. Eur J Immunol 31, 1674-1684. https://doi.org/10.1002/1521-4141(200106)31:6<1674::AID-IMMU1674>3.0.CO;2-2

Kraiczy, P., Skerka, C., Kirschfink, M., Zipfel, P.F., and Brade, V. (2001c). Mechanism of complement resistance of pathogenic Borrelia burgdorferi isolates. Int Immunopharmacol 1, 393-401.

Kraiczy, P., Hellwage, J., Skerka, C., Kirschfink, M., Brade, V., Zipfel, P.F., and Wallich, R. (2003). Immune evasion of Borrelia burgdorferi: mapping of a complement-inhibitor factor H-binding site of BbCRASP-3, a novel member of the Erp protein family. Eur J Immunol 33, 697-707. https://doi.org/10.1002/eji.200323571

Kraiczy, P., and Stevenson, B. (2013). Complement regulator-acquiring surface proteins of Borrelia burgdorferi: Structure, function and regulation of gene expression. Ticks Tick Borne Dis 4, 26-34. https://doi.org/10.1016/j.ttbdis.2012.10.039

Kraiczy, P. (2016a). Travelling between two worlds: complement as a gatekeeper for an expanded host range of Lyme disease spirochetes. Vet Sci 3. https://doi.org/10.3390/vetsci3020012

Kraiczy, P. (2016b). Hide and seek: how Lyme disease spirochetes overcome complement attack. Front Immunol 7, 385. https://doi.org/10.3389/fimmu.2016.00385

Kumar, D., Ristow, L.C., Shi, M., Mukherjee, P., Caine, J.A., Lee, W.Y., Kubes, P., Coburn, J., and Chaconas, G. (2015). Intravital Imaging of Vascular Transmigration by the Lyme Spirochete: Requirement for the Integrin Binding Residues of the B. burgdorferi P66 Protein. PLoS Pathog 11, e1005333. https://doi.org/10.1371/journal.ppat.1005333

Kurtenbach, K., De Michelis, S., Etti, S., Schafer, S.M., Sewell, H.S., Brade, V., and Kraiczy, P. (2002). Host association of Borrelia burgdorferi sensu lato--the key role of host complement. Trends Microbiol 10, 74-79.

Kuthejlova, M., Kopecky, J., Stepanova, G., and Macela, A. (2001). Tick salivary gland extract inhibits killing of Borrelia afzelii spirochetes by mouse macrophages. Infection and immunity 69, 575-578. https://doi.org/10.1128/IAI.69.1.575-578.2001

Labandeira-Rey, M., Seshu, J., and Skare, J.T. (2003). The absence of linear plasmid 25 or 28-1 of Borrelia burgdorferi dramatically alters the kinetics of experimental infection via distinct mechanisms. Infect Immun 71, 4608-4613.

Lagal, V., Portnoi, D., Faure, G., Postic, D., and Baranton, G. (2006). Borrelia burgdorferi sensu stricto invasiveness is correlated with OspC-plasminogen affinity. Microbes Infect 8, 645-652. https://doi.org/10.1016/j.micinf.2005.08.017

Larson, R.S., and Springer, T.A. (1990). Structure and function of leukocyte integrins. Immunol Rev 114, 181-217. https://doi.org/10.1111/j.1600-065x.1990.tb00565.x

Lawrence, M.B., and Springer, T.A. (1991). Leukocytes roll on a selectin at physiologic flow rates: distinction from and prerequisite for adhesion through integrins. Cell 65, 859-873.

Lawrence, M.B., Berg, E.L., Butcher, E.C., and Springer, T.A. (1995). Rolling of lymphocytes and neutrophils on peripheral node addressin and subsequent arrest on ICAM-1 in shear flow. European journal of immunology 25, 1025-1031.

Lawrenz, M.B., Kawabata, H., Purser, J.E., and Norris, S.J. (2002). Decreased electroporation efficiency in Borrelia burgdorferi containing linear plasmids lp25 and lp56: impact on transformation of infectious B. burgdorferi. Infection and immunity 70, 4798-4804.

Lawrenz, M.B., Wooten, R.M., Zachary, J.F., Drouin, S.M., Weis, J.J., Wetsel, R.A., and Norris, S.J. (2003). Effect of complement component C3 deficiency on experimental Lyme borreliosis in mice. Infection and immunity 71, 4432-4440.

Lawrenz, M.B., Wooten, R.M., and Norris, S.J. (2004). Effects of vlsE complementation on the infectivity of Borrelia burgdorferi lacking the linear plasmid lp28-1. Infection and immunity 72, 6577-6585. https://doi.org/10.1128/IAI.72.11.6577-6585.2004

Lee, W.Y., Moriarty, T.J., Wong, C.H., Zhou, H., Strieter, R.M., van Rooijen, N., Chaconas, G., and Kubes, P. (2010). An intravascular immune response to Borrelia burgdorferi involves Kupffer cells and iNKT cells. Nat Immunol 11, 295-302. https://doi.org/10.1038/ni.1855

Lee, W.Y., Sanz, M.J., Wong, C.H., Hardy, P.O., Salman-Dilgimen, A., Moriarty, T.J., Chaconas, G., Marques, A., Krawetz, R., Mody, C.H., et al. (2014). Invariant natural killer T cells act as an extravascular cytotoxic barrier for joint-invading Lyme Borrelia. Proceedings of the National Academy of Sciences of the United States of America 111, 13936-13941. https://doi.org/10.1073/pnas.1404769111

Leong, J.M., Morrissey, P.E., Ortega-Barria, E., Pereira, M.E., and Coburn, J. (1995). Hemagglutination and proteoglycan binding by the Lyme disease spirochete, Borrelia burgdorferi. Infection & Immunity 63, 874-883.

Li, X., Liu, X., Beck, D.S., Kantor, F.S., and Fikrig, E. (2006). Borrelia burgdorferi lacking BBK32, a fibronectin-binding protein, retains full pathogenicity. Infect Immun 74, 3305-3313. https://doi.org/10.1128/IAI.02035-05

Li, X., Pal, U., Ramamoorthi, N., Liu, X., Desrosiers, D.C., Eggers, C.H., Anderson, J.F., Radolf, J.D., and Fikrig, E. (2007). The Lyme disease agent Borrelia burgdorferi requires BB0690, a Dps homologue, to persist within ticks. Mol Microbiol 63, 694-710. https://doi.org/10.1111/j.1365-2958.2006.05550.x

Lin, B., Kidder, J.M., Noring, R., Steere, A.C., Klempner, M.S., and Hu, L.T. (2001a). Differences in synovial fluid levels of matrix metalloproteinases suggest separate mechanisms of pathogenesis in Lyme arthritis before and after antibiotic treatment. J Infect Dis 184, 174-180. https://doi.org/10.1086/322000

Lin, B., Short, S.A., Eskildsen, M., Klempner, M.S., and Hu, L.T. (2001b). Functional testing of putative oligopeptide permease (Opp) proteins of Borrelia burgdorferi: a complementation model in opp(-) Escherichia coli. Biochimica et biophysica acta 1499, 222-231.

Lin, T., Gao, L., Edmondson, D.G., Jacobs, M.B., Philipp, M.T., and Norris, S.J. (2009). Central role of the holliday junction helicase RuvAB in vlsE recombination and infectivity of Borrelia burgdorferi. PLoS Pathog 5, e1000679.

Lin, T., Gao, L., Zhang, C., Odeh, E., Jacobs, M.B., Coutte, L., Chaconas, G., Philipp, M.T., and Norris, S.J. (2012). Analysis of an ordered, comprehensive STM mutant library in infectious Borrelia burgdorferi: insights into the genes required for mouse infectivity. PLoS One 7, e47532. https://doi.org/10.1371/journal.pone.0047532

Lin, Y.P., Benoit, V., Yang, X., Martinez-Herranz, R., Pal, U., and Leong, J.M. (2014). Strain-specific variation of the decorin-binding adhesin DbpA influences the tissue tropism of the lyme disease spirochete. PLoS Pathog 10, e1004238. https://doi.org/10.1371/journal.ppat.1004238

Lin, Y.P., Chen, Q., Ritchie, J.A., Dufour, N.P., Fischer, J.R., Coburn, J., and Leong, J.M. (2015). Glycosaminoglycan binding by Borrelia burgdorferi adhesin BBK32 specifically and uniquely promotes joint colonization. Cell Microbiol 17, 860-875. https://doi.org/10.1111/cmi.12407

Lin, Y.P., Diuk-Wasser, M.A., Stevenson, B., and Kraiczy, P. (2020a). Complement Evasion Contributes to Lyme Borreliae-Host Associations. Trends in parasitology. https://doi.org/10.1016/j.pt.2020.04.011

Lin, Y.P., Frye, A.M., Nowak, T.A., and Kraiczy, P. (2020b). New Insights Into CRASP-Mediated Complement Evasion in the Lyme Disease Enzootic Cycle. Frontiers in cellular and infection microbiology 10, 1. https://doi.org/10.3389/fcimb.2020.00001

Lin, Y.P., Tan, X., Caine, J.A., Castellanos, M., Chaconas, G., Coburn, J., and Leong, J.M. (2020c). Strain-specific joint invasion and colonization by Lyme disease spirochetes is promoted by outer surface protein C. PLoS Pathog 16, e1008516. https://doi.org/10.1371/journal.ppat.1008516

Lipsett, S.C., Branda, J.A., and Nigrovic, L.E. (2019). Evaluation of the Modified Two-Tiered Testing Method for Diagnosis of Lyme Disease in Children. Journal of clinical microbiology 57. https://doi.org/10.1128/JCM.00547-19

Liveris, D., Mulay, V., Sandigursky, S., and Schwartz, I. (2008). Borrelia burgdorferi vlsE antigenic variation is not mediated by RecA. Infection and immunity 76, 4009-4018. https://doi.org/10.1128/IAI.00027-08

Lone, A.G., and Bankhead, T. (2020). The Borrelia burgdorferi VlsE Lipoprotein Prevents Antibody Binding to an Arthritis-Related Surface Antigen. Cell Rep 30, 3663-3670 e3665. https://doi.org/10.1016/j.celrep.2020.02.081

Lybecker, M.C., and Samuels, D.S. (2007). Temperature-induced regulation of RpoS by a small RNA in Borrelia burgdorferi. Molecular microbiology 64, 1075-1089.

Magnarelli, L.A., Anderson, J.F., and Barbour, A.G. (1989). Enzyme-linked immunosorbent assays for Lyme disease: reactivity of subunits of Borrelia burgdorferi. Journal of Infectious Diseases 159, 43-49.

Mahan, M.J., Slauch, J.M., Hanna, P.C., Camilli, A., Tobias, J.W., Waldor, M.K., and Mekalanos, J.J. (1993a). Selection for bacterial genes that are specifically induced in host tissues: the hunt for virulence factors. [Review]. Infectious Agents & Disease 2, 263-268.

Mahan, M.J., Slauch, J.M., and Mekalanos, J.J. (1993b). Selection of bacterial virulence genes that are specifically induced in host tissues [see comments]. Science (New York, NY 259, 686-688.

Marcinkiewicz, A.L., Dupuis, A.P., 2nd, Zamba-Campero, M., Nowak, N., Kraiczy, P., Ram, S., Kramer, L.D., and Lin, Y.P. (2019). Blood treatment of Lyme borreliae demonstrates the mechanism of CspZ-mediated complement evasion to promote systemic infection in vertebrate hosts. Cell Microbiol 21, e12998. https://doi.org/10.1111/cmi.12998

Margolis, N., Hogan, D., Tilly, K., and Rosa, P. (1994). Plasmid location of Borrelia purine biosynthesis gene homologs. Journal of bacteriology 176, 6427-6432.

Marques, A.R. (2018). Revisiting the Lyme Disease Serodiagnostic Algorithm: the Momentum Gathers. Journal of clinical microbiology 56. https://doi.org/10.1128/JCM.00749-18

Mason, C., Thompson, C., and Ouyang, Z. (2020). The Lon-2 protease of Borrelia burgdorferi is critical for infection in the mammalian host. Molecular microbiology. https://doi.org/10.1111/mmi.14460

McDowell, J.V., Sung, S.Y., Hu, L.T., and Marconi, R.T. (2002). Evidence that the variable regions of the central domain of VlsE are antigenic during infection with lyme disease spirochetes. Infection and immunity 70, 4196-4203.

McDowell, J.V., Hovis, K.M., Zhang, H., Tran, E., Lankford, J., and Marconi, R.T. (2006). Evidence that the BBA68 protein (BbCRASP-1) of the Lyme disease spirochetes does not contribute to factor H-mediated immune evasion in humans and other animals. Infect Immun 74, 3030-3034. https://doi.org/10.1128/IAI.74.5.3030-3034.2006

Medrano, M.S., Ding, Y., Wang, X.G., Lu, P., Coburn, J., and Hu, L.T. (2007). Regulators of expression of the oligopeptide permease A proteins of Borrelia burgdorferi. Journal of bacteriology 189, 2653-2659.

Metts, M.S., McDowell, J.V., Theisen, M., Hansen, P.R., and Marconi, R.T. (2003). Analysis of the OspE determinants involved in binding of factor H and OspE-targeting antibodies elicited during Borrelia burgdorferi infection in mice. Infect Immun 71, 3587-3596.

Mitchell, P.D., Reed, K.D., Vandermause, M.F., and Melski, J.W. (1993). Isolation of Borrelia burgdorferi from skin biopsy specimens of patients with erythema migrans. Am J Clin Pathol 99, 104-107. https://doi.org/10.1093/ajcp/99.1.104

Mohammed, D., Versaevel, M., Bruyere, C., Alaimo, L., Luciano, M., Vercruysse, E., Proces, A., and Gabriele, S. (2019). Innovative Tools for Mechanobiology: Unraveling Outside-In and Inside-Out Mechanotransduction. Front Bioeng Biotechnol 7, 162. https://doi.org/10.3389/fbioe.2019.00162

Moreth, K., Iozzo, R.V., and Schaefer, L. (2012). Small leucine-rich proteoglycans orchestrate receptor crosstalk during inflammation. Cell cycle (Georgetown, Tex 11, 2084-2091. https://doi.org/10.4161/cc.20316

Moriarty, T.J., Norman, M.U., Colarusso, P., Bankhead, T., Kubes, P., and Chaconas, G. (2008). Real-time high resolution 3D imaging of the Lyme disease spirochete adhering to and escaping from the vasculature of a living host. PLoS Pathog 4, e1000090. https://doi.org/10.1371/journal.ppat.1000090

Moriarty, T.J., Shi, M., Lin, Y.P., Ebady, R., Zhou, H., Odisho, T., Hardy, P.O., Salman-Dilgimen, A., Wu, J., Weening, E.H., et al. (2012). Vascular binding of a pathogen under shear force through mechanistically distinct sequential interactions with host macromolecules. Mol Microbiol 86, 1116-1131. https://doi.org/10.1111/mmi.12045

Narasimhan, S., Sukumaran, B., Bozdogan, U., Thomas, V., Liang, X., DePonte, K., Marcantonio, N., Koski, R.A., Anderson, J.F., Kantor, F., et al. (2007). A tick antioxidant facilitates the Lyme disease agent's successful migration from the mammalian host to the arthropod vector. Cell Host Microbe 2, 7-18. https://doi.org/10.1016/j.chom.2007.06.001

Nishimura, S.L. (2009). Integrin-mediated transforming growth factor-beta activation, a potential therapeutic target in fibrogenic disorders. Am J Pathol 175, 1362-1370. https://doi.org/10.2353/ajpath.2009.090393

Norman, M.U., Moriarty, T.J., Dresser, A.R., Millen, B., Kubes, P., and Chaconas, G. (2008). Molecular mechanisms involved in vascular interactions of the Lyme disease pathogen in a living host. PLoS Pathog 4, e1000169. https://doi.org/10.1371/journal.ppat.1000169

Norris, S.J. (2006). Antigenic variation with a twist--the Borrelia story. Mol Microbiol 60, 1319-1322. https://doi.org/10.1111/j.1365-2958.2006.05204.x

Norris, S.J. (2014). vls antigenic variation systems of Lyme disease Borrelia: eluding host immunity through both random, segmental gene conversion and framework heterogeneity. Microbiol Spectr 2. https://doi.org/10.1128/microbiolspec.MDNA3-0038-2014

Nowalk, A.J., Gilmore, R.D., Jr., and Carroll, J.A. (2006a). Serologic proteome analysis of Borrelia burgdorferi membrane-associated proteins. Infect Immun 74, 3864-3873. https://doi.org/10.1128/IAI.00189-06

Nowalk, A.J., Nolder, C., Clifton, D.R., and Carroll, J.A. (2006b). Comparative proteome analysis of subcellular fractions from Borrelia burgdorferi by NEPHGE and IPG. Proteomics 6, 2121-2134.

Obergfell, K.P., and Seifert, H.S. (2015). Mobile DNA in the Pathogenic Neisseria. Microbiol Spectr 3, MDNA3-0015-2014. https://doi.org/10.1128/microbiolspec.MDNA3-0015-2014

Ojaimi, C., Brooks, C., Casjens, S., Rosa, P., Elias, A., Barbour, A., Jasinskas, A., Benach, J., Katona, L., Radolf, J., et al. (2003). Profiling of temperature-induced changes in Borrelia burgdorferi gene expression by using whole genome arrays. Infect Immun 71, 1689-1705.

Onder, O., Humphrey, P.T., McOmber, B., Korobova, F., Francella, N., Greenbaum, D.C., and Brisson, D. (2012). OspC is potent plasminogen receptor on surface of Borrelia burgdorferi. J Biol Chem 287, 16860-16868. https://doi.org/10.1074/jbc.M111.290775

Ouyang, Z., Blevins, J.S., and Norgard, M.V. (2008). Transcriptional interplay among the regulators Rrp2, RpoN and RpoS in Borrelia burgdorferi. Microbiology (Reading, England) 154, 2641-2658.

Ouyang, Z., He, M., Oman, T., Yang, X.F., and Norgard, M.V. (2009a). A manganese transporter, BB0219 (BmtA), is required for virulence by the Lyme disease spirochete, Borrelia burgdorferi. Proceedings of the National Academy of Sciences of the United States of America 106, 3449-3454. https://doi.org/10.1073/pnas.0812999106

Ouyang, Z., Kumar, M., Kariu, T., Haq, S., Goldberg, M., Pal, U., and Norgard, M.V. (2009b). BosR (BB0647) governs virulence expression in Borrelia burgdorferi. Mol Microbiol 74, 1331-1343. https://doi.org/10.1111/j.1365-2958.2009.06945.x

Ouyang, Z., Deka, R.K., and Norgard, M.V. (2011). BosR (BB0647) controls the RpoN-RpoS regulatory pathway and virulence expression in Borrelia burgdorferi by a novel DNA-binding mechanism. PLoS Pathog 7, e1001272. https://doi.org/10.1371/journal.ppat.1001272

Pal, U., Wang, P., Bao, F., Yang, X., Samanta, S., Schoen, R., Wormser, G.P., Schwartz, I., and Fikrig, E. (2008). Borrelia burgdorferi basic membrane proteins A and B participate in the genesis of Lyme arthritis. J Exp Med 205, 133-141. https://doi.org/10.1084/jem.20070962

Parveen, N., and Leong, J.M. (2000). Identification of a candidate glycosaminoglycan-binding adhesin of the Lyme disease spirochete Borrelia burgdorferi. Molecular microbiology 35, 1220-1234.

Parveen, N., Cornell, K.A., Bono, J.L., Chamberland, C., Rosa, P., and Leong, J.M. (2006). Bgp, a secreted glycosaminoglycan-binding protein of Borrelia burgdorferi strain N40, displays nucleosidase activity and is not essential for infection of immunodeficient mice. Infection and immunity 74, 3016-3020.

Patton, T.G., Dietrich, G., Dolan, M.C., Piesman, J., Carroll, J.A., and Gilmore, R.D., Jr. (2011). Functional analysis of the Borrelia burgdorferi bba64 gene product in murine infection via tick infestation. PLoS ONE 6, e19536. https://doi.org/10.1371/journal.pone.0019536

Patton, T.G., Brandt, K.S., Nolder, C., Clifton, D.R., Carroll, J.A., and Gilmore, R.D. (2013). Borrelia burgdorferi bba66 gene inactivation results in attenuated mouse infection by tick transmission. Infect Immun 81, 2488-2498. https://doi.org/10.1128/IAI.00140-13

Paul, R., Lorenzl, S., Koedel, U., Sporer, B., Vogel, U., Frosch, M., and Pfister, H.W. (1998). Matrix metalloproteinases contribute to the blood-brain barrier disruption during bacterial meningitis. Annals of neurology 44, 592-600. https://doi.org/10.1002/ana.410440404

Perides, G., Noring, R., and Klempner, M.S. (1996). Inhibition of Borrelia burgdorferi-bound fibrinolytic enzymes by alpha2-antiplasmin, PAI-1 and PAI-2. Biochemical & Biophysical Research Communications 219, 690-695.

Perides, G., Tanner-Brown, L.M., Eskildsen, M.A., and Klempner, M.S. (1999). Borrelia burgdorferi induces matrix metalloproteinases by neural cultures. Journal of neuroscience research 58, 779-790. https://doi.org/10.1002/(sici)1097-4547(19991215)58:6<779::aid-jnr5>3.0.co;2-l

Phelan, J.P., Kern, A., Ramsey, M.E., Lundt, M.E., Sharma, B., Lin, T., Gao, L., Norris, S.J., Hyde, J.A., Skare, J.T., et al. (2019). Genome-wide screen identifies novel genes required for Borrelia burgdorferi survival in its Ixodes tick vector. PLoS Pathog 15, e1007644. https://doi.org/10.1371/journal.ppat.1007644

Phillips, Z.N., Tram, G., Seib, K.L., and Atack, J.M. (2019). Phase-variable bacterial loci: how bacteria gamble to maximise fitness in changing environments. Biochemical Society transactions 47, 1131-1141. https://doi.org/10.1042/BST20180633

Pickering, M.C., Cook, H.T., Warren, J., Bygrave, A.E., Moss, J., Walport, M.J., and Botto, M. (2002). Uncontrolled C3 activation causes membranoproliferative glomerulonephritis in mice deficient in complement factor H. Nature genetics 31, 424-428.

Pinne, M., Thein, M., Denker, K., Benz, R., Coburn, J., and Bergstrom, S. (2007). Elimination of channel-forming activity by insertional inactivation of the p66 gene in Borrelia burgdorferi. FEMS Microbiol Lett 266, 241-249. https://doi.org/10.1111/j.1574-6968.2006.00529.x

Pitzer, J.E., Sultan, S.Z., Hayakawa, Y., Hobbs, G., Miller, M.R., and Motaleb, M.A. (2011). Analysis of the Borrelia burgdorferi cyclic-di-GMP-binding protein PlzA reveals a role in motility and virulence. Infect Immun 79, 1815-1825. https://doi.org/10.1128/IAI.00075-11

Plasterk, R.H., Simon, M.I., and Barbour, A.G. (1985). Transposition of structural genes to an expression sequence on a linear plasmid causes antigenic variation in the bacterium Borrelia hermsii. Nature 318, 257-263.

Posey, J.E., and Gherardini, F.C. (2000). Lack of a role for iron in the Lyme disease pathogen. Science 288, 1651-1653.

Probert, W.S., and Johnson, B.J. (1998). Identification of a 47 kDa fibronectin-binding protein expressed by Borrelia burgdorferi isolate B31. Molecular microbiology 30, 1003-1015.

Probert, W.S., Kim, J.H., Hook, M., and Johnson, B.J.B. (2001). Mapping the ligand-binding region of the Borrelia burgdorferi fibronectin-binding protein BBK32. Infection & Immunity 69, 4129-4133.

Purser, J.E., Lawrenz, M.B., Caimano, M.J., Howell, J.K., Radolf, J.D., and Norris, S.J. (2003). A plasmid-encoded nicotinamidase (PncA) is essential for infectivity of Borrelia burgdorferi in a mammalian host. Mol Microbiol 48, 753-764.

Radolf, J.D., Goldberg, M.S., Bourell, K., Baker, S.I., Jones, J.D., and Norgard, M.V. (1995). Characterization of outer membranes isolated from Borrelia burgdorferi, the Lyme disease spirochete. Infection & Immunity 63, 2154-2163.

Raffel, S.J., Battisti, J.M., Fischer, R.J., and Schwan, T.G. (2014). Inactivation of genes for antigenic variation in the relapsing fever spirochete Borrelia hermsii reduces infectivity in mice and transmission by ticks. PLoS Pathog 10, e1004056. https://doi.org/10.1371/journal.ppat.1004056

Ramamoorthi, N., Narasimhan, S., Pal, U., Bao, F., Yang, X.F., Fish, D., Anguita, J., Norgard, M.V., Kantor, F.S., Anderson, J.F., et al. (2005). The Lyme disease agent exploits a tick protein to infect the mammalian host. Nature 436, 573-577. https://doi.org/10.1038/nature03812

Ramsey, M.E., Hyde, J.A., Medina-Perez, D.N., Lin, T., Gao, L., Lundt, M.E., Li, X., Norris, S.J., Skare, J.T., and Hu, L.T. (2017). A high-throughput genetic screen identifies previously uncharacterized Borrelia burgdorferi genes important for resistance against reactive oxygen and nitrogen species. PLoS Pathog 13, e1006225. https://doi.org/10.1371/journal.ppat.1006225

Revel, A.T., Talaat, A.M., and Norgard, M.V. (2002). DNA microarray analysis of differential gene expression in Borrelia burgdorferi, the Lyme disease spirochete. Proc Natl Acad Sci U S A 99, 1562-1567. https://doi.org/10.1073/pnas.032667699

Revel, A.T., Blevins, J.S., Almazan, C., Neil, L., Kocan, K.M., de la Fuente, J., Hagman, K.E., and Norgard, M.V. (2005). bptA (bbe16) is essential for the persistence of the Lyme disease spirochete, Borrelia burgdorferi, in its natural tick vector. Proc Natl Acad Sci U S A 102, 6972-6977. https://doi.org/10.1073/pnas.0502565102

Ribeiro, J.M., Weis, J.J., and Telford, S.R., 3rd (1990). Saliva of the tick Ixodes dammini inhibits neutrophil function. Exp Parasitol 70, 382-388. https://doi.org/10.1016/0014-4894(90)90121-r

Riley, S.P., Bykowski, T., Babb, K., von Lackum, K., and Stevenson, B. (2007). Genetic and physiological characterization of the Borrelia burgdorferi ORF BB0374-pfs-metK-luxS operon. Microbiology (Reading, England) 153, 2304-2311.

Ristow, L.C., Miller, H.E., Padmore, L.J., Chettri, R., Salzman, N., Caimano, M.J., Rosa, P.A., and Coburn, J. (2012). The β3-integrin ligand of Borrelia burgdorferi is critical for infection of mice but not ticks. Molecular microbiology 85, 1105-1118. https://doi.org/10.1111/j.1365-2958.2012.08160.x

Ristow, L.C., Bonde, M., Lin, Y.P., Sato, H., Curtis, M., Wesley, E., Hahn, B.L., Fang, J., Wilcox, D.A., Leong, J.M., et al. (2015). Integrin binding by Borrelia burgdorferi P66 facilitates dissemination but is not required for infectivity. Celular Microbiology 17, 1021-1036. https://doi.org/10.1111/cmi.12418

Rogers, E.A., Terekhova, D., Zhang, H.M., Hovis, K.M., Schwartz, I., and Marconi, R.T. (2009). Rrp1, a cyclic-di-GMP-producing response regulator, is an important regulator of Borrelia burgdorferi core cellular functions. Mol Microbiol 71, 1551-1573. https://doi.org/10.1111/j.1365-2958.2009.06621.x

Rogovskyy, A.S., and Bankhead, T. (2013). Variable VlsE is critical for host reinfection by the Lyme disease spirochete. PLoS One 8, e61226. https://doi.org/10.1371/journal.pone.0061226

Rogovskyy, A.S., Casselli, T., Tourand, Y., Jones, C.R., Owen, J.P., Mason, K.L., Scoles, G.A., and Bankhead, T. (2015). Evaluation of the Importance of VlsE Antigenic Variation for the Enzootic Cycle of Borrelia burgdorferi. PLoS One 10, e0124268. https://doi.org/10.1371/journal.pone.0124268

Rosa, P.A. (1997). Microbiology of Borrelia burgdorferi. Seminars in neurology 17, 5-10. https://doi.org/10.1055/s-2008-1040906

Rossmann, E., Kitiratschky, V., Hofmann, H., Kraiczy, P., Simon, M.M., and Wallich, R. (2006). Borrelia burgdorferi complement regulator-acquiring surface protein 1 of the Lyme disease spirochetes is expressed in humans and induces antibody responses restricted to nondenatured structural determinants. Infection and immunity 74, 7024-7028. https://doi.org/10.1128/IAI.01028-06

Ruoslahti, E. (1996). Integrin signaling and matrix assembly. [Review] [44 refs]. Tumour Biology 17, 117-124.

Ruoslahti, E. (1997). Integrins as signaling molecules and targets for tumor therapy. [Review] [44 refs]. Kidney international 51, 1413-1417.

Russell, T.M., Delorey, M.J., and Johnson, B.J. (2013). Borrelia burgdorferi BbHtrA degrades host ECM proteins and stimulates release of inflammatory cytokines in vitro. Molecular microbiology 90, 241-251. https://doi.org/10.1111/mmi.12377

Russell, T.M., and Johnson, B.J. (2013). Lyme disease spirochaetes possess an aggrecan-binding protease with aggrecanase activity. Mol Microbiol 90, 228-240. https://doi.org/10.1111/mmi.12276

Russell, T.M., Tang, X., Goldstein, J.M., Bagarozzi, D., and Johnson, B.J. (2016). The salt-sensitive structure and zinc inhibition of Borrelia burgdorferi protease BbHtrA. Molecular microbiology 99, 586-596. https://doi.org/10.1111/mmi.13251

Salman-Dilgimen, A., Hardy, P.O., Dresser, A.R., and Chaconas, G. (2011). HrpA, a DEAH-box RNA helicase, is involved in global gene regulation in the Lyme disease spirochete. PLoS ONE 6, e22168. https://doi.org/10.1371/journal.pone.0022168

Salman-Dilgimen, A., Hardy, P.O., Radolf, J.D., Caimano, M.J., and Chaconas, G. (2013). HrpA, an RNA helicase involved in RNA processing, is required for mouse infectivity and tick transmission of the Lyme disease spirochete. PLoS Pathog 9, e1003841. https://doi.org/10.1371/journal.ppat.1003841

Salo, J., Jaatinen, A., Soderstrom, M., Viljanen, M.K., and Hytonen, J. (2015). Decorin binding proteins of Borrelia burgdorferi promote arthritis development and joint specific post-treatment DNA persistence in mice. PLoS One 10, e0121512. https://doi.org/10.1371/journal.pone.0121512

Samuels, D.S. (2011). Gene regulation in Borrelia burgdorferi. Annu Rev Microbiol 65, 479-499. https://doi.org/10.1146/annurev.micro.112408.134040

Schlachter, S., Seshu, J., Lin, T., Norris, S., and Parveen, N. (2018). The Borrelia burgdorferi Glycosaminoglycan Binding Protein Bgp in the B31 Strain Is Not Essential for Infectivity despite Facilitating Adherence and Tissue Colonization. Infection and immunity 86. https://doi.org/10.1128/IAI.00667-17

Schopf, J.W. (2000). Solution to Darwin's dilemma: discovery of the missing Precambrian record of life. Proceedings of the National Academy of Sciences of the United States of America 97, 6947-6953. https://doi.org/10.1073/pnas.97.13.6947

Schuijt, T.J., Hovius, J.W., van Burgel, N.D., Ramamoorthi, N., Fikrig, E., and van Dam, A.P. (2008). The tick salivary protein Salp15 inhibits the killing of serum-sensitive Borrelia burgdorferi sensu lato isolates. Infect Immun 76, 2888-2894. https://doi.org/10.1128/IAI.00232-08

Schuijt, T.J., Coumou, J., Narasimhan, S., Dai, J., Deponte, K., Wouters, D., Brouwer, M., Oei, A., Roelofs, J.J., van Dam, A.P., et al. (2011). A tick mannose-binding lectin inhibitor interferes with the vertebrate complement cascade to enhance transmission of the lyme disease agent. Cell Host Microbe 10, 136-146. https://doi.org/10.1016/j.chom.2011.06.010

Schulze, R.J., and Zuckert, W.R. (2006). Borrelia burgdorferi lipoproteins are secreted to the outer surface by default. Mol Microbiol 59, 1473-1484. https://doi.org/10.1111/j.1365-2958.2006.05039.x

Schwan, T.G., Piesman, J., Golde, W.T., Dolan, M.C., and Rosa, P.A. (1995). Induction of an outer surface protein on Borrelia burgdorferi during tick feeding. Proc Natl Acad Sci U S A 92, 2909-2913.

Schwan, T.G., and Hinnebusch, B.J. (1998). Bloodstream- versus tick-associated variants of a relapsing fever bacterium. Science (New York, NY 280, 1938-1940.

Schwan, T.G., and Piesman, J. (2000). Temporal changes in outer surface proteins A and C of the lyme disease-associated spirochete, Borrelia burgdorferi, during the chain of infection in ticks and mice. Journal of clinical microbiology 38, 382-388.

Schwan, T.G., Raffel, S.J., and Battisti, J.M. (2020). Transgenic functional complementation with a transmission -associated protein restores spirochete infectivity by tick bite. Ticks and tick-borne diseases 11, 101377. https://doi.org/10.1016/j.ttbdis.2020.101377

Seiler, K.P., Vavrin, Z., Eichwald, E., Hibbs, J.B., Jr., and Weis, J.J. (1995). Nitric oxide production during murine Lyme disease: lack of involvement in host resistance or pathology. Infection & Immunity 63, 3886-3895.

Seshu, J., Boylan, J.A., Gherardini, F.C., and Skare, J.T. (2004). Dissolved oxygen levels alter gene expression and antigen profiles in Borrelia burgdorferi. Infection and immunity 72, 1580-1586.

Seshu, J., Esteve-Gassent, M.D., Labandeira-Rey, M., Kim, J.H., Trzeciakowski, J.P., Hook, M., and Skare, J.T. (2006). Inactivation of the fibronectin-binding adhesin gene bbk32 significantly attenuates the infectivity potential of Borrelia burgdorferi. Mol Microbiol 59, 1591-1601. https://doi.org/10.1111/j.1365-2958.2005.05042.x

Setubal, J.C., Reis, M., Matsunaga, J., and Haake, D.A. (2006). Lipoprotein computational prediction in spirochaetal genomes. Microbiology 152, 113-121. https://doi.org/10.1099/mic.0.28317-0

Shaw, D.K., Hyde, J.A., and Skare, J.T. (2012). The BB0646 protein demonstrates lipase and haemolytic activity associated with Borrelia burgdorferi, the aetiological agent of Lyme disease. Molecular microbiology 83, 319-334. https://doi.org/10.1111/j.1365-2958.2011.07932.x

Shi, Y., Xu, Q., McShan, K., and Liang, F.T. (2008). Both decorin-binding proteins A and B are critical for the overall virulence of Borrelia burgdorferi. Infect Immun 76, 1239-1246. https://doi.org/10.1128/IAI.00897-07

Showman, A.C., Aranjuez, G., Adams, P.P., and Jewett, M.W. (2016). Gene bb0318 Is Critical for the Oxidative Stress Response and Infectivity of Borrelia burgdorferi. Infection and immunity 84, 3141-3151. https://doi.org/10.1128/IAI.00430-16

Shrestha, B., Kenedy, M.R., and Akins, D.R. (2017). Outer Membrane Proteins BB0405 and BB0406 Are Immunogenic, but Only BB0405 Is Required for Borrelia burgdorferi Infection. Infection and immunity 85. https://doi.org/10.1128/IAI.00803-16

Siegel, C., Schreiber, J., Haupt, K., Skerka, C., Brade, V., Simon, M.M., Stevenson, B., Wallich, R., Zipfel, P.F., and Kraiczy, P. (2008). Deciphering the ligand-binding sites in the Borrelia burgdorferi complement regulator-acquiring surface protein 2 required for interactions with the human immune regulators factor H and factor H-like protein 1. The Journal of biological chemistry 283, 34855-34863.

Siegel, C., Hallstrom, T., Skerka, C., Eberhardt, H., Uzonyi, B., Beckhaus, T., Karas, M., Wallich, R., Stevenson, B., Zipfel, P.F., et al. (2010). Complement factor H-related proteins CFHR2 and CFHR5 represent novel ligands for the infection-associated CRASP proteins of Borrelia burgdorferi. PLoS ONE 5, e13519. https://doi.org/10.1371/journal.pone.0013519

Skare, J.T., Mirzabekov, T.A., Shang, E.S., Blanco, D.R., Erdjument-Bromage, H., Bunikis, J., Bergstrom, S., Tempst, P., Kagan, B.L., Miller, J.N., et al. (1997). The Oms66 (p66) protein is a Borrelia burgdorferi porin. Infect Immun 65, 3654-3661.

Skare, J.T., Shaw, D.K., Trzeciakowski, J.P., and Hyde, J.A. (2016). in vivo imaging demonstrates that Borrelia burgdorferi ospC is uniquely expressed temporally and spatially throughout experimental infection. PLoS One 11, e0162501. https://doi.org/10.1371/journal.pone.0162501

Skare, J.T., and Garcia, B.L. (2020). Complement Evasion by Lyme Disease Spirochetes. Trends in microbiology. https://doi.org/10.1016/j.tim.2020.05.004

Small, C.M., Ajithdoss, D.K., Rodrigues Hoffmann, A., Mwangi, W., and Esteve-Gassent, M.D. (2014). Immunization with a Borrelia burgdorferi BB0172-derived peptide protects mice against lyme disease. PLoS ONE 9, e88245. https://doi.org/10.1371/journal.pone.0088245

Smith, A.H., Blevins, J.S., Bachlani, G.N., Yang, X.F., and Norgard, M.V. (2007). Evidence that RpoS (sigmaS) in Borrelia burgdorferi is controlled directly by RpoN (sigma54/sigmaN). Journal of bacteriology 189, 2139-2144.

Stevenson, B., and Babb, K. (2002). LuxS-mediated quorum sensing in Borrelia burgdorferi, the lyme disease spirochete. Infection and immunity 70, 4099-4105. https://doi.org/10.1128/iai.70.8.4099-4105.2002

Stevenson, B., El-Hage, N., Hines, M.A., Miller, J.C., and Babb, K. (2002). Differential binding of host complement inhibitor factor H by Borrelia burgdorferi Erp surface proteins: a possible mechanism underlying the expansive host range of Lyme disease spirochetes. Infect Immun 70, 491-497.

Stewart, P.E., Hoff, J., Fischer, E., Krum, J.G., and Rosa, P.A. (2004). Genome-wide transposon mutagenesis of Borrelia burgdorferi for identification of phenotypic mutants. Applied and environmental microbiology 70, 5973-5979. https://doi.org/10.1128/AEM.70.10.5973-5979.2004

Stewart, P.E., Wang, X., Bueschel, D.M., Clifton, D.R., Grimm, D., Tilly, K., Carroll, J.A., Weis, J.J., and Rosa, P.A. (2006). Delineating the requirement for the Borrelia burgdorferi virulence factor OspC in the mammalian host. Infect Immun 74, 3547-3553. https://doi.org/10.1128/IAI.00158-06

Striz, I., and Costabel, U. (1992). The role of integrins in the immune response. Sarcoidosis 9, 88-94.

Sultan, S.Z., Pitzer, J.E., Miller, M.R., and Motaleb, M.A. (2010). Analysis of a Borrelia burgdorferi phosphodiesterase demonstrates a role for cyclic-di-guanosine monophosphate in motility and virulence. Mol Microbiol 77, 128-142. https://doi.org/10.1111/j.1365-2958.2010.07191.x

Sung, S.Y., McDowell, J.V., and Marconi, R.T. (2001). Evidence for the contribution of point mutations to vlsE variation and for apparent constraints on the net accumulation of sequence changes in vlsE during infection with Lyme disease spirochetes. Journal of bacteriology 183, 5855-5861.

Thome, S., Begandt, D., Pick, R., Salvermoser, M., and Walzog, B. (2018). Intracellular β2 integrin (CD11/CD18) interacting partners in neutrophil trafficking. European journal of clinical investigation 48 Suppl 2, e12966. https://doi.org/10.1111/eci.12966

Thompson, C., Mason, C., Parrilla, S., and Ouyang, Z. (2020). The Lon-1 protease is required by Borrelia burgdorferi to infect the mammalian host. Infection and immunity. https://doi.org/10.1128/IAI.00951-19

Tilly, K., Krum, J.G., Bestor, A., Jewett, M.W., Grimm, D., Bueschel, D., Byram, R., Dorward, D., Vanraden, M.J., Stewart, P., et al. (2006). Borrelia burgdorferi OspC protein required exclusively in a crucial early stage of mammalian infection. Infect Immun 74, 3554-3564. https://doi.org/10.1128/IAI.01950-05

Tilly, K., Bestor, A., and Rosa, P.A. (2013). Lipoprotein succession in Borrelia burgdorferi: similar but distinct roles for OspC and VlsE at different stages of mammalian infection. Mol Microbiol 89, 216-227. https://doi.org/10.1111/mmi.12271

Tokarz, R., Anderton, J.M., Katona, L.I., and Benach, J.L. (2004). Combined effects of blood and temperature shift on Borrelia burgdorferi gene expression as determined by whole genome DNA array. Infect Immun 72, 5419-5432. https://doi.org/10.1128/IAI.72.9.5419-5432.2004

Toledo, A., Coleman, J.L., Kuhlow, C.J., Crowley, J.T., and Benach, J.L. (2012). The enolase of Borrelia burgdorferi is a plasminogen receptor released in outer membrane vesicles. Infect Immun 80, 359-368. https://doi.org/10.1128/IAI.05836-11

Troxell, B., Xu, H., and Yang, X.F. (2012). Borrelia burgdorferi, a pathogen that lacks iron, encodes manganese-dependent superoxide dismutase essential for resistance to streptonigrin. The Journal of biological chemistry 287, 19284-19293. https://doi.org/10.1074/jbc.M112.344903

Troxell, B., and Yang, X.F. (2013). Metal-dependent gene regulation in the causative agent of Lyme disease. Frontiers in cellular and infection microbiology 3, 79. https://doi.org/10.3389/fcimb.2013.00079

Troxell, B., Ye, M., Yang, Y., Carrasco, S.E., Lou, Y., and Yang, X.F. (2013). Manganese and zinc regulate virulence determinants in Borrelia burgdorferi. Infection and immunity 81, 2743-2752. https://doi.org/10.1128/IAI.00507-13

Troy, E.B., Lin, T., Gao, L., Lazinski, D.W., Camilli, A., Norris, S.J., and Hu, L.T. (2013). Understanding barriers to Borrelia burgdorferi dissemination during infection using massively parallel sequencing. Infect Immun 81, 2347-2357. https://doi.org/10.1128/IAI.00266-13

Troy, E.B., Lin, T., Gao, L., Lazinski, D.W., Lundt, M., Camilli, A., Norris, S.J., and Hu, L.T. (2016). Global Tn-seq analysis of carbohydrate utilization and vertebrate infectivity of Borrelia burgdorferi. Mol Microbiol 101, 1003-1023. https://doi.org/10.1111/mmi.13437

van Burgel, N.D., Balmus, N.C., Fikrig, E., and van Dam, A.P. (2011). Infectivity of Borrelia burgdorferi sensu lato is unaltered in C3-deficient mice. Ticks Tick Borne Dis 2, 20-26. https://doi.org/10.1016/j.ttbdis.2010.10.003

van der Woude, M.W., and Baumler, A.J. (2004). Phase and antigenic variation in bacteria. Clin Microbiol Rev 17, 581-611, table of contents. https://doi.org/10.1128/CMR.17.3.581-611.2004

Verhey, T.B., Castellanos, M., and Chaconas, G. (2018a). Analysis of recombinational switching at the antigenic variation locus of the Lyme spirochete using a novel PacBio sequencing pipeline. Molecular microbiology 108, 461. https://doi.org/10.1111/mmi.13969

Verhey, T.B., Castellanos, M., and Chaconas, G. (2018b). Antigenic Variation in the Lyme Spirochete: Insights into Recombinational Switching with a Suggested Role for Error-Prone Repair. Cell Rep 23, 2595-2605. https://doi.org/10.1016/j.celrep.2018.04.117

Verhey, T.B., Castellanos, M., and Chaconas, G. (2019). Antigenic variation in the Lyme spirochete: detailed functional assessment of recombinational switching at vlsE in the JD1 strain of Borrelia burgdorferi. Mol Microbiol 111, 750-763. https://doi.org/10.1111/mmi.14189

Verma, A., Brissette, C.A., Bowman, A., and Stevenson, B. (2009). Borrelia burgdorferi BmpA is a laminin-binding protein. Infection and immunity 77, 4940-4946.

Verma, N.K., and Kelleher, D. (2017). Not Just an Adhesion Molecule: LFA-1 Contact Tunes the T Lymphocyte Program. J Immunol 199, 1213-1221. https://doi.org/10.4049/jimmunol.1700495

Vink, C., Rudenko, G., and Seifert, H.S. (2012). Microbial antigenic variation mediated by homologous DNA recombination. FEMS microbiology reviews 36, 917-948. https://doi.org/10.1111/j.1574-6976.2011.00321.x

von Lackum, K., Miller, J.C., Bykowski, T., Riley, S.P., Woodman, M.E., Brade, V., Kraiczy, P., Stevenson, B., and Wallich, R. (2005). Borrelia burgdorferi regulates expression of complement regulator-acquiring surface protein 1 during the mammal-tick infection cycle. Infection and immunity 73, 7398-7405.

von Lackum, K., Babb, K., Riley, S.P., Wattier, R.L., Bykowski, T., and Stevenson, B. (2006). Functionality of Borrelia burgdorferi LuxS: the Lyme disease spirochete produces and responds to the pheromone autoinducer-2 and lacks a complete activated-methyl cycle. Int J Med Microbiol 296 Suppl 40, 92-102.

Wager, B., Shaw, D.K., Groshong, A.M., Blevins, J.S., and Skare, J.T. (2015). BB0744 affects tissue tropism and spatial distribution of Borrelia burgdorferi. Infect Immun 83, 3693-3703. https://doi.org/10.1128/IAI.00828-15

Wagh, D., Pothineni, V.R., Inayathullah, M., Liu, S., Kim, K.M., and Rajadas, J. (2015). Borreliacidal activity of Borrelia metal transporter A (BmtA) binding small molecules by manganese transport inhibition. Drug Des Devel Ther 9, 805-816. https://doi.org/10.2147/DDDT.S77063

Walter, L., Surth, V., Rottgerding, F., Zipfel, P.F., Fritz-Wolf, K., and Kraiczy, P. (2019). Elucidating the Immune Evasion Mechanisms of Borrelia mayonii, the Causative Agent of Lyme Disease. Front Immunol 10, 2722. https://doi.org/10.3389/fimmu.2019.02722

Wang, D., Botkin, D.J., and Norris, S.J. (2003). Characterization of the vls antigenic variation loci of the Lyme disease spirochaetes Borrelia garinii Ip90 and Borrelia afzelii ACAI. Molecular microbiology 47, 1407-1417.

Wang, G., van Dam, A.P., Schwartz, I., and Dankert, J. (1999). Molecular typing of Borrelia burgdorferi sensu lato: taxonomic, epidemiological, and clinical implications. Clin Microbiol Rev 12, 633-653.

Wang, G., van Dam, A.P., and Dankert, J. (2001). Analysis of a VMP-like sequence (vls) locus in Borrelia garinii and Vls homologues among four Borrelia burgdorferi sensu lato species. FEMS microbiology letters 199, 39-45.

Wang, P., Lutton, A., Olesik, J., Vali, H., and Li, X. (2012). A novel iron- and copper-binding protein in the Lyme disease spirochaete. Molecular microbiology 86, 1441-1451. https://doi.org/10.1111/mmi.12068

Wang, X.G., Lin, B., Kidder, J.M., Telford, S., and Hu, L.T. (2002). Effects of environmental changes on expression of the oligopeptide permease (opp) genes of Borrelia burgdorferi. Journal of bacteriology 184, 6198-6206.

Wang, X.G., Kidder, J.M., Scagliotti, J.P., Klempner, M.S., Noring, R., and Hu, L.T. (2004). Analysis of differences in the functional properties of the substrate binding proteins of the Borrelia burgdorferi oligopeptide permease (Opp) operon. Journal of bacteriology 186, 51-60.

Weening, E.H., Parveen, N., Trzeciakowski, J.P., Leong, J.M., Hook, M., and Skare, J.T. (2008). Borrelia burgdorferi lacking DbpBA exhibits an early survival defect during experimental infection. Infect Immun 76, 5694-5705. https://doi.org/10.1128/IAI.00690-08

Winzer, K., and Williams, P. (2001). Quorum sensing and the regulation of virulence gene expression in pathogenic bacteria.PG - 131-43. Int J Med Microbiol 291.

Wood, E., Tamborero, S., Mingarro, I., and Esteve-Gassent, M.D. (2013). BB0172, a Borrelia burgdorferi outer membrane protein that binds integrin α3β1. Journal of bacteriology 195, 3320-3330. https://doi.org/10.1128/JB.00187-13

Woodman, M.E., Cooley, A.E., Miller, J.C., Lazarus, J.J., Tucker, K., Bykowski, T., Botto, M., Hellwage, J., Wooten, R.M., and Stevenson, B. (2007). Borrelia burgdorferi binding of host complement regulator factor H is not required for efficient mammalian infection. Infect Immun 75, 3131-3139. https://doi.org/10.1128/IAI.01923-06

Wormser, G.P., Molins, C.R., Levin, A., Lipsett, S.C., Nigrovic, L.E., Schriefer, M.E., and Branda, J.A. (2018). Evaluation of a sequential enzyme immunoassay testing algorithm for Lyme disease demonstrates lack of test independence but high diagnostic specificity. Diagnostic microbiology and infectious disease 91, 217-219. https://doi.org/10.1016/j.diagmicrobio.2018.02.006

Xie, J., Zhi, H., Garrigues, R.J., Keightley, A., Garcia, B.L., and Skare, J.T. (2019). Structural determination of the complement inhibitory domain of Borrelia burgdorferi BBK32 provides insight into classical pathway complement evasion by Lyme disease spirochetes. PLoS Pathog 15, e1007659. https://doi.org/10.1371/journal.ppat.1007659

Xu, H., Caimano, M.J., Lin, T., He, M., Radolf, J.D., Norris, S.J., Gherardini, F., Wolfe, A.J., and Yang, X.F. (2010a). Role of acetyl-phosphate in activation of the Rrp2-RpoN-RpoS pathway in Borrelia burgdorferi. PLoS Pathog 6, e1001104. https://doi.org/10.1371/journal.ppat.1001104

Xu, H., He, M., He, J.J., and Yang, X.F. (2010b). Role of the surface lipoprotein BBA07 in the enzootic cycle of Borrelia burgdorferi. Infection and immunity 78, 2910-2918. https://doi.org/10.1128/IAI.00372-10

Xu, Q., Seemanapalli, S.V., McShan, K., and Liang, F.T. (2006). Constitutive expression of outer surface protein C diminishes the ability of Borrelia burgdorferi to evade specific humoral immunity. Infection and immunity 74, 5177-5184.

Xu, Q., Shi, Y., Dadhwal, P., and Liang, F.T. (2012). RpoS regulates essential virulence factors remaining to be identified in Borrelia burgdorferi. PLoS ONE 7, e53212. https://doi.org/10.1371/journal.pone.0053212

Yang, X., Coleman, A.S., Anguita, J., and Pal, U. (2009). A chromosomally encoded virulence factor protects the Lyme disease pathogen against host-adaptive immunity. PLoS Pathog 5, e1000326.

Yang, X., Qin, J., Promnares, K., Kariu, T., Anderson, J.F., and Pal, U. (2013). Novel microbial virulence factor triggers murine lyme arthritis. J Infect Dis 207, 907-918. https://doi.org/10.1093/infdis/jis930

Yang, X., Smith, A.A., Williams, M.S., and Pal, U. (2014). A dityrosine network mediated by dual oxidase and peroxidase influences the persistence of Lyme disease pathogens within the vector. The Journal of biological chemistry 289, 12813-12822. https://doi.org/10.1074/jbc.M113.538272

Yang, X., Lin, Y.P., Heselpoth, R.D., Buyuktanir, O., Qin, J., Kung, F., Nelson, D.C., Leong, J.M., and Pal, U. (2016). Middle region of the Borrelia burgdorferi surface-located protein 1 (Lmp1) interacts with host chondroitin-6-sulfate and independently facilitates infection. Cell Microbiol 18, 97-110. https://doi.org/10.1111/cmi.12487

Yang, X.F., Alani, S.M., and Norgard, M.V. (2003). The response regulator Rrp2 is essential for the expression of major membrane lipoproteins in Borrelia burgdorferi. Proc Natl Acad Sci U S A 100, 11001-11006. https://doi.org/10.1073/pnas.1834315100

Yang, X.F., Lybecker, M.C., Pal, U., Alani, S.M., Blevins, J., Revel, A.T., Samuels, D.S., and Norgard, M.V. (2005). Analysis of the ospC regulatory element controlled by the RpoN-RpoS regulatory pathway in Borrelia burgdorferi. J Bacteriol 187, 4822-4829. https://doi.org/10.1128/JB.187.14.4822-4829.2005

Yang, X.F., Goldberg, M.S., He, M., Xu, H., Blevins, J.S., and Norgard, M.V. (2008). Differential expression of a putative CarD-like transcriptional regulator, LtpA, in Borrelia burgdorferi. Infection and immunity 76, 4439-4444.

Ye, M., Sharma, K., Thakur, M., Smith, A.A., Buyuktanir, O., Xiang, X., Yang, X., Promnares, K., Lou, Y., Yang, X.F., et al. (2016). HtrA, a Temperature- and Stationary Phase-Activated Protease Involved in Maturation of a Key Microbial Virulence Determinant, Facilitates Borrelia burgdorferi Infection in Mammalian Hosts. Infection and immunity 84, 2372-2381. https://doi.org/10.1128/IAI.00360-16

Zambrano, M.C., Beklemisheva, A.A., Bryksin, A.V., Newman, S.A., and Cabello, F.C. (2004). Borrelia burgdorferi binds to, invades, and colonizes native type I collagen lattices. Infect Immun 72, 3138-3146. https://doi.org/10.1128/IAI.72.6.3138-3146.2004

Zhang, J.R., Hardham, J.M., Barbour, A.G., and Norris, S.J. (1997). Antigenic variation in Lyme disease borreliae by promiscuous recombination of VMP-like sequence cassettes. Cell 89, 275-285.

Zhang, J.R., and Norris, S.J. (1998a). Genetic variation of the Borrelia burgdorferi gene vlsE involves cassette-specific, segmental gene conversion. Infection and immunity 66, 3698-3704.

Zhang, J.R., and Norris, S.J. (1998b). Kinetics and in vivo induction of genetic variation of vlsE in Borrelia burgdorferi. Infection and immunity 66, 3689-3697.

Zhang, K., Qin, Z., Chang, Y., Liu, J., Malkowski, M.G., Shipa, S., Li, L., Qiu, W., Zhang, J.R., and Li, C. (2019). Analysis of a flagellar filament cap mutant reveals that HtrA serine protease degrades unfolded flagellin protein in the periplasm of Borrelia burgdorferi. Molecular microbiology 111, 1652-1670. https://doi.org/10.1111/mmi.14243

Zhang, X., Yang, X., Kumar, M., and Pal, U. (2009). BB0323 function is essential for Borrelia burgdorferi virulence and persistence through tick-rodent transmission cycle. J Infect Dis 200, 1318-1330. https://doi.org/10.1086/605846

Zhao, Z., Chang, H., Trevino, R.P., Whren, K., Bhawan, J., and Klempner, M.S. (2003). Selective up-regulation of matrix metalloproteinase-9 expression in human erythema migrans skin lesions of acute lyme disease. J Infect Dis 188, 1098-1104. https://doi.org/10.1086/379039

Zhi, H., Weening, E.H., Barbu, E.M., Hyde, J.A., Hook, M., and Skare, J.T. (2015). The BBA33 lipoprotein binds collagen and impacts Borrelia burgdorferi pathogenesis. Mol Microbiol 96, 68-83. https://doi.org/10.1111/mmi.12921

Zhi, H., Xie, J., and Skare, J.T. (2018). The classical complement pathway Is required to control Borrelia burgdorferi levels during experimental infection. Front Immunol 9, 959. https://doi.org/10.3389/fimmu.2018.00959

Zipfel, P.F., and Skerka, C. (2009). Complement regulators and inhibitory proteins. Nature reviews Immunology 9, 729-740. https://doi.org/10.1038/nri2620

Zuckert, W.R., Lloyd, J.E., Stewart, P.E., Rosa, P.A., and Barbour, A.G. (2004). Cross-species surface display of functional spirochetal lipoproteins by recombinant Borrelia burgdorferi. Infect Immun 72, 1463-1469. https://doi.org/10.1128/iai.72.3.1463-1469.2004

Zuckert, W.R. (2019). Protein secretion in spirochetes. Microbiol Spectr 7. https://doi.org/10.1128/microbiolspec.PSIB-0026-2019

Chapter 14

Pathogenesis of Relapsing Fever

Job Lopez[1], Joppe W. Hovius[2] and Sven Bergström[3]

[1]Department of Pediatrics, Section of Tropical Medicine, Baylor College of Medicine and Texas Children's Hospital, Houston TX, USA
[2]Center for Experimental and Molecular Medicine, Amsterdam Medical centers, location Academic Medical Center, University of Amsterdam, 1105 AZ, Amsterdam, The Netherlands
[3]Department of Molecular Biology, Umeå Center for Microbial Research, Molecular Infection Medicine Sweden, Umeå University, Umeå, Sweden
*Corresponding author: sven.bergstrom@umu.se

DOI: https://doi.org/10.21775/9781913652616.14

Abstract
Relapsing fever (RF) is caused by several species of *Borrelia*; all, except two species, are transmitted to humans by soft (argasid) ticks. The species *B. recurrentis* is transmitted from one human to another by the body louse, while *B. miyamotoi* is vectored by hard-bodied ixodid tick species. RF *Borrelia* have several pathogenic features that facilitate invasion and dissemination in the infected host. In this chapter, we discuss the dynamics of vector acquisition and subsequent transmission of RF *Borrelia* to their vertebrate hosts. We also review taxonomic challenges for RF *Borrelia* as new species have been isolated throughout the globe. Moreover, aspects of pathogenesis including symptomology, neurotropism, erythrocyte and platelet adhesion are discussed. We expound on RF *Borrelia* evasion strategies for innate and adaptive immunity, focusing on the most fundamental pathogenetic attributes, multiphasic antigenic variation. Lastly, we review new and emerging species of RF *Borrelia* and discuss future directions for this global disease.

Introduction
Relapsing fever (RF), or what we believe to be RF, was historically known by different names including "Febris recurrentis", "recurrent fever", "famine fever", "spirillum fever", "spirochetal fever", "saddleback fever", "vagabond fever", "fowl nest fever", "Karlskrona fever", "gharib ghez", "Giesinger's bilious typhoid", and "kimputu." RF was recognized as a disease by physicians in ancient Greece at the time of Hippocrates (Felsenfeld, 1971). Later, between 1485 and 1551, five epidemics of "sweating sickness" that swept through England probably included outbreaks of RF. One of the best recorded descriptions of RF came from the physician John Rutty, who kept a detailed diary during his time in Dublin, where he described the weather and illnesses in the area during the mid-1700's (Rutty, 1770). Interestingly, the fatality rate was very low and most of the affected people did recover after two or three relapses.

RF symptoms also were described in detail by field medics during the 1788 Swedish-Russian war. The Swedish navy conquered the Russian 74-cannon battleship Vladimir and its 783 men crew at a battle in the Finnish bay near the Hogland Island. Following the victory, the ship was brought to Helsinki, but several of the Russian prisoners aboard were sick. The louse-borne RF spread rapidly in the Swedish Fortress Sveaborg and among Helsinki civilians. Five hundred sick soldiers were shipped to Karlskrona, the main Swedish naval base, which then also became heavily plagued by RF (Felsenfeld, 1971; Huldén, 2006). From the documentation, it is possible that many soldiers were afflicted by concomitant infection with *Rickettsia prowazekii* and possibly other microorganisms. Difficulty in clinically distinguishing RF from other vector-borne infections continues to this day. For example, RF and malaria are often mistaken in areas where both exist. Also, dengue and Colorado Tick Fever, present with similar symptoms to RF such as headache, constitutional symptoms, and "saddleback fever"-like (biphasic fever that may suggest a relapse).

The definite cause of RF was not established until 1868 when the German scientist Otto Obermeier identified a spirochete as the etiologic agent of an epidemic fever outbreak in Berlin. The organism causing this epidemic disease was first named *Spirocheta obermeieri* but later renamed *Borrelia recurrentis*. The largest outbreaks occurred during World War I over large areas in Europe, Asia and Africa. The epidemics among the soldiers and civilians were probably due to poor hygienic conditions and the forced migration of refugees (Felsenfeld, 1971).

RF is either tick-borne (TBRF) or louse-borne (LBRF) (Felsenfeld, 1971). LBRF is caused by *B. recurrentis*, and TBRF is caused by several different *Borrelia* species. Many of the RF outbreaks described by Rutty and others in Northern Europe were likely louse-borne. These outbreaks occurred when there were optimal conditions for the body louse - the weather was turning cooler and people were adding more layers of clothing, the preferred niche for the human body louse. Also, during wartime, people are more vulnerable to attack by lice because of reduced hygiene, including the lack of clean clothes and bed linen.

Since the initial descriptions of LBRF, there have been millions of human cases in large and small epidemics (Bryceson et al., 1970). However, endemic transmission of the agent continues to persist in the Horn of Africa and adjacent areas (Borgnolo et al., 1993; Mitiku and Mengistu, 2002; Sundnes and Haimanot, 1993). As expected for a disease with this degree of public health importance, there have been numerous clinical, epidemiologic, and entomologic studies of LBRF, its agent, and vector. Many of these reports date to the first half of the last century and provide a wealth of information about the clinical aspects of LBRF.

The reservoir of *B. recurrentis* is probably restricted to humans, and reports of experimental animal infections with *B. recurrentis* have been scarce (Barbour and Hayes, 1986). For example, *B. recurrentis* may cause a spirochetemia in weanling mice, newborn rabbits, and grivet monkeys (Judge et al., 1974a, b, c). More recently, Larsson and coworkers successfully developed a LBRF model in immunodeficient mice (Larsson et al., 2009). The findings demonstrated that, in addition to humoral immunity, innate immune responses could limit the pathogens in the mammalian host.

TBRF is a vector-borne zoonosis found in all regions of the world, apart from Oceania and Antarctica. TBRF accounts for thousands of cases of RF globally each year. Most of the studies on the pathogenesis and immunity of RF have been carried out on TBRF species, whose natural reservoirs are rodents and other mammals. However, the natural reservoir of the RF species *B. duttonii* has not yet been defined. For the TBRF *Borrelia* species, mice and rats serve as suitable experimental animal models.

This chapter by necessity emphasizes studies of TBRF, especially those of two Nearctic species, *B. hermsii* and *B. turicatae*, and two Afro-Tropical/ Palearctic species, *B. duttonii* and *B. crocidurae*. This chapter also covers important aspects of LBRF infection and its causative agent *B. recurrentis,* which is descended from the tick-borne species *B. duttonii* or a close relative (Lescot et al., 2008). Thus, the findings of experimental studies of tick-borne species might be applicable to *B. recurrentis* and LBRF. Furthermore, this chapter is an updated version of the chapter by Barbour and Guo (2010). In this second edition chapter, we have included a more comprehensive section on vector biology and an update on new and emerging hard tick-transmitted RF species, *Borrelia miyamotoi*.

Transmission-vector and reservoirs

Harry Hoogstraal, a medical acarologist and parasitologist, suggested that the *Borrelia* lineage began as symbionts of ticks rather than as parasites of vertebrates (Hoogstraal, 1979). This is an interesting hypothesis but, as Barbour and Hayes stated, "As it is metaphorically true for the chicken and the egg, it may not be known whether *Borrelia* were originally parasites of arthropods or of vertebrates" (Barbour and Hayes, 1986). Thus, spirochetes are found as endosymbionts in several invertebrates, e.g., termites and mollusks. From that viewpoint, infection of vertebrates is the main role for transmission of the spirochetes to another arthropod host, in this case soft-bodied (argasid) ticks of the genus *Ornithodoros* (Cooley and Kohls, 1944; Felsenfeld, 1971; Hoogstraal, 1979). On the other hand, one piece of evidence that favors an arthropod origin for the genus *Borrelia* is the finding of a high frequency of transovarial transmission observed for TBRF species, with between 35 - 100% of offspring, in most but not all species (Burgdorfer and Varma, 1967; Davis, 1943, 1952). When the *Borrelia* spirochete enters the ovary, it invades the developing oocyte-yolk complex before the impermeable shell is formed around the egg (Aeschlimann, 1958).

TBRF spirochetes, are transmitted either by argasid or ixodid ticks and their life cycles in each vector differ. In argasids, TBRF spirochetes enter the tick when feeding on an infected vertebrate host. While it is not completely clear if replication occurs in the vector, the spirochetes persistently colonize the tick midgut and salivary glands of argasids in the genus *Ornithodoros,* (Boyle et al., 2014; Krishnavajhala et al., 2017; Policastro et al., 2013; Schwan and Hinnebusch, 1998). Since *Ornithodoros* ticks are rapid feeders, completing the blood meal within 60 minutes, the population of TBRF spirochetes that colonized the salivary glands enters the host's skin and blood within seconds of attachment (Boyle et al., 2014; Burgdorfer, 1951). This was demonstrated when *Ornithodoros turicata* ticks engorged on the ears of mice, and after feeding the bite site was removed and placed into culture medium (Boyle et al., 2014). All the animals became infected, and spirochetes also were cultivated from ear biopsies. These results indicated that, while some spirochetes enter the bloodstream, others that remain at the bite site after feeding. While most species of TBRF are transmitted through the saliva of *Ornithodoros* ticks, transmission of *O. moubata* may occur through contamination of the bite site with infected coxal fluid (Varma, 1956).

Recent work with the ixodid tick vector of TBRF spirochetes suggests a slightly different life cycle compared to argasid ticks. Studies with *Ixodes scapularis,* the vector of *B. miyamotoi,* indicated that the bacteria persistently colonize the midgut and salivary glands of the tick (Breuner et al., 2017). Upon feeding infected ticks, spirochete DNA was detected in 10% of mice after 24 hours. This indicates that spirochetes in the salivary gland are the first to enter the vertebrate host. Interestingly, by the time ticks had completed feeding, several days later, 73% of the animals were infected. This suggests that unlike argasid-transmitted TBRF spirochetes, the population of *B. miyamotoi* that colonizes the midgut migrates to the salivary glands then plays a role in vertebrate infection as has been well established for the causative agent of Lyme disease.

The transmission dynamics of LBRF spirochetes are unique compared to the tick-transmitted species. After ingestion by the louse, LBRF spirochetes pass from the midgut to the hemolymph where they multiply and persist for the ~3-week lifespan of the louse. Subsequent transmission of LBRF spirochetes to humans does not occur through the bite or the saliva of the human body louse (*Pediculus humanus humanus).* LBRF spirochetes are transmitted mechanically and transovarial transmission of *B. recurrentis* does not occur (Bryceson et al., 1970; Felsenfeld, 1965). Only when the louse is crushed does contaminated hemolymph or feces enter the human (Houhamdi and Raoult, 2005). This occurs through contamination of the conjunctivae, abrasions in the mucous membranes of the mouth, or the louse bite site.

The vertebrate reservoir hosts of the TBRF *Borrelia* spp. are mostly different species of rodents (Nicolle and Anderson, 1927) but may also include animals like pigs, goats, sheep, rabbits, bats, opossums, armadillos, foxes, cats, dogs, and birds (Felsenfeld, 1971; Hoogstraal, 1985; Pavlovsky, 1963; Thomas et al., 2002). These animals usually become infected during the bite of an *Ornithodoros* or related soft-bodied tick. Accidental peroral transmission through consumption of livers and brains of infected animals also has been reported (reviewed in Barbour and Hayes, 1986). Transplacental transmission occurs also in mammals and may cause congenital infection with subsequent placental damage and stillbirths (Larsson et al., 2006a). However, the peroral and transplacental routes of RF transmission are rare, and their importance for transmission in nature is probably insufficient for the maintenance of the pathogen. Likewise, for humans, RF is nearly always acquired through contact with a tick or louse carrying the spirochetes. Rarely the disease is acquired through (i) accidental inoculation of infected blood; (ii) contact of abraded or lacerated skin, mucous membranes, or conjunctiva with infected patient or animal blood (Herms and Wheeler, 1935); or (iii) transplacental or perinatal transmission from mother to fetus (Fuchs and Oyama, 1969; Yagupsky and Moses, 1985). Additionally, pregnancy complications connected to RF infection have been reported, especially *B. duttonii* infections in the sub-Saharan region of Africa (Brasseur, 1985; Dupont et al., 1997; Goubau and Munyangeyo, 1983; Jongen et al., 1997; Melkert and Stel, 1991; van Holten et al., 1997). Importantly, transmission does not occur by aerosols, fomites, saliva, urine, feces, or sexual contact.

***Borrelia*-arthropod interactions: Life inside the tick!**
In the early and mid-twentieth century, there were many studies on the colonization and movement of *Borrelia* spirochetes in ticks and lice, reviewed in (Barbour and Hayes, 1986; Burgdorfer and Varma, 1967; Felsenfeld, 1965; Felsenfeld, 1971). An example was the comprehensive investigation by Willy Burgdorfer assessing *B. duttonii* colonization in its tick vector, *O. moubata* (Burgdorfer, 1951). In that study, he showed that a high density of spirochetes was found around ganglia and in the salivary glands of the tick (Burgdorfer, 1959). Interestingly, Burgdorfer also showed that the transmission of RF *Borrelia* occurred not only via a tick bite, but also through contamination from infected coxal fluid (Burgdorfer, 1951, 1959).

As efforts have focused on generating spirochete isolates from the field, we are gaining refined insights into vector competence of TBRF spirochetes. For example, multilocus sequencing of ~50 *B. hermsii* isolates distinguished two genomic groupings, GGI and GGII. A series of tick acquisition studies was performed to determine whether co-infection of GGI and GGII impacted vector colonization and subsequent transmission frequencies from *O. hermsi* (Policastro et al., 2013). Policastro and colleagues first infected cohorts of ticks with either GGI or GGII spirochetes. After molting, they fed ticks again to allow them to become infected with the opposite genomic group. Finally, in the third tick feeding, individual ticks were assessed to determine the impact of superinfection in transmission. Interestingly, ticks that were first infected with GGII then GGI transmitted only GGII spirochetes. However, prolonging the interval between the second and third feeding resulted in co-transmission of GGI and GGII spirochetes (Policastro et al., 2013). Thus, transmission of different populations of *B. hermsii* depends upon the order of tick acquisition and the interval between feedings.

Additional studies with *B. turicatae* further demonstrate that specific populations of TBRF spirochetes colonize disparate environments within the tick. *B. turicatae* producing GFP established a persistent infection for at least 18 months in the tick midgut and salivary glands (Krishnavajhala et al., 2017). While the population colonizing the salivary glands is the one responsible for transmission to the vertebrate host (Boyle et al., 2014; Schwan and Hinnebusch, 1998), the role of the midgut population remains unclear. Given the rapidity of *Ornithodoros* feeding and TBRF spirochete transmission, there is an insufficient time for the midgut population to be transmitted through the saliva. There are at least two potential roles for the midgut population. First, it may be involved with spirochete replication in the midgut after infected ticks feed. Alternatively, the midgut population may have a role in transovarial transmission to developing eggs.

Advances in high-throughput methodologies have begun to provide insight into the selective pressures and molecular mechanisms of vector colonization by TBRF spirochetes. Mans and colleagues seminal work produced the first salivary gland transcriptomes of *Ornithodoros* ticks, reporting the identification of immune regulators that may aid in tick feeding and TBRF spirochete transmission (Mans et al., 2008a; Mans et al., 2008b). Recent studies on the proteome of fed and unfed *O. erraticus* midguts indicated the production of proteins involved with antioxidant defenses (Oleaga et al., 2015). These findings were expanded further with the salivary gland transcriptome of *O. turicata* (Bourret et al., 2019). The expression of 57 genes involved in oxidant metabolism or antioxidant defenses were identified. Of these, the genes encoding glutathione peroxidase, thioredoxin peroxidase, manganese superoxide dismutase, copper-zine superoxide dismutase, and catalase were further evaluated in fed and unfed midguts and salivary glands. These findings indicated that the midguts and salivary glands of *O. turicata* were highly nitrosative and oxidative, respectively. Interestingly, the work also determined that *B. turicatae* was hyper-resistant to reactive oxygen species compared to Lyme disease (LD)-causing spirochetes (Bourret et al., 2019). Collectively, omics studies have begun to provide insight into TBRF adaptation within the tick vector; however, additional work is needed to understand the mechanisms utilized by the pathogens in the tick-mammalian infectious cycle.

As molecular tools have improved and become available for genetically manipulating TBRF spirochetes, the events of vector colonization are becoming clearer. The most significant advance on the molecular interactions between the arthropod vector and *Borrelia* was performed with *O. hermsi* and *B. hermsii*. Schwan and Hinnebusch began to unravel the intricacies of vector colonization and transmission with the identification of the variable tick protein (Vtp) (previously denoted plc, VmpC or Vsp33) (Schwan and Hinnebusch, 1998). They demonstrated that Vtp was preferentially produced by spirochetes in the salivary glands of unfed *O. hermsi* ticks and that its production could be induced *in vitro* by lowering the incubation temperature to 23° C. Interestingly, they reported that when spirochetes producing Vtp enter a new mouse host from the tick, they reverted back to the mammal-associated variant they last manifested in the mouse host (Schwan and Hinnebusch, 1998). Raffel and co-workers genetically inactivated the *vtp* locus and it became clear that the protein is involved with preadapting *B. hermsii* for mammalian entry (Raffel et al., 2014). Mutants lacking Vtp were able to colonize the salivary glands of *O. hermsi* ticks but failed to establish an infection in mice after tick feeding (Raffel et al., 2014). While the function of Vtp is unclear, the protein is homologous to the Vsp proteins of RF *Borrelia* spp. (see below) and OspC of Lyme borreliosis species (Carter et al., 1994; Marconi et al., 1993). Interestingly, the finding that the Vtp proteins from different *B. hermsii* strains are highly polymorphic suggests positive selection involving antibody responses or for the need to survive in different biological niches like rodents, canines, etc. While Vtp is essential in the establishment of early mammalian infection, less is known regarding the molecular mechanism of persistent vector colonization.

Transcriptional studies have been implemented to begin to understand how the TBRF spirochetes establish an infection in the tick vector. Work in *B. turicatae* suggested that the linear megaplasmid is primarily involved in vector colonization and/or initiating mammalian infection (Wilder et al., 2016). *B. turicatae* was grown under ambient temperature mimicking the tick and 67% of the genes localized on the linear megaplasmid were upregulated in these conditions compared to spirochetes isolated from infected murine blood. The expression of a cluster of genes on the 3′ end of the plasmid was investigated further in the tick because they were predicted to code for surface proteins. While the results demonstrated up-regulation in infected ticks compared to infected blood, the proteins are of

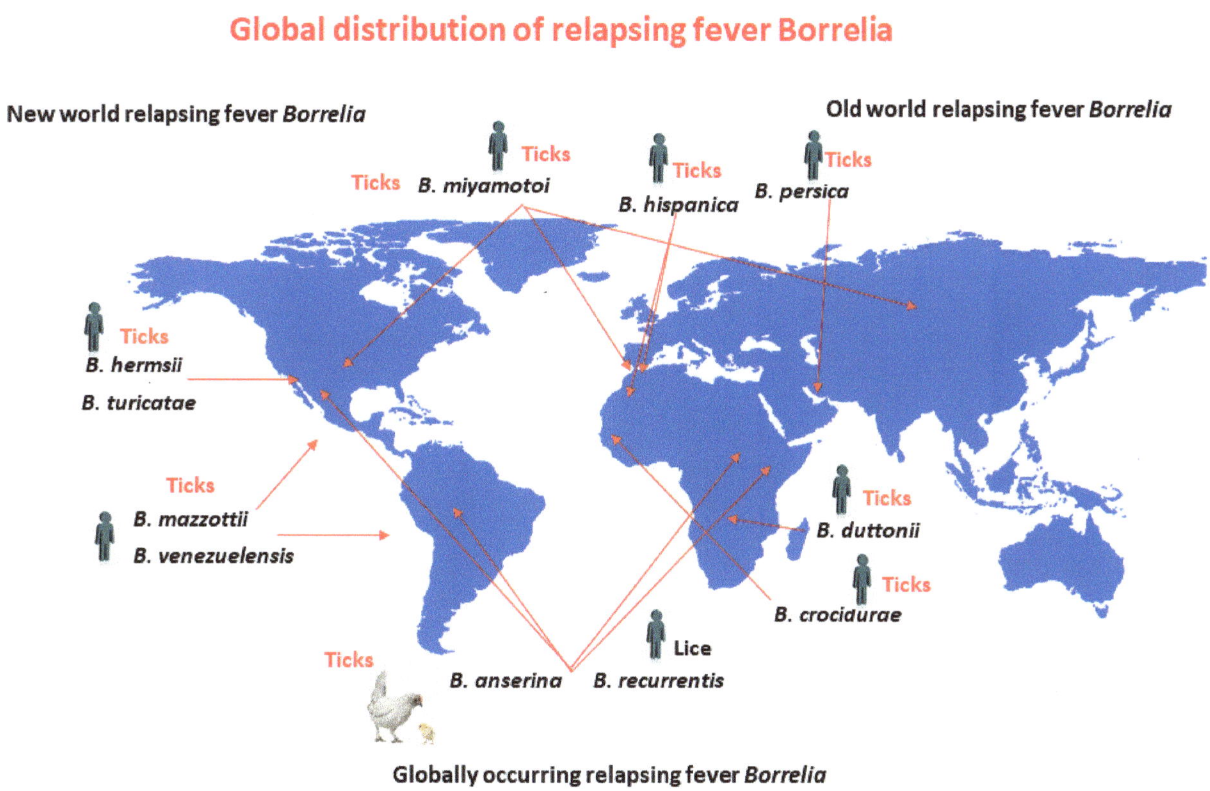

Figure 1. Global distribution of RF *Borrelia*. Only certain representative RF *Borrelia* species are included in the figure. A more detailed description of currently defined species is found in (Barbour and Schwan, 2018; Talagrand-Reboul et al., 2018). Drawing was generated by Haitham Elbir. The global map was created using R software version 4.0.1

unknown function. With advances in genetic systems for TBRF spirochetes, the identification of genes essential in vector colonization is now possible (Battisti et al., 2008; Lopez et al., 2013b).

RF *Borrelia* speciation: Old and New World borreliosis

Currently, there is no systematic way to classify RF *Borrelia*; nomenclature has been subject to change, and differences exist between the speciation of Old and New World pathogens (Figure 1). A comprehensive description of classified RF *Borrelia* species is found in the *Borrelia* chapter of Bergey's Manual of Systematics of Archaea and Bacteria as well as in the recent review by Talagrand-Reboul (Barbour and Schwan, 2018; Talagrand-Reboul et al., 2018). Well characterized Old World species include *B. duttonii* and *B. crocidurae,* which were named after Dr. Joseph Everett Dutton and the vertebrate host in the genus *Crocidura,* respectively. The Old World species, *B. recurrentis* (LBRF), was speciated based on its biology in the human host.

The speciation of New World RF spirochetes has been somewhat more systematic and primarily based on the species tick vector. For example, *B. hermsii, B. parkeri,* and *B. turicatae* were originally speciated according to their argasid vectors, *O. hermsi, O. parkeri,* and *O. turicata.* This method of speciation was based on the observed vector specificity of TBRF spirochetes. However, discrepancies do exist with other New World species. For example, *B. mazzottii*, which was detected in *O. talaje* from northern Mexico and likely distributed in the southern United States and other regions of Latin America, was named after Dr. Luis Mazzottii (Davis, 1956; Lopez et al., 2016).

Historically, confusion has surrounded the speciation of RF *Borrelia* and their tick vectors in Central and South America. For example, it was not clear whether *O. talaje, O. rudis,* or *O. venezuelensis* were independent species, and this led to the uncertainty in the speciation of their respective RF spirochetes (Davis, 1955). As the ambiguity regarding the tick vectors was clarified, it became apparent that *Borrelia neotropicalis* was synonymous with *B. venezuelensis* (Davis, 1955). Within the last 15 years, significant advances have been made toward identifying and speciating *Ornithodoros* ticks from Central and South America. Venzal and coworkers provided detailed morphological characteristics of larval *Ornithodoros* ticks and demonstrated the importance in characterizing this developmental stage for speciation (Venzal et al., 2008). Recently, Muñoz-Leal and colleagues identified *O. rudis* in Brazil and this resulted in the first isolation of a TBRF spirochete from the continent (Munoz-Leal et al., 2018). Given the historic association of *B. venezuelensis* with *O. rudis,* this novel *Borrelia* isolate was designated *B. venezuelensis.* Interestingly, the molecular characterization of this spirochete indicated that it was most closely related to the North American species, *B. turicatae.*

Speciating TBRF spirochetes in Eurasia has posed similar challenges. This is primarily because a few loci are sequenced for speciation, while genetic, geographical, ecological, and biological factors are not considered when designating a new species of TBRF spirochete. The need for a refined systematic classification of TBRF spirochetes from Eurasia and the Middle East was demonstrated recently with the first isolation of *B. caucasica* from Ukraine (Filatov et al., 2020). Filatov and co-workers performed a phylogenetic analysis of *B. caucasica* comparing several single loci. The results indicated that, depending on the locus sequenced (16S rRNA, flagellin, and DNA gyrase), this species could have been designated as *B. hispanica* or *B. persica.* However, the spirochete was speciated as *B. caucasica* given the tick vector from which it was isolated. Moreover, while the intragenic spacer sequence (IGS) was not evaluated in that study, there is accumulating evidence that sequencing this locus may provide the best resolution for speciation of TBRF spirochetes (Bunikis et al., 2004; Cutler et al., 2010; Nieto and Teglas, 2014; Schwan et al., 2007).

Clinical presentation- the crisis phenomenon and the Jarisch-Herxheimer reaction

The clinical hallmark of RF is periods of very high fever spaced by periods of relative well-being, i.e., the recurrent fever episodes. The incubation period is between 5-8 days for LBRF, but varies much more with TBRF (Southern and Sanford, 1969). The reported mortality rates for untreated RF differ (Barbour et al., 2006), with the largest variability seen in LBRF. The high mortality was probably caused by factors associated with wars, famine, and situations within refugee camps, i.e., conditions of malnutrition and insufficient health care. Patients with RF usually complain of headache, backache, muscle pain, arthralgia, and abdominal pain. Sometimes, the clinical examination reveals jaundice and tenderness over the liver and spleen. A gradual decline of general fitness, with extreme weakness and weight loss can occur if successive relapses occur without adequate treatment (Goodman et al., 1969). If death occurs during RF, it is most common during a crisis episode or within hours after the start of antibiotic therapy. The crisis phenomenon results from the rapid disintegration of spirochetes within the blood stream because of the increase of neutralizing antibodies. Antibiotic treatment of spirochetal infections provides a similar effect, which was termed the Jarisch-Herxheimer reaction (J-HR) after its first description by the dermatologists Adolf Jarisch and Karl Herxheimer. They both observed this reaction during antimicrobial treatment with mercury in patients with secondary syphilis. Interestingly, Balfour (1911) observed that chickens experimentally infected with *B. anserina* also produced an "artificial crisis" after treatment with Salvarsan (arsenic compound 606). Treatment of both LBRF and TBRF with antibiotics was shown to commonly result in the J-HR (Bryceson, 1976; Bryceson et al., 1972; Warrell et al., 1983; Webster et al., 2002). Several studies have tried to resolve if the type of antimicrobial used is important for the generation and severity of J-HR. It is logical to envision that a bactericidal antibiotic is more prone to induce a J-HR response as large amount of cell fragments are released from the bacteria. However, there is still no conclusive result differentiating the J-HR response of a bacteriostatic from a bactericidal antibiotic (Butler, 2017). With the appearance of either neutralizing antibodies or the administration of antibiotics, spirochetes in the blood are lysed, releasing proinflammatory bacterial cell fragments into the circulation. Within a couple of hours after initiation of antibiotic treatment, the patient often presents with acute symptoms, including headache, rigors, malaise, hypotension, and sweating. Subsequently, symptoms might even worsen with an increase of temperature by 1-2° C, tachycardia, and a rise in blood pressure. This period of worsening symptoms occurs over the next couple of hours, followed by diaphoresis, exhaustion, a decline in temperature, hypotension, and leukopenia. This apparent shock-like condition, with dilated vasculature and increased oxygen demands with fever (Warrell et al., 1983), may be exacerbated by myocardial dysfunction if the spirochetes have invaded the heart.

Given the similarity of J-HR to septic shock, early clinical observations predicted endotoxins to be the cause of the J-HR phenomenon (Bryceson et al., 1972). However, *Borrelia* spirochetes have an outer membrane architecture different from other Gram-negative bacteria and lacking endotoxin, i.e., LPS is not present (Ben-Menachem et al., 2003; Takayama et al., 1987). Instead, spirochetes contains ample amounts of glycolipids (Ben-Menachem et al., 2003; Hossain et al., 2001; Schroder et al., 2008) as well as a large amount of surface located lipoproteins. Another hypothesis about the crisis was that it was attributable to antigen-antibody complexes (Galloway et al., 1977), but changes in complement levels were not detected in patients with LBRF (Warrell et al., 1983). A more likely explanation for J-HR and the crisis is the action of liberated spirochetal lipoproteins on the toll-like 2 receptors (TLR2) of macrophages and other cells (Scragg et al., 2000; Vidal et al., 1998). In patients with RF, the levels of tumor necrosis factor (TNF-α), interleukin (IL)-6, and IL-8 rose several fold over pre-treatment levels during the J-HR in patients treated for LBRF (Negussie et al., 1992). Importantly, in a clinical trial, treatment with anti-TNF antibodies before the penicillin therapy reduced the severity of J-HR (Fekade et al., 1996). In mice experimentally infected with *B. turicatae* or *B. hermsii*, serum concentrations of the anti-inflammatory cytokine IL-10 reach high levels (Crowder et al., 2016; Londono et al., 2008). In patients with LBRF, IL-10 was highly elevated, however, treatments with recombinant IL-10 neither prevented J-HR nor inhibited the elevation of TNF, IL-6 or IL-8 (Cooper et al., 2000).

Mammalian infection: RF *Borrelia* pathogenesis

The different periods of time spent feeding by soft and hard ticks (minutes versus days, respectively) clearly separates the biology of transmission of RF *Borrelia* and LD *Borrelia*. Under these very restricted time periods of transmission, a single spirochete may be enough to initiate an RF infection in experimental animals (Beunders and Van Thiel, 1932; Schuhardt and Wilkerson, 1951; Stoenner et al., 1982). The bacteria multiply in the blood at an estimated rate of one cell division every ~6 hours until they number between 10^6 to 10^7, sometimes as many as 10^8, per milliliter of blood at their peak (Figure 2) (Cadavid et al., 1994; Crowder et al., 2016; Schuhardt and Wilkerson, 1951; Stoenner et al., 1982). Cultivable RF species achieve similar generation times in broth medium, such as BSK II (Barbour, 1986). The interval between relapses in experimental infections is usually 2 to 7 days. Untreated LBRF tends to have fewer relapses than does TBRF (Bodman and Stewart, 1948; Borgnolo et al., 1993; Brown et al., 1988; Bryceson et al., 1970; Garnham et al., 1947;

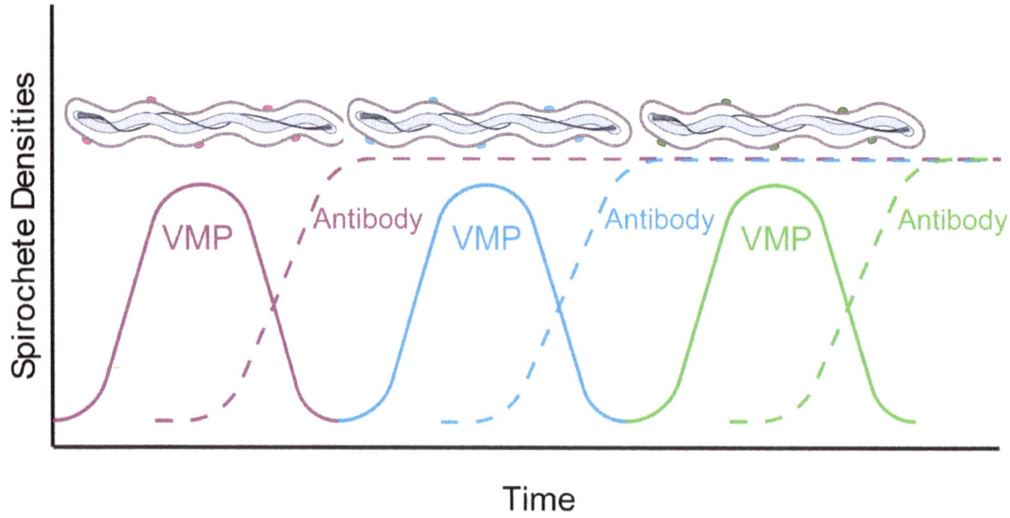

Figure 2. Antigenic variation and antibody response during RF borreliosis. Show are spirochete densities in the blood over time. As spirochetes replicate in the blood a predominant VMP variant emerges (pink line and spirochetes coated with pink proteins). The antibody response generated against this variant (pink dotted line) clears the population of spirochetes while a new population emerges (blue variant). Again, an antibody response is generated (blue dotted line) to clear the population, and yet another variant (green) emerges. The dynamics between antigenic variation and the host antibody response can continue for months. Drawing was generated using BioRender by Brittany A. Armstrong and Job Lopez.

Salih et al., 1977; Sundnes and Haimanot, 1993). The reason for this discrepancy in the number of consecutive relapses is not known, but the finding that *B. recurrentis* have a reduced genome compared to *B. duttonii,* suggests that the former could have lost genetic traits related to prolonged relapsing (Elbir et al., 2014; Lescot et al., 2008).

The incubation period between exposure and onset of fever or other signs of illness is from 3 to 12 days. During relapses, the number of spirochetes in the blood increases again, but usually not to the same density as during the first episode of spirochetemia. The lower concentration of cells in a relapse does not appear to be due to attenuated virulence of the relapse serotypes. Russell described the relapse populations as being as virulent as the "attack" serotype in naïve animals (Russell, 1936). Moreover, Crowder *et al.* observed that the relapse population of *B. hermsii* increased more slowly than the serotype that initiated the infection in mice; however, when each serotype was used to infect naïve mice, their growth rates were indistinguishable (Crowder et al., 2016). The existence of this growth retardation was corroborated by the finding that changes in host factors, involving the innate immune responses, could influence the growth rate during infection (Crowder et al., 2016). Further evidence for this was the observation that, during co-infection of experimental animals with a RF *Borrelia* species and an African trypanosome species, the population growth of the parasite was slower than during single infections (Felsenfeld and Wolf, 1973).

The growth rates and maximum cell densities of *Borrelia* species vary by species and environmental conditions. Most notably, TBRF species and *B. recurrentis* differ by a factor of 1,000 or more in the cell densities they achieve in the blood of immunocompetent as well as immunodeficient mice (Barbour and Bundoc, 2001; Crowder et al., 2016; Sadziene et al., 1993). However, during *in vitro* cultivation in the same type of medium, representative TBRF spirochetes and *B. recurrentis* have similar growth rates and reach stationary cell densities as high as ~10^8 per milliliter (Barbour, 1988; Cadavid et al., 1994; Crowder et al., 2016; Pennington et al., 1997). Also, expression profiles analyzed by genome-wide arrays revealed little difference between cells of *B. hermsii* harvested from BSK II broth medium or in the blood of mice, when the temperature and pH of the broth approximated that of blood, and cell densities at the time of collection were comparable (Zhong and Barbour, 2004). Given these observations, genetic differences in metabolic pathways or other inherent capabilities plausibly account for observed differences between TBRF and LBRF species in peak burdens in the

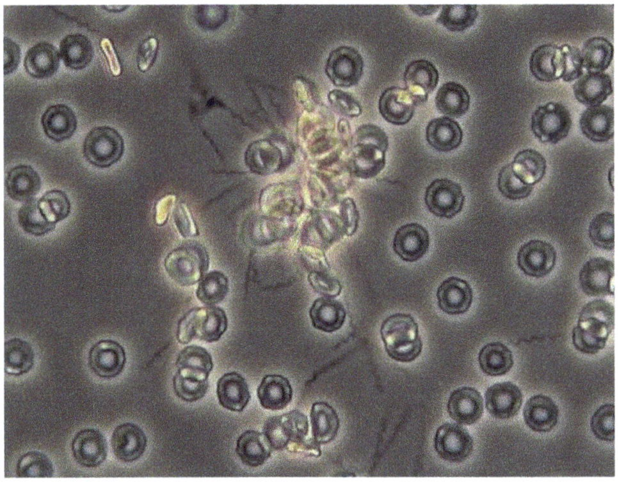

Figure 3. Phase contrast micrograph of rosette of spirochetes and erythrocytes formed during experimental RF. Blood was examined four days after inoculation of an adult C3H/HeN mouse with *Borrelia duttonii*. Magnification 400X. Photo courtesy of Marie Andersson, University of Umeå.

blood. Four pathways that differ between the two groups of species are in the acquisition of glycerol-3-phosphate (Bacon et al., 2004; Schwan et al., 2003; Schwan et al., 1996), pyrimidine biosynthesis (Zhong et al., 2006), purine salvage (Barbour et al., 2005; Lescot et al., 2008; Pettersson et al., 2007), and glutamate-arginine-proline biosynthesis (Lescot et al., 2008).

Greater persistence in the blood presumably increases the fitness of RF *Borrelia* for transmission from an infected vertebrate to a naïve set of tick vectors (Barbour and Restrepo, 2000). However, RF spirochetes also leave the vascular compartment through the endothelium to invade the central nervous system, the eye, the liver, testes, the placenta, and other organs and tissues (Cadavid and Barbour, 1998; Cadavid et al., 2001; Nordstrand et al., 2001; Shamaei-Tousi et al., 2001). The borreliae survive in these extravascular sites when they have been eliminated from the blood (Cadavid et al., 1993; Cadavid et al., 2006; Kazragis et al., 1996; Larsson et al., 2008). Persistence in extravascular niches may also serve to enhance transmissibility, because invasion of the blood from these sites can recur under conditions of immune suppression (Larsson et al., 2006b). Also, Larsson and coworkers suggest certain immune privileged sites can serve as a location for silent infections that provide a reservoir for reactivation of infection as a result of external signals (Larsson et al., 2006b; Larsson et al., 2008; Lundqvist et al., 2010).

In animal models, tissue invasion appears to be facilitated by recruitment of plasminogen to the borrelial cell surface. Once converted to enzymatically active plasmin, local degradation of extracellular matrix barriers in conjunction with efficient spirochetal motility enable spread to distant organs. Consequently, in plasminogen-deficient mice, *B. crocidurae* invasion of brain and kidney was delayed (Nordstrand et al., 2001). Another mechanism of tissue invasion occurs indirectly because of erythrocyte rosette formation and microemboli-mediated hemorrhage (Figure 3). In experimental rats, pressure-induced capillary rupture leads to leakage of aggregates of spirochetes and erythrocytes into adjacent tissue (Shamaei-Tousi et al., 2001).

When spirochetes are found in tissues, they generally are found in the interstitial spaces between cells, except when they have been engulfed by phagocytes. There is no compelling evidence that RF borreliae are facultative, let alone obligatory, intracellular pathogens. The pathologic findings and subsequent identification of RF spirochetes are mainly from autopsies of fatal human cases and from histologic examinations of experimentally-infected animals (Bryceson et al., 1970; Cadavid and Barbour, 1998; Southern and Sanford, 1969). Extracellular spirochetes were demonstrated in tissues, often in perivascular locations (Cadavid et al., 2001; Duray, 1987; Pennington et al., 1997). The liver and spleen were often enlarged and showed inflammation, with microabscesses in the spleen (Anderson and Zimmerman, 1955; Judge et al., 1974b; Thomas et al., 2002). The microabscesses, areas of necrosis, and petechiae may be the consequence, in part, of microemboli of spirochetes and blood cells, and transendothelial migration of neutrophils (Shamaei-Tousi et al., 2000; Shamaei-Tousi et al., 1999).

Intracranial hemorrhage has been observed in both LBRF and TBRF (Anderson and Zimmerman, 1955; Babes, 1916; Dewar and Walmsley, 1945; Judge et al., 1974d; Thomas et al., 2002). When spirochetes in the brain were noted, they were more common around vessels and adjacent to leptomeninges (Anderson and Zimmerman, 1955; Cadavid et al., 2001; Garcia-Monco et al., 1997). Although cardiac abnormalities are not usually associated with RF, myocarditis with the presence of numerous spirochetes has been demonstrated in experimental infections of both TBRF and LBRF (Breitschwerdt et al., 1996; Cadavid et al., 1994; Judge et al., 1974d; Londono et al., 2005; Wengrower et al., 1984). In experimental *B. turicatae* infection, the severity of illness correlated with the numbers of spirochetes in the blood (Pennington et al., 1997).

In the blood of infected patients, leukocyte counts are usually in the normal range or only slightly elevated (Parsons, 1947). Leukopenia and platelet counts below 50,000/mm^3 may occur during the crisis. Hemorrhagic complications, such as epistaxis, are common in LBRF (Dennis et al., 1976; Perine et al., 1971). The erythrocyte sedimentation rate often is moderately elevated. There may be laboratory evidence of hepatitis, such as elevated serum concentrations of aminotransferases and prolonged prothrombin and partial thromboplastin times (Barbour, 2011). In the cerebrospinal fluid there may be mildly elevated numbers of mononuclear cells and protein levels, but glucose levels are typically normal.

Neurotropism

Both RF and LD *Borrelia* affect the neurological system, resulting in various neurological and pathognomonic symptoms. LBRF often presents a more severe course, but TBRF, at least for some species of *Borrelia*, such as *B. duttonii* and *B. turicatae*, results in a higher frequency of neurologic complications, including symptoms like facial palsy, weakness, radiculopathy, and, occasionally, stupor or coma (Cadavid and Barbour, 1998). Also, neuropsychiatric symptoms with psychotic events, including confusion, anxiety and hallucination has been described (Aubin et al., 1947). Neurotropism has also been observed experimentally in mammals (Cadavid and Barbour, 1998; Cadavid et al., 1993; Dubois and Pearson, 1939; Kazragis et al., 1996; Larsson et al., 2006b; Shamaei-Tousi et al., 1999; Velu et al., 1930). RF spirochetes are seen in the brain of infected mice as early as the second day after inoculation (Andersson et al., 2007). *B. crocidurae* cells remain actively growing in the brain and persist there in numbers of ~1000 per mouse brain (Larsson et al., 2006b; Larsson et al., 2008). A similar phenomenon may also occur in the tick vector, as *B. duttonii* spirochetes accumulate around the central ganglion cells of *O. moubata* (Burgdorfer, 1951).

In both human cases and experimental infections of rodents, neurotropism is more common following infection by some species than by others (Cadavid and Barbour, 1998). For example, the greater incidence of neurologic involvement in cases of RF due to *B. duttonii* in Africa and *B. turicatae* in North America has long been noted (reviewed by Cadavid and Barbour, 1998). Also, *B. duttonii* appears to cause persistent brain infection that could be reactivated by immune suppression as well as with concomitant infection with malaria (Lundqvist et al., 2010). These findings also present the importance of humans as reservoir for certain RF *Borrelia* species and a relevant biological signal for reactivation of silent RF borreliosis. A newly described species of RF *Borrelia* in Spain appears to be particularly neurotropic (Garcia-Monco et al., 1997). This isolate was later taxonomically defined as *B. hispanica* (Toledo et al., 2010).

Neurotropism also varies between variants of a given strain. Cadavid et al. demonstrated in an experimental animal model that some variants of a *B. turicatae* strain were present in higher numbers in the brain compared to other variants of the same strain (Cadavid et al., 2001; Cadavid et al., 1997; Cadavid et al., 1994). However, neurotropism is not necessarily associated with greater virulence as is evident from the finding of residual brain infection up to several months for RF *Borrelia* such as *B. duttonii* (Larsson et al. 2006). A neuroinvasive variant of *B. turicatae* had a lower fatality rate among experimentally infected mice than an isogenic variant with a lower potential for invasion of the brain (Cadavid et al., 1994; Pennington et al., 1997). Moreover, the noticeable neurological disease symptoms in certain cases of RF is aggravated by the massive influx of immune cells, including T cells that are important for neuropathogenesis (Liu et al., 2010).

Not only do neurological complications impact treatment, they also can lead to a misdiagnosis of Lyme disease; consequently, caution should be exercised when diagnosing LD in historically nonendemic regions. For example, in Mexico, where *B. turicatae* is distributed (Cooley and Kohls, 1944; Donaldson et al., 2016), serological responses in patients presenting with neurological symptoms (facial palsy) were screened for LD (Gordillo-Perez et al., 2017; Gordillo-Perez et al., 2018). These tests used whole *B. burgdorferi* protein lysates or commercially available kits. Unfortunately, given the serological cross reactivity with patients infected with RF and LD, assays should have targeted antigens specific for TBRF species, like glycerophosphodiester phosphodiesterase (GlpQ) or the *Borrelia* immunogenic protein A (BipA) (Lopez et al., 2010; Schwan et al., 1996). Moreover, there is increasing evidence that serological responses generated during *B. turicatae* infections cross-react with commercially available LD assays (Gettings et al., 2019). Interestingly, when TBRF-specific antigens are utilized in Mexico, serologic studies point to ongoing circulation of TBRF spirochetes (Vazquez-Guerrero et al., 2019).

Adhesins

Adherence to host cells is an important bacterial virulence mechanism. This enables microbes to gain a foothold on surfaces and resist mechanical removal. Binding of RF *Borrelia* to cellular and soluble blood components, such as integrins and glycoaminoglycans, has been described (Alugupalli et al., 2001; Magoun et al., 2000). However, in the absence of knock-out mutants, the functional roles of these associations for infectivity and pathogenesis still remain to be confirmed. The function as well as mechanism for these host ligands in mediating adherence of spirochetes to host tissues and cells are likely, and would be consistent with, similar, more established pathogenetic mechanisms of other pathogens, including LD *Borrelia* (see chapter 13 by Coburn *et al.*).

Some species of RF spirochetes have been noted to form aggregates or rosettes with platelets or erythrocytes in the blood of infected animals (Alugupalli et al., 2003b; Burman et al., 1998; Mooser, 1958). Figure 2 shows a rosette of spirochetes and erythrocytes. Rosettes are best viewed when spirochete density in the blood is at its highest. For *B. crocidurae*, aggregation of spirochetes and erythrocytes into rosettes appears to delay immune clearance; antibodies appear and rise in the blood more slowly than is observed in experimental infection with *Borrelia* spp. that do not aggregate erythrocytes (Burman et al., 1998). Erythrocyte and spirochete rosettes also activate endothelial cells and cause microvascular thrombi and emboli that lead in turn to pathologic changes in tissues (Shamaei-Tousi et al., 2000; Shamaei-Tousi et al., 2001; Shamaei-Tousi et al., 1999). Erythrocyte aggregation has also been noted in experimental infections with *B. duttonii* (Takahashi et al., 2000).

B. hermsii attachment to platelets occurs via the integrin $\alpha_{IIb}\beta_3$ (Alugupalli et al., 2001; Alugupalli et al., 2003b), and *B. crocidurae* binds to erythrocytes via the neolactoglycan lacto-N-neotetraose (Guo et al., 2009). The bacterial ligands for these interactions have not yet been identified for RF species. One possible spirochete ligand is P66, or Oms66, an outer membrane protein porin of both LD and RF *Borrelia* species (Bunikis et al., 1998; Bunikis et al., 1995; Shang et al., 1998; Skare et al., 1997). P66 of *B. burgdorferi* has been reported to be an adhesin for the β_3-integrin and as such important for dissemination in the infected host (Coburn et al., 1999; Coburn and Cugini, 2003). Two other possible adhesins are products of open reading frames BH0512 and BH0553 of *B. hermsii*. Both proteins are characterized by multiple repeating domains. Indirect evidence for the adhesin function of these proteins was the finding that antisera to these proteins diminished attachment of *B. hermsii* to erythrocytes (Guyard et al., 2005). One of the surface lipoproteins of *B. turicatae*, Vsp1, binds to glycosaminoglycans to a greater degree than other Vsp proteins of the same species or of *B. hermsii* (Magoun et al., 2000; Zuckert et al., 2004).

Serum, complement, and phagocyte resistance

Early investigators noted that RF *Borrelia* were susceptible to non-immune sera from some species of animals but resistant to the sera from other species. For example, *B. duttonii* survived in fresh sera from mice, guinea pigs, rabbits, pigeons, chickens, and horses but not in fresh sera from cow, goat, pig, or sheep (Cuboni, 1929). If the latter sera were first heat-inactivated, viability of the spirochetes was retained, an indication that the alternative complement pathway was involved. This is similar to the differential serum resistance demonstrated by LD *Borrelia* species (see chapters 13 and 18 by Coburn *et al.* and Wooten *et al.*). Factor H is a major complement regulator in the serum that prevents C3b formation by accelerating decay of the C3 convertase and by facilitating the irreversible inactivation of C3b. Bacterial acquisition of factor H could enable complement evasion by inactivating C3b or preventing its deposition and, thus, phagocytosis. Some RF *Borrelia* species, including *B. hermsii*, but neither *B. anserina* nor *B. coriaceae*, display factor H binding activity (McDowell et al., 2003; Meri et al., 2006). This activity was attributed to plasmid-encoded proteins of ~20 kDa called FhbA proteins (Hovis et al., 2004). *B. hermsii* also binds to factor H-like protein 1, an alternative splice variant of factor H (Hovis et al., 2006; Lopez et al., 2008). FhbA elicits IgM and IgG responses during experimental infection

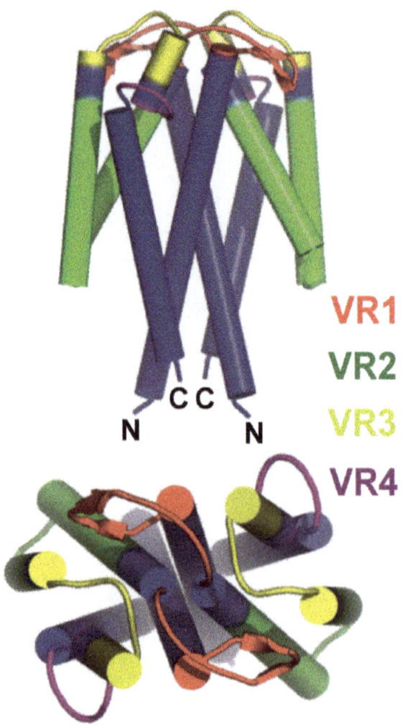

Figure 4. Schematic representation of structure of a Vsp lipoprotein dimer of *B. turicatae*. The four major alpha-helical chains in each monomer are depicted by cylinders. The first and fourth alpha-helix (blue) in each Vsp are relatively conserved in sequence between different Vsp proteins of a given species (Dai et al., 2006; Restrepo et al., 1992). The second and third alpha-helical chain and the loops between all the chains are variable in sequence and are divided into variable regions (VR) 1 (red), 2 (green), 3 (yellow), and 4 (purple). The greatest variability is at the top or dome of the protein. Adapted from Lawson et al. (Lawson et al., 2006).

of mice (Colombo and Alugupalli, 2008). The BhCRASP-1 protein of *B. hermsii* is a 20-kDa outer membrane lipoprotein that is orthologous to the CRASP protein BBA68 of *B. burgdorferi*. BhCRASP-1 appears to bind factor H as well as plasminogen (Rossmann et al., 2007). Finally, *B. duttonii* and *B. recurrentis* are reported to associate with C4b-binding protein in the serum (Meri et al., 2006). Several subsequent studies have identified additional complement factors, including FhbA, HcpA and CihC, which affect different regulators of complement activation. These factors make RF *Borrelia* resistant to killing by complement and provide other important virulence properties (Brenner et al., 2013; Fine et al., 2014; Grosskinsky et al., 2009; Grosskinsky et al., 2010; Schott et al., 2010; Woodman et al., 2009)

Antigenic variation

Vsp and Vlp proteins

The steady increase of spirochetes in the blood, followed by a rapid and deep drop in their numbers, and then their reappearance in the blood (Figure 2) are the direct result of the bacterium's unique capacity for antigenic variation. Investigators as early as 1918 noted that each new crop of spirochetes differed antigenically from the population it succeeded and the population to succeed it (Jancso, 1918; Meleney, 1928; Russell, 1936). The individual variants of a given strain are called serotypes. By definition, there is little or no antigenic cross-reactivity between serotypes with polyclonal antisera obtained either from infected patients or immunized animals (Coffey and Eveland, 1967; Stoenner et al., 1982). A serotype that had disappeared from an RF *Borrelia* population under the force of immunity could reappear again if even a single of the surviving population was transmitted to a naïve animal (Barbour, 1987); thus the hereditary information conferring serotype identity is not lost from a cell. This feature of RF indicated that antigenic variation was not simply an accelerated evolutionary process with selection for a succession of mutants, as typified by influenza A virus.

B. hermsii, a North American species that is most commonly encountered in the mountainous areas of the West (see chapter 10 by Telford and Goethert), has been the subject of the most extensive studies of antigenic variation (Barbour, 2002). Several serotypes of the HS1 strain of this species were derived from a single cell injected into a mouse (Barbour and Stoenner, 1985; Restrepo et al., 1992; Stoenner et al., 1982). Stoenner raised specific antisera in mice to each of 24 *B. hermsii* serotypes and used these reagents to study the relapses that occurred during experimental infections (Stoenner et al., 1982). With this battery of antisera in immunofluorescence assays of blood smears from infected animals, he identified the serotypes of ~80-90% of the variants that appeared during relapses in hundreds of mice. Examination of the bulk of *B. hermsii* chromosomal sequences and plasmids revealed additional *vsp* and *vlp* genes for 59 different serotypes (Dai et al., 2006).

Serotype identity is determined by expression of one of a repertoire of polymorphic lipoproteins (Barbour et al., 1982) (see chapter 7 by Zückert *et al.*). Similar to other bacterial lipoproteins, the processed N-terminus is acylated and anchors the protein in the outer membrane (Carter et al., 1994; Scragg et al., 2000). The known set of about 40 proteins, for which expression is documented, is divided about equally between two different families: variable large proteins (Vlp) of about 36 kDa and variable small proteins (Vsp) of about 20 kDa (Barbour et al., 1983; Burman et al., 1990; Carter et al., 1994; Restrepo et al., 1992). These are encoded by *vlp* and *vsp* genes, respectively. Full-length Vsp and Vlp proteins may have near-identical N-terminal sequences, but after the conserved signal peptide has been cleaved off, the processed proteins are highly divergent. Vsp and Vlp proteins are orthologs of the OspC proteins and VlsE proteins of LD-causing *Borrelia* species, respectively (Carter et al., 1994; Marconi et al., 1993; Zhang et al., 1997). Another ortholog of Vsp is the Vtp protein that is expressed in ticks (see above).

Figure 4 represents the structure of a Vsp protein dimer. The Vsp and Vlp proteins are not obviously homologous at the primary sequence level, but they resemble each other at the conformational level, being predominantly α-helical in secondary structure (Burman et al., 1990; Zuckert et al., 2004) and having similar folds (Lawson et al., 2006). Vsp proteins exist as dimers (Zuckert et al., 2001; Zuckert et al., 2004) and in this configuration resemble the deduced structure of a single Vlp protein (Eicken et al., 2002). The C-terminus is not anchored but is close to N-terminus near the cell's surface for both Vsp and Vlp proteins. The loops between the four α-helices of each monomer are exposed to the environment as ligands or epitopes. The first and fourth alpha helical chains are conserved in sequence, while the second and third chains and loops between the chains comprise four variable regions of the molecule. The Vlp family of proteins is furthered divided into four sub-families with less than 70% sequence identity among them: α, β, γ, and δ (Burman et al., 1990; Dai et al., 2006; Hinnebusch et al., 1998; Restrepo et al., 1992). Different strains of *B. hermsii* from distant geographic origins in western North America have the same four sub-families (Hinnebusch et al., 1998).

The considerable diversity among the sets of *vsp* and *vlp* genes contained in each *Borrelia* cell is the consequence of multiple rounds of intragenic recombination among the alleles within each cell (Rich et al., 2001b). This process was likely preceded by horizontal transfer of alleles between strains, for otherwise there would not be a fund of heterogeneous sequences upon which recombinational processes could operate. Horizontal transfer could be accomplished by the bacteriophage observed in association with *B. hermsii* (Barbour and Hayes, 1986), which may be homologous to the 32-kb circular plasmid prophages of *B. burgdorferi* (Eggers and Samuels, 1999; Stevenson et al., 2000; Zhang and Marconi, 2005). If an ancestor of RF species which possessed a single proto-*ospC/vsp* gene became merodiploid through acquisition of a second ortholog, then with subsequent rounds of recombination and gene duplications, the number of alleles might increase from two to several within a single lineage.

Genetic mechanisms for antigenic variation

Recognizing the cumulative studies of the biology, biochemistry, and immunology of RF, a modeler of the genetic mechanism of serotype-switching should account for the following features of the phenomenon, reviewed in (Barbour, 2002, 2003) (1) The variation is reversible; a serotype once cleared by an animal can appear again in another animal to whom the pathogen, even a single cell, is passed. (2) It is multiphasic, that is, involving three or more different antigens. (3) Serotype identity is conferred by a set of proteins, only one of which is expressed at a time. (4) Serotype-specific epitopes of the protein are located in different parts of the protein, not just in one region. (5) When there is an antigenic switch in a cell during infection, all or nearly all of the protein changes, not just a limited region of the protein, as in VlsE variation in *B. burgdorferi.* (6) Antigen switches are effectively random and spontaneous. In other words, they are not an effect of a change in the environment, such as inoculation into a mouse from culture medium or the appearance of neutralizing antibody in the blood (Stoenner et al., 1982).

Reversibility is achieved by archiving the antigen genes at locations where they are silent and, thus, presumably not under direct selection by immune systems and other host factors (Kitten and Barbour, 1990; Meier et al., 1985; Plasterk et al., 1985). For an antigen gene to be expressed, it must be copied and moved to a site where it can be transcribed (Barbour et al., 1991a; Burman et al., 1990; Sohaskey et al., 1999). This site may be altered by subsequent DNA rearrangements, leading to loss of the DNA that was located there, but a copy of that gene remains in the archive for future activation. The archived genes and the expression site for the duplicated gene are located on linear plasmids of 28 to 32 kb in *B. hermsii* (Ferdows et al., 1996; Kitten and Barbour, 1990;

Plasterk et al., 1985) and on plasmids of other sizes in other RF species (Lescot et al., 2008; Penningon et al., 1999; Tabuchi et al., 2002). There are multiple copies of each type of plasmid in each cell with the genomes distributed evenly within the approximately 20 µm long spirochetal cell (Kitten and Barbour, 1992). The number of copies varies due to growth and milieu, but the mean was calculated both *in vivo* and *in vitro* to be within the range of 16 to 22 genomes per cell (Kitten and Barbour, 1992).

Sequencing of the linear plasmids that bear *vsp* and *vlp* genes in *B. hermsii* further defined the organization and arrangement of the expression site and the several archival sites (Burman et al., 1990; Dai et al., 2006; Restrepo et al., 1992). There appears to be a complete or nearly complete *vsp* or *vlp* gene for each Vsp or Vlp protein that a spirochete is capable of expressing. The report by Dai et al. provides full physical maps of large regions of the plasmids (Dai et al., 2006). In brief, the archived *vsp* and *vlp* genes are, with a few exceptions, found in clusters, usually in the same orientation, i.e., head-to-tail, but head-to-head at some locations. Several of the silent genes are full-length, that is, with a start codon and the signal peptide. Most of these complete genes at silent sites also have consensus ribosomal binding sequences and transcriptional start sites, but not discernible promoters. Other archived genes are lacking some sequence at the 5′ end. Some *vsp*- and *vlp*-like sequences are likely pseudogenes; they are missing 3′ ends and/or have indels with frameshifts. The expression site for *vsp* and *vlp* genes is located near the telomere of a 28-kb plasmid (Kitten and Barbour, 1990); it has a σ^{70}-type promoter (Barbour et al., 1991a) and an upstream T-rich region that appears to enhance transcription (Barbour et al., 1991b; Sohaskey et al., 1999). Similar arrangements of the repertoires of *vsp* and *vlp* genes have been found on linear plasmids of various sizes in the RF species *B. duttonii* and *B. recurrentis* (Lescot et al., 2008; Tabuchi et al., 2002; Vidal et al., 2002).

In *B. hermsii*, two mechanisms can account for all the observed features of antigenic variation (Figure 5). The first and most common mechanism is a non-reciprocal recombination between a linear plasmid with an array of silent, archived *vsp* and *vlp* genes and another linear plasmid with a transcribed *vsp* or *vlp* gene (Kitten and Barbour, 1990; Meier et al., 1985; Plasterk et al., 1985). In *B. hermsii*, as well as in *B. turicatae*, what is effectively a gene conversion results in the replacement of the expressed gene for a variable major protein at a site downstream from a promoter at an expression site (Barbour et al., 1991a; Cadavid et al., 1997; Penningon et al., 1999; Pennington et al., 1999). An archived *vsp* gene can replace an expressed *vlp* gene or vice versa. The boundaries for the recombination are regions of sequence identity between silent and expression sites around and flanking the extreme 5′ and 3′ ends of the expressed and silent *vsp* or *vlp* genes (Barbour et al., 2000; Dai et al., 2006). In some cases, recombination at the expression site occurs within a *vsp* or *vlp* gene (Dai et al., 2006; Kitten et al., 1993). However, if the resultant chimeric protein retained one or more specific epitopes of the original protein, it presum ably would still be recognized by circulating antibodies elicited during the preceding spirochetemia (Barstad et al., 1985). In *B. hermsii*, the longest length of converted sequence that has been observed is about 2 kb (Kitten and Barbour, 1990; Restrepo et al., 1992). In *B. turicatae*, a more extensive gene conversion involving 10 or more kb downstream of the promoter on the expression-site plasmid occurs (Penningon et al., 1999). The mechanism of intermolecular recombination between plasmids first demonstrated in Nearctic species, like *B. hermsii*, also seems to occur in Afro-Tropical and Palearctic species. Rearrangements incorporating long lengths of plasmids have also been noted in *B. duttonii* (Takahashi et al., 2000).

The second mechanism for expressing a different variable major protein is an intramolecular DNA rearrangement of the plasmid with the expression site (Figure 5). This mechanism has only been noted when there is another variable major protein gene just downstream of an expressed *vlp* gene (Restrepo et al., 1994). In such instances, the expressed gene and the non-transcribed downstream gene are in the same orientation. A deletion between short direct repeats at the 5′ ends of each of the tandemly arrayed genes effectively excises the sitting *vlp* gene, and the downstream gene effectively takes the place of the deleted gene next to the promoter. After this deletion, the newly expressed gene is susceptible to small gene conversions that provide further diversity of variable membrane proteins (Restrepo and Barbour, 1994), but the significance of this phenomenon for further immune avoidance is not known.

The activation of the *vtp* gene when *B. hermsii* is in the tick occurs by a mechanism that does not require recombination or transposition. This is a third mechanism: modification of transcripts by expression site switching (Figure 5). The *vtp* gene is located at another locus on a different plasmid than the expression site. At that location, *vtp* has its own promoter (Barbour et al., 2000; Carter et al., 1994). There is only one copy of *vtp* per genome. The mechanism of differential transcription between the *vtp* locus and the expression site locus is unknown.

Programming of antigenic variation

The process of switching starts in one of the 16 to 22 genome equivalents of the spirochetes, which means that during a certain period the cell is expressing more than one Vmp. This hybrid spirochete would be considered recessive until it becomes homozygous for the new serotype gene because of the antibody response generated against the prior Vmp (Crowder et al., 2017). Moreover, the overall serotype switch rate for *B. hermsii* is an estimated 10^{-4} to 10^{-3} per cell per generation. That is, by the time a given serotype reaches a population size of 1,000 to 10,000 cells in an animal, at least one cell of a different serotype has likely appeared in the population (Barbour and Stoenner, 1985; Stoenner et al., 1982). However, the rates for individual switches, e.g., from serotype 3 to serotype 7 or from serotype 17 to serotype 19, vary

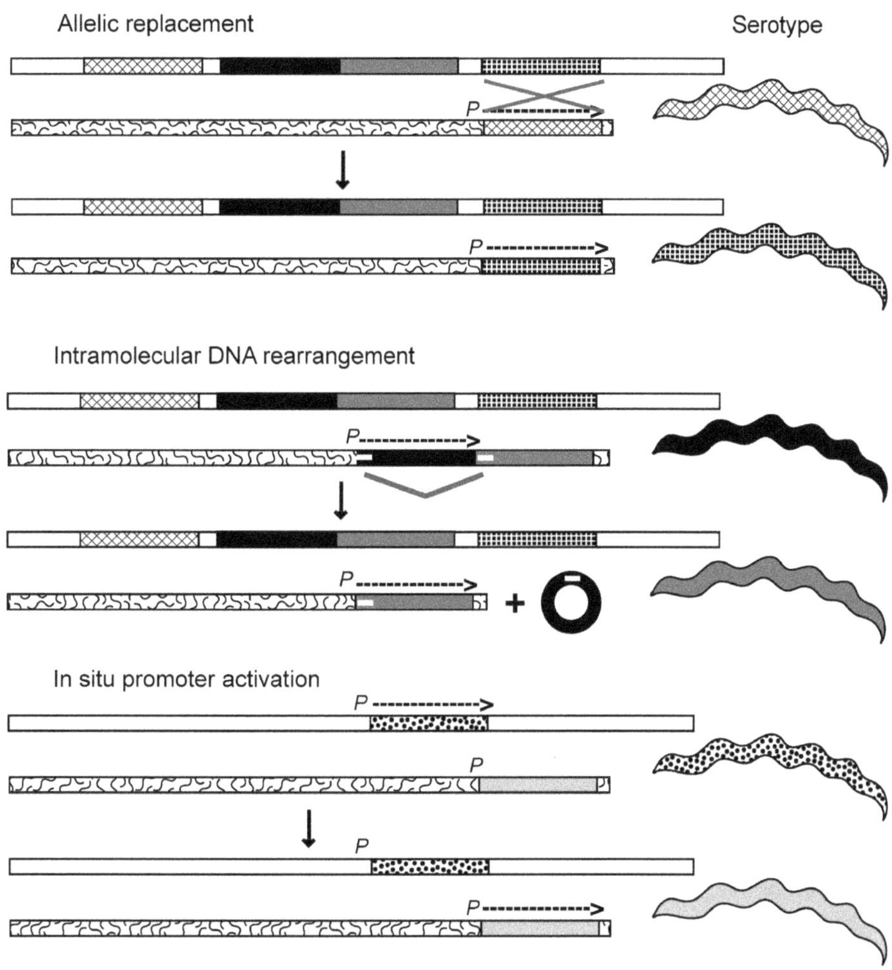

Figure 5. Three mechanisms for serotype switches by *B. hermsii*: allelic replacement, intramolecular DNA rearrangement, and in situ promoter activation. The different mechanisms are discussed in the text. *P* indicates the promoter and the location of the expression site near the right telomere of a linear plasmid; an overhead arrow shows whether the gene is active and the direction of transcription. *vsp* and *vlp* genes at the expression site location or in archival locations on the same (squiggly line background) or another linear plasmid (white background) are schematically represented by different patterns and grayscale values. The serotype of the cell, similarly indicated by pattern or grayscale, corresponds to *vsp* or *vlp* gene, at the expression site. In the second mechanism a deletion by direct repeats (small white horizontal bars) occurs and results in a non-replicative circle. (Courtesy of Alan Barbour, UC Irvine).

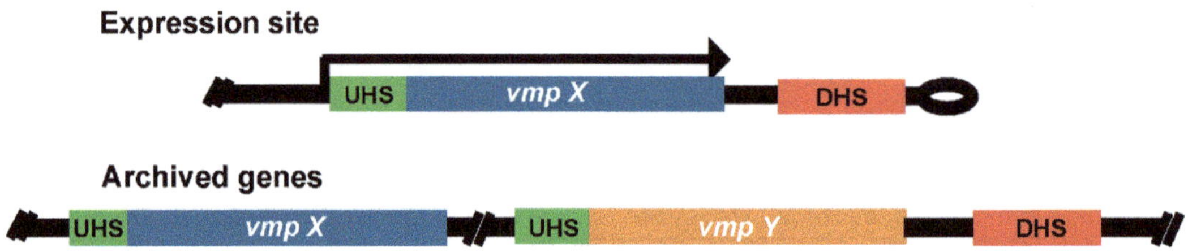

Figure 6. Organization of the expression plasmid (top) and one archival plasmid (bottom) in *B. hermsii*. Two examples of *vsp* or *vlp* genes are denoted by "X" and "Y" and different fill patterns. There is a single expression site on one plasmid, and silent variants of *vsp* and *vlp* genes are found on the same and other plasmids. The direction and extent of transcription of the duplicate gene at the expression site is indicated by the arrow. The UHS element at the expression site comprises 61 nt around the start codon of the variant gene (Barbour et al., 1991a; Dai et al., 2006; Kitten et al., 1993). Silent genes vary in the extent to which their UHS regions are identical to the expression site UHS; this is represented here by the relative length of the UHS block. There is a 214 nt noncoding DHS element downstream from the expression site and adjacent to the plasmid telomere and at various locations on the plasmids. The lengths of *vsp/vlp* gene examples and the UHS and DHS elements are not to scale. (Courtesy of Alan Barbour, UC Irvine).

over about a thousand-fold range (Barbour et al., 2006; Barbour and Stoenner, 1985). The rates in each direction are not equivalent. For instance, switches from serotype 7 to serotype 19 are much less frequent than a switch from serotype 19 to serotype 7. These differences in switch rates among all possible antigenic variation events could account for the rough ordering of the appearance serotypes during individual infections that has been observed (Coffey and Eveland, 1967; Frank and Barbour, 2006; Stoenner et al., 1982)

The organization of the *vsp* and *vlp* genes on the plasmids provided clues to the possible mechanisms for the observed differences in frequencies of serotypes during the first relapse (Dai et al., 2006). The more frequent *B. hermsii* serotypes had Vsp or Vlp proteins that were encoded by genes with certain characteristics in their silent or archival locations (Barbour et al., 2006). One important characteristic was the extent of identity between the 5′ end of archived *vsp* or *vlp* sequence and the 5′ end of the expression site (Figure 6). This region of identity of approximately 50 nucleotides is called the UHS region and is the usual site for the upstream crossover during recombination (Dai et al., 2006). The second major variable was the spatial relationship of a given archived gene to an extragenic repetitive sequence of ~200 nucleotides called the DHS element, which directly follows some but not all archived genes and, when present, is at various distances downstream of a silent allele. The DHS contains the downstream crossover site for the recombination (Dai et al., 2006).

The serotypes that were most frequently observed during the first relapse of infection in laboratory mice were specified by *vsp* or *vlp* genes which, in their archived locations, had the following properties: (1) a UHS sequence that was identical or near-identical to the UHS at the expression site, and (2) a DHS element 50-200 nucleotides downstream. Archived *vsp* or *vlp* alleles that lacked both properties were rarely expressed during the first or early relapses. Silent genes with either property were intermediate in frequency (Barbour et al., 2006). Presumably, possession of both (1) and (2) fostered recombination between the archived sequence and the expression site.

Immunity

Immunity in RF has been the subject of study since the late nineteenth century, as reviewed in (Barbour, 1987; Barbour and Restrepo, 2000). The relapse phenomena observed in experimental animals and in patients with neurosyphilis undergoing fever therapy with RF spirochetes provided evidence for early immunologists, such as Paul Ehrlich, of the fine specificity of the adaptive immune response. The role of immunity in bringing about clearance of spirochetes from the blood was demonstrated by findings that the initial spirochetemia of mice or rats can be prolonged by gamma irradiation (Barbour et al., 1983; Newman and Johnson, 1984), treatment with cyclophosphamide (Newman and Johnson, 1984; Stoenner et al., 1982), the severe combined immunodeficiency (SCID) phenotype (Cadavid et al., 1994), or splenectomy (Meleney, 1928).

Figure 2 shows the appearance of anti-*Borrelia* antibodies in response to infection and the proliferation of spirochetes. This process was thoroughly described in a study by Crowder and coworkers, which infected <10 cells of *B. hermsii* in a group of mice and obtained a spirochetal peak of ~10^8 followed by a clearance of spirochetes that coincides with the rise of variant-specific agglutinating antibodies (Crowder et al., 2016). This first immune response does not prevent the growth of a second wave of spirochetes. As Figure 2 suggests, adaptive immunity to RF species is largely, if not exclusively, variant-specific. If there is an immune response that provides protection against several antigenic variants of a strain or species of RF *Borrelia*, it has not to our knowledge been reported. This may be one reason why there has not been an effective vaccine against RF, although the agent has been known for over 120 years. Aristowsky and Wainstein immunized humans with one variant of a strain; this provided protection against homologous challenge, but not against heterologous variants (Aristowsky and Wainstein, 1929). The variant-specificity of immunity may also explain the paucity of useful serologic assays for confirmation of the diagnosis of RF or for seroepidemiologic surveys. The most promising subunit serologic assays for RF are based on the GlpQ and BipA proteins, which are found in RF *Borrelia* spp. but not in LD *Borrelia* spp. (Lopez et al., 2010; Lopez et al., 2013a; Porcella et al., 2000; Schwan et al., 1996).

In 1896, Gabritchewsky showed the importance of humoral immunity for clearance of spirochetes during RF when he demonstrated the lytic action of immune serum on *Borrelia* (Gabritchewsky, 1896). Novy and Knapp reported in 1906 that they passively protected animals against infection with immune serum (Novy and Knapp, 1906). Calabi subsequently found that serum taken at the time of clearance reduced the viable count of spirochetes in suspension by up to one million-fold (Calabi, 1959). The lesser importance of cellular immunity for resolution of infection was indicated by the findings that T cell deficient nude mice cleared spirochetemia as effectively as wild-type animals and that infection was cleared following reconstitution of lethally irradiated mice with B cells alone, but not T cells (Barbour and Bundoc, 2001; Newman and Johnson, 1984).

As the foregoing experiments would suggest, IgM antibodies from infected animals are alone sufficient to limit or eliminate infection in the blood (Alugupalli et al., 2003a; Arimitsu and Akama, 1973; Barbour and Bundoc, 2001; Connolly and Benach, 2001; Newman and Johnson, 1981, 1984). Within 24 to 48 hours after the spirochetes reach a peak of 10^6 to 10^7 organisms per milliliter of blood, they are undetectable by light microscopy in the blood of an immunocompetent animal. When antibodies are administered intravenously, the spirochetes are cleared from the blood within one hour (Barbour and Bundoc, 2001; Cadavid et al., 1993). As earlier empirical studies had suggested, the neutralizing antibody response is serotype-specific: the animal remains susceptible to other serotypes of the same strain or of different strains (Barbour, 1987; Barbour and Restrepo, 2000). Up to 10 different relapses have been recorded during experimental infections (Felsenfeld, 1965). Whether recovery is due to the development of antibodies to all or most serotypes during the infection's course or to the eventual rise of antibodies directed against antigens expressed by all serotypes is unknown. In endemic areas, newcomers or visitors appear to be at greater risk of symptomatic TBRF than long-term residents, an indication that long-lasting immunity to the local strain(s) may occur among humans.

Specific antibodies, even F_{ab} monomers, recognizing outer membrane proteins of *B. hermsii* and other RF species can kill the spirochetes in the absence of complement or phagocytes (Connolly and Benach, 2001; Connolly et al., 2004; Sadziene et al., 1994). The bactericidal action lies in the variable region (LaRocca et al., 2008), as it does for bactericidal antibodies directed to a surface protein of *B. burgdorferi* (Sadziene et al., 1994). Incubation of the antibody with *Borrelia* cells leads to bacterial lysis following outer membrane disruption and excessive bleb formation (LaRocca et al., 2008).

B cell memory commonly is dependent on stimulation by helper T cells. The fact that *Borrelia* clearance is not dependent on T cell function indicated that a different B cell type was responsible for generating bactericidal antibodies. B1b lymphocytes were shown to confer long-lasting immunity independent of a T cell response via rapid generation of *Borrelia*-specific IgM (Alugupalli et al., 2003a; Alugupalli et al., 2004) . Induction of this response occurs by way of toll-like receptor stimulation and B cell antigen receptor activation (Alugupalli et al., 2007).

During infection, both experimental animals and human patients develop antibody responses against several different spirochetal components, including the periplasmic flagella and heat shock proteins. However, only antibodies directed against variant-specific proteins appear to be effective in controlling infection (Barbour and Bundoc, 2001; Cadavid et al., 1993; Stoenner et al., 1982). Moreover, a host response that eradicates the spirochetes of the predominant serotype from the blood is often not successful in clearing spirochetes from the brain, cerebrospinal fluid, or eye (Barbour, 1987; Cadavid and Barbour, 1998; Larsson et al., 2006b). Unfortunately, given the mechanisms of antigenic variation, targeting the Vmps in a vaccine is not practical.

An alternative approach to generating a vaccine against RF spirochetes would be to neutralize them during early mammalian infection, as they enter the host during tick feeding. Since it takes several generations for antigenic variation to commence, this approach targets more conserved surface proteins that RF spirochetes produce while still colonizing the salivary glands. The efficacy of this approach was demonstrated initially in the *B. hermsii* model (Krajacich et al., 2015). Groups of mice were immunized with one of two antigenically distinct *vtp* variants and were subsequently challenged by tick bite (Krajacich et al., 2015). The animals were protected against homologous variant challenge, but not against heterologous challenge. Regardless, this study demonstrated that preventing the establishment of early mammalian infection is a promising approach toward generating a vaccine against RF spirochetes.

Other RF *Borrelia* species: Emerging diseases

Borrelia miyamotoi is a RF spirochete that, despite its discovery in 1995 (Fukunaga et al., 1995), has only recently been shown to be pathogenic to humans (Platonov et al., 2011). Unlike other RF *Borrelia*, *B. miyamotoi* is not transmitted by soft ticks or lice, which mostly reside in the Southern hemisphere, but instead by hard ticks of various *Ixodes* species that also serve as vectors for *B. burgdorferi* sensu lato (Wagemakers et al., 2015). The main *Ixodes* vectors for both *Borrelia* species live in the temperate climate zones of the Northern hemisphere: *Ixodes pacificus* and *Ixodes scapularis* (North-America), *Ixodes ricinus* (Europe and Asia) and *Ixodes pacificus* (Russia and Asia). In contrast to *B. burgdorferi* sensu lato, which is not transmitted from the female adult tick to her offspring (Steere et al., 2016), *B. miyamotoi* is efficiently transmitted transovarially (Han et al., 2019). Infection rates of nymphs, the *Ixodes* life stage that is most notorious to bite humans, differ for *B. miyamotoi* and *B. burgdorferi* sensu lato, with for instance average infection rates of *I. ricinus* nymphs with *B. burgdorferi* sensu lato of approximately 12% (Strnad et al., 2017) compared to approximately 2% for *B. miyamotoi* in Europe (Cutler et al., 2019; Wagemakers et al., 2015).

Similar to other *Borrelia* species, *B. miyamotoi* has a complex genome with one linear chromosome and a variable number of linear and circular plasmids (Hue et al., 2013; Kingry et al., 2017a; Kingry et al., 2017b; Kuleshov et al., 2018). Due to apparent genetic differences between isolates, the broader term *B. miyamotoi* sensu lato was proposed (Krause et al., 2015). As for other RF *Borrelia*, multiple plasmids encode different Vmps. Vsps and Vlps belonging to the α, β, γ, and δ sub-families, of which only one Vmp is present and transcribed at the expression site in the linear plasmid p41 in any moment in time in a given RF spirochete (Kuleshov et al., 2020). *B. miyamotoi*, similar to other RF *Borrelia* (Barstad et al., 1985; Plasterk et al., 1985) may be able to evade host immune responses by full Vmp gene switching (Wagemakers et al., 2016), in which the initial Vmp is replaced by a different archival full length Vmp gene that is subsequently expressed (Wagemakers et al., 2015). This could explain the occasional spirochetemia relapses in mice experiments (Wagemakers et al., 2016) and in humans (Platonov et al., 2011; Sarksyan et al., 2015a). This could also explain why SCID mice, lacking T and B cells, are able to develop continuous spirochetemia (Wagemakers et al., 2016). Phylogenetic analysis of the available (partial) genomes showed that there are genetic differences between clinical or tick isolates of *B. miyamotoi* sensu lato from Europe, Asia and North America, presumably related to differences in tick vectors in the respective geographical locations (Kuleshov et al., 2019). Importantly, *B. miyamotoi* sensu lato is resistant to killing by the human host complement system (Wagemakers et al., 2014). This is, at least in part, mediated through a protein designated complement binding and inhibitory protein A (CbiA), also called Factor H binding protein, that interacts with the complement regulators Factor H, C3, C3b, C4b, C5, and C9 (Rottgerding et al., 2017). CbiA inhibits the activation of the classical pathway and assembly of the terminal complement complex (Rottgerding et al., 2017) and can bind human plasminogen (Nguyen et al., 2018).

Although human exposure to, and infection with, *B. miyamotoi* appears to occur regularly (Jahfari et al., 2016; Kadkhoda et al., 2017; Krause et al., 2014; Sarksyan et al., 2015b; Sato et al., 2018; Smith et al., 2019; Wroblewski et al., 2017), human *B. miyamotoi* disease (BMD) is relatively infrequent. Nonetheless, case reports and case series of human BMD have been described in Europe, Asia and North America, with a considerable amount of cases originating from Russia and the United States (Chowdri et al., 2013; Crowder et al., 2014; Hoornstra et al., 2018; Jiang et al., 2018; Jobe et al., 2016; Karan et al., 2018; Koetsveld et al., 2017b; Krause et al., 2013; Krause et al., 2016; Molloy et al., 2015; Platonov et al., 2011; Sarksyan et al., 2015a; Sarksyan et al., 2015b; Sato et al., 2014; Sudhindra et al., 2016; Yamano et al., 2017). Generally, the clinical presentation is rather indistinct, but is characterized by fever a few days to weeks after a tick bite, accompanied by non-specific influenza-like symptoms, including chills, fatigue, headache, myalgia and arthralgia (Platonov et al., 2011). Relapses of fever, well known for other RF *Borrelia* infections, appear to only occur occasionally (Sarksyan et al., 2015a). Upon examination of blood, abnormalities, such as mild lymphocytopenia or thrombocytopenia and elevated liver enzymes and/or creatinine, can be observed. Thus, based on the clinical presentation and routine blood studies, it is difficult to differentiate BMD from other tick-borne diseases, or even from other non-tick bite related infectious or inflammatory conditions. In immunocompromised individuals, mainly those on B cell depleting therapy (e.g., anti-CD20 monoclonal antibody rituximab), BMD can present as a meningoencephalitis (Boden et al., 2016; Gugliotta et al., 2013; Hovius et al., 2013). However, a *B. miyamotoi*-induced meningoencephalitis also was described in a seemingly immunocompetent individual in Sweden (Henningsson et al., 2019).

B. miyamotoi disease requires a high level of suspicion among physicians, yet there is little public awareness and diagnostic tools are still experimental and scarcely available. Nonetheless, there are several diagnostic options. Firstly, although most likely of little clinical utility, it is possible to isolate *B. miyamotoi* from blood of infected individuals (Wagemakers et al., 2014). Secondly, and of greater clinical relevance, DNA from *B. miyamotoi* sensu lato can be detected by molecular methods in blood (plasma or serum) (Platonov et al., 2011) or cerebrospinal fluid (CSF) (Hovius et al., 2013) and, albeit less sensitive, spirochetes can be directly visualized by dark-field microscopy (Telford et al., 2019). The above-mentioned diagnostic modalities can be used only in febrile episodes, during which spirochetemia occurs. Outside of the spirochetemic episodes, especially with longer lasting symptoms, serology should be considered. Most studies have performed serological tests based on detection of antibodies against the (GlpQ) antigen expressed by RF *Borrelia*, but not by LD *Borrelia* (Schwan et al., 1996). Of note, GlpQ is also expressed in other gram-negative bacteria, although with low levels of similarity, and does not discriminate *B. miyamotoi* from other RF *Borrelia*. However, due to differences in the geographical spread of hard-bodied *Ixodes* ticks and soft ticks and their associated RF *Borrelia*, this is of little concern, except in specific areas such as California, USA (Krause et al., 2018). Another concern with GlpQ as a diagnostic marker is the large variation in the sensitivity to detect *B. miyamotoi* disease; depending on the population tested, the techniques used to determine seroreactivity results in whether or not patients were treated with antibiotics (Jahfari et al., 2017; Koetsveld et al., 2018a; Krause et al., 2013; Molloy et al., 2015). Therefore, additional antigens, including Vmps, have been investigated (Wagemakers et al., 2016). Interestingly, combining GlpQ with a selection of Vlps and Vsps, results in a high specificity (100% for IgM and 98.3% for IgG) and sensitivity (79% for IgM and 86.7% for IgG) (Koetsveld et al., 2018a). Preliminary and unpublished observations from follow-up studies, which included culture confirmed Lyme borreliosis patients as controls, indicate that Vmps suffer from cross-reactivity (personal communication A.E. Platonov]. Additionally, and vice versa, antibodies against Vmps from BMD patients cross-react with the so-called C6 ELISA (Koetsveld et al., 2020; Molloy et al., 2018), which is based on a conserved peptide within *B. burgdorferi* sensu lato and is a widely used serodiagnostic test for Lyme borreliosis. Interestingly, when using two-dimensional electrophoresis, several immunogenic *B. miyamotoi* proteins were identified, including a putative lipoprotein that specifically reacted with mouse anti-*B. miyamotoi* and not anti-*B. burgdorferi* antibodies (Harris et al., 2019). Another promising antigen is BipA, which not only can discriminate between LD *Borrelia* and RF *Borrelia* (Lopez et al., 2010), but also may differentiate between different RF *Borrelia* species (Lopez et al., 2013a). Future research will be needed to determine whether BipA can distinguish infection caused by *B. miyamotoi* and other RF *Borrelia*.

Based on seroprevalence studies, a considerable number of human *B. miyamotoi* infections can be asymptomatic. However, when *B. miyamotoi* disease is suspected or confirmed, antibiotic treatment should be initiated, although in reported cases from Europe and the United States, individuals recovered fully without antibiotic treatment (Hoornstra et al., 2018; Sudhindra et al., 2016). Additionally, *in vitro* analysis indicated that *B. miyamotoi* laboratory strains and Russian clinical isolates are susceptible to doxycycline, azithromycin, and ceftriaxone (Koetsveld et al., 2017a; Koetsveld et al., 2018b). In these studies, a (relative) resistance to amoxicillin was shown. Although, this has not been confirmed *in vivo* and human patients with *B. miyamotoi* disease have been successfully treated with amoxicillin (Molloy et al., 2015), caution might be warranted using this antibiotic until further studies have been performed. As with LD and RF caused by other *Borrelia*, when central nervous system involvement is suspected or proven, treatment with intravenous antibiotics, e.g., ceftriaxone, should be considered (Henningsson et al., 2019; Hovius et al., 2013).

Other RF *Borrelia* species

B. anserina and *B. coriaceae* are two species transmitted by soft ticks but not known to cause RF in humans. *B. anserina*, which is the type species of *Borrelia,* is the agent of avian or fowl spirochetosis. It is transmitted by *Argas persicus* and related species and is global in distribution in poultry flocks (DaMassa and Adler, 1979; Diab and Soliman, 1977; Zaher et al., 1977). The disease causes severe anemia, weight loss, and high mortality in affected poultry flocks and is the most widespread poultry disease in the world, so is of great economic significance. Heavy spirochetemia of the first episode of illness may extend for 4-6 days in chickens and is prolonged for weeks in bursectomized fowls (Bandopadhyay and Vegad, 1983; DaMassa and Adler, 1979). There are antigenic differences between strains of *B. anserina* and evidence of strain-specific immunity (DaMassa and Adler, 1979; Pachanian and Smith, 1970; Uppal and Rao, 1966). However, *B. anserina* has not been known for displaying antigenic variation during infection. The mortality of the early infection is often high, and many experimental infections were terminated after the first spirochetemia (Bandopadhyay and Vegad, 1983). Nevertheless, the occurrence of immune evasion is suggested by Balfour's report on "Spirochetosis of Sudanese fowls" (Balfour, 1911). He wrote that "in chicks relapses are common, and the disease therefore approaches mammalian spirochetosis in type." In experimental infections of chicks, Balfour noted that 2-3 day periods of heavy spirochetemia culminated in a "crisis." This was followed by disappearance of spirochetes from the blood, then reappearance of spirochetes in the blood, sometimes in "large numbers" after an interval of ~4 days. *B. anserina* has a Vsp-like lipoprotein of 22 kDa (Sambri et al., 1999). Chicks that were passively immunized with antisera to this protein were protected against subsequent homologous challenge with ~10^5 spirochetes, but antisera to a Vsp-like protein from another strain of *B. anserina* did not provide protection against heterologous challenge.

B. coriaceae occurs in *O. coriaceus* soft ticks in California (Hendson and Lane, 2000; Lane and Manweiler, 1988). These ticks feed on deer and cattle, but the role of these large mammals in maintenance of the infection remains to be determined. *B. coriaceae* was identified by PCR in the blood of a mule deer and a Columbian black-tailed deer in California (Lane et al., 2005; Yabsley et al., 2005). *O. coriaceus* has been associated with epizootic bovine abortion in California and Nevada (Schmidtmann et al., 1976). *B. coriaceae*'s etiological role in this disease has been suggested but not proven (Lane et al., 1985).

B. theileri is an RF-like *Borrelia* species that is transmitted by hard ticks of the metastriate genus *Rhipicephalus* rather than by argasid ticks. *B. theileri* produces a microscopically-detectable spirochetemia and weakness, loss of appetite, and anemia in cattle in Africa, Australia, and Europe (Brocklesby et al., 1963; Sharma et al., 2000; Smith et al., 1978). According to Theiler's original description of the infection, there may be two attacks of fever, suggesting the occurrence of antigenic variation (Theiler, 1902). Transovarial transmission of *B. theileri* in *Rhipicephalus* (*Boophilus) microplus* occurs (Smith et al., 1978). Another *Borrelia* sp. that clusters with *B. theileri* was identified in *Rhipicephalus (Boophilus) microplus* in Brazil (Yparraguirre et al., 2007).

In the last decade, other RF group species that are transmitted by hard rather than soft ticks have been identified. These all cluster with *B. theileri* in a monophyletic group distinct from species transmitted by soft ticks. One of the newly discovered species is *B. lonestari*, which was found in the U.S. in

Amblyomma americanum, another metastriate tick (Barbour et al., 1996; Rich et al., 2001a). The distribution in the southern and eastern US of *A. americanum* corresponds with the distribution of an illness characterized by a rash resembling erythema migrans of Lyme borreliosis, which has been called southern tick-associated rash illness or STARI. The clinical association between STARI and *B. lonestari* came from a positive PCR test from an erythema of a patient with an attached *A. americanum*, suggesting that this spirochete might be the causative agent of STARI (Campbell et al., 1995). Further clinical and serological studies have clearly separated the erythema skin lesions presented during infection with *B. burgdorferi* and *B. lonestari*, respectively (Philipp et al., 2006; Wormser et al., 2005). Interestingly, a subsequent study by the same research group showed that neither *B. lonestari* nor *B. burgdorferi* or any other spirochete was the causative agent of an erythema migrans like condition in a specific area in Missouri (Wormser et al., 2005). The etiology of the erythema migrans like disease found in that area is still unclear. Recent reports on cases of STARI infections in other areas in the US will aid in the understanding of the epidemiology of *B. lonestari* (Feder et al., 2011). Also, the successful culture isolation of *B. lonestari* will promote the understanding of this spirochete (Varela et al., 2004). To further study the infection biology of *B. lonestari* (Moyer et al., 2006) used a low-passage culture isolate to infect white-tailed deer (*Odocoileus virginianus*), a common host of *A. americanum* ticks, as well as dogs, mice, and cattle. Only deer developed spirochetemia, as evidenced by blood smear and PCR; the spirochetemia lasted for 6-9 days. Yabsley et al. subsequently found that raccoons were refractory to infection with *B. lonestari.* (Yabsley et al., 2008) The failure of *B. lonestari* to infect mice is attributable to the bactericidal activity of naïve mouse serum for this species (Grigery et al., 2005).

Future Directions

Many other species within the family of *Borreliaceae* probably have reservoir host-vector cycles that remain vague and their role in human infection is unknown. Consequently, these organisms have escaped attention. Of importance for the development of therapeutic, diagnostic, and preventative measures is the ability to isolate spirochetes. While isolation in culture medium can be challenging for RF spirochetes, once bacterial isolates are obtained the available tool-box of multiple omics techniques will enable further exploration to the unique life of these micro-organisms. Importantly, with the increasing number of tools for genetic manipulation of RF *Borrelia,* we will further unravel mechanisms of pathogenesis. Also, the accumulation of knowledge on RF *Borrelia* pathogenesis will aid in development of anti-virulence drugs that might alter the use of broad-spectrum antibiotics and, as such, decrease the burden on the pandemic of antibiotic resistance.

As additional species of *Borrelia* are identified, they may include organisms that, like *B. miyamotoi* and *B. lonestari*, share features with both the RF species and LD species or have other novel features. New species may not include features linking these two groups of spirochetes, but they could be useful for further defining features distinguishing the pathogenesis of RF and LD. Finally, especially for those RF *Borrelia* species with relatively short-lived and low levels of spirochetemia, e.g., *B. miyamotoi,* novel molecular and serological diagnostic antigens are warranted. Ideally those would be able to discriminate between spirochetes belonging to the *B. burgdorferi* sensu lato group and those that belong to RF *Borrelia* and even amongst different RF *Borrelia* species.

Acknowledgement

This chapter was based on a previous edition by Alan Barbour and Betty Guo, whose contributions are the foundation of the current chapter. JL was supported by the NIH grant AI137412,, JWH was supported by the European Union's regional development fund (INTERREG) as part of the NorthTick project and SB from the Swedish Research Council (VR-MH 01885).

References

Aeschlimann, A. (1958). Développement embryonnaire d'*Ornithodorus moubata* (Murray) et transmission transovarienne de *Borrelia duttoni.* Acta Tropica *15*, 15-64.

Alugupalli, K.R., Akira, S., Lien, E., and Leong, J.M. (2007). MyD88- and Bruton's tyrosine kinase-mediated signals are essential for T cell-independent pathogen-specific IgM responses. J Immunol *178*, 3740-3749. doi:10.4049/jimmunol.178.6.3740

Alugupalli, K.R., Gerstein, R.M., Chen, J., Szomolanyi-Tsuda, E., Woodland, R.T., and Leong, J.M. (2003a). The resolution of relapsing fever borreliosis requires IgM and is concurrent with expansion of B1b lymphocytes. J Immunol *170*, 3819-3827. doi:10.4049/jimmunol.170.7.3819

Alugupalli, K.R., Leong, J.M., Woodland, R.T., Muramatsu, M., Honjo, T., and Gerstein, R.M. (2004). B1b lymphocytes confer T cell-independent long-lasting immunity. Immunity *21*, 379-390. doi: 10.1016/j.immuni.2004.06.019

Alugupalli, K.R., Michelson, A.D., Barnard, M.R., Robbins, D., Coburn, J., Baker, E.K., Ginsberg, M.H., Schwan, T.G., and Leong, J.M. (2001). Platelet activation by a relapsing fever spirochaete results in enhanced bacterium-platelet interaction via integrin alphaIIbbeta3 activation. Mol Microbiol *39*, 330-340. doi:10.1046/j.1365-2958.2001.02201.x

Alugupalli, K.R., Michelson, A.D., Joris, I., Schwan, T.G., Hodivala-Dilke, K., Hynes, R.O., and Leong, J.M. (2003b). Spirochete-platelet attachment and thrombocytopenia in murine relapsing fever borreliosis. Blood *102*, 2843-2850. doi:10.1182/blood-2003-02-0426

Anderson, T.R., and Zimmerman, L.E. (1955). Relapsing fever in Korea; a clinicopathologic study of eleven fatal cases with special attention to association with Salmonella infections. Am J Pathol *31*, 1083-1109.

Andersson, M., Nordstrand, A., Shamaei-Tousi, A., Jansson, A., Bergstrom, S., and Guo, B.P. (2007). In situ immune response in brain and kidney during early relapsing fever borreliosis. J Neuroimmunol *183*, 26-32. doi:10.1016/j.jneuroim.2006.11.004

Arimitsu, Y., and Akama, K. (1973). Characterization of protective antibodies produced in mice infected with *Borrelia duttonii*. Jpn J Med Sci Biol *26*, 229-237. doi: 10.7883/yoken1952.26.229

Aristowsky, W.M., and Wainstein, A.B. (1929). Rekurrens-Schutzimpfungsversuche am Menschen. I. Zeitschrift für Immunitaetsforschung und Experimetelle Therapie *61*, 296-308.

Aubin, H., Gachkel, and et al. (1947). Les troubles mentaux dans la fièvre récurrente. Alger Medicale *50*, 408-416.

Babes, V. (1916). Hemorragies meningees et autres manifestations hemorragiques dans la fievre recurrente. Comptes Rendus des Seances Societe de Biologie (Paris) *79*, 855-857.

Bacon, R.M., Pilgard, M.A., Johnson, B.J., Raffel, S.J., and Schwan, T.G. (2004). Glycerophosphodiester phosphodiesterase gene (glpQ) of Borrelia lonestari identified as a target for differentiating Borrelia species associated with hard ticks (Acari:Ixodidae). J Clin Microbiol *42*, 2326-2328. doi:10.1128/JCM.42.5.2326-2328.2004

Balfour, A. (1911). Fourth Report of the Wellcome Tropical Research Laboratories at the Gordon Memorial College Khartoum (London: Wellcome Tropical Research Laboratories), pp. 76-107.

Bandopadhyay, A.C., and Vegad, J.L. (1983). Observations on the pathology of experimental avian spirochaetosis. Res Vet Sci *35*, 138-144.

Barbour, A.G. (1986). Cultivation of Borrelia: a historical overview. Zentralbl Bakteriol Mikrobiol Hyg A *263*, 11-14. doi:10.1016/s0176-6724(86)80095-5

Barbour, A.G. (1987). Immunobiology of relapsing fever. Contrib Microbiol Immunol *8*, 125-137.

Barbour, A.G. (1988). Laboratory aspects of Lyme borreliosis. Clin Microbiol Rev *1*, 399-414. doi:10.1128/CMR.1.4.399

Barbour, A.G. (2002). Antigenic variation of relapsing fever *Borrelia* spp. and other bacterial pathogens. In Mobile DNA II, N.L. Craig, R. Craigie, M. Gellert, and A. Lambowitz, eds. (Washington, D.C.: American Society for Microbiology), pp. 972-994. doi: 10.1128/9781555817954.ch41

Barbour, A.G. (2003). Antigenic variation in Borrelia: relapsing fever and Lyme borreliosis. In Antigenic Variation, A. Craig, and A. Scherf, eds. (London: Academic Press), pp. 319-356. doi:10.1016/B978-012194851-1/50040-8

Barbour, A.G. (2011). Relapsing fever and other Borrelia diseases. In Tropical Infectious Diseases Principles, Pathogens, & Practice, R.L. Guerrant, D.H. Walker, and P.F. Weller, eds. (Philadelphia: WB Saunders Co Ltd), pp. 295-302. doi:10.1016/B978-0-7020-3935-5.00044-6

Barbour, A.G., Barrera, O., and Judd, R.C. (1983). Structural analysis of the variable major proteins of Borrelia hermsii. J Exp Med *158*, 2127-2140. doi:10.1084/jem.158.6.2127

Barbour, A.G., and Bundoc, V. (2001). In vitro and in vivo neutralization of the relapsing fever agent Borrelia hermsii with serotype-specific immunoglobulin M antibodies. Infect Immun *69*, 1009-1015. doi: 10.1128/IAI.69.2.1009-1015.2001

Barbour, A.G., Burman, N., Carter, C.J., Kitten, T., and Bergstrom, S. (1991a). Variable antigen genes of the relapsing fever agent Borrelia hermsii are activated by promoter addition. Mol Microbiol *5*, 489-493. doi:10.1111/j.1365-2958.1991.tb02132.x

Barbour, A.G., Carter, C.J., Burman, N., Freitag, C.S., Garon, C.F., and Bergstrom, S. (1991b). Tandem insertion sequence-like elements define the expression site for variable antigen genes of Borrelia hermsii. Infect Immun *59*, 390-397. doi:10.1128/IAI.59.1.390-397.1991

Barbour, A.G., Carter, C.J., and Sohaskey, C.D. (2000). Surface protein variation by expression site switching in the relapsing fever agent Borrelia hermsii. Infect Immun *68*, 7114-7121. doi:10.1128/iai.68.12.7114-7121.2000

Barbour, A.G., Dai, Q., Restrepo, B.I., Stoenner, H.G., and Frank, S.A. (2006). Pathogen escape from host immunity by a genome program for antigenic variation. Proc Natl Acad Sci U S A *103*, 18290-18295. doi:10.1073/pnas.0605302103

Barbour, A.G., and Guo, B.P. (2010). Borrelia: Molecular Biology, Host Interaction and Pathogenesis (Caister Academic Press).

Barbour, A.G., and Hayes, S.F. (1986). Biology of Borrelia species. Microbiol Rev *50*, 381-400.

Barbour, A.G., Maupin, G.O., Teltow, G.J., Carter, C.J., and Piesman, J. (1996). Identification of an uncultivable Borrelia species in the hard tick Amblyomma americanum: possible agent of a Lyme disease-like illness. J Infect Dis *173*, 403-409. doi:10.1093/infdis/173.2.403

Barbour, A.G., Putteet-Driver, A.D., and Bunikis, J. (2005). Horizontally acquired genes for purine salvage in Borrelia spp. causing relapsing fever. Infect Immun *73*, 6165-6168. doi:10.1128/IAI.73.9.6165-6168.2005

Barbour, A.G., and Restrepo, B.I. (2000). Antigenic variation in vector-borne pathogens. Emerging Infectious Diseases *6*, 449-457. doi: 10.3201/eid0605.000502

Barbour, A.G., and Schwan, T.G. (2018). Borrelia. In Bergey's Manual of Systematics of Archaea and Bacteria, M.E. Trujillo, S. Dedysh, P. DeVos, B. Hedlund, P. Kämpfer, F.A. Rainey, and W.B. Whitman, eds. (John Wiley & Sons, Inc., in association with Bergey's Manual Trust), pp. 1-22. doi:10.1002/9781118960608.gbm01246.pub2

Barbour, A.G., and Stoenner, H.G. (1985). Antigenic variation of *Borrelia hermsii*. In Genome Rearrangement, M.I. Simon, and I. Herskowitz, eds. (New York: Alan R. Liss, Inc.), pp. 123-135.

Barbour, A.G., Tessier, S.L., and Stoenner, H.G. (1982). Variable major proteins of Borrellia hermsii. J Exp Med *156*, 1312-1324. doi:10.1084/jem.156.5.1312

Barstad, P.A., Coligan, J.E., Raum, M.G., and Barbour, A.G. (1985). Variable major proteins of Borrelia hermsii. Epitope mapping and partial sequence analysis of CNBr peptides. J Exp Med *161*, 1302-1314. doi:10.1084/jem.161.6.1302

Battisti, J.M., Raffel, S.J., and Schwan, T.G. (2008). A system for site-specific genetic manipulation of the relapsing fever spirochete *Borrelia hermsii*. Methods Mol Biol *431*, 69-84. doi: 10.1007/978-1-60327-032-8_6

Ben-Menachem, G., Kubler-Kielb, J., Coxon, B., Yergey, A., and Schneerson, R. (2003). A newly discovered cholesteryl galactoside from Borrelia burgdorferi. Proc Natl Acad Sci U S A *100*, 7913-7918. doi:10.1073/pnas.1232451100

Beunders, B., and Van Thiel, P. (1932). Untersuchungen ueber die persistenz von Spirochaeta duttoni in mausegehirnen bei experimenteller febris recurrens. Zeitschrift fuer Hygiene und Infektionskrankheiten *114*, 568-583.

Boden, K., Lobenstein, S., Hermann, B., Margos, G., and Fingerle, V. (2016). Borrelia miyamotoi-Associated Neuroborreliosis in Immunocompromised Person. Emerging Infectious Diseases *22*, 1617-1620. doi:10.3201/eid2209.152034

Bodman, R.I., and Stewart, I.S. (1948). Louse-borne relapsing fever in Persia. Br Med J *1*, 291-293. doi:10.1136/bmj.1.4545.291

Borgnolo, G., Hailu, B., Ciancarelli, A., Almaviva, M., and Woldemariam, T. (1993). Louse-borne relapsing fever. A clinical and an epidemiological study of 389 patients in Asella Hospital, Ethiopia. Trop Geogr Med *45*, 66-69.

Bourret, T.J., Boyle, W.K., Zalud, A.K., Valenzuela, J.G., Oliveira, F., and Lopez, J.E. (2019). The relapsing fever spirochete *Borrelia turicatae* persists in the highly oxidative environment of its soft-bodied tick vector. Cell Microbiol *21*, e12987. doi:10.1111/cmi.12987

Boyle, W.K., Wilder, H.K., Lawrence, A.M., and Lopez, J.E. (2014). Transmission dynamics of *Borrelia turicatae* from the arthropod vector. PLoS neglected tropical diseases *8*, e2767. doi:10.1371/journal.pntd.0002767

Brasseur, D. (1985). Tick-borne relapsing fever in a premature infant. Ann Trop Paediatr *5*, 161-162. doi:10.1080/02724936.1985.11748384

Breitschwerdt, E.B., Geoly, F.J., Meuten, D.J., Levine, J.F., Howard, P., Hegarty, B.C., and Stafford, L.C. (1996). Myocarditis in mice and guinea pigs experimentally infected with a canine-origin Borrelia isolate from Florida. Am J Vet Res *57*, 505-511.

Brenner, C., Bomans, K., Habicht, J., Simon, M.M., and Wallich, R. (2013). Mapping the ligand-binding region of Borrelia hermsii fibronectin-binding protein. PLoS One *8*, e63437. doi:10.1371/journal.pone.0063437

Breuner, N.E., Dolan, M.C., Replogle, A.J., Sexton, C., Hojgaard, A., Boegler, K.A., Clark, R.J., and Eisen, L. (2017). Transmission of *Borrelia miyamotoi* sensu lato relapsing fever group spirochetes in relation to duration of attachment by *Ixodes scapularis* nymphs. Ticks Tick Borne Dis *8*, 677-681. doi:10.1016/j.ttbdis.2017.03.008

Brocklesby, D.W., Scott, G.R., and Rampton, C.S. (1963). Borrelia theileri and transient fevers in cattle. Vet Rec *75*, 103-104.

Brown, V., Larouze, B., Desve, G., Rousset, J.J., Thibon, M., Fourrier, A., and Schwoebel, V. (1988). Clinical Presentation of Louse-Borne Relapsing Fever among Ethiopian Refugees in Northern Somalia. Ann Trop Med Parasit *82*, 499-502. doi:10.1080/00034983.1988.11812282

Bryceson, A., Parry, E., Perine, P., Warrel, D., Vukotich, D., and Leithead, C. (1970). Louse-borne relapsing fever: A clinical and laboratory study of 62 cases in Ethiopia and a reconsideration of the literature. Quarterly Journal of Medicine *153*, 129-170.

Bryceson, A.D. (1976). Clinical pathology of the Jarisch-Herxheimer reaction. J Infect Dis *133*, 696-704. doi:10.1093/infdis/133.6.696

Bryceson, A.D., Cooper, K.E., Warrell, D.A., Perine, P.L., and Parry, E.H. (1972). Studies of the mechanism of the Jarisch-Herxheimer reaction in louse-borne relapsing fever: evidence for the presence of circulating Borrelia endotoxin. Clin Sci *43*, 343-354. doi:10.1042/cs0430343

Bunikis, J., Luke, C.J., Bunikiene, E., Bergstrom, S., and Barbour, A.G. (1998). A surface-exposed region of a novel outer membrane protein (P66) of Borrelia spp. is variable in size and sequence. J Bacteriol *180*, 1618-1623. doi:10.1128/JB.180.7.1618-1623.1998

Bunikis, J., Noppa, L., and Bergstrom, S. (1995). Molecular analysis of a 66-kDa protein associated with the outer membrane of Lyme disease Borrelia. FEMS Microbiol Lett *131*, 139-145. doi:10.1111/j.1574-6968.1995.tb07768.x

Bunikis, J., Tsao, J., Garpmo, U., Berglund, J., Fish, D., and Barbour, A.G. (2004). Typing of *Borrelia* relapsing fever group strains. Emerging Infectious Diseases *10*, 1661-1664. doi:10.3201/eid1009.040236

Burgdorfer, W. (1951). Analyse des Infektionsverlaufes bei *Ornithodorus moubata* und der natürlichen Uebertragung von *Spirochaeta duttoni*. Acta Tropica *8*, 196-262.

Burgdorfer, W. (1959). Zum organotropismus der Spirochaete *B. duttoni* gegenuber der ubertragenden zecke. Acta Tropica *16*, 242-243.

Burgdorfer, W., and Varma, M.G. (1967). Trans-stadial and transovarial development of disease agents in arthropods. Annu Rev Entomol *12*, 347-376. doi:10.1146/annurev.en.12.010167.002023

Burman, N., Bergstrom, S., Restrepo, B.I., and Barbour, A.G. (1990). The variable antigens Vmp7 and Vmp21 of the relapsing fever bacterium Borrelia hermsii are structurally analogous to the VSG proteins of the African trypanosome. Mol Microbiol *4*, 1715-1726. doi:10.1111/j.1365-2958.1990.tb00549.x

Burman, N., Shamaei-Tousi, A., and Bergstrom, S. (1998). The spirochete Borrelia crocidurae causes erythrocyte rosetting during relapsing fever. Infect Immun *66*, 815-819. doi:10.1128/IAI.66.2.815-819.1998

Butler, T. (2017). The Jarisch-Herxheimer Reaction After Antibiotic Treatment of Spirochetal Infections: A Review of Recent Cases and Our Understanding of Pathogenesis. Am J Trop Med Hyg *96*, 46-52. doi:10.4269/ajtmh.16-0434

Cadavid, D., and Barbour, A.G. (1998). Neuroborreliosis during relapsing fever: review of the clinical manifestations, pathology, and treatment of infections in humans and experimental animals. Clin Infect Dis *26*, 151-164. doi:10.1086/516276

Cadavid, D., Bundoc, V., and Barbour, A.G. (1993). Experimental infection of the mouse brain by a relapsing fever Borrelia species: a molecular analysis. J Infect Dis *168*, 143-151. doi:10.1093/infdis/168.1.143

Cadavid, D., Pachner, A.R., Estanislao, L., Patalapati, R., and Barbour, A.G. (2001). Isogenic serotypes of Borrelia turicatae show different localization in the brain and skin of mice. Infect Immun *69*, 3389-3397. doi:10.1128/IAI.69.5.3389-3397.2001

Cadavid, D., Pennington, P.M., Kerentseva, T.A., Bergstrom, S., and Barbour, A.G. (1997). Immunologic and genetic analyses of VmpA of a neurotropic strain of Borrelia turicatae. Infect Immun *65*, 3352-3360. doi:10.1128/IAI.65.8.3352-3360.1997

Cadavid, D., Sondey, M., Garcia, E., and Lawson, C.L. (2006). Residual brain infection in relapsing-fever borreliosis. J Infect Dis *193*, 1451-1458. doi:10.1086/503367

Cadavid, D., Thomas, D.D., Crawley, R., and Barbour, A.G. (1994). Variability of a bacterial surface protein and disease expression in a possible mouse model of systemic Lyme borreliosis. J Exp Med *179*, 631-642. doi:10.1084/jem.179.2.631

Calabi, O. (1959). The presence of plasma inhibitors during the crisis phenomenon in experimental relapsing fever (*Borrelia novyi*). J Exp Med *110*, 811-825. doi:10.1084/jem.110.5.811

Campbell, G.L., Paul, W.S., Schriefer, M.E., Craven, R.B., Robbins, K.E., and Dennis, D.T. (1995). Epidemiologic and diagnostic studies of patients with suspected early Lyme disease, Missouri, 1990-1993. J Infect Dis *172*, 470-480. doi:10.1093/infdis/172.2.470

Carter, C.J., Bergstrom, S., Norris, S.J., and Barbour, A.G. (1994). A family of surface-exposed proteins of 20 kilodaltons in the genus Borrelia. Infect Immun *62*, 2792-2799. doi:10.1128/IAI.62.7.2792-2799.1994

Chowdri, H.R., Gugliotta, J.L., Berardi, V.P., Goethert, H.K., Molloy, P.J., Sterling, S.L., and Telford, S.R. (2013). Borrelia miyamotoi infection presenting as human granulocytic anaplasmosis: a case report. Ann Intern Med *159*, 21-27. doi:10.7326/0003-4819-159-1-201307020-00005

Coburn, J., Chege, W., Magoun, L., Bodary, S.C., and Leong, J.M. (1999). Characterization of a candidate Borrelia burgdorferi beta3-chain integrin ligand identified using a phage display library. Mol Microbiol *34*, 926-940. doi:10.1046/j.1365-2958.1999.01654.x

Coburn, J., and Cugini, C. (2003). Targeted mutation of the outer membrane protein P66 disrupts attachment of the Lyme disease agent, Borrelia burgdorferi, to integrin alphavbeta3. Proc Natl Acad Sci U S A *100*, 7301-7306. doi:10.1073/pnas.1131117100

Coffey, E.M., and Eveland, W.C. (1967). Experimental relapsing fever initiated by Borrelia hermsi. II. Sequential appearance of major serotypes in the rat. J Infect Dis *117*, 29-34. doi:10.1093/infdis/117.1.29

Colombo, M.J., and Alugupalli, K.R. (2008). Complement factor H-binding protein, a putative virulence determinant of Borrelia hermsii, is an antigenic target for protective B1b lymphocytes. J Immunol *180*, 4858-4864. doi:10.4049/jimmunol.180.7.4858

Connolly, S.E., and Benach, J.L. (2001). Cutting edge: the spirochetemia of murine relapsing fever is cleared by complement-independent bactericidal antibodies. J Immunol *167*, 3029-3032. doi: 10.4049/jimmunol.167.6.3029

Connolly, S.E., Thanassi, D.G., and Benach, J.L. (2004). Generation of a complement-independent bactericidal IgM against a relapsing fever Borrelia. J Immunol *172*, 1191-1197. doi:10.4049/jimmunol.172.2.1191

Cooley, R.A., and Kohls, G.M. (1944). The Argasidae of North America, Central America, and Cuba (South Bend: Notre Dame University Press). doi:10.5962/bhl.title.4511

Cooper, P.J., Fekade, D., Remick, D.G., Grint, P., Wherry, J., and Griffin, G.E. (2000). Recombinant human interleukin-10 fails to alter proinflammatory cytokine production or physiologic changes associated with the Jarisch-Herxheimer reaction. The Journal of infectious diseases *181*, 203-209. doi:10.1086/315183

Crowder, C.D., Carolan, H.E., Rounds, M.A., Honig, V., Mothes, B., Haag, H., Nolte, O., Luft, B.J., Grubhoffer, L., Ecker, D.J., *et al.* (2014). Prevalence of Borrelia miyamotoi in Ixodes ticks in Europe and the United States. Emerging Infectious Diseases *20*, 1678-1682. doi:10.3201/eid2010.131583

Crowder, C.D., Denny, R.L., and Barbour, A.G. (2017). Segregation Lag in Polyploid Cells of the Pathogen Genus Borrelia: Implications for Antigenic Variation. Yale J Biol Med *90*, 195-218.

Crowder, C.D., Ghalyanchi Langeroudi, A., Shojaee Estabragh, A., Lewis, E.R.G., Marcsisin, R.A., and Barbour, A.G. (2016). Pathogen and Host Response Dynamics in a Mouse Model of Borrelia hermsii Relapsing Fever. Vet Sci *3*. doi:10.3390/vetsci3030019

Cuboni, E. (1929). Sul potere spirocheticida del siero di sangue di alcuni animali. Bol Inst Sieroterap Milan *8*, 813-817.

Cutler, S., Vayssier-Taussat, M., Estrada-Pena, A., Potkonjak, A., Mihalca, A.D., and Zeller, H. (2019). A new Borrelia on the block: Borrelia miyamotoi - a human health risk? Euro Surveill *24*. doi: 10.2807/1560-7917.ES.2019.24.18.1800170

Cutler, S.J., Bonilla, E.M., and Singh, R.J. (2010). Population structure of East African relapsing fever *Borrelia* spp. Emerging Infectious Diseases *16*, 1076-1080. doi:10.3201/eid1607.091085

Dai, Q., Restrepo, B.I., Porcella, S.F., Raffel, S.J., Schwan, T.G., and Barbour, A.G. (2006). Antigenic variation by Borrelia hermsii occurs through recombination between extragenic repetitive elements on linear plasmids. Mol Microbiol *60*, 1329-1343. doi:10.1111/j.1365-2958.2006.05177.x

DaMassa, A.J., and Adler, H.E. (1979). Avian spirochetosis: natural transmission by Argas (Persicargas) sanchezi (Ixodoidea: argasidae) and existence of different serologic and immunologic types of Borrelia anserina in the United States. Am J Vet Res *40*, 154-157.

Davis, G.E. (1943). Relapsing fever: the tick *Ornithodoros turicata* as a spirochetal reservoir. Pub Health Rep *58*, 839-842.

Davis, G.E. (1952). Biology as an aid to the identification of two closely related species of ticks of the genus Ornithodoros. J Parasitol *38*, 477-480.

Davis, G.E. (1955). Relapsing Fever Spirochetes - the Present Status of Borrelia-Venezuelensis Brumpt and Borrelia-Neotropicalis Bates-and-St-John. Int B Bact Nomencl T *5*, 107-109. doi: 10.1099/0096266x-5-3-107

Davis, G.E. (1956). A relapsing fever spirochete, *Borrelia mazzottii* (sp. nov.) from *Ornithodoros talaje* from Mexico. American journal of hygiene *63*, 13-17. doi:10.1093/oxfordjournals.aje.a119787

Dennis, D.T., Awoke, S., Doberstyn, E.B., and Fresh, J.W. (1976). Bleeding in louse-borne relapsing fever in Ethiopia; clinical and laboratory features in 29 patients. East Afr Med J *53*, 220-225.

Dewar, H.A., and Walmsley, R. (1945). Relapsing fever with nephritis and subarachnoid haemorrhage. Lancet *2*, 630. doi:10.1016/s0140-6736(45)90763-x

Diab, F.M., and Soliman, Z.R. (1977). An experimental study of Borrelia anserina in four species of Argas ticks. 1. Spirochete localization and densities. Z Parasitenkd *53*, 201-212. doi:10.1007/BF00380465

Donaldson, T.G., Perez de Leon, A.A., Li, A.Y., Castro-Arellano, I., Wozniak, E., Boyle, W.K., Hargrove, R., Wilder, H.K., Kim, H.J., Teel, P.D., *et al.* (2016). Assessment of the Geographic Distribution of *Ornithodoros turicata* (Argasidae): Climate Variation and Host Diversity. PLoS neglected tropical diseases *10*, e0004383. doi: 10.1371/journal.pntd.0004383

Dubois, A., and Pearson, Y. (1939). Variabilite du neurotropisme d'une souche de Spirochaeta dutoni, immunite en l'abscence de neurotropisme. Comptes Rendus des Seances Societe de Biologie (Paris) *130*, 1379-1381.

Dupont, H.T., La Scola, B., Williams, R., and Raoult, D. (1997). A focus of tick-borne relapsing fever in southern Zaire. Clin Infect Dis *25*, 139-144. doi:10.1086/514496

Duray, P.H. (1987). The surgical pathology of human Lyme disease. An enlarging picture. Am J Surg Pathol *11 Suppl 1*, 47-60. doi: 10.1097/00000478-198700111-00005

Eggers, C.H., and Samuels, D.S. (1999). Molecular evidence for a new bacteriophage of Borrelia burgdorferi. J Bacteriol *181*, 7308-7313. doi:10.1128/JB.181.23.7308-7313.1999

Eicken, C., Sharma, V., Klabunde, T., Lawrenz, M.B., Hardham, J.M., Norris, S.J., and Sacchettini, J.C. (2002). Crystal structure of Lyme disease variable surface antigen VlsE of Borrelia burgdorferi. J Biol Chem *277*, 21691-21696. doi:10.1074/jbc.M201547200

Elbir, H., Abi-Rached, L., Pontarotti, P., Yoosuf, N., and Drancourt, M. (2014). African relapsing Fever borreliae genomospecies revealed by comparative genomics. Front Public Health *2*, 43. doi:10.3389/fpubh.2014.00043

Feder, H.M., Jr., Hoss, D.M., Zemel, L., Telford, S.R., 3rd, Dias, F., and Wormser, G.P. (2011). Southern Tick-Associated Rash Illness (STARI) in the North: STARI following a tick bite in Long Island, New York. Clin Infect Dis *53*, e142-146. doi:10.1093/cid/cir553

Fekade, D., Knox, K., Hussein, K., Melka, A., Lalloo, D.G., Coxon, R.E., and Warrell, D.A. (1996). Prevention of Jarisch-Herxheimer reactions by treatment with antibodies against tumor necrosis factor alpha. N Engl J Med *335*, 311-315. doi:10.1056/NEJM199608013350503

Felsenfeld, O. (1965). Borrelia, human relapsing fever, and parasite-vector-host relationships. Bacteriological Reviews *29*, 46-74.

Felsenfeld, O. (1971). Borrelia. Strains, Vectors, Human and Animal Borreliosis (St. Louis, MO: Warren H. Greene, Inc.). doi: 10.7326/0003-4819-75-5-822_4

Felsenfeld, O., and Wolf, R.H. (1973). Interference between borreliae and trypanosomes. I. Antigenic analysis, fluorescent antibody and immobilisine studies of Borrelia turicatae and Trypanosoma cruzi. Ann Trop Med Parasitol *67*, 335-340.

Ferdows, M.S., Serwer, P., Griess, G.A., Norris, S.J., and Barbour, A.G. (1996). Conversion of a linear to a circular plasmid in the relapsing fever agent Borrelia hermsii. J Bacteriol *178*, 793-800. doi:10.1128/jb.178.3.793-800.1996

Filatov, S., Krishnavajhala, A., Armstrong, B.A., Kneubehl, A.R., Nieto, N.C., Perez De Leon, A.A., and Lopez, J.E. (2020). Isolation and Molecular Characterization of Tick-Borne Relapsing Fever Borrelia Infecting Ornithodoros (Pavlovskyella) verrucosus Ticks Collected in Ukraine. J Infect Dis *221*, 804-811. doi:10.1093/infdis/jiz500

Fine, L.M., Miller, D.P., Mallory, K.L., Tegels, B.K., Earnhart, C.G., and Marconi, R.T. (2014). The *Borrelia hermsii* factor H binding protein FhbA is not required for infectivity in mice or for resistance to human complement in vitro. Infect Immun *82*, 3324-3332. doi:10.1128/IAI.01892-14

Frank, S.A., and Barbour, A.G. (2006). Within-host dynamics of antigenic variation. Infect Genet Evol *6*, 141-146. doi:10.1016/j.meegid.2004.10.005

Fuchs, P.C., and Oyama, A.A. (1969). Neonatal relapsing fever due to transplacental transmission of *Borrelia*. JAMA : the journal of the American Medical Association *208*, 690-692.

Fukunaga, M., Takahashi, Y., Tsuruta, Y., Matsushita, O., Ralph, D., McClelland, M., and Nakao, M. (1995). Genetic and phenotypic analysis of *Borrelia miyamotoi* sp. nov., isolated from the ixodid tick *Ixodes persulcatus*, the vector for Lyme disease in Japan. International Journal of Systematic Bacteriology *45*, 804-810. doi: 10.1099/00207713-45-4-804

Gabritchewsky, G.N. (1896). Les bases de la sérothérapie de la fievre récurrente. Annales Institut Pasteur *10*, 630-653.

Galloway, R.E., Levin, J., Butler, T., Naff, G.B., Goldsmith, G.H., Saito, H., Awoke, S., and Wallace, C.K. (1977). Activation of protein mediators of inflammation and evidence for endotoxemia in Borrelia recurrentis infection. Am J Med *63*, 933-938. doi: 10.1016/0002-9343(77)90548-4

Garcia-Monco, J.C., Miller, N.S., Backenson, P.B., Anda, P., and Benach, J.L. (1997). A mouse model of Borrelia meningitis after intradermal injection. J Infect Dis *175*, 1243-1245. doi: 10.1086/593681

Garnham, P., Davis, C., Heisch, R., and Timms, G. (1947). An epidemic of louse-borne relapsing fever in Kenya. Transactions of the Royal Society of Tropical Medicine and Hygiene *41*, 141-170. doi:10.1016/s0035-9203(47)90209-5

Gettings, J.R., Lopez, J.E., Krishnavahjala, A., Armstrong, B.A., Thompson, A.T., and Yabsley, M.J. (2019). Antibodies to Borrelia turicatae in Experimentally Infected Dogs Cross-React with Borrelia burgdorferi Serologic Assays. J Clin Microbiol *57*. doi:10.1128/JCM.00628-19

Goodman, R.L., Arndt, K.A., and Steigbigel, N.H. (1969). Borrelia in Boston. JAMA : the journal of the American Medical Association *210*, 722-724. doi:10.1001/jama.1969.03160300062020

Gordillo-Perez, G., Garcia-Juarez, I., Solorzano-Santos, F., Corrales-Zuniga, L., Munoz-Hernandez, O., and Torres-Lopez, J. (2017). Serological Evidence of Borrelia Burgdorferi Infection in Mexican Patients with Facial Palsy. Rev Invest Clin *69*, 344-348. doi:10.24875/RIC.17002344

Gordillo-Perez, G., Solorzano, F., Cervantes-Castillo, A., Sanchez-Vaca, G., Garcia-Ramirez, R., Diaz, A.M., Munoz, O., and Torres, J. (2018). Lyme Neuroborreliosis is a Severe and Frequent Neurological Disease in Mexico. Arch Med Res *49*, 399-404. doi:10.1016/j.arcmed.2018.11.007

Goubau, P.F., and Munyangeyo, C. (1983). Fièvre récurrente congénitale à Borrelia duttoni. A propos d'un cas au Rwanda [Congenital relapsing fever caused by Borrelia duttoni. Apropos of a case in Rwanda]. Ann Soc Belg Med Trop *63*, 367-369.

Grigery, C.N., Moyer, P., Little, S.E., and Masters, E.J. (2005). Bacteriocidal activity of lizard and mouse serum for Borrelia lonestari, putative agent of a Lyme-like illness (AKA STARI or Masters disease) in Missouri. Mo Med *102*, 442-446.

Grosskinsky, S., Schott, M., Brenner, C., Cutler, S.J., Kraiczy, P., Zipfel, P.F., Simon, M.M., and Wallich, R. (2009). *Borrelia recurrentis* employs a novel multifunctional surface protein with anti-complement, anti-opsonic and invasive potential to escape innate immunity. PLoS One *4*, e4858. doi:10.1371/journal.pone.0004858

Grosskinsky, S., Schott, M., Brenner, C., Cutler, S.J., Simon, M.M., and Wallich, R. (2010). Human complement regulators C4b-binding protein and C1 esterase inhibitor interact with a novel outer surface protein of *Borrelia recurrentis*. PLoS Negl Trop Dis *4*, e698. doi: 10.1371/journal.pntd.0000698

Gugliotta, J.L., Goethert, H.K., Berardi, V.P., and Telford, S.R., 3rd (2013). Meningoencephalitis from Borrelia miyamotoi in an immunocompromised patient. N Engl J Med *368*, 240-245. doi: 10.1056/NEJMoa1209039

Guo, B.P., Teneberg, S., Munch, R., Terunuma, D., Hatano, K., Matsuoka, K., Angstrom, J., Boren, T., and Bergstrom, S. (2009). Relapsing fever Borrelia binds to neolacto glycans and mediates rosetting of human erythrocytes. Proc Natl Acad Sci U S A *106*, 19280-19285. doi:10.1073/pnas.0905470106

Guyard, C., Chester, E.M., Raffel, S.J., Schrumpf, M.E., Policastro, P.F., Porcella, S.F., Leong, J.M., and Schwan, T.G. (2005). Relapsing fever spirochetes contain chromosomal genes with unique direct tandemly repeated sequences. Infect Immun *73*, 3025-3037. doi:10.1128/IAI.73.5.3025-3037.2005

Han, S., Lubelczyk, C., Hickling, G.J., Belperron, A.A., Bockenstedt, L.K., and Tsao, J.I. (2019). Vertical transmission rates of Borrelia miyamotoi in Ixodes scapularis collected from white-tailed deer. Ticks Tick Borne Dis *10*, 682-689. doi:10.1016/j.ttbdis.2019.02.014

Harris, E.K., Harton, M.R., de Mello Marques, M.A., Belisle, J.T., Molins, C.R., Breuner, N., Wormser, G.P., and Gilmore, R.D. (2019). Immunoproteomic analysis of Borrelia miyamotoi for the identification of serodiagnostic antigens. Sci Rep *9*, 16808. doi:10.1038/s41598-019-53248-5

Hendson, M., and Lane, R.S. (2000). Genetic characteristics of Borrelia coriaceae isolates from the soft tick Ornithodoros coriaceus (Acari: Argasidae). J Clin Microbiol *38*, 2678-2682. doi:10.1128/JCM.38.7.2678-2682.2000

Henningsson, A.J., Asgeirsson, H., Hammas, B., Karlsson, E., Parke, A., Hoornstra, D., Wilhelmsson, P., and Hovius, J.W. (2019). Two Cases of Borrelia miyamotoi Meningitis, Sweden, 2018. Emerging Infectious Diseases *25*, 1965-1968. doi:10.3201/eid2510.190416

Herms, W.B., and Wheeler, C.M. (1935). Tick transmission of California relapsing fever. J Econ Entom *28*, 846-855. doi:10.1093/jee/28.6.846

Hinnebusch, B.J., Barbour, A.G., Restrepo, B.I., and Schwan, T.G. (1998). Population structure of the relapsing fever spirochete Borrelia hermsii as indicated by polymorphism of two multigene families that encode immunogenic outer surface lipoproteins. Infect Immun *66*, 432-440. doi:10.1128/IAI.66.2.432-440.1998

Hoogstraal, H. (1979). Ticks and spirochetes. Acta Trop *36*, 133-136.

Hoogstraal, H. (1985). Argasid and nuttalliellid ticks as parasites and vectors. Adv Parasitol *24*, 135-238. doi:10.1016/s0065-308x(08)60563-1

Hoornstra, D., Koetsveld, J., Sprong, H., Platonov, A.E., and Hovius, J.W. (2018). Borrelia miyamotoi Disease in an Immunocompetent Patient, Western Europe. Emerging Infectious Diseases *24*, 1770-1772. doi:10.3201/eid2409.180806

Hossain, H., Wellensiek, H.J., Geyer, R., and Lochnit, G. (2001). Structural analysis of glycolipids from Borrelia burgdorferi. Biochimie *83*, 683-692. doi:10.1016/s0300-9084(01)01296-2

Houhamdi, L., and Raoult, D. (2005). Excretion of living *Borrelia recurrentis* in feces of infected human body lice. The Journal of infectious diseases *191*, 1898-1906. doi:10.1086/429920

Hovis, K.M., McDowell, J.V., Griffin, L., and Marconi, R.T. (2004). Identification and characterization of a linear-plasmid-encoded factor H-binding protein (FhbA) of the relapsing fever spirochete Borrelia hermsii. J Bacteriol *186*, 2612-2618. doi:10.1128/jb.186.9.2612-2618.2004

Hovis, K.M., Schriefer, M.E., Bahlani, S., and Marconi, R.T. (2006). Immunological and molecular analyses of the Borrelia hermsii factor H and factor H-like protein 1 binding protein, FhbA: demonstration of its utility as a diagnostic marker and epidemiological tool for tick-borne relapsing fever. Infect Immun *74*, 4519-4529. doi:10.1128/IAI.00377-06

Hovius, J.W., de Wever, B., Sohne, M., Brouwer, M.C., Coumou, J., Wagemakers, A., Oei, A., Knol, H., Narasimhan, S., Hodiamont, C.J., *et al.* (2013). A case of meningoencephalitis by the relapsing fever spirochaete Borrelia miyamotoi in Europe. Lancet *382*, 658. doi: 10.1016/S0140-6736(13)61644-X

Hue, F., Ghalyanchi Langeroudi, A., and Barbour, A.G. (2013). Chromosome Sequence of Borrelia miyamotoi, an Uncultivable Tick-

Borne Agent of Human Infection. Genome Announc *1*. doi:10.1128/genomeA.00713-13

Huldén, L. (2006). Den sexbente fienden fienden: leddjurens inverkan på västerländsk krigsföring (Helsingfors, Finland: Schildt).

Jahfari, S., Hofhuis, A., Fonville, M., van der Giessen, J., van Pelt, W., and Sprong, H. (2016). Molecular Detection of Tick-Borne Pathogens in Humans with Tick Bites and Erythema Migrans, in the Netherlands. PLoS neglected tropical diseases *10*, e0005042. doi:10.1371/journal.pntd.0005042

Jahfari, S., Sarksyan, D.S., Kolyasnikova, N.M., Hovius, J.W., Sprong, H., and Platonov, A.E. (2017). Evaluation of a serological test for the diagnosis of Borrelia miyamotoi disease in Europe. J Microbiol Methods *136*, 11-16. doi:10.1016/j.mimet.2017.02.013

Jancso, N. (1918). Experimentelle Untersuchung bezüglich der Pathogenese der Rezidive des Rückfallfiebers. Zentbl Bakt Orig *81*, 457-474.

Jiang, B.G., Jia, N., Jiang, J.F., Zheng, Y.C., Chu, Y.L., Jiang, R.R., Wang, Y.W., Liu, H.B., Wei, R., Zhang, W.H., *et al.* (2018). Borrelia miyamotoi Infections in Humans and Ticks, Northeastern China. Emerging Infectious Diseases *24*, 236-241. doi:10.3201/eid2402.160378

Jobe, D.A., Lovrich, S.D., Oldenburg, D.G., Kowalski, T.J., and Callister, S.M. (2016). Borrelia miyamotoi Infection in Patients from Upper Midwestern United States, 2014-2015. Emerging Infectious Diseases *22*, 1471-1473. doi:10.3201/eid2208.151878

Jongen, V.H., van Roosmalen, J., Tiems, J., Van Holten, J., and Wetsteyn, J.C. (1997). Tick-borne relapsing fever and pregnancy outcome in rural Tanzania. Acta Obstet Gynecol Scand *76*, 834-838. doi:10.3109/00016349709024361

Judge, D.M., La Croix, J.T., and Perine, P.L. (1974a). Experimental louse-borne relapsing fever in the grivet monkey, *Cercopithecus aethiops*. I. Clinical course. The American journal of tropical medicine and hygiene *23*, 957-961. doi:10.4269/ajtmh.1974.23.957

Judge, D.M., La Croix, J.T., and Perine, P.L. (1974b). Experimental louse-borne relapsing fever in the grivet monkey, *Cercopithecus aethiops*. II. Pathology. The American journal of tropical medicine and hygiene *23*, 962-968. doi:10.4269/ajtmh.1974.23.962

Judge, D.M., La Croix, J.T., and Perine, P.L. (1974c). Experimental louse-borne relapsing fever in the grivet monkey, *Cercopithecus aethiops*. III. Crisis following therapy. The American journal of tropical medicine and hygiene *23*, 969-973. doi:10.4269/ajtmh.1974.23.969

Judge, D.M., Samuel, I., Perine, P.L., Vukotic, D., and Ababa, A. (1974d). Louse-borne relapsing fever in man. Arch Pathol *97*, 136-170.

Kadkhoda, K., Dumouchel, C., Brancato, J., Gretchen, A., and Krause, P.J. (2017). Human seroprevalence of Borrelia miyamotoi in Manitoba, Canada, in 2011-2014: a cross-sectional study. CMAJ Open *5*, E690-E693. doi:10.9778/cmajo.20170070

Karan, L., Makenov, M., Kolyasnikova, N., Stukolova, O., Toporkova, M., and Olenkova, O. (2018). Dynamics of Spirochetemia and Early PCR Detection of Borrelia miyamotoi. Emerging Infectious Diseases *24*, 860-867. doi:10.3201/eid2405.170829

Kazragis, R.J., Dever, L.L., Jorgensen, J.H., and Barbour, A.G. (1996). In vivo activities of ceftriaxone and vancomycin against Borrelia spp. in the mouse brain and other sites. Antimicrob Agents Chemother *40*, 2632-2636. doi:10.1128/AAC.40.11.2632

Kingry, L.C., Replogle, A., Batra, D., Rowe, L.A., Sexton, C., Dolan, M., Connally, N., Petersen, J.M., and Schriefer, M.E. (2017a). Toward a Complete North American Borrelia miyamotoi Genome. Genome Announc *5*. doi:10.1128/genomeA.01557-16

Kingry, L.C., Replogle, A., Dolan, M., Sexton, C., Padgett, K.A., and Schriefer, M.E. (2017b). Chromosome and Large Linear Plasmid Sequences of a Borrelia miyamotoi Strain Isolated from Ixodes pacificus Ticks from California. Genome Announc *5*. doi:10.1128/genomeA.00960-17

Kitten, T., and Barbour, A.G. (1990). Juxtaposition of expressed variable antigen genes with a conserved telomere in the bacterium Borrelia hermsii. Proc Natl Acad Sci U S A *87*, 6077-6081. doi:10.1073/pnas.87.16.6077

Kitten, T., and Barbour, A.G. (1992). The relapsing fever agent Borrelia hermsii has multiple copies of its chromosome and linear plasmids. Genetics *132*, 311-324.

Kitten, T., Barrera, A.V., and Barbour, A.G. (1993). Intragenic recombination and a chimeric outer membrane protein in the relapsing fever agent Borrelia hermsii. J Bacteriol *175*, 2516-2522. doi:10.1128/jb.175.9.2516-2522.1993

Koetsveld, J., Draga, R.O.P., Wagemakers, A., Manger, A., Oei, A., Visser, C.E., and Hovius, J.W. (2017a). In Vitro Susceptibility of the Relapsing-Fever Spirochete Borrelia miyamotoi to Antimicrobial Agents. Antimicrob Agents Chemother *61*. doi:10.1128/AAC.00535-17

Koetsveld, J., Kolyasnikova, N.M., Wagemakers, A., Stukolova, O.A., Hoornstra, D., Sarksyan, D.S., Toporkova, M.G., Henningsson, A.J., Hvidsten, D., Ang, W., *et al.* (2018a). Serodiagnosis of Borrelia miyamotoi disease by measuring antibodies against GlpQ and variable major proteins. Clin Microbiol Infect *24*, 1338 e1331-1338 e1337. doi:10.1016/j.cmi.2018.03.009

Koetsveld, J., Kolyasnikova, N.M., Wagemakers, A., Toporkova, M.G., Sarksyan, D.S., Oei, A., Platonov, A.E., and Hovius, J.W. (2017b). Development and optimization of an in vitro cultivation protocol allows for isolation of Borrelia miyamotoi from patients with hard tick-borne relapsing fever. Clin Microbiol Infect *23*, 480-484. doi:10.1016/j.cmi.2017.01.009

Koetsveld, J., Manger, A., Hoornstra, D., Draga, R.O., Oei, A., Kolyasnikova, N.M., Toporkova, M.G., Sarksyan, D.S., Wagemakers, A., Platonov, A.E., *et al.* (2018b). In Vitro Antimicrobial Susceptibility of Clinical Isolates of Borrelia miyamotoi. Antimicrob Agents Chemother *62*. doi:10.1128/AAC.00419-18

Koetsveld, J., Platonov, A.E., Kuleshov, K., Wagemakers, A., Hoornstra, D., Ang, W., Szekeres, S., van Duijvendijk, G.L.A., Fikrig, E., Embers, M.E., *et al.* (2020). Borrelia miyamotoi infection leads to cross-reactive antibodies to the C6 peptide in mice and men. Clin Microbiol Infect *26*, 513 e511-513 e516. doi:10.1016/j.cmi.2019.07.026

Krajacich, B.J., Lopez, J.E., Raffel, S.J., and Schwan, T.G. (2015). Vaccination with the variable tick protein of the relapsing fever spirochete *Borrelia hermsii* protects mice from infection by tick-bite. Parasit Vectors *8*, 546. doi:10.1186/s13071-015-1170-1

Krause, P.J., Carroll, M., Fedorova, N., Brancato, J., Dumouchel, C., Akosa, F., Narasimhan, S., Fikrig, E., and Lane, R.S. (2018). Human Borrelia miyamotoi infection in California: Serodiagnosis is complicated by multiple endemic Borrelia species. PLoS One *13*, e0191725. doi:10.1371/journal.pone.0191725

Krause, P.J., Fish, D., Narasimhan, S., and Barbour, A.G. (2015). Borrelia miyamotoi infection in nature and in humans. Clin Microbiol Infect *21*, 631-639. doi:10.1016/j.cmi.2015.02.006

Krause, P.J., Narasimhan, S., Wormser, G.P., Barbour, A.G., Platonov, A.E., Brancato, J., Lepore, T., Dardick, K., Mamula, M., Rollend, L., *et al.* (2014). Borrelia miyamotoi sensu lato seroreactivity and seroprevalence in the northeastern United States. Emerging Infectious Diseases *20*, 1183-1190. doi:10.3201/eid2007.131587

Krause, P.J., Narasimhan, S., Wormser, G.P., Rollend, L., Fikrig, E., Lepore, T., Barbour, A., and Fish, D. (2013). Human Borrelia miyamotoi infection in the United States. N Engl J Med *368*, 291-293. doi:10.1056/NEJMc1215469

Krause, P.J., Schwab, J., Narasimhan, S., Brancato, J., Xu, G., and Rich, S.M. (2016). Hard Tick Relapsing Fever Caused by Borrelia miyamotoi in a Child. Pediatr Infect Dis J *35*, 1352-1354. doi:10.1097/INF.0000000000001330

Krishnavajhala, A., Wilder, H.K., Boyle, W.K., Damania, A., Thornton, J.A., Perez de Leon, A.A., Teel, P.D., and Lopez, J.E. (2017). Imaging of *Borrelia turicatae* producing the green fluorescent protein reveals persistent colonization of the *Ornithodoros turicata* midgut and salivary glands from nymphal acquisition through transmission. Appl Environ Microbiol *83*. doi:10.1128/AEM.02503-16

Kuleshov, K.V., Hoornstra, D., Sprong, H., Platonov, A.E., and Hovius, J.W. (2019). Draft Whole-Genome Sequences of Two Western European Borrelia miyamotoi Isolates. Microbiol Resour Announc *8*. doi:10.1128/MRA.01314-19

Kuleshov, K.V., Koetsveld, J., Goptar, I.A., Markelov, M.L., Kolyasnikova, N.M., Sarksyan, D.S., Toporkova, M.G., Kirdyashkina, N.P., Shipulin, G.A., Hovius, J.W., *et al.* (2018). Whole-Genome Sequencing of Six Borrelia miyamotoi Clinical Strains Isolated in Russia. Genome Announc *6*. doi:10.1128/genomeA.01424-17

Kuleshov, K.V., Margos, G., Fingerle, V., Koetsveld, J., Goptar, I.A., Markelov, M.L., Kolyasnikova, N.M., Sarksyan, D.S., Kirdyashkina, N.P., Shipulin, G.A., *et al.* (2020). Whole genome sequencing of *Borrelia miyamotoi* isolate Izh-4: reference for a complex bacterial genome. BMC Genomics *21*, 16. doi:10.1186/s12864-019-6388-4

Lane, R.S., Burgdorfer, W., Hayes, S.F., and Barbour, A.G. (1985). Isolation of a spirochete from the soft tick, Ornithodoros coriaceus: a possible agent of epizootic bovine abortion. Science *230*, 85-87. doi: 10.1126/science.3898367

Lane, R.S., and Manweiler, S.A. (1988). Borrelia coriaceae in its tick vector, Ornithodoros coriaceus (Acari: Argasidae), with emphasis on transstadial and transovarial infection. J Med Entomol *25*, 172-177. doi:10.1093/jmedent/25.3.172

Lane, R.S., Mun, J., Parker, J.M., and White, M. (2005). Columbian black-tailed deer (Odocoileus hemionus columbianus) as hosts for Borrelia spp. in northern California. J Wildl Dis *41*, 115-125. doi: 10.7589/0090-3558-41.1.115

LaRocca, T.J., Katona, L.I., Thanassi, D.G., and Benach, J.L. (2008). Bactericidal action of a complement-independent antibody against relapsing fever Borrelia resides in its variable region. J Immunol *180*, 6222-6228. doi:10.4049/jimmunol.180.9.6222

Larsson, C., Andersson, M., Guo, B.P., Nordstrand, A., Hagerstrand, I., Carlsson, S., and Bergstrom, S. (2006a). Complications of pregnancy and transplacental transmission of relapsing-fever borreliosis. J Infect Dis *194*, 1367-1374. doi:10.1086/508425

Larsson, C., Andersson, M., Pelkonen, J., Guo, B.P., Nordstrand, A., and Bergstrom, S. (2006b). Persistent brain infection and disease reactivation in relapsing fever borreliosis. Microbes Infect *8*, 2213-2219. doi:10.1016/j.micinf.2006.04.007

Larsson, C., Lundqvist, J., and Bergstrom, S. (2008). Residual brain infection in murine relapsing fever borreliosis can be successfully treated with ceftriaxone. Microb Pathog *44*, 262-264. doi:10.1016/j.micpath.2007.11.002

Larsson, C., Lundqvist, J., van Rooijen, N., and Bergstrom, S. (2009). A novel animal model of Borrelia recurrentis louse-borne relapsing fever borreliosis using immunodeficient mice. PLoS neglected tropical diseases *3*, e522. doi:10.1371/journal.pntd.0000522

Lawson, C.L., Yung, B.H., Barbour, A.G., and Zuckert, W.R. (2006). Crystal structure of neurotropism-associated variable surface protein 1 (Vsp1) of Borrelia turicatae. J Bacteriol *188*, 4522-4530. doi: 10.1128/JB.00028-06

Lescot, M., Audic, S., Robert, C., Nguyen, T.T., Blanc, G., Cutler, S.J., Wincker, P., Couloux, A., Claverie, J.M., Raoult, D., *et al.* (2008). The genome of Borrelia recurrentis, the agent of deadly louse-borne relapsing fever, is a degraded subset of tick-borne Borrelia duttonii. PLoS Genet *4*, e1000185. doi:10.1371/journal.pgen.1000185

Liu, H., Fitzgerald, D., Gran, B., Leong, J.M., and Alugupalli, K.R. (2010). Induction of distinct neurologic disease manifestations during relapsing fever requires T lymphocytes. J Immunol *184*, 5859-5864. doi:10.4049/jimmunol.0902737

Londono, D., Bai, Y., Zuckert, W.R., Gelderblom, H., and Cadavid, D. (2005). Cardiac apoptosis in severe relapsing fever borreliosis. Infect Immun *73*, 7669-7676. doi:10.1128/IAI.73.11.7669-7676.2005

Londono, D., Marques, A., Hornung, R.L., and Cadavid, D. (2008). IL-10 helps control pathogen load during high-level bacteremia. J Immunol *181*, 2076-2083. doi:10.4049/jimmunol.181.3.2076

Lopez, J.E., Krishnavahjala, A., Garcia, M.N., and Bermudez, S.E. (2016). Tick-borne relapsing fever spirochetes in the Americas. Vet Sci *3*, 1-18. doi:10.3390/vetsci3030016

Lopez, J.E., Schrumpf, M.E., Nagarajan, V., Raffel, S.J., McCoy, B.N., and Schwan, T.G. (2010). A novel surface antigen of relapsing fever spirochetes can discriminate between relapsing fever and Lyme borreliosis. Clin Vaccine Immunol *17*, 564-571. doi:10.1128/CVI.00518-09

Lopez, J.E., Schrumpf, M.E., Raffel, S.J., Policastro, P.F., Porcella, S.F., and Schwan, T.G. (2008). Relapsing fever spirochetes retain infectivity after prolonged in vitro cultivation. Vector Borne Zoonotic Dis *8*, 813-820. doi:10.1089/vbz.2008.0033

Lopez, J.E., Wilder, H.K., Boyle, W., Drumheller, L.B., Thornton, J.A., Willeford, B., Morgan, T.W., and Varela-Stokes, A. (2013a). Sequence analysis and serological responses against *Borrelia turicatae* BipA, a putative species-specific antigen. PLoS neglected tropical diseases *7*, e2454. doi:10.1371/journal.pntd.0002454

Lopez, J.E., Wilder, H.K., Hargrove, R., Brooks, C.P., Peterson, K.E., Beare, P.A., Sturdevant, D.E., Nagarajan, V., Raffel, S.J., and Schwan, T.G. (2013b). Development of genetic system to inactivate a *Borrelia turicatae* surface protein selectively produced within the salivary glands of the arthropod vector. PLoS neglected tropical diseases *7*, e2514. doi:10.1371/journal.pntd.0002514

Lundqvist, J., Larsson, C., Nelson, M., Andersson, M., Bergstrom, S., and Persson, C. (2010). Concomitant infection decreases the malaria burden but escalates relapsing fever borreliosis. Infect Immun *78*, 1924-1930. doi:10.1128/IAI.01082-09

Magoun, L., Zuckert, W.R., Robbins, D., Parveen, N., Alugupalli, K.R., Schwan, T.G., Barbour, A.G., and Leong, J.M. (2000). Variable small protein (Vsp)-dependent and Vsp-independent pathways for glycosaminoglycan recognition by relapsing fever spirochaetes. Mol Microbiol *36*, 886-897. doi:10.1046/j.1365-2958.2000.01906.x

Mans, B.J., Andersen, J.F., Francischetti, I.M., Valenzuela, J.G., Schwan, T.G., Pham, V.M., Garfield, M.K., Hammer, C.H., and Ribeiro, J.M. (2008a). Comparative sialomics between hard and soft ticks: implications for the evolution of blood-feeding behavior. Insect biochemistry and molecular biology *38*, 42-58. doi:10.1016/j.ibmb.2007.09.003

Mans, B.J., Andersen, J.F., Schwan, T.G., and Ribeiro, J.M. (2008b). Characterization of anti-hemostatic factors in the argasid, Argas monolakensis: implications for the evolution of blood-feeding in the soft tick family. Insect Biochem Mol Biol *38*, 22-41. doi:10.1016/j.ibmb.2007.09.002

Marconi, R.T., Samuels, D.S., Schwan, T.G., and Garon, C.F. (1993). Identification of a protein in several Borrelia species which is related to OspC of the Lyme disease spirochetes. J Clin Microbiol *31*, 2577-2583. doi:10.1128/JCM.31.10.2577-2583.1993

McDowell, J.V., Tran, E., Hamilton, D., Wolfgang, J., Miller, K., and Marconi, R.T. (2003). Analysis of the ability of spirochete species associated with relapsing fever, avian borreliosis, and epizootic bovine abortion to bind factor H and cleave c3b. J Clin Microbiol *41*, 3905-3910. doi:10.1128/jcm.41.8.3905-3910.2003

Meier, J.T., Simon, M.I., and Barbour, A.G. (1985). Antigenic variation is associated with DNA rearrangements in a relapsing fever Borrelia. Cell *41*, 403-409. doi:10.1016/s0092-8674(85)80013-1

Meleney, H.E. (1928). Relapse Phenomena of Spironema Recurrentis. J Exp Med *48*, 65-82. doi:10.1084/jem.48.1.65

Melkert, P.W., and Stel, H.V. (1991). Neonatal Borrelia infections (relapsing fever): report of 5 cases and review of the literature. East Afr Med J *68*, 999-1005.

Meri, T., Cutler, S.J., Blom, A.M., Meri, S., and Jokiranta, T.S. (2006). Relapsing fever spirochetes Borrelia recurrentis and B. duttonii acquire complement regulators C4b-binding protein and factor H. Infect Immun *74*, 4157-4163. doi:10.1128/IAI.00007-06

Mitiku, K., and Mengistu, G. (2002). Relapsing fever in Gondar, Ethiopia. East Afr Med J *79*, 85-87. doi:10.4314/eamj.v79i2.8908

Molloy, P.J., Telford, S.R., 3rd, Chowdri, H.R., Lepore, T.J., Gugliotta, J.L., Weeks, K.E., Hewins, M.E., Goethert, H.K., and Berardi, V.P. (2015). Borrelia miyamotoi Disease in the Northeastern United States: A Case Series. Ann Intern Med *163*, 91-98. doi:10.7326/M15-0333

Molloy, P.J., Weeks, K.E., Todd, B., and Wormser, G.P. (2018). Seroreactivity to the C6 Peptide in Borrelia miyamotoi Infections Occurring in the Northeastern United States. Clin Infect Dis *66*, 1407-1410. doi:10.1093/cid/cix1023

Mooser, H. (1958). Erytrozyten-adhäsion und hämagglomeration durch rückfallfieber-spirochäten. Tropenmedizin und Parasitologie *9*, 93-111.

Moyer, P.L., Varela, A.S., Luttrell, M.P., Moore, V.A.t., Stallknecht, D.E., and Little, S.E. (2006). White-tailed deer (Odocoileus virginianus) develop spirochetemia following experimental infection with Borrelia lonestari. Vet Microbiol *115*, 229-236. doi:10.1016/j.vetmic.2005.12.020

Munoz-Leal, S., Faccini-Martinez, A.A., Costa, F.B., Marcili, A., Mesquita, E., Marques, E.P., Jr., and Labruna, M.B. (2018). Isolation and molecular characterization of a relapsing fever *Borrelia* recovered from *Ornithodoros rudis* in Brazil. Ticks Tick Borne Dis *9*, 864-871. doi:10.1016/j.ttbdis.2018.03.008

Negussie, Y., Remick, D.G., DeForge, L.E., Kunkel, S.L., Eynon, A., and Griffin, G.E. (1992). Detection of plasma tumor necrosis factor, interleukins 6, and 8 during the Jarisch-Herxheimer Reaction of relapsing fever. J Exp Med *175*, 1207-1212. doi:10.1084/jem.175.5.1207

Newman, K., Jr., and Johnson, R.C. (1981). In vivo evidence that an intact lytic complement pathway is not essential for successful removal of circulating Borrelia turicatae from mouse blood. Infect Immun *31*, 465-469. doi:10.1128/IAI.31.1.465-469.1981

Newman, K., Jr., and Johnson, R.C. (1984). T-cell-independent elimination of Borrelia turicatae. Infect Immun *45*, 572-576. doi:10.1128/IAI.45.3.572-576.1984

Nguyen, N.T.T., Rottgerding, F., Devraj, G., Lin, Y.P., Koenigs, A., and Kraiczy, P. (2018). The Complement Binding and Inhibitory Protein CbiA of Borrelia miyamotoi Degrades Extracellular Matrix Components by Interacting with Plasmin(ogen). Frontiers in cellular and infection microbiology *8*, 23. doi:10.3389/fcimb.2018.00023

Nicolle, C., and Anderson, C. (1927). Etude compartive de quelques viru recurrents, pathogenes pour l'homme. Arch Inst Past de Tunis *16*, 123-206.

Nieto, N.C., and Teglas, M.B. (2014). Relapsing fever group Borrelia in Southern California rodents. J Med Entomol *51*, 1029-1034. doi:10.1603/me14021

Nordstrand, A., Shamaei-Tousi, A., Ny, A., and Bergstrom, S. (2001). Delayed invasion of the kidney and brain by Borrelia crocidurae in plasminogen-deficient mice. Infect Immun *69*, 5832-5839. doi:10.1128/iai.69.9.5832-5839.2001

Novy, F.G., and Knapp, R.E. (1906). Studies on *Spirillum obermeieri* and related organisms. J Infect Dis *3*, 291-293.

Oleaga, A., Obolo-Mvoulouga, P., Manzano-Roman, R., and Perez-Sanchez, R. (2015). Midgut proteome of an argasid tick, *Ornithodoros erraticus:* a comparison between unfed and engorged females. Parasit Vectors *8*, 525. doi:10.1186/s13071-015-1148-z

Pachanian, A., and Smith, J.B. (1970). Immunization of chicks against avian spirochetosis with vaccines of Borrelia anserina. Tex Rep Biol Med *28*, 287-301.

Parsons, L. (1947). Relapsing fever at Lake Tahoe, California-Nevada. Am J Clin Pathol *17*, 388-392.

Pavlovsky, Y.N. (1963). Tick-borne relapsing fever. In Human Disease with Natural Foci, Y.N. Pavlovsky, ed. (Moscow: Foreign Languages Publishing House), pp. 138-184.

Penningon, P.M., Cadavid, D., Bunikis, J., Norris, S.J., and Barbour, A.G. (1999). Extensive interplasmidic duplications change the virulence phenotype of the relapsing fever agent Borrelia turicatae. Mol Microbiol *34*, 1120-1132. doi:10.1046/j.1365-2958.1999.01675.x

Pennington, P.M., Allred, C.D., West, C.S., Alvarez, R., and Barbour, A.G. (1997). Arthritis severity and spirochete burden are determined by serotype in the Borrelia turicatae-mouse model of Lyme disease. Infect Immun *65*, 285-292. doi:10.1128/IAI.65.1.285-292.1997

Pennington, P.M., Cadavid, D., and Barbour, A.G. (1999). Characterization of VspB of Borrelia turicatae, a major outer membrane protein expressed in blood and tissues of mice. Infect Immun *67*, 4637-4645. doi:10.1128/IAI.67.9.4637-4645.1999

Perine, P.L., Parry, E.H.O., Vukotich, D., Warrell, D.A., and Bryceson, A.D.M. (1971). Bleeding in louse-borne relapsing fever I. Clinical studies in 37 patients. Transactions of the Royal Society of Tropical Medicine and Hygiene *65*, 776-781. doi:10.1016/0035-9203(71)90091-5

Pettersson, J., Schrumpf, M.E., Raffel, S.J., Porcella, S.F., Guyard, C., Lawrence, K., Gherardini, F.C., and Schwan, T.G. (2007). Purine salvage pathways among Borrelia species. Infect Immun *75*, 3877-3884. doi:10.1128/IAI.00199-07

Philipp, M.T., Masters, E., Wormser, G.P., Hogrefe, W., and Martin, D. (2006). Serologic evaluation of patients from Missouri with erythema migrans-like skin lesions with the C6 Lyme test. Clin Vaccine Immunol *13*, 1170-1171. doi:10.1128/CVI.00238-06

Plasterk, R.H., Simon, M.I., and Barbour, A.G. (1985). Transposition of structural genes to an expression sequence on a linear plasmid causes antigenic variation in the bacterium Borrelia hermsii. Nature *318*, 257-263. doi:10.1038/318257a0

Platonov, A.E., Karan, L.S., Kolyasnikova, N.M., Makhneva, N.A., Toporkova, M.G., Maleev, V.V., Fish, D., and Krause, P.J. (2011). Humans infected with relapsing fever spirochete Borrelia miyamotoi, Russia. Emerging Infectious Diseases *17*, 1816-1823. doi:10.3201/eid1710.101474

Policastro, P.F., Raffel, S.J., and Schwan, T.G. (2013). *Borrelia hermsii* acquisition order in superinfected ticks determines transmission efficiency. Infect Immun *81*, 2899-2908. doi:10.1128/IAI.00542-13

Porcella, S.F., Raffel, S.J., Schrumpf, M.E., Schriefer, M.E., Dennis, D.T., and Schwan, T.G. (2000). Serodiagnosis of Louse-Borne relapsing fever with glycerophosphodiester phosphodiesterase (GlpQ) from Borrelia recurrentis. J Clin Microbiol *38*, 3561-3571. doi:10.1128/JCM.38.10.3561-3571.2000

Raffel, S.J., Battisti, J.M., Fischer, R.J., and Schwan, T.G. (2014). Inactivation of genes for antigenic variation in the relapsing fever spirochete *Borrelia hermsii* reduces infectivity in mice and transmission by ticks. PLoS pathogens *10*, e1004056. doi:10.1371/journal.ppat.1004056

Restrepo, B.I., and Barbour, A.G. (1994). Antigen Diversity in the Bacterium B-Hermsii through Somatic Mutations in Rearranged Vmp Genes. Cell *78*, 867-876. doi:10.1016/S0092-8674(94)90642-4

Restrepo, B.I., Carter, C.J., and Barbour, A.G. (1994). Activation of a vmp pseudogene in Borrelia hermsii: an alternate mechanism of antigenic variation during relapsing fever. Mol Microbiol *13*, 287-299. doi:10.1111/j.1365-2958.1994.tb00423.x

Restrepo, B.I., Kitten, T., Carter, C.J., Infante, D., and Barbour, A.G. (1992). Subtelomeric expression regions of Borrelia hermsii linear plasmids are highly polymorphic. Mol Microbiol *6*, 3299-3311. doi:10.1111/j.1365-2958.1992.tb02198.x

Rich, S.M., Armstrong, P.M., Smith, R.D., and Telford, S.R., 3rd (2001a). Lone star tick-infecting borreliae are most closely related to the agent of bovine borreliosis. J Clin Microbiol *39*, 494-497. doi:10.1128/JCM.39.2.494-497.2001

Rich, S.M., Sawyer, S.A., and Barbour, A.G. (2001b). Antigen polymorphism in Borrelia hermsii, a clonal pathogenic bacterium. Proc Natl Acad Sci U S A *98*, 15038-15043. doi:10.1073/pnas.071042098

Rossmann, E., Kraiczy, P., Herzberger, P., Skerka, C., Kirschfink, M., Simon, M.M., Zipfel, P.F., and Wallich, R. (2007). Dual binding specificity of a Borrelia hermsii-associated complement regulator-acquiring surface protein for factor H and plasminogen discloses a putative virulence factor of relapsing fever spirochetes. J Immunol *178*, 7292-7301. doi:10.4049/jimmunol.178.11.7292

Rottgerding, F., Wagemakers, A., Koetsveld, J., Fingerle, V., Kirschfink, M., Hovius, J.W., Zipfel, P.F., Wallich, R., and Kraiczy, P. (2017). Immune evasion of Borrelia miyamotoi: CbiA, a novel outer surface protein exhibiting complement binding and inactivating properties. Sci Rep *7*, 303. doi:10.1038/s41598-017-00412-4

Russell, H. (1936). Observations on immunity in relapsing fever and trypanosomiasis. Trans R Soc Trop Med Hyg *30*, 179-190. doi:10.1016/S0035-9203(36)90086-X

Rutty, J. (1770). A Chronological History of the Weather and Seasons, and of the Prevailing Diseases in Dublin: With the Various Periods, Successions, and Revolutions, During the Space of Forty Years : with a Comparative View of the Difference of the Irish Climate and Diseases, and Those of England and Other Countries (Robinson and Roberts).

Sadziene, A., Barbour, A.G., Rosa, P.A., and Thomas, D.D. (1993). An OspB mutant of Borrelia burgdorferi has reduced invasiveness in vitro and reduced infectivity in vivo. Infect Immun *61*, 3590-3596. doi:10.1128/IAI.61.9.3590-3596.1993

Sadziene, A., Jonsson, M., Bergstrom, S., Bright, R.K., Kennedy, R.C., and Barbour, A.G. (1994). A Bactericidal Antibody to Borrelia-Burgdorferi Is Directed against a Variable Region of the Ospb Protein. Infection and Immunity *62*, 2037-2045. doi:10.1128/Iai.62.5.2037-2045.1994

Salih, S.Y., Mustafa, D., Abdel Wahab, S.M., Ahmed, M.A., and Omer, A. (1977). Louse-borne relapsing fever: I. A clinical and laboratory study of 363 cases in the Sudan. Trans R Soc Trop Med Hyg *71*, 43-48. doi:10.1016/0035-9203(77)90206-1

Sambri, V., Marangoni, A., Olmo, A., Storni, E., Montagnani, M., Fabbi, M., and Cevenini, R. (1999). Specific antibodies reactive with the 22-kilodalton major outer surface protein of Borrelia anserina Ni-NL protect chicks from infection. Infect Immun *67*, 2633-2637. doi:10.1128/IAI.67.5.2633-2637.1999

Sarksyan, D.S., Maleev, V.V., Platonov, A.E., Platonova, O.V., and Karan, L.S. (2015a). [Relapsing (recurrent) disease caused by Borrelia miyamotoi]. Ter Arkh *87*, 18-25. doi:10.17116/terarkh2015871118-25

Sarksyan, D.S., Platonov, A.E., Karan, L.S., Shipulin, G.A., Sprong, H., and Hovius, J.W. (2015b). Probability of Spirochete Borrelia miyamotoi Transmission from Ticks to Humans. Emerging Infectious Diseases *21*, 2273-2274. doi:10.3201/eid2112.151097

Sato, K., Sakakibara, K., Masuzawa, T., Ohnishi, M., and Kawabata, H. (2018). Case control study: Serological evidence that Borrelia miyamotoi disease occurs nationwide in Japan. J Infect Chemother *24*, 828-833. doi:10.1016/j.jiac.2018.06.017

Sato, K., Takano, A., Konnai, S., Nakao, M., Ito, T., Koyama, K., Kaneko, M., Ohnishi, M., and Kawabata, H. (2014). Human infections with Borrelia miyamotoi, Japan. Emerging Infectious Diseases *20*, 1391-1393. doi:10.3201/eid2008.131761

Schmidtmann, E.T., Bushnell, R.B., Loomis, E.C., Oliver, M.N., and Theis, J.H. (1976). Experimental and epizootiologic evidence associating Ornithodoros coriaceus (Acari:Agrasidae) with the exposure of cattle to epizootic bovine abortion in California. J Med Entomol *13*, 292-299. doi:10.1093/jmedent/13.3.292

Schott, M., Grosskinsky, S., Brenner, C., Kraiczy, P., and Wallich, R. (2010). Molecular characterization of the interaction of *Borrelia parkeri* and *Borrelia turicatae* with human complement regulators. Infect Immun *78*, 2199-2208. doi:10.1128/IAI.00089-10

Schroder, N.W., Eckert, J., Stubs, G., and Schumann, R.R. (2008). Immune responses induced by spirochetal outer membrane lipoproteins and glycolipids. Immunobiology *213*, 329-340. doi:10.1016/j.imbio.2007.11.003

Schuhardt, V.T., and Wilkerson, M. (1951). Relapse phenomena in rats infected with single spirochetes (*Borrelia recurrenis* var. turicatae). J Bacteriol *62*, 215-219. doi:10.1128/JB.62.2.215-219.1951

Schwan, T.G., Battisti, J.M., Porcella, S.F., Raffel, S.J., Schrumpf, M.E., Fischer, E.R., Carroll, J.A., Stewart, P.E., Rosa, P., and Somerville, G.A. (2003). Glycerol-3-phosphate acquisition in spirochetes: distribution and biological activity of glycerophosphodiester phosphodiesterase (GlpQ) among Borrelia species. J Bacteriol *185*, 1346-1356. doi:10.1128/jb.185.4.1346-1356.2003

Schwan, T.G., and Hinnebusch, B.J. (1998). Bloodstream- versus tick-associated variants of a relapsing fever bacterium. Science *280*, 1938-1940. doi:10.1126/science.280.5371.1938

Schwan, T.G., Raffel, S.J., Schrumpf, M.E., and Porcella, S.F. (2007). Diversity and distribution of *Borrelia hermsii*. Emerg Infect Dis *13*, 436-442. doi:10.3201/eid1303.060958

Schwan, T.G., Schrumpf, M.E., Hinnebusch, B.J., Anderson, D.E., Jr., and Konkel, M.E. (1996). GlpQ: an antigen for serological discrimination between relapsing fever and Lyme borreliosis. J Clin Microbiol *34*, 2483-2492. doi:10.1128/JCM.34.10.2483-2492.1996

Scragg, I.G., Kwiatkowski, D., Vidal, V., Reason, A., Paxton, T., Panico, M., Dell, A., and Morris, H. (2000). Structural characterization of the inflammatory moiety of a variable major lipoprotein of Borrelia recurrentis. J Biol Chem *275*, 937-941. doi:10.1074/jbc.275.2.937

Shamaei-Tousi, A., Burns, M.J., Benach, J.L., Furie, M.B., Gergel, E.I., and Bergstrom, S. (2000). The relapsing fever spirochaete, Borrelia crocidurae, activates human endothelial cells and promotes the transendothelial migration of neutrophils. Cell Microbiol *2*, 591-599. doi:10.1046/j.1462-5822.2000.00083.x

Shamaei-Tousi, A., Collin, O., Bergh, A., and Bergstrom, S. (2001). Testicular damage by microcirculatory disruption and colonization of an immune-privileged site during Borrelia crocidurae infection. J Exp Med *193*, 995-1004. doi:10.1084/jem.193.9.995

Shamaei-Tousi, A., Martin, P., Bergh, A., Burman, N., Brännström, T., and Bergström, S. (1999). Erythrocyte-aggregating relapsing fever spirochete Borrelia crocidurae induces formation of microemboli. J Infect Dis *180*, 1929-1938. doi:10.1086/315118

Shang, E.S., Skare, J.T., Exner, M.M., Blanco, D.R., Kagan, B.L., Miller, J.N., and Lovett, M.A. (1998). Isolation and characterization of the outer membrane of Borrelia hermsii. Infect Immun *66*, 1082-1091. doi:10.1128/IAI.66.3.1082-1091.1998

Sharma, S.P., Amanfu, W., and Losho, T.C. (2000). Bovine borreliosis in Botswana. Onderstepoort J Vet Res *67*, 221-223.

Skare, J.T., Mirzabekov, T.A., Shang, E.S., Blanco, D.R., Erdjument-Bromage, H., Bunikis, J., Bergström, S., Tempst, P., Kagan, B.L., Miller, J.N., *et al.* (1997). The Oms66 (p66) protein is a Borrelia burgdorferi porin. Infect Immun *65*, 3654-3661. doi:10.1128/IAI.65.9.3654-3661.1997

Smith, R.D., Brener, J., Osorno, M., and Ristic, M. (1978). Pathobiology of Borrelia theileri in the tropical cattle tick, Boophilus microplus. J Invertebr Pathol *32*, 182-190. doi:10.1016/0022-2011(78)90028-9

Smith, R.P., Jr., Elias, S.P., Cavanaugh, C.E., Lubelczyk, C.B., Lacombe, E.H., Brancato, J., Doyle, H., Rand, P.W., Ebel, G.D., and Krause, P.J. (2019). Seroprevalence of Borrelia burgdorferi, B. miyamotoi, and Powassan Virus in Residents Bitten by Ixodes Ticks, Maine, USA. Emerging Infectious Diseases *25*, 804-807. doi:10.3201/eid2504.180202

Sohaskey, C.D., Zuckert, W.R., and Barbour, A.G. (1999). The extended promoters for two outer membrane lipoprotein genes of Borrelia spp.

uniquely include a T-rich region. Mol Microbiol *33*, 41-51. doi:10.1046/j.1365-2958.1999.01443.x

Southern, P., and Sanford, J. (1969). Relapsing fever. A clinical and microbiological review. Medicine *48*, 129-150.

Steere, A.C., Strle, F., Wormser, G.P., Hu, L.T., Branda, J.A., Hovius, J.W., Li, X., and Mead, P.S. (2016). Lyme borreliosis. Nat Rev Dis Primers *2*, 16090. doi:10.1038/nrdp.2016.90

Stevenson, B., Porcella, S.F., Oie, K.L., Fitzpatrick, C.A., Raffel, S.J., Lubke, L., Schrumpf, M.E., and Schwan, T.G. (2000). The relapsing fever spirochete Borrelia hermsii contains multiple, antigen-encoding circular plasmids that are homologous to the cp32 plasmids of Lyme disease spirochetes. Infect Immun *68*, 3900-3908. doi:10.1128/iai.68.7.3900-3908.2000

Stoenner, H.G., Dodd, T., and Larsen, C. (1982). Antigenic variation of Borrelia hermsii. J Exp Med *156*, 1297-1311. doi:10.1084/jem.156.5.1297

Strnad, M., Honig, V., Ruzek, D., Grubhoffer, L., and Rego, R.O.M. (2017). Europe-Wide Meta-Analysis of Borrelia burgdorferi Sensu Lato Prevalence in Questing Ixodes ricinus Ticks. Appl Environ Microbiol *83*. doi:10.1128/AEM.00609-17

Sudhindra, P., Wang, G., Schriefer, M.E., McKenna, D., Zhuge, J., Krause, P.J., Marques, A.R., and Wormser, G.P. (2016). Insights into Borrelia miyamotoi infection from an untreated case demonstrating relapsing fever, monocytosis and a positive C6 Lyme serology. Diagn Microbiol Infect Dis *86*, 93-96. doi:10.1016/j.diagmicrobio.2016.06.015

Sundnes, K.O., and Haimanot, A.T. (1993). Epidemic of louse-borne relapsing fever in Ethiopia. Lancet *342*, 1213-1215. doi:10.1016/0140-6736(93)92190-5

Tabuchi, N., Mitani, H., Seino, S., and Fukunaga, M. (2002). The 44-kb linear plasmid molecule in the relapsing fever agent Borrelia duttonii strain Ly serve as a preservation of vmp genes. Microbiol Immunol *46*, 159-165. doi:10.1111/j.1348-0421.2002.tb02681.x

Takahashi, Y., Cutler, S.J., and Fukunaga, M. (2000). Size conversion of a linear plasmid in the relapsing fever agent Borrelia duttonii. Microbiol Immunol *44*, 1071-1074. doi:10.1111/j.1348-0421.2000.tb02605.x

Takayama, K., Rothenberg, R.J., and Barbour, A.G. (1987). Absence of lipopolysaccharide in the Lyme disease spirochete, Borrelia burgdorferi. Infect Immun *55*, 2311-2313. doi:10.1128/IAI.55.9.2311-2313.1987

Talagrand-Reboul, E., Boyer, P.H., Bergström, S., Vial, L., and Boulanger, N. (2018). Relapsing Fevers: Neglected Tick-Borne Diseases. Frontiers in cellular and infection microbiology *8*, 98. doi:10.3389/fcimb.2018.00098

Telford, S.R., Goethert, H.K., Molloy, P.J., and Berardi, V. (2019). Blood Smears Have Poor Sensitivity for Confirming Borrelia miyamotoi Disease. J Clin Microbiol *57*. doi:10.1128/JCM.01468-18

Theiler, A.J. (1902). Spirillosis of cattle. Compar Pathol Ther *17*, 47-55. doi:10.1016/S0368-1742(04)80003-1

Thomas, N.J., Bunikis, J., Barbour, A.G., and Wolcott, M.J. (2002). Fatal spirochetosis due to a relapsing fever-like Borrelia sp. in a northern spotted owl. J Wildl Dis *38*, 187-193. doi:10.7589/0090-3558-38.1.187

Toledo, A., Anda, P., Escudero, R., Larsson, C., Bergstrom, S., and Benach, J.L. (2010). Phylogenetic analysis of a virulent Borrelia species isolated from patients with relapsing fever. J Clin Microbiol *48*, 2484-2489. doi:10.1128/JCM.00541-10

Uppal, D.R., and Rao, S.B. (1966). Studies on the preparation of a spirochaete vaccine. I. Indian Vet J *43*, 191-195.

van Holten, J., Tiems, J., and Jongen, V.H. (1997). Neonatal Borrelia duttoni infection: a report of three cases. Trop Doct 27, 115-116. doi:10.1177/004947559702700229

Varela, A.S., Luttrell, M.P., Howerth, E.W., Moore, V.A., Davidson, W.R., Stallknecht, D.E., and Little, S.E. (2004). First culture isolation of Borrelia lonestari, putative agent of southern tick-associated rash illness. J Clin Microbiol *42*, 1163-1169. doi:10.1128/jcm.42.3.1163-1169.2004

Varma, M.G.R. (1956). Infections of *Ornithodoros* ticks with relapsing fever spirochaetes, and the mechanisms of their transmission. Ann Trop Med Parasitol *50*, 18-31. doi:10.1080/00034983.1956.11685735

Vazquez-Guerrero, E., Adan-Bante, N.P., Mercado-Uribe, M.C., Hernandez-Rodriguez, C., Villa-Tanaca, L., Lopez, J.E., and Ibarra, J.A. (2019). Case report: A retrospective serological analysis indicating human exposure to tick-borne relapsing fever spirochetes in Sonora, Mexico. PLoS Negl Trop Dis *13*, e0007215. doi:10.1371/journal.pntd.0007215

Velu, H., Balozet, L., and Zottner, G. (1930). Neurotropisme de Spirochaeta hispanicum (souche de Mansouriah) pour le lapin' forme nouvelle de l'infection inapparente. Comptes Rendus des Seances Societe de Biologie (Paris) *103*, 381-383.

Venzal, J.M., Estrada-Pena, A., Mangold, A.J., Gonzalez-Acuna, D., and Guglielmone, A.A. (2008). The *Ornithodoros (Alectorobius) talaje* species group (Acari: Ixodida: Argasidae): description of *Ornithodoros (Alectorobius) rioplatensis* n. sp. from southern South America. J Med Entomol *45*, 832-840. doi:10.1603/0022-2585(2008)45[832:toatsg]2.0.co;2

Vidal, V., Cutler, S., Scragg, I.G., Wright, D.J., and Kwiatkowski, D. (2002). Characterisation of silent and active genes for a variable large protein of Borrelia recurrentis. BMC Infect Dis *2*, 25. doi:10.1186/1471-2334-2-25

Vidal, V., Scragg, I.G., Cutler, S.J., Rockett, K.A., Fekade, D., Warrell, D.A., Wright, D.J., and Kwiatkowski, D. (1998). Variable major lipoprotein is a principal TNF-inducing factor of louse- borne relapsing fever. Nat Med *4*, 1416-1420. doi:10.1038/4007

Wagemakers, A., Koetsveld, J., Narasimhan, S., Wickel, M., Deponte, K., Bleijlevens, B., Jahfari, S., Sprong, H., Karan, L.S., Sarksyan, D.S., *et al.* (2016). Variable Major Proteins as Targets for Specific Antibodies against Borrelia miyamotoi. J Immunol *196*, 4185-4195. doi:10.4049/jimmunol.1600014

Wagemakers, A., Oei, A., Fikrig, M.M., Miellet, W.R., and Hovius, J.W. (2014). The relapsing fever spirochete Borrelia miyamotoi is cultivable in a modified Kelly-Pettenkofer medium, and is resistant to human complement. Parasit Vectors 7, 418. doi:10.1186/1756-3305-7-418

Wagemakers, A., Staarink, P.J., Sprong, H., and Hovius, J.W. (2015). Borrelia miyamotoi: a widespread tick-borne relapsing fever spirochete. Trends Parasitol *31*, 260-269. doi:10.1016/j.pt.2015.03.008

Warrell, D.A., Perine, P.L., Krause, D.W., Bing, D.H., and MacDougal, S.J. (1983). Pathophysiology and immunology of the Jarisch-Herxheimer-like reaction in louse-borne relapsing fever: comparison of tetracycline and slow-release penicillin. J Infect Dis *147*, 898-909. doi:10.1093/infdis/147.5.898

Webster, G., Schiffman, J.D., Dosanjh, A.S., Amieva, M.R., Gans, H.A., and Sectish, T.C. (2002). Jarisch-Herxheimer reaction associated with ciprofloxacin administration for tick-borne relapsing fever. Pediatr Infect Dis J *21*, 571-573. doi:10.1097/00006454-200206000-00020

Wengrower, D., Knobler, H., Gillis, S., and Chajek-Shaul, T. (1984). Myocarditis in tick-borne relapsing fever. J Infect Dis *149*, 1033. doi:10.1093/infdis/149.6.1033

Wilder, H.K., Raffel, S.J., Barbour, A.G., Porcella, S.F., Sturdevant, D.E., Vaisvil, B., Kapatral, V., Schmitt, D.P., Schwan, T.G., and Lopez, J.E. (2016). Transcriptional Profiling the 150 kb Linear Megaplasmid of *Borrelia turicatae* Suggests a Role in Vector Colonization and Initiating Mammalian Infection. PLoS One *11*, e0147707. doi:10.1371/journal.pone.0147707

Woodman, M.E., Cooley, A.E., Avdiushko, R., Bowman, A., Botto, M., Wooten, R.M., van Rooijen, N., Cohen, D.A., and Stevenson, B. (2009). Roles for phagocytic cells and complement in controlling relapsing fever infection. J Leukoc Biol *86*, 727-736. doi:10.1189/jlb.0309169

Wormser, G.P., Masters, E., Liveris, D., Nowakowski, J., Nadelman, R.B., Holmgren, D., Bittker, S., Cooper, D., Wang, G., and Schwartz, I. (2005). Microbiologic evaluation of patients from Missouri with erythema migrans. Clin Infect Dis *40*, 423-428. doi:10.1086/427289

Wroblewski, D., Gebhardt, L., Prusinski, M.A., Meehan, L.J., Halse, T.A., and Musser, K.A. (2017). Detection of Borrelia miyamotoi and other tick-borne pathogens in human clinical specimens and Ixodes scapularis ticks in New York State, 2012-2015. Ticks Tick Borne Dis *8*, 407-411. doi:10.1016/j.ttbdis.2017.01.004

Yabsley, M.J., Davidson, W.R., Stallknecht, D.E., Varela, A.S., Swift, P.K., Devos, J.C., Jr., and Dubay, S.A. (2005). Evidence of tick-borne organisms in mule deer (Odocoileus hemionus) from the western United States. Vector Borne Zoonotic Dis *5*, 351-362. doi:10.1089/vbz.2005.5.351

Yabsley, M.J., Murphy, S.M., Luttrell, M.P., Little, S.E., Massung, R.F., Stallknecht, D.E., Conti, L.A., Blackmore, C.G., and Durden, L.A. (2008). Experimental and field studies on the suitability of raccoons (Procyon lotor) as hosts for tick-borne pathogens. Vector Borne Zoonotic Dis *8*, 491-503. doi:10.1089/vbz.2007.0240

Yagupsky, P., and Moses, S. (1985). Neonatal Borrelia species infection (relapsing fever). Am J Dis Child *139*, 74-76. doi:10.1001/archpedi.1985.02140030076034

Yamano, K., Ito, T., Kiyanagi, K., Yamazaki, H., Sugawara, M., Saito, T., Ohashi, N., Zamoto-Niikura, A., Sato, K., and Kawabata, H. (2017). Case Report: Clinical Features of a Case of Suspected Borrelia miyamotoi Disease in Hokkaido, Japan. Am J Trop Med Hyg *97*, 84-87. doi:10.4269/ajtmh.16-0699

Yparraguirre, L.A., Machado-Ferreira, E., Ullmann, A.J., Piesman, J., Zeidner, N.S., and Soares, C.A. (2007). A hard tick relapsing fever group spirochete in a Brazilian Rhipicephalus (Boophilus) microplus. Vector Borne Zoonotic Dis *7*, 717-721. doi:10.1089/vbz.2007.0144

Zaher, M.A., Soliman, Z.R., and Diab, F.M. (1977). An experimental study of Borrelia anserina in four species of Argas ticks. 2. Transstadial survival and transovarial transmission. Z Parasitenkd *53*, 213-223. doi:10.1007/BF00380466

Zhang, H., and Marconi, R.T. (2005). Demonstration of cotranscription and 1-methyl-3-nitroso-nitroguanidine induction of a 30-gene operon of Borrelia burgdorferi: evidence that the 32-kilobase circular plasmids are prophages. J Bacteriol *187*, 7985-7995. doi:10.1128/JB.187.23.7985-7995.2005

Zhang, J.R., Hardham, J.M., Barbour, A.G., and Norris, S.J. (1997). Antigenic variation in Lyme disease borreliae by promiscuous recombination of VMP-like sequence cassettes. Cell *89*, 275-285. doi:10.1016/s0092-8674(00)80206-8

Zhong, J., and Barbour, A.G. (2004). Cross-species hybridization of a Borrelia burgdorferi DNA array reveals infection- and culture associated genes of the unsequenced genome of the relapsing fever agent Borrelia hermsii. Mol Microbiol *51*, 729-748. doi:10.1046/j.1365-2958.2003.03849.x

Zhong, J., Skouloubris, S., Dai, Q., Myllykallio, H., and Barbour, A.G. (2006). Function and evolution of plasmid-borne genes for pyrimidine biosynthesis in Borrelia spp. J Bacteriol *188*, 909-918. doi:10.1128/JB.188.3.909-918.2006

Zuckert, W.R., Kerentseva, T.A., Lawson, C.L., and Barbour, A.G. (2001). Structural conservation of neurotropism-associated VspA within the variable Borrelia Vsp-OspC lipoprotein family. J Biol Chem *276*, 457-463. doi:10.1074/jbc.M008449200

Zuckert, W.R., Lloyd, J.E., Stewart, P.E., Rosa, P.A., and Barbour, A.G. (2004). Cross-species surface display of functional spirochetal lipoproteins by recombinant Borrelia burgdorferi. Infect Immun *72*, 1463-1469. doi:10.1128/iai.72.3.1463-1469.2004

Chapter 15

Animal Models of Borreliosis

Monica E. Embers[1]*, Stephen W. Barthold[2] and Diego Cadavid[3]

[1]Division of Immunology, Tulane National Primate Research Center, Tulane University Health Sciences, Covington LA, USA
[2]Center for Immunology and Infectious Diseases, University of California, Davis CA, USA
[3]University of Massachusetts Medical School, Worcester MA, USA and Fulcrum Therapeutics, Cambridge MA, USA
*Corresponding author: members@tulane.edu

DOI: https://doi.org/10.21775/9781913652616.15

Abstract
Animal models have been extremely useful for understanding many features of infection with *Borrelia* species, including pathogenesis, host responses, host-vector interactions, *in vivo* bacterial gene expression, vaccine responses, and chemotherapy. In particular, the smaller animal models have been beneficial in understanding the molecular pathogenesis of Lyme and relapsing fever spirochetes, mechanisms of immune evasion and generation of arthritis. The larger animal models including dogs and primates, have been valuable for the development of vaccines and diagnostics. Each animal model brings its own set of benefits for studying aspects of borreliosis, which is what we aim to convey in this chapter. We begin with the laboratory mouse, with which the majority of borreliosis studies have been conducted. We also discuss wild mice and other rodent models, rabbits, dogs, horses and nonhuman primates. We conclude with a summary of results in animal models on the controversial topic of antimicrobial efficacy. Some degree of overlap may exist with these subjects that are covered in other chapters, but the goal of this chapter is to focus on the practical aspects of various animal models used for Lyme borreliosis and relapsing fever research.

Introduction to animal models of Lyme borreliosis
Ticks of the *Ixodes ricinus* complex (Charrier et al., 2019) feed upon a remarkable array of reservoir host species, in fact, more than any other group of ticks. Ixodid ticks are known to feed on more than 300 different kinds of mammals, birds, and reptiles, and *Borrelia burgdorferi* has been isolated or detected in 24 different species of mammals or birds (Anderson, 1991). Other than humans, incidental hosts for these zoonotic pathogens include domestic cattle, horses, dogs and cats (Burgess, 1988; Parker and White, 1992; Barthold et al., 1995b; Lischer et al., 2000; Krupka and Straubinger, 2010; Little et al., 2010). The broad host range of Lyme disease spirochetes lends well to experimental infection of a wide variety of species that can be used as animal models for research and disease characterization. The most common reservoir host in North America is the white-footed mouse (*Peromyscus leucopus*). In addition to wild mice, several rodent species are susceptible to experimental *B. burgdorferi* sensu lato (s.l.) infection, including hispid cotton rats (*Sigmodon hispidus*) (Burgdorfer and Gage, 1987), Mongolian gerbils (*Meriones unguiculatus*) (Preac Mursic et al., 1990), grasshopper mice (*Onychomys leucogaster*) (Czub et al., 1992), Syrian hamsters (*Mesocricetus auratus*) (Johnson et al., 1984; Goodman et al., 1991), guinea pigs (*Cavia porcellus)* (Sonnesyn et al., 1993b; Breitschwerdt et al., 1996), laboratory rats (*Rattus norvegicus)* (Barthold et al., 1988b; Moody et al., 1990b), and several strains of laboratory mice (*Mus musculus*) (Moody et al., 1990a; Czub et al., 1992). Laboratory rabbits (*Oryctolagus cuniculus*) also have been experimentally infected with Lyme disease spirochetes (Foley et al., 1995a). Larger animal models include nonhuman primates and dogs (England et al., 1997a; Straubinger et al., 1997a; Cadavid et al., 2000a).

The natural history of infection with *Borrelia burgdorferi* is to establish persistent infections with minimal harm to natural (reservoir) hosts. Humans and domestic animals are incidental hosts and clinical disease is an aberration from the norm. Thus, a major challenge to modelling human disease has been the ability to experimentally reproduce some aspects of the human disease in animal models. Generally, the larger animal models recapitulate various manifestations of human disease, and several strains and mutants of laboratory mice also favor susceptibility to disease. These models have provided essential tools for better understanding, preventing, diagnosing and treating Lyme borreliosis.

The laboratory mouse model

Effective use of the mouse model requires an awareness that infection may be influenced by mouse genotype, immune competence, and age, as well as *B. burgdorferi* species, isolate and strain; passage history, adaptive form, dose, route of inoculation, and site of inoculation. The mouse model shares many features in common with acute disseminated human Lyme borreliosis, including arthritis, carditis, vasculitis, myositis, peripheral neuritis, an intermittent course of disease remission and exacerbation, persistent infection, and immune response. Mice do not develop an EM rash and, despite recent evidence that *B. burgdorferi* can colonize the vasculature of the dura mater (Divan et al., 2018), mice do not develop neuroborreliosis. This lack of CNS invasion may be due to the paucity of collagenous tissue in their meninges and brain, and/or their inability of Lyme disease spirochetes to penetrate the blood-brain barrier in mice. The most commonly studied feature of the mouse model is arthritis, which is similar in character to that found in other rodents, dogs, nonhuman primates, and humans.

Basic features of the mouse model

Infection has two distinct phases in immunocompetent mice: the innate and acquired immune phases. In the early phase, spirochetes utilize their full armamentarium to evade host innate responses, proliferate, and disseminate broadly throughout all tissues. Spirochetes disseminate directly through tissue, as well as in the blood (bacteremia), and population dynamics reach their zenith in various tissues during this period. Arrival of spirochetes into joint and heart tissue results in an acute inflammatory response that is related to the presence of spirochetes, but inflammation is curiously minimal or absent in other tissues, such as the skin and urinary bladder, that are also infected during this phase (Barthold et al., 1991b).

As the acquired immune response evolves in the immune phase of infection, there is reduction (but incomplete clearance) of spirochetes in all tissues, with resolution of acute inflammatory lesions in joints and hearts. Disease resolution coincides with clearance of spirochetes from much of the heart tissue and the synovium, but persistence of spirochetes intercalated within collagen of connective tissue of the aortic wall, ligaments, tendons, joint capsules and dermis (Zimmer et al., 1990; Barthold et al., 1991b; Barthold et al., 1992a; Barthold et al., 1993; Pachner et al., 1995a; Barthold et al., 2006). Over the course of several months of infection, mice experience periodic exacerbations of acute oligoarticular arthritis, acute carditis, and bacteremia, associated with resurgence of spirochetes into the synovium. During the immune phase of infection, random foci of segmental perivasculitis, perineuritis and other chronic lesions are present (Barthold et al., 1991a; Barthold et al., 1992b; Barthold et al., 1993).

Kinetics of spirochete dissemination and tissue invasion

Initial studies were conducted to determine the susceptibility of multiple strains of laboratory mice to infection and disease caused by *B. burgdorferi* (Barthold et al., 1990). The C3H/He strain emerged as one in which both polyarthritis and carditis were observed consistently. In a comprehensive study utilizing C3H mice inoculated intradermally by syringe on the dorsal thorax, dissemination of spirochetes to various tissues was evaluated by culture, histochemistry (silver stain), immunohistochemistry, *in situ* DNA hybridization, PCR, and histology. Although that study involved syringe-inoculation of cultured non-clonal *B. burgdorferi* N40, similar events take place following infection with other *B. burgdorferi* isolates, by other means of infection, and with other adaptive states of spirochetes. Tissues were analyzed on days 1, 2, 3, 4, 7, 10, 15, 21, and 30 after inoculation. Blood and spleen of most mice were positive by day 3, and ear by day 10. Inflammation in joints and hearts coincided with invasion of spirochetes on days 4 through 7 in joints and days 7 through 10 in hearts. Spirochetes had a general predilection for collagenous tissues of joints, heart, arteries, nerves, muscle, skin and other tissues. Spirochete populations peaked at 15 days, after which bacterial burdens declined globally in all tissues, but persisted (Barthold et al., 1991a). Examination of mice at 30, 90, 180, and 360 days after inoculation confirmed persistence of spirochetes in multiple collagenous tissues for the duration of the study. At 30 days, hearts and many joints of all limbs were acutely inflamed, whereas at later intervals, both carditis and arthritis had generally resolved. However, some mice sampled at later intervals had acute carditis and oligoarticular arthritis, suggesting recurrent acute disease (Barthold et al., 1993).

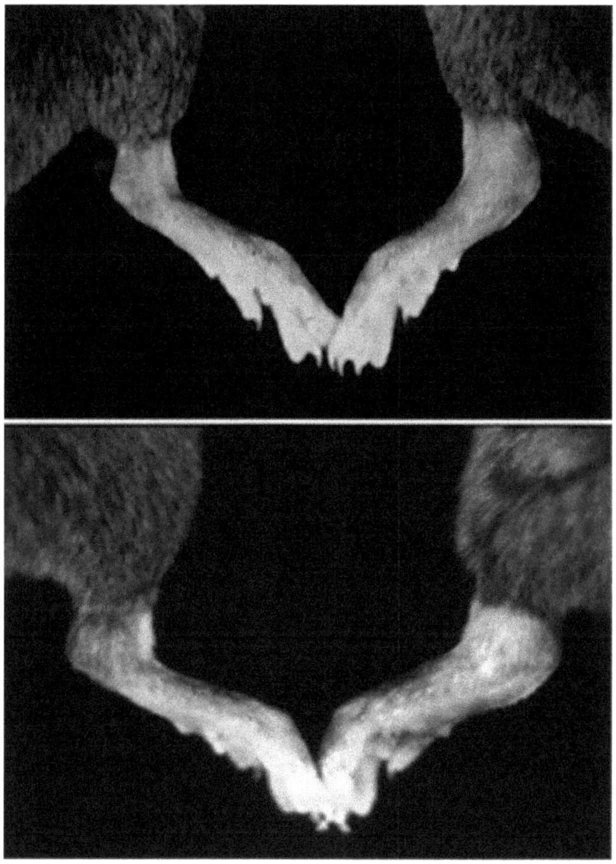

Figure 1. Tibiotarsal swelling in C3H mice infected with *B. burgdorferi*. The upper panel depicts an uninfected mouse (left), compared to an infected mouse (right) at around 2-3 weeks of infection. The lower panel depicts an immune-competent mouse (left) compared to a SCID mouse at around 60 days of infection.

Quantitative PCR analysis of spirochete populations in various tissues during the course of infection has demonstrated the early rise, peak at 2 weeks, and then decline but persistence of spirochetes at the inoculation site, distant skin, ear pinna, heart base, ventricular myocardium, tibiotarsal joint, and quadriceps muscle. The decline in spirochete populations during the immune phase of the infection is global in all tissues, including the bloodstream, but spirochetes persist in all tissues in significant numbers. In contrast, spirochete populations continue to rise in various tissues and bacteremia persists in C3H-*scid* mice at 4 and 6 weeks. However, spirochete populations plateau without massive overgrowth in immunodeficient mice, suggesting mediation by innate immunity and/or an unknown mechanism(s) for sensing and limiting bacterial population growth *in vivo* (Hodzic et al., 2003).

Musculoskeletal pathology

The "arthritis" in mice infected with *B. burgdorferi* consists of several different components. Joint disease arises in many joints, but is most evident as swelling of the tibiotarsal joint, and is particularly obvious in immunodeficient mice (Figure 1). It must be emphasized that caliper measurements of joints do not necessarily reflect arthritis. Grossly measurable joint swelling may be due to periarticular edema (with or without underlying synovitis/arthritis), proliferative synovitis, and/or fibrosis. In the initial pre-immune and early immune phases of infection, there is often grossly visible swelling of joints, particularly the tibiotarsal (ankle) joints, due to periarticular edema (Figure 2). The edema arises in the subcutis, and features sparse infiltration of neutrophils and macrophages in response to the presence of large numbers of spirochetes in the subcutaneous tissue (Barthold et al., 1993). This reaction may or may not be associated with underlying inflammation of joints. Notably, it has been shown that joint swelling and joint inflammation are differentially affected by separate genetic loci in the mouse (Roper et al., 2001). A major component of joint disease is polysynovitis (Figure 3), with synovial proliferation, infiltration of neutrophils, and exudation of neutrophils and fibrin into synovial lumina, involving many bursae, sheaths, and joints. The tibiotarsal region is optimal for examining this response, which peaks at 2 to 3 weeks, but many

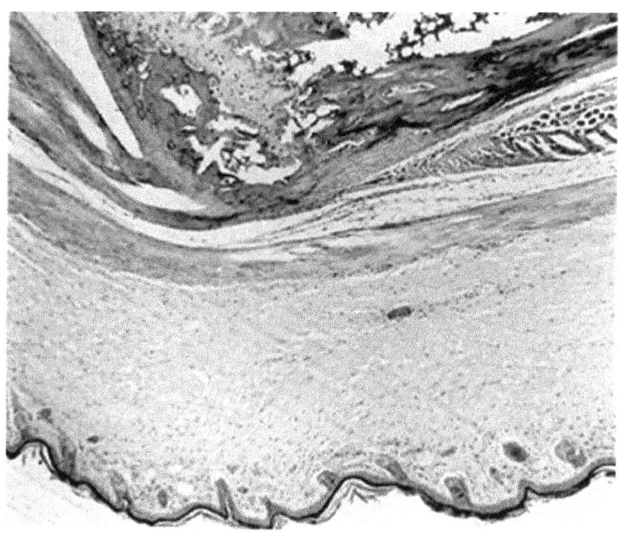

Figure 2. Tibiotarsal region of a C3H mouse during acute infection with *B. burgdorferi*. The dermis and subcutis are markedly edematous in the absence of inflammation of the underlying joints and tendon sheaths. This illustrates the concept that gross tibiotarsal swelling is not an accurate measure of arthritis.

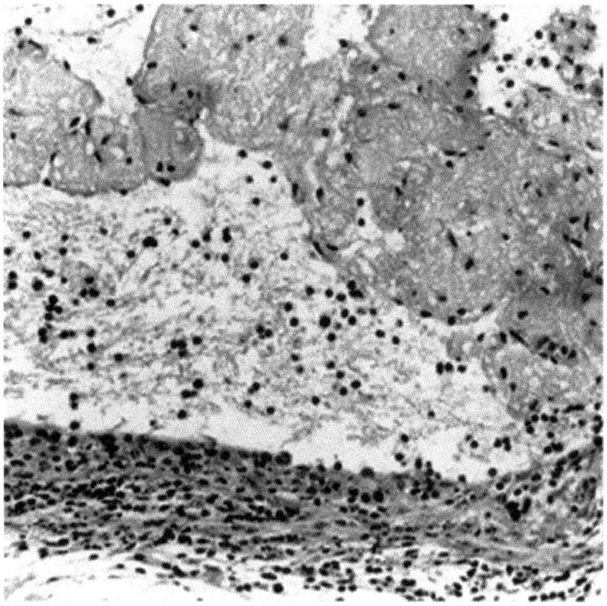

Figure 3. Synovium of a C3H mouse during acute infection with *B. burgdorferi*. The synovium is hypercellular due to synovial proliferation and leukocytic infiltration, and the lumen contains exudate consisting of leukocytes and fibrin.

joints of both the axial and appendicular skeleton may be involved. Spirochetes are present within the proliferating synovium and surrounding connective tissue but not in exudate within joint spaces. In addition, inflammation of attachment sites of ligaments and tendons to bone is a frequent finding (enthesopathy) in association with localization of associated with spirochetes among collagen fibers.

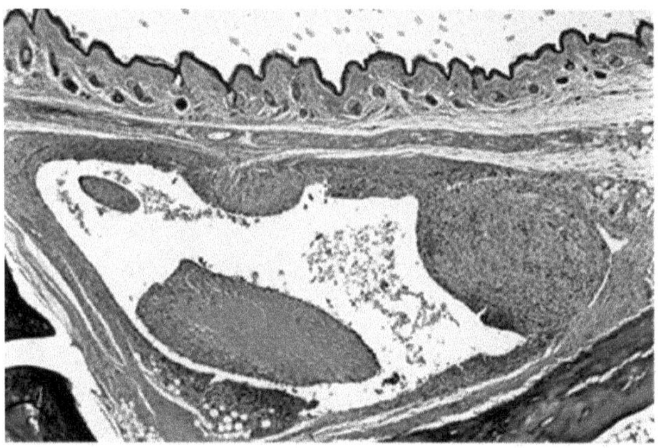

Figure 4. Tibiotarsal flexor tendon of a C3H mouse during acute infection with *B. burgdorferi*. The tendon sheath is thickened due to synovial hyperplasia and inflammation and the lumen contains exudate. This is a consistent site for examining synovial pathology in the mouse model.

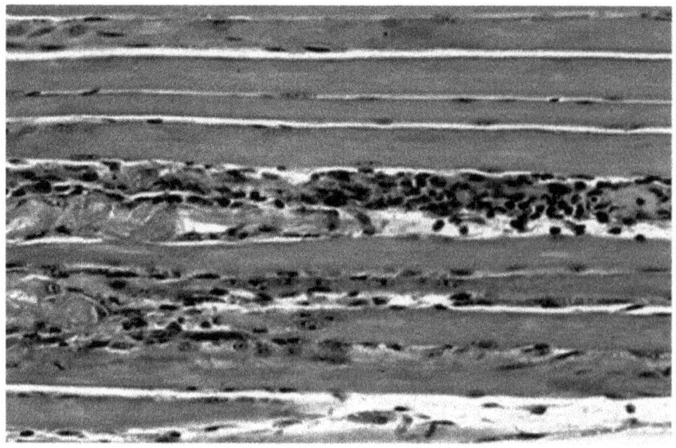

Figure 5. Myositis in a C3H mouse chronically infected with *B. burgdorferi*. There is infiltration of the endomesial connective tissue with mononuclear leukocytes, particularly in association with small vessels.

The tibial crest, Achilles tendon, and patellar ligaments are frequent sites for infection. Marked proliferative changes are consistently present around the tibiotarsal flexor tendon, its bursal sheath, and its fibrocartilaginous cushion on the dorsal aspect of the tibiotarsus (Barthold et al., 1990; Barthold et al., 1992b; Barthold et al., 1993; Ma et al., 1998) (Figure 4). If infection is allowed to progress in SCID mice, there is progressive pannus of joint spaces, with destruction of bone and articular cartilage (Schaible et al., 1990b; Barthold et al., 1992b). During the immune phase of infection in immunocompetent mice, the inflammatory and proliferative changes resolve, with residual capsular fibrosis and sparse mononuclear leukocyte infiltrates (Barthold et al., 1993). Tissue-specific analysis of spirochete burdens during the immune phase of infection has shown that spirochetes are eliminated from the proliferative synovium, but persist in adjacent collagenous tissues (capsule, tendons, ligaments) (Hodzic et al., 2003; Barthold et al., 2006).

Myositis of skeletal muscle is also a frequent feature of *B. burgdorferi* infection in the mouse and is quite apparent in immunodeficient mice. Lesions consist of infiltration of endomysial and perimysial connective tissue with macrophages and lymphocytes, particularly in perivascular regions (Schaible et al., 1990b; Barthold et al., 1991a; Museteanu et al., 1991; Barthold et al., 1992b; Defosse et al., 1992) (Figure 5).

Cardiovascular pathology

Evaluation of carditis requires optimal sampling and processing to include the heart base (including aortic

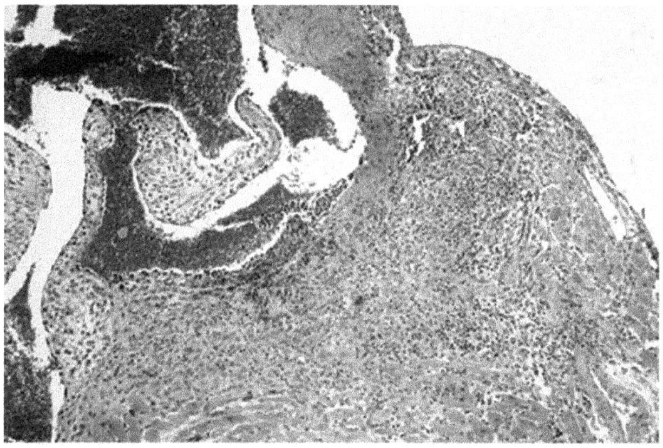

Figure 6. Acute carditis in a C3H mouse infected with *B. burgdorferi.* There is marked infiltration of the connective tissue at the heart base with macrophages, which extend into the adjacent myocardium (bottom of figure). The aortic root near the aortic valve has endothelial margination of neutrophils, with extension of inflammation into the vessel wall.

root, auricles, and ventricular myocardium). Syringe inoculation in the dorsal thoracic midline, in contrast to distant sites, such as tail base, results in more consistent cardiac involvement in immunocompetent mice. Tick-borne infection, which generally results from ticks feeding in the head and neck region, also favors development of carditis. Immunodeficient mice develop carditis, regardless of inoculation site. Although not as extensively studied as arthritis susceptibility, mouse genetic background is a major determinant of carditis susceptibility, which parallels arthritis susceptibility (Armstrong et al., 1992).

The temporal kinetics of cardiac infection have been studied in the carditis-susceptible C3H mouse following syringe inoculation. Tissues were examined by culture, PCR and immunohistochemistry for spirochete invasion at 2, 3, 4, 7, 10, 15, 21, and 30 days after inoculation. Spirochetes were first detectable in heart at day 7, which coincided with onset of inflammation (Barthold et al., 1991b). With the exception of the aortitis that occurs during acute infection, the inflammatory response of cardiac tissues to infection with *B. burgdorferi* is dominated by mononuclear leukocytes, and particularly macrophages (Ruderman et al., 1995). During the early, pre-immune phase of infection, there is acute inflammation of the aortic root above the aortic valve (Figure 6). The aortitis is characterized initially as segmental transmural arteritis, with inflammation extending from the endothelium through to the periarterial adventitia. The arteritis features infiltration of neutrophils throughout the vessel wall, and margination of neutrophils, thrombocytes, and fibrin on the endothelium. In addition, there is generalized infiltration of the atrial and ventricular myocardium with macrophages. Many of these lesions undergo resolution with the advent of host immunity, and are therefore most apparent in immunodeficient mice (Schaible et al., 1989b; Zimmer et al., 1990; Barthold et al., 1991a; Armstrong et al., 1992; Defosse et al., 1992; Barthold et al., 1993; Ruderman et al., 1995). During the pre-immune phase, lesions feature the presence of predominantly extracellular spirochetes, but occasional intracellular spirochetes have been visualized by electron microscopy within cardiac myofibers (Zimmer et al., 1990; Barthold et al., 1991b; Armstrong et al., 1992; Pachner et al., 1995a). Following the evolution of host immunity, carditis wanes, with residual infiltration of epicardium with lymphocytes (Figure 7), and segmental periaortic infiltration of lymphocytes and plasma cells (Figure 8). Spirochetes are present within all regions of the heart during the pre-immune phase (or in immunodeficient mice), but as the host immune response evolves, spirochetes are relatively cleared from myocardium and heart base tissue, but persist in the aortic wall at the base of the heart (Hodzic et al., 2002; Hodzic et al., 2003; Barthold et al., 2006).

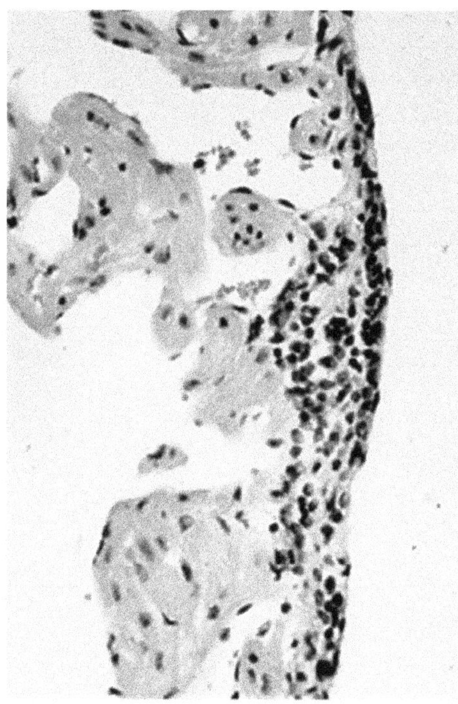

Figure 7. Epicardial infiltration of mononuclear leukocytes in a C3H mouse during chronic infection with *B. burgdorferi.*

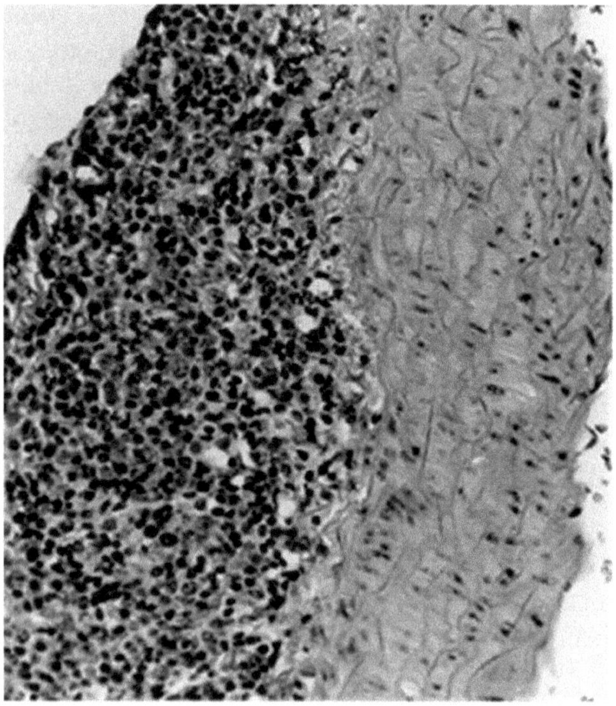

Figure 8. Segmental infiltration of periaortic adventitia with mononuclear leukocytes during chronic infection of a C3H mouse with *B. burgdorferi*. These lesions are commonly present around the base of the aorta during chronic infection.

Similar clearance occurs in SCID mice when induced by adoptive lymphocyte transfer, passive transfer of immune serum, or passive transfer of DbpA or Arp antisera (Barthold et al., 2006).

In addition to carditis, segmental arteritis of the coronary, femoral, plantar, saphenous, tibial, renal, pulmonary and other arteries may be present. During acute infection, these foci of arteritis are similar in character to the transmural aortitis that occurs at the base of the heart (Figure 9). During chronic infection, vessels frequently feature segmental periarterial lymphoplasmacytic infiltrates in the adventitia (Figure 10), similar to chronic lesions of carditis. These lesions are associated with the presence of visible spirochetes in silver-stained histosections (Barthold et al., 1992b; Barthold et al., 1993). Although vascular lesions are common in infected mice, their multifocal and variable locations are a challenge to experimental study.

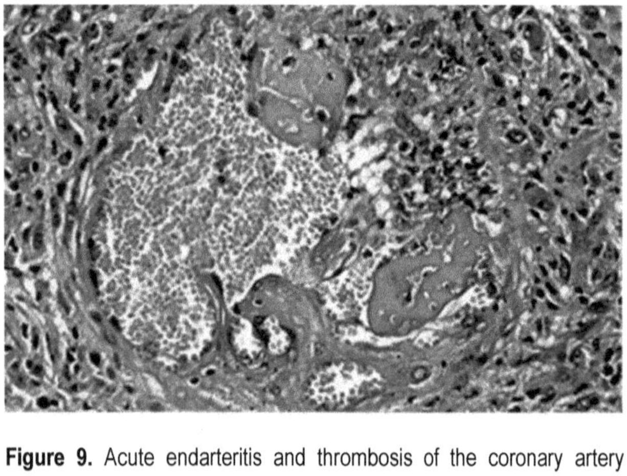

Figure 9. Acute endarteritis and thrombosis of the coronary artery during acute *B. burgdorferi* infection of a C3H mouse.

Despite the challenge of a heart rate of over 300 beats per minute, a limited number of studies have confirmed that experimentally infected mice develop cardiac conduction defects, which arise during the acute phase of infection in association with inflammation, and resolve with resolution of inflammation during the immune phase of infection (Armstrong et al., 1992; Saba et al., 2001).

Mouse genotype and age

All inbred strains and outbred stocks of laboratory mice are susceptible to *B. burgdorferi* infection. Strains of mice that have been shown to be susceptible include inbred A/J, A/WySn, AKR, BALB/c, BALB/cByJ, C.B-17, C3H/HeJ, C3H/HeN, C3H.OH, C3H.SW, C57BL/6, C.B-17, CAL.20, C57BL/10, DBA/2, PL, B10.A, B10.AQR, B10.A(5R), B10.BR, B10.D2, B10.WB, B10R111, B10.S, NZB, SWR, SJL, SWR, CRL:SKH(hr/hr)Br (hairless), and outbred ddY, NMRI, CD-1, ICR, and RML mice

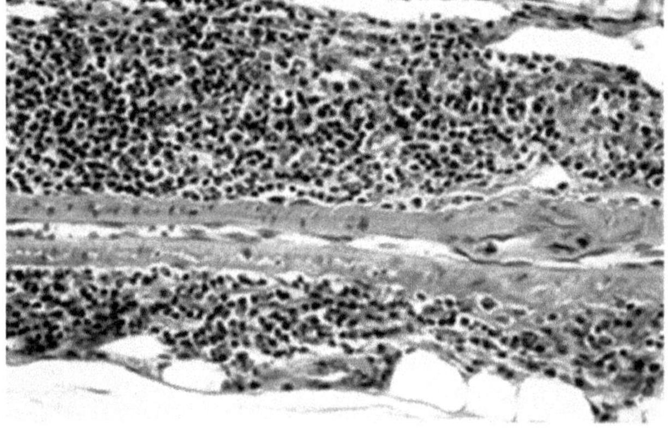

Figure 10. Segmental infiltration of periarterial (plantar artery) adventitia in a C3H mouse during chronic infection with *B. burgdorferi*. These lesions are randomly distributed in many tissues, but are most commonly found around major vessels of the extremities.

(Schaible et al., 1989b; Barthold et al., 1990; Schaible et al., 1991; Schwan et al., 1991; Masuzawa et al., 1992; Shih et al., 1992; Yang et al., 1992; Bockenstedt et al., 1993; Golde et al., 1993; Tuomi et al., 2002; Raveche et al., 2005). In addition, numerous mouse genetic mutants, both spontaneous and induced, have been used for Lyme disease studies (Table 1; also see chapter 18).

Susceptibility to arthritis (and probably carditis) is genetically determined but is also age dependent. Infection of several diverse genotypes of mice, representing major MHC haplotypes, including BALB/cByJ (H-2d), C3H/HeJ (H-2k), C57BL/6J (H-2b), SJL/J (H-2s), and SWR/J (H-2q) developed uniformly severe disease (based upon histopathology) at 30 days after intraperitoneal inoculation with 10^6 or 10^7 *B. burgdorferi* N40 at 3 days of age, but displayed distinctly different patterns of susceptibility when inoculated at 3 weeks of age. Among mice infected at 3 weeks of age, C3H and SWR mice were most susceptible to induction of arthritis, and C57BL/6, SJL and BALB/c were significantly more resistant to disease. C3H/He mice inoculated at 12 weeks of age were more resistant to disease compared with C3H mice inoculated at 3 weeks. Susceptible strains developed higher IgG ELISA titers to *B. burgdorferi* compared to resistant genotypes (Barthold et al., 1990). Disease-susceptible C3H mice weigh less than their age-matched control peers, particularly when infected as young mice, but body weight returns to normal during late, persistent infection (Saba et al., 2001).

A subsequent study examined arthritis (tibiotarsal enlargement) and antibody responses following *B. burgdorferi* ZS7 (10^8, tail base) inoculation of mice representing varying H-2 haplotypes, Igh allotypes, and genetic strain backgrounds, including: C57BL/6 (H-2b, Igh-b), BALB/c (H-2d, Igh-a), C.B-17 (H-2d, Igh-b), Cal-20 (H-2d, Igh-d), DBA/2 (H-2d, Igh-c), C3H/HeJ and C3H/HeN (H-2k, Igh-a), C3H.OH (H-2 dddk, Igh-a), AKR (H-2k, Igh-d), a number of C57BL/10 (B10) congenic lines with the same Igh allotype (b) but differing H-2 haplotypes, including B10.D2 (H-2d), B10.WB (H-2j), B10.R111 (H-2r), B10.S (H-2s), B10.A (H-2kkkd), B10.AQR (H-2qkkd), and B10.A(5R) (H-2 bbkd) mice. It was concluded that H-2k haplotype mice (AKR, C3H, B10.BR) developed chronic progressive arthritis, H-2 haplotypes H-2b (C57BL/6), H-2j (B10.WB), H-2r (B10.R111 and H-2s (B10.S) developed arthritis of variable duration and intensity that was not progressive, and mice of the H-2d haplotype (BALB/c, DBA/2, C.B-17, B10.D2 and Cal-20) developed minimal or no arthritis. All types of mice developed similar patterns of antibody reactivity to *B. burgdorferi* antigens on immunoblots (Schaible et al., 1991). The argument for H-2 linkage to arthritis susceptibility, however, was weakened by a study using C3H (susceptible) and B10 (resistant) congenic mice of the same genotype but differing haplotypes, including C3H/HeN (H-2k) compared to C3H/SW (H-2b), B10.BR (H-2k) and B10 (H-2b) mice. Arthritis was equally severe among C3H mice and equally mild among B10 mice, regardless of haplotype (Yang et al., 1992). In other studies, C3H congenic mice were equally susceptible to arthritis and DBA congenic mice were equally resistant, regardless of their H-2 haplotype (H-2k or H-2d) (Brown and Reiner, 2000). The differing conclusions from these MHC studies may reflect different *B. burgdorferi* isolates, doses, routes of inoculation, mouse suppliers, mouse genotypes, and methods of assessing arthritis. In another study using outbred ICR, BALB/c, B10, B10.BR and B10.D2 infected by tick, differences in immunoblot response to various *B. burgdorferi* proteins were noted to be H-2 restricted (Golde et al., 1993).

The C3H/He mouse has emerged as a prototype disease-susceptible genotype, and BALB/c or B6 (the standard abbreviation for C57BL/6) as resistant genotypes for disease modeling. Following intradermal inoculation of C3H/HeN and BALB/cAnN mice, C3H mice harbored 5-10 fold higher copy numbers of *ospA* and *flagellin* DNA in tissues at intervals up to 5 weeks, compared to BALB mice. F1 hybrids were disease-susceptible with high copy numbers of tissue DNA (Yang et al., 1994). The genetic resistance of BALB/c mice can be overcome with higher doses or relative virulence of inocula, resulting in severe disease and spirochete DNA levels equivalent to C3H mice (Weis et al., 1997a; Ma et al., 1998). Thus, the resistance of BALB/c mice was determined to be conditional, whereas B6 mice were resistant to disease at all doses of *B. burgdorferi*. Remarkably, levels of spirochetal DNA in tissues of B6 mice were equivalent to C3H (C3) mice. B6C3F1 hybrids developed intermediate disease severity (Ma et al., 1998). Susceptibility was evaluated in F2 intercross generation mice derived from C3H/HeN and B6 mice, revealing nine distinct quantitative trait loci (QTL) that were genetically linked to ankle swelling (loci on chromosomes 4 and 5), arthritis histopathology (loci on chromosomes 5 and 11), and humoral responses (loci on

Table 1. Mouse Genetic Mutations Utilized in Various Lyme Borreliosis Studies*

Mouse Strain - Gene Mutation	Reference
Natural Mutants	
BALB/c-*nude*	(Moody et al., 1990a)
C.B-17-*scid*	(Schaible et al., 1989a)
	(Zimmer et al., 1990)
	(Schaible et al., 1990b)
	(Schaible et al., 1991)
	(Barthold et al., 1992b)
	(Simon et al., 1996)
C3H-*scid*	(Barthold et al., 1992b)
B6-*scid*	(KeaneMyers et al., 1996)
	(Mbow et al., 2001)
B6-*beige* (NK deficient)	(Barthold and deSouza, 1995)
NIH-3 (*scid, beige*)	(Defosse et al., 1992)
	(Simon et al., 1996)
C3H/HeJ TLR4 deficient	(Barthold et al., 1991a)
	(Ma et al., 1998)
	(Yang et al., 1994)
A, AKR, B10.D2, DBA/2, SWR – complement 5 deficient	(Bockenstedt et al., 1993)
C3H/HeJ-*Tnfsf6*^{gld} (FasL deficient)	(Shi et al., 2006)
Targeted and Conditional Mutants**	
B6-*IgM* null (μ chain membrane exon)	(KeaneMyers et al., 1996)
B6-*Igh6* null	(McKisic et al., 2000b)
	(McKisic et al., 2000b)
	(Mbow et al., 2001)
B6-*Rag1*, B6-*Rag2* null	(Schaible et al., 1989a)
	(Schaible et al., 1991)
	(Simon et al., 1996)
	(McKisic and Barthold, 2000b)
	(McKisic et al., 2000a)
	(McKisic et al., 2000b)
B6-*Ncf1* null (NADPH oxidase *p47phox*)	(Crandall et al., 2005)
B6-*Cybb* null (NADPH oxidase *gp91phox*)	(Crandall et al., 2005)
B6-*Tcrβ* null	(Mbow et al., 2001)
B6-*Tcrδ* null	(Mbow et al., 2001)
	(Bockenstedt et al., 2003)
B6-*Tcrβ/δ* null	(Mbow et al., 2001)
	(McKisic and Barthold, 2000b)
	(McKisic et al., 2000b)
B6-*CIITA* null (MHC II transactivator)	(Fikrig et al., 1997a)
B6-*MHC II* null	(Mbow et al., 2001)
B6129-*CD40L* null	(Fikrig et al., 1996a)
MRL/MpJ-*Fcgrt* null (FcRn binds IgG Fc)	(Crowley et al., 2006)
B6-*CD1d* null	(Kumar et al., 2000)
C3H-*CD14* null	(Benhnia et al., 2005)
DBA-*IL4* null	(Brown et al., 1999)
BALB/c-*IL4* null	(Potter et al., 2000)
	(Satoskar et al., 2000)
B6-*IL4* null	(Potter et al., 2000)
BALB/c-*IL4R* null	(Potter et al., 2000)
BALB/c-*IL4-IL4R* null	(Potter et al., 2000)
B6-*IL6* null	(Anguita et al., 1998)
B6-*IL10* null	(Brown et al., 1999)
	(Lazarus et al., 2006)
C3H-*IFNγ* null	(Brown et al., 1999)
B6129-*IFNγ* null	(Christopherson et al., 2003)
129-*IFNγ Rα* null	(Anguita et al., 2001)
	(Glickstein et al., 2001)
C3H/HeJ-*Cmkar2* null (CXCR2)	(Brown et al., 2003)
DBA/2- *Cmkar2* null (CXCR2)	(Brown et al., 2003)

Mouse Strain - Gene Mutation	Reference
B6-CCR2 null	(Brown et al., 2003)
	(Montgomery et al., 2007)
C3H/HeJ-CCR2 null	(Brown et al., 2003)
	(Montgomery et al., 2007)
B6-*C3* null	(Woodman et al., 2007)
	(Lawrenz et al., 2003)
B6-*Bf* null (complement factor B)	(Woodman et al., 2007)
B6-*Cfh* null (complement factor H)	(Woodman et al., 2007)
B6-CD11a null	(Guerau-de-Arellano et al., 2005a)
B6-CD11c null	(Guerau-de-Arellano et al., 2005a)
B6-*Itgam* null (CD11b)	(Guerau-de-Arellano et al., 2005a)
B6-*Igb2* hypomorph (CD18)	(Guerau-de-Arellano et al., 2005b)
BALB/c-*Cr2* null (CD21/35 complement receptors)	(Jacobson et al., 2007)
B6-*Cr2* null (CD21/35 complement receptors)	(Jacobson et al., 2007)
B6-*TLR1* null and *TLR6* null	(Oosting, et al. 2011)
B6129-*TLR1* null	(Alexopoulou et al., 2002)
B6129-*TLR2* null	(Alexopoulou et al., 2002)
C3H-*TLR2* null	(Bolz et al., 2004)
	(Wang et al., 2004)
	(Wooten et al., 2002)
B6-IFNAR1 null	(Hastey, et al., 2014)
B6-MyD88 null × TRIF null	(Hastey, et al., 2014)
B6129-CD19.IFNAR null	(Hastey, et al., 2014)
B6129-*MyD88* null	(Liu et al., 2004)
	(Bockenstedt et al., 2006)
B6-*MyD88* null	(Bolz et al., 2004)
	(Behera et al., 2006)
B6129-FcεRγ null	(Belperron, et al., 2014)
B6129-*MKK3* null (regulator of p38 MAP kinase)	(Anguita et al., 2002)
C3H-*iNOS* null	(Brown and Reiner, 1999a)
DBA/2-*iNOS* null	(Brown and Reiner, 1999a)
C3H-*Stat1* null	(Brown et al., 2006)
DBA/2-*Stat1* null	(Brown et al., 2006)
B6129-*Plg* null (plasminogen)	(Coleman et al., 1997b)
BALB/c-*Dcn* null (decorin)	(Liang et al., 2004)
B6-*Dcn* null (decorin)	(Brown et al., 2001)
C3H/HeN-*Dcn* null (decorin)	(Brown et al., 2001)
B6129-*E- P-selectin* null	(Seiler et al., 1998)
B6129-*Spp1* null (osteopontin)	(Potter et al., 2002)
129-*Spp1* null (osteopontin)	(Potter et al., 2002)
B6-*Spp1* null (osteopontin)	(Craig-Mylius et al., 2005)
Transgenic Mutants	
B10.BR-dominant negative p38 MAP kinase	(Hedrick et al., 2006)
B6C3H-*SAF-1* (serum amyloid A-activating factor)	(Ray et al., 2004)
B10-*DR4Dw4* transgenic	(Feng et al., 1995)
B6-*ospA* transgenic	(Fikrig et al., 1996b)
C3H-*ospA* transgenic	(Fikrig et al., 1996b)
	(Fikrig et al., 1997b)
C3H-*flagellin* transgenic	(Fikrig and Barthold, 1995)

* Effects of mutations (phenotype) are varied and dependent upon experimental design, outcome measured, and methods of assessment.

** Mutations created by homologous recombination in 129 ES cells (129 strain not specified). Parental mouse strains are congenic to 129 (or insipient congenic at N5 or higher). In some cases, mutations were originally maintained on another parental strain, but fully backcrossed (N10) onto another parental strain, in which case the original parental strain is not indicated. If hybrid, the parental strain indicates such.

chromosomes 6, 9, 11, 12, and 17) (Weis et al., 1999). Reciprocal backcross populations derived from C3H/HeN and B6, and C3H/HeJ and BALB/cAnN mice have been used for composite interval mapping, which revealed 12 QTL, including seven previously reported loci. Further analysis revealed multiple QTL on chromosome 5 (Roper et al., 2001), 11 and 12 (Ma et al., 2009) that affected disease phenotypes (ankle swelling and histopathologic criteria of inflammation). Subsequent studies revealed a key role for type I interferon genes in regulation of arthritis severity (Miller et al., 2008; Ma et al., 2014) Comparative studies between arthritis-resistant DBA/2 and arthritis-susceptible C3H/HeJ mice (Barthold et al., 1990) confirmed studies with B6 and C3H mice, with equivalent levels of spirochetes in joint tissues, but markedly different disease severity (Brown and Reiner, 1998). These findings clearly illustrate that various aspects of Lyme borreliosis in the mouse model are under distinctly different and complex genetic control.

The genetic basis of arthritis susceptibility among inbred mouse strains remains manifest in globally immunodeficient mice. Arthritis is more severe in C3H-*scid* mice compared to BALB/c-*scid* mice (Barthold et al., 1992b). Infection of RAG-null mice on C3H, BALB/c, B6 or DBA/2 genetic backgrounds resulted in severe arthritis among C3H- and BALB-*Rag* null mice, whereas arthritis was mild or absent in the B6- or DBA-*Rag* null mice (Brown and Reiner, 1999b).

Infectivity of B. burgdorferi species, strains and isolates

Laboratory mice are susceptible to infection with numerous species, strains and isolates of *B. burgdorferi* sensu stricto (s.s.), *B. afzelii*, *B. garinii*, *B. japonica*, *B. bissettii*, and *B. lusitaniae* (Anderson et al., 1990; Masuzawa et al., 1992; Masuzawa et al., 1994a; Kaneda et al., 1998; Zeidner et al., 2001; Tuomi et al., 2002; Pachner et al., 2004; Yrjänäinen et al., 2006). Because different patterns of organotropism and disease are suspected in humans infected with different *B. burgdorferi* s.l. and different ribosomal DNA genotypes of *B. burgdorferi* s.s., it is tempting to compare patterns of tissue distribution and disease induction by various isolates.

With few exceptions, if *B. burgdorferi* s.l. isolates or strains are infectious in mice, they will cause disseminated and persistent infection and similar patterns of disease, but disease severity may vary. One study examined mice infected with *B. burgdorferi* s.s., *B. afzelii*, and *B. garinii* isolates. *B. afzelii* isolates tended to be cultured less often from urinary bladder but more often from heart and ear, compared to other species and isolates (Tuomi et al., 2002). A similar study suggested that *B. burgdorferi* s.s. isolates were more apt to infect heart, cause inflammation, and induce higher antibody responses, compared to *B. afzelii* and *B. garinii* isolates, but levels of *B. afzelii* and *B. garinii* were higher in skin compared to *B. burgdorferi* s.s. isolates (Pachner et al., 2004). The pathogenicity of ten clinical *B. burgdorferi* s.s. isolates, representing two distinct ribosomal DNA spacer restriction fragment length polymorphism genotypes (RST1 and RST3), revealed that RST1 isolates were more apt to be infectious in C3H mice following inoculation with high doses of cultured spirochetes, induced greater ankle swelling and arthritis scores, and were associated with higher levels of spirochetal *recA* DNA in tissues. These results correspond to a higher percentage of human patients with Lyme disease infected with RST1 isolates (Wang et al., 2001; Wang et al., 2002) (also see chapter 24). A similar study was performed with tick-borne exposure, revealing that bacteremia lasted longer in mice infected with RST1 isolates, but there were no differences in spirochetal flagellin DNA levels in tissues, or disease severity between RST1 and RST3 groups (Dolan et al., 2004). The most recently identified virulent strain of *B. burgdorferi s.l.* in North America, named *Borrelia mayonii,* exhibits higher spirochetemia in humans than other strains. When used to infect outbred CD-1 mice by tick bite, the kinetics of infection were found to be similar to that of *B. burgdorferi* sensu stricto (Dolan et al., 2017).

Infectious dose, adaptive state and route of inoculation.

The infectious dose of *B. burgdorferi* is lowest following intradermal inoculation. Three-week old C3H and BALB mice were equally susceptible to intradermal inoculation with 10 or fewer *B. burgdorferi* N40 spirochetes, compared to a hundredfold more spirochetes to establish infection following intraperitoneal inoculation (Barthold, 1991). The preferential infectivity by the intradermal route has been confirmed with other *B. burgdorferi* s.s., *B. afzelii*, and *B. garinii* isolates (Barthold et al., 1995a; Tuomi et al., 2002). Mice also are susceptible to infection following intragastric inoculation of very high doses of *B. burgdorferi* N40, but there is no evidence for contact transmission (Barthold, 1991), or *in utero* (placental) transmission, although maternal infection may cause fetal death (Silver et al., 1995; Weis et al., 1997b).

The site of intradermal inoculation can significantly influence the pattern of infection, disease, and host response. Inoculation in the dorsal midline thoracic region results in bacteremia, antibody response, and evolution of arthritis and carditis more rapidly than when mice are inoculated in the foot (Bockenstedt et al., 1996), observations that have been recapitulated (Motameni et al., 2005). Depending upon site of intradermal inoculation, peripheral blood and splenic lymphocytes manifest different proliferative responses to mitogens. Responses were impaired following inoculation in the shoulder region, whereas no effect occurred following inoculation of the foot (deSouza et al., 1993). Others (Matyniak and Reiner, 1995; KeaneMyers et al., 1996) have shown similar effects with cytokine responses, which could only be detected in regional lymph nodes draining sites of inoculation, but not the spleen. This phenomenon is related to direct tissue invasion of spirochetes. Mice inoculated at the base of the tail develop more severe arthritis in the rear legs, but mild or no carditis. Likewise, enlargement of regional lymph nodes is greatest near the site of inoculation, despite subsequent disseminated infection. Thus, purposeful targeting of specific sites can influence experimental outcome. Direct footpad inoculation with cultured spirochetes has also been used to favor induction of local arthritis (Masuzawa et al., 1994b; Christopherson et al., 2003).

Infecting dose and adaptive state of spirochetes used for inocula can significantly influence the host immune response. The OspA-OspC dichotomous expression serves as an example. Inoculation of high doses of culture-grown spirochetes stimulates an immune response to antigens within the input inoculum (reflecting *in vitro* conditions), as well as to the spirochetes that adapt and proliferate in the host (reflecting *in vivo* conditions). Seroconversion to OspA occurs with high-dose cultured inocula but does not occur when the inoculating dose is low (generally less than 10^4) or introduced by tick (Pachner et al., 1992; Barthold and Bockenstedt, 1993; Gern et al., 1993; Schaible et al., 1993; Golde et al., 1994; Barthold et al., 1995a). Seroconversion following low-dose or tick-borne infection reflects the antigenic stimulation that arises to host-adapted spirochetes during active infection, which includes reactivity to antigens that are expressed exclusively or preferentially *in vivo* (Simpson et al., 1991; Gern et al., 1993; Golde et al., 1993; Schaible et al., 1993; Golde et al., 1994; Barthold et al., 1995a; Suk et al., 1995; Wallich et al., 1995).

In addition to infection with cultured spirochetes or by tick-borne exposure, mice can be infected with host-adapted spirochetes. One means of inoculation with host-adapted spirochetes is subcutaneous transplantation of allografts from infected mice (Barthold and Bockenstedt, 1993; Barthold et al., 1995a). Spirochetes can be purified from peripheral blood of mice during the acute (spirochetemic) phase of infection, especially in SCID mice (Cassatt et al., 1998). Subcutaneously implanted metal net chambers (Jonsson et al., 1995) and perforated Teflon chambers (Hurtenbach et al., 1995), which allow spirochetes to enter from surrounding tissues, have successfully yielded spirochetes for study. In addition, these approaches have allowed harvest of tissue fluid for cytokine analysis (Hurtenbach et al., 1995). Subcutaneous or intraperitoneal placement of chambers equipped with 0.1 µm pore membrane filters has also been shown to support growth of host-adapted spirochetes (deSilva et al., 1998). Larger yields of host-adapted spirochetes have been harvested from dialysis membrane chambers implanted into the peritoneal cavities of rats (Akins et al., 1998).

More rapid dissemination, with earlier invasion of joint tissue and onset of disease, occurs when mice are infected with host-adapted spirochetes through allografts, compared with syringe or tick-borne infection, or with high doses of cultured spirochetes. In the case of tick-borne infection, it has been shown that spirochetes entering the mouse must undergo a brief period of adaptation before dissemination (Shih et al., 1992; Hodzic et al., 2002). Notably, dissemination following tick-borne infection is more rapid in genetically disease-susceptible C3H mice compared to resistant CD-1 mice (Shih et al., 1993). Cultured spirochetes must also undergo a period of local adaptation prior to dissemination, explaining their vulnerability to immune serum for up to 4 days after inoculation, whereas host adapted spirochetes are not vulnerable (Barthold et al., 1997; De Silva et al., 1998). High-passage infectious spirochetes remained vulnerable to passive immunization with immune serum for longer intervals after inoculation compared to low-passaged spirochetes, and were defective in up-regulation of genes that are typically preferentially expressed during infection (Anguita et al., 2000).

Tick-borne inoculation and xenodiagnosis

Both larval and nymphal stages of *Ixodes* spp. ticks readily feed upon laboratory mice. Larval ticks, which are not transovarially infected, can be fed upon laboratory mice that have been inoculated with any infectious *B. burgdorferi* species, strain, isolate or mutant of interest. Nymphs, infected as larvae with the *B. burgdorferi* isolate of interest, can then be used as a means of infection of experimental mice. Placement of ticks into the pelage of the dorsal neck and shoulder region will result in rapid attachment in the head and neck region (notably, ticks are more apt to attach locally when placed on pigmented strains of mice, and paradoxically tend to wander more when placed on albino mice). These sites are generally not accessible to grooming by the host mouse, but it is important that the mouse be maintained singly within the cage until the ticks have fed to repletion and detached to avoid inter-mouse grooming. Placement of mice in wire-bottomed cages over pans with water allows recovery of replete ticks for further analysis.

Xenodiagnosis is used as a highly sensitive method of detecting infection in rodents (Shih et al., 1995), including detection of low numbers of replication-suppressed spirochetes following antibiotic treatment of the host (Bockenstedt et al., 2002; Hodzic et al., 2008). Xenodiagnosis involves the placement of larval or nymphal ticks upon a host, allowing them to feed to repletion, molt and harden, and then testing the ticks for infection by culture, PCR, or transmission to other hosts. There is a burst of spirochete replication within the tick during feeding, which amplifies the ability to detect spirochetes (De Silva and Fikrig, 1995; Hodzic et al., 2002). Co-feeding of uninfected ticks with infected ticks has been shown to be an effective means of detecting very early infection in mice (Hua et al., 2003). Events (gene expression, population dynamics, etc.) that take place at the tick attachment/feeding site can be assessed by biopsy of the tick attachment site (Shih et al., 1992; Ohnishi et al., 2001a; Hodzic et al., 2002) Forceful removal of ticks results in attachment of a small fragment of mouse dermis to the tick hypostome, which has been successfully examined for spirochetes, gene, and protein expression (Ohnishi et al., 2001b).

Immune response to infection

Although T cells influence the host immune response to infection, key events in the host response are mediated by B cells or antibodies in the absence of cognate T cell help. Infection of immunodeficient SCID and NIH-3 (triple deficient) mice (Schaible et al., 1989b; Schaible et al., 1990a; Zimmer et al., 1990; Barthold et al., 1992b; Hodzic et al., 2003) or B6-*Rag1* (or *Rag2*) null mice (McKisic and Barthold, 2000b; McKisic et al., 2000b) results in progressive disease. A similar course occurs in B cell deficient (B6-*Igh6* null) mice, whereas T cell deficient BALB athymic nude or B6-*Tcr β/δ* null mice develop a disease course that parallels fully immunocompetent mice, with disease remission (Moody et al., 1990a; Kang et al., 1995; McKisic and Barthold, 2000b; McKisic et al., 2000b). Adoptive transfer of naïve T cells into infected B6-*Rag1* null mice exacerbated disease, whereas transfer of B cells induced disease remission. Immune serum from T cell-deficient mice, but not B cell-deficient mice, was protective against syringe challenge (McKisic et al., 2000b) and induced disease remission (Barthold et al., 2006). In addition, adoptive transfer of B cells, but not T cells, elicited protective immunity in SCID mice (Schaible et al., 1994). Collectively, these studies strongly implicate humoral immunity as a critical determinant of the host response to *B. burgdorferi* infection.

The host response to infection is more thoroughly covered in the chapter by Bockenstedt et al., but discussion of the mouse model is not complete without an overview of the role of the acquired humoral response in three major biologic events: protective immunity, disease-resolving immunity, and global reduction of spirochetes in tissues (with cessation of bacteremia) during the immune phase of infection. These critical events can be differentially dissected for experimental analysis. Protective and disease-resolving antibodies develop during active infection of mice with *B. burgdorferi* s.s., *B. afzelii* PKo or *B. garinii* PBi, indicating universality of the phenomena, but protection is species-specific (Barthold, 1999).

Serum from infected immunocompetent mice during the immune phase of infection is protective against tick- and syringe-borne cultured spirochetes but not against challenge with host-adapted spirochetes (Bockenstedt et al., 1996; Barthold et al., 1997; deSilva et al., 1998). When immune serum is passively transferred to naïve mice before or 1 day after tick attachment, relatively large, but physiological amounts of immune serum (100-500 µl) are protective (Shih et al., 1995; Shih and Liu, 1998; Bockenstedt et al., 2003). In contrast, as little as 0.0125 ml of immune serum is protective against challenge with low doses of syringe-inoculated cultured spirochetes if administered prior to, but not after, inoculation (Barthold and Bockenstedt, 1993; Barthold et al., 1997). Immune sera from human patients and naturally infected dogs are also protective against syringe challenge in mice and hamsters (Callister et al., 1993; Fikrig et al., 1994; Barthold et al., 1995b). Although cultured spirochetes do not reflect the natural state of spirochetes, the exquisite susceptibility of cultured spirochetes to minute amounts of immune serum containing antibody to *in vivo*-expressed antigens can be utilized in a passive protection assay of functional immunity that is amenable to titration. Using serial dilutions of immune serum and a standardized inoculum in such an assay, the passive protective activity in immune serum rises rapidly during early infection, peaks at 30 to 60 days, and then declines during the course of persistent infection. These kinetics do not correspond with the rising titers and expanding reactivity of antibodies that evolve during infection when measured by Western blots and ELISAs using antigens expressed by cultured spirochetes (Barthold and Bockenstedt, 1993; Barthold et al., 1997).

When mice are challenged with host-adapted spirochetes, immune serum is incapable of preventing dissemination, but prevents development of disease (Bockenstedt et al., 1996; Barthold et al., 1997; deSilva et al., 1998). Transfer of immune serum during ongoing arthritis induces resolution of arthritis in infected SCID mice, but does not eliminate infection (Barthold et al., 1996; Barthold et al., 1997; De Silva et al., 1998; Barthold et al., 2006). These effects require passive immunization with larger volumes of immune serum (150 µl or more) than those required to induce protective immunity against syringe challenge. Using serial dilutions of immune serum to passively immunize infected SCID mice, disease-resolving activity in immune serum follows similar kinetics as protective activity, with peak activity during early infection, followed by decline, despite persisting infection. These events may explain the recurrent bouts of acute disease during chronic infection.

Somewhat discordant observations have been reported regarding the effect of immune serum on carditis resolution. Early studies indicated that passive immunization of infected SCID mice with immune serum induced only partial resolution of carditis or had no effect (Barthold et al., 1996; McKisic and Barthold, 2000b). Other studies demonstrated that B cells were not necessary for carditis resolution, and that CD4+ helper T cells facilitated carditis resolution (Bockenstedt et al., 2001). However, recent studies have confirmed that immune serum alone is capable of inducing carditis resolution (Barthold et al., 2006). These discrepancies are likely due to the complex features of carditis, the fact that carditis is less sensitive to the effects of antibody, compared to arthritis, and may be impacted by the quality and quantity of antibody in immune sera, including the effects of T cell help.

Another important effect of immune serum is the reduction of spirochete numbers in all tissues following passive immunization of infected C3H-*scid* mice with immune serum from infected immunocompetent mice. This parallels the reduction of spirochete burdens that occurs during the evolution of host immunity in immunocompetent mice (Barthold et al., 1993; Hodzic et al., 2003), and also occurs following adoptive transfer of lymphocytes to infected SCID mice from immunocompetent mice (Barthold et al., 2006). Notably, IL-10 deficient mice exhibit up to tenfold greater clearance of spirochetes from tissues compared to wild type during the immune phase of infection, but their immune serum has similar spirochete-reducing properties as wild-type mice when passively transferred to infected SCID mice (Lazarus et al., 2006).

Based upon its protective, disease-resolving, and spirochete-reducing properties, immune serum has been used to screen *B. burgdorferi* genomic expression libraries to identify the responsible antigens. Antiserum against recombinant non-lipidated DbpA elicits protective immunity against syringe-borne challenge (Cassatt et al., 1998; Feng et al., 1998; Hanson et al., 1998) and induces disease resolution (Feng et al., 2000; Barthold et al., 2006), but DbpA immunity does not seem to protect against tick-borne infection (Hagman et al., 2000). Antiserum to recombinant non-lipidated Arp is not protective but induces disease resolution (Feng et al., 2000; Feng et al., 2003; Barthold et al., 2006). Despite their protective and disease-resolving activity, neither DbpA nor Arp antisera induced reductions in spirochete burdens when infected SCID mice were passively immunized (Barthold et al., 2006). Reduction of spirochete burdens in tissues and reduction in joint swelling (arthritis resolution was not confirmed by histopathology) can be induced by passive administration of antisera to recombinant BmpA and BmpB (Pal et al., 2008). Although these antigenic targets do not explain the whole story, they contribute toward dissecting the pathogenesis of Lyme borreliosis in the mouse model, using prototype antigens and antibodies with known biologic effect.

Using the ZS7 strain of *B. burgdorferi* s.s., it has been shown that passive immunization of infected SCID mice with polyclonal and monospecific antibody to recombinant OspC is capable of resolving carditis and arthritis, as well as clear infection, based upon culture of ear biopsies (Zhong et al., 1997; Zhong et al., 1999). In contrast, passive immunization of SCID mice infected with *B. burgdorferi* N40 with OspC antibody had no effect upon disease or infection status (Barthold et al., 1997). These results underscore the potential differences in OspC expression during infection among *B. burgdorferi* isolates, strains or species. It has been shown that passive immunization of *B. burgdorferi* B31-infected SCID mice with OspC antibody eliminated OspC-expressing spirochetes, but failed to clear infection (Liang et al., 2002a), but passive immunization of SCID mice infected with *B. burgdorferi* that constitutively expressed OspC resulted in clearance of infection (Xu et al., 2006). Results of these studies are influenced by the tissues examined and methods for detecting spirochetes.

Other rodent models of Lyme borreliosis

White-footed mouse model

Peromyscus leucopus received early attention as a model because of its importance as a natural reservoir host in North America and its availability as a laboratory animal. *Peromyscus* mice are readily susceptible to infection by tick, which transmit infection between 48 and 72 hours after attachment (Piesman et al., 1987), but do not become infected transplacentally from mothers (Mather et al., 1991). Studies in experimentally infected hamsters and naturally exposed *Peromyscus* mice revealed that ear tissue could be sampled as a non-invasive method for evaluating infection status (Sinsky and Piesman, 1989). This observation was soon capitalized upon with the advent of PCR for detection of *B. burgdorferi* DNA in museum specimens of rodents, confirming that *B. burgdorferi* was present in suitable reservoir hosts on Cape Cod in the 1890's (Persing et al., 1990). Likewise, this species revealed the high rate of infection of the urinary bladder, which is now generally tested to confirm disseminated infection in animal model studies, regardless of species (Schwan et al., 1988). The widely used method of xenodiagnosis, in which infection of the host can be evaluated with a high degree of sensitivity by feeding uninfected ticks upon infected hosts was also introduced with this model (Donahue et al., 1987). Finally, the phenomenon of a secondary wave of IgM isotype antibodies after *B. burgdorferi* infection was first noted in this reservoir species (Schwan et al., 1989). Microscopic studies in experimentally infected *Peromyscus* mice revealed no evidence of pathology in tissues in persistently infected animals, with spirochetes observed extracellularly within the interstitium of various tissues (Wright and Nielsen, 1990). However, when *Peromyscus* mice were infected as infants, they developed arthritis and carditis (Moody et al., 1994b).

The use of *Peromyscus* mice as a natural host upon which to study mechanisms of resistance and tolerance to the pathogen has been expounded by Barbour (Barbour, 2017). By infecting these wild mice with different strains of *B. burgdorferi,* variances in the antibody responses targeting specific antigens, which differentially reduce the spirochete burden, revealed that the white-footed mouse likely selects for diversity in the pathogen population, namely with respect to antigenic diversity (Baum et al., 2012). Following the sequencing of the genome for *Peromyscus leucopus*, RNA sequencing of infected mice revealed that differentially expressed genes were far more numerous in the skin than in the blood (Long et al., 2019).

Hamster model

In a remarkably insightful early study, Johnson and colleagues not only documented the susceptibility of Syrian hamsters to infection with *B. burgdorferi*, but also demonstrated that the animals developed early bacteremia, which was cleared after day 6, and then developed persistent disseminated infection of multiple organs (up to 52 days) with no overt clinical signs (Johnson et al., 1984). Furthermore, hamsters of both sexes and all ages (4, 10 and 52 weeks of age) were susceptible to disseminated infection. The important concept of loss of spirochete infectivity following multiple *in vitro* passage was also initially documented with the hamster model. Persistent infection was later confirmed for up to 14 months, with periodic recurrence of bacteremia in some animals after months of infection (Johnson et al., 1988).

The hamster model was used for examining efficacy of passive immunization, using antiserum from rabbits that had been hyperimmunized intravenously weekly with 10^8 viable spirochetes. Protective activity was effective when antiserum was administered 17-18 hours before, but not 17-18 hours after challenge. As little as 0.0125 ml of rabbit antiserum was effective against 10^8 intraperitoneal challenge (1,000 ID50). Furthermore, passive immunization was effective against challenge with homologous isolates, but not against geographically heterologous isolates (Johnson et al., 1986; Johnson et al., 1988). These pivotal passive immunization studies confirmed that *B. burgdorferi* was exquisitely susceptible to antibody, but only when antibody was present at the time of challenge, not once infection had been established. This concept of host adaptation of *B. burgdorferi* has been a significant subject of research interest, and likely underlies the critical mechanisms of persistence.

Inoculation of adult hamsters intraperitoneally or cutaneously with *B. burgdorferi* results in disseminated, persistent infection, but without lesions in internal organs (Duray and Johnson, 1986; Goodman et al., 1991). However, when inbred LSH/SsLak hamsters were inoculated in the hind paws with high doses (10^6) of spirochetes, acute transient synovitis was evident microscopically. Whole-body non-lethal irradiation prior to inoculation exacerbated the synovitis, with earlier onset, grossly evident swelling, and delayed resolution. Induction of joint swelling could also be induced in irradiated hamsters following intravenous and intraperitoneal inoculation, and with hind paw inoculation of fewer spirochetes. At one week, the tibiotarsal and intertarsal synovia of non-irradiated hamsters became hypertrophied and hyperplastic, with infiltration of neutrophils and fibrin. Numerous spirochetes were visible in the synovial and periarticular tissues. Inflammation subsided by 3 weeks, which coincided with seroconversion, and at 15 weeks, there was residual fibrosis, focal peri-neurovascular accumulations of lymphocytes and plasma cells, and spirochetes were no longer visible. Joints of irradiated hamsters had similar, but more severe inflammatory changes, with persistence of inflammation and spirochetes at 5 weeks. Inflammation of distant joints (fore paws) arose in irradiated hamsters, but lesions were less severe (Schmitz et al., 1989). A sequential study of arthritis in irradiated and nonirradiated hamsters reiterated the transient nature of the inflammation in nonirradiated hamsters, which was prolonged and more severe with irradiation. The abundance of extracellular spirochetes within the periarticular edematous tissue, which disappeared as inflammation resolved, and the tendency of spirochetes to align themselves along collagen bundles in tendons and fibrous connective tissue was noted (Hejka et al., 1989). Others have shown that hamsters inoculated as 3-day old infants intraperitoneally with 10^6 *B. burgdorferi* developed disseminated infections with polyarthritis and carditis at 30 days (Moody et al., 1990a).

Hamsters appear to require relatively high doses of spirochetes to establish infection following intraperitoneal or intradermal inoculation, but this may be a reflection of culture conditions or the strain of the inoculum (Roehrig et al., 1992; Obonyo et al., 2002). Despite the apparent need to use high doses of spirochetes when syringe inoculating, hamsters are readily susceptible to tick-borne infection and can transmit infection to feeding ticks (Piesman et al., 1987; Kurtti et al., 1996; Woodrum and Oliver, 1999). Infected nymphal ticks feeding upon hamsters or *Peromyscus* mice transmitted infection inefficiently at 24 and 48 h, but most animals became infected after 72 h of attachment (Piesman et al., 1987). Studies with tick-infected hamsters have shown no evidence of venereal, transplacental, or contact transmission among infected hamsters (Woodrum and Oliver, 1999).

The hamster model has been used to elucidate important concepts about biologically relevant host antibody responses during active infection. Serum from actively infected hamsters (immune serum) was protective when passively transferred to naïve irradiated hamsters prior to high dose foot challenge. Using serial dilutions of immune serum to titrate protective activity, protective activity was present within 7 days of infection, peaked between 3 and 5 weeks, and then declined, despite persistent infection. Protection was achieved with as little as 0.0125 ml of immune serum. The dynamics of the protective antibody response differed from the overall antibody reactivity against *B. burgdorferi* lysates, as measured by ELISA, which remained elevated throughout the course of infection (Schmitz et al., 1990; Schmitz et al., 1991b). The unusual dynamics of protective activity in immune serum were associated with complement (Schmitz et al., 1991a), IgG2 (Schmitz et al., 1992) and *in vitro* borreliacidal activity (Lovrich et al., 1991). Passive immunization with immune sera and challenge with homologous and heterologous isolates has also been used to demonstrate different seroprotective groups among isolates of *B. burgdorferi* and *B. garinii* (Lovrich et al., 1993; Lovrich et al., 1994).

The hamster model has been utilized for immunization studies, using inactivated *B. burgdorferi*, mutant *B. burgdorferi* and recombinant proteins (Hughes et al., 1993a; Johnson et al., 1995; Lovrich et al., 1995; Kurtti et al., 1996; Padilla et al., 1996; Rousselle et al., 1998). Immunization with a non-infectious *B. burgdorferi* mutant lacking OspA and OspB was shown to be protective against challenge with infectious wild-type spirochetes, demonstrating that host protective immunity can be elicited against antigens other than OspA/B (Hughes et al., 1993b). Hyperimmunization with formalin-inactivated whole *B. burgdorferi* prior to challenge with a homologous isolate in the hind paw may result in severe arthritis, but only if challenge took place prior to development of borreliacidal antibodies. Borreliacidal antibodies could be circumvented by challenge with a heterologous isolate (different seroprotective group) of *B. burgdorferi* (Lim et al., 1994b). Subsequent studies demonstrated that sensitized CD4+ T cells and macrophages were responsible for this effect (Lim et al., 1994a; Lim et al., 1995a; Lim et al., 1995b; Du Chateau et al., 1996; DuChateau et al., 1997; DuChateau et al., 1999). This phenomenon of destructive arthritis was later shown to occur when hamsters were pre-immunized with recombinant OspA (Croke et al., 2000). This observation provoked controversy, as it was published at a time when the human OspA vaccine (LYMErix) was being marketed (Parenti, 2000).

Guinea pig model

While guinea pigs are not natural hosts for *Ixodes scapularis,* larvae and nymphs readily attach and feed on these animals if they have not been pre-exposed to tick feeding. The two key features of the guinea pig model are their generation of anti-tick immunity and the higher susceptibility of younger animals to infection. A guinea pig model for Lyme disease was first investigated by Sonnesyn et al. (Sonnesyn et al., 1993a), which utilized a combination of outbred Duncan-Hartley and C4-deficient guinea pigs. For these initial studies, 10^7-to-10^8 spirochetes were administered either subcutaneously or intradermally via needle inoculation. Results demonstrated an age-dependent susceptibility to *B. burgdorferi*, with animals ≥6 months of age more resistant to infection. Infected animals displayed pathologic changes relevant to human Lyme disease and mounted a host immune response that was ultimately effective in clearing spirochetes from most affected tissues. Across all sample time points, ranging from 2 weeks to 6 months, *B. burgdorferi* was cultured from the following organs with decreasing frequency: stifle (79%), quadriceps muscle (67%) urinary bladder (56%), heart (48%), and spinal cord (10%). The stifle (knee joint) was the only tissue where spirochetes remained culture positive for ≥4 months P.I. No spirochetes were cultured from cerebrums, ears, or livers examined. In contrast, all cultures from guinea pigs ≥6 months of age at the time of inoculation were negative. Clinical evidence of arthritis or EM was not observed among infected animals. Positive cultures of tissue were not always accompanied by histologic changes. Additionally, inflammation often persisted following host clearance of spirochetes evidenced by a negative culture of inflamed tissues. This phenomenon has been observed across multiple animal models. Immunoblots utilizing *B. burgdorferi* whole cell extracts demonstrated an IgG antibody response in all infected animals by day 15 P.I., which became progressively more intense at days 23 and 30. The most abundant IgG antibodies noted were proteins migrating at 24, 31 (OspA), 39, and 41 (flagellin) kDa. Benefits of the guinea pig animal model taken from this study included utilization of an outbred immunocompetent host with pathologic changes in multiple tissues relevant to human disease including the nervous system, heart, joints, and urinary bladder. Preferential involvement of areas containing abundant conductive fibers in the heart is consistent with the cardiac conduction disturbances described clinically in humans with acute Lyme disease. Furthermore, the guinea pig is unique with respect to a predilection of spirochetal persistence in infected knee joints despite clearance elsewhere (Sonnesyn et al., 1993a).

In a later study by Breitschwerdt et al., outbred Hartley guinea pigs were inoculated intradermally with *Borrelia* spirochetes isolated from a canine clinical isolate (Breitschwerdt et al., 1994). These were distinctly not *Borrelia burgdorferi,* and were later recognized as relapsing fever spirochetes. As opposed to the previous study by Sonnesyn et al., spirochetemia was documented through positive blood cultures and direct observation of spirochetes in blood smears (Breitschwerdt et al., 1996). All inoculated guinea pigs seroconverted, with peak reciprocal antibody titers ranging from 128 to 256. Of the nine inoculated guinea pigs, two had focal lymphocytic epicarditis, so these may be another potential model for relapsing fever.

Nazario et al. were the first to establish transmission of *B. burgdorferi* through tick-bite to naïve outbred guinea pigs (Nazario et al., 1998). This was confirmed P.I. through concordant positive skin cultures and seroconversion in all inoculated guinea pigs. Furthermore, this work highlighted the fact that guinea pigs with repeated exposure to *Ixodes scapularis* not harboring *B. burgdorferi* develop anti-tick immunity and are subsequently resistant or highly resistant to challenge with *B. burgdorferi*-infected ticks. When comparing naïve and sensitized guinea pigs, indirect characteristics of tick sensitization included early detachment of nymphal ticks, decreased average weight of recovered ticks, and a robust localized basophilia in the dermis adjacent to tick bites. Direct features of anti-tick immunity included the absence of positive tissue cultures in the majority of guinea pigs (i.e., blood, urinary bladder, skin, and spleen) and complete absence of serologic conversion in any of the sensitized guinea pigs subsequently exposed to *B. burgdorferi*-infected nymphs. In summary, tick-mediated infection was restricted to the skin and anti-tick immunity, through prior sensitization, either eliminated or greatly hampered *B. burgdorferi* transmission in guinea pigs.

Rat model

Several tick isolates of *B. burgdorferi* (N8, N13, and N40) were shown to be infectious in 3-week-old inbred Lewis (LEW/SSNcr) rats. The rats developed transient bacteremia, disseminated infection and mild polyarthritis. Inoculation of infant (3-day-old) Lewis, Fischer 344 (F344/Ncr) and outbred Sprague Dawley (CR:RAR[SD]) rats revealed similar results, but arthritis was more severe in the younger rats, with grossly evident swelling of tibiotarsal joints and lameness. Spirochetes were visualized by silver stain in the periarticular edema, synovium and occasionally in the joint spaces (Barthold et al., 1988a; Barthold et al., 1988b). Infant rats developed periarticular tibiotarsal swelling by day 14, which subsided by day 60. Caliper measurements indicated significant persistent joint enlargement beyond 90 days. Bacteremia was evident at 14 days, but disappeared thereafter, and multiple tissues were infected for up to 1 year. Acute arthritis, with exudation of fibrin and neutrophils, subsided by 90 days, with subsequent fibrosis and infiltration of macrophages, lymphocytes and plasma cells in the perisynovial tissue. When rats were examined at 180, 270 and 360 days, some rats had acute oligoarticular arthritis in addition to chronic changes in other joints, suggesting disease exacerbation. In addition, infiltrates of plasma cells and lymphocytes were noted in the connective tissue at the base of the heart and aortic wall, as well as myocardium (Moody et al., 1990a). Rats were shown to be susceptible to infection with numerous different *B. burgdorferi* isolates from the northeast, Midwest and Pacific U.S, as well as Europe, with no apparent differences in patterns or ability to disseminate and induce disease (Barthold et al., 1990). An exception was the New York tick isolate *B. bissettii* 25015, which was infectious but non-arthritogenic (Anderson et al., 1990). Rats could be infected with significantly lower doses of spirochetes following intradermal inoculation, compared with intraperitoneal inoculation, but could not be infected through contact, oronasal inoculation, *in utero* transmission, or venereal transmission (Moody and Barthold, 1991). Intravenous inoculation of live *B. burgdorferi*, but not killed spirochetes, was found to increase the permeability of the blood-brain barrier, with mild pleocytosis and spirochetes noted in cerebrospinal fluid. These changes were less marked when rats were inoculated with long-term, *in vitro*-passaged organisms (Garcia-Monco et al., 1990). Intravenously inoculated radiolabeled spirochetes were cleared from the vasculature within 1 h of injection, with random invasion of multiple organs (Galbe et al., 1993).

The laboratory rat model also lends itself to intra-articular injection for the study of arthritogenesis, using an approach known as antigen-induced allergic arthritis in Wistar rats. Injection of butanol-extracted, water-soluble outer surface proteins into the knee joint was initially shown to be arthritogenic (Gondolf et al., 1994). More recent work demonstrated that lipidated recombinant OspA and OspC, but not OspB, were arthritogenic, whereas non-lipidated proteins were not, and synthetic lipid alone or coupled to bovine serum albumen was not. Thus, results indicated that the lipid and protein moieties were both needed for arthritogenesis (Batsford et al., 2004).

Finally, the larger body size of rats compared to other laboratory rodents has facilitated implantation of dialysis membrane chambers containing cultivated *B. burgdorferi* into the peritoneal cavity (Akins et al., 1998). The differences in gene expression between *in vitro*-cultured *B. burgdorferi* and those growing inside of the living mammalian host are significant (Liang et al., 2002b; Iyer et al., 2015). A major challenge to studying *B. burgdorferi* gene expression and antigen production in a host environment is their broad tissue distribution and seclusion in immune privileged sites. For analyses of phenotype (genomic and proteomic), the substantial contamination with host tissue can complicate experimentation and interpretation. The development of this *in vivo* culture method utilizing dialysis membrane chambers implanted into the peritoneal cavities of rats has been a major stride in solving this problem (Akins et al., 1998; Caimano, 2005). Here, pure populations of host-adapted spirochetes can be acquired for study. By growing the spirochetes in dialysis bags (8-10 kDa molecular weight cut off), they are able to adapt to the biochemical environment of the host by exchange of small molecules, but are protected from the immune response. Multiple studies have shown that the spirochetes alter gene expression in this system and that the phenotype is consistent with host adaptation (Akins et al., 1998; Brooks et al., 2003; Bugrysheva et al., 2003).

Lyme disease in horses

Infection of horses with the Lyme disease pathogen is common in endemic regions and this can lead to disease presentations including neuroborreliosis (James et al., 2010; Imai et al., 2011), uveitis, and cutaneous pseudolymphoma (Divers et al., 2018). Given the known susceptibility, equine experimental work in horses has employed the usage of ponies (*Equus ferus caballus*) due to economic feasibility and their ease of handling. These studies have illustrated the efficacy and safety of an equine Lyme vaccine utilizing recombinant OspA. (Chang et al., 1999). Following vaccination and challenge with *B. burgdorferi*-infected adult ticks collected from a highly endemic region (i.e., Westchester County, New York), the absence of additional bands by Western blot analysis following exposure suggested a productive infection was not established in vaccinated animals. Additionally, skin biopsies taken near the site of tick bites over the course of 3 months post-challenge were culture negative in all vaccinated animals; three of four control ponies challenged by tick exposure were skin culture positive. Significant histopathologic findings were confined to unvaccinated tick-inoculated control ponies and included nonsuppurative synovitis, mild to moderate subsynovial perivascular lymphocytic aggregates, and perivascular aggregates in the fibrous layer of the joint capsule or in the periarticular adipose tissue. Perivascular mononuclear inflammation was also observed around small arteries adjacent to the perineurium of peripheral nerves and spinal nerve roots, with light perivascular cuffs in the thalamus. It is of note, that clinical signs were not observed in any of these animals (Chang et al., 1999).

In another study, ponies were tick-inoculated with *B. burgdorferi* while being treated with dexamethasone over 5 consecutive days. Aside from localized skin inflammation associated with tick bites, no significant histopathologic changes were observed in the study. To confirm infection, *B. burgdorferi* was detected by culture and PCR performed on monthly skin biopsies and various tissues collected at postmortem examination. Interestingly, one animal exposed to clean laboratory-reared nymphs not harboring *B. burgdorferi* three weeks prior to *B. burgdorferi* tick-inoculation was negative by culture, PCR, and serology, suggesting a role of tick immunity in the prevention of *B. burgdorferi* transmission in horses (Chang et al., 2000). These results are similar to that previously described in guinea pigs and mice (Wikel, 1996; Nazario et al., 1998).

The canine model

Dogs manifest many of the signs seen in human Lyme borreliosis, but many dogs that are naturally infected in endemic areas are seropositive and clinically normal (Levy and Magnarelli, 1992). In contrast, some dogs develop fatal kidney disease (Littman et al., 2006). The pathogenesis of canine Lyme nephropathy is poorly understood, but it has features of an immune-complex glomerulonephritis. It is not associated with the direct presence of spirochetes in the kidney, and is tragically refractory to palliative clinical intervention. Dogs are at greater risk of infection than humans in endemic areas, and serve as a sentinel species for the presence of *B. burgdorferi* in a region (Lindenmayer et al., 1991).

Although dogs are susceptible to infection by sub-adult ticks, they tend to be preferentially and more efficiently infected by adult female ticks (Anderson and Magnarelli, 1993; Appel et al., 1993; Korshus et al., 2004). Exposure of pregnant bitches to infected ticks resulted in infection of the dams, but not their fetuses or pups (Appel et al., 1993), but in a separate study, intrauterine transmission was documented (Gustafson et al., 1993).

Disease manifestation

Dogs develop disseminated, persistent infections, with a course that closely mimics human disease, including intermittent bouts of oligoarthritis that arise several weeks after infection. Unlike humans, erythema migrans is inconsistently observed in dogs (or at least is difficult to discern). Because of their physical size, dogs offer the opportunity to collect clinical samples, including blood, synovial fluid, skin biopsies, and CSF. All ages of dogs are susceptible to infection, but 6-week-old puppies were more susceptible to development of clinical disease (lameness), compared with older dogs. Clinical illness is silent in some dogs, but in one study, over 60% developed one to six episodes of transient (3 or 4 days) lameness, with reluctance to move and fever commencing at 2 to 5 months after infection. Lameness was sudden in onset, and most often involved the elbow or shoulder joint on the side where ticks were placed. The episodes of lameness resolved spontaneously without therapy. Affected joints had slight swelling with marked exudation of fibrin and neutrophils into the joint space. Spirochetes were detectable in multiple tissues for 1 year or more, based upon culture and PCR, and could be isolated from joint capsules of both lame and clinically normal dogs. Bacteremia was variably detected, and transient. As in humans, total IgG levels were increased for up to 17 months. Immunoblot profiles resembled natural infection, with initial reactivity to 39- and 41- kDa antigens (Burgess, 1986; Appel et al., 1993; Barthold et al., 1995b; Summers et al., 2005).

Histopathology from dogs at several months after tick-borne infection revealed multifocal non-suppurative meningitis, encephalitis, perineuritis, periarteritis, polyarthritis, and perivascular infiltrates in skin near the site of initial tick attachment. All of these lesions were characterized by infiltration of lymphocytes, plasma cells, and macrophages. Most dogs, regardless of clinical status, had some joints with focal aggregation of lymphocytes, plasma cells, neutrophils, and fewer numbers of macrophages and eosinophils. Unlike humans, lymphoid follicles with germinal centers were not found, and synovial villous proliferation was rare. Furthermore, the dogs developed enlarged lymph nodes, particularly those near the site of tick attachment, with hyperplasia dominated by numerous plasma cells (Chang et al., 2001; Summers et al., 2005). Corticosteroid treatment at 420 days of infection appeared to reactivate arthritis (Straubinger et al., 1998b). The neurological aspects of canine Lyme borreliosis have not been well characterized or compared to the human condition, nor has cardiac disease. A naturally occurring case of complete heart block and myocarditis has been documented in a dog (Levy and Duray, 1988). Dogs would lend themselves well to electrocardiographic studies.

When dogs were infected by tick, spirochetes disseminated locally within tissues. Detection of spirochetes by culture and PCR was most apparent in tissues within the somatic quadrant of the site of tick attachment. Likewise, lymph node enlargement was greatest in the gland nearest to the tick site. Spirochetes were most apt to be detected in collagen-rich tissues, with synovia, fascia, and lymph nodes closest to the tick bite most often infected. Quantitative PCR of skin biopsies of infected dogs have indicated a significant drop in spirochete numbers that correlated with rising antibody levels (Straubinger et al., 1997b; Straubinger et al., 2000). Synovial fluids from dogs during acute lameness had elevated volume, cell count, cell viability, higher proportion of neutrophils, and chemotactic activity for neutrophils. At necropsy, IL-8 was found to be expressed in synovia, pericardium, and peritoneum, and TNF was elevated in enlarged lymph nodes draining affected joints. These indices were lower or absent during clinical silence or in joints with mild inflammation (Straubinger et al., 1997a). Canine synovial explants exposed to *B. burgdorferi* have also revealed up-regulation of pro-inflammatory cytokines, including IL-1 alpha, IL-1 beta, TNF and IL-8 in response to *B. burgdorferi* (Straubinger et al., 1998a).

As a species of veterinary importance, as well as for modeling the human condition, dogs have been utilized to demonstrate efficacy of vaccination and antibiotic treatment in a number of trials. Vaccination has been shown to be effective against tick-borne and syringe challenge using various formulations of whole-cell bacterins and recombinant OspA/B proteins (Chu et al., 1992; Chang et al., 1995; Coughlin et al., 1995; Straubinger et al., 1995; Topfer and Straubinger, 2007).

The rabbit model

Rabbits, also referred to as lagomorphs, were the first animal to be inoculated with the known cause of Lyme disease. In a seminal publication published in 1982, Burgdorfer and colleagues demonstrated that laboratory rabbits could be infected with treponema-like spirochetes (later to be named *B. burgdorferi*) following feeding by adult *Ixodes dammini* (later renamed *I. scapularis*) (Burgdorfer et al., 1982). Furthermore, the rabbits developed skin lesions reminiscent of EM. Other groups attempted to infect rabbits, yielding inconsistent results (Krinsky et al., 1982; Benach et al., 1984; Kornblatt et al., 1984a). In one early study, an EM-like lesion was shown to contain aggregates of lymphoid cells, and spirochetes were visualized with silver stain (Kornblatt et al., 1984a). These early efforts also confirmed bacteremia in rabbits (Benach et al., 1984; Kornblatt et al., 1984a; Kornblatt et al., 1984b), and acquisition of infection by ticks following feeding upon experimentally infected rabbits (Burgdorfer, 1984). Intradermal inoculation with high doses (10^6) of cultured spirochetes resulted in a high rate of PCR-reactivity in the CNS and visceral organs, but tissues were culture-negative. Following high dose (10^8) intraperitoneal inoculation of infant Dutch rabbits, spirochetes could be cultured from multiple organs and the rabbits developed carditis and arthritis at 30 days after inoculation (Moody et al., 1990a).

The rabbit model came into its own with the reproducible development of EM lesions following intradermal inoculation of adult New Zealand White rabbits with as few as 10^3 *B. burgdorferi* B31 spirochetes, or by tick-borne infection (Foley et al., 1995a). Lesions arose within 3-8 days, expanded to 7.5-79 mm diameter, and remained visible for 3-7 days, and then regressed. Similar lesions could be induced with a number of *B. burgdorferi* s.s. isolates. EM lesions were similar to EM in humans, with dermal aggregates of lymphocytes and plasma cells around adnexae, vessels and small nerves, appearing as inflammatory cuffing. In addition, spirochetes could be cultured from multiple organs, including spinal cord, but not cerebral cortex or CSF, for up to 8 weeks, but not thereafter, indicating that the spirochetes do not naturally persist in this model (Foley et al., 1995b). The EM lesions were associated with the presence of extracellular spirochetes in the dermal collagen by immunohistochemistry. Spirochetes remain in the skin following regression of lesions, but are eventually cleared (Chong-Cerrillo et al., 2001). The reproducibility of EM induction made possible the demonstration of resistance to reinfection with high challenge doses of the homologous *B. burgdorferi* strain (Foley et al., 1995b; Foley et al., 1997; Shang et al., 2000), but not with a heterologous strain (Shang et al., 2001). Similar results occurred when rabbits were challenged with host-adapted spirochetes, introduced by implantation of infected skin biopsies (Shang et al., 2000; Shang et al., 2001). As with rodent models, rabbits can be protected against infection (and development of EM) when immune rabbit serum is administered at multiple points before and after (-18, 0, +24, +48, +96 and +216 hours) challenge, but not when administered only after (+24, +48, +96, and +216 hours) challenge inoculation (Chong-Cerillo et al., 2000). The rabbit model has also been used to demonstrate efficacy of vaccination with recombinant OspA, outer membrane vesicles, and outer membrane protein Oms66 (P66) (Exner et al., 2000; Shang et al., 2000).

The rabbit model has been extremely valuable in understanding host-dependent protein expression by *B. burgdorferi*. The skin, which contains (albeit transiently) high numbers of spirochetes, has been used to examine expression of hydrophobic proteins between 5 and 21 days after infection, with maximal numbers of spirochetes at day 11. Using a method to extract and measure hydrophobic antigens from tissues, significant changes in the relative expression of *B. burgdorferi* proteins were found, including loss of detectable OspC by day 9, and increases in VlsE and other proteins (BmpA, OppA-2, DbpA, etc.) from days 7 through 21 (Crowley et al., 2004).

Multiple studies indicate that rabbits, unlike other species, are able to clear infection by 8 weeks. Unlike rodents, dogs, monkeys, and humans, which develop strong and early antibody responses to VlsE invariable region IR6, rabbits mount early antibody to the carboxyl-terminal domain of VlsE, but a very low or no response to IR6 during the first 6 to 11 weeks. Although *vlsE* gene sequence recombination occurs during infection of rabbits, a significant number of *B. burgdorferi* isolates (25 of 50 clones) obtained from rabbit skin were found to be devoid of lp-28, the B31 plasmid that encodes VlsE (among other genes). This rate of loss has not been observed in mice. Despite these findings, a few rabbits were found to seroconvert with high titers of antibody to IR6 after several weeks of infection, oscillation in antibody titers to the C-terminal region of VlsE, and the culture of spirochetes from one rabbit at 98 weeks (Embers et al., 2007). Interestingly, a recent study demonstrated that spirochetes retain the *vls* system, and that despite functionality of this antigenic variation system, the spirochetes fail to evade anti-*Borrelia* antibodies of NZW rabbits and do not persist (Batool et al., 2019). Thus, some rabbits, like other species, can be persistently infected, but most rabbits seem to mount an effective sterilizing immune response, which may reflect the unique immunophysiology of lagomorphs. The comparative

Figure 11. Erythematous lesions in primates infected with *B. burgdorferi* by tick bite. An expanding red erythema migrans is shown in the left panel and diffuse erythema is shown on the right. Arrows indicate the site of the initial tick bite and subsequent skin biopsy locations.

nuances of the rabbit immune response are well worth further investigation, especially as they may offer protective antibody responses.

Nonhuman primate (NHP) models

The variable and multi-organ disease caused by *B. burgdorferi* and the host responses to infection in rhesus macaques (*Macaca mulatta*) parallel human Lyme disease the closest of the animal models. The rhesus macaque taxonomic proximity with humans also favors this model, and this primate species is the one in which the majority of research has been conducted. Rhesus macaques exhibit signs that are characteristic of all three phases of human Lyme borreliosis, namely early localized, early disseminated and late-stage disease. Studies using this model have employed both Chinese and Indian origin rhesus macaques, several different strains of *B. burgdorferi* s.s., infection by needle or tick inoculation, coupled with the use of immunosuppressive drugs. Neuroborreliosis is an important aspect of this animal model, as neurological lesions are minimal or absent in rodents and lagomorphs.

Multi-organ character of Lyme borreliosis

Some variation has been observed when macaques are infected with different strains (B31, N40 and JD1), yet one common finding is that multiple organs become affected, and variation between individual animals exists. Spirochetes disseminate to multiple organs, including the central nervous system (cerebrum, brainstem and cerebellum, spinal cord, and dura mater) (Roberts et al., 1995; Cadavid et al., 2000b).

Early localized infection in humans is often characterized by erythema migrans (EM) in an estimated 70–80% of human cases (Steere and Sikand, 2003), yet the presentation can vary considerably (Berger, 1993). The proportion of macaques who produce EM following tick transmission of *B. burgdorferi* may vary with the strain, but typical and atypical features have been described, including diffuse erythema or a circular patch (Philipp et al., 1993). Images of an EM and small erythematous lesion on rhesus macaques are shown in Figure 11. Following tick-borne infection with *B. burgdorferi*, cultures of EM biopsies are almost always positive (Philipp et al., 1993; Embers et al., 2017). As with EM in humans (Berger, 1989), lesions in NHP feature monocytic and lymphocytic infiltrates of the superficial perivascular areas as well as the deep dermal adipose tissues and perivascular spaces within muscle (Philipp et al., 1993). Signs of early-disseminated disease included secondary EM (Philipp et al., 1993), lymphadenopathy (Embers et al., 2017), splenomegaly (Philipp et al., 1993; Roberts et al., 1995), fever (Roberts et al., 1995), conjunctivitis (Philipp et al., 1993) (a rare manifestation of human Lyme borreliosis) (Lesser, 1995), and cerebrospinal fluid (CSF) pleocytosis (with evidence of spirochetal DNA) (Pachner et al., 1995b).

Arthritis and carditis, signs of late Lyme borreliosis in humans, also occur in rhesus macaques, both with or without immunosuppression (Roberts et al., 1995; Cadavid et al., 2004; Crossland et al., 2018). In a study that involved five animals tick-inoculated with *B. burgdorferi* strain JD1 and one uninfected control, the joints most commonly affected were the knee and elbow joints, as with humans (Steere et al., 2004). Lesions included synovial hyperplasia and mild periarticular fibrosis. In severely affected joints, articular cartilage was marginally eroded, and in some cases pannus formation was observed. Synovial hyperplasia was accompanied by perisynovial fibrosis and perivascular accumulation of both B and T (primarily CD4) lymphocytes, plasma cells, and macrophages. The leukocyte subsets are similar to those described for human Lyme borreliosis (Steere et al., 1988). Evidence of spirochetal presence in synovial tissue was obtained by culture from joint tissue fragments (Roberts et al., 1995). In macaques persistently infected for a period of 12-14 months, three of ten animals exhibited minimal to mild lymphoplasmacytic inflammation that affected joint-associated structures in the stifle (knees), independent of prior antibiotic treatment (Crossland et al., 2018). Inflammation cuffing the vasculature (Figure 12B) was present within the synovium (Figure 12A) and perisynovial adipose tissue. These animals also had concurrent synovial epithelial hyperplasia ranging from minimal to mild, and spirochetes were identified within tissue sections (Figure 12C).

Lyme carditis has been reported in up to 10% of infected adults and up to 30% of children (Yeung and Baranchuk, 2019). The most common manifestation is atrioventricular block (AVB), where high degree AVB is seen in 80-90% of Lyme carditis patients, though varying degrees of AVB are associated with Lyme carditis. The pathophysiology of AVB involves myocardial invasion by the spirochetes and subsequent inflammatory foci which may be found both in the interstitium and in the pericardium. In an autopsy study of five patients who suffered sudden cardiac death from Lyme disease, the histopathology in all cases reflected a roadmap pattern of interstitial inflammatory infiltrates extending from the endocardium to the epicardium (Muehlenbachs et al., 2016). The infiltrate had a prominent perivascular distribution, and spirochetes were identified within the myocardial interstitial infiltrates, in the sub-endocardium, and occasionally in pericardial tissue.

In a study that involved 14 immune competent and 11 immune suppressed macaques that received a needle inoculum of *B. burgdorferi* strain N40, all but one of the animals developed carditis (Cadavid et al., 2004). Increased numbers of organisms and inflammatory cell infiltrates were observed in animals that were immunosuppressed with dexamethasone (2 mg/kg, i.m.,q.i.d.) either transiently, from days 7 to 28 post-inoculation, or for the duration of the experiment (Bai et al., 2004; Cadavid et al., 2004). Leukocytic infiltrates were composed primarily of plasma cells, monocytes and T-lymphocytes. In addition, carditis was evident in three of five rhesus macaques that were infected by tick-borne *B. burgdorferi* strain JD1 (Roberts et al., 1995). In a more recent study, the microscopic disease and presence of spirochetes was interrogated in 10 rhesus macaques infected with strain B31 by tick bite

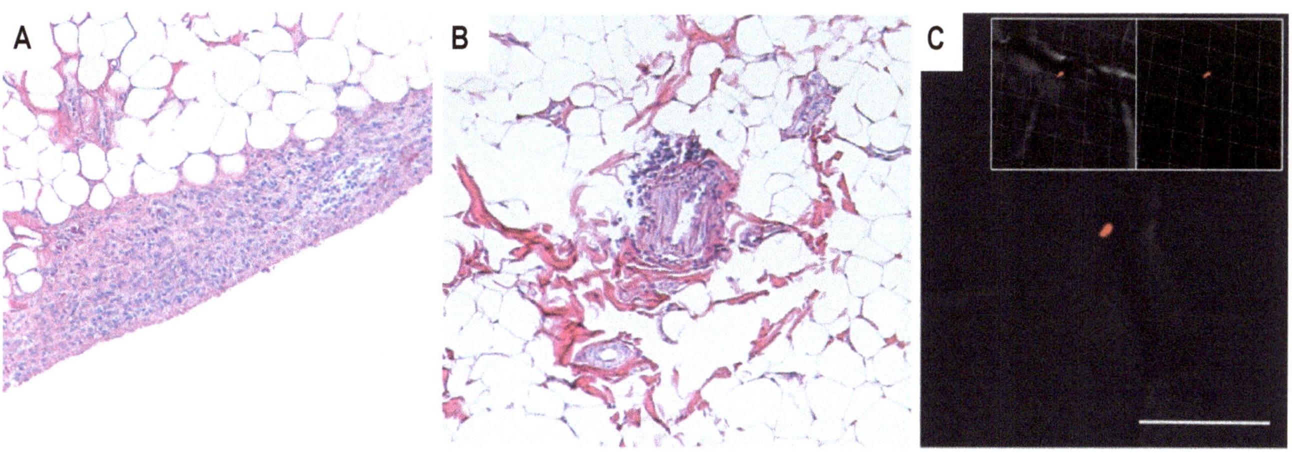

Figure 12. Pathology and *B. burgdorferi* detection in rhesus macaques infected with *B. burgdorferi*. A: Stifle (knee) synovium with mild lymphoplasmacytic inflammation infiltrating hyperplastic synovial epithelium. B: Stifle perisynovial adipose tissue with minimal localized lymphoplasmacytic perivascular cuffing. C: Immunofluorescent staining for *B. burgdorferi* shows localized red fluorescence in perisynovial adipose tissue. Three dimensional imaging in inset shows cylindrical shape of the structure, indicating that this is a cross section of the spirochete; 1 unit Z 6.17 mm. DIC image applied to enhance contrast (Crossland, et al. 2018, copyright permission-Elsevier).

(Embers et al., 2017; Crossland et al., 2018). The inflammatory pathology as described in human cases was very similar in some animals, albeit milder. Six of ten animals displayed mild myocardial inflammation, while one animal showed moderate pericarditis. As in the human cases, inflammation was not associated with histologic evidence of myocardial degeneration, necrosis, and/or fibrosis. Figure 13 shows inflammation observed in the interstitium and pericardium of macaques, along with a spirochete in the cardiac tissue identified by immunofluorescent antibody staining.

The possible correlation between Lyme carditis and heart function has not been studied in rhesus macaques. Continuous telemetric monitoring of heart function, a technique that has been used in studies of relapsing fever, is available for large animals and could be implemented for this purpose (Lopez et al., 2014; Markert et al., 2018).

Neuroborreliosis in primates.

Nervous system involvement in Lyme disease is known as Lyme neuroborreliosis (LNB), which manifests in approximately 15% of Lyme patients and may affect both the central and peripheral nervous systems (Halperin, 2019). Central nervous system (CNS) involvement may result in symptoms such as headache, fatigue, memory loss, learning disability, or depression. LNB affecting the peripheral nervous system (PNS) may result in facial nerve palsy, limb pain, sensory loss, and/or muscle weakness (Halperin, 2015; Marques, 2015). Neurogenic pain along the back, radiating into the legs and foot, accompanied by weakness, numbness, and tingling in the legs, attributed to radiculitis and/or radiculoneuritis, is the most common manifestation of peripheral LNB (Garcia-Monco and Benach, 2019; Rauer et al., 2020). Involvement of cranial or peripheral nerves occurs in approximately 10% of infected but untreated individuals and typically manifests quite early in infection (Murphree Bacon, 2008; Halperin, 2015). Though little human pathological data exist, lesions from human patients affected by Lyme disease have been characterized by inflammation of the nerve roots and/or dorsal root ganglia and multifocal axonal loss accompanied with epineural perivascular infiltrates or perineuritis (Halperin, 2015, 2017).

As mentioned, nonhuman primates are ideal models for LNB. Peripheral neuroborreliosis has been evaluated both functionally and histologically in rhesus macaques. In one study, nerve conduction was tested on eight animals inoculated with *B. burgdorferi* strain JD1 (five via tick bite, and three via injection) (England et al., 1997b). Motor and sensory nerve conduction studies (NCS) and conduction velocities were recorded on multiple nerves. Abnormal NCS values were further analyzed for evidence of axonal loss or demyelination. Three of the infected monkeys had normal NCS values. The remaining five infected animals had a mono-neuropathy or mononeuropathy multiplex nerve

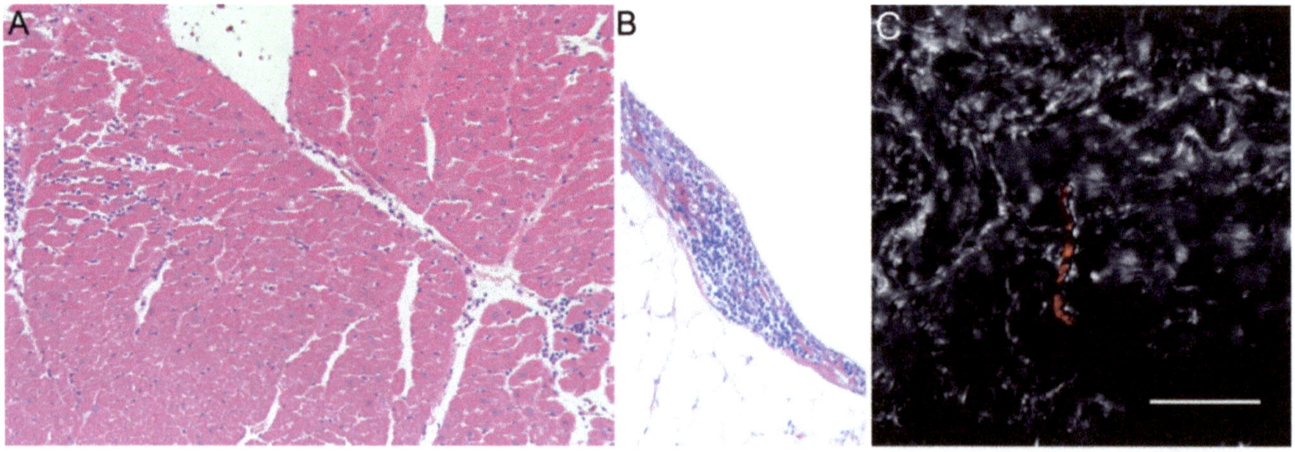

Figure 13. Histologic characterization and localization of *Borrelia burgdorferi* within heart tissue of rhesus macaques with late disseminated Lyme disease. A: Mild multifocal lymphoplasmacytic interstitial inflammation. B: Visceral pericardium with localized mild lymphoplasmacytic inflammation. C: Red fluorescence of a morphologically intact spirochete within myocardial interstitium; differential interference contrast (DIC) image applied to enhance contrast (Crossland, et al., 2018, copyright permission-Elsevier).

lesion pattern that was consistent with neuronal loss. Concentric needle electromyographic examinations performed on selected muscles showed confirmatory evidence of denervation, concordant with the distribution of nerve lesions in each affected monkey. Altogether, five of the infected monkeys demonstrated primarily axonal-loss-variety multifocal neuropathies. Pathologically, peripheral nerve specimens showed multifocal axonal degeneration and regeneration and occasional perivascular inflammatory cellular infiltrates without vessel wall necrosis (England et al., 1997b). In a study of five animals that had been infected by tick-borne *B. burgdorferi* strain JD1 for 3 months, the most prominent and consistent neurological manifestation observed was neuritis (Roberts et al., 1998). The neuritis was characterized by intraneural perivascular infiltration of mononuclear cells, including macrophages and B lymphocytes, nut not T cells.

Evidence for the presence of *B. burgdorferi* in the PNS was obtained in two separate studies. The earliest involved seven strain N40-infected rhesus macaques, three of which three were immune competent and four were immune suppressed. PCR was used to amplify the *ospA* gene to detect spirochete DNA present in PNS tissues (Cadavid et al., 2000a). In a more recent study involving 10 rhesus macaques infected with strain B31 by tick bite, peripheral nerve inflammation was observed in five of nine animals, characterized by minimal to moderate lymphoplasmacytic inflammation with a predilection for collagen-rich epineurium and perivascular spaces. Spirochetes were directly identified by immunofluorescent staining using polyclonal and monoclonal *Borrelia*-specific antibodies (Crossland et al., 2018).

Neuroborreliosis of the central nervous system

Patients with LNB often exhibit the neurologic triad of meningitis, cranial neuritis, and radiculoneuritis, commonly described as meningoradiculitis (Reik et al., 1979; Pachner and Steere, 1985). Lyme meningitis presents primarily as leptomeningitis, characterized by lymphocytic pleocytosis in the cerebrospinal fluid (CSF) (Garcia-Monco and Benach, 2019). LNB patients may experience encephalopathy, encephalitis, and encephalomyelitis concomitant with white matter inflammation in the brain and spinal cord (Garcia-Monco and Benach, 1995; Halperin, 2015; Garcia-Monco and Benach, 2019). Furthermore, cognitive impairment, behavioral changes in children, and depression can all result from *B. burgdorferi* infection affecting the brain (Bloom et al., 1998; Keilp et al., 2019).

Evidence that *B. burgdorferi* accesses the CNS of rhesus macaques was first provided by Pachner and co-workers (Pachner et al., 1995c). Their study included five animals that were given intradermal inoculations of *B. burgdorferi* strain N40. One of these animals was immune-compromised by dexamethasone injections. CNS invasion was documented by both culture and PCR of CSF specimens. Pleocytosis and spirochetal DNA were detected in the CSF of all of the animals, and two of the animals, one of which was immune-compromised, were CSF culture positive. This latter animal also showed evidence of CNS inflammation by magnetic resonance imaging. An important finding from this study was that both spirochetal DNA and pleocytosis were intermittently detected for as long as 18 weeks post-inoculation in all of the immune competent animals (Pachner et al., 1995c). This is an indication that *B. burgdorferi* infection may persist in the CNS for an extended time.

In a study designed to investigate *B. burgdorferi* localization in the CNS and in other organs, infected animals were immune-suppressed. Spirochetes were detected immunohistochemically chiefly in the leptomeninges, nerve roots, and dorsal root ganglia. By *ospA* PCR, spirochetal DNA was detected in the cerebrum, brainstem and cerebellum, spinal cord and dura mater, not only in immune-suppressed but also in immune competent macaques, albeit in much lower amounts (Cadavid et al., 2000a). Spirochetal DNA also has been detected in spinal leptomeninges, spinal roots, and dorsal root ganglia of cervical, thoracic, and lumbosacral spinal cord tissues, but only minimally in the spinal cord parenchyma of immunosuppressed animals (Pachner et al., 2001).

Inflammation related to the presence of *B. burgdorferi* occurs in the rhesus CNS, not only in the meninges, with associated pleocytosis, but also in the brain (Pachner et al., 2001; Crossland et al., 2018). One inflammatory mediator that was found to be consistently upregulated in skeletal muscle tissue of chronically-infected macaques is the B-cell chemokine CXCL13 (Pachner et al., 2002b), which also was detected by immunohistochemistry in the brain and spinal cord of *B. burgdorferi*-infected rhesus macaques (Narayan et al., 2005). Specifically, CXCL-13 was found to be produced by glial cells in response to *B. burgdorferi,* evidenced by an *ex vivo* brain explant model (Ramesh et al., 2008). In humans, CXCL13 has been found to be upregulated in the CSF of patients with neuroborreliosis and other CNS inflammatory diseases, but not in patients with non-inflammatory neurological diseases, suggesting its value as a diagnostic biomarker (Henningsson et al., 2018; Karrasch et al., 2018; Rupprecht et al., 2018).

Studies aimed at examining the pathophysiology and potential mitigation of CNS disease in LNB have utilized direct intrathecal inoculation of primates with *B. burgdorferi* (Ramesh et al., 2015; Martinez et al., 2017) One such study on 14 rhesus macaques compared the effect of treatment with anti-inflammatory drugs (i.e., dexamethasone and meloxicam) on indicators of inflammation, neurodegeneration, and demyelination following intrathecal inoculation of $1x10^8$ *B. burgdorferi* strain B31. CSF from untreated and meloxicam-treated animals displayed a greater lymphocytic and monocytic pleocytosis when compared to dexamethasone-treated animals, while specimens from sham-inoculated animals were nearly devoid of inflammatory cells. Multiple inflammatory cytokines and chemokines were investigated for their levels in CSF and serum amongst the different treatment groups; these included IL-6, IL-8, CCL2, and CXCL13. CCL2 (also known an monocyte chemoattractant protein-1 or MCP-1) and IL-8 are known to mediate inflammatory cell influx in the CNS compartment (Bose and Cho, 2013). IL-8 is a neutrophil and T-cell chemoattractant (Harada et al., 1994) and CCL2 attracts monocytes and T-cells and modulates monocyte immune functions (Bose and Cho, 2013) CXCL13, discussed previously, is the primary B-cell attractant into the CSF during neuroinflammation (Irani 2016). IL-6 is a pleiotropic cytokine that influences antigen-specific immune responses and inflammatory reactions (Kishimoto, 2006). Interestingly, both CCL2/MCP-1 and CXCL13 are also involved in the induction of neuropathic pain (White et al., 2009; Zhang et al., 2017). Following infection, CSF showed significantly elevated levels of IL-6, IL-8, CCL2, and CXCL13 and pleocytosis in all infected animals, except those treated with dexamethasone (Ramesh et al., 2015). Similarly, serum mediators CCL2, CXCL13, and IL-8 were elevated in untreated animals when compared to dexamethasone treated, but with later onset of peak values (approximately 4 weeks compared to 1-3 weeks for CSF). Treatment with meloxicam significantly diminished the concentrations of serum CXCL13 and CD40L levels, but retained other previously mentioned inflammatory cytokines described for untreated animals. Differential histopathologic findings among treatment groups revealed the absence of lymphocytic inflammatory lesions in the brain and spinal cord of dexamethasone and sham-inoculated controls. Consistent findings in untreated and meloxicam-treated animals included lymphocytic leptomeningitis in the brain and spinal cord with perivascular tropism, lymphocytic neuritis and ganglionitis of cranial nerves, and radiculoneuritis affecting dorsal root ganglia and both sensory and motor spinal nerve roots. The aforementioned lesions were not observed in either the dexamethasone or untreated sham-inoculated animals. Demyelination also was observed in the spinal nerve roots of all infected rhesus macaques except those treated with dexamethasone and sham-inoculated. Apoptosis was significantly reduced in both neurons and sustentacular satellite glial cells within the dorsal root ganglion (DRG) in dexamethasone-treated animals when compared to untreated and meloxicam groups. Furthermore, *B. burgdorferi*-specific antigen colocalized with areas of inflammation observed in the untreated and meloxicam-treated animals, supporting the role of *B. burgdorferi* in driving host inflammation. Dexamethasone treatment was thus effective in curbing inflammation and neuro-degeneration in the DRG and demyelination in the spinal nerve roots. In addition, dexamethasone significantly reduced the levels of Bb-induced neuronal and glial cell apoptosis in the DRG.

To more closely mimic natural infection and disease in humans, macaques have been infected by tick bite and evaluated at 12-14 months post-infection for pathologic signs of chronic disease (Embers et al., 2017; Crossland et al., 2018). Minimal and focal lymphoplasmacytic inflammation of the central nervous system was observed in two animals, both of which were treated with the antibiotic doxycycline. In one monkey, inflammation was observed in the leptomeninges overlying a section of temporal cerebral cortex (Figure 14A), whereas inflammation in the other macaque was characterized by perivascular inflammation of the brainstem adjacent to the fourth ventricle (Figure 14B). Green fluorescent *B. burgdorferi*-specific staining was detected free within the neuropil of both these animals (Figure 14 C and D) and was not directly associated with the minimal inflammation observed histologically. Of all the tissue sections examined in this study, the

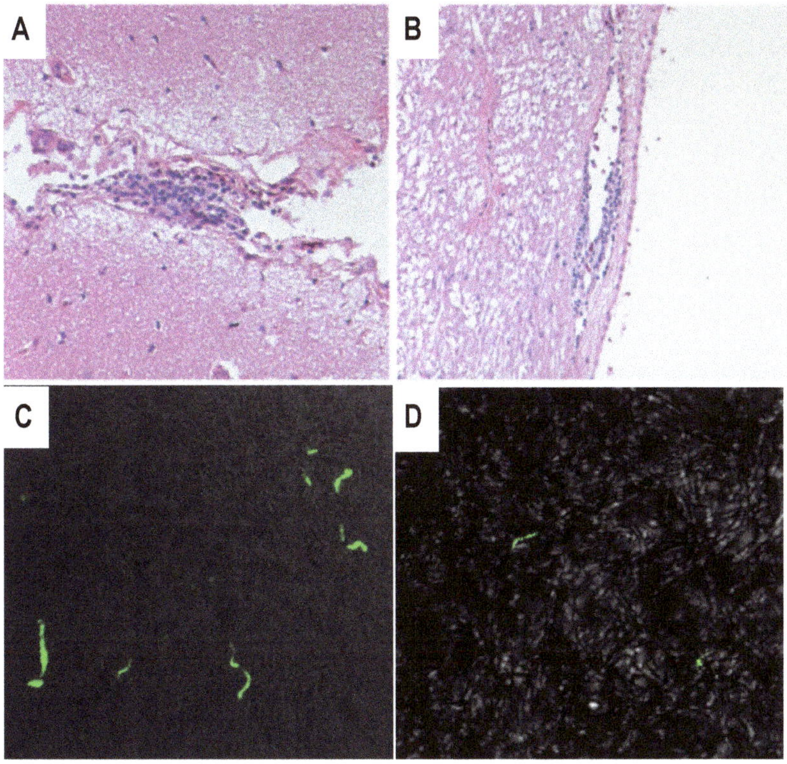

Figure 14. Inflammation and spirochetes in the CNS of chronically infected rhesus macaques. A: Cerebrum (doxycycline treated animal); leptomeninges are focally expanded by minimal localized lymphoplasmacytic inflammation. B: Brainstem (another doxycycline treated animal); minimal localized lymphoplasmacytic perivascular cuffing adjacent to the fourth ventricle. C: Green fluorescence shows six variably sized morphologically intact spirochetes within the cerebral neuropil. D: Green fluorescence reveals two morphologically intact spirochetes within the brainstem neuropil; differential interference contrast (DIC) applied to enhance contrast (Crossland, et al., 2018, copyright permission-Elsevier).

greatest density of morphologically intact spirochetes was observed in the cerebrum of the doxycycline-treated monkey showing leptomeningeal inflammation, with up to six spirochetes observed in a single high powered field (Figure 14C).

Collectively, these findings support the intrinsic role of inflammation, which is directly attributable to *B. burgdorferi* antigens, in the pathogenesis of LNB through induction of inflammatory mediators in glial and neuronal cells. Consistent with this notion, dexamethasone-treated animals displayed reduced CSF pleocytosis, inflammatory mediators in the CSF and serum, neuronal and glial cell apoptosis in the DRG, and inflammatory, neurodegenerative, and demyelinating lesions in the nervous system as a whole (Ramesh et al., 2015).However, it should be noted that inflammation in tissues is not always co-localized with intact spirochetes and vice versa.

Positron Emission Tomography (PET) imaging has been used in patients with chronic symptoms associated with Lyme disease to visualize inflammation in the CNS. (Coughlin et al., 2018). The diagnostic capability of PET imaging, especially for disease resolution, could similarly be evaluated in experimentally infected primates. Finally, cognitive impairment, behavioral changes in children, and depression can all result from *B. burgdorferi* infection affecting the brain (Bloom et al., 1998; Bransfield, 2018; Keilp et al., 2019). While this is difficult to assess in nonhuman primates, cognitive tests, telemetric measuring of physical activity, and behavioral monitoring could be used to improve the macaque model of LNB and offer insight into the inflammatory and biochemical changes in the brain associated with disease.

Preclinical assessment of vaccines

While initial experimental assessment of candidate vaccines is usually done in rodents, inbred and genetically modified mice often produce atypical responses to specific infections (Jameson and Masopust, 2018), and are inappropriate to study infection or vaccine responses to a variety of important human and veterinary infectious diseases (Griffin, 2002). The principal requirement for the use of an experimental animal model to test vaccines is authentication in that the experimental infection mimics the infection and disease pattern found in nature. For Lyme disease, preliminary testing in mice is valuable, but predictive immunogenicity and efficacy for a human vaccine are more accurately determined in primates. Safety, immunogenicity, and efficacy of outer surface protein A (OspA) vaccines were examined in rhesus macaques (Philipp et al., 1996; Philipp et al., 1997). The vaccines that were tested included one with the same formulation as that subsequently used in humans, namely, lipidated OspA combined with aluminum hydroxide (alum), as well as formulations that included non-lipidated OspA with and without the adjuvant monophosphoryl lipid A (Philipp et al., 1996). A multi-antigen vaccine, consisting of a sonicated preparation of *B. burgdorferi* strain N40, was also tested in rhesus macaques (Pachner et al., 1999). In this study the primary focus was on the prevention of disease, with emphasis on neuroborreliosis. In comparison to control animals, vaccinated animals did not show evidence for productive infection and invasion of the brain (Pachner et al., 1999).

Following the demise of the original OspA vaccine, several other vaccination strategies against Lyme disease have been tested (Embers and Narasimhan, 2013). Targeting the pathogen by transmission blockade (as with OspA) or targeting tick proteins themselves, the preclinical studies will ultimately be well-served by utilizing nonhuman primates, given their similarities to humans in immune response to tick-feeding and infection, as compared to rodents (Pachner et al., 2002a; Wikel, 2013).

Diagnostic test development

Animal models are also important for the development of diagnostic tests by their ability to generate clear "infected" and "uninfected" samples. The advantages of the nonhuman primate model for diagnostic test research are: (1) compared to mice, the disease course, including duration and quantity of *B. burgdorferi* in the blood is more similar to that of humans; (2) infection by the natural (tick) route will accurately represent antigens produced during early infection, including variable host responses (Embers et al., 2017); and (3) compared to human samples, the infection history (e.g., exact point of exposure) is known, allowing for the assessment of changes in the response over time (Embers et al., 2012b). The sensitivity of current serologic testing for Lyme disease varies by stage of disease, and is below 50% for early Lyme disease (Branda et al., 2017). Thus, interest in developing more sensitive serological tests and improved direct detection of Lyme disease spirochetes in blood or urine has grown, and several such tests are between the development and product stage (Schutzer et al., 2018).

Discovery of the antigenicity of the C6 peptide and development of the C6 ELISA represented a significant advance in Lyme disease serology (Liang et al., 1999a; Liang et al., 1999b). This test is based on the detection of serum antibodies to a peptide (C6) composed of a conserved amino acid sequence of a small fragment of VlsE, the antigenic variation protein of *B. burgdorferi* (Liang et al., 1999a). Three key features of the test were first described from rhesus macaques infected with *B. burgdorferi.* First, the ability of the test to detect antibody in serum early in the infection was precisely determined using serial samples (Liang et al., 2000). The absence of reactivity of the C6 ELISA with antibodies elicited by the OspA vaccine also was first uncovered with rhesus monkeys that had been vaccinated with the OspA vaccine used in humans (Liang et al., 1999b). The marked decline in C6 antibody titer that follows antibiotic treatment of patients with early Lyme borreliosis is the third feature of the C6 test that was first revealed with the rhesus model (Philipp et al., 2001). A longitudinal analysis of serum antibody levels revealed that C6 antibody declined at a much faster rate after treatment than did antibody to *B. burgdorferi* whole-cell antigen (standard Lyme ELISA).

The changes in longitudinal IgG responses during *Borrelia burgdorferi* infection and antibiotic treatment of rhesus macaques have been investigated by ELISAs utilizing recombinant antigens, including OspC, DbpA, BBK32, OspA, and OppA-2 (Embers et al., 2012b). Antibody profiles were derived from five macaques followed for 11 months subsequent to needle inoculation by *in vitro* cultured *B. burgdorferi* strain B31; two were untreated and three were treated with antibiotics. These specimens were further supplemented with 12 other serum samples from needle-inoculated *B. burgdorferi* strain JD1 infected and treated macaques as a means to assess responses to antibiotic administration. All NHP with either *B. burgdorferi* strain produced high level, sustained antibody responses to DbpA. This antigen has been shown to reflect a T-cell independent antigen in mice (Barthold et al., 2006). BBK32, an extracellular matrix binding protein, produced a more variable response and exhibited a lower potential for sensitivity. OspA antibodies have been associated with arthritis during disseminated infection and elevated in patients with post-treatment Lyme disease syndrome (Akin et al., 1999; Chandra et al., 2011). However, tick-infected animals did not generate anti-OspA responses, as expected, given its downregulation during the bloodmeal and transmission to the host. Anti-OspC antibodies arose early and steadily declined as expected as this antigen's expression is known to shut off after the advent of the humoral response (Liang et al., 2002a). Broad reactivity of macaques to OppA-2 with a slow and incomplete post-treatment decline, support the utility of this antigen as a diagnostic tool for evaluating serum before and after treatment. This was the first report of OppA-2 antigenicity in NHPs. A decline in response to the C6 peptide following antibiotic treatment was observed. Subsequent to this study, a five-antigen assay was developed and shown to be more sensitive than the two-tier test with human patient samples (Embers et al., 2016).

Finally, macaques have been utilized to identify potential *Borrelia* antigens that could be shed in the blood and thus used for direct detection of the pathogen. This technique, called *in vivo* Microbial Antigen Discovery (InMAD) utilized injection of mice with blood from primates at early time points of infection (Pflughoeft et al., 2019). By combining the antibodies generated with proteomic methods that include microbial biomarker enrichment, secreted proteins circulating during infection were identified. Using this strategy, six putative biomarkers from multiple samples were characterized. By integrating biomarkers identified from early infection with temporal patterns of expression, this platform allows for the data-driven selection of diagnostic targets for *B. burgdorferi* and other infections with a blood-borne phase.

Assessment of antimicrobial therapy

Lyme disease treated early with antimicrobial therapy is generally effective, but some patients, especially those whose treatment is delayed, may continue to have persistent signs such as multiple EM and persistent arthritis, or (more frequently) symptoms such as arthralgia, musculoskeletal pain, sensory disturbances, neuralgia, neurocognitive symptoms, and fatigue (Asch et al., 1994; Shadick et al., 1994). Due to the subjective nature of these symptoms, they are not easily measured in animal models and it remains unknown if similar symptoms affect animals, including NHPs. Persistent symptoms observed in humans are often referred to as post-treatment Lyme disease syndrome (PTLDS). The cause of PTLDS is currently unknown but long term antibiotic therapy does not seem to be helpful in terms of eradicating all of the symptoms (Klempner et al., 2001; Klempner, 2002), though repeated ceftriaxone treatment has shown some neurological benefit in patients (Fallon et al., 2008; Delong et al., 2012).

Continued symptomatology following antimicrobial therapy of Lyme disease may be attributed to several causes, including (1) spirochetes that persist in the tissues, tolerant to the antibiotic used; (2) inflammatory responses to residual antigens from dead organisms; or (3) autoimmune responses, possibly elicited by antigenic mimicry (Bolz and Weis, 2004). The etiology may be influenced by the duration of infection, genetic factors in the host, and a combination of the aforementioned causes (Please see chapter 23).

Mice

The mouse model has been utilized in several studies of antibiotic efficacy for treatment of early and established *B. burgdorferi* infections. Based upon culture of tissues after antibiotic treatment of infected mice, it has been shown that amoxicillin, ceftriaxone, and penicillin G could eliminate infection and disease, whereas oxytetracycline, doxycycline, chloramphenicol, erythromycin and azithromycin failed to cure infected mice (Malawista et al., 1994; Moody et al., 1994a; Kazragis et al., 1996; Konishi and Nakao, 1997; Pavia et al., 2002; Yrjänäinen et al., 2007) Doxycycline and vancomycin have been shown to be effective at preventing early infection (Kazragis et al., 1996; Zeidner et al., 2004).

The ability of *B. burgdorferi* to persist in immunologically competent hosts, a concept that has been demonstrated in various animal models, presents a challenge to complete eradication of infection with antibiotic therapy (also see chapter 23). "Persisters," a small subpopulation of bacteria that

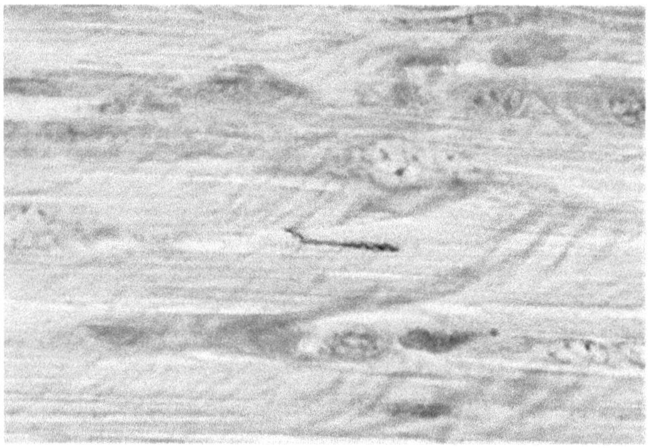

Figure 15. Immunohistochemical staining of a single *B. burgdorferi* spirochete in collagenous tendon tissue of a C3H mouse. Persisting spirochetes during chronic infection are typically associated with collagen bundles in various tissues, including tendons, ligaments, vessel walls, and dermis.

are dormant and non-dividing, have been documented since 1944 during early studies on the action of penicillin on *Staphylococcus* spp., and have subsequently been found among a wide variety of other bacteria. Persisters are tolerant (in contrast to resistant) to antibiotics because they have little or no cell wall synthesis, translation or topoisomerase activity (reviewed in (Lewis, 2007). Studies show that the spirochetes can form antibiotic-tolerant persisters *in vitro* (Caskey and Embers, 2015; Sharma et al., 2015). This may explain the reduced effectiveness of antibiotic therapy during late Lyme disease, and results from several studies in animal models (dogs, nonhuman primates and mice) in which spirochetes cannot be cultured after antibiotic treatment, but can be detected in tissues by PCR and RT-PCR, suggest persister forms. Furthermore, while intact and non-dividing populations of persisting spirochetes have not been detected by culture, they have been detected in antibiotic-treated mice by xenodiagnosis. These forms of spirochetes are acquired by feeding ticks, but they do not undergo the usual burst of replication in the ticks. The spirochetes, nevertheless, are transmissible by feeding infected ticks on naïve SCID mice. Although the spirochetes that infect these mice disseminate, they do not initially replicate to the levels of spirochetes from ticks that have fed upon antibiotic-untreated mice (Hodzic et al., 2008). However, a landmark study revealed that when mice were tested at 12 months post-treatment with ceftriaxone, the spirochete levels had resurged, comparable to the untreated mice (Hodzic et al., 2014). An earlier study in mice suggested that the spirochetes that were acquired by xenodiagnosis on antibiotic-treated mice were genetically attenuated and non-infectious (Bockenstedt et al., 2002), but this was not found in a subsequent study (Hodzic et al., 2008).

Under normal circumstances, culture and PCR are equally sensitive for detection of spirochetes in tissues, but this is not the case when spirochetes are in a non-dividing state. Another caveat that confounds most studies is that following antibiotic treatment, spirochetes are cleared from several of the tissues that are routinely cultured as read-outs, such as ear and urinary bladder. It is important to examine and test highly collagenous tissue of joints and heart base for maximal detection sensitivity of persisting spirochetes (Hodzic et al., 2008). These sites are preferential sites of spirochete persistence during the immune phase of chronic infection (Barthold et al., 2006) (Figure 15). The highly collagenous dermis is also a common site for persistence during chronic infection (Barthold et al., 1993).

Although evidence for non-dividing persister forms of *B. burgdorferi* is still incomplete, support for this concept has arisen from other studies. Both penicillin G and doxycycline have been shown to significantly alter the expression pattern of *B. burgdorferi* s.s. LW2 structural proteins, and most of the affected genes are chromosomal and involved in bacterial cell metabolism (Hunfeld et al., 2008). Other studies have shown that calprotectin, an abundant neutrophil cytoplasmic protein, is a potent bacteriostat, and physiologic levels of this protein induce *B. burgdorferi* to enter a non-dividing state without cell death. Calprotectin's effects are mediated through chelation of Zn, which is essential to growth of spirochetes, as well as activation of NADPH oxidase (Lusitani et al., 2003). A potential consequence of this suppression of division would be antibiotic tolerance, which indeed has been shown with *B. burgdorferi* exposed to penicillin G, but not doxycycline (Montgomery et al., 2006). It should also be noted that functional levels of calprotectin are present in synovial fluids of Lyme disease and other arthritides (Montgomery et al., 2006). Collectively, these results pose challenges to existing dogma about antibiotic treatment of Lyme borreliosis, and the mouse model offers opportunities for further investigation, but the model must be utilized in increasingly precise ways with full cognizance of these biologic features.

Horses
Three different 4-week antibiotic treatments including either doxycycline, ceftiofur, or tetracycline were evaluated for their efficacy in treating experimental Lyme disease in ponies. Animals were inoculated with *B. burgdorferi* via tick bite. Doxycycline and ceftiofur treatment did not appear to eliminate the infection (measured by elevations in specific antibody titers and positive necropsy tissue cultures). However, tetracycline seemed to eliminate persistent infection. The latter was supported by loss of seropositivity and negative cultures from multiple post-mortem tissues. Serum antibody levels to *B. burgdorferi* declined in all antibiotic treated ponies, but three of four doxycycline and two out of four ceftiofur treated ponies had elevated titers again three months after discontinuation of antibiotics (Chang et al., 2005).

Canines
Perhaps the most intriguing findings with the dog model have been the demonstration of persisting spirochetes after antibiotic treatment. Treatment of tick-infected dogs for 30 days with amoxicillin or doxycycline eliminated clinical disease, and reduced, but did not entirely eliminate infection, based upon PCR of tissues. Antibody titers to *B. burgdorferi* declined following treatment; in some cases to undetectable levels. When dogs were maintained for 6 months after treatment was completed, antibody levels rose again (Straubinger et al., 1997b). Studies with antibiotic-treated dogs showed a rapid decline in antibody response to C6 of VlsE, in contrast to persistence or a slower decline in overall antibody response to *B. burgdorferi* lysates, despite incomplete clearance of infection (Philipp et al., 2001). Evidence of persistence was found in dogs treated with amoxicillin, azithromycin, ceftriaxone, or doxycycline (Straubinger et al., 1998b; Straubinger, 2000). Notably, evidence of persistent infection in these studies has been based upon PCR, with no success at culturing viable spirochetes. In untreated dogs, PCR and culture have equal sensitivity. Similar results have been confirmed in the rhesus and mouse models.

Primates
In an initial effort to gain insight into the mechanism for antibiotic treatment failure in Lyme disease using the NHP model, two experiments were conducted to assess the efficacy of a prolonged regimen of antibiotic treatment administered 6.75 months (experiment 1) or 4 months (experiment 2) post-inoculation with *B. burgdorferi* in rhesus macaques (Embers et al., 2012a) To focus on the early disseminated phase of disease, five rhesus macaques were used for experiment 2; each animal was injected with *in vitro*-cultured late log-phase *B. burgdorferi* strain B31, isolate 5A19 for a total inoculum of $3x10^8$. Treatment was initiated at 4 months post inoculation, and animals were kept for a minimum of 6 months after treatment termination. At 4 months PI, 3 of 5 animals received one 50 mg doxycycline tablet BID for 28 consecutive days corresponding to >12 mg/kg/day, while the remaining two animals were untreated controls.

Antemortem and postmortem, tissue samples from multiple organs were collected for (1) *in vitro* culture of *B. burgdorferi;* (2) quantitative real-time PCR and reverse transcriptase (RT) PCR, for detection of *B. burgdorferi* DNA and RNA, respectively; (3) histopathology and immunohistochemistry or immunofluorescence to localize inflammatory lesions and spirochetes in tissues; (4) longitudinal serology for *B. burgdorferi*-specific C6 diagnostic peptide antibodies; and (5) detection of the presence of spirochetes by xenodiagnosis (experiment 2 only).

For experiment 1, all skin biopsies yielded positive cultures the first 4 weeks post-inoculation. Of tissues collected at necropsy, none of the treated animals yielded a positive culture and only one of the untreated animals was culture-positive (lung tissue). Of the postmortem tissues assessed for spirochetal DNA two untreated animals were positive, with spirochetal DNA found in the dorsal root ganglia for one animal, and in the heart for the other. One of the treated animals was PCR positive in multiple tissues including the meninges, bladder, spleen, and lungs. Two of the untreated animals were positive for *B. burgdorferi* RNA transcript, one in heart and the other in the brain. Two of the treated animals had detectable spirochetal RNA in the heart, with one additional treated animal positive for *B. burgdorferi* RNA in both the heart and brain. Three animals had moderate to severe inflammatory lesions in the heart. Interestingly, all of these animals were from the treated group. Positive immunofluorescence for *B. burgdorferi* appeared as fragments and was observed in either the heart and/or meninges in 7 of the treated animals and 9 of the untreated animals.

For experiment 2, PCR on skin biopsies targeting the *ospC* gene of *B. burgdorferi* was positive for all animals in at least 1 of the 4 tested time points. A few spirochetes grew in cultures of organ tissues collected postmortem from both treated and untreated animals, but subculture of spirochetes was not possible due to their slow growth. Cultures were pelleted and transcription of *ospA* was detected in tissues from treated and untreated animals via RT-PCR. Additionally, transcription of the gene *bbf26* located on the lp28-1 plasmid was verified in both untreated animals and one treated animal. For xenodiagnoses, all monkeys were fed upon by *Ixodes scapularis* nymphs for uptake of persisting spirochetes. Intact spirochetes were detected from cultured midgut contents or directly from tick midgut smears of two treated animals at 7 months PI. Additionally, spirochetal DNA was detected by amplification of *ospC* from midgut contents in an additional treated animal where organisms were not detected by culture or direct immunofluorescence. Antibody titers to C6 were evaluated longitudinally before, during and after treatment. For all untreated animals, the titers remained high, whereas the titers from the treated animals declined markedly, and remained at low level even at 34 weeks post-treatment. Taken together, these results demonstrate that *B. burgdorferi* can survive antibiotic treatment and persist in tissues post-dissemination in a primate host. This finding raised important questions about the pathogenicity of antibiotic-tolerant persisters and whether or not they can contribute to symptoms of PTLDS.

Subsequent to the publication of these results, a rebuke was written claiming "insufficient attention was placed on documenting the blood levels, pharmacokinetics, and pharmacodynamic parameters of the antibiotics" used in the NHP model and that it was impossible to conclude that the findings have validity in judging the efficacy of doxycycline or ceftriaxone for the treatment of *Borrelia burgdorferi* in primates (Wormser et al., 2012). A pharmacokinetic analysis of oral doxycycline in rhesus macaques was published shortly thereafter, validating the doses used for treatment (Embers et al., 2013). More recently, a comprehensive analysis of *B. burgdorferi* persistence in macaques following delayed treatment of tick-transmitted infection was performed (Embers et al., 2017; Crossland et al., 2018). The findings revealed that macaques exhibit variable manifestations in terms of tissues affected (as do humans) and differences in antibody responses. Persistent spirochetes were identified in multiple tissues, including the brain (Crossland et al., 2018), after a 28-day treatment with doxycycline and were shown to be transcriptionally active in a novel "*ex vivo*" culture system utilizing rats to culture infected primate tissues (Embers et al., 2017). This model could thus be further exploited to test novel treatments and to identify biomarkers to distinguish persistent infection from other causes of PTLDS.

Animal models of relapsing fever

Relapsing fever (RF) is a multisystemic spirochetal infection that occurs in two forms, epidemic and endemic, and is caused by a variety of *Borrelia* species. Epidemic RF is louse-borne (LBRF) and caused by *B. recurrentis*, whereas endemic RF is tick-borne (TBRF) and caused by different *Borrelia* (separated into Old World and New World species) in endemic areas throughout the world. Tick-borne RF is transmitted by soft ticks of the *Ornithodoros* genus. Among the several agents of endemic RF, the best studied have been the New World species *B. turicatae* and *B. hermsii* and the Old World species *B. duttoni* and *B. crocidurae* (Barbour and Hayes, 1986). The clinical and pathological manifestations of RF borreliosis are diverse, depending on the *Borrelia* species and the genetic background of the host. The relapsing course is mediated by antigenic variation of outer membrane lipoproteins, which allows infection to persist for several weeks with alternating periods of illness at times of high bacteremia and periods of relative well-being during low bacteremia (see chapter by Bergstrom, et al.). Untreated infection can result in multisystemic complications and even death. The greatest threat posed by RF is to the unborn, and more research is needed into the pathogenesis of this infection in humans. However, much has been learned of RF in experimental animal model systems. This section will present an overview of RF manifestations, pathology and host immune responses in experimental animals. For the epidemic form, the only available animal models are the nonhuman primate and immunodeficient mice (Judge et al., 1974a; Larsson et al., 2009). In contrast, mice and rats, in addition to primates, have proven to be excellent models to study the endemic form (Felsenfeld and Wolf, 1975; Cadavid and Barbour, 1998; Lopez et al., 2014; Boardman et al., 2019).

Relapsing fever in mouse models

Clinical disease

Significant heterogeneity of clinical manifestations of RF borreliosis in experimental animals exists. As a general rule, higher pathogen loads result in more severe disease. Therefore, animals inoculated with large numbers of organisms tend to fall ill and even die early (Wright and Woodrow, 1980). The clinical manifestations of infection in immunocompetent mice vary with their genetic background, in contrast to antibody-deficient mice that uniformly develop clinical disease independent of genetic background (Cadavid, 2006). Old World RF *Borrelia* species appear to be more pathogenic than the New World species: inoculation of BALB/c mice with *B. crocidurae* resulted in clinical disease, whereas *B. hermsii* did not (Shamaei-Tousi et al., 1999; Mehra et al., 2009). While inoculation with *B. turicatae* did not result in any clinical disease in any of several inbred mouse strains tested, including C3H/HeJ, SWR/J, or B6, severe vestibular dysfunction was observed in 1 in 4 outbred mice (Cadavid, 2006). Deficiency of TLR2, which plays a major role in recognition of *Borrelia* lipoproteins, did not result in observable clinical disease during infection with *B. turicatae* in mice of the B6 background, despite increased bacteremia (Alugupalli et al., 2007). The strong and transient production of IL-10 observed in immunocompetent mice during peak bacteremia likely explains the paucity of clinical manifestations (Cadavid and Londoño, 2009; Crowder et al., 2016). In mice, IgM antibodies produced by B1b cells have been shown to be important for clearance of the variant serotypes (Alugupalli et al., 2003a) (Connolly, 2001; Alugupalli et al., 2003a; Alugupalli et al., 2004).

Infection

All outbred stocks and inbred strains of mice that have been tested (B6, BALB/c, C3H/HeJ, SWR) develop relapsing bacteremia upon inoculation with endemic RF spirochetes, including *B. hermsii* (Cadavid et al., 1993), *B. turicatae* (Cadavid, 2006), *B. crocidurae* (Shamaei-Tousi et al., 1999), and an uncultivable strain from Spain (Gebbia et al., 1999). Peak bacteremia varies from 10^6/ml to 10^8/ml with higher counts usually observed during the first peak and with higher initial inocula (Alugupalli et al., 2003a; Larsson et al., 2006b). Spirochetes were observed free in the blood with the exception of erythrocyte-associated *B. crocidurae* and extravascularly in the kidney interstitium, lung and testis (Shamaei-Tousi et al., 1999). Dissemination to the kidneys was facilitated by plasminogen (Shamaei-Tousi et al., 1999; Nordstrand et al., 2001). Unlike *B. burgdorferi* (Coleman et al., 1997a), plasminogen plays no role in the bacteremia of RF (Gebbia et al., 1999; Nordstrand et al., 2001).

The subject of brain infection with RF spirochetes became an area of particular interest following the finding that brain homogenates were infectious after the blood had ceased to be infectious; a phenomenon called residual brain infection (Cadavid and Barbour, 1998). Cadavid *et al.* found residual brain infection after one month in 20% of mice inoculated with a neurotropic strain of *B. turicatae* (Cadavid et al., 2006). TLR2 deficiency did not increase susceptibility to residual brain infection. In the majority of cases, residual brain infection was caused by new serotypes, but in at least one case the originally inoculated serotype was responsible. Larsson et al. also studied residual brain infection with four different RF species and found that it was more prevalent with *B. duttoni* (Larsson et al., 2006b). Different serotypes of *B. duttoni* were capable of causing residual brain infection and the average pathogen load in the brain was around $2x10^3$ spirochetes/gram of brain. In some mice they observed recurrence of bacteremia with steroid treatment. At least two studies have shown that host plasmin facilitated brain infection (Gebbia et al., 1999; Nordstrand et al., 2001). There have been several studies to precisely localize RF spirochetes in the brain. Rare spirochetes were observed in the brain parenchyma by silver staining techniques in C3H/HeN but not in 129/Black Swiss mice infected with a Spanish strain (Garcia-Monco et al., 1997; Gebbia et al., 1999). In contrast, *B. crocidurae* was readily observed in the brain parenchyma early after inoculation, although extensive intravascular congestion, rosetting, and microthrombi made it difficult to establish whether these spirochetes were truly intrathecal (Shamaei-Tousi et al., 1999; Nordstrand et al., 2001). Others failed to find any spirochetes microscopically in the brain during the residual infection phase with *B. duttoni* (Larsson et al., 2006b).

With respect to maternal-fetal infection, a pregnant murine model of *Borrelia duttonii* infection was previously reported. Despite observations of intrauterine growth retardation, as well as placental damage and inflammation, impaired fetal circulation, and decreased maternal hemoglobin levels, the animals birthed live young and the study reported that infected pregnant mice were actually more resistant to illness (Larsson et al., 2006a). While this may not faithfully represent disease progression in pregnant women, the study showed that spirochetes frequently cross the maternal-fetal barrier, resulting in congenital infection.

Inflammation and pathology

During peak bacteremia, increased numbers of neutrophils, lymphocytes, and monocytes were observed in the blood of most rodent models, in addition to anemia, which worsened after the peak (Gebbia et al., 1999; Shamaei-Tousi et al., 1999). However, inoculation with large numbers of *B. duttoni* spirochetes resulted in a progressive decrease in the number of leukocytes that corresponded to the increase in bacteremia (Wright and Woodrow, 1980). There is a strong correlation between the magnitude of bacteremia and thrombocytopenia (Alugupalli et al., 2003a). Early and persistent thrombocytopenia was observed in mice infected with *B. duttoni* (Wright and Woodrow, 1980), *B. crocidurae* (Shamaei-Tousi et al., 1999), and an RF strain from Spain (Gebbia et al., 1999). RF spirochetes bind platelets *in vivo* and *in vitro* (Shamaei-Tousi et al., 1999; Alugupalli et al., 2001a; Alugupalli et al., 2001b; Alugupalli et al., 2003b). Inoculation with 10^6 *B. hermsii* spirochetes lowered platelet counts 5-20 fold and prolonged bleeding time 5 fold due to accelerated platelet clearance but without increasing D-dimers (Alugupalli et al., 2003b).

Marked enlargement of the spleen and liver is a consistent feature in small animal models of RF (Wright and Woodrow, 1980; Gebbia et al., 1999; Shamaei-Tousi et al., 1999). In grivet monkeys (discussed in the final section), *B. recurrentis* caused multiple microabscesses that replaced the white pulp in the spleen, and hepatitis with foci of midzonal necrosis (Judge et al., 1974b). Accumulation of neutrophils in and around microvessels was also observed in the kidneys with *B. crocidurae* but not with a Spanish strain (Gebbia et al., 1999; Shamaei-Tousi et al., 1999). BALB/c mice and rats inoculated with 1 million *B. crocidurae* spirochetes rapidly developed hemorrhages in the surface of the lungs, liver, and kidneys. Microscopic examination revealed intravascular erythrocyte aggregation with secondary extravasation and hemorrhage (Shamaei-Tousi et al., 1999; Shamaei-Tousi et al., 2001). Prominent vascular pathology occurred in mice inoculated with 5×10^7 *B. duttoni* spirochetes. This very large inoculum resulted in high mortality of 50% after only 4 days (Wright and Woodrow, 1980). The most conspicuous feature in infected tissues was the presence of blood vessel fibrin thrombi, most commonly in the kidney, that distended and occluded glomerular capillary loops. Intravascular fibrin thrombi were also seen in the liver, spleen, and lung. One of the mice showed a cortical infarct in the brain. Infection with *B. crocidurae* resulted in spirochetes associated with blood vessels and endothelial cell activation (Shamaei-Tousi et al., 2000; Nordstrand et al., 2001).

Myocarditis was also observed in rodents infected with Old World (Gebbia et al., 1999; Nordstrand et al., 2001), but not with New World RF species (Londoño et al., 2005). Infected hearts showed perivascular mononuclear leukocyte infiltrates in the pericardium and focal myocarditis with mononuclear leukocyte infiltrates, interstitial edema, and isolated cardiomyocyte necrosis (Gebbia et al., 1999).

Animal models that utilize large inocula have prominent tissue injury. Mice inoculated with 5×10^7 *B. duttoni* spirochetes developed focal areas of endothelial cell swelling in the kidneys and nuclear pyknosis and karyorrhexis in hepatocytes and Kupffer cells in the liver. Spleens were enlarged with large areas of necrosis and macrophages in the red pulp were markedly vacuolated (Wright and Woodrow, 1980). BALB/c mice inoculated with large numbers of *B. crocidurae* spirochetes developed changes suggestive of cell death in kidneys and lungs. These changes were thought to be the result of vascular occlusion from microthrombi formed by a combination of erythrocytes, spirochetes, and leukocytes (Shamaei-Tousi et al., 1999). TUNEL staining revealed the presence of apoptotic germ cells during peak bacteremia in rats infected with *B. crocidurae* (Shamaei-Tousi et al., 2001). In contrast, tissue damage is not a feature in mice inoculated with *B. hermsii* (Shamaei-Tousi et al., 2001), *B. turicatae* (Cadavid et al., 2006), a Spanish strain (Gebbia et al., 1999), or lower doses of *B. duttoni* (Larsson et al., 2006b; Andersson et al., 2007). No alterations in gene expression were found in the brains of mice with residual infection by *B. duttoni* (Larsson et al., 2006b).

There have been several studies of inflammation in the brain. Wild-type or TLR2-deficient B6 mice inoculated with *B. turicatae* developed leptomeningeal inflammation and brain parenchymal macrophage infiltration/activation with little if any lymphocytes (Cadavid et al., 2006). Inoculation of BALB/c mice with *B. crocidurae* resulted in increased perivascular macrophages/microgliosis as well as small assemblies of lymphocytes and granulocytes in the brain (Andersson et al., 2007). The inflammatory reaction was more pronounced in the choroid plexus and leptomeninges and consisted of lymphocytes and neutrophils. The meninges were congested and occasional petechial hemorrhages in the brain parenchyma were also found; activated macrophages/microglia were by far the most abundant inflammatory cells in the brain, followed by neutrophils with little if any increase in B or T cells. Brain inflammation during *B. crocidurae* infection decreased in the absence of plasminogen (Nordstrand et al., 2001). 129/Black Swiss, BALB/c, and C3H/HeN mice inoculated with an RF strain from Spain also developed perivascular mononuclear cell infiltration of the leptomeninges, accumulation of activated macrophages perivascularly, and plexitis (Garcia-Monco et al., 1997; Gebbia et al., 1999). In this study, the inflammatory infiltrate was strongly CD45R (B220) positive and there was submeningeal fibrin deposition that was not seen in the absence of plasminogen. No evidence of inflammation was found in the brain of mice experiencing residual brain infection with *B. duttoni* (Larsson et al., 2006b) or during any stage of infection with *B. hermsii* (Cadavid et al., 1993; Shamaei-Tousi et al., 1999).

Relapsing fever in immune deficient mice

Since serotype-specific antibodies rapidly clear RF spirochetes from blood and tissues, it has been possible to compare disease by specific serotypes over time. In the early 1990s, there was an outbreak of endemic RF in Texas with prominent neurological complications (Rawlings, 1995). Inoculation of spirochetes cultured from whole tick-homogenates into SCID mice allowed the identification of two *B. turicatae* serotypes that caused distinct differences in pathogen load in blood and brain (Cadavid et al., 1994). These two serotypes were originally named A and B, and subsequently renamed 1 and 2 (or Bt1 and Bt2). The only discernible difference between them is the variable major protein they express, 23 kDa Vsp1 in serotype 1 and 20 kDa Vsp2 in serotype 2 (Cadavid et al., 1997; Penningon et al., 1999; Pennington et al., 1999). Infection of B cell-deficient mice with serotypes 1 and 2 has been utilized to understand the basis of their different virulence and tissue infectivity.

Clinical manifestations

Since their original isolation, it became apparent that Bt2 was more virulent than Bt1. Both serotypes have similar abilities to establish infection with as little as 1 spirochete per mouse, but Bt2 causes more severe clinical manifestations. Clinical disease began in the first week by purulent discharge from the eyes and ruffled skin, was followed by tibiotarsal joint swelling and redness by the second week, and signs of vestibular dysfunction arose after 3 weeks (Cadavid et al., 1994). Tibiotarsal and metatarsal joint reddening and swelling began earlier and were more significantly more severe with Bt2 that also resulted in more functional impairment on an equilibrium bar. Despite the absence of specific antibody to control pathogen load, all the clinical manifestations either completely resolved (purulent eye discharge) or significantly decreased (skin and joints) over days to weeks (Cadavid et al., 1994). Mice co-infected with both Bt1 and Bt2 were compared with mice infected with either serotype alone. Although vestibular dysfunction began earlier with Bt2, only Bt1 caused severe vestibular dysfunction associated with early mortality. Interestingly, this was prevented by co-infection with Bt2. Unlike vestibular dysfunction, the severe arthritis and weight loss caused by Bt2 alone was not influenced by co-infection with Bt1 (Cadavid et al., 2007). In a series of recent studies that compared infection in Rag1 null (T and B cell deficient) and Igh6 null (B cell deficient) mice, it was shown that Rag1 null mice developed less clinical disease, despite higher pathogen load due to higher production of IL-10 (Gelderblom et al., 2007b; Londono et al., 2008).

Infection

The high number of spirochetes found in blood and tissues of B cell deficient mice inoculated with *B. turicatae* provided an opportunity to study their localization after systemic inoculation. Abundant spirochetes were observed microscopically in multiple tissues as early as 6 days after inoculation. In the brain, the localization was primarily leptomeningeal, although rarely spirochetes were also found in the brain parenchyma itself (Cadavid et al., 2001). The site of entry into the brain appeared to be the leptomeningeal microcirculation, where spirochetes were observed attached to and in the process of crossing brain-microvascular endothelial cells (Sethi et al., 2006). Spirochetes were also observed in the dura mater, middle and inner ear, endoneurium and perineurium of cranial and peripheral nerves, skin, and bone marrow (Cadavid et al., 2001). Clumps of spirochetes were observed only in the subarachnoid space and inner ear. Abundant spirochetes were present in the extracellular matrix of the skin, skeletal and cardiac muscle.

Bt2 causes higher bacteremia than Bt1. Pathogen load during co-infection was studied indirectly by comparing circulating levels of the B cell chemokine CXCL13 in SCID mice persistently infected with Bt2 alone or in combination with Bt1 (Cadavid et al., 2007; Gelderblom et al., 2007a; Gelderblom et al., 2007b). This approach was used as an alternative to directly counting spirochetes because the levels of this chemokine strongly correlate with pathogen load (Narayan et al., 2005). Results revealed that co-infected mice had significantly higher CXCL13 levels. When brain infection was studied by culture of whole brain homogenates, Bt1 was found in the brain much more frequently than Bt2. Furthermore, only Bt1 was apparent by Western blot on brain cultures from SCID mice that were co-infected with Bt1 and Bt2 (Cadavid et al., 1994). Since Bt2 readily infected the brain after intracerebral inoculation (Cadavid et al., 2001), this suggested that Bt1 is better than Bt2 at crossing the blood-brain barrier (BBB). Using immunohistochemistry, fivefold more Bt1 than Bt2 spirochetes were found in the leptomeninges in SCID mice examined 18 days after inoculation (Sethi et al., 2006).

Although it is firmly established that spirochetes can cause inflammation when infecting the central nervous system, in mice persistently infected with Bt1 spirochetes, widespread cerebral inflammation was consistently observed and out of proportion to the presence of spirochetes in the brain parenchyma itself (Cadavid et al., 2001). The possibility that *Borrelia* lipoproteins could cause brain inflammation by disseminating from the systemic circulation across the blood brain barrier was later examined. In a series of experiments using normal and Bt-infected mice injected intraperitoneally with lanthanide-labeled variable outer membrane lipoproteins, Vsp1 was found to be better than Vsp2 and Bt1 spirochetes at distributing from circulating blood to the brain parenchyma. The dissemination of Vsp1 from the periphery to the brain occurred in uninfected mice, but significantly increased in mice with ongoing Bt1 infection (Londoño and Cadavid, 2010). These results indicate that one potential mechanism of neuroinflammation in Borrelial infections is via dissemination of lipoproteins from the periphery.

Pathology

Examination of tissues from SCID mice persistently infected with Bt1 or Bt2 has confirmed significant differences in their pathogenicity. No inflammation was apparent for at least 5 days after inoculation. However, by 2 weeks arthritis was present and more severe with Bt2 than with Bt1: 19/20 joints from Bt2 infected mice had arthritis compared with 12/20 from Bt1 infected mice; it was severe only with Bt2. H&E-stained sections revealed mixed inflammation of the synovial lining of both joints and tendon sheaths with mononuclear and polymorphonuclear cells, with prominent thickening of the synovium and leukocyte exudate into the joint space. Infected SCID mice also showed prominent inflammation in the heart. Cardiac inflammation involved both the myocardium and pericardium and was characterized by prominent mononuclear leukocytic infiltrate between myocardial fibers with exudate in the pericardium. Cardiac inflammation was more severe in the atria and around the great vessels in the mediastinum (Londoño et al., 2005) similar to Lyme borreliosis (Barthold et al., 1993). Masson's trichrome staining showed loss of cardiomyocytes that was more prominent in the atrium; immunohistochemistry showed infiltration by activated macrophages in the atrium and periphery of the heart; and *in situ* hybridization revealed prominent upregulation of expression of IL-6 also more prominent in the heart base. Digital image analysis revealed that the loss of cardiomyocytes in the heart base was significantly worse with Bt2 than with Bt1. Morphological features of apoptosis, including fragmented nuclei and condensed chromatin, were readily apparent in areas of cardiomyocyte injury. Analysis of caspase activation by RNase protection assay showed significant upregulation of several caspases, most notably caspase 1 and caspase 11. The upregulation of caspase 1 was confirmed at the protein level (Londoño et al., 2005), with strong expression found in areas of injury that co-localized with activated macrophages. Cardiomyocyte apoptosis was also found in Rag1 and Igh6 deficient mice (Gelderblom et al., 2007b), with strong expression in areas of injury that co-localized with activated macrophages. Molecular analysis of cardiac tissue from Rag1 or Igh6 null mice showed prominent upregulation of both TNF and IL-10 of similar degree with either serotype.

Inflammation in the skin of SCID mice was equally severe with either serotype despite the marked differences in pathogen load (Cadavid et al., 2001). Most inflammatory cells in the skin were activated macrophages (Sethi et al., 2006). Significant inflammation was also found in the vestibular and auditory labyrinth, which likely explained the prominent vestibular dysfunction that characterizes this model (Gelderblom et al., 2007b). Mild inflammation was also found in the leptomeninges

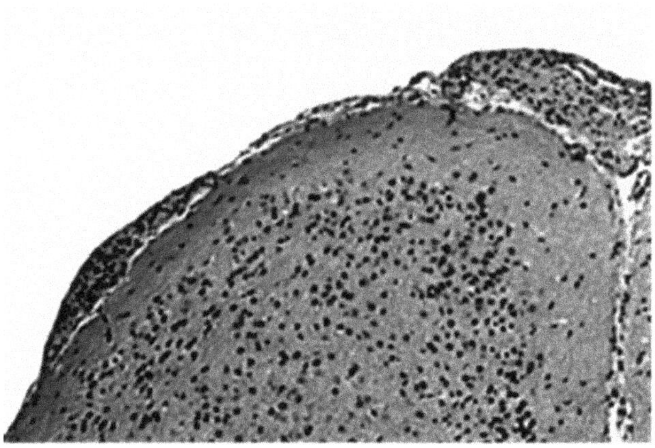

Figure 16. Mild leptomeningitis in a RAG2 deficient mouse persistently infected with serotype 1 of *B. turicatae.*

with no obvious difference between Bt1 and Bt2 (Figure 16). Although H&E-staining did not reveal any apparent abnormalities in the brain parenchyma, immunostaining for F4/80, a marker of activated microglia/macrophages revealed widespread infiltration by activated macrophages/microglia with a 30-fold increase in Bt1-infected mice compared to uninfected controls (Figure 17). Microgliosis was less apparent in the brain of Bt2 infected mice. The morphology of F4/80-stained cells was heterogeneous with some ameboid and some ramified. Examination of mice deficient in Rag1 or Igh6 confirmed that persistent infection with Bt1 results in severe cerebral microgliosis. Molecular analysis of the brain from Igh6 deficient mice showed that Bt1 infection resulted in significant upregulation of TNF and IL-10 by 2 weeks but, interestingly, by 4 weeks only upregulation of IL-10 persisted. Immunofluorescence showed that fully activated microglia were the main source of IL-10 in Bt1-infected brains. Oligoarray gene expression analysis in Igh6 deficient mice persistently infected with Bt1 compared with uninfected controls revealed 116 out of 14,109 well documented genes present on the Affymetrix 430A mouse gene chip were upregulated at least twofold (Gelderblom et al., 2007a). Many of them are genes known to be upregulated during microglia activation. CXCL13 was the most highly upregulated gene in infected brains and the most abundant of the chemokines in the blood (Gelderblom et al., 2007a). An analysis of functional categories within the gene ontology database showed that the majority of the 20 categories with the highest EASE scores refer to functions of the immune response, including proteins involved in antigen processing and presentation; chemotaxis, adhesion and migration; LPS/lipoprotein toll-like receptor ligation and signaling; and pathogen control and host defense (Gelderblom et al., 2007a).

The observation by Cooper et al. (2000) that humans infected with *Borrelia recurrentis* have extraordinarily high levels of circulating IL-10 at times when they appear to be doing relatively well prompted studies on its possible protective role in experimental animals (Cooper et al., 2000). High levels of IL-10 characterized persistent infection of B cell deficient mice with *B. turicatae* (Gelderblom et al., 2007a). In fact, IL-10 had the highest levels among all the cytokines measured with a strong positive correlation with pathogen load (Londoño et al., 2008; Londono et al., 2008). The increase in IL-10 levels was associated with a concomitant decrease in circulating levels of TNF and many other pro-inflammatory cytokines and chemokines. The upregulation of IL-10 was also found in tissues, including heart and brain. Exogenous IL-10 given to Igh6 null mice reduced bacteremia, clinical disease, cerebral microgliosis, and CXCL13.

Bt2-infected Rag2 null mice also deficient in IL-10 showed early mortality due to increased bacteremia, prominent apoptosis of leptomeningeal microvascular endothelial cells, and severe intracerebral hemorrhage (Figure 18). Bt1-infected IL-10 and Rag2 deficient mice did not die and only showed smaller areas of hemorrhage in the brain and other organs when necropsied 2 weeks after inoculation (Londoño

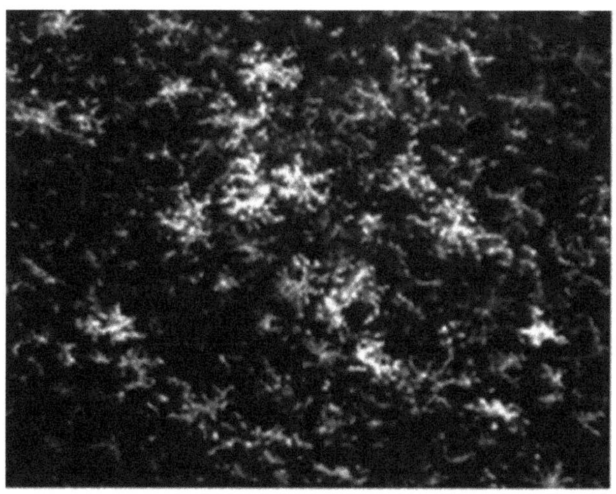

Figure 17. Severe microgliosis in the brain of a C.B-17-scid mouse persistently infected with serotype 1 of *B. turicatae*, depicted by immunohistochemical staining of F4/80. Note the prominence of ramified microglia.

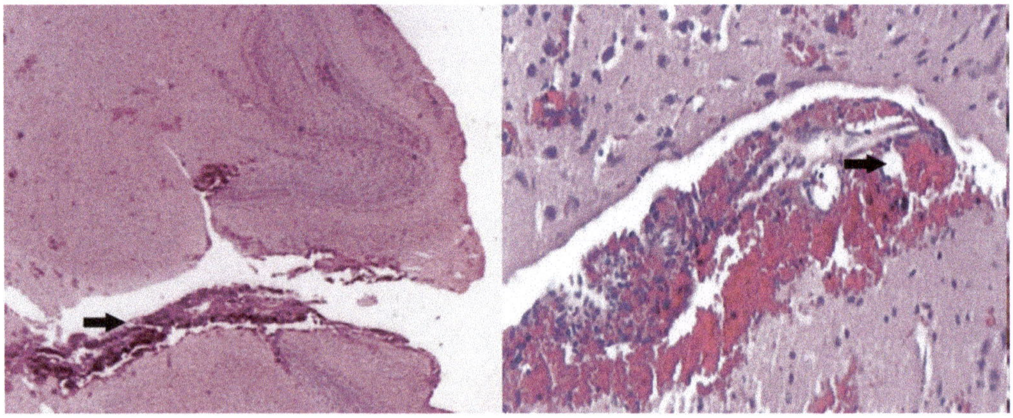

Figure 18. Pathological findings in the brains of RAG2/IL10j/j mice infected with *Borrelia turicatae*. Hematoxylin and eosin (H&E)-stained sections of the brain of a mouse examined on Day 5 after inoculation of serotype 2. Notice the extensive subarachnoid and intraparenchymal hemorrhages.

et al., 2008; Londono et al., 2008). These results revealed that a functional protective mechanism during RF borreliosis is the production of IL-10 systemically and locally.

Nonhuman primate models of relapsing fever

As mentioned, nonhuman primates are a sound model for epidemic LBRF caused by *B. recurrentis*. *Borrelia recurrentis* infection in grivet monkeys (*Cercopithecus aethiops*) uniformly caused fever as high as 40.9ºC and killed one in five animals, including two during a severe relapse (Judge et al., 1974a). Using this model, the therapeutic response and clinical pathology (leukopenia, tachycardia and fever) often seen in humans were recapitulated (Judge et al., 1974c). Finally, the pathology observed in these animals included histiocytic myocarditis and microabscesses of the spleen and liver, all of which have been noted in humans with LBRF.

Studies of TBRF in monkeys date back to 1938 when Francis first observed the longevity of *O. turicata* and the transmissibility of *B. turicatae* after the ticks had been starved for over five years (Francis, 1938). The ticks, obtained from caves in Texas, were placed on rhesus macaques for feeding. Multiple febrile episodes, with temperatures spiking at 41ºC and spirochetemic blood were observed, demonstrating the virulence of the spirochetes. More recently, sophisticated technology has been used to study RF borreliosis in primates, and these studies have further defined the intricacies of disease (Lopez et al., 2014). In a pilot study, four rhesus macaques were infected with *B. turicatae* by tick bite (Figure 19) and novel real-time telemetry was used to quantify disease progression. In all animals, blood samples were collected to quantify spirochete densities using

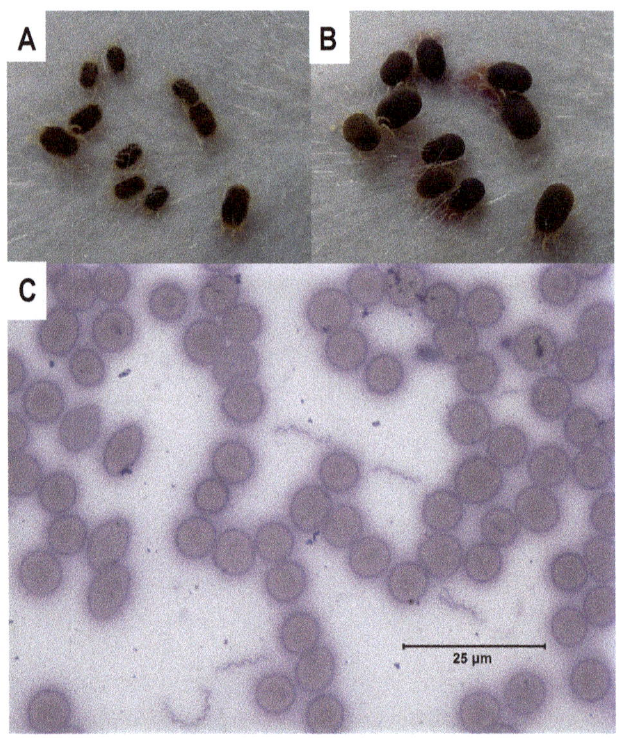

Figure 19. *Ornithodoros turicata* feeding and *Borrelia turicatae* spirochetemia in rhesus macaques. Ten ticks were placed on the abdomen of a rhesus macaque, quickly attached and fed to repletion. Shown are the ticks at 5 minutes (A) and 20 minutes (B) after placement. A blood smear from a day 13 spirochetemic episode is shown in (C).

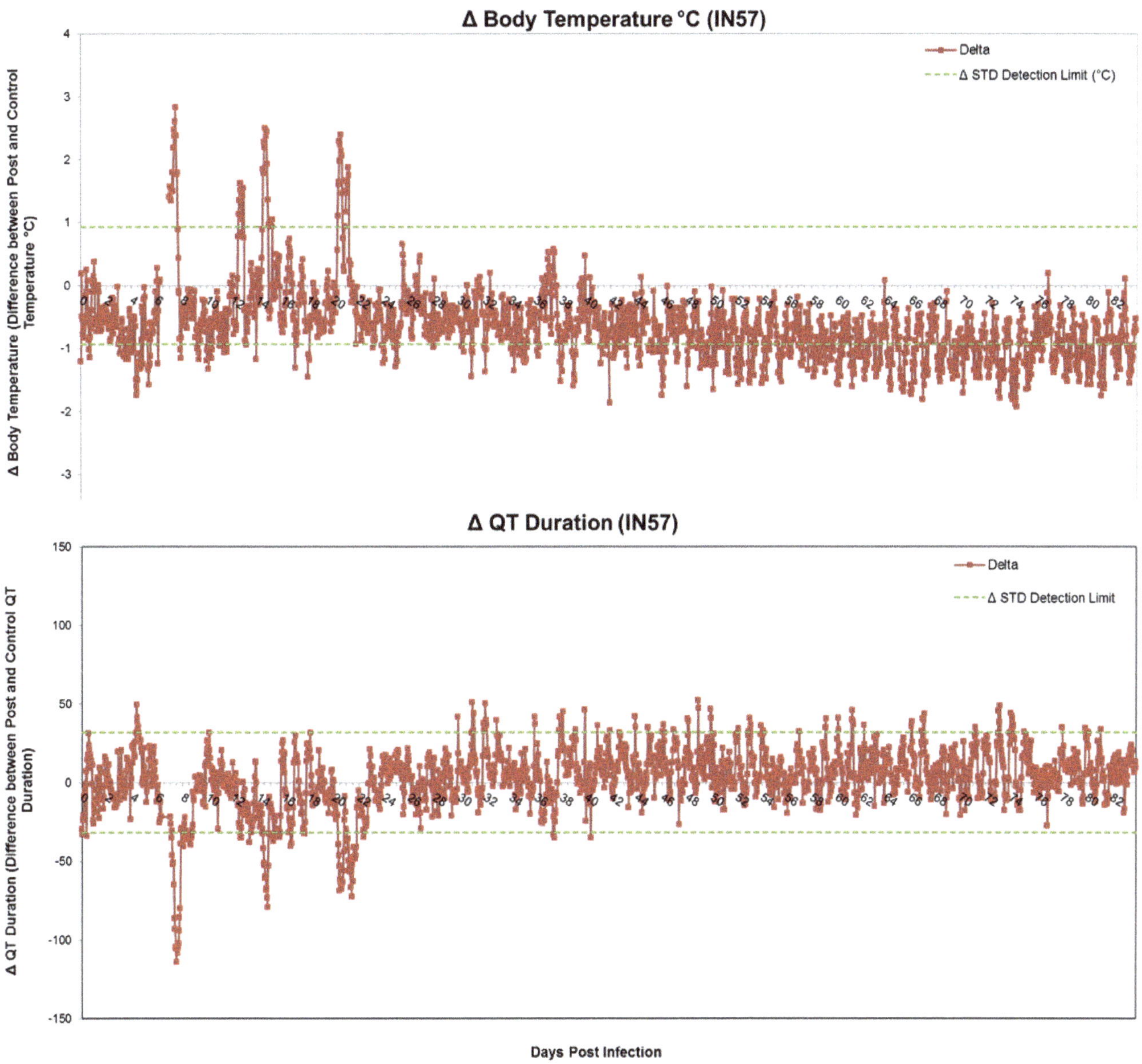

Figure 20. The inverse relationship between body temperature and QT interval in a macque infected with *Borrelia turicatae*. The body temperature and electrocardiograph were monitored in real time using radio telemetry.

Q-PCR and spirochetemia peaked at 10^7/ml; this is notable because a natural inoculum and route were used. Also, temperature readings were acquired from two animals in 24-hour increments for 14 consecutive days, while with the other two animals, temperature, heart and respiratory rate, and ECG intervals were recorded in real-time for 28 days using wireless telemetry devices. NHP infected with *B. turicatae* displayed symptoms associated with human illness including febrile episodes (Figure 20 with periods of hypothermia, and newly-described alterations to the ECG, characterized as decreases in the QT interval (Figure 21)(Lopez et al., 2014). With this study, immune responses unique to human infection were also observed. Inflammatory cytokines (IL-1β and TNF-α) linked to the Jarisch-Herxheimer reactions (Vidal et al., 1998) and miscarriages in rhesus macaques (Sadowsky et al., 2006) and human patients (Shaarawy and Nagui, 1997; Vitoratos et al., 2006) were identified. Distinct antibody responses to

B. turicatae surface antigens and early (14 day p.i.) reduction in peripheral B cell populations were also characterized in macaques (Embers et al., 2019). The rhesus macaque model would likely represent maternal-fetal transmission and pathogenesis quite accurately, so this may be the most relevant use in modern times for the animal model.

Concluding remarks

While alternatives to animal research are sought where possible, test tubes do not have the connected organ systems, such as the immune and neurological systems needed to truly understand disease and the host responses to infection. As zoonotic diseases, the adaptation to the host are of key importance. Each of the models discussed in this chapter has its strengths and weaknesses, and direct reference to human disease is not always possible. However, a great deal has been gleaned from the use of animal models in borreliosis research. It is readily apparent that a comparative commonality pervades all models of each type of borreliosis discussed, and to a certain extent, between different types of borreliosis, which allows reasonable assumptions for extrapolation to the human disease. Dr. Albert Sabin, best known for his role in the development of an oral polio vaccine, noted that "my own experience of over 60 years in biomedical research amply demonstrated that without the use of animals and of human beings, it would have been impossible to acquire the important knowledge needed to prevent much suffering and premature death not only among humans but also among [other] animals."

Acknowledgements

We gratefully acknowledge the contribution of Dr. Mario Phillip to the primate model of Lyme disease, and Dr. Nicholas Crossland for his input on the equine and guinea pig model sections.

References

Akin, E., McHugh, G.L., Flavell, R.A., Fikrig, E., and Steere, A.C. (1999). The immunoglobulin (IgG) antibody response to OspA and OspB correlates with severe and prolonged Lyme arthritis and the IgG response to P35 correlates with mild and brief arthritis. Infect Immun *67*, 173-181.

Akins, D.R., Bourell, K.W., Caimano, M.J., Norgard, M.V., and Radolf, J.D. (1998). A new animal model for studying Lyme disease spirochetes in a mammalian host-adapted state. J Clin Invest *101*, 2240-2250. https://doi.org/10.1172/jci2325

Alexopoulou, L., Thomas, V., Schnare, M., Lobet, Y., Anguita, J., Schoen, R.T., Medzhitov, R., Fikrig, E., and Flavell, R.A. (2002). Hyporesponsiveness to vaccination with *Borrelia burgdorferi* OspA in humans and in TLR1- and TLR2-deficient mice. Nature Medicine *8*, 878-884.

Alugupalli, K.R., Akira, S., Lien, E., and Leong, J.M. (2007). MyD88- and Bruton's tyrosine kinase-mediated signals are essential for T cell-independent pathogen-specific IgM responses. Journal of Immunology *178*, 3740-3749.

Alugupalli, K.R., Gerstein, R.M., Chen, J., Szomolanyi-Tsuda, E., Woodland, R.T., and Leong, J.M. (2003a). The resolution of relapsing fever borreliosis requires IgM and is concurrent with expansion of B1b lymphocytes. Journal of Immunology *170*, 3819-3827.

Alugupalli, K.R., Leong, J.M., Woodland, R.T., Muramatsu, M., Honjo, T., and Gerstein, R.M. (2004). B1b lymphocytes confer T cell-independent long-lasting immunity. Immunity *21*, 379-390.

Alugupalli, K.R., Michelson, A.D., Barnard, M.R., and Leong, J.M. (2001a). Serial determinations of platelet counts in mice by flow cytometry. Thrombosis & Haemostasis *86*, 668-671.

Alugupalli, K.R., Michelson, A.D., Barnard, M.R., Robbins, D., Coburn, J., Baker, E.K., Ginsberg, M.H., Schwan, T.G., and Leong, J.M. (2001b). Platelet activation by a relapsing fever spirochaete results in enhanced bacterium-platelet interaction via integrin alphaIIbbeta3 activation. Molecular Microbiology *39*, 330-340.

Alugupalli, K.R., Michelson, A.D., Joris, I., Schwan, T.G., Hodivala-Dilke, K., Hynes, R.O., and Leong, J.M. (2003b). Spirochete-platelet attachment and thrombocytopenia in murine relapsing fever borreliosis. Blood *102*, 2843-2850.

Anderson, J.F. (1991). Epizootiology of Lyme borreliosis. Scandinavian journal of infectious diseases Supplementum *77*, 23-34.

Anderson, J.F., Barthold, S.W., and Magnarelli, L.A. (1990). Infectious but nonpathogenic isolate of Borrelia burgdorferi. Journal of Clinical Microbiology *28*, 2693-2699.

Anderson, J.F., and Magnarelli, L.A. (1993). Epizootiology of Lyme disease-causing borreliae. Clin Dermatol *11*, 339-351. https://doi.org/10.1016/0738-081x(93)90088-t

Andersson, M., Nordstrand, A., Shamaei-Tousi, A., Jansson, A., Bergström, S., and Guo, B.P. (2007). In situ immune response in brain and kidney during early relapsing fever borreliosis. Journal of neuroimmunology *183*, 26-32. https://doi.org/10.1016/j.jneuroim.2006.11.004

Anguita, J., Barthold, S.W., Persinski, R., Hedrick, M.N., Huy, C.A., Davis, R.J., Flavell, R.A., and Fikrig, E. (2002). Murine Lyme arthritis development mediated by p38 mitogen-activated protein kinase activity. J Immunol *168*, 6352-6357.

Anguita, J., Rincaon, M., Samanta, S., Barthold, S.W., Flavell, R.A., and Fikrig, E. (1998). Borrelia burgdorferi-infected, interleukin-6-deficient mice have decreased Th2 responses and increased Lyme arthritis. J Infect Dis *178*, 1512-1515.

Anguita, J., Samanta, S., Revilla, B., Suk, K., Das, S., Barthold, S.W., and Fikrig, E. (2000). *Borrelia burgdorferi* gene expression in vivo and spirochete pathogenicity. Infect Immun *68*, 1222-1230.

Anguita, J., Thomas, V., Samanta, S., Persinski, R., Hernanz, C., Barthold, S.W., and Fikrig, E. (2001). *Borrelia burgdorferi*-induced inflammation facilitates spirochete adaptation and variable major protein-like sequence locus recombination. J Immunol *167*, 3383-3390.

Appel, M.J., Allan, S., Jacobson, R.H., Lauderdale, T.L., Chang, Y.F., Shin, S.J., Thomford, J.W., Todhunter, R.J., and Summers, B.A. (1993). Experimental Lyme disease in dogs produces arthritis and persistent infection. Journal of Infectious Diseases *167*, 651-664.

Armstrong, A.L., Barthold, S.W., Persing, D.H., and Beck, D.S. (1992). Carditis in Lyme disease susceptible and resistant strains of laboratory mice infected with Borrelia burgdorferi. Am J Trop Med Hyg *47*, 249-258.

Asch, E.S., Bujak, D.I., Weiss, M., Peterson, M.G., and Weinstein, A. (1994). Lyme disease: an infectious and postinfectious syndrome. Journal of Rheumatology *21*, 454-461.

Barbour, A.G. (2017). Infection resistance and tolerance in Peromyscus spp., natural reservoirs of microbes that are virulent for humans.

Semin Cell Dev Biol *61*, 115-122. https://doi.org/10.1016/j.semcdb.2016.07.002

Barbour, A.G., and Hayes, S.F. (1986). Biology of Borrelia species. Microbiological Reviews *50*, 381-400.

Barthold, S.W. (1991). Infectivity of *Borrelia burgdorferi* relative to route of inoculation and genotype in laboratory mice. J Infect Dis *163*, 419-420.

Barthold, S.W. (1999). Specificity of infection-induced immunity among *Borrelia burgdorferi* sensu lato species. Infect Immun *67*, 36-42.

Barthold, S.W., Beck, D.S., Hansen, G.M., Terwilliger, G.A., and Moody, K.D. (1990). Lyme borreliosis in selected strains and ages of laboratory mice. J Infect Dis *162*, 133-138.

Barthold, S.W., and Bockenstedt, L.K. (1993). Passive immunizing activity of sera from mice infected with Borrelia burgdorferi. Infection & Immunity *61*, 4696-4702.

Barthold, S.W., de Souza, M.S., Janotka, J.L., Smith, A.L., and Persing, D.H. (1993). Chronic Lyme borreliosis in the laboratory mouse. Am J Pathol *143*, 959-971.

Barthold, S.W., deSouza, M., and Feng, S. (1996). Serum-mediated resolution of Lyme arthritis in mice. Laboratory Investigation *74*, 57-67.

Barthold, S.W., and deSouza, M.S. (1995). Exacerbation of Lyme arthritis in beige mice. J Infect Dis *172*, in press.

Barthold, S.W., deSouza, M.S., Fikrig, E., and Persing, D.H. (1992a). Lyme borreliosis in the laboratory mouse. In Lyme Disease: Molecular and Immunological Approaches, S.E. Schutzer, ed. (Cold Spring Harbor, NY: Cold Spring Harbor Laboratory Press), pp. 223-242.

Barthold, S.W., Feng, S., Bockenstedt, L.K., Fikrig, E., and Feen, K. (1997). Protective and arthritis-resolving activity in sera of mice infected with Borrelia burgdorferi. Clinical Infectious Diseases *25 Suppl 1*, S9-17.

Barthold, S.W., Fikrig, E., Bockenstedt, L.K., and Persing, D.H. (1995a). Circumvention of outer surface protein A immunity by host-adapted Borrelia burgdorferi. Infection & Immunity *63*, 2255-2261.

Barthold, S.W., Hodzic, E., Tunev, S., and Feng, S. (2006). Antibody-mediated disease remission in the mouse model of lyme borreliosis. Infect Immun *74*, 4817-4825. https://doi.org/10.1128/iai.00469-06

Barthold, S.W., Levy, S.A., Fikrig, E., Bockenstedt, L.K., and Smith, A.L. (1995b). Serologic responses of dogs naturally exposed to or vaccinated against Borrelia burgdorferi infection. J Am Vet Med Assoc *207*, 1435-1440.

Barthold, S.W., Moody, K.D., Terwilliger, G.A., Duray, P.H., Jacoby, R.O., and Steere, A.C. (1988a). Experimental Lyme arthritis in rats infected with Borrelia burgdorferi. Journal of Infectious Diseases *157*, 842-846.

Barthold, S.W., Moody, K.D., Terwilliger, G.A., Jacoby, R.O., and Steere, A.C. (1988b). An animal model for Lyme arthritis. Ann N Y Acad Sci *539*, 264-273. https://doi.org/10.1111/j.1749-6632.1988.tb31860.x

Barthold, S.W., Persing, D.H., Armstrong, A.L., and Peeples, R.A. (1991a). Kinetics of Borrelia burgdorferi dissemination and evolution of disease after intradermal inoculation of mice. American Journal of Pathology *139*, 263-273.

Barthold, S.W., Persing, D.H., Armstrong, A.L., and Peeples, R.A. (1991b). Kinetics of *Borrelia burgdorferi* dissemination and evolution of disease following intradermal inoculation of mice. Am J Pathol *139*, 263-273.

Barthold, S.W., Sidman, C.L., and Smith, A.L. (1992b). Lyme borreliosis in genetically resistant and susceptible mice with severe combined immunodeficiency. Am J Trop Med Hyg *47*, 605-613.

Batool, M., Hillhouse, A.E., Ionov, Y., Kochan, K.J., Mohebbi, F., Stoica, G., Threadgill, D.W., Zelikovsky, A., Waghela, S.D., Wiener, D.J., *et al.* (2019). New Zealand White Rabbits Effectively Clear Borrelia burgdorferi B31 despite the Bacterium's Functional vlsE Antigenic Variation System. Infect Immun *87*. https://doi.org/10.1128/iai.00164-19

Batsford, S., Dunn, J., and Mihatsch, M. (2004). Outer surface lipoproteins of Borrelia burgdorferi vary in their ability to induce experimental joint injury. Arthritis Rheum *50*, 2360-2369.

Baum, E., Hue, F., and Barbour, A.G. (2012). Experimental Infections of the Reservoir Species Peromyscus leucopus with Diverse Strains of Borrelia burgdorferi, a Lyme Disease Agent. *3*, e00434-00412. https://doi.org/10.1128/mBio.00434-12 %J mBio

Behera, A.K., Hildebrand, E., Bronson, R.T., Perides, G., Uematsu, S., Akira, S., and Hu, L.T. (2006). MyD88 deficiency results in tissue-specific changes in cytokine induction and inflammation in interleukin-18-independent mice infected with Borrelia burgdorferi. . Infect Immun *74*, 1462-1470.

Benach, J.L., Bosler, E.M., Coleman, J.L., and Habicht, G.S. (1984). Experimental transmission of the Lyme disease spirochete to rabbits. J Infect Dis *150*, 786-787. https://doi.org/10.1093/infdis/150.5.786-a

Benhnia, M.R., Wroblewski, D., Akhtar, M.N., Patel, R.A., Lavezzi, W., Gangloff, S.C., Goyert, S.M., Caimano, M.J., Radolf, J.D., and Sellati, T.J. (2005). Signaling through CD14 attenuates the inflammatory response to Borrelia burgdorferi, the agent of Lyme disease. J Immunol *174*, 1539-1548.

Berger, B.W. (1989). Cutaneous manifestations of Lyme borreliosis. Rheumatic diseases clinics of North America *15*, 627-634.

Berger, B.W. (1993). Erythema migrans. Clin Dermatol *11*, 359-362. https://doi.org/https://doi.org/10.1016/0738-081X(93)90090-Y

Bloom, B.J., Wyckoff, P.M., Meissner, H.C., and Steere, A.C. (1998). Neurocognitive abnormalities in children after classic manifestations of Lyme disease. Pediatr Infect Dis J *17*, 189-196. https://doi.org/10.1097/00006454-199803000-00004

Boardman, K., Rosenke, K., Safronetz, D., Feldmann, H., and Schwan, T.G. (2019). Host Competency of the Multimammate Rat Mastomys natalensis Demonstrated by Prolonged Spirochetemias with the African Relapsing Fever Spirochete Borrelia crocidurae. Am J Trop Med Hyg *101*, 1272-1275. https://doi.org/10.4269/ajtmh.19-0590

Bockenstedt, L.K., Barthold, S., Deponte, K., Marcantonio, N., and Kantor, F.S. (1993). Borrelia burgdorferi infection and immunity in mice deficient in the fifth component of complement. Infection & Immunity *61*, 2104-2107.

Bockenstedt, L.K., Fikrig, E., Barthold, S.W., Flavell, R.A., and Kantor, F.S. (1996). Identification of a Borrelia burgdorferi OspA T cell epitope that promotes anti-OspA IgG in mice. Journal of Immunology *157*, 5496-5502.

Bockenstedt, L.K., Kang, I., Shanafelt, M.-C., Chang, C., Persing, D., Hayday, A., and Barthold, S.W. (2001). CD4+ Th1 cells facilitate regression of murine Lyme carditis. submitted.

Bockenstedt, L.K., Liu, N., Schwartz, I., and Fish, D. (2006). MyD88 defciiency enhances acquisition and transmission of Borrelia burgdorferi by Ixodes scapularis ticks. Infect Immun *74*, 2154-2160.

Bockenstedt, L.K., Mao, J., Hodzic, E., Barthold, S.W., and Fish, D. (2002). Detection of attenuated, non-infectious spirochetes after antibiotic treatment of *Borrelia burgdorferi*-infected mice. J Infect Dis *186*, 1430-1437.

Bockenstedt, L.K., Shanafelt, M.C., Belperron, A., Mao, J., and Barthold, S.W. (2003). Humoral immunity reflects altered T helper cell bias in Borrelia burgdorferi-infected gamma delta T-cell-deficient mice. Infection & Immunity *71*, 2938-2940.

Bolz, D.D., Sundsbak, R.S., Ma, Y., Akira, S., Kirschning, C.J., Zachary, J.F., Weis, J.H., and Weis, J.J. (2004). MyD88 plays a unique role in host defense but not arthritis development in Lyme disease. J Immunol *173*, 2003-2010.

Bolz, D.D., and Weis, J.J. (2004). Molecular Mimicry to Borrelia burgdorferi: Pathway to Autoimmunity? Autoimmunity *37*, 387-392. https://doi.org/doi:10.1080/08916930410001713098

Bose, S., and Cho, J. (2013). Role of chemokine CCL2 and its receptor CCR2 in neurodegenerative diseases. Archives of pharmacal research *36*, 1039-1050. https://doi.org/10.1007/s12272-013-0161-z

Branda, J.A., Body, B.A., Boyle, J., Branson, B.M., Dattwyler, R.J., Fikrig, E., Gerald, N.J., Gomes-Solecki, M., Kintrup, M., Ledizet, M., *et al.* (2017). Advances in Serodiagnostic Testing for Lyme Disease Are at Hand. Clinical Infectious Diseases *66*, 1133-1139. https://doi.org/10.1093/cid/cix943 %J Clinical Infectious Diseases

Bransfield, R.C. (2018). Neuropsychiatric Lyme Borreliosis: An Overview with a Focus on a Specialty Psychiatrist's Clinical Practice. Healthcare (Basel, Switzerland) *6*. https://doi.org/10.3390/healthcare6030104

Breitschwerdt, E.B., Geoly, F.J., Meuten, D.J., Levine, J.F., Howard, P., Hegarty, B.C., and Stafford, L.C. (1996). Myocarditis in mice and guinea pigs experimentally infected with a canine-origin Borrelia isolate from Florida. Am J Vet Res *57*, 505-511.

Breitschwerdt, E.B., Nicholson, W.L., Kiehl, A.R., Steers, C., Meuten, D.J., and Levine, J.F. (1994). Natural infections with Borrelia spirochetes in two dogs from Florida. Journal of clinical microbiology *32*, 352-357.

Brooks, C.S., Hefty, P.S., Jolliff, S.E., and Akins, D.R. (2003). Global analysis of Borrelia burgdorferi genes regulated by mammalian host-specific signals. Infect Immun *71*, 3371-3383. https://doi.org/10.1128/iai.71.6.3371-3383.2003

Brown, C.R., Blaho, V.A., Fritsche, K.L., and Loiacono, C.M. (2006). Stat1 deficiency exacerbates carditis but not arthritis during experimental Lyme borreliosis. J Interferon Cytokine Res *26*, 390-399.

Brown, C.R., Blaho, V.A., and Loiacono, C.M. (2003). Susceptibility to experimental Lyme arthritis correlates with KC and monocyte chemoattractant protein-1 production in joints and requires neutrophil recruitment via CXCR2. J Immunol *171*, 893-901.

Brown, C.R., and Reiner, S.L. (1998). Clearance of Borrelia burgdorferi may not be required for resistance to experimental Lyme arthritis. Infect Immun *66*, 2065-2071.

Brown, C.R., and Reiner, S.L. (1999a). Development of Lyme arthritis in mice deficient in inducible nitric oxide synthase. J Infect Dis *179*, 1573-1576.

Brown, C.R., and Reiner, S.L. (1999b). Genetic control of experimental Lyme arthritis in the absence of specific immunity. Infect Immun *67*, 1967-1973.

Brown, C.R., and Reiner, S.L. (2000). Genes outside the major histocompatibility complex control resistance and susceptibility to experimental Lyme arthritis. Med Microbiol Immunol *189*, 85-90.

Brown, E.L., Wooten, R.M., Johnson, B.J.B., Iozzo, R.V., Smith, A., Dolan, M.C., Guo, B.P., Weis, J.J., and Hook, M. (2001). Resistance to Lyme disease in decorin-deficient mice. J Clin Invest *107*, 845-852.

Brown, J.P., Zachary, J.F., Teuscher, C., Weis, J.J., and Wooten, R.M. (1999). Dual role of interleukin-10 in murine Lyme disease: regulation of arthritis severity and host defense. Infect Immun *67*, 5142-5150.

Bugrysheva, J., Dobrikova, E.Y., Sartakova, M.L., Caimano, M.J., Daniels, T.J., Radolf, J.D., Godfrey, H.P., and Cabello, F.C. (2003). Characterization of the Stringent Response and relBbu Expression in Borrelia burgdorferi. Journal of Bacteriology *185*, 957-965. https://doi.org/10.1128/jb.185.3.957-965.2003

Burgdorfer, W. (1984). The New Zealand white rabbit: an experimental host for infecting ticks with Lyme disease spirochetes. Yale J Biol Med *57*, 609-612.

Burgdorfer, W., Barbour, A.G., Hayes, S.F., Benach, J.L., Grunwaldt, E., and Davis, J.P. (1982). Lyme disease-a tick-borne spirochetosis? Science *216*, 1317-1319. https://doi.org/10.1126/science.7043737

Burgdorfer, W., and Gage, K.L. (1987). Susceptibility of the hispid cotton rat (Sigmodon hispidus) to the Lyme disease spirochete (Borrelia burgdorferi). Am J Trop Med Hyg *37*, 624-628. https://doi.org/10.4269/ajtmh.1987.37.624

Burgess, E.C. (1986). Natural exposure of Wisconsin dogs to the Lyme disease spirochete (Borrelia burgdorferi). Lab Anim Sci *36*, 288-290.

Burgess, E.C. (1988). Borrelia burgdorferi infection in Wisconsin horses and cows. Ann N Y Acad Sci *539*, 235-243. https://doi.org/10.1111/j.1749-6632.1988.tb31857.x

Cadavid, D. (2006). The mammalian host response to borrelia infection. Wiener klinische Wochenschrift *118*, 653-658. https://doi.org/10.1007/s00508-006-0692-0

Cadavid, D., Bai, Y., Hodzic, E., Narayan, K., Barthold, S.W., and Pachner, A.R. (2004). Cardiac involvement in non-human primates infected with the Lyme disease spirochete Borrelia burgdorferi. Laboratory Investigation *84*, 1439-1450.

Cadavid, D., and Barbour, A.G. (1998). Neuroborreliosis during relapsing fever: review of the clinical manifestations, pathology, and treatment of infections in humans and experimental animals. Clinical Infectious Diseases *26*, 151-164.

Cadavid, D., Bundoc, V., and Barbour, A.G. (1993). Experimental infection of the mouse brain by a relapsing fever Borrelia species: a molecular analysis. Journal of Infectious Diseases *168*, 143-151.

Cadavid, D., Garcia, E., and Gelderblom, H. (2007). Coinfection with Borrelia turicatae serotype 2 prevents the severe vestibular dysfunction and earlier mortality caused by serotype 1. The Journal of infectious diseases *195*, 1686-1693. https://doi.org/10.1086/516783

Cadavid, D., and Londoño, D. (2009). Understanding tropism and immunopathological mechanisms of relapsing fever spirochaetes. Clinical microbiology and infection : the official publication of the European Society of Clinical Microbiology and Infectious Diseases *15*, 415-421. https://doi.org/10.1111/j.1469-0691.2009.02785.x

Cadavid, D., O'Neill, T., Schaefer, H., and Pachner, A.R. (2000a). Localization of Borrelia burgdorferi in the nervous system and other organs in a nonhuman primate model of lyme disease. Laboratory investigation; a journal of technical methods and pathology *80*, 1043-1054. https://doi.org/10.1038/labinvest.3780109

Cadavid, D., O'Neill, T., Schaefer, H., and Pachner, A.R. (2000b). Localization of Borrelia burgdorferi in the nervous system and other organs in a nonhuman primate model of lyme disease. Laboratory Investigation *80*, 1043-1054.

Cadavid, D., Pachner, A.R., Estanislao, L., Patalapati, R., and Barbour, A.G. (2001). Isogenic serotypes of Borrelia turicatae show different localization in the brain and skin of mice. Infection & Immunity *69*, 3389-3397.

Cadavid, D., Pennington, P.M., Kerentseva, T.A., Bergstrom, S., and Barbour, A.G. (1997). Immunologic and genetic analyses of VmpA of a neurotropic strain of Borrelia turicatae. Infect Immun *65*, 3352-3360.

Cadavid, D., Sondey, M., Garcia, E., and Lawson, C.L. (2006). Residual brain infection in relapsing-fever borreliosis. The Journal of infectious diseases *193*, 1451-1458. https://doi.org/10.1086/503367

Cadavid, D., Thomas, D.D., Crawley, R., and Barbour, A.G. (1994). Variability of a bacterial surface protein and disease expression in a possible mouse model of systemic Lyme borreliosis. J Exp Med *179*, 631-642. https://doi.org/10.1084/jem.179.2.631

Caimano, M.J. (2005). Cultivation of Borrelia burgdorferi in dialysis membrane chambers in rat peritonea. Current protocols in microbiology *Chapter 12*, Unit 12C.13. https://doi.org/10.1002/9780471729259.mc12c03s00

Callister, S.M., Schell, R.F., Case, K.L., Lovrich, S.D., and Day, S.P. (1993). Characterization of the borreliacidal antibody response to Borrelia burgdorferi in humans: a serodiagnostic test. J Infect Dis *167*, 158-164. https://doi.org/10.1093/infdis/167.1.158

Caskey, J.R., and Embers, M.E. (2015). Persister Development by Borrelia burgdorferi Populations In Vitro. Antimicrobial Agents and Chemotherapy *59*, 6288-6295. https://doi.org/10.1128/aac.00883-15

Cassatt, D.R., Patel, N.K., Ulbrandt, N.D., and Hanson, M.S. (1998). DbpA, but not OspA, is expressed by Borrelia burgdorferi during spirochemia and is a target for protective antibodies. Infect Immun *66*, 5379-5387.

Chandra, A., Wormser, G.P., Marques, A.R., Latov, N., and Alaedini, A. (2011). Anti-Borrelia burgdorferi antibody profile in post-Lyme disease syndrome. Clin Vaccine Immunol *18*, 767-771. https://doi.org/10.1128/cvi.00002-11

Chang, Y.-F., Ku, Y.-W., Chang, C.-F., Chang, C.-D., McDonough, S.P., Divers, T., Pough, M., and Torres, A. (2005). Antibiotic treatment of experimentally Borrelia burgdorferi-infected ponies. Vet Microbiol *107*, 285-294. https://doi.org/https://doi.org/10.1016/j.vetmic.2005.02.006

Chang, Y., Novosol, V., McDonough, S.P., Chang, C.F., Jacobson, R.H., Divers, T., Quimby, F.W., Shin, S., and Lein, D.H. (1999). Vaccination against lyme disease with recombinant Borrelia burgdorferi outer-surface protein A (rOspA) in horses. Vaccine *18*, 540-548. https://doi.org/10.1016/s0264-410x(99)00187-5

Chang, Y.F., Appel, M.J., Jacobson, R.H., Shin, S.J., Harpending, P., Straubinger, R., Patrican, L.A., Mohammed, H., and Summers, B.A. (1995). Recombinant OspA protects dogs against infection and disease caused by Borrelia burgdorferi. Infection & Immunity *63*, 3543-3549.

Chang, Y.F., Novosel, V., Chang, C.F., Summers, B.A., Ma, D.P., Chiang, Y.W., Acree, W.M., Chu, H.J., Shin, S., and Lein, D.H. (2001). Experimental induction of chronic borreliosis in adult dogs exposed to Borrelia burgdorferi-infected ticks and treated with dexamethasone. Am J Vet Res *62*, 1104-1112.

Chang, Y.F., Novosol, V., McDonough, S.P., Chang, C.F., Jacobson, R.H., Divers, T., Quimby, F.W., Shin, S., and Lein, D.H. (2000). Experimental infection of ponies with Borrelia burgdorferi by exposure to Ixodid ticks. Vet Pathol *37*, 68-76. https://doi.org/10.1354/vp.37-1-68

Charrier, N.P., Hermouet, A., Hervet, C., Agoulon, A., Barker, S.C., Heylen, D., Toty, C., McCoy, K.D., Plantard, O., and Rispe, C. (2019). A transcriptome-based phylogenetic study of hard ticks (Ixodidae). Scientific Reports *9*, 12923. https://doi.org/10.1038/s41598-019-49641-9

Chong-Cerillo, C., Wu, S.-Y., Wang, Y.-P., Blanco, D.R., and Lovett, M.A. (2000). Immune serum from rabbits infected with Borrelia burgdorferi B31 confers complete passive protection against homologous challenge. J Spirochetal Tick-Borne Dis *7*, 3-9.

Chong-Cerrillo, C., Shang, E.S., Blanco, D.R., Lovett, M.A., and Miller, J.N. (2001). Immunohistochemical analysis of Lyme disease in the skin of naive and infection-immune rabbits following challenge. Infect Immun *69*, 4094-4102.

Christopherson, J.A., Munson, E.L., England, D.M., Croke, C.L., Remington, M.C., Molitor, M.L., DeCoster, D.J., Callister, S.M., and Schell, R.F. (2003). Destructive arthritis in vaccinated interferon gamma-deficient mice challenged with Borrelia burgdorferi: modulation by tumor necrosis factor alpha. Clin Diag Lab Immunol *10*, 44-52.

Chu, H.J., Chavez, L.G., Jr., Blumer, B.M., Sebring, R.W., Wasmoen, T.L., and Acree, W.M. (1992). Immunogenicity and efficacy study of a commercial Borrelia burgdorferi bacterin. J Am Vet Med Assoc *201*, 403-411.

Coleman, J.L., Gebbia, J.A., Piesman, J., Degen, J.L., Bugge, T.H., and Benach, J.L. (1997a). Plasminogen is required for efficient dissemination of B. burgdorferi in ticks and for enhancement of spirochetemia in mice. Cell *89*, 1111-1119.

Coleman, J.L., Gebbia, J.A., Piesman, J., Degen, J.L., Bugge, T.H., and Benach, J.L. (1997b). Plasminogen is required for efficient dissemination of *Borrelia burgdorferi* in ticks and for enhancement of spirochetemia in mice. Cell *89*, 1111-1119.

Connolly, S.E., J.L. Benach (2001). Cutting edge: the spirochetemia of murine relapsing fever is cleared by complement-independent bactericidal antibodies. J Immunol *167*, 3029-3032.

Cooper, P.J., Fekade, D., Remick, D.G., Grint, P., Wherry, J., and Griffin, G.E. (2000). Recombinant human interleukin-10 fails to alter proinflammatory cytokine production or physiologic changes associated with the Jarisch-Herxheimer reaction. J Infect Dis *181*, 203-209. https://doi.org/10.1086/315183

Coughlin, J.M., Yang, T., Rebman, A.W., Bechtold, K.T., Du, Y., Mathews, W.B., Lesniak, W.G., Mihm, E.A., Frey, S.M., Marshall, E.S., *et al.* (2018). Imaging glial activation in patients with post-treatment Lyme disease symptoms: a pilot study using [(11)C]DPA-713 PET. Journal of neuroinflammation *15*, 346-346. https://doi.org/10.1186/s12974-018-1381-4

Coughlin, R.T., Fish, D., Mather, T.N., Ma, J., Pavia, C., and Bulger, P. (1995). Protection of dogs from Lyme disease with a vaccine containing outer surface protein (Osp) A, OspB, and the saponin adjuvant QS21. Journal of Infectious Diseases *171*, 1049-1052.

Craig-Mylius, K., Weber, G.F., Coburn, J., and Glickstein, L. (2005). *Borrelia burgdorferi*, an extracellular pathogen, circumvents osteopontin in inducing an inflammatory cytokine response. . J Leukoc Biol *77*, 710-718.

Crandall, H., Ma, Y., Dunn, D.M., Sundsbak, R.S., Zachary, J.F., Olofsson, P., Holmdahl, R., Weis, J.H., Weiss, R.B., Teucher, C., *et al.* (2005). Am J Pathol. 167.

Croke, C.L., Munson, E.L., Lovrich, S.D., Christopherson, J.A., Remington, M.C., England, D.M., Callister, S.M., and Schell, R.F. (2000). Occurrence of severe destructive Lyme arthritis in hamsters vaccinated with outer surface protein A and challenged with *Borrelia burgdorferi*. . Infect Immun *68*, 658-663.

Crossland, N.A., Alvarez, X., and Embers, M.E. (2018). Late Disseminated Lyme Disease: Associated Pathology and Spirochete Persistence Posttreatment in Rhesus Macaques. The American Journal of Pathology *188*, 672-682. https://doi.org/https://doi.org/10.1016/j.ajpath.2017.11.005

Crowder, C.D., Ghalyanchi Langeroudi, A., Shojaee Estabragh, A., Lewis, E.R.G., Marcsisin, R.A., and Barbour, A.G. (2016). Pathogen and Host Response Dynamics in a Mouse Model of Borrelia hermsii Relapsing Fever. Vet Sci *3*, 19. https://doi.org/10.3390/vetsci3030019

Crowley, H., Alroy, J., Sproule, T.J., Roopenian, D., and Huber, B.T. (2006). The MHC class I-related FcRn ameliorates murine Lyme arthritis. International Immunol *18*, 409-414.

Crowley, T.R., Champion, C.I., Whitelegge, J.P., Aguilera, R., Wu, X.Y., Blanco, D.R., Miller, J.N., and Lovett, M.A. (2004). Temporal analysis of the antigenic composition of Borrelia burgdorferi during infection in rabbit skin. Infect Immun *72*, 5063-5072.

Czub, S., Duray, P.H., Thomas, R.E., and Schwan, T.G. (1992). Cystitis induced by infection with the Lyme disease spirochete, Borrelia burgdorferi, in mice. Am J Pathol *141*, 1173-1179.

De Silva, A.M., and Fikrig, E. (1995). Growth and migration of Borrelia burgdorferi in Ixodes ticks during blood feeding. Am J Trop Med Hyg *53*, 397-404. https://doi.org/10.4269/ajtmh.1995.53.397

De Silva, A.M., Fikrig, E., Hodzic, E., Kantor, F.S., Telford, S.R., 3rd, and Barthold, S.W. (1998). Immune evasion by tickborne and host-adapted Borrelia burgdorferi. Journal of Infectious Diseases *177*, 395-400.

Defosse, D.L., Duray, P.H., and Johnson, R.C. (1992). The NIH-3 immunodeficient mouse is a model for Lyme borreliosis myositis and carditis. Am J Pathol *141*, 3-10.

Delong, A.K., Blossom, B., Maloney, E.L., and Phillips, S.E. (2012). Antibiotic retreatment of Lyme disease in patients with persistent symptoms: a biostatistical review of randomized, placebo-controlled, clinical trials. Contemp Clin Trials *33*, 1132-1142. https://doi.org/10.1016/j.cct.2012.08.009

deSilva, A., Fikrig, E., Hodzic, E., TelfordIII, S.R., and Barthold, S.W. (1998). Immune evasion by tick-borne and host-adapted *Borrelia burgdorferi*. J Infect Dis *177*, 395-400.

deSouza, M.S., Smith, A.L., Beck, D.S., Kim, L.J., Hansen, G.M., Jr., and Barthold, S.W. (1993). Variant responses of mice to *Borrelia burgdorferi* depending on the site of intradermal inoculation. Infect Immun *61*, 4493-4497.

Divan, A., Casselli, T., Narayanan, S.A., Mukherjee, S., Zawieja, D.C., Watt, J.A., Brissette, C.A., and Newell-Rogers, M.K. (2018). Borrelia burgdorferi adhere to blood vessels in the dura mater and are associated with increased meningeal T cells during murine disseminated borreliosis. PloS one *13*, e0196893-e0196893. https://doi.org/10.1371/journal.pone.0196893

Divers, T.J., Gardner, R.B., Madigan, J.E., Witonsky, S.G., Bertone, J.J., Swinebroad, E.L., Schutzer, S.E., and Johnson, A.L. (2018). Borrelia burgdorferi Infection and Lyme Disease in North American Horses: A Consensus Statement. J Vet Intern Med *32*, 617-632. https://doi.org/10.1111/jvim.15042

Dolan, M.C., Breuner, N.E., Hojgaard, A., Boegler, K.A., Hoxmeier, J.C., Replogle, A.J., and Eisen, L. (2017). Transmission of the Lyme Disease Spirochete Borrelia mayonii in Relation to Duration of Attachment by Nymphal Ixodes scapularis (Acari: Ixodidae). Journal of medical entomology *54*, 1360-1364. https://doi.org/10.1093/jme/tjx089

Dolan, M.C., Piesman, J., Schneider, B.S., Schriefer, M., Brandt, K., and Zeidner, N.S. (2004). Comparison of disseminated and nondisseminated strains of Borrelia burgdorferi sensu stricto in mice naturally infected by tick bite. Infect Immun *72*, 5262-5266.

Donahue, J.G., Piesman, J., and Spielman, A. (1987). Reservoir competence of white-footed mice for Lyme disease spirochetes. Am J Trop Med Hyg *36*, 92-96.

Du Chateau, B.K., England, D.M., Callister, S.M., Lim, L.C., Lovrich, S.D., and Schell, R.F. (1996). Macrophages exposed to Borrelia burgdorferi induce Lyme arthritis in hamsters. Infect Immun *64*, 2540-2547. https://doi.org/10.1128/iai.64.7.2540-2547.1996

DuChateau, B.K., Jensen, J.R., England, D.M., Callister, S.M., Lovrich, S.D., and Schell, R.F. (1997). Macrophages and enriched populations of T lymphocytes interact synergistically for the induction of severe, destructive Lyme arthritis. Infect Immun *65*, 2829-2836. https://doi.org/10.1128/iai.65.7.2829-2836.1997

DuChateau, B.K., Munson, E.L., England, D.M., Lovrich, S.D., Callister, S.M., Jensen, J.R., and Schell, R.F. (1999). Macrophages interact with enriched populations of distinct T lymphocyte subsets for the induction of severe destructive Lyme arthritis. Journal of leukocyte biology *65*, 162-170. https://doi.org/10.1002/jlb.65.2.162

Duray, P.H., and Johnson, R.C. (1986). The histopathology of experimentally infected hamsters with the Lyme disease spirochete, Borrelia burgdorferi. Proceedings of the Society for Experimental Biology and Medicine Society for Experimental Biology and Medicine (New York, NY) *181*, 263-269. https://doi.org/10.3181/00379727-181-42251

Embers, M.E., Barthold, S.W., Borda, J.T., Bowers, L., Doyle, L., Hodzic, E., Jacobs, M.B., Hasenkampf, N.R., Martin, D.S., Narasimhan, S., *et al.* (2012a). Persistence of *Borrelia burgdorferi* in Rhesus Macaques following Antibiotic Treatment of Disseminated Infection. PLoS ONE *7*, e29914. https://doi.org/10.1371/journal.pone.0029914

Embers, M.E., Hasenkampf, N.R., Barnes, M.B., Didier, E.S., Philipp, M.T., and Tardo, A.C. (2016). A Five-Antigen Fluorescent Bead-based Assay for Diagnosis of Lyme Disease. Clinical and Vaccine Immunology. https://doi.org/10.1128/cvi.00685-15

Embers, M.E., Hasenkampf, N.R., Embers, D.G., and Doyle, L.A. (2013). Pharmacokinetic analysis of oral doxycycline in rhesus macaques. *42*, 57-61. https://doi.org/10.1111/jmp.12031

Embers, M.E., Hasenkampf, N.R., Jacobs, M.B., and Philipp, M.T. (2012b). Dynamic Longitudinal Antibody Responses during Borrelia burgdorferi Infection and Antibiotic Treatment of Rhesus Macaques. Clinical and Vaccine Immunology *19*, 1218-1226. https://doi.org/10.1128/cvi.00228-12

Embers, M.E., Hasenkampf, N.R., Jacobs, M.B., Tardo, A.C., Doyle-Meyers, L.A., Philipp, M.T., and Hodzic, E. (2017). Variable manifestations, diverse seroreactivity and post-treatment persistence in non-human primates exposed to Borrelia burgdorferi by tick feeding. PLoS ONE *12*, e0189071. https://doi.org/10.1371/journal.pone.0189071

Embers, M.E., Krishnavajhala, A., Armstrong, B., Curtis, M.W., Pahar, B., Wilder, H.K., Allen, M., Beare, P., Hasenkampf, N.R., and Lopez, J.E. (2019). Immunological responses to the relapsing fever spirochete <em>Borrelia turicatae</em> in infected Rhesus macaques: implications for pathogenesis and diagnosis. Infection and Immunity, IAI.00900-00918. https://doi.org/10.1128/IAI.00900-18

Embers, M.E., Liang, F.T., Howell, J.K., Jacobs, M.B., Purcell, J.E., Norris, S.J., Johnsonn, B.J.B., and Philipp, M.T. (2007). Antigenicity and recombination of VlsE, the antigenic variation protein of Borrelia burgdorferi, in rabbits, a host putatively resistannt to long-term infection with this spirochete. FEMS Immunol Med Microbiol *50*, 421-429.

Embers, M.E., and Narasimhan, S. (2013). Vaccination against Lyme disease: past, present and future. Frontiers in Cellular and Infection Microbiology *3*. https://doi.org/10.3389/fcimb.2013.00006

England, J.D., Bohm, R.P., Jr., Roberts, E.D., and Philipp, M.T. (1997a). Lyme neuroborreliosis in the rhesus monkey. Semin Neurol *17*, 53-56. https://doi.org/10.1055/s-2008-1040913

England, J.D., Bohm, R.P., Jr., Roberts, E.D., and Philipp, M.T. (1997b). Mononeuropathy multiplex in rhesus monkeys with chronic Lyme disease. Ann Neurol *41*, 375-384. https://doi.org/10.1002/ana.410410313

Exner, M.M., Wu, X., Blanco, D.R., and Miller, J.N. (2000). Protection elicited by native outer membrane protein Oms66 (P66) against host-adapted Borrelia burgdorferi: Conformational nature of bactericidal epitopes. Infect Immun *68*, 2647-2654.

Fallon, B.A., Keilp, J.G., Corbera, K.M., Petkova, E., Britton, C.B., Dwyer, E., Slavov, I., Cheng, J., Dobkin, J., Nelson, D.R., *et al.* (2008). A randomized, placebo-controlled trial of repeated IV antibiotic therapy for Lyme encephalopathy. Neurology *70*, 992-1003. https://doi.org/10.1212/01.WNL.0000284604.61160.2d

Felsenfeld, O., and Wolf, R.H. (1975). Reinfection of vervet monkeys (Cercopithecus aethiops) with Borrelia hermsii. Research Communications in Chemical Pathology & Pharmacology *11*, 147-150.

Feng, S., Barthold, S.W., Bockenstedt, L.K., Zaller, D.M., and Fikrig, E. (1995). Lyme disease in human DR4Dw4-transgenic mice. J Infect Dis *172*, 286-289.

Feng, S., Hodzic, E., and Barthold, S.W. (2000). Lyme arthritis resolution with antiserum to a 37-kilodalton Borrelia burgdorferi protein. Infection & Immunity *68*, 4169-4173.

Feng, S., Hodzic, E., Freet, K., and Barthold, S.W. (2003). Immunogenicity of Borrelia burgdorferi arthritis-related protein. Infection & Immunity *71*, 7211-7214.

Feng, S., Hodzic, E., Stevenson, B., and Barthold, S.W. (1998). Humoral immunity to *Borrelia burgdorferi* N40 decorin binding proteins during infection of laboratory mice. Infect Immun *66*, 2827-2835.

Fikrig, E., and Barthold, S.W. (1995). Lyme borreliosis in transgenic mice expression the *Borrelia burgdorferi* flagellin epitope implicated in human neuroborreliosis. Submitted.

Fikrig, E., Barthold, S.W., Chen, M., Chang, C.-H., and Flavell, R.A. (1997a). Protective antibodies develop, and murine Lyme arthritis regresses, in the absence of MHC class II and CD4+ T cells. J Immunol *159*, 5682-5686.

Fikrig, E., Barthold, S.W., Chen, M., Grewal, I.S., Craft, J., and Flavell, R.A. (1996a). Protective antibodies in murine Lyme disease arise independently of CD40 ligand. J Immunol *157*, 1-3.

Fikrig, E., Barthold, S.W., Chen, M., Tao, M., Ali-Salaam, P., Bockenstedt, L.K., and Flavell, R.A. (1997b). Lyme borreliosis in transgenic mice tolerant to OspA from *Borrelia burgdorferi* 25015. J Infect Dis *175*, 1000-1003.

Fikrig, E., Bockenstedt, L.K., Barthold, S.W., Chen, M., Tao, H., Ali-Salaam, P., Telford, S.R., and Flavell, R.A. (1994). Sera from patients with chronic Lyme disease protect mice from Lyme borreliosis. Journal of Infectious Diseases *169*, 568-574.

Fikrig, E., Tao, H., Chen, M., Barthold, S.W., and Flavell, R.A. (1996b). Lyme arthritis in transgenic mice tolerant to *Borrelia burgdorferi* OspA or OspB. J Clin Invest *96*, 1706-1714.

Foley, D.M., Gayek, R.J., Skare, J.T., Wagar, E.A., Champion, C.I., Blanco, D.R., Lovett, M.A., and Miller, J.N. (1995a). Rabbit model of Lyme borreliosis: erythema migrans, infection-derived immunity, and identification of Borrelia burgdorferi proteins associated with virulence and protective immunity. J Clin Invest *96*, 965-975. https://doi.org/10.1172/jci118144

Foley, D.M., Gayek, R.J., Skare, J.T., Wagar, E.A., Champion, C.I., Blanco, D.R., Lovett, M.A., and Miller, J.N. (1995b). Rabbit model of Lyme borreliosis: Erythema migrans, infection-derived immunity, and identification of *Borrelia burgdorferi* proteins associated with virulence and protective immunity. J Clin Invest *96*, 965-975.

Foley, D.M., Wang, Y.P., Wu, X.Y., Blanco, D.R., Lovett, M.A., and Miller, J.M. (1997). Acquired resistance to Borrelia burgdorferi infection in the rabbit. Comparison between outer surface protein A vaccine- and infection-derived immunity. J Clin Invest *99*, 2030-2035.

Francis, E. (1938). Longevity of the tick Ornithodoros turicata and of Spirochaeta recurrentis with this tick. . Publ Hlth Rep *53*, 2220-2241.

Galbe, J.L., Guy, E., Zapatero, J.M., Peerschke, E.I.B., and Benach, J.L. (1993). Vascular clearance of Borrelia burgdorferi in rats. Microbial Pathogenesis *14*, 187-201.

Garcia-Monco, J.C., and Benach, J.L. (1995). Lyme neuroborreliosis. Annals of Neurology *37*, 691-702.

Garcia-Monco, J.C., and Benach, J.L. (2019). Lyme Neuroborreliosis: Clinical Outcomes, Controversy, Pathogenesis, and Polymicrobial Infections. Annals of neurology *85*, 21-31. https://doi.org/10.1002/ana.25389

Garcia-Monco, J.C., FernandezVillar, B., CalvoAlen, J., and Benach, J.L. (1990). *Borrelia burgdorferi* in the central nervous system: experimental and clinical evidence for early invasion. J Infect Dis *161*, 1187-1193.

Garcia-Monco, J.C., Miller, N.S., Backenson, P.B., Anda, P., and Benach, J.L. (1997). A mouse model of Borrelia meningitis after intradermal injection. The Journal of infectious diseases *175*, 1243-1245. https://doi.org/10.1086/593681

Gebbia, J.A., Monco, J.C., Degen, J.L., Bugge, T.H., and Benach, J.L. (1999). The plasminogen activation system enhances brain and heart invasion in murine relapsing fever borreliosis. The Journal of clinical investigation *103*, 81-87. https://doi.org/10.1172/jci5171

Gelderblom, H., Londoño, D., Bai, Y., Cabral, E.S., Quandt, J., Hornung, R., Martin, R., Marques, A., and Cadavid, D. (2007a). High production of CXCL13 in blood and brain during persistent infection with the relapsing fever spirochete Borrelia turicatae. Journal of neuropathology and experimental neurology *66*, 208-217. https://doi.org/10.1097/01.jnen.0000248556.30209.6d

Gelderblom, H., Schmidt, J., Londono, D., Bai, Y., Quandt, J., Hornung, R., Marques, A., Martin, R., and Cadavid, D. (2007b). Role of interleukin 10 during persistent infection with the relapsing fever Spirochete Borrelia turicatae. American Journal of Pathology *170*, 251-262.

Gern, L., Schaible, U.E., and Simon, M.M. (1993). Mode of Inoculation of the Lyme Disease Agent Borrelia burgdorferi Influences Infection and Immune Responses in Inbred Strains of Mice. Journal of Infectious Diseases *167*, 971-975. https://doi.org/10.1093/infdis/167.4.971

Glickstein, L., Edelstein, M., and Dong, J.Z. (2001). Gamma interferon is not required for arthritis resistance in the murine Lyme disease model. Infect Immun *69*, 3737-3743.

Golde, W.T., Burkot, T.R., Sviat, S., Keen, M.G., Mayer, L.W., Johnson, B.J., and Piesman, J. (1993). The major histocompatibility complex-restricted response of recombinant inbred strains of mice to natural tick transmission of Borrelia burgdorferi. J Exp Med *177*, 9-17. https://doi.org/10.1084/jem.177.1.9

Golde, W.T., Kappel, K.J., Dequesne, G., Feron, C., Plainchamp, D., Capiau, C., and Lobet, Y. (1994). Tick transmission of Borrelia burgdorferi to inbred strains of mice induces an antibody response to P39 but not to outer surface protein A. Infect Immun *62*, 2625-2627. https://doi.org/10.1128/iai.62.6.2625-2627.1994

Gondolf, K.B., Mihatsch, M., Curscellas, E., Dunn, J.J., and Batsford, S.R. (1994). Induction of experimental allergic arthritis with outer surface proteins of Borrelia burgdorferi. Arthritis Rheum *37*, 1070-1077.

Goodman, J.L., Jurkovich, P., Kodner, C., and Johnson, R.C. (1991). Persistent cardiac and urinary tract infections with Borrelia burgdorferi in experimentally infected Syrian hamsters. J Clin Microbiol *29*, 894-896.

Griffin, J.F.T. (2002). A strategic approach to vaccine development: animal models, monitoring vaccine efficacy, formulation and delivery. Advanced Drug Delivery Reviews *54*, 851-861. https://doi.org/https://doi.org/10.1016/S0169-409X(02)00072-8

Guerau-de-Arellano, M., Alroy, J., Bullard, D., and Huber, B.T. (2005a). Aggravated Lyme carditis in CD11a-/- and CD11c-/- mice. Infect Immun *73*, 7637-7643.

Guerau-de-Arellano, M., Alroy, J., and Huber, B.T. (2005b). Beta2 integrins control the severity of murine Lyme carditis. Infect Immun *73*, 3242-3250.

Gustafson, J.M., Burgess, E.C., Wachal, M.D., and Steinberg, H. (1993). Intrauterine transmission of Borrelia burgdorferi in dogs. Am J Vet Res *54*, 882-890.

Hagman, K.E., Yang, X., Wikel, S.K., Schoeler, G.B., Caimano, M.J., Radolph, J.D., and Norgard, M.V. (2000). Decorin-binding protein A (DbpA) of Borrelia burgdorferi is not protective when immunized mice are challenged via tick infestation and correlates with the lack of DbpA expression by B. burgdorferi in ticks. Infect Immun *68*, 4759-4764.

Halperin, J.J. (2015). Nervous System Lyme Disease. Clin Lab Med *35*, 779-795. https://doi.org/https://doi.org/10.1016/j.cll.2015.07.002

Halperin, J.J. (2017). Neuroborreliosis. Journal of Neurology *264*, 1292-1297. https://doi.org/10.1007/s00415-016-8346-2

Halperin, J.J. (2019). Lyme neuroborreliosis. Current Opinion in Infectious Diseases *32*.

Hanson, M.S., Cassatt, D.R., Guo, B.P., Patel, N.K., McCarthy, M.P., Dorward, D.W., and Hook, M. (1998). Active and passive immunity against *Borrelia burgdorferi* decorin binding protein A (DbpA) protects against infection. Infect Immun *66*, 2143-2153.

Harada, A., Sekido, N., Akahoshi, T., Wada, T., Mukaida, N., and Matsushima, K. (1994). Essential involvement of interleukin-8 (IL-8) in acute inflammation. Journal of leukocyte biology *56*, 559-564.

Hedrick, M.N., Olson, C., Conze, D.B., Bates, T.C., Rincon, M., and Anguita, J. (2006). Control of *Borrelia burgdorferi*-specific CD4+-T-cell effector function by interleukin-12- and T-cell receptor-induced p38 mitogen-activated protein kinase activity. Infect Immun *74*, 5713-5717.

Hejka, A., Schmitz, J.L., England, D.M., Callister, S.M., and Schell, R.F. (1989). Histopathology of Lyme arthritis in LSH hamsters. Am J Pathol *134*, 1113-1123.

Henningsson, A.J., Lager, M., Brännström, R., Tjernberg, I., and Skogman, B.H. (2018). The chemokine CXCL13 in cerebrospinal fluid in children with Lyme neuroborreliosis. European journal of clinical microbiology & infectious diseases : official publication of the European Society of Clinical Microbiology *37*, 1983-1991. https://doi.org/10.1007/s10096-018-3334-3

Hodzic, E., Feng, S., Freet, K.J., and Barthold, S.W. (2003). Borrelia burgdorferi population dynamics and prototype gene expression during infection of immunocompetent and immunodeficient mice. Infection & Immunity *71*, 5042-5055.

Hodzic, E., Feng, S., Freet, K.J., Borjesson, D.L., and Barthold, S.W. (2002). Borrelia burgdorferi population kinetics and selected gene expression at the host-vector interface. Infection & Immunity *70*, 3382-3388.

Hodzic, E., Feng, S., Holden, K., Freet, K.J., and Barthold, S.W. (2008). Persistence of Borrelia burgdorferi following antibiotic treatment in mice. Antimicrob Agents Chemother *52*, 1728-1736.

Hodzic, E., Imai, D., Feng, S., and Barthold, S.W. (2014). Resurgence of Persisting Non-Cultivable *Borrelia burgdorferi* following Antibiotic Treatment in Mice. PLoS ONE *9*, e86907. https://doi.org/10.1371/journal.pone.0086907

Hua, C.M., Cheminade, Y., Perret, J.L., Weynants, V., Lobet, Y., and Gern, L. (2003). Early detection of *Borrelia burgdorferi* sensu lato infection of Balb/c mice by co-feeding *Ixodes ricinus* ticks. . Int J Med Microbiol *293*, 421-426.

Hughes, C.A., Engstrom, S.M., Coleman, L.A., Kodner, C.B., and Johnson, R.C. (1993a). Protective immunity is induced by a Borrelia burgdorferi mutant that lacks OspA and OspB. Infect Immun *61*, 5115-5122. https://doi.org/10.1128/iai.61.12.5115-5122.1993

Hughes, C.A.N., Engstrom, S.M., Coleman, L.A., Kodner, C.B., and Johnson, R.C. (1993b). Protective immunity is induced by a *Borrelia burgdorferi* mutant that lacks OspA and OspB. Infect Immun *61*, 5115-5122.

Hunfeld, K.P., Burg, S., Hanssen-Huber, C., Karas, M., Brade, V., and Kraiczy, P. (2008). Changes ion the expression pattern of structural proteins after exposure of Borrelia burgdorferi to penicillin G and doxycycline. Int J Med Microbiol *ePub, ahead of print*.

Hurtenbach, U., Museteanu, C., Gasser, J., Schaible, U.E., and Simon, M.M. (1995). Studies on early events of *Borrelia burgdorferi*-induced cytokine production in immunodeficient SCID mice by using a tissue chamber model for acute inflammation. Int J Exp Pathol *76*, 111-.

Imai, D.M., Barr, B.C., Daft, B., Bertone, J.J., Feng, S., Hodzic, E., Johnston, J.M., Olsen, K.J., and Barthold, S.W. (2011). Lyme neuroborreliosis in 2 horses. Vet Pathol *48*, 1151-1157. https://doi.org/10.1177/0300985811398246

Iyer, R., Caimano, M.J., Luthra, A., Axline, D., Jr., Corona, A., Iacobas, D.A., Radolf, J.D., and Schwartz, I. (2015). Stage-specific global alterations in the transcriptomes of Lyme disease spirochetes during tick feeding and following mammalian host adaptation. Mol Microbiol *95*, 509-538. https://doi.org/10.1111/mmi.12882

Jacobson, A.C., Ma, Y., Zachary, J.F., Weis, J.J., and Weis, J.H. (2007). Mice lacking CD21 and CD35 proteins mount effective immune responses against Borrelia burgdorferi infection. Infect Immun *75*, 2075-2078.

James, F.M., Engiles, J.B., and Beech, J. (2010). Meningitis, cranial neuritis, and radiculoneuritis associated with Borrelia burgdorferi infection in a horse. J Am Vet Med Assoc *237*, 1180-1185. https://doi.org/10.2460/javma.237.10.1180

Jameson, S.C., and Masopust, D. (2018). What Is the Predictive Value of Animal Models for Vaccine Efficacy in Humans? Reevaluating the Potential of Mouse Models for the Human Immune System. Cold Spring Harbor perspectives in biology *10*. https://doi.org/10.1101/cshperspect.a029132

Johnson, B.J., Sviat, S.L., Happ, C.M., Dunn, J.J., Frantz, J.C., Mayer, L.W., and Piesman, J. (1995). Incomplete protection of hamsters vaccinated with unlipidated OspA from Borrelia burgdorferi infection is associated with low levels of antibody to an epitope defined by mAb LA-2. Vaccine *13*, 1086-1094. https://doi.org/10.1016/0264-410x(95)00035-y

Johnson, R.C., Kodner, C., and Russell, M. (1986). Passive immunization of hamsters against experimental infection with *Borrelia burgdorferi*. Infect Immun *53*, 713-716.

Johnson, R.C., Kodner, C., Russell, M., and Duray, P.H. (1988). Experimental infection of hamsters with *Borrelia burgdorferi*. Ann NY Acad Sci *539*, 258-263.

Johnson, R.C., Marek, N., and Kodner, C. (1984). Infection of Syrian hamsters with Lyme disease spirochetes. J Clin Microbiol *20*, 1099-1101.

Jonsson, M., Elmros, T., and Berstrom, S. (1995). Subcutaneous implanted chambers in different mouse strains as an animal model to study genetic stability during infection with Lyme disease Borrelia. Microbial Pathogenesis *18*, 109-114.

Judge, D.M., La Croix, J.T., and Perine, P.L. (1974a). Experimental louse-borne relapsing fever in the grivet monkey, Cercopithecus aethiops. I. Clinical course. Am J Trop Med Hyg *23*, 957-961. https://doi.org/10.4269/ajtmh.1974.23.957

Judge, D.M., La Croix, J.T., and Perine, P.L. (1974b). Experimental louse-borne relapsing fever in the grivet monkey, Cercopithecus aethiops. II. Pathology. Am J Trop Med Hyg *23*, 962-968. https://doi.org/10.4269/ajtmh.1974.23.962

Judge, D.M., La Croix, J.T., and Perine, P.L. (1974c). Experimental louse-borne relapsing fever in the grivet monkey, Cercopithecus aethiops. III. Crisis following therapy. Am J Trop Med Hyg *23*, 969-973. https://doi.org/10.4269/ajtmh.1974.23.969

Kaneda, K., Masuzawa, T., Simon, M.M., Isogai, E., Isogai, H., Yasugami, K., Suzuki, T., Suzuki, Y., and Yanagihara, Y. (1998). Infectivity and arthritis induction of Borrelia japonica on SCID mice and immune competent mice: possible role of galactosylceramide binding activity on initiation of infection. Microbiol Immunol *42*, 171-175. https://doi.org/10.1111/j.1348-0421.1998.tb02268.x

Kang, I., Barthold, S.W., and Bockenstedt, L.K. (1995). Lyme borreliosis in B-cell deficient mice. Arthritis Rheum *38*, S345.

Karrasch, M., Fingerle, V., Boden, K., Darr, A., Baier, M., Straube, E., and Nenadic, I. (2018). Neuroborreliosis and acute encephalopathy: The use of CXCL13 as a biomarker in CNS manifestations of Lyme borreliosis. Ticks Tick Borne Dis *9*, 415-417. https://doi.org/10.1016/j.ttbdis.2017.12.008

Kazragis, R.J., Dever, L.L., Jorgensen, J.H., and Barbour, A.G. (1996). In vivo activities of ceftriaxone and vancomycin against Borrelia spp. in the mouse brain and other sites. Antimicrob Agents Chemother *40*, 2632-2636. https://doi.org/10.1128/aac.40.11.2632

KeaneMyers, A., Maliszewski, C.R., Finkelman, F.D., and Nickell, S.P. (1996). Recombinant IL-4 treatment augments resisitance to *Borrelia burgdorferi* infections in both normal susceptible and antibody-deficient susceptible mice. J Immunol *156*, 2488-2494.

Keilp, J.G., Corbera, K., Gorlyn, M., Oquendo, M.A., Mann, J.J., and Fallon, B.A. (2019). Neurocognition in Post-Treatment Lyme Disease and Major Depressive Disorder. Archives of clinical neuropsychology : the official journal of the National Academy of Neuropsychologists *34*, 466-480. https://doi.org/10.1093/arclin/acy083

Kishimoto, T. (2006). Interleukin-6: discovery of a pleiotropic cytokine. Arthritis Res Ther *8*, S2. https://doi.org/10.1186/ar1916

Klempner, M.S. (2002). Controlled trials of antibiotic treatment in patients with post-treatment chronic Lyme disease. Vector Borne & Zoonotic Diseases *2*, 255-263.

Klempner, M.S., Hu, L.T., Evans, J., Schmid, C.H., Johnson, G.M., Trevino, R.P., Norton, D., Levy, L., Wall, D., McCall, J., *et al.* (2001). Two controlled trials of antibiotic treatment in patients with persistent symptoms and a history of Lyme disease. New England Journal of Medicine *345*, 85-92.

Konishi, T., and Nakao, M. (1997). Antibiotic treatment in mice infected with Japanese Borrelia garinii: efficacy of ceftriaxone for eradicating the infection induced by Ixodes persulcatus tick bites. Microbiol Immunol *41*, 165-168. https://doi.org/10.1111/j.1348-0421.1997.tb01182.x

Kornblatt, A.N., Steere, A.C., and Brownstein, D.G. (1984a). Infection in rabbits with the Lyme disease spirochete. The Yale journal of biology and medicine *57*, 613-616.

Kornblatt, A.N., Steere, A.C., and Brownstein, D.G. (1984b). Infection in rabbits with the Lyme disease spirochete. Yale J Biol Med *57*, 613-614.

Korshus, J.B., Munderloh, U.G., Bey, R.F., and Kurtti, T.J. (2004). Experimental infection of dogs with Borrelia burgdorferi sensu stricto using Ixodes scapularis ticks artificially infected by capillary feeding. Medical Microbiology & Immunology *193*, 27-34.

Krinsky, W.L., Brown, S.J., and Askenase, P.W. (1982). Ixodes dammini skin lesions in guinea pigs and rabbits compared to erythema chronicum migrans in patients with Lyme arthritis. . Exp Parasitol *53*, 381-395.

Krupka, I., and Straubinger, R.K. (2010). Lyme Borreliosis in Dogs and Cats: Background, Diagnosis, Treatment and Prevention of Infections with Borrelia burgdorferi sensu stricto. Veterinary Clinics of North America: Small Animal Practice *40*, 1103-1119. https://doi.org/https://doi.org/10.1016/j.cvsm.2010.07.011

Kumar, H., Belperron, A., Barthold, S.W., Bendelac, B., and Bockenstedt, L.K. (2000). CD1d deficiency impairs murine host defense against the spirochete, Borrelia burgdorferi. J Immunol *165*, 4797-4801.

Kurtti, T.J., Munderloh, U.G., Hughes, C.A., Engstrom, S.M., and Johnson, R.C. (1996). Resistance to tick-borne spirochete challenge induced by Borrelia burgdorferi strains that differ in expression of outer surface proteins. Infect Immun *64*, 4148-4153. https://doi.org/10.1128/iai.64.10.4148-4153.1996

Larsson, C., Andersson, M., Guo, B.P., Nordstrand, A., Hagerstrand, I., Carlsson, S., and Bergstrom, S. (2006a). Complications of pregnancy and transplacental transmission of relapsing-fever borreliosis. Journal of Infectious Diseases *194*, 1367-1374. https://doi.org/10.1086/508425

Larsson, C., Andersson, M., Pelkonen, J., Guo, B.P., Nordstrand, A., and Bergstrom, S. (2006b). Persistent brain infection and disease reactivation in relapsing fever borreliosis. Microbes & Infection *8*, 2213-2219.

Larsson, C., Lundqvist, J., van Rooijen, N., and Bergström, S. (2009). A novel animal model of Borrelia recurrentis louse-borne relapsing fever borreliosis using immunodeficient mice. PLoS neglected tropical diseases *3*, e522-e522. https://doi.org/10.1371/journal.pntd.0000522

Lawrenz, M.B., Wooten, R.M., Zachary, J.F., Drouin, S.M., Weis, J.J., Wetsel, R.A., and Norris, S.J. (2003). Effect of complement component 3 deficiency on experimental Lyme borreliosis in mice. Infect Immun *71*, 4432-4440.

Lazarus, J.J., Meadows, M.J., Lintner, R.E., and Wooten, R.M. (2006). IL-10 deficiency promotes increased *Borrelia burgdorferi* clearance predominantly through enhanced innate immune responses. J Immunol *177*.

Levy, S.A., and Duray, P.H. (1988). Complete heart block in a dog seropositive for Borrelia burgdorferi. Similarity to human Lyme carditis. J Vet Intern Med *2*, 138-144.

Levy, S.A., and Magnarelli, L.A. (1992). Relationship between development of antibodies to Borrelia burgdorferi in dogs and the subsequent development of limb/joint borreliosis. J Am Vet Med Assoc *200*, 344-347.

Lewis, K. (2007). Persister cells, dormancy and infectious disease. Nature Reviews/Microbiology *5*, 48-56.

Liang, F.T., Aberer, E., Cinco, M., Gern, L., Hu, C.M., Lobet, Y.N., Ruscio, M., Voet, P.E., Jr., Weynants, V.E., and Philipp, M.T. (2000). Antigenic conservation of an immunodominant invariable region of the VlsE lipoprotein among European pathogenic genospecies of Borrelia burgdorferi SL. J Infect Dis *182*, 1455-1462. https://doi.org/10.1086/315862

Liang, F.T., Alvarez, A.L., Gu, Y., Nowling, J.M., Ramamoorthy, R., and Philipp, M.T. (1999a). An Immunodominant Conserved Region Within the Variable Domain of VlsE, the Variable Surface Antigen of <em>Borrelia burgdorferi</em>. The Journal of Immunology *163*, 5566.

Liang, F.T., Brown, E.L., Wang, T., Iozzo, R.V., and Fikrig, E. (2004). Protective niche for *Borrelia burgdorferi* to evade humoral immunity. Amer J Pathol *165*, 977-985.

Liang, F.T., Jacobs, M.B., Bowers, L.C., and Philipp, M.T. (2002a). An immune evasion mechanism for spirochetal persistence in Lyme borreliosis. J Exp Med *195*, 415-422. https://doi.org/10.1084/jem.20011870

Liang, F.T., Nelson, F.K., and Fikrig, E. (2002b). Molecular adaptation of Borrelia burgdorferi in the murine host. J Exp Med *196*, 275-280. https://doi.org/10.1084/jem.20020770

Liang, F.T., Steere, A.C., Marques, A.R., Johnson, B.J., Miller, J.N., and Philipp, M.T. (1999b). Sensitive and specific serodiagnosis of Lyme disease by enzyme-linked immunosorbent assay with a peptide based on an immunodominant conserved region of Borrelia burgdorferi vlsE. Journal of clinical microbiology *37*, 3990-3996.

Lim, L.C., England, D.M., DuChateau, B.K., Glowacki, N.J., Creson, J.R., Lovrich, S.D., Callister, S.M., Jobe, D.A., and Schell, R.F. (1994a). Development of destructive arthritis in vaccinated hamsters challenged with Borrelia burgdorferi. Infect Immun *62*, 2825-2833. https://doi.org/10.1128/iai.62.7.2825-2833.1994

Lim, L.C., England, D.M., DuChateau, B.K., Glowacki, N.J., and Schell, R.F. (1995a). Borrelia burgdorferi-specific T lymphocytes induce severe destructive Lyme arthritis. Infect Immun *63*, 1400-1408. https://doi.org/10.1128/iai.63.4.1400-1408.1995

Lim, L.C., England, D.M., Glowacki, N.J., DuChateau, B.K., and Schell, R.F. (1995b). Involvement of CD4+ T lymphocytes in induction of severe destructive Lyme arthritis in inbred LSH hamsters. Infect Immun *63*, 4818-4825. https://doi.org/10.1128/iai.63.12.4818-4825.1995

Lim, L.C.L., England, D.M., DuChateau, B.K., Glowacki, N.J., Creson, J.R., Lovrich, S.D., Callister, S.M., Jobe, D.A., and Schell, R.F. (1994b). Development of destructive arthritis in vaccinated hamsters challenged with *Borrelia burgdorferi*. Infect Immunq *62*, 2825-2833.

Lindenmayer, J.M., Marshall, D., and Onderdonk, A.B. (1991). Dogs as sentinels for Lyme disease in Massachusetts. Am J Public Health *81*, 1448-1455.

Lischer, C.J., Leutenegger, C.M., Braun, U., and Lutz, H. (2000). Diagnosis of Lyme disease in two cows by the detection of Borrelia burgdorferi DNA. Vet Rec *146*, 497-499. https://doi.org/10.1136/vr.146.17.497

Little, S.E., Heise, S.R., Blagburn, B.L., Callister, S.M., and Mead, P.S. (2010). Lyme borreliosis in dogs and humans in the USA. Trends Parasitol *26*, 213-218. https://doi.org/10.1016/j.pt.2010.01.006

Littman, M.P., Goldstein, R.E., Labato, M.A., Lappin, M.R., and Moore, G.E. (2006). ACVIM small animal consensus statement on Lyme disease in dogs: diagnosis, treatment, and prevention. J Vet Intern Med *20*, 422-434.

Liu, N., Montgomery, R.R., Barthold, S.W., and Bockenstedt, L.K. (2004). Myeloid differentiation antigen 88 deficiency impairs pathogen clearance but does not alter inflammation in *Borrelia burgdorferi*-infected mice. Infect Immun *72*, 3195-3203.

Londoño, D., Bai, Y., Zückert, W.R., Gelderblom, H., and Cadavid, D. (2005). Cardiac apoptosis in severe relapsing fever borreliosis. Infect Immun *73*, 7669-7676. https://doi.org/10.1128/iai.73.11.7669-7676.2005

Londoño, D., and Cadavid, D. (2010). Bacterial lipoproteins can disseminate from the periphery to inflame the brain. Am J Pathol *176*, 2848-2857. https://doi.org/10.2353/ajpath.2010.091235

Londoño, D., Carvajal, J., Arguelles-Grande, C., Marques, A., and Cadavid, D. (2008). Interleukin 10 protects the brain microcirculation from spirochetal injury. Journal of neuropathology and experimental neurology *67*, 976-983. https://doi.org/10.1097/NEN.0b013e318187a279

Londono, D., Marques, A., Hornung, R.L., and Cadavid, D. (2008). Relapsing fever borreliosis in interleukin-10-deficient mice. Infection & Immunity *76*, 5508-5513.

Long, A.D., Baldwin-Brown, J., Tao, Y., Cook, V.J., Balderrama-Gutierrez, G., Corbett-Detig, R., Mortazavi, A., and Barbour, A.G. (2019). The genome of Peromyscus leucopus, natural host for Lyme disease and other emerging infections. Sci Adv *5*, eaaw6441-eaaw6441. https://doi.org/10.1126/sciadv.aaw6441

Lopez, J.E., Vinet-Oliphant, H., Wilder, H.K., Brooks, C.P., Grasperge, B.J., Morgan, T.W., Stuckey, K.J., and Embers, M.E. (2014). Real-time monitoring of disease progression in rhesus macaques infected with Borrelia turicatae by tick bite. Journal of Infectious Diseases *210*, 1639-1648.

Lovrich, S.D., Callister, S.M., DuChateau, B.K., Lim, L.C., Winfrey, J., Day, S.P., and Schell, R.F. (1995). Abilities of OspA proteins from different seroprotective groups of Borrelia burgdorferi to protect hamsters from infection. Infect Immun *63*, 2113-2119. https://doi.org/10.1128/iai.63.6.2113-2119.1995

Lovrich, S.D., Callister, S.M., Lim, L.C.L., DuChateau, B.K., and Schell, R.F. (1994). Seroprotective groups of Lyme borreliosis spirochetes from North American and Europe. J Infect Dis *170*, 115-121.

Lovrich, S.D., Callister, S.M., Lim, L.C.L., and Schell, R.F. (1993). Seroprotective groups among isolates of *Borrelia burgdorferi*. Infect Immun *61*, 4367-4374.

Lovrich, S.D., Callister, S.M., Schmitz, J.L., Alder, J.F., and Schell, R.F. (1991). Borreliacidal activity of sera from hamsters infected with the Lyme disease spirochete. Infect Immun *59*, 2522-2528.

Lusitani, D., Malawista, S.E., and Montgomery, R.R. (2003). Calprotectin, an abundant cytosolic protein from human polymorphonuclear leukocytes, inhibits the growth of *Borrelia burgdorferi*. . Infect Immun *71*, 4711-4716.

Ma, Y., Bramwell, K.K.C., Lochhead, R.B., Paquette, J.K., Zachary, J.F., Weis, J.H., Teuscher, C., and Weis, J.J. (2014). Borrelia burgdorferi arthritis-associated locus Bbaa1 regulates Lyme arthritis and K/B×N serum transfer arthritis through intrinsic control of type I IFN production. Journal of immunology (Baltimore, Md : 1950) *193*, 6050-6060. https://doi.org/10.4049/jimmunol.1401746

Ma, Y., Miller, J.C., Crandall, H., Larsen, E.T., Dunn, D.M., Weiss, R.B., Subramanian, M., Weis, J.H., Zachary, J.F., Teuscher, C., *et al.* (2009). Interval-specific congenic lines reveal quantitative trait Loci with penetrant lyme arthritis phenotypes on chromosomes 5, 11, and 12. Infection and immunity *77*, 3302-3311. https://doi.org/10.1128/IAI.00396-09

Ma, Y., Seiler, K.P., Eichwald, E.J., Weis, J.H., Teuscher, C., and Weis, J.J. (1998). Distinct characteristics of resistance to Borrelia burgdorferi-induced arthritis in C57BL/6N mice. Infect Immun *66*, 161-168. https://doi.org/10.1128/iai.66.1.161-168.1998

Malawista, S.E., Barthold, S.W., and Persing, D.H. (1994). Fate of Borrelia burgdorferi DNA in tissues of infected mice after antibiotic treatment. Journal of Infectious Diseases *170*, 1312-1316.

Markert, M., Trautmann, T., Krause, F., Cioaga, M., Mouriot, S., Wetzel, M., and Guth, B.D. (2018). A new telemetry-based system for assessing cardiovascular function in group-housed large animals. Taking the 3Rs to a new level with the evaluation of remote measurement via cloud data transmission. Journal of pharmacological and toxicological methods *93*, 90-97. https://doi.org/10.1016/j.vascn.2018.03.006

Marques, A.R. (2015). Lyme Neuroborreliosis. *21*, 1729-1744. https://doi.org/10.1212/con.0000000000000252

Martinez, A.N., Burmeister, A.R., Ramesh, G., Doyle-Meyers, L., Marriott, I., and Philipp, M.T. (2017). Aprepitant limits in vivo neuroinflammatory responses in a rhesus model of Lyme neuroborreliosis. J Neuroinflammation *14*, 37. https://doi.org/10.1186/s12974-017-0813-x

Masuzawa, T., Beppu, Y., Kawabata, H., Yanagihara, Y., Iwamoto, Y., Shimizu, T., and Johnson, R.C. (1992). Experimental Borrelia burgdorferi infection of outbred mice. J Clin Microbiol *30*, 3016-3018. https://doi.org/10.1128/jcm.30.11.3016-3018.1992

Masuzawa, T., Kurita, T., Kawabata, H., Muramatsu, K., and Uchikawa, K. (1994a). Infectivity and early antibody response to *Borrelia burgdorferi* sensu lato isolated in Japan in outbred mice. Microbiol Immunol *38*, 621-.

Masuzawa, T., Kurita, T., Kawabata, H., Suzuki, H., and Yanagihara, Y. (1994b). Experimental induction of Lyme arthritis in outbred mice. Microbiol Immunol *38*, 263-268.

Mather, T.N., TelfordIII, S.R., and Adler, G.H. (1991). Absence of transplacental transmission of Lyme disease spirochetes from reservoir mice (Peromyscus leucopus) to their offspring. J Infect Dis *164*, 564-567.

Matyniak, J.E., and Reiner, S.L. (1995). T helper phenotype and genetic susceptibility in experimental Lyme disease. J Exp Med *181*, 1251-1254.

Mbow, M.L., Zeidner, N., Gilmore, R.D., Dolan, M., Piesman, J., and Titus, R.G. (2001). Major histocompatibility complex class II-independent generation of neutralizing antibodies against T-cell-dependent Borrelia burgdorferi antigens presented by dendritic cells: regulation by NK and gamma/delta T cells. Infect Immun *69*, 2407-2415.

McKisic, M.D., and Barthold, S.W. (2000a). T-cell-independent responses to *Borrelia burgdorferi* are critical for protective immunity and resolution of Lyme disease. Infect Immun *68*, 5190-5197.

McKisic, M.D., and Barthold, S.W. (2000b). T-Cell-Independent Responses to Borrelia burgdorferiAre Critical for Protective Immunity and Resolution of Lyme Disease. Infection and Immunity *68*, 5190-5197. https://doi.org/10.1128/iai.68.9.5190-5197.2000

McKisic, M.D., Redmond, W.L., and Barthold, S.W. (2000a). Cutting edge: T cell-mediated pathology in murine Lyme borreliosis. Journal of Immunology *164*, 6096-6099.

McKisic, M.D., Redmond, W.L., and Barthold, S.W. (2000b). Cutting edge: T cell-mediated pathology in murine lyme borreliosis. J Immunol *164*, 6096-6099.

Mehra, R., Londoño, D., Sondey, M., Lawson, C., and Cadavid, D. (2009). Structure-function investigation of vsp serotypes of the spirochete Borrelia hermsii. PLoS One *4*, e7597. https://doi.org/10.1371/journal.pone.0007597

Miller, J.C., Ma, Y., Bian, J., Sheehan, K.C.F., Zachary, J.F., Weis, J.H., Schreiber, R.D., and Weis, J.J. (2008). A critical role for type I IFN in arthritis development following Borrelia burgdorferi infection of mice. Journal of immunology (Baltimore, Md : 1950) *181*, 8492-8503. https://doi.org/10.4049/jimmunol.181.12.8492

Montgomery, R.R., Booth, C.J., Wang, X., Blaho, V.A., Malawista, S.E., and Brown, C.R. (2007). Recruitment of macrophages and polymorphonuclear leukocytes in Lyme carditis. Infect Immun *75*, 613-620.

Montgomery, R.R., Schreck, K., Wang, X., and Malawista, S.E. (2006). Human neutrophil calprotectin reduces the susceptibility of *Borrelia burgdorferi* to penicillin. . Infect Immun *74*, 2468-2472.

Moody, K.D., Adams, R.L., and Barthold, S.W. (1994a). Effectiveness of antimicrobial treatment against Borrelia burgdorferi infection in mice. Antimicrobial Agents & Chemotherapy *38*, 1567-1572.

Moody, K.D., and Barthold, S.W. (1991). Relative infectivity of *Borrelia burgdorferi* in Lewis rats by various routes of inoculation. Am J Trop Med Hyg *44*, 135-139.

Moody, K.D., Barthold, S.W., and Terwilliger, G.A. (1990a). Lyme borreliosis in laboratory animals: effect of host species and in vitro passage of Borrelia burgdorferi. Am J Trop Med Hyg *43*, 87-92.

Moody, K.D., Barthold, S.W., and Terwilliger, G.A. (1990b). Lyme borreliosis in laboratory animals: effect of host species and in vitro passage of Borrelia burgdorferi. Am J Trop Med Hyg *43*, 87-92. https://doi.org/10.4269/ajtmh.1990.43.87

Moody, K.D., Terwilliger, G.A., Hansen, G.M., and Barthold, S.W. (1994b). Experimental *Borrelia burgdorferi* infection in *Peromyscus leucopus*. J Wildlife Dis *30*, 155-161.

Motameni, A.R., Bates, T.C., Juncadella, I.J., Petty, C., Hedrick, M.N., and Anguita, J. (2005). Distinct bacterial dissemination and disease outcome in mice subcutaneously infected with Borrelia burgdorferi in the midline of the back and the footpad. FEMS Immunol Med Microbiol *45*, 279-284.

Muehlenbachs, A., Bollweg, B.C., Schulz, T.J., Forrester, J.D., DeLeon Carnes, M., Molins, C., Ray, G.S., Cummings, P.M., Ritter, J.M., Blau, D.M., *et al.* (2016). Cardiac Tropism of Borrelia burgdorferi: An Autopsy Study of Sudden Cardiac Death Associated with Lyme Carditis. The American Journal of Pathology *186*, 1195-1205. https://doi.org/https://doi.org/10.1016/j.ajpath.2015.12.027

Murphree Bacon, R.K., Kiersten J. ; Mead, Paul S. ; (2008). Surveillance for lyme disease--United States, 1992-2006.

Museteanu, C., Schaible, U.E., Stehle, T., Kramer, M.D., and Simon, M.M. (1991). Myositis in mice inoculated with Borrelia burgdorferi. Amer J Pathol *139*, 1267-1271.

Narayan, K., Dail, D., Li, L., Cadavid, D., Amrute, S., Fitzgerald-Bocarsly, P., and Pachner, A.R. (2005). The nervous system as ectopic germinal center: CXCL13 and IgG in lyme neuroborreliosis. Annals of neurology *57*, 813-823. https://doi.org/10.1002/ana.20486

Nazario, S., Das, S., de Silva, A.M., Deponte, K., Marcantonio, N., Anderson, J.F., Fish, D., Fikrig, E., and Kantor, F.S. (1998). Prevention of Borrelia burgdorferi transmission in guinea pigs by tick immunity. Am J Trop Med Hyg *58*, 780-785. https://doi.org/10.4269/ajtmh.1998.58.780

Nordstrand, A., Shamaei-Tousi, A., Ny, A., and Bergström, S. (2001). Delayed invasion of the kidney and brain by Borrelia crocidurae in plasminogen-deficient mice. Infect Immun *69*, 5832-5839. https://doi.org/10.1128/iai.69.9.5832-5839.2001

Obonyo, M., Munderloh, U.G., Sam, T.N., and Kurtti, T.J. (2002). Cultivation at 37 degrees C enhances Borrelia burgdorferi sensu stricto infectivity for hamsters. Medical microbiology and immunology *191*, 33-39. https://doi.org/10.1007/s00430-002-0116-3

Ohnishi, J., Piesman, J., and de Silva, A.M. (2001a). Antigenic and genetic heterogeneity of Borrelia burgdorferi populations transmitted by ticks. Proc Natl Acad Sci U S A *98*, 670-675. https://doi.org/10.1073/pnas.98.2.670

Ohnishi, J., Piesman, J., and deSilva, A. (2001b). Antigenic and genetic heterogeneity of Borrelia burgdorferi populations transmitted by ticks. Proc Natl Acad Sci USA *98*, 670-675.

Pachner, A.R., Basta, J., Delaney, E., and Hulinska, D. (1995a). Localization of *Borrelia burgdorferi* in murine Lyme borreliosis by electron microscopy. Am J Trop Med Hyg *52*, 128-133.

Pachner, A.R., Cadavid, D., Shu, G., Dail, D., Pachner, S., Hodzic, E., and Barthold, S.W. (2001). Central and peripheral nervous system infection, immunity, and inflammation in the NHP model of Lyme borreliosis. Annals of Neurology *50*, 330-338.

Pachner, A.R., Dail, D., Bai, Y., Sondey, M., Pak, L., Narayan, K., and Cadavid, D. (2004). Genotype determines phenotype in experimental Lyme borreliosis. Annals of Neurology *56*, 361-370. https://doi.org/10.1002/ana.20192

Pachner, A.R., Dail, D., Li, L., Gurey, L., Feng, S., Hodzic, E., and Barthold, S. (2002a). Humoral immune response associated with lyme borreliosis in nonhuman primates: analysis by immunoblotting and enzyme-linked immunosorbent assay with sonicates or recombinant proteins. Clin Diagn Lab Immunol *9*, 1348-1355.

Pachner, A.R., Dail, D., Narayan, K., Dutta, K., and Cadavid, D. (2002b). Increased expression of B-lymphocyte chemoattractant, but not pro-inflammatory cytokines, in muscle tissue in rhesus chronic Lyme borreliosis. Cytokine *19*, 297-307. https://doi.org/10.1006/cyto.2002.1973

Pachner, A.R., Delaney, E., and O'Neill, T. (1995b). Neuroborreliosis in the nonhuman primate: Borrelia burgdorferi persists in the central nervous system. Annals of Neurology *38*, 667-669.

Pachner, A.R., Delaney, E., O'Neill, T., and Major, E. (1995c). Inoculation of nonhuman primates with the N40 strain of Borrelia burgdorferi leads to a model of Lyme neuroborreliosis faithful to the human disease. Neurology *45*, 165-172.

Pachner, A.R., Delaney, E., and Ricalton, N.S. (1992). Murine Lyme borreliosis: route of inoculation determines immune response and infectivity. Regional immunology *4*, 345-351.

Pachner, A.R., Delaney, E., Zhang, W.F., O'Neill, T., Major, E., Frey, A.B., and Davidson, E. (1999). Protection from Lyme Neuroborreliosis in Nonhuman Primates with a Multiantigenic Vaccine. Clinical Immunology *91*, 310-313. https://doi.org/https://doi.org/10.1006/clim.1999.4703

Pachner, A.R., and Steere, A.C. (1985). The triad of neurologic manifestations of Lyme disease: meningitis, cranial neuritis, and radiculoneuritis. Neurology *35*, 47-53. https://doi.org/10.1212/wnl.35.1.47

Padilla, M.L., Callister, S.M., Schell, R.F., Bryant, G.L., Jobe, D.A., Lovrich, S.D., DuChateau, B.K., and Jensen, J.R. (1996). Characterization of the protective borreliacidal antibody response in humans and hamsters after vaccination with a Borrelia burgdorferi outer surface protein A vaccine. J Infect Dis *174*, 739-746. https://doi.org/10.1093/infdis/174.4.739

Pal, U., Wang, P., Bao, F., Yang, X., Samanta, S., Schoen, R., Wormser, G.P., Schwartz, I., and Fikrig, E. (2008). Borrelia burgdorferi basic membrane proteins A and B participate in the genesis of Lyme arthritis. J Exp Med *205*, 133-141.

Parenti, D.L. (2000). Outer surface protein A and arthritis in hamsters. Infect Immun *68*, 7212.

Parker, J.L., and White, K.K. (1992). Lyme borreliosis in cattle and horses: a review of the literature. The Cornell veterinarian *82*, 253-274.

Pavia, C., Inchiosa, M.A., Jr., and Wormser, G.P. (2002). Efficacy of short-course ceftriaxone therapy for Borrelia burgdorferi infection in C3H mice. Antimicrob Agents Chemother *46*, 132-134. https://doi.org/10.1128/aac.46.1.132-134.2002

Penningon, P.M., Cadavid, D., Bunikis, J., Norris, S.J., and Barbour, A.G. (1999). Extensive interplasmidic duplications change the virulence phenotype of the relapsing fever agent Borrelia turicatae. Molecular Microbiology *34*, 1120-1132.

Pennington, P.M., Cadavid, D., and Barbour, A.G. (1999). Characterization of VspB of Borrelia turicatae, a major outer membrane protein expressed in blood and tissues of mice. Infect Immun *67*, 4637-4645. https://doi.org/10.1128/iai.67.9.4637-4645.1999

Persing, D.H., Telford, S.R., III, Rys, P.N., Dodge, D.E., White, T.J., Malawista, S.E., and Spielman, A. (1990). Detection of *Borrelia burgdorferi* DNA in museum specimens of *Ixodes dammini* ticks. Science *249*, 1420-1423.

Pflughoeft, K.J., Mash, M., Hasenkampf, N.R., Jacobs, M.B., Tardo, A.C., Magee, D.M., Song, L., LaBaer, J., Philipp, M.T., Embers, M.E., *et al.* (2019). Multi-platform Approach for Microbial Biomarker Identification Using Borrelia burgdorferi as a Model. Frontiers in cellular and infection microbiology *9*, 179-179. https://doi.org/10.3389/fcimb.2019.00179

Philipp, M., Lobet, Y., Bohm, R., Conway, M., Dennis, V., Desmons, P., Gu, Y., Hanser, P., Lowrie, R., Roberts, E.J.J.O.S., *et al.* (1996). Safety and immunogenicity of recombinant outer surface protein A (OspA) vaccine formulations in the rhesus monkey. *3*, 67-79.

Philipp, M.T., Aydintug, M.K., Bohm, R.P., Jr., Cogswell, F.B., Dennis, V.A., Lanners, H.N., Lowrie, R.C., Jr., Roberts, E.D., Conway, M.D., Karaçorlu, M., *et al.* (1993). Early and early disseminated phases of Lyme disease in the rhesus monkey: a model for infection in humans. Infection and immunity *61*, 3047-3059.

Philipp, M.T., Bowers, L.C., Fawcett, P.T., Jacobs, M.B., Liang, F.T., Marques, A.R., Mitchell, P.D., Purcell, J.E., Ratterree, M.S., and

Straubinger, R.K. (2001). Antibody response to IR6, a conserved immunodominant region of the VlsE lipoprotein, wanes rapidly after antibiotic treatment of Borrelia burgdorferi infection in experimental animals and in humans. Journal of Infectious Diseases *184*, 870-878.

Philipp, M.T., Lobet, Y., Bohm, R.P., Jr., Roberts, E.D., Dennis, V.A., Gu, Y., Lowrie, R.C., Jr., Desmons, P., Duray, P.H., England, J.D., *et al.* (1997). The outer surface protein A (OspA) vaccine against Lyme disease: efficacy in the rhesus monkey. Vaccine *15*, 1872-1887. https://doi.org/10.1016/s0264-410x(97)00133-3

Piesman, J., Mather, T.N., Sinsky, R.J., and Spielman, A. (1987). Duration of tick attachment and *Borrelia burgdorferi* transmission. J Clin Microbiol *25*, 557-558.

Potter, M.R., Noben-Trauth, N., Weis, J.H., Teurscher, C., and Weis, J.J. (2000). Interleukin-4 (IL-4) and IL-13 signaling pathways do not regulate Borrelia burgdorferi-induced arthritis in mice: IgG1 is not required for host control of tissue spirochetes. Infect Immun *68*, 5603-5609.

Potter, M.R., Rittling, S.R., Denhardt, D.T., Roper, R.J., Weis, J.H., Teuscher, C., and Weis, J.J. (2002). Role of osteopontin in murine Lyme arthritis and host defense against Borrelia burgdorferi. Infect Immun *70*, 1372-1381.

Preac Mursic, V., Patsouris, E., Wilske, B., Reinhardt, S., Gross, B., and Mehraein, P. (1990). Persistence of Borrelia burgdorferi and histopathological alterations in experimentally infected animals. A comparison with histopathological findings in human Lyme disease. Infection *18*, 332-341. https://doi.org/10.1007/bf01646399

Ramesh, G., Borda, J.T., Dufour, J., Kaushal, D., Ramamoorthy, R., Lackner, A.A., and Philipp, M.T. (2008). Interaction of the Lyme disease spirochete Borrelia burgdorferi with brain parenchyma elicits inflammatory mediators from glial cells as well as glial and neuronal apoptosis. American Journal of Pathology *173*, 1415-1427.

Ramesh, G., Didier, P.J., England, J.D., Santana-Gould, L., Doyle-Meyers, L.A., Martin, D.S., Jacobs, M.B., and Philipp, M.T. (2015). Inflammation in the pathogenesis of lyme neuroborreliosis. Am J Pathol *185*, 1344-1360. https://doi.org/10.1016/j.ajpath.2015.01.024

Rauer, S., Kastenbauer, S., Hofmann, H., Fingerle, V., Huppertz, H.-I., Hunfeld, K.-P., Krause, A., Ruf, B., Dersch, R., and Consensus, g. (2020). Guidelines for diagnosis and treatment in neurology - Lyme neuroborreliosis. Ger Med Sci *18*, Doc03-Doc03. https://doi.org/10.3205/000279

Raveche, E.S., Schutzer, S.E., Fernandes, H., Bateman, H., McCarthy, B.A., Nickell, S.P., and Cunningham, M.W. (2005). Evidence of Borrelia autoimmunity-induced component of Lyme carditis and arthritis. J Clin Microbiol *43*, 850-856. https://doi.org/10.1128/jcm.43.2.850-856.2005

Rawlings, J.A. (1995). An overview of tick-borne relapsing fever with emphasis on outbreaks in Texas. Tex Med *91*, 56-59.

Ray, A., Kumar, D., Shakya, A., Brown, C.R., Cook, J.L., and Ray, B.K. (2004). Serum amyloid A-activating factor-1 (SAF-1) transgenic mice are prone to develop a severe form of inflammation-induced arthritis. J Immunol *173*, 4684-4691.

Reik, L., Steere, A.C., Bartenhagen, N.H., Shope, R.E., and Malawista, S.E. (1979). Neurologic Abnormalities of Lyme Disease. Medicine *58*.

Roberts, E.D., Bohm, R.P., Jr., Cogswell, F.B., Lanners, H.N., Lowrie, R.C., Jr., Povinelli, L., Piesman, J., and Philipp, M.T. (1995). Chronic lyme disease in the rhesus monkey. Laboratory investigation; a journal of technical methods and pathology *72*, 146-160.

Roberts, E.D., Bohm, R.P., Jr., Lowrie, R.C., Jr., Habicht, G., Katona, L., Piesman, J., and Philipp, M.T. (1998). Pathogenesis of Lyme neuroborreliosis in the rhesus monkey: the early disseminated and chronic phases of disease in the peripheral nervous system. J Infect Dis *178*, 722-732.

Roehrig, J.T., Piesman, J., Hunt, A.R., Keen, M.G., Happ, C.M., and Johnson, B.J. (1992). The hamster immune response to tick-transmitted Borrelia burgdorferi differs from the response to needle-inoculated, cultured organisms. J Immunol *149*, 3648-3653.

Roper, R.J., Weis, J.J., McCracken, B.A., Green, C.B., Ma, Y., Weber, K.S., Fairbairn, D., Butterfield, R.J., Potter, M.R., Zachary, J.F., *et al.* (2001). Genetic control of susceptibility to experimental Lyme arthritis is polygenic and exhibits consistent linkage to multiple loci on chromosome 5 in four independent mouse crosses. Genes Immun *2*, 388-397.

Rousselle, J.C., Callister, S.M., Schell, R.F., Lovrich, S.D., Jobe, D.A., Marks, J.A., and Wieneke, C.A. (1998). Borreliacidal antibody production against outer surface protein C of Borrelia burgdorferi. J Infect Dis *178*, 733-741. https://doi.org/10.1086/515382

Ruderman, E.M., Kerr, J.S., Telford, S.R., 3rd, Spielman, A., Glimcher, L.H., and Gravallese, E.M. (1995). Early murine Lyme carditis has a macrophage predominance and is independent of major histocompatibility complex class II-CD4+ T cell interactions. J Infect Dis *171*, 362-370. https://doi.org/10.1093/infdis/171.2.362

Rupprecht, T.A., Manz, K.M., Fingerle, V., Lechner, C., Klein, M., Pfirrmann, M., and Koedel, U. (2018). Diagnostic value of cerebrospinal fluid CXCL13 for acute Lyme neuroborreliosis. A systematic review and meta-analysis. Clin Microbiol Infect *24*, 1234-1240. https://doi.org/10.1016/j.cmi.2018.04.007

Saba, S., VanderBrink, B.A., Perides, G., Glickstein, L.J., Link, M.S., Homoud, M.K., Bronson, R.T., Estes3rd, M., and Wang, P.J. (2001). Cardiac conduction abnormalities in a mouse model of Lyme borreliosis. J Interv Card Electrophysiol *5*, 137-143.

Sadowsky, D.W., Adams, K.M., Gravett, M.G., Witkin, S.S., and Novy, M.J. (2006). Preterm labor is induced by intraamniotic infusions of interleukin-1beta and tumor necrosis factor-alpha but not by interleukin-6 or interleukin-8 in a nonhuman primate model. Am J Obstet Gynecol *195*, 1578-1589. https://doi.org/10.1016/j.ajog.2006.06.072

Satoskar, A.R., Elizondo, J., Monteforte, G.M., Stamm, L.M., Bluethmann, H., Katavolos, P., and TelfordIII, S.R. (2000). Interleukin-4-deficient BALB/c mice develop and enhanced Th1-like response but control cardiac inflammation following Borrelia burgdorferi infection. FEMS Microbiol Lett *183*, 319-325.

Schaible, U., Kramer, M., Museteanu, C., Zimmer, G., Mossman, H., and Simon, M.M. (1989a). The severe combined immunodeficient mouse: A laboratory model for the analysis of Lyme arthritis and carditis. J Exp Med *170*, 1427-1432.

Schaible, U.E., Gern, L., Wallich, R., Kramer, M.D., Prester, M., and Simon, M.M. (1993). Distinct patterns of protective antibodies are generated against Borrelia burgdorferi in mice experimentally inoculated with high and low doses of antigen. Immunology letters *36*, 219-226. https://doi.org/10.1016/0165-2478(93)90056-8

Schaible, U.E., Kramer, M.D., Eichmann, K., Modolell, M., Museteanu, C., and Simon, M.M. (1990a). Monoclonal antibodies specific for the outer surface protein A (OspA) of Borrelia burgdorferi prevent Lyme borreliosis in severe combined immunodeficiency (scid) mice. Proceedings of the National Academy of Sciences of the United States of America *87*, 3768-3772.

Schaible, U.E., Kramer, M.D., Justus, C.W., Museteanu, C., and Simon, M.M. (1989b). Demonstration of antigen-specific T cells and histopathological alterations in mice experimentally inoculated with Borrelia burgdorferi. Infect Immun *57*, 41-47. https://doi.org/10.1128/iai.57.1.41-47.1989

Schaible, U.E., Kramer, M.D., Museteanu, C., Zimmer, G., Mossmann, H., and Simon, M. (1990b). Lyme borreliosis in the severe combined immunodeficiency (scid) mouse manifests predominantly in the joints, heart and liver. Amer J Pathol *137*, 811-820.

Schaible, U.E., Kramer, M.D., Wallich, R., Tran, T., and Simon, M.M. (1991). Experimental Borrelia burgdorferi infection in inbred mouse strains: antibody response and association of H-2 genes with resistance and susceptibility to development of arthritis. European Journal of Immunology *21*, 2397-2405.

Schaible, U.E., Wallich, R., Kramer, M.D., Nerz, G., Stehle, T., Museteanu, C., and Simon, M.M. (1994). Protection against *Borrelia burgdorferi* infection in SCID mice is conferred by presensitized spleen cells and partially by B- but not T-cells alone. Internl Immun *6*, 671-681.

Schmitz, J.L., Lovrich, S.D., Callister, S.M., and Schell, R.F. (1991a). Depletion of complement and effects on passive transfer of resistance to infection with Borrelia burgdorferi. Infect Immun *59*, 1831-3818.

Schmitz, J.L., Schell, R.F., Callister, S.M., Lovich, S.D., Day, S.P., and Coe, J.E. (1992). Immunoglobulin G2 confers protection against Borrelia burgdorferi infection in LSH hamsters. Infect Immun *60*, 2677-2682.

Schmitz, J.L., Schell, R.F., Hejka, A.G., and England, D.M. (1990). Passive immunization prevents induction of Lyme arthritis in LSH hamsters. Infect Immun *58*, 144-148.

Schmitz, J.L., Schell, R.F., Hejke, A., England, D.M., Callister, S.M., and Schell, R.F. (1989). Induction of Lyme arthritis in LSH hamsters. Amer J Pathol *134*, 1113-1123.

Schmitz, J.L., Schell, R.F., Lovrich, S.D., Callister, S.M., and Coe, J.E. (1991b). Characterization of the protective antibody response to *Borrelia burgdorferi* in experimentally infected LSH hamsters. Infect Immun *59*, 1916-1921.

Schutzer, S.E., Body, B.A., Boyle, J., Branson, B.M., Dattwyler, R.J., Fikrig, E., Gerald, N.J., Gomes-Solecki, M., Kintrup, M., Ledizet, M., *et al.* (2018). Direct Diagnostic Tests for Lyme Disease. Clinical Infectious Diseases *68*, 1052-1057. https://doi.org/10.1093/cid/ciy614 %J Clinical Infectious Diseases

Schwan, T.G., Burgdorfer, W., Schrumpf, M.E., and Karstens, R.H. (1988). The urinary bladder: a consistent source of Borrelia burgdorferi in experimentally infected white-footed mice (Peromyscus leucopus). . J Clin Microbiol *26*, 893-895.

Schwan, T.G., Karstens, R.H., Schrumpf, M.E., and Simpson, W.J. (1991). Changes in antigenic reactivity of Borrelia burgdorferi, the Lyme disease spirochete, during persistent infection in mice. Canadian journal of microbiology *37*, 450-454. https://doi.org/10.1139/m91-074

Schwan, T.G., Kime, K.K., Schrumpf, M.E., Coe, J.E., and Simpson, W.J. (1989). Antibody response in white-footed mice (Peromyscus leucopus) experimentally infected with the Lyme disease spirochete (Borrelia burgdorferi). Infection and immunity *57*, 3445-3451. https://doi.org/10.1128/IAI.57.11.3445-3451.1989

Seiler, K.P., Ma, Y., Weis, J.H., Frenette, P.S., Hynes, R.O., Wagner, D.D., and Weis, J.J. (1998). E and P selectins are not required for resistance to severe murine Lyme arthritis. Infect Immun *66*, 4557-4559.

Sethi, N., Sondey, M., Bai, Y., Kim, K.S., and Cadavid, D. (2006). Interaction of a neurotropic strain of Borrelia turicatae with the cerebral microcirculation system. Infect Immun *74*, 6408-6418. https://doi.org/10.1128/iai.00538-06

Shaarawy, M., and Nagui, A.R. (1997). Enhanced expression of cytokines may play a fundamental role in the mechanisms of immunologically mediated recurrent spontaneous abortion. Acta Obstet Gynecol Scand *76*, 205-211. https://doi.org/10.3109/00016349709048142

Shadick, N.A., Phillips, C.B., Logigian, E.L., Steere, A.C., Kaplan, R.F., Berardi, V.P., Duray, P.H., Larson, M.G., Wright, E.A., Ginsburg, K.S., *et al.* (1994). The long-term clinical outcomes of Lyme disease. A population-based retrospective cohort study. Annals of Internal Medicine *121*, 560-567.

Shamaei-Tousi, A., Burns, M.J., Benach, J.L., Furie, M.B., Gergel, E.I., and Bergström, S. (2000). The relapsing fever spirochaete, Borrelia crocidurae, activates human endothelial cells and promotes the transendothelial migration of neutrophils. Cellular microbiology *2*, 591-599. https://doi.org/10.1046/j.1462-5822.2000.00083.x

Shamaei-Tousi, A., Collin, O., Bergh, A., and Bergström, S. (2001). Testicular damage by microcirculatory disruption and colonization of an immune-privileged site during Borrelia crocidurae infection. The Journal of experimental medicine *193*, 995-1004. https://doi.org/10.1084/jem.193.9.995

Shamaei-Tousi, A., Martin, P., Bergh, A., Burman, N., Brännström, T., and Bergström, S. (1999). Erythrocyte-aggregating relapsing fever spirochete Borrelia crocidurae induces formation of microemboli. The Journal of infectious diseases *180*, 1929-1938. https://doi.org/10.1086/315118

Shang, E., Wu, X.-Y., Lovett, M., Miller, J.N., and Blanco, D.R. (2001). Homologous and heterologous Borrelia burgdorfeir challenge of infection-derived immune rabbits using host-adapted organisms. Infect Immun *69*, 593-598.

Shang, E.S., Champion, C.I., Wu, X.-Y., Skare, J.T., Blanco, D.R., Miller, J.N., and Lovett, M.A. (2000). Comparison of protection in rabbits against host-adapted and cultivated Borrelia burgdorferi following infection-derived immunity or immunization with outer membrane vesicles or outer surface protein A. Infect Immun *68*, 4189-4199.

Sharma, B., Brown, A.V., Matluck, N.E., Hu, L.T., and Lewis, K. (2015). Borrelia burgdorferi, the causative agent of Lyme disease, forms drug-tolerant persister cells. Antimicrobial Agents and Chemotherapy. https://doi.org/10.1128/aac.00864-15

Shi, C., Wolfe, J., Russell, J.Q., Fortner, K., Hardin, N., Anguita, J., and Budd, R.C. (2006). Fas ligand deficiency impairs host inflammatory response against infection with the spirochete *Borrelia burgdorferi*. . Infect Immun *74*, 1156-1160.

Shih, C.M., and Liu, L.P. (1998). Differential efficacy of passive immunisation against infection by Lyme disease spirochaetes transmitted by partially fed vector ticks. J Med Microbiol *47*, 773-779. https://doi.org/10.1099/00222615-47-9-773

Shih, C.M., Pollack, R.J., Telford, S.R., and Spielman, A. (1992). Delayed dissemination of Lyme disease spirochetes from the site of deposition in the skin of mice. J Infect Dis *166*. https://doi.org/10.1093/infdis/166.4.827

Shih, C.M., Spielman, A., and Telford, S.R., 3rd (1995). Short report: mode of action of protective immunity to Lyme disease spirochetes. American Journal of Tropical Medicine & Hygiene *52*, 72-74.

Silver, R.M., Yang, L., Daynes, R.A., Branch, D.W., Salafia, C.M., and Weis, J.J. (1995). Fetal outcome in murine Lyme disease. Infect Immun *63*, 66-72. https://doi.org/10.1128/iai.63.1.66-72.1995

Simon, M.M., Wallich, R., and Kramer, M.D. (1996). Borrelia burgdorferi infection of inbred strains of mice provides insights into cellular and molecular parameters of pathogenesis and protection of Lyme disease: a viewpoint. J Spirochetal Tick Borne Dis *3*, 45-52.

Simpson, W.J., Burgdorfer, W., Schrumpf, M.E., Karstens, R.H., and Schwan, T.G. (1991). Antibody to a 39-kilodalton Borrelia burgdorferi antigen (P39) as a marker for infection in experimentally and naturally inoculated animals. J Clin Microbiol *29*, 236-243. https://doi.org/10.1128/jcm.29.2.236-243.1991

Sinsky, R.J., and Piesman, J. (1989). Ear punch biopsy method for detection and isolation of *Borrelia burgdorferi* from rodents. J Clin Microbiol *27*, 1723-1727.

Sonnesyn, S.W., Manivel, J.C., Johnson, R.C., and Goodman, J.L. (1993a). A guinea pig model for Lyme disease. Infection and immunity *61*, 4777-4784.

Sonnesyn, S.W., Manivel, J.C., Johnson, R.C., and Goodman, J.L. (1993b). A guinea pig model for Lyme disease. Infect Immun *61*, 4777-4784.

Steere, A.C., Coburn, J., and Glickstein, L. (2004). The emergence of Lyme disease. Journal of Clinical Investigation *113*, 1093-1101.

Steere, A.C., Duray, P.H., and Butcher, E.C. (1988). Spirochetal antigens and lymphoid cell surface markers in Lyme synovitis.

Comparison with rheumatoid synovium and tonsillar lymphoid tissue. Arthritis and rheumatism *31*, 487-495. https://doi.org/10.1002/art.1780310405

Steere, A.C., and Sikand, V.K. (2003). The Presenting Manifestations of Lyme Disease and the Outcomes of Treatment. N Engl J Med *348*, 2472-2474. https://doi.org/10.1056/nejm200306123482423

Straubinger, R.K. (2000). PCR-Based quantification of Borrelia burgdorferi organisms in canine tissues over a 500-Day postinfection period. Journal of Clinical Microbiology *38*, 2191-2199.

Straubinger, R.K., Chang, Y.F., Jacobson, R.H., and Appel, M.J. (1995). Sera from OspA-vaccinated dogs, but not those from tick-infected dogs, inhibit in vitro growth of Borrelia burgdorferi. Journal of Clinical Microbiology *33*, 2745-2751.

Straubinger, R.K., Straubinger, A.F., Härter, L., Jacobson, R.H., Chang, Y.F., Summers, B.A., Erb, H.N., and Appel, M.J. (1997a). Borrelia burgdorferi migrates into joint capsules and causes an up-regulation of interleukin-8 in synovial membranes of dogs experimentally infected with ticks. Infect Immun *65*, 1273-1285.

Straubinger, R.K., Straubinger, A.F., Summers, B.A., Erb, H.N., Harter, L., and Appel, M.J. (1998a). Borrelia burgdorferi induces the production and release of proinflammatory cytokines in canine synovial explant cultures. Infection & Immunity *66*, 247-258.

Straubinger, R.K., Straubinger, A.F., Summers, B.A., and Jacobson, R.H. (2000). Status of Borrelia burgdorferi infection after antibiotic treatment and the effects of corticosteroids: An experimental study. Journal of Infectious Diseases *181*, 1069-1081.

Straubinger, R.K., Straubinger, A.F., Summers, B.A., Jacobson, R.H., and Erb, H.N. (1998b). Clinical manifestations, pathogenesis, and effect of antibiotic treatment on Lyme borreliosis in dogs. Wien Klin Wochenschr *110*, 874-881.

Straubinger, R.K., Summers, B.A., Chang, Y.F., and Appel, M.J. (1997b). Persistence of Borrelia burgdorferi in experimentally infected dogs after antibiotic treatment. Journal of Clinical Microbiology *35*, 111-116.

Suk, K., Das, S., Sun, W., Jwang, B., Barthold, S.W., Flavell, R.A., and Fikrig, E. (1995). Borrelia burgdorferi genes selectively expressed in the infected host. Proc Natl Acad Sci U S A *92*, 4269-4273. https://doi.org/10.1073/pnas.92.10.4269

Summers, B.A., Straubinger, A.F., Jacobson, R.H., Chang, Y.F., Appel, M.J., and Straubinger, R.K. (2005). Histopathological studies of experimental lyme disease in the dog. J Comp Pathol *133*, 1-13.

Topfer, K.H., and Straubinger, R.K. (2007). Characterization of the humoral immune response in dogs after vaccination against the Lyme borreliosis agent A study with five commercial vaccines using two different vaccination schedules. Vaccine *25*, 314-326.

Tuomi, J., Rantamaki, L.K., and Tanskanen, R. (2002). Experimental infection of laboratory mice and rabbits with several isolates of Borrelia burgdorferi sensu lato; comparison of antigens from different genospecies in serological measurement of immune responses. Comp Immunol Microbiol Infect Dis *25*, 109-125. https://doi.org/10.1016/s0147-9571(01)00027-3

Vidal, V., Scragg, I.G., Cutler, S.J., Rockett, K.A., Fekade, D., Warrell, D.A., Wright, D.J., and Kwiatkowski, D. (1998). Variable major lipoprotein is a principal TNF-inducing factor of louse-borne relapsing fever. Nature medicine *4*, 1416-1420. https://doi.org/10.1038/4007

Vitoratos, N., Papadias, C., Economou, E., Makrakis, E., Panoulis, C., and Creatsas, G. (2006). Elevated circulating IL-1beta and TNF-alpha, and unaltered IL-6 in first-trimester pregnancies complicated by threatened abortion with an adverse outcome. Mediators Inflamm *2006*, 30485. https://doi.org/10.1155/MI/2006/30485

Wallich, R., Brenner, C., Kramer, M.D., and Simon, M.M. (1995). Molecular cloning and immunological characterization of a novel linear-plasmid-encoded gene, pG, of Borrelia burgdorferi expressed only in vivo. Infect Immun *63*, 3327-3335. https://doi.org/10.1128/iai.63.9.3327-3335.1995

Wang, G., Ma, Y., Buyuk, A., McClain, S., Weis, J.J., and Schwartz, I. (2004). Impaired host defense to infection and Toll-like receptor 2-independent killing of *Borrelia burgdorferi* clinical isolates in TLR2-deficient C3H/HeJ mice. FEMS Microbiol Lett *231*, 219-225.

Wang, G., Ojaimi, C., Iyer, R., Saksenberg, V., McCain, S.A., Wormser, G.P., and Schwartz, I. (2001). Impact of genotypic variation of Borrelia burgdorferi sensu stricto on kinetics of dissemination and severity of disease in C3H/HeJ mice. Infect Immun *69*, 4303-4312.

Wang, G., Ojaimi, C., Wu, H., Saksenberg, V., Iyer, R., Liveris, D., McClain, S.A., Wormser, G.P., and Schwartz, I. (2002). Disease severity in a murine model of Lyme borreliosis is associated with the genotype of the infecting Borrelia burgdorferi sensu stricto strain. J Infect Dis *186*, 782-791.

Weis, J.J., McCracken, B.A., Ma, Y., Fairbairn, D., Roper, R.J., Morrison, T.B., Weis, J.H., Zachary, J.F., Doerge, R.W., and Teuscher, C. (1999). Identification of quantitative trait loci governing arthritis severity and humoral responses in the murine model of Lyme disease. J Immunol *162*, 948-956.

Weis, J.J., Yang, L., PetriSeiler, K., and Silver, R.M. (1997a). Pathological manifestations in murine Lyme disease: association with tissue invasion and spirochete persistence. Clin Infect Dis *25(suppl)*, S18-S24.

Weis, J.J., Yang, L., Seiler, K.P., and Silver, R.M. (1997b). Pathological manifestations in murine Lyme disease: association with tissue invasion and spirochete persistence. Clin Infect Dis *25 Suppl 1*, S18-24. https://doi.org/10.1086/516172

White, F.A., Feldman, P., and Miller, R.J. (2009). Chemokine signaling and the management of neuropathic pain. Mol Interv *9*, 188-195. https://doi.org/10.1124/mi.9.4.7

Wikel, S. (2013). Ticks and tick-borne pathogens at the cutaneous interface: host defenses, tick countermeasures, and a suitable environment for pathogen establishment. *4*. https://doi.org/10.3389/fmicb.2013.00337

Wikel, S.K. (1996). Host immunity to ticks. Annual review of entomology *41*, 1-22. https://doi.org/10.1146/annurev.en.41.010196.000245

Woodman, M.E., Cooley, A.E., Miller, J.C., Lazarus, J.J., Tucker, K., Bykowski, T., Botto, M., Hellwege, J., Wooten, R.M., and Stevenson, B. (2007). *Borrelia burgdorferi* binding of host complement regulator factor H is not required for efficient mammalian infection. Infect Immun *75*, 3131-3139.

Woodrum, J.E., and Oliver, J.H. (1999). Investigation of venereal, transplacental, and contact transmission of the Lyme disease spirochete, Borrelia burgdorferi, in Syrian hamsters. J Parasitol *85*, 426-430.

Wooten, R.M., Ma, Y., Yoder, R.A., Brown, J.P., Weis, J.H., Zachary, J.F., Kirschning, C.J., and Weis, J.J. (2002). Toll-like receptor 2 is required for innate, but not acquired, host defense to *Borrelia burgdorferi*. J Immunol *168*, 348-355.

Wormser, G.P., Baker, P.J., O'Connell, S., Pachner, A.R., Schwartz, I., and Shapiro, E.D. (2012). Critical analysis of treatment trials of rhesus macaques infected with Borrelia burgdorferi reveals important flaws in experimental design. Vector borne and zoonotic diseases (Larchmont, NY) *12*, 535-538. https://doi.org/10.1089/vbz.2012.1012

Wright, D.J., and Woodrow, D.F. (1980). Terminal changes in mice experimentally infected with Borrelia duttoni. J Pathol *130*, 83-90.

Wright, S.D., and Nielsen, S.W. (1990). Experimental infecction of the white-footed mouse with Borrelia burgdorferi. Am J Vet Res *51*, 1980-1987.

Xu, Q., seemanapalli, S.V., McShan, K., and Liang, F.T. (2006). Constitutive expression of outer surface protein C diminishes the ability of Borrelia burgdorferi to evade specific humoral immunity. Infect Immun *74*, 5177-5184.

Yang, L., Ma, Y., Schoenfeld, R., Griffiths, M., Eichwald, E., Araneo, B., and Weis, J.J. (1992). Evidence for B-lymphocyte mitogen activity in

Borrelia burgdorferi-infected mice. Infect Immun *60*, 3033-3041. https://doi.org/10.1128/iai.60.8.3033-3041.1992

Yang, L.M., Weis, J.H., Eichwald, E., Kolbert, C.P., Persing, D.H., and Weis, J.J. (1994). Heritable susceptibility to severe *Borrelia burgdorferi*-induced arthritis is dominant and is associated with persistence of large numbers of spirochetes in tissues. Infect Immun *62*, 492-500.

Yeung, C., and Baranchuk, A. (2019). Diagnosis and Treatment of Lyme Carditis: JACC Review Topic of the Week. Journal of the American College of Cardiology *73*, 717-726. https://doi.org/https://doi.org/10.1016/j.jacc.2018.11.035

Yrjänäinen, H., Hytönen, J., Söderström, K.O., Oksi, J., Hartiala, K., and Viljanen, M.K. (2006). Persistent joint swelling and Borrelia-specific antibodies in Borrelia garinii-infected mice after eradication of vegetative spirochetes with antibiotic treatment. Microbes Infect *8*, 2044-2051. https://doi.org/10.1016/j.micinf.2006.03.008

Yrjänäinen, H., Hytönen, J., Song, X.Y., Oksi, J., Hartiala, K., and Viljanen, M.K. (2007). Anti-tumor necrosis factor-alpha treatment activates Borrelia burgdorferi spirochetes 4 weeks after ceftriaxone treatment in C3H/He mice. J Infect Dis *195*, 1489-1496. https://doi.org/10.1086/513873

Zeidner, N.S., Brandt, K.S., Dadey, E., Dolan, M.C., Happ, C., and Piesman, J. (2004). Sustained-release formulation of doxycycline hyclate for prophylaxis of tick bite infection in a murine model of Lyme borreliosis. Antimicrob Agents Chemotherapy *48*, 2697-2699.

Zeidner, N.S., Núncio, M.S., Schneider, B.S., Gern, L., Piesman, J., Brandão, O., and Filipe, A.R. (2001). A portuguese isolate of Borrelia lusitaniae induces disease in C3H/HeN mice. J Med Microbiol *50*, 1055-1060. https://doi.org/10.1099/0022-1317-50-12-1055

Zhang, Z.J., Jiang, B.C., and Gao, Y.J. (2017). Chemokines in neuron-glial cell interaction and pathogenesis of neuropathic pain. Cell Mol Life Sci *74*, 3275-3291. https://doi.org/10.1007/s00018-017-2513-1

Zhong, W., Gern, L., Kramer, M., Wallich, R., and Simon, M.M. (1997). T helper cell priming of mice to Borrelia burgdorferi OspA leads to induction of protective antibodies following experimental but not tick-borne infection. European Journal of Immunology *27*, 2942-2947.

Zhong, W., Gern, L., Stehle, T., Museteanu, C., Kramer, M., Wallich, R., and Simon, M.M. (1999). Resolution of experimental and tick-borne Borrelia burgdorferi infection in mice by passive, but not active immunization using recombinant OspC. Eur J Immunol *29*, 946-957.

Zimmer, G., Schaible, U.E., Kramer, M.D., Mall, G., Museteanu, C., and Simon, M.M. (1990). Lyme carditis in immunodeficient mice during experimental infection of Borrelia burgdorferi. Virchows Archiv A, Pathological anatomy and histopathology *417*, 129-135. https://doi.org/10.1007/bf02190530

Chapter 16

Genetic Manipulation of *Borrelia*

Patricia A. Rosa[1]* and Mollie W. Jewett[2]

[1]Rocky Mountain Laboratories, National Institute of Allergy and Infectious Diseases, National Institutes of Health, 903 S 4th St. Hamilton, MT 59840 USA
[2]Division of Immunity and Pathogenesis, Burnett School of Biomedical Sciences, University of Central Florida College of Medicine, 6900 Lake Nona Blvd, Orlando, FL 32827 USA
*Corresponding author: prosa@niaid.nih.gov

DOI: https://doi.org/10.21775/9781913652616.16

Abstract
Genetic studies in *Borrelia* require special consideration of the highly segmented genome, complex growth requirements and evolutionary distance of spirochetes from other genetically tractable bacteria. Despite these challenges, a robust molecular genetic toolbox has been constructed to investigate the biology and pathogenic potential of these important human pathogens. In this chapter we summarize the tools and techniques that are currently available for the genetic manipulation of *Borrelia,* including the relapsing fever spirochetes, viewing them in the context of their utility and shortcomings. Our primary objective is to help researchers discern what is feasible and what is not practical when thinking about potential genetic experiments in *Borrelia*. We have summarized published methods and highlighted their critical elements, but we are not providing detailed protocols. Although many advances have been made since *B. burgdorferi* was first transformed over 25 years ago, some standard genetic tools remain elusive for *Borrelia*. We mention these limitations and why they persist, if known. We hope to encourage investigators to explore what might be possible, in addition to optimizing what currently can be achieved, through genetic manipulation of *Borrelia*.

Introduction
Spirochetes in the genus *Borrelia* have been known to cause human disease for well over a century, but in-depth investigations of these pathogenic bacteria awaited many developments, including the advent of molecular genetics. Since transformation of *Borrelia burgdorferi* was first described in 1994 (Samuels et al., 1994b; Samuels, 1995), a molecular genetic approach has been applied with growing success to study the biology and pathogenic potential of these tick-borne spirochetes (Tilly et al., 2000; Cabello et al., 2001; Rosa et al., 2005; Battisti et al., 2008; Fine et al., 2011; Brisson et al., 2012; Raffel et al., 2014; Krishnavajhala et al., 2017; Raffel et al., 2018; Samuels et al., 2018). Both *in vitro* and *in vivo* analyses have thus been conducted to identify and describe the basic functions of many *Borrelia* genes and gene products and their contributions to the relationship of these important pathogens with their tick vectors and mammalian hosts. Direct applications of genetic tools developed for Gram-positive and Gram-negative bacteria have encountered limitations, however, due to the genetic distance and distinct codon usage of *Borrelia* relative to these bacteria (Paster et al., 1984; Fraser et al., 1997). Hence there is still a significant need to develop additional and alternative genetic tools for *Borrelia* in particular, and for spirochetes in general.

Over 30 years ago, Dr. Stanley Falkow, the father of the field of microbial pathogenesis, restated Koch's postulates as they could be applied in microbiology using the tools of molecular genetics (Falkow, 1988). In their simplest form, Falkow's molecular postulates entail the ability to create mutations, restore genes and analyze the phenotypes of an otherwise isogenic set of microbes. However, their interpretation and application continue to evolve as new technologies emerge with which to investigate the complex biology of pathogenic microbes (Falkow, 2004; Casadevall, 2017).

Molecular genetic analyses have provided insight into a wide array of regulatory and structural elements of *Borrelia*. However, limitations persist for genetic studies of both the Lyme disease (LD) and relapsing fever (RF) *Borrelia*, including low frequencies and efficiencies of transformation, complex and undefined growth media, and a variable ability to quantify or clone spirochetes as colonies in solid medium. In addition, the generation of isogenic wild-type, mutant and complemented clones present unique challenges in *Borrelia* due to the segmented and unstable nature of their genomes, and poorly defined, multifactorial barriers to the uptake and incorporation of exogenous DNA. Polar effects following insertional inactivation and allelic exchange can arise as a consequence of the compact chromosome of *B. burgdorferi*, which contains many genes with overlapping transcriptional and termination signals (Fraser et al., 1997; Adams et al., 2017b). The presence of paralogous gene families (Casjens et al., 2000) encoding related products with potentially redundant functions in *B. burgdorferi* can also complicate the interpretation of mutant phenotypes. Establishing the connection between a mutation and a phenotype through complementation is essential, but it can be particularly challenging. Isolation of nutritional mutants in *Borrelia* is limited by the complex and undefined nature of the growth media and a relatively long doubling time (>5 hours under optimal conditions) (Barbour, 1984). Despite these potential and real challenges, substantial progress has been made in utilizing genetic approaches to study factors involved in the biology and physiology of *Borrelia* and pathogenesis of the diseases they cause, as illustrated by a number of chapters in this book.

Clone and strain considerations

Lyme disease spirochetes

Most genetic studies to date have been conducted with *B. burgdorferi* sensu stricto (ss) strains, with only a few studies utilizing *B. afzelii* or *B. garinii* strains (Fingerle et al., 2007; Bontemps-Gallo et al., 2018; Cuellar et al., 2019; Strnad and Rego, 2020). Initially, the complete and annotated genomic sequence, including all plasmids, was only available for *B. burgdorferi* ss type strain B31 (Fraser et al., 1997; Casjens et al., 2000), which made it the obvious choice for genetic studies. However, this limitation no longer exists, and the sequences for the complete genomes, including plasmids, are either available or readily determined for virtually any *Borrelia* isolate (Casjens et al., 2017; Casjens et al., 2018). Sequencing advances that yield extremely long and accurate reads from single DNA molecules have been key to this development (Margos et al., 2017). The linear chromosome is highly conserved other than a short region at one telomere, whereas the plasmid component of the genome has undergone substantial rearrangement among strains. However, the total gene content of *B. burgdorferi* ss is relatively constant, with few absolute differences among the large number of isolates analyzed to date, despite widespread plasmid reorganization (Casjens et al., 2017). The variability in plasmid structure among strains suggests that the phenotype of a particular mutant (role/contribution of a particular gene) could vary with the context (strain background) in which it is analyzed. Ideally, mutations would be introduced and analyzed in more than a single clone or strain (Gilbert et al., 2007; Lybecker and Samuels, 2007). However, recognizing that experimental outcome may differ between strains and interpreting results accordingly is currently acceptable, given the limitations and challenges presented by genetic studies in *B. burgdorferi*,

Strain choice by individual labs generally reflects what has been used in previous experiments to facilitate comparison with existing data. There are no data indicating that a particular LD spirochete strain is inherently more suitable than another. However, the ease with which spirochetes can be grown and plated, the efficiency at which they can be transformed, and their phenotype in the mouse/tick infectious cycle are relevant traits that vary substantially among *B. burgdorferi* strains and between clones derived from a single strain (Elias et al., 2002). Hence a key consideration at the outset of any experiment is whether characteristics of the clone/strain/genospecies being used, such as those described above, are appropriate for the question(s) being asked.

Although differences in plasmid content (and corresponding differences in plasmid-borne restriction/modification systems) clearly contribute to the efficiency with which different LD strains and clonal derivatives can be transformed (Lawrenz et al., 2002; Kawabata et al., 2004; Rego et al., 2011), additional traits also affect transformability. These include surface properties that permit or restrict resuspension in electroporation solution (Tilly et al., 2000; Tilly et al., 2016). The highly segmented and potentially unstable structure of the *B. burgdorferi* genome mandates that plasmid content be continuously monitored to assure that mutant and wild-type clones remain isogenic except for directed genetic changes (Elias et al., 2002). Determining the plasmid content of strain B31 derivatives has been greatly facilitated by development of multiplex PCR methods (Bunikis et al., 2011; Norris et al., 2011). Complementation also represents a critical component of genetic experiments in *Borrelia*. Finally, it is important to remember that a wild-type *Borrelia* "clone" is neither clonal nor isogenic after passage through a mouse, due to rampant recombination at the *vlsE* and *vmp* expression loci of LD and RF spirochetes, respectively (Stoenner et al., 1982; Barbour, 1990; Restrepo and Barbour, 1994; Zhang et al., 1997; Zhang and Norris, 1998; Coutte et al., 2009; Verhey et al., 2019). Spirochetes isolated from a mouse comprise a complex population of genetically and phenotypically distinct bacteria expressing different alleles of these abundant and immunodominant surface proteins. This heterogenous population of spirochetes recovered from a mouse can serve as the source from which to derive a new infectious clone by plating or limiting dilution, but it should not be used directly as the starting material for genetic manipulation.

Transformation protocol

B. burgdorferi was first genetically transformed in 1994 by electroporation, following a previous study in which the effect of buffers and electroporation parameters on *B. burgdorferi* had been tested (Sambri and Lovett, 1990; Samuels et al., 1994a). A detailed protocol for electro-transformation of *B. burgdorferi* was published by Samuels et al. in 1995 (Samuels, 1995) and recently updated (Samuels et al., 2018). Although increasingly sophisticated genetic tools and techniques have become available, the only significant improvement to the original protocol is the addition of large amounts of DNA (10-50 μg) as electroporation substrate (Tilly et al., 2000). More recently, recognizing and avoiding barriers presented by plasmid-borne restriction/modification (R/M) genes (Lawrenz et al., 2002; Kawabata et al., 2004; Jewett et al., 2007a; Chen et al., 2008; Rego et al., 2011; Kasumba et al., 2015), and routinely monitoring total plasmid content of transformants by multiplex PCR (Bunikis et al., 2011; Norris et al., 2011) have permitted genetic manipulation of fully infectious *B. burgdorferi* clones, which is needed for *in vivo* studies.

Transformation frequency and efficiency

Two calculations are typically used to monitor how well a transformation has worked. Each provides important information for developing, optimizing or applying genetic tools to *Borrelia* and should be used routinely. The first, termed transformation frequency, describes what fraction of the population that survives electroporation (or other transformation method) has stably acquired the introduced DNA. This is determined by quantifying viable bacteria in comparable aliquots of the post-electroporation culture, with and without selection for the transforming DNA. The other standard calculation is transformation efficiency, which is determined by calculating the number of transformants obtained per microgram of DNA used in the electroporation. Transformation frequency and efficiency can be determined only if selection is imposed and viable bacteria are quantified shortly after recovery from electroporation, before the surviving bacteria have undergone substantial cell division. Accurate determinations of transformation frequency and efficiency also require verification that bacteria growing under selection are transformants and not spontaneous resistance mutants. The frequency and efficiency of transformation in *Borrelia* are notoriously poor, which means that a large number of spirochetes and many micrograms of DNA are needed per electroporation in order to recover a transformant. In general, high passage, non-infectious *B. burgdorferi* spirochetes are more readily transformed (higher frequency and efficiency) than low-passage infectious clones; this is due primarily to the loss of plasmids carrying R/M genes, but other poorly understood factors also contribute.

Preparation of competent cells

The preparation of electrocompetent *Borrelia* involves growing spirochetes to a suitable density, followed by a series of washes to concentrate cells and remove culture medium, especially salts, which cause arcing, cell death, and, of course, decreased transformation efficiency (Samuels et al., 1994a; Samuels, 1995; Samuels et al., 2018). Cell density and growth phase are important criteria to consider for successful electrotransformation of *Borrelia* (Samuels, 1995; Tilly et al., 2000; Hübner et al., 2001). Spirochetes in mid-log phase have the highest transformation frequency, but competent cell preparations from late-log phase cultures, which have a greater number of cells, yield the most transformants (Samuels, 1995; Tilly et al., 2000). Briefly, cells are harvested sterilely by centrifugation, washed several times, and finally resuspended in cold electroporation solution as an approximately 100-fold concentrated cell slurry. Some *B. burgdorferi* strains aggregate during the washes and cannot be resuspended in electroporation solution, which prevents or severely limits transformation (Tilly et al.,

2000); this undesirable trait is intensified when bacteria are grown to higher cell density. The presence of outer surface protein B (OspB) contributes to this clumping phenotype (Tilly et al., 2016); hence most strains selected for use in genetic studies are spontaneous mutants that lack OspB (Elias et al., 2002). Aliquots of competent cells can be transformed by electroporation immediately or they can be stored at –80°C, which decreases transformation efficiency by about 50% (Samuels, 1995; Samuels and Garon, 1997; Samuels et al., 2018).

Electrotransformation

Electroporation conditions for transformation of *B. burgdorferi* are similar to those used for other bacteria (Samuels et al., 1994a; Samuels, 1995; Samuels et al., 2018). A high-voltage electric pulse generates pores in the cellular membranes through which DNA can enter the cytoplasm (Shigekawa and Dower, 1988; Trevors et al., 1992; Nickoloff, 1995). Transmission electron microscopy of electroporated *B. burgdorferi* reveals darkly stained regions on the membrane along the length of the cell that are thought to be disruptions in the surface (Samuels and Garon, 1997). These putative pores are no longer visible after a 30 minute incubation in culture medium.

As mentioned above, a significant improvement to the original protocol was increasing by approximately 1000-fold the amount of DNA used per electroporation, from the nanogram levels typically used with other bacteria, to the microgram quantities needed to routinely transform *B. burgdorferi* (Bono et al., 2000; Tilly et al., 2000; Hübner et al., 2001; Stewart et al., 2001; Eggers et al., 2002; Elias et al., 2002). However, even when successful, relatively few *B. burgdorferi* transformants are typically recovered per electroporation, particularly when working with infectious clones. Aside from R/M systems, which can be eliminated, the basis for the extremely low transformation frequency and efficiency of *B. burgdorferi* remains a mystery. Perhaps very little of the transforming DNA actually gets into the cell, the DNA is rapidly targeted for degradation once it enters, or some other feature of the spirochete limits stable retention of transforming DNA. Whatever the basis, this unfortunate characteristic limits the routine construction of saturated libraries that are needed to apply powerful forward genetic screens in *Borrelia* (Lin et al., 2012; Ellis et al., 2013).

Transformation substrate

In addition to the amount of DNA, its source and form have an effect on transformation efficiency. Allelic exchange transformation was first demonstrated by introducing site-directed mutations that conferred antibiotic resistance, using PCR-generated linear DNA substrates (Samuels et al., 1994a). Short oligonucleotides also can be used as transformation substrates (Samuels and Garon, 1997), although this has limited utility because the site-directed mutation encoded by the oligonucleotide must confer a selectable advantage (currently limited to antibiotic resistance). A circular suicide plasmid was previously shown to recombine into the genome at the same frequency as linear DNA (Stevenson et al., 1998); hence linearized plasmid DNA or PCR-amplified DNA fragments are typically used when allelic exchange is the desired outcome. This prevents integration of suicide plasmids and generation of merodiploids, which are recovered more frequently when working with *Borrelia* clones lacking loci encoding restriction-modification (R/M) systems. As described in a later section, circular shuttle vectors have been constructed that replicate autonomously in *B. burgdorferi* and *E. coli* (Stewart et al., 2001; Eggers et al., 2002; Stewart et al., 2003; Byram et al., 2004). In addition, linear shuttle vectors have been constructed using replication functions from both native circular and linear plasmids (Chaconas et al., 2001; Beaurepaire and Chaconas, 2007). Transformation of autonomously replicating shuttle vectors into strains lacking R/M systems (see below) is generally more efficient than that of allelic exchange constructs (Elias et al., 2003), presumably because the latter require the additional step of recombination. Remarkably, intact native plasmids up to 36 kbp in length have been successfully transformed into *B. burgdorferi*. (Grimm et al., 2004; Jewett et al., 2007b).

The requirement for microgram quantities of substrate DNA per *Borrelia* transformation can present a problem in itself. Plasmids containing *Borrelia* sequences or *Borrelia* codon-optimized genes can limit *E. coli* growth, and rapid propagation of *E. coli* to high cell densities (as typically done for large scale plasmid preparations) can select for mutations in sequences conferring a growth phenotype. Propagating *E. coli* carrying recombinant plasmids at a lower temperature, such as 30°C, (Takacs et al., 2018), lessens the selection for mutations that confer a growth advantage, and plasmid DNA preparations should be sequenced to confirm the integrity of the insert.

DNA modification

One of the identified barriers to efficient transformation of *B. burgdorferi* is the presence of plasmid-encoded restriction-modification (R/M) systems (Lawrenz et al., 2002; Kawabata et al., 2004). Transformation of a set of closely related *B. burgdorferi* strain B31 clones with distinct plasmid complements demonstrated that loss of two linear plasmids, lp25 and lp56, correlated with higher transformation efficiency of a shuttle vector (Lawrenz et al., 2002). These two plasmids carry genes *(bbe02* and *bbq67)* that are predicted to encode type IV R/M enzymes. Disruption of *bbe02* increased shuttle vector transformation efficiency approximately 40-fold (Kawabata et al., 2004). Subsequent studies confirmed these initial observations and demonstrated that *bbe02* and *bbq67* of strain B31 encode distinct R/M enzymes that methylate endogenous DNA and restrict foreign DNA lacking sequence-specific modifications (Chen et al., 2008; Rego et al., 2011). The motifs modified by *bbe02* and *bbq67* were recently inferred by SMRT sequence analysis of genomic DNA from B31 clones that differed in R/M gene content (Casselli et al., 2018). This information should facilitate the design of optimized constructs for more efficient transformation of B31 clones containing intact *bbe02* and *bbq67* loci. Analysis of genomic sequences in the database indicates that some *B. burgdorferi* strains contain additional R/M loci that likely recognize and modify novel sequence motifs (Rego et al., 2011). However, a third plasmid-borne R/M locus in strain B31, *bbh09* on lp28-3, does not negatively impact shuttle vector transformation (Lawrenz et al., 2002; Kawabata et al., 2004), suggesting that the presence of R/M loci does not necessarily indicate a barrier to transformation in these strains.

In addition to the *bbe02* R/M locus, lp25 also carries genes required for mouse and tick infectivity (Purser and Norris, 2000; Labandeira-Rey and Skare, 2001; Purser et al., 2003; Grimm et al., 2005; Revel et al., 2005; Strother et al., 2005; Gilmore et al., 2014). The linkage of these disparate functions on lp25 is the primary reason that infectious B31 derivatives (containing lp25) are difficult to genetically transform. Transformation efficiency into low-passage infectious strains can be increased by first transforming the DNA into a strain carrying lp56 but lacking lp25 (Jewett et al., 2009). The resulting genomic DNA purified from *Borrelia* is expected to be partially modified and thus presents less of an obstacle when introduced in subsequent transformations into strains containing lp25. The transformation barrier imposed by putative restriction enzymes on lp56 and lp25 applies primarily to shuttle vector transformations rather than gene inactivation by allelic exchange (Elias et al., 2002; Lawrenz et al., 2002; Hübner et al., 2003; Kawabata et al., 2004; Tilly et al., 2004). This likely reflects the requirement for intact circular DNA for autonomous plasmid replication, whereas linearized DNA remains an effective template for allelic exchange. Consequently, *trans*-complementation on a shuttle vector presents a greater challenge than gene inactivation in infectious clones retaining lp25.

Fortunately, spirochetes do not require *bbe02* or *bbq67* for growth *in vitro* or at any stage of the experimental mouse/tick infectious cycle. Therefore, fully infectious B31 variants have been constructed in which the *bbe02* locus on lp25 has been deleted (Lawrenz et al., 2002; Kawabata et al., 2004; Rego et al., 2011). These modified strains typically lack the (non-essential) lp56 plasmid that carries *bbq67*, thereby further enhancing their suitability for genetic studies. As an added bonus, the presence of an antibiotic resistance cassette in *bbe02* provides selection for lp25 during *in vitro* growth and genetic manipulation. The *bbe02* locus on lp25 has been chosen as an optimal site for insertion of reporter genes, components of inducible systems, and *trans*-complementation (Gilbert et al., 2007; Chan et al., 2015; Kasumba et al., 2015), as targeted insertion at this site guarantees stable retention of introduced DNA at normal copy number throughout the mouse/tick infectious cycle. This is discussed more fully in subsequent **Complementation** and **Reporter gene** sections.

Alternative methods for transformation

In addition to electroporation, *B. burgdorferi* has been successfully transformed using a simple protocol that entails resuspension of bacteria and DNA together in a solution of PEG and DMSO (Chung et al., 1989; Elias et al., 2002). However, this method has not been widely used or optimized, and its general utility remains unexplored. Likewise, Eggers and colleagues have demonstrated that the endogenous prophage of *Borrelia* can mediate relatively efficient horizontal gene transfer between co-cultured spirochetes (Eggers et al., 2016), but phage transduction has not been developed as a standard genetic tool for *B. burgdorferi*. A substantial advantage of both protocols over electroporation is that they require fewer bacteria at the outset of an experiment because neither technique results in significant cell death, thereby greatly reducing (~100-fold) the initial culture volume needed per transformation.

Selectable markers

Antibiotic resistance markers are one of the most effective selective tools used to single out rare bacterial variants containing new genetic properties, as well as to select for and screen mutants deficient in the expression of specific genes (Pich et al., 2006). *B. burgdorferi* is exquisitely susceptible to many antibiotics (Hunfeld et al., 2002; Ruzic-Sabljic et al., 2005; Hunfeld and Brade, 2006), but the choice of antibiotic resistance genes as genetic markers is somewhat limited by biosafety considerations: markers encoding resistance to ampicillin and tetracycline are prohibited for use in *B. burgdorferi* because these antibiotics are clinically useful in the treatment of Lyme disease (Steere, 2001; Wormser, 2006). The selectable markers currently available for genetic studies in *B. burgdorferi* and their relevant features are listed in Table 1.

Endogenous genes

To date, only antibiotic resistance has been used to select *Borrelia* transformants (Samuels et al., 1994a; Bono et al., 2000; Sartakova et al., 2000; Elias et al., 2003; Frank et al., 2003). An allele of the endogenous *B. burgdorferi gyrB* gene (Samuels et al., 1994b), which encodes the B subunit of DNA gyrase, was the first selectable marker used in *Borrelia* (Samuels, 2006). However, the *gyrB301* marker, which confers resistance to the antibiotic coumermycin A_1, has limited utility for several reasons, including pleiotropic effects due to relaxed DNA supercoiling (Alverson and Samuels, 2002; Alverson et al., 2003). Nonetheless, the use of *gyrB301* was seminal in the development of transformation and genetic manipulation of *B. burgdorferi* (Samuels et al., 1994a; Rosa et al., 1996).

Hybrid genes

A major breakthrough in genetic analysis of *B. burgdorferi* was the construction of hybrid genes using native *B. burgdorferi* promoters fused to foreign open reading frames. This was first used to assay transcription by fusing various borrelial promoters to *cat*, encoding chloramphenicol acetyl transferase (Sohaskey et al., 1997). Bono et al. (Bono et al.,

Table 1. selectable markers for *Borrelia*

ANTIBIOTIC RESISTANCE CASSETTES						
gene	**promoter**	**origin**	**resistance**	**selection ($\mu g\ ml^{-1}$)**	**comments**	**reference**
gyrB301	native (*gyrB*)	*B. burgdorferi* chromosome	coumermycinA_1	0.2-5	high rate of recombination at endogenous *gyrB* locus; pleiotropic effects	(Samuels et al., 1994)
aphI	*flgB, flaB*	*E. coli* Tn*903*	kanamycin	200-400	first widely used selectable marker in *Borrelia*	(Bono et al., 2000; Battisti et al., 2008)
ermC	native	*Staphylococcus aureus*	erythromycin	0.03-0.06	some *Borrelia* strains are resistant to erythromycin	(Sartakova et al., 2000)
aacC1	*flgB, flaB*	*Pseudomonas aeruginosa* Tn*1696*	gentamicin	40	selection with this marker can be variable in *E. coli*	(Elias et al., 2003)
aadA	*flgB, flaB*	*Shigella flexneri* plasmid R100	streptomycin$_a$	50	$_a$confers resistance to both streptomycin and spectinomycin	(Frank et al., 2003; Jewett et al., 2007a)
bsd$_{Bb}$	*flgB*	*Aspergillus terreus*	blasticidin	10-160	codon-optimized for *Borrelia*	(Takacs et al., 2018)
hph$_{Bb}$	*flgB*	*E. coli*	hygromycin	0.2-0.3	codon-optimized for *Borrelia*	(Takacs et al., 2018)

$_a$Streptomycin is used for selection in *B. burgdorferi* because spontaneous resistance to spectinomycin is high, whereas spectinomycin is used *in E. coli*, because many strains are resistant to streptomycin.

2000) constructed the first efficient selectable marker for *Borrelia* by fusing the strong and constitutive *flgB* and *flaB* promoters of *B. burgdorferi* to the *aphI* gene from Tn*903*, which confers resistance to the aminoglycoside kanamycin; the resulting cassette confers kanamycin resistance in both *E. coli* and *Borrelia*. Several other chimeric genes were constructed in a similar fashion, notably selectable markers conferring resistance to gentamicin with the *aacC1* gene from Tn*1696* (Elias et al., 2003) and resistance to spectinomycin and streptomycin with the *aadA* gene from *Shigella flexneri* plasmid R100 (Frank et al., 2003). Most recently, this approach was applied to the *hph* gene of *E. coli* and the *bsd* gene of *Aspergillis terreus,* yielding codon-optimized selectable markers fused to the *flgB* promoter that confer resistance to hygromycin and blasticidin, respectively (Takacs et al., 2018). Although seemingly straightforward, not all antibiotic resistance cassettes constructed in this fashion function as selectable markers in *B. burgdorferi*. The *cat* and *pac* genes, fused to the *flgB* promoter, do not confer resistance to chloramphenicol and puromycin, respectively (Elias et al., 2003). Hybrid cassettes conferring resistance to various antibiotics and used as selectable markers in *B. burgdorferi* are summarized in Table 1.

Selection parameters

Transformants were originally isolated (Samuels et al., 1994a) as sub-surface colonies in semi-solid medium using a modification (Rosa and Hogan, 1992) of the plating protocol described by Kurtti *et al.* (Kurtti et al., 1987). Following electroporation, cells are allowed to recover in liquid culture for approximately 18 hours in the absence of selection before plating in semi-solid medium containing the selective agent and incubating in a humidified microaerobic atmosphere, typically a CO_2 incubator. Stably transformed bacteria form colonies in the semi-solid medium in five days to two weeks and can be isolated by aspirating a plug of agarose containing the colony to liquid medium. A PCR-screen for the presence of the selectable marker can be used to distinguish transformants from spontaneous antibiotic resistant mutants.

A method of selecting for clonal transformants by limiting dilution was developed by Yang *et al.* (Yang et al., 2004), who were unable to isolate *ospA* mutants in solid medium. Recovered transformants are diluted into 100 ml of liquid medium with selection and distributed among several 96-well plates. Again, the plates are incubated in a humidified atmosphere containing CO_2 and positive wells are detected by the change in color of the medium from red to yellow, which is due to the production of lactic acid by the bacteria during growth, as detected by phenol red in the medium. Putative transformants can be expanded in a larger volume of medium containing antibiotic for confirmation and subsequent analysis. Regardless of the method used to isolate transformants, there is the possibility that two spirochetes are intertwined and thus a colony in semi-solid medium or a positive well at the endpoint of limiting dilution may not be derived from a single cell; therefore, extensive mixing, vortexing, and/or pipetting is required to separate cells before plating.

Not currently available

Effective counter-selectable markers have not been widely applied to *B. burgdorferi*. The use of the *rpsL* gene for counter-selection in *Borrelia* has been described but not utilized in any genetic studies to date (Drecktrah et al., 2010). The lack of defined culture medium, coupled with an inherently limited metabolic capability, has impaired the development of auxotrophic selection as an *in vitro* genetic tool for *Borrelia*. However, Richards and colleagues recently demonstrated that *B. burgdorferi* mutants lacking genes encoding the initial enzymes in acetate utilization are non-infectious in mice and have an absolute requirement for the addition of mevalonate to culture medium, suggesting a means of auxotrophic selection *in vitro* and potential utility as selectable markers *in vivo* (Richards et al., 2015).

Gene inactivation

A number of reviews have outlined various strategies and provided detailed protocols for gene inactivation in *Borrelia* (Hyde et al., 2011b; Drecktrah and Samuels, 2018; Latham and Blevins, 2018; Samuels et al., 2018), and this information will not be reiterated in detail in this chapter. As described above, targeted gene inactivation by homologous recombination is straightforward but inefficient in *B. burgdorferi*, requiring microgram quantities of DNA and ranging in frequency from 10^{-5} to 10^{-8}, depending upon the recipient organism (Samuels et al., 1994a; Tilly et al., 2000; Elias et al., 2002). Fortunately, illegitimate recombination (insertion of transforming DNA at a non-homologous site) does not commonly occur in *B. burgdorferi* and thus does not impede the recovery of intended mutants. Simple gene disruption can be accomplished by integration (through a single crossover event) of a circular suicide vector carrying an appropriate selectable marker and a fragment of homologous DNA. Alternatively, if homologous sequences flank the selectable marker in the inactivation construct, allelic exchange can occur with the targeted gene through a double crossover event. Depending upon the design of the inactivation construct, allelic exchange can result in simple disruption or deletion of the targeted gene. A somewhat more elegant form of allelic exchange entails modifying the coding sequence of the targeted gene rather than deleting it, to assess the resulting mutant phenotype. This approach has been used for site-directed mutagenesis of a number of *B. burgdorferi* genes (Knight et al., 2000; Yang et al., 2003; Earnhart et al., 2010; Earnhart et al., 2011; Drecktrah et al., 2013).

As mentioned previously, a single crossover event can occur with allelic exchange constructs if they are introduced as circular DNA. This can result in the unintentional introduction into *B. burgdorferi* of other antibiotic resistance markers present on the cloning vector. As stated above, some selectable markers commonly used in *E. coli*, such as those conferring resistance to ampicillin and tetracycline, are not permitted for genetic studies in *B. burgdorferi* due to the clinical utility of these antibiotics in treating Lyme disease. At the minimum, cloning vectors carrying these markers should be linearized prior to transformation into *B. burgdorferi*. Preferably, allelic exchange constructs should be cloned into vectors that do not contain such restricted markers or the genes encoding them should be inactivated prior to transformation of *B. burgdorferi*.

Several approaches for engineering large deletions in *Borrelia* have been described. Chaconas and colleagues co-opted an endogenous cellular activity involved in the formation of telomeres in *Borrelia* to engineer staggered deletions from either end of a linear plasmid to targeted internal sites (Beaurepaire and Chaconas, 2005; Bankhead and Chaconas, 2007). Bestor et al. have adapted the Cre-*lox* site-specific recombination system to *Borrelia* and used it to generate large internal deletions on a linear plasmid (Bestor et al., 2010). Finally, a basic understanding of the genes/sequences required for autonomous replication and plasmid incompatibility has been applied to displace entire plasmids in *Borrelia* (Stewart et al., 2001; Eggers et al., 2002; Stewart et al., 2003; Byram et al., 2004; Stewart et al., 2005; Jewett et al., 2007a; Jewett et al., 2007b; Dulebohn et al., 2011; Dulebohn et al., 2013).

Conditional mutants, as described elsewhere in this chapter, can now be generated with any of several inducible systems available for *Borrelia*, (Blevins et al., 2007; Gilbert et al., 2007; Whetstine et al., 2009; Latham and Blevins, 2018) Conditional mutants provide an important means to assess the contributions of genes that encode essential functions or for which unmodulated expression is lethal. Finally, all forms of gene inactivation or modification in *Borrelia* entail insertion of a selectable marker at a targeted site, with potential polar effects on adjacent sequences. Interrupted or enhanced transcription of upstream or downstream sequences following insertion of a constitutively expressed antibiotic resistance cassette is not uncommon and should be considered when designing gene inactivation constructs and interpreting experimental outcome. Promoter-less resistance cassettes have been used successfully within constitutive operons of *B. burgdorferi* to target individual genes without impacting downstream genes (Motaleb et al., 2011; Showman et al., 2016).

Complementation

Complementation plays a crucial role in genetic studies of *B. burgdorferi* due to the extremely low frequency with which targeted gene inactivation occurs and the inherent instability of the segmented genome, as described above. Although measures can be taken to ensure that wild-type and derivative clones contain the same plasmids (Purser and Norris, 2000; Labandeira-Rey and Skare, 2001; Eggers et al., 2002; Elias et al., 2002; Parveen et al., 2006; Bunikis et al., 2011; Norris et al., 2011), screening for more subtle genomic alterations elsewhere in the genome is not routinely feasible. To control for unforeseen or unidentified second site mutations, or polar effects on flanking sequences, the goal of complementation is to leave the original mutation unaltered, but to re-introduce a wild-type copy of the targeted gene in *trans* on an autonomously replicating shuttle vector or elsewhere in the genome, or in *cis* by allelic replacement. As described below, various technical considerations, such as copy number, DNA supercoiling, genomic organization and placement of the antibiotic cassette, should be considered when selecting which complementation approach to use.

Endogenous plasmid shuttle vectors

A number of shuttle vectors have been derived from the endogenous plasmids of *B. burgdorferi*. As reviewed in the Chapters 3 and 4, the complete genome sequence of *B. burgdorferi* identified a set of paralogous gene families (PGFs) present on all plasmids, suggested to be involved in plasmid replication and maintenance, although lacking strict sequence homology to plasmid genes with comparable functions in other bacteria (Zückert and Meyer, 1996; Casjens et al., 2000). An early shuttle vector derived from a small, non-essential endogenous circular plasmid (cp9) of *B. burgdorferi* demonstrated that the region encompassing these PGFs did not support replication in *E. coli*, but was sufficient for autonomous replication in *B. burgdorferi* (Stewart et al., 2001). This shuttle vector, termed pBSV2, and its derivatives have become the workhorse for plasmid-based *trans*-complementation and other applications in *B. burgdorferi*. In the absence of strong counterselection (Tilly et al., 2006), pBSV2 is relatively stable both *in vitro* and *in vivo* (Stewart et al., 2001), and present at approximately 5-10 copies per cell (Tilly et al., 2006). However, sequences inserted on a shuttle vector (such as for complementation or fluorescent protein expression) can destabilize the shuttle vector, leading to its loss if antibiotic selection is not maintained. Depending on the mutation, the increased copy number of a wild-type gene on the shuttle vector has the potential for ineffective complementation due to deleterious effects of overproduction of the target protein (Sartakova et al., 2001; Eggers et al., 2002; Tilly et al., 2006; Ristow et al., 2012; Kasumba et al., 2015). Additional shuttle vectors have been derived from other endogenous plasmids (lp25, lp28-1, cp26, and a cp32) by cloning their PGF counterparts together with a functional selectable marker (Eggers et al., 2002; Stewart et al., 2003; Byram et al., 2004; Beaurepaire and Chaconas, 2005). Along with autonomous replication, the PGF regions also mediate incompatibility, resulting in displacement of the endogenous *B. burgdorferi* plasmid from which they were derived (Stewart et al., 2001; Eggers et al., 2002; Stewart et al., 2003; Beaurepaire and Chaconas, 2005), unless selection exists for the coexistence of both replicons (Byram et al., 2004). Displacement with incompatible shuttle vectors thus has been used to experimentally manipulate plasmid content in *B. burgdorferi* (Stewart et al., 2003; Byram et al., 2004; Grimm et al., 2004; Jewett et al., 2007a; Jewett et al., 2007b; Dulebohn et al., 2011; Dulebohn et al., 2013).

Genetic reconstitution at the original locus

An alternative complementation strategy is genetic reconstitution, in which the mutated copy of the gene is replaced with a wild-type copy and another closely linked marker (Hübner et al., 2001; Tilly et al., 2001; Li et al., 2007b), similar to the strategy used to introduce point mutations (Knight et al., 2000; Yang et al., 2003). This approach, however, does not leave the original mutation unaltered. A potential advantage to complementation by genetic reconstitution, at the site of the mutated locus or elsewhere on the same replicon, is that the reintroduced wild-type gene copy is at the same copy number and similar supercoiling state as the original gene. Some attempts to complement mutations in *trans* on a shuttle vector have only partially or unsuccessfully restored the wild-type phenotype (Sartakova et al., 2001; Ristow et al., 2012), raising questions about whether the context of the complementing gene has significantly altered its expression. Alternatively, such results could indicate the presence of an unrecognized second site mutation, which is responsible for the defective phenotype of the complemented mutant. Distinguishing among the possible causes of unsuccessful complementation is addressed below.

Trans-complementation in the genome

An improvement over *trans*-complementation on a shuttle vector is to insert the complementing gene elsewhere in the genome, such as on an essential plasmid (cp26, lp25, lp28-1 or lp36) (Byram et al., 2004; Kawabata et al., 2004; Grimm et al., 2005; Blevins et al., 2007; Gilbert et al., 2007; Jewett et al., 2007b; Caimano et al., 2019). This approach leaves the original mutation undisturbed and maintains *in vivo* selection for the reintroduced gene by virtue of insertion on an essential plasmid. A logical target for this approach is the *bbe02* locus on lp25 as the insertion site. Currently, many research labs employ *B. burgdorferi* clones in which *bbe02* has been inactivated as a means to increase transformation efficiency, yet maintain infectivity (Kawabata et al., 2004; Jacobs et al., 2006; Rego et al., 2011). Further, because lp25 is required for *B. burgdorferi* survival throughout the enzootic cycle, this provides biological selection for *in vivo* maintenance of the complementing gene. The allelic exchange suicide vectors pKBE-Strep and pKBE-Kan have been engineered with a multiple cloning site for insertion of the complementing gene upstream of a linked antibiotic cassette, and flanking DNA homologous to sequences upstrem and downstream of the *bbe02* R/M locus (Kasumba et al., 2015). Validation of this approach via complementation of an *ospC* mutant demonstrated copy number on lp25 consistent with the copy number of the native gene on cp26, as well as functional and stable maintenance of the *ospC* gene in *trans* on lp25 (Kasumba et al., 2015). Similarly, the intergenic region between oppositely transcribed *bb0445* and *bb0446* genes on the chromosome has been used successfully for *trans*-complementation (Li et al., 2007a; Zhang et al., 2009). Innocuous sites on lp25, lp28-1 and lp36 have been useful for insertion of an antibiotic resistance cassette to provide antibiotic selection to ensure *in vitro* maintenance (Blevins et al., 2007) or to restore to *B. burgdorferi* clones lacking each of the critical virulence plasmids (Grimm et al., 2004; Jewett et al., 2007b). Given that this approach effectively preserved or reestablished the wild-type infection phenotype, these sites like *bbe02* on lp25, provide preferable alternatives for stable *trans*-complementation for *in vivo* studies.

Limitations, failures and alternatives

Complementation, while critical to demonstrating the link between a mutant phenotype and gene, presents a substantial challenge when working with infectious *B. burgdorferi* clones retaining the full set of endogenous plasmids. As mentioned above, this is primarily due to the activities of restriction enzymes encoded by genes *bbe02* and *bbq67* on lp25 and lp56, respectively (Lawrenz et al., 2002; Kawabata et al., 2004). The DNA of *B. burgdorferi* contains N^6-methylated adenine only when *bbe02* and/or *bbq67* are present, indicating that both genes encode adenine methyltransferase activity (Rego et al., 2011), and recently the sequence motifs recognized by these restriction enzymes were identified (Casselli et al., 2018). Two genetic approaches to address the barrier that these R/M loci present for transformation of infectious *B. burgdorferi* are (i) to use a genetic background that lacks lp56 and gene *bbe02* (Kawabata et al., 2004; Jacobs et al., 2006; Rego et al., 2011; Jain et al., 2012; Showman et al., 2016; Aranjuez et al., 2019), or (ii) to perform genetic manipulations in a non-infectious, yet transformable, *B. burgdorferi* genetic background lacking lp25 and subsequently restore the *in vivo*-essential lp25-encoded *pncA* (*bbe22*) gene prior to mouse infection studies (Labandeira-Rey and Skare, 2001; Labandeira-Rey et al., 2003; Purser et al., 2003; Seshu et al., 2006; Weening et al., 2008). However, this second approach does not allow tick studies due to the requirement for lp25-encoded genes in addition to *pncA* for tick colonization and persistence (Grimm et al., 2005; Revel et al., 2005; Strother et al., 2005; Gilmore et al., 2014).

When repeated attempts to obtain an isogenic complemented mutant retaining essential plasmids are unsuccessful, an argument can be made linking a mutant phenotype and targeted gene if several independently derived mutants, obtained in separate transformation experiments, are compared with an isogenic wild-type clone. This argument is based on the low probability that second site mutations resulting in a similar mutant phenotype would have arisen independently in separate transformations. Occasionally, a quandary arises when a wild-type copy of the gene has been reintroduced successfully into the mutant strain, yet the defective phenotype persists, leading to the conclusion that complementation has failed. As discussed above, one explanation is that the complementing gene may not be appropriately expressed in *trans* due to changes in supercoiling, copy number or absence of key regulatory sequences. This issue may be addressed by employing one of the alternative complementation strategies. If persistence of the mutant phenotype in the complemented strain cannot be explained by dysregulated gene expression, the next possibility to investigate is polar effects of the original mutation on the transcription of adjacent sequences. If expression of adjacent transcripts appears to be altered by the mutation, then a different gene inactivation strategy should be employed that minimizes potential polar effects. Finally, if the complementing gene is intact and appropriately expressed, and there is no evidence that the mutation impacts adjacent sequences, then the most likely explanation for the defective phenotype of the complemented mutant strain is an unlinked and unrecognized mutation elsewhere in the genome. In this case, complementation has failed because the phenotype of the mutant strain does not derive from the introduced mutation, thus illustrating the essential role of complementation in genetic studies.

Use of published 5′ RNA-seq data to predict endogenous promoters

When designing a complementation construct it is important to incorporate sufficient sequence upstream of the complementing gene to ensure its appropriate regulation. Similarly, the genomic and transcriptional context of the gene should be taken into account, including whether or not the gene is transcribed alone or with other genes as a part of an operon. The *B. burgdorferi* 5′ end transcriptome has provided significant information for refined annotation of the *B. burgdorferi* genome, including a genome-wide map of transcription start sites and a comprehensive analysis of 5′ UTR length (Adams et al., 2017b). These data provide a valuable resource (https://lymedisease.med.ucf.edu/) for predicting endogenous promoters and putative regulatory sequence, critical for the design of effective mutagenesis and complementation approaches.

Surrogate genetics

Limitations on the genetic manipulation of *B. burgdorferi* have stimulated researchers to perform surrogate genetics in *E. coli* because of the well-developed genetic systems of the latter organism and the availability of extensively characterized mutants (Beckwith, 1991). Numerous *B. burgdorferi* genes have been expressed in *E. coli* from either their endogenous promoters or from cloning vector promoters, primarily as a source of purified recombinant protein. A number of *B. burgdorferi* genes have been used to complement *E. coli* or *Salmonella* mutants deficient in the homologous genes, which has permitted partial functional characterization of these borrelial genes (Tilly et al., 1993; Margolis et al., 1994; Knight and Samuels, 1999; Lin et al., 2001; Purser et al., 2003; Eggers et al., 2004; Liveris et al., 2004; Putteet-Driver et al., 2004; Bugrysheva et al., 2005; Jewett et al., 2011). Similarly, cloning of *B. burgdorferi* genes encoding phosphatidylcholine and phosphatidylglycerol-phosphate synthase in *E. coli* permitted the characterization of these enzymatic activities (Wang et al., 2004). Expression in *E. coli* of a phage display library led to the identification of a putative integrin-binding adhesin of *B. burgdorferi* (Coburn et al., 1999; Coburn and Cugini, 2003). Surrogate genetics in *E. coli* also permitted the characterization of *B. burgdorferi* promoter elements involved in transcriptional regulation (Alverson et al., 2003; Boylan et al., 2003; Caimano et al., 2004; Eggers et al., 2004; Burtnick et al., 2007; Medrano et al., 2010). However, differences in metal acquisition and homeostasis between *B. burgdorferi* and *E. coli* have been cited as a reason why *E. coli* may not be an appropriate surrogate system for studying *B. burgdorferi* gene expression and regulation (Ouyang et al., 2010).The advances in genetic manipulation of *B. burgdorferi* described in this chapter suggest that the need to perform surrogate genetic experiments in other bacterial hosts may decrease in the future. However, given its complex culture medium and slow growth rate, it is unlikely that *B. burgdorferi* will ever be as easy to manipulate as *E. coli.*

Engineered gene expression

Endogenous promoter fusions

The first tools employed to control gene expression in *B. burgdorferi* were its endogenous *flaB* and *flgB* promoters (Sohaskey et al., 1997; Bono et al., 2000; Sartakova et al., 2000). These two promoters have been used extensively to constitutively express endogenous and foreign genes in *B. burgdorferi,* and to develop new genetic markers and genetic tools, as described throughout this chapter. Enzootic stage-specific expression has been achieved using the *ospA* and *ospC* promoters, for tick stage and mammalian stage expression, respectively (Tilly et al., 2016; Adams et al., 2017b). Constitutive expression mediated by *flaBp* has also been utilized to study the role of *ospC*-repression in the ability of *B. burgdorferi* to evade mammalian host defenses (Xu et al., 2006). Similarly, *B. burgdorferi* containing *flaBp-ospAB* gene fusions were used to demonstrate that constitutive expression of these proteins led to rapid clearance of *Borrelia* in mice by the immune system (Strother et al., 2007). For many applications, strong gene expression driven by the *flaB* and *flgB* promoters is experimentally useful; however, overexpression of certain proteins can be limited by possible toxic effects, mis-localization and even altered function (Hayes et al., 2014). Recently, to address these challenges, constitutive promoters of low and intermediate strengths were identified from a global transcriptome data set that measured *B. burgdorferi* gene expression at multiple *in vitro* growth phases (Arnold et al., 2016; Takacs et al., 2018). This panel of cloned and validated promoters provides a range of levels of constitutively controlled expression of genes of interest (Takacs et al., 2018). Future analysis of the activities and utility of these promoters *in vivo* will be important.

Inducible gene expression

Inducible promoter systems that can regulate gene expression in a graded manner offer certain advantages over the total ablation of gene expression generated by null mutants, or constitutive expression generated by endogenous *B. burgdorferi* promoters. For example, information about the function of unknown but essential genes can be obtained through modulating gene expression and monitoring the phenotypic changes that occur when gene products become limiting. In *B. burgdorferi*, endogenous genetic switches have yet to be identified. However, well characterized genetic switches, such as the *E. coli* lactose operon (Jacob and Monod, 1961) and the *E. coli* tetracycline resistance gene of *Tn*10 (Gossen and Bujard, 1992; Bujard, 1999), have been adapted for use in *B. burgdorferi*.

lac *system*

The *lac* system has been adapted to experimentally regulate gene expression in *B. burgdorferi* (Blevins et al., 2007; Gilbert et al., 2007). Gilbert *et al.* (Gilbert et al., 2007) constructed the *flacp* inducible promoter by integrating the synthetic $lacO_{id}$ operator (Oehler et al., 1994) into the *B. burgdorferi flgB* promoter (Ge et al., 1997). This inducible promoter was fused to *ospC* and *rpoS* by directly incorporating *flacp* into the genome and replacing the native promoters of the respective genes (Gilbert et al., 2007). These investigators also generated *B. burgdorferi* strains that produce high levels of the LacI repressor by fusing the *flgB* promoter to *lacI* and inserting the hybrid into lp25, a plasmid required for infectivity (Purser and Norris, 2000; Labandeira-Rey and Skare, 2001). The expression of *flacp*-controlled genes in this background is tightly regulated in response to the inducer isopropyl β-D-thiogalactopyranoside (IPTG); genomic *flacp-ospC* and *flacp-rpoS* fusions bypass the normal regulation by environmental factors (Gilbert et al., 2007). In addition, *flacp-rpoS* was used to demonstrate that RpoS was not sufficient to downregulate OspA *in vitro* (Caimano et al., 2007). This *lac* system was used *in vivo* (Gilbert et al., 2007) to further test the hypothesis that OspC was required at an early time point in infection (Tilly et al., 2006; Tilly et al., 2007). This experiment indicated that the inducible system functions *in vivo* and can be used to test genes required for infectivity (Gilbert et al., 2007).

An alternative system of gene regulation based on the *E. coli lac* operon was also shown to be functional in *B. burgdorferi* (Blevins et al., 2007). Gene regulation by the *lac* operator was demonstrated using a codon-optimized luciferase reporter (luc_{Bb}) and through expression of the endogenous *B. burgdorferi* protein BptA. In this system, the *B. burgdorferi lac* expression system consisted of a single plasmid based on the pKFSS1 shuttle vector with the T5/*lac* hybrid promoter *pQE30* from plasmid QE30 (QIAGEN); this promoter contains two *lac* operators (Blevins et al., 2007). On the same plasmid, a codon-optimized *lacI* gene was present under the control of the *flaB* promoter (Blevins et al., 2007). Adding IPTG to *B. burgdorferi* carrying this plasmid with either luc_{Bb} or *bptA* cloned under the control of the *pQE30* promoter resulted in increased expression of these gene products. This *lac* system worked relatively well in *B. burgdorferi*, but induction was incomplete, which may result from either insufficient IPTG or excessive LacI inside the cell, or weak activity of the *pQE30* promoter in *Borrelia*.

tet *system*

Like the *lac* system, the tetracycline-inducible system has been adapted for controlling gene expression in *B. burgdorferi* (Cabello et al., 2006; Whetstine et al., 2009). These approaches use hybrid tetracycline-responsive *B. burgdorferi* promoters and constitutively expressed *tetR*. Placement of the *tet* operator upstream of the -10 and -35 sequences of the *bmpA* promoter resulted in inducible, but significantly leaky expression of the target gene (Cabello et al., 2006). In contrast, the hybrid P_{ost} promoter includes the *ospA* promoter and 5′ UTR sequence with the first 19 nucleotides of the 5′ UTR replaced with the *tetO* sequence (Whetstine et al., 2009). The P_{ost} promoter, along with the constitutively expressed *tetR* gene, are engineered on a single kanamycin-resistant shuttle vector, pCRW53 (Whetstine et al., 2009). This system demonstrates little to no background expression in the absence of anhydrotetracycline (ATc) and titratable expression up to 2 μg/ml ATc. Yet, maximal P_{ost} promoter activity was not achieved, perhaps due to read through-transcription of *tetR*. In its current design, the pCRW53 tet-inducible system is limited to environments in which *ospA* is expressed and further adaptation is required for its application during mammalian infection (Whetstine et al., 2009).

These results demonstrate that gene expression can be artificially regulated in *B. burgdorferi* using either the *lac* system or the *tet* system. Modifications yielding stronger and tighter induction of genes under the control of these inducible promoters should be possible in the future. The availability of inducible expression systems for *B. burgdorferi* has allowed isolation of conditional lethal mutants through regulated repression of a number of essential genes. (Lenhart and Akins, 2010; Groshong et al., 2012; Bandy et al., 2014; Ye et al., 2014; Chu et al., 2016; Groshong et al., 2017; Drecktrah et al., 2020). Conversely, increased expression of *rpoS* from an inducible *lac* promoter resulted in cell death, consistent with a failure to engineer over-expression of *rpoS* from the strong *flaB* promoter (Chen et al., 2013). Further optimization of these genetic switches, as well as development of endogenous genetic switches, will improve manipulation of *B. burgdorferi* gene expression *in vivo* and open a new avenue for studies of pathogenesis in Lyme borreliosis.

Random tagged mutagenesis

The use of transposon mutagenesis as a tool for bacterial genetics represents an important advance that has generally superseded chemical and radiation mutagenesis (Hayes, 2003; Bosse et al., 2006; Salama and Manoil, 2006; Burrack and Higgins, 2007). Mutagenized bacteria are easily isolated on the basis of antibiotic resistance or other selectable markers carried on the transposable element, and the mutated gene can be readily identified by routine molecular procedures (Judson and Mekalanos, 2000; Hayes, 2003; Bosse et al., 2006; Salama and Manoil, 2006; Burrack and Higgins, 2007).

Genome-wide transposon mutagenesis of *B. burgdorferi* was first demonstrated with a suicide plasmid that contained a modified *Himar1* element of the *mariner* family (Stewart et al., 2004). In this system, called pMarGent, the inverted repeats of the transposable element flank a ColE1 origin of replication and a gentamicin-resistance cassette that functions in both *E. coli* and *B. burgdorferi*, while the *Himar1* transposase is carried on a portion of the suicide plasmid that is not mobilized, and hence only transiently present in *B. burgdorferi* (Stewart et al., 2004; Stewart and Rosa, 2008). This approach yielded a saturated mutant library in a non-infectious *B. burgdorferi* clone and permitted the isolation and characterization of a large number of mutants, with random transposon insertions in the linear chromosome and all linear and circular plasmids (Stewart et al., 2004). These experiments demonstrated the strength and feasibility of transposon mutagenesis in *B. burgdorferi*.

The utility of transposon mutagenesis as a tool to study *B. burgdorferi* virulence was illustrated by using the pMarGent system to generate a limited number of transposon mutants in an infectious *B. burgdorferi* clone and subsequently assessing their ability to infect mice (Botkin et al., 2006a; Botkin et al., 2006b). However, this approach was limited due to the instability of the pMarGent plasmid in *E. coli* and the low transformation frequency of infectious *B. burgdorferi*. To reduce the possibility of recovering inactive plasmid from *E. coli* due to transposition of the gentamicin marker, a kanamycin resistance cassette was added adjacent to the transposase, creating the vector pGKT (Stewart and Rosa, 2008). However, this modification does not fully stabilize pGKT because both markers can be retained in the same cell at different genomic locations following transposition. The limitation on transposon mutagenesis imposed by *B. burgdorferi*'s low transformation frequency has been partially addressed, as in other genetic approaches, with *B. burgdorferi* clones lacking lp56 and gene *bbe02* on lp25, thereby avoiding the R/M systems they encode (Kawabata et al., 2004; Rego et al., 2011; Lin et al., 2012; Lin et al., 2014; Lin and Gao, 2018). Lin et al. further adapted the pGKT vector for signature tagged mutagenesis (STM) by inserting eleven distinct 7-bp tags in the region between the ColE1 and inverted terminal repeat 2 (Lin et al., 2012). This molecular genetic advance allowed for generation of a genome-scale STM library of 4,479 sequence-defined mutants, which can be interrogated in groups of clones to identify genes important for *in vivo* and *in vitro* phenotypes (Lin et al., 2012).

Isolation of a random, saturated mutant library requires an insertion every 200 base pairs on average, thus a completely saturated transposon library of *B. burgdorferi* would contain at least 7,500 independent mutants, since the *B. burgdorferi* genome is approximately 1,500,00 bp (Fraser et al., 1997; Casjens et al., 2000; Salama and Manoil, 2006). Generation of a genome-wide STM library necessitated development of a multiplex, high-throughput method for plasmid content analysis of the individual clones. The Luminex plasmid analysis assay employs a combination of three multiplex PCRs using 24 primer sets specific for each of the 21 *B. burgdorferi* B31 plasmids, and Luminex xMAP technology for microfluidic bead-based detection of the *B. burgdorferi* replicons (Norris et al., 2011). This approach allows simultaneous analysis of 30 *B. burgdorferi* clones in a 96-well plate format and has been used to assess the plasmid content of 44 B31 clones and 4,464 transposon mutants in an STM library (Norris et al., 2011). To date, 434 STM mutants from the library of 4,479 unique mutants have been screened for infectivity in mice (Lin et al., 2012), resulting in the identification of infection-relevant candidate genes (reviewed in Chapter 13). Initially, comparison of the input and output STM mutant pools to identify those mutants that were attenuated for infection involved a semi-quantitative Luminex-based multiplex PCR procedure (Lin et al., 2012). More recently, Tn-seq has been used to identify genes contributing to a phenotype. Using this approach, the fitness of STM mutants is determined

Table 2. reporter genes for *Borrelia*

REPORTER	APPLICATION(S)	LIMITATION(S)	REFERENCE(S)
Borrelia burgdorferi			
cat (chloramphenicol acetyl transferase)	*in vitro* gene expression	requires radioactivity or ELISA	(Sohaskey et al., 1997; Alverson et al., 2003)
lacZ (β-galactosidase)	*in vitro* gene expression; colorimetric detection in liquid and solid medium; easy blue/white screening	requires X-gal substrate	(Hayes et al., 2010)
FLUORESCENT REPORTERS (listed below)	*in vitro* and *in vivo* gene expression; cellular localization of fusion proteins; *in vivo* imaging of individual spirochetes in ticks and mice	fluorescence parameters vary among reporters and alleles; not all reporters tested for *in vivo* applications	
GFP, mEGFP, msfGFP (Green Fluorescent Protein)			(Carroll et al., 2003; Eggers et al., 2004; Moriarty et al., 2008; Dunham-Ems et al., 2009; Takacs et al., 2018)
YFP, mCitrine, msfYFP (Yellow Fluorescent Protein)			(Eggers et al., 2002; Takacs et al., 2018)
CFP, mCerulean, msfCFP (Cyan Fluorescent Protein)			(Eggers et al., 2002; Takacs et al., 2018)
mRFP, mCherry (Red Fluorescent Protein)			(Schulze and Zuckert, 2006; Takacs et al., 2018)
dTomato (Orange Fluorescent Protein)			(Lee et al., 2010)

by sequencing the genomic DNA flanking the transposon insertion sites in the entire library, before and after undergoing selection in the environment of interest (Troy et al., 2016). Tn-seq has been applied in *B. burgdorferi* to identify gene candidates important for mouse infection, survival in ticks, resistance to reactive oxygen and nitrogen species and carbohydrate utilization (Troy et al., 2016; Ramsey et al., 2017; Phelan et al., 2019), (see Chapter 13), as well as to identify when and where *B. burgdorferi* encounters barriers to mouse infection (Troy et al., 2016).

STM library screens are powerful molecular genetic tools to identify candidate genes of interest. It is important to note, however, that genetic screens of this nature do not assess the fitness of a single mutant alone in the challenge environment, rather the fitness of all mutants in the input pool relative to each other. For this reason, the phenotype of a single mutant alone may differ from that in the context of a pool of other clones. Further studies, including targeted inactivation and complementation, are required to validate the roles of the implicated genes in the phenotype of interest. Despite technical advancements in the transformation frequency of *B. burgdorferi,* the challenge of achieving a saturated transposon mutant library remains daunting. The increased technical feasibility and reduced cost of whole genome re-sequencing raises the possibility for the resurgence of chemical and radiation mutagenesis approaches as alternatives for genetic screens in *B. burgdorferi.*

Reporter genes

Reporter genes encode proteins whose function, such as enzymatic activity, fluorescence or selectable phenotype, can be readily assayed or detected; they have been used extensively to monitor gene expression, to track cells in complex environments, and to determine the subcellular locations of proteins. Different types of reporters are typically better suited for distinct applications. The reporter genes currently available for *Borrelia* and their pertinent features are summarized in Table 2.

In vitro *gene expression*

A number of reporter genes have been used successfully as transcriptional fusions with *B. burgdorferi* promoters to study gene expression, including *cat* (chloramphenicol acetyl transferase), *luc* (luciferase), *lacZ* (β-galactosidase), and *gfp* (green fluorescent protein) (Sohaskey et al., 1997; Sartakova et al., 2000; Carroll et al., 2003; Eggers et al., 2004; Blevins et al., 2007; Hayes et al., 2010). The initial studies in *B. burgdorferi* used promoter:*cat* fusions to assay gene expression (Sohaskey et al., 1997; Sohaskey et al., 1999), but *cat* has largely been replaced by other reporter genes that are easier to use or have broader utility. β-galactosidase, encoded by the *lacZ* gene, is a widely used reporter whose activity can be easily monitored in bacteria grown in liquid and solid media without specialized equipment or radioactivity (Sambrook et al., 1989). A codon-optimized *lacZ* reporter was developed for *B. burgdorferi*, and β-galactosidase activity measured in lysates of cultured spirochetes carrying *lacZ* transcriptional fusions with a standard Miller assay (Hayes et al., 2010). The omission of phenol red from BSK medium facilitates detection of *lacZ*-expressing (blue) *B. burgdorferi* colonies in plates containing X-Gal (Hayes et al., 2010). Use of a *lacZ* reporter in a blue-white colony screen confirmed the identity of a negative regulator of *ospC* encoded on linear plasmid lp17 (Sadziene et al., 1993; Sarkar et al., 2011; Hayes et al., 2014). The availability of *lacZ* further expands the repertoire of tools for investigating mechanisms of gene regulation in *B. burgdorferi*.

Whereas cat and *lacZ* reporters are routinely used to quickly measure gene expression in a population of cells, fluorescent protein reporters offer other advantages, such as directly monitoring gene expression at the individual cell level by microscopy or flow cytometry. Green fluorescent protein (GFP) was the first fluorescent protein reporter used to monitor gene expression and identify promoter elements in *B. burgdorferi* (Carroll et al., 2003; Eggers et al., 2004; Falkow, 2004; Valdivia et al., 2006; Srivastava and de Silva, 2008). GFP signal can be influenced by temperature, pH and oxygen concentration, but is relatively stable, which limits its utility for detecting rapid changes in gene expression (Valdivia et al., 2006). An expanded range of alternative fluorescent reporter genes that have been expressed in *B. burgdorferi* include yellow fluorescent protein, *yfp*; cyan fluorescent protein, *cfp* (Eggers et al., 2002); monomeric red fluorescent protein, *mRFP* (Schulze and Zuckert, 2006) and dTomato (Lee et al., 2010). Recently, a number of *B. burgdorferi* codon-optimized reporters have been developed, including CFP variants, mCerulean and msfCFP; GFP variants, mEGFP and msfGFP; YFP variants, mCitrine and msfYFP and a monomeric improved variant of mRFP1, mCherry (Takacs et al., 2018). In addition, an infrared fluorescent protein (iRFP) reporter was developed. iRFP is an extrinsically fluorescent protein that covalently binds biliverdin as the fluorophore (Filonov et al., 2011). Growth of *B. burgdorferi* expressing iRFP in medium supplemented with the biliverdin cofactor results in near-infrared fluorescence of the cells (Takacs et al., 2018). In addition to *in vitro* studies, iRFP may be useful for *in vivo* live imaging of infected mice and/or *ex vivo* imaging of infected tissue explants, as described elsewhere in this chapter, due to the high excitation light penetrance and low tissue auto-fluorescence in the infrared region of the spectrum. Further, the endogenous levels of biliverdin in animal tissues are predicted to be sufficient for maximal fluorescence of iRFP-expressing *B. burgdorferi* (Takacs et al., 2018), although this has yet to be tested experimentally. Combinatorial imaging studies demonstrated that CFP variants can be co-imaged with mCitrine, mCherry and iRFP, and GFP variants can be co-imaged with mCherry and iRFP, suggesting the possibility for simultaneous imaging of up to four proteins in the same spirochete (Takacs et al., 2018).

Subcellular protein localization

In addition to their utility as transcriptional fusions for gene expression studies, translational fusions of fluorescent proteins represent important tools for cell biology, as described in greater detail in Chapter 7. Monomeric red fluorescent protein (mRFP) has been used in this capacity to investigate the lipoprotein export pathway of *Borrelia* (Schulze and Zuckert, 2006) and to develop a FACS-based lipoprotein localization screen (Kumru et al., 2010). More recently, an mCherry fusion protein was employed in a microscopy-based approach, coupled with quantitative image analysis, to investigate the dynamic cellular localization of an envelope protein of *Borrelia* (Takacs et al., 2018). A key feature of mRFP and mCherry as fusion partners for protein localization studies is that they continue to fluoresce outside of the cytoplasm, which earlier variants of GFP did not allow (Schulze and Zuckert, 2006). However, this previous limitation has been overcome with super-folding variants of GFP (sfGFP) that maintain fluorescence in periplasmic and extracellular locations (Pedelacq et al., 2006; Dinh and Bernhardt, 2011). A shortcoming of fluorescent proteins as translational fusions is their relatively large physical size, which can hinder the function, folding or localization of the protein to which they are fused. This potential drawback can be avoided with an alternative fluorescent labeling approach that entails tagging a cellular protein with a small tetra-cysteine motif that can bind fluorescent biarsenical dyes (Griffin et al., 1998). This approach has been used to label membrane proteins in *B. burgdorferi* and while avoiding the size constraints of fluorescent proteins, it is considerably less sensitive, requires a labeling step and can be limited by non-specific binding of the fluorescent dye (Hillman et al., 2019).

In vivo *tracking and imaging*

Beyond applications with spirochetes grown *in vitro*, reporter genes are important tools for detecting *B. burgdorferi* in ticks and mice (see Chapter 17 for an in-depth review). GFP has permitted study of gene expression of individual spirochetes in ticks (Bykowski et al., 2006; Miller et al., 2006), and this reporter was used to visualize the interaction of the spirochete with the mouse vasculature in real time with high resolution imaging (Moriarty et al., 2008). Furthermore, GFP-expressing *B. burgdorferi* have been visualized in the intact skin of living mice and time-lapse images taken early after infection have allowed analysis of spirochete motility *in vivo* (Sultan et al., 2015; Novak et al., 2016). GFP-expressing *B. burgdorferi* were used to track spirochete migration from an infected mouse to a feeding tick (Bockenstedt et al., 2014) and spirochete dissemination within ticks (Dunham-Ems et al., 2009).

Further expansion of *in vivo* live imaging of *B. burgdorferi* during infection has come with adaptation of a codon-optimized luciferase gene (*luc*Bb) from the firefly *Photinus pyralis* for use in *B. burgdorferi* (Blevins et al., 2007). Luciferase activity of the firefly *luc*Bb reporter, which requires an exogenous substrate and is measured with a luminometer, was first used to monitor the activity of endogenous *Borrelia* promoters and an ITPG-induced *lac* expression system (Blevins et al., 2007), as described above. Subsequently, *in vivo* bioluminescent imaging of mice infected with *B. burgdorferi* constitutively synthesizing luciferase has been used for temporal and spatial tracking of spirochete infection at the population level (Hyde et al., 2011a; Chan et al., 2015; Wager et al., 2015; Skare et al., 2016; Adams et al., 2017b; Hyde and Skare, 2018). The *luc*Bb gene, driven by the strong constitutive *flaB* promoter and expressed on a multi-copy shuttle vector, allows detection of bioluminescent *B. burgdorferi* in live mice as early as three days post-inoculation (Adams et al., 2017b). This technique works well using both BALB/c and C3H/HeN mice (Hyde et al., 2011a; Adams et al., 2017b). The relative luminescence units (RLUs) detected by imaging mice infected with bioluminescent *B. burgdorferi* correlates with spirochete load in tissues of the same mice, as measured by quantitative PCR, allowing assessment of the ability of *B. burgdorferi* mutant clones to survive and disseminate in a live animal over time (Hyde et al., 2011a; Wager et al., 2015).

Quantifying in vivo *gene expression*

Application of *in vivo* bioluminescent imaging for quantitation of *B. burgdorferi* promoter activity as a means to measure expression of target genes during infection is less well developed. Currently this approach involves normalizing the luminescence signals of mice infected with *B. burgdorferi* carrying *luc*Bb fused to the target gene promoter, with those of other mice infected with *B. burgdorferi* constitutively expressing *luc*Bb, and subsequent qPCR analysis to demonstrate that all mice carry equivalent bacterial loads at the endpoint of the study (Skare et al., 2016). A dual reporter system that allows simultaneous measurement of the luminescence signal of the target promoter and a control constitutive promoter from the same population of spirochetes would provide a more straightforward and robust method to quantitate gene expression during infection. The first dual luciferase reporter system of *B. burgdorferi* was recently developed (Adams et al., 2017a). For this reporter system, constitutive expression of the *B. burgdorferi* codon-optimized *Renilla reniformis* (sea pansy) luciferase gene (P*flaB*-*rluc*Bb) allows normalization of the activity of a second promoter fused to the *B. burgdorferi* codon-optimized firefly luciferase (*fluc*Bb) gene on the same shuttle vector plasmid. The bioluminescence signals of the two reporters are distinguishable due to their respective specificities for different substrates, coelenterazine and luciferin. This dual luciferase system is a simple and robust approach for quantitative analysis of *B. burgdorferi* promoter activity under *in vivo* growth conditions and in infected ticks (Adams et al., 2017a). Application of this dual reporter system to *B. burgdorferi* promoter activity during mammalian infection remains a challenge, however, because coelenterazine demonstrates high background fluorescence in the mouse that was not circumvented through extensive modification of experimental parameters (Adams et al., 2017a). Future development of a dual reporter system for *in vivo* live imaging of *B. burgdorferi* infection may include a combination of green- and red-shifted luciferase enzymes (Branchini et al., 2005) or application of iRFP (Takacs et al., 2018), as discussed above.

Identifying in vivo *promoters*

Reporter genes are not limited to those whose expression can be measured as an enzymatic activity, fluorescence, luminosity or antibiotic resistance. A gene that is required for survival in a specific environment can also be used as a reporter of promoter activity. This concept is the basis of *in vivo* expression technology (IVET), a powerful and versatile method for genome-wide identification of promoters that are active in a host environment, by complementation of an auxotrophic mutation through gene fusion (Mahan et al., 1993). IVET has been adapted for *B. burgdorferi* using the *pncA* gene (Ellis et al., 2013; Casselli and Bankhead, 2015), which encodes a nicotinamidase activity that is required during infection of a mammalian host (Purser et al., 2003). The *pncA* promoter trap provides strong selection for *in vivo*-active promoters, as only those clones from the IVET library that express *pncA* during infection will survive in the mouse. Moreover, when expressed, *pncA* partially complements the phenotype of *B. burgdorferi* lacking virulence plasmid lp25 (on which *pncA* is encoded), allowing the IVET library to be generated in a more readily transformable genetic background, as described above (Lawrenz et al., 2002; Purser et al., 2003; Ellis et al., 2013). The initial IVET method in *B. burgdorferi* used the *pncA* reporter alone and identified promoters that were active both *in vitro* and during infection, as well as those that were specifically active during infection (Ellis et al., 2013; Adams et al., 2017b). This approach identified 233 unique infection-active promoters across a majority of the *B. burgdorferi* replicons (Ellis et al., 2013; Adams et al., 2017b). These sequences have led to the identification of novel genes critical for *B. burgdorferi* infectivity (Jain et al., 2015; Showman et al., 2016), as well as the discovery of previously unannotated intergenic, intragenic and antisense transcripts throughout the *B. burgdorferi* genome (Adams et al., 2017b).

In vivo phenotype

An aspect of genetic studies with *B. burgdorferi* that requires recognition (but is not unique to this bacterial pathogen) is the phenotype of the organism and the context in which gene function is addressed. Cultured spirochetes look and respond quite differently from genetically identical organisms present in infected ticks or mice (Iyer et al., 2015). Thus *in vitro* assays may not accurately reflect or predict the *in vivo* function of a particular *B. burgdorferi* gene product because other differentially synthesized bacterial proteins could mask or expose its function. There is no ready solution to this potential problem, but interpretation of results obtained from *in vitro* assays should include this caveat. Propagation of spirochetes in implanted chambers in the peritoneal cavity of rats results in bacteria exhibiting a partially host-adapted state, as assessed by altered patterns of protein synthesis and gene expression (Akins et al., 1998; Revel et al., 2002; Brooks et al., 2003; Caimano, 2005) and presents a somewhat more relevant source of bacteria for testing the *in vitro* properties of mutants.

Offsetting the inefficiencies and challenges of genetic studies in *B. burgdorferi* is the availability of relevant animal models in which to analyze mutants. Rodents are natural reservoir hosts, and both ticks and mice can be experimentally infected, making the entire infectious cycle amenable to investigation. Another chapter in this book (Chapter 15) describes the animal models of Lyme disease in greater detail. Several parameters of the in *vivo* model that merit attention with respect to mutant characterization are infectious dose, mode and route of challenge, mouse strain, number of animals and tissues analyzed, quantitation of tissue burden and gene expression, and the statistics used to evaluate the data. Competitive index may be a useful parameter for discovery and quantitation of more subtle phenotypes (Bestor et al., 2012; Dulebohn et al., 2013; Arnold et al., 2015). Experimental design typically is influenced by the mutant phenotype and the focus of the investigation.

When designing, executing and interpreting in *vivo* genetic screens that use a highly complex inoculum (such as a saturated library of mutants) in which the ability to be recovered (or not) is a defining trait (such as IVET or Tn-seq), it is important to keep in mind the population dynamics of *B. burgdorferi* infection. Population bottlenecks during infection have been experimentally demonstrated using both Tn-Seq and genetic bar-coding approaches (Troy et al., 2013; Rego et al., 2014). A significant population bottleneck was found to occur at the site of infection, due in part to early innate immune responses (Troy et al., 2013; Casselli and Bankhead, 2015). Population bottlenecks were also detected during persistent infection of mice, as well as during spirochete acquisition and transmission by ticks (Rego et al., 2014; Phelan et al., 2019). These data highlight the potential misinterpretation of a mutant phenotype in the context of a mixed infection (*in vivo* genetic screen or competition assay) because clones can undergo stochastic elimination at a bottleneck independently of their genotype.

Genetic manipulation of relapsing fever spirochetes
A significant advance in the *Borrelia* field has been the development of a genetic system for the relapsing fever (RF) spirochetes (Battisti et al., 2008; Fine et al., 2011; Guyard et al., 2013; Lopez et al., 2013; Fine et al., 2014; Raffel et al., 2014; James et al., 2016; Krishnavajhala et al., 2017; James et al., 2018; Jackson-Litteken et al., 2019). The availability of genetic tools and complete genomic sequences for multiple RF species (Lescot et al., 2008; Elbir et al., 2012; Miller et al., 2013; Barbour and Campeau Miller, 2014; Barbour, 2016; Wilder et al., 2016; Elbir et al., 2017; Kuleshov et al., 2020) has permitted investigation of genetically modified RF spirochetes in their respective tick vectors or rodent hosts, and informative comparisons with other *Borrelia*, as detailed in Chapter 3. As with LD spirochetes, there is considerable variation among RF strains in the efficiency with which they can be grown, transformed or recovered as colonies in solid medium (Raffel et al., 2018). A noteworthy feature of genetic studies with RF spirochetes is retention of infectivity following prolonged cultivation (Lopez et al., 2008), suggesting a more stable segmented genome or a stronger requirement for plasmid-encoded functions during *in vitro* growth, relative to the LD *Borrelia*. Additional research is needed to determine if this attribute is true for all RF strains and clones, and how it is influenced by media and culture conditions (Raffel et al., 2018). Another advantage of RF *Borrelia* for genetic studies is the relatively high density they reach in the blood of an infected mouse during a relapse, permitting direct visualization and quantitation of *in vivo* organisms, and sufficient material for downstream analyses.

The genetic approaches and tools developed for the LD spirochetes provided a strong foundation for molecular genetic manipulation of the closely related RF spirochetes. Generation of the first site-directed allelic-exchange deletion mutant in *Borrelia hermsii* was reported in 2008 (Battisti et al., 2008). Furthermore, this work included a method for genetic transformation by electroporation and isolation of individual *B. hermsii* clones in liquid medium by limiting dilution in 96-well plates. In addition, antibiotic resistance cassettes and a *B. hermsii* shuttle vector for *trans*-complementation and expression studies were described (Battisti et al., 2008). Further expansion of the genetic tools for RF spirochetes came with the generation of an infectious *B. hermsii* clone that constitutively expresses GFP, allowing visualization of fluorescent spirochetes both *in vitro* and *in vivo* (Fine et al., 2011). The methods for genetic manipulation of *B. hersmii* have been applied successfully to another RF spirochete, *B. turicatae* (Lopez et al., 2013; Krishnavajhala et al., 2017). Most recently, a method for reliable growth of *B. hermsii* and *B. turicatae* in solid medium has been developed, allowing isolation of single colonies following genetic manipulation, as well as enumeration of colony forming units from both *in vitro* and *in vivo* samples (Raffel et al., 2018). The application of reverse genetics to the RF spirochetes has opened the door to broad investigation of these tick-borne pathogens (Raffel et al., 2014; James et al., 2016, 2018; Jackson-Litteken et al., 2019; Schwan et al., 2020).

Conclusion
The genetic toolbox for *B. burgdorferi* continues to grow, expanding the spectrum of approaches available for investigating the genetic mechanisms driving the biology of this fascinating pathogen. An increased understanding of the R/M systems that limit transformation efficiency has resulted in *B. burgdorferi* clones with enhanced capacity for transformation while retaining infectivity. These *B. burgdorferi* clones have been critical for improved recovery of targeted mutant and complemented clones, as well as the generation of genome-wide mutant and promoter-trap libraries. The development of a wide array of new reporter genes have opened the door for novel microscopy, flow-cytometry, genetic screening, *in vivo* bioluminescence, and *in vivo* selection assays. Moreover, new constitutive and inducible promoters allow for various levels of engineered expression of genes of interest. Molecular genetic approaches pioneered in *B. burgdorferi* have facilitated the development of tools for genetic manipulation and investigation of the

closely related RF spirochetes. Transformation, plating in solid medium, reverse genetics, and florescence-tagging of *B. hermsii* and *B. turicatae* are critical advances for understanding these medically important spirochetes. In sum, although there are still challenges and obstacles to overcome, significant strides have been made in *Borrelia* genetics, and persistent innovation and creativity will continue to increase and enhance the genetic tools available for the study of LD and RF spirochetes.

Acknowledgements

The authors thank and acknowledge Drs. Felipe Cabello and D. Scott Samuels for their contributions as co-authors of the first edition of this chapter, substantial portions of which are retained or modestly revised in the current edition. Supported by the Intramural Research Program (PAR) and grant R01AI099094 (MWJ) of the National Institute of Allergy and Infectious Diseases, National Institutes of Health.

References

Adams, P.P., Flores Avile, C., and Jewett, M.W. (2017a). A Dual Luciferase Reporter System for *B. burgdorferi* Measures Transcriptional Activity during Tick-Pathogen Interactions. Frontiers in Cellular and Infection Microbiology 7, 225. 10.3389/fcimb.2017.00225

Adams, P.P., Flores Avile, C., Popitsch, N., Bilusic, I., Schroeder, R., Lybecker, M., and Jewett, M.W. (2017b). *In vivo* expression technology and 5' end mapping of the *Borrelia burgdorferi* transcriptome identify novel RNAs expressed during mammalian infection. Nucleic Acids Res *45*, 775-792. 10.1093/nar/gkw1180

Akins, D.K., Bourell, K.W., Caimano, M.J., Norgard, M.V., and Radolf, J.D. (1998). A new animal model for studying Lyme disease spirochetes in a mammalian host-adapted state. Journal of Clinical Investigation *101*, 2240-2250.

Alverson, J., Bundle, S.F., Sohaskey, C.D., Lybecker, M.C., and Samuels, D.S. (2003). Transcriptional regulation of the *ospAB* and *ospC* promoters from *Borrelia burgdorferi*. Mol Microbiol *48*, 1665-1677.

Alverson, J., and Samuels, D.S. (2002). *groEL* expression in *gyrB* mutants of *Borrelia burgdorferi*. J Bacteriol *184*, 6069-6072.

Aranjuez, G.F., Kuhn, H.W., Adams, P.P., and Jewett, M.W. (2019). *Borrelia burgdorferi bbk13* Is Critical for Spirochete Population Expansion in the Skin during Early Infection. Infect Immun *87*, e00887-00818. 10.1128/IAI.00887-18

Arnold, W.K., Savage, C.R., Antonicello, A.D., and Stevenson, B. (2015). Apparent role for *Borrelia burgdorferi* LuxS during mammalian infection. Infect Immun *83*, 1347-1353. 10.1128/IAI.00032-15

Arnold, W.K., Savage, C.R., Brissette, C.A., Seshu, J., Livny, J., and Stevenson, B. (2016). RNA-Seq of *Borrelia burgdorferi* in multiple phases of growth reveals insights into the dynamics of gene expression, transcriptome architecture, and noncoding RNAs. PLoS One *11*, e0164165. 10.1371/journal.pone.0164165

Bandy, N.J., Salman-Dilgimen, A., and Chaconas, G. (2014). Construction and characterization of a *Borrelia burgdorferi* strain with conditional expression of the essential telomere resolvase, ResT. J Bacteriol *196*, 2396-2404. 10.1128/JB.01435-13

Bankhead, T., and Chaconas, G. (2007). The role of VlsE antigenic variation in the Lyme disease spirochete: persistence through a mechanism that differs from other pathogens. Mol Microbiol *65*, 1547-1558.

Barbour, A.G. (1984). Isolation and cultivation of Lyme disease spirochetes. Yale J Biol Med *57*, 521-525.

Barbour, A.G. (1990). Antigenic variation of a relapsing fever *Borrelia* species. Annu Rev Microbiol *44*, 155-171.

Barbour, A.G. (2016). Chromosome and Plasmids of the Tick-Borne Relapsing Fever Agent *Borrelia hermsii*. Genome Announc *4*, e00528-00516. 10.1128/genomeA.00528-16

Barbour, A.G., and Campeau Miller, S. (2014). Genome Sequence of *Borrelia parkeri*, an Agent of Enzootic Relapsing Fever in Western North America. Genome Announc *2*, e00018-00014. 10.1128/genomeA.00018-14

Battisti, J.M., Raffel, S.J., and Schwan, T.G. (2008). A system for site-specific genetic manipulation of the relapsing fever spirochete *Borrelia hermsii*. Methods Mol Biol *431*, 69-84. 10.1007/978-1-60327-032-8_6

Beaurepaire, C., and Chaconas, G. (2005). Mapping of essential replication functions of the linear plasmid lp17 of *Borrelia burgdorferi* by targeted deletion walking. Mol Microbiol *57*, 132-142.

Beaurepaire, C., and Chaconas, G. (2007). Topology-dependent transcription in linear and circular plasmids of the segmented genome of *Borrelia burgdorferi*. Mol Microbiol *63*, 443-453.

Beckwith, J. (1991). Strategies for finding mutants. Methods Enzymol *204*, 3-18.

Bestor, A., Rego, R.O., Tilly, K., and Rosa, P.A. (2012). Competitive advantage of *Borrelia burgdorferi* with outer surface protein BBA03 during tick-mediated infection of the mammalian host. Infect Immun *80*, 3501-3511. 10.1128/IAI.00521-12

Bestor, A., Stewart, P.E., Jewett, M.W., Sarkar, A., Tilly, K., and Rosa, P.A. (2010). Using the Cre-*lox* recombination system to investigate lp54 gene requirement in the infectious cycle of *Borrelia burgdorferi*. Infect Immun *78*, 2397-2407.

Blevins, J.S., Revel, A.T., Smith, A.H., Bachlani, G.N., and Norgard, M.V. (2007). Adaptation of a luciferase gene reporter and *lac* expression system to *Borrelia burgdorferi*. Appl Environ Microbiol *73*, 1501-1513.

Bockenstedt, L.K., Gonzalez, D., Mao, J., Li, M., Belperron, A.A., and Haberman, A. (2014). What ticks do under your skin: two-photon intravital imaging of *Ixodes scapularis* feeding in the presence of the Lyme disease spirochete. Yale J Biol Med *87*, 3-13.

Bono, J.L., Elias, A.F., Kupko III, J.J., Stevenson, B., Tilly, K., and Rosa, P. (2000). Efficient targeted mutagenesis in *Borrelia burgdorferi*. J Bacteriol *182*, 2445-2452.

Bontemps-Gallo, S., Lawrence, K.A., Richards, C.L., and Gherardini, F.C. (2018). Genomic and phenotypic characterization of *Borrelia afzelii* BO23 and *Borrelia garinii* CIP 103362. PLoS One *13*, e0199641. 10.1371/journal.pone.0199641

Bosse, J.T., Zhou, L., Kroll, J.S., and Langford, P.R. (2006). High-throughput identification of conditionally essential genes in bacteria: from STM to TSM. Infect Disord Drug Targets *6*, 241-262.

Botkin, D.J., Abbott, A., Howell, J.K., Mosher, M., Stewart, P.E., Rosa, P.A., Kawabata, H., Watanabe, H., and Norris, S.J. (2006a). Transposon mutagenesis of infectious *Borrelia burgdorferi* B31: a pilot study. In Molecular Biology of Spirochetes, NATO Advanced Research Workshop Series, F.C. Cabello, and D. Hulinska, eds. (Amsterdam: IOS Press), pp. 13-24.

Botkin, D.J., Abbott, A., Stewart, P.E., Rosa, P.A., Kawabata, H., Watanabe, H., and Norris, S.J. (2006b). Identification of potential virulence determinants by *HimarI* transposition of infectious *Borrelia burgdorferi* B31. Infect Immun *74*, 6690-6699.

Boylan, J.A., Posey, J.E., and Gherardini, F.C. (2003). *Borrelia* oxidative stress response regulator, BosR: a distinctive Zn-dependent transcriptional activator. Proc Natl Acad Sci USA *100*, 11684-11689.

Branchini, B.R., Southworth, T.L., Khattak, N.F., Michelini, E., and Roda, A. (2005). Red- and green-emitting firefly luciferase mutants for bioluminescent reporter applications. Anal Biochem *345*, 140-148. 10.1016/j.ab.2005.07.015

Brisson, D., Drecktrah, D., Eggers, C.H., and Samuels, D.S. (2012). Genetics of *Borrelia burgdorferi*. Annu Rev Genet *46*, 515-536. 10.1146/annurev-genet-011112-112140

Brooks, C.S., Hefty, P.S., Jolliff, S.E., and Akins, D.R. (2003). Global analysis of *Borrelia burgdorferi* genes regulated by mammalian host-specific signals. Infect Immun *71*, 3371-3383.

Bugrysheva, J.V., Bryksin, A.V., Godfrey, H.P., and Cabello, F.C. (2005). *Borrelia burgdorferi rel* is responsible for generation of guanosine-3'-diphosphate-5'-triphosphate and growth control. Infect Immun *73*, 4972-4981.

Bujard, H. (1999). Controlling genes with tetracyclines. J Gene Med *1*, 372-374.

Bunikis, I., Kutschan-Bunikis, S., Bonde, M., and Bergstrom, S. (2011). Multiplex PCR as a tool for validating plasmid content of *Borrelia burgdorferi*. J Microbiol Methods *86*, 243-247. 10.1016/j.mimet. 2011.05.004

Burrack, L.S., and Higgins, D.E. (2007). Genomic approaches to understanding bacterial virulence. Curr Opin Microbiol *10*, 4-9.

Burtnick, M.N., Downey, J.S., Brett, P.J., Boylan, J.A., Frye, J.G., Hoover, T.R., and Gherardini, F.C. (2007). Insights into the complex regulation of *rpoS* in *Borrelia burgdorferi*. Mol Microbiol *65*, 277-293.

Bykowski, T., Babb, K., von Lackum, K., Riley, S.P., Norris, S.J., and Stevenson, B. (2006). Transcriptional regulation of the *Borrelia burgdorferi* antigenically variable VlsE surface protein. J Bacteriol *188*.

Byram, R., Stewart, P.E., and Rosa, P.A. (2004). The essential nature of the ubiquitous 26 kb circular replicon of *Borrelia burgdorferi*. J Bacteriol *186*, 3561-3569.

Cabello, F.C., Dubytska, L., Bryksin, A.V., Bugrysheva, J.V., and Godfrey, H.P. (2006). Genetic studies of the *Borrelia burgdorferi bmp* gene family. In NATO Advanced Research Workshop on Molecular Biology of Spirochetes, F.C. Cabello, Hulinska, D and Godfrey, H.P., ed. (Prague, Czech Republic: IOS Press, Amsterdam), pp. 235-249.

Cabello, F.C., Sartakova, M.L., and Dobrikova, E.Y. (2001). Genetic manipulation of spirochetes- light at the end of the tunnel. Trends Microbiol *9*, 245-248.

Caimano, M.J. (2005). Cultivation of Borrelia burgdorferi in dialysis membrane chambers in rat peritonea. Curr Protoc Microbiol *Chapter 12*, Unit 12C 13. 10.1002/9780471729259.mc12c03s00

Caimano, M.J., Eggers, C.H., Hazlett, K.R., and Radolf, J.D. (2004). RpoS is not central to the general stress response in *Borrelia burgdorferi* but does control expression of one or more essential virulence determinants. Infect Immun *72*, 6433-6445.

Caimano, M.J., Groshong, A.M., Belperron, A., Mao, J., Hawley, K.L., Luthra, A., Graham, D.E., Earnhart, C.G., Marconi, R.T., Bockenstedt, L.K., *et al.* (2019). The RpoS Gatekeeper in *Borrelia burgdorferi*: An Invariant Regulatory Scheme That Promotes Spirochete Persistence in Reservoir Hosts and Niche Diversity. Front Microbiol *10*, 1923. 10.3389/fmicb.2019.01923

Caimano, M.J., Iyer, R., Eggers, C.H., Gonzalez, C., Morton, E.A., Gilbert, M.A., Schwartz, I., and Radolf, J.D. (2007). Analysis of the RpoS regulon in *Borrelia burgdorferi* in response to mammalian host signals provides insight into RpoS function during the enzootic cycle. Mol Microbiol *65*, 1193-1217.

Carroll, J.A., Stewart, P., Rosa, P., and Garon, G. (2003). An enhanced GFP reporter system to monitor gene expression in *Borrelia burgdorferi*. Microbiology *149*, 1819-1828.

Casadevall, A. (2017). The Pathogenic Potential of a Microbe. mSphere *2*, e00015-00017. 10.1128/mSphere.00015-17

Casjens, S., Palmer, N., van Vugt, R., Huang, W.M., Stevenson, B., Rosa, P., Lathigra, R., Sutton, G., Peterson, J., Dodson, R.J., *et al.* (2000). A bacterial genome in flux: the twelve linear and nine circular extrachromosomal DNAs in an infectious isolate of the Lyme disease spirochete *Borrelia burgdorferi*. Mol Microbiol *35*, 490-516.

Casjens, S.R., Di, L., Akther, S., Mongodin, E.F., Luft, B.J., Schutzer, S.E., Fraser, C.M., and Qiu, W.G. (2018). Primordial origin and diversification of plasmids in Lyme disease agent bacteria. BMC Genomics *19*, 218. 10.1186/s12864-018-4597-x

Casjens, S.R., Gilcrease, E.B., Vujadinovic, M., Mongodin, E.F., Luft, B.J., Schutzer, S.E., Fraser, C.M., and Qiu, W.G. (2017). Plasmid diversity and phylogenetic consistency in the Lyme disease agent *Borrelia burgdorferi*. BMC Genomics *18*, 165. 10.1186/ s12864-017-3553-5

Casselli, T., and Bankhead, T. (2015). Use of *in vivo* Expression Technology for the Identification of Putative Host Adaptation Factors of the Lyme Disease Spirochete. J Mol Microbiol Biotechnol *25*, 349-361. 10.1159/000439305

Casselli, T., Tourand, Y., Scheidegger, A., Arnold, W.K., Proulx, A., Stevenson, B., and Brissette, C.A. (2018). DNA Methylation by Restriction Modification Systems Affects the Global Transcriptome Profile in *Borrelia burgdorferi*. J Bacteriol *200*, e00395-00318. 10.1128/JB.00395-18

Chaconas, G., Stewart, P.E., Tilly, K., Bono, J.L., and Rosa, P. (2001). Telomere resolution in the Lyme disease spirochete. Embo J *20*, 3229-3237.

Chan, K., Alter, L., Barthold, S.W., and Parveen, N. (2015). Disruption of *bbe02* by Insertion of a Luciferase Gene Increases Transformation Efficiency of *Borrelia burgdorferi* and Allows Live Imaging in Lyme Disease Susceptible C3H Mice. PLoS One *10*, e0129532. 10.1371/ journal.pone.0129532

Chen, L., Xu, Q., Tu, J., Ge, Y., Liu, J., and Liang, F.T. (2013). Increasing RpoS expression causes cell death in *Borrelia burgdorferi*. PLoS One *8*, e83276. 10.1371/journal.pone.0083276

Chen, Q., Fischer, J.R., Benoit, V.M., Dufour, N.P., Youderian, P., and Leong, J.M. (2008). In vitro CpG methylation increases the transformation efficiency of *Borrelia burgdorferi* strains harboring the endogenous linear plasmid lp56. J Bacteriol *190*, 7885-7891.

Chu, C.Y., Stewart, P.E., Bestor, A., Hansen, B., Lin, T., Gao, L., Norris, S.J., and Rosa, P.A. (2016). Function of the *Borrelia burgdorferi* FtsH Homolog Is Essential for Viability both In Vitro and In Vivo and Independent of HflK/C. MBio *7*, e00404-00416. 10.1128/mBio. 00404-16

Chung, C.T., Niemla, S.L., and Miller, R.H. (1989). One-step preparation of competent *Escherichia coli*: transformation and storage of bacterial cells in the same solution. Proc Natl Acad Sci USA *86*, 2172-2175.

Coburn, J., Chege, W., Magoun, L., Bodary, S.C., and Leong, J.M. (1999). Characterization of a candidate *Borrelia burgdorferi* ß3-chain integrin ligand identified using a phage display library. Mol Microbiol *34*, 926-940.

Coburn, J., and Cugini, C. (2003). Targeted mutation of the outer membrane protein P66 disrupts attachment of the Lyme disease agent, *Borrelia burgdorferi*, to integrin $a_v ß_3$. Proc Natl Acad Sci USA *100*, 7301-7306.

Coutte, L., Botkin, D.J., Gao, L., and Norris, S.J. (2009). Detailed analysis of sequence changes occurring during *vlsE* antigenic variation in the mouse model of *Borrelia burgdorferi* infection. PLoS Pathog *5*, e1000293.

Cuellar, J., Pietikainen, A., Glader, O., Liljenback, H., Soderstrom, M., Hurme, S., Salo, J., and Hytonen, J. (2019). *Borrelia burgdorferi* Infection in Biglycan Knockout Mice. J Infect Dis *220*, 116-126. 10.1093/infdis/jiz050

Dinh, T., and Bernhardt, T.G. (2011). Using superfolder green fluorescent protein for periplasmic protein localization studies. J Bacteriol *193*, 4984-4987. 10.1128/JB.00315-11

Drecktrah, D., Douglas, J.M., and Samuels, D.S. (2010). Use of *rpsL* as a Counterselectable Marker in *Borrelia burgdorferi*. Appl Environ Microbiol *76*, 985-987. 10.1128/AEM.02172-09

Drecktrah, D., Hall, L.S., Brinkworth, A.J., Comstock, J.R., Wassarman, K.M., and Samuels, D.S. (2020). Characterization of 6S RNA in the Lyme disease spirochete. Mol Microbiol *113*, 399-417. 10.1111/mmi.14427

Drecktrah, D., Hall, L.S., Hoon-Hanks, L.L., and Samuels, D.S. (2013). An inverted repeat in the *ospC* operator is required for induction in *Borrelia burgdorferi*. PLoS One *8*, e68799. 10.1371/journal.pone.0068799

Drecktrah, D., and Samuels, D.S. (2018). Genetic Manipulation of *Borrelia* Spp. Curr Top Microbiol Immunol *415*, 113-140. 10.1007/82_2017_51

Dulebohn, D.P., Bestor, A., Rego, R.O., Stewart, P.E., and Rosa, P.A. (2011). The *Borrelia burgdorferi* linear plasmid lp38 is dispensable for completion of the mouse-tick infectious cycle. Infect Immun *79*, 3510-3517.

Dulebohn, D.P., Bestor, A., and Rosa, P.A. (2013). *Borrelia burgdorferi* linear plasmid 28-3 confers a selective advantage in an experimental mouse-tick infection model. Infect Immun *81*, 2986-2996. 10.1128/IAI.00219-13

Dunham-Ems, S.M., Caimano, M.J., Pal, U., Wolgemuth, C.W., Eggers, C.H., Balic, A., and Radolf, J.D. (2009). Live imaging reveals a biphasic mode of dissemination of *Borrelia burgdorferi* within ticks. Journal of Clinical Investigation *119*, 3652-3665. 10.1172/JCI39401

Earnhart, C.G., Leblanc, D.V., Alix, K.E., Desrosiers, D.C., Radolf, J.D., and Marconi, R.T. (2010). Identification of residues within ligand-binding domain 1 (LBD1) of the *Borrelia burgdorferi* OspC protein required for function in the mammalian environment. Mol Microbiol *76*, 393-408. 10.1111/j.1365-2958.2010.07103.x

Earnhart, C.G., Rhodes, D.V., and Marconi, R.T. (2011). Disulfide-mediated oligomer formation in *Borrelia burgdorferi* outer surface protein C, a critical virulence factor and potential Lyme disease vaccine candidate. Clinical and vaccine immunology : CVI *18*, 901-906. 10.1128/CVI.05004-11

Eggers, C.H., Caimano, M.J., Clawson, M.L., Miller, W.G., Samuels, D.S., and Radolf, J.D. (2002). Identification of loci critical for replication and compatibility of a *Borrelia burgdorferi* cp32 plasmid and use of a cp32-based shuttle vector for expression of fluorescent reporters in the Lyme disease spirochaete. Mol Microbiol *43*, 281-295.

Eggers, C.H., Caimano, M.J., and Radolf, J.D. (2004). Analysis of promoter elements involved in the transcriptional initiation of RpoS-dependent *Borrelia burgdorferi* genes. J Bacteriol *186*, 7390-7402.

Eggers, C.H., Gray, C.M., Preisig, A.M., Glenn, D.M., Pereira, J., Ayers, R.W., Alshahrani, M., Acabbo, C., Becker, M.R., Bruenn, K.N., *et al.* (2016). Phage-mediated horizontal gene transfer of both prophage and heterologous DNA by phiBB-1, a bacteriophage of *Borrelia burgdorferi*. Pathog Dis *74*, ftw107. 10.1093/femspd/ftw107

Elbir, H., Gimenez, G., Robert, C., Bergstrom, S., Cutler, S., Raoult, D., and Drancourt, M. (2012). Complete genome sequence of *Borrelia crocidurae*. J Bacteriol *194*, 3723-3724. 10.1128/JB.00118-12

Elbir, H., Sitlani, P., Bergstrom, S., and Barbour, A.G. (2017). Chromosome and Megaplasmid Sequences of *Borrelia anserina* (Sakharoff 1891), the Agent of Avian Spirochetosis and Type Species of the Genus. Genome Announc *5*, e00018-00017. 10.1128/genomeA.00018-17

Elias, A.F., Bono, J.L., Kupko, J.J., Stewart, P.E., Krum, J.G., and Rosa, P.A. (2003). New antibiotic resistance cassettes suitable for genetic studies in *Borrelia burgdorferi*. J Mol Microbiol Biotechnol *6*, 29-40.

Elias, A.F., Stewart, P.E., Grimm, D., Caimano, M.J., Eggers, C.H., Tilly, K., Bono, J.L., Akins, D.R., Radolf, J.D., Schwan, T.G., *et al.* (2002). Clonal polymorphism of *Borrelia burgdorferi* strain B31 MI: implications for mutagenesis in an infectious strain background. Infect Immun *70*, 2139-2150.

Ellis, T.C., Jain, S., Linowski, A.K., Rike, K., Bestor, A., Rosa, P.A., Halpern, M., Kurhanewicz, S., and Jewett, M.W. (2013). *In vivo* expression technology identifies a novel virulence factor critical for *Borrelia burgdorferi* persistence in mice. PLoS Pathog *9*, e1003567. 10.1371/journal.ppat.1003567

Falkow, S. (1988). Molecular Koch's postulates applied to microbial pathogenicity. Rev Infect Dis *10 (suppl 2)*, S274-S276.

Falkow, S. (2004). Molecular Koch's postulates applied to bacterial pathogenicity - a personal recollection 15 years later. Nat Rev Microbiol *2*, 67-72.

Filonov, G.S., Piatkevich, K.D., Ting, L.M., Zhang, J., Kim, K., and Verkhusha, V.V. (2011). Bright and stable near-infrared fluorescent protein for in vivo imaging. Nat Biotechnol *29*, 757-761. 10.1038/nbt.1918

Fine, L.M., Earnhart, C.G., and Marconi, R.T. (2011). Genetic transformation of the relapsing fever spirochete *Borrelia hermsii*: stable integration and expression of green fluorescent protein from linear plasmid 200. J Bacteriol *193*, 3241-3245. 10.1128/JB.05037-11

Fine, L.M., Miller, D.P., Mallory, K.L., Tegels, B.K., Earnhart, C.G., and Marconi, R.T. (2014). The *Borrelia hermsii* factor H binding protein FhbA is not required for infectivity in mice or for resistance to human complement *in vitro*. Infect Immun *82*, 3324-3332. 10.1128/IAI.01892-14

Fingerle, V., Goettner, G., Gern, L., Wilske, B., and Schulte-Spechtel, U. (2007). Complementation of a *Borrelia afzelii* OspC mutant highlights the crucial role of OspC for dissemination of *Borrelia afzelii* in *Ixodes ricinus*. Int J Med Microbiol *297*, 97-107.

Frank, K.L., Bundle, S.F., Kresge, M.E., Eggers, C.H., and Samuels, D.S. (2003). *aadA* confers streptomycin resistance in *Borrelia burgdorferi*. J Bacteriol *185*, 6723-6727.

Fraser, C.M., Casjens, S., Huang, W.M., Sutton, G.G., Clayton, R., Lathigra, R., White, O., Ketchum, K.A., Dodson, R., Hickey, E.K., *et al.* (1997). Genomic sequence of a Lyme disease spirochaete, *Borrelia burgdorferi*. Nature *390*, 580-586.

Ge, Y., Old, I.G., Saint Girons, I., and Charon, N.W. (1997). Molecular characterization of a large *Borrelia burgdorferi* motility operon which is initiated by a consensus sigma70 promoter. J Bacteriol *179*, 2289-2299.

Gilbert, M.A., Morton, E.A., Bundle, S.F., and Samuels, D.S. (2007). Artificial regulation of *ospC* expression in *Borrelia burgdorferi*. Mol Microbiol *63*, 1259-1273.

Gilmore, R.D., Brandt, K.S., and Hyde, J.A. (2014). *pncA* and *bptA* are not sufficient to complement *Ixodes scapularis* colonization and persistence by *Borrelia burgdorferi* in a linear plasmid lp25-deficient background. Infect Immun *82*, 5110-5116. 10.1128/IAI.02613-14

Gossen, M., and Bujard, H. (1992). Tight control of gene expression in mammalian cells by tetracycline-responsive promoters. Proc Natl Acad Sci USA *89*, 5547-5551.

Griffin, B.A., Adams, S.R., and Tsien, R.Y. (1998). Specific covalent labeling of recombinant protein molecules inside live cells. Science *281*, 269-272. 10.1126/science.281.5374.269

Grimm, D., Eggers, C.H., Caimano, M.J., Tilly, K., Stewart, P.E., Elias, A.F., Radolf, J.D., and Rosa, P.A. (2004). Experimental assessment of the roles of linear plasmids lp25 and lp28-1 of *Borrelia burgdorferi* throughout the infectious cycle. Infect Immun *72*, 5938-5946.

Grimm, D., Tilly, K., Bueschel, D.M., Fisher, M.A., Policastro, P.F., Gherardini, F.C., Schwan, T.G., and Rosa, P.A. (2005). Defining plasmids required by *Borrelia burgdorferi* for colonization of tick vector *Ixodes scapularis* (Acari: Ixodidae). J Med Entomol *42*, 676-684.

Groshong, A.M., Dey, A., Bezsonova, I., Caimano, M.J., and Radolf, J.D. (2017). Peptide Uptake Is Essential for *Borrelia burgdorferi* Viability and Involves Structural and Regulatory Complexity of its Oligopeptide Transporter. MBio *8*, e02047-02017. 10.1128/mBio.02047-17

Groshong, A.M., Gibbons, N.E., Yang, X.F., and Blevins, J.S. (2012). Rrp2, a prokaryotic enhancer-like binding protein, is essential for viability of *Borrelia burgdorferi*. J Bacteriol *194*, 3336-3342. 10.1128/JB.00253-12

Guyard, C., Raffel, S.J., Schrumpf, M.E., Dahlstrom, E., Sturdevant, D., Ricklefs, S.M., Martens, C., Hayes, S.F., Fischer, E.R., Hansen, B.T., *et al.* (2013). Periplasmic flagellar export apparatus protein, FliH, is involved in post-transcriptional regulation of FlaB, motility and virulence of the relapsing fever spirochete *Borrelia hermsii*. PLoS One *8*, e72550. 10.1371/journal.pone.0072550

Hayes, B., Jewett, M., and Rosa, P. (2010). A *lacZ* reporter system for use in *Borrelia burgdorferi*. Appl Environ Microbiol *76*, 7407-7412.

Hayes, B.M., Dulebohn, D.P., Sarkar, A., Tilly, K., Bestor, A., Ambroggio, X., and Rosa, P.A. (2014). Regulatory protein BBD18 of the lyme disease spirochete: essential role during tick acquisition? MBio *5*, e01017-01014. 10.1128/mBio.01017-14

Hayes, F. (2003). Transposon-based strategies for microbial functional genomics and proteomics. Annu Rev Genet *37*, 3-29.

Hillman, C., Stewart, P.E., Strnad, M., Stone, H., Starr, T., Carmody, A., Evans, T.J., Carracoi, V., Wachter, J., and Rosa, P.A. (2019). Visualization of Spirochetes by Labeling Membrane Proteins With Fluorescent Biarsenical Dyes. Frontiers in cellular and infection microbiology *9*, 287. 10.3389/fcimb.2019.00287

Hübner, A., Revel, A.T., Nolen, D.M., Hagman, K.E., and Norgard, M.V. (2003). Expression of a *luxS* gene is not required for *Borrelia burgdorferi* infection of mice via needle inoculation. Infect Immun *71*, 2892-2896.

Hübner, A., Wang, X., Nolen, D.M., Popova, T.G., Cabello, F.C., and Norgard, M. (2001). Expression of *Borrelia burgdorferi* OspC and DbpA is controlled by a RpoN-RpoS regulatory pathway. Proc Natl Acad Sci USA *98*, 12724-12729.

Hunfeld, K.P., and Brade, V. (2006). Antimicrobial susceptibility of *Borrelia burgdorferi* sensu lato: what we know, what we don't know, and what we need to know. Wien Klin Wochenschr *118*, 659-668.

Hunfeld, K.P., Kraiczy, P., Kekoukh, K., Schafer, V., and Brade, V. (2002). Standardised *in vitro* susceptibility testing of *Borrelia burgdorferi* against well-known and newly developed antimicrobial agents - possible implications for new therapeutic approaches to Lyme disease. Int J Med Microbiol *291*, 125-137.

Hyde, J.A., and Skare, J.T. (2018). Detection of Bioluminescent *Borrelia burgdorferi* from *In Vitro* Cultivation and During Murine Infection. Methods Mol Biol *1690*, 241-257. 10.1007/978-1-4939-7383-5_18

Hyde, J.A., Weening, E.H., Chang, M., Trzeciakowski, J.P., Hook, M., Cirillo, J.D., and Skare, J.T. (2011a). Bioluminescent imaging of *Borrelia burgdorferi* in vivo demonstrates that the fibronectin-binding protein BBK32 is required for optimal infectivity. Mol Microbiol *82*, 99-113. 10.1111/j.1365-2958.2011.07801.x

Hyde, J.A., Weening, E.H., and Skare, J.T. (2011b). Genetic transformation of *Borrelia burgdorferi*. Curr Protoc Microbiol *Chapter 12*, Unit 12C 14. 10.1002/9780471729259.mc12c04s20

Iyer, R., Caimano, M.J., Luthra, A., Axline, D., Jr., Corona, A., Iacobas, D.A., Radolf, J.D., and Schwartz, I. (2015). Stage-specific global alterations in the transcriptomes of Lyme disease spirochetes during tick feeding and following mammalian host adaptation. Mol Microbiol *95*, 509-538. 10.1111/mmi.12882

Jackson-Litteken, C.D., Zalud, A.K., Ratliff, C.T., Latham, J.I., Bourret, T.J., Lopez, J.E., and Blevins, J.S. (2019). Assessing the Contribution of an HtrA Family Serine Protease During *Borrelia turicatae* Mammalian Infection. Frontiers Cellular Infection Microbiology *9*, 290. 10.3389/fcimb.2019.00290

Jacob, F., and Monod, J. (1961). Genetic regulatory mechanisms in the synthesis of proteins. J Mol Biol *3*, 318-356.

Jacobs, M.B., Norris, S.J., Phillippi-Falkenstein, K.M., and Philipp, M.T. (2006). Infectivity of the highly transformable BBE02⁻ lp56⁻ mutant of *Borrelia burgdorferi*, the Lyme disease spirochete, via ticks. Infect Immun *74*, 3678-3681.

Jain, S., Showman, A.C., and Jewett, M.W. (2015). Molecular dissection of a *Borrelia burgdorferi in vivo* essential purine transport system. Infect Immun *83*, 2224-2233. 10.1128/IAI.02859-14

Jain, S., Sutchu, S., Rosa, P.A., Byram, R., and Jewett, M.W. (2012). *Borrelia burgdorferi* harbors a transport system essential for purine salvage and mammalian infection. Infect Immun *80*, 3086-3093. 10.1128/IAI.00514-12

James, A.E., Rogovskyy, A.S., Crowley, M.A., and Bankhead, T. (2016). Characterization of a DNA Adenine Methyltransferase Gene of *Borrelia hermsii* and Its Dispensability for Murine Infection and Persistence. PLoS One *11*, e0155798. 10.1371/journal.pone.0155798

James, A.E., Rogovskyy, A.S., Crowley, M.A., and Bankhead, T. (2018). Cis-acting DNA elements flanking the variable major protein expression site of *Borrelia hermsii* are required for murine persistence. MicrobiologyOpen *7*, e00569. 10.1002/mbo3.569

Jewett, M.W., Byram, R., Bestor, A., Tilly, K., Lawrence, K., Burtnick, M.N., Gherardini, F., and Rosa, P.A. (2007a). Genetic basis for retention of a critical virulence plasmid of *Borrelia burgdorferi*. Mol Microbiol *66*, 975-990.

Jewett, M.W., Jain, S., Linowski, A.K., Sarkar, A., and Rosa, P.A. (2011). Molecular characterization of the *Borrelia burgdorferi in vivo*-essential protein PncA. Microbiology *157*, 2831-2840. 10.1099/mic.0.051706-0

Jewett, M.W., Lawrence, K., Bestor, A.C., Tilly, K., Grimm, D., Shaw, P., VanRaden, M., Gherardini, F., and Rosa, P.A. (2007b). The critical role of the linear plasmid lp36 in the infectious cycle of *Borrelia burgdorferi*. Mol Microbiol *64*, 1358-1374.

Jewett, M.W., Lawrence, K.A., Bestor, A., Byram, R., Gherardini, F., and Rosa, P.A. (2009). GuaA and GuaB are essential for *Borrelia burgdorferi* survival in the tick-mouse infectious cycle. J Bacteriol *191*, 6231-6241.

Judson, N., and Mekalanos, J.J. (2000). Transposon-based approaches to identify essential bacterial genes. Trends Microbiol *8*, 521-526.

Kasumba, I.N., Bestor, A., Tilly, K., and Rosa, P.A. (2015). Use of an endogenous plasmid locus for stable in *trans* complementation in *Borrelia burgdorferi*. Appl Environ Microbiol *81*, 1038-1046. 10.1128/aem.03657-14

Kawabata, H., Norris, S.J., and Watanabe, H. (2004). BBE02 disruption mutants of *Borrelia burgdorferi* B31 have a highly transformable, infectious phenotype. Infect Immun *72*, 7147-7154.

Knight, S.W., Kimmel, B.J., Eggers, C.H., and Samuels, D.S. (2000). Disruption of the *Borrelia burgdorferi gac* gene, encoding the naturally synthesized GyrA C-terminal domain. J Bacteriol *182*, 2048-2051.

Knight, S.W., and Samuels, D.S. (1999). Natural synthesis of a DNA-binding protein from the C-terminal domain of DNA gyrase A *in Borrelia burgdorferi*. Embo J *18*, 4875-4881.

Krishnavajhala, A., Wilder, H.K., Boyle, W.K., Damania, A., Thornton, J.A., Perez de Leon, A.A., Teel, P.D., and Lopez, J.E. (2017). Imaging of *Borrelia turicatae* Producing the Green Fluorescent Protein Reveals Persistent Colonization of the *Ornithodoros turicata* Midgut and Salivary Glands from Nymphal Acquisition through Transmission. Appl Environ Microbiol *83*, e02503-02516. 10.1128/AEM.02503-16

Kuleshov, K.V., Margos, G., Fingerle, V., Koetsveld, J., Goptar, I.A., Markelov, M.L., Kolyasnikova, N.M., Sarksyan, D.S., Kirdyashkina, N.P., Shipulin, G.A., *et al.* (2020). Whole genome sequencing of *Borrelia miyamotoi* isolate Izh-4: reference for a complex bacterial genome. BMC Genomics *21*, 16. 10.1186/s12864-019-6388-4

Kumru, O.S., Schulze, R.J., Slusser, J.G., and Zuckert, W.R. (2010). Development and validation of a FACS-based lipoprotein localization screen in the Lyme disease spirochete *Borrelia burgdorferi*. BMC microbiology *10*, 277. 10.1186/1471-2180-10-277

Kurtti, T.J., Munderloh, U.G., Johnson, R.C., and Ahlstrand, G.G. (1987). Colony formation and morphology in *Borrelia burgdorferi*. J Clin Microbiol *25*, 2054-2058.

Labandeira-Rey, M., Seshu, J., and Skare, J.T. (2003). The absence of linear plasmid 25 or 28-1 of *Borrelia burgdorferi* dramatically alters the kinetics of experimental infection via distinct mechanisms. Infect Immun *71*, 4608-4613.

Labandeira-Rey, M., and Skare, J.T. (2001). Decreased infectivity in *Borrelia burgdorferi* strain B31 is associated with loss of linear plasmid 25 or 28-1. Infect Immun *69*, 446-455.

Latham, J.I., and Blevins, J.S. (2018). Generation of Conditional Mutants in *Borrelia burgdorferi*. Methods Mol Biol *1690*, 225-239. 10.1007/978-1-4939-7383-5_17

Lawrenz, M.B., Kawabata, H., Purser, J.E., and Norris, S.J. (2002). Decreased electroporation efficiency in *Borrelia burgdorferi* containing linear plasmids lp25 and lp56: impact on transformation of infectious *Borrelia*. Infect Immun *70*, 4851-4858.

Lee, W.Y., Moriarty, T.J., Wong, C.H., Zhou, H., Strieter, R.M., van Rooijen, N., Chaconas, G., and Kubes, P. (2010). An intravascular immune response to *Borrelia burgdorferi* involves Kupffer cells and iNKT cells. Nat Immunol *11*, 295-302. 10.1038/ni.1855

Lenhart, T.R., and Akins, D.R. (2010). *Borrelia burgdorferi* locus BB0795 encodes a BamA orthologue required for growth and efficient localization of outer membrane proteins. Mol Microbiol *75*, 692-709. 10.1111/j.1365-2958.2009.07015.x

Lescot, M., Audic, S., Robert, C., Nguyen, T.T., Blanc, G., Cutler, S.J., Wincker, P., Couloux, A., Claverie, J.M., Raoult, D., *et al.* (2008). The genome of *Borrelia recurrentis*, the agent of deadly louse-borne relapsing fever, is a degraded subset of tick-borne *Borrelia duttonii*. PLoS Genet *4*, e1000185.

Li, X., Pal, U., Ramamoorthi, N., Liu, X., Desrosiers, D.C., Eggers, C.H., Anderson, J.F., Radolf, J.D., and Fikrig, E. (2007a). The Lyme disease agent *Borrelia burgdorferi* requires BB0690, a Dps homologue, to persist within ticks. Mol Microbiol *63*, 694-710. 10.1111/j.1365-2958.2006.05550.x

Li, X., Pal, U., Ramamoorthi, N., Liu, X., Desrosiers, D.C., Eggers, C.H., Anderson, J.F., Radolf, J.D., and Fikrig, E. (2007b). The Lyme disease agent *Borrelia burgdorferi* requires BB0690, a Dps homologue, to persist within ticks. Mol Microbiol *63*, 694-710.

Lin, B., Short, S.A., Eskildsen, M., Klempner, M.S., and Hu, L.T. (2001). Functional testing of putative oligopeptide permease (Opp) proteins of *Borrelia burgdorferi:* a complementation model in *opp(-) Escherichia coli.* Biochim Biophys Acta *1499*, 222-231.

Lin, T., and Gao, L. (2018). Genome-Wide Mutagenesis in *Borrelia burgdorferi*. Methods Mol Biol *1690*, 201-223. 10.1007/978-1-4939-7383-5_16

Lin, T., Gao, L., Zhang, C., Odeh, E., Jacobs, M.B., Coutte, L., Chaconas, G., Philipp, M.T., and Norris, S.J. (2012). Analysis of an ordered, comprehensive STM mutant library in infectious *Borrelia burgdorferi*: insights into the genes required for mouse infectivity. PLoS One *7*, e47532. 10.1371/journal.pone.0047532

Lin, T., Troy, E.B., Hu, L.T., Gao, L., and Norris, S.J. (2014). Transposon mutagenesis as an approach to improved understanding of *Borrelia* pathogenesis and biology. Frontiers in cellular and infection microbiology *4*, 63. 10.3389/fcimb.2014.00063

Liveris, D., Mulay, V., and Schwartz, I. (2004). Functional properties of *Borrelia burgdorferi recA.* J Bacteriol *186*, 2275-2280.

Lopez, J.E., Schrumpf, M.E., Raffel, S.J., Policastro, P.F., Porcella, S.F., and Schwan, T.G. (2008). Relapsing fever spirochetes retain infectivity after prolonged *in vitro* cultivation. Vector Borne Zoonotic Dis *8*, 813-820. 10.1089/vbz.2008.0033

Lopez, J.E., Wilder, H.K., Hargrove, R., Brooks, C.P., Peterson, K.E., Beare, P.A., Sturdevant, D.E., Nagarajan, V., Raffel, S.J., and Schwan, T.G. (2013). Development of genetic system to inactivate a *Borrelia turicatae* surface protein selectively produced within the salivary glands of the arthropod vector. PLoS neglected tropical diseases 7, e2514. 10.1371/journal.pntd.0002514

Lybecker, M.C., and Samuels, D.S. (2007). Temperature-induced regulation of RpoS by a small RNA in *Borrelia burgdorferi*. Mol Microbiol *64*, 1075-1089.

Mahan, M.J., Slauch, J.M., and Mekalanos, J.J. (1993). Selection of bacterial virulence genes that are specifically induced in host tissues. Science *259*, 686-688.

Margolis, N., Hogan, D., Tilly, K., and Rosa, P.A. (1994). Plasmid location of *Borrelia* purine biosynthesis gene homologs. J Bacteriol *176*, 6427-6432.

Margos, G., Hepner, S., Mang, C., Marosevic, D., Reynolds, S.E., Krebs, S., Sing, A., Derdakova, M., Reiter, M.A., and Fingerle, V. (2017). Lost in plasmids: next generation sequencing and the complex genome of the tick-borne pathogen *Borrelia burgdorferi*. BMC Genomics *18*, 422. 10.1186/s12864-017-3804-5

Medrano, M.S., Policastro, P.F., Schwan, T.G., and Coburn, J. (2010). Interaction of *Borrelia burgdorferi* Hbb with the p66 promoter. Nucleic Acids Res *38*, 414-427. 10.1093/nar/gkp1027

Miller, J.C., von Lackum, K., Woodman, M.E., and Stevenson, B. (2006). Detection of *Borrelia burgdorferi* gene expression during mammalian infection using transcriptional fusions that produce green fluorescent protein. Microb Pathog *41*, 43-47.

Miller, S.C., Porcella, S.F., Raffel, S.J., Schwan, T.G., and Barbour, A.G. (2013). Large linear plasmids of *Borrelia* species that cause relapsing fever. J Bacteriol *195*, 3629-3639. 10.1128/JB.00347-13

Moriarty, T.J., Norman, M.U., Colarusso, P., Bankhead, T., Kubes, P., and Chaconas, G. (2008). Real-time high resolution 3D imaging of the Lyme disease spirochete adhering to and escaping from the vasculature of a living host. PLoS Pathog *4*, e1000090.

Motaleb, M.A., Pitzer, J.E., Sultan, S.Z., and Liu, J. (2011). A novel gene inactivation system reveals altered periplasmic flagellar orientation in a *Borrelia burgdorferi fliL* mutant. J Bacteriol *193*, 3324-3331. 10.1128/JB.00202-11

Nickoloff, J.A. (1995). Electroporation Protocols for Microorganisms (Totowa, New Jersey: Humana Press).

Norris, S.J., Howell, J.K., Odeh, E.A., Lin, T., Gao, L., and Edmondson, D.G. (2011). High-throughput plasmid content analysis of *Borrelia burgdorferi* B31 by using Luminex multiplex technology. Applied and Environmental Microbiology *77*, 1483-1492. 10.1128/AEM.01877-10

Novak, E.A., Sekar, P., Xu, H., Moon, K.H., Manne, A., Wooten, R.M., and Motaleb, M.A. (2016). The *Borrelia burgdorferi* CheY3 response regulator is essential for chemotaxis and completion of its natural infection cycle. Cell Microbiol *18*, 1782-1799. 10.1111/cmi.12617

Oehler, S., Amouyal, M., Kolkhof, P., von Wilcken-Bergmann, B., and Müller-Hill, B. (1994). Quality and position of the three lac operators of *E. coli* define efficiency of repression. EMBO J *13*, 3348-3355.

Ouyang, Z., Haq, S., and Norgard, M.V. (2010). Analysis of the *dbpBA* upstream regulatory region controlled by RpoS in *Borrelia burgdorferi*. J Bacteriol *192*, 1965-1974. 10.1128/JB.01616-09

Parveen, N., Cornell, K., Bono, J., Chamberland, C., Rosa, P., and Leong, J. (2006). Bgp, a secreted GAG-binding protein of *Borrelia burgdorferi* strain N40, displays nucleosidase activity and is not essential for infection of immunodeficient mice. Infect Immun *74*, 3016-3020.

Paster, B.J., Stackebrandt, E., Hespell, R.B., Hahn, C.M., and Woese, C.R. (1984). The phylogeny of the spirochetes. Syst Appl Microbiol *5*, 337-351.

Pedelacq, J.D., Cabantous, S., Tran, T., Terwilliger, T.C., and Waldo, G.S. (2006). Engineering and characterization of a superfolder green fluorescent protein. Nat Biotechnol *24*, 79-88. 10.1038/nbt1172

Phelan, J.P., Kern, A., Ramsey, M.E., Lundt, M.E., Sharma, B., Lin, T., Gao, L., Norris, S.J., Hyde, J.A., Skare, J.T., *et al.* (2019). Genome-wide screen identifies novel genes required for *Borrelia burgdorferi* survival in its *Ixodes* tick vector. PLoS Pathog *15*, e1007644. 10.1371/journal.ppat.1007644

Pich, O.Q., Burgos, R., Planell, R., Querol, E., and Pinol, J. (2006). Comparative analysis of antibiotic resistance gene markers in *Mycoplasma genitalium*: application to studies of the minimal gene complement. Microbiology *152*, 519-527.

Purser, J.E., Lawrenz, M.B., Caimano, M.J., Radolf, J.D., and Norris, S.J. (2003). A plasmid-encoded nicotinamidase (PncA) is essential for infectivity of *Borrelia burgdorferi* in a mammalian host. Mol Microbiol *48*, 753-764.

Purser, J.E., and Norris, S.J. (2000). Correlation between plasmid content and infectivity in *Borrelia burgdorferi*. Proc Natl Acad Sci USA *97*, 13865-13870.

Putteet-Driver, A.D., Zhong, J., and Barbour, A.G. (2004). Transgenic expression of RecA of the spirochetes *Borrelia burgdorferi* and *Borrelia hermsii* in *Escherichia coli* revealed differences in DNA repair and recombination phenotypes. J Bacteriol *186*, 2266-2274.

Raffel, S.J., Battisti, J.M., Fischer, R.J., and Schwan, T.G. (2014). Inactivation of genes for antigenic variation in the relapsing fever spirochete *Borrelia hermsii* reduces infectivity in mice and transmission by ticks. PLoS Pathog *10*, e1004056. 10.1371/journal.ppat.1004056

Raffel, S.J., Williamson, B.N., Schwan, T.G., and Gherardini, F.C. (2018). Colony formation in solid medium by the relapsing fever spirochetes *Borrelia hermsii* and *Borrelia turicatae*. Ticks and tick-borne diseases *9*, 281-287. 10.1016/j.ttbdis.2017.11.001

Ramsey, M.E., Hyde, J.A., Medina-Perez, D.N., Lin, T., Gao, L., Lundt, M.E., Li, X., Norris, S.J., Skare, J.T., and Hu, L.T. (2017). A high-throughput genetic screen identifies previously uncharacterized *Borrelia burgdorferi* genes important for resistance against reactive oxygen and nitrogen species. PLoS Pathog *13*, e1006225. 10.1371/journal.ppat.1006225

Rego, R.O., Bestor, A., and Rosa, P.A. (2011). Defining the plasmid-encoded restriction-modification systems of the Lyme disease spirochete *Borrelia burgdorferi*. J Bacteriol *193*, 1161-1171. JB.01176-10 [pii] 10.1128/JB.01176-10

Rego, R.O., Bestor, A., Stefka, J., and Rosa, P.A. (2014). Population bottlenecks during the infectious cycle of the Lyme disease spirochete *Borrelia burgdorferi*. PLoS One *9*, e101009. 10.1371/journal.pone.0101009

Restrepo, B.I., and Barbour, A.G. (1994). Antigen diversity in the bacterium *B. hermsii* through "somatic" mutations in rearranged *vmp* genes. Cell *78*, 867-876.

Revel, A.T., Blevins, J.S., Almazan, C., Neil, L., Kocan, K.M., de la Fuente, J., Hagman, K.E., and Norgard, M.V. (2005). *bptA* (*bbe16*) is essential for the persistence of the Lyme disease spirochete, *Borrelia burgdorferi*, in its natural tick vector. Proc Natl Acad Sci USA *102*, 6972-6977.

Revel, A.T., Talaat, A.M., and Norgard, M.V. (2002). DNA microarray analysis of differential gene expression in *Borrelia burgdorferi*, the Lyme disease spirochete. Proc Natl Acad Sci USA *99*, 1562-1567.

Richards, C.L., Lawrence, K.A., Su, H., Yang, Y., Yang, X.F., Dulebohn, D.P., and Gherardini, F.C. (2015). Acetyl-Phosphate Is Not a Global Regulatory Bridge between Virulence and Central Metabolism in *Borrelia burgdorferi*. PLoS One *10*, e0144472. 10.1371/journal.pone.0144472

Ristow, L.C., Miller, H.E., Padmore, L.J., Chettri, R., Salzman, N., Caimano, M.J., Rosa, P.A., and Coburn, J. (2012). The beta(3)-integrin ligand of *Borrelia burgdorferi* is critical for infection of mice but not ticks. Mol Microbiol *85*, 1105-1118. 10.1111/j.1365-2958.2012.08160.x

Rosa, P., Samuels, D.S., Hogan, D., Stevenson, B., Casjens, S., and Tilly, K. (1996). Directed insertion of a selectable marker into a circular plasmid of *Borrelia burgdorferi*. J Bacteriol *178*, 5946-5953.

Rosa, P.A., and Hogan, D. (1992). Colony formation by *Borrelia burgdorferi* in solid medium: clonal analysis of *osp* locus variants. In Proceeding of the First International Conference on Tick Borne Pathogens at the Host-Vector Interface, U.G. Munderloh, and T.J. Kurtti, eds. (University of Minnesota, St. Paul, Minn.), pp. 95-103.

Rosa, P.A., Tilly, K., and Stewart, P.E. (2005). The burgeoning molecular genetics of the Lyme disease spirochaete. Nature Rev Microbiol *3*, 129-143.

Ruzic-Sabljic, E., Podreka, T., Maraspin, V., and Strle, F. (2005). Susceptibility of *Borrelia afzelii* strains to antimicrobial agents. Int J Antimicrob Agents *25*, 474-478.

Sadziene, A., Wilske, B., Ferdows, M.S., and Barbour, A.G. (1993). The cryptic *ospC* gene of *Borrelia burgdorferi* B31 is located on a circular plasmid. Infect Immun *61*, 2192-2195.

Salama, N.R., and Manoil, C. (2006). Seeking completeness in bacterial mutant hunts. Curr Opin Microbiol *9*, 307-311.

Sambri, V., and Lovett, M.A. (1990). Survival of *Borrelia burgdorferi* in different electroporation buffers. Microbiologica *13*, 79-83.

Sambrook, J., Fritsch, E.F., and Maniatis, T. (1989). Molecular Cloning. A Laboratory Manual, 2nd edn (Cold Spring Harbor, NY: Cold Spring Harbor Laboratory Press).

Samuels, D.S. (1995). Electrotransformation of the spirochete *Borrelia burgdorferi*. In Methods in Molecular Biology, J.A. Nickoloff, ed. (Totowa, NJ: Humana Press, Inc.), pp. 253-259.

Samuels, D.S. (2006). Antibiotic resistance in *Borrelia burgdorferi*: applications for genetic manipulation and implications for evolution. In NATO Advanced Research Workshop on Molecular Biology of Spirochetes, F.C. Cabello, D. Hulinska, and H.P. Godfrey, eds. (Prague, Czech Republic: IOS Press, Amsterdam), pp. 56-70.

Samuels, D.S., Drecktrah, D., and Hall, L.S. (2018). Genetic Transformation and Complementation. Methods Mol Biol *1690*, 183-200. 10.1007/978-1-4939-7383-5_15

Samuels, D.S., and Garon, C.F. (1997). Oligonucleotide-mediated genetic transformation of *Borrelia burgdorferi*. Microbiology *143*, 519-522.

Samuels, D.S., Mach, K.E., and Garon, C.F. (1994a). Genetic transformation of the Lyme disease agent *Borrelia burgdorferi* with coumarin-resistant *gyrB*. J Bacteriol *176*, 6045-6049.

Samuels, D.S., Marconi, R.T., Huang, W.M., and Garon, C.F. (1994b). *gyrB* mutations in coumermycin A_1-resistant *Borrelia burgdorferi*. J Bacteriol *176*, 3072-3075.

Sarkar, A., Hayes, B.M., Dulebohn, D.P., and Rosa, P.A. (2011). Regulation of the virulence determinant OspC by *bbd18* on linear plasmid lp17 of *Borrelia burgdorferi*. J Bacteriol *193*, 5365-5373. 10.1128/jb.01496-10

Sartakova, M., Dobrikova, E., and Cabello, F.C. (2000). Development of an extrachromosomal cloning vector system for use in *Borrelia burgdorferi*. Proc Natl Acad Sci USA *97*, 4850-4855.

Sartakova, M.L., Dobrikova, E.Y., Motaleb, M.A., Godfrey, H.P., Charon, N.W., and Cabello, F.C. (2001). Complementation of a nonmotile *flaB* mutant of *Borrelia burgdorferi* by chromosomal integration of a plasmid containing a wild-type *flaB* allele. J Bacteriol *183*, 6558-6564.

Schulze, R.J., and Zuckert, W.R. (2006). *Borrelia burgdorferi* lipoproteins are secreted to the outer surface by default. Mol Microbiol *59*, 1473-1484.

Schwan, T.G., Raffel, S.J., and Battisti, J.M. (2020). Transgenic functional complementation with a transmission -associated protein restores spirochete infectivity by tick bite. Ticks and tick-borne diseases *11*, 101377. 10.1016/j.ttbdis.2020.101377

Seshu, J., Esteve-Gassent, M.D., Labandeira-Rey, M., Kim, J.H., Trzeciakowski, J.P., Hook, M., and Skare, J.T. (2006). Inactivation of the fibronectin-binding adhesin gene *bbk32* significantly attenuates the infectivity potential of *Borrelia burgdorferi*. Mol Microbiol *59*, 1591-1601.

Shigekawa, K., and Dower, W.J. (1988). Electroporation of eukaryotes and prokaryotes: a general approach to the introduction of macromolecules into cells. BioTechniques *6*, 742-751.

Showman, A.C., Aranjuez, G., Adams, P.P., and Jewett, M.W. (2016). Gene *bb0318* Is Critical for the Oxidative Stress Response and Infectivity of *Borrelia burgdorferi*. Infect Immun *84*, 3141-3151. 10.1128/IAI.00430-16

Skare, J.T., Shaw, D.K., Trzeciakowski, J.P., and Hyde, J.A. (2016). In Vivo Imaging Demonstrates That *Borrelia burgdorferi ospC* Is Uniquely Expressed Temporally and Spatially throughout Experimental Infection. PLoS One *11*, e0162501. 10.1371/journal.pone.0162501

Sohaskey, C.D., Arnold, C., and Barbour, A.G. (1997). Analysis of promoters in *Borrelia burgdorferi* by use of a transiently expressed reporter gene. J Bacteriol *179*, 6837-6842.

Sohaskey, C.D., Zuckert, W.R., and Barbour, A.G. (1999). The extended promoters for two outer membrane lipoprotein genes of *Borrelia* spp. uniquely include a T-rich region. Mol Microbiol *33*, 41-51.

Srivastava, S.Y., and de Silva, A.M. (2008). Reciprocal expression of *ospA* and *ospC* in single cells of *Borrelia burgdorferi*. J Bacteriol *190*, 3429-3433.

Steere, A.C. (2001). Lyme Disease. N Engl J Med *345*, 115-125.

Stevenson, B., Bono, J.L., Elias, A., Tilly, K., and Rosa, P. (1998). Transformation of the Lyme disease spirochete *Borrelia burgdorferi* with heterologous DNA. J Bacteriol *180*, 4850-4855.

Stewart, P.E., Byram, R., Grimm, D., Tilly, K., and Rosa, P.A. (2005). The plasmids of *Borrelia burgdorferi*: essential genetic elements of a pathogen. Plasmid *53*, 1-13.

Stewart, P.E., Chaconas, G., and Rosa, P. (2003). Conservation of plasmid maintenance functions between linear and circular plasmids in *Borrelia burgdorferi*. J Bacteriol *185*, 3202-3209.

Stewart, P.E., Hoff, J., Fischer, E., Krum, J.G., and Rosa, P.A. (2004). Genome-wide transposon mutagenesis of *Borrelia burgdorferi* for identification of phenotypic mutants. Appl Environ Microbiol *70*, 5973-5979.

Stewart, P.E., and Rosa, P.A. (2008). Transposon mutagenesis of the Lyme disease agent *Borrelia burgdorferi*. Methods Mol Biol *431*, 85-95.

Stewart, P.E., Thalken, R., Bono, J.L., and Rosa, P. (2001). Isolation of a circular plasmid region sufficient for autonomous replication and transformation of infectious *Borrelia burgdorferi*. Mol Microbiol *39*, 714-721.

Stoenner, H.G., Dodd, T., and Larsen, C. (1982). Antigenic variation in *B. hermsii*. J Exp Med *156*, 1297-1311.

Strnad, M., and Rego, R.O.M. (2020). The need to unravel the twisted nature of the *Borrelia burgdorferi sensu lato* complex across Europe. Microbiology. 10.1099/mic.0.000899

Strother, K.O., Broadwater, A., and de Silva, A. (2005). Plasmid requirements for infection of ticks by *Borrelia burgdorferi*. Vector Borne Zoonotic Dis ***5***, 237-245.

Strother, K.O., Hodzic, E., Barthold, S.W., and de Silva, A.M. (2007). Infection of mice with Lyme disease spirochetes constitutively expressing outer surface proteins A and B. Infect Immun *75*, 2786-2794.

Sultan, S.Z., Sekar, P., Zhao, X., Manne, A., Liu, J., Wooten, R.M., and Motaleb, M.A. (2015). Motor rotation is essential for the formation of the periplasmic flagellar ribbon, cellular morphology, and *Borrelia burgdorferi* persistence within *Ixodes scapularis* tick and murine hosts. Infect Immun *83*, 1765-1777. 10.1128/IAI.03097-14

Takacs, C.N., Kloos, Z.A., Scott, M., Rosa, P.A., and Jacobs-Wagner, C. (2018). Fluorescent Proteins, Promoters, and Selectable Markers for Applications in the Lyme Disease Spirochete *Borrelia burgdorferi*. Appl Environ Microbiol *84*, e01824-01818. 10.1128/AEM.01824-18

Tilly, K., Bestor, A., Jewett, M.W., and Rosa, P. (2007). Rapid clearance of Lyme disease spirochetes lacking OspC from skin. Infect Immun *75*, 1517-1519.

Tilly, K., Bestor, A., and Rosa, P.A. (2016). Functional equivalence of OspA and OspB, but not OspC, in tick colonization by *Borrelia burgdorferi*. Infect Immun *84*, 1565-1573. 10.1128/IAI.00063-16

Tilly, K., Elias, A.F., Bono, J.L., Stewart, P., and Rosa, P. (2000). DNA exchange and insertional inactivation in spirochetes. J Mol Microbiol Biotechnol *2*, 433-442.

Tilly, K., Elias, A.F., Errett, J., Fischer, E., Iyer, R., Schwartz, I., Bono, J.L., and Rosa, P. (2001). Genetics and regulation of chitobiose utilization in *Borrelia burgdorferi*. J Bacteriol *183*, 5544-5553.

Tilly, K., Grimm, D., Bueschel, D.M., Krum, J.G., and Rosa, P. (2004). Infectious cycle analysis of a *Borrelia burgdorferi* mutant defective in transport of chitobiose, a tick cuticle component. Vector Borne Zoonotic Dis *4*, 159-168.

Tilly, K., Hauser, R., Campbell, J., and Ostheimer, J. (1993). Isolation of *dnaJ*, *dnaK*, and *grpE* homologues from *Borrelia burgdorferi* and complementation of *Escherichia coli* mutants. Mol Microbiol *7*, 359-369.

Tilly, K., Krum, J.G., Bestor, A., Jewett, M.W., Grimm, D., Bueschel, D., Byram, R., Dorward, D., Stewart, P., and Rosa, P. (2006). *Borrelia burgdorferi* OspC protein required exclusively in a crucial early stage of mammalian infection. Infect Immun *74*, 3554-3564.

Trevors, J.T., Chassy, B.M., Dower, W.J., and Blaschek, H.P. (1992). Electrotransformation of bacteria by plasmid DNA. In Guide to electroporation and electrofusion, D.C. Chang, B.M. Chassy, J.A. Saunders, and A.E. Sowers, eds. (San Diego: Academic Press), pp. 265-290.

Troy, E.B., Lin, T., Gao, L., Lazinski, D.W., Camilli, A., Norris, S.J., and Hu, L.T. (2013). Understanding barriers to *Borrelia burgdorferi* dissemination during infection using massively parallel sequencing. Infect Immun *81*, 2347-2357. 10.1128/IAI.00266-13

Troy, E.B., Lin, T., Gao, L., Lazinski, D.W., Lundt, M., Camilli, A., Norris, S.J., and Hu, L.T. (2016). Global Tn-seq analysis of carbohydrate utilization and vertebrate infectivity of *Borrelia burgdorferi*. Mol Microbiol *101*, 1003-1023. 10.1111/mmi.13437

Valdivia, R.H., Cormack, B.P., and Falkow, S. (2006). The uses of green fluorescent protein in prokaryotes. *Methods Biochem Anal 47*, 163-178.

Verhey, T.B., Castellanos, M., and Chaconas, G. (2019). Antigenic variation in the Lyme spirochete: detailed functional assessment of recombinational switching at *vlsE* in the JD1 strain of *Borrelia burgdorferi*. Mol Microbiol *111*, 750-763. 10.1111/mmi.14189

Wager, B., Shaw, D.K., Groshong, A.M., Blevins, J.S., and Skare, J.T. (2015). BB0744 Affects Tissue Tropism and Spatial Distribution of *Borrelia burgdorferi*. Infect Immun *83*, 3693-3703. 10.1128/IAI.00828-15

Wang, X.G., Scagliotti, J.P., and Hu, L.T. (2004). Phospholipid synthesis in *Borrelia burgdorferi: BB0249* and *BB0721* encode functional phosphatidylcholine synthase and phosphatidylglycerolphosphate synthase proteins. Microbiology *150*, 391-397.

Weening, E.H., Parveen, N., Trzeciakowski, J.P., Leong, J.M., Hook, M., and Skare, J.T. (2008). *Borrelia burgdorferi* lacking DbpBA exhibits an early survival defect during experimental infection. Infect Immun *76*, 5694-5705.

Whetstine, C.R., Slusser, J.G., and Zuckert, W.R. (2009). Development of a single-plasmid-based regulatable gene expression system for *Borrelia burgdorferi*. Appl Environ Microbiol *75*, 6553-6558. 10.1128/AEM.02825-08

Wilder, H.K., Raffel, S.J., Barbour, A.G., Porcella, S.F., Sturdevant, D.E., Vaisvil, B., Kapatral, V., Schmitt, D.P., Schwan, T.G., and Lopez, J.E. (2016). Transcriptional Profiling the 150 kb Linear Megaplasmid of *Borrelia turicatae* Suggests a Role in Vector Colonization and Initiating Mammalian Infection. PLoS One *11*, e0147707. 10.1371/journal.pone.0147707

Wormser, G.P. (2006). Clinical practice. Early Lyme disease. N Engl J Med *354*, 2794-2801.

Xu, Q., Seemanapalli, S.V., McShan, K., and Liang, F.T. (2006). Constitutive expression of outer surface protein C diminishes the ability of *Borrelia burgdorferi* to evade specific humoral immunity. Infect Immun *74*, 5177-5184.

Yang, X., Alani, S.M., and Norgard, M.V. (2003). The response regulator Rrp2 is essential for the expression of major membrane lipoproteins in *Borrelia burgdorferi*. Proc Natl Acad Sci USA *100*, 11001-11006.

Yang, X.F., Pal, U., Alani, S.M., Fikrig, E., and Norgard, M.V. (2004). Essential role for OspA/B in the life cycle of the Lyme disease spirochete. J Exp Med *199*, 641-648.

Ye, M., Zhang, J.J., Fang, X., Lawlis, G.B., Troxell, B., Zhou, Y., Gomelsky, M., Lou, Y., and Yang, X.F. (2014). DhhP, a cyclic di-AMP phosphodiesterase of *Borrelia burgdorferi*, is essential for cell growth and virulence. Infect Immun *82*, 1840-1849. 10.1128/IAI.00030-14

Zhang, J.R., Hardham, J.M., Barbour, A.G., and Norris, S.J. (1997). Antigenic variation in Lyme disease borreliae by promiscuous recombination of VMP-like sequence cassettes. Cell *89*, 275-285.

Zhang, J.R., and Norris, S.J. (1998). Kinetics and *in vivo* induction of genetic variation of *vlsE* in *Borrelia burgdorferi*. Infect Immun *66*, 3689-3697.

Zhang, X., Yang, X., Kumar, M., and Pal, U. (2009). BB0323 function is essential for *Borrelia burgdorferi* virulence and persistence through tick-rodent transmission cycle. J Infect Dis *200*, 1318-1330. 10.1086/605846

Zückert, W.R., and Meyer, J. (1996). Circular and linear plasmids of Lyme disease spirochetes have extensive homology: characterization of a repeated DNA element. J Bacteriol *178*, 2287-2298.

Chapter 17

Live Imaging

George Chaconas[1]*, Tara J. Moriarty[2], Jon Skare[3] and Jenny A. Hyde[3]

[1]Department of Biochemistry and Molecular Biology and Department of Microbiology, Immunology and Infectious Diseases, Snyder Institute for Chronic Diseases, University of Calgary, Calgary, Alberta, T2N 4N1, Canada
[2]Faculties of Dentistry and Medicine (Laboratory Medicine and Pathobiology), University of Toronto, Toronto, Ontario, M5G 1G6, Canada
[3]Department of Microbial Pathogenesis and Immunology, Texas A & M University Health Science Center, Bryan, Texas, 77807, USA
*Corresponding author: chaconas@ucalgary.ca

DOI: https://doi.org/10.21775/9781913652616.17

Abstract
Being able to vizualize a pathogen at a site of interaction with a host is an aesthetically appealing idea and the resulting images can be both informative as well as enjoyable to view. Moreover, the approaches used to derive these images can be powerful in terms of offering data unobtainable by other methods. In this chapter, we review three primary modalities for live imaging *Borrelia* spirochetes: whole animal imaging, intravital microscopy and live cell imaging. Each method has strengths and weaknesses, which we review, as well as specific purposes for which they are optimally utilized. Live imaging borriliae is a relatively recent development and the first edition of this book lacked a chapter to cover the area, which was in its infancy. Here, in addition to the methods themselves, we also review areas of spirochete biology that have been significantly impacted by live imaging and present a collection of images associated with the forward motion in the field driven by imaging studies.

Introduction
Imaging has become one of the primary methods for the analysis of pathogen-host interactions and imaging of living cells and organisms has recently made a large contribution to our knowledge of *B. burgdorferi* pathogenesis. The focus on imaging of live pathogens and or host cells/animals can provide powerful *in situ* results on dynamic interactions that are not obtainable using spirochetes grown in culture or analyzed *in vitro* or after tissue fixation. This chapter reviews currently available technologies for live imaging that have been used on *B. burgdorferi* and that have revealed significant new findings. We start with the view from 10,000 feet, which allows visualization of *B. burgdorferi* infection of the entire animal. This is followed by a lower altitude sighting of specific organ systems or anatomical sites by high resolution, real-time imaging of live spirochetes in a living animal. Finally, we focus on live imaging in cellular systems. Each approach can provide unique and important information. The major contributions that these methodologies have made to our current understanding of Lyme disease pathogenesis will be reviewed. This chapter cannot provide a comprehensive review of all the topics covered here, and non-imaging data from the papers cited below is not discussed; instead, the reader is referred to the original literature for complete details of the papers we examine.

The methodologies to be described here are highly varied, as are the processes studied and the resulting knowledge gained. Whole animal imaging makes use of bioluminescent probes and has been successfully used to study *Borrelia* pathogenesis including dissemination and colonization of the mouse using a variety of *B. burgdorferi* mutant strains. By coupling specific gene promoters to luminescent probes, the methodology has also proven useful to study the localized expression in the mouse of several spirochete genes following infection. Intravital microscopy has opened a new area of study in terms of *B. burgdorferi*-endothelial interactions, the innate immune response to *B. burgdorferi* infection, vascular transmigration, the localized effectiveness of antibiotic treatment in the mouse and *B. burgdorferi* motility and chemotaxis during infection. Intravital imaging allows the study of these processes at specific locations in a living mouse at high resolution and in real-time. Finally, live cell imaging provides the highest resolution analysis and has allowed for novel, in-depth of study of *B. burgdorferi*-tick and *B.* burgdorferi-mammalian cell interactions by the imaging of dissected whole mount infected tick midguts and salivary glands and *B. burgdorferi*-cell interactions in chambers mimicking vascular flow. Although widely diverse in technical methodology, the approaches described below are tethered to each other through their common use of light to image the desired processes.

Whole Animal Imaging

Bioluminescent *in vivo* imaging (BLI) is a widely utilized technique to investigate a variety of questions from tumor development, immune response and host-pathogen interactions to bacterial pathogenesis (Contag et al., 1995; Rehemtulla et al., 2000; Sadikot et al., 2002; Jenkins et al., 2003; Hutchens and Luker, 2007; Lienenklaus et al., 2009; Andreu et al., 2011; Xu et al., 2016). The first publication of bioluminescent bacteria was Contag *et al.* (Contag et al., 1995) in *Salmonella typhimurium* utilizing the luciferase gene from the firefly (*Photinus pyralis*) to spatiotemporally track infection in mice. This method provides a satellite view of infection as it progresses in mice and is not designed for an assessment of pathogenesis on a cellular level. Bioluminescence is the result of an enzymatic reaction by luciferase with the D-luciferin substrate, oxygen, and ATP (Figure 1). The robust signal is emitted from viable cells with relatively low background. Over the years, a variety of luciferase alleles from insects and marine organisms have been developed to capture light emission from deep tissues within mammalian hosts (Zhang et al., 1999; Waidmann et al., 2011; Xu et al., 2016). To capture images and quantify bioluminescence emission from mammalian tissue, a couple-charged device (CCD) camera is required. BLI was first implemented with *B. burgdorferi* using a constitutively expressed codon optimized *Photinus pyralis* luciferase gene on a multicopy shuttle vector (designated as Bb*luc*; (Blevins et al., 2007; Hyde et al., 2011). This work was followed by the development of single copy Bb*luc,* a borrelial codon optimized *Renilla* luciferase ($Rluc_{Bb}$), and several fluorescent proteins with potential use in the murine model (Chan et al., 2015; Adams et al., 2017; Takacs et al., 2018).

Methodology

Most bioluminescent borrelial strains employ a luciferase gene that is encoded on a shuttle vector and maintained through in vivo selective pressure (Hyde et al., 2011; Wager et al., 2015; Zhi et al., 2015; Skare et al., 2016; Adams et al., 2017). A multicopy plasmid-encoded form of luciferase is necessary for a high level of sensitivity in order to decipher subtle phenotypic differences. To facilitate this, constitutively expressed Bb*luc,* expressed from the strong *flaB* promoter (P_{flaB}), was introduced into pBBE22 within a strain background that used positive selection for *pncA* (*bbe22*) to maintain the presence of the *luc* gene during infection for up to 91 days post-infection without an added antibiotic (Figure 1; (Hyde et al., 2011; Wager et al., 2015; Zhi et al., 2015; Skare et al., 2016; Medina-Perez et al., 2020; Saputra et al., 2020). It has since been demonstrated that antibiotic supplementation in drinking water can maintain *luc*-based bioluminescence throughout a 10 day period of murine infection (Adams et al., 2016).

Experimentation using luciferase-based imaging with *B. burgdorferi* has been restricted to needle inoculation of mice although one study evaluated the acquisition of bioluminescent *B. burgdorferi* in crushed tick samples (Adams et al., 2017). Initially, Balb/c mice were used given the low skin content of melanin to maximize the acquired bioluminescent signal (Hyde et al., 2011; Wager et al., 2015; Skare et al., 2016; Saputra et al., 2020). More recently, however, C57 and C3H mice have been utilized, albeit at a lower level of sensitivity (Zhi et al., 2015; Medina-Perez et al., 2020). The advantage of needle inoculation is the ability to infect with a known amount of *B. burgdorferi.* To image, the mice are anesthetized using isoflurane and injected intraperitoneally with D-luciferin, the substrate for BbLuc, and then placed in an imager with a CCD camera that can detect light emission (Hyde and Skare, 2018). Following injection of the D-luciferin substrate, the mice can be imaged for up to 45 minutes; however, the signal does proportionally decrease over time. It is important to include a background control mouse in each image for every group of mice and time point for accurate analysis of the quantified bioluminescent signal. There are multiple acceptable approaches to control for background. One approach is to infect all the mice in the group with bioluminescent *B. burgdorferi* and not inject one mouse with D-luciferin (Hyde et al., 2011; Wager et al., 2015; Zhi et al., 2015; Hyde and Skare, 2018; Medina-Perez et al., 2020). Alternatively, one

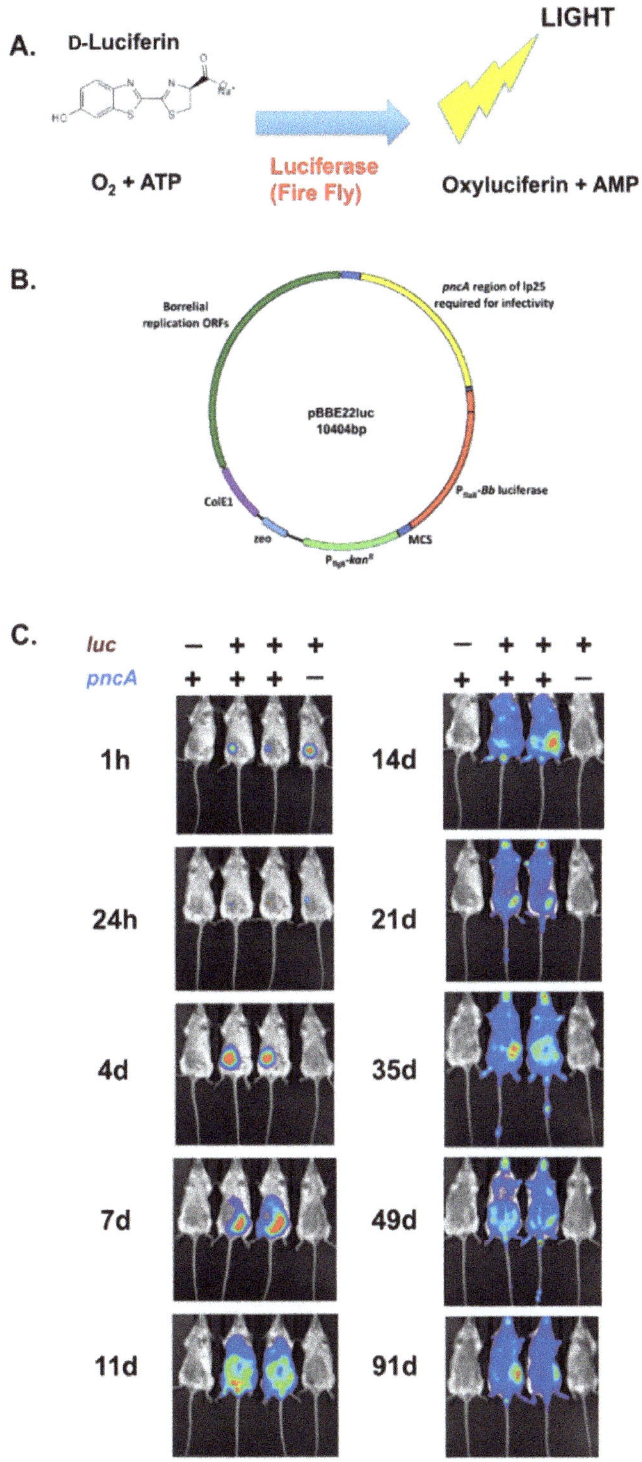

Figure 1. The implementation of bioluminescent imaging (BLI) in *Borrelia burgdorferi.* (A) Luciferase enzymology. Enzymology of luciferase that results in light formation. This reaction requires ATP and dissolved oxygen. (B) Construction of a luciferase containing plasmid for BLI following infection with *B. burgdorferi*. The Bb*luc* gene was cloned into the pBBE22 shuttle vector, which contains the 25 kb linear plasmid (lp25) *pncA* (*bbe22*) gene, to generate pBBE22*luc*. *B. burgdorferi* strains lacking the lp25 plasmid are non-infectious. The presence of *pncA* together with Bb*luc* promotes simultaneous positive selection for the maintenance of this plasmid during infection and the constitutive expression of Bb*luc*. (C) *In vivo* tracking of luciferase-expressing *B. burgdorferi*. A ventral view of Balb/C mice infected with 10^5 *B. burgdorferi* constitutively expressing Bb*luc*. The images are from the same 4 mice imaged at 1h out to 91 days (d). All mice imaged were given D-luciferin except the mouse on the left, which is infected with infectious *B. burgdorferi* with no Bb*luc* gene; the mouse on the right is infected with *B. burgdorferi* containing Bb*luc* but lacking the *pncA* (*bbe22*) gene required for persistence. The presence of the *pncA* and *luc* genes is indicated at the top left. For the images shown, the signal range averaged between 5.5×10^3 to 1.0×10^5 photons per second (p/s).

mouse in each group can be infected with *B. burgdorferi* lacking luciferase (Figure 1) or encoding a promoterless luciferase and then treated with D-luciferin as with the other mice in the group.

A challenge in experimental *B. burgdorferi* infection is determining tissue specific localization during later stages of disease. For deeper tissue colonization, mice can be treated with D-luciferin, sacrificed, organs harvested, incubated in additional substrate, and imaged (Skare et al., 2016; Saputra et al., 2020). Imaging of individual tissues allows the direct quantification of bioluminescence as a representation of bacterial burden or gene expression without further manipulation as would be necessary using real-time PCR.

Advantages

The greatest advantage of whole animal imaging is the ability to track infection of *B. burgdorferi* in a living animal over an extended period of time and observe quantifiable changes in infectivity, dissemination, and gene expression by non-invasive means. Of particular note, the *in vivo* bioluminescent evaluation of borrelial gene regulation involves environmental conditions that cannot be recapitulated under *in vitro* conditions or in dialysis membrane chambers incubated in rat peritoneal cavities. The low level of background luminescence, which often plagues fluorescence techniques, contributes to increased sensitivity and detection. Another distinguishing feature from fluorescence is that light emission is an indicator of bacterial viability (Figure 1). The requirement for oxygen and ATP as reaction co-factors for light production is found only in viable cells, whereas a decline in fluorescence indicates degradation of the protein and, therefore, diminished excitability. This sensitivity has proven to be particularly important in the comparison of mutants in various genes relative to the parental and complemented strains (Hyde et al., 2011; Wager et al., 2015; Zhi et al., 2015; Skare et al., 2016; Medina-Perez et al., 2020). Inherent to this, the number of animals required to get an assessment of infection is significantly reduced, and the same animals are repeatedly monitored, providing a less variable and more cohesive kinetic evaluation of infection.

Disadvantages

B. burgdorferi bioluminescent imaging is limited by the strength of the emitted signal, which is dependent on the number of bacteria present, strength of the promoter driving Bb*luc*, and the copy number of Bb*luc*. BLI in *B. burgdorferi* has been utilized primarily with the robust, constitutively expressed *flaB* promoter driving Bb*luc* expression encoded on a multicopy shuttle vector (Figure 1). Weaker promoters linked to Bb*luc* reduce the sensitivity of detection or possibly are not detectable above background levels. The quantifiable range of bioluminescence is 600 to 60,000 counts for valid analyses of bioluminescence imaging (Hyde and Skare, 2018). One other potential issue is the use of a multicopy Bb*luc* to visualize light emitting *B. burgdorferi*. To date, bioluminescence driven by *flaB* or other promoters on a multicopy plasmid has correlated well with borrelial burden or transcript levels, respectively, but this may not be the case for every gene of interest (Hyde et al., 2011; Skare et al., 2016; Saputra et al., 2020).

One issue with whole animal imaging is the lack of resolution of individual bacterial cells; that is, the signal only indicates localization within the skin in a broad sense as organisms disseminate intra-cutaneously (Waidmann et al., 2011). *In vivo* imaging of the whole animal or individual tissues provides limited insight into the molecular mechanisms of pathogenesis. Other techniques outlined in this Chapter involve individual *B. burgdorferi* cells interacting *in vivo* with the host vasculature and tissues. Another limitation of whole animal imaging is the inability to detect and quantify *B. burgdorferi* in deeper tissues; thus, signals in tissues such as the heart, bladder, tibiotarsal joint, and lymph nodes are not accurate. As indicated above, and outlined in more detail below, organs can be imaged *ex vivo* to determine borrelial localization, but this requires invasive techniques and more mice if several time points are needed (Skare et al., 2016; Saputra et al., 2020).

Phenotypic characterization of luciferase expressing **B. burgdorferi** *strains*

The ability to detect light-emitting *B. burgdorferi* has provided important insight into how Lyme disease spirochetes can establish and maintain an infection in a small cohort of mice both temporally and spatially (Hyde et al., 2011). This technology was used initially to assess the phenotype of a *B. burgdorferi bbk32* mutant that exhibited an apparent conflicting phenotype in the literature (Li et al., 2006; Seshu et al., 2006). The differences observed between these three week end point studies was attributed to the infectious dose used; specifically, when a high and non-physiologically relevant load was used (e.g., 10^5

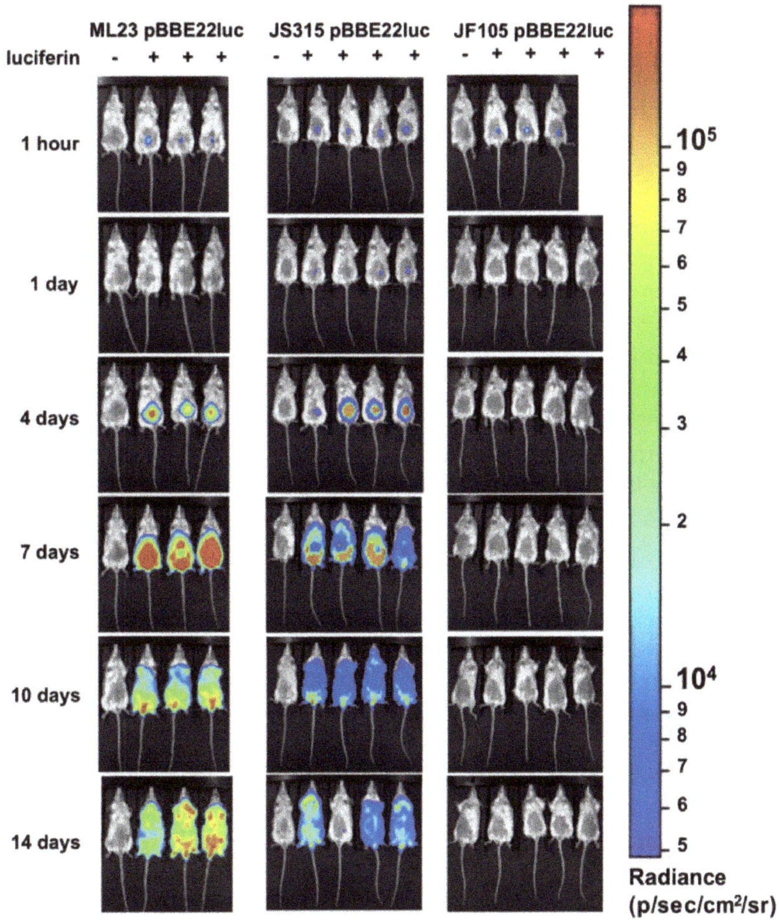

Figure 2. Tracking phenotypes of *B. burgdorferi* mutants using BLI. Infectivity of the wild type parent strain (ML23 pBBE22*luc*), a *B. burgdorferi bbk32* mutant (JS315 pBBE22*luc*), and a *B. burgdorferi dbpBA* deletion (JF105 pBBE22*luc*) were compared in Balb/c mice. Mice were infected intradermally at 10^3 *B. burgdorferi* and imaged. The mouse on the far left is infected equivalently as the others but was not given D-luciferin substrate (denoted with a "minus"). Mice given substrate are indicated with a "plus". All images for each time point and strain shown were standardized to the same photon per second range. This figure is reproduced from (Hyde et al., 2011; Zhi et al., 2015) with permission from John Wiley and Sons, Inc.

spirochetes), very little difference was observed between the infection of the *B. burgdorferi bbk32* mutant relative to its parent (Li et al., 2006; Seshu et al., 2006). However, when a lower and more physiological infectious dose was used (e.g., 10^3 spirochetes) a significant difference in infectivity was observed (Seshu et al., 2006). When the *B. burgdorferi bbk32* mutant was transformed with the Bb*luc* construct (Figure 2), the resulting data demonstrated an attenuated phenotype at both the lower and higher dose and provided additional quantitative infection kinetics of the *B. burgdorferi bbk32* mutant (Hyde et al., 2011). Specifically, the overall signal was significantly less starting at 7 days and was 4.1- to 7.7-fold less at all time points (Hyde et al., 2011). For the 10^5 dose, the light signal for the *B.burgdorferi bbk32* mutant was 3-fold less than that for the parent strain at day 7 (Hyde et al., 2011).

Imaging values also correlated well with qPCR. The *B. burgdorferi bbk32* mutant showed significantly reduced bacterial burdens by qPCR in the skin, lymph node, and joint, independent of dose (Hyde et al., 2011). Total genomic *B. burgdorferi* DNA from infected skin correlated well with bioluminescent imaging, indicating that the detection of light emitting *B. burgdorferi* is a valid approach for quantifying borrelial burdens during active infection (Hyde et al., 2011). This method also has the added advantage of deciphering subtle differences in infectivity not otherwise apparent without tracking genomic equivalents via qPCR at discrete time points. While a valuable and valid means to determine infectivity, qPCR at multiple time points has the disadvantage of being expensive, laborious, and, most importantly, requires a large number of animals to obtain relevant data.

The experiments discussed above used multicopy Bb*luc* to detect light emitting *B. burgdorferi* following infection. Another group integrated the Bb*luc* gene and a linked antibiotic resistance locus into the strain N40 *bbe02* restriction/modification system gene (Chan et al., 2015). By virtue of its construction, this strain was more competent for transformation and gained bioluminescence. When imaged, however, the signal detected was not uniform between the mice tested, presumably due to the low inoculum doses used (10, 100, and 10^3), which were, in some instances, at or below the 50% infectious dose for *B. burgdorferi* (ID_{50}).

Having established bioluminescent imaging as a method to characterize and quantify *B. burgdorferi* infectivity, numerous subsequent studies used this approach to track mutant phenotypes in live mice. In a manner similar to the *B. burgdorferi bbk32* mutant, the inactivation of *bb0744* (p83/100), as well as the non-coding RNA *ittA*, resulted in a significantly attenuated phenotype relative to the infectious parent strain (Wager et al., 2015; Medina-Perez et al., 2020). Other mutants that are severely impaired in their ability to cause experimental infection, including *B. burgdorferi dbpBA* and *bba33*, were vetted using bioluminescence, either confirming (for *dbpBA* mutants) or demonstrating (for *bba33*) their loss of virulence (Blevins et al., 2008; Shi et al., 2008; Weening et al., 2008; Hyde et al., 2011; Zhi et al., 2015). Typical images of "wildtype" (parent), subtle, or pronounced phenotypic differences are shown in Figure 2. Note that other borrelial genes have been targeted but did not yield an infectivity defect via imaging or other metrics of infection (unpublished). These analyses reinforced the strength of this approach, notably, that a phenotype, be it mild or profound, can be evaluated in a limited but significant pool of experimentally infected mice in a sensitive and timely manner.

Use of imaging to track the expression of B. burgdorferi *genes by transcriptional reporters*

B. burgdorferi undergoes complex gene regulation in the mammalian host and BLI enables spatiotemporal tracking of changes in gene expression. Promoters for *ospC* and *dbpBA* were linked to luciferase, P_{ospC}-*luc* and P_{dbpBA}-*luc,* respectively, to monitor the borrelial response to the distinct stages of pathogenesis (Figures 3 and 4) (Skare et al., 2016; Saputra et al., 2020). Bioluminescent reporter strains were normalized for changes in borrelial burden quantified from the constitutively expressed luciferase strain P_{flaB}-*luc*. These studies clearly

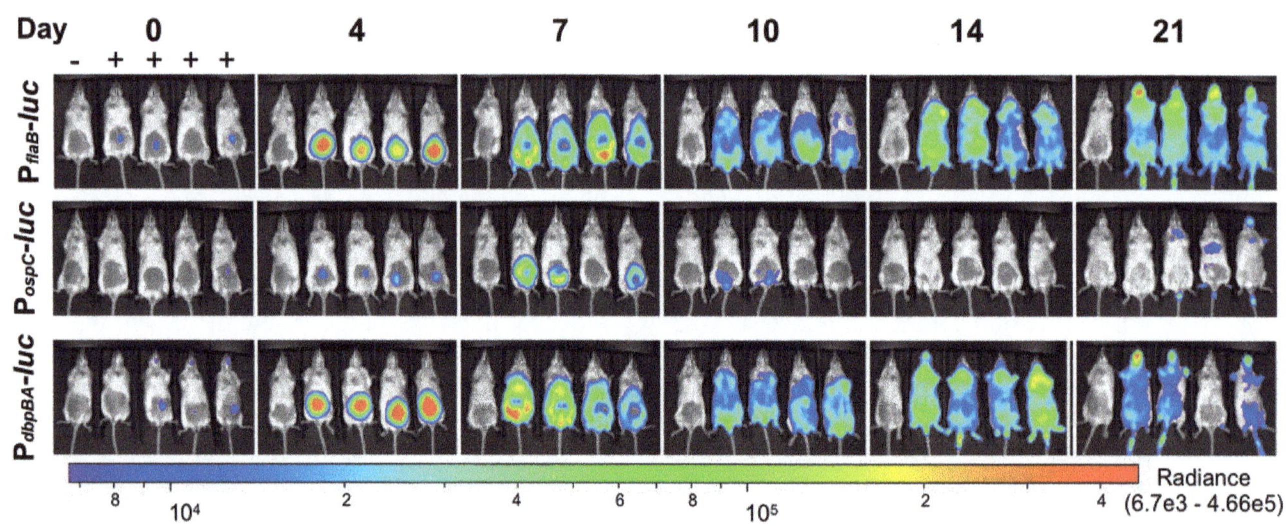

Figure 3. Tracking *B. burgdorferi* gene expression in infected mice using distinct reporter constructs. Imaging of infected mice with various reporters. Balb/c mice were infected with 10^5 infectious *B. burgdorferi* containing P_{flaB}-*luc*, P_{ospC}-*luc*, or P_{dbpBA}-*luc* reporters and imaged at the time points indicated. The control mouse ("-") was treated as described in Figure 2. All reporters were normalized for interstrain comparison to the spectral emission scale shown below the images. Images reproduced from (Skare et al., 2016; Saputra et al., 2020).

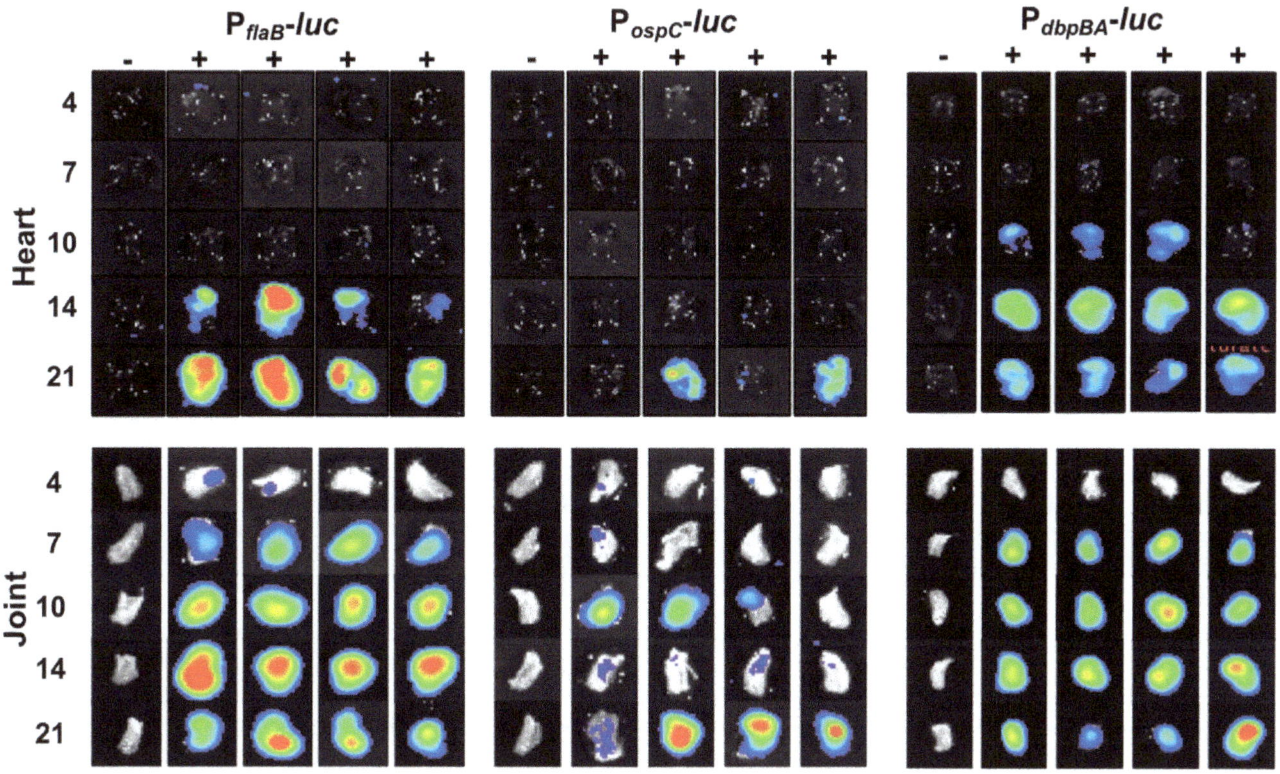

Figure 4. *Ex vivo* gene expression in tissues in *B. burgdorferi* infected mice. (A) Imaging of reporter constructs in infectious *B. burgdorferi* within heart and joint tissues. Heart and tibiotarsal joint from mice infected with infectious *B. burgdorferi* containing P_{flaB}-*luc*, P_{ospC}-*luc*, or P_{dbpBA}-*luc* were imaged at the time points indicated. Control mice were treated as described in Figure 2. Images for the P_{flaB}-*luc*, P_{ospC}-*luc*, and P_{dbpBA}-*luc* expression are set on individual scales to display the full spectrum of bioluminescence. Images reproduced from (Skare et al., 2016; Saputra et al., 2020).

demarcated the unique expression patterns of *ospC* and *dbpBA* despite both belonging to the RpoS regulon (Hubner et al., 2001; Caimano et al., 2004; Yang et al., 2005; Ouyang et al., 2012; Skare et al., 2016; Saputra et al., 2020). *dbpBA* was robustly expressed throughout infection, at times exceeding emission from P_{flaB}-*luc* (Figure 3) (Saputra et al., 2020). Given the long held paradigm that *ospC* is needed early in infection (Grimm et al., 2004; Tilly et al., 2006; Tilly et al., 2009), the observation that *ospC* expression was detectable 21 days post-infection, was surprising (Skare et al., 2016). Imaging of individual tissues demonstrated distinct tissue specific expression that changed over the course of infection. These results demonstrated that *ospC* and *dbpBA* are induced in the skin and heart early in infection after which expression declines at these sites (Figure 4). Consistent with the aforementioned atypical late expression of *ospC*, *ospC* levels were higher later in the infectious process in the bladder and joint, suggesting that OspC may be important in secondary colonization at these sites or persistence within these tissues. In the inguinal lymph node and tibiotarsal joint, *dbpBA* reached the highest level of expression after 21 days. Interestingly, *ospC* was not detectable in the lymph node. These studies highlight the utility of *in vivo* imaging to quantify the differential expression patterns of targeted *B. burgdorferi* genes in a tissue tropic fashion.

In addition to tracking the expression of known borrelial genes, luciferase based reporters have been used to characterize promoters that are induced following experimental infection of mice. Several of the promoters tested in this way were identified using the BbIVET "promoter trap" approach designed to identify promoters that function during infection (Ellis et al., 2013). This system uses randomly cloned fragments of *B. burgdorferi* DNA to drive expression of *pncA* (*bbe22*), an essential gene for borrelial infectivity. Several of the promoters isolated, including those that drive primary, internal, or antisense transcripts were re-tested using the Bb*luc* allele (Adams et al., 2017). These studies confirmed that *in vivo* reporters can be utilized in *B. burgdorferi* to study genes spatially and temporally expressed *in vivo* and, further, track quantifiable differences in live, experimentally infected mice.

Intravital Microscopy

Methodology

Intravital microscopy (IVM) allows the direct visualization of a pathogen at work in a living host, a goal sought for many years and that provides a gold standard for pathogen-host interactions. IVM has been used for decades to investigate a variety of biological processes in living animals (Secklehner et al., 2017). However, its use to investigate bacterial pathogen-host interactions is a relatively recent development with the first report appearing in 2005 (Laschke et al., 2005). An important feature of this methodology is being able to discriminate between the pathogen and the infected host. This is typically accomplished by fluorescent labelling of the pathogen, either by staining or, preferably, by expression of a fluorescent marker, such as GFP, coupled with staining of host cells with fluorescent antibodies or other fluorescent markers. Finally, accessibility of the tissue desired for imaging is an important feature of this methodology, which usually requires surgical preparation of the area of interest prior to imaging.

Early intravital imaging was performed using standard epifluorescence microscopy. More recently, however, more sophisticated microscopes providing superior resolution and imaging are available. Current microscopes of choice are either the spinning disk laser confocal microscope (SDLCM) (Oreopoulos et al., 2014) or a confocal multi-photon microscope (MPM) (Larson, 2011). The spinning disk allows imaging through multiple pinholes, thereby blocking out of focus light from both the excitation and emission light paths. The SDLCM allows high resolution imaging and rapid image acquisition. In contrast, MPM uses near infrared light where the sum of two or more photons reaching the sample provides the energy required to excite the fluorophore. The long wavelength light reduces light scattering and allows deep imaging into the sample and reduces photodamage. MPM also allows direct imaging of certain highly ordered molecular structures (eg collagen) without a fluorophore, through second harmonic generation. In this process, two near infrared photons merge their energies into a single emitted photon of visible light. Confocal microscopes of both SDLCM and MPM types also can be used for 3D reconstructions, allowing a more realistic analysis of processes in living organisms. Microscopes for intravital imaging need to be equipped with an appropriate stage, objectives and equipment for stabilizing an anesthetized living mouse for microscopic observation (Moriarty et al., 2008; Masedunskas et al., 2012; Belperron et al., 2018).

Advantages

The major advantage of IVM is that it allows for the direct visualization of a live pathogen in a living organism at high resolution and in real-time. Digital video images can be acquired and quantification is possible in terms of fluorescence intensity, velocity measurements and enumeration of spirochetes. Mechanistic studies are possible by coupling the imaging with genetics, injection of peptides, antibodies and small or macromolecules. Multichannel fluorescence experiments and 3D reconstructions are routinely performed. Combining intravital imaging with the above modalities offers an unprecedented glimpse of pathogenic mechanisms in a variety of locations.

Disadvantages

IVM is labor intensive and usually requires a high level of surgical skill with mice or other small animals and appropriate state-of-the-art microscopes specifically setup for intravital imaging. It relies on sites that can be effectively imaged (eg, skin, lung, liver, heart, spleen, brain, cremaster) (Secklehner et al., 2017) and the surgical expertise required to prepare each tissue for imaging. Intravital imaging typically requires a large number of pathogens to be present in order to locate and image the infecting organism. In the case of *B. burgdorferi*, IVM may require infection modalities that differ from tick inoculation and may therefore be subject to perturbations from a natural infection. Clarity and resolution can suffer sometimes from small movements due to respiration by the mouse. Finally, the methodology is definitely not high throughput.

B. burgdorferi-*endothelial interactions*

Hematogenous dissemination is an important, but poorly understood, facet of Lyme disease pathogenesis (Wormser, 2006). By disseminating to a variety of organ systems, *B. burgdorferi* can cause a wide variety of clinical manifestations and symptoms. But how do spirochetes manage to grab hold of the vascular endothelium in the presence of the extremely strong shear forces found in the

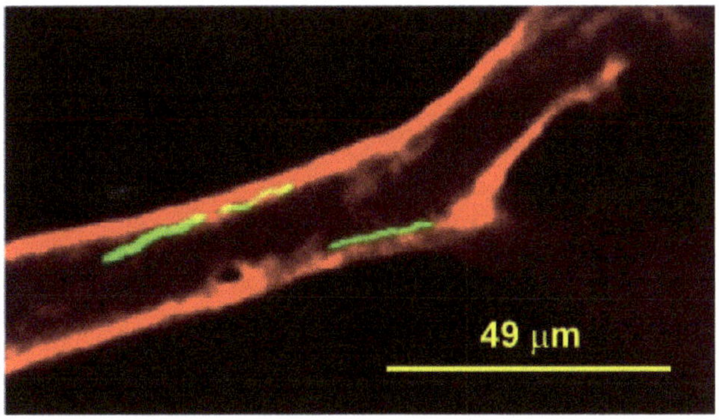

Figure 5. Intravital microscopy to study *B. burgdorferi*-microvascular interactions. Spinning disk confocal microscopy of GFP-expressing, infectious *B. burgdorferi* in a post-capillary venule of a living mouse. The vasculature was visualized with an AlexaFluor555-conjugated antibody to CD31 (PECAM-1). Image from (Norman et al., 2008).

bloodstream and then cross the endothelium to infect the parenchyma? This process is not effectively or realistically studied outside of the mouse, and it was the starting point for intravital imaging of *B. burgdorferi* by Moriarty and coworkers (Moriarty et al., 2008). The experimental system used continues to provide valuable information on *B. burgdorferi*-endothelial interactions and on the vascular transmigration process.

In the first *Borrelia* intravital imaging experiments, a high concentration of *B. burgdorferi* expressing GFP from a shuttle vector was injected directly into the bloodstream via the jugular or femoral veins of anesthetized mice (resulting in about 10^7 spirochetes/ml in the blood). Spirochetes were also visible in the ear at two weeks post-infection; however, this paper focused on *B. burgdorferi*-vascular interactions immediately after inoculation, when significant numbers of organisms could be observed. The vasculature was labelled with fluorescent CD31 (PECAM-1) antibody; a representative micrograph is shown in Figure 5. This work showed that, similar to leukocytes (Filippi, 2019), spirochetes engage in multi-stage interactions with the endothelium, including transient tethering interactions, short-lived dragging interactions and stationary adhesion to the vasculature (Figure 6). The vast majority of short-term interactions occurred on the surface of endothelial cells, while 75% of stationary adhesions were localized to endothelial junctions, suggesting that transmigration might be a paracellular process. Escaping spirochetes were present only at about 0.1% of the observed spirochete-endothelial interactions and were, therefore, difficult to reliably observe with any confidence. 3D reconstructions also were performed to accurately analyze placement of spirochetes within or outside the observed post-capillary venules. An average displacement velocity of 3.4 µm/min (range 1.1-8.1) was measured for *B. burgdorferi* escaping from post-capillary venules.

A second IVM study of B. burgdorferi-vascular interactions by the same group (Norman et al., 2008) focused on mechanistic issues involved in the pathogen-host endothelial interactions and coupled a variety of experimental approaches with IVM. These included the use of gain of function genetic

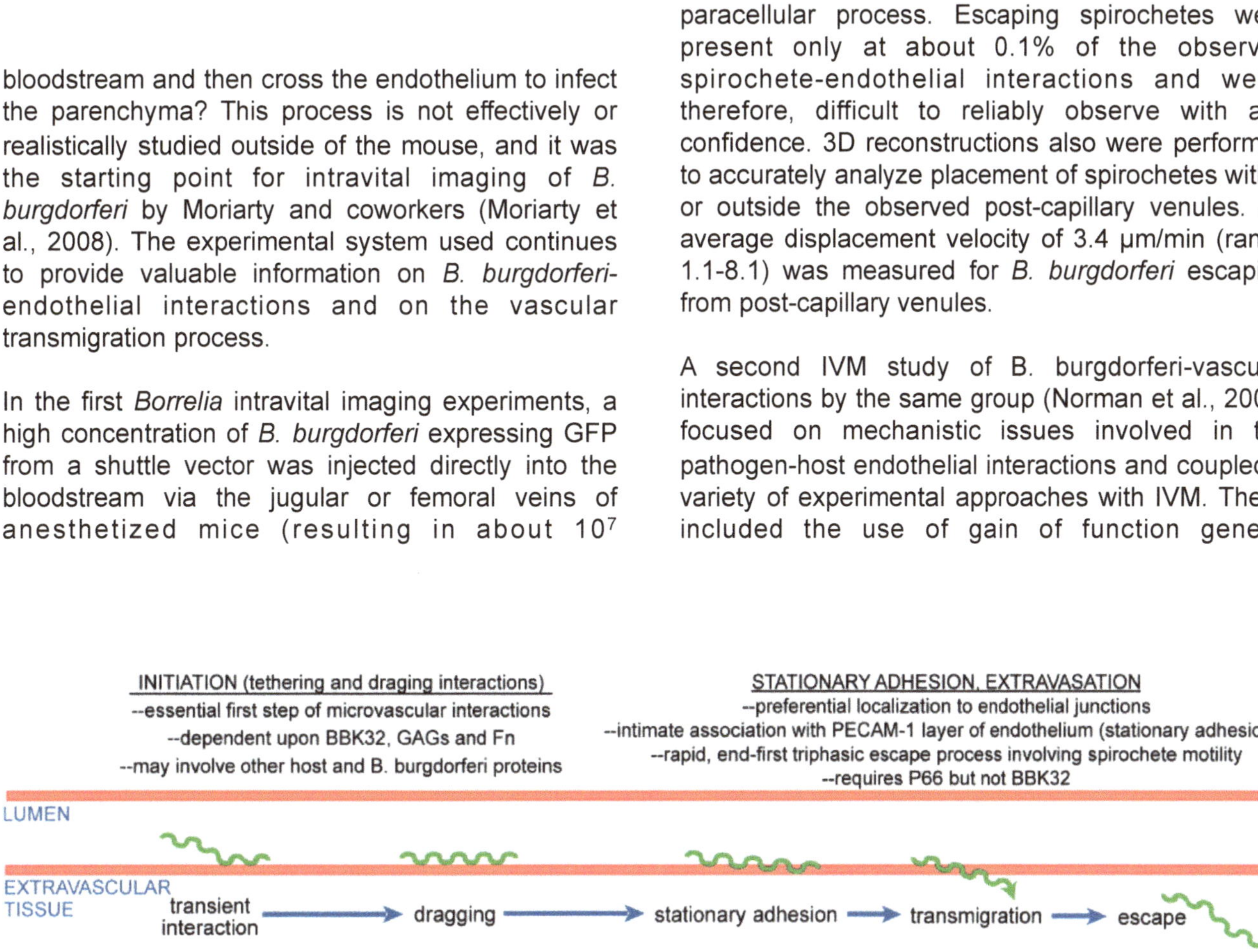

Figure 6. Schematic summarizing multi-stage interactions of *B. burgdorferi* with the mouse microvasculature based upon IVM (Moriarty et al., 2008; Norman et al., 2008). Tethering and dragging interactions comprise the essential first step of interactions with the microvasculature and are mechanistically distinct from each other regarding BBK32 interactions with Fn and GAGs, respectively (Moriarty et al., 2012). The tethering and dragging interactions are also mechanistically divergent from subsequent stationary adhesions. Stationary adhesion requires additional host and or spirochete factors and result in a redistribution of most spirochetes from cellular locations to endothelial junctions. Vascular transmigration requires P66 (Kumar et al., 2015). Figure modified from (Norman et al., 2008).

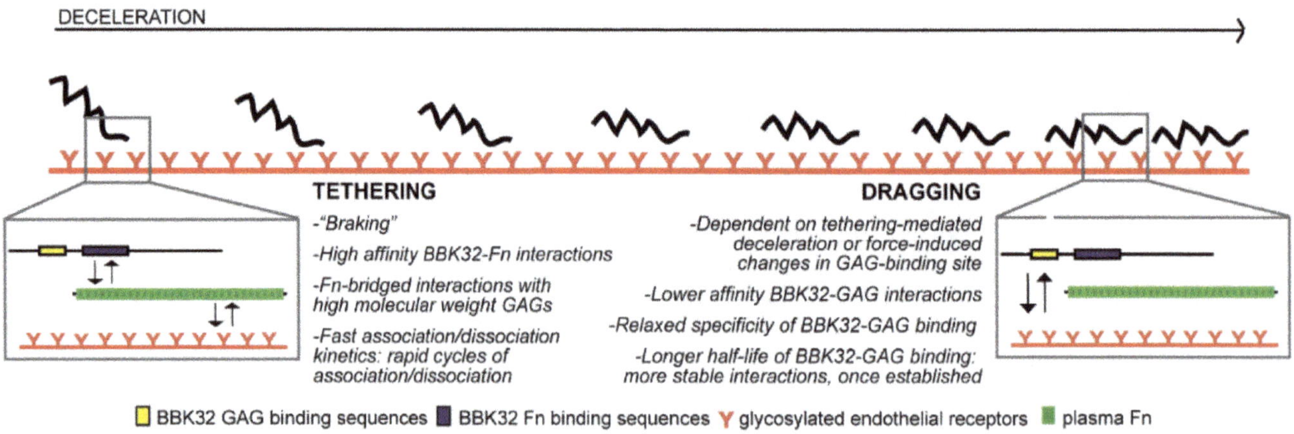

Figure 7. BBK-mediated tethering and dragging steps of vascular adhesion through distinct macromolecular interactions involving distinct Fn and GAG binding regions. Figure from (Moriarty et al., 2012), with permission from John Wiley and Sons, Inc.

constructs as well as antibodies, peptides and macromolecules to block *B. burgdorferi*-endothelial interactions. Tethering and dragging interactions were shown to be mechanistically distinct from stationary adhesions. Initial interactions were found to be mediated by host fibronectin (Fn) and glycoaminoglycans (GAGs), macromolecules known to interact with a variety of bacterial pathogens, and by the *B. burgdorferi* Fn and GAG binding protein BBK32. A subsequent IVM study further dissected the role of BBK32 by using BBK32 specifically reduced in either Fn or GAG binding (Moriarty et al., 2012). These studies showed that Fn binding by BBK32 is responsible for tethering interactions which provide the first point of contact for spirochetes careening through the vasculature under shear force. This molecular braking permits subsequent BBK32-GAG interactions that mediate spirochete dragging (Figure 7). This study also showed that RevA, RevB and BB0347 did not provide adhesion activity under shear stress in gain of function strains and that another unknown adhesin (Adhesin X) imparts strong adhesive properties to *B. burgdorferi* in the absence of BBK32. Further characterization of BBK32-endothelial interactions is discussed in the live cell imaging section.

The *B. burgdorferi* gain of function vascular adhesion system described above has also been used to assess the endothelial adhesion of Tp0751 (Pallilysin) from *Treponema pallidum* by expression of this protein in *B. burgdorferi* followed by IVM (Kao et al., 2017). Pallilysin was found to strongly interact with post-capillary venules in living mice and was also characterized by live cell imaging as noted in that section below.

The innate immune response to B. burgdorferi

IVM has also been used to study the hepatic intravascular immune response to *B. burgdorferi* (Lee et al., 2010). Imaging of the liver after intravenous injection of *B. burgdorferi* revealed that Kupffer cells ingested the spirochetes and induced iNKT cell clustering via CXCR3. Subsequently, direct contact between Kupffer cells and iNKT cells through the antigen-presenting molecule CD1D resulted in iNKT cell activation that reduced *B. burgdorferi* dissemination specifically into the knee joint and surrounding tissue.

A follow-up study by the same group (Lee et al., 2014) used multichannel spinning-disk intravital microscopy of joints and found iNKT cells adjacent to and surrounding the peripheral joint vasculature (Figure 8). The iNKT cells were positioned to ambush spirochetes at the vessel wall and interrupted transmigration of the spirochetes into the joints. *B. burgdorferi* that escaped iNKT cell immune surveillance and transmigrated into the joint tissue were subjected to attack by extravascular iNKT cells (Figure 9), which killed the spirochetes using a granzyme-dependent pathway. The combined results suggest that iNKT cells comprise a potent immune surveillance system and cytotoxic barrier for *B. burgdorferi* transmigration into joint tissue. This is further supported by the large increase in *B. burgdorferi* found specifically in joint tissue in *Cd1d-/-* mice, which are deficient in iNKT cells. This finding

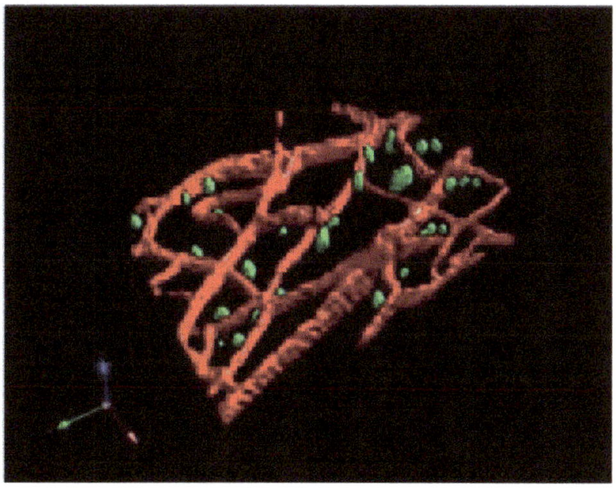

Figure 8. 3D reconstruction of iNKT cells in the mouse joint from IVM images. The iNKT cells are green (CXCR6-GFP, where most GFP-expressing extravascular cells were iNKT cells) and the vasculature red (stained with AlexaFluor555-conjugated antibody to PECAM-1. Image reprinted from (Lee et al., 2014).

was important for the development of an intravital assay to monitor the transmigration ability of *B. burgdorferi* (Kumar et al., 2015) as described in a subsequent section.

Antibiotic treatment of B. burgdorferi*-infected mice*

Two-photon confocal IVM has been used to visualize GFP-expressing *B. burgdorferi* in the dermis and calcaneal tendon of needle-inoculated and tick-infected *Myd88-/-* mice (Bockenstedt et al., 2012). The use of *Myd88-/-* mice deficient in the Toll-like receptor adaptor molecule MyD88 results in up to a 100-fold increase in spirochete burden (Barthold et al., 1992; Bolz et al., 2004; Liu et al., 2004; Wang et al., 2004) and, therefore, facilitates microscopic observation. Using real-time IVM, the authors reported that most *B. burgdorferi* were found moving along collagen fibers in the ear as can be seen in Figure 10. Treatment with ceftriaxone resulted in a dramatic disappearance of spirochetes in the dermis and tendons 24 hours after starting antibiotic treatment (Figure 11). Interestingly, at this time, two morphological transformations of elongated spirochetes into spherical forms were captured by video imaging. A very important finding from this study was that, although antibiotic treatment with ceftriaxone or doxycycline effectively killed *B. burgdorferi,* antigenic spirochete remnants were observed between 2 and 10 weeks after completion of treatment. Spirochete immunogenic and inflammatory antigens also were found in the joints of antibiotic-treated mice, suggesting that they might

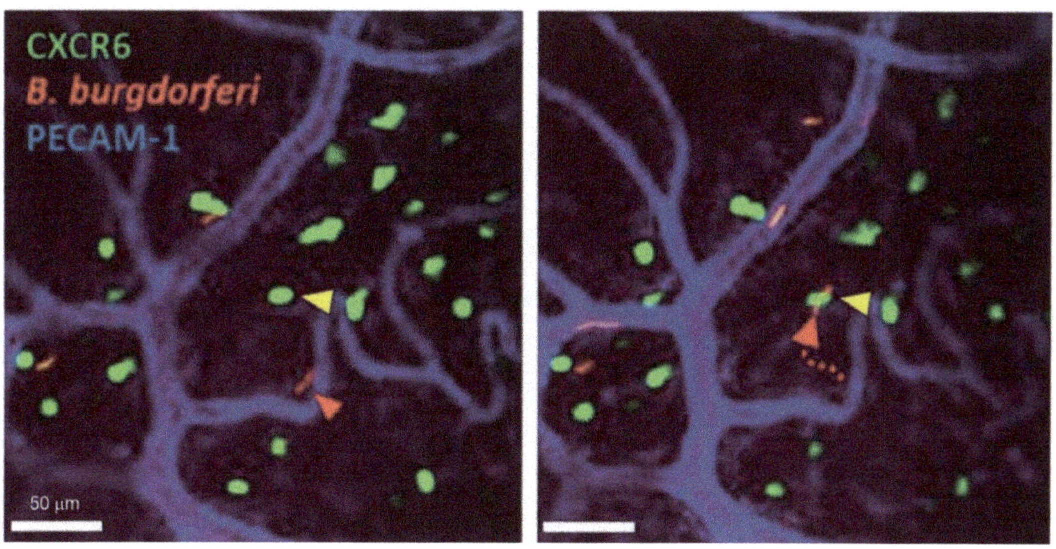

Figure 9. Spirochete attack by an extravascular iNKT cell in CXCR6-GFP mouse joint where most GFP-expressing extravascular cells were iNKT cells. The image on the left shows a motile tomato-labelled extravascular spirochete (red arrow) and a stationary iNKT cell (yellow arrow). The right image shows the path of spirochete migration (red dots) and the spirochete captured by the iNKT cell. The captured spirochete was rendered immobile for the remainder of the experiment (1 hour). Image reprinted from (Lee et al., 2014).

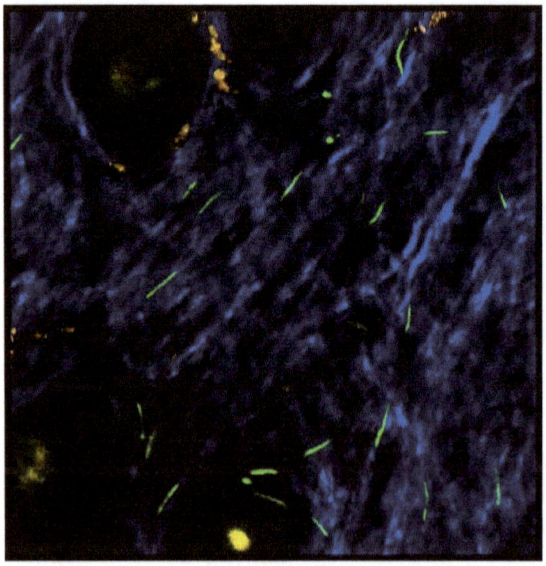

Figure 10. IVM of *B. burgdorferi* expressing GFP in the ear of an immunodeficient mouse 8 weeks post-infection, visualized by confocal two photon microscopy. Second harmonic generation of collagen results in a blue signal, melanin autofluoresces in all channels and appears white/orange, and the hair follicles auto fluoresce yellow (see Belperron et al., 2018). Image from Alexia Belperron and Linda Bockenstedt.

play a role in triggering the development of Lyme arthritis and/or other autoimmune sequelae.

Motility and chemotaxis of B. burgdorferi

The ability of Lyme disease spirochetes to disseminate in a vertebrate host is dependent upon their impressive flagella-driven motility (Sultan et al., 2013; Motaleb et al., 2015; Sultan et al., 2015). This motility has been studied by two-photon confocal IVM in the dermis of mouse ears following tick inoculation (Harman et al., 2012). Spirochete motility was observed to be similar to that previously reported in liquid and in gelatin. However, previously unobserved stationary but motile states were found. These were referred to as *wriggling* and *lunging* (Figure 12). The former remained at a fixed location but displayed movement as an undulating, traveling wave. The latter spirochetes were fixed at only one end while they underwent bending and relaxation in apparent attempts to free themselves from their attachment point. These stationary spirochetes are presumed to be tethered to the extracellular matrix by interactions strong enough to withstand the forces of spirochete motility. The IVM observations were compared with *B. burgdorferi* motility in gelatin and media to help establish *in vitro* conditions for studying spirochete motility that closely matched motility in the mouse dermis.

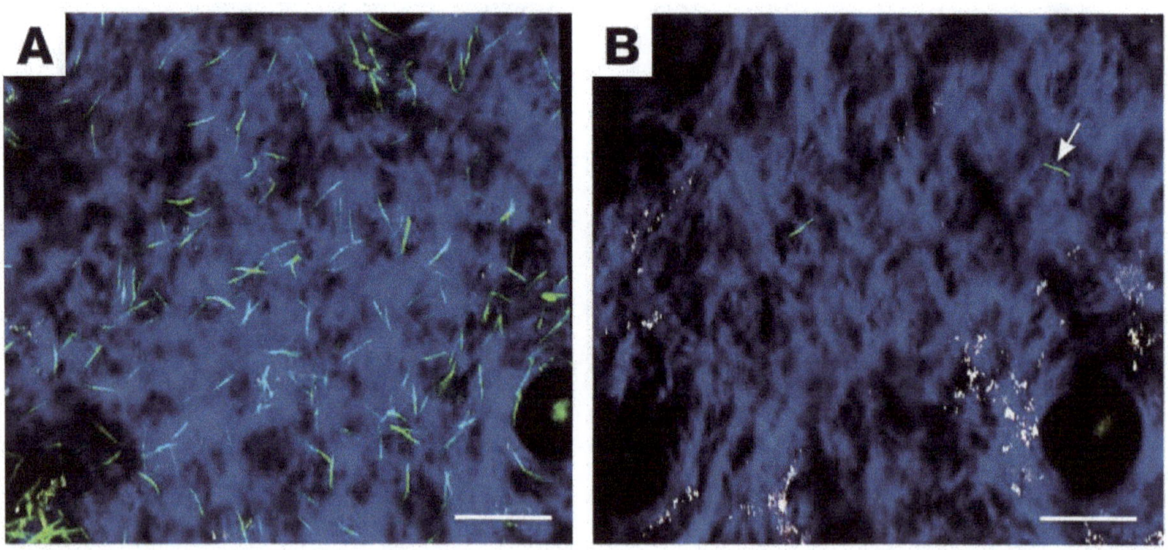

Figure 11. A) *B. burgdorferi* expressing GFP in skin of a *MyD88-/-* mouse 22 days post-infection. The blue signal is from second harmonics of skin collagen. **B)** A dramatic decrease in spirochete numbers is shown after one day of ceftriaxone treatment. The arrow denotes a spirochete that subsequently became a spherical form (not shown). Images from (Bockenstedt et al., 2012), reprinted with permission from the American Society for Clinical Investigation.

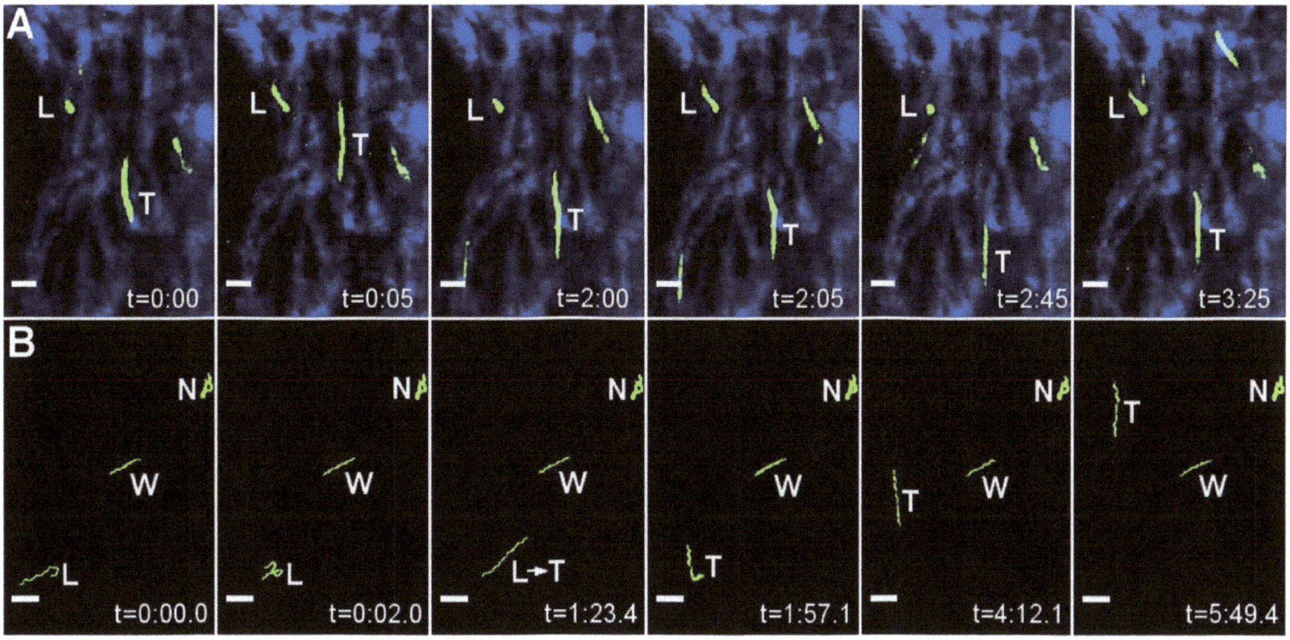

Figure 12. A time course of GFP-expressing spirochete motility patterns in mouse dermis and gelatin. A) Tick-inoculated mouse dermis. B) 3% gelatin. Lunging (L), translocating (T), wriggling (W) and non-motile (N) spirochetes. The blue signal is from second harmonics of skin collagen. The scale bar is 10μm. This figure is reprinted from (Harman et al., 2012).

Two photon IVM microscopy also has been used to visualize *Ixodes scapularis* feeding on mice in real time and to study the response of nearby *B. burgdorferi* to the feeding ticks (Bockenstedt et al., 2014). Either *rag-/-* mice, lacking an acquired immune response, or *MyD88-/-* mice were infected either by needle inoculation or by infestation with infected nymphs. For IVM, unfed nymphs were placed on the ears of mice infected for at least two weeks with GFP-expressing *B. burgdorferi*. The characteristic motility behavior of the observed spirochetes was the same in wild-type, *rag-/-* and *MyD88-/-* mice. Restructuring of collagen fibers proximal to the bite site was observed and, although some spirochetes seemed unaffected by the presence of a feeding tick, others were observed to migrate in the direction of the hypostome, suggesting a chemotactic response. This work also supports the currently held belief that spirochetes are acquired from the skin rather than from the blood during tick feeding.

IVM has also been used to study the role of the chemotaxis regulator CheY3 (Novak et al., 2016). *B. burgdorferi* (wild-type or spirochetes with a disrupted *cheY3* gene) were injected intradermally into the dorsal surface of the mouse ear. In comparison to wild-type *B. burgdorferi, ΔcheY3* spirochetes displayed little translational motility in the mouse skin at six hours post-infection, with a progressive loss in motility over 48 hours. The spirochetes that were motile at early times displayed translational motility velocities similar to wild-type spirochetes but were unable to reverse direction and continued to run in the same direction (Figure 13). The inability of spirochetes tethered to the ECM to reverse direction and free themselves may be responsible for their loss of motility. Δ*cheY3* spirochetes apparently were trapped near the inoculation site and, by 96 hours post-infection, all mutant spirochetes were cleared, while wild-type *B. burgdorferi* increased in numbers. The *cheY* spirochetes did not disseminate to other tissues. The IVM results clearly demonstrated an essential role for CheY3 in motility, viability and dissemination in the mouse.

Vascular transmigration by B. burgdorferi

Little is currently known about the vascular transmigration step of *B. burgdorferi* dissemination. Data suggesting a paracellular path whereby spirochetes cross the endothelium at endothelial cell junctions have been reported (Comstock and Thomas, 1989, 1991). In contrast, data suggesting that transmigration occurs via a transcellular pathway

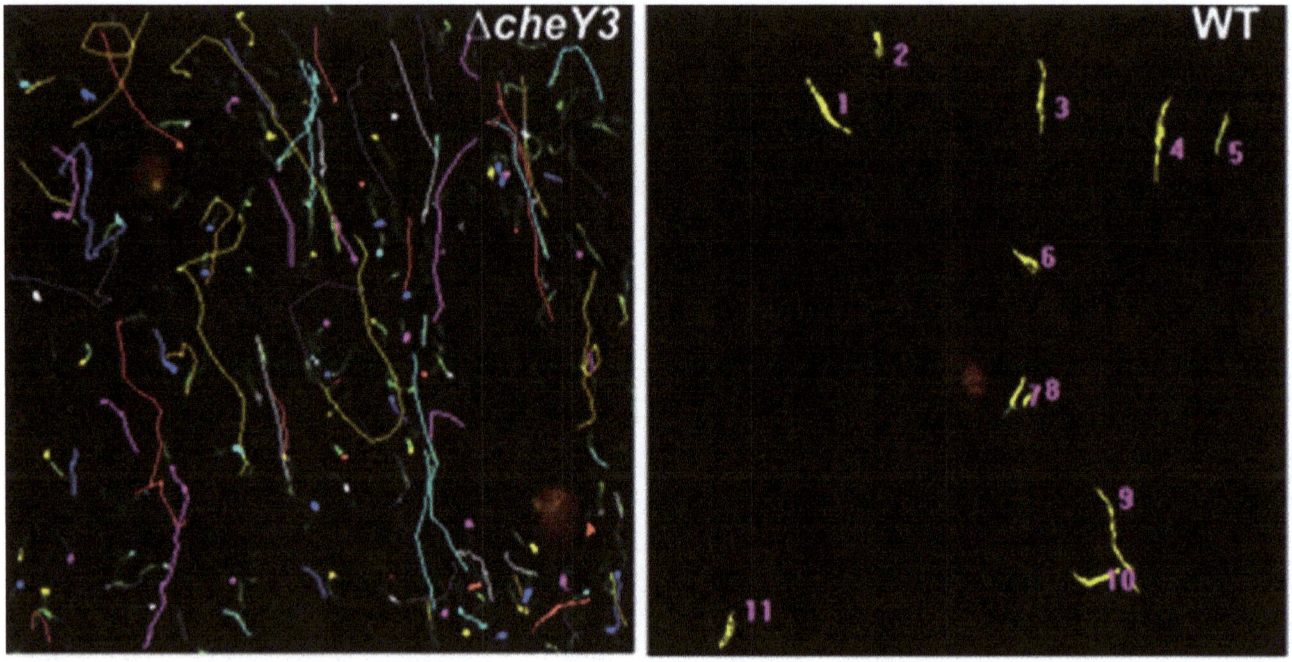

Figure 13. Contrasting motility behavior of *dcheY3 and* WT *B. burgdorferi* in mouse ear. Paths were tracked at six hours post-injection. Colored lines (except green) show the migration paths of the spirochete. Images from (Novak et al., 2016), with permission from John Wiley and Sons, Inc.

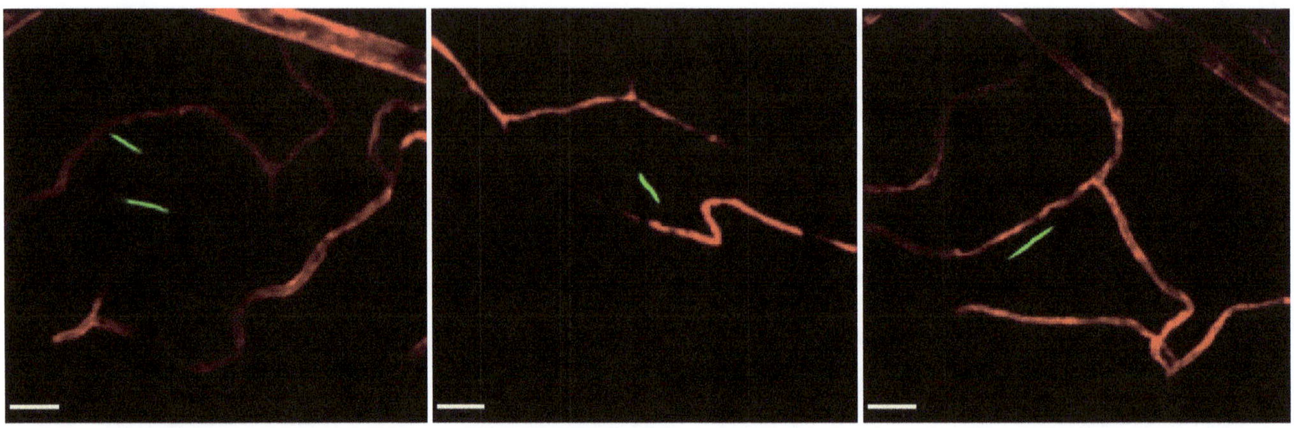

Figure 14. Intravital microscopy to study *B. burgdorferi* vascular transmigration. Spinning disk confocal microscopy of GFP-expressing, infectious *B. burgdorferi* that escaped from the vasculature in the knee joint-proximal tissue of a living mouse. The vasculature was visualized with an AlexaFluor555-conjugated antibody to CD31 (PECAM-1). Image from (Kumar et al., 2015).

have also been reported (Szczepanski et al., 1990). Moreover, little is known regarding the bacterial and host proteins and processes involved. The transmigration process was difficult to study by IVM using using the approach described above for vascular adhesion because looking for transmigrating spirochetes (0.1% of observed spirochetes) is akin to looking for a needle in a haystack. Moreover, transmigrated spirochetes rapidly leave the site of vascular escape in the skin, leaving no record. In addition, host defense systems can interfere with transmigration at several levels and *B. burgdorferi* genes involved in survival can be mistaken for genes playing a direct role in transmigration.

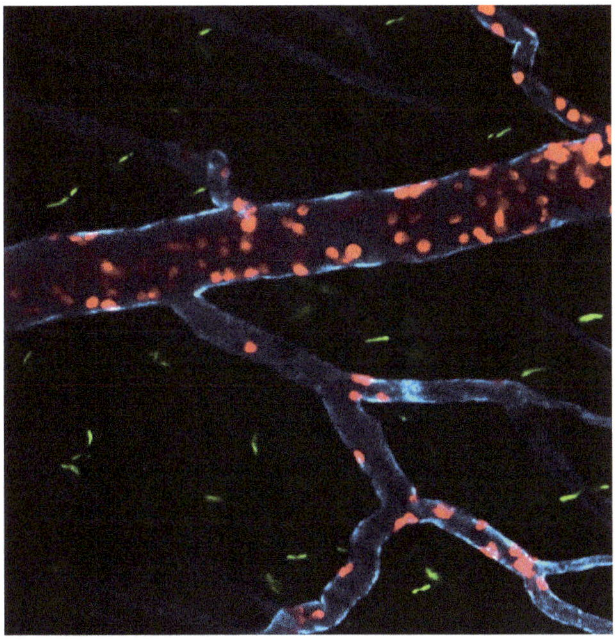

Figure 15. Spinning disk laser confocal micrograph showing endothelial activation in the murine cremaster microvasculature 24 hours after iv injection of GFP-expressing *B. burgdorferi.* Adherent and crawling neutrophils, physically distanced from spirochetes, were stained were with phycoerythin-conjugated anti-Ly6G (red) and venules were stained with Alexa Fluor 647-conjugated anti-PECAM1 (blue). See (Tan et al., 2020). Image from Björn Petri, Xi Tan and George Chaconas.

To circumvent these problems, an IVM transmigration assay was developed using information discovered in the study of the host innate immune response to *B. burgdorferi* described above (Lee et al., 2010; Lee et al., 2014). The assay uses the peripheral knee vasculature as the site of observation in *Cd1d-/-* mice, which lack iNKT cells. Use of *Cd1d-/-* mice is a crucial feature in this assay system as iNKT cells are the first line of defense against *B. burgdorferi* in mouse knee joint tissue (Lee et al., 2010; Lee et al., 2014). In wild-type mice, spirochetes are rapidly eliminated and unavailable for observation and enumeration. Moreover, the knee joint provides a stable environment where spirochetes remain in the local milieu and can be accurately counted after transmigration. Using this system, spirochetes were injected in the tail vein and visualized in knee joint tissue at 24 hours post-infection (Kumar et al., 2015); very little transmigration occurs before that time. Transmigrated spirochetes are clearly discernible from the vasculature (Figure 14) and display a reciprocating, translational motion outside the vessels. This IVM system was coupled with mutations in the genes encoding the Fn and GAG binding protein BBK32 and the integrin adhesin and porin, P66. Wild-type levels of transmigration were observed when *bbk32* was disrupted; however, disruption of *p66* resulted in a loss of transmigration. In particular, mutation of residues involved in integrin binding but not affecting the porin function abrogated transmigration, indicating a role for P66 and integrin binding in vascular transmigration of *B. burgdorferi.* In contrast, disruption of P66 did not influence vascular adhesion, suggesting that initial establishment of a borrelial-endothelial interaction is dependent upon BBK32 and perhaps other *B. burgdorferi* adhesins.

An important component of these IVM experiments was the monitoring of spirochete clearance from the vasculature. The removal of surface adhesins can affect the rate of spirochete clearance. This can result in an apparent decrease in transmigration that stems from a decrease in the number of spirochetes in the vasculature rather than an actual transmigration defect (Kumar et al., 2015).

More recently (Lin et al., 2020), another *Borrelia* adhesin has been shown to be required for vascular transmigration using IVM. The extracellular matrix binding activity of OspC is specific for fibronectin and dermatan sulphate, and ECM binding by OspC is required for extravasation. OspC does not appear to confer strong vascular adhesion, as BBK32 does, but likely interacts with the vascular endothelium, as P66 does, to promote transmigration subsequent to early tethering and dragging interactions that deaccelerate the spirochetes under the shear stress of bloodflow. The mechanism(s) by which P66 and OspC promote vascular transmigration is unknown at this time, however, it is tempting to speculate that they may induce endothelial signaling to promote cellular changes required for spirochetal transmigration.

Finally, a very recent IVM study (Tan et al., 2020) has revealed that the 24 hour lag observed for vascular transmigration of *B. burgdorferi* results from a requirement for endothelial activation to facilitate vascular escape. Extravasation is stimulated by a factor of three by host neutrophils, which do not recognize or clear the infecting spirochetes. The role of neutrophils appears to be the production of TNF-α, MCP-1 (CCL2) and IL-10, all of which can activate the endothelium for transmigration of *B. burgdorferi.* Figure 15 shows adherent and crawling neutrophils that do not recognize *B. burgdorferi* in the murine cremaster microvasculature.

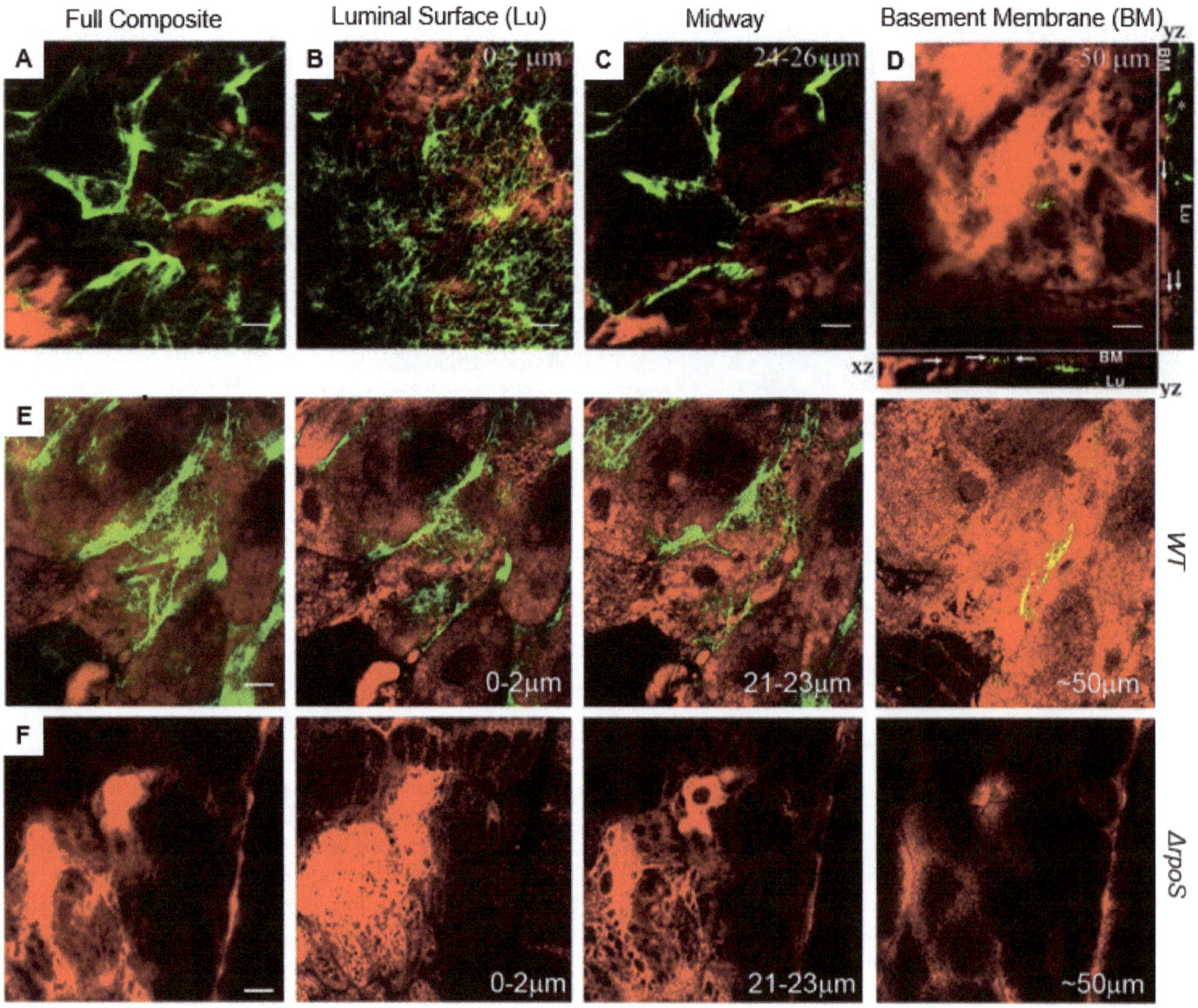

Figure 16. Live *ex vivo* 3D confocal microscopy z-series of dense, non-motile aggregates of GFP-expressing *B. burgdorferi* (green) in a hypertrophied region of the midgut epithelium of an infected feeding nymphal tick stained with FM4-64 membrane dye (red). Panel A depicts the composite (projection) image for all optical sections. Panels B-D show single slices obtained at different depths through the thickness of the midgut epithelial lining. xz and yz views of the composite stack are shown at far right/bottom. Panels E and F show similar regions from ticks infected with *wild-type* (*WT*) and *ΔrpoS* GFP-expressing *B.* burgdorferi. Images in A-D from (Dunham-Ems et al., 2009), reprinted with permission from the American Society for Clinical Investigation. Images in E-F from (Dunham-Ems et al., 2012).

Live Cell Imaging *Methodologies*

Ex vivo live cell imaging of *B. burgdorferi*-tick interactions

Direct observation of *B. burgdorferi* behaviour and biology in ticks is challenging because their cuticle is strongly autofluorescent (Hammer et al., 2001). Multiple approaches have been used to overcome this problem over the past 50 years, including whole organ dissection and cultivation (Grabowski et al., 2018). Current *ex vivo* imaging methods for studying *B. burgdorferi*-tick interactions visualize GFP-expressing *B. burgdorferi* in fluorescent counter-stained dissected whole mount tick midguts and salivary glands at different stages of the tick blood meal and transmission, using epifluorescence or confocal time lapse microscopy, and confocal optical sectioning for three-dimensional z-series imaging (Figure 16) (Dunham-Ems et al., 2009; Dunham-Ems et al., 2012; Caimano et al., 2015).

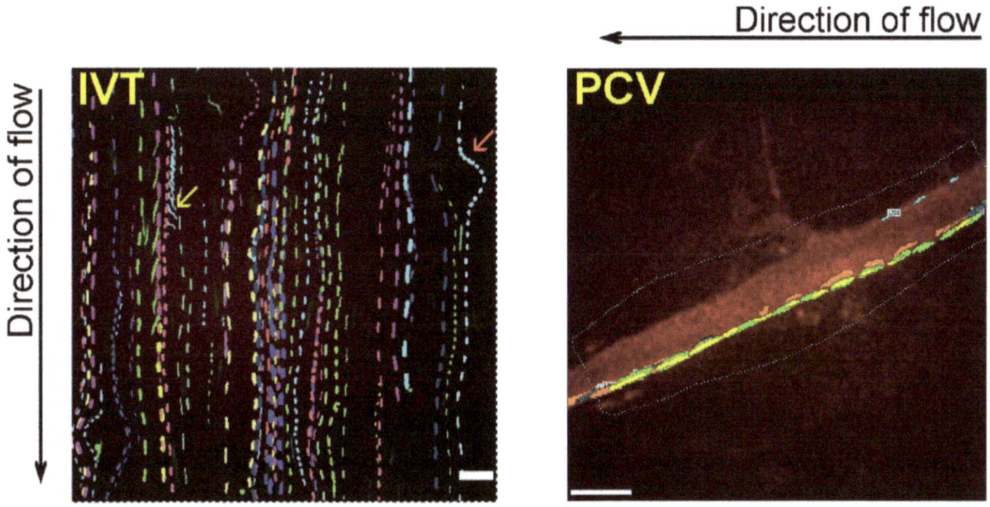

Figure 17. Trajectories of individual *B. burgdorferi* (different colours for each track) interacting with primary human endothelial cells in a flow chamber under postcapillary venule (PCV) shear stress conditions (IVT), and with a dermal PCV in a live mouse (PCV: IVM). Scale bars: IVT, 13 µm; PCV, 32 µm. Endothelial counterstain (red): plasma membrane dye (IVT), AlexaFluor 555-PECAM-1 antibody (PCV). Yellow and red arrows: *B. burgdorferi* interacting by one end only (yellow) and moving along edge of endothelial nucleus projecting above imaging plane (red). Image from (Ebady et al., 2016).

In vitro live cell imaging of *B. burgdorferi*-mammalian host interactions

IVM visualization of *B. burgdorferi* in murine skin and the dermal microvasculature in turn enabled the development and validation of *in vitro* fluorescent live cell imaging models of host-pathogen interactions in the skin dermis (fluorescent bacteria moving in gelatin matrices) (Figure 12) and dermal microvasculature (fluorescent bacteria interacting with fluorescent counterstained primary human endothelial cells in blood-vessel mimicking flow chambers) (Figure 17) (Harman et al., 2012; Ebady et al., 2016; Salo et al., 2016; Kao et al., 2017; Niddam et al., 2017). These methods typically rely on GFP-expressing *B. burgdorferi*, although bacteria labelled with a non-toxic live cell imaging membrane dye can also be used (Salo et al., 2016).

One of the biggest challenges for *in vitro* live imaging is keeping cells healthy and fully functional for the duration of experiments (Ettinger and Wittmann, 2014). Most live cell imaging experiments are performed in a medium that is hospitable to both bacteria and host cells (not always an easy task to find), and in a controlled physical environment where temperature, pH and sometimes humidity are stably maintained using specialized heaters, CO_2 and humidity controllers. Flow chamber live cell imaging also requires syringe or peristaltic pumps capable of precise output at low flow rates. All equipment must be regularly calibrated since small linear changes in variables such as shear stress can have log-scale effects on host-pathogen interactions in settings such as flow chambers.

Real-time time lapse imaging can be performed using epifluorescence, spinning disk confocal or resonant scanner confocal microscopes equipped with electron multiplying charge-coupled device (EMCCD) cameras or high sensitivity photomultiplier tube (PMT) sensors. As for all live cell microscopy, including *ex vivo* imaging and IVM, phototoxicity and photobleaching can pose significant challenges for acquiring repeated high frame rate time courses at the same position, particularly for confocal laser light sources (Ettinger and Wittmann, 2014). Visualizing thin bacteria, such as *B. burgdorferi*, in fast-moving settings such as blood vessels and flow chambers requires prioritization of image acquisition speed over image brightness and "beauty", especially for confocal microscopy, which excludes out of focus light. Prioritization of speed over beauty and brightness is especially important for measuring properties such as length and diameter of moving cells, which can appear larger when captured using EMCCD cameras and longer exposure times that produce more visually appealing, bright images (Jonkman et al., 2020). For this reason, accurate quantitative imaging for automated particle tracking and large-scale data analysis performed for fast-moving bacteria such as *B. burgdorferi* depends on non-saturated imaging conditions. For quantitative imaging, universal image acquisition conditions must also be pre-established and uniformly applied for all experimental conditions, and recalibrated over time as lasers age and become less powerful, and each time a new batch of fluorescent counterstain reagents is prepared or purchased (Jost and Waters, 2019).

Advantages

Ex vivo imaging in tick midguts and salivary glands after transmission to ticks or during transmission to mammals offers the only currently available method to directly observe *B. burgdorferi*-tick interactions at the single cell level. It has provided dynamic, localized insight into mechanisms of bacterial migration during tick transmission that would not be possible without direct observation at single cell resolution (Dunham-Ems et al., 2009; Dunham-Ems et al., 2012; Caimano et al., 2015).

Host-mimicking live cell imaging-based model systems support detailed, single cell resolution kinetic, biophysical, biochemical and genetic analysis of host-pathogen interaction mechanisms that are faster, less costly and more easily manipulated than IVM in animals (Harman et al., 2012; Ebady et al., 2016; Salo et al., 2016; Kao et al., 2017; Niddam et al., 2017). They facilitate interpretation of host-pathogen interaction data because immune clearance, which confounds IVM studies in blood vessels (Moriarty et al., 2012; Kumar et al., 2015), is largely absent. *In vitro* live cell imaging systems also support investigation of *B. burgdorferi* interactions with human cells, and rapid capture of large datasets (>10,000 datapoints) that can be analyzed by powerful segmentation and bioinformatics approaches.

Disadvantages

Ex vivo imaging in tick midguts and salivary glands requires delicate dissection, particularly if nymphal ticks are used, and separates these organs from the rest of the tick body, possibly disrupting some aspects of their physiology and function. Capturing optical sections through the full thickness of some tissues (e.g. late stage post-blood meal midguts) depends on physical sectioning or long working range objectives. Time-lapse imaging of live *B. burgdorferi* in tick midguts and salivary glands is performed for a maximum of 1.5 hours after dissection and fluorescent labeling of tick cells, necessitating the use of different ticks for each time point during the ~48-72 hour feeding period (Dunham-Ems et al., 2009).

In vitro live cell imaging model systems must be validated using *in vivo* methods such as whole body and IVM imaging, and/or qPCR and cultivation-based measurement of bacterial burden. *In vitro* live cell imaging depends on a stable, well-controlled physical environment. Quantitative live cell imaging also requires carefully controlled and calibrated imaging conditions, and substantial training and quality control monitoring. Efficient extraction of maximal amounts of information from high frame rate live cell imaging experiments can necessitate development and validation of automated methods for particle tracking and analyzing large volumes of data, and relies heavily on coding and bioinformatics skills that can be challenging for more conventionally trained biologists and microbiologists to quickly master if they are not working in collaborative teams.

Tick-B. burgdorferi interactions

Live *ex vivo* imaging methods developed for dissected tick guts and salivary glands (Dunham-Ems et al., 2009) have provided crucial insight into the processes of initial *B. burgdorferi* colonization of the tick midgut during mammalian host to tick transmission, and subsequent dissemination from the tick midgut to salivary glands during tick to mammalian host transmission. Live *ex vivo* imaging shows that these processes are controlled by the RpoS regulon (Dunham-Ems et al., 2012; Caimano et al., 2015). In *B. burgdorferi*-colonized unfed ticks, a mesh of non-motile spirochetes covers the epithelial lining of the tick midgut (Dunham-Ems et al., 2009). As colonized ticks feed, this mesh advances basally as the epithelium proliferates and differentiates in response to the blood meal. Basally positioned spirochetes then become motile and enter the hemocoel to migrate to and enter the salivary glands, from which they are transmitted to mammalian hosts (Dunham-Ems et al., 2009) (Figure 16). Spirochetes deficient in the RpoS regulon transform into a reversible round body morphotype as the blood meal progresses, and cannot migrate to salivary glands (Dunham-Ems et al., 2012). After naïve ticks feed on infected mammalian hosts, gene regulation switches to an RpoS-off state that promotes formation of the non-motile mesh of bacteria on epithelial surfaces that is typical of unfed, midgut colonization (Caimano et al., 2015).

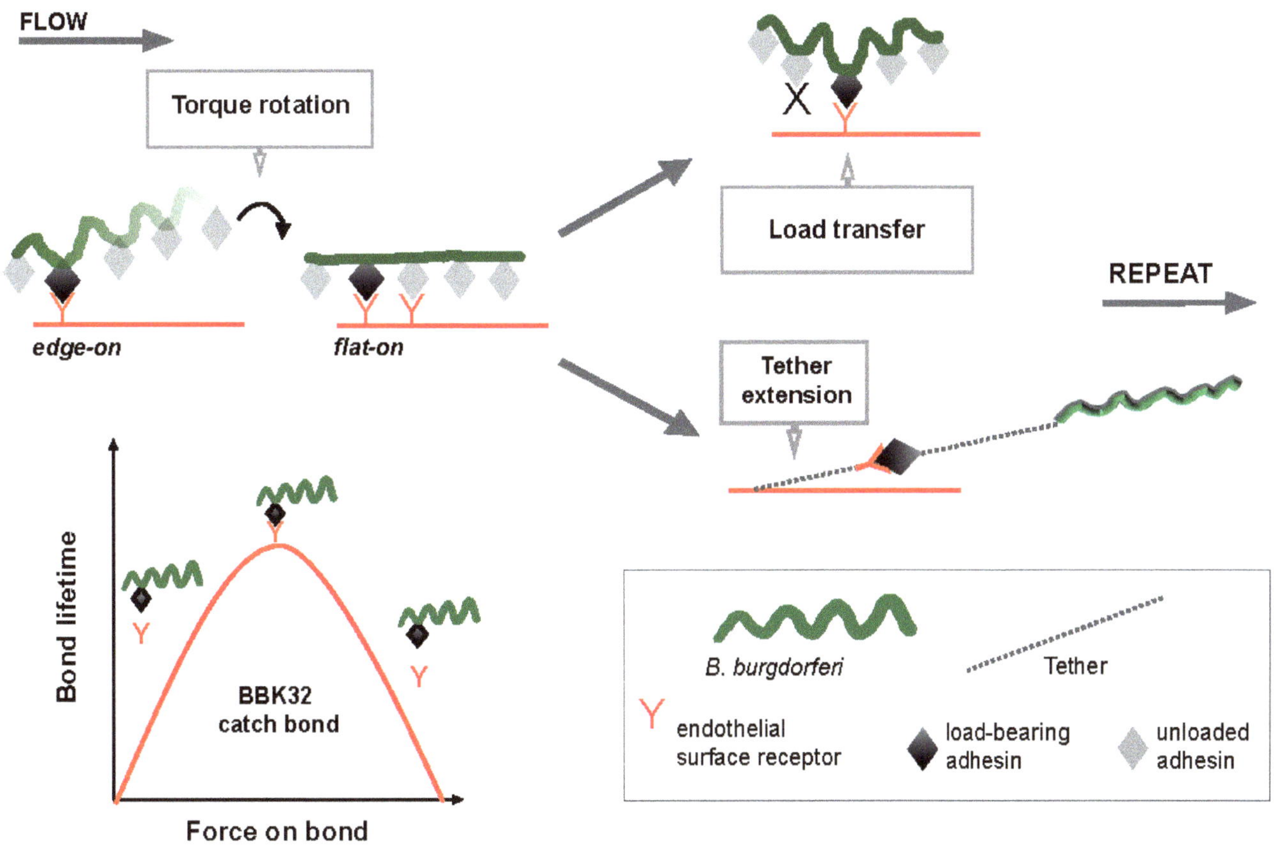

Figure 18. Under PCV shear stress conditions, slower moving "dragging" (untethered) *B. burgdorferi* is oriented edge-on to the endothelial surface and moves in a stepwise fashion, by transferring mechanical load from a single or coordinated, tightly clustered adhesion complex at the peak of the bacterial sine wave to another adhesion complex on the next sine wave peak. This process is facilitated by rotation of bacteria driven by flow-induced torque or bacterial motility. Faster-moving interactions are stabilized by tethers of unknown composition anchoring adhesion complexes and *B. burgdorferi* to the endothelial surface. Interactions are also stabilized by BBK32-dependent catch bonds, which are bonds that become longer lived as pulling force on the bond increases. Image from (Ebady et al., 2016).

Shared human and mouse B. burgdorferi-endothelial interaction mechanisms

Live cell imaging of *B. burgdorferi* interactions with primary human umbilical cord vein endothelial cells in flow chambers under postcapillary venule shear stress conditions reveals that basic interaction features and properties observed in dermal postcapillary venules (PCVs) of mice by IVM (Moriarty et al., 2008; Norman et al., 2008) are recapitulated with human endothelial cells *in vitro* (Ebady et al., 2016; Kao et al., 2017; Niddam et al., 2017). In both settings, *B. burgdorferi* exhibits tethering, dragging and stationary adhesion, tethering and dragging depend on Fn, GAGs, the vascular adhesin BBK32 and at least one additional unidentified vascular adhesin, and bacterial strain-specific differences in interaction properties are conserved. Conversely, *B. burgdorferi*-endothelial interaction mechanisms first identified in flow chambers, such as transfer of mechanical load along a series of single adhesion complexes, were subsequently confirmed *in vivo* (Ebady et al., 2016). Identification of a *T. pallidum* adhesin supporting human endothelial interactions in flow chambers (pallilysin) led to confirmation of a role for this adhesin in dermal PCV interactions in mice (Kao et al., 2017). Thus, data derived from flow chamber model systems and human cells are useful for dissecting and predicting mouse vascular interaction mechanisms *in vivo*.

Biomechanics of B. burgdorferi-endothelial interactions

High speed particle tracking performed using large flow chamber and IVM datasets reveals that *B. burgdorferi* interactions with murine skin

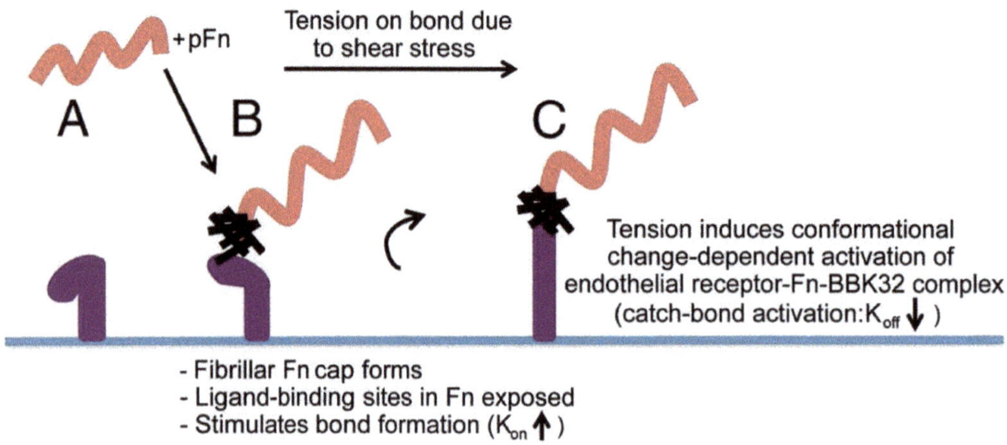

Figure 19. Proposed mechanism of plasma fibronectin (pFn)-dependent *B. burgdorferi*-vascular interactions under physiological shear stress. Conformational changes and polymerization of pFn exposed to a BBK32-expressing *bacterium* (*A*) exposes pFn binding sites for endothelial receptors and increases bond formation rate (K_{on}) (*B*). As bacteria are pushed/pulled by shear stress, stretching of the load-bearing adhesion complex induces additional force-induced conformational changes in adhesion complex components (BBK32, pFn, and endothelial receptors) to promote formation of longer-lived catch bonds and slower adhesion complex dissociation. Image from (Niddam et al., 2017).

microvessels and human endothelial cells in flow chambers occur by transfer of mechanical load along a series of single or tightly clustered adhesion complexes (Ebady et al., 2016), similar to the interaction mechanism used by leukocytes and platelets rolling along endothelial surfaces in PCVs (Begandt et al., 2017; Coenen et al., 2017) (Figure 17).

Leukocyte and platelet rolling interactions are stabilized by tethers, which are extendible membranous or filamentous structures that distribute (reduce) force imposed on load-bearing adhesion complexes (Begandt et al., 2017; Coenen et al., 2017). Particle tracking of *B. burgdorferi*-endothelial interactions in flow chambers and PCVs shows that these interactions are also stabilized by tethers, although the composition of these tethers remains unknown. These tethered interactions correspond to the faster-moving transient interactions referred to as "tethering" in prior observational IVM studies (Figure 18) (Ebady et al., 2016).

The slow interaction type identified as "dragging" in early IVM studies in fact corresponds to untethered bacteria that are oriented with their sine wave edge-on to endothelial surfaces, projecting up into flow (Figure 18) (Ebady et al., 2016). In untethered interactions, mechanical load is transferred from peak to peak of the sine wave, resulting in slow-moving step-wise incremental displacements equal to the distance between sine wave peaks (~2.83 μm). Load-bearing bonds supporting untethered interactions are subjected to considerably more pulling force, due to the projection of bacteria up into flow. For human umbilical vein endothelial cells and murine dermal PCVs, the adhesin BBK32 stabilizes higher force interactions by a catch bond mechanism. Catch bonds are adhesive bonds that become longer-lived (more stable) under increasing pulling force, and are important for stabilization of other interaction types in the vasculature, including leukocyte and platelet rolling (Begandt et al., 2017; Coenen et al., 2017).

Flow chamber and particle-tracking approaches also show that the *B. burgdorferi*-endothelial catch bond interaction mechanism depends on soluble Fn recruited from blood plasma (plasma Fn, pFn), and not on insoluble Fn found on endothelial surfaces (Figure 19) (Niddam et al., 2017). BBK32-expressing bacteria induce pFn polymerization to form pFn caps positioned at the mechanical load-bearing ends of *B. burgdorferi*, stimulating formation of adhesion complexes with receptors on endothelial surfaces, and slowing dissociation of these complexes by a catch bond mechanism.

Interestingly, multiple flow chamber and IVM studies have found that adhesive bond forces supporting bacterial-endothelial interactions under PCV shear stress are sufficiently small that propulsive forces generated by *B. burgdorferi* motility should be large enough to modulate the forces imposed on these bonds, or even to break them (Ebady et al., 2016; Kao et al., 2017; Niddam et al., 2017). It is also possible that *B. burgdorferi* motility contributes to maintaining untethered interactions in an edge-on position relative to endothelial surfaces or bacterial rotation during interactions (Figure 18), or could contribute to the pulling forces driving pFn polymerization to the load-bearing ends of cells (Figure 19).

Acknowledgements

The authors would like to thank Alexia Belperron and Linda Bockenstedt for providing the image for Figure 10 and Björn Petri and Xi Tan for the image for Figure 15. Work in the Chaconas aboratory is supported by the Canadian Institutes of Health Research (PJT-153336) and the National Institute of Health (R01AI18799) and work in the Moriarty laboratory is supported by the Canadian Institutes of Health Research (PJT-159466) and the Natural Sciences and Engineering Research Council of Canada (RGPIN-401938-17). Research in the Skare laboratory is currently funded by NIAID Public Health Service grants AI131656 and AI146930 and research in the Hyde laboratory is currently funded by NIAID Public Health Service grants AI101740 and AI131656.

References

Adams, P.P., Flores Avile, C., and Jewett, M.W. (2017). A Dual Luciferase Reporter System for *B. burgdorferi* Measures Transcriptional Activity during Tick-Pathogen Interactions. Frontiers in cellular and infection microbiology *7*, 225. https://doi.org/10.3389/fcimb.2017.00225

Adams, P.P., Flores Avile, C., Popitsch, N., Bilusic, I., Schroeder, R., Lybecker, M., and Jewett, M.W. (2016). In vivo expression technology and 5' end mapping of the *Borrelia burgdorferi* transcriptome identify novel RNAs expressed during mammalian infection. Nucleic Acids Res. https://doi.org/10.1093/nar/gkw1180

Andreu, N., Zelmer, A., and Wiles, S. (2011). Noninvasive biophotonic imaging for studies of infectious disease. FEMS Microbiol Rev *35*, 360-394. https://doi.org/10.1111/j.1574-6976.2010.00252.x

Barthold, S.W., Sidman, C.L., and Smith, A.L. (1992). Lyme borreliosis in genetically resistant and susceptible mice with severe combined immunodeficiency. The American journal of tropical medicine and hygiene *47*, 605-613. https://doi.org/https://doi.org/10.4269/ajtmh.1992.47.605

Begandt, D., Thome, S., Sperandio, M., and Walzog, B. (2017). How neutrophils resist shear stress at blood vessel walls: molecular mechanisms, subcellular structures, and cell-cell interactions. J Leukoc Biol *102*, 699-709. https://doi.org/10.1189/jlb.3MR0117-026RR

Belperron, A.A., Mao, J., and Bockenstedt, L.K. (2018). Two Photon Intravital Microscopy of Lyme *Borrelia* in Mice. Methods Mol Biol *1690*, 279-290. https://doi.org/10.1007/978-1-4939-7383-5_20

Blevins, J.S., Hagman, K.E., and Norgard, M.V. (2008). Assessment of decorin-binding protein A to the infectivity of *Borrelia burgdorferi* in the murine models of needle and tick infection. BMC microbiology *8*, 82. https://doi.org/10.1186/1471-2180-8-82

Blevins, J.S., Revel, A.T., Smith, A.H., Bachlani, G.N., and Norgard, M.V. (2007). Adaptation of a luciferase gene reporter and lac expression system to *Borrelia burgdorferi*. Appl Environ Microbiol *73*, 1501-1513. https://doi.org/https://doi.org/10.1128/AEM.02454-06

Bockenstedt, L.K., Gonzalez, D., Mao, J., Li, M., Belperron, A.A., and Haberman, A. (2014). What ticks do under your skin: two-photon intravital imaging of *Ixodes scapularis* feeding in the presence of the Lyme disease spirochete. Yale J Biol Med *87*, 3-13.

Bockenstedt, L.K., Gonzalez, D.G., Haberman, A.M., and Belperron, A.A. (2012). Spirochete antigens persist near cartilage after murine Lyme borreliosis therapy. J Clin Invest *122*, 2652-2660. https://doi.org/10.1172/JCI58813

Bolz, D.D., Sundsbak, R.S., Ma, Y., Akira, S., Kirschning, C.J., Zachary, J.F., Weis, J.H., and Weis, J.J. (2004). MyD88 plays a unique role in host defense but not arthritis development in Lyme disease. J Immunol *173*, 2003-2010. https://doi.org/https://doi.org/10.4049/jimmunol.173.3.2003

Caimano, M.J., Dunham-Ems, S., Allard, A.M., Cassera, M.B., Kenedy, M., and Radolf, J.D. (2015). Cyclic di-GMP modulates gene expression in Lyme disease spirochetes at the tick-mammal interface to promote spirochete survival during the blood meal and tick-to-mammal transmission. Infect Immun *83*, 3043-3060. https://doi.org/10.1128/IAI.00315-15

Caimano, M.J., Eggers, C.H., Hazlett, K.R., and Radolf, J.D. (2004). RpoS is not central to the general stress response in *Borrelia burgdorferi* but does control expression of one or more essential virulence determinants. Infect Immun *72*, 6433-6445. https://doi.org/10.1128/IAI.72.11.6433-6445.2004

Chan, K., Alter, L., Barthold, S.W., and Parveen, N. (2015). Disruption of *bbe02* by Insertion of a Luciferase Gene Increases Transformation Efficiency of *Borrelia burgdorferi* and Allows Live Imaging in Lyme Disease Susceptible C3H Mice. PLoS One *10*, e0129532. https://doi.org/10.1371/journal.pone.0129532

Coenen, D.M., Mastenbroek, T.G., and Cosemans, J. (2017). Platelet interaction with activated endothelium: mechanistic insights from microfluidics. Blood *130*, 2819-2828. https://doi.org/10.1182/blood-2017-04-780825

Comstock, L.E., and Thomas, D.D. (1989). Penetration of endothelial cell monolayers by *Borrelia burgdorferi*. Infect Immun *57*, 1626-1628. https://doi.org/https://doi.org/10.1128/IAI.57.5.1626-1628.1989

Comstock, L.E., and Thomas, D.D. (1991). Characterization of *Borrelia burgdorferi* invasion of cultured endothelial cells. Microb Pathog *10*, 137-148. https://doi.org/https://doi.org/10.1016/0882-4010(91)90074-k

Contag, C.H., Contag, P.R., Mullins, J.I., Spilman, S.D., Stevenson, D.K., and Benaron, D.A. (1995). Photonic detection of bacterial pathogens in living hosts. Mol Microbiol *18*, 593-603. https://doi.org/10.1111/j.1365-2958.1995.mmi_18040593.x

Dunham-Ems, S.M., Caimano, M.J., Eggers, C.H., and Radolf, J.D. (2012). *Borrelia burgdorferi* requires the alternative sigma factor RpoS for dissemination within the vector during tick-to-mammal transmission. PLoS Pathog *8*, e1002532. https://doi.org/https://doi.org/10.1371/journal.ppat.1002532

Dunham-Ems, S.M., Caimano, M.J., Pal, U., Wolgemuth, C.W., Eggers, C.H., Balic, A., and Radolf, J.D. (2009). Live imaging reveals a biphasic mode of dissemination of *Borrelia burgdorferi* within ticks. J Clin Invest *119*, 3652-3665. https://doi.org/https://doi.org/10.1172/JCI39401

Ebady, R., Niddam, A.F., Boczula, A.E., Kim, Y.R., Gupta, N., Tang, T.T., Odisho, T., Zhi, H., Simmons, C.A., Skare, J.T., *et al.* (2016). Biomechanics of *Borrelia burgdorferi* Vascular Interactions. Cell Rep. https://doi.org/10.1016/j.celrep.2016.08.013

Ellis, T.C., Jain, S., Linowski, A.K., Rike, K., Bestor, A., Rosa, P.A., Halpern, M., Kurhanewicz, S., and Jewett, M.W. (2013). In vivo expression technology identifies a novel virulence factor critical for

Borrelia burgdorferi persistence in mice. PLoS Pathog *9*, e1003567. https://doi.org/https://doi.org/10.1371/journal.ppat.1003567

Filippi, M.D. (2019). Neutrophil transendothelial migration: updates and new perspectives. Blood. https://doi.org/10.1182/blood-2018-12-844605

Grabowski, J.M., Offerdahl, D.K., and Bloom, M.E. (2018). The Use of Ex Vivo Organ Cultures in Tick-Borne Virus Research. ACS Infect Dis *4*, 247-256. https://doi.org/10.1021/acsinfecdis.7b00274

Grimm, D., Tilly, K., Byram, R., Stewart, P.E., Krum, J.G., Bueschel, D.M., Schwan, T.G., Policastro, P.F., Elias, A.F., and Rosa, P.A. (2004). Outer-surface protein C of the Lyme disease spirochete: a protein induced in ticks for infection of mammals. Proc Natl Acad Sci U S A *101*, 3142-3147. https://doi.org/https://doi.org/10.1073/pnas.0306845101

Hammer, B., Moter, A., Kahl, O., Alberti, G., and Gobel, U.B. (2001). Visualization of *Borrelia burgdorferi* sensu lato by fluorescence in situ hybridization (FISH) on whole-body sections of Ixodes ricinus ticks and gerbil skin biopsies. Microbiology *147*, 1425-1436. https://doi.org/10.1099/00221287-147-6-1425

Harman, M.W., Dunham-Ems, S.M., Caimano, M.J., Belperron, A.A., Bockenstedt, L.K., Fu, H.C., Radolf, J.D., and Wolgemuth, C.W. (2012). The heterogeneous motility of the Lyme disease spirochete in gelatin mimics dissemination through tissue. Proc Natl Acad Sci U S A *109*, 3059-3064. https://doi.org/https://doi.org10.1073/pnas.1114362109

Hubner, A., Yang, X., Nolen, D.M., Popova, T.G., Cabello, F.C., and Norgard, M.V. (2001). Expression of *Borrelia burgdorferi* OspC and DbpA is controlled by a RpoN-RpoS regulatory pathway. Proc Natl Acad Sci U S A *98*, 12724-12729. https://doi.org/https://doi.org10.1073/pnas.231442498

Hutchens, M., and Luker, G.D. (2007). Applications of bioluminescence imaging to the study of infectious diseases. Cell Microbiol *9*, 2315-2322. https://doi.org/10.1111/j.1462-5822.2007.00995.x

Hyde, J.A., and Skare, J.T. (2018). Detection of Bioluminescent *Borrelia burgdorferi* from In Vitro Cultivation and During Murine Infection. Methods Mol Biol *1690*, 241-257. https://doi.org/10.1007/978-1-4939-7383-5_18

Hyde, J.A., Weening, E.H., Chang, M., Trzeciakowski, J.P., Hook, M., Cirillo, J.D., and Skare, J.T. (2011). Bioluminescent imaging of *Borrelia burgdorferi* in vivo demonstrates that the fibronectin-binding protein BBK32 is required for optimal infectivity. Mol Microbiol *82*, 99-113. https://doi.org/https://doi.org/10.1111/j.1365-2958.2011.07801.x

Jenkins, D.E., Oei, Y., Hornig, Y.S., Yu, S.F., Dusich, J., Purchio, T., and Contag, P.R. (2003). Bioluminescent imaging (BLI) to improve and refine traditional murine models of tumor growth and metastasis. Clin Exp Metastasis *20*, 733-744. https://doi.org/10.1023/b:clin.0000006815.49932.98

Jost, A.P., and Waters, J.C. (2019). Designing a rigorous microscopy experiment: Validating methods and avoiding bias. J Cell Biol *218*, 1452-1466. https://doi.org/10.1083/jcb.201812109

Kao, W.A., Petrosova, H., Ebady, R., Lithgow, K.V., Rojas, P., Zhang, Y., Kim, Y.E., Kim, Y.R., Odisho, T., Gupta, N., *et al.* (2017). Identification of Tp0751 (Pallilysin) as a *Treponema pallidum* Vascular Adhesin by Heterologous Expression in the Lyme disease Spirochete. Sci Rep *7*, 1538. https://doi.org/10.1038/s41598-017-01589-4

Kumar, D., Ristow, L.C., Shi, M., Mukherjee, P., Caine, J.A., Lee, W.Y., Kubes, P., Coburn, J., and Chaconas, G. (2015). Intravital Imaging of Vascular Transmigration by the Lyme Spirochete: Requirement for the Integrin Binding Residues of the *B. burgdorferi* P66 Protein. PLoS Pathog *11*, e1005333. https://doi.org/10.1371/journal.ppat.1005333

Larson, A.M. (2011). Multiphoton microscopy. Nature Photonics *5*, 1. https://doi.org/https://doi.org/10.1038/nphoton.an.2010.2

Laschke, M.W., Kerdudou, S., Herrmann, M., and Menger, M. (2005). Intravital fluorescence microscopy: a novel tool for the study of the interaction of *Staphylococcus aureus* with the microvascular endothelium *in vivo*. J Infect Dis *191*, 435-443. https://doi.org/https://doi.org10.1086/427193

Lee, W.Y., Moriarty, T.J., Wong, C.H., Zhou, H., Strieter, R.M., van Rooijen, N., Chaconas, G., and Kubes, P. (2010). An intravascular immune response to *Borrelia burgdorferi* involves Kupffer cells and iNKT cells. Nat Immunol *11*, 295-302. https://doi.org/https://doi.org/10.1038/ni.1855

Lee, W.Y., Sanz, M.J., Wong, C.H., Hardy, P.O., Salman-Dilgimen, A., Moriarty, T.J., Chaconas, G., Marques, A., Krawetz, R., Mody, C.H., *et al.* (2014). Invariant natural killer T cells act as an extravascular cytotoxic barrier for joint-invading Lyme *Borrelia*. Proc Natl Acad Sci U S A *111*, 13936-13941. https://doi.org/https://doi.org/10.1073/pnas.1404769111

Li, X., Liu, X., Beck, D.S., Kantor, F.S., and Fikrig, E. (2006). *Borrelia burgdorferi* lacking BBK32, a fibronectin-binding protein, retains full pathogenicity. Infect Immun *74*, 3305-3313. https://doi.org/https://doi.org/10.1128/IAI.02035-05

Lienenklaus, S., Cornitescu, M., Zietara, N., Lyszkiewicz, M., Gekara, N., Jablonska, J., Edenhofer, F., Rajewsky, K., Bruder, D., Hafner, M., *et al.* (2009). Novel reporter mouse reveals constitutive and inflammatory expression of IFN-beta in vivo. J Immunol *183*, 3229-3236. https://doi.org/10.4049/jimmunol.0804277

Lin, Y.P., Tan, X., Caine, J.A., Castellanos, M., Chaconas, G., Coburn, J., and Leong, J.M. (2020). Strain-specific joint invasion and colonization by Lyme disease spirochetes is promoted by outer surface protein C. PLoS Pathog *16*, e1008516. https://doi.org/10.1371/journal.ppat.1008516

Liu, N., Montgomery, R.R., Barthold, S.W., and Bockenstedt, L.K. (2004). Myeloid differentiation antigen 88 deficiency impairs pathogen clearance but does not alter inflammation in *Borrelia burgdorferi*-infected mice. Infect Immun *72*, 3195-3203. https://doi.org/10.1128/IAI.72.6.3195-3203.2004

Masedunskas, A., Milberg, O., Porat-Shliom, N., Sramkova, M., Wigand, T., Amornphimoltham, P., and Weigert, R. (2012). Intravital microscopy: a practical guide on imaging intracellular structures in live animals. Bioarchitecture 2, 143-157. https://doi.org/10.4161/bioa.21758

Medina-Perez, D.N., Wager, B., Troy, E., Gao, L., Norris, S.J., Lin, T., Hu, L., Hyde, J.A., Lybecker, M., and Skare, J.T. (2020). The intergenic small non-coding RNA *ittA* is required for optimal infectivity and tissue tropism in *Borrelia burgdorferi*. PLoS Pathog *16*, e1008423. https://doi.org/10.1371/journal.ppat.1008423

Moriarty, T.J., Norman, M.U., Colarusso, P., Bankhead, T., Kubes, P., and Chaconas, G. (2008). Real-time high resolution 3D imaging of the lyme disease spirochete adhering to and escaping from the vasculature of a living host. PLoS Pathog *4*, e1000090. https://doi.org/https://doi.org/10.1371/journal.ppat.1000090

Moriarty, T.J., Shi, M., Lin, Y.P., Ebady, R., Zhou, H., Odisho, T., Hardy, P.O., Salman-Dilgimen, A., Wu, J., Weening, E.H., *et al.* (2012). Vascular binding of a pathogen under shear force through mechanistically distinct sequential interactions with host macromolecules. Mol Microbiol *86*, 1116-1131. https://doi.org/https://doi.org/10.1111/mmi.12045

Motaleb, M.A., Liu, J., and Wooten, R.M. (2015). Spirochetal motility and chemotaxis in the natural enzootic cycle and development of Lyme disease. Current Opinion in Microbiology *28*, 106-113. https://doi.org/http://dx.doi.org/10.1016/j.mib.2015.09.006

Niddam, A.F., Ebady, R., Bansal, A., Koehler, A., Hinz, B., and Moriarty, T.J. (2017). Plasma fibronectin stabilizes *Borrelia burgdorferi*-endothelial interactions under vascular shear stress by a catch-bond mechanism. Proc Natl Acad Sci U S A *114*, E3490-E3498. https://doi.org/10.1073/pnas.1615007114

Norman, M.U., Moriarty, T.J., Dresser, A.R., Millen, B., Kubes, P., and Chaconas, G. (2008). Molecular mechanisms involved in vascular interactions of the Lyme disease pathogen in a living host. PLoS Pathog *4*, e1000169. https://doi.org/https://doi.org/10.1371/journal.ppat.1000169

Novak, E.A., Sekar, P., Xu, H., Moon, K.H., Manne, A., Wooten, R.M., and Motaleb, M.A. (2016). The *Borrelia burgdorferi* CheY3 Response Regulator is Essential for Chemotaxis and Completion of its Natural Infection Cycle. Cell Microbiol. https://doi.org/10.1111/cmi.12617

Oreopoulos, J., Berman, R., and Browne, M. (2014). Spinning-disk confocal microscopy: present technology and future trends. Methods in Cell Biology *123*, 153-175. https://doi.org/10.1016/B978-0-12-420138-5.00009-4.

Ouyang, Z., Narasimhan, S., Neelakanta, G., Kumar, M., Pal, U., Fikrig, E., and Norgard, M.V. (2012). Activation of the RpoN-RpoS regulatory pathway during the enzootic life cycle of *Borrelia burgdorferi*. BMC microbiology *12*, 44. https://doi.org/https://doi.org/10.1186/1471-2180-12-44

Rehemtulla, A., Stegman, L.D., Cardozo, S.J., Gupta, S., Hall, D.E., Contag, C.H., and Ross, B.D. (2000). Rapid and quantitative assessment of cancer treatment response using in vivo bioluminescence imaging. Neoplasia *2*, 491-495. https://doi.org/10.1038/sj.neo.7900121

Sadikot, R.T., Wudel, L.J., Jansen, D.E., Debelak, J.P., Yull, F.E., Christman, J.W., Blackwell, T.S., and Chapman, W.C. (2002). Hepatic cryoablation-induced multisystem injury: bioluminescent detection of NF-kappaB activation in a transgenic mouse model. J Gastrointest Surg *6*, 264-270. https://doi.org/10.1016/s1091-255x(01)00064-6

Salo, J., Pietikainen, A., Soderstrom, M., Auvinen, K., Salmi, M., Ebady, R., Moriarty, T.J., Viljanen, M.K., and Hytonen, J. (2016). Flow-Tolerant Adhesion of a Bacterial Pathogen to Human Endothelial Cells Through Interaction With Biglycan. J Infect Dis *213*, 1623-1631. https://doi.org/10.1093/infdis/jiw003

Saputra, E.P., Trzeciakowski, J.P., and Hyde, J.A. (2020). *Borrelia burgdorferi* spatiotemporal regulation of transcriptional regulator *bosR* and decorin binding protein during murine infection. Sci Rep *10*, 12534. https://doi.org/10.1038/s41598-020-69212-7

Secklehner, J., Lo Celso, C., and Carlin, L.M. (2017). Intravital microscopy in historic and contemporary immunology. Immunol Cell Biol *95*, 506-513. https://doi.org/10.1038/icb.2017.25

Seshu, J., Esteve-Gassent, M.D., Labandeira-Rey, M., Kim, J.H., Trzeciakowski, J.P., Hook, M., and Skare, J.T. (2006). Inactivation of the fibronectin-binding adhesin gene *bbk32* significantly attenuates the infectivity potential of *Borrelia burgdorferi*. Mol Microbiol *59*, 1591-1601. https://doi.org/https://doi.org/10.1111/j.1365-2958.2005.05042.x

Shi, Y., Xu, Q., McShan, K., and Liang, F.T. (2008). Both decorin-binding proteins A and B are critical for the overall virulence of *Borrelia burgdorferi*. Infect Immun *76*, 1239-1246. https://doi.org/10.1128/IAI.00897-07

Skare, J.T., Shaw, D.K., Trzeciakowski, J.P., and Hyde, J.A. (2016). In Vivo Imaging Demonstrates That *Borrelia burgdorferi* ospC Is Uniquely Expressed Temporally and Spatially throughout Experimental Infection. PLoS One *11*, e0162501. https://doi.org/10.1371/journal.pone.0162501

Sultan, S.Z., Manne, A., Stewart, P.E., Bestor, A., Rosa, P.A., Charon, N.W., and Motaleb, M.A. (2013). Motility is crucial for the infectious life cycle of Borrelia burgdorferi. Infect Immun *81*, 2012-2021. https://doi.org/10.1128/IAI.01228-12

Sultan, S.Z., Sekar, P., Zhao, X., Manne, A., Liu, J., Wooten, R.M., and Motaleb, M.A. (2015). Motor rotation is essential for the formation of the periplasmic flagellar ribbon, cellular morphology, and Borrelia burgdorferi persistence within Ixodes scapularis tick and murine hosts. Infect Immun *83*, 1765-1777. https://doi.org/10.1128/iai.03097-14

Szczepanski, A., Furie, M., Benach, J., Lane, B., and Fleit, H. (1990). Interaction between *Borrelia burgdorferi* and endothelium *in vitro*. J Clin Invest *85*, 1637-1647. https://doi.org/https://doi.org/10.1172/JCI114615

Takacs, C.N., Kloos, Z.A., Scott, M., Rosa, P.A., and Jacobs-Wagner, C. (2018). Fluorescent Proteins, Promoters, and Selectable Markers for Applications in the Lyme Disease Spirochete *Borrelia burgdorferi*. Appl Environ Microbiol *84*. https://doi.org/10.1128/AEM.01824-18

Tan, X., Petri, B., and Chaconas, G. (2020). The Lyme disease spirochete hijacks the host immune system to extravasate through the microvasculature. *Manuscript submitted.*

Tilly, K., Bestor, A., Dulebohn, D.P., and Rosa, P.A. (2009). OspC-independent infection and dissemination by host-adapted *Borrelia burgdorferi*. Infect Immun *77*, 2672-2682. https://doi.org/10.1128/IAI.01193-08

Tilly, K., Krum, J.G., Bestor, A., Jewett, M.W., Grimm, D., Bueschel, D., Byram, R., Dorward, D., Vanraden, M.J., Stewart, P., *et al.* (2006). *Borrelia burgdorferi* OspC protein required exclusively in a crucial early stage of mammalian Infection. Infect Immun *74*, 3554-3564. https://doi.org/https://doi.org/10.1128/IAI.01950-05

Wager, B., Shaw, D.K., Groshong, A.M., Blevins, J.S., and Skare, J.T. (2015). BB0744 Affects Tissue Tropism and Spatial Distribution of *Borrelia burgdorferi*. Infect Immun *83*, 3693-3703. https://doi.org/10.1128/IAI.00828-15

Waidmann, M.S., Bleichrodt, F.S., Laslo, T., and Riedel, C.U. (2011). Bacterial luciferase reporters: the Swiss army knife of molecular biology. Bioeng Bugs *2*, 8-16. https://doi.org/10.4161/bbug.2.1.13566

Wang, G., Ma, Y., Buyuk, A., McClain, S., Weis, J.J., and Schwartz, I. (2004). Impaired host defense to infection and Toll-like receptor 2-independent killing of *Borrelia burgdorferi* clinical isolates in TLR2-deficient C3H/HeJ mice. FEMS Microbiol Lett *231*, 219-225. https://doi.org/10.1016/S0378-1097(03)00960-1

Weening, E.H., Parveen, N., Trzeciakowski, J.P., Leong, J.M., Hook, M., and Skare, J.T. (2008). *Borrelia burgdorferi* lacking DbpBA exhibits an early survival defect during experimental infection. Infect Immun *76*, 5694-5705. https://doi.org/10.1128/IAI.00690-08

Wormser, G.P. (2006). Hematogenous dissemination in early Lyme disease. Wien Klin Wochenschr *118*, 634-637. https://doi.org/https://doi.org/10.1007/s00508-006-0688-9

Xu, T., Close, D., Handagama, W., Marr, E., Sayler, G., and Ripp, S. (2016). The Expanding Toolbox of In Vivo Bioluminescent Imaging. Front Oncol *6*, 150. https://doi.org/10.3389/fonc.2016.00150

Yang, X.F., Lybecker, M.C., Pal, U., Alani, S.M., Blevins, J., Revel, A.T., Samuels, D.S., and Norgard, M.V. (2005). Analysis of the *ospC* regulatory element controlled by the RpoN-RpoS regulatory pathway in *Borrelia burgdorferi*. J Bacteriol *187*, 4822-4829. https://doi.org/https://doi.org/10.1128/JB.187.14.4822-4829.2005

Zhang, W., Contag, P.R., Madan, A., Stevenson, D.K., and Contag, C.H. (1999). Bioluminescence for biological sensing in living mammals. Advances in experimental medicine and biology *471*, 775-784. https://doi.org/10.1007/978-1-4615-4717-4_89

Zhi, H., Weening, E.H., Barbu, E.M., Hyde, J.A., Hook, M., and Skare, J.T. (2015). The BBA33 lipoprotein binds collagen and impacts *Borrelia burgdorferi* pathogenesis. Mol Microbiol *96*, 68-83. https://doi.org/10.1111/mmi.12921

Chapter 18

Immune Response to *Borrelia*: Lessons from Lyme Disease Spirochetes

Linda K. Bockenstedt[1]*, R. Mark Wooten[2] and Nicole Baumgarth[3]

[1]Department of Internal Medicine, Yale University School of Medicine, New Haven, CT 06520-8031, USA
[2]Department of Medical Microbiology and Immunology, University of Toledo Health Science Campus, Toledo, OH 43614, USA
[3]Center for Immunology and Infectious Diseases and Dept. Pathology, Microbiology and Immunology, University of California, Davis, Davis CA 95616, USA
*Corresponding author: linda.bockenstedt@yale.edu

DOI: https://doi.org/10.21775/9781913652616.18

Abstract

The mammalian host responds to infection with *Borrelia* spirochetes through a highly orchestrated immune defense involving innate and adaptive effector functions aimed toward limiting pathogen burdens, minimizing tissue injury, and preventing subsequent reinfection. The evolutionary adaptation of *Borrelia* spirochetes to their reservoir mammalian hosts may allow for its persistence despite this immune defense. This review summarizes our current understanding of the host immune response to *B. burgdorferi sensu lato*, the most widely studied *Borrelia* spp. and etiologic agent of Lyme borreliosis. Pertinent literature will be reviewed with emphasis on *in vitro*, *ex vivo* and animal studies that influenced our understanding of both the earliest responses to *B. burgdorferi* as it enters the mammalian host and those that evolve as spirochetes disseminate and establish infection in multiple tissues. Our focus is on the immune response of inbred mice, the most commonly studied animal model of *B. burgdorferi* infection and surrogate for one of this pathogen's principle natural reservoir hosts, the white-footed deer mouse. Comparison will be made to the immune responses of humans with Lyme borreliosis. Our goal is to provide an understanding of the dynamics of the mammalian immune response during infection with *B. burgdorferi* and its relation to the outcomes in reservoir (mouse) and non-reservoir (human) hosts.

Introduction

The mammalian immune system is critical for host defense against infection and for establishing memory responses that can respond rapidly to prevent or limit re-infection (Medzhitov, 2008). Sensor cells located in barrier tissues throughout the body play important roles in distinguishing pathogens as foreign (Murphy and Weaver, 2017). These cells orchestrate immune responses that evolutionarily have been selected as effective at controlling the perceived threat, either by its elimination or by limiting the pathogen burden and adverse effects on the host. Regulation of the inflammatory response is key to preventing excessive tissue injury as a consequence of this defense and restoring tissue homeostasis once the pathogen is controlled (Medzhitov, 2008; Feehan and Gilroy, 2019).

The goal of this chapter is to provide an understanding of the dynamics of the mammalian immune response during infection with *Borrelia burgdorferi* and how this response affects pathogen elimination and persistence. When visualized in tissues, *B. burgdorferi* is extracellular and has a predilection for extracellular matrix and connective tissue at all stages of infection (Duray and Steere, 1988; Barthold et al., 1990; Barthold et al., 1993). Although *B. burgdorferi* has been sighted rarely inside cells, both *ex vivo* and *in vitro* studies of its genome and its mechanisms for immune evasion (reviewed in Chapters 2 and 13), which are directed toward subverting phagocyte, antibody, and complement-mediated destruction, point overwhelmingly toward its residing in the host extracellularly (Hyde, 2017; Lin et al., 2020). *B. burgdorferi* has evolved to persist for extended periods of time in its reservoir hosts without associated inflammatory pathologies despite engagement of both innate and adaptive immunity (Moody et al., 1994; Baum et al., 2012). The adaptive immune response, while unable to eliminate all spirochetes from the host, is capable of protecting the host from re-infection with autologous strains (Bockenstedt et al., 1997; Nadelman et al., 2012; Khatchikian et al., 2014; Jacquet et al., 2015). The factors that allow for this state of "concomitant immunity" await a better understanding of the genetic and antigenic composition of spirochetes that persist within the immunocompetent vertebrate host.

Much of what has been learned about the innate response has been derived from *in vitro* studies of the interactions of cultured *B. burgdorferi* and its components with cells or cell lines. While providing much insight into the capacity of mammalian immune cells to recognize and respond to *B. burgdorferi* components, the ability of *B. burgdorferi* to rapidly change its gene expression program depending on its environment (culture, tick, or mammal) (see chapter 5) limits the extrapolation of *in vitro* studies to the unique *in vivo* microenvironments in which *B. burgdorferi* resides in the mammal (Radolf et al., 2012). A further distinction is made between *B. burgdorferi* infection in inbred mice, a surrogate reservoir host, and the clinical pathology known as "Lyme disease" or "Lyme borreliosis" resulting from *B. burgdorferi* infection of non-reservoir hosts, including humans. Studies in inbred mice provide important insights into immune components that are critical not only for limiting pathogen burden, resolving inflammation and preventing infection with *B. burgdorferi,* but also for identifying the spirochete's key immune evasion strategies. Mice, however, do not develop the pathology characteristic of Lyme borreliosis in humans (Duray and Steere, 1988; Barthold et al., 1992a; Steere et al., 2016) even though *in vitro* studies suggest that similar immune pathways are engaged by mouse and human cells exposed to *B. burgdorferi* and its components (discussed herein). Recent studies in humans have used advanced technologies to probe host responses in blood and tissues, with a goal toward understanding how these responses result in clinical pathology. In this chapter, we review the innate and acquired host responses to *B. burgdorferi* that control pathogen burden, prevent reinfection and limit the extent of tissue injury and clinical disease, as well as those that may allow for *B. burgdorferi* persistence.

Establishment of infection: The role of the tick vector in *B. burgdorferi* infection

B. burgdorferi infection begins with introduction of spirochetes into the mammalian skin during tick feeding. The skin constitutes the first barrier to tick-borne pathogens, and breaching of this barrier at the tick feeding site creates a critical portal for passage of the pathogen from the tick to the bloodmeal host (Frischknecht, 2007). As with other vector-borne pathogens, *B. burgdorferi* benefits from mechanisms that *Ixodes* ticks have evolved to obtain the bloodmeal and to prevent host immunity to tick refeeding. Blood feeding of *Ixodes* ticks is a highly ordered process that occurs over 3-10 days, with the majority of the blood ingested during the final 24 hours of attachment (Kemp et al., 1982). With the initiation of feeding, *B. burgdorferi* replicates in the tick midgut and then migrates through the hemocoel to the salivary glands prior to deposition in the deep epidermis or dermis of the bloodmeal host (Dunham-Ems et al., 2009). During this time, tick saliva bathes the feeding pit with pharmacologically active substances that impair hemostasis, host immunity and wound healing, and which can secondarily facilitate pathogen transmission to the bloodmeal host (Ribeiro et al., 1985)(see also Chapters 11 and 12). The composition of tick saliva changes during the course of tick feeding (Perner et al., 2018) and varies depending on the bloodmeal host species, possibly as a consequence of the vertebrate immune response (Narasimhan et al., 2019). Successful transmission of *B. burgdorferi* to the vertebrate host depends on its ability to exploit components of tick saliva as it moves from the tick salivary gland into the unique ecological niche of the tick feeding site (Nuttall and Labuda, 2004; Nuttall, 2019).

Several studies have shown that tick saliva can enhance infectivity of *B. burgdorferi.* Co-inoculation of *B. burgdorferi* with *I. scapularis* or *I. ricinus* salivary gland lysates into mice enhances pathogen burden in multiple organs in comparison to needle inoculation (Zeidner et al., 2002). This effect, however, is tick species- and spirochete strain-specific, as it was observed only when co-inoculating a spirochete strain isolated from the tick species from which the salivary gland lysates were obtained. Saliva and salivary gland extracts from *I. ricinus* promoted enhanced dissemination of *B. afzelii* when co-inoculated into mice, with an increase in pathogen burden in the skin and urinary bladder, but not the heart, by day 6 (Pechova et al., 2002; Horka et al., 2009). The early effect of tick saliva on pathogen survival is likely mediated by inhibition of cutaneous innate immune mechanisms. Consistent with this, *I. ricinus* salivary gland extracts impeded macrophage production of reactive oxygen species and killing of

B. afzelii by mouse macrophages *in vitro* (Kuthejlova et al., 2001). A more recent study demonstrated that *I. ricinus* salivary gland extracts inhibited mast cell activation *in vitro*. While mast cell-deficient mice did not exhibit increased replication of *B. burgdorferi* in the skin, dissemination to the joints occurred more quickly (Bernard et al., 2017).

The factors within tick saliva that facilitate *B. burgdorferi* transmission and dissemination are only partially defined. Analyses of the *I. scapularis* and *I. ricinus* sialotranscriptomes have identified differentially expressed genes during tick feeding that encode proteins predicted to have a broad range of pharmacologic activities (Narasimhan et al., 2002; Ribeiro et al., 2006; Kotsyfakis et al., 2015; Perner et al., 2018). The timing for production of some of these molecules varies with duration of tick feeding (Perner et al., 2018). These molecules have been catalogued into functional groups, such as the basic tail family proteins (e.g. salp14) that impair blood clotting (Narasimhan et al., 2002; Ribeiro et al., 2006), the tissue factor pathway inhibitors ixolaris and penthalaris (Francischetti et al., 2002; Francischetti et al., 2004), vasodilators such as prostacyclin, prostaglandin E2, and adenosine (Ribeiro et al., 1988; Sa-Nunes et al., 2007), and inhibitors of angiogenesis (Francischetti et al., 2005). More relevant to the survival of *B. burgdorferi* in the skin are factors that retard local inflammation at the tick bite site, especially those that interfere with the activity of phagocytes, antibody and complement. Tick saliva contains an abundance of protease inhibitors, including sialostatin L, ILS919, and ISL 1373, that impair neutrophil functions such as chemotaxis, integrin expression, and reactive oxygen generation, respectively (Kotsyfakis et al., 2006) (Guo et al., 2009). Sialostatins also modulate mouse dendritic cell (DC) responses to *B. burgdorferi in vitro*, reducing DC maturation and IFN-β production, as well as DC production of the chemokines MIP-1α and IP-10 that recruit mononuclear cells, T cells and mast cells (Lieskovska et al., 2015). Other factors inhibit macrophage activation and interfere with the alternate pathway of complement activation, including salp20, Isac and Isac paralogs found in *I. ricinus* (Valenzuela et al., 2000; Daix et al., 2007; Tyson et al., 2007; Couvreur et al., 2008). Additionally, the TSLP1 family inhibits the lectin complement cascade (Schuijt et al., 2011; Wagemakers et al., 2016). RNAseq analysis of single *I. ricinus* salivary glands revealed that the most abundant transcripts expressed within the first 24 hours of tick feeding were secreted metalloproteases, members of the basic tail superfamily proteins, and anti-complement Isac paralogs, including IRAC1, IRACII and IxACB1-5 (Daix et al., 2007; Couvreur et al., 2008; Perner et al., 2018). The actions of these molecules at the tick feeding site interfere with the initial skin immune defense and may help pathogens like *B. burgdorferi* establish infection in the mammal.

Components of tick saliva modulate adaptive immunity, skewing T cell differentiation toward a TH2 phenotype through actions on DCs and T cells. This could facilitate *B. burgdorferi* infection by reducing the actions of TH1-associated cytokines on phagocytes. PGE_2 in tick saliva increases IL-10 and reduces IL-12 and TNFα production by DCs, and suppresses DC-induced CD4+ T cell proliferation and IL-2 production *in vitro* (Sa-Nunes et al., 2007). Tick saliva also contains an IL-2 binding protein that inhibits T cell proliferation *in vitro* (Gillespie et al., 2001). Repeated tick infestation of C3H/HeN mice leads to reduced production of IL-2 and IFN-γ by splenocytes after concanavalin A stimulation and augmented production of IL-4 and IL-10 production. The increase in production of anti-inflammatory cytokines, especially IL-10, may help suppress phagocyte activation and clearance of *B. burgdorferi*. In support of this, treatment of C3H/HeJ mice with TNFα, or with IL-2 or IFN-γ for 10 days after onset of *B. burgdorferi*-infected tick feeding reduced mouse infection rates by 95% and 30-45%, respectively (Zeidner et al., 1996). Langerhans cells (LCs) are critical for tick-mediated suppression of the concanavalin A-induced TH1 response in lymph nodes draining the tick bite site, but their impact is attenuated when *B. burgdorferi* is introduced by feeding ticks (Vesely et al., 2009). Recently, a sphingomyelinase-like protein (IsSMase) has been identified in *I. scapularis* saliva that directly polarizes CD4+ T cells toward a TH2 phenotype and expression of IL-4 (Alarcon-Chaidez et al., 2009). *B. burgdorferi* also has exploited a tick salivary protein, salp15, to colonize the bloodmeal host (Ramamoorthi et al., 2005). The expression of this protein is upregulated in infected tick salivary glands and binds to outer surface protein C (OspC) on spirochetes as they enter the host. The binding of salp15 to OspC can shield *B. burgdorferi s.s.* from borrelicidal antibodies *in vitro*, suggesting that it could interfere with antibody-mediated killing *in vivo*. This effect appears to be species- specific as salp15 could not protect *B. garinii* or *B. afzelii* from being killed. Recombinant salp15 injected with spirochetes enhanced their survival in mice, and siRNA depletion of the protein in infected ticks reduced *B. burgdorferi* infectivity after vector transmission (Ramamoorthi et al., 2005). In addition, salp15 inhibited DC production of inflammatory cytokines through binding of DC-SIGN (Hovius et al., 2008), as well as T cell receptor signal transduction by binding to the T cell coreceptor CD4 (Garg et al., 2006; Juncadella et al., 2007). These diverse functions of salp15 not only aid *B. burgdorferi* infection and contribute to disease expression but also reduce the impact of host innate and adaptive immune responses induced by *B. burgdorferi* that could potentially interfere with tick feeding.

Initial host defense - the innate immune response

Both innate and acquired defenses control spirochetes in tissues, with the humoral response particularly important for lowering *B. burgdorferi* tissue burden, resolving disease and protecting from infection. Innate immune mechanisms are likely engaged to some extent at all stages of infection and wherever spirochetes are found in the mammal. Their greatest impact, however, would be during phases where *B. burgdorferi* initially enters the host, replicates and disseminates.

Early spirochetal dissemination into tissues and interactions with resident immune cells

During natural infection, ticks deposit *B. burgdorferi* within the epidermal or dermal layers of the skin, depending on the size/stage of the tick and the skin thickness of the new host. Studies of the earliest stages of tick-transmitted infection using intravital microscopy in living mice and GFP-expressing spirochetes have been unable to visualize spirochete deposition within the skin, consistent with the small numbers that are believed to be deposited in the feeding pit (Ohnishi et al., 2001; Bockenstedt et al., 2014). However, intradermal needle inoculation with a range of *B. burgdorferi* doses has been found to result in similar tissue burdens as achieved following tick transmission over time, with the absolute numbers varying depending on the mouse strain infected (Brown and Reiner, 1998b). After inoculation of cultured GFP-expressing spirochetes into mouse skin, intravital microscopy reveals that they immediately become motile within the skin, displaying a backward-forward motility along collagen fibers interspersed with longer runs, similar to patterns observed in tick-transmitted infection (Harman et al., 2012). Within 48 hours, they begin disseminating from the inoculation site through the dermis, particularly in areas possessing collagen (Barthold et al., 1991; Shih et al., 1992; Shih et al., 1993). During dissemination, spirochetes achieve velocities that average ~250 µm/min, while maintaining a primarily backward-forward motility interspersed with some longer runs in a single direction (Malawista and de Boisfleury Chevance, 2008; Harman et al., 2012). While some spirochetes access the bloodstream, the vast majority of bacteria appear to continue migrating through the dermis and lymphatic ducts (Tunev et al., 2011; Imai et al., 2013). After injection into one ear of a mouse, it consistently takes around 8 days for the spirochetes to reach the opposite ear, regardless of the inoculum dose; this also supports a mainly skin/lymphatics-based dissemination (Tunev et al., 2011). The first resident immune cells encountered during dissemination are LCs, which reside in the areas at the epidermal/dermal interface (Nestle et al., 2009), and constantly sample the environment via their long pseudopods that move at around 0.4 µm/min (Nguyen Hoang et al., 2014). Subsequently, the spirochetes come in contact with cells of the macrophage/DC lineage as they continue through the dermis (Nestle et al., 2009). These cells are capable of capturing and phagocytosing a subset of *B. burgdorferi* even though they move approximately 300-times slower than *B. burgdorferi* (Malawista and de Boisfleury Chevance, 2008). Within 4 hours of *B. burgdorferi* being injected into the skin, large numbers of neutrophils migrate specifically to the skin regions that contain spirochetes (Xu et al., 2007). These phagocytes travel at average speeds of 6 µm/min, which is around 40-times slower than spirochetes, capturing some of the disseminating *B. burgdorferi* (Park et al., 2018). The vast majority of spirochetes, however, continue to disseminate, with bacteria observed in the heart by ≤7 days and in the rear ankle joints by 7-10 days post-infection after inoculation into the back skin (Lazarus et al., 2006). The spirochetes can inhabit at least transiently almost any tissue in the host, including the meninges (Divan et al., 2018), residing in the extracellular spaces where they can interact with resident immune cells. During this persistent phase, *B. burgdorferi* may elicit inflammatory responses, though the exuberance of those responses is dependent largely on the host genetic make-up rather than the number of spirochetes that reside in those tissues (Ma et al., 1998; Brown et al., 1999; Miller et al., 2008b). Notably, the spirochetes continue to display the same backward-forward motility and velocities in skin tissues as demonstrated early in infection, even 2 years post-infection (Wooten et al., unpublished).

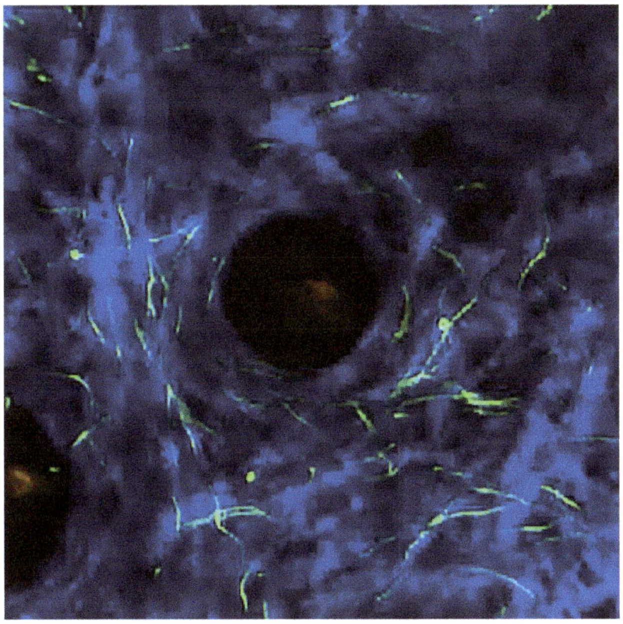

Figure 1. Image from a 2-photon intravital microscopy movie of GFP-expressing *B. burgdorferi* transformant 914 (derived from strain 297) in the ear skin of a 27-day infected C3H *myD88-/-* mouse. Infection was introduced by tick bite. Blue, second harmonics of skin collagen fibers. Note the numerous GFP spirochetes along the collagen fibers and surrounding the central hair follicle (See Bockenstedt, et al., 2012).

Aspects of the innate immune response appear to have the strongest effects on bacterial clearance.
A number of studies have helped delineate the relative importance of innate versus adaptive immunity in the control of *B. burgdorferi* numbers in host tissues. As outlined more below, SCID mice lacking B and T cells possessed similar spirochetal numbers as wild-type (WT) mice for several weeks post-infection. Over time, spirochetes reached higher numbers in several tissues of SCID mice compared to WT mice (Barthold et al., 1992b; Wang et al., 2005), with control of spirochete numbers mediated by B cells but not T cells (de Souza et al., 1993; Schaible et al., 1994). Mice lacking the pattern recognition receptor toll-like receptor 2 (TLR2) or myeloid differentiation antigen 88 (MyD88), the downstream signaling adaptor for numerous TLRs, including TLR2, quickly developed bacterial loads that were significantly higher than WT mice, and these differences were maintained even after the adaptive responses developed (Wooten et al., 2002b; Bolz et al., 2004) (Figure 1). Importantly, the antibody levels in these mice were similar to or significantly higher than those of infected WT mice, and passive transfer of their immune serum to naïve mice generated similar protection against *B. burgdorferi* challenge as immune serum from WT mice (Wooten et al., 2002b; Yoder et al., 2003; Bolz et al., 2004). These findings suggest that the absence of MyD88 reduces effector functions of the innate immune response important for controlling pathogen burden. Finally, massively parallel sequencing of a defined set of *B. burgdorferi* mutant strains following their transfer by needle-inoculation into WT and MyD88-deficient mice determined that the major bottleneck to infection occurs early at the site of infection, which is when innate immunity comprises the majority of immune responsiveness (Troy et al., 2013). Together with studies in SCID mice, these findings suggest that both the innate immune response and the production of antibodies are critical for controlling pathogen burden after *B. burgdorferi* infection in mice.

A caveat concerning most studies of *B. burgdorferi* interactions with immune cells is that they utilize assays performed *in vitro* or *ex vivo,* which are inadequate surrogates for the 3-dimensional tissues in which these spirochetal bacteria have evolved to reside. Some reasons for this include: 1) *B. burgdorferi* is a true obligate parasite that lacks many genes needed for growth outside a tick or vertebrate host; 2) *B. burgdorferi* is highly adaptable to the host or environment it encounters (i.e. vertebrate host vs. *in vitro* culture) and can rapidly change gene expression patterns; 3) we still do not know which gene products are expressed on the *B. burgdorferi* surface within a vertebrate host versus those expressed in an artificial environment; and 4) the relevance of *B. burgdorferi* motility and chemotaxis for immune evasion, particularly regarding the cellular responses, cannot be appreciated anywhere other than within the complex architecture of host tissues (Stevenson et al., 1995; Carroll et al., 1999; Charon and Goldstein, 2002; Rosa et al., 2005; Stewart et al., 2005). Thus, much of our understanding of how individual innate immune cell types respond to encounters with *B. burgdorferi* and/or its lipoproteins *in vitro* must be interpreted with caution. Some innate immune pathways have also been explored in mutant mice lacking relevant genes or after depletion of effector molecules. Table 1 summarizes those studies and the impact of interruption of specific innate immune pathways on pathogen burden, antibody responses and pathology.

Table 1. Effect of Innate Immune Mediators on *B. burgdorferi* Infection

Host gene or molecule affected	Pathogen burden relative to WT mice	Antibody (Ab) development	Disease development	References
CD1d/Jα18/Vα14 KO[1] (no iNKT cells)	Increased	Increased	Increased arthritis, carditis	(Tupin et al., 2008; Olson et al., 2009; Lee et al., 2010; Lee et al., 2014)
Complement component C3 KO	Increased	Decreased	Increased	(Lawrenz et al., 2003)
E-selectin/P-selectin KO	No change	N/A	No change	(Seiler et al., 1998)
Gp91phox (NADPH oxidase subunit) KO	No change	N/A	No change	(Crandall et al., 2005)
IFN Type I KO or depleted	No change	No change	Decreased	(Hastey et al., 2014; Ma et al., 2014; Paquette et al., 2017)
IFNγ/IFNγR KO or depleted[3]	Increased	N/A	Decreased	(Brown and Reiner, 1999b; Olson et al., 2009; Sonderegger et al., 2012)
IL-4/IL-13 KO	No change	Shift from IgG1 to IgG2b	No change	(Potter et al., 2000; Satoskar et al., 2000)
IL-10 KO	Decreased (skin, heart, joint)	Increased	Increased	(Brown et al., 1999; Lazarus et al., 2006; Sonderegger et al., 2012)
IL-12 depleted	Decreased (ear)	Decrease IgG2a only	Decreased arthritis	(Anguita et al., 1996; Anguita et al., 1997)
IL-17 KO	No change	No change	No change	(Lasky et al., 2015)
Mannose-binding lectin (MBL) KO	Increased early in skin; other tissues are WT	Increased	No change	(Coumou et al., 2019)
Myeloid-related protein (MRP) 8/14 KO	No change	No change	No change	(Mason et al., 2018)
MyD88 KO	Increased	No change	No change	(Bolz et al., 2004; Behera et al., 2006; Bockenstedt et al., 2006; Bolz et al., 2006)
Ncf1/p47phox (NADPH oxidase subunit) KO	No change	No change	No change	(Crandall et al., 2005)
Nitric oxide	No change	No change	No change	(Seiler et al., 1995; Brown and Reiner, 1999a)
Toll-like receptor 2 (TLR2) KO	Increased (ear, heart, joint, back skin)	Roughly similar to WT; lesser amounts of lipoprotein-specific Abs	No change	(Wooten et al., 2002; Yoder et al., 2003)

[1]KO – knock-out/gene deletion
[2]N/A – Not assessed
[3]Depleted – antibody-mediated depletion in mice

B. burgdorferi agonists and their innate pattern recognition receptors (PRRs)

B. burgdorferi populates the extracellular spaces of host tissues and, thus, is likely to initially interact with PRRs on cell surfaces. Early *in vitro* experiments indicated that *B. burgdorferi* is capable of stimulating a wide range of leukocytes and other host cells (Defosse and Johnson, 1992; Schoenfeld et al., 1992). Fractionation studies suggested that outer membrane lipoproteins were the primary pathogen-associated molecular patterns (PAMPs) through which innate immune cells sense *B. burgdorferi.* Prior to the identification of toll-like receptor (TLR) 4, the C3H/HeJ mouse line, a derivative of the C3H/HeN strain, was identified as resistant to septic shock after injection with lipopolysaccharide (LPS), but to be highly susceptible to infection with Gram-negative bacteria (Watson and Riblet, 1974; Weinstein et al., 1986); thus C3H/HeJ mice were presumed to lack an LPS receptor. Significantly, *B. burgdorferi* and/or their purified lipoproteins elicited similar responses by macrophages derived from either C3H/HeJ or C3H/HeN mice, and both mouse strains exhibited a similar course of infection with *B. burgdorferi* (Radolf et al., 1991; Yang et al., 1992; Barthold et al., 1993; Ma and Weis, 1993). This strongly suggested that *B. burgdorferi* lacked a classical LPS and that spirochetal lipoproteins activated macrophages through different innate receptors. The discovery of a family of TLRs (Medzhitov et al., 1997; Rock et al.,

1998) set the stage for subsequent work that identified TLR2 to be the receptor for *B. burgdorferi* lipoproteins.

In addition to its surface PAMPs, intracellular PAMPs of *B. burgdorferi* are likely exposed when the bacterial cells are damaged or after their phagocytosis and degradation within certain host cell types. These PAMPs would then be recognized by PRRs residing within the endosomal/phagosomal compartment of phagocytes, where they can either amplify or suppress particular immune/inflammatory pathways (Cervantes et al., 2014). Evidence for the existence of pathways involving *B. burgdorferi* agonists and PRRs other than TLR2 have come from reports demonstrating that a greater complexity of responses are triggered in human and mouse mononuclear cells cultured with living *B. burgdorferi* than with purified lipoproteins or with sonicated preparations of spirochetes (Benhnia et al., 2005; Moore et al., 2007; Cruz et al., 2008; Lazarus et al., 2008; Dennis et al., 2009; Salazar et al., 2009) (Wooten et al., unpublished). The many immune molecules and pathways involved in the innate response to *B. burgdorferi* are described below.

CD14

The first receptor identified to mediate *B. burgdorferi*-associated inflammatory signaling was CD14. This receptor can be utilized by cells that either possess CD14 directly linked to their surface via a glycosylphosphatidylinositol linkage (mCD14) on myeloid lineage immune cells (e.g. macrophages, neutrophils, dendritic cells) or as a soluble protein (sCD14) in serum. Both macrophages (via mCD14) and endothelial cells (via sCD14) were activated by *B. burgdorferi* recombinant lipoproteins or lipopeptides using pathways accentuated by CD14 (Sellati et al., 1998; Wooten et al., 1998). Prototypic *B. burgdorferi* lipoproteins OspA and OspC were directly bound to CD14 in a similar fashion as LPS (Wooten et al., 1998). Activation of macrophages by lipoproteins occurred in the absence of CD14, albeit only at higher concentrations, indicating a role for CD14 in innate immune cell monitoring for bacterial lipid-containing pathogen-associated molecular patterns (PAMPs) in host tissues (Wooten et al., 1998; Hirschfeld et al., 1999). The mechanism differed from LPS-mediated interactions with CD14, in that lipoprotein activities mediated by CD14 occurred independently of LPS-binding protein activity (Sellati et al., 1998; Wooten et al., 1998). Studies using CD14-deficient mice indicated that *B. burgdorferi* interactions with CD14 *in vivo* were important for lowering bacterial numbers in multiple target tissues and diminishing *B. burgdorferi*-associated inflammatory pathologies (Benhnia et al., 2005). CD14-mediated signaling increased neutrophil-associated activities in *B. burgdorferi*-infected tissues while promoting matrix metalloprotease 9 (MMP9) activities and collagen remodeling (Zhao et al., 2007; Sahay et al., 2011). CD14 can promote *B. burgdorferi* phagocytosis through interactions with complement receptor 3 (CR3; CD11b) and lipid raft microdomains of macrophages completely independent of the TLR2-signaling pathway (Hawley et al., 2012). These findings suggest that CD14 acts as a scavenger receptor to target *B. burgdorferi* for enhanced phagocytosis and/or interacts with TLR2 (see below) to activate CD14-associated host cells.

TLR2

Early publications identified both TLR2 (Kirschning et al., 1998; Yang et al., 1998) and TLR4 (Poltorak et al., 1998) as the LPS-receptor. However, work by Hirschfeld et al indicated that the apparent LPS-mediated signaling via TLR2 was actually due to contaminating lipoproteins in the crude LPS preparations used in those studies (Hirschfeld et al., 2000), confirming that TLR4 was the actual LPS receptor. Transfection of five different TLRs individually into 293 and U373 cell lines lacking some or all TLRs identified TLR2 as a major receptor for *B. burgdorferi*-mediated signaling, as only cells expressing TLR2 alone could become activated in response to recombinant OspA and OspB lipoproteins, synthesized triacylated lipopeptides, or sonicated *B. burgdorferi* (Hirschfeld et al., 1999; Lien et al., 1999). TLR2-mediated signaling in response to lipoproteins also activated NF-κB in THP-1 cells, and could elicit apoptosis and/or respiratory burst-associated products in those cells (Aliprantis et al., 1999). TLR2-mediated signaling controlled *B. burgdorferi* numbers *in vivo*, as multiple target tissues from TLR2-deficient mouse lines possessed up to 100-fold more spirochetes compared to WT mice months post-infection (Wooten et al., 2002a). Interestingly, TLR2-deficient mice produced similar or significantly greater levels of *B. burgdorferi*-specific antibodies than WT mice, and these antibodies recognized a similar range of bacterial antigens, suggesting that TLR2-mediated innate responses were most important for controlling *B. burgdorferi* numbers but are similarly unable to clear the infection (Wooten et al., 2002a)(Wooten et al., unpublished).

Immunization of TLR2-deficient and WT mice with rOspA elicited similar protection from challenge with *B. burgdorferi*, confirming that antibodies arising in the absence of TLR2 possessed indistinguishable protective capacities (Yoder et al., 2003). While TLR2-mediated signaling alone significantly contributes to the response to *B. burgdorferi*, the association of this receptor with TLR1 appears necessary for optimal production of antibodies against *B. burgdorferi* lipoproteins (Alexopoulou et al., 2002; Yoder et al., 2003). TLR2-activity is not required for phagocytosis of *B. burgdorferi* but does appear to be needed for sampling of spirochetal agonists after phagocytosis to allow optimal inflammatory signaling (Shin et al., 2008b; Salazar et al., 2009). In the case of human macrophages, phagocytosis of live *B. burgdorferi* elicits a more potent and varied upregulation of proinflammatory mediators compared to macrophages exposed to spirochetal lysates, suggesting that phagosomal signaling continues to be driven by subsequent exposure to TLR2-dependent and -independent *B. burgdorferi* agonists (Salazar et al., 2009). Interestingly, this TLR2-associated inflammatory signaling by murine macrophages is much stronger in response to killed *B. burgdorferi* or bacterial lysates compared to intact spirochetes, whereas human macrophages and DCs show much stronger signaling in response to live bacteria compared to bacterial lysates. The reason for these differences between mouse and human innate immune cell activation events is currently unknown (Lazarus et al., 2008; Salazar et al., 2009).

As noted above, *B. burgdorferi* infection of TLR2- or MyD88-deficient mice resulted in very high levels of bacteria in all target tissues tested compared to WT mice, despite similarity in their antibody responses (Bolz et al., 2004; Liu et al., 2004). Unlike TLR2-deficient macrophages, however, MyD88-deficient macrophages exhibited a defect in phagocytosis of *B. burgdorferi* and produced lower levels of inflammatory mediators in response to *B. burgdorferi* and its components than TLR2-deficient or WT macrophages (Shin et al., 2008b). These findings support a role for MyD88 in facilitating signals through innate receptors other than TLR2 after *B. burgdorferi* infection. Phagocytosis of *B. burgdorferi* also can be enhanced through signaling via the TIR-domain-containing adapter-inducing interferon-β (TRIF) adapter protein, which can be stimulated following TLR2 engagement (Shin et al., 2008b; Petnicki-Ocwieja et al., 2013; Petnicki-Ocwieja et al., 2015; Ullah et al., 2016). Together, these findings indicate that TLR2-signaling is critical for controlling *B. burgdorferi* numbers in tissues by affecting both innate and adaptive effector functions.

TLR1 and TLR6

TLR1 is expressed on the surface of many cell types and is known to dimerize with TLR2 to promote binding of bacterial triacylated lipoproteins, including those expressed by *B. burgdorferi*. TLR6 is also known to dimerize with TLR2 and to bind bacterial diacylated lipoproteins, which are presently mainly on Gram-positive bacteria. Thus, while one might predict that TLR1, but not TLR6, would have a major role in *B. burgdorferi*-signaling, there have been few studies assessing their function in controlling spirochete numbers independent of TLR2. TLR1 does appear necessary for optimal production of antibodies specific for *B. burgdorferi* lipoproteins (Alexopoulou et al., 2002). Co-cultures of *B. burgdorferi* lysates with different cell types isolated from the central nervous system elicited increased expression of TLR1 by astrocytes and microglia cells, but not neurons (Cassiani-Ingoni et al., 2006). In macrophages, TLR1 levels also were upregulated after exposure to *B. burgdorferi* (Izadi et al., 2007). Peritoneal macrophages from either TLR1- or TLR6-deficient mice exposed to *B. burgdorferi in vitro* produced a range of inflammatory cytokines at levels similar to those obtained from WT mice (Oosting et al., 2011a). Interestingly, splenocytes from TLR1-deficient mice produced significantly less IL-10 and significantly higher levels of IFNγ after exposure to *B. burgdorferi in vitro* compared to WT splenocytes. In contrast, TLR6-deficient mice exhibited a decreased IFNγ response. These findings indicate that the balance of TH1/TH2 associated cytokines is altered in the absence of TLR1 or TLR6 in mice. In contrast to TLR1-deficient mice, studies using PBMCs from human donors with polymorphisms in their TLR1 genes indicated that a subset of those mutations resulted in a significant inhibition of proinflammatory cytokines in response to *B. burgdorferi in vitro* (Oosting et al., 2011a), which could potentially impact the outcome of *B. burgdorferi* infection. Subsequent studies with a larger cohort of Lyme borreliosis patients identified a particular TLR1-polymorphism (i.e. TLR1-1805GG) that correlated with the development of persistent inflammatory arthritis after antibiotic treatment (post-antibiotic Lyme arthritis, discussed further below) (Strle et al., 2012). These patients developed higher levels of IFNγ and other inflammatory mediators than Lyme borreliosis patients who did not develop post-antibiotic Lyme arthritis, and PBMCs from these patients also produced higher levels of these mediators when exposed to *B. burgdorferi in vitro*. Together, these studies suggest that TLR1 can modulate inflammatory cytokine production in response to *B. burgdorferi*, and certain TLR1 polymorphisms may augment TH1-like inflammatory responses. While these effects may contribute to adverse outcomes from Lyme borreliosis, specific roles for TLR1 and TLR6 in controlling *B. burgdorferi* infection in vertebrate hosts have not yet been elucidated.

Other TLRs: TLR5/TLR7/TLR8/TLR9

TLR5 is a surface PRR that recognizes bacterial flagella. Although *B. burgdorferi* possess flagella, their location between the inner and outer membranes of the spirochete indicates they would not be exposed unless the spirochete has been damaged. While deficiency in TLR5 does not alter *B. burgdorferi* phagocytosis, it has been shown to affect production of inflammatory cytokines in macrophages and microglia. These observations suggest that phagocytosis and disruption of intact *B. burgdorferi* exposes flagella to TLR5 within endolysosomal compartments (Shin et al., 2008b; Parthasarathy and Philipp, 2015, 2017). However, macrophage responses to flagellin-deficient *B. burgdorferi* elicited similar levels of prototypic proinflammatory cytokines as produced in response to WT *B. burgdorferi*, indicating that TLR5 plays a minimal role in the inflammation caused by these spirochetes (Salazar et al., 2005). Disparate responses by certain TLRs are obtained depending on the host cell type studied. In contrast to human microglia, human monocytes stimulated with *B. burgdorferi* lysates or lipoproteins resulted in the upregulation of TLR2 and downregulation of TLR5, rendering the cells less responsive to TLR5 agonists (Cabral et al., 2006). TLR7 and TLR8 (PRRs for single-stranded RNA), as well as TLR9 (PRR for unmethylated CpG sequences in DNA), also respond to *B. burgdorferi* components that are likely exposed upon its degradation within host cells. Phagocytosis of *B. burgdorferi* has been shown to elicit production of Type I and III interferons from human PBMCs (Petzke et al., 2009; Love et al., 2014, 2015), monocytes, and DCs (Cervantes et al., 2011; Cervantes et al., 2013) via these TLR pathways, suggesting their involvement in the host response to *B. burgdorferi* infection. Interestingly, although type I IFN production was noted in *B. burgdorferi*-infected mice, production was shown to be independent of MyD88/TRIF, thus independent of TLR stimulation (Miller et al., 2008a; Hastey et al., 2014).

Nod-like receptors (NLRs)

Another group of intracellular PRRs is the NLRs that normally reside in the cytoplasm of leukocytes as well as other host cell types (Motta et al., 2015). Most of these possess a common C-terminal leucine-rich repeat segment that is associated with recognition of microbial agonists, as well as nucleotide-binding domains that allow protein-protein interactions. At least two different NLRs, NOD-1 and NOD-2, respond to muropeptides of bacterial peptidoglycan, with NOD-1 activated by muramyltripeptides of Gram-negative bacteria and NOD-2 by muramyl-dipeptides of Gram-positive bacteria. In those bacteria, inflammatory cytokines produced by macrophages are substantially reduced in the absence of NOD-2 but not NOD-1. *B. burgdorferi* produces a thin layer of peptidoglycan that elicits inflammatory events in mammals (Beck et al., 1990). As discussed elsewhere (see chapter 7), peptidoglycan of *Borrelia* spirochetes has a unique muramyldipeptide and, unlike other bacteria, these spirochetes do not recycle their peptidoglycan fragments during cell wall turnover (Jutras et al., 2019). This shedding of a highly inflammatory *B. burgdorferi* component may contribute to the disproportionate host inflammatory reaction in infected tissues in comparison to the numbers of spirochetes present and may explain the more exuberant responses elicited from innate immune cells *in vitro* by sonicated/disrupted spirochetes versus intact viable organisms (Jutras et al., 2019).

NOD2 activation induces the proinflammatory cytokines IL-1β and IL-18. The receptors for both of these cytokines require MyD88 to transmit intracellular signals, suggesting that these cytokines could play an important role in host defense. Absence of IL-18, however, does not alter host susceptibility to *B. burgdorferi* infection and disease (Behera et al., 2006a). Secretion of IL-1β requires the intracellular assembly of the inflammasome, comprised of NOD-, LRR- and pyrin domain-containing protein 3 (NLRP3), pro-caspase 1 and the adaptor apoptosis-associated speck-like protein ASC, through which caspase 1 cleaves pro-IL-β into its active form (Swanson et al., 2019). Upregulation of the inflammasome components occurs in response to perceived cell stress, including signals transmitted through pattern-associated molecular patterns (PAMPS; microbial agonists) and PRRs such as

NOD2. Several *in vitro* studies have shown that *B. burgdorferi* induction of IL-1β from mouse and human macrophages requires caspase 1 and ASC (Liu et al., 2009; Oosting et al., 2011b). Despite these *in vitro* observations, *in vivo* studies have not supported a critical role for caspase 1-dependent inflammasome activation in host defense against *B. burgdorferi* (Liu et al., 2009). No significant difference in pathogen burden or arthritis severity was noted in mice deficient in either caspase 1 or ASC when infection was introduced by intradermal inoculation. However, absence of caspase 1 or ASC does attenuate arthritis induced by direct inoculation of cultured *B. burgdorferi* into the joint (Oosting et al., 2012). Surprisingly, joint inflammation was not dependent on NLRP3, the principal NLR involved in the caspase 1-inflammasome, suggesting the involvement of other pathways.

Cellular mediators of the innate immune response to B. burgdorferi

Langerhans cells (LCs).

LCs are a subset of myeloid cells that reside in the epidermis near the dermal interface (Kaplan, 2017). These cells are identified by the cell surface receptor Langerin (CD207), a PRR. They possess a full complement of PRRs that allow them to respond to pathogens encountered near the skin surface. Their activation leads to the upregulation of MHC class II and subsequent migration to the draining lymph nodes to present antigens to lymphocyte populations in those tissues. Electron microscopy studies identified *B. burgdorferi* associated with LCs in the epidermis of the skin lesion erythema migrans (EM), an early manifestation of Lyme borreliosis in humans (Hulinska et al., 1994). An *ex vivo* model using human skin tissues showed that LCs become activated by *B. burgdorferi* via a TLR2-mediated mechanism and subsequently migrate out of these tissues (Mason et al., 2016). A study of human skin incubated with *B. burgdorferi in vitro* found that, in contrast to other DC populations, LCs were not able to phagocytose Lyme *Borrelia*, (Filgueira et al., 1996). Increased numbers of LCs were observed in biopsies of acrodermatitis chronica atrophicans (ACA) lesions, a late skin manifestation of untreated Lyme borreliosis, and these cells demonstrated a decrease in MHC class II expression compared to LCs in normal skin (Silberer et al., 2000). Transfer of LC-enriched epidermal cell populations, pulsed *in vitro* with *B. burgdorferi,* into naïve mice elicited significant antibody production that protected recipients from challenge infection with tick-transmitted bacteria, suggesting that LCs play a significant role in *B. burgdorferi*-directed adaptive immune responses (Mbow et al., 1997). LCs also appear to be involved in the suppression of TH1 responses in regional lymph nodes during and after tick-feeding on mice, though the mechanisms for this activity are unknown (Vesely et al., 2009). Altogether, these findings suggest that LCs are not effectors that phagocytose *B. burgdorferi.* Instead, LCs efficiently sense spirochetal agonists in the skin and then activate adaptive immune responses in secondary lymphoid tissues.

Macrophages/Dendritic cells

Macrophages and dendritic cells (DCs) are tissue-resident professional phagocytes that can be generated from several tissues, though the majority represent a progression from myeloid progenitors in the bone marrow to blood-resident monocytes (Geissmann et al., 2010).Upon receiving maturation signals, monocytes leave the vasculature and migrate into tissues, where they mature into either macrophages or DCs. Macrophages and DCs are both professional antigen-presenting cells capable of initiating adaptive immune responses via antigen presentation by MHC class I or II. DCs are specialized for migration to the T cell zones of lymphatic tissues after phagocytosis of microbes where they present processed antigens to initiate T cell responses. In contrast, macrophages remain mostly in tissues at the site of infection where they phagocytose invading microbes and initiate localized inflammatory responses that attract additional leukocytes and soluble immune mediators. Currently, the literature on macrophage interactions with *B. burgdorferi* far outnumber papers on DCs, though both have similar effector responses (Chung et al., 2013).

In vitro studies indicate that macrophages and DCs can efficiently phagocytose and kill *B. burgdorferi*, and these activities are enhanced by opsonization via complement and/or spirochete-specific antibodies that bind to CR3 or Fc receptors, respectively (Benach et al., 1984a; Peterson et al., 1984; Montgomery et al., 1994; Filgueira et al., 1996; Cinco et al., 1997; Montgomery et al., 2002; Suhonen et al., 2002; Chung et al., 2013). Macrophage activation also occurs in response to *B. burgdorferi* recombinant lipoproteins via interactions with TLR2, producing a similar elicitation of activation markers as reported after exposure to intact spirochetes (Radolf et al., 1995; Hirschfeld et al., 1999; Wooten et al.,

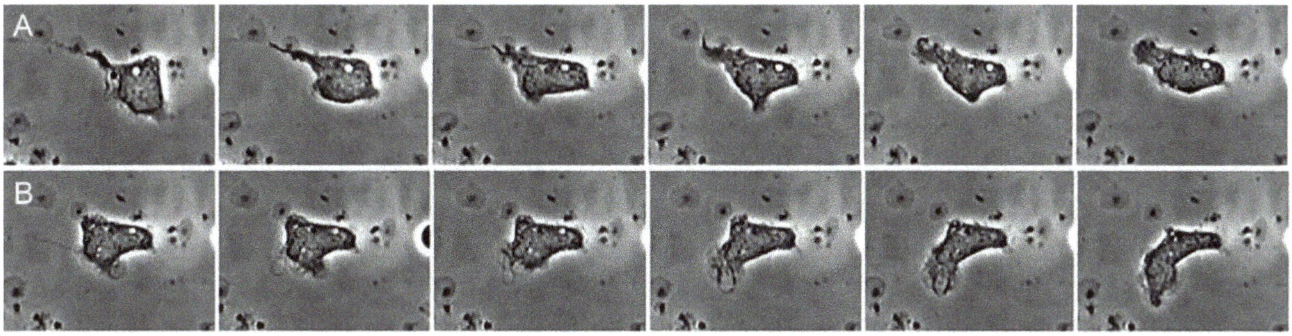

Figure 2. Serial images from a movie of a human neutrophil ingesting *B. burgdorferi* in vitro. Panel A shows engulfment of *B. burgdorferi* from one end by tube phagocytosis. Panel B shows the same neutrophil engulfing a spirochete initially attached at one end. The opposite end was subsequently captured, allowing the neutrophil to readily engulf the immobilized spirochete. From the unpublished archives of movies produced by Malawista (deceased) and de Boisfleury Chevance.

2002a; Chung et al., 2013). *B. burgdorferi* strains that have been passaged repeatedly *in vitro* are less infectious and more susceptible to phagocytosis by macrophages *in vitro* than low-passage strains, but both high and low passage strains equally elicit an oxidative burst after phagocytosis; the mechanisms behind these differences are unknown (Georgilis et al., 1991b).

Phagocytosis of *B. burgdorferi* also can be triggered by interaction of spirochetes with a number of surface PRRs, including integrin α3β1, MARCO, and TLRs, the latter capable of signaling via MyD88 and TRIF (Behera et al., 2006c; Behera et al., 2008; Petnicki-Ocwieja et al., 2013; Hawley et al., 2015; Killpack et al., 2017). Macrophages (and neutrophils) ingest *B. burgdorferi* primarily through both tube and coiling phagocytosis, in which a pseudopod of the cell wraps or "coils" around the entire length of the spirochete, resulting in its engulfment (Rittig et al., 1992; Montgomery and Malawista, 1996; Rittig et al., 1998; Linder et al., 2001; Naj et al., 2013; Naj and Linder, 2015) (Figures 2 and 3). Phagocytosis and degradation of *B. burgdorferi* releases intracellular PAMPs that macrophages then recognize via TLRs and other PRRs that are present in the phagolysosome and other endosomal compartments, as discussed above (Salazar et al., 2005; Shin et al., 2008b; Salazar et al., 2009; Cervantes et al., 2014; Petnicki-Ocwieja et al., 2015; Carreras-González et al., 2019). The cumulative effect of *B. burgdorferi* phagocytosis is a potent inflammatory response eliciting a wide range of pro-inflammatory mediators, including cytokines, chemokines, leukotrienes, and eicosanoids. These mediators prove to be a double-edged sword, in that they promote both spirochetal

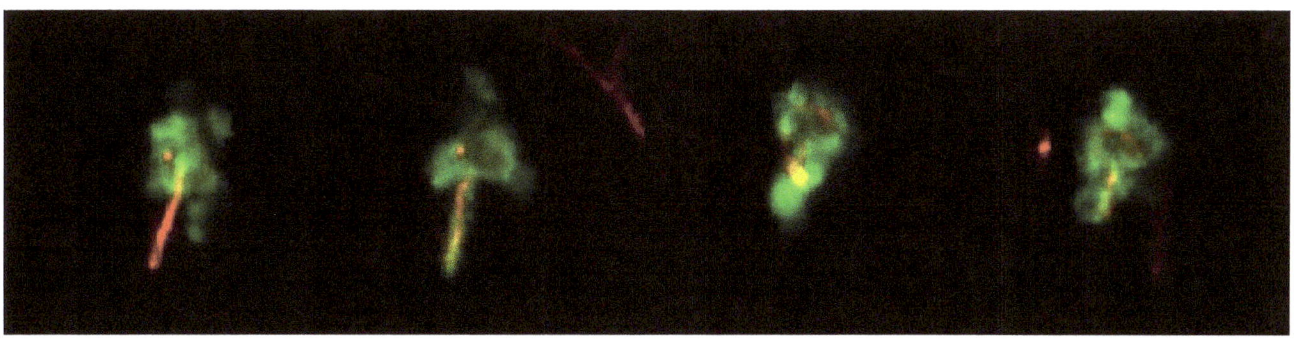

Figure 3. Phagocytosis of *B. burgdorferi* by a dermal professional antigen-presenting cell. Transgenic mice (I-Aβ-GFP expressing) possessing green MHC class II-expression were infected with DsRed-expressing *B. burgdorferi* intradermally into the ear. Intact ear tissues were imaged using intravital confocal microscopy on living mice at 24h post-infection. The four images were taken at 6-minute intervals (left-to-right). From the unpublished archives of movies produced by the R. Mark Wooten lab; images provided by John Presloid.

clearance, but also inflammatory pathology (Wang et al., 2008; Strle et al., 2009; Blaho et al., 2011; Pratt and Brown, 2014). This is reflected transiently in the joints and hearts of *B. burgdorferi*-infected mice, where macrophages have been shown to express phenotypic markers associated with both protective M1 and tissue remodeling M2 differentiation (Lasky et al., 2015b). Overall, these findings indicate a critical role for macrophages and DCs in initiating and mediating immune mechanisms necessary for controlling *B. burgdorferi* numbers in host tissues.

Neutrophils/Endothelial cells

Neutrophils are the most numerous phagocytic cells normally present in the circulation. When tissue-resident myeloid cells (e.g. macrophage and DCs) become activated, they secrete inflammatory mediators that activate endothelial cells lining local blood vessels. Activated endothelial cells upregulate the expression of adhesion molecules and chemokines that promote adherence of circulating immune cells and their subsequent migration from the bloodstream into the tissue. Chemotactic molecules secreted by activated tissue-resident immune cells create a gradient that directs leukocytes from the blood to the infected area, where they contribute to the phagocytosis and killing of extracellular pathogens. Since *B. burgdorferi* is known to persist long-term in the extracellular spaces in multiple tissues, endothelium and neutrophils must play an important role in controlling these infections.

In vitro studies using primary human umbilical vein endothelial cells (HUVECs) indicate they rapidly become activated after exposure to *B. burgdorferi* or their lipoproteins. Spirochetes bind to HUVECs via a mechanism involving fibronectin and then traverse the endothelial layers; this penetration was enhanced via binding of plasminogen (Szczepanski et al., 1990; Comstock and Thomas, 1991; Coleman et al., 1995). HUVECs co-cultured with *B. burgdorferi* or their lipoproteins elicit rapid NF-κB translocation to the nucleus, followed by upregulation of adhesion molecules critical for recruitment of leukocytes within 1-4 hours post-infection (Boggemeyer et al., 1994; Sellati et al., 1995; Wooten et al., 1996). Activation of HUVECs is greatly enhanced in the presence of sCD14 and activation is suppressed by *B. burgdorferi*-elicited IL-10 (Sellati et al., 1998; Wooten et al., 1998; Lisinski and Furie, 2002a). HUVEC secretion of pro-inflammatory cytokines and chemokines occurs within 4 hours of exposure, and neutrophils bind and traverse these endothelial layers by ≥4h post-exposure (Ma and Weis, 1993; Sellati et al., 1995; Sellati et al., 1996; Wooten et al., 1996; Ebnet et al., 1997).

In vitro studies indicate that neutrophils can efficiently phagocytose and kill *B. burgdorferi*, and these activities are enhanced in the presence of complement and/or spirochete-specific antibodies (Benach et al., 1984a; Peterson et al., 1984; Suhonen et al., 2000; Lusitani et al., 2002; Suhonen et al., 2002). Neutrophil activation also occurs in response to *B. burgdorferi* recombinant lipoproteins, producing a similar elicitation of activation markers as reported after exposure to intact spirochetes (Morrison et al., 1997; Cinco et al., 1998). Phagocytosis of *B. burgdorferi* by neutrophils occurs by tube phagocytosis in which pseudopods wrap around the entire 20-30μm length of these spirochetes to allow engulfment (Suhonen et al., 1998). Neutrophil-generated elastase, LL-37, and a number of other granular components can either directly kill *B. burgdorferi* or inhibit their growth *in vitro* (Garcia et al., 1998; Lusitani et al., 2002, 2003). Injection of *B. burgdorferi* or their lipoproteins into murine skin tissue results in a rapid influx of neutrophils (Norgard et al., 1995; Salazar et al., 2005). A number of chemokines and their receptors appear to be critical for recruitment of neutrophils into infected skin, heart, or joint tissues, including CXCR2 and KC (Georgilis et al., 1991a; Brown et al., 2003; Montgomery et al., 2007). Of significance, while neutrophils infiltrate *B. burgdorferi*-infected mouse skin efficiently within 6h, these numbers rapidly drop after 12h post-infection. Neutrophils are almost completely absent by day 7 post-infection, despite the continued presence of spirochetes (Xu et al., 2007). The downregulation of neutrophil influx may be a consequence of host adaptation of *B. burgdorferi* from the culture to the mammalian environment and appears to be important for the establishment of *B. burgdorferi* infection in this model. Of note, with tick-transmitted infection, inflammatory cells increased in numbers at the tick bite site during the first 3 days of tick attachment, but were excluded from the area immediately adjacent to the tick hypostome where spirochetes are deposited (Krause et al., 2009). Evidence that neutrophils can be deleterious to *B. burgdorferi* survival can be seen in studies using a *B. burgdorferi* transformant that expresses and secretes the murine neutrophil chemokine KC. Infection of mice with this transformant caused a significant increase and extended presence of neutrophils at the inoculation site, which correlated with a significantly increased clearance of bacteria in all tissues tested (Xu et al., 2007). Even though neutrophil presence highly correlated with more efficient control of *B. burgdorferi* numbers, mice deficient in genes that encode proteins within the NADPH oxidase complex (i.e. *Ncf1* or *Gp91phox*) showed similar numbers of spirochetes in joints, hearts, and ear skin at 4 weeks post-infection as WT mice (Crandall et al., 2005). On the surface, this appears at odds with the finding that mouse strains that develop arthritis after *B. burgdorferi* infection exhibit a short-lived neutrophil-mediated inflammation in the joints (Barthold et al., 1990). Although quantification of *B. burgdorferi* in mouse tissues by PCR does not distinguish live from dead spirochetes, these findings suggest that either neutrophils are not important for controlling spirochete numbers in those tissues, or that other, NADPH oxidase-independent mechanisms may be engaged. Neutrophils responding to *B. burgdorferi* can generate extracellular traps (NETs) formed by the release of nuclear and cytosolic proteins and chromatin that ensnare and kill spirochetes (Menten-Dedoyart et al., 2012). The formation of NETs has been shown to occur through both NADPH oxidase-dependent and -independent mechanisms, suggesting a mechanism in addition to phagocytosis whereby neutrophils responding to *B. burgdorferi* in infected joints may kill spirochetes (Douda et al., 2015). Overall, the data suggest that neutrophils initially control *B. burgdorferi* numbers, but that there is a lack of sustained neutrophil infiltration into infected skin and perhaps other tissues. Identifying the mechanisms behind this lack of continued neutrophil activation will be important to understanding how these spirochetes can establish persistence.

Natural killer cells

Natural killer (NK) cells are derived from cells of the lymphocytic lineage during hematopoiesis but do not subsequently undergo antigen-receptor rearrangement compared to T and B cells (Abel et al., 2018). Instead, they act as cytotoxic innate immune cells with invariant receptors that recognize alterations in expression of major histocompatibility complex proteins on multiple types of host cells. These specificities of NK cell receptors allow recognition of host cells that have either become infected by different classes of microbial pathogens or that have been transformed to become tumorigenic, even in the absence of MHC expression. NK cell binding to these "altered" cells also promotes binding to Fas ligand. The combined effects of these interactions lead to release of cytotoxic granules and formation of pores, leading to cell lysis and/or apoptosis. Within the host environment, their activities are enhanced by exposure to a subset of proinflammatory cytokines secreted by activated tissue-resident phagocytes, including IL-12 and IL-18, which promote their activation and secretion of IFNγ. This makes NK cells important for early detection of intracellular pathogens, such as viruses. The literature assessing the importance of NK cells in controlling *B. burgdorferi* during acute infection is limited. Patients with early or later stage manifestations of untreated Lyme borreliosis possessed significantly more $CD16^+$ NK cells in their blood compared to patients that had been treated with antibiotics, and these cells expressed different cell markers associated with activation (Golightly et al., 1988). Interestingly, exposure of isolated NK cells to *B. burgdorferi in vitro* caused the suppression of their cytotoxicity. Crude preparations of murine primary NK cells from lymph nodes suggest they become activated *in vitro* after exposure to *B. burgdorferi* or their recombinant lipoproteins, as evidenced by production of IFNγ (Brown and Reiner, 1998a). Antibody-depletion of NK cells in mice did not appear to alter host control of *B. burgdorferi* numbers in multiple tissues, as measured by tissue outgrowth in BSK medium cultures (Barthold and de Souza, 1995; Brown and Reiner, 1998a). However, synovial tissues from Lyme borreliosis patients possess increased numbers of NK and T cells which secrete high levels of IFNγ, suggesting they do promote the inflammatory lesions associated with human Lyme arthritis and could contribute to spirochete killing (Katchar et al., 2013) (Jutras et al., 2019). NK cells also express inhibitory receptors and help to maintain self tolerance and potentially could contribute to resolution of inflammation (Pegram et al., 2011). Overall, while NK cells represent a resident immune cell type that can be activated in response to *B. burgdorferi*, their importance in controlling infection or mediating the pathology seen in Lyme borreliosis is still poorly understood.

Mast cells

Mast cells are tissue-resident innate immune cells that originate from hemopoietic stem cells (Frossi et al., 2017). They are primarily found in the connective tissues or mucosal surfaces near smooth muscle, glands, nerves, and blood vessels. When these cells become activated in response to their surface IgE molecules binding particular ligands, they degranulate to release a plethora of preformed cytokines, proteases, heparin, and multiple vasoactive materials. This results in rapid effects on proximal tissues, such as local blood flow, smooth muscle contraction, vessel permeability, and secretion activities. Because *B. burgdorferi* is introduced into and persists within skin tissue, it is possible that interactions with skin-resident mast cells might influence the development of *B. burgdorferi* induced inflammation. *B. burgdorferi* infection of gerbils resulted in increased numbers of mast cells in most tissues (Preac Mursic et al., 1990). Purified *B. burgdo*rferi lipoproteins promoted the upregulation of CD28 on bone marrow-derived mast cells (Marietta et al., 1997), an important costimulatory receptor for T cells, and exposure of rodent mast cells to intact *B. burgdorferi* elicited degranulation and increased TNFα secretion *in vitro* (Talkington and Nickell, 1999). A more recent study found that exposure of C57BL/6 mouse mast cells to *B. burgdorferi* spirochetes *in vitro* resulted in degranulation and the production of the inflammatory cytokine IL-6 and upregulation of MCP1 mRNA; addition of *I. ricinus* salivary gland extract inhibited these responses (Bernard et al., 2017). Experiments in mast cell deficient (Kit^{wsh}*-/-*) mice, however, revealed no significant impact on host resistance to infection, although *B. burgdorferi* was detected in the joints slightly earlier than in WT mice. Taken together, although early mast cell activation may occur in response to *B. burgdorferi* infection, there is limited data indicating these cells play a significant role in host defense or disease.

Microglia

B. burgdorferi are known to cause both peripheral and central nervous system pathology, collectively termed Lyme neuroborreliosis (Halperin, 2019). The central nervous system contains a diverse repertoire of myeloid cells, including microglia found in the brain parenchyma and non-parenchymal macrophages that can be found in the perivascular areas, the meninges and the choroid plexus (Prinz et al., 2011). To date, the responses of microglia are the only CNS macrophage type that has been studied *in vitro*. Microglia phagocytose invading pathogens and elicit inflammatory responses that can recruit a range of infiltrating immune cell types (Forrester et al., 2018). *In vitro* studies with murine microglial cell lines and primary murine or human microglia reveal that they are able to efficiently phagocytose *B. burgdorferi*, even in the absence of complement, and that spirochetes were efficiently trafficked into phagolysosomes and digested, followed by production of various proinflammatory mediators (Rasley et al., 2002; Kuhlow et al., 2005; Greenmyer et al., 2018). Phagocytosis of *B. burgdorferi* by primary microglia cells from rhesus monkeys correlates with increases in TLR1, TLR2, and TLR5 expression (Bernardino et al., 2008). Production of proinflammatory mediators by primary microglial cells in response to *B. burgdorferi* is augmented in the presence of NOD2 (Chauhan et al., 2009). Killed or damaged *B. burgdorferi* elicit a greater magnitude of inflammatory mediators compared to intact spirochetes, and these inflammatory responses are associated with increased apoptosis of bystander cells (Myers et al., 2009; Greenmyer et al., 2018). These findings, albeit derived from studies with microglia, suggest that CNS macrophages have the capacity to respond vigorously to *B. burgdorferi* and likely play a role in the host response in neuroborreliosis.

Soluble mediators that significantly affect B. burgdorferi infection

Interleukin-10

The production of pro-inflammatory cytokines is an important outcome of phagocyte-mediated killing of microbes due to their ability to amplify innate host defenses and to activate the adaptive immune response (Liu et al., 2004; Salazar et al., 2005; Moore et al., 2007; Cruz et al., 2008; Shin et al., 2008a; Salazar et al., 2009). These responses must eventually be suppressed, however, to prevent damage to the host tissues. One major anti-inflammatory cytokine that provides this function is IL-10, which is usually produced at later times after infection to downregulate the inflammatory responses and prevent excessive tissue injury. However, some pathogens are able to elicit early IL-10 production or encode an IL-10 homolog, either of which can prematurely shut down inflammatory/immune responses that are needed to effectively clear the infection. Thus, IL-10 production provides an important counterbalance to productive inflammatory responses and serves to modulate those responses to prevent damaging the host.

In vitro studies determined that *B. burgdorferi* and/or their lipoproteins potently elicit IL-10 production from macrophages and DCs (Chung et al., 2013). Similar trends in IL-10 production were observed in macrophages from mouse, human, and rhesus macaques incubated with *B. burgdorferi* and its lipoproteins, with optimal signaling dependent on both TLR2 and CD14 (Giambartolomei et al., 1998; Brown et al., 1999; Wooten et al., 2002b; Benhnia et al., 2005; Dennis et al., 2006). Interestingly, it was shown recently that living *B. burgdorferi* induce greater production of IL-10 by murine macrophages than do equivalent numbers of non-viable organisms (Lazarus et al., 2008). Addition of IL-10 to cell cultures suppresses pro-inflammatory cytokine production by macrophages and B cells in response to *B. burgdorferi*. The *B. burgdorferi*-induced production of IL-10 appears dependent on the host genetic make-up, and is greater in macrophages from mildly arthritic C57BL/6 mice than severely arthritic C3H mice (Brown et al., 1999; Lazarus et al., 2006). Kinetic studies indicated that IL-10 is elicited earlier in response to *B. burgdorferi* than the production of pro-inflammatory mediators, and correlated with prematurely suppressed pro-inflammatory responses (Lazarus et al., 2008; Chung et al., 2013). These findings are supported by the observation of greatly enhanced production of inflammatory mediators by macrophages and DCs from IL-10-deficient mice, as well as suppressed phagocytosis of *B. burgdorferi* and reduced expression of markers indicative of efficient activation of adaptive immune responses (Lazarus et al., 2008; Chung et al., 2013). Intravital confocal microscopy supports both the early kinetics of IL-10 production in infected mouse skin, with the majority of IL-10-producing cells falling within the phenotype of MHC class II-expressing cells, which are presumed to be macrophages/DCs (Wooten et al., unpublished).

The role of IL-10 in modulating arthritis severity in *B. burgdorferi*-infected mice is apparent, as infection of either IL-10$^{-/-}$C57BL/6 mice or IL-10$^{-/-}$C3H mice resulted in more severe disease than was observed in the appropriate WT strain (Brown et al., 1999) (Brown et al., 2008). The suppression of arthritis by IL-10 is not due to better control of spirochetes in tissues, as IL-10$^{-/-}$ mice harbored reduced numbers of bacteria in their tissues (Brown et al., 1999). IL-10 in infected joints was produced primarily by cells of the macrophage and CD4^{+} T cell lineage (Sonderegger et al., 2012b). The enhanced control of bacteria *in vivo* was dependent on innate host defenses, not to an observed heightened production of IgG (Lazarus et al., 2006). Thus, although release of the tight control of inflammation by the lack of IL-10 resulted in more severe arthritis, this was coupled with more effective control of the bacteria. The role of IL-10 in controlling arthritis development is not well understood, as regulation of IL-10 production is complex and depends on the cell type producing it (Moore et al., 2001). Mechanistic insight has come from *in vitro* studies in which SOCS1 and SOCS3 have been implicated in the IL-10-mediated suppression of *B. burgdorferi*-induced inflammatory responses in macrophages (Dennis et al., 2006), while others have found that IL-10 effects on the endothelium can influence inflammatory cell infiltration (Lisinski and Furie, 2002b). Apoptosis also was observed in human monocytes following treatment with lipoproteins or after phagocytosis of *B. burgdorferi*, suggesting another means of limiting inflammatory responses (Aliprantis et al., 1999; Cruz et al., 2008).

Complement system

The complement system refers to a group of 30+ proteins and their fragments that comprise an essential immune monitoring system to identify and control invading microbes, particularly those residing in extracellular spaces (Lin et al., 2020; Skare and Garcia, 2020). Activation can be initiated via binding of specific antibodies (i.e. classical pathway), binding by innate pattern recognition receptors such as mannose binding lectins (i.e. lectin-binding pathway), or by the continuous small-scale activation of complement component C3, which will bind surfaces and initiate activation unless inactivated by regulatory proteins (i.e. alternative pathway). Activation by any of these three pathways results in the formation of a C3-convertase which will generate activation of the enzymatically-amplified cascade of complement factors to result in a panoply of immune effects that can kill microbial invaders. One is the production of a membrane attack complex (MAC), which is a self-assembling complex that forms a pore in the microbial membrane, leading to a loss of membrane potential and/or lysis. Microbes that can resist this killing are known as serum-resistant, which is essential for establishing infection in most host tissues. Another mechanism for killing is via the effect of covalently bound fragments of the C3 protein (e.g. iC3b) on the microbial surface, which opsonizes the microbe to be recognized by phagocytes possessing complement receptors, which potently promotes phagocytosis and killing by those host cells. Thus, *B. burgdorferi* must possess a number of mechanisms to evade the killing mechanisms of complement to establish persistent infection (see also Chapter 13).

Virtually all virulent *B. burgdorferi sensu lato* can grow/survive in the presence of human serum, a feature attributed to its ability to resist complement activation and complement-mediated lysis in the absence of specific antibodies (Kochi et al., 1991; Patarakul et al., 1999; Stevenson et al., 2002). This property is common to many human pathogens and is consistent with the need for this organism to survive extracellularly and disseminate hematogenously. Murine studies have described the involvement of a number of different complement components with *B. burgdorferi* during infection. *In vitro* studies indicated that *B. burgdorferi* can initiate complement activation via the classical and alternative pathways (Kochi and Johnson, 1988). While certain spirochete-specific antibodies have been reported to directly kill or inhibit growth of *B. burgdorferi in vitro* (Sadziene et al., 1993; Sadziene et al., 1994), in most cases complement is required for antibody-mediated killing of these bacteria (Kochi et al., 1993). Mice deficient in C3, which is central to all downstream complement-mediated immune mechanisms, harbored greater numbers of *B. burgdorferi* in some tissues and required five-fold fewer *B. garinii* spirochetes to establish infection (Lawrenz et al., 2003; Woodman et al., 2007). The fifth component of complement, C5, which is essential for formation of the membrane attack complex, is not required for antibody-mediated protection from infecting *B. burgdorferi* (Bockenstedt et al., 1993). Interestingly, serum-resistant *B. burgdorferi* strains express a CD59-like protein which can bind complement components C8 and C9, thus inhibiting assembly of the MAC (Pausa et al., 2003). *B. burgdorferi sensu lato* species also expresses three additional proteins (i.e. CspA, BGA66, and BGA71) that directly bind different components of the MAC, thus preventing the assembly of these lytic complexes (Hallstrom et al., 2013; Hammerschmidt et al., 2016). Together, these studies in mice suggest a major role for complement as an opsonin to promote clearance of the organism by phagocytes during early infection, rather than in direct lysis of the bacteria via the MAC, which is only seen at significant levels in the presence of *B. burgdorferi*-specific antibodies *in vitro* (Kochi et al., 1991).

B. burgdorferi is supremely adapted to evasion of complement mechanisms by producing a plethora of proteins capable of binding multiple host complement-regulatory proteins (see chapter 13) (Lin et al., 2020). These proteins include complement regulator-acquiring surface proteins (CRASPS) that bind factor H, OspC that binds C4b, BBK32 that binds the C1 complex and a 43 kD protein that binds the C4b binding protein C4bp. The combined effects of these interactions renders variable levels of resistance to clearance of spirochetes by the classical, alternative, and lectin-mediated pathways. Spirochete binding of host factor H prevents C3-convertases from accumulating on their surfaces, and thus renders them resistant to clearance by the alternative pathway. Overall, it is apparent that *B. burgdorferi* has invested much evolutionary currency to allow inactivation of complement immune activation. Collectively, these findings also suggest that therapies that can quell a subset of these activities might be able to allow the host innate responses to clear these infections. Recent examples of this are studies mutating the factor H-binding site of *B. burgdorferi* protein CspZ (i.e. CRASP-2) so that it is unable to bind factor H, and using this as a vaccine in mice (Marcinkiewicz et al., 2018; Marcinkiewicz et al., 2020). By preventing immediate complexing of this mutant protein with factor H within the mouse, antibodies are generated against multiple epitopes on the CspZ protein, which appears to confer resistance when challenged with *B. burgdorferi*.

Responses of acquired (adaptive) host defenses

T and B lymphocytes of the adaptive immune system and their secreted products are critical for controlling infections. These cells regulate anti-pathogen defenses following their highly antigen-specific, pathogen-induced activation, clonal expansion and eventual differentiation into various effector populations. These effectors control acute infections and generate memory responses that enable a host to more vigorously respond to repeat encounters with the same pathogen. The interaction of CD4 T cells with B cells is critical for the induction of optimal antibody responses, as well as the establishment of robust B cell memory. Experiments with SCID mice lacking T and B lymphocytes confirmed the importance of the adaptive immune system in controlling *B. burgdorferi* infections, as SCID mice cannot control *B. burgdorferi* spirochete burdens in blood and tissues and eventually succumb to uncontrolled *B. burgdorferi* expansion and the resulting inflammatory responses (Schaible et al., 1989; Schaible et al., 1990).

Yet, despite the fact that *B. burgdorferi* infection results in the generation of vigorous and specific antibody responses in their natural reservoir species, including small rodents such as *Peromyscus leucopus* and birds, infections are not cleared. Instead, as shown with studies on commonly used laboratory mouse strains, such as 129, BALB/c, C57BL/6 and C3H, tissue loads of *B. burgdorferi* are suppressed, but remain at levels that allow their acquisition by feeding ticks for many months. Clinical studies in dogs, horses, non-human primates and humans demonstrated that chronic, non-resolving infection can occur also in these incidental non-reservoir hosts (Appel et al., 1993; Barthold et al., 1993; Roberts et al., 1995; Straubinger et al., 1997; Radolf et al., 2012). Seroprevalence studies in humans as well as dogs residing in endemic areas have shown high exposure rates. Comparisons between IgG seroprevalence in individuals following a tick-bite and incidence of Lyme borreliosis suggests that only a fraction of exposed individuals who generate IgG will develop disease.(Fahrer et al., 1991; Rath et al., 1996; Huegli et al., 2011). Thus, while lifelong persistent infection in reservoir species is the likely outcome, in non-reservoir species outcomes vary from asymptomatic seroconversion to debilitating disease.

In persistent non-resolving infections of reservoir species, the clinical manifestations of disease are largely suppressed by the adaptive immune response. While spirochete burdens are reduced significantly, the pathogen is not eliminated. Importantly, the spirochete burden of disease-susceptible mice, such as the C3H mouse, which shows significant arthritis development after *B. burgdorferi* infection, is indistinguishable from that of disease-resistant laboratory mouse strains when using PCR to quantify Borrelia DNA (Barthold et al., 1993). Thus, disease associated with *B. burgdorferi* infection is not a manifestation of the pathogen causing damage, but rather the result of the inflammatory response to the pathogen. In C3H mice, for example, arthritis development is caused by a deficiency of lysosomal beta-glucoronidase. This leads to a failure to clear glycosaminoglycans in the joints of infected mice, which is a damage-associated molecular pattern (DAMP) that causes continued activation of inflammatory stimuli, resulting in tissue damage (Bramwell et al., 2014).

In sum, the adaptive immune response to *B. burgdorferi* infection is ineffective in clearing the pathogen from its natural reservoir species yet can protect the host against reinfection. In the infected host, it controls spirochete numbers, and thereby reduces the triggers of inflammatory responses that result in disease. In the clinical pathology characteristic of Lyme borreliosis, T cells may also become drivers of an ongoing inflammatory process. The features of adaptive immunity to *B. burgdorferi,* which have been studied mostly in mice, and which are summarized below, cannot be seen as reflecting an effective host defense that results in pathogen clearance. Rather, they represent the outcomes of a complex interplay of *B. burgdorferi* immune evasion strategies and the mammalian adaptive immune system that results in a state of concomitant immunity. This interplay allows *B. burgdorferi* to fulfill its complex lifecycle without inciting disease in the reservoir host.

B Cell Responses

B cells are critical for controlling *B. burgdorferi* tissue burden. Mice selectively lacking B cells (*Igh6-/-* mice) exhibited markedly elevated pathogen burdens in various tissues. B cells were also shown to be critical for preventing and/or resolving arthritis, carditis and myocarditis (McKisic et al., 2000). Their actions can be augmented by T cells, as carditis resolution was promoted by the adoptive transfer of IFNγ−producing T cells into T cell deficient mice, presumably through activation of macrophages, which are prominent in the diseased heart tissue (Bockenstedt et al., 2001). A role for antibodies as the main effectors produced by B cells was demonstrated in numerous passive transfer experiments, in which serum from previously infected mice or humans, but not non-infected controls, or the transfer of B and T cells, but not T cells alone, into SCID or T and B cell-deficient *rag1-/-* mice reduced pathogen burden and resolved arthritis (Schaible et al., 1989; Barthold et al., 1992b; McKisic and Barthold, 2000; Bockenstedt et al., 2001). Antibodies induced by *B. burgdorferi* infection are bactericidal, i.e. can kill the bacteria through both direct, complement-dependent and complement-independent spirochete lysis (LaRocca and Benach, 2008). They also can opsonize the spirochetes for uptake and degradation by macrophages and neutrophils following engagement of their surface-expressed FcγR or complement receptors (Benach et al., 1984b; Modolell et al., 1994).

Antigenic targets of the antibody response to *B. burgdorferi* infection.

Early clinical studies demonstrated strong serum antibody-antigen immune complex formation in patients with disseminated *B. burgdorferi* infection (Hardin et al., 1979; Hardin et al., 1984; Brunner and Sigal, 2001), indicating that robust antigen-specific humoral immune responses can be initiated among that group of patients. Indeed, strong early B cell responses have been associated recently with faster resolution of disease following antibiotic treatment (Blum et al., 2018). However, seroconversion takes time. Previous studies suggested that it is detectable in only about half of patients with a localized EM at the time of presentation (Aguero-Rosenfeld et al., 2005; Steere et al., 2008; Wormser et al., 2008; Pegalajar-Jurado et al., 2018; Nigrovic et al., 2019), although development of more sensitive modified-testing modalities have increased these numbers (Pegalajar-Jurado et al., 2018). Convalescent blood samples obtained after antibiotic treatment increases the percentage of patients testing positive for Lyme borreliosis. Lack of seroconversion has been observed in patients following rapid initiation of antibiotics, when the infection may have remained localized in the skin. In experimentally infected mice, seroconversion takes 10 -14 days, despite a robust and early B cell activation in the lymph nodes. Yet, by ELISPOT analysis, increased but transient local production of IgM and IgG was measurable in lymph nodes draining the site of infection by day 7. The data suggest that production of antibodies even in lymphoid tissues is not always reflected in the serum (Tunev et al., 2011; Hastey et al., 2012).

Comprehensive serological analysis of the immunogenic targets of *B. burgdorferi*-specific IgG from mice and humans, conducted with a genome-wide protein expression array, demonstrated the remarkable conservation of antibody targets between these species. Induction of antibodies was estimated to occur against ~15% of *B. burgdorferi* gene open reading frames (ORFs) (Barbour et al., 2008). This number is likely an underestimate, however, given the restrictions in this approach to screen effectively for all conformational epitopes. Nonetheless, the study provides the most comprehensive analysis yet on the antibody targets of *B. burgdorferi*. Previously identified surface-exposed lipoproteins such as OspC, p35, p37, p39, BmpA, BmpB and decorin-binding proteins (Dbp) A and B, which can induce antibodies with passive protective capacity (Fikrig et al., 1990; Fikrig et al., 1992; Fikrig et al., 1997b; Hanson et al., 1998; Pal et al., 2008), were strongly represented among *B. burgdorferi* peptide targets. In addition, the array analysis also identified strong IgG responses to previously unknown *Borrelia* proteins, including BBK07 among others, which were shown to be mostly absent from culture-grown *B. burgdorferi* but rapidly and strongly induced following mammalian infection (Barbour et al., 2008; Coleman et al., 2011). The ability of some of these antigens to elicit protective immunity after immunization appears to be *B. burgdorferi* strain-dependent. For example, neither OspC nor BmpA from *B. burgdorferi* N40 could elicit protective immunity, while they were protective to challenge with the B31 strain (Barthold et al., 1997; Bockenstedt et al., 1997).

The dynamic up- and down-regulation of surface proteins by *B. burgdorferi* affects the specificity of the antibody response induced to the infection and must be considered an immune evasion strategy. One prominent example is the expression of OspC, a protein required for mammalian infections (Grimm et al., 2004). Its expression is induced in the tick after initiation of the blood meal (Schwan et al., 1995; Pal et al., 2004), while OspA is downregulated (de Silva et al., 1996). Mutant *B. burgdorferi* with stabilized expression of OspC are rapidly cleared from the host (Embers et al., 2008). Once infection is established, however, expression of OspC is rapidly lost from *B. burgdorferi* in lymph nodes, where B cell responses are initiated (Skare et al., 2016). This suggests that the loss of this immunogenic surface protein reduces induction of OspC-specific antibodies, thereby avoiding antibody-mediated clearance during the earliest stages of *B. burgdorferi* infection, when Osp C is indispensable for the *B. burgdorferi* life-cycle (Tilly et al., 2006). Consistent with the expression pattern of OspC, IgG responses to this protein are overall low in infected mice compared to antibody responses to other surface proteins of *B. burgdorferi* (Barbour et al., 2008; Tunev et al., 2011).

Measurement of antibody responses to the surface-expressed VlsE protein of *B. burgdorferi* is used extensively for serological testing. The stalk region of this protein (the "C6" peptide) is highly invariable, immunogenic and appears to be constitutively expressed by *B. burgdorferi*. Most antibody responses to VlsE, however, are directed against the highly immunogenic protein head region, which is antibody accessible on the spirochete outer membrane (Eicken et al., 2002). That region undergoes extensive and continued antigenic variation, owing to the continued segmental recombination from a contiguous array of multiple *vlsE* silent-expression cassettes (Zhang et al., 1997; Zhang and Norris, 1998b, a; Bankhead and Chaconas, 2007; Coutte et al., 2009) (see chapter 13). The process of antigenic variation is considered a prominent mechanism through which *B. burgdorferi* evades immune clearance, as it results in the generation of strong, but ultimately ineffective antibody responses that are induced to antigens expressed only for short periods of time (Norris, 2006; Bankhead and Chaconas, 2007; Coutte et al., 2009; Rogovskyy and Bankhead, 2013; Magunda and Bankhead, 2016). Thus, although a large number of antibodies against a variety of surface proteins of *B. burgdorferi* are induced, and the antibodies engage with and eliminate large numbers of spirochetes over the course of infection, the pathogen uses a multitude of immune evasion strategies to reduce the functionality of the antibodies.

Characteristics of the antibody response to *B. burgdorferi*

Serum antibodies to *B. burgdorferi* are of the IgM, IgG and IgA isotypes in humans, with IgA responses most prevalent in early Lyme borreliosis in patients presenting with EM (Steere et al., 1979; D'Arco et al., 2017). IgM and IgG responses have been detected in the serum of 30-80% of patients at the initial visit (Aguero-Rosenfeld et al., 1996; Barbour et al., 2008; Tjernberg et al., 2009), which in the case of patients presenting with EM, is usually within the first 10-14 days after a tick bite. In experimentally-infected dogs, IgM responses were measurable as early as 7 days post infection and IgG by around day 10, similar to experimental infection of mice (Greene et al., 1988). As infection progresses beyond 6 weeks, IgM responses may remain elevated or may return to normal (Aguero-Rosenfeld et al., 1996). In untreated infection in humans, *B. burgdorferi*-specific IgG titers increase and are highest during the late phase of disease, when arthritis is present (Craft et al., 1986; Aguero-Rosenfeld et al., 2005). IgG titers may remain elevated for years after treatment despite clinical remission, or they may return to background levels following antibiotic treatment (Hammers-Berggren et al., 1994; Aguero-Rosenfeld et al., 1996; Nowakowski et al., 2003). Intrathecal IgM, IgG and IgA are also found in humans with neurological manifestations of the infection and are of high diagnostic value (Steere et al., 1990; Schutzer et al., 1997). Intrathecal antibodies to *Borrelia* are induced following direct intracerebral inoculation of spirochetes into mice but not when infection is introduced peripherally (outside the central nervous system) (Li et al., 2006).

A poorly-understood characteristic of the murine antibody response to *B. burgdorferi*, also reported for humans, is the continued presence of T-independent *Borrelia*-specific serum IgM, even months after initial infection (Steere, 2001; Liu et al., 2004; Gajovic et al., 2010; Hastey et al., 2020). Given that the half-life of serum IgM is short, on the order of 12 – 48 hours in mice, the continued presence of IgM in persistently-infected mice suggests either production of long-lived IgM secreting plasma cells or the continued *de novo* formation of plasmablasts. Because anti-*Borrelia* IgM responses are largely T-independent, the latter appears more plausible but would require the continued presence of *Borrelia* antigens. Of importance, the ratio of IgM and IgG therefore does not differ significantly between acute and persistent *B. burgdorferi* infection in mice. Ongoing IgM responses seem to control bacteremia, but do not appear to contribute to control of *Borrelia* tissue-burden, presumably because this very large molecule (850 kD) cannot gain access to tissues, except in situations of strong inflammation and associated vascular leakage (Blandino and Baumgarth, 2019; Hastey et al., 2020).

As stated above, hypergammaglobulinemia, often associated with chronic and persistent infections, and consisting of both *Borrelia*-specific and non-specific IgG, was found also in *B. burgdorferi*-infected mice (Hardin et al., 1979; Schutzer et al., 1990; Brunner and Sigal, 2001; Schutzer and Luan, 2003; Soulas et al., 2005). The mechanisms underlying the development of hypergammaglobulinemia, and the role this plays in effective immunity to *B. burgdorferi* is unknown. Studies on a chronic viral infection model in mice, the pathogen LCMV, suggested that hypergammaglobulinemia and strong production of self-reactive antibodies is dependent on CD4 T cell activation, and that this strong T cell induction paradoxically suppresses effective neutralizing antiviral antibody responses (Recher et al., 2004). Whether a similar effect might reduce the effectiveness of anti-*Borrelia* responses has not been tested.

B cell activation during *B. burgdorferi* infection

Following *B. burgdorferi* infection, B cells are activated and lymph nodes dramatically increase in size and cellularity, with much of the increase contributed by the accumulation of B cells (Sigal et al., 1988; Yang et al., 1992; Tunev et al., 2011). In contrast, B cell activation and induction of *B. burgdorferi*-specific antibody-secreting cells in the spleen are rarely observed in response to tick-borne infections or infection with host-adapted spirochetes via tissue-transplantation (Tunev et al., 2011). However, activation of marginal zone B cells resulting in secretion of IgM was reported in response to infection after intradermal inoculation of cultured *B. burgdorferi*, presumably in response to blood-borne spirochetes (Belperron et al., 2007; Malkiel et al., 2009). The extensive B cell accumulation observed in the lymph nodes was shown to depend on infection-induced, B cell extrinsic Type I IFN signaling, likely affecting lymph node stroma (Hastey et al., 2014). Co-cultures of B cells with culture-grown *B. burgdorferi* causes mitogenic B cell activation dependent largely on TLR2-mediated signals. However, the observed robust antibody response to *B. burgdorferi* infection of mice lacking either TLR2 or the adaptor protein MyD88, indicated that B cell activation and differentiation *in vivo* is not driven, or not driven solely, by these innate signals (Woods et al., 2008). Recent studies using transgenic mice expressing B cells specific for the hen egg lysozyme (HEL) demonstrated that HEL-reactive, *Borrelia* non-specific B cells are activated *in vivo* following infection (Soulas et al., 2005; Jung et al., 2016). These cells secreted IgM and could undergo class-switch recombination to IgG; however, the IgG-expressing cells were unable to differentiate into antibody-secreting cells. Thus, increases in transgene-induced, non-*Borrelia*-reactive antibodies following *B. burgdorferi* infection were largely of the IgM isotype, not considered to be pathogenic, and the HEL-specific B cells producing these antibodies contributed to the pool of activated B cells after infection (Soulas et al., 2005; Jung et al., 2016). This is consistent with another *in vivo* study, which concluded that much of the strong infection-induced IgG response in lymph nodes of B. *burgdorferi*-infected mice is *Borrelia*-specific, and, thus, the outcome of specific antigen-B cell interactions (Tunev et al., 2011). Paradoxically, mice lacking MyD88, the adaptor molecule of most TLRs, including TLR2, showed a dramatically enhanced hypergamma-globulinemia, which was shown to be a B cell-extrinsic effect (Woods et al., 2008), possibly triggered by the presence of increasing amounts of *Borrelia* antigens in these mice.

Some of the differences observed between B cell stimulation *in vivo* and *in vitro* may be explained by the extensive differences in *Borrelia* surface protein expression between culture-grown and mammalian-host adapted spirochetes (Samuels, 2011; Iyer and Schwartz, 2016), including expression of OspA, a mitogen for B cells (de Souza et al., 1992) that is rapidly downregulated after uptake of the bloodmeal and following transmission to the mammalian host (de Silva et al., 1996; Caimano et al., 2019). Thus, OspA is unlikely to drive early B cell responses *in vivo*, as it is not significantly expressed following infection of the mammalian host. (Schwan et al., 1995), consistent with a lack of antibody response development to OspA following natural infection or infection with host-adapted spirochetes. Taken together, *B. burgdorferi* infection typically results in the strong activation of lymph node B cells, including both *Borrelia*-specific and *Borrelia* non-specific, possibly self-reactive B cells. Their activation is affected by extrinsic and intrinsic inflammatory signals that act in concert with *Borrelia* antigens to drive strong B cell responses in the lymph nodes of infected mice. These responses result in strongly enhanced serum concentrations of both IgM and IgG that remain elevated throughout infection (Tran V.G. and Baumgarth, N., unpublished).

T-dependent follicular B cell responses to *B. burgdorferi*

Follicular B cell responses are classified into either extrafollicular or germinal center (GC) B cell responses. Extrafollicular responses result in the rapid formation of foci containing short-lived plasmablasts in the T-B border and the medullary cord area of lymph nodes, and can be induced in either a T-dependent or T-independent manner. In contrast, GC responses are dependent on cognate CD4 T cell help from T follicular helper cells (T_{FH}). GC responses are required for antibody somatic hyper-affinity maturation and result in the development of antibody-secreting long-lived plasma cells and recirculating memory B cells.

Despite the appreciated need for CD4 T cell help in the induction of strong B cell responses, antibodies that arise after *B. burgdorferi* infection in the absence of T cell help can support the resolution of arthritis and can, after passive transfer, protect mice from challenge infections (Fikrig et al., 1996; McKisic and Barthold, 2000). There are two possible explanations for these puzzling findings: First, most if not all T cell responses to *B. burgdorferi* are initiated in a T-independent manner through formation of extrafollicular foci; or second, CD4 T cell help is insufficient to enhance the functionality of the humoral response to *B. burgdorferi*. Support for induction of largely T-independent responses to *B. burgdorferi* was provided with studies in mice lacking CD40L, a critical co-stimulatory molecule expressed by T cells that induces Ig class-switch recombination and B cell activation (Fikrig et al., 1996). Antibodies generated to *B. burgdorferi* infection in these mice retained their ability to protect mice from a challenge infection. Similarly, serum from *B. burgdorferi*-infected mice lacking either all T cells, lacking CD4 T cells, or lacking MHCII, retained passive protective capacity (Fikrig et al., 1996; Fikrig et al., 1997a; McKisic and Barthold, 2000). Furthermore, strong extrafollicular plasmablast responses form in lymph nodes after *B. burgdorferi* infection (Tunev et al., 2011). However, IgG class-switching, with the exception of switching to IgG2b, was severely reduced in mice deficient in CD40L, MHCII, CD4 or all T cells (Fikrig et al., 1996; Fikrig et al., 1997a; McKisic and Barthold, 2000). Also, *B. burgdorferi* tissue burden was strongly increased in CD4 T cell-deficient mice (Elsner et al., 2015a); arthritis could not be resolved in SCID mice after passive-serum transfer (Fikrig et al., 1996; McKisic and Barthold, 2000; Elsner et al., 2015a); and carditis-resolution was delayed (Fikrig et al., 1997a). These data do not refute the notion that CD4 T cells are induced and likely important for the humoral response to *B. burgdorferi*. Rather, the passive protective capacity of antibody responses to *B. burgdorferi* is achieved even in the absence of IgG class switch recombination and the formation of T cell-dependent B cell responses. Thus, passive protection might not require antibody functionalities associated with antibody-tissue entry, which is needed for resolution of arthritis and carditis, as IgM is passively protective, or strongly hyper-affinity-matured antibodies, a process that requires T cell-dependent GC responses.

Taken together, these findings suggest that CD4 T cell helper responses to *B. burgdorferi* are insufficient to induce fully functional antibodies able to clear *B. burgdorferi* infection. In support of that finding, it was shown that functional GC responses do not develop after *B. burgdorferi* infection in mice (Hastey et al., 2012; Elsner et al., 2015a; Elsner et al., 2015b). Instead, they arise relatively late, around day 15 of infection, and then rapidly collapse by day 30, despite ongoing infection and despite the formation of CXCR5+ PD-1hi Bcl6+ CD4+ T_{FH} (Hastey et al., 2012; Elsner et al., 2015a; Elsner et al., 2015b; Hastey et al., 2016). This lack of sustained GC causes a failure to sustain antibody affinity maturation, and a failure to induce memory B cells and long-lived plasma cells (Hastey et al., 2012; Elsner et al., 2015b). A failure to induce memory B cell responses may explain why, following antibiotic treatment, *Borrelia*-specific serum antibodies are often lost and, as shown for experimental mice, hosts can become susceptible to reinfection (Piesman et al., 1997; Elsner et al., 2015b). The lack of effective T-dependent B cell response induction may contribute to the inability of the host to clear *B. burgdorferi* infection. It will be important to determine the mechanisms underlying the inability of *B. burgdorferi*-infected mice to induce robust T-dependent B cell responses and to assess whether their induction could result in clearance of *B. burgdorferi* from the mammalian host.

Pre-existing and innate-like B cell immunity to *B. burgdorferi*

Humoral defense begins prior to deposition of spirochetes within the host, when ticks ingest interstitial fluid containing IgG or hemorrhaged blood, which contains all Ig subclasses, at the feeding site (Belperron and Bockenstedt, 2001). Infected ticks that feed on B cell-deficient mice (B6.*Igh6-/-)* harbor more spirochetes than those that feed on WT mice. Moreover, passive transfer of normal mouse serum into B cell-deficient mice reduces spirochete burden in feeding ticks. Pre-existing antibodies might be generated through long-lived plasma cells, or the presence of circulating IgG following recent exposure to *B. burgdorferi*. Even individuals never exposed to *B. burgdorferi*, however, harbor so-called "natural" antibodies, mostly of the IgM isotype, that exist in all individuals. Natural antibodies are secreted in mice by a distinct fetal-derived B cell subset termed B-1, which have the capacity to generate long-lived antigen-specific immunity in the absence of T cell help (Alugupalli, 2008). They also can reduce early pathogen burden to various pathogens, likely because of their ability to bind structures common to many pathogens, thus functioning similar to innate soluble pattern recognition receptors (Baumgarth et al., 2005; Baumgarth, 2011). The presence of natural *B. burgdorferi*-binding IgM, generated in the absence of any previous exposure, can complicate IgM-based serodiagnosis for Lyme borreliosis. Natural IgM is sufficient to cause *B. burgdorferi* reductions in the ticks during blood uptake, whereas IgG antibodies have no effect (Belperron and Bockenstedt, 2001). IgG can pass into the tick haemolymph, where it is bound by immunoglobulin-binding proteins and is excreted in the tick saliva back into the host, a mechanism by which the tick may interrupt the host immune response to tick feeding (Wang and Nuttall, 1999).

T cell responses

In contrast to the B cell's ability to directly engage with *Borrelia* antigens, T cells require antigen-presentation by conventional or non-conventional MHC molecules. Conventional T cell activation entails the processing of protein antigens into peptides, followed by peptide loading into MHC I and II complexes, which allows their presentation by antigen-presenting cells (APC), such as dendritic cells, macrophages and B cells, to αβ T cell receptors (TCR) expressed on CD8 and CD4 T cells, respectively. The presentation of peptides by MHC II on B cells, following antigen binding and subsequent internalization by the BCR, is the hallmark of CD4 T-dependent B cell responses and results in the differentiation of CD4 T follicular helper cells (T_{FH}) during *B. burgdorferi* infection. It also causes the activation of various CD4 and CD8 effector T cells (Teff), and populations of CD4 T regulatory (Treg) cells, the latter considered essential in controlling overshooting inflammation and tissue-damage, which are discussed below.

"Unconventional" antigen presentation involves the presentation of lipids and glycolipids to T cells (Park and Kim, 2018; Godfrey et al., 2019). These types of antigens cannot be presented on MHC I and MHC II, yet are common antigens expressed by *B. burgdorferi*. These antigens are processed for presentation by non-classical MHC Ib molecules such as CD1d, as well as Qa-I in mice, the ortholog of the non-polymorphic HLA-E of humans, which activate "natural killer" T cells (Park and Kim, 2018). Another non-classical MHC Ib molecule is MR1, which can bind to and present small bacterial metabolites and activates so-called Mucosa-Associated Invariant T cells (Godfrey et al., 2019). While CD1-restricted T cell activation has been shown for *B. burgdorferi* infection, activation of MAIT cells has not been described.

T cells utilizing the γδ TCR are another type of "unconventional" T cell reported to play a role in *B. burgdorferi* infections. Identifying the antigens recognized by the γδ TCR remains the subject of ongoing investigations. In contrast to αβ TCR and other non-conventional T cells, γδ TCR antigen-binding appears to be more akin to that of the BCR. About 0.5% of γδ T cells in mice can directly bind to the inducible non-classical MHC Ib molecule T10 and the related T22, which lack an antigen-binding groove (Crowley et al., 2000). Recognition of lipid antigens in CD1 has also been observed (Divan et al., 2015). However, they can also bind directly to protein antigens (phycoerythrin) as well as haptens (Zeng et al., 2012; Chien et al., 2014; D'Souza et al., 2019). A recent study, using a tetramerized human γδ TCR expressed by a T cell clone derived from the synovial fluid of a patient with chronic Lyme arthritis, reported expression of IL-1β-induced cell surface ligand(s) on activated monocytes and T cells, although their exact identity remains to be revealed (Collins et al., 2019).

Responses to *B. burgdorferi* by nonconventional T cells

Antigens presented by CD1d to invariant TCR "natural killer" T (iNKT) cells, and by MR1 to MAIT cells, first demonstrated activation of T cell subsets with highly restricted TCR repertoires and are considered innate-like. The *B. burgdorferi*-derived glycolipid diacylglycerol (BbGL-II) was one of the first pathogen-derived antigens identified as being presented by CD1d to mouse and human invariant iNKT cells, causing their activation (Kinjo et al., 2006). iNKT cells were shown to be activated in response to the infection and to produce IFNγ (Tupin et al., 2008; Olson et al., 2009). Activation of iNKT cells appeared to be, at least in part, dependent on TLR2 signaling, as TLR2-deficient mice showed reduced accumulation of NKT cells in the arthritic joints, although their frequencies in the heart tissue was unchanged (Lasky et al., 2016). Production of IFNγ by iNKT cells may enhance macrophage-mediated *B. burgdorferi* uptake and thus could reduce *Borrelia* tissue burden; it can also drive the upregulation of CD1d as well as alter the chemokine milieu, thereby affecting the recruitment of macrophages and neutrophils (Sabino et al., 2011). However, IFNγ production by iNKT cells was not always observed (Whiteside et al., 2018). Lack of CD1d expression, and thus lack of iNKT cells, or deficiency in the major Vα14-expressing iNKT cell subset, resulted in enhanced arthritis development and increased spirochete tissue burden in mouse strains that are normally resistant to disease (C57BL/6 and BALB/c mice). Increased spirochete burden may underlie the observed increases in serum antibody titers against *B. burgdorferi* in these mice, including antibodies to the CD1d-ligand BbGLII (Kumar et al., 2000; Tupin et al., 2008). Together, the data suggest a protective role for CD1d-restricted iNKT cells; however, the mechanisms of that protective role remains to be revealed more fully.

In addition to CD1d-restricted invariant NKT cells, also termed "Type I NKT", more recent studies using tetramerized CD1a, CD1b and CD1c from humans (mice only express CD1d), identified CD1-restricted αβ T cells with a diverse TCR repertoire (dNKT, or

Type II NKT). These dNKT recognize antigens, including self-antigens, presented by CD1a, b or c, (Mori et al., 2016; Dhodapkar and Kumar, 2017). *B. burgdorferi* infection was shown to drive expression of CD1a-c on myeloid cells indirectly by stimulating the TLR2-induced secretion of IL-1β (Yakimchuk et al., 2011) and may drive the activation of CD1b-restricted NKT cells with self-antigen recognition potential in human patients (Reinink et al., 2019). The significance of these cells and their contributions to immunity to *B. burgdorferi* infection and/or infection-induced disease remain to be established.

T cells expressing the γδ TCR develop from the thymus throughout life, although their development includes early developmental waves that seed various tissues, including the skin, where they may interact with *B. burgdorferi* shortly after its release from the tick into the mammalian host (Divan et al., 2015). The main focus of studies on γδ T cells in infection has been on lymph tissue of mice and synovial fluid of patients with Lyme arthritis (Divan et al., 2015). Collectively, the results suggest that γδ T cells respond to cell stress and inflammatory responses triggered indirectly through TLR activation, as well as release of pro-inflammatory cytokines, such as IL-1β, that occurs during arthritis development. These signals seem to trigger the upregulation of γδ TCR ligands on numerous cell types, including monocytes, T cells, dendritic cells as well as fibroblasts (Collins et al., 2016; Collins et al., 2019). Similar to findings with CD1d-restricted iNKT cells, γδ T cells may contribute to production of IFNγ in tissues during *B. burgdorferi* infection (Bockenstedt et al., 2003).

Conventional T cell responses

The contributions of αβ T cells to host defense during *B. burgdorferi* infection remains incompletely defined. Studies with mice lacking all T cells (TCR β/δ -/-) or only αβ TCR-expressing cells, or with mice depleted of CD4 or CD8 T cells, demonstrated surprisingly modest effects of T cell-deficiency on the course of *B. burgdorferi* infection. These effects included some enhancements in spirochete tissue burden and increased arthritis severity (Keane-Myers and Nickell, 1995; Lim et al., 1995; Fikrig et al., 1997a; McKisic and Barthold, 2000; Elsner et al., 2015a). As discussed above, given that CD4 T cells are required for Ig class-switch recombination to most IgG subclasses, at least part of the effect on *Borrelia*-load can be explained by a failure to induce IgG in these animals. Antibody-independent mechanisms of host defense, however, cannot be ruled out (Keane-Myers and Nickell, 1995). These apparently modest effects of T cells on the course of *B. burgdorferi* infection are surprising, given the importance of CD4 T cells as crucial effectors and regulators of immune responses (Sakai et al., 2014; Tubo and Jenkins, 2014; Sallusto, 2016), and since effective B cell immunity usually requires T-dependent co-stimulatory signals that support Ig-class switch recombination, affinity-maturation, and the development of B cell memory and long-lived plasma cells. As outlined above, given that many of the features of a T-dependent B cell response are lacking in *B. burgdorferi* infected mice (Hastey et al., 2012; Hastey et al., 2014; Elsner et al., 2015a; Elsner et al., 2015b), together the data indicate that a fully-functional T cell response fails to develop or is suppressed during the infection, which could contribute to the development of *Borrelia* persistence.

Effector CD4 T cell responses are shaped by the inflammatory milieu in which they develop. Infections with extracellular pathogens, such as *B. burgdorferi,* are thought to be controlled by CD4 T cells polarized to express the transcriptional regulator RORγt and secreting the hallmark cytokine IL-17. Through secretion of cytokines, these cells regulate neutrophil accumulation at sites of infection, which then eliminate the pathogen via phagocytosis and enzyme-mediated killing. Exuberant and/or dysregulation of the pro-inflammatory IL-17-driven responses may also cause tissue damage (Chamoun et al., 2018). Remarkably, IL-17-deficiency on either the C57BL/6 or the C3H background did not significant impact *B. burgdorferi* tissue burden, course of disease, or the magnitude of the anti-*B. burgdorferi* antibody responses (Lasky et al., 2015a). As discussed in the innate immunity section of this chapter, the presence of *B. burgdorferi* in the skin of infected mice does not trigger continued accumulation of neutrophils or other effectors to the site of infection, further suggesting that TH17 responses are not significantly induced during *B. burgdorferi* infection in this model. However, the interpretation of the data is complicated by the fact that lack IL-17/RORγt can lead to reciprocal increases in T-bet expression and IFNγ production, which may mask some of the functions of IL-17 in *B. burgdorferi* infection by supporting a TH1-like pro-inflammatory response that contributes to *Borrelia* clearance.

Early studies suggested that induction of TH1 responses could drive the transient inflammatory pathology seen in mice infected with *B. burgdorferi*, and that a dominance of TH2 responses attenuated that response (Matyniak and Reiner, 1995). A pro-inflammatory role for IFNγ-producing T-bet+ CD4 TH1 cells in *B. burgdorferi* infection-induced inflammatory disease was suggested by the strong association between IFNγ−producing CD4 T cells with Lyme arthritis and carditis in mice and humans (Lim et al., 1995; Anguita et al., 1996; Gross et al., 1998b; McKisic et al., 2000). While adoptive transfer of T cells into *B. burgdorferi*-infected mice deficient in B cells led to significant inflammatory pathology, transfer of both B and T cells resolved inflammation, indicating that B cells are required for disease-ameliorating effects (McKisic et al., 2000). In another study, the transfer of IFNγ−producing CD4 T cells into T cell-deficient mice promoted carditis regression, not inflammation; this may be related to their effects on promoting macrophage-mediated clearance of *B. burgdorferi* at this site (Bockenstedt et al., 2001). Furthermore, treatment of mice with anti-IL-12 mAb, a cytokine that induces IFNγ production by T cells, reduced arthritis severity in C3H mice, but at the expense of an increased *B. burgdorferi* tissue-burden (Anguita et al., 1996). Others showed that the frequencies of activated $CD44^{hi}$ $CD11a^{hi}$ cells increased while total CD4 T cell numbers changed little in LN (or spleen) following *B. burgdorferi* infection (Hastey et al., 2012). No evidence of IFNγ production was found at those sites, despite the fact that *B. burgdorferi*-infection induces strong IFNβ responses, including in the LN (Miller et al., 2008a; Lochhead et al., 2012; Hastey et al., 2014; Ma et al., 2014). Moreover, C3H IFNγ-/- and DBA IL-4-/- mice developed arthritis of comparable severity and duration to their WT counterparts, as did disease-resistant 129/SvEv mice deficient in the IFNγ receptor, and C57BL/6 and BALB/c mice deficient in IL-4 or the IL-4 receptor alpha chain (Brown and Reiner, 1999; Potter et al., 2000; Glickstein et al., 2001).

Current studies on T cell responses to *B. burgdorferi* are hampered by inadequate tools to follow antigen-specific T cells. Recent evidence furthermore suggests that a significant proportion of T cells active in lymph nodes after *B. burgdorferi* infection do not react to *Borrelia* antigens (Whiteside et al., 2018). Mice deficient in the anti-inflammatory cytokine IL-10 develop robust arthritis on an otherwise disease-resistant background, which is associated with T cell infiltration of the synovium. A significant fraction of activated CD4 and CD8 T cells in the lymph nodes of these mice after *B. burgdorferi* infection were not induced by *Borrelia*-antigens but, rather, responded with strong “bystander” activation to inflammatory signals that required T cell-intrinsic TLR2-signaling. These bystander T cells, both CD4 and CD8, upregulated TLR2 in response to infection and secreted IFNγ, which supported inflammatory joint destruction (Whiteside et al., 2018). The data further highlight that control of the inflammatory milieu after *B. burgdorferi* infection can ameliorate the induction of potentially pathogenic T cell responses. They also highlight the fact that production of IFNγ is a double-edged sword, as it both supports reduction of *B. burgdorferi* tissue burden, but also drives inflammatory tissue destruction. Similarly, IL-10 is induced following *B. burgdorferi* infection (Sonderegger et al., 2012a) and appears to be a key regulatory factor that can reduce the amount of IFNγ produced following *B. burgdorferi* infection, and thereby the level of arthritis that develops in mice (Whiteside et al., 2018). However, as discussed above, greatly enhanced production of IL-10 may reduce macrophage activation and *Borrelia* clearance, which could cause secondary inflammatory responses due to excessive spirochete loads. In disease-resistant strains of mice, the balance between IFNγ and/or IL-17 and the production of IL-10 may support the failure to clear *B. burgdorferi* infections in the absence of overt manifestations of disease. In humans, who present with a spectrum of pathological manifestations following *B. burgdorferi* infections, a negative correlation was demonstrated between the levels of IL-10 produced by *in vitro* stimulated monocytes and reduced production of TNFα in response to TLR-signaling and disseminated Lyme borreliosis (Kisand et al., 2007).

Anti-inflammatory cytokines, such as IL-10 and TGFβ, are secreted in part by $CD4^{+}$ $CD25^{+}$ $Foxp3^{+}$ Treg cells. The balance of Teff populations with Treg may contribute to the regulation of Lyme arthritis (Sakaguchi, 2004; Belkaid and Rouse, 2005). In support, Treg are found in the synovium of patients with chronic antibiotic-refractory Lyme arthritis, where they were presented in higher numbers compared to control patients with antibiotic-responsive arthritis. Moreover, *in vitro* inhibitory activity of Treg was reduced when taken from antibiotic-refractory Lyme arthritis compared to Treg isolated from those responding to antibiotic therapy (Shen et al., 2010;

Vudattu et al., 2013). Activation of CD4+ Treg is observed also in the lymph nodes of *B. burgdorferi*-infected mice (Nardelli et al., 2004; Elsner et al., 2015a). Mice depleted of IFNγ and IL-17 developed no arthritis until mice were depleted also of CD25+ T cells (Nardelli et al., 2004). Furthermore, CD28-/- mice, which are deficient in natural Treg (Iliopoulou et al., 2007), had increased incidences, duration and severity of arthritis following *B. burgdorferi* infection in comparison to control mice. However, CD28-/- mice have impaired humoral immune responses (Shahinian et al., 1993) and lower pathogen-specific antibody titers after infection, which complicates interpretation of results.

Taken together, a diverse array of T cells is induced in response to *Borrelia* infection in both mice and humans. Production of proinflammatory cytokines such as TNFα and IFNγ supports the reduction in *Borrelia* tissue loads and inhibits the development of *Borrelia*-induced pathology, although it does not clear the infection. Conversely, induction of immune-regulatory cytokines, such as IL-10, may ameliorate manifestations of disease, while failing to support *Borrelia*-clearance. Altering the balance of pro- and anti-inflammatory stimuli to benefit stronger anti-pathogen responses can cause inflammatory disease, arthritis and carditis, either because *B. burgdorferi* infection is incompletely controlled, or because of the consequences of a dysregulated inflammatory response, as modeled in different strains of mice (Crandall et al., 2006; Ma et al., 2014; Paquette et al., 2017).

Host responses in human Lyme borreliosis

As discussed above, there are remarkable similarities between the immune molecules, cells, and signaling pathways that are engaged by *B. burgdorferi* and its components in mice and in humans. Without antibiotic treatment, spirochetes can persist in humans despite a panoply of immune responses, but most often these are associated with clinical disease. The outcomes in terms of pathology are quite different in untreated *B. burgdorferi* infection in mice and humans. Experimentally infected outbred *Peromyscus leucopus*, a main reservoir host for *B. burgdorferi*, exhibit no significant gross or microscopic pathology despite the presence of spirochetes in multiple tissues. An exception exists only when infection is introduced at a very young age (3 days), which can result in a transient arthritis (Moody et al., 1994). Inbred laboratory mice (*Mus musculus*) can exhibit inflammatory lesions in the hearts and joints, the appearance and severity of which are not only age- but also mouse strain-dependent (Barthold et al., 1990). In inbred mice, appearance and resolution of pathology follow the evolution of the immune responses to the spirochete as it multiplies in the skin and disseminates through the blood, lymphatics and tissues to establish foci of infection in other organ systems. The greatest burden of *B. burgdorferi* is detected during the period of peak disease, consistent with spirochete components driving the acute inflammatory response. Inbred mice also develop chronic inflammation in tissues, characterized by segmental perivascular and perineural infiltrates of lymphocytes and plasma cells that are often in association with visible spirochetes in adjacent tissues, although these pathologic changes do not give rise to clinical manifestations such as those seen in human Lyme borreliosis (Barthold et al., 1993).

B. burgdorferi infection of humans incites a broader range of tissue pathology, most often clinically apparent in the skin (EM, *Borrelia* lymphocytoma, and ACA), the heart (conduction system abnormalities and myocarditis), the nervous system (especially cranial nerve abnormalities, meningitis/meningo-encephalitis, and radiculoneuropathies) and in the joints (proliferative synovitis) (Steere et al., 2016) (also see chapter 24). These manifestations typically follow a characteristic timeline, with skin (except for ACA), heart and nervous system involvement in the first weeks to 1-2 months of infection, and arthritis and ACA occurring later. Clinical signs such as erythema migrans (EM) can remit without specific intervention, and others, like arthritis, can recur in the same or different site. Older studies of tissues from patients with Lyme borreliosis, many of whom had the infection for several years prior to diagnosis, have documented the extent of pathology that can occur in various organs during the course of untreated *B. burgdorferi* infection (Duray and Steere, 1988). It is important to note that lymphocytes dominate the cellular infiltrates of all human tissues examined by pathology, whereas innate immune cells are more prevalent in the transient inflammatory lesions in mice, with T and B lymphocytes infiltrating mouse organs as disease resolves. While these differences may reflect, in part, the duration of infection at which samples are evaluated for pathology, they also suggest that *B. burgdorferi* infection of reservoir hosts, such as mice, elicits an immune program much more permissive to silent persistent infection than the immune responses that arise in non-reservoir hosts exemplified by humans.

Host response in cutaneous manifestations of Lyme borreliosis

Experiments in laboratory mice have provided much information about the dynamics of the immune response to *B. burgdorferi* and its relationship to pathologic changes in the mouse, yet many features of human Lyme borreliosis remain to be explained. By the time patients present with clinical signs, cellular infiltrates in infected tissues reflect an adaptive immune response, with CD4+ T cells dominating the pathology (Steere et al., 2016). Typically, this presentation occurs within the first month of infection when one of the earliest clinical signs of Lyme borreliosis, the skin lesion EM, appears at the tick bite site (average 7-14 days after the bite). Not all humans present with EM, and mice do not develop EM after infection is introduced by tick bite or needle inoculation, even though the skin is a preferred site for *B. burgdorferi* infection and persistence (Barthold et al., 1992a). The reason for this is unclear, given that EM lesions are comprised of T lymphocytes, DCs, macrophages and rare plasma cells (Boer et al., 2007), and as described earlier, these cell subsets are engaged in mice within the first month of infection, when EM would be expected to appear. In humans, more spirochetes are found at the leading edge of the EM lesion than centrally, and smaller lesions harbor more spirochetes than larger ones (Liveris et al., 2002). As the size of the EM approximates the duration of infection, this suggests that the immune response is effective at reducing bacterial burdens in human skin, just as the bacterial burden at this site in mice declines with time, albeit without evidence of adaptive immune cell infiltration.

Characterization of EM by immunohistochemistry shows patchy perivascular and interstitial infiltrates, the density of which increase with the age of the lesion (Boer et al., 2007). It is presumed that skin-resident innate immune cells respond to the tick bite and *B. burgdorferi* deposited in the feeding pit as the effects of tick saliva wane, and their activation leads to release of inflammatory mediators that recruit additional immune cells to the infection site. In EM, perivascular infiltrates are dominated by CD4+ and CD8+ lymphocytes whereas interstitial infiltrates are comprised mainly of CD68+ histiocytes (Boer et al., 2007). CD1a+ DCs, which can drive TH1 differentiation, are less commonly found and are mainly perivascular. In one study, denser infiltrates had a greater abundance of B cells with rare CD138+ plasma cells, and this was associated with the age of the lesion (Boer et al., 2007). Flow cytometric analysis of dermal leukocytes extracted from EM lesions, using a blister technique, similarly revealed a predominance of T cells and also monocytes/ macrophages and DCs, with occasional neutrophils (Salazar et al., 2003). T cells had an effector phenotype, and DCs had upregulated HLA-DR, CD14, and TLRs 1, 2 and 4 expression, consistent with a mature phenotype capable of presenting antigens to T cells. The dominant cytokines extracted from the lesions by the blister method were IL-6 and IFNγ, and the levels were higher in patients with isolated EM in comparison to those with multiple EM lesions. Responses in the skin are reflected, at least in part, in the blood. Analysis of cytokines and chemokines of patients who present with EM revealed elevation of innate cytokines CCL2, IL-6, IL-10, TNF and IL-1β, as well as the TH1-associated cytokines and chemokines CXCL0, CXCL10, IFNγ, and CCL19. In addition, marked elevated in TH17 cell associated cytokines IL-23, IL-27, IL-25, IL-22, IL17F, IL-21, and IL-17A has been observed (Strle et al., 2017).

The findings above are consistent with a much earlier study comparing expression of select genes in EM lesions versus those expressed in the late skin manifestation ACA (Mullegger et al., 2000). ACA typically occurs months to years after *B. burgdorferi* infection, most often on a distal extremity, and is associated with *B. afzelii* infection. Both EM and ACA exhibit mononuclear cell infiltrates, although they are patchy and found throughout the dermis in EM lesions, whereas infiltrates in ACA are denser and concentrated in the upper dermis. Cytokine mRNA expressed within EM lesions, as detected by *in situ* hybridization, revealed prominence of TNFα and IFNγ, as well as the anti-inflammatory cytokine IL-10, whereas ACA lesions showed only TNFα and IL-4. The differences in the dominant cytokines in EM vs ACA led to the notion that a strong IFNγ-producing T cell response that promoted macrophage activation was important for controlling *B. burgdorferi* burden and tissue dissemination, and that development of a TH2 response early may set the stage for the later complication of ACA. Whether the initial reaction to the tick bite, which as noted earlier might be modulated by the introduction of tick salivary proteins that can skew immune responses toward a TH2 phenotype, is permissive to ACA has not yet been explored.

Transcriptomic analyses of RNA isolated from EM biopsies supported earlier findings of induction of innate and adaptive immune programs that favor TH1 differentiation and phagocyte activation (Marques et al., 2017). Of 254 genes differentially expressed in EM lesions versus healthy control skin, signatures for innate immune responses, cell migration and chemotaxis dominated among genes upregulated, whereas a prominent signature for tissue development and remodeling was found in those downregulated. In particular, two dominant clusters of induced genes were found: one with Type I and Type II IFN signaling, along with T cell chemoattractants CXCL9, CXCL10, and CXCL11; and the other related to innate immune cell functions, including phagosome formation, FcγR and TLRs 1 and 2. B cell chemoattractants, IFNγ, and genes involved in tryptophan catabolism, including the enzyme IDO1 (indoleamine 2,3-dioxygenase 1) were also found. Induction of CXCL9 and CXCL10 is consistent with the dominance of IFNγ−producing CD4+ TH1 cells in EM lesions that promote phagocyte activation, whereas CXCL11 promotes Treg formation. Phagocytic DC ingestion of spirochetes can lead to the induction of Type I IFNs *in vitro*, and this could drive the differentiation of IDO1-producing DCs that induce Treg cells, which can be found by immunohistochemistry in EM lesions. In addition to induction of Treg cells, tryptophan depletion can lead to suppression of the priming of CD8+ T cells, which are found in abundance in EM lesions. Taken together, these results suggest that the balance between proinflammatory responses and Treg cell induction may influence how permissive the host is to spirochete dissemination and ultimately to late complications of Lyme borreliosis.

Host response in the heart

Clinical involvement of the heart is rare in humans with Lyme borreliosis and most often occurs during the early, disseminated phase of *B. burgdorferi* infection. As a consequence, few studies have examined cardiac tissue specimens in Lyme carditis, which have largely been obtained either from endomyocardial biopsies or postmortem (Cary et al., 1990; Reimers et al., 1993; Tavora et al., 2008; Muehlenbachs et al., 2016). A recent autopsy case series evaluated the hearts of 5 patients who died of sudden cardiac death due to acute Lyme borreliosis (Muehlenbachs et al., 2016).The findings revealed significant inflammatory infiltrates in the atrioventricular and sinoatrial nodes, consistent with the clinical abnormalities typical of Lyme carditis, namely conduction system disease. There were also interstitial lymphoplasmacytic infiltrates in the myocardium in a curvilinear pattern with perivascular distribution extending from the endocardium to the epicardium, with infiltrates also involving the epicardial adipose tissue. Coronary arteries were not involved, nor was their evidence of vasculitis. Cellular infiltrates were predominantly T cells, histiocytes, and plasma cells, without significant recruitment of neutrophils or eosinophils. CD3+ T cells generally were in greater abundance than CD79a+ B cells, but lymphoid follicles were not observed. Cardiomyocyte complement components C4a and C9 were not detected. Cardiac tissue had increased collagen content and decorin was detected where collagen was present. These findings differ from the transient carditis seen in *B. burgdorferi*-infected mice, which arises in the absence of adaptive immune cells. In mice, inflammation is dominated by macrophages primarily located at the heart base, with T and B cells comprising less than 5% of the infiltrates (Ruderman et al., 1995). In all of the autopsy cases, spirochetes were detected in the heart tissues by Warthin-Starry stain and immunohistochemistry along the collagen fibers, consistent with their presence driving the inflammatory response. Despite involvement of the heart indicating disseminated *B. burgdorferi* infection, the only other site where spirochetes were found by immunohistochemistry was in the leptomeninges (2 patients) where they also were associated with mild perivascular inflammation. Spirochetes were not detected visually in other tissues available for analysis (lung, liver, kidney, spleen, prostate, nonlesional skin, and various soft tissues), including several that express decorin. It has been proposed from mouse studies that *B. burgdorferi* expression of DbpA is required for heart involvement and that absence of decorin diminishes the spirochete's ability to colonize this site (Brown et al., 2001; Lin et al., 2014). The results of autopsy studies demonstrate that in humans, tissue expression of decorin is not sufficient to allow for spirochete colonization of a site, and the development of pathology leading to Lyme carditis suggests *B. burgdorferi* has a tropism for infection at this site.

Host response in neuroborreliosis

Neurologic disease is a third example of how mice and humans differ in the pathology that arises after *B. burgdorferi* infection. Early studies revealed that while spirochetes could be visualized in the mouse perineurium, there was no associated inflammation. They also could only be cultured from mouse brains early in infection, during periods when blood cultures were also positive (Barthold et al., 1992a). More recent immunohistochemical studies, however, revealed that while *B. burgdorferi* did not involve the brain parenchyma in mice, it was detected in the meninges, most often intra- and peri-vascular, in disseminated infection (Divan et al., 2018). In humans, Lyme neuroborreliosis can manifest as peripheral or central nervous system involvement, with meningoencephalitis, cranial nerve palsies and peripheral radiculoneuropathies most commonly seen (Halperin et al., 1988; Pachner and Steiner, 2007). A characteristic feature of Lyme neuroborreliosis is the immune cell composition of cerebrospinal fluid (CSF), which is notable for a lymphocytic infiltrate with elevated protein levels and antibodies to *B. burgdorferi*. Studies conducted primarily in Europe, where Lyme neuroborreliosis is more prevalent, have partially characterized the immune response in the CSF. Induction of matrix metalloproteases (MMPs) may permit *B. burgdorferi* invasion into neural tissue, and production of chemokines, including CXCL11 and CXCL13, contributes to cell recruitment. The CSF immune cells in Lyme neuroborreliosis are comprised of >60%T cells and ~15% B cells, with NK cells and monocytes representing a much smaller fraction of the cellular response (Kowarik et al., 2014). The relatively high frequency of B cells, which is much higher than in other bacterial or viral CNS diseases, may be explained by the fact that *B. burgdorferi* can induce human monocytes to produce the B cell chemoattractant CXCL13 *in vitro*, and monocytes with increased expression of CXCL13 are found in the CSF (Narayan et al., 2005; Rupprecht et al., 2007). *B. burgdorferi*-specific antibodies to a number of *Borrelia* antigens can be detected in the CSF. An elevated ratio of pathogen-specific IgG in the CSF compared to the peripheral blood is indicative of local antibody production. C1q, C4, C3 and C3a levels were elevated in the CSF, without associated findings in the plasma, suggesting that the inflammatory response is localized to the CSF (Rupprecht et al., 2007). In early neuroborreliosis, CD8+ T cells were the predominant T cells in CSF, which then gave way to a dominant CD4+ TH1 type response (Jacobsen et al., 2003; Lunemann et al., 2007). Analysis of the TH1 cells isolated from a single US patient with acute neuroborreliosis identified peptides of *B. burgdorferi* heat shock protein 90, 30S ribosomal protein, and proteins involved with porphrine and lipoprotein biosynthesis as relevant T cell antigens (Lunemann et al., 2007). Interestingly, a number of T cells had specificities for human peptide sequences, including CNPase, a structural protein involved in synaptic transmission, but whether these self-reactive T cells play a role in the immune response or contribute to clinical signs and symptoms of neuroborreliosis remains unknown.

More recently, Eckman and colleagues examined the cytokines and chemokines in US patients with Lyme neuroborreliosis using multiplex assays (Eckman et al., 2018). Their study confirmed the dominant increase in CXCL13 in CSF specimens and also reported elevated levels of the chemokines CXCL12, CXCL10, CXCL8, CXCL1 and eotaxin as well as the cytokines IL-12, IFNγ, IL-1α, IL-6, TNFα, IL-10, IL-1 receptor antagonist (IL-1RA) and granulocyte-monocyte colony stimulating factor (GM-CSF). The expression of many of these cytokines and chemokines can be induced when human microglia (the predominant phagocytic cell of the brain) or astrocytes (the primary glial cell forming the blood-brain barrier) are exposed to *B. burgdorferi in vitro* (Casselli et al., 2017; Greenmyer et al., 2018). No studies have specifically detailed responses *in vitro* of non-parenchymal macrophages (resident cells of the meninges) to *B. burgdorferi*. Transcriptomic studies of astrocytes exposed to *B. burgdorferi in vitro* also revealed changes in expression of 38 miRNAs, and for about half, expression of their respective gene targets was altered. Interestingly, the miRNAs expressed, which included miR145 and miR146b, were different from those identified as influencing Lyme arthritis, supporting the concept that in humans, *B. burgdorferi* elicits a tissue – specific immune response.

Host response in Lyme arthritis

Lyme arthritis is considered a late manifestation of Lyme borreliosis, occurring several months (average 6 months) after the onset of infection (Steere et al., 1987). The late appearance of arthritis relative to the onset of infection distinguishes the arthritis seen in *B. burgdorferi*-infected mice from the affliction in humans. Age plays a role in the clinical presentation, with young children often presenting with an acute mono- or pauciarticular inflammatory arthritis involving the knee or hip that resembles septic arthritis (Willis et al., 2003; Thompson et al., 2009; Baldwin et al., 2016), whereas Lyme arthritis in teens and adults usually presents with significant swelling in a joint without much pain (Steere et al., 2016). Adults often report no history of a clinical syndrome compatible with acute Lyme borreliosis in the year preceding the onset of arthritis. In mice, arthritis is polyarticular and is mediated by PMNs in the joint space and surrounding tissues. Although synovial fluid from Lyme arthritis patients similarly has a predominance of PMNs, synovial tissue is notable for infiltration with CD4+ T cells, macrophages, B cells and plasma cells (Duray and Steere, 1988; Steere and Angelis, 2006). The degree of synovial proliferation is dependent upon the number of episodes of arthritis and the duration of arthritis prior to treatment. In acute arthritis, *B. burgdorferi* DNA can be detected in the synovial fluid and tissue (Li et al., 2011), consistent with infection driving the host response. In untreated patients, rare spirochetes and globular antigen deposits have been observed in perivascular areas where lymphocytes are also present (Duray and Steere, 1988). The spontaneous remission of an episode of acute arthritis, which can last for only a few days or for several months, is presumed to be immune-mediated, but organisms can persist in the absence of antibiotic therapy and give rise to recurrent episodes of acute arthritis. In one study of patients who did not receive antibiotic therapy for EM, arthritis developed in more than half, but the number of patients with recurrent episodes of arthritis declined by 10-20% each year even in the absence of therapy, suggesting that either the inflammatory response was eventually down-regulated or spirochetes driving that response had been eliminated (Steere et al., 1987).

The late appearance of arthritis raises the question as to where spirochetes reside during the asymptomatic phase before arthritis onset. Two-photon intravital imaging of *B. burgdorferi*-infected mice has revealed spirochetes adjacent to cartilage and in the tendon insertion sites (entheses) around joints (Bockenstedt et al., 2012) (see also Chapter 17). Entheses are relatively avascular areas rich in decorin, aggrecan, and other molecules *B. burgdorferi* is known to bind. Decorin in particular has been shown to be required for the development of *B. burgdorferi*-induced arthritis in mice (Brown et al., 2001). In both mice and humans, a specialized population of $CD3^{+}CD4^{-}CD8^{-}IL\text{-}23R^{+}ROR\gamma t^{+}$ T cells has been identified in entheses that is believed to be involved in the pathogenesis of other forms of inflammatory arthritis through the production of IL-17 (Sherlock et al., 2012; Reinhardt et al., 2016; Cuthbert et al., 2017). These cells express PRRs and respond to local or systemic administration of IL-23. Interestingly, $IL\text{-}23R^{+}ROR\gamma t^{+}$ T cells are found at sites associated with mechanical stress, like the enthesis and the aorta, which also are areas associated with *B. burgdorferi* pathology. For other forms of arthritis in which $IL\text{-}23R^{+}ROR\gamma t^{+}$ T cells are believed to play a role, mechanical stress damaging the enthesis is thought to be an initiating event for an episode of joint inflammation. Adjacent synovial tissue cells, including macrophages and synovial intimal fibroblasts that normally contribute to entheseal nourishment and repair, drive the inflammatory response that involves the joint. Whether $IL\text{-}23R^{+}ROR\gamma t^{+}$ T cells are involved in the host response in Lyme borreliosis has not been explored.

The initiation of an episode of Lyme arthritis in humans likely involves exposure of synovial innate immune cells expressing TLRs to *B. burgdorferi* or its components, and the induction of MMPs from synovial cells and other cells within the joint. *B. burgdorferi* elicits MMPs and aggrecanases from human chondrocytes *in vitro* and MMP-2, MMP-3, and MMP-9 are found in synovial fluid (Hu et al., 2001; Behera et al., 2006b). As noted above, MMP-9 expression is also linked to the development of *B. burgdorferi*-induced arthritis in mice (Heilpern et al., 2009). Both αβ- and γδ T cells have been isolated and characterized from synovial fluid and tissues of patients with Lyme arthritis. αβ T cells include TH1, TH17, and TH2 subsets, and the dominance of TH1 cells correlates with larger effusions. Immunohistochemistry reveals upregulation of adhesion molecules (P-selectin, vascular adhesion protein-1, intercellular adhesion molecule-1 and vascular cell adhesion molecule-1) on synovial lining cells, endothelial cells, and cell infiltrates (Akin et al., 2001). Analysis of synovial fluid chemokines have detected macrophage and neutrophil chemoattractants, as well as TH1 chemoattractants CXCL9 and CXCL10, the TH17 cell chemoattractant CCL2, and the B cell chemoattractant CXCL13 (Shin et al., 2007). Both pro-inflammatory (IL-6, TNFα, IL-1β, IFNγ) and anti-inflammatory cytokines (IL-10, IL-5 and IL-4) have been reported (Shin et al., 2007). NapA has been identified as one *B. burgdorferi* component that drives joint inflammation in Lyme arthritis through the induction of IL-6, IL-1β, and TGFβ, and NapA-specific TH17 cells have been found in synovial fluid. A feature of Lyme arthritis is that in some patients, resolution of joint inflammation may take weeks to a few months after completion of antibiotics. The reason for the slow resolution of arthritis is not clear, although some studies suggest that adjunctive use of intra-articular steroids or oral non-steroidal anti-inflammatory agents may aid resolution. Recently *B. burgdorferi* peptidoglycan has been detected in the joints of patients with Lyme arthritis both before and after treatment and may provide a stimulus for continued inflammation even after antibiotics have killed viable organisms (Jutras et al., 2019).

Return to homeostasis and immunoregulation

The resolution of inflammation and return to tissue homeostasis after an immune challenge involves timely control of both innate and adaptive immune responses to prevent excessive tissue damage and limit energy consumption that occurs with on-going immune activation (Paludan et al., 2020). A return to a new immune homeostasis after pathogen encounter depends on the degree of perceived threat to the host. In *B. burgdorferi* reservoir hosts such as mice, that threshold appears to be low as mice exhibit minimal pathology despite persistence of spirochetes. The human encounter with *B. burgdorferi*, in contrast, results in a range of outcomes, from asymptomatic seroconversion to persistence of inflammatory sequelae after treatment, as for example in post-antibiotic Lyme arthritis (discussed below). These differences in outcomes likely relate to individual variation in host responses to infectious and immune challenges. A full understanding of how variations in human immunity contribute to the expression and outcomes of Lyme borreliosis awaits a more comprehensive understanding of the human immune system, its regulation and relation to other biological systems within the human host.

Regulation of innate and adaptive immune responses begins virtually immediately after their engagement (Vigano et al., 2012; Cui et al., 2014). Examples include the induction of IL-1 receptor antagonist after IL-1 receptor signaling, downregulation of cytokine receptor expression, modulation of signaling through induction of suppressor of cytokine signaling (SOCS), induction of CTLA4 after T cell activation, and actions of Treg cells. In an effort to assess the evolution of the immune response in Lyme borreliosis, Bouquet et al conducted a longitudinal transcriptome analysis of PBMCs isolated from whole blood samples of Lyme borreliosis patients presenting with EM and followed for 6 months after antibiotic treatment (Bouquet et al., 2016). Because a known signature of *B. burgdorferi* infection is Type I IFN response, a comparison was made between differentially expressed genes in Lyme borreliosis with those found in other infectious and inflammatory conditions, some of which are noted for Type I IFN induction. In the acute phase at study entry, the Lyme borreliosis samples revealed upregulation of gene sets common to influenza infection, another Type I interferon-associated pathogen, as well as sets shared with the Gram positive bacterium *Streptococcus pneumoniae* bacteremia. These gene sets included significant enrichment for transcripts encoding interferon signaling as well as PRR-associated innate immune responses. The elevation in these genes was protracted and detected even at 3 weeks after antibiotic treatment (~5 weeks after presentation). By 6 months, the number of differentially expressed genes (DEGs) had diminished and the pattern had changed, showing evolution of the response from genes primarily upregulated to those downregulated. At this stage, 60% of the DEGs were shared with those seen in systemic lupus erythematosus, a Type I IFN-associated autoimmune disease, and 30% with rheumatoid arthritis (other infectious diseases were not reported at this time point). The most downregulated gene transcript was EIF2, which encodes for eIF2, a global regulator necessary for the translation initiation of mRNAs and also downregulated in systemic lupus erythematosus. Although the study cohort included patients who had not yet resolved subjective symptoms such as fatigue at the 6 month timepoint, there was no difference in DEGs between those who reported return to their pre-infection clinical baseline and those who did not. Taken together, these findings confirmed a robust inflammatory transcriptome during the early phase of Lyme borreliosis that was sustained for a period after antibiotic therapy for the infection. The significance of a protracted inflammatory signature should be considered in the light of new studies revealing the impact of antibiotics on host immunity. Specifically, antibiotics have global effects on the host microbiome and can alter host immunity acutely in response to an immune challenge, enhancing inflammatory signatures in the blood (Hagan et al., 2019). Moreover, doxycycline (the most common antibiotic used to treat EM) not only inhibits MMPs but also interferes with M2 (anti-inflammatory) macrophage differentiation (He and Marneros, 2014). Both of these effects could lead to prolonged inflammatory signatures in the blood even though patients are clinically improving. The subsequent evolution of the immune response to one that shares some features with a Type I IFN-associated autoimmune disorder may represent engagement of common pathways aimed at reducing innate immune activation during the elimination of cellular and pathogen inflammatory debris and the energy expenditure of excessive activation of adaptive immunity.

Post-antibiotic Lyme arthritis – the importance of immune regulation

About 10% of patients with Lyme arthritis develop chronic synovitis, usually in a single joint, that persists despite antibiotic therapy. In an early study, arthritis often had been present for many months to years in patients prior to antibiotic institution, and even then, joint inflammation generally responded to this intervention (Steere et al., 1987). With time, a second subgroup of patients has been identified in whom arthritis does not resolve within >3 months of completion of antibiotic therapy, and often patients have received multiple antibiotic courses (Steere et al., 2006). Inflammation persists even though *B. burgdorferi* can no longer be detected in synovial fluid or synovial tissue (Nocton et al., 1994). Development of this "post-infectious" form of Lyme arthritis (now termed "post-antibiotic Lyme arthritis") is associated with more invasive strains (RST1 type) of *B. burgdorferi sensu strico*, as well as genetic factors that may predispose to greater intensity and duration of inflammatory responses (Strle et al., 2011; Strle et al., 2012). Patients who develop post-antibiotic Lyme arthritis are more likely than those with antibiotic-responsive arthritis to bear rheumatoid arthritis-associated HLA-DRB1 molecules, especially HLA-DRB1*0401, HLA-DRB1*0101, and HLA-DRB1*0404 alleles. Because of the association of these HLA alleles with rheumatoid arthritis (Steere et al., 2006), much research has focused on autoreactivity in patients with post-antibiotic Lyme arthritis. A search for *B. burgdorferi* proteins that could incite autoreactivity through molecular mimicry initially identified an OspA peptide (aa163-175) as having some sequence similarity to human LFA-1, but subsequent studies have not supported its relevance to arthritis (Gross et al., 1998a; Trollmo et al., 2001). Linked T and B cell responses to several self-antigens have been identified subsequently, including endothelial cell growth factor (ECGF), MMP-10, apolipoprotein B 100, and annexin A2 (Drouin et al., 2013; Crowley et al., 2015; Crowley et al., 2016). Although antibodies to these proteins can be found in patients who resolve arthritis, they tend to be higher in those whose arthritis does not resolve. These antibodies to self-antigens are generally of the IgG2 and IgG4 subclasses, whereas those directed against *B. burgdorferi* antigens are usually IgG1 and IgG3 (Sulka et al., 2018). Interestingly, the magnitude of IgG4 to some of these self-antigens correlates with the degree of obliterative vasculopathy and fibrosis characteristic of post-antibiotic Lyme arthritis (Londono et al., 2014; Sulka et al., 2018).

There is increasing evidence that a heightened and dysregulated immune response contributes to the magnitude and duration of Lyme arthritis. A retrospective analysis compared cytokine and chemokine expression in archived synovial fluid samples from patients who resolved arthritis versus those who developed post-antibiotic arthritis (Shin et al., 2007). Samples from patients in whom arthritis persisted after antibiotic therapy had higher levels of TNFα, IL-1β and IFNγ in synovial fluid and increased levels of T cell and monocyte chemoattractants (CXCL9, CXCL10 and CCL4), indicating a more exuberant inflammatory response. Most of the TH1 cytokines and chemokines declined over time in the post-antibiotic Lyme arthritis group once PCR for spirochete DNA became negative, although patients continued to have elevated levels in comparison to controls. Resolution of inflammation may in part be mediated by FoxP3 Treg cells, as post-antibiotic Lyme arthritis has been associated with low numbers of Treg cells in synovial fluid (Shen et al., 2010). The magnitude of the TH17 response has been correlated with the severity of joint inflammation and to the development of autoreactivity that may contribute to persistent synovitis (Strle et al., 2017). Genetic factors that could contribute to more severe inflammation include a SNP of the CD14 gene (CD14-550 C/T), which is related to the levels of soluble CD14 (Inoue et al., 2007); soluble CD14 has been detected at elevated levels in patients with a more severe inflammatory response to *B. burgdorferi* infection (Lin et al., 2000). In addition, a Toll-like receptor 1 (1805GG) polymorphism has also been associated with heightened TH1 inflammatory responses and is more prevalent among patients with post-antibiotic Lyme arthritis (Strle et al., 2012). Certain microRNAs that modulate inflammatory gene expression, including miR-155, miR-223 and miR-146a, are elevated in patients who experience persistent synovitis in comparison to those whose arthritis resolves with antibiotic treatment (Lochhead et al., 2017).

A study of the transcriptome of synovium from patients with post-antibiotic Lyme arthritis has shed light on how infection-induced inflammation can evolve to an immune-mediated chronic inflammation after the pathogen is eliminated (Lochhead et al., 2019a). Synovial tissue of patients with post-antibiotic Lyme arthritis exhibits heightened Type I and Type II IFN responses that correlate with the degree of inflammatory cytokine expression at the protein level, especially IFNγ, and are inversely correlated with genes involved in tissue repair. A dominance of T cells that produced IFNγ, including CD4, CD8 and double negative populations, as well as a smaller population of IFNγ-producing NK cells were found. Fibroblast-like synoviocytes, which have been implicated in the pathogenesis of rheumatoid arthritis, were a main source of IFN-inducible genes and when exposed to IFNγ *in vitro*, became hyper-responsive to *B. burgdorferi* (Lochhead et al., 2019b). Dysregulation of fibroblast-like synoviocytes may favor the continued recruitment of IFNγ-producing T cells into the synovium and give rise to many of the observed pathologic changes in post-antibiotic-Lyme arthritis. These findings underscore the importance of host regulation of both innate and acquired immune responses in order to return to a new state of tissue homeostasis after pathogen encounter, control and elimination.

Concluding Remarks

From the above, it is apparent that the immune response to *B. burgdorferi* is both complex and nuanced and involves many cell types, soluble molecules and varies based on the organ system affected by infection with this pathogen. While the immune response to *B. burgdorferi* reduces and controls spirochete numbers, and prevents or resolves manifestations of infection-induced disease, *B. burgdorferi* ultimately establishes persistent infection in many vertebrate reservoir hosts, enabling its re-acquisition by tick vectors and transmission to other hosts. Thus, co-evolutionary adaptation allows *B. burgdorferi* to maintain its essential enzootic cycle while helping the vertebrate host to avoid disease. Persistent infections by extracellular microbes are rare, more often seen following infections of viruses, such as CMV and EBV rather than large and complex bacteria that are constantly exposed to cells of the immune system. The complexity of how the balance between persistence and a disease-free state is achieved would be nearly impossible to maintain across the genetic diversity of the entire spectrum of hosts susceptible to infection with B. *burgdorferi*. Indeed, most infected vertebrates can either clear *B. burgdorferi* efficiently or develop some level of inflammatory disease, the latter seen following infections of canine, equine, and human hosts. A better understanding of the mechanisms enabling disease-free persistence of *B. burgdorferi* versus the mechanisms causing inflammation and disease are critical for enhancing our ability to develop therapeutics that more directly target the mechanisms rather than the manifestations of *B. burgdorferi*-induced disease. The recent advances in research technologies, including intravital microscopy, single-cell transcriptomics, proteomics and metabolomics, to name but a few, are likely to result in a more nuanced understanding of immune response regulation to infections with *B. burgdorferi* that may have relevance not only to human Lyme borreliosis but to infection with other *Borrelia* spp.

Acknowledgements

This work was supported by NIH U19AI089992 and the Harold W. Jockers Professorship to LKB, NIH R01AI121970 to RMW, and DoD TB170139 to NB

References

Abel, A.M., Yang, C., Thakar, M.S., and Malarkannan, S. (2018). Natural Killer Cells: Development, Maturation, and Clinical Utilization. Front Immunol *9*, 1869. https://doi.org/10.3389/fimmu.2018.01869

Aguero-Rosenfeld, M.E., Nowakowski, J., Bittker, S., Cooper, D., Nadelman, R.B., and Wormser, G.P. (1996). Evolution of the serologic response to *Borrelia burgdorferi* in treated patients with culture-confirmed erythema migrans. J Clin Microbiol *34*, 1-9.

Aguero-Rosenfeld, M.E., Wang, G., Schwartz, I., and Wormser, G.P. (2005). Diagnosis of Lyme borreliosis. Clin Microbiol Rev *18*, 484-509. https://doi.org/10.1128/CMR.18.3.484-509.2005

Akin, E., Aversa, J., and Steere, A.C. (2001). Expression of adhesion molecules in synovia of patients with treatment-resistant Lyme arthritis. Infect Immun *69*, 1774-1780.

Alarcon-Chaidez, F.J., Boppana, V.D., Hagymasi, A.T., Adler, A.J., and Wikel, S.K. (2009). A novel sphingomyelinase-like enzyme in *Ixodes scapularis* tick saliva drives host CD4 T cells to express IL-4. Parasite Immunol *31*, 210-219. https://doi.org/10.1111/j.1365-3024.2009.01095.x

Alexopoulou, L., Thomas, V., Schnare, M., Lobet, Y., Anguita, J., Schoen, R.T., Medzhitov, R., Fikrig, E., and Flavell, R.A. (2002). Hyporesponsiveness to vaccination with *Borrelia burgdorferi* OspA in humans and in TLR1- and TLR2-deficient mice. Nat Med *8*, 878-884.

Aliprantis, A.O., Yang, R.B., Mark, M.R., Suggett, S., Devaux, B., Radolf, J.D., Klimpel, G.R., Godowski, P., and Zychlinsky, A. (1999). Cell activation and apoptosis by bacterial lipoproteins through toll-like receptor-2. Science *285*, 736-739.

Alugupalli, K.R. (2008). A distinct role for B1b lymphocytes in T cell-independent immunity. Curr Top Microbiol Immunol *319*, 105-130. https://doi.org/10.1007/978-3-540-73900-5_5

Anguita, J., Persing, D.H., Rincon, M., Barthold, S.W., and Fikrig, E. (1996). Effect of anti-interleukin 12 treatment on murine Lyme borreliosis. J Clin Invest *97*, 1028-1034.

Appel, M.J., Allan, S., Jacobson, R.H., Lauderdale, T.L., Chang, Y.F., Shin, S.J., Thomford, J.W., Todhunter, R.J., and Summers, B.A. (1993). Experimental Lyme disease in dogs produces arthritis and persistent infection. J Infect Dis *167*, 651-664.

Baldwin, K.D., Brusalis, C.M., Nduaguba, A.M., and Sankar, W.N. (2016). Predictive Factors for Differentiating Between Septic Arthritis and Lyme Disease of the Knee in Children. J Bone Joint Surg Am *98*, 721-728. https://doi.org/10.2106/JBJS.14.01331

Bankhead, T., and Chaconas, G. (2007). The role of VlsE antigenic variation in the Lyme disease spirochete: persistence through a mechanism that differs from other pathogens. Mol Microbiol *65*, 1547-1558. https://doi.org/MMI5895 [pii] 10.1111/j.1365-2958.2007.05895.x

Barbour, A.G., Jasinskas, A., Kayala, M.A., Davies, D.H., Steere, A.C., Baldi, P., and Felgner, P.L. (2008). A genome-wide proteome array reveals a limited set of immunogens in natural infections of humans and white-footed mice with *Borrelia burgdorferi*. Infect Immun *76*, 3374-3389. https://doi.org/IAI.00048-08 [pii] 10.1128/IAI.00048-08

Barthold, S.W., Beck, D.S., Hansen, G.M., Terwilliger, G.A., and Moody, K.D. (1990). Lyme borreliosis in selected strains and ages of laboratory mice. J Infect Dis *162*, 133-138.

Barthold, S.W., and de Souza, M. (1995). Exacerbation of Lyme arthritis in *beige* mice. J Infect Dis *172*, 778-784.

Barthold, S.W., de Souza, M., Fikrig, E., and Persing, D.H. (1992a). Lyme borreliosis in the laboratory mouse. In Lyme Disease: Molecular and Immunologic Approaches, S.E. Schutzer, ed. (Plainview (New York): Cold Spring Harbor Laboratory Press), pp. 223-242.

Barthold, S.W., de Souza, M.S., Janotka, J.L., Smith, A.L., and Persing, D.H. (1993). Chronic Lyme borreliosis in the laboratory mouse. Am J Pathol *143*, 959-971.

Barthold, S.W., Feng, S., Bockenstedt, L.K., Fikrig, E., and Feen, K. (1997). Protective and arthritis-resolving activity in sera of mice infected with *Borrelia burgdorferi*. Clin Infect Dis *25 Suppl 1*, S9-17.

Barthold, S.W., Persing, D.H., Armstrong, A.L., and Peeples, R.A. (1991). Kinetics of *Borrelia burgdorferi* dissemination and evolution of disease after intradermal inoculation of mice. Am J Pathol *139*, 263-273.

Barthold, S.W., Sidman, C.L., and Smith, A.L. (1992b). Lyme borreliosis in genetically resistant and susceptible mice with severe combined immunodeficiency. Am J Trop Med Hyg *47*, 605-613.

Baum, E., Hue, F., and Barbour, A.G. (2012). Experimental infections of the reservoir species *Peromyscus leucopus* with diverse strains of *Borrelia burgdorferi*, a Lyme disease agent. mBio *3*, e00434-00412.http s://doi.org/10.1128/mBio.00434-12

Baumgarth, N. (2011). The double life of a B-1 cell: self-reactivity selects for protective effector functions. Nat Rev Immunol *11*, 34-46.http s://doi.org/10.1038/nri2901

Baumgarth, N., Tung, J.W., and Herzenberg, L.A. (2005). Inherent specificities in natural antibodies: a key to immune defense against pathogen invasion. Springer Semin Immunopathol *26*, 347-362. https://doi.org/10.1007/s00281-004-0182-2

Beck, G., Benach, J.L., and Habicht, G.S. (1990). Isolation, preliminary chemical characterization, and biological activity of *Borrelia burgdorferi* peptidoglycan. Biochem Biophys Res Commun *167*, 89-95. https://doi.org/10.1016/0006-291x(90)91734-a

Behera, A.K., Durand, E., Cugini, C., Antonara, S., Bourassa, L., Hildebrand, E., Hu, L.T., and Coburn, J. (2008). *Borrelia burgdorferi* BBB07 interaction with integrin alpha3beta1 stimulates production of pro-inflammatory mediators in primary human chondrocytes. Cell Microbiol *10*, 320-331.http s://doi.org/10.1111/j.1462-5822.2007.01043.x

Behera, A.K., Hildebrand, E., Bronson, R.T., Perides, G., Uematsu, S., Akira, S., and Hu, L.T. (2006a). MyD88 deficiency results in tissue-specific changes in cytokine induction and inflammation in interleukin-18-independent mice infected with *Borrelia burgdorferi.* Infect Immun *74*, 1462-1470.

Behera, A.K., Hildebrand, E., Szafranski, J., Hung, H.H., Grodzinsky, A.J., Lafyatis, R., Koch, A.E., Kalish, R., Perides, G., Steere, A.C., *et al.* (2006b). Role of aggrecanase 1 in Lyme arthritis. Arthritis Rheum *54*, 3319-3329.

Behera, A.K., Hildebrand, E., Uematsu, S., Akira, S., Coburn, J., and Hu, L.T. (2006c). Identification of a TLR-independent pathway for *Borrelia burgdorferi*-induced expression of matrix metalloproteinases and inflammatory mediators through binding to integrin alpha 3 beta 1. J Immunol *177*, 657-664.

Belkaid, Y., and Rouse, B.T. (2005). Natural regulatory T cells in infectious disease. Nat Immunol *6*, 353-360.

Belperron, A.A., and Bockenstedt, L.K. (2001). Natural antibody affects survival of the spirochete *Borrelia burgdorferi* within feeding ticks. Infect Immun *69*, 6456-6462.

Belperron, A.A., Dailey, C.M., Booth, C.J., and Bockenstedt, L.K. (2007). Marginal zone B-cell depletion impairs murine host defense against *Borrelia burgdorferi* infection. Infect Immun *75*, 3354-3360.

Benach, J.L., Fleit, H.B., Habicht, G.S., Coleman, J.L., Bosler, E.M., and Lane, B.P. (1984a). Interactions of phagocytes with the Lyme disease spirochete: role of the Fc receptor. J Infect Dis *150*, 497-507.

Benach, J.L., Habicht, G.S., Gocinski, B.L., and Coleman, J.L. (1984b). Phagocytic cell responses to in vivo and in vitro exposure to the Lyme disease spirochete. Yale J Biol Med *57*, 599-605.

Benhnia, M.R., Wroblewski, D., Akhtar, M.N., Patel, R.A., Lavezzi, W., Gangloff, S.C., Goyert, S.M., Caimano, M.J., Radolf, J.D., and Sellati, T.J. (2005). Signaling through CD14 attenuates the inflammatory response to *Borrelia burgdorferi*, the agent of Lyme disease. J Immunol *174*, 1539-1548.

Bernard, Q., Wang, Z., Di Nardo, A., and Boulanger, N. (2017). Interaction of primary mast cells with *Borrelia burgdorferi (sensu stricto)*: role in transmission and dissemination in C57BL/6 mice. Parasit Vectors *10*, 313. https://doi.org/10.1186/s13071-017-2243-0

Bernardino, A.L., Myers, T.A., Alvarez, X., Hasegawa, A., and Philipp, M.T. (2008). Toll-like receptors: insights into their possible role in the pathogenesis of lyme neuroborreliosis. Infect Immun *76*, 4385-4395.http s://doi.org/10.1128/IAI.00394-08

Blaho, V.A., Zhang, Y., Hughes-Hanks, J.M., and Brown, C.R. (2011). 5-Lipoxygenase-deficient mice infected with *Borrelia burgdorferi* develop persistent arthritis. J Immunol *186*, 3076-3084. https://doi.org/10.4049/jimmunol.1003473

Blandino, R., and Baumgarth, N. (2019). Secreted IgM: New tricks for an old molecule. J Leukoc Biol *106*, 1021-1034. https://doi.org/10.1002/jlb.3ri0519-161r

Blum, L.K., Adamska, J.Z., Martin, D.S., Rebman, A.W., Elliott, S.E., Cao, R.R.L., Embers, M.E., Aucott, J.N., Soloski, M.J., and Robinson, W.H. (2018). Robust B Cell Responses Predict Rapid Resolution of Lyme Disease. Front Immunol *9*, 1634. https://doi.org/10.3389/fimmu.2018.01634

Bockenstedt, L.K., Barthold, S., Deponte, K., Marcantonio, N., and Kantor, F.S. (1993). *Borrelia burgdorferi* infection and immunity in mice deficient in the fifth component of complement. Infect Immun *61*, 2104-2107.

Bockenstedt, L.K., Gonzalez, D., Mao, J., Li, M., Belperron, A.A., and Haberman, A. (2014). What ticks do under your skin: two-photon intravital imaging of *Ixodes scapularis* feeding in the presence of the Lyme disease spirochete. Yale J Biol Med *87*, 3-13.

Bockenstedt, L.K., Gonzalez, D.G., Haberman, A.M., and Belperron, A.A. (2012). Spirochete antigens persist near cartilage after murine Lyme borreliosis therapy. J Clin Invest *122*, 2652-2660. https://doi.org/10.1172/JCI58813

Bockenstedt, L.K., Hodzic, E., Feng, S., Bourrel, K.W., de Silva, A., Montgomery, R.R., Fikrig, E., Radolf, J.D., and Barthold, S.W. (1997). *Borrelia burgdorferi* strain-specific Osp C-mediated immunity in mice. Infect Immun *65*, 4661-4667.

Bockenstedt, L.K., Kang, I., Chang, C., Persing, D., Hayday, A., and Barthold, S.W. (2001). CD4+ T helper 1 cells facilitate regression of murine Lyme carditis. Infect Immun *69*, 5264-5269.

Bockenstedt, L.K., Shanafelt, M.C., Belperron, A., Mao, J., and Barthold, S.W. (2003). Humoral immunity reflects altered T helper cell bias in *Borrelia burgdorferi*-infected gamma delta T-cell-deficient mice. Infect Immun *71*, 2938-2940.

Boer, A., Bresch, M., Dayrit, J., and Falk, T.M. (2007). Erythema migrans: a reassessment of diagnostic criteria for early cutaneous manifestations of borreliosis with particular emphasis on clonality investigations. Br J Dermatol *156*, 1263-1271.

Boggemeyer, E., Stehle, T., Schaible, U.E., Hahne, M., Vestweber, D., and Simon, M.M. (1994). *Borrelia burgdorferi* upregulates the adhesion molecules E-selectin, P-selectin, ICAM-1 and VCAM-1 on mouse endothelioma cells *in vitro*. Cell Adhes Commun 2, 145-157.

Bolz, D.D., Sundsbak, R.S., Ma, Y., Akira, S., Kirschning, C.J., Zachary, J.F., Weis, J.H., and Weis, J.J. (2004). MyD88 plays a unique role in host defense but not arthritis development in Lyme disease. J Immunol *173*, 2003-2010.

Bouquet, J., Soloski, M.J., Swei, A., Cheadle, C., Federman, S., Billaud, J.N., Rebman, A.W., Kabre, B., Halpert, R., Boorgula, M., *et al.* (2016). Longitudinal Transcriptome Analysis Reveals a Sustained Differential Gene Expression Signature in Patients Treated for Acute Lyme Disease. mBio *7*, e00100-00116. https://doi.org/10.1128/mBio.00100-16

Bramwell, K.K., Ma, Y., Weis, J.H., Chen, X., Zachary, J.F., Teuscher, C., and Weis, J.J. (2014). Lysosomal beta-glucuronidase regulates Lyme and rheumatoid arthritis severity. J Clin Invest *124*, 311-320. https://doi.org/72339 [pii] 10.1172/JCI72339

Brown, C.R., Blaho, V.A., and Loiacono, C.M. (2003). Susceptibility to experimental Lyme arthritis correlates with KC and monocyte chemoattractant protein-1 production in joints and requires neutrophil recruitment via CXCR2. J Immunol *171*, 893-901.

Brown, C.R., Lai, A.Y., Callen, S.T., Blaho, V.A., Hughes, J.M., and Mitchell, W.J. (2008). Adenoviral delivery of interleukin-10 fails to attenuate experimental Lyme disease. Infect Immun *76*, 5500-5507.

Brown, C.R., and Reiner, S.L. (1998a). Activation of natural killer cells in arthritis-susceptible but not arthritis-resistant mouse strains following *Borrelia burgdorferi* Infection. Infect Immun *66*, 5208-5214.

Brown, C.R., and Reiner, S.L. (1998b). Clearance of *Borrelia burgdorferi* may not be required for resistance to experimental Lyme arthritis. Infect Immun *66*, 2065-2071.

Brown, C.R., and Reiner, S.L. (1999). Experimental Lyme arthritis in the absence of interleukin-4 or gamma interferon. Infect Immun *67*, 3329-3333.

Brown, E.L., Wooten, R.M., Johnson, B.J., Iozzo, R.V., Smith, A., Dolan, M.C., Guo, B.P., Weis, J.J., and Hook, M. (2001). Resistance to Lyme disease in decorin-deficient mice. J Clin Invest *107*, 845-852.http s://doi.org/10.1172/JCI11692

Brown, J.P., Zachary, J.F., Teuscher, C., Weis, J.J., and Wooten, R.M. (1999). Dual role of interleukin-10 in murine Lyme disease: Regulation of arthritis severity and host defense. Infect Immun *67*, 5142-5150.

Brunner, M., and Sigal, L.H. (2001). Use of serum immune complexes in a new test that accurately confirms early Lyme disease and active infection with *Borrelia burgdorferi.* J Clin Microbiol *39*, 3213-3221.http s://doi.org/10.1128/jcm.39.9.3213-3221.2001

Cabral, E.S., Gelderblom, H., Hornung, R.L., Munson, P.J., Martin, R., and Marques, A.R. (2006). *Borrelia burgdorferi* lipoprotein-mediated TLR2 stimulation causes the down-regulation of TLR5 in human monocytes. J Infect Dis *193*, 849-859. https://doi.org/10.1086/500467

Caimano, M.J., Groshong, A.M., Belperron, A., Mao, J., Hawley, K.L., Luthra, A., Graham, D.E., Earnhart, C.G., Marconi, R.T., Bockenstedt, L.K., *et al.* (2019). The RpoS Gatekeeper in *Borrelia burgdorferi:* An Invariant Regulatory Scheme That Promotes Spirochete Persistence in Reservoir Hosts and Niche Diversity. Front Microbiol *10*, 1923. https://doi.org/10.3389/fmicb.2019.01923

Carreras-González, A., Barriales, D., Palacios, A., Montesinos-Robledo, M., Navasa, N., Azkargorta, M., Peña-Cearra, A., Tomás-Cortázar, J., Escobes, I., Pascual-Itoiz, M.A., *et al.* (2019). Regulation of macrophage activity by surface receptors contained within *Borrelia burgdorferi*-enriched phagosomal fractions. PLoS Pathog *15*, e1008163. https://doi.org/10.1371/journal.ppat.1008163

Carroll, J.A., Garon, C.F., and Schwan, T.G. (1999). Effects of environmental pH on membrane proteins in *Borrelia burgdorferi.* Infect Immun *67*, 3181-3187.

Cary, N.R., Fox, B., Wright, D.J., Cutler, S.J., Shapiro, L.M., and Grace, A.A. (1990). Fatal Lyme carditis and endodermal heterotopia of the atrioventricular node. Postgrad Med J *66*, 134-136. https://doi.org/10.1136/pgmj.66.772.134

Casselli, T., Qureshi, H., Peterson, E., Perley, D., Blake, E., Jokinen, B., Abbas, A., Nechaev, S., Watt, J.A., Dhasarathy, A., *et al.* (2017). MicroRNA and mRNA Transcriptome Profiling in Primary Human Astrocytes Infected with *Borrelia burgdorferi.* PLoS One *12*, e0170961.http s://doi.org/10.1371/journal.pone.0170961

Cassiani-Ingoni, R., Cabral, E.S., Lunemann, J.D., Garza, Z., Magnus, T., Gelderblom, H., Munson, P.J., Marques, A., and Martin, R. (2006). *Borrelia burgdorferi* Induces TLR1 and TLR2 in human microglia and peripheral blood monocytes but differentially regulates HLA-class II expression. J Neuropathol Exp Neurol *65*, 540-548. https://doi.org/10.1097/00005072-200606000-00002

Cervantes, J.L., Dunham-Ems, S.M., La Vake, C.J., Petzke, M.M., Sahay, B., Sellati, T.J., Radolf, J.D., and Salazar, J.C. (2011). Phagosomal signaling by *Borrelia burgdorferi* in human monocytes involves Toll-like receptor (TLR) 2 and TLR8 cooperativity and TLR8-mediated induction of IFN-beta. Proc Natl Acad Sci U S A *108*, 3683-3688. https://doi.org/10.1073/pnas.1013776108

Cervantes, J.L., Hawley, K.L., Benjamin, S.J., Weinerman, B., Luu, S.M., and Salazar, J.C. (2014). Phagosomal TLR signaling upon *Borrelia burgdorferi* infection. Front Cell Infect Microbiol *4*, 55.http s://doi.org/10.3389/fcimb.2014.00055

Cervantes, J.L., La Vake, C.J., Weinerman, B., Luu, S., O'Connell, C., Verardi, P.H., and Salazar, J.C. (2013). Human TLR8 is activated upon recognition of *Borrelia burgdorferi* RNA in the phagosome of human monocytes. J Leukoc Biol *94*, 1231-1241. https://doi.org/10.1189/jlb.0413206

Chamoun, M.N., Blumenthal, A., Sullivan, M.J., Schembri, M.A., and Ulett, G.C. (2018). Bacterial pathogenesis and interleukin-17: interconnecting mechanisms of immune regulation, host genetics, and microbial virulence that influence severity of infection. Crit Rev Microbiol *44*, 465-486. https://doi.org/10.1080/1040841X.2018.1426556

Charon, N.W., and Goldstein, S.F. (2002). Genetics of motility and chemotaxis of a fascinating group of bacteria: the spirochetes. Annu Rev Genet *36*, 47-73. https://doi.org/10.1146/annurev.genet.36.041602.134359 041602.134359 [pii]

Chauhan, V.S., Sterka, D.G., Jr., Furr, S.R., Young, A.B., and Marriott, I. (2009). NOD2 plays an important role in the inflammatory responses of microglia and astrocytes to bacterial CNS pathogens. Glia *57*, 414-423. https://doi.org/10.1002/glia.20770

Chien, Y.H., Meyer, C., and Bonneville, M. (2014). Gammadelta T cells: first line of defense and beyond. Annu Rev Immunol *32*, 121-155. https://doi.org/10.1146/annurev-immunol-032713-120216

Chung, Y., Zhang, N., and Wooten, R.M. (2013). *Borrelia burgdorferi* elicited-IL-10 suppresses the production of inflammatory mediators, phagocytosis, and expression of co-stimulatory receptors by murine macrophages and/or dendritic cells. PLoS One *8*, e84980. https://doi.org/10.1371/journal.pone.0084980

Cinco, M., Murgia, R., Perticarari, S., and Presani, G. (1998). Surface receptors of neutrophils towards *B. burgdorferi*. Wien Klin Wochenschr *110*, 866-869.

Cinco, M., Murgia, R., Presani, G., and Perticarari, S. (1997). Integrin CR3 mediates the binding of nonspecifically opsonized *Borrelia burgdorferi* to human phagocytes and mammalian cells. Infect Immun *65*, 4784-4789.

Coleman, A.S., Rossmann, E., Yang, X., Song, H., Lamichhane, C.M., Iyer, R., Schwartz, I., and Pal, U. (2011). BBK07 immunodominant peptides as serodiagnostic markers of Lyme disease. Clin Vaccine Immunol *18*, 406-413. https://doi.org/10.1128/CVI.00461-10

Coleman, J.L., Sellati, T.J., Testa, J.E., Kew, R.R., Furie, M.B., and Benach, J.L. (1995). *Borrelia burgdorferi* binds plasminogen, resulting in enhanced penetration of endothelial monolayers. Infect Immun *63*, 2478-2484.

Collins, C., Lui, Y., Santos, A.M., Ballif, B.A., Gogerly-Moragoda, A.M., Brouwer, H., Ross, R., Balagurunathan, K., Sharma, S., Wright, G.J., *et al.* (2019). Detection of Cell Surface Ligands for Human Synovial gammadelta T Cells. J Immunol *203*, 2369-2376. https://doi.org/10.4049/jimmunol.1900451

Collins, C.C., Bashant, K., Erikson, C., Thwe, P.M., Fortner, K.A., Wang, H., Morita, C.T., and Budd, R.C. (2016). Necroptosis of Dendritic Cells Promotes Activation of gammadelta T Cells. J Innate Immun *8*, 479-492. https://doi.org/10.1159/000446498

Comstock, L.E., and Thomas, D.D. (1991). Characterization of *Borrelia burgdorferi* invasion of cultured endothelial cells. Microb Pathog *10*, 137-148.

Coutte, L., Botkin, D.J., Gao, L., and Norris, S.J. (2009). Detailed analysis of sequence changes occurring during vlsE antigenic variation in the mouse model of Borrelia burgdorferi infection. PLoS Pathog *5*, e1000293. https://doi.org/10.1371/journal.ppat.1000293

Couvreur, B., Beaufays, J., Charon, C., Lahaye, K., Gensale, F., Denis, V., Charloteaux, B., Decrem, Y., Prevot, P.P., Brossard, M., *et al.* (2008). Variability and action mechanism of a family of anticomplement proteins in *Ixodes ricinus*. PLoS One *3*, e1400. https://doi.org/10.1371/journal.pone.0001400

Craft, J.E., Fischer, D.K., Shimamoto, G.T., and Steere, A.C. (1986). Antigens of *Borrelia burgdorferi* recognized during Lyme disease. Appearance of a new immunoglobulin M response and expansion of the immunoglobulin G response late in the illness. J Clin Invest *78*, 934-939.

Crandall, H., Dunn, D.M., Ma, Y., Wooten, R.M., Zachary, J.F., Weis, J.H., Weiss, R.B., and Weis, J.J. (2006). Gene expression profiling reveals unique pathways associated with differential severity of Lyme arthritis. J Immunol *177*, 7930-7942. https://doi.org/10.4049/jimmunol.177.11.7930

Crandall, H., Ma, Y., Dunn, D.M., Sundsbak, R.S., Zachary, J.F., Olofsson, P., Holmdahl, R., Weis, J.H., Weiss, R.B., Teuscher, C., *et al.* (2005). Bb2Bb3 regulation of murine Lyme arthritis is distinct from Ncf1 and independent of the phagocyte nicotinamide adenine dinucleotide phosphate oxidase. Am J Pathol *167*, 775-785.

Crowley, J.T., Drouin, E.E., Pianta, A., Strle, K., Wang, Q., Costello, C.E., and Steere, A.C. (2015). A Highly Expressed Human Protein, Apolipoprotein B-100, Serves as an Autoantigen in a Subgroup of Patients With Lyme Disease. J Infect Dis *212*, 1841-1850. https://doi.org/10.1093/infdis/jiv310

Crowley, J.T., Strle, K., Drouin, E.E., Pianta, A., Arvikar, S.L., Wang, Q., Costello, C.E., and Steere, A.C. (2016). Matrix metalloproteinase-10 is a target of T and B cell responses that correlate with synovial pathology in patients with antibiotic-refractory Lyme arthritis. J Autoimmun *69*, 24-37.http s://doi.org/10.1016/j.jaut.2016.02.005

Crowley, M.P., Fahrer, A.M., Baumgarth, N., Hampl, J., Gutgemann, I., Teyton, L., and Chien, Y. (2000). A population of murine gammadelta T cells that recognize an inducible MHC class Ib molecule. Science *287*, 314-316. https://doi.org/10.1126/science.287.5451.314

Cruz, A.R., Moore, M.W., La Vake, C.J., Eggers, C.H., Salazar, J.C., and Radolf, J.D. (2008). Phagocytosis of *Borrelia burgdorferi*, the Lyme disease spirochete, potentiates innate immune activation and induces apoptosis in human monocytes. Infect Immun *76*, 56-70.

Cui, J., Chen, Y., Wang, H.Y., and Wang, R.F. (2014). Mechanisms and pathways of innate immune activation and regulation in health and cancer. Hum Vaccin Immunother *10*, 3270-3285. https://doi.org/10.4161/21645515.2014.979640

Cuthbert, R.J., Fragkakis, E.M., Dunsmuir, R., Li, Z., Coles, M., Marzo-Ortega, H., Giannoudis, P.V., Jones, E., El-Sherbiny, Y.M., and McGonagle, D. (2017). Brief Report: Group 3 Innate Lymphoid Cells in Human Enthesis. Arthritis Rheumatol *69*, 1816-1822. https://doi.org/10.1002/art.40150

D'Arco, C., Dattwyler, R.J., and Arnaboldi, P.M. (2017). *Borrelia burgdorferi*-specific IgA in Lyme Disease. EBioMedicine *19*, 91-97.http s://doi.org/10.1016/j.ebiom.2017.04.025

D'Souza, M.P., Adams, E., Altman, J.D., Birnbaum, M.E., Boggiano, C., Casorati, G., Chien, Y.H., Conley, A., Eckle, S.B.G., Fruh, K., *et al.* (2019). Casting a wider net: Immunosurveillance by nonclassical MHC molecules. PLoS Pathog *15*, e1007567. https://doi.org/10.1371/journal.ppat.1007567

Daix, V., Schroeder, H., Praet, N., Georgin, J.P., Chiappino, I., Gillet, L., de Fays, K., Decrem, Y., Leboulle, G., Godfroid, E., *et al.* (2007). *Ixodes* ticks belonging to the *Ixodes ricinus* complex encode a family of anticomplement proteins. Insect Mol Biol *16*, 155-166. https://doi.org/10.1111/j.1365-2583.2006.00710.x

de Silva, A.M., Telford, S.R., 3rd, Brunet, L.R., Barthold, S.W., and Fikrig, E. (1996). *Borrelia burgdorferi* OspA is an arthropod-specific transmission-blocking Lyme disease vaccine. J Exp Med *183*, 271-275.

de Souza, M.S., Fikrig, E., Smith, A.L., Flavell, R.A., and Barthold, S.W. (1992). Nonspecific proliferative responses of murine lymphocytes to *Borrelia burgdorferi* antigens. J Infect Dis *165*, 471-478.

de Souza, M.S., Smith, A.L., Beck, D.S., Terwilliger, G.A., Fikrig, E., and Barthold, S.W. (1993). Long-term study of cell-mediated responses to *Borrelia burgdorferi* in the laboratory mouse. Infect Immun *61*, 1814-1822.

Defosse, D.L., and Johnson, R.C. (1992). *In vitro* and *in vivo* induction of tumor necrosis factor alpha by *Borrelia burgdorferi*. Infect Immun *60*, 1109-1113.

Dennis, V.A., Dixit, S., O'Brien, S.M., Alvarez, X., Pahar, B., and Philipp, M.T. (2009). Live *Borrelia burgdorferi* spirochetes elicit inflammatory mediators from human monocytes via the Toll-like receptor signaling pathway. Infect Immun *77*, 1238-1245.

Dennis, V.A., Jefferson, A., Singh, S.R., Ganapamo, F., and Philipp, M.T. (2006). Interleukin-10 anti-inflammatory response to Borrelia burgdorferi, the agent of Lyme disease: a possible role for suppressors of cytokine signaling 1 and 3. Infect Immun *74*, 5780-5789.

Dhodapkar, M.V., and Kumar, V. (2017). Type II NKT Cells and Their Emerging Role in Health and Disease. J Immunol *198*, 1015-1021.http s://doi.org/10.4049/jimmunol.1601399

Divan, A., Budd, R.C., Tobin, R.P., and Newell-Rogers, M.K. (2015). Gammadelta T Cells and dendritic cells in refractory Lyme arthritis. J Leukoc Biol *97*, 653-663. https://doi.org/10.1189/jlb.2RU0714-343RR

Divan, A., Casselli, T., Narayanan, S.A., Mukherjee, S., Zawieja, D.C., Watt, J.A., Brissette, C.A., and Newell-Rogers, M.K. (2018). *Borrelia burgdorferi* adhere to blood vessels in the dura mater and are associated with increased meningeal T cells during murine disseminated borreliosis. PLoS One *13*, e0196893. https://doi.org/10.1371/journal.pone.0196893

Douda, D.N., Khan, M.A., Grasemann, H., and Palaniyar, N. (2015). SK3 channel and mitochondrial ROS mediate NADPH oxidase-independent NETosis induced by calcium influx. Proc Natl Acad Sci U S A *112*, 2817-2822. https://doi.org/10.1073/pnas.1414055112

Drouin, E.E., Seward, R.J., Strle, K., McHugh, G., Katchar, K., Londono, D., Yao, C., Costello, C.E., and Steere, A.C. (2013). A novel human autoantigen, endothelial cell growth factor, is a target of T and B cell responses in patients with Lyme disease. Arthritis Rheum *65*, 186-196. https://doi.org/10.1002/art.37732

Dunham-Ems, S.M., Caimano, M.J., Pal, U., Wolgemuth, C.W., Eggers, C.H., Balic, A., and Radolf, J.D. (2009). Live imaging reveals a biphasic mode of dissemination of *Borrelia burgdorferi* within ticks. J Clin Invest *119*, 3652-3665. https://doi.org/10.1172/JCI39401

Duray, P.H., and Steere, A.C. (1988). Clinical pathologic correlations of Lyme disease by stage. Ann N Y Acad Sci *539*, 65-79.

Ebnet, K., Brown, K.D., Siebenlist, U.K., Simon, M.M., and Shaw, S. (1997). *Borrelia burgdorferi* activates nuclear factor-kappa B and is a potent inducer of chemokine and adhesion molecule gene expression in endothelial cells and fibroblasts. J Immunol *158*, 3285-3292.

Eckman, E.A., Pacheco-Quinto, J., Herdt, A.R., and Halperin, J.J. (2018). Neuroimmunomodulators in Neuroborreliosis and Lyme Encephalopathy. Clin Infect Dis *67*, 80-88. https://doi.org/10.1093/cid/ciy019

Eicken, C., Sharma, V., Klabunde, T., Lawrenz, M.B., Hardham, J.M., Norris, S.J., and Sacchettini, J.C. (2002). Crystal structure of Lyme disease variable surface antigen VlsE of Borrelia burgdorferi. J Biol Chem *277*, 21691-21696. https://doi.org/10.1074/jbc.M201547200

Elsner, R.A., Hastey, C.J., and Baumgarth, N. (2015a). CD4+ T cells promote antibody production but not sustained affinity maturation during *Borrelia burgdorferi i*nfection. Infect Immun *83*, 48 - 56. https://doi.org/IAI.02471-14 [pii] 10.1128/IAI.02471-14

Elsner, R.A., Hastey, C.J., Olsen, K.J., and Baumgarth, N. (2015b). Suppression of long-lived immunity following Borrelia burgdorferi induced Lyme disease. PLoS Pathog *11*, e1004976.

Embers, M.E., Alvarez, X., Ooms, T., and Philipp, M.T. (2008). The failure of immune response evasion by linear plasmid 28-1-deficient *Borrelia burgdorferi* is attributable to persistent expression of an outer surface protein. Infect Immun *76*, 3984-3991. https://doi.org/IAI.00387-08 [pii] 10.1128/IAI.00387-08

Fahrer, H., van der Linden, S.M., Sauvain, M.J., Gern, L., Zhioua, E., and Aeschlimann, A. (1991). The prevalence and incidence of clinical and asymptomatic Lyme borreliosis in a population at risk. J Infect Dis *163*, 305-310. https://doi.org/10.1093/infdis/163.2.305

Feehan, K.T., and Gilroy, D.W. (2019). Is Resolution the End of Inflammation? Trends Mol Med *25*, 198-214. https://doi.org/10.1016/j.molmed.2019.01.006

Fikrig, E., Barthold, S.W., Chen, M., Chang, C.H., and Flavell, R.A. (1997a). Protective antibodies develop, and murine Lyme arthritis regresses, in the absence of MHC class II and CD4+ T cells. J Immunol *159*, 5682-5686.

Fikrig, E., Barthold, S.W., Chen, M., Grewal, I.S., Craft, J., and Flavell, R.A. (1996). Protective antibodies in murine Lyme disease arise independently of CD40 ligand. J Immunol *157*, 1-3.

Fikrig, E., Barthold, S.W., Kantor, F.S., and Flavell, R.A. (1990). Protection of mice against the Lyme disease agent by immunizing with recombinant OspA. Science *250*, 553-556.

Fikrig, E., Barthold, S.W., Marcantonio, N., Deponte, K., Kantor, F.S., and Flavell, R.A. (1992). Roles of OspA, OspB, and flagellin in protective immunity to Lyme borreliosis in laboratory mice. Infect Immun *60*, 657-661.

Fikrig, E., Barthold, S.W., Sun, W., Feng, W., Telford, S.R., 3rd, and Flavell, R.A. (1997b). *Borrelia burgdorferi* P35 and P37 proteins, expressed in vivo, elicit protective immunity. Immunity *6*, 531-539.

Filgueira, L., Nestle, F.O., Rittig, M., Joller, H.I., and Groscurth, P. (1996). Human dendritic cells phagocytose and process *Borrelia burgdorferi*. J Immunol *157*, 2998-3005.

Forrester, J.V., McMenamin, P.G., and Dando, S.J. (2018). CNS infection and immune privilege. Nat Rev Neurosci *19*, 655-671. https://doi.org/10.1038/s41583-018-0070-8

Francischetti, I.M., Mather, T.N., and Ribeiro, J.M. (2004). Penthalaris, a novel recombinant five-Kunitz tissue factor pathway inhibitor (TFPI) from the salivary gland of the tick vector of Lyme disease, Ixodes scapularis. Thromb Haemost *91*, 886-898.

Francischetti, I.M., Mather, T.N., and Ribeiro, J.M. (2005). Tick saliva is a potent inhibitor of endothelial cell proliferation and angiogenesis. Thromb Haemost *94*, 167-174.

Francischetti, I.M., Valenzuela, J.G., Andersen, J.F., Mather, T.N., and Ribeiro, J.M. (2002). Ixolaris, a novel recombinant tissue factor pathway inhibitor (TFPI) from the salivary gland of the tick, *Ixodes scapularis:* identification of factor X and factor Xa as scaffolds for the inhibition of factor VIIa/tissue factor complex. Blood *99*, 3602-3612.

Frischknecht, F. (2007). The skin as interface in the transmission of arthropod-borne pathogens. Cell Microbiol *9*, 1630-1640.

Frossi, B., Mion, F., Tripodo, C., Colombo, M.P., and Pucillo, C.E. (2017). Rheostatic Functions of Mast Cells in the Control of Innate and Adaptive Immune Responses. Trends Immunol *38*, 648-656.https://doi.org/10.1016/j.it.2017.04.001

Gajovic, O., Todorovic, Z., Nesic, L., and Lazic, Z. (2010). [Lyme borreliosis--diagnostic difficulties in interpreting serological results]. Med Pregl *63*, 839-843.

Garcia, R., Gusmani, L., Murgia, R., Guarnaccia, C., Cinco, M., and Rottini, G. (1998). Elastase is the only human neutrophil granule protein that alone is responsible for in vitro killing of *Borrelia burgdorferi*. Infect Immun *66*, 1408-1412.

Garg, R., Juncadella, I.J., Ramamoorthi, N., Ashish, Ananthanarayanan, S.K., Thomas, V., Rincon, M., Krueger, J.K., Fikrig, E., Yengo, C.M., *et al.* (2006). Cutting edge: CD4 is the receptor for the tick saliva immunosuppressor, Salp15. J Immunol *177*, 6579-6583.

Geissmann, F., Manz, M.G., Jung, S., Sieweke, M.H., Merad, M., and Ley, K. (2010). Development of monocytes, macrophages, and dendritic cells. Science *327*, 656-661. https://doi.org/10.1126/science.1178331

Georgilis, K., Noring, R., Steere, A.C., and Klempner, M.S. (1991a). Neutrophil chemotactic factors in synovial fluids of patients with Lyme disease. Arthritis Rheum *34*, 770-775.

Georgilis, K., Steere, A.C., and Klempner, M.S. (1991b). Infectivity of *Borrelia burgdorferi* correlates with resistance to elimination by phagocytic cells. J Infect Dis *163*, 150-155.

Giambartolomei, G.H., Dennis, V.A., and Philipp, M.T. (1998). Borrelia burgdorferi stimulates the production of interleukin-10 in peripheral blood mononuclear cells from uninfected humans and rhesus monkeys. Infect Immun *66*, 2691-2697.

Gillespie, R.D., Dolan, M.C., Piesman, J., and Titus, R.G. (2001). Identification of an IL-2 binding protein in the saliva of the Lyme disease vector tick, Ixodes scapularis. J Immunol *166*, 4319-4326.

Glickstein, L., Edelstein, M., and Dong, J.Z. (2001). Gamma interferon is not required for arthritis resistance in the murine Lyme disease model. Infect Immun *69*, 3737-3743.

Godfrey, D.I., Koay, H.F., McCluskey, J., and Gherardin, N.A. (2019). The biology and functional importance of MAIT cells. Nat Immunol *20*, 1110-1128. https://doi.org/10.1038/s41590-019-0444-8

Golightly, M., Thomas, J., Volkman, D., and Dattwyler, R. (1988). Modulation of natural killer cell activity by *Borrelia burgdorferi*. Ann N Y Acad Sci *539*, 103-111. https://doi.org/10.1111/j.1749-6632.1988.tb31843.x

Greene, R.T., Levine, J.F., Breitschwerdt, E.B., Walker, R.L., Berkhoff, H.A., Cullen, J., and Nicholson, W.L. (1988). Clinical and serologic evaluations of induced Borrelia burgdorferi infection in dogs. Am J Vet Res *49*, 752-757.

Greenmyer, J.R., Gaultney, R.A., Brissette, C.A., and Watt, J.A. (2018). Primary Human Microglia Are Phagocytically Active and Respond to Borrelia burgdorferi With Upregulation of Chemokines and Cytokines. Front Microbiol *9*, 811. https://doi.org/10.3389/fmicb.2018.00811

Grimm, D., Tilly, K., Byram, R., Stewart, P.E., Krum, J.G., Bueschel, D.M., Schwan, T.G., Policastro, P.F., Elias, A.F., and Rosa, P.A. (2004). Outer-surface protein C of the Lyme disease spirochete: a protein induced in ticks for infection of mammals. Proc Natl Acad Sci U S A *101*, 3142-3147.

Gross, D.M., Forsthuber, T., Tary-Lehmann, M., Etling, C., Ito, K., Nagy, Z.A., Field, J.A., Steere, A.C., and Huber, B.T. (1998a). Identification of LFA-1 as a candidate autoantigen in treatment-resistant Lyme arthritis. Science *281*, 703-706. https://doi.org/10.1126/science.281.5377.703

Gross, D.M., Steere, A.C., and Huber, B.T. (1998b). T helper 1 response is dominant and localized to the synovial fluid in patients with Lyme arthritis. J Immunol *160*, 1022-1028.

Guo, X., Booth, C.J., Paley, M.A., Wang, X., DePonte, K., Fikrig, E., Narasimhan, S., and Montgomery, R.R. (2009). Inhibition of neutrophil function by two tick salivary proteins. Infect Immun *77*, 2320-2329.

Hagan, T., Cortese, M., Rouphael, N., Boudreau, C., Linde, C., Maddur, M.S., Das, J., Wang, H., Guthmiller, J., Zheng, N.Y., *et al.* (2019). Antibiotics-Driven Gut Microbiome Perturbation Alters Immunity to Vaccines in Humans. Cell *178*, 1313-1328 e1313. https://doi.org/10.1016/j.cell.2019.08.010

Hallstrom, T., Siegel, C., Morgelin, M., Kraiczy, P., Skerka, C., and Zipfel, P.F. (2013). CspA from *Borrelia burgdorferi* inhibits the terminal complement pathway. mBio *4*. https://doi.org/10.1128/mBio.00481-13

Halperin, J.J. (2019). A Neurologist's View of Lyme Disease and Other Tick-Borne Infections. Semin Neurol *39*, 440-447. https://doi.org/10.1055/s-0039-1692143

Halperin, J.J., Pass, H.L., Anand, A.K., Luft, B.J., Volkman, D.J., and Dattwyler, R.J. (1988). Nervous system abnormalities in Lyme disease. Ann N Y Acad Sci *539*, 24-34.

Hammers-Berggren, S., Lebech, A.M., Karlsson, M., Svenungsson, B., Hansen, K., and Stiernstedt, G. (1994). Serological follow-up after treatment of patients with erythema migrans and neuroborreliosis. J Clin Microbiol *32*, 1519-1525.

Hammerschmidt, C., Klevenhaus, Y., Koenigs, A., Hallstrom, T., Fingerle, V., Skerka, C., Pos, K.M., Zipfel, P.F., Wallich, R., and Kraiczy, P. (2016). BGA66 and BGA71 facilitate complement resistance of *Borrelia bavariensis* by inhibiting assembly of the membrane attack complex. Mol Microbiol *99*, 407-424. https://doi.org/10.1111/mmi.13239

Hanson, M.S., Cassatt, D.R., Guo, B.P., Patel, N.K., McCarthy, M.P., Dorward, D.W., and Hook, M. (1998). Active and passive immunity against *Borrelia burgdorferi* decorin binding protein A (DbpA) protects against infection. Infect Immun *66*, 2143-2153.

Hardin, J.A., Steere, A.C., and Malawista, S.E. (1984). The pathogenesis of arthritis in Lyme disease: humoral immune responses and the role of intra-articular immune complexes. Yale J Biol Med *57*, 589-593.

Hardin, J.A., Walker, L.C., Steere, A.C., Trumble, T.C., Tung, K.S., Williams, R.C., Jr., Ruddy, S., and Malawista, S.E. (1979). Circulating immune complexes in Lyme arthritis. Detection by the 125I-C1q binding, C1q solid phase, and Raji cell assays. J Clin Invest *63*, 468-477. https://doi.org/10.1172/JCI109324

Harman, M.W., Dunham-Ems, S.M., Caimano, M.J., Belperron, A.A., Bockenstedt, L.K., Fu, H.C., Radolf, J.D., and Wolgemuth, C.W. (2012). The heterogeneous motility of the Lyme disease spirochete in gelatin mimics dissemination through tissue. Proc Natl Acad Sci U S A *109*, 3059-3064. https://doi.org/10.1073/pnas.1114362109

Hastey, C.J., Elsner, R.A., Barthold, S.W., and Baumgarth, N. (2012). Delays and diversions mark the development of B cell responses to *Borrelia burgdorferi* infection. J Immunol *188*, 5612 - 5622.

Hastey, C.J., Elsner, R.A., Olsen, K.J., Mundigl, S., Barthold, S.W., and Baumgarth, N. (2020). *Borrelia burgdorferi* infection induces ongoing IgM production that controls bacteremia but not bacterial dissemination or tissue burden in mice. Submitted.

Hastey, C.J., Elsner, R.A., Olsen, K.J., Tunev, S.S., Escobar, E.D., Barthold, S.W., and Baumgarth, N. (2016). Compartmentalization of humoral immunity to *Borrelia burgdorferi i*nfection. In preparation.

Hastey, C.J., Ochoa, J., Olsen, K.J., Barthold, S.W., and Baumgarth, N. (2014). MyD88- and TRIF-independent induction of type I interferon drives naive B cell accumulation but not loss of lymph node architecture in Lyme disease. Infect Immun *82*, 1548-1558. https://doi.org/IAI.00969-13 [pii] 10.1128/IAI.00969-13

Hawley, K.L., Olson, C.M., Jr., Carreras-González, A., Navasa, N., and Anguita, J. (2015). Serum C3 Enhances Complement Receptor 3-Mediated Phagocytosis of *Borrelia burgdorferi*. Int J Biol Sci *11*, 1269-1271.http s://doi.org/10.7150/ijbs.13395

Hawley, K.L., Olson, C.M., Jr., Iglesias-Pedraz, J.M., Navasa, N., Cervantes, J.L., Caimano, M.J., Izadi, H., Ingalls, R.R., Pal, U., Salazar, J.C., *et al.* (2012). CD14 cooperates with complement receptor 3 to mediate MyD88-independent phagocytosis of *Borrelia burgdorferi*. Proc Natl Acad Sci U S A *109*, 1228-1232. https://doi.org/10.1073/pnas.1112078109

He, L., and Marneros, A.G. (2014). Doxycycline inhibits polarization of macrophages to the proangiogenic M2-type and subsequent neovascularization. J Biol Chem *289*, 8019-8028. https://doi.org/10.1074/jbc.M113.535765

Heilpern, A.J., Wertheim, W., He, J., Perides, G., Bronson, R.T., and Hu, L.T. (2009). Matrix metalloproteinase-9 plays a key role in Lyme arthritis but not dissemination of *Borrelia burgdorferi.* Infect Immun *77*, 2643-2649. https://doi.org/10.1128/IAI.00214-09

Hirschfeld, M., Kirschning, C.J., Schwandner, R., Wesche, H., Weis, J.H., Wooten, R.M., and Weis, J.J. (1999). Cutting edge: inflammatory signaling by *Borrelia burgdorferi* lipoproteins is mediated by toll-like receptor 2. J Immunol *163*, 2382-2386.

Hirschfeld, M., Ma, Y., Weis, J.H., Vogel, S.N., and Weis, J.J. (2000). Repurification of LPS eliminates signaling through both human and murine Toll-like receptor 2. J Immunol *165*, 618-622.

Horka, H., Cerna-Kyckova, K., Skallova, A., and Kopecky, J. (2009). Tick saliva affects both proliferation and distribution of *Borrelia burgdorferi* spirochetes in mouse organs and increases transmission of spirochetes to ticks. Int J Med Microbiol *299*, 373-380. https://doi.org/10.1016/j.ijmm.2008.10.009

Hovius, J.W., de Jong, M.A., den Dunnen, J., Litjens, M., Fikrig, E., van der Poll, T., Gringhuis, S.I., and Geijtenbeek, T.B. (2008). Salp15 binding to DC-SIGN inhibits cytokine expression by impairing both nucleosome remodeling and mRNA stabilization. PLoS Pathog *4*, e31. https://doi.org/10.1371/journal.ppat.0040031

Hu, L.T., Eskildsen, M.A., Masgala, C., Steere, A.C., Arner, E.C., Pratta, M.A., Grodzinsky, A.J., Loening, A., and Perides, G. (2001). Host metalloproteinases in Lyme arthritis. Arthritis Rheum *44*, 1401-1410.

Huegli, D., Moret, J., Rais, O., Moosmann, Y., Erard, P., Malinverni, R., and Gern, L. (2011). Prospective study on the incidence of infection by *Borrelia burgdorferi sensu lato* after a tick bite in a highly endemic area of Switzerland. Ticks Tick Borne Dis *2*, 129-136. https://doi.org/10.1016/j.ttbdis.2011.05.002

Hulinska, D., Bartak, P., Hercogova, J., Hancil, J., Basta, J., and Schramlova, J. (1994). Electron microscopy of Langerhans cells and *Borrelia burgdorferi* in Lyme disease patients. Zentralbl Bakteriol *280*, 348-359.

Hyde, J.A. (2017). Borrelia burgdorferi Keeps Moving and Carries on: A Review of Borrelial Dissemination and Invasion. Front Immunol *8*, 114. https://doi.org/10.3389/fimmu.2017.00114

Iliopoulou, B.P., Alroy, J., and Huber, B.T. (2007). CD28 deficiency exacerbates joint inflammation upon *Borrelia burgdorferi* infection, resulting in the development of chronic Lyme arthritis. J Immunol *179*, 8076-8082.

Imai, D.M., Samuels, D.S., Feng, S., Hodzic, E., Olsen, K., and Barthold, S.W. (2013). The early dissemination defect attributed to disruption of decorin-binding proteins is abolished in chronic murine Lyme borreliosis. Infect Immun *81*, 1663-1673. https://doi.org/10.1128/IAI.01359-12

Inoue, Y., Shimojo, N., Suzuki, Y., Campos Alberto, E.J., Yamaide, A., Suzuki, S., Arima, T., Matsuura, T., Tomiita, M., Aoyagi, M., *et al.* (2007). CD14 -550 C/T, which is related to the serum level of soluble CD14, is associated with the development of respiratory syncytial virus bronchiolitis in the Japanese population. J Infect Dis *195*, 1618-1624.

Iyer, R., and Schwartz, I. (2016). Microarray-Based Comparative Genomic and Transcriptome Analysis of *Borrelia burgdorferi.* Microarrays (Basel) *5*. https://doi.org/10.3390/microarrays5020009

Izadi, H., Motameni, A.T., Bates, T.C., Olivera, E.R., Villar-Suarez, V., Joshi, I., Garg, R., Osborne, B.A., Davis, R.J., Rincon, M., *et al.* (2007). c-Jun N-terminal kinase 1 is required for Toll-like receptor 1 gene expression in macrophages. Infect Immun *75*, 5027-5034. https://doi.org/10.1128/IAI.00492-07

Jacobsen, M., Zhou, D., Cepok, S., Nessler, S., Happel, M., Stei, S., Wilske, B., Sommer, N., and Hemmer, B. (2003). Clonal accumulation of activated CD8+ T cells in the central nervous system during the early phase of neuroborreliosis. J Infect Dis *187*, 963-973.

Jacquet, M., Durand, J., Rais, O., and Voordouw, M.J. (2015). Cross-reactive acquired immunity influences transmission success of the Lyme disease pathogen, *Borrelia afzelii.* Infect Genet Evol *36*, 131-140.http s://doi.org/10.1016/j.meegid.2015.09.012

Juncadella, I.J., Garg, R., Ananthnarayanan, S.K., Yengo, C.M., and Anguita, J. (2007). T-cell signaling pathways inhibited by the tick saliva immunosuppressor, Salp15. FEMS Immunol Med Microbiol *49*, 433-438.

Jung, S., Schickel, J.N., Kern, A., Knapp, A.M., Eftekhari, P., Da Silva, S., Jaulhac, B., Brink, R., Soulas-Sprauel, P., Pasquali, J.L., *et al.* (2016). Chronic bacterial infection activates autoreactive B cells and induces isotype switching and autoantigen-driven mutations. Eur J Immunol *46*, 131-146.http s://doi.org/10.1002/eji.201545810

Jutras, B.L., Lochhead, R.B., Kloos, Z.A., Biboy, J., Strle, K., Booth, C.J., Govers, S.K., Gray, J., Schumann, P., Vollmer, W., *et al.* (2019). *Borrelia burgdorferi* peptidoglycan is a persistent antigen in patients with Lyme arthritis. Proc Natl Acad Sci U S A *116*, 13498-13507. https://doi.org/10.1073/pnas.1904170116

Kaplan, D.H. (2017). Ontogeny and function of murine epidermal Langerhans cells. Nat Immunol *18*, 1068-1075. https://doi.org/10.1038/ni.3815

Katchar, K., Drouin, E.E., and Steere, A.C. (2013). Natural killer cells and natural killer T cells in Lyme arthritis. Arthritis Res Ther *15*, R183.http s://doi.org/10.1186/ar4373

Keane-Myers, A., and Nickell, S.P. (1995). T cell subset-dependent modulation of immunity to *Borrelia burgdorferi* in mice. J Immunol *154*, 1770-1776.

Kemp, D.H., Stone, B.F., and Binnington, K.C. (1982). Tick attachment and feeding: Role of the mouthparts, feeding apparatus, salivary gland secretions and the host response. In Physiology of Ticks, F.D. Obenchain, and R. Galun, eds. (Elmsford, NY: Pergamon Press), pp. 119-168.

Khatchikian, C.E., Nadelman, R.B., Nowakowski, J., Schwartz, I., Wormser, G.P., and Brisson, D. (2014). Evidence for strain-specific immunity in patients treated for early Lyme disease. Infect Immun *82*, 1408-1413.http s://doi.org/10.1128/IAI.01451-13

Killpack, T.L., Ballesteros, M., Bunnell, S.C., Bedugnis, A., Kobzik, L., Hu, L.T., and Petnicki-Ocwieja, T. (2017). Phagocytic receptors activate Syk and Src signaling during *Borrelia burgdorferi* phagocytosis. Infect Immun *85*. https://doi.org/10.1128/iai.00004-17

Kinjo, Y., Tupin, E., Wu, D., Fujio, M., Garcia-Navarro, R., Benhnia, M.R., Zajonc, D.M., Ben-Menachem, G., Ainge, G.D., Painter, G.F., *et al.* (2006). Natural killer T cells recognize diacylglycerol antigens from pathogenic bacteria. Nat Immunol *7*, 978-986.

Kirschning, C.J., Wesche, H., Merrill Ayres, T., and Rothe, M. (1998). Human toll-like receptor 2 confers responsiveness to bacterial lipopolysaccharide. J Exp Med *188*, 2091-2097.

Kisand, K.E., Prukk, T., Kisand, K.V., Luus, S.M., Kalbe, I., and Uibo, R. (2007). Propensity to excessive proinflammatory response in chronic Lyme borreliosis. APMIS *115*, 134-141. https://doi.org/10.1111/j.1600-0463.2007.apm_538.x

Kochi, S.K., and Johnson, R.C. (1988). Role of immunoglobulin G in killing of *Borrelia burgdorferi* by the classical complement pathway. Infect Immun *56*, 314-321.

Kochi, S.K., Johnson, R.C., and Dalmasso, A.P. (1991). Complement-mediated killing of the Lyme disease spirochete *Borrelia burgdorferi.* Role of antibody in formation of an effective membrane attack complex. J Immunol *146*, 3964-3970.

Kochi, S.K., Johnson, R.C., and Dalmasso, A.P. (1993). Facilitation of complement-dependent killing of the Lyme disease spirochete, *Borrelia burgdorferi,* by specific immunoglobulin G Fab antibody fragments. Infect Immun *61*, 2532-2536.

Kotsyfakis, M., Sa-Nunes, A., Francischetti, I.M., Mather, T.N., Andersen, J.F., and Ribeiro, J.M. (2006). Antiinflammatory and immunosuppressive activity of sialostatin L, a salivary cystatin from the tick Ixodes scapularis. J Biol Chem *281*, 26298-26307.

Kotsyfakis, M., Schwarz, A., Erhart, J., and Ribeiro, J.M. (2015). Tissue- and time-dependent transcription in *Ixodes ricinus* salivary glands and midguts when blood feeding on the vertebrate host. Sci Rep *5*, 9103.http s://doi.org/10.1038/srep09103

Kowarik, M.C., Grummel, V., Wemlinger, S., Buck, D., Weber, M.S., Berthele, A., and Hemmer, B. (2014). Immune cell subtyping in the cerebrospinal fluid of patients with neurological diseases. J Neurol *261*, 130-143.http s://doi.org/10.1007/s00415-013-7145-2

Krause, P.J., Grant-Kels, J.M., Tahan, S.R., Dardick, K.R., Alarcon-Chaidez, F., Bouchard, K., Visini, C., Deriso, C., Foppa, I.M., and Wikel, S. (2009). Dermatologic changes induced by repeated *Ixodes scapularis* bites and implications for prevention of tick-borne infection. Vector Borne Zoonotic Dis *9*, 603-610. https://doi.org/10.1089/vbz.2008.0091

Kuhlow, C.J., Garcia-Monco, J.C., Coleman, J.L., and Benach, J.L. (2005). Murine microglia are effective phagocytes for *Borrelia burgdorferi.* J Neuroimmunol *168*, 183-187.

Kumar, H., Belperron, A., Barthold, S.W., and Bockenstedt, L.K. (2000). Cutting edge: CD1d deficiency impairs murine host defense against the spirochete, *Borrelia burgdorferi.* J Immunol *165*, 4797-4801.

Kuthejlova, M., Kopecky, J., Stepanova, G., and Macela, A. (2001). Tick salivary gland extract inhibits killing of *Borrelia afzelii* spirochetes by mouse macrophages. Infect Immun *69*, 575-578.

LaRocca, T.J., and Benach, J.L. (2008). The important and diverse roles of antibodies in the host response to Borrelia infections. Curr Top Microbiol Immunol *319*, 63-103. https://doi.org/10.1007/978-3-540-73900-5_4

Lasky, C.E., Jamison, K.E., Sidelinger, D.R., Pratt, C.L., Zhang, G., and Brown, C.R. (2015a). Infection of Interleukin 17 Receptor A-Deficient C3H Mice with *Borrelia burgdorferi* Does Not Affect Their Development of Lyme Arthritis and Carditis. Infect Immun *83*, 2882-2888. https://doi.org/IAI.00533-15 [pii] 10.1128/IAI.00533-15

Lasky, C.E., Olson, R.M., and Brown, C.R. (2015b). Macrophage polarization during murine Lyme borreliosis. Infect Immun *83*, 2627-2635. https://doi.org/10.1128/iai.00369-15

Lasky, C.E., Pratt, C.L., Hilliard, K.A., Jones, J.L., and Brown, C.R. (2016). T Cells Exacerbate Lyme Borreliosis in TLR2-Deficient Mice. Front Immunol *7*, 468. https://doi.org/10.3389/fimmu.2016.00468

Lawrenz, M.B., Wooten, R.M., Zachary, J.F., Drouin, S.M., Weis, J.J., Wetsel, R.A., and Norris, S.J. (2003). Effect of complement component C3 deficiency on experimental Lyme borreliosis in mice. Infect Immun *71*, 4432-4440.

Lazarus, J.J., Kay, M.A., McCarter, A.L., and Wooten, R.M. (2008). Viable *Borrelia burgdorferi* enhances interleukin-10 production and suppresses activation of murine macrophages. Infect Immun *76*, 1153-1162.

Lazarus, J.J., Meadows, M.J., Lintner, R.E., and Wooten, R.M. (2006). IL-10 deficiency promotes increased *Borrelia burgdorferi* clearance predominantly through enhanced innate immune responses. J Immunol *177*, 7076-7085.

Li, L., Narayan, K., Pak, E., and Pachner, A.R. (2006). Intrathecal antibody production in a mouse model of Lyme neuroborreliosis. J Neuroimmunol *173*, 56-68. https://doi.org/10.1016/j.jneuroim.2005.11.019

Li, X., McHugh, G.A., Damle, N., Sikand, V.K., Glickstein, L., and Steere, A.C. (2011). Burden and viability of *Borrelia burgdorferi* in skin and joints of patients with erythema migrans or lyme arthritis. Arthritis Rheum *63*, 2238-2247. https://doi.org/10.1002/art.30384

Lien, E., Sellati, T.J., Yoshimura, A., Flo, T.H., Rawadi, G., Finberg, R.W., Carroll, J.D., Espevik, T., Ingalls, R.R., Radolf, J.D., *et al.* (1999). Toll-like receptor 2 functions as a pattern recognition receptor for diverse bacterial products. J Biol Chem *274*, 33419-33425. https://doi.org/10.1074/jbc.274.47.33419

Lieskovska, J., Palenikova, J., Langhansova, H., Campos Chagas, A., Calvo, E., Kotsyfakis, M., and Kopecky, J. (2015). Tick sialostatins L and L2 differentially influence dendritic cell responses to Borrelia spirochetes. Parasit Vectors *8*, 275. https://doi.org/10.1186/s13071-015-0887-1

Lim, L.C., England, D.M., DuChateau, B.K., Glowacki, N.J., and Schell, R.F. (1995). *Borrelia burgdorferi*-specific T lymphocytes induce severe destructive Lyme arthritis. Infect Immun *63*, 1400-1408.

Lin, B., Noring, R., Steere, A.C., Klempner, M.S., and Hu, L.T. (2000). Soluble CD14 levels in the serum, synovial fluid, and cerebrospinal fluid of patients with various stages of Lyme disease. J Infect Dis *181*, 1185-1188.

Lin, Y.P., Benoit, V., Yang, X., Martinez-Herranz, R., Pal, U., and Leong, J.M. (2014). Strain-specific variation of the decorin-binding adhesin DbpA influences the tissue tropism of the lyme disease spirochete. PLoS Pathog *10*, e1004238. https://doi.org/10.1371/journal.ppat.1004238

Lin, Y.P., Diuk-Wasser, M.A., Stevenson, B., and Kraiczy, P. (2020). Complement Evasion Contributes to Lyme Borreliae-Host Associations. Trends Parasitol. https://doi.org/10.1016/j.pt.2020.04.011

Linder, S., Heimerl, C., Fingerle, V., Aepfelbacher, M., and Wilske, B. (2001). Coiling phagocytosis of Borrelia burgdorferi by primary human macrophages is controlled by CDC42Hs and Rac1 and involves recruitment of Wiskott-Aldrich syndrome protein and Arp2/3 complex. Infect Immun *69*, 1739-1746.

Lisinski, T.J., and Furie, M.B. (2002a). Interleukin-10 inhibits proinflammatory activation of endothelium in response to *Borrelia burgdorferi* or lipopolysaccharide but not interleukin-1β or tumor necrosis factor α. J Leukoc Biol *72*, 503-511.

Lisinski, T.J., and Furie, M.B. (2002b). Interleukin-10 inhibits proinflammatory activation of endothelium in response to *Borrelia burgdorferi* or lipopolysaccharide but not interleukin-1beta or tumor necrosis factor alpha. J Leukoc Biol *72*, 503-511.

Liu, N., Belperron, A.A., Booth, C.J., and Bockenstedt, L.K. (2009). The caspase 1 inflammasome is not required for control of murine Lyme borreliosis. Infect Immun *77*, 3320-3327. https://doi.org/10.1128/IAI.00100-09

Liu, N., Montgomery, R.R., Barthold, S.W., and Bockenstedt, L.K. (2004). Myeloid differentiation antigen 88 deficiency impairs pathogen clearance but does not alter inflammation in *Borrelia burgdorferi*-infected mice. Infect Immun *72*, 3195-3203.

Liveris, D., Wang, G., Girao, G., Byrne, D.W., Nowakowski, J., McKenna, D., Nadelman, R., Wormser, G.P., and Schwartz, I. (2002). Quantitative detection of *Borrelia burgdorferi* in 2-millimeter skin samples of erythema migrans lesions: correlation of results with clinical and laboratory findings. J Clin Microbiol *40*, 1249-1253. https://doi.org/10.1128/jcm.40.4.1249-1253.2002

Lochhead, R.B., Arvikar, S.L., Aversa, J.M., Sadreyev, R.I., Strle, K., and Steere, A.C. (2019a). Robust interferon signature and suppressed tissue repair gene expression in synovial tissue from patients with postinfectious, *Borrelia burgdorferi*-induced Lyme arthritis. Cell Microbiol *21*, e12954. https://doi.org/10.1111/cmi.12954

Lochhead, R.B., Ordonez, D., Arvikar, S.L., Aversa, J.M., Oh, L.S., Heyworth, B., Sadreyev, R., Steere, A.C., and Strle, K. (2019b). Interferon-gamma production in Lyme arthritis synovial tissue promotes differentiation of fibroblast-like synoviocytes into immune effector cells. Cell Microbiol *21*, e12992. https://doi.org/10.1111/cmi.12992

Lochhead, R.B., Sonderegger, F.L., Ma, Y., Brewster, J.E., Cornwall, D., Maylor-Hagen, H., Miller, J.C., Zachary, J.F., Weis, J.H., and Weis, J.J. (2012). Endothelial cells and fibroblasts amplify the arthritogenic type I IFN response in murine Lyme disease and are major sources of chemokines in *Borrelia burgdorferi*-infected joint tissue. J Immunol *189*, 2488-2501. https://doi.org/jimmunol.1201095 [pii] 10.4049/jimmunol.1201095

Lochhead, R.B., Strle, K., Kim, N.D., Kohler, M.J., Arvikar, S.L., Aversa, J.M., and Steere, A.C. (2017). MicroRNA Expression Shows Inflammatory Dysregulation and Tumor-Like Proliferative Responses in Joints of Patients With Postinfectious Lyme Arthritis. Arthritis Rheumatol *69*, 1100-1110.http s://doi.org/10.1002/art.40039

Londono, D., Cadavid, D., Drouin, E.E., Strle, K., McHugh, G., Aversa, J.M., and Steere, A.C. (2014). Antibodies to endothelial cell growth factor and obliterative microvascular lesions in the synovium of patients with antibiotic-refractory Lyme arthritis. Arthritis Rheumatol *66*, 2124-2133. https://doi.org/10.1002/art.38618

Love, A.C., Schwartz, I., and Petzke, M.M. (2014). *Borrelia burgdorferi* RNA induces type I and III interferons via Toll-like receptor 7 and contributes to production of NF-kappaB-dependent cytokines. Infect Immun *82*, 2405-2416. https://doi.org/10.1128/IAI.01617-14

Love, A.C., Schwartz, I., and Petzke, M.M. (2015). Induction of indoleamine 2,3-dioxygenase by *Borrelia burgdorferi* in human immune cells correlates with pathogenic potential. J Leukoc Biol *97*, 379-390.http s://doi.org/10.1189/jlb.4A0714-339R

Lunemann, J.D., Gelderblom, H., Sospedra, M., Quandt, J.A., Pinilla, C., Marques, A., and Martin, R. (2007). Cerebrospinal fluid-infiltrating CD4+ T cells recognize *Borrelia burgdorferi* lysine-enriched protein domains and central nervous system autoantigens in early Lyme encephalitis. Infect Immun *75*, 243-251.

Lusitani, D., Malawista, S.E., and Montgomery, R.R. (2002). *Borrelia burgdorferi* are susceptible to killing by a variety of human polymorphonuclear leukocyte components. J Infect Dis *185*, 797-804.

Lusitani, D., Malawista, S.E., and Montgomery, R.R. (2003). Calprotectin, an abundant cytosolic protein from human polymorphonuclear leukocytes, inhibits the growth of *Borrelia burgdorferi*. Infect Immun *71*, 4711-4716.

Ma, Y., Bramwell, K.K., Lochhead, R.B., Paquette, J.K., Zachary, J.F., Weis, J.H., Teuscher, C., and Weis, J.J. (2014). *Borrelia burgdorferi* arthritis-associated locus Bbaa1 regulates Lyme arthritis and K/BxN serum transfer arthritis through intrinsic control of type I IFN production. J Immunol *193*, 6050-6060.http s://doi.org/jimmunol.1401746 [pii] 10.4049/jimmunol.1401746

Ma, Y., Seiler, K.P., Eichwald, E.J., Weis, J.H., Teuscher, C., and Weis, J.J. (1998). Distinct characteristics of resistance to *Borrelia burgdorferi*-induced arthritis in C57BL/6N mice. Infect Immun *66*, 161-168.

Ma, Y., and Weis, J.J. (1993). *Borrelia burgdorferi* outer surface lipoproteins OspA and OspB possess B-cell mitogenic and cytokine-stimulatory properties. Infect Immun *61*, 3843-3853.

Magunda, P.R., and Bankhead, T. (2016). Investigating the potential role of non-vls genes on linear plasmid 28-1 in virulence and persistence by *Borrelia burgdorferi*. BMC Microbiol *16*, 180. https://doi.org/10.1186/s12866-016-0806-4

Malawista, S.E., and de Boisfleury Chevance, A. (2008). Clocking the Lyme spirochete. PLoS One *3*, e1633. https://doi.org/10.1371/journal.pone.0001633

Malkiel, S., Kuhlow, C.J., Mena, P., and Benach, J.L. (2009). The loss and gain of marginal zone and peritoneal B cells is different in response to relapsing fever and Lyme disease Borrelia. J Immunol *182*, 498-506.

Marcinkiewicz, A.L., Lieknina, I., Kotelovica, S., Yang, X., Kraiczy, P., Pal, U., Lin, Y.P., and Tars, K. (2018). Eliminating Factor H-Binding Activity of *Borrelia burgdorferi* CspZ Combined with Virus-Like Particle Conjugation Enhances Its Efficacy as a Lyme Disease Vaccine. Front Immunol *9*, 181. https://doi.org/10.3389/fimmu.2018.00181

Marcinkiewicz, A.L., Lieknina, I., Yang, X., Lederman, P.L., Hart, T.M., Yates, J., Chen, W.H., Bottazzi, M.E., Mantis, N.J., Kraiczy, P., *et al.* (2020). The Factor H-Binding Site of CspZ as a Protective Target against Multistrain, Tick-Transmitted Lyme Disease. Infect Immun *88*. https://doi.org/10.1128/IAI.00956-19

Marietta, E.V., Weis, J.J., and Weis, J.H. (1997). CD28 expression by mouse mast cells is modulated by lipopolysaccharide and outer surface protein A lipoprotein from *Borrelia burgdorferi*. J Immunol *159*, 2840-2848.

Marques, A., Schwartz, I., Wormser, G.P., Wang, Y., Hornung, R.L., Demirkale, C.Y., Munson, P.J., Turk, S.P., Williams, C., Lee, C.R., *et al.* (2017). Transcriptome Assessment of Erythema Migrans Skin Lesions in Patients With Early Lyme Disease Reveals Predominant Interferon Signaling. J Infect Dis *217*, 158-167. https://doi.org/10.1093/infdis/jix563

Mason, L.M., Wagemakers, A., van 't Veer, C., Oei, A., van der Pot, W.J., Ahmed, K., van der Poll, T., Geijtenbeek, T.B., and Hovius, J.W. (2016). *Borrelia burgdorferi* Induces TLR2-Mediated Migration of Activated Dendritic Cells in an *Ex Vivo* Human Skin Model. PLoS One *11*, e0164040.http s://doi.org/10.1371/journal.pone.0164040

Matyniak, J.E., and Reiner, S.L. (1995). T helper phenotype and genetic susceptibility in experimental Lyme disease. J Exp Med *181*, 1251-1254. https://doi.org/10.1084/jem.181.3.1251

Mbow, M.L., Zeidner, N., Panella, N., Titus, R.G., and Piesman, J. (1997). *Borrelia burgdorferi*-pulsed dendritic cells induce a protective immune response against tick-transmitted spirochetes. Infect Immun *65*, 3386-3390.

McKisic, M.D., and Barthold, S.W. (2000). T-cell-independent responses to *Borrelia burgdorferi* are critical for protective immunity and resolution of lyme disease. Infect Immun *68*, 5190-5197.

McKisic, M.D., Redmond, W.L., and Barthold, S.W. (2000). Cutting edge: T cell-mediated pathology in murine Lyme borreliosis. J Immunol *164*, 6096-6099.

Medzhitov, R. (2008). Origin and physiological roles of inflammation. Nature *454*, 428-435. https://doi.org/10.1038/nature07201

Medzhitov, R., Preston-Hurlburt, P., and Janeway, C.A., Jr. (1997). A human homologue of the Drosophila Toll protein signals activation of adaptive immunity [see comments]. Nature *388*, 394-397.

Menten-Dedoyart, C., Faccinetto, C., Golovchenko, M., Dupiereux, I., Van Lerberghe, P.B., Dubois, S., Desmet, C., Elmoualij, B., Baron, F., Rudenko, N., *et al.* (2012). Neutrophil extracellular traps entrap and kill *Borrelia burgdorferi sensu stricto* spirochetes and are not affected by Ixodes ricinus tick saliva. J Immunol *189*, 5393-5401. https://doi.org/10.4049/jimmunol.1103771

Miller, J.C., Ma, Y., Bian, J., Sheehan, K.C., Zachary, J.F., Weis, J.H., Schreiber, R.D., and Weis, J.J. (2008a). A critical role for type I IFN in arthritis development following *Borrelia burgdorferi* infection of mice. J Immunol *181*, 8492-8503.

Miller, J.C., Ma, Y., Crandall, H., Wang, X., and Weis, J.J. (2008b). Gene expression profiling provides insights into the pathways involved in inflammatory arthritis development: murine model of Lyme disease. Exp Mol Pathol *85*, 20-27. https://doi.org/10.1016/j.yexmp.2008.03.004

Modolell, M., Schaible, U.E., Rittig, M., and Simon, M.M. (1994). Killing of *Borrelia burgdorferi* by macrophages is dependent on oxygen radicals and nitric oxide and can be enhanced by antibodies to outer surface proteins of the spirochete. Immunol Lett *40*, 139-146. https://doi.org/10.1016/0165-2478(94)90185-6

Montgomery, R.R., Booth, C.J., Wang, X., Blaho, V.A., Malawista, S.E., and Brown, C.R. (2007). Recruitment of macrophages and polymorphonuclear leukocytes in lyme carditis. Infect Immun *75*, 613-620.

Montgomery, R.R., Lusitani, D., Chevance Ad Ade, B., and Malawista, S.E. (2002). Human phagocytic cells in the early innate immune response to *Borrelia burgdorferi*. J Infect Dis *185*, 1773-1779.

Montgomery, R.R., and Malawista, S.E. (1996). Entry of *Borrelia burgdorferi* into macrophages is end-on and leads to degradation in lysosomes. Infect Immun *64*, 2867-2872.

Montgomery, R.R., Nathanson, M.H., and Malawista, S.E. (1994). Fc- and non-Fc-mediated phagocytosis of *Borrelia burgdorferi* by macrophages. J Infect Dis *170*, 890-893.

Moody, K.D., Terwilliger, G.A., Hansen, G.M., and Barthold, S.W. (1994). Experimental *Borrelia burgdorferi* infection in *Peromyscus leucopus*. J Wildl Dis *30*, 155-161. https://doi.org/10.7589/0090-3558-30.2.155

Moore, K.W., de Waal Malefyt, R., Coffman, R.L., and O'Garra, A. (2001). Interleukin-10 and the interleukin-10 receptor. Annu Rev Immunol *19*, 683-765.

Moore, M.W., Cruz, A.R., LaVake, C.J., Marzo, A.L., Eggers, C.H., Salazar, J.C., and Radolf, J.D. (2007). Phagocytosis of *Borrelia burgdorferi* and *Treponema pallidum* potentiates innate immune activation and induces gamma interferon production. Infect Immun *75*, 2046-2062.

Mori, L., Lepore, M., and De Libero, G. (2016). The Immunology of CD1- and MR1-Restricted T Cells. Annu Rev Immunol *34*, 479-510.http s://doi.org/10.1146/annurev-immunol-032414-112008

Morrison, T.B., Weis, J.H., and Weis, J.J. (1997). *Borrelia burgdorferi* outer surface protein A (OspA) activates and primes human neutrophils. J Immunology *158*, 4838-4845.

Motta, V., Soares, F., Sun, T., and Philpott, D.J. (2015). NOD-like receptors: versatile cytosolic sentinels. Physiol Rev *95*, 149-178.http s://doi.org/10.1152/physrev.00009.2014

Muehlenbachs, A., Bollweg, B.C., Schulz, T.J., Forrester, J.D., DeLeon Carnes, M., Molins, C., Ray, G.S., Cummings, P.M., Ritter, J.M., Blau, D.M., *et al.* (2016). Cardiac Tropism of Borrelia burgdorferi: An Autopsy Study of Sudden Cardiac Death Associated with Lyme Carditis. Am J Pathol *186*, 1195-1205. https://doi.org/10.1016/j.ajpath.2015.12.027

Mullegger, R.R., McHugh, G., Ruthazer, R., Binder, B., Kerl, H., and Steere, A.C. (2000). Differential expression of cytokine mRNA in skin specimens from patients with erythema migrans or acrodermatitis chronica atrophicans. J Invest Dermatol *115*, 1115-1123. https://doi.org/10.1046/j.1523-1747.2000.00198.x

Murphy, K., and Weaver, C. (2017). Principles of innate immunity. In Janeway's Immunobiology (New York, NY: Garland Science, Taylor & Francis Group, LLC), pp. 3-11.

Myers, T.A., Kaushal, D., and Philipp, M.T. (2009). Microglia are mediators of *Borrelia burgdorferi*-induced apoptosis in SH-SY5Y neuronal cells. PLoS Pathog *5*, e1000659. https://doi.org/10.1371/journal.ppat.1000659

Nadelman, R.B., Hanincova, K., Mukherjee, P., Liveris, D., Nowakowski, J., McKenna, D., Brisson, D., Cooper, D., Bittker, S., Madison, G., *et al.* (2012). Differentiation of reinfection from relapse in recurrent Lyme disease. N Engl J Med *367*, 1883-1890. https://doi.org/10.1056/NEJMoa1114362

Naj, X., Hoffmann, A.K., Himmel, M., and Linder, S. (2013). The formins FMNL1 and mDia1 regulate coiling phagocytosis of *Borrelia burgdorferi* by primary human macrophages. Infect Immun *81*, 1683-1695. https://doi.org/10.1128/iai.01411-12

Naj, X., and Linder, S. (2015). ER-Coordinated Activities of Rab22a and Rab5a Drive Phagosomal Compaction and Intracellular Processing of *Borrelia burgdorferi* by Macrophages. Cell Rep *12*, 1816-1830.http s:// doi.org/10.1016/j.celrep.2015.08.027

Narasimhan, S., Booth, C.J., DePonte, K., Wu, M.J., Liang, X., Mohanty, S., Kantor, F., and Fikrig, E. (2019). Host-specific expression of *Ixodes scapularis* salivary genes. Ticks Tick Borne Dis *10*, 386-397.http s://doi.org/10.1016/j.ttbdis.2018.12.001

Narasimhan, S., Santiago, F., Koski, R.A., Brei, B., Anderson, J.F., Fish, D., and Fikrig, E. (2002). Examination of the *Borrelia burgdorferi* transcriptome in Ixodes scapularis during feeding. J Bacteriol *184*, 3122-3125.

Narayan, K., Dail, D., Li, L., Cadavid, D., Amrute, S., Fitzgerald-Bocarsly, P., and Pachner, A.R. (2005). The nervous system as ectopic germinal center: CXCL13 and IgG in lyme neuroborreliosis. Ann Neurol *57*, 813-823.

Nardelli, D.T., Burchill, M.A., England, D.M., Torrealba, J., Callister, S.M., and Schell, R.F. (2004). Association of CD4+ CD25+ T cells with prevention of severe destructive arthritis in *Borrelia burgdorferi*-vaccinated and challenged gamma interferon-deficient mice treated with anti-interleukin-17 antibody. Clin Diagn Lab Immunol *11*, 1075-1084.

Nestle, F.O., Di Meglio, P., Qin, J.Z., and Nickoloff, B.J. (2009). Skin immune sentinels in health and disease. Nat Rev Immunol *9*, 679-691. https://doi.org/nri2622 [pii] 10.1038/nri2622

Nguyen Hoang, A.T., Chen, P., Bjornfot, S., Hogstrand, K., Lock, J.G., Grandien, A., Coles, M., and Svensson, M. (2014). Technical advance: live-imaging analysis of human dendritic cell migrating behavior under the influence of immune-stimulating reagents in an organotypic model of lung. J Leukoc Biol *96*, 481-489. https://doi.org/ 10.1189/jlb.3TA0513-303R

Nigrovic, L.E., Lipsett, S.C., Molins, C.R., Wormser, G.P., Bennett, J.E., Garro, A.C., Levas, M.N., Balamuth, F., Neville, D., Lingampalli, N., *et al.* (2019). Higher C6 enzyme immunoassay index values correlate with a diagnosis of noncutaneous Lyme disease. Diagn Microbiol Infect Dis *94*, 160-164. https://doi.org/10.1016/j.diagmicrobio. 2018.12.001

Nocton, J.J., Dressler, F., Rutledge, B.J., Rys, P.N., Persing, D.H., and Steere, A.C. (1994). Detection of *Borrelia burgdorferi* DNA by polymerase chain reaction in synovial fluid from patients with Lyme arthritis. N Engl J Med *330*, 229-234.

Norgard, M.V., Riley, B.S., Richardson, J.A., and Radolf, J.D. (1995). Dermal inflammation elicited by synthetic analogs of *Treponema pallidum* and *Borrelia burgdorferi* lipoproteins. Infect Immun *63*, 1507-1515.

Norris, S.J. (2006). Antigenic variation with a twist--the Borrelia story. Mol Microbiol *60*, 1319-1322. https://doi.org/MMI5204 [pii] 10.1111/j. 1365-2958.2006.05204.x

Nowakowski, J., Nadelman, R.B., Sell, R., McKenna, D., Cavaliere, L.F., Holmgren, D., Gaidici, A., and Wormser, G.P. (2003). Long-term follow-up of patients with culture-confirmed Lyme disease. Am J Med *115*, 91-96. https://doi.org/S0002934303003085 [pii]

Nuttall, P.A. (2019). Tick saliva and its role in pathogen transmission. Wien Klin Wochenschr. https://doi.org/10.1007/s00508-019-1500-y

Nuttall, P.A., and Labuda, M. (2004). Tick-host interactions: saliva-activated transmission. Parasitology *129 Suppl*, S177-189.

Ohnishi, J., Piesman, J., and de Silva, A.M. (2001). Antigenic and genetic heterogeneity of *Borrelia burgdorferi* populations transmitted by ticks. Proc Natl Acad Sci U S A *98*, 670-675.

Olson, C.M., Jr., Bates, T.C., Izadi, H., Radolf, J.D., Huber, S.A., Boyson, J.E., and Anguita, J. (2009). Local production of IFN-gamma by invariant NKT cells modulates acute Lyme carditis. J Immunol *182*, 3728-3734.

Oosting, M., Buffen, K., Malireddi, S.R., Sturm, P., Verschueren, I., Koenders, M.I., van de Veerdonk, F.L., van der Meer, J.W., Netea, M.G., Kanneganti, T.D., *et al.* (2012). Murine Borrelia arthritis is highly dependent on ASC and caspase-1, but independent of NLRP3. Arthritis Res Ther *14*, R247.http s://doi.org/10.1186/ar4090

Oosting, M., Ter Hofstede, H., Sturm, P., Adema, G.J., Kullberg, B.J., van der Meer, J.W., Netea, M.G., and Joosten, L.A. (2011a). TLR1/ TLR2 heterodimers play an important role in the recognition of Borrelia spirochetes. PLoS One *6*, e25998. https://doi.org/10.1371/ journal.pone.0025998

Oosting, M., van de Veerdonk, F.L., Kanneganti, T.D., Sturm, P., Verschueren, I., Berende, A., van der Meer, J.W., Kullberg, B.J., Netea, M.G., and Joosten, L.A. (2011b). Borrelia species induce inflammasome activation and IL-17 production through a caspase-1-dependent mechanism. Eur J Immunol *41*, 172-181. https://doi.org/ 10.1002/eji.201040385

Pachner, A.R., and Steiner, I. (2007). Lyme neuroborreliosis: infection, immunity, and inflammation. Lancet Neurol *6*, 544-552.

Pal, U., Wang, P., Bao, F., Yang, X., Samanta, S., Schoen, R., Wormser, G.P., Schwartz, I., and Fikrig, E. (2008). *Borrelia burgdorferi* basic membrane proteins A and B participate in the genesis of Lyme arthritis. J Exp Med *205*, 133-141.

Pal, U., Yang, X., Chen, M., Bockenstedt, L.K., Anderson, J.F., Flavell, R.A., Norgard, M.V., and Fikrig, E. (2004). OspC facilitates *Borrelia burgdorferi* invasion of *Ixodes scapularis* salivary glands. J Clin Invest *113*, 220-230.

Paludan, S.R., Pradeu, T., Masters, S.L., and Mogensen, T.H. (2020). Constitutive immune mechanisms: mediators of host defence and immune regulation. Nat Rev Immunol. https://doi.org/10.1038/ s41577-020-0391-5

Paquette, J.K., Ma, Y., Fisher, C., Li, J., Lee, S.B., Zachary, J.F., Kim, Y.S., Teuscher, C., and Weis, J.J. (2017). Genetic Control of Lyme Arthritis by *Borrelia burgdorferi* Arthritis-Associated Locus 1 Is Dependent on Localized Differential Production of IFN-beta and Requires Upregulation of Myostatin. J Immunol *199*, 3525-3534. https://doi.org/10.4049/jimmunol.1701011

Park, J.S., and Kim, J.H. (2018). Role of non-classical T cells in skin immunity. Mol Immunol *103*, 286-292. https://doi.org/10.1016/ j.molimm.2018.09.024

Park, S.A., Choe, Y.H., Park, E., and Hyun, Y.M. (2018). Real-time dynamics of neutrophil clustering in response to phototoxicity-induced cell death and tissue damage in mouse ear dermis. Cell Adh Migr *12*, 424-431. https://doi.org/10.1080/19336918.2018.1471322

Parthasarathy, G., and Philipp, M.T. (2015). Inflammatory mediator release from primary rhesus microglia in response to *Borrelia burgdorferi* results from the activation of several receptors and pathways. J Neuroinflammation *12*, 60. https://doi.org/10.1186/ s12974-015-0274-z

Parthasarathy, G., and Philipp, M.T. (2017). Receptor tyrosine kinases play a significant role in human oligodendrocyte inflammation and cell death associated with the Lyme disease bacterium *Borrelia burgdorferi*. J Neuroinflammation *14*, 110. https://doi.org/10.1186/ s12974-017-0883-9

Patarakul, K., Cole, M.F., and Hughes, C.A. (1999). Complement resistance in *Borrelia burgdorferi* strain 297: outer membrane proteins prevent MAC formation at lysis susceptible sites. Microb Pathog *27*, 25-41.

Pausa, M., Pellis, V., Cinco, M., Giulianini, P.G., Presani, G., Perticarari, S., Murgia, R., and Tedesco, F. (2003). Serum-resistant strains of *Borrelia burgdorferi* evade complement-mediated killing by expressing a CD59-like complement inhibitory molecule. J Immunol *170*, 3214-3222.

Pechova, J., Stepanova, G., Kovar, L., and Kopecky, J. (2002). Tick salivary gland extract-activated transmission of *Borrelia afzelii* spirochaetes. Folia Parasitol (Praha) *49*, 153-159.

Pegalajar-Jurado, A., Schriefer, M.E., Welch, R.J., Couturier, M.R., MacKenzie, T., Clark, R.J., Ashton, L.V., Delorey, M.J., and Molins, C.R. (2018). Evaluation of Modified Two-Tiered Testing Algorithms for Lyme Disease Laboratory Diagnosis Using Well-Characterized Serum Samples. J Clin Microbiol *56*. https://doi.org/10.1128/JCM.01943-17

Pegram, H.J., Andrews, D.M., Smyth, M.J., Darcy, P.K., and Kershaw, M.H. (2011). Activating and inhibitory receptors of natural killer cells. Immunol Cell Biol *89*, 216-224. https://doi.org/10.1038/icb.2010.78

Perner, J., Kropackova, S., Kopacek, P., and Ribeiro, J.M.C. (2018). Sialome diversity of ticks revealed by RNAseq of single tick salivary glands. PLoS Negl Trop Dis *12*, e0006410. https://doi.org/10.1371/journal.pntd.0006410

Peterson, P.K., Clawson, C.C., Lee, D.A., Garlich, D.J., Quie, P.G., and Johnson, R.C. (1984). Human phagocyte interactions with the Lyme disease spirochete. Infect Immun *46*, 608-611.

Petnicki-Ocwieja, T., Chung, E., Acosta, D.I., Ramos, L.T., Shin, O.S., Ghosh, S., Kobzik, L., Li, X., and Hu, L.T. (2013). TRIF mediates Toll-like receptor 2-dependent inflammatory responses to *Borrelia burgdorferi*. Infect Immun *81*, 402-410. https://doi.org/10.1128/iai.00890-12

Petnicki-Ocwieja, T., Kern, A., Killpack, T.L., Bunnell, S.C., and Hu, L.T. (2015). Adaptor Protein-3-Mediated Trafficking of TLR2 Ligands Controls Specificity of Inflammatory Responses but Not Adaptor Complex Assembly. J Immunol *195*, 4331-4340. https://doi.org/10.4049/jimmunol.1501268

Petzke, M.M., Brooks, A., Krupna, M.A., Mordue, D., and Schwartz, I. (2009). Recognition of *Borrelia burgdorferi*, the Lyme disease spirochete, by TLR7 and TLR9 induces a type I IFN response by human immune cells. J Immunol *183*, 5279-5292. https://doi.org/10.4049/jimmunol.0901390

Piesman, J., Dolan, M.C., Happ, C.M., Luft, B.J., Rooney, S.E., Mather, T.N., and Golde, W.T. (1997). Duration of immunity to reinfection with tick-transmitted *Borrelia burgdorferi* in naturally infected mice. Infect Immun *65*, 4043-4047.

Poltorak, A., He, X., Smirnova, I., Liu, M.Y., Huffel, C.V., Du, X., Birdwell, D., Alejos, E., Silva, M., Galanos, C., *et al.* (1998). Defective LPS Signaling in C3H/HeJ and C57BL/10ScCr Mice: Mutations in Tlr4 Gene. Science *282*, 2085-2088.

Potter, M.R., Noben-Trauth, N., Weis, J.H., Teuscher, C., and Weis, J.J. (2000). Interleukin-4 (IL-4) and IL-13 signaling pathways do not regulate *Borrelia burgdorferi*-induced arthritis in mice: IgG1 is not required for host control of tissue spirochetes. Infect Immun *68*, 5603-5609.

Pratt, C.L., and Brown, C.R. (2014). The role of eicosanoids in experimental Lyme arthritis. Front Cell Infect Microbiol *4*, 69.http s://doi.org/10.3389/fcimb.2014.00069

Preac Mursic, V., Patsouris, E., Wilske, B., Reinhardt, S., Gross, B., and Mehraein, P. (1990). Persistence of *Borrelia burgdorferi* and histopathological alterations in experimentally infected animals. A comparison with histopathological findings in human Lyme disease. Infection *18*, 332-341.

Prinz, M., Priller, J., Sisodia, S.S., and Ransohoff, R.M. (2011). Heterogeneity of CNS myeloid cells and their roles in neurodegeneration. Nat Neurosci *14*, 1227-1235. https://doi.org/10.1038/nn.2923

Radolf, J.D., Arndt, L.L., Akins, D.R., Curetty, L.L., Levi, M.E., Shen, Y., Davis, L.S., and Norgard, M.V. (1995). *Treponema pallidum* and *Borrelia burgdorferi* lipoproteins and synthetic lipopeptides activate monocytes/macrophages. J Immunol *154*, 2866-2877.

Radolf, J.D., Caimano, M.J., Stevenson, B., and Hu, L.T. (2012). Of ticks, mice and men: understanding the dual-host lifestyle of Lyme disease spirochaetes. Nat Rev Microbiol *10*, 87-99. https://doi.org/nrmicro2714 [pii] 10.1038/nrmicro2714

Radolf, J.D., Norgard, M.V., Brandt, M.E., Isaacs, R.D., Thompson, P.A., and Beutler, B. (1991). Lipoproteins of *Borrelia burgdorferi* and *Treponema pallidum* activate cachectin/tumor necrosis factor synthesis. Analysis using a CAT reporter construct. J Immunol *147*, 1968-1974.

Ramamoorthi, N., Narasimhan, S., Pal, U., Bao, F., Yang, X.F., Fish, D., Anguita, J., Norgard, M.V., Kantor, F.S., Anderson, J.F., *et al.* (2005). The Lyme disease agent exploits a tick protein to infect the mammalian host. Nature *436*, 573-577.

Rasley, A., Anguita, J., and Marriott, I. (2002). *Borrelia burgdorferi* induces inflammatory mediator production by murine microglia. J Neuroimmunol *130*, 22-31. https://doi.org/10.1016/s0165-5728(02)00187-x

Rath, P.M., Ibershoff, B., Mohnhaupt, A., Albig, J., Eljaschewitsch, B., Jurgens, D., Horbach, I., and Fehrenbach, F.J. (1996). Seroprevalence of Lyme borreliosis in forestry workers from Brandenburg, Germany. Eur J Clin Microbiol Infect Dis *15*, 372-377. https://doi.org/10.1007/BF01690092

Recher, M., Lang, K.S., Hunziker, L., Freigang, S., Eschli, B., Harris, N.L., Navarini, A., Senn, B.M., Fink, K., Lotscher, M., *et al.* (2004). Deliberate removal of T cell help improves virus-neutralizing antibody production. Nat Immunol *5*, 934-942. https://doi.org/10.1038/ni1102

Reimers, C.D., de Koning, J., Neubert, U., Preac-Mursic, V., Koster, J.G., Muller-Felber, W., Pongratz, D.E., and Duray, P.H. (1993). *Borrelia burgdorferi* myositis: report of eight patients. J Neurol *240*, 278-283.http s://doi.org/10.1007/BF00838161

Reinhardt, A., Yevsa, T., Worbs, T., Lienenklaus, S., Sandrock, I., Oberdorfer, L., Korn, T., Weiss, S., Forster, R., and Prinz, I. (2016). Interleukin-23-Dependent gamma/delta T Cells Produce Interleukin-17 and Accumulate in the Enthesis, Aortic Valve, and Ciliary Body in Mice. Arthritis Rheumatol *68*, 2476-2486. https://doi.org/10.1002/art.39732

Reinink, P., Souter, M.N.T., Cheng, T.Y., van Gorkom, T., Lenz, S., Kubler-Kielb, J., Strle, K., Kremer, K., Thijsen, S.F.T., Steere, A.C., *et al.* (2019). CD1b presents self and *Borrelia burgdorferi* diacylglycerols to human T cells. Eur J Immunol *49*, 737-746. https://doi.org/10.1002/eji.201847949

Ribeiro, J.M., Alarcon-Chaidez, F., Francischetti, I.M., Mans, B.J., Mather, T.N., Valenzuela, J.G., and Wikel, S.K. (2006). An annotated catalog of salivary gland transcripts from *Ixodes scapularis* ticks. Insect Biochem Mol Biol *36*, 111-129.

Ribeiro, J.M., Makoul, G.T., Levine, J., Robinson, D.R., and Spielman, A. (1985). Antihemostatic, antiinflammatory, and immunosuppressive properties of the saliva of a tick, *Ixodes dammini.* J Exp Med *161*, 332-344. https://doi.org/10.1084/jem.161.2.332

Ribeiro, J.M., Makoul, G.T., and Robinson, D.R. (1988). *Ixodes dammini:* evidence for salivary prostacyclin secretion. J Parasitol *74*, 1068-1069.

Rittig, M.G., Jagoda, J.C., Wilske, B., Murgia, R., Cinco, M., Repp, R., Burmester, G.R., and Krause, A. (1998). Coiling phagocytosis discriminates between different spirochetes and is enhanced by phorbol myristate acetate and granulocyte-macrophage colony-stimulating factor. Infect Immun *66*, 627-635.

Rittig, M.G., Krause, A., Haupl, T., Schaible, U.E., Modolell, M., Kramer, M.D., Lutjen-Drecoll, E., Simon, M.M., and Burmester, G.R. (1992). Coiling phagocytosis is the preferential phagocytic mechanism for *Borrelia burgdorferi*. Infect Immun *60*, 4205-4212.

Roberts, E.D., Bohm, R.F., Cogswell, F.B., NorbertLanners, H., Lowrie, R.C., Povinelli, L., Piesman, J., and Philipp, M.T. (1995). Chronic Lyme disease in the rhesus monkey. Lab Invest *72*, 146-160.

Rock, F.L., Hardiman, G., Timans, J.C., Kastelein, R.A., and Bazan, J.F. (1998). A family of human receptors structurally related to Drosophila Toll. Proc Natl Acad Sci U S A *95*, 588-593.

Rogovskyy, A.S., and Bankhead, T. (2013). Variable VlsE is critical for host reinfection by the Lyme disease spirochete. PLoS One *8*, e61226. https://doi.org/10.1371/journal.pone.0061226

Rosa, P.A., Tilly, K., and Stewart, P.E. (2005). The burgeoning molecular genetics of the Lyme disease spirochaete. Nat Rev Microbiol *3*, 129-143.

Ruderman, E.M., Kerr, J.S., Telford, S.R., 3rd, Spielman, A., Glimcher, L.H., and Gravallese, E.M. (1995). Early murine Lyme carditis has a macrophage predominance and is independent of major histocompatibility complex class II-CD4+ T cell interactions. J Infect Dis *171*, 362-370.

Rupprecht, T.A., Kirschning, C.J., Popp, B., Kastenbauer, S., Fingerle, V., Pfister, H.W., and Koedel, U. (2007). *Borrelia garinii* induces CXCL13 production in human monocytes through Toll-like receptor 2. Infect Immun *75*, 4351-4356.

Sa-Nunes, A., Bafica, A., Lucas, D.A., Conrads, T.P., Veenstra, T.D., Andersen, J.F., Mather, T.N., Ribeiro, J.M., and Francischetti, I.M. (2007). Prostaglandin E2 is a major inhibitor of dendritic cell maturation and function in I*xodes scapularis* saliva. J Immunol *179*, 1497-1505.

Sabino, G.J., Hwang, S.J., McAllister, S.C., Mena, P., and Furie, M.B. (2011). Interferon-gamma influences the composition of leukocytic infiltrates in murine Lyme carditis. Am J Pathol *179*, 1917-1928.http s://doi.org/10.1016/j.ajpath.2011.06.029

Sadziene, A., Jonsson, M., Bergstrom, S., Bright, R.K., Kennedy, R.C., and Barbour, A.G. (1994). A bactericidal antibody to *Borrelia burgdorferi* is directed against a variable region of the OspB protein. Infect Immun *62*, 2037-2045.

Sadziene, A., Thompson, P.A., and Barbour, A.G. (1993). In vitro inhibition of *Borrelia burgdorferi* growth by antibodies. J Infect Dis *167*, 165-172.

Sahay, B., Singh, A., Gnanamani, A., Patsey, R.L., Blalock, J.E., and Sellati, T.J. (2011). CD14 signaling reciprocally controls collagen deposition and turnover to regulate the development of lyme arthritis. Am J Pathol *178*, 724-734. https://doi.org/10.1016/j.ajpath.2010.10.025

Sakaguchi, S. (2004). Naturally arising CD4+ regulatory T cells for immunologic self-tolerance and negative control of immune responses. Annu Rev Immunol *22*, 531-562.

Sakai, S., Mayer-Barber, K.D., and Barber, D.L. (2014). Defining features of protective CD4 T cell responses to *Mycobacterium tuberculosis*. Curr Opin Immunol *29*, 137-142. https://doi.org/S0952-7915(14)00084-3 [pii] 10.1016/j.coi.2014.06.003

Salazar, J.C., Duhnam-Ems, S., La Vake, C., Cruz, A.R., Moore, M.W., Caimano, M.J., Velez-Climent, L., Shupe, J., Krueger, W., and Radolf, J.D. (2009). Activation of human monocytes by live *Borrelia burgdorferi* generates TLR2-dependent and -independent responses which include induction of IFNβ. PLoS Pathog *5*, e1000444. https://doi.org/10.1371/journal.ppat.1000444

Salazar, J.C., Pope, C.D., Moore, M.W., Pope, J., Kiely, T.G., and Radolf, J.D. (2005). Lipoprotein-dependent and -independent immune responses to spirochetal infection. Clin Diagn Lab Immunol *12*, 949-958.

Salazar, J.C., Pope, C.D., Sellati, T.J., Feder, H.M., Jr., Kiely, T.G., Dardick, K.R., Buckman, R.L., Moore, M.W., Caimano, M.J., Pope, J.G., *et al.* (2003). Coevolution of markers of innate and adaptive immunity in skin and peripheral blood of patients with erythema migrans. J Immunol *171*, 2660-2670.

Sallusto, F. (2016). Heterogeneity of Human CD4(+) T Cells Against Microbes. Annu Rev Immunol *34*, 317-334. https://doi.org/10.1146/annurev-immunol-032414-112056

Samuels, D.S. (2011). Gene regulation in *Borrelia burgdorferi.* Annu Rev Microbiol *65*, 479-499. https://doi.org/10.1146/annurev.micro.112408.134040

Schaible, U.E., Kramer, M.D., Museteanu, C., Zimmer, G., Mossmann, H., and Simon, M. (1990). Lyme borreliosis in the severe combined immunodeficiency (scid) mouse manifests predominantly in the joints, heart and liver. Amer J Pathol *137*, 811-820.

Schaible, U.E., Kramer, M.D., Museteanu, C., Zimmer, G., Mossmann, H., and Simon, M.M. (1989). The severe combined immunodeficiency (scid) mouse. A laboratory model for the analysis of Lyme arthritis and carditis. J Exp Med *170*, 1427-1432.

Schaible, U.E., Wallich, R., Kramer, M.D., Nerz, G., Stehle, T., Museteanu, C., and Simon, M.M. (1994). Protection against *Borrelia burgdorferi* infection in SCID mice is conferred by presensitized spleen cells and partially by B but not T cells alone. Int Immunol *6*, 671-681.

Schoenfeld, R., Araneo, B., Ma, Y., Yang, L.M., and Weis, J.J. (1992). Demonstration of a B-lymphocyte mitogen produced by the Lyme disease pathogen, *Borrelia burgdorferi*. Infect Immun *60*, 455-464.

Schuijt, T.J., Coumou, J., Narasimhan, S., Dai, J., Deponte, K., Wouters, D., Brouwer, M., Oei, A., Roelofs, J.J., van Dam, A.P., *et al.* (2011). A tick mannose-binding lectin inhibitor interferes with the vertebrate complement cascade to enhance transmission of the Lyme disease agent. Cell Host Microbe *10*, 136-146. https://doi.org/10.1016/j.chom.2011.06.010

Schutzer, S.E., Coyle, P.K., Belman, A.L., Golightly, M.G., and Drulle, J. (1990). Sequestration of antibody to *Borrelia burgdorferi* in immune complexes in seronegative Lyme disease. Lancet *335*, 312-315. https://doi.org/10.1016/0140-6736(90)90606-6

Schutzer, S.E., Coyle, P.K., Krupp, L.B., Deng, Z., Belman, A.L., Dattwyler, R., and Luft, B.J. (1997). Simultaneous expression of Borrelia OspA and OspC and IgM response in cerebrospinal fluid in early neurologic Lyme disease. J Clin Invest *100*, 763-767. https://doi.org/10.1172/JCI119589

Schutzer, S.E., and Luan, J. (2003). Early OspA immune complex formation in animal models of Lyme disease. J Mol Microbiol Biotechnol *5*, 167-171. https://doi.org/10.1159/000070267

Schwan, T.G., Piesman, J., Golde, W.T., Dolan, M.C., and Rosa, P.A. (1995). Induction of an outer surface protein on *Borrelia burgdorferi* during tick feeding. Proc Natl Acad Sci U S A *92*, 2909-2913.

Sellati, T.J., Abrescia, L.D., Radolf, J.D., and Furie, M.B. (1996). Outer surface lipoproteins of *Borrelia burgdorferi* activate vascular endothelium *in vitro*. Infect Immun *64*, 3180-3187.

Sellati, T.J., Bouis, D.A., Kitchens, R.L., Darveau, R.P., Pugin, J., Ulevitch, R.J., Gangloff, S.C., Goyert, S.M., Norgard, M.V., and Radolf, J.D. (1998). *Treponema pallidum* and *Borrelia burgdorferi* lipoproteins and synthetic lipopeptides activate monocytic cells via a CD14-dependent pathway distinct from that used by lipopolysaccharide. J Immunol *160*, 5455-5464.

Sellati, T.J., Burns, M.J., Ficazzola, M.A., and Furie, M.B. (1995). *Borrelia burgdorferi* upregulates expression of adhesion molecules on endothelial cells and promotes transendothelial migration of neutrophils *in vitro*. Infection and Immunity *63*, 4439-4447.

Shahinian, A., Pfeffer, K., Lee, K.P., Kundig, T.M., Kishihara, K., Wakeham, A., Kawai, K., Ohashi, P.S., Thompson, C.B., and Mak, T.W. (1993). Differential T cell costimulatory requirements in CD28-deficient mice. Science *261*, 609-612.

Shen, S., Shin, J.J., Strle, K., McHugh, G., Li, X., Glickstein, L.J., Drouin, E.E., and Steere, A.C. (2010). Treg cell numbers and function in patients with antibiotic-refractory or antibiotic-responsive Lyme arthritis. Arthritis Rheum *62*, 2127-2137. https://doi.org/10.1002/art.27468

Sherlock, J.P., Joyce-Shaikh, B., Turner, S.P., Chao, C.C., Sathe, M., Grein, J., Gorman, D.M., Bowman, E.P., McClanahan, T.K., Yearley, J.H., *et al.* (2012). IL-23 induces spondyloarthropathy by acting on ROR-gammat+ CD3+CD4-CD8- entheseal resident T cells. Nat Med *18*, 1069-1076.http s://doi.org/10.1038/nm.2817

Shih, C.M., Pollack, R.J., Telford, S.R., 3rd, and Spielman, A. (1992). Delayed dissemination of Lyme disease spirochetes from the site of deposition in the skin of mice. J Infect Dis *166*, 827-831.

Shih, C.M., Telford, S.R., 3rd, Pollack, R.J., and Spielman, A. (1993). Rapid dissemination by the agent of Lyme disease in hosts that permit fulminating infection. Infect Immun *61*, 2396-2399.

Shin, J.J., Glickstein, L.J., and Steere, A.C. (2007). High levels of inflammatory chemokines and cytokines in joint fluid and synovial tissue throughout the course of antibiotic-refractory Lyme arthritis. Arthritis Rheum *56*, 1325-1335.

Shin, O.S., Isberg, R., Akira, S., Uematsu, S., Behera, A.K., and Hu, L.T. (2008a). Distinct roles for MyD88, TLR2, TLR5, and TLR9 in phagocytosis of *Borrelia burgdorferi* and inflammatory signaling. Infect Immun *76*, 2341-2351.

Shin, O.S., Isberg, R.R., Akira, S., Uematsu, S., Behera, A.K., and Hu, L.T. (2008b). Distinct roles for MyD88 and Toll-like receptors 2, 5, and 9 in phagocytosis of *Borrelia burgdorferi* and cytokine induction. Infect Immun *76*, 2341-2351. https://doi.org/10.1128/IAI.01600-07

Sigal, L.H., Steere, A.C., and Dwyer, J.M. (1988). In vivo and in vitro evidence of B cell hyperactivity during Lyme disease. J Rheumatol *15*, 648-654.

Silberer, M., Koszik, F., Stingl, G., and Aberer, E. (2000). Downregulation of class II molecules on epidermal Langerhans cells in Lyme borreliosis. Br J Dermatol *143*, 786-794. https://doi.org/bjd3776 [pii]

Skare, J.T., and Garcia, B.L. (2020). Complement Evasion by Lyme Disease Spirochetes. Trends Microbiol. https://doi.org/10.1016/j.tim.2020.05.004

Skare, J.T., Shaw, D.K., Trzeciakowski, J.P., and Hyde, J.A. (2016). In Vivo Imaging Demonstrates That Borrelia burgdorferi ospC Is Uniquely Expressed Temporally and Spatially throughout Experimental Infection. PLoS One *11*, e0162501. https://doi.org/10.1371/journal.pone.0162501 PONE-D-16-28292 [pii]

Sonderegger, F.L., Ma, Y., Maylor-Hagan, H., Brewster, J., Huang, X., Spangrude, G.J., Zachary, J.F., Weis, J.H., and Weis, J.J. (2012a). Localized production of IL-10 suppresses early inflammatory cell infiltration and subsequent development of IFN-gamma-mediated Lyme arthritis. J Immunol *188*, 1381-1393. https://doi.org/10.4049/jimmunol.1102359

Sonderegger, F.L., Ma, Y., Maylor-Hagan, H., Brewster, J., Huang, X., Spangrude, G.J., Zachary, J.F., Weis, J.H., and Weis, J.J. (2012b). Localized production of IL-10 suppresses early inflammatory cell infiltration and subsequent development of IFNγ-mediated Lyme arthritis. J Immunol *188*, 1381-1393. https://doi.org/jimmunol.1102359 [pii] 10.4049/jimmunol.1102359

Soulas, P., Woods, A., Jaulhac, B., Knapp, A.M., Pasquali, J.L., Martin, T., and Korganow, A.S. (2005). Autoantigen, innate immunity, and T cells cooperate to break B cell tolerance during bacterial infection. J Clin Invest *115*, 2257-2267. https://doi.org/10.1172/JCI24646

Steere, A.C. (2001). Lyme disease. N Engl J Med *345*, 115-125. https://doi.org/10.1056/NEJM200107123450207

Steere, A.C., and Angelis, S.M. (2006). Therapy for Lyme arthritis: strategies for the treatment of antibiotic-refractory arthritis. Arthritis Rheum *54*, 3079-3086.

Steere, A.C., Berardi, V.P., Weeks, K.E., Logigian, E.L., and Ackermann, R. (1990). Evaluation of the intrathecal antibody response to *Borrelia burgdorferi* as a diagnostic test for Lyme neuroborreliosis. J Infect Dis *161*, 1203-1209. https://doi.org/10.1093/infdis/161.6.1203

Steere, A.C., Hardin, J.A., Ruddy, S., Mummaw, J.G., and Malawista, S.E. (1979). Lyme arthritis: correlation of serum and cryoglobulin IgM with activity, and serum IgG with remission. Arthritis Rheum 22, 471-483.http s://doi.org/10.1002/art.1780220506

Steere, A.C., Klitz, W., Drouin, E.E., Falk, B.A., Kwok, W.W., Nepom, G.T., and Baxter-Lowe, L.A. (2006). Antibiotic-refractory Lyme arthritis is associated with HLA-DR molecules that bind a *Borrelia burgdorferi* peptide. J Exp Med *203*, 961-971.

Steere, A.C., McHugh, G., Damle, N., and Sikand, V.K. (2008). Prospective study of serologic tests for Lyme disease. Clin Infect Dis *47*, 188-195.http s://doi.org/10.1086/589242

Steere, A.C., Schoen, R.T., and Taylor, E. (1987). The clinical evolution of Lyme arthritis. Ann Intern Med *107*, 725-731.

Steere, A.C., Strle, F., Wormser, G.P., Hu, L.T., Branda, J.A., Hovius, J.W., Li, X., and Mead, P.S. (2016). Lyme borreliosis. Nat Rev Dis Primers *2*, 16090. https://doi.org/10.1038/nrdp.2016.90

Stevenson, B., El-Hage, N., Hines, M.A., Miller, J.C., and Babb, K. (2002). Differential binding of host complement inhibitor factor H by *Borrelia burgdorferi* Erp surface proteins: a possible mechanism underlying the expansive host range of Lyme disease spirochetes. Infect Immun *70*, 491-497. https://doi.org/10.1128/iai.70.2.491-497.2002

Stevenson, B., Schwan, T.G., and Rosa, P.A. (1995). Temperature-related differential expression of antigens in the Lyme disease spirochete, *Borrelia burgdorferi*. Infect Immun *63*, 4535-4539.

Stewart, P.E., Byram, R., Grimm, D., Tilly, K., and Rosa, P.A. (2005). The plasmids of *Borrelia burgdorferi*: essential genetic elements of a pathogen. Plasmid *53*, 1-13.

Straubinger, R.K., Summers, B.A., Chang, Y.F., and Appel, M.J.G. (1997). Persistence of Borrelia burgdorferi in experimentally infected dogs after antibiotic treatment. J Clin Microbiol *35*, 111-116.

Strle, K., Drouin, E.E., Shen, S., Khoury, J.E.I., McHugh, G., Ruzic-Sabljic, E., Strle, F., and Steere, A.C. (2009). *Borrelia burgdorferi* Stimulates Macrophages to Secrete Higher Levels of Cytokines and Chemokines than *Borrelia afzelii* or *Borrelia garinii*. J Infect Dis *200*, 1936-1943.

Strle, K., Jones, K.L., Drouin, E.E., Li, X., and Steere, A.C. (2011). Borrelia burgdorferi RST1 (OspC type A) genotype is associated with greater inflammation and more severe Lyme disease. Am J Pathol *178*, 2726-2739.http s://doi.org/10.1016/j.ajpath.2011.02.018

Strle, K., Shin, J.J., Glickstein, L.J., and Steere, A.C. (2012). Association of a Toll-like receptor 1 polymorphism with heightened Th1 inflammatory responses and antibiotic-refractory Lyme arthritis. Arthritis Rheum *64*, 1497-1507. https://doi.org/10.1002/art.34383

Strle, K., Sulka, K.B., Pianta, A., Crowley, J.T., Arvikar, S.L., Anselmo, A., Sadreyev, R., and Steere, A.C. (2017). T-Helper 17 Cell Cytokine Responses in Lyme Disease Correlate With Borrelia burgdorferi Antibodies During Early Infection and With Autoantibodies Late in the Illness in Patients With Antibiotic-Refractory Lyme Arthritis. Clin Infect Dis *64*, 930-938. https://doi.org/10.1093/cid/cix002

Suhonen, J., Hartiala, K., Tuominen-Gustafsson, H., and Viljanen, M.K. (2000). Borrelia burgdorferi--induced oxidative burst, calcium mobilization, and phagocytosis of human neutrophils are complement dependent. J Infect Dis *181*, 195-202.

Suhonen, J., Hartiala, K., Tuominen-Gustafsson, H., and Viljanen, M.K. (2002). Sublethal concentrations of complement can effectively opsonize *Borrelia burgdorferi*. Scand J Immunol *56*, 554-560. https://doi.org/10.1046/j.1365-3083.2002.01171.x

Suhonen, J., Hartiala, K., and Viljanen, M.K. (1998). Tube phagocytosis, a novel way for neutrophils to phagocytize *Borrelia burgdorferi*. Infect Immun *66*, 3433-3435.

Sulka, K.B., Strle, K., Crowley, J.T., Lochhead, R.B., Anthony, R., and Steere, A.C. (2018). Correlation of Lyme Disease-Associated IgG4 Autoantibodies With Synovial Pathology in Antibiotic-Refractory Lyme Arthritis. Arthritis Rheumatol *70*, 1835-1846. https://doi.org/10.1002/art.40566

Swanson, K.V., Deng, M., and Ting, J.P. (2019). The NLRP3 inflammasome: molecular activation and regulation to therapeutics. Nat Rev Immunol *19*, 477-489. https://doi.org/10.1038/s41577-019-0165-0

Szczepanski, A., Furie, M.B., Benach, J.L., Lane, B.P., and Fleit, H.B. (1990). Interaction between *Borrelia burgdorferi* and endothelium *in vitro*. J Clin Invest *85*, 1637-1647.

Talkington, J., and Nickell, S.P. (1999). *Borrelia burgdorferi* spirochetes induce mast cell activation and cytokine release. Infect Immun *67*, 1107-1115.

Tavora, F., Burke, A., Li, L., Franks, T.J., and Virmani, R. (2008). Postmortem confirmation of Lyme carditis with polymerase chain reaction. Cardiovasc Pathol *17*, 103-107. https://doi.org/10.1016/j.carpath.2007.03.004

Thompson, A., Mannix, R., and Bachur, R. (2009). Acute pediatric monoarticular arthritis: distinguishing Lyme arthritis from other etiologies. Pediatrics *123*, 959-965. https://doi.org/10.1542/peds.2008-1511

Tilly, K., Krum, J.G., Bestor, A., Jewett, M.W., Grimm, D., Bueschel, D., Byram, R., Dorward, D., Vanraden, M.J., Stewart, P., *et al.* (2006). *Borrelia burgdorferi* OspC protein required exclusively in a crucial early stage of mammalian infection. Infect Immun *74*, 3554-3564. https://doi.org/10.1128/IAI.01950-05

Tjernberg, I., Sillanpaa, H., Seppala, I., Eliasson, I., Forsberg, P., and Lahdenne, P. (2009). Antibody responses to Borrelia IR(6) peptide variants and the C6 peptide in Swedish patients with erythema migrans. Int J Med Microbiol *299*, 439-446. https://doi.org/10.1016/j.ijmm.2008.10.006

Trollmo, C., Meyer, A.L., Steere, A.C., Hafler, D.A., and Huber, B.T. (2001). Molecular mimicry in Lyme arthritis demonstrated at the single cell level: LFA-1 alpha L is a partial agonist for outer surface protein A-reactive T cells. J Immunol *166*, 5286-5291. https://doi.org/10.4049/jimmunol.166.8.5286

Troy, E.B., Lin, T., Gao, L., Lazinski, D.W., Camilli, A., Norris, S.J., and Hu, L.T. (2013). Understanding barriers to *Borrelia burgdorferi* dissemination during infection using massively parallel sequencing. Infect Immun *81*, 2347-2357. https://doi.org/10.1128/IAI.00266-13

Tubo, N.J., and Jenkins, M.K. (2014). CD4+ T Cells: guardians of the phagosome. Clin Microbiol Rev *27*, 200-213. https://doi.org/27/2/200 [pii] 10.1128/CMR.00097-13

Tunev, S.S., Hastey, C.J., Hodzic, E., Feng, S., Barthold, S.W., and Baumgarth, N. (2011). Lymphoadenopathy during Lyme borreliosis is caused by spirochete migration-induced specific B cell activation. PLoS Pathog *7*, e1002066. https://doi.org/10.1371/journal.ppat.1002066 PPATHOGENS-D-10-00234 [pii]

Tupin, E., Benhnia, M.R., Kinjo, Y., Patsey, R., Lena, C.J., Haller, M.C., Caimano, M.J., Imamura, M., Wong, C.H., Crotty, S., *et al.* (2008). NKT cells prevent chronic joint inflammation after infection with *Borrelia burgdorferi.* Proc Natl Acad Sci U S A *105*, 19863-19868.

Tyson, K., Elkins, C., Patterson, H., Fikrig, E., and de Silva, A. (2007). Biochemical and functional characterization of Salp20, an *Ixodes scapularis* tick salivary protein that inhibits the complement pathway. Insect Mol Biol *16*, 469-479.

Ullah, M.O., Sweet, M.J., Mansell, A., Kellie, S., and Kobe, B. (2016). TRIF-dependent TLR signaling, its functions in host defense and inflammation, and its potential as a therapeutic target. J Leukoc Biol *100*, 27-45. https://doi.org/10.1189/jlb.2RI1115-531R

Valenzuela, J.G., Charlab, R., Mather, T.N., and Ribeiro, J.M. (2000). Purification, cloning, and expression of a novel salivary anticomplement protein from the tick, *Ixodes scapularis.* J Biol Chem *275*, 18717-18723.

Vesely, D.L., Fish, D., Shlomchik, M.J., Kaplan, D.H., and Bockenstedt, L.K. (2009). Langerhans cell deficiency impairs *Ixodes scapularis* suppression of Th1 responses in mice. Infect Immun *77*, 1881-1887.

Vigano, S., Perreau, M., Pantaleo, G., and Harari, A. (2012). Positive and negative regulation of cellular immune responses in physiologic conditions and diseases. Clin Dev Immunol *2012*, 485781. https://doi.org/10.1155/2012/485781

Vudattu, N.K., Strle, K., Steere, A.C., and Drouin, E.E. (2013). Dysregulation of CD4+CD25(high) T cells in the synovial fluid of patients with antibiotic-refractory Lyme arthritis. Arthritis Rheum *65*, 1643-1653. https://doi.org/10.1002/art.37910

Wagemakers, A., Coumou, J., Schuijt, T.J., Oei, A., Nijhof, A.M., van 't Veer, C., van der Poll, T., Bins, A.D., and Hovius, J.W. (2016). An *Ixodes ricinus* Tick Salivary Lectin Pathway Inhibitor Protects *Borrelia burgdorferi sensu lato* from Human Complement. Vector Borne Zoonotic Dis *16*, 223-228. https://doi.org/10.1089/vbz.2015.1901

Wang, G., Petzke, M.M., Iyer, R., Wu, H., and Schwartz, I. (2008). Pattern of proinflammatory cytokine induction in RAW264.7 mouse macrophages is identical for virulent and attenuated *Borrelia burgdorferi.* J Immunol *180*, 8306-8315.

Wang, H., and Nuttall, P.A. (1999). Immunoglobulin-binding proteins in ticks: new target for vaccine development against a blood-feeding parasite. Cell Mol Life Sci *56*, 286-295.

Wang, X., Ma, Y., Weis, J.H., Zachary, J.F., Kirschning, C.J., and Weis, J.J. (2005). Relative contributions of innate and acquired host responses to bacterial control and arthritis development in Lyme disease. Infect Immun *73*, 657-660.

Watson, J., and Riblet, R. (1974). Genetic control of responses to bacterial lipopolysaccharides in mice. I. Evidence for a single gene that influences mitogenic and immunogenic respones to lipopolysaccharides. J Exp Med *140*, 1147-1161.

Weinstein, D.L., Lissner, C.R., Swanson, R.N., and O'Brien, A.D. (1986). Macrophage defect and inflammatory cell recruitment dysfunction in *Salmonella*-susceptible C3H/HeJ mice. Cell Immunol *102*, 68-77.

Whiteside, S.K., Snook, J.P., Ma, Y., Sonderegger, F.L., Fisher, C., Petersen, C., Zachary, J.F., Round, J.L., Williams, M.A., and Weis, J.J. (2018). IL-10 Deficiency Reveals a Role for TLR2-Dependent Bystander Activation of T Cells in Lyme Arthritis. J Immunol *200*, 1457-1470. https://doi.org/10.4049/jimmunol.1701248

Willis, A.A., Widmann, R.F., Flynn, J.M., Green, D.W., and Onel, K.B. (2003). Lyme arthritis presenting as acute septic arthritis in children. J Pediatr Orthop *23*, 114-118.

Woodman, M.E., Cooley, A.E., Miller, J.C., Lazarus, J.J., Tucker, K., Bykowski, T., Botto, M., Hellwage, J., Wooten, R.M., and Stevenson, B. (2007). *Borrelia burgdorferi* binding of host complement regulator factor H is not required for efficient mammalian infection. Infect Immun *75*, 3131-3139.

Woods, A., Soulas-Sprauel, P., Jaulhac, B., Arditi, B., Knapp, A.M., Pasquali, J.L., Korganow, A.S., and Martin, T. (2008). MyD88 negatively controls hypergammaglobulinemia with autoantibody production during bacterial infection. Infect Immun *76*, 1657-1667.

Wooten, R.M., Ma, Y., Yoder, R.A., Brown, J.P., Weis, J.H., Zachary, J.F., Kirschning, C.J., and Weis, J.J. (2002a). Toll-like receptor 2 is required for innate, but not acquired, host defense to *Borrelia burgdorferi*. J Immunol *168*, 348-355.

Wooten, R.M., Ma, Y., Yoder, R.A., Brown, J.P., Weis, J.H., Zachary, J.F., Kirschning, C.J., and Weis, J.J. (2002b). Toll-like receptor 2 plays a pivotal role in host defense and inflammatory response to *Borrelia burgdorferi*. Vector Borne Zoonotic Dis 2, 275-278.

Wooten, R.M., Modur, V.R., McIntyre, T.M., and Weis, J.J. (1996). *Borrelia burgdorferi* outer membrane protein A induces nuclear translocation of nuclear factor-kB and inflammatory activation in human endothelial cells. J Immunol *157*, 4584-4590.

Wooten, R.M., Morrison, T.B., Weis, J.H., Wright, S.D., Thieringer, R., and Weis, J.J. (1998). The role of CD14 in signaling mediated by outer membrane lipoproteins of *Borrelia burgdorferi.* J Immunol *160*, 5485-5492.

Wormser, G.P., Nowakowski, J., Nadelman, R.B., Visintainer, P., Levin, A., and Aguero-Rosenfeld, M.E. (2008). Impact of clinical variables on *Borrelia burgdorferi*-specific antibody seropositivity in acute-phase sera from patients in North America with culture-confirmed early Lyme disease. Clin Vaccine Immunol *15*, 1519-1522. https://doi.org/10.1128/CVI.00109-08

Xu, Q., Seemanapalli, S.V., Reif, K.E., Brown, C.R., and Liang, F.T. (2007). Increasing the recruitment of neutrophils to the site of infection dramatically attenuates *Borrelia burgdorferi* infectivity. J Immunol *178*, 5109-5115.

Yakimchuk, K., Roura-Mir, C., Magalhaes, K.G., de Jong, A., Kasmar, A.G., Granter, S.R., Budd, R., Steere, A., Pena-Cruz, V., Kirschning, C., *et al.* (2011). *Borrelia burgdorferi* infection regulates CD1 expression in human cells and tissues via IL1-beta. Eur J Immunol *41*, 694-705.http s://doi.org/10.1002/eji.201040808

Yang, L., Ma, Y., Schoenfeld, R., Griffiths, M., Eichwald, E., Araneo, B., and Weis, J.J. (1992). Evidence for B-lymphocyte mitogen activity in *Borrelia burgdorferi*-infected mice. Infect Immun *60*, 3033-3041.http s://doi.org/10.1128/IAI.60.8.3033-3041.1992

Yang, R.B., Mark, M.R., Gray, A., Huang, A., Xie, M.H., Zhang, M., Goddard, A., Wood, W.I., Gurney, A.L., and Godowski, P.J. (1998). Toll-like receptor-2 mediates lipopolysaccharide-induced cellular signalling. Nature *395*, 284-288.

Yoder, A., Wang, X., Ma, Y., Philipp, M.T., Heilbrun, M., Weis, J.H., Kirschning, C.J., Wooten, R.M., and Weis, J.J. (2003). Tripalmitoyl-S-glyceryl-cysteine-dependent OspA vaccination of toll-like receptor 2-deficient mice results in effective protection from *Borrelia burgdorferi* challenge. Infect Immun *71*, 3894-3900.

Zeidner, N., Dreitz, M., Belasco, D., and Fish, D. (1996). Suppression of acute Ixodes scapularis-induced Borrelia burgdorferi infection using tumor necrosis factor-alpha, interleukin-2, and interferon-gamma. J Infect Dis *173*, 187-195.

Zeidner, N.S., Schneider, B.S., Nuncio, M.S., Gern, L., and Piesman, J. (2002). Coinoculation of Borrelia spp. with tick salivary gland lysate enhances spirochete load in mice and is tick species-specific. J Parasitol *88*, 1276-1278.

Zeng, X., Wei, Y.L., Huang, J., Newell, E.W., Yu, H., Kidd, B.A., Kuhns, M.S., Waters, R.W., Davis, M.M., Weaver, C.T., *et al.* (2012). Gammadelta T cells recognize a microbial encoded B cell antigen to initiate a rapid antigen-specific interleukin-17 response. Immunity *37*, 524-534. https://doi.org/10.1016/j.immuni.2012.06.011

Zhang, J.R., Hardham, J.M., Barbour, A.G., and Norris, S.J. (1997). Antigenic variation in Lyme disease Borreliae by promiscuous recombination of VMP-like sequence cassettes. Cell *89*, 275-285. https://doi.org/S0092-8674(00)80206-8 [pii]

Zhang, J.R., and Norris, S.J. (1998a). Genetic variation of the *Borrelia burgdorferi* gene vlsE involves cassette-specific, segmental gene conversion. Infect Immun *66*, 3698-3704.

Zhang, J.R., and Norris, S.J. (1998b). Kinetics and in vivo induction of genetic variation of vlsE in *Borrelia burgdorferi*. Infect Immun *66*, 3689-3697.

Zhao, Z., Fleming, R., McCloud, B., and Klempner, M.S. (2007). CD14 mediates cross talk between mononuclear cells and fibroblasts for upregulation of matrix metalloproteinase 9 by *Borrelia burgdorferi*. Infect Immun *75*, 3062-3069. https://doi.org/10.1128/IAI.00202-07

Chapter 19

Human and Veterinary Vaccines for Lyme Disease

Nathaniel S. O'Bier, Amanda L. Hatke, Andrew C. Camire and Richard T. Marconi*

Department of Microbiology and Immunology, Virginia Commonwealth University Medical Center, Richmond, VA 23298, USA
*Corresponding author: Richard.marconi@vcuhealth.org

DOI: https://doi.org/10.21775/9781913652616.19

Abstract
Lyme disease (LD) is an emerging zoonotic infection that is increasing in incidence in North America, Europe, and Asia. With the development of safe and efficacious vaccines, LD can potentially be prevented. Vaccination offers a cost-effective and safe approach for decreasing the risk of infection. While LD vaccines have been widely used in veterinary medicine, they are not available as a preventive tool for humans. Central to the development of effective vaccines is an understanding of the enzootic cycle of LD, differential gene expression of *Borrelia burgdorferi* in response to environmental variables, and the genetic and antigenic diversity of the unique bacteria that cause this debilitating disease. Here we review these areas as they pertain to past and present efforts to develop human, veterinary, and reservoir targeting LD vaccines. In addition, we offer a brief overview of additional preventative measures that should employed in conjunction with vaccination.

Introduction

The causative agents of Lyme disease

Borrelia burgdorferi, the primary causative agent of Lyme disease (LD) in North America, was cultivated from ticks and identified in 1981 (Burgdorfer et al., 1982). A direct linkage between *B. burgdorferi* and LD was established with its isolation from the blood of human LD patients (Burgdorfer et al., 1982; Benach et al., 1983). This previously uncharacterized pathogen was classified in the genus *Borrelia* based on its shared features with the causative agents of tick-borne relapsing fever (TBRF) (Benach et al., 1983). *B. burgdorferi* and TBRF spirochetes possess a unique segmented genome comprised of a series of linear and circular DNA plasmids and prophage (Barbour, 1988, 1993; Eggers and Samuels, 1999; Zhang and Marconi, 2005). Upon denaturation, the linear plasmids transition to a single-stranded circular DNA molecule. As a consequence of extensive complementarity, the single-stranded DNA molecules can reanneal upon themselves to form double-stranded linear DNA with covalently closed hairpin termini (Barbour and Garon, 1987). While linear DNA is rare in bacteria, genetic elements with similar structure are found in some viruses including the African Swine Fever Virus, an arbovirus (Ndlovu et al., 2020). *B. burgdorferi* and the TBRF *Borrelia* have a similar spiral ultrastructure, unique mode of motility and similar nutritional requirements (Barbour, 1984; Asbrink and Hovmark, 1985).

As the number of isolates available for study increased in the 1980's, it became apparent that *B. burgdorferi* strains varied in many of their genotypic and phenotypic properties. Comparative analyses of isolates from different biological and geographic sources revealed extensive diversity in plasmid and protein profiles (Barbour, 1988; Schwan et al., 1988; Stålhammar-Carlemalm et al., 1990). Genetic relationships among isolates were subsequently assessed through DNA-DNA hybridization (LeFebrve et al., 1989), restriction fragment length polymorphism analysis (Stålhammar-Carlemalm et al., 1990; Fukunaga et al., 1993; Postic et al., 1994; Liveris et al., 1995), S1 nuclease digestion (Baranton et al., 1992) and 16S rRNA sequencing (Marconi and Garon, 1992a, b). Shortly after its initial discovery, *B. burgdorferi* was divided into three distinct species: *B. burgdorferi* (Burgdorfer et al., 1982), *B. garinii* (Baranton et al., 1992), and *B. afzelii* (previously referred to as genomic group VS461) (Marconi and Garon, 1992b; Canica et al., 1993). At last count, twenty-one distinct species associated with LD have been identified (Eisen, 2020). To reflect the diversity among *B. burgdorferi*, the taxonomic designation sensu lato was applied. *B. burgdorferi, B. garinii,* and *B. afzelii* are responsible for the majority of LD cases in humans and companion animals (Eisen, 2020). All three of the established pathogenic species are found in Europe and Asia while *B. burgdorferi* is found in North America. *B. bavariensis* (Margos et al., 2013), *B. valaisiana* (Wang et al., 1997), *B. bissettii* (Postic et al., 1998), *B. kurtenbachii* (Margos et al., 2010), *B. speilmanii* (Maraspin et al., 2006), *B. lusitaniae* (Collares-Pereira et al., 2004) and *B. mayonii* (Pritt et al., 2016) are considered minor contributors to disease (Eisen, 2020). *B. mayonii*, the most recent species to be identified and characterized, was discovered through PCR analyses of human diagnostic specimens by the Mayo Clinic (Pritt et al., 2016; Johnson et al., 2017). While *B. mayonii* has been recovered from both humans and ticks, its overall contribution to LD case numbers appears to be minimal. The pathogenic potential of *B. americanum* (Rudenko et al., 2009), *B. carolinesis* (Foley et al., 2014), *B. californiensis* (Postic et al., 2007), *B. chilensis* (Verdugo et al., 2017), *B. japonica* (Kawabata et al., 1993), *B. lanei* (Margos et al., 2017), *B. sinica* (Masuzawa et al., 2001), *B. tanuki* (Fukunaga et al., 1996), *B. turdii* (Fukunaga et al., 1996), *B. andersonii* (Marconi et al., 1995), and *B. yangtzensis* (Margos et al., 2015) remains to be determined. The apparent inability of this latter group of species to infect and cause disease in humans and companion animals may be due to the feeding habits of the specific *Ixodes* tick species that serve as their vectors, as well as their sensitivity to complement. *B. andersonii*, for example, is vectored primarily by *Ixodes dentatus* ticks which do not commonly feed on humans and other large mammals (Anderson et al., 1989). In addition, *B. andersonii* is highly sensitive to human and canine complement, presumably due to the fact that it does not produce factor H binding proteins (McDowell et al., 2003b). The potential host range for many of the species listed above remains largely undefined. For convenience, henceforth we use the term LD spirochetes to refer collectively to species that cause disease in humans and companion animals.

In 2014, it was proposed that the *B. burgdorferi* sensu lato group be awarded separate genus status (Adeolu and Gupta, 2014). This reclassification was proposed based on epidemiological, biological, genotypic and clinical differences between the LD and tick-borne relapsing fever (TBRF) spirochetes. Species of the *B. burgdorferi* sensu lato complex were separated from the genus *Borrelia* and awarded a new genus designation; *Borreliella* (Adeolu and Gupta, 2014). As is often the case with taxonomic reclassification, this change was met with varying degrees of enthusiasm and reclassification has been rigorously debated at conferences and in the literature (Adeolu and Gupta, 2014; Barbour et al., 2017; Stevenson et al., 2018). Here, we do not seek to argue for or against reclassification. However, while the use of *Borreliella* is considered optional in the literature, it is important to note that this designation has been fully integrated into the sequence and protein structure databases. Hence, for practical purposes we have adopted this new genus designation.

The enzootic cycle of Lyme disease

The *Borreliella* are obligate host-dependent bacteria maintained in nature in an enzootic cycle involving diverse vertebrate reservoir species and *Ixodes* tick vectors (Barbour and Hayes, 1986; Loss et al., 2016). The strict requirement for cycling between reservoirs and ticks reflects the absence of transovarial and horizontal transmission. While there have been isolated reports of transovarial transmission in ticks (Hauck et al., 2020), it does not appear to be significant in terms of population maintenance in nature. Tick species that are competent vectors for one or more *Borreliella* species include *I. affinis, I. angustus, I. hexagonus, I. minor, I. muris, I. pacificus, I. ricinus, I. ovatus, I. scapularis,* and *I. spinipalpis* (Eisen, 2020). The identification of *Haemaphysalis longicornis* (Asian long horned tick) in North America initially raised concerns about the potential of this invasive and rapidly spreading species to serve as a vector for *Borreliella* species (Beard et al., 2018). To date, *Borreliella* species have not been recovered from *H. longicornis* and efforts to experimentally infect these ticks with *B. burgdorferi* were unsuccessful (Breuner et al., 2020).

Diverse reservoirs for the Borreliella: implications for vaccine development

While the list of identified reservoir species for the *Borreliella* is extensive, investigations into the competence of additional species of mammals and birds to serve as reservoirs is critically important, particularly in light of ongoing efforts to develop reservoir targeting vaccines and intervention strategies. *Peromyscus leucopus* (white-footed mouse) and *Apodemus sylvacticus* (wood mouse) are among the most abundant reservoir species in North America and Europe, respectively (Anderson, 1991). However, "inconspicuous" hosts including *Tamias striatus* (eastern chipmunk), *Sorex brevicauda* (short-tailed shrew), *Sorex cinereus* (masked shrew), *Myodes glareolus* (bank vole),

Microtus agrestis (field vole), *Sorex araneus* (common shrew) and *Sorex minutus* (pygmy shrew) may be equally important (Brisson et al., 2008; Coipan et al., 2018). Birds also serve as reservoir hosts but perhaps most important is the role that birds play in the transportation and geographic dispersal of ticks (Kurtenbach et al., 1998; Hanincova et al., 2003; Ogden et al., 2008; Scott et al., 2012; Scott et al., 2018). Ground foraging, passerine birds harbor heavy tick burdens and are considered "super-carriers" (Anderson et al., 1986; Hanincova et al., 2003; Ogden et al., 2008; Scott et al., 2012; Hasle, 2013; Lommano et al., 2014; Newman et al., 2015; Scott et al., 2018). Of the 162 species of birds assessed in one study for overall tick burden, 72 species were described as "tick infested" (Loss et al., 2016). Another recent study screened 205 tissue samples from the bone marrow or spleen of 169 individual upland game birds for *B. burgdorferi* DNA. Amplicons were obtained from 12% of the samples tested with the highest prevalence in wild turkeys (*Meleagris gallopavo*) (Cleveland et al., 2020).

Predators also influence or contribute directly or indirectly to the enzootic cycle (Ostfeld et al., 2006; Levi et al., 2012). Beyond their influence on small mammal reservoir populations through predation, they may also serve as reservoirs. In Europe, red foxes (*Vulpes vulpes*) have been demonstrated to be competent reservoirs (Mysterud et al., 2019) and it has been postulated that eastern coyotes (*Canis latrans*), an apex predator, may serve as reservoirs in the northeastern United States (Izac et al., 2020a). Robust antibody titers to several *B. burgdorferi* antigens were demonstrated in 64.8% of 128 eastern coyote plasma samples collected in Pennsylvania between 2015 and 2017 (Izac et al., 2020a). In the same study, 72.7% of the samples were antibody positive for the *Anaplasma phagocytophilum* antigens P44 and P130. Notably, 51% of the samples were positive for antibodies to both *B. burgdorferi* and *A. phagocytophilum* suggesting potential coinfection. While seropositivity is not a direct indication of reservoir competence, based on the behavioral habits of eastern coyotes it is possible that they may serve as a source of infection for feeding ticks. Knowledge of the full range of competent reservoir hosts will be essential to advance wildlife and reservoir targeting vaccines, which to date have been focused primarily on targeting mice (Tsao et al., 2004; Gomes-Solecki et al., 2006; Scheckelhoff et al., 2006; Bhattacharya et al., 2011; Meirelles Richer et al., 2011; Telford et al., 2011; Voordouw et al., 2013). It remains to be determined if an effective and practical reservoir targeting strategy can be developed that will be able to target the highly diverse reservoirs species of the *Borreliella*.

Justification for the development and use of Lyme disease vaccines

Human Lyme disease

The case for the development of a human LD vaccine is, first and foremost, supported by the failure of existing preventive strategies to reduce the incidence of disease. Current prevention strategies for tick-borne diseases, particularly in humans, are ineffective at the population level. In this section we discuss the justification for the development and use of LD vaccines and briefly review complementary, non-vaccination-based preventative strategies that are being applied in human and veterinary medicine. Prevention of tick-borne diseases will require a multi-pronged approach with vaccination as its center piece.

The human health toll and socioeconomic impact of LD cannot be understated. LD is the most common tick-borne disease in North America, Europe, and Asia. In the United States a total of 33,666 confirmed and probable cases of LD were reported to the CDC in 2018 (www.cdc.gov/lyme/index.html). While surveillance for LD and other tick-borne disease has provided important information about the incidence and spread of diseases, these approaches are almost entirely passive. It is widely recognized that passive surveillance data are compromised by low compliance, non-mandated reporting, misdiagnosis, and under and over reporting (Meek et al., 1996; Yound, 1998; Bacon et al., 2008). Studies suggest that the true incidence of LD in humans in some regions of the United States is 3 to 12 times greater than that inferred from the reported case numbers (Meek et al., 1996; Yound, 1998; Sykes and Makiello, 2017)). Between 1992 and 1998, a total of 88,967 cases of human LD were reported to the CDC by state departments of health (www.cdc.gov). Between 2008 and 2015, this number rose to 275,000 cases. A CDC based analysis of clinician diagnosed cases from 2005 through 2010 estimated the annual incidence of LD in the United States at 106.6 cases per 100,000 persons with 296,000 to 376,000 new cases each year (95% credible interval) (Nelson et al., 2015). A separate CDC study that analyzed insurance records and diagnostic codes also concluded that the actual number of cases is considerably higher than that inferred from passive surveillance (Tseng et al., 2015). Data from the National Notifiable Disease Surveillance System indicates that the geographic distribution of high-incidence areas for LD is expanding. The number of counties with an incidence of ≥10 confirmed cases per 100,000 persons increased from 324 in 2008 to 415 in 2018. It is clear that the LD threat is increasing.

Consistent with the geographic expansion of high-incidence zones for LD, the rapid and ongoing geographic spread of *I. scapularis* and *I. pacificus* ticks is well documented (Eisen et al., 2016; Eisen et al., 2017). *I. scapularis* and *I. pacificus* have been documented in 43 states and in 1,531 of the 3,110 continental United States counties (49.2%). This represents a 44.7% increase since an earlier assessment conducted in 1998. The number of counties in which *I. scapularis* is considered established (i.e., six or more individual ticks or one or more life stages identified in a single year) has more than doubled over the past two decades. The majority of county status changes have occurred in the North-Central and Northeastern states. It is anticipated that the two previously distinct foci for *I. scapularis* in the Northeast and North-Central states will merge in the upper Ohio River Valley resulting in the formation of a single contiguous "tick zone" encompassing much of the eastern half of the United States. Endemic regions for *Ixodes* ticks, and the incidence of LD, are also increasing in Canada, Europe, and Asia as well (Gasmi et al., 2017; Levy, 2017; Sykes and Makiello, 2017; Lloyd and Hawkins, 2018; Ogden et al., 2019).

The socioeconomic costs of LD in the United States are staggering. A study conducted in 1998, based on an incidence of 4.73 cases per 100,000 persons and year 1996 dollars, estimated costs at $500 million in the United States alone (Maes et al., 1998). A study published in 2015 estimated the direct costs of LD and post-treatment LD syndrome in the United States at approximately $712 million and $1.3 billion a year, respectively (Adrion et al., 2015). The most recent analysis of the inpatient costs of LD-related hospitalizations assessed data collected between 2005-2014 from the Truven Health Analytics MarketScan Commercial Claims and Encounters Databases (Schwartz et al., 2020). Of 20,983,165 admission records assessed, 2,823 (0.01%) met inclusion criteria for LD-related hospitalizations. Over half of the identified records contained an ICD-9-CM code for meningitis (n = 614), carditis (n = 429), facial palsy (n = 400) or arthritis (n = 377). The median cost per LD-related hospitalization was $11,688 (range: $140-$323,613). The median costs per stay based on clinical manifestations were $17,461 for carditis, $15,177 for meningitis, $13,012 for arthritis, and $10,491 for facial palsy. Extrapolation of the data to the U.S. population suggests that there are 2,196 LD-related hospitalizations in persons under 65 years of age annually with an estimated annual cost of $25,826,237. The estimated costs for vaccination of individuals at risk for LD is estimated to be approximately $100 a year per vaccinated individual. The potential cost-effectiveness of vaccination for LD is clear (Mac et al., 2019; Behler et al., 2020). The increasing incidence and expansion of endemic regions for LD, coupled with the real or perceived controversies surrounding diagnosis and treatment, highlight the need for aggressive prevention strategies. It follows that an effective and publicly accepted vaccine would address and lessen some of the most controversial aspects of LD diagnosis and treatment and dramatically reduce the costs associated with LD.

LD in companion animals

LD was first described in dogs in the mid 1980's (Lissman et al., 1984; Kornblatt et al., 1985; Magnarelli et al., 1985) and is now the most common tick-borne disease of companion animals. Point-of-care tests that allow for rapid serological screening of dogs for antibodies to *B. burgdorferi, Anaplasma phagocytophilum, Anaplasma platys, Ehrlichia canis* and *Ehrlichia ewingii* are widely used in veterinary medicine (Little et al., 2010; Littman et al., 2018). There is no equivalent screening strategy in place in human medicine. The data obtained from yearly screening informs our understanding of disease spread and exposure risk in humans (Mead et al., 2011; Yancey et al., 2014; Bowser and Anderson, 2018) . The Companion Animal Parasite Council (CAPC; https://capcvet.org/articles/parasite-prevalence-maps/), an independent non-profit organization that is focused on increasing awareness of infectious diseases in companion animals, gathers and reports antibody test results for tick-borne diseases that affect canines in the United States and Canada. In 2019, the results from 7,057,815 LD and tick-borne pathogen serological screening tests were catalogued by the CAPC. Approximately, 5.0% of the tests run were antibody positive for *B. burgdorferi.* The second most prevalent tick-borne disease in North America is anaplasmosis (*Anaplasma phagocytophilum*) with over 221,000 (3.1%) positive tests reported in 2019. It is important to note that the CAPC captures results for less than ~30% of the total number of serological tests conducted each year. Hence, the probable number of positive *B. burgdorferi* antibody tests in dogs may be three-fold greater and most likely exceeds 1,000,000. In Canada, serological screening of dogs suggests that endemic regions for LD currently exist as isolated foci. Defined risk areas with increasing incidence are found in Manitoba, Ontario, New Brunswick, Quebec, and Nova Scotia. In Nova Scotia, 19.5% of the serological tests for *B. burgdorferi* in dogs reported in 2019 were antibody positive. Significant case numbers have also been reported in British Columbia on Canada's western coast (Mak et al., 2000). With time, the now isolated and defined foci are expected to expand and merge. As in the United States, vaccination is widely employed in veterinary medicine in Canada. It is possible that with the implementation of an aggressive vaccination strategy, the growing threat of LD in dogs can be contained.

The exposure risk for LD in humans is associated with occupation and with the frequency and duration of outdoor activities (hiking, outdoor recreation, gardening etc.). Central tenets of the CDC's recommendations for prevention of tick-borne diseases are to minimize exposure to tick habitats (grassy, brushy and wooded areas) and to perform diligent tick checks (physical body inspection) after engaging in at risk activities. In both human and veterinary medicine, acaricides are available for use. The application of permethrin (0.5%) (Connally et al., 2018) to clothing is one preventative strategy recommended by the CDC. Permethrin is long-lasting and has been demonstrated to convey significant protection against bites from *I. scapularis* in laboratory studies and sustained protection against bites by *Amblyomma americanum* ticks in field tests. Recently a two-year randomized, placebo-controlled, double-blinded trial of 82 outdoor workers in Rhode Island and southern Massachusetts was completed (Mitchell et al., 2020). Study participants in the treatment-arm wore factory-impregnated permethrin clothing while the control group wore sham-treated clothing. Factory-impregnated clothing reduced tick bites by 65% in the first study year and by 50% in the second year. No significant treatment-related adverse outcomes were reported. Additional tick repellents registered with the Environmental Protection Agency are also available, but some, such as oil of lemon eucalyptus and para-methane-diol, should not be used on children under three years old. While the rationale behind the CDC recommendations listed above are obvious, the practicality or likelihood of widespread voluntary compliance with some of these practices at the population level seems unlikely.

The risk for tick-borne diseases is inherently greater for companion animals (Littman et al., 2018; Divers et al., 2019) than it is for humans due to their greater risk of exposure. As discussed above, veterinary medicine has several tools available to screen for and aid in the prevention of tick-borne diseases. The American College of Veterinary Internal Medicine 2018 consensus statement on LD (Littman et al., 2018) recommends screening of all healthy dogs that live in, live near, or travel to LD endemic areas in North America for antibody to *B. burgdorferi*. Furthermore, it is recommended that *B. burgdorferi* antibody positive dogs be tested for proteinuria due to concerns about the development of potentially fatal kidney disease (Lyme nephritis). Acaricides, including fluralaner and afoxolaner (isooxazoline family), are widely used on dogs and have good efficacy and safety profiles (Allen et al., 2020; Holzmer et al., 2020; Kryda et al., 2020; Petersen et al., 2020). While varying with the product, acaricides can be delivered orally as chewables, topically (so called "spot on" application) or through the use of treated collars. Tick checks should also be performed on dogs after outdoor activity but they can be difficult to thoroughly conduct due to the dense coat and skin tone of most common dog breeds. Several bacterin and subunit vaccines have been approved by the USDA and are commercially available. The composition, mechanism of action, safety, and efficacy of canine LD vaccines are discussed in detail below. For further information on the socioeconomic burden of human LD and the cost-effectiveness of treatment, general prevention and vaccination, readers are directed to a recent and comprehensive review of the existing literature (Mac et al., 2019).

Outer surface proteins and LD vaccine development

The *Borreliella* genome encodes a vast repertoire of surface-exposed proteins (Fraser et al., 1997; Casjens et al., 2000; Glockner et al., 2004; Schutzer et al., 2011). Many are differentially expressed in response to environmental variables (Fingerle et al., 2000; Revel et al., 2002; Roberts et al., 2002; Crandall et al., 2006; Caimano et al., 2007; Rogers et al., 2009b; Iyer et al., 2015; Caimano et al., 2016). To date, only two surface proteins have been successfully developed as subunit vaccine antigens in human or veterinary LD vaccines: outer surface protein A (OspA) and a laboratory designed outer surface protein C (OspC)-based recombinant chimeric epitope protein.

Outer surface protein A: transcriptional expression, biological function and diversity

Outer surface protein (Osp) A was among the first *B. burgdorferi* proteins to be identified and characterized (Howe et al., 1985). Early analyses of OspA were prompted in part by the simple observation that it was one of the most abundant *B. burgdorferi* proteins produced during *in vitro* cultivation. OspA is a 31 kDa surface lipoprotein encoded by a linear plasmid of 54 kb (lp54) (Barbour, 1988). It is cotranscribed as a bicistronic operon with *ospB* (Howe et al., 1986; Bergström et al., 1989). While *B.* burgdorferi strains that lack lp54 have been obtained upon prolonged *in vitro* cultivation (Hughes et al., 1993; Sadziene et al., 1995), lp54 is universal among isolates collected from nature (Casjens et al., 2012) and is essential for completion of the enzootic cycle. OspA is selectively produced during spirochete residence in ticks (Fikrig et al., 1990) with a transcriptional expression rank relative to all other detectable transcripts of 8th during cultivation and 18th in unfed larval ticks (Caimano et al., 2019). Upon ingestion of the bloodmeal, *ospA* transcription is downregulated (Schwan and Piesman, 2000) before terminating at the tick-tissue interface (Caimano et al., 2019). Gene deletion studies demonstrated that OspA is required for survival in ticks (Yang et al., 2004; Battisti et al., 2008). In the tick midgut, OspA functions as an adhesin and binds to the tick protein TROSPA (tick receptor for OspA) (Pal et al., 2000; Pal et al., 2004a). The crystal structure of recombinant OspA in complex with the Fab fragment of monoclonal antibody 184.1 was determined at a resolution of 1.9 A (PDB: 1OSP) (Li et al., 1997). A unique structural organization was revealed that consists of repetitive antiparallel beta sheets with a nonglobular region connecting the globular N- and C-terminal domains.

Antigenic determinants of OspA

Based on reactivity with monoclonal antibodies and comparative DNA sequence analyses, distinct variants or serotypes of OspA have been identified (discussed below) (Wilske et al., 1993a; Will et al., 1995; Wilske et al., 1996). OspA serotype 1 is the dominant variant produced by *B. burgdorferi* isolates from North America. OspA diversity is greater among *Borreliella* isolates from Europe where several serotypes have been identified (Wilske et al., 1993a; Wilske et al., 1995; Marconi et al., 1999). OspA serotype 2 is associated with *B. afzelii,* serotypes 3 and 5 with *B. garinii*, serotype 6 with *B. garinii* and serotype 4 with *B. bavariensis*. Serotype 4 has been postulated to correlate with a neuroinvasive phenotype, but since OspA is not produced in mammals, the potential correlation would be an indirect one (Marconi et al., 1999).

The protective efficacy of OspA was first demonstrated in the C3H/HeJ mouse model (Fikrig et al., 1990). Mice were vaccinated with either live *E. coli* expressing OspA (strain N40; 5x10^6 cells; three doses one week apart) or purified recombinant strain N40 OspA fused to GST. It is not indicated in the original publication if the formulation was adjuvanted. Both vaccine formulations provided short term protection. The live *E. coli* formulation expressing OspA protected against strains N40, B31 and CD16. The recombinant GST-OspA fusion protein was only tested for efficacy using strain N40. In a follow up study, mice were immunized with recombinant OspA (10 µg; 3 doses using Freund's complete and incomplete adjuvants) and protection against different spirochete challenge doses was assessed (Fikrig et al., 1992b). Mice were protected against 1x10^4 laboratory cultured *B. burgdorferi* cells but not 1x10^7. In the same study, the protective efficacy of OspB and flagellin were assessed. OspB protected mice from infection with an inoculum of 10^3 spirochetes but not a dose of 10^4. Consistent with its periplasmic localization, flagellin did not convey protection against any challenge dose. OspA has proven to be a protective vaccinogen in dogs, rhesus macaques (Probert and LeFebvre, 1994; Straubinger et al., 1995; Philipp et al., 1997) and humans (discussed in detail below). A caveat to OspA mediated protective antibody responses is that they are relatively short-lived and an aggressive booster schedule is required. The necessity for frequent boosts stems from the unique mechanism of action of OspA vaccinogens. Since OspA is produced only in ticks, antibody can only target *B. burgdorferi* prior to transmission into mammals. Hence, it is essential to maintain high levels of circulating antibody since an immune memory response will not develop following natural infection by tick inoculation.

Several studies have localized protective epitopes of OspA within its C-terminal half (Huang et al., 1998; Ding et al., 2000; Legros et al., 2000; Kharitonenkov et al., 2002; Izac et al., 2017; Shandilya et al., 2017). Using a protective and well-characterized monoclonal antibody designated as LA-2, the corresponding LA-2 epitope was identified (Huang et al., 1998; Ding et al., 2000). The LA-2 epitope is discontinuous and formed by residues contained within three surface-exposed loops present in the C-terminal domain of OspA (Ding et al., 2000). Based on this finding, and epitope mapping studies with other monoclonal antibodies, it has been suggested that maintenance of OspA structure is essential for maximizing productive antibody responses. Efforts have been taken to generate N-terminally truncated OspA proteins that still retain the native LA-2 epitope conformation (Koide et al., 2005). A lipidated OspA fragment consisting of residues 130-273 (*B. burgdorferi* B31; ORF designation, BBA_15) fused to the IgG binding domain of staphylococcal protein A (Koide et al., 2005) retained native conformation and bound monoclonal antibody LA-2. However, the protein proved to be unstable and it exhibited reduced efficacy in challenge studies in C3H-HeJ mice (Koide et al., 2005). To stabilize the fragment, residues involved in buried salt-bridge interactions were replaced with residues that promote hydrophobic interactions. Specifically, substitution of amino acids R139, E160 and K189 with M, Y and M, respectively, increased the stability of the fragment and restored protective efficacy. The vaccination protocol consisted of three doses of immunogen in alum at two-week intervals. Challenge was administered by feeding ten field collected ticks (infectivity percentage of 30%) on each mouse. To our knowledge, there have been no additional reports indicating that this protein has been pursued further as a vaccine antigen.

A recent study identified a continuous, linear epitope in OspA that is localized within amino acids spanning residues 221-240 (*B. burgdorferi* B31 OspA-serotype 1) referred to as the $OspA_{221-240}$ epitope (Izac et al., 2017). IFA assays using antibodies raised against $OspA_{221-240}$ peptide conjugated to keyhole limpet hemocyanin (KLH) surface labeled intact *B. burgdorferi.* The hyperimmune sera displayed potent complement-dependent bactericidal activity that killed *B. burgdorferi* at a level equivalent to that of full length OspA. The bactericidal activity of the antisera against *B. afzelii, B. bavariensis* and *B. garinii* was significant but reduced relative to that seen against *B. burgdorferi.* BLAST searches using the $OspA_{221-240}$ amino acid sequence revealed regions of conservation and variation within this epitope containing span of OspA. Interestingly, an amino acid segment of OspB spanning residues 244-260 displayed significant homology with the OspA 221-240 peptide. The corresponding OspB sequence was found to be well-conserved among *B. afzelii, B. bavariensis* and *B. garinii* isolates. Consistent with the decreased antibody mediated killing of *B. bavariensis* and *B. afzelii* isolates by the hyperimmune sera relative to that of *B. burgdorferi* isolates, the variable amino acid residues are centrally located within the $OspB_{244-260}$ sequence. The positioning of these mismatches most likely accounts for the lower level of bactericidal activity. Based on sequence similarity between the *B. burgdorferi* $OspA_{221-240}$ epitope and $OspB_{244-260}$, it is possible that antibody raised against variants of the $OspA_{221-240}$ domain may be able to target both OspA and OspB. It remains to be determined if OspA peptides or fragments consisting of the multiple variants of the 221-240 peptide can elicit protective antibodies in a tick-mouse challenge model.

Outer surface protein C (OspC) expression, structure and function

OspC is an approximate 22kDa surface lipoprotein encoded by a circular plasmid of 26 kb (cp26). Cp26 is genetically stable and ubiquitous among *Borreliella* species and isolates (Marconi et al., 1993a; Sadziene et al., 1993; Tilly et al., 1997). Orthologs of OspC have been identified in some TBRF species including *Borrelia hermsii* (Marconi et al., 1993b). *OspC* is the most highly expressed *B. burgdorferi* gene during early stage infection in mice and the 8th most abundant transcript in fed *I. scapularis* nymphs (Caimano et al., 2019). Antibody to OspC develops early and can be detected in the serum of mice within two weeks of syringe inoculation of *B. burgdorferi* (Oliver Jr et al., 2016). Consistent with this, non-targeted mass spectrometry of skin biopsies collected from mice seven days after delivery of *B. burgdorferi* by syringe inoculation also identified OspC as one of the most frequently detected *B. burgdorferi* proteins (Talagrand-Reboul et al., 2020). In contrast, during laboratory cultivation *ospC* is expressed at low and varying levels with a transcriptional ranking of 816 (Iyer et al., 2015). The transcriptional regulation of OspC is complex and has been detailed in several studies (Marconi et al., 1993a; Hubner et al., 2001; Alverson et al., 2003; Yang et al., 2005; Gilbert et al., 2007; Hayes et al., 2014).

The structure of OspC (PDB: 1GGQ, 1F1M, 1G5Z) offers significant insight into its potential functional domains as well as its antigenic determinants (Eicken et al., 2001; Kumaran et al., 2001a; Kumaran et al.,

2001b). In contrast to OspA, which is largely β-sheet and monomeric, OspC consists of five α-helices connected by loops. OspC forms a homodimer with an extensive buried interface suggesting that the biologically active form of the protein is the dimer. Gene deletion analyses demonstrated that OspC is required for *B. burgdorferi* to infect mice but not required to infect ticks or be transmitted by ticks (Grimm et al., 2004; Pal et al., 2004b; Dunham-Ems et al., 2012). Specific amino acids required for the *in vivo* function were identified through the introduction of amino acid substitution mutants into *B. burgdorferi* B31 *ospC* using allelic exchange replacement (Earnhart et al., 2010). The substitutions were targeted within a putative small ligand binding pocket (designated as LBD1) predicted using the ConCavity ligand analysis algorithm. LBD1 is formed near the membrane proximal end of the OspC dimer and lined by residues K60, E61, E63 and A64 of each monomer. The mutated proteins were expressed in *B. burgdorferi* and presented on the cell surface in a manner consistent with wild-type OspC. A single amino acid substitution in which E61 was replaced with Q rendered *B. burgdorferi* B31 non-infectious in C3H/HeJ mice (dose of 1×10^4 syringe administered spirochetes) (Earnhart et al., 2010). Substitution of K60 with Y did not affect infectivity but attenuated colonization of the heart relative to the parental B31 strain. The use of site-directed mutagenesis allelic exchange replacement is a powerful approach that can be used to identify the determinants of OspC that are involved in its different and distinct functions.

Evidence suggests that OspC is a multi-functional protein. Several tick or mammalian derived ligands have been demonstrated to bind to OspC including Salp15 (tick salivary protein) (Anguita et al., 2002; Ramamoorthi et al., 2005), plasminogen (Lagal et al., 2006; Onder et al., 2012) and complement C4b (Caine et al., 2017). Salp15 is a potential immune modulatory protein that may interfere with CD4+ T-lymphocyte function (Anguita et al., 2002). The binding of plasminogen has been postulated to facilitate tissue invasion (Lagal et al., 2006; Onder et al., 2012), while the binding of C4b has been hypothesized to inhibit the classical and lectin complement pathways and compete with complement protein C2 for binding to C4b (Caine et al., 2017). OspC has also been reported to have anti-phagocytic properties (Carrasco et al., 2015). Additional research is needed to demonstrate the biological significance of the individual ligand interactions summarized above.

OspC phylogeny

OspC is among the most variable LD spirochete proteins (Jauris-Heipke et al., 1993; Theisen et al., 1993; Wilske et al., 1993b). Phylogenetic analyses have delineated approximately 38 distinct OspC phyletic clusters or "types" (Seinost et al., 1999a; Wang et al., 1999; Brisson and Dykhuizen, 2004; Earnhart and Marconi, 2007c). Within a phyletic type, OspC sequences are highly conserved (>95% amino acid identity) while at the inter-type level, amino acid identity values range from 55 to 80% (Wang et al., 1999; Eicken et al., 2001; Kumaran et al., 2001b; Earnhart and Marconi, 2007c). OspC variants are referred to as OspC types; letter based, isolate of origin, or other designations are used to differentiate them (i.e., OspC type A, OspC type B, OspC type Pko, etc.,). In spite of extended regions of sequence conservation, antibody responses to OspC during natural infection or upon immunization with bacterin vaccines or recombinant OspC proteins are largely OspC type-specific indicating that the immuno-dominant epitopes are localized within variable domains of the protein (Earnhart et al., 2005; Earnhart et al., 2007; Oliver Jr et al., 2016; Izac et al., 2019). Considerable OspC diversity has been demonstrated among LD spirochete strains isolated from tightly defined geographic regions (Wang et al., 1999; Lin et al., 2002; Alghaferi et al., 2005; Earnhart et al., 2005). Twenty-one OspC types have been demonstrated in a single tick through deep sequencing (Di et al., 2018). *B. burgdorferi* isolates recovered from naturally infected animals (Seinost et al., 1999b; Oliver Jr et al., 2016; Izac et al., 2019) or from the tissues of dogs experimentally infected with field collected ticks (Rhodes et al., 2013) are highly heterogenous with regard to the OspC types represented. The existence of stable and distinct OspC types in nature may be essential for population maintenance and may facilitate the ability of the LD spirochetes to infect an immunologically primed, OspC antibody positive host (Seinost et al., 1999a; Wang et al., 1999). OspC diversity has long been a significant obstacle in the development of broadly protective OspC based vaccines. The recent development of recombinant proteins consisting of linear epitopes derived from numerous OspC types has effectively addressed the OspC diversity issue (Izac et al., 2020b). These unique proteins, referred to as chimeritopes, have been demonstrated to elicit potent bactericidal antibodies against a wide range of OspC types and to convey protection against the diverse strains presented during challenge with field-collected ticks (Ball, 2016; Marconi, 2020b). The construction and analysis of OspC chimeritopes is described in detail below.

Several studies have reported on the identification of conformationally defined OspC protective epitopes (Gilmore et al., 1996; Gilmore and Mbow, 1999; Norek and Janda, 2017). Gilmore et al (1996) vaccinated specific-pathogen-free outbred mice intradermally with recombinant non-denatured OspC excised from a gel and with denatured OspC (adjuvanted with alum). Three doses were delivered each two weeks apart. The mice vaccinated with non-denatured recombinant OspC were protected upon challenge with laboratory reared-ticks infected with *B. burgdorferi* strain B31 (ten nymphal ticks per mouse). In contrast, mice immunized with denatured OspC were not protected.

To further investigate the influence of conformation on protective efficacy, recombinant OspC preparations were denatured using heat or chemical treatment prior to animal immunization (Gilmore and Mbow, 1999). Additional groups of mice received soluble or insoluble non-denatured recombinant OspC. All proteins were adjuvanted with TiterMax (oil in immersion). After confirming seroconversion, the mice were challenged with *B. burgdorferi* B31 infected ticks. Mice immunized with soluble, nondenatured OspC were fully protected, whereas mice that received insoluble or denatured OspC, were not. Protective epitope localization experiments were then conducted. OspC fragments were tested for their ability to bind the protective OspC monoclonal antibody B5. The fragments did not bind prompting the suggestion that the amino or carboxy end of OspC is required to bind monoclonal antibody B5. Immunization of mice with truncated forms of OspC also failed to provide protection against tick challenge. The authors postulated that the N and C-terminal domains of OspC, which are in close physical proximity to one another in the OspC structure (Eicken et al., 2001; Kumaran et al., 2001b), interact to present a conformational epitope. A potential caveat of this study is that some of the recombinant OspC proteins were produced with a 38 kDa T7 gene 10 fusion at the N-terminus of OspC. It is possible that effects of denaturation, and the results obtained with truncated OspC proteins, were influenced by the large N-terminal tag.

The conserved C-terminal ten amino acids of OspC (amino acid sequence PVVAESPKKP) have also been reported to constitute an immunodominant, continuous linear epitope that elicits protective antibody responses (Rousselle et al., 1998; Jobe et al., 2003; Lovrich et al., 2005). However, this suggestion is difficult to reconcile since a single OspC protein elicits only strain-specific protection (Bockenstedt et al., 1993; Earnhart et al., 2007; Izac and Marconi, 2019). The conserved C-terminal motif of OspC has been referred to as the C7, C10, and pepC10 motifs (Mathiesen et al., 1998a; Mathiesen et al., 1998b; Porwancher et al., 2011; Earnhart et al., 2014; Izac and Marconi, 2019). It stands to reason that if this conserved C10 motif, which is shared by all OspC types, is in fact immunodominant, then antibody elicited by any OspC type protein should be immunoreactive with all OspC type proteins. To assess this further, the relative contribution of the C10 domain to OspC antibody responses and to OspC function was assessed by allelic exchange replacement of *B. burgdorferi* B31 wild-type *ospC* type A with an *ospC* gene truncated to eliminate the C10 domain (Earnhart et al., 2014). The truncated protein (OspCΔ10) was expressed and presented on the cell surface in a manner similar to that of the parental strain, and the B31::*ospC*Δ10 strain retained the ability to infect mice (syringe inoculation) and to elicit anti-OspC IgG titers equivalent to the B31 parental strain. Additional evidence that the C10 motif is not immunodominant came from a study in which OspCΔ10 truncations were generated for 20 different OspC type proteins (Izac et al., 2019). The OspC protein panel was screened by immunoblot and dot-blot with sera from experimentally infected C3H-HeJ mice and non-human primates (Rhesus macaques) and naturally infected dogs, horses and humans (Izac et al., 2019). The truncated proteins retained their type specific immunoreactivity. Finally, sera harvested from rats (Sprague-Dawley) immunized with recombinant OspC proteins (types I, T or F), and their corresponding OspCΔ10 proteins (Freund's adjuvant), displayed equivalent levels of complement-dependent bactericidal activity. It can be concluded that the C10 domain does not contribute to an essential *in vivo* function, is not an immunodominant epitope, and is not required for the presentation of epitopes that elicit protective antibody responses.

While conformational and discontinuous epitopes undoubtedly contribute to the antibody responses elicited by OspC (Pulzova et al., 2016; Norek and Janda, 2017), linear epitopes that elicit antibody responses have been identified (Buckles et al., 2006). Immunoscreening of overlapping OspC peptides and OspC fragments using sera from experimentally infected mice (C3H-HeJ) and naturally infected humans identified two distinct linear epitope containing sequences (Earnhart et al., 2005; Buckles et al., 2006). These epitopes, which reside within the C-terminal half of OspC, have been designated as the helix 5 (H5) and loop 5 (L5) epitopes. The L5 epitope maps within the C-terminal end of helix 3 and the N-terminal end of loop 5 while H5 epitope is contained within a region encompassing the 5th alpha helix and a few residues C-terminal to this helix. Both epitopes reside within membrane proximal surface-exposed segments of OspC. The L5 and H5 epitopes elicit OspC type-specific bactericidal antibody that kills through a complement dependent mechanism (Buckles et al., 2006). The specificity of the OspC antibody response elicited by the variable domain epitopes of OspC explains the narrow protective efficacy obtained upon vaccination with a single OspC type protein (Bockenstedt et al., 1997; Probert et al., 1997). It is noteworthy that type-specific antibody responses are also elicited upon vaccination with recombinant OspC proteins. Rabbits vaccinated individually with different recombinant OspC type proteins developed type-specific antibody responses (Oliver Jr et al., 2016). The membrane proximal location of the L5 and H5 epitopes may enhance the bactericidal activity of antibodies directed at these epitopes by fixing complement near the cell surface.

Important considerations in LD vaccine development

Approaches for assessing potential vaccine efficacy in vitro

Prior to advancing vaccine candidates to testing in animal models, *in vitro* assays can be used to assess correlates of protection. However, since a significant number of promising vaccine candidates are selectively expressed *in vivo* and not produced during laboratory cultivation, *in vitro* analyses can pose unique challenges. OspC serves as an example. The level of OspC expression during cultivation can vary considerably among strains (Xiang et al., 2017) and among cells in a population (Oliver Jr et al., 2016). IFA analyses in which *B. burgdorferi* B31 cells were screened with anti-OspC type A antisera revealed that less than 10% of cells produced detectable levels of OspC. The non-uniform production of OspC has in some cases compromised efforts to use wild-type strains in bactericidal assays. However, the expression levels of OspC and other antigens can be enhanced *in vitro* by several approaches. Adjusting the pH of the growth media or temperature-shifting cultures has been demonstrated to upregulate the expression of some *in vivo* antigens including OspC, OspE and OspF (Ramamoorthy and Philipp, 1998; Carroll et al., 1999; Ramamoorthy and Scholl-Meeker, 2001). However, it has proven difficult to achieve consistent levels of expression by manipulation of culture conditions. The dialysis membrane chamber implant model developed by Akins and colleagues was a key step forward in efforts to study *B. burgdorferi* in a host-adapted state (Akins et al., 1998). In the dialysis membrane chamber implant model, actively growing spirochetes are introduced into the peritoneal cavity of rats or mice within the confines of dialysis membrane chambers (5 kDa MW cutoff). After allowing for growth and adaptation to the mammalian environment, the chambers are removed from the peritoneal cavity and the host-adapted spirochetes harvested for analysis. Since cultures volumes as high as 10 mL can be recovered from a single rat, this approach allowed for the direct analysis of host-adapted without *in vitro* cultivation. The transcriptional and proteome profiles of host-adapted spirochetes are consistent with that reported for the expression of individual genes in mammals (Brooks et al., 2003; Tokarz et al., 2004; Caimano et al., 2007; Iyer et al., 2015). Perhaps the most efficient approach for mimicking the *in vivo* expression of an individual antigen is through the application of genetic manipulation methodologies (Samuels et al., 1994). The development of genetic manipulation approaches for *B. burgdorferi* was a significant breakthrough in LD research. Strains can be engineered to constitutively produce proteins such as OspC (Tilly et al., 1997) that are not expressed at steady levels in wild-type strains but that are dominant antigens *in vivo*. A *B. burgdorferi* B31 derived strain that constitutively produces OspC has been used in several studies to measure the antibody-dependent bactericidal activity of OspC derived recombinant vaccine antigens *in vitro* (Earnhart et al., 2014; Izac et al., 2019; Izac et al., 2020b).

Challenge studies: the approach matters

When conducting challenge studies, the route by which the challenge is delivered is important to consider. Studies of decorin binding protein A (DbpA) serve as an example as they demonstrated that protection is dependent upon route of administration (i.e., tick versus syringe). Vaccination of C3H/HeJ mice with recombinant DbpA in Freund's adjuvant induced high-titer antibody responses that were protective against syringe challenge with host-adapted *B. burgdorferi* (Cassatt et al., 1998).

However, when the challenge was delivered via tick bite, the mice were not protected (Hagman et al., 2000). While needle challenge is an important first step in efficacy studies, it is important to recognize its limitations and eventually progress to tick challenge. Naïve ticks or egg masses are now commercially available, and their availability has made an essential research tool more accessible. One approach to infect ticks with the desired strain is through immersion feeding (Policastro and Schwan, 2003). It is a simple approach whereby ticks (typically larval ticks) are immersed in a dense and actively growing culture. The ticks will imbibe the culture thus allowing the spirochetes to enter the midgut. The percentage of ticks that became infected was reported to be 65% after immersion for 45 minutes in a dense culture of *B. burgdorferi* B31. Ticks can also be infected by microinjection (Kariu et al., 2011; Smith et al., 2018). The process takes just a few hours and allows for the delivery of specific numbers of spirochetes into each tick. Microinjection can also be used to infect ticks with genetically modified strains that are unable to infect mice and thus cannot be introduced into ticks by the natural route, namely feeding of larvae on infected mice. It would be possible using either immersion feeding or microinjection to infect ticks with a mixture of strains. However, to our knowledge, this has not yet been reported. Lastly, large numbers of ticks can be infected by feeding on laboratory infected mice. While tick challenge studies are widely used for assessing vaccine efficacy, the overwhelming majority of studies published to date have used ticks infected in the laboratory with a single LD spirochete strain. As noted above, field collected ticks typically are infected with a heterogenous population of LD spirochetes and, as such, the use of field collected ticks in challenge studies most closely resembles natural challenge. A caveat of using field collected ticks is the potential for the ticks to be carrying multiple pathogens. With all of the approaches described above, it is critical to determine the percentage of infected ticks. This information is essential for determining the number of ticks that should be fed on an individual animal in order to ensure delivery of a robust challenge dose. In contrast to studies with mice, canine vaccine challenge studies have routinely used field collected ticks (Baum et al., 2014; Wagner et al., 2015; Stillman et al., 2019; Marconi, 2020b). Lastly, the mouse strain or dog breed used in challenge studies is also of great importance and consequence. Some dog breeds are more likely to develop symptoms of infection than others (Dambach et al., 1997; Goldstein et al., 2007; Littman, 2013; Preyß-Jägeler et al., 2020). Age is also an important variable and sex as a biological variable should be considered. These topics are covered elsewhere (Barthold et al., 1990; Moody and Barthold, 1998) and thus are not discussed here.

First generation human LD vaccines

Two investigational LD vaccines intended for use in humans (ImuLyme, Pasteur-Merieux-Connaught; LYMErix, SmithKline Beecham) were developed in the 1990's and advanced to clinical trials (Schoen et al., 1995; Sigal et al., 1998; Schoen et al., 2003). Both vaccines consisted of recombinant lipidated OspA derived from North American isolates of *B. burgdorferi* (Erdile et al., 1993; Sigal et al., 1998).

ImuLyme: a first generation OspA investigational vaccine

The safety of recombinant lipidated OspA (ImuLyme) in humans was investigated using a randomized, double-blinded, placebo-controlled trial (36 adult volunteers; aged 18 to 65 years) (Keller et al., 1994). Alum adsorbed and non-adsorbed OspA formulations were tested and both were found to induce bactericidal antibody responses that inhibited growth of *B. burgdorferi in vitro*. It is noteworthy that the authors reported that adsorption to alum did not influence antibody titers. Adverse health events (AHEs) after the first vaccine dose were minor and consisted primarily of local pain and tenderness at the injection site. An increase in AHEs was not observed following the second or third dose. It was concluded that a two or three dose series of the OspA Lyme vaccine was safe in adults.

The efficacy of the ImuLyme formulation was assessed in a double-blind trial (10,305 participants; 18 years of age or older) at 14 sites in endemic areas of the United States (Sigal et al., 1998). Participants were randomly assigned to receive either placebo (5149 subjects) or 30 µg of lipidated OspA vaccine (5156 subjects). Two doses of the non-adjuvanted vaccine were delivered 1 month apart with 7515 subjects receiving a booster at 12 months. The subjects were monitored for two seasons with the primary end point being the number of new clinically and serologically confirmed cases of LD. During the first year there were 49 confirmed cases of LD with 37 in the placebo group and 12 in the vaccine group (68% efficacy; 95 percent confidence interval - 36 to 85 %). The efficacy rate was highest among participants under the age of 60. In the second year, 35 cases of LD were confirmed with 28 in the placebo group and 7 in the vaccine group. Efficacy among those participants that received the 12-month booster was 92% (95 percent confidence interval - 69 to 97%). In the absence of the booster dose, there was little demonstrable effect of vaccination during the second year of the study. It is noteworthy that in this study, the percentage of subjects with definite LD was higher in the placebo group than in the vaccinates (1.3 percent vs. 0.4 percent, P<0.001). In addition, the percentage of participants that reported symptoms of LD was higher in the placebo group than in vaccinates (5.5% vs. 4.2 %, P=0.002). This was also the case for confirmed seroconversion (1.5% vs. 0.6%, P<0.001). While this study highlighted the efficacy of OspA in humans, it also revealed the need for booster vaccinations.

LYMErix: the only human LD vaccine to advance to market

Prior to the final formulation of the LYMErix vaccine, the safety, reactogenicity and immunogenicity of three precursor candidate OspA based vaccine formulations were assessed (Van Hoecke et al., 1996). Two different study designs were employed. The first was an open, randomized pilot study with 60 volunteers (20 per group; 18 to 50 years of age). The second, which was initiated 14 days after the first study began, was double-blinded and randomized with respect to vaccine formulation. Two-hundred and forty volunteers were enrolled and randomly assigned to three groups. The first formulation consisted of lipidated OspA (10 µg in alum) fused at its single cysteine residue to the N-terminal 81 amino acids of non-structural protein 1 (NS1) from influenza virus. The second formulation consisted of 10 µg of NS1-OspA in alum combined with the experimental adjuvant, 3-0-deacylated monophosphoryl lipid A (MPL). The MPL was derived from the lipid A moiety of *Salmonella minnesota* lipopolysaccharide (Ribi Immunochem Research). The third formulation consisted of lipidated OspA lacking the NS-1 sequence in alum. The vaccines were administered intramuscularly at monthly intervals. After each immunization, volunteers were monitored for AHEs. Only minor local and general reactions were noted. Serum was collected from each study participant prior to vaccination and then one month after each vaccine dose the sera were evaluated for anti-OspA IgG antibody levels. Of the three vaccine formulations tested the lipidated OspA in alum elicited the highest anti-OspA IgG levels and LA-2 equivalent geometric mean titers (GMTs). Volunteers that received this formulation remained seropositive for anti-OspA IgG for the longest period of time. Based on the results of this study, the lipidated OspA in alum formulation was pursued for further development.

The safety and efficacy of the recombinant lipidated OspA vaccine, LYMErix, was then assessed in 956 volunteers (17 to 72 years of age) at three LD endemic sites in the United States (Schoen et al., 2000). Anti-OspA IgG titers were measured in blood collected at 0, 2, 3, 12 and 13 months. The study compared the tolerability of a two-dose schedule (0 and 1 months; intramuscular administration) with a three-dose schedule (0, 1, and 2 months). AHEs were mild to moderate and transient. Three doses of the vaccine proved more immunogenic than two and conveyed a higher probability of protection. Based on the GMTs determined, an antibody kinetics model used to assess duration of immunity predicted that protection would only last for a single typical tick-transmission cycle.

Using an intention-to treat analysis, the efficacy of LYMErix in humans was measured in terms of its ability to protect against "definite" LD (Steere et al., 1998). Definite LD was defined as the presentation of erythema migrans or objective neurologic, musculoskeletal, or cardiovascular manifestations consistent with LD plus laboratory confirmation via culture, PCR positivity or seroconversion (immunoblot). The phase III-randomized controlled trial included 10,936 subjects (15 to 70 years of age) from endemic regions of the United States. A two-dose vaccination series (30 µg lipidated OspA; alum adjuvant) provided 49% efficacy (95% confidence interval (CI) = 15%-69%) while a three-dose series delivered over 13 months provided 76% efficacy (95% CI=58-68%). The minimum protective titer was determined by measuring anti-OspA antibody titers in a subset of the study participants at 2, 12, 13 and 20 months. The GMTs at one month and 10 months after the second vaccine dose were 1227 ELISA units/mL and 116 ELISA units/mL, respectively. One month after a third vaccine dose (month 13), the GMT rose to 6006 ELISA units/mL indicative of a vaccination induced amnestic response. The anti-OspA GMTs at 20 months declined to 1991 ELISA units/mL. The minimum protective titer was determined to be ≥1200 ELISA U/mL (CDC, 1999). This value serves as a comparator for measuring immune responses to new OspA vaccines in development. It should be noted that individuals or animals that are vaccinated with OspA, and then naturally exposed to infected ticks, do not develop an exposure induced amnestic response because OspA is not produced during infection.

The safety of LYMErix was assessed by following study participants for twenty months post-delivery of the first vaccine dose. Injection site soreness was the most frequent vaccination induced AHE and was reported by 24.1% of vaccine recipients (7.6% of placebo recipients; p <0.001). Vaccine recipients were more likely to report arthralgia and myalgia than placebo vaccinated individuals (p < 0.05). Potential AHEs were not reported in vaccine recipients with previous LD (Schoen et al., 1995). LYMErix was approved for use on December 21, 1998. The Advisory Committee on Immunization Practices (ACIP) suggested the LYMErix™ vaccine for persons who live in an endemic area and who engage in activities that result in frequent or prolonged exposure to ticks. The ACIP did not recommend Lyme vaccination for individuals in non-endemic areas or for those with low exposure risk or individuals under 15 years or older than 70 years of age. This rather lukewarm recommendation did not serve to further the cause for the widespread use of LYMErix.

LYMErix advanced to the market in 1998 and sales were initially robust. However, shortly after its introduction, reports emerged of potentially serious AHEs (Zundorf and Dingermann, 2008). The possible associations between the vaccine and putative AHEs were well-publicized and widely disseminated by some LD advocacy groups and the media. Class action lawsuits alleging vaccine-related harm were first filed in 1999. It was speculated that OspA may be associated with the development of antibiotic refractory arthritis with a reported linkage to particular MHC II alleles (HLA-DR4) (Steere et al., 1990; Kalish et al., 1993; Gross et al., 1998; Steere et al., 2011). Reports then emerged implicating an OspA T-cell epitope with sequence similarity to lymphocyte function antigen 1 as a contributor to autoimmune arthritis (Trollmo et al., 2001). In spite of these concerns, Phase IV studies were initiated by the manufacturer. These studies were designed to assess 25,000 vaccine recipients and 75,000 age and sex-matched unvaccinated controls. However, by November of 2000, only 10% of the planned enrollment had been met. In January of 2001, the results obtained up to that point were presented to an FDA advisory panel. A correlation between vaccination and autoimmunity was not identified. The CDC and FDA also evaluated data from the Vaccine Adverse Event Reporting System collected between December 1998 and July 2000. After the distribution of 1.4 million vaccine doses, 905 AHEs were reported and of these 102 were coded as arthritis, arthrosis, or rheumatoid arthritis and 12 were facial paralysis. However, the reported rate of both arthritis and facial paralysis were below what would be expected as background. This rigorous multi-agency review concluded that the reported AHEs were not related to the product and thus did not recommend removal of LYMErix from the market. Nonetheless, the commercial viability of the vaccine was irreparably damaged, and sales declined dramatically resulting in the voluntary removal of LYMErix from distribution in 2002.

Unsubstantiated safety issues aside, the vaccine was also compromised by evidence-based limitations including its booster schedule and moderate efficacy (Zundorf and Dingermann, 2008). The need for repeated boosters with LYMErix and most likely with OspA based vaccines that are currently in development, is a direct reflection of the mechanism of action of these vaccines. As noted above, OspA is not produced by the LD spirochetes during their residence in mammals. Hence, vaccination induced antibody decreases the risk of LD by targeting spirochetes in ticks upon ingestion of the bloodmeal (Fikrig et al., 1992a). Vaccination induced antibody is not effective at targeting spirochetes that have successfully transmitted into a vaccinated individual. Efficacy of OspA based vaccines is therefore strictly dependent on the maintenance of high circulating antibody titers at a level above a specific threshold. A second consequence of the absence of OspA expression in mammals is that when vaccinated individuals are naturally exposed to *B. burgdorferi* via tick bite they will not mount an amnestic response. It is important to point out that this does not negate the positive contributions that OspA can provide in a vaccine formulation, especially if coupled with an antigen that can lead to memory immune responses. As detailed below, the canine LD vaccine VANGUARD®crLyme, which consists of OspA and an OspC based chimeric-epitope protein (chimeritope) (Izac et al., 2020b) designated as Ch14, was designed with this in mind (Marconi, 2020b).

Next generation Lyme disease vaccines

Several efforts are underway to develop modified OspA based vaccines for use in humans. In a study supported by Sanofi, a prototype recombinant OspA-ferritin nanoparticle was generated by fusing non-lipidated OspA (serotype 1; *B. burgdorferi* B31) to the amino terminus of ferritin derived from *Helicobacter pylori* (Kamp et al., 2020). As detailed by the authors, the OspA-ferritin fusion was designed so as to allow for self-assembly of 24 ferritin units into a hollow spherical nanoparticle that presents the C-terminal domain of OspA outwards from the interior. The efficacy of this experimental vaccinogen was assessed in mice using the canine LD OspA based RECOMBITEK® Lyme vaccine (Boehringer-Ingelheim) as a comparator. The OspA-ferritin nanoparticles induced endpoint antibody titers 4.4-fold greater than the non-adjuvanted OspA of the comparator ($p < 0.001$). In addition, OspA-ferritin induced antibody persisted for a longer duration than antibody induced by the comparator vaccine lipidated OspA. Protective efficacy was assessed in C3H/HeN mice vaccinated intramuscularly with Addavax (Invivogen) adjuvanted OspA-ferritin nanoparticles in a two-dose series (four weeks apart). Mice were challenged two weeks after the final vaccine dose with 5 to 6 nymphal ticks infected with *B. burgdorferi* B31. The ticks were fed to repletion, collected, and efficacy assessed by culture of ear, ankle and heart biopsies. While sample size was relatively limited (4 mice per group), the OspA-ferritin formulation and the comparator vaccine protected all vaccinated mice from infection. The authors then tested an OspA-ferritin nanoparticle cocktail that included OspA serotypes 1, 2, 3, 4, 5 and 7. Robust antibody titers to each component OspA protein were demonstrated and the cocktail proved to be protective against challenge with *I. ricinus* ticks infected with *B. afzelii* (OspA serotype 2). The immunogenicity of the hexavalent formulation was compared with the comparator vaccine in non-human primates using the squalene-based adjuvant, AF03. The hexavalent formulation elicited an 11 to 200-fold higher antibody titer than the comparator. The durability of the antibody response, although relatively short (6 months), was also higher than RECOMBITEK® Lyme.

Baxter BioScience has also pursued a multi-valent OspA based vaccine that consists of three recombinantly engineered OspA based proteins (Wressnigg et al., 2013; Wressnigg et al., 2014). The three OspA antigens were designed to harbor protective epitopes from two different OspA serotypes. While the details on the molecular design of these proteins and the identity of the protective epitopes employed is not in the public domain, one protein consists of epitopes from OspA serotypes 1 (*B. burgdorferi*) and 2 (*B. afzelii*); the second of OspA serotypes 3 and 5 (*B. garinii*) and the third of OspA serotypes 6 (*B. garinii*) and 4 (*B. bavariensis*). For each recombinant protein, the proximal end of one OspA serotype was fused to the distal portion of another OspA serotype. The strategy was to retain the overall structure of native OspA so as to maintain potentially conformational epitopes. All three of the unique recombinant antigens were lipidated. A double blinded, randomized, dose escalation, phase1/2 study was conducted at four sites in Austria and Germany. Study participants (age 18-70; n=300) were administered 30, 60, or 90 ug of the antigen formulated with or without aluminum hydroxide. The dosing schedule consisted of three intramuscular vaccinations 28 days apart with a fourth dose delivered nine to twelve months after the initial immunization. The primary immunization endpoint was 28 days after the third dose, at which time the IgG titer to each OspA serotype was measured. Individual recombinant OspA proteins representing each serotype served as the detection antigens. A dose response was not evident in participants that received the adjuvanted formulation, yet an inverse dose response was noted after the fourth vaccine dose. A similar observation was noted in clinical trials of two viral vaccines (Ehrlich et al., 2008; Aichinger et al., 2011). It is also noteworthy that the adjuvanted formulation of the OspA cocktail was better tolerated than the non-adjuvanted formulation. The authors suggest that alum may mask or shield the lipidation motif thereby inhibiting its interaction with Toll-like receptors that are mediators of inflammatory responses (Hirschfeld et al., 1999). The 30 μg adjuvanted vaccine was determined to be the best dose and formulation. The bactericidal activity of vaccinal antibody was assessed by incubating hyperimmune sera with strains producing each OspA serotype. While vaccination induced bactericidal antibody, the killing activity dropped to near baseline prior to the booster vaccination. This is consistent with the relatively short-lived bactericidal activity elicited by other OspA based vaccines. In a follow up study, the safety of the OspA vaccine cocktail was assessed in an adult population that had been previously diagnosed with LD. The vaccine was well-tolerated and immunogenic (Wressnigg et al., 2014). Further development was not pursued by Baxter. The rights to the vaccine ultimately went to Shire which did not pursue further development.

Another engineered OspA based vaccine formulation, VLA15 (Comstedt et al., 2014), is currently in development by Valneva. VLA15 consists of three engineered recombinant proteins that harbor the C-terminal domain of OspA serotypes 1, 3, 4, 5, and 6. To stabilize the recombinant proteins and allow for heterodimerization, molecular modeling was conducted to identify amino acid positions that could be substituted with cysteines and thus allow for inter-monomer disulfide bond formation. Both the monomeric and heterodimeric proteins displayed increased thermal stability. A cocktail of the three resulting heterodimers (referred to in the original study as the "LB vaccine") was assessed for immunogenicity and protective efficacy using the mouse model for LD. LB-vaccine was delivered with or without aluminum hydroxide in a three-dose series. Ten-fold greater anti-OspA IgG titers were generated with the adjuvanted vaccine. A series of challenge experiments were conducted by sub-cutaneous injection of *B. burgdorferi* ZS7 (serotype 1), *B. garinii* PHei (serotype 5) or *B.* afzelii (serotype 2) individually. The challenge dose of spirochetes was stated to be between 5x and 10x of the ID_{50}. However, since the ID_{50} was not detailed in the study, the number of spirochetes in the challenge dose is not known. Challenge was delivered two weeks after the third dose of vaccine and efficacy assessed with cultivation, PCR, seroconversion, and bactericidal growth inhibition titers serving as readouts. For the tick challenge experiments, *I. ricinus* ticks infected with *B. afzelii* IS1 (serotype 2) were used. In contrast to most challenge analyses that feed 5 or more ticks on each animal, only 2 ticks were fed on each mouse. While the challenge inoculum might be considered low, the percentage of placebo mice that became infected in each experiment was at least 87%. The protective efficacy varied slightly for each experiment but was >85% in all cases. In a follow up study, protection against strains expressing OspA serotypes 1, 2, 4, 5 and 6 was demonstrated (Comstedt et al., 2017). Protection against infection with serotype 3 was inferred from *in vitro* bactericidal assays.

Phase 1 studies of VLA15 have been conducted in Europe. After initiating a phase 1 study in which immunogenicity and safety were evaluated, Valneva amended its Phase 1 study protocol to determine if a booster vaccination administered to a sub-cohort of the Phase 1 study population was warranted. VLA15 demonstrated a favorable safety profile and no significant AHEs were noted. In addition, the final Phase 1 immunogenicity results indicated that the alum-adjuvanted formulations elicit higher immune-responses at all time-points than the non-adjuvanted formulation. However, antibody titers declined post Day 84 across all groups, trending towards baseline at approximately one year post initial vaccination.

To evaluate the benefit of a booster dose, 64 subjects across the two higher-dose groups (48µg and 90µg; with and without alum) from the Phase 1 study received a booster 12 to 15 months after the primary immunization. The booster resulted in anti-OspA antibody titers that ranged from 2.7 to 5.8-fold higher than the titers observed at Day 84. In 2020, Valneva and Pfizer partnered to advance the VLA15 vaccine. The planned Phase 2 study will consist of 250 subjects that will receive one of two dosage levels of alum adjuvanted VLA15 (100 subjects each) or placebo (50 subjects). The investigational vaccine will be administered intramuscularly in three-doses on Days 1, 57 and 180. Study participants will be followed for 18 months with final immunogenicity readouts assessed on day 208. At the time of this writing, the initiation of the trial was pending. It remains to be determined if the dose vaccination schedule of this, and other potential OspA based vaccines, will be embraced by the public.

Chimeric epitope based OspC vaccinogens

As noted above, a significant challenge in the development of broadly protective OspC based vaccinogens is the type specificity of the antibody response (Wang et al., 1999; Brisson and Dykhuizen, 2004; Earnhart and Marconi, 2007c). With the identification of the antigenic determinants of OspC, recombinant proteins consisting of OspC linear epitopes derived from multiple OspC types were constructed (Earnhart et al., 2007; Earnhart and Marconi, 2007a, b). These unique epitope-based proteins are referred to as chimeritopes (Izac et al., 2020b). The original prototype chimeritope harbored L5 and H5 epitopes from OspC types A, B, K and D (Earnhart et al., 2007). The "ABKD" construct was immunogenic in mice and elicited antibodies that recognized all OspC types represented in the construct. Anti-ABKD antibodies readily surface-labeled intact spirochetes and displayed potent complement-dependent bactericidal activity. However, it was noted that the titers induced by immunization were lower for the C-terminal OspC type D epitope than observed for epitopes at the N-terminus of the chimeritope. The immunogenicity of the C-terminal epitopes was increased through epitope reiteration, epitope reordering, and by increasing the solubility of the construct (Earnhart and Marconi, 2007a). The tetravalent construct was then expanded to an octavalent construct through the addition of L5 and H5 epitopes derived from OspC types E, N, I and C (Earnhart and Marconi, 2007b). When delivered to mice, the OspC-A8.1 construct (adjuvanted with alum) elicited high-titer antibody to each component epitope that efficiently surface-labeled diverse *B. burgdorferi* strains. The antibody response was skewed to IgG2a and IgG2b isotypes, both of which efficiently fix complement. As efforts continued to introduce additional epitopes into the chimeritope so as to expand OspC type representation, the increasing molecular weight of the chimeritope hampered expression of 'clean' full-length protein. The chimeritopes that had been produced up to this point harbored linkers between the individual epitopes to maintain protein flexibility. However, the linkers and sequences shared by some H5 and L5 epitopes from different ospC types accounted for nearly 30% of the molecular weight of the proteins. To determine if the linkers and redundant sequences could be eliminated or minimized without adversely affecting immunogenicity, three chimeritopes designated as Chv1, Chv2, and Chv3, were constructed and comparatively assessed (Izac et al., 2020b). All three Chv proteins possess the same set of 18 linear epitopes derived from 9 OspC type proteins but differ in the physical ordering of epitopes and by the presence or absence of linkers. The Chv proteins were immunogenic in mice, rats and dogs, eliciting similar antibody titers. Immunoblot and ELISA assays demonstrated that the Chv proteins elicit IgG that recognizes a diverse array of OspC type proteins including North American and European strains of the LD spirochetes. This important observation indicates that some of the L5 and H5 epitope-containing domains share identity with OspC types that are not directly represented in their entirety in the Chv proteins. Thus, it appears that the 18 epitopes present in the Chv proteins collectively elicit antibodies that can recognize the majority of OspC types. Anti-Chv antisera surface-labeled intact *B. burgdorferi* demonstrating that vaccinal antibody can bind to epitopes as they are presented on the cell surface. Immunization induced antibody also displayed potent complement dependent-antibody mediated killing activity. This study demonstrated that high-valance OspC chimeritopes can potentially elicit broadly protective antibody responses. OspC chimeritopes have not yet advanced to a phase 1 clinical trial.

Other vaccine candidates (past and present)

A comprehensive assessment of all proteins that have been investigated as potential vaccine candidates is beyond the scope of this writing. Here we focus on those for which significant information was available at the time of submission. To begin, it is worthy of note that while numerous immunogenic proteins have been assessed as vaccine candidates, the majority do not induce bactericidal or protective antibody responses. Examples of immunogenic but non-protective antigens include OspD (Probert and LeFebvre, 1994), BmpA (also known as P39) (Gilmore et al., 2003), OspE (paralogs BBN38 and BBL39), CspA (McDowell et al., 2006), CspZ (Rogers et al., 2009a), and VlsE (Liang et al., 2001). Examples of immunogenic proteins that convey partial protective immunity include P35, P37 (Fikrig et al., 1997), BBA52 (Kumar et al., 2011), BB0405 (Kung et al., 2016), BBI39 (Singh et al., 2017), P66 (oms66) (Exner et al., 2000) and BB0172 (Small et al., 2014; Hassan et al., 2019). Studies conducted on a subset of these proteins are summarized below.

The outer membrane proteins BBA52, BBI39, and BBI36, have been reported to play possible roles in spirochete survival within fed ticks or during transmission to mammals (Kumar et al., 2010; Kumar et al., 2011; Kung et al., 2016; Singh et al., 2017). Each protein has been assessed for its potential to induce antibody that can block spirochete transmission. While each protein was immunogenic in mice (Freund's adjuvant), only limited protection against challenge by *B. burgdorferi* infected ticks was observed. Serum harvested from immunized mice also lacked significant *in vitro* bactericidal activity (Kumar et al., 2010; Kumar et al., 2011; Kung et al., 2016; Singh et al., 2017). Anti-BBI39 antisera was reported to have bacteriostatic properties against *in vitro* cultivated *B. burgdorferi* (Singh et al., 2017). The utility of a vaccine antigen that induces bacteriostatic immune responses remains to be determined.

The BB0172 protein is a von Willebrand factor A (vWFA) domain-containing protein that potentially functions as a metal-dependent, integrin binding protein (Wood et al., 2013). Proteinase K digestion and triton X-114 extraction and phase partitioning analyses demonstrated that BB0172 is an outer membrane protein (Wood et al., 2013). RT-PCR analyses did not detect BB0172 transcript in cells cultivated under standard conditions. However, transcript was detected in cells recovered from cultures maintained at pH of 6.8 and 37 °C. Based on this observation and the absence of an antibody to BB0172 during mammalian infection, it was postulated that BB0172 is produced specifically in fed ticks. Peptides derived from B0172 protein were subsequently tested as transmission blocking vaccine antigens (Small et al., 2014). Four BB0172 derived peptides, designated as peptides A, B, C and D, were synthesized, conjugated to Keyhole Limpet Hemocyanin, formulated in Titer Max® adjuvant, and used to immunize C3H/HeN mice (3 dose immunization series). Vaccinated mice were syringe-challenged 28 days after immunization with 10^3 or 10^5 *B. burgdorferi* cells. Peptide B, but not peptides A, C, and D, elicited protection against a challenge dose of 10^3 spirochetes but not against a dose of 10^5 spirochetes. In addition, immunized mice were only partially protected when challenged with infected ticks. In a follow up study, the 12 amino acid PepB peptide was conjugated to Cross-Reacting Material 197 (CRM197) or to Tetanus Toxoid heavy chain (TTHc). C3H/HeN mice were immunized with each vaccinogen in alum and challenged with 10^5 spirochetes or with infected ticks. While CRM197:PepB was not protective, TTHc:PepB conveyed induced protective responses in a subset of mice. Follow up studies on BB0172 or PepB have not been published.

Factor H binding proteins as vaccine antigens

Numerous microbial pathogens including several *Borrelia, Borreliella,* and *Treponema* species, evade complement-mediated killing in part by binding to the cell surface negative regulators of complement. The contribution of members of the factor H protein family to complement evasion has been intensively studied (Hellwage et al., 2001; Alitalo et al., 2002; Kraiczy et al., 2003; McDowell et al., 2003a; McDowell et al., 2003b; McDowell et al., 2009; Rogers et al., 2009a; McDowell et al., 2012). Factor H binding proteins have been investigated as potential vaccine candidates. The fHbp protein of *Neisseria meningitidis* (Beernink and Granoff, 2009), has been successfully developed as a vaccine for meningococcal disease caused by type B *Neisseria meningitidis* (Beeslaar et al., 2020a; Beeslaar et al., 2020b; Borrow et al., 2020). FHbp is the main vaccinogen in two FDA approved and licensed vaccines: Bexsero® (GlaxoSmithKline) and Trumenba®(Pfizer). Studies of fHbp demonstrated that the immunogenicity of the protein can be enhanced through the introduction of site-directed mutations that attenuate its ability to bind FH (Beernink et al., 2011; Beernink et al., 2012). The potential protective efficacy of two *Borreliella* factor H binding proteins, OspE and CspZ, have also been investigated (FHBP). While both proteins were immunogenic in mice (Hefty et al., 2002; Rogers et al., 2009a), they did not elicit bactericidal or protective antibody responses (Rogers et al., 2009a; Marcinkiewicz et al., 2018). To prevent masking of potentially protective epitopes by bound FH, CspZ was modified at two amino acid positions through site-directed mutagenesis to abolish factor H binding (Marcinkiewicz et al., 2020). The mutated protein was conjugated to virus-like particles and assessed for *in vitro* bactericidal activity and protective efficacy. Bactericidal activity was significantly enhanced using the mutated protein and immunization studies demonstrated protection in mice against both needle and tick delivered *B. burgdorferi* (Marcinkiewicz et al., 2020). FH binding proteins in general appear to be attractive candidates for vaccine development.

Veterinary Lyme disease Vaccines

Bacterin vaccines

The first USDA approved canine LD vaccines were formalin inactivated, *B. burgdorferi* cell lysate formulations (bacterins), produced by Fort Dodge Labs (Lyme Vax) and Schering-Plough Animal Health (Galaxy Lyme) (Chu et al., 1992; Levy et al., 1993). The licenses for these vaccines have since been transferred among companies over the past few decades and the original formulations modified to an unknown degree to decrease AHEs. The utility of one of the earliest canine LD vaccines was demonstrated in a study that compared the C6 seropositivity of client owned dogs vaccinated with Lyme Vax (Fort Dodge) with dogs that had not been vaccinated (Levy, 2002). Antibody responses to the C6 peptide of VlsE served as a readout of protective efficacy. Approximately 5% (8 of 163) of vaccinated dogs became C6 antibody positive overtime whereas 64% (25 of 39) of unvaccinated dogs seroconverted and were C6 antibody positive. These early studies established the health benefits and value of vaccination of dogs at risk for LD.

Currently, three LD bacterin vaccines are available on the North American market: NOVIBAC[7] Lyme (Merck; USDA approval, 03/19/2007) and Ultra™ Duramune® Lyme (Elanco; USDA approval 12/6/1995) and Duramune® Lyme (Elanco;

12/6/1995). Duramune® Lyme and Ultra™Duramune® Lyme differ primarily in the vaccine dose volume: 1.0 mL versus 0.5mL, respectively. It has not been demonstrated in the literature if the reduction in dose volume conveys added protection or decreases the potential for AHEs. All LD bacterin vaccines distributed in the United States consist of two strains of *B. burgdorferi*. Information provided by the manufacturers indicate that the rationale for the bivalent formulations is to include a strain that expresses OspA and a second strain that produces elevated levels of OspC. Since OspA is typically produced at high levels by wild-type laboratory cultivated strains, anti-OspA antibodies are indeed likely to be elicited upon vaccination and contribute to the neutralization of *B. burgdorferi* in ticks. However, claims regarding elevated OspC expression by the second strain in these vaccine formulations have not been rigorously substantiated by the manufacturers, or by independent peer reviewed publications. While NOVIBAC® Lyme has a USDA approved 12-month duration of immunity label, duration of immunity information is not available for Duramune® Lyme or Ultra™ Duramune® Lyme. Combination vaccines have been developed in which LD bacterin formulations were combined with *Leptospira* bacterins and antigens from parvovirus type b, adenovirus type 2, coronavirus, and canine distemper. There is limited information in the public domain regarding the efficacy of combination vaccines and hence, they are not discussed here. In Europe, a trivalent bacterin vaccine consisting of *B. afzelii, B. garinii,* and *B. burgdorferi* strains is available (MeriLym3; Merial). The rationale behind the trivalent formulation is to provide broad protective efficacy against the three primary *Borreliella* species associated with canine LD in Europe. While this appears conceptually sound, to our knowledge, evidence demonstrating specific protection against each species is not in the public domain.

While bacterin vaccines have played an important role in the ongoing development of LD vaccines for companion animals, a shift to vaccines of defined antigenic composition is desirable for several reasons. First, bacterins consist of all proteins produced by a LD spirochete during laboratory cultivation. Sequence analyses have revealed that the genome of a typical *Borreliella* strain harbors approximately 1500 open reading frames (Fraser et al., 1997; Casjens et al., 2000; Casjens, 2003; Casjens et al., 2011). Transcriptional profiling analyses have demonstrated that cultivated *B. burgdorferi* produce over 1000 different proteins (Iyer et al., 2015), the majority of which are not presented on the cell surface. This leads to non-productive antibody responses since intracellular proteins cannot be targeted by vaccinal antibody. Secondly, some *Borreliella* proteins that are selectively produced in the mammalian environment are absent from these vaccine formulations. Proteins expressed in mammals are advantageous as vaccinogens as they have the potential to elicit immune memory responses upon exposure of a vaccinated dog to *B. burgdorferi* infected ticks.

Recombinant protein-based subunit vaccines

The second general class of canine LD vaccines are subunit vaccines. LD subunit vaccines differ from bacterins in that they are of defined composition and contain well-characterized proteins that have been demonstrated to elicit protective antibody responses (reviewed in (Earnhart and Marconi, 2009; Izac and Marconi, 2019)). The use of vaccines with low complexity antigen content is important in the veterinary setting since multiple vaccines are typically delivered during a single visit. By eliminating extraneous antigens, the potential for vaccine induced AHEs is theoretically reduced. The first canine subunit vaccine to advance to market, RECOMBITEK® Lyme (Boehringer-Ingelheim; USDA approval; 03/27/1995), consists of lipidated OspA (90 µg) in a non-adjuvanted formulation. Efficacy and other important study data for RECOMBITEK® Lyme are not available on the USDA website because manufacturers were not required to compile and publish the data for vaccines evaluated by, and approved by USDA-APHIS, prior to 01/01/2007. However, post-licensure efficacy studies have been conducted (Conlon et al., 2000). In a blinded, controlled study using purpose-bred, mixed-breed dogs (10 to 12 weeks of age), vaccinate dogs received two-doses of vaccine, three weeks apart (Conlon et al., 2000). Sentinel dogs (not exposed to ticks, *B. burgdorferi,* or vaccine), which can provide information regarding the frequency with which random events may occur, were not included in the study. Three weeks after challenge with field collected ticks, the dogs were assessed for *B. burgdorferi* antibody by ELISA and immunoblot. Biopsies were also obtained for cultivation and PCR analyses. The vaccinate dogs were seronegative except for vaccinal antibody to OspA, and none were culture or PCR positive. Vaccinates did not present with clinical signs indicative of LD. In contrast, all placebo dogs were antibody and culture positive and nine of ten were PCR positive for *B. burgdorferi* DNA. Twenty percent of the placebo dogs developed symptoms indicative of LD. Vaccinates were protected from challenge delivered three weeks after the last boost. Long-term protective immunity was not assessed.

A subsequent study, conducted at a private veterinary hospital in Maine, assessed the ability of the vaccine to prevent reinfection in vaccine compliant dogs over a five year period (Eschner and Mugnai, 2015). The authors reported that 1% of the fully compliant dogs, defined as receiving all intended vaccine doses, developed serological responses indicative of LD. Of the dogs that were incompletely vaccinated (defined as missing one or more booster doses), 357 positive C6 antibody tests were reported (the number of dogs that this corresponds to was not indicated). The vaccination protocol used in this study did not follow the manufacturers recommended protocol. Label instructions call for a two-dose series, three weeks apart, and an annual booster thereafter. In the above referenced study, three vaccine doses were delivered during the first year of use (0, 1 and 6 months), followed by a single vaccine dose each year thereafter. The modified protocol was adopted based on a study that demonstrated enhanced antibody responses when a third dose of vaccine was delivered in year 1 (Topfer and Straubinger, 2007). This study, as well as the human OspA vaccines studies detailed above, speak to the utility of OspA as a vaccine component. However, a common underlying issue with all OspA alone based vaccines is the need to deliver frequent boosters.

VANGUARD®crLyme (Zoetis), the most recent canine LD vaccine to be approved by the USDA (12/23/2015), is a subunit vaccine consisting of OspA and an OspC chimeritope (Marconi, 2020b) designated as Ch14. The OspA component of VANGUARD®crLyme is non-lipidated and delivered at a 10µg antigen dose whereas the OspA in RECOMBITEK®crLyme is lipidated and delivered at a 90µg antigen dose. A recent study compared the IgG titers elicited by the OspA in these subunit vaccines in purpose bred dogs (Grosenbaugh et al., 2018). After a two-dose series the IgG titers elicited by lipidated OspA versus non-lipidated OspA were similar and differences were not statistically significant. Hence, lipidation and dose do not appear to influence anti-OspA IgG titer. As alluded to above, the OspC chimeritope in VANGUARD®crLyme, Ch14, is single recombinant protein that harbors at least 14 different linear epitopes derived from several antigenically distinct OspC proteins. The epitopes selected for inclusion included some that are predicted to elicit antibodies that can recognize multiple diverse OspC types. The efficacy of VANGUARD®crLyme was demonstrated in purpose bred dogs (7.1 to 7.7 weeks of age; 18 males; 18 females). Vaccinates received two-doses of vaccine (three weeks apart) by subcutaneous injection in the left dorsal scapular area. Three weeks after challenge, the vaccinates and placebo dogs were challenged with 30 ticks (15 males and 15 females; field collected from Rhode Island). Blood and tissue biopsies were collected at defined timepoints post-challenge. The dogs were euthanized approximately four months after challenge and necropsied for histopathology analyses. The serum samples collected over the course of the study were tested for antibody to the VlsE derived C6 peptide. Sentinel dogs, which were not exposed to *B. burgdorferi,* vaccine or ticks at any point during the study, were C6 antibody negative throughout. All placebo dogs (16/16) were C6 antibody positive on two or more occasions post-challenge. Fifteen of sixteen vaccinates were C6 antibody negative at all timepoints. A single vaccinate was C6 antibody positive 35 days after initiation of challenge but was antibody negative thereafter. Histopathological analyses of tissue from at or near the tick bite site identified nodular lymphoplasmacytic inflammation in the hypodermis/subcutis typical of *B. burgdorferi* infection in 14 of 16 placebo dogs, but not in the vaccinate or sentinel dogs. Within and around joints, histopathological changes observed in placebo dogs included synovial hypertrophy and hyperplasia (including papillary projections), mononuclear and neutrophilic inflammation within the joint and local subcutis, and fibrin deposition. Intra-articular lymphoplasmacytic inflammation was identified in 13 of 16 placebo dogs, 0 of 16 vaccinates, and 0 of 4 sentinel dogs. Inflammation was identified in more than one joint of each placebo dog. It can be concluded that vaccination resulted in a significant difference in the prevention of *B. burgdorferi* infection compared to placebo ($P < 0.0001$). Vaccination also prevented the synovitis and dermatitis typically associated with *B. burgdorferi* infection ($p<0.0001$). In 2017, VANGUARD®crLyme was approved for a 15-month duration of immunity label by the USDA. This is the longest duration of immunity label for any LD vaccine on the market (Littman et al., 2018).

A comprehensive field safety study using two serials of prelicensed VANGUARD®crLyme demonstrated an excellent safety profile (Marconi, 2020a). The vaccine was tested in its target population (dogs) under the conditions of its intended use. Six-hundred and twenty dogs, from three distinct geographic regions of the United States, were enrolled in this study with each receiving two-doses of vaccine 3 to 4 weeks apart by subcutaneous injection. Approximately, one-third of the dogs were ≤8 weeks of age as required to meet regulatory requirements. Safety was evaluated by observation of local and systemic reactions for at least 10 days after each vaccination. Minor AHEs occurred at low frequencies and serious AHEs were not observed. The study results demonstrated that VANGUARD®crLyme is safe for use in healthy dogs 8 weeks of age or older.

DNA based vaccine candidates

Two of the first studies to investigate DNA based LD vaccines assessed the protective potential of OspA encoded by different delivery plasmids (Simon et al., 1996; Luke et al., 1997). *OspA* was inserted into the pEF-BOS plasmid which carries the promoter/enhancer region of human elongation factor 1-alpha (EF-1alpha). A similar construct was generated that the lacked the EF-1alpha promoter enhancer region. One to four doses of each plasmid were delivered to BALB/c and AKR/N mice via intramuscular injection. Antibody to OspA was readily detected even after a single dose and the IgG titer increased in a dose dependent manner. Efficient antibody production was determined to not be dependent on the presence of the EF-1alpha promoter/enhancer element. The dominant IgG isotypes induced by vaccination were IgG2a, IgG2b and IgG1. Immunized mice were challenged with *B. burgdorferi* ZS7 (10^4 cells) and deemed to be fully protected. Passive transfer of the sera from vaccinated mice to SCID mice also provided protection.

An *ospA* based plasmid construct has also been tested. The DNA backbone of the plasmid was derived from the expression plasmid pVR2210. The construct carries *ospA* fused to the leader peptide of human tissue plasminogen activator (htpA) (Luke et al., 1997). A similar plasmid was constructed with *ospB* fused to the htpA leader peptide. Mice were administered three-doses of the constructs, two weeks apart. Sera were harvested from the vaccinated mice and assessed for OspA specific IgG antibody titer. However, the titer was determined to be 3 to 4-fold lower than that obtained using a comparator vaccine consisting of lipidated, recombinant OspA. *In vitro* growth inhibition assays demonstrated that sera from the *ospA* DNA vaccinated mice was inhibitory to growth. Mice that received the *ospA* DNA construct were protected against syringe challenge with 10^4 spirochetes delivered 76 days after the last vaccine dose. *OspB* was also investigated but it did not elicit protective antibody responses.

In a recent study, the efficacy of a proprietary DNA vaccine (pLD1; Inovio Pharmaceuticals) comprised of a synthetic consensus (SynConR) OspA combined with CELLECTRAR *in vivo* delivery technology was assessed. VANGUARD®crLyme served as a comparator (Guibinga et al., 2020). Results from both bacterial tissue burden and bacterial growth inhibition assays demonstrated equivalent protective efficacy of pLD1 and the comparator vaccine.

OspC based DNA vaccines also have been tested and demonstrated to elicit protective antibody responses (Scheiblhofer et al., 2003). *OspC* was fused to the htpA leader sequence to yield pCMV-TPA/ZS7. BALB/c and C3H/HeJ mice were immunized with the plasmid or with recombinant OspC as a comparator. Vaccine was administered by gene gun or syringe injection. In both mouse strains, purified protein as well as gene-gun delivered plasmid DNA skewed to Th2 type responses. In contrast, plasmid delivered by needle injection resulted in Th1 type antibody production. Mice challenged with 10^4 *B. burgdorferi* cells were protected with an efficacy range of 80% to 100%. This was the first study to demonstrate that a DNA vaccine encoding OspC induced protection against LD. OspC DNA constructs also have been delivered by DNA tattooing (Wagemakers et al., 2014). DNA tattooing is a novel vaccination approach that can be applied with an accelerated vaccination schedule (0, 2 and 4 weeks) (Wagemakers et al., 2014). The codon optimized *ospC* of *B. afzelii* PKo was assessed for efficacy with the corresponding recombinant protein serving as the comparator. The IgG responses induced by both approaches were comparable except that that IgG1/IgG2a ratio was lower when administration of the vaccine by DNA tattoo. Two weeks after syringe-challenge with 5×10^5 *B. afzelii* Pko spirochetes, most vaccinated mice were negative for *B. afzelii* DNA in tissues and all biopsies were culture negative. However, the DNA vaccine was not protective against heterologous strains including *B. burgdorferi* N40. Protection against tick challenge and long-term protective efficacy were not assessed. As described elsewhere in this review, the strain-specific protection reported in this study is not surprising in light of OspC diversity. Additional studies investigating DNA based LD vaccines have also been conducted, and varying levels of protection have been described (Wallich et al., 2001). As reported for vaccination studies using recombinant proteins, a multi-valent approach will most likely be required for DNA vaccines to elicit broadly protective antibody responses.

Reservoir targeting vaccines

Considerable research has focused on devising vaccination-based strategies to reduce the carriage rate of the LD spirochetes in specific reservoir populations. It follows that if carriage in reservoirs can be reduced, there will be a concomitant reduction in the percentage of ticks that are infected. In Europe, the yellow-necked mouse (*Apodemus flavicollis*) is an important reservoir for *Borreliella* species. In one study, yellow-necked mice were immunized with recombinant lipidated OspA (Kurtenbach et al., 1997) adjuvanted with ABM3. Three doses (10 μg protein) were delivered at 13-day intervals. Consistent with studies in laboratory mice, all immunized mice developed robust antibody responses to OspA including antibody to the epitope recognized by the protective monoclonal antibody LA-2. Infected *I. ricinus* ticks were cleared of infection after feeding on the vaccinated mice. This study was among the first to raise the possibility that natural mouse reservoir hosts are competent to generate transmission-blocking antibodies after vaccination that could potentially decrease carriage of *Borreliella* species in mice (Kurtenbach et al., 1997).

In another study, *Peromyscus leucopus* were infected with *B. burgdorferi* and then vaccinated with an OspA-GST fusion protein (Tsao et al., 2001). Using xenodiagnoses, it was demonstrated that ticks that fed on the vaccinated mice had a significantly reduced capacity to become infected with *B. burgdorferi*. The authors concluded that vaccination of *P. leucopus* with OspA could reduce transmission of *B. burgdorferi* to the tick vector, even in mice with prior infection.

In a controlled field study conducted in Connecticut, 1000 wild *P. leucopus* mice were captured, tagged, vaccinated with non-lipidated, recombinant OspA (N-terminal GST fusion) adjuvanted with Freund's and released. Follow up analyses at the defined field sites revealed that vaccination reduced the prevalence of *B. burgdorferi* in nymphal *I. scapularis* ticks. However, the study also revealed that non-mouse hosts played a more significant role than expected as a source for infecting ticks. The authors concluded that a wildlife vaccination strategy focused only on mice would be insufficient to decrease carriage rates overall and that vaccination of additional reservoir species would be required (Tsao et al., 2004). This is consistent with studies detailed above that revealed that inconspicuous hosts and birds are significant reservoirs for the LD spirochetes (Anderson et al., 1986; Brisson and Dykhuizen, 2004; Scott et al., 2018).

Parenteral vaccination of wildlife at a level sufficient enough to significantly reduce carriage of LD in reservoirs is arguably not a feasible and sustainable strategy. Hence, others have turned their efforts to vaccination using bait vaccines or reservoir targeting vaccines. An ideal vector would be one that is environmentally stable and able to elicit protection after a single dose. In addition, it would need to be non-toxic to both targeted and non-targeted wildlife species. Vaccinia virus has been suggested to meet these criteria and to be desirable due to its high levels of protein synthesis and ability to accept foreign DNA without losing infectivity. Precedent for the use of Vaccinia virus can be found in the oral rabies vaccine, Raboral™, which is targeted at raccoons and foxes (Pastoret et al., 1995). Oral gavage vaccination of C3H/HeN mice with Vaccinia virus encoding *ospA* resulted in protection of vaccinated mice against *B. burgdorferi* and significant clearance of *B. burgdorferi* from ticks that were fed on the vaccinated animals (Scheckelhoff et al., 2006). This approach was extended by delivery of the vaccinia virus-based vaccine to *P. leucopus* by oral gavage (Bhattacharya et al., 2011). The vaccine was found to be efficacious in decreasing the acquisition of *B. burgdorferi* by larval ticks and in decreasing transmission from ticks to *P. leucopus*. In another study, an OspA bait vaccine was developed using *E. coli* expressing OspA from a plasmid (Meirelles Richer et al., 2011). Lyophilized cells (200mg) were mixed with rolled oats and provided ad libitum. Mice that consumed the vaccine over periods of 1 or 4 months developed a neutralizing IgG response that persisted for 1 year. Based on the outcome of this study, the authors postulated that delivery of the vaccine twice a week for a period of 4 months from mid-April until mid-August could provide protection to young mice that are born throughout two seasons. In a prospective 5-year field trial, oral vaccination of wild white-footed mice resulted in reductions of 23% and 76% in the nymphal infection prevalence in a time-dependent manner (Richer et al., 2014). Significant decreases in infection prevalence were observed after 3 years of vaccine deployment. Hence, it is possible that a long-term bait vaccine strategy could be employed to reduce LD exposure risk. However, due to the limited migration range of mice and other small mammal reservoirs, wide and high-density distribution of bait vaccines over expansive geographic areas would be required. It remains to be determined if the costs of such an approach are feasible and sustainable.

Closing comments

Vaccination offers the most cost-efficient and reliable approach to address the expanding LD problem in both humans and companion animals. There have been advances made in the development of LD vaccines over the past 20 years that offer significant hope for those at risk for this debilitating disease. New technologies and innovative approaches have been employed that offer great promise and next generation LD vaccines for humans are undergoing clinical trials. However, as a cautionary note we must not forget or ignore the lessons learned from the rise and fall of LYMErix, the only human vaccine to advance to market. The perceived safety concerns that led to the voluntary removal of the vaccine from the market had a damaging effect on subsequent efforts to advance a human LD vaccine. Research on LD vaccines slowed and the interest of big pharma to take up LD vaccine development programs essentially came to a halt. Fortunately, interest in LD vaccines for humans has been renewed. It has become increasingly apparent that a vaccine that can recapture public interest and trust will need to be safe and convey long-lasting protection without the requirement for an aggressive booster schedule. As detailed above, it remains to be determined if a vaccine consisting only of OspA will meet these criteria. Vaccines that elicit protective antibodies that can work synergistically to attack multiple targets on the LD spirochete surface, and that have broad protective efficacy, are, in this authors opinion, our best, and potentially last opportunity to bring a successful vaccine to market.

Acknowledgements

This work was supported in part by grants to RTM from the National Institute of Allergy and Infectious Diseases (National Institutes of Health), The Stephen and Alexandra Cohen Foundation, The Christian Family Foundation, VCU Innovation Gateway Commercialization Award, and a Value and Efficiency Teaching and Research (VETAR) award from Virginia Commonwealth University.

References

Adeolu, M., and Gupta, R.S. (2014). A phylogenomic and molecular marker based proposal for the division of the genus Borrelia into two genera: the emended genus Borrelia containing only the members of the relapsing fever Borrelia, and the genus Borreliella gen. nov. containing the members of the Lyme disease Borrelia (Borrelia burgdorferi sensu lato complex). Antonie Van Leeuwenhoek *105*, 1049-1072.

Adrion, E.R., Aucott, J., Lemke, K.W., and Weiner, J.P. (2015). Health care costs, utilization and patterns of care following Lyme disease. PLoS One *10*, e0116767.

Aichinger, G., Ehrlich, H.J., Aaskov, J.G., Fritsch, S., Thomasser, C., Draxler, W., Wolzt, M., Müller, M., Pinl, F., Van Damme, P., *et al.* (2011). Safety and immunogenicity of an inactivated whole virus Vero cell-derived Ross River virus vaccine: a randomized trial. Vaccine *29*, 9376-9384. https://doi.org/10.1016/j.vaccine.2011.09.125

Akins, D.R., Bourell, K.W., Caimano, M.J., Norgard, M.V., and Radolf, J.D. (1998). A new animal model for studying Lyme disease spirochetes in a mammalian host-adapted state. J Clin Invest *101*, 2240-2250.

Alghaferi, M.Y., Anderson, J.M., Park, J., Auwaerter, P.G., Aucott, J.N., Norris, D.E., and Dumler, J.S. (2005). *Borrelia burgdorferi* ospC heterogeneity among human and murine isolates from a defined region of northern Maryland and southern Pennsylvania: lack of correlation with invasive and noninvasive genotypes. J Clin Microbiol *43*, 1879-1884.

Alitalo, A., T, M., Lankinen, H., Seppala, I., Lahdenne, P., Hefty, P.S., Akins, D., and Meri, S. (2002). Complement inhibitor factor H binding to Lyme disease spirochetes is mediated by inducible expression of multiple plasmid-encoded outer surface protein E paralogs. J Immunol *169*, 3847-3853.

Allen, K., Little, S., Petersen, M., Gruntmeir, J., Barrett, A., Herrin, B., Starkey, L., Sun, F., and Guerino, F. (2020). Evaluation of oral fluralaner (Bravecto(®)) for efficacy against nymphs of Amblyomma americanum and Rhipicephalus sanguineus (sensu lato). Parasit Vectors *13*, 315. https://doi.org/10.1186/s13071-020-04179-y

Alverson, J., Bundle, S.F., Sohaskey, C.D., Lybecker, M.C., and Samuels, D.S. (2003). Transcriptional regulation of the *ospAB* and *ospC* promoters from *Borrelia burgdorferi*. Mol Microbiol *48*, 1665-1677.

Anderson, J.F. (1991). Epizootiology of Lyme borreliosis. Scand J Infect Dis Suppl *77*, 23-34.

Anderson, J.F., Johnson, R.C., Magnarelli, L.A., and Hyde, F.W. (1986). Involvement of birds in the epidemiology of the Lyme disease agent Borrelia burgdorferi. Infect Immun *51*, 394-396.

Anderson, J.F., Magnarelli, L.A., LeFebvre, R.B., Andreadis, T.G., McAninch, J.B., Perng, G.C., and Johnson, R.C. (1989). Antigenically variable Borrelia burgdorferi isolated from cottontail rabbits and Ixodes dentatus in rural and urban areas. J Clin Microbiol *27*, 13-20.

Anguita, J., Ramamoorthi, N., Hovius, J.W., Das, S., Thomas, V., Persinski, R., Conze, D., Askenase, P.W., Rincon, M., Kantor, F.S., *et al.* (2002). Salp15, an ixodes scapularis salivary protein, inhibits CD4(+) T cell activation. Immunity *16*, 849-859.

Asbrink, E., and Hovmark, A. (1985). Successful cultivation of spirochetes from skin lesions of patients with erythema chronicum migrans Afzelius and acrodermatitis chronica atrophicans. Acta Pathol Microbiol Immunol Sect B *93*, 161-163.

Bacon, R.M., Kugeler, K.J., and Mead, P.S. (2008). Surveillance for Lyme disease--United States, 1992-2006. MMWR Surveill Summ *57*, 1-9.

Ball, E.C. (2016). Technical bulletin: VANGUARD crLyme: Chimeric recombinant vaccien technology for broad-spectrum protection against canine Lyme disease. Zoetis *SAB-00193*.

Baranton, G., Postic, D., Saint Girons, I., Boerlin, P., Piffaretti, J.C., Assous, M., and Grimont, P.A. (1992). Delineation of Borrelia burgdorferi sensu stricto, Borrelia garinii sp. nov., and group VS461 associated with Lyme borreliosis. Int J Syst Bacteriol *42*, 378-383. https://doi.org/10.1099/00207713-42-3-378

Barbour, A.G. (1984). Isolation and cultivation of Lyme disease spirochetes. The Yale journal of biology and medicine *57*, 521-525.

Barbour, A.G. (1988). Plasmid analysis of *Borrelia burgdorferi*, the Lyme disease agent. J Clin Microbiol *26*, 475-478.

Barbour, A.G. (1993). Linear DNA of *Borrelia* species and antigenic variation. Trends Microbiol *1*.

Barbour, A.G., Adeolu, M., and Gupta, R.S. (2017). Division of the genus Borrelia into two genera (corresponding to Lyme disease and relapsing fever groups) reflects their genetic and phenotypic distinctiveness and will lead to a better understanding of these two groups of microbes (Margos et al. (2016) There is inadequate evidence to support the division of the genus Borrelia. Int. J. Syst. Evol. Microbiol. doi: 10.1099/ijsem.0.001717). Int J Syst Evol Microbiol *67*, 2058-2067.

Barbour, A.G., and Garon, C.F. (1987). Linear plasmids of the bacterium *Borrelia burgdorferi* have covalently closed ends. Science *237*, 409-411.

Barbour, A.G., and Hayes, S.F. (1986). Biology of *Borrelia* species. Microbiol Rev *50*, 381-400.

Barthold, S.W., Beck, D.S., Hansen, G.M., Terwilliger, G.A., and Moody, K.D. (1990). Lyme borreliosis in selected strains and ages of laboratory mice. J Infect Dis *162*, 133-138.

Battisti, J.M., Bono, J.L., Rosa, P.A., Schrumpf, M.E., Schwan, T.G., and Policastro, P.F. (2008). Outer surface protein A protects Lyme disease spirochetes from acquired host immunity in the tick vector. Infect Immun *76*, 5228-5237.

Baum, E., Grosenbaugh, D.A., and Barbour, A.G. (2014). Diversity of antibody responses to Borrelia burgdorferi in experimentally infected beagle dogs. Clin Vaccine Immunol *21*, 838-846.

Beard, C.B., Occi, J., Bonilla, D.L., Egizi, A.M., Fonseca, D.M., Mertins, J.W., Backenson, B.P., Bajwa, W.I., Barbarin, A.M., Bertone, M.A., *et al.* (2018). Multistate Infestation with the Exotic Disease-Vector Tick Haemaphysalis longicornis - United States, August 2017-September 2018. MMWR Morb Mortal Wkly Rep *67*, 1310-1313.

Beernink, P.T., and Granoff, D.M. (2009). The modular architecture of meningococcal factor H-binding protein. Microbiol *155*, 2873-2883.

Beernink, P.T., Shaughnessy, J., Braga, E.M., Liu, Q., Rice, P.A., Ram, S., and Granoff, D.M. (2011). A meningococcal factor H binding protein mutant that eliminates factor H binding enhances protective antibody responses to vaccination. J Immunol *186*, 3606-3614.

Beernink, P.T., Shaughnessy, J., Pajon, R., Braga, E.M., Ram, S., and Granoff, D.M. (2012). The effect of human factor h on immunogenicity of meningococcal native outer membrane vesicle vaccines with over-expressed factor h binding protein. PLoS Pathog *8*, e1002688.

Beeslaar, J., Absalon, J., Anderson, A.S., Eiden, J.J., Balmer, P., Harris, S.L., Jones, T.R., O'Neill, R.E., Pregaldien, J.L., Radley, D., *et al.* (2020a). MenB-FHbp Vaccine Protects Against Diverse Meningococcal Strains in Adolescents and Young Adults: Post Hoc Analysis of Two Phase 3 Studies. Infect Dis Ther. https://doi.org/10.1007/s40121-020-00319-0

Beeslaar, J., Peyrani, P., Absalon, J., Maguire, J., Eiden, J., Balmer, P., Maansson, R., and Perez, J.L. (2020b). Sex, Age, and Race Effects on Immunogenicity of MenB-FHbp, A Bivalent Meningococcal B Vaccine: Pooled Evaluation of Clinical Trial Data. Infect Dis Ther. https://doi.org/10.1007/s40121-020-00322-5

Behler, R.P., Sharareh, N., Whetten, J.S., and Sabounchi, N.S. (2020). Analyzing the cost-effectiveness of Lyme disease risk reduction approaches. J Public Health Policy. https://doi.org/10.1057/s41271-020-00219-0

Benach, J.L., Bosler, E.M., Hanrahan, J.P., Coleman, J.L., Habicht, G.S., Bast, T.F., Cameron, D.J., Ziegler, J.L., Barbour, A.G., Burgdorfer, W., *et al.* (1983). Spirochetes isolated from the blood of two patients with Lyme disease. N Engl J Med *308*, 740-742.

Bergström, S., Bundoc, V.G., and Barbour, A.G. (1989). Molecular analysis of linear plasmid-encoded major surface proteins, OspA and OspB, of the Lyme disease spirochaete Borrelia burgdorferi. Mol Microbiol *3*, 479-486. https://doi.org/10.1111/j.1365-2958.1989.tb00194.x

Bhattacharya, D., Bensaci, M., Luker, K.E., Luker, G., Wisdom, S., Telford, S.R., and Hu, L.T. (2011). Development of a baited oral vaccine for use in reservoir-targeted strategies against Lyme disease. Vaccine *29*, 7818-7825. https://doi.org/10.1016/j.vaccine.2011.07.100

Bockenstedt, L.K., Fikrig, E., Barthold, S.W., Kantor, F.S., and Flavell, R.A. (1993). Inability of truncated recombinant Osp A proteins to elicit protective immunity to Borrelia burgdorferi in mice. J Immunol *151*, 900-906.

Bockenstedt, L.K., Hodzic, E., Feng, S., Bourrel, K.W., de Silva, A., Montgomery, R.R., Fikrig, E., Radolf, J.D., and Barthold, S.W. (1997). *Borrelia burgdorferi* strain-specific OspC mediated immunity in mice. Infect Immun *65*, 4661-4667.

Borrow, R., Taha, M.K., Giuliani, M.M., Pizza, M., Banzhoff, A., and Bekkat-Berkani, R. (2020). Methods to evaluate serogroup B meningococcal vaccines: From predictions to real-world evidence. J Infect. https://doi.org/10.1016/j.jinf.2020.07.034

Bowser, N.H., and Anderson, N.E. (2018). Dogs (Canis familiaris) as Sentinels for Human Infectious Disease and Application to Canadian Populations: A Systematic Review. Vet Sci *5*.

Breuner, N.E., Ford, S.L., Hojgaard, A., Osikowicz, L.M., Parise, C.M., Rosales Rizzo, M.F., Bai, Y., Levin, M.L., Eisen, R.J., and Eisen, L. (2020). Failure of the Asian longhorned tick, Haemaphysalis longicornis, to serve as an experimental vector of the Lyme disease spirochete, Borrelia burgdorferi sensu stricto. Ticks Tick Borne Dis *11*, 101311. https://doi.org/10.1016/j.ttbdis.2019.101311

Brisson, D., and Dykhuizen, D.E. (2004). ospC diversity in Borrelia burgdorferi: different hosts are different niches. Genetics *168*, 713-722.

Brisson, D., Dykhuizen, D.E., and Ostfeld, R.S. (2008). Conspicuous impacts of inconspicuous hosts on the Lyme disease epidemic. Proc Biol Sci *275*, 227-235.

Brooks, C.S., Hefty, P.S., Jolliff, S.E., and Akins, D.R. (2003). Global analysis of *Borrelia burgdorferi* genes regulated by mammalian host-specific signals. Infect Immun *71*, 3371-3383.

Buckles, E.L., Earnhart, C.G., and Marconi, R.T. (2006). Analysis of antibody response in humans to the type A OspC loop 5 domain and assessment of the potential utility of the loop 5 epitope in Lyme disease vaccine development. Clin Vaccine Immunol *13*, 1162-1165.

Burgdorfer, W., Barbour, A.G., Hayes, S.F., Benach, J.L., Grunwaldt, E., and Davis, J.P. (1982). Lyme disease-a tick-borne spirochetosis? Science *216*, 1317-1319.

Caimano, M.J., Drecktrah, D., Kung, F., and Samuels, D.S. (2016). Interaction of the Lyme disease spirochete with its tick vector. Cell Microbiol *18*, 919-927.

Caimano, M.J., Groshong, A.M., Belperron, A., Mao, J., Hawley, K.L., Luthra, A., Graham, D.E., Earnhart, C.G., Marconi, R.T., Bockenstedt, L.K., *et al.* (2019). The RpoS Gatekeeper in Borrelia burgdorferi: An Invariant Regulatory Scheme That Promotes Spirochete Persistence in Reservoir Hosts and Niche Diversity. Front Microbiol *10*, 1923. https://doi.org/10.3389/fmicb.2019.01923

Caimano, M.J., Iyer, R., Eggers, C.H., Gonzalez, C., Morton, E.A., Gilbert, M.A., Schwartz, I., and Radolf, J.D. (2007). Analysis of the RpoS regulon in Borrelia burgdorferi in response to mammalian host signals provides insight into RpoS function during the enzootic cycle. Mol Microbiol *65*, 1193-1217.

Caine, J.A., Lin, Y.P., Kessler, J.R., Sato, H., Leong, J.M., and Coburn, J. (2017). Borrelia burgdorferi outer surface protein C (OspC) binds complement component C4b and confers bloodstream survival. Cell Microbiol *19*.

Canica, M.M., Nato, F., du Merle, L., Mazie, J.C., Baranton, G., and Postic, D. (1993). Monoclonal antibodies for identification of *Borrelia afzelii* sp. nov. associated with late cutaneous manifestations of Lyme borreliosis. Scandinavian Journal of Infectious Disease *25*, 441-448.

Carrasco, S.E., Troxell, B., Yang, Y., Brandt, S.L., Li, H., Sandusky, G.E., Condon, K.W., Serezani, C.H., and Yang, X.F. (2015). Outer surface protein OspC is an antiphagocytic factor that protects Borrelia burgdorferi from phagocytosis by macrophages. Infect Immun *83*, 4848-4860.

Carroll, J.A., Garon, C.F., and Schwan, T.G. (1999). Effects of environmental pH on membrane proteins in *Borrelia burgdorferi*. Infect Immun *67*, 3181-3187.

Casjens, S. (2003). Prophage and bacterial genomics: what have we learned so far? Mol Microbiol *49*, 277-300.

Casjens, S., Palmer, N., van Vugt, R., Huang, W.M., Stevenson, B., Rosa, P., Lathigra, R., Sutton, G., Peterson, J., Dodson, R.J., *et al.* (2000). A bacterial genome in flux: the twelve linear and nine circular extrachromosomal DNAs in an infectious isolate of the Lyme disease spirochete *Borrelia burgdorferi*. Mol Microbiol *35*, 490-516.

Casjens, S.R., Mongodin, E.F., Qiu, W.G., Dunn, J.J., Luft, B.J., Fraser-Liggett, C.M., and Schutzer, S.E. (2011). Whole-genome sequences of two Borrelia afzelii and two Borrelia garinii Lyme disease agent isolates. J Bacteriol *193*, 6995-6996.

Casjens, S.R., Mongodin, E.F., Qiu, W.G., Luft, B.J., Schutzer, S.E., Gilcrease, E.B., Huang, W.M., Vujadinovic, M., Aron, J.K., Vargas, L.C., *et al.* (2012). Genome stability of Lyme disease spirochetes: comparative genomics of Borrelia burgdorferi plasmids. PLoS One *7*, e33280.

Cassatt, D.R., Patel, N.K., Ulbrandt, N.D., and Hanson, M.S. (1998). DbpA, but not OspA, is expressed by Borrelia burgdorferi during spirochetemia and is a target for protective antibodies. Infect Immun *66*, 5379-5387. https://doi.org/10.1128/iai.66.11.5379-5387.1998

CDC (1999). Recommendations for the use of Lyme disease vaccine. Recommendations of the Advisory Committee on Immunization Practices (ACIP). MMWR Recomm Rep *48*, 1-17, 21-15.

Chu, H.J., Chavez, L.G., Jr., Blumer, B.M., Sebring, R.W., Wasmoen, T.L., and Acree, W.M. (1992). Immunogenicity and efficacy study of a commercial Borrelia burgdorferi bacterin. J Am Vet Med Assoc *201*, 403-411.

Cleveland, C.A., Swanepoel, L., Brown, J.D., Casalena, M.J., Williams, L., and Yabsley, M.J. (2020). Surveillance for Borrelia spp. in Upland Game Birds in Pennsylvania, USA. Vet Sci *7*. https://doi.org/10.3390/vetsci7030082

Coipan, C.E., van Duijvendijk, G.L.A., Hofmeester, T.R., Takumi, K., and Sprong, H. (2018). The genetic diversity of Borrelia afzelii is not maintained by the diversity of the rodent hosts. Parasites & Vectors *11*, 454. https://doi.org/10.1186/s13071-018-3006-2

Collares-Pereira, M., Couceiro, S., Franca, I., Kurtenbach, K., Schäfer, S.M., Vitorino, L., Gonçalves, L., Baptista, S., Vieira, M.L., and Cunha, C. (2004). First isolation of Borrelia lusitaniae from a human patient. J Clin Microbiol *42*, 1316-1318. https://doi.org/10.1128/jcm.42.3.1316-1318.2004

Comstedt, P., Hanner, M., Schüler, W., Meinke, A., and Lundberg, U. (2014). Design and development of a novel vaccine for protection against Lyme borreliosis. PLoS One *9*, e113294. https://doi.org/10.1371/journal.pone.0113294

Comstedt, P., Schuler, W., Meinke, A., and Lundberg, U. (2017). The novel Lyme borreliosis vaccine VLA15 shows broad protection against Borrelia species expressing six different OspA serotypes. PLoS One *12*, e0184357. https://doi.org/10.1371/journal.pone.0184357

Conlon, J.A., Mather, T.N., Tanner, P., Gallo, G., and Jacobson, R.H. (2000). Efficacy of a nonadjuvanted, outer surface protein A, recombinant vaccine in dogs after challenge by ticks naturally infected with Borrelia burgdorferi. Vet Ther *1*, 96-107.

Connally, N.P., Rose, D.A., Breuner, N.E., Prose, R., Fleshman, A.C., Thompson, K., Wolfe, L., Broeckling, C.D., and Eisen, L. (2018). Impact of Wearing and Washing/Drying of Permethrin-Treated Clothing on Their Contact Irritancy and Toxicity for Nymphal Ixodes scapularis (Acari: Ixodidae) Ticks. J Med Entomol.

Crandall, H., Dunn, D.M., Ma, Y., Wooten, R.M., Zachary, J.F., Weis, J.H., Weiss, R.B., and Weis, J.J. (2006). Gene expression profiling reveals unique pathways associated with differential severity of lyme arthritis. J Immunol *177*, 7930-7942.

Dambach, D.M., Smith, C.A., Lewis, R.M., and Van Winkle, T.J. (1997). Morphologic, immunohistochemical, and ultrastructural characterization of a distinctive renal lesion in dogs putatively associated with Borrelia burgdorferi infection: 49 cases (1987-1992). Vet Pathol *34*, 85-96. https://doi.org/10.1177/030098589703400201

Di, L., Wan, Z., Akther, S., Ying, C., Larracuente, A., Li, L., Di, C., Nunez, R., Cucura, D.M., Goddard, N.L., *et al.* (2018). Genotyping and quantifying Lyme pathogen strains by deep sequencing of the outer surface protein C (ospC) locus. J Clin Microbiol.

Ding, W., Huang, X., Yang, X., Dunn, J.J., Luft, B.J., Koide, S., and Lawson, C.L. (2000). Structural identification of a key protective B-cell epitope in Lyme disease antigen OspA. J Mol Biol *302*, 1153-1164.

Divers, T.J., Chang, Y.F., Irby, N.L., Smith, J.L., and Carter, C.N. (2019). Leptospirosis: An important infectious disease in North American horses. Equine Vet J *51*, 287-292. https://doi.org/10.1111/evj.13069

Dunham-Ems, S.M., Caimano, M.J., Eggers, C.H., and Radolf, J.D. (2012). Borrelia burgdorferi requires the alternative sigma factor RpoS for dissemination within the vector during tick-to-mammal transmission. PLoS Pathog *8*, e1002532. https://doi.org/10.1371/journal.ppat.1002532

Earnhart, C., and Marconi, R.T. (2009). Lyme disease. In Vaccines for Biodefense and Emerging and Neglected Diseases, A.D. Barrett, and L.R. Stanberry, eds. (London: Elsevier), pp. 1032-1060.

Earnhart, C.G., Buckles, E.L., Dumler, J.S., and Marconi, R.T. (2005). Demonstration of OspC type diversity in invasive human Lyme disease isolates and identification of previously uncharacterized epitopes that define the specificity of the OspC murine antibody response. Infect Immun *73*, 7869-7877.

Earnhart, C.G., Buckles, E.L., and Marconi, R.T. (2007). Development of an OspC-based tetravalent, recombinant, chimeric vaccinogen that elicits bactericidal antibody against diverse Lyme disease spirochete strains. Vaccine *25*, 466-480. https://doi.org/10.1016/j.vaccine.2006.07.052

Earnhart, C.G., Leblanc, D.V., Alix, K.E., Desrosiers, D.C., Radolf, J.D., and Marconi, R.T. (2010). Identification of residues within ligand-binding domain 1 (LBD1) of the *Borrelia burgdorferi* OspC protein required for function in the mammalian environment. Mol Microbiol *76*, 393-408.

Earnhart, C.G., and Marconi, R.T. (2007a). Construction and analysis of variants of a polyvalent Lyme disease vaccine: approaches for improving the immune response to chimeric vaccinogens. Vaccine *25*, 3419-3427.

Earnhart, C.G., and Marconi, R.T. (2007b). An octavalent lyme disease vaccine induces antibodies that recognize all incorporated OspC type-specific sequences. Hum Vaccin *3*, 281-289.

Earnhart, C.G., and Marconi, R.T. (2007c). OspC phylogenetic analyses support the feasibility of a broadly protective polyvalent chimeric Lyme disease vaccine. Clin Vaccine Immunol *14*, 628-634.

Earnhart, C.G., Rhodes, D.V., Smith, A.A., Yang, X., Tegels, B., Carlyon, J.A., Pal, U., and Marconi, R.T. (2014). Assessment of the potential contribution of the highly conserved C-terminal motif (C10) of Borrelia burgdorferi outer surface protein C in transmission and infectivity. Pathog Dis *70*, 176-184.

Eggers, C., and Samuels, D.S. (1999). Molecular evidence for a new bacteriophage of *Borrelia burgdorferi*. J Bacteriol *181*, 7308-7313.

Ehrlich, H.J., Müller, M., Oh, H.M., Tambyah, P.A., Joukhadar, C., Montomoli, E., Fisher, D., Berezuk, G., Fritsch, S., Löw-Baselli, A., *et al.* (2008). A clinical trial of a whole-virus H5N1 vaccine derived from cell culture. N Engl J Med *358*, 2573-2584. https://doi.org/10.1056/NEJMoa073121

Eicken, C., Sharma, V., Klabunde, T., Owens, R.T., Pikas, D.S., Hook, M., and Sacchettini, J.C. (2001). Crystal structure of Lyme disease antigen outer surface protein C from Borrelia burgdorferi. J Biol Chem *276*, 10010-10015.

Eisen, L. (2020). Vector competence studies with hard ticks and Borrelia burgdorferi sensu lato spirochetes: A review. Ticks Tick Borne Dis *11*, 101359. https://doi.org/10.1016/j.ttbdis.2019.101359

Eisen, R.J., Eisen, L., and Beard, C.B. (2016). County-Scale Distribution of Ixodes scapularis and Ixodes pacificus (Acari: Ixodidae) in the Continental United States. J Med Entomol *53*, 349-386.

Eisen, R.J., Kugeler, K.J., Eisen, L., Beard, C.B., and Paddock, C.D. (2017). Tick-Borne Zoonoses in the United States: Persistent and Emerging Threats to Human Health. ILAR J *58*, 319-335. https://doi.org/10.1093/ilar/ilx005

Erdile, L.F., Brandt, M.A., Warkomski, D.J., Westrack, G.J., Sadziene, A., Barbour, A.G., and Mays, J.P. (1993). Role of attached lipid in the immunogenicity of *Borrelia burgdorferi* OspA. Infect Immun *61*, 81-90.

Eschner, A.K., and Mugnai, K. (2015). Immunization with a recombinant subunit OspA vaccine markedly impacts the rate of newly acquired Borrelia burgdorferi infections in client-owned dogs living in a coastal community in Maine, USA. Parasites & Vectors *8*, 92. https://doi.org/10.1186/s13071-015-0676-x

Exner, M.M., Wu, X., Blanco, D.R., Miller, J.N., and Lovett, M.A. (2000). Protection elicited by native outer membrane protein Oms66 (p66) against host-adapted Borrelia burgdorferi: conformational nature of bactericidal epitopes. Infect Immun *68*, 2647-2654.

Fikrig, E., Barthold, S.W., Kantor, F.S., and Flavell, R.A. (1990). Protection of mice against the Lyme disease agent by immunizing with recombinant OspA. Science *250*, 553-556. https://doi.org/10.1126/science.2237407

Fikrig, E., Barthold, S.W., Sun, W., Feng, W., Telford, S.R., 3rd, and Flavell, R.A. (1997). Borrelia burgdorferi P35 and P37 proteins, expressed in vivo, elicit protective immunity. Immunity *6*, 531-539. https://doi.org/10.1016/s1074-7613(00)80341-6

Fikrig, E., Telford, S.R., 3rd, Barthold, S.W., Kantor, F.S., Spielman, A., and Flavell, R.A. (1992a). Elimination of Borrelia burgdorferi from vector ticks feeding on OspA-immunized mice. Proc Natl Acad Sci U S A *89*, 5418-5421.

Fikrig, E.F., Barthold, S.W., Marcantonio, N., Deponte, K., Kantor, F.S., and Flavell, R.A. (1992b). Roles of OspA, OspB, and flagellin in protective immunity to Lyme borreliosis in laboratory mice. Infect Immun *59*, 553-559.

Fingerle, V., Laux, H., Munderloh, U.G., Schulte-Spechtel, U., and Wilske, B. (2000). Differential expression of outer surface proteins A and C by individual Borrelia burgdorferi in different genospecies. Med Microbiol Immunol *189*, 59-66.

Foley, J., Ott-Conn, C., Worth, J., Poulsen, A., and Clifford, D. (2014). An Ixodes minor and Borrelia carolinensis enzootic cycle involving a critically endangered Mojave Desert rodent. Ecol Evol *4*, 576-581. https://doi.org/10.1002/ece3.957

Fraser, C.M., Casjens, S., Huang, W.M., Sutton, G.G., Clayton, R., Lathigra, R., White, O., Ketchum, K.A., Dodson, R., Hickey, E.K., *et al.* (1997). Genomic sequence of a Lyme disease spirochaete, Borrelia burgdorferi. Nature *390*, 580-586.

Fukunaga, M., Hamase, A., Okada, K., and Nakao, M. (1996). *Borrelia tanukii* sp. nov. and *Borrelia turdae* sp. nov. found from ixodid ticks in Japan: rapid species identification by 16S rRNA gene-targeted PCR analysis. Microbiol Immunol *40*, 877-881.

Fukunaga, M., Sohnaka, M., and Yanagihara, Y. (1993). Analysis of *Borrelia* species associated with Lyme disease by rRNA gene restriction fragment length polymorphism. J Gen Microbiol *139*, 1141-1146.

Gasmi, S., Ogden, N.H., Lindsay, L.R., Burns, S., Fleming, S., Badcock, J., Hanan, S., Gaulin, C., Leblanc, M.A., Russell, C., *et al.* (2017). Surveillance for Lyme disease in Canada: 2009-2015. Can Commun Dis Rep *43*, 194-199.

Gilbert, M.A., Morton, E.A., Bundle, S.F., and Samuels, D.S. (2007). Artificial regulation of ospC expression in Borrelia burgdorferi. Mol Microbiol *63*, 1259-1273.

Gilmore, R.D., Bacon, R.M., Carpio, A.M., Piesman, J., Dolan, M.C., and Mbow, M.L. (2003). Inability of outer-surface protein C (OspC)-primed mice to elicit a protective anamnestic immune response to a tick-transmitted challenge of *Borrelia burgdorferi*. J Med Microbiol *52*, 551-556.

Gilmore, R.D., Jr., Kappel, K.J., Dolan, M.C., Burkot, T.R., and Johnson, B.J. (1996). Outer surface protein C (OspC), but not P39, is a protective immunogen against a tick-transmitted Borrelia burgdorferi challenge: evidence for a conformational protective epitope in OspC. Infect Immun *64*, 2234-2239.

Gilmore, R.D., Jr., and Mbow, M.L. (1999). Conformational nature of the Borrelia burgdorferi B31 outer surface protein C protective epitope. Infect Immun *67*, 5463-5469.

Glockner, G., Lehmann, R., Romualdi, A., Pradella, S., Schulte-Spechtel, U., Schilhabel, M., Wilske, B., Suhnel, J., and Platzer, M. (2004). Comparative analysis of the Borrelia garinii genome. Nucleic Acids Res *32*, 6038-6046.

Goldstein, R.E., Cordner, A.P., Sandler, J.L., Bellohusen, B.A., and Erb, H.N. (2007). Microalbuminuria and comparison of serologic testing for exposure to Borrelia burgdorferi in nonclinical Labrador and Golden Retrievers. J Vet Diagn Invest *19*, 294-297. https://doi.org/10.1177/104063870701900312

Gomes-Solecki, M.J., Brisson, D.R., and Dattwyler, R.J. (2006). Oral vaccine that breaks the transmission cycle of the Lyme disease spirochete can be delivered via bait. Vaccine *24*, 4440-4449.

Grimm, D., Tilly, K., Byram, R., Stewart, P.E., Krum, J.G., Bueschel, D.M., Schwan, T.G., Policastro, P.F., Elias, A.F., and Rosa, P.A. (2004). Outer-surface protein C of the Lyme disease spirochete: a protein induced in ticks for infection of mammals. Proc Natl Acad Sci U S A *101*, 3142-3147.

Grosenbaugh, D.A., De Luca, K., Durand, P.Y., Feilmeier, B., DeWitt, K., Sigoillot-Claude, C., Sajous, M.L., Day, M.J., and David, F. (2018). Characterization of recombinant OspA in two different Borrelia vaccines with respect to immunological response and its relationship to functional parameters. BMC Vet Res *14*, 312.

Gross, D.M., Forsthuber, T., Tary-Lehmann, M., Etling, C., Ito, K., Nagy, Z.A., Field, J.A., Steere, A.C., and Huber, B.T. (1998). Identification of LFA-1 as a candidate autoantigen in treatment-resistant Lyme arthritis. Science *281*, 703-706.

Guibinga, G.H., Sahay, B., Brown, H., Cooch, N., Chen, J., Yan, J., Reed, C., Mishra, M., Yung, B., Pugh, H., *et al.* (2020). Protection against Borreliella burgdorferi infection mediated by a synthetically engineered DNA vaccine. Hum Vaccin Immunother, 1-9. https://doi.org/10.1080/21645515.2020.1789408

Hagman, K.E., Yang, X., Wikel, S.K., Schoeler, G.B., Caimano, M.J., Radolf, J.D., and Norgard, M.V. (2000). Decorin-binding protein A

(DbpA) of Borrelia burgdorferi is not protective when immunized mice are challenged via tick infestation and correlates with the lack of DbpA expression by B. burgdorferi in ticks. Infect Immun *68*, 4759-4764.

Hanincova, K., Taragelova, V., Koci, J., Schafer, S.M., Hails, R., Ullmann, A.J., Piesman, J., Labuda, M., and Kurtenbach, K. (2003). Association of Borrelia garinii and B. valaisiana with songbirds in Slovakia. Appl Environ Microbiol *69*, 2825-2830.

Hasle, G. (2013). Transport of ixodid ticks and tick-borne pathogens by migratory birds. Front Cell Infect Microbiol *3*, 48. https://doi.org/10.3389/fcimb.2013.00048

Hassan, W.S., Giaretta, P.R., Rech, R., Ollivault-Shiflett, M., and Esteve-Gasent, M.D. (2019). Enhanced protective efficacy of Borrelia burgdorferi BB0172 derived-peptide based vaccine to control Lyme disease. Vaccine *37*, 5596-5606. https://doi.org/10.1016/j.vaccine.2019.07.092

Hauck, D., Jordan, D., Springer, A., Schunack, B., Pachnicke, S., Fingerle, V., and Strube, C. (2020). Transovarial transmission of Borrelia spp., Rickettsia spp. and Anaplasma phagocytophilum in Ixodes ricinus under field conditions extrapolated from DNA detection in questing larvae. Parasit Vectors *13*, 176. https://doi.org/10.1186/s13071-020-04049-7

Hayes, B.M., Dulebohn, D.P., Sarkar, A., Tilly, K., Bestor, A., Ambroggio, X., and Rosa, P.A. (2014). Regulatory protein BBD18 of the lyme disease spirochete: essential role during tick acquisition? MBio *5*, e01017-01014.

Hefty, P.S., Brooks, C.S., Jett, A.M., White, G.L., Wikel, S.K., Kennedy, R.C., and Akins, D.R. (2002). OspE-related, OspF-related, and Elp lipoproteins are immunogenic in baboons experimentally infected with Borrelia burgdorferi and in human lyme disease patients. J Clin Microbiol *40*, 4256-4265. https://doi.org/10.1128/jcm.40.11.4256-4265.2002

Hellwage, J., Meri, T., Heikkila, T., Alitalo, A., Panelius, J., Lahdenne, P., Seppala, I.J.T., and Meri, S. (2001). The complement regulator factor H binds to the surface protein OspE of *Borrelia burgdorferi*. J Biol Chem *276*, 8427-8435.

Hirschfeld, M., Kirschning, C.J., Schwandner, R., Wesche, H., Weis, J.H., Wooten, R.M., and Weis, J.J. (1999). Cutting edge: inflammatory signaling by Borrelia burgdorferi lipoproteins is mediated by toll-like receptor 2. J Immunol *163*, 2382-2386.

Holzmer, S., Kryda, K., Mahabir, S.P., and Everett, W. (2020). Evaluation of the speed of kill of a novel orally administered combination product containing sarolaner, moxidectin and pyrantel (Simparica Trio™) against induced infestations of Ixodes scapularis on dogs. Parasit Vectors *13*, 76. https://doi.org/10.1186/s13071-020-3953-2

Howe, T.R., LaQuier, F.W., and Barbour, A.G. (1986). Organization of genes encoding two outer membrane proteins of the Lyme disease agent *Borrelia burgdorferi* within a single transcriptional unit. Infect Immun *54*, 207-212.

Howe, T.R., Mayer, L.W., and Barbour, A.G. (1985). A single recombinant plasmid expressing two major outer surface proteins of the Lyme disease spirochete. Science *227*, 645-646. https://doi.org/10.1126/science.3969554

Huang, X., Yang, X., Luft, B.J., and Koide, S. (1998). NMR identification of epitopes of Lyme disease antigen OspA to monoclonal antibodies. J Mol Biol *281*, 61-67.

Hubner, A., Yang, X., Nolen, D.M., Popova, T.G., Cabello, F.C., and Norgard, M.V. (2001). Expression of *Borrelia burgdorferi* OspC and DbpA is controlled by a RpoN-RpoS regulatory pathway. Proc Natl Acad Sci USA *98*, 12724-12729.

Hughes, C.A., Engstrom, S.M., Coleman, L.A., Kodner, C.B., and Johnson, R.C. (1993). Protective immunity is induced by a Borrelia burgdorferi mutant that lacks OspA and OspB. Infect Immun *61*, 5115-5122.

Iyer, R., Caimano, M.J., Luthra, A., Axline, D., Jr., Corona, A., Iacobas, D.A., Radolf, J.D., and Schwartz, I. (2015). Stage-specific global alterations in the transcriptomes of Lyme disease spirochetes during tick feeding and following mammalian host adaptation. Mol Microbiol *95*, 509-538.

Izac, J.R., Camire, A.C., Earnhart, C.G., Embers, M.E., Funk, R.A., Breitschwerdt, E.B., and Marconi, R.T. (2019). Analysis of the antigenic determinants of the OspC protein of the Lyme disease spirochetes: Evidence that the C10 motif is not immunodominant or required to elicit bactericidal antibody responses. Vaccine *37*, 2401-2407. https://doi.org/10.1016/j.vaccine.2019.02.007

Izac, J.R., Camire, A.C., Schuler, E.J.A., Hatke, A.L., O'Bier, N.S., Oliver, L.D., Corondi, A., Plocinski, O.C., Desmond, R.P., Naimi, W.A., *et al.* (2020a). Serologic Evidence for the Exposure of Eastern Coyotes (Canis latrans)) in Pennsylvania to the Tick-Borne Pathogens Borreliella burgdorferi and Anaplasma phagocytophilum. mSphere *5*, e00544-00520. https://doi.org/10.1128/mSphere.00544-20

Izac, J.R., and Marconi, R.T. (2019). Diversity of the Lyme Disease Spirochetes and its Influence on Immune Responses to Infection and Vaccination. Vet Clin North Am Small Anim Pract. https://doi.org/10.1016/j.cvsm.2019.02.007

Izac, J.R., O'Bier, N.S., Oliver, L.D., Jr., Camire, A.C., Earnhart, C.G., LeBlanc Rhodes, D.V., Young, B.F., Parnham, S.R., Davies, C., and Marconi, R.T. (2020b). Development and optimization of OspC chimeritope vaccinogens for Lyme disease. Vaccine *38*, 1915-1924. https://doi.org/10.1016/j.vaccine.2020.01.027

Izac, J.R., Oliver, L.D., Jr., Earnhart, C.G., and Marconi, R.T. (2017). Identification of a defined linear epitope in the OspA protein of the Lyme disease spirochetes that elicits bactericidal antibody responses: Implications for vaccine development. Vaccine *35*, 3178-3185.

Jauris-Heipke, S., Fuchs, R., Motz, M., Preac-Mursic, V., Schwab, E., Soutschek, E., Will, G., and Wilske, B. (1993). Genetic heterogenity of the genes coding for the outer surface protein C (OspC) and the flagellin of Borrelia burgdorferi. Med Microbiol Immunol *182*, 37-50.

Jobe, D.A., Lovrich, S.D., Schell, R.F., and Callister, S.M. (2003). C-terminal region of outer surface protein C binds borreliacidal antibodies in sera from patients with Lyme disease. Clin Diagn Lab Immunol *10*, 573-578.

Johnson, T.L., Graham, C.B., Hojgaard, A., Breuner, N.E., Maes, S.E., Boegler, K.A., Replogle, A.J., Kingry, L.C., Petersen, J.M., Eisen, L., *et al.* (2017). Isolation of the Lyme Disease Spirochete Borrelia mayonii From Naturally Infected Rodents in Minnesota. J Med Entomol *54*, 1088-1092. https://doi.org/10.1093/jme/tjx062

Kalish, R.A., Leong, J.M., and Steere, A.C. (1993). Association of treatment-resistant chronic Lyme arthritis with HLA-DR4 and antibody reactivity to OspA and OspB of Borrelia burgdorferi. Infect Immun *61*, 2774-2779.

Kamp, H.D., Swanson, K.A., Wei, R.R., Dhal, P.K., Dharanipragada, R., Kern, A., Sharma, B., Sima, R., Hajdusek, O., Hu, L.T., *et al.* (2020). Design of a broadly reactive Lyme disease vaccine. NPJ Vaccines *5*, 33. https://doi.org/10.1038/s41541-020-0183-8

Kariu, T., Coleman, A.S., Anderson, J.F., and Pal, U. (2011). Methods for rapid transfer and localization of lyme disease pathogens within the tick gut. J Vis Exp. https://doi.org/10.3791/2544

Kawabata, H., Masuzawa, T., and Yanagihara, Y. (1993). Genomic analysis of *Borrelia japonica* sp. nov. isolated from *Ixodes ovatus* in Japan. Microbiol Immun *37*, 843-848.

Keller, D., Koster, F.T., Marks, D.H., Hosbach, P., Erdile, L.F., and Mays, J.P. (1994). Safety and immunogenicity of a recombinant outer surface protein A Lyme vaccine. JAMA *271*, 1764-1768.

Kharitonenkov, I.G., Pomelova, V.G., Bucher, D.J., Cozzolino, A., Bichenkova, T.A., Sadykbekova, R.K., and Osin, N.S. (2002). Epitope mapping of the outer surface protein A (OspA) of the spirochete Borrelia burgdorferi using a panel of monoclonal antibodies and lanthanide competition fluoroimmunoassay. Biochemistry (Mosc) *67*, 640-650.

Koide, S., Yang, X., Huang, X., Dunn, J.J., and Luft, B.J. (2005). Structure-based design of a second-generation Lyme disease vaccine based on a C-terminal fragment of *Borrelia burgdorferi* OspA. J Mol Biol *350*, 290-299.

Kornblatt, A.N., Urband, P.H., and Steere, A.C. (1985). Arthritis caused by Borrelia burgdorferi in dogs. J Am Vet Med Assoc *186*, 960-964.

Kraiczy, P., Hellwage, J., Skerka, C., Kirschfink, M., Brade, V., Zipfel, P.F., and Wallich, R. (2003). Immune evasion of *Borrelia burgdorferi*: mapping of a complement-inhibitor factor H-binding site of BbCRASP-3, a novel member of the Erp protein family. Eur J Immunol *33*, 697-707.

Kryda, K., Mahabir, S.P., Chapin, S., Holzmer, S.J., Bowersock, L., Everett, W.R., Riner, J., Carter, L., and Young, D. (2020). Efficacy of a novel orally administered combination product containing sarolaner, moxidectin and pyrantel (Simparica Trio™) against induced infestations of five common tick species infesting dogs in the USA. Parasit Vectors *13*, 77. https://doi.org/10.1186/s13071-020-3945-2

Kumar, M., Kaur, S., Kariu, T., Yang, X., Bossis, I., Anderson, J.F., and Pal, U. (2011). Borrelia burgdorferi BBA52 is a potential target for transmission blocking Lyme disease vaccine. Vaccine *29*, 9012-9019. https://doi.org/10.1016/j.vaccine.2011.09.035

Kumar, M., Yang, X., Coleman, A.S., and Pal, U. (2010). BBA52 facilitates Borrelia burgdorferi transmission from feeding ticks to murine hosts. J Infect Dis *201*, 1084-1095. https://doi.org/10.1086/651172

Kumaran, D., Eswaramoorthy, S., Dunn, J.J., and Swaminathan, S. (2001a). Crystallization and preliminary X-ray analysis of Borrelia burgdorferi outer surface protein C (OspC). Acta Crystallogr D Biol Crystallogr *57*, 298-300.

Kumaran, D., Eswaramoorthy, S., Luft, B.J., Koide, S., Dunn, J.J., Lawson, C.L., and Swaminathan, S. (2001b). Crystal structure of outer surface protein C (OspC) from the Lyme disease spirochete, Borrelia burgdorferi. EMBO J *20*, 971-978.

Kung, F., Kaur, S., Smith, A.A., Yang, X., Wilder, C.N., Sharma, K., Buyuktanir, O., and Pal, U. (2016). A Borrelia burgdorferi Surface-Exposed Transmembrane Protein Lacking Detectable Immune Responses Supports Pathogen Persistence and Constitutes a Vaccine Target. J Infect Dis *213*, 1786-1795. https://doi.org/10.1093/infdis/jiw013

Kurtenbach, K., Dizij, A., Voet, P., Hauser, P., and Simon, M.M. (1997). Vaccination of natural reservoir hosts with recombinant lipidated OspA induces a transmission-blocking immunity against Lyme disease spirochaetes associated with high levels of LA-2 equivalent antibodies. Vaccine *15*, 1670-1674.

Kurtenbach, K., Peacey, M., Rijpkema, S.G., Hoodless, A.N., Nuttall, P.A., and Randolph, S.E. (1998). Differential transmission of the genospecies of Borrelia burgdorferi sensu lato by game birds and small rodents in England. Appl Environ Microbiol *64*, 1169-1174.

Lagal, V., Portnoi, D., Faure, G., Postic, D., and Baranton, G. (2006). Borrelia burgdorferi sensu stricto invasiveness is correlated with OspC-plasminogen affinity. Microbes Infect *8*, 645-652.

LeFebrve, R.B., Perny, G.C., and Johnson, R.C. (1989). Characterization of *Borrelia burgdorferi* isolates ny restriction endonuclease analysis and DNA hybridization. J Clin Microbiol *27*.

Legros, V., Jolivet-Reynaud, C., Battail-Poirot, N., Saint-Pierre, C., and Forest, E. (2000). Characterization of an anti-Borrelia burgdorferi OspA conformational epitope by limited proteolysis of monoclonal antibody-bound antigen and mass spectrometric peptide mapping. Protein Sci *9*, 1002-1010.

Levi, T., Kilpatrick, A.M., Mangel, M., and Wilmers, C.C. (2012). Deer, predators, and the emergence of Lyme disease. Proc Natl Acad Sci U S A *109*, 10942-10947.

Levy, S. (2017). Northern Trek: The Spread of Ixodes scapularis into Canada. Environ Health Perspect *125*, 074002.

Levy, S.A. (2002). Use of a C6 ELISA test to evaluate the efficacy of a whole-cell bacterin for the prevention of naturally transmitted canine Borrelia burgdorferi infection. Vet Ther *3*, 420-424.

Levy, S.A., Lissman, B.A., and Ficke, C.M. (1993). Performance of a Borrelia burgdorferi bacterin in borreliosis-endemic areas. J Am Vet Med Assoc *202*, 1834-1838.

Li, H., Dunn, J.J., Luft, B.J., and Lawson, C.L. (1997). Crystal structure of Lyme disease antigen outer surface protein A complexed with an Fab. Proc Natl Acad Sci U S A *94*, 3584-3589. https://doi.org/10.1073/pnas.94.8.3584

Liang, F.T., Jacobs, M.B., and Philipp, M.T. (2001). C-terminal invariable domain of VlsE may not serve as target for protective immune response against *Borrelia burgdorferi*. Infect Immun *69*, 1337-1343.

Lin, T., Oliver, J.H., Jr., and Gao, L. (2002). Genetic diversity of the outer surface protein C gene of southern Borrelia isolates and its possible epidemiological, clinical, and pathogenetic implications. J Clin Microbiol *40*, 2572-2583.

Lissman, B.A., Bosler, E.M., Camay, H., Ormiston, B.G., and Benach, J.L. (1984). Spirochete-associated arthritis (Lyme disease) in a dog. J Am Vet Med Assoc *185*, 219-220.

Little, S.E., Heise, S.R., Blagburn, B.L., Callister, S.M., and Mead, P.S. (2010). Lyme borreliosis in dogs and humans in the USA. Trends Parasitol *26*, 213-218.

Littman, M.P. (2013). Lyme nephritis. J Vet Emerg Crit Care (San Antonio) *23*, 163-173. https://doi.org/10.1111/vec.12026

Littman, M.P., Gerber, B., Goldstein, R.E., Labato, M.A., Lappin, M.R., and Moore, G.E. (2018). ACVIM consensus update on Lyme borreliosis in dogs and cats. J Vet Intern Med *32*, 887-903.

Liveris, D., Gazumyan, A., and Schwartz, I. (1995). Molecular typing of Borrelia burgdorferi sensu lato by PCR-restriction fragment length polymorphism analysis. J Clin Microbiol *33*, 589-595.

Lloyd, V.K., and Hawkins, R.G. (2018). Under-Detection of Lyme Disease in Canada. Healthcare (Basel) *6*.

Lommano, E., Dvorak, C., Vallotton, L., Jenni, L., and Gern, L. (2014). Tick-borne pathogens in ticks collected from breeding and migratory birds in Switzerland. Ticks Tick Borne Dis *5*, 871-882.

Loss, S.R., Noden, B.H., Hamer, G.L., and Hamer, S.A. (2016). A quantitative synthesis of the role of birds in carrying ticks and tick-borne pathogens in North America. Oecologia *182*, 947-959. https://doi.org/10.1007/s00442-016-3731-1

Lovrich, S.D., Jobe, D.A., Schell, R.F., and Callister, S.M. (2005). Borreliacidal OspC antibodies specific for a highly conserved epitope are immunodominant in human Lyme disease and do not occur in mice or hamsters. Clin Diagn Lab Immunol *12*, 746-751.

Luke, C.J., Carner, K., Liang, X., and Barbour, A.G. (1997). An OspA-based DNA vaccine protects mice against infection with Borrelia burgdorferi. J Infect Dis *175*, 91-97. https://doi.org/10.1093/infdis/175.1.91

Mac, S., da Silva, S.R., and Sander, B. (2019). The economic burden of Lyme disease and the cost-effectiveness of Lyme disease interventions: A scoping review. PLoS One *14*, e0210280. https://doi.org/10.1371/journal.pone.0210280

Maes, E., Lecomte, P., and Ray, N. (1998). A cost-of-illness study of Lyme disease in the United States. Clin Ther *20*, 993-1008; discussion 1992. https://doi.org/10.1016/s0149-2918(98)80081-7

Magnarelli, L.A., Anderson, J.F., Kaufmann, A.F., Lieberman, L.L., and Whitney, G.D. (1985). Borreliosis in dogs from southern Connecticut. J Am Vet Med Assoc *186*, 955-959.

Mak, S., Morshed, M., and Henry, B. (2000). Ecological niche modeling of lyme disease in British Columbia, Canada. J Med Entomol *47*, 99-105.

Maraspin, V., Ruzic-Sabljic, E., and Strle, F. (2006). Lyme borreliosis and Borrelia spielmanii. Emerg Infect Dis *12*, 1177. https://doi.org/10.3201/eid1207.060077

Marcinkiewicz, A.L., Lieknina, I., Kotelovica, S., Yang, X., Kraiczy, P., Pal, U., Lin, Y.P., and Tars, K. (2018). Eliminating Factor H-Binding Activity of Borrelia burgdorferi CspZ Combined with Virus-Like Particle Conjugation Enhances Its Efficacy as a Lyme Disease Vaccine. Front Immunol *9*, 181. https://doi.org/10.3389/fimmu.2018.00181

Marcinkiewicz, A.L., Lieknina, I., Yang, X., Lederman, P.L., Hart, T.M., Yates, J., Chen, W.H., Bottazzi, M.E., Mantis, N.J., Kraiczy, P., *et al.* (2020). The Factor H-binding site of CspZ as a protective target against multi-strain, tick-transmitted Lyme disease. Infect Immun. https://doi.org/10.1128/iai.00956-19

Marconi, R.T. (2020a). Field Safety Study of VANGUARD®crLyme: a vaccine for the prevention of Lyme disease in dogs. Submitted.

Marconi, R.T. (2020b). VANGUARD®crLyme: A Next Generation Lyme Disease Vaccine That Prevents B. burgdorferi Infection in Dogs. Submitted.

Marconi, R.T., and Garon, C.F. (1992a). Development of polymerase chain reaction primer sets for diagnosis of Lyme disease and for species-specific identification of Lyme disease isolates by 16S rRNA signature nucleotide analysis. J Clin Microbiol *30*, 2830-2834.

Marconi, R.T., and Garon, C.F. (1992b). Identification of a third genomic group of Borrelia burgdorferi through signature nucleotide analysis and 16S rRNA sequence determination. J Gen Microbiol *138*, 533-536.

Marconi, R.T., Hohenberger, S., Jauris-Heipke, S., Schulte-Spechtel, U., LaVoie, C.P., Rossler, D., and Wilske, B. (1999). Genetic analysis of Borrelia garinii OspA serotype 4 strains associated with neuroborreliosis: evidence for extensive genetic homogeneity. J Clin Microbiol *37*, 3965-3970.

Marconi, R.T., Liveris, D., and Schwartz, I. (1995). Identification of novel insertion elements, restriction fragment length polymorphism patterns, and discontinuous 23S rRNA in Lyme disease spirochetes: phylogenetic analyses of rRNA genes and their intergenic spacers in *Borrelia japonica sp. nov.* and genomic group 21038 (*Borrelia andersonii sp. nov.*) isolates. J Clin Microbiol *33*, 2427-2434.

Marconi, R.T., Samuels, D.S., and Garon, C.F. (1993a). Transcriptional analyses and mapping of the ospC gene in Lyme disease spirochetes. J Bacteriol *175*, 926-932.

Marconi, R.T., Samuels, D.S., Schwan, T.G., and Garon, C.F. (1993b). Identification of a protein in several Borrelia species which is related to OspC of the Lyme disease spirochetes. J Clin Microbiol *31*, 2577-2583.

Margos, G., Chu, C.Y., Takano, A., Jiang, B.G., Liu, W., Kurtenbach, K., Masuzawa, T., Fingerle, V., Cao, W.C., and Kawabata, H. (2015). Borrelia yangtzensis sp. nov., a rodent-associated species in Asia, is related to Borrelia valaisiana. Int J Syst Evol Microbiol *65*, 3836-3840. https://doi.org/10.1099/ijsem.0.000491

Margos, G., Fedorova, N., Kleinjan, J.E., Hartberger, C., Schwan, T.G., Sing, A., and Fingerle, V. (2017). Borrelia lanei sp. nov. extends the diversity of Borrelia species in California. Int J Syst Evol Microbiol *67*, 3872-3876. https://doi.org/10.1099/ijsem.0.002214

Margos, G., Hojgaard, A., Lane, R.S., Cornet, M., Fingerle, V., Rudenko, N., Ogden, N., Aanensen, D.M., Fish, D., and Piesman, J. (2010). Multilocus sequence analysis of Borrelia bissettii strains from North America reveals a new Borrelia species, Borrelia kurtenbachii. Ticks Tick Borne Dis *1*, 151-158. https://doi.org/10.1016/j.ttbdis.2010.09.002

Margos, G., Wilske, B., Sing, A., Hizo-Teufel, C., Cao, W.C., Chu, C., Scholz, H., Straubinger, R.K., and Fingerle, V. (2013). Borrelia bavariensis sp. nov. is widely distributed in Europe and Asia. Int J Syst Evol Microbiol *63*, 4284-4288.

Masuzawa, T., Takada, N., Kudeken, M., Fukui, T., Yano, Y., Ishiguro, F., Kawamura, Y., Imai, Y., and Ezaki, T. (2001). Borrelia sinica sp. nov., a lyme disease-related Borrelia species isolated in China. Int J Syst Evol Microbiol *51*, 1817-1824.

Mathiesen, M.J., Christiansen, M., Hansen, K., Holm, A., Asbrink, E., and Theisen, M. (1998a). Peptide-based OspC enzyme linked immunosorbent assay for serodiagnosis of Lyme borreliosis. J Clin Microbiol *36*, 3474-3479.

Mathiesen, M.J., Holm, A., Christiansen, M., Blom, J., Hansen, K., Ostergaard, S., and Theisen, M. (1998b). The dominant epitope of Borrelia garinii outer surface protein C recognized by sera from patients with neuroborreliosis has a surface-exposed conserved structural motif. Infect Immun *66*, 4073-4079.

McDowell, J.V., Hovis, K.M., Zhang, H., Tran, E., Lankford, J., and Marconi, R.T. (2006). Evidence that the BBA68 protein (BbCRASP-1) of the Lyme disease spirochetes does not contribute to factor H-mediated immune evasion in humans and other animals. Infect Immun *74*, 3030-3034.

McDowell, J.V., Huang, B., Fenno, J.C., and Marconi, R.T. (2009). Analysis of a unique interaction between the complement regulatory protein factor H and the periodontal pathogen *Treponema denticola*. Infect Immun *77*, 1417-1425.

McDowell, J.V., Miller, D.P., Mallory, K.L., and Marconi, R.T. (2012). *Treponema denticola*: FhbB, Dentilisin, Complement Evasion and the Paradox of Factor H Cleavage. In The Pathogenic Spirochetes: strategies for evasion of host immunity and persistance, M.E. Embers, ed. (New York: Springer Science+Business Media), pp. 43-62.

McDowell, J.V., Tran, E., Hamilton, D., Wolfgang, J., Miller, K., and Marconi, R.T. (2003a). Analysis of the ability of spirochete species associated with relapsing fever, avian borreliosis, and epizootic bovine abortion to bind factor H and cleave c3b. J Clin Microbiol *41*, 3905-3910.

McDowell, J.V., Wolfgang, J., Tran, E., Metts, M.S., Hamilton, D., and Marconi, R.T. (2003b). Comprehensive analysis of the factor H binding capabilities of *Borrelia* species associated with lyme disease: delineation of two distinct classes of factor H binding proteins. Infect Immun *71*, 3597-3602.

Mead, P., Goel, R., and Kugeler, K. (2011). Canine serology as adjunct to human Lyme disease surveillance. Emerg Infect Dis *17*, 1710-1712.

Meek, J.I., Roberts, C.L., Smith, E.V., Jr., and Cartter, M.L. (1996). Underreporting of Lyme disease by Connecticut physicians, 1992. J Public Health Manag Pract *2*, 61-65. https://doi.org/10.1097/00124784-199623000-00017

Meirelles Richer, L., Aroso, M., Contente-Cuomo, T., Ivanova, L., and Gomes-Solecki, M. (2011). Reservoir targeted vaccine for lyme borreliosis induces a yearlong, neutralizing antibody response to OspA in white-footed mice. Clin Vaccine Immunol *18*, 1809-1816. https://doi.org/10.1128/cvi.05226-11

Mitchell, C., Dyer, M., Lin, F.C., Bowman, N., Mather, T., and Meshnick, S. (2020). Protective Effectiveness of Long-Lasting Permethrin Impregnated Clothing Against Tick Bites in an Endemic Lyme Disease Setting: A Randomized Control Trial Among Outdoor Workers. J Med Entomol. https://doi.org/10.1093/jme/tjaa061

Moody, K.D., and Barthold, S.W. (1998). Lyme borreliosis in laboratory mice. Lab Anim Sci *48*, 168-171.

Mysterud, A., Stigum, V.M., Jaarsma, R.I., and Sprong, H. (2019). Genospecies of Borrelia burgdorferi sensu lato detected in 16 mammal species and questing ticks from northern Europe. Scientific Reports *9*, 5088. https://doi.org/10.1038/s41598-019-41686-0

Ndlovu, S., Williamson, A.L., Heath, L., and Carulei, O. (2020). Genome Sequences of Three African Swine Fever Viruses of Genotypes IV and XX from Zaire and South Africa, Isolated from a Domestic Pig (Sus scrofa domesticus), a Warthog (Phacochoerus africanus), and a

European Wild Boar (Sus scrofa). Microbiol Resour Announc *9*. https://doi.org/10.1128/mra.00341-20

Nelson, C.A., Saha, S., Kugeler, K.J., Delorey, M.J., Shankar, M.B., Hinckley, A.F., and Mead, P.S. (2015). Incidence of Clinician-Diagnosed Lyme Disease, United States, 2005-2010. Emerg Infect Dis *21*, 1625-1631.

Newman, E.A., Eisen, L., Eisen, R.J., Fedorova, N., Hasty, J.M., Vaughn, C., and Lane, R.S. (2015). Borrelia burgdorferi sensu lato spirochetes in wild birds in northwestern California: associations with ecological factors, bird behavior and tick infestation. PLoS One *10*, e0118146. https://doi.org/10.1371/journal.pone.0118146

Norek, A., and Janda, L. (2017). Epitope mapping of Borrelia burgdorferi OspC protein in homodimeric fold. Protein Sci *26*, 796-806.

Ogden, N.H., Bouchard, C., Badcock, J., Drebot, M.A., Elias, S.P., Hatchette, T.F., Koffi, J.K., Leighton, P.A., Lindsay, L.R., Lubelczyk, C.B., *et al.* (2019). What is the real number of Lyme disease cases in Canada? BMC Public Health *19*, 849. https://doi.org/10.1186/s12889-019-7219-x

Ogden, N.H., Lindsay, L.R., Hanincova, K., Barker, I.K., Bigras-Poulin, M., Charron, D.F., Heagy, A., Francis, C.M., O'Callaghan, C.J., Schwartz, I., *et al.* (2008). Role of migratory birds in introduction and range expansion of Ixodes scapularis ticks and of Borrelia burgdorferi and Anaplasma phagocytophilum in Canada. Appl Environ Microbiol *74*, 1780-1790.

Oliver Jr, L.D., Earnhart, C.G., Virgina-Rhodes, D., Theisen, M., and Marconi, R. (2016). Antibody profiling of canine IgG responses to the OspC protein of the Lyme disease spirochetes supports a multivalent approach in vaccine and diagnostic assay development. The Veterinary Journal, 27-33.

Onder, O., Humphrey, P.T., McOmber, B., Korobova, F., Francella, N., Greenbaum, D.C., and Brisson, D. (2012). OspC Is Potent Plasminogen Receptor on Surface of Borrelia burgdorferi. J Biol Chem *287*, 16860-16868.

Ostfeld, R.S., Canham, C.D., Oggenfuss, K., Winchcombe, R.J., and Keesing, F. (2006). Climate, deer, rodents, and acorns as determinants of variation in lyme-disease risk. PLoS Biol *4*, e145. https://doi.org/10.1371/journal.pbio.0040145

Pal, U., de Silva, A.M., Montgomery, R.R., Fish, D., Anguita, J., Anderson, J.F., Lobet, Y., and Fikrig, E. (2000). Attachment of Borrelia burgdorferi within Ixodes scapularis mediated by outer surface protein A. J Clin Invest *106*, 561-569. https://doi.org/10.1172/jci9427

Pal, U., Li, X., Wang, T., Montgomery, R.R., Ramamoorthi, N., Desilva, A.M., Bao, F., Yang, X., Pypaert, M., Pradhan, D., *et al.* (2004a). TROSPA, an Ixodes scapularis receptor for Borrelia burgdorferi. Cell *119*, 457-468. https://doi.org/10.1016/j.cell.2004.10.027

Pal, U., Yang, X., Chen, M., Bockenstedt, L.K., Anderson, J.F., Flavell, R.A., Norgard, M.V., and Fikrig, E. (2004b). OspC facilitates Borrelia burgdorferi invasion of Ixodes scapularis salivary glands. J Clin Invest *113*, 220-230.

Pastoret, P.P., Boulanger, D., and Brochier, B. (1995). Field trials of a recombinant rabies vaccine. Parasitology *110 Suppl*, S37-42. https://doi.org/10.1017/s0031182000001475

Petersen, I., Chiummo, R., Zschiesche, E., Karas-Tecza, J., Rapti, D., Roepke, R., and Thomas, E. (2020). A European field assessment of the efficacy of fluralaner (Bravecto(®)) chewable and spot-on formulations for treatment of dogs with generalized demodicosis. Parasit Vectors *13*, 304. https://doi.org/10.1186/s13071-020-04159-2

Philipp, M.T., Lobet, Y., Bohm, R.P., Jr., Roberts, E.D., Dennis, V.A., Gu, Y., Lowrie, R.C., Jr., Desmons, P., Duray, P.H., England, J.D., *et al.* (1997). The outer surface protein A (OspA) vaccine against Lyme disease: efficacy in the rhesus monkey. Vaccine *15*, 1872-1887.

Policastro, P.F., and Schwan, T.G. (2003). Experimental infection of Ixodes scapularis larvae (Acari: Ixodidae) by immersion in low passage cultures of Borrelia burgdorferi. J Med Entomol *40*, 364-370.

Porwancher, R.B., Hagerty, C.G., Fan, J., Landsberg, L., Johnson, B.J., Kopnitsky, M., Steere, A.C., Kulas, K., and Wong, S.J. (2011). Multiplex immunoassay for Lyme disease using VlsE1-IgG and pepC10-IgM antibodies: improving test performance through bioinformatics. Clin Vaccine Immunol *18*, 851-859.

Postic, D., Assous, M.V., Grimont, P.A.D., and Baranton, G. (1994). Diversity of *Borrelia burgdorferi* sensu lato evidenced by restriction fragment length polymorphism of *rrf* (5S)-*rrl* (23S) intergenic spacer amplicons. Int J Sys Bacteriol *144*, 733-742.

Postic, D., Garnier, M., and Baranton, G. (2007). Multilocus sequence analysis of atypical *Borrelia burgdorferi* sensu lato isolates--description of *Borrelia californiensis* sp. nov., and genomospecies 1 and 2. Int J Med Microbiol *297*, 263-271.

Postic, D., Ras, N.M., Lane, R.S., Hendson, M., and Baranton, G. (1998). Expanded diversity among Californian *borrelia* isolates and description of *Borrelia bissettii* sp. nov. (formerly *Borrelia* group DN127). J Clin Microbiol *36*, 3497-3504.

Preyß-Jägeler, C., Hartmann, K., and Dorsch, R. (2020). Changes in renal parameters and their association with subclinical vector-borne infections in Bernese Mountain dogs. BMC Vet Res *16*, 285. https://doi.org/10.1186/s12917-020-02506-0

Pritt, B.S., Respicio-Kingry, L.B., Sloan, L.M., Schriefer, M.E., Replogle, A.J., Bjork, J., Liu, G., Kingry, L.C., Mead, P.S., Neitzel, D.F., *et al.* (2016). Borrelia mayonii sp. nov., a member of the Borrelia burgdorferi sensu lato complex, detected in patients and ticks in the upper midwestern United States. Int J Syst Evol Microbiol *66*, 4878-4880. https://doi.org/10.1099/ijsem.0.001445

Probert, W.S., Crawford, M., Cadiz, R.B., and LeFebvre, R.B. (1997). Immunization with outer surface protein (Osp) A, but not OspC, provides cross-protection of mice challenged with North American isolates of Borrelia burgdorferi. J Infect Dis *175*, 400-405.

Probert, W.S., and LeFebvre, R.B. (1994). Protection of C3H/HeN mice from challenge with *Borrelia burgdorferi* through active immunization with OspA, OspB, or OspC but not with OspD or the 83-kilodalton antigen. Infect Immun *62*, 1920-1926.

Pulzova, L., Flachbartova, Z., Bencurova, E., Potocnakova, L., Comor, L., Schreterova, E., and Bhide, M. (2016). Identification of B-cell epitopes of Borrelia burgdorferi outer surface protein C by screening a phage-displayed gene fragment library. Microbiol Immunol *60*, 669-677. https://doi.org/10.1111/1348-0421.12438

Ramamoorthi, N., Narasimhan, S., Pal, U., Bao, F., Yang, X.F., Fish, D., Anguita, J., Norgard, M.V., Kantor, F.S., Anderson, J.F., *et al.* (2005). The Lyme disease agent exploits a tick protein to infect the mammalian host. Nature Letters *436*, 573-577.

Ramamoorthy, R., and Philipp, M.T. (1998). Differential expression of Borrelia burgdorferi proteins during growth in vitro. Infect Immun *66*, 5119-5124.

Ramamoorthy, R., and Scholl-Meeker, D. (2001). Borrelia burgdorferi proteins whose expression is similarly affected by culture temperature and pH. Infect Immun *69*, 2739-2742.

Revel, A.T., Talaat, A.M., and Norgard, M.V. (2002). DNA microarray analysis of differential gene expression in Borrelia burgdorferi, the Lyme disease spirochete. Proc Natl Acad Sci U S A *99*, 1562-1567. https://doi.org/10.1073/pnas.032667699

Rhodes, D.V., Earnhart, C.G., Mather, T.N., Meeus, P.F., and Marconi, R.T. (2013). Identification of Borrelia burgdorferi ospC genotypes in canine tissue following tick infestation: Implications for Lyme disease vaccine and diagnostic assay design. Vet J *198*, 412-418.

Richer, L.M., Brisson, D., Melo, R., Ostfeld, R.S., Zeidner, N., and Gomes-Solecki, M. (2014). Reservoir targeted vaccine against Borrelia burgdorferi: a new strategy to prevent Lyme disease transmission. J Infect Dis *209*, 1972-1980.

Roberts, D.M., Caimano, M., McDowell, J., Theisen, M., Holm, A., Orff, E., Nelson, D., Wikel, S., Radolf, J., and Marconi, R.T. (2002). Environmental regulation and differential production of members of the Bdr protein family of Borrelia burgdorferi. Infect Immun *70*, 7033-7041.

Rogers, E.A., Abdunnur, S.V., McDowell, J.V., and Marconi, R.T. (2009a). Comparative analysis of the properties and ligand binding characteristics of CspZ, a factor H binding protein, derived from Borrelia burgdorferi isolates of human origin. Infect Immun *77*, 4396-4405.

Rogers, E.A., Terekhova, D., Zhang, H.M., Hovis, K.M., Schwartz, I., and Marconi, R.T. (2009b). Rrp1, a cyclic-di-GMP-producing response regulator, is an important regulator of *Borrelia burgdorferi* core cellular functions. Mol Microbiol *71*, 1551-1573.

Rousselle, J.C., Callister, S.M., Schell, R.F., Lovrich, S.D., Jobe, D.A., Marks, J.A., and Wieneke, C.A. (1998). Borreliacidal antibody production against outer surface protein C of Borrelia burgdorferi. J Infect Dis *178*, 733-741.

Rudenko, N., Golovchenko, M., Lin, T., Gao, L., Grubhoffer, L., and Oliver, J.H., Jr. (2009). Delineation of a new species of the Borrelia burgdorferi Sensu Lato Complex, Borrelia americana sp. nov. J Clin Microbiol *47*, 3875-3880.

Sadziene, A., Thomas, D.D., and Barbour, A.G. (1995). Borrelia burgdorferi mutant lacking Osp: biological and immunological characterization. Infect Immun *63*, 1573-1580.

Sadziene, A., Wilske, B., Ferdows, M.S., and Barbour, A.G. (1993). The cryptic *ospC* gene of *Borrelia burgdorferi* B31 is located on a circular plasmid. Infect Immun *61*, 2192-2195.

Samuels, D.S., Mach, K., and Garon, C.F. (1994). Genetic transformation of the Lyme disease agent *Borrelia burgdorferi* with coumarin-resistant *gyrB*. J Bacteriol *176*, 6045-6049.

Scheckelhoff, M.R., Telford, S.R., and Hu, L.T. (2006). Protective efficacy of an oral vaccine to reduce carriage of Borrelia burgdorferi (strain N40) in mouse and tick reservoirs. Vaccine *24*, 1949-1957.

Scheiblhofer, S., Weiss, R., Durnberger, H., Mostbock, S., Breitenbach, M., Livey, I., and Thalhamer, J. (2003). A DNA vaccine encoding the outer surface protein C from *Borrelia burgdorferi* is able to induce protective immune responses. Microbes and Infection *5*, 939-946.

Schoen, R.T., Deshefy-Longhi, T., Van-Hoecke, C., Buscarino, C., and Fikrig, E. (2003). An open-label, nonrandomized, single-center, prospective extension, clinical trial of booster dose schedules to assess the safety profile and immunogenicity of recombinant outer-surface protein A (OspA) Lyme disease vaccine. Clin Ther *25*, 210-224.

Schoen, R.T., Meurice, F., Brunet, C.M., Cretella, S., Krause, D.S., Craft, J.E., and Fikrig, E. (1995). Safety and immunogenicity of an outer surface protein A vaccine in subjects with previous Lyme disease. J Infect Dis *172*, 1324-1329. https://doi.org/10.1093/infdis/172.5.1324

Schoen, R.T., Sikand, V.K., Caldwell, M.C., Van Hoecke, C., Gillet, M., Buscarino, C., and Parenti, D.L. (2000). Safety and immunogenicity profile of a recombinant outer-surface protein A Lyme disease vaccine: clinical trial of a 3-dose schedule at 0, 1, and 2 months. Clin Ther *22*, 315-325. https://doi.org/10.1016/s0149-2918(00)80035-1

Schutzer, S.E., Fraser-Liggett, C.M., Casjens, S.R., Qiu, W.G., Dunn, J.J., Mongodin, E.F., and Luft, B.J. (2011). Whole-genome sequences of thirteen isolates of Borrelia burgdorferi. J Bacteriol *193*, 1018-1020.

Schwan, T.G., Burgdorfer, W., and Garon, C.F. (1988). Changes in infectivity and plasmid profile of the Lyme disease spirochete, *Borrelia burgdorferi*, as a result of in vitro cultivation. Infect Immun *56*, 1831-1836.

Schwan, T.G., and Piesman, J. (2000). Temporal changes in outer surface proteins A and C of the lyme disease-associated spirochete, Borrelia burgdorferi, during the chain of infection in ticks and mice. J Clin Microbiol *38*, 382-388.

Schwartz, A.M., Shankar, M.B., Kugeler, K.J., Max, R.J., Hinckley, A.F., Meltzer, M.I., and Nelson, C.A. (2020). Epidemiology and cost of Lyme disease-related hospitalizations among patients with employer-sponsored health insurance-United States, 2005-2014. Zoonoses Public Health *67*, 407-415. https://doi.org/10.1111/zph.12699

Scott, J.D., Anderson, J.F., and Durden, L.A. (2012). Widespread dispersal of Borrelia burgdorferi-infected ticks collected from songbirds across Canada. J Parasitol *98*, 49-59.

Scott, J.D., Clark, K.L., Foley, J.E., Bierman, B.C., and Durden, L.A. (2018). Far-Reaching Dispersal of Borrelia burgdorferi Sensu Lato-Infected Blacklegged Ticks by Migratory Songbirds in Canada. Healthcare (Basel) *6*.

Seinost, G., Dykhuizen, D.E., Dattwyler, R.J., Golde, W.T., Dunn, J.J., Wang, I.N., Wormser, G.P., Schriefer, M.E., and Luft, B.J. (1999a). Four clones of Borrelia burgdorferi sensu stricto cause invasive infection in humans. Infect Immun *67*, 3518-3524.

Seinost, G., Golde, W.T., Berger, B.W., Dunn, J.J., Qiu, D., Dunkin, D.S., Dykhuizen, D.E., Luft, B.J., and Dattwyler, R.J. (1999b). Infection with multiple strains of Borrelia burgdorferi sensu stricto in patients with Lyme disease. Arch Dermatol *135*, 1329-1333.

Shandilya, S., Kurt Yilmaz, N., Sadowski, A., Monir, E., Schiller, Z.A., Thomas, W.D., Jr., Klempner, M.S., Schiffer, C.A., and Wang, Y. (2017). Structural and molecular analysis of a protective epitope of Lyme disease antigen OspA and antibody interactions. J Mol Recognit *30*. https://doi.org/10.1002/jmr.2595

Sigal, L.H., Zahradnik, J.M., Lavin, P., Patella, S.J., Bryant, G., Haselby, R., Hilton, E., Kunkel, M., Adler-Klein, D., Doherty, T., *et al.* (1998). A vaccine consisting of recombinant *Borrelia burgdorferi* outer surface protein A to prevent Lyme disease. New Eng J Med *339*, 216-222.

Simon, M.M., Gern, L., Hauser, P., Zhong, W., Nielsen, P.J., Kramer, M.D., Brenner, C., and Wallich, R. (1996). Protective immunization with plasmid DNA containing the outer surface lipoprotein A gene of Borrelia burgdorferi is independent of an eukaryotic promoter. Eur J Immunol *26*, 2831-2840. https://doi.org/10.1002/eji.1830261206

Singh, P., Verma, D., Backstedt, B.T., Kaur, S., Kumar, M., Smith, A.A., Sharma, K., Yang, X., Azevedo, J.F., Gomes-Solecki, M., *et al.* (2017). Borrelia burgdorferi BBI39 Paralogs, Targets of Protective Immunity, Reduce Pathogen Persistence Either in Hosts or in the Vector. J Infect Dis *215*, 1000-1009. https://doi.org/10.1093/infdis/jix036

Small, C.M., Ajithdoss, D.K., Rodrigues Hoffmann, A., Mwangi, W., and Esteve-Gassent, M.D. (2014). Immunization with a Borrelia burgdorferi BB0172-derived peptide protects mice against lyme disease. PLoS One *9*, e88245. https://doi.org/10.1371/journal.pone.0088245

Smith, A.A., Yang, X., Fikrig, E., and Pal, U. (2018). Artificial Infection of Ticks with Borrelia burgdorferi Using a Microinjection Method and Their Detection In Vivo Using Quantitative PCR Targeting flaB RNA. Methods Mol Biol *1690*, 105-114. https://doi.org/10.1007/978-1-4939-7383-5_9

Stålhammar-Carlemalm, M., Jenny, E., Gern, L., Aeschlimann, A., and Meyer, J. (1990). Plasmid analysis and restriction fragment length polymorphisms of chromosomal DNA allow a distinction between *Borrelia burgdorferi* strains. Zentralblatt für Bakteriologie *274*, 28-39.

Steere, A.C., Drouin, E.E., and Glickstein, L.J. (2011). Relationship between immunity to Borrelia burgdorferi outer-surface protein A (OspA) and Lyme arthritis. Clin Infect Dis *52 Suppl 3*, s259-265.

Steere, A.C., Dwyer, E., and Winchester, R. (1990). Association of chronic Lyme arthritis with HLA-DR4 and HLA-DR2 alleles. N Engl J Med *323*, 219-223.

Steere, A.C., Sikand, V.K., Meurice, F., Parenti, D.L., Fikrig, E., Schoen, R.T., Nowakowski, J., Schmid, C.H., Laukamp, S., Buscarino, C., *et al.* (1998). Vaccination against Lyme disease with recombinant Borrelia burgdorferi outer-surface lipoprotein A with adjuvant. Lyme Disease Vaccine Study Group. N Engl J Med *339*, 209-215.

Stevenson, B., Fingerle, V., Wormser, G.P., and Margos, G. (2018). Public health and patient safety concerns merit retention of Lyme borreliosis-associated spirochetes within the genus Borrelia, and rejection of the genus novum Borreliella. Ticks Tick Borne Dis.

Stillman, B.A., Thatcher, B., Beall, M.J., Lappin, M., O'Connor, T.P., and Chandrashekar, R. (2019). Borrelia burgdorferi Antibody Test Results in Dogs Administered 4 Different Vaccines. Top Companion Anim Med *37*, 100358. https://doi.org/10.1016/j.tcam.2019.100358

Straubinger, R.K., Chang, Y.F., Jacobson, R.H., and Appel, M.J. (1995). Sera from OspA-vaccinated dogs, but not those from tick-infected dogs, inhibit in vitro growth of Borrelia burgdorferi. J Clin Microbiol *33*, 2745-2751.

Sykes, R.A., and Makiello, P. (2017). An estimate of Lyme borreliosis incidence in Western Europe. J Public Health (Oxf) *39*, 74-81. https://doi.org/10.1093/pubmed/fdw017.

Talagrand-Reboul, E., Westermann, B., Raess, M.A., Schnell, G., Cantero, P., Barthel, C., Ehret-Sabatier, L., Jaulhac, B., and Boulanger, N. (2020). Proteomic as an Exploratory Approach to Develop Vaccines Against Tick-Borne Diseases Using Lyme Borreliosis as a Test Case. Vaccines (Basel) *8*. https://doi.org/10.3390/vaccines8030463

Telford, S.R., 3rd, Cunningham, J.A., Waltari, E., and Hu, L. (2011). Nest box-deployed bait for delivering oral vaccines to white-footed mice. Ticks Tick Borne Dis *2*, 151-155. https://doi.org/10.1016/j.ttbdis.2011.06.001

Theisen, M., Frederiksen, B., Lebech, A.M., Vuust, J., and Hansen, K. (1993). Polymorphism in ospC gene of Borrelia burgdorferi and immunoreactivity of OspC protein: implications for taxonomy and for use of OspC protein as a diagnostic antigen. J Clin Microbiol *31*, 2570-2576.

Tilly, K., Casjens, S., Stevenson, B., Bono, J.L., Samuels, D.S., Hogan, D., and Rosa, P. (1997). The Borrelia burgdorferi circular plasmid cp26: conservation of plasmid structure and targeted inactivation of the ospC gene. Mol Microbiol *25*, 361-373.

Tokarz, R., Anderton, J.M., Katona, L.I., and Benach, J.L. (2004). Combined effects of blood and temperature shift on Borrelia burgdorferi gene expression as determined by whole genome DNA array. Infect Immun *72*, 5419-5432.

Topfer, K.H., and Straubinger, R.K. (2007). Characterization of the humoral immune response in dogs after vaccination against the Lyme borreliosis agent A study with five commercial vaccines using two different vaccination schedules. Vaccine *25*, 314-326.

Trollmo, C., Meyer, A.L., Steere, A.C., Hafler, D.A., and Huber, B.T. (2001). Molecular mimicry in Lyme arthritis demonstrated at the single cell level: LFA-1 alpha L is a partial agonist for outer surface protein A-reactive T cells. J Immunol *166*, 5286-5291. https://doi.org/10.4049/jimmunol.166.8.5286

Tsao, J., Barbour, A.G., Luke, C.J., Fikrig, E., and Fish, D. (2001). OspA immunization decreases transmission of Borrelia burgdorferi spirochetes from infected Peromyscus leucopus mice to larval Ixodes scapularis ticks. Vector Borne Zoonotic Dis *1*, 65-74.

Tsao, J.I., Wootton, J.T., Bunikis, J., Luna, M.G., Fish, D., and Barbour, A.G. (2004). An ecological approach to preventing human infection: vaccinating wild mouse reservoirs intervenes in the Lyme disease cycle. Proc Natl Acad Sci U S A *101*, 18159-18164.

Tseng, Y.J., Cami, A., Goldmann, D.A., DeMaria, A., Jr., and Mandl, K.D. (2015). Using Nation-Wide Health Insurance Claims Data to Augment Lyme Disease Surveillance. Vector Borne Zoonotic Dis *15*, 591-596.

Van Hoecke, C., Comberbach, M., De Grave, D., Desmons, P., Fu, D., Hauser, P., Lebacq, E., Lobet, Y., and Voet, P. (1996). Evaluation of the safety, reactogenicity and immunogenicity of three recombinant outer surface protein (OspA) lyme vaccines in healthy adults. Vaccine *14*, 1620-1626. https://doi.org/10.1016/s0264-410x(96)00146-6

Verdugo, C., Jiménez, O., Hernández, C., Álvarez, P., Espinoza, A., and González-Acuña, D. (2017). Infection with Borrelia chilensis in Ixodes stilesi ticks collected from Pudu puda deer. Ticks Tick Borne Dis *8*, 733-740. https://doi.org/10.1016/j.ttbdis.2017.05.007

Voordouw, M.J., Tupper, H., Önder, Ö., Devevey, G., Graves, C.J., Kemps, B.D., and Brisson, D. (2013). Reductions in human Lyme disease risk due to the effects of oral vaccination on tick-to-mouse and mouse-to-tick transmission. Vector Borne Zoonotic Dis *13*, 203-214. https://doi.org/10.1089/vbz.2012.1003

Wagemakers, A., Mason, L.M., Oei, A., de Wever, B., van der Poll, T., Bins, A.D., and Hovius, J.W. (2014). Rapid outer-surface protein C DNA tattoo vaccination protects against Borrelia afzelii infection. Gene Ther.

Wagner, B., Johnson, J., Garcia-Tapia, D., Honsberger, N., King, V., Strietzel, C., Hardham, J.M., Heinz, T.J., Marconi, R.T., and Meeus, P.F. (2015). Comparison of effectiveness of cefovecin, doxycycline, and amoxicillin for the treatment of experimentally induced early Lyme borreliosis in dogs. BMC Vet Res *11*, 163.

Wallich, R., Siebers, A., Jahraus, O., Brenner, C., Stehle, T., and Simon, M.M. (2001). DNA vaccines expressing a fusion product of outer surface proteins A and C from *Borrelia burgdorferi* induce protective antibodies suitable for prophylaxis but not for resolution of Lyme disease. Infect Immun *69*, 2130-2136.

Wang, G., van Dam, A.P., Le Fleche, A., Postic, D., Peter, O., Baranton, G., de Boer, R., Spanjaard, L., and Dankert, J. (1997). Genetic and phenotypic analysis of *Borrelia valaisiana* sp. nov. (*Borrelia* genomic groups VS116 and M19). Int J Syst Bacteriol *47*, 926-932.

Wang, I.N., Dykhuizen, D.E., Qiu, W., Dunn, J.J., Bosler, E.M., and Luft, B.J. (1999). Genetic diversity of *ospC* in a local population of *Borrelia burgdorferi* sensu stricto. Genetics *151*, 15-30.

Will, G., Jaruis-Heipke, S., Schwab, E., Busch, U., Rossler, D., Soutschek, E., Wilske, B., and Preac-Mursic, V. (1995). Sequence analysis of the *ospA* genes shows homogeneity within *Borrelia burgdorferi* sensu stricto and *Borrelia afzelii* strains but reveals major subgroups within the *Borrelia garinii* species. Med Microbiol Immunol *184*, 73-80.

Wilske, B., Busch, U., Fingerle, V., Jauris-Heipke, S., Preac Mursic, V., Rossler, D., and Will, G. (1996). Immunological and molecular variability of OspA and OspC. Implications for Borrelia vaccine development. Infection *24*, 208-212.

Wilske, B., Jauris-Heipke, S., Lobentanzer, R., Pradel, I., Preac-Mursic, V., Rossler, D., Soutschek, E., and Johnson, R.C. (1995). Phenotypic analysis of outer surface protein C (OspC) of Borrelia burgdorferi sensu lato by monoclonal antibodies: relationship to genospecies and OspA serotype. J Clin Microbiol *33*, 103-109.

Wilske, B., Preac-Mursic, V., Gobel, U.B., Graf, B., Jauris, S., Soutschek, E., Schwab, E., and Zumstein, G. (1993a). An OspA serotyping sustem for *Borrelia burgdorferi* based on reactivity with monoclonal antibodies and OspA sequence analysis. J Clin Microbiol *31*, 340-350.

Wilske, B., Preac-Mursic, V., Jauris, S., Hofmann, A., Pradel, I., Soutschek, E., Schwab, E., Will, G., and Wanner, G. (1993b). Immunological and molecular polymorphisms of OspC, an immunodominant major outer surface protein of *Borrelia burgdorferi*. Infect Immun *61*, 2182-2191.

Wood, E., Tamborero, S., Mingarro, I., and Esteve-Gassent, M.D. (2013). BB0172, a Borrelia burgdorferi outer membrane protein that binds integrin α3β1. J Bacteriol *195*, 3320-3330. https://doi.org/10.1128/jb.00187-13

Wressnigg, N., Barrett, P.N., Pollabauer, E.M., O'Rourke, M., Portsmouth, D., Schwendinger, M.G., Crowe, B.A., Livey, I., Dvorak, T., Schmitt, B., *et al.* (2014). A Novel multivalent OspA vaccine against Lyme borreliosis is safe and immunogenic in an adult population previously infected with Borrelia burgdorferi sensu lato. Clin Vaccine Immunol *21*, 1490-1499.

Wressnigg, N., Pöllabauer, E.M., Aichinger, G., Portsmouth, D., Löw-Baselli, A., Fritsch, S., Livey, I., Crowe, B.A., Schwendinger, M., Brühl, P., *et al.* (2013). Safety and immunogenicity of a novel multivalent OspA vaccine against Lyme borreliosis in healthy adults: a double-blind, randomised, dose-escalation phase 1/2 trial. Lancet Infect Dis *13*, 680-689. https://doi.org/10.1016/s1473-3099(13)70110-5

Xiang, X., Yang, Y., Du, J., Lin, T., Chen, T., Yang, X.F., and Lou, Y. (2017). Investigation of ospC Expression Variation among Borrelia burgdorferi Strains. Front Cell Infect Microbiol 7, 131.

Yancey, C.B., Hegarty, B.C., Qurollo, B.A., Levy, M.G., Birkenheuer, A.J., Weber, D.J., Diniz, P.P., and Breitschwerdt, E.B. (2014). Regional seroreactivity and vector-borne disease co-exposures in dogs in the United States from 2004-2010: utility of canine surveillance. Vector Borne Zoonotic Dis *14*, 724-732.

Yang, X.F., Lybecker, M.C., Pal, U., Alani, S.M., Blevins, J., Revel, A.T., Samuels, D.S., and Norgard, M.V. (2005). Analysis of the ospC regulatory element controlled by the RpoN-RpoS regulatory pathway in Borrelia burgdorferi. J Bacteriol *187*, 4822-4829.

Yang, X.F., Pal, U., Alani, S.M., Fikrig, E., and Norgard, M.V. (2004). Essential role for OspA/B in the life cycle of the Lyme disease spirochete. J Exp Med *199*, 641-648.

Yound, J.D. (1998). Underreporting of Lyme disease. N Eng J Med *338*, 1629.

Zhang, H., and Marconi, R.T. (2005). Demonstration of cotranscription and 1-methyl-3-nitroso-nitroguanidine induction of a 30-gene operon of Borrelia burgdorferi: evidence that the 32-kilobase circular plasmids are prophages. J Bacteriol *187*, 7985-7995.

Zundorf, I., and Dingermann, T. (2008). [Death of a vaccine--the fall of LYMErix]. Pharm Unserer Zeit *37*, 38-39.

Chapter 20

Lyme Borreliosis in Domestic Animals

Stephen W. Barthold*

Department of Pathology, Microbiology and Immunology, University of California Davis School of Veterinary Medicine, Davis, California, USA
*Corresponding author: swbarthold@ucdavis.edu

DOI: https://doi.org/10.21775/9781913652616.20

Abstract
Several domestic animals, including dogs, cats, horses, cows and small ruminants are frequent incidental hosts for infection with *Borrelia burgdorferi* sensu lato (s.l.). Subclinical infection is quite common, but all species covered in this chapter may develop clinical disease. Disease manifestations are variable and often transient, but arthritis is common. Neuroborreliosis has been described in dogs and is a particularly noteworthy syndrome in horses. The dog has served as an experimental model, and considerable information on *B. burgdorferi* pathogenesis, therapy and prevention is available for this species.

Introduction
As other chapters of this text have documented, *Borrelia burgdorferi* s.l. utilizes a remarkable array of hosts in its complex life cycle. Humans and their domestic animals are among the incidental hosts for these bacteria. Much of what is known, unknown, and speculated about Lyme disease in humans likewise pertains to the challenges of diagnosing and treating Lyme borreliosis in domestic animals. Understanding the manifestations of Lyme borreliosis in domestic animals has been extrapolated from experimental studies in animals, clinical case reports, and comparative clinical observations of human Lyme disease. However, animals do not express "symptoms" (**subjective** evidence of disease or physical disturbance observed by the patient), so discussion of Lyme borreliosis in animals is restricted to known clinical "signs" (**objective** evidence of a disease or physical disturbance recognized by the physician/veterinarian, nurse, or family member/owner). Animals, which do not understand illness, are by nature stoic and, thus, tend to obscure clinical signs that are readily recognized by their human counterparts. Accordingly, although exposure of domestic animals to *B. burgdorferi* is common in Lyme-endemic areas, clinical signs of Lyme borreliosis are relatively uncommon or unrecognized. Recognition of clinical signs is most likely to occur with animals that live in close association with their owners. Pet owners also should be aware that pet ownership increases human risk of encountering ticks, and possibly acquiring Lyme disease (Curran and Fish, 1989; Jones et al., 2018; Steere et al., 1978)

Domestic animals that are featured in this chapter include dogs, cats, horses, cattle and small ruminants. Naturally occurring, clinically apparent Lyme borreliosis has not been reported in other domestic species. As in humans with Lyme disease, the clinical manifestations of Lyme borreliosis in animals are highly variable, and depend upon age, immune status, species, and genetic background (breed) of the host, the strain or species of *Borrelia*, and co-infection with other tick-borne agents, particularly *Anaplasma phagocytophilum*. This chapter only covers *B. burgdorferi* infection.

Canine Lyme Borreliosis
Dogs are at high risk of acquiring tick infestations and, as a result, have a very high *B. burgdorferi* seroprevalence in Lyme disease-endemic areas in the US, Europe and Asia (Arashima, 1991; Barth et al., 2012; Bowman et al., 2009; Burgess, 1986b; Eng et al., 1988; Greene et al., 1988a; Levy, 2002; Levy and Magnarelli, 1992; Little et al., 2014; Magnarelli et al., 1984; Magnarelli et al., 1985; Magnarelli et al., 1990b; Magnarelli et al., 1987; Schulze et al., 1986). In Europe, dogs become naturally infected with one or more *B. burgdorferi* s.l. species, including *B. burgdorferi* s.s., *B. afzelii* and *B. garinii*, with *B. burgdorferi* s.s. and *B. garinii* being the most frequent (Barth et al., 2012; Hovius et al., 1999; Kybicova et

Figure 1. Acute arthritis in a spaniel from Connecticut with naturally acquired Lyme borreliosis. Note swelling of the rear leg and unwillingness to bear weight on the affected limb. Photo courtesy of Steven A. Levy.

al., 2009; Skotarczak, 2014; Speck et al., 2001; Zygner et al., 2009). In addition, dogs may be susceptible to other *Borrelia* spp. Spirochetemia due to a non-*B. burgdorferi* species was described in two Florida dogs with clinical signs resembling Lyme borrelosis (Breitschwerdt et al., 1994).

Seroconversion to *B. burgdorferi* is not predictive for development of clinical disease in dogs. In one clinical study involving 125 seropositive and 109 seronegative dogs, the development of limb/joint signs with lethargy, fever and anorexia (typical of clinical Lyme borreliosis in dogs) was equally prevalent among seropositive (4.8%) and seronegative (4.6%) dogs (Levy and Magnarelli, 1992). A higher prevalence of intermittent limb/joint disorders was noted in other reports, including 19 of 52 seropositive dogs (Magnarelli et al., 1985) and 127 of 175 seropositive dogs (Magnarelli et al., 1990b). This variance is likely due to a number of factors, but it is generally believed that less than 10% of infected (seropositive) dogs develop clinically recognized disease (Sykes, 2014). Regardless, dogs appear to be the most susceptible domestic animal to clinical Lyme borreliosis, perhaps because of their close affiliation with owners who are apt to recognize clinical illness in their pets. In addition, dogs are not only susceptible to *B. burgdorferi* infection, they also can serve as competent reservoir hosts that can transmit *B. burgdorferi* to feeding ticks, thereby increasing the risk of human exposure to infected ticks (Mather et al., 1994). Indeed, following experimental tick-borne infection, ear punch biopsies were consistently culture-positive for *B. burgdorferi* in all dogs as early as 6 days after tick feeding (Bosler et al., 1992). However, repeated infestation of dogs with *Ixodes scapularis* ticks has been shown to induce protective immunity against tick feeding (Gebbia et al., 1995).

Not surprisingly, there is a wide range of opinions among veterinarians and the lay public regarding canine Lyme borreliosis. Similar to the *Infectious Disease Society of America's* clinical guidelines for humans (Wormser et al., 2006), the *American College of Veterinary Internal Medicine* (ACVIM) has published consensus statements regarding diagnosis, treatment, and prevention (including vaccination) in dogs (Littman et al., 2018; Littman et al., 2006). In spite of differing opinions regarding diagnosis, treatment, and prevention of Lyme borreliosis in dogs, there is a foundation of experimental studies performed in dogs that offers considerable insight into *B. burgdorferi* infection and disease in dogs. The dog has been an important experimental model for studying pathogenesis and persistent infection in Lyme borreliosis, response to antibiotics, immune response, and vaccine response.

As in humans, clinical signs of canine Lyme borrelosis, when present, are quite variable. Erythema migrans has been reported in dogs, but is quite rare (Appel, 1990). The most frequent findings are "limb/joint disorders", including sudden onset of single or intermittent episodes of transient lameness, hot swollen joints, and pain involving the carpus, tarsus, digits, elbows, shoulders, stifles (knees) and cervical or lumbar spine (Figure 1). Episodes may be brief (lasting less than 10 days) or more prolonged (12 or more weeks). Episodes are typically accompanied by fever (102-105.8 ^{0}F), lethargy, anorexia, and lymphadenopathy (Appel, 1990; Bosler et al., 1984; Giudice et al., 2003; Kornblatt et al., 1985; Levy and Dressen, 1992; Levy and Magnarelli, 1992; Lissman et al., 1984; Madigan and Teitler, 1988; Magnarelli et al., 1985; Magnarelli et al., 1990b; Magnarelli et al., 1987; May et al., 1990; Schulze et al., 1986).

The interval between episodes has been reported to vary from 1 to 23 months (Kornblatt et al., 1985). Aspiration of fluid from acutely swollen joints may reveal an exudate that is acellular to rich in neutrophils, but spirochetes are only rarely found (Kornblatt et al., 1985). Some dogs may manifest intermittent non-specific lameness with no or low fever, mild anorexia and lethargy that is responsive to antibiotics (Levy and Dressen, 1992). Rheumatoid arthritis has been reported in two dogs following initial *B. burgdorferi* infection (Roush et al., 1989), but there is insufficient proof of such an association. A study in the Netherlands that performed PCR on multiple tissues of naturally infected dogs suggested that *B. burgdorferi* s.s. and *B. garinii* differed in their organotropism and that *B. burgdorferi* s.s. was most often associated with limb/joint signs (Hovius et al., 1999).

Because of the non-specific nature of clinical signs, diagnosis is determined by history of tick exposure in endemic areas (including serology), clinical signs, consideration of other differentials, and response to antibiotics (Littman et al., 2006). The caveat with this concept is that clinical signs are often, if not typically, transient and resolve spontaneously (despite continuing infection). A euphemism within the profession is "hurry up and treat the dog before it gets better." The spontaneous resolution of clinical signs and high rate of clinically normal seropositive dogs within endemic areas pose a challenge regarding the necessity of treatment (or vaccination) in such cases.

Neurologic signs may occur in dogs with Lyme borreliosis, but they are generally not well defined. Anecdotal accounts suggest that antibiotic treatment of some clinically normal seropositive dogs has resulted in beneficial behavioral changes. This observation may have merit. In two studies involving laboratory beagles, some of the dogs manifested aggressive behavior following infection (Chu et al., 1992; Evans et al., 1995a). Meningitis has been described in which there is sudden onset of fever with severe pain that is not localized to joints, anorexia, depression and reluctance to move. Manipulation of the head and neck elicits pain. Lymph nodes (popliteal and/or prescapular), which drain the site of the tick bite, may be enlarged (Appel, 1990; Levy and Dressen, 1992). Depression, disorientation, stumbling, head tilt and seizures have been noted in seropositive dogs, but cause and effect have not been proven (Cohen et al., 1990). A number of neurologic signs, including inability to stand (astasia), tonic convulsions, and hyperreflexia were noted in two seropositive dogs in Japan with a history of *Ixodes persulcatus* tick exposure. *Borrelia burgdorferi* was isolated from ticks removed from one of the dogs. Both dogs responded rapidly to antibiotic therapy, with complete resolution of clinical signs (Azuma et al., 1993). In Europe, *B. garinii* was detected by PCR in the blood of a dog with meningoencephalitis (Kybicova et al., 2009). Facial paralysis (Bell's palsy) involving the seventh cranial nerve is common in older dogs, but an association with *B. burgdorferi* has not been described in dogs.

Ophthalmic disorders have been reported in dogs with Lyme borreliosis, including conjunctivitis, corneal edema, anterior uveitis, chorioretinitis, retinal hemorrhage or detachment (Cohen et al., 1990; Dziezyc, 2000; Munger, 1990). The role of *B. burgdorferi* as the cause of these entities is not certain in dogs, but is well documented in human Lyme disease (Lesser, 1995). A case of orbital myositis, with exophthalmos, protrusion of the nictitating membrane and lateral strabismus was reported in a seropositive dog in Spain. The dog was euthanized and histopathology revealed myositis with argyrophilic spirochetal forms in the interstitium (Raya et al., 2010). Orbital myositis with similar features has been reported in human Lyme disease (Fatterpekar et al., 2002; Holak et al., 2006; Seidenberg and Leib, 1990).

Cardiac disease appears to be rare in dogs. Myocarditis with complete heart block was reported in a dog that presented with repeated syncope unrelated to exercise. Upon necropsy, the heart was grossly dilated and the lungs were filled with blood-tinged froth, indicative of heart failure. Histology revealed necrotizing myocarditis and infiltration of mixed leukocytes. Rare argyrophilic spirochetal forms were observed in the interstitium (Levy and Duray, 1988). Conduction defects and similar histologic features have been described in human cases of Lyme carditis (Marcus et al., 1985b; Muehlenbachs et al., 2016; Robinson et al., 2015).

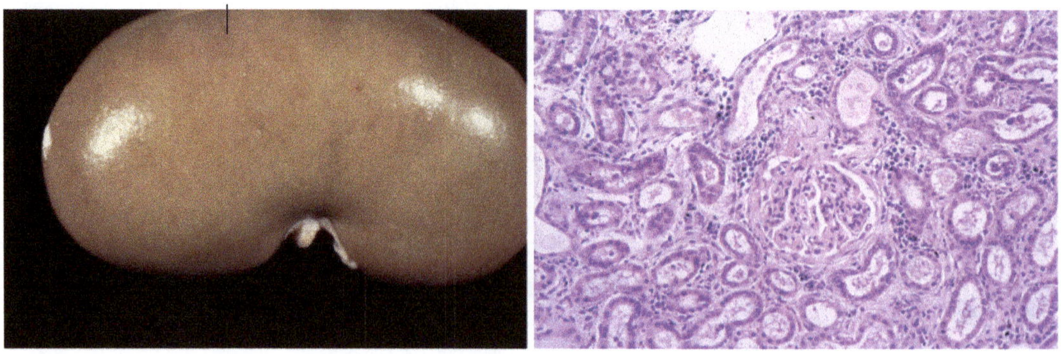

Figure 2. Lyme nephritis in a naturally infected dog from Pennsylvania. The kidney (left image) is pale and histopathology (right image) reveals glomerulonephritis, interstitial nephritis and tubular degeneration. Image courtesy of Donna M. Dambach.

Renal disease ("Lyme nephritis") is a clinically significant and widely recognized problem in a minority of canine Lyme borreliosis cases. Renal impairment was noted in 5 of 250 (2%) clinically ill seropositive dogs in the very first reports of canine Lyme borreliosis in the northeastern US (Grauer et al., 1988; Magnarelli et al., 1987). Affected dogs become acutely ill with severe, rapidly progressive protein-losing nephropathy, which may or may not be associated with concomitant limb/joint disease. The duration of clinical illness ranges from 24 hours to 8 weeks, with progressive deterioration despite treatment. There is a breed predisposition among Labrador retrievers and Golden retrievers, but a variety of other breeds may also develop Lyme nephritis. Age at onset (5.6 $\pm$ 2.6 years) is younger than age at onset for other types of glomerulonephritis (7.8 $\pm$ 3.5 years) in dogs (Dambach et al., 1997).

Because Lyme nephritis is relatively common and nearly always fatal, clinical chemistry and histopathology have been well described. Affected dogs are seropositive for *B. burgdorferi* and present with severe proteinuria, hypoalbuminemia, azotemia and other markers of nephrotic syndrome. The underlying cause is an immune complex membranoproliferative glomerulonephritis. At necropsy, kidneys are pale and histopathology reveals diffuse glomerulonephritis (Figure 2) with sub-endothelial *B. burgdorferi*-specific antigen-antibody deposition, including C3, IgG and IgM deposits. Acute tubular necrosis, nonsuppurative interstitial nephritis, arteriolar fibrinoid necrosis, and deposition of acellular hyaline material in arteriolar walls are variably present, as well as secondary lesions in distant tissues consistent with uremia (Chou et al., 2006; Dambach et al., 1997; Magnarelli et al., 1987). Ultrastructural changes consist of fusion of glomerular epithelial foot processes, expansion of the glomerular basement membrane with flocculent material, plexiform splitting, and subendothelial electron-dense deposits (Dambach et al., 1997). A number of *B. burgdorferi* antigens have been identified by immunohistochemistry of glomeruli, including p83, p34, OspA, OspB and lower molecular weight antigens (Chou et al., 2006; Grauer et al., 1988; Magnarelli et al., 1987). Elution of the antigen-antibody complexes identified p39, flagellin, p22 and small amounts of OspA (Littman, 2013). Few, if any, spirochetes have been found in association with Lyme nephritis in dogs, based upon silver stain, PCR, and fluorescence *in situ* hybridization (FISH) (Chou et al., 2006; Dambach et al., 1997; Grauer et al., 1988; Hutton et al., 2008). The clinical presentation of Lyme nephritis in dogs is typically severe, and it is uncertain if milder forms of the disease occur (Littman, 2013). In one study, 6 of 18 seropositive dogs with Lyme nephritis had been vaccinated with a commercial *B. burgdoreri* bacterin (Dambach et al., 1997).

The key diagnostic features for canine Lyme nephritis, in addition to the signalment of severe progressive renal disease, are *B. burgdorferi* seropositivitiy and proteinuria. Aggressive supportive therapy, including antimicrobial and immunosuppressive therapy has been recommended, but prognosis is consistently poor (Goldstein et al., 2013; Littman, 2013; Sykes, 2014). Lyme nephritis appears to be typically severe. Attempts to detect milder forms of the disease by measuring proteinuria in 268 subclinical, seropositive Labrador and Golden Retrievers found that microalbuminuria is not a consequence of *B. burgdorferi* infection (Goldstein et al., 2007). Bernese Mountain dogs are susceptible to development of a familial glomerulonephritis. An association with *B. burgdorferi* sensu lato infection was suspected, but later not confirmed (Gerber et al., 2009; Gerber et al., 2010; Gerber et al., 2007).

Although certainly more rare than Lyme nephritis in dogs, immune-complex membranoproliferative glomerulonephritis with renal failure has also been documented in humans with Lyme disease. Clinical and histologic features are similar to those described in dogs, but fortunately, the reported human cases appear to have been more responsive to therapy (Kelly et al., 1999; Kirmizis and Efstratiadis, 2013; McCausland et al., 2011; Papineni et al., 2010; Rolla et al., 2014; Schneider et al., 2013).

Dogs as Experimental Models of Lyme Borreliosis

Experimental studies in laboratory dogs have contributed significantly to understanding Lyme borreliosis in a non-rodent species, with very close similarity to human Lyme disease (Duray, 1989). Purpose-bred, specific pathogen free beagles have been utilized in nearly all experimental infection studies, with the exception of a single study that used purpose-bred mongrel dogs (Wasmoen et al., 1992). Recognizing the significant degree of breed predisposition to many diseases among dogs (including Lyme borreliosis), experimental studies in beagles provide a limited perspective of the full ramifications of infection in this species.

An initial study demonstrated that dogs were susceptible to infection with low-passage, cultured *B. burgdorferi*, but the infected dogs manifested no clinical signs. Spirochetemia could not be detected, except when infected dogs were treated with dexamethasone (Burgess, 1986a; Chang et al., 2001; Krimer et al., 2011; Straubinger et al., 2000). Based upon IFA seroconversion (*B. burgdorferi* lysate antigen), a contact control dog seroconverted, suggesting that contact transmission may have occurred between dogs (Burgess, 1986a). Another study involved inoculation of 4 beagle dogs with a cultured European isolate of *B. burgdorferi* (presumably s.s.). *Borrelia burgdorferi* was cultured from the blood and urine of one of the dogs, but no clinical signs were observed. Furthermore, that study also noted seroconversion of a contact control dog (Cerri et al., 1994), whereas another study found no evidence of contact transmission (Appel et al., 1993). The issue of contact transmission among dogs has not been further documented. Intrauterine transmission was shown among ten pregnant beagles inoculated multiple times with cultured *B. burgdorferi* s. s. throughout pregnancy. Eight of the inoculated dogs became infected (based upon culture and PCR of tissues and blood) and subsequently whelped. Spirochetes were cultured and detected by PCR in tissues of at least one neonatal or 6-week old pup from each of the eight litters (Gustafson et al., 1993). In contrast, no vertical transmission was found in beagle pups born to tick-inoculated dams (Appel et al., 1993). Dogs have endotheliochorial placentation, in contrast to humans with hemochorial placentation (Wooding and Burton, 2008).

Although the above studies failed to induce clinical signs, clinical signs were readily induced in a series of three studies utilizing mixed breed dogs inoculated with cultured *B. burgdorferi* obtained from a tick in Wisconsin. Dogs were inoculated by a combination of intraperitoneal, subcutaneous and intradermal routes on 7 consecutive days. Most of the dogs developed spirochetemia at various intervals after infection, and became febrile, depressed and anorectic. Most of the dogs developed transient and intermittent lameness, as well as milder signs of limb/joint disorder, including stiffness, limping and "walking on eggshells." One dog with relatively severe clinical illness was euthanized and necropsied at day 25. *B. burgdorferi* was isolated from a variety of tissues, including CSF, mesenteric lymph node, spleen, heart, meninges, kidney and vitreous fluid (Wasmoen et al., 1992).

A number of subsequent studies characterized the course of Lyme borreliosis in purpose-bred laboratory beagles infected by feeding naturally infected nymphal and adult *Ixodes scapularis* ticks. These studies, which utilized vector-borne infection, have proven to be an optimal model of Lyme borreliosis. Infected dogs developed disseminated persistent infection, intermittent limb/joint disorders and other manifestations that closely mimicked natural disease in dogs (Appel et al., 1993; Callister et al., 2000; Chang et al., 2001; Chang et al., 1996; Evans et al., 1995a; Evans et al., 1995b; Korshus et al., 2004; Krimer et al., 2011; Straubinger et al., 1997a; Summers et al., 2005; Susta et al., 2012). Furthermore, the well-characterized dog model has allowed investigation of immune response, antibiotic efficacy and vaccination protection studies (discussed below). Development of clinical signs was found to be related to the age of the dogs when

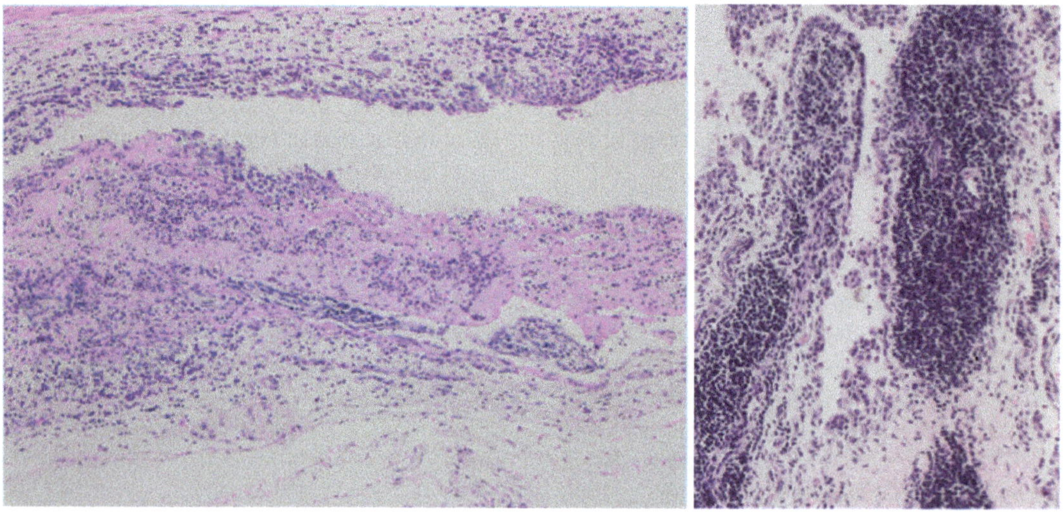

Figure 3. Acute synovitis (left image) with infiltration of leukocytes and fibrin. Chronic involvement (right image) with villous hyperplasia. Villous projections have intense nodular infiltration of mononuclear leukocytes. Images courtesy of Brian A. Summers.

exposed to infected ticks, with 6-week old pups being most susceptible to lameness compared to older dogs. Dogs infected at greater than 6 months of age became infected, but did not develop lameness. When observed, lameness arose at 2 to 7 months after tick exposure, and was intermittent, oligoarticular, and persisted for only around three to four days per episode. Most dogs had two episodes of lameness and a few had three (Appel et al., 1993; Callister et al., 2000; Chu et al., 1992). Affected joints were slightly swollen, with an increase in synovial fluid.

Histopathologic lesions that have been documented in dogs include acute fibrinopurulent synovitis (Figure 3) in dogs with acute lameness, whereas more chronic inflammatory changes were observed in some joints, consisting of villous proliferation and nodular infiltration of predominantly lymphocytes and plasma cells (Figure 3).

Lymphadenomegaly was observed most often in the axillary and popliteal lymph nodes that were located in the somatic quadrant of the arthritic joints, with marked cortical lymphoid hyperplasia, enlarged follicles and hypercellular parafollicular regions. Nonsuppurative inflammatory changes were observed in the skin near tick feeding sites and, to a lesser extent, distant sites. Dogs remained persistently infected for at least a year, and *B. burgdorferi* was cultured and detected by PCR in a wide variety of tissues, with a predilection for connective tissue. Furthermore, spirochetes were detected in tissues of the somatic quadrant nearest to the tick-feeding site (Appel et al., 1993; Chang et al., 2001; Chang et al., 1996; Straubinger et al., 1997a; Summers et al., 2005; Susta et al., 2012).

A similar study observed increased aggressive behavior, transient lameness and mild gait abnormalities, beginning approximately 6 months after tick-borne exposure. Minimal histopathologic lesions were found in the persistently infected dogs that were necropsied at 6 and 12 months after exposure. One lesion of note was renal glomerular mesangial proliferation, which was found in infected, but not control dogs (Evans et al., 1995a; Evans et al., 1995b). Other lesions that have been observed in experimentally infected beagles included mild focal meningitis, mild to moderate lymphoplasmacytic choroid plexitis, focal lymphoplasmacytic peripheral neuritis (Figure 4) and focal lymphoplasmacytic perivascululitis (Chang et al., 2001; Krimer et al., 2011; Straubinger et al., 1998; Summers et al., 2005; Susta et al., 2012).

Antibiotic therapy of dogs with Lyme borreliosis

Based upon the human literature, most veterinarians recommend treatment of canine Lyme borreliosis with tetracycline derivatives or amoxicillin. Doxycycline is often the drug of choice (10 mg/kg PO s.i.d. for a minimum of 1 month) for dogs because it is inexpensive, anti-inflammatory and also effective against other primary or co-infections that occur in

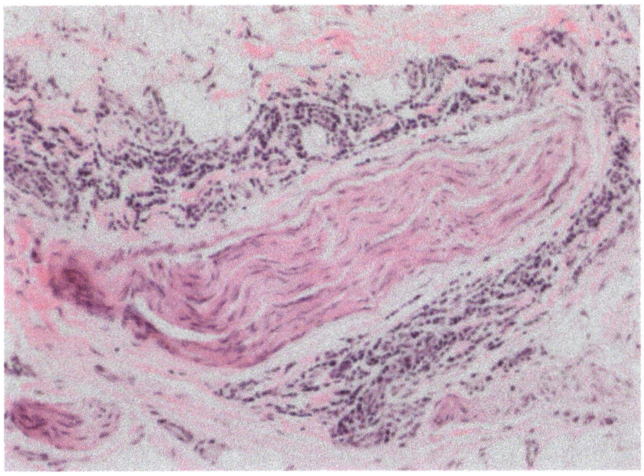

Figure 4. Peripheral neuritis in an experimentally infected beagle dog. The perineurium is infiltrated with mononuclear leukocytes. Similar infiltrates can be found in the adventitia of peripheral vessels. Image courtesy of Brian A. Summers.

dogs (*Anaplasma* spp., *Ehrlichia* spp., *Leptospira* spp., *Rickettsia* spp.) (Littman et al., 2006). Measurement of treatment success is challenging, as clinical signs are often transient, regardless of treatment. Some veterinarians follow a waning antibody response as indicative of treatment success (Wagner et al., 2011; Wagner et al., 2012; Wagner et al., 2013; Wagner et al., 2015). Treatment of subclinical, seropositive dogs (which are likely to be infected) is a subject of debate.

A number of studies evaluated the response to antibiotic treatment of both naturally infected and experimentally infected dogs. An early clinical study involved 69 naturally infected dogs of various breeds with limb/joint disorder that were treated with tetracycline, compared to 58 dogs with joint involvement that did not receive antibiotic. Limb/joint signs resolved in all dogs of both groups within 3 weeks. Furthermore, antibody titers (IgM and IgG) against *B. burgdorferi* lysate ELISA persisted following treatment or no treatment (Magnarelli et al., 1990b).

Experimental studies on the efficacy of antibiotic treatment have utilized purpose-bred, specific pathogen free beagles tick-inoculated with *B. burgdorferi* s.s. (Wagner et al., 2015). A series of seminal publications by Straubinger *et al.* (Straubinger, 2000; Straubinger et al., 2000; Straubinger et al., 1998; Straubinger et al., 1997b) examined the response of *B. burgdorferi*-infected beagles treated with 30 days of amoxicillin, azithromycin, ceftriaxone or doxycycline compared to untreated, infected controls. Response to treatment was followed clinically and by culture and quantitative *ospA* PCR of various tissues, histopathology of joints, and serology (*B. burgdorferi* lysate ELISA). With few exceptions, the results collectively found that regardless of antibiotic, treatment resulted in elimination of *B. burgdorferi* from all tissues based upon culture. Antibody titers declined over the course of 6 months, but began to rise again thereafter. Furthermore, the results demonstrated a general reduction in the amount of detectable *B. burgdorferi* DNA in tissues of treated dogs compared to untreated controls. However, DNA persisted at low levels in multiple tissues of treated dogs for months following treatment. Antibiotic-treated dogs that received corticosteroids prior to necropsy had DNA-positive tissues but did not become culture-positive.

Another study compared the effectiveness of cefovecin, doxycycline, and amoxicillin for the treatment of infected beagles. Based upon detection of *B. burgdorferi flaB* DNA by PCR of skin biopsy samples only, treated dogs became PCR-negative and antibody reactivity against *B. burgdorferi* C6, OspC and OspF declined following treatment (Wagner et al., 2015). It should be noted that skin biopsy PCR is an insufficient index for such trials, as skin biopsies are consistently negative by PCR following treatment of dogs with antibiotics, whereas other tissues often are positive. Furthermore, decline in antibody titers occurs in response to antibiotics, despite persisting *B. burgdorferi* DNA (Straubinger, 2000; Straubinger et al., 2000; Straubinger et al., 1998; Straubinger et al., 1997b). DNA is rapidly cleared following injection of killed spirochetes in dogs (Straubinger, 2000).

Studies in the dog model heralded similar findings of non-cultivable *B. burgdorferi* and persisting *B. burgdorferi* DNA following antibiotic treatment in laboratory mice (Barthold et al., 2010; Bockenstedt et al., 2002; Hodzic et al., 2008; Hodzic et al., 2014; Yrjanainen et al., 2010; Yrjanainen et al., 2007), horses (Chang et al., 2005), and rhesus macaques (Embers et al., 2012). The biological behavior of persisting *B. burgdorferi* DNA cannot be summarily dismissed as "DNA debris", but this issue is beyond the scope of this chapter.

Vaccination of dogs

There are, or have been, a number of USDA-licensed Lyme disease vaccines available for use in dogs in the United States. The reader is referred to the chapter on vaccines in this text. Canine vaccines include a monovalent bacterin (LymeVax, Fort Dodge Animal Health), bivalent bacterins (Galaxy Lyme, Schering-Plough Animal Health; Nobivac, Merial), a nonadjuvanted recombinant OspA vaccine (Recombitek, Merial), an adjuvanted recombinant OspA vaccine (ProLyme, Intervet), and a newly developed recombinant OspA/OspC chimeric recombinant vaccine (VANGARD crLyme, Zoetis) that contains multiple OspC antigenic epitopes in addition to OspA. In Europe, bacterins developed from *B. burgdorferi* s.l. (*B. burgdorferi* s.s., *B. garinii* and *B. afzelii* (Merilym3, Merial) or *B. garinii* and *B. afzelii* (Biocan B, Bioveta) also are available (Krupka and Straubinger, 2010; Pantchev et al., 2015). Natural infection, as well as both bacterins and recombinant OspA, elicit bacteriacidal, immobilizing, and growth inhibitory activity against cultured *B. burgdorferi* (Callister et al., 2000; Hovius et al., 2000; LaFleur et al., 2009; Ma et al., 1995; Straubinger et al., 1995; Straubinger et al., 2002).

Vaccination is not recommended in non-endemic areas and is controversial in endemic areas for a number of reasons. The majority of dogs do not develop clinical signs, and, when they do, signs generally are mild and transient. Furthermore, the common signs of Lyme borreliosis in dogs readily resolve in response to antibiotic therapy. Finally, there is risk of adverse effects in response to any vaccination, including development of immune complex Lyme nephritis (Dambach et al., 1997; Littman et al., 2006).

Experimental vaccination studies, nearly all of which have been performed in purpose-bred laboratory beagles, have shown varying degrees of protection against cultured or tick-borne *B. burgdorferi* challenge following immunization with *B. burgdorferi* s.s. bacterins, including bacterins prepared from adjuvanted OspA/OspB-negative *B. burgdorferi*, and OspA/OspB-negative *B. burgdorferi* that expresses high OspC (Chu et al., 1992; LaFleur et al., 2010; LaFleur et al., 2015; LaFleur et al., 2009; Straubinger et al., 2002; Topfer and Straubinger, 2007). A number of studies have also shown efficacy of adjuvanted or non-adjuvanted recombinant OspA and OspA/OspB vaccination against tick-borne challenge (Chang et al., 1995; Conlon et al., 2000; Coughlin et al., 1995; Grosenbaugh et al., 2016; Ma et al., 1996; Straubinger et al., 2002; Topfer and Straubinger, 2007; Wikle et al., 2006). Vaccination with either bacterin or recombinant OspA has been shown to provide protection for at least one year (LaFleur et al., 2010; Wikle et al., 2006). Clinical studies have also shown efficacy of bacterin and recombinant OspA vaccination in naturally exposed dogs (Eschner and Mugnai, 2015; Leschnik et al., 2010).

Serology

The serologic response, including Western blot patterns, of naturally infected dogs and of experimentally infected dogs with tick-borne infection is similar to that of humans and other species (Appel et al., 1993; Barthold et al., 1995; Greene et al., 1988a; Greene et al., 1988b) (Hovius et al., 2000; Jacobson et al., 1996; Korshus et al., 2004; Levy and Magnarelli, 1992; Magnarelli et al., 2001; Shin et al., 1993; Straubinger et al., 1995; Straubinger et al., 2000; Topfer and Straubinger, 2007; Wittenbrink et al., 1996). However, the specificity of canine serologic responses against *B. burgdorferi* lysates by ELISA or Western blot may be complicated by cross-reactivity against other pathogens that may infect dogs, including *Leptospira* serovars, *Salmonella typhimurium*, *Serpulina* spp., and other *Borrelia* spp. (Breitschwerdt et al., 1994; Magnarelli et al., 1987; Shin et al., 1993; Topfer and Straubinger, 2007; Wittenbrink et al., 1996). Furthermore, *B. burgdorferi* vaccination (bacterins and recombinant OspA) elicits a different serologic response that can be differentiated from natural infection. Naturally infected dogs may develop transient low levels of antibody to OspA, but Western blot patterns differ from the response of vaccinated animals, which develop a prominent OspA response (Barthold et al., 1995; Chang et al., 1995; Conlon et al., 2000; Greene et al., 1988b; Jacobson et al., 1996; LaFleur et al., 2009; Leschnik et al., 2010; Levy and Magnarelli, 1992; Ma et al., 1995; Ma et al., 1996; Magnarelli et al., 2001; Straubinger et al., 1995; Straubinger et al., 2002; Topfer and Straubinger, 2007).

Because of the availability and use of various bacterins and recombinant OspA/OspB/OspC vaccines, considerable effort has been devoted to differentiation of the serologic response of vaccinated vs. naturally infected dogs. As noted above, a prominent OspA antibody response is used to differentiate vaccinated from naturally infected dogs. *Borrelia burgdorferi* recombinant proteins have been examined as potential antigens for detecting specific antibody responses in infected dogs, including OspA, OspB, OspC, OspE, OspF, p22, p35 (fibronectin binding protein), p37, p39 and p41G (central fragment of flagellin) (Magnarelli et al., 1997) (Magnarelli et al., 2001). The serologic response of vaccinated dogs was found to include reactivity to OspA, OspB, p22, p37 and p41-G. In contrast, the serologic response to OspF, p35 and p39 reflected the response to natural infection (Magnarelli et al., 2001). The antibody response of beagles infected by tick-borne infection was followed sequentially. The authors concluded the potential utility of OspF antibody as a specific marker of natural infection (Callister et al., 2015). Other studies confirmed the utility of C6 (IR6 conserved region of VlsE protein) antibody response as a means of discriminating between natural infection and vaccination (Hebert and Eschner, 2010; Levy, 2002; Liang et al., 2000; O'Connor et al., 2004).

These studies have led to the use of recombinant OspA, OspC, OspF and C6 as antigens in commercially available multiplex assays (Accuplex4, Antech Diagnostics; Canine Multiplex Assay, Cornell University) or a quantitative C6 antibody assay (Lyme Quant C6, IDEXX) for discriminating between the antibody response of infected vs. vaccinated dogs (Moroff et al., 2015; Wagner et al., 2011; Wagner et al., 2012; Wagner et al., 2013). Studies in experimentally infected dogs, compared to naturally exposed seropositive dogs, have shown that early infection is often marked by low levels of transient antibody against OspA, an early and transient response to OspC, and rising titers of antibody to OspF and C6 over time, thereby allowing assessment of early vs. late infection responses, as well as confirming a vaccine response to OspA. It is further claimed that a decline in *B. burgdorferi* antibodies following antibiotic treatment reflects treatment success, but this must be interpreted with caution (see Antibiotic Treatment, above) (Wagner et al., 2011; Wagner et al., 2012; Wagner et al., 2013; Wagner et al., 2015). In addition, commercial multiplex assays are available for detection of antibody to *B. burgdorferi* (C6), *Anaplasma* spp. (*A. phagocytophilum*, *A. platys*), *Ehlichia canis*, and heartworm (*Dirofilaria immitis*) (SNAP4Dx, IDDEX) or *B. burgdorferi* (C6), *Ehlichia canis* and *Dirofilaria immitis* (SNAP3Dx, IDEXX) (Barth et al., 2014).

Feline Lyme Borreliosis

Cats seroconvert against *B. burgdorferi* under natural exposure conditions in endemic areas in the US and Europe (Levy et al., 2003; Linder and Bockel, 1995; Magnarelli et al., 1990a; May et al., 1994; Pantchev et al., 2016). Cats may be readily parasitized by all stages of *Ixodes scapularis* (larvae, nymphs and adults). Larvae removed from naturally infested cats were shown to be infected with *B. burgdorferi*, indicating that cats may serve as competent reservoir hosts (Magnarelli et al., 1990a). Furthermore, like dogs, natural infection is often subclinical, but some naturally infected cats may develop fever, fatigue, anorexia and lameness (Magnarelli et al., 1990a; Peterhans and Peterhans, 2010; Shaw et al., 2005). *Borrelia burgdorferi ospA* DNA was amplified from the blood of two naturally infected cats in the UK, but the genospecies was not identified (Shaw et al., 2005).

The susceptibility of cats to infection with cultured, needle-inoculated *B. burgdorferi*, as well as tick-borne *B. burgdorferi* has been confirmed experimentally. Clinical signs were frequently absent or minimal, but fever and lameness were noted (Burgess, 1992; Gibson et al., 1995a; Gibson et al., 1993; Gibson et al., 1995b; Lappin et al., 2015). In one study, histopathology revealed mild synovial hyperplasia with focal accumulations of lymphocytes in the stifle (knee) joint capsule, mild perivascular cuffing in the cerebrum, and lymphoid hyperplasia and plasmacytosis in a variety of lymph nodes. Focal nonsuppurative interstitial nephritis, chronic ischemic nephropathy, periglomerulitis and mild multifocal pyelitis were noted in kidneys. Mild alveolar emphysema with multifocal interstitial pneumonitis, plasmacytosis and lymphoid hyperplasia were found in the lungs of infected cats. In spite of these findings, direct cause-and-effect was not confirmed and infection status was based on serology, rather than culture or PCR (Gibson et al., 1995b). Nevertheless, at least some of the lesions that were observed were consistent with Lyme borreliosis in other species.

The C6 peptide is the most frequently used antigen for serodiagnosis of exposure in cats, with good agreement with IFA and Western blot results (Lappin et al., 2015; Levy et al., 2003) (Pantchev et al., 2016). One report in Europe found a lower prevalence of C6 antibody-positive cats compared to dogs but noted a correlation of cats with clinical signs of Lyme borreliosis and seroreactivity to C6 antigen (Pantchev et al., 2016). Because cats are quite stoic, clinical illness is generally not noted or clinically mild or transient when present. Serodiagnosis, antibiotic treatment and vaccination are rarely considered in cats. Cat ownership has been associated with increased risk of human Lyme disease (Steere et al., 1978), but the reasons for this relationship have not been defined.

Equine Lyme Borreliosis

Lyme borreliosis in equids has the same diagnostic and treatment challenges as in other species. The ACVIM recently published a consensus statement on *B. burgdorferi* and Lyme disease in horses (Divers et al., 2018). Serosurveys of equids in Lyme disease-endemic areas in the US and Europe indicate that horses are frequently exposed to *B. burgdorferi* infection but rarely manifest clinical signs (Burgess, 1988; Carter et al., 1994; Cohen et al., 1988; Dzierzecka and Kita, 2002; Egenvall et al., 2001; Funk et al., 2016; Magnarelli et al., 1988; Marcus et al., 1985a; Stefancikova et al., 2002). Some horses most certainly develop clinical signs of Lyme borreliosis, but accurate diagnosis can be a challenge. Diagnosis is based upon the environment in which the horse has been living or has travelled in (*i.e.*, endemic for *B. burgdorferi*), clinical signs (which often are absent, mild and highly variable and often non-specific), supplemental laboratory diagnostics, such as serology (which only confirms exposure), and ruling out other causes of disease. Treatment should be targeted to clinical signs of illness. Veterinarians in *B. burgdorferi* endemic areas are faced with the same challenges as physicians regarding diagnosis and treatment, often with strong opinions of horse owners about the appropriate course of action. In some cases, this has led to the off-label use of commercially available canine *B. burgdorferi* bacterins and recombinant OspA vaccines.

Typical of *B. burgdorferi* infection in humans, a wide variety of clinical signs have been observed in horses, but true attribution of infection as the cause of those signs has been elusive in this species. The veterinary community generally recognizes weight loss, low-grade fever, and a number of single or recurrent bouts of musculoskeletal signs, including stiffness, sporadic and shifting lameness in one or more limbs, muscle tenderness, hyperaesthesia, swollen joints, arthritis, and a variety of other orthopedic signs, including laminitis. Owners have complained that horses may exhibit chronically poor performance, lethargy and behavioral changes (Browning et al., 1993; Burgess, 1988; Butler et al., 2005; DeVilbiss et al., 2009; Divers, 2013; Egenvall et al., 2001; Gall and Pfister, 2006; Hahn et al., 1996; Levy, 1989; Magnarelli et al., 1988; Parker and White, 1992).

Dermatologic signs of Lyme borreliosis are generally undetectable in horses, but a case of cutaneous pseudo-lymphoma, with multiple dermal papules of the masseter region, was reported in which the lesions were confirmed to be positive for *B. burgdorferi* DNA and resolved with antibiotic treatment. The lesions contained mixed populations of T- and B-cells (Johnstone et al., 2016). Cutaneous lymphocytomas (lymphocytoma cutis) are a well-recognized manifestation of human Lyme borreliosis (Colli et al., 2004). Experimentally infected ponies developed lymphadenomegaly of the prescapular, axillary and popliteal lymph nodes with microscopic evidence of mild to moderate follicular hyperplasia (Chang et al., 2005).

Orthopedic signs, including recurrent lameness, swelling of various joints (and tendon sheaths, and laminitis are common clinical features of Lyme borreliosis in horses (Browning et al., 1993; Burgess, 1988; Burgess et al., 1986; Hahn et al., 1996; Magnarelli et al., 1988; Passamonti et al., 2015). There is only limited information regarding gross or microscopic joint lesions in horses, but the pathology resembles that found in human Lyme arthritis (Duray, 1989). Experimentally infected ponies had rare foci of mild to moderate perivascular lymphohistiocytic infiltrates in the perisynovial tissues (Chang et al., 2005). Microscopic findings in one case of Lyme arthritis in a horse revealed synovial villous hypertrophy, capillary proliferation, small granulomas, and focally fibrinous stroma that contained small aggregates of lymphocytes, histiocytes and few giant cells (Hahn et al., 1996). *Borrelia burgdorferi*-seropositive horses with high fever and limb edema are more likely infected (or co-infected) with *Anaplasma phagocytophilum* (Chang et al., 2000c; Pusterla et al., 2002).

Carditis is generally not clinically recognized in horses, but it probably occurs. In the retrospective survey of horses with neuroborreliosis, cardiac arrhythmia, epicarditis and myocarditis were found in some horses (Johnstone et al., 2016).

Horses of all breeds are prone to development of Equine Recurrent Uveitis (ERU), colloquially known as "moon blindness", which increases in prevalence with age (Allbaugh, 2016; Gilger and Deeg, 2011).

ERU shares similarities to recurrent uveitis in humans, as both have an underlying autoimmune pathogenesis and both are often associated with infectious agents, among other precipitating causes (Malalana et al., 2015). In horses, *Leptospira* spp. infection is a major underlying cause (Gilger and Deeg, 2011), but a number of reported cases have documented *B. burgdorferi*-associated uveitis in horses (Burgess et al., 1986; Hahn et al., 1996) (Johnstone et al., 2016; Priest et al., 2012), as well as humans (Lesser, 1995). Other ocular lesions, including vitreal inflammation, conjunctivitis, keratitis, retinitis and optic neuritis, have been observed in horses with neuroborreliosis (Johnstone et al., 2016). Although not definitively attributed to *B. burgdorferi*, these lesions likely illustrate that horses with Lyme borreliosis may potentially develop a much broader spectrum of ocular disease, as occurs in humans (Lesser, 1995; Rothermel et al., 2001).

Rarely, natural cases of neuroborreliosis have been reported in horses. In a recent retrospective review of northeastern US horses with presumptive clinical signs and postmortem lesions consistent with neuroborreliosis, horses presented with weight loss, cranial nerve deficits, ataxia, behavioral changes, dysphagia, laryngeal dysfunction, muscle fasciculations, neck stiffness, and fever, which were variably accompanied by uveitis, joint effusion, respiratory distress and cardiac arrhythmia (Johnstone et al., 2016). Upon necropsy, gross lesions included opacity, hyperemia, focal edema and yellow discoloration of the meninges. Microscopic findings in all of the horses included nonsuppurative leptomeningitis with segmental vasculitis, which often extended into Virchow-Robin spaces and rarely into the brain parenchyma. In addition, inflammation of cranial and spinal nerve roots, peripheral nerves, ganglia, and skeletal muscle were found in several horses. In all of the horses included in that study, clinical signs were sufficiently severe as to result in euthanasia, or the horses were euthanized because they responded poorly to antibiotic treatment. Clinical signs either continued to progress, did not improve, or recurred and progressed following treatment.

A small number of publications have definitively diagnosed neuroborreliosis in horses by culture or PCR amplification of *B. burgdorferi* DNA, as well as visualization of spirochetes in lesions by immunohistochemistry or silver stain. The lesions described in these cases were similar to those in the retrospective study described above. A clinical diagnosis of encephalitis, confirmed by culture of the brain but not histopathology, was documented in a Wisconsin horse that presented with flaccid paralysis of the tail, sweating, constant pacing and dysphagia (Burgess and Mattison, 1987). A horse in the United Kingdom presented with lethargy, anorexia, fever, and ataxia with hyperesthesia of the head. Clinical signs improved, but relapsed following treatment, at which time the mare was reluctant to move, had distention of several joints and tendon sheaths and developed bilateral uveitis. The ocular aqueous was *B. burgdorferi* PCR-positive. Necropsy revealed multiple focal raised meningeal nodules consisting of leukocytic infiltration, vasculitis and perivascular cuffing (Hahn et al., 1996). In another report (Imai et al., 2011), two US horses with progressive neurologic disease were found to have severe necrosuppurative or nonsuppurative perivascular to diffuse meningoradiculoneuritis and sclerosing meningeal vasculitis of the brain and spinal cord. The horses presented with pain along the back and neck, lumbar hyperesthesia, lethargy, anorexia, weight loss and muscle atrophy. Numerous spirochetes were identified in the dura mater of both horses with silver stain and with *B. burgdorferi*-specific immunohistochemistry. *Borrelia burgdorferi*-specific DNA was amplified and sequenced from the spinal cord of one horse.

Another report (James et al., 2010) described a horse with a similar severe clinical course of meningitis, cranial neuritis, and radiculoneuritis with positive *B. burgdorferi* DNA in its CSF. The horse was dull, depressed, and had diminished responses to external stimuli. It was reluctant to raise its head, and had hyperesthesia of the thoracic and lumbar regions, with muscle fasciculation of the hindquarters. At necropsy, severe inflammatory lesions were found, similar to the other cases described above. In contrast to these severe cases of neuroborreliosis, experimentally infected ponies developed no or only mild perivascular lymphocytic infiltrates in the meninges (Chang et al., 2005).

Notably, in the case described by James, et al. (James et al., 2010) the horse had only equivocal antibody titers against *B. burgdorferi*, and the authors demonstrated that the horse had severe B-cell lymphopenia and low serum IgM, consistent with common variable immunodeficiency of horses. Common variable immunodeficiency is characterized by progressive late-onset B cell lymphopenia, hypogammaglobulinemia, and B-cell depletion in lymphoid tissues. It occurs in various breeds of horses that become susceptible to repeated bacterial infections as adults (Flaminio et al., 2009). One of the horses with chronic neuroborreliosis described by Imai, et al. (Imai et al., 2011) was seronegative against *B. burgdorferi*, whereas its clinically normal pasture-mate was seropositive. In another report, three horses with common variable immunodeficiency were described with presumptive bacterial meningitis. The bacterial etiology was not determined, but the horses presented with musculoskeletal pain, dullness, ataxia and seizures (Pellegrini-Masini et al., 2005). A similar case was reported in a horse with chronic wasting, muscle atrophy, polyarthritis, panuveitis, and hyperaesthesia was confirmed to be infected with *B. afzelii* by PCR of synovial and eye fluids. As in other cases, it responded initially to therapy, but relapsed and deteriorated with sudden death 12 months later (Sommerauer et al., 2019). Recently, chronic neuroborreliosis was definitively documented in a horse with common variable immunodeficiency in which *B. burgdorferi* was confirmed by PCR, fluorescent in situ hybridization and immuno-histochemical staining (Pecoraro et al., 2019). These findings suggest that the rare, but unusually severe and refractory cases of equine Lyme borreliosis may be, as Robert Good described for human immunoglobulin deficiencies, "experiments in nature" (Good and Zak, 1956). Experimental studies with the mouse model of Lyme borreliosis have shown that *B. burgdorferi* antibody and B cells are singularly critical for suppressing inflammatory lesions in joints and heart (Barthold et al., 1997; Barthold et al., 2006; McKisic and Barthold, 2000). There are striking histologic similarities of severe equine neuroborreliosis with tertiary neurosyphilis in humans (Imai et al., 2011; Miklossy, 2012).

A limited number of studies have examined the course of experimental *B. burgdorferi* s.s. infections in horses (Bosler et al., 1988; Chang et al., 2000a; Chang et al., 2000b; Chang et al., 2005). As with natural infection, clinical signs of infection were usually mild or absent, and horses were infected persistently for months after initial exposure, based upon *ospA* and/or *23S rRNA* gene amplification and culture (Chang et al., 2000a; Chang et al., 2000b; Chang et al., 2005). Histopathology revealed generally mild lesions. Perivascular and perineural lymphocytic, histiocytic and plasma cell aggregates were observed in the deep dermis and panniculus of previous tick attachment sites nine months after tick feeding, and regional lymph nodes were reactive, with follicular hyperplasia, but lesions in other tissues were not found (Chang et al., 2000a). In another study that examined tissues at four months after tick exposure, similar lesions were found in the dermis, and one of four animals had evidence of mild to moderate perivascular lymphoplasmacytic infiltrates in subsynovial tissue of multiple joints, perivascular infiltrates around small arteries adjacent to peripheral nerves (including the facial nerve), thoracic and lumbar dorsal nerve roots, and skeletal muscle. A focus of perivascular lymphocyte cuffing was also observed in the thalamus (Chang et al., 2000b). Similar mild lesions were found in another study at eight months after tick infection. In that study, prescapular, axillary and popliteal lymph nodes were reactive, and perivascular infiltrates of lymphocytes were found in the meninges of two of four infected animals (Chang et al., 2005). Lesions, when present, were most often found in tissues of the somatic quadrant upon which ticks were attached (Chang et al., 2000b).

As in humans, serology is performed to confirm exposure of horses to *B. burgdorferi*, and although antibody titers decline following treatment, they are not accurate tests for treatment success. Various serology studies have used whole cell antigen, purified antigen or recombinant antigen (OspA, OspB, OspC, OspE, OspF, p41-G, p35, p37, p39, and VlsE) indirect immunofluorescent antibody assay, Enzyme-linked Immunosorbent Assay (ELISA), Western blot, serum growth inhibition assay, and agglutination assay (Burgess, 1988; Carter et al., 1994; Cohen et al., 1988; Egenvall et al., 2001; Magnarelli and Anderson, 1989; Magnarelli et al., 1988; Manion et al., 2001; Marcus et al., 1985a; Stefancikova et al., 2002). More recently, the veterinary community often relies on Cornell University to perform equine serology, which uses a fluorescent bead-based multiplex assay containing recombinant OspA, OspC and OspF antigens and Western blot analysis of whole cell lysate (Wagner et al., 2011). A commercially available in-clinic SNAP-4Dx (IDEXX Laboratories) test is also popular. The SNAP-4Dx test is an ELISA developed for dogs that includes recombinant *B. burgdorferi* C6 antigen, *Anaplasma phagocytophilum* p44 antigen, plus two antigens of *Ehrlichia canis*. Although developed for dogs, it has been found to be useful for equine *B. burgdorferi* and *A. phagocytophilum* serology (Chandrashekar et al., 2008; Johnson et al., 2008). If horses are manifesting particularly severe or treatment-refractory signs of neuroborreliosis, thorough analysis of immune competence should be pursued, including peripheral blood leukocyte phenotyping, lymphocyte proliferation assays, phagocyte oxidative burst activity, and serum IgG, IgG3, IgG5, IgM and IgA concentrations, as described (Flaminio et al., 2009).

PCR, generally targeting *ospA*, but also other sequences, has been shown to work well on tissues, blood, synovial fluid, ocular fluid and urine from both experimentally- and naturally-infected horses (Chang et al., 2000a; Hahn et al., 1996; Imai et al., 2011; James et al., 2010; Johnstone et al., 2016; Manion et al., 2001; Passamonti et al., 2015; Pecoraro et al., 2019; Priest et al., 2012; Sears et al., 2011; Sommerauer et al., 2019). Notably, *ospA* PCR testing of blood and urine sediment of naturally infected horses during early infection (based upon low antibody titers and acute clinical signs) revealed that 10 of 19 clinical cases had positive PCR results in blood and 2 of 10 clinical cases had positive PCR results in urine, as well as 1 positive blood sample among 21 clinically normal horses from the same endemic area (Manion et al., 2001). PCR was shown to be more sensitive than culture of tissues in experimentally infected ponies. In that study, 41 samples from 7 infected ponies were culture-positive, compared to 134 samples that were *ospA* PCR positive and 120 samples that were *23S rRNA* PCR positive (Chang et al., 2000a). The higher sensitivity of *ospA* PCR compared to the *23S rRNA* target correlates with the finding of "target imbalance" or disparity of *ospA* over other gene targets in the synovial fluid of human patients (Persing et al., 2000).

Treatment of horses for Lyme borreliosis is generally approached with intravenous tetracycline (which cannot be given orally because of low bioavailability in equids) or oral doxycycline (which cannot be given intravenously because of adverse effects in equids). Minocycline has better bioavailability than doxycycline in horse tissues, including higher levels in CSF and aqueous humor (Divers, 2013). In a single study, tick-infected horses were treated with recommended doses of tetracycline (intravenous), doxycycline (oral) or ceftiofur (intramuscular) for four weeks, and skin punch biopsies from the tick attachment sites were collected monthly, up to five months after treatment, at which time the horses were necropsied and multiple tissues were cultured. Three of four doxycycline-treated horses and two of four ceftiofur-treated horses had one or more culture-positive tissues at necropsy, indicating treatment failure, but none of the tetracycline-treated horses remained culture positive. Remarkably, however, skin biopsies taken at the site of original tick attachment remained *ospA* DNA positive by PCR in all antibiotic-treated and control horses for up to 5 months after treatment (Chang et al., 2005).

There are no commercially licensed *B. burgdorferi* vaccines for use in horses, but some veterinarians (and the lay public) take advantage of off-label use of commercially available canine *B. burgdorferi* vaccines in areas that are endemic for Lyme borreliosis. Considering the rarity, mildness, and transient nature of clinical signs, and resolution of signs with antibiotic treatment in most horses, vaccination of horses for prevention of Lyme borreliosis is controversial. It has no efficacy for treatment of active infection. An experimental study demonstrated efficacy of vaccination in preventing infection in horses. The study was performed with eight ponies that were vaccinated intramuscularly three times at day 0, 20 and 82 with 100 micrograms of recombinant OspA in 1% v/v aluminum hydroxide adjuvant, and four ponies that were non-vaccinated controls that received adjuvant alone. The ponies were challenged by exposure to *B. burgdorferi*-infected ticks at 112 days after the first vaccination. Protection was assessed by culture of three monthly skin biopsies near the site of tick attachment, and culture of multiple tissues at the end of the study. All of the vaccinated ponies were culture-negative, whereas the four control ponies were all culture-positive. The vaccinated ponies developed antibodies to OspA, but not other antigen bands on *B. burgdorferi* lysate Western blot (Chang et al., 2000b). In contrast, vaccination of horses with a variety of commercial canine vaccines resulted in only transient and low-level antibody responses, thereby questioning their efficacy for protection in horses (Guarino et al., 2017).

Lyme Borreliosis in Domestic Ruminants

Serologic surveys, using various methods, have found seroconversion of dairy and beef cattle to *B. burgdorferi* s.l. in the US, Canada, Europe, Japan, and China. Depending upon test method, interpretation of results may be difficult due to antigenic cross-reactivity with other related agents that infect cattle, including *B. coriaceae*, *B. theileri*, *Treponema* spp., and, to a lesser extent, *Leptospira* spp. (Blowey et al., 1994; Borko, 2000; Burgess, 1988; Hovmark et al., 1986; Ji and Collins, 1994;

Kasbohrer A and A, 1990; Magnarelli et al., 2004; Post et al., 1988; Stefancikova et al., 2008; Stefancikova et al., 2002; Wan et al., 1999; Wells et al., 1993). As in other species, antibody reactivity status may not necessarily correlate with clinical signs.

Clinical signs of Lyme borreliosis that have been reported in cattle include fever, stiff gait, joint swelling, lameness, decreased milk production, chronic weight loss, laminitis, abortion, and udder erythema (Parker and White, 1992). Most reports are anecdotal, but *B. burgdorferi* s.l. has been isolated or detected by PCR from blood, synovial fluid, colostrum, urine, and parenchymal organs in a limited number of reports (Burgess, 1988; Burgess et al., 1987; Burgess et al., 1993; Lischer et al., 2000). Spirochetal forms, identified with silver histochemistry, have been documented in hyperplastic synovial tissue in a *B. burgdorferi* seropositive cow with wasting and multiple swollen joints in Australia (Rothwell et al., 1989). Erythema migrans, involving the ventral udder and lower legs has been described in Connecticut dairy cows. Affected areas were warm, swollen, and hypersensitive, with decreased milk production. The animals also had lameness and swollen joints, but attempts at *B. burgdorferi* isolation failed (Post et al., 1988). However, *B. burgdorferi* sensu stricto and *B. afzelii* were definitively confirmed by PCR and gene sequencing in two Swiss dairy cows with similar signs, including udder erythema (Lischer et al., 2000).

A limited number of reports have documented necropsy results in cows with confirmed infection (Burgess, 1988; Burgess et al., 1987; Lischer et al., 2000), and similar findings were observed in a cow in Australia with unconfirmed infection (Rothwell et al., 1989). These cases represented unusually severe, chronic disease, which appears to be atypical in cattle, and merits investigation of such cases for underlying immune deficiencies.

Infection of sheep and goats with *B. burgdorferi* is suspected, but reports are scant. Arthritis was described in two 6-month old seropositive lambs, but *B. burgdorferi* infection was not confirmed (Fridriksdottir et al., 1992). Serologic surveys of small ruminants in China and Slovakia revealed that exposure to *B. burgdorferi* may be common (Travnicek et al., 2002; Yang et al., 2015), and *B. burgdorferi* s.l. DNA was documented in blood of sheep, goats, cattle and camels in Tunisia (BenSaid et al., 2016).

Sheep support the feeding of all three stages of *I. ricinus*, and can maintain the *B. burgdorferi* enzootic cycle in the absence of other hosts, can foster co-transmission between feeding ticks, but tend not to support systemic infections (Ogden et al., 1997). Notably, various ruminants, including cattle, sheep and wild cervids, appear to limit the prevalence of *B. burgdorferi* s.l. in ticks. Grazing has a negative effect on small rodents and on ticks (Gassner et al., 2008), and spirochetes appear to be eliminated from ticks feeding on domestic ruminants (Richter and Matuschka, 2010). This observation extends to hoofed game animals, among which it was found that few ticks retain infection by *B. burgdorferi* after feeding (Matuschka et al., 1993; Telford et al., 1988).

References

Allbaugh, R.A. (2016). Equine recurrent uveitis: a review of clinical assessment and management. Equine Veterinary Education 29:279-288. https://doi.org/10.1111/eve.12548

Appel, M.J.G. (1990). Lyme disease in dogs and cats. Compendium on Continuing Education for the Practicing Veterinarian *12*, 617-627.

Appel, M.J.G., Allan, S., Jacobson, R.H., Lauderdale, T.L., Chang, Y.-F., Shin, S.J., Thomford, J.W., Todhunter, R.J., and Summers, B.A. (1993). Experimental Lyme disease in dogs produces arthritis and persistent infection. Journal of Infectious Diseases *172*, 651-654. DOI: 10.1093/infdis/167.3.651

Arashima, Y. (1991). Anti-*Borrelia burgdorferi* antibody in dogs: Lyme disease as a zoonosis. Japanese Journal of Clinical Pathology *39*, 869-874.

Azuma, Y., Kawamura, K., Isogai, H., and Isogai, E. (1993). Neurologic abnormalities in two dogs suspected Lyme disease. Microbiology and Immunology *37*, 325-329. DOI: 10.1111/j.1348-0421.1993.tb03217.x

Barth, C., Straubinger, R.K., Krupka, I., Muller, E., Sauter-Louis, C., and Hartman, K. (2014). Comparison of different diagnostic assays for the detecton of *Borrelia burgdorferi*-specific antibodies in dogs. Veterinary Clinical Pathology *43*, 496-504. DOI: 10.1111/vcp.12213

Barth, C., Straubinger, R.K., Sauter-Louis, C., and Hartmann, K. (2012). Prevalence of antibodies against *Borrelia burgdorferi* sensu lato and *Anaplasma phagocytophilum* and their clinical relevance in dogs in Munich, Germany. Berliner Munchener tierarztl Wochenschrift *125*, 337-344.

Barthold, S.W., Feng, S., Bockenstedt, L.K., Fikrig, E., and Feen, K. (1997). Protective and arthritis-resolving activity in sera of mice infected with *Borrelia burgdorferi*. Clinical Infectious Disease *24*, S9-S19. https://doi.org/10.1086/516166

Barthold, S.W., Hodzic, E., Imai, D.M., Feng, S., and Luft, B.J. (2010). Ineffectiveness of tigecycline against persistent *Borrelia burgdorferi*. Antimicrobial Agents and Chemotherapy *54*, 643-651. doi: 10.1128/AAC.00788-09

Barthold, S.W., Hodzic, E., Tunev, S., and Feng, S. (2006). Antibody-mediated disease remission in the mouse model of Lyme borreliosis. Infection and Immunity *74*, 4817-4825. doi: 10.1128/IAI.00469-06

Barthold, S.W., Levy, S.A., Fikrig, E., Bockenstedt, L.K., and Smith, A.L. (1995). Serologic responses of dogs naturally exposed to or vaccinated against *Borrelia burgdorferi* infection. Journal of the American Veterinary Medical Association *207*, 1435-1440.

BenSaid, M., Belkahia, H., Alberti, A., Abdi, K., Zhioua, M., Daaloul-Jedidi, M., and Messadi, L. (2016). First molecular evidence of *Borrelia burgdorferi* sensu lato in goats, sheep, cattle and camels in Tunisia. Annals of Agricultural and Environmental Medicine *23*, 442-447. DOI: 10.5604/12321966.1219184

Blowey, R.W., Carter, S.D., White, A.G., and Barnes, A. (1994). *Borrelia burgdorferi* infections in UK cattle: a possible association with digital dermatitis. Veterinary Record *135*, 577-578.

Bockenstedt, L.K., Mao, J., Hodzic, E., Barthold, S.W., and Fish, D. (2002). Detection of attenuated, noninfectious spirochetes in *Borrelia burgdorferi*-infected mice after antibiotic treatment. Journal of Infectious Diseases *186*, 1430-1437. DOI: 10.1086/345284

Borko, C. (2000). Humoral immune response to bacteria *Borrelia burgdorferi* s.l. in cattle after natural and experimental exposure. Vet nov *26*, 120-123.

Bosler, E.M., Camay, H., Ormiston, B.G., and Benach, J.L. (1984). Spirochete-associated arthritis (Lyme disease) in a dog. Journal of the American Veterinary Medical Association *185*, 219-220.

Bosler, E.M., Cohen, D.P., Schultz, T.L., Olsen, C., Bernard, W., and Lissman, B. (1988). Host responses to *Borrelia burgdorferi* in dogs and horses. Annals of the New York Academy of Sciences *539*, 221-234. doi:10.1111/j.1749-6632.1988.tb31856.x.

Bosler, E.M., Evans, R.D., and Schneider, E.M. (1992). Biopsy of canine ear tissue for detection of *Borrelia burgdorferi* following *Ixodes dammini* feeding. FEMS Microbiology Letters *99*, 317-320. DOI: 10.1016/0378-1097(92)90047-r

Bowman, D., Little, S.E., Lorentzen, L., Shields, J., Sullivan, M.P., and Carlin, E.P. (2009). Prevalence and geographic distribution of *Dirofilaria immitis*, *Borrelia burgdorferi*, *Ehrlichia canis*, and *Anaplasma phagocytophilum* in dogs in the United States: results of a national clinic-based serologic survey. Veterinary Parasitology *160*, 138-148. doi: 10.1016/j.vetpar.2008.10.093.

Breitschwerdt, E.B., Nicolson, W.L., Kiehl, A.R., Steers, C., Meuten, D.J., and Levine, J.F. (1994). Natural infection with Borrelia spirochetes in two dogs from Florida. Journal of Clinical Microbiology *32f*, 352-357. doi: 10.1128/JCM.32.2.352-357.1994.

Browning, A., Carter, S.D., Barnes, A., May, C., and Bennett, D. (1993). Lameness associated with *Borrelia burgdorferi* infection in the horse. Veterinary Record *132*, 610-611. DOI: 10.1136/vr.132.24.610

Burgess, E.C. (1986a). Experimental inoculation of dogs with *Borrelia burgdorferi*. Zentralblatt Bakteriologie, Mikrobiologie und Hygiene A *263*, 49-54. DOI: 10.1016/s0176-6724(86)80102-x

Burgess, E.C. (1986b). Natural exposure of Wisconsin dogs to the Lyme disease spirochete (*Borrelia burgdorferi*). Laboratory Animal Science *36*, 288-290.

Burgess, E.C. (1988). *Borrelia burgdorferi* infection in Wisconsin horses and cows. Annals of the New York Academy of Sciences *539*, 235243. doi: 10.1111/j.1749-6632.1988.tb31857.x.

Burgess, E.C. (1992). Experimentally induced infection of cats with *Borrelia burgdorferi*. American Journal of Veterinary Research *53*, 1507-1511.

Burgess, E.C., Gendron-Fitzpatrick, A., and Wright, W.O. (1987). Arthritis and systemic disease caused by *Borrelia burgdorferi* infection in a cow. Journal of the American Veterinary Medical Association *191*, 1468-1470.

Burgess, E.C., Gillette, D., and Pickett, J.P. (1986). Arthritis and panuveitis as manifestations of *Borrelia burgdorferi* infection in a Wisconsin pony. Journal of the American Veterinary Medical Association *189*, 1340-1342.

Burgess, E.C., and Mattison, M. (1987). Encephalitis associated with *Borrelia burgdorferi* infection in a horse. Journal of the American Veterinary Medical Association *191*, 1457-1458.

Burgess, E.C., Wachal, M.D., and Cleven, T.D. (1993). *Borrelia burgdorferi* infection in dairy cows, rodents, and birds from four Wisconsin dairy farms. Veterinary Microbiology *35*, 61-77. DOI: 10.1016/0378-1135(93)90116-o

Butler, C.M., Houwers, D.J., Jongejan, F., and VanderKolk, J.H. (2005). *Borrelia burgdorferi* infections with special reference to horses. A review. Veterinary Quarterly *27*, 146-156.

Callister, S.M., Jobe, D.A., Schell, R.F., Lovrich, S.D., Onheiber, K.L., and Korshus, J.B. (2000). Detection of borreliacidal antibodies in dogs after challenge with *Borrelia burgdorferi*-infected *Ixodes scapularis* ticks. Journal of Clinical Microbiology *38*, 3670-3674. doi: 10.1128/JCM.38.10.3670-3674.2000.

Callister, S.M., LaFleur, R.L., Jobe, D.A., Lovrich, S.D., and Wasmoen, T.L. (2015). Antibody response to *Borrelia burgdorferi* outer surface proteins C and F in experimentally infected Beagle dogs. Journal of Veterinary Diagnostic Investigation *27*, 526-530. doi: 10.1177/1040638715583868.

Carter, S.D., May, C., Barnes, A., and Bennett, D. (1994). *Borrelia burgdorferi* infection in UK horses. Equine Veterinary Journal *26*, 187-190. doi: 10.1111/j.2042-3306.1994.tb04367.x.

Cerri, D., Farina, R., Andreani, E., Nuvoloni, R., and Pedrini, A. (1994). Experimental infection of dogs with *Borrelia burgdorferi*. Research in Veterinary Science *57*, 256-258. doi: 10.1016/0034-5288(94)90069-8.

Chandrashekar, R., Daniluk, D., Moffitt, S., Lorentzen, L., and Williams, J. (2008). Serologic diagnosis of equine borreliosis: evaluation of an in-clinic enzyme-linked immunosorbent assay (SNAP 4Dx). International Journal of Applied Research in Veterinary Medicine *6*, 145-150.

Chang, Y.-F., Appel, M.J.G., Jacobson, R.H., Shin, S.J., Harpending, P., Straubinger, R., Patrican, L.A., Mohammed, H., and Summers, B.A. (1995). Recombinant OspA protects dogs against infection and disease caused by *Borrelia burgdorferi*. Infection and Immunity *63*, 3543-3549. doi: 10.1128/IAI.63.9.3543-3549.1995.

Chang, Y.-F., McDonough, S.P., Chang, C.-F., Jacobson, R.H., Divers, T., Quimby, F.W., Shin, S., and Lein, D.H. (2000a). Experimental infection of ponies with *Borrelia burgdorferi* by exposure to Ixodid ticks. Veterinary Pathology *37*, 68-76. doi: 10.1354/vp.37-1-68.

Chang, Y.-F., Novosol, V., Chang, C.-F., Summers, B.A., Ma, D.-P., Chiang, Y.W., Acree, W.M., Chiu, H.-J., Shin, S., and Lein, D.H. (2001). Experimental induction of chronic borreliosis in adult dogs exposed to *Borrelia burgdorferi*-infected ticks and treated with dexamethasone. American Journal of Veterinary Research *62*, 1104-1112. DOI: 10.2460/ajvr.2001.62.1104

Chang, Y.-F., Novosol, V., McDonough, S.P., Chang, C.-F., Jacobson, R.H., Divers, T., Quimby, F.W., Sin, S., and Lein, D.H. (2000b). Vaccination against Lyme Disease with recombinant *Borrelia burgdorferi* outer-surface protein A (rOspA) in horses. Vaccine *18*, 540-548. doi: 10.1016/s0264-410x(99)00187-5. doi: 10.1016/s0264-410x(99)00187-5.

Chang, Y.F., Ku, Y.W., Chang, C.F., Chang, C.D., McDonough, S.P., Divers, T., Pough, M., and Torres, A. (2005). Antibiotic treatment of experimentally *Borrelia burgdorferi*-infected ponies. Veterinary Microbiology *107*, 285-294. doi: 10.1016/j.vetmic.2005.02.006.

Chang, Y.F., McDonough, S.P., Chang, C.F., and al., e. (2000c). Human granulocytic ehrlichiosis agent infection in a pony vaccinated with a *Borrelia burgdorferi* recombinant OspA vaccine and challenged by exposure to naturally infected ticks. Clinical and Diagnostic Laboratory Immunology *2000*, 68-71. doi: 10.1128/cdli.7.1.68-71.2000.

Chang, Y.F., Straubinger, R.K., Jacobson, R.H., and al., e. (1996). Dissemination of *Borrelia burgdorferi* after experimental infection in dogs. Journal of Spirochetal and Tick-Borne Diseases *3*, 80-86.

Chou, J., Wunschmann, A., Hodzic, E., and Borjesson, D.L. (2006). Detection of *Borrelia burgdorferi* DNA in tissues from dogs with presumptive Lyme borreliosis. Journal of the American Veterinary Medical Association *229*, 1260-1265. DOI: 10.2460/javma.229.8.1260

Chu, H.-J., Chavez, L.G., Blumer, B.M., Sebring, R.W., Wasmoen, T.L., and Acree, W.M. (1992). Immunogenicity and efficacy study of a commercial *Borrelia burgdorferi* bacterin. Journal of the American Veterinary Medical Association *201*, 403-418.

Cohen, D., Bosler, E.M., Bernard, W., Meirs, D., Eisner, R., and Schulze, T.L. (1988). Epidemiologic studies of Lyme disease in horses and their public health significance. Annals of the New York Academy of Sciences *539*, 244-257. doi: 10.1111/j.1749-6632.1988.tb31858.x.

Cohen, N.D., Carter, C.N., Thomas, M.A., Angulo, A.B., and Eugster, A.K. (1990). Clinical and epizootiologic characteristics of dogs seropositive for *Borrelia burgdorferi* in Texas; 110 cases (1988). Journal of the American Veterinary Medical Association *197*, 893-898.

Colli, C., Leinweber, B., Mullegger, R., Chott, A., Kerl, H., and Cerroni, L. (2004). *Borrelia burgdorferi*-associated lymphocytoma cutis: clinicopathologic, immunophenotypic, and molecular study of 106 cases. . Journal of Cutaneous Pathology *31*, 232-240. DOI: 10.1111/j.0303-6987.2003.00167.x

Conlon, J.A., Mather, T.N., Tanner, P., Gallo, G., and Jacobson, R.H. (2000). Efficacy of a nonadjuvanted, outer surface protein A, recombinant vaccine in dogs after challenge by ticks naturally infected with *Borrelia burgdorferi*. Veterinary Therapeutics *1*, 96-107.

Coughlin, R.T., Fish, D., Mather, N., Ma, J., Pavia, C., and Bulger, P. (1995). Protection of dogs from Lyme disease with a vaccine containing outer surface protein (Osp) A, OspB, and the saponin adjuvant QS21. Journal of Infectious Diseases *171*, 1049-1052. doi: 10.1093/infdis/171.4.1049.

Curran, K.L., and Fish, D. (1989). Increased risk of Lyme disease for cat owners. N Engl J Med *320*, 183.

Dambach, D.M., Smith, C.A., Lewis, R.M., and Winkle, T.J.V. (1997). Morphologic, immunohistochemical, and ultrastructural characterization of a distinctive renal lesion in dogs putatively associated with *Borrelia burgdorferi* infection. Veterinary Pathology *34*, 85-86. DOI: 10.1177/030098589703400201

DeVilbiss, B.A., Mohammed, H.O., and Divers, T.J. (2009). Perception of equine practitioners regarding the occurrence of selected equine neurologic diseases in the northeast over a 10-year period. Journal of Equine Veterinary Science *29*, 237-246.

Divers, T.J. (2013). Equine Lyme disease. Journal of Equine Veterinary Science *33*, 488-492. https://doi.org/10.1016/j.jevs.2013.03.187

Divers, T.J., Gardner, R.B., Madigan, J.E., Witonsky, S.G., Bertone, J.J., Swinebroad, E.L., Schutzer, S.E., and Johnson, A.I. (2018). *Borrelia burgdorferi* infection and Lyme disease in North American horses: A consensus report. Journal of Veterinary Internal Mediciine *32*, 617-632. doi: 10.1111/jvim.15042.

Duray, P.H. (1989). Histopathology of clincal phases of human Lyme disease. Rheumatic Disease Clinics of North America *15*, 691-709.

Dzierzecka, M., and Kita, J. (2002). The use of chosen serological diagnostic methods in Lyme disease in horses. Part I. Indirect immunoflourescence and enzyme-linked immunosorbent assay (ELISA). Polish Journal of Veterinary Science *5*, 71-77.

Dziezyc, J. (2000). Canine systemic bacterial infections. Veterinary Clinics of North America Small Animal Practice *30*, 1103-1117. doi: 10.1016/s0195-5616(00)05009-9.

Egenvall, A., Franzen, P., Gunnarsson, A., Egenvall, E., Vagsholm, I., Wikstrom, U.B., and Artursson, K. (2001). Cross-sectional study of the seroprevalence to *Borrelia burgdorferi* sensu lato and demographic, clinical and tick-exposure factors in Swedish horses. Prev Vet Med *49*, 191-208. doi: 10.1016/s0167-5877(01)00187-8.

Embers, M., Barthold, S.W., Borda, J.T., Bowers, L., Doyle, L., Hodzic, E., Jacobs, M.B., Hasenkampf, N.R., Martin, D.S., Narasimhan, S., *et al.* (2012). Persistence of *Borrelia burgdorferi* in rhesus macaques following antibiotic treatment of disseminated infection. PLoS ONE *7*, e29914. doi: 10.1371/journal.pone.0029914.

Eng, T.R., Wilson, M.L., Spielman, A., and Lastavica, C.C. (1988). Greater risk of *Borrelia burgdorferi* infection in dogs than in people. Journal of Infectious Diseases *158*, 1410-1411. DOI: 10.1093/infdis/158.6.1410

Eschner, A.K., and Mugnai, K. (2015). Immunization with a recominant subunit OspA vaccine markedly impacts the rate of newly acquired *Borrelia burgdorferi* infections in client-owned dogs living in a coastal community in Maine, USA. Parasites and Vectors *8*, 92. doi: 10.1186/s13071-015-0676-x.

Evans, R.D., Bosler, E.M., Orthel, F., Robertson, J.L., Schneider, E.M., LeFebvre, R.B., and Graham, M.D. (1995a). Canine Lyme borreliosis I. Gross clinical observations of laboratory beagles following exposure to ticks infected with *Borrelia burgdorferi*. Journal of Spirochetal and Tick-Borne Diseases 2, 28-32.

Evans, R.D., Robertson, J.L., Graham, M.D., Bosler, E.M., LeFebvre, R.B., and Schneider, E.M. (1995b). Canine Lyme borreliosis II. Minimal lesions in tissues of laboratory beagles following infection by exposure to Ixodid ticks infected with *Borrelia burgdorferi*. Journal of Spirochetal and Tick-Borne Diseases *2*, 33-36.

Fatterpekar, G.M., Gottesman, R.I., Sacher, M., and Som, P.M. (2002). Orbital Lyme disease: MR imaging before and after treatment: case report. Journal of Neuroradiology *23*, 657-659.

Flaminio, M.J.B.F., Tallmadge, R.L., Salles-Gomes, C.O.M., and Matychak, M.B. (2009). Common variable immunodeficiency in horses is characterized by B cell depletion in primary and secondary lymphoid tissues. Journal of Clinical Immunology *29*, 107-116. doi: 10.1007/s10875-008-9221-4.

Fridriksdottir, V., Overnes, G., and Stuen, S. (1992). Suspected Lyme borreliosis in sheep. Veterinary Record *130*, 323-324. DOI: 10.1136/vr.130.15.323

Funk, R.A., Pleasant, R.S., Witonsky, S.G., Reeder, D.S., Werre, S.R., and Hodson, D.R. (2016). Seroprevalence of *Borrelia burgdorferi* in horses presented for Coggins testing in southwest Virginia and change in positive results approximately 1 year later. Journal of Veterinary Internal Medicine *30*, 1300-1304. DOI: 10.1111/jvim.13973

Gall, Y., and Pfister, K. (2006). Survey on the subject of equine Lyme borreliosis. International Journal of Medical Microbiology *40*, 2724-2279. DOI: 10.1016/j.ijmm.2006.01.004

Gassner, F., Verbaarschot, P., Smallegange, R.C., Spitzen, J., VanWieren, S.E., and Takken, W. (2008). Variations in *Ixodes ricinus* density and Borrelia infections associated with cattle introduced into a woodland in The Netherlands. Applied and Environmental Microbiology *74*, 138-144. doi: 10.1128/AEM.00310-08.

Gebbia, J.A., Bosler, E.M., Evans, R.D., and Schneider, E.M. (1995). Acquired resistance in dogs to repeated infestation with *Ixodes scapularis* (Acari: Ixodidae) reduces tick viability and reproductive success. Experimental Applied Acarology *19*, 593-605. DOI: 10.1007/BF00048814

Gerber, B., Eichenberger, S., Haug, K., Wittenbrink, M.M., and Reusch, C.E. (2009). Association of urine protein excretion and infection with *Borrelia burgdorferi* sensu lato in Bernese Mountain dogs. Veterinary Journal *182*, 487-488. doi: 10.1016/j.tvjl.2008.08.022.

Gerber, B., Eichenberger, S., Joller-Jemelka, H.I., Wittenbrink, M.M., and Reusch, C.E. (2010). Complement C3 in Bernese Mountain Dogs. Veterinary Clinical Pathology *39*, 164-168. doi: 10.1111/j.1939-165X.2009.00205.x.

Gerber, B., Eichenberger, S., Wittenbrink, M.M., and Reusch, C.E. (2007). Increased prevalence of *Borrelia burgdorferi* infections in Bernese Mountain Dogs: a possible breed predisposition. BMC Veterinary Research *3*, 15. doi: 10.1186/1746-6148-3-15.

Gibson, M.D., Omran, M.T., and Young, C.R. (1995a). Experimental feline Lyme borreliosis as a model for testing *Borrelia burgdorferi* vaccines. Advances in Experimental Medicine and Biology *383*, 73-82. doi: 10.1007/978-1-4615-1891-4_9.

Gibson, M.D., Young, C.R., Omran, M.T., Edwards, J., Palma, K., Russell, L., and Rawlings, J. (1993). *Borrelia burgdorferi* infection of cats. Journal of the American Veterinary Medical Association *202*, 1786.

Gibson, M.D., Young, C.R., Omran, T., Palma, K., Edwards, J.F., Rawlings, J.A., and Lewis, D.H. (1995b). Lyme disease in an experimental cat model. International Journal of Angiology *4*, 155-159.

Gilger, B.C., and Deeg, C. (2011). Equine recurrent uveitis. In Equine Opththalmology, B.C. Gilger, ed. (St. Louis, Elsevier Saunders), pp. 317-349.

Giudice, E., Domina, F., Britti, D., DiPietro, S., and Pugliese, A. (2003). Clinical findings associated with *Borrelia burgdorferi* infection in the dog. Veterinary Research Communications *27 Suppl. 1*, 767-770. doi: 10.1023/b:verc.0000014267.25428.32.

Goldstein, R.E., Brovida, C., Palacio, M.J.F.-d., Littman, M.P., Polzin, D.J., Zatelli, A., and Cowgill, L.D. (2013). Consensus recommendations for treatment for dogs with serology positive glomerular disease. Journal of Veterinary Internal Medicine *27*, S60-S66. doi: 10.1111/jvim.12232.

Goldstein, R.E., Cordner, A.P., Sandler, J.L., Bellohusen, B.A., and Erb, H.N. (2007). Microalbuminuria and comparison of serologic testing for exposure to *Borrelia burgdorferi* in nonclinical Labrador and Golden Retrievers. Journal of Veterinary Diagnostic Investigation *19*, 294-297. doi: 10.1177/104063870701900312.

Good, R.A., and Zak, S.J. (1956). Disturbances in gamma globulin synthesis as experiments in nature. Pediatrics *18*, 109-149.

Grauer, G.F., Burgess, E.C., Cooley, A.J., and Hagee, J.H. (1988). Renal lesions associated with *Borrelia burgdorferi* infection in a dog. Journal of the American Veterinary Medical Association *193*, 237-238.

Greene, R.T., Walker, R.L., Burgess, E.C., and Levine, J.F. (1988a). Heterogeneity in immunoblot patterns obtained by using four strains of *Borrelia burgdorferi* and sera from naturally exposed dogs. Journal of Clinical Microbiology *26*, 2287-2291. DOI: 10.1128/JCM.26.11.2287-2291.1988

Grosenbaugh, D.A., Rissi, D.R., and Krimer, P.M. (2016). Demonstration of the ability of canine Lyme vaccine to reduce the incidence of histological synovial lesions following experimentall-induced canine Lyme borreliosis. Veterinary Immunology and Immunopathology *180*, 29-33. doi: 10.1016/j.vetimm.2016.08.014

Guarino, C., Asbie, S., Rohde, J., Glaser, A., and Wagner, B. (2017). Vaccination of horses with Lyme vaccines for dogs induces short-lasting antibody responses. Vaccine, 4140-4147. DOI: 10.1016/j.vaccine.2017.06.052

Gustafson, J.M., Burgess, E.C., Wachal, M.D., and Steinberg, H. (1993). Intrauterine transmission of *Borrelia burgdorferi* in dogs. American Journal of Veterinary Research *54*, 882-890.

Hahn, C.N., Mayhew, I.G., Whitwell, K.E., Smith, K.C., Carey, D., Carter, S.D., and Read, R.A. (1996). A possible case of Lyme borreliosis in a horse in the UK. Equine Veterinary Journal *28*, 84-88. DOI: 10.1111/j.2042-3306.1996.tb01595.x

Hebert, D., and Eschner, A. (2010). Seroprevalence of *Borrelia burgdorferi*-specific C6 antibody in dogs before and after implementation of a nonadjuvanted recombinant outer surface protein A vaccine in a Rhode Island small animal clinic. Veterinary Therapeutics *11*, E1-9.

Hodzic, E., Feng, S., Holden, K., Freet, K.J., and Barthold, S.W. (2008). Persistence of *Borrelia burgdorferi* following antibiotic treatment in mice. Antimicrobial Agents and Chemotherapy *52*, 1728-1736. doi: 10.1128/AAC.01050-07.

Hodzic, E., Imai, D., Feng, S., and Barthold, S.W. (2014). Resurgence of persisting non-cultivable *Borrelia burgdorferi* following antibiotic treatment in mice. PLoS ONE *9*, e86907. doi: 10.1371/journal.pone.0086907. eCollection 2014.

Holak, H., Holak, N., Huzarska, M., and Holak, S. (2006). Tick inoculation in an eyelid region: report on five cases with one complication of the orbital myositis associated with Lyme borreliosis. Klinikea Oczona *108*, 220-224.

Hovius, J.W.R., Hovius, K.E., Oei, A., Houwers, D.J., and vanDam, A.P. (2000). Antibodies against specific proteins of and immobilizing activity against three strains of *Borrelia burgdorferi* sensu lato can be found in symptomatic but not in infected asymptomatic dogs. Journal of Clinical Microbiology *38*, 2611-2621. doi: 10.1128/JCM.38.7.2611-2621.2000.

Hovius, K.E., Stark, L.A., Bleumink-Pluym, N.M., vandePol, I., Verbeek-deKruif, N., Rijpkema, S.G., Schouls, L.M., and Houwers, D.J. (1999). Presence and distribution of *Borrelia burgdorferi* sensu lato species in internal organs and skin of naturally infected symptomatic and asymptomatic dogs, as detected by polymerase chain reaction. Veterinary Quarterly *21*, 54-58. doi: 10.1080/01652176.1999.9694992

Hovmark, A., Asbrink, E., Schwan, O., Hederstedt, B., and Christensson, D. (1986). Antibodies to Borrelia spirochetes in sera from Swedish cattle and sheep. Acta Veterinaria Scandinavica *27*, 479-485.

Hutton, T.A., Goldstein, R.E., Njaa, B.L., Atwater, D.Z., Chang, Y.F., and Simpson, K.W. (2008). Search for *Borrelia burgdorferi* in kidneys of dogs with suspected "Lyme nephritis". Journal of Veterinary Internal Medicine *22*, 860-865. DOI: 10.1111/j.1939-1676.2008.0131.x

Imai, D.M., Barr, B.C., Daft, B., Bertone, J.J., Feng, S., Hodzic, E., Johnson, J.M., Olsen, K.J., and Barthold, S.W. (2011). Lyme neuroborreliosis in 2 horses. Veterinary Pathology *48*, 1151-1157. DOI: 10.1177/0300985811398246

Jacobson, R.H., Chang, Y.F., and Shin, S.J. (1996). Lyme disease: laboratory diagnosis of infected and vaccinated symptomatic dogs. Seminars in Veterinary Medicine and Surgery (Small Animal) *11*, 172-182. doi: 10.1016/s1096-2867(96)80030-2.

James, F.M., Engiles, J.B., and Beech, J. (2010). Meningitis, cranial neuritis, and radiculoneuritis associated with *Borrelia burgdorferi* infection in a horse. Journal of the American Veterinary Medical Association *237*, 1180-1185. DOI: 10.2460/javma.237.10.1180

Ji, B., and Collins, M.T. (1994). Seroepidemiologic survey of *Borrelia burgdorferi* exposure of dairy cattle in Wisconsin. American Journal of Veterinary Research *55*, 1228-1231.

Johnson, A.L., Divers, T.J., and Chang, Y.F. (2008). Validation of an in-clinic enzyme-linked immunosorbent assay kit for diagnosis of *Borrelia burgdorferi* infection in horses. Journal of Veterinary Diagnostic Investigation *20*, 321-324. DOI: 10.1177/104063870802000309

Johnstone, L.K., Engiles, J.B., Aceto, H., Buechner-Maxwell, V., Divers, T., Gardner, R., Levine, R., Scherrer, N., Tewari, D., Tomlinson, J., *et al.* (2016). Retrospective evaluation of horses diagnosed with neuroborreliosis on postmortem examination: 16 cases (2004-2015). Journal of Veterinary Internal Medicine *30*, 1305-1312. doi: 10.1111/jvim.14369.

Jones, E.H., Hinckley, A.F., Hook, S.A., Meek, J.I., Bachenson, B., Kugeler, K.J., and Feldman, K.A. (2018). Pet ownership increases human risk of encountering ticks. Zoonoses Public Health *65*, 74-79. DOI: 10.1111/zph.12369

Kasbohrer A, and A, S. (1990). Serologic studies of the occurrence of *Borrelia burgdorferi* in domestic animals in Berlin (West). Berliner Munchener tierartliche Wochenschrift *193*, 374-378.

Kelly, B., Finnegan, P., Cormican, M., and Callaghan, J. (1999). Lyme disease and glomerulonephritis. Irish Medical Journal *92*, 372.

Kirmizis, D., and Efstratiadis, G. (2013). MPGN secondary to Lyme disease: the role of cryoglobulins. Clinical Kidney Journal *6*, 349. doi: 10.1053/j.ajkd.2003.11.014.

Kornblatt, A.N., Urband, P.H., and Steere, A.C. (1985). Arthritis caused by *Borrelia burgdorferi* in dogs. Journal of the American Veterinary Medical Association *186*, 960-964.

Korshus, J.B., Munderloh, U.G., Bey, R.F., and Kurtti, T.J. (2004). Experimental infection of dogs with *Borrelia burgdorferi* sensu stricto using *Ixodes scapularis* ticks artificially infected by capillary feeding.

Medical Microbiology and Immunology *193*, 27-34. · DOI: 10.1007/s00430-003-0178-x

Krimer, P.M., Miller, A.D., Li, Q., Grosenbaugh, D.A., Susta, L., and Schatzberg, S.J. (2011). Molecular and pathological investigations of the central nervous system in *Borrelia burgdorferi*-infected dogs. Journal of Veterinary Diagnostic Investigation *23*, 757-763. · DOI: 10.1177/1040638711408281

Krupka, I., and Straubinger, R.K. (2010). Lyme borreliosis in dogs and cats: background, diagnosis, treatment and prevention of infections with *Borrelia burgdorferi* sensu stricto. Veterinary Clinics of North America Small Animal Practice *40*, 1103-1119. doi: 10.1016/j.cvsm.2010.07.011.

Kybicova, K., Schanilec, P., Hulinska, D., Uherkiva, L., Kurzova, Z., and Spejchalova, S. (2009). Detection of *Anaplasma phagocytophilum* and *Borrelia burgdorferi* sensu lato in dogs in the Czech Republic. Vector Borne Zoonotic Dis *9*, 655-661. doi: 10.1089/vbz.2008.0127.

LaFleur, R.L., Callister, S.M., Dant, J.C., Jobe, D.A., Lovrich, S.D., Warner, T.F., Wasmoen, T.L., and Schell, R.F. (2010). One-year duration of immunity induced by vaccination with a canine Lyme disease bacterin. Clinical and Vaccine Immunology *17*, 870-874. doi: 10.1128/CVI.00524-09

LaFleur, R.L., Callister, S.M., Dant, J.C., Wasmoen, T.L., Jobe, D.A., and Lovrich, S.D. (2015). Vaccination with the ospA- and ospB-negative *Borrelia burgdorferi* strain 50772 provides significant protection against canine Lyme disease. Clinical and Vaccine Immunology *22*, 836-839. doi: 10.1128/CVI.00193-15

LaFleur, R.L., Dant, J.C., Wasmoen, T.L., Callister, S.M., Jobe, D.A., Lovrich, S.D., Warner, T.F., Abdelmagid, O., and Schell, R.F. (2009). Bacterin that induces anti-OspA and anti-OspC borreliacidal antibodies provides a high level of protection against canine Lyme disease. Clinical and Vaccine Immunology *16*, 253-259. doi: 10.1128/CVI.00373-08

Lappin, M.R., Chandrashekar, R., Stillman, B., Liu, J., and Mather, T.N. (2015). Evidence of *Anaplasma phagocytophilum* and *Borrelia burgdorferi* infection in cats after exposure to wild-caught adult *Ixodes scapularis*. Journal of Veterinary Diagnostic Investigation *27*, 522-525. DOI: 10.1177/1040638715593598

Leschnik, M.W., Kirtz, G., Khanakah, G., Duscher, G., Leidinger, E., Thalhammer, J.G., Joachim, A., and Stanek, G. (2010). Humoral immune response in dogs naturally infected with *Borrelia burgdorferi* sensu lato and in dogs after immunization with a Borrelia vaccine. Clinical Vaccine Immunology *17*, 828-835. doi: 10.1128/CVI.00427-09

Lesser, R.L. (1995). Ocular manifestations of Lyme disease. American Journal of Medicine *98*, 60S-62S. https://doi.org/10.1016/S0002-9343(99)80045-X

Levy, S.A. (1989). Lyme borreliosis in domestic animals. Connecticut Medicine *53*, 352-353.

Levy, S.A. (2002). Use of a C6 ELISA test to evaluate the efficacy of a whole-cell bacterin for the prevention of naturally transmitted canine *Borrelia burgdorferi* infection. Veterinary Therapeutics *3*, 420-424.

Levy, S.A., and Dressen, D.W. (1992). Lyme borreliosis in dogs. Canine Practice 2, 5-13.

Levy, S.A., and Duray, P.H. (1988). Complete heart block in a dog seropositive for *Borrelia burgdorferi*. Similarity to human Lyme carditis. Journal of Veterinary Internal Medicine 2, 138-144.

Levy, S.A., and Magnarelli, L.A. (1992). Relationship between development of antibodies to *Borrelia burgdorferi* in dogs and the subsequent development of limb/joint borreliosis. Journal of the American Veterinary Medical Association *200*, 344-347.

Levy, S.A., O'Connor, T.P., Hanscom, J.L., and Shields, P. (2003). Evaluation of a canine C6 ELISA Lyme disease test for the determination of the infection status of cats naturally exposed to *Borrelia burgdorferi*. Veterinary Therapeutics *4*, 172-177.

Liang, F.T., Jacobson, R.H., Straubinger, R.K., Grooters, A., and Philipp, M.T. (2000). Characterization of a *Borrelia burgdorferi* VlsE invariable region useful in canine Lyme disease serodiagnosis by enzyme-linked immunosorbent assay. Journal of Clinical Microbiology *38*, 4160-4166.

Linder, A., and Bockel, K. (1995). Serological investigations into the occurrence of antibodies against *Borrelia burgdorferi* in dogs and cats. Praktische Tierarzt *76*, 177.

Lischer, C.J., Leutenegger, C.M., Braun, U., and Lutz, H. (2000). Diagnosis of Lyme disease in two cows by the detection of *Borrelia burgdorferi* DNA. Veterinary Record *146*, 497-499.

Lissman, B.A., Bosler, E.M., Camay, H., Ormiston, B.G., and Benach, J.L. (1984). Spirochete-associated arthritis (Lyme disease) in a dog. Journal of the American Veterinary Medical Association *185*, 219-220.

Little, S.E., Beall, M.J., Bowman, D.D., Chandrashekar, R., and Stamaris, J. (2014). Canine infection with *Dirofilaria immitis*, *Borrelia burgdorferi*, *Anaplasma* sp., and *Ehrlichia* sp. in the United States, 20110-2012. Parasitology Vectors *7*, 257. doi: 10.1186/1756-3305-7-257.

Littman, M.P. (2013). Lyme nephritis. Journal of Veterinary Emergency and Critical Care (San Antonio) *23*, 163-173. doi: 10.1111/vec.12026.

Littman, M.P., Gerber, B., Goldstein, R.E., Labato, M.A., Lappin, M.R., and Moore, G.E. (2018). ACVIM consensus update on Lyme borreliosis in dogs and cats. Journal of Veterinary Internal Mediciine *32*, 887-903. doi: 10.1111/jvim.15085.

Littman, M.P., Goldstein, R.E., Labato, M.A., Lappin, M.R., and Moore, C.E. (2006). ACVIM small animal consensus statement on Lyme disease in dogs: diagnosis, treatment, and prevention. Journal of Veterinary Internal Mediciine *20*, 442-434. doi: 10.1892/0891-6640(2006)20[422:asacso]2.0.co;2.

Ma, J., Bulger, P.A., Dante, S., davis, D.R., Perilli-Palmer, B., and Coughlin, R.T. (1995). Characterization of canine humoral immune responses to outer surface protein subunit vaccines and to natural infection by Lyme disease spirochetes. Journal of Infectious Diseases *171*, 909-915. doi: 10.1093/infdis/171.4.909.

Ma, J., Hine, P.M., Clough, E.R., Fish, D., Coughlin, R.T., Beltz, G.A., and Shew, M.G. (1996). Safety, efficacy, and immunogenicity of a recombinant Osp subunit to whole-cell and recombinant *Borrelia burgdorferi* antigens in cattle. Canadian Veterinary Journal *45*, 667-673.

Madigan, J.E., and Teitler, J. (1988). *Borrelia burgdorferi* borreliosis. Journal of the American Veterinary Medical Association *192*, 892-896.

Magnarelli, L.A., and Anderson, J.F. (1989). Class-specific and polyvalent enzyme-linked immunosorbent assays for the detection of antibodies to *Borrelia burgdorferi* in equids. Journal of the American Veterinary Medical Association *10*, 1365-1368.

Magnarelli, L.A., Anderson, J.F., Burgdorfer, W., and Chappell, W.A. (1984). Parasitism by *Ixodes dammini* (Acari:Ixodidae) and antibodies to spirochetes in mammals at Lyme disease foci in Connecticut. Journal of Medical Entomology *21*, 52-57. doi: 10.1093/jmedent/21.1.52.

Magnarelli, L.A., Anderson, J.F., Kaufman, A.F., Lieberman, L.L., and Whitney, G.D. (1985). Borreliosis in dogs from southern Connecticut. Journal of the American Veterinary Medical Association *186*, 955-059.

Magnarelli, L.A., Anderson, J.F., Levine, H.R., and Levy, S.A. (1990a). Tick parasitism and antibodies to *Borrelia burgdorferi* in cats. Journal of the American Veterinary Medical Association *197*, 63-66.

Magnarelli, L.A., Anderson, J.F., and Schreier, A.B. (1990b). Persistence of antibodies to *Borrelia burgdorferi* in dogs of New York and Connecticut. Journal of the American Veterinary Medical Association *196*, 1064-1068.

Magnarelli, L.A., Anderson, J.F., Schreier, A.B., and Ficke, C.M. (1987). Clinical and serologic studies of canine borreliosis. Journal of the American Veterinary Medical Association *191*, 1089-1094.

Magnarelli, L.A., Anderson, J.F., Shaw, E., Post, J.E., and Palka, F.C. (1988). Borreliosis in equids in northeastern United States. American Journal of Veterinary Research *49*, 359-362.

Magnarelli, L.A., Bushmich, S.L., Sherman, B.A., and Fikrig, E. (2004). A comparison of serologic tests for the detection of serum antibodiesMagnarelli, L.A., Flavell, R.A., Padula, S.J., Anderson, J.F., and Fikrig, E. (1997). Serologic diagnosis of canine and equine borreliosis: use of recombinant antigens in enzyme-linked immunosorbent assays. Journal of Clinical Microbiology *35*, 169-173. doi: 10.1128/JCM.35.1.169-173.1997.

Magnarelli, L.A., Levy, S.A., IJdo, J.W., Wu, C., Padula, S.J., and Fikrig, E. (2001). Reactivity of dog sera to whole-cell or recombinant antigens of *Borrelia burgdorferi* by ELISA and immunoblot analysis. Journal of Medical Microbiology *50*, 889-895. doi: 10.1099/0022-1317-50-10-889.

Malalana, F., Stylianides, A., and McGowan, C. (2015). Equine recurrent uveitis: human and equine perspectives. Veterinary Journal *206*, 22-29. doi: 10.1016/j.tvjl.2015.06.017.

Manion, T.B., Bushmich, S.L., Khan, M.I., Dinger, J., Werner, H., Mittel, L., Laurendeau, M., and Reilly, M. (2001). Suspected clinical Lyme disease in horses: serological and antigen testing differences between clinically ill and clinically normal horses from an endemic area. Journal of Equine Veterinary Science *21*, 229-234.

Marcus, L.C., Patterson, M.M., Gilfillan, R.E., and Urban, P.H. (1985a). Antibodies to *Borrelia burgdorferi* in New England horses: serologic survey. American Journal of Veterinary Research *46*, 2570-2571.

Marcus, L.C., Steere, A.C., Duray, P.H., Anderson, A.E., and Mahoney, E.B. (1985b). Fatal pancarditis in a patient with coexistant Lyme disease and babesiosis. Demonstration of spirochetes in the myocardium. Annals of Internal Medicine *103*, 374-376. DOI: 10.7326/0003-4819-103-3-374

Mather, T.N., Fish, D., and Coughlin, R.T. (1994). Competence of dogs as reservoirs for Lyme disease spirochetes (*Borrelia burgdorferi)*. Journal of the American Veterinary Medical Association *205*, 186-188.

Matuschka, F.R., Heiler, M., Eiffert, H., Fischer, P., Lotter, H., and Spielman, A. (1993). Diversionary role of hoofed game in the transmission of Lyme disese spirochetes. American Journal of Tropical Medicine and Hygiene *48*, 693-699. DOI: 10.4269/ajtmh.1993.48.693

May, C., Bennett, D., and Carter, S.D. (1990). Lyme disease in the dog. Veterinary Record, 293.

May, C., Carter, S.D., Barnes, A., Mclean, C., Bennett, M., Coutts, A., and Grant, C.K. (1994). *Borrelia burgdorferi* infection in cats in the UK. Journal of Small Animal Practice *35*, 517-520.

McCausland, F.R., Niedermaier, S., Bijol, V., Rennke, H.C., Choi, M.E., and Forman, J.P. (2011). Lyme disease-associated glomerulonephritis. Nephrology Dialysis Transplantation *26*, 3054-3056. DOI: 10.1093/ndt/gfr335

McKisic, M.D., and Barthold, S.W. (2000). T-cell-independent responses to *Borrelia burgdorferi* are critical for protective immunity and resolution of Lyme disease. Infection and Immunity *68*, 5190-5197. doi: 10.1128/iai.68.9.5190-5197.2000.

Miklossy, J. (2012). Chronic or late Lyme neuroborreliosis: analysis of evidence compared to chronic or late neurosyphilis. The Open Neurology Journal *6* 146-157. doi: 10.2174/1874205X01206010146.

Moroff, S., Woodruff, C., Woodring, T., Sokolchick, I., and Lappin, M.R. (2015). Multiple antigen target approach using the Accuplex4 BioCD system to detect *Borrelia burgdoreri* antibodies in experimentally infected and vaccinated dogs. Journal of Veterinary Diagnostic Investigation *27*, 581-588. doi: 10.1177/1040638715600196.

Muehlenbachs, A., Bollweg, B.C., Schultz, T.J., Forrester, J.D., DeleonCarnes, M., Molins, C., Rays, G.S., Cummings, P.M., Ritter, J.M., Blau, D.M., *et al.* (2016). Cardiac tropism of *Borrelia burgdorferi*: an autopsy study of suddent cardiac death associated with Lyme carditis. American Journal of Pathology *186*, 1195-1205. doi: 10.1016/j.ajpath.2015.12.027.

Munger, R.J. (1990). Uveitis as a manifestation of *Borrelia burgdorferi* infection in dogs. Journal of the American Veterinary Medical Association *197*, 811.

O'Connor, T.P., Esty, K.J., Hanscom, J.L., Shields, P., and Philipp, M.T. (2004). Dogs vaccinated with common Lyme disease vaccines do not respond to IR6, the conserved immunodominant region of the VlsE surface protein of *Borrelia burgdorferi.* Clinical and Diagnostic Laboratory Immunology *11*, 458-462. doi: 10.1128/CDLI.11.3.458-462.2004.

Ogden, N.H., Nuttall, P.A., and Randolph, S.E. (1997). Natural Lyme disease cycles maintained via sheep by co-feeding ticks. Parasitology *115*, 591-599. doi: 10.1017/s0031182097001868.

Pantchev, N., Pluta, S., Huisinga, E., Nather, S., Scheufelen, M., Vrhovec, M.G., Schweinitz, A., Hampel, H., and Straubinger, R.K. (2015). Tick-borne diseases (Borreliosis, Anaplasmosis, Babesiosis) in German and Austrian dogs: status quo and review of distribution, transmission, clinical findings, diagnostics and prophylaxis. Parasitology Research *114 (Suppl 1)*, S19-S54. doi: 10.1007/s00436-015-4513-0.

Pantchev, N., Vrhovec, M.G., Pluta, S., and Straubinger, R.K. (2016). Seropositivity of *Borrelia burgdorferi* in a cohort of symptomatic cats from Europe based on C6-peptide assay with discussion of implications in disease aetiology. Berliner und Munchen tierartzl Wochenschrift *129*, 333-339.

Papineni, P., Doherty, T., Pickett, T., Toth, T., and Boddana, P. (2010). Membranous glomerulonephritis secondary to *Borrelia burgdorferi* infection presenting as nephrotic syndrome. NDT Plus *3*, 105-106. DOI: 10.1093/ndtplus/sfp160

Parker, J.L., and White, K.K. (1992). Lyme borreliosis in cattle and horses: a review of the literature. Cornell Veterinarian *82*, 253-274.

Passamonti, F., Veronesi, F., Cappelli, K., Capomaccio, S., Reginato, A., Miglio, A., Vardi, D.M., Stefanetti, V., Coletti, M., Bazzica, C., *et al.* (2015). Polysynovitis in a horse due to *Borrelia burgdorferi* sensu lato infection - case study. Annals of Agriculture and Environmental Medicine *22*, 247-250. DOI: 10.5604/12321966.1152074

Pecoraro, H.L., Felippe, M.J.B., Miller, A.D., Divers, T.J., Simpson, K.W., Guyer, K.M., and Duhamel, G.E. (2019). Neuroborreliosis in a horse with common variable immunodeficiency. Journal of Veterinary Diagnostic Investigation *31*, 241-245. DOI: 10.1177/1040638718824146

Pellegrini-Masini, A., Bentz, A.I., Johns, I.C., Parsons, C.S., Beech, J., Whitlock, R.H., and Flaminio, M.J. (2005). Common variable immunodeficiency in three horses with presumptive bacterial meningitis. Journal of the American Veterinary Medical Association *227*, 114-122.

Persing, D.H., Rutledge, B.J., Rys, P.N., Podzorski, D.S., Mitchell, P.D., Reed, K.D., Liu, B., Fikrig, E., and Malawista, S.E. (2000). Target imbalance: disparity of *Borrelia burgdorferi* genetic material in synovial fluid from Lyme arthritis patients. Journal of Infectious Diseases *169*, 668-672. DOI: 10.1093/infdis/169.3.668 Peterhans, E., and Peterhans, E. (2010). "Lyme disease" as a possible cause for lameness in the cat. Schweizer Archiv fur Tierheilkunde *152*, 295-297. DOI: 10.1024/0036-7281/a000067

Post, J.E., Shaw, E.E., and Wright, S.D. (1988). Suspected borreliosis in cattle. Annals of the New York Academy of Sciences *539*, 488. DOI : 10.1111/j.1749-6632.1988.tb31916.x

Priest, H.L., Irby, N.L., Schlafer, D.H., Divers, T.J., Wagner, B., Glaser, A.L., Chang, Y.-F., and Smith, M.C. (2012). Diagnosis of Borrelia-associated uveitis in two horses. Veterinary Ophthalmology *15*, 398-405. DOI: 10.1111/j.1463-5224.2012.01000.x

Pusterla, N., Chae, J.-S., Kimsey, R.B., Pusterla, J.B., DeRock, E., Dumler, J.S., and Madigan, J.E. (2002). Transmission of *Anaplasma phagocytophilum* (human granulocytic ehrlichiosis agent) in horses using experimentally infected ticks (*Ixodes scapularis)*. Journal of Veterinary Medicine B *49*, 484-488. DOI: 10.1046/j.1439-0450.2002.00598.x

Raya, A.I., Afonso, J.C., Perez-Ecija, R.A., Blanco, B., Martin-de-LasMulas, J., Carrasco, L., Cortade, L.E., and Gomez-Laguna, J.

(2010). Orbital myositis associated with Lyme disease in a dog. Veterinary Record *167*, 663-664. DOI: 10.1136/vr.c4841

Richter, D., and Matuschka, F.-Z. (2010). Elimination of Lyme disease spirochetes from ticks feeding on domestic ruminants. Applied and Environmental Microbiology *76*, 7650-7652. DOI: 10.1128/AEM. 01649-10

Robinson, M.L., Kobayashi, T., Higgins, Y., Calkins, H., and Melia, M.T. (2015). Lyme carditis. Infectious Disease Clinics of North America *29*, 255-268. DOI: 10.1016/j.idc.2015.02.003

Rolla, D., Conti, N., Ansaldo, F., Panaro, L., and Lusenti, T. (2014). Post-infectious glomerulonephritis presenting as acute renal failure in a patient with Lyme disease. Journal of Renal Injury Prevention *3*, 17-20. DOI: 10.12861/jrip.2014.07

Rothermel, H., Hedges, T.R., and Steere, A.C. (2001). Optic neuropathy in children with Lyme disease. Pediatrics *108*, 477-481. DOI: 10.1542/ peds.108.2.477

Rothwell, J.T., Christie, B.M., Williams, C., and Walker, K.H. (1989). Suspected Lyme disease in a cow. Australian Veterinary Journal *66*, 296-298. DOI: 10.1111/j.1751-0813.1989.tb13956.x

Roush, J.K., Manley, P.A., and Dueland, R.T. (1989). Rheumatoid arthritis subsequent to *Borrelia burgdorferi* infection in two dogs. Journal of the American Veterinary Medical Association *195*, 951-953.

Schneider, C.A., Wiemer, J., Seibt-Meisch, S., Bruckner, W., Amman, K., and Scherberich, J.E. (2013). Borrelia and nephropathy: cryoglobulinemic membranoproliferative glomerulonephritis responsive to doxycycline in active Lyme disease. Clinical Kidney Journal *6*, 77-80. DOI: 10.1093/ckj/sfs149

Schulze, T.L., Bosler, E.M., Shisler, J.K., Ware, I.C., Lakat, M.F., and Parkin, W.E. (1986). Prevalence of canine Lyme disease from an endemic area as determined by serosurvey. Zentralblatt fur Bakteriologie and Hygiene A *263*, 427-434. doi: 10.1016/ s0176-6724(87)80104-9.

Sears, K.P., Divers, T.J., Neff, R.T., Miller, W.H., and McDonough, S.P. (2011). A case of Borrelia-associated cutaneous pseudolymphoma in a horse. Veterinary Dermatology *23*, 153-156. DOI: 10.1111/j. 1365-3164.2011.01013.x

Seidenberg, K.B., and Leib, M.L. (1990). Orbital myositis with Lyme disease. American Journal of Ophthalmology *109*, 13-16. · DOI: 10.1016/s0002-9394(14)75571-0

Shaw, S.E., Binns, S.H., Birles, R.J., Day, M.J., Smithson, R., and Kenny, M.J. (2005). Molecular evidence of tick-transmitted infections in dogs and cats in the United Kingdom. Veterinary Record *157*, 645-648. doi: 10.1136/vr.157.21.645.

Shin, S.J., Chang, Y.-F., Jacobson, R.H., Shaw, E., Lauderdale, T.-L., Appel, M.J., and Lein, D.H. (1993). Cross-reactivity between *Borrelia burgdorferi* and other spirochetes affects specificity of serotests for detection of antibodies to the Lyme disease agent in dogs. Veterinary Microbiology *36*, 161-174. doi: 10.1016/0378-1135(93)90137-v.

Skotarczak, B. (2014). Why are there several species of *Borrelia burgdorferi* sensu lato detected in dogs and humans? Infection, Genetics and Evolution *23*, 182-188. doi: 10.1016/j.meegid. 2014.02.014.

Sommerauer, S., Blohm, K.-O., Spergser, J., and Buchner, H.H.F. (2019). Arthritis, panuveitis and hyperaesthesia associated with *Borrelia afzelii* infection in a warmblood gelding. Veterinary Record Case Reports 7, e000911. doi:10.1136/vetreccr-2019-000911

Speck, S., Reiner, B., and Wittenbrink, M.M. (2001). Isolation of *Borrelia afzelii* from a dog. Veterinary Record *149*, 19-20. doi: 10.1136/vr. 149.1.19.

Steere, A.C., Broderick, T.F., and Malawista, S.E. (1978). Erythema chronicum migrans and Lyme arthritis: epidemiologic evidence for a tick vector. American Journal of Epidemiology *108*, 312-321. doi: 10.1093/oxfordjournals.aje.a112625.

Stefancikova, A., Adaszek, L., Petko, B., Winiarczyk, S., and Dudinak, V. (2008). Serologic evidence of *Borrelia burgdorferi* sensu lato in horses and cattle from Poland and diagnostic problems of Lyme borreliosis. Annals of Agricultural and Environmental Medicine *15*, 37-43.

Stefancikova, A., Stepanova, G., Derdakova, M., Petko, B., Kyselova, J., Ciganek, J., Strojny, L., Cislakova, L., and Travnicek, M. (2002). Serologic evidence for *Borrelia burgdorferi* infection assocated with clinical signs in dairy cattle in Slovakia. Veterinary Research Communications *26*, 601-611. doi: 10.1023/a:1020912618950.

Straubinger, R.K. (2000). PCR-based quantification of *Borrelia burgdorferi* organisms in canine tissues over a 500-day postinfection period. Journal of Clinical Microbiology *38*, 2191-2199.

Straubinger, R.K., Chang, Y.-F., Jacobson, R.H., and Appel, M.J.G. (1995). Sera from OspA-vaccinated dogs, but not those from tick-infected dogs, inhibit in vitro growth of *Borrelia burgdorferi.* Journal of Clinical Microbiology *33*, 2745-2751. doi: 10.1128/JCM. 33.10.2745-2751.1995.

Straubinger, R.K., Rao, T.D., Davidson, E., Summers, B.A., Jacobson, R.H., and Frey, A.B. (2001). Protection against tick-transmitted Lyme disease in dogs vaccinated with a multiantigenic vaccine. Vaccine *20*, 181-193. doi: 10.1016/s0264-410x(01)00251-1.

Straubinger, R.K., Straubinger, A.F., Harter, L., Jacobson, R.H., Chang, Y.-F., Summers, B.A., Erb, H.N., and Appel, M.J.G. (1997a). *Borrelia burgdoferi* migrates into joint capsules and causes an up-regulation of interleukin-8 in synovial membranes of dogs experimentally infected with ticks. Infection and Immunity *65*, 1273-1285. doi: 10.1128/IAI. 65.4.1273-1285.1997.

Straubinger, R.K., Straubinger, A.F., Summers, B.A., and Jacobson, R.H. (2000). Status of *Borrelia burgdorferi* infection after antibiotic treatment and the effects of corticosteroids: an experimental study. Journal of Infectious Diseases *181*, 1069-1081. doi: 10.1086/315340.

Straubinger, R.K., Straubinger, A.F., Summers, B.A., Jacobson, R.H., and Erb, H.N. (1998). Clinical manifestations, pathogenesis and effects of antibiotic treatment on Lyme borreliosis in dogs. Wiener Klinische Wochenschrift *110*, 874-881.

Straubinger, R.K., Summers, B.A., Chang, Y.-F., and Appel, M.J.G. (1997b). Persistence of *Borrelia burgdorferi* in experimentally infected dogs after antibiotic trearment. Journal of Clinical Immunology *35*, 111-116. doi: 10.1128/JCM.35.1.111-116.1997.

Summers, B.A., Straubinger, A.F., Jacobson, R.H., Chang, Y.-F., Appel, M.J.G., and Straubinger, R.K. (2005). Histopathological studies of experimental Lyme disease in the dog. Journal of Comparative Pathology *133*, 1-13. doi: 10.1016/j.jcpa.2004.11.006.

Susta, L., Uhl, E.W., Grosenbaugh, D.A., and Krimer, P.M. (2012). Synovial lesions in experimental canine Lyme borreliosis. Veterinary Pathology *49*, 453-261. DOI: 10.1177/0300985811424754

Sykes, J.E. (2014). Lyme Borreliosis. In Canine and Feline Infectious Diseases, J.E. Sykes, ed. (St. Louis, MO, Elsevier/Saunders), pp. 487-497.

Telford, S.R., Mather, T.N., Moore, S.I., Wilson, M.L., and Spielman, A. (1988). Incompetence of deer as reservoirs of the Lyme disease spirochete. American Journal of Tropical Medicine and Hygiene *39*, 105-109. doi: 10.4269/ajtmh.1988.39.105.

Topfer, K.H., and Straubinger, R.K. (2007). Characterization of humoral immune response in dogs after vaccination against the Lyme borreliosis agent. A study with five commercial vaccines using two different vaccination schedules. Vaccine *25*, 314-326. DOI: 10.1016/ j.vaccine.2006.07.031

Travnicek, M., Stefancikova, A., Nadzamova, D., Stanko, M., Cislakova, L., Petko, B., Mardzinova, S., and Bhide, M.R. (2002). Seroprevalence of anti-*Borrelia burgdorferi* antibodies in sheep and goats from mountainous areas of Slovakia. Annals of Agricultural and Environmental Medicine 9, 153-155.

Wagner, B., Freer, H., Rollins, A., Erb, H.N., and Lu, Z. (2011). Development of a multiplex assay for the detection of antibodies to *Borrelia burgdorferi* in horses and its validation using Bayesian and conventional statistical methods. Veterinary Immunology and Immunopathology *144*, 374-381. doi: 10.1016/j.vetimm.2011.08.005.

Wagner, B., Freer, H., Rollins, A., Garcia-Tapia, D., Erb, H.N., Earnhart, C., Marconi, R., and Meeus, P. (2012). Antibodies to *Borrelia burgdorferi* OspA, OspC, OspF, and C6 antigens as markers for early and late infection in dogs. Clinical Vaccine Immunology *19*, 527-535. doi: 10.1128/CVI.05653-11.

Wagner, B., Goodman, L.B., Rollins, A., and Freer, H.S. (2013). Antibodies to OspC, OspF and C6 antigens as indicators for infection with *Borrelia burgdorferi* in horses. Equine Veterinary Journal *45*, 533-537. doi: 10.1111/evj.12033.

Wagner, B., Johnson, J., Garcia-Tapia, D., Honsberger, N., King, V., Strietzel, C., Hardham, J.M., Heinz, T.J., Marconi, R.T., and Meeus, P.F.M. (2015). Comparison of effectiveness of cefovecin, doxycycline, and amoxicillin for the treatment of experimentally induced early Lyme borreliosis in dogs. BMC Veterinary Research *11*, 163. doi: 10.1186/s12917-015-0475-9.

Wan, K., Zhang, Z., Wang, H., and Hou, X. (1999). Preliminary investigation on reservoir hosts of *Borrelia burgdorferi* in China (in Chinese). Wei Sheng Yan Jiu *28*, 7-9.

Wasmoen, T.L., Sebring, R.W., Blumer, B.M., Chavez, L.G., Chu, H.-J., and Acree, W.M. (1992). Examination of Koch's postulates for *Borrelia burgdorferi* as the causative agent of limb/joint dysfunction in dogs with borreliosis. Journal of the American Veterinary Medical Association *201*, 412-418.

Wells, S.J., Trent, A.M., Robinson, R.A., Knutson, K.S., and Bey, R.F. (1993). Association between clinical lameness and *Borrelia burgdorferi* antibody in dairy cows. American Journal of Veterinary Research *54*, 398-405.

Wikle, R.E., Fretwell, B., Jarecki, M., and Jarecki-Black, J.C. (2006). Canine Lyme disease: one-year duration of immunity elicited with a canine OspA monovalent Lyme vaccine. International Journal of Applied Research in Veterinary Medicine *4*, 23-28.

Wittenbrink, M.M., Failing, K., and Kraus, H. (1996). Enzyme-linked immunosorbent assay and immunoblot analysis for detection of antibodies to *Borrelia burgdorferi* in dogs. The impact of serum absorption with homologous and heterologous bacteriae. Veterinary Microbiology *48*, 257-268. doi: 10.1016/0378-1135(95)00165-4.

Wooding, P., and Burton, G. (2008). Comparative Placentation: Structures, Functions and Evolution. (New York, Springer).

Wormser, G.P., Dattwyler, R.J., Shapiro, E.D., Halperin, J.J., Steere, A.C., Klempner, M.S., Krause, P.J., Bakken, J.S., Strle, F., Stanek, G., *et al.* (2006). The clinical assessment, treatment, and prevention of Lyme disease, human granulocytic anaplasmosis, and babesiosis: cliinical practice guidelines by the Infectious Disease Society of America. Clinical Infectious Disease *43*, 1089-1134. doi: 10.1086/508667.

Yang, J., Liu, Z., Guan, G., Li, Y., Cen, Z., Ma, M., Liu, A., Ren, Q., Wang, J., Luo, J., *et al.* (2015). Comprehensive surveillance of the antibody response to *Borrelia burgdorferi* s.l. in small ruminants in China. Annals of Agricultural and Environmental Medicine *22*, 208-211. DOI: 10.5604/12321966.1152066

Yrjanainen, H., Hytonen, J., Hartiala, P., Oksi, J., and Viljanen, M.K. (2010). Persistence of borrelial DNA in the joints of *Borrelia burgdorferi*-infected mice after ceftriaxone treatment. Acta Pathologica, Microbiologica, et Immunologica *118*, 665-673. doi: 10.1111/j.1600-0463.2010.02615.x.

Yrjanainen, H., Hytonen, J., Song, X.Y., Oski, J., Hartiala, K., and Viljanen, M.K. (2007). Anti-tumor necrosis factor-alpha activates *Borrelia burgdorferi* spirochetes 4 weeks after ceftriaxone treatment in C3H/He mice. Journal of Infectious Diseases *195*, 1489-1496. DOI: 10.1086/513873

Zygner, W., Gorski, P., and Wedrychowicz, H. (2009). Detection of the DNA of *Borrelia afzelii*, *Anaplasma phagocytophilum* and *Babesia canis* in blood samples from dogs in Warsaw. Veterinary Record *164*, 465-467. doi: 10.1136/vr.164.15.465.

Chapter 21

Epidemiology of Lyme Disease

Paul Mead* and Amy Schwartz

Bacterial Diseases Branch, Division of Vector-borne Diseases, National Center for Emerging and Zoonotic Infectious Diseases, Centers for Disease Control and Prevention (CDC), USA
*Corresponding author: pmead@cdc.gov

DOI: https://doi.org/10.21775/9781913652616.21

Abstract
Lyme disease affects populations throughout the northern hemisphere, with regional patterns reflecting differences in pathogen genospecies, vector biology, and human behavior. In the United States, reported cases increased from ~9,000 in 1992 to nearly 34,000 in 2018. The number of patients treated for Lyme disease is estimated at 300,000 annually. U.S. cases are focused in the Northeast and upper Midwest regions; however, these foci are expanding. Lyme disease cases have a distinct bimodal distribution relative to age, peaking among children aged 5 to 15 years and adults aged 60 years or older. Clinical features are influenced by both pathogen and host factors. A firm understanding of Lyme disease epidemiology is important for clinician and researcher alike.

Introduction
Lyme disease is a tickborne zoonosis caused by spirochetes of the *Borrelia burgdorferi* sensu lato complex. Common clinical features include erythema migrans rash, arthritis, and facial nerve palsy, along with general symptoms of fever and fatigue (see chapter 24 on Lyme disease in humans). Meningitis, radiculopathy and cardiac manifestations are less common but serious. The name "Lyme" is derived from a Connecticut town where Steere, Malawista and colleagues investigated an unusual cluster of arthritis cases in the mid-1970s (Steere et al., 1977b). They postulated a link to arthropod bites, and in 1982, Burgdorfer and colleagues reported isolating the presumed etiologic agent from *Ixodes* ticks (Burgdorfer et al., 1982). The spirochete was isolated from clinical samples the following year, and a serologic response was demonstrated in patients with erythema migrans (Benach et al., 1983; Steere et al., 1983). In retrospect, illness consistent with Lyme borreliosis was described in the upper Midwest in 1970 (Scrimenti, 1970) and in Europe as early as 1883 (Steere et al., 1977a; Matuschka et al., 1995). Today Lyme disease is considered the most common vector-borne disease in the northern hemisphere.

The epidemiology of Lyme disease is shaped by complex underpinnings. The organism is maintained in enzootic cycles involving both avian and mammalian hosts, with transmission by various tick species, each with its own geographic distribution and host preferences. These diverse enzootic cycles have given rise to multiple *Borrelia* genospecies and strains with variable capacity to cause human illness (Becker et al., 2016; Cerar et al., 2016; Gallais et al., 2018). Equally important are the habits and demographic features of the exposed human population. These determine the likelihood of tick bite as well as the care seeking practices required for diagnosis and reporting. Finally, surveillance practices and infrastructure differ greatly across public health jurisdictions, influencing both the apparent incidence of disease and the characteristics of reported cases (Ertel et al., 2012).

Agents and Relation to Human Illness
To date, 22 named genospecies of *Borrelia burgdorferi* sensu lato have been identified based on isolates from ticks and vertebrate hosts (Table 1) (NCBI, 2020). Three genospecies cause the majority of human infections: *B. afzelii*, *B. garinii/B. bavariensis*, and *B. burgdorferi* sensu stricto (hereafter *B. burgdorferi*). Genospecies isolated rarely from humans include *B. bissetii, B. lusitaniae, B. mayonii, B. spielmanii*, and *B. valaisiana*. (van Dam, 2002; Diza et al., 2004; da Franca et al., 2005; Maraspin et al., 2006; de Carvalho et al., 2008; Rudenko et al., 2008; Rudenko et al., 2009; Pritt et al., 2016). With the exception of *B. mayonii*, all nine

Table 1. Named genospecies of *Borrelia* by region, as of June 2020. Bolded names indicate geno-species with possible or confirmed links to human illness; shaded rows indicate common pathogens.

Genospecies	North America	South America	Europe	Asia
*B. afzelii**			X	X
B. americana	X			
B. andersonii	X			
*B. bavariensis**			X	
*B. bissettii**	X	?	X	X
*B. burgdorferi**	X		X	
B. californiensis	X			
B. carolinensis	X			
B. chilenensis		X		
B. finlandensis			X	
*B. garinii**			X	X
B. japonica				X
B. kurtenbachii	X			
B. lanei	X			
*B. lusitaniae**			X	
*B. mayonii**	X			
B. sinica				X
*B. spielmanii**			X	
B. tanukii				X
B. turdii				X
*B. valaisiana**			X	X
B. yangtzensis				X

*Detected in samples from humans

genospecies have been detected in Europe (van Dam, 2002) (Becker et al., 2016) (Table 1). In North America, human infections are mostly caused by *B. burgdorferi*, with several dozen documented cases of *B. mayonii* infection.

Tick Vectors Define the Geography of Risk

Borrelia burgdorferi sensu lato is transmitted among animals and to humans by *Ixodes* species ticks. Ticks of other genera sometimes harbor the organism; however, these have generally proven to be incompetent vectors infected incidentally by feeding on a reservoir host (Eisen, 2020). Over their 2 to 3-year lifespan, *Ixodes* ticks typically take three blood meals, one each as a larvae, nymph, and adult. Ticks acquire the spirochete by feeding on an infected reservoir host, typically a small mammal or bird, and can transmit the pathogen during subsequent blood meals. The spirochete is not transmitted vertically from adult ticks to offspring; thus larvae generally do not pose a risk of transmission. In some species, adult female ticks feed preferentially on large mammals, such as deer. These hosts are not reservoirs for the spirochete; however, their abundance affects tick reproductive success and, consequently, local risk of disease.

Four species of *Ixodes* ticks commonly bite humans; collectively their distribution covers temperate forests throughout the Northern hemisphere (Piesman and Gern, 2004). The principal vector in North America is *I. scapularis*, which is abundant in the mid-Atlantic, northeastern, and north central United States, with a range that extends into adjacent areas of Canada. A second, less-abundant species, *Ixodes pacificus*, occurs in pockets from the central California coast northward into British Columbia. The principal vector in Europe is *I. ricinus,* which is found at elevations below 1,300 meters throughout western, central, and eastern Europe, as well as northern Africa (Piesman and Gern, 2004; ECDC, 2020). The fourth species, *I. persulcatus,* is found in eastern Europe, where its distribution overlaps with that of *I. ricinus*, and in areas of Russia, Mongolia, China and Japan (Korenberg, 1994; Postic et al., 1997; Masuzawa, 2004; Hao et al., 2011; ECDC, 2020). Although numerous other *Ixodes* species transmit *Borrelia* and contribute to enzootic cycles (e.g., *I. affinis, I. minor, I spinipalpis*), these ticks rarely bite humans (Oliver et al., 2003; Harrison et al., 2010).

The four human biting *Ixodes* vectors differ with regard to carriage of *B. burgdorferi* sensu lato genospecies. *Ixodes ricinus* ticks carries all major pathogenic genospecies but with some regional differences in frequency (Estrada-Pena et al., 2018). Ticks in eastern Europe are more likely to carry *B. afzelii*, while those from western Europe are more often infected with *B. garinii* (Rauter and Hartung, 2005). Carriage of *B. burgdorferi* is less common and largely restricted to central Europe (Piesman and Gern, 2004; Estrada-Pena et al., 2018). In Asia, *I. persulcatus* carries *B. afzelii*, *B. garinii*, *B. bavariensis, and B. bissettii,* but it is not known to transmit *B. burgdorferi* (Korenberg, 1994; Korenberg et al., 2002 ; Masuzawa, 2004; Wang et al., 2019). In North America, *I. pacificus* transmits *B. burgdorferi*, and *I. scapularis* transmits both *B. burgdorferi* and *B. mayonii* (Steere, 2001; Piesman and Gern, 2004; Dolan et al., 2016). Animal studies and clinical observations indicate that *I. scapularis* ticks require at least 36 hours of attachment in order to transmit *B. burgdorferi*, supporting a possible preventive role for daily tick checks and showering after exposure (Vazquez, Muehlenbein et al. 2008, Connally, Durante et al. 2009). *Ixodes ricinus* ticks can transmit infection after much shorter periods of attachment, especially if infected with *B. afzelii* (Piesman and Gern 2004).

While the geographic distribution of vector ticks broadly defines areas of risk for Lyme disease, local risk can vary widely due to differences in tick abundance and infection prevalence. In the northeastern United States, infection prevalence can approach 40% among adult *I. scapularis,* the result of robust enzootic cycles. In contrast, *I. scapularis* in the southeastern United States are rarely infected with *B. burgdorferi*, possibly as a result of genetic and local ecologic factors (Piesman and Gern, 2004). It has been hypothesized that southern *I. scapularis* have evolved to quest below the leaf litter due to a greater risk of desiccation (Ginsberg et al., 2017). In the western United States, *I. pacificus* ticks are also rarely infected with *B. burgdorferi,* a consequence of their preferences for feeding on lizards that are not competent reservoirs for the spirochete (Lane and Loye, 1989; Piesman and Gern, 2004).

Distribution and Incidence in the United States

In the United States, Lyme disease has been a nationally notifiable condition since 1991. Surveillance case definitions have changed somewhat over time, as have the reporting practices of individual states. Nevertheless, a clear trend toward increasing incidence and a focal but expanding geographic distribution is evident. Reported cases have increased from ~9,000 in 1992 to nearly 34,000 in 2018 (Figure 1) (Bacon et al., 2008; Schwartz et al., 2017; CDC, 2020). The increase would likely have been greater but for recent cost-cutting measures enacted by some high incidence states. For example, the Massachusetts Department of Public Health recently discontinued routine follow-up of positive Lyme serology tests, resulting in a fall in reported cases from 2,933 in 2015 to just 13 in 2018 (CDC, 2020). Rather than a resolution of the Lyme disease problem, this decrease reflects the overwhelming burden that reporting places on providers and health departments in high incidence areas.

As with other nationally notifiable diseases, case counts reported to CDC capture but a portion of the true U.S. burden of Lyme disease. Under-reporting may be especially common with early forms of illness which rely upon clinical diagnosis in lieu of more readily captured laboratory data (Connecticut Department of Public Health, 2009; Ertel et al., 2012). One approach to quantifying the broader burden of disease is to assess coded claims for insurance reimbursement. An analysis of U.S. claims data for 2005-2010 yielded an estimate of between 296,000 and 376,000 clinician-diagnosed Lyme disease cases annually, roughly 8-fold greater than the number of cases reported to CDC through routine surveillance (Nelson, Saha et al. 2015). A similar analysis focusing on hospitalized cases yielded an

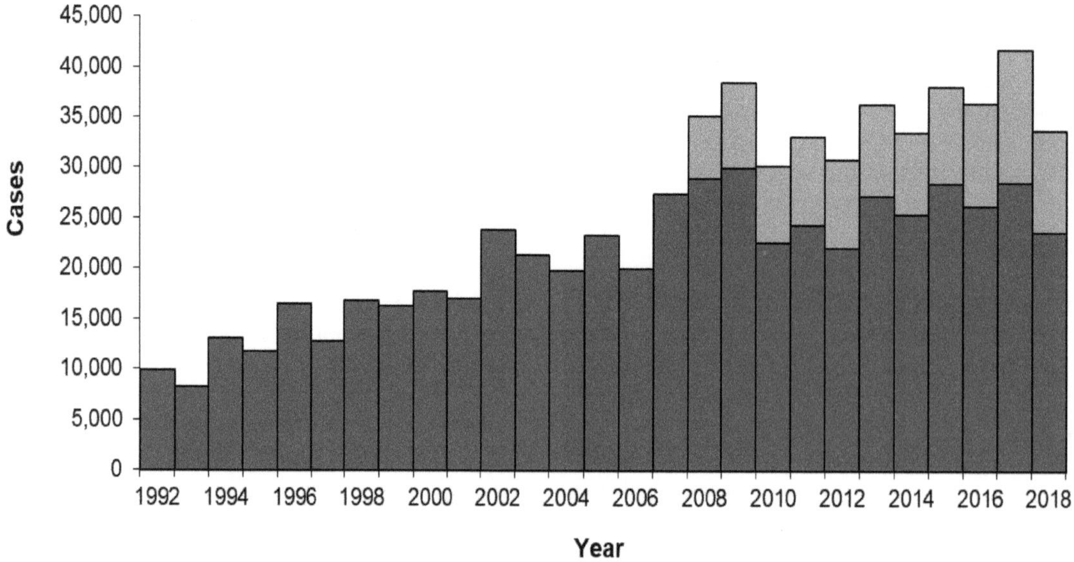

Figure 1. Reported Lyme disease cases by year, United States, 1992-2018.

annual estimate of ~2,200 Lyme disease-related hospitalizations among persons under age 65 years of age (Schwartz et al., 2020). While confirming that Lyme disease rarely requires hospitalization, the estimated cost of these hospitalizations nevertheless exceeded $25 million dollars annually. Claims data, although voluminous, are subject to the vagaries of coding practices and can still result in over and under counting of actual infections (Tseng et al., 2015; Rutz et al., 2018; Schiffman et al., 2018; White et al., 2018). At best they represent persons treated for a condition, some of whom may have been misdiagnosed. Analysis of more recent U.S. claims data is underway and likely to yield a larger estimate than the 296,000-376,000 annual cases derived for the years 2005-2010.

Within the United States, Lyme disease occurs in two major foci, one in the Northeast and mid-Atlantic region, and a second in upper Mid-west (Figure 2). Fourteen states in these two regions account for over 95% of all reported cases (Table 2) (Schwartz et al., 2017; CDC, 2020). Within these states county-level incidence can exceed 200 reported cases per 100,000 population per year (Bacon et al., 2008). This equates to an annual infection rate of approximately 1-2% after accounting for underreporting. Recent changes in reporting practices have produce a glaring dearth of cases in state of Massachusetts (Figure 2). Smaller foci of low-level transmission occur along the West Coast in California, Oregon, and Washington where *I. pacificus* is an established though less-commonly infected vector (Piesman and Gern, 2004).

Along with growing incidence, the geographic extent of the two major foci has expanded substantially over the last two decades. This expansion is not just northward, as might be presumed given climate change, but rather outward in all directions (Figure 2). The northeastern focus has expanded north into Maine, west through Pennsylvania and into Ohio, and south into west Virginia, Virginia, and North Carolina (Lantos et al., 2015a). Similar multidirectional expansion is evident in the upper Midwest focus (Stone et al., 2015; Lantos et al., 2017). In one assessment, the number of Northeastern counties meeting study specific high incidence criteria increased from 43 during 1993–1997 to 182 during 2008–2012, a gain of more than 320%. During the same time period, the number of high incidence counties in the upper Mid-west increased ≈250%, from 22 to 78 (Kugeler et al., 2015). Focal areas of intense transmission have recently been identified as far south as Buncombe County, North Carolina (Barbarin et al., 2020).

Occurrence Outside the United States

Expansion of U.S. Lyme disease foci extends into Canada, where Lyme disease is a limited but emerging problem (Ogden et al., 2019). *Ixodes scapularis* ticks are found in areas along the U.S.

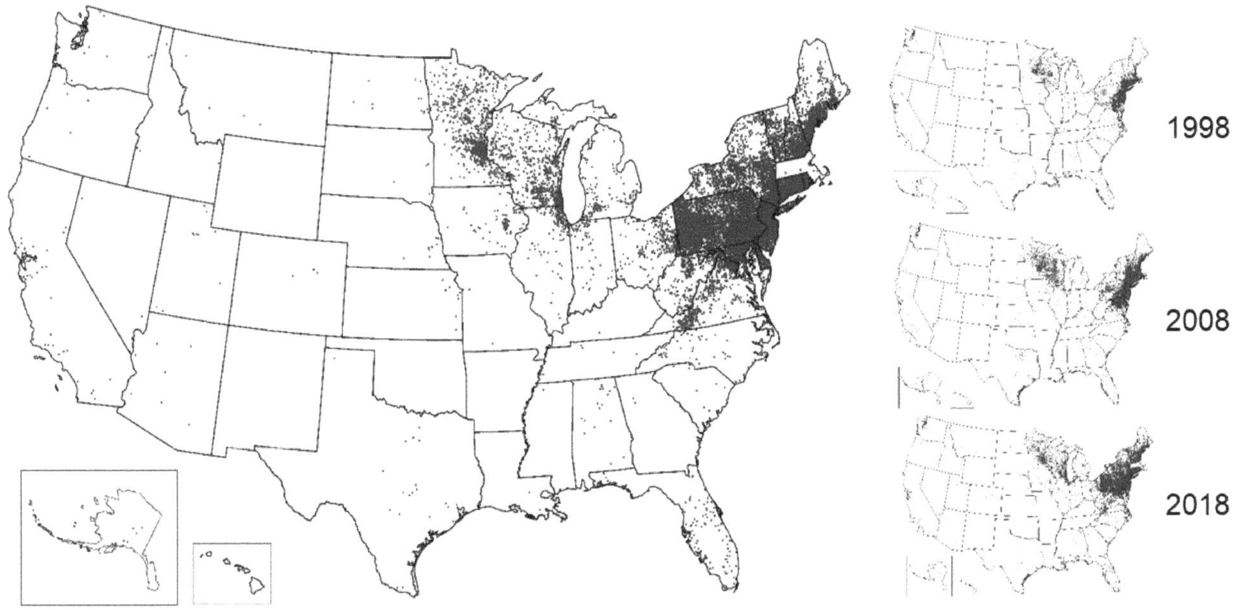

Figure 2. Geographic distribution of Lyme disease cases, United States, 1998, 2008, 2018. One dot per case placed randomly in county of patient residence.

border from eastern Manitoba to Nova Scotia; *I. pacificus* ticks are established in coastal areas of southern British Columbia (Ogden et al., 2009; Ogden et al., 2019). Although less common than in the United States, reports of human Lyme disease have increased substantially in recent years, from 144 in 2009 to 1,487 in 2018 (Public Health Agency of Canada, 2020). Ontario, Quebec and Nova Scotia account for the majority of reported cases. Incidence is especially high in Nova Scotia, which at 34 per 100,000 population in 2016 was nearly 13-fold greater than the national average (Public Health Agency of Canada, 2020). The high incidence rate in Nova Scotia is corroborated by a high sero-prevalence of anti-*B. burgdorferi* antibodies among domestic dogs in the province (Herrin et al., 2017; Ogden et al., 2019). There is concern that changing climatic conditions will allow northward expansion of *Ixodes* populations further into Canada, potentially increasing disease risk (Ogden et al., 2006).

Elsewhere in the Americas, the occurrence of human Lyme disease is uncertain, although clinically compatible illness has been reported occasionally (Neira et al., 1996). *Ixodes scapularis* is found in northeastern Mexico and *I. pacificus* in Baja (Piesman and Gern, 2004; Gordillo-Perez et al., 2009). *B. burgdorferi* has been detected by PCR in *I. scapularis* ticks collected in northern Mexico (Feria-Arroyo et al., 2014); however, genetic analysis suggests these results might reflect laboratory contamination (Norris et al., 2015). Gordillo-Perez and colleagues report that 28% of 606 Mexican adults and children with neurologic disease had serologic evidence of *Borrelia* infection (Gordillo-Perez et al., 2018). These results have been met with some skepticism, however, given the potential for serologic cross-reactions and the failure to date to isolate *Borrelia burgdorferi* from a Mexican patient (Faccini-Martlnez, 2019). Similarly, spirochetes have not been isolated from Brazilian patients with a "Lyme-like" illness known locally as Baggio-Yoshinari syndrome, nor do these patients have serologic evidence of *Borrelia* infection (de Oliveira et al., 2018). Nevertheless, spirochetes belonging to *Borrelia burgdorferi* sensu lato complex have been detected by molecular methods in Ixodid ticks collected in multiple South American countries, including Argentina, Brazil, Chile, and Uruguay (Carvalho et al., 2020). One of these, the newly named *B. chilensis,* was successfully cultured from *I. stilesi* ticks and long-tailed rice rats (Ivanova et al., 2014; Huang et al., 2015; Carvalho et al., 2020). The role of these and other spirochetes as a cause of human illness remains to be determine.

Table 2. Incidence of confirmed Lyme disease cases reported to the U.S. Centers for Disease Control and Prevention by state postal code and year, United States, 2015-2018.

State / District	2015	2016	2017	2018
AL	0.3	0.3	0.5	0.4
AK	0.1	0.8	1.1	1.1
AZ	0.1	0.1	0.3	0.1
AR	0	0	0.1	0
CA	0.2	0.2	0.2	0.2
CO	0	0	0.1	0.1
CT	52.2	34.6	38.5	35.5
DE	35.3	41.1	43.8	52.2
DC	11.6	9.7	8.9	8.1
FL	0.6	0.6	0.6	0.5
GA	0.1	0	0.1	0
HI	0	0	0	0
ID	0.2	0.5	0.9	0.3
IL	2.2	1.8	1.7	1.5
IN	1.5	1.9	1.4	1.1
IA	4.2	2.4	3.6	2.9
KS	0.4	0.6	0.4	0.3
KY	0.3	0.4	0.1	0.1
LA	0	0.1	0.2	0
ME	74.7	86.4	106.6	83
MD	20.8	21.2	19.7	14.8
MA	43	2.1	4.7	0.2
MI	1.3	1.6	2	1.8
MN	21.4	23.6	25.2	16.9
MS	0.1	0	0	0
MO	0	0	0	0

State / District	2015	2016	2017	2018
MT	0.2	1.2	0.4	0.5
NE	0.3	0.5	0.3	0.4
NV	0.2	0.1	0.2	0.2
NH	32.8	51.8	71.2	68.3
NJ	43.9	37.3	40.3	32.3
NM	0	0	0.1	0
NY	16.4	13.3	17.6	12.5
NC	0.4	0.3	0.7	0.5
ND	2	1.2	1.6	1.8
OH	1	1.1	2	2.1
OK	0	0	0	0
OR	0.1	0.3	0.3	0.2
PA	57.4	70.3	72.2	61.8
RI	53.4	50.6	56.2	62.5
SC	0.3	0.5	0.3	0.2
SD	0.6	0.9	0.5	0.5
TN	0.1	0.1	0.2	0.2
TX	0.1	0.1	0.1	0.1
UT	0.1	0.4	0.3	0.3
VT	78.4	78.1	103.6	55.6
VA	13.1	11.6	12.3	8.7
WA	0.2	0.2	0.4	0.1
WV	13.2	16.2	27.7	30.7
WI	22.7	26	31	19.3
WY	0	0.2	0.5	0
USA	**8.9**	**8.1**	**9.1**	**7.2**

In Europe, risk of Lyme disease is widespread (Hubalek, 2009; Rizzoli et al., 2011; Lorenc et al., 2017; Vandekerckhove et al., 2019), as are the key vectors *I. ricinus* and *I. persulcatus* (maps available at: https://www.ecdc.europa.eu/en/disease-vectors/surveillance-and-disease-data/tick-maps) (ECDC, 2020). Reported incidence rates vary by as much as four orders of magnitude, although comparisons are complicated by marked differences in surveillance policies and practices. The European Union recently designated neuroborreliosis a notifiable condition with a standard case definition, which should make future comparisons more meaningful (van den Wijngaard et al., 2017; Lancet, 2018). Overall, available data suggest that transmission is most intense in northeastern and central Europe and decreases moving west and southward into Spain and Italy (Lindgren and Jaenson, 2006; Rizzoli et al., 2011; Vandenesch et al., 2014; Vandekerckhove et al., 2019). Annual rates of 100 per 100,000 population or greater have been reported in areas of Austria, Estonia, Germany, Slovenia, Switzerland, Sweden, Norway, and the Netherlands (Huppertz et al., 1999; Bennet et al., 2006; WHO, 2012; Wilking and Stark, 2014; Hofhuis et al., 2015; Enkelmann et al., 2018; Vandekerckhove et al., 2019). Other countries report rates ranging from ~20-80 per 100,000 population, including Finland, France, Lithuania, and Poland (Sajanti et al., 2017; Septfons et al., 2019; Zbrzezniak and Paradowska-Stankiewicz, 2019; Petrulioniene et al., 2020). The 10-year mean annual rate for Bulgaria was 7 per 100,000 (Ermenlieva et al., 2019), and in the British Isles, rates are generally below 5, but with peaks as high as 44 per 100,000 in the Scottish Highlands (Mavin et al., 2015; Lorenc et al., 2017; Cairns et al., 2019; Tulloch et al., 2019). Rates less than 1 per 100,000 population have been rereported for Spain, Portugal, and Italy (Smith and Takkinen, 2006). In Iceland, all cases appear to be importations related to travel (Vigfusson et al., 2019).

Areas of Lyme disease risk extend across Eurasia, from eastern Europe through southern Russia, neighboring regions of Mongolia and China, and into Japan (Postic et al., 1997). Reported incidence in endemic areas of Russia generally ranges from 5 to 10 per 100,000 population (Hubalek, 2009). Higher rates have been reported in areas northeast of Moscow in Vologda oblast, in the Sverdlovsk (Urals) region, and in western Siberia (World Health Organization, 1995; EpiNorth, 2014). Although infected ticks have been found in Mongolia, information on human cases is scarce in the English literature (Jiang et al., 2019). At least six genospecies of *B. burgdorferi* sensu lato have been isolated from rodents and ticks in China, and over 2,000 human infections have been identified through serologic testing (Ai et al., 1990; Chu et al., 2008; Zhang et al., 2010; Hao et al., 2011; Fang et al., 2015). Based on a small number of clinical isolates, *B. garinii* appears to be the most common cause of human infection, followed by *B. afzelii* and a single isolate of a *B. valasiana*-related genospecies (Fang et al., 2015). Human cases have been reported from most areas but, most commonly, from the more northern provinces of Xinjiang, Gansu, Inner Mongolia, Jinlin and Heilongjiang (Fang et al., 2015). Although both *B. garinii* and *B. afzelii* have been isolated from patients in Japan, overall incidence is <0.1 per 100,000 population, with most cases occurring on Hokkaido Island and some from exposures in subalpine forested areas in central Japan (Nakama et al., 1994; Hashimoto et al., 2007). In a recent serosurvey, 10% of dogs in Sapporo, Japan, had antibodies to *B. burgdorferi* sensu lato, suggesting the existence of a robust, local enzootic cycle (Uesaka et al., 2016). Enzootic cycles are established in Korea and Taiwan, and *B. garinii* has been isolated from at least one patient from northern Taiwan (Chao et al., 2010).

Lyme-like illness has been reported on occasion from Australia (Russell, 1995) and South Africa (Stanek et al., 1987b). While the possibility of *Borrelia*-related illness with distinct enzootic cycles in these areas cannot be excluded, a great deal more information will be needed to determine the relationship, if any, between these reports and Lyme borreliosis as currently defined. Recent analyses have concluded that there was no credible evidence for Lyme disease in Australia (Collignon et al., 2016; Brown, 2018).

Seasonality

As with most vector-borne diseases, Lyme disease occurs most often in the warmer months. This temporal pattern reflects both tick questing habits and the recreational tendencies of humans (Ai et al., 1990; Piesman and Gern, 2004; Bacon et al., 2008). Nymphal *I. scapularis* quest in the spring and early summer, a behavior that correlates with a sharp increase in human cases (Figure 3). Adult ticks, which are much less abundant, typically quest in the fall, resulting in a subtle but discernable “tail” of human cases extending into October. Over half of all U.S. cases have onset in June or July (Bacon et al., 2008; Schwartz et al., 2017). In some reports from Europe, acute infections peak slightly later, in July and August, perhaps due to climatic differences, recreational patterns, or both (Berglund et al., 1995; Lindgren and Jaenson, 2006; Vandenesch et al., 2014; Wilking and Stark, 2014; Sajanti et al., 2017). Tick questing behavior is influenced by meteorological factors (Alekseev and Dubinina, 2000; Eisen et al., 2002), and the onset of human illness can vary from year to year based on climatic conditions (Bennet et al., 2006; Moore et al., 2014). The seasonal peak in cases is generally less pronounced for later stages of disease which have a longer and more variable incubation period (Stanek et al., 1987a; Berglund et al., 1995; Strle, 1999; Bacon et al., 2008; Vandenesch et al., 2014; Sajanti et al., 2017; Enkelmann et al., 2018).

Age and Sex

In the United States, Lyme disease cases have a distinct bimodal distribution relative to age. Rates

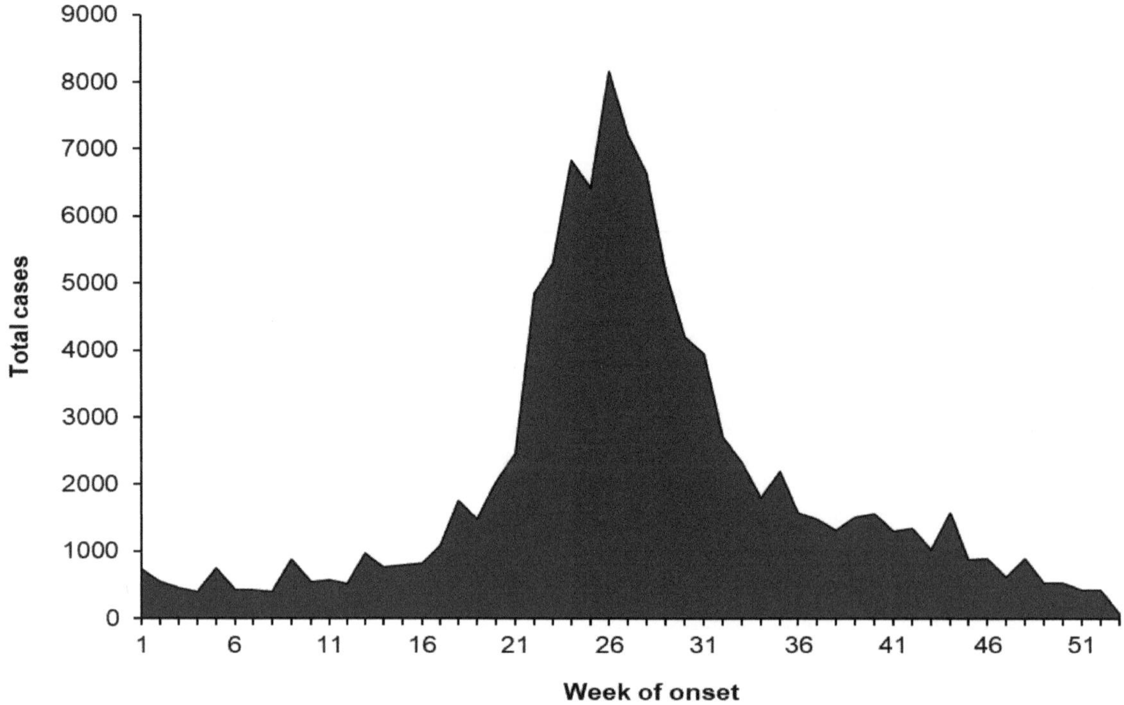

Figure 3. Seasonality of confirmed and probable Lyme disease cases, United States, 2015-2018.

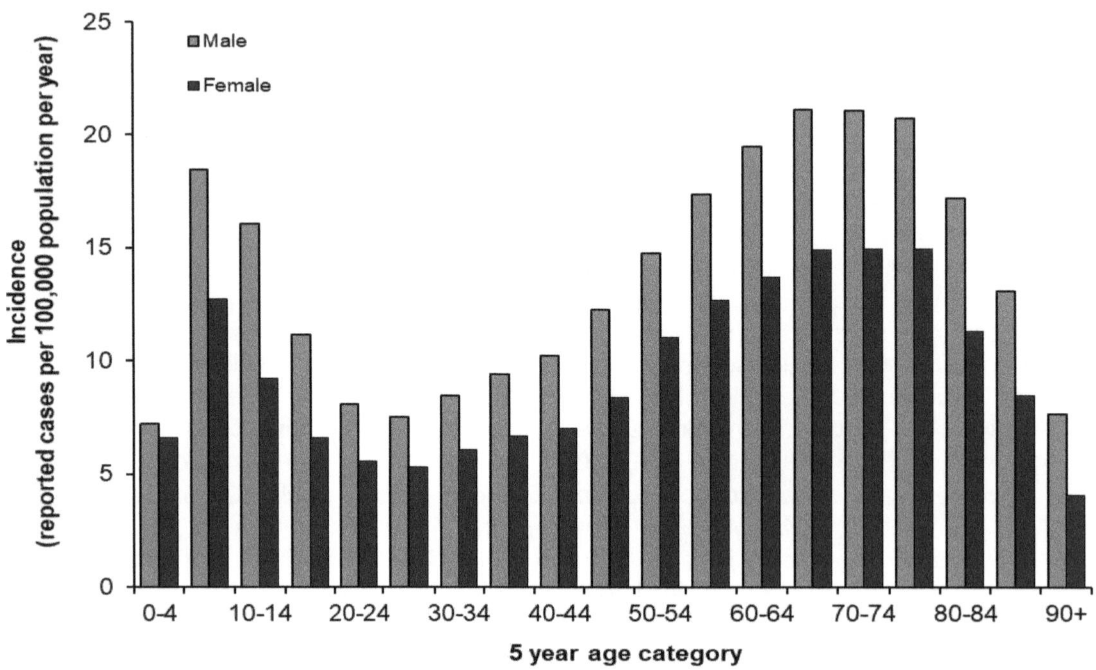

Figure 4. Age distribution of confirmed and probable Lyme disease cases by sex, United States, 2015-2018.

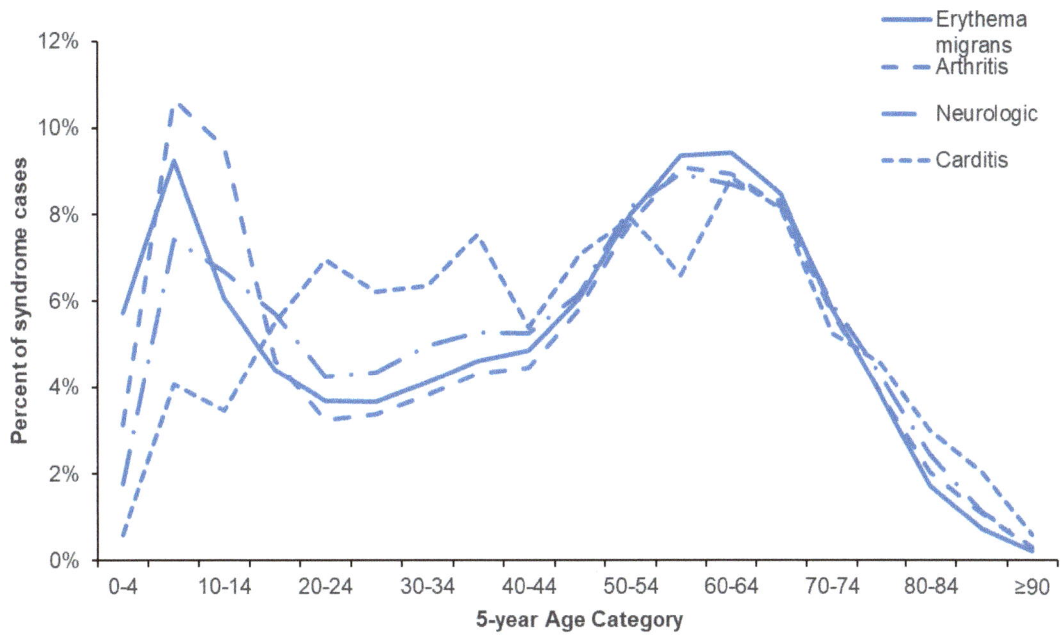

Figure 5. Age distribution of patients by clinical feature, reported confirmed Lyme disease cases, United States, 2014-2018.

peak among children between the ages of 5 and 15 years, decrease among 20- to 30-year-olds, and peak again among adults aged 60 years or older (Figure 5). Incidence for males is higher in all age categories, and the disparity with females has increased over time, shifting the overall sex ratio for confirmed cases from 51% male in 1992 to 53% in 2006 (Bacon et al., 2008) to 57% males during 2008-2015 (Schwartz et al., 2017). Although unexplained, this gender-specific increase has been most pronounced among children. Age and sex distributions also differ between areas with high and low reported incidence (CDC, 2004). During 2008-2015, the modal age for cases in high incidence states was 8 years with 57% of cases among males, as compared with 44 years with 45% of cases among males in low incidence states (Schwartz et al., 2017). Barring fundamental differences in risk factors for infection, this discrepancy suggests that some proportion of cases reported from low incidence areas may due to conditions other than Lyme disease. This is consistent with the lower predictive value of clinical and laboratory findings in the setting of low prior probability (Tugwell et al., 1997).

In Europe, a bimodal distribution is also observed with respect to age, with the important difference that the sex distribution shifts in adults to yield a higher incidence among women than men (Strle, 1999; Bennet et al., 2007; Sajanti et al., 2017; Enkelmann et al., 2018). Overall, females account for the majority of cases in series from Austria, France, Germany, Slovenia, Sweden, (Stanek et al., 1987a; Strle, 1999; Bennet et al., 2006; Wilking and Stark, 2014; Septfons et al., 2019). These patterns may reflect behavior-related differences in rates of exposure across populations, although they may also be influenced by age and sex specific differences in clinical illness or care seeking practices (Strle et al., 2013).

Clinical Features

Lyme disease is characterized by a range of dermatologic, rheumatologic, neurologic, and cardiac abnormalities (see chapter 24). Erythema migrans (EM) is the most common manifestation, accounting for 60-90% of cases in both North America and Eurasia (Ai et al., 1988; Steere, 2001; Stanek and Strle, 2003; Schwartz et al., 2017). Less common cutaneous features include lymphocytoma and acrodermatitis chronica atrophicans (ACA), seen almost exclusively in Europe (Busch et al., 1996). Dissemination of spirochetes from the skin to other organs can result in facial palsy, radiculopathy, arthritis, carditis (Feder et al., 1995; Steere, 2001; Stanek and Strle, 2003; Steere and Sikand, 2003; Nau et al., 2009; Steere et al., 2016). Readers should consult other sources for details of clinical disease; the purpose here is to briefly describe how clinical features differ across patient populations.

Epidemiologic data suggest different tissue tropisms for the principle genospecies of *B. burgdorferi* sensu lato. In European studies, isolates from patients with neuroborreliosis are most commonly *B. garinii (Ruzic-Sabljic et al., 2001b)*, while those from patients with EM -- and especially ACA -- are predominantly *B. afzelii* (van Dam et al., 1993; Busch et al., 1996; Ornstein et al., 2001; Ruzic-Sabljic et al., 2001a). *B. burgdorferi* is thought to have a particular capacity to cause arthritis (Steere, 2001; Steere and Glickstein, 2004). In the United States where nearly all cases are caused by *B. burgdorferi*, 27% of cases reported through national surveillance during 2008-2015 were associated with arthritis, while only 12% had neurological symptoms (Schwartz et al., 2017). Conversely, neuroborreliosis is more common that arthritis in several European series (Stanek et al., 1987a; Berglund et al., 1995; Letrilliart et al., 2005). In a large series from Sweden, 16% of patients had neuroborreliosis as compared to 7% with arthritis (Berglund et al., 1995). Among 873 Austrian patients, 24% had neurologic manifestations, as compared with only 2% with arthritis (Stanek et al., 1987a).

Patient age and sex also influence the clinical features of Lyme disease. In North America, Lyme arthritis is distinctly more common among children aged 5-15 years, while men aged 20-40 years account for a disproportionate number of carditis cases (Figure 5) (Forrester and Mead, 2014; Schwartz et al., 2017). In Europe, lymphocytoma and neuroborreliosis are diagnosed more commonly in children than adults (Stanek et al., 1987a; Berglund et al., 1995; Huppertz et al., 1999; Hubalek, 2009; Henningsson et al., 2010). ACA, in contrast, is primarily a disease of adult women, although a number of recent reports document cases in children (Asbrink et al., 1986; Stanek et al., 1987a; World Health Organization, 1995; Brzonova et al., 2002; Zalaudek et al., 2005; Hubalek, 2009). Dermatologic features in general appear to be more common among women in Europe, while non-cutaneous manifestations are more common among men (Fulop and Poggensee, 2008; Hubalek, 2009). In a large series from Slovenia, females accounted for 69% of patients with ACA, 59% with EM, 39% with neuroborreliosis, and only 25% of patients with Lyme arthritis (Strle et al., 2013).

Few deaths due to Lyme disease have been reported in the medical literature (Marcus et al., 1985; Kirsch et al., 1988; Waniek et al., 1995; Tavora et al., 2008; CDC, 2013). A review of U.S. death certificates for the years 1999-2003 identified 23 records listing Lyme disease as the underlying cause of death; however, careful review found that most of these reflected improper coding (Kugeler et al., 2011). In a retrospective study of over 120,000 U.S. patients diagnosed with Lyme disease, only 0.6% died within a year of diagnosis, a rate less than the expected, all-cause mortality for the population (Forrester et al., 2014). Nevertheless, sudden death due to Lyme carditis may go undetected, as shown by reports in which the diagnosis was recognized only because the patient was an organ donor (CDC, 2013). Sadly, some patients have died as a result of unproven treatments for Lyme disease, and some due to delayed treatment of other conditions misdiagnosed as Lyme disease. (Patel et al., 2000; Lantos et al., 2015b; Marzec et al., 2017; Strizova et al., 2019).

Risk factors

The fundamental risk factor for Lyme disease is, of course, exposure to infected ticks; however, this can occur in various ways in different settings. In the northeastern United States, homes are often located in heavily tick-infested areas, and exposure is believed to occur primarily around the home (Falco and Fish 1988, Maupin, Fish et al. 1991, Klein, Eppes et al. 1996). Factors that contribute to peridomestic risk include the amount of suitable tick habitat, landscaping practices, deer density, and outdoor activities such as gardening (Ley, Olshen et al. 1995, Orloski, Campbell et al. 1998, Smith, Wileyto et al. 2001, Rand, Lubelczyk et al. 2003). A Connecticut study found that risk of tick bite was associated with time spent in one's own yard rather than public recreational areas (Mead et al., 2018). In other areas, work and recreation away from the home may drive exposure. Increased risk of infection has been noted among forestry workers, farmers, hikers, soldiers, hunters, and orienteers in studies from the United States (Schwartz and Goldstein 1990) Asia (Ai, Zhang et al. 1994, Nakama, Muramatsu et al. 1994), and throughout Europe (Cinco, Barbone et al. 2004, Kaya, Parlak et al. 2008, Bilski 2009, Buczek, Rudek et al. 2009, Hubalek 2009, Rizzoli, Hauffe et al. 2011, Richard and Oppliger 2015).

Data gaps and misunderstandings

Two aspects of Lyme disease epidemiology have been recurring topics of confusion in the United States. The first involves the geographic distribution of risk. Lyme disease cases have been reported from all U.S. states, leading some to suggest that risk of infection exists throughout the country. There is no evidence to support this claim. By rule, U.S. notifiable disease cases are reported according to the state in which the patient officially resides. Infections acquired during travel or former residence in an endemic area are occasionally reported in states where there is no evidence of local transmission (Schwartz et al., 2017). For example, during 2015–2016, a total of two illnesses meeting the confirmed case definition for Lyme disease were reported to the Arkansas Department of Health; follow-up investigation revealed that both patients had exposures out of state in areas where Lyme disease is common (Kwit et al., 2017). Absence of risk is further supported by serologic testing of millions of dogs nationwide. Whereas 10% to 25% of dogs in Lyme endemic states of the Northeast are seropositive, less than 1% of dogs in the southeast and west are seropositive, a percentage consistent with the false positive rate of the assay (Little et al., 2014; Stillman et al., 2014; Dewage et al., 2019). Finally, the lack of appreciable enzootic cycles in many areas of the country has been affirmed through extensive acarological surveys (Diuk-Wasser et al., 2006).

A second topic of confusion concerns modes of transmission. While the epidemiology of Lyme disease is consistent with transmission by *Ixodes* ticks, alternate modes of transmission have been postulated and investigated. Laboratory-adapted strains of *B. burgdorferi* can survive under blood banking conditions, raising the possibility of transmission through blood transfusion (Johnson, Swaminathan et al. 1990). Nevertheless, no transfusion-associated cases of Lyme disease have ever been documented. Meanwhile, transmission through transfusion of less common *Ixodes*-transmitted pathogens, *Babesia microti* and *Anaplasma phagocytophilum*, has been demonstrated repeatedly (McQuiston, Childs et al. 2000, CDC 2008). Similarly, there is no credible evidence of transmission through sexual contact, semen, urine, or breast milk, despite a series of studies in animals (Moody and Barthold 1991, Woodrum and Oliver 1999). Intrauterine infection has been documented in rare reports of miscarriage and still birth in women infected during pregnancy (Schlesinger, Duray et al. 1985). A causal relationship to the miscarriage has not been established, however, as *B. burgdorferi* has also been identified in placentas of women with normal pregnancy outcomes (Figueroa, Bracero et al. 1996). Epidemiological studies have identified no definable pattern of teratogenicity and pregnant women who develop Lyme disease have good outcomes if they receive appropriate antimicrobial therapy (Markowitz, Steere et al. 1986, Walsh, Mayer et al. 2007). Nevertheless, a recent systematic review found significantly fewer adverse birth outcomes in women treated for gestational Lyme disease compared to those who were not treated during pregnancy (Waddell et al., 2018). These findings support prompt diagnosis and treatment of Lyme disease during pregnancy.

Conclusion

Along with other tickborne infections, Lyme disease is a substantial and growing health threat, both in the United States and abroad. The etiologic agent, principal vectors, and clinical manifestations vary by region, as does the magnitude of risk. An understanding of Lyme disease epidemiology is useful clinically and important for guiding the development of prevention measures. Although exposure to infected ticks can be reduced somewhat through environmental controls, tick checks and repellent use, additional practical and highly effective prevention measures are needed urgently.

The findings and conclusions in this article are those of the author and do not necessarily represent the views of the Centers for Disease Control and Prevention. This chapter is derived in part from previous chapters written by the first author.

References

Ai, C.X., Hu, R.J., Hyland, K.E., Wen, Y.X., Zhang, Y.G., Qiu, Q.C., Li, D.Y., Liu, X.D., Shi, Z.X., Zhao, J.H., *et al.* (1990). Epidemiological and aetiological evidence for transmission of Lyme disease by adult *Ixodes persulcatus* in an endemic area in China. Int J Epidemiol *19*, 1061-1065. https://doi.org/10.1093/ije/19.4.1061

Ai, C.X., Wen, Y.X., Zhang, Y.G., Wang, S.S., Qiu, Q.C., Shi, Z.X., Li, D.Y., Chen, D.Q., Liu, X.D., and Zhao, J.H. (1988). Clinical manifestations and epidemiological characteristics of Lyme disease in Hailin county, Heilongjiang Province, China. Ann N Y Acad Sci *539*, 302-313. https://doi.org/10.1111/j.1749-6632.1988.tb31864.x

Alekseev, A.N., and Dubinina, H.V. (2000). Abiotic parameters and diel and seasonal activity of *Borrelia*-infected and uninfected *Ixodes persulcatus* (Acarina: Ixodidae). J Med Entomol *37*, 9-15. https://doi.org/10.1603/0022-2585-37.1.9

Asbrink, E., Hovmark, A., and Olsson, I. (1986). Clinical manifestations of acrodermatitis chronica atrophicans in 50 Swedish patients.

Zentralblatt fur Bakteriologie Mikrobiologie und Hygiene (A) *263*, 253-261. https://doi.org/10.1016/s0176-6724(86)80128-6

Bacon, R.M., Kugeler, K.J., and Mead, P.S. (2008). Surveillance for Lyme disease--United States, 1992-2006. MMWR Surveill Summ *57*, 1-9.

Barbarin, A.M., Seagle, S.W., and Creede, S. (2020). Notes from the Field: Four Cases of Lyme Disease at an Outdoor Wilderness Camp - North Carolina, 2017 and 2019. MMWR Morb Mortal Wkly Rep *69*, 114-115. https://doi.org/10.15585/mmwr.mm6904a5

Becker, N.S., Margos, G., Blum, H., Krebs, S., Graf, A., Lane, R.S., Castillo-Ramirez, S., Sing, A., and Fingerle, V. (2016). Recurrent evolution of host and vector association in bacteria of the *Borrelia burgdorferi* sensu lato species complex. BMC Genomics *17*, 734. https://doi.org/10.1186/s12864-016-3016-4

Benach, J.L., Bosler, E.M., Hanrahan, J.P., Coleman, J.L., Habicht, G.S., Bast, T.F., Cameron, D.J., Ziegler, J.L., Barbour, A.G., Burgdorfer, W., *et al.* (1983). Spirochetes isolated from the blood of two patients with Lyme disease. N Engl J Med *308*, 740-742. https://doi.org/10.1056/NEJM198303313081302

Bennet, L., Halling, A., and Berglund, J. (2006). Increased incidence of Lyme borreliosis in southern Sweden following mild winters and during warm, humid summers. Eur J Clin Microbiol Infect Dis *25*, 426-432. https://doi.org/10.1007/s10096-006-0167-2

Bennet, L., Stjernberg, L., and Berglund, J. (2007). Effect of gender on clinical and epidemiologic features of Lyme borreliosis. Vector Borne Zoonotic Dis *7*, 34-41. https://doi.org/10.1089/vbz.2006.0533

Berglund, J., Eitrem, R., Ornstein, K., Lindberg, A., Ringer, A., Elmrud, H., Carlsson, M., Runehagen, A., Svanborg, C., and Norrby, R. (1995). An epidemiologic study of Lyme disease in southern Sweden. New Engl J Med *333*, 1319-1327. https://doi.org/10.1056/NEJM199511163332004

Brown, J.D. (2018). A description of 'Australian Lyme disease' epidemiology and impact: an analysis of submissions to an Australian senate inquiry. Intern Med J *48*, 422-426. https://doi.org/10.1111/imj.13746

Brzonova, I., Wollenberg, A., and Prinz, J.C. (2002). Acrodermatitis chronica atrophicans affecting all four limbs in an 11-year-old girl. Br J Dermatol *147*, 375-378. https://doi.org/10.1046/j.1365-2133.2002.04792.x

Burgdorfer, W., Barbour, A.G., Hayes, S.F., Benach, J.L., Grunwaldt, E., and Davis, J.P. (1982). Lyme disease-a tick-borne spirochetosis? Science *216*, 1317-1319. https://doi.org/10.1126/science.7043737

Busch, U., Hizo-Teufel, C., Bohmer, R., Fingerle, V., Rossler, D., Wilske, B., and Preac-Mursic, V. (1996). *Borrelia burgdorferi* sensu lato strains isolated from cutaneous Lyme borreliosis biopsies differentiated by pulsed-field gel electrophoresis. Scand J Infect Dis *28*, 583-589. https://doi.org/10.3109/00365549609037965

Cairns, V., Wallenhorst, C., Rietbrock, S., and Martinez, C. (2019). Incidence of Lyme disease in the UK: a population-based cohort study. BMJ Open *9*, e025916. https://doi.org/10.1136/bmjopen-2018-025916

Carvalho, L.A., Maya, L., Armua-Fernandez, M.T., Felix, M.L., Bazzano, V., Barbieri, A.M., Gonzalez, E.M., Lado, P., Colina, R., Diaz, P., *et al.* (2020). *Borrelia burgdorferi* sensu lato infecting *Ixodes auritulus* ticks in Uruguay. Exp Appl Acarol *80*, 109-125. https://doi.org/10.1007/s10493-019-00435-8

CDC (2004). Lyme disease--United States, 2001-2002. MMWR Morb Mortal Wkly Rep *53*, 365-369.

CDC (2013). Three sudden cardiac deaths associated with Lyme carditis - United States, November 2012-July 2013. MMWR Morb Mortal Wkly Rep *62*, 993-996.

CDC (2020). Lyme Disease Data and Surveillance. Available from: https://www.cdc.gov/lyme/datasurveillance/index.html?CDC_AA_refVal=https%3A%2F%2Fwww.cdc.gov%2Flyme%2Fstats%2Findex.html. Accessed: 06/02/2020.

Cerar, T., Strle, F., Stupica, D., Ruzic-Sabljic, E., McHugh, G., Steere, A.C., and Strle, K. (2016). Differences in Genotype, Clinical Features, and Inflammatory Potential of *Borrelia burgdorferi* sensu stricto Strains from Europe and the United States. Emerg Infect Dis *22*, 818-827. https://doi.org/10.3201/eid2205.151806

Chao, L.L., Chen, Y.J., and Shih, C.M. (2010). First detection and molecular identification of *Borrelia garinii* isolated from human skin in Taiwan. J Med Microbiol *59*, 254-257. https://doi.org/jmm.0.016626-0 [pii]10.1099/jmm.0.016626-0

Chu, C.Y., Jiang, B.G., Liu, W., Zhao, Q.M., Wu, X.M., Zhang, P.H., Zhan, L., Yang, H., and Cao, W.C. (2008). Presence of pathogenic *Borrelia burgdorferi* sensu lato in ticks and rodents in Zhejiang, south-east China. J Med Microbiol *57*, 980-985. https://doi.org/57/8/980 [pii] 10.1099/jmm.0.47663-0

Collignon, P.J., Lum, G.D., and Robson, J.M. (2016). Does Lyme disease exist in Australia? Med J Aust *205*, 413-417. https://doi.org/10.5694/mja16.00824

Connecticut Department of Public Health (2009). Lyme disease -- Connecticut, 2008. Connecticut Epidemiologist *29*, 14-16.

da Franca, I., Santos, L., Mesquita, T., Collares-Pereira, M., Baptista, S., Vieira, L., Viana, I., Vale, E., and Prates, C. (2005). Lyme borreliosis in Portugal caused by *Borrelia lusitaniae*? Clinical report on the first patient with a positive skin isolate. Wiener klinische Wochenschrift *117*, 429-432. https://doi.org/10.1007/s00508-005-0386-z

de Carvalho, I.L., Fonseca, J.E., Marques, J.G., Ullmann, A., Hojgaard, A., Zeidner, N., and Nuncio, M.S. (2008). Vasculitis-like syndrome associated with *Borrelia lusitaniae* infection. Clin Rheumatol *27*, 1587-1591. https://doi.org/10.1007/s10067-008-1012-z

de Oliveira, S.V., Faccini-Martinez, A.A., and Cerutti Junior, C. (2018). Lack of serological evidence for Lyme-like borreliosis in Brazil. Travel Med Infect Dis *26*, 62-63. https://doi.org/10.1016/j.tmaid.2018.04.005

Dewage, B.G., Little, S., Payton, M., Beall, M., Braff, J., Szlosek, D., Buch, J., and Knupp, A. (2019). Trends in canine seroprevalence to *Borrelia burgdorferi* and *Anaplasma spp.* in the eastern USA, 2010-2017. Parasit Vectors *12*, 476. https://doi.org/10.1186/s13071-019-3735-x

Diuk-Wasser, M.A., Gatewood, A.G., Cortinas, M.R., Yaremych-Hamer, S., Tsao, J., Kitron, U., Hickling, G., Brownstein, J.S., Walker, E., Piesman, J., *et al.* (2006). Spatiotemporal patterns of host-seeking *Ixodes scapularis* nymphs (Acari: Ixodidae) in the United States. J Med Entomol *43*, 166-176. https://doi.org/10.1603/0022-2585(2006)043[0166:spohis]2.0.co;2

Diza, E., Papa, A., Vezyri, E., Tsounis, S., Milonas, I., and Antoniadis, A. (2004). *Borrelia valaisiana* in cerebrospinal fluid. Emerg Infect Dis *10*, 1692-1693. https://doi.org/10.3201/eid1009.030349

Dolan, M.C., Hojgaard, A., Hoxmeier, J.C., Replogle, A.J., Respicio-Kingry, L.B., Sexton, C., Williams, M.A., Pritt, B.S., Schriefer, M.E., and Eisen, L. (2016). Vector competence of the blacklegged tick, *Ixodes scapularis*, for the recently recognized Lyme borreliosis spirochete Candidatus Borrelia mayonii. Ticks Tick Borne Dis *7*, 665-669. https://doi.org/10.1016/j.ttbdis.2016.02.012

ECDC (2020). European Centre for Disease Control. Tick species distribution maps: Ixodes rincinus/persulcatus - known distribution May 2020. Available from: https://ecdc.europa.eu/en/disease-vectors/surveillance-and-disease-data/tick-maps. Accessed: 2 June 2020.

Eisen, L. (2020). Vector competence studies with hard ticks and *Borrelia burgdorferi* sensu lato spirochetes: A review. Ticks Tick Borne Dis *11*, 101359. https://doi.org/10.1016/j.ttbdis.2019.101359

Eisen, L., Eisen, R.J., and Lane, R.S. (2002). Seasonal activity patterns of *Ixodes pacificus* nymphs in relation to climatic conditions. Med Vet Entomol *16*, 235-244. https://doi.org/10.1046/j.1365-2915.2002.00372.x

Enkelmann, J., Bohmer, M., Fingerle, V., Siffczyk, C., Werber, D., Littmann, M., Merbecks, S.S., Helmeke, C., Schroeder, S., Hell, S., *et al.* (2018). Incidence of notified Lyme borreliosis in Germany,

2013-2017. Sci Rep *8*, 14976. https://doi.org/10.1038/s41598-018-33136-0

EpiNorth (2014). A Co-operatation Project for Communicable Disease control in Northern Europe. EpiNorth Data: Lyme borreliosis. Available from: Available at: http://www.epinorth.org. Accessed: 10 July 2014.

Ermenlieva, N., Tsankova, G., and Todorova, T.T. (2019). Epidemiological study of Lyme disease in Bulgaria. Cent Eur J Public Health *27*, 235-238. https://doi.org/10.21101/cejph.a5007

Ertel, S.H., Nelson, R.S., and Cartter, M.L. (2012). Effect of surveillance method on reported characteristics of Lyme disease, Connecticut, 1996-2007. Emerg Infect Dis *18*, 242-247. https://doi.org/10.3201/eid1802.101219

Estrada-Pena, A., Cutler, S., Potkonjak, A., Vassier-Tussaut, M., Van Bortel, W., Zeller, H., Fernandez-Ruiz, N., and Mihalca, A.D. (2018). An updated meta-analysis of the distribution and prevalence of *Borrelia burgdorferi* s.l. in ticks in Europe. Int J Health Geogr *17*, 41. https://doi.org/10.1186/s12942-018-0163-7

Faccini-Martlnez, A.A. (2019). Call for Caution to Consider Lyme Neuroborreliosis as A Frequent Neurological Disease in Mexico. Arch Med Res *50*, 18. https://doi.org/10.1016/j.arcmed.2019.03.005

Fang, L.Q., Liu, K., Li, X.L., Liang, S., Yang, Y., Yao, H.W., Sun, R.X., Sun, Y., Chen, W.J., Zuo, S.Q., *et al.* (2015). Emerging tick-borne infections in mainland China: an increasing public health threat. Lancet Infect Dis *15*, 1467-1479. https://doi.org/10.1016/S1473-3099(15)00177-2

Feder, H.M., Jr., Gerber, M.A., Cartter, M.L., Sikand, V., and Krause, P.J. (1995). Prospective assessment of Lyme disease in a school-aged population in Connecticut. J Infect Dis *171*, 1371-1374. https://doi.org/10.1093/infdis/171.5.1371

Feria-Arroyo, T.P., Castro-Arellano, I., Gordillo-Perez, G., Cavazos, A.L., Vargas-Sandoval, M., Grover, A., Torres, J., Medina, R.F., de Leon, A.A., and Esteve-Gassent, M.D. (2014). Implications of climate change on the distribution of the tick vector *Ixodes scapularis* and risk for Lyme disease in the Texas-Mexico transboundary region. Parasit Vectors 7, 199. https://doi.org/10.1186/1756-3305-7-199

Forrester, J.D., and Mead, P. (2014). Third-degree heart block associated with lyme carditis: review of published cases. Clin Infect Dis *59*, 996-1000. https://doi.org/10.1093/cid/ciu411

Forrester, J.D., Meiman, J., Mullins, J., Nelson, R., Ertel, S.H., Cartter, M., Brown, C.M., Lijewski, V., Schiffman, E., Neitzel, D., *et al.* (2014). Notes from the field: update on Lyme carditis, groups at high risk, and frequency of associated sudden cardiac death--United States. MMWR Morb Mortal Wkly Rep *63*, 982-983.

Fulop, B., and Poggensee, G. (2008). Epidemiological situation of Lyme borreliosis in germany: surveillance data from six Eastern German States, 2002 to 2006. Parasitol Res *103 Suppl 1*, S117-120. https://doi.org/10.1007/s00436-008-1060-y

Gallais, F., De Martino, S.J., Sauleau, E.A., Hansmann, Y., Lipsker, D., Lenormand, C., Talagrand-Reboul, E., Boyer, P.H., Boulanger, N., Jaulhac, B., *et al.* (2018). Multilocus sequence typing of clinical *Borreliella afzelii* strains: population structure and differential ability to disseminate in humans. Parasit Vectors *11*, 374. https://doi.org/10.1186/s13071-018-2938-x

Ginsberg, H.S., Albert, M., Acevedo, L., Dyer, M.C., Arsnoe, I.M., Tsao, J.I., Mather, T.N., and LeBrun, R.A. (2017). Environmental Factors Affecting Survival of Immature *Ixodes scapularis* and Implications for Geographical Distribution of Lyme Disease: The Climate/Behavior Hypothesis. PLoS One *12*, e0168723. https://doi.org/10.1371/journal.pone.0168723

Gordillo-Perez, G., Solorzano, F., Cervantes-Castillo, A., Sanchez-Vaca, G., Garcia-Ramirez, R., Diaz, A.M., Munoz, O., and Torres, J. (2018). Lyme Neuroborreliosis is a Severe and Frequent Neurological Disease in Mexico. Arch Med Res *49*, 399-404. https://doi.org/10.1016/j.arcmed.2018.11.007

Gordillo-Perez, G., Vargas, M., Solorzano-Santos, F., Rivera, A., Polaco, O.J., Alvarado, L., Munoz, O., and Torres, J. (2009). Demonstration of *Borrelia burgdorferi* sensu stricto infection in ticks from the northeast of Mexico. Clin Microbiol Infect *15*, 496-498. https://doi.org/10.1111/j.1469-0691.2009.02776.x

Hao, Q., Hou, X., Geng, Z., and Wan, K. (2011). Distribution of *Borrelia burgdorferi* sensu lato in China. J Clin Microbiol *49*, 647-650. https://doi.org/10.1128/JCM.00725-10

Harrison, B.A., Rayburn, W.H., Jr., Toliver, M., Powell, E.E., Engber, B.R., Durden, L.A., Robbins, R.G., Prendergast, B.F., and Whitt, P.B. (2010). Recent discovery of widespread *Ixodes affinis* (Acari: Ixodidae) distribution in North Carolina with implications for Lyme disease studies. J Vector Ecol *35*, 174-179. https://doi.org/10.1111/j.1948-7134.2010.00044.x

Hashimoto, S., Kawado, M., Murakami, Y., Izumida, M., Ohta, A., Tada, Y., Shigematsu, M., Yasui, Y., Taniguchi, K., and Nagai, M. (2007). Epidemics of vector-borne diseases observed in infectious disease surveillance in Japan, 2000-2005. J Epidemiol *17 Suppl*, S48-55. https://doi.org/10.2188/jea.17.s48

Henningsson, A.J., Malmvall, B.E., Ernerudh, J., Matussek, A., and Forsberg, P. (2010). Neuroborreliosis--an epidemiological, clinical and healthcare cost study from an endemic area in the south-east of Sweden. Clin Microbiol Infect *16*, 1245-1251. https://doi.org/10.1111/j.1469-0691.2009.03059.x

Herrin, B.H., Peregrine, A.S., Goring, J., Beall, M.J., and Little, S.E. (2017). Canine infection with *Borrelia burgdorferi, Dirofilaria immitis, Anaplasma spp.* and *Ehrlichia spp.* in Canada, 2013-2014. Parasit Vectors *10*, 244. https://doi.org/10.1186/s13071-017-2184-7

Hofhuis, A., Harms, M., Bennema, S., van den Wijngaard, C.C., and van Pelt, W. (2015). Physician reported incidence of early and late Lyme borreliosis. Parasit Vectors *8*, 161. https://doi.org/10.1186/s13071-015-0777-6

Huang, W., Ojaimi, C., Fallon, J.T., Travisany, D., Maass, A., Ivanova, L., Tomova, A., Gonzalez-Acuna, D., Godfrey, H.P., and Cabello, F.C. (2015). Genome Sequence of Borrelia chilensis VA1, a South American Member of the Lyme Borreliosis Group. Genome Announc *3*. https://doi.org/10.1128/genomeA.01535-14

Hubalek, Z. (2009). Epidemiology of lyme borreliosis. Curr Probl Dermatol *37*, 31-50. https://doi.org/10.1159/000213069

Huppertz, H.I., Bohme, M., Standaert, S.M., Karch, H., and Plotkin, S.A. (1999). Incidence of Lyme borreliosis in the Wurzburg region of Germany. Eur J Clin Microbiol Infect Dis *18*, 697-703.

Ivanova, L.B., Tomova, A., Gonzalez-Acuna, D., Murua, R., Moreno, C.X., Hernandez, C., Cabello, J., Cabello, C., Daniels, T.J., Godfrey, H.P., *et al.* (2014). *Borrelia chilensis*, a new member of the *Borrelia burgdorferi* sensu lato complex that extends the range of this genospecies in the Southern Hemisphere. Environ Microbiol *16*, 1069-1080. https://doi.org/10.1111/1462-2920.12310

Jiang, Y., Hou, X., Zhang, L., Tan, Y., Lu, C., Xiao, D., Li, H., Hao, Q., and Wan, K. (2019). Case report: A patient coinfected by *Borrelia burgdorferi* sensu lato and spotted fever group Rickettsiae in Urumqi, China. Medicine (Baltimore) *98*, e17977. https://doi.org/10.1097/MD.0000000000017977

Kirsch, M., Ruben, F.L., Steere, A.C., Duray, P.H., Norden, C.W., and Winkelstein, A. (1988). Fatal adult respiratory distress syndrome in a patient with Lyme disease. JAMA *259*, 2737-2739.

Korenberg, E., Gorelova, N., and Kovalevskii, Y. (2002). Ecology of *Borrelia burgdorferi* sensu lato in Russia. In Lyme borreliosis: biology, epidemiology and control, K.O. Gray JS, Lane RS, Stanek G, ed. (New York: CABI Publishing), pp. 175-200.

Korenberg, E.I. (1994). Comparative ecology and epidemiology of lyme disease and tick-borne encephalitis in the former Soviet Union. Parasitol Today *10*, 157-160. https://doi.org/10.1016/0169-4758(94)90269-0

Kugeler, K.J., Farley, G.M., Forrester, J.D., and Mead, P.S. (2015). Geographic Distribution and Expansion of Human Lyme Disease,

United States. Emerg Infect Dis *21*, 1455-1457. https://doi.org/10.3201/eid2108.141878

Kugeler, K.J., Griffith, K.S., Gould, L.H., Kochanek, K., Delorey, M.J., Biggerstaff, B.J., and Mead, P.S. (2011). A review of death certificates listing Lyme disease as a cause of death in the United States. Clin Infect Dis *52*, 364-367. https://doi.org/10.1093/cid/ciq157

Kwit, N.A., Dietrich, E.A., Nelson, C., Taffner, R., Petersen, J., Schriefer, M., Mead, P., Weinstein, S., and Haselow, D. (2017). Notes from the Field: High Volume of Lyme Disease Laboratory Reporting in a Low-Incidence State - Arkansas, 2015-2016. MMWR Morb Mortal Wkly Rep *66*, 1156-1157. https://doi.org/10.15585/mmwr.mm6642a8

Lancet, T. (2018). Introducing EU-wide surveillance of Lyme neuroborreliosis. Lancet *392*, 452. https://doi.org/10.1016/S0140-6736(18)31738-0

Lane, R.S., and Loye, J.E. (1989). Lyme disease in California: interrelationship of *Ixodes pacificus* (Acari: Ixodidae), the western fence lizard (*Sceloporus occidentalis*), and *Borrelia burgdorferi*. J Med Entomol *26*, 272-278. https://doi.org/10.1093/jmedent/26.4.272

Lantos, P.M., Nigrovic, L.E., Auwaerter, P.G., Fowler, V.G., Jr., Ruffin, F., Brinkerhoff, R.J., Reber, J., Williams, C., Broyhill, J., Pan, W.K., *et al.* (2015a). Geographic Expansion of Lyme Disease in the Southeastern United States, 2000-2014. Open Forum Infect Dis *2*, ofv143. https://doi.org/10.1093/ofid/ofv143

Lantos, P.M., Shapiro, E.D., Auwaerter, P.G., Baker, P.J., Halperin, J.J., McSweegan, E., and Wormser, G.P. (2015b). Unorthodox alternative therapies marketed to treat Lyme disease. Clin Infect Dis *60*, 1776-1782. https://doi.org/10.1093/cid/civ186

Lantos, P.M., Tsao, J., Nigrovic, L.E., Auwaerter, P.G., Fowler, V.G., Ruffin, F., Foster, E., and Hickling, G. (2017). Geographic Expansion of Lyme Disease in Michigan, 2000-2014. Open Forum Infect Dis *4*, ofw269. https://doi.org/10.1093/ofid/ofw269

Letrilliart, L., Ragon, B., Hanslik, T., and Flahault, A. (2005). Lyme disease in France: a primary care-based prospective study. Epidemiol Infect *133*, 935-942. https://doi.org/10.1017/S0950268805004413

Lindgren, E., and Jaenson, T. (2006). Lyme borreliosis in Europe: influences of climate and climate change, epidemiology, ecology and adaptation measures. Available from: http://www.euro.who.int/__data/assets/pdf_file/0006/96819/E89522.pdf. Accessed: 1 October 2010.

Little, S.E., Beall, M.J., Bowman, D.D., Chandrashekar, R., and Stamaris, J. (2014). Canine infection with *Dirofilaria immitis, Borrelia burgdorferi, Anaplasma spp.*, and *Ehrlichia spp.* in the United States, 2010-2012. Parasit Vectors *7*, 257. https://doi.org/10.1186/1756-3305-7-257

Lorenc, T., Jones-Diette, J., Blanchard, L., Sutcliffe, K., Stansfield, C., Harden, M., Sowden, A., and Thomas, J. (2017). Incidence and surveillance of Lyme disease: Systematic review and policy mapping. (London: EPPI-Centre, Social Science Research Unit, UCL Institute of Education, University College London).

Maraspin, V., Ruzic-Sabljic, E., and Strle, F. (2006). Lyme borreliosis and *Borrelia spielmanii*. Emerg Infect Dis *12*, 1177. https://doi.org/10.3201/eid1207.060077

Marcus, L.C., Steere, A.C., Duray, P.H., Anderson, A.E., and Mahoney, E.B. (1985). Fatal pancarditis in a patient with coexistent Lyme disease and babesiosis. Demonstration of spirochetes in the myocardium. Ann Intern Med *103*, 374-376. https://doi.org/10.7326/0003-4819-103-3-374

Marzec, N.S., Nelson, C., Waldron, P.R., Blackburn, B.G., Hosain, S., Greenhow, T., Green, G.M., Lomen-Hoerth, C., Golden, M., and Mead, P.S. (2017). Serious Bacterial Infections Acquired During Treatment of Patients Given a Diagnosis of Chronic Lyme Disease - United States. MMWR Morb Mortal Wkly Rep *66*, 607-609. https://doi.org/10.15585/mmwr.mm6623a3

Masuzawa, T. (2004). Terrestrial distribution of the Lyme borreliosis agent *Borrelia burgdorferi* sensu lato in East Asia. Jpn J Infect Dis *57*, 229-235.

Matuschka, F.R., Ohlenbusch, A., Eiffert, H., Richter, D., and Spielman, A. (1995). Antiquity of the Lyme-disease spirochaete in Europe. Lancet *346*, 1367. https://doi.org/10.1016/s0140-6736(95)92382-9

Mavin, S., Watson, E.J., and Evans, R. (2015). Distribution and presentation of Lyme borreliosis in Scotland - analysis of data from a national testing laboratory. J R Coll Physicians Edinb *45*, 196-200. https://doi.org/10.4997/JRCPE.2015.304

Mead, P., Hook, S., Niesobecki, S., Ray, J., Meek, J., Delorey, M., Prue, C., and Hinckley, A. (2018). Risk factors for tick exposure in suburban settings in the Northeastern United States. Ticks Tick Borne Dis *9*, 319-324. https://doi.org/10.1016/j.ttbdis.2017.11.006

Moore, S.M., Eisen, R.J., Monaghan, A., and Mead, P. (2014). Meteorological influences on the seasonality of Lyme disease in the United States. Am J Trop Med Hyg *90*, 486-496. https://doi.org/10.4269/ajtmh.13-0180

Nakama, H., Muramatsu, K., Uchikama, K., and Yamagishi, T. (1994). Possibility of Lyme disease as an occupational disease--seroepidemiological study of regional residents and forestry workers. Asia Pac J Public Health *7*, 214-217. https://doi.org/10.1177/101053959400700403

Nau, R., Christen, H.J., and Eiffert, H. (2009). Lyme disease--current state of knowledge. Deutsches Arzteblatt international *106*, 72-81; I. https://doi.org/10.3238/arztebl.2009.0072

NCBI (2020). Taxonomy Browser. Available from: https://www.ncbi.nlm.nih.gov/Taxonomy/Browser/wwwtax.cgi?mode=Undef&id=1643685&lvl=3&lin=f&keep=1&srchmode=1&unlock. Accessed: 6/2/2020.

Neira, O., Cerda, C., Alvarado, M.A., Palma, S., Abumohor, P., Wainstein, E., Guzman, L., Juliet, C., Perez, C., Raggio, X., *et al.* (1996). [Lyme disease in Chile. Prevalence study in selected groups]. Rev Med Chil *124*, 537-544.

Norris, S.J., Barbour, A.G., Fish, D., and Diuk-Wasser, M.A. (2015). Response to Esteve-Gassent et al.: flaB sequences obtained from Texas PCR products are identical to the positive control strain *Borrelia burgdorferi* B31. Parasit Vectors *8*, 310. https://doi.org/10.1186/s13071-015-0899-x

Ogden, N.H., Bouchard, C., Badcock, J., Drebot, M.A., Elias, S.P., Hatchette, T.F., Koffi, J.K., Leighton, P.A., Lindsay, L.R., Lubelczyk, C.B., *et al.* (2019). What is the real number of Lyme disease cases in Canada? BMC Public Health *19*, 849. https://doi.org/10.1186/s12889-019-7219-x

Ogden, N.H., Lindsay, L.R., Morshed, M., Sockett, P.N., and Artsob, H. (2009). The emergence of Lyme disease in Canada. Can Med Assoc J *180*, 1221-1224. https://doi.org/180/12/1221 [pii] 10.1503/cmaj.080148

Ogden, N.H., Maarouf, A., Barker, I.K., Bigras-Poulin, M., Lindsay, L.R., Morshed, M.G., O'Callaghan C, J., Ramay, F., Waltner-Toews, D., and Charron, D.F. (2006). Climate change and the potential for range expansion of the Lyme disease vector *Ixodes scapularis* in Canada. Int J Parasitol *36*, 63-70. https://doi.org/10.1016/j.ijpara.2005.08.016

Oliver, J.H., Jr., Lin, T., Gao, L., Clark, K.L., Banks, C.W., Durden, L.A., James, A.M., and Chandler, F.W., Jr. (2003). An enzootic transmission cycle of Lyme borreliosis spirochetes in the southeastern United States. Proc Natl Acad Sci U S A *100*, 11642-11645. https://doi.org/10.1073/pnas.1434553100

Ornstein, K., Berglund, J., Nilsson, I., Norrby, R., and Bergstrom, S. (2001). Characterization of Lyme borreliosis isolates from patients with erythema migrans and neuroborreliosis in southern Sweden. J Clin Microbiol *39*, 1294-1298. https://doi.org/10.1128/JCM.39.4.1294-1298.2001

Patel, R., Grogg, K.L., Edwards, W.D., Wright, A.J., and Schwenk, N.M. (2000). Death from inappropriate therapy for Lyme disease. Clin Infect Dis *31*, 1107-1109. https://doi.org/10.1086/318138

Petrulioniene, A., Radzisauskiene, D., Ambrozaitis, A., Caplinskas, S., Paulauskas, A., and Venalis, A. (2020). Epidemiology of Lyme

Disease in a Highly Endemic European Zone. Medicina (Kaunas) *56*. https://doi.org/10.3390/medicina56030115

Piesman, J., and Gern, L. (2004). Lyme borreliosis in Europe and North America. Parasitology *129 Suppl*, S191-220. https://doi.org/10.1017/s0031182003004694

Postic, D., Korenberg, E., Gorelova, N., Kovalevski, Y.V., Bellenger, E., and Baranton, G. (1997). *Borrelia burgdorferi* sensu lato in Russia and neighbouring countries: high incidence of mixed isolates. Res Microbiol *148*, 691-702. https://doi.org/10.1016/S0923-2508(99)80068-0

Pritt, B.S., Mead, P.S., Johnson, D.K.H., Neitzel, D.F., Respicio-Kingry, L.B., Davis, J.P., Schiffman, E., Sloan, L.M., Schriefer, M.E., Replogle, A.J., *et al.* (2016). Identification of a novel pathogenic *Borrelia* species causing Lyme borreliosis with unusually high spirochaetaemia: a descriptive study. Lancet Infect Dis *16*, 556-564. https://doi.org/10.1016/S1473-3099(15)00464-8

Public Health Agency of Canada (2020). Surveillance of Lyme disease. Available from: https://www.canada.ca/en/public-health/services/diseases/lyme-disease/surveillance-lyme-disease.html#a3. Accessed.

Rauter, C., and Hartung, T. (2005). Prevalence of *Borrelia burgdorferi* sensu lato genospecies in *Ixodes ricinus* ticks in Europe: a metaanalysis. Appl Environ Microbiol *71*, 7203-7216. https://doi.org/10.1128/AEM.71.11.7203-7216.2005

Rizzoli, A., Hauffe, H., Carpi, G., Vourc, H.G., Neteler, M., and Rosa, R. (2011). Lyme borreliosis in Europe. Euro Surveill *16*.

Rudenko, N., Golovchenko, M., Mokracek, A., Piskunova, N., Ruzek, D., Mallatova, N., and Grubhoffer, L. (2008). Detection of *Borrelia bissettii* in cardiac valve tissue of a patient with endocarditis and aortic valve stenosis in the Czech Republic. J Clin Microbiol *46*, 3540-3543. https://doi.org/10.1128/JCM.01032-08

Rudenko, N., Golovchenko, M., Ruzek, D., Piskunova, N., Mallatova, N., and Grubhoffer, L. (2009). Molecular detection of *Borrelia bissettii* DNA in serum samples from patients in the Czech Republic with suspected borreliosis. FEMS Microbiol Lett *292*, 274-281. https://doi.org/10.1111/j.1574-6968.2009.01498.x

Russell, R.C. (1995). ?Lyme disease in Australia-still to be proven! Emerg Infect Dis *1*, 29-31.

Rutz, H., Hogan, B., Hook, S., Hinckley, A., and Feldman, K. (2018). Exploring an alternative approach to Lyme disease surveillance in Maryland. Zoonoses Public Health *65*, 254-259. https://doi.org/10.1111/zph.12446

Ruzic-Sabljic, E., Arnez, M., Lotric-Furlan, S., Maraspin, V., Cimperman, J., and Strle, F. (2001a). Genotypic and phenotypic characterisation of *Borrelia burgdorferi* sensu lato strains isolated from human blood. J Med Microbiol *50*, 896-901. https://doi.org/10.1099/0022-1317-50-10-896

Ruzic-Sabljic, E., Lotric-Furlan, S., Maraspin, V., Cimperman, J., Pleterski-Rigler, D., and Strle, F. (2001b). Analysis of *Borrelia burgdorferi* sensu lato isolated from cerebrospinal fluid. APMIS *109*, 707-713. https://doi.org/10.1034/j.1600-0463.2001.d01-136.x

Sajanti, E., Virtanen, M., Helve, O., Kuusi, M., Lyytikainen, O., Hytonen, J., and Sane, J. (2017). Lyme Borreliosis in Finland, 1995-2014. Emerg Infect Dis *23*, 1282-1288. https://doi.org/10.3201/eid2308.161273

Schiffman, E.K., McLaughlin, C., Ray, J.A.E., Kemperman, M.M., Hinckley, A.F., Friedlander, H.G., and Neitzel, D.F. (2018). Underreporting of Lyme and Other Tick-Borne Diseases in Residents of a High-Incidence County, Minnesota, 2009. Zoonoses Public Health *65*, 230-237. https://doi.org/10.1111/zph.12291

Schwartz, A.M., Hinckley, A.F., Mead, P.S., Hook, S.A., and Kugeler, K.J. (2017). Surveillance for Lyme Disease - United States, 2008-2015. MMWR Surveill Summ *66*, 1-12. https://doi.org/10.15585/mmwr.ss6622a1

Schwartz, A.M., Shankar, M.B., Kugeler, K.J., Max, R.J., Hinckley, A.F., Meltzer, M.I., and Nelson, C.A. (2020). Epidemiology and cost of Lyme disease-related hospitalizations among patients with employer-sponsored health insurance-United States, 2005-2014. Zoonoses Public Health *67*, 407-415. https://doi.org/10.1111/zph.12699

Scrimenti, R.J. (1970). Erythema chronicum migrans. Arch Dermatol *102*, 104-105.

Septfons, A., Goronflot, T., Jaulhac, B., Roussel, V., De Martino, S., Guerreiro, S., Launay, T., Fournier, L., De Valk, H., Figoni, J., *et al.* (2019). Epidemiology of Lyme borreliosis through two surveillance systems: the national Sentinelles GP network and the national hospital discharge database, France, 2005 to 2016. Euro Surveill *24*. https://doi.org/10.2807/1560-7917.ES.2019.24.11.1800134

Smith, R., and Takkinen, J. (2006). Lyme borreliosis: Europe-wide coordinated surveillance and action needed? Euro Surveill *11*, E060622 060621. https://doi.org/10.2807/esw.11.25.02977-en

Stanek, G., Flamm, H., Groh, V., Hirschl, A., Kristoferitsch, W., Neumann, R., Schmutzhard, E., and Wewalka, G. (1987a). Epidemiology of borrelia infections in Austria. Zentralbl Bakteriol Mikrobiol Hyg A *263*, 442-449. https://doi.org/10.1016/s0176-6724(87)80106-2

Stanek, G., Hirschl, A., Stemberger, H., Wewalka, G., and Wiedermann, G. (1987b). Does Lyme borreliosis also occur in tropical and subtropical areas? Zentralbl Bakteriol Mikrobiol Hyg A *263*, 491-495. https://doi.org/10.1016/s0176-6724(87)80117-7

Stanek, G., and Strle, F. (2003). Lyme borreliosis. Lancet *362*, 1639-1647. https://doi.org/10.1016/S0140-6736(03)14798-8

Steere, A.C. (2001). Lyme disease. N Engl J Med *345*, 115-125. https://doi.org/10.1056/NEJM200107123450207

Steere, A.C., and Glickstein, L. (2004). Elucidation of Lyme arthritis. Nat Rev Immunol *4*, 143-152. https://doi.org/10.1038/nri1267

Steere, A.C., Grodzicki, R.L., Kornblatt, A.N., Craft, J.E., Barbour, A.G., Burgdorfer, W., Schmid, G.P., Johnson, E., and Malawista, S.E. (1983). The spirochetal etiology of Lyme disease. N Engl J Med *308*, 733-740. https://doi.org/10.1056/NEJM198303313081301

Steere, A.C., Malawista, S.E., Hardin, J.A., Ruddy, S., Askenase, W., and Andiman, W.A. (1977a). Erythema chronicum migrans and Lyme arthritis. The enlarging clinical spectrum. Ann Intern Med *86*, 685-698. https://doi.org/10.7326/0003-4819-86-6-685

Steere, A.C., Malawista, S.E., Snydman, D.R., Shope, R.E., Andiman, W.A., Ross, M.R., and Steele, F.M. (1977b). Lyme arthritis: an epidemic of oligoarticular arthritis in children and adults in three connecticut communities. Arthritis Rheum *20*, 7-17. https://doi.org/10.1002/art.1780200102

Steere, A.C., and Sikand, V.K. (2003). The presenting manifestations of Lyme disease and the outcomes of treatment. N Engl J Med *348*, 2472-2474. https://doi.org/10.1056/NEJM200306123482423

Steere, A.C., Strle, F., Wormser, G.P., Hu, L.T., Branda, J.A., Hovius, J.W., Li, X., and Mead, P.S. (2016). Lyme borreliosis. Nat Rev Dis Primers 2, 16090. https://doi.org/10.1038/nrdp.2016.90

Stillman, B.A., Monn, M., Liu, J., Thatcher, B., Foster, P., Andrews, B., Little, S., Eberts, M., Breitschwerdt, E.B., Beall, M.J., *et al.* (2014). Performance of a commercially available in-clinic ELISA for detection of antibodies against *Anaplasma phagocytophilum, Anaplasma platys, Borrelia burgdorferi, Ehrlichia canis,* and *Ehrlichia ewingii* and *Dirofilaria immitis* antigen in dogs. J Am Vet Med Assoc *245*, 80-86. https://doi.org/10.2460/javma.245.1.80

Stone, B.L., Russart, N.M., Gaultney, R.A., Floden, A.M., Vaughan, J.A., and Brissette, C.A. (2015). The Western progression of Lyme disease: infectious and nonclonal *Borrelia burgdorferi* sensu lato populations in Grand Forks County, North Dakota. Appl Environ Microbiol *81*, 48-58. https://doi.org/10.1128/AEM.02422-14

Strizova, Z., Patek, O., Vitova, L., Horackova, M., and Bartunkova, J. (2019). Internet-based self-diagnosis of Lyme disease caused death in a young woman with systemic lupus erythematosus. Joint Bone Spine *86*, 650-651. https://doi.org/10.1016/j.jbspin.2018.12.011

Strle, F. (1999). Lyme borreliosis in Slovenia. Zentralbl Bakteriol Parasitenkd Infektionskr Hyg *289*, 643-652. https://doi.org/10.1016/s0934-8840(99)80023-1

Strle, F., Wormser, G.P., Mead, P., Dhaduvai, K., Longo, M.V., Adenikinju, O., Soman, S., Tefera, Y., Maraspin, V., Lotric-Furlan, S., *et al.* (2013). Gender disparity between cutaneous and non-cutaneous manifestations of Lyme borreliosis. PLoS One *8*, e64110. https://doi.org/10.1371/journal.pone.0064110

Tavora, F., Burke, A., Li, L., Franks, T.J., and Virmani, R. (2008). Postmortem confirmation of Lyme carditis with polymerase chain reaction. Cardiovasc Pathol *17*, 103-107. https://doi.org/10.1016/j.carpath.2007.03.004

Tseng, Y.J., Cami, A., Goldmann, D.A., DeMaria, A., Jr., and Mandl, K.D. (2015). Using Nation-Wide Health Insurance Claims Data to Augment Lyme Disease Surveillance. Vector Borne Zoonotic Dis *15*, 591-596. https://doi.org/10.1089/vbz.2015.1790

Tugwell, P., Dennis, D.T., Weinstein, A., Wells, G., Shea, B., Nichol, G., Hayward, R., Lightfoot, R., Baker, P., and Steere, A.C. (1997). Laboratory evaluation in the diagnosis of Lyme disease. Ann Intern Med *127*, 1109-1123. https://doi.org/10.7326/0003-4819-127-12-199712150-00011

Tulloch, J.S.P., Semper, A.E., Brooks, T.J.G., Russell, K., Halsby, K.D., Christley, R.M., Radford, A.D., Vivancos, R., and Warner, J.C. (2019). The demographics and geographic distribution of laboratory-confirmed Lyme disease cases in England and Wales (2013-2016): an ecological study. BMJ Open *9*, e028064. https://doi.org/10.1136/bmjopen-2018-028064

Uesaka, K., Maezawa, M., and Inokuma, H. (2016). Serological survey of Borrelia infection of dogs in Sapporo, Japan, where *Borrelia garinii* infection was previously detected. J Vet Med Sci *78*, 463-465. https://doi.org/10.1292/jvms.15-0392

van Dam, A.P. (2002). Diversity of *Ixodes*-borne *Borrelia* species--clinical, pathogenetic, and diagnostic implications and impact on vaccine development. Vector Borne Zoonotic Dis 2, 249-254. https://doi.org/10.1089/153036602321653833

van Dam, A.P., Kuiper, H., Vos, K., Widjojokusumo, A., de Jongh, B.M., Spanjaard, L., Ramselaar, A.C., Kramer, M.D., and Dankert, J. (1993). Different genospecies of Borrelia burgdorferi are associated with distinct clinical manifestations of Lyme borreliosis. Clin Infect Dis *17*, 708-717. https://doi.org/10.1093/clinids/17.4.708

van den Wijngaard, C.C., Hofhuis, A., Simoes, M., Rood, E., van Pelt, W., Zeller, H., and Van Bortel, W. (2017). Surveillance perspective on Lyme borreliosis across the European Union and European Economic Area. Euro Surveill *22*. https://doi.org/10.2807/1560-7917.ES.2017.22.27.30569

Vandekerckhove, O., De Buck, E., and Van Wijngaerden, E. (2019). Lyme disease in Western Europe: an emerging problem? A systematic review. Acta Clin Belg, 1-9. https://doi.org/10.1080/17843286.2019.1694293

Vandenesch, A., Turbelin, C., Couturier, E., Arena, C., Jaulhac, B., Ferquel, E., Choumet, V., Saugeon, C., Coffinieres, E., Blanchon, T., *et al.* (2014). Incidence and hospitalisation rates of Lyme borreliosis, France, 2004 to 2012. Euro Surveill *19*. https://doi.org/10.2807/1560-7917.es2014.19.34.20883

Vigfusson, H.B., Hardarson, H.S., Ludviksson, B.R., and Gudlaugsson, O. (2019). [Lyme disease in Iceland - Epidemiology from 2011 to 2015]. Laeknabladid *105*, 63-70. https://doi.org/10.17992/lbl.2019.02.215

Waddell, L.A., Greig, J., Lindsay, L.R., Hinckley, A.F., and Ogden, N.H. (2018). A systematic review on the impact of gestational Lyme disease in humans on the fetus and newborn. PLoS One *13*, e0207067. https://doi.org/10.1371/journal.pone.0207067

Wang, Y., Li, S., Wang, Z., Zhang, L., Cai, Y., and Liu, Q. (2019). Prevalence and Identification of *Borrelia burgdorferi* sensu lato Genospecies in Ticks from Northeastern China. Vector Borne Zoonotic Dis *19*, 309-315. https://doi.org/10.1089/vbz.2018.2316

Waniek, C., Prohovnik, I., Kaufman, M.A., and Dwork, A.J. (1995). Rapidly progressive frontal-type dementia associated with Lyme disease. J Neuropsychiatry Clin Neurosci *7*, 345-347. https://doi.org/10.1176/jnp.7.3.345

White, J., Noonan-Toly, C., Lukacik, G., Thomas, N., Hinckley, A., Hook, S., and Backenson, P.B. (2018). Lyme Disease Surveillance in New York State: an Assessment of Case Underreporting. Zoonoses Public Health *65*, 238-246. https://doi.org/10.1111/zph.12307

WHO (2012). Lyme disease - (CISID Table). Available from: http://data.euro.who.int/cisid/?TabID=434249. Accessed: 1 July 2020.

Wilking, H., and Stark, K. (2014). Trends in surveillance data of human Lyme borreliosis from six federal states in eastern Germany, 2009-2012. Ticks Tick Borne Dis *5*, 219-224. https://doi.org/10.1016/j.ttbdis.2013.10.010

World Health Organization (1995). Report of WHO workshop on Lyme borreliosis diagnosis and surveillance (who/cds/vph/95141-1) (Warsaw, Poland).

Zalaudek, I., Leinweber, B., Kerl, H., and Mullegger, R.R. (2005). Acrodermatitis chronica atrophicans in a 15-year-old girl misdiagnosed as venous insufficiency for 6 years. J Am Acad Dermatol *52*, 1091-1094. https://doi.org/10.1016/j.jaad.2005.01.125

Zbrzezniak, J., and Paradowska-Stankiewicz, I. (2019). Lyme disease in Poland in 2017. Przegl Epidemiol *73*, 317-320. https://doi.org/10.32394/pe.73.37

Zhang, F., Gong, Z., Zhang, J., and Liu, Z. (2010). Prevalence of *Borrelia burgdorferi* sensu lato in rodents from Gansu, northwestern China. BMC Microbiol *10*, 157. https://doi.org/10.1186/1471-2180-10-157

Chapter 22

Lyme Disease Diagnostics

Michael R. Mosel[1], John Aucott[2], Steve E. Schutzer[3], Adriana Marques[4], Paul M. Arnaboldi[5], Raymond Dattwyler[5] and Mark W. Eshoo[6*]

[1]Janus-I Science, Vista, CA 92081, USA
[2]Lyme Disease Research Center, Division of Rheumatology, Department of Medicine, Johns Hopkins University School of Medicine, USA
[3]Rutgers New Jersey Medical School, Newark NJ 07103, USA
[4]Laboratory of Clinical Microbiology and Immunology, National Institute of Allergy and Infectious Diseases, National Institutes of Health, 10 Center Drive, Bethesda, MD 20852, USA
[5]New York Medical College, 40 Sunshine Cottage Rd., Valhalla, NY 10595, USA
[6]BlueArc Biosciences, Del Mar, CA 92014, USA
*Corresponding author: mark.eshoo@bluearcbio.com

DOI: https://doi.org/10.21775/9781913652616.22

Abstract
Lyme disease, which is caused by *Borrelia burgdorferi* sensu lato infection from a tick bite, is growing with more than 300,000 estimated cases each year in the US alone. If diagnosed and treated early, antibiotics are most effective at eradicating infection and preventing sequelae. However, early diagnosis can be challenging in many instances. This chapter describes challenges regarding the clinical diagnosis of early Lyme disease and focuses on the current and future methodologies for detecting the presence of *B. burgdorferi* in clinical specimens. Both direct and indirect methods for pathogen detection are presented with regard to early and later-stage Lyme disease. In addition, potential alternative and evolving diagnostic methods such as detection of human host response and xenodiagnosis are discussed.

Introduction

Accurate diagnostic tests for Lyme disease are imperative. Lyme disease is caused by pathogenic members of the *Borrelia burgdorferi* sensu lato family. This tick-borne disease is the most common vector-borne infectious disease in the Northern Hemisphere. The U.S. Center for Disease Control and Prevention (CDC) estimates that there are approximately 300,000 Lyme disease cases per year in the United States, with over 3 million diagnostic tests performed each year (Hinckley et al., 2014; Nelson et al., 2015). Europe has a similar number of cases (Petrulioniene et al., 2020). The incidence of disease, the geographical distribution, and the number of diagnostic tests ordered are growing yearly in both North America and in Europe (Dessau et al., 2018; Petrulioniene et al., 2020).

Infection begins as an acute local infection in the skin at the tick bite site with *B. burgdorferi* proliferating locally and spreading out though the dermis. From there, *B. burgdorferi* can disseminate hematogenously to various organ systems including the heart, the nervous system, joints, and remote areas of the skin. Disseminated Lyme disease is associated with many severe symptoms and signs such as arthritis, cardiac manifestations, and neurological abnormalities (Steere et al., 2016). Outcomes in patients diagnosed and treated early in the course of infection are typically very good. Appropriate antibiotic treatment for 2 to 3 weeks duration is curative in the majority of patients, thus eliminating the risk of sequelae.

Most cases of early Lyme disease are clinically diagnosed by the presence of an EM lesion (Steere et al., 2003). However, other than the EM, no other clinical symptoms specifically point to early Lyme disease (Schoen, 2020). Moreover, EM is not present or recognized in up to 20% of individuals (Steere et al., 2003). Identification of EM is also difficult because the typical ring-within-a-ring (bull's eye) appearance is seen in only 20% of cases (Berger, 1984; Tibbles and Edlow, 2007). Thus, there is a great need for highly sensitive tests to aid in the diagnosis of active early Lyme disease.

Diagnostic tests can generally be divided into two categories, direct and indirect. Direct tests detect pathogen-specific nucleic acid or protein antigens, while indirect tests detect humoral or cellular responses to infection by identifying host response factors such as antibodies (i.e., serology-based testing) or other host response factors (*e.g.* mRNA, metabolites, T-cells, and others). The host humoral immune response to *B. burgdorferi* follows the classical pattern. The immunoglobulin M (IgM) response develops first, followed by IgG. Within 3 weeks following the onset of infection, IgM antibodies to *B. burgdorferi* can be detected in the majority of infected individuals. While serological testing is usually effective at detecting exposure to *B. burgdorferi* several weeks after infection, assays are highly insensitive during the first 2 weeks of early-localized infection with improved sensitivity in early disseminated disease (Wormser et al., 2008; Marques, 2015).

Direct diagnostic tests have a number of advantages over serologic assays. Direct detection tests can detect pathogens earlier in the course of infection and be used to monitor response to antibiotic treatment and the clearance of *B. burgdorferi*. Unlike serologic assays, direct detection tests can distinguish new from prior infection as they detect *B. burgdorferi* specific macromolecules rather than circulating antibodies that persist long after treatment or eradication of the pathogen by the host immune system. Although direct detection of *B. burgdorferi* in the clinical laboratory offers great promise, few methods are available for general use, with the exception of synovial fluid PCR which is widely used and reasonably sensitive (Li et al., 2011; Marques, 2015). One challenge is that infection in humans is characterized by low bacterial loads with an estimated 0.1 cultivable cells/mL of *B. burgdorferi* in whole blood during acute infection (Wormser et al., 2001). Further emphasizing the need for a direct test, is that unlike most bacterial infections, where the presence of the pathogen can be defined by direct observation or culture, *B. burgdorferi* is difficult to culture, requiring specialized media and long culture periods. Furthermore, some outer surface protein C (OspC) genotypes have been shown to be especially difficult to detect by culture (Xu et al., 2013).

This chapter focuses on the current and future methodologies and their challenges for the clinical diagnosis of Lyme disease from clinical specimens.

Clinical diagnosis

The clinical history and physical examination of the patient remain central to the diagnosis of Lyme disease. In the first 2-3 weeks of infection, antibody-based tests are extremely insensitive and clinical evaluation is the primary means of diagnosis. Patients with acute Lyme disease will usually present with EM, but 10-18% may present with non-specific symptoms of infection including fever, chills, fatigue, malaise and musculoskeletal pain. Respiratory symptoms such as rhinitis and cough are not typical for early Lyme disease. Data from the Lyme disease vaccine trial suggested that 7% of patients who are newly infected with *B. burgdorferi* may manifest no recognizable symptoms or signs of infection (Steere et al., 1998; Steere et al., 2003).

The EM skin lesion is the hallmark of early Lyme disease during the first few weeks of infection. While not all patients have or recognize the characteristic skin lesion, it is a reasonably sensitive diagnostic feature that in ideal circumstances is detected in approximately 80% of patients with acute Lyme disease (Steere et al., 2003). When the presence of a probable EM skin lesion is combined with other clinical and epidemiologic features the diagnosis is very likely. In Lyme-endemic areas, the combination of a solitary lesion of appropriate size, enlarging in size, a non-scaly perimeter, round or oval morphology, and in many cases the presence of systemic symptoms is relatively straightforward for making a treatment decision. Unfortunately, the accurate identification of the EM remains problematic. Only 30% of patients with EM recall a tick bite. EM can be highly variable in appearance, with only 20% of cases in North America exhibiting the stereotypical ring within a ring target appearance (Berger, 1984; Tibbles and Edlow, 2007). The more common EM lesions without central clearing may be missed by both patients and providers (Aucott et al., 2012). Lesions smaller than the current diagnostic cut off of 5cm in diameter also may represent false negatives. In a study of 118 cases of EM, in which *B. burgdorferi* infection was confirmed by culture or polymerase chain reaction, the lesion erythema was homogeneous in 59%, had central darker erythema in 32% and was a "bull's eye" with central clearing in only 9% (Smith et al., 2002). In another study of culture positive cases of EM central clearing was seen in 37% of cases (Nadelman and Wormser, 1995).

In another case series, the lack of central clearing, spontaneous resolution over several days, and tenderness led to missed diagnosis of EM (Feder and Whitaker, 1995). Reports of confusion with spider bites is a recurrent theme in case reports of misdiagnosed EM (Osterhoudt et al., 2002). Occasionally EM can be indurated, have central vesiculation or even become necrotic (Wetter and Ruff, 2011). The diagnosis of acute Lyme disease in Europe is complicated by the presence of several distinct Borrelia species, in addition to *B. burgdorferi*, that cause Lyme disease, particularly *B. garinii* and *B. afzelli*. In European EM lesions central clearing is more common than in the United States (79% vs. 19%) (Carlsson et al., 2003). *B. garinii*-associated EM formed more rapidly, were larger and were more likely to be homogeneous, compared with *B. afzelli*-associated lesions, which were slower to form, smaller and more likely to have the central clearing of the stereotypical target lesion (Carlsson et al., 2003).

Physician and patient surveys show that accurate EM identification remains elusive. When all the responses of the GPs are considered, only 72% of the cases of EM would have been recognized, and 23% of the cases of non-EM would have been diagnosed as EM (Lipsker et al., 2004). Of 3,104 non-physicians who accessed an online skin lesion identification survey, 72.7% of participants correctly identified the classic target EM commonly associated with Lyme disease. A mean of 20.5% of participants was able to correctly identify the four nonclassical EM, while 24.2% of participants incorrectly identified a tick bite reaction as EM (Aucott et al., 2012).

Unfortunately, the appearance of an EM-like lesion is also not specific as similar lesions occur in other circumstances, both infectious and noninfectious. Tick bite reactions and insect bites can be confused with EM (Feder and Whitaker, 1995). Tick bite reactions, which are due to an immediate reaction to tick bite proteins, are small (less than 1cm) and do not expand over time. Insect bites may be small or large but do not expand with observation over several days unless infected.

Bacterial cellulitis, most commonly caused by Group A *Streptococcus*, may occasionally be round or oval and therefore confused with EM (Imirzalioglu et al., 2010). *Staphylococcus aureus* skin infections can also occasionally be confused with EM, but typically have purulence, drainage or abscess formation which do not occur in EM. Primary dermatology conditions such as tinea, eczema, or rarely urticaria (Feder and Whitaker, 1995), erythema multiforme, and fixed drug eruption may also be confounding.

An increasingly recognized, and possibly important, condition called southern tick-associated rash illness (STARI) (Goddard, 2017) is transmitted by lone star ticks (*Amblyomma americanum*) and produces a lesion similar to that of EM (Logar et al., 2004; Wormser et al., 2005). However, no infectious etiology is known for STARI. The range of *A. americanum* tick has expanded into Lyme endemic areas of the Northeastern US as far north as Maine, resulting in a geographical overlap between the tick vectors of Lyme disease and STARI. As a result, the difficulty in distinguishing the STARI rash from the EM of Lyme disease is an increasing problem and will likely complicate future research, including diagnostics and vaccine development.

Clinical diagnosis of Lyme disease in the absence of a skin lesion is challenging. The differential diagnosis of the so called "summer flu" is a common clinical scenario. Summer flu-like illness in Lyme disease endemic regions may be due to tick-borne diseases including Lyme disease (Belongia et al., 2001). In endemic regions, such as the Upper Midwest (*e.g.*, northwestern Wisconsin), fever without skin lesion or localizing symptoms during May-Aug had evidence of tick-borne disease, primarily human anaplasmosis and Lyme disease, in 27% of patients (Belongia et al., 2001). Common symptoms include fatigue, malaise, myalgias, arthralgias, headache and neck pain. Important clues to other diagnoses such as rhinitis and cough are important to factor into a differential diagnosis, as respiratory symptoms and diarrhea would not be expected.

Dissemination of *Borrelia spp*. leads to more diverse clinical manifestations, which can include multiple EM skin lesions, meningitis, cranial neuritis (particularly facial palsy), radiculoneuritis, carditis (usually presenting as atrioventricular nodal block) and arthritis, that result from a combination of poorly understood host and pathogen related factors. In disseminated disease, serological assays are more sensitive (Wormser et al., 2008), though this is more likely attributable to the duration of infection rather than the fact of dissemination itself (Wormser et al., 2008). Awareness of the differing appearance of the

Table 1. Specific reasons why a direct diagnostic test is needed for Lyme disease

Cases are increasing >300,000 per year in the US.
Cases are spreading geographically.
Current antibody-based assays are insensitive in the first 1- 3 weeks of infection and a single test does not indicate active infection.
Inferences of new infection by antibody tests are challenged because individuals can get Lyme disease more than once and in an endemic area many persons are seropositive from past exposure.
Direct tests have potential to detect infection within days to first week.
Detection of circulating DNA or antigen is indication of an active infection.
Direct testing to diagnose active *B. burgdorferi* infection will reduce potential misdiagnoses, thus leading to prompt treatment minimizing long-term effects.

disseminated skin lesions of EM is underappreciated (Shah and Varatharaj Palraj, 2019). Clinical suspicion and awareness of subtle signs of infection such as radiculopathy and carditis and nonspecific symptoms are needed during this phase (Forrester and Mead, 2014; Steere et al., 2016).

Direct diagnostic assays for *B. burgdorferi*

The laboratory is an adjunctive tool to help clinicians make diagnoses, including infections. The major clinical laboratory tests used today to directly detect pathogens are cultures, microscopy, antigen detection, and nucleic acid detection of the offending pathogen (Table 1). In this section of the chapter, we will focus on those methods that have been or are near being translated to practical clinical use for Lyme disease. The tests will be of value to the clinician who is making a diagnosis with the intent to treat as well as researchers who want to enroll patients in a clinical trial that requires definitive evidence of the infecting agent. *B. burgdorferi* has several properties that challenge the usual diagnostic methods.

In contrast to common bacterial infectious agents, such as streptococci or *E. coli*, where there is an abundance of live organisms, their antigens, and nucleic acids for diagnostic detection, *B. burgdorferi* grows very slowly, thus, resulting in comparatively few cells available for detection and requires a complex growth media that is expensive and not readily available. Consequently, cultures take too long (weeks to months) to grow even if a detectable number of bacteria is present in a clinical sample. Similarly, it is difficult to search for the microbe with a microscope using staining methods even with specific probes. The low number of bacteria and their slow replication results in a lower concentration of specific antigens shed into the clinical sample, thus challenging common antigen detection tests. The low copy number of bacteria also means that there are few nucleic acid copies available for straight probing such as with a PCR assay. That is why a PCR blood test has been challenging with past methods. However, our knowledge of the genomics and consequently proteomics of *B. burgdorferi* has evolved markedly in the past few years. In addition, technology has advanced to permit more precise detection of infectious agents. These developments, coupled with some creative approaches, may someday permit us to overcome past hurdles in detecting the pathogen by antigen and DNA assays.

The collective data show that nucleic acid detection assays can detect *B. burgdorferi* with great specificity but with low sensitivity due to the low levels of organism in the bloodstream. However, the current methods may require enhancement or enrichment methods (Eshoo et al., 2012). Detection of the bacteria and/or it's DNA may occur only transiently and during a limited window of time.

Culture

Culture of blood and skin biopsies from EM is not a practical laboratory method for the diagnosis of early Lyme disease but remains an effective technique to confirm active infection by *B. burgdorferi* for research investigation or clinical trials. The disadvantages that limit culture to clinical settings are that it is slow, has low sensitivity with a risk of contamination, and requires specialized equipment (e.g., fluorescent or dark-field microscope) and expertise (Coulter et al., 2005; Johnson et al., 2013; Marques, 2015; Schutzer et al., 2019). *B. burgdorferi* replicates slowly, requiring complex growth media and a positive culture can take up to 12 weeks (Coulter et al., 2005; Schutzer et al., 2019). In general, sensitivities of 40% and 40-60% have been observed for high-volume plasma and skin biopsy cultures, respectively, from early Lyme disease patient samples (Marques, 2015). Lower sensitivities have also been reported in a study by Coulter et al., where *B. burgdorferi* was recovered from 27% of high-volume plasma cultures versus 32% from skin biopsy cultures requiring up to 56 days of cultivation (Coulter et al., 2005). Thus, without increasing speed, sensitivity, specificity, and simplicity, culture will never become part of routine diagnostic testing for early Lyme disease but will remain valuable in research settings as a gold standard for confirming *B. burgdorferi* positive specimens with active infection.

PCR

Past PCR assays of blood that did not have target enrichment or larger starting sample volumes than traditional PCR had low sensitivity (Nolte, 2012; Lee et al., 2014; Bil-Lula et al., 2015; Liu et al., 2016; Moore et al., 2016). The sensitivity of PCR tests of blood for Lyme disease is so low that CDC discouraged their use. However, Liveris *et al* showed that *B. burgdorferi* often was in the blood of untreated individuals with early Lyme disease and could be detected using culture enrichment methods (Liveris et al., 2011; Liveris et al., 2012). This was shown as well with more practical methods (Eshoo et al., 2012).

The use of PCR as a diagnostic tool for early Lyme disease is included in a variety of laboratory workflows. For example, amplified products can be detected and measured by gel electrophoresis, real-time PCR, next-generation sequencing NGS, or by PCR/ESI-MS (Eshoo et al., 2012; Snitkin et al., 2012; Eshoo et al., 2013; Schutzer et al., 2019). Sample types for analysis include whole blood, plasma, and EM-skin biopsies with or without a prior *B. burgdorferi* culture step. Some of the approaches are meant for clinical utility while others are intended for research use. PCR primers are typically designed to target *B. burgdorferi* gene target(s) such as *rpoC, leuS, OspC, gyrB, FlaB* and *OspA etc.* (Eshoo et al., 2013). While the sensitivity from PCR of skin biopsies may be as high as 80% (Liveris et al., 2002), the sensitivity from PCR of whole blood is approximately 20-60% (Eshoo et al., 2012; Liveris et al., 2012). The lower sensitivity observed from blood is most likely due to low levels of pathogen whole cells and free DNA during the time window in which samples are collected. Early Lyme disease patients are estimated to have 0.1 cultivable cells/mL of *B. burgdorferi* in whole blood (Wormser et al., 2001). To increase sensitivity, additional enrichment steps have been added to PCR workflows such as an upfront isothermal amplification to increase pathogen DNA along with using larger blood volumes (Eshoo et al., 2012) up to 20mLs (Mosel et al., 2019). PCR amplicon contamination is also a challenge, especially in nested PCR workflows, because a high level of amplicon is typically generated in the first round and is handled to setup the subsequent PCR reaction.

PCR using nucleic acid extracts from a small 2mm skin biopsy taken from the leading edge of an EM can have a high yield (Berger et al., 1985; Schwartz et al., 1992; Liveris et al., 2002). The biopsy is an easy to perform procedure by a primary care doctor and does not require sutures. However, because an EM lesion is not always present, PCR from blood would be useful specimen type if continued advancements in PCR sensitivity can be achieved.

PCR and electrospray Ionization mass spectrometry detection of B. burgdorferi

PCR followed by electrospray ionization mass spectrometry (PCR/ESI-MS) has been used to both directly detect and genotype *B. burgdorferi* from ticks and clinical specimens (Crowder et al., 2010; Eshoo et al., 2012; Eshoo et al., 2014; Eshoo et al., 2015; Honig et al., 2017; Mosel et al., 2019). PCR amplicons of about 100 base pairs are generated from gene-specific primers targeting *Borrelia* followed by desalting and ESI/MS. Basecount signatures are produced by counting the number of A's, T's, G's, and C's in each amplicon and a curated database is used to determine the presence of *B. burgdorferi* genotype(s). PCR/ESI-MS recently was utilized to measure response to antibiotic treatment using serially collected whole blood specimens from patients clinically diagnosed with early Lyme disease (Mosel et al., 2019). Two of four patients were found to be infected with more than one *B. burgdorferi* genotype and the ratio of these genotypes changed over time during antibiotic treatment. In one patient, the presence of *B. burgdorferi* DNA was still detected at Day 21, the latest time point of the study. In another study, PCR/ESI-MS was combined with an isothermal amplification step using a 1.25 mL volume of whole blood, to increase sensitivity and *B. burgdorferi* were detected in 13 of 21 (62%) of patients with clinically diagnosed and laboratory confirmed (seroconversion) early Lyme disease at the initial physician visit (Eshoo et al., 2012). Moreover, early Lyme disease was determined in 19 of 21 (90%) of these patients at the initial doctor's visit when PCR/ESI-MS results were combined with results of the two-tier test from visit one serology (Eshoo et al., 2012). These findings are quite comparable to the sensitivity for the detection of *B. burgdorferi* in blood by next generation sequencing and PCR combined with T2 magnetic resonance technology as discussed below, suggesting there may be a diagnostic limit for the molecular detection of *B. burgdorferi* in blood.

PCR/ESI-MS also has been used to detect *B. burgdorferi* in ticks used for the xenodiagnoses of early Lyme disease (Marques et al., 2014). While other direct molecular detection technologies have been shown to detect *B. burgdorferi* from clinical specimens such a whole blood, PCR/ESI-MS has an advantage of also providing genotyping capabilities, which may be useful in further understanding the complex nature of *B. burgdorferi* infection. Unfortunately, Abbott Laboratories in 2017 discontinued the commercial sales of the instrument, reagents and software to perform PCR/ESI-MS presumably based on financial, logistical and regulatory considerations (Ozenci et al., 2018).

Direct detection of B. burgdorferi by next generation sequencing

Metagenomic analysis by next generation sequencing (NGS) is a promising approach for the diagnosis of a wide range of infectious diseases. Importantly NGS combined with bioinformatics analyses has the potential to detect any pathogen including novel ones. The approach gained a great deal of enthusiasm in 2014 when next generation sequencing was used to diagnose a case of neuroleptospirosis (Wilson et al., 2014). The main challenge of this approach is the fact that clinical specimens contain an overwhelming amount of human DNA relative to pathogen DNA. For example, in the neuroleptospirosis study, only 475 of 3,063,784 sequencing reads were found to map to *Leptospira spp.* (Wilson et al., 2014). Despite this challenge NGS is being applied to a wide range of infectious diseases (Simner et al., 2018).

A Silicon Valley start-up, Karius Inc. (Redwood City, CA) has been developing an NGS-based analysis pipeline for the detection of cell-free infectious agent DNA from plasma specimens. Their technology sequences all the cell-free DNA found in plasma and the human DNA sequences are removed bioinformatically. Karius Inc., in collaboration with researchers at Stonybrook Children's Hospital in New York, started a clinical study in July 2018 to test up to 30 plasma specimens from pediatric patients with Lyme disease (ClinicalTrials.gov Identifier: NCT03505879). However, in October 2019 the group published a presentation abstract where they reported that NGS of cell-free DNA did not identify *B. burgdorferi* DNA in the plasma of five pediatric patients with active EM skin lesions. Based upon this finding, the authors concluded the approach is unlikely to be helpful in diagnosing early localized Lyme disease (Andrew S Handel, 2019). However, a second study that also used the Karius NGS assay found much better results. In this second study, the study authors used previously characterized plasma specimens collected from forty patients presenting with physician diagnosed EM (Branda et al., 2020). Of the 40, cell-free *B. burgdorferi* DNA was detected in 19 specimens (48%) by the Karius NGS diagnostic pipeline. Amongst the 40 patients, the diagnosis was laboratory confirmed in 28 either by recovering *B. burgdorferi* in cultures of an EM skin biopsy or by demonstrating seroconversion between acute- and convalescent-phase sera. Among this subset of specimens, *B. burgdorferi* cell-free DNA was detected in 18/28 patients (64%). These findings are consistent with previous studies for the detection of *B. burgdorferi* in blood using a multi-locus PCR and electrospray ionization mass spectrometry molecular detection assay where 13/21 patients (62%) with laboratory confirmed Lyme disease were found to have *B. burgdorferi* DNA in their blood (Eshoo et al., 2012).

NGS-based tests, like all Lyme disease diagnostics with the exception of culture, cannot confirm active infection by live bacteria. However, if *B. burgdorferi* DNA is found circulating in the blood in a suspected untreated early case, this is actionable evidence for treatment. This distinction could be especially important for highly sensitive molecular tests such as NGS as they may be detecting remnant DNA fragments from dead bacteria post treatment. A study looking at molecular detection from serial blood specimens collected from two patients with clinically and laboratory confirmed Lyme disease found *B. burgdorferi* DNA in the blood of two early Lyme disease patients after 8 days and 21 days of antibiotic treatment (Mosel et al., 2019). These finding suggest that the bacteria may reside in parts of the body that are not readily cleared, and bacterial remnants (DNA) may continue to leak into the circulatory system after antibiotic treatment (Wormser et al., 2012b).

T2 magnetic resonance detection of B. burgdorferi

A PCR-based assay in development for Lyme disease employs a T2 magnetic resonance (T2MR) assay to achieve a lower limit of detection than by standard PCR (Snyder et al., 2017). T2 Biosciences (Lexington, MA) has been working to develop an assay designed to detect Lyme *Borrelia* from whole blood using their FDA approved T2Dx Instrument. This automated cartridge-based platform provides a test result in 3 to 5 hours. The T2 Lyme test works by attaching a blood tube directly to the cartridge which processes a 2 mL specimen. The system concentrates the cells in the blood, lyses the cells and then performs PCR without a nucleic acid extraction step. Following multiplex PCR, the amplicons are hybridized to probe coated paramagnetic particles which cluster/aggregate in the presence of their cognate target DNA sequence. The clustering is then detected by T2 magnetic resonance within the T2 instrument. The main advantages of the system are that the test is a fully automated sample-to-answer system, and the Lyme disease test can be performed on an FDA-approved T2 platform which is already installed in hospitals and diagnostic testing laboratories. The company presented a poster on clinical performance of their Lyme assay at the 2018 International Conference on Lyme Borreliosis and Other Tick-Borne Diseases in Atlanta GA and published the poster on the company's website (Pierrette Andre, 2018). In this study, using clinical specimens from 21 subjects with early Lyme disease based on the presence of an EM, twelve patients were confirmed to have Lyme disease based upon one or more laboratory tests being positive: tissue PCR, serology, 2-tier test, culture and/or seroconversion. Amongst the laboratory confirmed Lyme disease positive specimens, the T2 Lyme test found 7/12 to be positive (58%). Though the study numbers are low, the sensitivity levels observed are consistent with other molecular diagnostic tests for early Lyme disease.

Xenodiagnosis

Xenodiagnosis is the use of a natural vector to detect the presence of a pathogen. This technique has been used to identify the etiology and provide evidence of some vector borne diseases, Chagas Disease being the classical example for human infections (Telford et al., 2014). For *B. burgdorferi*, xenodiagnosis using laboratory-reared *Ixodes sp* ticks has been frequently used to provide evidence of infection in animals (Donahue et al., 1987). Tick saliva contains a complex mixture of bioactive compounds that help ticks hide from the hosts and facilitate acquisition and transmission of microorganisms (Nuttall, 2019), and *Ixodes* tick saliva has been shown to be a chemoattractant for *B. burgdorferi*, promoting aggregation of bacteria at tick feeding sites (Shih et al., 2002).

Studies in dogs, mice and non-human primates have shown that *B. burgdorferi* DNA can be detected in tissues for extended periods after antibiotic therapy and that *Ixodes* ticks feeding on these animals were able to acquire *B. burgdorferi* during their blood meal (Straubinger et al., 1997; Straubinger et al., 2000; Bockenstedt et al., 2002; Yrjanainen et al., 2007; Hodzic et al., 2008; Barthold et al., 2010; Embers et al., 2012; Embers et al., 2017). Critiques to some of the animal studies include the dosage of the antibiotics used, infection by needle inoculation instead of using infected ticks, and the inability to cultivate the bacteria *in vitro* (Wormser and Schwartz, 2009; Wormser et al., 2012a). Xenodiagnosis using laboratory-reared, pathogen-free ticks on patients after antibiotic therapy could offer a minimally invasive method for determining the persistence of *B. burgdorferi* and help clarify the etiology of post-treatment Lyme disease syndrome.

The first study of the use of *I. scapularis* larva for the xenodiagnosis of *B. burgdorferi* infection in humans (Marques et al., 2014) enrolled 36 participants (10 patients with persistently high C6 antibody levels, 10 patients with post-treatment Lyme disease syndrome, 5 patients with erythema migrans after completion of antibiotic therapy, one patient with acute erythema migrans who started therapy at the same time of tick placement, and 10 healthy volunteers). The procedure was well tolerated (mild itching at the site was the most common adverse event) but labor and time intensive (larval ticks required 4–5 days to feed to repletion) (Marques et al., 2014; Turk et al., 2018). Twenty-three of the 26 patients with Lyme disease had at least 1 fed tick tested for *B. burgdorferi,* using one or more of multiple techniques that included culture, PCR, isothermal amplification followed by PCR and electrospray ionization mass spectroscopy (Eshoo et al., 2012), as well as feeding of ticks or injection of tick contents into severe combined immunodeficiency mice. Results were negative for 19 patients, indeterminate for 2 patients, and positive for

2 patients. One of the patients that tested positive had acute erythema migrans and initiated antibiotic therapy at the same time of tick placement (considered the positive control). The other patient had post-treatment Lyme disease syndrome and tested positive in two separate occasions. Because the evidence for the presence of *B. burgdorferi* in the xenodiagnostic ticks was based on DNA detection, questions have been raised regarding the interpretation of these findings (Bockenstedt and Radolf, 2014; Shapiro, 2015). A larger study is currently underway to assess the link between detection of *B. burgdorferi* by xenodiagnosis and persistence of symptoms in patients diagnosed with Lyme disease (NIAID, 2015).

Indirect diagnostic assays for *B. burgdorferi*

Indirect antibody-based tests (EIA/ELISA, IFA, Western blot, Immunoblot)

The mainstay of the laboratory diagnosis of Lyme disease remains the detection of anti-*B. burgdorferi* antibodies produced in response to infection. Despite issues with specificity and a lack of sensitivity in early infection, serologic-based assays are the only FDA approved laboratory diagnostics (Nowakowski et al., 2001; Bacon et al., 2003).

The modern era of Lyme serology began in 1995 with the establishment of the CDC recommended standard two-tier testing (STTT) paradigm. It consists of a first-tier assay (typically an EIA) that, if positive or equivocal, is followed by a second-tier western blot (WB) (Control., 1995). IgM and IgG WB are performed as the second step. The IgM WB results are used only for disease of less or equal to 30 days of duration (Figure 1). The WB is interpreted using standardized criteria, requiring at least two of three signature bands for a positive IgM WB (24 kDa (OspC), 39 kDa (BmpA), and 41 kDa (Fla)), and 5 of 10 signature bands for a positive IgG WB (18 kDa, 21 kDa (OspC) *, 28 kDa, 30 kDa, 39 kDa (BmpA), 41 kDa (Fla), 45 kDa, 58 kDa (not GroEL), 66 kDa, and 93 kDa).

The CDC's recommended STTT paradigm was necessitated primarily by the lack of specificity in laboratory diagnostics used in the late 1980's and early 1990's, which came to be known as first tier assays (Marques, 2018). Non-specificity of first-tier assays was principally driven by the antigen targets used. Early assays consisted of EIAs or IFAs that utilized whole *B. burgdorferi*, whole *B. burgdorferi* lysate, or mixtures of whole *B. burgdorferi* recombinant proteins as antigen targets. Whole *B. burgdorferi* protein antigens have the advantage of containing multiple linear and conformational epitopes but the disadvantage that some of these epitopes are not unique to *B. burgdorferi* and are homologous to epitopes found in unrelated antigens (Liang et al., 1999b; Aguero-Rosenfeld et al., 2005; Arnaboldi et al., 2013; Dattwyler and Arnaboldi, 2014). As a result, these antigen targets detect antibodies generated by immune responses against *B. burgdorferi*, but also may detect antibodies to unrelated bacteria or other antigens.

The introduction of the second-tier WB improved specificity but introduced other problems. Although STTT is sensitive in late infection, it lacks sensitivity in the first weeks of infection when the disease is most easily treated. WB are read manually or scanned by a machine that measures the density of the banding pattern and interprets the results (Binnicker et al., 2008). Scanning and manual scoring of bands can produce differing interpretations of the results between labs or medical personnel (Seriburi et al., 2012). Also, some clinicians have difficulty understanding the results or when to do additional testing.

If the first-tier test yields negative results and the illness duration is more than 30 days of duration, the clinician should consider an alternative diagnosis and no further testing is recommended. For patients with signs or symptoms consistent with Lyme disease, and illness duration equal or less than 30 days, the provider should consider treatment. A follow up convalescent serum can be obtained when the patient has had more time to develop an antibody response. If the first-tier test yields positive or equivocal results, an IgM and IgG WB is performed. If the pretest likelihood of infection is high, then either a positive IgM or IgG WB is reasonable evidence to assume recent infection with *B. burgdorferi*. If the infection is greater than four weeks, and a positive IgG result is present, this is good evidence of current or previous infection (Figure 1).

Recombinant or synthetic antigens and serological diagnostics

In the late 1990s, assays containing peptide-antigen targets were developed and marketed as a more specific alternative to existing whole antigen or whole bacteria-based assays. The C6-peptide, a 25-mer oligopeptide corresponding to the sixth invariable region (IR6) of the VlsE protein (Zhang and Norris,

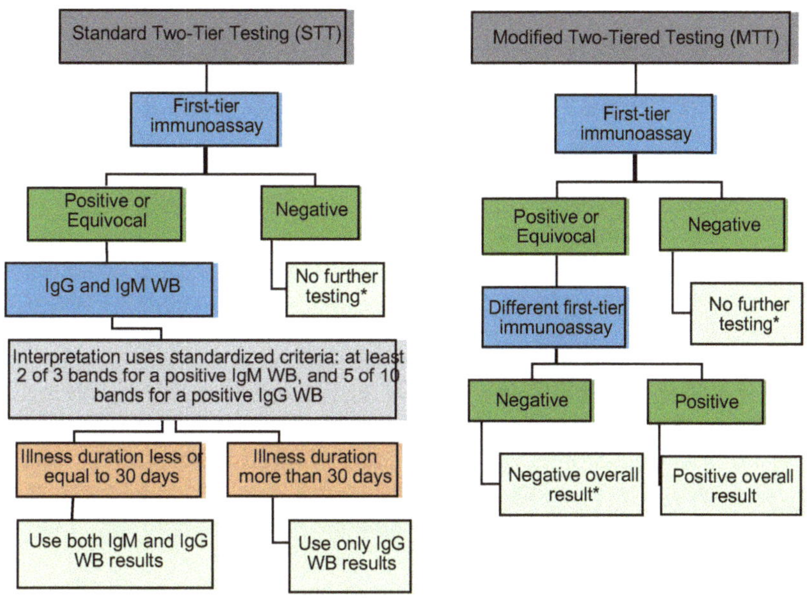

Figure 1. Standard two-tier testing and modified two-tier testing for serodiagnosis of Lyme disease. *For patients with signs or symptoms consistent with Lyme disease for less than or equal to 30 days, the provider may treat the patient and follow up with a convalescent serum. Patients with erythema migrans should receive treatment on the basis of the clinical diagnosis. Western blot (WB).

1998) was the target antigen for the C6 Peptide EIA (Liang et al., 1999b). This immunodominant epitope is partially conserved among *B. burgdorferi sensu lato* (Liang et al., 1999a; Gomes-Solecki et al., 2007). OspC is the major outer surface protein expressed by *B. burgdorferi* in the feeding tick and VlsE and C6 expression only develops after infection is establish in the skin. Recombinant VlsE and IR6 based EIAs are more specific than those using whole cell sonicates, but they are generally less sensitive (Ledue et al., 2008; Branda et al., 2011; Wormser et al., 2013; Molins et al., 2016; Pegalajar-Jurado et al., 2018b).

In any case, no first-tier assay is yet specific enough to be used by itself. One study purported that the C6 assay was robust enough to be used as a single tier assay, with higher sensitivity than STTT, with specificity of 98.9%, compared with 99.5% for STTT (Wormser et al., 2013). However, other studies using different serum sets, demonstrated that the specificity of the C6 assay was in the mid 90's. Branda reported a 94% specificity in endemic controls (Branda et al., 2011), Lipsett reported a 94.2% specificity and, interestingly, found that the STTT had a sensitivity of 81.6% while the sensitivity of C6 was 78.8% (Lipsett et al., 2016). Therefore, the C6 assay ultimately was found to be unsuitable as a single tier assay because it contained a single human epitope for which there is some heterogeneity and had slightly lower specificity than STTT, The C6 parent protein, VlsE is not expressed in the feeding tick and only expressed after infection was established (Ohnishi et al., 2003; Embers et al., 2007; Sillanpaa et al., 2007; Tjernberg et al., 2009). VlsE specificity is also in the mid 90's (Gomes-Solecki et al., 2007). Furthermore, it has become evident that both whole VlsE and C6 peptide cross-react with antibodies generated against similar antigens expressed by members of the relapsing fever *Borrelia*, including *Borrelia miyamotoi* (Koetsveld et al., 2020). Studies from the CDC reported a 4% false positive rate for the C6 EIA in non-endemic health controls obtained in Texas where relapsing fever is prevalent (Molins et al., 2017; Pegalajar-Jurado et al., 2018a). While C6 was not accepted as a single tier assay, it garnered further interest in the concept of peptide based sero-diagnostics, as it showed that even a one-epitope peptide assay could yield results similar to or better than the existing first tier assays (Liang et al., 1999b; Bacon et al., 2003; Heikkila et al., 2003; Ohnishi et al., 2003).

Most of the first tier sero-assays currently on the market are based on whole protein antigens, native or recombinant or peptides derived from VlsE, OspC or FlaB, or a combination of whole protein antigens and peptides. Among the leading assays, ZEUS ELISA Borrelia utilizes VlsE/pepC10; Diasorin LIAISON utilizes VlsE and recombinant OspC; BioRad multiplex uses recombinant OspC type B, recombinant OppA2 and a peptide containing an IR6-like epitope and a FlaB epitope; and the Quidel rapid assay has uses IR6- pepC10. None of these assays is specific enough to be used as a single tier assay and sensitivity in early infection remains an issue.

Until recently, the second tier of the two-tier paradigm consisted solely of a WB for the detection of either IgM or IgG binding. In 1995, at the time the STTT paradigm was instituted, *B. burgdorferi* antigens were poorly understood, and the proteins comprising the individual bands in the WB criteria were not well characterized. It is now recognized that some of the bands contain more than one antigen, and that some of these antigens, such as the p41 flagellin, are cross reactive with non-*B. burgdorferi* antigens. Furthermore, the WB criteria were developed utilizing antigens derived from cultured organisms. We now know that cultured *B. burgdorferi* do not express the full range of antigens expressed by *B. burgdorferi in vivo*. Thus, key *in vivo* expressed antigens, including VlsE, are not included in the WB criteria.

The initial immune response is primarily to *B. burgdorferi* antibodies first expressed in the feeding tick and when *B. burgdorferi* is first introduced into the skin, including OspC, FlaB and p66. FlaB and p66 antigens express epitopes that cross react with epitopes from other bacteria, while OspC has a high degree of variability and also contains epitopes with sequences similar to those of other bacteria (Liang et al., 1999b; Cugini et al., 2003; Arnaboldi and Dattwyler, 2015). Only after the infection is established does *B. burgdorferi* begin to express additional antigens, like VlsE, the parent antigen of C6 (Fallon et al., 2014; Marques, 2018; Pegalajar-Jurado et al., 2018b; Lipsett et al., 2019). Consequently, proper antigen target selection becomes a critical determinant for the development of a sensitive Lyme disease serodiagnostics test, especially in early disease. The use of cultured bacteria as a source of target antigens for the WB, which do not express the same spectrum of antigens expressed by *B. burgdorferi in vivo* contributed to the insensitivity of the STTT approach, particularly during early infection and a perception of a delayed immune response in Lyme disease.

As a result of the numerous drawbacks associated with the STTT paradigm, sensitivity of the combined assay is typically poor at the time many patients with EM first seek medical care, with sensitivity typically reported as less than 50% during early infection (Aguero-Rosenfeld et al., 2005). During early infection, while the antibody response is still developing, an individual will often not have a measurable antibody response against multiple antigens. Thus, even if a first-tier assay were positive at the time of the first clinical visit, second tier WB will often be negative necessitating a convalescent visit to obtain blood for a confirmatory serological test. This lack of sensitivity is typically not an issue in later disseminated disease as the patient has had additional time to generate a humoral response against a more diverse set of antigens.

An as an alternative to STTT, a new modified two-tier testing (MTTT) paradigm (Figure 1) recently was introduced which replaces the second tier WB with a second, distinct EIA (Mead et al., 2019). In comparison to STTT, MTTT is more sensitive in early disease and has similar sensitivity for later infection (Molins et al., 2016; Marques, 2018; Pegalajar-Jurado et al., 2018b; Lipsett et al., 2019). MTTT approaches are also less labor intensive, more cost efficient, and eliminates subjective interpretation of results.

Part of understanding the limitations of *B. burgdorferi* serological testing is understanding the limitations of diagnostic serology all together. No serologic test can be used to assess disease activity or the response to treatment. As with most infections, once a mature humoral immune response develops, antibodies can remain detectable for years. With the lack of specificity associated with current FDA approved first tier serologic assays, generally in the 2-4+% range, great care must be taken interpreting test results. They should only be interpreted in light of the patient's clinical picture.

In all laboratory tests, predictive value must be considered to place the test results in context for each individual patient. The predictive value of a test is determined by the sensitivity, specificity and the prevalence of the disease in the population to be tested. To appropriately interpret any test result, one must understand the false positive and false negative rates and the pretest likelihood that the patient has the illness being tested for. Lyme disease is not different. Positive predictive value is the ability of a positive serology to correctly identify someone who has or had Lyme disease; negative predictive value is the ability for a negative serology to correctly identify that an individual does not have Lyme disease. Pretest likelihood of Lyme disease directly relates to both positive and negative predictive values. Individuals who present with only protracted nonspecific complaints and no objective clinical abnormalities associated with Lyme disease have a low pretest likelihood of having the illness. For these patients, a negative Lyme disease serology is very highly predictive that that patient does not have Lyme disease, but a positive serology is likely to be a false positive result. In an individual with clear objective, later clinical manifestations compatible with Lyme disease, a positive Lyme disease serology has a high positive predictive value. In contrast, a patient with a compatible exposure history and a suspected localized EM skin lesion have a high probability of illness based on the clinical suspicion alone. As the sensitivity of serological tests are about 50% for this stage of the infection, a negative result DOES NOT exclude the disease. Therefore, these patients should be diagnosed and treated based on the clinical evidence alone.

Nonspecific symptoms, including fatigue, stiff neck, arthralgia, myalgia, palpitations, abdominal pain, sleep disturbance, poor concentration, irritability, depression, back pain, headache, dizziness, are reported by 20%–25% of the population. In people with these complaints for more than 4 weeks, the positive predictive value is poor and certainly not diagnostic. A study of "Lyme specialty laboratories" found that the use of unvalidated WB criteria by one of them produced a high rate of positive results in healthy controls. Of the 40 healthy controls, 15 of 40 (37.5%) met that laboratory's criteria for IgM positivity, 11 of the 40 (27.5%) for IgG positivity, and 23 (57.5%) met one or the other (Fallon et al., 2014). False positive results frequently lead to the prescription of unnecessary antibiotics (Nelson et al., 2019).

Other diagnostic approaches

T cell-based diagnostic assays for Lyme disease

Interest in cellular-based laboratory diagnostics for Lyme disease can be traced to the late 1980s and early 1990's when lymphoproliferative assays were first investigated (Dattwyler et al., 1988; Dressler et al., 1991). Data suggested that T cell responses could be detected prior to the development of detectable antibody responses, but ultimately, these assays did not prove sensitive enough for diagnostic use. This was followed by an effort to develop T cell diagnostics based on cytokine secretion rather than cell proliferation. Two basic methodologies have been employed, the Enzyme Linked Immunospot (ELISPOT) assay (Forsberg et al., 1995; Wang et al., 1995; Ekerfelt et al., 1997; Ekerfelt et al., 1999; Bauer et al., 2001; Ekerfelt et al., 2001; Skogman et al., 2012; Jin et al., 2013; Gyllemark et al., 2017; van Gorkom et al., 2018; van Gorkom et al., 2020), which measures the number of cells producing a particular cytokine of interest in response to *ex vivo* antigen stimulation, and a whole blood assay (Sikand et al., 1999; Callister et al., 2016), which measures the concentration of cytokine secreted in a given time period in a proscribed amount of whole blood mixed with target antigens. Though a number of diagnostic studies examined the production of different cytokines produced by T cells in Lyme disease (Forsberg et al., 1995; Wang et al., 1995; Ekerfelt et al., 1997; Ekerfelt et al., 1999; Bauer et al., 2001; Ekerfelt et al., 2001; Skogman et al., 2012; Jin et al., 2013; Gyllemark et al., 2017; van Gorkom et al., 2018; van Gorkom et al., 2020), all assays investigating the diagnostic potential of T cell cytokine secretion have ultimately focused on interferon gamma (IFNγ). Many of the ELISPOT studies have been performed in Europe and have focused on secretion of IFNγ by patients with either neuroborreliosis or asymptomatic Lyme disease patients (defined as serologically positive individuals with no history of past disease) (Ekerfelt et al., 1999; Ekerfelt et al., 2001; Skogman et al., 2012; Gyllemark et al., 2017). The majority of ELISPOT studies have utilized either whole sonicated *Borrelia* as an antigen target or select outer membrane proteins/preparations. In general, most studies demonstrated more IFNγ spot forming units (SFU) in patients with current or prior Lyme disease compared to seronegative control patients with no history of Lyme disease (Forsberg et al., 1995; Wang et al., 1995; Ekerfelt et al., 1999; Bauer et al., 2001; Ekerfelt et al., 2001; Jin et al., 2013; Gyllemark et al., 2017; van Gorkom et al., 2018; van Gorkom et al., 2020), though, few carefully designed prospective studies have been performed. Like serological assays, IFNγ ELISPOTS have been reported to be unable to distinguish between patients with active disease from those with prior disease (Wang et al., 1995; Ekerfelt et al., 1999; Ekerfelt et al., 2001; van Gorkom et al., 2018; van Gorkom et al., 2020). This includes patients with treated early manifestations and neuroborreliosis (Wang et al., 1995; van Gorkom et al., 2018; van Gorkom et al., 2020), as well as so-called asymptomatic patients that were seropositive in the absence of a clinical history of Lyme disease (Ekerfelt et al., 1999; Ekerfelt et al., 2001). It is unclear how many of the asymptomatic patients may have been positive due to cross-reactive target antigens in the assays used vs. actual disease, but the presence of both seropositivity and IFNγ secretion to different target antigens implies that at least some of these patients were exposed to *Borrelia*. In contrast, Skogman *et al.* were unable to distinguish patients with prior treated Lyme disease, supposed asymptomatic Lyme disease patients, and healthy controls in a cohort of children using an IFNγ ELISPOT (Skogman et al., 2012). Thus, the inability to distinguish active from prior disease could also be linked to study design and the antigens used. van Gorkom *et. al.* recently conducted a large, prospective trial evaluating the utility of an IFNγ ELISPOT using peripheral blood mononuclear cells in the laboratory diagnosis of Lyme neuroborreliosis, concluding that the assay was unable to distinguish active from treated prior disease and that it had poor diagnostic performance (van Gorkom et al., 2018; van Gorkom et al., 2020). However, it is important to note that in other studies evaluating neuroborreliosis patients, IFNγ+ T cells were present in much higher abundance in cerebrospinal fluid than peripheral blood (Forsberg et al., 1995; Wang et al., 1995; Ekerfelt et al., 1997). Carefully designed prospective studies are needed to define the role, if any, that IFNγ ELISPOT assays could play in Lyme disease. Although their diagnostic utility is unclear and no IFNγ ELISPOT assays are approved by either the U.S. FDA or European regulatory agencies, at least two IFNγ ELISPOT assays are commercially available in specialty laboratories. A lymphocyte transformation assay has been developed and was shown to have high sensitivity (89.4%) and specificity (98.7%) for the laboratory diagnosis of Lyme disease, and demonstrated the ability to monitor treatment (von Baehr et al., 2012); however, issues with the methodology of this study have been raised (Dessau et al., 2014), and it is not currently cleared by U.S. or European regulatory authorities.

Whole blood IFNγ release assays also have been evaluated for Lyme disease. In 1999, Sikand et. al. demonstrated that overnight incubation of whole blood with whole *B. burgdorferi* sonicate induced secretion of IFNγ in 67% of samples from Lyme disease patients and 4% of controls (Sikand et al., 1999). More recently, Callister *et. al.* demonstrated similar results, finding that positive IFNγ secretion was observed in 69% of early Lyme disease patients assessed, but only 3.2% of controls (Callister et al., 2016). Furthermore, a sharp decrease in IFNγ was observed in blood collected after treatment, suggesting that the assay might be used to monitor treatment efficacy. Further study is required to confirm this, and it is unclear if this assay has the same limitations as current ELISPOTs. No whole blood IFNγ release assays is currently available for clinical use.

Other host response biomarkers of Lyme disease

Establishment of bacterial infections by pathogens such as *B. burgdorferi* are the result of highly complex interactions between the host immune system and the pathogen. Measuring host response biomarkers for the diagnosis and prognosis of infection offers a novel way to detect infection by potentially any agent without the need for culture or a pathogen specific identification test. Circulating host immune cells alter their gene expression patterns in response to infection, and this response can be measured by detecting changes in the relative levels of mRNA within these cells. These gene expression changes are the body's first response to infection and can potentially be used to detect an infection prior to the onset of symptoms when treatments are the most effective. Recently, several studies have shown that bacterial and viral infections can be detected by looking at host biomarkers (Zaas et al., 2013; Langley et al., 2014; Tsalik et al., 2014; Ko et al., 2015; Tsalik et al., 2015; McClain et al., 2016a; McClain et al., 2016b; Tsalik et al., 2016; Yang et al., 2016).

Untreated *B. burgdorferi* in hosts and humans can cause chronic infections and thus has evolved unique interactions with the immune system and elicit a distinctive immune response profile (Bouquet et al., 2016). Transcriptomic profiling by RNA sequencing (RNA-Seq) has been used to identify unique transcriptional signatures in the blood of patients with acute Lyme disease (Bouquet et al., 2016). Using bioinformatics analysis of gene expression data from blood, the authors identified a set of 20 classifier genes that can discriminate Lyme disease from other bacterial and viral infections. In another study DNA microarrays were used to identify transcriptional gene expression patterns in EM skin lesions (Marques et al., 2017). This study found 254 differentially regulated genes (180 induced and 74 repressed) characterized by the induction of chemokines, cytokines, Toll-like receptors, antimicrobial peptides, monocytoid cell activation markers, and numerous genes annotated as interferon (IFN)-inducible.

Another area of host-response based diagnostics is focused on metabolic biosignatures associated with Lyme disease and clinically similar illness. In 2015 Molins *et al* identified a panel of 44 metabolic biomarkers (called molecular features) in serum that could distinguish early Lyme disease from healthy controls (Molins et al., 2015). The same approach was also used to identify 228 molecular features that could distinguish early Lyme disease from healthy controls using urine specimens (Pegalajar-Jurado et al., 2018a). Host metabolic response also has been used to identify biosignatures to distinguish early localized Lyme disease vs early disseminated Lyme disease and healthy controls (Fitzgerald et al., 2020). Perhaps most important, a panel of 261 molecular features was developed that could distinguish early Lyme disease from STARI. The capability to distinguish early Lyme disease from STARI could prove to be especially important to establish the clinical sensitivity and specificity of Lyme diagnostics where there could be a significant number of STARI cases among the study patients.

Though host response-based diagnostics using gene expression or metabolomics are still early in their development and present many technical challenges for their commercial implementation, the use of the host immune response and its products offers an especially useful tool for the very early detection of Lyme disease before antibodies have had time to develop.

Conclusions

What is desperately needed is a sensitive and specific direct diagnostic test for early Lyme disease that can diagnose the infection at the first sign of symptoms to assure prompt and appropriate treatment. Despite the discovery of the *B. burgdorferi* as the causative agent of Lyme disease in 1982 from the tick vector Ixodes scapularis (Burgdorfer et al., 1982) and subsequently from skin biopsy, cerebrospinal fluid, or blood specimens of patients with Lyme disease in the United States (Benach et al., 1983; Steere et al., 1983) there has not been an FDA-approved direct diagnostic test available for early Lyme disease.

Unfortunately, there are many barriers for the development of direct diagnostic technologies for the sensitive detection of early Lyme disease. Efforts to date have been challenged due to the low levels of bacteria present in the blood or other specimens of patients with Lyme disease. Biopsy of the leading edge of an untreated EM is a relatively good specimen for both culture and PCR-based methodologies, but EM is not present in many cases of Lyme disease and the acquisition of this specimen type is not practical for routine diagnostic testing. However, direct identification of *B. burgdorferi* in skin biopsies plays an important role for the confirmation of early Lyme disease in clinical studies. Furthermore, technical advances in specimen collection *(e.g.* microsampling), together with improvements in processing and detection of *B. burgdorferi*, have the potential to overcome some of the obstacles of skin biopsy-based diagnostics.

There are also challenges for the design and interpretation of clinical trials for early Lyme disease diagnostics. Most early Lyme disease patients are treated by their primary care physicians making the enrollment of large numbers of patients difficult. Furthermore, the clinical diagnosis of early Lyme disease can be challenging. As patients are treated immediately upon clinical diagnosis there is a higher proportion of Lyme disease patients that fail to seroconvert (Trevejo et al., 1999; Aguero-Rosenfeld et al., 2005). With serological tests having low sensitivity for early Lyme disease and many patients failing to seroconvert, there is not a good comparator diagnostic to show sensitivity and specificity for newer more sensitive tests for early Lyme disease. Early Lyme disease is also both seasonal and regional which both limits the commercial opportunity and creates further challenges for large clinical studies.

In summary, it is important to consider each current or prospective assay in terms of its purpose. There is a great distinction between use for research alone and use for clinical diagnosis. It is likely that clinical demands will continue to require the presence of a battery of tests that are sensitive and specific and applicable at different times during the infection (Branda et al., 2018; Schutzer et al., 2019).

Acknowledgements and Disclaimer

Funding for this study was provided in part by the Division of Intramural Research, National Institute of Allergy and Infectious Diseases, National Institutes of Health. The content of this publication does not necessarily reflect the views or policies of the

Department of Health and Human Services, nor does the mention of trade names, commercial products, or organizations imply endorsement by the U.S. Government.

References

Aguero-Rosenfeld, M.E., Wang, G., Schwartz, I., and Wormser, G.P. (2005). Diagnosis of lyme borreliosis. Clin Microbiol Rev *18*, 484-509. https://doi.org/18/3/484 [pii] 0.1128/CMR.18.3.484-509.2005

Andrew S Handel, M., Carine Ho, MD, Desiree D Hollemon, MSN, MPH2 David K Hong, MD, and Christy Beneri, DO (2019). 231. Microbial cell-free DNA Sequencing to Detect Borrelia burgdorferi DNA in the Plasma of Pediatric Patients with Lyme Disease. Open forum infectious diseases *6*, 1. https://doi.org/10.1093/ofid/ofz360.306

Arnaboldi, P.M., and Dattwyler, R.J. (2015). Cross-Reactive Epitopes in Borrelia burgdorferi p66. Clin Vaccine Immunol *22*, 840-843. https://doi.org/10.1128/CVI.00217-15

Arnaboldi, P.M., Seedarnee, R., Sambir, M., Callister, S.M., Imparato, J.A., and Dattwyler, R.J. (2013). Outer surface protein C peptide derived from Borrelia burgdorferi sensu stricto as a target for serodiagnosis of early lyme disease. Clin Vaccine Immunol *20*, 474-481. https://doi.org/10.1128/CVI.00608-12

Aucott, J.N., Crowder, L.A., Yedlin, V., and Kortte, K.B. (2012). Bull's-Eye and Nontarget Skin Lesions of Lyme Disease: An Internet Survey of Identification of Erythema Migrans. Dermatology research and practice *2012*, 451727. https://doi.org/10.1155/2012/451727

Bacon, R.M., Biggerstaff, B.J., Schriefer, M.E., Gilmore, R.D., Jr., Philipp, M.T., Steere, A.C., Wormser, G.P., Marques, A.R., and Johnson, B.J. (2003). Serodiagnosis of Lyme disease by kinetic enzyme-linked immunosorbent assay using recombinant VlsE1 or peptide antigens of Borrelia burgdorferi compared with 2-tiered testing using whole-cell lysates. J Infect Dis *187*, 1187-1199. https://doi.org/JID20849 [pii] 0.1086/374395

Barthold, S.W., Hodzic, E., Imai, D.M., Feng, S., Yang, X., and Luft, B.J. (2010). Ineffectiveness of tigecycline against persistent Borrelia burgdorferi. Antimicrob Agents Chemother *54*, 643-651. https://doi.org/10.1128/AAC.00788-09

Bauer, Y., Hofmann, H., Jahraus, O., Mytilineos, J., Simon, M.M., and Wallich, R. (2001). Prominent T cell response to a selectively in vivo expressed Borrelia burgdorferi outer surface protein (pG) in patients with Lyme disease. Eur J Immunol *31*, 767-776. https://doi.org/10.1002/1521-4141(200103)31:3<767::aid-immu767>3.0.co;2-m

Belongia, E.A., Reed, K.D., Mitchell, P.D., Mueller-Rizner, N., Vandermause, M., Finkel, M.F., and Kazmierczak, J.J. (2001). Tickborne infections as a cause of nonspecific febrile illness in Wisconsin. Clin Infect Dis *32*, 1434-1439. https://doi.org/10.1086/320160

Benach, J.L., Bosler, E.M., Hanrahan, J.P., Coleman, J.L., Habicht, G.S., Bast, T.F., Cameron, D.J., Ziegler, J.L., Barbour, A.G., Burgdorfer, W., *et al.* (1983). Spirochetes isolated from the blood of two patients with Lyme disease. N Engl J Med *308*, 740-742.

Berger, B.W. (1984). Erythema chronicum migrans of Lyme disease. Arch Dermatol *120*, 1017-1021.

Berger, B.W., Kaplan, M.H., Rothenberg, I.R., and Barbour, A.G. (1985). Isolation and characterization of the Lyme disease spirochete from the skin of patients with erythema chronicum migrans. J Am Acad Dermatol *13*, 444-449.

Bil-Lula, I., Matuszek, P., Pfeiffer, T., and Wozniak, M. (2015). Lyme Borreliosis--the Utility of Improved Real-Time PCR Assay in the Detection of Borrelia burgdorferi Infections. Advances in clinical and experimental medicine : official organ Wroclaw Medical University *24*, 663-670. https://doi.org/10.17219/acem/28625

Binnicker, M.J., Jespersen, D.J., Harring, J.A., Rollins, L.O., Bryant, S.C., and Beito, E.M. (2008). Evaluation of two commercial systems for automated processing, reading, and interpretation of Lyme borreliosis Western blots. J Clin Microbiol *46*, 2216-2221. https://doi.org/10.1128/JCM.00200-08

Bockenstedt, L.K., Mao, J., Hodzic, E., Barthold, S.W., and Fish, D. (2002). Detection of attenuated, noninfectious spirochetes in Borrelia burgdorferi-infected mice after antibiotic treatment. J Infect Dis *186*, 1430-1437. https://doi.org/10.1086/345284

Bockenstedt, L.K., and Radolf, J.D. (2014). Xenodiagnosis for posttreatment Lyme disease syndrome: resolving the conundrum or adding to it? Clin Infect Dis *58*, 946-948. https://doi.org/10.1093/cid/cit942

Bouquet, J., Soloski, M.J., Swei, A., Cheadle, C., Federman, S., Billaud, J.N., Rebman, A.W., Kabre, B., Halpert, R., Boorgula, M., *et al.* (2016). Longitudinal Transcriptome Analysis Reveals a Sustained Differential Gene Expression Signature in Patients Treated for Acute Lyme Disease. mBio *7*, e00100-00116. https://doi.org/10.1128/mBio.00100-16

Branda, J.A., Body, B.A., Boyle, J., Branson, B.M., Dattwyler, R.J., Fikrig, E., Gerald, N.J., Gomes-Solecki, M., Kintrup, M., Ledizet, M., *et al.* (2018). Advances in Serodiagnostic Testing for Lyme Disease Are at Hand. Clin Infect Dis *66*, 1133-1139. https://doi.org/10.1093/cid/cix943

Branda, J.A., Lemieux, J.E., Blair, L., Ahmed, A.A., Hong, D.K., Bercovici, S., Blauwkamp, T.A., Hollemon, D., Ho, C., Strle, K., *et al.* (2020). Detection of Borrelia burgdorferi Cell-free DNA in Human Plasma Samples for Improved Diagnosis of Early Lyme Borreliosis. Clin Infect Dis. https://doi.org/10.1093/cid/ciaa858

Branda, J.A., Linskey, K., Kim, Y.A., Steere, A.C., and Ferraro, M.J. (2011). Two-tiered antibody testing for Lyme disease with use of 2 enzyme immunoassays, a whole-cell sonicate enzyme immunoassay followed by a VlsE C6 peptide enzyme immunoassay. Clin Infect Dis *53*, 541-547. https://doi.org/10.1093/cid/cir464

Burgdorfer, W., Barbour, A.G., Hayes, S.F., Benach, J.L., Grunwaldt, E., and Davis, J.P. (1982). Lyme disease-a tick-borne spirochetosis? Science *216*, 1317-1319.

Callister, S.M., Jobe, D.A., Stuparic-Stancic, A., Miyamasu, M., Boyle, J., Dattwyler, R.J., and Arnaboldi, P.M. (2016). Detection of IFN-gamma Secretion by T Cells Collected Before and After Successful Treatment of Early Lyme Disease. Clin Infect Dis *62*, 1235-1241. https://doi.org/10.1093/cid/ciw112

Carlsson, S.A., Granlund, H., Jansson, C., Nyman, D., and Wahlberg, P. (2003). Characteristics of erythema migrans in Borrelia afzelii and Borrelia garinii infections. Scand J Infect Dis *35*, 31-33. https://doi.org/10.1080/0036554021000026978

Control., C.f.D. (1995). Recommendations for test performance and interpretation from the Second National Conference on Serologic Diagnosis of Lyme Disease. MMWR Morb Mortal Wkly Rep *44*, 590-591.

Coulter, P., Lema, C., Flayhart, D., Linhardt, A.S., Aucott, J.N., Auwaerter, P.G., and Dumler, J.S. (2005). Two-year evaluation of Borrelia burgdorferi culture and supplemental tests for definitive diagnosis of Lyme disease. J Clin Microbiol *43*, 5080-5084. https://doi.org/10.1128/JCM.43.10.5080-5084.2005

Crowder, C.D., Matthews, H.E., Schutzer, S., Rounds, M.A., Luft, B.J., Nolte, O., Campbell, S.R., Phillipson, C.A., Li, F., Sampath, R., *et al.* (2010). Genotypic variation and mixtures of Lyme *Borrelia* in *Ixodes* ticks from North America and Europe. PLoS ONE *5*, e10650. https://doi.org/10.1371/journal.pone.0010650

Cugini, C., Medrano, M., Schwan, T.G., and Coburn, J. (2003). Regulation of expression of the Borrelia burgdorferi beta(3)-chain integrin ligand, P66, in ticks and in culture. Infect Immun *71*, 1001-1007. https://doi.org/10.1128/iai.71.2.1001-1007.2003

Dattwyler, R.J., and Arnaboldi, P.M. (2014). Comparison of lyme disease serologic assays and lyme specialty laboratories. Clin Infect Dis *59*, 1711-1713. https://doi.org/10.1093/cid/ciu705

Dattwyler, R.J., Volkman, D.J., Luft, B.J., Halperin, J.J., Thomas, J., and Golightly, M.G. (1988). Seronegative Lyme disease. Dissociation of specific T- and B-lymphocyte responses to Borrelia burgdorferi. N Engl J Med *319*, 1441-1446. https://doi.org/10.1056/NEJM198812013192203

Dessau, R.B., Fingerle, V., Gray, J., Hunfeld, K.P., Jaulhac, B., Kahl, O., Kristoferitsch, W., Stanek, G., and Strle, F. (2014). The lymphocyte transformation test for the diagnosis of Lyme borreliosis has currently not been shown to be clinically useful. Clin Microbiol Infect *20*, O786-787. https://doi.org/10.1111/1469-0691.12583

Dessau, R.B., van Dam, A.P., Fingerle, V., Gray, J., Hovius, J.W., Hunfeld, K.P., Jaulhac, B., Kahl, O., Kristoferitsch, W., Lindgren, P.E., *et al.* (2018). To test or not to test? Laboratory support for the diagnosis of Lyme borreliosis: a position paper of ESGBOR, the ESCMID study group for Lyme borreliosis. Clin Microbiol Infect *24*, 118-124. https://doi.org/10.1016/j.cmi.2017.08.025

Donahue, J.G., Piesman, J., and Spielman, A. (1987). Reservoir competence of white-footed mice for Lyme disease spirochetes. Am J Trop Med Hyg *36*, 92-96. https://doi.org/10.4269/ajtmh.1987.36.92

Dressler, F., Yoshinari, N.H., and Steere, A.C. (1991). The T-cell proliferative assay in the diagnosis of Lyme disease. Ann Intern Med *115*, 533-539.

Ekerfelt, C., Ernerudh, J., Bunikis, J., Vrethem, M., Aagesen, J., Roberg, M., Bergstrom, S., and Forsberg, P. (1997). Compartmentalization of antigen specific cytokine responses to the central nervous system in CNS borreliosis: secretion of IFN-gamma predominates over IL-4 secretion in response to outer surface proteins of Lyme disease Borrelia spirochetes. J Neuroimmunol *79*, 155-162. https://doi.org/10.1016/s0165-5728(97)00118-5

Ekerfelt, C., Forsberg, P., Svenvik, M., Roberg, M., Bergstrom, S., and Ernerudh, J. (1999). Asymptomatic Borrelia-seropositive individuals display the same incidence of Borrelia-specific interferon-gamma (IFN-gamma)-secreting cells in blood as patients with clinical Borrelia infection. Clin Exp Immunol *115*, 498-502.

Ekerfelt, C., Masreliez, C., Svenvik, M., Ernerudh, J., Roberg, M., and Forsberg, P. (2001). Antibodies and T-cell reactivity to Borrelia burgdorferi in an asymptomatic population: a study of healthy blood donors in an inland town district in the south-east of Sweden. Scand J Infect Dis *33*, 806-808.

Embers, M.E., Barthold, S.W., Borda, J.T., Bowers, L., Doyle, L., Hodzic, E., Jacobs, M.B., Hasenkampf, N.R., Martin, D.S., Narasimhan, S., *et al.* (2012). Persistence of Borrelia burgdorferi in rhesus macaques following antibiotic treatment of disseminated infection. PLoS One *7*, e29914. https://doi.org/10.1371/journal.pone.0029914

Embers, M.E., Hasenkampf, N.R., Jacobs, M.B., Tardo, A.C., Doyle-Meyers, L.A., Philipp, M.T., and Hodzic, E. (2017). Variable manifestations, diverse seroreactivity and post-treatment persistence in non-human primates exposed to Borrelia burgdorferi by tick feeding. PLoS One *12*, e0189071. https://doi.org/10.1371/journal.pone.0189071

Embers, M.E., Jacobs, M.B., Johnson, B.J., and Philipp, M.T. (2007). Dominant epitopes of the C6 diagnostic peptide of Borrelia burgdorferi are largely inaccessible to antibody on the parent VlsE molecule. Clin Vaccine Immunol *14*, 931-936. https://doi.org/10.1128/CVI.00075-07

Eshoo, M.W., Carolan, H.E., Massire, C., Chou, D.M., Crowder, C.D., Rounds, M.A., Phillipson, C.A., Schutzer, S.E., and Ecker, D.J. (2015). Survey of Ixodes pacificus Ticks in California Reveals a Diversity of Microorganisms and a Novel and Widespread Anaplasmataceae Species. PLoS One *10*, e0135828. https://doi.org/10.1371/journal.pone.0135828

Eshoo, M.W., Crowder, C.C., Rebman, A.W., Rounds, M.A., Matthews, H.E., Picuri, J.M., Soloski, M.J., Ecker, D.J., Schutzer, S.E., and Aucott, J.N. (2012). Direct molecular detection and genotyping of *Borrelia burgdorferi* from whole blood of patients with early Lyme disease. PLoS ONE *7*, e36825. https://doi.org/10.1371/journal.pone.0036825 PONE-D-12-01579 [pii]

Eshoo, M.W., Crowder, C.D., Carolan, H.E., Rounds, M.A., Ecker, D.J., Haag, H., Mothes, B., and Nolte, O. (2014). Broad-range survey of tick-borne pathogens in Southern Germany reveals a high prevalence of Babesia microti and a diversity of other tick-borne pathogens. Vector Borne Zoonotic Dis *14*, 584-591. https://doi.org/10.1089/vbz.2013.1498

Eshoo, M.W., Schutzer, S.E., Crowder, C.D., Carolan, H.E., and Ecker, D.J. (2013). Achieving molecular diagnostics for Lyme disease. Expert Rev Mol Diagn *13*, 875-883. https://doi.org/10.1586/14737159.2013.850418

Fallon, B.A., Pavlicova, M., Coffino, S.W., and Brenner, C. (2014). A comparison of lyme disease serologic test results from 4 laboratories in patients with persistent symptoms after antibiotic treatment. Clin Infect Dis *59*, 1705-1710. https://doi.org/10.1093/cid/ciu703

Feder, H.M., Jr., and Whitaker, D.L. (1995). Misdiagnosis of erythema migrans. Am J Med *99*, 412-419.

Fitzgerald, B.L., Molins, C.R., Islam, M.N., Graham, B., Hove, P.R., Wormser, G.P., Hu, L., Ashton, L.V., and Belisle, J.T. (2020). Host Metabolic Response in Early Lyme Disease. J Proteome Res *19*, 610-623. https://doi.org/10.1021/acs.jproteome.9b00470

Forrester, J.D., and Mead, P. (2014). Third-degree heart block associated with lyme carditis: review of published cases. Clin Infect Dis *59*, 996-1000. https://doi.org/10.1093/cid/ciu411

Forsberg, P., Ernerudh, J., Ekerfelt, C., Roberg, M., Vrethem, M., and Bergstrom, S. (1995). The outer surface proteins of Lyme disease borrelia spirochetes stimulate T cells to secrete interferon-gamma (IFN-gamma): diagnostic and pathogenic implications. Clin Exp Immunol *101*, 453-460. https://doi.org/10.1111/j.1365-2249.1995.tb03134.x

Goddard, J. (2017). Not All Erythema Migrans Lesions Are Lyme Disease. Am J Med *130*, 231-233. https://doi.org/10.1016/j.amjmed.2016.08.020

Gomes-Solecki, M.J., Meirelles, L., Glass, J., and Dattwyler, R.J. (2007). Epitope length, genospecies dependency, and serum panel effect in the IR6 enzyme-linked immunosorbent assay for detection of antibodies to Borrelia burgdorferi. Clin Vaccine Immunol *14*, 875-879. https://doi.org/10.1128/CVI.00122-07

Gyllemark, P., Forsberg, P., Ernerudh, J., and Henningsson, A.J. (2017). Intrathecal Th17- and B cell-associated cytokine and chemokine responses in relation to clinical outcome in Lyme neuroborreliosis: a large retrospective study. J Neuroinflammation *14*, 27. https://doi.org/10.1186/s12974-017-0789-6

Heikkila, T., Huppertz, H.I., Seppala, I., Sillanpaa, H., Saxen, H., and Lahdenne, P. (2003). Recombinant or peptide antigens in the serology of Lyme arthritis in children. J Infect Dis *187*, 1888-1894. https://doi.org/10.1086/375371

Hinckley, A.F., Connally, N.P., Meek, J.I., Johnson, B.J., Kemperman, M.M., Feldman, K.A., White, J.L., and Mead, P.S. (2014). Lyme disease testing by large commercial laboratories in the United States. Clin Infect Dis *59*, 676-681. https://doi.org/10.1093/cid/ciu397

Hodzic, E., Feng, S., Holden, K., Freet, K.J., and Barthold, S.W. (2008). Persistence of Borrelia burgdorferi following antibiotic treatment in mice. Antimicrob Agents Chemother *52*, 1728-1736. https://doi.org/AAC.01050-07 [pii] 0.1128/AAC.01050-07

Honig, V., Carolan, H.E., Vavruskova, Z., Massire, C., Mosel, M.R., Crowder, C.D., Rounds, M.A., Ecker, D.J., Ruzek, D., Grubhoffer, L., *et al.* (2017). Broad-range survey of vector-borne pathogens and tick host identification of Ixodes ricinus from Southern Czech Republic. FEMS Microbiol Ecol *93*. https://doi.org/10.1093/femsec/fix129

Imirzalioglu, C., Hain, T., Hossain, H., Chakraborty, T., and Domann, E. (2010). Erythema caused by a localised skin infection with Arthrobacter mysorens. BMC Infect Dis *10*, 352. https://doi.org/10.1186/1471-2334-10-352

Jin, C., Roen, D.R., Lehmann, P.V., and Kellermann, G.H. (2013). An Enhanced ELISPOT Assay for Sensitive Detection of Antigen-Specific T Cell Responses to Borrelia burgdorferi. Cells *2*, 607-620. https://doi.org/10.3390/cells2030607

Johnson, B.J., Pilgard, M.A., and Russell, T.M. (2013). Assessment of New Culture Method to Detect Borrelia species in Serum of Lyme Disease Patients. J Clin Microbiol. https://doi.org/JCM.01674-13 [pii] 0.1128/JCM.01674-13

Ko, E.R., Yang, W.E., McClain, M.T., Woods, C.W., Ginsburg, G.S., and Tsalik, E.L. (2015). What was old is new again: using the host response to diagnose infectious disease. Expert Rev Mol Diagn *15*, 1143-1158. https://doi.org/10.1586/14737159.2015.1059278

Koetsveld, J., Platonov, A.E., Kuleshov, K., Wagemakers, A., Hoornstra, D., Ang, W., Szekeres, S., van Duijvendijk, G.L.A., Fikrig, E., Embers, M.E., *et al.* (2020). Borrelia miyamotoi infection leads to cross-reactive antibodies to the C6 peptide in mice and men. Clin Microbiol Infect *26*, 513 e511-513 e516. https://doi.org/10.1016/j.cmi.2019.07.026

Langley, R.J., Tipper, J.L., Bruse, S., Baron, R.M., Tsalik, E.L., Huntley, J., Rogers, A.J., Jaramillo, R.J., O'Donnell, D., Mega, W.M., *et al.* (2014). Integrative "omic" analysis of experimental bacteremia identifies a metabolic signature that distinguishes human sepsis from systemic inflammatory response syndromes. Am J Respir Crit Care Med *190*, 445-455. https://doi.org/10.1164/rccm.201404-0624OC

Ledue, T.B., Collins, M.F., Young, J., and Schriefer, M.E. (2008). Evaluation of the recombinant VlsE-based liaison chemiluminescence immunoassay for detection of Borrelia burgdorferi and diagnosis of Lyme disease. Clin Vaccine Immunol *15*, 1796-1804. https://doi.org/10.1128/CVI.00195-08

Lee, S.H., Vigliotti, J.S., Vigliotti, V.S., Jones, W., and Shearer, D.M. (2014). Detection of borreliae in archived sera from patients with clinically suspect Lyme disease. Int J Mol Sci *15*, 4284-4298. https://doi.org/10.3390/ijms15034284

Li, X., McHugh, G., Damle, N., Sikand, V.K., Glickstein, L., and Steere, A.C. (2011). Burden and viability of Borrelia burgdorferi in skin or joints, of patients with erythema migrans or lyme arthritis. Arthritis Rheum. https://doi.org/10.1002/art.30384

Liang, F.T., Alvarez, A.L., Gu, Y., Nowling, J.M., Ramamoorthy, R., and Philipp, M.T. (1999a). An immunodominant conserved region within the variable domain of VlsE, the variable surface antigen of Borrelia burgdorferi. J Immunol *163*, 5566-5573.

Liang, F.T., Steere, A.C., Marques, A.R., Johnson, B.J., Miller, J.N., and Philipp, M.T. (1999b). Sensitive and specific serodiagnosis of Lyme disease by enzyme-linked immunosorbent assay with a peptide based on an immunodominant conserved region of Borrelia burgdorferi vlsE. J Clin Microbiol *37*, 3990-3996.

Lipsett, S.C., Branda, J.A., McAdam, A.J., Vernacchio, L., Gordon, C.D., Gordon, C.R., and Nigrovic, L.E. (2016). Evaluation of the C6 Lyme Enzyme Immunoassay for the Diagnosis of Lyme Disease in Children and Adolescents. Clin Infect Dis *63*, 922-928. https://doi.org/10.1093/cid/ciw427

Lipsett, S.C., Branda, J.A., and Nigrovic, L.E. (2019). Evaluation of the Modified Two-Tiered Testing Method for Diagnosis of Lyme Disease in Children. J Clin Microbiol *57*. https://doi.org/10.1128/JCM.00547-19

Lipsker, D., Lieber-Mbomeyo, A., and Hedelin, G. (2004). How accurate is a clinical diagnosis of erythema chronicum migrans? Prospective study comparing the diagnostic accuracy of general practitioners and dermatologists in an area where lyme borreliosis is endemic. Arch Dermatol *140*, 620-621. https://doi.org/10.1001/archderm.140.5.620

Liu, W., Liu, H.X., Zhang, L., Hou, X.X., Wan, K.L., and Hao, Q. (2016). A Novel Isothermal Assay of Borrelia burgdorferi by Recombinase Polymerase Amplification with Lateral Flow Detection. Int J Mol Sci *17*. https://doi.org/10.3390/ijms17081250

Liveris, D., Schwartz, I., McKenna, D., Nowakowski, J., Nadelman, R., Demarco, J., Iyer, R., Bittker, S., Cooper, D., Holmgren, D., *et al.* (2012). Comparison of five diagnostic modalities for direct detection of Borrelia burgdorferi in patients with early Lyme disease. Diagn Microbiol Infect Dis *73*, 243-245. https://doi.org/10.1016/j.diagmicrobio.2012.03.026

Liveris, D., Schwartz, I., McKenna, D., Nowakowski, J., Nadelman, R.B., Demarco, J., Iyer, R., Cox, M.E., Holmgren, D., and Wormser, G.P. (2011). Quantitation of cell-associated borrelial DNA in the blood of Lyme disease patients with erythema migrans. Eur J Clin Microbiol Infect Dis. https://doi.org/10.1007/s10096-011-1376-x

Liveris, D., Wang, G., Girao, G., Byrne, D.W., Nowakowski, J., McKenna, D., Nadelman, R., Wormser, G.P., and Schwartz, I. (2002). Quantitative detection of Borrelia burgdorferi in 2-millimeter skin samples of erythema migrans lesions: correlation of results with clinical and laboratory findings. J Clin Microbiol *40*, 1249-1253. https://doi.org/10.1128/jcm.40.4.1249-1253.2002

Logar, M., Ruzic-Sabljic, E., Maraspin, V., Lotric-Furlan, S., Cimperman, J., Jurca, T., and Strle, F. (2004). Comparison of erythema migrans caused by Borrelia afzelii and Borrelia garinii. Infection *32*, 15-19. https://doi.org/10.1007/s15010-004-3042-z

Marques, A., Schwartz, I., Wormser, G.P., Wang, Y., Hornung, R.L., Demirkale, C.Y., Munson, P.J., Turk, S.P., Williams, C., Lee, C.R., *et al.* (2017). Transcriptome Assessment of Erythema Migrans Skin Lesions in Patients With Early Lyme Disease Reveals Predominant Interferon Signaling. J Infect Dis *217*, 158-167. https://doi.org/10.1093/infdis/jix563

Marques, A., Telford, S.R., 3rd, Turk, S.P., Chung, E., Williams, C., Dardick, K., Krause, P.J., Brandeburg, C., Crowder, C.D., Carolan, H.E., *et al.* (2014). Xenodiagnosis to detect *Borrelia burgdorferi* infection: a first-in-human study. Clin Infect Dis *58*, 937-945. https://doi.org/10.1093/cid/cit939

Marques, A.R. (2015). Laboratory diagnosis of Lyme disease: advances and challenges. Infect Dis Clin North Am *29*, 295-307. https://doi.org/10.1016/j.idc.2015.02.005

Marques, A.R. (2018). Revisiting the Lyme Disease Serodiagnostic Algorithm: the Momentum Gathers. J Clin Microbiol *56*. https://doi.org/10.1128/JCM.00749-18

McClain, M.T., Henao, R., Williams, J., Nicholson, B., Veldman, T., Hudson, L., Tsalik, E.L., Lambkin-Williams, R., Gilbert, A., Mann, A., *et al.* (2016a). Differential evolution of peripheral cytokine levels in symptomatic and asymptomatic responses to experimental influenza virus challenge. Clin Exp Immunol *183*, 441-451. https://doi.org/10.1111/cei.12736

McClain, M.T., Nicholson, B.P., Park, L.P., Liu, T.Y., Hero, A.O., 3rd, Tsalik, E.L., Zaas, A.K., Veldman, T., Hudson, L.L., Lambkin-Williams, R., *et al.* (2016b). A Genomic Signature of Influenza Infection Shows Potential for Presymptomatic Detection, Guiding Early Therapy, and Monitoring Clinical Responses. Open forum infectious diseases *3*, ofw007. https://doi.org/10.1093/ofid/ofw007

Mead, P., Petersen, J., and Hinckley, A. (2019). Updated CDC Recommendation for Serologic Diagnosis of Lyme Disease. MMWR Morb Mortal Wkly Rep *68*, 703. https://doi.org/10.15585/mmwr.mm6832a4

Molins, C.R., Ashton, L.V., Wormser, G.P., Hess, A.M., Delorey, M.J., Mahapatra, S., Schriefer, M.E., and Belisle, J.T. (2015). Development of a metabolic biosignature for detection of early Lyme disease. Clin Infect Dis *60*, 1767-1775. https://doi.org/10.1093/cid/civ185

Molins, C.R., Delorey, M.J., Replogle, A., Sexton, C., and Schriefer, M.E. (2017). Evaluation of bioMerieux's Dissociated Vidas Lyme IgM II and IgG II as a First-Tier Diagnostic Assay for Lyme Disease. J Clin Microbiol *55*, 1698-1706. https://doi.org/10.1128/JCM.02407-16

Molins, C.R., Delorey, M.J., Sexton, C., and Schriefer, M.E. (2016). Lyme Borreliosis Serology: Performance of Several Commonly Used Laboratory Diagnostic Tests and a Large Resource Panel of Well-Characterized Patient Samples. J Clin Microbiol *54*, 2726-2734. https://doi.org/10.1128/JCM.00874-16

Moore, A., Nelson, C., Molins, C., Mead, P., and Schriefer, M. (2016). Current Guidelines, Common Clinical Pitfalls, and Future Directions for Laboratory Diagnosis of Lyme Disease, United States. Emerg Infect Dis *22*. https://doi.org/10.3201/eid2207.151694

Mosel, M.R., Carolan, H.E., Rebman, A.W., Castro, S., Massire, C., Ecker, D.J., Soloski, M.J., Aucott, J.N., and Eshoo, M.W. (2019). Molecular Testing of Serial Blood Specimens from Patients with Early Lyme Disease during Treatment Reveals Changing Coinfection with Mixtures of Borrelia burgdorferi Genotypes. Antimicrob Agents Chemother *63*. https://doi.org/10.1128/AAC.00237-19

Nadelman, R.B., and Wormser, G.P. (1995). Erythema migrans and early Lyme disease. Am J Med *98*, 15S-23S; discussion 23S-24S. https://doi.org/10.1016/s0002-9343(99)80040-0

Nelson, C.A., Saha, S., Kugeler, K.J., Delorey, M.J., Shankar, M.B., Hinckley, A.F., and Mead, P.S. (2015). Incidence of Clinician-Diagnosed Lyme Disease, United States, 2005-2010. Emerg Infect Dis *21*, 1625-1631. https://doi.org/10.3201/eid2109.150417

Nelson, R.E., Burganowski, R.P., Colton, L., Escobar, J.D., Pathak, S.R., Gambino-Shirley, K.J., and Webber, B.J. (2019). Evaluation of serological testing for Lyme disease in Military Health System beneficiaries in Germany, 2013-2017. MSMR *26*, 22-26.

NIAID (2015). Xenodiagnosis After Antibiotic Treatment for Lyme Disease. In https://clinicaltrialsgov/ct2/show/NCT02446626 (Clinicaltrials.gov), pp. Clincal study of Lyme Disease after treatment by Xenodiagnosis.

Nolte, O. (2012). Nucleic Acid Amplification Based Diagnostic of Lyme (Neuro-)borreliosis - Lost in the Jungle of Methods, Targets, and Assays? Open Neurol J *6*, 129-139. https://doi.org/10.2174/1874205X01206010129

Nowakowski, J., Schwartz, I., Liveris, D., Wang, G., Aguero-Rosenfeld, M.E., Girao, G., McKenna, D., Nadelman, R.B., Cavaliere, L.F., Wormser, G.P., *et al.* (2001). Laboratory diagnostic techniques for patients with early Lyme disease associated with erythema migrans: a comparison of different techniques. Clin Infect Dis *33*, 2023-2027. https://doi.org/10.1086/324490

Nuttall, P.A. (2019). Tick saliva and its role in pathogen transmission. Wien Klin Wochenschr. https://doi.org/10.1007/s00508-019-1500-y

Ohnishi, J., Schneider, B., Messer, W.B., Piesman, J., and de Silva, A.M. (2003). Genetic variation at the vlsE locus of Borrelia burgdorferi within ticks and mice over the course of a single transmission cycle. J Bacteriol *185*, 4432-4441. https://doi.org/10.1128/jb.185.15.4432-4441.2003

Osterhoudt, K.C., Zaoutis, T., and Zorc, J.J. (2002). Lyme disease masquerading as brown recluse spider bite. Ann Emerg Med *39*, 558-561. https://doi.org/10.1067/mem.2002.119509

Ozenci, V., Patel, R., Ullberg, M., and Stralin, K. (2018). Demise of Polymerase Chain Reaction/Electrospray Ionization-Mass Spectrometry as an Infectious Diseases Diagnostic Tool. Clin Infect Dis *66*, 452-455. https://doi.org/10.1093/cid/cix743

Pegalajar-Jurado, A., Fitzgerald, B.L., Islam, M.N., Belisle, J.T., Wormser, G.P., Waller, K.S., Ashton, L.V., Webb, K.J., Delorey, M.J., Clark, R.J., *et al.* (2018a). Identification of Urine Metabolites as Biomarkers of Early Lyme Disease. Sci Rep *8*, 12204. https://doi.org/10.1038/s41598-018-29713-y

Pegalajar-Jurado, A., Schriefer, M.E., Welch, R.J., Couturier, M.R., MacKenzie, T., Clark, R.J., Ashton, L.V., Delorey, M.J., and Molins, C.R. (2018b). Evaluation of Modified Two-Tiered Testing Algorithms for Lyme Disease Laboratory Diagnosis Using Well-Characterized Serum Samples. J Clin Microbiol *56*. https://doi.org/10.1128/JCM.01943-17

Petrulioniene, A., Radzisauskiene, D., Ambrozaitis, A., Caplinskas, S., Paulauskas, A., and Venalis, A. (2020). Epidemiology of Lyme Disease in a Highly Endemic European Zone.. Medicina (Kaunas) *56*. https://doi.org/10.3390/medicina56030115

Pierrette Andre, R.S., Nitin Damle, Clayton Green, Michelle Perry, Beck Jacobson, Anna Schotthoefer, Robert Shivers, John Branda, Tom Lowery (2018). A Prospective Clinical Evaluation of the T2Lyme® Diagnostic Demonstrated a High Positive Predictive Value for Borrelia Infection. Paper presented at: International Conference on Lyme Borreliosis and Other Tick-Borne Diseases (Atlanta, Georgia.).

Schoen, R.T. (2020). Challenges in the Diagnosis and Treatment of Lyme Disease. Curr Rheumatol Rep *22*, 3. https://doi.org/10.1007/s11926-019-0857-2

Schutzer, S.E., Body, B.A., Boyle, J., Branson, B.M., Dattwyler, R.J., Fikrig, E., Gerald, N.J., Gomes-Solecki, M., Kintrup, M., Ledizet, M., *et al.* (2019). Direct Diagnostic Tests for Lyme Disease. Clin Infect Dis *68*, 1052-1057. https://doi.org/10.1093/cid/ciy614

Schwartz, I., Wormser, G.P., Schwartz, J.J., Cooper, D., Weissensee, P., Gazumyan, A., Zimmermann, E., Goldberg, N.S., Bittker, S., Campbell, G.L., *et al.* (1992). Diagnosis of early Lyme disease by polymerase chain reaction amplification and culture of skin biopsies from erythema migrans lesions. J Clin Microbiol *30*, 3082-3088.

Seriburi, V., Ndukwe, N., Chang, Z., Cox, M.E., and Wormser, G.P. (2012). High frequency of false positive IgM immunoblots for Borrelia burgdorferi in clinical practice. Clin Microbiol Infect *18*, 1236-1240. https://doi.org/10.1111/j.1469-0691.2011.03749.x

Shah, A.S., and Varatharaj Palraj, B.R. (2019). Multiple Erythema Migrans Rashes Characteristic of Early Disseminated Lyme Disease, Before and After Therapy. Mayo Clin Proc *94*, 172-173. https://doi.org/10.1016/j.mayocp.2018.11.001

Shapiro, E.D. (2015). Repeat or persistent Lyme disease: persistence, recrudescence or reinfection with Borrelia Burgdorferi? F1000Prime Rep *7*, 11. https://doi.org/10.12703/P7-11

Shih, C.M., Chao, L.L., and Yu, C.P. (2002). Chemotactic migration of the Lyme disease spirochete (Borrelia burgdorferi) to salivary gland extracts of vector ticks. Am J Trop Med Hyg *66*, 616-621. https://doi.org/10.4269/ajtmh.2002.66.616

Sikand, V.K., Rothel, J.S., and Martin, R.M. (1999). Diagnosis of Lyme borreliosis by a whole-blood gamma interferon assay for cell-mediated immune responses. Clin Diagn Lab Immunol *6*, 445.

Sillanpaa, H., Lahdenne, P., Sarvas, H., Arnez, M., Steere, A., Peltomaa, M., and Seppala, I. (2007). Immune responses to borrelial VlsE IR6 peptide variants. Int J Med Microbiol *297*, 45-52. https://doi.org/10.1016/j.ijmm.2006.09.001

Simner, P.J., Miller, S., and Carroll, K.C. (2018). Understanding the Promises and Hurdles of Metagenomic Next-Generation Sequencing as a Diagnostic Tool for Infectious Diseases. Clin Infect Dis *66*, 778-788. https://doi.org/10.1093/cid/cix881

Skogman, B.H., Hellberg, S., Ekerfelt, C., Jenmalm, M.C., Forsberg, P., Ludvigsson, J., Bergstrom, S., and Ernerudh, J. (2012). Adaptive and innate immune responsiveness to Borrelia burgdorferi sensu lato in exposed asymptomatic children and children with previous clinical Lyme borreliosis. Clin Dev Immunol *2012*, 294587. https://doi.org/10.1155/2012/294587

Smith, R.P., Schoen, R.T., Rahn, D.W., Sikand, V.K., Nowakowski, J., Parenti, D.L., Holman, M.S., Persing, D.H., and Steere, A.C. (2002). Clinical characteristics and treatment outcome of early Lyme disease in patients with microbiologically confirmed erythema migrans. Ann Intern Med *136*, 421-428. https://doi.org/200203190-00005 [pii]

Snitkin, E.S., Zelazny, A.M., Thomas, P.J., Stock, F., Group, N.C.S.P., Henderson, D.K., Palmore, T.N., and Segre, J.A. (2012). Tracking a hospital outbreak of carbapenem-resistant Klebsiella pneumoniae with whole-genome sequencing. Sci Transl Med *4*, 148ra116. https://doi.org/10.1126/scitranslmed.3004129

Snyder, J.L., Giese, H., Bandoski-Gralinski, C., Townsend, J., Jacobson, B.E., Shivers, R., Schotthoefer, A.M., Fritsche, T.R., Green, C., Callister, S.M., *et al.* (2017). T2 Magnetic Resonance Assay-Based Direct Detection of Three Lyme Disease-Related Borrelia Species in Whole-Blood Samples. J Clin Microbiol *55*, 2453-2461. https://doi.org/10.1128/JCM.00510-17

Steere, A.C., Grodzicki, R.L., Kornblatt, A.N., Craft, J.E., Barbour, A.G., Burgdorfer, W., Schmid, G.P., Johnson, E., and Malawista, S.E. (1983). The spirochetal etiology of Lyme disease. N Engl J Med *308*, 733-740.

Steere, A.C., Sikand, V.K., Meurice, F., Parenti, D.L., Fikrig, E., Schoen, R.T., Nowakowski, J., Schmid, C.H., Laukamp, S., Buscarino, C., *et al.* (1998). Vaccination against Lyme disease with recombinant Borrelia burgdorferi outer-surface lipoprotein A with adjuvant. Lyme Disease Vaccine Study Group. N Engl J Med *339*, 209-215. https://doi.org/10.1056/NEJM199807233390401

Steere, A.C., Sikand, V.K., Schoen, R.T., and Nowakowski, J. (2003). Asymptomatic infection with Borrelia burgdorferi. Clin Infect Dis *37*, 528-532. https://doi.org/10.1086/376914

Steere, A.C., Strle, F., Wormser, G.P., Hu, L.T., Branda, J.A., Hovius, J.W., Li, X., and Mead, P.S. (2016). Lyme borreliosis. Nat Rev Dis Primers *2*, 16090. https://doi.org/10.1038/nrdp.2016.90

Straubinger, R.K., Straubinger, A.F., Summers, B.A., and Jacobson, R.H. (2000). Status of Borrelia burgdorferi infection after antibiotic treatment and the effects of corticosteroids: An experimental study. J Infect Dis *181*, 1069-1081. https://doi.org/10.1086/315340

Straubinger, R.K., Summers, B.A., Chang, Y.F., and Appel, M.J. (1997). Persistence of Borrelia burgdorferi in experimentally infected dogs after antibiotic treatment. J Clin Microbiol *35*, 111-116.

Telford, S.R., 3rd, Hu, L.T., and Marques, A. (2014). Is there a place for xenodiagnosis in the clinic? Expert Rev Anti Infect Ther *12*, 1307-1310. https://doi.org/10.1586/14787210.2014.966084

Tibbles, C.D., and Edlow, J.A. (2007). Does this patient have erythema migrans? JAMA *297*, 2617-2627. https://doi.org/297/23/2617 [pii] 10.1001/jama.297.23.2617

Tjernberg, I., Sillanpaa, H., Seppala, I., Eliasson, I., Forsberg, P., and Lahdenne, P. (2009). Antibody responses to borrelia IR(6) peptide variants and the C6 peptide in Swedish patients with erythema migrans. Int J Med Microbiol *299*, 439-446. https://doi.org/10.1016/j.ijmm.2008.10.006

Trevejo, R.T., Krause, P.J., Sikand, V.K., Schriefer, M.E., Ryan, R., Lepore, T., Porter, W., and Dennis, D.T. (1999). Evaluation of two-test serodiagnostic method for early Lyme disease in clinical practice. J Infect Dis *179*, 931-938. https://doi.org/10.1086/314663

Tsalik, E.L., Henao, R., Nichols, M., Burke, T., Ko, E.R., McClain, M.T., Hudson, L.L., Mazur, A., Freeman, D.H., Veldman, T., *et al.* (2016). Host gene expression classifiers diagnose acute respiratory illness etiology. Sci Transl Med *8*, 322ra311. https://doi.org/10.1126/scitranslmed.aad6873

Tsalik, E.L., Langley, R.J., Dinwiddie, D.L., Miller, N.A., Yoo, B., van Velkinburgh, J.C., Smith, L.D., Thiffault, I., Jaehne, A.K., Valente, A.M., *et al.* (2014). An integrated transcriptome and expressed variant analysis of sepsis survival and death. Genome Med *6*, 111. https://doi.org/10.1186/s13073-014-0111-5

Tsalik, E.L., McClain, M., and Zaas, A.K. (2015). Moving toward prime time: host signatures for diagnosis of respiratory infections. J Infect Dis *212*, 173-175. https://doi.org/10.1093/infdis/jiv032

Turk, S.P., Williams, C., and Marques, A. (2018). Xenodiagnosis Using Ixodes scapularis Larval Ticks in Humans. Methods Mol Biol *1690*, 337-346. https://doi.org/10.1007/978-1-4939-7383-5_26

van Gorkom, T., Sankatsing, S.U.C., Voet, W., Ismail, D.M., Muilwijk, R.H., Salomons, M., Vlaminckx, B.J.M., Bossink, A.W.J., Notermans, D.W., Bouwman, J.J.M., *et al.* (2018). An Enzyme-Linked Immunosorbent Spot Assay Measuring Borrelia burgdorferi B31-Specific Interferon Gamma-Secreting T Cells Cannot Discriminate Active Lyme Neuroborreliosis from Past Lyme Borreliosis: a Prospective Study in the Netherlands. J Clin Microbiol *56*. https://doi.org/10.1128/JCM.01695-17

van Gorkom, T., Voet, W., Sankatsing, S.U.C., Nijhuis, C.D.M., Ter Haak, E., Kremer, K., and Thijsen, S.F.T. (2020). Prospective comparison of two enzyme-linked immunosorbent spot assays for the diagnosis of Lyme neuroborreliosis. Clin Exp Immunol *199*, 337-356. https://doi.org/10.1111/cei.13393

von Baehr, V., Doebis, C., Volk, H.D., and von Baehr, R. (2012). The lymphocyte transformation test for borrelia detects active lyme borreliosis and verifies effective antibiotic treatment. Open Neurol J *6*, 104-112. https://doi.org/10.2174/1874205X01206010104

Wang, W.Z., Fredrikson, S., Sun, J.B., and Link, H. (1995). Lyme neuroborreliosis: evidence for persistent up-regulation of Borrelia burgdorferi-reactive cells secreting interferon-gamma. Scand J Immunol *42*, 694-700. https://doi.org/10.1111/j.1365-3083.1995.tb03713.x

Wetter, D.A., and Ruff, C.A. (2011). Erythema migrans in Lyme disease. CMAJ *183*, 1281. https://doi.org/10.1503/cmaj.101533

Wilson, M.R., Naccache, S.N., Samayoa, E., Biagtan, M., Bashir, H., Yu, G., Salamat, S.M., Somasekar, S., Federman, S., Miller, S., *et al.* (2014). Actionable diagnosis of neuroleptospirosis by next-generation sequencing. N Engl J Med *370*, 2408-2417. https://doi.org/10.1056/NEJMoa1401268

Wormser, G.P., Baker, P.J., O'Connell, S., Pachner, A.R., Schwartz, I., and Shapiro, E.D. (2012a). Critical analysis of treatment trials of rhesus macaques infected with Borrelia burgdorferi reveals important flaws in experimental design. Vector Borne Zoonotic Dis *12*, 535-538. https://doi.org/10.1089/vbz.2012.1012

Wormser, G.P., Bittker, S., Cooper, D., Nowakowski, J., Nadelman, R.B., and Pavia, C. (2001). Yield of large-volume blood cultures in patients with early Lyme disease. J Infect Dis *184*, 1070-1072. https://doi.org/JID010331 [pii] 10.1086/323424

Wormser, G.P., Masters, E., Nowakowski, J., McKenna, D., Holmgren, D., Ma, K., Ihde, L., Cavaliere, L.F., and Nadelman, R.B. (2005). Prospective clinical evaluation of patients from Missouri and New York with erythema migrans-like skin lesions. Clin Infect Dis *41*, 958-965. https://doi.org/10.1086/432935

Wormser, G.P., Nadelman, R.B., and Schwartz, I. (2012b). The amber theory of Lyme arthritis: initial description and clinical implications. Clin Rheumatol *31*, 989-994. https://doi.org/10.1007/s10067-012-1964-x

Wormser, G.P., Nowakowski, J., Nadelman, R.B., Visintainer, P., Levin, A., and Aguero-Rosenfeld, M.E. (2008). Impact of clinical variables on Borrelia burgdorferi-specific antibody seropositivity in acute-phase sera from patients in North America with culture-confirmed early Lyme disease. Clin Vaccine Immunol *15*, 1519-1522. https://doi.org/10.1128/CVI.00109-08

Wormser, G.P., Schriefer, M., Aguero-Rosenfeld, M.E., Levin, A., Steere, A.C., Nadelman, R.B., Nowakowski, J., Marques, A., Johnson, B.J., and Dumler, J.S. (2013). Single-tier testing with the C6 peptide ELISA kit compared with two-tier testing for Lyme disease. Diagn Microbiol Infect Dis *75*, 9-15. https://doi.org/10.1016/j.diagmicrobio.2012.09.003

Wormser, G.P., and Schwartz, I. (2009). Antibiotic treatment of animals infected with Borrelia burgdorferi. Clin Microbiol Rev *22*, 387-395. https://doi.org/10.1128/CMR.00004-09

Xu, G., Wesker, J., White, C., Campbell, J., Reich, N.G., and Rich, S.M. (2013). Detection of heterogeneity of Borrelia burgdorferi in ixodes ticks by culture-dependent and culture-independent methods. J Clin Microbiol *51*, 615-617. https://doi.org/10.1128/JCM.03009-12

Yang, W.E., Suchindran, S., Nicholson, B.P., McClain, M.T., Burke, T., Ginsburg, G.S., Harro, C.D., Chakraborty, S., Sack, D.A., Woods, C.W., *et al.* (2016). Transcriptomic Analysis of the Host Response and Innate Resilience to Enterotoxigenic Escherichia coli Infection in Humans. J Infect Dis *213*, 1495-1504. https://doi.org/10.1093/infdis/jiv593

Yrjanainen, H., Hytonen, J., Song, X.Y., Oksi, J., Hartiala, K., and Viljanen, M.K. (2007). Anti-tumor necrosis factor-alpha treatment activates Borrelia burgdorferi spirochetes 4 weeks after ceftriaxone treatment in C3H/He mice. J Infect Dis *195*, 1489-1496. https://doi.org/10.1086/513873

Zaas, A.K., Burke, T., Chen, M., McClain, M., Nicholson, B., Veldman, T., Tsalik, E.L., Fowler, V., Rivers, E.P., Otero, R., *et al.* (2013). A host-based RT-PCR gene expression signature to identify acute respiratory viral infection. Sci Transl Med *5*, 203ra126. https://doi.org/10.1126/scitranslmed.3006280

Zhang, J.R., and Norris, S.J. (1998). Genetic variation of the Borrelia burgdorferi gene vlsE involves cassette-specific, segmental gene conversion. Infect Immun *66*, 3698-3704. https://doi.org/10.1128/IAI.66.8.3698-3704.1998

Chapter 23

The Widening Gyre: Controversies in Lyme Disease

Adriana Marques[1]*, Jacob E. Lemieux[2] and Linden T. Hu[3]

[1]Laboratory of Clinical Microbiology and Immunology, National Institute of Allergy and Infectious Diseases, National Institutes of Health, 10 Center Drive, Bethesda, MD 20852, USA
[2]Department of Medicine, Massachusetts General Hospital and Harvard Medical School, Boston, 52 Oxford Street, Cambridge, MA 02138, USA
[3]Tufts University, 145 Harrison Avenue Boston, MA 02111, USA
*Corresponding author: amarques@niaid.nih.gov

DOI: https://doi.org/10.21775/9781913652616.23

"Turning and turning in the widening gyre
The falcon cannot hear the falconer;
Things fall apart; the centre cannot hold;
Mere anarchy is loosed upon the world,
The blood-dimmed tide is loosed, and everywhere
The ceremony of innocence is drowned;
The best lack all conviction, while the worst
Are full of passionate intensity."
William Butler Yeats, The Second Coming

Abstract
In this chapter we explore controversies surrounding the issues of chronic Lyme disease (CLD) and post-treatment Lyme disease (PTLD) syndrome. We describe each of the terms as they are used in the literature and then will look at the data supporting (or not supporting) some of the most prominent hypotheses for their etiology. Among the areas we will discuss are roles in PTLD for autoimmunity, immune dysregulation or maladaptive host responses, the persistence of bacterial remnants that cause chronic immune stimulation; and chronic infection by *Borrelia burgdorferi*. Our examination of the data shows that there still remains much to be known about CLD and PTLD, and we provide recommendations for future research to try to resolve the controversies.

Introduction

Since its initial description in 1980, the Lyme disease field has seen its share of controversies, both scientific and societal. Some of these controversies have been addressed and "resolved"; others remain flashpoints for discussion. In this chapter, we will discuss some of the more high-profile controversies, the evidence supporting the different perspectives on the controversy and the remaining issues that require further investigation.

The most controversial topics revolve around the concept of "chronic Lyme disease"—does it exist and if so, what is the cause. We will examine some of the more contentious approaches and theories that have been raised over the years regarding chronic Lyme disease, including autoimmunity, inappropriate immune activation, persistence of spirochetes, including survival of alternative forms and the use of long-term antibiotics.

Chronic Lyme: What is actually going on here?

The course of untreated Lyme disease has been very well described and is largely non-controversial. Lyme disease usually begins with the characteristic skin lesion, erythema migrans. After several days or weeks, the spirochete may disseminate to other sites and patients may develop neurologic, cardiac and rheumatologic involvement (see chapter 24).

Patients with a confirmed diagnosis of Lyme disease who remain symptomatic despite appropriate medical therapy are referred to as having Post-Treatment Lyme Disease (PTLD) syndrome or symptoms. Yet patients with PTLDS are a small subset of the spectrum of what is called "chronic Lyme disease" (CLD), one of the most vexing problems in the Lyme disease field. CLD is a highly significant problem, with a 2011 national survey performed by the US Centers for Disease Control and Prevention (CDC) showing 0.5% of the participants saying they had CLD and 10.5% saying they knew someone else with CLD (Hook et al., 2015). Nevertheless, CLD

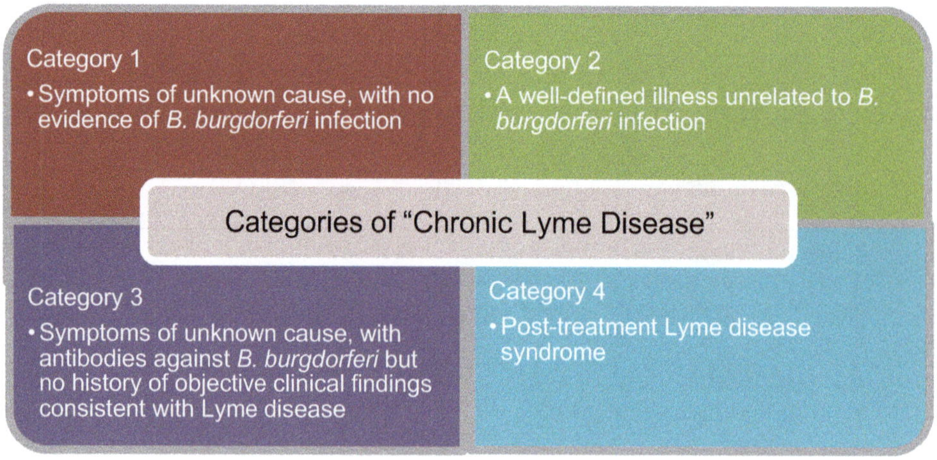

Figure 1. Categories of "Chronic Lyme Disease". Adapted from (Feder et al., 2007).

lacks a clear definition, and the term is used in various ways to refer to highly heterogeneous groups of patients. For example, patients with clearly defined objective manifestations of late Lyme disease (for example, arthritis, encephalomyelitis, or peripheral neuropathy) have sometimes been included as CLD. Late manifestations of *B. burgdorferi* infection should be treated with antibiotics as recommended (Wormser et al., 2006).

Patients with CLD have been classified into four categories (Figure 1): 1) patients with medically unexplained symptoms (MUS) who receive the CLD diagnosis based on unproven and/or non-validated laboratory tests and clinical criteria; 2) patients with other medical conditions whose symptoms are misdiagnosed as CLD; 3) patients with signs and symptoms of unknown etiology and laboratory evidence of exposure to *B. burgdorferi*; and 4) patients with PTLDS. Therefore, the term CLD has largely fallen out of use among mainstream medical providers and in the academic literature, where it has been replaced by PTLDS and other more-specific labels.

The majority of patients under the CLD umbrella fall into categories 1 and 2. These patients received the diagnosis of CLD by practitioners using alternative clinical criteria and non-validated laboratory tests and/or interpretation criteria that have been shown to have low specificity (Marques et al., 2009a; Fallon et al., 2014; CDC, 2015). Studies from referral centers have shown that 50 to 80% of the patients seeking further evaluation for Lyme disease had no evidence of *B. burgdorferi* infection (Sigal, 1990; Steere et al., 1993; Rose et al., 1994; Reid et al., 1998; Aucott et al., 2012; Kobayashi et al., 2019). Patients with MUS accounted for 19% to 54% of the cases, while an alternative diagnosis was made in 15 to 55% of the evaluations. PTLD symptoms accounted for 3 to 15% of the cases (Figure 2). This was reflected in the difficulty of accruing PTLDS patients into the placebo-controlled trials of antibiotic treatment, with less of 10% of the screened individuals being eligible for the studies (Marshall, 1999; Klempner et al., 2001; Krupp et al., 2003; Fallon et al., 2007).

MUS, or persistent physical symptoms, are a common and important problem across general medicine (Nimnuan et al., 2001; Deary et al., 2014; van Westrienen et al., 2019; van Westrienen et al., 2020); the management and treatment of these patients is one of the most difficult tasks facing healthcare services. The term refers to a diverse range of conditions including chronic pain, fibromyalgia, myalgic encephalomyelitis/chronic fatigue syndrome (ME/CFS), irritable bowel syndrome, functional dyspepsia, overactive bladder syndrome as well as persistent physical symptoms such as dizziness and chest pain. The etiology and pathogenesis of these disorders are unknown, but it is thought that symptoms are caused, maintained and/or exacerbated by complex interactions between biological, psychological, and social mechanisms. Because symptoms significantly overlap between these patients (Persson et al., 2015; Chalder and Willis, 2017), there has been discussion that these syndromes may not be distinct entities but a product

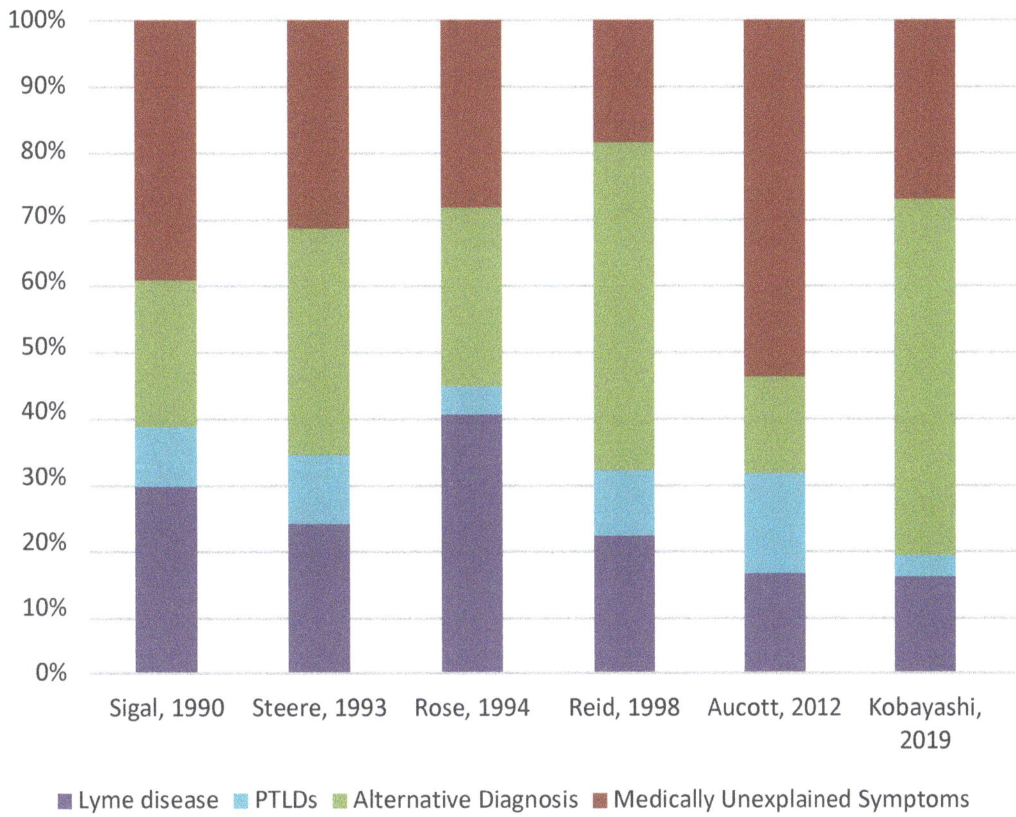

Figure 2. Categorization of patients referred for evaluation of possible Lyme disease. Post treatment Lyme disease symptoms (PTLDs).

of medical specialization and differing diagnostic approaches (Wessely et al., 1999). It is interesting that patients with "alternatively diagnosed chronic Lyme syndrome" (ADCLS), a suggested label for CLD, were phenotypically similar to patients with ME/CFS. The different label was most likely due to false-positive results from alternative criteria or laboratory tests (for example, a positive band at 31 or 34 kDa on a western blot or a decreased CD57 cell count) (Patrick et al., 2015). There were no significantly differential gene expression, B-cell/T-cell receptor profiles and potential viral infections when ME/CFS and ADCLS were compared to matched controls (Bouquet et al., 2017).

In recent years, the definition of CLD has further evolved to include polymicrobial infectious syndrome (another term used is Lyme-Multiple Systemic Infectious Disease Syndrome), with many (if not most) patients being diagnosed with co-infections with multiple organisms (*Babesia*, *Ehrlichia*, *Anaplasma*, *Rickettsia*, *Bartonella*, *Mycoplasma*, *Chlamydia*) (Horowitz and Freeman, 2018; Johnson et al., 2018; Horowitz and Freeman, 2019) as well as suffering from nutritional and enzyme deficiencies, food allergies/sensitivities, leaky gut, adrenal fatigue, metabolic and hormonal imbalances, immune dysfunction, damage from toxins, free radical/oxidative stress, and liver and mitochondrial dysfunction. The diagnosis of polymicrobial infectious syndrome is based on clinical judgment since standard laboratory tests are not helpful (Citera et al., 2017). Tests sent to specialty laboratories that focus on these diseases often utilize unapproved tests to support the diagnosis. The physicians diagnosing and treating CLD typically treat for prolonged periods, usually for months to years, with multiple antibiotics and supplements (Cameron et al., 2004; Stricker et al., 2010; Shor et al., 2019). There is very little formal, high quality, research evaluating this large, heterogeneous group of patients or the multiple treatments recommended (Auwaerter et al., 2011; Auwaerter and Melia, 2012; Lantos and Wormser, 2014; Lantos et al., 2015; Kobayashi et al., 2019).

Table 1. Post Treatment Lyme Disease Symptoms - Possible mechanisms

• Part of the expected resolution of symptoms after treatment
• Post-infective fatigue syndrome
• Other tick-borne infections
• Intercurrent conditions
• Host immune response/autoimmune mechanisms
• Persistence of remnants of spirochetes that cause inflammation
• Persistent infection with *B. burgdorferi*

Our verdict:
The lack of consensus definitions for CLD, together with the use of low-quality testing and the lack of standardization of treatment offered to these patients, have made it hard to make conclusive statements about CLD from the literature. As such, we will focus below on studies of PTLD. However, this is not intended as a dismissal of patients who do not meet the definition of PTLD. Indeed, there is a critical need for better definitions and research into diverse mechanisms of etiopathogenesis for all patients who fall under the heading CLD, as well as properly controlled, prospective randomized trials of the myriad reatments for which the evidence of efficacy is poor to non-existent.

What is post-treatment Lyme disease syndrome?
The Infectious Diseases Society of America proposed a research case definition of post treatment Lyme disease syndrome (PTLDS) in 2006 (Wormser et al., 2006). Patients with PTLDS must (i) have had a documented episode of Lyme disease, (ii) received an accepted antibiotic treatment with resolution or stabilization of objective manifestation(s) of Lyme disease, and (iii) have persistent or relapsing non-specific symptoms for at least 6 months after completion of antibiotic therapy. The non-specific symptoms must have started within 6 months of the diagnosis of Lyme disease and have caused a substantial reduction in levels of activity. There should be no other condition that could explain the symptoms (Wormser et al., 2006).

There have been efforts to develop an operational model for patients with symptoms after treatment of Lyme disease to better describe this clinically defined population, so that outcomes and other data can be compared between studies (Aucott et al., 2013; Turk et al., 2019; Wormser et al., 2020). A standardized approach allows for quantitation of the severity level of symptoms in PTLDS patients. Depending of the severity of impairment, patients can then be classified within PTLDS (when symptoms are severe enough to cause a substantial reduction in previous levels of activity), versus a less severe category of symptoms, when patients have not returned to health but do not have substantial functional impairment. Also, information about the timing of preceding Lyme disease, along with the absence of an alternate explanation for symptoms, is important for arriving at a diagnosis of PTLDS.

The majority of patients who present with clearly defined manifestations of Lyme disease will have a good health outcome after treatment with currently recommended regimens, and objective evidence of treatment failure is quite uncommon. The frequency of persisting or relapsing subjective complaints of mild to moderate intensity after therapy has varied between 0 to 23% of patients with erythema migrans that were evaluated 6 to 24 months after completion of treatment (Smith et al., 2002; Nowakowski et al., 2003; Wormser et al., 2003; Wormser et al., 2006; Cerar et al.; Kowalski et al.; Tory et al.; Nizic et al., 2012; Stupica et al., 2012; Arnez and Ruzic-Sabljic, 2015; Bechtold et al., 2017; Borsic et al., 2018). These symptoms included fatigue, myalgias, arthralgias, headache, paresthesias, neck stiffness, irritability, sleep problems and difficulty with word finding, memory, and concentration (Mac et al., 2020). Risk factors for PTLDS include symptom severity at the time of treatment and presentation with multiple erythema migrans skin lesions (Nowakowski et al., 2003; Borsic et al., 2018).

The pathogenesis of these nonspecific symptoms is unknown, and it is likely that the underlying cause will differ among patients (Table 1). A post-infective fatigue syndrome might be triggered by Lyme disease, as has been described with other infections (Hickie et al., 2006; Katz et al., 2009; Carter et al., 2019; Furberg et al., 2019), but prolonged fatigue attributed to early Lyme disease is uncommon. In a study of 100 patients with erythema migrans evaluated at 11 to 20 years after presentation, only 3% of patients appeared to have persistent fatigue possibly due to Lyme disease (Wormser et al., 2015b). Co-infection with other tick-borne pathogens is uncommon in PTLDS patients (Wang et al., 2000; Klempner et al., 2001; Ramsey et al., 2002; Fallon et al., 2007; Lantos and Wormser, 2014), though patients with concurrent Lyme disease and untreated *Babesia* infection had an increased number, severity, and duration of symptoms as compared to *B. burgdorferi* mono-infection (Krause et al., 1996).

A major issue in PTLDS (and CLD in general) is that symptoms may be due to previous or new co-morbidities unrelated to Lyme disease. This is particularly significant in the clinical setting, as it can be difficult for practitioners to definitely attribute a nonspecific symptom to a specific condition, or to absolutely exclude a condition as causing that symptom. There is substantial background prevalence of nonspecific symptoms in the general population. Musculoskeletal pain is a very common complaint, and chronic pain, fatigue and sleep disturbances are often reported together (Rohrbeck et al., 2007). Studies in patients with Lyme disease presenting with erythema migrans that followed prospectively both patients and controls showed no difference between patients and controls regarding the presence of new or increased symptoms at follow up (Cerar et al., 2010; Stupica et al., 2012; Stupica et al., 2018a; Stupica et al., 2018b; Wormser et al., 2020). Similar results were found in a study of children after treatment of Lyme neuroborreliosis (Skogman et al., 2008) which did not detect any significant differences in long-term outcome between patients and healthy controls. A prospective study of 101 patients with diverse manifestations of Lyme disease showed that patients' quality of life scores increased to just above the US national average after 3 years of follow-up time, regardless of disease stage or severity at diagnosis, while the presence of co-morbidities was significantly associated with long-term symptoms or lower quality of life scores (Wills et al., 2016). Patients with erythema migrans evaluated for a median of 16 years after initial diagnosis showed similar results (Wormser et al., 2015a).

Another facet of this issue is the possible misattribution to Lyme disease of nonspecific symptoms due to other conditions. These patients would not be expected to respond well to antibiotic therapy, or at least not as favorably as patients where the diagnosis of Lyme disease is accurate. This is shown on studies that include both definite and possible cases. A study of patients with disseminated Lyme disease showed that 90% of the patients with definite diagnosis had excellent or good outcomes, compared with only 47% of possible cases (Oksi et al., 2007). A systematic review of residual symptoms after Lyme neuroborreliosis found that studies of definite or probably Lyme neuroborreliosis had a lower prevalence of residual symptoms than studies including possible cases (Dersch et al., 2016).

Our verdict:
Based on existing published evidence, PTLDS is a distinct entity that occurs in a small group of patients with Lyme disease who have received appropriate antibiotic treatment (Wormser et al., 2006). This fits with the experiences of the authors and others (Wormser et al., 2020) who feel that there is a small number of patients whose persistent symptoms are related to their Lyme disease diagnosis.

Controversies regarding the potential mechanisms of PTLDS

Of the group of patients who fit the definition of PTLDS, several hypotheses for the cause of persistent symptoms in Lyme disease have been advanced, including but not limited to 1) immune dysregulation or maladaptive host responses, 2) chronic infection by *B. burgdorferi* bacteria, including drug-tolerant "persister" cells, and 3) the persistence of bacterial remnants that cause chronic immune stimulation. These mechanisms do not need to be exclusive, i.e., they all could play a role in PTLDS.

Is immune dysregulation the cause of PTLDS?
Antibiotic refractory Lyme arthritis is another post-antibiotic Lyme disease complication (Steere, 2020), but its clinical presentation is very different from PTLDS. Patients with antibiotic refractory Lyme arthritis continue to show signs of joint inflammation after antibiotic treatment due to persistent inflammatory, proliferative synovitis, usually affecting a knee. Antibiotic refractory Lyme arthritis responds to treatment with disease-modifying anti-rheumatic drugs, such as methotrexate and tumor necrosis factor (TNF) inhibitors, but does not respond to additional antibiotic therapy (Steere, 2019). One

hypothesis for the etiology of antibiotic refractory Lyme arthritis is an autoimmune response initiated by molecular mimicry between self-antigens and *B. burgdorferi* antigens. The disease is associated with certain HLA-DR alleles that bind to peptides derived from *B. burgdorferi* outer-surface protein A (OspA) (Kalish et al., 1993; Steere et al., 2006; Drouin et al., 2008b; Iliopoulou et al., 2009; Steere et al., 2011). Patients with antibiotic refractory Lyme arthritis have higher cellular and humoral immune responses to OspA (Kalish et al., 1993; Lengl-Janssen et al., 1994; Chen et al., 1999), which is not usually expressed by *B. burgdorferi* in mammalian hosts. It was proposed that cross reactive T-cell recognition of an OspA epitope with human lymphocyte function–associated antigen-1 (LFA-1) may be the cause of the prolonged inflammation (Gross et al., 1998). Further work showed that other factors were involved, as some patients had weak or partial responses to the OspA peptide presumptive mimitopes and the responses also could be found in patients that responded to antibiotic therapy (Trollmo et al., 2001; Kannian et al., 2007; Drouin et al., 2008a). Since then, other epitopes of *B. burgdorferi* proteins have been studied as potential molecular mimics (Drouin et al., 2008b), as well as responses against autoantigens (endothelial cell growth factor (ECGF), matrix metalloproteinase-10, apolipoprotein B-100, annexin A2) (Drouin et al., 2013; Londono et al., 2014; Crowley et al., 2015; Pianta et al., 2015; Crowley et al., 2016; Sulka et al., 2018).

More recent studies exploring mechanisms of antibiotic refractory Lyme arthritis have focused on immune dysregulation rather than autoimmunity. Studies of patients with antibiotic refractory Lyme arthritis have shown alterations in T cell populations (Shen et al., 2010; Vudattu et al., 2013) and changes in cytokine production that may alter the differentiation and function of synoviocytes (Strle et al., 2017; Lochhead et al., 2019a; Lochhead et al., 2019b). An inactivating polymorphism in the toll-like receptor 1 (TLR1) gene, which is an important receptor in controlling innate immune responses, has been linked to the development of antibiotic refractory Lyme arthritis (Strle et al., 2012) when associated with infection with 16s–23s rRNA intergenic spacer type 1 (RST1) *B. burgdorferi* strains (Strle et al., 2011).

B. burgdorferi factors are also very likely to play a role in the outcome of the infection. Studies have shown that variants of *B. burgdorferi* have differing capacities for dissemination and induction of inflammatory responses (Wormser et al., 1999; Wang et al., 2001; Wang et al., 2002; Jones et al., 2006; Dykhuizen et al., 2008; Strle et al., 2011; Cerar et al., 2016). In particular, *B. burgdorferi* RST1 strains are associated with greater inflammation, dissemination and more severe disease than RST2 or RST3 strains (Wormser et al., 1999; Jones et al., 2006; Wormser et al., 2008a; Strle et al., 2011) *B. burgdorferi* RST1 strains are also associated with antibiotic refractory Lyme arthritis (Jones et al., 2009).

The clinical features of PTLDS are quite different from antibiotic refractory Lyme arthritis, and the pathogenesis is likely to be distinct. Nevertheless, the investigation of possible mechanisms has followed similar lines for the two conditions. Autoimmunity also has been proposed as a possible mechanism for PTLDS. Different studies have shown increases in antibodies to *B. burgdorferi* protein such as specific parts of VlsE (Chandra et al., 2011a) or cross reacting antibodies to human proteins such as ECGF (Strle et al., 2014) or to neural tissues in general (Chandra et al., 2010). In the case of ECGF, a subsequent study did not confirm the association of anti-ECGF antibodies and PTLDS (Tang et al., 2015). One concern about an autoimmune cause for PTLDS is that it does not behave like other autoimmune diseases—i.e. worsening without anti-immune therapy.

There are now multiple studies that have examined the question of immune dysregulation in PTLDS. One study of European patients used a panel with 23 cytokines (measured by Luminex) to analyze serum samples from patients with erythema migrans (mostly infected with *B. afzelli*) who were prospectively followed from diagnosis to 1 year after antibiotic therapy (Strle et al., 2014). The study compared results from 45 patients with at least 1 post-Lyme symptom after antibiotic therapy, and 41 recovered patients (who were randomly selected from 296 patients). The number of patients with symptoms after Lyme disease decreased over time (32 at 2 months, 22 at 6 months and 12 at 12 months). Patients with persistent symptoms had higher interleukin (IL)-23 levels before therapy, and the levels remained significantly higher at each posttreatment visit, up to 12-month follow-up period. IL-23 is a key mediator of the T-helper 17 response and the authors hypothesized that T-helper 17 response could play a role in disease-associated pathology. C-C motif chemokine ligand 19 (CCL19, see below) was measured in this study, and not found to be significantly different between the groups. There were no differences between the two groups regarding the presence and the decline of the antibody response against the C6 peptide of *B. burgdorferi*. There was a trend towards patients with post-Lyme symptoms being more likely to have antibodies against ECGF. The increase in antibodies levels against ECGF in PTLD patients was not found in a separate study (Tang et al., 2015).

There have been three studies examining diverse immunological and inflammatory mediators in a cohort of patients with erythema migrans followed prospectively from the time of acute infection for a period of at least 12 months after antibiotic treatment, the Study of Lyme disease Immunology and Clinical Events (SLICE) (Soloski et al., 2014; Aucott et al., 2016; Bouquet et al., 2016; Blum et al., 2018). These patients were classified into 3 outcome groups: return-to health, symptoms only, and PTLDS at either the 6-month or 1-year time points. The first study measured the levels of 58 immune mediators and 7 acute phase proteins in 44 patients with Lyme disease and 23 controls (Soloski et al., 2014). Patients with acute, untreated Lyme disease had increased levels of inflammatory markers, including C-reactive protein, serum amyloid A, IL-6, C-X-C motif chemokine 10 (CXCL10), CXCL9, and CCL19. Hierarchical clustering divided the patients into mediator-high and mediator-low groups. As expected, patients in the high mediator group were more likely to be more symptomatic, had higher rates of seroconversion, lymphopenia and elevated liver enzymes. Interestingly, the groups did not differ with respect to the presence of multiple erythema migrans lesions, an unexpected finding, as patients with multiple erythema migrans (a sign of hematogenous dissemination of *B. burgdorferi*) are more likely to be symptomatic, be seropositive and have higher markers of inflammation (Wormser et al., 2005; Wormser et al., 2008b; Liveris et al., 2012).

In a follow up study, the authors investigated associations between PTLDS and immune mediator levels during acute illness and at several time points following treatment in the SLICE cohort (Aucott et al., 2016). They analyzed the results of 58 cytokines and chemokines (including IL-23), and six acute-phase markers on 76 patients with erythema migrans for 1 year after antibiotic therapy. These patients included 11 PTLDS patients at either 6 or 12 months, 29 patients with symptoms-only, and 36 recovered (return-to-health) patients. Only CCL19, ferritin, fibrinogen, CXCL10, CXCL9, C-reactive protein, and serum amyloid A were found to be differentially regulated between Lyme disease and healthy controls at the pretreatment visit. Using these 7 markers, generalized logistical regression was used to predict Lyme disease clinical outcome at each of the study time points. None of the markers was significantly associated with an increased odds of PTLDS at the pretreatment visit. An increase in CCL19 of 0.5 standard deviation at posttreatment visits was associated with PTLDS. The serum CCL19 levels were similar between the Lyme disease groups (and different from healthy controls) at the pretreatment visit, but CCL19 serum levels for the PTLDS group remained significantly elevated at the posttreatment visits, while CCL19 serum levels for both patients who recovered and symptoms-only patients were not different from controls at posttreatment visits. Age, gender, single versus multiple erythema migrans, serostatus (positive or negative) were not significantly associated with PTLDS, symptoms-only or return to health.

Also analyzing patients from the SLICE cohort, Blum at al. (Blum et al., 2018) showed that blood plasmablasts were increased in 32 untreated Lyme patients (7 PTLDS and 25 recovered patients) and returned to baseline levels by 6 months following antibiotic treatment. Patients with disseminated EM (10 recovered and 1 PTLDS) had higher plasmablasts numbers compared with those with single EM (15 recovered and 6 PTLDS). Interestingly, PTLDS patients had significantly lower blood plasmablasts than patients who returned to health at the pretreatment timepoint, and less seroreactivity to surface proteins and peptides from *B. burgdorferi*. Of note, other studies have found that PTLDS patients had higher antibody reactivity frequencies against certain *B. burgdorferi* antigens than recovered patients (Chandra et al., 2011a; Chandra et al., 2011b; Jacek et al., 2013). However, a cause and effect between these antibody levels and PTLDS has not been established.

The longitudinal transcriptomic response was investigated in a subset of the SLICE cohort (Bouquet et al., 2016). This study included 29 individuals with Lyme disease (15 who returned to health, 9 with symptoms only, and 4 with PTLD) and 13 matched controls. Transcriptional responses were evaluated by RNA-seq at the time of Lyme disease diagnosis, at the end of the 3-week course of doxycycline treatment, and at 6 months after therapy. There were no significantly differentially expressed genes at any single time point between the 15 who returned to health and the 13 patients with persistent symptoms. In an analysis comparing patients with resolved Lyme disease (n=15) to those meeting defined PTLDS criteria (n=4), a single gene (GPR15) differentially expressed at the time of diagnosis was upregulated in PTLD. To date, there are no studies of host genetic or *B. burgdorferi* factors involved in PTLDS.

Our verdict:
More research is needed. There is no confirmation of a role for autoimmunity in PTLDS. Although potential cross reacting epitopes between *B. burgdorferi* and human tissue have been proposed, the chance for false discovery is very high, and none have been confirmed to be important in the persistence of PTLD symptoms. For the more recent studies of inflammatory markers, it is notable that no marker has been identified in more than one study. There are many potential reasons for this. First, the number of patients with early Lyme disease who go on to suffer from PTLD symptoms in prospective studies is small, and, given the large numbers of variables examined in each study, the risk of type I error is elevated. The lack of identification of similar markers in different studies could reflect false discovery and/or a weak or partial role in PTLDS. Other possible reasons include differences in the patient groups studied, differences between the distribution of strains of *B. burgdorferi* infecting European and U.S. patients, and differences in the technologies used to measure the markers. Until there is confirmation of these markers by independent groups, we reserve judgement on the utility of any of them either as a diagnostic tool or for understanding disease mechanisms.

Are long term antibiotics useful in the treatment of PTLDS?

A key concern has been that the symptoms of PTLDS may be due to persistent infection with *B. burgdorferi*. Because this has been difficult to prove, investigators have taken an alternative method for testing this hypothesis by examining the response to additional antibiotic therapy. Five randomized, placebo-controlled, double-blind studies of antibiotic treatment (four studying PTLDS patients and one study of patients with symptoms attributed to Lyme disease, including PTLDS) have been reported. Briefly, the consensus viewpoint is that prolonged treatment has no proven benefit but has substantial risks.

Two studies were published together, as they used the same intervention (ceftriaxone 2 g intravenous daily for 30 days, followed by oral doxycycline, 200 mg daily for 60 days, or matching intravenous and oral placebos) and outcomes (improvement in the 36-item Short-Form General Health Survey (SF-36) score on day 180 of the study) (Klempner et al., 2001). Patients in one study were IgG seropositive for *B. burgdorferi* at enrollment, while patients in the other study were seronegative. All patients had well-documented Lyme disease. In the seropositive study, there were 39 patients in each treatment group, while the seronegative study had 25 patients randomized to antibiotics and 26 to placebo. A planned interim analysis, using the data on 107 patients who had completed 180 days, showed that demonstrating a difference between treatment groups was unlikely (1.4% chance in the seropositive study and a 4% in the seronegative study, if full enrollment had been reached), so the studies were stopped. Intention-to-treat analyses at 30, 90, and 180 days showed no significant differences in the SF-36 scores between the patients in the antibiotic groups and those in the placebo groups in the seropositive study, the seronegative study, or both studies combined. There were 2 severe adverse events related to treatment.

The STOP-LD (Study and Treatment Of Post Lyme Disease) trial enrolled 55 PTLDS patients who had significant fatigue (Krupp et al., 2003). These patients were randomized to ceftriaxone 2 g (28 patients) or placebo (27 patients, but only 24 started treatment) intravenously daily for 28 days. Twenty-six patients in the ceftriaxone group and 22 in the placebo group completed the 6-month follow-up. The intention-to-treat analysis (assuming that two patients in the ceftriaxone group and five patients in the placebo group had no improvement) showed improvement of fatigue with ceftriaxone therapy at 6 month, with similar results for the 48 patients who received therapy and completed follow up. Patients who had a positive immunoblot, had not received previous intravenous antibiotic therapy and had less pain were more likely to improve. There was no improvement in mental speed or other neurocognitive measures. Three patients in each group discontinued therapy due to side effects, and 4 had to be hospitalized.

The next study enrolled patients with PTLDS who had been treated with at least 3 weeks of intravenous antibiotic therapy, were seropositive by IgG western blot, and presented objective memory impairment (Fallon et al., 2007). Patients were randomized 2:1 to receive 10 weeks of intravenous ceftriaxone (23 patients) or intravenous placebo (14 patients). The primary analysis was chosen by a data-driven selection process, which complicated the interpretation of the results (Marques et al., 2009b). The authors reported a small improvement at 12 weeks in the ceftriaxone group of borderline significance. Both the placebo and the ceftriaxone groups had similar improvements from baseline when analyzed at 24 weeks. Nine patients discontinued therapy due to side effects.

Berende et al (Berende et al., 2016) conducted the Persistent Lyme Empiric Antibiotic Study Europe (PLEASE) trial. This trial enrolled 281 participants with symptoms attributed to Lyme disease who were treated with 2 weeks of open label ceftriaxone and then randomly assigned to treatment with 12 weeks of oral doxycycline, 12 weeks of placebo, or 12 weeks of clarithromycin-hydroxychloroquine. Patients were eligible if they had persistent symptoms temporally related to an erythema migrans or a positive *B. burgdorferi* IgM or IgG test. The median duration of symptoms was more than 2 years, and about 90% of the patients had received previous antibiotic treatment. Only 96 (34%) patients had objective evidence of prior Lyme disease. All groups improved when compared with their initial scores, but there were no significant differences between groups at the end of treatment (week 14) or at 26, 40, and 52 weeks. The authors concluded that prolonged antibiotics beyond two weeks are not helpful for reducing persistent symptoms attributed to Lyme disease. As there was no full placebo arm, one cannot access if the initial improvement was due to the ceftriaxone therapy or a placebo effect. Patients had improvement in fatigue similar to the intravenous placebo arms in 2 other studies (Wormser, 2016). Another criticism of this study is that the lenient enrollment criteria may have included patients with untreated Lyme disease as well as patients who did not have Lyme disease, causing further confusion in interpreting the results. In any event, like its predecessors, this study saw no benefit from prolonged antibiotic therapy (Melia and Auwaerter, 2016).

Our verdict:
Additional treatment with the same antibiotics used in the initial treatment of Lyme disease does not affect symptoms of PTLDS. This issue has been resolved by multiple studies from different groups and should not be studied further. Given the difficulties in defining PTLDS, it is possible that there may be subgroups of patients who may benefit from antibiotic therapy. Until markers to identify different subgroups of PTLDS have been discovered, there would not be a justification for repeating these interventional studies. It is possible that other antibiotics with different mechanisms of action may affect symptoms of PTLDS, but, based on the negative results from the prior studies, a strong rationale for possible efficacy of a different antibiotic or combination would be needed before embarking on additional clinical trials.

Does *B. burgdorferi* or its components persist after antibiotic therapy?

The question at the heart of the antibiotic trials, whether there are residual *B. burgdorferi* after antibiotic treatment, has been examined in multiple ways by different groups. In studies of animals (including mice, dogs and monkeys) after antibiotic treatment, *B. burgdorferi* DNA and RNA can be detected for months to a year after antibiotic treatment (Straubinger et al., 1997; Straubinger et al., 2000; Bockenstedt et al., 2002; Yrjanainen et al., 2007; Hodzic et al., 2008; Barthold et al., 2010; Bockenstedt et al., 2012; Embers et al., 2012; Hodzic et al., 2014). Levels of DNA and RNA may even increase long after antibiotic therapy (Hodzic et al., 2014). However, after antibiotic therapy, it is very rare to be able to culture *B. burgdorferi* from tissue using standard growth media.

Some of the animal studies used xenodiagnosis to try to identify *B. burgdorferi* after antibiotic therapy (Bockenstedt et al., 2002; Hodzic et al., 2008; Barthold et al., 2010; Bockenstedt et al., 2012; Embers et al., 2012; Hodzic et al., 2014). *B. burgdorferi* DNA was detected by PCR in the ticks fed on these animals, but again, it has not been possible to recover the organism from xenodiagnostic ticks. Two studies (from the same research group) showed that *B. burgdorferi* DNA could be detected in severe combined immunodeficiency mice after the feeding of xenodiagnostic ticks that had fed on an infected, antibiotic treated animal; however, even here, no live organisms could be cultured (Hodzic et al., 2008; Hodzic et al., 2014). Xenodiagnosis has been tried on humans and was well tolerated; however the rate of positivity of xenodiagnostic ticks for *B. burgdorferi* DNA was low, and cultures were negative (Marques et al., 2014).

Criticism of the studies showing persistence of DNA/RNA in tissues and detection by xenodiagnoses have focused on the dosing schedules used and documentation that antibiotic exposure was similar to that expected in humans, as well as technical issues (Wormser and Schwartz, 2009; Wormser et al., 2012a; Bockenstedt and Radolf, 2014; Wormser et al., 2018). A major point of discussion is how to interpret the detection of DNA or mRNA, while being unable to culture the bacteria from tissues or xenodiagnostic ticks (Wormser and Schwartz, 2009; Barbour, 2012; Wormser et al., 2012a; Bockenstedt and Radolf, 2014; Wormser et al., 2018). Do these results represent viable *B. burgdorferi* or remnants of the infection? If the bacteria are alive, why are they not cultivable? If the bacteria are not alive, how did they reach the xenodiagnostic ticks? It is difficult to explain how ticks that sample about 1 microliter of fluid during feeding, could so consistently acquire *B. burgdorferi* DNA in the absence of replicating organisms. It has been suggested that the DNA may be carried by inflammatory cells attracted to the tick bite site, but this has not been proven. It is even more difficult to explain how DNA could be transferred to another animal in sufficient amounts to be detected by xenodiagnostic ticks without replication in the new host. Interestingly, the criticism regarding antibiotic dosage in the animal studies presupposes that the source of the DNA is bacteria that survive due to incomplete killing by the antibiotics. If this were true, given large differences in compliance with timely administration of antibiotics, differences in drug distribution due to size and differences in serum drug levels due to metabolism/excretion, one might expect that survival of *B. burgdorferi* is likely in some individuals. A criticism of the human xenodiagnosis study is that further investigation in untreated patients was needed in order to properly interpret the negative results in PTLDS subjects (Bockenstedt and Radolf, 2014).

The inability to culture bacteria from animals after antibiotic therapy despite the long-term detection of DNA/RNA has led to an alternative hypothesis that the surviving bacteria exist in a modified form. Some authors have raised the possibility that phenotypic changes in a small subset of cells may allow persistence despite antibiotic pressure. In this case, repeated or prolonged therapy with the same antibiotics used to treat initially would not be expected to have an effect as the form may be resistant or require different types of antibiotics for eradication. *In vitro* evidence of a subset of cells that may survive and regrow after antibiotic therapy has been shown to occur for *B. burgdorferi*, and these have been termed "persister cells" or just "persisters" (Mursic et al., 1996; Feng et al., 2014; Feng et al., 2015; Sharma et al., 2015).

Persister bacteria are the proportion of a bacterial population that survives antibiotic exposure, but, as opposed to classically antibiotic resistant bacteria, do not differ genetically from the antibiotic susceptible population. Persisters have been identified for almost all bacteria where they have been investigated (Fisher et al., 2017; Wilmaerts et al., 2019). Their ability to survive in the face of antibiotics initially was thought to be due to target inactivity (Lewis, 2010; Fisher et al., 2017), because of slow growth or growth arrest, usually due to stress (Fisher et al., 2017). For example, for antibiotics that are dependent upon bacterial growth for their activity (e.g. b-lactam antibiotics), this may mean that persisters consist of the dormant fraction of a bacterial population. Populations of bacteria may be heterogeneous in their growth rate as a "bet-hedging" strategy for survival in different conditions, including antibiotic exposure (Kussell et al., 2005). Interestingly, *B. burgdorferi* itself does not show the same degree of increased persister formation during stationary phase as other bacteria with exposure to beta-lactam antibiotics, remaining sensitive to ceftriaxone even as growth slows. This may be due to the continued turnover of peptidoglycan even during stationary phase (Wu et al., 2018). Although dormancy originally was thought to be the mechanism for most bacterial persisters, subsequent research has suggested that there are other mechanisms for persister formation (Nguyen et al., 2011; Wakamoto et al., 2013; Volzing and Brynildsen, 2015). These include, but are not limited to, heterogeneity in cellular energy levels, membrane potential, translational rate and toxin-anti-toxin systems.

For *B. burgdorferi*, persister bacteria have been identified by multiple groups of investigators *in vitro* after treatment with different classes of antibiotics used to treat human disease (Feng et al., 2014; Caskey and Embers, 2015; Sharma et al., 2015; Pothineni et al., 2016) and associated with cell density and bacterial growth (Caskey and Embers, 2015). The largest fraction of persisters *in vitro* appears after treatment with doxycycline, but some persisters also appear after *in vitro* treatment of *B. burgdorferi* with ceftriaxone, ampicillin and erythromycin (Sharma et al., 2015). *In vitro*, removal of the antibiotic by exchanging with fresh media is sufficient to allow the persister cells to reactivate and create a new population of bacteria. The new population contains a mixture of antibiotic sensitive cells and persisters, in the same proportion as the original population (Sharma et al., 2015). Although for other bacteria, persister cells also have favored the development of truly genetically resistant bacteria, this does not appear to occur with *B. burgdorferi*.

B. burgdorferi undergoes phenotypic changes when grown in culture to late stationary phase, as well as when exposed to a multitude of adverse conditions (prolonged cultivation, lack of serum, extreme pH values, high temperature, H2O2, antibiotics) *in vitro* (Brorson and Brorson, 1997; Murgia et al., 2002; Murgia and Cinco, 2004). These structures have been called cystic forms, spheroplasts, L-forms, round bodies, starvation forms, morphologic variants, propagules, and cell wall–deficient forms, but they appear to have an intact cell wall (Merilainen et al., 2015). Some studies have suggested that these forms of *B. burgdorferi* may be more resistant to some antibiotics (Mursic et al., 1996; Feng et al., 2014; Feng et al., 2016a; Feng et al., 2016b). However, again, there is no evidence for the existence of these forms during the mammalian part of the *Borrelia* lifecycle (Lantos et al., 2014).

Bacterial persistence has been postulated to play an important role in recrudescence of infection after treatment in diseases such as biofilm-related catheter infections and in tuberculosis. The causative agent typically can be recovered by culture in relapses due to persisters. For Lyme disease, the role of persisters remains unknown. There is no convincing data that *B. burgdorferi* form biofilms *in vivo*. The term "biofilm-like colonies" or "aggregated biofilm-like microcolonies" has been used to describe aggregate forms *of B. burgdorferi* seem in old stationary-phase cultures (Sapi et al., 2012; Feng et al., 2014; Feng et al., 2016b). While multiple laboratories have reported that *B. burgdorferi* do aggregate into clusters when grown *in vitro* or in ticks (Burgdorfer et al., 1982; Gilmore and Piesman, 2000; Srivastava and de Silva, 2009), it is unclear if bacterial aggregates seen *in vitro* fulfill criteria for a biofilm. Biofilms are a surface-associated complex community of bacteria, that secrete and surround themselves with an extracellular polymer matrix that consists of polysaccharides, extracellular DNA, lipids and proteins (Flemming and Wingender, 2010). Evidence that *B. burgdorferi* form biofilms is mainly from one laboratory (Sapi et al., 2012; Sapi et al., 2016; Sapi et al., 2019) and is based on microscopy images of *B. burgdorferi* aggregates and immunofluorescence using an anti-alginate antibody. Alginate is an exopolysaccharide important in forming the biofilm architecture in *Pseudomonas aeruginosa* (Flemming and Wingender, 2010). A query of the *B. burgdorferi* B31 genome using the genes associated with all five synthase-dependent exopolysaccharide systems (alginate, cellulose, acetylated cellulose, poly-β-1,6-N-acetyl-D-glucosamine, and the Pel polysaccharide), described in Table S3 of reference (Bundalovic-Torma et al., 2020), showed no clear orthologs in *B. burgdorferi*. Other laboratories, using newer live imaging techniques, demonstrated that *B. burgdorferi* remain highly motile and is not clustered in mouse tissue (Belperron et al., 2018).

The inability to recover live bacteria from animals after courses of antibiotic therapy is difficult to explain if persisters do remain in the tissues of treated animals. *In vitro*, *B. burgdorferi* persisters are able to grow back easily after removal of the antibiotic. One possibility to explain why live *B. burgdorferi* cannot be recultured in standard media after exposure to antibiotics is that they require a specific signal for reactivation and growth *in vitro*. This has been reported for *M. tuberculosis* where the addition of pre-conditioned culture filtrate greatly enhanced recovery of persister bacteria (Chengalroyen et al., 2016; McAulay et al., 2018). However, no similar signal that can restart growth of *B. burgdorferi* from antibiotic treated animals has been found.

Regardless of whether *B. burgdorferi* persisters exist *in vivo*, it is important to note that **no evidence** points to their involvement in PTLD. There is currently no animal model of PTLD. Xenodiagnosis in humans has shown that the detection rate of *B. burgdorferi* DNA detection in PTLDS is quite low (Marques et al., 2014). Studies to compare the rate of xenodiagnostic positivity between PTLDS patients and recovered patients have not been reported to date.

The difficulty in detecting live bacteria from animals after antibiotic therapy has raised the question as to whether detection of DNA/RNA represents the remnants of dead bacteria (Barbour, 2012; Iyer et al., 2013; Bockenstedt and Radolf, 2014). An "amber" theory of persistence of *B. burgdorferi* products, trapped and preserved in tissue after infection has been proposed (Wormser et al., 2012b). Remnants of *B. burgdorferi,* rather than live bacteria, could persist after antibiotic treatment, causing inflammation, and playing a role in PTLD. In mouse studies, deposits of antigens (with inflammatory potential) persisted near the cartilage and within joint entheses, after antibiotic treatment with doxycycline in MyD88 knockout (deficient in an important innate immune adaptor molecule) and wild-type mice infected with green fluorescent protein-tagged *B. burgdorferi* (Barbour, 2012; Bockenstedt et al., 2012). Using intravital microscopy, persistent antigen deposits, but no spirochetes, were found adjacent to the cartilage after antibiotic therapy. This material was capable of stimulating macrophages to produce TNF-alpha *in vitro* and also of provoking an immune response (as measured by *B. burgdorferi* IgG) when injected into naive mice. The presence of *B. burgdorferi* peptidoglycan in Lyme arthritis (Jutras et al., 2019) provides support for the hypothesis that residual antigen may drive immune responses in some patients. No studies in PTLDS have been performed to date.

Our verdict:
Alternative forms of *B. burgdorferi*, including persisters, cystic forms and aggregates, have been reported by multiple laboratories and are well documented *in vitro*. True biofilms have not been documented and it does not appear that *B. burgdorferi* form classically defined biofilms. However, the existence of any of these forms *in vivo* and their relevance to PTLDS has not been established and further research is warranted.

It is becoming clear that components of *B. burgdorferi* including DNA and peptidoglycan are detectable in humans and/or animals for prolonged periods of time after antibiotic therapy. What remains controversial is whether these components are detectable because they are being produced by live organisms or because they are the remnants of dead organisms.

Summary

Despite decades of studies, questions about the etiology of PTLDS persist with few accepted answers. In our review of the literature, we believe the following facts to be well established by studies from multiple independent groups: 1) PTLDS is much less common than CLD; 2) the majority of symptoms associated with Lyme disease improve with time; 3) additional treatment with antibiotics used for the treatment of acute Lyme disease are not effective in resolving symptoms of PTLDS; and 4) *B. burgdorferi* DNA, but not cultivable organisms, can be detected in animal tissue by PCR or xenodiagnoses after completion of antibiotic therapy.

The lack of support for high-quality studies required to provide answers to open questions in Lyme disease is, in the authors' opinion, the principal reason for the number of inconclusive, non-generalizable studies. While results from basic and animal models are very important, the significance of the findings for human Lyme disease is uncertain and requires additional clinical investigation before their medical relevance is established. This area, sometimes referred to as translational research, is greatly neglected in Lyme disease, as there are many technical challenges of conducting clinical studies on Lyme disease. Subjects present with heterogeneous manifestations and are variously seen by primary care providers, infectious disease specialists, neurologists, dermatologists, and numerous other clinicians. Thus, an effort to recruit such individuals requires costly and time-consuming collaboration among multiple specialists. Furthermore, microbiological support for the diagnosis of a fastidious organism requires additional time and cost compared to standard diagnostic methods for conventional bacteria or viruses. Together, these challenges mean that it is extremely hard to conduct research studies in the typical timeframe of 2-5 years supported by most research grants. This issue is likely to have an outsized effect on junior researchers who are typically discouraged from tackling projects that cannot lead to publication within the promotion cycle of academic medicine.

These challenges have been recognized in other areas of medicine and resulted in, for example, development of the Aids Clinical Trials Group and the Rare Diseases Network. This points to a critical need to develop networks to enroll Lyme disease subjects at different stages of disease using unified definitions. Because of the relatively small number of patients affected annually (compared to, for example, cardiac disease) and nature of the distribution of Lyme disease, with most patients seen in primary care settings where it is difficult to conduct studies, it would be hard for any single center to accumulate sufficient numbers of subjects.

The importance of identifying markers and criteria that would allow differentiation of PTLDS patients from those suffering from other diseases cannot be overstated. Without this critical piece, treatment studies are unlikely to succeed due to heterogeneity in the enrolled subjects. In addition, such markers may give insights into the potential pathways for treatment of PTLDS patients. While there remains much work to be done, we remain hopeful that with cooperation between researchers working on PTLDS we will have much more clarity by the next edition of this book.

An added challenge to the Lyme disease field is the toxic effect of long-simmering controversies (Auwaerter et al., 2011). This, along with a lack of investment in fundamental, translational, and clinical science in *B. burgdorferi* infection, have greatly reduced interest in the field for young researchers. The Lyme disease field is in dire need of experimental approaches and new funding that will help bring evidence-based answers to open questions, recruit new clinical and basic researchers into the field, and improve the care of patients.

Acknowledgements and Disclaimer

Funding for this study was provided in part by the Division of Intramural Research, National Institute of Allergy and Infectious Diseases, National Institutes of Health. The content of this publication does not necessarily reflect the views or policies of the Department of Health and Human Services, nor does the mention of trade names, commercial products, or organizations imply endorsement by the U.S. Government.

References

Arnez, M., and Ruzic-Sabljic, E. (2015). Azithromycin Is Equally Effective as Amoxicillin in Children with Solitary Erythema Migrans. Pediatr Infect Dis J *34*, 1045-1048. https://doi.org/10.1097/INF.0000000000000804

Aucott, J.N., Crowder, L.A., and Kortte, K.B. (2013). Development of a foundation for a case definition of post-treatment Lyme disease syndrome. Int J Infect Dis *17*, e443-449. https://doi.org/10.1016/j.ijid.2013.01.008

Aucott, J.N., Seifter, A., and Rebman, A.W. (2012). Probable late lyme disease: a variant manifestation of untreated *Borrelia burgdorferi* infection. BMC Infect Dis *12*, 173. https://doi.org/10.1186/1471-2334-12-173

Aucott, J.N., Soloski, M.J., Rebman, A.W., Crowder, L.A., Lahey, L.J., Wagner, C.A., Robinson, W.H., and Bechtold, K.T. (2016). CCL19 as a Chemokine Risk Factor for Posttreatment Lyme Disease Syndrome: a Prospective Clinical Cohort Study. Clin Vaccine Immunol *23*, 757-766. https://doi.org/10.1128/CVI.00071-16

Auwaerter, P.G., Bakken, J.S., Dattwyler, R.J., Dumler, J.S., Halperin, J.J., McSweegan, E., Nadelman, R.B., O'Connell, S., Shapiro, E.D., Sood, S.K., *et al.* (2011). Antiscience and ethical concerns associated with advocacy of Lyme disease. Lancet Infect Dis *11*, 713-719. https://doi.org/10.1016/S1473-3099(11)70034-2

Auwaerter, P.G., and Melia, M.T. (2012). Bullying Borrelia: when the culture of science is under attack. Trans Am Clin Climatol Assoc *123*, 79-89; discussion 89-90.

Barbour, A. (2012). Remains of infection. J Clin Invest *122*, 2344-2346. https://doi.org/10.1172/JCI63975

Barthold, S.W., Hodzic, E., Imai, D.M., Feng, S., Yang, X., and Luft, B.J. (2010). Ineffectiveness of tigecycline against persistent *Borrelia burgdorferi*. Antimicrob Agents Chemother *54*, 643-651. https://doi.org/10.1128/AAC.00788-09

Bechtold, K.T., Rebman, A.W., Crowder, L.A., Johnson-Greene, D., and Aucott, J.N. (2017). Standardized Symptom Measurement of Individuals with Early Lyme Disease Over Time. Arch Clin Neuropsychol *32*, 129-141. https://doi.org/10.1093/arclin/acw098

Belperron, A.A., Mao, J., and Bockenstedt, L.K. (2018). Two Photon Intravital Microscopy of Lyme Borrelia in Mice. Methods Mol Biol *1690*, 279-290. https://doi.org/10.1007/978-1-4939-7383-5_20

Berende, A., ter Hofstede, H.J., Vos, F.J., van Middendorp, H., Vogelaar, M.L., Tromp, M., van den Hoogen, F.H., Donders, A.R., Evers, A.W., and Kullberg, B.J. (2016). Randomized Trial of Longer-Term Therapy for Symptoms Attributed to Lyme Disease. N Engl J Med *374*, 1209-1220. https://doi.org/10.1056/NEJMoa1505425

Blum, L.K., Adamska, J.Z., Martin, D.S., Rebman, A.W., Elliott, S.E., Cao, R.R.L., Embers, M.E., Aucott, J.N., Soloski, M.J., and Robinson, W.H. (2018). Robust B Cell Responses Predict Rapid Resolution of Lyme Disease. Front Immunol *9*, 1634. https://doi.org/10.3389/fimmu.2018.01634

Bockenstedt, L.K., Gonzalez, D.G., Haberman, A.M., and Belperron, A.A. (2012). Spirochete antigens persist near cartilage after murine Lyme borreliosis therapy. J Clin Invest *122*, 2652-2660. https://doi.org/10.1172/JCI58813

Bockenstedt, L.K., Mao, J., Hodzic, E., Barthold, S.W., and Fish, D. (2002). Detection of attenuated, noninfectious spirochetes in *Borrelia burgdorferi*-infected mice after antibiotic treatment. J Infect Dis *186*, 1430-1437. https://doi.org/10.1086/345284

Bockenstedt, L.K., and Radolf, J.D. (2014). Xenodiagnosis for posttreatment Lyme disease syndrome: resolving the conundrum or adding to it? Clin Infect Dis *58*, 946-948. https://doi.org/10.1093/cid/cit942

Borsic, K., Blagus, R., Cerar, T., Strle, F., and Stupica, D. (2018). Clinical Course, Serologic Response, and Long-Term Outcome in Elderly Patients with Early Lyme Borreliosis. J Clin Med *7*. https://doi.org/10.3390/jcm7120506

Bouquet, J., Gardy, J.L., Brown, S., Pfeil, J., Miller, R.R., Morshed, M., Avina-Zubieta, A., Shojania, K., McCabe, M., Parker, S., *et al.* (2017). RNA-Seq Analysis of Gene Expression, Viral Pathogen, and B-Cell/T-Cell Receptor Signatures in Complex Chronic Disease. Clin Infect Dis *64*, 476-481. https://doi.org/10.1093/cid/ciw767

Bouquet, J., Soloski, M.J., Swei, A., Cheadle, C., Federman, S., Billaud, J.N., Rebman, A.W., Kabre, B., Halpert, R., Boorgula, M., *et al.* (2016). Longitudinal Transcriptome Analysis Reveals a Sustained Differential Gene Expression Signature in Patients Treated for Acute Lyme Disease. mBio *7*, e00100-00116. https://doi.org/10.1128/mBio.00100-16

Brorson, O., and Brorson, S.H. (1997). Transformation of cystic forms of *Borrelia burgdorferi* to normal, mobile spirochetes. Infection *25*, 240-246. https://doi.org/10.1007/BF01713153

Bundalovic-Torma, C., Whitfield, G.B., Marmont, L.S., Howell, P.L., and Parkinson, J. (2020). A systematic pipeline for classifying bacterial operons reveals the evolutionary landscape of biofilm machineries. PLoS computational biology *16*, e1007721. https://doi.org/10.1371/journal.pcbi.1007721

Burgdorfer, W., Barbour, A.G., Hayes, S.F., Benach, J.L., Grunwaldt, E., and Davis, J.P. (1982). Lyme disease-a tick-borne spirochetosis? Science *216*, 1317-1319. https://doi.org/10.1126/science.7043737

Cameron, D., Gaito, A., Harris, N., Bach, G., Bellovin, S., Bock, K., Bock, S., Burrascano, J., Dickey, C., Horowitz, R., *et al.* (2004). Evidence-based guidelines for the management of Lyme disease. Expert Rev Anti Infect Ther *2*, S1-13. https://doi.org/10.1586/14789072.2.1.s1

Carter, B.L., Stiff, R.E., Elwin, K., Hutchings, H.A., Mason, B.W., Davies, A.P., and Chalmers, R.M. (2019). Health sequelae of human cryptosporidiosis-a 12-month prospective follow-up study. Eur J Clin Microbiol Infect Dis *38*, 1709-1717. https://doi.org/10.1007/s10096-019-03603-1

Caskey, J.R., and Embers, M.E. (2015). Persister Development by *Borrelia burgdorferi* Populations In Vitro. Antimicrob Agents Chemother *59*, 6288-6295. https://doi.org/10.1128/AAC.00883-15

CDC, C.f.D.C.a.P. (2015). Laboratory tests that are not recommended.

Cerar, D., Cerar, T., Ruzic-Sabljic, E., Wormser, G.P., and Strle, F. (2010). Subjective symptoms after treatment of early Lyme disease. Am J Med *123*, 79-86. https://doi.org/10.1016/j.amjmed.2009.05.011

Cerar, T., Strle, F., Stupica, D., Ruzic-Sabljic, E., McHugh, G., Steere, A.C., and Strle, K. (2016). Differences in Genotype, Clinical Features, and Inflammatory Potential of *Borrelia burgdorferi* sensu stricto Strains from Europe and the United States. Emerg Infect Dis *22*, 818-827. https://doi.org/10.3201/eid2205.151806

Chalder, T., and Willis, C. (2017). "Lumping" and "splitting" medically unexplained symptoms: is there a role for a transdiagnostic approach? J Ment Health *26*, 187-191. https://doi.org/10.1080/09638237.2017.1322187

Chandra, A., Latov, N., Wormser, G.P., Marques, A.R., and Alaedini, A. (2011a). Epitope mapping of antibodies to VlsE protein of *Borrelia burgdorferi* in post-Lyme disease syndrome. Clin Immunol *141*, 103-110. https://doi.org/10.1016/j.clim.2011.06.005

Chandra, A., Wormser, G.P., Klempner, M.S., Trevino, R.P., Crow, M.K., Latov, N., and Alaedini, A. (2010). Anti-neural antibody reactivity in patients with a history of Lyme borreliosis and persistent symptoms. Brain Behav Immun *24*, 1018-1024. https://doi.org/10.1016/j.bbi.2010.03.002

Chandra, A., Wormser, G.P., Marques, A.R., Latov, N., and Alaedini, A. (2011b). Anti-*Borrelia burgdorferi* antibody profile in post-Lyme disease syndrome. Clin Vaccine Immunol *18*, 767-771. https://doi.org/10.1128/CVI.00002-11

Chen, J., Field, J.A., Glickstein, L., Molloy, P.J., Huber, B.T., and Steere, A.C. (1999). Association of antibiotic treatment-resistant Lyme arthritis with T cell responses to dominant epitopes of outer surface protein A of Borrelia burgdorferi. Arthritis Rheum *42*, 1813-1822. https://doi.org/10.1002/1529-0131(199909)42:9<1813::AID-ANR4>3.0.CO;2-0

Chengalroyen, M.D., Beukes, G.M., Gordhan, B.G., Streicher, E.M., Churchyard, G., Hafner, R., Warren, R., Otwombe, K., Martinson, N., and Kana, B.D. (2016). Detection and Quantification of Differentially Culturable Tubercle Bacteria in Sputum from Patients with Tuberculosis. Am J Respir Crit Care Med *194*, 1532-1540. https://doi.org/10.1164/rccm.201604-0769OC

Citera, M., Freeman, P.R., and Horowitz, R.I. (2017). Empirical validation of the Horowitz Multiple Systemic Infectious Disease Syndrome Questionnaire for suspected Lyme disease. Int J Gen Med *10*, 249-273. https://doi.org/10.2147/IJGM.S140224

Crowley, J.T., Drouin, E.E., Pianta, A., Strle, K., Wang, Q., Costello, C.E., and Steere, A.C. (2015). A Highly Expressed Human Protein, Apolipoprotein B-100, Serves as an Autoantigen in a Subgroup of Patients With Lyme Disease. J Infect Dis *212*, 1841-1850. https://doi.org/10.1093/infdis/jiv310

Crowley, J.T., Strle, K., Drouin, E.E., Pianta, A., Arvikar, S.L., Wang, Q., Costello, C.E., and Steere, A.C. (2016). Matrix metalloproteinase-10 is a target of T and B cell responses that correlate with synovial pathology in patients with antibiotic-refractory Lyme arthritis. J Autoimmun *69*, 24-37. https://doi.org/10.1016/j.jaut.2016.02.005

Deary, V., Metcalfe, L., and Wilson, J.A. (2014). Persistent (unexplained) physical symptoms: evidence-based highlights. Br J Hosp Med (Lond) *75*, 564-567. https://doi.org/10.12968/hmed.2014.75.10.564

Dersch, R., Sommer, H., Rauer, S., and Meerpohl, J.J. (2016). Prevalence and spectrum of residual symptoms in Lyme neuroborreliosis after pharmacological treatment: a systematic review. J Neurol *263*, 17-24. https://doi.org/10.1007/s00415-015-7923-0

Drouin, E.E., Glickstein, L., Kwok, W.W., Nepom, G.T., and Steere, A.C. (2008a). Human homologues of a Borrelia T cell epitope associated with antibiotic-refractory Lyme arthritis. Mol Immunol *45*, 180-189. https://doi.org/10.1016/j.molimm.2007.04.017

Drouin, E.E., Glickstein, L., Kwok, W.W., Nepom, G.T., and Steere, A.C. (2008b). Searching for borrelial T cell epitopes associated with antibiotic-refractory Lyme arthritis. Mol Immunol *45*, 2323-2332. https://doi.org/10.1016/j.molimm.2007.11.010

Drouin, E.E., Seward, R.J., Strle, K., McHugh, G., Katchar, K., Londono, D., Yao, C., Costello, C.E., and Steere, A.C. (2013). A novel human autoantigen, endothelial cell growth factor, is a target of T and B cell responses in patients with Lyme disease. Arthritis Rheum *65*, 186-196. https://doi.org/10.1002/art.37732

Dykhuizen, D.E., Brisson, D., Sandigursky, S., Wormser, G.P., Nowakowski, J., Nadelman, R.B., and Schwartz, I. (2008). The propensity of different *Borrelia burgdorferi* sensu stricto genotypes to cause disseminated infections in humans. Am J Trop Med Hyg *78*, 806-810.

Embers, M.E., Barthold, S.W., Borda, J.T., Bowers, L., Doyle, L., Hodzic, E., Jacobs, M.B., Hasenkampf, N.R., Martin, D.S., Narasimhan, S., *et al.* (2012). Persistence of *Borrelia burgdorferi* in rhesus macaques following antibiotic treatment of disseminated infection. PLoS One *7*, e29914. https://doi.org/10.1371/journal.pone.0029914

Fallon, B.A., Keilp, J.G., Corbera, K.M., Petkova, E., Britton, C.B., Dwyer, E., Slavov, I., Cheng, J., Dobkin, J., Nelson, D.R., *et al.* (2007). A randomized, placebo-controlled trial of repeated IV antibiotic therapy for Lyme encephalopathy. Neurology.

Fallon, B.A., Pavlicova, M., Coffino, S.W., and Brenner, C. (2014). A comparison of lyme disease serologic test results from 4 laboratories in patients with persistent symptoms after antibiotic treatment. Clin Infect Dis *59*, 1705-1710. https://doi.org/10.1093/cid/ciu703

Feder, H.M., Jr., Johnson, B.J., O'Connell, S., Shapiro, E.D., Steere, A.C., Wormser, G.P., Ad Hoc International Lyme Disease, G., Agger, W.A., Artsob, H., Auwaerter, P., *et al.* (2007). A critical appraisal of "chronic Lyme disease". N Engl J Med *357*, 1422-1430. https://doi.org/10.1056/NEJMra072023

Feng, J., Auwaerter, P.G., and Zhang, Y. (2015). Drug combinations against *Borrelia burgdorferi* persisters in vitro: eradication achieved by using daptomycin, cefoperazone and doxycycline. PLoS One *10*, e0117207. https://doi.org/10.1371/journal.pone.0117207

Feng, J., Shi, W., Zhang, S., Sullivan, D., Auwaerter, P.G., and Zhang, Y. (2016a). A Drug Combination Screen Identifies Drugs Active against Amoxicillin-Induced Round Bodies of In Vitro *Borrelia burgdorferi* Persisters from an FDA Drug Library. Front Microbiol *7*, 743. https://doi.org/10.3389/fmicb.2016.00743

Feng, J., Wang, T., Shi, W., Zhang, S., Sullivan, D., Auwaerter, P.G., and Zhang, Y. (2014). Identification of novel activity against *Borrelia burgdorferi* persisters using an FDA approved drug library. Emerg Microbes Infect *3*, e49. https://doi.org/10.1038/emi.2014.53

Feng, J., Weitner, M., Shi, W., Zhang, S., and Zhang, Y. (2016b). Eradication of Biofilm-Like Microcolony Structures of *Borrelia burgdorferi* by Daunomycin and Daptomycin but not Mitomycin C in Combination with Doxycycline and Cefuroxime. Front Microbiol *7*, 62. https://doi.org/10.3389/fmicb.2016.00062

Fisher, R.A., Gollan, B., and Helaine, S. (2017). Persistent bacterial infections and persister cells. Nat Rev Microbiol *15*, 453-464. https://doi.org/10.1038/nrmicro.2017.42

Flemming, H.C., and Wingender, J. (2010). The biofilm matrix. Nat Rev Microbiol *8*, 623-633. https://doi.org/10.1038/nrmicro2415

Furberg, M., Anticona, C., and Schumann, B. (2019). Post-infectious fatigue following Puumala virus infection. Infect Dis (Lond) *51*, 519-526. https://doi.org/10.1080/23744235.2019.1605191

Gilmore, R.D., Jr., and Piesman, J. (2000). Inhibition of *Borrelia burgdorferi* migration from the midgut to the salivary glands following feeding by ticks on OspC-immunized mice. Infect Immun *68*, 411-414. https://doi.org/10.1128/iai.68.1.411-414.2000

Gross, D.M., Steere, A.C., and Huber, B.T. (1998). T helper 1 response is dominant and localized to the synovial fluid in patients with Lyme arthritis. J Immunol *160*, 1022-1028.

Hickie, I., Davenport, T., Wakefield, D., Vollmer-Conna, U., Cameron, B., Vernon, S.D., Reeves, W.C., Lloyd, A., and Dubbo Infection Outcomes Study, G. (2006). Post-infective and chronic fatigue syndromes precipitated by viral and non-viral pathogens: prospective cohort study. BMJ *333*, 575. https://doi.org/10.1136/bmj.38933.585764.AE

Hodzic, E., Feng, S., Holden, K., Freet, K.J., and Barthold, S.W. (2008). Persistence of *Borrelia burgdorferi* following antibiotic treatment in mice. Antimicrob Agents Chemother *52*, 1728-1736. https://doi.org/10.1128/AAC.01050-07

Hodzic, E., Imai, D., Feng, S., and Barthold, S.W. (2014). Resurgence of persisting non-cultivable *Borrelia burgdorferi* following antibiotic treatment in mice. PLoS One *9*, e86907. https://doi.org/10.1371/journal.pone.0086907

Hook, S.A., Nelson, C.A., and Mead, P.S. (2015). U.S. public's experience with ticks and tick-borne diseases: Results from national HealthStyles surveys. Ticks Tick Borne Dis *6*, 483-488. https://doi.org/10.1016/j.ttbdis.2015.03.017

Horowitz, R.I., and Freeman, P.R. (2018). Precision Medicine: The Role of the MSIDS Model in Defining, Diagnosing, and Treating Chronic Lyme Disease/Post Treatment Lyme Disease Syndrome and Other Chronic Illness: Part 2. Healthcare (Basel) *6*. https://doi.org/10.3390/healthcare6040129

Horowitz, R.I., and Freeman, P.R. (2019). Precision medicine: retrospective chart review and data analysis of 200 patients on dapsone combination therapy for chronic Lyme disease/post-treatment Lyme disease syndrome: part 1. Int J Gen Med *12*, 101-119. https://doi.org/10.2147/IJGM.S193608

Iliopoulou, B.P., Guerau-de-Arellano, M., and Huber, B.T. (2009). HLA-DR alleles determine responsiveness to *Borrelia burgdorferi* antigens in a mouse model of self-perpetuating arthritis. Arthritis Rheum *60*, 3831-3840. https://doi.org/10.1002/art.25005

Iyer, R., Mukherjee, P., Wang, K., Simons, J., Wormser, G.P., and Schwartz, I. (2013). Detection of *Borrelia burgdorferi* nucleic acids after antibiotic treatment does not confirm viability. J Clin Microbiol *51*, 857-862. https://doi.org/10.1128/JCM.02785-12

Jacek, E., Fallon, B.A., Chandra, A., Crow, M.K., Wormser, G.P., and Alaedini, A. (2013). Increased IFNalpha activity and differential antibody response in patients with a history of Lyme disease and persistent cognitive deficits. J Neuroimmunol *255*, 85-91. https://doi.org/10.1016/j.jneuroim.2012.10.011

Johnson, L., Shapiro, M., and Mankoff, J. (2018). Removing the Mask of Average Treatment Effects in Chronic Lyme Disease Research Using Big Data and Subgroup Analysis. Healthcare (Basel) *6*. https://doi.org/10.3390/healthcare6040124

Jones, K.L., Glickstein, L.J., Damle, N., Sikand, V.K., McHugh, G., and Steere, A.C. (2006). *Borrelia burgdorferi* genetic markers and disseminated disease in patients with early Lyme disease. J Clin Microbiol *44*, 4407-4413. https://doi.org/10.1128/JCM.01077-06

Jones, K.L., McHugh, G.A., Glickstein, L.J., and Steere, A.C. (2009). Analysis of *Borrelia burgdorferi* genotypes in patients with Lyme arthritis: High frequency of ribosomal RNA intergenic spacer type 1 strains in antibiotic-refractory arthritis. Arthritis Rheum *60*, 2174-2182. https://doi.org/10.1002/art.24812

Jutras, B.L., Lochhead, R.B., Kloos, Z.A., Biboy, J., Strle, K., Booth, C.J., Govers, S.K., Gray, J., Schumann, P., Vollmer, W., *et al.* (2019). *Borrelia burgdorferi* peptidoglycan is a persistent antigen in patients with Lyme arthritis. Proc Natl Acad Sci U S A *116*, 13498-13507. https://doi.org/10.1073/pnas.1904170116

Kalish, R.A., Leong, J.M., and Steere, A.C. (1993). Association of treatment-resistant chronic Lyme arthritis with HLA-DR4 and antibody reactivity to OspA and OspB of *Borrelia burgdorferi*. Infect Immun *61*, 2774-2779. https://doi.org/10.1128/IAI.61.7.2774-2779.1993

Kannian, P., Drouin, E.E., Glickstein, L., Kwok, W.W., Nepom, G.T., and Steere, A.C. (2007). Decline in the frequencies of *Borrelia burgdorferi* OspA161 175-specific T cells after antibiotic therapy in HLA-DRB1*0401-positive patients with antibiotic-responsive or antibiotic-refractory lyme arthritis. J Immunol *179*, 6336-6342. https://doi.org/10.4049/jimmunol.179.9.6336

Katz, B.Z., Shiraishi, Y., Mears, C.J., Binns, H.J., and Taylor, R. (2009). Chronic fatigue syndrome after infectious mononucleosis in adolescents. Pediatrics *124*, 189-193. https://doi.org/10.1542/peds.2008-1879

Klempner, M.S., Hu, L.T., Evans, J., Schmid, C.H., Johnson, G.M., Trevino, R.P., Norton, D., Levy, L., Wall, D., McCall, J., *et al.* (2001). Two controlled trials of antibiotic treatment in patients with persistent symptoms and a history of Lyme disease. N Engl J Med *345*, 85-92. https://doi.org/10.1056/NEJM200107123450202

Kobayashi, T., Higgins, Y., Samuels, R., Moaven, A., Sanyal, A., Yenokyan, G., Lantos, P.M., Melia, M.T., and Auwaerter, P.G. (2019). Misdiagnosis of Lyme Disease With Unnecessary Antimicrobial Treatment Characterizes Patients Referred to an Academic Infectious Diseases Clinic. Open Forum Infect Dis *6*. https://doi.org/10.1093/ofid/ofz299

Kowalski, T.J., Tata, S., Berth, W., Mathiason, M.A., and Agger, W.A. (2010). Antibiotic treatment duration and long-term outcomes of patients with early lyme disease from a lyme disease-hyperendemic area. Clin Infect Dis *50*, 512-520. https://doi.org/10.1086/649920

Krause, P.J., Telford, S.R., 3rd, Spielman, A., Sikand, V., Ryan, R., Christianson, D., Burke, G., Brassard, P., Pollack, R., Peck, J., *et al.* (1996). Concurrent Lyme disease and babesiosis. Evidence for increased severity and duration of illness. JAMA *275*, 1657-1660.

Krupp, L.B., Hyman, L.G., Grimson, R., Coyle, P.K., Melville, P., Ahnn, S., Dattwyler, R., and Chandler, B. (2003). Study and treatment of post Lyme disease (STOP-LD): a randomized double masked clinical trial. Neurology *60*, 1923-1930. https://doi.org/10.1212/01.wnl.0000071227.23769.9e

Kussell, E., Kishony, R., Balaban, N.Q., and Leibler, S. (2005). Bacterial persistence: a model of survival in changing environments. Genetics *169*, 1807-1814. https://doi.org/10.1534/genetics.104.035352

Lantos, P.M., Auwaerter, P.G., and Wormser, G.P. (2014). A systematic review of *Borrelia burgdorferi* morphologic variants does not support a role in chronic Lyme disease. Clin Infect Dis *58*, 663-671. https://doi.org/10.1093/cid/cit810

Lantos, P.M., Shapiro, E.D., Auwaerter, P.G., Baker, P.J., Halperin, J.J., McSweegan, E., and Wormser, G.P. (2015). Unorthodox alternative therapies marketed to treat Lyme disease. Clin Infect Dis *60*, 1776-1782. https://doi.org/10.1093/cid/civ186

Lantos, P.M., and Wormser, G.P. (2014). Chronic coinfections in patients diagnosed with chronic lyme disease: a systematic review. Am J Med *127*, 1105-1110. https://doi.org/10.1016/j.amjmed.2014.05.036

Lengl-Janssen, B., Strauss, A.F., Steere, A.C., and Kamradt, T. (1994). The T helper cell response in Lyme arthritis: differential recognition of *Borrelia burgdorferi* outer surface protein A in patients with treatment-resistant or treatment-responsive Lyme arthritis. J Exp Med *180*, 2069-2078. https://doi.org/10.1084/jem.180.6.2069

Lewis, K. (2010). Persister cells. Annu Rev Microbiol *64*, 357-372. https://doi.org/10.1146/annurev.micro.112408.134306

Liveris, D., Schwartz, I., McKenna, D., Nowakowski, J., Nadelman, R., Demarco, J., Iyer, R., Bittker, S., Cooper, D., Holmgren, D., *et al.* (2012). Comparison of five diagnostic modalities for direct detection of *Borrelia burgdorferi* in patients with early Lyme disease. Diagn Microbiol Infect Dis *73*, 243-245. https://doi.org/10.1016/j.diagmicrobio.2012.03.026

Lochhead, R.B., Arvikar, S.L., Aversa, J.M., Sadreyev, R.I., Strle, K., and Steere, A.C. (2019a). Robust interferon signature and suppressed tissue repair gene expression in synovial tissue from patients with postinfectious, *Borrelia burgdorferi*-induced Lyme arthritis. Cell Microbiol *21*, e12954. https://doi.org/10.1111/cmi.12954

Lochhead, R.B., Ordonez, D., Arvikar, S.L., Aversa, J.M., Oh, L.S., Heyworth, B., Sadreyev, R., Steere, A.C., and Strle, K. (2019b). Interferon-gamma production in Lyme arthritis synovial tissue promotes differentiation of fibroblast-like synoviocytes into immune effector cells. Cell Microbiol *21*, e12992. https://doi.org/10.1111/cmi.12992

Londono, D., Cadavid, D., Drouin, E.E., Strle, K., McHugh, G., Aversa, J.M., and Steere, A.C. (2014). Antibodies to endothelial cell growth factor and obliterative microvascular lesions in the synovium of patients with antibiotic-refractory lyme arthritis. Arthritis Rheumatol *66*, 2124-2133. https://doi.org/10.1002/art.38618

Mac, S., Bahia, S., Simbulan, F., Pullenayegum, E.M., Evans, G.A., Patel, S.N., and Sander, B. (2020). Long-Term Sequelae and Health-Related Quality of Life Associated With Lyme Disease: A Systematic Review. Clin Infect Dis *71*, 440-452. https://doi.org/10.1093/cid/ciz1158

Marques, A., Brown, M.R., and Fleisher, T.A. (2009a). Natural killer cell counts are not different between patients with post-Lyme disease syndrome and controls. Clin Vaccine Immunol *16*, 1249-1250. https://doi.org/10.1128/CVI.00167-09

Marques, A., Shaw, P., Schmid, C.H., Steere, A., Kaplan, R.F., Hassett, A., Shapiro, E., and Wormser, G.P. (2009b). Re: A randomized, placebo-controlled trial of repeated IV antibiotic therapy for Lyme encephalopathy. Prolonged Lyme disease treatment: enough is enough. Neurology *72*, 383-384; author reply 384. https://doi.org/10.1212/01.wnl.0000343855.45704.7c

Marques, A., Telford, S.R., 3rd, Turk, S.P., Chung, E., Williams, C., Dardick, K., Krause, P.J., Brandeburg, C., Crowder, C.D., Carolan, H.E., *et al.* (2014). Xenodiagnosis to detect *Borrelia burgdorferi* infection: a first-in-human study. Clin Infect Dis *58*, 937-945. https://doi.org/10.1093/cid/cit939

Marshall, E. (1999). Lyme disease. Patients scarce in test of long-term therapy. Science *283*, 1431. https://doi.org/10.1126/science.283.5407.1431

McAulay, K., Saito, K., Warrier, T., Walsh, K.F., Mathurin, L.D., Royal-Mardi, G., Lee, M.H., Ocheretina, O., Pape, J.W., Fitzgerald, D.W., *et al.* (2018). Differentially Detectable Mycobacterium tuberculosis Cells in Sputum from Treatment-Naive Subjects in Haiti and Their Proportionate Increase after Initiation of Treatment. mBio *9*. https://doi.org/10.1128/mBio.02192-18

Melia, M.T., and Auwaerter, P.G. (2016). Time for a Different Approach to Lyme Disease and Long-Term Symptoms. N Engl J Med *374*, 1277-1278. https://doi.org/10.1056/NEJMe1502350

Merilainen, L., Herranen, A., Schwarzbach, A., and Gilbert, L. (2015). Morphological and biochemical features of *Borrelia burgdorferi* pleomorphic forms. Microbiology *161*, 516-527. https://doi.org/10.1099/mic.0.000027

Murgia, R., and Cinco, M. (2004). Induction of cystic forms by different stress conditions in *Borrelia burgdorferi*. APMIS *112*, 57-62. https://doi.org/10.1111/j.1600-0463.2004.apm1120110.x

Murgia, R., Piazzetta, C., and Cinco, M. (2002). Cystic forms of *Borrelia burgdorferi* sensu lato: induction, development, and the role of RpoS. Wien Klin Wochenschr *114*, 574-579.

Mursic, V.P., Wanner, G., Reinhardt, S., Wilske, B., Busch, U., and Marget, W. (1996). Formation and cultivation of *Borrelia burgdorferi* spheroplast-L-form variants. Infection *24*, 218-226. https://doi.org/10.1007/BF01781096

Nguyen, D., Joshi-Datar, A., Lepine, F., Bauerle, E., Olakanmi, O., Beer, K., McKay, G., Siehnel, R., Schafhauser, J., Wang, Y., *et al.* (2011). Active starvation responses mediate antibiotic tolerance in biofilms and nutrient-limited bacteria. Science *334*, 982-986. https://doi.org/10.1126/science.1211037

Nimnuan, C., Hotopf, M., and Wessely, S. (2001). Medically unexplained symptoms: an epidemiological study in seven specialities. J Psychosom Res *51*, 361-367. https://doi.org/10.1016/s0022-3999(01)00223-9

Nizic, T., Velikanje, E., Ruzic-Sabljic, E., and Arnez, M. (2012). Solitary erythema migrans in children: comparison of treatment with clarithromycin and amoxicillin. Wien Klin Wochenschr *124*, 427-433. https://doi.org/10.1007/s00508-012-0194-1

Nowakowski, J., Nadelman, R.B., Sell, R., McKenna, D., Cavaliere, L.F., Holmgren, D., Gaidici, A., and Wormser, G.P. (2003). Long-term follow-up of patients with culture-confirmed Lyme disease. Am J Med *115*, 91-96. https://doi.org/10.1016/s0002-9343(03)00308-5

Oksi, J., Nikoskelainen, J., Hiekkanen, H., Lauhio, A., Peltomaa, M., Pitkaranta, A., Nyman, D., Granlund, H., Carlsson, S.A., Seppala, I., *et al.* (2007). Duration of antibiotic treatment in disseminated Lyme borreliosis: a double-blind, randomized, placebo-controlled, multicenter clinical study. Eur J Clin Microbiol Infect Dis *26*, 571-581. https://doi.org/10.1007/s10096-007-0340-2

Patrick, D.M., Miller, R.R., Gardy, J.L., Parker, S.M., Morshed, M.G., Steiner, T.S., Singer, J., Shojania, K., Tang, P., and Complex Chronic Disease Study, G. (2015). Lyme Disease Diagnosed by Alternative Methods: A Phenotype Similar to That of Chronic Fatigue Syndrome. Clin Infect Dis *61*, 1084-1091. https://doi.org/10.1093/cid/civ470

Persson, R., Wensaas, K.A., Hanevik, K., Eide, G.E., Langeland, N., and Rortveit, G. (2015). The relationship between irritable bowel syndrome, functional dyspepsia, chronic fatigue and overactive bladder syndrome: a controlled study 6 years after acute gastrointestinal infection. BMC Gastroenterol *15*, 66. https://doi.org/10.1186/s12876-015-0296-0

Pianta, A., Drouin, E.E., Crowley, J.T., Arvikar, S., Strle, K., Costello, C.E., and Steere, A.C. (2015). Annexin A2 is a target of autoimmune T and B cell responses associated with synovial fibroblast proliferation in patients with antibiotic-refractory Lyme arthritis. Clin Immunol *160*, 336-341. https://doi.org/10.1016/j.clim.2015.07.005

Pothineni, V.R., Wagh, D., Babar, M.M., Inayathullah, M., Solow-Cordero, D., Kim, K.M., Samineni, A.V., Parekh, M.B., Tayebi, L., and Rajadas, J. (2016). Identification of new drug candidates against *Borrelia burgdorferi* using high-throughput screening. Drug Des Devel Ther *10*, 1307-1322. https://doi.org/10.2147/DDDT.S101486

Ramsey, A.H., Belongia, E.A., Gale, C.M., and Davis, J.P. (2002). Outcomes of treated human granulocytic ehrlichiosis cases. Emerg Infect Dis *8*, 398-401. https://doi.org/10.3201/eid0804.010222

Reid, M.C., Schoen, R.T., Evans, J., Rosenberg, J.C., and Horwitz, R.I. (1998). The consequences of overdiagnosis and overtreatment of Lyme disease: an observational study. Ann Intern Med *128*, 354-362. https://doi.org/10.7326/0003-4819-128-5-199803010-00003

Rohrbeck, J., Jordan, K., and Croft, P. (2007). The frequency and characteristics of chronic widespread pain in general practice: a case-control study. Br J Gen Pract *57*, 109-115.

Rose, C.D., Fawcett, P.T., Gibney, K.M., and Doughty, R.A. (1994). The overdiagnosis of Lyme disease in children residing in an endemic area. Clin Pediatr *33*, 663-668.

Sapi, E., Bastian, S.L., Mpoy, C.M., Scott, S., Rattelle, A., Pabbati, N., Poruri, A., Burugu, D., Theophilus, P.A., Pham, T.V., *et al.* (2012). Characterization of biofilm formation by *Borrelia burgdorferi* in vitro. PLoS One *7*, e48277. https://doi.org/10.1371/journal.pone.0048277

Sapi, E., Kasliwala, R.S., Ismail, H., Torres, J.P., Oldakowski, M., Markland, S., Gaur, G., Melillo, A., Eisendle, K., Liegner, K.B., *et al.* (2019). The Long-Term Persistence of *Borrelia burgdorferi* Antigens and DNA in the Tissues of a Patient with Lyme Disease. Antibiotics (Basel) *8*. https://doi.org/10.3390/antibiotics8040183

Sapi, E., Theophilus, P.A., Pham, T.V., Burugu, D., and Luecke, D.F. (2016). Effect of RpoN, RpoS and LuxS Pathways on the Biofilm Formation and Antibiotic Sensitivity of *Borrelia Burgdorferi*. Eur J Microbiol Immunol (Bp) *6*, 272-286. https://doi.org/10.1556/1886.2016.00026

Sharma, B., Brown, A.V., Matluck, N.E., Hu, L.T., and Lewis, K. (2015). *Borrelia burgdorferi*, the Causative Agent of Lyme Disease, Forms Drug-Tolerant Persister Cells. Antimicrob Agents Chemother *59*, 4616-4624. https://doi.org/10.1128/AAC.00864-15

Shen, S., Shin, J.J., Strle, K., McHugh, G., Li, X., Glickstein, L.J., Drouin, E.E., and Steere, A.C. (2010). Treg cell numbers and function in patients with antibiotic-refractory or antibiotic-responsive Lyme arthritis. Arthritis Rheum *62*, 2127-2137. https://doi.org/10.1002/art.27468

Shor, S., Green, C., Szantyr, B., Phillips, S., Liegner, K., Burrascano, J.J., Jr., Bransfield, R., and Maloney, E.L. (2019). Chronic Lyme Disease: An Evidence-Based Definition by the ILADS Working Group. Antibiotics (Basel) *8*. https://doi.org/10.3390/antibiotics8040269

Sigal, L.H. (1990). Summary of the first 100 patients seen at a Lyme disease referral center. Am J Med *88*, 577-581. https://doi.org/10.1016/0002-9343(90)90520-n

Skogman, B.H., Croner, S., Nordwall, M., Eknefelt, M., Ernerudh, J., and Forsberg, P. (2008). Lyme neuroborreliosis in children: a prospective study of clinical features, prognosis, and outcome. Pediatr Infect Dis J *27*, 1089-1094. https://doi.org/10.1097/INF.0b013e31817fd423

Smith, R.P., Schoen, R.T., Rahn, D.W., Sikand, V.K., Nowakowski, J., Parenti, D.L., Holman, M.S., Persing, D.H., and Steere, A.C. (2002). Clinical characteristics and treatment outcome of early Lyme disease in patients with microbiologically confirmed erythema migrans. Ann Intern Med *136*, 421-428. https://doi.org/10.7326/0003-4819-136-6-200203190-00005

Soloski, M.J., Crowder, L.A., Lahey, L.J., Wagner, C.A., Robinson, W.H., and Aucott, J.N. (2014). Serum inflammatory mediators as markers of human Lyme disease activity. PLoS One *9*, e93243. https://doi.org/10.1371/journal.pone.0093243

Srivastava, S.Y., and de Silva, A.M. (2009). Characterization of *Borrelia burgdorferi* aggregates. Vector Borne Zoonotic Dis *9*, 323-329. https://doi.org/10.1089/vbz.2008.0148

Steere, A.C. (2019). Treatment of Lyme Arthritis. J Rheumatol *46*, 871-873. https://doi.org/10.3899/jrheum.190320

Steere, A.C. (2020). Posttreatment Lyme disease syndromes: distinct pathogenesis caused by maladaptive host responses. J Clin Invest *130*, 2148-2151. https://doi.org/10.1172/JCI138062

Steere, A.C., Drouin, E.E., and Glickstein, L.J. (2011). Relationship between immunity to *Borrelia burgdorferi* outer-surface protein A (OspA) and Lyme arthritis. Clin Infect Dis *52 Suppl 3*, s259-265. https://doi.org/10.1093/cid/ciq117

Steere, A.C., Klitz, W., Drouin, E.E., Falk, B.A., Kwok, W.W., Nepom, G.T., and Baxter-Lowe, L.A. (2006). Antibiotic-refractory Lyme arthritis is associated with HLA-DR molecules that bind a *Borrelia burgdorferi* peptide. J Exp Med *203*, 961-971. https://doi.org/10.1084/jem.20052471

Steere, A.C., Taylor, E., McHugh, G.L., and Logigian, E.L. (1993). The overdiagnosis of Lyme disease. JAMA *269*, 1812-1816.

Straubinger, R.K., Straubinger, A.F., Summers, B.A., and Jacobson, R.H. (2000). Status of *Borrelia burgdorferi* infection after antibiotic treatment and the effects of corticosteroids: An experimental study. J Infect Dis *181*, 1069-1081. https://doi.org/10.1086/315340

Straubinger, R.K., Summers, B.A., Chang, Y.F., and Appel, M.J. (1997). Persistence of *Borrelia burgdorferi* in experimentally infected dogs after antibiotic treatment. J Clin Microbiol *35*, 111-116. https://doi.org/10.1128/JCM.35.1.111-116.1997

Stricker, R.B., Green, C.L., Savely, V.R., Chamallas, S.N., and Johnson, L. (2010). Safety of intravenous antibiotic therapy in patients referred for treatment of neurologic Lyme disease. Minerva Med *101*, 1-7.

Strle, K., Jones, K.L., Drouin, E.E., Li, X., and Steere, A.C. (2011). *Borrelia burgdorferi* RST1 (OspC type A) genotype is associated with greater inflammation and more severe Lyme disease. Am J Pathol *178*, 2726-2739. https://doi.org/10.1016/j.ajpath.2011.02.018

Strle, K., Shin, J.J., Glickstein, L.J., and Steere, A.C. (2012). Association of a Toll-like receptor 1 polymorphism with heightened Th1 inflammatory responses and antibiotic-refractory Lyme arthritis. Arthritis Rheum *64*, 1497-1507. https://doi.org/10.1002/art.34383

Strle, K., Stupica, D., Drouin, E.E., Steere, A.C., and Strle, F. (2014). Elevated levels of IL-23 in a subset of patients with post-lyme disease symptoms following erythema migrans. Clin Infect Dis *58*, 372-380. https://doi.org/10.1093/cid/cit735

Strle, K., Sulka, K.B., Pianta, A., Crowley, J.T., Arvikar, S.L., Anselmo, A., Sadreyev, R., and Steere, A.C. (2017). T-Helper 17 Cell Cytokine Responses in Lyme Disease Correlate With *Borrelia burgdorferi* Antibodies During Early Infection and With Autoantibodies Late in the Illness in Patients With Antibiotic-Refractory Lyme Arthritis. Clin Infect Dis *64*, 930-938. https://doi.org/10.1093/cid/cix002

Stupica, D., Lusa, L., Ruzic-Sabljic, E., Cerar, T., and Strle, F. (2012). Treatment of erythema migrans with doxycycline for 10 days versus 15 days. Clin Infect Dis *55*, 343-350. https://doi.org/10.1093/cid/cis402

Stupica, D., Maraspin, V., Bogovic, P., Ogrinc, K., Blagus, R., Cerar, T., and Strle, F. (2018a). Comparison of Clinical Course and Treatment Outcome for Patients With Early Disseminated or Early Localized Lyme Borreliosis. JAMA Dermatol *154*, 1050-1056. https://doi.org/10.1001/jamadermatol.2018.2306

Stupica, D., Veluscek, M., Blagus, R., Bogovic, P., Rojko, T., Cerar, T., and Strle, F. (2018b). Oral doxycycline versus intravenous ceftriaxone for treatment of multiple erythema migrans: an open-label alternate-treatment observational trial. J Antimicrob Chemother *73*, 1352-1358. https://doi.org/10.1093/jac/dkx534

Sulka, K.B., Strle, K., Crowley, J.T., Lochhead, R.B., Anthony, R., and Steere, A.C. (2018). Correlation of Lyme Disease-Associated IgG4 Autoantibodies With Synovial Pathology in Antibiotic-Refractory Lyme Arthritis. Arthritis Rheumatol *70*, 1835-1846. https://doi.org/10.1002/art.40566

Tang, K.S., Klempner, M.S., Wormser, G.P., Marques, A.R., and Alaedini, A. (2015). Association of immune response to endothelial cell growth factor with early disseminated and late manifestations of Lyme disease but not posttreatment Lyme disease syndrome. Clin Infect Dis *61*, 1703-1706. https://doi.org/10.1093/cid/civ638

Tory, H.O., Zurakowski, D., and Sundel, R.P. (2010). Outcomes of children treated for Lyme arthritis: results of a large pediatric cohort. J Rheumatol *37*, 1049-1055. https://doi.org/10.3899/jrheum.090711

Trollmo, C., Meyer, A.L., Steere, A.C., Hafler, D.A., and Huber, B.T. (2001). Molecular mimicry in Lyme arthritis demonstrated at the single cell level: LFA-1 alpha L is a partial agonist for outer surface protein A-reactive T cells. J Immunol *166*, 5286-5291. https://doi.org/10.4049/jimmunol.166.8.5286

Turk, S.P., Lumbard, K., Liepshutz, K., Williams, C., Hu, L., Dardick, K., Wormser, G.P., Norville, J., Scavarda, C., McKenna, D., *et al.* (2019). Post-treatment Lyme disease symptoms score: Developing a new tool for research. PLoS One *14*, e0225012. https://doi.org/10.1371/journal.pone.0225012

van Westrienen, P.E., Pisters, M.F., Toonders, S.A.J., Gerrits, M., de Wit, N.J., and Veenhof, C. (2020). Quality of life in primary care patients with moderate medically unexplained physical symptoms. Qual Life Res *29*, 693-703. https://doi.org/10.1007/s11136-019-02358-8

van Westrienen, P.E., Pisters, M.F., Veenhof, C., and de Wit, N.J. (2019). Identification of patients with moderate medically unexplained physical symptoms in primary care with a five years follow-up. BMC Fam Pract *20*, 66. https://doi.org/10.1186/s12875-019-0950-7

Volzing, K.G., and Brynildsen, M.P. (2015). Stationary-Phase Persisters to Ofloxacin Sustain DNA Damage and Require Repair Systems Only during Recovery. mBio *6*, e00731-00715. https://doi.org/10.1128/mBio.00731-15

Vudattu, N.K., Strle, K., Steere, A.C., and Drouin, E.E. (2013). Dysregulation of CD4+CD25(high) T cells in the synovial fluid of patients with antibiotic-refractory Lyme arthritis. Arthritis Rheum *65*, 1643-1653. https://doi.org/10.1002/art.37910

Wakamoto, Y., Dhar, N., Chait, R., Schneider, K., Signorino-Gelo, F., Leibler, S., and McKinney, J.D. (2013). Dynamic persistence of antibiotic-stressed mycobacteria. Science *339*, 91-95. https://doi.org/10.1126/science.1229858

Wang, G., Ojaimi, C., Iyer, R., Saksenberg, V., McClain, S.A., Wormser, G.P., and Schwartz, I. (2001). Impact of genotypic variation of *Borrelia burgdorferi* sensu stricto on kinetics of dissemination and severity of disease in C3H/HeJ mice. Infect Immun *69*, 4303-4312. https://doi.org/10.1128/IAI.69.7.4303-4312.2001

Wang, G., Ojaimi, C., Wu, H., Saksenberg, V., Iyer, R., Liveris, D., McClain, S.A., Wormser, G.P., and Schwartz, I. (2002). Disease severity in a murine model of lyme borreliosis is associated with the genotype of the infecting *Borrelia burgdorferi* sensu stricto strain. J Infect Dis *186*, 782-791. https://doi.org/10.1086/343043

Wang, T.J., Liang, M.H., Sangha, O., Phillips, C.B., Lew, R.A., Wright, E.A., Berardi, V., Fossel, A.H., and Shadick, N.A. (2000). Coexposure to *Borrelia burgdorferi* and *Babesia microti* does not worsen the long-term outcome of lyme disease. Clin Infect Dis *31*, 1149-1154. https://doi.org/10.1086/317465

Wessely, S., Nimnuan, C., and Sharpe, M. (1999). Functional somatic syndromes: one or many? Lancet *354*, 936-939. https://doi.org/10.1016/S0140-6736(98)08320-2

Wills, A.B., Spaulding, A.B., Adjemian, J., Prevots, D.R., Turk, S.P., Williams, C., and Marques, A. (2016). Long-term Follow-up of Patients With Lyme Disease: Longitudinal Analysis of Clinical and Quality-of-life Measures. Clin Infect Dis *62*, 1546-1551. https://doi.org/10.1093/cid/ciw189

Wilmaerts, D., Windels, E.M., Verstraeten, N., and Michiels, J. (2019). General Mechanisms Leading to Persister Formation and Awakening. Trends Genet *35*, 401-411. https://doi.org/10.1016/j.tig.2019.03.007

Wormser, G.P. (2016). Longer-Term Therapy for Symptoms Attributed to Lyme Disease. N Engl J Med *375*, 997. https://doi.org/10.1056/NEJMc1608044

Wormser, G.P., Baker, P.J., O'Connell, S., Pachner, A.R., Schwartz, I., and Shapiro, E.D. (2012a). Critical analysis of treatment trials of rhesus macaques infected with *Borrelia burgdorferi* reveals important flaws in experimental design. Vector Borne Zoonotic Dis *12*, 535-538. https://doi.org/10.1089/vbz.2012.1012

Wormser, G.P., Brisson, D., Liveris, D., Hanincova, K., Sandigursky, S., Nowakowski, J., Nadelman, R.B., Ludin, S., and Schwartz, I. (2008a). *Borrelia burgdorferi* genotype predicts the capacity for hematogenous dissemination during early Lyme disease. J Infect Dis *198*, 1358-1364. https://doi.org/10.1086/592279

Wormser, G.P., Dattwyler, R.J., Shapiro, E.D., Halperin, J.J., Steere, A.C., Klempner, M.S., Krause, P.J., Bakken, J.S., Strle, F., Stanek, G., *et al.* (2006). The clinical assessment, treatment, and prevention of lyme disease, human granulocytic anaplasmosis, and babesiosis: clinical practice guidelines by the Infectious Diseases Society of America. Clin Infect Dis *43*, 1089-1134. https://doi.org/10.1086/508667

Wormser, G.P., Liveris, D., Nowakowski, J., Nadelman, R.B., Cavaliere, L.F., McKenna, D., Holmgren, D., and Schwartz, I. (1999). Association of specific subtypes of *Borrelia burgdorferi* with hematogenous dissemination in early Lyme disease. J Infect Dis *180*, 720-725. https://doi.org/10.1086/314922

Wormser, G.P., McKenna, D., Carlin, J., Nadelman, R.B., Cavaliere, L.F., Holmgren, D., Byrne, D.W., and Nowakowski, J. (2005). Brief communication: hematogenous dissemination in early Lyme disease. Ann Intern Med *142*, 751-755. https://doi.org/10.7326/0003-4819-142-9-200505030-00011

Wormser, G.P., McKenna, D., Karmen, C.L., Shaffer, K.D., Silverman, J.H., Nowakowski, J., Scavarda, C., Shapiro, E.D., and Visintainer, P. (2020). Prospective Evaluation of the Frequency and Severity of Symptoms in Lyme Disease Patients With Erythema Migrans Compared With Matched Controls at Baseline, 6 Months, and 12 Months. Clin Infect Dis. https://doi.org/10.1093/cid/ciz1215

Wormser, G.P., Nadelman, R.B., and Schwartz, I. (2012b). The amber theory of Lyme arthritis: initial description and clinical implications. Clin Rheumatol *31*, 989-994. https://doi.org/10.1007/s10067-012-1964-x

Wormser, G.P., Nowakowski, J., Nadelman, R.B., Visintainer, P., Levin, A., and Aguero-Rosenfeld, M.E. (2008b). Impact of clinical variables on *Borrelia burgdorferi*-specific antibody seropositivity in acute-phase sera from patients in North America with culture-confirmed early Lyme disease. Clin Vaccine Immunol *15*, 1519-1522. https://doi.org/10.1128/CVI.00109-08

Wormser, G.P., O'Connell, S., Pachner, A.R., Schwartz, I., Shapiro, E.D., Stanek, G., and Strle, F. (2018). Critical analysis of a doxycycline treatment trial of rhesus macaques infected with *Borrelia burgdorferi*. Diagn Microbiol Infect Dis *92*, 183-188. https://doi.org/10.1016/j.diagmicrobio.2018.06.007

Wormser, G.P., Ramanathan, R., Nowakowski, J., McKenna, D., Holmgren, D., Visintainer, P., Dornbush, R., Singh, B., and Nadelman, R.B. (2003). Duration of antibiotic therapy for early Lyme disease. A randomized, double-blind, placebo-controlled trial. Ann Intern Med *138*, 697-704. https://doi.org/10.7326/0003-4819-138-9-200305060-00005

Wormser, G.P., and Schwartz, I. (2009). Antibiotic treatment of animals infected with *Borrelia burgdorferi*. Clin Microbiol Rev *22*, 387-395. https://doi.org/10.1128/CMR.00004-09

Wormser, G.P., Weitzner, E., McKenna, D., Nadelman, R.B., Scavarda, C., Molla, I., Dornbush, R., Visintainer, P., and Nowakowski, J. (2015a). Long-term assessment of health-related quality of life in patients with culture-confirmed early Lyme disease. Clin Infect Dis *61*, 244-247. https://doi.org/10.1093/cid/civ277

Wormser, G.P., Weitzner, E., McKenna, D., Nadelman, R.B., Scavarda, C., and Nowakowski, J. (2015b). Long-term assessment of fatigue in patients with culture-confirmed Lyme disease. Am J Med *128*, 181-184. https://doi.org/10.1016/j.amjmed.2014.09.022

Wu, X., Sharma, B., Niles, S., O'Connor, K., Schilling, R., Matluck, N., D'Onofrio, A., Hu, L.T., and Lewis, K. (2018). Identifying Vancomycin as an Effective Antibiotic for Killing Borrelia burgdorferi. Antimicrob Agents Chemother *62*. https://doi.org/10.1128/AAC.01201-18

Yrjanainen, H., Hytonen, J., Song, X.Y., Oksi, J., Hartiala, K., and Viljanen, M.K. (2007). Anti-tumor necrosis factor-alpha treatment activates *Borrelia burgdorferi* spirochetes 4 weeks after ceftriaxone treatment in C3H/He mice. J Infect Dis *195*, 1489-1496. https://doi.org/10.1086/513873

Chapter 24

Lyme Disease in Humans

Justin D. Radolf[1-5]*, Klemen Strle[6], Jacob E. Lemieux[7] and Franc Strle[8]

Departments of [1]Medicine, [2]Pediatrics, [3]Genetics and Genome Sciences, [4]Molecular Biology and Biophysics, and [5]Immunology, UConn Health, Farmington, CT 06030, USA
[6]Division of Infectious Diseases, Wadsworth Center, NY Department of Health, Albany NY, 12208, USA
[7]Department of Medicine, Massachusetts General Hospital, Harvard Medical School, Boston, MA 02114, USA
[8]Department of Infectious Diseases, University Medical Center Ljubljana, Ljubljana, Slovenia
*Corresponding author: JRadolf@uchc.edu

DOI: https://doi.org/10.21775/9781913652616.24

Abstract
Lyme disease (Lyme borreliosis) is a tick-borne, zoonosis of adults and children caused by genospecies of the *Borrelia burgdorferi* sensu lato complex. The ailment, widespread throughout the Northern Hemisphere, continues to increase globally due to multiple environmental factors, coupled with increased incursion of humans into habitats that harbor the spirochete. *B. burgdorferi* sensu lato is transmitted by ticks from the *Ixodes ricinus* complex. In North America, *B. burgdorferi* causes nearly all infections; in Europe, *B. afzelii* and *B. garinii* are most associated with human disease. The spirochete's unusual fragmented genome encodes a plethora of differentially expressed outer surface lipoproteins that play a seminal role in the bacterium's ability to sustain itself within its enzootic cycle and cause disease when transmitted to its incidental human host. Tissue damage and symptomatology (*i.e.*, clinical manifestations) result from the inflammatory response elicited by the bacterium and its constituents. The deposition of spirochetes into human dermal tissue generates a local inflammatory response that manifests as erythema migrans (EM), the hallmark skin lesion. If treated appropriately and early, the prognosis is excellent. However, in untreated patients, the disease may present with a wide range of clinical manifestations, most commonly involving the central nervous system, joints, or heart. A small percentage (~10%) of patients may go on to develop a poorly defined fibromyalgia-like illness, post-treatment Lyme disease (PTLD) unresponsive to prolonged antimicrobial therapy. Below we integrate current knowledge regarding the ecologic, epidemiologic, microbiologic, and immunologic facets of Lyme disease into a conceptual framework that sheds light on the disorder that healthcare providers encounter.

Introduction

Lyme disease is the prototype of an emerging infectious disease (Steere et al., 2004; Paules et al., 2018). The isolation of its etiologic agent, *Borrelia burgdorferi*, from humans in 1983 (Benach et al., 1983; Steere et al., 1983a; Barbour and Benach, 2019) capped an intensive hunt for a pathogen that just a short time before had been cultured from a black legged (deer) tick (Burgdorfer et al., 1982), initially named *Ixodes dammini* (Spielman et al., 1979) but subsequently found to belong to a species, *I. scapularis*, whose range had been expanding in the U.S. since it was first recognized since the 1920s (Burgdorfer and Gage, 1986; Eisen and Eisen, 2018). Critical to the chain of events that led to the discovery of the Lyme disease spirochete was the observation that many patients involved in an outbreak of oligoarthritis in Southeastern Connecticut also had a skin rash, erythema chronicum migrans (ECM; now erythema migrans, EM) (Steere et al., 1977a; Steere et al., 1977b), previously associated in Europe with the bite of the sheep tick *Ixodes ricinus* (Afzelius, 1910; Lipschütz, 1913; Afzelius, 1921; Lipschütz, 1923). The isolation of *B. burgdorferi* (Benach et al., 1983; Steere et al., 1983a; Barbour and Benach, 2019) sparked an explosive increase in our knowledge of the bacterium, the disease it causes, and the enzootic cycle that sustains and creates risk to humans who intrude upon it. We now know that Lyme disease (Lyme borreliosis) is the most prevalent tick-borne illness in the Palearctic region of the Northern Hemisphere and that its incidence continues to increase globally due to myriad demographic and environmental factors, including climate change (Mead, 2015; Ostfeld and Brunner, 2015; Schotthoefer and Frost, 2015; Semenza and Suk, 2018; Sharareh et al., 2019). Although the clinical manifestations of Lyme disease continue to be a source of considerable controversy, it is generally accepted that a relatively small number of syndromes dominate the clinical picture and that the vast majority of patients present with treatment-responsive acute illness (Steere et al., 2016; Stanek and Strle, 2018). Serologic surveys conducted in high prevalence areas indicate that asymptomatic infection also is relatively common (Hanrahan et al., 1984; Steere et al., 2003; Wilhelmsson et al., 2016; Carlsson et al., 2018); thus, despite the bacterium's notorious reputation, benign outcomes often occur. The genomic sequence of *B. burgdorferi* revealed that the spirochete lacks genes encoding known toxigenic molecules as well as the secretory apparatus required to deliver them to the extracellular milieu it inhabits within its mammalian host (Fraser et al., 1997; Casjens et al., 2000). Whereas reservoir hosts are unaffected by lifelong infection with Lyme disease spirochetes (Oliver et al., 2003; Hersh et al., 2014) due to a poorly understood form of immunologic tolerance (Barbour, 2017), infected humans often mount local and systemic inflammatory responses that make them ill (Steere et al., 2016; Stanek and Strle, 2018). From this perspective, one can regard clinical Lyme disease in humans as an evolutionary "mismatch" between pathogen and the intolerant immune system of its incidental host. Beyond this reductionist view, however, we still have only a limited understanding of the microbial factors, pathogenic mechanisms, and immunologic responses that determine outcomes following the adventitious encounter of humans with this zoonotic microorganism.

Over the years, a number of excellent clinical reviews of Lyme disease have been published in journals and textbooks, and there is no need to reiterate all of this information herein (Stanek et al., 2012; Steere et al., 2016; Stanek and Strle, 2018). In addition, medical societies in both the United States and Europe have issued comprehensive guidelines for the diagnosis and management of this infection (Wormser et al., 2006; Halperin et al., 2007; Eldin et al., 2019). Rather, the primary objective of this chapter is to integrate current knowledge regarding the ecologic, epidemiologic, microbiologic, and immunologic facets of Lyme disease into a conceptual framework that sheds light on the disorder practitioners see and manage. As will be seen, many of the principal factors that determine the level of risk for populations and individuals lie outside the sphere of human activity. Since our goal is to develop a mechanistic picture of the human disorder, we intend to rely as much as possible upon data obtained from human studies; extrapolation from *in vitro* and animal models is necessary, indeed, unavoidable, given the constraints of human experimentation. At the same time, this chapter attempts to grapple with a vexing but fundamental issue—the extent to which infection in humans deviates from the infectious process in nature and that observed in experimental animal models.

Historical overview

The history of what we now call Lyme disease, dating back to the early part of the twentieth century, is instructive for contemporary understanding (Burgdorfer, 1986, 1993). However, the discoveries that ushered in our current understanding of the illness began in 1981, when Willy Burgdorfer, Jorge Benach and Alan Barbour identified a new *Borrelia* species from ticks collected on eastern Long Island (Burgdorfer et al., 1982; Barbour and Benach, 2019). As Burgdorfer (Burgdorfer, 1993) and, most recently, Barbour and Benach (Barbour and Benach, 2019) note in their colorful first-hand accounts, the finding was serendipitous and well illustrates Louis Pasteur's famous dictum that "chance favors only the prepared mind". Burgdorfer and Benach were seeking a vector to explain an outbreak of Rocky Mounted spotted fever on Eastern Long Island but stumbled across spirochetes when they dissected midguts from *Ixodes scapularis* ticks, recently recognized as the vector for *Babesia microti* (Spielman et al., 1979). Burgdorfer was aware that a possible spirochetal etiology for Lyme disease had been "in the air" since the late 1940s. Two other strokes of fortune contributed to the discovery: Benach possessed a bank of sera from convalescent Lyme disease patients, while Barbour, at the time studying relapsing fever at the Rocky Mountain Laboratory, had developed an improved medium for cultivation of *Borrelia*. They found that antibodies in the sera of Lyme disease patients reacted intensely with spirochetes in dissected tick midguts and spirochetes isolated from tick midguts using Barbour's newly formulated (BSKII) medium (Burgdorfer et al., 1982). Within a year, groups separately led by Benach and Allen Steere isolated spirochetes from blood, skin, and cerebrospinal fluid (CSF), thereby establishing it as the etiologic agent (Burgdorfer et al., 1982;

Burgdorfer, 1986). Using DNA-DNA hybridization, two groups (Hyde and Johnson, 1984; Schmid et al., 1984) subsequently showed that the spirochete was a new species of *Borrelia*, subsequently named *B. burgdorferi* (Johnson et al., 1984).

It soon became apparent that a variety of clinical syndromes described in the European medical literature were, in fact, manifestations of a tick-transmitted disorder caused by members of what eventually came to be known as the *B. burgdorferi* sensu lato complex (Belfaiza et al., 1993; Wang et al., 1999a; Cutler et al., 2017). EM, the classic skin lesion associated with this infectious disease, was first described by Afzelius, a Swedish dermatologist, in 1910 (Afzelius, 1910). Afzelius also correctly hypothesized that EM resulted from the tick-borne transmission to humans of a zoonotic pathogen (Afzelius, 1921); in 1921, Lipschütz identified *Ixodes ricinus* as the vector (Lipschütz, 1923). By the beginning of World War II, European investigators knew that EM was associated with several neurologic and dermatologic disorders and with musculoskeletal complaints (Garin, 1922; Bannwarth, 1941, 1944). After the war, Lenhoff (Lenhoff, 1948), a Swedish pathologist, described what he believed were spirochetes in biopsies of EM lesions, while Hollström (Hollstrom, 1951) demonstrated that penicillin was effective for its treatment.

In 1970, Rudolph J. Scrimenti, a Wisconsin dermatologist, reported the first case of Lyme disease acquired in the United States. The patient, by serendipity a physician, was bitten by a tick above his right iliac crest while grouse hunting in North Central Wisconsin. Three months later, he presented with an enormous EM rash extending from his right mid-chest to mid-back, encircling his right axilla and iliac crest, accompanied by hyperesthesia of the T12 and L1 dermatomes. Fortunately, Scrimenti knew of this "curious condition" from the European literature and of its responsiveness to penicillin; incredibly, the patient was symptom-free within 48 h of receiving what by today's standards is considered a miniscule dose (1.2 MU) of intramuscular benzathine penicillin G. In 1976, Mast and Burrows (Mast and Burrows, 1976) reported the first cluster of cases from Southeastern Connecticut. Soon afterwards, Yale rheumatologists Allen Steere and Steven Malawista began investigating cases of arthritis in patients, many of whom were children, in and around Old Lyme, Connecticut. Mothers of afflicted children, skeptical of the diagnosis of juvenile rheumatoid arthritis made by local physicians, had informed the State Health Department about the outbreak and called it to the attention of Steere and Malawista. In their initial reports, they called the mysterious ailment Lyme arthritis (Steere et al., 1977a; Steere et al., 1977b). However, with the realization that most arthritis patients previously had EM and that non-arthritic manifestations (heart block, facial nerve palsy and/or meningitis) were associated with the rash, they subsequently changed the name to "Lyme disease" (Steere and Malawista, 1979). Shortly thereafter, they "closed the loop" by correlating cases of Lyme disease with the distribution of *I. scapularis* in the Northeast and *I. pacificus* in California and Oregon (Steere and Malawista, 1979). Of note, European authorities prefer Lyme borreliosis because, in their view, U.S. patients diagnosed with Lyme disease do not always have a disorder with a clear-cut infectious etiology (Stanek and Strle, 2018). Detailed narratives of the medical sleuthing that led to the discovery of *B. burgdorferi* can be found in chapter one in this volume, in Edlow's entertaining book Bull's Eye, Unraveling the Medical Mystery of Lyme Disease (Erdlow, 2003), and the gripping narrative recently published by Barbour and Benach (Barbour and Benach, 2019).

Epidemiology

Please see chapter 21 for a complete discussion of the epidemiology of Lyme disease. Lyme disease became a notifiable condition in the U.S. in 1991. Since 2008, a confirmed case has been defined as either (i) EM in a person with possible or known tick exposure in an endemic area or laboratory evidence of infection (almost always serological) or (ii) at least one recognized clinical manifestation other than EM along with confirmatory laboratory evidence (Schwartz et al., 2017). From 2008 to 2015, 208,834 confirmed cases were reported to the CDC with the highest number (29,959) in 2009 (Figure 1) (Schwartz et al., 2017). During this period, Lyme disease accounted for 82% of all tick-borne diseases and 63% of all vector-borne disease reported in the US, making it by far the most prevalent vector-borne illness in the United States and *Ixodes scapularis* the most important vector (Hamer et al., 2010; Mead, 2015; Rosenberg et al., 2018). Nationwide studies of health insurance claims (Nelson et al., 2015) and commercial laboratory diagnostic tests (Hinckley et al., 2014) suggest that underreporting is common and that the actual number of cases is closer to 300,000 per year (*i.e.*, about tenfold higher than reported). Fourteen states, all located in the

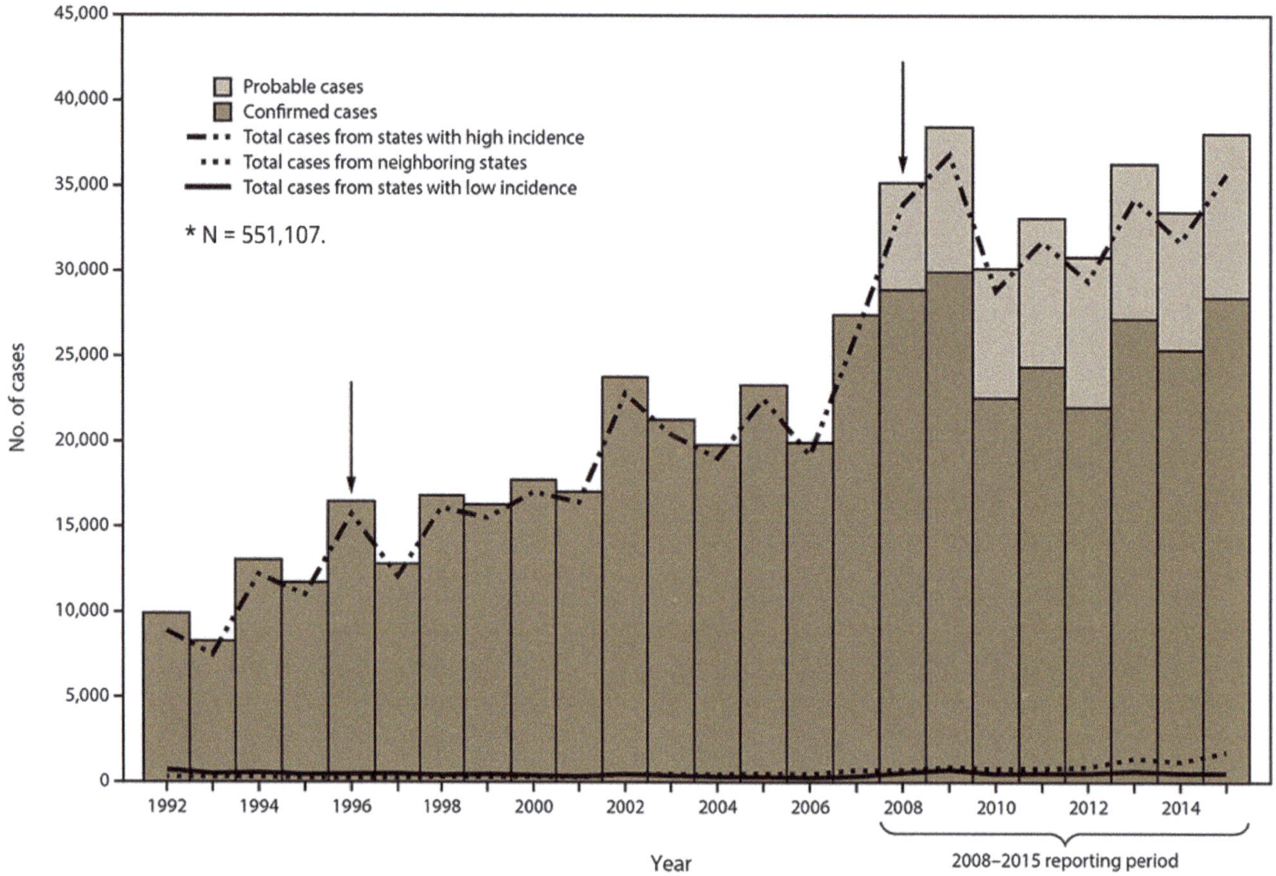

Figure 1. Number of confirmed and probable Lyme disease cases in the United States, 1992-2015. Arrows indicate notable changes in case definitions. The case definition was revised in 1996 to recommend a two-step testing method and in 2008 to increase specificity of laboratory evidence of infection and to include provision for report of probable cases (reproduced from Schwartz *et al.*, 2017).

Northeast, mid-Atlantic, and upper Midwest (Connecticut, Delaware, Maine, Maryland, Massachusetts, Minnesota, New Hampshire, New Jersey, New York, Pennsylvania, Rhode Island, Vermont, Virginia, and Wisconsin), accounted for 95.7% of confirmed cases; Delaware, Connecticut, and Vermont had the highest incidences (~65-69 cases per 100,000 population) (Figure 2) (Schwartz et al., 2017). For all years, confirmed and probable cases peaked during the first week in July, consistent with nymphs being the principal stage for transmission. The age distribution was bimodal, with peaks between 5-9 years and 50-55 years, a slight male predominance (56%), and whites representing the overwhelming majority (~90%) of cases (Schwartz et al., 2017). EM was the most common clinical manifestation, accounting for nearly three-fourths (72.2%) of patients, and carditis least common (1.5%); 27.5% had arthritis and 12.5% had a neurologic manifestation. Serologic surveys in the U.S. and Europe have revealed substantial rates of asymptomatic or subclinical infections among persons living in endemic areas (Hanrahan et al., 1984; Steere et al., 2003; Wilhelmsson et al., 2016; Carlsson et al., 2018). Not surprisingly, risk is proportional to time spent outdoors, whether recreationally or occupationally, in or near tick-infested woods and vegetation (Hengge et al., 2003; Finch et al., 2014).

The last three decades have witnessed not only an impressive increase in the incidence of Lyme disease in North America but also a relentless expansion of its geographic range. Although historically associated with incursion into deciduous forests (Dennis and Hayes, 2002), Lyme disease now poses a threat to

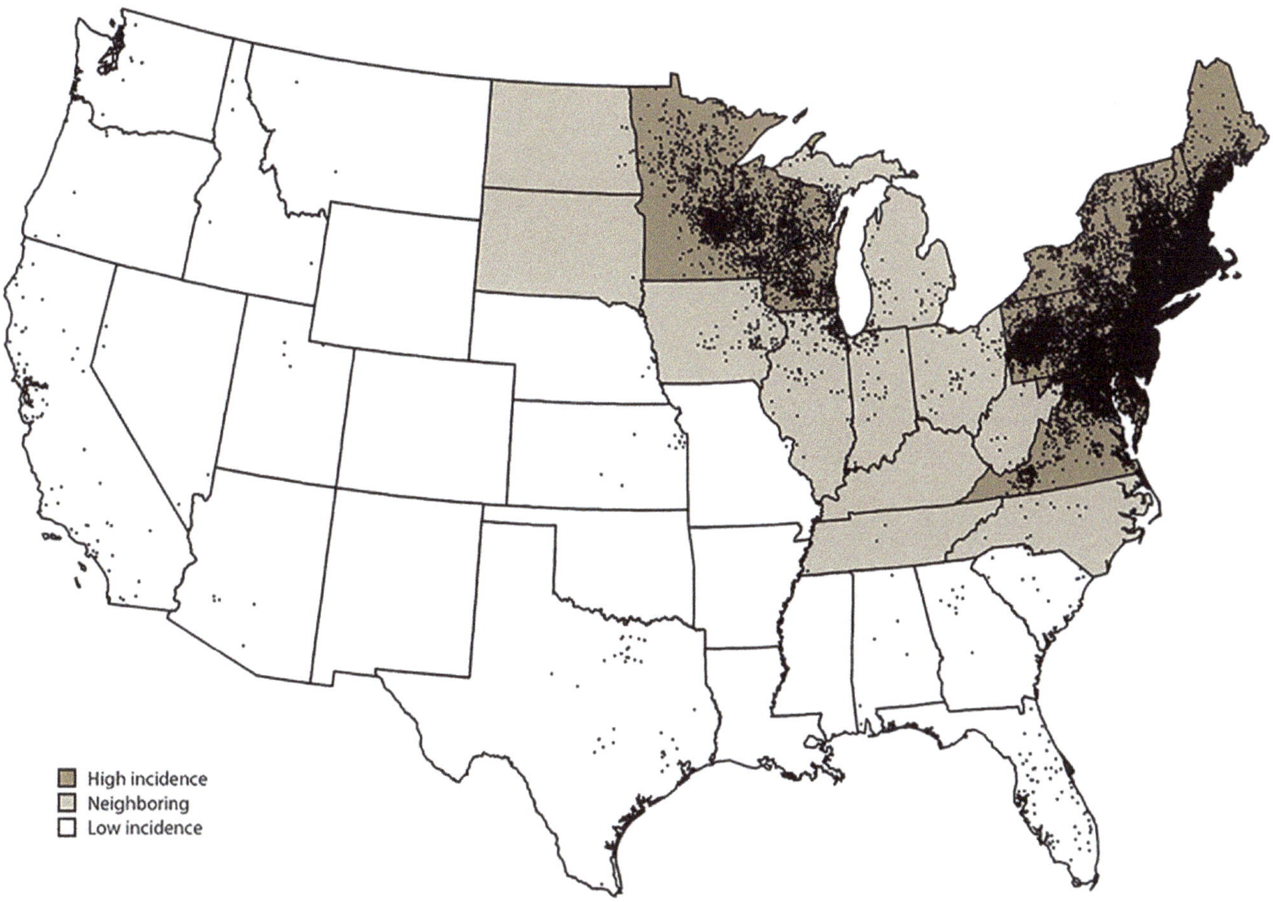

Figure 2. Average annual number of confirmed Lyme disease cases by county of residence in the United States, 2008-2015. Each dot represents one confirmed case (reproduced from Schwartz *et al.*, 2017).

urban dwellers, as evidenced by cases acquired in New York City (VanAcker et al., 2019) and identification of B. *burgdorferi*-infected ticks in Chicago (Hamer et al., 2012). Moreover, the disease has expanded beyond the confines of the continental United States. It has emerged as a health threat in Southern Canada (Ogden et al., 2009; Gasmi et al., 2019), where the number of reported cases increased from 144 in 2005 to more than 2000 in 2017, and multiple surveys have reported identification of *B. burgdorferi*-infected *I. scapularis* ticks (Bouchard et al., 2015; Gasmi et al., 2016; Ogden et al., 2019). Although improved reporting and increased awareness are likely contributory factors (Orloski et al., 1998; Aenishaenslin et al., 2016), there is a strong consensus among entomologists that these epidemiologic trends reflect the collective impact of environmental drivers that increase the likelihood of human encounters with infected ticks (Eisen et al., 2016; Stone et al., 2017). Among the most important of these are (i) climate-mediated expansion of tick habitats (Ostfeld and Brunner, 2015; Dumic and Severnini, 2018), (ii) dispersal of infected *I. scapularis* by migratory birds (Olsen et al., 1995; Brinkerhoff et al., 2011; Hasle et al., 2011), (iii) increased densities of vertebrate reservoirs and deer populations upon which *I. scapularis* feed and mate as lands cleared for agriculture become reforested (Eisen and Eisen, 2018), and (iv) the increased risk of transmission associated with decreased biodiversity in endemic areas (LoGiudice et al., 2003; Granter et al., 2014; Ruyts et al., 2016). In short, one cannot divorce environmental factors fueling the proliferation of *B. burgdorferi* sensu lato in the wild (Gern, 2008) from their cumulative effects on human populations (Schwartz et al., 2017; Rosenberg et al., 2018).

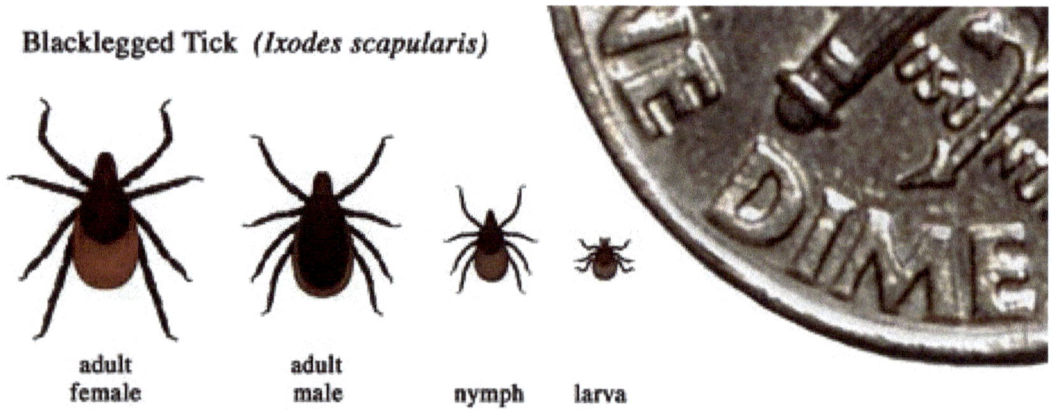

Figure 3. ***Ixodes scapularis*** **stages.**

Lyme disease also is the most prevalent vector-borne illness in Europe, where it is widely, though nonuniformly, distributed (Hubalek, 2009; Rizzoli et al., 2011; Sykes and Makiello, 2017). Remarkably, Lyme disease is not a mandatory notifiable disease in many European countries, complicating country by country comparison of epidemiologic data (van den Wijngaard et al., 2017). Although methods used to acquire surveillance and laboratory data vary greatly (van den Wijngaard et al., 2017), an estimated 85,000 cases occur each year throughout Europe (Sykes and Makiello, 2017). In Europe, as in the U.S., new cases peak in the summer months of June through August (Hubalek, 2009), and underreporting is believed to be common (van den Wijngaard et al., 2017). In Northern Europe, disease rates are highest in the Baltic states and Southern Sweden; in Central Europe, highest incidences are in Austria and Slovenia (Hubalek, 2009; Rizzoli et al., 2011; Sykes and Makiello, 2017). At the southern limits of the disease range (*e.g.*, Italy and the Balkans), incidence decreases rapidly from north to south (Hubalek, 2009). Disease rates across the Continent parallel the densities of *I. ricinus* ticks affected with the pathogenic species most frequently detected in patients, *B. afzelii* and *B. garinii* (Coipan et al., 2016; Strnad et al., 2017; Estrada-Pena et al., 2018). Lyme disease rates are increasing in Europe for the same reasons as in North America—increased awareness (Smith and Takkinen, 2006), coupled with increasing distribution and abundance of *I. ricinus* due to the same environmental drivers, with climate change probably a major culprit (Medlock et al., 2013; Semenza and Suk, 2018).

Ecology

Although ticks capable of vectoring Lyme disease spirochetes often take their blood meals from humans, humans are not required for perpetuation of either ticks or spirochetes in nature. Humans are incidental, presumably "dead-end," hosts that become infected when their lifestyles or activities intersect with habitats harboring spirochetes (Gern, 2009; Radolf et al., 2012; Eisen and Eisen, 2018). Only ticks belonging to the hard tick genus *Ixodes* are vector competent, that is, capable of acquiring and transmitting spirochetes (Lane et al., 1991; Gern, 2009; Eisen and Eisen, 2018). *B. burgdorferi* sensu lato is transmitted mainly by ticks of the *Ixodes ricinus* complex (Burgdorfer et al., 1991; Piesman and Gern, 2004) (also see chapters 10 and 11), *I. scapularis* in the Northeastern and Upper Midwestern United States; *I. pacificus* on the Pacific Coast; *I. ricinus* in Europe, Western Asia, and North Africa; and *I. persulcatus* in Eastern Europe and Asia (Gern, 2009; Mannelli et al., 2012; Franke et al., 2013). *Ixodes* ticks have a two-year life cycle with four life stages: egg, larva, nymph and adult (Figure 3). Ticks are born uninfected. Larvae acquire the spirochete by feeding on an infected reservoir host, and, after molting to the nymphal stage, transmit the pathogen when they feed on an uninfected reservoir or incidental host (Figure 4) (Gern, 2009; Radolf et al., 2012; Eisen and Eisen, 2018). The dependence of *B. burgdorferi* sensu lato on efficient transstadial transmission for long-term survival is an important distinction from relapsing fever spirochetes, which can be maintained within their argasid vectors by vertical or transovarial transmission (Barbour and Hayes, 1986; Rollend et al., 2013) (also see Chapter 10).

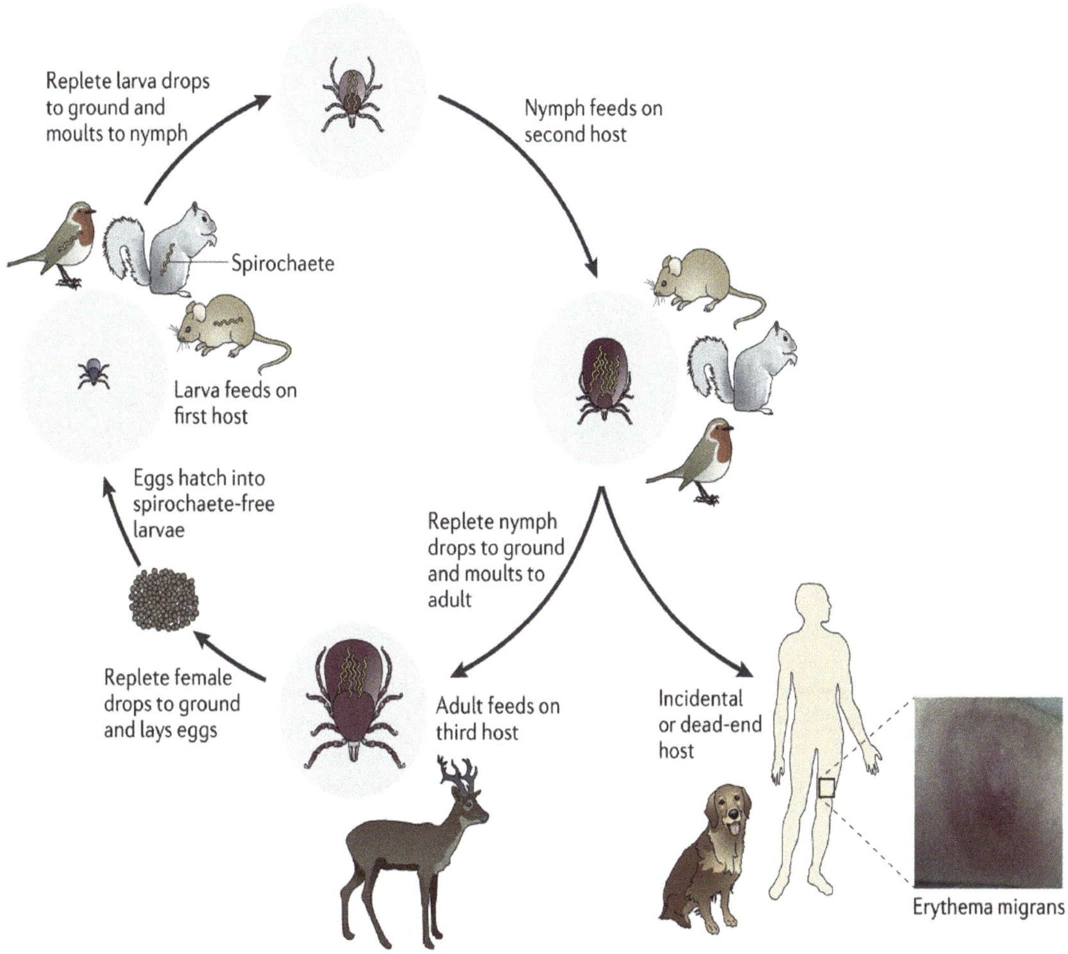

Figure 4. The enzootic cycle of *Borrelia burgdorferi*. *Ixodes* ticks undergo a three-stage life cycle (larva, nymph and adult, with one blood meal per stage). Larval ticks acquire spirochetes by feeding on an infected reservoir animal, and the bacterium is retained during the subsequent stages. Transmission of spirochetes to a competent reservoir host by a feeding nymph perpetuates the enzootic cycle for the next generation of larval ticks. Adult ticks are not important for maintenance of *B. burgdorferi* in the wild; however, deer are important for maintenance of the tick population because adult ticks mate on them. Nymphs are responsible for the vast majority of spirochete transmission to humans, generally considered dead-end hosts, as are dogs. (Reprinted with permission from Radolf *et al.* 2012.)

Nymphal ticks are responsible for the majority of human infections (Piesman et al., 1987a; Falco et al., 1996; Dennis and Hayes, 2002). In addition to having high infection rates, nymphs quest during the summer months when humans are most likely to be outdoors and, because of their small size (approximately that of a poppy seed), are difficult to detect on body surfaces, clothing, and pets (Dennis and Hayes, 2002). Not surprisingly, the risk of infection for humans correlates with infection rates in vectors and reservoir hosts as well as tick density (Mather et al., 1996; Stafford et al., 1998; Falco et al., 1999; Pepin et al., 2012).

Immature ticks (larvae and nymphs) have a broad host range, including rodents, insectivores, birds, lagomorphs, and ungulates (LoGiudice et al., 2003; Ogden et al., 2008). Besides explaining how humans acquire infection, the aggressive feeding behaviors of these non-nidicolous (openly host-seeking) generalists enhance the opportunities for transmission of spirochetes amongst infection-competent vertebrates, linkage of ecological niches, and expansion of the geographic range of the disease (Kurtenbach et al., 2006). Although adult stage ticks have a twofold greater prevalence of infection than nymphs (Schwartz et al., 1997), they are much less important as vectors of human disease because adult males do not feed, and adult females usually feed on large reservoir-incompetent animals, typically white-tailed deer (Anderson, 1988; Kurtenbach et al., 2006). Furthermore, adults quest in late autumn through early spring when humans are less apt to encounter them and more apt to be wearing protective clothing (Falco et al., 1999; Dennis and Hayes, 2002). The western black-legged tick, *Ixodes pacificus*, is the primary vector of Lyme disease on the Pacific Coast (Campagna et al., 1983; Lane et al., 2007). Because *I. pacificus* larvae and nymphs preferentially feed on *B. burgdorferi*-refractory lizards (Lane and Loye, 1989; Lane and Quistad, 1998), infection rates in *I. pacificus* ticks are low (Lane et al., 2013; Rose et al., 2019), with a corresponding decrease in regional prevalence of human disease. Natural transmission cycles exist in non-endemic regions of the United States but are of lesser importance for humans because they involve reservoir hosts in remote geographic areas, less vector-competent *Ixodes* species, tick species with narrow host ranges that tend not to bite humans, and/or *Borrelia* species with limited infectivity for humans (Maupin et al., 1994; Dolan et al., 1997; Norris et al., 1999; Oliver et al., 2003; Franke et al., 2013).

Animals are reservoir competent if they become infected following the bite of an infected tick and can re-transmit the pathogen to a naïve vector (Mather et al., 1989; Hanincova et al., 2006; Brunner et al., 2008). In the case of Lyme disease, infection in a reservoir host must be of long enough duration to serve as a blood meal source for more than one tick life stage. As eloquently stated by Barbour (Barbour, 2017), what this (*i.e.*, reservoir competence) "effectively means is usually a combination of resistance to and tolerance of infection in reservoir hosts of long-standing." *Peromyscus leucopus*, the white-footed mouse, which thrives in habitats ranging from pristine forest to degraded woodlots, is a principal reservoir in the Northeast and North Central United States (Donahue et al., 1987; LoGiudice et al., 2003; Barbour, 2017). Once infected with *B. burgdorferi*, *P. leucopus* can remain infected for life without end-organ pathology (*i.e.*, inflammatory response) or decreased longevity (Moody et al., 1994; Oliver et al., 2003; Schwanz et al., 2011; Voordouw et al., 2015) – dual indicators of a high degree of tolerance. Despite *B. burgdorferi*'s reputation as a "generalist" pathogen (Hanincova et al., 2006), not all *I. scapularis* blood meal hosts are equally competent reservoirs; moreover, evidence exists that tick hosts other than *P. leucopus* (*e.g.*, shrews and chipmunks) can contribute to the maintenance of enzootic cycles in endemic areas (LoGiudice et al., 2003; Kurtenbach et al., 2006; Brisson et al., 2008; Franke et al., 2013). The degree of biodiversity in a given locale, more specifically, the relative proportions of competent and incompetent species, is a major determinant of the transmission risk for humans (LoGiudice et al., 2003; Granter et al., 2014; Ruyts et al., 2016). As noted earlier, in recent years, there has been growing appreciation of passerine birds as both reservoir hosts and vehicles for dissemination of infected ticks (Richter et al., 2000; Hasle et al., 2011; Norte et al., 2013). *I. ricinus*, the sheep tick, is the principal vector for Lyme disease spirochetes isolated from European patients (Gern, 2009). This tick is widely distributed throughout Europe with a range extending from Ireland to the Urals and from Southern Sweden to North Africa (McCoy et al., 2013; Cull et al., 2018; Estrada-Pena et al., 2018). The spatial prevalence of spirochete-infected *I. ricinus* ticks differs considerably and can be quite patchy even in areas with high overall densities (Estrada-Pena et al., 2018). According to a recent meta-analysis (Strnad et al., 2017), the highest rate of infected ticks was found in Central Europe and the lowest in the British Isles. *B. afzelii* and *B. garinii*, the genospecies most frequently associated with disease in European patients (Stanek and Strle, 2018), are the most commonly identified in questing *I. ricinus* nymphs (Rauter and Hartung, 2005; Estrada-Pena et al., 2018). The vector ecology of Lyme disease in Europe is even more complicated than in North America because of the greater diversity of European *Borrelia* populations (discussed below) and differences in reservoir host preferences (Gern, 2009; Mannelli et al., 2012; Franke et al., 2013). Small mammals, ground-foraging birds, and reptiles are common hosts for the larval and nymphal stages of *I. ricinus,* while adults (females) feed mostly on large mammals such as ungulates; both immature and adult stages will attach to humans (Gern, 2002). Only a small number of the more than 300 vertebrates serving as blood meal hosts for questing *I. ricinus* ticks are reservoir competent (Gern, 2009; Mannelli et al., 2012). Whereas both rodents and birds can serve as reservoirs for *B. burgdorferi* (Kurtenbach et al., 2006), *B. afzelii* depends mainly on rodents (mice and voles), while *B. garinii* depends on birds (Comstedt et al., 2011; Mannelli et al., 2012). Thus, *I.*

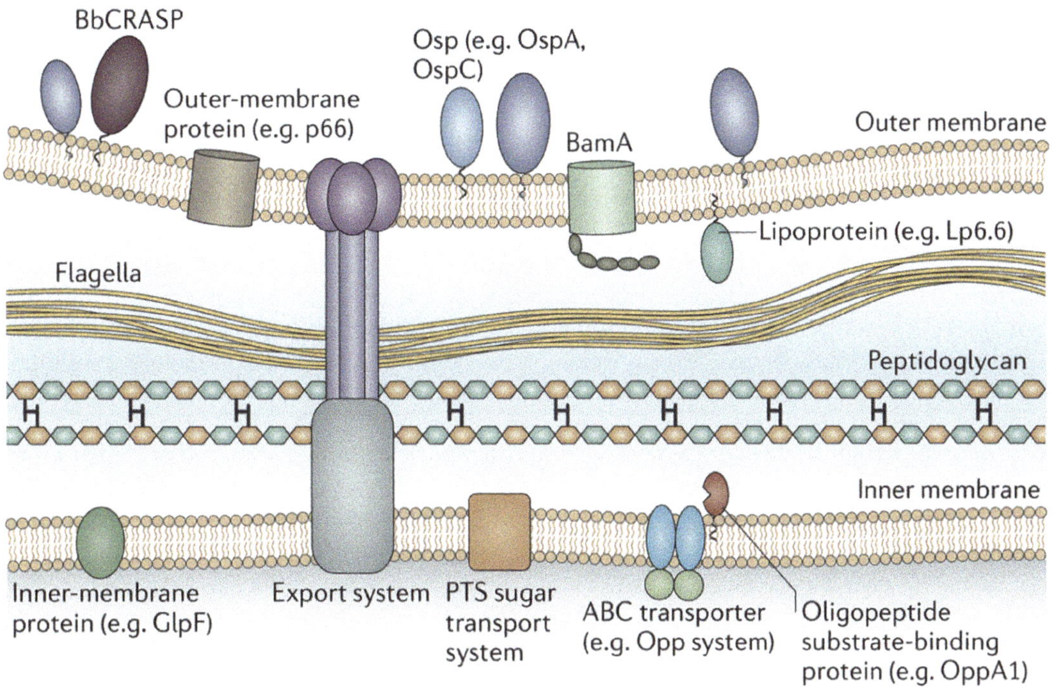

Figure 5. The borrelial cell envelope. The outer membrane contains outer-surface lipoproteins (Osps) in high density and β-barrel outer-membrane-spanning proteins such as BamA in low density. The inner membrane is rich in integral membrane proteins, many of which are transporters. BbCRASP, complement regulator-acquiring surface protein; OppA1, oligopeptide permease A1; PTS, phosphotransferase system. Reprinted with permission from Radolf *et al.* 2012.

ricinus can be likened to a "mixing vessel" for different *Borrelia* strains and species, with host associations driven by the filtering effect of spirochete selectivity for particular vertebrates, rather than adaptation of the bacterium to a vector with narrow feeding preferences (Margos et al., 2011). Two additional tick species, *I. hexagonus* and *I. uriae*, maintain the spirochete in transmission cycles separate from those of *I. ricinus*. *I. hexagonus* has a more restricted host range than *I. ricinus*, feeding primarily on carnivores, such as foxes and mustelids, occasionally on lagomorphs, and rarely birds (Gern, 2002). Transmission cycles involving *I. uriae*, a seabird specialist that feeds on a range of avian marine species, have been implicated in global dispersal of *B. garinii* (Olsen et al., 1993; Olsen et al., 1995; Comstedt et al., 2011; Munro et al., 2019).

The spirochete

The spirochete-host interface - the outer membrane

Like all spirochetes (Holt, 1978), *B. burgdorferi* is a diderm consisting of an outer membrane (OM) that surrounds the periplasmic space, the peptidoglycan, the cytoplasmic membrane, and the protoplasmic cylinder (Figure 5) (Barbour and Hayes, 1986; Charon et al., 2009). The organelles of motility, the flagella, are contained entirely within the periplasmic compartment (Charon et al., 2012). In addition to propagating a planar wave that enables the spirochete to penetrate collagen matrices in connective tissue and endothelial junctions (Norman et al., 2008; Charon et al., 2012; Harman et al., 2012; Harman et al., 2013), the flagellar filaments also serve a cytoskeletal function (Charon et al., 2012). As they wind around the protoplasmic cylinder, they push against the elastic peptidoglycan sacculus, bending it to create the cell's distinctive flat-wave morphology (Motaleb et al., 2000; Charon et al., 2009) (also see Chapter 8).

The *B. burgdorferi* OM comprises the host-pathogen interface in all milieus through which the spirochete transits or in which it takes up final residence; it is not surprising, therefore, that this structure has attracted great interest over the years (Barbour and Hayes, 1986; Kenedy et al., 2012; Radolf et al., 2012; Zuckert, 2019) (also see Chapter 7). Because of its double-membrane architecture, *B. burgdorferi* often has been likened to Gram-negative bacteria. This analogy is inaccurate from the standpoints of phylogenetics (Paster et al., 1991; Daubin et al., 2002), ultrastructure (Radolf et al., 2012; Zuckert, 2019), composition (LaRocca et al., 2010; Radolf et al., 2012; LaRocca et al., 2013), and genomics (Fraser et al., 1997; Casjens et al., 2000; Stewart et al., 2005; Qiu and Martin, 2014). The OM of Gram-negative bacteria is an asymmetric bilayer composed of glycerophospholipids in the inner leaflet and the highly inflammatory glycolipid lipopolysaccharide (LPS) in the outer (Konovalova et al., 2017). Early reports that *B. burgdorferi* contains LPS (Beck et al., 1985; Habicht et al., 1986) were disproved, initially by chemical and immunological analysis (Takayama et al., 1987; Radolf et al., 1991), and subsequently by genomic sequencing (Fraser et al., 1997). The absence of LPS has important clinical ramifications inasmuch as spirochetemic Lyme disease patients rarely, if ever, manifest sepsis-like pathophysiology comparable to that seen in patients with Gram-negative bacteremia (Wormser et al., 2005; Wormser, 2006). Consequently, the presence of a sepsis syndrome in a Lyme disease patient should prompt a search for co-infections, such as babesiosis and anaplasmosis, also transmitted by ixodid ticks (Sanchez et al., 2016).

The OM of *B. burgdorferi* differs in other important respects from its Gram-negative counterparts: (i) It is much more easily damaged during routine laboratory manipulations (*e.g.*, centrifugation and resuspension) and is far more susceptible to detergent solubilization (Brusca et al., 1991; Cox et al., 1996). (ii) It contains a much lower density of proteins with membrane-spanning domains, as assessed by freeze-fracture electron microscopy (Walker et al., 1991; Radolf et al., 1994). (iii) Although proteins with porin-like properties and function have been identified in the *B. burgdorferi* OM (Pinne et al., 2004; Pinne et al., 2007; Barcena-Uribarri et al., 2013; Kenedy et al., 2014), the bacterium does not contain orthologs for well-characterized Gram-negative porins (Fraser et al., 1997; Nikaido, 2003; Kenedy et al., 2016). (iv) The spirochete, however, does contain orthologs for TolC and the other components of an ATP-dependent efflux pump shown to contribute to the bacterium's inherent antimicrobial resistance (Bunikis et al., 2008). (v) Though lacking LPS, *B. burgdorferi* OMs contain three abundant (comprising 50-60% of total lipids), immunogenic, but non-inflammatory, lower molecular weight glycolipids – cholesteryl-β-D-galactopyranoside, cholesteryl 6-*O*-acyl-β-D-galactopyranoside, and mono-α-galactosyl-diacyl-glycerol (Wheeler et al., 1993; Norgard et al., 1996; Ben-Menachem et al., 2003; Kinjo et al., 2006; Schroder et al., 2008; Huang et al., 2016). The cholesterol glycolipids spontaneously phase partition from the other lipids, forming lipid rafts or microdomains into which segregates a subset of outer surface lipoproteins (LaRocca et al., 2010; Toledo et al., 2014; Huang et al., 2016). Surprisingly, *B. burgdorferi* contains an ortholog for LptD, the outer membrane protein in Gram-negative bacteria that inserts newly exported LPS into the outer membrane (Botos et al., 2016); it is tempting to speculate that *Borrelia* appropriated the Gram-negative LPS transport pathway to serve its own needs – localization of glycolipids to the OM. (vi) Arguably, the most notable difference is the number and variety of lipoproteins that adorn the borrelial surface (Kenedy et al., 2012; Radolf et al., 2012; Dowdell et al., 2017). In Gram-negative microorganisms, lipoproteins typically are anchored to the inner leaflet of the outer membrane or the periplasmic leaflet of the cytoplasmic membrane and not exported to the bacterial surface (Zuckert, 2014; Konovalova et al., 2017). Although the tertiary structures of borrelial outer surface lipoproteins differ considerably (Li et al., 1997; Eicken et al., 2001; Kumaran et al., 2001; Eicken et al., 2002; Brangulis et al., 2018), their membrane topologies are identical. They are soluble polypeptides tethered to the external leaflet of the outer membrane by *N*-terminal lipids (Jones et al., 1995). This topological configuration presumably enables *Borrelia* to differentially express the enormous number of surface structures needed to sustain its dual-host lifestyle and helps explain its ability to infect a wide variety of vertebrate hosts (Wywial et al., 2009; Radolf et al., 2012; Brisson et al., 2013; Brissette and Gaultney, 2014; Caine and Coburn, 2016; Tufts et al., 2019).

Differential gene expression – tick transmission and back again

Because animals syringe-inoculated with *in vitro*-cultivated organisms develop manifestations indistinguishable from animals inoculated with ticks (Barthold et al., 2010), the arthropod phases of the bacterial life cycle might be considered irrelevant to pathogenesis. However, in the real world, Lyme disease spirochetes are transmitted by ticks and, therefore, the tick-mammal interface must be regarded as the starting point for the infectious process (Tilly et al., 2008; de Silva et al., 2009; Radolf et al., 2012; Steere et al., 2016) (also see

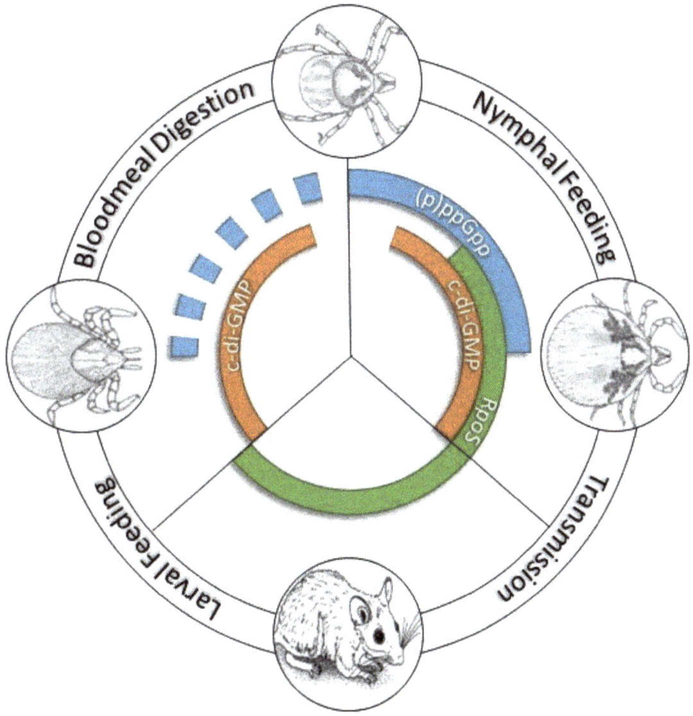

Figure 6. *B. burgdorferi* gene regulatory programs throughout the enzootic cycle. The second messengers (p)ppGpp and cyclic di-GMP regulate gene expression during the tick phases of the cycle. RpoS transcribes genes required for transmission (nymphal blood meal) and infection of the vertebrate reservoir. RpoS is OFF in unfed ticks and feeding larvae. (Figure courtesy of Dr. Ashley Groshong).

Chapters 11 and 12). Indeed, there is now overwhelming evidence that spirochetes in feeding ticks undergo complex alterations in their transcriptional and protein profiles that are not reproduced by *in vitro* culture conditions (Cugini et al., 2003; Iyer et al., 2015; Caimano et al., 2016; Stevenson and Seshu, 2018). Collectively, these changes enable the spirochete to adapt physiologically to the feeding midgut environment (He et al., 2011; Pappas et al., 2011; Dunham-Ems et al., 2012; Caimano et al., 2015; Bontemps-Gallo et al., 2016; Caimano et al., 2016), while, at the same time, preparing it for challenges looming during the mammalian phase (Grimm et al., 2004; Fisher et al., 2005; Dunham-Ems et al., 2012). They also promote binding to the bacterial surface of serum proteins that exploit mammalian proteolytic systems to facilitate dissemination from tick to mammal (Hu et al., 1995; Coleman et al., 1997; Onder et al., 2012) and protect against complement-mediated lysis (Kraiczy, 2016b; Zhi et al., 2018; Xie et al., 2019). The obligatory time course for this programmatic sequence explains why spirochetes are infrequently transmitted to mice (Piesman et al., 1987b; Ohnishi et al., 2001; Dunham-Ems et al., 2009) or humans (Berger et al., 1995; Falco et al., 1996; Sood et al., 1997; Nadelman et al., 2001) when ticks are attached for less than 48 h. It is critical for physicians to be aware of this time frame because it establishes the window of opportunity for antimicrobial prophylaxis (Nadelman et al., 2001; Wormser, 2006) and can be used to provide re-assurance to individuals who removed recently attached ticks that their risk of infection is low.

Our understanding (albeit still rather limited) of the genetic regulatory mechanisms that control the infectious process (Figure 6) dates back to the seminal discovery by Schwan and co-workers (Schwan et al., 1995) that OspC, an outer surface lipoprotein *B. burgdorferi* requires to establish mammalian infection (Grimm et al., 2004; Tilly et al., 2006; Dunham-Ems et al., 2012), is upregulated within the nymphal midgut during the blood meal. They also noted that this phenomenon can be mimicked by shifting from ambient to mammalian body temperature during *in vitro* cultivation. The Norgard group's description of the RpoN/RpoS master regulatory pathway in a series of landmark papers (Hubner et al., 2001; Yang et al., 2003; Ouyang et al., 2009) provided mechanistic insight into these observations. Following a temperature shift *in vitro*, or at the outset of the nymphal blood meal in nature (Caimano et al., 2007), the spirochete's alternative sigma factor RpoN works in concert with the response regulator Rrp2 and the Fur/PerR ortholog, BosR, to transcribe *rpoS*, the downstream "effector" sigma factor (for a complete discussion see Chapter 5). Although RpoS was shown originally to transcribe just *ospC* and the *dbpA* gene encoding decorin-binding protein A (DbpA), another virulence determinant (Guo et al., 1995; Hagman et al., 1998; Fischer et al., 2003), we now know that the RpoS regulon encompasses approximately 10% of the *B. burgdorferi* genome, and includes many genes of unknown function (Caimano et al., 2007; Caimano et al., 2019). In *Escherichia coli*, activation of RpoS induces a complex, coordinated adaptive response involving a large cohort of genes that enables the bacterium to resist abiotic, physiological and environmental stresses (Chiang and Schellhorn, 2010; Hengge, 2011). The *B. burgdorferi* RpoS regulon, in contrast, contains only a handful of genes with discernible roles in physiology and stress responses (Caimano et al., 2007; Caimano et al., 2019). Organisms lacking RpoS are avirulent by tick- as well as needle-inoculation (Caimano et al., 2004; Fisher et al., 2005; Hyde et al., 2009; Ouyang et al., 2009; Xu et al., 2010; Dunham-Ems et al., 2012), implying that the RpoN/RpoS pathway regulates genes that promote dissemination within the tick as well as genes that function within the mammal. In contrast to the RpoS regulon in mammals, the cohort of genes controlled by the RpoN/RpoS pathway within feeding ticks is poorly defined. Gilmore's group has identified two RpoS-dependent genes, *bba64* and *bba66*, required for tick transmission (Gilmore et al., 2010; Patton et al., 2013).

Following inoculation into the dermis of its naïve vertebrate host, the spirochete must solidify its foothold at the bite site via a poorly understood program for differential gene expression with multiple regulatory layers designated by the umbrella term "mammalian host adaptation" (Barthold et al., 1995; Montgomery et al., 1996; Akins et al., 1998) (see Chapter 5). One involves turning OFF pathways that enable spirochetes to survive the many noxious aspects of the blood meal (Bontemps-Gallo et al., 2016); principal among these is the Hk1/Rrp1 two-component system that signals via the pleiotropic effector molecule cyclic-di-GMP (Hengge, 2009; He et al., 2011; Kostick et al., 2011; Caimano et al., 2015). Whether the absence of a tick-specific environmental signal or the appearance of a new mammalian host-derived cue(s) turns OFF synthesis of c-di-GMP by the diguanylate cyclase Rrp1 remains a matter of conjecture. The RpoN/RpoS pathway plays two critical, inter-dependent roles in mammalian host adaptation. First, it represses σ^{70}-dependent tick-phase genes (Caimano et al., 2005; Caimano et al., 2007; Caimano et al., 2019), the midgut colonization factor OspA (de Silva et al., 1996; Pal et al., 2004a; Yang et al., 2004) being the prototype. This so-called "gatekeeper" repressor function of RpoS appears to work in tandem with the loss of c-di-GMP signaling to terminate expression of tick-phase genes (Caimano et al., 2019). For reasons that have yet to be determined, not all tick-phase genes (*e.g., ospA* and *lp6.6*) repressed by RpoS are upregulated by c-di-GMP. Although it is often stated that downregulation of OspA occurs within the tick in response to the blood meal (Pal et al., 2004a; Tilly et al., 2008; Caine et al., 2017), multiple lines of evidence argue that it is delayed until spirochetes reach the mammal (Belperron and Bockenstedt, 2001; Ohnishi et al., 2001; Mulay et al., 2009; Adams et al., 2017; Caimano et al., 2019). In other words, the reciprocal relationship between the expression of OspA and OspC is a mammalian host phase, not tick phase, phenomenon (Montgomery et al., 1996). RpoS-mediated downregulation of OspA and other immunogenic tick-phase lipoproteins explains why infection fails to generate antibodies against them (Gern et al., 1993; Golde et al., 1993; Brunet et al., 1995; Piesman et al., 1997; Vaz et al., 2001).

As noted already, the RpoN/RpoS pathway also upregulates expression of gene products required for infectivity. Of these, the lipoproteins OspC, DbpA/B, and BBK32 are by far the most extensively explored (Radolf et al., 2012; Groshong and Blevins, 2014; Caine and Coburn, 2016). The X-ray crystal structure of OspC, solved independently by two groups nearly twenty years ago, revealed an elongated, α-helical homodimer with putative ligand-binding sites at the distal "crown"of the dimer and at the interface between two opposing helices of the monomers (Eicken et al., 2001; Kumaran et al., 2001; Earnhart et al., 2010). The relationship(s) between this structure and OspC's reported biological functions remains enigmatic. Along with directly recruiting the immunosuppressive tick salivary protein SALP15 to the bacterial surface (Anguita et al., 2002;

Ramamoorthi et al., 2005; Hovius et al., 2008a), OspC interferes with innate clearance mechanisms (Stewart et al., 2006), putatively preventing phagocytosis by macrophages (Carrasco et al., 2015), and serves as a surface receptor for plasminogen (Lagal et al., 2006; Onder et al., 2012) and complement inhibitors (Caine et al., 2017) (see Chapters 13 and 18). DbpA and B are adhesins for decorin, heparan, dermatan sulfate, heparan sulfate, and glycosoaminoglycans (GAGs) (Guo et al., 1998; Brown et al., 2001; Fischer et al., 2003; Pikas et al., 2003; Morgan and Wang, 2013). BBK32 is an adhesin, binding fibronectin and GAGs (Probert et al., 2001; Seshu et al., 2006; Moriarty et al., 2012; Lin et al., 2015), and an inhibitor of the classical complement pathway (Garcia et al., 2016; Xie et al., 2019). Whereas OspC functions primarily or exclusively at the bite site and is downregulated several weeks after infection due to the immune pressure exerted by the appearance of OspC antibodies (Liang et al., 2002; Tilly et al., 2006), DbpA/B and BBK32 function "downstream", facilitating hematogenous dissemination and spirochete tropisms for heart and joints (Brown et al., 2001; Norman et al., 2008; Weening et al., 2008; Hyde et al., 2011; Moriarty et al., 2012; Fortune et al., 2014; Lin et al., 2014; Caine and Coburn, 2015; Lin et al., 2015; Ebady et al., 2016). Elegant studies by Moriarty's group (Moriarty et al., 2012) revealed that BBK32 exerts its fibronectin and GAG binding activities in a sequential manner. Tethering to fibronectin on the endovascular surface recruits spirochetes from the circulating blood compartment, while binding to GAGs stabilizes interactions with endothelial cells, setting the stage for transmigration between endothelial cells (Szczepanski et al., 1990; Coleman et al., 1995). DbpA/B and BBK32 are just two elements of *B. burgdorferi*'s complicated adhesin story. Lyme disease spirochetes express a bewildering array of surface molecules with redundant ligand-binding activities; whether they work cooperatively or preferentially depending on the milieu and host infected is simply not understood (Caine and Coburn, 2016). This wide assortment of adhesins may help to determine the range of vertebrate hosts, including humans, that a borrelial species and, even strains within a species, can parasitize (Tufts et al., 2019).

Correct expression of adhesins in time and space is just one facet of the pathogenic process. Spirochetes deploy an arsenal to persist long enough in the reservoir to transit back to the vector when opportunity "knocks" in the form of a larva taking a blood meal. Such "persistence functionalities" include (i) directed motility to locate and penetrate dermal blood vessels following deposition and negotiate endovascular and tissue barriers at metastatic sites, all in response to chemotactic signals not even remotely understood (Charon et al., 2012; Motaleb et al., 2015; Hyde, 2017); (ii) expression of the proper combination of outer and inner membrane transporters to appropriate the huge spectrum of nutrients "generously" supplied by the vertebrate host to support the bacterium's extremely limited biosynthetic capacity (Fraser et al., 1997; Gherardini et al., 2010; Corona and Schwartz, 2015); (iii) adjustment of central metabolism to take maximal advantage of the nutrients available within the various micro-environments in which it takes up residence (Corona and Schwartz, 2015; Iyer et al., 2015; Groshong et al., 2017), all while fending off innate and subsequently adaptive defenses, particularly complement (Kraiczy, 2016b; Tracy and Baumgarth, 2017) and antibodies (Norris, 2014; Stone and Brissette, 2017), as an "exposed" parasite within the extracellular milieu (Radolf et al., 2012). (Please see Chapters 8 and 13 for comprehensive reviews of motility/chemotaxis, virulence, and immune evasion mechanisms.)

Upon larval acquisition, the RpoN/RpoS pathway rapidly shuts OFF, the HK1/Rrp1 pathway rapidly turns ON, and expression of tick-phase genes resumes, enabling successful colonization of the vector (Donahue et al., 1987; Yang et al., 2004; Caimano et al., 2015; Iyer et al., 2015; Caimano et al., 2019). How often do spirochetes execute this complex maneuver in humans? Although there is no question that Lyme disease spirochetes can persist in untreated humans and give rise to disease manifestations well after inoculation (Steere et al., 2016; Stanek and Strle, 2018), there are no reliable data as to whether persistence in humans is a common or rare occurrence. Are humans dead-end hosts because they are incompetent reservoirs or poorly accessible targets of opportunity for larvae? The biology underlying this question has clinical relevance. The technique called xenodiagnosis – allowing naïve larvae to feed on subjects to assess infection status (Telford et al., 2014) - is being used to determine whether individuals who remain symptomatic following treatment for Lyme disease (see below) harbor spirochete "persisters" (Marques et al., 2014). The rationale for xenodiagnosis as a diagnostic tool becomes cloudy if untreated individuals cannot infect ticks (Bockenstedt and Radolf, 2014).

Phylogenetic diversity and human disease

Taxonomy and disease

Borrelia species fall into two major phyletic clusters, each with considerable heterogeneity (Bunikis et al., 2004; Cutler et al., 2017; Stone et al., 2017). One contains the relapsing fever spirochetes. With the notable exceptions of the louse-borne agent of endemic relapsing fever, *B. recurrentis*, and the *I. ricinus* complex-borne *B. miyamotoi*, relapsing fever spirochetes are transmitted by soft-bodied, argasid ticks (see Chapters 10, 11, and 14). The other cluster contains the agents of Lyme disease. The recognition by the early 1990s that multiple *Borrelia* species cause Lyme disease led to the designation of this cluster as the *Borrelia burgdorferi* sensu lato complex and the original isolate as *B. burgdorferi* sensu stricto (herein *B. burgdorferi*) Presently, the sensu lato complex comprises at least 20 proposed or confirmed species worldwide, nine of which have been found to cause human disease: *B. burgdorferi*, *B. afzelii*, *B. garinii*, *B. bavariensis*, *B. spielmanii*, *B. lusitaniae*, *B. bissettii*, *B. valaisiana*, and *B. mayonii* (Belfaiza et al., 1993; Wang et al., 1999a; Schotthoefer and Frost, 2015; Pritt et al., 2016; Cutler et al., 2017). However, the great majority of human Lyme disease cases are due to four pathogenic species, namely *B. burgdorferi* sensu stricto, *B. afzelii*, and *B. garinii* and the closely related *B. bavariensis*. Other species have been detected in specimens from only a few patients or, in single cases, which raises questions regarding their importance in the pathogenesis of human disease. In contrast to the United States, Lyme borreliosis in Europe is caused predominantly by *B. afzelii*, *B. garinii*, *B. bavariensis* (formerly *B. garinii* OspA type 4) and only rarely by *B. burgdorferi* (Stanek and Strle, 2018). As noted earlier, *B. afzelii* depends mainly on rodents as reservoir hosts, while *B. garinii* (as well as *B. bavariensis*) preferentially parasitizes birds (Comstedt et al., 2011; Mannelli et al., 2012). *In vitro* susceptibility to rodent versus avian complement has been invoked to explain these differences (Kraiczy, 2016b).

B. burgdorferi was thought to be the sole genospecies in the United States until 1995 when Marconi and associates (Marconi et al., 1995) isolated *B. andersonii* from cottontail rabbits and *I. dentatus* ticks. Subsequently, it became evident that enzootic cycles exist involving non-sensu stricto species and "specialist" ixodid ticks with strong, selective host preferences and little or no proclivity to bite humans. Perhaps the best characterized of these is *B. bissettii* (Bissett and Hill, 1987), transmitted by *I. pacificus*, *I. affinis* and *I. spinipalpis* and recovered and/or detected by PCR throughout the United States in ticks and a variety of vertebrates (Postic et al., 1998; Picken and Picken, 2000; Piesman, 2002; Oliver et al., 2003; Margos et al., 2016) and, in rare instances, humans (Picken et al., 1996; Girard et al., 2011; Golovchenko et al., 2016; Rudenko et al., 2016). Molecular evidence for infection of humans by *B. americana* and *B. andersonii*, particularly in Southern states, also has been reported (Clark et al., 2014). Moreover, *B. mayonii* was recently isolated from a few human specimens in upper midwestern U.S. (Pritt et al., 2016). However, in the United States, *B. burgdorferi* remains the primary agent of disease.

Recently, an already confusing taxonomic situation has become controversial. A comparative genomic search for molecular signatures to distinguish relapsing fever from Lyme disease spirochetes spawned a proposal that the two clusters be placed into separate genera (Adeolu and Gupta, 2014). The genus *Borrelia* would contain just the agents of relapsing fever, along with *B. miyamotoi*, while a new genus, *Borreliella*, would contain the agents of Lyme disease. The phylogenetic significance of these differences, and whether they justify the division, has been hotly debated (Barbour et al., 2017; Margos et al., 2017b; Margos et al., 2018; Estrada-Pena and Cabezas-Cruz, 2019). Opponents of this reclassification argue that it entails risks to public health and patient safety (Stevenson et al., 2019).

Comparison of *B. burgdorferi* sensu lato genospecies in United States and Europe

The most obvious clues that microbial genetics may be responsible for the differences in human disease have come from studies of the clinical presentation of Lyme borreliosis caused by different *Borrelia* genospecies and subspecies: in North America (*B. burgdorferi*) or Europe (namely *B. afzelii, B. garinii, and B. bavariensis)*. With all four species, the first sign of infection is usually an expanding EM skin lesion (Nadelman et al., 1996; Strle et al., 1996; Steere, 2001). However, compared with *B. afzelii* or *B. garinii* infection in Europe, EM caused by *B. burgdorferi* in the U.S. is associated with a greater number of symptoms and more frequent hematogenous dissemination (Strle et al., 1999; Carlsson et al., 2003; Logar et al., 2004; Wormser et al., 2005; Wormser, 2006; Jones et al., 2008; Cerar et al., 2016). More pronounced differences are observed with later manifestations of disease that demonstrate *B. burgdorferi* is considerably more arthritogenic than *B. afzelii*, which usually remains localized to the skin, or *B. garinii* and *B. bavariensis,* which are predominantly associated with neurologic complications (van Dam et al., 1993; Balmelli and Piffaretti, 1995; Coipan et al., 2016; Jahfari et al., 2017; Gallais et al., 2018; Stanek and Strle, 2018; Grillon et al., 2019). An appreciation of these regional differences may help clinicians with the diagnosis and treatment of Lyme disease specific to a region. Differences in clinical presentation in North America and Europe also are observed by comparing patients infected on the two continents by *B. burgdorferi*. Infection with *B. burgdorferi* in the U.S. is associated with more symptomatic early infection compared to that in Europe, which resembles the milder infection seen with *B. afzelii* and *B. garinii*. Moreover, European *B. burgdorferi* strains appear to be more neurotropic than *B. burgdorferi* genotypes from North America and are associated with certain clinical manifestations, such as acrodermatitis chronica atrophicans, seldomly, if ever, found in the U.S. (Jungnick et al., 2015). These findings raise the intriguing possibility that strains in a region accrue similar characteristics by sharing genetic information; if so, then the feeding nymphal tick, where actively replicating strains can encounter each other, is the likely venue for such exchange.

Intraspecies genotypes are associated with distinct clinical phenotypes

B. burgdorferi can be divided into three genotypes, 16S-23S rRNA intergenic spacer types 1-3 (RST1-3), based on restriction fragment length polymorphisms of the 16S-23S rRNA intergenic spacer (IGS) (Wang et al., 1999a). OspC typing divides *B. burgdorferi* strains into ~30 genotypes (Liveris et al., 1999; Wang et al., 1999a; Wormser et al., 1999; Wang et al., 2001; Wang et al., 2002; Jones et al., 2006; Wormser et al., 2008a; Hanincova et al., 2013). The high degree of sequence variability among *ospC* genes is believed to reflect the selection of allelic variants by immunologic pressure exerted during infection of vertebrates (Wang et al., 1999b; Barbour and Travinsky, 2010; Baum et al., 2013). Individual vertebrate species serve as reservoir hosts for only a subset of OspC genotypes (Brisson and Dykhuizen, 2004; Hanincova et al., 2013; Vuong et al., 2014). A similar situation pertains to humans – only a subset of *ospC* genotypes circulating in ticks cause human disease (Seinost et al., 1999; Dykhuizen et al., 2008; Wormser et al., 2008a; Hanincova et al., 2013). These results imply that the tick serves as an incubator for *ospC* diversity, while the vertebrates upon which infected ticks feed exert a "filtering" effect, selecting for OspCs that "match" putative ligands in the blood meal host.

OspC and RST typing systems have been particularly useful in stratifying *B. burgdorferi* strains according to clinical presentation of disease (Liveris et al., 1999; Seinost et al., 1999; Wang et al., 1999a; Wormser et al., 1999; Bunikis et al., 2004; Alghaferi et al., 2005; Jones et al., 2006; Hanincova et al., 2008; Wormser et al., 2008a; Hanincova et al., 2013). All three RST types and 24 (of over 30) OspC types of *B. burgdorferi* have been recovered from patients with Lyme disease (Liveris et al., 1999; Wang et al., 1999a; Jones et al., 2006; Wormser et al., 2008b; Jones et al., 2009; Barbour and Travinsky, 2010). More recently, multilocus sequence typing (MLST) studies have been used to further sub-stratify the strains and provide additional insights into human infection that ultimately will require comparative genomics strategies to elucidate (Jungnick et al., 2015). According to the *Borrelia* MLST Database (https://pubmlst.org/borrelia/), >900 MLST sequence types have been identified, of which 220 are associated with disease in humans. Use of multiple typing systems together presumably can provide more information for clinical correlations than either system alone.

Several studies have now demonstrated differential pathogenicity among various genotypes within *B. burgdorferi* (Seinost et al., 1999; Wormser et al., 1999; Wang et al., 2001; Wang et al., 2002; Jones et al., 2006; Wormser et al., 2008b; Jones et al., 2009; Strle et al., 2011b). As a generalization, RST1 genotypes contain *ospC* alleles associated with invasive disease, whereas RST3 genotypes contain predominantly noninvasive *ospC* alleles (Hanincova et al., 2008; Wormser et al., 2008a). For example, RST1 strains are more often detectable in blood in mice and humans (Wormser et al., 1999; Wang et al., 2001; Wang et al., 2002; Jones et al., 2006; Dykhuizen et al., 2008), suggesting that they disseminate more readily and/or reach higher numbers in blood. Moreover, RST1 OspC type A strains are associated with more symptomatic early infection in patients with EM, and they more frequently cause antibiotic-refractory Lyme arthritis than other strains (Jones et al., 2009; Strle et al., 2011b). The greater disease severity and propensity for dissemination of RST1 vs RST3 genotypes was corroborated experimentally in the murine model (Wang et al., 2001; Wang et al., 2002). The data also suggest that RST1 strains are likely a contributing factor in the prevalence and severity of Lyme disease in the Northeastern U.S. (Derdakova et al., 2004; Margos et al., 2008; Hoen et al., 2009). In the mouse model, RST1 OspC type A strains have higher transmission efficiency from mice to ticks than other strains (Derdakova et al., 2004). In addition, they appear to be a recently evolved clonal lineage that may be an important factor in the emergence of the Lyme disease in the northeastern U.S. (Margos et al., 2008; Qiu et al., 2008; Hoen et al., 2009).

Like *B. burgdorferi*, only certain subsets of *B. afzelii* or *B. garinii ospC* types could be linked to human infection. Of the 14 *ospC* groups in *B. afzelii*, nine were found in human isolates and only two associated with invasive disease. Gallais *et al.* (Gallais et al., 2018) as well as Coipan *et al.* (Coipan et al., 2016) found a small number of *B. afzelii* sequence types associated with localized or disseminated infection. Similarly, only nine of the 22 *ospC* groups identified in *B. garinii* were isolated from humans, and four contained all invasive isolates. In a PCR-based comparison of tick and EM *B. afzelii* strains, Tijsee-Klasen *et al.* (Tijsse-Klasen et al., 2013) also found a correlation between specific IGS and *ospC* haplotypes in the patient samples. These results were extended by MLST analyses that indicated that only a small number of *B. afzelii* sequence types associated with localized or disseminated infection (Derdakova et al., 2004; Margos et al., 2008; Qiu et al., 2008; Hoen et al., 2009; Gallais et al., 2018).

Plasmid-encoded variable lipoproteins and host specificity

The *B. burgdorferi* genome encodes over 120 lipoproteins (Setubal et al., 2006), the large majority of which make their way to the spirochete's surface at some point during the enzootic cycle via a poorly understood secretory pathway (Kenedy et al., 2012; Dowdell et al., 2017; Zuckert, 2019); many exhibit high degrees of sequence polymorphisms presumably driven by adaptive pressures exerted by vertebrates (Roberts et al., 1998; Wywial et al., 2009; Haven et al., 2011; Casjens et al., 2012; Brisson et al., 2013; Mongodin et al., 2013). Among these sequence variable lipoproteins are the lp56-encoded DbpA/B paralogs; the cp32-endoded OspE/OspF/Elp, Mlp, and RevA paralogous families; and the PFam54 paralogs, many, but not all, encoded on lp54. In addition to sequence variability at specific loci, for some of these families, the mix of paralogs varies between strains (Wywial et al., 2009; Brisson et al., 2013; Caimano et al., 2019). Importantly, variants of proteins within a family can differ functionally. For example, isogenic *B. burgdorferi* mutants expressing strain-specific DbpA variants exhibit pronounced differences in binding to GAGs and cultured kidney epithelial cells (Benoit et al., 2011) and marked differences in tissue tropisms in the murine model (Lin et al., 2014). Despite the high degree of structural similarity among Pfam54 paralogs, only one, BBA68 (CspA), binds complement inhibitory factors (Wywial et al., 2009; Brangulis et al., 2019). Although a challenging hypothesis to test experimentally, it is now widely believed that variability in the repertoires of these host-interactive Osps is a principal determinant of differential infectivity for vertebrate species, including humans (Tufts et al., 2019). As noted above, work with OspC establishes a paradigm for this line of thinking. Another, pioneered by Kurtenbach (Kurtenbach et al., 1998; Kurtenbach et al., 2002), is that serum resistance/susceptibility determines host-range; this appealing concept has prompted investigation of an ever expanding array of complement inhibitory proteins (Garcia et al., 2016; Kraiczy, 2016a, b; Caine et al., 2017; Marcinkiewicz et al., 2019; Tufts et al., 2019; Xie et al., 2019). One can integrate these paradigms by positing that dissemination and long-term survival of *B. burgdorferi* in a particular host requires an "appropriate" combination of early survival/host adaptation Osps (*e.g.,* OspC and anticomplementary lipoproteins) and adhesins/invasins that function during later stages of infection. It then follows that (i) whether an inoculated strain has the correct assortment of Osps to cause local or systemic human disease is a chance event and (ii) variations in Osp repertoires may give rise to different clinical manifestations. It is even conceivable that some clinical manifestations without known counterparts in nature (*e.g.*, neuroborreliosis) are the inadvertent consequence of adaptive changes at loci functionally downstream of OspC.

Novel sequencing approaches to characterize *Borrelia* genetic diversity and its impact on phenotypic heterogeneity

Recent advances in sequencing technologies raise the possibility that deep genetic characterization of clinical isolates may elucidate strain- and species-specific differences that promote virulence, immunogenicity, organ tropism, or persistent infection. Second-generation technologies, particularly those commercialized by Illumina, excel at "resequencing" and can produce very high-quality genomic sequences. However, the sequence reads are relatively short (36-250 bases) and cannot resolve longer genomic fragments or areas with extensive diversity. Thus, while short read approaches have been used with success to study the genome sequence of *B. burgdorferi* chromosome and conserved plasmids cp26 and lp54 (Castillo-Ramirez et al., 2016; Margos et al., 2017a; Tyler et al., 2018), they are poorly suited to the study of the less-conserved plasmids. Recently, technologies that produce longer reads have been introduced by Oxford Nanopore Technologies (ONT) and Pacific Biosciences. These methods can generate much longer sequence reads, albeit at lower sequence quality, making them unsuitable for assembly in the absence of additional high-accuracy sequence (Bashir et al., 2012; Antipov et al., 2016; Wick et al., 2017). However, the combination of these short- and long-read sequencing approaches is poised to open the full landscape of *B. burgdorferi* genomic analyses for clinical correlations. Early successes include demonstration of the utility of long-read approaches for obtaining finished sequence of plasmids (Kingry et al., 2016; Margos et al., 2017a; Jabbari et al., 2018) and a detailed study, carried out by Chaconas and colleagues, of the mutation rate and switching kinetics at the *vlsE* locus (Verhey et al., 2018).

Infection of humans

The bite site

The seminal event in the natural history of Lyme disease is, of course, the deposition of spirochetes into the skin when an infected tick, usually a nymph, feeds on a human instead of a vertebrate in the wild. Although spirochetes undergo dramatic expansion ($\geq$ 300-fold) within the midgut during the blood meal, only a remarkably small number manage to complete the journey from tick to vertebrate (De Silva and Fikrig, 1995; Coleman et al., 1997; Dunham-Ems et al., 2009). Along the way, they encounter myriad anatomical, biochemical, and immunological barriers that passively and actively reduce their numbers (Pal et al., 2004b; Fisher et al., 2005; Bontemps-Gallo et al., 2016; Sonenshine and Macaluso, 2017; Shaw et al., 2018). qPCR and immunofluorescence analysis yielded values of approximately 20 spirochetes per salivary gland at peak infectivity (Ohnishi et al., 2001; Piesman et al., 2001). While this number is well below published ID_{50} values determined using *in vitro*-cultivated organisms (Sadziene et al., 1993; Xu et al., 1996), it must be remembered, per above, that spirochetes delivered by ticks undergo dramatic transcriptomic and proteomic changes during feeding (Cugini et al., 2003; Pal and Fikrig, 2003; Drecktrah et al., 2015; Iyer et al., 2015; Caimano et al., 2016; Bernard et al., 2018; Stevenson and Seshu, 2018) in addition to exploiting the battery of pharmacologic, hemostatic, anticomplementary, and immuno-suppressive factors in the tick salivary cocktail ("saliva-assisted transmission") (Hovius, 2009; Chmelar et al., 2016; Simo et al., 2017; Nuttall, 2019) (also see Chapters 11 and 12). In recent years, tick salivary proteins that contribute to pathogen transmission have been identified (Fikrig and Narasimhan, 2006; Hovius et al., 2008b; Murfin and Fikrig, 2017; Simo et al., 2017). Of these, the multi-functional Salp15 has been the most extensively studied. In addition to inhibiting activation of naïve $CD4^+$ T cells by binding to CD4 (Anguita et al., 2002; Tomas-Cortazar et al., 2017) and dendritic cells through interactions with DC-SIGN (Hovius et al., 2008a), Salp15 binds to OspC on spirochetes within saliva, protecting against antibody-mediated killing (Ramamoorthi et al., 2005). Even so, the spirochetes that do complete the journey from midgut to dermis initially have a precarious existence. In mice, dissemination does not occur if the inoculation site is excised within the first several days of tick detachment (Shih et al., 1992). Moreover, spirochetes inoculated into skin are antigenically heterogeneous and do not appear to have a uniform ability to establish infection (Ohnishi et al., 2001). At some point, they overcome the local bottleneck to colonization of their new mammalian host and dissemination (Troy et al., 2013; Rego et al., 2014). The importance of the RpoN/RpoS-regulated genes (Figure 6) and anti-complementary surface lipoproteins for early survival in the mammal was discussed above (also see Chapters 5 and 13).

With their foothold established, spirochetes begin to replicate and migrate outwards along the plane of the skin and downward towards the dermal microvasculature (Figure 7) (Skare et al., 2016; Hyde, 2017). In *ex vivo* gelatin matrices that mimic the extracellular collagenous matrix spirochetes traverse *in vivo* (Zambrano et al., 2004; Dunham-Ems et al., 2009), the pathogen moves at a rapid clip, ~4 microns per second (Harman et al., 2012), far faster than pursuing phagocytes (Vig and

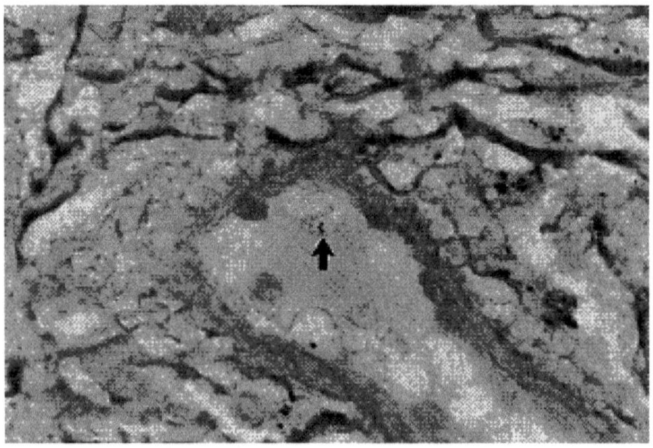

Figure 7. Silver stained biopsy from an erythema migrans lesion showing a spirochete that has penetrated a dermal venule. (Reproduced with permission from Duray, 1989b).

Wolgemuth, 2014). Motility, however, is not uniform; spirochetes transition between a variety of motility states determined by transient adhesions, although the specific adhesins involved have yet to be identified (Harman et al., 2012). Presumably, these movements, along with PAMPs shed by live and dead organisms, trigger local danger signals that result in the accumulation of resident innate immune cells, primarily macrophages and dendritic cells, as well as the recruitment of circulating immune cells with skin-homing capacity (Mullegger et al., 2000; Salazar et al., 2003; Petzke and Schwartz, 2015; Marques et al., 2017). The ensuing inflammatory response gives rise to the hallmark skin lesion, erythema migrans (EM) (Wormser, 2006; Nadelman, 2015; Stanek and Strle, 2018). Eventually, organisms reach densities easily detected by culture, PCR (Nowakowski et al., 2001; Aguero-Rosenfeld et al., 2005; Ruzic-Sabljic and Cerar, 2017), and even histopathology (Duray, 1989b). Not surprisingly, detection of spirochetes by culture or PCR correlates with lesion size (Stupica et al., 2015). The rate of expansion of EM lesions (~20 cm^2 per day) is thought to reflect the speed at which spirochetes migrating away from the bite site are trailed by the local inflammatory response (Dandache and Nadelman, 2008). Along these lines, a mathematical model predicted that the rates of bacterial replication and dissemination within the dermis are the primary determinants of the rate of EM progression (Vig and Wolgemuth, 2014). A cardinal difference between Lyme disease in humans and mice is that the latter do not develop EM (Barthold, 1996); in other words, development of EM can be considered an indication of the lack of local tolerance of humans to the presence of *B. burgdorferi.* Like humans, tick or needle-inoculated non-human primates develop EM (Philipp et al., 1993; Pachner et al., 2001; Embers et al., 2017), as do rabbits (Wheeler et al., 1989; Foley et al., 1995), another non-reservoir host species.

Asymptomatic infection

Clinical manifestations are not an inevitable outcome of tick inoculation with *B. burgdorferi.* As noted previously, seroepidemiologic studies in Lyme disease endemic areas have shown that substantial proportions of persons with antibodies to *B. burgdorferi* are asymptomatic. In Europe, asymptomatic or subclinical infection may be as common as clinically apparent disease, whereas in the U.S. only a small subset of infected patients are asymptomatic (Hanrahan et al., 1984; Steere et al., 2003; Wilhelmsson et al., 2016; Carlsson et al., 2018). In the OspA vaccine trial, 30 of the 269 patients who met the criteria for Lyme disease were classified as having asymptomatic IgG seroconversion to *B. burgdorferi* (Steere et al., 1998). Upon subsequent investigation, 14 of these individuals were found to have had symptoms and/or a rash compatible with EM that was not appreciated during their participation in the study. Eight patients, however, were truly asymptomatic and went untreated. Surprisingly, only one of these eight individuals developed a late complication – arthritis (Steere et al., 2003). Wormser and co-workers (Wormser et al., 2001b) proposed that some asymptomatic infections may be attributed to non-invasive strains. This is another way of stating that at least some uneventful outcomes reflect the general non-permissiveness of humans as hosts for Lyme disease spirochetes. Unfortunately, methodologies to distinguish patients who have cleared "benign" asymptomatic infection from those who may be persistently infected and at risk for subsequent late complications do not exist. The management of asymptomatic persons found to be seroreactive for Lyme disease is an important unresolved issue for practitioners in endemic areas (Wormser et al., 2006). A comparison with syphilis, caused by *Treponema pallidum*, a spirochete that has evolved to persist in humans, is instructive: It is well recognized that patients with latent syphilitic infection are at risk of recrudescent disease, and, therefore, must be treated (Radolf et al., 2019).

Hematogenous dissemination and organ system invasion

That spirochetes disseminate hematogenously during early infection has been known since *B. burgdorferi* was first isolated in the early 1980s (Benach et al., 1983; Steere et al., 1983a). Wormser and colleagues (Wormser et al., 2001a) showed that spirochetemia occurs in approximately 40% of patients with EM but that the extremely low spirochete concentrations in blood (estimated to be ~1 bacterium per 10 ml) necessitate culturing large volumes of plasma (Wormser et al., 2001a) (also see Chapter 22). The low spirochete burdens in blood during early Lyme disease also explains why PCR analysis of blood, a volume-limited technique, has poor diagnostic yield (Aguero-Rosenfeld et al., 2005; Lohr et al., 2018). Since untreated patients cannot be followed prospectively, it is not known whether spirochetemia occurs constantly at low levels or at varying levels intermittently. In a large retrospective review of spirochetemic EM patients evaluated using the large-volume culture technique, hematogenous dissemination was not associated with duration or size of EM (Wormser et al., 2005); these results are in accord with data that borrelial genotypic factors are the predominant determinants of whether invasive infection occurs. Approximately 20% of EM patients have secondary EM-like skin lesions, a clinical indicator of hematogenous dissemination analogous to the rash of secondary syphilis (Wormser et al., 2005). Interestingly, blood cultures were positive in only five of 26 patients with extracutaneous manifestations of Lyme disease, four of whom had concomitant erythema migrans (Nowakowski et al., 2009); these results suggest that spirochetemia occurs predominantly during early infection. Even though spirochetemia may be difficult to detect, the high proportion of constitutional symptoms strongly points to the systemic nature of Lyme disease even when it appears microbiologically localized (Wormser, 2006; Steere et al., 2016). The discovery of metabolic (Molins et al., 2017) and proteomic (Zhou et al., 2020) signatures in the blood of early Lyme disease patients further indicates the systemic nature of earlier illness regardless of whether clinical signs of dissemination are present.

Little is known about the mechanisms by which circulating spirochetes recognize and invade target organs. Studies with cultured human umbilical vein endothelial cells have shown that organisms rapidly attach to vascular endothelium and negotiate their way through intercellular junctions (transmigration), subsequently attaching to subendothelial matrix components (Comstock and Thomas, 1989; Szczepanski et al., 1990). Real-time intravital confocal microscopy revealed that transmigration through capillaries and post-capillary venules *in vivo* is a multi-step process engaged in by only a small percentage of GFP-expressing spirochetes introduced intravenously into mice (Moriarty et al., 2008; Norman et al., 2008). The role of the adhesin BBK32 in trans-endothelial migration was described above.

Aside from skin, the heart, joints, and nervous system are the most affected metastatic sites (see below). How frequently do blood-borne spirochetes gain access to these organ systems? Unlike with experimental animal models, obtaining precise values for the incidence rates of disseminated manifestations of Lyme disease in humans is a major challenge. For example, while some authorities cite incidence rates for carditis as high as 10% (Scheffold et al., 2015), these values are almost certainly overestimates, perhaps reflecting the era when early Lyme disease often went unrecognized and, therefore, untreated. Based on surveillance data obtained from 2001-2010, the CDC determined that cardiac involvement occurs in approximately 1% of reported cases (Forrester et al., 2014). On the other hand, cardiac involvement occurs in diverse strains of laboratory mice and in most inoculated animals (Barthold et al., 1990) (also see Chapter 15). Thus, the comparatively low percentage of carditis in humans suggests that *B. burgdorferi* does not have as strong a tropism for human cardiac tissue. This conclusion is supported by studies with rhesus macaques that found that carditis was absent or mild unless animals were immunosuppressed (Philipp et al., 1993; Cadavid et al., 2004).

In an early report describing 12 U.S. patients with acute disseminated Lyme disease, six with acute cranial neuritis, three with multiple EM, and three with both cranial neuritis and multiple EM, eight patients had spirochetal DNA in their cerebrospinal fluid (CSF) detectable by PCR (Luft et al., 1992). More recent, large European studies involving patients infected with neurotropic *Borrelia* genospecies indicate that these findings greatly overestimated the incidence of CNS invasion during early Lyme disease. During a 5.5-year period, Strle and colleagues (Ogrinc et al., 2013) found neurologic symptoms in only 161 of 2751 (6%) patients with erythema migrans; of these, only 31 (19%) had CSF

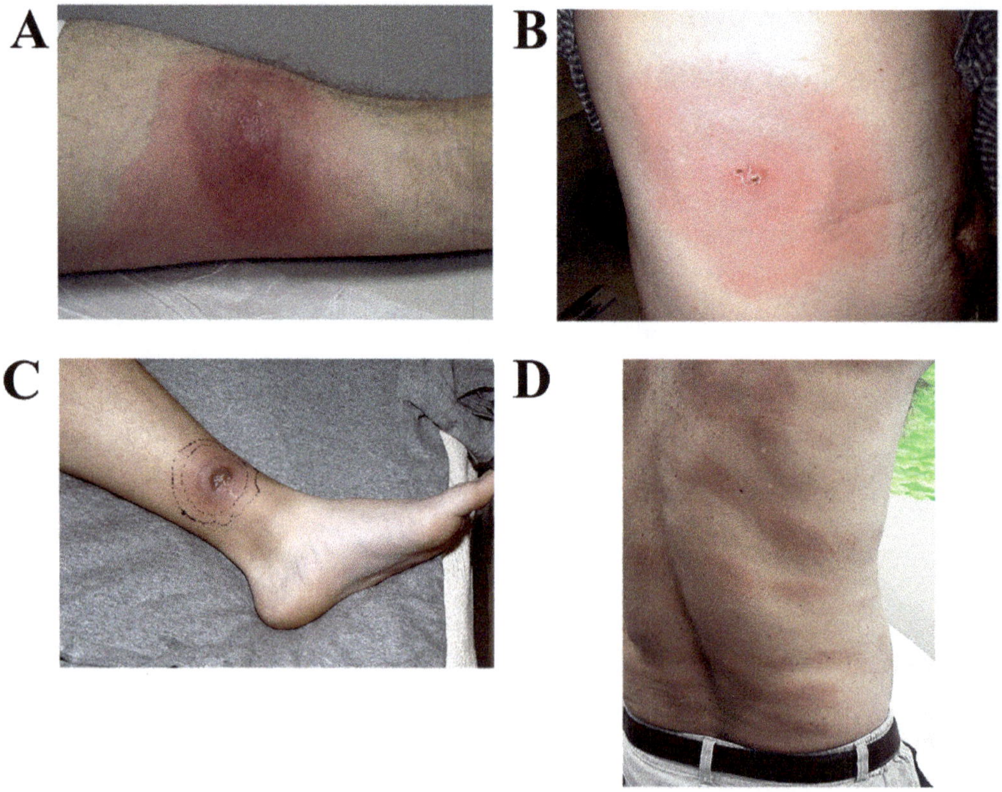

Figure 8. Erythema migrans (EM) lesions: (A) Diffuse EM expanding centripetally over the right quadriceps. (B) EM lesion on the right torso forming a classic "bulls-eye" pattern; note the central eschar (the tick bite site). (C) Ulcerated necrotic and EM lesion. Black lines demarcate area of receding erythema following antibiotic therapy. (D) Multiple EM lesions of disseminated early Lyme disease over the posterior trunk and lower back.

abnormalities, and spirochetes were isolated from only 6 of 127 untreated individuals (Ogrinc et al., 2013). In a subsequent study, they isolated spirochetes from only 12 of 177 persons with Bannwarth Syndrome (15%) (Ogrinc et al., 2016). These latter results bear out Halperin's contention that neurologic complications in Lyme disease often result from inflammation of blood vessels supplying nerve roots and peripheral nerves as opposed to penetration of the blood-brain barrier and/or invasion of brain parenchyma (Halperin, 2008).

Clinical Manifestations

Early studies divided Lyme disease into three distinct stages analogous to those of syphilis (Steere, 2001): Stage 1, erythema migrans; Stage 2, neurologic or cardiac involvement, and Stage 3, arthritis. This staging system is still in use today (Steere et al., 2016). It is particularly important for caregivers on the front lines to think in terms of localized infection, EM, and early and late consequences of dissemination (Trayes et al., 2018; Schoen, 2020). However, only a minority of untreated patients develops all clinical manifestations, although in some patients signs of early and disseminated Lyme disease are present at the same time.

Skin

Numerous clinical series have established that EM is the most common manifestation of *B. burgdorferi* infection in the United States and Europe (Berglund et al., 1995; Steere and Sikand, 2003; Wormser et al., 2006; Enkelmann et al., 2018; Stanek and Strle, 2018). Most patients presenting with this rash have primary EM, the lesion that develops at the site of tick inoculation, typically within 7 to 14 days after detachment. Primary EM can be located anywhere but in adults occurs most commonly below the waist. In North American studies, only 14-32% of individuals presenting with EM recall a tick bite. To distinguish EM from the transient and more localized inflammatory reactions that often develop following

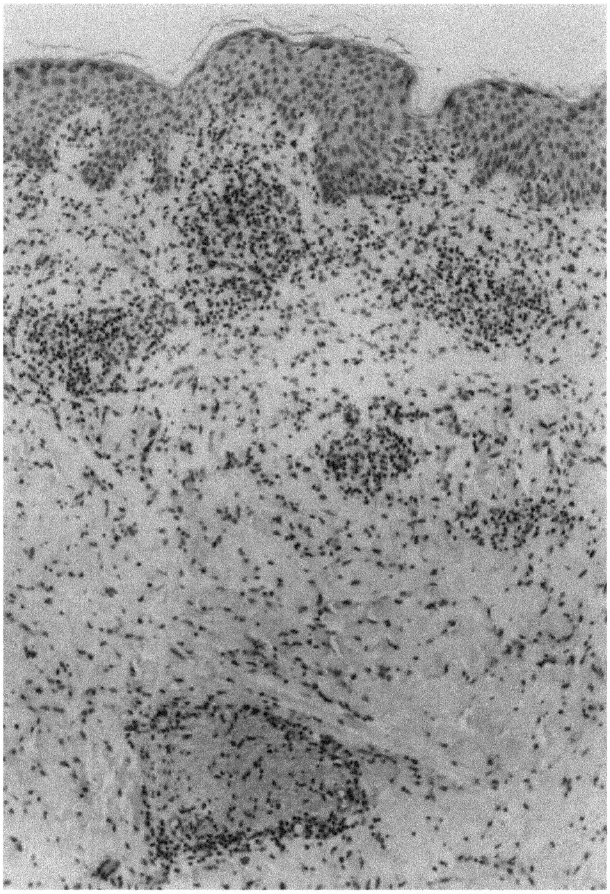

Figure 9. Biopsy taken from the center of an EM lesion showing a superficial and deep perivascular and interstitial infiltrate of lymphocytes and histiocytes. H&E, x 100 (Reprinted with permission from de Koning et al.)

an insect bite, the Centers for Disease Control and Prevention designated 5 cm as the minimum lesion diameter required for clinical diagnosis (Figure 8). While this size criterion improves the specificity of clinical diagnosis, it must be emphasized that *bona fide* EM lesions can be less than 5 cm, particularly when detected shortly after tick detachment. EM is traditionally described as an expanding, annular, erythematous skin lesion with central clearing, the so-called classic "bull's eye rash" (Steere, 2001). Central clearing, however, is not as characteristic as was once thought: two large North American studies found only 37% (Nadelman et al., 1996) and 9% (Smith et al., 2002) of patients had central clearing. Central clearing appears to be more common in Europe, where most cases are due to *B. afzelii* and have a longer duration prior to treatment (Strle et al., 1999; Dandache and Nadelman, 2008). EM-associated systemic illness can vary from none to moderate constitutional symptoms consisting of arthralgia, malaise, fatigue, headache, and low-grade fever and chills (Berglund et al., 1995; Wormser et al., 2006; Dandache and Nadelman, 2008; Stanek and Strle, 2018). Studies of EM patients have reported a difference in the incidence of extracutaneous manifestations in America and Europe. Less than 50% of European patients have extracutaneous manifestations as opposed to more than 75% of patients in the United States (Strle et al., 1999; Dandache and Nadelman, 2008). Regional lymphadenopathy is the most common physical finding associated with EM in both Europe and North America. Differences in EM and associated symptoms caused by *B. garinii* and *B. afzelii* also have been noted (Logar et al., 2004; Strle et al., 2011a). Patients with *B. garinii* had shorter incubation periods; faster evolution of their EM, and more symptomatic lesions (burning, itching, and pain) as well as modestly increased systemic symptomatology.

The most common histological pattern of EM is a superficial and deep dermal infiltrate consisting mostly of lymphocytes, but also containing neutrophils, macrophages, and plasma cells, often in a perivascular distribution (Figure 9) (Duray, 1989b; de Koning, 1993). Over the years, efforts to dissect this cellular response have been predicated on the assumption that it plays a pivotal role in determining the outcome of infection, possibly helping to initiate the ill-defined, treatment-recalcitrant symptomatic state known as post-treatment Lyme disease syndrome (PTLDS) (Crowder et al., 2014; Weitzner et al., 2015; Bouquet et al., 2016) (also see Chapter 23). Using an epidermal blister fluid suction technique to extract dermal cells for immunophenotypic analysis by flow cytometry, Salazar and co-workers (Salazar et al., 2003) showed that EM infiltrates contain diverse cellular elements derived from both the innate and adaptive arms of the immune system, including plasmacytoid dendritic cells and memory-effector T cells. Greater than 80% of lesional T cells expressed CCR5 and/or CXR3, suggesting a strongly polarized Th1 response, confirmed by high levels of IFN-γ in the blister (interstitial) fluids. Th1 polarization of skin-recruited, antigen-sensitized T cells was supported by Glickstein *et al* (Glickstein et al., 2003), who found IFN-γ to be the predominant cytokine produced by peripheral blood leukocytes from culture-confirmed EM incubated with borrelial lysates, and by Jones *et al.* (Jones et al., 2008), who found high levels of

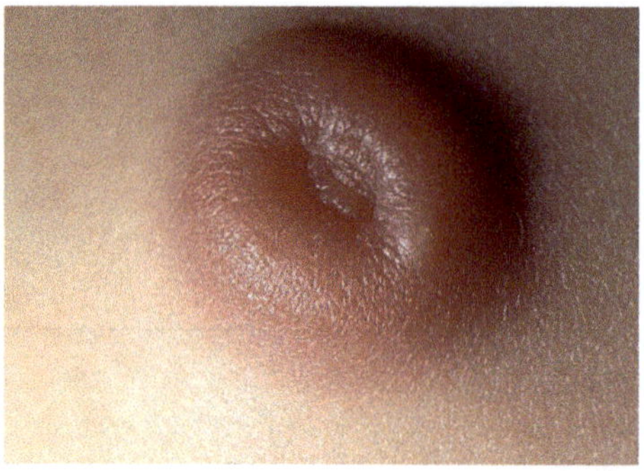

Figure 10. Borrelial lymphocytoma. Image shows a firm, indolent swelling of the left nipple of an eight-year-old boy with prior history of a tick bite and serological evidence of Lyme disease. Photograph courtesy of Ulrich Heininger, University Children's Hospital, Basel, Switzerland.

transcripts for IFN-γ and Th1-associated chemokines in RNAs extracted from EM lesions. The latter study also was noteworthy for its comparison of the chemokine and cytokine mRNA profiles of EM lesions from European and U.S. patients infected with *B. afzelii* and *B. burgdorferi*, respectively. Consistent with the generally milder course of EM in Europe (Strle et al., 1999; Dandache and Nadelman, 2008), they found that lesions from U.S. patients had significantly higher mRNA levels for chemokines associated with macrophage activation and Th1 cell recruitment. Similarly, studies in serum from EM patients, or in macrophage cell cultures stimulated with *Borrelia* isolates from these patients, demonstrated that infection with U.S. *B. burgdorferi* strains elicits higher levels of cytokines and chemokines associated with innate and Th1 adaptive responses than European *B. afzelii* or *B. garinii* strains (Strle et al., 2009).

A recent comparative microarray analysis of mRNAs extracted from 18 EM biopsies and 27 controls has provided an unprecedented global assessment of the transcriptional response elicited by *B. burgdorferi* infection of human skin (Marques et al., 2017). CXCL9, CXCL10, and CXCL11, chemokines strongly implicated in early and late Lyme disease pathogenesis (Lepej et al., 2005; Mullegger et al., 2007; Shin et al., 2010), were the most highly induced genes. Interferon signaling was the top activated pathway, followed by pathways for pattern recognition receptors of bacteria and viruses and DC maturation. Based upon this comprehensive dataset, Marques *et al.* (Marques et al., 2017) proposed a model for the immunopathogenesis of cutaneous *B. burgdorferi* infection that explains the genesis and amplification of the local inflammatory response along with a novel mechanism spirochetes deploy for evading it. Briefly, TLR-mediated activation of local and recruited phagocytes upon ingestion and degradation of spirochetes elicits numerous inflammatory mediators, including Th1-polarizing chemokines. Production of IFN-γ by T cells and type I IFNs by macrophages and plasmacytoid dendritic cells bolsters phagocytosis and elicits myriad interferon-responsive genes; the latter include genes encoding enzymes involved in tryptophan catabolism. The authors hypothesized that spirochetes exploit the inhibition of T cell priming and enhanced development of FoxP3$^+$ Tregs resulting from tryptophan depletion.

In Europe, but not North America, Lyme disease is associated with two additional dermatologic disorders: borrelial lymphocytoma (BL) and acrodermatitis chronica atrophicans (ACA) (Steere et al., 2016; Stanek and Strle, 2018). BL may occur concurrently with EM or a few days before or up to several weeks after EM (Strle et al., 1992; Brehmer-Andersson et al., 1998; Glatz et al., 2015; Maraspin et al., 2016), generally manifesting as solitary, bluish-red nodule on the ear lobe or the nipple area (Figure 10) (Strle et al., 1992; Colli et al., 2004). Histopathologically, BL is characterized by a dense polyclonal lymphocytic infiltrate in the dermis or subcutaneous tissue; differentiation from lymphoma is usually not very difficult since lymphocytic infiltration in BL is polyclonal, while in lymphoma it is oligoclonal or monoclonal (Colli et al., 2004). Maraspin *et al.* (Maraspin et al., 2016) reported an isolation rate of one in three from BL tissue and that *B. afzelii* is the main causative agent; of 13 typed strains, 11 were *B. afzelii*, one was *B. garinii,* and one was *B. bissettii.* Although systemic symptoms are rarely associated with lymphocytoma, most patients report mild symptoms localized to the site of the lesion.

ACA is an extremely late manifestation, typically developing months to years after the onset of untreated infection (Brehmer-Andersson et al., 1998; Brandt et al., 2015). Occurring throughout Northern, Central, and Eastern Europe, ACA is associated primarily with *B. afzelii* infection. Initially the skin lesions are distinguished by inflammation and edema. With time, the lesions become increasingly atrophic. ACA typically involves the distal extensor part of extremities and, less commonly the trunk, sparing the face. Polyneuropathy, small joint arthritis with subluxation, arthritis of the large joints, and periosteal thickening of the bones may occur in the same extremity as ACA.

Lyme neuroborreliosis

The neurologic manifestations of Lyme disease, often referred to as Lyme neuroborrelisosis, reflect the capacity of *B. burgdorferi* sensu lato to invade diverse targets within the peripheral and central nervous system and to cause neurological complications weeks to months after infection (Halperin, 2003; Koedel et al., 2015; Halperin, 2018; Garcia-Monco and Benach, 2019). Early nervous system involvement is usually manifested by the involvement of cranial and/or peripheral nerves or nerve roots typically associated with lymphocytic meningitis. Classical Bannwarth syndrome (BS) is defined as meningoradiculoneuritis; cranial neuritis may be or may not be present (Halperin, 2015; Ogrinc et al., 2016; Shah et al., 2018; Garcia-Monco and Benach, 2019). BS is the most typical manifestation of Lyme neuroborreliosis in adult European populations (Ogrinc et al., 2016); although rare in the U.S., a cluster of five cases recently was reported from the upper Midwest (Shah et al., 2018). Currently, peripheral facial palsy and an isolated aseptic (viral-like) meningitis picture are the most common early neurologic manifestation in the U.S. (Garcia-Monco and Benach, 2019). Initial studies of North American patients with untreated EM found that approximately 15% developed meningitis or cranial neuritis within the first three months after presentation (Steere et al., 1983b; Halperin et al., 1989). It is now believed that early neurologic involvement is much more common in Europe than in the United States, reflecting the prevalence of neurotropic strains of *B. garinii* (Koedel et al., 2015; Steere et al., 2016; Stanek and Strle, 2018).

Although late disseminated *B. burgdorferi* infection has been linked to a variety of nervous system disorders over the years, the neurologic entities now generally accepted as causally associated with late Lyme disease – encephalomyelitis, encephalopathy, and axonal polyneuropathy (Logigian et al., 1990; Logigian and Steere, 1992; Wormser et al., 2006) – are quite rare. As documented in the 2006 IDSA Lyme disease guidelines (Wormser et al., 2006), only one case of encephalomyelitis, nine patients with peripheral neuropathy and seven patients with encephalopathy were seen by the panel members in the 5 years preceding their publication. Peripheral nerve involvement in late Lyme disease is an axonopathy that may be distributed symmetrically or asymmetrically, and in some cases, resemble polyneuritis multiplex (Halperin et al., 1990; Logigian et al., 1990). Patients with this entity complain of paresthesias and hyperthesias and have abnormal nerve conduction. In Europe, peripheral neuropathy may be found in association with ACA but rarely, if ever, occurs in patients without ACA or other Lyme disease manifestations (Kristoferitsch et al., 1988; Mygland et al., 2006). Halperin (Halperin, 2008) has voiced considerable skepticism that Lyme disease is a significant cause of stroke.

Not surprisingly, the inflammatory mechanisms underlying CNS neuroborreliosis are complex, involving both innate and adaptive immune systems. Building upon their *in vivo* work with the primate model, Philipp and co-workers (Ramesh et al., 2003; Bernardino et al., 2008; Ramesh et al., 2008) demonstrated in an elegant series of studies that spirochetes induce profound proinflammatory and apoptotic changes in rhesus macaque glial cells and astroctyes and, most recently, isolated brain slices, that involve diverse chemokines, cytokines, and TLR-mediated signals. Analysis of cerebrospinal fluids from neuroborreliosis patients has identified borrelial antigen-specific $CD4^+$ T cells, $CD8^+$ T cells, and B cells (Cepok et al., 2003; Jacobsen et al., 2003; Rupprecht et al., 2005; Lunemann et al., 2007), as well as unusual subsets of dendritic cells (Pashenkov et al., 2001). Collectively, these results underscore how greatly CNS infection by Lyme disease spirochetes differs from meningitis caused by pyogenic bacteria.

Lyme carditis

As with all forms of disseminated Lyme disease, the incidence of symptomatic cardiac involvement in the United States appears to have decreased dramatically and is now considered to be approximately 1% (Forrester et al., 2014). Available evidence indicates that carditis does not occur if early Lyme disease patients without cardiac manifestations are appropriately treated (Robinson et al., 2015). When first described, concurrent EM and other forms of early Lyme disease was the rule (van der Linde, 1991); in recent years, however, one-half or less of patients present with concurrent EM (Forrester and Mead, 2014; Marcos et al., 2019). In their recent review, Forrester and Mead (Forrester and Mead, 2014) noted a striking male predominance (84%), a finding supported by a retrospective chart review

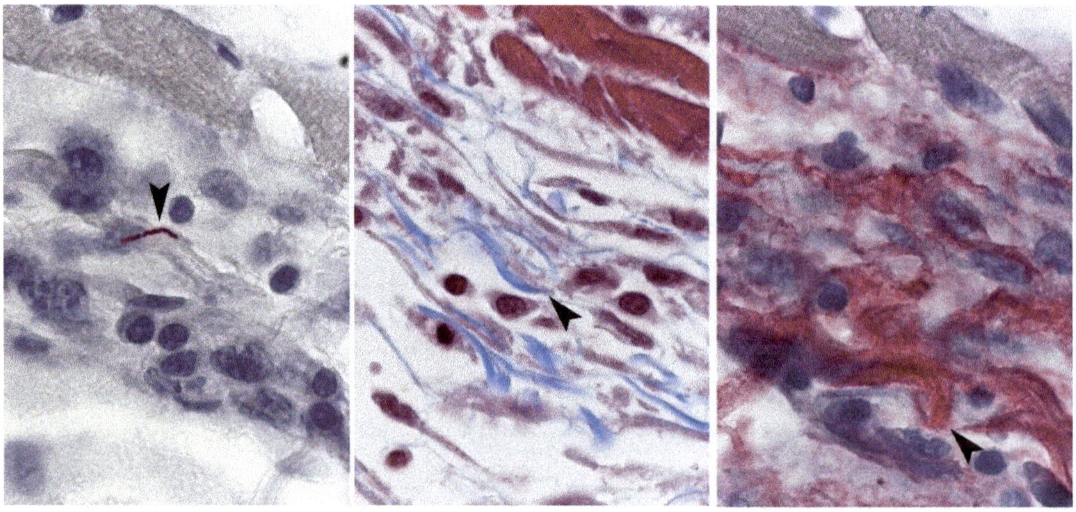

Figure 11. Localization of spirochetes (left panel, immunohistochemistry, arrowhead) with collagen (middle panel, blue, trichrome, arrowhead) and decorin (right panel, red, immunohistochemistry) in the myocardium of a patient with fatal myocarditis. (Reproduced with permission from Muehlenbachs et al., 2016).

conducted at Stony Brook, Long Island (Marcos et al., 2019).

The principal manifestation of Lyme carditis is atrioventricular block proximal to the bundle of His, which can fluctuate and progress rapidly (typically within hours, occasionally within minutes) (Fish et al., 2008; Robinson et al., 2015). Other forms of conduction system disease, such as atrial fibrillation (Zainal et al., 2019) and sick sinus syndrome (Gazendam et al., 2020), also have been reported. Although diffuse myocardial involvement has been seen in cases that have come to autopsy (Duray, 1989a; Tavora et al., 2008; Muehlenbachs et al., 2016), clinically apparent acute myocarditis is uncommon (Robinson et al., 2015). Valvular heart disease also occurs but is extremely rare, with only a handful of microbiologically confirmed cases reported (Nikolić et al., 2020). Two PCR-confirmed cases occurred in the United States (Paim et al., 2018; Haddad et al., 2019), while one European case was attributed to *B. bissettii* (Rudenko et al., 2008). Early European studies reported an etiologic association between Lyme disease and chronic dilated cardiomyopathy, but these have not subsequently been reproduced on either side of the Atlantic (Robinson et al., 2015). Even without therapy, Lyme carditis may be self-limited (Steere et al., 1980). With appropriate therapy, the prognosis is excellent, with a medium time of 6 days to improvement or restoration of normal sinus rhythm.

With only 11 cases in the literature, sudden death is an extremely rare, though, tragically, now well-recognized outcome of Lyme carditis (Marx et al., 2019). In 2013, the CDC called attention to the potential for Lyme disease to cause cardiac fatalities when it reported three cases occurring from November, 2012 to July, 2013 in young individuals (26 to 38 years) from Connecticut, Massachusetts, and New Hampshire (Centers for Disease and Prevention, 2013). Muehlenbachs *et al.* (Muehlenbachs et al., 2016) evaluated autopsy tissue samples from these, plus two additional cases, by light microscopy, immunohistochemistry (IHC), and PCR. Pancarditis with a characteristic "road map" distribution of infiltrates consisting of lymphocytes (T cell predominant but also B cells), histiocytes, and plasma cells were seen in all five. The conduction system was specifically evaluated in two cases; the presence of severe inflammation in the sinoatrial and AV nodes, including intense necrotizing inflammation of the AV node in one, led them to speculate that involvement of the conduction system caused the deaths. However, arrhythmias also seem likely in some. In all five, spirochetes were widely detected by Warthin-Starry silver stain, IHC, and PCR and co-localized with collagen and decorin (Figure 11). The ease of detecting spirochetes in heart samples

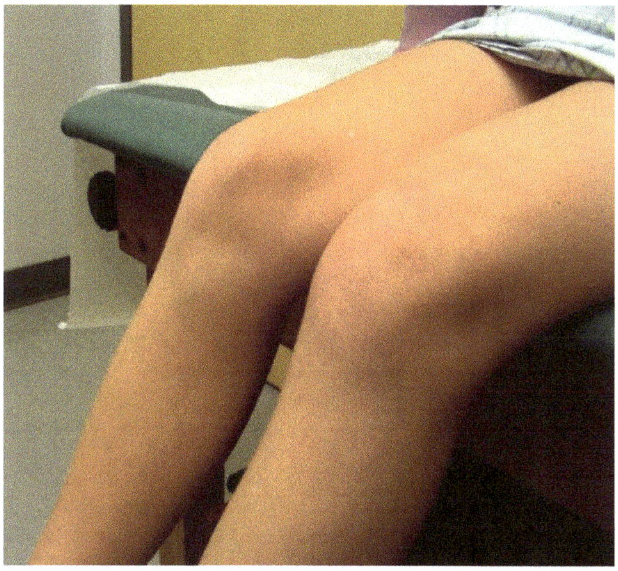

Figure 12. Lyme arthritis manifested as left knee swelling in a young adult. Clinically there was no redness, few constitutional symptoms, and little pain. The patient recovered after a 4-week course of oral antibiotics. Photgraph courtesy of Dr. Jeffrey Thompson, Connecticut Children's Medical Center.

contrasted with the paucity of bacteria detected by IHC and PCR in numerous other organs, indicating that in these patients *B. burgdorferi* clearly possessed a tropism for cardiac tissue.

Lyme arthritis

Arthritis is the most common and serious rheumatologic consequence of spirochetal infection and dissemination. Although Lyme arthritis was once believed to be a North American phenomenon because of the defining epidemic in the 1970s (Steere et al., 1977b) and the paucity of reports of tick-related arthritis in the European literature of the day (Stanek and Strle, 2018), this entity also occurs in European patients without a clear genospecies predominance (Herzer, 1993; Steere, 1997; Haugeberg et al., 2014; Enkelmann et al., 2018). The incubation period for joint involvement can vary from days to years, usually 3 to 6 months, from the time of infection (Steere et al., 1987; Herzer, 1991), and it is presumed that spirochetes typically invade joints well before the onset of arthritic symptoms (Bockenstedt and Wormser, 2014). In some patients, episodes of periarticular pain may precede frank arthritis by weeks to months (Stanek and Strle, 2018). The presence of high-titer antibodies to multiple *B. burgdorferi* antigens in the sera of Lyme arthritis patients (Akin et al., 1999; Wormser et al., 2006) is consistent with the concept of a period of quiescence within the joint prior to the development of clinically evident arthritis (Bockenstedt and Wormser, 2014).

Lyme arthritis is a monoarticular or oligoarticular large joint arthritis characterized by episodes of inflammation with swelling, large effusions and surprisingly little pain (Arvikar and Steere, 2015). The knee is the most commonly affected joint, but other large or small joints, such as the ankle, shoulder, elbow or wrist, may be affected (Figure 12); exclusive involvement of small joints is unusual (Arvikar and Steere, 2015; Stanek and Strle, 2018). In the original study defining the arthritic complications of Lyme disease, prior to use of antibiotics for treatment of the infection, 34 of 55 patients who presented with EM went on to develop arthritis (Steere et al., 1987). Twenty-eight of the 34 patients experienced intermittent arthritis with the episodes of joint swelling lasting from a few days to several months before spontaneously resolving. Over time, the interval between episodes gradually increased before the arthritis spontaneously remitted. Six patients, however, had a year or more of unremitting arthritis. According to a more recent report by the CDC, of ~150,000 newly reported cases of Lyme disease in the U.S. (reported on from 1992-2006) up to 30% present as Lyme arthritis. In contrast, in Europe, the percentage of *B. burgdorferi*-infected individuals who develop arthritis is less than 10% (Enkelmann et al., 2018; Stanek and Strle, 2018). Synovial biopsy of untreated Lyme arthritis shows synovial hypertrophy with vascular proliferation, a mixed infiltrate including T cells, B cells, and follicular dendritic cells (Steere et al., 1979; Duray and Steere, 1986). Consistent with findings in tissue, joint fluid in Lyme arthritis patients contains markedly elevated levels of cytokines and chemokines associated with innate and adaptive immune responses, with particularly high levels of IFNγ-inducible chemokines CXCL9 and CXCL10. The levels of these mediators are nominally lower in serum supporting the idea that these immune responses are occurring locally in joints (Jones et al., 2008; Strle et al., 2017). In addition to inflammatory mediators, matrix-degrading enzymes such as matrix metalloproteinases (MMPs), which degrade both collagen and proteoglycans (Murphy and Nagase, 2008), are believed to contribute to the erosion of cartilage and bone. MMPs have been detected by zymography in the synovial fluids of Lyme arthritis patients (Hu et al., 2001), and in *ex vivo* experiments, *B. burgdorferi* elicits their production by cartilage explants (Lin et al., 2001), chondrocytes (Behera et al., 2004) and monocytes (Gebbia et al., 2004).

The pathogenesis of Lyme arthritis is enigmatic from several standpoints. Though undoubtedly arising from hematogenous dissemination of spirochetes into joints or periarticular tissues, with only two documented exceptions (Snydman et al., 1986; Schmidli et al., 1988), efforts to recover live spirochetes from synovial fluid as a rule have been unsuccessful. In 1994, Nocton and co-workers (Nocton et al., 1994) made two seminal observations: (i) *B. burgdorferi* DNA can be detected in synovial fluids from the large majority (75 of 88) of Lyme arthritis patients and (ii) PCR-positive patients usually respond well to antimicrobial therapy. In PCR-positive patients, resolution of arthritis with antibiotics confirms that spirochete-elicited inflammation drives the arthritic process (Bockenstedt and Wormser, 2014). However, where the spirochetes producing the DNA are hiding in these individuals and what triggers the inflammation after a presumptive period of quiescence remain mysteries (Bockenstedt and Wormser, 2014). The murine model offers some help with the first conundrum. In mice, spirochetes do not infect the joint spaces per se but are visualized in collagen-rich, periarticular structures (Barthold et al., 1991; Bockenstedt and Wormser, 2014).

Although Lyme arthritis usually resolves following appropriate antibiotic therapy, some patients (estimated to be ~10%) have a persistent proliferative synovitis, characterized by synovial hyperplasia, vascular damage, intense inflammation, and fibrosis, termed post-infectious, antibiotic-refractory Lyme arthritis (Steere and Angelis, 2006; Chan and Pollock, 2015). Patients diagnosed with this entity are strongly seropositive but usually lack a PCR signal for *B. burgdorferi* DNA in joints (Wormser et al., 2006; Li et al., 2011). The central mechanistic question is how spirochetes initiate a localized, dysregulated inflammatory process that persists after they have been eradicated (Singh and Girschick, 2004; Shin et al., 2007; Shen et al., 2010; Strle et al., 2012). The increased prevalence of certain HLA alleles that are also associated with rheumatoid arthritis (Steere et al., 1990) spawned the hypothesis that antibiotic-refractory Lyme arthritis is a form of infection-induced autoimmunity caused by molecular mimicry (Klempner and Huber, 1999; Bolz and Weis, 2004). Subsequent studies identified a sequence in OspA as the molecular mimic for a sequence in hLFA-1 (Gross et al., 1998; Chen et al., 1999) and reported that the HLA-DR alleles overrepresented in antibiotic-refractory arthritis patients bind the OspA mimotope (Steere et al., 2006). The public health consequences of these publications were enormous since OspA was the antigenic component of the Lyme disease vaccine released at approximately the same time (Sigal et al., 1998; Steere et al., 1998). Unsubstantiated concerns about autoimmunity due to vaccination with OspA contributed to the commercial failure of the vaccine and its withdrawal from the market (Plotkin, 2011; Willyard, 2014). In subsequent publications, Steere's group backed away from the association of persistent synovitis with T cell reactivity to either OspA or LFA-1 (Kannian et al., 2007; Drouin et al., 2013). However, in support of the concept of spirochete-induced autoimmunity, the Steere group has now identified four novel self-antigens, ECGF, ApoB-100, Annexin A2, and MMP10, which are targets of T and B cell responses in patients with refractory Lyme arthritis (Drouin et al., 2013; Pianta et al., 2015; Crowley et al., 2016; Arvikar et al., 2017; Strle et al., 2017; Steere, 2020).

Although the etiology remains unclear, much has been learned in recent years about the dysregulated inflammatory processes that underlie post-infectious Lyme arthritis (Steere, 2020). For example, patients with a particular single nucleotide polymorphism in TLR1, particularly those who are infected with *B. burgdorferi* RST1 strains, appear to be predisposed to excessive Th1-associated inflammation (Strle et al., 2012). Compared with antibiotic-responsive Lyme arthritis patients, persons with post-infectious arthritis had lower levels of $CD4^{+}CD25^{+}$ regulatory T cells in their synovial fluids and their regulatory T cells were less effective at downregulating Th1 effector responses (Vudattu et al., 2013). High throughput RNA sequencing revealed a robust IFNγ response in synovium that correlates inversely with expression of genes involved in tissue repair and drives differentiation of fibroblast-like synoviocytes into immune effector cells that are thought to amplify and perpetuate these dysregulated Th1 responses (Lochhead et al., 2019a; Lochhead et al., 2019b). An alternative, or perhaps complementary, mechanism for persistent inflammation based on innate immunity recently has been proposed. Jutras *et al.* (Jutras et al., 2019) reported that *B. burgdorferi* cannot recycle peptidoglycan and sheds peptidoglycan fragments (muropeptides) into its environment. They postulated that peptidoglycan fragments shed into synovial fluid prior to the eradication of bacteria causes chronic activation of NOD2-dependent arthritogenic pathways. Thus, the infection with certain *Borrelia* strains in genetically predisposed individuals appears to set the stage for dysregulated immune responses with particularly high levels of IFNγ that promote synovial hypertrophy, inhibit tissue repair, and are permissive to the development of autoreactive T and B cell responses in joints. Microbial remnants such as peptidoglycan, which can persist in the absence of active spirochete infection, may further perpetuate these responses.

Lyme Disease in Children

As noted earlier, Lyme disease began as an affliction of children (Steere et al., 1977b) and it continues so to this day. The clinical manifestations of Lyme disease in children, as in adults, mainly involve the skin, central nervous system, joints, and heart (Feder, 2008; Sood, 2015). The primary EM lesion in children is most commonly found on the head and neck (26%), arms or legs (25%), or lower back (24%) (Salazar et al., 1993; Gerber et al., 1996). Awareness of EM in the head and neck is important because such lesions can be hidden by the hairline. Less frequently lesions can be found over the abdomen (9%), the axilla (8%), the groin (5%), or the chest (3%) (Gerber et al., 1996). Most single EM lesions in children are uniformly erythematous; however, they also can have central erythema or clearing and, in rare instances, painful central ulcerations (Figure 8C). Children also may present with constitutional symptoms, including fever (24%), fatigue (58%), headache (42%) and arthralgias (33%) (Gerber et al., 1996). The presence of multiple EMs has been documented in up to 40% of children (Gerber et al., 1996; Arnez et al., 2003) and most have associated fever (45%), fatigue (80%), and headache (70%). Similar to adults with Lyme disease, ~20% of children may present with constitutional symptoms as the only manifestation (Feder et al., 1993; Gerber et al., 1996). Practitioners in endemic areas need to have a high index of suspicion to avoid missing the diagnosis in such children.

Of course, arthritis also may be the presenting manifestation of Lyme disease in children (Steere et al., 1977b; Szer et al., 1991). As in adults, the incubation period is highly variable with a mean of about 4 months (Szer et al., 1991). In contrast to the early era, most children who present with Lyme arthritis today do not have a history of EM. In these children, the rash was either unseen, or perhaps they never had it at all. Lyme arthritis in children usually involves single large joints, most commonly the knee (Gerber et al., 1998). However, series describing monoarticular arthritis of the hips and elbows in children have appeared of late (Cruz et al., 2017; Gendelberg and Hennrikus, 2018). Fever occurs in 25-50% of children with Lyme arthritis, as opposed to adults who are usually afebrile. Joint fluid obtained from these children may have large numbers of leukocytes (> 50,000 cells/mm^3) with a predominance of neutrophils. If not suspected, the disease can be mistaken for septic arthritis caused by pyogenic bacteria (*e.g.*, *Staphylococcus aureus*). A recent series comparing Lyme arthritis of the hip to septic arthritis emphasized that the former is more likely to occur in children who are afebrile and without leukocytosis (Cruz et al., 2017). If treated appropriately, the prognosis is excellent and with rare exceptions a full recovery is the outcome. Antibiotic-refractory Lyme arthritis is unusual in children (Daikh et al., 2013; Sood, 2015).

Lyme disease in children also may present with neurologic manifestations (Cook et al., 1997; Eppes et al., 1999; Sood, 2015). Most children with neuroborreliosis do not provide a history of EM. The most common neurologic presentations in children include facial nerve palsy and lymphocytic meningitis (Cook et al., 1997; Eppes et al., 1999; Sood, 2015). Children with Lyme meningitis also may have increased intracranial pressure, and some have optic neuritis and papilledema (Kan et al., 1998); increased intracranial pressure may occur in the absence of meningitis (Kan et al., 1998; Sood, 2006). Though rare, BS should be considered in children in Europe or who have traveled to Europe who present with characteristic radicular pain (Sood, 2015). The CSF pleocytosis is predominantly lymphocytic; in one study, a CSF containing 10% or more neutrophils had a negative predictive value of 99% for Lyme disease compared to viral meningitis (Shah et al., 2005). More severe neurologic manifestations, such as myelitis and severe encephalitis, are exceedingly rare in young children. Though rare, Lyme carditis may also occur in children (Bolourchi et al., 2019). When it occurs, the most common manifestation in children is transient high-grade AV block (Silver et al., 2007). Less commonly, children may present in heart failure because of myocarditis. The prognosis of both neurologic and cardiac Lyme disease in children is excellent following appropriate therapy (Wormser et al., 2006).

Maternal-fetal transmission of *B. burgdorferi* has never been definitively documented, and epidemiologic investigations have failed to establish an association between infection and adverse outcomes of pregnancy (Strobino et al., 1993; Gerber and Zalneraitis, 1994; Sood, 2015).

Reinfection

Recurrent episodes of Lyme disease can theoretically be due to either relapse or reinfection (Nadelman and Wormser, 2007). For EM to be due to relapse, the lesion must occur at the same site as the original infection and within a relatively short time interval. Reinfection is defined as EM occurring at a site distant from the initial lesion, often with a punctum or eschar indicative of a recent tick bite, months to years after the initial treated episode. Evidence for the relapse scenario is scant. Two groups have reported their inability to re-isolate spirochetes from the site of EM following therapy, although administration of antibiotics at the time of re-culture confounds interpretation of these data (Berger et al., 1992; Nadelman et al., 1993). In contrast, longitudinal studies conducted in highly endemic areas, Westchester, NY and Block Island, RI, found that up to 20% of patients treated for EM met criteria for reinfection (Nowakowski et al., 2003; Krause et al., 2006). An impressive report provided compelling molecular evidence based on *ospC* genotyping of isolates from 17 patients with 22 paired episodes of EM that recurrences are caused by different *B. burgdorferi* strains (Nadelman and Wormser, 2013). Using statistical modeling, this group subsequently argued that elicitation of durable strain-specific immunity explains why patients were reinfected with different strains (Khatchikian et al., 2014). Without serological data, however, this hypothesis remains unproven, and there is evidence from the mouse model that protective immunity following tick inoculation wanes within a year (Piesman et al., 1997). Furthermore, Barthold and Bockenstedt (Barthold and Bockenstedt, 1993) showed that levels of antibodies capable of conferring passive protection fall off much more rapidly than antibodies detected by ELISA with whole cell lysates. The implication of this finding is that high reactivity in serodiagnostic tests may not be an accurate indicator of immune status. One also must remember that a substantial percentage of persons with EM do not have detectable antibodies, even after treatment (Wormser, 2006). Such individuals will be susceptible to reinfection within a relatively short time if they live in an endemic area or frequently engage in high-risk activities.

Serologic response

The need for accurate serologic tests for diagnosis of Lyme disease has been a major driving force behind efforts to elucidate the humoral response to the spirochete. Consequently, the antibody response to *B. burgdorferi* sensu lato has been extensively dissected over the years with increasingly sophisticated methodologies in order to identify borrelial polypeptides that are not only highly immunogenic in a large percentage of patients but also specific for *B. burgdorferi* (Craft et al., 1986; Dressler et al., 1993; Engstrom et al., 1995; Aguero-Rosenfeld et al., 2005; Nowalk et al., 2006; Lohr et al., 2018) (also see Chapter 22). This combination has not been easy to find. Cross-reactivity between *B. burgdorferi* FlaB, the major flagellar sheath protein, and flagellar antigens of commensal bacteria is the primary reason why this polypeptide, which induces a robust antibody response early in infection (Dressler et al., 1993), fell out of favor as a diagnostic antigen (Aguero-Rosenfeld et al., 2005; Lohr et al., 2018). OspC also induces a strong antibody response during early infection (Fung et al., 1994), consistent with the murine studies demonstrating its critical role for the establishment of infection following tick inoculation, but immunodominant epitopes of OspC proteins from different isolates are extremely heterogeneous (Lohr et al., 2018). A number of other borrelial proteins, many, but not all, lipoproteins, are strong immunogens but are hypervariable, expressed relatively late in the course of disease, and/or recognized by an insufficient percentage of patients (Aguero-Rosenfeld et al., 2005; Wilske et al., 2007; Lohr et al., 2018). Ironically, the antigenically variable VlsE protein, which in the murine model facilitates evasion of antibody responses, has a region with a conserved epitope that is highly immunogenic in early as well as late Lyme disease patients and has been exploited for improved serodiagnosis in Europe as well as North America (Liang et al., 1999) (Liang et al., 1999; Wormser et al., 2013; Lohr et al., 2018; Stanek and Strle, 2018; Zannoli et al., 2020).

Because approximately one-half of patients with EM do not mount detectable antibody responses to the pathogen, lack of seroreactivity cannot be used to rule out the diagnosis of EM (Wormser et al., 2006; Dandache and Nadelman, 2008). There is evidence that serodiagnostic sensitivity is improved by employing borrelial antigens known to be expressed *in vivo* during early infection (Brandt et al., 2019). Seroreactivity increases substantially following therapy for EM (Vaz et al., 2001; Dandache and

Nadelman, 2008), indicating that killing of organisms causes liberation of spirochetal antigens and enhances processing for antibody production. Most untreated patients have detectable antibodies within a month of infection. Accordingly, clinical manifestations caused by the dissemination of spirochetes, are for the most part, accompanied by seroreactivity (Wormser et al., 2006; Lohr et al., 2018; Stanek and Strle, 2018). The flip side of this coin is that IgG antibodies can be present for life and, therefore, cannot be used by themselves as indicators of active infection. While assays using whole cell lysates are less affected by strain variability and the heterogeneity in patient responses than those using single antigens, they can yield false-positives due to the presence of cross-reactive, background antibodies. This problem has been circumvented by the development of the two-tiered testing algorithm in which immunoblotting is used to confirm serologic reactivity by ELISA (Aguero-Rosenfeld et al., 2005; Wormser et al., 2006; Lohr et al., 2018; Stanek and Strle, 2018). The success of the two-tiered approach rests upon the implementation of strict criteria for the interpretation of immunoblots based on reactivity with a small subset of borrelial polypeptides defined by their SDS-PAGE mobilities (*i.e.*, apparent molecular masses) (Dressler et al., 1993). It is noteworthy, considering the T cell reactivity studies discussed below, that reactivity with OspA is not a criterion for immunoblot reactivity.

The general picture that has emerged over the years is that IgG and IgM antibodies to the spirochete develop slowly and are directed against an increasingly diverse array of proteins as infection progresses (Craft et al., 1986; Dressler et al., 1993; Lohr et al., 2018). The earliest responses are to flagellin B (FlaB), OspC (25 kDa) and BmpA (39 kDa) with responses to a number of additional antigens, such as VlsE, fibronectin-binding protein (BBK32), and decorin-binding protein A (DbpA), developing as *B. burgdorferi* disseminates (Bacon et al., 2003; Aguero-Rosenfeld et al., 2005; Wilske et al., 2007; Lohr et al., 2018). This temporal pattern is consistent with the notion that the bacterium draws upon an expanding repertoire of differentially expressed proteins once within its vertebrate host, including the phased expression of paralogous surface-exposed lipoproteins (Caimano et al., 2019). Carroll and co-workers (Nowalk et al., 2006) completed an extensive proteomics-based survey of murine and human antibody responses to *B. burgdorferi* proteins, concluding that the contours of the two antibody profiles share many similarities during early infection.

What role, if any, do antibodies play in control of the disease process in humans? This question is far more difficult to answer for humans than in mice because humans cannot be followed serially without treatment and they cannot be passively immunized prior to experimental infection. The presence of intense and diverse antibody responses in patients with late manifestations of disease (Steere, 2001) is clear-cut evidence that, as in mice, spirochetes can persist despite high titers of circulating antibodies (see below). Nevertheless, there is *in vitro* evidence that the antibodies produced during human infection have bactericidal activity (Pavia et al., 1997; Callister et al., 2002) and can passively protect animals against inoculation with *in vitro* cultivated spirochetes (Fikrig et al., 1994). Robinson and coworkers (Blum et al., 2018) dramatically advanced this notion using single-cell paired-chain and bulk heavy-chain antibody repertoire sequencing to investigate circulating B cell responses during early Lyme disease. They demonstrated expanded memory B cell and plasmablast populations that produce antibodies with specificities for several *B. burgdorferi* antigens (*e.g.*, VlsE, DbpA, DbpB, and OspC) and that recombinant mAbs recognizing several of these inhibit spirochete growth *in vitro*. They also noted that robust plasmablast responses encoding *B. burgdorferi*-inhibitory antibodies were associated with more rapid disease resolution. As stated earlier, infection with less invasive borrelial strains has been proposed as one explanation for the high prevalence of asymptomatic *B. burgdorferi* infection in endemic areas (Wormser et al., 2001b). The capacity of humoral responses to contain and even eliminate spirochetes provides a second, non-mutually exclusive explanation, which also is in accord with the idea that humans lack reservoir-competence for the bacterium (Bockenstedt and Radolf, 2014).

Chronic Lyme disease and post-treatment Lyme disease syndrome

Please see the comprehensive discussion by Marques and Hu (Chapter 23). At one time, the term "chronic Lyme disease" (CLD) referred to late stage manifestations of untreated *B. burgdorferi* sensu lato infection, such as encephalomyelitis, encephalopathy, neuropathy, arthritis, and ACA (Koedel et al., 2015). Over the years, however, this term has come to denote persons with chronic, subjective complaints (*e.g.*, neurocognitive syndromes, mood disorders, fibromyalgia-type symptoms, and chronic fatigue), in the absence of a documented history of Lyme disease or objective physical or laboratory evidence of infection, who some practitioners maintain are afflicted by treatment-recalcitrant Lyme disease spirochetes (Cameron et al., 2004; Feder et al., 2007; Marques, 2008; Baker, 2010; Stricker and Johnson, 2014; Lantos, 2015). Few "mainstream" authorities believe there is convincing evidence for persistent infection (or any infection, for that matter) in these individuals (Feder et al., 2007; Koedel et al., 2015; Baker and Wormser, 2017; Shapiro et al., 2017). Clinicians who routinely diagnose CLD contend that the conventional laboratory tests mainstream practitioners rely upon are notoriously unreliable (Cameron et al., 2004; Stricker and Johnson, 2014). To further support their position, these self-designated "Lyme literate" physicians and their staunch supporters in advocacy groups cite myriad studies purporting to show that during chronic infection *B. burgdorferi* assumes unusual morphologies (variously termed "L forms", "round bodies", "L forms", etc.) and/or metabolic states resistant to standard treatment regimens (Stricker and Johnson, 2011; Lantos, 2015; Sharma et al., 2015; Timmaraju et al., 2015; Feng et al., 2016b; Baker and Wormser, 2017). Even so, the pathophysiologic rationale is difficult to grasp (Halperin, 2008; Baker and Wormser, 2017). As noted by Halperin, for spirochetes to persist and cause disease without provoking discernable inflammation or an immunologic response manifested by detectable serum antibodies, *B. burgdorferi* would have to differ from virtually every other known chronic, systemic bacterial pathogen. The outcome of this questionable line of thinking is prolonged antimicrobial therapy of uninfected (though truly unwell) individuals, sometimes involving unusual regimens and agents of unproven efficacy for Lyme disease (Feder et al., 2007; Lantos et al., 2015). The potential dangers of these treatment regimens have been well documented (Lantos, 2015). While historically a North American phenomenon (Barbour and Fish, 1993), CLD appears to be gaining traction in Europe and, understandably, generating increasing concern among European investigators (Koedel et al., 2015; Gentilini and Bricaire, 2019; Peri et al., 2019).

Almost 30 years ago, clinicians began to report that some Lyme disease patients have persistent nonspecific symptoms after receiving an adequate course of therapy (Dinerman and Steere, 1992; Asch et al., 1994; Shadick et al., 1994). In its 2006 guidelines, the Infectious Diseases Society of America created a working definition for this entity, now called "post-treatment Lyme disease syndrome" or PTLDS: clinical symptoms persisting at least six months after treatment for Lyme disease in persons who lack objective evidence of treatment failure, reinfection, or relapse (Wormser et al., 2006; Aucott, 2015; Lantos, 2015; Strle and Strle, 2020). PTLDS differs from CLD in one particularly important respect – patients with PTLDS have unequivocal documentation for appropriately treated Lyme disease. PTLDS occurs in only a small percentage of treated patients. As noted by Lantos in an excellent review (Lantos, 2015), only 222 (3.8%) of 5846 patients screened to participate in clinical trials for PTLDS had credible evidence for past Lyme disease, while less than 10% of patients in ten prospective studies of EM and early disseminated Lyme disease described persistent symptoms such as myalgias or fatigue 9 or more months following treatment. Fortunately, multiple longitudinal studies in the U.S. and Europe report that functional impairment by PTLDS diminishes over time (Cerar et al., 2010; Wormser et al., 2015; Stupica et al., 2018a; Stupica et al., 2018b; Wormser et al., 2020).

As with CLD, the central questions surrounding PTLDS are the underlying mechanism(s) of persistent symptomatology and whether extended antimicrobial therapy, or indeed any antimicrobial treatment in addition to that initially administered, is beneficial (Delong et al., 2012; Fallon et al., 2012; Klempner et al., 2013). Convincing evidence for *Borrelia* infection in PTLDS patients has not been obtained using PCR, culture, or xenodiagnosis of blood and/or spinal fluid (Klempner et al., 2001; Fallon et al., 2008; Marques et al., 2014). In the absence of direct microbiologic evidence, response to extended therapy has come to be regarded as an alternative means of addressing the persistence question (Klempner et al., 2013). Five double-blind, placebo-controlled treatment trials in the U.S. and Europe have failed to demonstrate lasting benefit for extended therapy (Klempner et al., 2001; Kaplan et al., 2003; Krupp et al., 2003; Fallon et al., 2008; Berende et al., 2016). "Lyme literate" skeptics reject these results on methodologic grounds (Delong et al., 2012; Fallon et al., 2012; Klempner et al., 2013), also pointing to the evidence, per above, that spirochetes *in vivo* become "persisters" refractory to conventionally used agents (Stricker and Johnson, 2011; Sharma et al., 2015; Timmaraju et al., 2015;

Feng et al., 2016b). Proponents of the persister theory were heartened by a highly publicized study maintaining, based on scant data for recovery of viable organisms, that spirochetes persisted in treated rhesus macaques (Embers et al., 2012). Agents that purportedly kill persister spirochetes *in vitro* have been identified (Sharma et al., 2015; Theophilus et al., 2015; Feng et al., 2016a; Feng et al., 2016c), while one group has reported that the combination of daptomycin plus doxycycline eradicated persister (aggregated, stationary phase) organisms from mice, whereas doxycycline plus ceftriaxone did not (Feng et al., 2019). One can safely predict that at some time in the future clinical trials of regimens believed capable of eliminating persister forms will be done.

In the absence of compelling evidence for bacterial persistence, other investigators lean towards the notion that spirochetes initiate an inflammatory process that is not turned off once the organisms have been eradicated. Using intravital microscopy in mice, Bockenstedt *et al.* (Bockenstedt et al., 2012) showed that *B. burgdorferi* antigens, but not infectious spirochetes, remain adjacent to cartilage for extended periods after antibiotic treatment; they proposed that this residual debris is the cause of ongoing inflammation. Investigations in humans also provide some, but not entirely consistent, support for a post-infectious inflammatory process. Strle *et al.* (Strle et al., 2014) reported that high IL-23 levels at presentation with early Lyme disease correlated with development of persistent symptoms, while Aucott *et al.* (Aucott et al., 2016) found elevated levels of the T cell chemokine CCL19 in PTLDS patients. Wormser's group found that PTLDS patients have significantly higher C-reactive protein (CRP) levels than treated patients who become asymptomatic (Uhde et al., 2016) or persons with chronic fatigue/fibromyalgia (Uhde et al., 2018). RNA-seq analysis of early Lyme disease patients revealed a differential gene expression signature that continued for at least six months post-treatment in symptomatic, but not asymptomatic, treated individuals (Bouquet et al., 2016). In contrast, another group recently reported that transcriptional signatures associated with early disseminated Lyme disease normalized within six months of antibiotic treatment (Petzke et al., 2020). And so it goes....

Future Directions

Lyme disease is now regarded as one of the first in a series of explosive infectious disease phenomena that gave rise to the contemporary concept of emerging infectious diseases - the notion that the infectious agents that impact human welfare are dynamic and strongly influenced by factors both within and outside the normal sphere of human activity. In fact, since the first description of the disease, we have come to appreciate the global dimensions of the threat to human health posed by *B. burgdorferi* sensu lato and its arthropod vectors. Over the years, the notoriety and novelty of the disease have attracted the attention of numerous investigators; as this chapter and the volume that contains it attest, our knowledge of the disorder, its etiologic agent, and the natural world that shape them has expanded at a meteoric pace. So, what remains to be done from the perspective of disease in humans? Vaccine development is an obvious answer, although pharmaceutical companies have shown at best muted enthusiasm for the development of second-generation vaccines after the commercial failure and precipitous withdrawal of the first-generation vaccine based on OspA. Clearly, alternative prophylactic strategies that focus on preventing enzootic transmission and spread of the pathogen warrant great emphasis. But the major thrust needs to be in the realms of cellular and molecular pathogenesis.

While genetic and bioinformatics breakthroughs undoubtedly will be made, the ultimate question is how to interpret this information in the context of human disease. As any syphilologist will affirm, Lyme disease researchers are blessed not only with a cultivatable organism but an excellent and extremely versatile animal model whose full potential has not yet been realized. However, to quote a now clichéd maxim, "humans are not mice", and extrapolation is always fraught with danger. Nor are all the major features of human disease recapitulated by the murine model. As examples one need only point to erythema migrans and central nervous system complications, both of which underscore the existence of unique interactions between the spirochete and human cells. With the exception of AIDS, and, perhaps, COVID-19 most recently, no infectious disease presents a stronger paradigm for translational research. With the advent of systems biology and related genomics-based technologies, investigators need to build upon findings made with surrogate systems rather than regarding them as ends in themselves. The stakes are high, judging by the intensity of the Lyme disease wars currently being waged on the medical and legal fronts (see

Chapter 23). We not only need to understand how *B. burgdorferi* accomplishes its parasitic strategy once introduced into humans, but also how the outcomes of this interaction differ from those seen in experimental systems and how engagement of the human immune system by the bacterium appears to leave debilitating footprints after its progress has been arrested.

Acknowledgements

This work was supported by grants R01 AI-29735 and R21 AI-139940 from the National Institute of Allergy and Infectious Diseases to JDR, by funds generously provided by the Department of Research of Connecticut Children's Medical Center to JDR, by grants R21 AI14491 and R21 AI149278 from the National Institute of Allergy and Infectious Diseases to KS and grant P3-0296 from the Slovenian Research Agency to FS.

References

Adams, P.P., Flores Avile, C., and Jewett, M.W. (2017). A dual luciferase reporter system for *B. burgdorferi* measures transcriptional activity during tick-pathogen interactions. Front Cell Infect Microbiol *7*, 225. https://doi.org/10.3389/fcimb.2017.00225

Adeolu, M., and Gupta, R.S. (2014). A phylogenomic and molecular marker based proposal for the division of the genus *Borrelia* into two genera: the emended genus *Borrelia* containing only the members of the relapsing fever *Borrelia*, and the genus *Borreliella* gen. nov. containing the members of the Lyme disease *Borrelia* (*Borrelia burgdorferi* sensu lato complex). Antonie Van Leeuwenhoek *105*, 1049-1072. https://doi.org/10.1007/s10482-014-0164-x

Aenishaenslin, C., Bouchard, C., Koffi, J.K., Pelcat, Y., and Ogden, N.H. (2016). Evidence of rapid changes in Lyme disease awareness in Canada. Ticks Tick Borne Dis *7*, 1067-1074. https://doi.org/10.1016/j.ttbdis.2016.09.007

Afzelius, A. (1910). Verhandlungen der dermatologischen Gesellschaft zu Stockholm. Arch Dermatol Syph, 405-406.

Afzelius, A. (1921). Erythema chronicum migrans. Acta Derm Venereol, 120-125.

Aguero-Rosenfeld, M.E., Wang, G., Schwartz, I., and Wormser, G.P. (2005). Diagnosis of Lyme borreliosis. Clin Microbiol Rev *18*, 484-509. https://doi.org/10.1128/CMR.18.3.484-509.2005

Akin, E., McHugh, G.L., Flavell, R.A., Fikrig, E., and Steere, A.C. (1999). The immunoglobulin (IgG) antibody response to OspA and OspB correlates with severe and prolonged Lyme arthritis and the IgG response to P35 correlates with mild and brief arthritis. Infect Immun *67*, 173-181.

Akins, D.R., Bourell, K.W., Caimano, M.J., Norgard, M.V., and Radolf, J.D. (1998). A new animal model for studying Lyme disease spirochetes in a mammalian host-adapted state. J Clin Invest *101*, 2240-2250. https://doi.org/10.1172/JCI2325

Alghaferi, M.Y., Anderson, J.M., Park, J., Auwaerter, P.G., Aucott, J.N., Norris, D.E., and Dumler, J.S. (2005). Borrelia burgdorferi ospC heterogeneity among human and murine isolates from a defined region of northern Maryland and southern Pennsylvania: lack of correlation with invasive and noninvasive genotypes. J Clin Microbiol *43*, 1879-1884. https://doi.org/10.1128/JCM.43.4.1879-1884.2005

Anderson, J.F. (1988). Mammalian and avian reservoirs for *Borrelia burgdorferi*. Ann N Y Acad Sci *539*, 180-191. https://doi.org/10.1111/j.1749-6632.1988.tb31852.x

Anguita, J., Ramamoorthi, N., Hovius, J.W., Das, S., Thomas, V., Persinski, R., Conze, D., Askenase, P.W., Rincon, M., Kantor, F.S., *et al.* (2002). Salp15, an ixodes scapularis salivary protein, inhibits CD4(+) T cell activation. Immunity *16*, 849-859.

Antipov, D., Korobeynikov, A., McLean, J.S., and Pevzner, P.A. (2016). hybridSPAdes: an algorithm for hybrid assembly of short and long reads. Bioinformatics *32*, 1009-1015. https://doi.org/10.1093/bioinformatics/btv688

Arnez, M., Pleterski-Rigler, D., Luznik-Bufon, T., Ruzic-Sabljic, E., and Strle, F. (2003). Solitary and multiple erythema migrans in children: comparison of demographic, clinical and laboratory findings. Infection *31*, 404-409. https://doi.org/10.1007/s15010-003-4007-3

Arvikar, S.L., and Steere, A.C. (2015). Diagnosis and treatment of Lyme arthritis. Infect Dis Clin North Am *29*, 269-280. https://doi.org/10.1016/j.idc.2015.02.004

Arvikar, S.L., Crowley, J.T., Sulka, K.B., and Steere, A.C. (2017). Autoimmune arthritides, rheumatoid arthritis, psoriatic arthritis, or peripheral spondyloarthritis following Lyme disease. Arthritis & Rheumatology *69*, 194-202. https://doi.org/10.1002/art.39866

Asch, E.S., Bujak, D.I., Weiss, M., Peterson, M.G., and Weinstein, A. (1994). Lyme disease: an infectious and postinfectious syndrome. J Rheumatol *21*, 454-461.

Aucott, J.N. (2015). Posttreatment Lyme disease syndrome. Infect Dis Clin North Am *29*, 309-323. https://doi.org/10.1016/j.idc.2015.02.012

Aucott, J.N., Soloski, M.J., Rebman, A.W., Crowder, L.A., Lahey, L.J., Wagner, C.A., Robinson, W.H., and Bechtold, K.T. (2016). CCL19 as a chemokine risk factor for posttreatment Lyme disease syndrome: a prospective clinical cohort study. Clin Vaccine Immunol *23*, 757-766. https://doi.org/10.1128/CVI.00071-16

Bacon, R.M., Biggerstaff, B.J., Schriefer, M.E., Gilmore, R.D., Jr., Philipp, M.T., Steere, A.C., Wormser, G.P., Marques, A.R., and Johnson, B.J. (2003). Serodiagnosis of Lyme disease by kinetic enzyme-linked immunosorbent assay using recombinant VlsE1 or peptide antigens of Borrelia burgdorferi compared with 2-tiered testing using whole-cell lysates. J Infect Dis *187*, 1187-1199. https://doi.org/10.1086/374395

Baker, P.J. (2010). Chronic Lyme disease: in defense of the scientific enterprise. FASEB J *24*, 4175-4177. https://doi.org/10.1096/fj.10-167247

Baker, P.J., and Wormser, G.P. (2017). The clinical relevance of studies on *Borrelia burgdorferi* persisters. Am J Med *130*, 1009-1010. https://doi.org/10.1016/j.amjmed.2017.04.014

Balmelli, T., and Piffaretti, J.C. (1995). Association between different clinical manifestations of Lyme disease and different species of Borrelia burgdorferi sensu lato. Res Microbiol *146*, 329-340. https://doi.org/10.1016/0923-2508(96)81056-4

Bannwarth, A. (1941). Chronische lymphocytare Meningitis, entzundliche Polyneuritis und "Rheumatismus". Ein Beitrag zum Problem "Allergie und Nervensystem". Arch Psychiatri, 284-376.

Bannwarth, A. (1944). Zur klinik und pathogenese de "chronischen lymphocytaren meningitis". Arch Psychiatr Nervenkr, 161-185.

Barbour, A.G., and Hayes, S.F. (1986). Biology of *Borrelia* species. Microbiol Rev *50*, 381-400.

Barbour, A.G., and Fish, D. (1993). The biological and social phenomenon of Lyme disease. Science *260*, 1610-1616.

Barbour, A.G., and Travinsky, B. (2010). Evolution and distribution of the *ospC* gene, a transferable serotype determinant of *Borrelia burgdorferi*. MBio *1*. https://doi.org/10.1128/mBio.00153-10

Barbour, A.G. (2017). Infection resistance and tolerance in *Peromyscus* spp., natural reservoirs of microbes that are virulent for humans. Semin Cell Dev Biol *61*, 115-122. https://doi.org/10.1016/j.semcdb.2016.07.002

Barbour, A.G., Adeolu, M., and Gupta, R.S. (2017). Division of the genus Borrelia into two genera (corresponding to Lyme disease and relapsing fever groups) reflects their genetic and phenotypic distinctiveness and will lead to a better understanding of these two groups of microbes (Margos et al. (2016) There is inadequate evidence to support the division of the genus Borrelia. Int. J. Syst. Evol. Microbiol. doi: 10.1099/ijsem.0.001717). Int J Syst Evol Microbiol *67*, 2058-2067. https://doi.org/10.1099/ijsem.0.001815

Barbour, A.G., and Benach, J.L. (2019). Discovery of the Lyme Disease Agent. MBio *10*. https://doi.org/10.1128/mBio.02166-19

Barcena-Uribarri, I., Thein, M., Maier, E., Bonde, M., Bergstrom, S., and Benz, R. (2013). Use of nonelectrolytes reveals the channel size and oligomeric constitution of the Borrelia burgdorferi P66 porin. PLoS One *8*, e78272. https://doi.org/10.1371/journal.pone.0078272

Barthold, S.W., Beck, D.S., Hansen, G.M., Terwilliger, G.A., and Moody, K.D. (1990). Lyme borreliosis in selected strains and ages of laboratory mice. J Infect Dis *162*, 133-138.

Barthold, S.W., Persing, D.H., Armstrong, A.L., and Peeples, R.A. (1991). Kinetics of *Borrelia burgdorferi* dissemination and evolution of disease after intradermal inoculation of mice. Am J Pathol *139*, 263-273.

Barthold, S.W., and Bockenstedt, L.K. (1993). Passive immunizing activity of sera from mice infected with *Borrelia burgdorferi*. Infect Immun *61*, 4696-4702.

Barthold, S.W., Fikrig, E., Bockenstedt, L.K., and Persing, D.H. (1995). Circumvention of outer surface protein A immunity by host-adapted *Borrelia burgdorferi*. Infect Immun *63*, 2255-2261.

Barthold, S.W. (1996). Lyme borreliosis in the laboratory mouse. J Spirochet Tick-Borne Dis, 22-44.

Barthold, S.W., Cadavid, D., and Philipp, M.T. (2010). Animal Models of Lyme Borreliosis. In *Borrelia*: Molecular Biology, Host Interaction, and Pathogenesis, D.S. Samuels, and J.D. Radolf, eds. (Norfolk, UK: Caister Academic Press), pp. 359-411.

Bashir, A., Klammer, A., Robins, W.P., Chin, C.S., Webster, D., Paxinos, E., Hsu, D., Ashby, M., Wang, S., Peluso, P., *et al.* (2012). A hybrid approach for the automated finishing of bacterial genomes. Nat Biotechnol *30*, 701-707. https://doi.org/10.1038/nbt.2288

Baum, E., Randall, A.Z., Zeller, M., and Barbour, A.G. (2013). Inferring epitopes of a polymorphic antigen amidst broadly cross-reactive antibodies using protein microarrays: a study of OspC proteins of *Borrelia burgdorferi*. PLoS One *8*, e67445. https://doi.org/10.1371/journal.pone.0067445

Beck, G., Habicht, G.S., Benach, J.L., and Coleman, J.L. (1985). Chemical and biologic characterization of a lipopolysaccharide extracted from the Lyme disease spirochete (*Borrelia burgdorferi*). J Infect Dis *152*, 108-117.

Behera, A.K., Thorpe, C.M., Kidder, J.M., Smith, W., Hildebrand, E., and Hu, L.T. (2004). *Borrelia burgdorferi*-induced expression of matrix metalloproteinases from human chondrocytes requires mitogen-activated protein kinase and Janus kinase/signal transducer and activator of transcription signaling pathways. Infect Immun *72*, 2864-2871. https://doi.org/10.1128/iai.72.5.2864-2871.2004

Belfaiza, J., Postic, D., Bellenger, E., Baranton, G., and Girons, I.S. (1993). Genomic fingerprinting of *Borrelia burgdorferi* sensu lato by pulsed-field gel electrophoresis. J Clin Microbiol *31*, 2873-2877.

Belperron, A.A., and Bockenstedt, L.K. (2001). Natural antibody affects survival of the spirochete *Borrelia burgdorferi* within feeding ticks. Infect Immun *69*, 6456-6462. https://doi.org/10.1128/IAI.69.10.6456-6462.2001

Ben-Menachem, G., Kubler-Kielb, J., Coxon, B., Yergey, A., and Schneerson, R. (2003). A newly discovered cholesteryl galactoside from *Borrelia burgdorferi*. Proc Natl Acad Sci U S A *100*, 7913-7918. https://doi.org/10.1073/pnas.1232451100

Benach, J.L., Bosler, E.M., Hanrahan, J.P., Coleman, J.L., Habicht, G.S., Bast, T.F., Cameron, D.J., Ziegler, J.L., Barbour, A.G., Burgdorfer, W., *et al.* (1983). Spirochetes isolated from the blood of two patients with Lyme disease. N Engl J Med *308*, 740-742. https://doi.org/10.1056/NEJM198303313081302

Benoit, V.M., Fischer, J.R., Lin, Y.P., Parveen, N., and Leong, J.M. (2011). Allelic variation of the Lyme disease spirochete adhesin DbpA influences spirochetal binding to decorin, dermatan sulfate, and mammalian cells. Infect Immun *79*, 3501-3509. https://doi.org/10.1128/IAI.00163-11

Berende, A., ter Hofstede, H.J., Vos, F.J., van Middendorp, H., Vogelaar, M.L., Tromp, M., van den Hoogen, F.H., Donders, A.R., Evers, A.W., and Kullberg, B.J. (2016). Randomized Trial of Longer-Term Therapy for Symptoms Attributed to Lyme Disease. N Engl J Med *374*, 1209-1220. https://doi.org/10.1056/NEJMoa1505425

Berger, B.W., Johnson, R.C., Kodner, C., and Coleman, L. (1992). Failure of *Borrelia burgdorferi* to survive in the skin of patients with antibiotic-treated Lyme disease. J Am Acad Dermatol *27*, 34-37. https://doi.org/10.1016/0190-9622(92)70152-6

Berger, B.W., Johnson, R.C., Kodner, C., and Coleman, L. (1995). Cultivation of *Borrelia burgdorferi* from human tick bite sites: a guide to the risk of infection. J Am Acad Dermatol *32*, 184-187.

Berglund, J., Eitrem, R., Ornstein, K., Lindberg, A., Ringer, A., Elmrud, H., Carlsson, M., Runehagen, A., Svanborg, C., and Norrby, R. (1995). An epidemiologic study of Lyme disease in southern Sweden. N Engl J Med *333*, 1319-1327. https://doi.org/10.1056/NEJM199511163332004

Bernard, Q., Smith, A.A., Yang, X., Koci, J., Foor, S.D., Cramer, S.D., Zhuang, X., Dwyer, J.E., Lin, Y.P., Mongodin, E.F., *et al.* (2018). Plasticity in early immune evasion strategies of a bacterial pathogen. Proc Natl Acad Sci U S A *115*, E3788-E3797. https://doi.org/10.1073/pnas.1718595115

Bernardino, A.L., Myers, T.A., Alvarez, X., Hasegawa, A., and Philipp, M.T. (2008). Toll-like receptors: insights into their possible role in the pathogenesis of lyme neuroborreliosis. Infect Immun *76*, 4385-4395. https://doi.org/10.1128/IAI.00394-08

Bissett, M.L., and Hill, W. (1987). Characterization of *Borrelia burgdorferi* strains isolated from *Ixodes pacificus* ticks in California. J Clin Microbiol *25*, 2296-2301.

Blum, L.K., Adamska, J.Z., Martin, D.S., Rebman, A.W., Elliott, S.E., Cao, R.R.L., Embers, M.E., Aucott, J.N., Soloski, M.J., and Robinson, W.H. (2018). Robust B cell responses predict rapid resolution of Lyme disease. Front Immunol *9*, 1634. https://doi.org/10.3389/fimmu.2018.01634

Bockenstedt, L.K., Gonzalez, D.G., Haberman, A.M., and Belperron, A.A. (2012). Spirochete antigens persist near cartilage after murine Lyme borreliosis therapy. J Clin Invest *122*, 2652-2660. https://doi.org/10.1172/JCI58813

Bockenstedt, L.K., and Radolf, J.D. (2014). Xenodiagnosis for posttreatment Lyme disease syndrome: resolving the conundrum or adding to it? Clin Infect Dis *58*, 946-948. https://doi.org/10.1093/cid/cit942

Bockenstedt, L.K., and Wormser, G.P. (2014). Unraveling Lyme disease. Arthritis Rheumatol *66*, 2313-2323. https://doi.org/10.1002/art.38756

Bolourchi, M., Silver, E.S., and Liberman, L. (2019). Advanced heart block in children with Lyme disease. Pediatr Cardiol *40*, 513-517. https://doi.org/10.1007/s00246-018-2003-8

Bolz, D.D., and Weis, J.J. (2004). Molecular mimicry to Borrelia burgdorferi: pathway to autoimmunity? Autoimmunity *37*, 387-392. https://doi.org/10.1080/08916930410001713098

Bontemps-Gallo, S., Lawrence, K., and Gherardini, F.C. (2016). Two different virulence-related regulatory pathways in *Borrelia burgdorferi* are directly affected by osmotic fluxes in the bllood meal of feeding *Ixodes* ticks. PLoS Pathog *12*, e1005791. https://doi.org/10.1371/journal.ppat.1005791

Botos, I., Majdalani, N., Mayclin, S.J., McCarthy, J.G., Lundquist, K., Wojtowicz, D., Barnard, T.J., Gumbart, J.C., and Buchanan, S.K.

(2016). Structural and functional characterization of the LPS transporter LptDE from Gram-negative pathogens. Structure *24*, 965-976. https://doi.org/10.1016/j.str.2016.03.026

Bouchard, C., Leonard, E., Koffi, J.K., Pelcat, Y., Peregrine, A., Chilton, N., Rochon, K., Lysyk, T., Lindsay, L.R., and Ogden, N.H. (2015). The increasing risk of Lyme disease in Canada. Can Vet J *56*, 693-699.

Bouquet, J., Soloski, M.J., Swei, A., Cheadle, C., Federman, S., Billaud, J.N., Rebman, A.W., Kabre, B., Halpert, R., Boorgula, M., *et al.* (2016). Longitudinal transcriptome analysis reveals a sustained differential gene expression signature in patients treated for acute Lyme disease. MBio *7*, e00100-00116. https://doi.org/10.1128/mBio.00100-16

Brandt, F.C., Ertas, B., Falk, T.M., Metze, D., and Boer-Auer, A. (2015). Histopathology and immunophenotype of acrodermatitis chronica atrophicans correlated with *ospA* and *ospC* genotypes of *Borrelia* species. J Cutan Pathol *42*, 674-692. https://doi.org/10.1111/cup.12550

Brandt, K.S., Ullmann, A.J., Molins, C.R., Horiuchi, K., Biggerstaff, B.J., and Gilmore, R.D. (2019). Evaluation of in vivo expressed *Borrelia burgdorferi* antigens for improved IgM serodiagnosis of early Lyme disease. Diagn Microbiol Infect Dis *93*, 196-202. https://doi.org/10.1016/j.diagmicrobio.2018.09.012

Brangulis, K., Akopjana, I., Petrovskis, I., Kazaks, A., Kraiczy, P., and Tars, K. (2018). Crystal structure of the membrane attack complex assembly inhibitor BGA71 from the Lyme disease agent *Borrelia bavariensis*. Sci Rep *8*, 11286. https://doi.org/10.1038/s41598-018-29651-9

Brangulis, K., Akopjana, I., Petrovskis, I., Kazaks, A., and Tars, K. (2019). Crystal structure of *Borrelia burgdorferi* outer surface protein BBA69 in comparison to the paralogous protein CspA. Ticks Tick Borne Dis *10*, 1135-1141. https://doi.org/10.1016/j.ttbdis.2019.06.009

Brehmer-Andersson, E., Hovmark, A., and Asbrink, E. (1998). Acrodermatitis chronica atrophicans: histopathologic findings and clinical correlations in 111 cases. Acta Derm Venereol *78*, 207-213.

Brinkerhoff, R.J., Folsom-O'Keefe, C.M., Tsao, K., and Diuk-Wasser, M.A. (2011). Do birds affect Lyme disease risk? Range expansion of the vector-borne pathogen *Borrelia burgdorferi*. *9*, 103-110. https://doi.org/10.1890/090062

Brissette, C.A., and Gaultney, R.A. (2014). That's my story, and I'm sticking to it--an update on B. burgdorferi adhesins. Front Cell Infect Microbiol *4*, 41. https://doi.org/10.3389/fcimb.2014.00041

Brisson, D., and Dykhuizen, D.E. (2004). *ospC* diversity in *Borrelia burgdorferi*: different hosts are different niches. Genetics *168*, 713-722. https://doi.org/10.1534/genetics.104.028738

Brisson, D., Dykhuizen, D.E., and Ostfeld, R.S. (2008). Conspicuous impacts of inconspicuous hosts on the Lyme disease epidemic. Proc Biol Sci *275*, 227-235. https://doi.org/10.1098/rspb.2007.1208

Brisson, D., Zhou, W., Jutras, B.L., Casjens, S., and Stevenson, B. (2013). Distribution of cp32 prophages among Lyme disease-causing spirochetes and natural diversity of their lipoprotein-encoding *erp* loci. Appl Environ Microbiol *79*, 4115-4128. https://doi.org/10.1128/AEM.00817-13

Brown, E.L., Wooten, R.M., Johnson, B.J., Iozzo, R.V., Smith, A., Dolan, M.C., Guo, B.P., Weis, J.J., and Hook, M. (2001). Resistance to Lyme disease in decorin-deficient mice. J Clin Invest *107*, 845-852. https://doi.org/10.1172/JCI11692

Brunet, L.R., Sellitto, C., Spielman, A., and Telford, S.R., 3rd (1995). Antibody response of the mouse reservoir of *Borrelia burgdorferi* in nature. Infect Immun *63*, 3030-3036.

Brunner, J.L., LoGiudice, K., and Ostfeld, R.S. (2008). Estimating reservoir competence of *Borrelia burgdorferi* hosts: prevalence and infectivity, sensitivity, and specificity. J Med Entomol *45*, 139-147. https://doi.org/10.1603/0022-2585(2008)45[139:ercobb]2.0.co;2

Brusca, J.S., McDowall, A.W., Norgard, M.V., and Radolf, J.D. (1991). Localization of outer surface proteins A and B in both the outer membrane and intracellular compartments of *Borrelia burgdorferi*. J Bacteriol *173*, 8004-8008.

Bunikis, I., Denker, K., Ostberg, Y., Andersen, C., Benz, R., and Bergstrom, S. (2008). An RND-type efflux system in *Borrelia burgdorferi* is involved in virulence and resistance to antimicrobial compounds. PLoS Pathog *4*, e1000009. https://doi.org/10.1371/journal.ppat.1000009

Bunikis, J., Garpmo, U., Tsao, J., Berglund, J., Fish, D., and Barbour, A.G. (2004). Sequence typing reveals extensive strain diversity of the Lyme borreliosis agents *Borrelia burgdorferi* in North America and *Borrelia afzelii* in Europe. Microbiology *150*, 1741-1755. https://doi.org/10.1099/mic.0.26944-0

Burgdorfer, W., Barbour, A.G., Hayes, S.F., Benach, J.L., Grunwaldt, E., and Davis, J.P. (1982). Lyme disease-a tick-borne spirochetosis? Science *216*, 1317-1319.

Burgdorfer, W. (1986). Discovery of the Lyme disease spirochete: a historical review. Zentralbl Bakteriol Mikrobiol Hyg A *263*, 7-10.

Burgdorfer, W., and Gage, K.L. (1986). Susceptibility of the black-legged tick, *Ixodes scapularis*, to the Lyme disease spirochete, *Borrelia burgdorferi*. Zentralbl Bakteriol Mikrobiol Hyg A *263*, 15-20.

Burgdorfer, W., Anderson, J.F., Gern, L., Lane, R.S., Piesman, J., and Spielman, A. (1991). Relationship of *Borrelia burgdorferi* to its arthropod vectors. Scand J Infect Dis Suppl *77*, 35-40.

Burgdorfer, W. (1993). How the discovery of *Borrelia burgdorferi* came about. Clin Dermatol *11*, 335-338.

Cadavid, D., Bai, Y., Hodzic, E., Narayan, K., Barthold, S.W., and Pachner, A.R. (2004). Cardiac involvement in non-human primates infected with the Lyme disease spirochete *Borrelia burgdorferi*. *84*, 1439-1450. https://doi.org/10.1038/labinvest.3700177

Caimano, M.J., Eggers, C.H., Hazlett, K.R., and Radolf, J.D. (2004). RpoS is not central to the general stress response in *Borrelia burgdorferi* but does control expression of one or more essential virulence determinants. Infect Immun *72*, 6433-6445. https://doi.org/10.1128/IAI.72.11.6433-6445.2004

Caimano, M.J., Eggers, C.H., Gonzalez, C.A., and Radolf, J.D. (2005). Alternate sigma factor RpoS is required for the in vivo-specific repression of *Borrelia burgdorferi* plasmid lp54-borne ospA and lp6.6 genes. J Bacteriol *187*, 7845-7852. https://doi.org/10.1128/JB.187.22.7845-7852.2005

Caimano, M.J., Iyer, R., Eggers, C.H., Gonzalez, C., Morton, E.A., Gilbert, M.A., Schwartz, I., and Radolf, J.D. (2007). Analysis of the RpoS regulon in *Borrelia burgdorferi* in response to mammalian host signals provides insight into RpoS function during the enzootic cycle. Mol Microbiol *65*, 1193-1217. https://doi.org/10.1111/j.1365-2958.2007.05860.x

Caimano, M.J., Dunham-Ems, S., Allard, A.M., Cassera, M.B., Kenedy, M., and Radolf, J.D. (2015). Cyclic di-GMP modulates gene expression in Lyme disease spirochetes at the tick-mammal interface to promote spirochete survival during the blood meal and tick-to-mammal transmission. Infect Immun *83*, 3043-3060. https://doi.org/10.1128/IAI.00315-15

Caimano, M.J., Drecktrah, D., Kung, F., and Samuels, D.S. (2016). Interaction of the Lyme disease spirochete with its tick vector. Cell Microbiol *18*, 919-927. https://doi.org/10.1111/cmi.12609

Caimano, M.J., Groshong, A.M., Belperron, A., Mao, J., Hawley, K., Luthra, A., Graham, D.E., Earnhart, C.G., Marconi, R.T., Bockenstedt, L.K., *et al.* (2019). The RpoS gatekeeper in *Borrelia burgdorferi*: an invariant regulatory scheme that promotes spirochete persistence in reservoir hosts and niche diversity. Front Microbiol *in press*.

Caine, J.A., and Coburn, J. (2015). A short-term *Borrelia burgdorferi* infection model identifies tissue tropisms and bloodstream survival conferred by adhesion proteins. Infect Immun *83*, 3184-3194. https://doi.org/10.1128/IAI.00349-15

Caine, J.A., and Coburn, J. (2016). Multifunctional and redundant roles of *Borrelia burgdorferi* outer surface proteins in tissue adhesion, colonization, and eomplement evasion. Front Immunol *7*, 442. https://doi.org/10.3389/fimmu.2016.00442

Caine, J.A., Lin, Y.P., Kessler, J.R., Sato, H., Leong, J.M., and Coburn, J. (2017). *Borrelia burgdorferi* outer surface protein C (OspC) binds complement component C4b and confers bloodstream survival. Cell Microbiol *19*. https://doi.org/10.1111/cmi.12786

Callister, S.M., Jobe, D.A., Agger, W.A., Schell, R.F., Kowalski, T.J., Lovrich, S.D., and Marks, J.A. (2002). Ability of the borreliacidal antibody test to confirm Lyme disease in clinical practice. Clin Diagn Lab Immunol *9*, 908-912.

Cameron, D., Gaito, A., Harris, N., Bach, G., Bellovin, S., Bock, K., Bock, S., Burrascano, J., Dickey, C., Horowitz, R., *et al.* (2004). Evidence-based guidelines for the management of Lyme disease. Expert Rev Anti Infect Ther *2*, S1-13. https://doi.org/ERI0201S [pii]

Campagna, J., Lavoie, P.E., Birnbaum, N.S., and Furman, D.P. (1983). Lyme disease in northern California. West J Med *139*, 319-323.

Carlsson, H., Ekerfelt, C., Henningsson, A.J., Brudin, L., and Tjernberg, I. (2018). Subclinical Lyme borreliosis is common in south-eastern Sweden and may be distinguished from Lyme neuroborreliosis by sex, age and specific immune marker patterns. Ticks Tick Borne Dis *9*, 742-748. https://doi.org/10.1016/j.ttbdis.2018.02.011

Carlsson, S.A., Granlund, H., Jansson, C., Nyman, D., and Wahlberg, P. (2003). Characteristics of erythema migrans in Borrelia afzelii and Borrelia garinii infections. Scand J Infect Dis *35*, 31-33. https://doi.org/10.1080/0036554021000026978

Carrasco, S.E., Troxell, B., Yang, Y., Brandt, S.L., Li, H., Sandusky, G.E., Condon, K.W., Serezani, C.H., and Yang, X.F. (2015). Outer surface protein OspC is an antiphagocytic factor that protects *Borrelia burgdorferi* from phagocytosis by macrophages. Infect Immun *83*, 4848-4860. https://doi.org/10.1128/IAI.01215-15

Casjens, S., Palmer, N., van Vugt, R., Huang, W.M., Stevenson, B., Rosa, P., Lathigra, R., Sutton, G., Peterson, J., Dodson, R.J., *et al.* (2000). A bacterial genome in flux: the twelve linear and nine circular extrachromosomal DNAs in an infectious isolate of the Lyme disease spirochete Borrelia burgdorferi. Mol Microbiol *35*, 490-516.

Casjens, S.R., Mongodin, E.F., Qiu, W.G., Luft, B.J., Schutzer, S.E., Gilcrease, E.B., Huang, W.M., Vujadinovic, M., Aron, J.K., Vargas, L.C., *et al.* (2012). Genome stability of Lyme disease spirochetes: comparative genomics of *Borrelia burgdorferi* plasmids. PLoS One *7*, e33280. https://doi.org/10.1371/journal.pone.0033280

Castillo-Ramirez, S., Fingerle, V., Jungnick, S., Straubinger, R.K., Krebs, S., Blum, H., Meinel, D.M., Hofmann, H., Guertler, P., Sing, A., *et al.* (2016). Trans-Atlantic exchanges have shaped the population structure of the Lyme disease agent Borrelia burgdorferi sensu stricto. Sci Rep *6*, 22794. https://doi.org/10.1038/srep22794

Centers for Disease, C., and Prevention (2013). Three sudden cardiac deaths associated with Lyme carditis - United States, November 2012-July 2013. MMWR Morb Mortal Wkly Rep *62*, 993-996.

Cepok, S., Zhou, D., Vogel, F., Rosche, B., Grummel, V., Sommer, N., and Hemmer, B. (2003). The immune response at onset and during recovery from Borrelia burgdorferi meningoradiculitis. Arch Neurol *60*, 849-855. https://doi.org/10.1001/archneur.60.6.849

Cerar, D., Cerar, T., Ruzic-Sabljic, E., Wormser, G.P., and Strle, F. (2010). Subjective symptoms after treatment of early Lyme disease. Am J Med *123*, 79-86. https://doi.org/10.1016/j.amjmed.2009.05.011

Cerar, T., Strle, F., Stupica, D., Ruzic-Sabljic, E., McHugh, G., Steere, A.C., and Strle, K. (2016). Differences in genotype, clinical features, and inflammatory potential of *Borrelia burgdorferi* sensu stricto strains from Europe and the United States. Emerg Infect Dis *22*, 818-827. https://doi.org/10.3201/eid2205.151806

Chan, S.S., and Pollock, A.N. (2015). Lyme arthritis. Pediatr Emerg Care *31*, 680-681. https://doi.org/10.1097/PEC.0000000000000576

Charon, N.W., Goldstein, S.F., Marko, M., Hsieh, C., Gebhardt, L.L., Motaleb, M.A., Wolgemuth, C.W., Limberger, R.J., and Rowe, N. (2009). The flat-ribbon configuration of the periplasmic flagella of *Borrelia burgdorferi* and its relationship to motility and morphology. J Bacteriol *191*, 600-607. https://doi.org/10.1128/JB.01288-08

Charon, N.W., Cockburn, A., Li, C., Liu, J., Miller, K.A., Miller, M.R., Motaleb, M.A., and Wolgemuth, C.W. (2012). The unique paradigm of spirochete motility and chemotaxis. Annu Rev Microbiol *66*, 349-370. https://doi.org/10.1146/annurev-micro-092611-150145

Chen, J., Field, J.A., Glickstein, L., Molloy, P.J., Huber, B.T., and Steere, A.C. (1999). Association of antibiotic treatment-resistant Lyme arthritis with T cell responses to dominant epitopes of outer surface protein A of *Borrelia burgdorferi*. Arthritis Rheum *42*, 1813-1822. https://doi.org/10.1002/1529-0131(199909)42:9<1813::AID-ANR4>3.0.CO;2-0

Chiang, S.M., and Schellhorn, H.E. (2010). Evolution of the RpoS regulon: origin of RpoS and the conservation of RpoS-dependent regulation in bacteria. J Mol Evol *70*, 557-571. https://doi.org/10.1007/s00239-010-9352-0

Chmelar, J., Kotal, J., Kopecky, J., Pedra, J.H.F., and Kotsyfakis, M. (2016). All for one and one for all on the tick-host battlefield. Trends Parasitol *32*, 368-377. https://doi.org/10.1016/j.pt.2016.01.004

Clark, K.L., Leydet, B.F., and Threlkeld, C. (2014). Geographical and genospecies distribution of *Borrelia burgdorferi* sensu lato DNA detected in humans in the USA. J Med Microbiol *63*, 674-684. https://doi.org/10.1099/jmm.0.073122-0

Coipan, E.C., Jahfari, S., Fonville, M., Oei, G.A., Spanjaard, L., Takumi, K., Hovius, J.W., and Sprong, H. (2016). Imbalanced presence of *Borrelia burgdorferi* s.l. multilocus sequence types in clinical manifestations of Lyme borreliosis. Infect Genet Evol *42*, 66-76. https://doi.org/10.1016/j.meegid.2016.04.019

Coleman, J.L., Sellati, T.J., Testa, J.E., Kew, R.R., Furie, M.B., and Benach, J.L. (1995). *Borrelia burgdorferi* binds plasminogen, resulting in enhanced penetration of endothelial monolayers. Infect Immun *63*, 2478-2484.

Coleman, J.L., Gebbia, J.A., Piesman, J., Degen, J.L., Bugge, T.H., and Benach, J.L. (1997). Plasminogen is required for efficient dissemination of *B. burgdorferi* in ticks and for enhancement of spirochetemia in mice. Cell *89*, 1111-1119.

Colli, C., Leinweber, B., Mullegger, R., Chott, A., Kerl, H., and Cerroni, L. (2004). *Borrelia burgdorferi*-associated lymphocytoma cutis: clinicopathologic, immunophenotypic, and molecular study of 106 cases. J Cutan Pathol *31*, 232-240.

Comstedt, P., Jakobsson, T., and Bergstrom, S. (2011). Global ecology and epidemiology of *Borrelia garinii* spirochetes. Infect Ecol Epidemiol *1*. https://doi.org/10.3402/iee.v1i0.9545

Comstock, L.E., and Thomas, D.D. (1989). Penetration of endothelial cell monolayers by *Borrelia burgdorferi*. Infect Immun *57*, 1626-1628.

Cook, S.P., Macartney, K.K., Rose, C.D., Hunt, P.G., Eppes, S.C., and Reilly, J.S. (1997). Lyme disease and seventh nerve paralysis in children. Am J Otolaryngol *18*, 320-323.

Corona, A., and Schwartz, I. (2015). *Borrelia burgdorferi*: carbon metabolism and the tick-mammal enzootic cycle. Microbiol Spectr *3*. https://doi.org/10.1128/microbiolspec.MBP-0011-2014

Cox, D.L., Akins, D.R., Bourell, K.W., Lahdenne, P., Norgard, M.V., and Radolf, J.D. (1996). Limited surface exposure of Borrelia burgdorferi outer surface lipoproteins. Proc Natl Acad Sci U S A *93*, 7973-7978.

Craft, J.E., Fischer, D.K., Shimamoto, G.T., and Steere, A.C. (1986). Antigens of *Borrelia burgdorferi* recognized during Lyme disease. Appearance of a new immunoglobulin M response and expansion of the immunoglobulin G response late in the illness. J Clin Invest *78*, 934-939. https://doi.org/10.1172/JCI112683

Crowder, L.A., Yedlin, V.A., Weinstein, E.R., Kortte, K.B., and Aucott, J.N. (2014). Lyme disease and post-treatment Lyme disease syndrome: the neglected disease in our own backyard. Public Health *128*, 784-791. https://doi.org/10.1016/j.puhe.2014.06.016

Crowley, J.T., Strle, K., Drouin, E.E., Pianta, A., Arvikar, S.L., Wang, Q., Costello, C.E., and Steere, A.C. (2016). Matrix metalloproteinase-10 is a target of T and B cell responses that correlate with synovial pathology in patients with antibiotic-refractory Lyme arthritis. J Autoimmun *69*, 24-37. https://doi.org/10.1016/j.jaut.2016.02.005

Cruz, A.I., Jr., Aversano, F.J., Seeley, M.A., Sankar, W.N., and Baldwin, K.D. (2017). Pediatric Lyme arthritis of the hip: The great imitator? J Pediatr Orthop *37*, 355-361. https://doi.org/10.1097/BPO.0000000000000664

Cugini, C., Medrano, M., Schwan, T.G., and Coburn, J. (2003). Regulation of expression of the *Borrelia burgdorferi* b_3-chain integrin ligand, P66, in ticks and in culture. Infect Immun *71*, 1001-1007.

Cull, B., Pietzsch, M.E., Hansford, K.M., Gillingham, E.L., and Medlock, J.M. (2018). Surveillance of British ticks: An overview of species records, host associations, and new records of *Ixodes ricinus* distribution. Ticks Tick Borne Dis *9*, 605-614. https://doi.org/10.1016/j.ttbdis.2018.01.011

Cutler, S.J., Ruzic-Sabljic, E., and Potkonjak, A. (2017). Emerging borreliae - Expanding beyond Lyme borreliosis. Mol Cell Probes *31*, 22-27. https://doi.org/10.1016/j.mcp.2016.08.003

Daikh, B.E., Emerson, F.E., Smith, R.P., Lucas, F.L., and McCarthy, C.A. (2013). Lyme arthritis: a comparison of presentation, synovial fluid analysis, and treatment course in children and adults. Arthritis Care Res (Hoboken) *65*, 1986-1990. https://doi.org/10.1002/acr.22086

Dandache, P., and Nadelman, R.B. (2008). Erythema migrans. Infect Dis Clin North Am *22*, 235-260, vi. https://doi.org/10.1016/j.idc.2007.12.012

Daubin, V., Gouy, M., and Perriere, G. (2002). A phylogenomic approach to bacterial phylogeny: evidence of a core of genes sharing a common history. Genome Res *12*, 1080-1090. https://doi.org/10.1101/gr.187002

de Koning, J. (1993). Histopathologic patterns of erythema migrans and borrelial lymphocytoma. Clin Dermatol *11*, 377-383.

De Silva, A.M., and Fikrig, E. (1995). Growth and migration of *Borrelia burgdorferi* in Ixodes ticks during blood feeding. Am J Trop Med Hyg *53*, 397-404.

de Silva, A.M., Telford, S.R., 3rd, Brunet, L.R., Barthold, S.W., and Fikrig, E. (1996). Borrelia burgdorferi OspA is an arthropod-specific transmission-blocking Lyme disease vaccine. J Exp Med *183*, 271-275.

de Silva, A.M., Tyson, K.R., and Pal, U. (2009). Molecular characterization of the tick-*Borrelia* interface. Front Biosci (Landmark Ed) *14*, 3051-3063.

Delong, A.K., Blossom, B., Maloney, E.L., and Phillips, S.E. (2012). Antibiotic retreatment of Lyme disease in patients with persistent symptoms: a biostatistical review of randomized, placebo-controlled, clinical trials. Contemp Clin Trials *33*, 1132-1142. https://doi.org/10.1016/j.cct.2012.08.009

Dennis, D.T., and Hayes, E.B. (2002). Epidemiology of Lyme Borreliosis. In Lyme Borreliosis Biology, Epidemiology, and Control, J.S. Gray, O. Kahl, R.S. Lane, and G. Stanek, eds. (New York: CABI Publishing), pp. 251-280.

Derdakova, M., Dudioak, V., Brei, B., Brownstein, J.S., Schwartz, I., and Fish, D. (2004). Interaction and transmission of two *Borrelia burgdorferi* sensu stricto strains in a tick-rodent maintenance system. Appl Environ Microbiol *70*, 6783-6788. https://doi.org/10.1128/AEM.70.11.6783-6788.2004

Dinerman, H., and Steere, A.C. (1992). Lyme disease associated with fibromyalgia. Ann Intern Med *117*, 281-285.

Dolan, M.C., Maupin, G.O., Panella, N.A., Golde, W.T., and Piesman, J. (1997). Vector competence of *Ixodes scapularis*, *I. spinipalpis*, and *Dermacentor andersoni* (Acari:Ixodidae) in transmitting *Borrelia burgdorferi*, the etiologic agent of Lyme disease. J Med Entomol *34*, 128-135. https://doi.org/10.1093/jmedent/34.2.128

Donahue, J.G., Piesman, J., and Spielman, A. (1987). Reservoir competence of white-footed mice for Lyme disease spirochetes. Am J Trop Med Hyg *36*, 92-96.

Dowdell, A.S., Murphy, M.D., Azodi, C., Swanson, S.K., Florens, L., Chen, S., and Zuckert, W.R. (2017). Comprehensive spatial analysis of the *Borrelia burgdorferi* lipoproteome reveals a compartmentalization bias toward the bacterial surface. J Bacteriol *199*. https://doi.org/10.1128/JB.00658-16

Drecktrah, D., Lybecker, M., Popitsch, N., Rescheneder, P., Hall, L.S., and Samuels, D.S. (2015). The *Borrelia burgdorferi* RelA/SpoT homolog and stringent response regulate survival in the tick vector and global gene expression during starvation. PLoS Pathog *11*, e1005160. https://doi.org/10.1371/journal.ppat.1005160

Dressler, F., Whalen, J.A., Reinhardt, B.N., and Steere, A.C. (1993). Western blotting in the serodiagnosis of Lyme disease. J Infect Dis *167*, 392-400.

Drouin, E.E., Seward, R.J., Strle, K., McHugh, G., Katchar, K., Londono, D., Yao, C., Costello, C.E., and Steere, A.C. (2013). A novel human autoantigen, endothelial cell growth factor, is a target of T and B cell responses in patients with Lyme disease. Arthritis Rheum *65*, 186-196. https://doi.org/10.1002/art.37732

Dumic, I., and Severnini, E. (2018). "Ticking Bomb": the impact of climate change on the incidence of Lyme disease. Can J Infect Dis Med Microbiol *2018*, 5719081. https://doi.org/10.1155/2018/5719081

Dunham-Ems, S.M., Caimano, M.J., Pal, U., Wolgemuth, C.W., Eggers, C.H., Balic, A., and Radolf, J.D. (2009). Live imaging reveals a biphasic mode of dissemination of *Borrelia burgdorferi* within ticks. J Clin Invest *119*, 3652-3665. https://doi.org/10.1172/JCI39401

Dunham-Ems, S.M., Caimano, M.J., Eggers, C.H., and Radolf, J.D. (2012). *Borrelia burgdorferi* requires the alternative sigma factor RpoS for dissemination within the vector during tick-to-mammal transmission. PLoS Pathog *8*, e1002532. https://doi.org/10.1371/journal.ppat.1002532

Duray, P.H., and Steere, A.C. (1986). The spectrum of organ and systems pathology in human Lyme disease. Zentralbl Bakteriol Mikrobiol Hyg A *263*, 169-178.

Duray, P.H. (1989a). Clinical pathologic correlations of Lyme disease. Rev Infect Dis *11 Suppl 6*, S1487-1493.

Duray, P.H. (1989b). Histopathology of clinical phases of human Lyme disease. Rheum Dis Clin North Am *15*, 691-710.

Dykhuizen, D.E., Brisson, D., Sandigursky, S., Wormser, G.P., Nowakowski, J., Nadelman, R.B., and Schwartz, I. (2008). The propensity of different *Borrelia burgdorferi* sensu stricto genotypes to cause disseminated infections in humans. Am J Trop Med Hyg *78*, 806-810.

Earnhart, C.G., Leblanc, D.V., Alix, K.E., Desrosiers, D.C., Radolf, J.D., and Marconi, R.T. (2010). Identification of residues within ligand-binding domain 1 (LBD1) of the Borrelia burgdorferi OspC protein required for function in the mammalian environment. Mol Microbiol *76*, 393-408. https://doi.org/10.1111/j.1365-2958.2010.07103.x

Ebady, R., Niddam, A.F., Boczula, A.E., Kim, Y.R., Gupta, N., Tang, T.T., Odisho, T., Zhi, H., Simmons, C.A., Skare, J.T., *et al.* (2016). Biomechanics of *Borrelia burgdorferi* vascular interactions. Cell Rep *16*, 2593-2604. https://doi.org/10.1016/j.celrep.2016.08.013

Eicken, C., Sharma, V., Klabunde, T., Owens, R.T., Pikas, D.S., Hook, M., and Sacchettini, J.C. (2001). Crystal structure of Lyme disease antigen outer surface protein C from *Borrelia burgdorferi*. J Biol Chem *276*, 10010-10015. https://doi.org/10.1074/jbc.M010062200

Eicken, C., Sharma, V., Klabunde, T., Lawrenz, M.B., Hardham, J.M., Norris, S.J., and Sacchettini, J.C. (2002). Crystal structure of Lyme disease variable surface antigen VlsE of *Borrelia burgdorferi*. J Biol Chem *277*, 21691-21696. https://doi.org/10.1074/jbc.M201547200

Eisen, R.J., Eisen, L., Ogden, N.H., and Beard, C.B. (2016). Linkages of weather and climate with *Ixodes scapularis* and *Ixodes pacificus* (Acari: Ixodidae), enzootic transmission of *Borrelia burgdorferi*, and Lyme disease in North America. J Med Entomol *53*, 250-261. https://doi.org/10.1093/jme/tjv199

Eisen, R.J., and Eisen, L. (2018). The blacklegged tick, *Ixodes scapularis* : an increasing public health concern. Trends in Parasitology *34*, 295-309. https://doi.org/10.1016/j.pt.2017.12.006

Eldin, C., Raffetin, A., Bouiller, K., Hansmann, Y., Roblot, F., Raoult, D., and Parola, P. (2019). Review of European and American guidelines for the diagnosis of Lyme borreliosis. Med Mal Infect *49*, 121-132. https://doi.org/10.1016/j.medmal.2018.11.011

Embers, M.E., Barthold, S.W., Borda, J.T., Bowers, L., Doyle, L., Hodzic, E., Jacobs, M.B., Hasenkampf, N.R., Martin, D.S., Narasimhan, S., *et al.* (2012). Persistence of *Borrelia burgdorferi* in rhesus macaques following antibiotic treatment of disseminated infection. PLoS One *7*, e29914. https://doi.org/10.1371/journal.pone.0029914

Embers, M.E., Hasenkampf, N.R., Jacobs, M.B., Tardo, A.C., Doyle-Meyers, L.A., Philipp, M.T., and Hodzic, E. (2017). Variable manifestations, diverse seroreactivity and post-treatment persistence in non-human primates exposed to *Borrelia burgdorferi* by tick feeding. PLOS ONE *12*, e0189071. https://doi.org/10.1371/journal.pone.0189071

Engstrom, S.M., Shoop, E., and Johnson, R.C. (1995). Immunoblot interpretation criteria for serodiagnosis of early Lyme disease. J Clin Microbiol *33*, 419-427.

Enkelmann, J., Bohmer, M., Fingerle, V., Siffczyk, C., Werber, D., Littmann, M., Merbecks, S.S., Helmeke, C., Schroeder, S., Hell, S., *et al.* (2018). Incidence of notified Lyme borreliosis in Germany, 2013-2017. Sci Rep *8*, 14976. https://doi.org/10.1038/s41598-018-33136-0

Eppes, S.C., Nelson, D.K., Lewis, L.L., and Klein, J.D. (1999). Characterization of Lyme meningitis and comparison with viral meningitis in children. Pediatrics *103*, 957-960. https://doi.org/10.1542/peds.103.5.957

Erdlow, J.A. (2003). Bull's Eye. Unraveling the Medical Mystery of Lyme Disease (New Haven: Yale University Press).

Estrada-Pena, A., Cutler, S., Potkonjak, A., Vassier-Tussaut, M., Van Bortel, W., Zeller, H., Fernandez-Ruiz, N., and Mihalca, A.D. (2018). An updated meta-analysis of the distribution and prevalence of *Borrelia burgdorferi* s.l. in ticks in Europe. Int J Health Geogr *17*, 41. https://doi.org/10.1186/s12942-018-0163-7

Estrada-Pena, A., and Cabezas-Cruz, A. (2019). Phyloproteomic and functional analyses do not support a split in the genus *Borrelia* (phylum Spirochaetes). BMC Evol Biol *19*, 54. https://doi.org/10.1186/s12862-019-1379-2

Falco, R.C., Fish, D., and Piesman, J. (1996). Duration of tick bites in a Lyme disease-endemic area. Am J Epidemiol *143*, 187-192.

Falco, R.C., McKenna, D.F., Daniels, T.J., Nadelman, R.B., Nowakowski, J., Fish, D., and Wormser, G.P. (1999). Temporal relation between *Ixodes scapularis* abundance and risk for Lyme disease associated with erythema migrans. Am J Epidemiol *149*, 771-776.

Fallon, B.A., Keilp, J.G., Corbera, K.M., Petkova, E., Britton, C.B., Dwyer, E., Slavov, I., Cheng, J., Dobkin, J., Nelson, D.R., *et al.* (2008). A randomized, placebo-controlled trial of repeated IV antibiotic therapy for Lyme encephalopathy. Neurology *70*, 992-1003. https://doi.org/10.1212/01.WNL.0000284604.61160.2d

Fallon, B.A., Petkova, E., Keilp, J.G., and Britton, C.B. (2012). A reappraisal of the U.S. Clinical trials of post-treatment lyme disease syndrome. Open Neurol J *6*, 79-87. https://doi.org/10.2174/1874205X01206010079

Feder, H.M., Jr., Gerber, M.A., Krause, P.J., Ryan, R., and Shapiro, E.D. (1993). Early Lyme disease: a flu-like illness without erythema migrans. Pediatrics *91*, 456-459.

Feder, H.M., Jr., Johnson, B.J., O'Connell, S., Shapiro, E.D., Steere, A.C., Wormser, G.P., Ad Hoc International Lyme Disease, G., Agger, W.A., Artsob, H., Auwaerter, P., *et al.* (2007). A critical appraisal of "chronic Lyme disease". N Engl J Med *357*, 1422-1430. https://doi.org/10.1056/NEJMra072023

Feder, H.M., Jr. (2008). Lyme disease in children. Infect Dis Clin North Am *22*, 315-326, vii. https://doi.org/10.1016/j.idc.2007.12.007

Feng, J., Shi, W., Zhang, S., Sullivan, D., Auwaerter, P.G., and Zhang, Y. (2016a). A drug combination screen identifies drugs active against amoxicillin-induced round bodies of *in vitro Borrelia burgdorferi* persisters from an FDA drug library. Front Microbiol *7*, 743. https://doi.org/10.3389/fmicb.2016.00743

Feng, J., Weitner, M., Shi, W., Zhang, S., and Zhang, Y. (2016b). Eradication of biofilm-like microcolony structures of *Borrelia burgdorferi* by daunomycin and daptomycin but not mitomycin C in combination with doxycycline and cefuroxime. Front Microbiol *7*, 62. https://doi.org/10.3389/fmicb.2016.00062

Feng, J., Zhang, S., Shi, W., and Zhang, Y. (2016c). Ceftriaxone pulse dosing fails to eradicate biofilm-like microcolony *B. burgdorferi* persisters which are sterilized by daptomycin/doxycycline/cefuroxime without pulse dosing. Front Microbiol *7*, 1744. https://doi.org/10.3389/fmicb.2016.01744

Feng, J., Li, T., Yee, R., Yuan, Y., Bai, C., Cai, M., Shi, W., Embers, M., Brayton, C., Saeki, H., *et al.* (2019). Stationary phase persister/biofilm microcolony of Borrelia burgdorferi causes more severe disease in a mouse model of Lyme arthritis: implications for understanding persistence, Post-treatment Lyme Disease Syndrome (PTLDS), and treatment failure. Discov Med *27*, 125-138.

Fikrig, E., Bockenstedt, L.K., Barthold, S.W., Chen, M., Tao, H., Ali-Salaam, P., Telford, S.R., and Flavell, R.A. (1994). Sera from patients with chronic Lyme disease protect mice from Lyme borreliosis. J Infect Dis *169*, 568-574.

Fikrig, E., and Narasimhan, S. (2006). *Borrelia burgdorferi*--traveling incognito? Microbes Infect *8*, 1390-1399. https://doi.org/10.1016/j.micinf.2005.12.022

Finch, C., Al-Damluji, M.S., Krause, P.J., Niccolai, L., Steeves, T., O'Keefe, C.F., and Diuk-Wasser, M.A. (2014). Integrated assessment of behavioral and environmental risk factors for Lyme disease infection on Block Island, Rhode Island. PLoS One *9*, e84758. https://doi.org/10.1371/journal.pone.0084758

Fischer, J.R., Parveen, N., Magoun, L., and Leong, J.M. (2003). Decorin-binding proteins A and B confer distinct mammalian cell type-specific attachment by *Borrelia burgdorferi*, the Lyme disease spirochete. Proc Natl Acad Sci U S A *100*, 7307-7312. https://doi.org/10.1073/pnas.1231043100

Fish, A.E., Pride, Y.B., and Pinto, D.S. (2008). Lyme carditis. Infect Dis Clin North Am *22*, 275-288, vi. https://doi.org/10.1016/j.idc.2007.12.008

Fisher, M.A., Grimm, D., Henion, A.K., Elias, A.F., Stewart, P.E., Rosa, P.A., and Gherardini, F.C. (2005). *Borrelia burgdorferi* s[54] is required for mammalian infection and vector transmission but not for tick colonization. Proc Natl Acad Sci U S A *102*, 5162-5167. https://doi.org/10.1073/pnas.0408536102

Foley, D.M., Gayek, R.J., Skare, J.T., Wagar, E.A., Champion, C.I., Blanco, D.R., Lovett, M.A., and Miller, J.N. (1995). Rabbit model of Lyme borreliosis: erythema migrans, infection-derived immunity, and identification of Borrelia burgdorferi proteins associated with virulence and protective immunity. J Clin Invest *96*, 965-975. https://doi.org/10.1172/JCI118144

Forrester, J.D., and Mead, P. (2014). Third-degree heart block associated with lyme carditis: review of published cases. Clin Infect Dis *59*, 996-1000. https://doi.org/10.1093/cid/ciu411

Forrester, J.D., Meiman, J., Mullins, J., Nelson, R., Ertel, S.H., Cartter, M., Brown, C.M., Lijewski, V., Schiffman, E., Neitzel, D., *et al.* (2014). Notes from the field: update on Lyme carditis, groups at high risk, and frequency of associated sudden cardiac death--United States. MMWR Morb Mortal Wkly Rep *63*, 982-983.

Fortune, D.E., Lin, Y.P., Deka, R.K., Groshong, A.M., Moore, B.P., Hagman, K.E., Leong, J.M., Tomchick, D.R., and Blevins, J.S. (2014).

Identification of lysine residues in the *Borrelia burgdorferi* DbpA adhesin required for murine infection. Infect Immun *82*, 3186-3198. https://doi.org/10.1128/IAI.02036-14

Franke, J., Hildebrandt, A., and Dorn, W. (2013). Exploring gaps in our knowledge on Lyme borreliosis spirochaetes--updates on complex heterogeneity, ecology, and pathogenicity. Ticks Tick Borne Dis *4*, 11-25. https://doi.org/10.1016/j.ttbdis.2012.06.007

Fraser, C.M., Casjens, S., Huang, W.M., Sutton, G.G., Clayton, R., Lathigra, R., White, O., Ketchum, K.A., Dodson, R., Hickey, E.K., *et al.* (1997). Genomic sequence of a Lyme disease spirochaete, *Borrelia burgdorferi*. Nature *390*, 580-586. https://doi.org/10.1038/37551

Fung, B.P., McHugh, G.L., Leong, J.M., and Steere, A.C. (1994). Humoral immune response to outer surface protein C of *Borrelia burgdorferi i*n Lyme disease: role of the immunoglobulin M response in the serodiagnosis of early infection. Infect Immun *62*, 3213-3221.

Gallais, F., De Martino, S.J., Sauleau, E.A., Hansmann, Y., Lipsker, D., Lenormand, C., Talagrand-Reboul, E., Boyer, P.H., Boulanger, N., Jaulhac, B., *et al.* (2018). Multilocus sequence typing of clinical *Borreliella afzelii* strains: population structure and differential ability to disseminate in humans. Parasit Vectors *11*, 374. https://doi.org/10.1186/s13071-018-2938-x

Garcia-Monco, J.C., and Benach, J.L. (2019). Lyme Neuroborreliosis: clinical outcomes, controversy, pathogenesis, and polymicrobial infections. Ann Neurol *85*, 21-31. https://doi.org/10.1002/ana.25389

Garcia, B.L., Zhi, H., Wager, B., Hook, M., and Skare, J.T. (2016). *Borrelia burgdorferi* BBK32 inhibits the classical pathway by blocking activation of the C1 complement complex. PLoS Pathog *12*, e1005404. https://doi.org/10.1371/journal.ppat.1005404

Garin, C., and Bujadoux, C (1922). Paralysie par les tiques. J Med Lyon *3*, 765-767.

Gasmi, S., Ogden, N.H., Leighton, P.A., Lindsay, L.R., and Thivierge, K. (2016). Analysis of the human population bitten by *Ixodes scapularis* ticks in Quebec, Canada: Increasing risk of Lyme disease. Ticks Tick Borne Dis *7*, 1075-1081. https://doi.org/10.1016/j.ttbdis.2016.09.006

Gasmi, S., Ogden, N.H., Ripoche, M., Leighton, P.A., Lindsay, R.L., Nelder, M.P., Rees, E., Bouchard, C., Vrbova, L., Rusk, R., *et al.* (2019). Detection of municipalities at-risk of Lyme disease using passive surveillance of *Ixodes scapularis* as an early signal: A province-specific indicator in Canada. PLoS One *14*, e0212637. https://doi.org/10.1371/journal.pone.0212637

Gazendam, N., Yeung, C., and Baranchuk, A. (2020). Lyme carditis presenting as sick sinus syndrome. Journal of Electrocardiology *59*, 65-67. https://doi.org/10.1016/j.jelectrocard.2020.01.007

Gebbia, J.A., Coleman, J.L., and Benach, J.L. (2004). Selective induction of matrix metalloproteinases by *Borrelia burgdorferi* via toll-like receptor 2 in monocytes. J Infect Dis *189*, 113-119. https://doi.org/10.1086/380414

Gendelberg, D., and Hennrikus, W.L. (2018). Lyme arthritis of the pediatric elbow: A case series. Orthopedics *41*, e511-e515. https://doi.org/10.3928/01477447-20180511-01

Gentilini, M., and Bricaire, F. (2019). Chronic Lyme disease: A scam that should be condemned! Med Mal Infect *49*, 83-84. https://doi.org/10.1016/j.medmal.2019.01.001

Gerber, M.A., and Zalneraitis, E.L. (1994). Childhood neurologic disorders and Lyme disease during pregnancy. Pediatr Neurol *11*, 41-43. https://doi.org/10.1016/0887-8994(94)90088-4

Gerber, M.A., Shapiro, E.D., Burke, G.S., Parcells, V.J., and Bell, G.L. (1996). Lyme disease in children in southeastern Connecticut. Pediatric Lyme Disease Study Group. N Engl J Med *335*, 1270-1274. https://doi.org/10.1056/NEJM199610243351703

Gerber, M.A., Zemel, L.S., and Shapiro, E.D. (1998). Lyme arthritis in children: clinical epidemiology and long-term outcomes. Pediatrics *102*, 905-908.

Gern, L., Schaible, U.E., and Simon, M.M. (1993). Mode of inoculation of the Lyme disease agent *Borrelia burgdorferi* influences infection and immune responses in inbred strains of mice. J Infect Dis *167*, 971-975. https://doi.org/10.1093/infdis/167.4.971

Gern, L. (2008). *Borrelia burgdorferi* sensu lato, the agent of lyme borreliosis: life in the wilds. Parasite *15*, 244-247.

Gern, L. (2009). Life cycle of *Borrelia burgdorferi* sensu lato and transmission to humans. Curr Probl Dermatol *37*, 18-30. https://doi.org/10.1159/000213068

Gern, L., and Humair,P.-F (2002). Ecology of *Borrelia burgdorferi* in Europe. In Lyme Borreliosis Biology, Epidemiology, and Control, J.S. Gray, Kahl,O., Lane, R.S., and Stanek, G., ed. (New York, NY: CABI Publishing), pp. 149-174.

Gherardini, F., Boylan, J., Lawrence, K., and Skare, J. (2010). Metabolism and Physiology of *Borrelia.* In Borrelia Molecular Biology, Host Interaction and Pathogenesis, D.S. Samuels, and J.D. Radolf, eds. (Norfolk, UK: Caister Academic Press), pp. 103-138.

Gilmore, R.D., Jr., Howison, R.R., Dietrich, G., Patton, T.G., Clifton, D.R., and Carroll, J.A. (2010). The *bba64* gene of *Borrelia burgdorferi*, the Lyme disease agent, is critical for mammalian infection via tick bite transmission. Proc Natl Acad Sci U S A *107*, 7515-7520. https://doi.org/10.1073/pnas.1000268107

Girard, Y.A., Fedorova, N., and Lane, R.S. (2011). Genetic diversity of *Borrelia burgdorferi* and detection of *B. bissettii*-like DNA in serum of north-coastal California residents. J Clin Microbiol *49*, 945-954. https://doi.org/10.1128/JCM.01689-10

Glatz, M., Resinger, A., Semmelweis, K., Ambros-Rudolph, C.M., and Mullegger, R.R. (2015). Clinical spectrum of skin manifestations of Lyme borreliosis in 204 children in Austria. Acta Derm Venereol *95*, 565-571. https://doi.org/10.2340/00015555-2000

Glickstein, L., Moore, B., Bledsoe, T., Damle, N., Sikand, V., and Steere, A.C. (2003). Inflammatory cytokine production predominates in early Lyme disease in patients with erythema migrans. Infect Immun *71*, 6051-6053. https://doi.org/10.1128/iai.71.10.6051-6053.2003

Golde, W.T., Burkot, T.R., Sviat, S., Keen, M.G., Mayer, L.W., Johnson, B.J., and Piesman, J. (1993). The major histocompatibility complex-restricted response of recombinant inbred strains of mice to natural tick transmission of *Borrelia burgdorferi.* J Exp Med *177*, 9-17. https://doi.org/10.1084/jem.177.1.9

Golovchenko, M., Vancova, M., Clark, K., Oliver, J.H., Jr., Grubhoffer, L., and Rudenko, N. (2016). A divergent spirochete strain isolated from a resident of the southeastern United States was identified by multilocus sequence typing as *Borrelia bissettii.* Parasit Vectors *9*, 68. https://doi.org/10.1186/s13071-016-1353-4

Granter, S.R., Bernstein, A., and Ostfeld, R.S. (2014). Of mice and men: Lyme disease and biodiversity. Perspect Biol Med *57*, 198-207. https://doi.org/10.1353/pbm.2014.0015

Grillon, A., Scherlinger, M., Boyer, P.H., De Martino, S., Perdriger, A., Blasquez, A., Wipff, J., Korganow, A.S., Bonnard, C., Cantagrel, A., *et al.* (2019). Characteristics and clinical outcomes after treatment of a national cohort of PCR-positive Lyme arthritis. Semin Arthritis Rheum *48*, 1105-1112. https://doi.org/10.1016/j.semarthrit.2018.09.007

Grimm, D., Tilly, K., Byram, R., Stewart, P.E., Krum, J.G., Bueschel, D.M., Schwan, T.G., Policastro, P.F., Elias, A.F., and Rosa, P.A. (2004). Outer-surface protein C of the Lyme disease spirochete: a protein induced in ticks for infection of mammals. Proc Natl Acad Sci U S A *101*, 3142-3147. https://doi.org/10.1073/pnas.0306845101

Groshong, A.M., and Blevins, J.S. (2014). Insights into the biology of *Borrelia burgdorferi* gained through the application of molecular genetics. Adv Appl Microbiol *86*, 41-143. https://doi.org/10.1016/B978-0-12-800262-9.00002-0

Groshong, A.M., Dey, A., Bezsonova, I., Caimano, M.J., and Radolf, J.D. (2017). Peptide uptake is essential for *Borrelia burgdorferi* viability andiInvolves structural and regulatory complexity of its oligopeptide transporter. mBio *8*. https://doi.org/10.1128/mBio.02047-17

Gross, D.M., Forsthuber, T., Tary-Lehmann, M., Etling, C., Ito, K., Nagy, Z.A., Field, J.A., Steere, A.C., and Huber, B.T. (1998). Identification of LFA-1 as a candidate autoantigen in treatment-resistant Lyme arthritis. Science *281*, 703-706.

Guo, B.P., Norris, S.J., Rosenberg, L.C., and Hook, M. (1995). Adherence of *Borrelia burgdorferi* to the proteoglycan decorin. Infect Immun *63*, 3467-3472.

Guo, B.P., Brown, E.L., Dorward, D.W., Rosenberg, L.C., and Hook, M. (1998). Decorin-binding adhesins from *Borrelia burgdorferi*. Mol Microbiol *30*, 711-723.

Habicht, G.S., Beck, G., Benach, J.L., and Coleman, J.L. (1986). *Borrelia burgdorferi* lipopolysaccharide and its role in the pathogenesis of Lyme disease. Zentralbl Bakteriol Mikrobiol Hyg A *263*, 137-141.

Haddad, O., Gillinov, M., Fraser, T., Shrestha, N., and Pettersson, G.B. (2019). Mitral valve endocarditis: A rare manifestation of Lyme disease. Ann Thorac Surg *108*, e85-e86. https://doi.org/10.1016/j.athoracsur.2018.12.046

Hagman, K.E., Lahdenne, P., Popova, T.G., Porcella, S.F., Akins, D.R., Radolf, J.D., and Norgard, M.V. (1998). Decorin-binding protein of *Borrelia burgdorferi* is encoded within a two-gene operon and is protective in the murine model of Lyme borreliosis. Infect Immun *66*, 2674-2683.

Halperin, J., Luft, B.J., Volkman, D.J., and Dattwyler, R.J. (1990). Lyme neuroborreliosis. Peripheral nervous system manifestations. Brain *113 (Pt 4)*, 1207-1221. https://doi.org/10.1093/brain/113.4.1207

Halperin, J.J., Luft, B.J., Anand, A.K., Roque, C.T., Alvarez, O., Volkman, D.J., and Dattwyler, R.J. (1989). Lyme neuroborreliosis: central nervous system manifestations. Neurology *39*, 753-759.

Halperin, J.J. (2003). Lyme disease and the peripheral nervous system. Muscle Nerve *28*, 133-143. https://doi.org/10.1002/mus.10337

Halperin, J.J., Shapiro, E.D., Logigian, E., Belman, A.L., Dotevall, L., Wormser, G.P., Krupp, L., Gronseth, G., Bever, C.T., Jr., and Quality Standards Subcommittee of the American Academy of, N. (2007). Practice parameter: treatment of nervous system Lyme disease (an evidence-based review): report of the Quality Standards Subcommittee of the American Academy of Neurology. Neurology *69*, 91-102. https://doi.org/10.1212/01.wnl.0000265517.66976.28

Halperin, J.J. (2008). Nervous system Lyme disease. Infect Dis Clin North Am *22*, 261-274, vi. https://doi.org/10.1016/j.idc.2007.12.009

Halperin, J.J. (2015). Nervous system Lyme disease. Infect Dis Clin North Am *29*, 241-253. https://doi.org/10.1016/j.idc.2015.02.002

Halperin, J.J. (2018). Neuroborreliosis. Neurol Clin *36*, 821-830. https://doi.org/10.1016/j.ncl.2018.06.006

Hamer, S.A., Tsao, J.I., Walker, E.D., and Hickling, G.J. (2010). Invasion of the Lyme disease vector *Ixodes scapularis*: implications for *Borrelia burgdorferi* endemicity. Ecohealth *7*, 47-63. https://doi.org/10.1007/s10393-010-0287-0

Hamer, S.A., Goldberg, T.L., Kitron, U.D., Brawn, J.D., Anderson, T.K., Loss, S.R., Walker, E.D., and Hamer, G.L. (2012). Wild birds and urban ecology of ticks and tick-borne pathogens, Chicago, Illinois, USA, 2005-2010. Emerg Infect Dis *18*, 1589-1595. https://doi.org/10.3201/eid1810.120511

Hanincova, K., Kurtenbach, K., Diuk-Wasser, M., Brei, B., and Fish, D. (2006). Epidemic spread of Lyme borreliosis, northeastern United States. Emerg Infect Dis *12*, 604-611. https://doi.org/10.3201/eid1204.051016

Hanincova, K., Liveris, D., Sandigursky, S., Wormser, G.P., and Schwartz, I. (2008). *Borrelia burgdorferi* sensu stricto is clonal in patients with early Lyme borreliosis. Appl Environ Microbiol *74*, 5008-5014. https://doi.org/10.1128/AEM.00479-08

Hanincova, K., Mukherjee, P., Ogden, N.H., Margos, G., Wormser, G.P., Reed, K.D., Meece, J.K., Vandermause, M.F., and Schwartz, I. (2013). Multilocus sequence typing of *Borrelia burgdorferi* suggests existence of lineages with differential pathogenic properties in humans. PLoS One *8*, e73066. https://doi.org/10.1371/journal.pone.0073066

Hanrahan, J.P., Benach, J.L., Coleman, J.L., Bosler, E.M., Morse, D.L., Cameron, D.J., Edelman, R., and Kaslow, R.A. (1984). Incidence and cumulative frequency of endemic Lyme disease in a community. J Infect Dis *150*, 489-496.

Harman, M., Vig, D.K., Radolf, J.D., and Wolgemuth, C.W. (2013). Viscous dynamics of Lyme disease and syphilis spirochetes reveal flagellar torque and drag. Biophys J *105*, 2273-2280. https://doi.org/10.1016/j.bpj.2013.10.004

Harman, M.W., Dunham-Ems, S.M., Caimano, M.J., Belperron, A.A., Bockenstedt, L.K., Fu, H.C., Radolf, J.D., and Wolgemuth, C.W. (2012). The heterogeneous motility of the Lyme disease spirochete in gelatin mimics dissemination through tissue. Proc Natl Acad Sci U S A *109*, 3059-3064. https://doi.org/10.1073/pnas.1114362109

Hasle, G., Bjune, G.A., Midthjell, L., Røed, K.H., and Leinaas, H.P. (2011). Transport of *Ixodes ricinus* infected with *Borrelia* species to Norway by northward-migrating passerine birds. *2*, 37-43. https://doi.org/10.1016/j.ttbdis.2010.10.004

Haugeberg, G., Hansen, I.J., Skarpaas, T., Noraas, S., and Kjelland, V. (2014). Lyme arthritis in southern Norway--an endemic area for Lyme borreliosis. BMC Infect Dis *14*, 185. https://doi.org/10.1186/1471-2334-14-185

Haven, J., Vargas, L.C., Mongodin, E.F., Xue, V., Hernandez, Y., Pagan, P., Fraser-Liggett, C.M., Schutzer, S.E., Luft, B.J., Casjens, S.R., *et al.* (2011). Pervasive recombination and sympatric genome diversification driven by frequency-dependent selection in *Borrelia burgdorferi*, the Lyme disease bacterium. Genetics *189*, 951-966. https://doi.org/10.1534/genetics.111.130773

He, M., Ouyang, Z., Troxell, B., Xu, H., Moh, A., Piesman, J., Norgard, M.V., Gomelsky, M., and Yang, X.F. (2011). Cyclic di-GMP is essential for the survival of the Lyme disease spirochete in ticks. PLoS Pathog *7*, e1002133. https://doi.org/10.1371/journal.ppat.1002133

Hengge, R. (2009). Principles of c-di-GMP signalling in bacteria. Nat Rev Microbiol *7*, 263-273.

Hengge, R. (2011). Stationary-phase gene regulation in *Escherichia coli* EcoSal Plus *4*. https://doi.org/10.1128/ecosalplus.5.6.3

Hengge, U.R., Tannapfel, A., Tyring, S.K., Erbel, R., Arendt, G., and Ruzicka, T. (2003). Lyme borreliosis. Lancet Infect Dis *3*, 489-500.

Hersh, M.H., LaDeau, S.L., Previtali, M.A., and Ostfeld, R.S. (2014). When is a parasite not a parasite? Effects of larval tick burdens on white-footed mouse survival. Ecology *95*, 1360-1369.

Herzer, P. (1991). Joint manifestations of Lyme borreliosis in Europe. Scand J Infect Dis Suppl *77*, 55-63.

Herzer, P. (1993). Rheumatic manifestations in Lyme borreliosis. Clin Dermatol *11*, 401-406. https://doi.org/10.1016/0738-081x(93)90096-u

Hinckley, A.F., Connally, N.P., Meek, J.I., Johnson, B.J., Kemperman, M.M., Feldman, K.A., White, J.L., and Mead, P.S. (2014). Lyme disease testing by large commercial laboratories in the United States. Clin Infect Dis *59*, 676-681. https://doi.org/10.1093/cid/ciu397

Hoen, A.G., Margos, G., Bent, S.J., Diuk-Wasser, M.A., Barbour, A., Kurtenbach, K., and Fish, D. (2009). Phylogeography of *Borrelia burgdorferi* in the eastern United States reflects multiple independent Lyme disease emergence events. Proc Natl Acad Sci U S A *106*, 15013-15018. https://doi.org/10.1073/pnas.0903810106

Hollstrom, E. (1951). Successful treatment of erythema migrans Afzelius. Acta Derm Venereol *31*, 235-243.

Holt, S.C. (1978). Anatomy and chemistry of spirochetes. Microbiol Rev *42*, 114-160.

Hovius, J.W., de Jong, M.A., den Dunnen, J., Litjens, M., Fikrig, E., van der Poll, T., Gringhuis, S.I., and Geijtenbeek, T.B. (2008a). Salp15 binding to DC-SIGN inhibits cytokine expression by impairing both nucleosome remodeling and mRNA stabilization. PLoS Pathog *4*, e31. https://doi.org/10.1371/journal.ppat.0040031

Hovius, J.W., Levi, M., and Fikrig, E. (2008b). Salivating for knowledge: potential pharmacological agents in tick saliva. PLoS Med *5*, e43. https://doi.org/10.1371/journal.pmed.0050043

Hovius, J.W. (2009). Spitting image: tick saliva assists the causative agent of Lyme disease in evading host skin's innate immune response. J Invest Dermatol *129*, 2337-2339. https://doi.org/10.1038/jid.2009.202

Hu, L.T., Perides, G., Noring, R., and Klempner, M.S. (1995). Binding of human plasminogen to *Borrelia burgdorferi*. Infect Immun *63*, 3491-3496.

Hu, L.T., Eskildsen, M.A., Masgala, C., Steere, A.C., Arner, E.C., Pratta, M.A., Grodzinsky, A.J., Loening, A., and Perides, G. (2001). Host metalloproteinases in Lyme arthritis. Arthritis Rheum *44*, 1401-1410. https://doi.org/10.1002/1529-0131(200106)44:6<1401::AID-ART234>3.0.CO;2-S

Huang, Z., Toledo, A.M., Benach, J.L., and London, E. (2016). Ordered nembrane domain-forming properties of the lipids of *Borrelia burgdorferi*. Biophys J *111*, 2666-2675. https://doi.org/10.1016/j.bpj.2016.11.012

Hubalek, Z. (2009). Epidemiology of Lyme borreliosis. Curr Probl Dermatol *37*, 31-50. https://doi.org/10.1159/000213069

Hubner, A., Yang, X., Nolen, D.M., Popova, T.G., Cabello, F.C., and Norgard, M.V. (2001). Expression of *Borrelia burgdorferi* OspC and DbpA is controlled by a RpoN-RpoS regulatory pathway. Proc Natl Acad Sci U S A *98*, 12724-12729. https://doi.org/10.1073/pnas.231442498

Hyde, F.W., and Johnson, R.C. (1984). Genetic relationship of Lyme disease spirochetes to *Borrelia*, *Treponema*, and *Leptospira* spp. J Clin Microbiol *20*, 151-154.

Hyde, J.A., Shaw, D.K., Smith Iii, R., Trzeciakowski, J.P., and Skare, J.T. (2009). The BosR regulatory protein of *Borrelia burgdorferi* interfaces with the RpoS regulatory pathway and modulates both the oxidative stress response and pathogenic properties of the Lyme disease spirochete. Mol Microbiol *74*, 1344-1355. https://doi.org/10.1111/j.1365-2958.2009.06951.x

Hyde, J.A., Weening, E.H., Chang, M., Trzeciakowski, J.P., Hook, M., Cirillo, J.D., and Skare, J.T. (2011). Bioluminescent imaging of *Borrelia burgdorferi* in vivo demonstrates that the fibronectin-binding protein BBK32 is required for optimal infectivity. Mol Microbiol *82*, 99-113. https://doi.org/10.1111/j.1365-2958.2011.07801.x

Hyde, J.A. (2017). *Borrelia burgdorferi* keeps moving and carries on: a review of borrelial dissemination and invasion. Front Immunol *8*, 114. https://doi.org/10.3389/fimmu.2017.00114

Iyer, R., Caimano, M.J., Luthra, A., Axline, D., Jr., Corona, A., Iacobas, D.A., Radolf, J.D., and Schwartz, I. (2015). Stage-specific global alterations in the transcriptomes of Lyme disease spirochetes during tick feeding and following mammalian host adaptation. Mol Microbiol *95*, 509-538. https://doi.org/10.1111/mmi.12882

Jabbari, N., Glusman, G., Joesch-Cohen, L.M., Reddy, P.J., Moritz, R.L., Hood, L., and Lausted, C.G. (2018). Whole genome sequence and comparative analysis of Borrelia burgdorferi MM1. PLoS One *13*, e0198135. https://doi.org/10.1371/journal.pone.0198135

Jacobsen, M., Zhou, D., Cepok, S., Nessler, S., Happel, M., Stei, S., Wilske, B., Sommer, N., and Hemmer, B. (2003). Clonal accumulation of activated CD8+ T cells in the central nervous system during the early phase of neuroborreliosis. J Infect Dis *187*, 963-973. https://doi.org/10.1086/368131

Jahfari, S., Krawczyk, A., Coipan, E.C., Fonville, M., Hovius, J.W., Sprong, H., and Takumi, K. (2017). Enzootic origins for clinical manifestations of Lyme borreliosis. Infect Genet Evol *49*, 48-54. https://doi.org/10.1016/j.meegid.2016.12.030

Johnson, R.C., Schmid, G.P., Hyde, F.W., Steigerwalt, A.G., and Brenner, D.J. (1984). *Borrelia burgdorferi* sp. nov: Etiologic agent of Lyme disease. Int J Syst Bacteriol *34*, 496-497.

Jones, J.D., Bourell, K.W., Norgard, M.V., and Radolf, J.D. (1995). Membrane topology of *Borrelia burgdorferi* and *Treponema pallidum* lipoproteins. Infect Immun *63*, 2424-2434.

Jones, K.L., Glickstein, L.J., Damle, N., Sikand, V.K., McHugh, G., and Steere, A.C. (2006). *Borrelia burgdorferi* genetic markers and disseminated disease in patients with early Lyme disease. J Clin Microbiol *44*, 4407-4413. https://doi.org/10.1128/JCM.01077-06

Jones, K.L., Muellegger, R.R., Means, T.K., Lee, M., Glickstein, L.J., Damle, N., Sikand, V.K., Luster, A.D., and Steere, A.C. (2008). Higher mRNA levels of chemokines and cytokines associated with macrophage activation in erythema migrans skin lesions in patients from the United States than in patients from Austria with Lyme borreliosis. Clin Infect Dis *46*, 85-92. https://doi.org/10.1086/524022

Jones, K.L., McHugh, G.A., Glickstein, L.J., and Steere, A.C. (2009). Analysis of *Borrelia burgdorferi* genotypes in patients with Lyme arthritis: High frequency of ribosomal RNA intergenic spacer type 1 strains in antibiotic-refractory arthritis. Arthritis Rheum *60*, 2174-2182. https://doi.org/10.1002/art.24812

Jungnick, S., Margos, G., Rieger, M., Dzaferovic, E., Bent, S.J., Overzier, E., Silaghi, C., Walder, G., Wex, F., Koloczek, J., *et al.* (2015). *Borrelia burgdorferi* sensu stricto and *Borrelia afzelii*: population structure and differential pathogenicity. Int J Med Microbiol *305*, 673-681. https://doi.org/10.1016/j.ijmm.2015.08.017

Jutras, B.L., Lochhead, R.B., Kloos, Z.A., Biboy, J., Strle, K., Booth, C.J., Govers, S.K., Gray, J., Schumann, P., Vollmer, W., *et al.* (2019). *Borrelia burgdorferi* peptidoglycan is a persistent antigen in patients with Lyme arthritis. Proc Natl Acad Sci U S A *116*, 13498-13507. https://doi.org/10.1073/pnas.1904170116

Kan, L., Sood, S.K., and Maytal, J. (1998). Pseudotumor cerebri in Lyme disease: a case report and literature review. Pediatr Neurol *18*, 439-441. https://doi.org/10.1016/s0887-8994(97)00215-4

Kannian, P., Drouin, E.E., Glickstein, L., Kwok, W.W., Nepom, G.T., and Steere, A.C. (2007). Decline in the frequencies of *Borrelia burgdorferi* OspA161 175-specific T cells after antibiotic therapy in HLA-DRB1*0401-positive patients with antibiotic-responsive or antibiotic-refractory lyme arthritis. J Immunol *179*, 6336-6342.

Kaplan, R.F., Trevino, R.P., Johnson, G.M., Levy, L., Dornbush, R., Hu, L.T., Evans, J., Weinstein, A., Schmid, C.H., and Klempner, M.S. (2003). Cognitive function in post-treatment Lyme disease: do additional antibiotics help? Neurology *60*, 1916-1922.

Kenedy, M.R., Lenhart, T.R., and Akins, D.R. (2012). The role of Borrelia burgdorferi outer surface proteins. FEMS Immunol Med Microbiol *66*, 1-19. https://doi.org/10.1111/j.1574-695X.2012.00980.x

Kenedy, M.R., Luthra, A., Anand, A., Dunn, J.P., Radolf, J.D., and Akins, D.R. (2014). Structural modeling and physicochemical characterization provide evidence that P66 forms a b-barrel in the *Borrelia burgdorferi* outer membrane. J Bacteriol *196*, 859-872. https://doi.org/10.1128/JB.01236-13

Kenedy, M.R., Scott, E.J., 2nd, Shrestha, B., Anand, A., Iqbal, H., Radolf, J.D., Dyer, D.W., and Akins, D.R. (2016). Consensus computational network analysis for identifying candidate outer membrane proteins from *Borrelia* spirochetes. BMC Microbiol *16*, 141. https://doi.org/10.1186/s12866-016-0762-z

Khatchikian, C.E., Nadelman, R.B., Nowakowski, J., Schwartz, I., Wormser, G.P., and Brisson, D. (2014). Evidence for strain-specific immunity in patients treated for early Lyme disease. Infect Immun *82*, 1408-1413. https://doi.org/10.1128/IAI.01451-13

Kingry, L.C., Batra, D., Replogle, A., Rowe, L.A., Pritt, B.S., and Petersen, J.M. (2016). Whole genome sequence and comparative genomics of the novel Lyme borreliosis causing pathogen, *Borrelia mayonii*. PLoS One *11*, e0168994. https://doi.org/10.1371/journal.pone.0168994

Kinjo, Y., Tupin, E., Wu, D., Fujio, M., Garcia-Navarro, R., Benhnia, M.R., Zajonc, D.M., Ben-Menachem, G., Ainge, G.D., Painter, G.F., *et al.* (2006). Natural killer T cells recognize diacylglycerol antigens from pathogenic bacteria. Nat Immunol *7*, 978-986. https://doi.org/10.1038/ni1380

Klempner, M.S., and Huber, B.T. (1999). Is it thee or me?--autoimmunity in Lyme disease. Nat Med *5*, 1346-1347. https://doi.org/10.1038/70907

Klempner, M.S., Hu, L.T., Evans, J., Schmid, C.H., Johnson, G.M., Trevino, R.P., Norton, D., Levy, L., Wall, D., McCall, J., *et al.* (2001). Two controlled trials of antibiotic treatment in patients with persistent symptoms and a history of Lyme disease. N Engl J Med *345*, 85-92. https://doi.org/10.1056/NEJM200107123450202

Klempner, M.S., Baker, P.J., Shapiro, E.D., Marques, A., Dattwyler, R.J., Halperin, J.J., and Wormser, G.P. (2013). Treatment trials for post-Lyme disease symptoms revisited. Am J Med *126*, 665-669. https://doi.org/10.1016/j.amjmed.2013.02.014

Koedel, U., Fingerle, V., and Pfister, H.W. (2015). Lyme neuroborreliosis-epidemiology, diagnosis and management. Nat Rev Neurol *11*, 446-456. https://doi.org/10.1038/nrneurol.2015.121

Konovalova, A., Kahne, D.E., and Silhavy, T.J. (2017). Outer membrane biogenesis. Annu Rev Microbiol *71*, 539-556. https://doi.org/10.1146/annurev-micro-090816-093754

Kostick, J.L., Szkotnicki, L.T., Rogers, E.A., Bocci, P., Raffaelli, N., and Marconi, R.T. (2011). The diguanylate cyclase, Rrp1, regulates critical steps in the enzootic cycle of the Lyme disease spirochetes. Mol Microbiol *81*, 219-231. https://doi.org/10.1111/j.1365-2958.2011.07687.x

Kraiczy, P. (2016a). Travelling between two worlds: complement as a gatekeeper for an expanded host range of Lyme disease spirochetes. Vet Sci *3*. https://doi.org/10.3390/vetsci3020012

Kraiczy, P. (2016b). Hide and seek: how Lyme disease spirochetes overcome complement attack. Front Immunol *7*, 385. https://doi.org/10.3389/fimmu.2016.00385

Krause, P.J., Foley, D.T., Burke, G.S., Christianson, D., Closter, L., Spielman, A., and Tick-Borne Disease Study, G. (2006). Reinfection and relapse in early Lyme disease. Am J Trop Med Hyg *75*, 1090-1094.

Kristoferitsch, W., Sluga, E., Graf, M., Partsch, H., Neumann, R., Stanek, G., and Budka, H. (1988). Neuropathy associated with acrodermatitis chronica atrophicans. Clinical and morphological features. Ann N Y Acad Sci *539*, 35-45. https://doi.org/10.1111/j.1749-6632.1988.tb31836.x

Krupp, L.B., Hyman, L.G., Grimson, R., Coyle, P.K., Melville, P., Ahnn, S., Dattwyler, R., and Chandler, B. (2003). Study and treatment of post Lyme disease (STOP-LD): a randomized double masked clinical trial. Neurology *60*, 1923-1930.

Kumaran, D., Eswaramoorthy, S., Luft, B.J., Koide, S., Dunn, J.J., Lawson, C.L., and Swaminathan, S. (2001). Crystal structure of outer surface protein C (OspC) from the Lyme disease spirochete, *Borrelia burgdorferi*. EMBO J *20*, 971-978. https://doi.org/10.1093/emboj/20.5.971

Kurtenbach, K., Sewell, H.S., Ogden, N.H., Randolph, S.E., and Nuttall, P.A. (1998). Serum complement sensitivity as a key factor in Lyme disease ecology. Infect Immun *66*, 1248-1251.

Kurtenbach, K., De Michelis, S., Etti, S., Schafer, S.M., Sewell, H.S., Brade, V., and Kraiczy, P. (2002). Host association of *Borrelia burgdorferi* sensu lato--the key role of host complement. Trends Microbiol *10*, 74-79.

Kurtenbach, K., Hanincova, K., Tsao, J.I., Margos, G., Fish, D., and Ogden, N.H. (2006). Fundamental processes in the evolutionary ecology of Lyme borreliosis. Nat Rev Microbiol *4*, 660-669. https://doi.org/10.1038/nrmicro1475

Lagal, V., Portnoi, D., Faure, G., Postic, D., and Baranton, G. (2006). *Borrelia burgdorferi* sensu stricto invasiveness is correlated with OspC-plasminogen affinity. Microbes Infect *8*, 645-652. https://doi.org/10.1016/j.micinf.2005.08.017

Lane, R.S., and Loye, J.E. (1989). Lyme disease in California: interrelationship of *Ixodes pacificus* (Acari: Ixodidae), the western fence lizard (*Sceloporus occidentalis*), and *Borrelia burgdorferi*. J Med Entomol *26*, 272-278. https://doi.org/10.1093/jmedent/26.4.272

Lane, R.S., Piesman, J., and Burgdorfer, W. (1991). Lyme borreliosis: relation of its causative agent to Its vectors and hosts in North America and Europe. *36*, 587-609. https://doi.org/10.1146/annurev.en.36.010191.003103

Lane, R.S., and Quistad, G.B. (1998). Borreliacidal factor in the blood of the western fence lizard (*Sceloporus occidentalis*). J Parasitol *84*, 29-34.

Lane, R.S., Mun, J., Peribanez, M.A., and Stubbs, H.A. (2007). Host-seeking behavior of *Ixodes pacificus* (Acari: Ixodidae) nymphs in relation to environmental parameters in dense-woodland and woodland-grass habitats. J Vector Ecol *32*, 342-357.

Lane, R.S., Fedorova, N., Kleinjan, J.E., and Maxwell, M. (2013). Eco-epidemiological factors contributing to the low risk of human exposure to ixodid tick-borne borreliae in southern California, USA. Ticks Tick Borne Dis *4*, 377-385. https://doi.org/10.1016/j.ttbdis.2013.02.005

Lantos, P.M. (2015). Chronic Lyme disease. Infect Dis Clin North Am *29*, 325-340. https://doi.org/10.1016/j.idc.2015.02.006

Lantos, P.M., Shapiro, E.D., Auwaerter, P.G., Baker, P.J., Halperin, J.J., McSweegan, E., and Wormser, G.P. (2015). Unorthodox alternative therapies marketed to treat Lyme disease. Clin Infect Dis *60*, 1776-1782. https://doi.org/10.1093/cid/civ186

LaRocca, T.J., Crowley, J.T., Cusack, B.J., Pathak, P., Benach, J., London, E., Garcia-Monco, J.C., and Benach, J.L. (2010). Cholesterol lipids of *Borrelia burgdorferi* form lipid rafts and are required for the bactericidal activity of a complement-independent antibody. Cell Host Microbe *8*, 331-342. https://doi.org/10.1016/j.chom.2010.09.001

LaRocca, T.J., Pathak, P., Chiantia, S., Toledo, A., Silvius, J.R., Benach, J.L., and London, E. (2013). Proving lipid rafts exist: membrane domains in the prokaryote *Borrelia burgdorferi* have the same properties as eukaryotic lipid rafts. PLoS Pathog *9*, e1003353. https://doi.org/10.1371/journal.ppat.1003353

Lenhoff, C. (1948). Spirochaetes in aetiologically obscure diseases. Acta Dermatologica Venereologica 295-324.

Lepej, S.Z., Rode, O.D., Jeren, T., Vince, A., Remenar, A., and Barsic, B. (2005). Increased expression of CXCR3 and CCR5 on memory CD4+ T-cells migrating into the cerebrospinal fluid of patients with neuroborreliosis: the role of CXCL10 and CXCL11. J Neuroimmunol *163*, 128-134. https://doi.org/10.1016/j.jneuroim.2005.03.005

Li, H., Dunn, J.J., Luft, B.J., and Lawson, C.L. (1997). Crystal structure of Lyme disease antigen outer surface protein A complexed with an Fab. Proc Natl Acad Sci U S A *94*, 3584-3589.

Li, X., McHugh, G.A., Damle, N., Sikand, V.K., Glickstein, L., and Steere, A.C. (2011). Burden and viability of *Borrelia burgdorferi* in skin and joints of patients with erythema migrans or lyme arthritis. Arthritis Rheum *63*, 2238-2247. https://doi.org/10.1002/art.30384

Liang, F.T., Steere, A.C., Marques, A.R., Johnson, B.J., Miller, J.N., and Philipp, M.T. (1999). Sensitive and specific serodiagnosis of Lyme disease by enzyme-linked immunosorbent assay with a peptide based on an immunodominant conserved region of *Borrelia burgdorferi* vlsE. J Clin Microbiol *37*, 3990-3996. https://doi.org/10.1128/JCM.37.12.3990-3996.1999

Liang, F.T., Jacobs, M.B., Bowers, L.C., and Philipp, M.T. (2002). An immune evasion mechanism for spirochetal persistence in Lyme borreliosis. J Exp Med *195*, 415-422. https://doi.org/10.1084/jem.20011870

Lin, B., Kidder, J.M., Noring, R., Steere, A.C., Klempner, M.S., and Hu, L.T. (2001). Differences in synovial fluid levels of matrix metalloproteinases suggest separate mechanisms of pathogenesis in Lyme arthritis before and after antibiotic treatment. J Infect Dis *184*, 174-180. https://doi.org/10.1086/322000

Lin, Y.P., Benoit, V., Yang, X., Martinez-Herranz, R., Pal, U., and Leong, J.M. (2014). Strain-specific variation of the decorin-binding adhesin DbpA influences the tissue tropism of the lyme disease spirochete.

PLoS Pathog *10*, e1004238. https://doi.org/10.1371/journal.ppat.1004238

Lin, Y.P., Chen, Q., Ritchie, J.A., Dufour, N.P., Fischer, J.R., Coburn, J., and Leong, J.M. (2015). Glycosaminoglycan binding by *Borrelia burgdorferi* adhesin BBK32 specifically and uniquely promotes joint colonization. Cell Microbiol *17*, 860-875. https://doi.org/10.1111/cmi.12407

Lipschütz, B. (1913). Uber eine seltene Erythemform (Erythema chronicum migrans) Arch Dermatol Syph *118*, 349-356.

Lipschütz, B. (1923). Weiter Beitgrag zur Kenngtnis des "Erythema chronicum migrans". Arch Dermatol Re *143*, 365-374.

Liveris, D., Varde, S., Iyer, R., Koenig, S., Bittker, S., Cooper, D., McKenna, D., Nowakowski, J., Nadelman, R.B., Wormser, G.P., *et al.* (1999). Genetic diversity of *Borrelia burgdorferi* in lyme disease patients as determined by culture versus direct PCR with clinical specimens. J Clin Microbiol *37*, 565-569.

Lochhead, R.B., Arvikar, S.L., Aversa, J.M., Sadreyev, R.I., Strle, K., and Steere, A.C. (2019a). Robust interferon signature and suppressed tissue repair gene expression in synovial tissue from patients with postinfectious, *Borrelia burgdorferi*-induced Lyme arthritis. Cell Microbiol *21*, e12954. https://doi.org/10.1111/cmi.12954

Lochhead, R.B., Ordonez, D., Arvikar, S.L., Aversa, J.M., Oh, L.S., Heyworth, B., Sadreyev, R., Steere, A.C., and Strle, K. (2019b). Interferon-gamma production in Lyme arthritis synovial tissue promotes differentiation of fibroblast-like synoviocytes into immune effector cells. Cell Microbiol *21*, e12992. https://doi.org/10.1111/cmi.12992

Logar, M., Ruzic-Sabljic, E., Maraspin, V., Lotric-Furlan, S., Cimperman, J., Jurca, T., and Strle, F. (2004). Comparison of erythema migrans caused by Borrelia afzelii and Borrelia garinii. Infection *32*, 15-19. https://doi.org/10.1007/s15010-004-3042-z

Logigian, E.L., Kaplan, R.F., and Steere, A.C. (1990). Chronic neurologic manifestations of Lyme disease. N Engl J Med *323*, 1438-1444. https://doi.org/10.1056/NEJM199011223232102

Logigian, E.L., and Steere, A.C. (1992). Clinical and electrophysiologic findings in chronic neuropathy of Lyme disease. Neurology *42*, 303-311.

LoGiudice, K., Ostfeld, R.S., Schmidt, K.A., and Keesing, F. (2003). The ecology of infectious disease: effects of host diversity and community composition on Lyme disease risk. Proc Natl Acad Sci U S A *100*, 567-571. https://doi.org/10.1073/pnas.0233733100

Lohr, B., Fingerle, V., Norris, D.E., and Hunfeld, K.P. (2018). Laboratory diagnosis of Lyme borreliosis: Current state of the art and future perspectives. Crit Rev Clin Lab Sci *55*, 219-245. https://doi.org/10.1080/10408363.2018.1450353

Luft, B.J., Steinman, C.R., Neimark, H.C., Muralidhar, B., Rush, T., Finkel, M.F., Kunkel, M., and Dattwyler, R.J. (1992). Invasion of the central nervous system by Borrelia burgdorferi in acute disseminated infection. JAMA *267*, 1364-1367.

Lunemann, J.D., Gelderblom, H., Sospedra, M., Quandt, J.A., Pinilla, C., Marques, A., and Martin, R. (2007). Cerebrospinal fluid-infiltrating CD4+ T cells recognize Borrelia burgdorferi lysine-enriched protein domains and central nervous system autoantigens in early lyme encephalitis. Infect Immun *75*, 243-251. https://doi.org/10.1128/IAI.01110-06

Mannelli, A., Bertolotti, L., Gern, L., and Gray, J. (2012). Ecology of *Borrelia burgdorferi* sensu lato in Europe: transmission dynamics in multi-host systems, influence of molecular processes and effects of climate change. FEMS Microbiol Rev *36*, 837-861. https://doi.org/10.1111/j.1574-6976.2011.00312.x

Maraspin, V., Nahtigal Klevisar, M., Ruzic-Sabljic, E., Lusa, L., and Strle, F. (2016). Borrelial Lymphocytoma in Adult Patients. Clin Infect Dis *63*, 914-921. https://doi.org/10.1093/cid/ciw417

Marcinkiewicz, A.L., Dupuis, A.P., 2nd, Zamba-Campero, M., Nowak, N., Kraiczy, P., Ram, S., Kramer, L.D., and Lin, Y.P. (2019). Blood treatment of Lyme borreliae demonstrates the mechanism of CspZ-mediated complement evasion to promote systemic infection in vertebrate hosts. Cell Microbiol *21*, e12998. https://doi.org/10.1111/cmi.12998

Marconi, R.T., Liveris, D., and Schwartz, I. (1995). Identification of novel insertion elements, restriction fragment length polymorphism patterns, and discontinuous 23S rRNA in Lyme disease spirochetes: phylogenetic analyses of rRNA genes and their intergenic spacers in Borrelia japonica sp. nov. and genomic group 21038 (Borrelia andersonii sp. nov.) isolates. J Clin Microbiol *33*, 2427-2434.

Marcos, L.A., Castle, P.M., Smith, K., Khoo, T., Morley, E.J., Bloom, M., and Fries, B.C. (2019). Risk factors for Lyme carditis: A case-control study. European Journal of Preventive Cardiology, 204748731987604. https://doi.org/10.1177/2047487319876046

Margos, G., Gatewood, A.G., Aanensen, D.M., Hanincova, K., Terekhova, D., Vollmer, S.A., Cornet, M., Piesman, J., Donaghy, M., Bormane, A., *et al.* (2008). MLST of housekeeping genes captures geographic population structure and suggests a European origin of *Borrelia burgdorferi.* Proc Natl Acad Sci U S A *105*, 8730-8735. https://doi.org/10.1073/pnas.0800323105

Margos, G., Vollmer, S.A., Ogden, N.H., and Fish, D. (2011). Population genetics, taxonomy, phylogeny and evolution of *Borrelia burgdorferi* sensu lato. Infect Genet Evol *11*, 1545-1563. https://doi.org/10.1016/j.meegid.2011.07.022

Margos, G., Lane, R.S., Fedorova, N., Koloczek, J., Piesman, J., Hojgaard, A., Sing, A., and Fingerle, V. (2016). *Borrelia bissettiae* sp. nov. and *Borrelia californiensis* sp. nov. prevail in diverse enzootic transmission cycles. Int J Syst Evol Microbiol *66*, 1447-1452. https://doi.org/10.1099/ijsem.0.000897

Margos, G., Hepner, S., Mang, C., Marosevic, D., Reynolds, S.E., Krebs, S., Sing, A., Derdakova, M., Reiter, M.A., and Fingerle, V. (2017a). Lost in plasmids: next generation sequencing and the complex genome of the tick-borne pathogen *Borrelia burgdorferi.* BMC Genomics *18*, 422. https://doi.org/10.1186/s12864-017-3804-5

Margos, G., Marosevic, D., Cutler, S., Derdakova, M., Diuk-Wasser, M., Emler, S., Fish, D., Gray, J., Hunfeldt, K.P., Jaulhac, B., *et al.* (2017b). There is inadequate evidence to support the division of the genus *Borrelia.* Int J Syst Evol Microbiol *67*, 1081-1084. https://doi.org/10.1099/ijsem.0.001717

Margos, G., Gofton, A., Wibberg, D., Dangel, A., Marosevic, D., Loh, S.M., Oskam, C., and Fingerle, V. (2018). The genus *Borrelia* reloaded. PLoS One *13*, e0208432. https://doi.org/10.1371/journal.pone.0208432

Marques, A. (2008). Chronic Lyme disease: a review. Infect Dis Clin North Am *22*, 341-360, vii-viii. https://doi.org/10.1016/j.idc.2007.12.011

Marques, A., Telford, S.R., 3rd, Turk, S.P., Chung, E., Williams, C., Dardick, K., Krause, P.J., Brandeburg, C., Crowder, C.D., Carolan, H.E., *et al.* (2014). Xenodiagnosis to detect *Borrelia burgdorferi* infection: a first-in-human study. Clin Infect Dis *58*, 937-945. https://doi.org/10.1093/cid/cit939

Marques, A., Schwartz, I., Wormser, G.P., Wang, Y., Hornung, R.L., Demirkale, C.Y., Munson, P.J., Turk, S.P., Williams, C., Lee, C.R., *et al.* (2017). Transcriptome assessment of erythema migrans skin lesions in patients with early Lyme disease reveals predominant interferon signaling. J Infect Dis *217*, 158-167. https://doi.org/10.1093/infdis/jix563

Marx, G.E., Leikauskas, J., Lindstrom, K., Mann, E., Reagan-Steiner, S., Matkovic, E., Read, J.S., Kelso, P., Kwit, N.A., Hinckley, A.F., *et al.* (2019). Fatal Lyme carditis in New England: Two case reports. Annals of Internal Medicine. https://doi.org/10.7326/l19-0483

Mast, W.E., and Burrows, W.M., Jr. (1976). Erythema chronicum migrans in the United States. JAMA *236*, 859-860.

Mather, T.N., Wilson, M.L., Moore, S.I., Ribeiro, J.M., and Spielman, A. (1989). Comparing the relative potential of rodents as reservoirs of the Lyme disease spirochete (*Borrelia burgdorferi*). Am J Epidemiol *130*, 143-150. https://doi.org/10.1093/oxfordjournals.aje.a115306

Mather, T.N., Nicholson, M.C., Donnelly, E.F., and Matyas, B.T. (1996). Entomologic index for human risk of Lyme disease. Am J Epidemiol *144*, 1066-1069.

Maupin, G.O., Gage, K.L., Piesman, J., Montenieri, J., Sviat, S.L., VanderZanden, L., Happ, C.M., Dolan, M., and Johnson, B.J. (1994). Discovery of an enzootic cycle of *Borrelia burgdorferi* in *Neotoma mexicana* and *Ixodes spinipalpis* from northern Colorado, an area where Lyme disease is nonendemic. J Infect Dis *170*, 636-643. https://doi.org/10.1093/infdis/170.3.636

McCoy, K.D., Leger, E., and Dietrich, M. (2013). Host specialization in ticks and transmission of tick-borne diseases: a review. Front Cell Infect Microbiol *3*, 57. https://doi.org/10.3389/fcimb.2013.00057

Mead, P.S. (2015). Epidemiology of Lyme disease. Infect Dis Clin North Am *29*, 187-210. https://doi.org/10.1016/j.idc.2015.02.010

Medlock, J.M., Hansford, K.M., Bormane, A., Derdakova, M., Estrada-Pena, A., George, J.C., Golovljova, I., Jaenson, T.G., Jensen, J.K., Jensen, P.M., *et al.* (2013). Driving forces for changes in geographical distribution of *Ixodes ricinus* ticks in Europe. Parasit Vectors *6*, 1. https://doi.org/10.1186/1756-3305-6-1

Molins, C.R., Ashton, L.V., Wormser, G.P., Andre, B.G., Hess, A.M., Delorey, M.J., Pilgard, M.A., Johnson, B.J., Webb, K., Islam, M.N., *et al.* (2017). Metabolic differentiation of early Lyme disease from southern tick-associated rash illness (STARI). Sci Transl Med *9*. https://doi.org/10.1126/scitranslmed.aal2717

Mongodin, E.F., Casjens, S.R., Bruno, J.F., Xu, Y., Drabek, E.F., Riley, D.R., Cantarel, B.L., Pagan, P.E., Hernandez, Y.A., Vargas, L.C., *et al.* (2013). Inter- and intra-specific pan-genomes of *Borrelia burgdorferi* sensu lato: genome stability and adaptive radiation. BMC Genomics *14*, 693. https://doi.org/10.1186/1471-2164-14-693

Montgomery, R.R., Malawista, S.E., Feen, K.J., and Bockenstedt, L.K. (1996). Direct demonstration of antigenic substitution of Borrelia burgdorferi ex vivo: exploration of the paradox of the early immune response to outer surface proteins A and C in Lyme disease. J Exp Med *183*, 261-269.

Moody, K.D., Terwilliger, G.A., Hansen, G.M., and Barthold, S.W. (1994). Experimental *Borrelia burgdorferi* infection in *Peromyscus leucopus*. J Wildl Dis *30*, 155-161. https://doi.org/10.7589/0090-3558-30.2.155

Morgan, A., and Wang, X. (2013). The novel heparin-binding motif in decorin-binding protein A from strain B31 of *Borrelia burgdorferi* explains the higher binding affinity. Biochemistry *52*, 8237-8245. https://doi.org/10.1021/bi401376u

Moriarty, T.J., Norman, M.U., Colarusso, P., Bankhead, T., Kubes, P., and Chaconas, G. (2008). Real-time high resolution 3D imaging of the Lyme disease spirochete adhering to and escaping from the vasculature of a living host. PLoS Pathog *4*, e1000090. https://doi.org/10.1371/journal.ppat.1000090

Moriarty, T.J., Shi, M., Lin, Y.P., Ebady, R., Zhou, H., Odisho, T., Hardy, P.O., Salman-Dilgimen, A., Wu, J., Weening, E.H., *et al.* (2012). Vascular binding of a pathogen under shear force through mechanistically distinct sequential interactions with host macromolecules. Mol Microbiol *86*, 1116-1131. https://doi.org/10.1111/mmi.12045

Motaleb, M.A., Corum, L., Bono, J.L., Elias, A.F., Rosa, P., Samuels, D.S., and Charon, N.W. (2000). *Borrelia burgdorferi* periplasmic flagella have both skeletal and motility functions. Proc Natl Acad Sci U S A *97*, 10899-10904. https://doi.org/10.1073/pnas.200221797

Motaleb, M.A., Liu, J., and Wooten, R.M. (2015). Spirochetal motility and chemotaxis in the natural enzootic cycle and development of Lyme disease. Curr Opin Microbiol *28*, 106-113. https://doi.org/10.1016/j.mib.2015.09.006

Muehlenbachs, A., Bollweg, B.C., Schulz, T.J., Forrester, J.D., DeLeon Carnes, M., Molins, C., Ray, G.S., Cummings, P.M., Ritter, J.M., Blau, D.M., *et al.* (2016). Cardiac tropism of *Borrelia burgdorferi*: An autopsy study of sudden cardiac death associated with Lyme carditis. Am J Pathol *186*, 1195-1205. https://doi.org/10.1016/j.ajpath.2015.12.027

Mulay, V.B., Caimano, M.J., Iyer, R., Dunham-Ems, S., Liveris, D., Petzke, M.M., Schwartz, I., and Radolf, J.D. (2009). *Borrelia burgdorferi* bba74 is expressed exclusively during tick feeding and is regulated by both arthropod- and mammalian host-specific signals. J Bacteriol *191*, 2783-2794. https://doi.org/JB.01802-08 [pii]
10.1128/JB.01802-08

Mullegger, R.R., McHugh, G., Ruthazer, R., Binder, B., Kerl, H., and Steere, A.C. (2000). Differential expression of cytokine mRNA in skin specimens from patients with erythema migrans or acrodermatitis chronica atrophicans. J Invest Dermatol *115*, 1115-1123. https://doi.org/10.1046/j.1523-1747.2000.00198.x

Mullegger, R.R., Means, T.K., Shin, J.J., Lee, M., Jones, K.L., Glickstein, L.J., Luster, A.D., and Steere, A.C. (2007). Chemokine signatures in the skin disorders of Lyme borreliosis in Europe: predominance of CXCL9 and CXCL10 in erythema migrans and acrodermatitis and CXCL13 in lymphocytoma. Infect Immun *75*, 4621-4628. https://doi.org/10.1128/IAI.00263-07

Munro, H.J., Ogden, N.H., Mechai, S., Lindsay, L.R., Robertson, G.J., Whitney, H., and Lang, A.S. (2019). Genetic diversity of *Borrelia garinii* from Ixodes uriae collected in seabird colonies of the northwestern Atlantic Ocean. Ticks Tick Borne Dis *10*, 101255. https://doi.org/10.1016/j.ttbdis.2019.06.014

Murfin, K.E., and Fikrig, E. (2017). Tick bioactive molecules as novel therapeutics: beyond vaccine targets. Front Cell Infect Microbiol *7*, 222. https://doi.org/10.3389/fcimb.2017.00222

Murphy, G., and Nagase, H. (2008). Progress in matrix metalloproteinase research. Mol Aspects Med *29*, 290-308. https://doi.org/10.1016/j.mam.2008.05.002

Mygland, A., Skarpaas, T., and Ljostad, U. (2006). Chronic polyneuropathy and Lyme disease. Eur J Neurol *13*, 1213-1215. https://doi.org/10.1111/j.1468-1331.2006.01395.x

Nadelman, R.B., Nowakowski, J., Forseter, G., Bittker, S., Cooper, D., Goldberg, N., McKenna, D., and Wormser, G.P. (1993). Failure to isolate *Borrelia burgdorferi* after antimicrobial therapy in culture-documented Lyme borreliosis associated with erythema migrans: report of a prospective study. Am J Med *94*, 583-588.

Nadelman, R.B., Nowakowski, J., Forseter, G., Goldberg, N.S., Bittker, S., Cooper, D., Aguero-Rosenfeld, M., and Wormser, G.P. (1996). The clinical spectrum of early Lyme borreliosis in patients with culture-confirmed erythema migrans. Am J Med *100*, 502-508.

Nadelman, R.B., Nowakowski, J., Fish, D., Falco, R.C., Freeman, K., McKenna, D., Welch, P., Marcus, R., Aguero-Rosenfeld, M.E., Dennis, D.T., *et al.* (2001). Prophylaxis with single-dose doxycycline for the prevention of Lyme disease after an Ixodes scapularis tick bite. N Engl J Med *345*, 79-84. https://doi.org/10.1056/NEJM200107123450201

Nadelman, R.B., and Wormser, G.P. (2007). Reinfection in patients with Lyme disease. Clin Infect Dis *45*, 1032-1038. https://doi.org/10.1086/521256

Nadelman, R.B., and Wormser, G.P. (2013). Reinfection versus relapse in Lyme disease. N Engl J Med *368*, 1063-1064. https://doi.org/10.1056/NEJMc1215615

Nadelman, R.B. (2015). Erythema migrans. Infect Dis Clin North Am *29*, 211-239. https://doi.org/10.1016/j.idc.2015.02.001

Nelson, C.A., Saha, S., Kugeler, K.J., Delorey, M.J., Shankar, M.B., Hinckley, A.F., and Mead, P.S. (2015). Incidence of clinician-diagnosed Lyme disease, United States, 2005-2010. Emerg Infect Dis *21*, 1625-1631. https://doi.org/10.3201/eid2109.150417

Nikaido, H. (2003). Molecular basis of bacterial outer membrane permeability revisited. Microbiol Mol Biol Rev *67*, 593-656.

Nikolić, A., Boljević, D., Bojić, M., Veljković, S., Vuković, D., Paglietti, B., Micić, J., and Rubino, S. (2020). Lyme endocarditis as an emerging infectious disease: A review of the literature. Frontiers in Microbiology *11*. https://doi.org/10.3389/fmicb.2020.00278

Nocton, J.J., Dressler, F., Rutledge, B.J., Rys, P.N., Persing, D.H., and Steere, A.C. (1994). Detection of *Borrelia burgdorferi* DNA by polymerase chain reaction in synovial fluid from patients with Lyme arthritis. N Engl J Med *330*, 229-234. https://doi.org/10.1056/NEJM199401273300401

Norgard, M.V., Arndt, L.L., Akins, D.R., Curetty, L.L., Harrich, D.A., and Radolf, J.D. (1996). Activation of human monocytic cells by *Treponema pallidum* and *Borrelia burgdorferi* lipoproteins and synthetic lipopeptides proceeds via a pathway distinct from that of lipopolysaccharide but involves the transcriptional activator NF-kB. Infect Immun *64*, 3845-3852.

Norman, M.U., Moriarty, T.J., Dresser, A.R., Millen, B., Kubes, P., and Chaconas, G. (2008). Molecular mechanisms involved in vascular interactions of the Lyme disease pathogen in a living host. PLoS Pathog *4*, e1000169. https://doi.org/10.1371/journal.ppat.1000169

Norris, D.E., Johnson, B.J., Piesman, J., Maupin, G.O., Clark, J.L., and Black, W.C.t. (1999). Population genetics and phylogenetic analysis of Colorado *Borrelia burgdorferi*. Am J Trop Med Hyg *60*, 699-707. https://doi.org/10.4269/ajtmh.1999.60.699

Norris, S.J. (2014). *vls* antigenic variation systems of Lyme disease *Borrelia*: eluding host immunity through both random, segmental gene conversion and framework heterogeneity. Microbiol Spectr 2. https://doi.org/10.1128/microbiolspec.MDNA3-0038-2014

Norte, A.C., Ramos, J.A., Gern, L., Nuncio, M.S., and Lopes de Carvalho, I. (2013). Birds as reservoirs for *Borrelia burgdorferi* s.l. in western Europe: circulation of *B. turdi* and other genospecies in bird-tick cycles in Portugal. Environ Microbiol *15*, 386-397. https://doi.org/10.1111/j.1462-2920.2012.02834.x

Nowakowski, J., Schwartz, I., Liveris, D., Wang, G., Aguero-Rosenfeld, M.E., Girao, G., McKenna, D., Nadelman, R.B., Cavaliere, L.F., Wormser, G.P., *et al.* (2001). Laboratory diagnostic techniques for patients with early Lyme disease associated with erythema migrans: a comparison of different techniques. Clin Infect Dis *33*, 2023-2027. https://doi.org/10.1086/324490

Nowakowski, J., Nadelman, R.B., Sell, R., McKenna, D., Cavaliere, L.F., Holmgren, D., Gaidici, A., and Wormser, G.P. (2003). Long-term follow-up of patients with culture-confirmed Lyme disease. Am J Med *115*, 91-96.

Nowakowski, J., McKenna, D., Nadelman, R.B., Bittker, S., Cooper, D., Pavia, C., Holmgren, D., Visintainer, P., and Wormser, G.P. (2009). Blood cultures for patients with extracutaneous manifestations of Lyme disease in the United States. Clin Infect Dis *49*, 1733-1735. https://doi.org/10.1086/648076

Nowalk, A.J., Gilmore, R.D., Jr., and Carroll, J.A. (2006). Serologic proteome analysis of *Borrelia burgdorferi* membrane-associated proteins. Infect Immun *74*, 3864-3873. https://doi.org/10.1128/IAI.00189-06

Nuttall, P.A. (2019). Wonders of tick saliva. Ticks Tick Borne Dis *10*, 470-481. https://doi.org/10.1016/j.ttbdis.2018.11.005

Ogden, N.H., Lindsay, L.R., Hanincova, K., Barker, I.K., Bigras-Poulin, M., Charron, D.F., Heagy, A., Francis, C.M., O'Callaghan, C.J., Schwartz, I., *et al.* (2008). Role of migratory birds in introduction and range expansion of *Ixodes scapularis* ticks and of *Borrelia burgdorferi* and *Anaplasma phagocytophilum* in Canada. Appl Environ Microbiol *74*, 1780-1790. https://doi.org/10.1128/AEM.01982-07

Ogden, N.H., Lindsay, L.R., Morshed, M., Sockett, P.N., and Artsob, H. (2009). The emergence of Lyme disease in Canada. CMAJ *180*, 1221-1224. https://doi.org/10.1503/cmaj.080148

Ogden, N.H., Bouchard, C., Badcock, J., Drebot, M.A., Elias, S.P., Hatchette, T.F., Koffi, J.K., Leighton, P.A., Lindsay, L.R., Lubelczyk, C.B., *et al.* (2019). What is the real number of Lyme disease cases in Canada? BMC Public Health *19*. https://doi.org/10.1186/s12889-019-7219-x

Ogrinc, K., Lotric-Furlan, S., Maraspin, V., Lusa, L., Cerar, T., Ruzic-Sabljic, E., and Strle, F. (2013). Suspected early Lyme neuroborreliosis in patients with erythema migrans. Clin Infect Dis *57*, 501-509. https://doi.org/10.1093/cid/cit317

Ogrinc, K., Lusa, L., Lotric-Furlan, S., Bogovic, P., Stupica, D., Cerar, T., Ruzic-Sabljic, E., and Strle, F. (2016). Course and outcome of early European Lyme neuroborreliosis (Bannwarth Syndrome): clinical and laboratory findings. Clin Infect Dis *63*, 346-353. https://doi.org/10.1093/cid/ciw299

Ohnishi, J., Piesman, J., and de Silva, A.M. (2001). Antigenic and genetic heterogeneity of *Borrelia burgdorferi* populations transmitted by ticks. Proc Natl Acad Sci U S A *98*, 670-675. https://doi.org/10.1073/pnas.98.2.670

Oliver, J.H., Jr., Lin, T., Gao, L., Clark, K.L., Banks, C.W., Durden, L.A., James, A.M., and Chandler, F.W., Jr. (2003). An enzootic transmission cycle of Lyme borreliosis spirochetes in the southeastern United States. Proc Natl Acad Sci U S A *100*, 11642-11645. https://doi.org/10.1073/pnas.1434553100

Olsen, B., Jaenson, T.G., Noppa, L., Bunikis, J., and Bergstrom, S. (1993). A Lyme borreliosis cycle in seabirds and *Ixodes uriae* ticks. Nature *362*, 340-342. https://doi.org/10.1038/362340a0

Olsen, B., Duffy, D.C., Jaenson, T.G., Gylfe, A., Bonnedahl, J., and Bergstrom, S. (1995). Transhemispheric exchange of Lyme disease spirochetes by seabirds. J Clin Microbiol *33*, 3270-3274.

Onder, O., Humphrey, P.T., McOmber, B., Korobova, F., Francella, N., Greenbaum, D.C., and Brisson, D. (2012). OspC is potent plasminogen receptor on surface of *Borrelia burgdorferi*. J Biol Chem *287*, 16860-16868. https://doi.org/10.1074/jbc.M111.290775

Orloski, K.A., Campbell, G.L., Genese, C.A., Beckley, J.W., Schriefer, M.E., Spitalny, K.C., and Dennis, D.T. (1998). Emergence of Lyme disease in Hunterdon County, New Jersey, 1993: a case-control study of risk factors and evaluation of reporting patterns. Am J Epidemiol *147*, 391-397. https://doi.org/10.1093/oxfordjournals.aje.a009462

Ostfeld, R.S., and Brunner, J.L. (2015). Climate change and *Ixodes* tick-borne diseases of humans. Philos Trans R Soc Lond B Biol Sci *370*. https://doi.org/10.1098/rstb.2014.0051

Ouyang, Z., Kumar, M., Kariu, T., Haq, S., Goldberg, M., Pal, U., and Norgard, M.V. (2009). BosR (BB0647) governs virulence expression in *Borrelia burgdorferi*. Mol Microbiol *74*, 1331-1343. https://doi.org/10.1111/j.1365-2958.2009.06945.x

Pachner, A.R., Gelderblom, H., and Cadavid, D. (2001). The rhesus model of Lyme neuroborreliosis. Immunol Rev *183*, 186-204.

Paim, A.C., Baddour, L.M., Pritt, B.S., Schuetz, A.N., and Wilson, J.W. (2018). Lyme endocarditis. Am J Med *131*, 1126-1129. https://doi.org/10.1016/j.amjmed.2018.02.032

Pal, U., and Fikrig, E. (2003). Adaptation of *Borrelia burgdorferi* in the vector and vertebrate host. Microbes Infect *5*, 659-666.

Pal, U., Li, X., Wang, T., Montgomery, R.R., Ramamoorthi, N., Desilva, A.M., Bao, F., Yang, X., Pypaert, M., Pradhan, D., *et al.* (2004a). TROSPA, an Ixodes scapularis receptor for Borrelia burgdorferi. Cell *119*, 457-468. https://doi.org/10.1016/j.cell.2004.10.027

Pal, U., Yang, X., Chen, M., Bockenstedt, L.K., Anderson, J.F., Flavell, R.A., Norgard, M.V., and Fikrig, E. (2004b). OspC facilitates *Borrelia burgdorferi* invasion of Ixodes scapularis salivary glands. J Clin Invest *113*, 220-230. https://doi.org/10.1172/JCI19894

Pappas, C.J., Iyer, R., Petzke, M.M., Caimano, M.J., Radolf, J.D., and Schwartz, I. (2011). *Borrelia burgdorferi* requires glycerol for maximum fitness during the tick phase of the enzootic cycle. PLoS Pathog 7, e1002102. https://doi.org/10.1371/journal.ppat.1002102
PPATHOGENS-D-11-00446 [pii]

Pashenkov, M., Huang, Y.M., Kostulas, V., Haglund, M., Soderstrom, M., and Link, H. (2001). Two subsets of dendritic cells are present in human cerebrospinal fluid. Brain *124*, 480-492.

Paster, B.J., Dewhirst, F.E., Weisburg, W.G., Tordoff, L.A., Fraser, G.J., Hespell, R.B., Stanton, T.B., Zablen, L., Mandelco, L., and Woese,

C.R. (1991). Phylogenetic analysis of the spirochetes. J Bacteriol *173*, 6101-6109.

Patton, T.G., Brandt, K.S., Nolder, C., Clifton, D.R., Carroll, J.A., and Gilmore, R.D. (2013). *Borrelia burgdorferi* bba66 gene inactivation results in attenuated mouse infection by tick transmission. Infect Immun *81*, 2488-2498. https://doi.org/10.1128/IAI.00140-13

Paules, C.I., Marston, H.D., Bloom, M.E., and Fauci, A.S. (2018). tickborne diseases - confronting a growing threat. N Engl J Med *379*, 701-703. https://doi.org/10.1056/NEJMp1807870

Pavia, C.S., Wormser, G.P., and Norman, G.L. (1997). Activity of sera from patients with Lyme disease against *Borrelia burgdorferi*. Clin Infect Dis *25 Suppl 1*, S25-30.

Pepin, K.M., Eisen, R.J., Mead, P.S., Piesman, J., Fish, D., Hoen, A.G., Barbour, A.G., Hamer, S., and Diuk-Wasser, M.A. (2012). Geographic variation in the relationship between human Lyme disease incidence and density of infected host-seeking Ixodes scapularis nymphs in the Eastern United States. Am J Trop Med Hyg *86*, 1062-1071. https://doi.org/10.4269/ajtmh.2012.11-0630

Peri, F., Nistico, D., Morabito, G., Occhipinti, A., Ventura, A., Barbi, E., and Cozzi, G. (2019). Somatic symptom disorder should be suspected in children with alleged chronic Lyme disease. Eur J Pediatr *178*, 1297-1300. https://doi.org/10.1007/s00431-019-03416-6

Petzke, M., and Schwartz, I. (2015). *Borrelia burgdorferi* pathogenesis and the immune response. Clin Lab Med *35*, 745-764. https://doi.org/10.1016/j.cll.2015.07.004

Petzke, M.M., Volyanskyy, K., Mao, Y., Arevalo, B., Zohn, R., Quituisaca, J., Wormser, G.P., Dimitrova, N., and Schwartz, I. (2020). Global transcriptome analysis identifies a diagnostic signature for early disseminated Lyme disease and its resolution. mBio *11*. https://doi.org/10.1128/mBio.00047-20

Philipp, M.T., Aydintug, M.K., Bohm, R.P., Jr., Cogswell, F.B., Dennis, V.A., Lanners, H.N., Lowrie, R.C., Jr., Roberts, E.D., Conway, M.D., Karacorlu, M., *et al.* (1993). Early and early disseminated phases of Lyme disease in the rhesus monkey: a model for infection in humans. Infect Immun *61*, 3047-3059.

Pianta, A., Drouin, E.E., Crowley, J.T., Arvikar, S., Strle, K., Costello, C.E., and Steere, A.C. (2015). Annexin A2 is a target of autoimmune T and B cell responses associated with synovial fibroblast proliferation in patients with antibiotic-refractory Lyme arthritis. Clin Immunol *160*, 336-341. https://doi.org/10.1016/j.clim.2015.07.005

Picken, R.N., Cheng, Y., Strle, F., and Picken, M.M. (1996). Patient isolates of *Borrelia burgdorferi* sensu lato with genotypic and phenotypic similarities of strain 25015. J Infect Dis *174*, 1112-1115.

Picken, R.N., and Picken, M.M. (2000). Molecular characterization of *Borrelia* spp. isolates from greater metropolitan Chicago reveals the presence of *Borrelia bissettii*. Preliminary report. J Mol Microbiol Biotechnol *2*, 505-507.

Piesman, J., Mather, T.N., Dammin, G.J., Telford, S.R., 3rd, Lastavica, C.C., and Spielman, A. (1987a). Seasonal variation of transmission risk of Lyme disease and human babesiosis. Am J Epidemiol *126*, 1187-1189.

Piesman, J., Mather, T.N., Sinsky, R.J., and Spielman, A. (1987b). Duration of tick attachment and *Borrelia burgdorferi* transmission. J Clin Microbiol *25*, 557-558.

Piesman, J., Dolan, M.C., Happ, C.M., Luft, B.J., Rooney, S.E., Mather, T.N., and Golde, W.T. (1997). Duration of immunity to reinfection with tick-transmitted *Borrelia burgdorferi* in naturally infected mice. Infect Immun *65*, 4043-4047.

Piesman, J., Schneider, B.S., and Zeidner, N.S. (2001). Use of quantitative PCR to measure density of *Borrelia burgdorferi* in the midgut and salivary glands of feeding tick vectors. J Clin Microbiol *39*, 4145-4148. https://doi.org/10.1128/JCM.39.11.4145-4148.2001

Piesman, J. (2002). Ecology of *Borrelia burgdorferi* sensu lato in North America. In Lyme Borreliosis Biology, Epidemiology, and Control, J.S. Gray, O. Kahl, R.S. Lane, and G. Stanek, eds. (New York: CABI Publishing), pp. 223-250.

Piesman, J., and Gern, L. (2004). Lyme borreliosis in Europe and North America. Parasitology *129 Suppl*, S191-220.

Pikas, D.S., Brown, E.L., Gurusiddappa, S., Lee, L.Y., Xu, Y., and Hook, M. (2003). Decorin-binding sites in the adhesin DbpA from Borrelia burgdorferi: a synthetic peptide approach. J Biol Chem *278*, 30920-30926. https://doi.org/10.1074/jbc.M303979200

Pinne, M., Ostberg, Y., Comstedt, P., and Bergstrom, S. (2004). Molecular analysis of the channel-forming protein P13 and its paralogue family 48 from different Lyme disease *Borrelia* species. Microbiology *150*, 549-559. https://doi.org/10.1099/mic.0.26728-0

Pinne, M., Thein, M., Denker, K., Benz, R., Coburn, J., and Bergstrom, S. (2007). Elimination of channel-forming activity by insertional inactivation of the p66 gene in *Borrelia burgdorferi*. FEMS Microbiol Lett *266*, 241-249. https://doi.org/10.1111/j.1574-6968.2006.00529.x

Plotkin, S.A. (2011). Correcting a public health fiasco: The need for a new vaccine against Lyme disease. Clin Infect Dis *52 Suppl 3*, s271-275. https://doi.org/10.1093/cid/ciq119

Postic, D., Ras, N.M., Lane, R.S., Hendson, M., and Baranton, G. (1998). Expanded diversity among Californian borrelia isolates and description of *Borrelia bissettii* sp. nov. (formerly *Borrelia* group DN127). J Clin Microbiol *36*, 3497-3504.

Pritt, B.S., Mead, P.S., Johnson, D.K.H., Neitzel, D.F., Respicio-Kingry, L.B., Davis, J.P., Schiffman, E., Sloan, L.M., Schriefer, M.E., Replogle, A.J., *et al.* (2016). Identification of a novel pathogenic *Borrelia* species causing Lyme borreliosis with unusually high spirochaetaemia: a descriptive study. Lancet Infect Dis *16*, 556-564. https://doi.org/10.1016/S1473-3099(15)00464-8

Probert, W.S., Kim, J.H., Hook, M., and Johnson, B.J. (2001). Mapping the ligand-binding region of *Borrelia burgdorferi* fibronectin-binding protein BBK32. Infect Immun *69*, 4129-4133. https://doi.org/10.1128/IAI.69.6.4129-4133.2001

Qiu, W.G., Bruno, J.F., McCaig, W.D., Xu, Y., Livey, I., Schriefer, M.E., and Luft, B.J. (2008). Wide distribution of a high-virulence *Borrelia burgdorferi* clone in Europe and North America. Emerg Infect Dis *14*, 1097-1104. https://doi.org/10.3201/eid1407.070880

Qiu, W.G., and Martin, C.L. (2014). Evolutionary genomics of *Borrelia burgdorferi* sensu lato: findings, hypotheses, and the rise of hybrids. Infect Genet Evol *27*, 576-593. https://doi.org/10.1016/j.meegid.2014.03.025

Radolf, J.D., Norgard, M.V., Brandt, M.E., Isaacs, R.D., Thompson, P.A., and Beutler, B. (1991). Lipoproteins of *Borrelia burgdorferi* and *Treponema pallidum* activate cachectin/tumor necrosis factor synthesis. Analysis using a CAT reporter construct. J Immunol *147*, 1968-1974.

Radolf, J.D., Bourell, K.W., Akins, D.R., Brusca, J.S., and Norgard, M.V. (1994). Analysis of *Borrelia burgdorferi* membrane architecture by freeze-fracture electron microscopy. J Bacteriol *176*, 21-31.

Radolf, J.D., Caimano, M.J., Stevenson, B., and Hu, L.T. (2012). Of ticks, mice and men: understanding the dual-host lifestyle of Lyme disease spirochaetes. Nat Rev Microbiol *10*, 87-99. https://doi.org/10.1038/nrmicro2714

Radolf, J.D., Tramont, E.C., and Salazar, J.C. (2019). Syphilis (*Treponema pallidum*). In Mandell, Douglas and Bennett's Principles and Practice of Infectious Diseases, G.L. Mandell, R. Dolin, and M.J. Blaser, eds. (Philadelphia: Churchill Livingtone Elsevier), pp. 2865-2892.

Ramamoorthi, N., Narasimhan, S., Pal, U., Bao, F., Yang, X.F., Fish, D., Anguita, J., Norgard, M.V., Kantor, F.S., Anderson, J.F., *et al.* (2005). The Lyme disease agent exploits a tick protein to infect the mammalian host. Nature *436*, 573-577. https://doi.org/10.1038/nature03812

Ramesh, G., Alvarez, A.L., Roberts, E.D., Dennis, V.A., Lasater, B.L., Alvarez, X., and Philipp, M.T. (2003). Pathogenesis of Lyme neuroborreliosis: *Borrelia burgdorferi* lipoproteins induce both proliferation and apoptosis in rhesus monkey astrocytes. Eur J Immunol *33*, 2539-2550. https://doi.org/10.1002/eji.200323872

Ramesh, G., Borda, J.T., Dufour, J., Kaushal, D., Ramamoorthy, R., Lackner, A.A., and Philipp, M.T. (2008). Interaction of the Lyme disease spirochete *Borrelia burgdorferi* with brain parenchyma elicits inflammatory mediators from glial cells as well as glial and neuronal apoptosis. Am J Pathol *173*, 1415-1427. https://doi.org/10.2353/ajpath.2008.080483

Rauter, C., and Hartung, T. (2005). Prevalence of *Borrelia burgdorferi* sensu lato genospecies in Ixodes ricinus ticks in Europe: a metaanalysis. Appl Environ Microbiol *71*, 7203-7216. https://doi.org/10.1128/AEM.71.11.7203-7216.2005

Rego, R.O., Bestor, A., Stefka, J., and Rosa, P.A. (2014). Population bottlenecks during the infectious cycle of the Lyme disease spirochete Borrelia burgdorferi. PLoS One *9*, e101009. https://doi.org/10.1371/journal.pone.0101009

Richter, D., Spielman, A., Komar, N., and Matuschka, F.R. (2000). Competence of American robins as reservoir hosts for Lyme disease spirochetes. Emerg Infect Dis *6*, 133-138. https://doi.org/10.3201/eid0602.000205

Rizzoli, A., Hauffe, H., Carpi, G., Vourc, H.G., Neteler, M., and Rosa, R. (2011). Lyme borreliosis in Europe. Euro Surveill *16*.

Roberts, W.C., Mullikin, B.A., Lathigra, R., and Hanson, M.S. (1998). Molecular analysis of sequence heterogeneity among genes encoding decorin binding proteins A and B of *Borrelia burgdorferi* sensu lato. Infect Immun *66*, 5275-5285.

Robinson, M.L., Kobayashi, T., Higgins, Y., Calkins, H., and Melia, M.T. (2015). Lyme carditis. Infect Dis Clin North Am *29*, 255-268. https://doi.org/10.1016/j.idc.2015.02.003

Rollend, L., Fish, D., and Childs, J.E. (2013). Transovarial transmission of *Borrelia* spirochetes by *Ixodes scapularis*: a summary of the literature and recent observations. Ticks Tick Borne Dis *4*, 46-51. https://doi.org/10.1016/j.ttbdis.2012.06.008

Rose, I., Yoshimizu, M.H., Bonilla, D.L., Fedorova, N., Lane, R.S., and Padgett, K.A. (2019). Phylogeography of *Borrelia* spirochetes in *Ixodes pacificus* and *Ixodes spinipalpis* ticks highlights differential acarological risk of tick-borne disease transmission in northern versus southern California. PLoS One *14*, e0214726. https://doi.org/10.1371/journal.pone.0214726

Rosenberg, R., Lindsey, N.P., Fischer, M., Gregory, C.J., Hinckley, A.F., Mead, P.S., Paz-Bailey, G., Waterman, S.H., Drexler, N.A., Kersh, G.J., *et al.* (2018). Vital signs: trends in reported vectorborne disease cases - United States and territories, 2004-2016. MMWR Morb Mortal Wkly Rep *67*, 496-501. https://doi.org/10.15585/mmwr.mm6717e1

Rudenko, N., Golovchenko, M., Mokracek, A., Piskunova, N., Ruzek, D., Mallatova, N., and Grubhoffer, L. (2008). Detection of *Borrelia bissettii* in cardiac valve tissue of a patient with endocarditis and aortic valve stenosis in the Czech Republic. *46*, 3540-3543. https://doi.org/10.1128/jcm.01032-08

Rudenko, N., Golovchenko, M., Vancova, M., Clark, K., Grubhoffer, L., and Oliver, J.H., Jr. (2016). Isolation of live *Borrelia burgdorferi* sensu lato spirochaetes from patients with undefined disorders and symptoms not typical for Lyme borreliosis. Clin Microbiol Infect *22*, 267 e269-215. https://doi.org/10.1016/j.cmi.2015.11.009

Rupprecht, T.A., Pfister, H.W., Angele, B., Kastenbauer, S., Wilske, B., and Koedel, U. (2005). The chemokine CXCL13 (BLC): a putative diagnostic marker for neuroborreliosis. Neurology *65*, 448-450. https://doi.org/10.1212/01.wnl.0000171349.06645.79

Ruyts, S.C., Ampoorter, E., Coipan, E.C., Baeten, L., Heylen, D., Sprong, H., Matthysen, E., and Verheyen, K. (2016). Diversifying forest communities may change Lyme disease risk: extra dimension to the dilution effect in Europe. Parasitology *143*, 1310-1319. https://doi.org/10.1017/S0031182016000688

Ruzic-Sabljic, E., and Cerar, T. (2017). Progress in the molecular diagnosis of Lyme disease. Expert Rev Mol Diagn *17*, 19-30. https://doi.org/10.1080/14737159.2016.1246959

Sadziene, A., Barbour, A.G., Rosa, P.A., and Thomas, D.D. (1993). An OspB mutant of *Borrelia burgdorferi* has reduced invasiveness in vitro and reduced infectivity in vivo. Infect Immun *61*, 3590-3596.

Salazar, J.C., Gerber, M.A., and Goff, C.W. (1993). Long-term outcome of Lyme disease in children given early treatment. J Pediatr *122*, 591-593.

Salazar, J.C., Pope, C.D., Sellati, T.J., Feder, H.M., Jr., Kiely, T.G., Dardick, K.R., Buckman, R.L., Moore, M.W., Caimano, M.J., Pope, J.G., *et al.* (2003). Coevolution of markers of innate and adaptive immunity in skin and peripheral blood of patients with erythema migrans. J Immunol *171*, 2660-2670. https://doi.org/10.4049/jimmunol.171.5.2660

Sanchez, E., Vannier, E., Wormser, G.P., and Hu, L.T. (2016). Diagnosis, treatment, and prevention of Lyme disease, human granulocytic anaplasmosis, and babesiosis: a review. JAMA *315*, 1767-1777. https://doi.org/10.1001/jama.2016.2884

Scheffold, N., Herkommer, B., Kandolf, R., and May, A.E. (2015). Lyme carditis--diagnosis, treatment and prognosis. Dtsch Arztebl Int *112*, 202-208. https://doi.org/10.3238/arztebl.2015.0202

Schmid, G.P., Steigerwalt, A.G., Johnson, S.E., Barbour, A.G., Steere, A.C., Robinson, I.M., and Brenner, D.J. (1984). DNA characterization of the spirochete that causes Lyme disease. J Clin Microbiol *20*, 155-158.

Schmidli, J., Hunziker, T., Moesli, P., and Schaad, U.B. (1988). Cultivation of *Borrelia burgdorferi* from joint fluid three months after treatment of facial palsy due to Lyme borreliosis. J Infect Dis *158*, 905-906. https://doi.org/10.1093/infdis/158.4.905

Schoen, R.T. (2020). Challenges in the diagnosis and treatment of Lyme disease. Curr Rheumatol Rep *22*, 3. https://doi.org/10.1007/s11926-019-0857-2

Schotthoefer, A.M., and Frost, H.M. (2015). Ecology and epidemiology of Lyme borreliosis. Clin Lab Med *35*, 723-743. https://doi.org/10.1016/j.cll.2015.08.003

Schroder, N.W., Eckert, J., Stubs, G., and Schumann, R.R. (2008). Immune responses induced by spirochetal outer membrane lipoproteins and glycolipids. Immunobiology *213*, 329-340. https://doi.org/10.1016/j.imbio.2007.11.003

Schwan, T.G., Piesman, J., Golde, W.T., Dolan, M.C., and Rosa, P.A. (1995). Induction of an outer surface protein on *Borrelia burgdorferi* during tick feeding. Proc Natl Acad Sci U S A *92*, 2909-2913.

Schwanz, L.E., Voordouw, M.J., Brisson, D., and Ostfeld, R.S. (2011). *Borrelia burgdorferi* has minimal impact on the Lyme disease reservoir host *Peromyscus leucopus*. Vector Borne Zoonotic Dis *11*, 117-124. https://doi.org/10.1089/vbz.2009.0215

Schwartz, A.M., Hinckley, A.F., Mead, P.S., Hook, S.A., and Kugeler, K.J. (2017). Surveillance for Lyme Disease - United States, 2008-2015. MMWR Surveill Summ *66*, 1-12. https://doi.org/10.15585/mmwr.ss6622a1

Schwartz, I., Fish, D., and Daniels, T.J. (1997). Prevalence of the rickettsial agent of human granulocytic ehrlichiosis in ticks from a hyperendemic focus of Lyme disease. N Engl J Med *337*, 49-50. https://doi.org/10.1056/NEJM199707033370111

Seinost, G., Dykhuizen, D.E., Dattwyler, R.J., Golde, W.T., Dunn, J.J., Wang, I.N., Wormser, G.P., Schriefer, M.E., and Luft, B.J. (1999). Four clones of *Borrelia burgdorferi* sensu stricto cause invasive infection in humans. Infect Immun *67*, 3518-3524.

Semenza, J.C., and Suk, J.E. (2018). Vector-borne diseases and climate change: a European perspective. FEMS Microbiol Lett *365*. https://doi.org/10.1093/femsle/fnx244

Seshu, J., Esteve-Gassent, M.D., Labandeira-Rey, M., Kim, J.H., Trzeciakowski, J.P., Hook, M., and Skare, J.T. (2006). Inactivation of the fibronectin-binding adhesin gene bbk32 significantly attenuates the infectivity potential of *Borrelia burgdorferi*. Mol Microbiol *59*, 1591-1601. https://doi.org/10.1111/j.1365-2958.2005.05042.x

Setubal, J.C., Reis, M., Matsunaga, J., and Haake, D.A. (2006). Lipoprotein computational prediction in spirochaetal genomes. Microbiology *152*, 113-121. https://doi.org/10.1099/mic.0.28317-0

Shadick, N.A., Phillips, C.B., Logigian, E.L., Steere, A.C., Kaplan, R.F., Berardi, V.P., Duray, P.H., Larson, M.G., Wright, E.A., Ginsburg, K.S., *et al.* (1994). The long-term clinical outcomes of Lyme disease. A population-based retrospective cohort study. Ann Intern Med *121*, 560-567.

Shah, A., O'Horo, J.C., Wilson, J.W., Granger, D., and Theel, E.S. (2018). An unusual cluster of neuroinvasive Lyme disease cases presenting with Bannwarth syndrome in the Midwest United States. Open Forum Infect Dis *5*, ofx276. https://doi.org/10.1093/ofid/ofx276

Shah, S.S., Zaoutis, T.E., Turnquist, J., Hodinka, R.L., and Coffin, S.E. (2005). Early differentiation of Lyme from enteroviral meningitis. Pediatr Infect Dis J *24*, 542-545. https://doi.org/10.1097/01.inf.0000164767.73746.6e

Shapiro, E.D., Baker, P.J., and Wormser, G.P. (2017). False and misleading information about Lyme disease. Am J Med *130*, 771-772. https://doi.org/10.1016/j.amjmed.2017.01.030

Sharareh, N., Behler, R.P., Roome, A.B., Shepherd, J., Garruto, R.M., and Sabounchi, N.S. (2019). Risk factors of Lyme disease: an intersection of environmental ecology and systems science. Healthcare (Basel) *7*. https://doi.org/10.3390/healthcare7020066

Sharma, B., Brown, A.V., Matluck, N.E., Hu, L.T., and Lewis, K. (2015). *Borrelia burgdorferi*, the causative agent of Lyme disease, forms drug-tolerant persister cells. Antimicrob Agents Chemother *59*, 4616-4624. https://doi.org/10.1128/AAC.00864-15

Shaw, D.K., Tate, A.T., Schneider, D.S., Levashina, E.A., Kagan, J.C., Pal, U., Fikrig, E., and Pedra, J.H.F. (2018). Vector immunity and evolutionary ecology: the harmonious dissonance. Trends Immunol *39*, 862-873. https://doi.org/10.1016/j.it.2018.09.003

Shen, S., Shin, J.J., Strle, K., McHugh, G., Li, X., Glickstein, L.J., Drouin, E.E., and Steere, A.C. (2010). Treg cell numbers and function in patients with antibiotic-refractory or antibiotic-responsive Lyme arthritis. Arthritis Rheum *62*, 2127-2137. https://doi.org/10.1002/art.27468

Shih, C.M., Pollack, R.J., Telford, S.R., 3rd, and Spielman, A. (1992). Delayed dissemination of Lyme disease spirochetes from the site of deposition in the skin of mice. J Infect Dis *166*, 827-831.

Shin, J.J., Glickstein, L.J., and Steere, A.C. (2007). High levels of inflammatory chemokines and cytokines in joint fluid and synovial tissue throughout the course of antibiotic-refractory Lyme arthritis. Arthritis Rheum *56*, 1325-1335. https://doi.org/10.1002/art.22441

Shin, J.J., Strle, K., Glickstein, L.J., Luster, A.D., and Steere, A.C. (2010). *Borrelia burgdorferi* stimulation of chemokine secretion by cells of monocyte lineage in patients with Lyme arthritis. Arthritis Res Ther *12*, R168. https://doi.org/10.1186/ar3128

Sigal, L.H., Zahradnik, J.M., Lavin, P., Patella, S.J., Bryant, G., Haselby, R., Hilton, E., Kunkel, M., Adler-Klein, D., Doherty, T., *et al.* (1998). A vaccine consisting of recombinant *Borrelia burgdorferi* outer-surface protein A to prevent Lyme disease. Recombinant outer-surface protein A Lyme disease vaccine study consortium. N Engl J Med *339*, 216-222. https://doi.org/10.1056/NEJM199807233390402

Silver, E., Pass, R.H., Kaufman, S., Hordof, A.J., and Liberman, L. (2007). Complete heart block due to Lyme carditis in two pediatric patients and a review of the literature. Congenit Heart Dis *2*, 338-341. https://doi.org/10.1111/j.1747-0803.2007.00122.x

Simo, L., Kazimirova, M., Richardson, J., and Bonnet, S.I. (2017). The essential role of tick salivary glands and saliva in tick feeding and pathogen transmission. Front Cell Infect Microbiol *7*, 281. https://doi.org/10.3389/fcimb.2017.00281

Singh, S.K., and Girschick, H.J. (2004). Lyme borreliosis: from infection to autoimmunity. Clin Microbiol Infect *10*, 598-614. https://doi.org/10.1111/j.1469-0691.2004.00895.x

Skare, J.T., Shaw, D.K., Trzeciakowski, J.P., and Hyde, J.A. (2016). in vivo imaging demonstrates that *Borrelia burgdorferi ospC* is uniquely expressed temporally and spatially throughout experimental infection. PLoS One *11*, e0162501. https://doi.org/10.1371/journal.pone.0162501

Smith, R., and Takkinen, J. (2006). Lyme borreliosis: Europe-wide coordinated surveillance and action needed? Euro Surveill *11*, E060622 060621.

Smith, R.P., Schoen, R.T., Rahn, D.W., Sikand, V.K., Nowakowski, J., Parenti, D.L., Holman, M.S., Persing, D.H., and Steere, A.C. (2002). Clinical characteristics and treatment outcome of early Lyme disease in patients with microbiologically confirmed erythema migrans. Ann Intern Med *136*, 421-428.

Snydman, D.R., Schenkein, D.P., Berardi, V.P., Lastavica, C.C., and Pariser, K.M. (1986). Borrelia burgdorferi in joint fluid in chronic Lyme arthritis. Ann Intern Med *104*, 798-800.

Sonenshine, D.E., and Macaluso, K.R. (2017). Microbial invasion vs. tick immune regulation. Front Cell Infect Microbiol *7*, 390. https://doi.org/10.3389/fcimb.2017.00390

Sood, S.K., Salzman, M.B., Johnson, B.J., Happ, C.M., Feig, K., Carmody, L., Rubin, L.G., Hilton, E., and Piesman, J. (1997). Duration of tick attachment as a predictor of the risk of Lyme disease in an area in which Lyme disease is endemic. J Infect Dis *175*, 996-999.

Sood, S.K. (2006). What we have learned about Lyme borreliosis from studies in children. Wien Klin Wochenschr *118*, 638-642. https://doi.org/10.1007/s00508-006-0689-8

Sood, S.K. (2015). Lyme disease in children. Infect Dis Clin North Am *29*, 281-294. https://doi.org/10.1016/j.idc.2015.02.011

Spielman, A., Clifford, C.M., Piesman, J., and Corwin, M.D. (1979). Human babesiosis on Nantucket Island, USA: description of the vector, *Ixodes* (Ixodes) *dammini*, n. sp. (Acarina: Ixodidae). J Med Entomol *15*, 218-234.

Stafford, K.C., 3rd, Cartter, M.L., Magnarelli, L.A., Ertel, S.H., and Mshar, P.A. (1998). Temporal correlations between tick abundance and prevalence of ticks infected with *Borrelia burgdorferi* and increasing incidence of Lyme disease. J Clin Microbiol *36*, 1240-1244.

Stanek, G., Wormser, G.P., Gray, J., and Strle, F. (2012). Lyme borreliosis. Lancet *379*, 461-473. https://doi.org/10.1016/S0140-6736(11)60103-7

Stanek, G., and Strle, F. (2018). Lyme borreliosis-from tick bite to diagnosis and treatment. FEMS Microbiol Rev *42*, 233-258. https://doi.org/10.1093/femsre/fux047

Steere, A.C., Malawista, S.E., Hardin, J.A., Ruddy, S., Askenase, W., and Andiman, W.A. (1977a). Erythema chronicum migrans and Lyme arthritis. The enlarging clinical spectrum. Ann Intern Med *86*, 685-698.

Steere, A.C., Malawista, S.E., Snydman, D.R., Shope, R.E., Andiman, W.A., Ross, M.R., and Steele, F.M. (1977b). Lyme arthritis: an epidemic of oligoarticular arthritis in children and adults in three connecticut communities. Arthritis Rheum *20*, 7-17.

Steere, A.C., Gibofsky, A., Patarroyo, M.E., Winchester, R.J., Hardin, J.A., and Malawista, S.E. (1979). Chronic Lyme arthritis. Clinical and immunogenetic differentiation from rheumatoid arthritis. Ann Intern Med *90*, 896-901.

Steere, A.C., and Malawista, S.E. (1979). Cases of Lyme disease in the United States: locations correlated with distribution of *Ixodes dammini*. Ann Intern Med *91*, 730-733.

Steere, A.C., Batsford, W.P., Weinberg, M., Alexander, J., Berger, H.J., Wolfson, S., and Malawista, S.E. (1980). Lyme carditis: cardiac abnormalities of Lyme disease. Ann Intern Med *93*, 8-16.

Steere, A.C., Grodzicki, R.L., Kornblatt, A.N., Craft, J.E., Barbour, A.G., Burgdorfer, W., Schmid, G.P., Johnson, E., and Malawista, S.E.

(1983a). The spirochetal etiology of Lyme disease. N Engl J Med *308*, 733-740. https://doi.org/10.1056/NEJM198303313081301

Steere, A.C., Pachner, A.R., and Malawista, S.E. (1983b). Neurologic abnormalities of Lyme disease: successful treatment with high-dose intravenous penicillin. Ann Intern Med *99*, 767-772.

Steere, A.C., Schoen, R.T., and Taylor, E. (1987). The clinical evolution of Lyme arthritis. Ann Intern Med *107*, 725-731.

Steere, A.C., Dwyer, E., and Winchester, R. (1990). Association of chronic Lyme arthritis with HLA-DR4 and HLA-DR2 alleles. N Engl J Med *323*, 219-223. https://doi.org/10.1056/NEJM199007263230402

Steere, A.C. (1997). Diagnosis and treatment of Lyme arthritis. Med Clin North Am *81*, 179-194. https://doi.org/10.1016/s0025-7125(05)70510-1

Steere, A.C., Sikand, V.K., Meurice, F., Parenti, D.L., Fikrig, E., Schoen, R.T., Nowakowski, J., Schmid, C.H., Laukamp, S., Buscarino, C., *et al.* (1998). Vaccination against Lyme disease with recombinant Borrelia burgdorferi outer-surface lipoprotein A with adjuvant. Lyme Disease Vaccine Study Group. N Engl J Med *339*, 209-215. https://doi.org/10.1056/NEJM199807233390401

Steere, A.C. (2001). Lyme disease. N Engl J Med *345*, 115-125. https://doi.org/10.1056/NEJM200107123450207

Steere, A.C., and Sikand, V.K. (2003). The presenting manifestations of Lyme disease and the outcomes of treatment. N Engl J Med *348*, 2472-2474. https://doi.org/10.1056/NEJM200306123482423

Steere, A.C., Sikand, V.K., Schoen, R.T., and Nowakowski, J. (2003). Asymptomatic infection with *Borrelia burgdorferi*. Clin Infect Dis *37*, 528-532. https://doi.org/10.1086/376914

Steere, A.C., Coburn, J., and Glickstein, L. (2004). The emergence of Lyme disease. J Clin Invest *113*, 1093-1101. https://doi.org/10.1172/JCI21681

Steere, A.C., and Angelis, S.M. (2006). Therapy for Lyme arthritis: strategies for the treatment of antibiotic-refractory arthritis. Arthritis Rheum *54*, 3079-3086. https://doi.org/10.1002/art.22131

Steere, A.C., Klitz, W., Drouin, E.E., Falk, B.A., Kwok, W.W., Nepom, G.T., and Baxter-Lowe, L.A. (2006). Antibiotic-refractory Lyme arthritis is associated with HLA-DR molecules that bind a *Borrelia burgdorferi* peptide. J Exp Med *203*, 961-971. https://doi.org/10.1084/jem.20052471

Steere, A.C., Strle, F., Wormser, G.P., Hu, L.T., Branda, J.A., Hovius, J.W., Li, X., and Mead, P.S. (2016). Lyme borreliosis. Nat Rev Dis Primers 2, 16090. https://doi.org/10.1038/nrdp.2016.90

Steere, A.C. (2020). Posttreatment Lyme disease syndromes: distinct pathogenesis caused by maladaptive host responses. J Clin Invest. https://doi.org/10.1172/JCI138062

Stevenson, B., and Seshu, J. (2018). Regulation of gene and protein expression in the Lyme disease spirochete. Curr Top Microbiol Immunol *415*, 83-112. https://doi.org/10.1007/82_2017_49

Stevenson, B., Fingerle, V., Wormser, G.P., and Margos, G. (2019). Public health and patient safety concerns merit retention of Lyme borreliosis-associated spirochetes within the genus *Borrelia*, and rejection of the genus novum *Borreliella*. Ticks Tick Borne Dis *10*, 1-4. https://doi.org/10.1016/j.ttbdis.2018.08.010

Stewart, P.E., Byram, R., Grimm, D., Tilly, K., and Rosa, P.A. (2005). The plasmids of *Borrelia burgdorferi*: essential genetic elements of a pathogen. Plasmid *53*, 1-13. https://doi.org/10.1016/j.plasmid.2004.10.006

Stewart, P.E., Wang, X., Bueschel, D.M., Clifton, D.R., Grimm, D., Tilly, K., Carroll, J.A., Weis, J.J., and Rosa, P.A. (2006). Delineating the requirement for the *Borrelia burgdorferi* virulence factor OspC in the mammalian host. Infect Immun *74*, 3547-3553. https://doi.org/10.1128/IAI.00158-06

Stone, B.L., and Brissette, C.A. (2017). Host immune evasion by Lyme and relapsing fever *Borreliae*: findings to lead future studies for *Borrelia miyamotoi*. Front Immunol *8*, 12. https://doi.org/10.3389/fimmu.2017.00012

Stone, B.L., Tourand, Y., and Brissette, C.A. (2017). Brave new worlds: the expanding universe of Lyme disease. Vector Borne Zoonotic Dis *17*, 619-629. https://doi.org/10.1089/vbz.2017.2127

Stricker, R.B., and Johnson, L. (2011). Lyme disease: the next decade. Infect Drug Resist *4*, 1-9. https://doi.org/10.2147/IDR.S15653
idr-4-001 [pii]

Stricker, R.B., and Johnson, L. (2014). Lyme disease: call for a "Manhattan Project" to combat the epidemic. PLoS Pathog *10*, e1003796. https://doi.org/10.1371/journal.ppat.1003796

Strle, F., Pleterski-Rigler, D., Cimperman, J., Pejovnik- Pustinek, A., Ruzic, E., and Stanek, G. (1992). Solitary borrelial lymphocytoma: Report of 36 cases. Infection *20*, 20`-206. https://doi.org/10.1007/BF02033059

Strle, F., Nelson, J.A., Ruzic-Sabljic, E., Cimperman, J., Maraspin, V., Lotric-Furlan, S., Cheng, Y., Picken, M.M., Trenholme, G.M., and Picken, R.N. (1996). European Lyme borreliosis: 231 culture-confirmed cases involving patients with erythema migrans. Clin Infect Dis *23*, 61-65.

Strle, F., Nadelman, R.B., Cimperman, J., Nowakowski, J., Picken, R.N., Schwartz, I., Maraspin, V., Aguero-Rosenfeld, M.E., Varde, S., Lotric-Furlan, S., *et al.* (1999). Comparison of culture-confirmed erythema migrans caused by *Borrelia burgdorferi* sensu stricto in New York State and by *Borrelia afzelii* in Slovenia. Ann Intern Med *130*, 32-36.

Strle, F., Ruzic-Sabljic, E., Logar, M., Maraspin, V., Lotric-Furlan, S., Cimperman, J., Ogrinc, K., Stupica, D., Nadelman, R.B., Nowakowski, J., *et al.* (2011a). Comparison of erythema migrans caused by *Borrelia burgdorferi* and *Borrelia garinii*. Vector Borne Zoonotic Dis *11*, 1253-1258. https://doi.org/10.1089/vbz.2010.0230

Strle, K., Drouin, E.E., Shen, S., El Khoury, J., McHugh, G., Ruzic-Sabljic, E., Strle, F., and Steere, A.C. (2009). *Borrelia burgdorferi* stimulates macrophages to secrete higher levels of cytokines and chemokines than *Borrelia afzelii* or *Borrelia garinii*. J Infect Dis *200*, 1936-1943. https://doi.org/10.1086/648091

Strle, K., Jones, K.L., Drouin, E.E., Li, X., and Steere, A.C. (2011b). *Borrelia burgdorferi* RST1 (OspC type A) genotype is associated with greater inflammation and more severe Lyme disease. Am J Pathol *178*, 2726-2739. https://doi.org/10.1016/j.ajpath.2011.02.018

Strle, K., Shin, J.J., Glickstein, L.J., and Steere, A.C. (2012). Association of a Toll-like receptor 1 polymorphism with heightened Th1 inflammatory responses and antibiotic-refractory Lyme arthritis. Arthritis Rheum *64*, 1497-1507. https://doi.org/10.1002/art.34383

Strle, K., Stupica, D., Drouin, E.E., Steere, A.C., and Strle, F. (2014). Elevated levels of IL-23 in a subset of patients with post-Lyme disease symptoms following erythema migrans. Clin Infect Dis *58*, 372-380. https://doi.org/10.1093/cid/cit735

Strle, K., Sulka, K.B., Pianta, A., Crowley, J.T., Arvikar, S.L., Anselmo, A., Sadreyev, R., and Steere, A.C. (2017). T-Helper 17 cell cytokine responses in Lyme disease correlate with *Borrelia burgdorferi* antibodies during early infection and with autoantibodies late in the illness in patients with antibiotic-refractory Lyme arthritis. Clin Infect Dis *64*, 930-938. https://doi.org/10.1093/cid/cix002

Strle, K., and Strle, F. (2020). Posttreatment symptoms in Lyme borreliosis. Clin Infect Dis. https://doi.org/10.1093/cid/ciz1218

Strnad, M., Hönig, V., Růžek, D., Grubhoffer, L., and Rego, R.O.M. (2017). Europe-wide meta-analysis of *Borrelia burgdorferi* sensu lato prevalence in questing *Ixodes ricinus* ticks. Applied and Environmental Microbiology *83*. https://doi.org/10.1128/aem.00609-17

Strobino, B.A., Williams, C.L., Abid, S., Chalson, R., and Spierling, P. (1993). Lyme disease and pregnancy outcome: a prospective study of two thousand prenatal patients. Am J Obstet Gynecol *169*, 367-374. https://doi.org/10.1016/0002-9378(93)90088-z

Stupica, D., Lusa, L., Maraspin, V., Bogovic, P., Vidmar, D., O'Rourke, M., Traweger, A., Livey, I., and Strle, F. (2015). Correlation of culture positivity, PCR positivity, and burden of *Borrelia burgdorferi* sensu lato in skin samples of erythema migrans patients with clinical findings.

PLoS One *10*, e0136600. https://doi.org/10.1371/journal.pone.0136600

Stupica, D., Maraspin, V., Bogovic, P., Ogrinc, K., Blagus, R., Cerar, T., and Strle, F. (2018a). Comparison of clinical course and treatment outcome for patients with early disseminated or early localized Lyme borreliosis. JAMA Dermatol *154*, 1050-1056. https://doi.org/10.1001/jamadermatol.2018.2306

Stupica, D., Veluscek, M., Blagus, R., Bogovic, P., Rojko, T., Cerar, T., and Strle, F. (2018b). Oral doxycycline versus intravenous ceftriaxone for treatment of multiple erythema migrans: an open-label alternate-treatment observational trial. J Antimicrob Chemother *73*, 1352-1358. https://doi.org/10.1093/jac/dkx534

Sykes, R.A., and Makiello, P. (2017). An estimate of Lyme borreliosis incidence in Western Europe. J Public Health (Oxf) *39*, 74-81. https://doi.org/10.1093/pubmed/fdw017

Szczepanski, A., Furie, M.B., Benach, J.L., Lane, B.P., and Fleit, H.B. (1990). Interaction between *Borrelia burgdorferi* and endothelium *in vitro*. J Clin Invest *85*, 1637-1647. https://doi.org/10.1172/JCI114615

Szer, I.S., Taylor, E., and Steere, A.C. (1991). The long-term course of Lyme arthritis in children. N Engl J Med *325*, 159-163. https://doi.org/10.1056/NEJM199107183250304

Takayama, K., Rothenberg, R.J., and Barbour, A.G. (1987). Absence of lipopolysaccharide in the Lyme disease spirochete, *Borrelia burgdorferi*. Infect Immun *55*, 2311-2313.

Tavora, F., Burke, A., Li, L., Franks, T.J., and Virmani, R. (2008). Postmortem confirmation of Lyme carditis with polymerase chain reaction. Cardiovasc Pathol *17*, 103-107. https://doi.org/10.1016/j.carpath.2007.03.004

Telford, S.R., 3rd, Hu, L.T., and Marques, A. (2014). Is there a place for xenodiagnosis in the clinic? Expert Rev Anti Infect Ther *12*, 1307-1310. https://doi.org/10.1586/14787210.2014.966084

Theophilus, P.A., Victoria, M.J., Socarras, K.M., Filush, K.R., Gupta, K., Luecke, D.F., and Sapi, E. (2015). Effectiveness of *Stevia rebaudiana* whole leaf extract against the various morphological forms of *Borrelia Burgdorferi in vitro*. Eur J Microbiol Immunol (Bp) *5*, 268-280. https://doi.org/10.1556/1886.2015.00031

Tijsse-Klasen, E., Pandak, N., Hengeveld, P., Takumi, K., Koopmans, M.P., and Sprong, H. (2013). Ability to cause erythema migrans differs between *Borrelia burgdorferi* sensu lato isolates. Parasit Vectors *6*, 23. https://doi.org/10.1186/1756-3305-6-23

Tilly, K., Krum, J.G., Bestor, A., Jewett, M.W., Grimm, D., Bueschel, D., Byram, R., Dorward, D., Vanraden, M.J., Stewart, P., *et al.* (2006). *Borrelia burgdorferi* OspC protein required exclusively in a crucial early stage of mammalian infection. Infect Immun *74*, 3554-3564. https://doi.org/10.1128/IAI.01950-05

Tilly, K., Rosa, P.A., and Stewart, P.E. (2008). Biology of infection with *Borrelia burgdorferi*. Infect Dis Clin North Am *22*, 217-234, v. https://doi.org/S0891-5520(07)00131-6 [pii]
10.1016/j.idc.2007.12.013

Timmaraju, V.A., Theophilus, P.A., Balasubramanian, K., Shakih, S., Luecke, D.F., and Sapi, E. (2015). Biofilm formation by *Borrelia burgdorferi* sensu lato. FEMS Microbiol Lett *362*, fnv120. https://doi.org/10.1093/femsle/fnv120

Toledo, A., Crowley, J.T., Coleman, J.L., LaRocca, T.J., Chiantia, S., London, E., and Benach, J.L. (2014). Selective association of outer surface lipoproteins with the lipid rafts of *Borrelia burgdorferi*. MBio *5*, e00899-00814. https://doi.org/10.1128/mBio.00899-14

Tomas-Cortazar, J., Martin-Ruiz, I., Barriales, D., Pascual-Itoiz, M.A., de Juan, V.G., Caro-Maldonado, A., Merino, N., Marina, A., Blanco, F.J., Flores, J.M., *et al.* (2017). The immunosuppressive effect of the tick protein, Salp15, is long-lasting and persists in a murine model of hematopoietic transplant. Sci Rep *7*, 10740. https://doi.org/10.1038/s41598-017-11354-2

Tracy, K.E., and Baumgarth, N. (2017). *Borrelia burgdorferi* manipulates innate and adaptive immunity to establish persistence in rodent reservoir hosts. Front Immunol *8*, 116. https://doi.org/10.3389/fimmu.2017.00116

Trayes, K.P., Savage, K., and Studdiford, J.S. (2018). Annular lesions: diagnosis and treatment. Am Fam Physician *98*, 283-291.

Troy, E.B., Lin, T., Gao, L., Lazinski, D.W., Camilli, A., Norris, S.J., and Hu, L.T. (2013). Understanding barriers to *Borrelia burgdorferi* dissemination during infection using massively parallel sequencing. Infect Immun *81*, 2347-2357. https://doi.org/10.1128/IAI.00266-13

Tufts, D.M., Hart, T.M., Chen, G.F., Kolokotronis, S.O., Diuk-Wasser, M.A., and Lin, Y.P. (2019). Outer surface protein polymorphisms linked to host-spirochete association in Lyme borreliae. Mol Microbiol *111*, 868-882. https://doi.org/10.1111/mmi.14209

Tyler, S., Tyson, S., Dibernardo, A., Drebot, M., Feil, E.J., Graham, M., Knox, N.C., Lindsay, L.R., Margos, G., Mechai, S., *et al.* (2018). Whole genome sequencing and phylogenetic analysis of strains of the agent of Lyme disease Borrelia burgdorferi from Canadian emergence zones. Sci Rep *8*, 10552. https://doi.org/10.1038/s41598-018-28908-7

Uhde, M., Ajamian, M., Li, X., Wormser, G.P., Marques, A., and Alaedini, A. (2016). Expression of C-reactive protein and serum amyloid A in early to late manifestations of Lyme disease. Clin Infect Dis *63*, 1399-1404. https://doi.org/10.1093/cid/ciw599

Uhde, M., Indart, A., Fallon, B.A., Wormser, G.P., Marques, A.R., Vernon, S.D., and Alaedini, A. (2018). C-reactive protein response in patients with post-treatment Lyme disease symptoms versus those with myalgic encephalomyelitis/chronic fatigue syndrome. Clin Infect Dis *67*, 1309-1310. https://doi.org/10.1093/cid/ciy299

van Dam, A.P., Kuiper, H., Vos, K., Widjojokusumo, A., de Jongh, B.M., Spanjaard, L., Ramselaar, A.C., Kramer, M.D., and Dankert, J. (1993). Different genospecies of *Borrelia burgdorferi* are associated with distinct clinical manifestations of Lyme borreliosis. Clin Infect Dis *17*, 708-717.

van den Wijngaard, C.C., Hofhuis, A., Simoes, M., Rood, E., van Pelt, W., Zeller, H., and Van Bortel, W. (2017). Surveillance perspective on Lyme borreliosis across the European Union and European Economic Area. Euro Surveill *22*. https://doi.org/10.2807/1560-7917.ES.2017.22.27.30569

van der Linde, M.R. (1991). Lyme carditis: clinical characteristics of 105 cases. Scand J Infect Dis Suppl *77*, 81-84.

VanAcker, M.C., Little, E.A.H., Molaei, G., Bajwa, W.I., and Diuk-Wasser, M.A. (2019). Enhancement of risk for Lyme disease by landscape connectivity, New York, New York, USA. Emerg Infect Dis *25*, 1136-1143. https://doi.org/10.3201/eid2506.181741

Vaz, A., Glickstein, L., Field, J.A., McHugh, G., Sikand, V.K., Damle, N., and Steere, A.C. (2001). Cellular and humoral immune responses to *Borrelia burgdorferi* antigens in patients with culture-positive early Lyme disease. Infect Immun *69*, 7437-7444. https://doi.org/10.1128/IAI.69.12.7437-7444.2001

Verhey, T.B., Castellanos, M., and Chaconas, G. (2018). Antigenic Variation in the Lyme spirochete: insights into recombinational wwitching with a suggested role for error-prone repair. Cell Rep *23*, 2595-2605. https://doi.org/10.1016/j.celrep.2018.04.117

Vig, D.K., and Wolgemuth, C.W. (2014). Spatiotemporal evolution of erythema migrans, the hallmark rash of Lyme disease. Biophys J *106*, 763-768. https://doi.org/10.1016/j.bpj.2013.12.017

Voordouw, M.J., Lachish, S., and Dolan, M.C. (2015). The Lyme disease pathogen has no effect on the survival of its rodent reservoir host. PLoS One *10*, e0118265. https://doi.org/10.1371/journal.pone.0118265

Vudattu, N.K., Strle, K., Steere, A.C., and Drouin, E.E. (2013). Dysregulation of CD4+CD25(high) T cells in the synovial fluid of patients with antibiotic-refractory Lyme arthritis. Arthritis Rheum *65*, 1643-1653. https://doi.org/10.1002/art.37910

Vuong, H.B., Canham, C.D., Fonseca, D.M., Brisson, D., Morin, P.J., Smouse, P.E., and Ostfeld, R.S. (2014). Occurrence and transmission efficiencies of *Borrelia burgdorferi* ospC types in avian and mammalian wildlife. Infect Genet Evol *27*, 594-600. https://doi.org/10.1016/j.meegid.2013.12.011

Walker, E.M., Borenstein, L.A., Blanco, D.R., Miller, J.N., and Lovett, M.A. (1991). Analysis of outer membrane ultrastructure of pathogenic Treponema and Borrelia species by freeze-fracture electron microscopy. J Bacteriol *173*, 5585-5588.

Wang, G., van Dam, A.P., Schwartz, I., and Dankert, J. (1999a). Molecular typing of *Borrelia burgdorferi* sensu lato: taxonomic, epidemiological, and clinical implications. Clin Microbiol Rev *12*, 633-653.

Wang, G., Ojaimi, C., Iyer, R., Saksenberg, V., McClain, S.A., Wormser, G.P., and Schwartz, I. (2001). Impact of genotypic variation of *Borrelia burgdorferi* sensu stricto on kinetics of dissemination and severity of disease in C3H/HeJ mice. Infect Immun *69*, 4303-4312. https://doi.org/10.1128/IAI.69.7.4303-4312.2001

Wang, G., Ojaimi, C., Wu, H., Saksenberg, V., Iyer, R., Liveris, D., McClain, S.A., Wormser, G.P., and Schwartz, I. (2002). Disease severity in a murine model of Lyme borreliosis is associated with the genotype of the infecting Borrelia burgdorferi sensu stricto strain. J Infect Dis *186*, 782-791. https://doi.org/10.1086/343043

Wang, I.N., Dykhuizen, D.E., Qiu, W., Dunn, J.J., Bosler, E.M., and Luft, B.J. (1999b). Genetic diversity of ospC in a local population of *Borrelia burgdorferi* sensu stricto. Genetics *151*, 15-30.

Weening, E.H., Parveen, N., Trzeciakowski, J.P., Leong, J.M., Hook, M., and Skare, J.T. (2008). *Borrelia burgdorferi* lacking DbpBA exhibits an early survival defect during experimental infection. Infect Immun *76*, 5694-5705. https://doi.org/10.1128/IAI.00690-08

Weitzner, E., McKenna, D., Nowakowski, J., Scavarda, C., Dornbush, R., Bittker, S., Cooper, D., Nadelman, R.B., Visintainer, P., Schwartz, I., *et al.* (2015). Long-term assessment of post-treatment symptoms in patients with culture-confirmed early Lyme disease. Clin Infect Dis *61*, 1800-1806. https://doi.org/10.1093/cid/civ735

Wheeler, C.M., Coleman, J.L., Habicht, G.S., and Benach, J.L. (1989). Adult *Ixodes dammini* on rabbits: development of acute inflammation in the skin and immune responses to salivary gland, midgut, and spirochetal components. J Infect Dis *159*, 265-273.

Wheeler, C.M., Garcia Monco, J.C., Benach, J.L., Golightly, M.G., Habicht, G.S., and Steere, A.C. (1993). Nonprotein antigens of *Borrelia burgdorferi*. J Infect Dis *167*, 665-674.

Wick, R.R., Judd, L.M., Gorrie, C.L., and Holt, K.E. (2017). Completing bacterial genome assemblies with multiplex MinION sequencing. Microb Genom *3*, e000132. https://doi.org/10.1099/mgen.0.000132

Wilhelmsson, P., Fryland, L., Lindblom, P., Sjowall, J., Ahlm, C., Berglund, J., Haglund, M., Henningsson, A.J., Nolskog, P., Nordberg, M., *et al.* (2016). A prospective study on the incidence of *Borrelia burgdorferi* sensu lato infection after a tick bite in Sweden and on the Aland Islands, Finland (2008-2009). Ticks Tick Borne Dis *7*, 71-79. https://doi.org/10.1016/j.ttbdis.2015.08.009

Willyard, C. (2014). Resurrecting the 'yuppie vaccine'. Nat Med *20*, 698-701. https://doi.org/10.1038/nm0714-698

Wilske, B., Fingerle, V., and Schulte-Spechtel, U. (2007). Microbiological and serological diagnosis of Lyme borreliosis. FEMS Immunol Med Microbiol *49*, 13-21. https://doi.org/10.1111/j.1574-695X.2006.00139.x

Wormser, G.P., Liveris, D., Nowakowski, J., Nadelman, R.B., Cavaliere, L.F., McKenna, D., Holmgren, D., and Schwartz, I. (1999). Association of specific subtypes of *Borrelia burgdorferi* with hematogenous dissemination in early Lyme disease. J Infect Dis *180*, 720-725. https://doi.org/10.1086/314922

Wormser, G.P., Bittker, S., Cooper, D., Nowakowski, J., Nadelman, R.B., and Pavia, C. (2001a). Yield of large-volume blood cultures in patients with early Lyme disease. J Infect Dis *184*, 1070-1072. https://doi.org/10.1086/323424

Wormser, G.P., Nadelman, R.B., Nowakowski, J., and Schwartz, I. (2001b). Asymptomatic *Borrelia burgdorferi* infection. Med Hypotheses *57*, 435-438. https://doi.org/10.1054/mehy.2001.1338

Wormser, G.P., McKenna, D., Carlin, J., Nadelman, R.B., Cavaliere, L.F., Holmgren, D., Byrne, D.W., and Nowakowski, J. (2005). Brief communication: hematogenous dissemination in early Lyme disease. Ann Intern Med *142*, 751-755.

Wormser, G.P. (2006). Clinical practice. Early Lyme disease. N Engl J Med *354*, 2794-2801. https://doi.org/10.1056/NEJMcp061181

Wormser, G.P., Dattwyler, R.J., Shapiro, E.D., Halperin, J.J., Steere, A.C., Klempner, M.S., Krause, P.J., Bakken, J.S., Strle, F., Stanek, G., *et al.* (2006). The clinical assessment, treatment, and prevention of lyme disease, human granulocytic anaplasmosis, and babesiosis: clinical practice guidelines by the Infectious Diseases Society of America. Clin Infect Dis *43*, 1089-1134. https://doi.org/10.1086/508667

Wormser, G.P., Brisson, D., Liveris, D., Hanincova, K., Sandigursky, S., Nowakowski, J., Nadelman, R.B., Ludin, S., and Schwartz, I. (2008a). *Borrelia burgdorferi* genotype predicts the capacity for hematogenous dissemination during early Lyme disease. J Infect Dis *198*, 1358-1364. https://doi.org/10.1086/592279

Wormser, G.P., Liveris, D., Hanincova, K., Brisson, D., Ludin, S., Stracuzzi, V.J., Embers, M.E., Philipp, M.T., Levin, A., Aguero-Rosenfeld, M., *et al.* (2008b). Effect of *Borrelia burgdorferi* genotype on the sensitivity of C6 and 2-tier testing in North American patients with culture-confirmed Lyme disease. Clin Infect Dis *47*, 910-914. https://doi.org/10.1086/591529

Wormser, G.P., Schriefer, M., Aguero-Rosenfeld, M.E., Levin, A., Steere, A.C., Nadelman, R.B., Nowakowski, J., Marques, A., Johnson, B.J., and Dumler, J.S. (2013). Single-tier testing with the C6 peptide ELISA kit compared with two-tier testing for Lyme disease. Diagn Microbiol Infect Dis *75*, 9-15. https://doi.org/10.1016/j.diagmicrobio.2012.09.003

Wormser, G.P., Weitzner, E., McKenna, D., Nadelman, R.B., Scavarda, C., Molla, I., Dornbush, R., Visintainer, P., and Nowakowski, J. (2015). Long-term assessment of health-related quality of life in patients with culture-confirmed early Lyme disease. Clin Infect Dis *61*, 244-247. https://doi.org/10.1093/cid/civ277

Wormser, G.P., McKenna, D., Karmen, C.L., Shaffer, K.D., Silverman, J.H., Nowakowski, J., Scavarda, C., Shapiro, E.D., and Visintainer, P. (2020). Prospective evaluation of the frequency and severity of symptoms in Lyme disease patients with erythema migrans compared with matched controls at baseline, 6 months, and 12 months. Clin Infect Dis. https://doi.org/10.1093/cid/ciz1215

Wywial, E., Haven, J., Casjens, S.R., Hernandez, Y.A., Singh, S., Mongodin, E.F., Fraser-Liggett, C.M., Luft, B.J., Schutzer, S.E., and Qiu, W.G. (2009). Fast, adaptive evolution at a bacterial host-resistance locus: the PFam54 gene array in *Borrelia burgdorferi*. Gene *445*, 26-37. https://doi.org/10.1016/j.gene.2009.05.017

Xie, J., Zhi, H., Garrigues, R.J., Keightley, A., Garcia, B.L., and Skare, J.T. (2019). Structural determination of the complement inhibitory domain of *Borrelia burgdorferi* BBK32 provides insight into classical pathway complement evasion by Lyme disease spirochetes. PLoS Pathog *15*, e1007659. https://doi.org/10.1371/journal.ppat.1007659

Xu, H., Caimano, M.J., Lin, T., He, M., Radolf, J.D., Norris, S.J., Gherardini, F., Wolfe, A.J., and Yang, X.F. (2010). Role of acetyl-phosphate in activation of the Rrp2-RpoN-RpoS pathway in Borrelia burgdorferi. PLoS Pathog *6*, e1001104. https://doi.org/10.1371/journal.ppat.1001104

Xu, Y., Kodner, C., Coleman, L., and Johnson, R.C. (1996). Correlation of plasmids with infectivity of *Borrelia burgdorferi* sensu stricto type strain B31. Infect Immun *64*, 3870-3876.

Yang, X.F., Alani, S.M., and Norgard, M.V. (2003). The response regulator Rrp2 is essential for the expression of major membrane lipoproteins in *Borrelia burgdorferi*. Proc Natl Acad Sci U S A *100*, 11001-11006. https://doi.org/10.1073/pnas.1834315100

Yang, X.F., Pal, U., Alani, S.M., Fikrig, E., and Norgard, M.V. (2004). Essential role for OspA/B in the life cycle of the Lyme disease spirochete. J Exp Med *199*, 641-648. https://doi.org/10.1084/jem.20031960

Zainal, A., Hanafi, A., Nadkarni, N., Mubasher, M., Lingutla, D., and Hoefen, R. (2019). Lyme carditis presenting as atrial fibrillation. BMJ Case Rep *12*. https://doi.org/10.1136/bcr-2018-228975

Zambrano, M.C., Beklemisheva, A.A., Bryksin, A.V., Newman, S.A., and Cabello, F.C. (2004). *Borrelia burgdorferi* binds to, invades, and colonizes native type I collagen lattices. Infect Immun *72*, 3138-3146. https://doi.org/10.1128/IAI.72.6.3138-3146.2004

Zannoli, S., Fantini, M., Semprini, S., Marchini, B., Ceccarelli, B., Sparacino, M., Schiavone, P., Belgrano, A., Ruscio, M., Gobbetti, M., *et al.* (2020). Multicenter evaluation of the C6 Lyme ELISA kit for the diagnosis of Lyme disease. Microorganisms *8*, 457. https://doi.org/10.3390/microorganisms8030457

Zhi, H., Xie, J., and Skare, J.T. (2018). The classical complement pathway Is required to control *Borrelia burgdorferi* levels during experimental infection. Front Immunol *9*, 959. https://doi.org/10.3389/fimmu.2018.00959

Zhou, Y., Qin, S., Sun, M., Tang, L., Yan, X., Kim, T.K., Caballero, J., Glusman, G., Brunkow, M.E., Soloski, M.J., *et al.* (2020). Measurement of organ-specific and acute-phase blood protein levels in early Lyme disease. J Proteome Res *19*, 346-359. https://doi.org/10.1021/acs.jproteome.9b00569

Zuckert, W.R. (2014). Secretion of bacterial lipoproteins: through the cytoplasmic membrane, the periplasm and beyond. Biochim Biophys Acta *1843*, 1509-1516. https://doi.org/10.1016/j.bbamcr.2014.04.022

Zuckert, W.R. (2019). Protein secretion in spirochetes. Microbiol Spectr *7*. https://doi.org/10.1128/microbiolspec.PSIB-0026-2019

Index

D

E

F

G

H

I

L

M

N

O

P

R

S

T

U

V

W

www.ingramcontent.com/pod-product-compliance
Lightning Source LLC
LaVergne TN
LVHW072141210926
848482LV00026B/860
* 9 7 8 1 9 1 3 6 5 2 6 1 6 *